Global Energy Market Trends

Global Energy Market Trends

Anco S. Blazev

Routledge
Taylor & Francis Group

LONDON AND NEW YORK

Published 2020 by River Publishers

River Publishers

Alsbjergvej 10, 9260 Gistrup, Denmark

www.riverpublishers.com

Distributed exclusively by Routledge

4 Park Square, Milton Park, Abingdon, Oxon OX14 4RN

605 Third Avenue, New York, NY 10017, USA

First issued in paperback 2023

Library of Congress Cataloging-in-Publication Data

Names: Blazev, Anco S., 1946- author.

Title: Global energy market trends / Anco S. Blazev.

Description: Lilburn, GA : Fairmont Press, Inc., 2016. | Includes
 bibliographical references and index.

Identifiers: LCCN 2016014316 (print) | LCCN 2016016131 (ebook) | ISBN
 0881737542 (alk. paper) | ISBN 9788770223348 (electronic) | ISBN
 9781498786577 (alk. paper) | ISBN 9788770223348 (Electronic)

Subjects: LCSH: Energy industries. | Power resources. | Renewable energy
 sources.

Classification: LCC HD9502.A2 B63 2016 (print) | LCC HD9502.A2 (ebook) | DDC
 333.79--dc23

LC record available at https://lccn.loc.gov/2016014316

Global Energy Market Trends by Anco S. Blazev

First published by Fairmont Press in 2016.

Routledge is an imprint of the Taylor & Francis Group, an informa business

Publisher's Note
The publisher has gone to great lengths to ensure the quality of this reprint but points out that
some imperfections in the original copies may be apparent.

ISBN 13: 978-87-7022-936-4 (pbk)

ISBN 13: 978-1-4987-8657-7 (hbk)

ISBN 13: 978-8-7702-2334-8 (online)

ISBN 13: 978-1-0031-5201-9 (ebook master)

While every effort is made to provide dependable information, the publisher, authors, and editors
cannot be held responsible for any errors or omissions.

The views expressed herein do not necessarily reflect those of the publisher.

Table of Contents

Foreword

Everything is energy and that's all there is to it.
—Albert Einstein

The above statement is very deep and all encompassing, so there is nothing we can add to it. It only confirms the genius of the man and today's energy market is the key witness. Now days energy is many times more important than at the time when Einstein concluded that energy is all that there is to it. This makes his statement even more true than ever.

Energy in all its forms is the lifeblood of the modern economies. No energy, no economic development, no conveniences, and no comfort at home. It is that simple! The big problem is that some countries have a lot of energy resources, while others have less, or none. Countries with excess energy resources (or products) will export these to countries that need them. This is one of the important functions of the global energy markets, where energy sources, products and services are sold, bought, and traded among countries and companies. While this is the primary activity in the energy markets, it is only a part of the entire global energy market scheme, as we will see in this text.

Energy market activities usually start with somebody needing energy. This somebody then directly or indirectly commissions somebody else to find, produce, and deliver the energy products and services to the customers.

This is one of the important functions of the global energy markets, where energy sources and products and services are sold, bought, or traded among countries and companies. But this is only a small part of the global energy markets in their entirety. Energy markets activities start with somebody needing energy. This somebody then directly or indirectly commissions somebody else to find, produce, and deliver the energy products and services to the customers.

The complex energy production-distribution-use cycle, repeated thousands of times daily, involves thousands of different businesses, millions of transactions, affecting billions of people around the globe.

*These non-stop activities—dynamic, ever changing chains of events all over the world—form the **global energy markets**.*

On top of the following figure we see the very important energy producers—energy producing countries and companies—that provide different types of energy sources (coal, natural gas, and crude oil) to the intermediaries, the electric power generators and the transportation fuels producers. These entities complete the process by getting the product and delivering it to the consumers.

These are many and very complex activities that are done with the direct and indirect involvement of a great number of different companies and enterprises—peripheral to the energy sector companies—that are usually not directly involved in energy production.

The peripheral entities assist in, and otherwise support, the activities of the principals in one way or another.

And, of course, the energy consumers—the customer base—is the largest group, since it includes every single country, region, company, entity, and person on Earth.

Generating electric power, for example, starts with obtaining energy sources (coal, natural gas, and crude oil), building the power plants, and generating power. This process involves many equipment manufacturers,

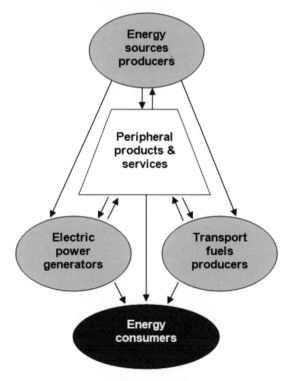

The global energy markets (top tier view)

construction companies, engineering firms, technical and logistics specialists, consultants, transport fleets, maintenance crews, and service companies, involved during different stages.

The transportation sector, which is the main consumer of fossil fuels (crude oil and natural gas), also starts with obtaining the related energy sources (crude oil and natural gas), making the equipment for, and constructing refineries, storage, and handling installations. This process also involves many different equipment manufacturers, construction companies, engineering firms, technical and logistics specialists, consultants, transport fleets, maintenance crews, and service companies in the different stages and levels of the overall cycle.

These enterprises are responsible for a number of important tasks: from selecting sites; construction of mines, gas and oil wells and power plants; manufacturing, delivering and installing specialized equipment; operation of the mines, wells, power plants and refineries, to managing the respective supply and delivery channels and the production operations. These principals also deliver the final products (chemicals, energy products, etc.) to the final consumers.

Many of these business enterprises are "peripheral" to the core energy business (since they do not handle or produce energy products), but are still an integral part of it by providing needed equipment, services, and know-how. The energy markets could not operate properly without even the smallest of these businesses, which is why we take a close look at each and every one of them—producers, users, and everyone and everything in between—all through the global energy markets.

We describe above the physical aspect of energy markets. On the financial side of the energy markets, no physical products are handled at all. Instead, commodities and futures are bought and sold by millions of traders, speculators, and investors on the financial markets.

These are also part of the peripheral energy markets, where most of the people and entities usually do not participate in the actual physical aspects of the energy markets (as described above). Instead, they are only interested in making a profit from the financial transactions involving energy products and services. Nevertheless, this activity has a huge effect on the overall global energy markets. The unprecedented increase of crude oil prices in 2008, for example, was attributed to speculative manipulation of the energy markets via massive stock and futures transactions.

The energy market as a whole is considered to be the largest business enterprise on Earth.

The energy market's size varies from area to area and from country to country, with the developed countries leading the pack in numbers, transaction size, and dollar value. There are, for example, over 25,000 top executives in the U.S. energy sector alone, which represent the core of the energy market. This number multiplied by thousands of people in each energy company would give us a good idea of the size of the entire U.S. energy market.

The size of the peripheral energy markets—these that ensure indirectly the proper function of the energy markets proper—is even larger. These businesses—equipment manufacturers, consultants, lawyers, insurance and security agents, etc., etc, cover almost all aspects of the U.S. and global economies.

As a matter of fact, it is difficult to find an economic sector that is not involved in some aspect of the energy markets.

The energy market, therefore, is like a huge octopus, holding the entire world from corner to corner in its powerful embrace.

The goal of this text is to review and analyze as closely as possible, all sides of the energy markets in their physical, technological, economic, political, regulatory, environmental, financial, and legal aspects. These are the many arms of the global energy octopus that are very long and strong, making the situation very complicated and fluid.

Since all aspects of the energy markets are interwoven, leaving one out does not do justice to the beast, and might lead to misunderstandings. Covering the entire spectrum, however, sounds like an impossible task, so we attempt to achieve the impossible by compromising between the complexity and the immenseness of the energy markets, letting the esteemed reader be the judge.

In this text, the subject of energy is narrowed down to its role in the function of the global energy markets. Here, all practical types and forms of energy, from the production of energy sources—coal, natural gas, crude oil, hydro, solar and wind power—to their delivery and use, are discussed from technical, logistics, marketing, political, legal, and financial points of view.

The emphasis here is on the physical energy markets, as the most important part of the equation. The financial energy markets are peripheral to the global physical energy market, and are thus reviewed only briefly in this text. The financial energy markets are detailed in other books on the subject.

Chapter 1

Energy Today

*Energy—in its different forms of production, distribution,
and utilization—is the engine of today's global economy.*
—Anco Blazev

BACKGROUND

The concept of energy in this text encompasses anything, everything, anybody and everybody related to or involved in the production, distribution, and use of anything and everything energy.

The energy markets basically include the fossil energy sources and fuels (coal, natural gas, and crude oil) and the related petrochemicals, as well as all energy aspects of transportation, electric power and, of course, the different types of alternative and renewable energy sources (nuclear, hydro-, ocean-power, geo-power, bio-fuels, solar, and wind).

The facilities, equipment, processes, procedures, and people involved in these steps and processes are also integral parts of the energy markets, and are therefore presented and discussed in some detail in this text as well.

Note: For additional technical details, please refer to our other books on the subjects at hand: *Power Generation and the Environment* (2014), and *Energy Security for the 21st Century* (2015), also published by The Fairmont Press.

The long and complex cradle-to-grave cycle of the different energy sources and their location, production, distribution, and use determines and drives the energy markets.

It is a long and complex process, with many different aspects and complex dynamics. We cannot claim that we have covered all aspects of the energy markets in this text, but the esteemed reader will get a good idea of their present and future.

Since in this text we discuss and analyze the key aspects of the energy markets, we will start with listing the major categories, thus exposing the broad spectrum of subjects we are covering.

Table 1-1
The Energy Markets Components

Energy Source Production
- Identifying the need for energy
- Making a request for energy
- Initial investment and legal work
  — Private banks and investors
  — Government finance
- Energy source locating and discovery
  — Equipment and services
  — Operations
- Energy sources production
  — Mines construction
  — Equipment and services
  — Operations
  — Risks

Fossil Energy Sources
- Coal
- Natural gas
- Crude oil

Electric Power Generation
- Energy source procurement
- Energy generation
- Coal, gas, crude oil, nuclear, hydro power.
  — Power plants construction
  — Equipment and services
  — Power plants operation
  — Power plants decommissioning
- Energy distribution
  — The National Electric Grid
  — Equipment and services
  — Operations
  — The smart grid
- Risks

The Power Grid
- Operation and maintenance
- Smart Grid
- The challenges

Energy Storage
- Importance
- Market development

Petrochemicals
- Production facilities construction
- Raw materials procurement
- Petrochemicals processing
- Final products transport
  — Pipelines
  — Trains, trucks, barges
- Retail distribution and sales
- Risks

Transportation
- Energy source procurement
- Energy source processing
  — Refineries construction
  — Equipment and services
  — Operations
- Fuels transport
  — Pipelines
  — Trains, trucks, barges
- Retail distribution and sales
- Risks

Other Energy Sources
- Firewood
- Nuclear power
- Hydropower
- Clean coal products

The Renewables
- Solar
- Wind
- Biofuels
- Geo- and ocean energy

Energy market transactions
- Domestic energy markets

— Production
— Transport
— Distribution
- International energy markets
  — Trade
  — Global shipping
- Energy market security issues
- Risks

Financial Aspects
- Private investments
- Government finance and subsidies
- Energy stocks, options, and futures trading
- Risks

Political and Regulatory Aspects
- Government direction
- Political decisions
- Regulatory compliance

Energy Security
- Products
- Services

Legal Aspects
- Policy challenges
- Lobbying issues
- Lawsuits

Environmental Aspects
- Environmental markets
- CO_2 trading
- Government and legal action
- Risks

Future Energy Concerns
- Fossil-less lifestyle options
- The role of the renewables
- New technologies and approaches

We will be looking at many of these subjects and sub-subjects in this text. Since we cannot cover every single aspect, we will focus on the key elements and issues of each subject, that are essential for the functioning of the energy markets. Even then, this is not an easy task, but an important one, as far as our present and future energy well-being and survival are concerned.

Energy Today...

Energy, in all its forms, is at the forefront of societal and economic development. Successful economic decisions today deal—in addition to the issues of the existing energy sources—with the use of renewable energy, environmental pollution, global warming, climate change, etc. phenomena related to energy production and use.

Developed and developing countries alike are increasingly dependent on electricity, which takes a major chunk of the budgets of households and firms. This development creates new needs and opportunities for efficient and smart usage of electricity and transportation in our daily lives.

The electrical industries were the first to deregulate and liberalize, at which point more room was given for new and better market mechanisms.

We are now finally able to fully evaluate and analyze the energy industry developments and its markets functions.

The serious technological, political, and logistics constraints that existed until recently (and restricted the use of efficient market mechanisms) are mow being lifted, which enables the use of new models, such as incentive oriented real-time pricing, and many other new market approaches.

We must emphasize, however, that the cycle is still developing. For example, while power generation has been deregulated, transmission and distribution are still mostly regulated and driven by monopolies. Nevertheless, we are witnessing significant structural changes in the power network systems. They are brought upon us by the need to change to more efficient and intelligent networks (smart grids and their components). The key advantage of these intelligent systems is that they change the one way traffic in our dumb networks to a new, two way, dynamic, and interactive, traffic.

The future smart systems will allow the consumers to change from passive takers of whatever the utilities have to offer, to active users and optimizers of the energy systems. They will be given many choices and some control over the overall system function, which creates new challenges and possibilities for the entire the power generation and distribution system.

The dynamics in the transportation sector of late are even more spectacular. Huge oil and gas discoveries and production in the U.S., followed by drastic fall of oil prices brought up a number of unexpected and unprecedented events. We will discuss these in detail later on in this text.

We must understand and support the goal of the new energy systems designers to bring us to a new and energy worry-free future. All parts of the power system are important and must be considered and improved at the same time. The flow of electricity from generation through transmission and distribution to the final consumer, and the roles of all the players in this chain of events, present challenges and opportunities from energy marketing point of view.

In this text, we intent to paint a comprehensive picture—from technical, political, economic, and logistics point of view—of the global energy markets, as a central field of our 21st century societies.

Take a quick around you and try to figure out what percentage of the objects in the house or office need some sort of energy to run. Our bet is that you will see dozens of house or office appliances plugged in the wall, or running on batteries. Computers, TVs, refrigerators, heaters, air conditioners, coffee makers, lights, fans, cell phones, and many, many other gadgets and appliances make our lives easy and comfortable.

A close look outside reveals a multitude of cars, trucks, boats, trains, planes and other moving vehicles zooming all over the place. We rely on these vehicles to take us from place to place, as well as to deliver goods to the groceries stores and many, many other equally important applications.

All of these moving and non-moving objects have one important thing in common; they all need some sort of energy to keep their motors running, and their batteries full. The energy they use could be in the form of electricity, natural gas, gasoline, diesel, jet fuel, steam, or some other form and must be delivered and used on time for the proper function of these moving or non-moving objects.

If we could imagine for a split second a world without energy we would see very little motion, and a lot of suffering and death.

So what is this thing we call energy, that is so important? Where does it come from and what does it do? Let's take a close look at the different types of energy and the related concepts.

ENERGY SCIENCE

What is energy and what are its various forms and properties. Where does it come from and how we use it? The intuitive answer is that it is the stuff we need to do physical work such as walking, running, heating and cooling a home, powering a TV set, or fueling a car. And this thinking is correct, because all these activities require energy, but it only conveys what energy is used for. It does not tell us where this energy came from, what it is made of, and how it behaves.

People usually ask questions like: Is energy a thing, or condition of a thing? What are its properties? This subject could be confusing to some, due to its complexity, where, for example, light is a thing (photons), which can be seen, counted and even manipulated. Most other types of energy, however, are a condition, or state of a thing or matter.

So, generally speaking, energy is not a thing, but a condition that accompanies an object or process. It can be generated, transported, bought and sold, but unlike anything around us, energy is never lost; it only changes from one to another state.

Energy is never created anew and never destroyed; it only changes states. -This gives it a very unique place among all things and conditions on Earth.

When speaking of energy we must always remember that the greatest source of energy—which as a matter of fact is the origin and sustenance of ALL life on Earth is sun's energy. We could live without most other energies and energy sources, at least for some time, but without sunlight life on Earth would cease to exist within a very short time period.

The Sun's Energy

We all know our good 'ol friend the sun is up there every day, greeting us with its abundant light and heat. But many of us have never even imagined that *everything* around us is arranged so, and/or controlled, by the sun's energy. The clouds, the wind, the global weather and most natural events around us depend on the sun's energy. Vegetation, animals, and humans cannot survive too long if the sun stops shining upon our Earth. At such a time, the Earth would become a dark, frozen solid mass of ice, where all life will gradually slow down and eventually become totally extinct.

And yet, we don't know much about the sun. Where did it come from, how does it function?. Oh, yes, there are a number of scientific theories, but no one has ever been on the sun, nor even close to it, so all we know about it are hypothesis and theories—some plausible, and some less so.

As a matter of fact, it is amazing how little we know about this source of life on Earth, as well as many other events of our past and present. We don't know how and when IT all started…if IT even has a beginning? We don't know how big IT is and if IT has an end? But even beyond that, we don't even understand some of the Earth's phenomena life depends on.

The Earth's rotation that gives as day and night time and the seasons, the gravity that holds everything together, the molten Earth's core, the magnetic fields around us, etc., etc. How do these things work? Theories abound, yes, but most of them have big holes, and we know that any theory with a hole (even a very small one) is a suspect.

And so, with all due respect to modern science and the wise scientists, we still don't know many of the basics of life on Earth and the Universe. This fact alone must humble us, as we analyze and speculate about events and developments of global or Cosmic character.

With this sort of humble awareness, we approach the subjects in this text too—including everything related to energy and the respected energy markets. We constantly keep in mind that although we know quite a bit about what energy is, where it comes from, and how the different types of energy are generated, transported and used, we must always remember that we don't know EVERYTHING about it. That there are materials, processes, and developments, we can only guess about. And that many "facts" and variables, part of every scientific equations related to life on Earth and energy, are not fully understood or resolved.

So, before we get to the subject of energy, let's take one of the paths, scientists believe we have been following to get to today's world order. During the Big Bang (or whatever initiated the birth of our Universe) the protons of the matter in the primordial mix got so close together during the initial collision that the extraordinary short-range nuclear forces took over and dominated the following processes. These strong nuclear forces pulled protons together to fuse them into new nucleuses.

When this happened, fusion took place, so not only was the original kinetic energy of the colliding protons released, but some of the mass of the particles themselves was converted into energy. Such an encounter between nuclear particles is called a "thermonuclear reaction," triggered by high temperatures and pressures, and maintained by a nuclear chain reaction.

The sequence of nuclear reactions that provides the enormous energy of a star, such as our sun, is a multistep process, the net accomplishment of which is to convert four hydrogen nuclei (protons) into a helium nucleus. The helium nucleus is slightly less massive than the total of the four protons. This extra mass is converted into energy—mostly heat and light.

It is these thermonuclear reactions that account for the enormous energy of our sun. They turned on when gravitational contraction produced sufficient heat. This new source of energy so increased the outward pressure in the sun's interior that gravitational contraction was halted and the sun stayed at its present size.

Our star, Earth, was probably not formed of the primordial hydrogen from the Big Bang, which occurred 14 billion years ago, since the sun is only about 4.5 billion years old. It was instead, and most probably formed out of hydrogen gas enriched with elemental debris from earlier stars that had exploded at the end of several cycles of more complicated nuclear reactions, in which not only helium, but heavier elements, had been formed. From this richer stuff around the early sun, the various planets, including our home planet, number three from the sun, were probably formed.

Too many "probably" and "most likely" suppositions to be comfortable with these theories, but this is all

Figure 1-1. The sun is what powers everything on Earth.

we have now. As time goes by, we will get more knowledge and understanding of our brilliant beginnings, our present status, and what to expect in the future.

What we know for sure now is that the sun has enormous potential energy storage and exhibits amazingly great kinetic energy every second of the day. It is emitting energy at an enormous rate, where approximately 4.5 million tons of mass per second are reacted and converted into energy. While this is a lot of mass and energy, it is very little compared with the overall mass of the sun and its energy supply. It has been estimated that would take over 50 billion years to use up just 50% of the sun's energy.

Sunlight arrives on Earth at power of about 1,000 W/m3, as measured at a cloudless summer noontime in the desert. Approximately this much sun power is available all over the world, but clouds and other Earth phenomena obstruct its path. Because of that, we measure much less solar energy during some seasons and regions of the world.

The sun's rays have different length and frequency, which determine their unique behavior within the solar spectrum.

The different wavelengths of the solar spectrum also have different energies, with the most important to us humans being in the 400 to 700 nm range, which is actually the visible part of the sunlight. Some wavelengths, like UV and IR, which are at both ends of the solar spectrum, are invisible and even harmful to humans and some materials.

Earth's energy balance

The Earth's energy balance formula is simple:

Energy in = Energy out + Energy captured

where:
 Energy in is the sun's energy + energy from Earth's core + stored energy
 Energy out is all reflected and radiated back into space energy
 Energy captured is the energy retained* and stored in soil, air, and water
*This includes solar, wind, bio-, geo-, etc. energy converted into electricity

Generally speaking, an energy balance model is an accounting of all energy coming into and going out of a system. The Earth's energy balance can be modeled as a whole, which is very hard thing to do, or breaking it into latitudinal bands with more predictable energy behavior. In all cases, we need to account for how much energy is coming from all energy sources, and how much is being reflected due to the Earth's albedo, how much infrared energy is being radiated back to space, etc.

No matter how we do this, there are many unknowns and variables, such as storms, melting of ice, changes in the albedo, etc. And as a result, all the information we get from the different sources has a margin of error, which depends on the size of the study, its length

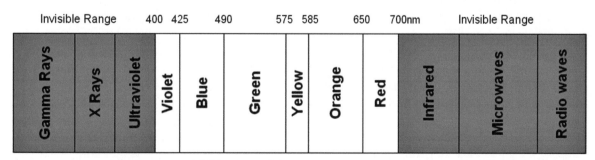

Figure 1-2. The solar spectrum

and precision.

The most amazing, almost unbelievable fact is that sun's energy is radiated into space in all directions—it is not focused on Earth, as some of us might imagine. Nope; we are not the center of the Solar system—the sun is. And as such, at the great distance of 93 million miles away from Earth, the sun spreads its energy in form of light and heat all around its periphery to places that are as distant as infinity.

What a waste of energy, you may say. And we'd agree in general, but we should not complain, because any more than what we get now is going to burn us alive. On the other hand, if we get less of our daily al- location, life on Earth would be different, and we just might freeze to death.

Actually, what we get is just the right amount, al- though it is only a very small portion of the overall sun energy, because the Earth is only a small dot—a golf ball size—compared with the football stadium size of the sun. And even smaller when the circumference of the sun rays is considered.

And as such, we get much less than a billionth of sun's radiation. So, from the 4.5 million tons of mass converted into energy every second, we get approxi- mately 12 lbs of energy. This small. 12 lbs sun matter (energy), however, is enough to send us enough sunlight to keep life on Earth going and provides a comfortable climate over most of the Earth's surface.

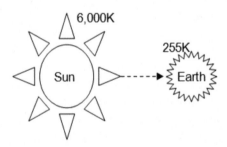

Figure 1-3. Sun and Earth radiation levels

The sun's emission temperature is 6,000K (10,340 F), with most of its energy in the shorter, visible, wave- lengths. The Earth's emission temperature is about 255K (-1 F) degrees, which is mostly infrared (invisible heat) energy. This differential is due to the minute amount of sunlight we receive in one single day (as compared to the total amount of energy emitted by the sun during the same time).

The amount of sunlight (energy) falling on Earth is equal to the total amount of energy all Earth's citizens use in an entire year.

The solar energy flowing from the sun to Earth dwarfs all other energy sources known to man put together. This is an awesome though, isn't it? We are, however, just now capable and willing to use the sun's energy, albeit still in very small amounts.

It takes sunlight (photons) about 6-8 minutes to reach the Earth' surface. Some of it is reflected back into space (30%), some is absorbed and reradiated (47%), some is used in evaporation, precipitation and wind/waves creation, while some provides sunlight for photosynthesis of plants.

The practical measure of energy is in terms of kilowatts (kW), megawatts (MW), gigawatts (GW), or terawatts (TW) of electric power generated, or used.

Just think…the sun sends us close to 173 trillion kilo- watts (TkW) of energy at any single moment of the day. It can power the entire world, if and when we find a way to capture and use all this energy.

During summer days in the world's desert and semi-desert areas (10% of the world's surface) we get approximately 10 hours of uninterrupted sunlight. This means that—given the appropriate equipment and unobstructed sunlight—the world can generate:

17.3 TkW x 10 hrs = 173 TkWh

Keeping in mind that the world uses about 150 TkW annually, this means that we can have an enor- mous surplus of electric generated from the sunlight hitting the desert areas alone. But, of course, we are very far from such reality. There are too many obstacles to overcome before we can get even close to it. We will take a look at these later on in the text.

If we only had a way to capture and convert enough of the sunlight falling on Earth into useful en- ergy, we would have eliminated most of the problems related to energy scarcity, environmental pollution, and climate change. Alas, this is not possible under the present economic and politics status-quo and the immature state-of-the-art of the energy generating technologies. So here we are; looking for a way out of the energy dependence and environmental mess we got ourselves in.

But the sun's energy is not fully wasted. It creates the winds, moves the oceans, and basically drives the major weather events on Earth. It changes the local weather and shapes the global climate. It is a great power that we depend on every day of lives.

OK, we are convinced; the sun is a great source of energy, but we also have many other types of energy.

Today's Energy Basics

What is energy, as we know it and use it in our daily lives? Generally speaking, energy is what makes everything move, change, grow, and even exist. More specifically, translated from Greek energy means "activity," and/or "operation." It is a quantity that exists, but is indirectly observed (except light), for it is generally speaking invisible and immeasurable in its purest form.

In layman's terms, we only see the effects of energy on objects and events. We seldom see energy itself.

In practical terms, energy is represented as the ability of a physical system to do work on other physical systems, or on itself. Work, on the other hand, is defined as the force acting through a distance (a length of space), where energy is always equivalent to the ability to exert pulls or pushes against the basic forces of nature, along a path of a certain length.

Here we come to the two basic types of energy: *potential energy* and *kinetic energy*.

The important facts here are that:

- *Potential energy* has the potential to do work, but it does nothing and remains idle until activated by its components; force and distance. It is *"indirectly observed"* entity when any of its two main constituents "force" or "distance" are missing.

- *Kinetic energy*, on the other hand, is observable (at times) and *measurable* (at other times) entity, if and when both of the variables (*force and distance*) are present. In this case the potential energy has been "activated," and it goes into action, by becoming *kinetic* energy in the process. Then, and only then, energy does work, and can be observed and/or measured.

- *Internal energy* is a concept that most of us are unfamiliar with, and not that interested in. It is the interactions of the atoms and molecules within a substance. It is different from the above (kinetic and potential) energy types in that it is always present in both types. For example, and object laying on a table might be at its potential state, but the molecules and atoms within it are always in motion. As it falls down from the table, its energy is now converted into kinetic type, while the molecules and atoms within it are still in motion as if

nothing happened. Internal energy, therefore, is independent from the other energy states the object might be at the moment.

The other two very important conditions to remember about energy:

- Energy that is visible and measurable in an object, or a person, can be identified with, and measured by, its mass, and

- Energy cannot be created or destroyed.

And finally, in this text we observe, describe, and measure the impact of energy—including power generation, fuels creation, and their use—on our lives and the environment, where we distinguish three different types of energy generation:

- Conventional energy (power) generators, such as coal, oil, natural gas, and crude oil, also called *fossils*,

- Renewable energy (power) generators, such as solar, wind, ocean-, geo-, and bio-fuels, and

- Nuclear and hydro are a third category of energy (power) generators, since they don't fit exactly in the above energy systems.

We will discuss and describe these in great detail in the following chapters.

Energy

Energy in general (as a physical entity) is measured in joules, but in many fields (power generation included) other units, such as kilowatt-hours and kilocalories, are used too. All of these units translate to units of work, which are always defined in terms of forces and the distances that the forces act through. In the electric field, we use the word power to replace energy, but for all practical purposes it is one and the same. And so we do use these terms interchangeably in this text.

When ordinary material particles are changed into energy (such as energy of motion, or radiation), the mass of the system does not change through the transformation process. However, there may be mechanistic limits as to how much of the matter in an object may be changed into other types of energy and thus into work, on other systems. Energy, like mass, is a scalar physical quantity.

A system can also transfer energy to another system by simply transferring matter to it (since matter is equivalent to energy, in accordance with its mass). However, when energy is transferred by means other than matter-transfer, the transfer produces changes in the second system, as a result of work done on it.

This work manifests itself as the effect of force(s) applied through distances within the target system. For example, a system can emit energy to another system by transferring (radiating) electromagnetic energy, which creates forces upon the particles that absorb the radiation. Similarly, a system may transfer energy to another by physically impacting it, but in this case the energy of motion in an object, or the kinetic energy, results in forces acting over distances (new energy) to appear in another object that is struck.

Transfer of thermal energy by heat occurs by both of these mechanisms: heat can be transferred by electromagnetic radiation, or by physical contact in which direct particle-particle impacts could transfer the energy from system to system.

Energy may be stored in systems without being present as matter, as kinetic or electro-magnetic energy. Stored energy is created whenever a particle has been moved through a field it interacts with (requiring a force to do so), but the energy to accomplish this is stored as a new position of the particles in the field. This new configuration must be "held" or fixed by a different type of force, otherwise, the new configuration would resolve itself by the field pushing or pulling the particle back toward its previous position.

This type of energy "stored" by force—fields and particles that have been forced into a new physical configuration in the field by doing work on them by another system—could be referred to as potential energy. A simple example here is the work needed to lift an object in a gravity field, up to a support, or the spinning rotor of a mechanical energy storage system.

Each of the basic forces of nature is associated with a different type of potential energy, and all types of potential energy (like all other types of energy) appear as system mass, whenever present. For example, a compressed spring will be slightly more massive than before it was compressed, albeit it would be difficult to measure the change in mass. Likewise, whenever energy is transferred between systems by any mechanism, an associated mass is transferred with it, which is also difficult to measure.

Any form of energy may be transformed into another form. For example, all types of potential energy could be converted into kinetic energy when the objects are given freedom to move to different position (as for example, when an object falls off a support). When energy is in a form other than thermal energy, it may be transformed with good or even perfect efficiency, to any

other type of energy, including electricity or production of new particles of matter. With thermal energy, however, there are often limits to the efficiency of the conversion to other forms of energy, as described by the second law of thermodynamics.

In any and all such energy transformation processes, the total energy remains the same, and a transfer of energy from one system to another, results in a loss to compensate for any gain. Although the total energy of a system does not change with time, its value may depend on the frame of reference. For example, a seated passenger in a moving airplane has zero kinetic energy relative to the airplane, but significant kinetic energy (and higher total energy) relative to the Earth below.

The Energy Laws

The concept of energy and its transformations is useful in explaining and predicting most natural phenomena. The direction of transformations in energy (what kind of energy is transformed to what other kind) is often described by entropy (equal energy spread among all available degrees of freedom) considerations, as in practice all energy transformations are permitted on a small scale.

Certain larger transformations, however, are not permitted because it is statistically unlikely that energy or matter will randomly move into more concentrated forms or smaller spaces. Confused? Wait, this is just the beginning… This is, nevertheless, needed exercise due to the complexity of the subject discussed below.

We know that energy in a closed system, such as our Universe, can be neither created nor destroyed, since it is one of the building blocks of the physical universe. We call this condition *the first law of thermodynamics*, which is also known as the "law of conservation of energy," and which states that:

> *In a closed system, energy can*
> *be neither created nor destroyed.*

We must remember that the ultimate closed system is the Universe, which is isolated and self-contained, and so it is these facts that we base all our assumptions and calculations on. And so, none of the energy we use here on Earth can be destroyed, or created. It can only be inverted from one form to another, in the Universal sense of the matter at hand. On a more practical; level, a steam turbine, for example, undergoes many energy transfers and conversion processes to or from heat energy, but the overall energy balance is preserved.

In this case:

Energy in = energy out + energy losses

Energy in is measured by the amount and temperature of the incoming steam, while the energy out is measured in amount of electricity generated. The difference (energy in – energy out) is energy that has been lost due to inefficient equipment or processes. The lost energy is difficult to measure, so we can only calculate it indirectly.

One amazing fact is that at the end of a long chain of energy conversions, i.e. radiant energy conversion to chemical energy in the plants' photosynthesis process, the burning of the resulting biomass usually ends up in heat energy (and other energy forms) which is equal to the sum of all other energies (including energy losses) involved in the process.

This is a preliminary evidence of another and even more important law, called *the second law of thermodynamics*, which also rules the energy conversions. It is evidenced in the conversion of the stored chemical energy in coal to electricity through the operations in an electric power plant.

In coal power plant operations there are many instances of the conversion of different energy forms into heat energy. Basically all of the chemical energy in coal can be converted to heat energy. On the other hand all of the electrical energy carried through a wire at the end of the power plant conversion process can be converted back to heat energy as well (just like in electric hot water heaters).

The problem occurs when we try to convert heat energy back to mechanical energy. This type of conversion is controlled and limited by different laws of nature. It is here that the second law of thermodynamics is illustrated best and can be stated in several ways.

For example:

No device can operate in a cycle (like an engine), extracting energy from some source, and then converting it <u>completely</u> into mechanical energy.

And so, we can rephrase the first law as, "You can't get something for nothing," while the second law would chime in to complete the energy cycle by stating, "You can't even get all you see, or think you can get." The second law of thermodynamics accounts for the energy losses we see in all energy systems.

The second law of thermodynamics provides broad and far reaching consequences. It not only categorically denies the possibility of converting *more than a fraction* of the energy in the dam water or the steam (as in the steam turbine) into work, but it also assures us that the energy that escapes in the form of heat at all other conversion points (as waste heat, or waste mechanical energy) cannot ever be completely reclaimed for work. Energy losses cannot be avoided…and there are no exceptions.

This might seem confusing, but it simply means that the second law claims that in any transfer of energy or conversion of one to another form, the energy becomes less useful after every step and sometimes with every second. It is not lost, nor created, but is rather converted into other forms that have other functions—and not necessarily the form we expected, or wanted.

Note: Please note that energy becomes 'less useful' with every step, because part of it is converted into something we cannot use. For all practical purposes, this energy is 'lost' to us, because we cannot use it.

The energy is not lost, but has become useless to us. This is the foundation of our discussion on energy efficiency all through this text.

Practical Types of Energy

There are several major types of energy that we need to take a close look at, especially these that can be used to generate power on Earth, in order to understand them well, and see if we can do something to improve them for our use.

Energy on Earth can be divided into three major categories:

* *Gravitational energy.*
  This is the energy (or force) that attracts all objects, humans and animals to the Earth' center. Without it we would not be able to step on the ground, and would most likely be floating around in space…if we ma

* *Electromagnetic energy.*
  This is the energy emitted and absorbed by charged particles. It exhibits wave-like behavior as it travels through space. It has both electric and magnetic field components in equal proportions and force. In a vacuum it moves at the speed of light. sunlight is type of electromagnetic energy.

* *Nuclear energy.*
  This is the energy that is hidden in the atoms and molecules of the elements and substances that make this world. This energy plays a major role in chemical and other reactions on the molecular level.

We can also divide energy, as briefly mentioned before, into two important types, according to their state and the work they do, or intend to do:

- Potential energy is any energy that is stored and is idle; waiting to be released. A good example of potential energy is a large tank of water perched on a tall hill. The water in the tank is stored and idle, but has "potential" to do some work if and when released, at which point it will obtain kinetic energy. But until then its energy is "potential," or idle, for all practical purposes.

- Kinetic energy is energy in action. It is what moves things. Any object that is in motion has some amount of kinetic energy. The amount of kinetic energy at a certain time can be expressed as power.

In the more practical terms, there are several forms of energy that drive our lives, the main of which are:

- Physical (or mechanical) energy is the energy that moves things

- Chemical energy is actually a number of energy types that drive chemical reactions

- Electrical energy makes our lights, computers, TVs, etc. electrical appliances work

- Thermal energy, or heat, is used to elevate temperature of objects

- Biomass energy is generated by plants

- Heat energy is what we get when burning coal and other fuels

- Geothermal energy comes from hot springs deep in the ground

- Fossil fuels are the old energy sources; coal, oil and natural gas

- Solar energy is the energy we obtain by capturing and converting sunlight

- Hydropower is obtained from water turning a wheel, which generates energy

- Ocean energy is the energy generated by ocean waves and tides

- Nuclear energy generates heat by nuclear reactions

- Solar energy comes from the sun and can be converted into heat or electricity

- Wind energy makes the wings of a wind mill rotate to generate electricity

- Transportation energy facilitates moving large loads

- Magnetic energy is created by permanent magnets or electromagnets

- Sound energy is created by increasing the noise level

- Cosmic energy is the different types of energy contained in the Universe

Let's take a look at some of the major types that would help us clarify the concepts of energy and its use on Earth:

Physical Energy

Energy exists in its physical forms, albeit it the physical realm it is invisible and immeasurable in its purest form. When another component is added to it, however, such as force, mass, distance, speed, etc., it comes to life, and is very easily measurable. In conjunction with its other components (force and distance) it is truly a physical entity that we use in our daily lives. In those cases, energy becomes physical work, and can be measured as a physical unit. The work then could be observed, measured and expressed as heat, electric power, mass, speed, etc. variables.

For example, a large body of water sitting behind the secure walls of a dam contains energy which is hidden from view while the water is just sitting in the lake-reservoir. The energy is released immediately when the intake gate is opened and the water is allowed to run down the penstocks. At the end of its travel it hits the turbine blades with a great force to generate electricity.

Water hitting the turbine blades of a hydropower generator converts the potential energy stored in the lake into kinetic energy (work) and eventually into electric power. Thus generated power is measured in kilowatts (kW) and megawatts (MW).

Chemical Energy

Energy can be changed by rearranging the atoms in a molecule or by combining free atoms, at which point chemical energy becomes available. The reverse of the photosynthesis process that created the huge forests discussed above is a good example of that process. When the carbon, hydrogen, and oxygen molecules in carbohydrates and sugars (as in the fossils) are oxidized, either by burning or by the process of digestion, CO_2 and H_2O are formed and energy is released (such as when burning fossil fuels). Free carbon (in coal, or oil) combines with oxygen during the burning process to form CO_2 and release heat energy too.

Chemical energy can be generated also when strips of lead (Pb) are dipped into sulfuric acid (H_2SO_4), where the Pb replaces the H_2 in the H_2SO_4 to form $PbSO_4$ and H_2O (like in a lead battery). This process goes in reverse during charge and releases energy during discharge, which can be used to generate electrons in a wire connected between the electrodes, which are then extract in a form of electric current. This is the process of conversion of chemical potential energy into electrical energy, which takes place in a car battery. Many other combinations of metals, liquids, and gases can be arranged and forced to release energy in a similar way.

Chemical potential energy is one of our most important forms of energy known to man, since it is the energy of fossil fuels formation and use, as well as many other chemical reactions, which provide our food and power our society.

The fossil fuels are becoming too expensive to burn, however, so we must find other ways to generate electric energy, since it consumes the largest amount of fossils.'

Solar Energy

We now know how solar energy is created at the sun, where immense amounts of matter are converted to light and heat via fusion reactions of hydrogen converted to helium. The sunlight then traverses the 93 million miles to Earth in a matter of minutes, and arrives at our atmosphere as radiant energy contained in particles called photons. This is pure energy, riding on its own electromagnetic fields through space. The photons also contain potential energy which can be released upon impact.

Radiant energy (the photons) has properties that are wavelike, but at the same time it has particle-like properties. These wavelike, particle-like units are delivered in discrete chunks that have two distinct properties; wavelength and frequency.

The wavelength is measured by the distance between the peaks of the waves, while the frequency is the speed of its vibration, which, in turn determines its energy. Higher the frequency of the photons, higher the amount of energy they carry. For example, X-ray photons are vibrating at very frequencies, and carry enough energy to do physical damage to body tissues and organs. At the same token, radio waves vibrate several times slower, and do no damage.

Sunlight radiation waves have a wide range of vibrations contained within, and carry enough energy to cause chemical changes in molecules that they encounter in the way. Most of the incoming radiation is visible light, and life forms have developed in response to those wavelengths. The reflected, or converted solar radiation leaves Earth at the much longer infrared (IR, or heat) radiation.

Radiation Energy

Radiant energy interacts with matter by, 1.) reflection from surfaces it encounters on its way, 2.) absorption, where it sets molecules or atoms into vibration, which could cause re-radiation in the same or at longer wavelengths, 3.) absorption and then dissipation within a substance as heat, and 4.) absorption and production of a chemical change.

Heat absorbed by land provides the major heat input into the lower atmosphere. It is this converted solar energy that provides the kinetic energy for the great trade winds and all the other winds of the "atmospheric engine." Windmills are driven, indirectly, by solar energy which creates wind conditions. The combination of unequal heating, the Earth's gravity, and the spin about its axis are responsible for
the atmospheric and oceanic motion.

Much of the solar technology depends on this radiant-to-heat energy conversion. Solar collectors for space and water heating, and other solar collectors rely on this interaction.

Radiant energy absorbed by the ocean and converted to heat ultimately provides humankind with another useful form of energy. If water molecules receive enough energy to become hot enough, they can escape from the liquid. This is the process we call evaporation. This water vapor is then lifted by rising heated air and is carried by the winds until it falls as rain. By this lifting of great masses of water away from the surface of the Earth, some of the solar energy is turned to gravitational potential energy.

Some amount of this gravitational potential energy is converted to mechanical energy when water is forced to run through a turbine on its way back to the sea. Thus, hydropower is a second indirect consequence of solar energy.

The selective absorption of radiant energy in the atmosphere (interaction 2 above) also has important consequences. Without the absorption at the top of the atmosphere of most of the ultraviolet radiation and other short-wavelength radiation life on Earth would be impossible. If this energetic radiation were not stopped by the thin layer of ozone there, no living thing could exist in the sunlight.

Nuclear Energy

The nuclear energy we discuss here is the energy that is stored in the nuclei of atoms and can be released by rearrangements of the protons and neutrons in it. One example of this energy are the thermonuclear reactions in the sun (fusion), as well as the atom bomb (fission).

In fusion reactions, the lighter elements (with nucleus made up of a few protons and neutrons) can combine to make heavier ones, during which process great amount of energy is released, as four protons fuse to make a helium nucleus, which contains two protons and two neutrons. Since the helium nucleus has less mass than the combined mass of the four protons, the mass difference is converted to energy.

Energy can be released in fission nuclear reactions too, by elements with heavy nucleus, such as uranium. It can be split into two medium-mass nuclei, during which a lot of energy is released. The two medium-mass nuclei have less mass than the original uranium and in this reaction the missing mass is also converted Into energy.

For our everyday energy generating purposes we must be able to control the nuclear reactions. Thus far we have learned how to control the fission nuclear reactions, which are used in our many nuclear power plants. We, however, still don't know how to efficiently initiate and safely control the fission nuclear reactions. This challenge is going be left for the next generations of the 21st and 22nd century to resolve. And no doubt they will, since it might be the only choice for energy production they'd have by then.

Gravitational Energy

Energy is contained in and released during moon's travel around the Earth. Potential energy, in form of gravitational energy, is stored in the space between the Earth, moon and sun. At the same time all three bodies have a lot of kinetic energy, due to their rotation around each other and around their own axis. Nevertheless, there is a lot we still don't know about gravity as a force on Earth. There are thousands of books published and that many theories expressed on the subject, but gravity is still a mystery for the most part.

We observe the effects and can calculate the results of its presence, the most obvious of which is our presence and movement on Earth, driven by gravity forces. Without gravity, we, and everything around us, would fly in orbit around the Earth…and yet we don't even know how this magical force, that keeps our feet to the ground and cars on the roads, works.

We also see the effect of the Earth-moon gravitational forces, which are converted into awesome mechanical energy by the mechanism of ocean tides creation. Tides are formed as a results of the gravitational push-pull mechanism of the moon, and to a lesser extend by the sun, during Earth's rotation. Since the oceans' surface is flexible, the waters beneath it can move up and down according to the moon's position. The converted energy, created by the interaction between the Earth and moon gravitational forces, causes the rise and fall of the tides, while some is dissipated as heat energy through friction.

The possibility of converting waves and tidal energy into useful electric energy has been on the agenda for many decades. Turbines can be turned by letting seawater run through them as it comes and goes in and out with the tides, or up and down with the waves. Such projects are feasible in a number of places, but technical difficulties and expense have not allowed their wide deployment.

This is just another unexplored, and yet very plausible, alternative for the next generations to work on. And they better be ready for it, because they won't have most of the less complicated and less expensive energy choices we have now days.

Geothermal Energy

Geothermal energy is one of the most interesting and elegant solutions to energy generation. The nucleus of the Earth consists of molten lava and some of the heat leaks through cracks in the Earth' core and crust and escapes to the surface by forming hot springs, geysers, or volcanoes. This is free heat in its basic form—lots of it and ready to use.

The Earth's heat can be captured as hot water, steam, or high-pressure steam. Different energy generators are used in each case, with the overall result of using this free (waste) energy to generate useful electric energy. In some places of the world, the hot water or gasses are used to heat homes, or in commercial applications, but these are exceptions.

Geologists are not exactly sure how the Earth's molten core was formed, but we now know that the inner heat is maintained by the slow decay of radioactive materials in it. The concentration of radioactive material in the Earth is small, but yet enough to balance the heat loss through the surface.

As we, however, remove significant amounts of the radioactive materials in the Earth's crust, and as we use more and more of its heat, it remains to be seen if geothermal energy would remain renewable energy for very long. Increasing the use of geothermal energy complicates the environmental picture, and seems to be a knife with two edges, so we must be careful using it.

Figure 1-4. Geothermal energy wells

The next generation will be forced to take a closer look at the inner and outer heat energy balance of the Earth and decide if this energy source can be expanded to serve their daily needs.

Wind Energy

Wind power is another form of solar energy, where billions of photons arriving from the sun hit the Earth' surface and the air around it and the energy in them is converted into mechanical and heat energy. Thus heats the air molecules to the point where they start rising up. When enough air is warmed and moved up, an air current is created, which interacts with other environmental elements and surface structures and creates turbulence. Increased turbulence gives rise to wind currents, which can travel at a high speed across great distances.

Wind energy is a major electric power generator in the world now days. Large scale wind farms are built in many areas of the world, and provide significant portion of the global commercial electric power.

Cosmic Energy

There is a lot of radiation in the Universe that bombards everything in its way. Celestial bodies, such as stars, gas giants, and black holes are the best examples of massive radiation. Stars, Planets, and Black holes are continuously producing high energy radiation, which in most cases comes from nuclear fusion. The combining of separate atoms produce high levels of energy that are dispersed into space in the full electromagnetic spectrum. Gas giants can also produce large fields of radiation. This is because most gas giants are celestial bodies that failed to fully become stars not having sufficient

mass to complete fusion.

There is also radiation from the universe as a whole, which is called cosmic microwave background radiation. Luckily, each of the cosmic radiation sources is far beyond the safety of Earth's atmosphere. Unfortunately, in order to be of use to people, the cosmic energy must be captured and controlled, something we are simply unable to do at this point.

Nevertheless, cosmic energy is available in immense quantities in high space, so the next generations will be charged with the task of figuring out how capture it, and to generate useful power for their use.

Energy Effects

The atmosphere does not absorb very much radiation at the visible wavelengths, since it is transparent to them. When this radiant energy is absorbed at the Earth's surface, however, it is reradiated at longer wavelengths, in the infrared region.

CO_2 and other gasses act as a barrier which traps some of the wavelengths, which helps keep the Earth warmer at night. Without this effect, the Earth would radiate back out into space most of the energy gathered during the day, thus the nights would be very cold. Instead, the absorbed energy is reradiated so that about half of it comes back to Earth, thus maintaining a much milder night temperature.

This "green house effect" is quite sensitive to the content of the type and amount of gasses in the barrier layer, and we'll discuss it in more detail in the following chapters.

The process of photosynthesis is another interaction that provides not only food and fossil fuels, but also the resource of biomass in general, and the life-giving things; wood, water, grains, etc., which we increasingly count on to provide food and energy.

The photovoltaic reactions are another example of this interaction. In the text below, we will take a very close look at this process, as well as some of the new ways to get usable energy from sunlight in general.

A number of other effects are important, when painting a full picture of the energy cycles. Some of these are the shielding action of ozone (O_3), where it is formed when photons strike O_2 molecules on top of the atmospheric layer to make an additional layer of O_3. where it is beneficial to the Earth's environment. O_3 is also produced in the lower atmosphere by a similar reaction, but there it is harmful to humans.

Power, Force, Work, and Heat

Matter can have energy in many different forms.

The three major forms of potential energy are gravitational, nuclear, and electrical. The kinetic or transient forms of energy are the mechanical forms of mass in motion, moving along straight or curved paths, and its other forms; heat and radiation.

Energy in its potential form is latent; it is there but it isn't going anywhere, nor doing anything. In order to do something with this energy, to use it to move something, change the temperature of a substance, provide light or some other form of radiation, etc., the potential energy has to be converted into a kinetic energy form, which is then expressed as power and can be measured.

There are many energy conversion mechanisms. Burning (or oxidation) of coal in oxygen rich atmosphere is a major conversion mechanism. Here solid coal burns to generate heat and CO_2. This heat can then be used for power generation, or heating in domestic of industrial applications.

$$C + O_2 = CO_2 + heat$$

The process is similar when burning natural gas, or crude oil, except that their molecules are much more complex. The final products, heat and CO_2, however, are similar too.

We have already mentioned a number of processes and devices (i.e. steam turbines) that convert the energy contained in a substance (steam) into mechanical energy (turning turbine blades and generator rotor), where the mechanical energy is converted into electric power.

Energy can also be transferred from one place to another. Electric energy moves long distance via wires. Heat energy heating a substance changes its state (cold to warm) and/or shape (melting iron or ice). It can also move, or lift substances by means of a physical contact with steam (as in a steam turbine), or with wind (as in a wind turbine).

Energy is transferred from its potential form into kinetic form as work (mechanical energy), heat, or radiation. The practical energy exchange and the corresponding changes of state, or transformation into another type of energy, are best represented by the variables of power, work and heat.

Power

In the real world, power is the rate at which energy is transferred, used, or transformed. The unit of power is the joule per second (J/s), known as the watt (in honor of James Watt, the eighteenth-century developer of the steam engine). For example, the rate at which a light bulb transforms electrical energy into heat and light is measured in watts—the more wattage, the more power, or equivalently the more electrical energy is used per unit time.

Energy transfer can be used to do work, so power is also the rate at which this work is performed. The same amount of work is done when carrying a load up a flight of stairs whether the person carrying it walks or runs, but more power is expended during the running because the work is done in a shorter amount of time.

The output power of an electric motor is the product of the torque the motor generates and the angular velocity of its output shaft. The power expended to move a vehicle is the product of the traction force of the wheels and the velocity of the vehicle.

The average power P_{avg} over a period of time is given by the formula

$$P_{avg} = \frac{\Delta W}{\Delta t}$$

where:
ΔW is the amount of work performed
Δt is the period of time of duration

It is the average amount of work done or energy converted per unit of time. The average power is often simply called "power" when the context makes it clear. The instantaneous power is then the limiting value of the average power as the time interval Δt approaches zero.

$$P = \lim_{\Delta t \to 0}{}^* Pavg = \lim_{\Delta t \to 0} \frac{\Delta W}{\Delta t} = \frac{dW}{dt}$$

In the case of constant power P, the amount of work performed during a period of duration T is given by:

$$W = PT$$

In the context of energy conversion it is more customary to use the symbol E rather than W.

Note: The rate of energy conversion is critical. As an example a medium size coal power plant produces as much heat as that released in an instant by a small a-bomb. The difference here is the rate of release. Because the a-bomb instantaneous reaction releases energy much more quickly, it delivers far more power than the coal power plant.

The basic unit for power is the Watt, which subdivisions are:

Table 1-2. Power denominations and equivalents

Units		Equivalent to power in:
Picowatt	$pW = 10^{-12}W$	human cell
Nanowatt	$nW = 10^{-9}W$	Microchip
Microwatt	$\mu W = 10^{-6}W$	quartz wrist watch
Milliwatt	$mW = 10^{-3}W$	laser beam in a CD player
Watt	W	light bulb, house appliances in general
Kilowatt	$kW = 10^{3}W$	propelling engines in general
Megawatt	$MW = 10^{6}W$	power of train engines and power plants in general
Gigawatt	$GW = 10^{9}W$	large hydropower plants capacity, average power consumption in a country
Terawatt	$TW = 10^{12}W$	average world power consumption, global annual energy production by photosynthesis in the world
Petawatt	$PW = 10^{15}W$	solar power received by the Earth

The integral of power over time defines the work done. Because this integral depends on the trajectory of the point of application of the force and torque, this calculation of work is said to be "path dependent."

Note: Although in theory there is a significant difference between energy and power, at times we use them interchangeably in this text, realizing that this might cause some learned folks headache and confusion. We do emphasize, however, that energy is a measure of the potential to do work, while power is the expression of that work.

For example, we say, "The sun's energy is converted into electric power by the PV effect. Both, the incoming solar energy and the generated electric power are measurable entities, but are in different forms and shapes." Nevertheless, we can replace power with energy as the end result, thus saying, "The sun's energy is converted into electric energy by the PV effect." While theoretically incorrect, electric energy is a popular term, which is more convenient for use at times.

Force

Force is an act that causes an object to undergo a change in any of its possible conditions; be it its movement, direction, shape, or structure. Force is what causes an object with a certain mass to start moving. It can also cause it to change its velocity if it is already moving—or to accelerate. In simple term this could be described as push and pull.

In physics, the Newton's second law is used to clarify that the net force acting upon an object is equal to the rate at which its momentum changes. Or, acceleration of an object is directly proportional to the net force acting on it; is in the direction of the net force, and is inversely proportional the mass of the object.

So, force has both magnitude and direction, making it a vector quantity. It can be measured with the SI unit of Newtons (N) and represented by the symbol F.

Mathematically force in motion is expressed as:

$$\vec{F} = ma$$

where
   m is mass
   a is acceleration
   → implies a vector quantity with both magnitude and direction.

Force includes also the concepts of thrust and drag (increase and decrease in the velocity of an object respectively), and torque (or change in rotational speed of an object).

Force acting in an un-uniform way on parts of a physical body causes mechanical stresses, which can also move or deform the body. Force, or mechanical stress in a liquid or gas causes changes in their pressure, volume, and temperature.-

Force is also a defining variable in calculating the amount of work done by, and/or within, a system.

Gravitational Force

Gravitational force, or gravity, is a natural phenomenon, where physical bodies attract each other with a force proportional to their masses. It is the invisible force that keeps our solar system together, where the sun's large mass attracts all celestial bodies, making them rotate around it. Gravity is also responsible for the moon's rotation around the Earth, which in turn create tides, natural convection, and a number of other phenomena.

It is also the force that gives weight to objects and people on Earth. It keeps them on the Earth's surface, instead of floating somewhere in space, and causes them to fall to the ground when elevated above the surface and dropped.

The gravitational force acts upon all objects—small and large. It itself, however, is very small, and since the mass of the object has a lot to do with the overall interaction, we can notice the gravity force of large bodies, like the Earth, moon and the sun. Which is also why, the smaller objects are attracted by the larger ones, as is the case of the Earth being attracted by the sun and forced to rotate around it at infinitum. It is also why, all objects and people on Earth are attracted by it and stick to its surface like little pins onto a large magnet.

Gravity is also the force that causes dispersed matter to coalesce, and coalesced matter to remain intact. This accounts for the existence of planets, stars, galaxies on one hand, and most of the microscopic and macroscopic objects around us, on the other. A truly awesome force that is all around, and constantly acting upon, us and everything around us. Yet, we don't fully understand it, and seldom see it, or its effects.

Mass vs. Weight

This might be a good time to point out that although we use mass and weight interchangeably, there is a big difference between them. While mass represents only the amount of matter (stuff) that is contained in an object, weight takes into account also it gravitational state and the related forces acting upon it.

As an example, a bowling ball with a weight of 10 kg means that it has that much mass in its structure as it sits idle. As a matter of fact, the true name for mass is slug. Really!? But by adding the gravity variable to the concept of idle mass of the bowling ball, we get its actual weight, which is also its potential energy. Since gravity acceleration constant is 9.8 m/s^2, that multiplied by the 10 kg of the ball, we get 98 kg.m/s^2.

Or our bowling ball with 10 kg of mass, therefore, actually weighs 98 N (Newtons) on Earth. This is useful to know, because when we get to the moon someday, our 10 kg bowling ball, will actually be 1/6 as heavy, and will weigh only 11.3 N, or 1/6th of its Earth weight—although its original mass (shape) remains the same.

This way we can distinguish between 10 kg mass on Earth and 10 kg mass on the moon, and thus make accurate calculations between our lives on different places with different gravity forces. This will get increasingly important as we move to live on the moon or Mars someday. BTW, you can obtain tickets for flights to the moon or Mars by calling us. Soon we will have tickets for flights to the sun too.

But seriously, and in more practical terms, the bowling ball with 10 kg mass will also weigh more (albeit not that much more) on the bottom of the Grand Canyon, than on top of mount Everest. This is because on the bottom of the Grand Canyon it is 4-5 miles closer to the Earth' center than on top of Everest. Gravity woks! Although we still don't know how ...embarrassing

Work

Work is defined as the act of transfer of energy that is accomplished by the application of a force over a distance. I.e., a car engine does work by moving the car down the road, or pumping water up into a tank, and also the water doing work when it flows back down.

Note: The difference between work and power is that:

a) Work is the act of energy transfer, while

b) Power is the rate of that same energy transfer process.

And so, our steam turbine does work (spinning) when hit by the steam blast, which determines the rate of electric power it will generate.

We can distinguish two kinds of work, processing work and storage work. In processing work, the initial energy is completely converted to kinetic energy. In this case matter is moved about, changed, and rearranged.

In storage work, some potential energy is converted to kinetic and then reconverted to potential energy. The automobile is doing processing work when running down the highway, while the pumped water in a dammed lake is an example of storage work.

In all these examples work is accomplished by the application of a force, a push or a pull. The force may not be obvious (like sunlight falling on solar panels), but we know that it is there when we see the evidence of the work done (electricity generated in this case).

Work is done by one part of a system on another part of the same or a different system. It is a measure of the change of energy of a part of the system. Work is done by that part of a system which exerts a force. Work is done on that part of a system on which a force acts. Work is often also done against a force.

Processing work is done against the resisting forces such as friction or air resistance. It is also done against inertial forces, the resistance of mass to being accelerated or decelerated.

The important part is that work is a measure of the change in energy when energy is transferred from one part of a system to another by the action of a force.

Heat

By definition, heat is the form of energy that flows between two bodies due to their temperature differential. Heat is basically energy always in transit between a warmer and cooler body, such as hot gases in the power plant furnace turning the water in the boiler to hot steam. This phenomena is due to the higher energy of the hotter molecules. It is also called "internal energy" and can have many forms.

Internal energy is in fact the energy stored in a system at a molecular level. It has no practical way of measurement, so it is usually calculated by measuring macroscopic variables, such as volume, temperature, pressure and composition. It cannot be absolutely calculated as a function, and is only used in relation to an initial state of reference.

Heal energy, as internal energy, is also described in terms of a random motion. It can be a more complicated vibratory motion of the molecules or atoms of the substance and thus a mixture of kinetic and potential energy. A broader definition of internal energy includes all forms of energy a substance can have at a certain time and state.

Heat Types

Heat comes in different types and forms. Some of the major types are:

- Heat conduction
  During heating, the "hotter" molecules flow towards and collide with the "cooler" ones and immediately transfer energy to them. Thus, some energy leaves the hotter substance (or part of it), thus leaving it with less energy—or cooling it—as it enters the cooler substance and warming it. This is the process of heat conduction.

- Heat convection
  The other two mechanisms of heat transfer are convection and radiation. In convection matter actually moves while carrying heat energy with it, as when hot water rises from the bottom of a kettle on a hot stove.

- Radiation process is when heat energy is changed into its electromagnetic form (light, infrared, etc.)

and is carried away. In all these types of energy transfers, however, the net flow of energy must still be from a hot body to cooler surroundings.

One very important substance in the heat process is the gas that is generated in most cases during heating, It is the hot gas, such as the steam in the steam engine, or the hot gases in a burning coal furnace, that do the actual work. Gas is also the simplest state of matter to understand, so its most important properties can be deduced from a model that treats the gas as a substance made up of identical, non-interacting (no forces other than contact forces between them) spheres. These spheres collide elastically with each other, with the container walls, or with another matter, and thus transfer their energy during impact.

In a simple gas, the most important form of internal energy is this random molecular (fire) or atomic (plasma discharge) motion of the heat energy. The hotter a substance is, the more energy its molecules contain. The hotter it is, the faster its molecules move, because the kinetic energy of motion increases with the velocity or speed of the moving matter. Heat energy, therefore, in the broad definition that we are using here, is related to the average kinetic energy of the moving molecules of the substance.

Heat is often created by combustion of matter, and is defined as a chemical exothermic reaction in which oxygen is combined with some other element, releasing heat in the process. Since natural fuels, such as coal, oils, natural gas, and biomass contain carbon. The combustion, or oxidation of the carbon molecules occurs during the reaction,

$$C + O_2 = CO_2 + 94.03 \text{ kcal of heat.}$$

Carbon has atomic weight of 12, so during burning 7.8 kcal per gram of carbon are produced.

Natural fuels contains carbon and hydrogen molecules, the additional combustion, or oxidation, of which is $H_2 + O = H_2O + 68.37$ kcal. Each gram of hydrogen produces 34.2 kcal of heat, which is four times the heat produced by the same amount of carbon.
This is one of the reasons natural gas and oil products have higher energy content than coal. This gives another important advantage of these fuels—they are more compact and easier to transport.

Note: Some natural fuels are partially oxygenated, which means that they contain oxygen in their molecule, so they normally lose (waste) part of the energy that they could have produced during combustion.

Table 1-3. Fuels heat content

Heat source	Heat (kcal/g)	Btu
Natural gas	13.2	52.4
Oil	10.0	39.7
Fat	9.1	36.1
Coal	7.8	31.0
Ethyl alcohol	7.1	28.2
Proteins	5.7	22.6
Sugars	4.1	16.3
Wood	3.3	13.1

Table 1-3 shows the heat content, in kilocalories and Btus, obtained when burning just one gram of these natural fuels. The data tells us that natural gas has the highest heat value (or heat energy released during burning). Natural gas is the only natural fuel sold in a gas form, and one gram of liquid natural gas (LNG) creates a lot of energy and some exhaust gases.

Wood, on the other hand, has a fairly high heat content, but in its natural, freshly cut, form it contains 50-80% water. Burning such wood lowers its heat value significantly, because a lot of the heat is used to boil and evaporate the water content. Oven dry wood, which is sold at a premium, has a much higher heat value; close to that of coal, but the drying process requires a lot of energy, which in effect reduces the overall heat value of the final product, which in turn increases its price significantly.

A different form of energy is that generated and used by humans and animals. Fats, proteins, sugars and starches are contained in most of the foods we eat. When they are ingested and 'burnt' in our body, they are converted into CO_2 and water, and generate energy which we can use for our bodily functions. The conversion process is different, but the final result is energy and waste products, all of which can be measured.

And speaking of measuring energy…

Energy Measurements

The key types of energy used frequently in this text are mechanical, thermal, and electric energy.

Mechanical Energy

The universal energy unit is the joule (J), named after Sir James Prescott Joule. It was derived from his hands-on experiments and calculations on the mechanical equivalent of heat energy, where 1 joule is equal to 1 Newton-meter and can be expressed as:

$$1J = 1kg \left(\frac{m}{s}\right)^2 = 1\frac{kg/m^2}{s^2}$$

where
  kg is in kilograms
  m is in meters, and
  s is in seconds

The US units for both energy and work are: the foot-pound force (or 1.3558 J), the British thermal unit (Btu, or about 1055 J), and the horsepower-hour (or 2.6845 MJ).

Mechanical energy of an object consists of two primary components; kinetic and potential energy.

Or,

Em = U + K

Where
  Em = mechanical energy
  U = potential energy
  K = kinetic energy

This is simply the energy that propels, or is contained, in objects and which can take many diverse forms, such as, collisions, gravity, jet planes, ocean waves, rivers water, rolling and sliding, rotation, pendulum motion, projectile motion, pulling and pushing, steam turbine, sound energy, walking.

There are also a number of different measurement units for the different types of mechanical energy, most of which are not part of this text, so we will take just a look at the major ones.

Units of Work and Mechanical Energy

Work is defined as force times a distance. In SI units the work unit is Newtons times meters or joules. A force of one Newton operating over a distance of one meter produces one joule of work.

1 N x 1 m = 1J

In the English system, the force is measured in pounds and the distance in feet, so the unit of work is the foot-pound (ft-lb).

1 ft x 1 lb = 1 ft-lb.

This unit is important in the engineering practice, but it is not used much in energy calculations.

Work is act of energy change. According to the first law of thermodynamics, the change in total internal energy of a system equals the added heat minus the work performed by the system.

$\delta E = \delta Q - \delta W$

Also, from Newton's second law for rigid bodies it can be shown that work on an object is equal to the change in kinetic energy of that object,

$W = -\Delta KE$

The work of forces generated by a potential function is known as potential energy and the forces are said to be conservative. Therefore work on an object moving in a conservative force field is equal to minus the change of potential energy of the object,

$W = - \Delta PE$

This shows that work is the energy associated with the action of a force, and so has the physical dimensions and units of energy.

The work energy principles apply to mechanical, chemical, and electrical types of work.

Thermal Energy

The SI *heat energy* unit is the calorie (cal) or kilocalorie (kcal), or the amount of heat energy needed to raise the temperature of 1 kg of water (at a specific temperature and pressure), by 1°C (degree Celsius). It is also the traditional measure of the energy value of foods.

In the *English system* the amount of water is measured in pounds (lbs) and the temperature rise in degrees Fahrenheit (°F), while the heat energy is the British thermal unit (Btu.). Btu stands for British Thermal Unit and is the amount of energy required to raise the temperature of 1 lb of water (at standard temperature and pressure) one degree Fahrenheit. In practical terms, we can think of a Btu as approximately equivalent to the heat given off by burning one match head. A Btu is equivalent to 1055 Joules, and from this follows that a therm is about 105,500,000, or 105.5 million Joules.

Home Heating

The basic unit for thermal energy in home heating applications is the "therm," which is defined to be 100,000 Btus:

1 therm = 100,000 Btus.

To get a feeling for how much energy a therm is, a home furnace is typically rated at somewhere around one therm per hour. Typical annual heating loads, for a house with 1,800 ft2 is about 400 therms in Phoenix, Arizona, 1,500 in Santa Fe, New Mexico, and 1,800 in Great Falls, Montana. This is because Phoenix has no winter weather per se, New Mexico has some, while Montana

as a northern state experiences very cold winters.

Comparing this with the annual energy from the sun falling on the same surface area in these cities, we see that in Phoenix, Arizona, it is about 13,000 therms, Santa Fe, New Mexico, 12,000, while in Great Falls, Montana, it is about 8,000-9,000 therms.

It can be seen that there is relatively abundant solar energy for heating homes, even in northern state like Montana. Thus, we also see that a homes in Phoenix and Santa Fe should be designed to capture about 10% of the available solar insolation, while most homes in Montana should capture much more.

Buildings that are built to capture just the right amount of solar energy without overheating or other negative effects, have deployed successful passive solar design.

Fuels Energy

The thermal energy value of fuels such as coal and oil are usually given in Btus. As a result, most of the energy data in English-speaking countries are presented in Btus. The calorie, watt, horse power, ton of oil equivalent, and joule units are used also in some scientific measurements and for simplicity in some conversions and calculations.

Table 1-3. Key energy conversion units

Unit	Measure
1 joule (J)	1N.m
	1kg.m/s^2
	0.2388 cal
	9.4782^{-4} Btu
	2.7778^{-4} Wh
	107 ergs
1 calorie	4.1868 J
1 kWh	3.6 x 10^{13} ergs
	3600 kJ
	860 kcal
	8.6-5 toe *
1 toe*	1010 cal
	41.8 GJ
	11.63 MWh
	1.28 tons coal equivalent
	39.68 million Btu
1 million Btu	1.0551 GJ
	2.52^{-2} toe
	0.2931 MWh
1 Watt	1 J/s
1 HP	746 W
1 GWh	86 toe

Note: toe means tons of oil equivalent

Electric Energy (power)

The International System of Units, or metric system, also form the basis for the electrical units we use in the U.S. and most other countries around the globe. When we talk about electric power, used for powering appliances in our home, for example, we are not usually interested in how much energy an appliance uses per se, but rather the rate of energy use. With other words, we need to know how much energy per unit time the appliance draws. This quantity is called the "power."

Power is thus defined:

Power = Energy / Time.

In the metric International units, the unit of power corresponding to 1 joule per second is called a watt:

1 watt = 1 joule per second.

Because the voltage V tells us the number of joules per coulomb, and the current I tells us the number of coulombs per second, all we have to do to get the current is to multiply them:

Power = number of joules/second = joules/coulomb x coulombs/second = I V,

or

P = I V = W

For measuring electrical power we use the unit measurement of "Watt"

1 Watt = 1 Joule/Second

Electric power is usually calculated in terms of amps and volts. Voltage, measures how much electrical energy is delivered if a certain charge, or the number of electrons transmitted through an electric circuit. The number of electrons is measured with the unit of a Coulomb, which consists of 1.6×10^{19} electrons.

Amps are a measure of how many coulombs per second are being transmitted, which is call the current. Thus, a current of one amp in a wire means exactly 1.6×10^{19} electrons per second are flowing past any given point in the wire.

A voltage of 1 volt means that 1 joule of energy is delivered for each coulomb of charge that flows through the circuit. Since current is the number of coulombs per second, and power is the number of joules per second

(watts), we see that

Power = Joules/Second = (Joules/Coulomb) x (Coulomb/Second) = volts x amps = number of Watts, or:

P = J/s = J/C = J/C x J/s = V x A = W

For example, a hair dryer rated to draw 1000 watts of power from an U.S. 110 V wall outlet will produce:

A = P/V 1000 watts/110 volts = 9.1 A

The power of an electric circuit of 110 volts and a current of 5 amps expressed in watts units of power would be:

110 V x 5 A = 550 W

Note: We must distinguish between power and energy. Although closely related, they refer to different things. While power is the rate at which energy is delivered, it is not the amount of energy itself.

So,

Energy = Power x Time.

For example, a 100 watt light bulb converts 100 joules of electrical energy into 100 joules of electromagnetic radiation (light) every second it is on. 100 watt light turned on for one hour (3600 seconds) gives us

(100 Joules/Second) x (3600 Seconds) = 360,000 Joules

Watt is a convenient unit to use when working with electrical devices, but there are also times when we are want to know the total energy use. Such as when calculating the monthly utility bill. Working with Joules in this case would be awkward, because we get such large numbers. So, another unit, called "kilo-watt hour," or kWh, is used instead.

1 kWh = 1000 watts of power per hour

This is the amount of energy you would use to run a typical hair dryer for one hour. Or in the above example it is 100 watts per hour, 100 Wh, or 0.1 kWh.

The unit used to represent everyday electricity, as in utility bills, is the kilowatt-hour (kWh), where one kWh is equivalent to 1,000 Watts used during one hour, or about 3600 kJ. Another useful unit is that of kilowatt-hours per year (kWh/yr), which is actually the average electric energy (power) consumption, or the average rate at which energy is used during the period of one year.

Thermodynamic Systems

A short description of some key energy concepts follows, in order to bring us all under a common denominator.

Exothermic and Endothermic Heat Processes

In thermodynamics (the science of heat), the term exothermic, from the Greek's outside heating, describes a process that releases energy from the system, usually in the form of heat, but also in the form of light (e.g. a spark, flame, or explosion). It can also produce electricity (as in a battery), or sound (such as when burning hydrogen). Explosions are some of the most violent and most noticeable exothermic reactions.

Exothermic refers to a transformation of energy in which a system releases energy (heat) to the surroundings: $Q < 0$, when the transformation occurs at constant pressure, $\Delta H < 0$, and constant volume $\Delta U < 0$.

In a closed system that does not give off heat to the surroundings, such as water boiling in a pressure cooker, the exothermic process results in an increase in internal temperature, and in this case in rise of pressure as well.

The opposite of an exothermic process is an endothermic process, one that absorbs energy in the form of heat as needed to complete a reaction or work activity.

In chemical reactions, the heat that is absorbed is in the form of electromagnetic energy. The loss of kinetic energy via reacting electrons causes light energy to be released. This amount of light is equivalent in energy to the stabilization energy of the energy that is needed to complete the chemical reaction, i.e. the bond energy. The light that is released can be absorbed by other molecules in solution to give rise to molecular vibrations or rotations, which results in heat generation—and in the end represents an exothermic reaction, which has negative enthalpy ($\Delta H < 0$), see enthalpy below.

In endothermic reactions, in order for the process to be completed, energy is absorbed to place an electron in a higher energy state, such that the electron can associate with another atom to form another chemical complex, thus completing the process. The loss of energy within solution is absorbed by the endothermic reaction and therefore is a loss of heat. Therefore in an exothermic reaction the energy needed for the reaction to occur is less than the total energy released, and the system enthalpy at that point is positive ($\Delta H > 0$), see enthalpy below.

Enthalpy

Enthalpy is another important variable, which is basically a measure of the total energy of a thermodynamic system. It is a measure of the thermodynamic potential of the system. It includes the internal energy, which is the energy required to create a system, and the amount of energy required to make room for it by displacing its environment and establishing its volume and pressure. This can be figuratively seen reflected in the behavior of water vapor in a water boiler. As the vapor goes through the boiler's piping, it carries its energy to another place; be it a space heater or a steam turbine. At this point there are no transfers of energy and no chemical reactions. It is simply a physical transport of a heated mass to point of use.

The total enthalpy, H, of a system cannot be measured directly. Thus, change in enthalpy, ΔH, is a more useful quantity than its absolute value. The change ΔH is positive in endothermic reactions, and negative in heat-releasing exothermic processes. ΔH of a system is equal to the sum of non-mechanical work done on it and the heat supplied to it.

Enthalpy is the preferred expression of the energy changes in chemical, biological, and physical measurements, because it simplifies the description of different energy transfer mechanisms. This is so, because a change in enthalpy takes account of energy transferred to the environment through the expansion of the system under study.

Supposing a closed system, we then see the system's enthalpy combining the internal energy of the vapor and variations of pressure and volume. As with internal energy, it does not have an absolute value and only its variations (ΔH) can be calculated in relation to a state of reference.

In exothermic reactions, those that generate heat, the enthalpy is negative ($\Delta H < 0$). A practical example is the formation of water from hydrogen combustion. In an endothermic reaction, in turn, there is heat absorption, and thus with positive enthalpy ($\Delta H > 0$). A practical example is that of water that vaporizes at the base of a waterfall. It turns into a gas with heat absorption generated by the kinetic energy of the falling water at the moment of impact.

Energy consumed as heat during a chemical transformation under constant pressure may be defined as reaction enthalpy and may be measured by a device called a 'calorimeter,' or calculated by the difference between the formation heats of products and of chemical reagents. However, understanding and interpreting the reaction heat (enthalpy) for a particular transformation requires understanding how chemical energy accumulates in matter.

Enthalpy is also measures by the joule, and/or the other units (Btu, calorie, etc.) as discussed above.

Energy Disorder

There is a tendency in all energy conversions to change from some order to increasing disorder. Looking at hot water molecules in a boiling water tank and steam column we can see them constantly bumping in and colliding with each other, and with the walls of the container, and moving chaotically in all directions. Their motion is totally randomized. to where we draw the conclusion that it is disordered, with no apparent preference for moving in one direction over any other. Heat energy is a disordered type of energy, and so the water molecules in the steam move randomly.

On the other hand, there are examples of much ordered forms of energy? One of these is the flow of electric current in a wire, or water running in a pipe. Under ideal conditions (no resistance in the wire and no roughness in the pipe walls) the electrons and molecules are lined up and moving in a perfect order.

In the real world, however, both of these cases some disordered motion is expected and usually observed. The electrons in the wires and the water molecules in the pipe would be bumping in the walls, bouncing in whichever way and off their neighbors. In both cases, however, the ordered motion is evident and is superimposed over the disorder.

As time goes on, the order gets more disordered. In the wire example, the wire gets heated, the resistance increases and the electron flow becomes much harder and almost random, until it is used in an electric light or motor, where it is converted into much more disordered heat, light, or mechanical work states of energy.

The water in the pipe cannot stay forever in also, and usually ends up splashing in a sink, where the molecules enter a totally disordered state and bounce whichever way gravity or impact directs them.

In all cases, state of disorder is easier to achieve, thus it is more probable under most circumstances. The process usually goes from order to disorder, and not the other way around.

Entropy

In the examples above (the electrical wire and the water pipe) the application of work provides the orderly motion; an electric force forced the electrons to move and a gravitational forced the water molecules to line up and flow in orderly manner. Until in both cases the orderly state eventually changed and/or disappears.

The electricity is eventually converted into more disordered heat energy. The water in the water pipe eventually runs into the sink, where it is splashing all over the place. Disorder seems more probable to exist than order, and probability of this happening is the key to understanding the broader statement of the second law of thermodynamics.

The evidence that order is less probable than disorder is all around us. Disorder is more probable than order. If a sample of matter is in an ordered state, spontaneous change will take it in the direction of disorder.

In all changes of state within a closed system the total energy remains unchanged. That is a statement of the first law of thermodynamics. But something changes in the system, and it has to do with order and disorder in the state of the system. We call that something "entropy."

The "entropy" of the state of a system is a measure of the probability of its occurrence. States of low probability have low entropy; states of high probability have high entropy. With this definition we see. from the previous discussion and examples, that in any transfer or conversion of energy, because the spontaneous direction of the change of slate of a closed system is from a less to a more probable state, the entropy of the system must increase. That is also a broader statement needed to describe the second law of thermodynamics.

In any energy transfer or conversion within a closed system, the entropy (or disorder) of that system increases. Energy conversions can proceed so that the entropy of a part of a system is decreased too, i.e. charging a storage battery, or freezing ice cubes. In each of these examples, order has been won from disorder and entropy has decreased during the input of energy.

If the total system is considered, however, the total effect has been an increase in disorder. To charge a battery we must provide energy above and beyond that necessary to re-form the chemical combinations in the battery plates. Some of this low entropy electrical energy (ordered state) is changed into high-entropy heat energy (disordered state) in the current-carrying wires and in the output devices.

In freezing ice we increase the order and thus decrease the entropy of the water in the ice cube trays by removing heat from it. The heat energy removed, however, has lo flow into a substance (the expanded Freon gas) that is at a much lower temperature than its surroundings. Thus, the entropy and the disorder of the gas is increased.

Another way of stating the consequence of the second law of thermodynamics is to say that all energy transfers or conversions are irreversible. They will go spontaneously in the direction of increasing entropy (disorder). They will not go spontaneously toward a state of lower entropy (order). Thus, all the losses we described in the power plant conversions are necessary to fulfill the nature's laws. Some of them can be minimized, but none of hem can be eliminated. Entropy must increase. And it always does.

Energy vs. Power

As we mentioned above, the terms "energy" and "power" are essentially synonyms, although most scientists and engineers would disagree, since they prefer to distinguish between them. This is because the specific conditions in each case are different and in precise technical and engineering calculations and designs we need to have a clear understanding of their origin, nature, and behavior. And we must agree that technically speaking, power is not the same as energy.

*Energy is the potential of doing work, while power is the **rate** at which energy is converted and work is performed.*

This energy we can be expressed and measured in units of power, such as Watt, kW MW, GW, TW etc. as used in the electric industry, which we are most interested in.

As we also mentioned before, energy is invisible and immeasurable until its components activate it. And so, for example, (potential) energy sits in our car's gas tank until it is pumped into the engine and forced to do work—at which point we feel the power of the fuel taking the car ahead. The power at that point (the kinetic energy) can be measured in units of horse power in the engine torque, feet per second in road speed, and/or degrees Celsius in the engine cylinders.

At the same time, a hydroelectric plant generates power by allowing the water above the dam to pass through turbines, where it actually converts the water's potential energy into kinetic energy and ultimately into electric energy. So, the amount of electric energy that is generated per unit of time is the electric power generated and sent into the grid for use at homes and businesses.

In everyday discussions, however, the terms electricity, electric energy, and electric power are interchangeable.

We allow ourselves to do this in this text as well, simply because they refer to the same final activity—the flow of energized electrons on their way to do work somewhere, sometime. And so we use these terms interchangeably and somewhat indiscriminately. Yes, it is incorrect from a technical point of view, but is accepted by the masses. So we, being part of the masses—and ignoring our technical background—will use the prevailing language.

Energy Transfers

The transfer of energy between a "system" and adjacent regions is expressed as work. One example is mechanical work, which in the simplest of cases is expressed in the following equation:

$\Delta E = W$ (if there are no other energy-transfer processes involved.)

where
ΔE is the amount of energy transferred, and
W represents the work done on the system.

In this case the ΔE is the difference of the energy before and after the transfer. For example, burning a log generates 100 Btus of heat. ΔE here would be 100 Btu, because we started from zero.

Energy transfer can be split into several categories:

$\Delta E = W + Q$

where
Q represents the heat flow into the system.

There are other ways in which an open system can gain or lose energy. In chemical systems, for example, energy can be added to a system by means of adding substances with different chemical potentials. These potentials can then be extracted as by fueling an auto; a system which gains in energy without addition of either work or heat. Winding a clock would be adding energy to a mechanical system (the spring).

These terms may be added to the above equation, or they can generally be subsumed into a quantity called "energy addition term, E, which refers to any type of energy carried over the surface of a control volume or system volume.

As an example, the kinetic energy of a stream of particles entering a system, or energy from a laser beam adds to system energy, without either being either work-done or heat-added, in the classic senses.

$$\Delta E = W + Q = Ea$$

where

Ea represents other (additional) energy not covered by work done on a system, or heat added to it.

Energy is also transferred from potential energy, E_p, to kinetic energy E_k and then back to potential energy constantly. This is referred to as conservation of energy, although there are always losses in the transfer processes.

In a closed system, energy cannot be created or destroyed; therefore, the initial energy and the final energy will be equal to each other. In this case the "losses" would be accounted as energy that is transferred into other types of energy, but yet remaining in the system.

This transfer can be represented by:

$$E_{pi} + E_{ki} = E_{pF} + E_{kF}$$

where:

E_{pi} is the initial potential energy,
E_k is the initial kinetic energy
E_{pF} is the final potential energy,
E_{kF} is the final kinetic energy

The equation can then be simplified further since

$$E_p = mgh$$

where

E_p is the potential energy
m is the mass
g is gravity, and
h is the height

and

$$E_k = 1/2mv^2$$

where

E_k is the kinetic energy
m is the mass, and
v is the velocity

Then the total amount of energy can be found by adding

$$E_p + E_k = E_{total}$$

Energy Conversions

Energy in the universe is sometimes quickly and easily changed from one state (or type) into another. At times the conversion process is not that quick, nor easy, depending on the type of materials and processes involved.

Here are some examples of the major energy conversion cycles:

- Radiative to/from thermal energy conversion:
  — Incandescence is an example of radiative to thermal energy conversion
  — the solar collector is an example of thermal to radiative energy conversion
- Radiative to/from chemical energy conversion:
  — photosynthesis is an example of radiative to chemical energy conversion
  — Chemi-luminescence is an example of chemical to radiative energy conversion
- Radiative to/from electric energy conversion:
  — The solar (PV) cell is an example of radiative to electric energy conversion,
  — The fluorescent light is an example of electric to radiative energy conversion,
- Thermal to/from electrical energy conversion:
  — The thermo-cell is an example of thermal to electric energy conversion
  — The electrical resistance of wires is an example of electric to thermal energy conversion
- Thermal to/from mechanical energy conversion
  — Heated steam is used to turn the turbines in power stations.
  — The mechanical energy from rubbing two sticks together is converted into heat
- Thermal to/from chemical energy conversion:
  — Endothermic reactions are example of thermal to chemical energy conversion
  — Exothermic reactions are example of chemical to thermal energy conversion
- Chemical to/from electric energy conversion:
  — The car battery is an example of chemical to electric energy conversion
  — Electrolysis is an example of electric to chemical energy conversion
- Chemical to mechanical energy conversion process is observed in human muscles, where the chemical energy stored in the body is converted into mechanical, as needed to activate the different body muscles.
- Mechanical to/from electric energy conversion:
  — The car alternator is an example of mechanical

to electric energy conversion
— The electric car engine is an example of electric to mechanical energy conversion
- Mechanical to/from thermal energy conversion:
— Car brakes are example of mechanical to thermal energy conversion
— The locomotive is an example of thermal to mechanical energy conversion
- Mechanical to/from sound energy conversion:
— A running car engine is an example of mechanical to sound energy conversion
— Aretha Franklin breaking a glass with the high pitch in her voice is an example of sound to mechanical energy conversion
- Nuclear energy, stored in nuclear materials, is converted to thermal energy in the nuclear power plants, and the resulting heat is used to generate electricity.

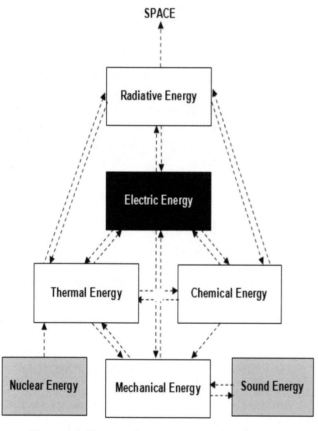

Figure 1-5. Universal energy conversion diagram

Energy transformations in the universe over time are characterized by various kinds of potential energy that has been available since the Big Bang, and later being "released" (transformed to more active types of energy such as kinetic or radiant energy), when a trig-

gering mechanism is available.

Familiar examples of such processes include nuclear decay, in which energy is released that was originally "stored" in heavy isotopes (such as uranium and thorium), by nucleosynthesis, a process ultimately using the gravitational potential energy released from the gravitational collapse of supernovae, to store energy in the creation of these heavy elements before they were incorporated into the solar system and the Earth. This energy is triggered and released in nuclear fission bombs. In a slower process, radioactive decay of these atoms in the core of the Earth releases heat.

This thermal energy drives plate tectonics and may lift mountains, via orogenesis. This slow lifting represents a kind of gravitational potential energy storage of the thermal energy, which may be later released to active kinetic energy in landslides, after a triggering event. Earthquakes also release stored elastic potential energy in rocks, a store that has been produced ultimately from the same radioactive heat sources.

Thus, according to present understanding, familiar events such as landslides and Earthquakes release energy that has been stored as potential energy in the Earth's gravitational field or elastic strain (mechanical potential energy) in rocks. Prior to this, they represent release of energy that has been stored in heavy atoms since the collapse of long-destroyed supernova stars created these atoms.

In another similar chain of transformations beginning at the dawn of the universe, nuclear fusion of hydrogen in the sun also releases another store of potential energy which was created at the time of the Big Bang. At that time, according to theory, space expanded and the universe cooled too rapidly for hydrogen to completely fuse into heavier elements.

This means that hydrogen represents a store of potential energy that can be released by fusion. Such a fusion process is triggered by heat and pressure generated from gravitational collapse of hydrogen clouds when they produce stars, and some of the fusion energy is then transformed into sunlight. Sunlight sun may again be stored as gravitational potential energy after it strikes the Earth, as (for example) water evaporates from oceans and is deposited upon mountains (where, after being released at a hydroelectric dam, it can be used to drive turbines or generators to produce electricity).

Sunlight also drives many weather phenomena, save those generated by volcanic events. An example of a solar-mediated weather event is a hurricane, which occurs when large unstable areas of warm ocean, heated over months, give up some of their thermal energy suddenly to power a few days of violent air movement.

Sunlight is also captured by plants as chemical potential energy in photosynthesis, when carbon dioxide and water (two low-energy compounds) are converted into the high-energy compounds carbohydrates, lipids, and proteins. Plants also release oxygen during photosynthesis, which is utilized by living organisms as an electron acceptor, to release the energy of carbohydrates, lipids, and proteins.

Release of the energy stored during photosynthesis as heat or light may be triggered suddenly by a spark, in a forest fire, or it may be made available more slowly for animal or human metabolism, when these molecules are ingested, and catabolism is triggered by enzyme action.

Through all of these transformation chains, potential energy stored at the time of the Big Bang is later released by intermediate events, sometimes being stored in a number of ways over time between releases, as more active energy. In all these events, one kind of energy is converted to other types of energy, including heat.

ENERGY APPLICATIONS

Energy, in all its forms and variations, is widely used in the sciences. For example:

- *In physics*, energy is considered a quantity that exists, but is indirectly observed (or invisible and immeasurable in its purest form. It comes to life, and is measurable when its other components (force and distance) are considered. In that case energy becomes work, and can be measured as a physical entity. The work then could be observed, measured and expressed as heat, electric power, mass, speed, etc. variables.

    For example, photons traveling from the sun through space have potential energy stored, which upon impact onto a solar panel is released and converted into heat (heating the panels) and electricity, which can be extracted and used.

- *In chemistry*, energy is an attribute of a substance as a consequence of its atomic, molecular or aggregate structure. Since a chemical transformation is accompanied by a change in one or more of these kinds of structure, it is invariably accompanied by an increase or decrease of energy of the substances involved. Some energy is transferred between the surroundings and the reactants of the reaction in the form of heat or light; thus the products of a reaction may have more or less energy than the reactants.

Chemical reactions are invariably not possible unless the reactants overcome an energy barrier known as the activation energy. The speed of a chemical reaction (at given temperature T) is related to the activation energy E, by the Boltzmann's population factor e−E/kT, which represents the probability of a molecule to have energy greater than or equal to E at the given temperature T. This exponential dependence of a reaction rate on temperature is known as the Arrhenius equation. The activation energy necessary for a chemical reaction can be in the form of thermal energy.

- *In biology*, energy is an attribute of all biological systems from the biosphere to the smallest living organism. Within an organism it is responsible for growth and development of a biological cell or an organelle of a biological organism. Energy is thus often said to be stored by cells in the structures of molecules of substances such as carbohydrates (including sugars), lipids, and proteins, which release energy when reacted with oxygen in respiration. In human terms, the human equivalent (H-e) (Human energy conversion) indicates, for a given amount of energy expenditure, the relative quantity of energy needed for human metabolism, assuming an average human energy expenditure of 12,500kJ per day and a basal metabolic rate of 80 watts.

    Our bodies run on average at 80 watts, then a light bulb running at 100 watts is running at 1.25 human equivalents (100 ÷ 80) i.e. 1.25 H-e. For a difficult task of only a few seconds duration, a person can put out thousands of watts, many times the 746 watts in one official horsepower. For tasks lasting a few minutes, a fit human can generate perhaps 1,000 watts. For an activity that must be sustained for an hour, output drops to around 300; for an activity kept up all day, 150 watts is about the maximum. The 'human equivalent' helps us understand the energy flows in physical and biological systems by expressing energy units in human terms. This provides a "feel" for the use of a given amount of energy.

- *In geology*, continental drift, mountain ranges, volcanoes, and Earthquakes are phenomena that can be explained in terms of energy transformations in the Earth's interior, while meteorological phenomena like wind, rain, hail, snow, lightning, tornadoes and hurricanes, are all a result of energy transformations brought about by solar energy on the atmosphere of the planet Earth.

For example, an erupting volcano releases its energy to create land slides, start fires etc. phenomena related to release of its thermal and mechanical energies stored in it through the millennia

- *In cosmology and astronomy* the phenomena of stars, nova, supernova, quasars and gamma ray bursts are the universe's highest-output energy transformations of matter. All stellar phenomena (including solar activity) are driven by various kinds of energy transformations.

Energy in such transformations is either from gravitational collapse of matter (usually molecular hydrogen) into various classes of astronomical objects (stars, black holes, etc.), or from nuclear fusion (of lighter elements, primarily hydrogen).

For the purposes of this book, the term "energy" is used with the understanding that it is contained in fuels, which can generate some type of energy, such as electric power, or move vehicles. The energy contained in these fuels is converted from one form to another—usually from potential to kinetic—for example by burning coal, oil, and natural gas. The chemical energy contained in these is converted into mechanical movement and changes in their particles, which results into heat. Thus produced heat boils water into steam, which drives steam turbines and generators that produce another form of energy; electric power, which can be measured and used to power our lives.

Similarly, energy (electric power) can also be generated by collecting sunlight, or wind energy and converting them into a flow of useful electrons, which can be used as electric energy (power).

In more practical terms, energy is all around us and can be observed and measured in its different forms.

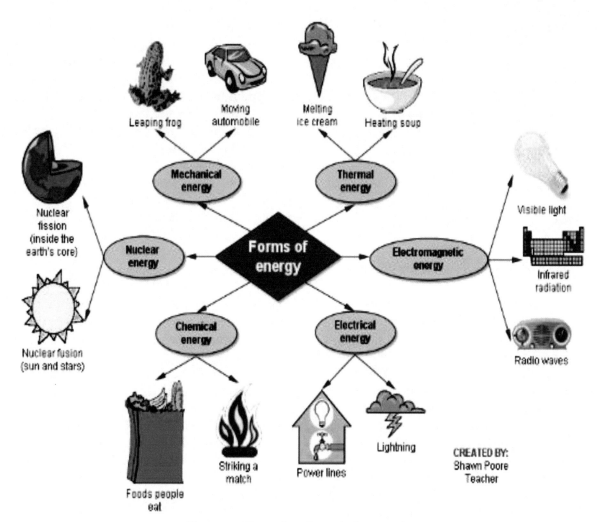

Figure 1-6. Examples of practical energy forms.

In this text, the subject of energy is narrowed down to its role in the proper function of the global energy markets. Here, all practical types and forms of energy; from the production of energy sources—coal, natural gas, crude oil, hydro, solar and wind power—to their delivery and use in different ways by different types of customers, are discussed from technical, logistics, legal, political, financial, and marketing points of view.

THE NEW ENERGY ECONOMY

Energy is the foundation of our lives. It drives us around in our cars, lights our houses, runs the appliances, and in general drives our economies. Energy today is obtained from burning fossil fuels—coal and natural gas to generate electricity, and crude oil to power our vehicles and to produced precious chemicals and products.

The Energy Revolutions

Energy, in its different material forms, has gone through several stages during the ages.

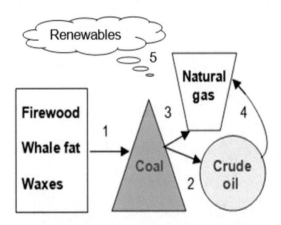

Figure 1-7. The Energy transformation cycles

Figure 1-7 shows the several major energy transformation cycles. The world's energy history started with firewood, used for many centuries as the only, or at least major, source of energy in form of heat and for cooking. As a matter of fact, some areas of the world are still using firewood as the primary energy source.

Later on, during the *first energy transformation cycle*, firewood was replaced by coal, which also had other uses, like train and car engine fuel. Coal was (and still is) also used in mass quantities for electric power generation.

Then came crude oil, during the *second energy transformation cycle*, taking the transportation sector by a storm and revolutionizing the world by making it very mobile.

The *third energy transformation cycle* came recently,

when natural gas was recognized as a cleaner, more efficient, and cheaper alternative to coal power generation. As a result, many coal power plants in the U.S., and other developed countries, have been gradually replaced by natural gas.

In parallel with the third, we are now going through a *fourth energy transformation cycle*—that of replacing crude oil as a vehicular fuel with natural gas. The reasons are the same; natural gas is cleaner and cheaper. This transformation would take some time, due to the expenses related to building the transportation infrastructure. So we expect crude oil to dominate the transportation sector until its last drop.

The *fifth energy revolution*, that of the renewables (wind and solar in particular) taking a greater part in power generation and transportation, is also ongoing. It is, however, still controversial and its future is not well defined, let alone certain. Nevertheless, these energy sources will be the major, and maybe the only, hope of the future generations, since we are mercilessly burning as much fossils as we want and can.

What is being steadily depleted today are the low-entropy (perhaps the lowest possible) forms of energy; the fossil fuels. These energy resources lie deep underground in their primordial state of rest until we dig or pump them out; just to burn them. At that time, they jump from a state of lowest (dormant) entropy to enter (albeit temporarily) into a state of highest entropy during the burning process. At that time they also change their physical state; from that of liquid or solid into gas (smoke) and solid residue (ash and other particles). Since there is no way to get back to their original state of low entropy, they are gone up in smoke forever; never to be seen again in their original form or quantity!

So when we use the fossils we do not really use up just their energy; we use up also the entire usefulness of the energy contained in these, as well as their original shape and form. This is what the present day energy crisis consists of. It is the state of low entropy that Mother Nature demanded and built through the millennia, which we managed to destroy and convert into highest form of entropy. By burning and eliminating large part of the fossils in the process, we are entering a new era; that of extreme and increasing pollution and approaching the day when the fossils will be no more.

During the golden age of the fossil fuels (1920-1970), we didn't worry about any of this because useful fossil energy was plentiful and cheap. We also didn't have all the facts in those days, nor was there a sign of increased pollution, and/or global warming talk.

We also thought the fossils bliss would last forever,

and caught in chasing and living the American Dream, we used it carelessly. Large 8 cylinder cars were the norm and we drove them up and down the highway just because it was fun, and just because we could. We actually still do, but are now somewhat more aware of the fact that time has come to begin the practice of "the new economy"—that of planned conservation of our precious low-entropy energy forms.

We know that action has to be taken, and that we only delay the inevitable by half-way measures and lots of talk. The high gas pumps prices and the heavy smog in the large cities are a constant reminder of the new reality, and reminds us constantly how foolish and selfless we have been.

The second law of thermodynamics tells us that because entropy increases in any transfer or conversion of energy, we should try to find energy transfer and/or conversion mechanisms (and products) in which the entropy of the source and the end use are as close as possible, thus minimizing the entropy increase.

This is still a form of delaying the inevitable, but at least it will buy us more time to find some alternative (renewable) energy sources for the long run. We should not leave it for the new generations to worry about. But as things are going today, we cannot even take care of our own problems and messes, so it is quite likely that the upcoming generations will inherit a mess, which they have to fix.

This new way of thinking and living leads us to a new way of measuring efficiency in terms of "increased entropy." In the long run, the entropy of the universe will increase no matter what we do. Entropy is the "arrow of time" which increases the chaos level in the direction of the future.

In any actual decision about the use of energy, of course, we have other things besides the entropy increase to consider. We must take into account the present cost, time, and environmental impact, for instance. But although we have considered these in the past, we have not paid much attention to the entropy involved. This may be just the right time to begin considering it as a reliable measurement.

We must begin to apply this concept as a type of a new thermodynamic economy and make certain that each use of low entropy is for a necessary end and is accomplished with as small an increase in entropy as possible.

Fossils then...

250-300 million years ago, in the middle part of the Late Paleozoic Era, plants emerged from the oceans, and the tropical marshlands were covered by vegetation of different types. Giant forests of ferns, horsetails, and mosses grew in enormous wooded areas, while human life was still work in progress. Human genes were just beginning their evolution in some of the creatures in the newly formed jungles and oceans.

The abundant sun energy was driving the life processes of the huge fern trees and other vegetation, which were used as food by equally giant animals. The radiant energy of sunlight was converted into actual life-giving energy, the third great form of potential energy (and the source of life and energy on Earth).

The Global Energy Storage

In the energy-converting process of photosynthesis, the energy of sunlight was used by plants to break up molecules of carbon dioxide (CO_2) and water (H_2O) and to rearrange their atoms into carbohydrates (sugars and starches). Large quantity of oxygen was produced as a byproduct at the time as well.

A form of potential energy storage was created in the huge vegetation areas and the animals of the time. The carbohydrates formed with the help of the sun's energy are either food or fuel, depending on the use we make of them. The jungle of 300 million years ago grew in a fertile environment and much of the world at that time had an almost tropical climate, wet and warm. The atmosphere was also rich in CO_2, which was carried up from the depths of the Earth by many active volcanoes and hot springs and used by the vegetation.

The huge trees grew and died, some falling into shallow waters from which they had emerged initially. Some of them decayed and were buried in their watery graves. Larger disasters killed more trees and animals with time, and more plants and dead animals were covered with water or dirt. Some remains decayed, some were buried deeper into the soil, where oxygen-requiring bacteria could not destroy them completely.

Over tens of thousands of years, huge amounts of this un-decayed plant material built up into the spongy mass we call peat. In peat, the anaerobic bacteria (those that do not require oxygen) broke down the carbohydrates without forming CO_2. Pressure and accompanying heat caused by the overbearing material drove off some of the water, oxygen, nitrogen, and miscellaneous products so that the percentage of carbon was increased. These chemical changes and the increases in density also increased the potential energy per pound of the buried material.

More and more of the water, nitrogen, oxygen, and other materials were driven out so that the percentage of carbon content increased—from an original 50% by weight up to 80%. The energy density, energy per pound, was significantly up also. The peat was now compacted into a hard, black mineral; and a coal basin was formed.

Through a similar processes, we get crude oil and natural gas deposits all over the world. These are packed with energy, which is actually sunlight collected and converted into living matter thousands of years ago, and which is available to us today.

Fossils Today and Tomorrow…

The fossil sources (coal, oil, natural gas) contain stored (potential) energy. This is one of the several primary forms of energy available to our industrious species. Bio- and nuclear energy also fit quite nicely in this classification as well. The other key energy forms are in their kinetic form (in our broad definition), since they are constantly on the move and can always be used for producing other forms of energy. These are: solar, wind, geo-, bio-, hydro-, and tidal energy.

*The renewables represent the mix of energy resources
available to our beloved, unique, and irreplaceable
'Spaceship Earth', They are best equipped to power
it in the not so distant future.*

Solar energy comes from the "mother ship," so to speak; our sun, and is responsible for many life-giving activities on our planet. As a matter of fact, the renewables are directly dependent on sun's energy. It creates wind currents, ocean waves, rain and snow, and makes plants grow.

Simply put, if there was no sunlight, there would be no renewables on Earth. As a matter of fact, if the sunlight was not at the exact quantities it is now days, life for all of us here would be quite different, and may even cease to exist, due to extreme (too hot, or too cold) temperatures.

*The most amazing fact about solar energy (sunlight)
is how many of us don't even notice, let alone use, it.*

The fossils—in all their types, forms and shapes—are also here due to sunlight, which created their origins millions of years ago. It took all this time to convert the useless mass of dead vegetation and animals into the major energy source in our world. We have been using the fossils for a long time at an ever increasing rate, and there is no slowing us down. We simply don't seem bothered by the fact that very soon—as soon as 20-30 years from today—many of them would be depleted and gone forever.

As we burn through the fossils, we are also and even more importantly, destroying the energy and physicochemical balance of Mother Earth.

*What took Mother Nature thousands of centuries
to create, we would eliminate in just over a century.*

Once gone, the fossils will not return…ever! The conversion process (vegetation and animals decay to fossils) is very slow and requires such special conditions, so that the combination of these simply cannot be reproduced under the present situation. Or ever.

The sad truth is that the "fossil fuels" that were buried deep into the ground millions and billions of years ago are all we have now. They are also all we are going to have ever. There are no new "fossil fuel" reserves in development stages that we know of, nor are there any expected to be created ever.

And as if this is not bad enough, burning the fossils in such huge amounts is creating great environmental pollution and climate changes that we (or the future generations) may not even survive. We might have already caused so much damage to Mother Earth that it would be unlikely to return to normal even if we stopped all fossil related activities today.

But with the ever increasing population growth, and the accompanying demand for better life (which requires lots of energy) we don't expect slowing down of these activities. Oh, yes, there is a lot of talk about it, but as a matter of fact China, India and other major economies have turned back to fossil fuels for most of their energy needs. Drill baby, drill is echoing all over the world. Climate change? Oh, yea, but it can wait…

Energy Materials and Fuels

*The foundation of the energy markets are
the raw materials, fuels, chemicals, and all
other energy related products and services.*

These key components are involved in one or another way in the energy life-cycle of; production, manufacturing, processing, distribution, and use of energy products.

Note: In this text we refer to the energy materials and fuels as 'energy sources.'

In this chapter we take a brief look at the major energy materials (or energy sources), and will expand in the following chapters on their use and function in the energy markets.

All power plants around the world use some type of fuel, be it primary, or secondary, renewable or not. The fossil fuels (coal, oil, and natural gas) account for more than 80% of the total amount of materials used in the global electric power generation. They have been and still are the most important primary energy generating sources around the world. The rest of the total power generation is contributed by hydropower, nuclear energy, and in much smaller amounts by solar, wind, and geothermal power.

Note: Firewood, which is also used in enormous quantities around the world, falls in a different category all together and we will review it as such in this text. It is a very little known fact that firewood is one of the major energy sources; both in developing and developed countries, so be prepared to be amazed as you read along.

Solar and wind energy have a special place here too, because they have been involved in the creation of most of the primary and secondary energy materials and fuels since the beginning of time. This is another complex process, with many combinations and permutations, which requires entire book of its to review in detail. For our purposes, in this text we review both; solar energy coming from the sun, followed by solar power generated on Earth by humans, and specifically the electric power generated by photovoltaic and thermal solar power devices.

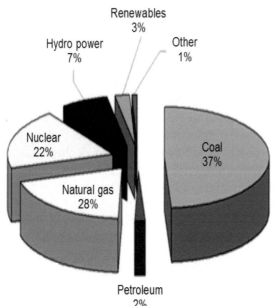

Figure 1-8. Energy use in the U.S.

Obviously, it follows from Figure 1-8 that coal rules in the energy sector, and the power generation in particular…and will continue so for a long time to come. Coal, due to its enormous reserves stored in the Earth's bosom, might be the last fossil fuel left, when all the rest are depleted—even as the predictions call for increased use of coal in the future. The other fuel types are also increasing in use, but coal is still on top, and is foreseen to be there for a very long time.

Coal

Coal has been, and still is, the energy engine of the world. During the last two centuries, the U.S. coal industry grew slowly, starting in the early nineteenth century. By mid-century, the superiority of coal over wood burning (which was until then the primary fuel) on an energy-per-volume basis, was undisputable. The growing scarcity of fire wood in some parts of the country caused the first great energy replacement cycle, which established coal as the fuel of the past century.

Anthracite is a type of coal, which contains almost 100% carbon. This quality, and the fact that it is cleaner burning made it most popular at first. By the end of the century, however, bituminous coal with lower carbon content but comparable heat value, took its place as the preferred fuel. Until then more than half of the coal came from the great "Pittsburgh seam," which lay exposed along the rivers above the city of Pittsburgh, PA.

Coal's History

The world has gone through several transformation (changes) of types of energy sources used during different periods of time. The first energy transformation—that of wood to coal use—was dramatic. In the U.S. wood fell from 90% of the energy input in 1850 to less than 20% by the end of the century. At the same time coal use grew to 70% throughout the first part of the 20th century.

One of the reasons for the swiftness of the replacement were the rich seams of coal in the East, where high population density and industrial concentration created an energy demand that the vanishing forests could not satisfy. Changes in industry and transportation fed the demand for coal.

Before 1900, less than 800 Mt (million tons) of hard coals and lignites accounted for about 95% of the world's total annual primary energy supply. The coal production went up and down during that part of the century, but doubled by 1949, to nearly 1.3 Gt (giga tons) of hard coals, and about 350 Mt of lignites, mostly because of the expansion of traditional underground manual mining. By 1988 the world produced and consumed 4.9 Gt, with 3.6 Gt of hard coal and 1.3 Gt of lignites. 40% of the coal at the time came from Eastern Germany and Russia, and was of very low quality.

Figure 1-9. Surface coal mining

During the 20th century, the U.S. coal production increased to the point that it not only energizes our own industry, heat our homes, and power our trains, but it enabled us to become a major exporter as well. During the early peak in coal production, in 1918, 678 million tons were taken from the ground, and 27 million tons were exported overseas. During the next peak of the wartime 1944-1945, 688 million tons of coal were mined, while exports amounted to 94 million tons. This amounted to nearly half of the world's coal production at the time.

U.S. coal exports grew again as coal production increased during the early 1980s, when coal production reached a new peak of approximately 800 million tons annually, with some 100 million tons of exports. At that time, however, the world's coal production was four times larger than the U.S.

Coal's customers changed as radically as the peaks and valleys of its production. In the early twentieth century industrial and home heating accounted for 60%, while railroads consumed 20% of the coal burned. Electric utilities consumed only 8%. By the middle of the 20th century, the railroad's share had been cut to 15%, industry's share was 40%, and 20% was used for home and commercial heating. The number of electric power plants grew at the time too, so they consumed 15% of all coal produced in the country.

Gradually, the use of coal changed even more drastically. Diesel powered railroads consumed only 2% and the retail coal market had shrunk to less than 10%. Industry, where natural gas and oil competed strongly, used only 36% of the total. Most of the rest, well over 50%, was consumed by the electric utilities.

By the 1980s the electric utilities' share has climbed to 82% of the total coal consumed in the U.S. Less than 1% was used for heating.

In 2011 90% of the coal produced in the U.S. was used by the nation's electric power generators.

Although, coal use has been reduced lately, as discussed in this text, coal is not going to give up easily it domination of the electric power markets. It is still the most abundant, most accessible, and relatively cheap energy source on Earth. As such, it is available in many countries, most of which don't have other alternatives and/or cannot afford them. Coal use is actually expected to increase in Asia and other places in the years to come. This means that it will be around for a long time, and due to its huge reserves. coal may be around even longer than any other fossils.

Recent estimates place the total amount of coal in the Earth's crust at over 20 trillion tons, but only 1.0 trillion is economically exploitable for now.

This means that the times for getting coal the easy way are nearly over. The rest of the coal will be much harder and expensive to dig out.

There are reports and claims of vast coal supplies in the U.S. and abroad, which would be able to fuel the global economies for 150 years or more. Several other studies suggest, however, that coal may not be nearly as abundant as once believed. At least not the coal that is practical to mine.

A 2007 National Academy of Sciences report concluded that the a more realistic number is about 100 years, or less, considering the fact that coal's consumption is expected to grow beyond the current rate. Even more disturbingly, data from EIA shows that the total heat content of U.S. coal has been declining since the late 1990s while production level has been steady. This means that coal grades are dropping, so that the actual energy we get from domestic coal is less and the waste from coal burning is increasing.

Worldwide reserves are probably half of what is currently stated, simply because, just like with oil reserves, coal reserve estimates have been steadily dropping over time down to what is actually minable, and not necessarily to what might be available.

In 2007, the Germany's Energy Watch Group suggests that China, which has some of the largest coal resources in the world, has supposedly already mined more than 20% of the existing coal in the country. Keeping in mind that China's industrial; revolution is only a couple of decades old, and that coal use is still increasing steadily, it is easy to see that China will run out of coal by the end of this century…or even sooner

In the U.S. the electric utilities still dominate the coal market, with coal gasification, new coal converting technologies, carbon sequestration and conversion to natural gas taking over the industry. The new developments in the energy sector, however, make the future of coal here uncertain. New and large natural gas reserves are found and exploited daily, and new renewable technologies are taking over—all of them trying to replace the old reliable (but dirty) coal as a fuel in critical power generation operations.

The new trend of converting coal burning plants to natural gas, and adding renewables to the energy mix, are quickly growing phenomena which would aid and extend the energy transition. This might allow us (and the future generations) to adapt to the dwindling global energy resources and find new, more sustainable way to generate and use energy.

New technologies of "cleaning" coal are in the development stages, so we would not be surprised if they become practical and cheap enough soon. At that time, the old and tired coal, with its new facelift, may once again take over the electric energy markets. This transition, however, would take some time, since the related technologies are not fully developed and are too expensive.

Crude Oil

Crude oil was also created many millions of years ago by decomposition of organic materials in the Earth's bosom. The oil-based fossils, however, are in liquid form, and are much younger than coal. They are usually found in sedimentary basins, but the exact fossilization mechanism is still a bit foggy, and intensive research is underway.

The fossils, during the period of their initial formation, consisted of large and very complex organic (hydrocarbon) molecules. At that stage they were very similar to these in many living cells. And not surprisingly so, since they were formed by the gradual transformation of living cells (vegetation, animals, etc.) into fossils.

During their "gestation" period, which also lasted many millions of years, they were also subjected to extreme temperatures and pressures. Under these conditions, and with time, their molecules "cracked" into lighter and more mobile such, thus creating different types and quality of crude oil.

When extracted and refined, crude oil reproduces the natural cracking process, but run in reverse at a much higher speed. When done in a precisely controlled

way, many different fuels, such as light oils, gasoline, diesel, jet fuel, etc. are obtained. On the other side of the spectrum, refining also produces heavy oils, greases, lubricants, asphalt, etc.

Figure 1-10. Crude oil pumping

Oil and natural gas tend to migrate through the rock layers toward the Earth's surface, pushed by the high pressure in their graves. Some reach the surface and produce oil pools and gas flares that are the first evidence of the presence of these fuels. However, most of the crude oil and natural gas are trapped in rock-solid, impermeable formations. Here, they form the pools and pressure chambers, which are then explored and pumped out by humans.

Presently, gasoline is the major product of the refinery operations, followed by the distillate fuel oil used in furnaces and the residual fuel oil that industry and utilities burn. Diesel, and kerosene in jet fuels are also important petrochemical products.

The switch from coal to petroleum products, during the second of the energy transformation cycles, saw coal go from 76% of total consumption in 1910 to less than 30% by 1955. During the same time, petroleum products increased their share from 10% to 65%. And production doubled just about every eight years.

Oil has improved our economies and personal lives significantly, but the problem is that we have become dependent on oil imports. These peaked to 3.21 billion barrels in 1977, and at times during that year crude oil imports accounted for half of the total crude oil input to U.S. refineries. That level of oil imports—with some fluctuations—remains until today.

In 2000 USGS estimated that the entire global supply of oil is approximately 3.0 trillion barrels, of which

24% has been already pumped out and used, while 29% was in discovering stages. While this suggest that there is still a lot of oil left in the Earth's crust, it is getting clearer that it is much harder and expensive to reach and extract the remaining oil.

In 2006 the oil imports peaked at 12.2 million barrels per day, which is equal to 4.5 billion barrels annually. Amazing numbers, so one has to wonder how much longer can we do this at this rate?

In 2011, the U.S. crude oil imports leveled to 8.7 million barrels of oil daily, which is close to the 1977 imports number. So, we are back where we started… Progress? Not much!

It has been estimated that the average global production in existing oilfields has been declining by about 4% annually. This means that if no new fields, or new capacities are added, the world will run out of oil in 25 years. That also means that the industry is forced to develop new oil production capacity constantly. The size of each of these developments is enormous—equivalent to the current capacity of the North Sea oil field, which is one of the largest ever. In order to grow production the oil industry must exceed this amount. This, however, is unlikely to happen due to increasing difficulties in extracting oil.

Luckily, the ever-inventive American spirit was put in action, and discovered hydrofracking. Thanks to this new technology and its hi-tech processes and equipment, we now have much more oil than before. We don't know how long the bliss is going to last, but presently, we are in the best shape we have ever been… at least as the amount of potential oil and gas reserves is concerned.

Being the most advanced (technologically) nation on Earth pays off…big time… most of the time

There is also lots of oil in the tar sands in Canada, additional heavy oil deposits in Venezuela, and huge oil shale reservoirs in Colorado and many other places. While many of these are real and feasible sources of oil, they also represent the most difficult to extract oil types. The easy oil is mostly gone. The days when poking a stick in the sand ended up with a burst of crude oil are over, Now we have to dig miles upon miles under the Earth's surface to get to the oil. And in many cases, we have to break it out of the rocks and pump it out at a great expense and serious environmental damage.

What we have to deal with in the conventional oil wells now days, is increasingly difficult oil types that are already proving to be a great challenge to the oil indus-

try. This will certainly result in future price increases. High-cost, and capital-intensive production methods will slow down the rate of production; it will become slower to ramp up and more difficult to maintain. The types of heavy and dirty hydrocarbons from the old wells must be heavily processed at high cost using enormous amounts of energy and chemicals, and so they are much more expensive. And will get much more so too as time goes on.

And so, yes we have a lot of oil, and oil-like substances left in the Earth' crust, but they are getting difficult to get, their quality is decreasing, and so the oil industry must figure out a way to get enough oil at an acceptable price. Basically speaking, the big issue for the 21st century is the rate of production, regardless of the size of the available resources. And equally importantly, these developments will determine the price of all oil products on the market place.

Oil's derivatives are convenient to use fuels and products, which we all use on daily basis, and cannot even imagine our life without them. Oil become crucial to the economic development of the U.S. and most other countries thanks to the abundance of inexpensive supplies from the Middle East early in the 20th century. Since World War II the world was has "hooked" on crude oil and the benefits it brought us. This has led to a number of negative effects—with environmental changes and climate warming as the main problems that we are now forced to deal with.

Reducing the petroleum imports dependence is a major goal of the U.S. government and it seems that recently the decision to proceed with the increase of domestic oil and natural gas production is seen as a solution. Yes, it might be so, but for how long? What are the possible scenarios and feasible alternatives?

Amazingly enough, we are all aware of the issues at hand, and yet oil is still numero uno on our energy agenda. It is the evil we can't do with or without. We have learned to just put up with the problems it brings and don't worry about the future. Time is running, however; crude oil reserves are limited and won't last forever.

At the very least, we need to take a very close look at our personal use of oil and oil derivatives, and get involved in the campaign to reduce use of oil in our daily lives. This is one thing we surely can do on personal level, in addition to electing government officials that will help us in the battle to reduce oil use.

Reliance on government, however, is not the best approach. It had not worked thus far, and we are sure it won't work in the future. We need to take the initiative in our hands and bring it on personal level. Although this is easier said than done, it is the best, if not the only, way to prevent a disastrous end of the fossils era.

Natural Gas

Natural gas was the most popular fuel in this country in the beginning of the 20th century, booming from about 5% of the market in 1925 to a peak of 33% in 1970, but then dropping to 26% in 1982. Natural gas is a normal component of petroleum production, and come through the same formation process. It is also often found in the same formations and extracted from them in a similar way.

Natural gas contains mostly methane ($CH4$), a very simple hydrocarbon, and smaller amounts of heavier and more complex hydrocarbon chains. Some of these can be liquefied at atmospheric pressure, and are known as natural gas liquids (NGL). One of them is a natural gasoline that is usually added to refined gasoline. The fuels sold as bottled gas, such as propane and butane, are also NGLs.

In the near past, the difficulty of storage and shipment of gas products resulted in much of it being burned (flared) at the well. Over 90% of the natural gas produced in Oklahoma, Texas, and California in the early 1900s was burned, or vented in the atmosphere, onsite.

Later on a pipeline network connected the rich Southwestern and Western fields with the populations of the Midwest and East, thus increasing the use of natural gas in homes and businesses. Then, large underground storage systems (some of them in natural caverns) were built to store excess gas for emergencies. At that time natural gas began to compete with the others, and many American homes and businesses were equipped with gas lines and meters. Natural gas soon became a daily energy source similar and equal to electricity.

The biggest boost came from two federally financed pipelines. The "Big Inch," a 24-inch diameter pipeline, 1250 miles long, was the first, followed by the "Little Big Inch," pipeline made out of 20 inches in diameter and 1475 miles long steel pipes. These pipelines were built during World War II in order to connect the Southwestern oil fields to the East Coast in peace, war, and emergency situations.

During the "pipeline boom" period of 1935 to 1955, the total pipeline length increased from 167,400 miles to 448,770 miles. Over a third of these sections are long-distance transmission lines. The overall length of the gas pipelines in 1980 was 1,029,800 miles. 263,500 miles of these were transmission lines, while 688,500 miles were distribution lines. The aging distribution pipelines are

reminding us other their age constantly by increasing incidence of leaks and explosions. The release of raw natural gas (via leaks) and the emission of GHGs (via natural gas explosions) is also the beginning of a new, growing, and very serious environmental hazard.

In 2000, USGS estimated that there are 15.4 quadrillion cubic feet of natural gas in the world (or the energy equivalent of about 2.5 trillion barrels of oil), of which 11% has been already pumped out and used, while over 31% is in the process of discovery.

Recently, there were a number of additional large natural gas supplies discovered in North America, which brings the total number of available gas to the energy equivalent of over 2.25 trillion barrels of oil, or a guaranteed minimum of 100 years supply of natural gas at the present rate of use.

But that was then—using rates of consumption for that time. Natural gas has now been promoted to a much more important—if not the most important role—of "transition fuel," which will power our society with the least environmental damage and lowest possible cost, while we figure out ways to deploy renewable and other clean energy types.

This, however, means that the rate of use and production of natural gas will increase in parallel with growth of the population and commerce. Replacing coal and oil is a big job, and natural gas is tasked with it. Simple calculations show that at just 2% annual growth, the previously suggested 100-year U.S. domestic natural gas supply will be exhausted in about 50 years, or less.

The picture gets much worse if the rate of gas production growth rises drastically, as needed to meet the economic growth expected by the middle of the 21st century, and by rising exports. An estimated 3% growth rate, exhaustion is expected in just over 45 years, while a 5% production growth rates will cause exhaustion of the national gas supplies in just over 35 years.

The good news is that predictions for the U.S. are for a gas production growth rate of just 0.4% per year through 2035, so this would extend the natural gas time line significantly. Also, there were some major gas reserves discoveries lately, which increases the gas supply estimate significantly.

The bad news is that according to U.S. Geological Survey assessment of recoverable natural gas in one of the largest of the shale deposits, the Marcellus Shale, the estimate was reduced from the previously estimated 410 tcf to just 84 tcf. This is almost 80% reduction of the reserves in one of our largest reservoirs. If this is a sign of things to come, then this is not the only estimate that has to be revised.

Also, the 100 year gas supply estimates were done before natural gas was promoted as a "transitional fuel," the fuel that will take us into the 22nd century. This is a great task, for the proper execution of which great quantity of gas will be needed, so the estimates must be recalculated from that vintage point.

And there are also a number of additional, and equally serious, questions to answer as far as gas production and use are concerned:

1.   Can the natural gas of the future be efficiently and economically recovered?, and

2.   Are the obvious, and many of the hidden, physical and chemical damages caused by gas exploration (fracking and emissions) going to kill the local environment and some of us in the process?

Note: Raw natural gas (methane gas) is over 25 times more potent and dangerous than carbon dioxide as a greenhouse gas. It is released into the environment during the entire lifecycle of natural gas production and use—from the first drilling, through delivery, and actual use. Shale gas, which is the source of the future, is much worse in this respect too; releasing more greenhouse gasses during production and use than any other fossils.

We cannot predict the natural gas future, but we see it burning bright (no pun intended) at least for awhile, which will help a lot. Natural gas is a cleaner burning fuel, so the environment will take a breath of relieve at least for the duration. Assuming that the fracking issues are resolved, then natural gas would be a true disrupter in the energy sector.

In conclusion, natural gas has become a premium fuel; preferred by all consumers for a number of reasons. It provides (at least it did until the removal of federal price controls) cheap Btus; it is clean and requires a very simple and efficient furnace to generate electricity. These advantages make it very popular in both residential and industrial heating and has made inroads into the electric utility market.

The popularity of natural gas as a fuel is undeniable, and at this point unavoidable. All signs of late point in that direction. It is used in the energy generating sectors, and recently increasingly so in LPG-powered truck, bus, and automobile fleets. It also enjoys a wide use in the residential market, which accounted for 40% of natural gas consumption until recently.

Natural gas is not only making inroads into the coal and crude oil markets, but is also challenging the electricity production sector. And in some areas, like space heating, where it has the advantage of high efficiency (60-70% efficiency), it is irreplaceable.

Its efficiency stems from the fact that natural gas needs only one single conversion, rather than the several conversions (as in coal and oil electricity production), which reduce the efficiency of electrical heating below 30%.

Natural gas as vehicle fuel is also entering the energy markets in a big way. New vehicles and infrastructure are planned to replace the gasoline and diesel cars on the nation's highways and byways. Some time is needed for all of this to be designed, financed and built, but the trend has been accepted by most consumers. We already see huge fleets of busses and transport vehicles converted to natural gas, so it is not going to take very long before we see major conversion to natural gas across the board.

In the near future, fuel cell (presently technology under development) would give natural gas an even greater efficiency advantage. This technology is also under intensive development, and we expect it to be made practical very soon. Due to its flexibility and other qualities, natural gas combined with solar and wind power would find place in this and many other future energy technologies.

With one word, natural gas is the king of the energy markets now, and is the fuel of the future. More specifically, of the near-term future, since its long term future ends in a dead end. As a finite commodity the natural gas supplies will dwindle by mid-, end-century and its kingdom will end. We better be ready to replace it with comparable and long-lasting energy source.

Nuclear Power

As a major power generator for decades, nuclear power is well established, and most countries depend on its reliable, clean, efficient, and cheap power. We cannot even imagine what this world would be without nuclear power. Yet, it has a major problem, and its future depends on a number of factors mostly out of our control. Nuclear power has a poor safety record due to several major nuclear disasters. Nuclear power fits in a very special category of energy sources; these in peril and on trial. The circumstances around these incidents and their consequences have added a high level of uncertainty and fear, which threatens to curtail global nuclear power development. And things could get much worse, if another major nuclear accident happens anytime soon. Such an event could mark the end of nuclear power as we know it.

Like the genie in the bottle, nuclear power brings miraculously huge quantities of useful energy to its masters.

When released by mistake or malicious intent, however, it becomes a monster bringing instant and merciless misery, pain, and devastation to thousands of innocent and unsuspecting victims.

Since the tragic Chernobyl nuclear disaster, nuclear power has been the Cinderella of the energy industry. Like Cinderella, it does good most of the time, but from time to time, things turn out badly. Usually it is the fault of the masters—their negligence or ignorance, that create the disasters. Nuclear power, however, is always blamed for the masters' mistakes.

Like Cinderella, it is never thanked for the good things it does, but is always blamed for anything bad that happens around it. As a result, some love, some envy, and some even hate her. In the end, most of us don't know for sure what to think and or do about our nuclear Cinderella.

The recent Fukushima nuclear accident elevated suspicion and fear surrounding nuclear power. Its veil of mystery, complexity, and danger became even more intimidating for the average person. Because of that, most of us don't know what to say about it for fear of being technically or politically incorrect. So, the controversy on high levels continues.

Nuclear power is a truly mysterious and misunderstood force. One frightening thing we have learned the hard way is that its ominous power is hard to control even with our advanced technologies and methods. This is a big problem, which, regardless of the assurances of energy companies and governments, puts nuclear power on the guilty list, where it will remain until proven innocent.

Nuclear power is unequalled in its fierce responses. No other disaster—natural or manmade—can get even close in magnitude and cruelty. Just whispering the word nuclear sends shivers down our spines and conjures images of explosions, fires, smoke, twisted metal, buildings with blown off roofs, and death.

This evokes respect and terror, inevitably reminding us of Hiroshima, Nagasaki, Three Mile Island, Chernobyl, and Fukushima. These are symbols of the fierce power, frightful widespread devastation, and merciless revenge of a wicked, out-of-this-world force. It is much more than we can even imagine, until we experience the effects it ourselves.

There are awful signs of nuclear power damage around the world, and there will be more before we find similarly useful, but safer, form of large-quantity electric power generation. For now we have no choice, if we want to continue living in comfort provided by unlimit-ed quantities of electric power...compliments of nuclear power generation.

Many people see nuclear power as a serial killer, a mass murderer that can strike any moment and kill thousands in seconds. Yet, most of us cannot live without it. We are, therefore, willing to gamble.

Nuclear power generates a significant portion of our electricity, so we need the beast, which provides us with lots of reliable and cheap electric power on demand. We just have to tame and befriend this necessary evil to the best of our abilities, to control and use it efficiently and safely.

This is not an easy chore, and we are not there yet. So, from time to time, we have problems handling the overwhelming and finicky power contained in nuclear reactors...and sometimes we pay for it with our health and lives.

Nuclear power pushes our technical abilities to the limits, keeping us on our toes—afraid that we might make a mistake to which the beast would respond with unspeakable fury, widespread destruction, and deadly devastation.

Yet, we are fully engaged in the nuclear power generation game, and although many countries are making serious efforts to reduce their dependence on nuclear power, it will not happen easily or anytime soon. As a matter of fact, we need nuclear power now more than ever before, because the world population is exploding and the fossil fuels are becoming too expensive.

We are also acutely aware that the fossils are coming close to their end, and that nuclear is the perfect fuel to help us transition to a post-fossils future. A real catch 22, no? Giving up on nuclear power now, and before we have lined up alternative energy sources, will cause unimaginable hardship during the transition to fossil-less energy reality in the future.

Nuclear Nightmare…

Before we get far into the future, a quick glimpse at the present energy situation in Japan provides a good picture of what nuclear power means to us, and what can be expected with and without it. It is the latest proof that we cannot live with it or without it.

During several tragic days in March 2011, we witnessed the dangers of nuclear power after the Fukushima accident, and now we are witnessing what it is to live without it. The entire Japanese nuclear power fleet was shut down after the 2011 accident, and the country entered, and is still meandering in, a period of energy and environmental crisis.

To replace the lost power, Japan started importing

and burning more coal, natural gas, and crude oil than ever before. This makes it vulnerable to the risks and fluctuations of the global energy markets, so Japan's energy security is on a limb for the duration. As a result of the increased fossil consumption, Japan now also emits more GHGs than ever before. The Kyoto Protocol agreements can wait...now is more important than tomorrow.

Now, imagine all global nuclear power generation shut down for a long while. Not possible? Yes, possible! One more Fukushima-like accident in the near future anywhere in the world would seriously cripple the nuclear industry for a long time. Perhaps forever...

The global nuclear industry would not be able to survive the resulting panic and massive nuclear plants shutdowns around the globe that would most likely follow another major nuclear disaster. This is because people have not forgotten the Chernobyl and Fukushima tragedies, and another one on top of those would be too much to bear. Unprecedented chaotic responses and decisions might just make nuclear power unfeasible for a long time.

There are countries, like France and others, however, where nuclear power provides more than half of the daily power generation. What would France and the other nuclear power-dependant countries do if their power fleets were shut down? Hard to tell, but their energy situation would be even worse than Japan's is now. That, in turn, would increase the risks and compromise their energy security, in addition to increasing significantly the global GHG emissions.

Nevertheless, at least for the foreseeable future, we have to live with nuclear power, keeping our fingers crossed and dealing with its problems, trying to avoid another big surprises at all cost.

Germany and Japan tried to reduce and even eliminate nuclear power for awhile, but are now reconsidering their previous decisions. Germany's nuclear power is back up 100%, while Japan is considering doing the same soon, but gradually.

Although we all are afraid of nuclear power, we need it, since we simply cannot replace it easily with the other sources. U.S. politicians and regulators were not impressed, let alone scared, by the Fukushima events, most likely thinking that such a thing cannot happen here. This is the U.S., after all, with the best nuclear technology and nuclear safety in place. So what can go wrong... go wrong... go wrong...?

We hope they're right, but we must remember that the Fukushima power plant was equipped with U.S. hardware and expertise, and look what happened to it. This cannot happen again, they say. But this is what they said before Fukushima too, so whom are we to believe? Several months after the Fukushima disaster, when the damages and the victims were being still counted, the nuclear industry watchdog, the U.S. Nuclear Regulatory Commission, approved the applications for several new nuclear power plants for the first time since 1978.

In defiance of the Fukushima disaster and its aftermath, the U.S. will add to its fleet of nuclear plants. We will not let one accident stop us in our drive for more and cheaper energy. No sir! But it wouldn't hurt to keep our fingers crossed, because one Fukushima-like disaster in the U.S. would certainly shove a big rod in the nuclear power industry's spokes. The final outcome would be anyone's guess. Now at least two new nuclear reactors are back on the schedule to be built in Georgia. Other states may follow soon.

Is this the first of the last batch of nuclear power plants to be built in the U.S., or is this a new wave of nuclear power increase? Is this the beginning of the nuclear renaissance, we have been talking about for decades? We don't know what will happen, because it could go either way fast. We just have to be very cautious with using nuclear power. Period!

Nuclear Facts

Obviously, nuclear power is very important, so let's see what it is and what it offers in terms of convenient, reliable, and plentiful energy generation, as compared to its cost, safety, and environmental effects.

> *Amazing fact: the specific energy of one kilogram coal (2.2 lbs.) is 24 MJ, while that of uranium-235 is 80,620,000 MJ...or 3.4 million times greater.*

While one kilogram of coal can theoretically produce 8 kWh of electric power, 1 kg. of uranium-235 produces over 25 million kWh of electric energy. This is about 3,400 tons of coal, or 34 train cars full of coal, 100 tons each, generating as much power as one single lump of 1 kg. uranium.

It's hard to imagine, yet this is only the theoretical side of the equation. When all other factors and variables—such as labor, transport, power generating efficiency, and all other losses—are included, nuclear looks even better.

Nuclear power generation also uses less labor, and less transport, it is more efficient in operation, cheaper, and much less polluting. These are huge advantages that should not be overlooked when comparing the different power sources. No matter how one looks at it, however, nuclear contains a tremendous amount of power, which if properly used and controlled, can generate a lot of electricity—much more than any other sources.

If not properly controlled, however, it could do a lot of damage…also much more than any other energy source! One single picture from Chernobyl or Fukushima tells the whole story. Most of us feel lucky that we were not nearby when those nuclear accidents happened, and hope that we will never witness such monstrous events in our lifetime.

Table 1-4. Specific energy of fuels (MJ per kilogram)

Energy Source	MJ/kg
Brown coal (lignite)	24
Firewood (dry)	16-18
TNT	4-5
Compressed hydrogen	142
Natural Gas	40-45
Crude Oil	45-50
Uranium (nuclear grade)	80,620,000

Table 1-4 shows the energy content of different materials (in mega joules per kilogram) that can be used as fuel for different applications, including heat and electric power generation. The incredible energy content of uranium, results in a huge difference, as compared to the energy content of the fossils. This is an enormous advantage for the nuclear power generation, in addition to other benefits and conveniences.

- Nuclear power offers the convenience of transporting and handling much smaller amounts of raw material; a truck load of uranium, for example, would keep producing power for many months and even years. This, compared to endless lines of coal cars loaded and unloaded 24/7 at the mines and coal-fired power plants respectively for the duration.

- And then, once the nuclear fuel is spent, only one truck load of nuclear waste is driven away from the nuclear plant. This, vs. several hundred train cars loaded with ash, soot, and slurry at the coal-fired power plant dumped periodically nearby, or transported to some distant, hazardous dump site.

- The truck full of nuclear waste, however, presents a number of specific and significant dangers too, but in a smaller package. This smaller package is more difficult to handle, because while coal byproducts can be just dumped in specially designated areas and left there unattended, nuclear wastes have to

be packaged in special containers, deposited in special storage sites, maintained, and guarded ad infinitum.

This is one of the big problems the nuclear power industry has not been able to resolve thus far. In addition to the safety of nuclear operations, the disposal of nuclear waste remains unresolved and very controversial issue.

One very important advantage of nuclear power from environmental point of view is the fact that nuclear plants produce less harmful gasses or liquids. At the same time, thousands of tons of toxic smoke, ash, and soot are produced at the coal-fired plants. Not to mention the acres of liquid and solid waste contamination around coal-fired plants and at remote dump sites.

Yet, nuclear is not 100% clean as we will see below. Since there is no perfect solution, we need to pick and chose carefully how we generate and use energy, keeping in mind that in all cases there would be some problems. Because of that, we must be well informed and educated on the issues, in order to make the right conclusions and take appropriate decisions. We dive deeper into the world of nuclear power, and its effect on the global energy markets later on in this text.

Firewood

As we mentioned before, firewood is another widely used energy source, used by millions of people in the underdeveloped countries. Many of these people simply have no other choice. Due to their poverty or isolation from civilization, they have access only to one fuel source—firewood—which is abundant and free in most cases. People in developed countries also use wood, but in different forms and for different purposes, as we will see below.

Figure 1-11. Firewood gathering is a big business

For most of us in the developed world, however, firewood is hidden behind images of poor people gathering sticks and branches in bare lands, as the only source of energy for heating and cooking. Although these images are correct, wood is much more than that. Although it is also much less famous of glorious than its fossil or renewable energy cousins, it has been, and still is, one of the major energy sources worldwide.

The use of firewood in the underdeveloped world is exceptionally important for a number of reasons. In addition to the fact that many of these people simply have no choice, the daily reliance on firewood by the locals have created several phenomena. Excessive deforestation, and desertification are the present day symptoms of these activities. And just recently, we have come to the realization, that firewood is creating health problems at home, and emitting huge amounts of GHG in the atmosphere, thus worsening the already worse climate conditions.

The amount of firewood used world wide is staggering. It is hard to measure, but is thought of as one of the major energy sources now days. The greatest use is by people in the developing countries, although many in the developed countries use firewood to, but for different reasons and in smaller amounts.

Until the middle of the nineteenth century, the major source of energy around the world was wood. The same energy source that has been used since the beginning of time. Even when coal became a major energy source, firewood was the next in line, mostly due to its almost unlimited supply from forests, and its renewable nature.

Wood accounted for over 90% of U.S. energy generation, mostly for home heating and cooking, until just about 100 years ago.

An average of 18 cords (each 4 x 4 x 8 feet pile) of wood per year were needed to heat an average American house in the mid 1800. This is 128 x 18 = 2.304 cubic feet of wood per house. Quite a pile of wood, and a lot of wasted energy.

The heat energy equivalent of this wood pile is actually 2.5-3 times the amount of heat energy needed to heat a typical home today. This was mostly due to the inefficiency of the old fireplaces, which had less than 10% efficiency rating (at today' standards), and which swallowed three quarters of all wood used at the time (approximately 75 million cords of wood).

Stoves are about four times as efficient, but were not popular in the Americas at that time, mostly for convenience reasons. Wood stoves became fashionable in North America in the mid 20th century, and are still quite fashionable, which is one of the reasons wood consumption has increased with time.

Firewood stoves are quite fashionable now days, and are selling at a rate of 200,000 per year in the U.S. alone.

Wood also powered the growing industries of the 19th century too. This included steamships and trains running on wood fire, and burned nearly 8 million cords of wood a year at the time. The iron and steel industry were the major industrial users of wood, although there was a twist. Most of the energy for iron smelting came from charcoal, which was made out of 1.5 million cords of wood per year.

The charcoal (made from wood) used in iron smelting totaled some 700,000 to 750,000 tons. This, compared to estimated 750,000 tons of charcoal are used every year by outdoor chefs in the back yard barbeques and tailgate parties in this country alone.

Charcoal is still important worldwide. Brazil uses charcoal in 45% of its iron smelting. 3.6 million tons of charcoal are used in Brazil annually, and consumption is still increasing. Charcoal is also a major fuel in less developed countries such as Ghana and Kenya, where 250-300,000 tons are used every year.

Over 50 million cords of firewood were used annually for home heating alone, and wood and wood residue provided almost 2 quads of total energy per year, most of it in the wood products and paper industries. In these industries these "biofuels" served about half of the total energy needs too.

Consumption of fuel wood in this country reached a peak of almost 150 million cords in 1870, since at that time wood was the primary fuel. In comparison, the 100 million cords used now days in the US account for only a few percent of the total primary energy input.

Even if the use of wood and biomass increases to a maximum, the total annual potential of biomass energy generation, could provide only about 1/4 of the U.S. primary energy input.

The situation is different in the developing countries, of course. Firewood is, and will remain for a long time, a primary fuel especially in rural areas.

The gasses, which are produced during the wood fuel burning process are serious pollutants, and account for a large portion of the world's environmental and climate change problems. We take a close look at the environmental problems of the energy sources in our book on the subject, *Power Generation and the Environment*, published by Fairmont Press in 2014.

The Renewables

It would be unfair to complete this section without mentioning the role of solar and wind power generators in the global energy markets. The problem here is that these energy sources do not use any energy materials that can be considered part of the energy markets, since sun and wind are free. Free products are not part of any market, so we must look at these from a different point of view.

How to put a price on sun and wind is a question for the next generations, because most of us now days take them for granted. Many people also think of them as useless and gutless attempt to emulate the real power generators. Others are even frustrated by their presence, for it complicates the function of the energy markets, and at times affects negatively their profit margins.

The reality, however, is that solar and wind power is all we and the future generations have to rely on when the fossils are gone. And so, like it or not, the renewables are here to stay for a long, long time. Very possibly until the end of life on Earth as we know it. So we at least must consider, and get ready for the transition.

But since there are no energy materials to speak of in the renewable energy markets, we need to cut this section short. We must agree that this is a good thing, since there are no additional fuel expenses to worry about, as is the case with the fossil, or nuclear power sources.

Instead, there are technical difficulties stemming mostly from the fact that sunlight and wind are intermittent. When sun is behind a cloud, or at night, even the most efficient solar power generators quit generating power. Instead, they just sit idle until the sun shines again…whenever that might be. Period. With so many variables, this is not a good way to run a profitable business, is it?

The same is true for wind power. No wind, no power generation. It is that simple, and overly frustrating. It puts a big gap in the energy equation too, which has to be resolved sooner or later. We will take a close look at the variability phenomena and the possible solutions in the following chapters of this text

Renewables' Dilemma

Energy in all its forms, has always been, and still is, the major force that created, and drives, everything in the world. The sun's energy was the first energy source that supported life on Earth. Then, over 100,000 years ago, our predecessors started using energy materials for a first time by burning firewood. This was the first man-made energy conversion process, which they developed and managed successfully to survive. It provided warmth and made their food more digestible. It eventually led to their industrial progress.

Wood being a primary source of solar energy conversion was used extensively by our primates. As a matter of fact, wood burning (for cooking and home heating) was the primary energy source until the mid 19th century. After that coal became the fuel that provided most energy and contributed most to the development and industrialization of the European and American civilizations. Coal was what people used to warm their homes, and what turned the wheels of transportation and industry.

Coal replaced wood, which is inconvenient for use in most of the then advanced applications. Wood is also much lower energy fuel per volume than coal, so its use plummeted as the industrial progress advanced.

By the end of the nineteenth century the first of the great energy source replacement cycles was under way, and coal was providing 90% of this country's primary energy. It was abundant in the industrialized East, and had more Btus per pound than wood. Coal became the fuel for the fledgling steel industry and for the trains that ran on the steel rails. This went on until 1950, when oil pushed coal out of the picture.

Oil was even more convenient fuel, especially to transport and use in vehicles. It offered much more energy (Btus) per pound. It replaced coal in the railroads and pushed the nation into the automobile age, which we are still in love with. By 1970 oil and oil products provided 75% of the nation's energy, with coal less than 20%. But coal found its niche with the electric utility industry and made an impressive comeback. It is now the fuel of choice in most electric power generating stations.

Natural gas was wasted early on by burning it at the oil wells, but became the fuel of choice for home heating and industrial furnaces when new gas pipelines were constructed to the Midwest and the East of the country.

The U.S. gas production peaked in 1970, and its use increased, while most of the oil was, and still is, imported. In 1977, imported crude oil was almost 50% of the total oil used, and almost 25% of our total primary energy consumption. Oil imports continued to increase through the years to over 60% total crude oil imports from 80 countries.

Slowly but surely, the United States and much of the world built a fossil fuels dependent industry and society. This period, however, is coming to an end, because the fossils have problems, as described above, and we are now looking for other energy production and use.

No one knows what exactly will happen in the future. but we know now that there is a problem that needs to be addressed and solved. We are presently in the "addressing" stage. It is an important stage, because in order to solve a problem one must recognize that there is a problem.

We are just now agreeing that there is a problem, and that we need to address it fully—which is the goal of this book as well. Herein we intend to present a complete and unadulterated account of the different energy sources; their technical characteristics, as well as their effect on the socio-political and environmental aspects of life in the U.S. and the world.

The solutions will come only after we have reached an agreement, for which we need to get prepared by complete understanding the issues at hand—which is the other key goal of this book

The fossil fuel industry and utilities executives have their agenda, geared to sell their products. Period. Everything else is secondary. Any talk about environment, climate warming, environmental disasters etc., is secondary to their purposes—which is making a buck. Often this talk is in conflict with their interests, so we must view everything they say and do with a grain of salt. This is not to say that these are bad people, nope!, but in the capitalist society everyone must defending their own turf—especially if it is what they rely on making a living—and it is truly their turf we are invading, and their livelihood we are attacking with such talk, so we must be considered and proceed carefully.

Government officials, politicians, regulators and utilities officials have their own agendas too. Although the intentions are good, the results from the latest energy markets development show that there are complex interactions and numerous interwoven interests that are not easily understood, which makes things very complicated and the issues not easily solved.

The U.S. energy policy has been that of, "Drill baby, drill," supplemented by large quantity of crude imports. Recently, due to the new crude oil and natural gas discoveries, the U.S. strategy was modified to that of, "Good-To-The-Last-Drop." Pumping huge amounts of oil and gas means a great windfall for oil and gas producers, and the government. And of course, we hear lectures about the benefits of these fuels and their miraculous benefits to our society.

We must recognize that we depend heavily on coal, oil and gas in our daily lives. They are the evil that we cannot do without.

So we must treat these energy sources with respect, and do our best to work towards the best outcome possible for the sake of our children and their children. The immediate issue is that our industrial society—now and for the foreseeable future—depends almost totally on the fossil fuels for its proper functioning. The continuous input of high-grade energy resources is critical to our progress, and if they decline, the system will falter.

We saw the effects of high oil prices during 2008 oil price hikes and constrained oil supplies at the time, and the way they affected the markets. They actually suppressed our economic activity and pressured our already fragile financial system into a defense pose. High oil prices were the straw that broke the camel's back, and marked the start of the great 2008-2012 economic slow down in the U.S. and the world.

The unexpected effect of the economic slow down was the renaissance of the renewable energy sector. Wind and solar benefitted most from the generous and abundant government subsidies during that time and they increased their share in the energy market very fast. The U.S. public supports the renewable technologies, so their growth is assured for now.

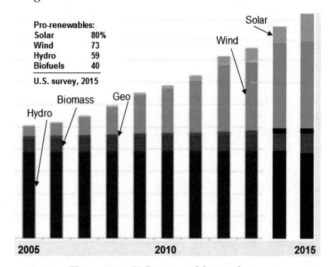

Figure 1-12. U.S. renewables to date

But now we are at the peak of our technological development, which has brought a lot of oil and gas onto the U.S. and global energy markets. But this is the bottom of the barrel, since more efficient, cheaper, and cleaner, fossil fuel technologies, although feasible are not very likely to emerge in the near future.

And so, we should not rely on a technological miracle coming our way any time soon in this field. The known trends and data tell us clearly that we should not wait any longer, but instead we must start looking into the most promising on the long run renewable energy technologies. We must also look into new and more efficient ways to use energy at home and in the industry.

The end of the fossils is near, and the transition to renewables is inescapable. It will have to happen some day, since there is no other way on the long run. The only question is whether we, as a responsible society, will consider starting the trend now, or if we would chose to let somebody in the distant future take care of it.

Do we have a feasible and focused plan for a rapid transition to renewable energy technologies, and new ways to use energy? If we don't, we will just let the global energy chaos prevail, and will just have to endure the economic up and downs, as we go from one energy-, or environment-induced crisis to another.

This is not a good or smart way for responsible people to behave in the face of the inevitable. Nor a good way to account for our actions, or lack thereof, to the upcoming generations.

EFFICIENCY OF ENERGY CONVERSION AND USE

There are few rare instances in which energy (or the fuel it is contained into) is used without conversion, such as heating of buildings using geothermal power (steam coming from the Earth's core), or burning crude oil in some industrial operations. In most other instances, energy is converted from one form to another, as needed for more efficient or more convenient delivery and use. In some cases, several conversions are necessary before we obtain energy where we want it and in the way we want it. This creates a number of inefficiencies and energy losses during each step of the long process.

For example, nuclear energy generated by the radioactive elements (chemical energy) in the nuclear plant is converted into steam (heat energy), which then turns a turbine and a generator (mechanical energy) to generate flow of electrons in the wires (electric energy), which (as the final energy product) can then be used whenever and where ever we need it. Similarly so, solar energy is captured by solar panels to be converted into electricity before sent into the electric grid.

Each energy conversion type and cycle has advantages and disadvantages, and is accompanied by significant losses of energy during the different steps of the transfer process. These losses are usually caused by inefficiencies in the conversion process and related equipment.

The key measurement of energy efficiency is then expressed by a number, which we assign to the completeness of these conversions in percentage—100% being total and complete conversion.

100% efficient conversion is not possible with today's technologies. Instead, 10%, 20%, and 30% are the usual numbers.

For example, if we assume that the efficiency of geothermal or solar power as 100% BEFORE the conversion process, then during the conversion, losses at every step and every inch of the process sequence bring losses—be it heat or electrical power.

Geothermal power is used directly to power steam turbines, where it arrives with its 100% efficiency potential, only to undergo a series of losses, starting at the well mouth. Here the steam enters long pipes leading to the turbine, or boiler, and its temperature (or potential efficiency) drops by several degrees. In the turbine, mechanical resistance and inefficiency of the turbine blades cause it to lose more of its efficiency.

After that, mechanical resistance in the electric generator and electric resistance in the wires cause additional losses. In the end, by the time thus generated electric power arrives at its final destination, the initial 100% efficiency has dropped to 20-40%, depending on location and type of equipment used.

As another example, solar energy (sunlight) arriving with 100% of its potential efficiency is converted into electricity by the solar panels, where more than 85% is usually lost. This is due to the inefficient conversion efficiency of the solar cells, electrical resistance in the wires, etc. factors. The solar equipment and the related processes, need a lot of improvement before achieving higher conversion efficiency.

Coal, on the other hand, undergoes a long, long process from the mines, to transport vehicles, pre-burn processing, and burning. It, however, does not lose any of its efficiency during these steps and arrives at the power plant with its original 100% potential efficiency. From here on, the efficiency of the final product—electricity—will drop during each step of the process. At the end, the coal-fired power plant would be lucky to be able to claim 30-40% energy conversion efficiency.

Energy Content

The energy content (or 100% potential efficiency) of the different fuels is different, and is determined by their heat value and burning properties, as in Table 1-5.

In more practical terms, we can generate 1 kW/h of electric energy by burning:

0.00052 tons, or 1.03 pounds, of coal
0.01003 Mcf, or 1,000 cubic feet, of natural gas, or
0.0016 barrels, or 0.2 lbs. of fuel oil

Table 1-5. Energy generated by different fuels

Fuel	Energy
Natural Gas	1,030 Btu/cu ft
Propane	2,500 Btu/cu ft
Methane	1,000 Btu/cu ft
Landfill gas	500 Btu/cu ft
Butane	3,200 Btu/cu ft
Kerosene	135,000 Btu/gal
Biodiesel oil	120,000 Btu/gal
Gasoline	125,000 Btu/gal
Anthracite coal	26,000,000 Btu/ton
Bituminous coal	24,000,000 Btu/ton
Electricity	3,412 Btu/kW
Softwood	15,000,000 Btu/cord
Hardwood	20,000,000 Btu/cord
Corn, shelled	15,000,000 Btu/ton

Or, the number of kW/h generated per unit of fuel used are:

2,000 kWh per ton of coal, or 0.1 kWh per pound of coal

100 kWh per Mcf (million cubic feet) of natural gas, or

610 kWh per barrel, or 14.5 kWh per gallon, of fuel oil

Obviously, the different fuels have different energy (or heat) value, which determines the efficiency of the burning process too. It all depends on the type and quality of the different fuels, so the fuel's state (gas, liquid, or solid) and heat values must be considered, in order to chose the proper fuel for each power plant design.

Energy Efficiency

Energy conversion efficiency depends on many factors, the major of which are, a.) the completeness of the conversion process, b.) the quality and the degree of usefulness of the output product as compared to the input.

For example, part of the heat produced from burning coal in a furnace is dissipated as waste heat and so the process would not be 100% efficient. As a matter of fact, we would be lucky to get 30-40% efficiency from any and all energy conversion processes we use now days. This means, alarmingly so, that more than half of the fuels we use go up in smoke...which is another big problem we review in this text.

The coal furnace in this example is an energy converter and the burning process is an example of an energy transformation of energy in the fuel from its potential to kinetic state.

In all cases, the efficiency of a process can be expressed, ad measured, as:

$$\eta = \frac{P_{out}}{P_{in}}$$

where:

P_{out} is the energy that is produced (energy kcal produced)

P_{in} is the energy input (coal kcal burned)

Case Study: Practical Coal Efficiency Calculations

In the coal-fired power plant example, coal comes with its 100% energy content (100% potential energy) and the potential of operating (burning and generating power) at 100% efficiency. This, however, is an ideal case scenario...if and when coal is burned and converted to electricity in 100% efficient energy conversion equipment and processes. Since in real life we don't have such processes and equipment now days, we have to deal with the inefficiencies as part of the final efficiency calculations.

So, when the coal trains deliver coal to the power plant, it is crushed into smaller (almost powder) particles, transported to, and injected in, the boilers. So far so good. Coal still has its 100% potential energy and the potential of providing electric power at 100% efficiency. But here is where reality hits, things don't go as in the ideal case, and so the efficiency start dropping fast.

- During burning in the boilers, some of the coal is used to heat the boiler's walls, resulting in wasted energy just to keep the process going. Some of the coal is also not completely burned, and is removed as smoldering ash. Let's say 15%, so the original 100% efficiency of the incoming coal now drops to 85%.

- After the water is converted to steam by the burning coal in the boilers, it is sent at high speed to the turbine, but on the way it heats the cooler pipes, where it loses more of its energy to heat them and to overcome the resistance of their walls. Another 15% efficiency loss, so we are now down to 70% efficiency.

- In the turbine, the incoming steam has to fight with the resistance of the blades, which are trying to stay stationary, the bearings, which do not want to turn, and the turbine chamber's steam injection orifice and its walls, which constantly grab and pull the steam back. Another 10% are lost here to bring the efficiency down to 60%.

- With its work done, and cooled down during its path, the steam returns back in the boiler to be heated again. On the way back it loses more of its potential energy, which affects the next cycle it goes through.

- The spinning turbine's blades are attached to an electric power generator, which generates electricity. Here, the generator's bearings and winding exert mechanical and electrical resistance respectively, and more efficiency is lost. Maybe another 15%, or we are now down to 45% of efficiency.

- Thus generated electric power is sent into the power grid by first going through electrical transformers and conditioners, where the efficiency is dropped another 5%, or down to 45% of final efficiency. The last stage is not related to the coal's efficiency at all, but affects the final efficiency of the power plant, so it cannot be ignored.

So, coal comes ready to play, with 100% potential energy at our disposal, but we can use only 45% of its potential energy. Due to our inefficient equipment and process, we waste more than half the coal we use. The situation is quite similar with the other energy sources and fuels.

Effectiveness Measurements

Efficiency is a technical and physical term, which can be translated in degree of usefulness of an energy conversion process. It can also be expressed in terms of its effectiveness and efficacy, which are related and are used in some engineering calculations.

Energy conversion efficiency is usually expressed in numbers between 0 and 1.0, or in percentage 0 to 100% with the maximum being 1.0 or 100% respectively. Actually, strictly technically speaking, 100% is assigned only to the perpetual motion machine, so efficiency of 0.95, or 95% would be considered an efficiency ceiling in our modern society. In practice, we are happy (and lucky) to get 20, 30, or 40% efficiency.

Examples of the measurement of efficiency:

Pump = [(kg.m²/s²)/(kWh)]
Turbine = [(kg.m²/s²)/(kg.m²/s²)]
Engine = [(kW)/(kg/s) x (kJ/kg)]
Steam gen. = [(kg/s) x (kJ/kg)/(kg/s) x (kJ/kg)]
Boiler = [(kg/s) x (kJ/kg)/(kg/s) x (kJ/kg)]

where
kg is in kilograms
kW is in kilowatts

kWh is in kilowatt/hours
m is in meters
kJ is in kilojoules

In some exceptions, effectiveness measures can exceed 1.0, or 100%, but these usually apply to heat pumps and other devices that move heat rather than convert it.

Table 1-7. Efficiency of energy convertors

Energy Use	Btu	kWh
Human daily activity	6000	1.8
Gallon of gasoline	10^5	30
Mfc of natural gas	3×10^5	90
Lightning bolt	10^5	300
Barrel fuel oil	2×10^6	600
Making one car	10^7	3,000
Ton of coal*	2×10^7	6,000
Rocket launch	10^9	300,000
Boiling Lake Michigan	10^{13}	3×10^9
H bomb	10^{14}	3×10^{10}
World energy use	10^{17}	3×10^{13}
Photosynthesis, annual	10^{19}	3×10^{15}
Solar energy, annual**	10^{23}	3×10^{19}
Earth rotation	10^{25}	3×10^{21}

Note: The high efficiency of the electric motors and electric heaters above has to be taken with a grain of salt, because they use secondary energy that has already been converted from fossil to electric energy, where it has lost over 50% of its efficiency. So the conversion of electric energy into motion (by the electric motors) and heat (by the heaters is just another, final, and much more efficient step in the overall energy conversion process.

The efficiency of power generating plants varies according to fuel and technology used. For example a wind power generating station could be 50% efficient, if the wind blows most of the time. This means that half of the energy of the incoming wind was converted into useful power.

Similarly, a 15% efficient solar module means that 85% of the solar energy coming in at 1,000W/m2 is wasted—either by reflection, or conversion to waste heat, or is simply not converted due to the cell's inefficiency.

If our solar module produces 150 Watts of power when exposed to 1,000 Watts of solar power, then its efficiency in this case would be:

$$\eta = \frac{P_{out}}{P_{in}} = \frac{150W}{1000W} = 15\% \ (0.15)$$

But there is another variable that has to be considered when calculating the efficiency of power plants. It is the capacity factor, which denotes the overall capacity of a power plant or a single solar module to produce power during a certain time period (usually 24 hrs, week, month, or a year).

Capacity Factor

The capacity factor of a power plant is the ratio of the actual output of a power plant over a period of time and its potential output if it had operated at full nameplate capacity the entire time. The name plate is the maximum possible output, or the output determined by the manufacturer.

To calculate the capacity factor, we take the total amount of energy the plant produced during a period of time and divide it by the nameplate capacity.

As an example a large solar power plant rated at 1,000 MWp (megawatt peak) produces 200,000 MW/h (megawatt hours) in a month. The number of MW/h that would have been produced had the plant been operating at full capacity is determined by multiplying the plant's nameplate (maximum) capacity by the number of hours in the time period. 1,000 MW × 30 days × 24 hours/day is 720,000 MW/h.

The capacity factor is then determined by dividing the actual output with the maximum possible output. In this case, the capacity factor is 0.28, or 28% of the total potential output.

$$Cf = \frac{200,000 \text{ MW/h}}{1,000 \text{ MWp x 30 days x 24 hrs/day}} = 0.28$$

This particular power plant has a capacity factor of 0.28, or 28%, which is actually quite good for a solar power plant now days. It is low, compared with fossil fuel plants which operate with 80-90% capacity factors, the reason for which is that solar power cannot be generated at night and/or during cloudy or rainy weather.

Note: The capacity factor is different from the availability factor, which is the actual operating time minus repairs and such. It is also different from the conversion efficiency, which we reviewed above. The capacity factors vary greatly depending on the type of fuel that is used, the technology, and the design of the plant. In the case of renewables, location is also a great factor determining capacity.

The capacity factor also determines the overall effectiveness of the energy sources and/or the installations that use them to generate power.

Lifetime Energy Balance

Life time energy balance (LEB) is the energy used during the energy sources production, transport, installation, and operation, compared with the power they generate during their useful lifetime. LEB is an important factor, which needs to be kept in mind when analyzing and calculating the energy benefits from a PV power plant. LEB basically tell us how much energy we save by using PV energy generating sources.

The solar cells and modules manufacturing process starts with melting and purifying sand in huge, dirty, dangerous, and energy-guzzling furnaces. The produced metallurgical grade (MG) silicon is crushed in huge mills and further purified in a complex network of chemical and electric equipment, again using enormous amounts of natural resources and energy. The product at this point is solar grade (SG) silicon of varying purity, the quality of which depends on the raw materials and process quality.

The SG silicon is crushed again and melted again in large furnaces at high temperatures for a very long time. Thus produced ingots of mono or poly silicon are sliced into thin wafers onto which the solar cells will be built. The wafers are cleaned, baked, fired several times, and coated several more times until finally a solar cell emerges at the end of the line. They are then sorted, lined up, and vacuum-bake sealed into the module frames, then transported to the location. The next time these PV modules are transported will be at the end of their life cycle, going to the crusher.

All this requires energy, materials, and resources. Fortunately, the production equipment and processes are very efficient these days, but even so, making a solar cell and module is an energy-consuming undertaking.

The LEB Equation

Consider this equation for a lifetime, life energy balance (LEB):

$$LEB = \frac{E_{prod} + E_{trans} + E_{inst} + E_{use} + E_{decom}}{E_{gen}}$$

Where:

E_{prod}	=	energy used during production of materials, wafers, cells, and modules
E_{trans}	=	energy used to transport materials, modules, and BoS to PV site
E_{inst}	=	energy used to assemble and install the PV power plant
E_{use}	=	energy used to operate the PV power plant
E_{recyc}	=	energy used to decommission, transport, and

recycle the PV field

E_{gen} = energy generated during the life of the PV power plant

In all cases, Egen must be much higher than the sum of the other sources of energy used in order for the system to be an effective energy source.

The term "energy payback" describes this "energy in-energy out" ratio and is what we will have to consider when designing, pricing, and justifying a PV system. Or in other words, how long do we need to operate a PV system before we recover the energy used, and pollution generated, during its manufacture, transport, and operation before it is decommissioned?

There are energy payback estimates ranging from 1 to 4 years for different technologies; more specifically:

- 3-4 years for systems using current multi-crystalline silicon PV modules

- 2-3 years for current thin-film PV modules

These estimates, however, vary from product to product and from manufacturer to manufacturer. Cost of energy (crude oil in particular) also has a great effect on the estimates, so these numbers must be adjusted periodically. The ever changing socio-economic situation in different countries is another great factor and we expect that the present-day economic slow-down and worldwide financial difficulties will drastically reshape these and most other estimates as well.

Lifetime CO₂ Balance

Similarly, a system's lifetime CO_2 balance (LCB) is a factor that takes into consideration the CO_2 emitted during the manufacturing and use of energy components vs. the amount of CO_2 saved by using the PV components instead of coal- or oil-fired power generation.

Note: This is most adequate for use with renewable energy sources—solar, wind, biomass and the like.

Solar wafers, cells, and modules manufacturing processes, for example, generate significant amounts of CO_2 and other harmful gasses, which must be taken into consideration when talking about the advantages of PV technologies over the conventional energy sources.

The LCB Equation

Consider this equation for lifetime CO_2 balance (LCB):

$$LCB = \frac{C_{prod} + C_{trans} + C_{inst} + C_{use} + C_{decom}}{C_{save}}$$

Where:

C_{prod} = energy used during production of materials, wafers, cells, and modules

C_{trans} = energy used to transport materials, modules, and BoS to PV site

C_{inst} = energy used to assemble and install the PV power plant

C_{use} = energy used to operate the PV power plant

C_{recyc} = energy used to decommission, transport, and recycle the PV field

C_{save} = energy generated during the life of the PV power plant

In all cases, C_{save} must be much higher than the sum of the CO_2 generation sources in order for the system to be an effective energy source. Nevertheless, and regardless of the LCB ratio, the PV, or wind power plant will receive carbon credits for the CO_2-free power generated during its lifetime.

Here again, we have seen estimates that the total quantity of CO_2 emitted during the cradle-to-grave cycle of PV components is compensated within 2-4 years of CO_2-free PV power generation, by reducing CO_2 emissions as compared to conventional power generators.

In all cases, solar and wind power generation are much less energy hungry, polluting, and damaging to the environment and the living things in it, than the conventional energy sources (fossils, nuclear, etc.). Nevertheless, we must always consider the fact that they are not totally pollution free, and to watch out for the problems along their life cycle.

PRACTICAL ENERGY USE

Energy comes in different forms, some of which are not easy to measure. We, however, need to have a good idea of what it is that it is and what it does, in order to be able to comprehend and compare its different states and properties.

Table 1-7 is a list of energy used in different areas of our activities and under different conditions:

*Note: The theoretical heat content in coal is in the 3,000-6,000 kWh per ton range. When burned, converted into steam and then into rotational energy of the turbine blades, however, the coal's energy efficiency drops significantly, so that in practice we get approximately 2,000-3,000 kWh electric energy from each ton of coal—depending on the type of coal and the efficiency of the power plant's equipment.

At the same time, solar, wind, hydro and biofuels

Table 1-7. Comparative energy use.

Energy Use	Btu	kW/h
Human daily activity	6000	1.8
Gallon of gasoline	10^5	30
Mfc of natural gas	3×10^5	90
Lightning bolt	10^6	300
Barrel fuel oil	2×10^6	600
Making one car	10^7	3,000
Ton of coal*	2×10^7	6,000
Rocket launch	10^9	300,000
Boiling Lake Michigan	10^{13}	3×10^9
H bomb	10^{14}	3×10^{10}
World energy use	10^{17}	3×10^{13}
Photosynthesis, annual**	10^{23}	3×10^{19}
Earth rotation	10^{25}	3×10^{21}

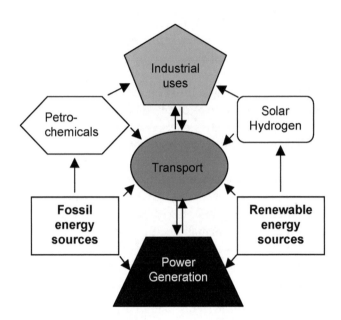

Figure 1-13. Energy sources and their use.

energy is difficult to present in the conventional energy terms, and even harder to compare with the conventional fuels. Their importance in the world energy generation is gradually increasing, however, so in the following chapters we will review in great detail their contribution to the world energy balance, as well as the environmental consequences of their production and use.

The overall energy use in the U.S. can be presented approximately as:

Table 1-8. Energy use in the U.S.

Energy Use	%
Space heating	15
Water heating	5
Cooking	2
Air conditioning	5
Refrigeration	2
Raw materials	7
Process steam	6
Direct heat	8
Electric motors	5
Cars	13
Trucks	6
Airplanes	2
Trains	1
Other	23

Figure 1-13 shows the main energy sources—fossils and renewables—and their use in the transportation and power generation sectors, as well as in different industrial applications. While this looks like a solid and stable way to handle things, it is far from it. This is because the present balance, as shown in Figure 1-13,

The fossils, which represent the majority of the energy markets will not be around forever. So we must be aware and take measure for the time when they become extinct. Yes, this is not a priority right now, but we should not forget this fact and act accordingly. Unfortunately, this is not happening on global level, so the future balance of the energy sources remains uncertain.

Energy Types

As we saw above, there are many energy types, materials, and fuels. We can divide these into several categories, according to their specific characteristics and practical applications.

The major practical division is that between primary and secondary energy sources, as follow:

1. *Primary* energy sources, which include coal, oil, natural gas, solar, wind, water falls, tidal, biomass and geothermal. These are called primary because they are found in nature and can be used one or another way to generate power (heat and/or electricity).

2. *Secondary* energy sources, which include electricity and hydrogen, since these are not found in nature, but instead are obtained by a conversion of the primary energy sources. I.e., burning coal to generate electricity, or using solar energy to decompose water for hydrogen production. Refined fuels, such as gasoline and diesel are also secondary sources, because they are obtained by processing a primary fuel, crude oil.

Another classification of the major energy sources is that of renewable and non-renewable types:

1. Non-renewable energy sources include coal, oil, natural gas, and nuclear. They are non-renewable because the quantity of their deposits in the Earth's crust are limited, slow to replace and will be eventually exhausted at the present rate of use.

   Note: Some of the non-renewable energy sources (coal, oil, and natural gas) are also called fossil fuels, because they were made in primordial times by anaerobic decomposition of fossils (vegetation and animals).

2. Renewable energy sources include solar energy, wind energy, hydro energy, tidal energy, biomass energy and geothermal energy. As the name suggests, these are constantly renewed by natural processes and will never be exhausted under normal use. Exceptions are hydro and biomass, because they depend on other (natural and man made) conditions that might limit their supply.

A third classification of energy sources depends on their initial cost:

1. *Commercial* energy sources are these that are usually sold or traded as entity (like selling a ton of coal), or their equivalents (such as selling one kilowatt of electricity made by burning some amount of coal in the coal-fired power plant).

   Note: At a cost of $100/bbl in the complex 2012 world energy, economic, and financial situations, the energy-related commercial (monetary) transactions in the global energy markets were estimated at about $2.0 trillion annually. Later on, during the 2014-2015 drop of oil prices to below $50/bbl, this number went way down, bringing unprecedented chaos and pain to most of the participants in the global energy markets.

2. *Non-commercial* energy sources are these that are freely obtained, such as sunlight and water. They can be freely used as is, for example the case of solar cooking using the sun's heat, or turning a water wheel at a stream to power an irrigation pump. Although they don't have a commercial value in their "as is" form—primarily because the oil companies have not figured out how to patent and include them in their portfolios as yet—they can, nevertheless, be used commercially and at a profit.

Table 1-9 gives us a good idea of the potential use of the different energy technologies, based on their initial cost. Coal, natural gas, nuclear power, hydropower, biomass, geothermal power, and onshore wind are consistently cheaper than the rest. Solar (PV) follows close by and is expected to join the ranks of the cheaper technologies sometime soon. This trend is expected to continue for the next decade, or more.

The primary energy sources (fossil fuels mostly) are used extensively in the actual process of generating practical and useful type of energy (as in electric energy, or heat) for use in our daily lives. These are complex, expensive and often dirty and dangerous processes. There are also a number of steps needed to get the different types of energy sources ready for generating secondary energy, like electricity or gasoline.

To start this process, we need to get the raw materials, which the fuels are made out of their original location (usually deep underground). These are mined, dug out, pumped up, or otherwise removed from where they have been sitting thousands and millions of years undisturbed. These places are most often deep under the ground, or even deeper under the ocean floor.

Once we get them out, we must transport them to a processing plant, followed by more transport to, and storage at, power plants. Finally, we burn them to generate heat or electricity, which is then sent into the grid for distribution to homes and businesses.

In most cases, the primary energy sources (the fossils especially) cannot be used in their "as is" state, so they have to be converted into secondary energy sources (heat or electricity). They can also be processed into other materials, like gasoline, paints, fertilizers, medicines, and different chemicals. These are long, complex, and expensive processes, which require a lot of equipment and people.

During each step of this process, huge amount of different types of energy are used, and a lot of gaseous, liquid, and solid pollution is created in return. The balance sheet of effort spent, energy used, expenses paid, environmental pollution generated and other negatives, vs. the positive aspects of these processes is also long and complex. It is part of the energy markets, so we take a close look at the key components in this text as well.

All things considered, now days the energy sources are used primarily for generating electric power and for transportation.

Power Generation

Power generation starts with the production of energy sources, such as coal, oil, natural gas, nuclear

Table 1-9. Cost of power generation in $/MWh (World Energy Council, 2014)

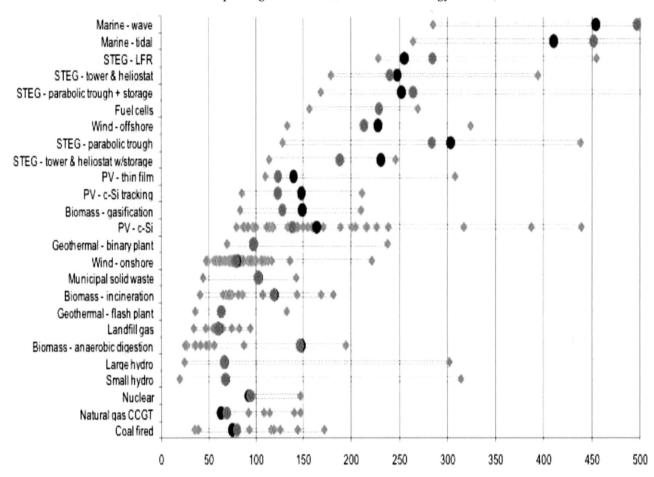

materials, biomass, etc. materials, which we also call fuels. And, of course, we have to consider the renewable energy sources, which, unlike the fossils, do not use any fuels or energy materials, but instead generate power by converting the natural energy of sunlight, wind, ocean waves, etc.

All power plants use some type of fuel, be it primary, or secondary, renewable or not. The fossil fuels (coal, oil, and natural gas) account for more than 90% of the total amount of materials used in energy generation. They have been and still are the most important primary energy generating sources around the world. The rest of the total global power generation is contributed by hydropower, nuclear energy, and in much smaller amounts by solar, wind, and geothermal power.

Solar energy has a special place here, because it has been involved in the creation of most of the primary and secondary fuels since the beginning of time. This is another complex process, which requires another book to review. For our purposes, we will consider solar power generation only, which is the actual production of energy by photovoltaic or thermal solar conversion devices.

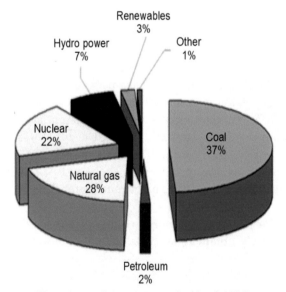

Figure 1-14. Power generation in the U.S.

It is obvious from Figure 1-14 that coal rules the power generation sector. The other fuel types are creeping up in use, but coal is still on top, and is foreseen to be there for a very long time.

The Costs

The LCOE numbers in Table 1-10 show that there is a big difference in constructing and operating a conventional combined cycle (CCC) gas-fired power plant vs. a solar CSP plant. This is because coal power is well established technology, using readily available conventional equipment, while CSP technology is relatively new technology, using specially designed and made equipment. There are a also a great number of other factors at play as reviewed in this text, and as detailed in our earlier books on the subject, see reference 18 and 19.

Table 1-10. Cost of LCOE (in $/MWh)

Energy source	Capital cost	O&M cost	LCOE cost
CCC gas	18	2	68
Natural gas	40	3	80
Hydro	77	4	90
Coal	65	4	100
Wind*	83	10	100
Geo	77	11	100
Nuclear	90	10	115
CCS coal	95	10	140
Solar PV	145	8	160
Solar CSP	210	40	250

*Note. As a clarification,

1. While coal power plant is much cheaper to construct than any solar plant, it uses millions of tons of coal, which makes it expensive to operate. Also, the new EPA GHG standards make it impossible to meet by any coal technology, so no new coal plants will be build any time soon in the U.S., regardless of the price and anything else.

2. There is another important factor that deserves a special attention and evaluation: it is the variability of the renewable energy sources. It is important to understand that the conventional technologies (coal, gas, and oil) operate at 85-90% capacity, or basically they are available 24/7/365. At the same time, the renewables (solar and wind) operate at 20-30% capacity, or they are available only 1/4 to 1/3 of the time.

3. It is also important to note here that:
    a. Solar is most efficient during the day, when electric power is most needed, and
    b. combining solar and wind at some location provides much higher capacity factor, thus their combined availability rate can be increased up to 75%.

If and when the above factors are managed to the full advantage of the variable power sources, and if and when energy storage becomes more efficient, wind and solar will then, and only then, become major sources of base load power in the U.S. and the world.

Presently coal and gas energy prices are the lowest (with externalities* excluded) and are the standard of the power generating industry today. If we assume coal prices to be 1.0, as in Table 7-2, then the other technologies are much higher, with the rest wind and solar usually more expensive…for now.

*Note: *Externalities* in this case are the negative effects of coal use on the economic activity on a third party. When coal is mined, transported and burned to generate power, external costs include the impacts of; air, soil and water pollution, toxic coal waste, the long-term damage to local and global ecosystems, and most importantly; their effect on human health and life.

If the total cost of externalities is included in the energy prices, however, then coal, oil and gas-fired, as well as hydro and nuclear power prices would be increased greatly—several fold in most cases.

We take a very close look at the power generation sector and the related energy markets in the following chapters. There we analyze the different process steps of the cradle-to-grave life cycle of all energy sources from technical and marketing point of views.

Transportation

The economies of many countries are experiencing unprecedented rise, which is expected to accelerate during the next decade or two. As of the summer of 2014, at least part of the progress can be attributed to the low crude oil prices. This only shows how important oil is. And since it is used mostly for transportation, this also points to the importance of transport vehicles in our modern lives.

As a consequence of the developments of late, the global energy markets are changing quickly and promise even greater fundamental changes in the future.

Transportation is a major part of the rise of economic prosperity and global trade. In addition to the already busy trade routes, new logistics passageways are under development between Asia and Africa, between Asia and South America, as well as with, and within, remote areas of Asia and Africa.

These activities mean movement of cargo and people from place to place, which requires huge amount of equipment and never ending flow of different types of energy. This part of the energy markets is the international energy trade. Here, energy in form of crude oil and its derivatives, natural gas, coal, solar panels, and to a smaller extent bio-fuels and other renewables, is what drives the global trade.

But most countries—including the U.S.—simply don't have enough fuel to drive all the moving parts of the huge national and international transportation sectors. And so, we, and most other countries, import crude oil and other transportation fuels.

Because of the huge amount of imports, and since this is the only energy product we import in such large quantities, crude oil is becoming the Achilles Heel of our economy. We simply have no practical ways—even with the immense new oil and gas field developments—to provide enough fuels for the transportation sector.

This is an extremely important commodity for the present and future economic development in the U.S. and the world, which is why we spend so much time on discussing it here and in the following chapters.

On top of that we have another book, titled, Energy Security for the 21st Century, published by The Fairmont Press in 2015, which details the subject from the point of view of our energy and national security.

Crude Oil Again...

The world transportation sector consumes more oil than all other sources combined. In the U.S. alone more than 13 million barrels of oil-equivalent are used every day to fuel cars, trucks, trains, boats, and planes. This is over two-thirds of total U.S.' daily oil use.

> *Gasoline, diesel, and jet fuels account for nearly 95% of all fossil fuels used for transportation in the U.S.*

Oil products power virtually every mile we drive, and we drive a lot of miles in our gas guzzling cars and trucks. And then, millions of gallons of oil are used every year for motor oil and lubricating oils, greases, and many other industrial products.

Figure 1-15 summarizes the global picture. While the developed countries are successfully reducing the use of crude oil, the developing world is increasing it. This is the case with the other fossil fuels as well. At this rate of use of the natural resources, we must expect major shortages and price hikes in the near future.

But blaming the others and pointing fingers is easy. Regardless of the trends, we are still the major user of crude oil, and will remain so for the duration. Probably until the last drop of oil, which we most likely would be the last to be able to afford.

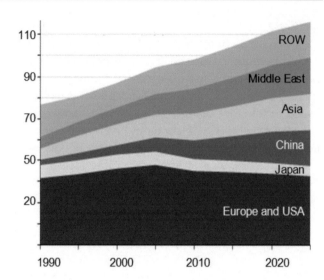

Figure 1-15. Future global crude oil demand (in million barrels/day)

There are more than 250 million cars and trucks on the U.S. roads alone today, traveling a whopping 3 trillion miles annually. This is enough miles for one car to make more than 14,000 round-trip voyages to the sun, and enough oil to fill a lake with oil used for lubrication.

The U.S.' transportation system requires about 5 billion barrels of crude oil to meet the annual needs of all cars, trucks, trains, and boats on the road, and planes in the sky. At the same time, the number of vehicles and miles driven are expected to rise in the future.

Oil companies and a handful of foreign countries benefit from this trend, while the consumers are usually the losers.

There's, however, a better, cheaper, and cleaner way to power America's transportation system. By increasing the use of clean biofuels and creating the next generation of advanced vehicles that no longer rely exclusively on oil, we can decrease our reliance on petroleum for fuel.

By improving the fuel efficiency of our cars and trucks, we can dramatically reduce the amount of oil we need in the first place.

Clean vehicle and fuel technologies are discussed and some designs are implemented, but much more work is needed to reduce the amount of oil we use for transportation.

> *If we act now, we can cut America's projected crude oil used for transportation in half during the next 20 years.*

This would save us billions of dollars at the gas pump, will slash oil consumption, move us toward a cleaner environment, and ensure our energy security on the long run. Yes, this is just a pipe-dream for now, but it will certainly become a reality when crude oil hits $250/bbl. The time is coming…

Transportation Efficiency

Crude oil is the life blood of the transportation sector. High oil prices and reduced oil volumes put pressure on the transport sector, and affect negatively the consumers and the entire economy. One way to put some order in the unstable oil supply situation is by optimizing the transport sector efficiency.

For that purpose, the 2007 Energy Independence and Security Act (EISA) amended the "corporate average fuel efficiency" (CAFE) standards.

CAFE mandates that by 2020, a manufacturer's combined fleet of passenger and non-passenger vehicles must achieve an average efficiency of 35 mpg.

According to the American Council on an Energy Efficient Economy, when implemented, this standard will save 2.4 million bbl/d of oil by 2030. This is a significant number, equivalent to nearly 15% of the current U.S. refinery output.

The EPA and the National Highway Transportation Safety Administration (NHTSA) recently published final rules to implement the first phase of these new standards. They also announced their goal, in addition to improving fuel efficiency, to reduce greenhouse gas (GHG) emissions for commercial trucks. Later on, they intend to adopt the second phase of GHG and fuel economy standards for light-duty vehicles.

Usually, improving energy efficiency releases an economic reaction that partially offsets the original energy savings, by causing a "rebound effect. This means that improving a vehicle's fuel economy reduces its fuel cost per mile driven.

This is the good thing. The bad thing is that reduced per-mile cost of driving will encourage some drivers to increase the amount of driving they do, just because it is cheaper to do so. The magnitude of such response is uncertain, but most experts agree that imposing stricter fuel economy standards will eventually increase the total miles driven. In such case, the increased fuel efficiency is not only not doing any good, but is actually harming the environment by increase of total GHG emissions.

Research on the magnitude of the rebound effect in light-duty vehicles done in the early 1980s concluded that a statistically significant rebound effect—due to increased driving habits—usually occurs with every increase of vehicle fuel efficiency.

Recent evidence shows that the rebound effect declines over time, and may decline even further if and when income rises faster than gasoline prices. In light of the various study results NHTSA elected to use a 10% rebound effect in its analysis of fuel savings and the overall benefits from higher CAFE standards for MY2012-MY2016 vehicles. The EPA prefers to use a more conservative 5% effect.

The legal mandate for increased ethanol content in motor fuels further complicates the effort to improve vehicle fuel efficiency. This is because, based on the energy content, it would take roughly 1.4 gallons of E85 (85% ethanol, 15% gasoline mix) to move a vehicle the same distance as one gallon of pure gasoline.

EPA's partial Clean Air Act waiver allows the sale of E15 (15% ethanol, 85% gasoline mix) which might cause further decreases in vehicles' advertised mile-per-gallon ratings. Obviously, great brains are at work in the U.S. and some other countries, analyzing and designing the ways and methods to reduce oil use. How fast and how far these efforts will take us into the fossil-less future remains to be seen.

Unfortunately, this is not the case in most developing countries, where with the number of cars and trucks on the roads is increasing daily. In parallel with that, the use of crude oil in these areas is increasing, as is the pollution from millions of vehicles and thousands of power plants. And there is no stopping this trend, which, as a matter of fact, is expected to increase exponentially in the near future. Most likely until fuel prices rise to unsupportable levels and/or until there are no more fossil fuels.

In summary, we must note here and agree that energy—in form of electricity and different fuels—drives our personal and business lives. Energy is more important than ever in the proper and efficient function of our society. If there is no electricity, homes would be dark and cold, and many business activities will decrease and some cease all together. If there are no fuels, we can't even go to the grocery store, or drive to work. And the vicious circle continues and expands ad infinitum.

Of course, such dark and extreme scenario is not very likely any time soon, but we must be aware of the possibility—as small and distant as it might be today. We must also always keep in mind the huge industrial progress we have experienced in one single century, which was made possible exclusively by the increased use of all types of energy. No energy, no progress.

Life on Earth will continue countless centuries to come, so we must wonder what would the people in the distant future do, if they have no access to the energy sources we have today? And such scenario is very likely.

In the following chapters we present the actual situation in the U.S. and global energy markets—today and in the future—viewed from a technical, logistics, political, and financial points of view. We base our observations and theories on the fact that the energy laws are set in cement and do not change, while human behavior changes often.

Human activities affect the end results of using different types of energy for different purposes and under different scenarios. In discussing these important subject matters, we must always keep in mind the possibility of drastic, and unwelcomed, developments in the energy sectors and the related energy markets. Better safe now than sorry later, is the best approach.

Notes and References

1. U.S. IEA, Medium Term Oil and Gas Markets 2010, OECD/ IEA, Paris.
2. http://www.worldenergyoutlook.org/resources/energydevelopment/energyforallfinancingaccessforthepoor/
3. Energy and Environment, http://www.multi-science.co.uk/ ee.htm
4. Energy and Climate Change, http://www.whitehouse.gov/ energy/
5. UN Development Program, http://www.undp.org/content/ undp/en/home/ourwork/environmentandenergy/overview.html
6. International Energy Agency, http://www.iea.org/
7. Energy + Environment, http://www.energyandenvironment. com/
8. U.S. Energy Information Agency, http://www.eia.gov/countries/
9. China Energy Group, LBNL, http://china.lbl.gov/
10. Europe's Energy Portal, http://www.energy.eu/
11. *Photovoltaics for Commercial and Utilities Power Generation*, Anco S. Blazev. The Fairmont Press, 2011.
12. *Solar Technologies for the 21st Century*, Anco S. Blazev. The Fairmont Press, 2013.
13. *Power Generation and the Environment*, Anco S. Blazev, The Fairmont Press, 2014
14. *Energy Security for the 21st Century*, Anco S. Blazev. The Fairmont Press, 2015

Chapter 2

Global Energy Balance

Energy, in form of electric power and vehicle fuels, runs this world.
Without sufficient, reliable, and affordable energy supplies, darkness,
chaos, uncertainties, and armed conflicts would prevail on Earth.
—Anco Blazev

BACKGROUND

From modern scientific point of view, the term *"energy balance"* has several practical meanings:

1. *In Biology and Physiology, energy balance* means the type and amount of calories one takes in order to support a healthy body and spirit. We have a *neutral energy balance* when taking in equal amount of calories to the calories expended. This is the ideal situation, where weight and health are properly maintained—this is the best, albeit most difficult to achieve, body and soul energy balance.

2. *In Ecology, energy balance* determines how the natural organisms affect and live with each other. *Energy balance* in nature is a state where the interactions between organisms and their environment produce a steady, healthy, and balanced ecosystem. *Neutral energy balance* means that the natural system is in a state of equilibrium. Due to man's reckless activities, the natural energy balance has been destroyed in many places around the globe. This has brought devastation of local environments and death to thousands of species.

3. *In Physics and the natural sciences, energy balance* refers to the energy types and quantity coming from the Sun vs. the portion of energy that is emitted back into space, and that which is stored in the Earth surface. This balance determines the temperature of the Earth's surface, and affects all natural climate events. Here too, *neutral energy balance* would be best, because we can then count on stable and predictable weather and climate conditions.

4. *In Chemistry, energy balance* is a state of a system, where the internal activities of molecules and atoms are in balance inside, and outside, the system.

This state of chemical reactions and the related substances is also known as *stoichiometry*, which is based on the law of conservation of mass, where the total mass of the reactants equals the total mass of the products. With other words, in any chemical reaction, the amount of the reactants is equal, and/or corresponds, to the amount of the final product.

5. *In the energy (power generation and transportation fuels) sector, energy balance* refers to the available amount of energy in its different forms (coal, oil, gas, solar, wind, etc.) vs. the amount used at the same location. On global level, it means that the amount of energy used is equal to the energy produced to replenish it.

Global energy balance in this text refers to the availability and production, vs. the use, of energy resources around the world. A global *"neutral energy balance"* is an ideal scenario, where we produce as much energy as we use. This is unachievable condition, since 98% of the energy used now days is not replenished, and is instead gone forever.

> *Unfortunately, the global **energy balance** has been seriously and irreparably violated by the uncontrolled use of fossil energy sources.*

On the long run, the global energy balance is increasingly driven by increased demand, rising energy prices and the threat of depletion of the fossils, since their natural reserves are limited. This will bring increasing volatility in the energy markets, followed by socio-political chaos and violence.

> *Half of the fossils that took millions of years to create, have been burned in a little more than a century.*

In this text, we take a close look at the *global energy balance*, focusing on *man's activities* and their consequences, and how these affects the global energy market and economy. We focus on the different energy products and technologies that accompany and drive the energy markets—keeping in mind their volatility and the possibility of pending gradual and abrupt changes.

A good and quite relevant example of this is the newly developed situation in the U.S., where less than a decade ago we were in panic; the sky was falling, we were running out of oil and gas. There were plans for increase of imports and dire straights scenarios abounded in the U.S. energy market. All of a sudden, almost overnight, we now produce more gas than we need, and over 50% of the crude oil we use too. But wait; this is almost as much as the entire Saudi Arabian oil production capacity. Amazing! The U.S. is suddenly and very quickly becoming the world's largest producer of oil and gas. Who knew?

But this new found wealth is not guaranteed solution to our energy problems. This was demonstrated in 2014-2015 by the drastic reduction of oil prices, forced by Saudi Arabia. At global price of $40-50 per barrel, the U.S. oil and gas producing wells went silent. The new wealth was worthless. Although temporary (Saudis cannot keep prices low for very long) it shows how volatile and even dangerous the global energy sector could be to any economy.

This and other developments are changing the global energy markets dynamics and shifting the energy balance. The present oil glitch won't last long and then the dominance of the present day major energy exporters (especially that of the major crude oil producers) is expected to decrease, while the U.S. is becoming ever more important factor in the global energy markets. Not only as importer, but as exporter too.

The U.S. has always been an important player in the global energy markets, but now it is becoming a controlling factor.

The game has changed. Large quantities of coal, natural gas, and petrochemicals exported by the U.S. have increasingly profound effect on the global energy balance. In response, Saudi Arabia—our best friend in sheep's clothing—dropped the global oil prices to the very minimum possible in an attempt to suffocate the burgeoning U.S. gas and oil industry. That was a clear attempt to shift of balance in the global energy market. The Saudi effort almost succeeded by hurting many oil and gas drillers, but did not achieve the goal of killing the U.S. energy sector.

Would be nice to think of the world as a place where all nations are united in pursuing and achieving

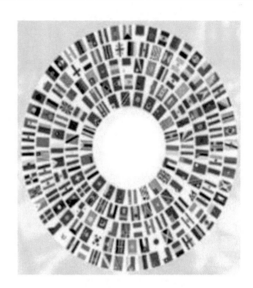

Figure 2-1. Global unity utopia

the ultimate goals of welfare and wellbeing for their populations. Instead, what we see more and more, is each nation going in its own direction. There are many reasons for this trend; some are tired of waiting for solutions and take matters in their own hands, while others take advantage of the situation to benefit at the expense of other nations. In all cases, this disunity is causing, or further complicating, the already complicated situation in the energy markets.

These, and many other changes of late are shaking the foundation of the global energy markets, some of which we take a close look in this text.

Note: For different people the term "energy markets" mean different things, so let's clarify and sort out the major subjects first.

The Energy Markets

The foundation of the global energy markets is the sales, trade, transport, processing, and use of all kinds of energy products and the related services. Coal mining, hydrofracking for oil and gas, hydropower, geo- and bio-energy, solar, and wind power, and all other forms of producing, transporting, and using energy are the foundation of the global energy markets.

In addition, there are dozens and hundreds additional products and services that originate from, or support, the different energy market segments. We call these *peripheral*, for obvious reasons, and review them in this text from the point of view of their utmost importance for the proper and efficient function of the global energy markets.

The major segments of the global energy markets are:

- Mining, rigging and power plant equipment manufacturing
  — Construction equipment
  — Transportation vehicles
  — Production and power generation
  — Safety equipment
- Construction of mines, drill sites, and support facilities
- Raw materials production
  — Exploration of fossil reserves
  — Engineering and technical support
- Production process (mining and gas/oil wells operations)
- Raw energy materials transport
- Refining and processing of raw energy materials
- Finished energy materials and fuels transport and distribution
- Power generation, conditioning, transmission and distribution, and
- Renewables development, production, installation, and use

The basic peripheral products and services, related to the energy markets, consist of:

- Consulting and contracting services
- General service providers
  — Engineering services
  — Risk-management services
  — Financial services
- Electro-mechanical services
- Security and risk avoidance
- Transportation services
- Energy efficiency
- Political and regulatory
  — Energy production support
  — Import/export
  — Strategic reserves
- Investing and finance
- Insurance services
- Legal services
- Environmental services
- Energy commodities and futures trading

In this text we take a closer look at the key energy markets segments; their products, services, and the related principle players in the global energy sector.

To start with, we would like to make a clear distinction between the producing, or "exporting," and "importing" countries. As we will see below, most countries are producers of energy products and services.

Depending on the type and quantity of energy products available in, and used by, the particular country, it could be classified as a net energy exporter or net energy importer.

Most countries do both; import and export different quantities of different energy materials and services. Very few, however, are blessed with enormous supplies of one or other type of energy sources, which makes them *major* net exporters of the type available on their territory. The rest of the countries are *net importers* of energy products and services. Saudi Arabia, for example, is one of the world's major *net crude oil exporters*, while the U.S. is one of the *major net importers* of *crude oil*.

Note: Although the U.S. is the world's largest importer of crude oil, it actually produces 50% of its oil. The in-country oil and gas production is expected to increase significantly in the future, when the U.S. might become a net zero importer of crude oil. The only problem then would be how long the fossils will last at the projected rate of exploitation.

In more detail:

The Energy Activities

Most countries fall into the category of "energy producers," since there is some (even if very limited) production of energy sources (coal, gas, oil), hydropower, solar and wind, etc. types of energy. These usually feed, albeit partially, the domestic economies, and shape the domestic energy markets. The rest of the energy needs are filled by imports, which affect directly the global energy markets by creating a demand (since not enough energy is produced domestically to satisfy the total national demand).

And so, the energy input-output formula for any energy producing country goes like this:

$$Et = Ed + Ei + Es + Ee$$

where:
Et is the total energy activities of a country or a region
Ed is the energy produced and used domestically (net energy balance)
Ei is the energy imports needed to supplement the domestic energy production, and
Es is the energy stored for rainy days (the U.S. has 90 days emergency oil supplies)
Ee is the exports of surplus energy and energy products

Even though most countries are producing some type and quantities of energy, we could divide them into net-energy exporting and net-importing countries.

Here, with a few exceptions, the exporting countries export more energy that they use, while the importing countries import more (or at least some) of the energy products they use.

What makes a country net importer or exporter is the difference between the amount of energy produced vs. the amount of energy used during the same period of time.

$$Et = (Ed + Es + Ee) - Ei$$

Note: It might seem strange that net-importers would export energy, but it happens for a number of reasons, as we will see below.

The above equation can be viewed in terms of the sum of all energy sources, i.e. coal, gas, oil, hydro-, nuclear-power, renewables, etc., but it is much easier and clearer if the different energy sources are analyzed separately. The resulting balances then can be used to project the total energy picture of the country.

For example, the U.S. produces a lot of coal and gas. It does not import any of these commodities, but imports about 50% of its oil needs, while exporting some petrochemicals. So, the approximate crude oil energy balance of the U.S. today can be represented as:

$$Et = (Ed + Es + Ee) - Ei$$

Or,

$$Et = (40\% + 5\% + 5\%) - 50\% = 0\%$$

With other words, in this example, we produce 40% of all our oil domestically (Ed), store 5% (Es), and export about 5% (Ee) of the total oil in form of petrochemicals, and import about half of the oil we use (Ei). This leaves as with a net zero balance. This, however, means that we need to import minimum 50% of the oil we use domestically in order to keep our energy balance at zero %.

Still, 50% oil imports make the U.S.
a major net oil importing country.

Considering the other energy sources, however, we see that we have abundant quantities of coal, natural gas, nuclear, and hydro-power. Due to reduction in domestic use of coal, and new gas discoveries, we are now exporting significant amount of these commodities, including products and electricity produced by them. There are also plans for increasing of these exports in the future.

The U.S. now is a major exporter
of energy products (except oil)

If we consider the total U.S. energy balance, we will see that the total energy exports (coal and gas) almost equal the amount of oil imports.

With nearly equal total energy imports and exports,
the U.S. today is nearly oil energy neutral country.

Figure 2-2 shows a schematic of a net-energy exporting county, where the energy exports are much greater than the imports. In this case, the domestic energy production is much greater than the imports. The difference between the domestic and imported energy content usually represent an excess of energy products that are exported. Such countries benefit financially by receiving huge payments from the importers.

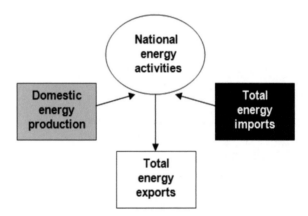

Figure 2-2. Net energy exporting countries

Figure 2-3 shoes an extreme case, where some countries depend totally on importing fossil energy sources and petrochemicals. These countries depend exclusively on energy imports for every energy need. The are extremely vulnerable to the energy markets fluctuations in availability and prices.

Many European countries, for example, depend exclusively on Russian natural gas imports. This is a dire situation to be in, especially during the winter months. The Kremlin magicians are known from time to time, with one wave of the magic wand, to stop the life-supporting gas flow to freezing European homes and businesses.

Most countries are somewhere in between, where the energy imports are only part of the total national energy budget.

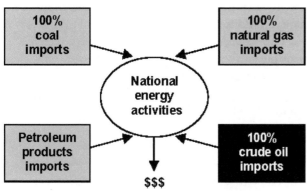

Figure 2-3. Net energy importing countries

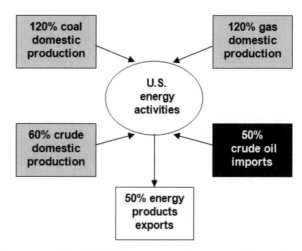

Figure 2-4. The U.S. (almost neutral) energy balance (near future)

Energy Neutrality

A number of countries do both; import and export energy products; with the U.S. posed to become a leader in this field. For example, we import negligible quantities of coal and natural gas, and with the surplus created by the new gas and oil bonanza we are planning to increase significantly the volume of exports of surplus coal and natural gas in the future.

At the same time, however, we import about half of the crude oil we use (not all of which is needed per se). We then turn around and export some of the imported crude oil—AFTER processing it into an array of petrochemicals.

In Figure 2.4, the U.S. has enough coal and natural gas supplies to satisfy 100% of the domestic demand for these energy sources and some more. So, from all energy sources we are deficient only on crude oil, thus importing about half of what we need.

And so, for the U.S. the above energy formula has to be modified as follow:

$$Et = Ed + Ei + Ee + Ep$$

Et is the total energy activities (or balance) of a country or a region
Ed is the energy produced and used domestically
Ei is the energy imports needed to supplement the local energy production, and
Ee is the exports of surplus energy and products.
Ep is the export of refined petroleum products

Note that the additional variable, Ep above, refers to export of petroleum products refined in the U.S. These products are made from crude oil, since not all imported crude oil—which we pay top dollar for—is needed domestically. As a matter of fact, about half of the imported oil (or 25% of the total crude oil we use), is

processed into different petrochemicals, some of which goes for export.

Got that? We buy billions barrels of Arab oil, transport it through thousands of miles of dangerous ocean routes, only to process and sell it on the global energy markets, sending it again on another long and dangerous trip around the world. Who knew?

One reason for this anomaly is to keep American refineries and chemical companies running, making money and keeping their people employed. All these activities are done at the cost of increased energy use and GHG emissions during different steps of the process. But even more importantly, by exporting precious crude oil derivatives, we are getting much faster to the inevitable fossil-less reality.

Nevertheless, for now at least, we are nearly oil energy neutral, which is the dream of all countries in this world. There are many benefits to be energy neutral, but there are also many detractors. In the U.S. case, we are still importing huge amounts of crude oil, which is a major part of the energy "net-neutral" equation.

Also, we are exporting huge quantity of coal and natural gas, which create several new problems:

— We are cutting GHG emissions by reducing the use of coal in our power plants, but are exporting these same quantities (and more) to be burned in Asia and Europe. There, these fossils would produce even more pollution on the short term, due to the lower efficiency of the equipment in these areas.

— But even more importantly, on the long run, we are reducing drastically our national natural energy supplies (which are still limited) by exporting them overseas. Just imagine what the 22nd century generations would think of that…if they even survive the fossil-less future we are preparing for them.

Global Energy Market Structure

The global energy markets start at the exporting countries, which have a lot of energy (mostly fossils) resources, and are able to extract and export them successfully at a profit. If there is enough surplus, after allocating enough energy for domestic use, it (the surplus energy and products) can be exported to countries all over the world—which is where the global market activities start.

According to Figure 2-5, the key elements in the energy markets, as follow:

Exporting Countries

The energy markets start at the producing countries who provide energy of some kind to those who

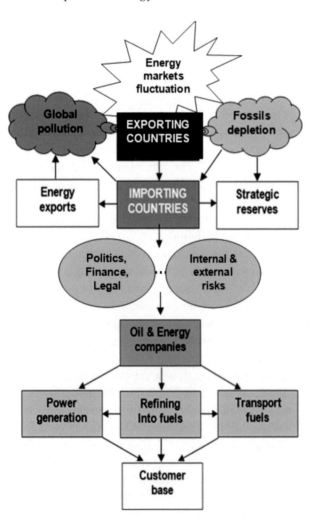

Figure 2-5. The global energy markets

need it. If there is no energy production, there are no energy markets. Simple as that!

Importing Countries

The importing counties generate the demand for energy, which is supplied either by internal means or by imports. Most countries import one or another type of energy.

Energy Companies

The energy companies produce, buy, process, convert, and sell all kinds of energy, products and services. This category includes mainly coal mining, crude oil and natural gas producing companies. Recently, some of these have been getting involved in wind and solar power generation as well.

Oil Refineries

Oil refining is conducted by the energy companies and often by third parties, such as chemical and other companies.

The Utilities

Power generation and distribution is the business of the local and regional utilities, since they manage the electric power generation and distribution.

Customer Base

The residential and commercial customers, of course, drive the energy markets by determining the overall energy demand in the different countries.

We address the energy markets' fluctuations, the related pollution, and fossils depletion later on in this text. It suffices to say here that the energy markets are responsible for the progress of the world's economies, which is a good thing. At the same time, however, many of the participants are acting irresponsibly in the way they treat the accompanying pollution, and the increasing threat of fossils depletion, which are very bad things.

These controversial dynamics are a product of the capitalist system, which drives the energy market activities.

Making a quick buck, and living comfortably TODAY, are the objectives of most energy markets participants.

The mighty dollar is the major objective, which controls the key activities in the energy markets. There are talks, and even some positive actions, that address the major concerns, but the overall effect of all energy activities is still negative. The global environment and the future generations are ultimately the victims of our actions today.

Imports vs. Exports

Considering all this, we see that the global energy balance now days is tilting towards the overall contribution of the importing countries, since they dictate the demand side of the global energy balance equation. Although they are stuck with paying enormous import bills, they still influence (and in some cases dictate) the global energy markets by controlling the demand side of the equation.

If there are no demands, there would be no imports. If there are no imports, there would be no exports, and the energy market would consist of fragmented entities contained within individual countries. The global energy market as we know it today, would not exist in the absence of energy imports and exports.

On the other side of the scale, we see the exporting countries, which have a say so in everything to do with energy, since they hold the key to the fossil reserves. To a greater extend, the exporting countries dictate the type, amount, and price of energy exports to different areas. Most of the exporting countries depend on energy exports as a primary driver of their economies, so it is a critical component of their daily activities. As such, they are very careful and extremely protective of their energy activities. Without enough crude oil, the Arab oil oligarchy will have to move from the palaces into tents and riding on camel back.

Basically speaking, the importing countries want an abundant and cheap source of crude oil, while the exporters tend to control the amount and price, in order to obtain maximum benefits.

Energy exports is what keeps many exporting countries afloat economically, and some are doing exceptionally well. They export millions of barrels of crude oil daily and pocket billions of dollars in return, which ensures an exceptional life of luxury for the leaders and most of the population. Entire generations in some countries do not know what real work is; living on government subsidies instead.

On several occasions, we witnessed the difficulties these countries encounter when the crude oil prices or export volume fell. This is because their economies are built on oil production and supported by oil exports. Without steady crude oil exports, their economies would simply come to a grinding halt.

It is quite obvious that without oil exports, many Arab sheikhs would be forced to leave their palaces and go back to living in tents in the deserts. They would also have to abandon their Rolls Royces too, and learn how to navigate the deserts on camel backs, just like their grandfathers did for many centuries. For now and for the foreseeable future, however, they control the crude oil energy market.

Crude Oil Market Dynamics

Crude oil imports and exports are major contributor to the overall global energy market dynamics. Crude oil is an important component in our daily lives. It drives (literally) the vehicles we use to transport ourselves and everything we produce, or use. The U.S., for example, imports 50% of the oil used in the country, while many other countries import much larger percentage.

The U.S. has been able to reduce the oil volume and price fluctuations by taking special measures and precautions. The national strategic reserves play important buffer role here, smoothening the sharp market fluctuations. In an extreme case of energy emergency, we can depend on them to provide fuel for our military and other emergency services. The U.S. has up to 6 months supply of crude oil, stored as reserve for such cases, which, with some efficiency measures could be extended

The economic activities drive the demand segment, which puts the energy markets in motion and keeps them going. As the crude oil demand increases, it dictates the energy markets direction. The demand is accompanied (and in some cases caused) by furious tech-

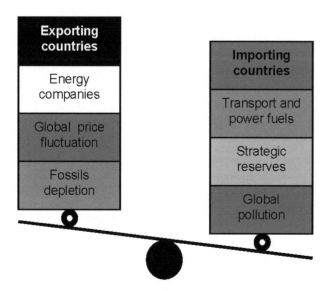

Figure 2-6. The global energy balance (or rather imbalance)

nological developments and unprecedented economic growth, which lead to excess energy use and pollution on the short term. On the long run, this will lead to permanent depletion of the fossils.

Although we experienced a short period of low crude oil prices recently, crude oil prices will continue to increase, while the production volumes are decreasing. A day will come—most likely before the end of the 21st century—when the gap would become so great that the world will enter the "post-fossils" era with a great bang. At that time, without enough crude oil to propel the world economies, many will collapse.

At the post-fossils era, the exporters will have nothing to export, so their economies will gradually transition back to the pre-fossils levels. The importers will be the first and most affected by the lack of fossils during the transition period. Since the exporters control the amount and price of fossils on the global markets, they will make sure that they have enough as long as possible by cutting the export volumes and jacking up prices at a whim. So, the importing nations will have to find a way to live and work without fossils, when the global supplies they depend on dwindle and the energy prices skyrocket.

There is no way to avoid the price and availability gaps, as the world is steadily headed into fossil-less future.

The other key component of the energy balance are the energy companies, who are also enjoying life of luxury based on excess crude oil and other energy exports and imports. Many of the multi-nationals make money at both; the import and export side of the oil cycle.

As we contemplate the pending fossil-less society of the 22nd century, we must consider the fact that the life of many people would be drastically changed, when the end of the fossils finally arrives. And make no mistake; it is coming sooner that you might think!

Before we get any deeper into the global markets, we need to understand the energy cycle; the cradle-to-grave life path of the different energy sources and the related market activities.

The Energy Cycle

The entire *cradle-to-grave* energy* production, distribution, and use of energy products determines and drives the energy markets. This is long and complex process, with many different steps, which we intent to present in as much detail as possible herein.

*Note: The concept of *energy* in this text encompasses everything and everybody involved in the pro-

duction, distribution, and use of energy sources (coal, gas, oil, renewables, etc.), electric power generation and distribution, transportation fuels, the different types of petrochemicals, as well as all other energy products and services on the energy markets today and in the future.

The energy market's cradle-to-grave life cycle could be presented as follow:

Energy Source Production
- Need and request for energy
  — Nation-to-nation
  — Company-to-company
- Initial investment and legal work
  — Private banks and investors
  — Government finance
  — Negotiations and contracts
- Energy source locating and discovery
  — Equipment and services
  — Exploration activities
- Energy source production
  — Mines construction
  — Well drilling
  — Equipment and services
  — Production operations
  — Health and safety
  — Energy sources transport
  — Pipelines
  — Trains, trucks, barges, oil tankers

Electric Power Generation
- Energy sources (fuels) procurement
- Power generation
  — Power plants construction
  — Equipment and services
  — Power plants operation
  — Power plants decommissioning

Power Transmission and Distribution
  — Equipment and services
  — Operation and maintenance
  — The Smart Grid effects

Energy Storage
  — Equipment and services
  — Market development
  — Reliability challenges

Energy Security
  — Physical security
  — Cyber security
  — Future threats

Transportation
- Energy sources (fuels) procurement
  — Fossil fuels
  — Renewable fuels
- Energy source processing
  — Refineries construction
  — Equipment and services
  — O&M processes
- Fuels transport
  — Pipelines
  — Trains, trucks, barges
  — Retail distribution and sales

Petrochemicals
  — Energy source procurement
  — Petrochemicals processing
  — Production facilities construction
  — O&M processes
  — Final products transport
  — Pipelines
  — Trains, trucks, barges
- Retail distribution and sales

Unconventional Energy Sources
  — Firewood
  — Clean coal products
  — Methane hydrates
  — New hi-tech technologies

Energy Market Transactions
- Domestic energy markets
  — Production issues
  — Transport and distribution
  — International energy markets
  — Trade negotiations and contracts
  — Global shipping
- Security

Political and Regulatory Aspects
- Government policies
- Political decisions
- Regulatory compliance

Financial Aspects
- Private investments
- Government finance and subsidies
- Energy stocks, options, and futures trading
- Risk management

Insurance
  — Negotiations and contracts
- Risk management
  — Case handling

Legal Aspects
- Policy challenges
- Lobbying
- Risk management
- Lawsuits and mitigation

Environmental Aspects
- Environmental markets
- CO_2 trading
- Government and legal action
- Risks and challenges

Risk Management Services
  — Energy production
  — Energy transport and distribution
  — Energy services

Future Energy Concerns
- Fossil-less lifestyle options
- The role of the renewables in the 22nd century
- New technologies and approaches in the 22nd century

This is a very long, and still incomplete, list, which we cannot possible cover thoroughly in a single book. Several volumes must be written to cover all aspects of these important subjects. Because of that, we will cover the topics that are most important and pertinent to the energy markets today.

Note: For additional information on the impact of the energy market activities on the environment and the energy security, we would direct the esteemed reader to our books on the subjects at hand: *Power Generation and the Environment*, and *Energy Security for the 21st Century*, published by The Fairmont Press, in 2014 and 2015 respectively.

The Energy Costs (in brief)
Energy in its most basic and practical forms is used mostly as fuel for:
a) Electric power generation, and
b) Transportation vehicles.

Electric Power Generation Costs
The cost of energy—electric power generated at power plants—vary from location to location and from plant to plant, and we take a closer look later on in this text at the different energy sources and applications. For now, however, we just need to mention that the cost of energy is calculated (and depends) mostly on:
a) the initial investment, and

b) the operations and maintenance (O&M) expenses.

There are other costs too, such as taxes, insurance, waste disposal, decommissioning, etc. and we discuss these in more detail elsewhere in this text as well. These are usually included in the price reflected in the monthly electric bill and the gas pump.

The total O&M costs of electric power generation include: fuel costs, the actual cost to operate and maintain the plant, including fuel, equipment, labor, waste management, and other costs.

— *Fuel Cost* is the total annual cost associated with the burning of fuel as needed to properly operate the power plant. This cost is based on the amortized costs associated with the purchasing of fuels (coal, natural gas, oil, uranium, conversion, enrichment, and fabrication services along with storage and shipment costs, and inventory charges—less any expected salvage value. This cost also included interest charges paid for any of the products and activities.

In a typical 1.0 GWe BWR or PWR nuclear power plant, the approximate cost of replacing one third of the core during a partial reload is about $40 million, based on an 18-month refueling cycle. The entire load of this reactor nuclear fuel is replaced within a 4-5 year cycle at a total cost of $120 million. Thus, the average *fuel cost* at a nuclear power plant is about 0.80 cents/kWh. Not bad!

One advantage of nuclear plants is that they need to be refueled only every 12-18 months. This allows advanced planning, which reduces the fuel price volatility—something that natural gas and oil power plants have to deal with daily.

In contrast, solar and wind power plants have no fuel cost, since sun and wind are free energy sources (for now). We can easily see a day coming when we will be obligated to pay sun and wind taxes, on top of all other taxes we pay now days too.

— *O&M Cost* is the cost associated with the operation, maintenance, administration, and support of a power plant. Included here are all costs related to labor, material, supplies, consumables, contractor services, licensing fees, insurance, employee expenses and regulatory fees.

The average non-fuel O&M cost for a nuclear power plant is about 1.5 cents per kWh.

— *Production Costs* are the combined O&M and fuel costs at a power plant. Nuclear power plants have the lowest production costs among their cousins; the coal, natural gas, and oil power plants. Fuel costs make up 30% of the overall production costs of nuclear power plants. Fuel costs for coal, natural gas and oil range between 60 and 80% of the total production costs.

— *Waste Management Costs* are the costs that are allocated to remove and dispose of the materials left over by the production processes. All power plants produce some sort of a waste. Coal power plants leave a huge pile of ashes and sludge in the back yards, in addition to the gas emissions. Natural gas and oil power plants emit large amounts of GHGs too, but very little solid or liquid waste.

The nuclear power industry has its own waste fund, which in the U.S. alone has accumulated to over $40 billion since the 1980s. This amounts to about 0.1 cent per kWh generated electricity by the nuclear industry.

— *Carbon economy* is the new kid in the energy markets. It is undeveloped, misunderstood, resisted, and even mistreated at times. The new carbon economy's goal is to reduce the GHG emissions by punishing the emitters and giving credit to the "green" products and services. For example, a coal-fired power plants will be forced to pay "carbon tax" proportional to the tons GHGs it emits, while a solar or wind power plant nearby will be given credit for NOT emitting GHGs proportional to the amount of power it generates. There is no global agreement on how to implement the carbon economy, so each country and location improvise as they see fit. This is causing confusion and making the process inefficient.

— *Decommissioning Cost* is the amount of money needed to decommission a power plant; to shut down and remove all production equipment, demolish the facilities, and bring the land to pre-project condition. This amounts to $300 to $500 million per plant, which in the case of nuclear plants includes estimated radiological and used fuel activities, and site restoration costs. This amounts to less than 0.001 cents per kWh generated. Coal, natural gas, and oil power plants usually cost much less than that to decommission.

The total daily production costs of a coal, oil, or gas power plant operation (not everything included) are:

Table 2-1. Production costs of power plants

• Fuel costs,	60% of total cost
• Day-to-day O&M	10%
• Capital costs	10%
• Depreciation	5%
• Miscellaneous	15% (taxes, insurance, etc.)

The total of all O&M costs of such plants is in the $0.03 to $0.08 per kWh range, depending on the type of fuel, equipment, and methods used. Solar, and wind power plants have the lowest O&M costs, since they do not use any fuel and require very little maintenance. Their fuel is free, or nearly free, and the reduced O&M eliminates the major expenses. They, however, have other problems, that reduce their efficiency, which in turn affects their usefulness, and reduces their profitability, as we discuss in detail later on in this text as well.

Hydropower plants also use free fuel—water in the storage reservoirs—and are the most efficient and cheapest *base load* generators now days.

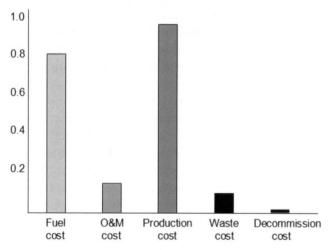

Figure 2-7. Expenses of a typical 1.0 GW nuclear plant (cents/kWh)

Nuclear plants are the most powerful and reliable (when properly functioning) power generators now days. The problems of safety and waste disposal are throwing a shadow of doubt on the entire industry, but for now we rely on them to generate almost 20% of pour power.

The costs in Figure 2-7 do not include the initial investment of plant site design, construction, and equipment purchase and setup, as well as insurance, interest and other financing charges. Some of these costs are subsidized by the governments, so they are hidden and not easy to figure out.

The initial costs for a new nuclear plant are in the range of $6-$7 per watt installed. So, a typical 1.0 GW nuclear power plant would require $6-$7 billion to start operations. The initial cost for coal, gas, and oil power plants is about 20-40% of the nuclear power plants initial cost.

Nuclear fuel price is unaffected by fluctuations in the energy markets.

The independent nuclear fuel flow is an important characteristic of the nuclear power industry, and is a significant factor in maintaining constant electric power prices. Other, non-production costs and expenses, however, such as accident insurance and nuclear waste disposal are partially covered by government subsidies and transactions, so the true (and final) cost of nuclear power is unclear. In the end, barring any major nuclear accidents, the public enjoys consistently lower prices of power generated at nuclear plants.

There are also other costs that we don't hear much about and which are hard to calculate. These are the so called *externalized* costs of power generation. A coal mine, for example, could make miners and locals sick, which damage includes insurance payments, medical payments, lawsuits, etc. expenses paid by different institutions, including the government. We take a close look at these later on in this text as well.

P.S. An even more detailed look at these and all other expenses related to power generation is covered in our book, Power Generation and the Environment, published in 2014 by Fairmont/CRC Press.

Transportation Fuels Costs

Transportation fuels are primarily based on, and derived from, crude oil. There are some other sources introduced in the market lately, such as bio-fuels, LP gas, hydrogen, etc., but their quantity is still quite small compared to that of crude oil derivatives.

Crude oil based fuels—gasoline, diesel, and jet fuels—drive the global economy (literally), thus they are critical to our present and future economic development.

No transportation fuels; no economic development, it is that simple.

In case of fuels shortage, all cars, trucks, trains, planes, and ships would be grounded, and all economic activities would just stop.

The lack of fuels would be especially hard felt by the military, since all military operations require huge amounts of different (mostly fossil based) fuels. If and when such a dark day comes, where the military vehicles can no longer run, fly, or float around the world, we would be faced with the greatest challenges of our lives.

Hard to imagine, but it is a fact, since vehicles cannot run without fuel.

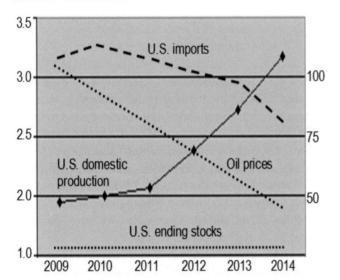

Figure 2-8. U.S. domestic crude oil production vs. imports (in billion barrels annually) vs. global oil price (in $/bbl, right hand side)

The recent crude oil phenomena—sharp increase of domestic production, reduced oil imports, combined with sharp decrease in global prices—is represented graphically in Figure 2-8. The newly created situation introduced several new variables in the global energy markets.

The new variables of, 1.) unprecedented high volume of oil production in the U.S. and other countries, 2.) temporarily reduced oil demand, and 3.) artificially lowered oil prices, are now the foundation of the crude oil market.

The low crude oil price was forced upon us and the world by our best friend, Saudi Arabia, whose sheikhs are looking to stifle the increased global oil production in order to extend their domination of the oil markets a little bit longer. We, and the world, were not expecting such a move and were not ready for it. The score now is 2:0 for Saudi Arabia.

Although the Saudi machinations succeeded in slowing down the world's oil production during 2014-2015, the situation is unsustainable. They cannot enforce the low oil prices for a long periods of time, because they are equally hurt by them.

One thing the new situation did, however, is to show us a clear picture of the global crude oil markets. It proved the Saudi's undisputed dominance and almost complete control, while focusing on the emerging production capacities in the U.S. and other major non-OPEC oil producers.

Another, even more important, fact emerged during this time too. It is the increasing importance of crude oil in the global economic system.

Most economies can run fairly well without some energy sources, but none of them can do well without enough crude oil supplies.

This is a truly scary scenario. We depend on oil for everything we do every moment of the day, but we do not stop for a second to think that this is unsustainable situation that cannot go on forever.

Crude oil is a finite energy resource. It took Mother Nature millions of years to create it, but in one century we burned half of it. It will take us less than that to burn the other half.

Yet, very few of us are asking what we are going to do tomorrow when the oil wells run dry, which they undoubtedly will do some day soon. And, unfortunately, even fewer of us are doing something positive about it.

So, is talking about oil price that important? Does it matter if it is $10, or $100 per barrel? It actually does matter, because only rich countries can afford high prices. Unfortunately, on the long run, the only way to reduce the amount of oil used around the globe is to increase its price to the point where we use it only for essential needs and emergencies. This would force us to look for, and find, crude oil replacements before the entire global oil supply is completely exhausted.

Nevertheless, the opposite is happening now days, as we discuss in detail in this text.

THE ENERGY BALANCE

As we saw in the previous chapter, energy comes in different shapes, forms, and quantities. And none of this is equally, let alone fairly, distributed around the world. Even the Sun, which seems to be the most fair and generous energy provider is not fair to all places and people on Earth. While countries in the equatorial zone receive much more sun shine than they can use, or want—every day, day after day, those living close to the poles get very little at short time periods. How fair is this? And how do you transition to renewable technologies in the future without sun and wind?

The same is true for energy sources on the surface of our Earth. Large rivers deliver a lot of water, which can be turned into electricity at some places, while others have not seen a rain drop in decades. While some places are drowning periodically and systematically in floods caused by rain storms and hurricanes, other places do not get more than a cup full of rain for years.

But the most unfair distribution of energy wealth is in what lays underground. There are countries that abound in crude oil, others in coal, and third in natural gas. But most countries have very little, or none, of the life-giving energy sources. They all, however, have sunlight and wind, so all they have to do is figure out a way to use what they have.

Sunlight, wind, waters, and the fossils are the major components of the natural Earth energy budget.

Obviously, the natural Earth's budget varies from place to place and from time to time. Here again, we see inequality. Some places have abundant solar energy, others not so much. And there are places and entire countries that very little of all energy sources. What can we do to equalize this energy inequality?

Here is where the global energy markets play a significant role.

The major (practical) role of the energy markets is distributing energy sources around the globe by export from energy rich to energy poor countries.

That then generates a number of direct and indirect activities, which make the energy market a dynamic place. An important part of this dynamic are the *peripheral energy markets*, which are usually indirectly related to the energy production, transport, and use. They support the market activities, thus playing an important role in the day-to-day energy market dynamics. We focus on the peripheral energy markets later on in this text.

A close look at the global energy balance reveals that we basically rely on,

a)   natural energy resources, and
b)   man-made energy sources.

The *natural energy resources*, from a point of view of this text, are these that we find in Nature, and before we temper with them in any way. These are the fossils (coal, natural gas, crude oil, and uranium).

The *man-made energy sources* are usually devices and/or processes that capture the natural resources (solar, wind, hydro, geo, nuclear, and bio energy) and convert them into useful energy or products. All these are reviewed in some detail in this text.

Below we review the major energy resources and their daily use for power generation, transport, commodities, and other practical purposes. The one and only most important energy source, which is in the foundation of all energy products, comes from our old friend, the Sun. Life on Earth depends on it, and energy markets—like it or not—are directly and indirectly affected by it.

The fossils (coal, natural gas, and crude oil), like it or not, are the major energy sources used both in power generation and transportation now days. Contrary to logic, which dictates reduced fossils use, many countries are increasing the use of fossils in their energy mix.

Fossil Energy

Our oldest energy friends, the fossils (coal, natural gas. and crude oil) have been essential in bringing us, and our advanced society, to where we are now. They have been providing non-stop abundant and cheap heat, electricity, and vehicle fuels to every area of the world. We have been using these dinosaur remains at will in our homes and businesses for over a century. We are still using them and they are still ready and willing to serve us until the end.

They are, however, victims of their own success, since more we use them more obvious their problems become. So now we have a dilemma; a huge problem to resolve, where the fossils' benefits are weighed against the damages they cause.

There are many reasons why this is so, and we review these issues in more detail in the following chapters. But try as we may, we just cannot cover all the details in this text, so we would recommend one of our other books on the subject, *Power Generation and the Environment*, published by Fairmont/CRC Press in 2014. In it, all types of fossils, and their properties, production, use, and the problems they bring are discussed in great detail.

We address the fossil fuels (coal, natural gas, and crude oil) first in this chapter, because they are the most important, controversial in the short term, and troublesome on the long run of all energy sources. In this text we discuss their availability, production, transport, processing, use, costs, prices, and the related problems. All this information is needed

Coal

Starting with our best friend—the oldest fossil known to man, coal—we must mention that it has long been and still remains the primary solid fuel used all over the world to produce electricity and heat through combustion.

Presently the world uses over 7.0 billion tons of coal annually, which is expected to increase to over 9.0 billion tons by 2030 and double by 2050 as the world population increases.

Got that? 9 billion tons of coal burned every year… this is 22.5 trillion lbs of coal going up in smoke every 12 months, or 62 billion lbs of coal belching millions of tons of pollution into the sky every day.

The burning coal emits poison gasses at the rate of 15 trillion tons annually now and 20 trillion by 2030.

China and India produce and use nearly half of the world's coal; over 4.0 billion tons annually now days. This amount is projected to increase to well over 6.0 billion tons by 2030, and doubling by 2050. And, of course, this is accompanied by proportionally increasing GHG emissions.

One doesn't have to be a scientist to figure out that at this rate several things will happen:

1.   Belching such huge amounts of poisonous gasses will choke our atmosphere and all that live in it to death,

2.   The most painful part will start in the near future, when the coal (and the other fossil fuels) are near depletion. At that point of no return, coal price will rise sky-high, so that it would be unaffordable; first in the poor countries, and then affecting the richer ones

3.   And finally, no matter how much coal there is in the Earth's depth, sooner or later coal will be depleted. No more coal, no more cheap electricity, no more warm homes.

We don't know exactly how and when exactly these events will develop and culminate, but at the end of this irresponsible coal-burning frenzy, the world will wake up one day choked by smoke and with no more coal for the basics of life. We can only guess what life is going to be then?

Table 2-2. The largest coal-fired power plants

Name	Power MW	Country
Taichung	7,424	Taiwan
Tuoketuo	5,400	China
Bełchatów	5,354	Poland
Guodian Beilun	5,000	China
Waigaoqiao	5,000	China
Guohoa Taishan	5,000	China
Jiaxing	5,000	China
Paiton	4,870	Indonesia
Yingkou	4,840	China
Shengtou	4,600	China

Presently, nearly 70% of China's electricity comes from burning coal. The U.S. is next with almost 50% of the total power generation done by burning coal. This is a lot of coal, a lot of smoke, and all of these activities are on the rise.

Coal used for electricity generation is usually pulverized and then burned in a furnace equipped with a boiler. Here the heat converts water into steam, which is then used to spin turbines. The turbine then turns electric generators which create electricity.

The operational efficiency of the coal-fired power generation process has been improved over time; from of 25% in the near past, to over 50% of the newest supercritical and "ultra-supercritical" steam cycle turbines. In these processes, obtaining temperatures over 600 °C and pressure over 3900 psi are responsible for the efficiency increase. They, of course, come at the higher cost of equipment and O&M expenses. Further efficiency improvements have been achieved by optimizing the pre-drying and cooling steps of the burning process, which adds to the complexity and expenses.

Even then, only 50% of the coal reaching the coal power plant is converted into electric power, while 100% of it goes up in smoke. While these achievements might impress 19th century engineers, they are not to be applauded by the 21st century generations. We must do much better than this!

Coal Today

Looking at the way coal has been produced and used in the past, and the way it is used even more today, one would think that we have unlimited quantities available all over the place. And while we can justify the unlimited use of coal in the past with the ignorance of our grandparents, the future increasing trend of coal use is totally unjustifiable, dangerous, and even scary. I*t is just like cutting the branch we are sitting on. Are we that ignorant, dumb, and selfish to do this day after day?

Table 2-3. World's coal reserves (in million tons)

Country	Total (MT)	% of World Total
United States	237,295	22.6
Russia	157,010	14.4
China	114,500	12.6
Australia	76,400	8.9
India	60,600	7.0
Germany	40,699	4.7
Ukraine	33,873	3.9
Kazakhstan	33,600	3.9
South Africa	30,156	3.5
Serbia	13,770	1.6
Colombia	6,746	0.8
Canada	6,528	0.8
Poland	5,709	0.7

The world's coal resources are estimated at a grand total of 950 billion tons.

So, taking 9 billion tons of annual coal use as an average, we end up with slightly over 100 years of coal reserves left in the world. This also means that we have used over half of all available coal reserves in one single century, and are planning to eliminate coal from the face of the Earth in another century.

Coal reserves decrease even faster today, due to increased use of coal in some countries. So, granted, we have a lot of coal now and so the coal markets will thrive for another century (maybe). But how about after that? How do we justify using all the coal in the world, which took billions of years to form, in two short centuries?

But business is business, and in the capitalist society making a buck and living comfortably today is the utmost priority. And so, China, the U.S., India, Russia, and most other countries will continue to burn coal—as much coal as they want until it is all gone. But giving credit where credit is due, we must acknowledge the fact that the U.S.—the leader of the world in energy use and environmental protection—is actually reducing the use of coal by replacing it with natural gas.

This theoretically does several things:

1. It reduces the use of coal in the country's coal-fired power plants by retrofitting them for using natural gas, thus reducing the GHG emitted by the U.S. coal fired power plants, and

2. Increases the time coal would be available in the U.S. for use by the future generations.

But wait; this is only a theory. The reality is different. The U.S. is actually planning to increase coal production...for export. So here goes our theory! While the U.S. is decreasing the use of coal in the country, it is in fact increasing its use in other countries. The exports to EU, China, and many other places are planned to increase with time.

This new situation does three things:

1. While reducing the GHGs emitted in the U.S., the global emissions will be increased proportionally by the U.S. coal exports, which will be burned in China, India and other countries.

2. The damage to the global environmental will continue at even higher rate, because of the inferior equipment in these countries, which emits more GHGs, and

3. The increased exports would bring the U.S. sooner to the fatal fossil-less future.

Nevertheless, the U.S. internal coal market is changing by the reduced use of coal in the country, while opening the global markets for exporting millions of tons of U.S. coal. New coal export terminals are planned in several areas of the country in order to ramp up the coal exports with time.

Natural Gas

Natural gas, just like coal (and often together with coal) is found in deep underground rock formations. It is usually found in coal and crude oil beds and also exists as large deposits of methane clathrates in the depth of the permafrost. Although the amount of the latter is enormous, they cannot exploited, due to technical complexity and economic barriers...for now.

Natural gas was created over time by two mechanisms: biogenic and thermogenic. Biogenic gas is created by methanogenic organisms in marshes, bogs, landfills, and shallow sediments. Thermogenic gas, on the other hand, is created from buried organic material deeper in the Earth at greater temperature and pressures.

Natural gas is used as a fuel in electric power generation or heating and cooking in homes and businesses. In all cases, fresh from the wells, it is transported to refineries to be processed as needed to remove impurities and water. This step is needed, in order to meet the specifications of commercial natural gas before being sent to the users.

The by-products of the natural gas processing include ethane, propane, butanes, pentanes, and higher molecular weight hydrocarbons, hydrogen sulfide, carbon dioxide, water vapor, helium and nitrogen. All these find application, and are used in large quantities, in different industries.

Note: Natural gas, often called "gas" is not to be confused with the abbreviation of "gas" when referring to "gasoline." In this text we use the term "gas" to mean natural gas and the term "gasoline" for gasoline.

Table 2-4. The largest gas-fired power plants

Name	Power MW	Country
Surgut-2	5,597	Russia
Jebel Ali	5,163	UAE
Futtsu	5,040	Japan
Kawagoe	4,802	Japan
Higashi-Niigata	4,600	Japan
Tatan	4,272	Taiwan
Himeji	3,992	Japan
Chita	3,966	Japan
Chiba	3,882	Japan
Anegasaki	3,606	Japan

Natural Gas Markets

Recent estimates place the amount of global natural gas reserves at about 900 trillion cubic meters of "unconventional" gas such as shale gas. Of this amount, however, only about 190 trillion may be easily recoverable with today's technologies and prices. Nevertheless, the experts at MIT and the U.S. DOE foresee natural gas to account for much larger, and ever increasing, portion of electricity generation and residential and commercial heat production in the future.

Natural gas quality varies as the well pressure drops with time and non-associated gas is extracted from a field under supercritical (pressure/temperature) conditions. Under these conditions, the higher molecular weight components may partially condense upon isothermic depressurizing. This effect is called "retrograde" condensation.

Thus formed liquid may get trapped when the pores of the gas reservoir get depleted. At this time dried gas is re-injected to maintain the underground pressure and to allow re-evaporation and extraction of condensates. Often the liquid condenses at the surface, at which point the gas production plant starts collecting the condensate, also called natural gas liquid (NGL). NGLs have special commercial applications and can be processed into different, and equally valuable, commercial products.

> *Natural gas is not a pure product,*
> *so in all cases it has to be refined.*

The largest gas field in the world is the offshore

Table 2-5. Proven natural gas reserves (in trillion m^3)

Country	In-country Reserves
Russia	48.7
Iran	33.6
Qatar	25.1
Turkmenistan	17.5
United States	9.5
Saudi Arabia	8.2
Venezuela	5.5
Nigeria	5.2
Algeria	4.5
Australia	4.3
Iraq	3.6
China	3.1
Indonesia	3.0
Total World	**187.3**

South Pars/North Dome Gas-Condensate field, in the waters of Iran and Qatar. This field alone is estimated at over 50 trillion cubic meters of natural gas and 50 billion barrels of natural gas condensates.

As a result of new discoveries and new extraction technologies, the U.S. is stepping up as the world's leader in the field. The new boom in the natural gas exploitation area have led to major changes in the energy sector. In addition to major conversion from coal-fired to gas-fired power generation, the U.S. is making grandiose plans to export ever increasing quantities of natural gas overseas. EU, some Asian countries, Japan and China seem to be the beneficiaries of these newly developed exports.

How good or bad this move might be depends on the view point of the participants, but several things are for sure;

1. The U.S. companies will profit from exporting this vital energy source,

2. The increased amount of natural gas used around the world will increase the total amount of GHGs emitted in the atmosphere,

3. The increased exploitation, use, and export volumes in the U.S. will ultimately bring the depletion of this precious commodity.

Another natural gas product, compressed natural gas (CNG), is used in rural homes without connections to piped-in public utility services. It is also used in great quantities in portable grills. CNG costs more than LPG, so LPG (propane) is the dominant source of rural gas.

LPG (liquefied petroleum gas or liquid petroleum gas) is also referred to as simply propane or butane. It is a flammable gas mixture, which includes some of the natural gas liquids (NGL) components ethane, propane, butanes, and marketed refinery olefins.

LPG is used as fuel in residential and commercial heating appliances, cooking equipment, and vehicles. It is also increasingly used as an aerosol propellant and a refrigerant, replacing chlorofluorocarbons as needed to reduce environmental damage.

LPG is made by refining petroleum or "wet" natural gas during the refining of crude oil, or it could be extracted from crude oil or natural gas streams. It was first produced as a commercial product in 1912, and currently provides about 3% of all energy consumed globally. LPG burns relatively cleanly with no soot and very few sulfur emissions. In its gaseous form it does not pose ground or water pollution hazards, but causes air pollution.

LPG has a typical specific calorific value of 46.1 MJ/kg compared with 42.5 MJ/kg for fuel oil and 43.5 MJ/kg for gasoline. LPG energy density per volume unit of 26 MJ/L is lower than either that of petrol or fuel oil, as its relative density is lower (about 0.5–0.58, compared to 0.71–0.77 for gasoline).

In most cases, powerful odorants (bad smelling) substances, such as ethanethiol, tetrahydrothiophene (thiophane) or amyl mercaptan are added to LPG to help detect leaks and prevent accidental suffocation, or explosions.

In 2012, the U.S. became a net exporter of liquefied petroleum gases (LPG) for the first time. According to EIA, the U.S. will continue to be increasing the exports of LPG at least through 2040, due to continued increases in natural gas and oil production. EIA does not say what will happen after that…

There are plans to add more infrastructure to support the LPG exports, since the net exports of LPG are projected to grow, but there are some 'ifs' and 'maybes' in these plans.

Figure 2-9 represents the three possible scenarios for U.S. exports of LPG, according to EIA. Starting with 2010, when the U.S. was a net-importer of LPG, the exports volume increased exponentially. Now the plans are to continue increasing the exports, where, according to EIA there are three different scenarios:

• In the *Worst case* scenario, we see reduction in exports through 2030, with slight increase by 2040. This might be due to a number of factors, the most likely of which are:

— Reduction in production due to technical and logistic glitches, and/or

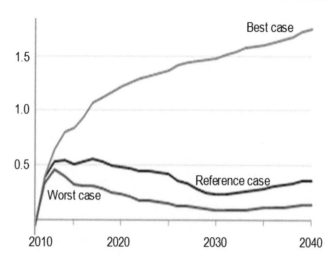

Figure 2-9. U.S. LPG exports (in million barrels per day)

— Increase of natural gas use domestically, and/or

— Drop in global LNG prices due to increased competition, and/or

— Drastic change in the global political and socio-economic structures.

• In the *Reference case* scenario, LPG exports are projected to grow by more than a half-million barrels per day until 2017. After that, however, the exports decline as wet gas (containing liquids) production declines, resulting in lower NGL production from natural gas processors.

• In the *Best case* scenario, we expect higher levels of long-term net exports for two reasons:

— Natural gas production is 36% higher in the High Oil and Gas Resource case than in the Reference case, and most of the difference is in shale gas production, which is heavy with liquids, and

— Tight oil production in the High Oil and Gas Resource case is projected to be more than double the level in the Reference case. Refinery processing of crude oil also contributes to the LPG supply. Industrial demand for LPG in the U.S. is not projected to keep pace with supply despite the number of ethylene crackers and other chemical projects under construction and planned through 2017. As a result, net LPG exports in the Best case scenario are 1.4 million barrels per day higher than in the Reference case by 2040.

Barring unexpected disasters, for the short term—at least until 2017—the U.S. will be a net exporter of LNG. The official plan right now is to increase the ex-

Table 2-6. Natural gas exports (in billion m³/year)

Country	Gas Exports
Russia	173.0
Arab League	144.2
Canada	107.3
Norway	85.7
Algeria	59.4
Netherlands	55.7
Turkmenistan	49.4
Qatar	39.3
Indonesia	32.6
Malaysia	31.6
United States	23.3
World total	929.9

ports on the long run too, but the 'ifs' are still there, so it is uncertain by how much.

Note: 23.3 billion m³/year is approximately 4.0 million barrels of LPG annual exports from the U.S. today. This number is going to increase to over 182 million barrels of annual LPG exports by 2017. Over 45 times increase in less than a decade. Not a peanut operation...

Considering the huge LPG export numbers, we must question the wisdom and practicality of increasing LPG, and any other fossils exports.

Crude Oil

Every single day we hear, "The price of a barrel of oil went up (or down) today" on the media. To most of us this means only one thing; higher gasoline and diesel prices and increase in the price of many other commodities.

Unlike most other commodities, however, the price of crude oil (even in the mighty U.S. of A. is determined by bearded sheikhs wrapped in sheets in Saudi Arabia, uncivilized revolutionary leaders in Iran, jungle lords in Nigeria, and confused presidents in Venezuela, bent on communism.

So how did it happen that the most powerful country in the world depends on questionable regimes?

Why did we allow them to set the rules of the most important energy commodity—crude oil—and rely on them to determine its volume and price?

Crude oil comes out of the ground in many places around the world, including the U.S. It was made from remains of animals and plants that existed millions of years ago, thus its classification as fossil fuel. It was then stored in large underground pools for us to discover, pump up, and use as we wish.

Crude oil consists of a variety of chemical elements, like carbon, hydrogen and sulfur, mixed with some inorganic impurities. In its purest form it is just a stinky goo, of little use to anyone. Crude oil must be refined in order to make it into a number of useful commodities.

The refining process splits the dirty, useless, crude oil mass into fractions, such as gasoline, diesel fuel, kerosene, heating oil, asphalt, and other products. These products can be further processed into plastics, lubricants, medications, fertilizers, etc. precious commodities. These final products are then sent to gas stations and factories all over the world to be used as is, or made into other oil-based products.

Crude oil brings a number of serious problems:

— Most nations—including the U.S.—don't have enough oil reserves and so are forced to import large quantities of oil from the above mentioned questionable regimes, as needed by their growing populations and energy needs,

— As the global demand for more and more crude oil increases, so do prices. Except for occasional glitches, the oil price is expected to rise steadily with time,

— As the global demand increases, the amount of oil in the ground decreases. In less than a century (actually mostly during the last 50 years) we have already used more than half of the total global crude oil reserves.

— Crude oil cannot be created, so as the reserves dwindle, the global competition will increase, which will bring further price increases, shortages and possibly even wars.

For now, however, we are concerned mostly, if not solely, with the short term price increases. The present trend of crude oil exploitation and use indicates that we are not worried about the future.

True to the capitalist ideal, we are focused on making a quick buck and ensuring our comforts and conveniences today.

We seem to have decided to start worrying about the future of crude oil when we get close to crossing that bridge. OK, so forget about tomorrow. What causes the price of oil to go up and down today? Why doesn't the cost of gasoline, diesel, and most other oil-based commodities stay at a constant level? Why does the price of a gallon of milk go up every time crude oil prices do?

Crude oil is a "commodity," a product that is generally the same no matter who, how, or what produces it. It is just like the other commodities, like wheat, corn, coffee beans, gold, iron, and copper.

The prices of all these commodities fluctuate from time to time and from place to place, because they all depend on worldwide supply and demand and many other factors. The fluctuation of one commodity, however, does not usually affect the other commodities. At least not in a big way. The price of crude oil, on the other hand, can make or a break a commodity. This is how easily affected the entire global commodity market is on the price and availability of crude oil. Much more than any other commodity, for sure.

The commodities prices are affected in different ways. For example, when ethanol fuel started becoming a popular alternative fuel option in vehicles during the mid-2000s, the price of fuel corn, from which ethanol is usually produced in the U.S., spiked. One commodity seriously affected by another (energy-related) commodity. From this example we could extrapolate and expand the role of crude oil to include all energy related commodities—coal, natural gas, bio-fuels, etc. This is simply because crude oil (and the other energy products) are in the foundation of our personal lives and our economy.

Crude oil is used in many aspects of our industrial complex, so without crude oil (used as raw material and/or fuel) many factories, and entire sectors, will have to simply shut down. Even more importantly, our entire economy depends on transportation, which is fueled mostly by crude oil derivatives.

No crude oil, no transport. Without transport, people cannot get to work, factories cannot power their production equipment, and finished products cannot get to market. This anomaly in turn would shut down many companies, and even entire economies.

Crude oil's importance in the industrial complex can be compared to the blood circulating in our vanes. If the vital blood flow stops, our bodily functions will stop too.

There are many reasons why crude oil global availability and prices fluctuate. The most important are:

- The crude oil production and export volumes are controlled by self-serving dictators, who in many cases are enemies to anything to do with the U.S.

This method has been used in two different ways:

— Reducing the flow of oil, as it happened during the 1973 Arab oil embargo, when we saw unprecedented high prices and long lines at the gas pumps, and

— Increasing the flow of oil, as it happened during 2014-2015, when Saudi Arabia caused a global chaos by dropping the oil prices below the production costs. The low prices threw the U.S. oil industry in a limbo, forcing it to shrink.

- The domestic crude oil production affects the internal crude oil and finished products prices too. More oil we produce domestically, more control we have over its availability and prices. And yet, the low global oil prices during 2014-2015 showed who is the global oil boss. And it ain't the U.S. Yet*!!!*

- Bad weather and refinery problems also affect the availability and price of the finished products (gasoline, diesel, etc.) Such events will also cause the price of oil to increase.

- Another factor in crude oil pricing is the international commodities market. Here investors and speculators hedge bets on how much they think the price of oil will increase or decrease in the near future. The volume of speculators' bets affects the global prices of crude oil.

Note: The 2008-2009 amazing jump of global crude oil prices to $180/bbl was thought to be caused, at least partially, by extreme number of speculators, betting on crude oil.

This all is very complex and even confusing. And we have only scratched the surface, since there are many, many other factors that determine how much crude is available now and tomorrow, and what its price is going to be before it ends up in your car or truck.

Crude Oil Markets Today

Some countries are blessed with a lot of oil, while others have very little or none. The oil-poor countries are majority, and so are forced to import from the few oil-rich countries.

And amazingly enough crude oil is also used in many oil-fired power generating plants in the world—including the U.S. Here we have oil-fired power plants in DC, Alaska, Hawaii and some other remote locations.

The use of crude oil for power generation is even more common in the oil-rich countries. Not just Saudi

Table 2-7. Oil exports (in million bbl/day)

Country/Region	Oil Exports
Saudi Arabia	8.9
Russia	7.2
United Arab Emirates	2.6
Nigeria	2.5
Kuwait	2.4
Iraq	2.2
Iran	1.8
Qatar	1.8
Angola	1.7
Venezuela	1.7
Norway	1.7
Canada	1.6
Algeria	1.6
Mexico	1.5
Kazakhstan	1.4
Libya	1.3
Azerbaijan	0.8
Colombia	0.8

Arabia and other OPEC nations, but developed nations, like the UK, Japan, and Germany.

The reason for using crude oil for power generation usually is lack of other fuels locally. No matter how you look at it, however, this is a huge waste of a precious commodity of limited availability. Nevertheless, this situation won't change any time soon and maybe never… or at least until there is oil to burn.

Table 2-8. The largest oil-fired power plants

Name	Power MW	Country
Shoaiba	5,600	Saudi Arabia
Kashima	4,400	Japan
Hirono	3,800	Japan
Surgut-1	3,268	Russia
Peterhead	2,177	United Kingdom

Table 2-9. The largest oil shale-fired power plants

Station	Power MW	Country
Eesti	1,615	Estonia
Balti	765	Estonia
Huadian	100	China
Mishor Rotem	13	Israel
Dotternhausen	9.9	Germany

And so, it is clear that the use of crude oil and the related commodities is on the rise…until it is fully depleted. This means that we will be going as if there is no tomorrow, and dealing with this unpredictable situation as if it did not exist for the remainder of this century. Or

Table 2-10. The largest peat oil-fired power plants

Station	Power MW	Country
Shatura	1,500	Russia
Kirov	300	Russia
Keljonlahti	209	Finland
Toppila	190	Finland
Haapavesi	154	Finland
West Offaly	150	Ireland
Edenderry	120	Ireland
Lough Ree	100	Ireland
Väo	25	Estonia

at least until the oil prices increase several fold, at which point some changes will have to be made.

We cannot even predict what exactly these changes would be, since there are not many substitutes for large volume transportation fuels to be use in all areas of the economy. So we just have to wait and see, hoping that it is not late.

But wait, there might be a solution or two to our gloomy fossil-less future. Could solar energy be one of them? Let's see…

Solar Energy

The sun is the only external energy coming to, and going away from, the Earth's surface in great quantities every day. A huge amount of solar energy reaches different areas of the Earth non-stop, day after day. Sunlight creates and supports life on our planet, drives our weather and for the most part determines the global climate.

The primary global energy balance on Earth, therefore, is the balance between the incoming energy from the sun, its activities here, and the waste energy leaving the Earth.

It has been determined that a major part—almost half—of the incoming solar energy is reflected, or absorbed by clouds, and the atmosphere even before it hits the Earth's surface. Some of it that reaches the Earth's surface is absorbed by oceans and land. Another third of the sunlight is reflected back into space.

Figure 2-10 reflects what happens to the solar radiation on its way to, and after it reaches, the Earth's surface. All in all, sunlight brings huge amounts of light, heat, and other energy types to Earth, all of which are in one or another way the foundation of our lives. Sunlight makes vegetation grow, brings rain water to the fields, and wind to the oceans. Without sunlight, life on Earth will slow down and eventually cease. Vegetation and

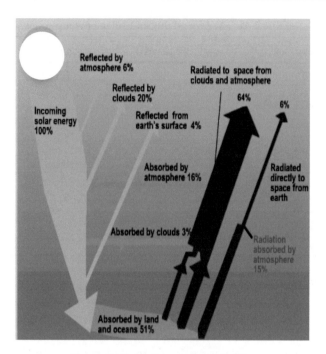

Figure 2-10. Sun's energy on Earth

animals would slowly disappear and humans will eventually freeze in the constant darkness and temperatures plummeting several hundred degrees below zero.

But for now, sunlight is generously bathing the Earth every day, and it seems that it would be doing so for many centuries to come. So, we can and should use it as we please without causing any harm to the Sun, the Earth or any other global eco-system.

Solar- and wind-power generation technologies are the most sustainable, and promising technologies today.

Solar and wind power generation technologies are also the most environmentally friendly of all energy technologies known to man. They cause minimal pollution and other environmental disturbance during their lifetime cycle.

So what is solar energy and how does it (and the related solar power generation in particular) affect the global energy balance now days?

Solar Power Today

Man saw that the Sun is his best and most reliable friend, and that solar energy is abundant and free, so he decided to use it. In primitive times, sunlight was quite useful for drying fruits, vegetables, fish, and meat, and other activities critical to sustaining life and wellbeing during the winter months.

Later on, he used the sunlight to heat water for domestic needs, and more advanced societies used

sunlight for cooking and other needs too. But it was not until the middle of the 20th century, when sunlight was introduced and used in large scale water heating and electric power generation schemes.

Solar energy falls on Earth in form of sunlight, the intensity and angle of which varies with location, seasons, and local climate. Because of these variations, some areas on our planet are more adaptable to solar energy generation than others. Alaska, for example would be a poor choice for a commercial solar plant, because of its 6 months darkness and serious weather problems over most of the year. Many areas around the equator and in the world's deserts, on the other hand, enjoy plentiful sunlight and would be a good choice for solar power generation.

It is also well known that some solar technologies can be used successfully in some areas, but not all can be used successfully all over the globe. Concentrated solar power (CSP) technology, for example requires intense sunlight to be reflected by mirrors in order to provide high temperature heating. This technology, therefore, can be used successfully only in desert areas.

We will take a closer look at the major solar technologies as they are manufactured, deployed, and used today. At the present, we have two major types of solar technologies:

— Solar Thermal (CSP), and
— Solar Photovoltaic (PV)

In more detail:

Concentrated Solar Power (CSP)

Large scale solar thermal power, also called concentrated solar power (CSP) generation was introduced in the U.S. in the 1980s, after the U.S. government was shaken by the Arab oil crisis, poured millions of dollars into solar research and development. Several large solar power plants were constructed in a hurry. These consisted on thousands of mirrors reflecting sunlight onto huge boilers to heat water and other liquids. These were then sent to make steam and turn the rotors of electric generators. Thus generated electricity was sent into the grid.

Nine Solar Energy Generating Systems (SEGS) were installed in California's deserts around that time, with the combined capacity of 354 megawatts. This is still the world's second largest solar thermal energy generating facility, after the brand new Ivanpah thermal solar power plant close by.

The SEGS use a total of 936,384 parabolic mirrors,

spread over 1,600 acres of desert land. If lined up in a row, the mirrors would extend over 230 miles.

Figure 2-11. SEGS thermal solar power plant in California desert

The total electric output from the nine SEGS plants is around 70-80 MWe at a production capacity factor of 21%. Thus generated electricity powers nearly 250,000 homes and displaces about 4,000 tons of GHG emission annually. As an added bonus, the steam turbines can also generate electric power at night by burning natural gas to propel them.

The installation cost at the time was about $3.00 per peak watt generated power. Plus $3 million annual operation and maintenance cost, or about $0.04 per kWh. Averaged to lifetime of 20 years, the entire cost comes to about 14-15 cents per kWh generated power.

Not bad...but there are problems that slowed down the rapid expansion of this technology in the U.S. and around the world. We take a closer look at these factors in this text.

Presently there are several large CSP power plants around the world:

No doubt, looking at Table 2-11, the Spaniards lead the world in thermal solar power installations. All these sites were completed in the 2007-2012 time period, mostly as a result of generous subsidies and other help from the government. Now days things have changed, and so the plant owners have fallen on hard times. In addition, CSP power plants require huge amount of water for cooling the steam, but there is not much water in the deserts, which adds to the CSP technology vows.

We will take a close look at this situation later on in this text as well.

Table 2-11. The largest thermal solar power plants.

Name	Power MW	Country
Ivanpah	392	USA
SEGS	354	USA
Solana	280	USA
Genesis	250	USA
Solaben	200	Spain
Andasol	150	Spain
Extresol	150	Spain
Solnova	150	Spain
Aste Solar	100	Spain
Helioenergy	100	Spain
Helios	100	Spain
Manchasol	100	Spain

Solar PV

Solar photovoltaic (PV) technology had a much less successful beginning in the U.S. Starting in the in the 1970s and 80s most of the money and effort went into building several solar thermal plants (SEGS). These were quite successful and are still in operation in Southern California.

The U.S. SEGS are the only large solar project in the world with a full (over 30 years) life cycle track record.

The amazing thing here is that with all the developments of late, no other technology, project, or company can boast such major achievement. This is a big problem for the solar industry in general, because millions of shiny solar panels have been already installed in the deserts without assurance that they can last 30 years under the extreme heat.

Significant funds were also directed at the time towards developing efficient and reliable PV technologies. The R&D work in this area went on for several years, and a lot of work was done, but no large scale PV power plants were built to speak of until recently.

During the last financial/energy crisis of 2008-2012, the U.S. and most EU countries again poured many billions of dollars in developing—and this time in following through the development of—solar PV technologies and projects.

As a result, there are many PV power plants around the world, some small and some quite large in size. The larger PV power plants around the world are showm in Table 2-12.

The first place in solar (PV) installations is shared among the U.S., China, and Germany, but China seems to be winning lately. Their future plans far exceed these of the competing countries, so we see China rising as

Figure 2-12. Solar (PV) power plant

Table 2-12. The largest solar (PV) power plants

Name	Power MW	Country
Desert Sunlight	500	USA
Topaz Solar Farm	500	USA
Longyangxia Dam	320	China
California Valley	292	USA
Agua Caliente	290	USA
Antelope Valley	266	USA
Charanka Solar	224	India
Mesquite Solar	207	USA
Mount Signal	206	USA
Huanghe Golmud	200	China
Imperial Valley	200	USA
Copper Mountain	192	USA

the world's number one producer and user of PV power generating technologies. This was confirmed by the new plans to double and triple the solar installations in the country.

These are the known facts. What remains unknown is the quality of the Chinese solar energy generating equipment. We have seen examples of sloppy workmanship and poor quality of solar panels, which deteriorate and fail quickly, so we would not be surprised if 10-15 years down the road, the Chinese discover that their mega solar fields have become large junk yards.

Nevertheless, at least for the immediate future, the development, production, and installation of PV technologies will continue to grow exponentially. China will lead the world in installed capacity, and we will watch carefully the developments there, since they will pave the road to the future…or not.

CPV

Concentrated PV (CPV) is another type of PV technology that uses optics to concentrate sunlight on very efficient type of solar cells. The combination of optics and efficient cells is putting this technology in a category of its own.

Figure 2-13. CPV tracker producing 50 kW DC power

CPV solar cells now days are over 45% efficient, which is 2-3 times higher than the regular silicon solar cells.

CPV and HCPV technologies use all three of the important methods available to increase efficiency;

1) Multi-junction cells, which are the highest performing cells in the industry, with over 45% conversion efficiency. This, vs. 15-18% efficiency of regular solar cells,

2) Concentrating photovoltaics, which provides the highest possible sunlight concentration possible onto the solar cells all throughout the daylight hours, and

3) Dual-axis tracking, which allows the most efficient and longest direct exposure to sunlight possible. This, vs. 20-30% sunlight exposure of non-tracking systems,

The combination of these superior functions ensures that the total amount of electric power generated per unit area is much greater than any of the solar competitors. As a result, the number of CPV installations is also growing, albeit at a much slower pace.

Table 2-13. The largest solar (CPV) power plants

Name	Power MW	Country
Alamosa HCPV	37.0	USA
Navarra	7.8	Spain
Hatch	5.0	USA
Casa Quemada	1.9	Spain
Sevilla	1.2	Spain
Victor Valley	1.0	USA
Questa	1.0	USA

CPV and HCPV (high concentration PV) technologies are not fully implemented as yet, due mostly to their complexity and added cost. They, however, hold the greatest promise for large scale deployment in the world's deserts in the near future.

We will take a closer look at the solar energy markets in the following chapters.

Wind Energy

Wind currents have been used directly by people since the beginning of time. Wind was used to sail the world and for other practical purposes. For thousands of years, wind-powered mills have been used to grind grain and to pump water for irrigation and home use. Wind-powered pumps were used to drain the huge polders of the low lying areas of the Netherlands. In the American mid-West and the Australian outback, wind pumps provided water for live stock and for powering steam engines.

The first windmill used for the production of electricity was built in Scotland in 1887. It was a 30 foot high, cloth-sailed wind turbine that was used to charge batteries for lighting a cottage. Later on the same Scott built a wind turbine that supplied power to an asylum. The association of wind and power was misinterpreted and so the wind power technology did not caught on. It was considered "crazy"; the works of the devil, while on more practical terms it did not appear to be economically viable for the times.

In the U.S., the first large commercial wind mill was designed and constructed In Ohio in 1988. It had a 56 foot wing span, mounted on a 60 foot tower. It produced only 12 kW of electric power, which was used for lighting and powering motors in a work shop- a humble beginnings of a mighty industry.

Throughout the 20th century, many wind mills were installed in small wind plants suitable for farms or residences. Later on, larger utility-scale wind generators connected to electricity grids for remote use of power started to appear as well. The wind industry "exploded"

during the 2007-2012 financial and energy boon as an alternative to expensive and dirty coal-fired power generators. As a result, wind power plants operate in almost every state and country, varying in size from several kW to many GW.

Wind Power Today

Wind currents are widely available across the nation and the globe, and wind power generation does not require any fuel. Wind energy, just like solar, however, comes to Earth equally free but in variable quantities and time frequencies. This unpredictable variability is perhaps the biggest obstacle to be resolved in the future, in order for wind to become a main-stream power generator.

Large wind farms are, nevertheless, of great interest to the power industry now days, and it is where a major part of their efforts in the renewable area are focused. This is mostly due to the fact that wind power plants produce large quantity of electricity, which translates into equally large share of revenue.

The large-scale wind power plants consist of dozens and hundreds of individual wind turbines which generate electricity when the wind blows and which are connected directly to the electric power transmission network. Small land-based wind farms can also provide electricity to remote locations and the local utility companies could buy surplus electricity produced by these as well.

Offshore (installed in the ocean) wind farms can harness more frequent and powerful winds than are available to land-based installations. They also have less visual impact on the landscape, but their construction costs are considerably higher. Offshore installations are not readily accessible for maintenance, and the violence of the sea is a determining factor in the daily O&M. Waiting for a long time for calm weather can result in delays in installation and maintenance schedules. In these cases offshore turbines could be idling for long periods of time, thus reducing energy production and profits.

Large land based wind farms don't have these problems, but face other obstacles, such as permitting delays, land availability, local opposition, right of way (for accessing the power grid), inconsistent wind velocity, etc.

Both, on- and off-shore wind power plants have basic advantages and disadvantages, which vary with type of technology, location around the globe, and the particular application.

Wind power is an alternative to fossil fuels, but unlike fossils it is renewable, widely distributed, and clean energy source. It produces no greenhouse gas emissions during operation and uses little land. Although there are some environmental effects, these are much less problematic than those from the fossil based power sources.

Figure 2-14. Wind power plant (wind mill farm).

Wind power made significant progress in the U.S. lately. It is expected to continue growing, but at a slower pace.

Worldwide, Denmark is generating 25% of its electricity from wind, and other 83 countries are using wind power on commercial basis. In 2010, wind energy production was over 2.5% of total worldwide electricity usage, and growing rapidly at more than 25% per annum.

Wind power levels are consistent on year-to-year basis, but have significant variation on hour-to-hour, daily, and seasonal basis. This condition creates problems when wind power is a major contributor in a local electrical system. And as the proportion of wind contribution increases, new approaches for using, and storing, wind power need to be implemented.

Power management techniques such as having excess capacity energy storage, geographically distributed turbines, dispatchable backing sources, exporting and importing power to neighboring areas, or reducing demand when wind production is low, can greatly mitigate these problems.

Mixing wind and solar generation is another feasible approach, for the peaks of wind and solar compliment each other. This approach could be used to level the renewable energy production and provide power in peak periods. Weather forecasting and management also permits the electricity network to be readied for the predictable variations in renewable energy production that might occur.

One of the biggest problems wind power faces, despite of its general acceptance by the public, is that the construction of new wind farms is not welcomed in most neighborhoods, due to aesthetics and safety concerns. The NIMBY (not in my back yard) sentiment this brings is hindering the spread of the gentle giants, with their slowly swinging long arms, around the U.S. and most of the developed world.

This issue can be resolved by locating large wind mill farms away from populated centers, which usually increases the final cost of wind power.

Table 2-14. The world largest wind power plants

Onshore wind	Power MW	Country
Gansu Wind	6,000	China
Oak Creek-Mojave	1,320	USA
Jaisalmer Wind	1,064	India
Shepherds Flat	845	USA
Roscoe Wind	782	USA
Horse Hollow	736	USA
Capricorn Ridge	662	USA
Fântânele-Cogealac	600	Romania
Fowler Ridge	600	USA
Whitelee	539	UK
Offshore wind	*Power MW*	*Country*
London Array	630	UK
Greater Gabbard	504	UK
Anholt	400	Denmark
BARD Offshore 1	400	Germany

Presently, there are over two hundred thousand wind turbines operating worldwide. The estimated total nameplate capacity of the global wind-power generating fleet is about 350 GW. This represents over 500 TWh, or nearly 3.0% of the global annual electric power generation.

The countries in the European Union had over 100 GW nameplate capacity installed and operational in 2013. The U.S. had about half that much, or about 50 GW, similar to China's grid connected capacity.

All in all wind is becoming, and fast, the world's most important renewable energy source. We will take a closer look at wind power generation in the following chapters.

Hydropower

Flowing water is another important and abundant energy source. It is available around the world in variable quantities, where it can be captured to be turned into electricity. The most common type of hydroelectric power plant uses a dam on a river to store water in a reservoir. Water released from the reservoir flows through a turbine, spinning its blades, which in turn activates a generator to produce electricity.

Hydroelectric power doesn't necessarily require a large dam. There are several other technologies for hydropower generation.

Some hydroelectric power plants, for example, use a small canal to channel the river water through a turbine. Another type of hydroelectric power plant, a pumped storage plant, can even store energy for later use. Here, excess electric power is sent into the electric generators, which spin the turbines backward, which causes the turbines to pump water from a river or lower reservoir to an upper reservoir, where the water (and its potential energy) are stored. When electric power is needed, the water is released from the upper reservoir back down into the river or lower reservoir. This spins the turbines forward, activating the generators to produce electricity.

Hydro Power Today

There were many GW or hydropower added around the world during the last part of the 20th century. That trend continues on paper, but the actual rise of hydropower capacity is uncertain for a number of reasons, as we will see below in this text.

Table 2-15. The largest hydropower dams

Name	Power GW	Country
Three Gorges	20.3	China
Itaipu Dam	14.0	Brazil
Guri Dam	10.2	Venezuela
Tucurui Dam	8.4	Brazil
Kashiwazaki	8.0	Japan
Grand Coulee	6.8	USA
Longtan	6.4	China
Sayano	6.4	Russia
Bruce	6.3	Canada
Krasnoyarsk	6.0	Russia
Hanui	5.9	S. Korea
Hanbit	5.8	S. Korea

Although hydropower is viewed by some as "renewable" power source, it is officially not recognized as such. At least not in the U.S. The battle on the subject continue, but whatever the outcome, we must agree that water (quality and quantity) is becoming a major issue

Table 2-16. The largest run-of-the-river hydropower plants

Name	Power MW	Country
Chief Joseph	2,620	USA
John Day	2,160	USA
Beauharnois	1,903	Canada
The Dalles	1,780	USA
Nathpa Jhakri	1,500	India

Table 2-17. The largest pumped storage hydropower plants

Name	Power MW	Country
Bath County	3,003	USA
Huizhou	2,448	China
Guangdong	2,400	China
Okutataragi	1,932	Japan
Ludington	1,872	USA

Table 2-18. The largest ocean tide hydropower dams

Name	Power MW	Country
Sihwa Lake	254	South Korea
Rance	240	France
Annapolis Royal	20	Canada
Jiangxia	3.9	China
Kislaya Guba	1.7	Russia

in many parts of this world. There are many examples of long lasting severe draughts in many U.S. states and countries, which reflect on the present and potential use of hydropower.

Thought of as a consequence from changing global climate patterns, these draughts are capable of bringing social and economic disasters to large areas. The amount of hydroelectric power generated is strongly affected by changes in precipitation and surface runoff, so when draughts hit, the local dams dry up and electric power generation suffers.

Nevertheless, the installed hydropower capacity is about 100 GW, which makes hydroelectric power plants are the second largest producer of 'renewable' energy in the U.S. after biomass, and the largest 'renewable' source of electricity. In recent years, hydroelectric power provided over 50% of the total renewable electricity in the U.S., and nearly 7% of the total U.S. electric power generation. Hydroelectricity production reached nearly 300 TWh annually, which is about 9% of the world total hydropower generation.

Until recently, the U.S. was the 4th in the hydroelectricity power in the world after China, Canada and

Brazil. Hydroelectric stations exist in at least 34 US states. The largest concentration of hydroelectric generation is in the Columbia River basin, which is the source of nearly 45% of the nation's hydroelectricity.

Large hydroelectricity projects such as Hoover Dam, Grand Coulee Dam, and the Tennessee Valley Authority are considered iconic large construction projects of national size and importance.

Bio-Energy

Bioenergy refers to many things, but for our purposes it is the energy produced from materials that have been derived from biological sources. Biomass is the general term for any organic material which has stored chemical energy during its lifetime. Biomass used for fuels include wood, wood waste, straw, manure, sugarcane, corn stalks, grass, and many other organic products and byproducts from a variety of agricultural and industrial processes.

Biofuels, on the other hand, are fuels derived from similar biological sources. Biomass, therefore, is the material used to make biofuels out of.

Note: There is some confusion about the meaning of bioenergy, biomass and biofuels, so we'd like to clarify it as follow:

> *Bioenergy is energy extracted from biomass, and biomass is the raw material biofuels are made from.*

Furthermore, while biomass and biofuels are physical entities, bioenergy is solely the energy contained in the biofuels. The confusion is further complicated by the fact that the word "bioenergy" is used in Europe when referring to biofuels.

Bioenergy Today

Bioenergy products are produced and used in large quantities around the world. Electric power can be, and is, generated by bioenergy sources as well. There is about 40 GW of bioenergy electricity generation around the globe, with about 7 GW in the U.S.

A little know fact is that biomass is the largest components of the renewable energy market in the U.S. and many other countries. One example of this is the use of biomass in Brazil. Here the production process of sugar and ethanol from sugarcane takes full advantage of the energy stored in it.

Once the juice is squeezed out of the sugarcane stalks, the remaining bagasse is usually burned at the mill to provide heat and electricity for the process. This allows ethanol plants to be energetically self-sufficient

Table 2-19. The largest biofuel producing plants

Name	Gallons (million)	Country
Dynoil, LLC	1,500	USA
SE Energy	320	USA
Dominion	300	Canada
Brazil Eco	220	Brazil
Energen	120	Jamaica
Agri-Source	120	USA
Imperium	100	USA
L. Dreyfus	80	USA
Green Fuels	64	Canada
Oilsource	60	USA

Table 2-20. The largest biomass power plants

Name	Power MW	Fuel	Country
Tilbury B	750	wood pellets	UK
Drax	660	wood pellets	UK
Ironbridge B	600	wood pellets	UK
Alholmens Kraft	265	forest residue	Finland
Maasvlakte 3	220	Biomass	Netherlands
Połaniec	205	Woodchips	Poland
Atikokan	205	Biomass	Canada
Rodenhuize	180	wood pellets	Belgium
Ashdown Paper	157	black liquor	USA
Wisapower	150	Wood	Finland

and even sell surplus electricity to utilities. This secondary activity is expected to boom now that utilities have been induced to pay full price for 10 year contracts.

This electricity is especially valuable to utilities because it is produced mainly in the dry season when hydroelectric dams are getting low. The potential power generation from bagasse is estimated at close to 10 GW. Higher estimates assume different processes, such as: gasification of biomass, replacement of current low-pressure steam boilers and turbines by high-pressure ones, and use of harvest trash currently left behind in the fields.

Geo-power; Earth's Thermal Energy

Geothermal (geo-) energy is thermal energy usually generated and stored deep under the Earth's surface. In scientific terms it is the energy that provides heat from inside the Earth, which in turn determines the temperature of matter.

Its history dates back to the original formation of the planet, which contributed 20% of the available geo-energy. Most of it, or the remainder of 80%, is generated by ongoing radioactive decay of minerals in the Earth's core. The difference in temperature between the core of the planet and its surface, the geothermal gradient, is a dynamic variable. It creates a continuous flow of thermal energy in the form of heat from the core to the surface with many interactions and activities in between.

The thermal gradient is huge, since the temperature in the Earth's core reach over 7,000 °F, while the surface area is in the 40-60°F range. The high temperature and pressure in Earth's core melts adjacent rocks, which then force the solid mantle to behave plastically. This sometimes result in portions of mantle flowing upward since it is lighter than the surrounding rock. During this process, rock and water in the vicinity heat and emerge at the crust at up to 700 °F.

This process creates hot springs, which is the oldest form of geothermal energy known to man, Hot springs have been used for bathing and medical purposes since Paleolithic times. Later on the Romans used the natural hot water for space heating.

Now days, hot springs are still used for bathing purposes, but geothermal energy is used more intensively for electricity generation, with nearly 12 GW of geo-power plants online in 24 countries. Another 30 GW of geothermal heating capacity is used for space heating, spas, industrial processes, desalination and agricultural applications.

Geo-Power Today

The Earth's thermal energy can be converted into two distinctly different energy forms:

a) *geothermal power,* which means generating electricity by means of a power plant, which uses hot water and steam pumped from deep underground. The steam is then fed into a steam turbine to rotate its blades, which are attached to a electric power generator, and

b) *geothermal heating and cooling* uses hot water and steam from beneath the ground to heat homes and businesses directly. The same hot water and steam can be in industrial processes, and/or indirectly for cooling buildings using special air conditioning systems.

Geothermal electricity generation technologies in use today include dry steam power plants, flash steam power plants and binary cycle power plants. This type of power is generated in about 25 countries around the globe. At the same time, geothermal heating and cooling is in use in more than 70 countries, since it is much simpler to setup and cheaper to operate.

The global geo-electric power generating potential is estimated at over 1,000 GW, but current installed capacity is only about 12 GW. The largest geo-electricity capacity is in the U.S.; about 3.5 GW. Other countries, like Iceland, El Salvador, Kenya, the Philippines, and Costa Rica generate 15-20% of their total electricity from geothermal sources.

The geo-thermal power generation capacity is substantial and sustainable, because the final heat extraction is minute, compared with the Earth's total heat content which is immense.

Table 2-21. Global geo-power generation

#	*Power MW*	*Country*
1	3,400	USA
2	1,900	Philippines
3	1,350	Indonesia
4	1,000	Mexico
5	925	Italy
6	910	New Zealand
7	675	Iceland
8	550	Japan
9	225	El Salvador
10	220	Kenya
11	210	Costa Rica
12	175	Turkey

In 2013, the United States led the world in geothermal electricity production with 3,400 MW of installed capacity from 77 power plants across the country. The largest concentration of geothermal power plants in the world is located at The Geysers geothermal field in California. The Philippines is the second highest producer of geothermal power in the world, with over 1,900 MW, where geothermal power makes up approximately 27% of the country's electricity generation.

The region of greatest geo-thermal potential is the Canadian Cordillera, stretching from British Columbia to the Yukon, where estimates of generating output are in the 3,000-5,000 MW. Unfortunately, and regardless of the great potential, Canada has not made any moves in this field as yet.

Geothermal power is cost effective, reliable, sustainable, and environmentally friendly, but has historically been limited to areas near tectonic plate boundaries. Recent technological advances have dramatically expanded the range and size of viable resources, especially for applications such as home heating, opening a potential for widespread exploitation.

The Earth's geothermal resources are theoretically

more than adequate to supply the entire globe with electric energy. Nevertheless, only a very small fraction may be profitably exploited. Drilling, exploration for deep resources, and their transport from remote areas, are very expensive undertakings.

Forecasts for the future of geothermal power are difficult, because it all depends on solving technological and logistics issues. Here we also must consider the competition from the other fuels which determine the energy prices, the available subsidies, and other money matters like interest rates.

Pilot programs show that customers would be willing to pay a little more for a renewable energy source like geothermal. Also, as a result of government assisted research and industry experience, the cost of generating geothermal power has decreased by 25% over the past two decades.

Geothermal wells release greenhouse gases trapped deep within the earth. At 45 grams of CO_2 emitted per kWh of electric generation, geo-power emits about 5% of the GHG amount emitted by conventional coal-fired plants. This and other facts mean that we must count on geo-power as one of the promising power-generating technologies in the future.

Nuclear Energy

Nuclear electricity is produced by splitting uranium atoms in a nuclear fission process. This process releases energy that is used to make steam, which drives a turbine attached to a generator to generate electricity. Presently nuclear power accounts for nearly 20% of the United States' electricity production. More than 100 nuclear generating units are currently in operation in the U.S.

Uranium is extracted from the earth through traditional mining techniques or chemical leaching. Once mined, the uranium ore is sent to a processing plant to be concentrated into enriched fuel (i.e., uranium oxide pellets). Enriched fuel is then transported to the nuclear power plant.

Uranium is a nonrenewable resource that cannot be replenished on a human time scale.

Nuclear power generation also produces a number of radioactive by-products, including tritium, cesium, krypton, neptunium, and forms of iodine.

Although nuclear power plants are regulated by federal and state laws to protect human health and the environment, there is a wide variation of environmental impacts associated with their operation.

Nuclear power plants do not emit carbon dioxide, sulfur dioxide, or nitrogen oxides as part of the power gener-

ation process. Nevertheless, there are serious amounts of GHG emitted during the uranium mining, its enrichment process and the transport of equipment and uranium fuel and waste to and from the nuclear plants.

Nuclear reactors also use excessive amounts of water for steam production and for cooling. Nuclear power plants remove large quantities of water from water bodies, which affect fish and other aquatic life. Heavy metals and salts build up in the water used in all power plant systems, including nuclear ones. These water pollutants, as well as the higher temperature of the water discharged from the power plant, can negatively affect water quality and aquatic life.

Nuclear power plants sometimes also discharge small amounts of tritium and other radioactive elements as allowed by their individual wastewater permits. Waste generated from uranium mining operations and rainwater runoff can contaminate groundwater and surface water resources with heavy metals and traces of radioactive uranium.

Every 18 to 24 months, nuclear reactors must be shut down to remove and replace the "spent" uranium fuel, which has released most of its energy as a result of the fission process and has become radioactive waste. The spent fuel is usually stored at the nuclear plants in steel-lined, concrete vaults filled with water or in aboveground steel or steel-reinforced concrete containers with steel inner canisters. Unfortunately, there are no permanent storage facilities for the safe disposal of spent nuclear fuel.

Nuclear Power Today

Now days, nuclear power plants generate about 6% of the world's energy, and 13% of the world's electricity. IAEA reported in 2013 that there are 437 operational nuclear power reactors in 31 countries.

Table 2-22. Global nuclear power generation

#	*Power MW*	*Country*
1	99,081	United States
2	63,130	France
3	42,388	Japan
4	23,643	Russia
5	20,721	South Korea
6	17,978	China
7	13,538	Canada
8	13,107	Ukraine
9	12,068	Germany
10	9,474	Sweden
11	9,243	United Kingdom
12	7,121	Spain

Not every one of the mentioned reactors produce commercial electricity, to which we must add the nuclear power generated by non-commercial naval nuclear fission reactors and many R&D facilities. This is not a small number, since, for example, there are approximately 140 naval vessels using nuclear propulsion in operation, powered by some 180 nuclear reactors. The other non-commercial, R&D and other such reactors are close in numbers too.

Nuclear power is reliable and cheap, but has problems that are too many and too complex to even mention here. We cover some of these in the next chapters, while the technical; details are covered in our book on the subject, *Power Generation and the Environment*, published in 2013 by Fairmont Press.

Generating electricity from sustained *nuclear fusion* reactions has been a dream of nuclear scientists'. Excluding the natural fusion power of the sun, and what we know about the fusion process, however, we are fairly sure that commercial *fusion* nuclear power generation will remain a dream for the remainder of this century. For now we have our old nuclear fission to deal with—enjoying its benefits while avoiding major disasters.

Most governments are fully involved in managing the destiny of their energy sectors—including the use of fossils and renewable forms of energy. Their role is getting more important by the day…

So what is government's role in energy markets?

THE GOVERNMENT'S ROLE

The U.S. government designs and enforces (some of) the energy policy of the country, which in turn determines the path and scope of the domestic, and (in part, and directly or indirectly) the international energy markets. The U.S. energy policy is determined by federal, state, and local entities, and addresses all issues related to energy production, distribution, and consumption. This includes; building codes, gas mileage standards, transport rules and regulations, energy related legislation, international treaties, energy subsidies and incentives to investment, energy conservation guidelines, pollution and waste treatment, energy taxation, and other such activities.

The U.S., however, does not have a solid,
*all encompassing **national** energy policy.*

The history of the U.S. energy policy making abounds of good and bad examples of what energy is,

how it is to be produced, and used. Several policies were proposed over the years, one of the most famous being President Nixon's directive that gasoline will never exceed \$1.00/gallon. Another one is President Carter's that the U. S. will never import as much oil as it did in 1977. This only shows that it is one thing to wish, and another to bring into practice. Unfortunately, these (proposed) policies were just that; a wish, and no comprehensive long-term energy policies were implemented.

One of the most revealing cases, pointing to the complete lack of direction on part of the U.S. government was the installation of solar panels on top of the White House by one President, only to be removed and junked by the next one. While one President saw solar energy as the solution to the energy problems of the country, the next one saw it as nonsense and nuisance

On-today, off-tomorrow energy policies, especially
in the renewables energy sector, have been the
MO of the U.S. government since the early 1970s.

Everybody from the President, Congress, DOD, and to the last citizen have been, and still are, concerned about energy issues, and the failure of our energy policies, and yet, they still remain inconsistent in nature. The U.S. energy policies, it seems, are still *work in progress*; a play toy in the hands of government and energy companies officials. As a result, we are lucky to get gasoline at \$3.00/gallon today, and oil imports will continue to be a major part of our energy reality ad infinitum.

The major (and persisting) problem that reflects the semi-functionality of the U.S. energy policies is the fact that in most cases they (the federal energy policies) have been dominated by a reactive thinking, induced by the crisis-mentality that started with the 1970s Arab oil embargo. This inconsistency has attributed to expensive quick fixes and single-shot (in the dark at times) solutions that disregard the established energy market realities.

Time after time the U.S. administrations have put forward politically expedient policies, which did not have much chance of success. In many cases, following an imaginary target, these policies disregarded the high costs, the environment, and even our national security. For example, we still lack realistic and stable rules in support of basic research, while letting the American enterprise free to implement and develop its entrepreneurship and innovation.

Nevertheless, we must acknowledge some of the successes in the energy sector, such as: year-round Daylight Saving Time that was imposed and successfully implemented; the United States Strategic Petroleum Reserve was created; and the National Energy Act of 1978 was introduced. Recently, three Energy Policy Acts were passed; (in 1992, 2005, and 2007), which deal directly with energy conservation, the best examples of which are the "Energy Star" program, and grants and tax incentives for all (or most) types of energy production, generation, and distribution. While these actions are major parts of the energy picture, and the lowest hanging fruit on the tree of our energy security, they are only a part of the entire energy picture.

Now days, hydrofracking plays a major role in the U.S. energy markets. It brought unprecedented natural gas and crude oil bonanza, which rule the energy game in the country today. With that, the U.S. energy policies have been undergoing another major revision. The U.S. is now among the top gas and oil producer in the world, and with that the energy policy is turning away from supporting the renewables to producing, using, and even exporting as much gas and oil as we possibly can.

These short-term gains are going to bring us prosperity today, but at the same time promise to limit sharply the availability of fossil fuels in the future. They will also delay the development and implementation of the very much needed renewable technologies. Good and bad...what is a government to do?

On top of that, the U.S. refused to endorse the Kyoto Protocol, letting the market drive CO_2 reductions as needed to mitigate global warming. Self-mitigation is not the best point of the capitalist enterprise, which is geared to making a quick buck today and at all cost. A real mitigation will require carbon emissions taxation, which can be done only with government's supervision.

The energy markets cannot control themselves efficiently in the capitalist society, due to a number of major factors (in addition to making a quick buck).

Ignorance, negligence, greed, and self-interests protection on part of corporate principals and governments are some of the major drivers (and obstacles) in the global energy markets.

And so, the world governments, who watch the energy markets very carefully, must exert most serious and long lasting pressure by establishing and enforcing rules and guidelines for the energy markets to operate within. If not, we will continue the wild ride to the proverbial energy cliff in a runaway train of the global energy markets.

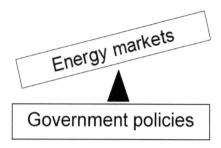

Figure 2-15. The governments' balancing role

Government and Energy

Without the government supervision and direction, cancers usually grow in the markets and bad things happen with or without a notice. The bad things come in the forms of shady operations, unfair competition, and the formation of cartels and monopolies, which develop and grow fast to overtake market segments, and even entire economies. The governments' role of judge and executioner, in most cases is important and fair. At times, however, the governments' intervention has negative results too. In all cases, the governments' actions—subsidies included—are always criticized by people on one or other side of the question.

The U.S. energy sector has had its share of government intervention, which did not help much the renewables development until the mid-2000. At that time, overly-generous government subsidies were largely responsible for the explosive growth of renewable and other energies during the last decade. Although the U.S. government's intentions and proper use of funds at the time are still under question, the entire energy industry benefited tremendously.

The energy sector experienced several cancerous growths in the recent past, and it was only after a government intervention of late that the market balance was restored. Like it or not, the government has powerful tools that, if properly used, provides a path for the markets. This is especially true for the energy markets in the 21st century. It was proven during the last financial crisis that proper and timely decisions always bring fruits.

For example, the Obama administration was crucial in assisting and encouraging the alternative energies during the 2008-2012 time period. This resulted in unprecedented progress in solar and wind installations, which were helped by bending rules and regulations as needed to favor the quick development of the sector. But even more importantly, there were billions of dollars thrown at fledgling alternative energy companies. Many of these companies seized the moment and are now thriving, along with the entire sector.

The French, German, Italian, and Spanish governments were even more helpful during that time. They also bent the rules and poured billions of Euros into the development of alternative energy companies and projects. The results of these efforts are undeniable, and can be clearly seen in the solar and wind power plants churning electric power around Europe. Was all this money spent wisely? Hardly, but the overall results are still good.

Between 2002 and 2008 the U.S. government spent:

— Over $79 billion on subsidies for oil, gas, and coal activities,

— Untold billions in support of nuclear power generation and development

— Over $12 billion on (one time) subsidies for renewable energy such as wind, solar, and hydroelectric power,

— Nearly $17 billion on subsidies for corn ethanol production,

Post-2008, the subsidies continued to flow, but while these for the fossils, hydro, and nuclear continue with no end in sight, some of the renewables' subsidies have ended and some are ending soon.

Are all these subsidies necessary, and do they achieve the goals at hand? Some do, some do not is the simplistic answer. The subsidies in themselves are not evil, and the government is not always responsible if and when some of them get wasted due to incapable, or crooked company officials.

Some of the subsidies that go to the energy sector actually also benefit regular people. They provide jobs and cheap energy for home and business owners. A good example is the Low-Income Home Energy Assistance Program that helps poor people pay their electric bills. The Oil Spill Liability Trust Fund, on the other hand, is controversial from the get go, simply because it is unfair to make poor tax-payers and customers pay for the mistakes of the rich oil industry.

The general formula that drives the flow of money in U.S. energy market is as follow:

Total energy spending = Consumer spending
+ Government Subsidies

While the consumer (home and business owners) spending is critical for day to day operations, the government subsidies drive the energy markets to higher levels. The truth is that with the monetary rewards come other benefits, not the least of which is the good feeling of being under government's protection. Some industries could not function properly without them. The nuclear industry, for example, would have never been able to achieve the progress it enjoys today without the aggressive government intervention in all aspects of its work. It could not afford the great expense of waste disposal without government's help either. And it surely cannot afford the quickly rising accident insurance rates without government's guarantees.

Billions of tax payers dollars have been spent on transferring the nuclear know-how from military to civilian usage. More billions were then spent on designing and building a number of nuclear reactors and nuclear plants. Additional billions of dollars were spent—and are still being spent—on transport and storage of waste nuclear materials. And now days, the nuclear industry would not be able to cover the costs of a nuclear disaster (a la Fukushima) without government subsidies in form of unlimited liability insurance. What a cozy place to be; under Mother Hen's wing.

And let's not forget that the government built, operates, and at least partially owns most large hydropower complexes, each of which is considered a National project and is heavily subsidized from the beginning to the very end.

The renewable sector is equally dependent on government subsidies. A number of companies were created by the generous subsidies of the last financial crisis. Many of them prospered by virtue of their ingenuity and continue to be successful. Others were propped up by millions and billions of dollars thrown at them and their projects.

One example of this is the U.S. company First Solar, which received over $3 billion of loan guarantees (which is as good as cash) from the U.S. government. That money helped the company to develop several large projects and establish itself as one of the largest in the world. Another $1.0 billion was thrown at First Solar by the U.S. Ex-Im bank, which helped it to develop its international projects.

Many companies sputter, slow down, and even fail if and when the government subsidies are discontinued.

So without the government "intervention" we simply would not have nuclear, hydro, wind, and solar power...or at least they would not be at today's level. Since the sum of these industries provide over 35% of our electricity, we must admit that the government has a stake in the energy sector. A big one!

And even more importantly for today's conditions, the government (the U.S. military) ensures safe and timely passage of the crude oil carriers across the world's oceans. This is not a direct subsidy to the oil sector, but without that support, the oil sector would not function properly. The U.S. Navy patrols the transport routes day and night, without which some of the oil headed to the U.S. would never get here. Imagine that...

Note: We take a closer look at the government subsidies in the following chapters, so here we are only mentioning some, and some of their effects.

The Renewables

After pouring billions of dollars in an effort to revive the renewable energy sector (solar and wind mostly), the EU and U.S. governments are now changing direction abruptly. After years of active support, these governments are pulling the rug from under the alternative energy industry. Awash in newly found energy sources, they consider the renewable energy sector mature and want it to run on its own.

The counterpoint here is that the nuclear and hydro industries are even more mature, and yet they still get full and almost unlimited technical and financial support. Without that support, these almost century old industries could not survive, or at the very least would be quite different.

And yet, when push comes to shove, the renewables are what gets pushed off of the vital government subsidies list. For that purpose, most European countries have eliminated any and all assistance to the fledgling renewables sector, and the U.S. government is following in their steps. It only remains to be seen if these industries could survive on their own on the long run. There is a 50:50 chance for that, and we would not bet on things going either way.

The worst, and most unfair, part now days is the fact that the U.S. government is making natural gas officially the king of the energy sector. Not by excess subsidies, but by government support; policies and regulations, which favor heavily the development of oil and natural gas exploitation via hydrofracking. This, in fact, is a different form of subsidy, which encourages unlimited and unrestricted natural gas production and use as electric power generator to grow exponentially.

There are also plans to *export* huge quantities of this precious commodity in the near future, which gives additional boost to natural gas' rule over the global energy market. King Natural Gas rules and flourishes, with substantial political and regulatory support. The

rest of the energy market segments—especially the renewables—watch and wonder about their short term future.

The U.S. Military

The global energy markets do not stay in balance by themselves for very long. There are areas in the global energy activities, where every once in a while a disruption of the routine occurs. The order is often re-established by enforcing existing rules and regulations, but quite often the problems are hidden in far away places and not under the usual control of the respective entities. In such cases the iron fist of the U.S. military is needed to restore the order.

Case in point are the crude oil transport routes, as discussed in detail in the following chapters, and especially the potential problems at number of choke points around the world's oceans. The U.S. Navy and Air Force are busy day and night patrolling the transport routes, making sure that the flow of oil is uninterrupted...at all cost! They know very well that if the oil doesn't get to its destination, the consequences would be catastrophic.

It takes energy—a lot of energy—to produce and protect the energy we need daily.

The military itself could not perform efficiently (if at all) without the millions of barrels of crude oil it needs to power its road vehicles, armament, ships, and planes. And so, the daily activities include non-stop protection of the ocean routes and many other areas that affect our energy supplies. Its a never ending 24/7/365 job that requires a lot of expertise, effort and energy (human sweat and fossil fuels) to complete successfully day after day.

The Crude Oil Dilemma

With other words, it takes a lot of energy to move and operate the military machine, as needed to ensure our energy and national security. Energy, energy, energy. It is all about energy... actually, mostly about crude oil, since it moves everything and everybody—including the military equipment and solders.

Figure 2-16 is an oversimplified schematic representation of the mighty U.S. military machine. The most remarkable thing here is that crude oil is in its very foundation. Without sufficient oil supplies, the entire military structure will collapse like a house of cards. Or it would rather sit still and helpless, like the proverbial duck on ice.

Really!

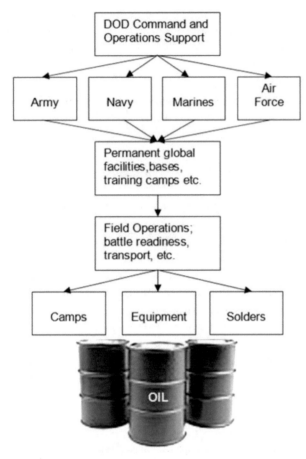

Figure 2-16. The U.S. military pyramid.

The military is driven by oil and its derivatives.
No oil; no movement and no activities.

How is a military, equipped with the latest equipment and gadgets—most of them crude oil based fuels gabbling beasts—going to function without crude oil? How would the equipment and soldiers move without gasoline, diesel, jet fuel, and lubricants? A picture of a military base and its equipment covered by dust comes to mind in such fossil-less future.

Yes, this is a simplistic view of a huge and complex system in distress. Also true that we are far from such a bleak destiny, but one doesn't need to be an expert, or a genius, to figure out what could happen if and when the crude oil and the fuels supply is reduced. A grim picture emerges, with Army trucks parked, F-18 fighter jets and Apache helicopters grounded, battleships and submarines anchored, and M1A3 Abrams tanks sitting still in the parking lots.

Only God knows what could happen in case of oil shortage, since only crude oil can move the U.S. military machinery.

What one needs to imagine are the thousands of people, equipment, weaponry, and vehicles that are represented in Figure 2-16. Some of the machinery is as large as a skyscraper, while some are as fast as lightening, and some can blow up an entire town out of existence in a split second. No other army, or natural force on this Earth, can do as much damage at a whim as the U.S. military machine.

Just think "huge" and "awesome," since these two words summarize the enormity of the U.S. military machine, and represent the essence of what we are talking about here.

Huge and awesome = U.S. Army + crude oil.
Without crude oil these terms have no meaning.

It is not easy to imagine the whole thing no matter how hard we try… It is even harder to imagine the immense amount of fuel that is needed to roll, fly, and float the millions of pounds of equipment, weaponry, vehicles, ammunition, and people around the world every day, day and night.

For that purpose, there are thousands of specialized trucks, trains, airplanes, and ships that haul non-stop millions of gallons of gasoline, diesel, jet fuels, nuclear materials, lubricants and other energy products around the country and the globe every day. These and all other pieces of equipment and vehicles count on the prompt delivery of fuels every day. It has to be done on schedule, else they won't budge, or may fall from the sky

It is truly amazing, but a fact, that without the daily (and hourly) fossil fuel refills, the military machine will come to a grinding halt. In such a case, all war operations will be severely constricted, and most will just fizzle and cease. All military plans and programs will sit idle until a compromise solution is found. Nothing in the military machine can move, or operate properly, without timely (fossil) energy input. Crude oil is the main energy source (fuel and lubricants)) used by all DoD branches, so it is what allows the daily operations to proceed as planned.

Since we import over half of the crude oil we
use, our energy security, and in extension our
national security, are less than 50% ensured.

What does that mean to the future of our energy (and national) security. The DoD generals will have to answer this question, for it is too much for us to even grasp. Why worry about fixing it, if it ain't broken? You say? Maybe so…for now. But crude oil is getting scarce, prices are going up, and a day is coming soon when oil will be extremely expensive and then it will be no more. Crude oil is a finite commodity that cannot and will not last forever.

The DoD generals see these problems clearly, since they pay the increasing energy bills and are responsible for battle readiness, and their soldiers' lives. They take the real and perceived problems seriously and have been taking decisive action for awhile now, as summarized in their energy plans. Is this enough, is another question that we cannot answer, but which they are surely working on.

We take a closer look at the U.S. military use of energy in our book, *Energy Security of the 21st Century*, published by The Fairmont Press/CRC. The overwhelming conclusion of that investigation is the fact that the DoD is looking very seriously at the energy and environmental problems it is faced with. Some important measures have been already taken and some are planned, as needed to reduce the use of fossils and the related pollution emissions.

While these efforts change the way the military machine operates, they are designed to not hinder with the principal goals of ensuring our energy and national security.

These are important concepts, which deserve a closer look:

ENERGY SECURITY, INDEPENDENCE, AND SURVIVAL

There is a lot of talk now days in Washington DC and other high places about our energy present and future. Some of the issues are obscured by political, some by financial, and some by technical misunderstanding. This then is a good time and place to clarify the meaning of,

a) energy security,
b) energy independence, and
c) energy survival (this one is quite important, but usually ignored).

In brief:
- *Energy security* is ensuring an uninterrupted energy supply today and tomorrow, by employing efficient supply methods, complete with accident and crime prevention measures,
- *Energy independence* is simply the idea of producing

our own energy, so that we don't have to depend on anyone else to import any fuels and energy products,

- *Energy survival* is ensuring that we have enough energy now AND in the **_near future_**, so that we now, and/or the future generations later on, don't end up in an energy blackout. This is an obscure subject, somewhere on the back burner for now, and most people are not ready or willing to talk about it.

Energy security, energy independence, and energy survival, while meaning different things, are interwoven in an integral way. They must be addressed as equal in importance and weight. One cannot exist without the other, especially on the long run. If one is ignored, our present and/or future energy balance and energy security would be at risk.

Our national security depends on maintaining the proper balance of these three concepts.

We know very well that we are using way too much energy and that at the rate we are using fossil fuels today, they will not last long. And yet we continue the fossils extermination at an accelerated pace, silently ignoring the grim reality of pending fossils doom.

At the same time, the implementation of the renewable energy sources around the globe is going at a snail's pace due to political hesitation, corporate grid, and many other socio-economic reasons. So what are we to do?

Let's take an even closer look at these three vitally important concepts:

Energy Security
There are a number of reasons why energy security is so important for providing normal (or rather opulent) life in our society. Without energy, or in a case of temporary interruptions of the energy supply system, we would be vulnerable to a number of internal and external threats.

Energy security is one of the pillars of the national security and efficient economic development of our modern society.

Without energy, the industrial machine stands still and does not produce anything. The little amount of products that can be produced, cannot be moved to the markets. Armies and their vehicles also stand still and cannot even get to the battle field. So, it is imperative for each nation to have enough energy to function properly. Interruption of the proper energy sector function brings a number of undesirable consequences, external problems, and internal turmoil.

We can easily divide the present day risks to our energy sector and our energy security into internal and external:

Internal Risks

There are a number of internal risks to our energy security. They are *internal*, because they are based, and conducted. on our national territory. The internal risks can also be divided into natural (caused by natural disasters and such) and man-made (caused intentionally or unintentionally by human activities).

Unforeseen natural events and man-made accidents in mines, transport routes, refineries, and power plants threaten to disrupt the energy supply chain. Mine accidents are still happening in the U.S. and around the world. They take lives and shut down mining operations temporarily or permanently. Railroad cars and transport ships accidents and spillage can close a transport route indefinitely and do significant damage to the environment at the same time. Power plants are vulnerable to internal sabotage and terrorism, and the risks of this happening in the U.S. are increasing. Some of the accidents of the past have resulted in loss of human life as well. Hurricanes damage and shut down refineries and power plants every year, which cause another set of problems.

As the state-of-the-art of the energy technologies matures, and we learn how to handle the natural and man-made accidents, some of the internal risks get more manageable. With that, the number of mine, railroad and transport ships accidents is significantly reduced. Refineries and power plants, because of their complex infrastructure, on the other hand, are vulnerable to the force of the hurricanes and other natural events and not much can be done to improve that situation.

The worst and most dangerous risks of all in the energy sector are these caused by nuclear power plant incidents, accidents, and failures. Just mentioning Chernobyl and Fukushima brings gruesome memories and forces us to imagine what would happen if a similar accident happens nearby. Nuclear accidents are an extremely rare occurrence but when they happen, the local devastation is total and permanent, and their effects are felt over broad areas—even across continents.

This is a critical component in considering our energy security and safety. One single nuclear plant accident can not only disrupt the energy supply, but even destroy the entire area and kill and sicken thousands. This is a serious issue that needs to be taken into consideration as we consider our energy options for the 21st century.

Internal terrorism is one way to gain control and damage a refinery, or a power plant…and God forbid that it might be a nuclear power plant. There are a number of reported incidents, and even a greater number of unreported such of such incidents. Now days computers run everything—including our energy production, power generation, and the entire energy infrastructure. Every step of our energy life cycle, from mining operations to transporting, processing, refining, and using energy sources is monitored and/or controlled by computers. This creates a serious problem; that of computer hacking and terrorism.

In 2012, for example, a backdoor in a piece of industrial software used to control power plants, allowed hackers to illegally access a New Jersey power company's internal heating and air-conditioning system. One of the viruses was accidentally discovered after an employee called in IT technician to troubleshoot the USB drive. A simple virus check discovered sophisticated malware, capable of doing a lot of damage to the plant equipment. Case solved, and the plant operation was safe again

Since the computer safety technology is fairly new, the defense mechanisms have not been fully developed and/or deployed, as was in this and many other cases. The workstations lacked backup systems, so they were lucky in discovering the malware before it was activated…this time!

Another intentional malware attack, spread by a USB drive, affected 10 computers in a steam turbine control system of a power plant. The incident resulted in downtime for repair of the impacted systems, which delayed the plant restart by three weeks.

The Stuxnet worm and the Flame malware were developed by the US and Israel to spy on, and even control, critical systems in power plants. In the summer of 2012, these programs successfully disabled an entire enrichment facility in Iran. These programs relied on USB drives to store the commands, propagate attack codes, and carry intercepted communications over computer networks. Microsoft has patched these vulnerabilities on Windows computers, but there are additional steps to be taken by power plant and other energy sector operators.

And of course, we cannot leave out the humbling experience of the massive computer security breach in

December 2013 by foreign hackers who stole the identity and other private information of several hundred thousand Target and other U.S. companies' customers.

If hackers can get into one of America's largest and most prestigious chain-stores' computer systems, getting into the utilities' control systems is only a step away.

As a matter of fact during 2012-2013, critical control systems in two US power plants were found infected with computer malware, spread intentionally by computer virus brought in the plants via USB drives. The infected computers controlled critical systems controlling power generation equipment.

Intentionally planted malware poses a real threat by allowing the attackers to disable key equipment, thus disrupting its normal operation, or destroying it all together. What a disaster that would be if allowed to shut down the cooling system of a nuclear plant!

But it is not just computers that are the internal terrorists' target. In 2013, for example, a well organized and executed sniper attack on power grid components shut down the power in a large part of the Silicon Valley area, and threw residences and businesses into darkness for several hours and even days in some areas.

The sniper shot at several transformers in a local high voltage substation which malfunctioned as a result and caused the power grid in a large part of the San Jose-Fremont area to shut down. The repairs were started almost immediately, but it took some time to figure out what is happening and to repair the damages.

These examples are cautionary tales and lessons in safety procedures, as well as a call to action by owners and operators of critical energy infrastructure. New, 21st century security policies must be developed and implemented, in order to maintain up-to-date antivirus mechanisms, manage system patching and the use of removable media.

Other, site-specific protection measures must be implemented for ensuring the energy infrastructure's physical safety. There are weak points all through the power system—from power generation to delivery—which need to be evaluated and protected from physical and cyber attacks. This is not simple, not cheap task, but it has to be done, if we don't want to risk waking up some day in darkness, or even worse; in the midst of a nuclear disaster.

On the overall, internal risks are unavoidable, but we are well positioned to anticipate and respond in most cases. The internal security situation is basically under control for the most part, and we just have to add the

necessary cyber and physical protection to key components, be more careful and not let our guard down.

The external risks are more serious and complicated.

External Risks

The U.S. has many foreign enemies. Traveling around the world, one can see the signs of animosity towards the U.S. in almost every country. Some hate us, because we have done bad things in the past (and God know we have). Others hate us, because they are jealous, and yet others hate us for the sake of hating. Either way, each of these groups has their own reasons and ways to inflict damage and pain to the mighty US of A.

Figure 2-17. Terrorist attack on French oil tanker Limburg in 2002

Terrorist attacks targeting oil facilities, pipelines, tankers, refineries, and oil fields are referred to as "industry risks," because they are part of daily life in the energy sector. The energy infrastructure is extremely vulnerable to external sabotage, with oil transportation, and its exposure at the five ocean chokepoints being on top of the risks list.

The Iranian controlled Strait of Hormuz is a prime example of a chokehold, where one attack on a Saudi oil field, or on tankers in the Strait of Hormuz, could disturb the global oil supply. A prolonged conflict in the area, would surely throw the entire world energy market into a chaos.

External risks have always been, and will continue being a problem in the U.S.' energy security.

Foreign terrorists spend a lot of time planning attacks on U.S. properties and personnel around the world and at home. This is the price we pay for being the world's greatest power and a shining example of a functioning democracy.

International terrorism affecting the world's energy reserves is of great concern lately as well, as evidenced by NATO leaders meeting in Bucharest in 2008, where international terrorism against energy resources was one of the key subjects. The group discussed the possibility of using military force to ensure the energy security of the region. One of the possibilities discussed include strategic placement of NATO troops in the Caucasus energy fields to police and protect oil and gas pipelines from terrorist attacks.

The U.S. energy supply and energy infrastructure are in the terrorists eyesight too, and attacks are, no doubt, planned daily in the terrorists' dens around the world. But how would they do that? We are too far away, and too powerful, to invade by sea or air. Instead, they do it by terrorism, which could be a physical attack on people or structures, but that doesn't work very well either, so recently they have been choosing different weapons and changing the battlefield tactics

The Risk Levels

So let's take a look at the risk levels of the different energy sources.

Figure 2-18 is a rough summary of the level of risk the different energy sources present to our energy security. Going from left to right, we see a decreasing level of *risk-free* operation of the different energy sources.

- Solar, wind, and geo-power are the most *risk-free* energy sources, simply because the fuels (sunshine and wind) are readily available and so they, and the related technologies, do not depend on any external factors. There are, of course, some inherited problems, such as the variability of sunshine and wind currents, and internal risks, such as materials availability and cost, but these are few and between and are easily manageable for the most part.

This makes these energy sources virtually risk-free and highly reliable ad infinitum. We foresee them becoming major power generators in the future, as their technologies mature and the fossils get more expensive. They are the best (if not the only) bet for the future generations.

- Coal is also relatively risk-free as far as its production, distribution, and use are concerned. Apart from its excess pollution, and mine accident problems, coal is, and will be, readily available and cheap enough for reliable and risk-free operation for a long time.

Its depletion in the more distant future, however, is imminent, so it is only a temporary solution to the present-day energy issues. Let's not forget that some day soon—most likely during the next century—there will be no more coal on this Earth.

- Natural gas presents an increased level of risks, due to its nature and the related production and distribution issues. The transport of natural gas, via pipelines and different land and ocean based vehicles increases the risks of natural accidents and criminal activities.

But even more importantly, the U.S. natural gas distribution infrastructure—the thousands of miles of gas pipelines running under our streets and homes—is old. In some places the pipelines are over 100 years old, and are developing leaks and creating all sorts of problems. The incidents of gas line failure are increasing and getting more violent. The last and most revealing example is the gas explosion that destroyed two apartment complexes is New York city and killed 8 people.

Natural gas is also a temporary solution to our energy problems. No matter how exiting the latest natural gas bonanza in the U.S. might be, we should not forget that Its quantity is limited and that sooner of later—and surely sometime this century—it will be completely depleted.

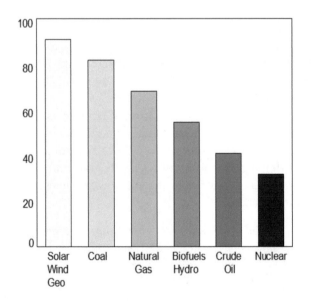

Figure 2-18. Energy security risks and dependencies (100 is risk-free)

- Biofuels and hydro power share the same unfortunate dependency on natural events—extended draughts and uncontrolled climate changes and weather patterns threaten to the reliability of these energy sources. Biofuels also compete with the world's food supply and as the global population increases, the competition between crops for biofuels and food will grow too.

Cellulosic biomass has a great potential as a future energy source, but it is too expensive for now. It, however, will become more reliable and cheaper in the future, as the technology is improved and optimized.

- Crude oil, and its imports in particular, present a major risk to our energy security. 50% of the oil used in the U.S. is imported. It comes from some of the most volatile and unfriendly areas of the world, and is transported through some of the most dangerous chokepoints now days. There is a constant threat to the production levels in some countries, in addition to the risks of the distribution channels, thus its import is a risky proposition.

But most importantly, let's not forget that crude oil production is entering a period of decline. The prices will keep rising, and the quantities will diminish steadily. One thing is for sure—there will be no more crude oil some day—and surely sometime this century.

- Nuclear power presents serious dangers and risks to our energy security and personal safety. Computer warfare is the name of the game now days, so a terrorist sitting in a cave in Afghanistan, or in a high rise in Beijing, could gain access to a the computer controls at a power plant or refinery and simply put them out of commission, or worse.

Nuclear reactors are potential terrorist targets, since they are not designed to withstand attacks by large aircraft, rockets and other air-born weapons. A well-coordinated attack, using powerful weapons, could damage the reactors in a nuclear plant, which will in turn have severe consequences for human health and the environment.

A recent study concluded that such an attack on the Indian Point Reactor in Westchester County, New York, could result in high radiation within 50 miles of the reactor, and cause 44,000 deaths from acute radiation sickness. Additional 500,000 long-term deaths from cancer and other radiation-caused illnesses would be

expected as well.

Terrorists could also target a spent fuel storage facility by using high explosives delivered by ground or air vehicles. This would also results in radiation contamination of the immediate area. A terrorist group may infiltrate the personnel of a nuclear plant and sabotage it from inside. They could, for example, disable the cooling system of the reactor core, or drain water from the cooling storage pond. An internal attack is perhaps the most likely, and most dangerous, terrorist attack on a nuclear-power reactor.

There is also an inextricable link between nuclear energy and nuclear weapons, which pose the greatest danger related to nuclear power. The problem is that the same process used to manufacture low-enriched uranium for nuclear fuel, also can be used for the production of highly enriched uranium for nuclear weapons.

Expansion of nuclear power generation could, therefore, lead to an increase in the number of rogue states with nuclear weapons, produced at their "civil" nuclear programs. This is exactly what we are seeing being played out in Iran and possibly North Korea. The use of nuclear power would, at the same time, increases the risk that commercial nuclear technology will be used to construct clandestine weapons facilities, as was done by Pakistan in the recent past.

Today, the majority in Japan is against nuclear power. As a result, Japan's nuclear plants are shut down and the GHG emissions increased by 15% as more fossils are used for power generation in Japan since Fukushima.

One more, even much smaller, nuclear accident on the island of Japan will surely put an end to nuclear in Japan forever. In order to prevent that from happening, Japan's nuclear plants spent billions of dollars for upgrades with the latest hardware and security measures, as needed to fight Mother Nature and eliminate any man-made accidents.

But even without any additional accidents, nuclear power in Japan will be on the back burner for a long time. Its destiny around the globe is also uncertain, and will be determined mostly by its safety record.

Environmental Impact

The environment is part of the external risks related to energy security simply because it is the media we live in. It is the air we breath, the water we drink, and the land we walk on. When we talk about energy security we must always, and before anything else, consider the environment and the impact on it our actions could bring.

It won't do us any good to have plentiful energy resources if we cannot breath the air, drink the water, or have safe land to live and walk on.

We see a glimpse of these abnormalities reflected in the dense Beijing smog, the contaminated with fracking liquids water in some U.S. states, and the disappearing land mass in the Maldives and other places. These phenomena—smog from coal fired power plants, increased fracking activities, and rising ocean levels respectively in this case—are hurting millions of people.

Coal-fired power plants belch millions of tons of GHG gasses, hydrofracking activities pollute the fresh water supplies in large areas, and rising ocean levels threaten to submerge portions of inhabited land mass. These activities are part of the energy production cycle, which is part of our modern society, but which results in property damage, make people sick and even kill many of us.

Energy security obtained at the expense of humans' safety and wellbeing is a partial security.

This 'partial security' benefits some people (the majority), while hurting others (the minority). As such, the minority in many cases is sacrificed for the benefit of the majority. We all are well aware of that fact, but it become important only when we become the minority. This situation is unsustainable and must be corrected before we reach a point of no return, which is much easier said than done.

The problem is that the majority always wins. In the capitalist system, there are many interests to protect and a lot of money to be made by maintaining the status quo. And because of that, the talks continue, and many volumes are written on the subject, but the actions do not follow in most cases. For now…

Fossils Depletion

Another external factor of great importance, and even greater consequences, is the pending depletion of global fossil reserves. For over a century now we have been relentlessly digging and pumping, day and night, non-stop, huge amounts of coal, crude oil, and natural gas. It is now quite obvious that any and all measures discussed and planned will not be enough to reduce the onslaught of fossils.

The rate of exploitation of the fossils is actually increasing in most areas of the world. This is especially true in the developing countries, many of which are determined to achieve Western-like lifestyle in the near

future. This effort requires a lot of energy and they are determined to get it one way or another.

The present day approach to achieving energy security à la "drill baby, drill" MO promises energy poverty, followed by total lack of fossil reserves in the 22ⁿᵈ century and beyond.

One doesn't have to be a genius or a specialist on the matter to figure out that no matter how much fossils we have, they are of limited supply. This means that by the end of this century, and even before that, there will be no more crude oil. The sucking sound of empty natural gas reservoirs will follow soon thereafter. The demise of the coal reserves will complete the fossils life cycle, thus bringing us to a fossil-less reality and energy debacle of unprecedented proportions. Not possible? Yes, possible!

Energy Independence

Depending for energy (oil in particular) on the whims of the dictators of politically unstable countries is the major U.S. energy security risk at the moment. Our economy is still exposed to manipulation of energy supplies, such as the OPEC-orchestrated oil crisis of 1973, the extraordinary jump of oil prices in 2008, and the even more surprising forceful drop in oil prices during 2014-2015.

These were key events that showed how much we depend on 'our friends', the oil producing and exporting countries. We also saw the negative effect of these events on our economy and personal lives. Time after time, we are reminded that crude oil is the engine that drives the national and global economies. It follows then that more oil we import more vulnerable we are to the whims of the exporters.

Presently, we import about 50% of the crude oil we use. Does this make us 50% vulnerable? Maybe, but in all cases, our energy security is threatened by the 50% crude oil we don't have, under which conditions energy independence is simply impossible.

Energy independence is impossible without 100%, cheap, domestic energy production.

The logical conclusion is that we cannot hope for energy independence while importing such large quantity of crude oil every day.

In 1973 the American government realized that oil imports are a big problem and started a feverish effort to develop new energy sources in order to obtain energy independence...or rather to ensure an acceptable level of energy security. Soon after the crisis was averted, the energy security drive gave way to making a quick buck by importing cheap Arab oil again.

The lesson of 1973 was not learned, and as a consequence we find ourselves in the same embarrassing and difficult situation today! We see the global energy prices dictated by the same Arab sheikhs all over again...and there is nothing we can do. The proverbial duck on ice comes to mind here too.

The efforts to implement new, renewable energy sources fizzled away as soon as OPEC lifted the embargo in 1974 and the oil price dropped. The energy security effort was shelved until the next energy crises. The saddest part is that this vicious cycle was repeated again and again.

It has been proven time after time that oil imports and prices are vulnerable to intentional or unintentional disruptions of oil supplies. These could be due to in-state conflicts, exporters' interests, and/or non-state players (terrorists and others) targeting the supply and transportation of oil resources.

The 1973 oil embargo is a good example of how oil supplies can be used against the U.S. The Arab nations decided to punish the U.S. for its support of Israel during the Yom Kippur War and turned the spigot off.

This was also the case during the economic negotiations during the Russia-Belarus energy dispute in 2007-2008. As a result, Putin shut down the Druzhba pipeline, which supplies oil to Germany, Poland, Ukraine, Slovakia, the Czech Republic, and Hungary, which caused several days extreme anguish and led to power failures and misery in these countries during severe winter cold.

In 2014-2015 we were punished again by—of all things—lower oil prices this time. The victim this time was intentional and well orchestrated interruption of our path to energy independence. As soon as our gas and oil production rose above the likes of the Arab sheikhs, they dropped the global oil prices to the point where it made no financial sense to produce additional quantities of oil and gas in the U.S. And so, the U.S. gas and oil industry shrunk, thus delaying its progress by several years and bankrupting many companies as a consequence.

One good thing about this situation is the fact that producing less gas and oil today means that more will be available to be used by the next generations—small consolation in the midst of the devastation, but we have no choice.

Wars, political conflicts and many other factors, such as strikes, can also prevent the proper flow of energy supplies. Venezuela is a prime example of everything that can go wrong with the oil supply chain. First it was the nationalization of the oil industry, followed by strikes and protests. Several years after the fact, Venezuela's oil production is yet to recover fully.

Since the nationalization, Hugo Chávez threatened to cut off oil supplies to the United States time after time. He was holding American oil exports hostage and used them extensively to further his ideas and political agendas. His death in the spring of 2013 marks a new era in the relations between our two countries, but the long term results remain predictably fluid and fluctuating.

Exporters like Venezuela are driven by economic and political incentives, which at times force them to limit their exports. In other cases, export revenues are used to finance terrorist groups like Hamas and Hezbollah.

And amazingly enough we (the U.S.) are paying the expenses on both sides of the conflicts. Saudi Arabia gets $150 billion annually from the U.S. oil exports, about $4 billion from which goes to the Wahhabis to teach their children to hate us, and train them to fight us. And then we pay billions of dollars and thousands of lives to fight these children when they grow up. Fair?

There is also an increased competition for energy resources around the world now days, due to the increase of population and standard of living in India, China, and other developing countries. This creates energy price rises, which can get to high extremes like the unprecedented and unforgettable jump of oil prices to $180/bbl in 2008.

Thus increased competition over energy resources may eventually lead to formation of security pacts between the major powers, in order to enable an equitable distribution of oil and gas. If and when this happens, however, it will affect positively the developed economies, while the developing countries will continue the daily struggle to supply fuels to power their homes and fledgling economies.

The concerns with the pending *oil peak* are lurking on the horizon too. This means reduced amounts of oil and higher prices, which will create another wave of competition among the major importers. This will introduce more inequality to the energy markets, and would make achieving energy independence difficult and even impossible for many countries, including the U.S.

One big problem with the energy independence debate is that it is basically misunderstood, and had even become a mantra of environmentalists and others interested in pushing their agendas ahead; renewable energy at all cost, climate change threats, etc. The energy independence debate is also used by interest groups as a weapon for defending technological and economic reasons for using expensive and unreliable energy sources that are unable to compete on the market with oil and gas, in order to justify subsidizing them.

A close look at the energy independence debate reveals that all interests groups think that theirs is the only way of getting close to it. One group insists that the fossils are indispensible for providing reliable and cheap power, so they must be supported at all cost. Another group promotes wind and solar energy as the only solution for the future, while a third group insists that nuclear power is the best solution to our energy problems.

The problem is that the support structure is limited and crumbling, so the different energy sources must find a way to support themselves while taking the country closer towards energy independence. This, however, requires a thorough understanding of the issues on both sides of the debate and coordinated effort—both of which are missing today.

We can use energy independence only as a measure of the degree of our dependence on foreign fossils imports.

The Energy Independence and Security Act of 2007

For the purpose of ensuring our energy security and independence, the U.S. Congress passed The Energy Independence and Security Act of 2007. It is an Act of Congress addressing concerns with the energy policy of the United States. It was originally named the Clean Energy Act of 2007.

In brief, the Act consists of:

- *Title I-Energy Security Through Improved Vehicle Fuel Economy*, which requires increase in fuel economy standards for passenger cars, and established the first efficiency standard for medium-duty and heavy-duty commercial vehicles.

It is estimated to save Americans a total of $22 billion, bn the year 2020, and to show a significant reduction in emissions equivalent to removing 28 million cars from the road. Title I is responsible for 60% of the estimated energy savings of the bill.

- *Title II: Energy Security Through Increased Production of Biofuels*, which contains the first legislation that specifically requires the addition of renewable biofuels to diesel fuel.

Biomass-based Diesel, fuel must be able to reduce emissions by 50% when compared to petroleum diesel.

Biodiesel is the only commercial fuel that meets this requirement thus far.

- Title III: Energy Savings Though Improved Standards for Appliance and Lighting, which contains standards for ten appliances and equipment.

When fully implemented, the Title III modifications will void the burning of millions of barrels of oil, which will save millions of dollars, and contribute to cleaner environment.

- Title IV: Energy Savings in Buildings and Industry, which establishes new initiatives for promoting energy conservation in buildings and the industry.

When fully implemented, Title IV modifications will reduce the energy used in Federal buildings by 30% by the year 2015.

- The Act also provides: awards for developing a hydrogen economy; funding of R&D of renewable technologies; expanding research of carbon sequestration technologies; new training programs for "Energy efficiency and renewable energy workers"; new initiatives for highway, sea and railroad infrastructure; small businesses loans toward energy efficiency improvements; modernization of the electricity grid to improve reliability and efficiency via Smart Grid technologies; new federal standards for drain covers and pool barriers; and exclusions for people who have UV sensitivity that can be triggered by the higher UV radiation of the CFs.

A solid step, or two, forward, no doubt. There are already some good results to be reported from the above measures. And yet, energy independence is far from being a reality in the U.S. and most other countries.

Energy Survival

This is the hardest component of the energy dilemma. It is the skeleton in the closet, which no one wants to talk, let alone do something, about.

Energy survival refers to the continuance of energy supplies in the distant future.

Implementing energy survival measures is the one and only way to ensure that the coming generations do not experience sudden and irreversible lack of fossil fuels. Such plunge in the dark is absolutely unacceptable, and so it is our responsibility to make everything possible to avoid it.

This subject, however, is foreign to most government leaders. Instead, excessive use and abuse of fossil fuels now days is bringing them closer to the end, while increasing adverse climate and environmental effects. Continuing at this rate of use of fossils will put the world's population in increasing danger of land loss, excessive air pollution, rapid global warming, and many other negative effects. The combination of these effects puts human wellbeing and health in danger, so we cannot continue doing nothing to solve the problems.

It is obvious that not much is done now days to ensure the energy survival of the next generations, which is a serious issue that deserve our full attention.

Here is why:

1. We know and agree that *energy security and safety* are a must, for they affects us directly. Because of that all parties are working together and doing everything possible towards those goals.

2. We also know that *energy independence* is important for maintaining our way of life, so we are doing everything possible to ensure it too.

3. *Energy survival*, however, is not on the list of our priorities. It does not affect us directly today, because it is a long term problem. It does not generate profits, and since we don't have to deal with its consequence directly it is kept hidden in the closet of our guilty collective sub consciousness. This makes the subject quite uncomfortable for discussion, because our behavior now days is similar to the way irresponsible parents spend their children's inheritance on frivolous and expensive life style, totally disregarding the future generations needs.

As a matter of fact, most of our actions in ensuring energy independence, such as the increased exploitation and use of fossils (increased use of coal in China, and fracked oil and gas in the U.S.), are severely damaging to the integrity of the long term energy survival of the affected countries and the entire world. These increased activities and fossils use deplete the limited resources, and pollute the environment excessively, thus leaving deadly inheritance to the next generations.

We are actually only postponing the inevitable transition to non-fossil economy, doing it simply because we are too comfortable with the way things are now, and are not willing to sacrifice for the sake of the

future generations. So as things are going now days, we will be out of oil, gas, and coal by mid-end of this century. What are the people living then going to do in a fossil-less society? What would they think of us?

There is a chance that by then some fossils will be replaced by new and renewable energy technologies. Using solar, wind, biomass, and nuclear energy might be enough to substitute for some of the energy the fossils are providing us today. These replacements, combined with more efficient use of residential and commercial energy, might be just enough to provide society's needs. If that happens, then our sin of indiscriminately abusing the precious energy sources will be washed away and maybe even forgotten. This is a big 'maybe', but we still have time and a chance to do the right thing.

Presently, however, we are entering a period of 'immediate gratification', during which most nations are rushing into excess energy use frenzy. The result is a 'delayed energy transition', during which period we will see significant shortages of fossil energy sources and sharply increased prices. Then, and only then, we will start thinking about real energy transition (from fossils to renewables) and implementing the necessary measures.

It is difficult to predict when, and what exactly can be expected at that time. Looking around us today and seeing fossils in everything we do, however, we doubt that the life style of those who follow in our steps would be similar to ours.

One could argue that the energy transition might take longer than a century, but whatever it takes, the fossil supplies are limited, so they will be gone eventually. If not this century, for sure during some time next century to come they will be no more.

Sooner or later, one of the next generations
is going to wake up in a fossil-less world.

Not a drop of oil, a whiff of natural gas, or a piece of coal will be left on the face of the Earth. Just imagine what a very cold, dark and sad day that would be.

The to-be-affected people will try to figure out what's happening with their energy sources and reserves. And they better find a solution quickly, and well before the last piece of coal and the last drop of oil go up in smoke. If they have not found energy substitutes by then, their lives would be quite miserable, and even in great jeopardy of extinction.

We, the present generations, are obviously not concerned very much with the approaching end times of fossil energy. Presently, a great part of the U.S. energy economy is based on fossils. The world as an entity is mining more coal than ever, pumping more oil than ever, and increasing gas and oil fracking to the highest levels imaginable. The use of all these precious fuels is increasing dramatically, especially in the developing countries, and the trend is growing.

In just one century, we have pumped up and
dug out and more than half of the fossils on Earth.

Now we are getting ready to dig and pump out the rest within this century. That will leave the next generations with next to nothing in ready to use fossil reserves. This makes our energy survival short lived enterprise of ensuring enough energy now, but with no provisions for the future.

Is there any way to change that? If there is, we don't see it clearly and very plausible plans for doing anything about it exist. Capitalism dictates living good today, for tomorrow will take care of itself...somehow.

THE ENVIRONMENTAL EFFECTS

The balance of Earth's energy and its environment is maintained constant by the natural forces. Or, it is relatively constant, with some notable changes, most of which are accounted for. Some of the changes are caused by complex natural events, while others are caused by humans and their daily activities. The interaction of these two make the situation much more dynamic and complex. Due to its importance to life on Earth, it is a subject to numerous studies.

One of these suggests that "radiative forcing" accounts for the balance of solar energy as a major force which affects all aspects of our lives. It is basically the overall change in the balance between the solar radiation coming into our atmosphere and the radiation in form of heat and other energy states going out.

A positive radiative forcing (heat retention)
is responsible for warming the surface of the
Earth, while negative forcing tends to cool it.

Note: The causes for environmental changes are too numerous and too complex to analyze in this text, so we would recommend again our book, *Power Generation and the Environment*, published by Fairmont Press in 2014, where these subjects are discussed and analyzed in greatest of detail.

There are recorded changes of the global annual averages of anthropogenic radiative forcing due to changes in concentrations of greenhouse gases and aerosols from pre-industrial times to present day, as well as these due to natural changes in solar output. There are, however, many of these and other factors that cannot be accounted in the overall models.

One of these unknowns is the forcing caused by stratospheric aerosols resulting from natural disturbances, like hurricanes, volcanic eruptions, etc. major events, which vary wildly between regions and over time periods. Due to their random occurrence, their influence on the overall environment, although great, is hard to calculate and even estimate. The worst part is that many of these events are unpredictable, so we are left in the dark as far as what could happen at any place on Earth tomorrow, or next month, let alone next year.

One predictable factor is global pollution, which has a major affect on the global balance of energy.

Table 2-23 lists some of the worst greenhouse gases (GHG) used today and their concentrations in parts per million in pre-industrial and around 2000. Note the individual contribution of the different pollutants to the global warming potential (GWP), which is defined as the cumulative radiative forcing between the present and some chosen time. The GWP effect is calculated as volume of gas emitted now relative to a reference gas, CO_2 in this case, some time in the future.

Table 2-23. The worst atmospheric pollutants

GHG Gas	1800	2000	Life (years)	GWP
Carbon dioxide	250	380	10-100	1
Methane	0.7	2.0	10-15	21
CFC-12	0	0.1	100	1500
Nitrous oxide	0.2	0.3	120	6500

The GWP of each gas provides a measure of its relative radiative effects. From this number we can then calculate the future global warming potential of any GHG species over 100 years time span by multiplying the GWP number by the amount of gas particular emitted. This is not an exact number, since there are many variables and unknowns to be considered during the time period in question. Its most important value is to compare the polluting potential of different GHGs species using the same unit measures.

As Table 2-23 suggests, methane is over 20 times worse as GHG than CO_2, but much less damaging than CFC-12 (which was widely used in car and home air conditioning systems until recently) and even less so than nitrous oxides (NO_x), which are emitted by many industrial processes and car emissions.

There are many other "bad" guys in the GHG lineup, and one can spend a life time chasing them. The important truth here is that we are inundated by bad gasses, coming from all areas of the economy: chemical industry, power generation, agriculture, animal husbandry, waste disposal sites, households, commercial transport, and personal cars. Everywhere we turn, there is some bad gas around us that we are forced to breath and live with. There is no escaping it, and so some of us get sick.

The time that different GHG agents can live in the atmosphere is also important, because the cumulative effect allows the concentration of pollutants to increase with time. It is obvious from the data in Table 2-23 that NO_x and CFC-12 are long-lived, and can remain air-born 100 years or more, during which time they multiply and increase quickly.

According to EPA, the U.S. emits about 5 billion tons of GHGs annually. Worldwide this amount grows to about 35 billion (with B) tons of GHGs annually.

Over 35 billion tons of GHGs are added into our atmosphere every year.

This is 350 billion tons in 10 years and over 3,500 billion tons of GHGs by the end of this century. A lot of pollution during one single century, spread over a confined area; circulating in the closed system of our Earth's environment. And the energy use and GHG emissions around the world are increasing steadily with no end in sight. So we are really getting close to a saturation point. A time when the polluting gasses would rain over us like dirty poisonous shower. We see some of that already happening in places like Beijing and Los Angeles, as well as in the vicinity of large power plants.

We are in the midst of a century-long process of non-stop GHG emissions, which are accumulating up high above us, and all around us. They are choking the Earth's atmosphere and all living things in it to death; causing all kinds of havoc in the air, water, and soil. If this trend continues, and if there are enough fossils available then, the present amount of GHGs in the atmosphere might double during the next century, reaching and exceeding the saturation point.

We are now basically paying the consequences of what our forefathers started doing in the beginning of the 20th century. And as things are going, our grand-children will pay the consequences of what we are doing now. They will be surprised at the high price they have to pay, and will be quite frustrated about our ways…and who can blame them?

We have learned some of the lessons of the past and are fully aware of what's going on. As a result, there is an ongoing effort in the U.S. and other developed countries to control the energy use and abuse practices in order to prevent an environmental debacle. The developing world, however, is not willing to follow our example, and understandably feels that it is unfair to deprive their growing population of its increasing energy demand.

The developing world sees that we did not restrain from using excess energy, so why should they?

As things look today, there is no turning back. According to the experts, we have reached the proverbial 'point of no return', so that whatever we do and whatever energy technology or methods we use, there would be some added cumulative negative impact on our lives and the environment.

Please note here the key word *'cumulative'*, which means that most of the GHG emitted in the atmosphere stay there for dozens and hundreds of years as their quantities increase daily. This suggests that even if we reduce the use of fossils now, their concentration in the atmosphere will still increase.

We are quite familiar with the damages done by nuclear, coal, and natural gas power generation. We also know the negative effects of gas and oil exploitation, transportation, and use. It is less obvious that hydropower, geo-power and bio-fuels also add to the complexity of the situation and contribute to the energy-environmental conundrum.

What is even less obvious, and what most of us are not aware of, however, is that even solar and wind power generators, which are considered by some 100% pollution-free, are actually far from that, since they leave a long trail of pollution during their life-cycle. This also adds to the cumulative effects of our activities on the Earth. The answer to how good or bad this all is depends on whom you ask. In this text, we try to be as unbiased as possible by presenting the facts from both sides of the respective arguments.

Humans and CO_2

One of the greatest debates now days evolves around the causes and causers of the environmental changes we see increasing lately. On one hand, we see people who think that all that is happening has happened before, therefore it is part of Earth's natural ways. Others, however, show facts that human activities are the major reason for the negative changes, such as increased air pollution and climate warming.

Power plants and vehicles have been blamed for excess GHG emissions. We humans are also CO_2 generators. Even when at rest, we breath at a different rate, depending on our age, body mass, health condition, etc. Similarly so, live stock and other industries also emit a lot of GHGs.

The question is; how much GHGs do humans produce and what is their effect on the global environment?

After all, the world's population exceeds 6 billion people, so, no doubt, there must be some effect.

EPA estimates that the average person produces approximately 2.3 pounds of CO_2 per day, or about 840 lbs annually. With 6.9 billion people, this is nearly 2.7 billion (with a B) metric tons of CO_2 emitted by humans just by breathing.

Putting this into perspective, the estimated emissions of CO_2 in the U.S. resulting from the generation of electric power is about 2.2 billion (with a B) metric tons. This includes all coal, petroleum, natural gas, and even the non-fossil fuel burning power plants. So, the entire world population produces about 115% as much CO_2 as all power plants in the U.S. This is also almost twice (185%) the amount of CO_2 emitted by the passenger vehicles in the U.S. Who knew…

Now, considering the other animals on earth, which outnumber the humanity many times over, we could come with another set of numbers that show their effect on the environment by emitting huge amounts of CO_2. And they surely do. Cattle in the U.S., for example, generates more CO_2 and other GHG gasses, via breathing and other functions, than all people in the country.

So what do we do? stop breathing? An extreme solution, but we are used to it. There was a protest in the spring of 2015 in Portland, OR, for example, where people were protesting against capitalism. They did not say what system they would like to replace it with, so we could only guess.

The next step in this process is a protest against people breathing too much. It is killing us and must be stopped. What we can replace it with is still unknown, but we are sure that somebody will come up with a solution. So not to worry; we are in good hands. Just imagine what a wonderful world this would be, once capitalism in America is eliminated, and people stop breathing. But the majority disagrees, so the debate continues…

THE GLOBAL ENERGY BALANCE

The global energy balance—everything to do with "energy" and the way energy is affecting the global economy and socio-political systems—is changing rapidly. These changes have profound effect on the way every country and region functions. And even more importantly, the way they relate to one another now and in the future.

The energy issues are increasing in importance today, and will dominate the global developments in the future.

Life in the *global village* has a lot of different twists and turns, many of which are unprecedented and some are unpredictable. Energy issues complicate further many areas of life in our already complex global cohabitation.

One of the most unexpected and welcomed news of this century was the Financial Times report in the fall of 2014 claiming that the U.S. is to overtake Saudi Arabia in oil production in the very near future. Who would've ever thought that Saudi Arabia, with its immense oil reserves, would be surpassed by its biggest and most important customer. This way, the U.S. would become officially the world's largest producer of crude oil. The only difference is that Saudi Arabia exports 80% of the oil they produce, while we still import 50% of the oil we use.

The funny part here is that we will continue importing 50% of the oil we use even after becoming the world's largest oil producer. This is the irony of the energy markets, which determine the global energy balance and our energy and national security.

Note: The difference between the U.S. and Saudi Arabia is due to the fact that, although both countries produce a lot of oil, Saudi Arabia uses much less, so it has a lot of oil to export. On the other hand, the U.S. is a very large country and uses a lot of oil, thus it needs much more than it produces. This has been changing quickly lately, where the U.S. oil production can be (and will be some day) ramped up to reduce the oil imports. We, however, do not see a day when the U.S. would be able to produce enough oil to meet 100% of its needs.

Note: Late 2015, the lawmakers lifted the four-decade ban on U.S. crude oil exports. The fear was that this would bring scarcity and high prices at the pump. Wrong! The opposite is happening. The crude oil exports have actually gone down 5-7%. Yes, this is right; the U.S. oil companies have always exported oil—ban, or no ban. The only difference was that they exported it under different names and packages.

And now the sky is the limit, but low global oil prices and oversupply will keep the exports down for a long while. This is good news, since less we export, more would be left for the future generations.

In other words, the energy situation in the U.S. is changing and we can expect some unprecedented and even unexpected developments in this area in the near future.

Until recently, the estimates (based on conventional methods for the recovery of oil and natural gas), were that the U.S. would remain a middle level producer, thus requiring significant imports in the future. This is no longer so, and as a matter of fact the production levels of oil, gas, and related petrochemical products of both the U.S. and Saudi Arabia were at an equal level; about 12 million barrels a day in 2014. The energy balance will continue to grow in U.S. favor, as discussed above.

The other big news of late was that Nigeria, which previously was one of the top six crude oil suppliers for a long time, is no longer on our oil suppliers list. Who knew?

Perhaps the greatest news in the changing nature of the global supply and demand of crude oil, was that the price for Brent Crude; the global benchmark for trading crude oil, fell to about $75/bbl in 2014, or the lowest price since 2011. Only a month or two later, the price was below $50, where it remained for the entire 2014 and part of 2015. It bottomed at $25/bbl.

Note: Please keep in mind that $75/bbl is at the low end of profitability of oil production for most countries. If oil prices go below that level, most countries, including the U.S., cannot produce oil at a profit, since their production costs are high. Which is exactly what happened in 2015, as we discuss here.

On the other hand, low oil prices help the U.S. to purchase large quantities of oil and move quickly towards having an energy surplus. It also helps the economy and the public with low prices at the gas pumps. Weather we would ever get to a complete energy independence is another question that only time will answer, but at the very least we now have partially ensured our energy security enough to sleep well at night. For now!

And to top off the good news, the U.S. production of crude oil is estimated to continue increasing steadily until 2020…after the oil drillers recover from the low oil prices of 2014-2015. The US natural gas production is also estimated to significantly exceed domestic consumption by the same time. This means that the U.S. officially have enough natural gas to become a large net exporter of this commodity. As the gas volumes increase, so will its exports. Amazing turn of events!

The main reason for this energy prosperity is the unprecedented technological advance in the oil and gas exploration and exploitation areas. The new hydrofracking technologies are primarily responsible for the new energy wealth in the U.S. These technological advances have allowed the discovery and extraction of oil and gas that previously was not possible. Another factor is the change in environmental regulations and controls, which has allowed a significant increase in production via hydraulic fracking and horizontal drilling.

At the same time, the energy consumption in the U.S. is in decline as a result of energy efficiency measures, and changes in fuel economy requirements for vehicles, which reduce the overall amount of fuels used. Liquefied natural gas is also slowly offsetting coal-fired power generation and diesel fuel consumption. Solar and wind power generation are increasing at a fast pace, and in some locations are exceeding the use of fossil fuels. All this is reducing the imbalance of domestic production vs. imports of crude oil

There are estimates that by 2030 the U.S. could be a major net exporter of oil and LNG. As energy prices in the country get lower, many offshore manufacturing operations and jobs could be sent back home in the U.S., where new forms of high-tech based industrialization would blossom, based on lower cost domestic energy.

These developments will have significant geopolitical consequences too, which will allow the U.S. to dictate (at least in large part) the distribution of energy around the world. Actually, the U.S. is already a major factor in the energy markets, but mostly as a major importer.

In the near future U.S. will drive the global energy markets as a major exporter of energy products.

The future will bring many other changes in the global energy markets too. We see Europe gradually freeing itself from Russia's gas supply hegemony, and China becoming the biggest energy consumer of the Middle East oil. At the same time, Saudi Arabia will be increasingly forced to re-evaluate its actions and relationships by choosing carefully how much oil to pump, when, and where to send it, if it is to retain a significant global influence. But in all cases, the glory days of Saudi Arabia's oil dominance are over…almost.

This was how we saw the global energy markets in the summer of 2014...and then things changed again.

The Oil Bust

And then came the low global oil prices of 2014. Saudi Arabia saw the increased activities in the energy sectors of the U.S. and other countries as threatening, and put its foot down. By manipulating its oil production volume, the Saudi sheikhs were able to lower the global oil prices down to $70-80, and then even further to $25-30 per barrel; well below the production costs of most countries. The increasing oil production in the U.S. was reduced to a trickle.

> *In 2015, the U.S. oil and gas drilling industry came to a screeching halt.*

More than half of the new drill sites were abandoned by the summer of 2015. So now we know who is the boss. The Saudi Arabian sheikhs still control the energy markets...or at least the crude oil global market.

Our crystal ball is not very reliable, as it can be seen from our predictions in the previous segment. Having learned the lesson, we will not try to predict what exactly will happen tomorrow, let alone 20 years down the road, but several things are quite clear today. Some of these will drive the future global energy markets:

— Saudi Arabia cannot keep prices too low for very long, since oil exports are its principal income. The sheikhs, however, know the importance of oil to the world economies and will surely continue to attempt to control the global oil markets by regulating production levels and prices as long as they can.

— No matter what happens, the huge oil and gas deposits in the U.S. remain and can be exploited whenever prices go up. This is inevitable, so the U.S. will be able to keep some control over the volume and price of its energy production. At least until their supplies last...which is another painful subject.

— The energy (hydrofracking driven) revolution underway in the U.S. and other countries (which was temporarily stifled by artificially low prices during 2014-2015) is quite significant and is promising to reshape the world's energy markets.

— As the energy wars continue, they will most certainly affect and reshape the global relationships during the next several decades. This will shift the global energy balance with time.

— The major players in the 21st century global energy markets (per volume of energy production and use) are shaping to be: the U.S., OPEC countries, China, India, Europe, and Russia.

— Most importing countries will be forced to adapt to the changes, like it or not. The new, shifting global energy balance has been imposing unfavorable conditions on some of them, so some more blood and tears are to be expected in the near future on the energy front in the affected areas.

— All said and done, *global energy balance* is a misnomer, since there is no balance in the energy markets. Instead, what we have is a temporary (day-to-day) titter-tatter of negotiations and compromises by exporting and importing countries. As the differences between these groups grow with time, the negotiations will become more complicated and the compromises would be more difficult to reach. Because of that, serious (even armed) conflicts among the different groups are to be expected.

— One of the most surprising facts is that as the world goes through the daily ups and downs of the energy markets, very few of us keep an eye on the global crude oil level gauge. Crude oil is the single most important component of the energy markets, and one can clearly see that its level is quickly diminishing and getting closer to the *empty* mark. Yet, we continue burning this very important, but finite, commodity as if there is no tomorrow. But tomorrow will come, and it will be oil-less... What would the energy markets look at that point in time?

Let's take a closer look at some of the key players, who determine the present day dynamics, and promise to shape the future, of the global energy markets:

China

China is slowly becoming the world's largest producer and consumer of all kinds of energy. That growth, driven by the increasing needs of a growing population, comes with a price. It is reflected in the quick depletion of natural resources, combined with severe air pollution in some Chinese cities. This extraordinary, and somewhat chaotic development is also causing rising prices of the available energy resources. At the same time, the rapidly growing Chinese economy is putting additional pressure on the national and global energy markets.

As a net crude oil importer since the early 1990s, China's dependence on foreign oil and other fossil fuels has been steadily and rapidly growing. The Chinese middle class has seen the major growth, and millions of people—over 300 million to be exact—are now in the middle class category. This is a population of the size of the U.S., which insists on having a similar lifestyle to that enjoyed by the Americans. This means all kinds of luxuries, including millions of cars replacing the bicycles and mopeds on the Chinese streets.

As a matter of fact, cars in China are now much more than means of transportation; they are a status symbol. Many Chinese in major cities can afford cars, but only a few are permitted. Yes, there is a lottery in many cities, where people bid on car permits. $10-15, and even 20 thousand are what one has to pay for the right to own a car—if they win the bid. The car does not come with the winning ticket.

BMW in China also has a different meaning. In addition to being a very desirable (and very expensive) vehicle, it is also used in marriage negotiations. Yes, many marriages are negotiated, where negotiators evaluate the potential spouses. The men are evaluated on financial basis, where having a job is number one priority. Having an apartment comes second, and having a car comes third, and yet very important requirement.

"Be my wife," is what any Chinese man who owns a BMW can tell the girl of his choice and win her hand in marriage. At least this is how the saying goes, so there must be some truth to it. Even more amazing is the fact that this same young man driving the BMW, was riding a bicycle on his way to work just a few years back. If this is not progress…

Figure 2-19. BMW is the best marriage negotiator in China

Yet, there are no huge natural resources in China, so the leadership has put in place ambitious goals for deriving a significant percentage of its power from renewable sources. So now, China's long-term strategic energy goals, in addition to increased domestic production and the imports of fossils, include ramping up the renewable energy industry.

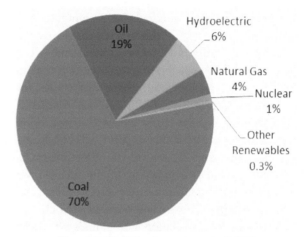

Figure 2-20. China's energy mix (2013)

As Figure 2-20 shows, China is a coal country. There are some plausible developments in the other areas of the energy sector recently and more are planned in the future, but coal has been, is, and will always be China's major energy generator.

Coal's share is expected to fall to 62% by 2035, down from 70% presently, mostly due to increased energy efficiency and use of renewables and other energy sources.

*Nevertheless, the **total** coal consumption will double by 2035, together with the energy use.*

What this means, is that the total energy use will continue increasing, and while coal's total presence in the energy mix might decrease as a result of introducing renewables and such, its total use will double from the present levels to meet the ever increased energy demand. Not good for China's populace or the global environment. Not to mention that such huge exploitation of the limited coal and other fossils resources will bring them much faster to their eventual and unavoidable depletion.

One of the major goals of the Chinese government is to reduce carbon emissions per unit of GDP by at least 40% by 2020, as compared to 2005 levels. China has also announced plans to reduce its energy intensity levels (energy consumed per unit of GDP) by 31% by 2020, as compared to 2010 levels.

There are also plans to increase non-fossil fuel energy consumption to 15% of the energy mix during the same time period. This is more than double the present level of non-fossil use (hydropower and renewables), which is possible, but easier said than done, simply because there are many other factors that hinder the quick development of these resources.

But even if China succeeds in implementing these plans on time, coal use is still on the rise (in step with increasing population and energy needs) and will continue so for the foreseeable future. So much that in 20 years, China will use twice as much coal as it uses today.

China's 12th five-year plan was released following the plenum and approved by the National People's Congress on March 14, 2011, with the goals of addressing rising inequality and creating an environment for more sustainable growth by prioritizing more equitable wealth distribution, increased domestic consumption, and improved social infrastructure and social safety nets.

The plan is representative of China's efforts to rebalance its economy, shifting emphasis from investment towards consumption and development from urban and coastal areas toward rural and inland areas, initially by developing small cities and greenfield districts to absorb coastal migration. The plan also continues to advocate objectives set out in the 11th Five-Year Plan to enhance environmental protection, accelerate the process of opening and reform, and emphasize Hong Kong's role as a center of international finance.

The main targets for the 12th Five-Year Plan were to:
— Grow of GDP by around 8%,
— 7% annual growth of per capita income,
— Spend 2.2% of GDP on research and development by 2015,
— Bring the population below 1.39 billion by 2015,
— Readjust income distribution to stop the yawning gap,
— Firmly curb excessive rise of housing prices,
— Implement prudent monetary policy,
— Intensify anti-corruption efforts,
— Accelerate economic restructuring, and
— Deal with the complex situations in present development

Among the other highlights of the plan are also:
— Urbanization rate reaching 51.5%
— Value-added output of emerging strategic industries accounting for 8% of GDP
— Inviting of foreign investment in modern agriculture, high-tech, and environment protection industries
— Moving coastal regions from being the "world's factory" to hubs of research and development, high-end manufacturing, and the service sector
— More efficient development of nuclear power under the precondition of ensured safety
— Increased momentum for large-scale hydropower plants in southwest China
— Length of high-speed railways reaching 45,000 km
— Length of highway networks reaching 83,000 km
— A new airport to be built in Beijing
— 36 million new affordable apartments for low-income people

The plan is also heavy on energy developments and improvements:
— Reduce fossil energy consumption,
— Restructure China's energy sector
— Promote low-carbon energy sources, and
— Gradually establish a carbon trade market.
— 16% reduction in energy intensity (energy consumption per unit of GDP);
— Increasing non-fossil energy to 11.4% of total energy use; and
— 17% reduction in carbon intensity (carbon emissions per unit of GDP).

Although there is significant progress noticeable in some of these areas, the final results of the 12th five-year Plan are to be made known in the future. A new 5 year plan was introduced in 2015 in which China's demand for industrial commodities will probably be hurt as the biggest energy and metals user intensifies efforts to rein in pollution, according to Citigroup Inc.

China's leaders are planning to focus on the environment as never before during the new Five-Year Plan of 2016-2021. There are plans for the initiation of an intensive carbon trading program by 2017. Costs for energy-intensive industries will increase, with steel and electricity likely to be the first to implement the scheme, and suffer the consequences.

China is the world's biggest carbon emitter, thus the move to address the environmental damage that's been a byproduct of its breakneck economic expansion and excess coal-fired power generation.

Environmental initiatives are expected to be among the most emphasized portions of the new 5-year plan. They are likely to carry some of the most aggressive targets, which is likely to increase costs for emissions heavy industries including coal power plants, metals refineries, oil and petrochemical plants.

The New China

There are lots of changes in China's economy and political life. The economy is tittering dangerously between growth and decline, while the government is headed into uncharted territory. Never before had a communist regime being in charge of such an economic growth. Unlike the conventional Lenin-Stalin type of communism, which kept the masses in poverty, the Chinese leaders are encouraging the economic growth and are implementing other changes that promise to keep them in power for awhile longer.

Relying on their good fortunes thus far, and determined to see the trend through, the Chinese government continues to promote positive social changes, the most significant of which is allowing Chinese couples to have more children. Another very significant change is allowing religion in the country.

Here again, the Chinese Politburo is breaking away with the conventional Soviet-era communism of ridiculing and punishing religion. Instead, it is all of a sudden decided to invite Buddhist monks to bring "opium for the masses" directly into the broad society without any government control or influence.

Amazing! Could it be that the wolf in sheep's clothing is actually a sheep? It is surely behaving as such, so it remains to be seen how long it could keep up the appearances.

Figure 2-21. China Buddhism and Communism hand-in-hand.

Note: The author was born and grew up in an Eastern European communist system, the stupidity and inequality of which forced him to escape to the West. 25 years life in a communist system would make anyone a specialist on the subject of communism, but a closer look into the Chinese *whatever system*, reveals little resemblance of communism. Instead, it functions like a giant mafia, to some sort of organized crime, entity under the cover of communist ideas.

The amazing progress of the Chinese economy during the last 20-30 years can be credited to anything but communism. The actions and policies of the "communist-in-words-only" government are on the other side of the spectrum of all we know about communism. So the overall conclusion is that what is happening in China is either, a.) not communism at all, or b.) it is some sort of new and totally different from of communism. In any case it is not anything like the Soviet-era communism we experienced until 1990.

In any case, the economic, financial, and social progress in China—communism or no communism—is simply remarkable.

Aside of the mafia-type activities of the central and local governments, there are remarkable and very progressive moves recently in the China energy field as well. China's government is determined (at least on paper) to reduce energy use via energy efficiency measures and through the introduction of new technologies—including renewables. It is also making some progress towards reducing the country's killer GHG emissions.

This is a more urgent, complex, and very expensive effort, which still needs to be defined and implemented. As the China economy falters (as it does from time to time), we see this effort put on the back burner. Yet, China and its government must be given credit for the good intentions and excellent first steps towards a more efficient and clean energy future for the largest country in the world.

Taking a closer look at China's energy sector we see that presently the Chinese power plants burn over two times more coal than these in the U.S., and almost as much as the rest of the world combined. The country's economy is reliant on domestic coal and coal imports for 75% of its total energy consumption and over 80% of its electricity generation.

China's coal reserves are estimated at 55 billion tons of anthracite and 45 billion tons of lignite quality coal. China's coal mines produce about 4.0 billion tons of coal annually, so at the present level of use, China will run out of domestic coal by the end of the century.

For over 15 consecutive years now, China's economy has seen rising thermal coal consumption, with some new changes finally taking a hold of, and modifying, the trend.

Chinese coal production increased 400% since 2000. It is also expected to continue to increase through 2050.

There is, however, a new trend in China; that of reducing coal use, which is driven by China's war on pollution. It is finally affecting the coal prices and is forcing China to implement a number of new and unprecedented measures. Lots of talk and negotiations started in 2014, some rules were changed and others implemented, buy so far the effect has been negligible.

The goal of reducing coal mining output and consumption to less than 4.0 billion tons per year as part of China's "war on pollution" is still a reality only in the dreams of the policy makers. New curbs on consumption were enacted, some bans on new coal-fired power plants were imposed, and limits on the total energy consumption were put in place recently. And yet, the results unfortunately are close to zero.

The air in Shanghai and many other cities is still untreatable and the soil in many area is still badly contaminated with hazardous waste materials. A national emission trading scheme will be initiated in 2016, in addition to the seven regional schemes already in place.

If successful, this measure will create the largest ever carbon emissions market in the world.

One positive result of all this now days is that the Chinese coal imports declined slowly during 2013, and are kept at the lowest level since 2012. Still too high for comfort, but lower, nevertheless. Domestic production, however, is still at its all time peak. As a matter of fact, the large mines are a national treasure and are thus untouchable. At the same time, efforts of the authorities to clamp down on small scale mines, in order to reduce coal production, have been largely unsuccessful too. And so, the domestic production is continuing to ramp-up unaffected by the policy changes.

China is expected to impose a cap on coal use when it issues its next five-year plan in 2016, but the scope and size is unclear. The new five-year plan is expected to contain such a cap, but the question is whether it will be legally binding and enforceable. The experts' answer to this question is, *maybe.*

There are also planned, but yet unpublished and unenforced, regulations that ban the imports of low-quality coal, while at the same time placing restrictions on production at the 14 largest domestic coal mining operations. If these regulations are implemented successfully someday soon, then things might change… somewhat. At least as far as reducing the amount of GHG emissions is concerned.

Note: China also imports a lot of coal, but it amounts to only 7% of its total coal consumption. Al-though coal imports are usually cheaper than domestic production, China doesn't increase imports, because of wild price swings driven by the exporting countries. Since some of these countries are not that friendly, or well-meaning, China's government is very careful with the imports. Presently, the U.S. is shaping as one of the best exporters to the Chinese coal market. This is a controversial development, so we need to watch it carefully.

There have been unofficial and undisclosed plans to increase the coal imports of cheaper coal from the U.S. and other countries. The new regulations limiting high-ash and high-sulfur coal imports, however, could slash overseas cargoes by 80 million tons annually… maybe. The coal production output cuts of domestic mines could take an additional 200 million tons off the market…yet another maybe.

The increasing energy demand is most likely to increase coal prices by 10% and so the entire recovery plan might collapse in favor of the imports, unless significant government support and subsidies are introduced to avoid export increase. Coal prices have started to show a recovery since the moment the government decided to step in. The recovery may accelerate the going forward with domestic production trend, according to the experts.

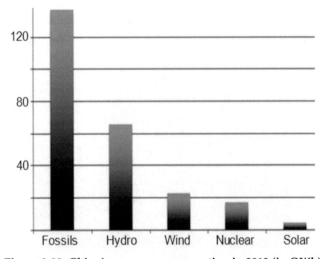

Figure 2-22. China's new power generation in 2013 (in QWh)

The Chinese coal market is full of good and bad news, but the overwhelming fact remains; China's growing economy and increasing population with increasing needs, will need more and more electric power. It is the sign of progress and wellbeing. Only coal can deliver so much power, so quickly in China at least for now, so it is not clear how well and efficient the new reforms would be on reducing coal use, coal imports, and coal emissions.

The Intergovernmental Panel on Climate Change said recently that demand for coal power led to a spike in global carbon dioxide emissions between 2000 to 2010. Much of the new demand for coal seen over the past 10 years has come from China, which is building new coal plants at a rapid pace to keep up with electricity consumption.

At the same time, China has been praised for installing huge quantities of solar installations. As a matter of fact, it is about to become the number one in the world in "green" power installs. Unfortunately solar and wind power plants, albeit impressive in number and size, are a minute portion of the total energy portfolio. As the need for energy in the country grows, coal is the fuel of choice. It is the answer to the energy problems in China—clean environment or not.

It is easy to see from Figure 2-22 that solar in China (despite of the loud media noise) is almost irrelevant as far as the total of its new power generation is concerned as compared with the fossils and the other energy sources. While wind and nuclear take a more respectful place in the energy mix, it is obvious that fossils play a huge role in the national power generation cycle. Is it possible, then, for solar, wind or any of the renewables to become major power sources in this huge and growing economy? The obvious answer is, *maybe later.*

On the positive side, some of the new fossil power plants replaced older facilities with outdated and much more polluting equipment. In 2014, China gave green energy advocates some hope when its energy administration announced that 1,725 small-scale coal mines would be closed by the end of the year. This effort, however, has proven to be not that successful in practice and most of these mines are still open. These closures are part of China's effort to reduce CO_2 levels in the atmosphere and cut massive pollution problems in cities like Beijing. But that was more of a PR move than a step towards a greener future.

China hopes to cap its total coal production at 4.1 billion metric tons by 2015, which is actually up 10% from the 3.7 billion metric tons produced in 2013. The picture is clear enough: While solar and wind are making dramatic strides in China, fossil fuels dominate by making even more dramatic strides. Fossils—coal in particular—will continue to dominate the Chinese energy markets for a long time.

What does this do to China's image as the world's largest emitter of GHG pollutants? While it is a major discussion point in all climate change debates, and especially as far as coal-burning issues are concerned, the issues are wide reaching:

- China is planning to continue burning dirty coal for a long time, since there is no other choice to generate enough electricity to power its growing population.

- Awareness of the fact that air pollution is choking and killing people in the major cities is increasing.

- Discussions on the subject are intensifying, but the actions are few and far between.

- There are some efforts by the government to reduce pollution, and though solar, wind, nuclear and hydroelectric dams are making some impressive gains, all this is small potatoes in the huge borsch pot of Chinese energy.

- At the same time, the country's economy is slowing down and getting more efficient, with a tendency to shift away from heavy, less energy demanding, industries.

- All this considered, experts estimate that China's coal use will peak by 2015, and might stabilize and even decrease after that.

- But there is another possibility: that of China's increasing use of coal to continue indefinitely. In such case, even if coal-plant construction is reduced, the thousands of existing power plants are going to increase emitting their poisons for many decades—at least until the middle of the 21st century.

- Even the most optimistic low-coal power generation forecasts show that coal will provide about 60% of China's electric power by 2020. This, in the best of cases, will still be many times more than all clean technologies put together.

So, while Europe, India and some other countries are foreseen to hit "peak coal use" in the near future, this is not the case in China. Instead, China is financially capable of, and is surely planning to, ride the coal use band wagon for a long time—even though its domestic coal reserves will not last but a few decades.

China's coal consumption is expected to increase during the next few decades to over half of the total world's coal use.

This, in turn, makes tackling climate change impossible, since as a result of China's increased coal use, the global GHG emissions will also continue to increase at least for the next two-three decades. This comes at a time when emissions need to start declining, if we want to avoid 2°C or more of global warming. Yet, China's priority is power generation at all cost and the rest is just words…

Decades of excess carbon emissions from China's increasing coal fleet will nullify all international efforts. So, short of implementing some ingenious carbon capture and storage technology for coal plants emissions in China, our environment is doomed and uncontrolled global warming will become a reality in the very near future.

The coal-rush in China is on, and only a miracle can stop it. We see no such miracle on the horizon, now or in the near future. So, in China the GHG gasses will continue flowing, negatively affecting the global environment.

How much would the U.S. help China and the local environment by exporting millions of tons of coal there as planned? How much would that help us on the long run? Or is it just the old, established priority of the capitalist system of making money today before anything else?

Europe

The EU community is an unenviable place of depending on huge foreign imports. It imports about 85% of the crude oil, and about 60% of the natural gas, it uses. These are very large quantities of liquids and gasses, which make the EU the world's leading importer of fossil fuels. At the same time, only 3% of the uranium used in European nuclear reactors was mined in Europe. So now, Russia, Canada, Australia, Niger and Kazakhstan supply about 75% of the total EU nuclear material needs. The rest 11-12%, as well as most of the oil and gas imports, come from unstable and unfriendly nations around the world.

And so, most European countries have a huge energy problem to solve. Poor on fossil resources, they depend on energy imports from other countries—and Russia in particular. Europe is importing most of its natural gas from Russia. Not a good place to be in. Dealing with Putin is like confronting a polar bear with a toothache; you never know what her next move is going to be. And most often than none, the move is not very friendly.

But things might be changing. The Russian bear needs US dollars, so the nationalistic sentiment has to be shelved from time to time, like it or not. It seems that this is exactly what happened in the fall of 2014, when Europe and Ukraine were getting ready for another severe winter cold hitting their homes and businesses.

Despite the problems in Ukraine and the worsening relations in the region, the Ukrainian energy problems for the winter of 2014 were resolved surprisingly fast, starting with an agreement with the Russian bear, despite its huge toothache. An agreement for natural gas purchases was signed in October, 2014, in the form of a Binding Protocol regarding the conditions for gas delivery from the Russian Federation to Ukraine for the period from November 2014 until 31st of March 2015. This helps Europe too, so maybe the fuel crisis is avoided. The official wording of the Agreement is as follow:

In this context the signing Parties of this Protocol agree on the following:

1. Purchase of natural gas
    1.1 Purchase Price
    For the period from November 2014 until 31st of March 2015 the price of natural gas supplied by Gazprom to Naftogaz as per Contract No. KP shall be decreased by the amount of export duty decrease provided by the Government of the Russian Federation. The decrease of the export duty shall be calculated according to the following formula: at a price of $333.33 US, and higher per 1000 cubic meters of natural gas the reduction is $100 US, at a price lower than $333.33 US dollars the reduction is 30 percent from such price.
    1.2 Payment and Delivery conditions
    The payments for natural gas purchased during the period between November 2014 until 31st of March 2015 shall be carried out before delivery takes place. Gazprom commits to deliver all paid volumes in accordance with Contract No. KP and Addendum to it dated 30 October 2014. Gazprom would forgo invoking the take or pay clause of the Contract No. KP for the period from 1st November 2014 until 31st March 2015.

2. The payment for transit services
    The payment for transit services shall be made by Gazprom in full to the bank account of Naftogaz in accordance with the Transit Contract.

3. Schedule of Payments for unpaid invoices
    3.1. Without prejudice to the decision on the eventual debt settlement to be made by the Arbitration Institute of the Stockholm Chamber of Commerce, the Parties agree that Naftogaz will pay to Gazprom 3.1 billion US dollars of the unpaid invoices for volumes delivered for the months November and December 2013 as well as April, May and June 2014 in the following order:
    — 1.45 billion US dollars (first tranche) before the

first delivery of natural gas in 2014;
— 1.65 billion US dollars (second tranche) before the 31st of December 2014.

3.2. The payment of the first tranche as mentioned in section 3.1 is a condition for the start of gas supplies according to section 1. of this Protocol. The payment of the first and second tranches as mentioned in section 3.1 is a condition for gas supply after 31st December 2014.

3.3. The total amount of 3.1 billion US dollars to be paid under this schedule has been set aside for this purpose on a separate account.

3.4 For Naftogaz, the payment of 3.1 billion US dollars under the schedule of payments constitutes the full payment of the unpaid invoices (calculation based on a price of 268.5 US dollars per 1000 cubic meters).

For Gazprom, the payment of 3.1 billion US dollars under the schedule of payments constitutes a partial payment of the unpaid invoices of Naftogaz for the gas supplied in November and December 2013 as well as in April, May and June 2014 in accordance with the Contract No. KP (calculation based on the relevant conditions of the Contract No. KP).

4.   Implementation of the Protocol
Simultaneously with the signing of this Protocol, Gazprom and Naftogaz shall sign the Addendum to the Contract No. KP outlining how the Protocol shall be implemented with regard to all issues falling under the responsibility of the two Companies.

5.   Miscellaneous
This Protocol is executed in three (3) original copies, in English, in Ukrainian and in Russian languages one copy for each Party. In case of discrepancies, the English text shall prevail. Done in Brussels on 30 October, 2014.

Wow! This looks like a great achievement after all the hoopla and animosities in Crimea and East Ukraine, where Putin confronted not only the locals, but the entire Western World with his nationalistic ambitions and threats for retaliation. At that time, it sounded like Russia will turn off the natural gas spigot to Ukraine and its major European customers, letting them freeze in the coming winter.

It was not to be! The Russian bear had a moment of sanity and signed a contract with its Ukrainian and European foes. Capitalism in action! Putin would not have done this during the Cold War years, but had to swallow his ego this time. *This time* is the key word, because there may not be a next time, if it is up to him.

Yet, the agreement with Ukraine and the EU countries is not a done deal until it is a done deal. It might last for awhile, but how long? The Russian bear has a lot of tricks up its sleeve, and we do not even doubt that they will use some of them somehow, someday.

The 2014-2015 gas imports agreement is a good step in the right direction, but will Putin sign it next year, and the next? The EU is not out of the woods and cannot count on uninterrupted energy supplies while dealing with Putin's KGB-like, self-proclaimed semi-democracy. So, EU member countries are looking in the alternatives; renewable energies, nuclear power, etc.

India

India is a huge country, with one of the world's largest and poorest population. It has a large human potential but limited resources and a number of serious problems. Energy, or lack thereof, is only one of the problems plaguing the country. It is, however, an important one, since it drives the economy and determines people's wellbeing.

The Challenges

The Indian Planning Commission has identified "Twelve Strategy Challenges" that need to be addressed and included in the five-year plans. According to the Commission, they are the main core areas that require addressing and re-evaluation to produce the desired positive results in the future. The challenges are:

GDP Growth

India has been sustaining an average of 8% annual GDP growth. Increasing it to 9-10% requires aggressive mobilization of investment resources and much better allocation of the available resources via much more efficient capital markets. Increased investment in infrastructure, education, and health care, as well as more efficient use of the available public resources, are also needed to increase the GDP.

Employment

India's economy is stagnant in the area of jobs generation. This is limiting the livelihood improvement of the majority of the population. And yet, there are manpower shortages in many sectors of the economy. One way of solving this problem is through improving the education and professional training systems. In parallel, efficient and accessible labor markets for all skill categories must be created, in addition to aggressively encouraging faster growth of small and micro enterprises.

Efficiency and Inclusion

Many sectors of the Indian economy are non-existent or incomplete at best, with those dominated by public provisioning suffering the most. Open, integrated, and well-regulated markets for land, labor, and capital and for goods and services are the mark of advanced economy, which are essential for the overall growth of the economy. This can be achieved by increased efficiency of industrial and agricultural operations, and inclusion of the populace in the economic activities.

Decentralization

Informed participation of all citizens in the decision-making processes, open accountability, and allowing them to exercise their rights and entitlements is necessary element in achieving the economic goals of the country. Self-determination of the course of the individual people's lives is critical to faster growth and sustainability.

Technology

Technological improvements and advancing organizational innovation are the keys to higher productivity and competitive enterprises. Encouraging and incentivizing innovation and its diffusion in academia, government, and the enterprises is one key to success.

Transport Infrastructure

India also suffers from inadequate transport infrastructure, which results in lower efficiency and productivity in all areas of the economy. Infrastructure inadequacies also increase the products and transactions costs, and insufficient access to the different segments of the national markets. The creation of an efficient and widespread multi-modal national transport network would solve all these, and many other, problems.

Agriculture

Rural India's infrastructure is well below the very minimum world standards, and suffers from lack of, or inadequate, amenities. This and other issues have hindered agricultural growth, which perpetuates food shortages and nutritional insecurities all through the country and also reduces rural incomes. Innovative ways are absolutely and urgently needed to encourage and support the rural population in improving their working and livelihood conditions.

Urbanization

India's populated centers, metros and cities especially, suffer from inadequate social and physical infrastructure. This, coupled with the worsening pollution in many areas, puts the populace under increased stress. With that migration pressures are becoming a serious problem, which is expected to increase. Making the large cities more livable, and ensuring that smaller cities and towns do not follow their example tomorrow is one of the goals ahead of us.

Education

India's educational system have improved in recent years, as a result of increase of educational and training facilities and the related opportunities for the people to attend. They all, however, suffer from inadequate access, affordability, and quality, which are the main concerns in the sector. Employability of the educated people also remains an unresolved issue. So improving the quality and the usefulness of the India's educational system, while ensuring equity and affordability, need to be addressed.

Health Care

India's health indicators are also well below the world standards, and are well below many other socio-economic indicators of the country in general as well. Good healthcare is unavailable in many areas, and is also unaffordable to many people. Improving the healthcare system; and its curative and preventive areas, especially relating to women and children, is another goal of the future plans.

The Environment

The environmental conditions in parts of the country, and the pronounced and increasing ecological degradation, are growing problems of serious local and global implications. And as usually, the most affected are the poor and other vulnerable classes. A new way of thinking and protecting the environment must be found, as needed to encourage responsible behavior of all sectors of the industry, agriculture, and governance—without hindering the economic development of the country.

Energy Security

Energy security is an important element in economic development of any country, including India. Energy drives faster and more inclusive economic growth, which requires rapid increase in energy resource consumption. But India has limited domestic natural resources, so it can not meet its energy needs without huge imports. This, however, has to be done equitably and affordably without compromising the economy or the environment. Not an easy task!

The Planning Commission of India has also developed an energy scenario building tool; the India Energy Security Scenarios 2047 (IESS 2047). It explores a range of potential future energy scenarios for India, for diverse energy demand and supply sectors leading up to the year 2047. See Figure 2-23.

IESS 2047 evaluates India's possible energy futures across energy supply sectors such as solar, wind, biofuels, oil, gas, coal, and nuclear, and energy demand sectors such as transport, industry, agriculture, cooking, and lighting and appliances. The overwhelming result of all this is that India's energy needs are increasing quickly, while the domestic power generation and fuels production is lingering, so the energy import imbalance is increasing.

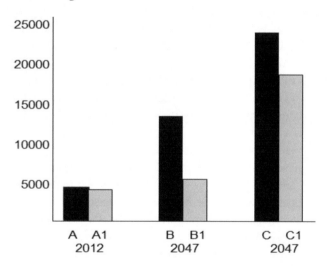

Figure 2-23. Energy demand in India 2012-2047 (in TWh)

Legend:
A is the actual energy demand in 2012 (4,905 TWh)
A1 is the actual energy production in 2012 (4,826 TWh)
B is the least possible energy demand in 2047 (14,732 TWh)
B1 is the least possible energy production in 2047 (5,684 TWh)
C is the most possible energy demand for 2047 (23,679 TWh)
C1 is the most possible energy production in 2047 (18,129 TWh)

Figure 2-23 depicts the increasing discrepancy between energy demand and supply in the country. Here *energy* includes any and all energy types produced and used by different energy sources: gas, coal, oil, renewables, hydropower, nuclear power, carbon capture and storage, electricity imports, bioenergy, as well as energy from municipal waste.

This means that in all cases, India would have energy deficit all through the middle of this century, and that it will rely on heavy energy imports. Because of that, many of the issues listed above will not be solved and so the Indian government has to find compromises in many areas.

The bright spot in India's energy conundrum is their effort in developing a new type of nuclear power generation, using thorium as source material. India has large deposits of thorium and a successful development of the new technology might bring an unprecedented energy bonanza. This, however, won't happen any time soon, so India still has to work on resolving its present energy problems in other ways.

Large amounts of cheap fossils are available today around the globe, while imported coal, oil, or gas are 60 to 80% more expensive than locally produced such. That is Finance 101, and while many countries are getting serious about getting a handle on this problem, India is not one of them.

Even though India is home to one of the world's largest coal deposits, the Indian coal sector is in shambles.

Figure 2-24. Coal transport in India

The country can't mine coal fast enough, and then it lacks the infrastructure to transport it to the points of use. While in the U.S., for example, unit trains (100 cars loaded with 100 tons of coal each) shuttle back and forth between mines and the power plants, in India coal transport is still done by horse carts and trucks.

Cheap coal imports have been blamed as a major part of the energy problems in India. Because of their high cost, the Indian government relies heavily on them to plug the faltering domestic supply. And the discrepancy is growing, with official government estimates placing coal imports to around 50-60% of total coal demand by 2030.

Putting these numbers in perspective; China's record-breaking coal imports account for only slightly over 7% of the country's total coal consumption. Imagine an increase to 50-60% imports, as in India, which has a similarly huge demand for energy. No wonder India's economy is broken.

One reason China doesn't increase imports to such high levels is that power plants and the economy they power are subjected to wild price swings driven by the exporting countries. Since some of these countries are not that friendly, or well-meaning, China's government is very careful with the imports.

Note: The events in India promise to affect the global thermal coal markets. For example, recently India's highest court ruled that many of the 218 coal licenses issued to companies since 1993 were awarded illegally, so penalties and mines shut downs are expected. In 2014 India's government mooted plans to let state-run Coal India, the world's largest producer of thermal coal, to increase imports and pool it with domestic supply. This was as a result from the revelation that nearly half of the country's coal power plants only have enough stocks to last a week. Other estimates show that India's imports could rise by and additional 38 million tons annually. According to a separate study the expected rise in imports would absorb nearly 90% of the 2014 seaborne coal surplus. These are, however, only speculations. The energy situation in India is too complex and dynamic for any reasonable estimates to be expected.

In all cases, India is past the point of no return in its energy security, so for now (and for the foreseeable future) it must import huge amounts of fossils from abroad. Coal imports, in particular, pose a real threat to India's energy sector and the current deficit levels, which could run up to 15% of total GDP by 2030. Fifteen percent of GHP is a big chunk of money for India or any other country.

Even worst-case scenarios estimate that the deficit could increase as high as nearly 40%. The rupee has fallen precipitously already, so such a huge drain on foreign exchange is putting India in a precarious and totally unsustainable position from energy and budget sustainability standpoints.

International advisers have suggested that India start planning for the oncoming "peak coal" too. Remember "peak oil?" It is when oil extraction around the world reaches a maximum and then starts declining. It is the same with coal. It will eventually reach a peak point and then dwindle into oblivion...unless something

drastic is done soon.

We have not considered "peak coal" as a serious problem until now, because it is still far in the distant future, at least for the U.S. and most other countries. Not so in India. And not because coal is running out per se (although in reality it is, but very slowly.) The "peak coal" for India is coming from the fact that the current path of increasing coal imports is simply unsustainable economically. This also means that—since increasing domestic coal production is not considered feasible—the government should dramatically ramp down the planned coal-fired power plants construction.

The current five-year plan proposes 51 GW of new coal-fired power plant construction within this decade, which would bring India into the world's first "peak coal" reality. The experts consider that the only sustainable path is for new coal-fired power plant construction to be reduced to about 10 GW by 2030.

In such case, replacing the potentially lost coal capacity would require 350-400 GW of new (alternative) power generation additions. This would require a dramatic ramp up of solar energy power generation: a jump from the estimated 2 GW solar power added annually today to 20-25 GW annually by 2020. This is quite unlikely due to "confusing" policies that change with time and differ from state to state.

The wind manufacturing also must be at least doubled and tripled, during the same time from the current 10 GW production level. But the existing policies are as confusing as in the case of solar power.

A drastic energy transition makes economic sense too, since new coal power generation is estimated at 25-29 cents/kWh, while new wind projects, for example, are estimated 8-10 cents/kWh, and solar is roughly 10-12 cents/kWh.

Such transition, however, requires providing both energy sectors with stable political support and enough low-cost finance, which is particularly important. Amidst the political confusion and scarcity of capital for clean energy in India, however, this is not very likely to happen soon.

So, in India the problem is clearly visible and solvable; imported coal is expensive and unavailable, clean energy is available, but equally expensive. And equally importantly, the entire energy situation in the country is in a state of confusion.

If and when India's government and regulators get a good grasp on the problems at hand, they have clear choices to make: continue with unsustainable fossils imports and throw the country down the "peak coal" cliff (the first country to get there), or take the "green" path. Not an easy, quick, or cheap choice to be sure, but one that cannot be postponed…for the sake of the Indian people!

Continuing to count on huge coal imports increases will solve the immediate energy supply problem, but it does not guarantee the overall energy security of the country. It might even bankrupt the already struggling economy.

Taking the "green" path, on the other hand, would be harder in the short term, but promises to provide long-term energy survival, at least as far as India's power generation is concerned. In the best of cases, there is still the crude oil imports issue, which is as serious, and getting even more serious with time, but which is, unfortunately, not easily solvable. So what will the Indian government do?

It is obvious from the above listed issues that there are serious gaps in the economic and social development of the country. All aspects of India's economy—including energy—are full of uncertainties and difficulties, some of which are growing in parallel with the growing population and its needs. There is no single answer to any of the issues at hand, since they are too complex and interrelated, so we can just wish that the politicians and industry leaders will find the best way to proceed.

Australia

Australia's new government made a pledge during 2013 national election to revoke the unpopular tax on mining companies' profits. The 30% tax on profits on coal and iron ore mining was seen as one of the primary reasons behind the downfall of the previous government. This shows the power of coal and its supporters.

This victory, however, was possible only after the leaders-to-be struck a deal with mining tycoon-turned-politician Clive Palmer's United Party, which holds the balance of power in the upper house of Parliament. This victory also comes at the expense of some Australians, since it includes freezing the government's contributions to state pensions, while benefitting other programs, like the bonus program for school children.

Australia's economy is heavily dependent on mining, which has contributed to it being one of the world's largest per capita greenhouse gas emitters. The laws had sought to put a price on greenhouse-gas emissions, but generated years of political debate, since the carbon tax has been blamed for higher energy bills and living cost.

The decade-long mining boom in Australia also underpinned a glut of spending on new cars and household durable goods, such as furniture, appliances, TVs,

cameras and computers. The estimates show that purchases in those categories were 30% and 20% respectively higher than what would have been without the so-called boom. Everything considered, the mining boom of late is believed to have substantially increased Australian living standards.

There were two main factors driving the spending trend. First, a surge in household disposable income per person, which was 13% higher than if there had been no mining boom. Second, the increased demand for Australian minerals pushed up the exchange rate, making imported goods such as cars, TVs and computers cheaper for local consumers.

But even more importantly, without the mining boom, Australia's jobless rate would be about 1.25% higher, according to the researchers. And the boon is just now starting. The $15 billion Carmichael coal mine, which is one of the world's largest, will hire over 5,000 people for the construction phase beginning in 2016. It will be then hiring many other workers for coal exploration, rail construction and operations, infrastructure construction, port expansion and operations, and coal mining.

The new mine is designed to produce over 60 million tons of thermal coal annually, and consists of six open-cut pits and five underground mines. Significant part of the coal produced here will be exported to India for power generation. It is estimated that the new Australian coal will generate enough electricity for over 100 million people.

There is a great controversy brewing over the expansion of port Adani on Abbot Point, since the project would dredge up over 3 million cubic meters of sand from the shallow waters and dump it near the Great Barrier Reef. This might place the Great Barrier Reef on UNESCO's endangered list, so the battle is just now starting.

Russia

Russia is another large country, but in contrast with India, it has been blessed with huge natural resources. Russia is a large exporter of all kinds of commodities, including energy products. As such, however, it has proven to be unreliable and even controversial player in the global energy markets.

Recently, Russia under Putin's misguided leadership is shaping as a wild card—the biggest security risk to the world energy markets, driven by nationalistic and self-serving interests of Putin's ruling class. Using old KGB tactics they don't hesitate to confront the entire world, while the Russian economy is in slump and the ruble dropped down 30% in the last 6 months of 2014.

Falling oil prices usually play a major role in shaping the geo-political landscape. Over 50% of the experts see Russia as the biggest loser from the lower price of oil, followed by Venezuela with over 20%, and Saudi Arabia and Iran with 12% and 6% respectively.

Russia and some smaller producers will be the biggest losers because they depend on a high oil price to finance the government budget and external trade, while the Middle Eastern oil exporters would like a higher price, but can afford, and live with, a lower price too. Russia cannot!

Russia is the biggest loser when crude oil prices drop, while. the U.S. is the most likely beneficiary from lower oil prices.

During 2014, the greatest risks to the global financial markets, with over 50% of the experts' votes, was the Russia-Ukraine conflict, followed by the Islamic State with 25%. Ebola was way down with 5% of a chance to affect the financial markets.

Russia is becoming a rogue state, which will be the biggest loser in the changing geo-politics of the energy markets—regardless of the global crude oil prices.

Russia is enduring a barrage of blows of sanctions from the Western world led by the U.S. This, and the oil-market volatility threaten to destroy its economy. And yet, the invasion of Ukraine was a priority for Putin and his lackeys.

The Russia-Ukraine situation is very dangerous for it is setting a precedent of a sovereign state, trying to increase its power by creating chaos by leading war activities against a neighboring country. This might impact the common thinking of how civilized and developed we are today, which might impact the financial markets and increase the risk premium of its products and services.

Putin global ambitions got out of hand when Russia moved warships toward Australia on the eve of a Group of 20 summit. And to top it off, at the same time Putin also announced plans to extend its long-range bomber patrols as far as the Gulf of Mexico.

According to the same experts, the world financial outlook is darkening, with about 90% see Europe's deflation risk increasing, and while Russia is bracing for catastrophic oil-price slump. This might lead to more complications in the global energy markets too, as Putin continues to cover its internal problems with blatant actions of global aggression.

All this said, we must not give up on Russia as yet.

Things might change rather quickly, if and when His Majesty Emperor Putin steps down from the throne. But all indications are that this won't happen soon, so we must brace for many more Kremlin surprises a la old Putin style.

Japan

Japan is a small country—island to be exact—but has the world's 5th largest economy. This is almost unbelievable from the energy markets perspective, since Japan has less natural resources than any of the developed, and many developing, countries.

How can such a small country with almost no natural resources, which was practically destroyed after WWII, progress so much in such a short time period. We can cite the Japanese discipline, Japanese ingenuity, and a number of other virtues of the Japanese culture, but it is most likely that it is some combination of all these.

Nevertheless, Japan is a brilliant example of a small nation rising wounded from the ashes of its ruined infrastructure in 1945 to become a world technological and economic leader only 30-40 years later. It still holds its place as such today, but things are slowly changing in Japan, where its future is uncertain.

It all started in 2011, when one of its largest nuclear power plants blew up and poisoned the air, soil, and water in the area. That event triggered immediate shut down of all nuclear power plants in the country, followed by similar actions in other countries around the world.

For a country with limited natural resources, Japan depends on nuclear power for 1/3rd of its electric power generation. Without it, Japan's energy sector is back where it started 40-50 years ago—burning huge amounts of coal and natural gas. How sustainable this is, from economic and environmental points of views, depends on whom you ask. Is Japan going to recover from this blow of bad luck? Probably, but not very easily; history repeats itself for Japan in this case.

Japan is the world's largest importer of liquefied natural gas.

Japan has undertaken an ambitious renewable energy sector expansion, expecting rapid built-up of renewable energy; solar energy, wind, geo-thermal, and other expected to follow. The government is also working on liberalization of Japan's energy markets, to where the Japan's energy sector is in the midst of rapid changes, which present large opportunities for innovation and unprecedented developments.

Table 2-24 shows that the pronounced trend of decreasing foreign imports from 2004 until 2010 was interrupted in 2011, after the Fukushima disaster, so now most (almost 4/5th) of the energy used in Japan is imported.

Table 2-24. Energy in Japan (in Mtoe)

	Natural resources TWh	Foreign Imports TWh	Electricity Generation TWh	CO_2 Emissions MT
2004	1,125	5,126	1,031	1,215
2007	1,052	5,055	1,083	1,236
2008	1,031	4,872	1,031	1,151
2009	1,091	4,471	997	1,093
2010	1,126	4,759	1,070	1,143
2012	1,013	5,532	851	1,207

Most developing countries have a similar ratio of domestic vs. imported energy sources. We also see that in 2012—a year after Fukushima—Japan increased the energy imports to where now full 4/5th are imported. This is also accompanied by a significant increase in CO_2 emissions in the home of the Kyoto protocol.

This is an exceptional case that deserves a closer look, which we will take later in this text. It suffices to say here that there are a number of inconsistencies in Japan's energy sector, hard to understand and justify, because of the extraordinary situation, which required an instantaneous response.

For example, in addition to being the largest natural gas importer in the world, Japan's crude oil use contributes to about 50% of the total energy consumption in the country. Natural gas is second with 24%, followed by coal at 23%. Over 90% of these fossils are imported. Nuclear and the renewables provide about 3% of the total energy mix.

Prior to 2011, declaring climate change and environment were top priority for the Kyoto protocol host country, Japan, which demonstrated its commitment to pressing ahead in these domains. For awhile there, Japan was a world leader in advancing energy technology transfer and environmental policies. It was determined to further improve its domestic policies, moving it towards to a more sustainable and secure energy pathway for the long term. Along with other priorities and accomplishments, government support for energy R&D were to be optimized and policies to enhance the efficiency of appliances–both for domestic consumption and export–were planned to be made into models for

other countries to follow.

Japan was speeding ahead on the global arena by introducing a greater reliance on market forces throughout the system leading customers to choices that enhance security, raise economic efficiency and promote environmental protection. In the climate change goals areas, Japan was the world's fifth-largest greenhouse gas emitter, so the goal was to strengthen the value of greenhouse gas emissions in order to give the appropriate signals that customers and all involved need to make the right choices.

Enhancing energy savings through efforts aimed at particular sectors were to be a major part of the overall policy mix, along with ongoing leadership in promoting energy efficiency and environmental responsibility. The government was committed to work to complement existing voluntary instruments with stronger ones, including such that rely more on market incentives, standards and requirements.

What a wonderful picture of global concern, dedication, and compliance a la Japanese style. What a wonderful world Japan was attempting to lead us into...But that was then...

All the plans and promises stopped in one single split second in March 2011, when Fukushima nuclear power plant went up with explosions, smoke and radiation. All plans were put on hold, and from that moment everything in Japan—at least everything in the energy sector—changed drastically.

Figure 2-25 shows the sharp drop of nuclear power generation down to zero in the beginning of 2011, accompanied by large increase of fossil energy sources imports.

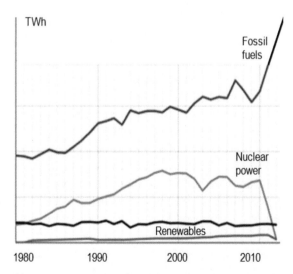

Figure 2-25. Japan's electricity generation 1980-2011

According to IEA, only 10% of the total primary energy used in Japan in 2012 was produced domestically.

All of a sudden—moments after the Fukushima disaster was confirmed—all the niceties about energy efficiency, environmental protection, basic R&D support etc. dropped in priorities and most dropped from the list all together. Japan went into survival mode. Keeping the energy sector going—including providing enough electric power and limiting the Fukushima damages—became the utmost, if not only, priority of the government and many companies and organizations.

An unprovoked, unforeseen, and unprecedented incident put one of the most advanced technologically, developed economically, richest, and most powerful nations in the world on its knees.

Figure 2-26. Fukushima and the environment

Not surprisingly, after the Fukushima disaster, Japan is headed to a near-term energy future powered mostly by coal. The Japanese government released a new energy plan in 2014, which intends to increase coal's use in the energy matrix at home. The new energy plan puts coal in the role of an "important long-term electricity source" while the costlier "green energy" is conspicuously absent from the new plan.

The plan also gives nuclear power the same prominence as coal, despite ongoing concerns regarding nuclear fall-out from the 2011 meltdown at the Fukushima plant. Japan's nuclear reactors have been idled since 2011 for safety checks and upgrades. As a result, 10 power companies consumed a record breaking 5.6 million metric tons of coal in January 2014. This is 12% more than the same time a year earlier.

Japan—and its energy future—are stuck between a hammer and a hard place now. Its huge nuclear power fleet, which was responsible for providing 30% of the nation's electricity (with plans for growth to over 40% by 2020), was completely shut down after the Fukushima accident in 2011.

Overnight, literally, Japan went from a leader in clean environment to a major polluter, along with China and the developing countries. After trying the renewables route for awhile, Japan realized that it will require a lot of time and money to achieve its energy goals with solar and wind, so they're back to using a lot of coal as a long-term (maybe permanent) solution to the lost 30% power generation capacity.

Yet, being the good steward of Mother Earth it is, Japan is now looking at the possibility of implementing "clean coal" technologies by improving on the latest

gasification and carbon storage technologies that make carbon emit less carbon into the atmosphere. One step forward, two steps back…
Solar and wind are still considered, but are far down on the list of priorities per the new 2014 energy plan. This new situation draws a totally new path for Japan, the Kyoto Protocol instigator and host.

After making real progress in GHG emissions reduction until 2011, Japan is again a major polluter for the foreseeable future. The Japanese government seems to have decided to put the Protocol in the closet, where it will stay wait for better days.

Until then, King Coal rules the energy sector in the land of the Rising Sun and the Kyoto Protocol will have to wait.

Japan will rise from the Fukushima ashes just like it rose from those of World War Two. No doubt! It will take a while and it will be a painful process, but by the end of this decade, Japan will be back where it stopped on that faithful day of March 2011.

What will remain forever, however, is the unsettled feeling, and a reminder, of just how temporary and fragile everything around us is. Not just nuclear power, but many other things—including some major forces—are uncontrollable and could wipe out a town, country, and even the entire world in one split second. Although it is not very likely, the possibilities of such event exist, and we must be fully aware of our vulnerabilities and be as prepared as we can be to face them.

Summary

Energy has been, and still is the force that created and moves the world. 100,000 years ago our predecessors started using energy for a first time by burning firewood. This was the first energy conversion process they developed and managed successfully. It provided warmth and made their food more digestible. It eventually led to their industrial progress.

Wood, being a major example of solar energy conversion, was used extensively by our primates. Wood burning was also the primary energy source until the mid 19th century. After that, coal became the fuel that provided most energy and contributed most to the development of the European and American civilizations. Coal was what people used to warm their homes with, and what turned the wheels of transportation and industry; replacing wood, which is inconvenient and low-energy fuel.

By the end of the nineteenth century the first of the great replacement cycles was under way, and coal was providing 90% of this country's primary energy. It was abundant in the industrialized East, and had more Btu's per pound than wood. Coal became the fuel for the fledgling steel industry and for the trains that ran on the steel rails. This went on until 1950, when oil pushed coal out of the picture.

Oil was even more convenient fuel, especially to transport and use in vehicles. It offered much more energy (Btu's) per pound. It replaced coal in the railroads and pushed the nation into the automobile age, which we are still in love with. By 1970 oil and oil products provided 75% of the nation's energy, with coal less than 20%. But coal found its niche with the electric utility industry and made an impressive comeback. It is now the fuel of choice in most electric power generating stations.

Natural gas was wasted early on by burning it at the oil wells, but became the fuel of choice for home heating and industrial furnaces when new gas pipelines were constructed to the Midwest and the East of the country.

US gas production peaked in 1970, and the use increased. Most of the oil was, and still is, imported. In 1977, imported crude oil was almost 50% of the total oil used, and almost 25% of our total primary energy consumption. Oil imports continued to increase through the years to over 60% total crude oil imports from 80 countries.

Slowly but surely, the United States and much of the world built a fossil fuels dependent industry and society. This period, however, is coming to an end, because the fossils have problems, as described above, and we are now looking for other energy production and use.

No one knows the future. but we know now that there is a problem that needs to be addressed and solved.

We are presently in the "addressing" stage. It is an important stage, because in order to solve a problem one must recognize that there is a problem.

We are just now agreeing that there is a problem, and that we need to address it fully—which is the goal of this book as well. Herein we intend to present a complete and unadulterated account of the different energy sources; their technical characteristics, as well as their effect on the socio-political and environmental aspects of life in the U.S. and the world.

The solutions will come only after we have reached an agreement, for which we need to get prepared by complete understanding the issues at hand—which is the other key goal of this book

The fossil fuel industry and utilities executives have their agenda, geared to sell their products. Period. Everything else is secondary. Any talk about environment, climate warming, environmental disasters etc., is secondary to their purposes—which is making a buck. Often this talk is in conflict with their interests, so we must view everything they say and do with a grain of salt. This is not to say that these are bad people, nope!, but in the capitalist society everyone must defending their own turf—especially if it is what they rely on making a living—and it is truly their turf we are invading, and their livelihood we are attacking with such talk, so we must be considered and proceed carefully.

Government officials, politicians, regulators and utilities officials have their own agendas too. Although the intentions are good, the results from the latest energy markets development show that there are complex interactions and numerous interwoven interests that are not easily understood, which makes things very complicated and the issues not easily solved.

The U.S. energy policy has been that of, "Drill baby, drill," and recently it was modified to that of, "Good-To-The-Last-Drop." This means a huge windfall for oil and gas producers, and to a certain extent coal producers. And of course, we hear about the benefits of these fuels and miraculous benefits to our society.

We must recognize that, no matter what we think about all this, we depend on coal, oil and gas in our daily lives. They are the evil that we cannot do without. And as such, we must treat it with respect and do our best to find and resolve the differences and work towards the best outcome possible—for the sake of our children and their children.

The immediate issue is that our industrial society—now and for the foreseeable future—depends on the fossil fuels for its proper functioning. The continuous input of high-grade energy resources is critical to our progress, and if they decline, the system will falter. As a matter of fact, we saw the effects of high oil prices (during 2008 price hikes) and constrained oil supplies, and the way they affected the markets and suppressed economic activity and pressured our already fragile financial system. High oil prices were the straw that broke the camel's back, and marked the start of the great 2008-2012 economic slow down in the U.S. and the world.

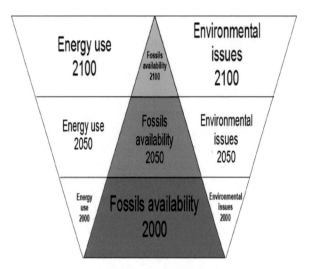

Figure 2-27. The global energy balance

What Figure 2-27 tells us that the inevitable progression, which is now largely ignored or acknowledged in passing, is quite simple and clear. It will become even more so by mid 21st century, when serious urgencies and emergencies will arise in the energy markets. The fossils are quickly depleted, the energy use is increasing rapidly, and the environment is polluted beyond measure. How can things not go wrong if we continue unhindered in our way?

The global energy markets are affected, and driven, by increased energy use vs. fossils availability and environmental degradation.

In conclusion: We are at the peak of our technological development, and so any future miracles of more efficient, cheaper, and cleaner, fossil fuel technologies

although feasible are not very likely. Because of that, we should not rely on another technological breakthrough coming our way any time soon. The known trends and data tell us clearly that we should not wait any longer, but instead start looking into the cleaner, flexible, more reliable and long lasting (renewable) energy technologies, and into new, more efficient, ways to consume energy.

The transition to renewables is inescapable. It will have to happen, since there is no other way on the long run. The only question is whether we, as a responsible society, will do it now by doing some changes and sacrifices, or let somebody in the distant future take care of it at a great price and pain.

Do we have a feasible and focused plan for a rapid transition to renewable energy technologies, and new ways to use energy? Yes, there are plans, but their focus is on using a lot of energy today with little worry about tomorrow.

If we don't plan and act properly now, however, the global energy balance would be disturbed and chaos will prevail on the energy markets eventually. We will then have to endure the economic up and downs, as we go from one energy-, or environment-induced, crisis to another. Not a good or smart way for responsible people to behave in the face of the inevitable. Not a good way to account for our actions, or lack thereof, to the upcoming generations.

Notes and References

1. U.S. Energy Information Administration. http://www.eia.gov
2. World Coal Association http://www.worldcoal.org/coal/use-sof-coal/coal-electricity
3. http://toxtown.nlm.nih.gov/text_version/chemicals.php?id=73
4. USGS. http://woodshole.er.usgs.gov/project-pages
5. http://www.globalgreenhousewarming.com/solar-parabolic-trough.html
6. *Photovoltaics for Commercial and Utilities Power Generation,* Anco S, Blazev. The Fairmont Press, 2011
7. *Solar Technologies for the 21st Century,* Anco S Blazev. The Fairmont Press, 2013
8. *Power Generation and the Environment,* Anco S. Blazev. The Fairmont Press, 2014
9. *Energy Security for the 21st Century,* Anco S. Blazev. The Fairmont Press, 2015

Chapter 3

Energy Marketing

The different energy sources and fuels have advantages and disadvantages. Presenting these in a way that appeals to the public, thus ensuring broad customer support, is what energy marketing is all about today.

—Anco Blazev

BACKGROUND

The times are changing fast, and with that the emphasis of the marketing efforts is shifting from selling a product by primarily pointing to its usefulness and qualities, to keeping the customers interested, involved, and happy. And this has to be done seamlessly; regardless of who provides the services and/or where the products were made, or come from.

Perfect customer service is the new flavor of the 21st century energy marketing.

Among many modifications and improvements, 21st century style energy marketing also includes a brand new concept; that of ensuring *positive* and satisfying *customer experience* as an important ingredient in the different marketing strategies. It is an expansion of the old *customer satisfaction* approach, which has worked for a long time, but is no longer good enough. The difference is subtle, but with a great effect. This is now the new game in town, so those who are good at it benefit tremendously, while the rest are trying to find a way to catch up.

Energy marketing is a huge and very important subject, so where shall we start? It might be best to start by a brief look at the subject of *marketing* in general, since it is in play at, and often dictates, the energy markets function.

MARKETING

Marketing involves an incredibly broad scope of ideas and ever increasing in complexity activities. So much so that asking several people what marketing is will result in several totally different answers. And

of course, asking any of the *markets experts,* we will be promptly informed that the rest are simply wrong.

The simplified classic definition of traditional marketing is quite simple,

Marketing is the art of finding buyers for products and services at a profit.

This is still the core of the modern marketing process, but the simplicity and the agreements on the substance end with the last word of the above definition. Many different combinations and permutations of sophisticated marketing approaches have been added through the years, so the marketing "art" is now developing into a complex science. The definition evolves with time and morphs into different forms and shapes for the different sectors, industries, and the companies within these.

Something we all must agree on, is that marketing in general is one of the major driver of economic progress in the 21st century.

Since energy drives the economies of the 21st century, it follows that the energy markets are some of the most important factors in the progress of today's society too.

We also agree that energy is NOT the one and only factor, but that it is surely one of the major ones in the chain of events and priorities that determine the speed and direction of economic development.

The Evolution

The marketing process usually starts a long time before any products or services hit the actual marketplace. It involves intensive analysis of the present customers needs, superimposing these into the short- and long-term future. Then, equally intensive research and information gathering follows, as needed to design and produce successful products and services—all geared towards satisfying the customers' preferences. This then is coupled with creating and maintaining a solid relationship with the supply chain and the customers.

Complex endeavor, no doubt, that can be as brilliant and well received as Steve Jobs' presentations of new Apple gadget, or as infamous and full of customers anger as defending the actions of BP and Exxon during the respective oil spills and the resulting environmental damages.

In the energy sector we seldom, if ever, see the glamour of Steve Job's marketing brilliance in pushing innovative technologies and shiny gadgets. This is because most of the energy products and services are old, less than brilliant, and most are even stinky, dirty, and even dangerous.

Nevertheless, equally appropriate, innovative, and successful marketing approaches are applied here too. Some think that marketing in the energy sector is not needed, and sure enough; when did you see a TV commercial advertising Saudi Arabian crude oil, or Pennsylvania coal? OK, so maybe the Saudis don't need open public marketing and advertisement, and yet they use (and very successfully so) marketing methods that are invisible to the untrained eye. The same is true for the coal mining operations. Their products too have to be sold to somebody, which is usually done by well trained marketing people or organizations. Their activities are not as obvious as Apple's, and yet they are as sophisticated and successful.

Marketing professionals (of different types and shapes) coordinate and direct the global energy markets.

Governments buy and sell energy and energy products just like any salesman would do. The only difference is that the transactions are, a.) done in secret, or at least are not widely publicized, and, b.) the transaction amounts are usually huge and of long term duration.

Energy companies and utilities are much more conspicuous about marketing their products and services, and very good in covering their mistakes and damages. BP is an example of an intensive and quite successful *after-the-fact* marketing approach. Only days after the Deep Horizon oil spill that flooded the Gulf with millions of gallons of crude oil and other toxic chemicals, BP started flooding the North American TV channels and the virtual media with commercials about their swift response to the crisis, and the benefits of their

products and services to the U.S. economy.

A year or two after the disaster, BP started showing the beautiful Gulf coast beaches, making it sound as if the unprecedented large scale disaster has contributed to making the Gulf even better, more beautiful, and surely much safer than ever before. BP's ads campaign, of course, does not show the long lasting and permanent damage to parts of the Gulf coast and its environment, wild life, and people. And the damage is significant…

Trying to give us a warm and fuzzy feeling about the events and the company itself, it uses a set of misleading maneuvers and cover-ups. These are wrapped up and presented in a relentless public brain-washing program via TV and radio. This is an example of a most aggressive, if not blatant, *after-the-fact* marketing of products and services at its best.

On the other hand, the now bankrupt Solyndra used open and very aggressive *before-the-fact* marketing methods, when approaching private investors and the U.S. government. The marketing techniques used by Solyndra were equally aggressive and at the end proved to be exceptionally successful; raising nearly $750 million from private parties and the government.

Solyndra's marketing genius raised millions for a miracle product that existed on paper only; it was fatally flawed from a technical point of view and had no chance for success on the energy market.

Solyndra's tube-shaped solar receivers look good and would make a good art exhibit, but this is where their usefulness ends. Solyndra's managers knew very well that their product had no chance on the global marketplace, so how did they succeed in raising almost a billion dollars for useless technology? Was this marketing excellence, blatantly criminal behavior, or combination of these? The jury is still out on this, but right or wrong, it shows that marketing in its different forms works.

The marketing campaign reached the highest U.S. government levels too. Solyndra's lobbyists had full and unobstructed access to the U.S. Congress, DOE management, and even the White House. Why and how is not that important, as are the results of these blatant actions. Because of the lobbyists' marketing success, Solyndra's management was able to sell their useless technology and raise 3/4 billion dollars, which was needed to survive (lavishly, we must add) at least for a short while.

Another example of energy marketing, this one of the *during-the-fact* varieties, and on a much larger scale, is the natural gas commercials from different companies in the U.S., marketing natural gas to the public through a flood of media ads. We see the warmth and comfort, natural gas provides, thus making it indispensible for getting all the benefits for our daily lives. Some of the commercials are selling the public on the good sides of the hydrofracking process as the next miracle that will save us from energy annihilation. Hydrofracking, according to the oil companies marketing gurus, is the safest, cleanest, and cheapest way to get energy in form of abundant natural gas. It, by the way, also creates a lot of jobs.

They, of course, do not mention the bad sides of the process, which is increasing in size and severity. And, of course, most of the people—those who are not personally affected—are unaware of the truth and believe the commercials. But there are presently thousands of Americans who are suffering the consequences of hydrofracking. They are the minority, so their voice is not heard.

On the more mundane side of things, most energy companies use different marketing techniques to promote their usual commercial products and services. Petrochemical refineries, for example, market their products to the private and commercial customers that could use them, using conventional marketing approaches. Wind and solar companies, on the other hand, use contemporary marketing tools to sell their products and services to different segments of the population and businesses. Both of these entities need their customers' approval to sell their products and services.

On the other extreme of things, electric power companies and utilities also advertise their products and services via new multimedia channels, even though they don't need to, since they have a captive market. They really do not need to move a finger to retain their customers, because the customers are stuck in their local grid like flies in a spider net. There is nothing one can do if s/he does not like the local utility, short of selling the house and moving to another location, serviced by another utility. And then, what guarantee is there that the new utility won't have the same problems as the old one?

Yet, the utilities put a significant effort in marketing their services in order to keep the customers happy. New services are offered periodically via TV, radio, emails, and other media channels. These provide the customers with all kinds of information and energy buying options, including equipment preventive maintenance and efficiency enhancement types of marketing approaches.

But back to marketing as a science.

The Marketing Process

The conventional marketing, as defined by the American Marketing Association, is '…the process of planning and executing the conception, pricing, promotion, and distribution of ideas, goods, and services to create exchanges that satisfy individual and organizational objectives…'

This is a broad definition of the way we sell and buy goods and services, where marketing in general aims to:

a) Assess the needs of prospective buyers, and
b) Satisfy these needs, whatever they might be and whatever it takes.

The spectrum of responsibilities, and the focus, of the marketing experts is on getting a solution and response on the following:

- Goods (come in a package)
- Services (intangible, a process)
- Experiences (feel good factor; wow factor)
- Events (exhibitions; tradeshows; road shows)
- Places (locations, channels, market space)
- Persons (staff, stakeholders, suppliers, media, government)
- Properties (real estate, land, etc.)
- Organizations (groups of people e.g. 4-2-1 member families)
- New ideas/information on future products and services development

While the modern marketing enterprise uses some (or most) of these approaches, its priorities have shifted somewhat. This is because today's enterprise has changed significantly, which requires a different marketing approach.

Today's industrial enterprise consists of several systems (departments). These usually are as follow:
- Strategic system
- HR system
- Finance and Accounting
- Production/Operations
- Marketing system
- R&D system

The structure and function of these, in addition to other departments such as Legal, Maintenance, Environmental, Safety, Training, Contracting and Consulting services, etc. determine the overall structure and functionality of the enterprise. Marketing is just one of the management systems, but it usually determines to a great extend the success or failure of a product/service line. That, in most cases determines the success of the entire enterprise.

Marketing, in short, drives the overall development, and determines the success of the enterprise on the long run.

Companies with capable and flexible marketing department are usually leaders in their fields.

Figure 3-1. The marketing details in brief...or not.

There were several basic marketing philosophies in the past:

- *Production oriented.* Mass produce and distribute, easily available and cheap, expand markets e.g. fast moving consumer goods (FMCG), such as toothpastes, toilet paper, baby food, etc.
- *Product oriented.* Design and improve, not always acceptable by customers, e.g. Eight Track players, which had a very short life s[pan.
- *Selling oriented.* Push sales; no customer care; little input from customers; inside-out thinking; products not in demand by most customers e.g. fancy cars, very expensive restaurants
- *Marketing oriented.* Cares and listens to customers, delivers promises, outside-in approach, customer-centered, produce superior performance and value. Apple is a good example of this approach.
- *Holistic oriented.* This is an individual marketing concept, comprised of the following key segments
  - *Relationship marketing* seeks long-term, "win-win" transactions between marketers and the stakeholders (suppliers, customers, distributors, government etc...)
  - *Integrated marketing* is a marketing mix that

involves recognition and use of the four P's (product, price, place, and promotion) and the four C's (customer solution, customer cost, convenience, and communication) in all stages of the product and service development.

- *Internal marketing* is where each person within the organization knows how their activity relates to the customer and contributes to the effort.
- *Performance marketing* is based on financial accountability, where marketing personnel is responsible for cheaper investments, dealing with smaller budgets, and greater requirements for returns from investments.
- *Social Responsibility* is a new trend—especially in the energy sector—where the marketing personnel is required to be fully aware of the role of the company's products in ensuring the overall social welfare.
- *Green marketing* is a variation of the latter type, where safety and non-polluting products and programs are used to help people. This is a growing trend in the energy sector, which every company is expected to respond with new, green, products in services.

The marketing professionals today employ all the above strategies, but the most successful companies focus on the latter two major approaches; the marketing and holistic oriented methods. There are many reasons for this, but the major one is the fact that the customers today are part of and (directly or indirectly) participate in the marketing process.

Marketing Today

Marketing today starts with choosing the level of involvement and the overall direction the company is to take with a particular product, or service. Based on that, and projected in time, the marketing effort can be divided into several strategies:

- *Market introduction strategies*, which are used during the initial design and introduction of new products. Here the marketing gurus have only two principle strategies to choose from:
  - Penetration of, and fighting on, *existing* market territory, which involves a great risk from encountering unproportional resistance from the competition, or
  - Penetration of, or creating a new, niche market, which is much easier to do, but offers narrower customer base and limited returns—especially in the initial stages of product exposure.

- *Market growth strategies are used in* the early growth stages, when the marketing effort needs a push. Here the choice is also between two strategic alternatives:
  — Segment expansion, where one part of the product or service is considered most likely to succeed in reviving the market activities and is given an aggressive market push.
  — Brand expansion is chosen when the entire brand with all its components, advantages and disadvantages is considered worthy of a market push.

- *Market maturity strategies* are used later on, when sales growth slow, and/or stagnate. In this case, a market maintenance strategies are deployed, where the firm has a choice to maintain or hold a stable marketing mix. This decision depends on the maturity of the product and the changes in the marketplace.

- *Market decline* strategies are used much later in the life of a product or service, when sales decline to the point of equaling or exceeding the product costs—including all hidden costs. At this point, companies try a harvesting strategy, designed to get the maximum possible from the product before it becomes totally unprofitable. This effort is usually followed by a divesting move, where the company tries to get rid of the product or service by selling, or shutting down, a brand.

Once a path is chosen, and the product shows promise in time, the company have to chose a growth strategy.

The growth strategy choices are:
- *No changes*, where the company prefers to keep the present (hopefully most successful) level of operating. No growth is projected, nor expected. This model is unusual for American enterprises larger than hot dog vendors' carts, but is common in many other, mostly developing, countries.

- *Horizontal integration*, or horizontal growth, is where a firm grows by acquiring other businesses in the same line of work. This is the most common business model in developed countries as well.

- *Vertical integration* can be forward or backward.
  — *Forward integration* is where the company grows towards its customers.
  — *Backward integration* is where the company grows towards its source of supply.

Vertical integration in either direction is more expensive and harder to achieve, but is usually more successful too. It is usually undertaken by large companies, or leaders in the field.

- *Diversification* is where a company enters into other lines of products or services. Although most difficult, this approach has been the most successful by far, and is preferred by most large U.S. enterprises. It pays on the long run, where companies can switch products or services according to the market changes and demands.

- *Intensification* is where a company intensifies its preference in a product or service and pushes them aggressively on the marketplace. This happens in large operations, where seasonal or emergency products find wider applications on the market at one time or another. The intensification effort could be temporary or permanent, depending on the market conditions. Here is where diversified companies cash in by ramping up product lines or services as and when needed.

- *Exit strategy* is an important strategy that could determine the final destiny—hence the final success or failure—of the enterprise. *Exit strategy* is simply a future plan of what to do, if and when things change, or after the predetermined objective for the product or service on hand has been achieved. This is usually a strategy to transfer ownership to another party, diversify, and/or mitigate the failure. Very often lack of exit strategy means lack of forward thinking and a possibility of a pending quagmire at a most inconvenient time.

In its very foundation the marketing process is split into two major segments (phases): the customer base and the market environment

The Customer Base

This is the most important part of the marketing effort, which often determines the difference between success and failure. Here, the focus is on figuring out the customers' needs and wants, and developing strategies to market the products and services to the broadest section of the customer base as possible.

The work at the beginning stages of a product/service marketing usually proceeds as follow:

- Collecting data about the customers and their particular needs and wishes
- Processing the data and outlining marketing strategies,
- Evaluating the responses to each of these strategies, and
- Designing a product, or service that brings profits and customer satisfaction

In parallel with that effort, the marketing departments take a close look at the market space itself, and the competition in particular.

The Market Environment

This is the other side of the marketing equation, where the majority of the work is dedicated to collecting data on the market place—the particular product or service as related to the competitions and other market peculiarities.

The work in this sector, or during the effort in this phase, usually proceeds as follow:

- Collect data on the specific marketplace and the competition,
- Design competitive marketing strategies,
- Evaluate the marketplace responses, and
- Design exit strategies

The key here is collecting enough and accurate data on the specific marketplace and the competition working in it. More accurate the data, more efficient the marketing effort would be. The key here is to develop a marketing strategy that is capable of generating and increasing sales (and profit) while achieving a competitive advantage.

The rule of thumb here (especially for small and new entrants) is to avoid entering into a field that is dominated by a major and capable competitor.

Such a move might prove fatal, although there are many examples of companies successfully breaking that rule. In all cases, entry into such area, however successful, is difficult, dangerous and expensive.

Strategic Marketing

In the recent past, companies used the conventional marketing approach as depicted in the left side column of Figure 3-2. It was perfectly plausible for developing marketing strategy under the existing market

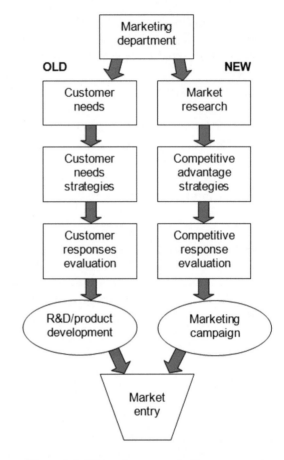

Figure 3-2. Old and new marketing processes

conditions, but it was lacking in dimension. As things kept changing with time, that type of marketing caused heavy losses to some companies, who were late in adapting to the new marketing methods.

Marketing, in the corporate context, was traditionally *tactical*. Under this method, the different resources were rallied in response to changes in the sales landscape on current (even day-to-day) basis. Tactical marketing was competitor-driven and market-reactive, where promotional strategies and advertising campaigns were designed to capitalize on perceived weaknesses of the competition in a certain market, at a certain time, and/or to emphasize a product's comparative strengths at the time.

Today there are other, new and more sophisticated, marketing methods and approaches, but the strategic marketing concept remains in the very foundation of the marketing science, including that of the energy sector.

Strategic marketing is a consumer-based approach to product promotion that a) identifies the different market segments, and b) attempts to make them profitable by providing superior value.

Figure 3-3. The strategic marketing scheme

The concept of strategic marketing is actually based on the same theories and approaches as *strategic management*. *Strategic marketing* takes a careful look at the best (cheapest, most effective and profitable) ways to get the products and services to market and put them in the hands of the largest demographic group possible. And then, the 21st century addition is the necessity to keep the group involved and satisfied. This is the mark of the new marketing strategies.

The concept of strategic marketing is a direct time value approach to profitable product and services planning.

A *strategic marketing* plan differs from the tactical, or any other, marketing methods by its focus on precisely and successfully *combining the overall market (customer) situation with the company's direction*. The company's capabilities and strengths are thus aligned for successful campaign, bringing satisfied customers and lots of profits.

For *consumer marketing*, this means using geographic and demographic segmentation, as well as psychographic segmentation (i.e., values, attitudes, lifestyles), and product usage motivations and particulars. For example, the aging population bubble creates a general increase in demand for a wide range of products (not only medications). It also creates market niches that are large enough to make product development and marketing worthwhile. The same approach can be used for marketing products and services to the young generations, according to the swings of the trends.

For *business-to-business* marketing, this means com-bining industry sector segmentation and product use with other factors related to purchasing decisions. These include the purchase criteria and decision motivations that affect large (enterprise size) purchases. The trend toward increased use of outsourcing to both domestic and global vendors, for example, creates new markets for some suppliers who do not have in-house manufacturing capabilities. They, however, still need to have a *strategic marketing vision* in order to see and penetrate the new markets early enough and grow fast, in order to take advantage of each opportunity.

The strategic marketing process typically consists of three stages:
- Segmentation of the target market
  — Geographic
  — Demographic
  — Psychographic
  — Behavioral

- Profiling the target market's segments
  — Revenue potential
  — Market share potential
  — Profitability potential

- Developing a marketing strategy for each segment
  — Market leader or product line extension
  — Mass marketing or targeted marketing
  — Direct or indirect sales

The process starts with a team of marketing specialists coming from various professional backgrounds, who analyze the different market segments, research the targeted customers interests and the related buy/sell processes during the product and services life cycle. They then go through the rest of the steps of the initial marketing research and planning process. The general object is to take a wide and long view of the market place, before finalizing the plan. When all this work is done, then a *strategic marketing plan* is finalized and issued, which usually includes:

- Situational Analysis. Present status of the company
  — Market Characteristics
  — Key Success Factors
  — Competition and Product Comparisons
  — Technology Considerations
  — Legal Environment
  — Political environment
  — Social Environment
  — Problems and Opportunities

- Marketing Objectives. The direction of the company
  — Product Profile
  — Target Market(s)
  — Target Volume (estimate in $)

- Marketing Strategies. Path to success.
  — Product Strategy
  — Pricing Strategy
  — Promotion Strategy
  — Distribution Strategy
  — Marketing Strategy Projection
  — Exit plan(s)

This process of developing a strategic marketing plan is one way to insure that the marketing programs take into account, are compatible with, and fully support, the company's capabilities, goals and objectives. And equally importantly; the company's strengths and product qualities are thoroughly and consistently conveyed to the existing and potential customers. This way, the company efficiency in the critical areas are optimized, in order to maximize the revenue by ensuring successful market penetration, market share growth, and multitude of satisfied customers

There are also many new and "innovative" marketing strategies today, which are applied and tried in some cases. Examples of these are: *Experimental marketing, Undercover marketing, Alternative marketing, Ambush marketing, Presence marketing, Presume marketing, Viral marketing, Drip marketing, Ambient marketing, and many more.* The names hint to what these methods refer to and use, but most exist on paper only and have been tried on limited basis.

Companies who can successfully and efficiently incorporate the most appropriate marketing strategy of the time would have significant advantage. This, however, is complex and expensive undertaking, so most companies compromise by adapting a modified marketing plan—either taking a short cut or modifying it to an extent that would render it useless.

Successful companies focus on satisfying the customers' needs, while sustaining a competitive edge.

The combination of these two factors—customer satisfaction and competitive advantage—is the only direct and surest path to long term profitability. It is also the most difficult, and expensive to achieve in providing successful products or services.

And speaking of services…

Service Economy

The U.S. economy is increasingly service oriented. There are many reasons for this, but the major one is that—for a number of reasons—we prefer not to manufacture many things and have consequently exported a large part of U.S. manufacturing overseas. Another major effect is the amazing growth of Chinese manufacturing, where today most items we purchase are made in China, simply because it is much cheaper and most convenient that way.

As a consequence of the developments of late, the U.S. is now predominantly a service economy.

This new MO is bringing increased importance of the service sector in our, and most other industrialized economies. As a matter of fact, most Fortune 500 companies today are service oriented companies, with fewer manufacturers than in the previous several decades. The situation is similar in most developed countries

Globally, the service economies in developing countries are mostly in the area of financial services, hospitality, retail, health, human services, information technology and education. The manufacturing sector in these countries, on the other hand, is flourishing.

The overall effect of this is that today most products sold in the developed countries—domestic or imported—have a much higher service component than in previous decades. We have in fact servitized (increased service content of) the products we use, where every product today has a service component attached to it. Some have been less, some more, servitized. Many "products" have been being transformed 100% into services, while very few have little or no service content.

Servitization usually adds value to the products, but in many cases it is just a marketing gimmick.

In all cases, 'service first, product later' marketing is now accepted and rewarded in the U.S. It is part of life; an illusion of simplified, uniform, modern, service-product continuum that we have embraced as the new 21st century standard. For example, many electronic products manufacturers today refer to their business as a "service" business, where the one and only goal is to serve the customer and make/keep them happy. Many companies do not even mention their manufacturing operations. And some have got rid of the manufacturing all together, in order to focus on the service side. The end goal is providing something that the customer likes—and it is all that matters.

*Many Western companies do not focus on, nor do they showcase, their manufacturing operations because they simply **have none**.*

This is important, so don't miss the irony here.

From a manufacturing powerhouse, the U.S. has become a country of service oriented businesses. Today most companies are acting as "distributors," selling products made by someone else. Apple and most other small and large U.S. companies, for example, do not manufacture their products directly. Instead, they use cheap labor provided by companies operating in China, India, Malaysia, and other such dark corners of the world. That way they cannot even show, let alone be proud of, their involvement in the manufacturing operations. And yet, they are actively, albeit indirectly, involved in, and responsible for, every step of the process.

The problem is that often the managers are closing their eyes at some short cuts and injustices. As a matter of fact, many companies—even the major they are often embarrassed by abnormalities, such as child labor, workers abuse, etc. These abnormalities are actually the MO in the manufacturing countries, but we hear about the problems only if and when a major disaster occurs, or some journalist stumbles accidentally upon them.

So the service component is the main focus for most companies. It is used, abused, polished, and pushed hard to cover up, and in some cases totally hide, the presence of physical products and their origins. It is, after all, only the final results that matter.

Apple, and most other U.S. companies using foreign labor, stand behind their products 100%, and this is all that matters to them and should matter to their customers. Who cares where all this stuff was made, and how many people out there (including children) were involved, exploited, mistreated, and/or hurt. Yes, there are numerous examples of life-threatening accidents in Asia and the surrounding areas, where entire substandard buildings collapsed, or caught on fire, and dozens of under aged employees were hurt and killed.

And so, these manufacturing-less companies do have full lines of products, with the physical goods being projected by the marketing schemes as a fairly small part of the "business solutions" offered by them. Excellent service and customer satisfaction is what they sell, and what is most visible on the surface.

As a matter of fact, there is an entire 'business solutions' industry, which is based on the fact that price elasticity of the demand for 'successful business solutions' is much less than that for hardware. Because of that, the entire process is easier to maintain on the short term and

more profitable on the long run.

For examples, buying a TV, or a PC, usually comes with a warranty period, during which you don't care what is inside the box. And even less you care who made it—after all, a reputable U.S. company is standing behind every bolt and nut in this product. You know for sure that no matter what happens somebody will somehow take care of the problem—any problem—so the rest is irrelevant and even non-existent for all practical purposes. Hardware in this case is irrelevant; it is the final result of a satisfied customer that is important and pushed forward by today's marketing gurus.

Very few people take a close look at Apple's operations, for example. All they see is shiny, brightly lit store fronts, usually in crowded malls or bustling business centers, where Apple products are showcased and serviced. Serviced, yes, not manufactured products. 'Apple provides excellent service' is what the marketing people want the customer to remember. The rest—thousands of low paid workers in China and other countries, the injustice and pollution the manufacturing process create—do not matter much.

And this brings us to the broader aspect of *product stewardship* in energy marketing:

Product Stewardship

'Product stewardship' is a fairly new concept; a specific requirement or measure in which a special service, for example product waste disposal, or some other special treatment of, is included in the life cycle and distribution of a product.

Product stewardship is built in the marketing of today's successful business operations.

It is an additional service that is marketed as a good thing for the customers, or the environment. It is, in some cases, also one thing that separates the company from the rest. This service is usually charged to the customer, who has been convinced by the marketing campaign that this is the right thing to do. And feeling happy to be included and part of the solution, the satisfied customer doesn't mind paying for it. S/he then can proudly talk about his or her participation in the process and contribution to the wellbeing of the masses.

Buying tires, batteries, solar panels etc. items today, for example, is accompanied by a bill for their disposal at end of life (EOL). The waste disposal charges are usually paid in advance; often at time of purchase. And so, we are paying for the disposal service in advance, yet relying on the seller for the safe and proper disposal days, weeks, months, and even years after the fact. Many products are sold that way today, although this type of "service" component, and the results thereof, are not heavily advertised.

This type of service is becoming a pre-requisite to a strict service economy, and the proper interpretation of the product/service continuum. It applies to most chemicals, paints, tires, car batteries, plastic bottles, solar panels, and many other goods that become toxic waste, and are dangerous if not disposed of properly.

In some countries, the law requires full and uninterrupted attention to development and outcome of the entire chain of events (cradle to grave) during extraction, production, distribution, use, and disposal of the products. This includes most of the energy products and services, but we will get deeper into this later on in this text.

Disposal of solar panels at the end of their useful life is (or should be) part of this service. It is approached differently by different companies and promises will become a major factor in the near future. As the number of solar panels installed on house roofs and in the desert increases, their disposal is becoming a key issue and a major plan of the exit plans of solar projects.

Note: Some solar panels contain toxic substances, lead, tin, cadmium, arsenic, etc., and must be disposed of safely and properly at the end of their useful life. This includes hazardous waste disposal services, which could be very, very expensive. If the exit plan of the solar project in question does not include funds and procedures for proper waste disposal, the panels will end up littering the fields for a long time and the U.S. taxpayers will pay the cleanup bills.

All parties who profit from the products and services during their lifetime are held legally responsible for any negative outcome along the way. This includes EOL handling and disposal. There is a trend in the UK and the U.S., where we have witnessed many class action suits related to proper and effective product stewardship liability. In many cases, the lawsuits hold companies responsible for damages, and/or for product performance that is different from the advertised.

This is the foundation of many accounting reform efforts, which are focused on achieving full cost accounting (including the entire cradle-to-grave) life time of the product or service. This is thought to be much better and fairer approach than letting the liability for these problems be assigned to the public sector, or to be handled as an afterthought by expensive lawsuits,

This is basically the financial side of the comprehensive outcome, which considers and assigns the gains and losses to all parties involved; not just those investing or purchasing. This is also the foundation of "moral purchasing," making it more practical and attractive, since it removes the liability and potential for future lawsuits.

This is a good step ahead towards a more fair markets, but we are just in the beginning stages of this development. Because of that, we now must evaluate some products and services of the past that have deviated from the "moral purchasing" path and the 'right the wrongs' of yesterday. This will provide a baseline of developments, responsibilities, and related appropriate actions.

Then we can proceed with taking a close look at the present situation, comparing the results, and mitigating today's problems as much as possible. We must learn from the past in order to design fail-proof methods of operation in the future.

For the U.S. energy markets, the U.S. Environmental Protection Agency (EPA) advocates product stewardship in order to "reduce the life-cycle environmental effects of products." The ideal of product stewardship, according to the EPA, "taps the shared ingenuity and responsibility of businesses, consumers, governments, and others." And so, this is the direction the U.S. and most developed countries are headed into during the 21st century.

Although, this is not a well travelled road yet, it makes a big difference when every individual and company are held responsible (and usually accept the responsibility) for their actions and so is judged according to the level of damage and degree of intent. This effort is still work in progress, but at least it is becoming accepted as a standard for future behavior and actions.

This, however, is not the case in most developing countries, where the socio-political infrastructure is much less established and in many cases is dysfunctional, or non-existent. And because of that, we see lakes of crude oil covering large portion of Nigeria, which contaminated ground and water there and in many other countries.

The *product stewardship* and *corporate responsibility* concepts have not arrived there yet, and it will take many years and decades to be understood and accepted. We just have to do our best to help these countries in their struggles, because we are indirectly paying for, and suffering the consequences, of their ignorance and negligence.

Marketing Ethics

Energy marketing, and marketing in general, changed significantly in the 1980-1990s, due to a realization on part of the corporate world that marketing their products and services include elements of *ethical responsibility* to the customers. While this was a known fact since day one, a number of developments brought a deeper meaning of the concept of *marketing ethics*.

Marketing ethics is based on the full responsibility of the market place to the society as a whole.

This concept—intentionally or unintentionally—was ignored for a long time in the past, in order to maximize profits. At the very least, it was not a priority and was not a subject of discussion. The ethical and societal responsibility of marketing was not reflected in the marketing strategies themselves and was not a visible part of the marketing concept per se. At least not as a visible (and official) expression of adherence to the normal rules of ethics goes.

For over a century, marketing was geared to sell products and services at a profit. That was the bottom line. Period! Today, due mostly to global socio-economic and political changes, as well as media globalization, a new element is introduced; that of satisfying the customer needs through sound (and ethical) marketing strategies.

Ethical treatment of products, services, and customers is all of a sudden of utmost importance, hence it is emphasized in business and marketing plans, and all through the companies' activities. It is intended to address the need of the customer to trust the product and/or service in a most sincere and personal way. This set of issues concern the basic ethics of corporate marketing gurus and is essential in meeting their responsibilities to the customers.

This is actually a priority emphasized in most companies' Mission Statements. Actually, *Mission Statements* were the big thing of the 20th century that has grown in status to be now viewed as a mirror into the quality of the products and services of the company in question.

As if a paragraph or two of words, usually copied from some marketing wizard's book, can determine the way a company operates, and the way its people work and behave. Yet, we (customers and investors alike) have bought into it, and expect to see on colorful, heart-breaking statements dripping of ethics and honesty. These words make us feel good, so if a company dares not to include them in the Missions Statement—or God forbid does not have a clearly visible such—then we would not thrust its products.

The problem is that this all the Missions Statements hoopla is nothing but loud words on paper only. Clichés that mean little and seldom reflect the true nature of the

company. These words are even more seldom fully implemented in practice, since doing is much harder than saying.

Enron's motto, for example, was "Respect, Integrity, Communication and Excellence." Their "Vision and Values" Mission Statement declared, "We treat others as we would like to be treated ourselves. We do not tolerate abusive or disrespectful treatment. Ruthlessness, callousness and arrogance don't belong here."

Really?! Stop and think for a second what the same mission statement would've sounded like in the ears of the judges and jury during the Enron court proceedings, "We are the rulers of the world. We treat others as if they were stupid idiots. We are the bastion of abusive or disrespectful treatment. Ruthlessness, callousness and arrogance is what we stand for here. And all of this is because we feel that we are mighty and untouchable."

Well, Enron, its Mission Statement, and misdeeds are history now. It was a lesson, which some learned not to repeat, but the bad taste in the mouths of millions of customers and employees will last forever. There are, however, many other company executives out there who are hiding behind the corporate mission statements and exercising Enron-like corporate behavior. We don't need to look very far; just several years back during the 2008 financial crisis, when we saw an entire barrage of Enron-like disasters. The Enron implosion, however, was more violent, expensive, and memorable. It will not prevent, but at the very least will serve as a measuring stick for, future debacles.

But how can you identify the unethical actions behind a given company's products and services?

Somebody said, "Unethical marketing uses false advertising to deceive the public, while ethical advertising uses the truth to deceive the public."

How true, and how wide and deep this goes in our corporate world is anyone's guess, but there is something that we must be aware of and watch for in all sectors of our economy, including energy. We saw in 2008 an army of financial and other corporate executives hiding behind the glass walls of their offices, feeling protected by the corporate Mission Statements of honesty and integrity, only to be eventually exposed like the crooks and thieves they were.

It didn't take long for many of these crooks and thieve to start falling off their thrones one by one. And yet many of them are still enjoying the fruits of their misdeeds in luxurious retirement, or doing what they did before somewhere else. Once a crook, always a crook, the saying goes. Well passed unethical, the behavior of the Wall Street gang of 2008-2009 is hard to summarize in one sentence. They were acting like czars in their little kingdoms, using and abusing government and investors alike.

Energy companies were not the exception at the time, and many of their names are written on the wall of shame of the time. Solyndra is only one—the most obvious and visible—of them. Many others followed Solyndra's example and received what they deserve. All this in the middle of the bastion of American democracy. Lesson learned? Maybe, but let's not drop our guards.

Marketing Blunders

The new marketing concepts work, but have several problems. For example, just because a company says that it subscribes to the new customer-oriented marketing concept, does not always mean that the company will ensure ethical behavior in all its activities. Often, official marketing plans and ads appear to be satisfying customer needs (on paper), while in reality the needs have been ignored, or subverted, at the expense of product quality or other problems.

There are so many examples of faulty products entering the marketplace after aggressive marketing campaigns. One needs to take a close look in the distant past, where all this started. The national 20th century debacles involving lead pipes, nicotine, and asbestos materials have left significant scars on the American customer experience. They are examples of manufacturers ignoring the ethics rules all together, while chasing the mighty dollar instead.

Today we see failing medications, toys, car parts, etc., etc. faulty products coming from reputable and not-so-reputable companies and countries. As recently as the summer of 2015, we had to deal with a number of major automotive debacles. How could anyone imagine that Japanese manufacturers would be as negligent to install millions of faulty air bags in their cars? Isn't the quality and sophistication of Japanese technology the world's golden standard?

Or what master mind would dare tinker with the pollution controls in millions of VW vehicles? In the summer of 2015, VW was the world's largest and most respected car manufacturer, with a century old tradition of unrivaled excellence. So, where did the wicked idea to turn the emission controls of the vehicles off when in operation, come from? Was this a marketing gimmick brewed and implemented by overzealous marketing wizards, or was it just a misguided engineering cover up?

VW is an example of a blatant mismatch of marketing words and corporate deeds

How fast the mighty fall... Just one mistake and you are out. We may never find out what exactly happened, and VW is not anxious talking about it in the face of multiple lawsuits, so we just have to guess. Taking a close look, we see that, amazingly enough, VW Group doesn't have an official corporate Mission Statement. The closest to it is the VW company goal statement, "The VW Group's goal is to offer attractive, safe and environmentally sound vehicles which can compete in an increasingly tough market and set world standards in their respective class."

Wow, 'environmentally sound vehicles' is one of their goals. Sounds very attractive, all right, until you become aware that your VW is environmentally safe in the garage, but which is designed to switch to a polluting monster on the streets and highways. How could anyone even think of such blatant and unsustainable solution is beyond us.

In the energy sector, Solyndra's case is the closest we get to a VW-like blunder. The California solar company received half a billion dollars from the U.S. government on the faulty premises that its products are the most practical, reliable, and efficient. Lie after lie, that went over the heads of the administration bureaucrats and DOE engineers.

Solyndra's marketing campaign was exceptional. It was well designed and executed to the tee. Its product, however, was not even close to what the marketing bulletins said. It was not even tested properly, and had no market share to speak of, which eventually forced the company into bankruptcy. And yet, Solyndra spent $733 million (including $535 million obtained with U.S. government loan guarantee) on a 300,000 square feet building; the equivalent of five full-size football fields.

The metal and glass building (looking more like a space ship than a manufacturing plant) was state-of-the-art with robots that whistled Disney tunes to keep the workers happy, spa-like showers with liquid-crystal displays of the water temperature, and glass-walled conference rooms.

A Taj Mahal in the midst of the otherwise dreary, and gloomy at the time, Silicon Valley reality. A symbol of useless technology dressed up in space-age attire, surrounded by mystery and misinformation, it was totally void of substance. A symbol of millions of tax payers dollars wasted on self-indulgence, covered by thin veil of arrogance and lies.

Why the White House and DOE officials were unable to see through it is a serious question that is now buried in complex bureaucratic procedures and thousands of pages of documentation and hundreds of law suites. And as in many such cases, nobody was held responsible for half a billion dollars wasted on something that did not exist.

Figure 3-4. Solyndra's fancy, but totally useless building

Solyndra's motto was "What we do here will someday change the world." Maybe, but "someday" never came. Instead, Solyndra's end came much sooner. Even the largest and fanciest building in the industry, built with tax payers money and accompanied by even more grandiose motto, could not hide the fact that Solyndra had no feasible product and that it did not deserve even a penny of tax payers money.

What Solyndra had was a bunch of excellent engineers who produced shiny prototypes and convincing product brochures, which the marketing gurus and lobbyists took to Washington DC. These people, with the blessing of the top management, were able to confuse and convince some (but not all) of us. In the end, they were able to trick those who matter; the U.S. DOE and the government officials to give them money to manufacture this imaginary product that would change the world. Marketing at its best. Getting paid for making the "Emperor's New Clothes?" Sarcasm aside, good or bad, Solyndra will be forever alive in our minds...

Note: The author was a manager of a company that provided outside technical services to Solyndra during its hey days. The original Solyndra building was equipped with specially designed and manufactured equipment that was suitable for producing the final product. That product, however, had a number of problems, which needed to be resolved, and some effort was directed in that area.

As soon as millions of dollars became available,

however, the company top hats lost the vision and focused on building a futuristic buildings with even more futuristic equipment. As if that was the solution to the myriad of problems the final product had. Without a successful marketable product, and with money running out, Solyndra could not pay the bills and was forced to close its doors.

Not one single top hat was held properly responsible for this treason. Yes, that was an example of a treason of public interest, resulting in a national size disaster.

Solyndra is not the proverbial *dead horse* that we keep on beating. It is, instead, a monument; a symbol of how things must NOT be done in the energy sector. It is a text-book example of how good people and ideas could turn bad. It also shows how personal ego-mania and greed could trump corporate and even national interests. We must never forget Solyndra!

We can give dozens of additional examples of faulty products and services, leading to company failures, that have been misrepresented in marketing campaigns. As a matter of fact, we have detailed hundreds of these in our book, *Solar Technologies for the 21st Century*, published in 2013 by the Fairmont Press. The list is too long—over 50 pages—so we cannot include it herein.

Note: One similarity failing companies of the period had was the fact that most of the bankrupted U.S. and EU hi-tech companies received government subsidies. Was that what caused the failures is still debated, but one cannot ignore that fact. Easy money, easy go, as they say...

But even when marketing properly reflects and legitimately satisfies the customers' needs, the final results may infringe on the rights and needs of other customers. Hydrofracking is the best example in this case, since it satisfies the need of the majority for more and cheaper fuel. At the same time, however, it hurts the minority by poisoning their soil and water.

The same is true for solar panels manufacturers. They advertise 20 year performance guarantee for their products, but these products have never been exposed to 20 years of non-stop operation under extreme desert conditions. As a matter of fact, most manufacturers, have not even seen a desert, let alone know what it can do to their products during 20 years.

The ends justify the means is not a good marketing method, and yet it is often used. It, however, has limits, which have to be drawn when the "means" companies use hurt people. Mining, hydrofracking, oil drilling, nuclear plants operation are some examples of activities which hurt people, and yet continue, and will most likely continue, ad infinitum.

But in most cases the companies involve justify the means any way possible. Today we see convincing ads on TV about how clean, efficient, and beneficial natural gas is, which is true for most of us, but not for those who end up in the hospital with toxic exposure, even though their number is quite small.

> *This set of issues often go beyond the company's direct (legal) responsibilities, and pushes the limits of their ethical responsibilities.*

So what is unethical marketing practice? The marketing people are not monsters, but in all cases the interest of the company comes first and everything else is distant second. Solyndra had good technical and marketing people, who did their job properly (at first) but then intentionally took a short cut—just because they could. And because they were given an opportunity backed by millions of easy money. This was, and usually is, in direct conflict with the key values of the marketplace and goes against customers' interests.

> *The U.S. taxpayers lost over half a billion dollars. Not one single person was punished for the crime.*

The conclusion then is that in the U.S. a theft of $100 would land you in jail, but stealing $535 million is not a punishable offense...if backed up by a skillful marketing campaign with efficient White House and DOE lobbying effort.

Marketing vs. Ethics

How do marketing people deal with the pressures of upholding their ethics vs. defending the company's interests at all cost. Solyndra is an example of a case where marketing people either did not have good understanding of the overall situation, or ignored their better judgment. They were clearly (intentionally or unintentionally) stepping over the ethical line limits... time after time, day after day. The line was blurred in this case either by technical misinformation, by pressure from the top management, or a combination of both.

There are usually grey areas and gaps in any project, which could be used, or misused whichever way by the marketing people. To start with, each marketing team relies on its own value system to determine what is ethical. They usually recognize consumers' rights to safety, and the need for full information in assessing the value of the product and service they pay for. But then, they (the marketing people) have to coordinate their actions with the company's interests and top management direction, which always boils down to "making the biggest profit possible."

There are some official guidelines and rules that are used when making ethical decisions. For example, the American Marketing Association (AMA) has established a code of ethics, which provide guidelines for ethical conduct. According to AMA, " Marketers shall uphold and advance the integrity, honor and dignity of the marketing profession by being honest in serving consumers, clients, employees, suppliers, distributors, and the public." The code then outlines responsibilities for each component of the marketing mix. For products and services, it says that marketers have the responsibility to ensure product safety, to disclose all product risks, and to identify any factor that might affect the product safety and proper operation.

Looking at the Solyndra's case one last time, we can't help it but wonder, if the marketing people were thinking about upholding the company's and their own integrity, honor, and dignity. Or were they simply concerned with their jobs and making profit for the company—under any circumstance and at all cost. This, we will also never know, so our guess is as good as yours.

And finally, there are sets of rules issued by federal, state, and local governments that deal with misinformation, misrepresentation, misleading statements, etc. intentional or unintentional intents to twist the truth, which might cause problems to the customers. This, however, has not stopped misinformation and distorted truth in marketing campaigns. We often see TV ads that simply make no sense from a number of points of view, and at times blatantly cross the truth lines. How do we determine the boundaries, for example, in insisting that hydrofracking is the safest and best for us all, if there is a new brown-black puddle bubbling of poisons in your back yard, which happened to be close to a recently started hydrofracking site? Or, how do we interpret BP's claims that they have made the Gulf coast "even better than before?"

These are unresolved issues that the U.S. and the world's energy industries, and their marketing professionals, are tackling presently. Some of these are very complicated and require an entire army of engineers, lawyers, and financial experts to resolve. This is a lengthy and expensive process, which we will be witnessing for many years to come. The marketing genius of the 21st century is at the front lines of this process, filling the gaps and cracks, and figuring out the next steps towards success.

21st Century Marketing

Generally speaking, marketing is the art of developing, advertising, and distributing goods and services to all kinds of consumers. During the 20th century, *availability and quality* of the products manufactured and offered on the marketplace was the primary goal of most reputable companies. Later on, quality customer service was added as a measure of the companies' compatibility and suitability on the global markets.

The 21st century also introduced a *third dimension* in marketing; it is the advent of a new economy, based on technological innovation and strategic development:

- The 21st century economy is driven by the digital revolution. Consumers have access to all types of information for products and services offered by all competing companies. A significant result of that is the fact that *standardization* has been replaced by *customization* of the type and number of products and services offered to the customer base.

- 21st century companies are making marketing decisions with the help of various computer systems and computer simulations. This, combined with the increased speed of communication via mobile phones, e-mail, SMS, etc. is making marketing a sophisticated enterprise that helps modern companies to make decisions and implement strategies much more precisely and quickly than ever before.

- The digital revolution also brought changes to the purchasing experience of the customer base too. Now day, online purchasing can be done 24/7 and includes products and services offered by many competitors all over the world and delivered to the customers globally. The customer base is also much more educated, informed, and sophisticated than ever before, so the marketing methods have to take all these changes seriously if their products and service are to be successful.

- Marketing is no longer limited to goods and services; instead, it is extended to everything from places to ideas, and everything in between. This makes choosing proper and effective strategy decisions much more complicated, and the answer to these challenges is determined by the particular market. In order to stay competitive and avoid problems, the global marketing people have to consider culture diversity, international trade laws, trade agreement, and the regulatory requirements of the individual market.

- The other side of the 21st century marketing dimension is the companies' sensitivity and responsibility to (or lack thereof) *environmental and social changes*. Any company that offers products and service to the U.S. and EU customers is required to show, one way or another, its concern with protecting the environment and the customers' rights. This is reflected in new packaging designs, the introduction of new labeling systems, and company involvement in the local social life.

- The organizational hierarchy of the 20th century was top-heavy, where the top management gave out instructions to be enforced by the middle managers, who directed the efforts of the work force. Today, companies have to consider, and take advantage of, immense amount of information available, which allows (or forces) them to use the *team approach* more efficiently by including all employees in the marketing effort. This leads to the design of more efficient marketing programs across different consumers and various distribution channels.

All these changes have forced companies to fine tune their modus operandi. Today there are several trends that determine the companies' products and services success.

- There is a trend of streamlining companies' processes and systems, focusing on cost reduction. This is most often done by swift internal process optimization, combined with outsourcing.

- Another trend is encouragement of the entrepreneur style of work environment across the company hierarchy with emphasis on global approach, usually built on long term relationship with consumers. This requires thorough understanding of consumer needs and preference, and so marketing today is geared to treat the marketing and distribution channels as life-long partners; not as mere customers.

The Customer Experience

Summarizing all this, we come to the 21st century marketing model. Today marketing is no longer just selling some product or service.

*The successful companies market the **entire customer experience**, which is packaged in different ways.*

The customer experience is actually a complete package that starts with educating the customer about the product or service, making him or her to like and want the product or service, completing the order, delivering the product or service, ensuring satisfaction in using it, and getting good feedback and recommendations for the experience from the customer. The customer is drawn in to participate in the process.

When people buy an iPhone, for example, it is not just the product they buy and use. Not even close. It (the phone) comes with an amazing experience that people talk for years. It starts with receiving the news of this new, shiny, slick, and amazing item that does almost everything short of serving your dinner. Then you end up in a long line of interesting people, waiting hours and days for the Apple store, so they can be the first to buy the technological miracle.

You feel proud to be part of this bunch of advanced society representatives. Then it is your turn at the counter, where you get a $600 gadget for mere $200 (and a two year contract with your phone company.) What a glorious day! And then the real fun begins! You can show off your new possession to the less enlightened family members and coworkers.

You can dazzle and amaze people with the things you can do with the new gadget. This amazing experience makes you feel so grown up and important. And the experience never ends, because every time you have a problem, you run to the Apple store, where they wait for you and put you on a throne of respect and super-customer service. It is also here, where you can meet like-minded people and strike some awesome relationships.

What Steve Jobs have sold you is a life-time experience. Brilliant, to say the least. It is an example of true 21st century market genius, which everyone is trying to emulate, but very few have been this successful.

After all, not all products can be packaged and sold the way Apple does it. How do you package the products and services of a coal-fired power plant? Nevertheless, the goal of the successful marketers is to always find a way to provide an unprecedented in its completeness experience, during which the customer is kept interested, satisfied and happy, while feeling like being part of the entire process throughout the entire time they are in contact with, and under the spell of, the marketing genius.

One car company even states in their ads that, "We believe the experience should start before the car does." We don't know if their belief has materialized, but we know that it was actually Steve Jobs who initiated the trend of this type of a "pre-experience" experience by personally promoting the new gadget his company offered way before it was even a reality.

The "pre-experience" of the anxious potential owners of shiny gadgets started as soon as the products were advertised, which in some cases was weeks or even months before any of them were available in the stores. Then the pre-experience continued when the gadgets hit the stores. It then culminated in endless lines of excited buyers waiting for hours and even days for the stores to open. And then the real experience started, when the

customers got the new gadgets in their hands.

That experience is still kept alive all over the world as Apple's marketing genius puts new gadgets on the market from time to time. The *product-service* offering campaigns make us all believe that iPhone 6 is a miracle; much different and superior to its older cousin, iPhone 5. We are led to believe that it would change our lives, so thousands and millions of us are lining up in front of Apple stores to shed cash for a new gadget and trash the old one, which was equally good.

The customer experience fire is still kept burning by Apple's marketing genius, but it is showing signs of decay. All good things end with time, so Apple's time might have come too, as other similarly good products and services hit the market.

Subscription Pricing

Another, somewhat more practical marketing trend of this century is that of *subscription pricing*, which ensures a true connection with, and more opportunities to influence the customers. The "subscription" means that instead of receiving a single payment for a piece of manufactured equipment (and be done with it), many manufacturers are now receiving a steady stream of revenue for ongoing service related contracts *after* the sale. They often lower the initial price of the product in order to secure the sale and this way the customer is hooked for the duration. This way, in addition to the initial price of the hardware, the sellers receive payments for additional services they may (or may not) provide for the duration of the contracts.

For example, AT&T and most other phone companies offer drastically reduce the prices of their fancy phones to customers who sign a two year contract. Under these contracts, a $600 iPhone can be purchased for $200, or less. The contract then obligates the new customer to abide by the company rules and pay whatever the monthly bill requires.

Captive Customers

Another very successful marketing approach today is expressed as increased attention paid to captive customers. The captive customers are those who have no other choice but to use one product or service in their geographical area or area of expertise. Many local utilities, for example, are providing extra services to all their customers at no additional charge.

Their captive customers cannot change vendors, unless they sell the house and move out of the area, so they are stuck with that local utility. And yet, the utilities are increasingly paying attention to the customers' needs via periodic emails, reminders, meetings, and other means, where they disclose information on new services, display energy usage data, etc.

This represents a new and much higher level of customer support, while selling us the same, over a century old, product. The old product is now dressed up and primed up as new and brilliant service aimed to please you; the customer.

This all boils down to the fact that today full cost accounting can be achieved much better (if not solely) by the inclusion of a good service economy model in the corporate modus operandi.

Service rules the markets today.

Brand Awareness

And of course, we need to mention the *brand awareness* here. It is actually a 20th century phenomena, which marketing executives emphasized and encouraged. They still do, but are now dressing it in new and more attractive vestments.

A stroll down main street Anytown, USA, we see people walking around with company logos on their hats, shirts, pants, hand bags, and shoes. Like little voluntary advertising boards, they proudly carry all kind of company logos everywhere they go. The amazing thing is that instead of the companies paying the people to carry their logos, the people pay (sometime astronomical prices) for the privilege to carry the vendor companies logos. What a bonanza for the marketing people of these companies…

Using all forms of digital and other media, the marketing focus is on providing "content" via "content marketing." Today this is done via the help of Twitter, Facebook, LinkedIn, and other such Internet services, which if properly used could be converted into successful marketing tools.

Figure 3-5. 21st century social (content marketing) icons

The final goal of this effort is to create "brand loyalty," which is the jewel in any company's crown of marketing activities.

Securing a significant level of loyal followers is the ultimate goal of, and the greatest reward for, marketing executives.

Apple is an example of such brand loyalty success. There are millions of (otherwise intelligent) people all around the world blindly following every step Apple makes. They are willing to spend night after night in a freezing cold, or fierce heat, just to be one of the first in line to get the latest gadget that Apple's marketing genius has come up with.

Nike is another very successful brand, with millions of people—from football players to babies—displaying proudly the famous logo everywhere they go. Many of these people don't even know what Nike stands for, let alone using their sports products, and yet they are compelled to be part of the crowd and adorn their attire with Nike's logo.

Not many companies can claim such a faithful customer base. Those who get even close to Apple's and Nike's following are quite successful, thank you.

Although many energy companies have brand recognition, it is far from the level of Apple's success. As a matter of fact, it is often held against them. Because of that, we seldom see Shell of BP logo on sports or fashion wear. This is why the marketing methods used by energy companies are somewhat different from those used by Steve Jobs.

Marketing Decision Making

Figure 3-6 shows the most logical, practical, and efficient way of making marketing decisions today. So let's take a closer look at this process. We believe that this is important because the energy markets, and as extension we all, depend on the decisions of politicians, company executives, and the related enterprises. In case of government decisions, the participants are many more, and some have a veto power for whatever reason they can come up with. This creates an overwhelming inefficiency and grid lock in the decision making channels, as we can see happening today. Washington is badly constipated, so most of the decisions are either not coming out, or come out half baked.

So, instead, we will take a look at the more reasonable and functional world of corporate (and other functional capitalist system enterprises) decision making process. How do we approach a new situation that requires a new product?

Let's look at an example of designing and developing a new product—a solar cell with improved performance.

Case study—
Solar Cell Improvement Project

Below are the steps in the decision making process

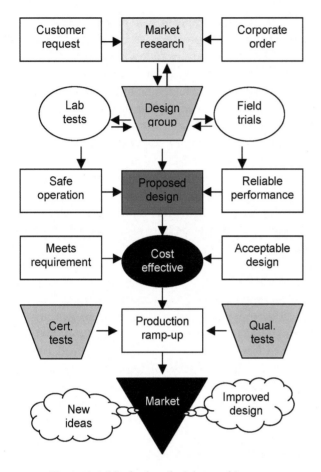

Figure 3-6. Marketing decision making tree

in response to the need for a new, more reliable, more efficient, and cheaper solar cell.

Day 1

Customer requests for more efficient and cheaper solar cells are constantly on the Sales and Marketing (S&M) groups' meeting agendas, and at certain point corporate issues an order to take a serious look into it and present a Proposal. At that point, a vigorous marketing research is undertaken by the S&M departments in order to find out what exactly the customers want, what the competition is doing to meet the demands, and what can be done in this case to take advantage of the situation.

Day 30

Preliminary work results are evaluated and a Design Group is formed in order to evaluate further the market research data and come up with a working model and a Project Proposal. The Design Group consists of S&M managers, engineers from different key disciplines, and other specialists in different areas of the solar cell industry.

Day 60

After a period of data collection and intense communication and collaboration among the different groups, a new and improved solar cell model is chosen and is sent to the R&D team for evaluation. The R&D team produces a prototype, which is tested and sent to manufacturing.

Day 120

Another group of professionals has been working on developing the necessary processes for optimizing the new solar cell manufacturing process. Finally the new cell is manufactured, tested, and pronounced good.

Day 180

The new and improved solar cell is ready for field testing. A batch of cells is manufactured and sent out for field tests. Here, normal performance tests are conducted in order to verify the efficiency and other characteristic of the new device. These are followed by accelerated tests, which are conducted with special equipment and under special conditions, designed to simulate long term exposure to the sun and the elements.

At this stage, the new cells could be sent to an independent test lab for verification of the test results and preliminary certification of the new product.

Day 360

Six months later, the accelerated field tests show that the new prototype is working well. The results are verified by the independent test lab, and show that the new cell is safer, more efficient, reliable, and cheaper than the original models. The data is collected, processed, analyzed, and included in a final Report.

The final Report and a Work Proposal are submitted to the top management for evaluation and authorization of the new solar cell project. It includes all that is needed to retrofit the existing, or construct new, production lines to manufacture the new cell in mass quantities. This includes the necessary equipment and process changes, complete with time and cost estimates

Day 420

The Report and Work Proposal are reviewed by the top management during a series of meeting with the different working groups. Finally, it is decided that the new solar cell is cost effective (the bottom lines speaks the loudest at this point), that it is acceptable, and meats all requirements. At this time, the Proposal is approved and given a green light. Time schedule and budget are agreed upon, and the work on a new production line, or retrofitting the existing one, is underway.

Day 480

Now the new solar cell production line is in full operation. It, however, cannot start shipping product until the new solar cells (encapsulated in solar panels) are officially qualified and certified by the respective national and international agencies. This process might take additional 6 to 12 months.

Day 660

After 6 months of qualification and certification tests, the new solar cells (in solar panels) are ready for the market. At this point, the production line is ramped up and the new and improved product is shipped to the happy customers.

Ongoing.

Based on information from the new customers and continuing field tests, new ideas are submitted to the S&M and Design teams for evaluation. New and improved designs are produced and tested. Some of these are submitted to the top management for further review and approval.

If and when approval is obtained, the new products or services undergo the above process step by step. Eventually, the newer and even-more-improved solar cells are sent through R&D, field tests and so on, all the way to the market. Here they will compete with its older cousin and the other solar technologies.

Note: This is logical and most practical way to introduce a new product (solar cells and panels in this case) into the marketplace. What we saw in the last 2007-2012 boom-bust cycle of the solar industry, however, was quite different. The race to the markets was on, so new solar cells and panels were introduced routinely (mostly by Chinese companies) with minimal, if any, testing. In some cases even the qualification and certification tests were bypassed, or somehow modified to speed up the process.

The most dangerous consequence of this rush to market is the fact that in most cases the new solar cells and panels were not put through long term field tests. Because of that, we now have millions upon millions of solar panels installed and operating in the world's deserts that have not been proven efficient, reliable, or safe for long term operation under extreme conditions. This means that we might see a number of serious surprises in these fields in the not so distant future.

New Technology Development and Finance

How are new technologies and products developed? By whom, and who pays for all this work?

New technologies, and/or improvement of existing ones, are worked on and developed every day. If a high-tech company has as a goal to stay competitive, it must continuously improve its technological tool box, and add new tools to it. This is why we see announce-

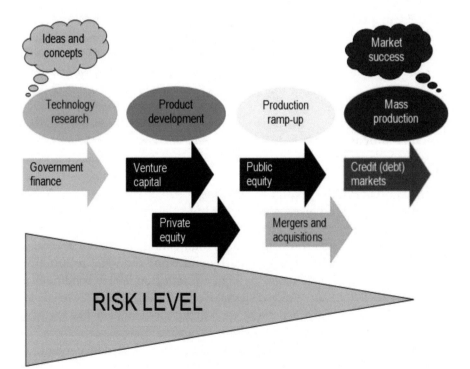

New technology development risks

Ideas and concepts   – 90%
Technology research -- 80%
Product development   – 60%
Product ramp-up   -- 40%
Mass production   -- 20%
Market failure   -- 10%
Field failures   -- 5%

Figure 3-7. New technology development and finance processes

ments of new achievements in efficiency, or reliability on daily basis.

Every improvement or new technology development goes through a complex series of events before being offered on the market. There are different risks at every step of the overall process of developing a new technology, which usually determines the type and level of financing. The greatest risk, and questionable expense, is during the initial stages of the new technology research and product development.

There were many energy related (solar mainly) companies that stumbled and fell on their faces during the renewables boom and bust cycle of 2008-2012 during this first step of technology research. Even though the initial stage of product research was fully or partially financed by government grants and subsidies, success was not guaranteed. As a matter of fact, very few of the recipients of government money succeeded during this initial steps of their development.

Large failure rates of solar companies were observed during the transition steps from research to development, and even larger to production ramp-up. In many cases, what worked in the lab, did not work on the production floor, and even less during field operations. This discrepancy is easy to understand, since the conditions in the lab are very different from these in the production line, and even more so from these in the field.

Companies that were able to push their product to the production ramp-up stage usually had some level of success. Those who reached the mass production stage were on their way to success...if they could only outsmart and outdo the competition. Which is where marketing takes over...

Once on the market, however, the product is beyond marketing's ability to control its behavior and function. This is where a series of unknowns in field performance and the related abnormalities and failures usually appear. The new solar cells and modules look good and shows good results in lab tests. They also performs well during short term field tests. But a month, or a year after installation, the PV modules efficiency drops and even total failures start showing their ugly head.

We saw a number of such cases during the quick ramp up of solar technologies and projects. The manufacturers were in a hurry to get as much of the ARRA (subsidies) money as possible, and produce a product with better performance characteristics than the competition. Time was of the essence. This left very little time for extensive—let alone long term—tests, so we ended up many marginal products...just think Solyndra's out of this world PV panels.

Even today, when things are much more stable in this area, we hear daily about new processes and products, and new performance barriers exceeded...in the lab. Can this new process, or improvement, be applied to the mass production line is another question, which needs more time and money to answer.

And even more importantly, the question of long term efficiency and reliability cannot be answered until the product has been in field operation for months and even years. This is where we see a lot of companies and customers disappointed by the unsatisfactory performance of new technologies in the near future.

Bypassing key steps in the product development process might cause millions of PV panels in the deserts to fail.

In Summary

Product development is a long, complex, expensive, and full of technical obstacles and financial risks process. Even in best of cases, the new product is not guaranteed to perform as specified, nor could the top management expect trouble-free market acceptance and/or profits. Any gaps in the decision making process complicate the situation even further. Some of them are as a result of making assumptions about availability, quality, and pricing of the raw materials and chemicals, which could change with time and throw a monkey wrench in the works.

There are also always problems with new design, and the related R&D, testing and manufacturing processes. Here, making one prototype under controlled Lab conditions is not the same as ramping up a production line that produces thousands up thousands of these devices.

The Lab-to-Production gap accounts for the largest amount of failures in the solar industry.

Note: Every time we hear that company X has achieved a new record of solar cell efficiency, we must remember the Lab-to-Production gap. The R&D departments function under totally different—more controlled and flexible—work conditions than the production lines. Achieving high efficiency under lab conditions is many times easier than on the production line. As a matter of fact, the product that comes from any line varies in quality even under best of conditions.

An average production line can produce solar cells and panels with variable efficiency. The difference is due to different batches of raw materials and supplies, varying equipment condition, and varying technical personnel abilities. Raw materials and supplies coming from company A, for example, would produce different quality cells that these coming from company B. Production equipment changes its performance with time. Night shift operators are usually more prone to making mistakes than their day shift counterparts, etc., etc. All this affects the quality of the final product.

And then, even if all goes well with the product development and ramp-up stages, the market acceptance of the new and improved product s not guaranteed. It does little good to the bottom line if a well made and superior product is not accepted by the customers for one or another reason. For example, even the best U.S. made solar cells and PV modules could not compete with the cheap Chinese competition.

Engineering-Marketing Team Work

During the first years of the 21st century, the author was a part of an *integrated **engineering-marketing*** team at a hi-tech company. This was a new approach, born in response to increasing number of problems with products in the field. A special team was formed for the purpose of addressing and solving the issues at hand and developing new products and services.

The new team consisted of key engineers, representing the different departments, working in tandem with key sales and marketing personnel. It was to lead the effort, which was a new MO for the company; something that had never been tried before. To everybody's amazement, the new approach showed exceptional results within very short period of time.

The different marketing and engineering teams across the company were communicating directly and cooperating at every step of the projects at hand. Although still functioning separately in the areas of their expertise and responsibilities, they coordinated their work load with the *Engineering-Marketing* team and took directions from it.

In the new *Engineering-Marketing* team approach, the close daily interaction between key engineers and marketing people proved to be a very powerful tool. It didn't take long to realize that the major factor in the success of this approach was a drastic change in the thinking and behavior of the team members. Now the entire large company was thinking and acting as one unit.

In the days prior to the formation of the *Engineering-Marketing* team, all parties were specialized in, and focused on, their specific areas of the business. They worried and focused on their specific areas of responsibility only. This left huge gaps in the project execution and led to misunderstandings and conflicts; many ending with angry finger pointing.

On one hand, the engineering personnel did not understand, nor did it care, about the marketing aspects of the projects, and often blamed marketing for their failure to research or present the projects properly. On the other hand, the marketing people did not understand fully the engineering aspects of the projects and blamed the engineering personnel for not giving them the appropriate information. That opened a gap that swallowed many opportunities.

After the creation of the official *Engineering-Marketing* team, the entire product development and project execution spectrum was covered solidly, and no gaps in understanding or communication were allowed. Success was much easier to achieve working that way.

This type of approach is easy to implement, and would be quite applicable for use by the energy companies and the utilities engineering and marketing teams, where gaps of understanding and communication still exist. In many cases this means breaking old habits and forcing new approaches on people, who are comfortable with the way things are. Once passed this barrier, however, the integrated *Engineering-Marketing* team approach offers many advantages.

THE ENERGY MARKETS

Energy marketing starts with thorough and honest analysis of the corporate side of the equation—the company capabilities and the different product or service characteristics. Then the marketing folks must look at the other side of the marketplace; the competition, the customers, and all related issues. This is not an easy task, keeping in mind that one is usually much more familiar with one side of the equation and not that much with the other side.

To do this right, the energy marketing professionals must be familiar with, and take a close look at all sides of the energy markets. In addition to the products and services they offer, they must be familiar with the middle-men, the regulators, the utilities, the myriad of service providers, and all other companies and individuals—including the customer base—as well as all other aspects of the energy markets and their function.

Taking such an all-encompassing approach, however, is seldom done, since it requires a large team of specialists. Instead, market analysts (of the past) prefer to look at the markets from one single point of view;

usually the one they are most familiar and comfortable with. This is not a productive or profitable way to do business, since many things can go wrong with other parts of the market. Today things are changing, and energy marketing has been forced to evolve into a much more complex and all encompassing enterprise.

In this text we try to do the impossible; taking a close look at *all* sides of the energy markets. Since we don't have a team of experts, but instead rely on the experience of one person only, we may fall short of the mark…hopefully not very short.

Generally speaking, the energy markets, have, a) physical, and b) financial sides, which are quite different. They can be easily distinguished by the products and services they offer, as well as by the intentions and goals of the market participants involved.

- The *physical energy markets* deal with the natural resources and the related infrastructure and institutions involved in producing energy sources (fuels) and electricity, and delivering them to consumers. This includes the trading and transport of, and payment for, the physical energy commodities (coal, natural gas, crude oil, solar panels, wind turbines, and electricity).

Physical products are basically those whose contracts involve the physical change of hands and delivery of energy sources or electricity. Physical market participants are those who are in the market to make or take delivery of the actual physical product or service.

The physical markets can be further differentiated by:
- Location: regions, nodes, zones or hubs;
- Time frames: hourly, daily, monthly, quarterly or yearly;
- Types of products: natural gas molecules or electricity electrons, pipeline or wire transmission capacity and storage; and
- Nature of sales: retail sales, which involve most sales to end-use customers; or wholesale sales, which involve buying and selling of energy in large quantity, usually on long term basis.

- The *financial energy markets* do not involve the actual delivery of physical energy sources or electricity; instead, they only involve the exchange of money via complex financial transactions. The financial markets include the buying and selling of financial products based on, or derived from, physical energy sources and electricity. The financial markets

also include market structures and institutions, market participants, products and trading, and have their own drivers of demand and supply.

Financial markets are where companies and individuals go if they need to raise or invest money too. They are important to the energy markets in two key ways.
- They provide access to the capital needed for operations, and
- Some energy-related products may trade in commodity markets or, as derivative products, in financial markets.

Energy sources (fossils, solar, wind, etc.) and electricity are traded like commodities, such as metals, corn, wheat or oil. These are not visible commodities, but can be turned on and off, and even measured. Commodity markets began as ways for farmers to sell their products, or a portion of their production, even before it was harvested. This provides them with capital to continue operations and mitigate some of the risks.

The commodities markets are powerful tools to manage loss and wealth in the energy sector.

In this book, we present all sides of the energy markets; loosely adhering to the above official designations and regardless of the impact our analysis would have on any particular product, companies, and/or individuals. This is not easy to do, so we hope that the critics would keep in mind that our only goal is thorough and honest representation of the important matter of energy, the energy markets, and all that is related to them.

Although the day to day function of the energy markets is somewhat different from that of most other consumer markets, some key rules of the new 21st century marketing science apply to an extent here too. Yet, the methods and application (as well as the results) of energy marketing are quite different.

Energy companies have many products and services to offer to a broad segment of the population and the business community. How they approach the marketing of their products and services determines the difference between a successful and failing product and/or service.

There is a degree of difficulty added to the proper marketing of energy products and services. It comes from the fact that the customer base wants these products and services, but is not willing to pay the high prices, and is even less willing to put up with the damage

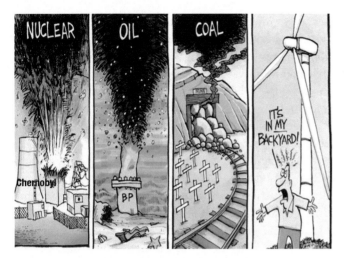

Figure 3-8. Some of the serious issues.

and harm done by some of these.

We can't help but wonder how Steve Jobs would've approach iPad and iPhone marketing if these products had some negative environmental impact. Would he be able to justify the damages resulting from using an iPhone that emits tons of GHGs, or harmful radiation? How would he explain and justify the damage and the remediation efforts? Would he be able to include these in, and sell them with, the remediation efforts, as part of the overall Apple *experience*? These are some of the problems energy marketing must to deal with today.

Note: One little known, or talked about, fact is that the manufacturing of iPhones and their different components use a lot of toxic compounds. There are lakes of fuming and bubbling liquids behind some of the factories involved in the parts manufacturing of iPhones and other electronic devices. The GHG emissions during each step of the manufacturing process, including transport, are also significant.

The customers, however, are mostly unaware of, or could care less, about these issues, because the products are manufactured in China, which is where all the bad stuff stays. All we care about is how shiny and functional our new devices are. So, Steve Jobs dealt with these (and many other negative) issues by not talking about them. Instead, he mind boggled us all with his new shiny devices that could do so many amazing, interesting and practical things.

Most energy company executives wish they could do the same; not talk about the issues at hand. Instead, today's energy companies are being blamed and held accountable for very serious damages right here at home; something even Steve Jobs could not have been able to defend. For example, while hydrofracking is adding a

level of security to our energy supplies, many people are against it because of the real and perceived damage it is doing to the environment and people's health. The same is true for coal mining and coal-fired power generation, nuclear power, and building new hydropower dams.

Nuclear power has an added dimension of deadly danger, which cannot be hidden by any marketing gurus. So it has to be dealt with openly and frankly. This fact was emphasized and even exaggerated by the 2011 Fukushima nuclear accident. While we all know that nuclear power is unprecedented as an efficient and cheap power generator, it is now labeled as a fierce mass killer for the whole world to know and ponder about. Most people find it hard to decide what to think or do with this fact. The same problem plagues the Japanese government, which is trying to decide what to do with the shut down nuclear fleet in the country. Not an easy decision...

These, and many other, discrepancies affect the function of the electric utilities, which must generate power that is needed for daily life and business operations. Electric power is absolutely necessary in all aspects of life, so the utilities must generate it one way or another day in and day out. So, they use all that is available to them to complete their task. While doing this, in many cases they cause some damage to the environment, which affect negatively human health and wellbeing.

So the utilities are stuck between a hammer and a hard place. On one hand they must generate enough electric power non-stop, whatever it takes. On the other hand, while trying their best, they encounter expenses that cause price increases, and/or use polluting technologies, all of which make the customers unhappy. Damn if you do, damn if you don't!

And so, what are the utilities to do? In order to reduce the negative impact of their products and services, some of them get into creative marketing mode. They use advertising campaigns, and introduce alternative programs that are geared to educate the customers, in an attempt to get on their good side. This is because their motto today is, "Informed customer is a happy customer." This is quite different from the past, when information was limited and was on the back burner of the energy marketing scheme.

The Customer Divisions

A number of studies in recent years have gauged consumers' attitudes toward the different segments of the energy markets; energy efficiency, new energy programs, the renewables, climate change, and the future of energy use at home and business in general. The researchers use different models and methods, but the results are quite similar. Predictably, customers who are more likely to be early adopters of green technology and programs also tend to be tech-savvy and more concerned about climate change and the introduction of renewable technologies. The rest follow according to the level of their understanding and interests.

This is a common finding across these studies, where the largest segment of consumers follow (hesitantly) the early adopters. This is a more diverse group of the so called "early majority" or "pragmatics." They are aware and somewhat concerned about the energy and environmental issues at hand, but are mainly motivated by cost, comfort, and value.

Simplicity is another common denominator in this group. On the other side of the dial are those suspicious, or critical of new energy products and services. Some people in this group are indifferent about, while some are even resisting, the upcoming changes.

The utilities are becoming aware of the growing spread of customer interests, likes, and dislikes. They understand the problems associated with their products, and are careful to not present conflicting messages, or complicated menus of energy services. A "happy meal" style packages with "set it and forget it" options geared for a specific group of customers is the best, most sustainable, and successful for the targeted customers.

The search for strategies for finding the right messages and messengers for the different consumer groups continues. Advertising the different brands with added

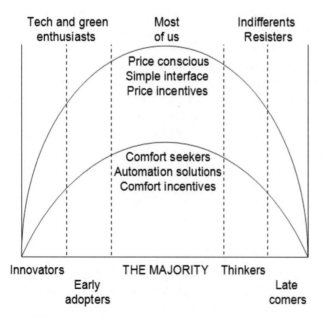

Figure 3-9. The U.S. energy customers segments

sense of emotion that appeals to the majority is one of the preferred approaches. Another method preferred by some energy specialists is the use of more practical, flexible, community-based programs, delivered by "trusted messengers." These approaches work well in some cases and not so much in others.

For example, a power company selling only renewable energy products and services, markets to the high-end, tech and green-savvy consumers, who are identified through sophisticated analysis of local and national statistics. The power company marketing scheme doesn't have to reach the broad market; just select individuals and groups of people and companies. This is a type of focused marketing that is used by data-driven marketers in other industries too.

The focused marketing approach is becoming the model for some energy services as well.

Most utilities, however, have a blend of technologies and services, so focusing on a group of customers is difficult. Instead, they are forced to find a happy medium in the way they present these to the different customer groups. Not an easy task. To overcome this problem, some utilities use periodic "perception trackers," that query customers on their awareness of the technological changes and willingness to participate in certain programs.

Some energy companies use online questioners, where panels of hundreds of customers are tested for their perception of certain energy products and services. Another surging trend is the use of periodic communications and customer testing that allows for what some call "fearless failures," combined with innovation and new approaches that lead to successful results.

Basically speaking the energy markers have been trying different things since the beginning of time. The big difference today is that random, general (wide circle of customers) testing is a thing of the past. Instead clearly defined test groups are chosen and tested periodically, using specially designed methods tuned to the interests of each group.

Finding new and better approaches is a priority today, since customer satisfaction is becoming a priority in the energy markets. A number of disasters in different areas of energy production have had a negative effect on the energy sector and the public perception of its function and goals. Energy production related disasters: oil spills, hydrofracking leaks, refineries' fire and storm damage, and the fear of nuclear accidents are playing a major role in the function of energy companies. This is creating a new situation, which is twisting the conventional marketing models towards increased attention to customers' preferences and ensuring their acceptance and satisfaction.

Note: The example letter bellow is from a large U.S. utility company, which has been increasing the interactive side of their business by online programs, emails and letters, such as this one:

Dear Customer,

We thank you, our valued customer, for your patience and understanding during this very active monsoon season. The summer of 2014 will be remembered for a series of powerful monsoon storms that caused record rainfalls and damaging winds that, at times, left tens of thousands in the Valley without power.

In fact, the most recent storm on Sept. 27 was one of the most damaging monsoon storms in history, resulting in more than 30 downed power poles, numerous downed power lines and damaged transformers, and more than 67,000 customers without electricity in our service territory.

Some of our customers were without electricity for only a few minutes, while others living near the downed poles and lines had to wait much longer until their power was restored. Our crews worked around the clock, safely restoring power to customers as quickly as possible.

We work hard to avoid outages by consistently maintaining and improving our strong electric grid to ensure reliability for all our customers. The fierceness of this storm was unusual, to say the least. We will be further strengthening our system.

Mother Nature is unpredictable in the desert, but our customers can be assured that we will respond quickly when storms hit to maintain the level of service our customers have come to expect.

Sincerely,
Chief Customer Executive

This utility is one of the largest in the country. It is unregulated entity and has a *captive market*, where people who live in the area serviced by it do not have any other choice. The utility is not obligated, nor needs to offer any special products or services, since the customers will get whatever it serves them. There is no other way. And yet, the utility has special customer satisfaction programs with Chief Customer Executive managing them. The management obviously goes through great lengths to ensure its customers' satisfaction. Good example of modern marketing, or what? Does that keep the customers satisfied? Maybe not, but this gesture

shows them that the utility is their friend and cares for their well-being.

The competitive advantage, which is less visible in the classic marketing model, is now also on top of the list of priorities of sales and marketing departments of energy products and services providers. The competition is increasing as the energy sector is now being geared towards providing higher quality customer services. The service models are increasingly designed to compliment the related products by emphasizing or supporting their quality, price, and benefits. A new dawn is breaking through the clouds of mistrust and dissatisfaction in the U.S. energy markets.

Energy Advertising

What does advertising have to do with the energy markets? Funny you should ask, because believe it or not, it is a very important vehicle used by all players in the energy sector. BP, for example, is still spending millions upon millions of dollars on a long term nationwide TV and radio campaign. The good looking people with BP badges on their shirts paint a beautiful picture of the serene coast line, while trying to convince us how concerned BP is with our land and people's health, and how much money they have spend on fixing the problems.

Similarly so, gas companies pour millions of dollars every day in the TV networks' pockets in trying to convince us how clean, cheap, and useful their product is. And, of course, how it generates a lot of good jobs.

OK, we are convinced...except for those of us whose health and livelihoods were ruined by the BP oil spill in the Gulf of Mexico, and the hydrofracking fluids in their back yards. But the negatively affected are still the minority, so forward we go drilling and pumping as much as we can.

Nevertheless, the advertising campaigns are shaping as an important tool, and the millions of dollars keep on pouring down the TV networks pockets. For example, several million dollars were spent in Albany, NY during the fight to drill for gas in the area, for TV advertising, lobbying and campaign contributions. The media war intensified as the state administration entered the final phase of deciding when and where to allow a controversial form of natural gas extraction that is opposed by environmental groups and many of the locals.

Natural gas drillers spent more than $3.2 million in 2011 alone in this area, lobbying the state government. Additional money came from the national natural gas industry, which was, and still is, advertising heavily using all kinds of media outlets to convince the public that hydrofracking is safe and economically beneficial. In the process they totally ignore and try to nullify the negative effects. And there are serious negative effects, which if not resolved soon might grow into a large disaster of National proportions.

The nuclear industry is also in the media. In 2014, several energy organizations pursued a *Future of Energy* expensive advertising campaign, focusing on nuclear energy as a clean source of electricity that provides jobs, contributes to a more secure energy future and enhances energy diversity.

A number of prominent voices were behind the campaign, which involved advertising in print and on television. This, in addition to radio and TV ads aired in the Washington, D.C. area, together with web-based and social media outreach. Famous people like Bill Gates and Jose Reyes are behind some of the campaigns for promoting nuclear energy options that are scalable and safe approaches. The new nuclear reactors designs—which have not been yet tested properly—are already promoted as one of the options to power safely the future generations. They are hailed as one of the ways to solve global warming issues, provide power for water desalination, and abundant cheap electricity.

While this is so, it is only one side of the equation. The other side is the suffering and death of thousands at Fukushima and Chernobyl. Environmental groups and local protestors don't have much money, so they cannot win the media war. Instead, they have to resort to homespun campaigns, warning that hydrofracking is threatening the local water supply, and that nuclear has its own problems, all of which could cause untold environmental disasters, including wild life and human health problems.

How far and wide the wars will go depends on the strategy chosen by the involved, and will be also driven by the developments in the energy sector. For example, in the summer of 2015 hydrofracking in the U.S. is shrinking significantly, due to low oil prices, so the media war is being moved to other arenas. Nuclear power is most likely to be on the front lines during the next several years.

The New Twists

Today, many companies are adding new approaches to the old and proven marketing methods. Twists like *Social Responsibility, Commitment to Sustainability, Environmentally Friendly*, etc., etc. moralistic terms are constantly tossed around, showing the companies' awareness of the issues of today. In the process, they also try to make us feel good, by feeling protected and well taken care of by such good and responsible neighbors.

So, what do the *Social Responsibility, Commitment to Sustainability, and Environmentally Friendly* labels mean?

Social Responsibility

By definition, *social responsibility* is a form of ethical framework, or behavior, reflecting the company's willingness to accept the obligation to act for the benefit of the entire society. In practical terms, *Social Responsibility* means sustaining the equilibrium between economic development and the welfare of the society for the good of the environment.

Social Responsibility can be expressed in avoiding engaging in socially harmful acts (passive), or by getting involved in activities geared to advance social and environmental goals (active). In order to show that they are *socially responsible*, corporations usually use ethical approaches (albeit most often on paper only) to secure their business success.

These touching socially responsible activities are often designed to prevent government agencies from getting involved with the corporate business affairs. Staying on the good side of EPA, for example, is a major goal of most energy companies. Most companies officially follow the EPA guidelines (on paper) for emissions of dangerous pollutants, but to be on the safe side, they add an extra-curriculum activity or two.

This is usually done by going the extra step of getting involved in the community in order to address its concerns. This is done officially, by adding the *Social Responsibility* spiel to the company mantra, thus showing EPA and the public that they are aware of the issues at hand and are doing everything possible to solve them. This way, the customers are happy to be protected, while at the same time the company reduces the chance of EPA investigation.

Corporate *Social Responsibility* has also become a shield behind which most atrocious acts are hidden. BP oil spill is a good, albeit extreme, example of social responsibility taken seriously (on paper).

Since it is not part of the basic economic role of most businesses, social responsibility is seen by most of us as a superficial *window-dressing*. It looks good on the very top of the corporate website and is always included in the business plans as an integral part of the business activities.

Commitment to Sustainability

The term *sustainability* in itself is usually used as an indication of the way biological systems remain versatile and productive. It also reflects the level of health and long-term endurance of these systems. The driving force of sustainable development are the interconnected domains of: ecology, economics, politics, society, and culture. *Commitment to Sustainability*, therefore, must include awareness and care of each of these domains.

In practical terms, *Commitment to Sustainability* means finding ways to reduce the negative impact of human activities on the local society and the global environment. This is done by developing environmentally friendly methods of conducting daily operations, and offering similarly friendly products and services.

This often involves complex and expensive processes of chemical engineering, environmental resources management, and environmental protection measures. These must be coordinated with the available information obtained from green chemistry, earth and environmental sciences, conservation biology, ecological and human economies, as they affect the related natural ecosystems.

So, what would be the *Commitment to Sustainability* of a coal-fired power plant? And how do we measure the level of its implementation. Let's assume that the health of the environment, and the species (including humans) in the plant's neighboring area averaged 90% before the power plant went into operation. The question would then be, what level of *Commitment to Sustainability* is to be expected the first year, the second, the 20th after the plant has been in operation? How does the *sustainability* concept fits in the dark picture of dead trees in the surrounding area? Or miners with lung disease?

Obviously, a report of no negative environmental effects whatsoever would mean that the plant has achieved 100% *Commitment to Sustainability*. But this is never the cases, so how do we label the level of commitment in case of 1% increase of cancer rates, or lung disease in the area? How about 5%, 10%, 30%? What if the local streams and water table are contaminated? What would be the level *Commitment to Sustainability* in these cases? These concepts and issues are usually left out of the daily discussions, and are mentioned only under special situations, or in lawsuit cases.

Environmentally Friendly

Environmentally friendly is another elusive label many companies use to identify themselves and their products and services with. *Environmentally friendly*, which has many other names, such as: environment-friendly, eco-friendly, nature-friendly, society-friendly, and/or "green," are a set of *marketing* terms used by companies in their fight to gain popular acceptance and win some additional marketplace share.

Environmentally Friendly usually refers to products and services, the design, production, and use of which also includes many laws, regulations, guidelines and policies. These are designed to reduce, or eliminate the harm inflicted upon the environment and the humans in it.

Environmentally Friendly is one of the most frequently used terms in the energy industry. It is a term without a measure, and supported by a few regulations, so it has been abused to the very maximum possible. It is actually one of the most ambiguous terms in the industry. It is widely used to promote products and services, by adding usually very broad, and sometimes more specific certifications, such as product specific eco-labels.

The term is referred by some as green-washing, with which any product can be made to seem cleaner and greener. Standards ISO 14020 and ISO 14024, developed by the International Organization for Standardization, establish some principles and procedures for environmental labels and certification. These eco-standards call (quite broadly and vaguely) for the use of "sound scientific methods and accepted testing procedures." The standards do not specify what "sound" and "accepted" testing procedures are, so the enforcement of these standards is lagging.

This is just another gap in our socio-economic system that allows the discrepancy in the industry to grow. Because of these gaps, *Environmentally Friendly* is the term of choice of any company, trying to promote their products and services—often regardless of how *Environmentally Friendly* these really are. It is our (the customers') responsibility to get involved in the process and voice our opinion on the matter.

In summary, *Social Responsibility, Commitment to Sustainability, and Environmentally Friendly* labels sound good and are becoming integral part of marketing campaigns and doing business in the energy industry today. Oil and gas companies show us how clean and beneficial fracking is. Solar companies show us the benefits of solar as far as saving the environment is concerned. Nuclear power companies…well, they are mostly silent. Why wake up the memories?, most likely their marketing gurus think.

"A sense of responsibility is an integral part of our corporate ethos. We aspire to be socially and environmentally responsible in every decision we make. To this end, we strive to reduce our energy consumption and carbon emissions throughout the manufacturing process," we read on one energy company's website. "We understand that the legacy we leave as a corporate citizen is not simply based on the types of products we make, but also the methods we employ in our own manufacturing facilities," the website's moralistic pitch continues.

It seems as if they are trying to convince us that they have invented these approaches from scratch. That these are brand new things, that we must be aware of, and that they are the ones responsible for awaking us to the reality of today. The pitch also conveys the thought that this company is the only one fully aware of the importance of these values and the only one putting them into practice—as if suddenly the old, dirty coal has been cleaned up, or as if natural gas had just been invented and were the cleanest, safest and best form of energy we ever saw.

In reality, we usually get the same old products and services, hidden behind the new shiny ads. Because how can a TV ad about the safety and usefulness of oil and gas drilling guarantee us that the hydrofracking site close to our neighborhood won't poison the underground water? Especially today, when we already have many examples of people and entire population centers being poisoned by the fracking chemicals as we speak?

How can the millions of cubic feet of natural gas burned every day be good for us and the environment, when in fact the sharp increase in natural gas use may be doing more harm to the environment than the coal it is replacing?

How can a solar company, making solar panels loaded with hazardous materials, hide behind the *Social Responsibility, Commitment to Sustainability, and Environmentally Friendly* labels when in reality, by virtue of its products and activities it is causing harm to the environment of the future. How can you hide thousands of tons of sand and dirt that have to be dug out, transported, melted, crushed, dissolved in acids, melted again, and again, to be made into silicon wafers? And then, these wafers undergo another long sequence of process steps, using a lot of energy, harsh acids, and other chemicals.

Amazing amount of energy, in form of electricity, heat, and vehicle fuels is used during these steps. Each step of the process is also accompanied by huge amounts of gas, liquid, and solid waste. Each step of the long cradle-to-grave life cycle of solar cells and modules leaves a significant footprint of energy use and pollution as well. When you combine all steps of all companies involved in this business, you get an ocean of liquid and solid waste, and millions of tons of gaseous pollution.

But none of that is reflected in the *Social Responsibility, Commitment to Sustainability, Environmentally Friendly* labels and ads. These titles are like the top of the iceberg that shows above the water. Nice, white and friendly. The bottom part, and its violent and polluting past is hidden from the public, which in its majority is not able to, or is not interested in, diving below the surface to see the total reality.

With all fairness, we must repeat that these terms cannot be qualified, nor quantified, so the fossil and solar industries alike can still find the gaps and hide behind the clean and green titles. In the future, however, we will have to come up with more exact ways to determine the level of compliance in all these areas. It is our social responsibility to the future generations.

The Energy Lobby

In addition to media advertising, many energy companies have an army of lobbyists on local, state, and federal levels. The energetic and well supported lobbying efforts of the oil industry are geared to influence lawmakers and regulators to twist the rules just enough as to allow them access to land for their drilling operations and relax the drilling restrictions in the shale rich formations.

In 2015, the oil companies lobbyists' dream came true; the US oil export ban was removed. Sure enough, the oil export ban which ruled the land for over 40 years is gone. The energy lobby cannot take full credit for this change, but it paved the road to it. The question now is how much oil do we have and how smart is it to export this precious commodity?

Texas, Pennsylvania, Ohio, New York, and other states have seen many millions of dollars spent in recent years by drilling companies on lobbying and campaign contributions. There is a lot at stake as the state and federal administrations are reviewing the situation and seeking to develop hydrofracking regulations. This is a critical time, so the game is played at the highest levels with no time, money, or effort to spare.

The energy industry promises to create many jobs and increase spending in some of the most economically struggling parts of the country. This is good, but the environmentalists warn that the risks to water quality and damage to roads will affect negatively and permanently a great portion of the U.S. population. This twists the balance unfairly in favor of the powerful minority, while hydrofracking puts billions in their pockets.

So today the 'hydrofracking vs. no-hydrofracking' game is driven by substantial amount of money in an effort to shift public policy towards pro-fracking as beneficial and profitable undertaking. Money talks; the lobbyists know that and count on the tremendous pressure this puts on government and regulatory agencies.

The oil drilling and recently the hydrofracking

issue have created a cottage industry for paid lobbyists, which until recently had little presence in the government offices. For example, Chesapeake Energy, a major gas producer, spent more than $1.6 million on lobbying over the past three years, employing 11 lobbyists and three private companies. It spent less than $40,000 in the previous three years—or 40 times increase in lobbying activities. This, in addition to increase in paid TV advertisements and roadside billboards around the state.

The crude oil and natural gas activities of late have also created a large cottage industry for legal and financial firms, which is much larger, so we will take a closer look at these developments in the following chapters.

Expensive national advertising campaigns are being waged over the issue by deep-pocketed industry groups, including the American Petroleum Institute. Companies, Exxon Mobil, and ConocoPhillips, who are paying for TV ads that promote the safety and the benefits of natural gas. These efforts are supported not only by the drilling companies but also by distributors and pipeline companies, who are making steady campaign contributions to elected state and federal officials. This represents a massive industry effort to influence public opinion on natural gas drilling and to persuade lawmakers in gas-rich states to support it.

The renewables are not much behind. One of the examples of the recent past is the way Solyndra's lobbyists were able to penetrate deep into the U.S. government all the way to the White House. As a result, they were able to secure over half a billion dollars of tax payers money in grants for non-existing technology. And, no; there were no consequences. No trials, no jail term for any of the perps, and the entire case went away as if nothing happened.

Similarly so, First Solar lobbyists were able to also access, befriend, and mislead high level government decision makers. Eventually, they were able to obtain last minute approval of several billion dollars in government loan guarantees. Over $3.0 billion to be exact, plus a billion or so from the U.S. Import-Export Bank, which future also seems questionable.

Such enormous amount of subsidies given to a single company—any company—must ring the alarm bells. Not in this case; First Solar continues to finance huge projects with taxpayers money, although its technology is less efficient than the competition. And despite the fact that its solar panels are not proven reliable for long term operation in the U.S. deserts. Not to mention the often ignored fact that its CdTe PV panels contain dangerous, hazardous, and cancerous materials. It remains to be seen if First Solar can stand on its feet without gov-ernment subsidies.

Nope, there is nothing we, the taxpayers, could do in these cases, except keep on paying the bills. Money talks in our capitalist society…and the battle on the energy fronts continue with lobbyists pulling the strings behind the scenes.

Energy Market Divisions

Most companies, including energy producers, generators and providers, fall within the following categories:
- Leaders
- Followers
- Challengers, and
- Niche hangers.

- It is obvious who the *leaders* in the energy markets are. On the production side these are the producing countries; Saudi Arabia and its OPEC cohorts. On the distribution side, the leaders are the large energy companies which control the refining and resale of energy sources; electricity, gasoline, diesel, jet fuel, and other petroleum products.

- The followers are manufacturing and service companies that work in parallel with, or provide services to, the major energy providers. Most equipment manufacturers, as well as other supply chain companies, have no other choice but to follow their customer base (the energy market leaders) in supplying whatever and whenever needed materials and services.

- The *challengers* come and go, but lately the renewables are shaping as the major challengers to the existing energy sector hierarchy. Solar and wind power technologies are coming in face-to-face battles with their coal and gas power leaders in the field. Although the fight is unfairly biased in favor of the status-quo that supports the fossils, the future is pointing towards a major shift.

It is inescapable for the renewables to become the primary energy sources in the distant future.

For now, however, the renewables are faced with long and difficult battles in order to stay alive and well. They will continue to challenge the leaders, and win in some cases, but

- The *niche hangers* are mostly smaller companies, or divisions of large companies that are providing unconventional products and services. Here we see a number of solar companies developing new and untested products, such as specialized thin film and nanoparticles based solar products. These are geared towards capturing existing small niche markets, or developing new such for their products.

Presently a new phenomena is growing in the U.S. oil and gas field. The large companies continue to dominate the market, while many small companies—leaders wanna-bes—also fall under the niche (or cliff) hangers category of oil and gas drilling and producers. Many are stumbling and even failing due to recent events in the crude oil market. Most of them started several years back, showing a great potential in delivering large quantities of oil and gas from rich natural reservoirs. How can you lose with so much black gold in the ground. And sure enough, many of them were able to borrow a lot of money, started operations full of hope, and did well for awhile.

Just like the good old days of the Gold Rush, however, this black gold lost 50% of its value lately, due to price manipulations by Saudi Arabia. At such low prices, drilling and fracking all of a sudden became unprofitable, so what can the small guys do? They went the same way most unprofitable businesses go; they reduced production, laid off people, and went into a survival mode, waiting for the bad times to pass. The lucky ones were able to endure the low oil price storm. Some of the smaller guys were not fortunate, and burdened by debt, which the falling oil prices made even worse, were forced out of business.

Large vs. Small

The differences among the energy companies are great and usually determine their paths. It is quite easy to make marketing plans and talk about competition, customer service, etc. fine things in the energy markets, when you are an official of a large energy company. It is quite the opposite for smaller energy products or services companies, struggling to stay in business. A multi-billion dollar company can undertake a large project and usually make money on it, but it wouldn't be a big deal if it loses money on a project or two. A small company, on the other hand, would go bankrupt if its one major project fails.

There is, therefore, a huge discrepancy between the marketing strategies of large vs. small companies operating in different areas of the energy sector. Under the capitalist democracy, the rich have much more latitude and can survive even large disasters. Just take a look at

Figure 3-10. FERC vs. the banks

BP: after polluting half of the Gulf of Mexico, they are now more famous and more profitable than ever. How many small companies could survive even a minute replica of the 2010 Deep Horizon oil spill and live to tell about it? This is because BP is one of few international conglomerates with undisputed monopoly in the crude oil business. How could a small company compete? The best it could hope for is getting a small piece of the action serving the giant mother BP, or tiptoeing around it.

Similarly so, First Solar—the largest U.S. solar company—is trying to dominate the U.S. and global solar markets. This is another large (financially speaking) company, which can afford losing money on a project or two. And after receiving several billion (yes, with capital B) in loan guarantees from the U.S. government recently, they are still doing very well, thank you, and advancing their mission. Things are much easier with that much cash and credit, while at the same time hundreds of their competitors went bankrupt during the 2011-2013 solar energy bust. Capitalist reality…

Another large U.S. company, Solar City, has been quite successful in cornering the solar roof installation market in the country. With millions of dollars in investments and loan guarantees, Solar City can implement new marketing strategies, including innovative payment options, financing and such, that the small business competition cannot even dream of. In the process of "getting the low hanging fruit first," Solar City at times forgets the rules of proper and efficient customer service. The race is on, and they have decided to win… takes whatever it takes…running over the competition not excluded.

What will happen on the long run with these companies is anyone's guess, but we need to be aware that an aggressive drive to monopolize a market niche, although legal, is unethical, and not always successful enterprise in today's global business environment. Such predatory behavior must be discouraged at all cost and in all cases.

Fairness in the marketplace is a nicety that people like to talk about in meetings and conferences, but we all know that this is not what capitalism is all about. Dog eat dog, is what capitalism calls for, so unless there are clearly established, and thoroughly enforced rules, the big dogs will eat, or chase out, the small ones. Always! It never fails.

The U.S. government is the only entity that could provide some balance—a level of fairness—in the energy markets. It has the power, but is not using it properly thus far, so fairness in the energy markets is still a dream. The large oil and gas companies make sure that this remains so.

The Energy Sector Function

Slowly, but surely, the large oil companies—the leaders of the energy markets—have been changing. The major shift has been in going from pure production of petroleum compounds to adding a long list of new products and services. Most of the large companies can do this with their own investment capital. Also, many of them are vertically integrated, which allows them to handle all stages of the process on their own; from initial exploration to final marketing. This gives them enough latitude to diversify and add or remove any type of new product or service in their portfolios.

Since all the steps of the oil process—exploration, production, transportation, refining, and marketing—are independent activities, crude oil has a defined (but variable) price at each stage. This is complicated by the fact that the major oil companies also use third parties to fill the gaps. Thus, today, Shell gasoline might come from a BP refinery, and vice versa. Quite confusing scenario, but nobody seems bothered by it.

And to top the confusion up, the major U.S. oil companies import major part of the crude oil they process and sell as different products. This way, a major part of the profits go the oil producing nations, usually through their government-controlled oil companies. As world's oil production varies in quantity and price, governments and oil companies alike are equally affected. This is creating some interesting scenarios around the globe.

And so the price of the final products—gasoline at the pump, for example—is a complex mixture of different charges, not a small part of which are the different taxes imposed (by the governments) on each gallon we

put in the gas tank. In order to decipher all this, we venture into a broad analysis of the energy sector and the related energy markets.

The energy markets today can be subdivided into major categories by energy source as follow:

Table 3-1. The major energy markets sectors

FOSSILS	RENEWABLES	OTHER
Coal	Solar power	Nuclear power
Natural gas	Wind power	Hydropower
Crude oil		Fire wood
		Ocean power
		Geo-power
		Bioenergy

Table 3-1 shows the major sectors of the global energy markets, as reviewed in this text. There are many other ways to classify and categorize these, but this three-prong division seems to be the simplest and clearest, at least for the purposes of this text.

Below is a complete list of the products and services provided by, or to, the different branches of the global energy sector. This is a very long list of complex and expensive products and services. We will not be able to cover all items in this text in detail, so instead, we will focus on the key elements and aspects of the energy sector and the global energy markets.

In more detail, the energy sector—broken down by energy sources—consists of:

Fossil Energy Sources
Coal
- Exploration and location services
- Engineering design services
- Mine equipment manufacturing
- Mine construction
- Mine operation and maintenance
- Transport services

Natural Gas and Crude Oil
- Exploration and location services
- Engineering design services
- Well equipment manufacturing
- Well construction
- Well operation and maintenance
- Transport services

Renewable Energy Sources
Solar and Wind Power
- Production equipment manufacturing
- Manufacturing of solar and wind power plant equipment

- Power plant exploration and location
- Power plant engineering design
- Power plant construction
- Power plant operation and maintenance
- Power transmission and distribution

The Other Energy Sources
Wood
- Global resources
- Production
- Residential markets
- Commercial markets

Nuclear Power
- Raw materials markets
- Uranium
- Transport services

Hydropower
- Exploration and design services
- Hydropower dam construction
- Hydropower power plant construction

Ocean- and Geo-power
- Production equipment manufacturing
- Manufacturing of ocean and geo power plant equipment
- Power plant exploration and location
- Power plant engineering design
- Power plant construction
- Power plant operation and maintenance

Biofuels
- Production equipment manufacturing
- Processing plant exploration and location
- Processing plant engineering design
- Processing plant construction
- Agricultural field operations
- Processing plant operation and maintenance
- Fuels transport and distribution

Energy Services
Electric Power Generation
- Exploration and siting services
- Engineering design services
- Power plant equipment manufacturing
- Power plant construction
- Transport services
- Power plant operation and maintenance
- Energy storage

Transportation Sector
- Vehicles and equipment manufacturing
- Rail, road, and water transport
- Fuel transport and storage
- Mass transport
- Personal transport

Politics, Regulation, Legality, and Finance
- Political direction
- Regulations
- Legal aspects
- Financial aspects

Government/Military
- Contracting services
- Consulting services
- Technical services
- Equipment manufacturing
- Civilian transport services

Peripheral Products and Services
- Commodity trading
- Physical and cyber security
- Political and regulatory services
- Waste storage and disposal
- Risk, accident, and safety management
- Environmental services
- Legal and financial services
- Power transmission and distribution
- Pipe lines
- Power lines
- Land, water, and air transport
- Energy storage

Transportation fuels
- Fuels processing
- Fuels delivery
- Fuels use

Specialty Energy Materials
- Silicon production; mining and processing
- Uranium mining and processing
- Rare metals mining and processing
  — Silver
  — Strontium
  — Cadmium

This is a long list, a close look at which would require several books. Because of that we cover only the most important characteristics and issues of the energy markets in this book. We, however, do take closer look at the specific energy products and services in our other books on the matter; *Photovoltaics for Commercial and Utilities Power Generation, Solar Technologies for the 21st Century, Power Generation and the Environment, and Energy Security for the 21st Century*, published by The Fairmont Press.

The Energy Markets

The energy sector's function depends on developments in its different branches and interdependencies. At the same time, it also depends heavily on the peripheral products and services of these enterprises for its proper, efficient, and cost-effective function. A quite complex picture, which is actually only the top of the iceberg, since there are many other—less obvious and not as essential—products and services, which complete the entirety of the energy markets. We will take a close look at most of the major, and some of the lesser, elements of the energy markets in this text.

We must first clarify the confusion surrounding the energy markets by dividing them into two major categories:

1. Physical energy market, and
2. Financial energy market

- The physical energy markets, where actual energy and energy products—in solid, liquid, or gaseous form—are sold and bought, depend on the following key factors:
- Geographic location of energy sources: regions, nodes, zones or hubs;
- Timeframes of power generation and use: hourly, daily, monthly, quarterly or yearly,
- Types of energy products: power generation, motor fuels, pipelines, storage, etc., and
- The energy sales, namely:
- Wholesale, which involve large transactions of energy sources and fuels.
- Retail sales, which involve smaller size sales geared for different customers.
- *The financial energy markets* are quite different in that only currencies change hands on the world financial markets—in form of investments, stock and futures trading, hedging, and other financial instruments—and where no physical transfer of energy products takes place.

Looking at Figure 3-11, one can easily see the importance of energy—electric power, heat, vehicle fuels, etc.—for the proper function and further development of the economy. All types of energy are absolutely need-ed every second and in every corner of our lives. Not having enough cheap energy is not an option, and the energy markets are responsible for delivering it without delay and at affordable prices.

In more detail:

Physical Energy Markets

The overall function of the energy markets is determined by a number of factors, some of which are:

- *The geographic location,* which determines the availability of energy sources in the region, which in turn creates either an energy-rich, or energy-poor area, or country.
- *The level of availability* of natural energy resources determines the economic development of the affected area. Countries with a lot of energy reserves are usually rich countries with booming economies. Those with limited energy resources struggle to keep up with the demand and the economic growth by importing energy sources. This usually restricts their economic development.

There are, of course, exceptions to this rule, with Japan being the best example of an energy-poor country with booming economy. This, however, is accompanied by a number of internal struggles and abnormalities—some of which are quite serious—as we will see later on in this text.

- *The timeframe of power generation and use* determines the amount of energy used at a specific time. There are times of the day and the seasons when energy is not needed, while at other times it is indispensible and even a life saver. A good example of this is Arizona, where during the very hot summer months the electricity demand peaks to the maximum of the generating capacities. During the rest of the year, most of the generators are idle, since the electric demand is very low. Another good example is the difference of power use during the day hours, when people in homes and businesses use electric appliances and equipment, vs. night time, when all activities are suspended and power demand nears zero.
- *The retail energy markets* are limited to the purchase and sale of electricity, heating oil, and vehicle fuels; all of which are secondary energy sources, since they usually are a result from wholesale markets activities and processing of the primary energy sources (coal, natural gas, crude oil, and renewable energies).

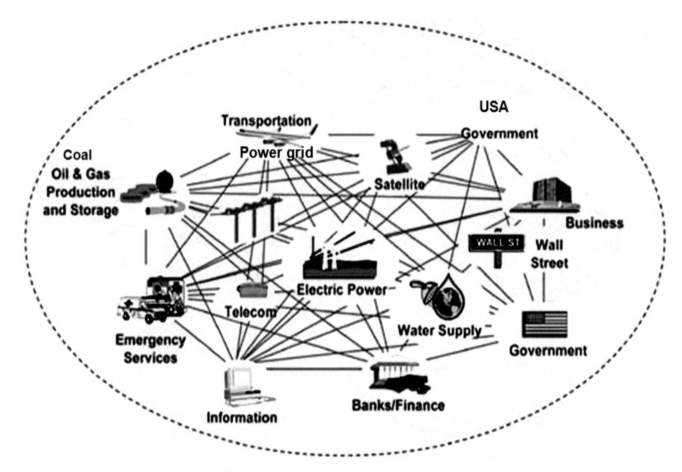

Figure 3-11. Key economic and energy sectors and their interdependencies

The physical energy markets exhibit some unique characteristics, and are driven by factors, such as:

- *Necessity* is a major driver of the energy markets. Unlike most other products, natural gas, electric power, gasoline, diesel, and heating oil fuels are bare necessities today. Lack of heating oil, for example, could mean that millions of customers in NE USA and Europe would be left without heat, the ability to cook food, and/or run their businesses. Lack of electricity during a hot summer day in Arizona could put many people in the hospital or the morgue.

- *Blackouts and other service disruptions* create operating problems and hazards. Consumers cannot stop or postpone the purchase of electricity or natural gas. They may be able to turn up or down their thermostats to reduce the energy use, but cannot eliminate consumption altogether for an extended period of time. Electricity and fossil fuels are basic necessities. They cannot be replaced easily, or at all, under the present energy structure.

- *Variable (seasonal) demand in some regions* deter-

mines the demand response. In the Arizona desert, the summer heat tests the limits of the power generating and transmission networks. During the other times of the year, these same facilities are idling most of the time, with some rare exceptions in the winter months.

- *Few options exist for storing electricity, where* for natural gas, many vendors and consumers are able to store gas for later use. Consumers may install batteries for electricity storage, but it cannot drive refrigeration and air conditioning units. Without energy storage, consumers cannot buy when prices are low and use their stored product when prices rise, nor can they use their powerful air conditioners when needed. This limits consumers' choices and response to changes in prices.

- *Few substitutes for natural gas, electricity, and vehicle fuels* exist, which means that if natural gas or gasoline prices go up, consumers cannot quickly switch to a different product. Longer term, they may be able to switch to gas from electricity for heating, for example. Or, they may choose to insulate, or install new, windows or take other steps to reduce their consumption of energy. Substituting gasoline, however, is virtually impossible for 99% of the people in this world.

- *Demand-response programs* offered by some utilities can provide benefits to those who can control their energy needs for at certain times of cheaper energy rates. This might include some inconveniences like turning off air conditioning and refrigeration during the hottest part of a day in order to help reduce electric load during periods of higher energy rates.
- *The wholesale energy markets* in the U.S. trade all energy sources (primary and secondary) competitively. Here, the energy markets have been established through administrative processes based on the cost of providing service. In competitive markets, prices are largely driven by the economic concepts of meeting basic needs, and supply and demand.
- Physical fundamentals are underlying the supply and demand for energy sources and electricity. These are, in fact, the physical realities of how markets produce and deliver energy to consumers, and how the prices are calculated.
- *Wholesale energy markets* differ from other competitive markets in critical ways, but the demand in most cases is ultimately determined at the retail (consumer) level.
- *Retail electric power and fuels prices and use* fluctuate slightly in the short-term, with some more pronounced long term variations, most of which are predictable in time and size. Meeting the customers needs and demands is a primary driver of the retail energy markets. This in turn usually determines the activities at the wholesale energy market level.

Retail Electricity Markets

The retail electric energy market is part of the physical energy markets, but it is so large and important that it deserves a closer look. The electricity markets are established through administrative processes based on the need and cost of providing service. In competitive markets, prices are largely driven by the economic concepts of supply and demand. Underlying the supply and demand for energy sources and electricity are physical fundamentals. These are the physical realities of how markets produce and deliver energy to consumers and how they form prices.

The wholesale energy markets differ from other competitive (consumables trading) markets somewhat, but not fundamentally. We cover these differences in more detail in this text, so we will not waste time with them here.

While the function of the wholesale energy markets is very important, the demand is ultimately and primarily determined at the retail level. Retail use is relatively inelastic in the short-term, although this may be less so with some larger customers, and changes with locations and seasons. The excess demand at certain points of peak energy use determines some of the risks associated with these markets.

The retail use of energy sources and electricity exhibits some unique characteristics, the most important of which are:

- *Necessity* is a major driver in the energy markets. We need gasoline and electricity now and any time when we have to use them. Lack, delay, or shortage of these products is not an option in our way of life. Thus, we are willing (usually unwillingly) to pay whatever is required to obtain them.
- *Limited substitutes* are another characteristic of the retail energy markets, since consumers cannot switch easily to a different product if and when they need or want.
- *Limited customer energy storage options* are another limitation, since retail consumers have few (or usually none) options for storing and using the energy they need. They just get whatever is offered to them by the local gas pump, or utility company; whenever the seller can or wants to. Period.
- There is a growing problem today, due to slowly increasing awareness and discontent of the retail customers with their limited participation in the energy markets and choice of products and services. The growing discontent has contributed to a number of positive changes in the industry. The regulators, utilities, and energy companies are becoming fully aware of that problem and are responding by offering a variety of products and services.

We cover these developments in more detail in this text, so it suffices to say here that the customers are looking for ways to control the way energy is produced, delivered, and priced. Solar and wind power, different energy storage options, different fuels, etc. give the energy customers choices that were not available until recently. These choices are changing the way customers and energy providers look at the energy markets and each other. This is changing the way the energy markets operate and we foresee more and greater changes in the future.

Financial Energy Markets

The *financial energy markets* are where companies and individuals go if they need to raise, invest, make, or lose money. These activities are a resemblance of a gambling game, but unlike Las Vegas casinos, the financial energy markets are under strict government and energy industry control. They play a significant role in the day-to-day operations of the energy markets, and the entire economy, in a number of ways:

- The Financial energy markets provide access to capital needed for energy related operations, and
- Energy products trade in large ($$$) numbers in the commodity markets and, as derivative products, in the financial markets.

Natural gas, crude oil, and electricity are traded (on the financial markets) like commodities, such as metals, corn, wheat, or corn oil. The commodities themselves may not be present or visible, as in a physical trade, but one could them as a financial assets, and could also measure them in financial terms (at least on paper).

Commodity markets began as ways for farmers to sell their products, or a portion of their production, before it was harvested. The object of this activity was/is to provide them with capital to continue operations without major losses. Commodity markets evolved to provide tools for farmers (and other commodity producers) to manage their risk, notably the risk of adverse changes in price due to bad weather and other disasters. These financial products were derived from the physical natural gas and electric products, and are known as derivatives.

Since their inception, trading in physical commodities and derivatives has attracted another group, not directly involved in the physical markets; the speculators. These are people and companies hoping to make some profit from changes in energy commodity prices. The market for natural gas derivatives, for example, has grown enormously within the past decade, as competitive natural gas and electric markets matured and investors came to see energy commodities as investments, not just a source of power.

The problem is that there are some dark and ugly sides of the financial energy markets. This is because there are always those who attempt to abuse and manipulate the system for their own benefit. These people use methods and practices that undermine the market's ability to operate efficiently, reduce other market participants' confidence in the markets, and distort market outcomes; including prices.

A good example of that was seen during the 2008 economic crises, when investors and speculators took advantage of the newly created chaos and caused an avalanche of events to raise the price of crude oil, albeit temporarily, to unheard of levels. A similar method, and another example of the power of the financial markets, is the formation known as a "contango," which was used during the 2014-2015 oil price drop. Brent crude prices for delivery in March 2015 were $10 a barrel cheaper than those for March 2016. This made it attractive to buy oil in 2015 and put it into storage for sale later. Large scale use of this strategy has the potential of increasing the oil sales, thus lifting the prices further and keeping them high longer.

This example comes as a proof that the financial markets are significant driver, which is not to be ignored, when considering our energy future.

We take a closer look at these speculative and lucrative (or not) markets later on in this text.

U.S. ENERGY

Energy is the driver of the U.S. economy. It is also largely responsible for our comfortable and opulent life style. Living in huge houses, driving large SUVs, flying all over the world, and other luxuries would not be possible if we didn't have abundant and cheap energy. How long this is going to last is the big question, which we are just now beginning to acknowledge and starting to discuss. It will take a long time before we find all the answers, and even longer before we implement the corrective measures, but it is not late…yet.

Background

The U.S. learned a bitter lesson in 1973, when OPEC cut down the oil supplies to the West. It took awhile for the good, 'ol, mighty USA to get off of its knees and get back on its (oil) feet. And just then, as we were dusting off our bruised energy knees, a new disaster hit the world energy markets. The 1979 Iranian Revolution started the second oil crisis in the United States. It all began with massive protests in Iran, which forced the Shah of Iran, Mohammad Reza Pahlavi, to flee his country in early 1979. The Ayatollah Khomeini assumed power as the new authoritarian leader of Iran.

The internal turmoil disrupted the Iranian oil sector, production was greatly reduced and finally all exports were suspended. Oil exports resumed under the new regime soon thereafter, but they were inconsistent and at a much lower volume. The new reality contributed to drastically increased world oil prices. The Iran-Iraq War later on in the 1980s reduced world oil exports once again when Iran nearly stopped oil production, and Iraq's oil production was severely cut as a result of the war.

The U.S. government was anticipating another 1973 gas shortage, and printed coupons which were to be used in an attempt to reduce the gas line problems. The coupons were never used, because the crisis did not last long enough to affect the motoring public significantly.

Nevertheless, this shows that the only weapon the U.S. government had at the time was reduction of gasoline available to the public. Is resorting to gasoline coupons possible today? Next year? Not likely, but we must be aware that, although we fully depend on, we do not control the global oil supplies.

Figure 3-12. 1980 gasoline coupons

At the time, OPEC nations compensated the reduction in exports from Iran by increasing production, so the overall drop in worldwide production was only about 4 per cent. Our Arab friends to the rescue… Nevertheless, the world was in panic resulting from the political chaos in Iran culminating with the taking of American hostages. All this contributed to oil prices rising beyond what would be expected under normal circumstances.

In response, President Carter began a phased deregulation of oil prices in the U.S., which allowed the U.S. oil output to rise sharply from a number of oil fields. Those efforts did not prevent long lines from appearing once again, and several states even implemented odd-even gas rationing, as in 1973, but the issue never reached the levels it did in '73.

The average 1970s vehicle—mostly V-8 large sedans—consumed over a gallon of gas per hour during idle.

With millions of gas-guzzlers idling for hours in the long lines at the gas pumps, during several months, American cars wasted over 200,000 barrels of oil per day just idling.

What choice did we have? Americans at the time believed the rumors that oil companies were responsible for artificially created oil shortages, intended to drive up prices. Although the amount of oil sold in the U.S. in 1979 was only 3.5% less than the previous years, 54% of Americans believed the energy shortages were created by the oil companies. Only 37% of Americans thought the energy shortages were real, and 9% had no opinion.

President Carter outlined his plans to reduce oil imports and improve energy efficiency in his "Crisis of Confidence" speech (also known as the "malaise" speech). Carter encouraged citizens to do what they could to reduce their use of energy. He had already installed solar hot water panels on the roof of the White House and a wood-burning stove in the building. A small gesture of a great man, in an attempt to lead the world by example. Unfortunately, the world did not chose to follow then is still hesitating.

Sure enough, the solar panels were removed in 1986 by another great man, Ronald Reagan. The excuse for the removal was reportedly for routine roof maintenance. The panels were never reinstalled. The incident remains as a symbol of the teeter-totter of government's energy policies (fossils vs. solar) that still persist today.

In January 1980, Carter issued the Carter Doctrine, which declared that any interference with U.S. oil interests in the Persian Gulf would be considered an attack on the vital interests of the United States. Carter also proposed removing price controls (imposed during the administration of Richard Nixon before the 1973 crisis), as part of the deregulation effort. He suggested removing price controls in phases, an effort finally dismantled in 1981 by President Reagan.

Carter also suggested imposing a windfall profit tax on oil companies. The internally regulated price of domestic oil was kept at $6 a barrel, while the world market price at the time was $30 a barrel. The oil companies were not happy, and this unsustainable situation did not last very long.

President Carter called the 1979 oil crisis "the moral equivalent of war," as a justification of his proposals and mandates. The critics argue that his proposals did not serve any practical purpose on the international front and only made the bad situation worse.

In 1980, the U.S. government established the Synthetic Fuels Corporation to produce an alternative to imported fossil fuels. A number of other similar measures were introduced and some implemented. Millions of dollars were also poured into solar, wind and other alternative energy technologies. Many new companies started work in these areas, and some showed amazing progress within a short time. Examples of the solar technologies of those days are still operating in the U.S. deserts.

The solar euphoria, however, did not last long. As soon as oil prices went down, all work on these technologies was abandoned. A trend we have seen repeated numerous times throughout the last several decades. Detroit was also affected by the sudden rise in fuel prices, with ridiculously looking small cars imported from Japan, Italy, and Germany starting to dominate the domestic car market.

The imported cars used novelty fuel-saving devices such as fuel injection and multi-valve engines replacing the conventional carburetor used in most American cars at the time. American automakers learned from the competition, and made some changes, which contributed to overall increase in fuel economy. This also was one of the factors leading to the subsequent 1980s oil glut.

Thirty-three years after President Carter warned against "outside" control of the oil-rich Persian Gulf, we see that the U.S. effectively enforced a corollary to that doctrine. This prevented control by a regional power during the Iraq wars of 1991 and 2003. Today, however, the Carter Doctrine must be revisited in light of Iran's nuclear ambitions and threats in the Gulf. This is perhaps the greatest immediate threat to U.S. energy and national security and a menace to U.S. economic interests, the least result of which could be long-term spike in oil prices.

The Carter Doctrine is as relevant today as it was thirty years ago, especially as the Persian Gulf's oil production capacity and transport activities are concerned. The global demand for crude oil is roughly the same today, about 30%, as it was in 1980. In case of a hostile takeover of, or serious interference with the Gulf activities, the U.S. and China would be immediately and seriously affected, and crude oil prices would rise accordingly.

A major threat to the global oil market is Iran's bazaar behavior and its nuclear weapons ambition. The threat is taken seriously and so, U.S.' Fifth Fleet and other military assets in the region are busy keeping the free flow of oil through the Persian Gulf day after day. A nuclear-armed Iran, however, would change the balance. It would gain de facto immunity from a conventional attack from another country, significantly limiting the

effectiveness of U.S. forces influence in the region. What would happen then? Hard to tell, but nothing good could be expected for the Gulf crude oil supply, and Iran's neighbors safety.

Figure 3-13. U.S. Navy warships patrol in the Gulf.

The overwhelming conclusion here is that price fluctuations, and subsequent anomalies, usually started with some type of crude oil manipulation; be it by foreign governments, by companies, by public frenzy, or by stock trading abuse. Often, one anomaly feeds the other, so the problem grows bigger with or without a root cause.

We need to disentangle and control these major and interwoven external and internal political, social and economic drivers to avoid similar energy and economic disasters in the future. The key to achieving this is optimizing our energy security and achieving total energy independence, which starts with understanding the issues at hand.

The result of the 1970s crisis was that oil-rich countries in the Middle East benefitted tremendously from increased prices, while slowing down oil production in other areas of the world. Some countries, like Norway, Mexico, and Venezuela, benefitted as well by becoming world class exporters of this crucial commodity.

In the U.S., Texas and Alaska, as well as some other oil-producing areas, experienced major economic booms too, due to high demand and soaring oil prices. The above mentioned foreign nations and U.S. states thrived while most of the rest of the nation and most of the world struggled with oil shortages and stagnant economies. The economic boon of the oil producing nations and U.S. states, however, was short lived and came to a abrupt halt as soon as prices stabilized and dropped in the 1980s.

The 1970s oil deficit and other oil-related problems were all but forgotten in the 1980s, when suddenly we saw ourselves floating in oil. Reduced oil demand and overproduction resulted in a glut on the world market. Oil consumption in major countries was down about 15% from 1978 to 1982, due in part the large increases in oil prices by OPEC and other oil exporters. This caused a six-year-long decline in oil prices, culminating with a 46% price drop in 1986.

Several years of peace were enough to rebuild the American public's confidence in continuous oil supplies and reasonable prices...until another war put our energy security and oil supplies in jeopardy. In 1990, Iraq invaded Kuwait, which started a 7-month armed conflict ending with occupation of Kuwait by U.S. military. Iraq officially claimed that Kuwait was stealing oil, but its true motives seem to have been much more complicated. Iraq owed Kuwait $14 billion in debt, loaned during the Iraq-Iran war. Iraq also felt that Kuwait was producing too much oil, thus lowering market prices, which hurt Iraqi oil profits.

The war caused interruption of oil production and exports in the area, which coupled with threats to Saudi Arabian oil fields, led to a rise in prices. The price went from $21 per barrel to $28 per barrel within a month after the invasion. By the end of the invasion, oil prices rose to nearly double that, and went as high as $55 per barrel on the world energy markets. Soon after, however, prices dropped significantly, albeit temporarily. The war and its energy consequences did not last long. After less than a year, the world oil supplies were back to normal, but this time oil prices remained high relative to pre-war levels.

There were no long lines at the gas pumps in the U.S. this time either, but oil prices kept rising steadily, as a result of persisting political turmoil and economic problems worldwide. Until recently, when the U.S. hit an unprecedented oil and gas bonanza. This new situation was threatening the Saudi Arabia's reign over the global energy markets, so the response was swift and significant. Oil production was reduced or kept steady in the summer of 2014, thus controlling the export volume in order to reduce, and keep low, the global oil prices.

The artificially low oil prices had a serious effect on the global energy markets. The gasoline prices at the pump dropped, which was good for the consumers, but at the same time many U.S. oil drillers were forced out of business. At global prices of $40-50 per barrel, oil drilling and pumping was not profitable.

The global oil prices tumbled from over $100/bbl to as low as $40/bbl. Over 50% price drop is a serious market anomaly, which usually has serious consequences. This time, however, the U.S. consumer and the economy were not hurt in any significant way. Except for reduction in fracking activities, the consequences for the U.S. economy were actually positive.

Dow Jones kept on rising, the gasoline prices kept falling, and Americans thought that they are getting a new chance of achieving the American Dream. Not so fast...Saudi Arabia won't allow that. All the sheikhs have to do is turn the oil spigot off again, and we will plunge back into the depths of the $100/bbl energy markets where we came from.

At the same time, however, most oil-producing countries were badly hurt. Among them, Russia suffered the greatest consequences of relying heavily on oil exports. At global oil price at $50/bbl, the Russian oil did not bring any profits, which exposed the entire economy to another recession

Russia has not even recovered from the 2008 economic disaster, which brought misery to most people, and now the country is faced with another economic cliffhanger. The Russian economy—half of which depends on oil and gas exports—is expected to contract by nearly 1% annually during the next several years. This budget is based on estimates of crude oil trading at $80/bbl. As a break-even point. If oil prices stay lower, then the Russian economy might enter into a state of a free fall, starting with economy falling 4-5% annually. How high and fast could it fall depends on the oil prices. Not an envious state of affairs for the mighty Russian bear.

Other countries, like Venezuela, Nigeria, Iran, and others oil-exporting countries are in the same situation as Russia. Low oil prices will hinder their economic development, and if the crisis continues too long, might bring them to their knees.

U.S. Energy Policy

The U.S. has a patchwork of energy policies and laws, which have never been cohesive and all-encompassing enough to cover the entire energy market and provide a solid direction for its future development. Recently, some more intensive effort is noticeable, but even then, it does not go far enough to solve all energy and environment related issues.

The U.S. Energy Policy Act, 42 USC §13201 et seq.

(2005) replaced the 1992 Energy Policy Act by addressing a number of new situations and issues in the nation's energy production. This includes; energy efficiency; renewable energy (solar and wind in particular); coal, oil and gas, nuclear power issues and security; motor vehicle fuels (ethanol production in particular); hydrogen; hydro-, geo-, and ocean-power energy; electricity; energy tax incentives; Tribal energy use; and climate change issues and technologies.

The Act provides loan guarantees for entities that develop or use innovative technologies that prove successful in avoiding GHG emissions. It also increases the amount of biofuel additive that must be mixed with gasoline sold in the U.S. It also seeks to increase coal as an energy source, while reducing air pollution at the same time, via $200 million annually for clean coal initiatives.

The Act also authorized subsidies for solar, wind, and other alternative energy producers and added ocean energy sources, including wave and tidal power for the first time as separately identified, renewable technologies. $50 million annually were allocated for biomass development grants for the life of the law, and includes provisions aimed at making geothermal energy more competitive with fossil fuels in generating electricity.

In addition, it requires all public electric utilities to offer net metering on request to their customers, and sets federal reliability standards regulating the electrical grid, in response to the 2003 North America blackout.

Additional incentives were provided to companies drilling for oil in the Gulf of Mexico, while exempting oil and gas producers from certain requirements of the Safe Drinking Water Act, but bans drilling for gas or oil in or underneath the Great Lakes.

On the energy efficiency front, the Act extended the daylight saving time by four to five weeks, depending on the year, and requires that all federal vehicles, capable of operating on alternative fuels be operated on these fuels exclusively. Special tax breaks were also allowed for people and businesses making energy conservation improvements to their homes or offices.

The Act authorized the following tax reductions:

$4.3 billion for nuclear power
$2.8 billion for fossil fuel production
$2.7 billion to extend the renewable electricity production credit
$1.6 billion in tax incentives for investments in "clean coal" facilities
$1.3 billion for energy conservation and efficiency
$1.3 billion for alternative motor vehicles and fuels (bioethanol, biomethane, liquified natural gas, propane)

$500 million Clean Renewable Energy Bonds (CREBS) for government agencies for renewable energy projects.

Not bad for starters. The problem is that as we started on the right leg, we are still standing on it. A lot of work and words were put into it, but little progress has been made. The U.S. also went back and forth on any of the Act's points, and presently there is a lack of direction, and many gaps in the function of the energy sector are quite obvious.

The current U.S. energy policy is a mixed bag of disconnected policies designed for specific constituencies, but no clear national energy direction in sight.

The U.S. energy policy is in reactive mode.
We do what seems best for us today only.
No meaningful mid-, or long-term plans.

We have subsidies for fossil fuels, nuclear power, wind and solar, and for insulating and retrofitting buildings. Energy standards for energy using appliances and miles per gallon standards for automobiles were added recently too. These subsidies fill the immediate large gaps, but do not address the main question of the policy debate.

We have not decided on the ultimate goal, and the course the country is to take, and how and when we are to achieve it. Instead, we are reacting to present (urgent) national needs and global events.

Presently, the U.S. energy sector is going through changes, driven by:

- The cheapest price for energy possible, without considering the future of the fossils and the environmental consequences as part of the final cost.
- A transition to renewable energy with a tentative goal to replace the fossils (partially at best),
- Short-term transition from burning coal to burning natural gas,
- Long-term transition from fossils to renewables and nuclear power, with no set direction and vanishing support,
- Developing *all* sources of energy to make sure we have enough at reasonable prices. This is the so called "all-of-the-above energy strategy."

So, looking at this, our energy policy is a hodgepodge of policies. But why does energy policy matter at all? In reality, the energy market runs normally by itself, and everybody is going around happy, or at least worried about things that are not energy related. So why the fuss about energy policy?

There are only a few important issues related to energy policies, or lack thereof. The most important are:

- First, fossils provide over 80% of the world's energy, and they are in finite quantities. As the world's economy improves and developing countries increasingly seek better life, so does the amount of fossils they use. But it is simply illogical, and even stupid, to count on the fossils to supply us with more and more energy indefinitely.

- Secondly, but equally importantly, due to ever increasing use of fossils, climate change, induced by their use is proceeding faster than any models have predicted. This does not leave us much time to reduce our carbon emissions enough to avert the risk of catastrophic climate consequences. But this is just not happening.

In addressing these issues—*managing energy supplies and mitigating climate risks*—the U.S. has open the national and the global discussions, but has done little in terms of implementation. Given the gravity of the challenges we and the world face, serious government policies are needed.

The increasing risk is not only a probability, but a product of probability multiplied by severity.

The problem is that we have no reliable ways to assess these variables in a strictly numerical and believable way, due to the complexity of the energy and climate systems. We can clearly see, however, that—not if, but when—even the most pessimistic scenarios for both; the demise of energy supplies, and run away climate change become reality, we now and the future generations later on, are in deep trouble. It may become so deep, that there would be no way out of it for some of us, and for most of those behind us.

This is a risk management on steroids. Since we deal with forecasts to estimate the risk, we must remember that the accuracy of any forecast decays quickly with time. This is even more so when we deal with long-term forecasts. In this case, in addition to time, the wide range of possible outcomes become utmost important. A forecast that doesn't include a range of outcomes is practically worthless for any practical purposes, and even less for believable scenario planning purposes. Such forecasts, more often than none, are designed to deceive rather than inform.

We have no precedents, to compare many potential upcoming energy and climate disasters, but it does not matter how frequently something happens, if the potential failure (or damage) is too costly. Here Chernobyl and Fukushima come to mind. The fossils, nuclear, climate, and other uncertainties are a signal to spring into action. We must not wait to see how bad things will get before doing something drastic to change the course we have taken.

So, the general consensus is that the U.S.—as the world's greater producer and user of energy—must lead the entire world in solving the biggest problems of the day; fossils depletion and climate change. We do not see this reflected in the government energy policy thus far, so some of us will be rudely awakened some day soon.

There is a chance that a bi-partisan energy plan will be submitted in the near future (2016-2017), which would address these issues, but we cannot count on any significant results coming from Washington during this administration. So it might be 2020 before we see any major policies enacted in the U.S. energy sector.

Energy Policy and Strategy

The link between policy and strategy in the energy markets is a critical aspect of developing new methods for managing the energy sources, the electrical grid, and our energy security in general. Through various directives, government policy has the power to regulate the types of energy sources utilized in the market as well as the different characteristics produced by those sources. Policies—if and when following the proper strategy—can also influence consumer preferences based on different directives and initiatives.

Efficient long term energy strategy, supported by effective government policies, are badly needed to bring the U.S. into the 21st century.

Government policy and other political factors significantly affect energy regulation and energy-related industries.

Policies can be negatively or positively enforced to help prevent or promote different results. For example, taxes and regulations could be used to reduce pollution and encourage the development of new technologies by shifting behaviors. Money talks!

The different policies and directives can be inspired by various events and influences, and are usually changed with change in the overall strategic course. No change means inflexibility, while too many changes could bring volatility, so:

*A **proper and steady strategic course** must be established first, in order to achieve the goals.*

The political landscape certainly influences the types of energy used within a country as well as any advancements to the grid through economic incentives. These advancements, due to economic incentivizing, are called 'induced innovation.' These play a substantial role in reducing costs for energy and increasing research into different sources of energy production, delivery, and use.

In general terms, government policy has serious consequences on pricing and production, since changes in the capacity and generation of energy are driven by price. These changes are ultimately shaped by government policy and initiatives that greatly influence not only electricity producers, but also affect the consumer base.

The impacts of significant government policy and directives such as the Renewable Portfolio Standards and the NetZero program are significant and drive social and corporate behaviors, change business practices, and bring new energy initiatives for the national energy grid. The energy sector can also help to influence government policy, by creating an interactive rather than a reactive relationship among the entities and the consumers.

Considering the looming energy crisis, depleting natural resources, pending threats of natural disasters, and terrorist attacks, the focus should be on altering the energy production and power grid to mitigate these concerns. There are several alternatives to developing answers to these pressing questions with the objective of building a stronger, smarter, and cleaner, more energy-efficient electric grid.

The basis of pricing as a means to influence consumer preferences and energy source investment is one of the factors that play a significant role in effective government policy. By influencing behavior in the social and business sectors it helps to reinforce government policies focused on advancing the energy supplies and the national electric grid.

The economic inefficiencies in the energy market may put pressure on politicians to alter their ideology that significantly influences the strategic path and the overall energy policies. Pricing policy and strategy can also provide incentives for producers to invest in research and development of new technology, which in turn alters buying practices for consumers and adds choices to meet their preferences.

The relationship between government policies, initiatives, directives, and the energy markets is a critical factor in developing a methodology to optimize efficiency and mitigate many of the concerns facing the U.S. With pending energy crisis and other threats to the power grid, energy is and will continue to be an important government policy issue.

Understanding the relationships between energy strategy and policy, the energy sources, and energy outputs and use is imperative to building economic solutions on national and global levels. Proper and timely implementation of these solutions should be a priority. It should not be hindered by third-party (personal and/or corporate) interests, which might twist the decision makers' thinking and influence their actions.

Unfortunately, new Solyndra-type managers and lobbyists are born every day. These people have a serious, but unpredictable, effect on the way the energy markets operate, and how the energy policies are made and implemented. This is the capitalist way, so we have to be aware of the good and bad around us. We, the consumers, like it or not, are heavily involved in the system, and must try to change it, and the way we are affected by it.

The U.S. Energy Market
The U.S., and most other economies, have four main energy market sectors:

- *Primary sector*, which involves the extraction and production of raw materials, such as coal, crude oil, wood, etc. Here, a coal miner and a fisherman are both workers in the primary sector.

- *Secondary sector*, which involves the transformation of raw or intermediate materials into goods, such as processing crude oil into gasoline, making steel into cars, or textiles into clothing. Here, a refinery worker and a car assembly worker, are both workers in the secondary sector.

- *Tertiary sector*, which involves the provision of services to consumers and businesses, such as providing electricity to homes and businesses, or banking services. Here, a utility worker and a bank teller are both workers in the tertiary sector.

- *Quaternary sector*, which involves the research and development (R&D) needed to produce products from natural resources. Here, a technician developing new products in a solar, or coal company R&D lab, work in this sector.

The U.S. primary sector employs only 2% of all workers in the country. Its size and activities vary with national and international developments. For example, coal production is dropping due to EPA restrictions in power generation. Fishing is also on the decrease, mostly due to overfishing, climate changes, and environmental damage.

The secondary sector represents about 23% of the U.S. employment, and is also in the decline, because during the 80s and 90s we exported many jobs overseas. As the global financial and environmental situations change, some of these jobs are coming back to the U.S., but we have become a consumer society, so the growth in this sector is not clear.

The tertiary sector, which for all practical purposes also includes the quaternary sector, represents the majority—about 75%—of the U.S. employment. It is growing fast too, as we are getting increasingly comfortable as a service oriented society.

The best thing here is the fact that the U.S. is still *numero uno* in developing new products, technologies, and services. This great nation is bringing more new products, techniques, and ideas to the world than all other nations combined. And the energy market drives all this progress.

The U.S. energy market is spread over all sectors of the U.S. economy. Combined, it is the third largest market segment in the U.S. economy.

The U.S. energy market encompasses huge spectrum of different products and services, spread over all sectors of the U.S. economy. It is not even possible to think of one single activity or business enterprise nthat does not use some sort of energy in its daily activities.

The U.S. energy market is expected to obtain over $700 Billion in private investment over the next few decades. This includes a large amount of advancement in technological development, IP gains, and other energy undertakings in the near future.

There are also many federal resources enticing both domestic and foreign companies to invest in the U.S. energy industry and the related markets. These federal resources include the Department of Energy Loan Guarantee, the American Reinvestment and Recovery Act (ARRA), the Smart Grid Stimulus Program, the Executive Order on Industrial Energy Efficiency, and others. All these programs allow for a very lucrative investment for U.S. and many foreign companies wishing to compete in the U.S. energy markets.

The unprecedented technological advancement in recent years—harnessing the power of wind, solar, bio-, and hydro-power in the U.S.—marks a shift in focus towards alternative energy. This is also driving down the prices of oil due to a decrease in demand and reduced imports. There are more incentives now than ever before to develop these new technologies and bring them into large scale use.

Energy Break Down

The U.S. uses a lot of energy. From a lot of different sources. The energy consumption in the U.S. by the different sectors of the economy is as follow:

Electric power	40%
Transportation	29
Industrial	21
Residential & commercial	10

Presently, the majority of the energy used in the U.S. comes from fossils (coal and gas), which are mostly burned as fuel for power generation, while crude oil is mostly used to fuel transport vehicles. A significant portion of these fossils are also used for making chemicals, medications, fertilizers, cosmetics and other consumables, some of which are exported.

The U.S. energy markets can be further broken into:

- Transportation—a $535 billion business where:
  94% is powered by crude oil products, and
  6% is powered by natural gas and other fuels
  Note: This is the Achilles Heel of our energy security since we import about 50% of the crude oil we need from unstable countries, transporting it through some of the most dangerous areas of the world.

- Electricity generation—a $357 billion business of which:
  48% is powered by coal,
  21% is powered by nuclear power,
  19% is powered by natural gas, and
  12% is powered by renewables (hydro and alternative technologies)
  Note: This ratio is changing dramatically, as coal is being replaced by natural gas, and the renewables' contribution increases.

- Industrial uses—a $217 billion business in which:
  52% is powered by crude oil products,
  34% is powered by natural gas, and
  14% is powered by renewables
  Note: Some crude oil, natural gas, and coal are used for production of consumables and in many different industrial processes and applications.

- Heating and cooling—a $125 billion business in which:
  58% is powered by natural gas,
  21% is powered by electricity,
  11% is powered by renewables, and
  10% is powered by crude oil products.

The U.S. uses over 4.0 trillion kilowatt-hours of electric power per year as needed to run its economy. There has been a steady growth in energy usage since the early 1990s, starting with 3.0 trillion kilowatt hours of energy annually around that time. Traditionally, the energy sources used to meet our energy needs have been oil, coal, nuclear, renewable energy, and natural gas.

The breakdown of these major fuels as a percentage of the overall electric power generation in the 1990s and today is as follow;

Table 3-2. Electric power in the U.S.

Resource	1995	2015
Coal	53%	37%
Natural gas	13	28
Nuclear	19	22
Hydro	9	7
Renewables*	2	4
Crude oil	4	2

Table 3-2 shows how the power generation ratio in the country is changing in favor of natural gas. Almost 50% drop in coal use since the 1990s...not a small change for huge industry. This is affecting other areas of the energy markets and the nation's economy, as we will see later on in the text.

The global coal consumption, however, has grown faster than any other fuel since the 2000s. The five largest coal users today are; China, USA, India, Russia and Japan account for over 75% of the total global coal use.

In the U.S. we are witnessing a dramatic drop in power generation from coal, and a significant increase in both natural gas and renewable energy sources. Hydroelectric power accounts for most of the renewable energy production in the U.S., but increasing government funding, grants, and incentives have been contributing towards the introduction of many new biofuel, wind, and solar energy production industries.

We will take a close look at the different energy sources that drive the energy markets in the following chapters.

ENERGY MARKETING CHALLENGES

Marketing energy products is easy, if you have enough of them and a captive customer base. This is the case with most U.S. electricity customers, who totally depend on the local utility for the power they use. Most of us are stuck in our homes and depend on the local utility company to provide (or not) electric power at whatever price it (the utility and its associates) see fit. What choice do we have? Pickup and move to a different area? Not possible, since our lives in general are locked in a house built on unmovable cement foundation, or a multi-story apartment complex. The local utility is well aware of that situation, and has been playing cat and mouse game with us for a long time.

As a matter of fact, most utilities don't even have a marketing department. What do they need to market something that has 100% demand, and the supply of which they control 100%. There is no better example of a lopsided market than the electricity supply.

The utilities have it made. At least until now... Their top tier organizational structure is quite simple:

Director
Assistant Director(s)
Utilities Analyst
O&M Manager
Engineering Manager
Project Manager(s)
Lead Generator(s)
Customer Service

Simple, no pressure setup. Customer Service department provides token services that deals with emergencies and some day-to-day routines. Nothing significant...thus far. Today, however, things are changing. Slowly. The customers all of a sudden have been awaken from their century long slumber, and are having lots of questions. They also have more say so and the utilities are paying attention, as we discuss in this text. Although most utilities don't see a need for a full size marketing department, marketing is becoming a significant part of their daily work. In most cases, the additional work load is divided among the existing departments, with a full size marketing inevitably in the future of most U.S. utilities.

The other aspects of the energy markets are much more complex, diversified, and volatile. So how do energy marketing professionals deal with the myriad of problems plaguing the energy sector and the global economies? The problems are obvious; the fossils are dirty and in limited quantities, the renewables are unreliable and expensive, nuclear is dangerous, hydropower is suffering from water shortages, etc., etc. How do they justify all these problems and sell their products? In this text we take a look at the major problems marketing people have, and how they get around in order to sell their products and services on the national and global energy markets.

Let's start with the energy market foundation; the energy sources used for power generating and transportation. The power generation sources (coal, natural gas, hydro, nuclear, and renewables) are the materials and fuels used to generate electric power, which is essential for the proper function of our daily activities.

The transportation energy sources (crude oil, natural gas, and biofuels) are the other key aspect of the U.S. and global energy markets.

There are also some additional energy sources, some of which are futuristic, and some too complex or costly, to be counted today. They, however, have to be considered for full development and use in the future. We take a look at these as well in the next chapters.

The Energy Sources

All power plants and vehicles use some type of fuel, be it primary, or secondary, renewable or not. Fossil fuels (coal, oil, and natural gas) account for more than 90% of the total amount of materials used in energy generation and vehicle fuels.

The fossils have been, and are still, the most important primary energy generating sources in the world. The rest of the power generation is contributed by hydropower, nuclear energy, and in much smaller amounts by solar, wind, and geothermal power. The global transportation is dominated by crude oil and its derivatives. Natural gas and bio-fuels are also used, but in much smaller amounts.

The use of fossils in the U.S. has been decreasing, but fossils use around the world is increasing. This means that no matter what the U.S. does, the total energy use and pollution around the world will continue to increase ad infinitum. At least until their prices become unaffordable. At that time, the global energy crisis will morph into a war among the nations for enough and affordable energy.

We see from Figure 3-14, that coal rules in the power generation sector! Still. The other fuel types are creeping up in use, but coal is still on top, and is foreseen to be there for a very long time.

The fossils are mostly burned as fuel for power generation or transport.

These ratios change with time in favor of one or other energy source, depending on the global and national political and economic conditions. In all cases, however, the fossils provide the major share of our electric power, vehicle fuels, and other materials for the entire industry.

A significant portion of fossil sources is also used

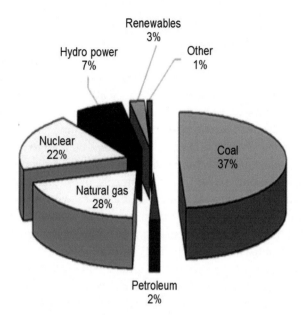

Figure 3-14. Power generation in the U.S. (2014)

for making chemicals, medications, fertilizers, cosmetics and other consumables, some of which are exported. Yes, the fossils supply thousands of materials—raw and secondary—for many areas of the U.S. economy. They are also used to make fertilizers. medicines, cosmetics, plastics, synthetic fabrics, and lubricants. Toothpaste, toothbrushes, shoes, sunglasses, car and truck tires, tennis racquets, basketballs, TVs, computers, etc., etc. are all products using some type of fossils. The fact is that a huge number of products that we use every day, wouldn't be available without the coal, oil, and gas extracted and processed in different U.S. states.

Table 3-3. Key energy sources in the U.S.

Energy Type	Use in the U.S.
Crude oil	37% (mostly used for transportation and production of consumables)
Natural gas	26% (mostly used for electric power generation and transportation)
Coal	21% (mostly used for electric power generation and in industrial processes)
Nuclear	8% (mostly used for electric power generation)
Hydro	5% (mostly used for electric power generation)
Renewables	3% (mostly used for electric power generation)

The fossils are the foundation of our economy. They provide personal comforts and power businesses.

Houses will not be comfortable places to live without enough power for lighting, heating, and cooling. Businesses will not be able to operate efficiently, or at all, without electric power, natural gas, and crude oil products. People and goods would not be able to move freely around without gasoline and diesel. Medications, plastic products, fertilizers, chemicals, paints, and many commodities would not be available, or would be extremely expensive, if we didn't have the fossils. We can go on and on with the gloom and doom of impossibilities in a fossil-less world, but this is enough to give us a good picture of what could happen. And we assure you; it will happen sooner or later. There is o way around it.

We have to agree that it is a serious problem, and yet, we don't have complete control over the supply chain of these materials—and especially that of crude oil. Lack, or even temporary disruption, of crude oil supplies would have a chain-reaction effect on the entire economy with disastrous results. Airplanes, trucks, trains, and most other vehicles would have to be parked for the duration. This would affect our personal lives too, since we would not be able to move around—even going to work would be a challenge. The economy would slow down, causing increasingly serious damage to companies and entire industrial sectors over time.

And on the fringes of extreme dangers, we see the U.S. military machine grounded too in case of oil shortage. Its 900 or so bases around the world would become parking lots of expensive equipment, frozen in time and useless for all practical purposes. Solders would be stuck in their barracks without means of transportation. The consequences of such a doomsday scenario are hard to even imagine, let alone estimate accurately.

This apocalyptic event, however, is not very likely today, or tomorrow, and yet it shows a (potentially) serious gap in our national energy security. It provides a level of uncertainty and vulnerability that can be attributed to a number of internal and external risks. Ever changing domestic and international political, regulatory, and economic developments make it hard to predict and control events, thus adding to the energy uncertainty.

There are also threats of natural disasters, internal and external terrorist attacks, and other maladies that could disrupt the energy supply chain at any time.

Most importantly, there is an eminent danger of running

out of the key energy sources (coal, oil, and gas) in the near future.

This is a key issue that has been a topic of discussion in energy circles since the 1970s. There have been many conferences, discussions, negotiations, and suggestions since then, but the results show that the world is still going in the wrong direction. As a matter of fact, most people (and entire countries) tend to turn a blind eye and forget about the pending energy doom. Others think that we are already too close to it and that it is too late to reverse the events no matter what we do. So all we, and they all, can do is predict different doomsday scenarios.

Until recently, the predominant thinking in the U.S. was that we will run out of fossil fuels in the very near future. But lo and behold, suddenly things have changed drastically for the good, making all previous doubts, prediction, and plans irrelevant. Instead of importing massive amounts of fossils, we now see the U.S. getting ready to export huge quantities of all kinds of energy products: coal, oil, natural gas, and refined crude oil products. Even crude oil, which has been an export no-no for half a century, is now being considered by Congress as an export commodity.

This all is an unexpected and very welcome bonanza for the oil companies, from which the consumers will benefit too. But the nagging question is still here, "How long will the new supplies last?"

Unfortunately, even if we discover 2-3 time more fossils in the U.S. and abroad, the fact of their limits remains. We clearly see glimpses of doomsday scenarios in the midst of all this newly acquired energy bliss too. We see the energy prosperity lasting for awhile but not forever! This is the key: fossils are a finite resource, so they are here today but gone tomorrow. We only don't know when that "tomorrow' will come. We also see that the short-term benefits do not translate into long-lasting energy bliss. Everything has limits—and so do fossils. The gains of late do not assure us that we will have fossils forever.

Blinded by the new energy boon, we are failing to consider a reasonable balance between:

• The amount of fossils we are using,
• The total amount available,
• When will they be depleted, and
• What to do when they are depleted.

And, yes, they will be depleted someday soon. Not today, but soon enough to worry about it. There are signs of the end, reflected as difficulties in the new oil and gas drilling operations. In the past we saw oil wells producing high quality oil non-stop for 30-50 years. This is because they stood on top of large lakes of liquid crude oil. Easy money!

Today, the new hydrofracking rigs sit on top of a rock that has some oil impregnated in it. To get to it, the drillers need some fancy and expensive procedures. But even the,, the amount of oil in the rock, and the access to it, are limited. Breaking oil from a rock is a far cry from pumping it from a lake.

Hence, the difference. The new hydrofracking wells do not produce enough high quality oil, and within a short period of time, they run dry. A new drill site has to be chosen and explored with the same overall results. This approach, of course, is much better than not having any oil, but is quite different from what we are used to. This new situation requires a new strategy, which must be well though of, properly developed, and efficiently implemented.

And we have not even mentioned the harm that many new hydrofracking wells cause to the local environment and people. Regardless of all this, the energy frenzy in the U.S. is underway unrestricted. Not a good thing, because unless we manage to put it under control, it might lead to some unexpected negative consequences affecting us and tomorrow's generations.

These are the pertinent issues of today's energy and energy security, and is what we discuss from a technical point of view, and resolve (at least on paper), in this text.

Energy Sources Availability

Ensuring availability of energy sources; primarily coal, natural gas, and crude oil is a major responsibility and every day preoccupation of governments, private companies, and individuals around the world. Alternative energy sources, wind and solar mostly, are also entering in the picture in a big way, which makes it much more complex.

The importance of securing energy is actually growing, as the world population, and its needs, increase. Governments cannot run their armies and economies without energy. Companies cannot conduct their business efficiently without energy. And individuals cannot heat or cool their homes and drive their cars without energy.

Energy is in the forefront of our daily lives.

Energy availability and pricing are of utmost importance to all of us, so let's take a closer look at what all that means.

The Energy Divisions

The large pictures reflects a world divided into haves and have-nots. On one side of the world table of goods are the fossil energy-rich countries, which have huge reserves of coal, natural gas, and crude oil. Some of them have so much that they don't know what to do with it. Saudi Arabia, for example, is burning more crude oil for electric power generations daily than most countries use in a month to run their economies.

On the other side of the energy table sit the fossil energy-poor countries. Being energy-poor does not necessarily mean that the entire country would be poor too. Take Japan for example; an island country with very limited natural energy supplies and almost no fossil energy supplies. It is, nevertheless, one of the largest global economies. Even though Japan imports 90+ percent of its energy sources, it manages to sustain a significant economic progress.

Most of the energy-poor countries, however, are poor. Period. It is hard to run an efficient business, let alone entire national economy, by importing expensive energy every day. India's economy, for example, is struggling mostly due to lack of enough and cheap energy. The situation is similar in most countries in Africa too.

And amazingly enough, even some of the energy-rich countries are poor, which is a consequence of either poor management of their resources, or extreme dependence on these. In the latter case, when energy prices are low, these countries suffer. This is exactly what happened in 2014-2015, when global oil prices fell 50%. At those prices, Russia, Venezuela, and most other small oil producers could not make ends meet and things got from bad to worse for them.

The U.S. is blessed with huge natural energy resources. We have coal for at least a century at the present rate of use. We also have a lot natural gas, which would last for another half a century, and crude oil for the same time period. What will happen after that? Silence! This is the best guarded secret of the energy sector. We discuss this issue in detail in this text, so we will only touch on it here.

Even though we have a lot of energy in the U.S., we use a lot too. In some cases, we use more than what we can produce. Crude oil is the prime example of this inequality, since we import nearly 50% of the oil we use.

This anomaly is a daily worry of the U.S. government, because this country runs on crude oil. The national economy would fall flat on its face if crude oil becomes unavailable or too expensive. On the other hand, it always benefited when oil prices fell. This was confirmed in 2014-2015, when crude oil prices fell, thus saving billions of dollars to consumers, and providing a significant boost to the national economy.

Dow Jones kept on climbing and shot up above 18,000 by the end of the year, the economy grew an estimated 5% at the time, gasoline prices fell below $2.00/gallon for awhile, and the consumer confidence returned and grew quickly as never before. All this was reflected in an optimistic outlook for the years to come. The euphoria did not last long, and the U.S. oil industry suffered a setback because of very low oil prices, but the bumps in the road were not due to crude oil availability or prices alone.

Now imagine what would happen if the crude oil prices rise 50% over the average of $100/bbl. At $150/bbl, as it happened in 2008, gasoline will shoot up again to nearly $5.00/gallon, and people would stop driving, shopping, going out and many other activities. The economy would slow down and stop growing all together. And this scenario is if *only* crude oil prices rise. Now imagine what would happen if natural gas prices follow the trend and rise by 50% or more. This would most likely bring coal prices up too, and the country would most likely enter a new depression-like era.

Figure 3-15 clearly shows that our national security (and the economic progress it represents) is directly and indirectly dependent on a number of internal and external factors and risks. Energy is one of the most important of these, which is why it is on the forefront of all governments' agendas.

What happens around the world affects us here at home one way or another. Sometimes we see these effects as increases in the price at the gas pumps, and sometimes as increases in the military budget. Sometimes we feel them in a more direct and painful way, like the 911 attacks and their results.

What we went through during the 19703 Arab oil embargo was shocking, but it made us more aware and prepared for things to come. It woke us up from the American Dream and pointed its weak points—mainly our excess use of energy and our country's energy deficiency. What we saw on the global energy market in 2014-2015 was very unique, but not unexpected development; that of oil-rich countries manipulating the oil market. This time they caused global oil prices to fall to very low levels. This was in response to the rapidly increasing crude oil and natural gas production in the U.S. and other countries, which jeopardizes Saudi Arabia's dominance and control of the oil market.

The oil-rich countries made an attempt to control the world's oil supplies by simply lowering the oil prices to levels that made it unprofitable for domestic production.

This affected negatively the U.S. well drilling and hydrofracking operations, forcing some operators to reduce their production, while other shut down production all together. Russia, Venezuela and other producer countries suffered most, since their economies depend fully on oil exports, which the low oil prices made unprofitable.

Energy Security and Independence

The availability of energy sources is a function of our *energy security*. Energy security is ensuring an uninterrupted energy supply today and tomorrow, by employing efficient supply methods complete with accident and crime prevention measures. Since we depend on importing some fossils (mostly crude oil) we are concerned with, and make a considerable effort to eliminate the risks, which represent clear and present danger to our energy and national security.

We also talk about *energy independence*, which means producing our own energy, so that we do not rely on imports. Obviously, we would not be able to achieve a complete energy independence if we continue to depend on energy imports of any sort and at any time. Period!

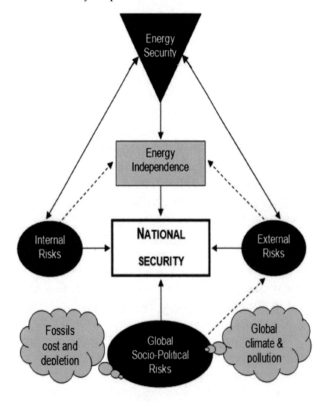

Figure 3-15. Energy-related dependencies

Another factor in the energy availability equation is our *energy survival*, which is ensuring that we have enough energy now **AND** in the near future, so that neither we nor future generations end up in an energy blackout. This is an obscure subject, kept on the back burner…for now, since we are preoccupied with making sure that we have enough energy TODAY. This is all that matters for now! Would things change any time soon? Probably not, because capitalism requires making a buck, and living comfortably, today. And today only; since the future does not figure in the capitalist model.

We see a split in the general population and the scientific community's opinions and actions on the issues at hand. Some people clearly see and acknowledge the new energy security risks around us—including increased terrorist activities, pending energy sources depletion, increasing energy demand, unreasonable energy waste, and disastrous environmental damage.

Others, however, don't see any of this (or pretend not to), while yet others just don't care. This is normal for a population as diverse and opulent as that in the U.S. It is also normal to see the glass as half empty or half full, depending on which side of the issue you sit.

So, oil company executives see the new situation one way, while the unemployed oil worker sees it totally differently. The executives' primary concern is to increase corporate profits—at all cost—while the unemployed struggle to pay ever-increasing energy bills.

Energy Prices

Americans are concerned basically by two different energy pricing systems; these of, 1) vehicle fuels (gasoline and diesel), and 2) electric power. Gasoline and diesel prices are—with very few exceptions—dependent on the price of crude oil. Because of that, they fluctuate in direct proportion with the global oil prices. In some exceptional cases, such as natural disasters or economic difficulties, the fuel prices might change course away from that of crude oil. These events, however, are of a temporal nature, driven by other factors, like natural disasters, refinery shut downs, etc.

Electricity prices also depend to a great extent on fossils prices, but the U.S. enjoys almost 100% dependence on local energy sources (coal and natural gas) for of its electricity generation. Nevertheless, there are a number of additional factors to be considered, some of which we have little, or no control over.

Vehicle Fuel Pricing

The price of vehicle fuels; gasoline, diesel, and jet fuel is determined by a number of factors, but the price

of crude oil is almost ¾ of the price at the gas pump.

Table 3-4. Fuels price structure

Service	Gas	Diesel
Crude oil	67%	61%
Taxes	13	14
Oil refining	11	15
Distribution	9	10

We see in Table 3-4 that the price at the gas pump is driven mostly by the crude oil prices. Oil prices vary with time, which makes the gas pump prices to go up and down accordingly. Some times the change in price is insignificant, and sometimes it is great. Good example of this dependence is the extremely high price of crude oil in 2008 of over \$150/barrel, which drove gasoline and diesel prices to nearly \$5.00/gallon.

On the other extreme, we saw very low oil prices in 2015, below \$50/barrel, which was reflected at a gas pump price of \$1.80/gallon for awhile.

We dedicate more space on this issue in the following chapters.

Electric Power Pricing

The pricing of electric power is much more complex. It is getting even more so, due to the introduction of new technologies, such as the smart grid, micro grids, and renewable energy production. This forces new pricing strategies to be developed as needed to capture the dynamics of supply and demand in this market. The new electricity pricing strategy development is one of the most important and complex aspects of the present-day energy debate.

The dynamic pricing strategies that are used to offer customers shifting prices depend on several internal and external factors. Dynamic pricing makes value and cost of energy use transparent to consumers which enables them to determine when cost exceeds value, which in turn enables the adjustment of energy consumption and production needs.

As smart grid and other technologies emerge, the necessity of new and efficient means of energy pricing become urgent. There has been an array of attempted pricing and forecasting models by utility companies and regulators—usually designed to increase end use efficiency, energy conservation, capacity utilization, savings, in the short term, and to ensure energy security, in the long run.

A price based energy network architecture is

based on marginal costs, consumer supply and demand, and the effects of deregulation and competition within a market. There are many different factors that impact electricity pricing, and which change with advances in technological capacity. These factors affect pricing schemes no matter which pricing model is implemented.

Prices fluctuate by location—city, state, country, region—and by input source (petroleum, oil, coal, solar, etc.).

Key factors that affect the price of electricity and energy in general include:

- Peak and demand rates are vehicles that help the power companies pay their bills and make capital investments to meet the present and anticipated peak demands (by charging the customers as and when needed).

- Alternative energy technologies can help reduce the rates (especially the peak rates), but despite advances in recent years their use is limited due to high implementation costs, uncertainty in their long-term operation, legal constraints, and recently due to conflicts with utilities operations (it costs too much to operate the grid with many renewables hooked into it).

- The state and local taxes imposed on utilities change from state to state and from time to time, affecting energy rates.

- The power generation or input source determines the price of energy delivered to the customers. A mix of energy sources is ideal for trimming peak rates, where, for example, base-load (gas-fired power plant) is assisted by a solar power plant, which delivers maximum power during the peak hours.

- Environmental considerations complicate the situation and limit the choice of the utilities, forcing them to use technologies with a smaller carbon footprint (wind and solar), which are more expensive to use.

- The capital expense and maintenance of the transmission network is of great importance, as is the distance from the energy source. For example, greater losses are experienced over long-distance power transmission, also making rates higher.

Many factors affect energy rates. The overall picture is complex, and the expenses for producing and maintaining energy supplies are great. Utilities are caught in the middle between the customers (who demand lower prices) and the regulators (who regulate prices). So, the utilities are forced to kick back by increasing prices in whichever area of their activities they can.

Recently, the utilities have been trying to increase the prices charged to residential solar power owners to compensate for the free use of the power grid by the thousands of new solar installations. The utilities claim that the increased use of the power grid by thousands of new residential solar installations is forcing new capital investment and additional maintenance.

True, but the customers do not like the utilities' dictatorship and are rebelling against it. Nevertheless, the utilities have the last word, and the customers are stuck with the local utility, since they cannot move their houses to another location. And so, we will witness many more price battles in the future.

Energy Efficiency

Energy efficiency is important when discussing energy use, because we can reduce the energy we use by increased energy efficiency of its generation and use. So, let's see first how much energy we waste and how. This is a very important subject, because we cannot hope to achieve any degree of energy efficiency, if we continue wasting large amounts of it.

Here are a few examples of energy waste:

Packaging

What does packaging have to do with energy efficiency? It is actually one of the most visible and noticeable symbols of energy and environmental impact of the 21st century. Plastic bags, containers, wrappers of all sorts and sizes are all around us. We see them around the house, the park, in the rivers and the oceans. Packaging is credited with heavily contributing to our economic development and raising our standard of living. One look at the items in a U.S. supermarket, as compared with the street market in most African and Asian countries, gives you a good idea of the stark difference that packaging makes.

In the U.S., with our sophisticated packaging, storage, and distribution systems, only 2-3% of the food is wasted before it gets to the consumer. In developing countries, on the other hand, 40-50% of food shipments are spoiled because of inadequate packaging, storage, and distribution systems.

On the consumer side, however, this ratio is reversed; 30-50% of the food from U.S. consumers ends in the waste baskets, while in the developing countries only 2-5% of the food total is wasted by the consumers. So how can people without efficient cooling and packaging of food items be so efficient in their use? How can they do without packaging, and why are we so wasteful having good packaging options?

Packaging plays a growing role in the global economy, especially in the developed countries where it is not only for protection of goods from contamination, but also used to convey information about contents, preparation, and use, but in most cases it also keeps would-be tamperers from poisoning or otherwise hurting the customers. At the same time, however, packaging—its manufacture, transport, use, and disposal—uses enormous amounts of energy and is causing an environmental disaster.

Packaging is the largest and fastest growing contributor to energy waste and environmental problems.

In the U.S., packaging accounted for more than 30% of the municipal solid waste stream in 1990. Where is all this packaging going? In this country, most packaging and other waste is buried in landfills. But even with its abundance of open land, America is running out of room for its garbage. One quarter of the country's municipalities have already exhausted their landfill capacities, so that more than half the population lives in regions with lack of landfill capacity.

Even though the environmentally sound alternatives to burying garbage—recycling, reuse, and energy recovery—are well developed, our disposable and throwaway society is still generating more garbage from packaging than the entire rest of the world. Packaging is not the only culprit in the solid waste crisis, but it is the most highly visible component in a garbage dump. It also directly involves the customers, who use it for a very short time. Its short lifetime exacerbates the problem, since the useful lives of most packages are very limited. While some packaging materials may last as long as several years (i.e. paint and food cans, and reusable canisters), the useful lives of most others can be as fleeting as a few minutes, or hours.

Think of the hamburger wrapper, or a new gadget in a large cardboard box received in the mail. We deal with a huge volume of different types of packaging daily that goes into the solid waste stream. This is a big problem, which also presents an opportunity, since even very small improvements in packaging, due to its huge amount, can make a real difference in the magnitude of the garbage crisis. So packaging offers this unique opportunity for individuals and companies to assume leadership roles in environmental responsibility.

Industry's response to the environmental chal-

lenge of excess packaging has been focused on recycling and packaging materials reduction. But this is not enough, because these are complex issues that demand a systemic, integrated approach based on comprehensive analysis. Long-term vision and implementation of innovative solutions are also some of the tools we must use. Life-cycle analysis is another tool, which needs to be applied to every product we consume. Think of millions of dirt-cheap gadgets coming from China enclosed in large packages made of heavy-duty plastics. How much material and energy was use to make these packages? How much energy was used to manufacture, use, and dispose of them? What is their overall benefit to our well-being?

Life-cycle analysis is an important step toward understanding the full energy and environmental implications of packaging choices.

Using this technique, we consider the energy use and environmental implications integral parts of each product. Considering the energy used during the entire product cycle—from raw material acquisition, design, manufacturing, transportation, to final use and disposal of the packaging—will allow us to see the other side of packaging. That's the side that affects our economy and contributes to great energy waste. This will allow us to make better packaging choices.

Our energy problems will not be resolved by solving the packaging problems, but present-day packaging is one of the elements of our way of life that is indicative of our societal problems in need of fixing. Less waste means more and cheaper energy for all.

And one of the greatest wastes of energy and materials…

Food Waste

Because we cannot live without food, it is on our government's priority list, and a lot of effort (and energy) goes into ensuring the food supply on the American tables. This comes at a heavy price.

Getting food from the farm to our dinner table eats up 10% of the total U.S. energy budget, uses 50% of U.S. land, and swallows 80% of all freshwater consumed in the country.

Yet, about 40% of all food in the U.S. today is wasted.

After spending so much effort and energy to produce and get it on the table, almost half of the food we buy is wasted. This is over 20 pounds of food per person every month ending up in waste baskets and eventually in municipal garbage dumps. Amazing facts and shocking numbers…

Get this…today, the average American consumer wastes 10 times as much food as people in Southeast Asia and Africa, and 50% more than we wasted in the 1970s. This not only means that Americans are throwing out the equivalent of $165 billion of different food items each year, but that the uneaten foods must be collected, transported, and left rotting in landfills. All this uses precious resources, which is another waste.

Food waste is the single largest component of U.S.' municipal solid waste, where it also accounts for a large portion of all U.S. methane emissions.

Organic matter (mostly food leftovers) in municipal waste dumps accounts for 16% of U.S. methane emissions and contamination of 25% of all fresh water supplies in the country. Not to mention the huge amounts of energy wasted in this process too.

There is work underway in Europe in search for better understanding of the drivers of the food waste problems, and identifying potential solutions for it. In 2012, the European Parliament adopted a resolution to reduce food waste by 50% by 2020. It also designated 2014 as the "European year against food waste."

An extensive U.K. public awareness campaign called "Love Food Hate Waste" has been conducted over the past five years too. As a result, over 50 of the leading food retailers and brands have adopted a resolution to reduce waste in their own operations, as well as upstream and downstream in the supply chain. As a result of these initiatives, during the last five years, household food waste in the United Kingdom has been reduced by 18%.

The situation is somewhat different in various countries, where America is in a class of its own. Here, food represents a small portion of most Americans' budgets, which makes the financial cost of wasting food low enough to be considered a regular, more convenient, mode of life by most of us.

Then there is the basic economic truth of advanced capitalist society, that if consumers waste more, industry sells more. This affects the entire supply chain—from the fields to the supermarkets. Here, any waste downstream translates to higher sales for the upstream companies.

Overcoming these and other related challenges is not easy, and will require the total cooperation of the government, consumers, and businesses. But this effort can start only after everyone understands and agrees that this is a problem. Only then, can the problem of reducing food waste be raised to a higher level of priority. Reducing waste food losses in the U.S. by just 15% could feed 25-30 million people in Africa every year. It could also save millions of barrels of oil wasted during food production, storage, transportation, preparation, and disposal.

One simple solution (one of many that the government can undertake) is to standardize and clarify the meaning of expiration date labels on food packages. All foods have recommended use periods stamped on their packages. These stamps are hard to read and are easy to misunderstand. This is causing tons of foods to be thrown out due to misinterpretation. Expiration date clarification on food packages could prevent about 15-20% of wasted food in U.S. households. Businesses must understanding the extent of their own waste streams and adopt best practices. For example, a U.S. food chain saves over $100 million annually after an extensive analysis of the freshness, shrink, and customer satisfaction in the perishable foods department. The consumers can reduce their waste by getting better educated on the basics of foods: shelf life, when and how the different types of foods go bad, as well as buying, storing and cooking food with waste reduction in mind.

Implementing efficient food waste reduction strategies can bring tremendous social and economic benefits. It can reduce hunger, save energy, and reduce environmental pollution. Are we ready for it?

The Energy Efficiency Game

Energy efficiency is the new game in town. It is expressed in reducing the amount of energy required to provide products and services. Insulating a home, for example, reduces the energy use by requiring less heating and cooling to maintain a comfortable temperature inside. Changing old habits by reducing driving is another energy saving method. Compact fluorescent lights (CFL) or natural skylights, reduce the amount of energy required to light a room by 75%, as compared to using traditional incandescent light bulbs. CFL also last 5-10 times longer than their old incandescent cousins, so both money and energy are saved by their use.

The key motivations to improving energy efficiency are reducing energy use and cost saving.

Great energy efficiency increase is attributed to new and more efficient technologies and production processes, and/or the application of proven methods of eliminating energy losses. Energy efficiency is also one of the solutions to reducing carbon dioxide emissions.

According to the experts, improved energy efficiency in buildings, industrial processes, and transportation could reduce the world's energy needs by 1/3 in 2050, and help control global emissions of greenhouse gases by an equal amount. Energy efficiency is also seen as a pillar in national energy security, because it reduces the energy imports from foreign countries. And very importantly, it also slows down the rate at which domestic energy resources are used and depleted.

And speaking of energy efficiency:

The U.S. Energy Efficiency Syndrome

Although the energy debate in the U.S. is escalating, and there are some good results to show, we don't expect any major developments on that front in the near future. Because of that, we will continue using huge amount of energy, mostly crude oil and electricity.

The excess use of energy use in the U.S. due to two main factors:

- *Infrastructure.* The way the U.S. roads and transport infrastructure are designed and built is very different from those in most other countries. The vast territory of the 48 continental states is scarcely populated, so public transport is not the best option. Although there are trains, planes, and buses running all over the place, they account for a very small percentage of human transport. Instead, most people prefer to drive from place to place, and even from state to state.

Mass transport systems exist in a few major cities, but in most U.S. cities and rural areas it is non-existent. Because of that, people drive everywhere around town in their own cars. The interstate highway system is so well developed that one can make 100-200 miles drive in a car faster than taking a train, bus, or even a plane in many cases (not to mention that this is a much more convenient and pleasant way to travel).

So millions of people do just that—drive and drive! Nevertheless, it would not be that difficult or expensive to modify the national mass transportation system, so that more people use it. Trains and busses crisscross the cities and the country half-empty, so filling them would boost our energy efficiency. This, however, is unlikely to happen anytime soon, because of the other, even more important factor—our life style.

- *Our Way of Life* is very different from that of any other humans on Earth. We are used to having

our cake and eating it too. Like spoiled children, we want everything to be done our way, and done now or sooner. Amazingly, we usually get it that way too.

These high expectations and the resulting high standard of living come at the cost of long hours of stressful work. However, the culture has become so entrenched in this system that most people feel there is no alternative. After all, it is a small price to pay for our almost unlimited freedoms and comforts.

And so, we drive large gas guzzling cars and trucks for no other reason than just because. We also live in huge houses with many lights, and large cooling and heating units. These houses are also the most comfortable places n the world. The temperature inside is kept at the most comfortable levels no matter if it is freezing cold or boiling hot outside.

This is how it has been for almost a century, this is how it is now and not much can be done in the future to change our way of life. Oh, yes, we will talk about it, we will discuss the alternatives, but then we will slam the doors of our gas guzzlers and drive to our comfortable houses…unless and until something drastic forces us to change.

Practical Energy Efficiency

An example of energy efficiency in motion is Executive Order (E.O.) 13514, Federal Leadership in Environmental, Energy, and Economic Performance, which was signed on October 5, 2009, by President Obama. It expanded upon the energy reduction and environmental performance requirements of the previous E.O. 13423. Surely, the Administration sees a strong relationship between energy production and use, and the environment. This is a good step ahead toward an energy efficient and environmentally friendly future.

E.O. 13514 states that the U.S. government is the largest consumer of energy in the country, with over 500,000 large, old, and mostly energy-inefficient buildings. E.O. 13514 sets numerous federal energy requirements, starting with a mandate for 15% of existing federally owned or leased buildings to meet Energy Efficiency Guiding Principles by 2015.

This means that at least 75,000 buildings are to be converted to energy efficient structures by 2015—a tall order. The annual progress is to be made toward 100% conformance of all federal buildings, where 100% of all new federal buildings are to achieve zero-net-energy status by 2030.

A "zero-net-energy building" according to E.O.

13514 is "a building that is designed, constructed, and operated to require a greatly reduced quantity of energy to operate, meet the balance of energy needs from sources of energy that do not produce greenhouse gases, and therefore result in no net emissions of greenhouse gases and be economically viable." All-in-one package. The best solution to our energy and environmental problems. The only problem is that it is very difficult and expensive to do, so it will be done slowly and partially.

The executive order also states the general direction the Administration is taking the country in, "… the Federal Government must lead by example … increase energy efficiency; measure, report, and reduce their greenhouse gas emissions from direct and indirect activities… design, construct, maintain, and operate high performance sustainable buildings in sustainable locations; strengthen the vitality and livability of the communities in which Federal facilities are located; and inform Federal employees about and involve them in the achievement of these goals."

In October 2013, President Obama also signed a Memorandum under his Climate Action Plan, directing the federal government to consume 20% of its electricity from renewable sources by 2020. This is more than double the current level. Another tall order.

As part of this effort, federal agencies will also identify formerly contaminated lands, landfills, and mine sites that might be suitable for renewable energy projects, in order to return those lands to productive use with minimal fossil energy use.

To improve the federal agencies' ability to manage energy consumption and reduce costs, the Memorandum also directs them to use Green Button (a tool developed by the industry) which provides utility customers with easy and secure access to their energy usage information in a consumer-friendly format. This will provide transparency and will help further in managing the effort.

This is a great step ahead for the U.S. energy policymakers. Unfortunately, it is relevant only to the government complex. Similar patchworks of energy and environmental guidelines exist here and there across the national economy too, but they are much more sporadic, and not as clearly defined.

The final decision in executing the civilian energy policies is left to the individual states.

Since different states have different types and levels of energy production and use, which determines their preferences, it would be hard to predict how the entire program will develop. A more unified, country-wide federal program is needed for putting some order in our energy present and future, but there are no signs for that thus far, so we will not hold our breath. For the time being our energy markets will continue developing according to global events and partial federal and state energy policies.

Although this approach usually works well for other industries, it is not the best for the energy sector. It is especially detrimental for the renewable energies, which are struggling to survive the competition with their older and better supported competitors—the fossil energy sources.

And yet, this is capitalism, so letting the markets do what markets usually do, may work in the end for the energy markets too. At least we hope so…

Major Change? Naahh…

For generations, Americans have enjoyed living in large houses with large appliances, and large brightly lit rooms maintained at constant temperature throughout the year. Since the 1950s, Americans have been driving large cars, trucks, and SUVs without regard to availability, price, or anything else. The automobile is a symbol of freedom and independence which are two of the highest values for Americans.

Because of that, any thought of altering our electrical or transportation system is viewed as a threat to our most cherished values. It is a violation of our rights to achieving the American Dream.

Americans insist on doing whatever, whenever at a whim. It is unreasonable to expect Americans to live any other way unless something drastic happens to force a major change. But what major changes could we expect? Even during the Great Depression many people drove cars, so we see cars as a key indicator of our way of life. It is how it is. This means that we will hang onto the car driving habit as a necessity until the last drop of oil is gone.

Major change in energy use in the U.S. will come only when crude oil is totally depleted.

The same is true for electricity. Also during the Depression, FDR authorized electric power subsidies, which indicates the importance of this energy source in American life. Today we depend even more, and can't even imagine life without electricity, so it is to be expected that we will burn coal and natural gas until they too are gone. Only then will we take steps towards implementing major change in our energy use.

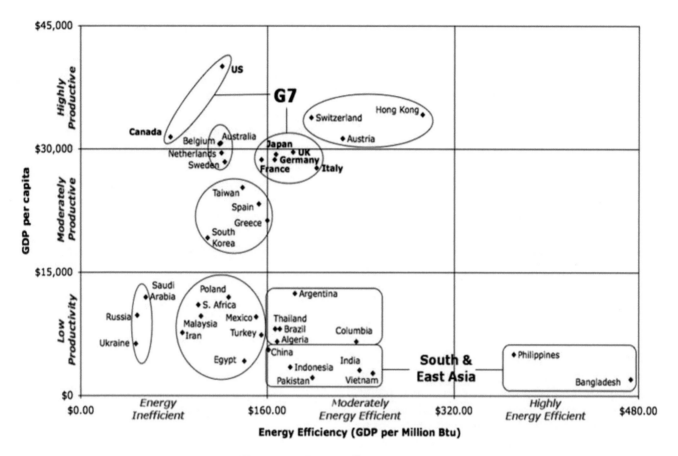

Figure 3-16. Energy efficiency vs. GDP

Figure 3-16 needs some clarification. The U.S. is perched high on top of the world as far as its GDP is concerned. We are a rich country, no doubt! At the same time we are near the bottom of global energy efficiency. Could it be that Bangladesh is more energy efficient than the U.S.? Yes; about three times more efficient.

Why? It is the way of life in the U.S? We are very smart and productive, but do not know how (or don't care) to conserve energy at home, or at work. Bangladesh, on the other hand, has no energy reserves, and the people are not accustomed to wasting energy, nor can they afford to, so they simply don't use it. This is not an option in the U.S.

Europeans and Asians are accustomed to living in small, crowded, dimly lit rooms, with shoebox-size refrigerators, air-dry clothes lines all over the rooms and balconies, and many ride bicycles to work. Most have accepted the idea of living with ever-increasing energy sacrifices, while others are even contemplating life in strictly regulated eco-villages.

Well, that's perfectly OK, if this is what they want to do. We will let them. This, however, is not going to happen on any large scale in the U.S. Not now, not ever…at least not until the major changes are forced upon us.

The conservative thinkers of today must understand that "our way of life" cannot be changed, and work around it. It matters not what our opinions of the status quo are, or how much better the alternatives might be; it is what it is and making any significant changes is not an option. It is the American way of life and people do not like change.

While a small number of free thinkers in Portland, Oregon, might consider living in an eco-village, going to Tyler, Texas with that idea might get you in big trouble with the locals. Just think of a Texas farmer living on a 500-acre farm in a 5,000-square-foot house, with walk-in refrigerator/freezer and two air conditioning units on the roof. Then take a look at his equipment barn full of huge tractors and other machinery—F-150 truck, Escalade luxury SUV, a Softail Harley Davidson in the three car garage—and his twin-engine fishing boat tied to the lake pier.

Now imagine this same guy riding a bicycle to the local bar at night, or to church on Sunday. How about moving out of the 3-bedroom home into a studio apartment with a shoebox-size refrigerator, 25 W lights, no air conditioner, and air-dry clothes line in the bathroom.

Not a chance! This is not going to happen. Not in America. Not now, not later, not ever… unless and until the big major change comes. So perhaps we should focus on greater energy efficiency instead? Yes, it is good to talk about it, so don't stop. You, however, cannot take my four-bedroom 3,000-square-foot home with two full-size refrigerator/freezers and 5-ton air conditioner on the roof away from me. And I will also continue driving my 8-cylinder SUV, my 1100 cc. motorcycle, and 30-foot diesel motor home anytime and everywhere I want. Our way of life encourages this type of behavior and only a major change could make a difference.

But let's continue the energy efficiency talk; it is a good thing, if it is not pushed down our throats. Instead, it must be somehow woven in the American fabric without sudden changes of our daily routines. The politicians know this very well, since they live in even bigger houses and drive even bigger cars and campaign busses. They are very careful not to step on any toes too, because one wrong step could cost them their political future, together with their large homes and cars.

So, the energy efficiency, environmental protection and the need for major change debate in this country continues, but the changes that are planned or expected will have to be carefully crafted and even more carefully implemented. They must come slowly and without much sacrifice on the part of the people and their way of life. How far we can go that way is uncertain, but there are few alternatives. And that's the honest (albeit politically incorrect) truth!

Energy Intensity

Energy intensity is another concept of importance to the energy markets. It measures the energy efficiency as compared to the economy of a nation, calculated as units of energy per unit of GDP. High energy intensities indicate a high cost of converting energy into GDP, while low energy intensity means lower cost of converting energy into GDP.

For example, 1 million Btus consumed with an energy intensity of 8,553 produced $116.92 of GDP for the U.S. between 2000 and 2010. At the same time, 1 million Btus of energy consumed in Bangladesh with an Energy Intensity of 2,113 produced $473 of GDP. This is more than four times the effective US rate. This, however, is an extreme case that does not set a precedent.

Energy intensity can be used as a comparative measure between countries, whereas the change in energy consumption required to raise GDP in a specific

country over time is described as its energy elasticity. Energy intensity is different in different world regions. For example, it is three times higher in CIS than in European countries. Energy intensity in OECD Asia and Latin America are about 15% above those in Europe, while North America stands 40% higher but remains below the world average. India is on a par with the world average, with energy intensity levels 60% higher than in Europe. The high energy intensity in the CIS, the Middle East, China and other Asian developing countries is due to the predominant use of energy-intensive industries and relatively low energy prices.

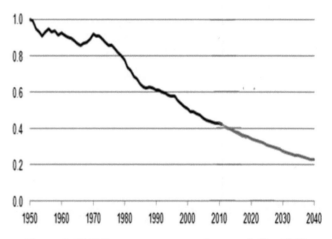

Figure 3-17. U.S. energy consumption per dollar GDP

Figure 3-17 shows that the U.S. energy intensity is in decline (a good thing). From 1950 to 2010, energy intensity in the U.S. decreased by almost 60% per GDP dollar. There was a period of sharp increase around the 1970s, although energy prices fluctuated only about 3% from year to year.

After the 1973 Arab oil embargo, energy prices rose significantly above the previous years, which lead to changes in the national energy policy. New vehicle efficiency standards were established, and consumer attitudes began to change as well. Since 1973, U.S. energy intensity has declined at a rate closer to 2% per year, and is expected to continue through 2040.

Surge of the knowledge-based economy, expressed in the rise of computer hardware, software and digital technology from 1991 to 2000, global economic productivity increased without parallel increases in energy use. New energy production technologies and use improved energy efficiency in almost every aspect of the economy. At the same time, the energy-intensive industries in the U.S. declined as well, with continuing structural changes in the economy.

Global energy intensity declined at an average annual rate of almost 1% in the 1980s and 1.40% in the 1990s. From 2001 to 2010, the rate dropped to 0.03%. This was a period of decline, when the developed countries restructured their economies with energy-intensive heavy industries accounting for a shrinking share of production. Global energy intensity increased 1.35% in 2010, reversing a broader trend of decline over the last 30 years.

In contrast, China's energy intensity declined 4.52% annually between 1981 and 2002, and a staggering 15.37% between 2005 and 2010. But that fell short of the government's incredible goal of 20%. The main reason for this shortfall was that over half of China's $630 billion stimulus plan was invested in infrastructure development, which drove up energy consumption.

The latest economic recovery has led to an increase in total energy consumption per unit of GDP, for the first time in more than 20 years (+0.5%) in the developed countries.

The European Union accounts for the highest increase in energy intensity, about 2.5% as compared to a 1.7% average annual decrease before the crisis. The poor EU performance was due to the industrial sector, where energy consumption did not decrease at the same pace as the value-added segment due to lower efficiency.

The key drivers of the present and expected decline patterns in the U.S. include:

- Residential energy intensity, measured as delivered energy used per household, is expected to decline about 27% by 2040.
- Commercial energy intensity, measured as delivered energy used per square foot of commercial floor space, is expected to decline about 17% by 2040.
- Industrial sector's energy intensity, measured as delivered energy per dollar of industrial sector shipments, rises above its 2005 level initially owing to the 2007-09 recession but is expected to decrease 25% below its 2005 level by 2040.
- Transportation sector's energy intensity is more difficult to measure because of the multiple modes of transportation. Light-duty vehicles are by far the largest energy consuming part of the sector. Light-duty vehicle energy intensity, which is measured as their consumption divided by the number of vehicle-miles traveled, is projected to decline by more than 47% by 2040 from their 2005 value.

Things are looking good in the U.S., from that perspective at least.

Energy Use Intensity

Energy use intensity (EUI) is a measure of energy use per unit area of a building or business. As one of the key metrics in the efficiency mix, EUI basically expresses a building's energy use as a function of its size, function, and other characteristics.

Note: EUI is not to be confused with the (national) energy intensity we discussed above. EUI is a measure of a single unit—building, business, or a home.

For most property types, the EUI is expressed as energy per square foot per year. It's calculated by dividing the total energy consumed by the building in one year (measured in kBtu or GJ) by the total gross floor area of the building. This can also be calculated from the energy use information in the monthly utility bills.

Generally speaking, a low EUI signifies good energy performance. A church campus, for example, uses much less energy compared to a hospital. The intensity of energy use, and the time periods are what makes the difference. Certain property types, however, will always use more energy than others.

It is obvious from Figure 3-18 that supermarkets and hospitals (usually 24 hrs./day operation) are using much more energy than churches with their 2-hour weekly use, or warehouses with dim and partial lighting. Because of that, the efforts to increase energy efficiency must be focused on the large users.

Note: Why a bank branch would use that much energy is a mystery to us, but the figures are right, so we must assume that banks have some operations (we don't know about) that require a lot of energy.

Energy and GDP

The amount of energy that a nation uses is related to its productivity, as expressed in its GDP and its population, which of course drives GDP. Therefore, to compare nations with different outputs and populations, it seems wise to divide both consumed energy and GDP by population.

Figure 3-19 shows unequivocally the wide range in power use (in kW per capita) between the lowest consumer (India) and the highest (USA). The GDP per capita is directly proportionate to the energy use per capita. In other words, the more money people make, the more energy they use.

Note: Japan is the exception in this case, as it is in many other cases. While its GDP per capita is even higher than the U.S., its energy use per capita is much less. Two times less, to be exact. Japanese are different people with different habits. They have accepted the fact that Japan is poor in energy sources, and are dealing with

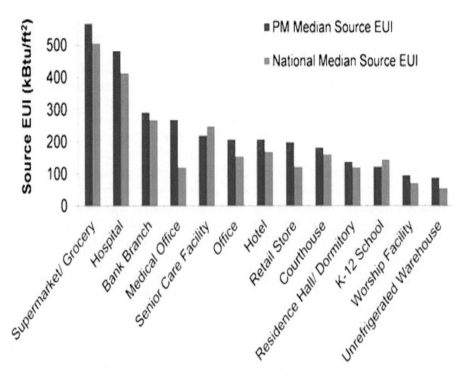

Figure 3-18. EUI of different building/business types

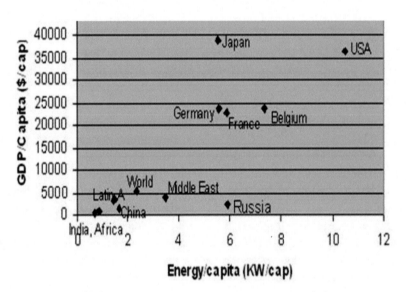

Figure 3-19. Energy use vs. GDP per capita

the restrictions and the high energy prices by using less energy. Could this happen in the U.S.? Not now and not any time soon.

India is an extreme case. Half of the rural areas in India don't have access to electric power, so the locals do with whatever they can, burning wood and coal, and many businesses generate electricity via diesel generators. In stark contrast, on the other side of the scale, each U.S. citizen needs an average more than 10 kW at any

moment to keep our way of life going. This includes all types of energy use in our daily lives: transportation, work, food, housing, leisure, etc.

This begs the question: How much of this energy is wasted?

There is a rough overall correlation between GDP/capita and energy/capita. This is because higher productivity per individual will cause a higher energy need for each. But comparing a developing country with a highly developed one is like comparing apples with oranges.

A more fair way to do this is to compare the U.S. and Japan, where the productivity per capita is comparable and are the best in the world from a productivity point of view. Yet Japan uses only half of the power used in the U.S.

Because of that, Japan emits 9.5 tons of CO_2/capita annually, vs. 19.7 tons per capita in the U.S., which roughly corresponds to the 2:1 energy consumption ratio of the two countries. Remember that the global per-capita CO_2 emission is only 4 tons per year—2.5 times less than the Japanese and almost 5 times less than the U.S. For some developing countries this ratio goes up to 20-30 times less than the U.S.

If we use energy as frugally as the Japanese, we would not need any crude oil imports. This is one (but not the only) way to obtain energy security and energy independence, while significantly lowering the total GHG emissions at the same time.

Note: After the 2011 Fukushima disaster, however, Japan shut down its nuclear plants and replaced the lost power by coal and natural gas-fired such. This, of course, increased the air pollution significantly, so the emission ratio has changed upward, but is still lower than the U.S.

The U.S. energy use per capita was constant from 1990 to 2007 and began to fall after 2007. Energy use per capita continues to decline as a result of improvements in energy efficiency, reflected in the use of new more-efficient appliances and CAFE standards. We have been also slowly introducing changes in the way we use energy in the form of energy conservation. As a result, under the present trend, the U.S. population is expected to increase 21% by 2040, but the energy use will increase only 12%. The energy use per capita is expected to decline 8% during the same time too.

The U.S. CO_2 emissions (in 2005 dollars of GDP) have tracked closely with energy use. With lower-carbon fuels accounting for a growing share of total energy use, CO_2 emissions per dollar of GDP are expected to decline more rapidly than energy use per dollar of GDP, or about 2.3% per year for a total of 56% below their 2005 level by 2040.

This is good and probably the most that can be expected from U.S. consumers, which is actually much better than the rate of energy use and CO_2 emissions expected from other, especially the developing, countries. This, however, is an unfair comparison, due to the much different (even opulent) lifestyle in the U.S., which makes a huge difference in every area of daily life.

Note: There is no official measure of the level of opulence (or extreme life styles) for different countries, but we don't have to dig too deep to see the striking difference. Just one example would illustrate the great life style chasm that exists between life in the States and other parts of the world.

During the 2014 Ebola epidemic events in Africa, thousands of people, including doctors, were infected and many died. In the best cases, the infected people—the lucky ones—got a bed and some food, but not much more than that. There were simply no adequate facilities and services to help them. They were basically left to fight the vicious Ebola on their own with little outside help. Almost half of the infected died, and the rest were left struggling in their misery for weeks and months.

At the same time, three American doctors were infected by the virus. With that news, the world stopped and held back its collective breath. Literally. The global media was brimming with concerns about the infected American doctors' status and following their progress step by step. An intense and very expensive effort started, to save the lives of the Americans at all cost. No amount of effort or expense was going to delay, let alone stop, the operation.

A private jet was equipped with special isolation tents and specialized equipment, attended by dozens of specialists onboard and on the ground. One by one, the doctors were carefully isolated in special suites, placed in the isolation tents in the jet and flown to specialized hospitals in the U.S., which are equipped for, and specialized in, handling acute infectious cases. Several doses of a new experimental vaccine were administered with the hope of stopping the virus and speeding the recovery period. The evacuation and treatment operations took several weeks at a cost of millions of dollars and ended successfully with saving the Americans' lives, who were up and well within a week or two.

Looking at the two different scenarios we clearly see that Americans have enormous advantages over the locals in any African village. Money, expertise, technology, and preferential treatment are always available when our wellbeing is at stake. This includes the advantages we have in the energy sector, where many African villages don't even have access to electricity.

During the 2014 Ebola disaster in Africa, many (mostly make-shift) hospitals didn't have the basic necessities to keep the patients alive. Some didn't even have electricity and/or running water. Thousands of people died from this lack of the most basic patient care needs.

At the same time, most Americans have access to unlimited electric power, water, health care and all the benefits that come with life in America, including unlimited resources in efforts to save American lives overseas. We also have unlimited right to complain about everything and everyone we wish. We are entitled…

No wonder the developing nations are trying to catch up with us at all cost. Who can blame them?

The Entropy Dilemma

Today the fossils represent the major energy sources, which our power generation and transportation sectors totally depends on.

The fossils (coal, oil, and gas) represent the lowest entropy — the lowest possible state of energy — on Earth.

The fossils' low entropy state is a blessing and a curse! Because of it, they are very convenient and cheap for power generation and transportation. And also because of that, we are using and abusing them by steadily, systematically, and uncompromisingly depleting and gradually eliminating them from the face of the Earth.

The fossils lie there in their primordial state of rest until we dig or pump them out. Then they jump from a state of lowest (dormant) entropy to enter (albeit temporarily) into a state of highest entropy, in order to deliver the maximum amount of energy for our needs.

During burning in the furnaces of power plants, or in cars engines, they also change their physical state, from that of liquid or solid to gas and solid residue. Since there is no way to return to their original state of low entropy, they are gone up in smoke. Literally! Forever! Only their smoke and sludge is left behind for us to deal with for hundreds of years.

When we use the fossils, we use up their energy, which depletes their entire usefulness by converting them in waste. What was created billions of years ago is gone in a split moment, never to be seen again. All that is left of the once glorious energy sources is a puff of hazardous smoke, and a pile, or a puddle of hazardous materials This is the shape and significance of the present-day energy crisis.

Burning and eliminating a large part of the fossils is changing the energy state of Earth; from the lowest to the highest possible form of entropy.

The transition to high entropy means that the total energy reserves on Earth—stored in the fossil energy resources—is reduced by the day. It is not hard to deduct, therefore, that some unpredictable changes are to be expected in the future.

During the golden age of fossil fuels (1920-1970) we didn't worry about any of this because useful fossil energy was plentiful and cheap. We thought it would last forever, and used it carelessly. Large 8-cylinder cars were the norm, and we drove them non-stop all over the place just for fun and because we could.

We still use and waste energy at a high rate, but now we are more careful, mostly because of its high price. We are also more acutely aware that time has come to begin the practice of "the new energy economy"; that of the conservation of our precious low-entropy energy forms. Even in best of cases, however, we only delay the inevitable with half-way measures and lots of talk. The high gas-pump prices are a constant reminder of the new reality.

The temporary low oil price trend of 2014-2015 only confirmed the volatility and un-sustainability of the energy markets. Unfortunately, even Saudi Arabia— the largest and richest oil producer—cannot afford such low prices, so prices will steadily increase with time.

The real long term problem is that even Saudi Arabia with estimated 265 billion barrels of oil reserves in the ground cannot produce crude oil forever. With production of about 10 million barrels a day (as in the fall of 2014), Saudi Arabia is number one oil producer, with the second oil producer coming at 3.25 million barrels daily production.

At this level, Saudi Arabia has oil (per some estimates) for about 45-50 years. After that oil will get much harder to pump out and with its quality diminishing, the prices will rise even faster and further than the present estimates. Until one day the pumps run dry…for ever!

With oil approaching its finite end, we must be thinking about the alternatives. They are many, but are in a disadvantageous situation (including the renewables) as far as competitive pricing is concerned. Because of that, things will move slowly (very slowly) for them, until some distant day when they would dominate the energy markets.

But even before we get to the end of the fossils, we still have to consider the present day consequences of using so much of them. The second law of thermodynamics tells us that because entropy increases in any transfer or conversion of energy, we should try to find energy transfer and/or conversion mechanisms (and products) in which the entropy of the source and the end use are as close as possible, thus minimizing the entropy increase. This is still a form of delaying the inevitable (the pending depletion of available energy sources), but at least it will buy us more time to find some alternative

Figure 3-20. The energy bridge to the future

energy sources for the short- and mid- terms.

This new way of thinking and living will lead us to a new way to measure efficiency in terms of "increased entropy." In the long run, the entropy of the universe will increase no matter what we do. Entropy is the "arrow of time" which increases the chaos level in the direction of the future.

In any actual decision about energy use, we have to consider other things affecting the entropy increase. Most importantly for our short term survival, we must also take into account the present energy costs and the environmental impacts caused by energy production and use, for instance.

Although we have considered these in the past, we have not paid much attention to the entropy involved in these processes. So, this may be just the right time to begin considering it as a reliable measurement. We must begin to apply this concept as a type of new thermodynamic economy, and a new structure of energy marketing, making certain that each use of low entropy is for a necessary end to be accomplished with as small an increase in entropy level and speed as possible. But doing this at the present rates of production and use of fossil energy sources makes the tasks difficult now, and becoming even impossible in the future.

We must always remember that the pending fossils' depletion is the most serious problem of the 21st century.
If unresolved soon, it could bring utmost devastation.

So what can we do? Since the world population increases, together with its need for more energy, increased energy efficiency in all steps of energy sources production and use seems the most plausible short term solution. The implementation of the renewable energy technology might be another solution (albeit partial) of the long term energy problems.

It all depends on how we—citizens of Mother Earth—look at all this and what we say and do about it. So let's see what people think.

The Energy Issues

To clarify the situation, we show in Figure 3-21, what the experts expect to see by the mid-21st century in the global energy field. And especially in the power generation sector. Obviously, as previously discussed, energy use around the globe is increasing due to growing demand. The rate of increase is expected to keep up with the demand; with coal as the dominant power generator and major polluter. Natural gas' role globally will increase too, approaching, and in some cases surpassing, that of coal.

Hydropower is also expected to grow, as are nuclear and renewables power generation, but at a moderate (and unpredictable) pace in a limited number of countries.

Here we must emphasize the fact that the future of the fossils (coal and natural gas) is quite clear and we don't expect anything to get on their way on the short

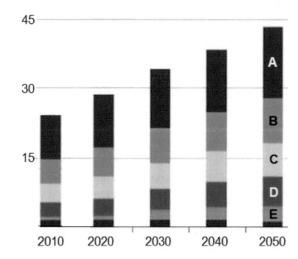

Figure 3-21. Global power generation by source (in trillion kWh/annum)
Legend:
A Coal
B Natural gas
C Hydropower
D Nuclear
E Renewables

run. Their long term future, however, is also very clear, but not that bright for the fossils.

At this pace of exploitation, the fossils are doomed to a complete extinction by the end of this century…or sooner!

The situation with hydro, nuclear and the renewables is more volatile and many good or bad surprises are expected:

- Hydropower depends heavily on water availability, which is getting scarcer. Periodic draughts of extended duration and severity are expected to aggravate the water scarcity around the globe, so hydropower might become an unwilling victim of the global warming.
- Nuclear power's future, as predicted in Figure 3-21 is possible ONLY IF no new Fukushima accidents occur any time soon. One more such accident could put the entire global nuclear power fleet on standby and the entire nuclear industry on its knees for undetermined amount of time, or in worst case scenario, forever.
- The progress of the renewables (solar and wind especially(depend heavily on the above developments and many other factors. While coal and natural gas are abundant and cheap, the renewables will be treated as an expensive novelty and will not be given a chance for mass increase. Only if and when the fossils get scarce and expensive (most likely by the end of the century) will solar and wind be able to dominate the energy sector.

Government support and subsidies are also critical for the renewables; especially at this early stage of their development. Without government assistance, the renewables will stagnate until the conditions change.

- The other renewables (biofuels, ocean wave energy, etc.) will continue growing modestly too, but only to pacify the environmentalists and meet the minimum government mandates. Their future also depends heavily on the availability and pricing of the fossil fuels. Government assistance is of utmost importance as well.

The above points are heavily debated on all levels. The debate is quite interesting and simply shows how deeply divided our society is. We, the U.S., are truly a nation divided. The energy debate is only one of the divisions, and quite small in comparison…for now. As the energy and environmental issues grow, and the public gets more and more informed, the debate will grow in parallel.

The Energy Debate

As discussed throughout this text, the fossils rule for now in the U.S. and around the globe, but a closer look shows that their end is approaching. As a result, an energy debate is underway about fossils vs. renewables. Why, when, who, how much, etc., etc. questions and issues prevail in the debate.

Our energy security, the environmental issues, economic development, etc. variables, are also included in the debate, so it is a quite complex mixture of science, technology, politics, economics, and legal issues.

The major portion of the debate is about the present and future role of the renewable technologies. The number of interested parties in this subject is growing, and when it comes to power generation—fossils vs. renewables—they are now cleanly divided into three major groups.

The major groups in the present day energy debate are:

1. Those who are strongly against renewables,
2. Those who are strongly pro renewables, and
3. Those who don't care, or don't want to get involved.

In more detail:

- *Against renewables.* The most striking case against renewables we've ever seen, was quite eloquently expressed in an article titled, *"Big Green, Not Big Oil, is the Enemy"* published in the summer of 2013 in "the commentator." In it, the authors provide provocative ideas, information, and misinformation showing the renewables as a cancer growing within our society. Cancer ready to devour us, our values, and our society as we know it today.

"…because of angst-ridden theoretical speculation—note: not empirical science—the modern green agenda has effected an intellectual disconnect. It is a disconnect that has seen eco-theories eclipse energy realities such that national leaders, industry executives and even reasonable people are not engaging in rational debate let alone action," the authors claim.

"There is no contest from an economic point of view. Solar, wind and other alternatives favored by the greens are not and will never be viable. From a thermodynamic point of view they will never amount to much more than one percent of world energy demand without

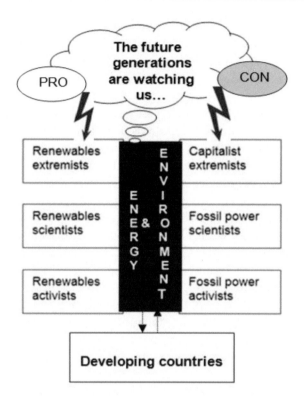

Figure 3-22. The energy and environment battlefield

massive and unsustainable government subsidies," is the authors' opinion, so forget about renewables and focus on oil and gas.

And they continue, "The reality is, however, that it's the well-funded lobby of Big Green that has the greatest influence on government and government policy—and presents the greatest threat to the economy."

And sure enough, they make it quite clear, "Let's get it straight, Big Green is Big Business. Next time you are moved to pitch your 'hard-earned' cash into a tin shaken before you while being regaled with a nonsense message—e.g. that carbon dioxide is a 'pollutant'—you might like to bear that in mind.

"As it stands, today's prevailing cultural 'green agenda' is nothing less than a case of anti-intellectualism, costing the earth, in pursuit of the illusory."

In a broader attack, the authors of the article, *Fossil Fuels Will Save the World (Really)*, published in the Wall Street Journal in March, 2105, argues, "That fossil fuels are finite is a red herring. The Atlantic Ocean is finite, but that does not mean that you risk bumping into France if you row out of a harbor in Maine. The buffalo of the American West were infinite, in the sense that they could breed, yet they came close to extinction. It is an ironic truth that no nonrenewable resource has ever run dry, while renewable resources—whales, cod, forests, passenger pigeons—have frequently done so."

Brilliant, ain't it? Pretty shallow as a scientific argument, as far as the definition of "finite" goes. The Atlantic Ocean is not finite, but constantly replenishing water body. And nobody ever thought that the buffalo were infinite, since there is a finite number of them.

But the most brilliant conclusion the author draws is a classic case of blind leading the blind. " Wind power, for all the public money spent on its expansion, has inched up to—wait for it—1% of world energy consumption in 2013. Solar, for all the hype, has not even managed that: If we round to the nearest whole number, it accounts for 0% of world energy consumption."

With one stroke of the pen, the genius author was able to eliminate (and equate to zero) all the achievements of the solar industry during the last decade. The billions of dollars spent on solar equipment, presently operating on thousands of acres around the globe, and growing, were reduced to ashes.

OK, we must agree that the renewables are not much (compared to the fossils) today, and will never amount to much, if the author's point of view is accepted as the status quo. But—wait for it—solar and wind have progressed that much in less than a decade, while the fossils have been around for centuries (literally speaking) and have been widely used for over a century.

But the first authors' conclusion, "The future of oil and gas is not solar and wind; the future of oil and gas is oil and gas," while the latter, blaming the rich for denying the poor the right to get rich through fossils, are void of any practical and scientific truths. They only muddy further the already muddy water of the energy markets.

Both of these writings represent the views of anti-renewables extremists. A blatant denial of the fact that the fossils—regardless of their potential, quantity, and present worth—are here today and gone tomorrow; a temporary solution to our energy problems. It is not possible for anyone to argue successfully that the fossils will last forever, or even more than a century. It is physically impossible for this to happen, because their quantities are limited and they are not replenished. It is the nature of things.

It took Mother Nature billions of years to create and store the fossils in its bosom. It took us—people—only a century to use half of them. How long would it take us to use the rest? Not very long, for sure!

If we do not have enough renewables by the time the inevitable "fossil-less" reality arrives, we and the future generations would be faced with an unprecedented challenges. Horrific fights for survival would erupt around the world, the outcome of which is unpredictable.

Maintaining that using oil and gas indiscriminately is the only way to go, is a good example of *capitalist extremism* (making a buck today, heck with tomorrow), and a blatant refusal of the fact that oil and gas cannot last forever. It is self-centered egoism thinly veiled in economic justification.

No matter how better oil and gas are, and/or how much of them we have, they are a finite energy source. They will be imminently depleted some day soon. 20, 30, 50, or 100 years from today at the present level of exploitation. And then what?

In contrast, sun and wind will always be here. If, God forbid, someday they are gone, life on Earth will be gone too. So it follows from a practical point of view—forget about politics and scientific debates—that we must be careful when considering using fossils for ever. It simply cannot happen, regardless of how much we like that idea, so the authors are barking at the wrong tree.

•   Pro-renewable advocates can also be divided in several groups:
    —   The extreme tree-huggers,
    —   The caution-driven realists, and
    —   The "let's wait and see" pragmatists

This is a large group in the U.S., where 80% of the people poled in 2014 have approved of renewables, while at the same time, 80% are against use of fossils.

Each of these affiliations has its problems, stemming from misunderstandings of the technical details. So the overwhelming conclusion is that we must be well educated on all issues at hand, and be very careful when discussing the different aspects of energy and energy security.

This is especially important for those who are given the privilege of making decisions on behalf of all of us—namely politicians and big company executives.

•   The agnostics; those who don't care and/or don't want to get involved in any of these issues, can also be divided into several groups;
    —   Those who are badly misinformed,

    —   Those who don't have access to information, and
    —   Those who refuse to even consider the issues at hand

This might be the largest group of the energy debate, since it includes hundreds of millions of people in third-world countries who have not even heard of the issues, or are too busy with daily survival tasks to worry about the pending energy issues.

Each of the sub-groups has its own problems that cannot be resolved overnight. In all cases, reliable information, delivered in a believable way, is critical for educating these people. Armed with it, they can then make their conclusions and decisions.

Without taking part in the debate, we must only add that the future of the renewables in the U.S. and the world is closely related to the future of,

a) power grid expansion and upgrades, and

b) new energy storage options.

Without these two factors, the renewable energy sources—solar and wind in particular—will never reach the level of mainstream, base load, power generators. There is no debate on these issues for now, but the discussions are starting and the issues will become more relevant and important as time goes on.

Although the future of the energy sources and the related markets is not determined by the debate, it is somewhat affected by the disagreements. To avoid that becoming a real barrier, more reliable scientific information must be provided to the general public in order to educate us all on the details of the energy technologies and their effects on the energy markets and the environment.

Notes and References

1.   http://www.energymarketers.com/
2.   ExxonMobile Perspectives
3.   https://www.energymarketingconferences.com/
4.   http://www.strategicenergymarketing.com/
5.   http://www.renewableenergymarketing.net/
6.   *Photovoltaics for Commercial and Utilities Power Generation*, Anco S, Blazev. The Fairmont Press, 2011
7.   *Solar Technologies for the 21st Century*, Anco S Blazev. The Fairmont Press, 2013
8.   *Power Generation and the Environment*, Anco S. Blazev. The Fairmont Press, 2014
9.   Energy Security for the 21st Century, Anco S. Blazev. The Fairmont Press, 2015

Chapter 4

Fossil Energy Markets

The conventional energy sources (the fossils) dictate the global energy markets. Continuing this trend will eventually create more and deeper conflicts around the world, throwing it into darkness and untold disparity on the long run.

—Anco Blazev

BACKGROUND

In the energy sector, the term *fossil energy sources, also known as conventional, or traditional energy sources* was used until recently to refer to a number of products and services related to coal, natural gas, and crude oil, and their use in power generation and transportation. Now days, these terms still stand in the foundation of the *conventional energy* concept, but are expanded significantly by the inclusion of additional products and services, a la 21st century, as needed to make (in reality or just on paper) the old *conventional energy sources* safer, cheaper, more efficient, and even more functional.

These are new types of products and services that add to the qualities of the fossils in an attempt to present them in a new and more favorable light. This new phenomena is forced upon us by the new world order we live in, and reflected in the relentless media attacks we endure daily. The sudden explosion of the renewables on the energy stage throws more fuel in the fire. This augments the competition battles and the need for the fossils energy sector to add new products and services.

The fossil energy sources, and the new products and services their principals offer, are one of the key subjects in this text, so let's start with taking a close look at our old friends, the fossils.

The fossil energy markets, we review in this chapter, are the markets that deal directly or indirectly with the major conventional energy sources: coal, crude oil, and natural gas. The electricity markets are also part of this discussion, since the fossils are heavily involved in them.

For the purposes of this text, we subdivide the fossils, and their respective markets, in the following categories.

Coal
— Exploration and location services
— Engineering design services
— Mine equipment manufacturing
— Mine construction
— Mine operation and maintenance
— Coal exports
— Transport services

Natural Gas and Crude Oil
— Exploration and location services
— Engineering design services
— Well equipment manufacturing
— Well construction
— Well operation and maintenance
— Imports and exports
— Transport services

Future Developments
— Undiscovered fossil reserves
— Methane hydrates
— New technologies

Please note that we have grouped Natural Gas and Crude Oil in one single category. This is because the entire life cycle of these energy sources—from the initial creation to the final use—is quite similar, if not identical. They are usually found and exploited together as well. The big difference comes in their use. Once produced, refined, they are transported to different areas, where we use crude oil mostly for transportation, while natural gas is used mostly for power generation.

Another significant difference here is that we import half of the crude oil we use, while at the same time we have excess of natural gas and even export some of it. This is a big difference that we take a close look herein as well.

Very importantly, let's remember that the term *Fossil Energy Markets*—at least for the purposes of this text—includes anything and everything related to the different fossil materials; mining; drilling and pumping; processing; power generation, as well as the distribution and sale of fuels and electricity coming from fossils-burning plants. Here we include anything and anybody participating in this long process.

Let's start with our oldest and most reliable fossil friend, coal.

COAL

The undeniable fact is that coal has powered the world during the last century, and is solely responsible for the success of the industrial revolution. If there was not enough coal during the 19th century, we still might be riding in horse carts. Now days coal continues to be the most important fuel around the world. Without coal, most of the world's leading economies would sputter and some would even stall. Most of our houses would be cold and dark places too.

The other undeniable, but not as pleasant fact is that coal emissions are killing us and are damaging the global environment in the process of generating electricity. As a matter of fact, this abnormality is getting more and more alarming as time goes by.

Coal is one of these things in life that
we love and hate at the same time.

We simply cannot live with and without it. No matter what we think or do about it, however, the facts are that coal is going to be around for a long time to come, so we just have to learn to live with it. In order to do that, we must first learn all there is to learn about coal; its properties, behavior, etc. In order to cover all there is about coal, however, we need to write another book, so in this text we will focus only on its basics as related to the energy markets.

Note: For more information and data on the fossils, we would refer the esteemed reader to our books on the subject, *Power Generation and the Environment*, and *Energy Security for the 21st Century*, published by the Fairmont Press in 2014 and 2015 respectively.

The U.S. alone mines, transports, and
burns over 1 trillion tons of coal annually.

Producing and using 1 trillion tons of coal every year is a lot of coal to burn in a short period of time. Looking at the official data, the U.S. has about 19 trillion tons of coal reserves in the existing mines. What this means is that in 19-20 years from today our present coal producing mines will be depleted and we have to dig somewhere else for coal.

The U.S. also has over 250 quadrillion tons of
proven (mostly unexploited) coal reserves.

Considering the fact that the entire world's proven coal reserves are about 950 quadrillion, the U.S. has nothing to worry, as far as the future of coal is concerned. With over 1/4 of the world's coal deposits on our territory, the U.S. is the Saudi Arabia of coal. At the present levels of exploitation, the 250 quadrillion tons coal reserves would ensure at least 200-250 years of non-stop coal availability in the U.S. This is great news, of course, but we must also consider the negatives; price increases with time, and the additional millions of tons of GHGs coal-fired power plants will inject in the atmosphere as time goes on.

We produce and use a lot of coal, and that situation will not change much in the future, so let's see what is done with these mountains of coal and ash, accompanied by clouds of emissions.

Coal's Cradle-to Grave Process

The U.S. mines produce 1 trillion tons of coal annually, which means that the U.S. miners dig out and load 10 million train cars in 100,000 unit train compositions every year. These 100,000 unit trains transport the coal to power plants all over the country. About 275 unit trains, hauling about 100 cars each, where each car is loaded with 100 tons of coal, travel every single day from the mines to the power plants.

At the power plants, the coal is unloaded, processed, and sent to be burned in a special furnace. Water boiling in the furnace generates steam, which is then injected into turbines. The turbine's blades are attached to an axle, which is in turn attached to an electric generator, which generates electric power when the blades turn. The electric power is finally sent into the national grid for distribution to residences and businesses across the country.

The only thing left of the trillion tons of coal after the furnace are mountains of dust, ash, and sludge at the mines and the power plants' back yards. Additionally, millions of tons of toxic and green house gasses (GHGs) are emitted in the atmosphere from each power plant's smoke stacks.

This process has been repeated non-stop every single day all across the country and the world for over a century. Taking a closer look at the different steps of the process, we see a huge number of people, machinery, and activities that make all this happen. Millions of miners, train engineers, power plant workers, utility technicians, contractors, consultants and many other people and companies are involved in the coal production, transport, processing, burning, and waste disposal process daily.

There are about 90,000 miners in the U.S. mines, which means that each miner is responsible for digging over 11,000 tons annually, or over 30 tons daily. Of course, the digging is no longer done with hand tools, but with huge machinery, and yet, this is a lot of coal to dig out, move, and burn non-stop every single day, 365 days every year.

Upon a closer look, we see also a layer of people and companies, that are indirectly involved in the cradle-to-grave process of coal production and use. These are people involved in locating coal and constructing the mines, as well as people who manufacture, transport and install the mining equipment in the mines. There is also an army of consultants, contractors, and subcontractors, who make sure the work is done properly and safely from day one, and who then assist in the day to day operations until the end of the mine's life cycle.

And there is yet another layer of people and companies that are indirectly involved in the process. Here we see thousands of government officials, bankers, insurance agents, doctors, restaurant and hotel workers, and an army of lawyers. These people get involved in the different stages of process from time to time, thus completing the huge and versatile market supporting the coal industry.

Let's take a closer look at our old friend, Mr. Coal. Where did it come from, what are his properties and behavior?

In the Beginning…

Coal is a part of the fossils group, named so because they (coal, natural gas, and crude oil) came from fossil remains. Coal was formed millions of years ago mostly from plant material that accumulated in some areas, where complete decay of organic matter was delayed or entirely prevented by the lack of oxygen.

Multitude of plants and animals that died as a result of a catastrophic event would be covered quickly with water, silt, sand, and other sediments. This way they were protected from reacting with oxygen and decomposing to carbon dioxide and water, as is normal for all organic materials.

Note: Coal deposits have been formed during many periods in the millennia, but the records show that the best conditions for coal conversion were during the Carboniferous Age, about 300-350 million years ago, and later on during the Upper Cretaceous Age, about 100-150 million years ago.

Anaerobic bacteria attacked the dead plant debris and converted it to simpler forms, such as pure carbon and a mixture of other simple compounds of carbon and hydrogen. Thus created compounds and links in the mass are what we now call hydrocarbons.

At the initial stages, the decaying matter is soft, woody material, which we call "peat." Peat is still found in some areas of the world and used as a fuel, although it is a low quality fuel. It burns poorly and emits huge amounts of dense smoke.

Most of the peat was allowed to remain in the ground for millions of years, during which it eventually became buried deeper and compacted as layers of sediment (overburden) piled above it. Thus created tremendous pressure, which also created heat, slowly converted the fragile and soggy peat into coal. At this point of the process, lignite or brown coal is formed.

More time goes by and the continued compaction and heat exerted by the overburden converted the lignite into bituminous (or soft) coal. After more time, the soft coal became hard coal, or anthracite.

Coals, from different types and times of formation, exhibit different quality and properties.

Today, coal is often found in "seams," huddled between layers of sedimentary rock. Often, the coal deposits have been pushed up and may be found on, or very near, the earth's surface. Most often, however, they are buried deep underground; hundred and thousands of feet below the surface.

The actual coal seams vary between 2 and 200 ft in thickness. Because of the variability of depth and thickness, we now use different methods for mining coal, as we will see later on in this text.

Coal Properties

Coal has a large and complex molecule. It consists mostly of carbon combined with hydrogen, oxygen, nitrogen and sulfur atoms. The actual numbers of different atoms in each molecule vary, but basically speaking its chemical formula can be expressed as

$C_{100}H_{73}O_8NS$.

What this formula tells us is that for every 100 atoms of carbon in the coal molecule, there are 73 atoms of hydrogen, 8 oxygen, one nitrogen, and one sulfur atom. This ratio may vary some for the different coal types. The distribution of the atoms and links to each other in this molecular structure are too complex for this text, so we won't get into it for obvious reasons. Only organic chemists, and other scientists specialized in this subject understand exactly, and can deal properly with, the coal structure and the relations within, so we will leave this task to them.

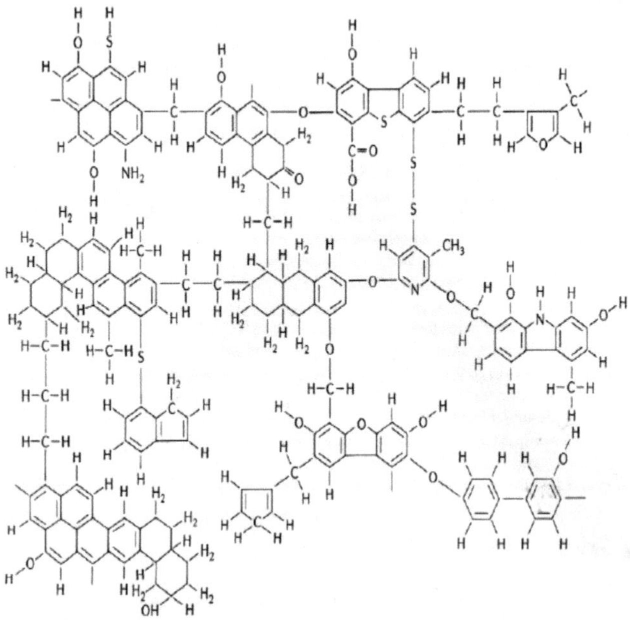

Figure 4-1. Segment of a coal molecule.

Figure 4-1 could prove too confusing for most of us, but we could not resist showing how complicated our old, and seemingly simple, friend Mr. Coal is. For our purposes, it suffices to say that coal' chemical structures consist of a large number of carbon, hydrogen and oxygen atoms, which are different for each type of coal type. The molecules in each lump of coal are interlocked into long chains, which, in addition to the quantity of carbon, hydrogen and oxygen atoms, determine the overall properties and qualities of the different types of coals.

Although the different types of coals have somewhat similar formulas, they have different qualities, which are determined by the ratio and arrangement of the atoms in the molecules. This arrangement varies with coal type and location and is usually an indication of the time frame, and the specific way, the coal layers were formed.

One important thing to remember here is that the hydrogen, oxygen, and other atoms in the coal molecule

determines its qualities. Basically speaking, coals that were formed in more recent past tend to have more hydrogen and oxygen atoms in their molecules. They are also of lower commercial quality and monetary value, mostly due to their lesser heating value.

There are four types or coal ranks: lignite or brown coal, bituminous coal or black coal, anthracite, and graphite. Each of these has a certain set of physical characteristics which are expressed by the molecule structure, in addition to moisture and volatile compounds content (in terms of aliphatic or aromatic hydrocarbons) and overall carbon content in the coal.

Note: Graphite is type of coal, but his properties are totally different. It does not burn, therefore it is not part of the energy sources subject of this text.

A closer look of coal's properties:

Energy Content
The most important (from commercial point of view) property of coal is its energy content. It is expressed in Btu/lb of coal, which is basically the amount of heat in Btus that a pound of coal can generate.

Coal Grade	(Btu/lb)
Anthracite	12910
Semi-Anthracite	13770
Low-volatile bituminous	14340
Medium-volatile bituminous	13840
High-volatile bituminous A	13090
High-volatile bituminous B	12130
High-volatile bituminous C	10750
Sub-bituminous B	9150
Sub-bituminous C	8940
Lignite	6900

Figure 4-2. Heating values of coal

The *amount of energy or fuel content* is the amount of potential energy contained in the coal that can be easily converted into actual heating ability. This heating ability (or heating value) is the factor that determines what coal is going to be used for, how much, and at what price. Anthracite and its sub-categories are the most valuable from that perspective and bring the highest price per unit as compared with most of the other coal types.

Another way of evaluating the heat content in coal is looking at its calorific value Q, which is the actual heat liberated by its complete combustion at different oxygen levels. The Q value is determined experimentally by using special test equipment, such as calorimeters.

The approximate formula for determining Q at an oxygen content of less than 10% is:

$$Q_{kJ/kg} = 337C + 1442(H - O/8) + 93S$$

Where:

C is the mass percent of carbon in the coal sample,
H is the mass percent of hydrogen,
O is the mass percent of oxygen, and
S is the mass percent of sulfur.
Q in this case is given in kilojoules per kilogram.

Physical Properties

In addition to the basic structure, *moisture* is one of the most important property of coal, which determines in part its commercial value. All coals contain some amount of moisture. Moisture may occur in four possible forms within coal:

1. *Surface moisture*, or water held on the surface of the coal particles;
2. *Hydroscopic moisture*, or water held by capillary action within the microfractures of the coal pieces;
3. *Decomposition moisture*, or water held within the coal's decomposed organic compounds; and
4. *Mineral moisture*, or water as part of the crystal structure of hydrous silicates such as clays.

External moisture, known as *adventitious* (or surface) moisture (as in 1 above) is readily evaporated during the mining and transport. Moisture that is held within the coal itself (as in 2, 3, and 4 above) is known as *inherent moisture*. It is analyzed and treated differently, but is, generally speaking, much more difficult to deal with. Its quantity usually determines the final price of the coal batch.

Volatile matter are all other components of coal, except for moisture which can be freed (evaporated) at high temperature in absence of air. These are usually mixtures of short and long chain hydrocarbons, aromatic hydrocarbons and often sulfur. The volatile matter of coal is determined under rigidly controlled standards. In Europe, for example, this involves heating the coal sample to 1650 ±10 °F in a muffle furnace and measuring the residue. The U.S. procedures involves heating the coal samples to 1740 ± 45 °F in a vertical platinum crucible and weighing the before and after samples. The difference is the volatile matter, which number is used to determine coal's quality and its price.

Fixed carbon is the carbon found in the material which is left after all volatile materials has been evaporated. This is different from the *ultimate* carbon content of the coal, because some carbon is lost with the volatiles. Fixed carbon is an estimate of the amount of coke that will be yielded from this type of coal. Fixed carbon is determined by removing the mass of volatiles determined by the volatility test from the original mass of the coal sample.

Ash is the non-combustible residue which is left after coal is burnt. It is the bulk mineral matter left behind after all carbon, oxygen, sulfur and water has been evaporated during combustion. Ash content is determined by burning the coal thoroughly and the remaining ash material is weighed and expressed as a percentage of the original weight.

Coal is sorted and classified by rank, according to its physical properties. This is basically a measure of the process of its formation and its present state, which represents the amount of alteration (the amount of heat and pressure) that the coal has undergone during its formation. The increase in rank describes an increase in temperature and pressure which results in the coals having a lower volatile content, therefore increased carbon content. That also determines its heat content, which in turn determines its commercial value, with anthracite having the highest heating and commercial value.

Coal is also classified according to the content of its contaminants; sulfur, phosphorous, volatile and ash contents, which generally vary according to, and determines, its rank and price as well. Consecutive stages in evolution of rank, from an initial peat stage, are brown coal (or lignite), sub-bituminous coal, bituminous coal, and anthracite.

In addition to being burned in coal-fired power plants for electricity generation, coal (in form of cocking coal) is also used in the steel making industry. This type of coal must have specific qualities such as low sulfur and phosphorous content.

Approximately 630 kg of cocking coal are used for every ton of steel produced.

Electricity generation normally uses lignite (thermal) coal, which is ground to a fine powder prior to combustion.

Coal Types

Coal is classified into four general categories, or "ranks." They range from lignite to sub-bituminous, bituminous, and anthracite. These ranks reflect the formation process in terms of time and the progressive response of individual deposits of coal to increasing heat and pressure.

The carbon content of coal determines its heating value (or most of it), but other factors also influence the amount of energy it contains per unit of weight.

Note: The amount of energy in coal is expressed in British thermal units per pound. A Btu is the amount of heat required to raise the temperature of one pound of water one degree Fahrenheit.

About 90% of the coal in the U.S. falls in the bituminous and subbituminous categories, which rank below anthracite and, for the most part, contain less energy per unit of weight. Bituminous coal predominates in the Eastern and Mid-continent coal fields, while subbituminous coal is generally found in the Western states and Alaska.

Lignite ranks the lowest in value and is the youngest of the coals. Most lignite is mined in Texas, but large deposits also are found in Montana, North Dakota, and some Gulf Coast states.

The coal types are priced by their carbon and energy content, as determined during testing procedures.

Commercial Classification

The commercial classification of coal is generally based on the content of volatiles. However, the exact classification varies between countries.

Class	Volatile matter weight in %
Anthracites	< 6.1
Dry steam coals	9.1 - 13.5
Cooking steams coal	15.1 - 17.0
Low volatile steam coal	19.1 - 19.5
Prime cooking coal	19.6 - 32.0
Heat altered coal	19.6 - 32.0
Strong coking coal	32.1 - 36.0
Medium coking coal	32.1 - 36.0
Weak coking coal	32.1 - 36.0
Very weak coking coal	32.1 - 36.0
Non-coking coal	32.1 - 36.0

Figure 4-3. Coal types per volatile matter content

So, depending on the different variables, specific uses and application we now have several types of coal, as follow:

Anthracite

Anthracite is coal with the highest carbon content, between 86 and 98%, and a heat value of nearly 15,000 Btus-per-pound. Most frequently associated with home heating, anthracite is a very small segment of the U.S.

coal market. There are 7.3 billion tons of anthracite reserves in the United States, found mostly in 11 northeastern counties in Pennsylvania.

Steam Coal

Steam coal is a grade between bituminous coal and anthracite, which was widely used as a fuel for steam locomotives in the past. In this specialized use, it is sometimes known as "sea-coal" in the US. Small steam coal, or dry small steam nuts, was used as a fuel for domestic water heating in the past as well. Now days steam coal is used in a number of applications that can afford its somewhat higher cost.

Bituminous Coal

Bituminous coal has a carbon content ranging from 45 to 86% carbon and a heat value of 10,500 to 15,500 Btus-per-pound. The most plentiful form of coal in the United States, bituminous coal, is used primarily to generate electricity and make coke for the steel industry. The fastest growing market for coal, though still a small one, is supplying heat for industrial processes.

Subbituminous Coal

Subbituminous coal contains 42 to 52% carbon (on a dry, ash-free basis), and its heat (calorific) values ranges between 19 and 26 MJ per kilogram, or about 8,200 to 11,200 Btus per pound. It is also characterized by greater compaction than lignite and has greater brightness and luster. The lower quality, woody-like structure lignite is not found in subbituminous coal, which exhibits alternating dull and bright maceral bands composed of vitrinite in patterns similar to those found in bituminous coals. Some subbituminous coal is macroscopically indistinguishable from bituminous coal.

Subbituminous coal is the type of coal used most extensively in coal-fired power plants, so it deserves a closer look. It is also called black lignite, and has dark brown to black color. Its qualities (and heating value) are between lignite and bituminous coal according to the coal classification used in the United States and Canada. In many other countries subbituminous coal is considered to be a brown coal and is used for domestic and commercial heating and cooking purposes.

There are reliable estimates that nearly half of the world's proven coal reserves are made up of subbituminous coal and lignite. There are large deposits of these coal types in Australia, Brazil, Canada, China, Russia, Ukraine, Germany, other European countries, and the United States.

Most subbituminous coal is considered to be relatively young from a geological point of view, dating from the Mesozoic and Cenozoic eras, or about 250 million years ago. Although age is important, the quality of coal is determined primarily by the pressure and temperature reached during the "cooking" cycle.

Subbituminous coal usually also contains less water, around 10 to 25% and is harder than lignite. This makes it more valuable and easier to transport, store, and use.

One important characteristic of great importance to the energy generation sector is the fact that although subbituminous coal has lower heat value than bituminous coal, it is lower in sulfur content, usually less than 1%, so it is preferred in some cases. Since it has a lower heat value, however, more subbituminous coal has to be burned in order to obtain equal amount of energy.

Recently a number of coal-fired power generating plants have switched from burning bituminous coal to subbituminous coal and lignite. The main reason is the relatively low sulfur content.

Lignite

Lignite is a geologically young coal which has the lowest carbon content, 25-35%, and a heat value ranging between 4,000 and 8,300 Btus-per-pound. Sometimes called brown coal, it too is mainly used for electric power generation, but its low energy content makes it less desirable than any of the other types.

Peat

Peat is a fairly young coal, actually considered to be a precursor of coal. It has some industrial uses as a fuel for some special applications. One of the wide applications is its dehydrated form. Since dehydrated peat is a highly effective absorbent, it is used in great quantities for fuel and oil spills on land and water. It is also used as a conditioner for soil to make it more able to retain and slowly release water.

Graphite

Graphite is technically the highest rank of coal with the highest carbon content, but its physical structure and chemical properties make it difficult (even impossible) to ignite, so it is not commonly used as fuel. It is instead mostly used for making pencils, electrodes and, when powdered and/or mixed with other chemicals, is widely used in lubricants.

Commercial Uses of Coal

Somebody said that the present-day role of coal is to help fill the gaps in our power generation, and to alle-

viate our energy poverty. This is true, but in addition to its wide use in electric power generation and domestic heating and cooking, coal has a number of other, quite versatile and useful applications. It is an amazingly versatile product, but most people are unaware of its variety of important applications.

There are many important commercial uses of coal, in addition to power generation. These form entire industries, or support the function of other industries. So take a look at the facts below before deciding to erase humble Mr. Coal, from the list of our good friends:

Sea Coal

Sea coal is finely ground bituminous coal, known in this application as sea coal, is a constituent of foundry sand. While the molten metal is in the mold, the coal burns slowly, releasing reducing gases at pressure, and so preventing the metal from penetrating the pores of the sand. It is also contained in 'mould wash', a paste or liquid with the same function applied to the mould before casting.

Sea coal can be mixed with the clay lining (the "bod") used for the bottom of a cupola furnace. When heated, the coal decomposes and the bod becomes slightly friable, easing the process of breaking open holes for tapping the molten metal.

Chemicals Production

Coal is used extensively as feedstock to produce chemicals using processes which require substantial quantities of water, and which release a number of toxic gasses and liquids. Because of that, presently most of the coal to chemical production is concentrated in China, where environmental regulation and water management policies are weak to non-existent.

In coal-to-chemicals, synthesis gas (syngas), which is a gaseous mixture of primarily carbon monoxide and hydrogen, gas is produced by gasification of coal. The syngas can then be used as a chemical building blocks in a number of chemical processes, such as in making methanol or acetyls.

Ammonia and urea are significant products of coal-to-chemicals for use in fertilizers. The syngas composition, or the ratio of hydrogen to carbon monoxide, is important for some downstream processes, so a water-gas shift reactor is sometimes used to change this balance.

Coking Coal

Coking coal (coke) is a solid carbonaceous residue derived from low-ash, low-sulfur bituminous coal from which the volatile constituents are driven off by baking in an oven in the absence of oxygen at 1,832°F in order to fuse the fixed carbon and residual ash together.

Coke from coal is grey, hard, and porous and has a heating value of 24.8 million Btu/ton (29.6 MJ/kg). Some coke-making processes produce valuable byproducts, including coal tar, ammonia, light oils, and coal gas.

Metallurgical coke is used as a fuel and as a reducing agent in smelting iron ore in a blast furnace. The result is pig iron, and is too rich in dissolved carbon, so it must be treated further to make steel. The coking coal is low in sulfur and phosphorus, so that they do not migrate into the metal to deteriorate its properties.

The coke is also strong enough to resist the weight of overburden in the blast furnace, which is why coking coal is so important in making steel using the conventional route. An alternative route is direct reduced iron, where any carbonaceous fuel can be used to make sponge or pelletized iron.

Petroleum coke is the solid residue obtained in oil refining, which resembles coke, but contains too many impurities to be useful in metallurgical applications.

Gasification

Coal gasification can be used to produce syngas, which basically a mixture of carbon monoxide (CO) and hydrogen (H_2) gas. This syngas can then be converted into transportation fuels, such as gasoline and diesel, through the Fischer-Tropsch process.

This technology is currently used by the Sasol chemical company of South Africa to make motor vehicle fuels from coal and natural gas. Alternatively, the hydrogen obtained from gasification can be used for various purposes, such as powering a hydrogen economy, making ammonia, or upgrading fossil fuels.

During gasification, the coal is mixed with oxygen and steam while also being heated and pressurized. During the reaction, oxygen and water molecules oxidize the coal into carbon monoxide (CO), while also releasing hydrogen gas (H_2).

This process has been conducted in underground coal mines and in the production of town gas.

$$C(\text{as Coal}) + O_2 + H_2O \rightarrow H_2 + CO + O_2$$

If the refiner wants to produce gasoline, the syngas is collected at this state and routed into a Fischer-Tropsch reaction. If hydrogen is the desired end-product, however, the syngas is fed into the water gas shift reaction, where more hydrogen is liberated.

$$CO + H_2O \rightarrow CO_2 + H_2$$

In the past, using these methods, coal was converted into *coal gas* (town gas), which was piped to customers to burn for illumination, heating, and cooking. This method is no longer used for these purposes, although town gas is used in some industrial applications.

Liquefaction

Coal can also be converted into synthetic fuels equivalent to gasoline or diesel by several processes. In the direct liquefaction processes, the coal is either hydrogenated or carbonized. Hydrogenation processes are the Bergius process, the SRC-I and SRC-II (solvent refined coal) processes and the NUS Corporation hydrogenation process. In the process of low-temperature carbonization, coal is coked at temperatures between 680 and 1,380°F. These temperatures optimize the production of coal tars richer in lighter hydrocarbons than normal coal tar. The coal tar is then further processed into fuels.

Alternatively, coal can be converted into a gas first, and then into a liquid, by using the Fischer-Tropsch process. Coal liquefaction methods involve carbon dioxide (CO_2) emissions in the conversion process. If coal liquefaction is done without employing either carbon capture and storage (CCS) technologies or biomass blending, the result is life-cycle greenhouse gas footprints that are generally greater than those released in the extraction and refinement of liquid fuel production from crude oil.

If CCS technologies are employed, reductions of 5–12% can be achieved in Coal to Liquid (CTL) plants and up to a 75% reduction is achievable when co-gasifying coal with commercially demonstrated levels of biomass (30% biomass by weight) in coal/bomass-to-liquids plants. For future synthetic fuel projects, carbon dioxide sequestration is proposed to avoid releasing CO_2 into the atmosphere.

Sequestration, though, adds to production cost. Currently, all US and at least one Chinese synthetic fuel projects, include sequestration in their process designs.

Refined Coal

Refined coal is the product of a coal-upgrading technology that removes moisture and certain pollutants from lower-rank coals such as sub-bituminous and lignite (brown) coals. It is one form of several pre-combustion treatments and processes for coal that alter coal's characteristics before it is burned.

The goals of pre-combustion coal technologies are to increase efficiency and reduce emissions when the coal is burned. Depending on the situation, pre-combustion technology can be used in place of or as a supplement to post-combustion technologies to control emissions from coal-fueled boilers.

The above uses of coal are only a small part of the overall coal industry, which is a huge business. Here we see millions of people working in thousands of different enterprises; mines construction, operation and maintenance; trains driving and maintenance; power plants operation and maintenance; equipment manufacturing, installation, and maintenance professions and related support industries.

The Coal Industry

A close look at the coal industry reveals a number of different segments, directly related to the coal production and commercial use. There are, however, also a number of segments that are not directly related to coal, *per se*. Yet, they have a direct impact on the different stages of coal's production and use.

Direct products and service:
- Raw materials (coal)
  — Exploration and location of coal reserves
  — Construction of mines and support facilities
  — Mining operations
  — Coal transport
  — Cleaning and processing
- Power generation and distribution
- Production of commercial products

Indirect products and services in the coal industry consist of:
- Consulting and contracting services
  — Engineering services
  — Electro-mechanical services
  — Energy efficiency consulting
  — Misc. service providers
- Transportation services
- Political and regulatory
- Energy production
- Import/export
- Strategic reserves management
- Investing and finance services
- Insurance services
- Legal services
- Mining equipment manufacturers
- Power plant equipment manufacturers
- Consumables supply chain
- Security, safety, and risk-management services
- Environmental issues and risks
- Energy commodities and futures trading

Note: The above services are similar to these used in the other energy sectors, so we dedicate a special chapter on Peripheral Services in this text.

Below, we will take a close look at some of the key segments of coal's life cycle.

COAL PRODUCTION

Coal is found in many places around the U.S. and the globe. Not all coal deposits are physically or economically feasible to exploit, and so many are sitting deep underground awaiting their day under the sun. The most economical method of coal extraction from coal seams vary, depending on the depth and quality of the seams, the local geology, and a number of environmental factors.

Coal mining processes are mainly differentiated by whether they operate *on the surface* or *underground*. As the name suggests, *surface mining* entails gathering coal from, or close to, the surface. Underground mining, on the other hand, is what we are more familiar with since it is the oldest method of coal mining.

For obvious reasons, surface mining is cleaner and safer than underground mining, but it causes serious visual and environmental damages. Underground mining is equally harmful to the environment, but the scars and damage are much less visible in most cases. It is also one of the most dangerous human undertakings and jobs on Earth.

The entire coal production process undergoes the following steps:

Planning

In order to determine the technical and economic feasibility of a new mine, a number of pieces of information are needed, such as:

- *Market analysis*. Potential customers, contract agreements, size and location of the markets, etc. data needed to figure out the size and other peculiarities of the potential markets.

- *Transportation*. Size and type of loads, property access, elevation of the mine and the customer sites, road and rail systems, etc.

- Utilities. Availability and distance to electric power and substations, location, right-of-way and the related costs for new transmission lines.

- Water. Type, quantity, and quality of potable and process water, and source location, type, and the cost of aqueducts etc.

- Labor. Local availability, type and quality, rates and trends, and the local labor history.

The technical and economic feasibility of coal mining are evaluated according to the following criteria:
- Regional geologic conditions
- Overburden characteristics
- Coal seam continuity, thickness, structure, quality, and depth
- Condition of the ground above and below the seam for use as roof and floor
- Local topography, especially altitude and slope
- Local climate and weather conditions
- Land ownership as it affects the availability of land for mining and access
- Surface drainage patterns
- Ground water conditions
- Availability of labor and materials
- Coal purchaser' requirements in terms of tonnage, quality, and destination; and
- Capital investment requirements.

Based on these analysis, the experts can determine if mining coal in the particular location is feasible and profitable. If yes, then they have to decide what type of mining process to use.

Surface, or strip, mining is usually preferred for extracting coal that is less than 200 feet under the surface. And so, today we have two distinct methods of mining coal; a) surface, or strip, mining and, b) underground mining.

For example, coal that is found at depths of 180 to 300 ft. is usually mined in underground mines, but in some cases surface mining techniques can be used. For example, some western U.S. coal that occur at depths in excess of 200 ft are mined by the open pit methods, due to thickness of the seam of 60–90 feet at places.

Coals deposits below 300 ft. are usually mined in underground mines. Although there are open pit mining operations working on coal seams up to 1000–1500 feet below ground level, in some regions in Germany, they are exceptions.

Below we will take a close look at the cradle-to-grave coal mining and use operations.

Note: "Cradle-to-grave" in this case means every single step of the process, and every single material and effort used from the very beginning of the coal mining to the very end of its existence and burial. All these materials and efforts have to be accounted for, because they have some impact on the overall process.

So let's see. The cradle-to-grave coal mining, cleaning, transport and burning process basically consists of:

Mine Planning
- Locating, testing and exploring the coal reserves in the proposed mine location
- Estimating the construction and operating cost and the related issues
- Applying and obtaining the necessary federal, state and local permits

Mine design
- Mine area design,
- Equipment design, and
- Process design

Mine construction
- Mine construction
- Surface facilites
- Infrastructure

Daily operations
- Coal digging and surface transport
- Coal preparation and waste treatment
- Coal and waste transport

Mine decommissioning and land reclamation
- Mine shutdown and evacuation
- Mine reconstruction
- Surface land reclamation

In addition, a major portion of thus mined coal is transported to a coal-fired power plant to be burned into furnaces. The heat of these furnaces is used to boil water and generate steam, which is run through turbines attached to electricity generators. Thus generated electric power is sent into the national grid for use at residences and businesses.

Let's start with a close look at coal mines' development and operation:

Surface Mining in Brief

When coal seams are near the surface, it may be economical to extract the coal using surface (also referred to as open cut, open cast, open pit, or strip) mining methods. Surface coal mining recovers a greater proportion of the coal deposit than underground methods, as more of the coal seams in the strata may be exploited.

Large surface mines can cover an area of many square kilometers and use very large pieces of equipment. This equipment can include: draglines which operate by removing the overburden, power shovels, large trucks in which transport overburden and coal, bucket wheel excavators, and conveyors.

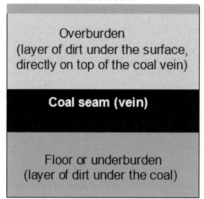

Figure 4-4. Coal deposit, cross section

In surface mining, explosives are first used in order to break through the surface soil and the overburden, of the mining area. The overburden is then removed by draglines or by power shovel and truck. Once the coal seam is exposed, it is drilled, fractured and thoroughly mined in strips. The coal is then loaded on to large trucks or conveyors for transport to either the coal preparation plant or directly to where it will be used.

Most open cast mines in the U.S. extract bituminous coal. In Australia and South Africa open cast mining is used for both thermal and metallurgical coals. In New South Wales open casting for steam coal and anthracite is practiced. Surface mining accounts for around 80% of production in Australia, while in the U.S. it is used for about 67% of production. Globally, about 40% of coal production involves surface mining.

Surface mining is the most widely spread mining method, with most obvious—in your face—environmental damages. Underground mining has its problems too, but they are hidden and harder to observe.

Underground Mining in Brief

Most coal seams are deep underground, and since opencast mining is impractical, underground mining is used instead. It accounts for about 60% of world coal production. Underground mining offers a number of totally different challenges extracting coal from the Earth. Since coal is very deep in some locations, and strip mining would not work there, deep holes (shafts) are dug into the ground down to the level of the coal veins. Miners are lowered down into the hole, where they dig the coal and bring it to the surface.

Figure 4-5. Underground mining operations

Let's follow the miners in their daily work. They arrive early morning by the shaft' opening and get into an elevator, which lowers them down; hundreds and sometimes thousands feet under the ground. The cage stops at the bottom of the mine shaft, and the minors spill out to their assigned places for the daily toil.

Some of them dig with picks, some drill holes, some drive coal trains, but all are covered with, and breath, great amounts of black coal dust. Most miners wear tiny lamps on their caps, since it is dark in there and not many lamps are allowed. Now days large machines do most of the digging and loading, but manual work in different areas of the mine operations is still a major part of the miners daily effort.

Looking around you'll see a low roof, usually held up by pillars of wood and coal. It is noisy and dusty in there. From time to time even greater noise approaches and a light moving towards signals the approaching coal train, dragging many cars loaded with coal. It soon rushes back with empty cars, taking them back to be refilled. And explosions would thunder at certain periods, when new veins are to be exploited.

There are small rooms and larger chambers everywhere you go. They were created by digging out the coal and taking it up to the surface. You can only imagine the amount of earth and rock above the ceilings of these rooms; hundreds of feet in thickness. And you also start imagining what would happen, if the roof in one of these rooms caves in. The rooms are usually small enough to not allow caving of the roofs, but you never know. Stuff happens…

Note: In many cases the "room-and-pillar" mining methods are used, where "pillars" of coal are created by digging the coal around the columns—pillars that serve as supports for the rooms' ceilings. This method accounts for a significant portion of the total mineral production in the United States. Well in excess of $6 billion worth of mineral commodities are produced each year by this method. A substantial portion ($3.55 billion) of coal production still comes from room-and-pillar mining.

Coal mines experience large-scale, catastrophic pillar collapses, if and when the strength of a pillar in a room-and-pillar mine is exceeded. When one pillar collapses, the load that it supported will be transferred to neighboring pillars. The additional load on these pillars may lead to increased stress of their structures and the ceiling above them.

This mechanism of pillar overloading, load transfer, and continuing pillar failures can lead to the rapid collapse of very large areas of a mine. In some cases, only a few pillars might fail; however, in extreme cases, hundreds, even thousands, of pillars can fail.

This kind of failure has many names, such as: progressive pillar failure, massive pillar collapse, domino-type failure, or pillar run. A special term, "cascading pillar failure" or CPF has been coined to describe and study these rapid pillar collapses. There are over 21 instances of large-scale pillar collapses in room-and-pillar mines, mainly in the United States, in the recent past.

There are many other dangers of underground mining, which we will review in the next chapters.

Coal Mines Logistics

Large-scale industrial process operations are expensive, and most of the cause significant environmental damage—usually negative—which has to be taken into consideration and calculated in terms of economic and health terms. Coal mining is no exception.

When all costs are considered, the process of locating the coal, applying for and getting the necessary permits, designing the mine structure (including facilities, equipment, process, and labor), hiring and training engineers, technicians and engineers, and the mining operations mining process (its daily operation and maintenance) is overwhelming and expensive. There are a number of steps in the overall mining process that cause significant air, soil or water table damage.

Let's take a close look at the entire cradle to grave, mining process and estimate the cost and environmental and other damages at each step of the process.

It all starts with an idea, "Hey, there is lots of coal in this area. Let's go and get it" When the principals agree and decide that a mine is justifiable in the reference area, they start the process, which consists of:

The cradle-to-grave coal mining, cleaning, transport and burning process basically consists of number of consecutive steps, which vary in type and magnitude from state to state and from country to country.

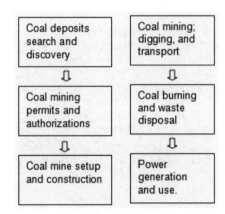

Figure 4-6. Cradle-to-grave coal power generation

In more detail:

Preliminary Mine Design Considerations

Coal mine sites vary in type and size, but typically involve large land areas, especially when surface mining methods are used. Coal seams vary in size, depth and quality, and the mining methods is chosen on the basis of physical feasibility, economical viability and safety. The size of the mining operation will depend on the site characteristics, the coal reserve, and the mining method.

For coal mined by surface methods, the mine plant should be located off the outcrop (as in a visible exposure of coal deposits on a hill), if possible.

The mine plant itself consists of coal handling and storage facilities, offices, shops and laboratories, equipment storage buildings, and waste disposal areas.

Access to coal deposits at a surface operation involves the use of large equipment such as bucket-wheel excavators, draglines, and shovels to remove overburden from the coal so extraction can begin. As mining progresses, development consists mainly of extending paved roads and power lines, and constructing new roads for access to the coal deposit.

The coal mined in the coal mine typically goes on a conveyor belt and on small cars to a preparation plant that is usually located close to the mining site. At the plant the coal gets cleaned and otherwise processed to remove dirt, rocks, ashes, sulfur, and other unwanted materials. After that the coal is sorted by quality and size, according to the customer's needs. This is needed to increase the heating value of the coal.

Once the coal is processed, it is shipped typically by rail, but also by truck or barge or even a coal-slurry pipeline, to the coal burning power plant. Transportation methods depend on the distance to be traveled, as well as the access to existing transportation systems.

Coal is delivered to coal-fired power plants and burnt to boil water. Thus produced steam is injected in steam turbines that turn generators to produces electricity. The electricity is sent into transmission lines system that consists of electric transmission lines, towers, substations and other components. Coal accounts for over 50% of the electricity produced in the US.

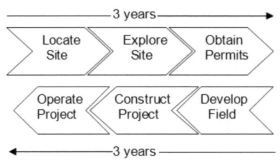

Figure 4-7. Project time line

The process, however, starts with finding the coal seam, or vein.

Coal Reserves Locating, Tests and Estimates

Before a coal mine can be planned and built, a number of tests and surveys must be conducted to estimate the type and quantity of coal in the particular location. Then engineering and financial estimates and calculations must be made in order to ensure the project's technical and economical feasibility. The environmental aspects of the entire cradle-to-grave process are plugged in as well, to complete the picture.

These activities include historical research, mapping, drilling (to obtain geological samples), and geophysical exploration. The latter include a number of field and lab tests, such as: aerial photography, airborne geophysical surveys (magnetic, radiometric, and electromagnetic); on-the-ground geophysical surveys (drill-hole logging; electrical, magnetic, electromagnetic, radiometric, gravimetric, and refraction-seismic surveys; induced polarization surveys using exposed electrodes). Field surveys for identifying cultural resources, paleontological resources, and ecological resources (habitats, species, etc.) in the project area should also be conducted during this phase.

Reclamation of exploratory areas that will not be part of the project area would occur during this phase as well.

A number of engineering disciplines and scientific methods are applied when evaluating all activities relat-

ed to the mine design, and estimating its technological and economic potential. The cradle-to-grave cost and environmental impact are major parts of, and considered during, every step of the process.

At the end, the obtained data and information about the different aspects of the potential mine are provided to the mine engineers, managers and investors as needed to make decisions on the plant design, financing. Some of the information needed for such decisions is as follow

Geology

A complete geological information on the soil, overburden and coal deposits, obtained from historical data and from the preliminary exploration steps. This includes the overall geological structure and the relevant physical properties of the proposed mine; the thickness and variability of the different layers, the quality and quantity of the coal seams, etc. vital information.

Geometry and Geography

A complete picture of the location, its climate and earthquake zoning, as well as the size shape, continuity, attitude, and drainage patterns of the different layers.

Hydrology

A complete picture of the permeability and porosity of the overburden and coal layers, as well as location of the aquifers.

Toxicology

A complete picture of the hazardous, toxic, reactive, and radioactive species found on the surface and in the ground. Additionally, best and worst case scenarios are developed for dealing with industrial accidents and hazmat situations during operation, as well as during the mine's decommissioning and land reclamation stages.

Surface Mines

The planning and design stages of a large size surface or pit mine in the U.S. is a complex and specialized undertaking, with many different and equally complex variables. The information collected in the evaluation stage, as well as all available project-related information, are reviewed as a first step, in order to develop the most appropriate for the particular case coal extraction plan.

Various mining methods and equipment set combinations and permutations are considered by the team of mining engineers, geologists, environmentalists, and economists. After the technology, process and equip-

Figure 4-8. Surface mine

ment set are determined, an economic and market analysis are performed of the best case scenario, as needed to determine the project viability from a financial point of view.

The preliminary mine plan usually goes through a number of iterations, which finally result in a technically feasible and economical operating plan, subject to the specific contractual, legal, environmental, and other constraints of the specific property and the related equipment and operating procedures.

The process of strip mining is quite simple, and self-explanatory. It basically consists of, a) carefully clearing the topsoil (mountaintop removal in some cases) and stockpiling it for reclamation, b) removing the overburden and stockpiling it for the same reasons, and then c) mining of the coal seam.

The coal is removed from the ground layers by a number of methods, depending on type of formation etc. It is the loaded on dump trucks or conveyor belt for transport to the preparation plant. Here it is processed as required by the customer, and is loaded on railroad cars for transport to the coal burning power plant.

As the mining of the surface seam moves forward, the mined area is simultaneously reclaimed by replacing and re-contouring the overburden and replacing the topsoil.

In open-pit mines, the mining begins by drilling and blasting waste rock and clearing the overburden and debris to expose the coal seam. The removing of the coal is done one layer (bench) at a time, which forms terraces. The mined bench continues to get wider as the mining continues, and goes deeper with each bench.

This is, nevertheless, a long and complex process, which, together with obtaining the necessary permits, etc., could take in 10-15 years, or more, in addition to

millions of dollars of related expenses...long before a single lump of coal is removed from the mine.

Here are some of the lengthy, complex, and expensive steps in this process:

Environmental impact package
• Initial site evaluation
• Scope of work program
• Environmental impact report
• Environmental monitoring program
• Reclamation bonds

Coal mining package
• Lease or buyout rights acquisition
• Mapping the area
• Site drilling and sample analysis
• Coal type and quantity evaluation
• Drilling coal samples and analysis
• Development drilling and sample analysis

Land rights
• Land requirements, surface land and minerals ownership, oil and gas wells location and ownership, etc.

Taxes and royalties
• Federal, state and local taxation, royalties payments, zoning, and operating and reclamation requirements.

Process design and equipment set planning
• Concept mine design
• Mining process design
• Equipment selection and ordering
• Economic evaluation
• Overall mining plan development

NEPA procedures
• Lead EIS agency communications
• EIS draft and reviews
• EIS hearing and federal review
• CEQ filing and approvals

Permits
• Surface drilling rights acquisition
• Federal permits
• USFS land use
• State water use and mining
• State industrial siting
• Local permits

Mining preparation
- Stripping equipment setup
- Loaders and conveyors setup
- Support equipment setup
- Labor hiring and training
- Production ramp-up procedures
- Full production specs and documentation

A quick look at these steps reveals that:
- The environmental studies of the proposed site, combined with post-test monitoring, are a major part of the process, and could take a long time; sometimes as long as 10 years, or more.

- The initial process design and equipment ordering are strictly engineering disciplines, the completion is also of long time duration and might take 2-3 years as a very minimum,

- The design and construction of the site facilities and support structures are specialized tasks that would take another 2-3 years, or more

- Completing the National Environmental Policy Act (NEPA) requirements and the related procedures would take 5-6 years,

- The necessary federal, state, local, tribal and other permits and negotiations would take the project well over the 10 year mark, and might even extend it to ad infinitum.

- If all goes well with the above tasks and sub-projects, and if their design and implementation are well coordinated and properly and efficiently executed, the initial setup and limited coal production could be started around year 10 from the beginning of the project.

It is important to point out that most of the tasks and sub-projects mentioned above can be conducted in parallel, thus saving time.

There are a number of additional complex details that go into the mine development and exploitation process. The design and equipment selection, for example, are very special and vary with individual sites. In some cases, these are considered 'exceptional.'

Some of these exceptions are:
- Equipment limitations
  In many cases the size and type of equipment limits the full exploration of the coal seams. Land instability might also limit the activities in some areas. Some of these limitations cannot be foreseen, and so they must be dealt with by planning on eventual equipment

or strategy changes in order to obtain the maximum amount of coal from the area.

- Mining losses
  As in any process, mining also suffers losses during full operation. Some mines are different than others. The most common losses are these in the top, bottom, and rib coal layer. These are losses, where the loaders, for a number of reasons, cannot grab all the coal from the top, bottom and the sides of the coal seam.

  There are also fly rock losses from blasting, and transportation losses from dump trucks and conveyor belts. These are so called barrier losses, and can amount to over 5% of the mine gross, which is significant number and which must be taken into consideration in the preliminary designs.

  When finally a green light is given, a year or more will be needed for final setup, operators training, and ramp up of the production cycle. Another year will be needed to establish the final, full production cycle and all its steps, specs, procedures, training, and the related documentation. And then the mine is in full production until there is enough coal to dig out at full speed.

  A day comes many years later, when the coal deposits in the mine seams will diminish and the mine production volume will be reduced. This requires special, albeit temporary, measures, which eventually lead to reduced profits. Eventually the mine has to be shut down, and special procedures have to be followed to accomplish that process successfully and safely.

  After shutdown of mining operations, the mine area has to be reconstructed as close as possible to its original state. This is seldom possible, so there a number of mines that have been simply abandoned, while others are partially reconstructed.

  Note: As part of the permitting process, new mine constructions in the US are required to purchase Reclamation Bonds, as required by the Bureau of Land Management (BLM) and various state environmental agencies. These are long term surety obligations, which are like insurance that the site will be returned to its original condition upon termination of the mining operations.

  Once issued, the Bonds cannot be canceled. Adequate performance can be highly subjective, and bond penalties and losses in some cases can be large.

Surface Mine Operations
  Surface mining, "strip mining" is used where large coal deposits are near to the surface. This way the coal can be reached and extracted easily by digging deep trenches and loading it directly on large trucks. This

method has its advantages; it is much safer and efficient than underground mining. It, however, has its disadvantages too; it damages the Earth's surface and changes the environment and life in it for miles at a time.

Surface mining is simply working on the surface of a mountain or a hill, where huge machines, called power shovels, scoop large amounts of soil just above the coal seam. The seam, also called overburden is then dug out by the same machines in similarly large quantities.

This technique has vast economic advantage since manual labor is replaced by machines. Some of these power shovels have buckets that hold over 200 cubic yards, or over 300 tons with each scoop, which is enough to fill 15-20 regular dump trucks.

This level of mechanization and automation of strip-mined coal is responsible for the low price of coal, which is estimated at almost two times lower cost than underground mined coal. And the gap is increasing due to increasing labor costs and new mine safety regulations for underground mines.

There are several types of surface mining:
*   *Area mining* is the preferred technique in most areas,
*   *Open pit mining* is used on very thick coal seams,
*   *Contour mining* is used in mostly hilly terrain, and
*   *Auger mining* usually accompanies contour mining.
*   *Mountain top removal mining* involves removal of large land areas.

Area mining is a surface mining method where the overburden is removed in a mile long and 100 feet wide strips. Holes are drilled in the exposed coal seam, filled with explosives and blasted. The coal is scooped and loaded onto dump trucks or conveyors for transport to the coal preparation (or wash) plant.

Once this strip is empty of coal, the process is repeated with a new strip being created next to it. At the same time, but with some delay, a parallel trench is dug close to the first trench and the overburden from it (the new trench) is dumped in the old trench as the machines move forward simultaneously.

The coal from the new trench is also removed, and another, third, parallel trench is dug. This process is repeated until all the coal of the surface is removed.

If the vein is deep enough, the process might be taken to another, deeper level. The depth at which coal can be profitably reached has been increasing as the equipment has become larger and more efficient. The maximum overburden that could be handled at first was about 70 feet, increasing to 125 feet with time, and now it is close to 200 feet.

The equipment used in strip mining depends on the local geological conditions. For example, to remove overburden that is loose or unconsolidated, a bucket wheel excavator might be used as the most efficient and productive piece of equipment.

Some area mines may be productive for more than 50 years.

The problem with area mining method is that flat farmland is replaced by a series of ridges and gullies, which brings the usefulness of the land and its value for crops, recreation, or anything else, to practically zero. Elaborate reclamation process can be used to restore the land to its original shape and quality, but this is a very expensive, and rarely completely successful, undertaking.

Area mining is wide spread in the Midwest USA, and the huge coal reserves in Montana, Wyoming, the Dakotas, and the Southwest. About 1/3 of the estimated 440 billion tons of U.S. coal reserves lies in beds that are less than 100 feet below the surface, thus suitable for strip mining.

Open pit mining is similar to area mining except that larger and often deeper trenches of up to 1000 feet wide, are usually dug out. This method is used on thick beds of coal, and similarly when the coal is removed from the first trench, it is filled with overburden from the second trench. The land is left in a similarly useless shape and quality for any practical purpose.

Contour mining is used in mountainous coal deposits, and consists of removing overburden from the seam in a pattern following the contours along a ridge or around a hillside. It is most commonly used in areas with rolling to steep terrain. It is quite different from the area strip mining, and is much more destructive.

The process begins with the power shovel cutting at the area where the coal seam reaches the surface. The resulting overburden is pushed down the mountainside and the coal is removed in long strips, in a way very similar to paring an apple.

The haul-back or lateral movement method is widely used and consist of an initial cut with the overburden (or spoil) deposited down slope or at some other site. Spoil from the second cut usually refills the first. A ridge of undisturbed natural material 15 to 20 ft (5–6 m) wide is often intentionally left at the outer edge of the mined area. This barrier adds stability to the reclaimed slope by preventing spoil from slumping or sliding downhill

The shovel moves toward the center of the mountain, with each cut removing more overburden. A high wall of over 100 feet high is created and is too thick to remove, and an auger is used at that time to drill out more coal.

The overburden is stacked on the edge or thrown down the slope, and this what causes the major damage to the surrounding area. The loose soil in an area with no trees or grass to anchor it is easily eroded and washed down the hill and into the streams below. Heavily traveled access roads, with heavy dump trucks and other equipment roaring up and down day and night, add even more to the erosion and overall area damage.

But erosion is not the only problem. Loose boulders and landslides are responsible for even larger and permanent damage. Entire towns in Wales and Virginia have been buried by such landslides.

The remaining high walls circle the mountains, making them useless and inaccessible. There are over a million such disturbed acres in the U.S., with additional 30,000 acres added every year. Most of this land remains unreclaimed.

Adding insult to the injury, rainwater floods the abandoned coal seams and reacts with the sulfur-containing pyrites and other minerals in them. Leachate, a corrosive and destructive pollutant such as sulfuric acid is formed in many areas runs down the hills killing all vegetation, It then runs into the streams where it kills aquatic life by increase in acidity.

Mine drainage also contributes to increased amounts of sediments, sulfates, iron, and hardness in the streams and lakes, which also changes their environment and affects the life forms living in them.

Surveys in some post-contour stripping areas in the U.S. show that about 5% of the hill sides had a pH less than 3 (highly acidic), and 80% with pH of 3-5— which is also unacceptable. Over 6000 miles in of the Appalachian streams are affected by acid mine drainage, and over 11,000 miles of streams are affected by other mine pollutants as well.

The limitations on contour strip mining are economic and technical. When the operation reaches a predetermined stripping ratio (tons of overburden/tons of coal), it becomes unprofitable. Also, depending on the equipment used, it may not be technically feasible to reach a certain height of high wall. Producing more coal with the auger method is possible today.

Auger mining is a method for coal extraction by boring into a coal seam at the base of strata exposed by excavation. Augering is usually associated with contour strip-mining, recovering coal for a limited depth (up to 1,000 feet) beyond which stripping becomes uneco-

nomical because the seam of coal lies so far beneath the surface. It is also limited to horizontal or slightly pitched seams that have been exposed by geologic erosion.

In this process an auger drills mounted with cutter heads cut and fracture through both overburden and coal, operating very similar to a drill machine. The augering differs from other types of coal cutting machines such as continuous miners in that it tends to exploit the lower tensile strength of coal rather than trying to over compensate for its high compressive strength.

The power of the auger as well as diameter of the cutterhead are the two features that govern an auger drill's performance. The greater the power of the machine, the greater the depth of coal seam into which it is able to bore down, producing a higher rate of coal.

Augers drills used in auger mining can range from 60 to 200 feet (18 to 61 m) in length, and two to seven feet (0.6 to 2.1 m) in diameter. The cutter head on the auger bores a number of openings into the seam, similar to how a wood drill produces wood shavings. The coal is then extracted and transported up to the surface.

As the depth of the bored hole is extended, coal production is most likely to decrease. The auger drill will continue to penetrate into a high wall until the maximum torque of an auger is reached, usually at a depth of 492 feet (150 m). Once the coal arrives at the surface, it is lifted up to a dump truck for hauling by a conveyor or front end loader.

A recent auger drill technology have led to the introduction of a new type of auger drilling machine called the thin-seam miner (TSM). It is actually a type of continuous miner that can cut an entry up to eight feet (2.4 m) wide and up to five feet (1.5 m) high into a coal seam situated under a high wall in surface mines.

One of the drawbacks of this method is that once the cutter head enters the coal seam, the operator is unable to view the cutting action directly and must rely more on a sense of feel for the machine, as needed to control it, and its performance, as well as to detect potential problems.

Mountaintop removal mining is a surface mining method that relies on the removal of entire mountaintops, in order to expose large coal seams. Mountaintop removal is a combination of area and contour strip mining methods. In areas with rolling or steep terrain with a coal seam occurring near the top of a ridge or hill, the entire top is removed in a series of parallel cuts, with the overburden is deposited in nearby valleys and hollows.

This method usually leaves ridge and hill tops as flattened plateaus. The process is highly controversial since it creates drastic changes in local topography. It is accompanied by the creation of head-of-hollow-fills, or filling in valleys with mining debris, and for covering streams and disrupting ecosystems.

In preparation for filling the overburden disposal area, vegetation and soil are removed and a rock drain is constructed down the middle of the area to be filled, thus replacing natural drainage. Upon completion of the fill, the under-drain forms a continuous water runoff system from the upper end of the valley to the lower end of the fill. Typical head-of-hollow fills are graded and terraced to create permanently stable slopes.

In all cases, the change of many acres of the local area is dramatic and permanent.

Underground Mines

Until recently mining was a fully manual job, where miners cut into the walls and ceilings of the rooms with special pickaxes. They reached as far as the pickaxe would reach, and then a hole was bored into the top of the coal, and a cartridge of dynamite was exploded to release a ton or two of coal. The coal was then shoveled into a car and pushed out of the room to join the long string of cars going to the surface. The digging and exploding work continued all day long.

Today, miners cut and dig the coal with machines that are as sophisticated and safe as the state of the art and economics allow. The machines grind the coal, and leave a deep cut all along the side of the room. Then another group of miners bore holes in the walls for the blasting. The holes are made with powerful, compressed air-driven drills, and dynamite charges are placed in the holes.

After the explosions are set off, and the dust has settled, the coal is loaded manually or with machines into the cars. Then it is taken to the surface and made ready for market.

Modern, fully automated operations are the most efficient and safest in deep mining. This type of mining is done along the seam with machines, and pillars and timbers are left standing to support the mine roof.

When the seam mining is complete, for reasons of geology formations or economics, a supplementary version of the room and pillar mining (second mining) takes over. It consists of removal of the coal in the support pillars in the dugout, thus recovering the maximum amount of coal possible from the seam.

Modern methods for coal pillar sections removal use remote-controlled equipment, which includes large hydraulic machines that are used to support the roof during the pillar removal process. The mobile roof supports look like a large dining-room table with hydraulic jacks for legs. After the coal pillars are removed, the legs of the table are shortened and it is withdrawn to a safe area.

This efficient and safe method is used to prevent cave-ins until the miners and their equipment have left a work area, because the unsupported roof of the room usually collapses when the roof supports are removed.

If the coal is to be mined by underground methods, the mine plant is constructed near the main portal or entrance. Access to coal deposits at an underground operation is provided by drifts, slopes, or shafts. The coal bed is developed for further operations by driving entries. Although terminology varies, the following system of entries is universal in the industry.

Main entries are extensions of access openings and often run several miles in one direction. Three or more parallel entries, 12 to 22 feet wide and 40 to 100 feet between centers, are driven in a given direction and connected at intervals by crosscuts to provide proper air circulation. These are the major routes of underground transport and access, and serve for the life of the mine.

Panel entries are driven from the main entries, resulting in a subdivision of the coal bed into blocks or panels having dimensions that may be as much as 1 by 1/2 mile. Panel entries serve as routes from the main entries to the working places, and for air circulation. Although coal is removed during the driving of the main and panel entries, the production cycle begins upon completion of the panel entries.

Underground Mine Planning and Design

The main activities during planning and design of mine construction and during operation phases of either surface or underground mining are focused on the efficient and safe function of the proposed mine as well as any auxiliary facility (e.g., shaft construction) and coal transport system (e.g., access roads, rail lines, pipelines, conveyor systems).

As with the surface mine, we need to go through a number of steps, some of which are:

1. Coal mine
   - Planning and design
   - Permitting
   - Land preparation
   - Facilities construction
   - Equipment purchase, delivery and installation
2. Daily operations
   - Coal digging
   - Onsite transport
   - Coal preparation (on-site)

3. Coal transport
    • Loading
    • Unit train transport
    • Unloading
4. Coal preparation (off-site)
    • Sorting
    • Washing
    • Treating
5. End of life decommissioning and waste disposal
    • Mine or plant shutdown and disabling
    • Equipment disassembly and demolition
    • Facilities demolition
    • Waste disposal
    • Land decontamination and reconstruction

Here, as in the surface mines' initial steps, we need to go through the different steps of site location, testing, and exploration. Then many months and thousands of dollars will be spent obtaining the necessary federal, state and local permits and approvals. Design of the actual mine, the mining equipment, and the support infrastructure (buildings, roads, railroad, etc.) follow. When all these tasks are completed, the miners go to work.

Underground Mine Development

The process of underground mining includes cutting into the coal deposit and removing it from the coal face via room-and-pillar methods using a continuous mining machine, or through longwall methods using a longwall cutting machine. In either method, once the coal is removed, the supports or pillars can be removed and the roof of the mine is allowed to collapse. The mined area is then abandoned and later the land around them is reconstructed, if and as much as possible.

Underground mine planning and design is a unique engineering discipline. It involves the development of infrastructure and working conditions that are very sophisticated, highly specialized, and quite different from other industrial processes.

This type of mine design consists of the three basic engineering phases, conceptual, preliminary, and final design. While the second and third steps are common for many industrial processes, the development of the conceptual design is different for mining and is the key to the success of the entire operation. An error in interpreting the results from preliminary tests, for example, could lead the entire process in the wrong direction, cause technical and financial difficulties, and even disaster.

The goal of underground mine planning and design is integrated with mine systems design, with the fi-

nal result being efficient and safe extraction of coal. The coal is then prepared to desired market requirements, at a minimum cost, while meeting social, legal and regulatory constraints.

A number of engineering disciplines are needed for successful mine planning and design process. Mining is a complex undertaking, so proper planning leads to the correct selection and implementation of all subsystems. Proper design, by the same token, ensures the implementation of traditional engineering subsystems.

An underground mining operation is a system which, due to the diversity of the technological processes, facilities, personal skills, and large capital investment, must consider and coordinate the behavior of, and interactions between the different subsystems.

Advances in several fields used in mining operations have the potential for making a significant impact on mining. These, therefore, must be taken into account during the planning and design process as well.

First Things First

The initial process of evaluation and exploration of the particular area considered for underground mining is similar to that used for surface mining, although the objectives and the results are different.

After this step, the planning and design process is also similar to that of surface mining, although additional steps and safety precautions are added for obvious reasons.

The following is a list of steps taken in the initial evaluation of underground mine sites:

Baseline Assessment

This is an essential process, and encompasses the evaluation of all available data, prior to starting the actual planning efforts. This is a comprehensive review of all available information gathered through historical materials and by actual site tests.

This process is somewhat more complicated than the surface mine evaluation, where one can dig a shallow hole to find some of the results. A number of specialized tests and measurements have to be taken and properly evaluated when looking for coal deep underground.

Underground mine data evaluation also includes the review of all geographic, geologic, environmental, technical, economic, and other data available. Hopefully, the available data contain enough accurate historical and present-day information for proper planning and mine design.

Preliminary Planning

Most plans start with a feasibility study—an overview of the project, making reasonable assumptions and estimates of the physical and other key operating factors of the mine. The intent is to figure out as quickly as possible if the project justifies further effort.

A life-of-mine plan must be developed to determine the reserve's type and size, and other mining parameters, including the costs of site reclamation. Reclamation costs are often great, so those could put the project's feasibly in question.

Regulatory and Legal Factors

The planning process must also review the state of current regulatory affairs. These play a significant role in the overall mining operation, and these must be faced in the very beginning and attacked in a proactive manner, rather than addressing them after the fact.

Each sub-system and step of the process is subject to compliance, and often needs to be submitted to the various agencies for inspection and approval. At a minimum, these may include mine layout, strata and roof control plan, shaft ventilation plan, fan stoppage plan, medical and emergency evacuation plan, fire control evacuation plan, and escape route plan. It is of utmost importance to ensure that the latest regulations, policies, and proposed rulemaking have been incorporated.

Geologic/Geotechnical Factors

The most important part of the data collection is the information on the coal deposits. Coal variability can be mathematically defined, if enough accurate information is available.

Exploration permits are needed to start the data collection, and must be first on the plan schedule.

Understanding the regional geology and features of the deposit are of utmost importance. Potentially adverse geologic conditions, such as faults, wants, rolls, low cover, or water inflow must be located and well defined. Seam or horizon conditions are important also.

A thorough review of the land lease is needed to determine the requirements and compliance provisions, which may be excessive for a profitable and safe mine operation.

Reserves Data

A complete and accurate coal reserves inventory is needed. Since composing the geological model is a free interpretation of the available data, it depends heavily on the experience of the geologists working on it. Exploration efforts do not rely on actual core recovery, but on indirect conclusions.

Changes in the data interpretation are usually made as new data come in, and might change the entire geologic model. Geophysical logging, core photography, and petrographic identification are used with the most success. There are different approaches for geologic modeling: accepting the geology and developing the plan around it, or considering the geologic model incomplete and incorporating flexibility in the planning for potential changes.

The evaluation and calculation of the coal reserve is one of the most crucial factors to the long-term success of a mine. The reserve type, magnitude, grade, depth, inclination, geometry, etc. are key to proper mine design.

There are a number of methods applied in proper estimation of the variables and the overall mine design characteristics. Mathematical methods are of utmost importance. This involves taking data, such as drilled samples, and extrapolating the data into blocks or grids to make the appropriate calculations. Mapping, determining reserve classification, leasehold boundaries, etc. are required for the final calculations. The data is then processed via different mathematical techniques, such as polygonal, inverse distance weighing, and others.

Geographic and Economic Factors

Geographic factors include the location, transportation infrastructure, type, size, and skill level of the local work force, private and public facilities available locally, local climate, local power availability, etc.

Economic factors include the local political and tax environment, government stability (if a foreign country), socio-economic conditions, and availability of support networks. Economics usually favor starting with the "lowest hanging fruit" approach, which in this case means extraction of the best-grade material, or starting with the lowest mining cost areas. While this approach might maximize the return on investment in the short term and shorten the payback period, it might also create a compromise in the mine's design and operation.

Environmental Factors

Environmental data gathering is very important from feasibility and economic points of view. Because of that, sometimes as much as 5 years' data are necessary, especially if an environmental impact statement is required. The minimum necessary baseline environmental data required for planning include a) topsoil, subsoil, and overburden analysis, b) hydrologic studies, c) vegetation and land use surveys, d) air quality analyses, e) wildlife surveys, and f) archaeological survey.

Tasks such as sealing core holes, and site reclamation, and their impacts on local environment must be considered and included in planning stages. Some of the impacts are aesthetics, noise, air quality, vibration, water discharge and runoff, subsidence, and process wastes. Surface and groundwater quality during operation and through the remedial and treatment stages must be developed to meet supply and discharge standards.

Planning is basically responsible for environmental protection, from the initial exploration to final reclamation. It is to alleviate or mitigate potential impacts of mining to a) minimize the cost of environmental protection by proper steps in the overall design, thus eliminating remedial measures, and b) minimize negative publicity or poor public relations which may have severe economic consequences.

Technical Factors

From an engineering point of view, the technical aspects of mining operation planning and design are the most extensive and detailed. Data from regulatory, geologic, and environmental analysis must be evaluated and translated into technical specifications. This information is used to determine which process to use, and then to outline and develop each step.

The layout of the mine is determined by the size and shape of the coal deposits, and these features are used to calculate the mine reserves and determine the best way to extract them. Access to the reserves can be by vertical shafts, inclined slopes and drifts, or horizontal entries, and the production levels will determine the number and size of the access openings made.

The technical parameters will form the conceptual basis for the plan, from which the detailed plan will be drawn. The larger and more extensive the area, the more complicated the plan.

Each item and step of mine construction and the production process are defined by the available key assumptions, the physical factors, the equipment, mine facilities and infrastructure, and transportation. These are detailed as much as possible in the beginning, and then modified as each item is executed, according to the new data and the "best-fit" technical models.

Equipment

The type and size of coal deposits, including their hardness, will determine the types of equipment to use. The seam and working height, mining dilution limits, production rates, and property extent, ventilation, size constraints, regulations, and floor pressures may impact the choice of equipment. For example, a large flat-lying coal seam may allow the use of longwall mining equipment. Floor condition plays a big part in the equipment type. Equipment productivity is also a factor that might prove essential in the final decision.

Maintenance, equipment overhaul and replacement schedules must be developed accordingly, to ensure continuous production. Transportation of the product may be by rail, truck or a combination.

Transportation

Movement of materials, personnel, and equipment into and out of a mine is a critical part of mine operation. Workers must reach their designated work area in an expeditious manner. Supplies must get to their points of use before need becomes critical. The equipment itself must be transported through the mine to the working area. Then the coal must be transported from the working face to the processing facility.

A smooth flow of people and materials is critical to the efficient operation of a mine, so the various transport vehicles and pathways must be properly selected for efficient operation.

Underground Mine Operations

Upon obtaining the necessary permits, and after a series of inspections and modifications, the mine is ready for exploitation. There are basically several methods of underground mining:

Room and Pillar Mining

Here coal deposits that are mined by cutting a network of rooms into the coal seam. Pillars of coal are left behind to hold up the roof. The pillars can make up

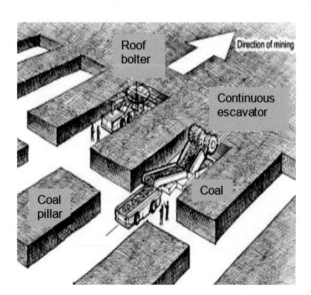

Figure 4-9. Room and pillar mining operation

to 30-40% of the total coal in the seam, as needed to provide space for head and floor coal.

There is evidence from recent open cast excavations that 18th century operators used a variety of room and pillar techniques to remove 92% of the in situ coal. The coal in the pillars can be extracted at a later stage by the retreat mining method, where a large machine with a self-supporting roof digs out the coal in the pillars and lets the roof collapse, as it pulls out of the area.

Longwall Mining

This process accounts for about 50% of underground production. The longwall method utilizes a large machine, Shearer, with a cutting face of 1,000 feet (300 m) or more. It is a sophisticated machine with a rotating drum that moves mechanically back and forth across a wide coal seam. The loosened coal falls onto a pan line that takes the coal to the conveyor belt for removal from the work area.

Longwall systems have their own hydraulic roof supports which provide safety, and can advance with the machine as mining progresses.

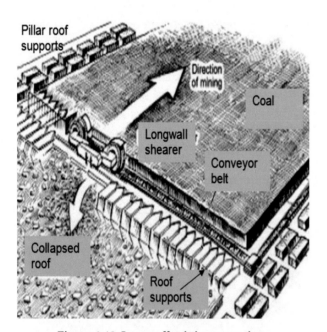

Figure 4-10. Longwall mining operations

As the longwall mining equipment moves forward, overlying rock that is no longer supported by coal is allowed to fall behind the operation in a controlled manner. Sensors detect how much coal remains in the seam, while robotic controls enhance the efficiency of the process. Longwall systems allow a 60-100% coal recovery rate when surrounding geology allows the use of this method. Once the coal is removed from approx-

imately 75% of a section, the roof is allowed to collapse in a safe manner.

Continuous Mining

This process utilizes a continuous-miner machine with a large rotating steel drum equipped with tungsten carbide teeth that scrape coal from the seam. Operating in a "room and pillar" (also known as "board and pillar") system—where the mine is divided into a series of 20- to 30-foot (5-10 m) "rooms" or work areas cut into the coal bed—a mine produces as much as five tons of coal a minute. This is more than a non-mechanized mine of the 1920s would produce in an entire day.

Conveyors transport the removed coal from the seam. Remote-controlled continuous miners are used to work in a variety of difficult seams and conditions, and robotic versions controlled by computers are becoming increasingly common. Continuous mining is truly a misnomer, room and pillar coal mining is very cyclical. In the US, one can generally cut 20 ft (or a bit more with MSHA permission, or 40 ft in South Africa) before the continuous miner goes out and the roof is supported by the roof bolter. After this, the face must be serviced, before it can be advanced again.

During servicing, the continuous miner moves to another face. Some continuous miners can bolt and dust the face (two major components of servicing) while cutting coal. A trained crew must be able to advance ventilation to truly earn the "continuous" label. However, very few mines are able to achieve it.

Most continuous mining machines in use in the U.S. lack the ability to bolt and dust. This may be partly because incorporation of bolting makes the machines wider, and therefore, less maneuverable.

Blast, or Conventional, Mining

This is an older practice that uses explosives such as dynamite to break up the coal seam, after which the coal is gathered and loaded onto shuttle cars or conveyors for removal to a central loading area. This process consists of a series of operations that begins with "cutting" the coal bed so it will break easily when blasted with explosives. This type of mining accounts for less than 5% of total underground production in the US today.

Shortwall Mining

This method currently accounts for less than 1% of deep coal production. It involves the use of a continuous mining machine with movable roof supports, similar to those used in the longwall method. The continuous miner shears coal panels 150 to 200 feet (40 to 60 m) wide and more than a half-mile (1 km) long, keeping in mind the local geological strata and other factors.

Retreat Mining

This is a method in which the pillars, or coal ribs, used to hold up the mine roof are extracted, allowing the mine roof to collapse as the mining works back towards the entrance. This is one of the most dangerous forms of mining, owing to the unpredictability of ceiling behavior, and possibility of collapse, which could crush or trap miners.

In all cases, and no matter what type of mining is used, underground mining is complex, expensive and dangerous work, where experience counts. That often determines the difference between loss and profit—even life and death.

Major Coal Mines

Coal us in the U.S. is decreasing, due to tightening EPA regulations, but the worldwide demand is increasing by the day. There are presently nearly 1,200 proposed coal plant projects in 59 countries around the globe, with most targeted for the Pacific market. China and India top the list with over 800 coal-fired plants planned for implementation in the next several years.

So while new coal-fired plant construction has slowed somewhat and is under increasing fire in the U.S., the rest of the world is poised to go all-in with a dirty coal future. The expert conclusion is that coal demand could rise over 20% by 2035.

This will increase the coal mining activities proportionally and we expect that the US will be a major player in this new game by increasing coal exports.

Coal production is growing fastest in Asia, while Europe has declined lately. The top coal mining nations in Table 4-1 (in millions of tons) are:

Table 4-1. Coal mining, 2010

Country Mining	
China	3,050
United States	973
India	557
Australia	409
South Africa	250
Russia	298
Indonesia	252
Poland	135
Kazakhstan	101
Colombia	72
TOTAL	6,097

Usually, a major part of the coal production is used in the country of origin, with only around 15% of hard coal production being exported. The U.S. is the only country that is planning to increase drastically its coal exports in the future. This controversial move promises to bring interesting developments, so we will watch carefully.

Global coal production is expected to more than double to over 13,000 Mt/yr by 2030. At that time, steam coal production is projected to reach around 5,200 Mt/yr; coking coal 620 Mt/yr; and brown coal 1,200 Mt/yr.

There were 1,325 mines in the U.S. in 2011, spread all over the U.S.

Table 4-2. Major U.S. mines production (in million tons/annum)

Mine	Type	State	Annual
North Antelope	Surface	WY	98,279,377
Black Thunder	Surface	WY	81,079,043
Cordero Mine	Surface	WY	39,380,964
Antelope Coal Mine	Surface	WY	33,975,524
Jacobs Ranch Mine	Surface	WY	29,021,485
Belle Ayr Mine	Surface	WY	28,395,952
Enlow Fork Mine	Underground	PA	11,092,684
Bailey Mine	Underground	PA	10,232,360
Mcelroy Mine	Underground	VA	9,863,588
Foidel Creek Mine	Underground	CO	7,827,079

Coal mining produced over 1.1 billion tons in 2011, with Northeastern Wyoming contributing the largest amount produced in any state in the US. It also presently produces more coal than any other region in the world.

Note the larger amount of coal produced at surface mines. This shows the relative ease of extraction by this method, which ultimately determines the total amount and price of thus produced coal. But this is accompanied by destruction of the local environment by leaving huge scars in the surface. While this is justifiable to an extent by the need of coal for base-load power generation, it is totally unjustifiable from the point of view of digging massive amounts of coal for export.

Exports simply punish the future generations by depriving them of this valuable commodity. They also contribute to increase of environmental pollution somewhere else around the globe. All this for the sake of making a quick buck today… Fair? Hardly!

Cost of Mining Operations

Underground mining currently accounts for about 60% of world coal production, although surface is prevalent in some regions of the US and some countries. For example, surface mining accounts for around 80% of production in Australia, while in the USA it is used for about 67% of production.

The cradle-to-grave mining operation is an expensive undertaking. The cost of a large mine—from concept to exploitation—is in the many millions and even billions of dollars.

- The site location, exploration and data collection alone is a major task, consisting of a number of sub-tasks, involving a number of professionals and specialized firms.
- The construction of the surface facilities, the local infrastructure, and mine structure is another grand undertaking, worth many millions of dollars and involving huge equipment and hundreds of people.
- The equipment procurement and installation is a considerable effort and expense as well,
- Mine exploitation is, of course, the main goal of this undertaking and its day-to-day operations are a never ending stream of expensive equipment, project and sub-projects, and finally;
- The mine decommissioning and land reclamation are also major, and equally lengthy, labor intensive and expensive, undertakings.

The initial process of mine location, development and construction is estimated on the average at about $150 million per each million tons of annual production, Or, a 5 million tons per annum coal mine would have an initial cost estimate of $750 million. This is almost a billion dollars needed, if everything goes well, to plan, design, setup, and start operations.

And then the mining operations start; equally expensive and complex process, the cost of which is determined mostly by the initial cost and coal cleaning. In the Eastern USA, for example this cost ranges from $15 to $45 per ton of clean coal (in 2010 dollars).

Assuming that the overall weighted average cost per ton of clean coal is $25, our 5 million tons of annual production would spends about $125 million annually for the daily operations—not including emergencies, accidents, etc.

The retail value of coal depends on the type, quality, quantity, the season, and the location of the order, and so it varies up and down accordingly. Due to increasing transport charges, taxes, broker fees, etc., the retail prices could be twice the cost of mining, and even more. They also vary with the type of coal, the period of the year, etc, and fluctuate in the $25-$125/ton range.

Mining Labor

As in any industry, labor pay in the mines is different from person to person and from mine to mine. Labor remuneration in the mines is categorized according to the level of expertise and experience. It is also dependent on the mine location. Different states have different pay scales. And of course, underground workers are paid higher wages, due to the increased level of difficulty and danger, which requires additional training and expertise.

Table 4-3. Labor rates in different states

Worker	Pay $/hour
Laborer—surface mine	7.00-22.00
Laborer—underground mine	14.00-27.00
Mill equipment operator	12.00-30.00
Stationary equipment operator	19.00-26.00
Mechanic—surface mine	11.00-30.00
Electrician—underground mine	14.00-32.00
Equipment operator	11.00-32.00
Production truck driver	9.00-28.00
Heavy equip. operator	9.00-30.00

Miners are not the best paid workers in the world. Considering the dangers, the dirt and misery in underground mines, and the other hazards these people are exposed on daily basis, they might be the most underpaid workers in the world.

Equipment Cost

Mine operations require a number of specialized pieces of equipment, as needed for digging, loading and transport of coal from the mining site to the surface and beyond.

Mining equipment is usually very large, really large, and not cheap. When all additional expenses related to equipment transport, assembly, maintenance, disassembly, EOL disposal, etc. are added, we get many more billions of dollars spent on mining equipment (after the initial purchase) through its long life in the mines.

Mine Decommissioning

After many years of exploitation, the coal in any mine would get depleted. The law requires that the land the mine occupies is brought back, as close as possible, to its original condition. Restoring the land 100% is very seldom possible, so enforcing the law is subjective at best.

Table 4-4. Mining equipment specs and prices

Equipment	Specifications	Weight in lbs.	Cost in $
Dragline	55 cu yd bucket, 250 ft. dump height	16 million	$100 million
Shovel, hydraulic	5.2 cu yd bucket 23.6 ft (7.2 m) dump height	131,000	$925,000
Loader, wheel	9.0 cu yd bucket, 12'1" dump height	114,000	$720,000
Truck, rear-dump	60 ton, 46 cu yd, mechanical drive	22,000	$120,000
Drill, rotary (crawler)	5.13" to 7.88" hole, 25 ft drill length	30,000	$600,000
Tractor, crawler (dozer)	13.7' maximum blade width	39,100	$300,000
Grader, road	14 ft blade width	52,200	$450,000
Truck, water	5,000 gallon water tank	20,000	$255,000
Truck, service	Off-road tire service truck	15,000	$55,000
Truck, shot loader	1,000 per minute capacity	15,000	$75,000

Figure 4-11. Mining equipment...try it for size.

Land reclamation is the preferred post-strip mining method for bringing it back to original conditions. It, however, is expensive and at a cost of about $10,000 per acre, it would cost $2-3 billion to reclaim all damaged land in the U.S. This is possible but improbable to happen any time soon.

And so, what happened in the past more often than none is that mine companies would just abandon the mine, file for bankruptcy, or find another way to evade the expensive land restoration process.

There are, however, examples of great success in this area. For example, in the Rhineland, Germany, coal fields they store both the topsoil and the subsoil during excavation, to be replaced later on. The land is fully refilled, graded and then fertilized and seeded. Drains carry the water away before it can get contaminated by forming sulfuric acid. The acid in the leachate is sometimes neutralized with limestone, as part of a long term land management process.

The acid can be also neutralized with ashes, which are a serious solid waste from coal burning. Using these ashes to fill the trenches and neutralize the leachate acidity would, however, be practical only at mine mouth

plants (power plants operating close to the mine). Transport of large quantity of ashes to mines that are far away from the power plant would be prohibitively expensive and has never been tried.

In other cases, sewage sludge and even liquid sewage from the water reclamation plants can be spread over strip-mined areas to help restore the fertility and soil texture. The land then can be re-vegetated, thus slowly returned to its original state.

The most difficult reclamation is in the North Central U.S. regions, where grassy ranch land is being destroyed in large numbers. Restoring this arid land to usable status would be almost impossible from economics point of view. As a result, no coal company has ever gotten its reclamation performance bond (which is deposited at the beginning of the mine process) returned to a Montana coal company.

Note: Meeting the requirements of the federal surface mining controls adds to the cost of the coal. The cost of reclaiming Western surface mines is $1,000 to $5,000 per acre. A more useful comparison is provided by the additional cost per million Btu (MBtu) of energy obtained from coal.

In the mid-1980s coal energy was available to electric utilities at an average cost of about $1.50 per MBtu. It is estimated now that federally mandated reclamation adds about $0.02/MBtu to Western coal (where the seams are deep, but the energy content is relatively low) about $0.05/MBtu to Midwestern coal, and $0.11/MBtu to Appalachian coal.

Existing requirements already cost an additional $0.10 MBtu for Appalachian coal. Added to these costs is a tax, equivalent to $0.02/MBtu, assessed on the coal companies for the reclamation of abandoned strip-mined land.

Thus, the highest total of additions, Appalachia's $0.23/MBtu, raises the cost of coal to the utility, and ultimately of electricity to the consumer, by over 15%.

> *About 1.1 million acres of coal mined land currently needs reclamation and new land is being disturbed at a rate of about 65,000 acres/year. At this pace, in 20 years we would have an additional 1.2 million acres of pit mining land in need of surface restoration and reclamation.*

Reclamation laws need careful enforcement are not up to par and cannot keep with the fast pace of proliferation of coal draglines around the country. The situation is even worse in many other countries.

The Impacts

Typical activities during the decommissioning and site reclamation phase include removing infrastructure, such as structures, conveyors, rail lines; filling in the mined area or shafts; recontouring the surface; and revegetation. Potential impacts from these activities are presented below, by the type of affected resource. Depending on the mining method, some reclamation activities occur while the coal mining continues, such as in strip mining.

The following potential impacts may result from decommissioning and site reclamation:

- *Acoustics (Noise)* sources during decommissioning would be similar to those during construction and mining, and would include equipment (rollers, bulldozers, and diesel engines) and vehicular traffic. Whether the noise levels exceed guidelines established by the U.S. Environmental Protection Agency (EPA) or local ordinances would depend on the distance to the nearest residence. If near a residential area, noise levels could exceed the EPA guideline, but would be intermittent and occur for a limited time.

- *Air Quality*, affecting Global Climate Change and Carbon Footprint during decommissioning activities include vehicle tailpipe emissions; diesel emissions from large construction equipment and generators; and fugitive dust from many sources such as backfilling, dumping, restoration of disturbed areas (grading, seeding, planting), and truck and equipment traffic. Permitting would be required (as during construction and mining), and therefore these emissions would not likely exceed air quality standards or impact climate change.

- *Cultural Resources* would be unlikely to be affected during decommissioning because these resources would have been removed professionally prior to mining, or would have been already disturbed or destroyed by prior activities. Collection of artifacts could be a problem if access roads were left in place and the area was not monitored.

- *Visual impact* of the coal mine would be mitigated if the site were restored to its preconstruction state. However, despite the physical removal of any surface facilities, the impact of a scarred landscape on an area would likely remain.

- *Ecological Resources* impacted by the decommissioning activities would be similar in nature to impacts from construction and mining, with a reduction or elimination of blasting activities. Negligible to no reduction in wildlife habitat would be expected, and injury and mortality rates of vegetation and wildlife could be lower than they would be during mining. Impacts resulting from acid mine drainage could continue if not properly managed. Restoration of the mine site would reduce habitat fragmentation. Following site reclamation, the ecological resources at the project site could return to preproject conditions.

- *Environmental Justice* could result from significant impacts occurred in any resource areas, and when these impacts disproportionately affect the populations. The environmental justice impact. Issues that could be of concern during decommissioning are noise, air quality, water quality, loss of employment and income, and visual impacts from the project site.

- *Hazardous Materials* and Waste Management of industrial wastes, such as lubricating oils, hydraulic fluids, coolants, solvents, and cleaning agents would be treated similarly to wastes generated during mining activities (that is, put in containers, characterized and labeled, possibly stored briefly, and transported by a licensed hauler to an appropriate permitted off-site disposal facility). Impacts could result if these wastes were not properly handled and were released to the environment. Additional solid and industrial waste would be generated during the dismantling of any ancillary facilities. Much of the solid material from dismantling facilities could be recycled and sold as scrap or used in road building or bank re-stabilization projects; the remaining nonhazardous waste would be sent to permitted disposal facilities.

- *Human Health and Safety* are potential impacts to worker and public health and safety during the decommissioning and reclamation of a coal mine, and would be similar to those from any construction-type project with earthmoving, crushing, large equipment, and transportation of overweight and oversized materials. Added risk may be involved with the reclamation of underground mines due to the potential for mine subsidence. In addition, health and safety issues include working in potential weather extremes and possible contact with natural hazards, such as uneven terrain and dangerous plants, animals, or insects.

- *Land Use*, upon decommissioning of the mine site and rectifying the impacts from coal mining, would be largely reversed and made ready for use. Future subsidence of underground mines could be a long-term issue. Open pit mines could have lasting land-use impacts; the land may be irreversibly altered if reclamation to pre-development condition is not possible. Alternate land uses may be established.

- *Paleontological Resources* during decommissioning activities would not be impacted, because these resources would have been removed professionally prior to mining, or would have been already disturbed or destroyed by prior activities. Fossil collection could be a problem if access roads were left in place and the area was no longer periodically monitored.

- *Socioeconomics impacts* of decommissioning of the mine and reclamation would include the impacts resulting from the cessation of mining activities, including job loss and revenue loss, and also the creation of new jobs for workers during reclamation activities and the associated income and taxes paid. Indirect impacts would occur from both the loss of economic development created by the loss of mining jobs and new economic development that would include things such as new jobs at businesses that support the reclamation workforce or that provide project materials and associated income and taxes.

No adverse effect to property values is anticipated as a result of decommissioning. Site reclamation could result in economic values of residential properties adjacent to the coal mine becoming equivalent to similarly developed residential areas that were not affected by the coal mine. The loss of royalty and tax revenue could adversely impact the local and regional economies.

- *Soils and Geologic Resources* (including Seismicity/ Geo Hazards) activities during the decommissioning/reclamation phase, include removal of access and on-site roads and heavy vehicle traffic. Surface disturbance, heavy equipment traffic, and changes to surface runoff patterns can cause soil erosion. Impacts of soil erosion include soil nutrient loss and reduced water quality in nearby surface water bodies. Disturbed areas would be contoured and revegetated to minimize the potential for soil erosion.

- *Transportation impact* is reflected in short-term increase in the use of local roadways, occurring during the reclamation period. Heavy equipment would remain at the site until reclamation is completed. Overweight and oversized loads, when removing the heavy equipment, could cause temporary disruptions to local traffic.

- *Visual Resources* during decommissioning would be similar to those from construction and mining. Restoring a decommissioned site to preproject conditions would entail recontouring, grading, scarifying, seeding and planting, and perhaps stabilizing disturbed surfaces. Newly disturbed soils would create visual contrasts that would persist at least several seasons before revegetation would begin to disguise past activity. Restoration to preproject conditions may take much longer. Invasive species may colonize newly and recently reclaimed areas. Nonnative plants that are not locally adapted could produce contrasts of color, form, texture, and line.

- *Water Resources* (Surface Water and Groundwater) might be trucked in from off-site or obtained from local groundwater wells or nearby surface water bodies, depending on availability. It would be used for dust control for road traffic and mine filling and for consumptive use by the decommissioning/site reclamation crew.

- *Water Quality* could be affected by continued acid mine drainage if not effectively managed, activities that cause soil erosion, weathering of newly exposed soils leading to leaching and oxidation that could release chemicals into the water, discharges of waste or sanitary water, and pesticide applications. Upon completion of decommissioning, disturbed areas would be contoured and re-vegetated to minimize the potential for soil erosion and water-quality-related impacts.

- *Water Flow* would be affected by withdrawals made for water use, wastewater and storm water discharges, and the diversion of surface water flow for access road reclamation or storm water control systems. The interaction between surface water and groundwater could also be affected if the two resources are hydrologically connected, potentially resulting in unwanted dewatering or recharging of any of these water resources.

Thus produced coal cannot be used as is, so instead it has to be treated for its final journey. Once coal is dug out and loaded on tracks or trucks, it is transported to a treatment facility.

Below we take a look at the different steps of coal-fired power generation.

COAL-POWER GENERATION

Coal straight from the ground, known as *run of mine* (ROM) coal, cannot be used in its *'as is'* state. It usually contains unwanted impurities such as rocks and dirt. It also comes in a mixture of different-sized fragments, which are not easy to use as is in most commercial operations.

Coal users, with very few exceptions, need coal of consistent quality and size, so coal preparation—also known as coal beneficiation or coal washing—and/or other treatment of ROM coal is performed to ensure consistent quality and to enhance its suitability for particular end uses.

Coal can be transported to a special pretreatment processing facility, but that requires an additional step of loading and unloading, so instead, so it is cheaper to process it at, or close to, the mine.

Coal Treatment Process

The mined coal deep down in the mine shafts, is loaded on special cars and transported to an elevator.

The cars are then hoisted to the surface, where the coal is dumped in storage bins or loading areas. Surface mines, of course, omit this step since the coal is near the surface. Once on the surface, the coal is piled in great piles, awaiting its turn in the preparation process.

The preparation work is done in large buildings close to the mine, or at separate locations. These buildings, called *breakers,* reach heights of 150 feet or more. Here the coal is taken to the top of the breaker by a conveyor belt system and undergoes several transformations in different pieces of large equipment on its way down.

The coal is crushed between rollers, and is then sifted over sorting screens into different shapes and sizes, as required by the different customers. On the way down, the smaller pieces fall through the sifters and are sold as a lower quality coal. The coal that remains in the sorters is what gets sold to the different customers. Remember that coal is used not only for power generation, so some of it is processed and destined for industrial uses.

One efficient way of sorting coal is putting it, and the accompanying slate (dirt), into moving water. The slate is heavier than the coal, and sinks or dissolves in the water, while the coal floats on the surface and can easily be separated and carried away.

The different sizes and types of coal have different names. For example, an "egg" must be between 2 inch, and 2 and 5/8 inch in diameter, a "nut" is usually between 3/4 and 1 and 1/8 inches; and a "pea" is between one 1/2 and 3/4 of an inch. Then there is powdered coal, which is used mostly for power generation.

Crushing and cleaning of mine-run coal is also referred to as *beneficiation or preparation*. Often, crushing and sizing is all that is required, but many coal seams, especially those in eastern and mid-western states, contain enough impurities to necessitate further cleaning. Whether the cleaning process is wet or dry, it is commonly referred to as 'coal washing.'

The dry washing method uses high-pressure, pulsating airflow to blow dust and dirt from the coal. Wet washing starts with breaking and screening the coal to remove the large, hard pieces of impurities.

Larger material is usually treated using *dense medium separation,* where the coal is separated from other impurities by floating in a tank containing a liquid of specific gravity, usually a suspension of finely ground magnetite.

As the coal is lighter, it floats and can be separated off, while heavier rock and other impurities sink and are removed as waste. The smaller size fractions are treated in a number of ways, usually based on differences in mass, such as in centrifuges. A centrifuge (machine which turns a container around very quickly) is used to cause solids and liquids inside it to separate.

Equipment can include any of the following pieces: jigs, screens, landers, heavy-medium cyclones, tricone separators, concentrating tables, froth flotation, cells, filters, and driers. Most of these are huge, complex, dangerous, and very expensive pieces of equipment, which require skilled labor and a lot of energy to operate.

Additional cleaning depends on the amount, size, and nature of impurity, how it is dispersed in the coal, and how the coal is to be used. The treatment depends on the properties of the coal and its intended use. It may require only simple crushing or it may need to go through a complex treatment process to reduce impurities.

Alternative preparation/washing methods use the different surface properties of coal and waste. In *froth flotation*, for example, coal particles are removed in a froth form, which is produced by blowing air into a water bath containing chemical reagents. The bubbles attract the coal but not the waste, and is skimmed off to recover the coal fines.

Recent technological developments have helped increase the recovery of ultra fine coal material too, which reduces waste and increases profit.

Once processed, the coal is loaded on unit trains for transport to the power plants, or other commercial enterprises.

Coal Transport

After the coal is dug out from the ground and taken up to the surface, it is processed per consumer specifications of size, cleanliness, etc. before being shipped to the customer for burning. Transportation is a major step in the overall coal production-use process.

In all cases, the coal transport system depends on site-specific and project-specific factors. So it could be a conveyor system within the mine site to the coal preparation plant or a rail system outside the mine site. A system of haul roads is also likely to be present in coal transport schemes.

Note: Often, coal from the mines is transported *as is* to a processing facility far away. Here it is processed, and loaded back on a train for transport to the power plants. Transporting coal off-site may be accomplished by rail, truck, barge, or some combination thereof. A coal-slurry pipeline also may be used to send coal off-site.

Some mines are very large and produce hundreds of tons of coal annually. Most of the 4,325 coal mines

in the U.S., however, produce less than 50,000 tons per year. Coal is processed at the mine site in many cases. There are also many smaller mines in the U.S., but it is impractical, and/or economically unfeasible, to process coal on-site at the smaller mines, so it is usually trucked to a central processing facility. Upon arrival, the coal is unloaded, cleaned, sorted by size and otherwise processed to customer specs before shipping to the point of use (POU)—the coal-fired power plants.

If the mined coal is to be processed at a dedicated facility, away from the mine, then coal transport includes: loading the coal on railroad cars at the mine and transporting it to a processing plant for unloading, treatment, and upgrading (cleaning, sorting, refining, etc.). Once the coal has been processed, it is loaded and transported again to a power plant or other industrial site. Here it is unloaded again and burnt for electricity generation or used in other industrial processes.

A close look at the coal transportation systems shows that coal mines are growing larger and their production increases exponentially. There were 316 mines in the mid-1970s with an annual output of above 500,000. These large mining complexes execute all coal preparation operations onsite. From there the coal moves, usually by train, to the end use site—utility power plant or a large industrial consumer.

Most of the coal in the U.S. is shipped to the processing facilities and power plants by *unit trains*. These are special train compositions, containing 100 or more coal hopper cars. The unit trains shuttle back and forth between large coal mines and coal-fired power plants non-stop, often on dedicated railroads lines

The typical coal train is 100-120 hopper cars long, with each of the cars holding 100-115 tons. This is almost a mile of coal, which can feed a large coal burning power plant operating about a day, or two maximum. The larger surface mines load two or three unit trains of coal a day.

Note: Two coal trains a day is 100 cars x 100 tons x 2 trains = 20,000 to 30,000 tons of coal daily. This is the daily coal production of a smaller mine. The large coal mines produce and transport several times this amount. Just think of the noise, dust, stink, human labor, and expense that accompanies the loading, transport, and unloading operations.

Remember that the U.S. mines, transports,
and burns over 1 trillion tons of coal annually.

There are approximately eighty trains leaving Wyoming mines every day, or about 26,000 trains annually.

This is 26,000 miles of coal, or more than the circumference of the earth. This is only a quarter of the U.S. annual coal production. If the unit trains from the other coal producing states are added, they can be wrapped around the Earth several times.

Figure 4-12. Unit trains

There is a trend of building *mine mouth* plants, in order to avoid the transportation costs. These power plants are built and operated right at, or very close to, the mines, because the coal is cheap at the mine (about $5 per ton), since it requires no transport.

Coal transportation makes 60-80% of the total cost of coal, depending on location, type, quantity and other variables.

Environmental and capital expense considerations, however, make the mine mouth plants option impractical in most cases, and unprofitable in some locations, such as the West Coast. There are, however, several large complexes in the Four Corners area (where Colorado, New Mexico, Arizona, and Utah join) and in the Dakotas and Montana, which are good example of *mine mouth* plants.

Coal can be also shipped by water—by barge on the nation's inland waterways, or by ocean freighters, to coal's export customers. Overseas transport (coal for export) is done by large ocean cargo ships. A huge transoceanic coal export facility is located at Norfolk, Virginia. Coal is currently shipped also from sea ports of Baltimore, Philadelphia, New York City, New Orleans, and Los Angeles.

Coal can be shipped by another water method; *slur-ry pipelines*. In this method, a mixture of powdered coal and water is mixed at the mine and pumped through a long, large diameter, pipeline to the coal burning power plant.

An example of that is/was the 273-mile-long pipeline from the Peabody mine on the "Black Mesa" in Arizona to the 1500-MW Mohave power plant near Page, Arizona. The Black Mesa slurry pipeline delivered about 8 tons of coal per minute along with 2700 gallons or 11 tons of water. At the plant the coal was dried and burned.

Note: This slurry pipeline is an example of lack of consideration for the natural resources and the local environment. Built on Indian land in the harshest and driest desert on Earth, it used 2700 gallons of water a minute, or close to 4 million gallons of water per day. This is 1.5 billion gallons—an entire lake—taken from one area of the desert and pumped to another, where it is discharged as waste water after the coal has been filtered out.

What brilliant mind conceived such a monstrosity? And where did they get these lakes of water to waste in the middle of one of the most arid deserts in the world? Did they think of the consequences?

Similar slurry pipeline are proposed for the North Central coal fields of the Dakotas, Montana, and Wyoming to feed the power plants of the Midwest (Chicago, St. Louis, etc.). Although studies show that slurry transport is less expensive than rail—after the pipeline is installed—environmental and economic controversies have so far frustrated this pipeline projects.

Transportation Cost

The cost of transport via unit train today is approximately $0.020 per ton/mile. 1,000 mile trip would cost approximately $20.00 per ton of coal transported from the mine to the power plant. This multiplied by 100 tons per car and 100 cars per unit train gives us the grand total of $200,000, which is the amount power plants would pay. Remember that at the mine, coal started with a humble price of $5 per ton at some mines.

The transportation cost is even higher when considering other transportation methods, such as barges, trucks, etc. This, compared with the cost of mining the coal, which is approximately $5-15 per ton in the U.S. (but much higher in other countries), is a true highway robbery. There is, however, no way around it; if you want it, you have to ship it. Except in the case of "mine mouth" versions, where the power plant is located at the mine site, so the coal goes directly into the furnace to be converted into electricity.

Another loss of power occurs during the transmission of the generated electric power from the remote power station at the *mine mouth* to the populated centers in this case, but this is a small price to pay, compared with shipping millions of tons of coal across country.

The coal delivered to the power plant is stored in large piles, where it awaits the final journey to the steam furnaces.

Coal Use

Coal has been going through a rough spot in the U.S. recently, but a new hope is emerging in the (thermal) coal market. After a dire run for a number of years here and abroad, which has seen prices crash to near six-year lows, the beginning of the stabilization of the coal market, as far as quantity and price are concerned, is on the horizon.

A correction in the supply imbalance is noticeable, still, the tough times for coal are not over yet, and may continue for some time to come, but the cycle will turn. For most thermal coal operations, aggressive cost-cutting is allowing the miners to keep their heads above water, as coal has been gradually replaced by natural gas.

All in all, about 200 million tons of coal production had fallen out of the global market, including in China, and there are still more casualties to come. This, however, has triggered a new era in coal exports, so the global coal market is now more dynamic than it has ever been. The U.S. and China are leaders in coal operations and in one or another way (import and export mostly) dictate the global coal market presently.

A major characteristic in any market is the availability of the products and services. The quantity and price drive the markets in most cases. Coal is the exception in the energy game today, because it is still available in large quantities all over the world. This contributes to its price to remain fairly stable from day to day and from location to location. The feared *coal peak* is still far away in the future, so coal is not threatened with extinction during this century as the other fossils are.

Obviously, as Figure 4-13 shows, coal is the most abundant fossil fuel, available in many places around the world. Yet, it is definitely a commodity of limited quantities. Yes, there is a lot of coal for now, but we are using a lot of it too. A lot minus a lot, equals zero. The only difference and unknown variable here is time. So no matter how full the coal bucket is now, given enough time, it will eventually get empty. That would be an ugly, sad, cloudy, dark, and very dangerous day, so we hope we can delay it until we are prepared for the new inevitable coal-less future.

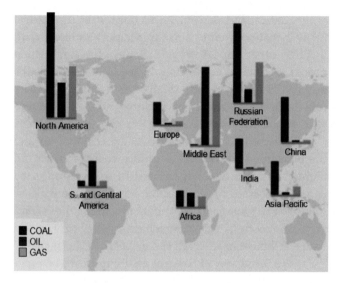

Figure 4-13. Global fossil reserves

Coal-Fired Power Generation

The coal-fired power plants daily operations consist of:

- Coal unloading and storage
- Coal preparation
- Coal burning,
- Power generation and conditioning
- Solid and liquid coal waste disposal

A coal-fired power plant operates primarily by burning coal to generate electricity. There are a number of different steps which allow this process to proceed.

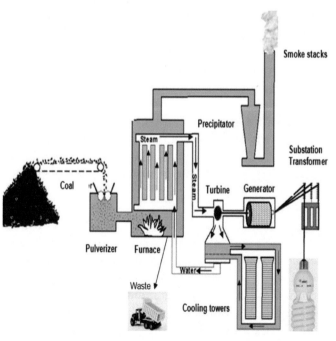

Figure 4-14. Coal burning power plant

Coal cannot be efficiently burned in its natural form, because it comes in large chunks mixed with dirt and other impurities, which make the combustion process inefficient and dirtier. Instead, the coal must be pulverized into extremely small particles, as fine as baby powder in most cases. This ensures most efficient and cleaner burning

Thus produced powder is mixed with hot air and the mixture is blown into a furnace called a "firebox." Here the coal powder is burned in suspension; long before is allowed to settle on the bottom or the walls of the furnace. This *space burning* results in the most complete combustion of the coal, which produces the hottest flame and most heat which can be obtained from coal.

The furnace walls are covered with rows of pipes with continuous flow of water during the burning process. The heat from the burning coal heats the water quickly and turns it into a high-temperature (1,000°F) and high-pressure (3,000 PSI) steam.

The superheated steam is injected into the blades of a turbine, where the extreme pressure is enough to turn the turbine blades fast. When the turbine blades turn fast enough, they engage a generator, which produces electric power by rotating magnets on its axles into wire coils. Thus generated electric power is sent to a transformer in the substation, where it is conditioned and sent into the transmission lines for use at a distant location.

After the steam exits the turbine compartment, it is sent into a condenser in the basement of a power plant, where the steam is cooled by running it between rows of pipes in which flows cool water that which can be returned to the boiler where the entire process is repeated.

The cooling water, upon exiting the condenser, is very hot and could be used for heating in industrial processes, or is discharged into a local river, lake, or the ocean.

The daily operation of the different stages of coal burning and power generation are complex, but well under control. It is done by standardized operations executed by highly skilled personnel, in controlled and documented at all times manner, with safety as a priority number one.

Now days, however, many coal-fired power plants in the U.S. and other developed countries are being replaced with, or retrofitted to, gas burning such. This trend seems irreversible, and is even increasing, so the future of the U.S. coal power industry is uncertain. Coal mining in the country, however, will not be affected much on the long run, because we plan to export most of the coal we cannot burn. Climate change, or no climate change, the U.S. coal industry will continue to produce the maximum amount of coal.

U.S. Coal Market

In the not-so-distant North American past, legend has it, Saint Nick visited children at Christmas to give them presents, if they were good. If there were bad and naughty, however, all they got was a lump of coal. Since these children did not meet the minimal qualifications for receiving gifts, they got only a lump of coal to have and burn to keep them warm.

This way even the naughty children were provided with enough comfort, while given a second chance to become good and get real presents next year. In this case, coal represents a less-than-desirable child's present, but is a symbol of a practical necessity that would keep people warm and comfortable

It is not much of a stretch of the imagination to equate the lump of coal which the naughty children got from Saint Nick, with the coal we get and use today. It is not the best thing to get or deal with, but it provides us with much needed heat, light, and comfort. Since we need a lot of comfort, we use a lot of coal. One lump won't do. Fortunately for coal and for all of us, we have a lot of coal and its use has been diminishing lately…at least in the U.S. This, however, is not so in other parts of the world, as we will see later in this text.

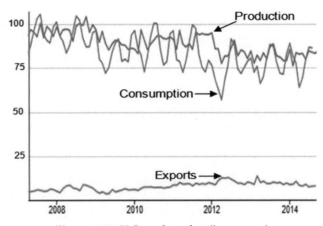

Figure 4-15. U.S. coal market (in percent)

Figure 4-15 tells the story of coal production and use reduction in the U.S. The overall production volumes are getting lower, and the consumption is even lower. Exports, on the other hand have a tendency to fill the gap by gradual increase in volume. And so it goes in the U.S. with coal: future reductions in consumption are planned, while the production volumes are expected to remain stable, thanks to gradual increase in coal exports.

The average price for coal in the recent past has been around $60 per ton, but since mid 2011, it has stabilized at an average of $50-55 per ton. This means that an entire unit train of 100 cars full of coal, about 100 tons each (about 10,000 tons of coal) would cost the power plant operators $55/ton x 10,000 tons= $550,000. Including the transport charges.

One ton of coal generates about 2,000 kWh of electric power

At 2,000 kWh per ton, the 100 cars of 100 tons each in the unit train contain 10,000 tons x 2,000 kWh = 20,000,000 kWh of electric energy, or 20,000 GWh total (after considering all losses and inefficiencies of the power plant equipment). The unit train, therefore, would be able to provide about 20 hours of electric power non-stop at 100% capacity for an average power plant. Since the power demand diminishes during some parts of the day, and is near zero at night, this unit train load might last day and a half, or even 2 days.

The power plant owners (the utilities) charge their customers about $0.10 per kWh electric power used, so the unit train would bring them about $2 million in gross proceeds. Minus $500,000 for the coal, and that much for operational expenses, the utility might end up with a million dollars gross profit. Not bad for a day or two of work as usual. There are, of course, a number of other expenses, such as insurance, plant and transmission lines maintenance and upgrades, etc., so at the end of the day, the actual profit to the utility is usually much less than that.

In 2012, the use of coal for electricity generation declined substantially, or about 21% from 2011 levels. 27 GW of capacity from coal-fired generators were retired, or substituted by natural gas, which use increased correspondingly by 30% recently.

Coal's share of electricity generation in the U.S. dropped to just over 36% from its high of over 50% a decade ago.

And now, since that much coal is not needed here, lot of our coal goes, and even more is planned to go, to China and other countries. This new situation is quite controversial, however, because our coal will contribute to significant increase in GHG emitted during its burning in less efficient Chinese power plants. The battles on this front have started and will flare with time.

U.S. Coal Market Development

A number of railroad transportation problems and increased rail traffic for other commodities were common in 2014, and yet the year-to-date coal railcar loadings showed 1% increase over 2013. The weekly coal carloads in December 2014 was 120,914 and kept close to this number in 2015.

Nearly half a million train cars full of coal are driven to, and burned at, the U.S. coal-fired power plants every month.

Total electric power sector coal stocks were over 12 million short tons (MMst) in October 2014, which was the largest stock build since a 12.5-MMst build during the same month in 2011. The increase in coal inventories follows typical seasonal pattern where coal plants build stocks during the autumn months in preparation for increased coal consumption during the winter.

Despite the increase, 2014 coal stocks of 136.3 MMst were 17 MMst (11%) lower than the previous year and 19% lower than the previous four-year average for the same month. The large year-over-year decrease in stocks reflects high levels of coal-fired electricity generation during the winter of 2013-14 across a large portion of the country and subsequent decrease in coal deliveries because of rail transportation issues.

The EIA estimates that total coal production in the U.S. for 2014 was 994 MMst, or 1% (10 MMst) higher than in 2013. The annual coal production has been going down, and is expected to decline further with time.

Regional shifts in production are as follow: Appalachian coal production averaged 272 MMst in 2014, declined 3.6% in 2015 and 2.9% in 2016 as a result of higher mining costs, weak demand from domestic and export markets, and a shift to higher-sulfur, lower-cost interior region coal.

Interior region coal production averaged 187 MMst in 2014, grew by 1.0% in 2015, and 1.3% in 2016. Many power generators have recently installed sulfur dioxide scrubbers in response to environmental regulations, allowing them to switch from Appalachian and Western region coal to Interior region coal.

Western region coal production averaged 535 MMst in 2014, and remains largely unchanged in both 2015 and 2016.

All in all, coal use has decreased significantly during this decade. Coal production has also fallen some, but disproportionately so, due to compensation from increased exports.

The Coal Export Issues

Thus far, most coal produced in the U.S. has been consumed domestically at a fairly steady pace. At the same time, the U.S. coal exports have been going slowly up, but now with decreased domestic demand for coal, there are serious plans to export huge amounts of coal to China and other countries.

New and ever increasing emission regulations are further forcing the coal companies to look at exporting their production as needed to keep the mines open. As a result, we foresee export volumes of coal on the increase. The proportion of coal production going toward exports has also increased, doubling from 5% in 2009 to 10% in 2011, sharply increasing for awhile in 2012, and projected to take off in the near future.

Several new shipping terminals have been planned in the U.S. and the coal companies are gearing up for exporting as much coal as they can produce—to whomever wants to buy the stuff. But what does that mean to the future generations? What would exporting so much coal do to their energy supplies? Isn't that similar to the proverbial "shooting oneself in the foot?"

The majority of present-day coal production in the U.S. consists of thermal (steam coal) used mostly for electricity generation.

Metallurgical (coking coal), although in smaller quantities, is also produced in large quantities, It is very important for the iron and steel production industries. Seventy percent of global steel production depends on this type of coal. In 2011, metallurgical coal exports accounted for 77% of U.S. coal exports in terms of trade dollars and 65% in terms of volume. In 2010, exports to Asian markets increased 176% from 2009 levels, primarily because of a surge in exports of metallurgical coal to China, Japan, and South Korea. Metallurgical coal exports accounted for 83% of the growth in 2010 export volumes.

U.S. coal producers increased both export volumes and prices in 2010, because there were huge demands from China and India, due to reduced world coal supply caused by heavy rains and flooding in major coal exporting countries. In 2012, export coal prices turned down sharply, due to decreased domestic demand, a slowdown in economic growth in both China and India. These, and some European, countries promise to be a bottomless coal market.

Oil and natural gas prices in Europe are, and will remain, quite high for a long time, so relatively inexpensive U.S. coal will be in even greater demand by European utility companies. This, combined with the decline of carbon emissions permits in the European Union, makes coal even more attractive to the end users there.

Today U.S. coal export volume and prices are influenced by the difference between domestic and international demand, the natural gas prices, the coal use in developing countries, and by the Asian coal production levels. While all indicators point to favorable trade winds for U.S. coal exports, the nagging question persist, "Is this the best we can do with such an important energy commodity, that is in limited supply?" Is this the best for our long term energy survival? The answer is also blowing in the wind…

Summary
- Coal fuels almost half of the electricity generated in the U.S.
- Coal also emits half of the CO_2 gasses that harm our environment.
- Electricity generated from coal costs about 5-6 cents per kWh, which is among the lowest in the world.
- The U.S. national average price of electricity from all fuel sources—with 50% from coal—is about 10 cents per kWh, which is also one of the lowest in the world.
- Each person in the U.S. uses on the average over 7,500 pounds of coal annually.
- Coal accounts for 90% of America's fossil fuel reserves, and is enough to power the U.S. for the next 150-200 years, but this needs to be adjusted as coal exports are increasing.

Note: The question that begs an answer here is, why are we in such a hurry to get rid of this precious commodity by exporting it? Is "selling the branch you are sitting on" (which is exactly what we are doing by exporting coal and other fossils), the smart and responsible thing to do? This is a very serious question that nobody is even trying to answer.

- Coal is a key element of our national security, since it is the lifeblood of our domestic energy supply, which depends on reliable electricity. It allows us to avoid further dependence on imports of energy from abroad.
- Coal provides thousands of long-term, well paid jobs. Each mining job also creates 3-4 additional jobs somewhere in the U.S. economy, or about 500,000 people (including over 125,000 coal miners) are involved in the U.S. coal industry. Many times more people are involved in the global coal industry.
- Coal mining jobs average $65,000 per year, or

about twice as much as the average wage for other industrial jobs at the same technical level.

- Coal mining, however, is a dirty, hard, and dangerous job, so we must thank our miners and their families, who stand by them in good and bad times, every time we turn the lights on.
- 65-70% of U.S. coal production is via surface mining, which alone is responsible for over $5.0 billion of business activities every year and provides fuel to power more than 25 million American homes.
- The U.S. coal exports are projected to increase significantly during the next several decades. China, EU, and other countries will eventually benefit from the U.S. exports. How good or bad this is remains to be seen, but the scene is set and the wheels are in motion, so there is no turning back...

Conclusion

There are dozens and hundreds of companies, factories, and organizations involved in the coal's life cycle. Thousands of skilled, unskilled, and specialized personnel are involved in all stages of the mines, transportation, power plants operation, equipment manufacturing, installation, and operation, etc. Many other thousands are involved in the production, transport, and processing of raw materials and the manufacturing and transport of primary and secondary materials, supplies, and consumables. All these products and services are needed for the day-to-day operations of mines, power plants, and other industries.

And, of course, we must not forget the undeniable importance of the utility companies which are involved in most of these steps, and which are in charge of delivering enough quality electric power to our homes and businesses day and night.

Designing, directing, and coordinating some of the critical steps of the process are the marketing people of the different companies. They are responsible for ensuring the uninterrupted and profitable market for their materials, components, supplies, fuels, electricity, services, etc. All these separate markets put together make the huge global energy market.

All companies, factories, organizations—as well as the thousands of people working in them—involved in this process are elements of the national and global energy markets. Without even one of these elements, the entire energy market structure won't function properly. If several of these elements are missing or fail, then the entire structure could collapse. At the very least, it won't be the same as we know it today! Thankfully, the structure is well established and the coal industry, as a

principal element of the energy markets, is functioning properly and efficiently today.

CRUDE OIL

Crude oil is second to none in importance in a number of key areas—but primarily the transportation sector—which in turn affects the energy independence and the national security of the U.S. and all importing nations.

The importing nations represent 95% of the global population, so the importance of crude oil is obvious.

Crude oil is widely used in the U.S. and around the world, as follow:
— Primarily, as a transportation fuel,
— Secondarily, as an important raw material in a number of industries, and
— Partially, as a convenient power generation fuel in some remote areas where other fuels are not available, and

Because of this unrivaled importance, we need to take a very closer look at this important and irreplaceable commodity.

Background

Our dependence on crude oil's magical qualities during the last century has been total and unprecedented, albeit unsustainable on the long run. The bad part is that within a single century, we have managed to burn, and otherwise use, a major part of the oil reserves on Earth, which it took Nature millions of years to create.

There is no possible way to create more crude oil, so we must be very careful how we handle it now and in the near future. Obviously, the long term situation is very serious and deserves our full attention and utmost care.

Using less oil now, will leave more for the basic needs of future generations.

Unfortunately, reducing the use of crude oil now would mean significantly slowing down the economic development in many countries, and most of them are not willing to accept that. Overuse, at the present rates, will exhaust the already semi-depleted oil supply within several decades and force us into an era of oil-less existence. This is truly a catch 22 situation...

Crude oil is the lifeblood of our nation's economy, and drives our technological, and social progress.

Just like blood flows in our veins, so does the vehicular traffic flow across the country, non-stop all day and all night, all through the year. Vehicles of all types and sizes travel the roads and highways, the waterways and the skies, carrying important products and people at work or play. A large fleet of large trucks, boats, trains, and airplanes is carrying millions of tons of goods across the country. From the agricultural fields of Imperial Valley, California, to the streets of New York, they carry fruits and vegetables, bread, and equipment. From Main to Arizona they haul people and goods non-stop, day and night.

Cars, trucks, boats, trains, and planes fill their tanks with oil products—millions of gallons of gasoline, diesel, and jet fuel—every day. This continuous flow—zillions of tons of vehicles and cargo—must keep going, fueled exclusively by crude oil, in order to keep the economy going and the people happy. This is something only crude oil can do...for now

Got that? No crude oil: no vehicles on the roads. No vehicles on the roads: no people at work and no vegetables and fish on the dinner tables.

Electric and hybrid vehicles, fuel cells, natural gas, and biofuels powered vehicles, and all such hi-tech wanna-be-substitutes cannot compete with the power, speed, reliability, and flexibility of gas and diesel powered cars and trucks. And even less can they compete with the fossils for driving our 18-wheelers, and large boats, trains, and planes. Not for now, not for a long time, and very likely never!

With the finality of fossil fuels, we see a simple and scary reality, that of no crude oil and no traffic on the roads! Simple and final...the lifeblood flow stops. Period. End of story, and end of the American Dream. The consequences to the economy would be disastrous, even if the vehicular flow is slightly reduced for lack of crude oil. Remember the horror of 1973...? If not, then you need to learn all about it, to see what could happen.

Here's hoping that our political and economic powers consider the importance of oil today and what could happen, if and when we run out of it. The consequences of a fossil-less future are very serious and we all must think again before deciding to drill another oil well or exporting another drop of crude oil products abroad...as planned by our politicians

In the Beginning...

Immediately after the Big Bang there were no coal, oil, or natural gas deposits on Earth. It took millions of years to complete the conversion of organic matter into coal, crude oil, and natural gas. As a matter of fact, the process is not fully completed yet. This is evidenced by the different coal layers which show different stages of coal development—from peat (the most recent) to anthracite (the oldest) coal products.

The oil and natural gas formation process, in most cases, were somewhat different from that of coal. The origins of oil and natural gas could be traced to the much shallower and warmer oceans, where dead animals, vegetation, and other organic matter was continuously falling and piling on the bottom. There, the bulk of the organic matter were tiny, surface-dwelling organisms called plankton, which includes several animal species (zoo plankton) and plant species (phytoplankton). The phytoplankton uses photosynthesis to capture the solar energy that is the basis of marine food chains.

The organic materials would sink to floor eventually, to join with the inorganic matter, consisting of sand and dirt, carried in by the rivers and land runoff. The different materials slowly mixed up on the ocean floor, resulting in a thick layer of ocean-floor sediments. As the ocean levels increased, the pressure of the overlying water compacted the sediments. A number of natural processes, such as crystallization and cementation of inorganic minerals eventually converted the soft sediments into solid material.

As time went by, the ocean-floor sediments got buried deeper and deeper, and the water got deeper, so the sediment layers were subjected to ever-increasing, extremely high pressures and temperatures. This resulted in the formation of the layered solids under the ocean floor, called sedimentary rock.

During the long exposure to high pressure and elevated temperature, organic material in the sedimentary layers got cooked, very similar to preparing food in pressure cookers, but at much higher pressure and temperature. Of course, the process took a very long time, which is the key to producing the large quantities of fossil materials with proper quality that we use today.

Initially, the sediments cooked to a waxy substance called kerogen. Kerogen, as in tar and shale oil deposits, is similar to the peat segment of the coal formation process. It represents a not-yet completed process of conversion to crude oil—sort of half-cooked crude oil.

Like peat, it has different (inferior) heating properties than the respective final products, crude oil and coal. As the cooking processes continued, the sediment was eventually converted into a combination of organic liquids and gases which we call petroleum.

Note: Although the term *petroleum* is usually used for liquids (i.e. crude oil), *petroleum products* refers to both liquids and gases. So, when we talk about *petroleum*, we refer to a group of fossils which contain both crude oil and natural gas and their derivatives and by-products.

Petroleum products contain some of the energy (sunlight) that was captured many, many years ago by the plankton on the ocean surface. This energy (created millions of years ago) is what we need badly and use today, and which our comfort and progress depend on. To that end, we now extract millions of gallons from the large petroleum deposit areas, which were formed by migration and pooling of the liquid oils and gaseous natural gas into large underground reservoirs.

There are also small petroleum deposits in many sedimentary rock formations around the world, but reaching them, due to the insufficient quantities and the great effort and expense involved, is technologically difficult and economically infeasible with the present state of the art. Many of them, however, could and will be exploited at the right time—most likely when petroleum prices increase enough to make the process economically feasible.

Note: Remember, when we talk about petroleum, we do refer to a group of fossils, which contain both crude oil and natural gas. Below we will take a close look at the present-day crude oil production and use, followed by natural gas.

In the Recent Past…

We know that oil in the large deposits comes mainly from strata of coarse sandstone, which was formed by sand deposited by water ages and ages ago. We've also noticed that there is no oil in flat strata, so we must deduce that the oil flowed from the sediment layers into the sandstone strata, where it was held as in a sponge, and where it remained throughout the millennia.

After the initial Big Bang and the following overheating at the very beginning, the Earth mass grew cooler. At that time and later on, it went though a number of processes, such as folding, during which horizontal movements press inward and move the rock layers upward into a fold or anticline.

Note: Faulting resulted where the layers of rock crack and one side shifts upward or downward. Pinching out happened where a layer of impermeable rock was squeezed upward into the reservoir rock.

While undergoing these processes, the Earth's surface layer—the crust—twisted and wrinkled, so that oil and gas are now found in the large sandstone wrinkles—with the gas on top and oil on the bottom.

If cracks develop in the strata, the oil would usually run away by gravity to find another place to settle. This is how large reservoirs are formed. There is also trapped gas in the mix, which is usually under pressure, and is constantly looking for a way out. If we know where it is, we could easily drill to it and pipe it for use in a house or business.

Not very long ago, there was a gas deposit close to Niagara Falls that constantly released gas. One day somebody drove a pipe down in the ground and lit the gas. The flame was quite high and was visible from far away. This, however, is a rare occurrence, since the oil and gas deposits are usually much deeper in the ground.

The presence of gas is often an indication that there is oil underneath too. This is how the first colonists discovered petroleum in New York. They, however, didn't know what to do with it, while the native Americans nearby used oil found in the local springs as medicine. Later on they figured out that oil can be sold, so they would soak their blankets with oil from the springs and wring them out to sell the oil. This was the first commercial oil extraction process.

Later on, the first "gusher" was discovered by accident. Workmen had drilled 500 feet down when a high pressure stream of crude oil burst forth, hurling tools and people high up into the air. They just let the oil gush out until the pressure was reduced enough. They did not know how to stop it, nor what to do with it, so a lot was wasted that way.

A large gusher in Lakeview, California, threw up 50,000 barrels of oil a day, to a height of 350 feet for a long time. The steady oil column sprayed the country for a mile around. The oil flowed away as a river, and no one could stop it or store it. Finally, a large storage tank was built around the well with stones and sand bags.

Yet, oil could not be efficiently collected, processed, or used for a long time.

The high pressure that oil comes up with sometimes is due mostly to the high pressure of natural gas in the deposit mix. This is confirmed by the fact that the pressure is reduced with time, after the oil is pumped out for awhile, and/or when the gas pressure is reduced.

So in the past, the lucky land owner who thought that there was oil underneath would build a derrick over it with the intention of collecting some of the oil and selling it. The derrick principle is used today too,

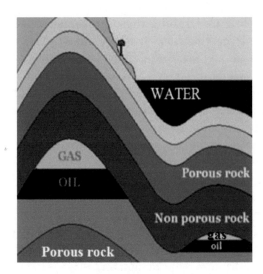

Figure 4-16. Crude oil deposits

and it usually consists of four wooden or metal beams, 30-100 feet high, firmly held together by crossbeams. It is positioned over where the oil well is to be dug. An engine house nearby provides power for the drilling machine.

An iron pipe eight or ten inches in diameter is driven slowly down in the ground until it comes to the oil-containing rock. At this point regular drilling operations begin. A pulley mounted on top of the derrick holds a heavy-duty rope to one end of which are attached the drilling tools. The drilling tools go down in the pipe and drill a hole in the rock day and night. While the drill is busy making the hole, a sand pump sucks out water and loose bits of stone from the drilling hole.

As soon as the drill reaches the bottom of the strata, the sides of the bore hole are cased to keep the water out, and the drilling continues. This time the drill penetrates the oil-bearing sandstone and if there is oil in it, it will gush out and the lucky land owner will join the ranks of oil millionaires.

Often, however, drill as much as you may, oil is nowhere to be found. That's because although it is underground, it is near the surface in some places, but more often than not it is deep down to 1,000-5,000 feet depth. To reach these depths, special and very expensive equipment and advanced procedures are usually needed.

In some cases, the oil is not under much pressure, so it is hard to determine where the oil deposit is even if it is hit directly. And this is where the modern oil drilling industry comes in handy.

Modern drilling rigs with their sophisticated drilling tools, computerized controls, and GPS navigation can find and reach oil and gas many thousands of feet under the surface—be it land or sea—with amazing precision and speed.

A new, more sophisticated and efficient, method of horizontal drilling is promising extraction from a depth of several miles, and from locations that were impossible to even imagine accessing a few years back. Hydrofracking is a new process which has revolutionized the oil and gas industry. Later, we will take a closer look at this process, its advantages and issues.

Crude Oil Properties

Crude oil is actually a mixture of many (dozens and even hundreds) of different species of organic chemical compounds, most of which are classified as hydrocarbons.

Hydrocarbons are organic chemicals that contain both hydrogen (H-hydro) and carbon (C-carbon)atoms.

Table 4-5. Hydrocarbon species in crude oil and natural gas

Formula	Name	# C atoms	MP	BP	State
CH_4	Methane	1	-183	-162	Gas
C_2H_6	Ethane	2	-172	-89	Gas
C_3H_8	Propane	3	-187	-42	Gas
C_4H_{10}	Butane	4	-138	0	Liquid
C_5H_{12}	Pentane	5	-130	36	Liquid
C_6H_{14}	Hexane	6	-95	69	Liquid
C_7H_{16}	Heprane	7	-91	98	Liquid
C_8H_{18}	Octane	8	-57	126	Liquid
C_9H_{20}	Nonane	9	-54	151	Liquid
$C_{10}H_{22}$	Decane	10	-30	174	Liquid

Methane and ethane (one and two carbons respectively) are the main components in natural gas.

Propane (three carbons) is most commonly used as gas, but can be liquefied under modest pressure. Its different forms are used in countless applications, where petroleum products are used for energy; cooking, heating, and transportation.

Butane (four carbons) is addedto motor fuels in winter because its high vapor pressure helps with cold starts of car and truck engines. It is also used to fill cigarette lighters at slightly above atmospheric pressure. It is also a main fuel source in many developing countries.

Octane, a main component of crude oil, has 8 carbon molecules, and is one of the most useful components of the motor fuels. It so important that the quality of gasoline is expressed in *octane* units. So now we have

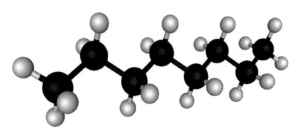

Figure 4-17.
Octane is a complex hydrocarbon found in crude oil.

98, 91, etc. *octane gas* at the gas pumps. It is part of the alkanes group, which start with pentane (five carbons). The alkanes are refined into gasoline, while hydrocarbons with a higher carbon number from nonane (nine carbons) to hexadecane (sixteen carbons) are refined into diesel, kerosene, and jet fuel.

Alkanes with more than 16 carbon atoms are not used for fuel, but are refined into fuel oil and lubricating oils. At the high end of the alkanes range is paraffin wax which has 25 carbon atoms, and asphalt with 35 and more carbons. Most of these fractions are usually cracked (refined) by modern refineries into lighter and more valuable products.

Some other types found in crude oil are paraffins, naphthenes, aromatics, or combinations of these, such as alkyl naphthenes and aromatics, and the polycyclic compounds. Mercaptan, thiophene, and others are also present in high concentration as well. There are also olefins, which usually are not present in crude oil, but are formed during decomposition of the crude oil components during high temperature distillation processes.

The shortest molecule alkanes, those with four or fewer carbon atoms, are the petroleum gasses, which remain in a gaseous state at room temperature. Depending on demand and the cost of recovery, these gases are either flared off, sold as liquefied petroleum gas under pressure, or used to power the refinery's own processes.

The cycloalkanes, also known as naphthenes, are saturated hydrocarbons which have one or more carbon rings to which hydrogen atoms are attached according to the formula CnH2n. Cycloalkanes have similar properties to alkanes but have higher boiling points.

The aromatic hydrocarbons are unsaturated hydrocarbons which have one or more planar six-carbon rings called benzene rings, to which hydrogen atoms are attached with the formula CnHn. They tend to burn with a sooty flame, and many have a sweet aroma. Some are carcinogenic.

To complete the soup mix, a number of other chemicals, some of which are inorganic (non carbon and hydrogen containing) are found in raw crude oils as well.

Sulfur produces the most troublesome byproducts of the oil refining process, in the form of free (elemental) sulfur (S) and hydrogen sulfide (H_2S). They cause corrosion and produce sulfur dioxide (SO_2) which is toxic and creates acid rain when burned. The other major sulfur compound, H_2S, is a vicious poison, which paralyzes the olfactory nerves so that its victim is unaware of its presence and can choke to death.

On the other hand, the petroleum industry produces large quantities of sulfur by converting the H_2S byproduct to elemental sulfur, which is quite useful in other industries. A lot of nitrogen is produced also, which also can be captured and used in other applications.

Salts, such as chlorides, sulfates, and carbonates (compounds of sodium, calcium, and magnesium) are found in crude oil in liquid or small particles of these salts, which are expressed in pounds of salt (NaCl equivalent) per thousand barrels of crude.

These salts cause corrosion and deposits in the heating and heat transfer equipment, by adhering to the surfaces of boilers and pipes. Worse, when in the presence of H_2S and H_2O, they are extremely corrosive and can damage all metal equipment, even stainless steel.

Of course, and very importantly, crude oil and all its organic components are highly flammable. They have different boiling point ranges. It can be easily seen in Table 4-6 that the boiling point increases as the compounds get heavier—their MW increases.

Table 4-6. Crude oil fractions

Compound	Boil°F	MW
LPG	-44–31	44–58
Gasoline	31–400	100–110
Jet fuel	380–520	160–190
Diesel fuel	520–650	245
Gas oil	650–800	320
Residuals	800–1000	—
Vacuum gas	800–1000	430
Crude coke	2,000	2,500

where:
Boil°F is the boiling point in degrees F, and
MW is the molecular weight of the different compounds

Most crude oil compounds can be ignited with open flame and at low temperatures, which can be good, but could also be quite dangerous. For this purpose, we use a standard called auto-ignition temperature (AIT), which is used for proper use and safety purposes.

AIT is basically a measure of the temperature at which a vapor from a particular compound will ignite spontaneously (in the absence of a flame).

Note: AIT should not be confused with the boiling point of the respective substances. Gasoline and naphtha, for example, boil at 100°F or less, but could ignite spontaneously on coming into contact with a hot surface at 600°F or above, even if in the absence of an open flame.

Classification

An important crude oil measure is the American Petroleum Institute gravity, or API gravity. API is a measure of how heavy or light a petroleum liquid is compared to water. American Petroleum Institute has an inverted scale for denoting the lightness or heaviness of crude oils and other liquid hydrocarbons. Calibrated in API degrees (or degrees API), it is used universally to expresses a crude's relative density in an inverse measure. The lighter the crude, the higher the API gravity, and vice versa, because lighter crude has a higher market value. Wow. Who knew…

If its API gravity is greater than 10, it is lighter and floats on water; if less than 10, it is heavier and sinks. API gravity is thus an inverse measure of the relative density of a petroleum liquid and the density of water, but it is used to compare the relative densities of petroleum liquids. For example, if one petroleum liquid floats on another and is therefore less dense, it has a greater API gravity.

Although mathematically, API gravity has no units (see the formula below), it is nevertheless referred to as being in "degrees." API gravity is gradated in degrees on a hydrometer instrument. The API scale was designed so that most values would fall between 10 and 70 API gravity degrees.

$$API = (141.5 / S.G.) - 131.5$$

where:

S.G. is the specific gravity, which is the ratio of the density of the material at 60°F to the density of water (accepted as 1) at that temperature.

For example, water at 60°F has

$$API = (141.5 / 1) - 131.5 = 10° \ API.$$

Crudes can be classified as "light" or "heavy," a characteristic which refers to the oil's relative density based on the American Petroleum Institute (API) Gravity. This measurement reflects how light or heavy a crude oil is compared to water. If an oil's API Gravity is greater than 10, it is lighter than water and will float on it. If an oil's API Gravity is less than 10, it is heavier than water and will sink.

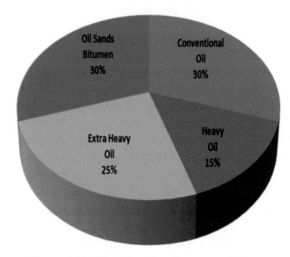

Figure 4-18. Global crude oil reserves by type

Light crude oil is defined as having an API gravity higher than 31.1°API. Brent Crude has an of 35.5°, and gasoline's API is 50°.

*   Medium oil is defined as having an API gravity between 22.3°API and 31.1°API.
*   Heavy oil is defined as having an API gravity below 22.3°API.
*   Asphalt is the heaviest of the petrochemicals with average API gravity of 8° or less.

Lighter crudes are easier and less expensive to produce and bring the highest prices. They generally have a higher percentage of light hydrocarbons that can be recovered with simple distillation at a refinery.

Heavy crudes cannot be produced, transported, and refined by conventional methods because they have high concentrations of sulfur and several metals, particularly nickel and vanadium. Heavy crudes have densities well above that of water. Some heavy crude oils are also known as "tar sands" because of their high bitumen (tar) content.

Crude oil quality is also measured and qualified by their sulfur content. Crude oils with low sulfur content are classified as "sweet." Those with a higher sulfur content are classified as "sour." Sulfur content is generally considered an undesirable characteristic with respect to both processing and end-product quality. Sweet crudes are usually more desirable and valuable than sour crudes.

Unfortunately, the light and sweet crude oil supplies are being exhausted at a high pace, and will be eliminated the fastest. What we have left now are the heavy, sour, and much dirtier oils, which also will be gone soon enough.

Toxicity

Toxicity of crude oil and its derivatives refers to how harmful these materials might be to humans and other living organisms. Generally, the lighter the oil, the more toxic it is. EPA has classified crude oils in four categories that reflect how the oils would behave in a spill and its aftermath:

- Class A crude oils are light and highly liquid, these clear and volatile oils can spread quickly on impervious surfaces and on water. Their odor is strong and they evaporate quickly, emitting volatiles. Usually flammable, these oils also penetrate porous surfaces such as dirt and sand and may remain in areas into which they have seeped. Humans, fish, and other biota face the danger of toxicity to Class A oils. These high quality light crudes and the products produced from them are in this class.
- Class B oils are considered less toxic than Class A, these oils are generally non-sticky but feel waxy or oily. The warmer it gets, the more likely Class B oils can be to soak into surfaces and they can be hard to remove. When volatile components of Class B oils evaporate, the result can be a Class C or D residue. Class B includes medium to heavy oils.
- Class C oils are heavy, tarry oils (which include residual fuel oils and medium to heavy crudes). They are slow to penetrate into porous solids and are not highly toxic. However, Class C oils are difficult to flush away with water and can sink in water, so they can smother or drown wildlife.
- Class D are non-fluid, thick oils that are comparatively non-toxic and don't seep into porous surfaces. Mostly black or dark brown, Class D oils tend to dissolve and cover surfaces tightly when they get hot, which makes cleanup much harder. Heavy crude oils, such as the bitumen found in tar sands, fall into this class.

The new planned Keystone Pipeline is scheduled to transport tar sands oil from Canada through several U.S. states to Gulf Coast refineries. The battle between pipeline operators and the locals (supported by environ-

mental groups) is presently centered on the danger of soil and water table contamination resulting from spills of these Class C and D oils.

In any case, all types of crude oils and their derivatives are considered hazardous compounds that can harm living things and humans under some circumstances. Because of that, special care must be exercised when handling, or otherwise dealing with, all types of oils and petrochemicals.

Combustion Properties

An important measure of the oil quality and safety is its vapor pressure, also called raid vapor pressure (RVP). Different oils have different RVP, which determines the volatility, or how fast the substance evaporates, and the related explosion risks. Crude oils with higher RVP are usually more prone to explosions during storage and transport than those with lower RVP.

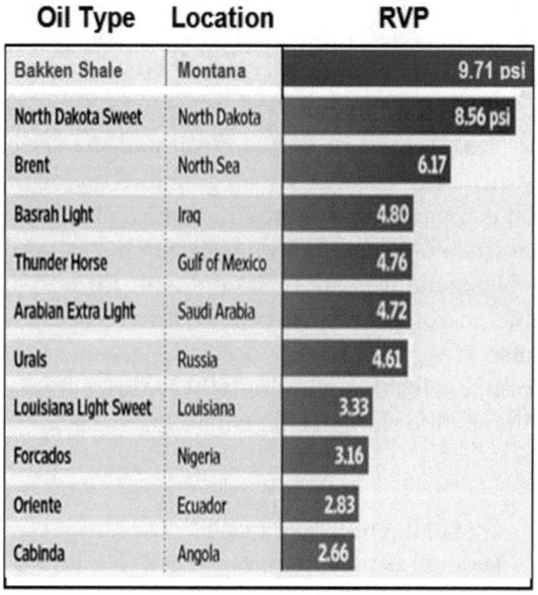

Oil Type	Location	RVP
Bakken Shale	Montana	9.71 psi
North Dakota Sweet	North Dakota	8.56 psi
Brent	North Sea	6.17
Basrah Light	Iraq	4.80
Thunder Horse	Gulf of Mexico	4.76
Arabian Extra Light	Saudi Arabia	4.72
Urals	Russia	4.61
Louisiana Light Sweet	Louisiana	3.33
Forcados	Nigeria	3.16
Oriente	Ecuador	2.83
Cabinda	Angola	2.66

Figure 4-19. RVP of different oils per location

Crude oil from the Bakken Shale formation, for example, contains several times the combustible gases as average crude oil from other places. This raises questions about the safety of producing and shipping it by rail or pipes across the U.S.

The volume of oil moving by rail from the Bakken Shale had soared to nearly a million barrels daily at the end of 2013; way up from about 300,000 barrels a day in 2010.

The rapid growth of oil production has increased the transport of oil by rail from other locations too, and with that the number of accidents have increased.

In the summer of 2013 an oil-tanker train loaded with 72 cars of crude oil exploded near Lac-Mégantic, Quebec, leveling the downtown area and killing 47 people. Later that year, trains derailed and exploded in Alabama and North Dakota, creating huge fires, complete with giant fireballs in the sky.

Tanker car derailments are usually caused by track problems or some other equipment failure, so the crude oil cannot be blamed for the accidents. But crude oil is considered hazardous, and although it usually does not explode, it did on several occasions recently.

Bakken crude is a mixture of oil, ethane, propane and other gaseous liquids, which are in a higher concentration than is usually found in conventional crude oils. Unlike conventional oil, which sometimes looks like black syrup, Bakken crude tends to be very light in color.

Some people claim that it smells like gasoline, so you can just fill your tank with it and run the engine with no problems. How true this is depends on whom you are talking to, so we just need to remember that crude oil presents a number of dangers, which we must be aware of and account for.

We also should know that it is hard to separate the liquids from the gasses in the raw oil, and since there are no particulate federal or state regulations on the matter, the oil companies just fill the railroad tanker cars with whatever comes out of the oil well spigot.

The standards are upcoming, but for now it is a gamble, so you don't always get what you see or want. Because of that, oil tanker accidents accompanied by oil spills, explosions, and fires will continue. Properties and human life along the oil tanker railroads are also in danger.

Crude Oil Cradle-to-Grave Cycle

Crude oil is usually pumped out in a liquid from underground deposits, where it could vary in density and other physical and chemical properties. There are also other forms and shapes that crude oil can be found in. These, of course, require different production and processing methods, some of which we will take a closer look later on in this text.

According to OPEC, at the end of 2011, world proven crude oil reserves stood at 1,481,526 million barrels, of which 1,199,707 million barrels, or 81%, were in OPEC Member Countries. The picture today is changing, however, and a number of non-OPEC member countries—including the U.S.—are increasing their production levels significantly, mostly due to new discoveries and more efficient technologies and processes.

One major change in the oil equation today is the fact that the U.S. now produces domestically over 50% of the oil used across the country on daily basis. Other countries are following our example.

What do we need to find, extract, and use crude oil? In general terms, the cradle-to-grave oil production and use process consists of:
- Oil search and location
- Proposed area survey
- Environmental studies
- Oil rights and agreements
- Permits and legal issues
- Drill area preparation
- Drilling rig construction
- Daily drill and well O&M
- Oil transport to refinery
- Oil processing
- Finished product distribution
- Oil products sales and use
- Oil well decommissioning

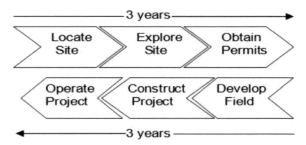

Figure 4-20. Crude oil well time line

Figure 4-20 shows the steps that each new oil well site goes through. It takes several years before a site is rendered feasible for operation and the appropriate permits are obtained. That much time is then needed to get equipment to the site, start construction, and finally produce the expected quantity and quality of oil.

Many such projects hit the jackpot and deliver enormous quantities of high quality oil. Many others, however, are not that lucky. In most cases where hydrofracking is used, as a matter of fact, the wells produce enough oil for awhile, but soon after the flow diminishes and the well is considered unprofitable. This is because there is no reservoir (a lake) of oil under the ground, but instead oil is extracted from rocks in the vicinity. The operators, therefore, do not know how much oil is contained in these rocks and how much can be extracted.

We'll take a close look at these steps and conditions, to determine the total energy cost and environmental impact of the cradle-to-grave oil production and use process.

Oil Search and Locating

The first step of the oil extraction process is locating significant oil deposits that are worth the time and expense to drill and extract the oil from them. Geologists employed by oil companies or under contract from a private firm, are the ones that look for and find oil. Their efforts are focused on recognizing the signs and indications of oil deposits. Some of the signs could be a source rock, reservoir rock, or entrapment.

In the past, geologists had to read the surface features, surface rock and soil types, and then drilled small core samples to confirm their findings. A variety of other methods were applied in different cases too.

Sensitive gravity meters were used to measure small deviations in the Earth's gravitational field, which could be translated into gaps of oil deposits. Magnetometers were also used to measure changes in the Earth's magnetic field that might be caused by flowing oil.

Detecting the smell of escaping gas by sensitive electronic sniffers is another way, but most commonly seismology is used. In this method shock waves are created close to the surface and travel through the rock layers, where some are reflected by the different materials.

Such shock waves are created by a compressed-air gun which shoots pulses of air into the water (for exploration over water). Thumper trucks are used to slam heavy plates into the ground (for exploration over land). Explosives can be detonated in holes drilled into the ground (for exploration over land) or thrown overboard (for exploration over water) to create shock waves.

The waves reflected back to the surface travel at different speeds depending on the type and/or density of rock layers through which they must pass. Sensitive microphones or vibration detectors detect the reflections (hydrophones over water, and seismometers over land are used). Seismologists interpret the readings, looking for signs of oil and gas reservoirs. Once a prospective oil strike is found, they mark the location using GPS coordinates on land or by marker buoys on water.

Today, satellite images and other sophisticated, computerized technologies do most of the job of detecting, interpreting and locating oil deposits.

Oil drilling is the next step of the process, and it can be divided into on land (on-shore) and into the ocean (off-shore) drilling. On-shore drilling is done on solid ground, while off-shore drilling is done miles from the shore, sometimes at great depths in the ocean.

Crude Oil Production

Once oil has been located—be it as a pool or as a rich content in the rocks underground—the entire area must be surveyed, in order to determine the exact boundaries. Then, the exact drilling site has to be selected, and environmental impact studies are conducted. If everything checks out and the decision is made to proceed with drilling at this site, then the legal work starts. Land lease agreements, or land titles, rights-of-way access, legal jurisdiction determination and other conditions must be determined and met.

After the legal issues are settled and the appropriate permits and documents have been signed, the work crew goes about preparing the land and the drilling location, The land is cleared and leveled as needed to accommodate the drilling derrick erection, pumping equipment, and support structures construction. Access roads also need to be constructed for transporting equipment, products, and people to and from the drill site.

A lot of water is used during the drilling process, so a source of water must be located nearby as well. A water well has to be drilled in many cases prior to initiating oil drilling. In many cases, local water is just not enough, so truck-tanks are used to deliver water to the site instead.

A reserve water and water disposal pits must be dug nearby as well, to deposit rock cuttings and drilling mud during the drilling process. The pits are temporary structures, which are usually lined with heavy plastic to prevent ground contamination and the related cleanup expense. In ecologically sensitive areas, such as a marsh or wilderness, a pit might not be allowed, so the waste materials must be dumped in barrels or trucks and disposed of by trucking off site.

The next step is to dig several holes in the place of the oil rig and the main hole. A rectangular pit, or cellar, is dug around the drilling hole area. This is needed to provide a level and safe work space around the hole for the workers and drilling accessories.

The main hole drilling can be then started, first with a small drill truck, since the top part of the hole is larger and shallower than the main portion. The drill hole is then lined with a large-diameter metal pipe, and additional holes are dug off to the side for temporary equipment storage.

The main rig components can be set up now, starting with the power system, which usually is a diesel engine coupled with a generator, to provide electric power for the drilling rig and the auxiliary equipment and structures. The components of the mechanical system, a hoisting system and a turntable are setup next.
The hoisting setup is used for lifting heavy loads and consists of a mechanical winch with a large steel cable

spool, a block-and-tackle pulley, and a receiving storage reel for the cable. The turntable, which is part of the drilling mechanism is installed next.

The main part of the drilling mechanism is the rotating equipment, which consists of a swivel, which is a large handle that holds the weight of the drill string and allows it to rotate. It also makes a pressure-tight seal at the hole. A kelly is four- or six-sided pipe that transfers rotary motion to the turntable and drill string, while the turntable drives the rotating motion, using power from the electric motors. The drill string consists of a drill pipe (connected sections of about 30 feet each) and drill collars consisting of larger-diameter, heavier pipe that fits around the drill pipe and places weight on the drill bit.

A special drill bit is mounted at the end of the drill that actually cuts into the rock formation deep below. The drill bits come in many shapes and materials, such as tungsten carbide steel, diamond, etc. Different drill bits are usually specifically designed for the different drilling tasks and rock formations. Finally, the casing, a large-diameter concrete pipe that lines the drill hole and prevents it from collapsing and leaking is placed in the hole. It also allows the drilling mud that is pumped into the smaller drill pipe, to circulate to the surface and deposited in the pit.

Thus constructed oil rig, or derrick, is tall enough to allow new sections of drill pipe to be added to the drilling apparatus as drilling progresses and the drill bit goes farther down into the ground. A blowout preventer is added to prevent explosions. It consists of several high-pressure valves that seal the high-pressure drill lines and relieve pressure when necessary to prevent a blowout. This prevents uncontrolled gush of gas or oil to the surface and eliminates fires.

The circulation system, when activated, pumps the drilling mud through the kelly, the rotary table, the drill pipes and collars. A mixture of water, clay, and other materials and chemicals are pumped down into the drill hole at a high pressure, and is also used to carry rock cuttings from the drill bit on its way back to the surface.

The pump sucks mud from the mud pits and pumps it to the drilling apparatus and down the hole. Pipes and hoses connect the pump to the drilling apparatus, while the mud-return line returns mud from the hole to the surface and into a shale shaker, which separates rock cuttings from the mud.

The shale slide pumps send the waste rock cuttings into the reserve pit, where they are separated from the mud. The cleaned mud is sent into a mud pit to be mixed and recycled in a mud-mixing hopper for another use.

Proven Oil Reservoir Drilling

Drilling into a proven oil reservoir starts by lowering the drill bit into the ground and rotating it. The drill bit goes down and is stopped when is very close to hitting the oil deposit. Technical and safety checks are completed at this point, and a surface hole is then drilled from the starter hole to a pre-set depth, just above the oil deposit. The surface hole is lined with a casing pipe, which is cemented in place in the center of the hole, as needed to prevent it from collapsing in on itself. The casing pipe is centered by spacers around the outside to keep it centered in the hole. The cement is allowed to harden and is then tested for hardness, alignment and a proper seal.

Drilling in the main hole continues for awhile, and then stops again for the new surface hole area to be lined and cemented. Then the drilling continues farther down, repeating the lining and cementing of the surface hole until the rock cuttings in the mud show oil sand, which is a signal that the well's final depth is very near.

Now the drilling mechanism is removed from the hole and a number of tests are performed to confirm the oil presence. These tests are done by lowering different sensors in the main drill hole. Some of tests are well logging, measuring the rock formations, drill-stem testing

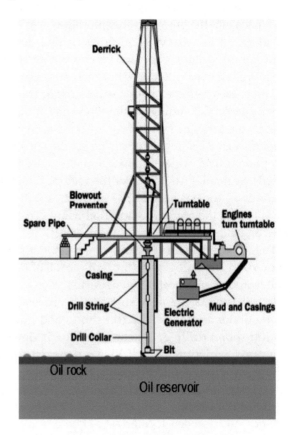

Figure 4-21. Land oil drilling rig

for measuring the pressures in the hole, and core samples for taking samples of rock to look for characteristics of oil deposit bedrock.

If the tests look good, the hole is drilled farther until oil flows into the casing in a controlled manner.

A perforating gun is lowered into the well to the depth of the oil deposits, using explosive charges to create holes in the casing through which oil can flow. A smaller-diameter pipe is then run in the hole as a conduit for the oil and gas to flow up through the well and to the surface.

A packer is then run down the outside of the tubing and is allowed to expand to form a seal around the outside of the tubing. A multi-valve structure called a Christmas tree is mounted on the top of the tubing and is cemented to the top of the casing. It allows control of the flow of oil from the bottom of the well to the surface.

The flow of oil into the well is started by acids pumped down into the well and out the perforations. Thus dissolved channels in the limestone allow oil into the well. If the rock is sandstone, then a specially blended fluid containing a mixture of sand, walnut shells, and aluminum pellets is pumped down the well. The pressure makes small fractures in the sandstone that allow oil to flow into the well, while the mixture holds these fractures open.

At this point the oil rig is removed from the site and a set of production equipment is installed for continuous extraction of the oil from the well to the surface. Several methods are used for this. The pump system uses an electric motor to drive a gear box that moves a lever. The lever pushes and pulls the pump rod, attached to a pump. The up and down action creates a suction that draws oil up through the well.

If the oil is too heavy or too deep (or the pressure inside the well is too low or none) the oil cannot flow freely, so an enhanced oil recovery method is used, where a second hole is drilled into the oil reservoir, and steam is injected into it under pressure. The heat thins the oil and the pressure helps push it up the well.

Offshore Oil Drilling

While oil extracted by the methods described above is usually done on solid ground, there is a lot of oil under the deep waters of rivers, lakes and the oceans. Extracting oil deposits covered by deep water is a much more complex and dangerous effort. Done correctly, it can be efficient, safe and profitable, but if things go wrong, the results can be deadly for the oil workers and devastating for the surrounding environment. We only need to mention Deep Horizon to get a feel of the conse-

quences from such a scenario.

The search, location, and exploration of potential oil deposits is similar to those described for land use. When all preliminaries are completed, a mobile offshore drilling unit (MODU) is brought in and installed over the potential oil hole. It is then used to dig the initial well and sometimes is converted into production rigs. More often, however, the MODU rigs are replaced by permanent oil production rigs for long-term oil extraction.

Figure 4-22. Offshore oil rig

There are several different types of MODUs:
- Submersible MODU is used in shallow and calm waters. It is a barge supported on the sea floor, and on the deck are steel posts that extend above the water line. A drilling platform rests on top of the steel posts and is used to drill the oil hole, similar to the methods described above.
- Jackup MODU is a rig that sits on the deck of a floating barge, which is towed to the drilling site. The jackup can be used in depths of up to 525 feet, by extending its legs down to the sea floor and resting them on it without penetrating the floor. The jackup is then ratcheted up so that the platform is kept above the water level to keep it safe from high waves. Drilling can commence and proceed in a similar fashion as described above
- Drill ships are special ships, designed for deep water oil drilling. They are equipped with a drilling rig mounted on the top deck, with a drill setup operating through a hole in the hull. Once the drilling

starts, the ship uses a combination of anchors and propellers to maneuver as needed to correct for waves and currents.

- Semi-submersible drilling rigs float on the surface of the ocean on top of huge, submerged pontoons, using propulsion systems to navigate to drilling sites and to maneuver over the hole. Computers control the anchor chain tension and engine power to correct for waves and currents.

During the drilling process, a blowout preventer (BOP) is installed at the ocean floor. It is equipped with a pair of hydraulically powered clamps that close off the pipe to the rig in the case of a blowout.

When the hole is drilled and ready for production, the well is sealed by a pair of plugs. The bottom plug sits near the oil deposit, and drilling mud or seawater keeps it in place, while the top plug is placed to cap the oil well. Then the well is hooked to a production rig, which operates in a similar way to the land-based oil rigs.

The most promising oil exploration technology today is hydrofracking.

The Hydrofracking Process

Hydraulic fracturing, hydrofracking, or fracking, is a major process that has played an increasingly important role in the development of America's oil and natural gas resources for over half a century. Until recently, few wells were developed and exploited by hydrofracking, simply because the technology was not fully developed.

Advances in the hydrofracking state of the art have allowed the technology to be used safely, cheaply and efficiently where it was unthinkable just a few years ago. Presently, there are over 35,000 wells processed and operated with the hydraulic fracturing method. Estimates are that over one million such wells have been developed during the last half century.

There are differences in each well development and operation, and with each well the industry takes a step ahead in its progress. This allows the development of new best practices, to increase safety and minimize the environmental and societal impacts associated with new well development.

At the present levels of oil and gas extraction, up to 80% of oil and natural gas wells drilled in the near future will require hydraulic fracturing. This process is essential for oil and gas production from hard-to-reach formations and for production optimization of existing wells.

Horizontal drilling is a key component in the hydraulic fracturing process.

Horizontal Drilling vs. Vertical Drilling

In traditional drilling, a well is drilled more or less vertically downward. When the target formation is reached, drilling continues for some distance into the target formation. The operator then uses a special tool to create perforations in a portion of the wellbore that is within the target formation. Oil or gas can then flow into the well through the perforations.

The longer the length of perforated pipe, the faster oil or gas can flow into the well, but with vertical drilling the length of pipe that can be perforated is limited by the vertical height of the target rock formation.

On the other hand, a formation that may be only a couple of hundred feet or less in height might extend horizontally for miles. Horizontal drilling takes advantage of this. In horizontal drilling, the operator drills vertically downward toward the target formation, then turns the drill bit at an angle to drill in a horizontal direction. The drilling might then proceed horizontally for a mile or more within the target formation, providing a long horizontal "leg" that can be perforated and exploited.

Hydraulic fracturing makes it feasible to produce oil and gas from shale and other low-permeability formations from which production would otherwise not be feasible. Such production and the activity associated with it are beneficial for several reasons.

- First, the activity has substantial economic benefits. Economists have estimated that shale gas development has created more than 200,000 jobs—direct and indirect—in the United States. Some of the new jobs are in the oil and gas industry itself, while other jobs are with companies that supply products, materials or services to the oil and gas industry.

These include companies that mine the sand or manufacture the ceramic particles used as proppants; transport water, sand, and equipment to drilling sites; manufacture the high-pressure pumps used in fracturing; operate pipelines; perform construction; and operate the hotels, restaurants and caterers that house and feed workers.

State and local governments benefit from increased tax revenue, mineral royalty, and other revenue streams. In northwest Louisiana, where Haynesville Shale is located, for example, some local governments have seen their sales tax revenue double over the course of a few years, enabling those governments to pay cash for the construction of numerous capital improvements, even while state and local governments elsewhere are struggling. State and local governments in other areas, including Texas, Pennsylvania, and North Dakota, also have benefitted.

— Second, the increased production of oil and gas bolster our energy security by reducing our country's dependence on imported crude oil, some of which comes from areas that are politically unstable, and unfriendly to the U.S.

— Third, hydraulic fracturing can have environmental benefits because it is often used to facilitate production of natural gas, which is the cleanest-burning of all fossil fuels. For a given amount of energy production, the combustion of natural gas produces only half as much carbon dioxide as coal, and about one-third less than oil. The combustion of natural gas also produces smaller, albeit significant, amounts of other pollutants.

The glass is always half empty or half full, depending on your point of view and your personal interests. Fracking is no exception, since in addition to the benefits, it has also introduced some most controversial issues and debates on very important socio-economic and political subjects to date.

On one hand we are happy that we now have plenty of clean, affordable natural gas, which can replace coal burning, thus killing two birds with one stone—providing cheap energy and cleaning the environment at the same time.

On the other hand, however, there are documented reports of environmental disasters caused by large-area fracking wells—water table contamination, ground cave-ins, earthquakes, toxic gas emission, etc. People and animals have been hurt by the liquid and gas emissions, and property values have been decreased because of them as well.

The Issues

In addition to causing local traffic congestion, dust and other problems for the locals, fracking can contaminate large land areas with waste water from wells and chemical storage ponds. At times the process goes wrong—both below and above ground. In some cases, the drilling wells are not built sturdily enough, or a piece of equipment malfunctions, causing leaks that contaminate the adjacent soil and the groundwater in the area.

"Flowback" water (process water used during the production cycle) has been documented at a number of sites to travel to the surface or into water sources, thus contaminating water tables, lakes, and streams.

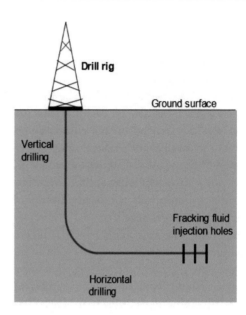

Figure 4-23. Horizontal (directional) drilling and fracking

Once the contamination enters the fresh water supplies, it shows up at people's dinner tables. In some cases, it can be seen bubbling out in water wells, and even coming out of kitchen faucets. These, albeit rare, occurrences are reason enough for us to consider the consequences of large-scale water contamination.

Luckily, at least for now, such incidents have been recorded in remote locations, affecting people living in small towns. Such contamination would be disastrous in a densely populated centers.

The fracking frenzy is increasing and proceeding quickly—well ahead of appropriate environmental safeguards. For example, toxic fracking fluids, including known cancer-causing chemicals like benzene and toluene, are not controlled and are exempt from federal regulation, according to the old Safe Drinking *Water Act.*

More importantly, there are no requirements for the drilling companies to disclose the contents of their fracking fluid, which are proprietary in most cases.

This regulatory gap lets millions of gallons to contaminate seep into the ground around each drilling site and eventually the water table. Because of that, property values in many areas near fracking sites have fallen, and human health has been adversely affected.

This is a precursor of a large-scale disaster, and since no one knows what is in the contaminated areas, except the drilling owners, nobody knows what can happen next. The drill operators are not going to reveal the toxic components for fear of competition and lawsuits…unless forced to.

At the same time, for state and local governments,

oil and gas are cash cows so they are not going to make too many changes, let alone stop, or even limit, the hydrofracking activities anytime soon.

Another danger of the fracking over-development is the fact that, just as happened in the Rust Belt in the recent past, large industries coming into unsuspecting areas, use the local resources and leave gross toxic waste, and eventually high unemployment behind. This has happened thousands of times; everyone is aware of it, and yet such disasters happen time after time.

Short-term solutions to the key points of the hydrofracking controversy are to:

• Open the communication channels between producers and locals,

• Accelerate the R&D and testing of new proppants, chemicals and procedures for use in fracking, and

• Introduce new legislation that is acceptable to all parties involved.

The latter option might be easier said than done, but it is the only fair and most efficient solution. If properly designed and executed, said legislation could be the best long-term solution, where the interests of all parties could be protected.

For now, however, crude oil and natural gas—and by association the hydrofracking process—rule over the energy sector in the U.S. and many other countries. The positive economic effect of this new development is so great that the entire U.S. economy is starting to depend on it and its products—abundant crude oil and cheap natural gas.

While nobody can deny the short-term benefits of the hydrofracking process, it is extremely likely that as the fracking sites multiply in numbers, so will the number and intensity of long-term negative effects. It is also very likely that eventually they will get so big (and so many people will be negatively affected) that the state and federal governments will need to step in and take appropriate—even drastic—measures to protect people and their properties.

Until then, it will be "Drill, Baby, drill." Oil and gas drilling are the most important (albeit short-term) solutions to the national energy security issues, after all…

Case Study: The Energy Country of Texas

Today Texas produces more energy products—fossils and lately wind and solar energy—than most countries in the world. Recently hydrofracking and other new technologies have increased the energy production to unprecedented levels. The Eagle Ford Shale deposits are bringing a new oil bonanza to Texas and the U.S. in general. The massive production coming out of Eagle Ford and other shale deposits in Texas is significantly enhancing our energy security.

With that in mind, we pronounce Texas the energy capital of the world, with Houston as its HQ. It is the nerve center of the U.S. energy sector too, with more energy-related jobs than most other places on Earth. At least 200,000 people are directly involved in the energy business, with thousands additional oil- and gas-field service industry jobs. Most oil and gas companies are headquartered in Houston, each with hundreds and thousands employees in the city.

Factoring all indirect and induced jobs that service the industry (hotels, restaurants, gas stations, and supermarkets) the dollar numbers are in the millions. Putting all this together, we see that Texas' economy is larger than that of many countries too.

If Texas were a country, it would rank as the 14th largest oil and 3rd largest natural gas producing nation on Earth.

Table 4-7. Annual gas production

Country	Thousand m^3
United States	681,400,000,000
Russia	669,700,000,000
State of Texas	225,000,000,000
European Union	164,600,000,000
Iran	162,600,000,000
Canada	143,100,000,000
Qatar	133,200,000,000
Norway	114,700,000,000
China	107,200,000,000

Texas produces about 30% of U.S. natural gas and about 30% of our oil production. By volume of production it is surpassed only by Russia and the combined production of the other U.S. states. The Eagle Ford Shale in East Texas alone tripled natural gas production in the state. It is estimated to be the largest oil field ever discovered in the lower 48 states. If this were not enough, there are now estimates that the Cline Shale formation in West Texas is even larger than the Eagle Ford. Who knew?

The implications of the Texas shale revolution are equally important for the entire country and the world. There are estimates now that due to these and other developments the U.S. will surpass Saudi Arabia in total oil production by 2020. And that was before the potential of the Cline Shale was added to the total. It is now also projected that, if things go as planned, the U.S. could become completely independent of imports by 2020-2025.

One amazing fact is that there are about 400 major rigs in the US drilling for natural gas, while four years ago, there were over 1,600 such rigs. While the rig count has dropped by 75%, the total natural gas production has continued to rise steadily. This is because technologies and rig operators are more efficient and because the major oil shales—the Eagle Ford, the Bakken, and the Permian Basin deposits—contain a lot of associated natural gas. So, natural gas is becoming more available whether the companies are drilling for it or not. There is also a huge amount of excess drilling capacity in the system, waiting to come on line as soon as natural gas and crude oil prices stabilize (and preferably rise some more).

This true-to-life energy revolution started only 6-7 years ago, when most of the major new shale plays today were discovered. The best part is that new shale plays are still being discovered on a daily basis, and we are far from having discovered all of them. In addition, new technologies promise to bring even more deposits, and larger quantities will be extracted from each.

This is one of the best news for the U.S. oil companies and partially for the consumers. It is, however, bad news for the future generations. If we proceed as planned, we will discover and pump out all the oil and gas in the U.S. within a short time. That would leave those who come after us with a serious energy deficit, and we clearly see them scratching their heads, saying, "What were those people thinking?"

During 2014-2015, however, drastic changes in the global oil and gas markets caused the U.S. drilling industry to slow down. The low crude oil prices, forced upon the world by Saudi Arabia, made drilling for new capacity unprofitable. As a result, many rigs were idle and some shut down all together. This was a temporary situation, of course, but it showed how vulnerable the energy markets, and the entire world economy, are.

Crude Oil Transport

Crude oil is the life blood of the global transportation and petro-chemical sectors. It is also vital to the international energy market. Since most of it is found in areas that are far away from the point of use, it has to be transported—sometimes at great distances. The transport is done via pipelines, or transport vehicles—tankers, trains, and trucks.

After crude oil is pumped out of the ground, it has to be transported to a refinery for conversion into useful final products. The different oil transport methods include pipelines, marine vessels, tank trucks, and rail tank cars to transport crude oils, compressed and liquefied hydrocarbon gases, liquid petroleum products and other chemicals from their point of origin to pipeline terminals, refineries, distributors and consumers.

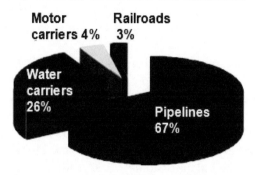

Figure 4-24. Crude oil transport methods and ratios

Oil is usually transported from the well to the refinery by a pipeline, which can be hundreds, even thousands, of miles long. The pipelines are either connected directly to the oil wells, or are supplied by storage tanks at tanker ports and terminals.

The U.S. pipeline system totals about 223,000 miles.

Following is a closer look at the main components of this important stage of crude oil's cradle-to-grave life cycle.

Pipeline Transport

Transporting oil via pipes is the safest, albeit not 100% safe, method of moving oil from one location to another—even if these are thousands of miles a part. The U.S. pipeline system totals about 223,000 miles. From the refineries, gasoline, fuel oil, and such, go by truck or train tank car to wholesalers or large consumers. The U.S. tank truck fleet numbers about 160,000, and the number of rail tank cars is slightly larger, over 165,000.

The risks related to pipe transport are few, but serious in nature if and when the pipe system fails. Oil spills are the most common result from pipeline failure, and result in contamination of large land areas. These spills are difficult to clean and are the cause of a number of law suites and protests from locals and environmental organizations—some of which have contributed to shutting down pipeline sections and prevented others from being built.

Background

Pipelines are widely used for transport of crude oil from the oil wells to the refineries for processing. Thus processed products are also often delivered via pipelines to the point of use.

There are aboveground, underwater and underground pipelines, varying in size from several inches to several feet in diameter. These pipelines move vast amounts of crude oil, natural gas, LHGs and liquid petroleum products across the US and most other countries.

The first successful crude-oil pipeline, a 2-inch-diameter wrought iron pipe, 6 miles long with a capacity of about 800 barrels a day, was built in Pennsylvania in 1865.

During WWII large pipe networks were built in the US, to move oil from coast to coast. The Keystone XL pipeline, planned to deliver Canadian syncrude oil to Gulf Coast refineries, is stalled in the permitting and political debate stages. If and when completed, it will transport millions of gallons of crude oil and petrochemicals across the U.S.

Many locals and a number of environmental groups are fighting the project, because of oil spill dangers, but it is unlikely that they can stop it. Oil is too important to our energy and national securities; so, if not this one, the next administration for sure, will make the Keystone XL pipeline happen.

The lower 48 states receive 1.5 to 1.6 million barrels of oil per day, courtesy of the Alaskan pipeline. This pipeline, 789 miles of 48-inch steel pipe, curving up and down and around hills and valleys, carries great quantities of oil from the rich North Slope field to the port of Valdez in southern Alaska. Here the oil is loaded on tankers for shipment south. The Alaskan pipeline was built in controversy which is still unresolved and is a subject of high-level debates and legal battles.

Today, liquid petroleum products are moved long distances through pipelines at speeds of up to 6 miles per hour, assisted by large pumps and compressors along the way, at intervals ranging from 60 miles to 200 miles.

Pipeline pumping pressures and flow rates are controlled throughout the system to maintain a constant movement of product within the pipeline. Sensors, control mechanisms, and safety devices are installed along the length of the major pipelines to prevent spills, fires, and other possible disasters. Pipelines run from the frozen tundra of Alaska and Siberia to the hot deserts of the Middle East, across rivers, lakes, seas, swamps and forests, over and through mountains and under cities and towns.

There is a network of over 95,000 miles of petroleum product pipelines in the United States. This network delivers finished petroleum products to the end customers. It is separate from the network of crude oil pipelines, and balances the demand and supply conditions in each region.

The initial construction of pipelines is difficult and expensive, but once they are built, properly maintained and operated, they provide one of the safest and most economical means of transporting these products.

Types of Pipelines

There are four basic types of pipelines in the oil and gas industry—flow lines, gathering lines, crude trunk pipelines and petroleum product trunk pipelines.

- Flow pipelines move crude oil or natural gas from producing wells to producing field storage tanks and reservoirs. Flow lines may vary in size from 5 cm in diameter in older, lower-pressure fields with only a few wells, to much larger lines in multi-well, high-pressure fields. Offshore platforms use flow lines to move crude and gas from wells to the platform storage and loading facility. Lease lines carry all of the oil produced on a single lease to a storage tank.
- Gathering and feeder pipelines collect oil and gas from several locations for delivery to central accumulating points, such as from field crude oil tanks and gas plants to marine docks. Feeder lines collect oil and gas from several locations for delivery directly into trunk lines, such as moving crude oil

Figure 4-25. Section of the Alaska pipeline

from offshore platforms to onshore crude trunk pipelines. Gathering lines and feeder lines are typically larger in diameter than flow lines.

- Crude trunk pipelines move natural gas and crude oil long distances, from producing areas or marine docks to refineries and from refineries to storage and distribution facilities by 1- to 3-m-diameter or larger trunk pipelines.
- Petroleum product trunk pipelines move liquid petroleum products such as gasoline and fuel oil from refineries to terminals and from marine and pipeline terminals to distribution terminals. Product pipelines may also distribute products from terminals to bulk plants and consumer storage facilities, and occasionally from refineries direct to consumers. Product pipelines are used to move LPG from refineries to distributor storage facilities or large industrial users.

Batch Intermix and Interface

Although pipelines originally were used to move only crude oil, they evolved into carrying all types and grades of liquid petroleum products. Because petroleum products are transported by pipelines in successive batches, there is co-mingling or mixing of the products at the interfaces.

The product intermix is controlled by one of three methods: downgrading (derating), using liquid and solid spacers for separation, or reprocessing the intermix. Radioactive tracers, color dyes and spacers may be placed into the pipeline to identify where the interfaces occur. Radioactive sensors, visual observation, or gravity tests are conducted at the receiving facility to identify different pipeline batches.

Petroleum products are normally transported through pipelines in batch sequences with compatible crude oils or products adjoining. One way to maintaining product quality and integrity, downgrading or derating, is by lowering the interface between the two batches to the level of the least affected product. For example, a batch of high-octane premium gasoline is typically shipped immediately before or after a batch of lower-octane regular gasoline. The small quantity of the two products which has intermixed will be downgraded to the lower octane rating regular gasoline.

When shipping gasoline before or after diesel fuel, a small amount of diesel interface is allowed to blend into the gasoline, rather than blending gasoline into the diesel fuel, which could lower its flash point. Batch interfaces are typically detected by visual observation, gravitometers or sampling.

Liquid and solid spacers or cleaning pigs may be used to physically separate and identify different batches of products. The solid spacers are detected by a radioactive signal and diverted from the pipeline into a special receiver at the terminal when the batch changes from one product to another. Liquid separators may be water or another product that does not co-mingle with either of the batches it is separating and is later removed and reprocessed. Kerosene, which is downgraded (derated) to another product in storage or is recycled, can also be used to separate batches.

A third method of controlling the interface, often used at the refinery ends of pipelines, is to return the interface to be reprocessed. Products and interfaces which have been contaminated with water may also be returned for reprocessing.

Oil-Tank Trucks

From the wells, or refineries, oil, gasoline, fuel oil, and other petrochemical products go by taker trucks or train tank cars to wholesalers, or to large consumers. The U.S. tank truck fleet numbers about 160,000 and the number of rail tank cars is over 165,000.

Tank trucks are normal heavy-duty trucks, modified to carry large metal oil tanks. The oil tanks are typically constructed of carbon steel, aluminum, or plasticized fiberglass material, and vary in size from 1,900 liter tank wagons to jumbo 53,200 liter capacity. The optimum capacity of tank trucks is governed by regulatory agencies, and usually is dependent upon highway and bridge capacity limitations and the allowable weight per axle or total amount of product allowed along the scheduled routes.

There are pressurized and non-pressurized tank trucks, which may be non-insulated or insulated depending on their service and the products transported. Pressurized tank trucks are usually single-compartment, and non-pressurized tank trucks may have single or multiple compartments.

Regardless of the number of compartments on a tank truck, each compartment must be treated individually, with its own loading, unloading and safety-relief devices. Compartments may be separated by single or double walls. Regulations may require that incompatible products and flammable and combustible liquids carried in different compartments on the same vehicle be separated by double walls. When pressure testing compartments, the space between the walls is also tested for liquid or vapor.

Tank trucks have either hatches which open for top loading, valves for closed top- or bottom-loading and unloading, or both. All compartments have hatch entries for cleaning and are equipped with safety relief devices to mitigate internal pressure when exposed to abnormal conditions. These devices include safety relief valves held in place by a spring which can open to relieve pressure and then close, and hatches on non-pressure tanks which pop open if the relief valves fail and rupture discs on pressurized tank trucks.

A vacuum relief valve is provided for each non-pressurized tank truck compartment to prevent vacuum when unloading from the bottom. Non-pressurized tank trucks have railings on top to protect the hatches, relief valves, and vapor recovery system in case of a rollover. Tank trucks are usually equipped with breakaway, self-closing devices installed on compartment bottom loading and unloading pipes and fittings to prevent spills in case of damage in a rollover or collision.

Railroad Tank Cars

Railroad tank cars are constructed of carbon steel or aluminium and may be pressurized or unpressurized. Modern tank cars can hold up to 171,000 liters of compressed gas at pressures up to 600 psi. Non-pressure tank cars have evolved from small wooden tank cars of the late 1800s to jumbo tank cars which transport as much as 1.31 million liters of product at pressures up to 100 psi.

Non-pressure tank cars may be individual units with one or multiple compartments, or a string of interconnected tank cars, called a tank train. Tank cars are loaded individually, and entire tank trains can be loaded and unloaded from a single point. Both pressure and non-pressure tank cars may be heated, cooled, insulated and thermally protected against fire, depending on their service and the products transported.

Figure 4-26. Burning oil-tanker train

All railroad tank cars have top- or bottom-liquid or vapor valves for loading and unloading and hatch entries for cleaning. They are also equipped with devices intended to prevent the increase of internal pressure when exposed to abnormal conditions. These devices include safety relief valves held in place by a spring which with rupture discs that burst open to relieve pressure but cannot reclose; or a combination of the two devices.

A vacuum relief valve is provided for non-pressure tank cars to prevent vacuum formation when unloading from the bottom. Both pressure and non-pressure tank cars have protective housings on top surrounding the loading connections, sample lines, thermometer wells and gauging devices. Platforms for loaders may or may not be provided on top of cars.

Older non-pressure tank cars may have one or more expansion domes. Fittings are provided on the bottom of tank cars for unloading or cleaning. Head shields are provided on the ends of tank cars to prevent puncture of the shell by the coupler of another car during derailments.

Oil Tanker Transport

A great portion of the world's crude oil and about half of the oil headed to the U.S. is transported via ocean tankers. Oil transport is very big business. The world tank ship fleet numbers more than 5000. On any given day as much as 750 million barrels of crude oil and products may be in tankers on the world's oceans.

Oil tankers and barges are vessels designed with the engines and quarters at the rear of the vessel and the remainder of the vessel divided into special compartments (tanks) to carry crude oil and liquid petroleum products in bulk. Cargo pumps are located in pump rooms, and forced ventilation and inerting systems are provided to reduce the risk of fires and explosions in pump rooms and cargo compartments.

Modern oil tankers and barges are built with double hulls and other protective and safety features required by the United States Oil Pollution Act of 1990 and the International Maritime Organization (IMO) tanker safety standards. Some new ship designs extend double hulls up the sides of the tankers to provide additional protection.

Generally, large tankers carry crude oil and small tankers and barges carry petroleum products.

• Oil tankers are ocean traveling vessels, which in addition to ocean travel can navigate restricted passages such as the Suez and Panama Canals, shallow coastal waters and estuaries. Large oil tankers, which range from 25,000 to 160,000

SDWTs, usually carry crude oil or heavy residual products. Smaller oil tankers, under 25,000 SDWT, usually carry gasoline, fuel oils and lubricants.

- Barges carrying oil products operate mainly in coastal and inland waterways and rivers, alone or in groups of two or more, and are either self-propelled or moved by tugboat. They may carry crude oil to refineries, but more often are used as an inexpensive means of transporting petroleum products from refineries to distribution terminals. Barges are also used to off-load cargo from tankers offshore whose draft or size does not allow them to come to the dock.

- Supertankers are the least expensive way for long-distance shipment of oil and oil products. These modern ocean-going vessels are huge floating oil tanks that make their slow cumbersome way from the giant oil fields of the Middle East to ports in the industrial countries. But there are no supertanker ports in this country, so they are usually unloaded off-shore into smaller tankers. Supertankers are huge.

  The largest ones presently in service have a DWT (deadweight tonnage, i.e., cargo and fuel capacity) of more than 546,000 tons. They are up to 400 meters long (the length of about five football fields laid end to end) and can carry about 40 million barrels of oil.

- Ultra-large and very large crude carriers (ULCCs and VLCCs) are restricted to specific routes of travel by their size and "draft" (depth of water to which a ship sinks according to its load). ULCCs are vessels whose capacity is over 300,000 SDWTs, and VLCCs have capacities ranging from 160,000 to 300,000 SDWTs. Most large crude carriers are not owned by oil companies, but are chartered from transportation companies which specialize in operating these super-sized vessels.

Their draft is very important for their navigation. A fully loaded draft of over 90 feet prevents such a large ship from going into any of the U.S. ports. Our deepest port, Los Angeles, cannot handle ships of more than 100,000 deadweight tons. As a result supertankers now unload offshore in the Caribbean and transfer their cargo to smaller tankers for delivery to the United States.

The ocean transfer method costs a lot of additional energy and money, and for this purpose an alternative LOOP facility was built. LOOP is the Louisiana Offshore Oil Port (LOOP), which is a deep-water port in the Gulf of Mexico off the coast of Louisiana, near the town of Port Fourchon. It provides offloading and temporary storage services for crude oil transported on some of the largest tankers in the world, since most of them cannot enter US ports.

LOOP presently handles 13% of the nation's oil imports, or about 1.2 million barrels a day. It is connected by pipeline to the refineries, thus feeding 50% of the U.S. refining capability.

Tankers offload at LOOP by pumping crude oil through hoses connected to a single point mooring (SPM) base. Three SPMs are located 8,000 feet from the marine terminal. The SPMs are designed to handle ships of up to 700,000 deadweight tons. The crude oil then moved to the marine terminal via a 56-inch diameter submarine pipeline.

The marine terminal consists of a control platform and a pumping platform. The control platform is equipped with a helicopter pad, living quarters, control room, vessel traffic control station, offices and life-support equipment. The pumping platform contains four 7,000-hp pumps, power generators, metering and laboratory facilities. Crude oil is handled only on the pumping platform where it is measured, sampled, and boosted to shore via a 48-inch-diameter pipeline.

The Oil Transport Issues

The issues related to security of global oil transport could be summarized as follow:

- Global problems, such as political changes and turmoil, can affect the oil routes and the safety of the tankers and the people working on them.
- Weather related problems, such as storms and hurricanes on the tanker route could cause serious damage to the vessels and their personnel.
- Terrorist attacks are the most threatening part of global oil transport routes. This is the worst form of damage to the oil transport tankers and their personnel. There are several "choke points" around the world, where the ocean transports are most vulnerable.

The Choke Points

The total world oil production amounts to approximately 80-90 million barrels per day. About one-half of this quantity is moved by oil tankers on fixed maritime routes. The global economic downturn of 2008 reduced the world oil demand, and the volumes of oil shipped to markets via pipelines and along maritime routes. Nevertheless, a large number of oil tankers crisscross the world's oceans daily. These undergo major risks every step of the way as they traverse the world's oceans, and especially in the most vulnerable places—choke points.

There are several choke points around the world, through which large quantities of oil float daily and which are under constant threat of terrorism. By volume of oil transit, the Strait of Hormuz leading out of the Persian Gulf, the Strait of Malacca linking the Indian and Pacific Oceans, Babel-Mandeb at the entrance to the Red Sea, and the Strait of Gibraltar at the exit from the Mediterranean Sea, are the world's most strategic and dangerous choke points.

Oil tankers crossing those choke points are exposed to political unrest in the form of wars or hostilities and are vulnerable to theft from pirates and terrorist attacks. All this can lead to shipping accidents, resulting in disastrous oil spills and or blockage of the waterways. The blockage of a choke point, even temporarily, can lead to substantial increases in total energy costs.

Over half of America's oil is imported, so terror organizations like al-Qaeda and its affiliates see the disruption of oil transportation routes as one of the ways to hurt the U.S. and its allies. And sure enough, disruption of scheduled oil flow through any of those routes could impact negatively the global oil prices.

Several attacks against oil tankers in the Arabian Gulf and Horn of Africa have been planned in the past, starting with June 2002, when a group of al-Qaeda operatives suspected of plotting raids on British and American tankers passing through the Strait of Gibraltar were arrested by the Moroccan government. In October of the same year, a boat packed with explosives rammed and badly damaged a French supertanker off Yemen. al-Qaeda claimed responsibility for that attack, and for plans to seize a ship and crash it into another vessel or into a refinery or port.

The terrorists' goal is to cut the economic lifelines of the world's industrialized societies, but such attacks would also weaken and perhaps topple some of the Gulf oil monarchies. This would have even greater and long lasting effect on the world's oil supplies and transport routes.

There are approximately 4,000 tankers crossing the world's oceans every day. Each of those slow and vulnerable giants can be attacked at the narrow passages, where a single disabled or burning oil tanker and its spreading oil slick could block the route for other tankers.

This constant threat and the increased risks contribute to further increase in oil prices. Insurance carriers are raising the premiums to cover tankers in risky waters, as evidenced by the fact that the insurance for oil tankers entering Yemeni waters tripled since the attack in Yemen a decade ago.

A typical supertanker with two million barrels of oil on board would pay about \$450,000 instead of the usual \$150,000 for insurance of the ship alone. The cargo requires a separate and equally expensive insurance policy each trip. This adds about 15-25 cents a barrel to the cost of the oil.

Even the best insurance cannot ensure complete safety and reliability of our energy supply under those conditions. The only way to achieve energy independence, therefore, is to eliminate the need to import and transport oil across the globe. This, however, is not going to happen anytime soon, so the new situation has created a new niche market for individuals and companies that provide insurance and ensure the physical safety of the oil tankers.

Some oil tank owners are resorting to placing security personnel on board, especially when crossing pirate-infested international waters. This measure, in addition to equipping the tankers with special equipment, is one of the services that is expected to grow. The number of complexities of those services, as well as their costs, will increase as the pirate attacks are on the rise, at least until we figure a way around it.

Crude Oil Processing

Crude oil is the stuff that comes from the ground when a crude oil deposit is found. It is the unprocessed oil, which is also known as petroleum, or liquid fossil fuel (coal is the solid, and natural gas is the gaseous form of the fossil fuels).

Crude oil was made naturally a long time ago from decaying plants and animals (or fossils) living in ancient seas millions of years ago. In most cases, crude oil is found under the oceans, or in places which were once sea beds.

Crude oils vary in color, from clear to tar-black, and their viscosity varies from water-like, to molasses-like, to almost solid. Crude oil is an extremely valuable material. It is for making many different substances, which also contain hydrocarbons.

Hydrocarbons (HC) are organic compounds that contain hydrogen and carbon atoms in their molecules. They come in various sizes and structures;, from single atoms to straight molecular chains, to branching chains, and rings.

The smallest hydrocarbon molecule is methane (CH_4). With its one carbon atom, it, like all HCs with up to 4 C atoms, is a gas. Longer HC chains with 5 or more C atoms are liquids, like gasoline and diesel. Very long HC chains, with 6 or more carbon atoms are solids, or

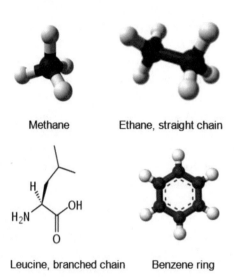

Methane   Ethane, straight chain

Leucine, branched chain   Benzene ring

Figure 4-27. Different HC configurations

semi-solids, like wax and tar.

The two most important characteristics of crude oil and its derivatives are:

- Their high energy content. Many of the things derived from crude oil, like gasoline, diesel fuel, paraffin wax and many others, have very large energy content.

- Hydrocarbon chains are very versatile and can take on many different forms and properties, which makes them very useful.

- Via special chemical reactions we can react and cross-link hydrocarbon chains to obtain consumer goods—anything from synthetic rubber, to nylon, to different types of plastics.

The major classes of hydrocarbon types contained in crude oil include:

- Paraffins with general formula $CnH2n+2$, where n is a whole number, usually from 1 to 20. These are straight- or branched-chain molecules, which can be gasses or liquids at room temperature depending upon the molecules. Some of these are methane, ethane, propane, butane, isobutane, pentane, hexane—all of which are gasses at room temperature.

- Aromatics with general formula $C6H5$ are ringed structures with one or more rings. Each ring contains six carbon atoms, with alternating double and single bonds between the carbons. These hydrocarbons are typically liquids, like benzene and napthalene.

- Napthenes or Cycloalkanes with general formula: $CnH2n$, where n is a whole number usually from 1 to 20 are also ringed structures with one or more rings. Their rings contain only single bonds

between the carbon atoms. These compounds are typically liquids at room temperature, like cyclohexane and methyl cyclopentane.

- Alkenes with general formula $CnH2n$, where n is a whole number, usually from 1 to 20, are linear or branched chain molecules containing one carbon-carbon double-bond. These hydrocarbon materials can be liquid or gas, such as ethylene, butene, isobutene

- Dienes and Alkynes with general formula $CnH2n-2$, where n is a whole number, usually from 1 to 20, are linear or branched chain molecules, containing two carbon-carbon double-bonds. These hydrocarbons can be liquid or gas, such as acetylene and butadienes.

Now that we know what's in crude oil, let's see what we can make from it.

Crude Oil Refinery

We refine crude oil to produce a variety of fuel products, ranging from the heaviest oils for use in industrial boilers, to fuel oil used in home heating, to diesel oil that powers most trucks and some cars, to jet aircraft fuel, to gasoline for our cars. Each of these products contains a mix of different molecules, so no single chemical formula describes a given fuel. Molecules typically found in gasoline, for example, include heptane, octane, nonane, and decane. These molecules consist of long chains of carbon with attached hydrogen molecules.

Higher grades of gasoline have a greater ratio of octane to heptane which makes them burn better in high-performance engines. This is the origin of the term high octane at the gas pump. The energy content of refined fuels varies somewhat, but a rough figure is about 45 MJ/kg for any product derived from crude oil.

Oil refineries are huge developments—entire cities of buildings, towers, piping, electrical substations, wires, storage tanks, railroad stations and ports. The life of a refinery is that of never-ending loading and unloading, processing and re-processing of oil products in gaseous, liquid and solid form.

Oil refining is an energy-intensive process; some 8% of the total U.S. energy consumption is used to run the fleet of oil refineries around the country.

The actual quantitative breakdown of the different fractions produced in a refinery varies with demand. For example, the percentage of crude oil refined into heavier fuel oil for heating increases significantly in the winter months.

Figure 4-28. Oil refinery

The Oil Refining Process

The crude oil is pumped from the storage tanks by charge pumps located at one of the oil storage tanks. Each of them can pump 350 gallons of oil per minute at 240 PSI outlet pressure.

The crude is then mixed with "make-up water," which is fresh water pumped from a well into an automated water desalination system, to purify it prior to mixing. After mixing with "make-up water," the crude is pumped through two heat exchangers, connected in series and mounted about twenty feet above the ground, in close proximity to the crude distillation tower.

The crude goes through the tubes of the heat exchangers and gets heated to over 400°F by indirect contact with the hot liquids coming from different stages of the process. It is then pumped into a condenser unit to separate one of the low-boiling fractions contained in the oil product.

Then the crude is mixed with more "make-up water" to form a water-in-oil emulsion. Salts, of the type of magnesium chloride and calcium chloride are in this emulsion and must be removed. The goal is to obtain salt concentration less than 1 pound of salt per 1000 barrels of crude. This is achieved by an electrostatic device and by adding demulsifying chemicals to the mix.

Thus purified crude is then pumped into the atmospheric distillation unit where the crude oil is distilled (separated) into low boiling point fractions. It is then pumped into a vacuum distillation unit, which distills

the higher boiling point fractions remaining after the atmospheric distillation.

The most important processes take place in the distillation column, or vacuum distillation units. This is a method of separating the different fractions by reducing the pressure in the vessel, which together with the high temperature of the oil causes evaporation of the volatile liquids. Vacuum distillation can be used also without heating the mixture, but in such case the high boiling point fractions will not be separated.

Some of the fractions go through a naphtha hydro-treater unit where hydrogen gas is used to de-sulfurize the naphtha fraction left from the previous atmospheric distillation. The naphtha is then hydro-treated to separate other fractions and impurities, before sending it to a catalytic reformer unit, where the naphtha fractions are converted into higher octane products. Hydrogen is released during the catalyst reaction, and is used either in the hydrotreating or the hydrocracking process steps.

The different fractions go through a number of additional steps and pieces of equipment, such as fluid catalytic cracker, hydrocrackers, visbreaking, merox, coking, alkylation units, dimerization, isomeration, steam reforming, amine gas treater, claus unit, and other units and steps during the long, complex refining process.

The final products, obtained and separated during the different process steps are petroleum based products, that can be grouped into light distillates (LPG, gasoline, naphtha); middle distillates (kerosene and diesel); heavy distillates (misc. products); and residuum (heavy fuel oil, lubricating oils, wax, asphalt).

In more practical terms, the refinery produces are liquefied petroleum gas, gasoline, naphtha, kerosene, jet aircraft fuel mixes, diesel fuel, fuel oils, lubricating oils, paraffin wax, asphalt and tar, petroleum coke, and sulfur. A number of gasses, like propane and others, are also produced.

All products are stored and eventually shipped for use to different customers in different industries.

Oil Products Types

Crude oil contains hundreds of different types of hydrocarbons and impurities, all mixed together in different proportions in the bulk. To be useful, the different fractions must be separated by type of hydrocarbons and impurities. This is done by refining the crude oil in special refineries.

Different hydrocarbon chain lengths have progressively higher boiling points, so they can all be separated by distillation. This is what happens in an oil refinery—in one part of the process, crude oil is heated and the different chains are pulled out by their vaporization temperatures and separated for later use. Each different chain length has a different property that makes it useful for different purposes.

There is great diversity contained in crude oil, which is why refining it is so important to our society.

The following is a list of some key products that can be refined from crude oil:

• Petroleum gas is used for heating, cooking, and making a number of plastics. It contains small alkanes molecules (1 to 4 carbon atoms), commonly known by the names methane, ethane, propane, butane with a boiling range of less than 104 degrees Fahrenheit. These can be liquefied by compressing them to a high pressure to create LPG (liquefied petroleum gas). This process uses a lot of energy and requires special container and transport, so it is used only in special circumstances usually for export purposes.

• Naphtha or Ligroin is an intermediate product that will be further processed to make gasoline. It is a mix of 5 to 9 carbon atom alkanes with a boiling range of 140 to 212°F.

• Gasoline is a liquid motor fuel, which is a mix of alkanes and cycloalkanes, containing 5 to 12 carbon atoms. Its boiling range is 104 to 401°F.

• Kerosene is fuel for jet engines and tractors and is also used as a starting material for making other products. It is liquid mix of alkanes, containing 10 to 18 carbons, and aromatics. Its boiling range is 350 to 617°F.

• Gas oil or Diesel distillate is used for diesel fuel and heating oil, as well as a starting material for making other products. It is liquid mix of alkanes, containing 12 or more carbon atoms. The boiling range is 482 to 662°F.

• Lubricating oil is used for motor oil, grease, other lubricants. It is a liquid long chain, containing 20 to 50 carbon atoms, alkanes, cycloalkanes, and aromatics. Its boiling range is 572 to 700°F.

• Heavy gas or fuel oil is used for industrial fuel, as well as a starting material for making other products. It is a liquid long chain, containing 20 to 70 carbon atoms, alkanes, cycloalkanes, and aromatics. Its boiling range is 700 to 1112°F.

• Residuals are materials like coke, asphalt, tar, waxes. They are also used as a starting material for making other products. These materials are usually solid multiple-ringed compounds with 70 or more carbon atoms. Their boiling range is greater than 1112°F.

All these products have different sizes and boiling ranges, which is the basis for their refining. They can be easily separated by heating, boiling and condensing each fraction, as we will see below.

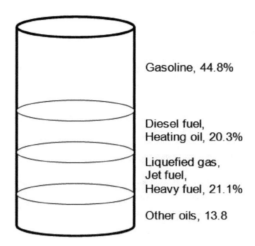

Gasoline, 44.8%

Diesel fuel,
Heating oil, 20.3%

Liquefied gas,
Jet fuel,
Heavy fuel, 21.1%

Other oils, 13.8

Figure 4-29. Crude oil products

As seen in Figure 4-29, most of the oil fractions produced from each barrel of crude oil are fuels (gasoline, diesel and such) intended to be burned in a number of applications, while about 7-8% of the crude oil is turned into road-making products, lubricants, and other petrochemical feedstocks.

A smaller amount of the crude is converted into oil derivatives, many of which go into making a vast array of the commonplace products we use today. Plastics are of the largest quantity of these products, followed by medicines, contact lenses, insecticides, toothpastes, cosmetics, fertilizers, paint, food preservatives, etc.

What a waste of a precious commodity—to be burned with no trace left except for deadly, climate warming gasses. What will future generations think of us, burning their inheritance without remorse or thought about a tomorrow without fossil reserves?!

Oil-Fired Power Plants

Crude oil-fired power plants are the exception to the rule today. They are few and far between and are used in some special circumstances; remote locations and such lack other energy resources. Washington, DC, Alaska, and Hawaii, for example, are powered mostly by crude oil, which is also the case, albeit to a lesser extent, in some other states.

The furnaces of these types of power plants are powered directly by unrefined crude oil, or more often by some of its derivatives. The oil-fuel is transported to the power plants by ship, pipelines, truck, or train, where several methods can be used to generate electricity from the oil.

One way is to burn the oil in boilers to produce heat and steam, which is used by a steam turbine to generate electricity. Another common method is to inject the oil in combustion turbines, which are similar in operation to jet engines.

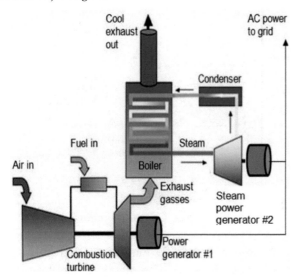

Figure 4-30. Combined cycle power plant

A more efficient method, or "combined cycle" technology, where oil is burn in a combustion turbine, but the hot exhaust gasses is then used to make steam and drive a steam turbine and generator to generate electricity. This technology is much more efficient because it uses the same fuel source twice.

What makes this setup efficient is that the waste steam generated in the combustion turbines is send to a boiler to generate steam, which drives the turbine and generator sets.

Another use of crude oil derivatives (gasoline and diesel) is to fuel internal combustion engines, which then drive the turbine-gen sets to generate electricity.

Saudi Arabia's Shoaiba is one of the largest oil-fired power plants in the world. Following completion of stage 1 in 2003, stage 2 consisting of six additional 400 MW units was completed, bringing the total output of the plant's stage 1 and 2 to 4400 MW.

Stage 3 was completed recently, adding 1.2 GW of electric power generation, bringing the total power generation to 5.6 GW. The plant uses oil furnaces to burn crude oil and generate steam, which drives the steam turbines and generators to produce electric power.

This monster crude-oil-wasting plant alone provides over 40% of the country's western region's current power requirements. It is also a record breaker in terms

of size for any oil-fired plant in the world. During the construction of stage 2, several records were broken in terms of construction and commissioning time for a steam power plant.

A multi-stage flash distillation water desalination plant was constructed as part of the plant complex. It has desalination capacity of 50 million cubic meters of water per year. A second desalination plant with another capacity of 50 million cubic meters of water per year was constructed later on as well.

The clever feature of this setup is that the waste steam generated in the power plant turbines is send to the desalination plant to be used to heat sea water. This provides cheap(er) clean water, while at the same time cools the steam before reuse, thus reducing the waste heat cooling losses.

The crude oil-fired power plant life-span cycle consists of several phases, very similar to those of a gas- or coal-fired power plant, as follow:

Plant site and structure planning and design
• Selecting, testing and exploring the proposed power plant location
• Engineering design of plant facilities, equipment and process
• Estimating the construction and operating cost and the related tasks
• Applying for and obtaining the necessary federal, state and local permits

Plant construction and setup
• Plant building construction
• Plant equipment procurement and setup
• Labor training

Power plant daily operation
• Petroleum fuel oil delivery
• Oil burning and energy generation
• Electric power distribution

Power plant decommissioning and land reclamation
• Plant shutdown and disassembly
• Plant waste disposal
• Surface land reclamation

Since the different steps of the power production process are similar to those used for coal-fired plants construction and use, we'd refer the reader to the section on COAL for more details.

In addition to power generation, crude oil is used for transportation (which is its major use) and in the production of different chemicals and products (its secondary use).

In more detail:

Heating Oil

Home and business oil-fired furnaces are a major use of heating oil, which is derived from refining crude oil. Home oil heating is widely used in developed countries, especially during the winter months.

Heating oil is usually delivered by tank truck to residential, commercial and municipal buildings and stored in above-ground storage tanks. These are located either outside, near the building, or in empty areas like basements and garages. When used in larger quantities the oil is stored in underground storage tanks. Heating oil is sometimes used as a fuel in industrial applications, and for small or remote power generation.

Heating oil's properties are very similar to that of diesel fuel, and consist of a mixture of petroleum-derived hydrocarbons in the 14- to 20-carbon atom range. During oil distillation, heating oil condenses at between 482 and 662°F. It condenses at a lower temperature than the heavy (C20+) hydrocarbons such as petroleum jelly, bitumen, candle wax, and lubricating oil, but it condenses at a higher temperature than kerosene, or over 500°F.

Heating oil has a high heating value, producing 138,500 Btus per gallon—close to the same heat per unit mass of diesel fuel—and is known in the U.S. as No. 2 heating oil. It must conform to ASTM standard D396, which is somewhat different from that of diesel and kerosene.

Heating oil is widely used during the winter months in parts of the United States and Canada where natural gas is not available and propane is priced higher. The northeastern United States and Canada are the most likely users for heating coming mostly from Irving Oil's refinery located in Saint John, New Brunswick, which is the largest oil refinery in Canada.

In addition to the toxic gases emitted during the burning process, leaks from tanks and piping are a well known environmental concern. Various federal and state regulations are in place regarding the proper transportation, storage and burning of heating oil, which are classified as hazardous material by U.S. federal regulators. With about 4% of the U.S. total consumption of crude oil going into heating oil, this is a significant issue, which is on the radar of the environmentalists, awaiting prompt resolution.

Consumable products, transportation fuels, and heating oil are only some of the many products derived from petroleum. Americans consume petroleum products at a rate of 3½ gallons of oil and more than 250 cubic feet of natural gas per day—per person!

We in the U.S. use two times more oil than we produce. The effect is obvious; we are spending a lot of money to support sheikhs and their artificial kingdoms, instead of using the money to build our own economy. This unreasonable and unjustifiable imbalance has been going on for almost a century, during which time we made several nations very rich, and have pumped and used most of the oil, which took billions of years to form. How long would this uncontrolled pumping and burning of a limited resource continue?

Petroleum is not only used as a fuel; there are many other uses of this precious material. One 42-gallon barrel of oil creates 19.4 gallons of gasoline and 9-10 gallons of diesel and heating oil. The rest (over half of the total content) is used to make things like:

- ammonia, ammonia based fertilizers, anesthetics, antifreeze, antihistamines, antiseptics, artificial limbs, artificial turf, aspirin, awnings, balloons, ballpoint pens, bandages, basketballs, bearing grease, bicycle tires, boats,
- cameras, candles, car battery cases, car enamel, car tires, cassettes, caulking, CD player, CDs & DVDs, clothes, clothesline, cold cream, combs, cortisone, crayons, curtains,
- dashboards, denture adhesive, dentures, deodorant, detergents, dice, diesel fuel, dishes, dishwasher parts, dresses, drinking cups, dyes,
- electric blankets, electrician's tape, enamel, epoxy, eyeglasses,
- fan belts, faucet washers, fertilizers, fishing boots, fishing lures, fishing rods, floor wax, folding doors, food preservatives, football cleats, football helmets, footballs,
- gasoline, glycerin, golf bags, golf balls, guitar strings,
- hair coloring, hair curlers, hand lotion, heart valves, house paint,
- ice chests, ice cube trays, ink, insect repellent, insecticides,
- life jackets, linings, linoleum, lipstick, luggage,
- model cars, mops, motor oil, motorcycle helmet, movie film,
- nail polish, nylon rope,
- oil filters, oils and lubricants
- paint, paint, brushes, paint rollers, panty hose, parachutes, percolators, perfumes, petroleum jelly, pillows, plastic wood, purses, putty,
- refrigerants, refrigerators, roller skates, roofing, rubber cement, rubbing alcohol,

- safety glasses, shag rugs, shampoo, shaving cream, shoe polish, shoes, shower curtains, skis, slacks, soap, soft contact lenses, solvents, speakers, sports car bodies, sun glasses, surf boards, sweaters, synthetic rubber,
- telephones, tennis rackets, tents, toilet seats, tool boxes, tool racks, toothbrushes, toothpaste, transparent tape, trash bags, TV cabinets,
- umbrellas, upholstery,
- vaporizers, vitamin capsules,
- water pipes, wheels,
- yarn, and many, many more.

This is only a partial list of products made from petroleum—only 150 of the over 6000 different items made every day from this miraculous liquid. Our lives are so wrapped in it that it is absolutely impossible to even imagine living without it. Just think…no gas for the car, no heat for the house, no plastic bags, no cosmetics…

Yet, this is exactly what is going to happen by the mid 21st century when most petroleum deposits in the ground will be pumped up, refined, used and otherwise wasted. Just think…

Crude Oil Markets

Figure 4-31 depicts the versatility and enormity of the crude oil related markets.

The world of oil is dark in some places and at some times, but is also quite cheerful in others. It could be most precisely described as full of controversies, misinformation, and disparity in comprehension.

We all know that without oil our entire existence, as we know it today, would be quite different. Although for decades now we have been worried about the pending doom of oil (the dreaded oil peak) and have had countless discussions about the end of the oil era, crude oil is still a primary energy source. It still drives (literally) our economies and is indispensable in our personal lives. And we still use huge amounts of it as if there is no tomorrow. According to the leading global indicators,

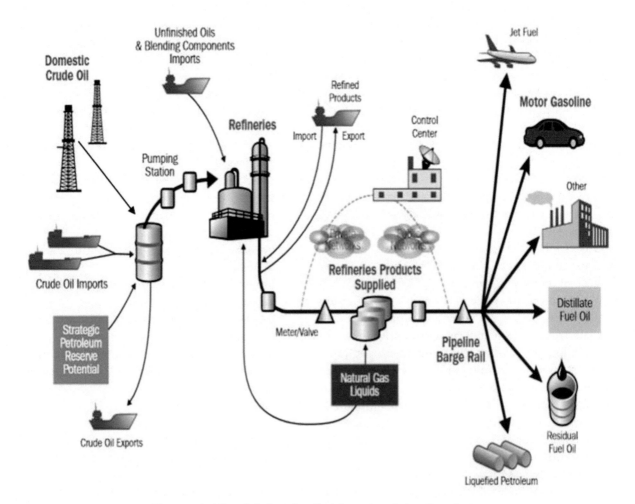

Figure 4-31. The global crude oil and some peripheral markets

the proportion of oil as a source of energy is still 34% vs. 26% of coal and 21% of gas.

We ignore the fact that crude oil it is a non-renewable energy source, which means that its days, in the long term, are inevitably numbered.

On an intellectual level, we see crude oil as an important material that shapes, changes, models, directs and configures the past, present, and future history of the world. Just like the era of the Roman worlds is still alive in our minds—although the Romans have been gone for centuries.

Crude oil has given the world economic progress, unspeakable comforts, and social and geopolitical models which will last many centuries to come. But just like the Roman empire, oil's empire is going to end soon. We, however, are concerned only with two things: is oil available and how much it costs. The rest (including the future) seems unimportant and is thus absolutely ignored.

What has such a powerful influence on our lives is not oil itself, but the combination of forces related to; its availability, costs, the decisions regarding investments and expenses, use of oil reserves, risks management, oil futures speculations, the policies of contracting companies and nations, the actions of oil companies, and any such activities that drive the global crude oil markets.

This is a huge picture, which we can only glance at and analyze in a small way in this text The multitude of complex elements that make the energy markets, create a long chains of events and actions that influence the variables of fundamental importance for the entire global economy. Here we see oil affecting the variations of the dollar and the euro, GDP ups and downs, the balance of payments, the price of gasoline and diesel, the cost of electricity, the rate of inflation, employment rates etc.

We as a matter of fact take a much closer look at crude oil and its issues in our books on the subject; *Power Generation and the Environment,* and *Energy Security for the 21st Century*, published by the Fairmont Press in 2014 and 2015 respectively.

With one word; oil (its availability and price mostly) has a vast and ever present impact on our lives. And despite all this, the issues are mentioned, hinted and explained, but beyond the inner circle of specialists no one talks much about them. The circle of specialists, unfortunately, is very restricted and limited in its influence. It consists of people educated and trained in certain areas of the energy markets, where even smaller number are familiar with the entire chain of events and all aspects of the energy sector. This adds to the disparity of understanding and limits the effectiveness of the actions taken in the energy sector, if any of importance.

Crude oil is a major force that drives the economic progress, and yet its issues are not fully understood and properly addressed by the media. There is also deficiency in the technical degree courses. How many economists know how an oil refinery operates, and/or what products it makes. Let alone describing their chemical and physical properties and behavior.

And so the ignorance propagates, and combined with the indifference of the customers, leaves the decisions in the hands of negligent government bureaucrats and greedy oil companies officials, while the customers watch and scratch their heads. What can we expect from this mix is evident in the present day situation with crude oil and the other energy sources. In the end, we follow the established capitalist model of making a quick buck today for our companies and ourselves, without worrying or planning for tomorrow. And the consumer pays the bills… Fair?

Oil Pricing

The most popular traded grades are Brent North Sea Crude (Brent Crude) and West Texas Intermediate (WTI).

- *Brent Crude* refers to oil produced in the Brent oil fields and other sites in the North Sea, and is the (price) benchmark for African, European and Middle Eastern crude. Brent Crude pricing mechanism determines about two-thirds of the world's crude oil production.

Brent Crude is considered "sweet" crude oil (sweet oil has sulfur content below 5%). Brent Crude actually has sulfur content of less than 0.4%. The lower the sulfur content the easier and cheaper it is to refine into products, such as gasoline, lower the emissions and higher the profits for the producers.

- *WTI* is the other major crude oil traded on the global energy market. It is the (price) benchmark crude for North America produced crude oil. One important thing here is that WTI is even "sweeter" than Brent Crude oil, with sulfur content of less than 0.25%. WTI is a better overall crude oil for the production of gasoline, while Brent oil is better for producing diesel fuels.

Note: The NYMEX (New York Mercantile Exchange) division of the CME (Chicago Mercantile Exchange) lists futures contracts of WTI crude oil, while the delivery of WTI crude futures is executed in Cushing, Oklahoma. Brent crude oil futures trade on the Intercontinental Exchange (ICE). Most Asian countries use both, Brent and WTI, benchmark prices for their crude oil.

It costs approximately $3-$4 per barrel to ship crude oil via supertankers from the U.S. to Europe. There are also different charges for storing crude oil in the European and North American storage and trading hubs. The price difference between Brent and WTI is usually around $2.50-$4.00 in favor of WTI crude, which is due mostly to the lower sulfur content. At times, however, the difference is reversed in favor of Brent Crude, which is most often due to abnormalities arising from political and/or other abnormalities.

The prices, and sometimes the spread between these two major crude oils moves drastically up or down for long periods of time. For example, in 2011, the Brent and WTI prices were almost equal. Then the Arab Spring came to blossom causing the spread to widen by the end of 2011 with Brent trading higher than WTI. The threat of closing the Suez Canal and a low oil supply raised Brent crude prices much higher than WTI. That, however, did not last very long and when tensions in the region eased, the price difference started to disappear. In late 2011, the Iranian government threatened to close the Straits of Hormuz through which 20% of the world's oil flows each day. For a second time in one year the Brent Crude prices went up; this time with a $25 premium per barrel over WTI.

Recent Oil Price Fluctuations

The local price of crude oil is just one factor involved in setting the prices in the global crude oil market. The spread between Brent and WTI is a good example of how quality, location spreads, and global developments affect the market structure. These factors dictate the pricing of crude oil, and ultimately that of all petroleum products (including gasoline and diesel fuels) around the globe.

The developments during the last several years are a good example of how oil prices fluctuate, driven by a number of factors:

2008. Oil skyrocketed to its all-time high of $143.68/bbl in July, sending gas prices to over $4/gallon. The reasons for this abnormality are still unclear, but commodity and futures trading is believed to be one of the major reasons. A gas station in Bakersfield, California sold diesel at $6.50/gallon, which is the highest fuel prices we've ever seen in the U.S.

2009. Oil fell to $39.41/barrel in February, when, due to the emerging financial crisis, investors stayed away from any investments except the ultra-safe U.S. Treasuries. Gas prices fell to $1.67/gallon by the end of the year.

2010. Oil prices stayed within the range of $70-$80/bbl all through the year, until December when they jumped to $90/barrel. Gas prices followed suit, staying below $3.00/gallon for the year.

2011. The price of oil reached a peak of $126.64/bbl in May. Gas prices peaked at the same time, at over $4.00/gallon. Gas prices stayed above $3.50/gallon all summer due to fears about refinery closures from the Mississippi River floods.

2012. Iran threatened to close the Straits...again, sending oil prices to their peak of $128.14/bbl in March. Gas peaked to nearly $4.00/gallon, but both returned to normal until August. Then commodities traders began bidding up oil prices sending oil at $117.48. The traders were hedging against the Federal Reserve's QE3 program, which would have lowered the value of the dollar and force oil prices higher, since oil is traded in dollars). Then, Hurricane Isaac closed some Gulf refineries, sending gas prices again close to $4.00 in September. Gas prices kept rising to over $4.50 a gallon in California, due to local distribution shortages.

2013. Oil rose again to $118.90/bbl in February, thanks to Iran's threatening war games near the Straits of Hormuz. Gasoline prices followed by rising to $3.85

2014. Prices remained around $100/bbl for most of the year, propped by the U.S. shale oil bonanza. By the end of the year, Saudi Arabia started playing a game of cat and mouse, which caused the global oil market to collapse at below $60/bbl. Gasoline prices in Texas dropped below $2.00/gallon for the duration.

Oil Production Costs

During the last half of 2014, crude oil prices collapsed over 50%, and later on recovered, but we still don't know where the bottom of the oil market is. The bottom price estimates for the different global oil producers vary from $85 a barrel to the low $30s, depending on whom you talk to. The bottom price is also called break-even price. It is the price at which producers no longer see any incentive to produce a barrel of oil.

The U.S. shale producers' break-even point is about $55/boe. The Canadian break-even point is $60/boe, while Russia is at $85/boe.

Note: The abbreviation *'boe'* means barrels of oil equivalent.

So, if it costs a producer $55 to produce a barrel of oil, then the bottom, or break even price for this producer is $55; everything included. When the global oil prices go lower than the bottom $55, s/he better stop producing oil, because it will be at a loss.

During 2014-2015, with global oil prices at $40-50 per barrel, many producers in the U.S. and abroad simply seized operations rather than continuing to lose money.

Unfortunately, there is no clear consensus on what the bottom break-even point number is, since different producers have different production costs. And yet, we see a number of common denominators that underline the domestic and global oil costs.

The major driver, and reason for the collapse in the price of oil during 2014-2015 was the intentional increase of oil supply over the last 5 years. This anomaly in the global supply and demand scenario was dictated by the producing nations, and Saudi Arabia in particular. This is what started the oil prices drop, and which then brought into the open what makes or breaks the oil prices today.

So, in the U.S., oil producers unintentionally pulled the rug from under themselves. By producing a lot of oil, they flooded the market, which contributed to lowering the prices. The production increase in the U.S. and Canada, augmented by that in Saudi Arabia, caused a sharp drop in global oil price.

To confirm the abnormality, on November 27, 2014 OPEC ministers got together and stood firm on not-cutting production. They simply refused to cut even a single barrel to support prices. This, despite the fact that lower prices hurt them too.

This is how OPEC, and Saudi Arabia in particular, punished the U.S. oil producers, who were ramping up their oil and gas production as if these was no tomorrow.

Does this remind you of the 1973 Arab oil embargo? Isn't Saudi Arabia saying again, "Hey, all mighty USA; we can squish you and your economy like a bug any time we want. And there is nothing you can do."

Chilling reality, we must add! Just another lesson in international politics we need to learn.

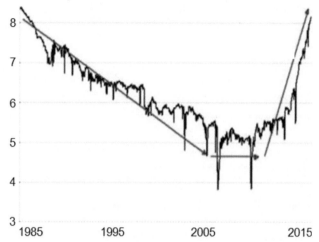

Figure 4-32. The U.S. shale oil revolution, 2010-2015.

The oil production increase was even more significant on national level, where production increased from 5 million barrels to over 8.5 million barrels since 2010. This is almost an 80% increase. See Figure 4-32. This is unprecedented increase of oil production for the U.S., or any other country. And so, at this level of production increase, the market was quickly oversupplied and prices dropped.

The low prices forced the U.S. oil companies to reduce their production and the prices eventually went up. So, it all boils down to the global price of oil, which is determined by the supply and demand ratios. The global oil prices then determine which oil company will make money and which will not. It all depends on the actual production costs of each company. If they are higher than the oil price, then the company cannot make profit and must find a way to survive by cutting operational expenses.

Oil Production Costs Breakdown

Since oil overproduction was the biggest contributor to the price wars of 2014-2015, and since the survival of the oil producers depends on the break-even point, we will take a close look at it.

The costs and expenses of oil producers can be broken into several major categories:

Development Costs

The development costs (also called fixed development costs) are incurred during the time when the well is in construction and in pre-production tests and preparations. In the U.S. it is about $20 per boe. Initial well development expenses have been falling in recent years too. There are reports of drastic reductions from $7.5 million per well, to $5.0 million. Some companies have even much lower fixed cost due to the fact that they use wells and infrastructure build in the past.

Dry Well Costs

Not all wells produce as projected, and some do not produce at all. Heavy losses might occur if there is no cost factored for drilling low-producing or dry wells. Technological improvements have reduced the risks in this area, but such anomalies still happen from time to time, at a heavy loss for the company and its investors.

Drop in Production Costs

The production levels of new oil (especially shale) wells are expected to drop significantly during the first years of operation. Since the development and operating cost (and the corresponding investments) are based on projected production curves, a steeper than expected drop in the production curve could result in higher per barrel costs.

Operating Costs

Operating costs are fairly consistent, and range between $10 and $12 barrel of oil equivalent (boe) for most oil companies, with some exceptions at $15-$18 (boe).

The operating costs depend mostly on:

+ Volume, where a well producing 1,000 boe per day would have a significantly higher per barrel operating cost than a well producing 100,000 boe per day.

- Production decline for shale oil wells is about 75% within the first year, so the production costs will gradually increase with time as productions decline.

Transport and Processing Costs

Once oil is pumped up, it has to be loaded on trucks, or trains, and taken to a refinery. That cost could be significant and varies with location and time periods. The Bakken sites, for example, located in North Dakota lack transport infrastructure and pipeline for the most part, and transport costs about $8-$9/boe. At the same time, Eagle Ford wells enjoy much lower transport costs of about $2/boe. These costs vary daily, and usually reflect the spread between the oil price at the well and the actual daily WTI price.

SGA Costs

Oil wells need management and workers to operate properly and efficiently. These people are paid salaries, and incur general costs, including stock options

and higher management salaries. In the U.S., these amount to an average of $3.50-$4.50/boe. These costs are different for each company, and vary according to the financial strength of the company.

Interest Costs

The cost of production is also affected by the interest rates a company is paying. This amount is significant, given the fact that a most U.S. shale drillers have an EBITDA-to-debt ratio below 2. High interest rates could reduce the profit margin significantly.

Royalties

Percentage of the revenue is paid to the state and title holders, which averages around 7% in the U.S. Canada has a much higher royalty tax around 14%. The rate could increase in some cases to up to 22%. In 2015, however, some Bakken sites did not pay any taxes when oil prices went below $50/bbl.

Profit Margins

Since most wells require investments, they must show profit. Without profit, and/or operating at a loss will chase the investors away and cause the oil company to eventually shut down. In the US a profit margin of $10/boe can be expected if and when the company's gross profit before tax is about $16/boe. So, if the company's break-even point is $45, then its oil selling price must be at least $61/boe. If it is not, the company would not have many investors, and if the low oil price situation persists, it may even go out of business. This is exactly what most U.S. oil producers were struggling with during the low global oil prices of 2014-2015.

Hedging

Many oil drillers were clever enough to hedge future prices, so that they can operate existing wells for awhile and still make money. Some drillers claim that hedging has helped them to drill and pump at much less than a $50.00/boe throughout the low oil price crisis. Long term, however, reality will hit depending on how long this goes on.

Hedges range from zero for some companies, to 100% at $90 per barrel for other companies, and there are many in between. Hedging became a major driver in determining who will survive and who won't during the low oil price debacle of 2014-2015.

Natural Gas to the Rescue

Most oil wells have excess natural gas. Some well owners are flaring it, but others have found a way to capture and sell it, which increases their profit per barrels of oil. Many companies at Bekken sites are doing much better job at capturing the associated high-Btu natural gas at their wells, and getting it to market, as compared to some of their competitors. This helps their bottom line significantly.

The U.S. Energy (Oil) Status

The U.S. is at a historic junction as far as its energy future is concerned. Recently, energy production (oil and gas mainly) has surged across the continent, with the U.S. poised to become the world's top oil producer. Canada is also scaling output to unprecedented levels, and Mexico is forging landmark energy reform.

As recently as 2008, the global energy gurus projected a very bleak future for North America. The expected increase in oil and gas imports, coupled with growing susceptibility to supply side shocks, foretold higher energy prices and increased pressure on economic growth. But all of a sudden, today, the North America's energy outlook has dramatically improved.

In addition to substantial increases in oil production, natural gas and renewable energy production are also on the rise. As a result, the U.S. energy trade deficit has shrunk, while jobs in the energy sector are growing faster than the private sector as a whole.

The opportunities presented by North American energy resources are tremendous. Maximizing energy efficiency and integration has the potential to support job growth in key sectors, including transportation, power generation, manufacturing, and petrochemicals, while also reducing harmful emissions.

The Big Dilemma…Oil Imports

Granted, coal and natural gas abound in the U.S., so we will erase them from the list of our worries and energy security priorities. But crude oil…? Unfortunately, oil is the Achilles Heel of the U.S. economy, which affects everything from our daily driving to the U.S. military operations around the globe. Oil drives everything and is therefore an indispensable commodity, which we just don't have enough of.

As Figure 4-33 shows, the crude oil imports have been steadily increasing for the last half a century. With the new oil discoveries in the U.S. we have been able to reduce the imports, but that won't last long. Our oil deposits are not as many, and are not of high quality, as we would wish them to be in order to reduce the imports for good.

As a matter of fact, new research estimates that the U.S. imports will start increasing slowly with time, at

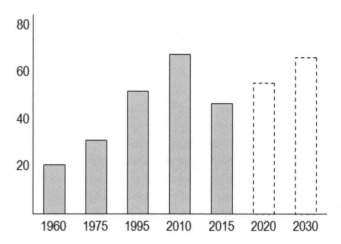

Figure 4-33. U.S. crude oil imports (in % of total use)

least until 2030, because:

1) The domestic oil production will level, and start decreasing due to cost increases, and
2) The domestic demand will increase with time as the population increases.

Note: Although oil prices fell over 50% by the end of 2014 and part of 2015, this drop cannot be used as a measuring stick for things to come, because it is a temporary situation imposed by Saudi Arabia. Sooner or later, things usually go back to normal—with oil prices steadily rising being the normal. There is simply no way around it, so the question here is not IF but WHEN oil prices will go sky high again, and what are we going to do then?

NATURAL GAS

Natural gas is (indirectly) shaping the "shale revolution" in the U.S. and other countries, where large quantities of this amazing commodity were found, are presently exploited, and used across the economy in large quantities. The size of the new natural gas reserves and their potential to change the direction the energy sector has been following lately, is nothing short of "energy revolution."

Who would've thought just several short years back that the U.S. would become a major fossil fuels exporter? Yet, this is exactly what is happening today. We are now planning to export large quantities of surplus natural gas and crude oil products to other countries.

In addition to being the world's largest importer (of crude oil), we would now become a major exporter of these precious commodities to the global energy market.

Maybe even grabbing the title of the largest natural gas and crude oil producer and exporter someday soon... How wise this is, is another question.

Natural Gas Properties

Raw natural gas is a mixture consisting of mostly methane (up to 96%), higher hydrocarbons (primarily ethane), and some noncombustible gases. Some other constituents, principally water vapor, hydrogen sulfide, helium, and other petroleum gases are present in different ratios. These are usually removed from the mix prior to distribution and use by the public.

Table 4-8. Natural gas components

Gas species	Content
Methane, CH_4	70 to 96%
Ethane, $C2H_6$	1 to 14%
Propane, C_3H_8	up to 4%
Butane, C_4H_{10}	up to 2%
Pentane, C_5H_{12}	up to 0.5%
Hexane, C_6H_{14}	up to 2%
Carbon dioxide, CO_2	up to 2%
Oxygen, O_2	up to 1.2%
Nitrogen, N_2	4 to 17%

Natural Gas Units

Natural gas has specific measurement units:
- m^3—cubic meter (used on the international market)
- CF, or ft^3—cubic foot
- CCF—one hundred cubic feet
- Mcf—one thousand cubic feet of natural gas
- Mmcf—one million cubic feet of natural gas
- Bcf—one billion cubic feet of natural gas
- Tcf—one trillion cubic feet of natural gas
- Mmcf/d—millions of cubic feet of gas per day

The energy equivalents of natural gas are:
- Boe (barrel of oil equivalent)—equal to 6,000 ft^3 natural gas
- Mboe—one thousand barrels of oil equivalent
- Mmboe—one million barrels of oil equivalent
- Mmcfe—one million cubic feet of natural gas equivalent
- Bcfe—one billion cubic feet of natural gas equivalent
- Tcfe—one trillion cubic feet of natural gas equivalent

Some of the practical natural gas measurements are as follow:
- 1 cubic foot (CF) natural gas averages 1,000 Btu

A typical gas spec allows 950 to 1,100 Btu per CF
- 1 hundred cubic feet (CCF) average 100,000 Btus
- 1 CCF is about 1 therm
- 1 therm averages 100,000 Btus
- 1 dekatherm (10 therms) averages 1,000,000 Btus
- 1 MCF (10 CCF) = 1,000 CF = 1,000,000 Btus

The composition of natural gas depends on its location and source. Commercial gas is usually a mixture of gas drawn from various sources, so its composition can vary some. Nevertheless, a fairly constant heating value is usually maintained for control and safety. Heating values of natural gases vary from 900 to 1,200 Btu/ft3; but the usual commercial range is 1,000 to 1,050 Btu/ft3 at sea level.

Natural gas in Nature is a nearly odorless and colorless gas that usually accumulates in the upper parts of oil and gas wells. For safety purposes, odorants (such as mercaptans) are added to natural gas and LPG before distribution to consumers

Wet Gas

Wet gas is any gas with a small amount of liquid present, which includes a range of conditions ranging from a *humid* gas which is gas saturated with liquid vapor, to a *multiphase* flow with a 90% volume of gas.

Wet gas is a particularly important concept in the field of flow measurement, as the varying densities of the constituent material present a significant problem. A typical example of wet gas flows are in the production of natural gas in the oil and gas industry.

Natural gas is a mixture of hydrocarbon compounds with quantities of various non hydrocarbons. It exists in either a gaseous or liquid phase or in solution with crude oil in porous rock formations. The amount of hydrocarbons present in the liquid phase of the wet gas extracted depends on the reservoir temperature and pressure conditions, which change over time as the gas and liquid are removed.

Changes in the liquid and gas content also occur when a wet gas is transported from a reservoir at high temperature and pressure to the surface where it experiences a lower temperature and pressure. The presence and changeability of this wet gas can cause problems and errors in the ability to accurately measure the gas phase flow rate.

It is important to be able to measure the wet gas

flow accurately, in order to quantify production from individual wells and to maximize the use of equipment and resources which in turn would reduce the final costs.

Natural Gas Cradle-to-Grave Cycle

Natural gas and crude oil are often found and exploited together, using the same equipment and procedures. In order to avoid repetitions, we would refer the esteemed reader to the previous section for details on these processes.

The refining process, transport, and distribution of natural gas, however, are somewhat different from that for crude oil, so here are some of these details:

The Natural Gas Refining Process

There are a great many ways to configure the various steps in the processing and refining of raw natural gas. The usual refining process flow (from well to market) can be described as follow:

To start with, the raw natural gas is pumped up from a well or a group of adjacent wells. It is then processed on-site usually, for removal of free liquid water and natural gas condensate. The condensate is trucked to a petroleum refinery and the water is disposed of as wastewater.

Processing raw natural gas yields several byproducts, such as natural gas condensate, sulfur, ethane, and natural gas liquids (NGL), such as propane, butanes and C5+ (commonly used term for pentanes plus higher molecular weight hydrocarbons). All these fractions must be separated and otherwise treated in separate streams, until the final product is pure enough.

The somewhat cleaned raw gas is sent from the wellhead via pipeline to a gas processing plant where it is treated first to remove the acid creating gases; hydrogen sulfide and carbon dioxide, using Amine gas treating. Polymeric membranes can be used also to dehydrate and separate the carbon dioxide and hydrogen sulfide from the natural gas stream.

Thus removed acid gases are sent to a sulfur recovery unit, where they are converted into elemental sulfur, using the Claus process in most cases. The residual gas from the Claus process is processed in a tail gas treating unit (TGTU) to recover and recycle residual sulfur-containing compounds.

The final residual gas from the TGTU is incinerated, so that the carbon dioxide in the raw natural gas

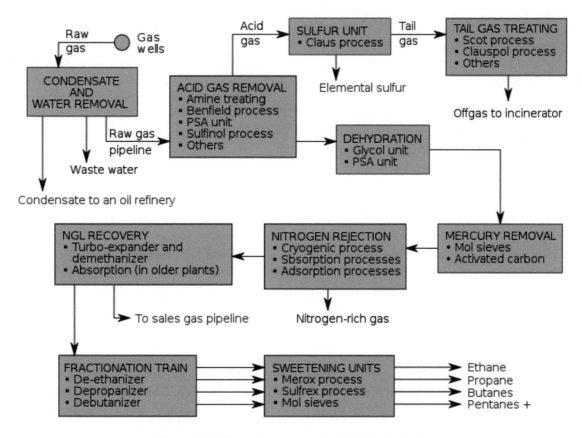

Figure 4-34. Natural gas processing (refining) plant

ends up in the incinerator flue gas stack and is disposed by venting in the atmosphere. Removal of water vapor from the gas is an important step, which is done by using either regenerable absorption in liquid triethylene glycol (TEG), also called glycol dehydration, or by using a pressure swing adsorption (PSA) unit using a solid adsorbent. Mercury is then removed by using adsorption processes using activated carbon or regenerable molecular sieves.

Nitrogen is removed next and rejected using one of these processes:

- Cryogenic process, using low temperature distillation.
- Absorption process, using lean oil or a special solvent as the absorbent, or
- Adsorption process, using activated carbon or molecular sieves as the adsorbent.

The natural gas liquids (NGL) are recovered next, via cryogenic low-temperature distillation process involving expansion of the gas through a turbo-expander, followed by distillation in a demethanizing fractionating column. Lean oil absorption process can be used here also, rather than the cryogenic turbo-expander process. The residue gas from the NGL recovery section is the final, purified "sales" gas which is pipelined to end-user markets.

The recovered NGL stream is processed through a fractionation system consisting of three distillation towers in series: de-ethanizer, de-propanizer, and debutanizer. The overhead product from the de-ethanizer is ethane and the bottoms are fed to the de-propanizer. The overhead product from the depropanizer is propane and the bottoms are fed to the debutanizer. The overhead product from the debutanizer is a mixture of normal and isobutane, and the bottoms product is a C5+ mixture.

The recovered streams of propane, butanes and C5+ are each "sweetened" in a Merox process unit to convert undesirable mercaptans into disulfides and, along with the recovered ethane, are the final NGL by-products from the gas processing plant.

Once pumped out of the ground, natural gas has to be transported to the refinery. From the refinery it is transported to storage or point of use facilities.

Natural Gas Transport and Distribution

The transport and distribution of natural gas starts at the wellhead, where it has to be transported from the origination point to storage, refinery, or a customer. This involves a number of pipelines, combined with several physical transfers of custody, and multiple processing steps along the way.

Once the gas leaves the well, a pipeline gathering system directs the flow either to a natural gas processing plant or directly to the mainline transmission grid, depending upon the initial quality of the wellhead product.

The processing plant produces pipeline-quality natural gas, which meets the federal and states standards as needed for use in a variety of residential and commercial applications. This gas is then transported by pipeline to consumers or is put into underground storage for future use. Storage helps to maintain pipeline system operational integrity and to meet customer requirements during peak-usage periods.

Transporting natural gas from wellhead to market involves a number of pieces of equipment, a series of processes, and an array of physical facilities.

The list of these is long, but the key pieces are:

- Gathering Lines are small-diameter pipelines which move natural gas from the wellhead to the natural gas processing plant or to an interconnection with a larger mainline pipeline.
- Processing Plant is the operation that extracts natural gas liquids and impurities from the natural gas stream. Thus produced plant gas is then sent to the customers via mainline pipelines.
- Mainline Transmission System is a large-diameter, long-distance pipelines that transport natural gas from the producing area, or the refinery to the market areas.
- Market Hubs/Centers are locations where pipelines intersect and gas flows are combined or transferred.
- Underground Storage Facilities are special places around the U.S. where some large quantity natural gas is stored for later use. These could be depleted oil and gas reservoirs, aquifers, and salt caverns, where the excess gas is pumped in for future use.
- Peak Shaving is a system design methodology permitting a natural gas pipeline to meet short-term surges in customer demands with minimal infrastructure. Peaks can be handled by using gas from storage or by short-term gas line-packing.
- Significant quantities of natural gas are compressed and liquefied for ease of transport. This requires special equipment (compressors) and sturdy tanks, which are mounted on special trucks, train cars, of ships. Natural gas is usually exported in its compressed (liquid) form, known as LNG

Externalized Cost of Natural Gas

Natural gas is the new priority of many governments because of its low cost and abundance. Low prices and abundance, however, cannot change the fact that gas drilling is a hazardous operation, from both environmental and human health points of view.

Although safety is paramount in gas drilling and production operations (especially in the US), there are still problems which often result in accidents and injuries—even in the US. In addition to the large accidents that make news, there are smaller ones daily, in which people get hurt or sick as a result of the effects of gas drilling and related operations.

The damages could be classified as:

- Burn injuries occur at the site of a gas drilling operation, where the victims are likely to suffer burn injuries from gas or chemicals, accompanied pain, suffering and loss of limbs or life.
- Chemical spills and leaks of toxic chemicals can cause injury to people close to a spill by direct exposure, or contact or inhalation of fumes, which could lead to chemical burns and respiratory illness. If the water supply becomes contaminated, it could lead to long-term health problems such as cancer.
- Drill site fires at oil and gas rigs are a constant threat. The source could be negligence, lack of experience, or defective equipment, but the consequences can be grave.
- Explosions could occur too, although the oil and natural gas industry is tightly regulated to ensure the safety of workers. An explosion in a gas drilling rig is a major event that could hurt and kill people, so precautions are the norm, but even then, accidents do happen.
- Gas field accidents—small and large—are not unheard of. They are rare, but when they do happen, those affected can be badly injured and even killed.
- Gas truck and train accidents are exceedingly dangerous, given the combustible nature of the cargo. Injuries can vary from broken bones to severe burns and death when a tank is ignited.
- Property damage happens during gas drilling accidents. Homeowners and individuals who live close to a drilling rig could be indirectly hurt by damage to their property, caused by fire, chemical spill or air contamination.
- Groundwater contamination is one of the most pervasive forms of property damage from gas drilling today.
- Toxic vapors and chemical exposure might cause damages that appear immediately or are not evident for months or years. Common examples include chemical burns, respiratory complications, and various forms of cancer related to contaminated air or water at the drill site. In these cases, gas rigs and processing workers, as well as local people, are sickened by pollution from the gas emissions and liquids escaping from the work sites.

Hydrofracking is a main cause of contamination of soil and the water table, causing permanent property damages, and where fish, wild life, and people are poisoned by harsh chemicals, dumped intentionally or unintentionally into their habitat.

The externalized costs of gas production and use are estimated to be in the billions. Since the gas drilling business is growing daily, estimates vary, but it seems that soon costs will be even larger than those of coal and oil.

As a result (of the "externalities"), the future generations will be heavily impacted by global warming from the exhaust gases and liquid wastes that gas drilling, production and use spew into the air and spill into the soil.

We would like to emphasize here that the complex, expensive, and dangerous gas drilling, exploitation and natural gas use (fuel for power generators and cars) have some components that are very complex and difficult to understand, sort, and estimate accurately.

Nevertheless, while we must be aware of the problems and work towards their solution, we must also keep in mind that natural gas is a necessary evil. It is something we need badly. And so, we need to treat natural gas and the people who work in the industry with respect and admiration, because their efforts and sacrifices make our lives comfortable and prosperous.

Note: In the summer of 2015 EPA came up with a report from a five-year investigation on the effects of hydrofracking on the environment. From EPA's point of view, only 1-2% of the drilling sites are contaminated, so not to worry. The contaminated sites, according to EPA do not pose a real threat to the water supply and the environment in general.

Yes, true! When looking at 1-2% of the total from a desk in Washington DC may not be much when considering the entire enormous picture of thousands of drilling sites. It is, however, quite different for the people who live near the contaminated fracking sites. In addition to the traffic congestion in the area, they have to deal with the threat of water table, soil, and air contamination.

1-2% of the total is not much for most of us, but is everything to those—thousands of people—who are affected. And sure enough, EPA admits that there are risks of water contamination in some areas. These incidents, however, are small compared to the total of fracking wells in the country.

So let's see, the EPA study estimates that 25,000 to 30,000 new wells were drilled and hydraulically fractured every year from 2011 to 2014. Hydrofracking wells were in operation in 25 states from 1990 to 2013, and so approximately 9.4 million people lived, or still live, within one mile of a hydrofracked well.

At the same time, about 6,800 public drinking water systems were located within one mile of at least one hydraulically fractured well during the same period. These water system provide drinking water to more than 8.6 million people today.

So, the conclusion is that no new hydrofracking regulations are needed. But wait, the report's authors were sitting on their desks in the EPA headquarters on Pennsylvania Avenue in Washington DC, relying on information and data supplied by oil and gas companies, while rarely visiting drilling sites. But even a day or two on a drilling site is not enough to see all that is going on around and below the rig.

And so, EPA wrote that limitations in the supplied data "preclude a determination of the frequency of impacts with any certainty." So, we now have a guesswork document, in which the only certainty is that oil and gas company will use it to claim that new regulations are not needed. This is like saying that 1+1 is either 2 or 3, admitting that there is uncertainty in the result, but yet insisting that no correction is needed.

Nevertheless, the new EPA study confirms the fact that millions of Americans live close to, and are in danger of, getting dirty contaminates water running out of their wells and faucets. This is a fact that most of us will ignore, but not those who live near the contaminated sites. For them, the danger hanging over their families will turn the American Dream into nightmare.

The Solutions

Natural gas production is quickly expanding across the nation—the new bonanza allowed by new technologies that make it easier to extract gas from previously unknown and/or inaccessible sites.

Recently, the industry has drilled thousands of new wells in the Central and South USA, and is expanding operations in the Eastern side of the country. The most promising being a 600-mile-long rock formation called the Marcellus Shale, which stretches from West

Virginia to western New York.

90% of all natural gas production today is done via hydrofracking—a technique that was perfected recently. Here, many (some dangerous) chemicals are mixed with large quantities of water and sand and are then injected into the drilling wells at extremely high pressure.

Hydrofracking is suspected to have polluted drinking water in Arkansas, Colorado, Pennsylvania, Texas, Virginia, West Virginia and Wyoming. In many of these paces, residents have reported unusual events and water quality deterioration. In most cases, these findings coincide with the commencement and expansion of hydrofracking operations.

Natural gas producers have been ignoring the complaints of the communities across the country with their extraction and production activities for too long. In the meanwhile, the number of cases of contaminated water supplies, dangerous air pollution, destroyed streams, and devastated landscapes is increasing daily.

The industry deals with a number of problems related to weak safeguards and inadequate oversight, which fail to protect the locals from harm. We shouldn't have to accept unsafe drinking water just because we have a lot of natural gas, and because it burns more cleanly than coal.

The rules of the game are unclear at best for now, and many companies ignore even the few rules set thus far. The industry in general prefers to use its political power to escape accountability for its actions, which leaves the locals unprotected.

New safeguards are needed, as follow:

Figure 4-35. Waste water and chemicals storage at a hydrofracking site

- The sensitive lands, including critical watersheds, must be off limits to fracking
- A lot of independent, in-depth research still remains to be done to understand and determine exactly the effects of the different chemical mixtures on vegetation, wild life, and humans.
- New clean air standards are needed to ensure that natural gas leaks from wells are under control, and as needed to prevent local air pollution and reduce global warming.
- New and safer well drilling and construction standards must be implemented to ensure strongest well siting, casing and cementing, as well as other best practices during drilling and operation of the gas wells.
- Closing Clean Air, Clean Water and Safe Drinking Water loopholes is needed to reduce toxic waste, and hold toxic oil and gas waste to the same standards as other types of hazardous waste.
- Robust inspections and enforcement of the safety programs, including full disclosure of fracking chemicals, are needed
- The affected communities close to fracking sites must be allowed to protect their environment and themselves by having a voice in comprehensive zoning, planning, and management of fracking sites.

There is no doubt that having enough natural gas is a good thing, which is critical to our energy future. We cannot and should not stop the developments, but we must make sure that they are not hurting and killing us in the process. Efficient regulations are needed to make the process fully transparent and protect the environment, and property and lives of the locals.

Natural Gas Markets

Natural gas is the fuel of the future and a leading fuel in ensuring our energy security. It powers large numbers of chemical and refining processes in the U.S., and is also the fuel of choice for our electric power generating network. It is a major feedstock for hydrogen production, hydrofracking, hydro-desulfurization, ammonia and other chemicals production.

Note: There is some confusion about the abbreviations of the different types of commercial gasses on the market today.

LPG (liquefied petroleum gas) is the common household name, whereas LNG and CNG are the specific energy terms for this commodity.

LNG (liquefied natural gas) is natural gas (mostly methane) in its liquid form, which is achieved by lowering the temperature and/or compressing the gas.

CNG (compressed natural gas) is natural gas (mostly propane and butane) in compressed (but not yet liquefied) form. It is often called "propane."

All these are natural gas in different physical forms and slightly different composition, which are used in different situations, depending on transport mode and final destination.

Natural gas is also used for making other commodities, such as syngas, methanol, and its derivatives—MTBE, formaldehyde, and acetic acid. The condensate derivatives, ethane and propane, are used as an advantageous raw material in a number of processes and products. Via ethylene and propylene, natural gas is also used in much of the organic chemicals production industries today.

Since at times there is much more natural gas produced than can be used, it is then stored underground inside depleted gas reservoirs from previous gas wells, salt domes, or in liquefied natural gas tanks. In most cases, the gas is injected in a time of low demand and extracted when demand picks up. Storage near end users helps to meet volatile demands, but due to limited capacity and the volatility of the gas, storage may not always be practical.

The world's top 15 producing countries account for 85% of the global gas extraction, so access to natural gas has become an important issue in international politics, and countries vie for control of the gas pipelines. As an example, in the 2000s, the Russian Gazprom started disputes with Ukraine and Belarus over the price of natural gas and on several occasions shut off the supplies in the midst of freezing winter weather.

The increasingly provocative Russian control of natural gas supplies has also increased the concerns of the largest European economies, which also depend heavily on Russian gas. The Europeans are looking for alternatives, but there are few. Using larger quantities of U.S. coal and natural gas is one possibility, and is under discussion on different levels of the European governments.

For now, however, Europe depends on Russia to open the gas spigot in the winter months. If supplies were reduced or shut down, a lot of Europeans could freeze to death, and lots of businesses would shut down.

Natural gas is mainly used for power generation, but it has many other uses as well. There are some important variations and derivatives of natural gas, which are also part of the energy markets, as follow:

NGLs

Natural gas liquids (NGLs) are hydrocarbons, other than methane, that are separated from raw natural gas during the refining process at the refineries. These liquids include ethane, propane, butane and pentane in variable amounts. Natural gas for an average well contains 11% ethane, 5% propane, 2% butane and about 2% of natural gasoline or drip gas (a low-octane fuel used mostly as a solvent.)

Other natural gas sources contain much higher percentage of methane and correspondingly smaller percentages of NGL—7% ethane, 4% propane, 1% butane, and other components including carbon dioxide and pentanes.

In most cases, ethane makes up about half of the NGL total, propane makes up about a quarter, and butane makes up 5-10%. Unfortunately, ethane cannot be used as is for vehicle fuel; only propane and butane can be used as such.

The problem here is that they are a small component (25-35%) of the total NGL volume, and since most of the as-is volume of NGL cannot easily be used as vehicle fuel, propane and butane will not become major vehicles fuels, due to their small fraction of the total natural gas volume. This makes their total availability limited by the total amount of natural gas extracted.

Within the last decade, the industry began to count the NGLs as part of our energy supply, and there is increasing talk about using some of these as vehicle fuel. Some NGLs can also be used as feedstocks for chemical production, just as petroleum is. Nevertheless, although NGLs would not be able to displace crude oil in this market, as it is currently configured, they play an important role in it.

So, in the future NGLs will most likely be used as a fuel in gas-fired power plants and other heat-producing applications. Their use in the transportation sector as replacement for crude oil, however is most important to our economy, since it is one key element of our energy security and future energy independence that no other fuel can provide.

The more natural gas and NGLs we use for transportation fuels, the closer we get to our goal of reaching energy independence, so the efforts to include NGLs in the fuel mix will continue.

Associated Petroleum Gas

Associated petroleum gas (APG), or associated gas, is a form of natural gas which is found with deposits of petroleum, either dissolved in the oil or as a free "gas cap" above the oil in the reservoir. Historically, this type of gas was released as a waste product from the petroleum extraction industry. Due to the remote location of many oil fields, either at sea or on land, this gas is simply burnt off in gas flares. When this occurs the gas is referred to as flare gas.

The flaring of APG is controversial as it is a pollutant, a source of global warming and is a waste of a valuable fuel source. APG is flared in many countries where there are significant power shortages. In the United Kingdom, gas may not be flared without written consent from the UK Government in order to prevent unnecessary wastage and to protect the environment. Russia is the world leader in the flaring of APG, and flares 30 per cent of the total APG flared globally.

The World Bank estimates that over 150 billion cubic meters of natural gas are flared or vented annually. This amount of gas is worth over $30 billion dollars and is equivalent to 25% of the U.S.' yearly gas consumption, or 30% of the EU's annual gas consumption.

APG gas can be utilized in a number of ways after processing. It can be sold and included in the natural gas distribution networks, or used for on-site electricity generation with engines or turbines, re-injected for enhanced oil recovery, or used as feedstock for the petrochemical industry.

Coal Bed Methane

Coal bed methane (CBM), coal-bed gas, coal seam gas (CSG), or coal-mine methane (CMM) is a form of natural gas, which, as the name suggests, is extracted from coal beds. In recent decades it has become an important source of energy in U.S., Canada, Australia, and other countries.

The term refers to methane adsorbed into the solid matrix of the coal. It is also called 'sweet gas' because of its lack of hydrogen sulfide. The presence of this gas is well known from its occurrence in underground coal mining, where it presents a serious safety risk. Coal-bed methane is distinct from a typical sandstone or other conventional gas reservoir, as the methane is stored within the coal by a process called adsorption. The methane is in a near-liquid state, lining the inside of pores within the coal (coal's matrix). The open fractures in the coal (coal cleats) can also contain free gas, or can be saturated with water.

Unlike much natural gas from conventional reservoirs, coal-bed methane contains very little heavier hydrocarbons such as propane or butane, and no natural-gas condensate. It often contains up to a few percent carbon dioxide. Some coal seams, such as those in certain areas of the Illawarra Coal Measures in New South Wales, contain little methane, with the predominant coal seam gas being carbon dioxide.

Landfill Gas

Landfill gas is produced from chemical reactions and microbes acting upon the waste as the putrescible materials begin to break down in the landfill. The rate of production is affected by waste composition and landfill geometry, which in turn influence the microbial populations within it, chemical make-up of waste, thermal range of physical conditions, and the biological ecosystems co-existing simultaneously within most sites. This heterogeneity, together with the frequently unclear nature of the contents, makes landfill gas production more difficult to predict and control than standard industrial bioreactors for sewage treatment.

Due to the continual production of landfill gas, the increase in pressure within the landfill (together with differential diffusion) causes the gases release into the atmosphere. Such emissions lead to important environmental, hygiene and security problems in the landfill.

Due to the risk presented by landfill gas there is a clear need to monitor gas produced by landfills. In addition to the risk of fire and explosion, gas migration in the subsurface can result in contact of landfill gas with groundwater. This, in turn, can result in contamination of groundwater by organic compounds present in nearly all landfill gas.

Landfill gas is approximately forty to sixty percent methane, with the remainder being mostly carbon dioxide. Landfill gas also contains varying amounts of nitrogen and oxygen gas, water vapor, hydrogen sulphide, and other contaminants. Most of these other contaminants are known as "non-methane organic compounds" or NMOCs.

Some inorganic contaminants, such as mercury, are also present in the gas of some landfills. There are sometimes also radioactive contaminants, such as tritium, found in landfill gas. The non-methane organic compounds usually make up less than one percent of landfill gas.

General options for managing landfill gas are: flaring, boiler heating, power generation, converting the methane to methyl alcohol, cleaning it to use in other industries or into the natural gas lines.

The U.S. EPA identified ninety-four non-methane organic compounds, including toxic chemicals like benzene, toluene, chloroform, vinyl chloride, and carbon tetrachloride. At least 41 of the non-methane organic compounds are halogenated compounds (chemicals containing halogens, such as chlorine, fluorine, or bromine).

There are approximately 6,000 landfills in the U.S., most of which are composed of municipal waste, and, therefore, produce methane. These landfills are the largest source of anthropogenic methane emissions in the U.S. These landfills will contribute an estimated four hundred and fifty to six hundred and fifty billion cubic feet of methane annually.

LPG

Liquid petroleum gas (LPG) is a form of natural gas, also called propane (C_3H_8) or butane (C_4H_{10}) on the commercial markets. When used as a vehicle fuel, it is referred to as autogas. LPG is derived from fossil fuel sources; i.e., during the refining of crude oil, or extracted from petroleum or natural gas streams as they emerge from the ground. It is used primarily for heating, cooking, and as an aerosol propellant and a refrigerant, replacing chlorofluorocarbons which damage to the ozone layer.

LPG's heating value is approximately 10% lower than natural gas (methane vs. propane). There is about 4% propane in the commercial natural gas used today. The major differences between LPG and natural gas are significant.

Natural gas is:
- Very safe as it is lighter than air, and when leaked it will float up and dissipate quickly.
- Ready to use as it is in gas form.
- Its ignition point is 593 degree C.
- The concentration at which natural gas would ignite or explode is 5-15% gas in air.
- Colorless and odorless, and burns completely, without emissions of soot or sulfur. Storage tanks are usually not needed for storing natural gas.

LPG is:
- Less safe, as it is heavier than air, and when leaked it will pool on the ground.
- A liquid, so it needs to be converted to gas.
- Colorless and odorless, but odor is normally added for safety reasons.
- It burns completely, without emissions of soot or sulfur.
- Storage tanks and advance ordering are needed in most cases.
- Its ignition point is lower than natural gas; 410-580 degree C.

• The concentration at which LPG gas would ignite or explode is 2.0-9.5% gas in air.

Note: The great variation of ignition points in LPG is due to the fact that there are a number of different petroleum compounds in it, each of which has a different ignition point. Basically, LPG is more dangerous to use than natural gas. It is easier to ignite or explode LPG gas than natural gas, since its ignition point is significantly lower. LPG is also heavier than air, so it tends to settle down and concentrate in corners, causing death by suffocation, and explosions in the presence of a spark or open flame.

U.S. Natural Gas Markets

Natural gas markets have an increasing impact on the global economy and the companies and individuals who rely on the fuel for electric generation, manufacturing, heating, cooking and other purposes.

The U.S. Energy Information Administration (EIA) estimates that natural gas supplies 25 percent of all the energy used in the United States. This amounts to about 24 trillion cubic feet (Tcf) of gas is used every year in U.S.

Under the Natural Gas Act (NGA), the Federal Energy Regulatory Commission (FERC) has jurisdiction over the transportation and sale of natural gas, and the companies engaged in those activities.

As in all energy markets, the natural gas market is an amalgamation of a number of subsidiary and peripheral markets. At the top tier, there is the physical market, in which natural gas is produced, transported, stored, and used. And then, there is also the financial component that plays a major part in the natural gas markets. Here, natural gas is bought and sold only as a financial product derived from physical natural gas activities. Only money changes hands, but which usually affects the actual physical markets activities and prices.

Table 4-9. Natural gas use in the U.S.

GAS USE	M ft³
Electric power	9,100,000
Industrial	7,225,000
Residential	4,150,000
Commercial	2,900,000
Vehicles	30,100
TOTAL	23,410,000

Table 4-9 shows that residential home use of different types of natural gas, delivered by pipeline, (LNG and CNG), is almost half the gas used in the U.S. for power generation, and almost 20% of all gas delivered to customers in the country.

Note: This gas was used mostly for heating in winter and cooking by U.S. households.

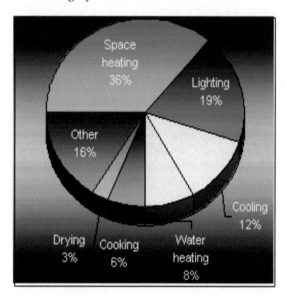

Figure 4-36. Residential natural gas use in the U.S.

When the amount of natural gas used by commercial and industrial enterprises is added, we come up with 60% of the total natural gas delivered to customers being used by these businesses. The amount of natural gas used by U.S. homes, commercial, and industrial enterprises has kept steady during the last decade.

At the same time, natural gas used for vehicle fuel has increased by about 20% and is expected to continue increasing as these vehicles are becoming widely spread around the U.S. and the globe. The largest increase of natural gas use is for electricity generation—a 30% jump, from 6,800,000 to 9,100,000 M ft3 annually in 2012.

The total use of natural gas increased by another 10% from 2012 to 2013 to 26,034,000 M ft3. This was due mostly to the increase of natural gas use for electric power generation. That trend is expected to continue for the foreseeable future, as natural gas is quickly replacing coal for electric power generation in the country.

Market Segments

The U.S. natural gas industry has three major segments;
• *The supply sector*, which includes exploration and development of natural gas resources and reserves,

and their production, which includes drilling, extraction and gas gathering.

- The midstream sector is where small-diameter gathering pipeline systems transport the gas from the wellhead to natural gas processing facilities for processing. Here, the impurities and other hydrocarbons are removed from the gas to create pipeline-quality dry natural gas,
- Transportation and distribution sector includes intrastate and interstate pipeline systems that move natural gas through large-diameter pipelines to storage facilities and a variety of consumers, including power plants, industrial facilities and local distribution companies, which deliver the gas to retail consumers.

The U.S. physical energy markets can be further differentiated by:

- Location: regions, nodes, zones or hubs;
- Time frames: hourly, daily, monthly, quarterly or yearly;
- Types of products: natural gas (as gas or electrons),
- Delivery by pipeline or transmission lines,
- Possibility of energy storage, and
  — Nature of sales; retail sales vs. wholesale.

The U.S. natural gas markets are regional, with prices for natural gas varying with location and seasons, the demand characteristics of the local market, as well as the region's access to different supply basins, pipelines and storage facilities. Prices are usually lowest in areas with low-cost production, ample infrastructure and limited demand, which is the case in gas producing states. At the same time, prices are highest where production or transportation is limited and demand is high, such as in some North Eastern states.

The U.S. natural gas markets are divided into several regions: the West, Midwest, Gulf Coast, Northeast and Southeast regions. These regions have differing supply, transportation and demand characteristics, resulting in different prices.

These regions are subdivided into hubs, where two or more major pipelines interconnect. These physical hubs are also market hubs for buying and selling gas. The key hub used to reflect the U.S. natural gas market as a whole is the Henry Hub, in Louisiana. Prices at oth-

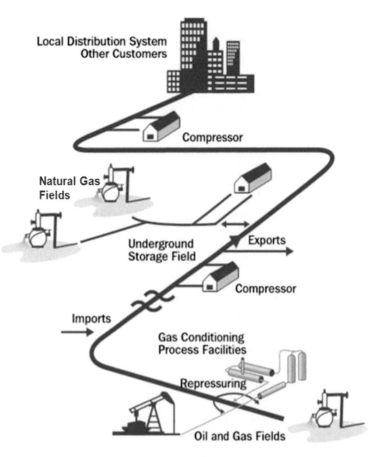

Figure 4-37.
The natural gas sector (production and distribution)

er locations are frequently shown as Henry Hub plus or minus some amount.

The U.S. natural gas markets are undergoing a major period of transition. Recently, various factors shifted the dynamics of supply and demand. These include:

- Development of new technologies, like hydraulic fracturing and horizontal drilling, which enable access to unconventional resources such as those in shale formations. This has vastly expanded supply and is increasing the amount of natural gas produced to unprecedented levels.
- Natural gas is now investment opportunity in addition to being a traded commodity. There are now two distinct markets for physical natural gas: a) the cash market, where natural gas is bought and sold for immediate delivery; and b) the forward market, where natural gas is bought and sold under contract months in the future.
- Natural gas demand for power generation is rising and is expected to increase significantly in the future, years. Natural gas-fired power plants are cheaper to build and operate and emit less air pol-

lution than coal- or oil-fired power

• Interstate pipeline expansions have brought gas to the Northeast and Mid-Atlantic, and have contributed to lowering the energy prices. Excess natural gas supplies have also tempered extreme price movements during periods of peak demand.

Natural Gas Exports

During the last decade, the U.S. natural gas production increased to unprecedented levels. So much, in fact, that there are now serious plans for increasing the natural gas exports to countries all over the world.

The U.S. DOE suggests four scenarios of export-related increases in natural gas demand:

• 6 billion cubic feet per day (Bcf/d), phased in at a rate of 1 Bcf/d per year (low/slow scenario),

• 6 Bcf/d phased in at a rate of 3 Bcf/d per year (low/rapid scenario),

• 12 Bcf/d phased in at a rate of 1 Bcf/d per year (high/slow scenario), and

• 12 Bcf/d phased in at a rate of 3 Bcf/d per year (high/rapid scenario).

The total (averaged) marketed natural gas production in the county is about 75 Bcf/d. The two ultimate levels of increased natural gas demand due to additional exports in the DOE scenarios represent roughly 7 and 16% (low and high scenarios respectively) of current production. So although that is still a huge amount of natural gas going overseas, it is not a major factor for the U.S. energy markets.

DOE then requested that EIA consider the four scenarios of increased natural gas exports in the context of four cases from the EIA's 2011 Annual Energy Outlook (AEO2011) that reflect varying perspectives on the domestic natural gas supply situation and the growth rate of the U.S. economy.

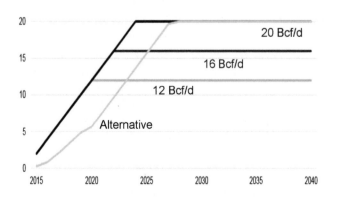

Figure 4-38.
U.S. LNG export scenarios (in billion cubic feet/day)

It is obvious that the U.S. is fixing to flood the global energy markets with very much needed natural gas. There are a number of uncertainties and factors that would influence the export rates, so several scenarios are under consideration:

• The High Shale Estimated Ultimate Recovery (EUR) case, reflecting more optimistic assumptions about domestic natural gas supply prospects, with the EUR per shale gas well for new, undrilled wells, assumed to be 50% higher than in the Reference case,

• The Low Shale EUR case, reflecting less optimistic assumptions about domestic natural gas supply prospects, with the EUR per shale gas well for new, undrilled wells assumed to be 50% lower than in the Reference case, and

• The High Economic Growth case, assuming the U.S. gross domestic product will grow at an average annual rate of 3.2% to 2035, compared to 2.7% in the Reference case, which increases domestic energy demand.

Either way, significant quantity of natural gas exports are well within the U.S.' capabilities for now, and so the plans are underway. But there are problems…

• On the long run; we will export our liquid gold while it is available. This is making a quick buck, without considering the long term energy security of the country. Its the energy economy, stupid! The LNG export plans in Figure 4-38 project to year 2040, and we cannot help it but wonder what will happen after that? Would there be enough LNG to continue exporting it much longer? How longer is another question that we need to consider, because we know that all fossils are finite commodities with beginning and end. We know the beginnings, and it would be smart and prudent to consider the end too.

• On the short term; we export LNG to some countries that do not have a free-trade agreement (FTA) with the U.S. The LNG export procedures call for the DOE to review each application, but it can only deny it only if the exports are not in the "public interest." This is a vague definition and requires added clarity in the process. But the U.S. has a FTA only with South Korea, which means that the DOE must have some way to determine which country is suitable for LNG exports and act upon it.

DOE relies on the Natural Gas Act to make these decisions, but the Act presumes that exports are in the national interest unless it is proven otherwise. This raises a question about our national security, which especially today with political turmoil around the world must include a better defined criteria as of whom and why we export our energy. These decisions also affect the cost of constructing export facilities, and changes in the LNG prices according to changes in the gas market.

In order to ramp-up of LNG exports, DOE gave conditional approval in 2013 to the Freeport LNG export terminal to ship LNG to countries that do not have a free-trade agreement with the U.S. This, in addition to the Sabine Pass terminal is the second export point for such dubious exports. What is next in the rush to make a quick buck, at the price of depriving the future generations of this precious and irreplaceable commodity, and endangering our energy security. Money first?

THE U.S. OIL AND GAS REVOLUTION

During the past several decades, the oil and gas production in the U.S. was insufficient, and even in decline. Thus, we were forced to import a lot of oil. All of a sudden during the last decade, the U.S. oil industry snapped into action and within several short years discovered new ways to look for and produce oil.

This resulted in branching from the old way of producing oil by pumping it from large underground pools, wherever these were located by Mother Nature, to producing oil from rock formation. What was considered impossible and unprofitable until now, was all of a sudden producing gold (errr, oil).

And why not? We are the most advanced technologically nation in the world, so sucking oil from rocks should not be that difficult. And ahead we went with new equipment and processes, using hydrofracking and other techniques to get oil from places where oil was not even suspected before.

So we have a lot of oil and gas in the U.S. and the production levels are increasing. The U.S. has some of the lowest energy prices in the world too, and reinvestment rates in energy-intensive manufacturing (those that create high-value jobs) lag those of Asia and the Middle East, by an impressive ratio of 15:1.

In 2013, North America spent nearly $200 billion on producing oil and gas, and in the process attracted over 50% of global upstream investment. This was more investment (in dollars) than Russia and Saudi Arabia combined…not 2 or 3 times more, but by an astonishing factor of 10:1 more.

But now we see another phenomena lurking on the horizon. The "demand response" phase of the oil revolution, which was growing fast is all of a sudden stalling.

Encouraged by this influx of money, the supply-side investment surges ahead, while the demand side investment is beginning to lag.

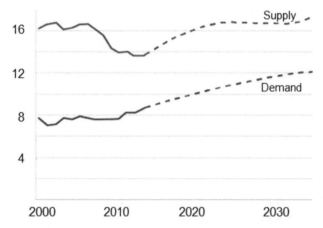

Figure 4-39. U.S. natural gas supply/demand (in Bcf/day)

We are now awakening to the reality that having a lot of oil and gas requires the corresponding infrastructure. It is needed to ensure proper use of the abundant energy supplies, without which (infrastructure) the new energy supplies cannot be delivered to the end users.

When during the winter of 2013 temperatures plunged well below zero in the North East USA, the heating gas could not be delivered to the homes and businesses where it was needed. This created regional price spikes, outages and an array of other customer problems. This anomaly brought up discomfort and illness to many people.

Also, many new drilling sites are far away from the existing infrastructure—pipelines, railroads, and roads—which often cause production delays and interruptions.

These problems will continue to hinder the development of new energy sources, and will make us unable to harness the benefits from the new oil revolution. If the problems are not corrected, we will also forego more than 2 million new jobs over the next decade. There is also a risk to 1.0 % of additional GDP growth and at least a 5% incremental reduction in greenhouse-gas emissions.

Nevertheless, the U.S. oil revolution is underway, although its progress was delayed for awhile during

Table 4-10. U.S. natural gas infrastructure

Production[b]	478,562 gas and condensate wells
Processing[c]	More than 500 gas processing plants (lower 48)
Storage[d]	401 underground storage facilities 8.5 trillion cubic feet capacity 109 LNG storage facilities
Pipeline[e]	
Gathering	20,215 miles of gathering pipeline
Transmission	298,993 miles of interstate pipeline
Distribution	1.2 million miles of intrastate pipeline

2014-2015. Extremely low global oil prices forced many drillers to reduce or stop production.

There are certain economic advantages created by the oil revolution, but they were based on legacy infrastructure rather than resource availability. Many other countries have similar resources, with China and Russia as extreme examples of enormous energy resources, which lack the infrastructure needed to exploit, transport, and deliver them to the customer.

Yet, it is only a matter of time before China and other nations catch up with the U.S., so the solution is simple: stable and well-defined energy, environmental and transportation policies, which only the government can establish.

Thus far the U.S.' "oil revolution," or as some call it *'The New Oil Order'* has been developing without a well defined energy policy, accompanied by provision of short-term, quick-turnaround investments. The demand-side investments that are needed today are very large and require decades to recoup the investment. Because of that, they also require a high level of confidence in sector's performance which requires stable future policies—which, again, only the government can provide.

Such policies would establish long-term confidence and an opportunity for business and government to work together to support the shared goal of cheap energy, clean environment, strong economy and sustainable job creation.

All the hoopla about natural gas' benefits for some people, comes at a price for others.

Case Study: New York State

Energy companies spend millions of dollars for television advertising, lobbying and political campaign contributions. The battle has been most violent in the state of New York, while the state was deciding when and where to allow a controversial form of natural gas extraction that is opposed by many locals and the environmental groups. The battle is over now that the state of New York decided NOT to allow hydrofracking on its territory.

Natural gas companies spent millions of dollars lobbying state government and the broader natural gas industry has been giving hundreds of thousands of dollars to the campaign accounts of lawmakers. Many national energy companies are also advertising heavily in an effort to convince the public that the extraction method, commonly known as hydrofracking, is safe and economically beneficial.

The locals and most environmental groups don't have money for ads, so they are mounting a more homespun campaign, warning the public of the dangers of hydrofracking, which could taint the water supply and cause serious environmental ruin.

The energy sector lobby is pushing hard, in an attempt to influence the public, the lawmakers, and regulators in states rich in shale formations. Several states, like Texas, Pennsylvania and Ohio, have also seen millions of dollars spent by drilling companies on both; lobbying, and campaign contributions.

New York state tried to develop hydrofracking regulations that please all involved, since the industry could create jobs and provide prosperity to the region by spending in some of the most economically struggling parts of the state, especially its Southern Tier. On the other hand, the environmentalists warned of risks to water quality and damage to roads.

The bottom line, however—like in all capitalist enterprises—is that hydrofracking could generate billions in revenue for energy companies. And so, there was a concerted effort of substantial amount of money to try to move public policy into a pro-fracking mood.

Money talks, so the gas-drilling companies' lobbying and TV ads dwarfed the efforts of the locals and the environmental groups. The money those companies have spent in New York state alone in 2014 is more than four times the roughly $800,000 that the environmental groups have spent.

That created a new peripheral market (a cottage industry), where paid lobbyists make a lot of money, helping the gas industry, which previously was not present in the U.S. capital. For example, one single major gas driller, spent over $1.6 million on lobbying over a 3-year period, while spending only $40,000 in the three years before that. This company had 11 lobbyists to advocate on its behalf in DC, and also hired 3 private companies to promote hydrofracking by all means possible.

Expensive national advertising campaigns were underway, financed by deep-pocketed industry groups and companies, including the American Petroleum Institute, Exxon Mobil, and ConocoPhillips. Their slick TV ads show how invisible, safe, and beneficial natural gas is. They also held numerous community meetings to answer questions from residents worried about hydrofracking. With a straight faces, they tell all that were willing to listen, that hydrofracking is as safe as a morning walk in the forest.

In the end, however, they always admitted that there are 'some' risks involved. As any other fuel source, hydrofracking too has risks attached to it, but, they always concluded that the rewards outweigh the risks. This is true for some people, but is of little consolation to those with bubbling green toxins in their back yards, or who cannot use the local water any longer. So, the battle of frack or not to frack in the state of New York went on for a long time.

Fifteen New York towns were upset with Democratic Gov. Andrew Cuomo's decision to ban fracking, and have threatened to secede from the state and join neighboring Pennsylvania, where fracking is allowed.

Finally in early 2015, the state banned all fracking activities on its territory. Now 15 towns, all members of the Upstate New York Towns Association, are threatening secession. The town association is compiling a report to assess the feasibility of joining the state of Pennsylvania. The new Greece of the United States. Is this even possible…?

In their defense, these towns are in the Southern Tier of New York, where there are no jobs and the economy has been terrible for some time. But the area is rich in natural gas. Permitting drilling in the region would provide an avenue for new jobs and a way to raise money for local schools and governments, just like it has across the border in Pennsylvania.

Leaving New York and joining the Keystone State would also mean lower taxes for businesses and lower insurance payments. Secession, however, is far-fetched. It take much more than job creation, and would require the approval of the New York legislature, the Pennsylvania legislature and the federal government to do such a thing. So raising the idea only helps highlight the region's discontent with New York's government's decision.

OK, we get the job situation in upper New York state, but what is the big deal with hydrofracking, and what are the risks everybody is talking about? None, if you live far away from the drilling sites, but it is all, if you happen to live nearby. Let's see why:

Fracking Chemistry

Proppants are chemicals, water, oil, and a large number of other organic and inorganic compounds that are used in large quantities during hydrofracking operations. These chemicals are pumped at a high volume and under high pressure in the oil and natural gas wells to increase their production.

The proppants work miracles in releasing huge quantity of gas and oil trapped in rocks and other underground formations. This was simply impossible just several years back, and is the main reason for the present day bonanza in the U.S. and many other countries.

Intentionally or unintentionally, however, proppants have leaked up into pastures, playgrounds, and people's back yards. Water tables have been contaminated by escaping from their containment proppants, and have hurt vegetation, animals, and humans as well.

Even the best of intentions and the best methods of operation cannot prevent occasional spills and leaks. It is the nature of the beast. And so, properties, the environment, wild life, and even humans have been hurt by this process. All in the name of natural gas—pumping it as much and as fast as possible out of the ground.

Human life and wellbeing should not be sacrificed in the name of energy security. We should not pay for our energy security and independence with human suffering. And yet, this is exactly what is increasingly happening today…

Over 700 chemicals have been identified and listed as proppant additives for hydraulic fracturing in a report to the US Congress in 2011. But the manufacturers are not required to officially disclose the contents of their proppant liquids. And so, they basically use whatever amount and combination of chemicals they want, regardless of what the final effect of these might be on the local environment and human life

The following is a partial list of some of the chemical constituents used in fracturing operations, which have been identified by the New York State Department of Environmental Conservation.

Table 4-11. List of chemicals and substances in proppants

1,2-Benzisothiazolin-2-one/1,2-benzisothiazolin-3-one

1,2,4-trimethylbenzene

1,4-Dioxane

1-eicosene

1-hexadecene

1-octadecene

1-tetradecene

2,2 Dibromo-3-nitrilopropionamide, a biocide

2,2'-azobis-{2-(imidazlin-2-yl)propane}-dihydrochloride

2,2-Dibromomalonamide

2-Acrylamido-2-methylpropane sulphonic acid sodium salt
    polymer

2-acryloyloxyethyl (benzyl) dimethylammonium chloride

2-Bromo-2-nitro-1,3-propanediol

2-Butoxy ethanol

2-Dibromo-3-Nitriloprionamide (2-Monobromo-3-nitriilopro-
    pionamide)

2-Ethyl Hexanol

2-Propanol/Isopropyl Alcohol/Isopropanol/Propan-2-ol

2-Propen-1-aminium, N,N-dimethyl-N-2-propenyl-chloride,
    homopolymer

2-propenoic acid, homopolymer, ammonium salt

2-Propenoic acid, polymer with 2 p-propenamide, sodium
    salt

2-Propenoic acid, polymer with sodium phosphinate (1:1)

2-propenoic acid, telomer with sodium hydrogen sulfite

2-Propyn-1-ol/Propargyl alcohol

3,5,7-Triaza-1-azoniatricyclo[3.3.1.13,7]decane, 1-(3-chloro-2-
    propenyl)-chloride,

3-methyl-1-butyn-3-ol

4-Nonylphenol Polyethylene Glycol Ether Branched/Nonyl-
    phenol ethoxylated/Oxyalkylated Phenol

Acetic Anhydride

Acetone

Acrylamide—sodium 2-acrylamido-2-methylpropane sulfon-
    ate copolymer

Acrylamide—Sodium Acrylate Copolymer or Anionic Poly-
    acrylamide

Acrylamide polymer with N,N,N-trimethyl-2[1-oxo-2-prope-
    nyl]oxy Ethanaminium chloride

Acrylamide-sodium acrylate copolymer

Aliphatic Hydrocarbon/Hydrotreated light distillate/Petro-
    leum Distillates/Isoparaffinic solvent/Paraffin Solvent/
    Napthenic Solvent

Aliphatic acids

Aliphatic alcohol glycol ether

Alkyl Aryl Polyethoxy Ethanol

Alkylaryl Sulfonate

Alkenes

Alkyl (C14-C16) olefin sulfonate, sodium salt

Alkylphenol ethoxylate surfactants

Amines, C12-14-tert-alkyl, ethoxylated

Amines, Ditallow alkyl, ethoxylated

Amines, tallow alkyl, ethoxylated, acetates

Ammonia

Ammonium acetate

Ammonium Alcohol Ether Sulfate

Ammonium bisulfate

Ammonium bisulfite

Ammonium chloride

Ammonium Cumene Sulfonate

Ammonium hydrogen-difluoride

Ammonium nitrate

Ammonium Persulfate/Diammonium peroxidisulphate

Ammonium Thiocyanate

Aromatic hydrocarbons

Aromatic ketones

Aqueous ammonia

Bentonite, benzyl(hydrogenated tallow alkyl) dimethylam-
    monium stearate complex/organophilic clay

Benzene

Benzene, 1,1'-oxybis, tetratpropylene derivatives, sulfonated,
    sodium salts

Benzenemethanaminium, N,N-dimethyl-N-[2-[(1-oxo-2-pro-
    penyl)oxy]ethyl]-, chloride, polymer with 2-propenam-
    ide

Boric acid

Boric oxide/Boric Anhydride

Butan-1-ol

Ethoxylated Alcohol

Alcohol, Ethoxylated

Carboxymethylhydroxypropyl guar

Cellulase/Hemicellulase Enzyme

Chlorine dioxide

Citrus Terpenes

Cocamidopropyl betaine

Cocamidopropylamine Oxide

Coco-betaine

Copper(II) sulfate

Crissanol A-55

Crystalline Silica (Quartz)

Cupric chloride dihydrate

Decyldimethyl Amine

Decyl-dimethyl Amine Oxide

Dibromoacetonitrile

Diethylbenzene

Diethylene glycol

Diethylenetriamine penta (methylenephonic acid) sodium salt

Diisopropyl naphthalenesulfonic acid

Dimethylcocoamine, bis(chloroethyl) ether, diquaternary ammonium salt

Dimethyldiallylammonium chloride

Dipropylene glycol

Disodium Ethylene Diamine Tetra Acetate

D-Limonene

Dodecylbenzene

Dodecylbenzene sulfonic acid

Dodecylbenzenesulfonate isopropanolamine

D-Sorbitol/Sorbitol

Endo-1,4-beta-mannanase, or Hemicellulase

Erucic Amidopropyl Dimethyl Betaine

Erythorbic acid, anhydrous

Ethanaminium, N,N,N-trimethyl-2-[(1-oxo-2-propenyl)oxy]-, chloride, homopolymer

Ethane-1,2-diol/Ethylene Glycol

Ethoxylated 4-tert-octylphenol

Ethoxylated alcohol

Ethoxylated branch alcohol

Ethoxylated C11 alcohol

Ethoxylated Castor Oil

Ethoxylated fatty acid, coco

Ethoxylated hexanol

Ethoxylated octylphenol

Ethoxylated Sorbitan Monostearate

Ethoxylated Sorbitan Trioleate

Ethyl alcohol/ethanol

Ethyl Benzene

Ethyl lactate

Ethylene Glycol-Propylene Glycol Copolymer (Oxirane, methyl-, polymer with oxirane)

Ethylene oxide

Ethyloctynol

Fatty alcohol polyglycol ether surfactant

Ferric chloride

Ferrous sulfate, heptahydrate

Formaldehyde

Formaldehyde polymer with 4,1,1-dimethylethyl phenol-methyl oxirane

Formamide

Formic acid

Fumaric acid

Glassy calcium magnesium phosphate

Glutaraldehyde

Glycerol/glycerine

Guar Gum

Heavy aromatic petroleum naphtha

Hemicellulase

Hydrochloric Acid/muriatic acid

Hydrogen peroxide

Hydroxy acetic acid

Hydroxyacetic acid ammonium salt

Hydroxyethyl cellulose

Hydroxylamine hydrochloride

Hydroxypropyl guar

Isomeric Aromatic Ammonium Salt

Isoparaffinic Petroleum Hydrocarbons, Synthetic

Isopropanol

Isopropylbenzene (cumene)

Isoquinoline, reaction products with benzyl chloride and quinoline

Kerosene

Lactose

Light aromatic solvent naphtha

Light Paraffin Oil

Magnesium Silicate Hydrate (Talc)

methanamine, N,N-dimethyl-, N-oxide

Methanol

Methyloxirane polymer with oxirane, mono (nonylphenol) ether, branched

Mineral spirits/Stoddard Solvent

Monoethanolamine

N,N,N-trimethyl-2[1-oxo-2-propenyl]oxy Ethanaminium chloride

Naphtha (petroleum), hydrotreated heavy

Naphthalene

Naphthalene bis(1-methylethyl)

Naphthalene, 2-ethoxy-

N-benzyl-alkyl-pyridinium chloride

N-Cocoamidopropyl-N,N-dimethyl-N-2-hydroxypropylsulfobetaine

Nitrogen, Liquid form

Nonylphenol Polyethoxylate

Organophilic Clays

Oxyalkylated alkylphenol

Petroleum distillate blend

Petroleum Base Oil

Petroleum naphtha

Phosphonic acid, [[(phosphonomethyl)imino]bis[2,1-ethanediylnitrilobis(methylene)]]tetrakis-, ammonium salt

Pine Oil

Polyethoxylated alkanol

Polymeric Hydrocarbons

Poly(oxy-1,2-ethanediyl), a-[3,5-dimethyl-1-(2-methylpropyl) hexyl]-w-hydroxy-

Poly(oxy-1,2-ethanediyl), a-hydro-w-hydroxy/Polyethylene Glycol

Poly(oxy-1,2-ethanediyl), α-tridecyl-ω-hydroxy-

Polyepichlorohydrin, trimethylamine quaternized

Polyethlene glycol oleate ester

Polymer with 2-propenoic acid and sodium 2-propenoate

Polyoxyethylene Sorbitan Monooleate

Polyoxylated fatty amine salt

Potassium acetate
Potassium borate
Potassium carbonate
Potassium chloride
Potassium formate
Potassium Hydroxide
Potassium metaborate
Potassium sorbate
Precipitated silica/silica gel
Propane-1,2-diol, or Propylene glycol
Propylene glycol monomethyl ether
Quaternary Ammonium Compounds
Quinoline,2-methyl-, hydrochloride
Quinolinium, 1-(phenylmethl),chloride
Salt of amine-carbonyl condensate
Salt of fatty acid/polyamine reaction product
Silica, Dissolved
Sodium 1-octanesulfonate
Sodium acetate
Sodium Alpha-olefin Sulfonate
Sodium benzoate
Sodium bicarbonate
Sodium bisulfate
Sodium bromide
Sodium carbonate
Sodium Chloride
Sodium chlorite
Sodium chloroacetate
Sodium citrate
Sodium erythorbate/isoascorbic acid, sodium salt
Sodium Glycolate
Sodium Hydroxide
Sodium hypochlorite
Sodium Metaborate .8H$_2$O
Sodium perborate tetrahydrate
Sodium persulfate
Sodium polyacrylate
Sodium sulfate
Sodium tetraborate decahydrate
Sodium thiosulfate
Sorbitan Monooleate
Sucrose
Sulfamic acid
Surfactants
Tall Oil Fatty Acid Diethanolamine
Tallow fatty acids sodium salt
Tar bases, quinoline derivs., benzyl chloride-quaternized
Terpene and terpenoids
Terpene hydrocarbon byproducts
Tetrahydro-3,5-dimethyl-2H-1,3,5-thiadiazine-2-thione (a.k.a. Dazomet)

Tetrakis (hydroxymethyl) phosphonium sulfate (THPS)
Tetramethyl ammonium chloride
Tetrasodium Ethylenediaminetetraacetate
Thioglycolic acid
Thiourea
Thiourea, polymer with formaldehyde and 1-phenylethanone
Toluene
Tributyl tetradecyl phosphonium chloride
Triethanolamine hydroxyacetate
Triethylene glycol
Trimethylolpropane, Ethoxylated, Propoxylated
Trisodium Ethylenediaminetetraacetate
Trisodium Nitrilotriacetate
Trisodium orthophosphate
Urea
Vinylidene Chloride/Methylacrylate Copolymer
Xylene
Misc. fillers and additives

The list in Table 4-11, albeit partial, is the most extensive and elaborate mixture of vicious chemicals for use in one single operation ever concocted. Taking a close look at the list, we see so many problems with using these chemicals that it is hard to even figure out where to start.

Here are the problems with any mixture containing any of these chemicals, in no particular order of importance or gravity:

1. Most of the proppant chemicals, in varying concentrations and exposure time, can, and usually do, damage the environment, and hurt wild life and humans, when given a chance.
2. Many of these chemicals are toxic, and some are carcinogenic, and yet
3. Often, there is not enough information as of the degree of harm these could cause and under what circumstances.
4. In most cases, the information available in the MSDS sheets has not undergone human toxicity and carcinogenic tests, so the exposure levels are uncertain.
5. Whatever information can be extracted fro the MSDS sheets, most of it cannot be applied directly in any particular case, because the manufacturers do not specify the concentrations of the chemicals in the mixtures, and finally,
6. Even if we knew the toxicity levels and the concentration of each chemical, however,
    a. We do not know what the inter-reaction among

the chemicals in the vicious soup would be, since there is no previous experience, and/or any research done on the matter, and

b. In best of cases we cannot possibly know the reactions occurring underground between the different chemicals and the varying rock formations.

Because of these, and many other reasons we simply would not be able to predict the effect the different proppant formulations might have on the environment and humans as well

Note: We would like to officially challenge anyone who can explain and demonstrate from technical—chemistry, geology, and biology—point of view, what exactly interactions and results can be expected from a mixture of any of these chemicals used as proppants.

What exactly do they do to the underground layers? How can the chemical reactions and mechanical (pressure) action be controlled in all directions and at all times? How to prevent chemical over-reaction, over-pressurizing certain areas, and the related unwanted cracks creation, and excess penetration of the soup into the surface layers as a result of the chemicals' reactivity under high pressure and increased temperature?

What reactions to expect in/with the upper soil levels? What would these chemicals do to the local vegetation and agricultural crops? What effects to expect from the chemical mixture, or parts of it, entering the local water supply? What would that do to wild life and humans who depend on it daily?

And finally, the most important question, 'How to prevent all possible negative effects?' Anyone?

The Failures

And so, what we have here (the proppants) is magic, the effects of which nobody can even predict. It is magic, because it does miracles underground. It is also a miracle that we don't have more incidents and accidents of proppant damage to property and people' health.

The chemicals used in hydrofracking create serious problems by leaking into places where they do not belong. Since some of the chemicals' properties are not well known, and the different manufacturers use different chemicals in different concentrations, we simply cannot even imagine how we will ever be able to put this chemical debacle under complete control.

Since the number of negatively affected people is not that great (yet), but the number is steadily increasing, this problem is becoming a scenario of many possi-

bilities and of very long duration.

While this might sound abstract to most people, it is a scary proposition to deal with if it affects you personally. Just imagine waking up one morning to find out a green-brown pool full of unknown chemicals bubbling in your back yard. What would you think and do?

Or even worse yet, imagine turning the faucet to get a glass of drinking water, but what you see instead is a yellow blub floating in it. And then you smell natural gas smell all over the house. What is the next step you'd take? Call the owner of the nearby drill? Or the state authorities? EPA?

Many people have gone through these experiences and none of them is pleasant. Many people have lost their property values and even their health in similar situations. We can only hope that the number doesn't grow much faster...although we are sure that it will grow some in the future.

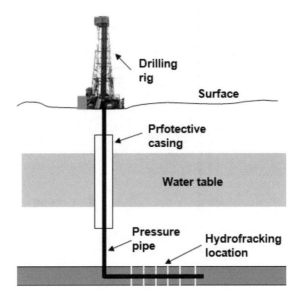

Figure 4-40. Normal hydrofracking setup

In most cases of normal daily operation, there are no mechanical problems, nor is there any presence of ground or surface abnormalities. All of the drilling rig components do their job as planned and no problems are encountered. The proppants are pumped into the pressure pipe and send deep down under to initiate the hydrofracking process as they are designed to do deep and safe bellow the surface.

After many days, weeks and even months of pumping huge amount of pressurized proppants in the formation and shuttering it piece by piece, enough gas is released, at which point the proppant flow is stopped and the gas production starts.

'So what can go wrong?' you ask. A number of

things. Contrary to what we are told, no production material, or process setup is perfect.

So, there is always a chance of:

1. Mechanical failure and breakage of the piping system, the well casings, pipe connections, etc. critical components, and

2. Cracks, pores etc. imperfections in the ground could expand, extend, and propagate uncontrollably in whichever direction under the high pressure and chemical action of the proppants.

Any of these failures and malfunctions could cause unpredictable, and sometime hazardous, below or above ground incidents and accidents.

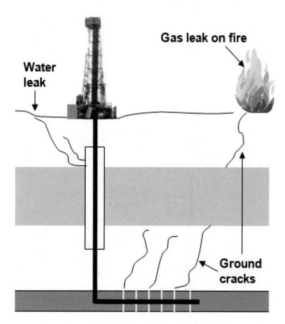

Figure 4-41. Hydrofracking failures

Figure 4-41 shows what we could see in the end after many days of pushing proppants into the rock formation. Proppant fluids and/or gas under high pressure could be forced to flow all the way into the water table and to the surface through the existing or newly formed cracks.

Thousands and millions of gallons of proppants are pushed down the pipe and into the rock formation day after long day. The pressure and the chemical action of the proppants is tremendous and there is no way to control what it does deep under the ground.

With time, the proppants could penetrate any existing pore and crack in their way, or create new ones and travel into the ground water and onto the surface areas around the drilling rig.

If the well casing fails too, then the water table will be surely contaminated and there is no man-made process to stop a disaster from happening in such case.

These are not planned events, mind you, and don't happen all the time, but when they happen, the consequences vary from the nuisance of dealing with small spills and gas leaks, to major property damage and health problems related to the chemicals in proppants.

These events might be quite small in the overall scheme of things, but they are huge for the locals that are affected by them. There are now a number of cases, where environment quality has been destroyed, properties have been made uninhabitable, and wild life and people have been made sick and killed.

What we see often, as result of the damages caused by hydrofracking, is one, or a combination, of the following:

*   Ground water and aquifer contamination
*   Air quality degradation
*   Migration of gases and hydraulic fracturing chemicals to the surface
*   Risk of gases in potable water wells, creating "flammable water"
*   Tremors and earthquakes within surrounding geographic areas
*   Potential waste water spillage risks
*   Loss of land and agricultural production
*   Property damage
*   Loss of land value
*   Human and animal health risks
*   Birth defects,
*   Miscarriages,
*   Cancers,
*   Attention deficit disorder,
*   Hyperactivity,
*   Learning deficiencies
*   Neurological damages

The chemical mixture in the proppants is blamed for a number of these anomalies, so controlling the mixture' contents and use would require a number of things to happen. Such an effort, if ever undertaken, will take many, many years, and billions of dollars, so we don't see it happening any time soon.

*One of the principals of our energy security efforts is to ensure the **safe** production and use of energy for all people involved. Hydrofracking violates this principle for some of us because the proppants used in the process have not been proven safe for use in the way they are presently used.*

For now we have to assume that we are on our own and depending on the good will of the manufacturers and well operators. We also trust that the U.S. regulators will become more active in this area and come up with an acceptable standardized way to deal with the issues at hand.

Our hope and prayers are that this will happen before the vicious underground chemical attack has a chance spread beyond the point of return for some people.

Land Damage

Hydraulic fracturing is a controversial process to say the least. On the positive side of things, oil and gas companies are now able to reach previously inaccessible reserves, which bring financial prosperity to many states and increase the domestic energy production.

The downside is the creation of huge waste chemicals ponds, which damage local lands and water table. Millions of gallons of salty, chemical-infused wastewater, or brine, is used during drilling and fracking at each well site. Drillers inject the brine deep underground, but some of it doesn't make it that far.

In N. Dakota alone there are more than 1,000 accidental releases of oil, drilling wastewater, and other fluids annually. Many more illicit releases go unreported, according to the state regulators. These are instances when companies dump truckloads of toxic fluid along the road or drained waste pits illegally for different reasons.

Brine laced with carcinogenic chemicals and heavy metals, dumped in rivers and streams have wiped out aquatic life and damaged wetlands and sterilized farmland. These effects are not temporary; they can last years and decades.

The enforcement of drilling and fracking operations in the U.S. is inadequate. In some cases there are no regulations to enforce.

Most companies clean up spills voluntarily, but not all and not in all cases. As the activities increase, so does the number of these 'exceptions.' Over 1,000 wells are drilled every month across the country, where millions of gallons of brine are pumped into them, and that much is stored in waste brine pits. Who can figure out and keep track of the number and size of these 'exceptions.'

Oil companies report accidental spills to the Department of Mineral Resources, which then works with the Health Department to investigate, assess, and eventually mitigate the incidents. The Department of Mineral Resources requires companies to report brine use and how much they dispose of as waste, but usually do not

audit the numbers.

And so, short of catching bad operators in the act, the regulators have no way to detect, let alone stop illegal dumping. As a result, the states have no real estimate for how much fluid spills out accidentally from waste pits, storage tanks, pipes, trucks and other fracking equipment.

About 30% of all spill reports involve hydrofracking brine, where in most cases the operators were not able to contain a leak and some waste materials leaked into the ground or waterways.

There are also many unofficial and uninvestigated reports of truckers often dumping their wastewater load at the roadside instead of waiting in line at the injection wells.

Extensive damage to pastures, cattle watering holes, rivers, streams, and roads have been reported. Amazingly enough, even when the authorities catch a well operator in illegal action—such as dumping large quantity of brine—they either can't, or don't bother to, impose sanctions. And so, the intense contamination of the U.S. land and water continues unhindered.

Pollution Footprint

A number of large and small pieces of equipment are used during gas drilling, production, and use. Starting with the gas rig and power plant design and construction, we see pollution taking place in the shape of heavy plumes of smoke from large trucks and bulldozers all around the gas site. There are also trains and trucks used for transporting the production equipment to the site, the exhausts of which are significant.

Don't forget that all this machinery was made somewhere, sometime in the past, where and when their manufacturing processes and transport created a significant pollution footprint as well. Since most of these pieces of equipment are used only temporarily, we cannot assign the entire footprint to them, so we would consider assigning 5-10% of it to the manufacture and transport of gas rigging and power plant equipment (A).

The equipment arrives in wooden crates on enormous trailers or railroad cars. The different pieces are assembled with the help of other equipment and put to work after creating another set of pollution problems (B).

In addition, there are many other items and small pieces of equipment, tools and consumables—some for personal use, and some for specialized operations. These are shovels, helmets, computers, power drills, boots, shoes, overalls, goggles, first aid kits, chemicals... the list is long. All of these were also made sometime in the past, and transported to the gas site. During their manufacture and transport, they also left a significant pollution footprint. (C).

To complete the pollution footprint picture, we must add the footprint of the people involved in the entire process—engineers and technicians involved in equipment manufacturing, transport, installation, set-up, and operation. These people drive cars, or travel by air, trains and buses, all of which leave another significant set of pollution footprints (D).

So, the total cradle-to-grave pollution footprint of the well drilling and exploitation equipment is expressed by:

$$Pf = A + B + C + D$$

where

Pf is the total pollution footprint

A is the gas rigging equipment manufacturing and transport to the drill site

B is gas production equipment assembly, testing and installation at the drill site

C is the manufacturing and transport of tools and consumables to the new well

D is the pollution footprint of the equipment personnel during the design, setup, drilling, and production phases of the operation.

Combining the pollution footprints of A (equipment manufacturing and transport), B (production equipment assembly, test and installation), C (manufacturing and transport of tools and consumables), and D (pollution footprint of equipment personnel) gives us the environmental footprint of the gas rigging and production equipment before even starting drilling operations.

This pollution footprint is significant, to be sure! It goes deep and wide, stretching from one part of the world to the other.

To put a value on it, we must take inventory of all items delivered to the site, as needed for its design, construction and operation. We must take a close look at the manufacturing of a gas rig, a bulldozer, a dump truck and a number of other pieces of equipment used at the gas well site.

Keep in mind that some of these pieces are quite large, and that it takes many tons of metals to make them.

Each of these pieces started as an iron or aluminum ore dug from a mine, and which was transported to a smelter. The molten metal was then shaped in different forms and shipped to various parts manufacturers, who drilled, welded and otherwise constructed parts for the equipment.

Again, remember that we are talking about huge pieces of metal—some of the gas well pipes are 20-30 feet long, and longer. One single bolt from the gas rig might be over a foot long, and the drill bore might need 5,000 feet of large- and small-diameter tubing for casing.

After the equipment is manufactured, which is usually an energy-intensive and polluting process, these parts are packed and loaded on railroad cars or trucks to be shipped to the assembly plant where they are assembled into major components, or entire vehicles. After one more loading and unloading operation, they finally arrive at the gas well site, or the power plant, for final assembly and testing, before being put to work.

This is a long and winding process with many loading and unloading steps, all of which require a lot of energy and generate a lot of toxic gasses.

Putting a dollar value to all this would be too complex for our purposes, so it suffices to say that the energy use and pollution effects during and after these steps are significant. We will take a closer look at these steps and their environmental effects in the following chapters.

Production Equipment Environmental Impact

Now, we have started the drilling operation and see another set of significant pollution footprints that are somewhat hidden. As an example, at the remote gas rig, a number of large diesel generators provide power for drilling operations and personnel accommodations. During an average work day and under a full load, these will burn 1,000 gallons of diesel...every day. Multiply this number by several thousand such rigs around the U.S. and many more around the world and you get the picture. Or at least part of it...

Several bulldozers, water carriers, and dump trucks are serving the rig for many days and months during the drilling of the gas well. They also burn a lot of fuel in their never-ending runs, so the entire area is fogged with dust and soaked in diesel fumes and lubricating oils, leaving their pollution imprint day after day, hour after hour—for the duration!

Spare parts for equipment periodically arrive from the parts manufacturers, after leaving their footprint around the country and the globe too. This is to be added to the pollution footprint left during the manufacturing of all well drilling and operation equipment.

There are a number of work related activities require the use of different pieces of equipment during the 20-30 years of non-stop operation of the gas well. People driving to and from work in cars and trucks, pumps running day and night, water and chemicals hauled to and from the site in huge trucks, etc.

After most of the gas has been produced, the well is finally declared exhausted. At this point it must be shut down and decommissioned. Decommissioning and land reclamation are another set of major, expensive, and polluting undertakings. The equipment and materials used during the decommissioning and waste disposal process create additional air, soil, and water table pollution which is not to be ignored.

Specialized equipment and personnel (environmental specialists and inspectors) are usually brought in to coordinate and supervise the effort. Some of the equipment and chemicals residue must be disposed of as hazardous waste, so it is loaded on trucks and transported to waste disposal facility. With that, more pollution footprints are left at the area around the well site.

When the final tally is made, a significant part of the emissions and overall pollution are to be attributed to the gas drilling and production equipment, chemicals, and support personnel. This effect is significant and needs to be well understood and included in all calculations of gas well design, drilling, and exploitation.

Summary

Crude oil and natural gas markets have a significantly increasing impact on the economy and on the individuals who rely on the fuel for vehicle fuels, electric generation, manufacturing, heating, cooking and other purposes. Natural gas supplies 25% of the energy used in the U.S., or over 24 trillion cubic feet (Tcf) of natural gas are produced, transferred, and used every year. Crude oil drives the transportation sector and everything that needs to be moved, and without it the economy will go into a standstill mode.

Note: Under the Natural Gas Act (NGA), the Federal Energy Regulatory Commission (FERC) has jurisdiction (only) over the transportation and sale of natural gas and the companies engaged in those activities. Section 1(b) of the NGA exempts production and gathering facilities from FERC jurisdiction.

Also, the Wellhead Decontrol Act of 1989, Pub. L. No. 101-60 (1989); 15 U.S.C. § 3431(b)(1)(A), completely removed federal controls on new natural gas, except sales for resale of domestic gas by interstate pipelines, LDCs or their affiliates. In Order No. 636, FERC required interstate pipelines to separate, or unbundle, their sales of gas from their transportation service, and to provide comparable transportation service to all shippers whether they purchase gas from the pipeline or another gas seller.

This brings a number of questions to mind; all boiling down to how good or bad it is to regulate the energy sector. There is no one answer to this question, so we attempt to provide some answers to these questions in different parts of this text.

The crude oil and natural gas markets are an amalgamation of a number of subsidiary markets. As discussed above, here we also have a physical market and a financial market. In the physical market, crude oil and natural gas are produced, transported, stored and consumed as actual physical products. In the financial market, physical oil and natural gas are bought and sold (on paper) as financial products derived from physical oil and natural gas.

The oil and natural gas markets can also be separated into regions, where the availability and price for natural gas vary with the demand characteristics of the specific market, the region's access to different supply basins, and available pipelines and storage facilities.

Generally speaking, the crude oil and natural gas industry has three segments.

- The supply segment, which includes exploration and development of the natural reserves, and production, which includes drilling, extraction and gas gathering.

- The midstream segment includes small-diameter gathering pipeline systems that transport the gas from the wellhead to natural gas processing facilities. Here the incoming gas is processed to remove impurities and other hydrocarbons, as needed to create pipeline-quality, dry, natural gas.

- The transportation, which includes intrastate and interstate pipeline systems that move natural gas through large-diameter pipelines to storage facilities and a variety of consumers, including power plants, and industrial facilities. The local distribution companies distribute and deliver the gas to retail consumers via smaller diameter pipes and special infrastructure.

Each component of the supply chain is critical in a number of ways. The quantity of reserves usually affect production that can in turn affect market participants' expectations about current and future supply, and thus can affect prices. Similarly, the availability of pipeline and storage capacity determines which supply basins are used and the amount of gas that can be transported from producers to consumers. All of these factors affect the supply chain, but they also affect the supply-demand balance, both nationally and regionally.

Crude oil and natural gas markets in the U.S. are generally divided into the West, Midwest, Gulf Coast, Northeast and Southeast regions. These regions have differing supply, transportation and demand characteristics, resulting in different prices. Within these regions are hubs—the interconnection of two or more pipelines—that also become market hubs for buying and selling oil and gas.

The key hub used by the U.S. natural gas market as a whole is the Henry Hub, in Louisiana. Prices at other locations are frequently shown as "Henry Hub" plus or minus some amount. These regional differences in supply and demand result in different prices for natural gas at various locations.

Prices are lowest in areas with low-cost production, ample infrastructure and limited demand, such as the Opal Hub in Wyoming. Prices are highest where production or transportation is limited and demand is high, as in the Algonquin citygate, in Massachusetts.

In the short-term, demand is usually driven by seasonal and weather factors, economic activity, oil availability, and changing relationships between coal and natural gas prices. Over the long term, crude oil and natural gas use and prices are driven by a number of factors, such as: economic and population growth, environmental policy, energy efficiency, technological changes and prices of the substitute energy sources such as, coal and electricity.

After reviewing the conventional energy sources; their production and use, we have a clear picture of the wide spectrum of activities and different market segments these create. We know about the many products and services that are needed for their proper function, as well as the many products and services they create.

We also met the thousands of people involved in the daily activities that make all this possible. We can clearly see thousands of engineers and technicians searching for new reserves, designing and constructing mines, drilling platforms, and power plants. There are also many others assisting as consultants and contractors, and many other services.

All these people put together are entire army of dedicated professionals, who are busy producing, transporting, and delivering electric power, fuels, products and services non-stop.

Just think of the huge number of activities, which the conventional energy markets need for their proper operation, and the products and services they create in the process:

The key activities are:

- Exploration and location services
- Engineering siting and design services
- Mines and oil/gas wells equipment manufacturing
- Mines and oil/gas wells construction
- Mines and oil/gas wells operation and maintenance
- Transport services for raw materials and finished goods
- Power plants design and construction
- Power plants equipment manufacturing
- Power plants operation and maintenance
- Power grid construction, operation and maintenance
- Electric power and fuels distribution
- Pipelines construction, operation and maintenance
- Fuels distribution and sales
- Environmental services
- Political direction and regulations
- Legal and financial services
- Contracting and consulting services
- Physical and cyber security
- Commodities and futures trading markets

Coal is still the leading power generation fuel, and will remain so for a long time to come. Natural gas is moving up the rankings by taking increasingly larger share in the power generation sector. The battles between coal and natural gas for market domination—on the national and international fronts—will continue for the foreseeable future. Both are expected to do well on both fronts…at least for a while.

Crude oil rules the transportation sector, with the oil market's fluctuations of late marking the beginning of a new era; that of increasing uncertainties and manipulations by the different players. Crude oil has no sizable competitors, so its importance (and price) will continue to drive the energy markets.

Not a pretty picture, but one that can be tolerated if it was not for the other sides of the fossil energy sources; a) environmental pollution and b) eventual depletion. We discussed both of these issues in quite a detail in this section, so there is no point in beating a dead horse, as they say.

We will, therefore, only summarize the subject by reminding the esteemed reader that these issues are not well understood, efficiently discussed, nor sufficiently corrected at the time of this writing. Nor are they expected to be addressed and solved fully any time soon.

As such, they are becoming more and more urgent. Turning our head the other direction has worked thus far, but the train is coming full speed at us, so looking the other way will certainly have very tragic consequences. The conventional fuels have been helping us until now, but now they promise to bring us hardship and tragedy…unless we wake up and do something about it, quickly! If not, the energy markets will become a wreck in the very near future.

The prolonged, artificially low crude oil prices of 2014-2015, which forced shutting down half of the gas drilling sites, confirmed the lack of national energy policy. Instead, gas and all other energy sources depend on international events, most of which are driven by volatile dictators.

What is needed from the U.S. government immediately is:

- Reducing the uncertainty in the energy markets through durable, effective, and enforceable policies and regulations.
- Optimizing energy and emissions costs along the entire energy market value chain, and
- Focusing on standardization, scalability, and diversification of the different energy technologies.

In the gas drilling sector, we need to ask, and get answers to, the following questions:

- What are the best and safest fracking practices?
- What are the chemicals used at each site?
- What are the water usage rules?
- How can pipeline rules and regulations be improved?
- What are optimal strategies for capturing and storing fugitive methane?
- How can LNG and electric fueled vehicles be encouraged?
- What further reforms in the power generation sector should be instituted now and later?
- How do the renewables figure in today's energy markets?
- How much LNG can we export now, and plan to export later, and why?

And finally, there is another extremely important question that is insufficiently and even improperly addressed by the government. It is the fact that the natural resources will not last forever, and we need to prepare for the transition to the upcoming fossil-less economy. This is a complex and expensive task, which involves reducing the use of natural energy resources, increasing the use of renewable energy, and other energy related measures.

Although we see efforts in all these areas, they seem sporadic and almost happening by a chance; usually driven by national or international events. There are no coherent, long term, government policies geared to address this unique, unprecedented, cataclysmic event. If we continue going full speed on the present route of unlimited energy use, we will hit the proverbial fossil-less future like a brick wall. Wow to those who happen to be in the speeding train at the time.

Notes and References

1. EIA, Crude Oil Production. http://www.eia.gov/dnav/pet/pet_crd_crpdn_adc_mbbl_m.htm
2. EIA. Natural Gas Production. http://www.eia.gov/natural-gas/weekly/
3. BP and Oil Production. http://www.bp.com/en_us/bp-us/what-we-do/exploration-and-production.html?utm_source=bing&utm_medium=CPC&utm_term=oil%20production&utm_campaign=NonBranded+-+Exploration+%26+Production+-+Exact
4. The New Oil Order. http://www.goldmansachs.com/our-thinking/pages/the-new-oil-order/index.html?CID=PS_02_11_07_00_00_15_01
5. Barnett Shale News. http://www.energyfromshale.org/
6. World Oil. http://www.worldoil.com/topics/production
7. Coal Power. http://www.power-eng.com/coal.html
8. World Coal Association. http://www.worldcoal.org/coal/uses-coal/coal-electricity
9. Power-Gen International. http://www.power-gen.com/index.html
10. IEA. http://www.iea.org/ciab/papers/power_generation_from_coal.pdf
11. API, http://www.api.org/oil-and-natural-gas-overview/crude-oil-and-product-markets
12. IHS, https://www.ihs.com/products/crude-oil-market-forecast-analysis.html
13. EIA, http://www.eia.gov/finance/markets/financial_markets.cfm
14. *Photovoltaics for Commercial and Utilities Power Generation*, Anco S. Blazev. The Fairmont Press, 2011
15. *Solar Technologies for the 21st Century*, Anco S Blazev. The Fairmont Press, 2013
16. *Power Generation and the Environment*, Anco S. Blazev. The Fairmont Press, 2014
17. Energy Security for the 21st Century, Anco S. Blazev. The Fairmont Press, 2015

Chapter 5

Renewable Energy Markets

*Solar and wind energy are fighting an uphill battle today
on their way to dominate the future global energy markets.*

—Anco Blazev

BACKGROUND

A single paragraph says it all. The fight is on, and all that is needed is a fair and level playing-field, clear rules, and no tricks. A simple concept at a first sight, but a closer look reveals many turns and twists, loaded with prejudice, unfair practices, self-interests, corporate protectionism, and borderline illegal operations—on both sides of the battle lines. In reality, however, there are many problems, created by ignorant, negligent, selfish, and incompetent managers and referees, including government officials and entities.

As if to complicate things further, the conditions and drivers of the global energy markets change from time to time. The results from these changes, often make all previous efforts—good or bad—useless, or at least in need of additional adaptive changes to deal with the new situations. And so, we go stumbling from change to change, one step at a time.

This is precisely the case of the historic development of the U.S. renewable energy markets since the 1970s. The solar and wind industries had several false starts since then. After desperate attempts to revive these industries, usually driven by rising oil prices, they were shoved back in the closet by government strategists and oil corporate interests, as soon as the global oil prices went down.

Starting in 2007-2008, however, the revival effort exceeded all expectations. The inertia provided by huge government subsidies propelled the renewables into high orbit, and there is no stopping, and putting them back in the closet this time. The billions upon billions of dollars poured into the energy sector by the U.S. and EU governments, helped the renewable industry to step on its feet. Unfortunately, it still needs help to start walking before being able to run on its own. This is especially true for the solar energy sector. Subsidies in form of loan guarantees and tax credits still drive the solar projects in the country, and without these solar would stumble again.

A lot of good things were accomplished since 2007, and significant progress was achieved, but a lot of bad situations were exposed as well. In the end, the renewable energy sector survived the turmoil and after a short hesitation during 2012-2013 is headed again in the right direction towards long-term progress. Unfortunately, the long-term success is not guaranteed, since there are several obstacles in the way.

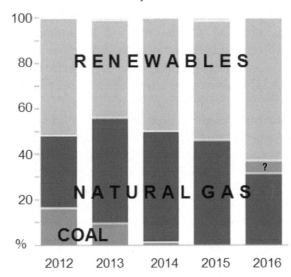

Figure 5-1. U.S. new energy capacity additions to date

Figure 5-1 shows the new phenomena—that of the renewables taking the energy markets by a storm. It is obvious from it that the renewables account for over 50% of all new *energy capacity additions* since 2012. Here we also see that coal is being replaced quickly by natural gas, mostly by retrofitting old coal-fired power plants to gas-fired. This trend is due mostly to the fact that the renewables got a boost during the 2008-2012 boom era, when government subsidies and various programs helped them step on their feet. How long this trend will last depends on a number of factors, the greatest of which is the amount of government support.

Note: The question mark (?) in Figure 5-1 represents (to be) added nuclear power generation capacity

273

in the country. The uncertainty is due to government restrictions and hesitation on part of regulators and investors. Although there are several nuclear plants under construction, their future is not guaranteed. One more Fukushima-like accident could put them, and the entire U.S. nuclear power fleet on hold.

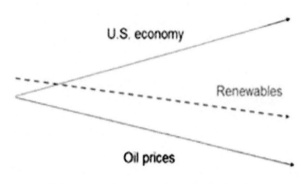

Figure 5-2. The renewables dependencies

Figure 5-2 shows the dependence of renewables on national and global events and developments. The usual scenario is that of decline of the renewables development when oil prices go down. The need for them, which is high during oil price rise, seems to evaporate in thin air when the prices go down.

While the global economy was in decline during the last economic crises, and oil prices went through the roof, huge amount of money was channeled into the renewable energy technologies. This demonstrated one more time that they are an important part of our energy mix. Billions of government dollars were spent on renewable technologies during the solar and wind energy boom of 2008-2012. This helped these industries to reach new highs, that were unimaginable before.

As soon as the economic forecasts improved, however, the government assistance all but dried up. The renewables, solar in particular, is again an orphan that nobody wants. And things were looking even worse, with the approaching December 2016 deadline, threatening to eliminate the vital ITC credit for new solar installations. This would've been a disaster for the solar industry, and would've put it on the back burner again…as it happened so many times before.

Luckily, the U.S. Congress authorized another 5 year extension of the critical ITC support.

Now solar has a chance to achieve unprecedented $0.04-0.05/kWh.

New strong wind is blowing in the U.S. solar's sails now…at least for another 5 years. The only thing

limiting its growth is lack of energy storage for large-scale power generation and use. If and when this limitation is removed, solar might become a major base-load power generator.

Although the renewables represent a large portion of the energy markets additions of late in the U.S. and Europe, this is not the case in most other countries. As a matter of fact, the renewables are in a slight decline now that many governments can no longer afford to subsidize them. China might be an exception here, because there are now ambitious solar and wind projects popping all over the country. Even more ambitious plans are made for the future of the renewables and the entire national power infrastructure.

The *other* renewable energy technologies—hydropower, biofuels, ocean and geo-power—are also affected by the changes of late and their future also depends on a number of factors. Being classified as renewable and green gives them a leg up over the other energy sources, and yet we don't see them growing very fast.

For this, and a number of other good reasons, we refer to them as *other* in this text and review them in the following chapter. In this chapter we take a closer look at the key renewable technologies—solar and wind—and their markets.

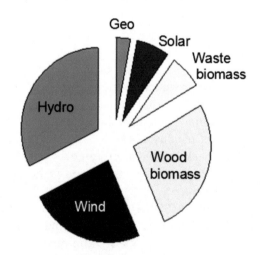

Figure 5-3. The renewables' share

Figure 5-3 shows the size of each renewable technology as a percentage of the total U.S. renewable energy market. Wind and solar are growing fast, but their share is still small. Their potential, however, is unlimited, while the other renewables have, or are close to, reaching their limits.

Solar and wind technologies and their markets can be broken down into the following sub-categories:

Solar and Wind Power
- Production equipment manufacturing
- Solar panels and wind turbines manufacturing
- Power plant exploration and location
- Power plant engineering design
- Power plant construction
- Power plant operation and maintenance
- Power transmission and distribution

We need to emphasize here that the solar and wind power markets include anything and everything related to: the raw materials used for making the equipment (solar panels, wind mills, etc.); different parts and components manufacturing; transportation; power plants design, construction, and O&M; as well as the distribution and sale of electricity coming from the solar and wind power plants. This also includes anything and anybody participating in this long process too.

Here we start with a close look at one of the major renewable technologies; solar power generation.

SOLAR POWER

We allocate significant amount of space to solar and wind technologies in this text, because they are the most promising and feasible chances for ensuring sustainable long-term energy future.

More we understand of the renewable technologies, more chances they have of achieving their full potential of bringing us a bright and sustainable future.

Today solar power can be divided into:
- Thermal solar, which consists mainly of,
  1. flat plate water heaters and
  2. concentrated solar power (CSP), and

- Photovoltaic (PV) solar technologies, which consist of:
  1. Silicon based, and
  2. Thin film based solar technologies

We will start with a quick glance at the solar thermal technologies, since they are viable, but still limited by technical, environmental, and financial reasons. We will then dedicate more space to, and take a much closer look at, the solar PV technologies because we see these as the most promising of all renewable technologies now and on the long run.

SOLAR THERMAL TECHNOLOGIES

Solar thermal equipment uses sunlight to convert its energy directly into heat which is then used for heating, or for electric power generation. The heat can be used for heating homes, or as a heat source in commercial processes. Most often, however, the heat generated by modern is converted into electric power which is then sent into the electric grid.

The conversion of sunlight is fairly straight forward process, while the conversion of heat to electricity is somewhat more complex and requires expensive equipment and large installations to make it cost effective. This conversion is usually done at the so-called "utility scale power plants," using concentrated solar power (CSP) equipment which is the technology of choice today.

There are currently four different types of solar thermal systems.
- The first type, *flat plate water heater* solar thermal energy generator is in its own category, because it generates heat only and is the simplest and cheapest of the bunch. It can be mounted on the roofs of houses and businesses and is used only to heat water (or other liquids) to a moderate temperature. This technology has been successfully used by commercial operations, such as restaurants, laundromats, canning facilities, etc. for several decades.

Smaller size, roof-mounted, parabolic troughs were also popular in the Southwest USA in the 1980s, and are making a slow comeback today, while generally speaking, the major CSP technologies are large, ground-mounted, grid-connected systems.
- The other types of thermal solar systems; *the power tower (central receiver), the parabolic trough tracker, and the Stirling engine-dish* are in the category of concentrated solar power (CSP) technologies, because they do capture the sunlight and concentrate (focus) it onto a receiver. The heat is then normally used to make electricity by several different methods which we will review below as well.

We will review the solar thermal and thermo-electric technologies, focusing on the CSP technologies, since they are most suitable for commercial and utilities power generation.

The solar thermal technologies of today can be described as follow:

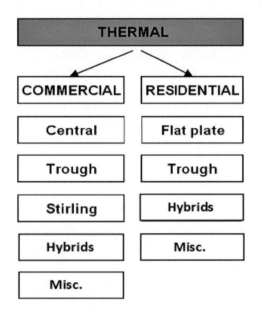

Figure 5-4. Major thermal solar technologies for the 21st century

Flat Plate Water Heater

Flat plate water heating systems have been, and still are, used in residential, commercial and industrial applications, primarily for heating water in laundromats, restaurants, public parks, car washes, and canning and bottling facilities. These heating systems could be used practically anywhere where low temperature hot water is needed during the day. Adding a storage tank could provide water for use during the night and/or cloudy days. In all cases, they are truly "thermal" systems designed to provide hot water.

These are the simplest and cheapest energy conversion devices today, consisting of a frame onto which a heat exchanger plate (or some modification of) is mounted. Water runs through the heat exchanger plate and absorbs the sunlight energy, thus heating the plate and the water (or other heat absorbing liquid) running through it.

The materials, as well as the manufacturing, installation and operation procedures are straightforward and relatively inexpensive. The return on investment (ROI) is one of the highest in the industry, if the systems are properly designed, installed and operated. There are also a number of incentives today which make it even more feasible and desirable to own and operate such a renewable energy system.

CSP Technologies

With the flat plate water heater covered as much as needed for our purposes and filed in the category of thermal (heat) generation for residential and small commercial applications, we will now concentrate on the three major types of thermo-electric systems presently used engine dish, the parabolic trough, and the power tower.

These three technologies have one thing in common: they all use trackers and optics of some type or another to optimize their efficiency. They also require relatively flat land with slopes not exceeding three percent to accommodate the solar collectors. The area of land required depends on the type of plant, but it is about five acres per installed megawatt (MW). It is anticipated that a commercial scale CSP facility of any type, would be in the range of 100 MW or larger and will require in excess of 500 acres plus whatever else is needed for their installation and proper infrastructure.

Unlike solar photovoltaic technologies which use semiconductors to convert sunlight directly into electricity, CSP plants generate electricity by converting sunlight into heat first. Much like a reflective mirror their reflectors focus sunlight onto a receiver. The heat

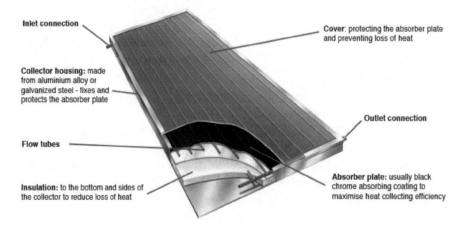

Figure 5-5. Flat plate solar water heater

absorbed by the receiver is used to move an engine piston (Sterling engine), or generate steam that drives a turbine to produce electricity (parabolic troughs and power tower). Power generation after sunset is possible also by storing excess heat in large, insulated tanks filled with molten salt during the day and using it at night. Since CSP plants require high levels of direct solar radiation to operate efficiently, deserts make ideal locations. As a matter of fact, these types of systems cannot operate efficiently in any other environment.

A study by Ausra, a solar energy company based in California, indicates that more than 90% of fossil fuel-generated electricity in the U.S. and the majority of U.S. oil usage for transportation could be eliminated by using solar thermal power plants and will cost less than it would cost to continue importing oil.

90% of U.S. electricity can be generated by about 15,000 square miles of CSP plants in the deserts.

This is approximately 15% of the total area of the state of Nevada, which may sound like a large tract of land. In fact, it is less land per equivalent electrical output than large hydroelectric dams, when flooded and wasted land is included. This is even less than the conventional coal-fired plants when factoring in the land used for mining and waste disposal.

Another study, published in Scientific American, proposes using CSP and PV plants to produce 69% of U.S. electricity and 35% of total U.S. energy including transportation fuels by 2050. While theoretically this is true, and a lot of effort and money was put in CSP technologies, in reality it is impractical mostly because of a number of issues, such as environmental conservation and the need for a lot of cooling water by the CSP power plants.

The deserts, where these power plants are most efficient, have a delicate environment and wild life structure. Developing large areas would harm the structure, so there are limited number of sites available for large-scale development.

Another big problem for the CSP technologies is their need for huge quantity of water for cooling. There is not much water in the deserts, and periodic draughts hit even non-desert areas, such as these in California, so the future of the CSP technologies is uncertain. Nevertheless, the technology is here and ready to go, awaiting solution of all its problems.

The major CSP technologies today are:
- The power tower (central receiver)
- The parabolic trough tracker, and
- The Stirling engine-dish tracker

We will take a close look and discuss each of these below, focusing on their technological advancements and use in large-scale solar installations.

The Solar Power Tower

The power tower (or central receiver) power generation uses methods of collection and concentration of solar power based on a large number of sun-tracking mirrors (heliostats) reflecting the incident sunshine to a receiver (boiler) mounted on the top of a high tower, usually in the middle of the collection field. 80 to 95% of the reflected energy is absorbed into the working fluid which is pumped up the tower and into the receiver. The heated fluid (or steam) returns down the tower and is fed into a thermal electrical power plant, steam turbine, or an industrial process that uses the heat.

The difference between the central receiver concept of collecting solar energy and the trough or dish collectors discussed previously, is that in this case all of the solar energy to be collected in the entire field is transmitted optically to a relatively small central collection region rather than being piped around a field as hot fluid. Because of this central receiver systems are characterized by large power levels (100 to 500 MW) and higher temperatures (540 to 840°C) of the working fluids which allows the creation of high quality superheated steam which is more efficient for electricity generation.

Power tower technology for generating electricity has been demonstrated in the Solar One pilot power plant at Barstow, California, since 1982. This system consists of 1818 heliostats, each with a reflective area of 39.9 m^2 (430 ft2) covering 291,000 m^2 (72 acres) of land. The receiver is located at the top of a 90.8 m (298 ft) high tower and produces steam at 516°C (960°F) at a maximum rate of 42 MW (142 MBtu/h).

The reflecting element of a heliostat is typically a thin, back (second) surface, low-iron glass mirror. This heliostat is composed of several mirror module panels rather than a single large mirror. The thin glass mirrors are supported by a substrate backing to form a slightly concave mirror surface. Individual panels on the heliostat are also canted toward a point on the receiver. The heliostat focal length is approximately equal to the distance from the receiver to the farthest heliostat. Subsequent "tuning" and optimization of the closer mirrors is done upon installation.

Another heliostat design concept, not so widely developed, uses a thin reflective plastic membrane stretched over a hoop. This design must be protected from the weather but requires considerably less expenditure in supports and the mechanical drive mechanism because of its light weight. Membrane renewal and cleaning appear to be important considerations with this design. In all cases, the reflective surface is mounted on a pedestal that permits movement about the azimuth and elevation axis. Movement about each axis is provided by a fractional-horsepower motor through a gearbox drive. These motors receive signals from a central control computer that accurately points the reflective surface normal halfway between the sun and the receiver.

System design and evaluation for a central receiver application is performed in a manner similar to that when other types of collectors are used. Basically, the thermal output of the solar field is found by calculating collection efficiency and multiplying this by the solar irradiance falling on the collector (heliostat) field, minus some optical, transmission and other losses.

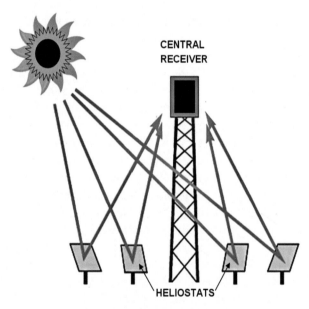

Figure 5-6. Power Tower Field

The major power tower system components are:

Tracking and Positioning

The heliostats must follow the sun all day long in order to focus the sunlight on the tower receiver. This is achieved by means of two electric motor-drive assemblies on each unit. In order to keep parasitic energy low, fractional horsepower motors with high gear rations are used to move the heliostat about its azimuth and elevation axes. This produces a powerful, slow, steady and accurate tracking motion. Under emergency conditions, however, rapid movement of the heliostats to a safe or

stow position is an important design criterion. A typical minimum speed requirement would be that the entire field defocus to less than 3% of the receiver flux in 2 minutes. Higher speed is desired in case of impending disasters, such as high wind, hail and such, in order to protect the mirrors from mechanical damage.

Since it is currently considered best to stow the heliostats face-down during high wind, hail storms, and at night, an acceptable time to travel to this position from any other position would be a maximum of 15 minutes. The requirement for inverted stow is being questioned since it requires that the bottom half of the mirror surface be designed with an open slot so that it can pass through the pedestal. This space reduces not only the reflective surface area for a given overall heliostat dimension, but also the structural rigidity of the mirror rack. However, face-down stow does keep the mirror surface cleaner and safer.

Receivers

The receiver, placed at the top of a tower, is located at a point where reflected energy from the heliostats can be intercepted most efficiently. The receiver absorbs the energy being reflected from the heliostat field and transfers it into a heat transfer fluid. Taking a closer look at the receivers, we see that there are two basic types of receivers, external and cavity receivers.

- *External receivers* normally consist of panels of many small (20-56 mm) vertical tubes welded side-by-side to approximate a cylinder. The bottoms and tops of the vertical tubes are connected to headers that supply heat transfer fluid to the bottom of each tube and collect the heated fluid from the top of the tubes. The tubes are usually made of Incoloy 800 and are coated on the exterior with high-absorption black paint. External receivers typically have a height to diameter ratio of 1:1 to 2:1. The area of the receiver is kept to a minimum to reduce heat loss. The lower limit is determined by the maximum operating temperature of the tubes and hence the heat removal capability of the heat transfer fluid.

- Cavity receivers are an attempt to reduce heat loss from the receiver by placing the flux absorbing surface inside an insulated cavity, thereby reducing the convective heat losses from the absorber. The flux from the heliostat field is reflected through an aperture onto absorbing surfaces forming the walls of the cavity. Typical designs have an aperture area of about one-third to one-half of the internal absorbing surface area.

Cavity receivers are limited to an acceptance angle of 60 to 120 degrees. Therefore, either multiple cavities are placed adjacent to each other, or the heliostat field is limited to the view of the cavity aperture. The aperture size is minimized to reduce convection and radiation losses without blocking out too much of the solar flux arriving at the receiver. The aperture is typically sized to about the same dimensions as the sun's reflected image from the farthest heliostat, giving a spillage of 1-4%.

Heat Transfer Fluids

The choice of the heat transfer fluid to be pumped through the receiver is determined by the application. The primary choice criterion is the maximum operating temperature of the system followed closely by the cost-effectiveness of the system and safety considerations. The heat transfer fluids with the lowest operating temperature capabilities are heat transfer oils.

Both hydrocarbon and synthetic-based oils may be used, but their maximum temperature is around 425°C (797°F). However, their vapor pressure is low at these temperatures, thus allowing their use for thermal energy storage. Below temperatures of about -10°C (14°F), heat must be supplied to make most of these oils flow. Oils have the major drawback of being flammable and thus require special safety systems when used at high temperatures. Heat transfer oils cost about $0.77/kg ($0.35/lb).

Water has been studied for many central receiver applications and is the heat transfer fluid used in many power tower plants. Maximum temperature applications are around 540°C (1000°F) where the pressure must be about 10 MPa (1450 psi) to produce a high boiling temperature. Freeze protection must be provided for ambient temperatures less than 0°C (32°F). The water used in the receiver must be highly deionized in order to prevent scale buildup on the inner walls of the receiver heat transfer surfaces. However, its cost is lower than that of other heat transfer fluids. Use of water as a high-temperature storage medium is difficult because of the high pressures involved.

Nitrate salt mixtures can be used as both a heat transfer fluid and a storage medium at temperatures of up to 565°C (1050°F). However, most mixtures currently being considered freeze at temperatures around 140 to 220°C (285 to 430°F) and thus must be heated when the system is shutdown. These mixtures have good storage potential because of their high volumetric heat capacity. The cost of nitrate salt mixtures is around $0.33/kg ($0.15/lb), making them an attractive heat transfer fluid candidate.

Liquid sodium can also be used as both a heat transfer fluid and storage medium, with a maximum operating temperature of 600°C (1112°F). Because sodium is liquid at this temperature, its vapor pressure is low. However, it solidifies at 980°C (208°F), thereby requiring heating on shutdown. The cost of sodium-based systems is higher than the nitrate salt systems since sodium costs about $0.88/kg ($0.40/lb).

For high-temperature applications such as Brayton cycles, the use of air or helium as the heat transfer fluid and operating temperatures of around 850°C (1560°F) at 12 atm. pressure are being proposed. Although the cost of these gases would be low, they cannot be used for storage and require very large diameter piping and expensive compressors to transport them through the system.

The Parabolic Trough Technology

Parabolic trough solar systems consist of frame in a parabolic trough shape in which glass, metal or plastic reflectors are mounted to focus the sun's energy onto a receiver pipe running above and in parallel to the trough's length. The receiver pipe, or heat collection element (HCE), is centered at the focal point of the reflectors and is heated by the reflected sunlight to very high temperatures. Liquid of some sort is pumped through the receiver pipe and is heated in the process.

The HCE of the parabolic trough units is usually composed of a metal pipe with a glass tube surrounding it, and with the space between these evacuated to provide low thermal losses from the pipe. The pipe is coated with a material that improves the absorption of solar energy. Several improvements have been made or are underway to improve performance, the most significant of which is the seal between the glass and the pipe, which seal has not been as reliable as desired and development of better seal materials/seal configuration is still underway.

Parabolic troughs can focus the sunlight many times its normal intensity on the receiver pipe, where heat transfer fluid (HTF—usually mineral or synthetic oil) flowing through the pipe is heated. This heated fluid is then used to generate steam which powers a turbine that drives an electric generator. The collectors are aligned on an east-west axis and the trough is rotated north-south, following the sun as needed to maximize the sun's energy input to the receiver tube.

Parabolic trough power plants, also called solar electric generating systems (SEGS), represent the most mature CSP technology, with the most installed capacity of all CSP technologies. The first SEGS solar trough

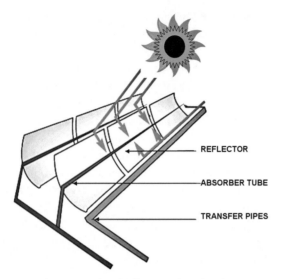

Figure 5-7. Parabolic trough technology

plant started operating in 1984, with the last one coming on line in 1991. Altogether, nine such plants were built, SEGS I–VII at Kramer Junction and VIII and IX at Harper Lake and Barstow respectively. In February 2005, all but two (I and II) of the Kramer Junction SEGS plants were acquired by FPL Energy and Carlyle/Riverstone and are still operating.

A natural gas system added to the plant "hybridizes" it and contributes up to 25% of the output. This feature also allows operation later at night or on cloudy days to meet the requirements of the grid. FPL now runs these systems, making it the largest solar power generator in the United States. All of the power generated from the SEGS projects is sold to Southern California Edison under long-term contracts negotiated by Luz back in the 1980s. There are a number of such plants operating around the world and many others are planned.

One big advantage of CSP systems is their ability to generate power after the sun has gone down. In these cases, the HTF fluid going through the receiver pipe is routed through a thermal storage system which permits the plant to keep operating for several hours after sunset while the electrical demand is still relatively high. The thermal storage system consists of a "hot" storage tank equipped with heat exchanger where HTF circulates and gives up a portion of its heat to heat the storage solution in the tank during the day.

At night, the hot storage solution flows through the same heat exchanger heating up HTF which is sent to the steam turbines for generating power. The cooled-down storage solution flows from the heat exchanger to a "cold" storage tank where it stays until daytime when it is reheated and returned to the "hot" storage tank.

And the cycle is repeated every night.

The greatest limitation of the use of this technology is the need of huge quantity of fresh water for cooling the steam. Since water is the gold of the deserts, and there is not much of it, this CSP technology is faced with a big problem.

Linear Fresnel Reflector

Linear Fresnel reflectors (LFR) systems are similar to the parabolic trough, but use an array of nearly flat Fresnel reflectors instead. These reflectors concentrate solar radiation onto elevated inverted linear receivers. Water, or other liquid, flows through the receivers and is converted into steam. This system is also line-concentrating with the advantages of low costs for structural support and reflectors, fixed fluid joints, a receiver separated from the reflector system and long focal lengths that allow the use of flat mirrors.

The LFR technology is seen as a potentially lower-cost alternative to trough technology for the production of solar process heat. Planned commercial applications are estimated at a size from 50 to 200 MW. Linear Fresnel applications are mostly at the experimental stage for now. Companies working in the field claim higher efficiency and lower costs per kWh than its direct competitor, parabolic trough, due to high density of mirrors. Fresnel mirror is available at little more than EUR 7.00 per m². According to Ausra, this technology can generate electricity for EUR 0.10 per kWh now and under EUR 0.08 per kWh within next 3 years.

The Fraunhofer Institute has contributed greatly in making the key components such as the absorber pipe, the secondary reflectors, primary reflector array and their control ready for operation. Based on theoretical investigations and the specific conditions found in sunny climates, Fraunhofer researchers have calculated that the electricity production costs will not rise above EUR 0.12 per kWh.

The linear Fresnel CSP technology derives its name from a type of optical system that uses a multiplicity of small flat optical faces, invented by the French physicist Augustin-Jean Fresnel who, while Commissioner for Lighthouses, invented the segmented lighthouse lens. Flat moving reflectors follow the path of the sun and reflect its radiation to the fixed pipe receivers above. Molten salt or other operating liquid powers a steam turbine, or is stored for night use. The technology itself is simple; the biggest challenge is setting mirrors to track the sun and reflect rays effectively. Flat mirrors are much cheaper to produce than parabolic ones, so this is a bonus.

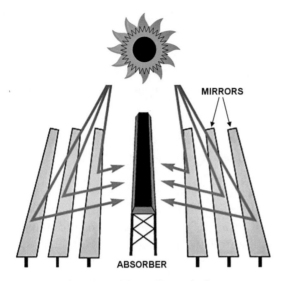

Figure 5-8. Linear Fresnel plant

Another advantage of the compact linear Fresnel reflector CLFR is that it allows for a greater density of reflectors in the array. In addition, Fresnel technology is less sensitive to wind loads and allows parallel land use to a large extent.

The LFR technology is more competitive economically due to:
• More effective land use than rival technologies;
• Low visual impact on landscape;
• Lower infrastructure costs due to its design;
• Lighter base, less steel used, flat instead of curved mirrors.

The Stirling Engine

One of the most elegant and flexible solar thermal power conversion technologies today is the Stirling engine dish system. It consists of mirrors mounted on a frame which is continuously tracking the sun all through the day. The mirrors focus the reflected sunlight onto the receiver of the Stirling engine mechanism which is activated by the heat and turns on a shaft, connected to the rotor of an electric generator similar to that of the alternator of your car. The generator rotor turns with the engine shaft and generates electric power while the sun is shining and the receiver is hot enough to activate the engine and rotate the shaft.

A Stirling engine system is actually a solar electricity generator because the heat produced by the mirrors attached to it is converted into electric energy on the spot so small installations of a few units are possible; something that is just not practical with the other CSP technologies. The Stirling engine needs cooling just like a car engine for more efficient operation and in order to

cool the engine walls, bearings and other moving parts.

The mirror, or mirrors, are mounted on a metal frame which is driven by two motor-gear assemblies (x-y drives), programmed to move the frame in such a fashion that it follows the sun's movement precisely all day long, thus providing accurate focusing of the sunlight onto the heating plate of the Stirling engine. When the plate gets hot enough, the air in one of the cylinders in it is compressed and forces the piston in it to move up. This action forces the piston in the other cylinder (which is simultaneously cooled) to move down. Eventually, the compression in the second cylinder increases to the point that its piston is forced to go back, thus forcing the piston in the first cylinder to assume its initial position. The cycle repeats over and over while there is enough heat to maintain the process.

The Stirling engine function, under ideal operating conditions, can be represented by four cycles, or thermodynamic process segments, of interaction between the working gases, the heat exchanger, pistons and the cylinder walls.

The Stirling engine is a very ingenious and efficient piston engine, without the noise and exhaust of internal combustion engines. As a matter of fact, it can be classified as an "external combustion" engine. The gasses inside the cylinders are not exhausted, so there is no pollution and there are few moving parts with very little noise, so it can be used virtually anywhere.

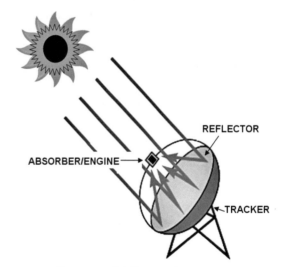

Figure 5-9. Stirling engine system

Since its invention in 1816 by the Scottish inventor Dr. Robert Stirling, the Stirling engine has been considered and proposed for use in many different applications. Presently it is used in some specialized applications, where quiet and clean (no exhaust) operation is required. Some fancy and special purpose submarines and vessels use Stirling engines part of the time under special conditions.

Unfortunately, mass-market application for the Stirling engine is not found as yet, although many scientists and inventors are working on it. Its use in solar power generating equipment might be a good start in that direction. There are currently a number of installations using this technology, but most of them are smaller, demo type systems.

No large CSP power plant using Stirling engine technology is in operation today, to our knowledge, and as a matter of fact two of the largest (and maybe only) Stirling engine power plants planned for installation in the California deserts were cancelled in 2011 for a number of reasons, The major problems for these failures were related to technological complications, as well as the unproven reliability, and high cost of the technology.

In conclusion, although solar thermal technologies are very efficient and cheap, their future is limited, because:

• The flat plate solar heaters will continue to expand around the globe, but in limited quantities, serving single families and small businesses,

• The Stirling engine technology was shelved recently, due to technological, reliability, and cost issues.

• The other CSP technologies—power tower and parabolic trough—will also be a part of the energy mix of the future but in reduced quantity, due mostly to the fact that they require large quantities of water for cooling. But water is just not available in the deserts—where these technologies are most efficient—and periodic draughts make the situation even worse. Because of that, we see reduction in the present and future progress of these, otherwise capable, technologies.

And now, we will be looking into the most promising, photovoltaic (PV), solar energy technology. The major PV technologies are, a) silicon PV based, and b) thin films based.

Let's start with a close look at the silicon (Si) solar technologies.

SILICON PV TECHNOLOGIES

The silicon based photovoltaic (PV) technology is a major—and the largest—branch of the global solar PV industry. It deals with the design, manufacturing and use of photovoltaic devices made out of silicon,

and the related materials and processes. Silicon (Si), or crystalline silicon (c-Si) PV technology is quite different from the above discussed CSP solar technologies. Here, obtaining heat is not the goal, and whatever heat is generated is wasted. It is even harmful to the system and its performance.

The goal of PV devices is direct and efficient conversion of solar energy into electricity.

The greater advantage of PV systems is their flexibility, since they can be used in many applications; in different location, and implemented in different sizes and configurations. There are many different types of PV based systems, with limitless feasibility for use in different applications. There is no doubt now that the PV technologies are most promising on the long run too—as far as efficiency, reliability and cost are concerned—and will continue to grow with time.

Silicon PV devices depend on the ability of silicon semiconductor materials to capture and convert sunlight into electricity.

A large number of variations and combinations of materials and properties are in use today, so we will start from the very beginning of the cradle-to-grave lifetime cycle of the silicon devices.

The life cycle of all silicon based PV devices basically consists of those items shown in Table 5-1.

We will start with the production of the silicon raw materials.

SG Silicon

Silicon PV devices are made out of solar grade silicon (SG Si), which is materials of special qualities, made specifically for PV device manufacturing. SG silicon can be further divided into *single crystal and multi-crystal silicon,* as we will see further in this text.

In general terms, solar grade (SG) silicon is a refined version of sand. The raw material—silica sand—is found in large quantities in nature in the form of special sand, or silicon oxide ($SiO2$). Most sand, however, contains so much dirt (like the sand in the deserts), or rocks like beach sand, that it is useless for all practical purposes.

Silicon is environmentally friendly material, although the waste products of the SG production process create some problems in the local areas. SG can be easily melted, shaped and formed into mono-crystalline or poly-crystalline ingots and wafers. Devices made from

Table 5-1. Life-cycle of PV devices

- Silicon raw material production
  Mining silicate sands
  Melting the sands into metallurgical grade (MG) silicon,
  Refining MG silicon into solar grade (SG) silicon, and
  Sorting and testing the final SG silicon material

- Solar wafers manufacturing
  Mono- and poly-crystalline silicon ingots production
  Slicing the ingots into wafers,
  Testing and sorting, and
  Preparing the wafers for processing into solar cells

- Solar cells manufacturing
  Wafer surface preparation
  p-n junction creation
  Metallization
  Testing and sorting, and

- Solar modules manufacturing
  Layout of components
  Interconnecting the solar cells,
  Lamination, and
  Final testing and sorting
  PV modules transport

- Solar (PV) arrays installation
  PV field design
  PV modules transport to the field
  Field preparation and installation
  PV power plant certification
  PV power plant O&M

silicon can operate at up to 125°C air temperatures (with some loss of efficiency), which allows their use in some harsh environments, although there are some exceptions and restrictions.

The purity of silicon depends on its use, and is usually qualified as follow:
- Metallurgical silicon, 98%
- Semiconductor grade, 99.9999999% (nine nines), and
- Solar grade, 99.9999% (six nines) pure silicon, or better, is used in today's solar wafers and cells manufacturing operations. Today, leading manufacturers provide 99.99999% or 99.999999% (seven or eight nines) purity.

The actual number of 9s refers to silicon's quality, which is extremely important, and usually determines the final quality, performance, and longevity of the end product (solar cells and modules).

Since it is impossible to change the initial silicon material quality at a later stage, it is critical that a high quality production starts with high quality (with minimum possible impurities) silicon material—seven or eight nines is best.

The SG Production Process

The basic process for making SG silicon as needed in the manufacturing of c-Si PV cells is as follows:
- High purity silica sand found in large quantities in several areas around the world is dug out, sifted and transported for further processing. This process is very similar to the coal surface mining operations we reviewed in the previous chapters, so we won't spend much time on it.
- Mountains of sand are loaded in huge furnaces, where it is melted for removal of the oxygen in the silica molecule through a reaction with carbon (coal or charcoal) added to the mix while heating it to 1500-2000°C in huge electrode-arc type furnaces. The goal here is to remove the oxygen from the silicon dioxide molecule, in order to obtain pure and simple silicon molecules.

$$SiO_2 + C \rightarrow Si + CO_2$$
$$SiO_2 + 2C \rightarrow Si + 2CO$$

- The resulting, molten metallurgical grade (MG) Si material is usually treated further for impurities removal by the addition of different additives:

$$SiO_2 + 2SiC + additives \rightarrow 3Si + 2CO$$

Note that CO_2 and CO gasses are emitted during these processes. Then imagine mountains of sand being melted in the presence of coal and coal-like materials. Day and night the huge furnaces are fed sand and coal, which are melted by equally huge amounts of electric power or natural gas burners. In either case, significant quantity of CO and CO_2 are emitted in the atmosphere.

So, yes, it is unarguable that solar technologies are cleaner than the rest.

> *The energy used, and pollution emitted, at different stages of the Si PV devices life cycle are significant.*

We must not close our eyes at this fact, but should take it into consideration when discussing the benefits of the solar technologies.

The final silicon material coming out of the furnaces is metallurgical grade (MG) silicon, still containing a lot of impurities—well over 2-3%—and as such it cannot be used for solar cell manufacturing. At this state it is good for use as an additive in the metallurgical industry only. Additional purification must take place if it is to be used for solar cells, and even more purification is needed if it is to be used in the production of semiconductor devices.

After solidification, the chunks of MG silicon are crushed and shipped to a processing plant for refining as needed to bring its quality to that needed for manufacturing efficient and reliable solar cells.

• At the refining facility, the MG silicon chunks are reacted with hydrochloric acid (HCl) to make silicon containing gases:

$$Si + 4HCl \rightarrow SiCl4 + 2H2$$

And/or

$$Si + 3HCl \rightarrow SiHCl3 + H2$$

The process chemicals go through the distillation and purification processes in order to obtain and isolate the main silicon containing gas SiHCl3 trichlorosilane (TCS) which is then purified and sent for preparation of solar grade (SG) silicon material.

The TCS gas is processed in plasma reactors and solidified in the shape of large ingots of SG multi-silicon (poly) as follow:

$$SiCl4 + 2Zn \rightarrow Si + 2ZnCl2 \text{ (DuPont process)}$$

or

$$2SiHCl3 \rightarrow Si + 2HCl + SiCl3 \text{ (Siemens process)}$$

or

$$4HSiCl3 \rightarrow 3SiCl4 + SiH4, \text{ where } SiH4 \rightarrow$$
$$Si + 2H2 \text{ (REC process)}$$

The final product from the refining process is chunks of solar grade (SG) silicon, which are much purer than the MG silicon, but are still not good enough for making solar cells.

• At this point the chunks are crushed, packed and shipped to be melted into high purity doped ingots from which solar wafers will be sliced and processed into solar cells.

SG Silicon Quality

The quality of the SG grade silicon material quality is extremely important, because it will determine the final quality of the solar cells made from it. So basically speaking, SG silicon has to be pure enough, and the type and quantity of each impurity must be known and controlled, because excess impurities of some types will invariably result in lower efficiency and potentially shorter life span of the cells made from it. This in turn will shorten the life of the PV modules in the field, so a deep understanding of the differences and strict adhesion to quality control procedures by manufacturers is paramount for ensuring the quality of the final product.

On the other hand, however, the best materials cost too much, so a balance must be struck, where quality is good enough at the best price possible. Semiconductor silicon, for example is orders of magnitude purer and more controlled than solar grade. It is also much more expensive.

Table 5-2 offers a close look at the types and amounts of impurities contained in semiconductor grade silicon (left column), vs. those in MG silicon (right column), and reveals huge differences, which translate into drastic differences in quality as well. Semiconductor grade Si is several times purer than MG Si, which means that a lot of energy, time and effort must be spent on complex processes, in order to purify it to this degree.

Table 5-2. Quality of different silicon types

Chemical Impurity	Semiconductor Grade Si (ppm)	MG-Si (ppm)
B	4.157	14.548
C	14.264	107.565
O	17.554	66.706
Mg	<0.001	8.204
Al	<0.005	520.458
Si	Matrix	Matrix
P	6.801	21.762
S	<0.044	0.096
K	<0.007	<0.036
Ca	<0.007	44.849
Ti	<0.001	47.526
V	<0.001	143.345
Cr	<0.001	19.985
Mn	<0.001	19.938
Fe	<0.005	553.211
Co	<0.002	0.763
Ni	<0.002	22.012
Cu	<0.001	1.724
Zn	<0.002	0.077
As	<0.002	0.007
Sr	<0.0003	0.353
Zr	<0.0003	2.063
Mo	<0.001	0.790
I	<0.0002	<0.001
Ba	<0.0002	0.266
W	<0.0003	0.024

Solar grade (SG) c-Si could also benefit from such high purity as that of the semiconductor grade silicon, but the benefits do not justify the steep price increase. Because of that, SG Si is somewhat less pure than its semiconductor grade cousin and is somewhere between the MG and semiconductor grade silicon quality in Table 5-2.

Some of the impurities play a much larger role in the final quality of solar cells and modules, so when we say that 6 nines SG silicon is suitable for solar cell manufacturing we must also qualify the actual types and amounts of the different impurities; i.e., if the amount of some harmful metal impurities is too high, the final product might be of questionable quality as well.

Once the SG silicon has passed the final tests it is ready to be melted and shaped into ingots, and then sliced into mono- or poly-crystalline wafers suitable for making silicon solar cells.

SG Silicon Wafers

Thus produced SG silicon is in large chunks that are useless as is. The chunks are then loaded into another furnace, melted, solidified, and sliced into thin wafers. The properties and manufacturing process of solar grade wafers, which are used for making solar cells can be divided into two major groups:

Mono-crystalline and multi-crystalline (poly) silicon solar wafers.

The major types of silicon solar wafers are mono-crystalline, also called single crystalline, and multi-crystalline, also called poly silicon wafers. Figure 5-10a is a graphic representation of silicon material processed by "pulling" as a single crystal (mono-crystalline) material via the so-called Czochralski process, while 5-10,b is silicon wafer produced by melting silicon chunks in a crucible, thus the asymmetry and imperfections in the wafer can be easily seen.

Mono-crystalline silicon wafers

Mono-crystalline silicon (also called mono-silicon, single-crystal, mc-Si, or sc-Si) wafers are produced by melting purified solar grade (SG) silicon in special furnaces, called *ingot pullers*, where a long, cylindrical silicon ingot is "pulled" up from the melt as a single crystal silicon. This process of mono-silicon production is known as the Czochralski (CZ) method. The CZ ingot, 4, 5, or 6" in diameter, is sliced into thin wafers which are then processed into solar cells.

This, and similar mono-silicon ingot production methods (like float zone, or FZ) are sophisticated and expensive production processes, so the single crystal wafers and cells tend to be more efficient and more ex-

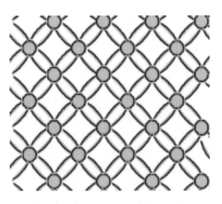

a) Mono-crystalline silicon wafer

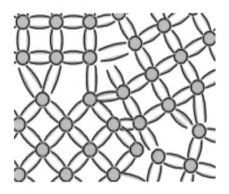

b) Multi-crystalline silicon wafer

Figure 5-10. Silicon wafers

pensive than their close relative, the polycrystalline solar wafers and solar cells. The efficiency increase is due to the uniform structure of the single crystal materials which facilitates the photoelectric effect and provides a good level of reliability and longevity.

mc-Si wafers are cut from cylindrical ingots and then squared, a process which creates substantial waste of silicon material. Their uneven shape also leaves unused gaps in the PV modules' surface which lowers their overall efficiency.

Like most other types of solar cells these also suffer from a reduction in output at temperatures over 25°C. A significant drop of output can be expected, especially at very high (desert) temperatures, but which drop is actually much lower than that of poly-crystalline silicon cells and some other types of solar cells. For this and other reasons we consider mono-crystalline silicon cells to be the most efficient and reliable PV devices available today.

Mono-crystalline solar panels are first-generation solar technology and have been around a long time, providing evidence of their durability and longevity. Several of the early mc-Si PV modules installed in the 1970s are still producing electricity today, albeit at reduced output levels, and some have even withstood the rigors of space travel.

Note: The mono-silicon materials and processes are well understood, since they have been used for over half a century in the semiconductor industry.

Solar wafers, cells and modules manufacturers benefit tremendously from the wealth of knowledge and practical experience of the semiconductor manufacturers and in some cases work together with them to advance the state-of-the-art. This is a great advantage over most other types of solar cells and modules which have to be developed without specialized outside help.

Multi-crystalline silicon (poly) wafers

Multi-crystalline silicon (poly) material is produced by melting solar grade (SG) silicon in a large cube-like crucible. The solidified cube of poly material is split into 4-, 5-, or 6-inchwide and 18- to 36-inch-long bars which are then sliced into thin wafers. Since it is not "pulled" as a single-crystal but rather as randomly mixed columns, the resulting material and the wafers sliced from it, look non-uniform with visible boundaries in the bulk which play a significant role in the solar cells' performance.

Large randomly oriented grains (in asymmetric rows and columns) crisscross the silicon bulk giving the poly wafers their distinguished and esthetically pleasing multi-color look. This segmentation, on the other hand, significantly reduces the efficiency of the solar cells made from it. The segmentation (grains) can be thought of as separate pieces of single crystal silicon, scattered randomly in the bulk and separated from each other by boundaries.

These are actual physical and electrical barriers which create obstacles in the operation of these types of solar cells. The boundaries prevent electrons and holes from freely moving across, and contribute to the recombination process, which minimizes the photoelectric effect, thus reducing the efficiency of the solar cells.

The grains are usually so large that they extend all through the wafers when cut from the solidified silicon block. The incorporation of hydrogen during device processing plays an important role in passivating the inter-grain boundaries which contributes to improving the efficiency and operating stability of the solar cells. Other techniques are also used to boost the effi-

ciency and reliability of poly-silicon solar cells, but they increase the cost of the final product, so their use is limited according to the specific needs.

The grain boundaries create a number of effects manifested as sub-grain boundaries, slip deformations, and twinning. These are basically additional lower level boundaries in the form of single dislocations or web of dislocations with different orientation, Burger's vectors and similar abnormalities, all of which complicate the photoelectric process, reduce the efficiency of the solar cells, and contribute to latent (delayed) failure mechanisms.

Advantages of using multi-crystalline growth over the Czochralski (single crystal silicon) method include:
- Lower capital costs,
- Higher throughput,
- Less sensitivity to the quality of the silicon feedstock used, and
- Higher packing density of cells to make a module and less material waste, due to the precise square shape of the material prior to final wafering and processing.

Silicon Solar Cells Manufacturing

Silicon solar wafers are made out of solar grade (SG) silicon, via melting the chunks and "pulling" or "casting" the molten material, thus creating mono, or poly Si materials respectively. The resulting material is cooled down and sliced into very thin, 200-300 microns (0.007"-.0.012") slices (wafers). These wafers are then processed into solar cells.

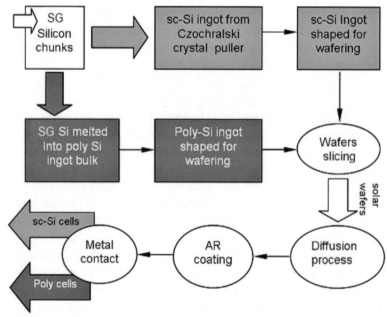

Figure 5-11. Si solar cells manufacturing process

The solar cells manufacturing process sequence below was used by Alpha Solarco, Inc., during the late 1990s at their solar cells manufacturing facility in Phoenix, Arizona. Variations of this process are used presently.

Multi-crystalline Silicon Wafers Manufacturing

The multi-crystalline (poly, or mc-Si) wafers are easier and cheaper to manufacture, and as such represent the majority of solar cells on the global markets.

The poly-Si wafers manufacturing sequence is as follows:

Sorting, inspection and cleaning the SG silicon

Silicon chunks, purchased directly from the manufacturers or indirectly from dealers, are received, documented, inspected, sorted by size and tested for resistivity, metal and other contaminants.

Bulk silicon cleaning and surface prep

The sorted Si chunks which pass the initial QC procedures are sent into the wet chemistry room for cleaning with detergents and etching with acids. This is needed in order to remove all dirt, oils and impurities from the material's surface. The chunks are then rinsed, dried and taken to the wafer process room for melting in the poly Si ingot HEM (heat exchange method) furnace.

Prepare Ingot Crucible

The HEM furnace uses a special crucible to melt the silicon in. Before loading the crucible, its inner surface is spray coated with a proprietary solution. This eliminates possible chemical reactions between the silicon material and the silica crucible in the HEM furnace at high temperatures. After application of the coating, the crucible is baked in a kiln for 24 hours to dry the sprayed solution. It is then ready for loading in the HEM furnace.

Load HEM Furnace

Assemble graphite support plates around the sides of the crucible before loading it into the furnace. Load crucible by adding 80 KG of silicon meltstock and calculated amount of dopant (Boron salt). After loading, clean exterior of furnace, control module, and area adjacent to furnace from any residue or overspray.

Grow Ingot

Apply vacuum to furnace and backfill with Argon gas. Re-apply vacuum to ensure that all moisture has been eliminated inside the furnace chamber. The control instrumentation on the HEM furnace is designed to automatically control the ingot growth process. No manual intervention is allowed during the entire process, except in emergency.

Cool and Remove Ingot

Ingot is annealed prior to cool down. Annealing is achieved by maintaining a furnace temperature slightly below the melting temperature of silicon. Stress relief of the ingot is achieved over a 12-hour period. Ingot is then removed from the crucible and ready for cutting. Cutting is done in two steps, block sawing and wafer slicing.

Block Sawing

After growing an ingot, its surfaces must be shaved square using a special ingot saw. The smooth squared block of silicon is then sectioned into bars that measure 10 cm x 10 cm or as desired and variable length. Some additional proprietary processing of the ingot sides follow. It is intended to reduce some of the surface damage of the ingot and the resulting wafers.

Wafer Slicing

Wire saws are used to slice the wafers from the bars. Each bar is attached to a glass sheet with a proprietary epoxy adhesive which is dried a minimum of 24 hours. This glass/silicon bar is then attached to a metal work piece (holder) and placed under the cutting wires of the wire saw. The cutting oil and abrasive mix must be prepared and properly mixed in the mixer of the slurry tank at the bottom of the wire saw. Wafers are sliced to the desired thickness and removed from the wire saw with the holder.

Pre-cleaning

After slicing, the wafers are separated from the holder and loaded into cassettes. They are then taken to a special degreasing area for gross removal of surface contamination, prior to cleaning. Cutting oils and abrasive residues left on the wafers' surfaces from the slicing process are properly and thoroughly removed via proprietary cleaning formulations. The wafers are stored in a water tank in order to keep them wet.

Wafers Uniformity Test

Some wafers are taken out of the batch, cleaned thoroughly and tested at five different locations on the surface of each wafer to ensure that the desired degree of thickness uniformity and resistivity have been achieved by the slicing process. Additional inspection and testing procedures are carried out as needed to ensure the quality of the wafers prior to processing into solar cells.

Wafer Storage

The wafers are loaded in cassettes for ease of handling during the rest of the processing sequence. The cassettes loaded with wafers are stored in water tanks where they are kept until taken for final cleaning.

Special instrumentation and procedures are used at each process step to ensure proper processing and to check tight adherence to the specs.

A special deionized (DI) water generating station is used to make DI water for rinsing the wafers. Proper rinsing of the silicon material (chunks) and the resulting wafers is accomplished, leaving the wafers with clean surfaces.

The critical steps of the solar cells wafer and cells manufacturing process are executed in a specially designed area, called a "clean room." It is built and operated to keep any environmental disturbances such as wind, humidity, heat and cold, particles, etc. isolated from the work environment. Special HEPA filters deliver purified and deionized air, and this way the solar cells are processed under near-ideal conditions and without any outside contamination.

Mono-crystalline Silicon Wafers

The mono-crystalline (or single crystal) silicon wafers manufacturing sequence is similar to the above, with the addition of several complex and expensive steps. The major difference is the use of a different type of ingot furnace, where the equipment and processes are quite different, as follow:

- Clean the ingot puller furnace and process crucible.
- Fill crucible with silicon material and add exact amount of doping chemical. The crucible surface tends to dissolve so it has to be made of pure silica and kept at maximum cleanliness possible. The process equipment and the execution of each step must also be kept up to spec as well to avoid any excess contamination.
- Melt the silicon chunks in the crucible and keep the temperature close to the melting point. Convection in the silicon melt is suppressed by correct application of magnetic fields around the crucible. The strength and duration of these fields is a proprietary know-how.
- Insert the silicon seed crystal into the melt and stabilize the temperature
- Start withdrawing the seed crystal slowly by rotating the seed crystal and the crucible. This process step is proprietary for each operation, but basically, fast withdrawal during the initial stages is needed

to reduce the diameter of the growing crystal a few mm. This is needed to ensure that most seed-crystal induced dislocations will be removed.

- The withdrawal rate must be decreased now in order to increase the ingot diameter slightly above the desired size. The impurity concentration (including dopants and oxygen) in the melt will change due to segregation, and the resulting crystal properties will change too, usually by increasing impurities concentration from top to bottom of the ingot. The temperature profile will vary at this stage and has to be adjusted. The homogeneity of the crystal depends on the correct temperature and speed of rotation and withdrawal regimes. These parameters are also proprietary and not much actual information can be found for free.
- When almost all silicon has left the crucible, the ingot growth is complete. Some molten silicon must be left in the crucible, because it is where the impurities are concentrated due to their low segregation coefficients.
- Pull the ingot out, but beware because the thermal shock introduces temperature gradients which cause stress gradient, and dislocations are nucleated into the crystal. The new dislocations will interact with previous dislocations, causing serious damage. The withdrawal rate at this point is critical and is usually gradually increased. This leads to a reduced diameter in cone-like shape of the tail end of the ingot.
- Remove the ingot slowly and place it on a suitable clean surface for cooling.
- Shape the ingot on a special lathe to give it more uniform surface and bring it to the exact outside diameter—4, 5 or 6" diameter.
- Inspect and test the ingot for mechanical defects and chemical impurities.
- Clean and slice it into wafers of the desired thickness.
- Inspect, test, sort and pack the wafers.

The resulting mono-crystalline silicon wafers are used for in-house solar cells manufacturing, or are shipped to solar cells and modules manufacturing facilities around the world.

The best quality PV cells made out of mono-crystalline silicon have 5-8% higher efficiency than those made of multi-crystalline silicon, but cost approximately 15-30% more to produce.

Silicon Solar Cells

Solar cells have been mass manufactured since the 1950s, starting with 1.0" diameter silicon wafers and 2-3% efficiencies. The energy crisis of the 1970s shifted the attention of the US government and public to PV technologies which allowed the quick development of more efficient solar cells and modules, and the related production equipment and process. Still, the progress was seriously hindered by technological issues such as the small size of the silicon wafers, lack of adequate processing equipment and unresolved issues at the different process steps.

Today there are almost unlimited combinations and permutations of different materials, equipment options, processing sequences and techniques which allow cost-effective and efficient manufacturing of solar cells and modules. The manufacturers keep most of their process specifics secret, but the overall equipment configuration, process scheme, and different procedures are unchanged since the 1970s.

The manufacturing processes for mono- and multi-silicon solar cells are virtually identical.

The different manufacturers have, of course, made modifications to the base process outlined below, as needed to reduce cost and increase the efficiency of their products.

Solar Cells Manufacturing Process Considerations

Keeping all functional and operational parameters of the solar cells in mind, design engineers must consider a large number of process and application factors in order to come up with a practical, efficient, reliable and profitable final product with long lifetime.

Some of the main considerations are as follow:

- The solar cell's material is of primary consideration. Its electro-mechanical and chemical properties, impurities, availability, price, ease and cost of processing, etc. factors are thoroughly evaluated and compared with competing materials and technologies. Starting with high quality silicon, and considering all process materials and related issues is the first and most important job of the design team.
- Actual solar cell structure and process sequence design considerations include:
  - Evaluating the quality of cleaning, texturing, doping, edge etch chemicals, gasses, and related equipment and process steps specs
  - Optical losses calculations and AR coating type and process specs
  - Selection of materials and processes for metallization of the front and back contacts
  - Quality control specs and inspection procedures for all process steps
  - In-process and final tests and sorting procedures selection
- The actual solar cells and modules manufacturing process is done best by using the most appropriate combination of materials, process equipment, chemicals and other consumables. Adequate knowledge and understanding of the cradle-to-grave process steps is paramount.

Silicon Solar Cell Manufacturing

The basic Si solar processing sequence, as used in the late 1990s by an associate company is outlined below. This process, or a variation of it, is used by most world class Si PV cells and modules manufacturers today.

The major process steps are as follow:

Wafers Inspection and Sorting

Wafers are placed on inspection tables and are inspected visually and with optical equipment. Any wafers with visible mechanical defects are rejected. The wafers are then tested with a 4-point probe, and are sorted according to their resistivity. Wafers, or wafer samples are sent to an outside lab for metal and organic contamination analysis. Results from these tests determine the level of quality of the finished cells.

Wafers Cleaning and Etch

The wafers that pass all initial inspections and tests go to the wet cleaning line and are chemically processed in special chemicals where they are cleaned and etched to remove damage and oxide formed on the surface. The wafers are then rinsed with de-ionized (DI) water and dried via spin dryer.

Surface Etch (Chemical Etch)

This process is used only on single-crystal silicon with 1-0-0 orientation. (Polycrystalline wafers cannot be textured, because the different strings of silicon have different orientation and the resulting surface is only partially and unevenly textured, if at all.) A controlled chemical solution (composition, concentration, temperature and time) etches the pyramid-like structures in the wafer surface and the surface takes on a dark-gray appearance. The pyramids blend into each other and block excess light reflection. Each pyramid is approximately 4-10 microns high. This step has critical process parameters. The wafers are then rinsed, spin-dried, and stored in special containers for processing.

Note: In a variation of this process, wafers are loaded in a fixture two-by-two with the backs of each pair touching, so that the pyramid structure is formed only on their front surfaces.

Diffusion for P-N Junction Formation

The clean wafers are oven dried, placed in the diffusion furnace at 900-950°C in reactive carrier gasses to impregnate the wafers. POCl3 gas is used for n-type diffusion, which diffuses phosphorous atoms in the wafer's surfaces. This creates a p-n junction in the lightly boron-doped wafers.

Note: This method is easier to control and has more uniform distribution of dopant than using the spray-on diffusion liquid and belt furnace diffusion process used by many companies today. This step has critical process parameters, so any compromise will be reflected in the cells' overall performance and longevity. In a variation of this process, wafers are loaded in a fixture two-by-two with the backs of each pair touching, so that the diffusion layer is formed only, or mostly, on their front surfaces. This facilitates the processing of the back surface later on.

Mass production solar cell operations use a different method in which the wafers are sprayed with a dopant chemical and run through a conveyor belt type furnace, where the dopant is diffused in the wafers' surface. Both methods have advantages and disadvantages.

Plasma Etch for Removal of Edge Layer (diffusion etc.)

The diffusion process implants P dopant in the wafers' side edges, causing an electrical short circuit between the top and bottom (negative and positive) surfaces of the cell, so it is necessary to remove the dopant with a wet chemistry or plasma etch. Wafers are coin-stacked and etched for a brief period in an RF plasma etch reactor; only the edges of the wafer are exposed to the plasma which removes the diffusion coating.

Wafers are then etched gently in a bath of dilute hydrofluoric acid to remove any oxides formed during the plasma etch step, rinsed with DI water, and finally spin dried.

Anti-reflective Coating

AR coating is deposited on the front surface of the wafers. The purpose of the AR coating is to reducing the amount of sunlight reflected from the finished cell surface. The AR coating is deposited via chemical vapor deposition (CVD) or by spraying the chemicals on the wafers and then baking. Both methods achieve similar outcome of enhancing the solar cells' output and giving them the distinctive dark blue color (for poly solar cells).

Note: Different manufacturers deposit and fire the AR coating using different process parameters and sequence order. This is an important process, nevertheless, so its proper design and execution will determine the final, most important aesthetic and performance aspects of the cells.

Printing (Metallization)

Several screen printing steps are used to apply the metallization on the front and back of the wafers. First, silver paste is printed on the top surface which then becomes the front metal pattern (top contacts, or fingers). The paste is dried and the wafers are flipped for printing the back surface with aluminum paste. After drying, silver paste is printed in special slots in the dried aluminum and the wafers are then transported into the firing furnace.

Metal Firing

Thus metalized on both sides, wafers are run slowly through an IR-heated furnace where the metal pastes on the top and bottom sides of the wafers diffuse into the substrate, to make an electric contact with the p-n junction and the back surface. This step has critical process parameters.

Note: The firing of the front contacts is a very delicate process, where time and temperature are controlled to achieve the desired depth of penetration of the metal into the silicon surface. The depth of penetration determines the electro-mechanical properties of the finished cell. Specially designed automated printing-firing equipment is available for more precise and consistent process control.

Inspection and Quality Control

Solar wafers and cells are inspected and tested at several stages of the process sequence. This is done by eye inspection, using magnification and other instrumentation. Electrical tests are also performed at some steps of the process. The final inspection is the most important step and must be performed by well trained and experienced operators.

Cell Flash Testing and Sorting

A certain percent of the completed wafers are placed on a test stand in the solar simulator and are illuminated for a period of time. I/V curve is generated for each cell and the output data are used to sort the cells into groups according to the I/V curve characteristics, prior to soldering and lamination into modules.

Cell Storage

The cells are finally loaded into cassettes, or coin-stacked (with protective material in between) and packed for ease of handling, transportation, or storage prior to laminating into solar modules or shipping to another location.

Si Solar Cells Structure and Function

Crystalline silicon (c-Si) solar cells are relatively simple in design at first glance, but their physical and chemical composition, structure, and electrical properties are quite complex.

We will take a closer look at the c-Si solar cell's structure in order to get a good understanding of the different aspects of its function. We will also review the major issues and failure mechanisms related to its manufacturing, installation and operation.

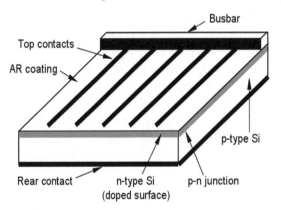

Figure 5-12. Silicon solar cell

The p-n Junction

The p-n junction is the engine of the silicon solar cells. It is an integral part of the solar cell operation, since it is where the photoelectric process starts and where the electric current is generated, so let's take a close look at it:

Figure 5-12 shows a cross section of a solar cell with its top surface (N-type region) on the right-hand side and the bottom (P-type region) on the left.

The p-n junction is in the middle and represents an electro-chemical boundary between the P and N regions. The N region and the resulting p-n junction are created during the diffusion process by doping the N side (usually the top of the wafer) with phosphorous or other such element, thus saturating a very shallow area (less than a micron) below the top surface of the cell with phosphorous atoms. These atoms have excess electrons,

which are loosely attached and are readily knocked out to facilitate the electric power generation, if and when energized by sunlight.

The p-n junction is located at the border line between the phosphorus saturated area and the original silicon bulk (P type region) which was very lightly doped with boron (or other similar chemical atoms) during the final silicon ingot melting process. The P region will provide the holes (+ charge) when the electron-hole pairs are broken by the incoming sunlight and the electrons extracted during the photoelectric effect.

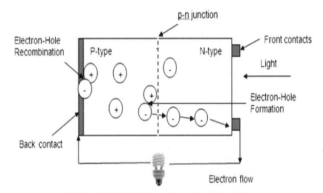

Figure 5-13. The p-n junction function

When at rest, the p-n junction and the areas around it are static with no meaningful activities in or across it. When photons from sunlight with proper energy levels impinge onto the solar cell surface they penetrate into the cell material and impact onto the electrons (in the electron-hole pairs) thus transferring energy to them and breaking the electron-hole pairs.

This creates free electrons (-) which start moving around, while the holes (+) from the corresponding electron-hole pairs are mostly left in their original place. Thin areas on both sides of the p-n junction, called the "depletion" region effectively separate the holes from the electrons which are forced towards the N-type region and are finally extracted from the cell as electric current.

The constant movement of electrons back and forth across the p-n junction and through the layers is quite complex, but the final and practical result of it is the creation of DC electric current which flows in an outside circuit attached to the cell. A large part of the electrons do leave the cell through the metal fingers and bars on the N-type side (usually on top of the solar cell), and if we close the circuit, electric current will flow through it. The electrons in the outside circuit will re-enter the cell via the back side metallization on the P-type side of the cell. Here they will recombine with excess holes in the region and the process repeats indefinitely, or at least while there is enough sunlight to keep it going.

The electron flow provides the current (I) in Amperes and the cell's electric field creates a potential, or voltage (V) in Volts. With both current and voltage above zero we have power (P) in Watts, which is a product of the two and which is what we use to define the PV cells and modules power output.

$$P_{Watts} = I_{Amperes} * V_{Volts}$$

When an external load (such as an electric bulb or a battery) is connected between the front and back contacts of the cell, DC electricity flows through the cell and the external circuit, and powers the load connected to the closed external circuit.

Note: Remember that the actual electric current of the external circuit flows in the opposite direction of the electron flow.

Power Generation

Solar cells are characterized by the power they generate, which is at maximum when the current and voltage are at their maximum levels. This can happen only when the cell is fully operational (no defects), and receives maximum solar insolation.

*Pmax = Imax * Vmax*

With these terms at zero, the conditions

V = Voc/I = 0 and V = 0/I = Isc

also represent zero power.

A combination of maximum current and maximum voltage maximizes the generated power and is called the "maximum power point" (MPP).

*MPP = Imax * Vmax*

So the MPP of a solar cell with 3.0 A current and 0.5 V voltage would be 1.5 W. A 100 pc. PV module made out of these solar cells connected in series would generate 150 Wp under full solar insolation of 1000 W/m².

The solar conversion efficiency η of a PV cell or module is used most commonly to express and compare performance.

The efficiency is given by:

η = Voc*Isc*FF/P

where:

Voc is the open circuit voltage (the voltage generated when the load resistance is infinite, or there is no resistance), Isc is the short circuit current (the current generated when the load resistance is zero, FF is the fill-factor, calculated as follows:

$$FF\% = \frac{\text{Imax * Vmax (actual measurements)}}{\text{MPP (maximum obtainable power)}}$$

This is simply the ratio of the actual measurements (Voc and Isc generated by the cell under the specific testing conditions) divided by the maximum power the cell can generate. In other words, the efficiency is basically the ratio of the amount of power a solar cell or module could produce vs. the total amount of power contained in the incoming sunlight and how efficiently the cell converts it into electric power.

So, if a PV module with 1.0 m² active surface area is rated at 15% efficiency which is average for c-Si PV modules, we can quickly deduct that it theoretically could produce 150 Watts DC power under 1000 W/m² solar insolation (its maximum power). So if it generates 50 V and 3 A then its FF will be 1. If it produces 50 V and only 2 A (or 100 W), then its FF will be (50 * 2)/150 W, or FF = 0.67.

It's a mouthful of concepts, but a good simplified description of the major practical effects of solar power generation which are the foundation of solar cells' and modules' ability to convert sunlight into useful DC power.

Table 5-3. Typical silicon solar cell specifications

Material:	Mono-Crystalline Silicon
Size:	4" diameter
Voltage:	0.55V
Current:	0.275-0.33 A
Voltage (load):	0.484V
Current (load):	0.250-0.275 A
Light Level:	1 Full Sun, based on STC (standard test conditions)
STC:	1,000 W/m², 25°C temp.
Heat Effect:	0.5% efficiency drop/°C)
Degradation:	1% drop of output per year

Note that:

• The voltage output is a constant 0.5-06V per unit, regardless of the type and size of the silicon solar cells.

• The amperage will vary depending on the type

and size of the cell. I.e., mono-Si solar cells have 5-10% higher amperage output.

- Elevated cell temperature affect all solar cells (not just silicon) and cause drop in power output according to the type of the cell. I.e., poly-crystalline-Si solar cells exhibit higher drop of efficiency as compared to their first cousins; mono-crystalline Si solar cells.
- All solar cells (not just silicon) lose efficiency with time due to material degradation. I.e., poly-crystalline solar cells are less reliable and do tend to degrade more and faster than the mono-crystalline type..

Solar Cell Types

We looked at silicon material and wafer types above, so we just have to remind the reader that the category "crystalline silicon," c-silicon, or c-Si is the general designation of all types of crystalline silicon-based products, including c-Si wafers, solar cells and PV modules.

CRYSTALLINE Si	THIN FILMS	OTHER
Mono-Crystalline	Amorphous Si	GaAs
Multi-Crystalline	Epitaxial Si	InP
Micro-Crystalline	CdTe	Other III-V
Super c-Si	CIGS	Germanium
Si Ribbon	Organic/Polymer	CPV Cells

Figure 5-14. Types of solar cells

The major types of silicon solar cells therefore are as follow:

Mono-crystalline Solar Cells

Single crystal silicon, also called mono-crystalline, mono-silicon, or mono-Si, or mc-Si, or sc-Si is a type of silicon that was grown by the very special and expensive Czochralski (or CZ) method, or via the float zone (FZ) method. Both methods use similar equipment and production methods and end up with the best, most efficient, and much superior silicon material for semiconductor and solar cells device manufacturing.

Solar cells and panels made out of CZ or FZ silicon material have the highest efficiency and longevity of all silicon based PV devices, primarily due to the uniform, stable and predictable nature of the bulk material.

Multi-crystalline Silicon PV Cells

Multi-crystalline silicon is the most widely used silicon material. It is most often called "poly," "poly-silicon," or "polycrystalline" silicon (which is what we will call it in this text too) because it consists of many (poly) strings instead of one single crystal.

Poly is made by melting and casting silicon chunks into large blocks, splitting the blocks into smaller rectangular blocks and slicing these into thin, square-shaped wafers.

Poly-Crystalline Silicon Solar Cells

Poly crystalline (note the difference between poly- and multi-crystalline) silicon, which is also called poly silicon, or poly, or pc-Si, is actually a thin film of silicon, deposited via CVD, or LPCVD processes on semiconductor type wafers, to be used as a gate material in MOSFET transistors and CMOS microchips. The solar industry uses similar equipment and processes to deposit very thin layers of silicon (pc-Si and a-Si) onto polysilicon or other substrates. The resulting devices are of lower efficiency, as compared to sc-Si or mc-Si.

Note: There is a confusion created by the term "poly" as it is used widely to identify PV cells modules made out of multi-crystalline silicon, instead of its actual use in the semiconductor thin film. Since we cannot change the decades-long use of the term "poly" in the solar industry to identify multi-crystalline silicon products, we will continue using it too with a certain degree of caution and with due clarification when needed.

Amorphous Silicon Solar Cells

Amorphous silicon is also thin film silicon, alpha silicon, or a-Si is used in p-i-n type solar cells. Typical a-Si modules include front side glass, TCO film, thin film silicon, back contact, polyvinyl butyral (PVB) encapsulant and back side glass. a-Si has been used to power calculators for some time now, mostly because it is easily and cheaply deposited on any substrate.

Micro-Crystalline Silicon Solar Cells

Micro crystalline silicon, also called nano-crystalline silicon, uc-Si, or nc-Si, is a form of silicon in its allotropic form, very similar to a-Si. nc-Si has small grains of crystalline silicon within the amorphous phase, and if grown properly can have higher electron mobility due to the presence of the silicon crystallites. It also shows increased absorption in the red and infrared wavelengths.

Super Mono-crystalline Silicon

This is a new purer type of silicon with more perfect crystalline structure, which exhibits reduced

294 Global Energy Market Trends

phonon-phonon and phonon-electron interactions. This phenomenon increases certain transport properties, resulting in 60% better room temperature thermal conductivity than natural silicon.

In this text we focus on mono- and poly-crystalline silicon solar cells and modules, their structure, function and the related properties and issues. This is because the bulk of the production today, and that planned for the future, involves these types of solar cells.

Silicon PV Modules

A PV module consists of a number of interconnected solar cells (typically 36 to 72 connected in series for battery charging), and many more for large-scale applications. Individual solar cells are soldered in strings and encapsulated into a single, hopefully long-lasting unit simply because PV modules cannot be disassembled for repairs. The main purpose for encapsulating a set of electrically connected solar cells into a module is to protect them from the harsh environment in which they are going to operate.

Solar cells are relatively thin and fragile and are prone to mechanical damage due to vibration or impact unless well protected. In addition, the metal grid on the top and bottom surfaces of the solar cells, the wires interconnecting the individual solar cells, as well as the soldered junctions can be corroded by moisture or water vapor entering the module, if the protecting materials are damaged or absent. So the encapsulation: a) provides a manageable package that can be installed in the field, b) prevents mechanical damage to the solar cells, and c) prevents water or water vapor from penetrating the module and corroding the electrical contacts and junctions.

Many different types of PV modules exist, and the module structure is often different for different types of solar cells or for different applications. For example, amorphous silicon, and other thin film solar cells are often encapsulated in a flexible array, while crystalline silicon solar cells are usually mounted in rigid metal frames with a glass front surface. Module lifetimes and warranties on bulk silicon PV modules are often 20-25 years, which assumes robust and durable encapsulation of the PV modules. Failing encapsulation will cause quick performance degradation and failure with time.

Most PV bulk silicon PV modules consist of:
• Glass top surface, which protects the top of the module,
• Top encapsulant, which protects the top side of the solar cells,
• String of PV cells, which provide the DC power,

• Back encapsulant, which protects the back side of the solar cells,
• Back cover, which protects the back of the module
• Metal frame, which is wrapped around the outer edge of the modules to protect the sides of the module.

In most modules, the top surface is glass, but plastics are used in some cases. The top encapsulants is usually EVA (ethyl vinyl acetate), while the back encapsulant layer is usually Tedlar, or a number of similar plastic and thermo-plastic materials.

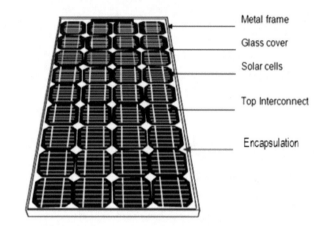

Figure 5-15. Standard silicon PV module, top view

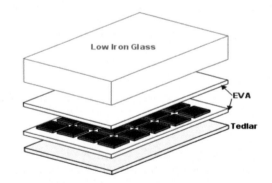

Figure 5-16. Standard silicon PV Module, side view

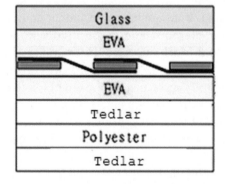

Figure 5-17. PV Modules cross section

Typical module components are:

Cover Glass

The cover of the front surface of a PV module is usually glass with high transmission in the wavelengths which can be used by the solar cells, usually in the range of 350 nm to 1200 nm. In addition, the reflection from the front surface should be low too. While theoretically the reflection could be reduced by applying an anti-reflection coating to the glass surface, in practice these coatings are not robust enough to withstand some of the conditions in which most PV systems are used.

An alternative technique used to reduce reflection is to "roughen," or texture the top glass surface. In this case dust and dirt are very likely to adhere to the top surface and less likely to be dislodged by wind or rain. These glass surfaces are not "self-cleaning" and the advantages of reduced reflection are quickly outweighed by losses incurred due to increased top surface soiling. Texturing the inside of the glass is also practiced by some manufacturers, but there are some disadvantages in doing this as well, so the proper glass has to selected according to the module type and designation.

In addition, the top surface should have good safety properties and good impact resistance, should be stable under prolonged UV exposure and should have a low thermal resistivity. There are several choices for a top surface material including acrylic, polymers and glass. Tempered, low iron-content glass is most commonly used as it is low cost, strong, stable, highly transparent, mostly impervious to water and gases, and has good self-cleaning properties. This type of glass is the most stable and trouble-free component of the entire module assembly. It doesn't deteriorate easily regardless of the harshness of the elements, and unless is broken, it will withstand the test of time for 25-30 years largely unaffected. Once the glass is broken, however, the module must be removed or put on a special maintenance schedule.

Encapsulant

The top encapsulant is a transparent, plastic, material used to provide adhesion between the solar cells, the top surface and the rear surface of the PV module. The encapsulant should remain chemically stable at elevated temperatures and high UV exposure. It should also be optically transparent and should have a low thermal resistance.

EVA (ethyl vinyl acetate) is the most commonly used encapsulant material. EVA comes in thin sheets which are inserted between the solar cells and the top surface and the rear surface. This sandwich is then heated to 150°C to polymerize the EVA and permanently bond the module together. Thus formed structure cannot be disassembled without destroying the bond, and with that encapsulated components.

EVA is responsible for protecting the cells from moisture and reactive species entering the module. Long exposure to UV and IR radiation tends to damage the EVA and it becomes yellow, which reduces its transmittance and reduces the module efficiency. Cracks and pores created in it under long exposure will allow the elements to enter the module and destroy the cells.

The back side encapsulant can be made of different materials that provide good mechanical, and electrical resistance, as needed to keep water out and prevent short circuiting the cells' interconnects by touching the back cover. EVA/Tedlar layers are placed under the cell strings in some modules to provide the needed protection.

Cell Strings

A number of solar cells are interconnected and sealed (laminated) between plastic materials, which insulate them from each other, from the interconnecting wires and from the elements. The cells can be arranged and wired in a number of ways, one of which is shown in Figure 5-18.

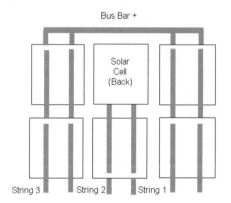

Figure 5-18. Cell stringing

Thus generated DC power is extracted from the module via two wires protruding from the module and routed into a junction (or terminal) box which is fitted with connectors for quick interconnect within the other array components.

Back Cover

Most c-Si modules have a thin sheet of aluminum for back cover, which is screwed into the frame, with the terminal box attached to it. A key characteristic of the back cover is that it must have low thermal resistance and that it must prevent the ingress of water or water vapor. In most modules, a thin polymer sheet, typically Tedlar, is used as the rear surface, which provides electrical and environmental protection for the solar cells.

Some PV modules, known as bifacial modules, are designed to accept light from either the front or the rear of the solar cell. In bifacial modules both the front and the rear must be optically transparent, so glass is the preferred material for these.

Note: The c-Si module's standard configuration is solid and strong enough to withstand transportation bumps, handling, and high winds during operation. Nevertheless, it is not intended to be used for support, or to be stepped on as some installers do. Even if the glass doesn't break in such case, the weight and impact puts enough stress on the cells to cause micro-cracks and other interruptions, which eventually grow into much bigger problems.

Side Frame

A final structural component of the module is the edging or framing, which provides additional mechanical strength and isolation from the elements. Module edges are sealed by the encapsulant layers in it and by additional adhesive materials for additional protection against the elements. An aluminum frame is then fastened around the edges of the module. The frame structure should be free of projections or pockets which could trap water, dust or other foreign matter.

Some PV modules are sold without side frames, as needed to reduce weight and cost, but this leaves the sides unprotected and exposed to moisture penetration with time. When that happens, the affected modules would fail.

PV Module Design Considerations

There are a large number of variables to keep in mind when designing or evaluating PV modules. Some of these are listed below:

Cell Packing Density

The packing density of solar cells in a PV module refers to the area of the module that is covered with solar cells compared to that which is blank (between the cells). The packing density affects the output power of the module as well as its operating temperature. The packing density depends on the shape of the solar cells used. For example, single crystalline solar cells are round or semi-square, while multi-crystalline silicon wafers are usually square. Therefore, if single-crystalline solar cells are not cut squarely, the packing density of the resulting module will be lower than that of a multi-crystalline module, because of excess wasted space between the cells.

Sparsely packed cells in a module with a white rear surface can also provide marginal increases in output via the "zero depth concentrator" effect. Some of the light striking regions of the module between cells and cell contacts is scattered and channeled to active regions of the module.

Power Output

While the voltage from the PV module is determined mostly by the number of solar cells, the current from the module depends primarily on the size of the solar cells and also on their efficiency. At AM 1.5 and under optimum tilt conditions, the current density from a commercial solar cell is approximately between 30 to 36 mA/cm^2. Single-crystal solar cells are often 100cm^2, giving a total current of about 3.5 A from each cell.

Poly-crystalline silicon modules have comparatively larger size individual solar cells but a lower current density. However, there is a large variation in the size of poly-crystalline silicon solar cells, and therefore the current will vary. Current from a module is not affected by temperature in the same way that voltage is, but instead depends heavily on the tilt angle of the module and the sunlight intensity reaching its surface.

If all the solar cells in a module have identical electrical characteristics, and they all experience the same insolation and temperature, then all the cells will be operating at exactly the same current and voltage. In this case, the IV curve of the PV module has the same shape as that of the individual cells, except that the voltage and current are increased proportionally to the number of cells in the module. If one single cell in the module, however, has different electrical characteristics (i.e. higher resistivity) then the entire module is affected, and will most likely underperform down to the level of the failing cell.

The different cell might also overheat and fail under the increased load, thus making the entire module fail. And because not all cells are made equal and do not perform exactly the same, they should be tested and sorted before stringing and encapsulating into finished PV modules.

This, however, is a tricky operation, because cells tested under "normal" or "standard" test conditions in-house, often perform totally differently in the field. This anomaly could be caused by difference in materials quality and/or improper process execution. In all cases, the established manufacturers who have long experience with solar cells, testing and field applications, have proven ways to sort and assemble the cells in order to eliminate or reduce this type of field problem.

In all cases, proper design calculations and tests must be executed in order to come up with an efficient cell and module design.

The equation for the module power under normal operating conditions used to evaluate the different cells and modules is:

$$I_T = M * I_L - M * I_0 \left[esp \left(\frac{q \frac{Vt}{N}}{nkT} \right) - 1 \right]$$

where:

- I_T is the total current from the circuit;
- N   is the number of cells in series;
- M   is the number of cells in parallel;
- V_T is the total voltage from the circuit;
- I_0 is the saturation current from a single solar cell;
- I_L is the short-circuit current from a single solar cell;
- n   is the ideality factor of a single solar cell; and
- q, k,   and T are respective constants

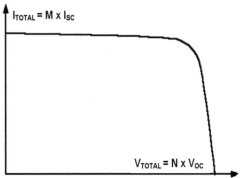

Figure 5-19. I-V curve for N cells in series x M cells in parallel.

This formula can be used also to predict the behavior of different cells in modules in the field. The situation changes drastically under not-so-normal field conditions, such as operation under high desert sunlight. Although there are formulas to calculate the temperature effect (as discussed in more detail in this writing), experience shows that elevated temperatures and humidity create havoc through individual cells affecting them differently and causing them to behave differently over time. This in turn results in unpredictable behaviors of the affected modules, and these anomalies often lead to reduced power output and short lifetime.

Mismatch of Series Connected Cells

As most PV modules are series-connected, series mismatches are the most common type of mismatch encountered. Of the two simplest types of mismatch considered (mismatch in short-circuit current or in open-circuit voltage), a mismatch in the short-circuit current is more common, as it can easily be caused by shading part of the module. This type of mismatch is also the most severe.

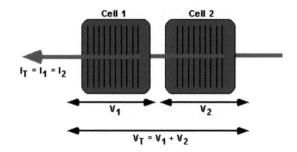

Figure 5-20. Cells in series

For two cells connected in series, the current through the two cells is the same. The total voltage produced is the sum of the individual cell voltages. Since the current must be the same, a mismatch in current means that the total current from the configuration is equal to the lowest current.

Open circuit voltage mismatch

A mismatch in the open-circuit voltage of series-connected cells is a relatively benign form of mismatch. So, at short circuit current, the overall current from the PV module is unaffected. At the maximum power point, the overall power is reduced because the poor cell is generating less power. As the two cells are connected in series, the current through the two solar cells is the same, and the overall voltage is found by adding the two voltages at a particular current.

Short-Circuit Current Mismatch

A mismatch in the short-circuit current of series connected solar cells can, depending on the operating point of the module and the degree of mismatch, have a drastic impact on the PV module. As shown in Figure 5-20, at open-circuit, the impact of a reduced short-circuit current is relatively minor. There is a minor change in the open-circuit voltage due to the logarithmic dependence of open-circuit voltage on short-circuit current. However, as the current through the two cells must be the same, the overall current from the combination cannot exceed that of the poor cell.

Therefore, the current from the combination cannot exceed the short-circuit current of the poor cell. At low voltages where this condition is likely to occur, the extra current-generating capability of the good cells is not dissipated in each individual cell (as would normally occur at short circuit), but instead is dissipated in the poor cell.

Overall, in a series-connected configuration with current mismatch, severe power reductions are experienced if the poor cell produces less current than the maximum power current of the good cells. Also, if the combination is operated at short circuit or low voltages, the high power dissipation in the poor cell can cause irreversible damage to the module.

Hot-spot Heating

This condition occurs when there is one low current solar cell in a string of at least several high short-circuit current solar cells, as shown in Figure 5-21.

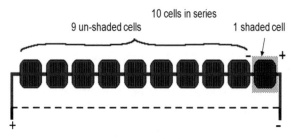

Figure 5-21. Cell shading

If the terminals of the module are connected (module Isc), the power from all unshaded cells is dissipated across the shaded cell, and might result in overheating.

One shaded cell in a string reduces the current through the good cells, causing the good cells to produce higher voltages that can often reverse bias the bad cell. If the operating current of the overall series string approaches the short-circuit current of the "bad" cell, the overall current becomes limited by the bad cell. The extra current produced by the good cells then forward biases the good solar cells. If the series string is short circuited, then the forward bias across all of these cells reverses the bias of the shaded cell. Hot-spot heating occurs when a large number of series connected cells cause a large reverse bias across the shaded cell, leading to large dissipation of power in the poor cell.

Essentially the entire generating capacity of all the good cells is dissipated in the poor cell. The enormous power dissipation occurring in a small area results in local overheating, or "hot-spots," which in turn leads to destructive effects, such as cell or glass cracking, melting of solder or degradation of the solar cell.

Bypass Diodes

The destructive effects of hot-spot heating may be circumvented through the use of a bypass diode. A bypass diode is connected in parallel, but with opposite polarity, to a solar cell. Under normal operation, each solar cell will be forward biased and therefore the bypass diode will be reverse biased and will effectively be an open circuit. However, if a solar cell is reverse biased due to a mismatch in short-circuit current between several series connected cells, then the bypass diode conducts, thereby allowing the current from the good solar cells to flow in the external circuit rather than forward biasing each good cell.

The maximum reverse bias across the poor cell is reduced to about a single diode drop, thus limiting the current and preventing hot-spot heating. In practice, however, one bypass diode per solar cell is generally too expensive and instead bypass diodes are usually placed across groups of solar cells. The voltage across the shaded or low-current solar cell is equal to the forward bias voltage of the other series cells which share the same bypass diode plus the voltage of the bypass diode.

The voltage across the unshaded solar cells depends on the degree of shading on the low-current cell. For example, if the cell is completely shaded, then the unshaded solar cells will be forward biased by their short circuit current and the voltage will be about 0.6V. If the poor cell is only partially shaded, some of the current from the good cells can flow through the circuit, and the remainder is used to forward bias each solar cell junction, causing a lower forward bias voltage across each cell.

The maximum power dissipation in the shaded cell is approximately equal to the generating capability of all cells in the group. The maximum group size per diode, without causing damage, is about 15 cells/bypass diode, for silicon cells. For a normal 36-cell module, therefore, 2 bypass diodes are used to ensure the module will not be vulnerable to "hot-spot" damage.

Within Module Heat Generation

A PV module exposed to sunlight generates heat as well as electricity. For a typical commercial PV module operating at its maximum power point, only 10 to 15% of the incident sunlight is converted into electricity, with much of the remainder being converted into heat.

Factors which affect the heating of the module are:

a.   Reflection from the top surface.

Light reflected from the front surface of the module does not contribute to the electrical power generated. Such light is considered an electrical loss mechanism which needs to be minimized. Neither does reflected light contribute to heating of the PV module. The maximum temperature rise of the module is therefore calculated as the incident power multiplied by the reflection. For typical PV modules with a glass top surface, the reflected light contains about 4% of the incident energy.

b.   Electrical operating point.

The operating point and efficiency of the solar cell determine the fraction of the light absorbed by the solar cell that is converted into electricity. If the solar cell is operating at short-circuit current or at open-circuit voltage, then it is generating no electricity and hence all the power absorbed by the solar cell is converted into heat.

c.   Absorption of sunlight in areas not covered by solar cells.

The amount of light absorbed by the parts of the module other than the solar cells will also contribute to the heating of the module. How much light is absorbed and how much is reflected is determined by the color and material of the rear backing layer of the module.

d.   Absorption of low-energy (infrared) light.

The amount of light absorbed by the parts of the module other than the solar cells will also contribute to the heating of the module. How much light is absorbed and how much is reflected is determined by the color and material of the rear backing layer of the module.

e.   Packing density of the solar cells.

Solar cells are specifically designed to be efficient absorbers of solar radiation. The cells will generate significant amounts of heat, usually higher than the module encapsulation and rear backing layer. Therefore, a higher packing factor of solar cells increases the generated heat per unit area.

Nominal Operating Cell Temperature (NOCT)

A PV module will generally be rated at 25°C under 1 kW/m^2. However, when operating in the field, they typically operate at higher temperatures and at somewhat lower insolation conditions. To determine the power output of the solar cell, it is important to determine the expected operating temperature of the PV module.

The nominal operating cell temperature (NOCT) is defined as the temperature reached by open circuited cells in a module under the conditions listed below:
• Irradiance on cell surface = 800 W/m^2
• Air temperature = 20°C

• Wind velocity = 1 m/s
• Mounting = open back side.

The equations for solar radiation and temperature difference between the module and air show that both conduction and convective losses are linear with incident solar insolation for a given wind speed, provided that the thermal resistance and heat transfer coefficient do not vary strongly with temperature. The best case includes aluminum fins at the rear of the module for cooling which reduces the thermal resistance and increases the surface area for convection.

$$E = \frac{1}{2}mv^2 = \frac{1}{2}(Avtp)v^2 = \frac{1}{2}Avtpv^3$$

where:
S is the insolation in mW/cm^2

Thermal Expansion and Stress

Thermal expansion is another important temperature effect which must be taken into account when modules are designed.

The spacing between cells tries to increase an amount δ given by:

$$\delta = (\alpha_G C - \alpha_c D)\Delta T$$

where:
$\alpha_G \alpha_c$ are the expansion coefficients of the glass and the cell respectively;
D is the cell width; and
C is the cell center to center distance.

Typically, interconnections between cells are looped to minimize cyclic stress. Double interconnects are used to protect against the probability of fatigue failure caused by such stress.

In addition to interconnect stresses, all module interfaces are subject to temperature-related cyclic stress which may eventually lead to delamination.

Other Module Design and Evaluation Considerations

A bulk silicon PV module consists of multiple individual solar cells connected nearly always in series as needed to increase the power and voltage to the desired level. The voltage of a PV module is usually chosen to be compatible with a 12V battery, if used for automotive or battery charging purposes, and any other voltage and current combination as needed for each particular application. An individual silicon solar cell has a voltage around 0.6V under 25°C and AM1.5 illumination.

Take into account the expected reduction in PV module voltage due to temperature and the fact that a battery may require 15V or more to charge the modules containing 36 solar cells in series. This gives an open-circuit voltage of about 18-21V under standard test conditions, and an operating voltage at maximum power and operating temperature of about 17 or 18V.

The remaining excess voltage is included to account for voltage drops caused by other elements of the PV system, including operation away from maximum power point and reductions in sunlight intensity.

The same principle is in effect for modules used in commercial or large-scale PV installations. These, however, contain a much larger number of cells in order to generate higher voltage, in the 150-300Wp range. These modules are also much larger in size in order to accommodate the greater number of cells.

A number of additional factors are considered when designing or evaluating different modules for different applications. The source and quality of the materials and components used in building the PV modules in question must be addressed, although this is easier said than done.

Taking a close look at the certification documents may provide a good estimate of the modules' performance and longevity. We also recommend discussing some of the certification test points with the manufacturers.

Practical Pv Modules Characteristics

This is a list of practical factors and issues related to PV modules selection, purchase and use.

There are a lot of modules to choose from but there is also uncertainty about the survival of many PV module manufacturers because of the large supply-versus-demand imbalance that currently exists, along with the challenge for many manufacturers to produce their products at today's cheap module pricing. These factors—combined with the usual considerations, such as module output ratings, power tolerance, efficiency, and pricing, along with new inverter, mounting, and module-level electronics options—make module selection more complex.

This is a list of top considerations that will be helpful for any array design exercise. Additionally, a module selection example is provided, given our potential roof space and energy generation goals.

- Power tolerance is a measurement of how close a module's actual output will be to its rated output

under standard test conditions (STC: cell temperature = 77°F and irradiance = 1,000 watts/m^2). For example, if a 200-watt module has a power tolerance of +/-3%, its actual output (under STC conditions) can vary from 194 W to 206 W. Some modules have a positive-only (such as "+5/-0") power tolerance, which means that these modules should be able to produce at least rated power under STC, and possibly more.

- PVUSA test conditions (PTC) calculate module output using an ambient air temperature of 68°F (at 1,000 watt/m^2 irradiance), which typically causes cell temperatures to be about 113°F to 122°F (36°F to 45°F higher than STC).

PTC-to-STC ratios specify module power output for settings that more closely represent real-world conditions, which makes them lower than STC ratings. The STC temperature of 77°F for a module's cells is often not a very realistic temperature for these dark cells exposed to direct sunlight; their temperature will commonly be much higher. As cell temperature increases, voltage drops, which reduces module power output.

However, modules are sold based on their STC-rated power output rather than by PTC ratings, making it more difficult to compare realistic performance between modules. A PTC-to-STC ratio is included in the table for all modules. The closer the PTC rating is to STC, the higher the module output is under more common conditions. For example, if a "200-watt" module has a PTC-to-STC ratio of 0.9 or better, then its PTC rating should be 180 W or higher; if the ratio is 0.85, then its PTC rating will be only 170 W. Although that difference may seem negligible, when you add the power up for an entire array, it can be significant.

- *Module voltage and string inverter input window* need to be considered for any grid-tied PV project that uses a string inverter. Each module has a specific maximum power point and open-circuit voltage, and each site has specific temperature ranges it will experience, which will determine the actual voltage each module will operate at.

Additionally, each inverter has its own input voltage limitations, which will dictate string size for module models being considered. Many string inverter manufacturers have online sizing calculators to help find string configurations that work for each PV module, considering local climate.

- *Power density* of a module is dependent on module efficiency and is given in watts per square foot. The greater the density, the more power the array can generate per square foot. But higher module efficiency also means more dollars per watt, so before you assume you need a high-efficiency module, check the amount of space you have compared to the total power you want (see the "Selecting Modules for My Garage" sizing example).

- *Module dimensions* need to be considered, especially if you're working with limited mounting space but trying to maximize array capacity. Often, you'll have to compare layouts, including both portrait and landscape configurations, to find the appropriate array layout for a rooftop. When using a string inverter, layout options may need to consider the required number of modules in series (and the number of parallel strings) to make sure the array layout is compatible with the inverter's input and output limitations.

- *PV manufacturer location* can be an important factor. First, some production-based incentives, such as Washington state's RE System Cost Recovery program, pay a higher per-kWh incentive for systems with locally manufactured equipment. Systems funded by the American Recovery and Reinvestment Act (ARRA) and installed on public buildings must use domestically manufactured modules (or foreign modules that use 100% domestic cells). The less distance a module has to travel to its ultimate destination, the less embodied energy that module has. Finally, many people want to support local manufacturing jobs over foreign jobs and imports, and yet most of the modules installed in the U.S. during the 2008-2012 solar boom-bust period were Chinese made. Go figure....

- *Module frames and back-sheets* are important to mounting technique and aesthetics. Options include frameless modules, module frames with mounting grooves for rail-less mounting, modules that allow some light to pass through (popular for awning systems), and dark back-sheets (black), which provide a more uniform look within the array.

PV wire leads are required for ungrounded arrays. Transformerless inverters are becoming increasingly popular because of increased inverter efficiency and enhanced safety (see "Ungrounded PV Systems" in HP150). However, they do require the array to be ungrounded, and the modules selected must have "PV-wire" cables for these installations. (PV-wire has specific benefits over standard USE-2 conductors including thicker insulation, higher voltage ratings, better UV resistance and flexibility in extreme cold)

- *Warranty* is important, and while most PV manufacturers offer 25-year power output warranties, material warranties can range from two to 10 years. A warranty is only helpful if the company offering it sticks around to service a future claim. With the PV manufacturing industry undergoing so much change right now, and many companies merging or exiting the market, some manufacturers are offering noncancellable warranties serviced by third-party insurance companies.

- *Cost* is always a factor and budget can dictate the array you ultimately purchase. Module pricing has been on a downward trend over the last few years. A brief online search shows many modules are available for $1.50 per W; some even below $1. Online module shopping—to try get the cheapest deal—instead of buying from a local module supplier/installer has drawbacks. While the array may cost less up-front, you may be without support should problems arise. In certain areas, installing a grid-tied system without a licensed installer means forfeiting some incentives. While the modules table lists more than 900 modules, no matter whether you buy online or through your local PV installer, available options will be limited to those modules currently offered by that supplier.

- *Compromises Are Inevitable.* Weighing all of these factors can be a time-consuming process that inevitably ends up in compromise. And each system designer or homeowner will have to establish their own priorities.

c-Si PV Module Manufacturing

A number of potential issues are encountered during the cell and module manufacturing processes, all of which must be taken into consideration, if we are to have reliably performing PV cells and modules, lasting 25-30 years.

The major issues to keep in mind when designing or planning to use c-Si PV modules are:

- Quality of silicon material, chemicals and consumables
- Cell type and design parameters
- Quality of the cells' manufacturing equipment and process
- Module type and design parameters
- Modules' manufacturing equipment and process
- Possible cell malfunctions within this type and make of module

• Possible module malfunctions within the particular array

Once the materials—PV cells, laminates, glass, back cover, wiring etc.—have been received and gathered at the module production site, the module is assembled in the following sequence (see Figure 5-22).

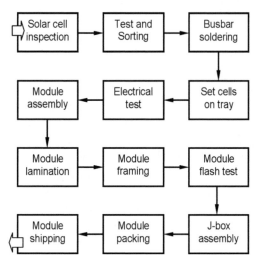

Figure 5-22. PV module assembly sequence

Cell Sorting, Arranging and Soldering

Finished solar cells are flash tested and sorted by their I-V characteristics and power output. Cells that pass the test are placed in bins according to their performance and stored or taken to the module assembly area.

Wiring and Assembly

Cells are connected in a series circuit manually, or by a semi-automated soldering machine using solder coated metal ribbon (usually two in parallel) soldered to the top of one cell and to the bottom of the next cell. This process forms a string of cells which could be as long as desired but usually it is shaped to fit in the respective PV modules tray. Electrical continuity and resistivity tests are performed on some modules to make sure that the bonds are good. "Pull" tests are done sometimes, to check the mechanical strength of the bonds. Thus, connected cells make a complete circuit (string), which is ready for lamination into a completed module.

Lamination

A lay-up for lamination is prepared with clean top glass, and EVA film, and strings of wired cells are placed on it. Sometimes the backing materials (Tedlar and back cover) are placed on top too, forming a complete module. Several lay-ups, each consisting of the above components are lined up in a large cabinet called

a laminator. Using silicone vacuum blankets, the batch of lay-ups is heated and vacuum laminated at one time. After cooling, the modules are ready for use.

This method of laminating is much cheaper than laminating one or two units at a time. The excess lamination is trimmed and terminal wiring is attached. In most cases, an aluminum extruded frame is assembled around the material and the unit is ready for shipping.

Basically, module laminators consist of a large-area heated metal platen mounted in a cabinet-like vacuum chamber. The top of the cabinet opens for loading and unloading modules. A flexible diaphragm is attached to the top of the chamber, and a set of valves allows the space above the diaphragm to be evacuated during the initial pump step and backfilled with room air during the press step. A pin lift mechanism is sometimes used to lift modules above the heated platen during the initial pump step, but most standard modules don't require it.

Temperature uniformities of ±5°C at the lamination point are sufficient for obtaining good laminations with acceptable gel content and adhesion across the module. While more uniform temperatures are available from some laminator suppliers, there is no real benefit to the module manufacturer.

Laminators are available with two types of cover opening systems: clamshell and vertical post. In the clamshell design, the cover is mounted on a hinge at the back of the laminator, which opens like the hood of a car. This leaves the laminator wide open on three sides, making it easy for an operator to load and unload modules manually. Automated belt-fed laminators, on the other hand, use the vertical post method, which lifts the cover horizontally above the process chamber.

Because the cover does not need to travel much for belt loading, the chamber opening and closing times, and resulting process steps (heating and vacuum pump down) are reduced. As a result, most high throughput module lines use belt-fed laminators with vertical cover lifts.

Note: Fully automated cell assembly and lamination lines exist today, but most low- to mid-volume assembly operations, especially those in Asia, still prefer manual lay-up and stringing operations, combined with low-throughput clamshell type laminators. This is due mostly to the availability of cheap labor, though automating labor-intensive processes is no guaranty of a high quality product.

Modules Flash Testing and Sorting

After completing the assembly process by adding edge sealers, side frame, terminal box, etc., the modules are placed on a test stand in the solar simulator (flasher) and are illuminated with special type of light that resembles the solar spectrum at STC, for a period of time. The temperature of the modules is kept at 25°C during the test by active or passive cooling. I-V curve is then generated for each module. The output data are used to identify, sort and label the modules according to their output. The modules that pass the test are packed and shipped to the customers.

Table 5-4. Example of Si PV modules specs

Specs \ Power	20 Watt	30 Watt	45 Watt
Dimension, mm L x W x H	656x306 x18	680x426 x18	665x537 x30
Dimension, in L x W x H	25.8x12.04 x0.7	26.8x16.8 x0.7	26.1x21.1 x1.18
Voc, V	21.2V	21.6V	21.9V
Vmp, V	16.8V	17.2V	17.6V
Isc, A	1.32A	1.93A	2.7A
Imp, A	1.19A	1.74A	2.56A
Pm (at STC), Wp	20Wp	30Wp	45Wp

Types of PV Modules

PV modules come in different makes, types, shapes, sizes and designations, but their ultimate purpose is to produce DC power. The more electricity a PV module generates from a given active surface area, the more efficient it is.

Different types of modules, using different solar cells, materials and manufacturing techniques have different efficiencies and longevity, but for the most part:

• The efficiency of the solar cells in each module determines its overall efficiency. One not-so-efficient cell can reduce the efficiency of an entire module. As a matter of fact, one "bad" cell can ruin the performance of an entire array of modules—or at least it can lower the output significantly. The cell's slow degradation, will also affect the module's performance over time.

• The proper structure of the module, and in particular its ability to keep the elements from penetrating inside and attacking the cell components, determines its durability, degree of accelerated degradation and overall longevity.

This is why product quality and quality control procedures during the entire production process are such an important aspect of PV cells and modules manufacturing and use.

Efficiency

PV modules come in different types and sizes. The efficiency of the different types of modules is one of the most understood and discussed properties (Table 5-5).

Clearly, sc-Si is the most efficient of the single junction PV cells for all practical purposes (III-V semiconductor cells are usually of the multi-junction type). It is, in our opinion, also most reliable for long-term use as well, simply because its structure is simple, its crystal is perfect and not as easily affected by different electro-mechanical abnormalities.

Polysilicon, however, is most widely used, mostly because of its pleasant aesthetics, ease of manufacturing and lower cost, even though its efficiency is lower and the imperfections of its crystal bring some unwanted effects in the long run as well.

Generally speaking, we must be aware of the behavior patterns, variables, performance and longevity

Table 5-5. Performance of different cells and PV modules

Solar cell material	Cell efficiency η_z (laboratory) (%)	Cell efficiency η_z (production) (%)	Module efficiency η_M (series production) (%)
Monocrystalline silicon	24.7	21.5	16.9
Polycrystalline silicon	20.3	16.5	14.2
Ribbon silicon	19.7	14	13.1
Crystalline thin-film silicon	19.2	9.5	7.9
Amorphous silicon[a]	13.0	10.5	7.5
Micromorphous silicon[a]	12.0	10.7	9.1
CIS	19.5	14.0	11.0
Cadmium telluride	16.5	10.0	9.0
III-V semiconductor	39.0[b]	27.4	27.0
Dye-sensitized call	12.0	7.0	5.0[c]
Hybrid HIT solar cell	21	18.5	16.8

issues of the different types of PV cells and modules, if we are to design and build efficient and long-lasting PV power plants.

Key Module Issues; Detailed Technical Discussion

Keeping in mind all possible behavior conditions and problems we discussed above, we will now take a look at the possible problems at each process step of PV cells and modules manufacturing. These issues can be traced to materials, equipment, processes and labor-related problems, and are usually demonstrated during testing or long-term exposure to the elements.

Now we will examine possible defects and failures which can occur during the final module test or, more importantly, during long-term exposure to harsh climatic conditions.

So, the key issues and all possible effects caused by long-term on-sun exposure, following the PV module elements shown in Figure 5-23, are detailed as follows:

ONE—Glass Cover (Top Cover)

The cover glass is basically designed to provide protection to the fragile solar cells and module components from mechanical damage, such as vibration, impact, etc., as well as to prevent the elements (rain, moisture, dust) from entering the module. If the glass itself is stressed, cracked and broken during processing, handling, transport, installation and operation stages, then the insides of the module might already be damaged as well. Subsequent damage could occur in such cases, by means of mechanical stress and chemical destruction of the cells and other active components.

This is even more important for thin film PV modules because the active layers are deposited directly on the front or back covers, thus any stress or breakage of the glass will directly, and immediately, affect the performance and longevity of the damaged modules. The advantage of c-Si PV modules in this area is that the EVA layer is enveloping the active components (solar cells), so that even if the top or bottom covers are damaged, the EVA envelope is given a chance to protect the cells from an invasion of the elements and chemical attack.

Modules with damaged cover glass—be it cracked or with hazy appearance—should not be installed and must be removed from service immediately. Repairing damaged glass is not an option in most cases, so the entire module must be properly disposed of, or sent back to the manufacturer for credit or for replacement. Periodic testing of the voltage/current output of modules with partially cracked cover glass, if acceptable at all,

should be part of the PM schedule.

Note: In all cases damaged modules should be handled with utmost care.

Cover glass surface soiling is another serious issue during normal operation. Rain and dust will deposit layers of contaminants and water spots on the surface, which will prevent sunlight from reaching the solar cells underneath. With time the layers might grow so thick that very little light goes through them, significantly reducing the output of the modules. Washing the top surface with a soapy water solution, followed by proper rinsing with DI water will restore the modules' efficiency, so it should be part of the PM schedule and carefully executed. This, however, is an expensive undertaking, especially in the deserts. Periodic use of water and chemicals also causes concerns with water table contamination, which has to be kept in mind when deciding on a PV power field location and O&M procedures.

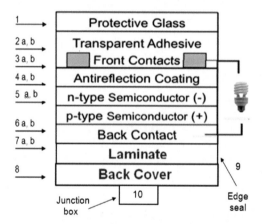

Figure 5-23. Standard silicon PV module cross section

TWO—EVA Encapsulant Deterioration and Delamination

The layer of plastic encapsulation (EVA usually) and the cover glass are in intimate contact and partially bonded. These two, however, have different elasticity, and coefficients of expansion and friction. With long-time exposure to the elements (heating, freezing, excess UV radiation, mechanical stress, fatigue) the EVA plastic will eventually change. Its physical and chemical properties change slowly and we can see it turning a yellowish color. Yellowing of the EVA results in reduction of its transmissivity and causes a decrease of power output.

Yellowing is one of the few changes occurring in the modules that can be observed with the naked eye and which gives us a clear indication of the changes within the module. As the changes continue, air bubbles, cracks, pits and cavities could form in the EVA material, which could cause it to separate from the top glass. This contributes further to increasing optical losses due to reflection and poor transmission, ultimately resulting in serious performance deterioration. Once the EVA material damage and the accompanying delamination processes have started they cannot be stopped. At this point the damaged modules are suspect and must be put under periodic observation as part of the O&M schedule.

EVA Delamination from The AR Layer

Some of the above-mentioned damage and defects in the EVA material will affect also the adhesion of the EVA material to the AR coating to which it is bonded. EVA delamination from the glass and AR coating surface is caused by the never ending UV bombardment, expansion and contraction of the layers and the resulting friction between them. This action might result in an air gap which will cause another (second) optical barrier which will further decrease the power output from the affected cells.

The newly created gaps and cracks might allow moisture, air and reactive gasses to enter in the module which might attack the other modules components.

Changes at the EVA-AR film interface (air gap, color change, cracking, etc.) are another of the few visually observable changes in PV modules. Depending on the type of cells (mono or poly crystalline) and type of AR coating (many different formulations are available), the visual effects might change the cell's surface color from black to faded black, and from dark blue to light blue (respectively), as well as a number of shades in between.

THREE—Top Contacts and Interfaces

The top contacts (fingers) consist of silver metal with metal ribbons soldered on top of the contacts to provide connection to the other cells and to the outside circuit. This creates several separate contact points (interfaces) as follow:

a. Silicon to silver metal of the top contact,
b. Silver metal to the soldered metal strip, and
c. Soldered metal strip to the EVA encapsulant.

The different interfaces provide good adhesion to their components and are quite stable under normal operating conditions. In abnormal situations, however, such as excess heat, overcurrent conditions, mechanical or thermal stress, and/or exposure to the elements (water vapor, chemicals and reactive gasses), the metals undergo serious changes which are usually initiated in the interfaces. The damage could increase with time and cause significant changes in the contacts' structure. In all cases these events will result in electrical changes or

failures (high resistivity or open circuit). Interrupting the contact with the adjacent cells will lead to a failure of the module.

In some cases, one or more of the affected contact layers might burn, chemically disintegrate, or undergo other destructive processes, which could cause the device to fail partially, or completely, in which case the entire solar module might stop generating power. In all cases, there will be noticeable reduction in power from the module, due to mismatch caused by the affected solar cell(s).

Top Contact Adhesion into the Silicon Substrate

Under normal operating conditions, the adhesion between the front contacts (silver metal) and the silicon material which they are fused into is good enough to keep them together and properly conducting electricity for the life of the module. Abnormally, however, mechanical stress, excess heat, and moisture and air penetrating into the module, could affect the front contacts. Mechanical forces could induce stress in the materials and their interfaces, while ingress of water and chemicals might corrode and separate them from the substrate.

These processes would negatively affect the bond at the interface between these key components (silicon substrate and silver contact). In such cases:

a. The diffusion area might be affected and undergo internal changes which might affect the cell's performance, and/or
b. The resistivity at the affected area might increase, which could result in overheating, and eventually a breakdown (open circuit), caused by partial or complete separation of the contact from the substrate.

These phenomena could bring partial or total failure of the solar cell and/or the entire module.

FOUR—AR Coating Surface Damage

A number of distractive mechanisms on the cell's surface, caused by mechanical stress and chemical attacks could affect the AR layer. These changes could occur under different operating conditions and cause the AR coating top surface to change mechanically or chemically. This could cause a change in the AR layer's optical properties, decreasing the cell's efficiency.

In addition, blistering of large areas of the AR coating might also stress the cell and the top contacts, causing further increase of resistance and heat generation. Overheating could damage the contacts and/or their interface. The warped AR surface might also reflect some of the incoming sunlight at a greater rate than designed. In all cases, the final result is reduction in affected cells' efficiency.

These effects (AR layer delamination, blistering etc.) also result in color change of the AR coating which represents another of the few visually observable changes in the solar cell and module. Depending on the type of cells (mono or poly crystalline) and type of AR coating, the visual effects might change the cell's surface appearance to different variations of the original colors.

AR Coating Adhesion to the Substrate

The antireflection coating (AR) is a fairly thin film of inorganic material, i.e., TiO2, Si3N4 and such, which is very thin, semi-transparent and fragile under certain conditions. Very thin and lightly bonded to the cell surface, it is easily damaged by mechanical and chemical means. There are several mechanisms that can contribute to changes of the AR layer and its adhesion to the substrate. One is contamination of the silicon substrate surface prior to AR coating, which would result in poor adhesion over time. The surface contamination could be caused by insufficient cleaning and rinsing, improper handling, or an out-of-spec process during the AR coating deposition. All of these inadequacies would compromise the adhesion integrity of the thin AR film to the silicon substrate.

If the AR layer is not fully adhered to the silicon substrate, there would be physical and optical gaps which would cause excess reflection and/or obstruct sunlight transmission into the cell. Combinations of these conditions will contribute greatly to reduction of the cell's efficiency.

Further, and more seriously, changes in the AR film properties could occur quicker if the damaged area is exposed to the elements via moisture and air leaking into the module (see above). The adhesion forces in the interface between the AR film and the substrate in such cases would weaken even more under the outside attacks, and the AR film might disintegrate or separate from the surface, amplifying the negative effects.

Most changes in the AR coating's adhesion and other changes of its properties are also visible on the module's surface, but equally impossible to correct. When these changes occur, future cell/module behavior is impossible to predict without destructive tests, so the module must be checked periodically as part of the O&M schedule.

Note: All visually observable changes mentioned above start, and might even continue, as microscopic imperfections which eventually grow bigger and more

visible. A trained eye is needed to observe the changes in many cases.

FIVE—P-N Junction and Diffusion Process Issues

The p-n junction is the most critical area of solar cells' function. It has the greatest impact on cell/module efficiency, performance and longevity. The p-n junction is formed by diffusing (injecting) a specified concentration of foreign atoms in the silicon bulk, which then initiate and drive the photoelectric effect. Unacceptable amounts of impurities in the bulk silicon, such as Fe, Cu, K, Na and any other contaminants in the process chemicals and gasses will affect the diffusion concentration and depth, ultimately changing the behavior of the p-n junction. Changes in the concentration of the foreign atoms in the junction, due to parasitic effects during its field operation would produce similar results.

If the wafer surface was not properly prepared, cleaned and rinsed, or if the diffusion process was not properly executed (time, gas concentration and temperature), the p-n junction might not be stable enough. Thus affected solar cells might operate well in the beginning, but eventually the diffusion layer will start changing (decreased concentration and diffusing further into the substrate) causing a drop in output. These effects sometimes occur months or years after the module has been in operation, and are called "latent" effects. They are most dangerous for the large-scale power plant's long-term success, because if they happen after the warranty period has expired and if the failures are great enough they might cause serious technical and financial difficulties.

The most common negative field effect of the p-n junction is change in the diffusion depth (and the resulting dopant concentration reduction at the junction) over time. While this phenomenon is well understood and the p-n formation parameters are controlled during the manufacturing sequence, certain material and process conditions force greater changes, over time, than allowed by design. In such cases, the cells start losing power quicker than expected and eventually fail altogether.

Diffusion processes techniques vary from manufacturer to manufacturer. Some use the old-fashioned, but most reliable, diffusion furnace process, where the wafers are placed into a glass tube and heated to over 1000°C in a controlled environment. Doping gases are passed through the tube and by precise control of temperature, process time and gas volume, precise deposition concentration and depth are obtained. A high-volume, but less precise, diffusion process has been widely used lately. It consists of spraying or printing the dopant liquid on the wafer surface and then baking it in a conveyor belt type of furnace. This process has more uncontrollable variables and is basically inferior from process control and product quality points of view than the old diffusion furnace method. Nevertheless, it is cheaper and faster, thus we will see it more and more in the future. And due to its less-than-precise and not-so-easy-to-control process parameters, we expect to see more diffusion and p-n junction changes issues in the field as well.

Diffusion layer depth and concentration changes cannot be visually detected, nor can they be measured under field conditions, so at this point we are looking at a black box and relying on the manufacturer's experience and his properly and efficiently executed quality control procedures.

P-N Junction Field Performance

So, what if the diffusion process were less than perfect? First, and most important, if the materials and process specs were even slightly out of control, the solar cells and modules would show low efficiently during the final tests. After sorting, the less efficient would go into the group of "cheaper" cells and modules to be shipped to a customer who hopes they will work well. But if the diffusion process were somewhat out of spec, thus the reason for the lower quality of some cells, then there would be a good chance that the affected cells and modules might deteriorate much quicker than the rest.

Exposure to extreme temperature is especially testy, and is the reason for many malfunctions and failures encountered in c-Si modules. It is possible for the species in the diffusion layer to start migrating into the bulk under abnormal conditions brought upon by extreme temperature regimes and/or any imperfections in the silicon material close to the p-n junction area. These will accelerate the changes in the electrical characteristics of the cells, thus causing additional efficiency decrease.

In summary, as soon as the newly processed solar cells are flash-tested they are ready for encapsulation into a module. Provided that the materials are of high quality and all process steps have been properly executed, the p-n junction and the cells will operate properly for a long time. But if we had materials quality problems or the diffusion process were not properly executed, then we might be looking for a surprise in our power field, over time.

Note: Here we need to mention the great effect of silicon bulk material on the efficiency and other properties of the solar cells that were made from it. We now know that the quality of the solar wafers determines the quality of the solar cells made from them. The silicon wafers are made from 99.9999% pure solar grade silicon material. Even at that purity, however, a slight increase in one of the harmful contaminants (usually metals) might have serious effects on the solar cells' field performance and longevity. Over time, and when operating at extreme temperatures, parasitic metals could start diffusing through the cell and alter p-n junction properties, with the diffusion speed increasing.

SIX—Rear Contact Damage

The rear contact also consists of several interfaces:
a. Silicon bulk to aluminum BSF
b. Silicon to silver back contact, and
c. Silver to soldered metal band

This metal structure and its interfaces are exposed to the same electro-mechanical, thermal and chemical attacks and changes as the front contacts we reviewed above. Although these are somewhat more protected from and less affected by the UV and IR radiation than the front contacts, their long-term quality is as critical. Any elemental impurities, chemical contaminants, mechanical damage, such as cracks, pits and other imperfections in the structure will have profound effect on the cell's efficiency and longevity.

The quality of the metal deposition is of great importance here too, because if it is defective in any way (high resistivity, cracks and pores, poor adhesion, oxidation at the interface, etc.), it will cause overheating, delamination and ultimately reduced efficiency of the cell, and even total failure. Nevertheless, the rear contacts are much less problematic and affect the cells' function less than the front contacts, but failure of the encapsulants and edge sealers could damage them seriously enough to cause performance degradation and failures.

Rear Contact Adhesion

In case of poor contact quality due to process control inadequacies (time, temperature, or metal quality) the metal film can partially separate from the substrate, eventually leading to decreased efficiency due to higher resistivity and overheating.

Excess overheating at the back surface might contribute to further erosion and delamination of the metal film and total failure of the cell and module.

SEVEN—Laminate Issues

Laminate materials (EVA, PET, PVB, Tedlar, etc.) are organic (plastic) materials, which have been in use and remain basically unchanged in composition and use, for the last 30+ years. Also unchanged is their vulnerability to exposure to the elements, where EVA and other organic compounds in the module do undergo mechanical and chemical changes and degradation over time.

Continuous expansion during hot days and shrinking during freezing nights, extreme IR and UV bombardment, and chemicals' ingress will affect the EVA and other organic (plastic) materials. This will eventually result in mechanical changes—discoloration, mechanical stress and disintegration of the lamination materials—creating cracks, bubbles, pits, voids, etc.

Ingress of moisture and gasses into the module via the above formed cracks and voids will contribute further to the degradation process by decomposing the module materials (solar cells, wiring, contacts etc.).

The laminate structure on the back of the module usually consists of the EVA envelope with thin Tedlar and/or other plastic materials backing, laid onto the rear metal cover. The rear EVA-on-Tedlar structure is protected from the elements (UV and IR radiation) so it lasts much longer without damage. Its yellowing and even delamination from the back cover don't have such dramatic effects on the module performance.

The adhesion of the EVA envelope to the back of the cells, however, can be affected with time. Excess heating, freezing, and moisture penetrating from the sides might eventually debilitate the adhesion to the solar cells. In that case, moisture and air penetrating the laminate material and reaching the solar cells will cause rapid oxidation of the back surface metallization (aluminum and silver metals), which will also affect the interface between these metals and the back surface of the solar cell. This deterioration of the cells' adhesion at interface might cause increased resistivity and overheating, and eventually delamination of the metal contacts from the substrate, which will cause performance issues and eventually failure of the affected cells and modules.

EIGHT—Rear Cover and Frame

The purpose of the frame and back cover (aluminum sheet or glass plate) is to protect the module's insides from attack by the elements. It is highly unlikely for the frame and back cover to degrade significantly with time under normal use, but a severe mechanical or chemical attack could compromise its integrity, forming cracks and voids which might allow moisture and air to penetrate the module. At that point, any of the above-described events might cause the cells and module to decrease in efficiency or fail.

NINE—Side Edge Seal

c-Si PV modules usually have side protection, as an extension of their edge protection. Ingress of harmful elements into the module is usually initiated through the sides (the edge seal) and could be slowed by a well-installed and sealed metal frame around the edges. A lot of effort is dedicated to finding better sealing materials, and assembly processes have improved significantly lately as a result. Edge sealing—its effects and weaknesses—are well understood.

No matter how good the sealing materials are, however, they are made of plastic (organic) materials, which are affected by the elements—IR and UV radiation and environmental chemicals. Thirty years under the blistering Arizona desert sun would challenge any organic material or compound.

Edge seals are prone to accelerated changes with time which usually lead to the formation of voids, cracks, pores, and bubbles. These imperfections eventually allow moisture and air to penetrate the modules and cause damage as described time after time above. Therefore, new types of frames and edge seal protection should be developed to provide additional protection.

Note: Edge seal quality and protection issues are even more important with some thin film PV modules, which are frameless. The modules consist of two glass plates with no sides, so that the edge seal is exposed to the elements. This makes the thin films inside the modules more vulnerable to the elements, which could be detrimental to their performance and longevity.

TEN—Junction Box

The junction box is just that—a small box where the wiring "junction" or connection is made. It is a metal container intended for easy, safe, and reliable electrical connections. It is also intended to conceal them from the elements and prevent tampering. The box is attached to the back cover of the module by means of screws and/or glue and contains connectors for wires coming from the module and those connecting it to the external circuitry.

Corrosion of the contacts in the boxes is the most frequent problem encountered during long-term operation in harsh climates. This might cause increased resistance, reducing output power and eventually resulting in fire or an open circuit. This problem, however, is the only one that can be fixed by replacing the defective parts without tearing the module apart.

Summary

As a summary of issues affecting solar cells and modules performance and longevity, if we represent the actual efficiency and longevity of a solar module using the above described conditions and issues, we get:

Field efficiency:
$\eta a = \eta t - (1 + 2a + 2b + 3a + 3b + 4a + 4b + 5a + 5b + 6a + 6b + 7a + 7b + 8 + 9 + 10)$

where:
ηa is the actual field efficiency of the module
ηt is the theoretical (optimum) efficiency of the module, and
Numbers 1 through 10 are the sequence of conditions and issues discussed above.

Longevity:
$La = Lt - (1 + 2a + 2b + 3a + 3b + 4a + 4b + 5a + 5b + 6a + 6b + 7a + 7b + 8 + 9 + 10)$

where:
La is the actual longevity of the module
Lt is the theoretical (optimum) longevity of the module
Numbers 1 through 10 are the sequence of conditions and issues discussed above.

It is impossible to predict, let alone put a numerical value on most of the conditions and issues in 1 through 10 above, so these formulas are good only to show roughly the qualitative dependence of the efficiency and longevity of the modules to the members of this long chain of events and how they could affect (usually in a negative way) the cells and modules during their long-term on-sun operation.

As we can clearly see, any discrepancy in the chain of ten events can only reduce the efficiency and/or longevity of the cells and modules, according to the seriousness of the deviation. Or as we mentioned, "The highest quality of the finished device is determined by the lowest quality of any process step," or event in this case.

Applying the related manufacturing process steps, sub-steps and procedures to the each of the members (1-10) of the above formula will result in a long string of variables (literally hundreds of them). Each of these additional sub-routines and variables would have an equally negative effect on the modules' performance and longevity, if not properly executed.

This is why quality of materials, quality of design, and process control, combined with know-how and experience, are so important in ensuring acceptable quality of the final product. This is not impossible, but we fear that very few people are fully aware of the complexity of these matters. Our hope is that now, with all

the issues on the table, we will be able to start an open discussion about their resolution. Standardization of materials quality and process controls is the ultimate solution to the issues at hand, so until that occurs we must use great care when designing and evaluating solar cells and modules for large-scale power generation, especially in harsh climates.

PV MODULES CERTIFICATION AND FIELD TESTS

Testing PV modules before installation in the field for long-term operation is a necessary step, but the results are not always indicative of actual lifetime field performance. Many modules qualified and certified as required, and operating for several years in the field, have been found malfunctioning after the abuse of Mother Nature.

Once the solar modules are manufactured, one of the best batches goes to US or European labs for testing and final certification. This is a very important test, which determines the future of this set of modules, and possibly that of the company that manufactures them.

Thus, selected modules undergo standard test procedures. Some pass the testing program, but start deteriorating prematurely and at times even fail completely. Sometimes they fail at the pre-test and pre-certification stages—even before the testing procedure has started.

Pre-test Failures

So let's take a look at an unusual trend of failures, which were discovered even before the actual testing was began. These failures tell us a lot about the manufacturer and don't give us too much confidence in their products. See for yourself:

According to Intertek, a certified test facility in California, a number of test modules fail this pre-test screening. Here are five typical reasons why PV modules coming for testing at their facility fail initial inspection—even before the testing can be started. The reasons are shocking, for a supposedly hi-tech business, such as PV cells and module manufacturing operations:

- Inappropriate/incomplete installation instructions
- Models provided for testing do not accurately represent the entire production model scheme being listed (largest module must be submitted for test)
- Testing requested without prior construction evaluation being performed
- Complete bill of materials with ratings and certification information not provided prior to start of the project

- Lack of back-sheet panel RTI rating

Note: The author's extensive experience with engineering and quality control operations in the world's semiconductor and solar industries shows that the phenomena described above is quite unusual from a quality control point of view. It points to major weakness in the QC/QA systems of the manufacturers in question.

Keep in mind that most of the above mentioned abnormalities came from Chinese made PV modules during the hay days of the solar boom and bust cycle of 2009-2013. At that time, hundreds of Chinese companies hastily switched from what they were doing to making solar cells and PV modules. Under pressure to produce as much as possible, many took short cuts

The above listed failures show weaknesses in the production cycle, which are among the very few that can be readily recognized *before* the PV modules are packed, shipped and installed in the field. There are many other gaps and issues in the production process, which cannot be discovered and corrected on time.

If this is do, then, how are we going to catch the possible failures of the thousands of modules we need for our 100MWp power plant during and after installation in the field? This is a question that we must answer in the daily quest for products needed for large-scale PV installations. How do we make sure that the PV modules we are ordering will perform as specified during next 20-30 years of non-stop operation under extreme conditions?

The quick answers:
- Find out as much as you can about the "sand-to-module" history of the product, before placing a large order.
- Take your time to thoroughly check all engineering and quality control procedures used during the processing and testing of your batch of PV modules. The manufacturer must provide some of this information. If not, then go to someone who will. Hands-on help from qualified PV specialist is highly recommended at this step of the process.
- Even better, have a team of qualified engineers visit the manufacturer and inspect the process line your modules go through. This is a difficult task, but short of that, you are working with a black box that may or may not work properly. The choice is yours.

Note: Of course, hands on help and team of engineers is possible only when designing large projects, so the small project owner has no such options. The best, if not the only thing s/he can do is buy PV modules and

BOS equipment from reputable companies and hope for the best.

Certification Test Conditions

All modules go through a minimum internal testing, and only these that pass are shipped. The rest are either downgraded and sold as a lower quality product, or are rejected and scrapped. A batch of modules (usually the best of the batch, hand-picked by QC engineers) is usually sent to third party testing and certifications. Here, the modules are exposed to a series of tests and judged according to the outcome. The modules passing the qualification tests are considered good and ready for mass production.

The qualification and certification tests usually require each sample to meet the following criteria as a minimum:
- Degradation of the maximum power output at standard test conditions (STC) does not exceed 5% after each test or 8% after each test sequence;
- The requirements of tests 10.3 and 10.20 are met;
- There is no major visible damage (broken, cracked, torn, bent or misaligned external surfaces, cracks in a solar cell which could remove a portion larger than 10% of its area, bubbles or delamination, loss of mechanical integrity;
- No sample has exhibited any open circuit or ground fault during the tests;
- For IEC 61646 only: the measured maximum output power after final light-soaking shall not be less than 90% of the minimum value specified by the manufacturer.

These qualification test conditions are simple but stringent, and as a result, the quality of both c-Si and thin film modules tested until recently was absolutely unacceptable by today's standards. Quality has improved in some areas during the last several years, as evidenced in some tests, but is still lower than practicable and profitable for use in large-scale PV power generating installations. Large failure rates are possible in some regions of high humidity and excess heat, which would bankrupt any installation.

No PV module manufacturer has completed successfully a full life-cycle (30 years) field test.

Today's solar industry is quite new and immature, with most companies being in existence only 5-10 years. During this time, they usually go through several changes and improvements of their production process, which means that they just don't have a product that can show successful, long-term, non-stop field operation.

Performance Tests and Failure Rates

Most manufacturers have in-house test facilities where they test their cells and modules prior to certification and during production. This approach shows immediate problems and allows for correction of process deviations, so it is indispensible for ensuring high quality of the product.

This, however, is only part one of two parts solution. Part two, which is many times more important, is long-term tests of modules under actual operating conditions. It is one thing to test a module for several minutes, or hours, in the comfort of a clean, air-conditioned, and dry test lab. It is totally different when this same module is installed in the field and is exposed to several long years of non-stop sunlight, sand and rain storms, etc.

And there is another level of tests that PV modules must undergo before being accepted as reliable—the long-term desert tests. This is the ultimate test for any solar technology. In the desert, the merciless extreme heat during the day and instant freeze at night, combined with UV radiation, sand storms, rain and snow challenge any material and structure.

PV modules are most efficient in the deserts, and countless millions have been already installed in the world's deserts without appropriate long-term desert tests. Albeit too late to correct the problems and failures these will experience in the decades to come, we must not ignore this problem. It is a key to the progress of the global PV industry, because if the modules in the deserts start failing at great numbers after 5-10 years of operation, the industry will be put on its knees and have to look for new solutions.

Below we show results from the largest, long-term field test with many PV modules from different manufacturers.

Long-term Desert Field Tests

As confirmation of the facts and issues discussed in previous chapters, a research study published in the fall of 2010 confirms the presence of major issues in PV modules operating under harsh desert conditions. The study describes the results from long-term scientific tests done with PV modules operating in an official desert test field for a number of years (ranging from 10 to 17 years).

Note: To our knowledge, this study is the largest-ever evaluation of long-term performance of a large number of production-grade PV modules under desert conditions.

The study was done at one of the most prestigious PV test sites in the world; APS's STAR solar test facility in Phoenix, Arizona, where most world-class PV modules manufacturers send their modules for long-term testing, so we would not be surprised if the tests include some famous names.

The work was supervised by one of the world's most reputable PV specialists, Dr. Govindasamy Tamizhmani (Mani) from the Arizona College of Technology and Innovation, and President of TUV Rheinland, Tempe, Arizona.

Dr. Mani's team carefully and thoroughly visually inspected, tested electrically, and IR measured each module in the nearly 1,900 PV modules test batch.

The final results from the long-term tests in the Arizona desert are shown in Table 5-6.

Amazingly, almost 100% of the nearly 1,900 PV modules tested in that study showed some sort of visual deterioration. Some were more impacted than others, but the overall conclusion was that a large percent of PV modules exposed to constant sunlight and heat in the deserts deteriorate beyond the manufacturer's warran-

Table 5-6. Long-term desert field test results

Module	Module Count	Modules Affected
A—Mono-Si (17 years; glass/polymer	384	
Browning in cell center	384	100.00%
Frame seal deterioration	384	100.00%
Hot spot (IR Scan)	4	1.0%
B—Mono-Si (12.3 years; glass/polymer)	1092	
Encapsulated delamination	2	0.2%
Browning in cell center	1092	100.0%
Hot spot (IR Scan)	2	0.6%
C—Poly-Si (10.7 years; glass/glass)	171	
Broken cells	47	27.5%
Encapsulated delamination	55	32.2%
Hot spot (IR Scan)	26	15.2%
White material near edge cells	2	1.2%
White material browning	56	32.8%
D—Poly-Si (10.7 years; glass/polymer)	48	
Browning in cell center	37	77.1%
Hot spot (IR Scan)	4	8.3%
E—Mono-Si (10.7 years; glass/polymer)	50	
Bubbling substrate	33	66.0%
Browned substrate near Jbox	50	100.0%
F—Poly-Si (10.7 years; glass/polymer)	120	
Discolored cell patches	6	5.0%
Backsheet bubbling	1	0.8%
Browned spots on backsheet	2	1.7%
Metalization discoloration	22	18.3%
Frame seal deterioration	15	12.5%
Hot spot (IR Scan)	4	3.3%
Total Modules:	**1865**	

ty within 10-17 years of operation. Many of the tested modules show average 1.3 to 1.9% annual power degradation, which translates into 23-33% total power loss for the duration (some only 10 years). This, extrapolated for the extended time of use, is well above the manufacturer's warranty of maximum 20% degradation in 20 years.

Interestingly enough, significant EVA encapsulation discoloration and delamination was observed on many modules, even those that were exposed to the sun only 10 years. Solar cell discoloring (due to a number of factors) was observed on most cells too. Since the oldest modules in this study were made in the mid-1990s we must assume that encapsulation and cell manufacturing processes have improved some during the last 10- 15 years to reduce the defect ratio. However, we have no proof of such progress, so this becomes another unknown that needs our attention when dealing with PV modules destined for desert applications.

In summary, browning in cell center was almost 100%, encapsulation issues followed as a distant second, followed by frame seal deterioration, hot spots, broken cells and glass, etc. Many of these defects are serious enough to cause power loss and eventually lead to total failure, be it due to optical interference, electrical resistance, overheating, mechanical disintegration, or any combination of these.

This is simply not supposed to happen; not in such large numbers at such an important test site, so we must ask, "What caused the problems? Was it the materials quality (silicon, metals, EVA, glass), or the manufacturing process (diffusion, metallization, encapsulation), or was it some special combination of these that was primarily responsible for such a large number of defects and failures?"

In any case, this study confirms the unfortunate reality that PV modules of any type and size are prone to rapid deterioration and failure under harsh desert conditions. Although we cannot predict future performance and overall condition of modules during years 20, 25 and 30, the logical assumption is that the deterioration processes will continue at the same rate, and even faster in some cases.

Because of that, we must conclude that each module type and size must be very well designed and properly manufactured with high quality materials and proper processes in order to be given a chance to survive the desert elements.

Note: The reasons for the above described defects and the related explanations and justifications are too numerous to launch into at this point. Detailed references and explanation to many of the problems found in these modules can be found in our previous book on the subject, *Solar Technologies for the 21st Century*, published by The Fairmont/CRC Press in 2013.

An obvious conclusion from the above tests is that some modules perform better than others. Is this due to different quality materials, and/or different processes used? The answer is less than obvious, so this means that the issues at hand need to be brought out in the open, discussed and resolved ASAP.

A counter-argument that we hear on the above matter is that the PV modules in this test are old, thus not made with contemporary materials and procedures.

This is a valid argument, but please note that:

1. PV modules materials and manufacturing process change slowly, but there is no way to identify and quantify the changes,

2. The above point emphasizes our point that the solar manufacturers are constantly looking for new, better, and cheaper materials and processes, so we never know which generation product is to be trusted and why, and

3. The failures shown in the above study are disastrous, so we must assume that the materials and processes have improved some. We just don't know how and how much has changed.

These points point to a set of uncertainties that we must deal with and resolve, as needed for solar installers and owners to have 100% assurance that their PV modules will last 30 years and perform as specified. The PV industry has a bright future, but is still learning how to walk, and we need to help it get on its feet sooner. The first step in that direction is to make the manufacturing process transparent, standardize it, and implement unquestionable state-of-the-art QC/QA procedures. Not an easy task.

The semiconductor industry went through a similar learning curve, so it took many years of hard work and billions of dollars to get where it is today. Due to global standardization and implementation of modern processing and QC/QA procedures, every semiconductor product coming from the major manufacturers assembly lines is proven and guaranteed to perform as specified, thus removing all uncertainties in its longterm use.

If and when would the global solar industry would get to this point is the big question today, but we are sure that it will happen some day in the future. Until then, however, we need to be aware of the problems at hand and be prepared for surprises in the solar fields.

Catastrophic Failures

Those of us who live in the desert are well aware of its destructive nature, and the fierce forces that are in play non-stop in its vast territory day after day, year after year.

It is a fact of life that the desert has no friends and shows no mercy to anyone or anything!

Anything left on the desert floor for a long period of time will be damaged or destroyed sooner or later. Usually it is much sooner than later. There are very few exceptions to this reality, but unfortunately PV modules are not one of them.

Because of that, we must be prepared for the worst when installing PV fields in the desert. Excess stifling heat, freezes, highly destructive UV radiation, severe sand storms, strong winds, monsoon rains, golf ball-size hail, and brush fires, just to name a few of the surprises that must be expected at any time in the deserts.

The extreme conditions, added to the high temperature in, and high voltage current flowing through, the PV panels create scenarios on the extreme end of most materials' tolerance limits. Plastics crack and disintegrate with time, the active structures inside the modules (silicon or thin films) experience excessive thermal and mechanical abuse, shrinking and expanding with the temperature changes, and all this can create defects in the structure which lead to unpredictable failures.

The stresses and defects caused by mechanical, electric, electronic, or chemical abnormalities in the modules and the support infrastructure could lead to partial or total failures at any time.

- *Partial failures* result in power degradation which results in power loss, but the modules are still operational. This is the case with most modules in the ASU field test. Most of them show some visible or otherwise measurable damage, defects and power loss, but are still functioning. Partial failure could also be caused by sand and rain storms that temporarily disable the power production by soiling the modules, or damage their support structures. After proper inspection, troubleshooting, and maintenance, the field could be up and running within hours. Excess decrease in power output might require replacing a number of modules and/or support equipment.
- *Total failures* result in permanent damage, be it mechanical damage to the modules, electrical failure causing open or closed circuits, or chemical disintegration of the active structure inside the modules. These abnormalities could result in total shutdown of modules and strings of modules, and in some cases even cause fires. These are then classified as catastrophic failures.
- *Catastrophic failures* are not an everyday occurrence in PV power plants, but when they happen they cause a lot of damage. The most dangerous catastrophic module failure, DC arcing, occurs when the modules lose their dielectric properties after long-time operation under the elements (due to lamination break-down) and start leaking electric current across the module elements. DC arcing is an example of premature module failure and is the worst catastrophic field failure.

The problem with DC arcing is that it cannot be easily foreseen or prevented, and once started it cannot be stopped. The arc will burn out only when there is no more material to burn, and in worst cases it could ignite the module, burning it and the entire array. This is a catastrophic failure, which could cause damage to the entire project.

Proper O&M procedures (including using IR sensors to detect hot spots and overheated modules), albeit expensive and time-consuming might detect the precursors of DC arc creation, so they are highly recommended.

There is simply no way to predict, let alone prevent, the desert elements and related surprises, but properly designed and manufactured modules, using highest quality materials and processes are a good start in that direction. Proper installation, again using highest quality materials and procedures, is the next best thing we can do to extend the useful life of the modules. Using proper O&M procedures for the duration will close the circle of our responsibilities.

In addition, an adequate manufacturers' warranty, including long-term performance warranty, is needed to provide protection against product defects. And finally, insurance against natural disasters, complete with long-term performance insurance, will bridge the gaps between the manufacturers' warranties and what Mother Nature has in store for our modules during their 30 years field operation. This way nothing is left to chance, and only now are we ready for anything the desert is going to send our way for the duration.

The Desert Phenomena

In order to wrap-up this discussion and put it in the right perspective, we need to take a long and serious look at the actual exposure of PV modules to long-term desert operation. This leads us to emphasize again and again the importance of well designed and constructed modules, always keeping in mind that they are not disposable toys to be stored in a dark closet after two weeks of use. Instead, they are high-tech devices destined to experience 30 years of non-stop torture in the world's harshest areas—the deserts.

The desert does not pick and choose, and does not have friends; it attacks decisively and violently anything and anybody in its midst or in its path. It destroys anything that dares challenge its reputation as an unforgiving and cruel tyrant. Blistering heat, freezing cold, hail, sand and monsoon storms (accompanied by high humidity and violent lightning strikes) are some of its tricks, which we who live in the desert know well and have learned to respect.

We cannot help but wonder, how many module manufacturers are even aware of the extreme conditions their modules are expected to endure in the deserts. We wonder if they have taken all this into consideration in their design process and manufacturing operations. And since the PV boom of late brought so many new companies in the field, we seriously doubt that any of them have any long-term test results to show.

PV modules are exposed to this harsh treatment non-stop, day after day, and for very long time. They heat up in the morning and overheat at noon, which not only causes a drop in efficiency, but actually creates a minuscule damage in the cell and modules' structures (micro-cracks, electrical contact deterioration, etc.) with every cycle.

Plastics are especially vulnerable under the high IR and UV radiation of the desert sun. With time, they start deteriorating and developing cracks and gaps. Moisture could then enter the module cause damage to the fragile solar cells inside. The daily minuscule damage is compounded for the duration and can at anytime grow larger and more noticeable, eventually destroying the active PV structure inside out..

In many cases, the damage becomes noticeable by visual inspection or electrical measurements (as in the ASU-APS field test study mentioned above).

When damage of any sort is detected, be it visually, or by instrumentation, it means that:

- The modules have deteriorated significantly, and/or
- The power output is reduced above the specified values, and/or
- There is a good chance of further power reduction, and/or
- There is increased chance of partial or total failure with time.

No type or amount of lab testing can accurately reproduce the different types of field failures. There is no test available today that can reproduce 30 years of non-stop desert torture, and so no one can accurately predict the types, level and duration of the different cycles of torture the desert will choose to put the modules through for the duration. Underestimating the power of the desert, however, will surely lead to certain field failure.

On top of that, the modules in our large-scale power field are connected together in long strings that are continuously exposed to the internal torture of several hundred DC volts of electric current running through them—non-stop, day after day, all day long. The electric current also generates significant amounts of heat in the cells and modules in its path to the grid. The sun heats the modules from the outside, while the produced electricity heats them from inside.

Any flaw in cells, modules, and electrical support structures will eventually create even more heat due to increased resistance or sunlight (IR and UV) absorption, or moisture penetration. Things only get worse from here on. Due to the complexity of the external and internal factors at play, no type or amount of lab testing to date can reproduce and predict accurately the effects of internal heat generation under desert conditions.

During many years of PV cells and module testing we have seen different variations of field behaviors, but one thing remains constant in the desert: it is the excess internal overheating of the modules that peaks around the noon hour. Internal mid-summer temperatures of up to 185°F and higher have been measured regularly in silicon PV modules in the Arizona and Nevada deserts. Temperatures abnormalities are exaggerated by any small defect in the solar cell/module design or construction. They can then be amplified and lead to a disaster.

In conclusion, the proper PV module design and choice for large-scale desert operation requires:

- Understanding and respect of the harsh desert conditions and related behavioral abnormalities of PV materials and structures. We must always remember that the desert kills anything left there for a long time, and PV modules (especially the plastic materials in them) are no exception.
- Understanding the long-term behavior of solar cells and active structures, and choosing the best and most appropriate materials and manufacturing processes to fit the extreme operating conditions.
- Understanding the long-term behavior of PV modules operating in the desert.

Caution: Remember: every new module is expected to last a long time under hellish conditions, so it better be made properly and with high quality materials. Any mistake or compromise might bring doom to it, the batch of modules and the entire PV project.

- Designing and executing module construction that is suitable for desert operation by
  — Providing best possible edge sealing to protect the active layers
  — Providing well designed side frame to cover the exposed module edges
  — Avoiding glass/glass module construction, since the edges are unprotected
- Proper and careful field installation as needed to avoid mechanical damage, corrosion and overheating.
- Proper operation and thorough maintenance, starting with careful watch for signs of early mortality and power fluctuations. This is needed to identify failing modules or strings and take immediate action, hopefully before the warranty period has expired.
- There is no way to predict with any certainty what will happen, when and where in the desert, so we need to have contingency plans for all possible circumstances, so
  — Manufacturer's and long-term performance warranties must be well designed to account for, and take care of, any materials, labor and performance deviations,
  — Performance insurance must be well designed to take care of excess power drop, module failures, and any accidents and incidents—man-made or caused by Mother Nature.
- And finally, pray. Pray unceasingly, because the desert is one place where bad things happen on a regular basis; sometimes very quickly and in an awesome fashion.

THIN FILM PV TECHNOLOGY

Thin film photovoltaic (TFPV) technologies are a relatively new branch of the solar industry, which has grown much faster in popularity and size than the other PV technologies during the last several years. Since the active layers in TFPV cells and modules are deposited in the form of thin films, via thin film deposition methods, we refer to them as "thin film" PV products.

Thin films of special photovoltaic materials can produce solar cells or modules with relatively high conversion efficiencies, while at the same time using much less semiconductor material than c-Si cells. In addition, thin film equipment and manufacturing methods allow efficient, cheap, fully automated mass production which is the reason for their success lately.

This, however, comes at the expense of reduced efficiency (average 6-9%), which is not expected to increase much in the future (in mass production mode). On the other end of the efficiency spectrum, multi-layer thin film CPV cells have reached efficiencies over 40% and are getting higher by the day with theoretical efficiency limits in the 80% range.

Due to their versatility, TFPV products have become very popular for use in a number of applications. Recently they have gained a share in large-scale installations as well. Some TFPV modules also show better efficiency under reduced solar radiation than the c-Si competition. This is very useful in many regions with cloudy climates, and could account to their quick rise in European and other world energy markets.

The major types of TFPV technologies considered for commercial and large-scale installations are:
- Cadmium telluride thin films
- CIGS thin films
- Amorphous silicon thin films
- Silicon ribbon
- Epitaxial silicon thin films
- Light absorbing dyes thin films
- Organic/polymer thin films
- Ink thin films
- Nano-crystalline cells
- Indium phosphide
- Single-junction III-V cells
- Multi-junction cells
  — Gallium arsenide based cells
  — Germanium based cells
  — CPV solar cells

Thin Films PV Structure

As the name implies, "thin films" are just that; very thin films (layers) of organic or inorganic materials. Thin film PV (TFPV) modules consist of several very thin layers (thin films) of different materials piled up on top of each other, to form a structure that is suitable for trapping and converting sunlight into electricity. The thickness of each film is usually several microns (1 micron is 0.001mm, or 0.0004 inch). Visualize the thickness of a human hair (avg. 100 microns) and you'll get a good idea of what 100 microns is—50-100 times the thickness of an individual thin film.

Now visualize layering these super-thin films until there are 3, 4, 8 or 10 layers. This is how TFPV structures are made and it is what they look like. A better visualization might be using a piece of Scotch tape as an example. Standard Scotch tape is ~0.060mm thick, or 60 microns, which is at least 10-20 times thicker than most thin films.

Yes, the entire thin film structure—all different layers combined in your TFPV modules is many times thinner than a Scotch tape strip. And the different layers (different chemical compounds) are even thinner. They are stacked on top of each other, held together by weak electro-mechanical forces of complex nature and behavior. We'll take a closer look at these forces and the related interactions in this text.

Thin films of any type and size are affected by chemical, mechanical, thermodynamic and electric forces and changes in, between, and around them. TFPV structures also depend heavily on the surrounding materials, glass, laminates, etc., and components in the PV module for protection. Complex picture, yes, but well understood by design and process engineers and research scientists, thanks to the broad experience we've gained from the semiconductor industry, which is based on thin films and processes.

Note: Remember this picture, because we will revisit it later on in this chapter, to explain the behavior of thin films in TFPV modules under different environmental conditions.

Thin Film Manufacturing Process

So let's see how these thin films and TFPV modules are made today:

Thin Film Deposition Processes

Thin film PV cells and modules manufacturing processes, similar to thin film processes used in the semiconductor industry, are well controlled. The level of process control depends only on the quality requirements and budget restrictions. The actual thin films deposition is usually done on a substrate (glass, plastic or such) which has been thoroughly inspected, cleaned and otherwise properly prepared for the deposition step.

- Substrate inspection and preparation is a key factor in maintaining process control, and determines the overall quality, performance and longevity of the final product. Basically speaking, dirty substrate will not only produce defective product, but will also contaminate the equipment, forcing lengthy and expensive clean-up procedures. The pre-deposition process starts with a thorough visual inspection of the glass substrate (glass panes)

- Pre-deposition cleaning of large glass substrates (panes) is done in automated washers, where brushes, soap and/or high pressure water solutions remove all particles and organic material from both surfaces. The glass panes are then rinsed with DI water and dried with forced air and heat applied to both surfaces. This step is also critical, because moisture retained on the large surfaces is a great enemy of vacuum/plasma processes and could seriously affect product quality.

- The substrate is then placed in a large vacuum chamber (usually on a horizontal or vertical conveyor belt that moves the material along the process path), where powerful vacuum pumps suck out all the air, to clear any mechanical (dust) and chemical contaminants (reactive gasses and water). The substrates are then usually heated, to remove any residual moisture and to heat the surface close to the temperature of the deposited species, reducing potential thermal disequilibrium and stress of the thin films to be deposited on it. Argon gas-based DC or RF plasma is ignited and maintained at a proper pressure and power density during the process, facilitating the deposition process and the related reactions.

- The material to be deposited as thin film is then evaporated (by actually melting the material and directing the resulting vapor clouds onto the deposit surface), or it is sputtered (by dislodging small clusters of the material via high voltage-generated ion bombardment) onto the substrate. These processes are also called chemical vapor deposition (CVD) and physical vapor deposition (PVD), respectively.

In both cases thin film material particles impinge on and adhere to the heated substrate surface on impact. The strength of the adhesion between film and substrate, or between the individual films, depends on the design and execution quality of the entire process sequence—quality of all materials, cleanliness of substrate and chamber interior, vacuum integrity, actual process temperature, forward and reverse plasma and substrate bias power levels, time duration, partial and total gas pressures, speed, and other process variables.

- Coated substrate with the films deposited on top of it is taken out of the chamber and exposed to a number of additional operations, such as wet chemistry treatment, rinsing, drying, annealing, and wire attachment. The above CVD, PVD and wet chem processes are repeated several times for some devices, following strict process and quality control procedures all through the sequence. Upon completion of the PV structure creation, the substrate with the deposited thin films is joined to similar substrate (glass usually) with the help of adhesives and encapsulants. Thus, TFPV modules are tested, sorted and packaged for shipment to lucky customers.

Process and Device Issues

No process is perfect, and a defect could be generated at any step of the manufacturing process, including the tightly controlled thin film deposition processes in the semiconductor fabs. Starting with the basics, materials selection is paramount for obtaining quality product.

"Garbage in, garbage out," goes the wise man saying. So the quality of the substrates, process chemicals and gasses, and all consumables must meet and exceed specifications—not a simple task. The different pieces of process equipment must be well taken care of, tuned and qualified periodically at the very minimum at the beginning of every shift and after each maintenance procedure. This is necessary to make sure all components are operating within spec.

- Step one of the thin film process is cleaning the substrate materials. Contaminated cleaning materials or equipment, or improper procedures will render the glass surface unsuitable for proper adhesion of deposited films. Cleaning the substrate with poorly selected chemicals, or shoddy procedures, might even introduce fatal impurities, such as Cu, K, Na, Fe metal ions, and phosphates. These impurities would eventually start their own demolition processes from within the thin film structure enclosed in the PV module, reducing its efficiency and longevity.

Dirt, particles, fingerprints, residual moisture, and strong electrostatic charge on the glass surface entering the vacuum chamber will have a profound, usually negative, effect on process integrity and final product quality. These are only few of the things to watch for at this stage of the process.

- The plasma deposition process could also introduce impurities and defects in film structures if improper parameters or dirty hardware are used. Impurities in carrier gasses, contaminated vacuum chamber walls, back-streaming vacuum pumps, unplanned drops or increases of total or partial pressure, air leaks, process time or belt speed discrepancies are process abnormalities which could

introduce other impurities or create defects with immediate or latent problems.

Out-of-spec process materials and consumables could be blamed for many process failures. Poor quality substrates, contaminated gasses and chemicals (overlooked during incoming quality control procedures) could cause slight or very great defects in the finished product. But most often, it is the process itself that creates problems in thin film manufacturing.

Generally speaking, there are a number of known, unknown, controlled and uncontrolled manufacturing process factors and variables in the thin film deposition sequence:

- Human error (ignorance and/or negligence) is the #1 variable in high-volume thin film manufacturing. People tend to cut corners, improvise, push the wrong buttons, rush through operation, maintenance and inspection steps, etc. These kinds of workplace behaviors cause defects of unpredictable proportions and consequences. Handling substrate materials and consumables, as well as operating the complex process equipment requires highly trained engineers, technicians and operators, who in many cases have bad habits that are hard to break even with the best of training.

- Equipment's proper operation, without any deviations or malfunctions, is #2 on the list of process-related variables in sophisticated thin film manufacturing operations. Thin film process equipment is complex in its design, operation and maintenance. Most production equipment made now is of good quality, but even the best equipment can malfunction and create headaches.

Key components and process control instrumentation slowly drift out of spec over time, causing product quality variation within batch, or from batch to batch. One serious drift in the multi-step process could cause serious defects in one or several batches. Anomalies during processing also occur unexpectedly and quickly. If not handled promptly and properly, they could cause final product quality problems.

- Poor quality supplies, materials and chemicals purchased from third-party manufacturers are another serious problem that we encounter in the thin film process. Since high volume operations use a lot of outsourced materials, the incoming quality is hard to verify 100% of the time, so any problem at the third-party vendor's plant will have negative effects in the final PV product quality.

• And of course, the operators, engineers and management have a lot to do with the proper, efficient and quality execution of each step of the process.

A number of additional factors and variables affect the quality of thin film PV modules, but it suffices to say that the TFPV manufacturing process consists of dozens of different materials and complex process steps, each of which must be immaculately designed, planned, executed and controlled.

Basically, the lowest quality of any process step (or material) in the process sequence determines the highest quality of the final product. One mishandled step, or one bottle of contaminated chemical or gas, or one negligent operator, could lead to rejection of a batch of modules, or worse—to their failure in the field, where things get very expensive…and embarrassing.

TFPV Functional Considerations

Even though the electro-mechanical and chemical properties of PV thin films are thoroughly studied and well understood, improper design and manufacturing procedures are still encountered at times. We already discussed the TFPV manufacturing process and the related issues, so now we'll take a closer look at the behavior of thin films in TFPV modules. (Refer to Figure 5-23.)

Remember our discussion at the beginning of this chapter that thin films are 1 micron thin, or thinner, and that this is less than 1/100 the thickness of human hair. Recall, too, that the entire thin film structure in TFPV modules is several times thinner than a piece of Scotch tape.

Some of these thin films are so thin in places that you could actually count the molecules across the film thickness, if you had a way to do that. And as you look at thin film very closely, you'll see all kinds of imperfections as well; interruptions, distortions, breaks, cracks, splits, slips, pits, voids, bubbles, lose particles, and all kind of other sub-structures in/on the thin film structure. In other words, the films, although they look perfectly smooth and uniform to the naked eye, are anything but.

And you will also notice in Figure 5-23 a clear delineation between the films—the boundary (interface) between each pair. These are very special areas, which play a huge role in the performance of the PV devices made out of these films. They are also the weakest points in most thin film structures, because it is where the electrical resistivity, mechanical stress and chemical degradation are usually initiated, stored, executed and amplified. These boundaries represent the time and

place where the process was interrupted and switched from one material deposition to another; and/or from one deposition step to another. This involves moving the substrate, which could be accompanied by longer than specified delays, switching gasses, changing vacuum and plasma power levels, and a number of additional inter-step process modifications.

The inter-step changes do cause momentary out-of-control conditions which could cause a number of process and product abnormalities, such as contamination of the surface by carrier gases or pump oil, and temporary process destabilization (total and partial pressures, gas mixing, power fluctuations, temperature imbalance, etc.).

All the combinations and permutations of possible process abnormalities and extraordinary scenarios is too much to discuss here, but you get the idea; this is a most complex set of parameters and conditions that must be executed perfectly all the time at all process steps, to obtain the best quality of final product.

For the purposes of this writing, however, we'll just agree that the inter-layer boundaries (interfaces) are extremely critical areas, which have different properties than the parent materials. These areas are more fragile, so there is a limit as to how much abuse they can handle during long-term field operation.

Figure 5-24. Cross section of a typical TFPV structure

Note in Figure 5-24 that the entire thin film structure is less than two microns. Also note that there are gaps within and between the layers. This is normal condition, which is determined by the materials, the process parameters, and their execution. No two thin films would be alike, due to slight variations in the quality of materials and the process execution.

Each material pair and its interface in the thin film stack has different properties. The thin films have well defined mechanical, chemical, electric and thermal stress limits. Equally so, the interfaces also have different properties, depending on the critical de-bonding energy and chemical inertness levels specific for each interface between two different thin film materials.

How much mechanical stress, heat, electrical charge, chemical reactivity or combination of these will be needed to cause delamination, adhesion problems, mechanical, electrical or chemical changes and/or disintegration of the bond between the materials, and eventually the materials themselves?

Amazingly enough, the energy levels and forces acting in these areas are very small, relative to the mass of the materials and strength of the films. So, without getting into much technical detail, just looking at such extremely small numbers. one can deduct that it doesn't take much effort to damage an interface which will affect the entire thin film structure.

Additionally, in most cases, many of these forces are inter-acting and counter-acting constantly—on and off many times daily for the duration (30 years or more) of the module's on-sun operation. Worst of all, once any destructive mechanical (stress) or chemical (oxidation) processes have started, they cannot be stopped. As a matter of fact, things usually only get worse with every daily cycle, and can accelerate quickly.

Taking a closer look at the major destructive processes acting upon thin film PV structures, we see:

Mechanical Stress

Mechanical stress. Thin film modules are exposed to mechanical stress from the moment they leave the production line. Never ending vibration, hitting, rubbing, pushing, pressing and squeezing of the modules, affect the layers in the thin film structure. Stress is induced during handling, packing, transport (long truck and train rides are the worst), installation (careless handling during installation is a major problem) and operation (high winds, hail and storms are some of the worst enemies).

Each of these actions and interactions increases the mechanical stress in the film structure.

Because of the films' non-uniformity, and the different coefficients of expansion of the different adjacent films, there is a lot of stress and friction within each film and among the adjacent films. As with other materials stuck together, they will be slipping and sliding, expanding and contracting, thus creating never ending bending, pulling, pushing and tugging against each other and the materials around them, be it thin films or encapsulants, glass etc.

In most cases these activities will produce some usually unpredictable changes. The question is what and how bad will these changes be? Stress, cracks, voids and general weakening of the thin film system is expected with time, followed at times by partial or full

disintegration of the films, depending on operating conditions. Results from these changes would be expressed as gradual loss of power, intermittent power, and finally complete failure.

Electrical and Heat Stress

Generation of electric energy is the primary purpose of any PV module. The photoelectric process and the accompanying extraction and transmission of electrons (electric current) through the different thin film layers, their interfaces and contacts is usually accompanied by heat generation. Parasitic (excess) heat is generated when the resistivity of the materials increases, which inevitably happens when the internal temperature of the module goes up. And it starts going up from the second the sun hits the module. The higher the sun, the more electricity is generated, and the higher the resistivity goes.

On a bright sunny day in the desert, the module interior could see temperatures as high as 180°F, or more. The heat build-up is a combination of the simultaneous increase of air temperatures and internal resistivity of the thin films and their interfaces. This heat is not enough to damage or destroy the films, since each one can withstand much higher temperatures. The excess heat, however, forces the materials to expand and shift in one direction. They then shrink at night and move in the opposite direction, which creates friction at the interfaces and stresses the materials in each layer. The more temperature differential the layers are exposed to, the more they stress and shift.

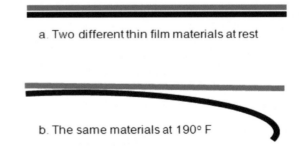

a. Two different thin film materials at rest

b. The same materials at 190° F

Figure 5-25. Thin film behavior under elevated temperature and humidity

Now imagine these films, packed tightly in the module, getting very hot during the day in the desert (measured inside module temperatures exceed 190°F in Arizona deserts), and freezing at below –20°F at night in winter. This process goes non-stop, 365 times a year. This translates into ~3.650 min-max temperature cycles in 10 years, and close to 11,000 cycles in 30 years.

Could any thin films in the PV modules withstand the constant, 30 years worth, of stress and overheating?

After awhile—one, five, ten years—one of them, or its interface, could give up and break down mechanically, or disintegrate chemically? Would that cause an untimely power drop, or complete failure, of the modules?

The results from one, two, or even hundreds, of these cycles may go unnoticed, but the non-stop (30 years and over 10,000 cycles) push and pull in and on the layers will fatigue thin film structure to an extent, thus shifting it into a different performance mode. As the module gets older, the resistivity of the films and their interfaces increases proportionally, according to the internal heat build-up and dissipation within the module. Electrical output will drop by ~1.0% every year, partially due to the above effects, or combinations of them. In some cases, the power output drop is even larger, and tends to increase with time.

There is no getting away from the harmful effects of stress, excess heat and moisture in field operations. They are variables in the PV generation equation, which cannot be ignored, nor can they be efficiently controlled.

Highest quality materials and processes are the only solution to long-term success of PV modules.

Chemical Interactions

The thin films in TFPV modules are well protected from the elements by the glass-glass structure. The problem is that the two glass panes are 'glued' together and the edges sealed with a plastic resin. With time, the never-ending UV and IR radiation attack the edge sealer (plastic resin) and eventually it would develop cracks and gaps. This cannot be avoided, especially in the deserts Once that happens, the thin films inside are easy pray to moisture and chemical gasses in the atmosphere around them.

All unprotected thin films—without exception—react with many chemicals and water. It is usually the interface (the boundary between the films) that is affected first, and is where problems can be observed. Some films and interfaces will disintegrate instantly with a simple touch (human sweat contains salts), while others will withstand weak chemicals, and some will require strong acids or bases in order to dissolve or decompose them or their interface. Nevertheless, all thin films are subject to mechanical and chemical changes under certain conditions.

CdTe, CIGS, a-Si and other thin film structures, as well as interfaces between layers, are affected in a similar fashion. Some of them are more chemically resistive than others, but prolonged exposure to weak acids (brought on by moist air, or contained in rain water) penetrating the module lamination (or entering through cracked glass) will eventually cause the films to react, delaminate and decompose. The type and speed of the decomposition process, and the newly created chemical species, will be determined by the types of thin films, the active chemicals species and the types of reactions these invoke.

Again, we count on the encapsulants and glass or metal frame to keep moisture, air and related chemicals out of the module for 30 years. How many modules will survive the test of time? What would be the total failure rate? These are questions to which we simply have no answers because there are no precedents.

Moisture ingress is the culprit of pronounced power output degradation of CIS and CdTe modules in Figure 5-27, reducing their output significantly within 6 months of exposure. This example also leads to the conclusion that good encapsulation is paramount, but since it is never perfect, harsh climate conditions will force moisture to penetrate thin film structure and change its composition and behavior, resulting in power loss and, eventually, failure.

Environmental effects initiate and accelerate the evolution of defects in TFPV modules. Damaged encapsulation allows moisture to penetrate modules and attack layers, resulting in their delamination or separation from the substrate, as shown in Figure 5-27.

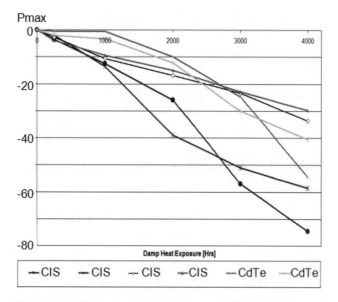

Figure 5-26. Damp heat degradation of CIS and CdTe modules

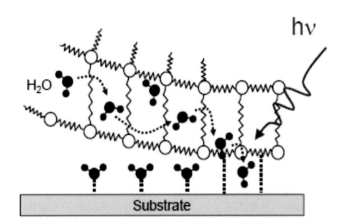

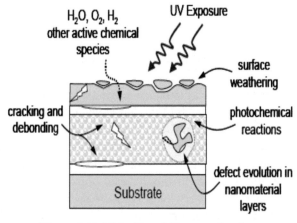

Figure 5-27. The chemical invasion process on molecular level

Figure 5-29. Thin films failure mechanisms

Just as in tooth decay, once moisture or environmental acids and gasses find a gap in the encapsulation they start the decomposition process which cannot be stopped without outside intervention. A dentist has the option to mechanically separate (drill) decay out of the tooth surface, but drilling decomposition damage out of a TFPV module is impossible. This means that modules will start losing power at a rate proportional to the inflicted damage, usually much quicker than the standard 1.0% power loss per annum.

Eventually—when the internal decay has affected large parts of the critical areas of the thin film structure—the affected modules will fail completely and must be replaced. The process is accelerated when affected modules are exposed to extreme heat and humidity (see Figure 5-28).

The sum of effects depicted in Figure 5-28 play a significant role in compromising thin film structure integrity. Some of these are vicious and fast acting, while some are slow and cause less damage. The actions of

each effect are hard to predict, because different operating conditions have their own peculiarities. It is even harder to predict the combined effect of these processes during non-stop, 30-year, on-sun operation.

The inevitable exposure to excess UV and IR radiation, thermal cycling, mechanical stress, moisture ingress, and chemically active environmental species leads to unpredictable degradation, and unreliable kinetics and reliability models, because of the ever changing types and numbers of forces acting on the materials in inhospitable regions.

In summary, we don't know what to expect of fragile thin film layers in TFPV modules exposed to unending mechanical, chemical, electrical and thermal action during on-sun operation for 30 years or more. Still, there is enough evidence to conclude that PV modules made with quality materials via proper processing would have a greater chance to survive over time than those made of low-quality materials, and/or using poor design and flawed manufacturing processes.

Below we review each of these technologies and their specific structure and function, focusing on their use in large-scale PV power generation and related issues.

Thin Film PV Module Tests

Thin film PV (TFPV) modules are tested in a fashion similar to that used for testing Si-based PV modules, described above. We will, therefore, skip the testing details, except the one that is very important, especially in the case of TFPV modules.

Note: One key difference to keep in mind here is that while Si-based PV modules have a solid metal frame covering the back and sides of the structure, TFPV modules often (in most cases) are without a metal frame all together.

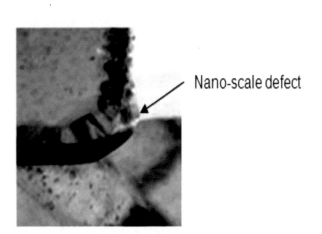

Figure 5-28. The invasion process has started

TFPV modules usually consist of two glass plates, between which the thin film structure is encapsulated. The encapsulation is, therefore, the only thing separating the fragile thin film components from destruction by moisture and the elements.

Since the encapsulation consists of plastic compounds, it is no match to the ferocious, non-stop exposure to excess heat, moisture, and UV radiation. Even a small defect or crack in the encapsulation could allow the elements to destroy the entire thin film structure inside the module. In the deserts, plastic encapsulation does not last very long. We have seen some types, used by world class manufacturers, disintegrate within months, while other types last several years. In no case, however, can we guarantee any plastic encapsulation to last 100% intact for 5, or 10; let alone 20 or 30 years.

No plastic is match for the desert elements.

This is why, special attention and is paid to the so called *damp test*, which, albeit partially, tests the integrity of the encapsulation of TFPV modules.

Damp Heat (moisture) Tests

This test seems to be the weakest point of TFPV technology, so a closer look is justified. Damp heat studies of CIGS and CdTe cells have been conducted by subjecting cells and mini-modules to an environment of 60°C/90% RH (relative humidity).

Two key conclusions can be made:
* Both CIGSS and CdTe cells degrade rapidly under 60°C/90% RH, if not adequately protected, and
* At the very least, the damp heat stress will cause changes in junction transport properties and minority carrier transport characteristics of the cell absorber, thus significantly reducing the cell's output.

The damp tests in Figure 5-30 show fatal degradation for the un-encapsulated thin film's structure, which confirms that:
* Thin film structures will be attacked, disintegrated and decomposed if and when exposed to the elements, and
* Thin film structures depend exclusively on encapsulation for protection from the elements.

A series of tests performed by the author, and other researchers, with different metal and non-metal TF structures (solar cells and other applications) exposed to hot and humid conditions ("sweat box" tests under dif-

ferent temperature and humidity regimes) show clearly that bare thin film structures (any type and combination thereof) will degrade if exposed to damp heat. See Figure 5-30.

Even the most resistant and inert to chemicals (Ti, Cr, etc.) do succumb and disintegrate after constant heat and moisture attacks. One of the explanations is that moisture in ambient air carries all kinds of contaminants, which create acids and other harmful chemical mixtures. Upon contact, these mixtures attack the thin films and can damage them.

Even if the chemical mixtures do not affect the thin film itself, they can get among and underneath the thin film layers, and cause separation of the films from each other, or from the substrate. There is no exception to this rule, so the only protection the otherwise fragile thin films in TFPV modules have is the encapsulant under the glass cover. If and when it cracks due to excess heat or age, moisture could enter the module and the thin films in it would be damaged and destroyed.

The destruction process usually starts slowly at the films' interfaces, and then accelerates quickly, causing delamination and/or destruction of the entire TF structure. TF structures enveloped in impermeable materials have a better chance to survive this test. An important conclusion here is that the modules' encapsulation is the only thing separating them from destruction and total failure.

This raises a serious question, "What type, and how much, protection is needed to ensure proper operation of TFPV modules, for 30 years in excessively hot and humid areas?" It is obvious that encapsulation determines performance and longevity of modules, because once moisture and other environmental chemicals and gasses reach the thin film structure, it undergoes unpredictable changes. These changes depend on the thin film composition and stability, and also on the nature of the attacking species, but most often the attacks will

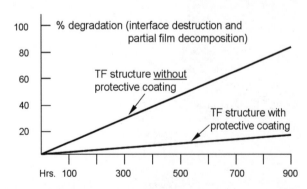

Figure 5-30. Damp test of thin film structure with and without protective coating

result in reduced power output and eventually lead to total failure.

In summary, thin-film PV modules are different from conventional c-Si modules, as follow:

- The thin film structure is deposited directly on the front or rear glass plate, thus it is exposed to outside elements in case of delamination of the encapsulant from the glass surface.
  Note: In Si-based modules, even if the EVA delaminates from the glass, the cells are still enveloped in it, thus they are still protected from the outside elements. This is a huge difference in longevity and reliability.

- Most TFPV modules, unlike Si PV modules, usually have no side frame (metal frame around the edges of the module), which leaves the fragile thin films inside fully dependent on the edge sealers, which are exposed to the elements and could be easily damaged or destroyed by heat and UV light.

This is an important differences, because leakage via unframed module's edges can easily reach the active thin film structure, thus damaging it and causing gradual power decrease and eventually total failure.

We must emphasize also that even the best damp-heat test scenarios and exceptionally good test results cannot guarantee 20-30 years damage-free encapsulation under excess heat and humidity conditions.

This, in fact, is one of the best kept secret of the industry. Damage of TFPV modules by the elements is something that has been consistently and systematically downplayed by the major manufacturers. The public and the governments have been buying it, but we have seen field failures caused by intense heat and damp heat phenomena already, and expect to see many more in the future.

The large TFPV installations, with millions of CdTe and SIGS PV modules in each are still brand new and the failures are not that obvious. We will revisit this problem in the large-scale TFPV fields 10-15 years down the road, in order to calculate the failure rates due to heat and moisture—especially in the desert regions of the U.S. and the world.

Major Thin Film PV Technologies

We classify the PV technologies reviewed below as "major" for the purposes of this text, because they are presently considered feasible for use in PV power generation projects. A number of additional thin film PV technologies, some of which we review below are also

feasible, but have not been used in large-scale projects.

A third group of thin film PV technologies are still in the planning or R&D stages. We have reviewed these in a previously published book, Solar Technologies for the 21st Century, by The Fairmont/CRC Press.

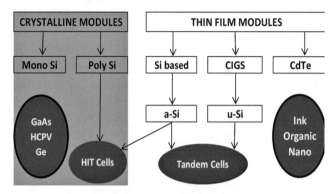

Figure 5-31. Key PV technologies 2010

The major thin film technologies are the: cadmium telluride, CIGS, and amorphous silicon thin films:

Cadmium Telluride (CdTe)

Cadmium telluride (CdTe) is a type of solar cell and module based on thin films of the heavy metal cadmium and its compounds, cadmium telluride (CdTe) and cadmium sulfide (CdS). CdTe is an efficient light-absorbing material, quite adaptable for the manufacture of thin-film solar cells and modules. Compared to other thin-film materials, CdTe is easier to deposit in mass production environments and more suitable for large-scale production.

CdTe bandgap is 1.48 eV, which makes it almost perfect for PV conversion purposes. At 16.5% demonstrated efficiency in the lab, it is a candidate for a major role in the energy future. Mass production modules are sold with 8-9% efficiency. No significant increase is expected with the present production materials and methods, although manufacturing costs are down—well below \$1.0/Wp. This price drop, however, is in response to pressure from low-cost c-Si manufacturers (mostly Chinese), so it might be a temporary situation. Prices in the \$1.00-\$1.50 seem most reasonable for TFPV modules.

With a direct optical energy bandgap of 1.48 eV and high optical absorption coefficient for photons with energies greater than 1.5 eV, only a few microns of CdTe are needed to absorb most of the incident light. Because only very thin layers are needed, material costs are minimized, and because a short minority diffusion length (a few microns) is adequate, expensive materials processing time and costs can be avoided.

The structure of a CdTe TFPV module, as shown in Figure 5-32, consists of a front contact, usually a transparent conductive oxide (TCO), deposited onto a glass substrate. The TCO layer has a high optical transparency in the visible and near-infrared regions and high n-type conductivity. This is followed by the deposition of a CdS window layer, the CdTe absorber layer, and finally the back contact.

Figure 5-32. CdTe/CdS thin-film solar cell

For high-volume devices, the CdS layer is usually deposited using either closed-space sublimation (CSS) or chemical bath deposition, although other methods have been used to investigate the fundamental properties of devices in the research laboratory. In all cases, mass production and automation is possible, which is the greatest advantage of this technology.

The CdTe p-type absorber layer, 3-10 μm thick, can be deposited using a variety of techniques including physical vapor deposition (PVD), CSS, electrodeposition, and spray pyrolysis. To produce the most efficient devices, an activation process is required in the presence of $CdCl_2$ regardless of the deposition technique. This treatment is known to recrystallize the CdTe layer, passivating grain boundaries in the process, and promoting inter-diffusion of the CdS and CdTe at the interface.

Forming an ohmic contact to CdTe, however, is difficult because the work function of CdTe is higher than all metals. This can be overcome by creating a thin p+ layer by etching the surface in bromine methanol or HNO_3/H_3PO_4 acid solution and depositing Cu-Au alloy or ZnTe:Cu. This creates a thin, highly doped region that carriers can tunnel through.

However, Cu is a strong diffuser in CdTe and causes performance to degrade with time. Another approach is to use a very low bandgap material, e.g. Sb_2Te_3, followed by Mo or W. This technique does not require a surface etch and the device performance does not degrade with time.

CdTe PV modules manufacturing is a sophisticated process; much more sophisticated than that of the conventional c-Si modules process, which uses simple 1970s manufacturing equipment, materials and processes. CdTe TFPV modules are manufactured with the help of modern, complex and expensive semiconductor type equipment and processes. Because of that, the precision and accuracy of the resulting process steps, and ergo the quality of the final product, are limited only by the quality of the materials and supplies, and the capabilities of the engineers, technicians and operators on the production lines.

CdTe thin-film solar modules are now being mass produced very cheaply, and it is expected with economies of scale that they will achieve the cost reduction needed to compete directly with other forms of energy production in the near future. Since CdTe thin film PV devices still have a long way to go to achieve maximum efficiencies, it will be interesting to see which materials and methods are most successful.

The most efficient CdTe/CdS solar cells (efficiencies of up to 16.5%) have been produced using a Cd_2SnO_4 TCO layer which is more transmissive and conductive than the classical SnO_2-based TCOs, and including a Zn_2SnO_4 buffer layer which improves the quality of the device interface.

CdTe PV research, done by manufacturers, universities and R&D labs focuses on some of these challenges:

- Boosting efficiencies by exploring innovative transparent conducting oxides that allow more light into the cell to be absorbed and at the same time more efficiently collect the electrical current generated by the cell.
- Studying mechanisms such as grain boundaries that can limit voltage.
- Understanding the degradation some CdTe devices exhibit at the contacts and redesigning the devices to minimize this phenomenon.
- Designing module packages that minimize any outdoor exposure to moisture.
- Engaging aggressively in both indoor and outdoor cell and module stress testing.

These efforts are geared to address the main problems with CdTe PV modules: a) the relatively low efficiency which contributes to using more land and mounting hardware, b) temperature power degradation, c) annual power degradation, and other negative long-term effects.

Availability of the rare metals used in CdTe TFPV technology and their toxicity are other serious issues which manufacturers and regulators have put on the back burner. Let's hope that we won't have to wait for a serious accident before bringing these issues out in the open and discussing possible solutions.

CIGS Solar Cells

Early solar cells and modules of this type were based on the use of CuInSe2 (CIS). However, it was rapidly realized that incorporating Ga to produce Cu(In,-Ga)Se2 (CIGS) structure, results in widening the energy bandgap to 1.3 eV and an improvement in material quality, producing solar cells with enhanced efficiencies. CIGS have a direct energy bandgap and high optical absorption coefficient for photons with energies greater than the bandgap, such that only a few microns of material are needed to absorb most of the incident light, with consequent reductions in material and production costs.

The best performing CIGS solar cells are deposited on soda lime glass in the sequence—back contact, absorber layer, window layer, buffer layer, TCO, and then the top contact grid. The back contact is a thin film of Mo deposited by magnetron sputtering, typically 500-1000 nm thick.

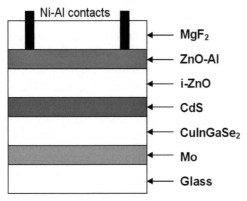

Figure 5-33. CIGS cell cross section

The CIGS absorber layer is formed mainly by the co-evaporation of the elements either uniformly deposited, or using the so-called three-stage process, or the deposition of the metallic precursor layers followed by selenization and/or sulfidization. Co-evaporation yields devices with the highest performance while the latter deposition process is preferred for large-scale production.

Both techniques require a processing temperature >500°C to enhance grain growth and recrystallization. Another requirement is the presence of Na, either directly from the glass substrate or introduced chemically by evaporation of a Na compound. The primary effects of Na introduction are grain growth, passivation of grain boundaries, and a decrease in absorber layer resistivity.

The junction is usually formed by the chemical bath deposition of a thin (50–80 nm) window layer. CdS has been found to be the best material, but alternatives such as ZnS, ZnSe, In2S3, (Zn,In)Se, Zn(O,S), and Mg-ZnO can also be used.

The buffer layer can be deposited by chemical bath deposition, sputtering, chemical vapor deposition, or evaporation, but the highest efficiencies have been achieved using a wet process as a result of the presence of Cd2+ ions. A 50 nm intrinsic ZnO buffer layer is then deposited and prevents any shunts. The TCO layer is usually ZnO:Al 0.5–1.5 μm. The cell is completed by depositing a metal grid contact Ni/Al for current collection, then encapsulated.

CIGS solar cells have been produced under lab conditions with efficiencies of 19.5%, and lately modules with efficiencies of 15.7% were verified as well. Commercial, mass produced, CIGS PV module efficiency, however, is still lower than CdTe PV modules—and this will have a major impact on their future unless ways to increase their efficiency and reduce their costs are found soon.

CIGS TFPV modules have similar problems as those plaguing CdTe TFPV technologies. They have low efficiency, require larger mounting infrastructure, exhibit power loss under excess heat and have a significant annual degradation rate. Scarcity of materials and related toxicity issues are, as in the CdTe PV case, on the back burner for now.

These issues must be evaluated from the point of view of large-scale installations, where thousands and millions of these modules will be installed. In such cases, minute amount of toxic materials in each module are multiplied mega times and become a substantial threat to the environment. Also, special measures must be taken for proper disposal or recycling of these modules.

CIGS research is focused on several of today's challenges of this promising technology:

- Pushing efficiencies even higher by exploring the chemistry and physics of the junction formation and by examining concepts to allow more of the high-energy part of the solar spectrum to reach the absorber layer.
- Dropping costs and facilitating the transition to a commercial stage by increasing the yield of CIS modules—which means increasing the percentage of modules and cells that make it intact through the manufacturing process.

- Decreasing manufacturing complexity and cost, and improving module packaging.

Note: At a meeting of PV specialists in February 2011, called PV Module Reliability Workshop (PVM-RW), the degradation and longevity of PV technologies and products were discussed by representatives of several solar products manufacturing companies. The susceptibility to moisture of SIGS thin film modules was addressed as one of the major concerns, and packaging solutions were presented.

Location-specific reliability tests and evaluations was also one of the topics, which is a step in the right direction. We are glad that such open discussions are underway, since this is the fastest way to resolve the issues and put the promising technologies on the energy market.

Amorphous Silicon

Amorphous silicon (a-Si) is produced via thin film processes, based on depositing thin layers of silicon films on different substrates. Silicon thin-film cells are mainly deposited by chemical vapor deposition (CVD), typically plasma-enhanced (PE-CVD), using silane and hydrogen reactive and carrier gasses for the actual deposition. Depending on the deposition parameters and the stoichiometry of the process, this reaction can yield different types of thin film structures, such as amorphous silicon (a-Si, or a-Si:H), protocrystalline silicon or nanocrystalline silicon (nc-Si or nc-Si:H), also called microcrystalline silicon.

These types of silicon feature dangling and twisted bonds, which result in deep defects (energy levels in the bandgap) as well as deformation of the valence and conduction bands (band tails), which lead to reduced efficiency. Proto-crystalline silicon mixed with nano-crystalline silicon is optimal for high, open-circuit voltage.

Solar cells and modules made from these materials tend to have lower energy conversion efficiency than those made from bulk silicon, but have some operating advantages (such as lower temperature degradation). They are also less expensive to produce, although the capital equipment expense is greater, due to equipment complexity.

a-Si has a somewhat higher bandgap (1.7 eV) than crystalline silicon (c-Si) (1.1 eV), which means that it absorbs the visible part of the solar spectrum more efficiently than the infrared portion. nc-Si has about the same bandgap as c-Si, so nc-Si and a-Si can advantageously be combined in thin layers, creating a layered cell called a "tandem cell," where the top a-Si cell ab-

sorbs the visible light and leaves the infrared part of the spectrum for the bottom cell in nc-Si.

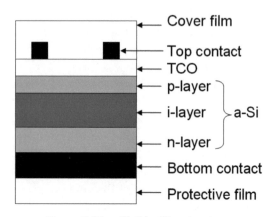

Figure 5-34. a-Si thin film structure

The biggest problem with a-Si TFPV technology and a barrier to its success, however, is its low efficiency. Today's best cell efficiencies are about 12% in the lab, which is almost 50% lower than other PV technologies. Mass produced a-Si cells and modules are in the 8% efficiency range today.

A second problem with a-Si is its manufacturing cost as related to initial capital investment, which is quite high as compared with the competing PV technologies. Two proposed solutions to this problem are higher manufacturing rates, and batch (simultaneous) processing of multiple modules. Good progress has been made in rates that are 3-10 times higher than those being used in production, but all this is still on a lab scale and is to be proven in reality and on a large scale.

On the positive side, while some of the more efficient cells and modules lose about 20-30% of their output in the field, due to excess heat exposure, a-Si loses only about 5-10%, due to its lower temperature coefficient.

Also, the active thin film structure is composed mainly of silicon films which have inert and homogeneous natures that show better chemical and mechanical stability than some of the competing thin films—in case of an encapsulation failure. a-Si modules are also more resistive to the negative effects of shading in the field.

Of equal importance is the fact that a-Si modules do not contain any hazardous materials, which is paramount where large-scale PV installations are concerned. These qualities put a-Si on the top of the list of PV technologies suitable for large-scale power generation in deserts and other inhospitable areas.

Even with low efficiency (well under 10%), a-Si thin film technology is being successfully developed for building-integrated photovoltaics (BIPV) in the form of semi-transparent solar cells which can be applied as window glazing. These cells function as window tinting while generating electricity. It remains to be seen if the amount of generated electricity covers initial and operating expenses.

A triple-junction a-Si TFPV power system has been operating near Bakersfield, CA, for several years, and is providing proof of the excellent performance of this technology. The 500 kW grid-connects system has been performing well, meeting or exceeding its design goals. Performance data from this larger-scale installation confirms data obtained from smaller a-Si systems and proves that this thin film PV technology can be successfully used in large-scale power plants, if the low efficiency can be justified.

Great research effort is underway at universities and R&D labs around the world, geared towards solving efficiency and cost issues and obtaining a-Si that is truly competitive in the energy market. a-Si manufacturers need to work on understanding the key areas of this technology and the processes, and focus on their optimization by:

- Improving the light-stabilized electronic quality of a-Si and low-gap a-Si:H cells to achieve broader spectrum conversion, and increased and stable overall efficiency.
- Increasing the growth rates of a-Si, a-SiGe, etc. layers while maintaining high electronic quality, to obtain increased throughput and reduced capital cost.
- Developing high-growth-rate methods for nanocrystalline silicon while maintaining high electronic quality as needed for increased efficiency, stability and reduced cost.
- Understanding and controlling light-induced degradation in a-Si:H as needed for increased efficiency and understanding of the intrinsic limits of the efficiency.
- Developing in-situ in-line process monitoring for increased yield.
- Improving light-management to obtain maximum efficiency.
- Improving stability and conversion efficiency of a-Si modules in actual use by addressing the Staebler-Wronski negative effects, where the conversion efficiency of the a-Si module decreases when it is first exposed to sunlight.
- Reducing capital equipment costs for manufacturing a-Si panels by improved manufacturing processes that include increasing the deposition rates.

- Improving module-packaging designs to make them more resilient to outdoor environments and less susceptible to glass breakage or moisture ingress.
- Developing new module designs for building-integrated applications.

These and other similarly complex issues must be addressed, because the future of a-Si PV products depends on their proper and timely resolution.

Note: The untimely exit of Applied Materials from the a-Si equipment manufacturing business in the summer of 2010 cast a shadow of doubt over the a-Si technology. Applied Materials is an example of a new-comer in the PV field with great potential and ambition, but inexperienced with the energy market's wants, needs, and overall peculiarities. Applied Materials had its own reasons to leave the field, but a-Si TFPV is here to stay. It is a promising technology that has already found niche markets, and will become even more popular with time, until it finds its place in the large-scale energy markets.

PERC

Passivated emitter rear contact (PERC) solar cells are fairly new development that opens a new door into the optimization of silicon solar cells. During 2013-2014, progress in the manufacturing of PERC solar cells has begun to reveal the full technological and commercial potential of this new cell architecture.

High efficiency levels in standard cell architectures are limited by the tendency for photo-generated electrons to recombine shortly after their liberation. PERC cells maximize the electrical gradient across the p-n junction, which allows for a steadier flow of electrons, reduction in electron recombination, and higher efficiency levels. PERC cells also enjoy increased conversion efficiency by the added dielectric passivation layer at the rear side of the cell.

There have been developments in equipment and manufacturing processes, as well as research, summarized in the Figures 5-35a and 5-35b.

Although the structure of the PERC cell is basically similar in the front surface, the back end is structurally and functionally different. The 100% back surface field (BSF) coverage in a standard cell is replaced by a local (partial) BSF, adjacent to a passivation layer, assisted by a capping layer.

Although PERC solar cells boast over 20% efficiency today, their price is prohibitive due to additional and complex process steps, and materials.

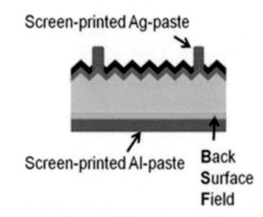

Figure 5-35a. Standard solar cell

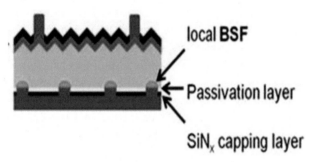

Figure 5-35b. PERC solar cell

The increased demand for PERC is being spurred by its potential to provide better efficiency at a competitive cost. As noted in the graph below, p-type mono PERC cell efficiency claims in the lab range from 19.8% to 20.4%, while standard p-type multi production line efficiency is in the range of 17.2% to 17.8%.

The upper-edge range for mass produced PERC cells will depend on line execution as PERC capacity ramps up. Despite that caveat, reported PERC commercial module efficiency claims still trump multi-module levels; current values are 17% and 15%, respectively.

The higher efficiency claims have piqued the interest of the downstream market, especially in Japan and the EU, with high-efficiency-focused suppliers ramping up capacity. A majority of capacity additions are taking place in Taiwan (Neo Solar, WINAICO, Gintech), though there have been recent commercial manufacturing announcements from JA Solar and SunEdison, as well.

Due to its high price and optimized efficiency, a majority of near-term demand will likely continue to be in residential, high-ASP markets. JA Solar is launching its PERC product, PERCIUM, in the Japanese, British, Israeli, Chinese, and German markets.

Moving forward, the key challenge for this cell technology will be reducing production costs. Currently, p-type mono PERC cells are on average 1.5 times the price of standard p-type multi cells. The price differential reflects the advanced cell architecture's higher cost. This premium is on par with that of standard mono cells, which are priced ~1.3 times as much as multi cells.

However, PERC cells currently provide a higher level of efficiency gain, about 15-20% higher than standard multi-crystalline cells and 7-9 % higher than standard mono-crystalline cells. This means that with incremental improvements in cost structure and efficiency, PERC mono-crystalline solar cells will be well on their way to being competitive with the standard mono-crystalline technology, and yet providing more power per unit area.

This is a great advantage, but one thing we need to warn the PERC enthusiasts, is that the additional structural complexity, and increased number of steps and layers of different materials in the back of the PERC cells would make them quite more expensive than their simpler, albeit less efficient, silicon cousins. Then, the battle would be decided by the ratio of added expense vs. efficiency gains.

Another significant problem is that the new complex structure opens a door for additional defects in the cells and more field failures. Especially during long-term exposure to excess heat in the deserts, PERC cells have more opportunities to develop defects in the back surface than standard solar cells. Because of that, extensive long-term tests are needed before we can be sure that mass produced PERC technology is reliable enough.

Emerging Thin Film Technologies

As mentioned above, we divide the thin film PV technologies into three different groups. Below, we discuss the not widely used, but promising, thin film PV technologies of: silicon ribbon, epitaxial thin film silicon, DSSC, organic, ink, nano-crystalline, perovskite, and III-V PV technologies.

Silicon Ribbon

Called EFG ("edge defined film fed growth"), this method, is not exactly a "thin film" process, as we know it, but the resulting material is thin enough, so it belongs in this category. Here, a graphite dye is immersed into molten silicon, making it rise into the dye by capillary action. It is then pulled as a self-supporting very thin sheet of silicon which hardens in the air above the dye. It can then be cut in different shapes and sizes for processing into solar cells and modules.

This method is more efficient than conventional c-Si ingot and wafer processes in terms of producing c-Si substrates of exact thickness and avoiding slicing it into wafers. Conventional processes waste 20-40% of the silicon material, use a lot of energy, and produce tons of hazardous waste materials.

Another similar process we need to mention here is called "dendritic web growth process." It consists of two dendrites, which are placed into molten silicon and withdrawn quickly, causing the silicon to exit and solidify as a thin sheet. A modification of this method now in use is called the "string ribbon method," where two graphite strings are used (instead of the dendrites) to draw the silicon sheet, which makes process control much easier. Again, the silicon sheet can be cut into different shapes and sizes.

In all cases the silicon produced by these methods is multi-crystalline with a quality approaching that of the directionally solidified material. Although lab tests show efficiencies in the 17-18% range, solar modules made using silicon ribbons and produced via these methods, generally have efficiencies in the 10-12% range.

After the initial hoopla, that silicon ribbon technologies would dominate the market, their share is quite small—less than 1% of total sales today—and does not seem to be growing. This is mostly due to the fact that the process is not easy to control, and the wafers' surface is not uniform enough, thus resulting in breakage, processing defects, and performance inefficiencies. The silicon sheets forming process is also complex and uses a lot of energy, which makes it comparably more expensive than some of the other mass produced TFPV technologies.

Epitaxial Thin Film Silicon

The high cost of silicon material accounts for about half of the production cost of current conventional, industrial-type silicon solar cells. In order to reduce the amount of consumed silicon, the photovoltaics (PV) industry is counting on a number of options presently being developed. The most obvious is to move to thinner silicon substrates by producing thinner Si wafers, or shaving the thicker wafers, but this is proving hard to do for a number of reasons. A more feasible approach is the so-called epitaxial deposition of a thin film of silicon on a cheap substrate, thus creating efficient but cheap solar cells.

There are several approaches that can be used to create such a thin film cell:

Epitaxial Single Crystal sc-Si

To create an epitaxial thin-film solar cell on a cheap substrate we start with highly doped sc-Si wafers (e.g., from low-grade silicon or scrap Si material), and deposit

an epi layer of Si by chemical vapor deposition (CVD). The resulting mix of a high quality epi layer and a cheap substrate is a compromise between high cost and efficiency, and yet offers a solution to gradual transition from a wafer-based (heavy material dependence) to a thin-film technology (less material and more sophisticated processing). This process is easier to implement than most other thin-film technologies today, but it remains to be seen if its efficiency and cost will be able to compete in the energy market.

Epitaxial Polysilicon Thin Film

To produce thin-film polysilicon solar cells, a thin layer (only a few microns) of polysilicon Si is deposited on a cheap foreign substrate, such as ceramic or high-temperature glass. These seed layers are then epitaxially thickened into absorber layers several microns thick using high-temperature CVD with a deposition rate exceeding 1 m/min.

Polycrystalline silicon films with grain sizes between 1-100 m appear to be particularly good candidates. Good polycrystalline silicon solar cells can be obtained using aluminum-induced crystallization of amorphous silicon. This process leads to very thin layers with an average grain size around 5 m. This technology is still in R&D stages, but shows high cost-reduction potential and might become very important, especially in case of silicon shortage, or very high prices in the future.

Light-absorbing Dyes (DSSC)

These are special types of dye-sensitized solar cells, where a ruthenium metalorganic dye (Ru-centered) is used as a monolayer of light-absorbing material. The dye-sensitized solar cell depends on a mesoporous layer of nanoparticulate titanium dioxide to greatly amplify the surface area (200-300 m^2/g $TiO2$, as compared to approximately 10 m^2/g of flat single crystal).

Photogenerated electrons from the light-absorbing dye are passed on to the n-type $TiO2$, and the holes are passed to an electrolyte on the other side of the dye. The circuit is completed by a redox couple in the electrolyte, which can be liquid or solid. This type of cell allows a more flexible use of materials, and is typically manufactured by screen printing and/or use of ultrasonic nozzles, with the potential for lower processing costs than those used for bulk solar cells.

However, the dyes in these cells also suffer from degradation under heat and UV light, and the cell casing is difficult to seal due to the solvents used in assembly. In spite of these problems, this is a popular emerging technology with special applications and significant commercial impact forecast within this decade.

The first commercial shipment of DSSC solar modules was recorded in July 2009 from G24i Innovations.

Organic/Polymer Solar Cells

Organic and polymer solar cells are built from thin films (typically 100 nm) of organic semiconductors such as small-molecule compounds like poly-phenylene vinylene, copper phthalo-cyanine (a blue or green organic pigment), and carbon fullerenes and fullerene derivatives, such as PCBM.

Energy conversion efficiencies achieved to date using conductive polymers are low compared to inorganic materials. However, they were improved in the last few years and the highest NREL certified efficiency has reached 6.77%. In addition, these cells could be beneficial for some applications where mechanical flexibility and disposability are important.

These devices differ from inorganic semiconductor solar cells in that they do not rely on the large built-in electric field of a p-n junction to separate the electrons and holes created when photons are absorbed. Instead, the active region of an organic device consists of two materials, one which acts as an electron donor and the other as an acceptor. When a photon is converted into an electron hole pair, typically in the donor material, the charges tend to remain bound in the form of an exciton, and are separated when the exciton diffuses to the donor-acceptor interface.

The short exciton diffusion lengths of most polymer systems tend to limit the efficiency of such devices. Nanostructured interfaces, sometimes in the form of bulk hetero-junctions, can improve performance. Instability of the films, especially under harsh environmental effects is a major problem, which needs to be resolved, before full-scale implementation. Even with its advantages, this technology still has far to go to full market acceptance and serious deployment.

Ink PV Cells

A fairly new development, this light-activated power generating product is based on a unique and patented solvent-based silicon nanomaterial platform that can be applied like ink on any substrate. Developers claim that this approach has cost savings over traditional silicon products by using less silicon and having a more efficient manufacturing process as well as unique optical advantages.

This new technology consists of processing the quantum dots in the silicon "ink" in a way that makes it possible to use the old "roll-to-roll" printing technology

used for printing on paper or film. Applying ink directly on any substrate (including a flexible one) allows applications such as tagless printing for clothing labels and portable chargers for consumer and military customers.

By controlling the sizes of the dots from 2 to 10 nm, the absorption or emission spectra of the resulting film can be controlled. This allows capture of everything from infrared to ultraviolet and the visible spectrum in between which is not possible with conventional technology.

The technology is also used as an efficient light source. By controlling particle size, you can produce light of any color or a combination of particle sizes that will give off white light. This application might provide additional, and possibly larger markets for this technology in the near future—at least in some specialized areas.

Nano-crystalline Solar Cells

These structures make use of some of the usual thin-film light absorbing materials, but are deposited as a very thin absorber on a substrate (supporting matrix) of conductive polymer or mesoporous metal oxide having a very high surface area to increase internal reflections. Hence, the probability of light absorption increases.

Using nanocrystals allows one to design architectures on the length scale of nanometers, the typical exciton diffusion length. In particular, single-nanocrystal ('channel') devices, an array of single p-n junctions between the electrodes and separated by a period of about a diffusion length, represent a new architecture for solar cells and potentially high efficiency.

We envision the development of this type of photo-conversion to be in the R&D labs for a while yet, but it opens new possibilities in areas where other technologies simply cannot compete, thus opening promising niche markets for these cells.

Perovskite Solar Cells

Perovskite solar cells are made out of perovskite; a special calcium titanate mineral $CaTiO3$. Perovskie mineral is predominant in the Ural Mountains of Russia, and lends its name to the class of compounds which have the same type of crystal structure as $CaTiO3$. This unusual structure ($^{XII}A^{2+VI}B^{4+}X^{2-}_3$) is known as the 'perovskite structure'.

Here, 'A' and 'B' are two cations of different sizes, while X is an anion that bonds to both. The 'A' atoms are larger than the 'B' atoms, and the ideal cubic-symmetry structure has the B cation in 6-fold coordination, surrounded by an octahedron of anions, and the A cation in 12-fold cuboctahedral coordination. Slight buckling and distortion of the structure can produce several lower-symmetry distorted versions, in which the coordination numbers of A cations, B cations or both are reduced.

While this special crystal structure and its applications have been well known since the 1940s, it was not until recently when it was discovered that it could also be successfully used in making efficient and cheap solar cells. The special properties of the perovskite material are under intense research, and the perovskite, according to the scientists, could cause a breakthrough in cheaper PV energy generation.

Synthetic perovskites materials have been created and identified as possible inexpensive base materials for high-efficiency commercial photovoltaic cells and modules. Some of these cells show conversion efficiency of 15-16%. The key advantage is that these cells can be manufactured by the usual thin-film manufacturing techniques, which are used for making thin film solar cells.

A group of methyl-ammonium tin and lead halides are of the greatest interest for use in dye-sensitized solar cells, since they have reached efficiencies of over 20%. Another approach uses organic-inorganic perovskite-structured semiconductors, the most common of which is the triiodide ($CH_3NH_3PbI_3$). These cells exhibit high charge carrier mobility and charge carrier lifetime that allow light-generated electrons and holes to move far enough to be extracted as current, instead of losing their energy as heat within the cell. Their effective diffusion lengths are about 100 nm for both electrons and holes, which promises increased efficiency and current output.

The perovskite materials are deposited by low-temperature solution methods, typically applied by spin-coating techniques. Experiments with nanostructured cells using a combination of methyl-ammonium iodide and lead iodide with a small amount of chloride substitution show nearly 12% conversion efficiency.

Planar heterojunction perovskite solar cells can be manufactured in simplified device architectures (without complex nanostructures) using only vapor deposition. This technique has produced cells with over 15% solar-to-electrical power conversion as measured under simulated full sunlight. The technique offers the potential of low cost because of the low temperature solution methods and the absence of rare elements.

The problem is that finished cell reliability is still insufficient for commercial use. Several teams of scientists are conducti8ng studies with perovskite in Alta Floresta (Brazil), Frenchman Flat, (USA), Granada (Spain), Beijing (China), Edinburgh (UK) and Solar Village (Saudi Arabia). They have confirmed its efficiency in converting light to power, not only under sunlight, but in a range of atmospheric conditions. Once the reliability issues are solved, this new type of very cheap cell has the potential to revolutionize the solar energy markets.

III-V PV Cells

These are very special PV technologies also based on the deposition of thin films made of special materials of the III-V groups of the Periodic Table. The major types of solar power generating devices in this group are as follow.

Gallium Arsenide Based Multi-Junction Cells

High-efficiency multi-junction cells were originally developed for special applications such as satellites and space exploration, but at present, their use in terrestrial concentrators might be the lowest cost alternative in terms of $/kWh and $/W. These multi-junction cells consist of multiple thin films produced via metal-organic vapor phase epitaxy. A triple-junction cell, for example, may consist of the semiconductors GaAs, Ge, and GaInP2.

Each type of semiconductor will have a characteristic band gap energy which, loosely speaking, causes it to absorb light most efficiently at a certain color, or more precisely, to absorb electromagnetic radiation over a portion of the spectrum. Semiconductors are carefully chosen to absorb nearly all of the solar spectrum, thus generating electricity from as much of the available solar energy as possible.

GaAs-based multi-junction devices are some of the most efficient solar cells to date, reaching a record high of 40.7% efficiency under "500-sun" solar concentration and laboratory conditions.

This technology is currently being utilized mostly in powering spacecrafts. Demand for tandem solar cells based on monolithic, series-connected, gallium indium phosphide (GaInP), gallium arsenide GaAs, and germanium Ge p-n junctions is rapidly rising. Prices are rising dramatically as well.

Twin-junction cells with indium gallium phosphide and gallium arsenide can be made on gallium arsenide wafers. Alloys of In.5Ga.5P through In.53Ga.47P may be used as the high band gap alloy. This alloy range allows band gaps in the range of 1.92eV to 1.87eV. The lower GaAs junction has a band gap of 1.42eV.

In spacecraft applications, cells have a poor current match due to a greater flux of photons above 1.87eV vs. those between 1.87eV and 1.42eV. This results in too little current in the GaAs junction, and hampers the overall efficiency since the InGaP junction operates below MPP current and the GaAs junction operates above MPP current. To improve current match, the InGaP layer is intentionally thinned to allow additional photons to penetrate to the lower GaAs layer.

In terrestrial concentrating applications, the scatter of blue light by the atmosphere reduces photon flux above 1.87eV, better balancing junction currents. GaAs was the material of the highest-efficiency solar cell, until recently, when Germanium-based MJ cells capped the world record at 41.4% efficiency.

Indium Phosphide Based Cells

Indium phosphide is used as a substrate to fabricate cells with band gaps between 1.35eV and 0.74eV. Indium phosphide has a band gap of 1.35eV. Indium gallium arsenide (In0.53Ga0.47As) is lattice matched to indium phosphide with a band gap of 0.74eV. A quaternary alloy of indium gallium arsenide phosphide can be lattice matched for any band gap in between the two.

Indium phosphide-based cells are being researched as a possible companion to gallium arsenide cells. The two differing cells may be either optically connected in series (with the InP cell below the GaAs cell), or through the use of spectra splitting using a dichroic filter.

The presence of varying quantities of toxic materials in these devices must be considered when planning their use in large quantities.

CPV Solar Cells

Concentrating photovoltaics (CPV) is a branch of the PV industry, using special cells, optics and tracking mechanisms, developed in the 1970s by several companies under contracts and financing from U.S. Departments of Energy.

The key differences between CPV solar cells and their competitors are:

- CPV solar cells are very small, about 1 cm^2 in size,
- About 120 cells can be made from a single 4" wafer; the size of a normal Si solar cell.
- Each small CPV cell can generate 20-30 W of power under 500x-600x concentration.
- The CPV solar cells are manufactured via sophisticated and extremely precise semiconductor equipment and processing conditions, thus their quality is unmatched.
- CPV type solar cells today are 45% efficient, with 60% efficiency expected by 2020,
- CPV solar cells do not degrade with temperature or time, as the competing technologies do.

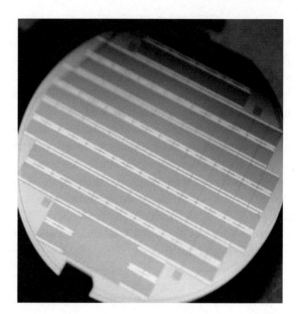

Figure 5-36. 4" Ge wafer yields over 120 CPV cells

Early CPV systems used silicon-based CPV cells, which had a problem with elevated temperatures. These were later replaced by GaAs-based multi-junction cells, which have much higher efficiency, but still suffer from the effects of high temperatures. At first GaAs CPV cells were made by using straight gallium arsenide in the middle junction. Later cells utilized In(0.015)Ga(0.985)As, due to the better lattice match to Ge, resulting in a lower defect density.

Today, the high efficiency CPV cells are fabricated by depositing a combination of thin film materials onto germanium substrate.

Germanium (0.86eV band gap) is a semiconductor material, with properties far superior to other substrate materials used for PV cells and modules. It is ~40-50% more efficient than silicon and has a much lower temperature coefficient. It is several times more expensive than silicon, too, but with new superior slicing techniques, it can be cut into very thin wafers, saving a lot of material. This, combined with its higher efficiency and less degradation than silicon, could put it on the competitors' list within the next few years.

Germanium-based solar cells have been used mostly for space applications, but a number of manufacturers have geared up for mass producing them for high concentration HCPV and other high efficiency applications (the record is currently 45% efficiency).

As you can see in Figure 5-37, CPV cells are complex structures, consisting of many layers (some deposited, some diffused) in, or piled on top of, Germanium semiconductor material, which has more superior process and performance characteristics than silicon.

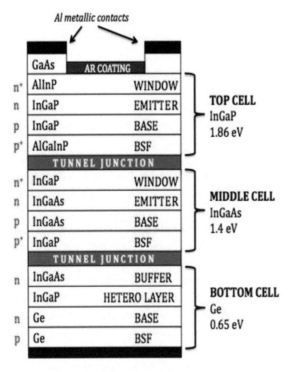

Figure 5-37. CPV cell

CPV cells are usually mounted under lenses, which concentrate sunlight falling on the cells 100 to 1000 times. This allows high efficiency and reliability, better land utilization, and other benefits. Cell-lens assemblies are mounted on trackers, which track the sun precisely through the day, providing the most power possible. Efficiencies of over 35%, measured in the grid are obtainable with these devices.

We foresee CPV technologies as the primary choice for installation and use in large-scale power plants, especially those in desert regions. And since we are already discussing CPV components, let's take a close look at the CPV technology itself. It is the most efficient and promising solar technology, so we will allocate some additional space in this text to emphasize its advantages and benefits.

CONCENTRATING PV TECHNOLOGY

Concentrating PV (CPV) and high concentration PV (HCPV) technologies have been around for over 30 years. They were developed by several US companies in the 1970s and 1980s with the help of DOE and its satellite national labs.

HCPV is the most efficient PV technology in the world, but its commercialization has been bogged down by a number of factors unrelated to its efficiency.

CPV systems concentrate the sunlight up to about 100 times (100x), while HCPV systems can concentrate it over 1,000x. Our experience, and the discussion below, is based on 500x concentration.

The design, installation, and O&M of these systems is somewhat different and more complex than other PV technologies. In contrast with 'flat-plate' PV modules, where a large area of photovoltaic material (usually crystalline silicon) is exposed to the maximum sunlight, CPV systems, as the name suggests, use lenses or mirrors to focus (or concentrate) sunlight onto a small amount of non-silicon photovoltaic material, thus using less land area..

General Description

HCPV systems operate by concentrating sunlight through Fresnel lenses onto specially designed, and very efficient (over 40%) CPV solar cells. Additional components, such as secondary optics, insulators, heat sinks, and trackers are needed to keep the light focused precisely onto the solar cells, to dissipate the generated heat and conduct the generated electricity to the electric grid, or user's site.

CPV solar cells mounted into CPV assemblies are installed in modules, which are mounted on a steel frame—a frame that pivots on a pedestal, Figure 5-38. GPS controllers send a signal to the x-y drive motors which drive actuators to position the frame and modules exactly perpendicular to the sun all day long with a .01% degree margin of error.

The CPV solar cells are made out of GaAs, Ge or other more efficient and more durable semiconductor

Figure 5-38. HCPV tracker

materials with thin films deposited on them to create a multi-junction device of very high efficiency—over 45% presently. CPV cells are sophisticated semiconductor devices that have followed Moore's law to an extent, as far as increase of efficiency per active area is concerned. At the very least, they are much closer to it than any competing PV technology to date

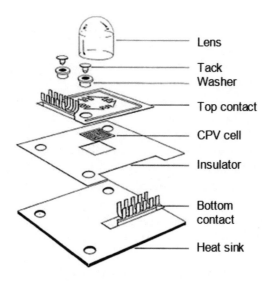

Figure 5-39. CPV cell assembly

The CPV cell assembly is the engine of the CPV systems. Here, the CPV cells are arranged in a special way, which, together with the other elements of the system, ensure proper orientation, efficient collection of sunlight and cooling of cells during operation, as well as the proper conduction of electric power into the wiring harness.

The CPV cells are mounted into an aluminum heat sink block, which is designed to absorb most of the heat generated at the cell, thus keeping the cells cooler and more efficient during operation. The heat sink is attached to the bottom of the module, with the ribs protruding to the outside, where they are most efficiently cooled by occasional breeze.

Each cell is configured with a blocking diode, which protects it from reverse bias currents and surges, and also isolates it from the circuit in case of total failure. The cell and diode are usually mounted on a heat sink, which extracts heat from the cell during operation and dissipates it in the surrounding environment outside the module.

A Fresnel lens, mounted directly on top of the CPV assembly is used to capture a larger area of sunlight and focus it onto the small CPV cell. The Fresnel lens is basically flat (or slightly curved) plastic or glass lens that uses a miniature sawtooth design on its bottom to redirect and focus the incoming light onto the small area cell, several inches away from the lens. When the sawtooth teeth on the lens are arranged in concentric circles, light is focused at a central point, and CPV system. When the teeth run in straight rows, the lenses act as line-focusing concentrators, and the resulting equipment is called the linear-focus CPV system.

The concentration ratio of each CPV device can vary as well, depending on lens and cell design. For example, if sunlight falling onto 100 cm² Fresnel lens is focused onto 1 cm² CPV cell, the ratio is considered to be 100 suns, or 100 x. If the light from a 1000-cm² Fresnel lens is focused onto a 1-cm² CPV cell, then the ratio is 1000 suns, or 1000 x. If thus concentrated sunlight light falls onto a well designed CPV cell, it will produce 100, or 1000 times the electricity respectively that a normal c-Si PV cell of the same size would produce under the same operating conditions.

Commercial concentration ratios are between 100 and 500 suns, and as much as 1000 suns, but there are theoretical and practical limits of min.- max. sun concentration levels, with the mid-range 400- 600 considered the most efficient and practical for now.

Most CPV systems use only direct solar radiation, so these installations operate best under direct sunlight (such as in desert areas) and always involve trackers, forcing the modules to rotate and follow the sun all day, which keeps the lenses and CPV cells looking right at the sun at all times, thus generating maximum possible electric energy while the sun is shining.

The cells and cell assemblies are placed in modules, which are mounted on a large frame, which rotates around its two axes to follow sun movement all day. Tracking is achieved with one or two gear-motor assemblies, which get a signal from a photo-detector or GPS controller, which knows where the sun is at all times and runs the motors to move the frame accordingly.

Precise tracking is needed, to keep the sun constantly at exactly 90 degrees (+/-0.1 degree) with respect to the Fresnel lens and CPV cell, so that sunlight is focused onto the solar cell at all times. If the sunlight comes at an angle, or is diffused (as on a cloudy or foggy day), some, if not most, sunlight will get de-focused, reflected, or otherwise fall away from the cells, resulting in very low light-to-power conversion efficiency.

Tracking systems add cost in terms of motor and controller maintenance, but this cost is relatively small compared with other cost savings that trackers provide. A single tracker motor, for example, controls more than 50 kWp of PV power generation, and requires only annual lubrication.

The operating and maintenance (O&M) cost of a utility-scale tracking system ends up being less than $0.01/kWh more than that of a fixed configuration. And this calculation does not factor in the savings from increased energy production and reduces land use, which is a different subject all together.

Nevertheless, mostly due to its complexity and higher cost, the CPV technology has not been widely accepted and used today. We, however, are sure that it will be a leading solar energy technology by mid-end of this century.

Development Trends

Current research on high-performance multi-junction thin film devices, and the entire operation of CPV systems, focuses on several major challenges:

CPV cells
- Determining high-bandgap alloys based on I-III-VI and II-VI compounds and other novel materials for the top cell.
- Considering low-bandgap CIS and its alloys, thin-film silicon, and other novel approaches for the bottom cell.
- Studying the difficult task of integrating the thin-film tunnel junction (interconnect) with the top cell. This work includes understanding the role of defects, how they affect the transport properties of this junction, and the diffusion of impurities into the bulk material.
- Fabricating a monolithic, two-terminal tandem cell based on polycrystalline thin-film materials that requires low-temperature deposition for several layers.
- Avoiding deterioration of the top cell when fabricating the bottom cell if a low-bandgap cell is fabricated after a high-bandgap cell with a superstrate structure, such as CdZnTe.

Avoiding temperatures and processes that could damage the CIS bottom cell if a high-bandgap cell is fabricated on top.

CPV System
- Developing efficient heat transfer materials for the components in the cell assembly
- Developing light and UV resistant plastic materials for the module cover lenses
- Developing more precise x-y controls, as needed to maintain 0.01 degree accuracy
- Developing better gear-drives for stable operation under wind conditions

Note: The efficiency of CPV solar cells has been rising steadily since the 1990s. Starting with 15% efficiency in 1993 of CPV solar cells made by Spectrolab, the efficiency of CPV cells made by Sharp was 44.4% as of the summer of 2013 and is now over 45%.

At this rate, we expect to see efficiency of 60% or higher within the next decade. This is possible because the efficiency of the materials involved in making CPV cells is not limited as is that of silicon, which theoretical maximum efficiency cannot exceed 28.5%.

Note: The "theoretical efficiency maximum" of solar cells is based on the fact that sunlight is made up of a bunch of photons with wavelengths. A lot of the total energy is stored in the infrared band, but silicon is transparent to photons with near infrared wavelengths, so all the energy of these wavelengths goes right through the silicon solar cell unconverted.

A second problem with silicon is that it readily absorbs ultraviolet light. Its thirst for UV light is so strong that a thin layer of silicon can absorb all UV falling on it. The combination of these factors reduces the upper efficiency limit of silicon simply because many photons of different wavelengths cannot reach down to the p-n junction where the actual energy conversion takes place.

And so, if we assume that all wavelengths in sunlight are 100% of the light hitting the solar cell, then silicon can trap and convert to electricity the absolute maximum of 28.5% of all these wavelengths, and so we end up with about the maximum possible efficiency for silicon of 28.5%. Or this means that only 28.5% of all sunlight illuminating a silicon solar cell can be converted into electricity under best of conditions.

CPV cells have the advantage of being made of a number of layers of different materials (Ge, Ga, In, As, etc.), which are superior in terms of efficiency. In addition, when sunlight goes through the array of different materials of the CPV cell, different wavelengths get absorbed by the different layers at each step of the process. This way we get to the 44.4% maximum efficiency today, which can easily go over 60% in the near future.

In conclusion, although there is still some work to do in the optimization of the CPV systems, we see this technology playing major role in the near future. Not immediately, but in the more distant future, when the technology will be developed to the level needed for reliable operation in large-scale desert installations. This will allow these more complex and more efficient systems to compete with the cheaper and simpler fixed mount Si and TFPV installations.

HYBRID PV TECHNOLOGIES

Hybrid PV technologies are these that combine the use of photovoltaics with other solar types, wind, or fossil fuels. This is done in special cases for the sake of practicality and/or efficiency. The combination between photovoltaic and thermal (PV-T) hybrid technologies is most useful in applications that require both electricity and hot water at the same time. Thus created hybrids can be divided into several types, according to their components and applications.

The present day hybrid solar technologies involving photovoltaics and heat generation can be divided into:
1. Photovoltaic/solar thermal hybrid, or PV/T,
2. Concentrating PV/solar thermal (CPV-T) hybrid, and
3. High-concentration PV/solar thermal hybrid, or HCPV-T
4. Solar/wind power cogeneration

A Word of Explanation and Clarification

PV-T is a combination of photovoltaic devices, which incorporate both: photovoltaics and thermal heating. These can operate at 1 x (1 sun) to generate electricity, while running water through the system for use somewhere else.

CPV-T therefore, is a combination (hybrid) of PV concentrators which concentrate the sunlight 10-100 times onto a receiver tube covered with solar cells mounted on a receiver tube, through which water or cooling liquid flow. The trough and receiver track the elevation of the sun all day long (one-axis tracking), and thus the device generates more DC power per active area, and the water/coolant are at much higher temperatures than their PV/T cousin.

HCPV-T is a combination of PV concentrators (mirrors or Fresnel lenses), which operate on higher magnification (100-1000 x) and track the sun in x-y direction (two-axis tracking) all through the day. These devices produce more power per area than their CPV-T cousin, while the water/liquid temperature is controlled as needed to optimize the efficiency of the device.

We'll start with a close look at the mechanics and thermodynamics of the simplest PV-T system—that of a fixed mounted flat panel equipped for generating both; PV electricity and hot water capabilities, which makes it into a PV-T hybrid.

These systems combine photovoltaic with thermal solar. The advantage is that the thermal solar part carries heat away and cools the photovoltaic cells; keeping temperature down lowers the resistance and improves the cell efficiency. Modified CPV systems have been tested, where the CPV cells are cooled by active flow of liquid, thus generating both heat and electricity.

Below we are taking a very close look at the PV/T device operation and their thermo-dynamics and static behavior, both of which could be used with some translation and extrapolation for use with CPV-T and HCPV-T devices. These calculations are absolutely necessary for the proper design of the devices and the power fields these are supposed to operate in.

The hybrid photovoltaic-thermal (PV/T) system basically consists of an array of PV cells positioned directly on top of the absorber plate of a conventional forced circulation type solar water heater—similar to these we see on roofs of houses.

PV-T Hybrid System

A PV/T solar collector is composed of:
- Transparent cover allowing sunlight to pass towards the absorber and to create an effect of greenhouse. It is composed by one or more glass or plastic panes.
- PV cells for the production of electricity.
- Absorber plate for transferring heat to the water or in the tubes built into it.
- Frame, for protection of the whole of these elements.

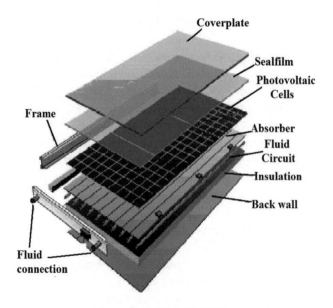

Figure 5-40. Hybrid PV/T

- Insulator, allowing limiting the losses by conduction through the walls back and side.

The schematics of the PV/T collector are shown in Figure 5-40. The top cover is represented by a glass sandwich that includes PV cells. The cell area can cover the entire active surface or can be distributed in a grid where the spacing between adjacent columns and rows can allow a direct gain of solar radiation to the absorber plate.

Different configurations of PV/T collector can be created by changing the cell area density in order to balance electricity and thermal energy output of the system.

Dynamic model of a PV/T system

Below, an explicit dynamic model suitable for PV/T system simulation is introduced. (7)

The effectiveness of PV/T system compared to a photovoltaic panel (PV) is underlined. The model provides the thermal state of various collector components and generates results for hourly and transient performance analysis (thermal/electrical gain).

The dynamic thermal model of the PVT collector is built upon the finite-difference control-volume technique. The PV/T collector is composed of four major components which represent the different nodes: a) Glass cover, b) solar cell, c) absorber plate, and d) water in channels and in storage tank. See below detailed calculations.

The energy and fluid flow equations are developed on the base of the four nodes. All sub-parts in each node are considered lumping together in proportion to give the average properties of the representing major component, (Figure 5-41.)

Where

T_g, T_c, T_p, T_f, T_a are respectively: the temperature of glazing, the solar cell, the absorber plate, the water circulation and the ambient temperature.
W is the wind speed,
G is the incident solar irradiation,
m is the mass flow rate of fluid,
P is the electric power output, and
Q is the thermal gain.

Since the PV/T system consists of several different elements working in unison, we need to consider the separate dynamics by analyzing separately, a) the glass cover, b) the solar cells, c) the absorber plate, and d) the heated water, and e.) the generated electric power.

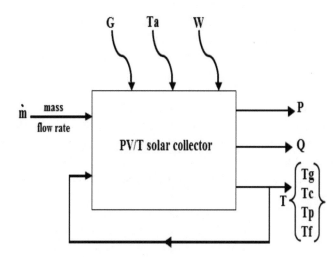

Figure 5-41. Modeling of PV/T system

These are elaborate thermodynamic and electro-mechanical operations, so we will refer the reader to our book, *Solar Technologies for the 21st Century*, published in 2013 by The Fairmont/CRC Press, where we take a very close look at the different PV-T technologies—from silicon, to thin films, and their most advanced and futuristic cousins.

In summary, PV-T technology is quite simple, but very versatile, cheap and efficient for use in certain locations, where both; hot water and electricity are needed. It is particularly practical for use in remote locations, such as vacation cabins, field hospitals, military installations, immigration camps, etc.

CPV-T Technology

The concentrating photovoltaic-hot water, hybrid (CPV/T) technology is fairly new development, and al-

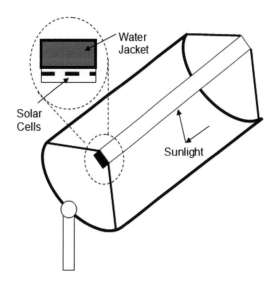

Figure 5-42. CPV-T one-axis tracker

though a number of companies are working on this concept, full commercialization and large-scale deployment are still way off for a number of reasons.

The CPV/T equipment usually consists of a parabolic trough lined with a mirror surface, focused on a bank of solar cells mounted on a heater tube through which flows water or some liquid. When the sun hits the trough, its mirror surface reflects the sunlight directing it to the cells, which generate DP power at 15-20% efficiency. When the cells heat up under the sun rays, they transmit the heat to the tube they are mounted on and heat the liquid running thru the tube.

The above dynamic and static calculations for the generic PV/T device could be used here too, with because the basic glass-absorber-cells-liquid configuration exists here in almost identical fashion. Their properties and behavior are similar, except for different results, due to much higher solar illumination, DC voltage and temperatures.

Several companies use this technology—by combining concentrated (low concentration) PV electricity production and heat collection to deliver simultaneously electricity and hot water for commercial, industrial and institutional applications.

Mirrors and single-axis trackers are used to focus light on a row of PV cells in a special CPV/T assembly. Heat, which is normally lost in a typical PV-only system is used to heat water that runs through the CPV/T assembly and cools the solar cells.

Cooler solar cells generate more electrical energy and thus generated hot water is used for heating or process water at the same time. Some of these are turnkey systems designed for easy assembly and rooftops or ground installations.

Although the electricity-hot water hybrid cogeneration equipment has a PV element, and thus competes with the conventional PV technologies, the dollar-per-watt metric is somewhat less applicable. The metric, in addition to the electricity generated within a certain time, includes gallons of hot water generated during the same time period.

The advantage here is that this type of technology does not compete head-to-head with solar panel makers but rather with natural gas. And since natural gas in the U.S. is cheap (roughly $0.02 to $0.03 per kilowatt-hour), this technology is most suited for locations with higher gas prices such as Japan and Europe.

Of course the climate has to be suitable as well, because for this technology to function properly and efficiently, it needs direct unobstructed (desert-like) sunlight. Appropriate (niche market) customers are needed

as well, these with demand for hot water. This includes the food and beverage industry, schools, hospitals, manufacturing facilities, Army bases, etc.

Another advantage of the system is the capability to store hot water for later use. The ability to generate energy in off-grid applications is also advantageous, and creates another niche market. In addition, thus generated heat can also be used for cooling applications, which opens the opportunity for supplying electricity and hot water to buildings with large "chilling loads," as well.

These CPV/T systems produces water at 70°C, and require no special permitting or roof modifications to support the additional load. They also come complete with integrated hydronics, controls, and inverters.

Another great advantage is the fact that if the CPV/T system is SRCC- (Solar Rating and Certification Corporation) and IEC- (International Electrotechnical Commission) certified, it qualifies for both thermal and electric rebates. This includes the California Solar Initiative (CSI), which offers an incentive for solar hot water systems.

CSI thermal rebates, currently offered in the first-tier level of $12.82 per year for one displaced gas, complement the CSI photovoltaic rebates. This results in faster paybacks than traditional solar hot water or solar electric systems alone.

On top of that, the installations are also eligible for the 30% Investment Tax Credit. With the CSI and federal rebates, CPV/T system customers can yield a return on investment in five years, while hedging against future utility price hikes and gas price volatility.

PV-T systems produces 3-5 times more energy (electricity and heat combined) and three times the greenhouse gas reductions of the competing solar technologies. And when compared watt-per-watt usage, solar water heating is 3 to 5 times more efficient than PV, which ensures a much faster payback, and essentially doubles the value of a CPV/T system.

Since this is an accepted way of doing things in China, Europe, and Austria, the technology will be widely accepted there.

The value of hot water production is, however, somewhat compromised by the plumbing aspect of hot water generation, so this places this technology in a number of somewhat limited in size niche energy markets.

HCPV-T Technology

High concentration photovoltaic/thermal (HCPV-T) hybrid technology is even newer and more undeveloped than its CPV-T cousin. As a matter of fact we know of only one company (SolarTech of Arizona) that is working on its optimization and commercialization.

The function of the HCPV-T system is similar to that of its older cousin, the CPV-T energy system, which we review above. It also requires tracking and operates at very high—over 60% overall—combined efficiency.

The same dynamic and static calculations for the generic PV-T device could be used here too, because the basic glass-absorber-cells-liquid configuration exists here in almost identical fashion. Their properties and behavior are similar, except for different results, due to the high concentration of sunlight onto the modules, which results in much higher solar illumination, DC voltage and temperatures.

In summary:

*   HCPV/T hybrid systems can be used (with some modifications) without liquid cooling; using air cooling of the heat sinks only, instead
    —   The system can be built in different sizes; from 0.5 kWp to 500 kWp as needed to fit different locations and applications—including large-scale ground-mounted installations.

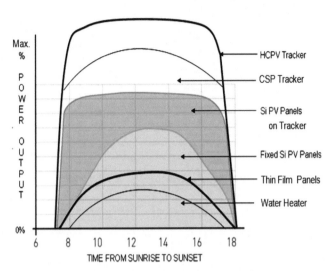

Figure 5-45. Efficiency of solar technologies

*   As another option, 6 regular silicon solar cells are added to each HCPV cell assembly in a special arrangement, so that some of the sunlight that is reflected by the HCPV cell illuminates the 6 regular solar cells around it, thus producing additional DC power. I.e., this configuration increases a 3.0 kWp power output of a standard 7 modules HCPV/T system to nearly 5.0 kWp.
    —   Different combinations of these options allow unprecedented efficiency, product flexibility, and offer a number of different application choices.

Total Energy Management

Figure 5-46 is a graphic representation of a functional, all-purpose combination power generator, consisting of a combo (hybrid) PV power/hot water/gas generator. This is a complete self-contained, self controlled, hybrid PV-T system for use in remote locations.

Here, the CPV tracker array in Figure 5-46 generates DC power and hot water. These are fed into, a) inverter, and b) hot water storage tank, respectively. A gas-fired heater and power generation set is on stand-by for use as and when needed to provide un-interrupted heat or electric power.

Thus generated electricity and hot water by the CPV array is fed to the respective systems, and are used as needed by the local facility. When the sun goes down, the power generator kicks in and keeps the entire setup running by providing both power and hot water.

This system provides power on 24/7 basis and covers the best of both worlds—DC/AC power and hot water for all day and night uninterrupted use. Such a system would be quite convenient for use in field hos-

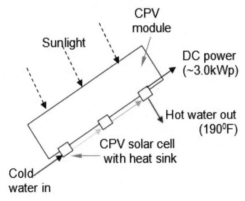

Figure 5-43. HCPV-T module cross section

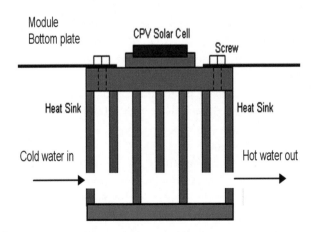

Figure 5-44. Solar cell heat sink and water heating assembly

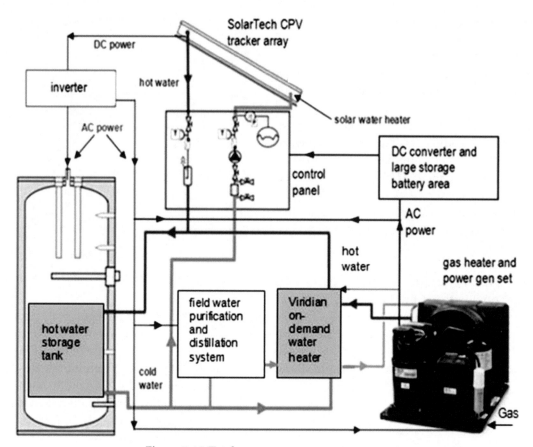

Figure 5-46. Total energy management system.

pitals, refugee camps and other such operations, where power and hot water are inaccessible.

This system would be exceptionally efficient in areas with lots of sunlight—desert areas are best, of course. It can be configured with or without batteries, and with or without a gas generator. In all cases, the solar collector(s) will provide much needed DC power and hot water during the day, where the only limiting factor is the amount of sunlight available to the collector(s).

By adjusting the controls and the different modus operandi of the system, we could obtain maximum efficiency under different climatic conditions. Such a system is what the future holds for a number of critical applications, so we foresee a bright future for these types of energy systems in the 21st century.

Hybrid CPV-T System Financial Analysis

CPV-T energy systems generate both DC electricity and hot water at the same time, at a very high efficiency—let's assume 40% and 20% efficiency respectively—totaling close to 60% overall utilization of solar power.

The amount of energy produced depends on the size of the system, its actual setup and operation, and—very importantly—the location and the amount of sunlight available for its operation. We reviewed the design and function of CPV-T hybrid power generating systems above, so here we will present an example of the operation of a small- to medium-size concentrating PV (CPV) and thermal (T), or CPV-T hybrid system in US deserts.

This system consists of a CPV tracker equipped with highly efficient solar cells (over 40% efficiency at the cell level), with cooling water lines, which feed a water heating loop with hot water as a final product.

Since CPV technology requires direct beam sunlight, it is best suited for use in the U.S. Southwest and a number of sunny locations around the country and the globe.

For the present performance analysis we'll pick a location in the Arizona deserts with the following average performance criteria:
- 320 sunny days annually
- Average 900W/m^2 full solar insolation
- Average full solar insolation of 8 hrs. per day
  — Summer time full solar insolation for CPV use is ~10 hrs,
  — Winter time solar insolation for CPV use is ~6 hrs.

So, under ideal conditions 900 W/m² x 8 hrs = 7.2 kW/h energy is received daily per m² Each CPV-T unit is over 40% efficient and occupies ~6.0 m² area. Or 7.2 kW/h x 6.0m² x 40% = ~17.3 kW/h daily generation (vs. 5.5 kW/h for PV panels*) 300 CPV-T units (1800 m²) would generate 1.0 MWp or ~8.0 MW/h daily averaged on annual basis.

At $0.20 avg. PPA (including carbon credits, federal and local subsidies and incentives**) this represents $1,600/day, or $512,000 annual income from DC/AC power generation alone.

In addition, each CPV/T unit generates hot water at ~20% efficiency (180W/m² = 614 BTU/m²) Thus, 614 BTU x 8 hrs x 6.0m² x 320 days = 9.43 million BTUs are generated annually. Or, our 1.0 MWp plant would generate additional ~$180,000 annually from hot water generation, or equivalent energy use needed to heat similar amount of hot water.

Useful power generation (considering DC/AC and other conversion losses and O&M inefficiencies) would be ~20% less. Therefore, we can expect gross income during the first year of operation from our 1.0 MWp CPV/T plant to be ~$550,000***.

Total capital expense for building the CPV-T power plant is estimated at: $1.30/Wp for equipment and $2.50W/p for BOS, land and administrative expenses, or total of $3.8 million are needed to install and start operation of the 1.0 MWp plant. Additional expenses of approx. 15% of gross income must be assumed and allocated for annual O&M operations, which brings total net operating income to ~468,000 annually.

Therefore, our 1.0 MWp CPV-T plant will be paid for in approx. 8 years, providing clean, green energy and reducing global CO_2 generation by over 1 million lbs. annually. The number of years will vary as other factors, such as government and local subsidies and incentives, carbon credits, taxes and initial investment repayment, are considered.

The drawback of this type of system, and where the uncertainty of this technology lies, in addition to sun's variability, is the maintenance of the complex key components; positioning controls, x-y gear assemblies, and the related motors, bearings, etc.. In best of cases, the maintenance would be 2-5% higher than fixed-mount modules, but in reality we expect 5-10 % higher

maintenance costs. We must hope that technological advances will improve the system, in order to reduce the O&M costs.

Once more, we need to emphasize the enormous potential of the CPV, HCPV, and hybrid technologies, which time, unfortunately, has not come yet. Nevertheless, we insist that they will be the leading solar power generating technologies by the end of this century.

Co- and Tri-generation

Cogeneration, or combined heat and power (CHP) uses a heat engine fed by renewable or fossil energy to generate electric power and useful heat at the same time

Tri-generation, or combined cooling, heat, and power (CCHP) is the simultaneous generation of electricity and useful heating and cooling from the combustion of a fuel or a solar heat collector.

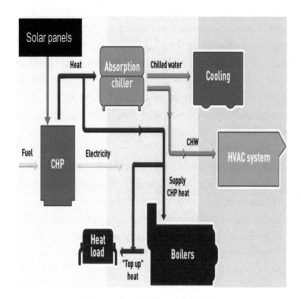

Figure 5-47. Tri-generation

This is a good example of how different energy technologies and fuels can work together. The renewables can increase the efficiency of CHP and CCHP systems to the maximum possible. CHP and CCHP hybrid systems can be used in small residential and commercial applications to provide 100% uninterrupted heat and electric power 24/7/365 at full capacity

This approach is very flexible and reliable. It is finding wide use in its different combinations and permutations in developing countries and in remote residential and commercial applications.

Solar/Wind Power Cogeneration

Solar energy generation could be combined with other energy sources when constant output is the goal.

---

*CPV is several times more efficient than fixed mounted PV panels. It also tracks the sun all day, thus getting maximum power possible from morning until evening.

**Federal, state and local subsidies might cover 30-40% of the initial cost.

***Conventional PV panels lose 1% efficiency annually. CPV-T systems do not.

Addition of solar power to conventional power sources (power plants) is one approach. Using energy storage devices (batteries or water storage) is another. A more natural and most efficient way, however, is combining PV power with wind at locations especially chosen for this purpose. At such locations PV power is complementary to the output of wind generation, since it is usually produced during the peak load hours when wind energy production may not be available.

Variability around the average demand values for the individual characteristic wind and solar resources can fluctuate significantly on a daily basis. However, as illustrated by Figure 5-49, solar and wind plant profiles—when considered in aggregate—can be a good match to the load profile and hence improve the resulting composite capacity value for variable generation.

In the Figure 5-48 example, the average load (upper line) is closely followed during the day by the average output from the combined wind and solar generators (the second from top line) during the same time. This average is created regardless and because of the fluctuations of the individual wind and solar power generators. This is a marriage made in heaven, and this combined power generating combination will work quite well if wind and solar power outputs can be matched as closely as this one.

Although there are areas in the US and abroad that match this wind and solar profile, the combined effort is usually hard to execute, because the best places for wind and solar are at different locations, often miles apart, and also because of lack of infrastructure at the most suitable locations.

Truthfully, it will require a great effort and expense to implement large-scale "wind-solar load matching" schemes anytime soon with existing technologies.

Still, having as a goal matching wind and solar power outputs will force us to seek the most suitable locations and appropriate technologies for this match. This won't happen overnight, but if we approach the solution intelligently, we will have a large-scale power output—nationally—that matches the grid power loads.

Conclusion

The solar energy industry, in general, and the related markets went through tremendous changes recently, and a lot of progress was documented. Although it seems likely that this trend will continue, the level of progress is uncertain, since it depends on a number of political and socio-economic factors.

The Winners

There are a number of major solar companies that have grown significantly in size and market share during the last several years. The battles for market share in the world's solar fields is increasing in size and intensity.

Every day we hear about new deals and projects announced on all continents. China seems to lead the pack in planned capacity, so we will be watching carefully the developments.

Company	Technology
Trina Solar, China	c-Si
Yingli Green, China	c-Si
Canadian Solar, China	c-Si
Hanwha Solar, China	a-Si, c-Si
Jinko Solar, China	c-Si
JA Solar, China	c-Si
Sharp, Japan	c-Si, Thin Film Si
ReneSola, China	c-Si
First Solar, USA	CdTe
Kyocera, Japan	c-Si

Figure 5-49. The leaders in the solar field, 2014

The solar energy market in the U.S. has been growing exponentially since 2008, achieving a record year in 2014 by growing 34% over 2013. Nearly 7,000 MW of new solar electric capacity was installed in one single year. This is the equivalent of building 7 new coal-fired power plants, without the emissions.

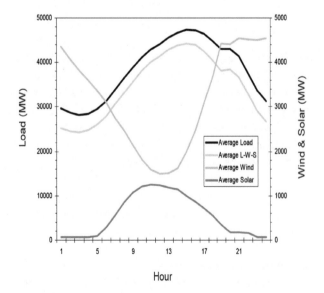

Figure 5-48. Simultaneous wind and solar generation

The photovoltaic (PV) sector installed over 6,200 MW of new capacity during the same time. Residential and utility market segments led, growing by 51% and 38%, respectively. Concentrating solar power segment had its largest year ever too. Nearly 800 MW of new capacity was installed.

All solar installations put together represented 32% of all new electricity generating capacity installed in the U.S.. This volume is second only to natural gas, where many new gas-fired power plants are added or retrofitted from coal-fired such.

Over 25 GW of cumulative solar electric capacity is operating in the U.S.; enough to power more than 4.5 million homes.

There are over 220,000 installations on nearly 700,000 U.S. homes and businesses. During 2014 and 2015, a new solar project was installed about every 2-3 minutes. The total solar capacity is expected to increase to 30 GW by the end of 2016.

The price of solar has been falling steadily since the implementation of the ITC in 2006. The overall cost to install solar has dropped by more than 80% since then. Residential solar costs have dropped by 45% since 2010, while utility-scale costs have dropped even more. Recent utility size contracts were issued at prices below $0.06/kWh.

In 2015, the 30% ITC was extended for another 5 years, so the U.S. solar industry is now breathing a collective sigh of relief. By 2020, residential installations would become the norm, and utilities will get solar power for less than $0.05/kWh. Hopefully by then energy storage would be perfected and made cheap enough too, at which point solar would become a major power generator.

WIND POWER

Wind is a powerful force that has been used through the centuries to provide mostly mechanical power to a number of agricultural and commercial operations. These include powering sail ships, running corn and wheat grinding mills etc. The sophistication of the wind power equipment has been increasing with time, and today these amazing wind giants—the wind turbines—can be seen all over the place.

They can be seen from far away, swinging their giant arms as if saying, "Here we are, don't you see how important we are?" And important they are, no doubt.

There is a lot of wind available in many areas of the U.S. and the world, where the wind energy is waiting to be harnessed efficiently, economically, and safely—for wind power is one of the cleanest and most efficient technologies around. How clean and efficient it is, we'll see below, but first we'll take a look at wind power and its present state-of-the-art.

Background

Wind power is basically the conversion of the free-willing wind currents and the energy contained in them, into useful form of energy. This way, huge amount of wind energy can be captured and converted into electric power, which is used locally, or sent into the national grid for use at distant places.

In parallel with this development, the importance of wind for direct water pumping or drainage, and propelling ships is still considerable, but it seems to be decreasing with time.

Figure 5-50. Wind power farm, or power plant.

Large wind farms are the focus of the power industry, and it is where a major part of their efforts and money goes today. This is mostly due to the fact that the large wind power plants produce large amount of electricity, which translates into equally large profits. The large-scale power plants consist of dozens and hundreds of individual wind turbines which are usually connected to the electric power transmission network.

Offshore wind farms can harness more frequent and powerful winds than are available to land-based installations and have less visual impact on the landscape, but their construction costs are considerably higher. There are also permitting problems, navigational risks and other obstacles that have slowed down their development to a snail's speed.

Offshore installations are also not readily accessible, and the violence of the sea is a determining factor during installation and daily O&M. Waiting for a long time for calm weather to repair a problem in the middle of the ocean can result in delays in installation and maintenance schedules. In these cases offshore turbines could be idling for long periods of time, thus reducing energy production and profits.

Large land based wind farms don't have these problems, but still have to fight a number of obstacles, such as permitting, land availability, right of way (for accessing the power grid), inconsistent wind velocity, etc. Small land-based wind farms can provide electricity to remote locations and the local utility companies could buy surplus electricity produced by these as well. Because of these advantages, land based wind farm dominate the global wind power market.

Wind power is an alternative to fossil fuels, and unlike fossils it is renewable, widely distributed, and clean. It produces no greenhouse gas emissions during operation, and uses little land. Although there are some environmental effects, these are much less problematic than those from other power sources.

Wind power made significant progress in the U.S. lately, but its growth has been slowing down and its future hard to predict for a number of reasons.

Worldwide, Denmark is generating 25% of its electricity from wind, and other 83 countries are using wind power on a commercial basis. In 2015 wind energy production was over 3.5% of total worldwide electricity usage, and growing rapidly at more than 25% per annum.

Wind power (potential) levels are consistent on year to year basis, but it has significant variation on hour-to-hour basis. No wind, no power. And since we cannot control when wind blows, or how hard, this creates problems in the supply-demand cycle, especially when wind power is a major contributor in a local electrical system.

As the proportion of wind contribution increases, new approaches for using wind power are being considered. Power management techniques such as having excess capacity energy storage, geographically distributed turbines, dispatchable backing sources, exporting and importing power to neighboring areas, or reducing demand when wind production is low, can greatly mitigate the variability problems. Mixing wind and solar generation is another feasible approach, for the peaks of wind and solar compliment each other. This could be used to level the energy production and provide power in peak periods. Weather forecasting and management also permits the electricity network to be readied for the predictable variations in production that might occur.

Another big problem wind power faces, despite of its general acceptance by the public, is that the construction of new wind farms is not welcomed by the locals in most neighborhoods, due to aesthetics, safety and other issues.

The NIMBY (not in my back yard) sentiment is hindering the spread of the gentle giants around the U.S. and most of the developed world. This issue can be resolved only by locating large wind mill farms away from populated centers, which usually increases the final cost of wind power.

Wind Energy

Wind energy is the kinetic energy of air in motion. Total wind energy flowing through specific area A during the time t is:

$$E = \frac{1}{2}mv^2 = \frac{1}{2}(Avtp)v^2 = \frac{1}{2}Avtpv^3$$

where
ρ is the density of air
v is the wind speed
Avt is the volume of air passing through A (which is considered perpendicular to the direction of the wind)
$Avt\rho$ is therefore the mass m passing per unit time.

Note that ½ ρv^2 is the kinetic energy of the moving air per unit volume.

Power is energy per unit time, so the wind power incident on A (e.g. equal to the rotor area of a wind turbine) is:

$$P = \frac{E}{t} = \frac{1}{2}A\rho v^3.$$

Wind is defined as movement of air across the surface of the Earth, affected by changes in temperature of the Earth's surface. It is usually directed from high to low pressure areas. The surface of the Earth is heated unevenly by the Sun, depending on factors such as the angle of incidence of the sun's rays at the surface (which differs with latitude and time of day) and whether the land is open or covered with vegetation. Large bodies of water, such as the oceans, heat up and cool down slower than the land.

The heat energy absorbed at the Earth's surface is transferred to the air directly above it and, as warmer air is less dense than cooler air, it rises above the cool air to form areas of high pressure and thus pressure differentials. The rotation of the Earth drags the air around with it causing turbulence. All these effects combine to cause a constantly varying, and largely unpredictable, pattern of air movement (winds) across the surface of the Earth.

The total amount of economically extractable power available from the wind is considerably more than present human power use from all sources. Some estimates show that around 50 TW of useful power could be extracted from wind turbines. Other estimate show that 1,700 TW of wind power are available worldwide at an altitude of 100 meters over land and sea, from which over 100 TW could be extracted in a practical and cost-competitive manner.

Practical wind power generation, however, is usually from winds that are very close to the surface of the earth. This is because wind turbines are limited as to how high the blades can be located for safe operation. The maximum practical altitude is 200-300 feet. But winds are usually much stronger and more consistent higher up we go. This is forcing technological developments as needed to generate electricity from high altitude winds.

So, if the wind turbines cannot be higher than 300 feet, then we can install them on high elevations. The only problem here is that these areas are usually far from populated centers and grid connections, so new power lines must be installed, which becomes a financial problem.

Wind Power Equipment

We saw some large equipment in use at coal and nuclear plants, so looking at a wind mill won't impress us much. And yet, they are most impressive mechanical units we see around. Hundreds of feet tall, they are noticeable from a great distance. And when these large giants are working side by side, they look like something out of this world. Looking at their large arms swinging in the air—200 feet above—whistling and whooshing nonstop, one gets an out-of-this-world sensation.

The Wind Turbine

A wind turbine is an electro-mechanical device that is designed to convert the kinetic energy of the wind first into mechanical energy, which can be used to drive machinery. Most often, however, it is used to generate electricity.

Today's wind turbines are manufactured in a wide range of vertical and horizontal axis types. Small wind turbines are used for water pumping, battery charging, auxiliary power on boats, and such. The large size wind turbines are usually used for grid-connected electric power generation, and their use is becoming increasing-

ly important for the commercial electricity generation.

Wind turbines have a number of key components:

- *The structural support* of the wind turbine includes the tower and rotor yaw mechanism.
- The *rotor* supports and controls the turbine blades for efficient and safe operation, in order to catch the maximum amount of wind and rotate the main axle.
- A gearbox or continuously variable transmission are attached to the main axle and convert its low speed (but powerful) rotation to high speed (low resistance) rotation as needed for generating electricity.
- An *electric generator* and control electronics convert the rotational energy of the axle into electricity, which is sent into the grid

The turbine blades are designed to rotate about either a horizontal or a vertical axis, with the horizontal type being much older and more common.

Figure 5-51. Wind turbine blades in transport

Wind Turbine Manufacturing

Prior to 2005, only one wind turbine original equipment manufacturer (OEM) assembled utility-scale turbines in the U.S. There were 28 wind turbine manufacturers of wind turbines in 2011. In 2012, 8 new facilities came online in 8 different states, so a total of 12 nacelle assembly facilities were online by the end of 2012.

The growth in the U.S. wind-related manufacturing has occurred across the entire value chain, so there are currently 13 facilities producing blades and 12 tower manufacturing facilities, while the wind-related manufacturing facilities are spread across 44 states and employ more than 25,000 people across the country. At least 10 R&D facilities conduct advanced wind turbine research.

Growing domestic production has resulted in a significant trickle-down effect as OEMs establishing a manufacturing presence in the U.S. have attracted supply chain companies capable of supplying the thousands of components comprising a wind turbine. As a result, there are now at least 559 wind-related manufacturing facilities spread across 44 states supplying components to the wind energy industry.

The growing manufacturing capacity here is reflected in the increasing domestic content of turbines, with 67% of U.S.-deployed turbines' value being manufactured domestically in recent years, up from less than 25% prior to 2005.

Wind turbines assembly consists of about 8,000 different components, some as strong and large as a skyscraper, and some as fragile and small as a semiconductor chip. The different components of the wind turbine system form sub-systems, as follow:

- Materials: turbines are primarily composed of large amounts of steel, but other materials, such as composites, ductile iron, concrete, aluminum, copper and adhesives, are also used.

- Equipment: a variety of components, such as fall protection, turbine lighting and other systems are needed. Turbines also require unique construction and on-site equipment.

- Structural: turbines use a huge number of fasteners, castings and other steel products.

- Power transmission: wind turbines have a sizable and complex power transmission system, requiring bearings, couplings, gears, hydraulic systems, brakes, machined and fabricated components and shafts, among other components.

- Electrical: the electrical system is a critical part of a wind turbine. Common components include power converters, controls, sensors and generator components.

There are several major wind turbine manufacturers world wide; Gamesa, GE Energy, Goldwing, Siemens Power, Nordex, and Vestas, to mention a few. The wind turbine manufacturing, transport, assembly, and O&M process consists of the following steps (22)

The Components

The major components in a wind turbine are:
- Rotor comprising four principal components
  — Blades,
  — Blade extender,
  — Hub, and
  — Pitch drive system;

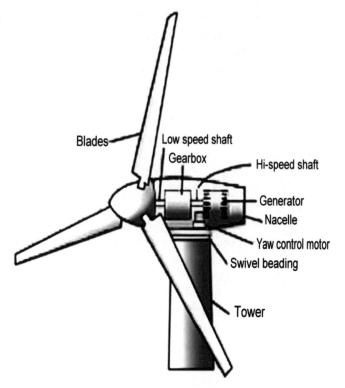

Figure 5-52. Wind turbine

- Nacelle, which is the external shell or structure resting atop the tower containing and which houses the major components and sub-systems;
  — Gearbox,
  — Bearing assemblies,
  — Connecting shafts,
  — Electric generator,
  — Controller, and

- All electronic components that allow the turbine to monitor changes in wind speed and direction;

- Tower, normally made of rolled steel tube sections that are bolted together to provide the support system for the blades and nacelle; and,

- Other components, including transformers, circuit breakers, fiber optic cables, and ground-mounted electrical equipment. Beyond the major components, there are also many subcomponents in a wind turbine as well. The tower, for example, is over 26% of the total cost of a wind turbine, rotor blades 22%, the gearbox 13%, and the other components 5% or less.

Components Transport

Wind turbines have some of the largest components in the world. Their blades are as long as half a football field and because of that they, and some other components, which require special and very expensive transport with complex logistics procedures.

Manufacturers use specialized means of transport to convey the wind turbines to their destination on the wind farm. These means of transport allow for access to any kind of terrain, including the most difficult, with a minimal impact on the environment.

The transport procedures could be classified as:
- *Road transport* uses ready access to rural and remote areas that are common locations for wind farms; however, limited carrier capacity, height, weight and length restrictions, in addition to rising fuel costs, state and local permitting issues, driver shortages and other issues, all pose challenges.
- *Rail transport* is cheaper than transport by road, particularly for long distances; however, height, weight, and length restrictions can be problematic, and rail lines don't run directly to wind farm site, which necessitates road transport to the final destination.
- *Water transport* via barge is comparatively inexpensive; however, access to waterways for wind components is limited, and transport time can be slow. International transport is almost always done via ocean freight.

There are a number of major issues facing the industry, such as varying state and local permit rules and restrictions, as well as varying engagement by state and local transportation officials, create significant challenges.

There is also a limited supply of trailers and railcars capable of hauling wind turbine components, and a limited number of companies willing to make the investment to increase the supply. This is due to cost and the fact that the specialized equipment used to haul wind turbine components often cannot be used for other hauling, which limits its usefulness.

Field Assembly

Once on site, an experienced team of operators carry out the installation of the turbines. Finally, the manufacturers perform the start up operations and takes responsibility for the operation and maintenance of the wind turbine throughout its working life.

The field assembly of the wind turbines and the wind power plant consist of:

Construction and Foundation

Before the wind turbine is transported and assembled, various tasks are performed to prepare the land,

such as concrete work and constructing the assembly platform. The latter requires compaction able to support weights of around $4\,\mathrm{kg/cm^2}$.

Tower Erection

The tower sections are placed one on top of the other using lattice cranes. They may be either caterpillar or hydraulic jack types. Caterpillar-type cranes, with widths between 8.5 and 10 m., are able to change position easily. 5 m.-wide jack types are suitable for working on difficult terrain, due to their narrow width.

Once the sections are in place, the field personnel connect and assemble the parts. The placement and height of the wind turbine are studied beforehand to guarantee maximum wind performance.

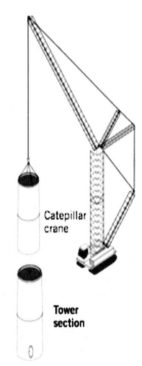

Figure 5-53

Nacelle installation

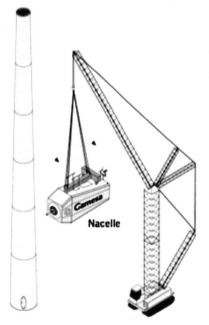

Figure 5-54a

Once the tower is assembled, the next step is to install the nacelle, which is connected to the top tower section. The electrical connection is established then to all components, parallel to the wind turbine assembly.

Rotor installation

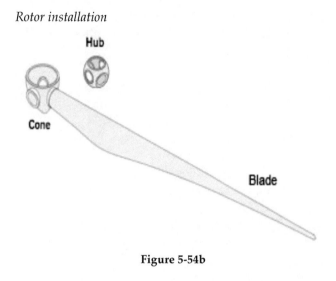

Figure 5-54b

This assembly may be performed on the ground, either connecting all three blades to the hub, or by connecting them blade by blade. The second method requires less space for maneuvering and makes the assembly process quicker.

Once the nacelle is installed, the hub and cone are raised, followed by the blades, which are lifted horizontally one by one and installed in that order.

Keeping in mind the out-of-this-world sizes of the different components, we can easily that the installation and assembly of a wind turbine is an extra-terrestrial experience. A few highly qualified and experienced engineers and technicians at each site are able to complete these operations successfully, but not without difficulties and a lot of sweat.

Wind Turbine Specs

Just to give the esteemed reader an idea of what we are talking about, here are some specs of these gentle giants—the largest among the renewable power generators:

Tower:

Purpose:	supports the entire assembly
Type:	3-section, bolted together steel tube
Material:	1.5 inch thick welded steel sheets.
Height:	230 ft
Bottom diameter:	14.0 ft
Top diameter:	6.0 ft
Weight:	100 tons.
Total height:	325 ft
Hub height:	240 ft

Rotor

Purpose:	rotates with the wind
Diameter:	175 ft
Type:	Upwind, Stall regulated.

Hub

Purpose	holds and controls the blades
Type	spherical
Hub diameter:	6.0 ft
Weight:	4.0 tons

Blades

Purpose:	catches the wind and rotate
Type:	stall regulated
Blade length:	75 ft
Blade weight:	4.0 tons

Nacelle

Purpose:	houses main shaft, gearbox, and generator
Type:	fiberglass housing on a steel platform
Length:	20.0 ft
Height:	10.0 ft
Width:	7.0 ft
Weight:	25 tons

Main shaft

Purpose:	connects gear box and generator
Type:	almost solid, painted steel
Diameter:	3.0 ft
Length:	7.0 ft
Weight:	2.5 tons

Gearbox

Purpose:	connects the blades to the generator
Type:	helical-planetary
Ratio	1:81
Length:	6.0 ft
Width:	6.0 ft
Weight:	5.5 tons

Generator

Purpose:	power generation
Type:	asynchronous
Stator:	water-cooled stator
Windings:	air cooled
Power rating:	two 200 kW and 900 kW generators
Power produced:	2,500 MWh annually.

Operation
- Low wind-speeds (<4 m/s) 15 rpm blade rotation is converted to 1200 rpm at the 200 kW generator
- High wind-speeds (> 4 m/s) 22.4 rpm at the blade is converted to 1800 rpm and 900 kW generator
- Turbines operate at maximum 1000 rpm.
- Air brakes slow blade rotation and stop rotation at high wind speeds.
- Hydraulics are deployed at power down or unresolved problems.
- Computer 'reads' wind speed and variability
- Computer selects generator based on weather data analysis
- At the selected blade RPMs, the generator makes a contact with the grid.
- The generator 'follows' the sine-wave and synchronizes with the grid power.
- The generator shuts down when disconnected from the grid

Approximately 75% of the total cost of a wind farm is related to upfront costs such as the cost of the turbine, foundation, electrical equipment, grid-connection, etc.. Fluctuating fuel costs have no impact on wind power generation costs. A wind turbine is capital-intensive compared to conventional fossil fuel fired technologies such as a natural gas power plant, where as much as 40-70% of costs are related to fuel and operations and maintenance. In turn, a wind farm does not need any fuel to operate. It is also one of the cleanest and greenest power generating technologies available today.

There are number of types of wind turbines, the major of which are the horizontal and vertical turbines:

Horizontal-Axis Wind Turbines

Horizontal-axis wind turbines (HAWT) are the most common types used today, and the one we review in more detail. These turbines are designed with the main rotor shaft and electrical generator at the top of a tower, and operate by pointing into the wind. Small turbines are pointed by a simple wind vane, while large turbines generally use a wind sensor coupled with a servo motor. Larger turbines have a gearbox, which turns the slow rotation of the blades into a quicker rotation that is more suitable to drive an electrical generator.

Since a tower and the casing produce turbulence, the turbine blades are usually positioned upwind. Turbine blades are stiff, which prevents mechanical stress and fatigue of the material and also keep the blades from bending and smashing into the tower during high winds.

Downwind versions are used for small turbines, since they don't need an additional mechanism for keeping them in line with the wind, and because in high winds the blades can be allowed to bend which reduces their swept area and thus their wind resistance.

Turbines used in wind farms for commercial production of electric power are usually three-bladed and pointed into the wind by computer-controlled motors. The high tip speeds of large turbines is over 320 km/h (200 mph), high efficiency, and low torque ripple, which contribute to good reliability.

The blades are usually colored white for daytime visibility by aircraft and range in length from 60 to 130 ft. and more. The tubular steel towers range from 200 to 300 ft. of height. The blades rotate at 10 to 22 revolutions per minute, where the blade tip speed exceeds 300 ft./s.

A gear box is used to step up the speed of the generator, but some designs use direct drive of an annular generator. Some models operate at constant speed, but more energy can be collected by variable-speed turbines which use a solid-state power converter to interface to the transmission system.

All turbines are equipped with protective features to avoid damage at high wind speeds, by feathering the blades into the wind which ceases their rotation, which can be supplemented by brakes for even safer operation.

Vertical-Axis Wind Turbines

Vertical-axis wind turbines (or VAWTs) are different from the above discussed horizontal turbines, in that the main rotor shaft is arranged vertically. The advantages here is that the turbine does not need to be pointed into the wind to be effective. This is an advantage on sites where the wind direction is highly variable.

The key disadvantages include the low rotational speed with the consequential higher torque and hence higher cost of the drive train, and the inherently lower power coefficient. The 360 degree rotation of the aerofoil within the wind flow during each cycle and hence the highly dynamic loading on the blade, the pulsating torque generated by some rotor designs on the drive train, and the difficulty of modeling the wind flow accurately and hence the challenges of analyzing and designing the rotor prior to fabricating a prototype.

With a vertical axis, the generator and gearbox can be placed near the ground, using a direct drive from the rotor assembly to the ground-based gearbox, hence improving accessibility for maintenance. When a turbine is mounted on a rooftop, the building generally redirects wind over the roof and this can double the wind speed at the turbine.

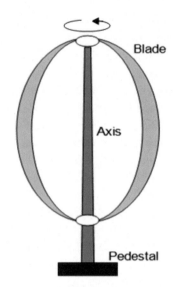

Figure 5-55. Vertical axis wind turbine

If, for example, the height of the rooftop mounted turbine tower is approximately 50% of the building height, this is near the optimum for maximum wind energy and minimum wind turbulence. However, we should remember that wind speeds close to structures are generally much lower than at exposed rural sites. Unpredictable turbulence and noise may be a concern in such cases. Another fact that should not be ignored is that old structures exposed to turbulent wind loads may not be able to adequately resist the additional stress.

There are also two major types location-based wind power generation: onshore and offshore.

Onshore Wind Power

Onshore wind power has been, is, and will remain a power source of significance and preference. The quantity of wind power generation is still relatively small, but the importance of generating power from wind in locations where other technologies are unavailable, is increasing.

Since the 1980s The United States has led the world in installed capacity, but in the late 1990s Germany surpassed the U.S., and after titter-tatter of leadership, Germany is now the leader with 100 GW, vs. the U.S. 50 GW wind installations. China has been rapidly expanding its wind installations since the late 2000s and now has 50 GW of installed wind capacity.

Presently there are over two hundred thousand wind turbines operating worldwide, with a total nameplate capacity of about 250 GW. World wind power generation increased four-fold between 2000 and 2006, doubling about every three years.

Wind power generates over 400 TWh annually, or about 2.5% of worldwide electricity. And the numbers are growing fast. Wind power generation is expected to reach over 10% of the total world's electricity by 2020. A number of countries are already close to that mark. Denmark produces 28% of its grid electricity via wind farms. Portugal follows with 20%, Spain with 16%, Ireland with 15, and Germany with 9%. 83 countries around the world use significant amount of wind power on a commercial basis.

Europe accounts for nearly half of the world total wind power generation capacity. The financial crisis of the 2008 changed the dynamics of the wind generation scenario, but progress is still evident in many countries, including the U.S..

Offshore Wind Power

Offshore wind power refers to the construction of wind farms in large bodies of water to generate electricity. These installations are a new development, intended to utilize the more frequent and powerful winds that are available in these locations and have less aesthetic impact on the landscape than land based projects.

However, the overall design, construction, operation and maintenance of off-shore wind power plants is much more difficult and complex, so the costs are considerably higher.

As of 2011, offshore wind farms were at least 3 times more expensive than onshore wind farms of the same nominal power but these costs are expected to fall as the industry matures.

Siemens and Vestas are the leading turbine suppliers for offshore wind power. DONG Energy, Vattenfall and E.ON are the leading offshore operators. As of October 2010, 3.16 GW of offshore wind power capacity was operational, mainly in Northern Europe. According to BTM Consult, more than 16 GW of additional capacity will be installed before the end of 2014 and the UK and Germany will become the two leading markets. Offshore wind power capacity is expected to reach a total of 75 GW worldwide by 2020, with significant contributions from China and the US.

There are 33 offshore wind projects in varying stages of development in the U.S., and while there are nearly 4 GW of installed offshore wind capacity in Europe and China, the U.S. has no operational offshore projects as yet. There are, however, nine advanced-stage plans, representing nearly 4.0 GW of potential wind capacity.

The U.S. offshore wind power generation is scheduled for commercial operation by 2017, so the DOE awarded $28 million per year in Advanced Technology Demonstration funding to seven projects. Is that enough, and what else is needed, remains to be seen, but progress is made, nevertheless.

Wind Power Markets

Wind energy, like other power technologies based on renewable resources, is widely available throughout the world. It uses no fuel, so there is no fuel price risk or any fuel-related constraints, which is critical to improving the security of domestic power supply.

Note: Wind uses no fuel, but is dependent on the weather condition, so it is considered a variable power generator, which brings a set of problems, as we will see later on in this text

Furthermore, and very importantly, wind power emits no direct greenhouse gas (GHG) emissions or any other pollutants (such as oxides of sulfur and nitrogen) during operation and consumes no water. Air pollution and extensive use of fresh water for cooling of thermal power plants are serious concerns in hot or dry regions, so wind power development in these regions becomes increasingly important.

Wind power plants do not emit GHGs directly, but there is a significant indirect pollution created during the equipment manufacturing, transport, and installation phases. There is also indirect pollution created during the O&M phase of the power plants, by service personnel's activities, which affect the local environment negatively.

Nevertheless, wind power ads to and enhances the energy diversity of the energy sources, and hedges against price volatility of fossil fuels. This has a stabilizing effect on the cost of electricity generation in the long term.

Types of Wind Power Markets

There are basically two major types of commercial wind turbines; on-shore and off-shore.

- On-shore, or land-based, wind refers to the energy generated by wind turbines deployed in the mainland. Land-based wind power is a proven and mature renewable energy technology that is being deployed globally on a mass scale. Wind turbines extract kinetic energy from moving air flow (wind) and convert it into electricity via an aerodynamic rotor, which is connected (often via a transmission system) to an electric generator.

Today's standard turbine has three blades rotating on a horizontal axis, upwind of the tower, with a synchronous or asynchronous generator connected to the grid using power electronics. Two-blade and di-

rect-drive (without a gearbox) turbines are also available.

The electricity output of a turbine is roughly proportional to the rotor area; therefore, fewer, larger rotors (on taller towers) use the wind resource more efficiently than more numerous, smaller machines.

The typical commercial type wind turbines used today are these of 1.0-2.0 MW size. The largest wind turbines are 5.0-6.0 MW in size, with a rotor diameter of up to 415 feet.

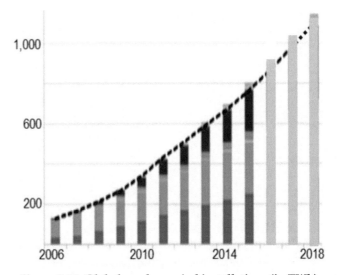

Figure 5-56. Global on-shore wind installations (in TWh)

- Offshore wind energy refers to the energy generated by wind turbine deployed in the sea. Depending on the depth of the sea, and climate conditions, the turbines can be installed tens of miles off the shoreline.

Deploying turbines in the sea takes advantage of better wind resources than at land-based sites. Offshore turbines, therefore, achieve significantly more full-load hours. Offshore wind farms can be located near large coastal demand centers, often avoiding long transmission lines to get power to demand. This can make them particularly attractive for countries with coastal demand areas and land-based resources located far inland, such as China, several European countries and the US.

While needing to satisfy environmental stakeholders, offshore wind farms generally face less public opposition and, to date, less competition for space compared with developments on land. As a result, projects can be large, with 1.0 GW power plants likely to be the norm in the future.

Large-scale offshore deployment has started, more slowly than initially hoped, mostly in Europe. By the end of 2012, 5.4 GW had been installed (up from 1.5 GW in 2008), mainly in the United Kingdom (3 GW) and Denmark (1 GW), with large offshore wind power plants installed in Belgium, China, Germany, the Netherlands and Sweden. Additional offshore turbines are operating in Norway, Japan, Portugal and Korea, while new projects are planned in France and the United States. In the United Kingdom, 46 GW of offshore projects are registered, of which around 10 GW have been progressing to consenting, construction or operation.

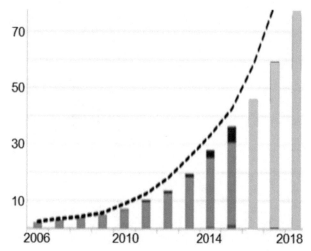

Figure 5-57. Global off-shore wind installations (in TWh)

As Figure 5-57 show, on-shore wind power installations predominate the wind market by a factor of 10 today, which difference is expected to increase in the future. This is because, generally speaking, off-shore wind power installations are more complex technologically, and more expensive than their on-shore cousins. They are expected to double from the present level by 2020, but that would still represent less than 10% of the on-shore installed wind power capacity around the world.

A handful of wind turbine manufacturers supply more than 70% of the U.S. market, but the diversification continues. Until recently, In 2012, GE Wind led the U.S. market with more than 5 GW of wind turbines newly installed in 2012, representing nearly 40% market share. Siemens held second place with a 20% market share, followed by Vestas with 14%, and Gamesa with 10%.

There was a notable increase in the number of wind turbine manufacturers serving the U.S. market. The number of these installing more than 1 MW increased from just five in 2005 to 25 in 2012. The large turbine suppliers; GE Wind, Vestas, and Siemens actually gained market share since 2008.

Worldwide, GE Wind and Vestas are the top supplier of turbines worldwide. A number of Chinese tur-

bine manufacturers entered the global wind market at the time and continue to occupy top positions in the global ratings. None of them is on the top five, and their growth has been based almost entirely on sales to the Chinese market. Chinese and South Korean wind turbine manufacturers include Goldwind, China Creative Wind Energy, Guodian United Power, Sinovel, Hyundai, HZ Windpower, and Sany Electric.

Global Wind Power

Annual wind power generation capacity additions in the U.S. achieved record levels during the 2008-2012 boom period. The main reason, especially during the last phase of this period, was the then-planned expiration of federal tax incentives at the end 2012, in addition to improvements in the cost and performance of wind power technology.

Today, the uncertainty of federal tax incentives, low natural gas and crude oil prices, modest electricity demand growth, and limited near-term demand from state renewable portfolio standards, have put a damper on the wind industry growth These trends continue to impact the wind industry, and its long-term future remains uncertain—totally dependent on natural gas prices, fossil plant retirements, and political decisions.

Since 2000, cumulative installed capacity has grown at an average rate of 24% per year. In 2012, about 45 GW of new wind power capacity were installed in more than 50 countries, bringing global onshore and offshore capacity to a total of 282 GW at the time. New investment in wind energy in 2012 was USD 76.6 billion. Among the largest clean energy projects financed in 2012 were four offshore wind sites of 216 MW to 400 MW in the German, UK and Belgian waters of the North Sea, with investments of $1.1 billion to $2.1 billion.

Progress made since 2008 shows a positive trend: in 2012, wind power generated about 2.6% of global electricity while capacity and production information for wind resources around the globe show steady expansion.

Repowering, i.e. replacing "old" wind turbines with more modern and productive equipment, is on the rise too. Repowering is shown to increase wind power while reducing its footprint.

Today, a 2.0 MW wind turbine with a 260 feet diameter rotor generates 4-6 times more electricity than a 500 kW turbine with a 130 feet diameter rotor built in 1995.

Wind power additions hit a record in 2012, with 13.1 GW of new installations and $25 billion Invested. Cumulative wind power capacity in the country grew to a total of 60 GW. Wind power generation represented the largest source of U.S. electric power generating at the time.

The U.S. wind market represented roughly 29% of global installed capacity in 2012, a steep rise from the 16% registered in 2011. The competition is growing fast, where about 30% of Denmark's electricity demand is fed by wind, 18% for Portugal and Spain, 16% for Ireland, and 10% for Germany.

The U.S. cumulative wind power capacity in 2012 was estimated to roughly 4.4% of the total electricity demand.

Note: An increasing number of wind turbines are being installed in cold climates, where they are exposed to icy conditions and/or low temperatures outside the design limits of standard wind turbines. At the end of 2012, nearly 69 GW of installed capacity were estimated to be located in cold climate areas in Scandinavia, North America, Europe and Asia, of which 19 GW were in areas with temperatures below 20°C and the rest subject to icing risks. Between 45 GW and 50 GW of additional capacity are likely to be installed in cold climates before end 2017.

Denmark, had about half of global markets in 2005, but Danish companies represented only 20% of operating turbines in 2012—still a huge amount for a country that has slightly more than 1% of global installed wind capacity. In addition to Denmark, strong manufacturers in Spain and Germany make Europe a large exporter of wind technology; in 2010, net exports were about $6.5 billion. The United States and India are also among the large manufacturing countries.

The wind industry has contributed substantially to the socio-economic development of several regions. A clear example is significant job creation in Spain during the first decade of the century, where a sound support scheme attracted several foreign industrial companies across the value chain for wind projects, together with a strong local industry.

The United Kingdom is currently attracting industry because of its thriving offshore wind market: between 2007 and 2010, jobs in the sector grew by nearly 30%. Jobs in the wind industry (both direct and indirect) reached approximately 265 000 in both China and the European Union (of which 118 000 in Germany), 81 000 in the United States, 48 000 in India and 29 000 in Brazil. Employment figures are not easy to compare across technologies, but wind generally provides more jobs per investment

Most wind turbine manufacturers are concentrated in six countries (the United States, Denmark, Germany, Spain, India and China), with components supplied from a wide range of countries. Market shares have changed in the past five years. New players from China are growing and have started exporting; the six largest Chinese companies (among the top 15 manufacturers globally) together have exceeded 20% of market share in recent years. Countries with emerging manufacturers include France and Korea, while Brazil has an increasing number of manufacturing facilities.

Despite many challenges, a growing percentage of the equipment used in the U.S. wind power installations of late has been sourced domestically. U.S. trade data show that the U.S. was a large importer of wind power equipment in 2012, but that growth in installed wind power capacity has outpaced the growth in imports in recent years,

The U.S. wind market consists of over 600 wind-related manufacturing facilities. The domestic wind power content is now over 75%, up from 25% before 2006.

The majority participants in the U.S. wind power market are local equipment manufacturers, complete with the product and services providers in their supply chain. Then, we have a large group of installers—small and large companies—who are responsible for the design, siting, transport, and construction of the power fields. On the peripheral side of the wind markets—but playing very important role—are many contractors and consultants, legal and insurance firms, transportation providers, government entities and a many other.

As a result, a growing percentage of the equipment (in dollar-value terms) used in wind power projects are now sourced domestically. Focusing on selected trade categories, and when presented as a fraction of total equipment-related wind turbine costs, the overall import fraction is estimated to have declined considerably, from 75% in 2006 to 25% in 2012. Conversely, domestic wind power equipment content has increased from 25% in 2006 to 75% now and the exports have also grown proportionally. These trends continue today and are expected to grow significantly in the future too.

China

China has large land mass and long coastline, all of which provide excellent wind resources. The estimates of late are that China has over 3,000 GW of exploitable wind power capacity on land and almost unlimited at sea.

By 2013, there were over 80 GW of wind electricity generating capacity installed in China, which is more than the total capacity of China's nuclear power stations. The wind power plants produced 115,000 GW-hours of wind electricity annually.

The original plan was to have 100 GW of grid-connected wind power capacity by the end of 2015 and to generate 190 terawatt-hours of wind power annually. In 2014, the government pledged to produce 20% of the country's total energy through non-fossil fuels by 2030, doubling its current level, while capping growth in its carbon emissions by the same year, it not earlier.

Wind power as a key growth component of the country's energy policy and its overall economy, and researchers claim that China could meet all of their electricity demands from wind power by the same time. However, this is just a wish, because in practice, wind energy generation in China has not kept up with the fast construction of wind power capacity in the country.

Just like in all other sectors of their construction industry, the Chinese are good at building things, but half as good in efficient use of the new creations. Because of that, and mostly due to poor planning, many wind power plants sit idling, because their power cannot be connected to the grid, or is used improperly.

Experts estimate the U.S. wind industry to be 40% more efficient than the Chinese, mostly due to inefficient grid connection.

By 2010, nearly 20 Chinese companies were producing wind turbines and many more were producing different components. Turbine sizes of 1.5 MW to 3 MW were manufactured routinely by leading wind power companies Goldwind, Dongfang Electric, and Sinovel.

The production of smaller size turbines was also increased to over 100,000 turbines for the same time. The most amazing thing at the time was the fact that the Chinese wind power industry seemed unaffected by the global financial crisis of 2008-20012.

At the same time period, China had the largest wind power installations in the world, with over 40 GW. Unfortunately, 1/4 to 1/3 of this total was not connected to the grid. By 2013, nearly 80 GW were installed of which almost 20 GW sat idle. Even after the National People's Congress permanent committee passed a law that requires the Chinese energy companies to purchase all the electricity produced by the renewable energy sector, the wind power plants could not be connected to the grid.

Table 5-7. Wind power in the China (in MW)

Year	Capacity	Production
2005	1,260	1,927
2006	2,599	3,675
2007	5,912	5,710
2008	12,200	14,800
2009	16,000	26,900
2010	31,100	44,622
2011	62,700	74,100
2012	75,000	103,000
2013	91,424	134,900
2014	114,763	153,400

The *off-shore* wind energy in China, just like that in the U.S., is developing much slower than expected. In 2013, China had only two operational *off-shore* wind farms, with the total of 250 MW wind power generation. By 2015, China doubled the off-shore wind capacity, but still had less than 600 MW offshore wind capacity, far from the target goal of 5 GW by the end of 2015 and 30 GW by 2020.

The largest wind power generation project in China, and maybe the world, is the Gansu Wind Farm Project. It is under construction in western Gansu province, as part of the six national wind power mega-projects approved by the Chinese government. Several Gansu Wind farms are expected to generate 20 GW of wind electricity by 2020. The project's total cost is estimated at nearly $18 billion, or an amazing $0.90 per installed watt of wind capacity.

In 2008 construction began on a dedicated 750 kV AC power line to carry electricity from the wind farm to populated centers. This is part of the massive power grid expansion underway in China. Thousands of miles of high voltage lines are strung all over the country, in an attempt to bring China's inefficient power grid to the 21st century standards.

When completed, the new power grid would be able to accommodate hundreds of new wind and solar power plants spread all over the country. This is an impressive and very expensive undertaking, which lays the foundation of China's renewable future.

Japan

Since 2011, Japan has been going through a rough patch in the energy sector. All nuclear power plants in Japan were shut down immediately aᵒfter the Fukushima disaster and the race for alternative energy sources started.

Increasing the coal-fired power generation capacity was a no-brainer and the coal imports were increased by almost 1/4. The solar was promoted to an urgency status and work started to install new solar capacity. But recently, a number of challenges to large-scale solar in Japan, including disputes over grid connection, are having a negative effect on the industry. This has led to poor financial results for a number of major PV companies.

The major Japanese thin-film manufacturer, Solar Frontier and the Norwegian solar energy developer, Scatec Solar, reported in 2015 that the 2014 decision by five utility companies to stop considering large PV projects for connection approval had had damaging consequences. If the situation is not improved, the entire solar energy sector in Japan might be in big trouble.

Several solar companies have been forced to stop work on a number of projects in Japan due to the utilities' decisions. Many more plans for new solar installations are presently on hold too, due to overall policy uncertainty. Most companies have been withdrawing their applications for connection to the grid even discontinue all activities in Japan activities. This means that a substantial pipeline of projects planned or under development in Japan will be stopped indefinitely postponed.

It is highly uncertain when and how the Japanese government will move forward with utility scale solar, so most foreign companies are abandoning the Japanese solar market place. Solar Frontier, the vertically integrated Japanese CIGS manufacturer, that the delays to its projects in Japan had caused it to reduce the production of its module shipment volumes. Nevertheless, the residential domestic demand is unaffected and remains robust, and for awhile in 2014 and 2015 represented nearly 90% of Solar Frontier's total module sales.

Over 130 solar companies complained about the state of solar in Japan. About 60% of these companies claimed to be directly affected in a negative way by both the grid connection issue and measures introduced to resolve them. This includes revisions by government entities of the rules governing Japan's feed-in tariff.

The saga started in the fall of 2014, with the suspension of all new grid applications by Kyushu Electric, one of the largest Japanese utilities. This action was followed almost immediately by four other of Japan's 10 utility companies. The solar industry was not consulted and the large-scale PV sector was put on its knees.

As in many other countries, each of the Japanese utilities has a monopoly of the area it serves. This includes full responsibility for the transmission and distribution infrastructure. The grid connection issue has been, and still is, a major concern for the utilities. It, however, is not the only reason for this bottleneck of solar projects. There have been various other issues for

awhile now, starting with the rising electricity prices. On top of it all, there is an added burden placed on ordinary consumers to pay for the solar FiT, which is quite generous. This has led to complaints of high prices and accusations of profiteering by different entities and officials.

Because of that, the Japanese government undertook a wide-ranging review of the FiT at the end of 2014. Coming with a complete solution is unlikely, and the solar industry in the country is in limbo since companies find it increasingly difficult to conduct proper and profitable business in the country in the newly created vacuum. As a result, solar companies cannot make business plans for the medium term future, which means that it would be awhile before the solar industry is back on track.

Strangely enough, since the Fukushima disaster, renewable energy capacity in Japan tripled to 25 GW, with solar accounting for more than 80% of that. And now, nearly 2.5 GW of expensive and polluting oil-fired energy plants are to be shut down in 2016 and switched to alternative fuels.

This means that solar is one of the major solutions to Japan's energy wows, so we just have to hope that the regulatory conditions will improve with time in order to allow solar to take the place it deserves in the country's energy sector.

Denmark

For a small country, tucked in a corner of the European continent, Denmark has a very interesting and dynamic energy history. It started in the mid-1970s, when in response to the Arab oil embargo, Denmark, and many other European countries, added large quantity of coal-fired power generation. By the 1980s, the carbon foot-print of the country, and the region, had grown noticeably high. At that point the Danes decided to go down the unproven renewables path with wind as a leader.

Around that time, a year before the Chernobyl nuclear disaster, Denmark adopted a law forbidding the construction of nuclear power plants. The home-grown anti-nuclear movement is credited for this achievement. With their logo, *Nuclear Power, No Thanks* became popular in the country and spread around Europe and the world.

Renewable and energy alternatives were in, and were heavily promoted by the Danish Organization for Renewable Energy (OVE). With the government on-board, planning to add more wind power was agreed upon and deliberately streamlined by the authorities in order to minimize the hurdles along the way.

Although Denmark has modest average wind speeds in the range of 5-6 m/s at 30 feet height, it is a major energy source and had to be used. Onshore wind is available mostly in the western part of the country, and also on the eastern islands. The offshore wind resources are even larger, with large areas of shallow (15-30 feet) seas, where wind power is most feasible. Many of the off-shore sites also offer higher wind speeds, in the range of roughly 8–9 m/s at 150 feet height.

Starting in 2005, Denmark had installed wind capacity of 3.2 GW, which produced 6.6 TWh of electricity. The actual energy produced, however, was 755 MW at a capacity factor of 24%. The reduction or generated power of nearly 75% is due to the variability of wind (no wind—no energy).

By 2010, Denmark's capacity grew to 3.5 GW, when the new 209 MW Horns Rev 2 offshore wind farm was put in operation. In 2010, another wind off-shore farm, Rødsand-2, increased the nation's wind capacity to 3.75 GW. And in 2014, the 400 MW Anholt wind farm, in addition to other smaller wind projects, brought the nation's wind capacity to today's level of 4.8 GW. Not bad for a country of nearly 6 million people.

This puts Denmark at the enviable position of having the highest proportion of wind power in its energy mix in the world. On the average, it is nearly 40% of the total power consumption in the country. During January 2014, for example, the wind power share was over 61% of the total electricity use. During some days in December, the wind share was over 102%, and at one point reached 135% of the total energy use.

During the summer moths, however, the wind power plants are not very productive, due to seasonal changes. In response, the Danes have also installed nearly 600 MW of solar power, which helps bring up the renewables' share in the base load during the summer months. So, solar compensates wind's inefficiency during the summer months, when the sun is high. A good combination of renewable resources, which, in the end, reduces energy spot market prices in general, and causes a net reduction of pre-tax electricity prices.

But the Danes have other tricks up their sleeves too. Denmark's power grid is connected by transmission lines to other European countries. This way, Denmark does not need to install additional peak-load plants to balance its wind and solar power. Instead, it purchases additional power from its neighbors when necessary. The plan now is to expand the national grid and increase the wind power share to over 50% of total energy use by 2020, and up to nearly 85% by 2035.

The efficient wind and solar power use, in addition to some energy efficiency measures introduced lately, are a remarkable achievement for any nation. Denmark shines, when compared to many countries who also tried to introduce wind and solar power in their energy mix. Most of these efforts did not succeed as planned for a number of reasons. And so, the world needs to take a closer look, and learn the lesson the Danes are offering us.

The Danish wind power industry is an example of a successful execution of a renewables energy plan. The government role in this progress is undeniable, since it provided over 30% of the initial capital cost, and many incentives, during the early years of the wind industry conception. The government help was gradually reduced to near zero today, but the feed-in tariff system is still in place.

The question here is, how can a country as big and diversified as the U.S. follow this Danes' example, but... the answer is still in the wind, my friend.

Europe

The other countries in the European Union (EU) had variable success in the attempt to introduce wind power into their energy mix. Although not nearly as successful as Denmark, by 2015, the installed wind power capacity in EU countries totaled 130 GW, with a compound annual growth rate (CAGR) of 10% during 2000-2013 time period.

In 2014 alone, a total of nearly 12 GW of wind power was installed on the continent, which represents about 32% of all new power capacity. On the average, the wind power capacity installed thus far would produce 275 TWh of electricity, which is nearly 8% of the entire EU electricity consumption.

Wind has 80% approval rate among the EU people, and the plan now is to introduce 230 GW of new wind capacity by 2020. 190 GW of this capacity would be onshore wind farms, and 40 GW off-shore. The total wind capacity would produce 15-18% of the EU's electricity. This would also avoid emitting over 330 million tons of CO_2 per year, while saving the EU over $40 billion a year in avoided fuel costs.

In Germany, wind power plays an important role in the country's energy mix. By 2015, the installed domestic capacity was nearly 40 GW, of which offshore contributed about 650 MW. Germany presently leads the EU countries by the amount of wind installations. And yet, wind is still a fairly small percentage of the electricity consumption in the country.

On a sunny and windy spring day in 2014, Germany's renewable energy sector generated 75% of the electricity used in the country's. Obviously, Sundays are not a high energy demand days, and yet, a record is a record. During the entire month, the renewables met about 27% of the nation's power demand, thanks in part to *favorable weather conditions*.

Table 5-8. Wind power in the EU (2014)

Country	MW
Germany	39,165
Spain	22,986
UK	12,440
France	9,285
Italy	8,663
Sweden	5,425
Portugal	4,914
Denmark	4,845
Poland	3,834
Romania	2,954
Netherlands	2,805
Ireland	2,272
Austria	2,095
Greece	1,980
Belgium	1,959

Summary

Note the "favorable weather conditions" remark above. The usefulness of wind farms is often exaggerated by talking in terms of their installed, or total, "capacity," which is impressive. This, however, hides the fact that the actual output from any wind farm wavers between 100% actual power generation to absolute zero... when the wind is not blowing. zero.

As reflected in the above discussion, wind and solar power plants are idling significant part of the time. This is due to the natural weather phenomena, which dictates when the wind blows and when the sun shines. Since we have no control over these weather conditions, we must take them into consideration every time we discuss or design renewable power systems.

The more we depend on wind and solar, the more we will encounter the three major technical problems with these energy sources.

First, it is very difficult, and even impossible, to maintain a consistent supply of power to the grid, when the weather is not "favorable enough."

When the wind doesn't blow and the sun doesn't shine the wind mills and solar panels sit idle.

Second, the wildly fluctuating renewable output from wind and solar power plants has to be balanced by input from conventional power stations. To keep that back-up power constant requires fossil-fuel power plants (peakers) to idle constantly and fire up on and off as needed. The renewables, therefore, introduce a new— very inefficient and expensive—way of maintaining constant power levels. They waste a lot of energy (while idling), which forces the emission of much more carbon pollution than normal.

Peaker power plants are only 20-40% efficient, waste a lot of energy and emit a lot of pollution.

It is obvious that the free lunch we were hoping to get by using renewable energy sources—wind and solar in particular—is not free at all. It comes with all kinds of conditions and complications. So, while including the renewables in the total energy mix, we must take into consideration their variability and immaturity.

We have ways to reduce the variable power generation by adding energy storage capacity, but this adds another set of technical and financial problems to the already complicated energy equation. And so, it seems that for now we must be very careful to consider all possible energy sources and use them in a most efficient way possible.

The renewables—wind and solar in particular—affect the energy markets significantly in a complex way, as we saw above. They, however, are part of the energy market, and will remain so for the foreseeable future, so we must figure out the best ways of integrating them in our energy mix. Our future depends on them...

There are several major trends in the renewable markets, that need to be acknowledged:

- The U.S. renewable energy is growing, but quite unevenly. Renewable energy represented 61% of all new U.S. electricity capacity installations in 2013, reaching 14.8% of total electric capacity and 13.1% of actual power generation at the time. This, considering the fact that renewables were just 9.5% of total generation in 2004—less than a decade before. The future growth, however, is uncertain and we already saw the renewables new capacity additions going down to 41% during the first three quarters of 2014. We do expect the growth to continue varying according to the prevailing national and global socio-political conditions and the related changes.

- Globally, the renewables are nearly 25% of the total power generation. Cumulative installed global renewable electricity capacity grew 108% from

2000 to 2013. It contributed 23% of all global power generation in 2013. Wind and solar grew fastest, with wind generation growing by a factor of 18 and solar by a factor of 68 between 2000 and 2013. Please note that hydropower is part of the renewables, and remains the largest source of renewable energy.

- Wind power generation is the winner in the Chinese renewables. At the end of 2013, China had 91.4 GW of cumulative wind power capacity, while the U.S. had 61.1 GW, Spain 23.0 GW, India 20.2 GW. Many other countries had wind power in the 5-10 GW range. In 2014 China added another 20.7 GW of wind power, while the U.S. added only 4.7 GW. This gap was narrowed in 2015 but China is setting new records by adding large quantity of installed wind capacity to the tune of 120 GW total installed capacity in 2015. The plan now is to add another 100 GW of wind power by 2020.

- Solar power generation is also growing very fast. The lag time from industry assessments to government reports is most evident with solar power. The U.S. solar photovoltaic installed capacity jumped from 7.3 GW in 2012 to 12.1 GW in 2013, and 18.3 in 2014. Solar's growth over time looks truly exponential. Yet, the U.S. lags in global solar capacity. Germany was the world's undisputed solar leader with 35.9 GW at the end of 2013, but only about 40.0 MW were added in 2014. China is second with 19.9 GW and Japan fourth overall with 13.6 GW at the end of 2013. 15.0 GW of new solar in China and 10.0 GW in Japan were added during 2014.

- Energy efficiency is the biggest hope for the future of energy. Yet, the U.S. is lagging in this area. Buildings are the largest energy consumers in America, consuming over 40% of the total national energy supply. Residential buildings consume just a bit higher than commercial buildings. The big problem is the fact that 52.1% of commercial energy consumption and 46% of residential energy consumption is due to electrical system energy losses. What? Roughly half of all U.S. power is wasted! This presents a huge opportunity for energy efficiency measures. Here is where the smart grid technologies will shine, and where a shift toward distributed generation, away from centralized power supplies, will take place.

- Biofuels are another growing energy business in the U.S. and the biofuels dominate U.S. alternative fuel consumption. We led the worldwide biofuels production with 13.3 billion gallons ethanol and

1.8 billion gallons biodiesel in 2013. This is more than twice Brazil's output of 6.2 billion gallons ethanol, 766 million gallons biodiesel. The renewable and alternative fuels (including electric vehicles) are growing fast, and yet they're still minuscule as compared to petroleum-based fuels. The U.S. fuel consumption is a daunting challenge, but at the same time it is also a huge opportunity, for clean energy technologies.

- Fuel cells are the new kid on the block and as such they dominate the headlines at times. The sobering fact is that hydrogen and fuel cells are very small of the energy mix. Stationary fuel cells, used primarily for backup power, grew 25% in 2013, but still only totaled 160 MW of installed capacity in 1,137 total systems. Fuel cell vehicles are an even fewer. The lack of growth is clearly shown by the low number of operational U.S. hydrogen fueling stations. There were only 53 of these in 2013, and not many more are planned.

- Electric and hybrid vehicles are another dimension in the energy puzzle that needs to be considered. Although they are not the solution, they are part of it. Especially as their numbers and popularity increase. There are thousands of these vehicles on the roads, and plugged in the electric grid. They are sucking electric power, but can be also used to store and release it as needed. The potential is great, but the time for their full utilization is yet to come.

Conclusions

During the transition to fossil-less future, we will try to switch to running the entire economy on renewables. This, however, is simply impossible. The renewables—all of them taken together—do not have even a fraction of the immense power and flexibility of the fossil fuels.

A significant fact to remember here is that the renewable technologies—solar, wind, and the rest—require huge amounts of fossil energy during their manufacturing, transportation, installation, operation, and maintenance stages.

Melting silicon, aluminum, and glass needed for making solar panels and wind mills, uses many gigawatts of coal and oil fuels. These non-stop, 24/7, operations cannot be interrupted, so it is quite hard to imagine running a huge silicon melting furnace on solar or wind power. It is also hard to imagine transporting a 200-foot, many tons wind mill blade using electric vehicles.

Different components of the renewable power generating equipment have to be replaced every 10-20-30 years.

The 'renewables' are actually 'recyclables', which requires more energy to replace the old hardware.

Note: Germany is hailed as the most advanced in solar energy. This praise has a high price tag. The 40 GW of heavily subsidized solar installations around the country, cost the German taxpayers about $200 billion. Additional $15-20 billion public money is to be spent every year on FiT subsidies.

Now get this; within the next 15-20 years the German solar fields would be so inefficient (solar loses 1% efficiency annually) that all of them has to be demolished and sent to dump sites. This entire operation would cost between $50 and $100 billion to complete properly.

Then, the cycle (of recycling the renewable solar energy technologies) starts again. New solar hardware has to be manufactured and installed, for which German taxpayers will have to pay another $200 billion to rebuild the old solar power fields. And another $200 billion again 25-30 years down the line...and again and again. During one century, over a trillion dollars would be spent on recycling solar power plants, and that much for doing the same with wind power plants.

All this expense provides 5-6% of the energy load.

A never ending cycle of making new hardware, dumping it in the waste dumps, replacing it with new hardware and doing this over and over again. And equally importantly, the amount of energy to continue the cycle cannot be provided by the renewables themselves. Remember that 24 hr/day energy demand by huge metal, silicon, and glass melting furnaces cannot be derived from renewable sources. Not having enough fossil energy sources would limit the amount of metal, silicon, and glass needed to manufacture new renewable power generators.

So, eventually, in the complete absence of fossils, the solar and wind farms would be shut down and the only energy left on Earth would be that delivered directly by the Sun. Back to the 19th century we go—poking the land by hand, cooking with firewood and using candles for light.

Not possible? Yes possible. If we don't change our ways soon enough.

On the bright side of things, however, we must also always remember, and count on, the American ingenuity and entrepreneurial spirit. We have been close to the edge of the cliff before, and have managed—time after time—to pull out of the danger zones unscathed. This time is no exception. As time goes on and the energy disaster gets more and more obvious and unavoidable, we will roll our sleeves and get to work. We will figure out how to deal with it and build a new energy future based on renewables, and/or whatever it takes.

As our good friend Churchill said, 'You can always count on Americans to do the right thing—after they've tried everything else.' How true! So now we are in the 'trying everything first' stage. Then will come the hard part, but we will overcome...or else.

Notes and References

1. SEIA http://www.seia.org/research-resources/solar-industry-data
2. Solar Markets. http://solarcellcentral.com/markets_page.html
3. World Solar Energy Market. https://www.alliedmarketresearch.com/solar-energy-market
4. IEA Solar Programs. http://www.iea-shc.org/
5. Wind Market Reports. http://www.reportlinker.com/report/best/keywords/Wind%20Power?utm_source=bing&utm_medium=cpc&utm_campaign=Energy_And_Environment&utm_adgroup=Wind_Power
6. AWEA, http://www.awea.org/
7. IHS, https://www.ihs.com/products/renewable-energy-market-analysis.html
8. IEA, http://www.iea.org/topics/renewables/
9. DOE, http://energy.gov/eere/renewables
10. EERE, http://energy.gov/eere/renewables
11. EU Commission, http://ec.europa.eu/energy/en/topics/renewable-energy
12. CA.gov, http://www.energy.ca.gov/renewables/renewable_links.html
13. REN21, http://www.ren21.net/status-of-renewables/global-status-report/
14. *Photovoltaics for Commercial and Utilities Power Generation*, Anco S. Blazev. The Fairmont Press, 2011
15. *Solar Technologies for the 21st Century*, Anco S Blazev. The Fairmont Press, 2013
16. *Power Generation and the Environment*, Anco S. Blazev. The Fairmont Press, 2014
17. Energy Security for the 21st Century, Anco S. Blazev. The Fairmont Press, 2015

Chapter 6

The Other Energy Markets

The fossils and the renewables make 2/3 of the global energy markets.
The "other" energy sources contribute only 1/3, which is expected
to increase to 1/2 by the end of the century, unless...
—Anco Blazev

BACKGROUND

Unless in the above statement is the sixty four thousand dollar question we discuss in detail and attempt to clarify in an unbiased way in this text. Its meaning is becoming more important by the day, because *unless* we change the way we live and work, and focus on the priorities of today, the global energy markets will swerve in a path we would not like. In order to avoid the disastrous consequences of such deviation, we need to emphasize on energy efficiency, the renewables, and the *other* energy sources.

So what are the *other* energy sources? For the purposes of this text, these are:

- Firewood
- Nuclear
- Hydropower
- Ocean-power
- Geo-power
- Bio-energy

We must first clarify, why we put these energy sources in a separate category. What does firewood has to do with nuclear power? Or ocean- with bio-energy? The main reason for this classification is that these technologies are different from the rest—the fossils and renewables—for a number of reasons. And this, in our humble opinion, that puts them in a class of their own. We will explain these reasons in detail as we go along.

For now, it suffices to say that the 'other' energy sources are quite different from the fossils and somewhat different from the renewables. They are not fossils, per se, although nuclear power plants use fossils-like materials. They are not truly renewable either, because they heavily depend on uncontrollable variable and rapidly changing natural events, as we will see below.

*All **other** energy sources have one thing in common; it is their uncertain future.*

There are many reasons for this inequality, as we will see in this text, and yet these energy sources are of utmost importance for the present, and even more important for the future, energy markets. By putting them in a different class, we are not minimizing, but rather re-emphasizing, their importance. We review them as an integral and very important part of the energy markets, emphasizing and re-emphasizing their importance and the fact that they need special attention.

In more detail, the different energy sources in this section, and their special conditions, are as follow:

- *Firewood* used as an energy source is the best kept secret of the energy sector. There is a lot of wood worldwide and much of it is used every day by millions and for different purposes. Using firewood as an energy source is only one of these purposes. The problem is the rate at which we use it today, and the increasing effects of natural phenomena (fires, draughts, floods, etc.) which will list wood as an endangered species by the end of this century. Wood also has a number of other characteristics, uses, and problems that distinguish it from other energy sources, thus it deserves special attention and a special place in the *other* category.

- *Nuclear power* uses energy materials (uranium, plutonium thorium, etc.) that are neither fossils nor renewable. They are not fossils, because they were created in a different manner and different time period than the fossils. They also have very different properties and behavior than the fossils. They are not renewables either, since their quantity on Earth is limited, albeit large enough to last for centuries.

361

Another significant characteristic of the nuclear fuels is that they are recyclable. As the recycling technology advances, a day may come when most of the used nuclear fuels can be recycled and reused. This would ensure their future for many centuries to come.

And even more importantly, nuclear power is not your typical "conventional" power generator.

Nuclear energy sources contain power thousands and millions of times greater than any energy source known to man.

The immense power contained in nuclear fuels is the greatest advantage of nuclear power generation and the reason for its wide spread use. It is, however, also its biggest fault.

The immense power contained in nuclear fuel is a blessing and a curse.

This is because it provides huge amounts of energy when properly done, but it can devastate its masters and the local surroundings if mishandled.

There is also the belief that nuclear power is clean and green, which on the surface it is. Looking deeper into it, however, we see environmental problems created by normally operating nuclear plants, as well as unspeakable devastation of huge land areas and many people hurt and killed during and after nuclear accidents. Nuclear power, therefore, is a blessing to most of us, but is a curse to some who have suffered the consequences of nuclear accidents. Because of that, its future is uncertain, which, with all due respect, is another reason for it to be in a class of its own.

• *Hydropower* has a number of special characteristics that distinguish it from the rest of the energy sources too. The most important of these is the fact that, due to land limitations and climate change, its future is limited to some locations, and uncertain in others. Most of the favorable for hydropower generation sites have been already used, which limits its future expansion. Most importantly, hydropower is faced with increasing water shortages in the future, which will further limit its use. While it has been a major energy source in the past, and is still growing today, albeit slowly, its future is uncertain. Thus, its special place and need for special attention and treatment.

• *Ocean energy* is renewable in its foundation, but its progress is hindered by technological limitations, changing climate, high installation costs, and rising O&M expenses. Ocean power generation technologies are also still quite immature, so the actual power generation around the world is limited to a few prototypes and demo projects. It is hard, therefore, to imagine large power plants, using waves or ocean currents technologies in the world's oceans any time soon. Even if some emerge, they would be the exception, not the rule. This uncertainty is what brings ocean energy into the *other* category.

• *Geo-thermal energy* sources are renewable, as long as the Earth's magma continues to generate heat and brings it close to the surface. They are, however, of limited importance, since the most promising sites for this type of power generation usually are far away from potential power users. This and other barriers make geothermal technology isolated and too expensive for large scale use in most cases.

At the same time, we must give it credit for its wide deployment in some places, such as Iceland. Here, due to unique conditions, geo-thermal power supplies over 26% of the nation's energy. Yet, the future of geo-power is uncertain for these and other reasons in most areas around the world.

• *Bioenergy* sources and biofuels can also be considered renewable to a point, but they depend on natural conditions—rain and snow—which is not readily available in many areas around the globe. Biofuels also depend heavily on fossils—fertilizers and diesel fuel for the agricultural and processing equipment used during their life cycle. Extensive land cultivation and switching to bio-energy crops are also causing enormous environmental and human suffering problems around the world. The bio-fuel crops are also affected by climate changes and periodically changing economic conditions. So, in order to keep the global balance, bioenergy sources' future is limited to producing the most essential fuels in limited quantities.

Putting all these facts, pros, and cons together, we see that the energy sources in the 'other' category are unique in some or other aspects of their production and/or application. What all they have in common is technological limitations and other uncertainties in their future development that limit their use. They are, nevertheless, an important part of the energy markets at this

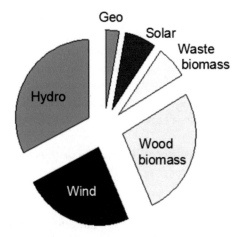

Figure 6-1. Renewable and *other* energy sources in the U.S.

stage of development and as such, they deserve special attention.

The *other* energy sources, as reviewed in this text are:

Firewood
- Huge global resources
- Uncontrollable use in poor countries
- Used mostly as residential energy source
- Limited use in commercial applications

Nuclear Power
- Huge future potential, if...
- Technological advances expected
- Limited use now for fear of accidents

Hydro Power
- Cleanest form of power generation
- Huge expense in hydropower dam construction
- Limited expansion, due to location and water scarcity

Ocean- and Geo-power
- Huge potential, limited by:
- Technological barriers and
- Financial and location challenges

Bio-energy and Bio-fuels
- Huge (potential) global resources
- Marginal environmental benefits
- Limited expansion, due to food market competition and climate change.

Let's start with a closer look at the oldest known energy source: firewood.

FIREWOOD ENERGY MARKETS

Firewood is the best kept secret of the energy markets. It is one of the major energy sources used around the world, but very few of us are aware of its importance.

Most of us in the U.S. do not use wood for heating, or any other practical purposes. Nor do we think of it as an energy source of any significance to us personally, or our economy. Let alone thinking of it as a major energy source, competing for recognition on global level. But guess what; it is! Regardless of what we know or think of it, wood—and wood materials in all their shapes and forms—is one of the world's largest sources of energy. Believe it, or not.

The largest amount of renewable fuel used in Europe, Asia, and Africa is firewood. Who knew?

Just think; the only competition Putin, and his natural gas exports to Europe, has is firewood. Many households and businesses around the continent can switch instantly to burning wood and coal, if and when the flow of natural gas from Russia stops. Not the best, or practical, choice or, but is there for the taking. Although not as convenient and clean as natural gas, firewood does a good job in cooking meals and heating buildings. And predicting what Putin is going to do next winter is much harder than collecting a cord or two of wood from the nearby forest.

Firewood is even more important in other places of the globe. There are thousands of villages and even small cities in Africa and Asia, which inhabitants depend partially, and in many cases solely, on firewood for all their energy needs. Firewood is essential for heating and cooking, and even running small businesses. Firewood, in many cases makes the difference between taking care of the basic necessities, and living miserable life in darkness, cold, and hunger.

Some of us also think that wood is a renewable resource, which in theory it is. The way people have been using and misusing it lately, however, is putting it in a new and special category; that of the *endangered species*. Just like them, much more wood is harvested globally than is created, which eventually would lead to complete depletion of the global forests and wooded areas.

Wood is a very important material and energy source packed in one package. This is another distinction of wood, since it is used in many areas of our economy and daily lives too. As a matter of fact, we are surrounded by wood; entire buildings are made out of wood, and it is in our offices, living and bed rooms. So, let's see the *"how and why"* of wood energy, and the function of the wood markets, in light of the new global energy order.

The major use of wood we are concerned in this text is that for energy generation.

Firewood Types

First, we need to clarify here that there are many types of *wood,* and they all have many uses. It is important to clarify also that what we are talking about here is not *wood* in general, but *firewood.*

> *Firewood is a special commodity made of wood and wood products, which people use for heating and cooking at home and also for running small businesses.*

Wood is also used on a large scale for making steam for heating large buildings and even entire cities. It is also used for generating electric energy in some parts of the world, as few and far between these might be.

Wood is the oldest energy source known to man. It might not be obvious that wood is an important part of the energy markets, but there are billions of people in the underdeveloped world, who depend on wood exclusively for their energy needs. There are also some people in the developing countries that use a lot of wood for cooking and heat during winter months. Of course, there are entire industries involved in providing wood products and services. Wood, as a matter of fact, is one of the most (if not the most) used energy source in the world…

Wood is used mostly in the rural areas of the world, with the developing countries using the bulk of wood and wood materials globally.

> *Hardwood is preferred over softwood because it creates less smoke and burns much longer.*

A woodstove or fireplace in a home is often considered a luxury that adds ambiance, warmth, and safety, so nowadays wood and wood pellets have become one of the most important heating fuels for homes in the U.S. and Canada. This is confirmed by the significant increase of fire wood in the region use since the 1970s.

A common hardwood, red oak, has an energy content, or heat value, of about 15.5 MJ/kg, or the equivalent of about 4.25 kWh per kg. of wood This is also equivalent to about 6,500 Btus per pound of wood. During burning, about 70% (or about 11.0 MJ) of the heat

content in wood can be recovered and used as direct heat.

In practical terms, the energy content of wood is more closely related to its moisture content than its species. The energy content improves as moisture content decreases. This is simply because it takes more energy (wasted) to burn wet wood.

> *Firewood usually refers to timber or trees unsuitable for building or construction.*

Most of the wood fuel (firewood) comes from native forests around the world. Plantation wood is rarely used for firewood, as it is more valuable as timber or wood pulp. Some wood fuel is gathered from trees planted amongst crops, which activity is known as agro-forestry.

The harvesting of wood can have serious environmental implications for the collection area. These concerns include all the problems created by regular logging activities. The removal of large quantities of wood from forests can cause habitat destruction and soil erosion. In many countries, for example in Europe and Canada, the forest residues are being collected and turned into useful wood fuels with minimal impact on the environment. Consideration is given to soil nutrition and erosion.

The environmental impact of wood as a fuel depends on how it is grown and burned. Some tree (wood types) are more energy efficient and less polluting than others. Higher efficiency firewoods generate higher temperatures during burning. This results in more complete combustion and less noxious gases emission, as a result of the pyrolysis process.

Note: Burning wood from a fully sustainable source is considered carbon-neutral. The justification is that a tree, or a crop, over the course of its lifetime, absorbs as much carbon (carbon dioxide) as it releases when burnt. The problem, however, is that the trees are usually grown in one place; usually high in the mountains, but are burned far away; usually close to populated centers. The instantaneous environmental impact, therefore, is felt at the place of burning, which is usually already heavily polluted.

Some firewood is harvested in "woodlots" designed and managed for that purpose. In heavily wooded areas, however, wood is more often harvested as a byproduct of natural forests. Deadfall that has not started to rot is preferred, since it is already partly seasoned. Standing dead timber is considered better still, as it is both seasoned, and has less rot. Harvesting this form of timber reduces the speed and intensity of bushfires.

Harvesting timber for firewood is normally carried out by hand, where people use chainsaws and heavy duty equipment to cut and load the trees for transport. Longer pieces require less manual labor and are less expensive, with the only limitation being the size of the transport and the firebox.

Wood prices vary considerably with location and quality. The distance from the wood lots determines the cost of transportation, while the wood quality determines its final use.

Fuel wood costs ~$15/1 million Btus, or ~$ 0.05/kWh.

This is not bad, as compared to most other energy sources, since most of them cost twice that much. However, there are other factors that influence the final prices. Availability, distance, and transporting firewood to urban centers, for example, could double its price.

Wood (as Biomass)

Wood is a very versatile product, used in many sectors of the global economy. It is also very important for the energy markets, where it is considered humankind's very first source of energy. Historically, wood has been used since the beginning of time for heating and cooking. Later on in the 17th and 18th century it was used commercially in some industrial applications as a heat source and steam generation. Wood used for powering locomotives and other stationary and mobile equipment dominated the transportation and other economic sectors well into the 19th century. Wood was later on replaced by other fuels, like crude oil, natural gas, and nuclear energy.

Today it still is one of the most important single source of renewable energy. Wood-fuels are obtained from many sources, including forests, other wooded land and trees outside forests. It is also obtained from many co-products from wood processing, post-consumer recovered wood, and processed wood-based fuels.

Wood and wood products provide about 9-10% of all global primary energy supply. This is more than all other renewables put together.

Wood and wood products are also used for power generation, transport (in some countries), and making bio-fuels. This makes wood a versatile; multi-use and multi-purpose raw material. As a matter of fact, some European countries produce a significant fraction of their electricity needs from wood or wood wastes. In

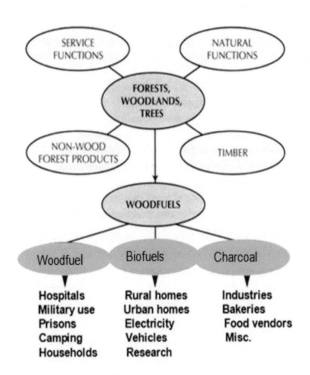

Figure 6-2. Major wood products and services

Scandinavian countries, for example, the cost of manual labor to process firewood is very high. Therefore, it is common to import firewood from countries with cheap labor and natural resources. The wood exports to Scandinavian countries come mostly from neighboring Estonia, Lithuania, and Latvia.

Some countries use wood waste as fuel for home and industrial heating, in the form of compacted wood pellets. Different types of wood are used in the process of pellet making; firewood, woodchips, wood shavings, etc. The biggest pellet plant in the EU is Võrumaa, Sõmerpalu, in the Baltic region of Estonia. It has a total output of 250,000 tons of wood pellets annually. Most of its output is exported to Europe.

The main pellet consumers in Europe are the UK, Denmark, the Netherlands, Sweden, Germany and Belgium. In Denmark and Sweden, pellets are used by power plants, and by households and medium size businesses for heating. In Austria and Italy, pellets are mainly used for heating small scale private residential and industrial boilers.

Over 1.5 million households in Australia use firewood as the main form of domestic heating. Approximately 2.0 million cubic meters of firewood is used annually in the province of Victoria alone. This amount is close to the total amount of wood consumed by all of Victoria's saw-log and pulp-log forestry operations. Half of this quantity is consumed in Melbourne proper.

More than two billion people in many countries around the globe depend on wood energy for lighting and heating their homes. Millions of households in developing countries use food for cooking too, since It represents the only domestically available and affordable source of energy.

Residential cooking and heating with wood fuels represents over 1/3 of the total global renewable energy consumption.

Just think, 1/3 of all renewables generated energy comes from firewood. This makes it the largest and most decentralized energy type in the world. *Who knew?* These are amazing facts about a fuel most people in the U.S. don't even consider a fuel.

Wood fuel is also often used as important energy source for emergency backup. People in most countries—even the most developed—switch quickly back to wood energy during economic difficulties, natural disasters, conflict situations, or temporary fossil energy supply shortages. Even in the U.S., a cord or two of firewood can be seen side by side with a natural gas, or heating oil, tank in most back yards in the North East part of the country.

Wood fuels are also a very important forest product, or rather byproduct. Even as a byproduct, the global production of fuelwood exceeds the production of industrial roundwood in volume. Fuelwood and charcoal production is often the predominant use of woody biomass in developing countries and some economies in transition.

Today wood energy has entered into a new phase of high importance and visibility with climate change and overall energy concerns. Wood energy is considered as a climate neutral and socially viable source of renewable energy, but only when meeting certain conditions.

Some of the conditions for carbon-neutral wood considerations are:
- Wood arising from sustainably managed resources (forests, trees outside forests, etc.).
- Appropriate fuel parameters (water content, calorific value, shape, etc.).
- Efficient incineration or gasification minimizing indoor and outdoor emissions.
- Cascade use of wood fibers, favoring material use, re-use and recycling before energy use.

Even the mighty, all developed and sophisticated USA is using a lot of firewood. Not surprisingly, the amount has increased dramatically since the 1970s, in

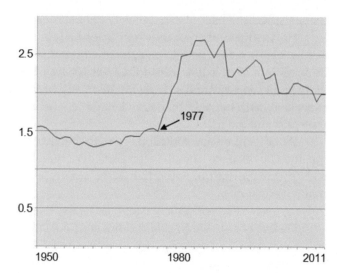

Figure 6-3. U.S. U.S. wood fuel consumption (in quadrillion Btu/annum)

parallel with the increase of fossil prices. Note the large jump in use since the end of the 1970s in Figure.6-3. Was this in response to the Arab oil embargo and a wake-up call of the American energy genius? Even now, with all the abundant oil and gas production in the U.S., wood fuel production is expected to increase gradually until 2020, then fluctuate up and down before leveling by 4040.

Wood fuel is the second-leading form of renewable energy after hydropower in the U.S.

Who knew? Wood in the most modern and technologically advanced country in the world/. Yet, in the U.S. wood is used as a main heating source in many homes during the winter months. It has actually gained popularity in many areas of the country in recent years. The increase, of course, is most notable in the Northeast, where the winters are most severe. Wood use in Arizona and New Mexico, for example, is unknown to most locals.

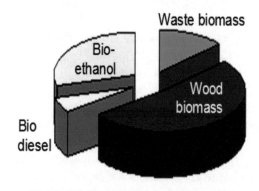

Figure 6-4. Biomass share in renewables

All nine states in the New England and the Middle Atlantic divisions saw at least a 50% jump from 2005 to 2012 in the number of households that rely on wood as the main heating source. Note the 'main' classification of wood as a heating source. Not so much fossil fuels for these people. No delayed deliveries and increased prices of heating oil and LPG during the freezing winter months. The use of fuel oil, LPG, and kerosene in this region has declined in recent years, as many households have turned to lower-cost alternatives, including solar and wood.

In total, about 2.5 million households across the country use wood as the main fuel for home heating, up from 1.9 million households in 2005. An additional 9 million households use wood as a secondary heating fuel. This combination of main and secondary heating accounts for about 500 trillion British thermal units (Btu) of wood consumption per year in the residential sector, or about the same as propane consumption and slightly less than fuel oil consumption.

Heating stoves are the most common equipment used by households that rely on wood as the main source of heat, and fireplaces are the most common choice for secondary wood heating. Most households still burn split logs, although wood pellet use has risen in recent years. While most households in higher income brackets are likely to use wood as a fashion statement, those at lower income levels who burn wood, on the average consume more of the stuff.

About 20% of households with $20,000 annual income use firewood for heating in the winter. At the same time, over twice that many (50%) of households with $100-120,000 income use firewood. The difference is that the poor families use firewood as a primary heating and cooking energy source in wood stoves. The richer families, use less firewood from time to time as a comfort, entertainment, or ornamental media.

No matter how it is used, the environmental impact of burning wood fuel is becoming a serious issue lately. Several localities have moved towards setting standards of use and/or bans of wood burning fireplaces. For example, the city of Montréal, Québec passed a resolution to ban wood fireplace installation in new construction. Phoenix, Arizona, and the surrounding cities are issuing fireplace burning bans almost on daily basis during the winter months. Hefty fines are imposed on violators, and the ban is becoming so frequent that it is almost impossible to use fireplace in many places like the Phoenix Metro area and other metropolis around the country.

Wood stoves and fireplaces are getting so common,

and so many have been used, so that The Environment Protection Agency (EPA) recently proposed updated emissions standards for new wood-burning stoves and other biomass heating equipment. Although these proposed rules address health concerns associated with the release of fine particulates from burning wood, the standards would also result in increased efficiency levels of new wood-burning equipment.

Here we can only imagine the number of industries and jobs created, and to be created, by using *wood* in the energy sector. Starting with growing, cutting, loading and transporting wood to processing facilities, processing or burning the delivered wood, and finally removing and disposing the waste materials, represent several large market segments. It is easy to see here a lot of facilities, equipment, and thousands of people busy every day working with wood in the energy sector.

All this is good, but the supply and demand balance is tilted towards demand, due to increased use of wood and the deforestation of large land areas around the world.

It takes minutes to chop down and burn a large tree, but it takes decades to grow it.

Since the supply and demand balance is out of whack, the shortage of wood in some parts of the world is obvious, and its increase in the coming decades is inevitable. The U.S. has good forests use policies, so we don't foresee such a problem here any time soon. Unless unforeseen events change the situation. In many places around the world, however, wood is truly an endangered species.

Wood Resources

Recent estimates put the wood (trees in the forests mostly) on Earth at one trillion tons. This includes all forests from one end of the Earth to the other. One can't help it but wonder how can the number and weight of trees in the forests be calculated to derive even close estimate of the total world's wood mass. This is such huge, diverse, and dynamic heterogeneous entity that boggles the mind with its immenseness.

The wood situation is further complicated by the never ending natural processes. Forest fires, floods, draught, etc. events consume millions of acres every year in the U.S. alone. The number of forest disasters is multiplied several times over in other countries. Man-made deforestation of the jungles in South America and other places is another unpredictable and incalculable event.

Yet, we need to trust the experts who give us these

numbers, at least until we find a better and most believable source. According to these same experts, part of the total wood amount is chopped down and used every day at variable amounts, with an average of about 40 billion tons annual use globally. At the same time, some new trees grow at a rate of about 10 billion tons per year. This is 4:1 ratio in favor of the wood use.

Or, with other words, the world uses 30 billion tons of wood every year more than is grown. This amount, out of a trillion tons total every year, means that at this rate of use, the world has wood for only 30-40 years.

The world is using 4 times more wood that is re-grown. At this rate, we would run out of wood before century's end.

Are we entering a wood-peak period? Remember the oil-peak discussions? Are we going to run out of wood at the same time we run out of the fossils? This is a new and very important question, which we need to address. The future generations should not be forced to deal with so may problems at once.

For now, however, wood is abundant, and considered carbon-neutral renewable resource. As such, its use will increase, as is already happening in Europe. Even more important here is the fact that the dominant uses of wood is for furniture and building construction. These industries use the best part of the wood—the high quality part of the tree trunk—for their purposes. The leftover of low quality wood and woody materials (branches, bushes, etc.)—most of which is considered waste—is used in other industries, like pulp and paper making, and only what is left over from these is used as fuel in millions of homes, businesses, and wood-fired power generation.

Today, some utilities are looking into adding wood to their fuels arsenal as a coal replacement in electric power co-generation plants. This might bring upon us another forests devastation movement, so we need to watch for the trends in the wood markets.

Some of the ways wood and wood products are used for energy today are:

Sawdust

Sawdust, or wood dust, is a by-product of cutting, grinding, drilling, sanding, or otherwise pulverizing wood, and is therefore composed of fine particles of wood. It is also the byproduct of certain animals, birds and insects which live in wood, such as the woodpecker and carpenter ant. It can present a hazard in manufacturing industries, especially in terms of its flammability.

Sawdust is the main component of particleboard.

A major use of sawdust is for particleboard; coarse sawdust may be used for wood pulp. Sawdust has a variety of other practical uses, including serving as a mulch, as an alternative to clay cat litter, or as a fuel. Until the advent of refrigeration, it was often used in icehouses to keep ice frozen during the summer. It has been used in artistic displays, and as scatter. It is also sometimes used to soak up liquid spills, allowing the spill to be easily collected or swept aside. As such, it was formerly common on barroom floors, and used to make Cutler's resin. Mixed with water and frozen, it forms pykrete, a slow-melting, much stronger form of ice, which finds some special applications.

Sawdust is also used in the manufacture of charcoal briquettes, the invention of which is credited to Henry Ford who made some from the wood scraps and sawdust produced by his automobile factories.

Burning sawdust to generate electricity is not very common, but is still used in some special situations, mostly in developing countries.

In addition to the pollution problems described above, at sawmills, sawdust burners generate a lot of toxic gasses. Sawdust also may be stored in large outdoor piles causing harmful leachates into local water systems, thus creating an environmental hazard. This has become a serious problem for small sawyers and environmental agencies, putting them in a deadlock.

Questions about the science behind the determination of sawdust piles being an environmental hazard remain to be answered by sawmill operators, who compare wood residuals to dead trees in a forest. Technical advisors have reviewed some of the environmental studies encounter lack of standardized methodology, and/or evidence of a direct impact on wildlife. The analysis don't take into account large drainage areas around the globe, so the amount of material that is getting into the water from these sites in relation to the total drainage area is minuscule.

Other scientists have a different view, saying the "dilution is the solution to pollution" argument is no longer accepted in environmental science. The decomposition of a tree in a forest is similar to the impact of sawdust pile, but the difference is of scale. Sawmills may be storing thousands of cubic meters of wood residues in one place, so the issue becomes one of concentration.

Of larger concern are substances such as lignins and fatty acids that protect trees from predators while they are alive, but can leach into water and poison wildlife. Those types of things remain in the tree and, as the tree decays, they slowly are broken down. But when sawyers are processing a large volume of wood and large concentrations of these materials from the piles permeate into the runoff, the toxicity they cause could be harmful to a broad range of organisms.

Charcoal

Charcoal is a type of man-made fuel, usually made from wood or vegetation or animal matter, via special process, where wood is burned in a controlled presence of oxygen. The main reason for producing this type of fuel was, and still is, to convert unusable materials (even waste materials) into useful fuels that are easy to transport and use.

Charcoal is soft, brittle, light, with a dark grey to black appearance, very similar to coal. It consists mainly of carbon and ash, which remain after the water and other volatile compounds are removed from the raw materials. Its combustion properties are also close to these of coal.

In the past, charcoal was produced by piling wood logs, or other combustible materials, in a conical pile, which was covered to restrict contact with the air. A small openings at the base was used to control the intake of air, which was directed through a central air shaft serving as a flue. The fire was started at the bottom of the flue, and let to gradually spread upwards, which process took days at a time to complete, since slow combustion is the key to producing quality charcoal. Small scale production of this type yields 50% by volume, or 25% by weight, of charcoal, which produces heat equivalent to the heat produced by the starting raw materials.

Today charcoal is produced by a special carbonization process, where small pieces of wood, sawdust, or other waste materials are burned under controlled conditions in cast iron retorts, which produce charcoal of different quality and for different purposes.

When charcoal is made at 300°C it is brown, soft, friable, and readily inflames at 380 °C. When it is made at higher temperatures it is hard and brittle, and does not fire until heated to about 700°C. Charcoal production is extensively practiced for the recovery of byproducts that have some useful heat content, such as tree branches, wood shavings, sawdust, etc.

There was massive production of charcoal during the last several centuries, which supported a large industry employing hundreds of thousands of workers in Central Europe and the UK. Most of the energy for iron smelting in the 1800s came from charcoal, which was made out of 1.5 million cords of wood per year. The charcoal used in iron smelting totaled some 700,000 to 750,000 tons annually. This compared to estimated

750,000 tons of charcoal used every year by 'outdoor and tailgate chefs' in the barbeques in this country alone.

Thousands of acres of forests were cut, which created a major deforestation in those areas. Large wooded areas were cut and re-grown cyclically, in order to provide a steady supply of charcoal at all times. Over-exploitation and lack of new supplies, and the increased demand for charcoal created a supply and demand un-equilibrium, which facilitated the switch to coal, and later on to fossil fuels for domestic and industrial use.

Charcoal is still important energy source worldwide. Brazil uses charcoal in 45% of its iron smelting. 3.6 million tons of charcoal are used annually in Brazil alone, and consumption is increasing there and other countries. Charcoal is also a major fuel in less developed countries such as Ghana and Kenya, where 250-300,000 tons are used every year for home heating and cooking.

Biochar

Biochar is a type of charcoal, which is also created by pyrolysis of some types of biomass (agricultural waste is of special interest here). It is presently considered as an approach to carbon sequestration to produce negative carbon dioxide emissions. Biochar thus has the potential to help mitigate climate change via carbon sequestration in the future.

For now, it is used in special occasions, such as a soil additive and amendment. It is also a possible source of carbon sequestration, so it has the potential to help mitigate climate change, via carbon sequestration.

When mixed with soils, biochar can increase soil fertility, increase agricultural productivity and provide protection against some foliar and soil-borne diseases. Biochar is a stable, solid, rich in carbon material that can endure in the soil for thousands of years.

Biochar made from agricultural waste can substitute wood charcoal. As wood stock becomes scarce, this alternative is gaining ground. In some parts of Africa, for example, biomass briquettes are being marketed as an alternative to charcoal to prevent deforestation, driven by increased charcoal production.

Bagasse Fuel

Bagasse is the dry, fibrous matter that remains after crushing sugarcane, blue agave, or sorghum stalks, and extracting their juice. It is mostly used in the production of biofuels, paper products, and some special building materials.

Sucrose accounts for little more than 30% of the chemical energy stored in the mature sugarcane plant, while about 35% is in the leaves and stem tips, which are left in the fields during harvest. Another 35% of the energy in the plant is in the fibrous material (bagasse) left over from crushing and pressing the sugarcane stalks, which can be used for making fuel gas.

Bagasse is often used as a primary fuel source for sugar mills, where it is burned in large quantity to produce enough heat to run an entire sugar mill. The energy in biogas can be also used to provide both; heat energy, used in the mill, and electricity, which is typically sent into the electricity grid. This allows the plants to be energetically self-sufficient and even sell surplus electricity to utilities.

An average sugar- or ethanol-producing plant could produce 500 MW electricity for self-use, and 100 MW for sale by burning bagasse. The sale of power is expected to boom as new regulations force the utilities to pay "fair price " for it. This type of power is also especially valuable to utilities because it is produced mainly in the dry season when hydroelectric dams, and the electricity produced by them, are running low.

Estimates of the potential power generation from bagasse in Brazil range from 2.0 to 9.0 GW, depending on time, technology used, and locations of the generators. The higher estimates assume gasification of biomass, replacement of current low-pressure steam boilers and turbines by high-pressure ones, and use of harvest trash currently left behind in the fields. The location of the power plant determines how much power, if any, can be sent into the utility grid.

Presently, an average sugar processing plant can generate about 300 MJ of electricity from the residues of one ton of sugarcane, of which about 200 MJ would be used by the plant itself. Thus a medium-size distillery processing 1 million ton of sugarcane per year could generate and sell about 5 MW of its surplus electricity for the duration. At current prices, it would earn $18 million from sugar and ethanol sales, and about $1 million from surplus electricity sales. Not bad for a small operation.

With advanced boiler and turbine technology, the electricity yield of the same distillery could be more than doubled to 650 MJ per ton of sugarcane. However, the current electricity prices do not justify the necessary investment for such upgrade for it is quite expensive.

Note: Presently the World Bank would finance investments in bagasse power generation only if the price is at least $0.07/kWh. This is not much of incentive for most such power plants, so they prefer to limit their power generation for local use, since it is much more profitable.

In many other countries (such as Australia), sugar factories contribute significant amount of 'green' power to the electricity supply. In the U.S., Florida Crystals Corporation, one of America's largest sugar companies, owns and operates the largest biomass power plant in North America. The 140 MW facility uses bagasse and urban wood waste as fuel to generate enough energy to power its large milling and refining operations as well as supply enough renewable electricity for nearly 60,000 homes.

Researchers are also working with cellulosic ethanol, in order to optimize the extraction of ethanol from sugarcane bagasse and other plants viable on an industrial scale. The cellulose-rich bagasse is being widely investigated for its potential for producing commercial quantities of cellulosic ethanol. For example, Verenium Corporation is building a cellulosic ethanol plant based on cellulosic by-products like bagasse, in Jennings, Louisiana.

Bagasse is being sold for use as a fuel (replacing heavy fuel oil) in various other industries too, including citrus juice concentrate, vegetable oil, ceramics, and tire recycling. The state of São Paulo, Brazil uses 2 million tons of bagasse as fuel, thus saving about $35 million in fuel oil imports.

Bagasse burning is environmentally friendly compared to other fuels like oil and coal. Its ash content is only 2.5%, vs. 30–50% of coal-fired power plants, and it also contains very little sulfur. Since it burns at relatively low temperatures, it produces little nitrous oxides, which has a dual effect of;
- Reducing some of the worst pollutants and their aftereffects (acid rain especially), and
- Allowing to introduce ways to reduce nitrous oxides generation (which is not possible at high sulfur levels).

The resulting CO_2 emissions are equal to the amount of CO_2 that the sugarcane burnt in the power plant absorbed from the atmosphere during its growing phase, which makes the process of cogeneration greenhouse gas-neutral. The controversy of this process is based on the different locations of the respective activities. Since the sugar processing plants are located far away from the fields, the area around them is polluted, while that in the agricultural areas is not. So, the argument goes, one location benefits while others suffer from the bagasse burning.

All in all, sugarcane and bagasse are shaping as an integral part of our energy future. The only problem with sugar cane is the need of lots and lots of water during the growing process. With increasing threat of global draughts, this might become a stumbling block for this promising energy source.

Wood Gas

Wood gas is a synthetic fuel in gaseous form, which can be used to fuel furnaces and stoves. After special processing, it can also be used as vehicle fuel instead of gasoline, diesel or other fuels. Wood and other biomass materials are gasified within oxygen-limited environment in a wood gas generator to produce hydrogen and carbon monoxide.

These gases can then be burnt as a fuel within an oxygen rich environment to produce carbon dioxide, water and heat. In some cases, this process is preceded by pyrolysis, where the wood or biomass is first converted to char, releasing methane and tar, rich in polycyclic aromatic hydrocarbons.

The quality of wood gas varies a great deal, depending on the raw materials, and the processes used. Staged gasifiers, for example, where pyrolysis and gasification occur separately (instead of in the same reaction zone), can be made to produce essentially tar-free gas (less than 1 mg/m³). Single-reactor fluid-bed gasifiers, on the other hand, may exceed 50,000 mg/m³ tar.

The heating value of wood is typically 15-18 MJ/kg, but these values vary from sample to sample. The heat of combustion of *producer gas* (wood gas used for car engine fuel) is rather low compared to other fuels. Producer gas has a heating value of 5.7 MJ/kg versus ten times less than the 55.9 MJ/kg energy content of natural gas, and 44.1 MJ/kg for gasoline.

Gas	Content
N_2	50.9%
CO	27.0%
H_2	14.0%
CO_2	4.5%
CH_4	3.0%
O_2	0.6%.

Figure 6-5. Producer gas content (by volume)

The chemical composition of producer gas, shown in Figure 6-5, can also vary with type of raw materials and process. We see here that its low heating value is due to low content of hydrogen and methane (or total of about 17% of the total), which are the only combustible gasses in the mix.

Nevertheless, producer gas has been used quite successfully in the past, and a day may come soon, when it might be one of the major vehicular fuels in the world.

The questions waiting for an answer today are:

- Can wood be considered a renewable energy source? If so, it could benefit from a number of government subsidies and incentives.
- What could be done to ensure that it qualifies for government support?
- Can wood be considered for use as replacement fuel for coal, oil, and natural gas?
- What steps should be taken to make sure that it fulfils its role of a substitute when the time comes?

With other words, humanity is fully determined to extract and burn the fossil fuels as quickly as possible. Since the fossils do not grow on trees and cannot be replenished, we need to think of their replacement by some other fuels.

Wood fuels are the ideal replacement for the fossils…if and when we figure out efficient and sustainable methods of doing so.

Can wood play this role successfully and fulfill the need for renewable fuels in the future? Is it possible that people in the 22nd century and beyond will go back to using wood as their primary energy sources? History repeats itself, they say. So maybe we will make a full circle back to the 17-18th century, when wood was the major, if not the only, reliable fuel.

But if that happens, then are we going to be able to grow enough trees and biomass? Are we going to be able to control excess deforestation? How would we make sure that the forests are not wasted in a short time period? What about the great GHG pollution coming from wood processing and burning?

We don't have the answers to these questions, so all we can do is and hope that the future generations will eventually find the answers before it is too late.

Domestic and Agricultural Refuse

Also called garbage, domestic refuse is a waste type consisting of everyday items that are discarded by the public and end up in the municipal dumps, where they rot and emit all kinds of gasses.

Wood waste is a large part of the solid waste mass. Wood and woody residue; grass, leaves, brush and branches from back yard maintenance comprise a significant amount of the domestic waste, which end up in the municipal dumps to add to the emissions with time.

Agricultural waste, consisting of dead vegetation and small animals is added in large amounts to the pollution emitters on daily basis. Manure, of course is inevitable sign of large animal herding and can be found in large amounts in the fields, and in the cow, pig, and chicken feeding farms. Agricultural refuse, mostly in form of crop (woody) residue, manure, and feed scraps is also mostly organic matter that can be burned for heat or power, or can be used as organic fertilizer in agriculture, where they contribute to the fertility of the soil by adding organic matter and nutrients to it.

Large part of thus generated refuse can be used for biogas and/or electricity generation. Several technologies have been developed that make the processing of solid waste for energy generation cleaner and more economical than ever before. This includes landfill gas capture, combustion, pyrolysis, gasification, and plasma arc gasification methods.

Older waste incineration plants emit high levels of pollutants, recently new technologies have significantly reduced this type of pollution. For example, EPA regulations in 1995 and 2000 under the Clean Air Act reduced emissions of dioxins from waste-to-energy facilities by more than 99% from 1990 levels. At the same time mercury emissions have been reduced by over 90% as well.

These improvements allow waste-to-energy sources to have less environmental impact than almost any other source of electricity. At the same time, however, they require more complex and expensive technologies. This reduces the incentive of waste site operators to invest in energy technologies.

Wood Use in Europe

Globally, developing countries use most of the firewood. There is, however, a growing trend in the developed world too. It is driven by the increase of daily wood use in most North European countries, where a large and lucrative firewood energy market is thriving.

In its various forms—from sticks to pellets to sawdust—wood (and biomass in general) accounts for about half of Europe's renewable-energy consumption.

In countries like Poland and Finland, wood meets more than 80% of renewable-energy demand. In Germany, energy transformation is the major project of the 21st century, and for that purpose huge government subsidies went into wind and solar power development. Yet, almost 40% of non-fossil fuel consumption comes from the old wood stuff.

After spending billions, and boasting about their high-tech achievements in the low-carbon energy revolution, the main reason (and beneficiary) of the energy hoopla is still the old firewood of the pre-industrial era. So, now wood has also been put on a pedestal as an environmental savior.

The idea that the rotting, smoky wood log smoldering in the old fireplace is low in carbon sounds bizarre, to say the least.

But contrary to what we might believe or not, firewood is included in the EU's official list of renewable-energy supplies. The idea is that IF wood coming from properly managed forests is used in a properly designed and managed wood-fired power plant, then the carbon that billows out of its chimney is (potentially) offset by the carbon that is captured and stored by the trees, the firewood comes from. Carbon in and carbon out equal zero...at least on paper. Here we see again a controversy related to location. While the forests where wood comes from benefit from the CO_2 absorption, the areas where wood is burnt are polluted; hence the discrepancy and the complaints.

Wood can be carbon-neutral if the entire process from seeding to burning is done right. This is hard to do, so whether it actually turns out to be so is a different matter that depends on many complex conditions and processes. Nevertheless, the decision was made in the EU, and wood is now called renewable energy source. Due to numerous renewables benefits and subsidies, firewood usage soared in most European countries recently.

And no doubt, firewood has a number of advantages in many areas, including the energy sector. It is a very convenient fuel for burning in fireplaces, where it brings heat and pleasant atmosphere. Who would not enjoy curling with a good book by a slow-burning log in the fireplace. This alone makes wood preferred fuel (especially in the winter months) by most Europeans.

The fact that the air quality in some large cities is affected by excess wood smoke during the months is becoming an issue that different governments tackle in a different ways. In some U.S. cities, for example there is a fireplace burning law. Phoenix, Arizona, is one of these cities. Here, fireplace burning ban is issued almost every day, and surely every weekend, during most winter days. Luckily, there are not many cold days in the desert, so fireplace burning ban is not causing any major difficulties or harms.

In the electricity generating sector, wood also has various advantages. Convenience and cost makes wood better than planting huge fields of windmills or solar panels. At the same time, most coal-burning power plants can be modified to burn a mixture of 90% coal and 10% wood in a co-firing process. This is a fairly simple modification that requires no new investment, since the equipment is in place and the power plant is already connected to the grid. This modification can bring power prices down, in addition to bringing carbon savings from using renewable fuels in the old coal-fired power plants.

Biomass Co-firing

Biomass is used for electric power generation by co-firing with coal, per the methods shown in Figure 6-6a, b, and c.

But even more importantly, biomass and firewood energy is one of the non-intermittent renewable energy sources. Unlike sun and wind power generation, firewood power generation does not require backup power at night, or on calm and cloudy days. For these and other reasons, wood is quickly becoming popular with the utilities.

Wood can be used in coal-fired power plants to meet the emission standards and prevent shut downs.

In this role, wood is seen as a major competitor to natural gas, which presently is the only replacement of coal-fired power generation. As a result, an alliance of pro-firewood enthusiasts formed in EU countries. It goal was to explore ways to get more subsidies for biomass use. Here, unusual developments caused unlikely teams to be formed. Green advocates, promoting wood as carbon-neutral fuel, joined with utilities, who saw wood-coal co-firing as a cheap way to save their coal plants from closing.

And then the governments who saw wood as the only way to meet the short term renewable-energy targets, joined the alliance. The EU community has set a target to get 20% of its energy from renewable sources by 2020, but this cannot be achieved by relying on solar and wind power alone.

Firewood to the rescue! A new sort of energy business is on the making now in Europe.

Actually, firewood is the oldest form of man-made energy since our ancestors have been using firewood since the invention of the fire. It was later on replaced by coal and was put on the back burner. Until recently, electricity generated from wood was a rare, usually small-scale waste-recycling operation. First, pulp and paper mills built small power stations near the plants to burn branches and sawdust, thus generating power for their own needs. Later on, co-firing was introduced in some power plants with a marginal interest by the utilities. And finally in 2011, RWE, a large German utility, decided to convert its Tilbury B power station in eastern England to run 100% on wood pellets. This was the beginning of the new trend in large-scale electric power generation.

Soon thereafter, Drax, a large British coal-fired power plant, converted three of its six boilers to burn wood. The work is underway, and when finished, these units will generate 12.5 TWh electric power annually. Now, here is the twist. The old Drax was a dirty coal-fired clunker, doomed to pay millions in penalties for excess carbon emissions, and eventually forced to shut down. Drax got a new makeup, with the shiny title of *renewable* power generator added to its title. The new co-fired power will also get a significant subsidy, called a renewable obligation certificate.

According to the present rules, Drax would get $68 for each MWh on top of the market price for electricity, because of the *renewable wood fuel* used in the three turbines. This is about $650 million annual income in form of subsidies for using renewable biomass and other woody materials. This is a significant—threefold—increase of Drax' pretax profit of $225 million. Not bad,

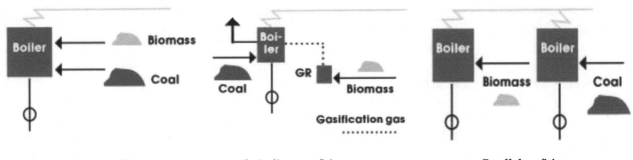

Figure 6-6.a. Direct co-firing   **b. Indirect co-firing**   **c. Parallel co-firing**

for burning tree branches and stumps. Following Drax' example, many other utilities, lured by the potential incentives, are scouring the Earth for wood.

Europe consumes 13-15 million tons of wood pellets annually, which amount is expected to double by 2020.

Even more demand could be expected if and when more utilities get into the coal-wood co-firing game. Whoever thought that wood is irrelevant to the energy markets, better think again. Wood is back and here to stay…in Europe at least.

We will not see this trend duplicated in the U.S., at least not in the near future. This is because we have too many other options, so wood would be the last to enter the U.S. energy markets, but It will some day!

On the other end of the energy spectrum from wood and wood products, is the most powerful energy source known to man; nuclear power. Just hearing the words *nuclear power* brings to mind horrible images from the devastation of Hiroshima and Nagasaki during WWII, and then Chernobyl and Fukushima more recently.

What is nuclear power, and why is it so special? Let's see…

NUCLEAR POWER

Nuclear power, like the Genie in the bottle, does miracles when under control. It, however, is fierce, unrelenting, and truly devastating force when out of control.

One question many people ask is if nuclear materials, of the likes of uranium, plutonium, thorium, etc., are fossils. The short answer is that these materials were not created as the fossils, so they don't fit directly in this category. They, however, just like the fossils, were created millions of years ago—actually many millions of years before the fossils—shortly after the Big Bang.

Also, just like the fossils, the nuclear materials are laying deep underground, awaiting the time to be dug out and transported to power plants. Very much like the fossils, they are not renewable energy sources, but instead are in limited quantities. As such, they are quite similar to the fossils, and yet due to their special properties and limitations, we put them in the *other energy sources* category of this text.

Nevertheless, the United Nations classifies a subset of presently operating nuclear fission reactors and the related technologies as part of the renewable sector.

Specifically, some reactors that produce more fissile fuel than they consume, such as breeder and nuclear fusion reactors (whenever they become a reality) are classified by the UN as renewable energy sources; in the same category with solar and wind power.

We can argue the pros and cons of this classification, but it won't change the fact that—with due respect to the thousands of nuclear power plant engineers, technicians, and operators—nuclear is a controversial energy source, with a questionable future.

The only problem with nuclear power is that we, humans, have not learned how to design, build, and use nuclear reactors properly and safely.

The operators' mistakes at Chernobyl's reactor, and the poor design of Fukushima power plant led to huge disasters. The victims of these accidents with horrific results are witnesses to the fact that we have a lot to learn about what nuclear power is and what it is not.

So, whether the United Nations classifies nuclear power as renewable energy source, or not, is irrelevant. What matters is that nuclear power has to be proven safe by not allowing any major accidents in the future. This, in our opinion, is not an easy task, because any technology can fail for some reason. Especially when humans are involved.

An equipment manufacturer had a plaque attached to his equipment, which read, 'This equipment is 100% reliable. It is only 50% reliable if installed by humans, and only 10% if operated by them.' How true!

The difference here is that when nuclear equipment fails, the reliability goes down to zero, and thousands of people living nearby are hurt and killed. The entire area around the plant is a waste land for hundreds of year, the entire world shudders in horror. In such cases, the nuclear industry suffers major setbacks and takes it awhile to recover.

We are looking into the future of nuclear power with the fear that another major nuclear accident might confirm that we cannot handle this awesome beast. In such case, we might see all nuclear plants around the world shut down for good.

In the Beginning…

There are a number of theories about the creation and availability of uranium and other nuclear (radioactive) materials on Earth. Some of these theories claim that the radioactive materials were produced in one or more supernovae; exploding stars in which the radiated energy increases several billion times within, as a result of the catastrophic collapse of the star's core. Such events are hard to describe exactly, because of the enormity and complexity of the different energies generated and released for the duration.

Supernovae events are extremely luminous and noisy, and cause a burst of radiation that often briefly outshines an entire galaxy, before fading from view over several weeks or months. During this short interval a supernova can radiate as much energy as the Sun is expected to emit over its entire life span.

The initial explosion expels much or all of a star's material at high velocity of up to 20,000 miles per second, thus driving a shock wave into the surrounding interstellar medium. This shock wave sweeps up an expanding shell of gas and dust called a supernova remnant, accompanied by a flood of free neutrons. The main process that leads to the creation of uranium and other radioactive isotopes can be explained by the rapid capture of these free neutrons on seed nuclei at rates greater than disintegration through radioactivity during the explosive supernovae event.

For the sake of simplification, we will assume that the Earth's uranium and some of the other radioactive materials we find here were produced through similar processes in one or more supernovae, which must've occurred billions of years ago. This is actually a crude oversimplification, since there were many other extraordinary events with even more spectacular effects around the time of the Big Bang and shortly thereafter. I.e., there is evidence that more than ten separate and distinctly different stellar sources were involved in the genesis of the solar system material.

Thus, the relative abundance of U-235 and U-238 at the time of formation of the solar system is most likely as a result of the explosive debris of many supernovae and the related interactions. In any and all cases, the final result of all this is that we now have large deposits of uranium, thorium, and other radioactive materials here on Earth that allow us to extract and use them for generating electric power via nuclear power reactors.

The average abundance of uranium in meteorites is about 0.008 parts per million (ppm, or gram/ton), while the abundance of uranium in the Earth's "primitive mantle" (prior to the extraction of the continental crust) was 0.021 ppm. Allowing for the extraction of a core-forming iron-nickel alloy with no uranium (because of the characteristic of uranium which makes it combine more readily with minerals in crustal rocks rather than iron-rich ones), this still represents a roughly two-fold enrichment in the materials forming the proto-Earth compared with average meteoritic materials.

The present-day abundance of uranium in the "depleted" mantle exposed on the ocean floor is about 0.004 ppm. The continental crust, on the other hand, is relatively enriched in uranium at some 1.4 ppm. This represents a 70-fold enrichment compared with the primitive mantle. In fact, the uranium lost from the "depleted" oceanic mantle is mostly sequestered in the continental crust. The processes which transferred uranium from the mantle to the continental crust are complex and consist of many consequent steps over long time period.

It took over 2 billion years to go through the:
- Formation of oceanic crust and lithosphere through melting of the mantle at mid-ocean ridges,
- Migration of the oceanic lithosphere laterally to a site of plate consumption (this is marked at the surface by a deep-sea trench),
- Production of fluids and magmas from the down-going (subducted) lithospheric plate and overriding mantle "wedge" in these subduction zones,
- Transfer of these fluids/melts to the surface in zones of "island arcs" (such as the Pacific's Ring of Fire),
- Production of continental crust from the island arc protoliths, through re-melting, granite formation and intracrustal recycling.
- In nature, uranium ore is found as uranium-238 (99.27%), uranium-235 (0.72%), and a very small amount of uranium-234 (0.006%).

All through the crust-forming cycle, the lithophile character of uranium is manifested in the constancy of the potassium to uranium ratio in the containing rock, ranging from peridotite to granite. Keeping track of how uranium is distributed in the Earth, we see the abundance and isotopic characteristics of lead (a relative of U-235 and U-238) as a useful parameter.

There is relatively low abundance of lead in the Earth's mantle and high uranium to lead ratio, compared with meteorites, which can be explained by lead's volatile nature and its tendency to combine with iron. Thus, lead is being lost during terrestrial accretion and core separation.

Note: Uranium-233 can also be presented as U-233, or 233U. This nomenclature is also valid for the other nuclear elements

One of the consequences of these high ratios is the comparatively high radiogenic/non-radiogenic content of Pb-207/Pb-204, and conversely Pb-206/Pb-204 in the Earth's crust and mantle compared with meteorites or the Earth's core. Pb-207 is the final stable decay product of U-235, and Pb-206 is that of U-238, while Pb-204 is non-radioactive.

Uranium decays slowly by emitting alpha particles. The half-life of U-238 is about 4.47 billion years and that of U-235 is 704 million years. Therefore, the decay from strongly radioactive U-235 to its final state of inert and harmless Pb-207 takes several hundred million years. This is fairly uniform process, which makes it useful in dating the age of the Earth and the materials contained in it.

Nuclear Power Life Cycle

All of the actions of mining, transporting, refining, purifying, using, recycling, reprocessing, reusing, and ultimately disposing of the nuclear fuel materials put together make up the nuclear power life cycle. It starts deep underground, where uranium and other radioactive ores are dug out for transport to a processing facility. Here the ore is converted into nuclear fuel, loaded in special canisters and transported to the nuclear plant to fuel the reactors, where it generates steam and electricity non-stop for several months.

When the fuel is exhausted, the reactor is shut down, the fuel bundles are removed, packed in special containers and taken to temporary storage. This process sequence forms the life-cycle of the nuclear fuel. New batch of bundles are loaded in the reactor and the power generating cycles starts anew with fresh or recycled uranium.

Nuclear Materials

One question to ask here is, "Do the nuclear power (its materials, like uranium, plutonium, thorium, etc.) belong in the fossils category?" The answer is categorically *maybe*… These materials were not created as the fossils, so they don't fit directly in this category. They, however, just like the fossils, were created millions of years ago—actually many millions of years before the fossils—shortly after the Big Bang.

Like the fossils, the nuclear materials are laying deep underground, awaiting to be dug out and transported to power plants. And also, very much like the fossils, they are not renewable energy sources, and just like them are in limited quantities.

One thing that separates them from most other energy sources is the fact that some of the key nuclear materials can be recycled and reused. This is a great advantage, which gets them close to the classification of 'renewable' technologies. Because of these border-line similarities to both fossils and renewables, we put the

nuclear materials in the *'other'* energy sources category.

Generally speaking, materials that can sustain a chain nuclear reaction and produce a nuclear explosion are known as nuclear, or *fissile* materials, such as uranium. Some examples of the nuclear fuels are; ^{233}U, ^{235}U, ^{239}Pu, ^{237}Np, ^{243}Am, ^{232}Th etc.

Energy is released when heavy elements like thorium, uranium, or plutonium undergo fission, during which process the heavy nuclide with atomic mass of 235 for ^{235}U, is split into two lighter nuclides like ^{90}Sr, or ^{137}Cs, several neutrons, and occasionally a hydrogen atom.

Uranium

Uranium is the preferred fuels for most nuclear reactors today, and presently generates over 16% of all electricity worldwide. This makes it an important fuel, and a key chemical element, which also played an important role in the evolution of the Earth.

Uranium is found in a number of places around the world, since it is as common as some more common metals, such as tin and germanium. As a matter of fact, uranium is a constituent of most rocks, dirt, and of the oceans—albeit in very small quantities. And this is a problem, because mining is economically feasible only in deposits of somewhat larger concentration of uranium.

The estimates show that, at the present rate of use there are enough uranium deposits around the globe to last about 100 years. Today, uranium is economically recovered at a price of $100 to 150/kg. This higher level of assured Uranium (U) is a very heavy and hard, silvery-white metallic chemical element, of the rare earth/actinides series in the periodic table. Its chemical symbol is U, and corresponds to atomic number 92. Each uranium atom has 146 neutrons, 92 protons and 92 electrons in its atom. Six of the electrons are in the outer shell and are its valence electrons. It has a high melting point, of 2070°F and is found in three major allotropic forms:

α (orthorhombic) stable to up to 660°C
β (tetragonal) stable from 660°C to 760°C
γ (body-centered cubic) from 760°C to melting point.

This is the most malleable and ductile state.

Uranium is malleable, ductile, slightly paramagnetic, strongly electropositive and is a poor electrical conductor. It has very high density, being approximately 70% denser than lead, and only slightly less dense than gold. Unlike gold, it reacts readily with almost all non-metallic elements and their compounds, with reactivity increasing with temperature.

Hydrochloric and nitric acids dissolve uranium, but non-oxidizing acids (other than hydrochloric acid) attack the element very slowly. When finely ground, it can react with cold water, and in air it gets coated with a dark layer of uranium oxide. Because of that, uranium in ores is extracted chemically and converted into uranium dioxide or other chemical forms usable in industry.

Naturally occurring uranium ores contain one or more of the three U isotopes: uranium 234, 235 and 238. All three isotopes are radioactive, but only one of them, uranium 235 with 143 neutrons, is fissionable and can be used in the generation of nuclear power.

Pure uranium is slightly radioactive, and has the second highest atomic weight of the primordially occurring elements, lighter only than plutonium. Its density is about 70% higher than that of lead, but not as dense as gold or tungsten. It occurs naturally in low concentrations of a few parts per million in soil, rock and water, and is commercially extracted from uranium-bearing minerals such as uraninite.

The uranium atom decays slowly by emitting an alpha particle. The half-life of uranium-238 is about 4.47 billion years and that of uranium-235 is 704 million years, which makes them useful in dating the age of the Earth.

The uranium nuclear fuel is usually based on the metal oxides of uranium metal. The oxides are used rather than the pure metal itself simply because the oxide melting point is much higher than that of the metal, thus they are safer in case of reactor overheating and meltdown. Another benefit of the oxides is that they cannot burn, being already in the final oxidized state of matter of the uranium metal. This is another important operational and safety consideration

Table 6-2. Specific energy of fuels (in MJ/kg)

Energy source	MJ / kg
Brown coal (lignite)	24
Firewood (dry)	16-18
TNT	4-5
Compressed hydrogen	142
Natural Gas	40-45
Crude Oil	45-50
Uranium (nuclear grade)	80,620,000

Table 6-2 shows the energy content of different materials (in mega joules per kilogram) that are used as fuel for different applications, including heat and electric power generation. Note the incredible energy content of uranium in Table 6-2, which results in a huge difference, as compared to the energy content of the fossils. This is an enormous advantage for the nuclear power generation, in addition to other benefits and conveniences.

According to the numbers in Table 6-2, nuclear fuel is 3.3 million times more powerful than coal. Considering all the conversions and losses during the power generation cycle, we see that we need many tons of coal to generate the power of just one lump of uranium. This mind boggling fact reveals the greatest advantage of nuclear power.

Note: Another amazing fact in Table 6-2 is that TNT contains 5-6 times less energy per unit mass than coal. So then why TNT is such a fierce explosive and coal is not? It is due mostly to the fact that coal burns slowly, while TNT releases its energy instantaneously. Because of that, even small amounts of TNT explode with a violent power, accompanied by a lot of light and noise.

Uranium Mining and Processing

In nature, uranium is found as uranium-238 (99.2742% U-238) and uranium-235 (0.7204% U-235). All U-based isotopes are radioactive, and pose serious health danger. They can be deadly upon exposure, when improperly handled.

At the uranium mine, the uranium ore is dug out via mechanized tools, similar to those used in coal mines. In addition to the usual safety measures used at mine operations, special precautions are taken to pre-

Table 6-1. Uranium isotopes

Isotope	Half Life
U-230	20.8 days
U-231	4.2 days
U-232	70.0 years
U-233	159000.0 years
U-234	247000.0 years
U-235	7.0004xE8 years
U-236	2.34xE7 years
U-237	6.75 days
U-238	4.47xE9 years
U-239	23.5 minutes
U-240	14.1 hours

vent the radiation from harming the workers.

The uranium ore is then loaded on special trucks or trains and transported to a processing facility. Here the ore is first crushed to a fine powder by crushers and grinders. The "pulped" ore is further processed by a treatment with concentrated acid, alkaline, and/or peroxide solutions, which dissolve and extract the uranium from the mix. The resulting solution is further refined, filtered and dried to yield the final product, called yellowcake, or urania, which is actually brown or black in color.

Note: The yellowcake name is a remainder of the color and texture of the final product in the past, when it was considered to be ammonium or sodium di-uranate, and was quite un-uniform and unstable, depending on the refining process conditions. The natural uranium, yellowcake, is sold on the uranium market as U_3O_8. It is then processed into UO_2, as required for making fuel rods.

Yellowcake is a coarse powder which has a pungent odor, is insoluble in water and contains about 80% uranium oxide (U_3O_8), which melts at approximately 5212.4°F. It contains, among other things; uranyl hydroxide, uranyl sulfate, sodium para-uranate, and uranyl peroxide, and some uranium oxides.

Today yellowcake's quality is tightly controlled to contain about 90% triuranium octoxide (U_3O_8) by weight. It is produced by all countries in which uranium ore is mined. Yellowcake is used predominantly in the preparation of uranium fuel for nuclear reactors, for which it is smelted into purified UO_2 for use in fuel rods for pressurized heavy-water reactors and other systems that use natural un-enriched uranium.

Purified, pure uranium, metal (not UO_2) can be enriched into the isotope U-235, by combining the pure uranium metal with fluorine to form uranium hexafluoride gas (UF_6). The gas is then processed via gaseous diffusion, or through a gas centrifuge, where it undergoes isotope separation. This process produces low-enriched uranium containing up to 20% U-235, or the type used in most large civilian electric-power generating nuclear reactors.

Further processing produces highly enriched uranium, containing over 20% U-235, which is used in smaller reactors to power naval warships and submarines. Even further processing can yield weapons-grade uranium, which contains over 90% U-235, which is used for making nuclear weapons.

Since the official end of the Cold War in 1990, there is a worldwide surplus of highly-enriched uranium, some of which is often diluted for use in nuclear reactors.

Isotope Enrichment

Isotope separation is designed to concentrate (enrich) the fissionable uranium-235 for nuclear weapons and most nuclear power plants, except for gas cooled reactors and pressurized heavy water reactors. Most neutrons released by a fissioning atom of uranium-235 must impact other uranium-235 atoms to sustain the nuclear chain reaction. The concentration and amount of uranium-235 needed to achieve this is called a "critical mass."

To be considered "enriched," the uranium-235 fraction should be between 3% and 5%. This process produces huge quantities of uranium that is depleted of uranium-235 and with a correspondingly increased fraction of uranium-238, called depleted uranium (DU).

To be considered "depleted," uranium-235 isotope concentration should be no more than 0.3%. The price of uranium has risen since 2001, so enrichment tailings containing more than 0.35% uranium-235 are being considered for re-enrichment, driving the price of depleted uranium hexafluoride.

The gas centrifuge process, where gaseous uranium hexafluoride (UF_6) is separated by the difference in molecular weight between $235UF_6$ and $238UF_6$ using high-speed centrifuges, is the cheapest and leading enrichment process. The gaseous diffusion process had been the leading method for enrichment for a long time. Here, uranium hexafluoride is repeatedly diffused through a silver-zinc membrane, and the different isotopes of uranium are separated by diffusion rate (since uranium 238 is heavier it diffuses slightly slower than uranium-235).

Today, however, gas diffusion is becoming an obsolete technology that is steadily being replaced by the later generations of technologies, as the diffusion plants reach their ends-of-life. The molecular laser isotope separation method employs a laser beam of precise energy to sever the bond between uranium-235 and fluorine. This leaves uranium-238 bonded to fluorine and allows uranium-235 metal to precipitate from the solution.

An alternative laser method of enrichment, known as atomic vapor laser isotope separation (AVLIS), employs visible tunable lasers such as dye lasers to achieve the desired results.

The uranium enrichment facilities are usually designed, built, and operated under strict security, to which very few outsiders have access. Because of that, we will limit our discussion on uranium processing and handling to the very minimum.

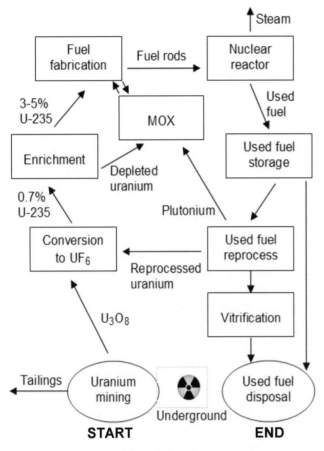

Figure 6-7. The nuclear fuel life cycle

escaping the fuel into the coolant and contaminating it.

The fuel rod assemblies are then shipped to the nuclear power plant for installation in the reactors. Since there are a number of different types of reactors, the fuel rod assemblies are also different in type and size. Several nuclear fuel bundles are submerged in water inside a pressure vessel, where the water acts as a coolant.

If the bundles are left unattended in the reactor, the uranium fission would accelerate beyond control and would eventually overheat the bundles, evaporate the water and melt the containment vessel. To prevent such uncontrolled conditions and overheating, a number of "control rods" made of a material that absorbs neutrons are inserted between the active rods in the uranium bundle.

While the uranium rods are stationary, and usually attached to the bottom of the reactor, the control rods are attached to a special mechanism in the ceiling of the reactor, that can be raised or lowered at will with great accuracy. By raising and lowering the control rods, operators can control the rate of the nuclear reaction and the overall temperature in the reactor.

In order for the uranium core to produce more heat, the control rods are slowly lifted out of the uranium bundle. This reduces the surface area of the uranium rods in contact with the control rods, so that they absorb fewer neutrons, which results in greater heat generation.

To reduce the speed of the reaction and lower the generated heat, the control rods are lowered into the uranium bundle. This increases their surface area in con-

Nuclear Fuel Production Process

During production of the reactor fuel rods, the final uranium dioxide (UO_2) product, in the form of a fine powder, is compacted into cylindrical pellets and sintered at high temperatures. The objective is to produce ceramic nuclear fuel pellets with a high density and well defined physical properties and chemical composition. The pellets are machined to give them uniform and precise cylindrical shape. Thus obtained fuel pellets are then stacked into metallic tubes—fuel rods.

The metal tubes type, size, and shape depends on the design of the reactor. Stainless steel used in the past is now replaced by zirconium alloy, which is highly corrosion-resistant and has lower neutron absorption. The metal tubes with the fuel pellets inside (fuel rods) are sealed, and grouped into fuel assemblies which comprise the core of a power reactor.

The outer layer of the fuel rods (cladding) is made of a corrosion-resistant material with low absorption cross section for thermal neutrons. It also protects the tubes from reacting with the surrounding media. Cladding also prevents radioactive fission fragments from

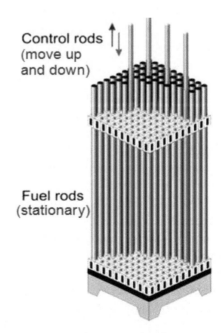

Figure 6-8. Uranium control rods arrangement

tact with the uranium rods, which leads to the absorption of more neutrons, which slows the nuclear reaction and reduces the heat in the reactor.

The lower the rods, the more area to absorb neutrons, less power produced. If the rods are lowered completely into the uranium bundle, then all emitted neutrons are absorbed and the entire nuclear reaction stops. This is how a nuclear reactor is normally shut down for maintenance, or in the event of a malfunction in an attempt to prevent an accident.

If the rods are not lowered in time, however, the reaction can get out of control, causing overheating and even explosion in the reactor. This is one of the ways nuclear accidents start, and what follows could be anyone's guess. In worst-case scenarios, we end up with a Chernobyl-type nuclear accident.

Nuclear Fuel Assemblies

The type, quality and use of radioactive materials, as well as those of the nuclear fuel assemblies, is regulated by the federal Nuclear Regulatory Commission (NRC), as well as different state regulating agencies

The major fuel rod assembly for the different types of reactors are:

- *Pressurized water reactors'* (PWR) nuclear fuel assemblies consist of cylindrical rods put into bundles. Here, uranium oxide ceramic is formed into pellets and inserted into the Zircaloy tubes that are bundled together. The uranium oxide is dried before inserting into the tubes to try to eliminate moisture in the ceramic fuel that can lead to corrosion and hydrogen embitterment. The Zircaloy tubes are pressurized with helium to try to minimize pellet-cladding interaction which can lead to fuel rod failure over long periods.

The fuel bundles are usually enriched to several percent of U-235, with tubes of about 1 cm in diameter. The fuel cladding gap is filled with helium gas to improve the conduction of heat from the fuel to the cladding. There are between 179 and 264 fuel rods per fuel bundle. Between 121 to 193 fuel bundles are loaded into each reactor core. Generally, the fuel bundles consist of fuel rods bundled in groups of 14×14 to 17×17 and about 4 meters long. In PWR fuel bundles, control rods are inserted through the top directly into the fuel bundle.

- *Boiling water reactors'* nuclear fuel, as the name suggests, is used in boiling water (BWR) reactors. It is similar to PWR fuel, except that the bundles are "canned" by means of a thin metal tube of ap-

propriate diameter enveloping each bundle. This is primarily done to prevent local density variations (difference between different tubes) from affecting neutronics and thermal hydraulics of the reactor core.

Modern BWR fuel bundles consist of either 91, 92, or 96 fuel rods per assembly, depending on the manufacturer and reactor type. The reactor core contains a total of 368 assemblies for the smallest, and 800 assemblies for the largest U.S. based BWR reactors. Each fuel rod is back filled with helium to a pressure of about three atmospheres, and is used similarly to the PWR rods.

- *CANada Deuterium Uranium* (CANDU) nuclear fuel bundles are about a half meter long and 10 cm in diameter. They consist of sintered (UO_2) pellets in zirconium alloy tubes, welded to zirconium alloy end plates. Each bundle weighs about 20 kg, and a typical core loading is in the order of 4500-6500 bundles, depending on the design. Modern types typically have 37 identical fuel pins radially arranged about the long axis of the bundle. Several different configurations and numbers of pins have been used in the past to figure out the ultimate, most efficient and safe design.

- *The CANDU FLEXible* fueling (CANFLEX) nuclear fuel bundle has 43 fuel elements, with two element sizes. It is also about 4 inches in diameter, 20 inches long and weighs about 44 lbs. It replaces the 37-pin standard bundle in use in some reactors. This configuration was designed specifically to increase fuel performance by utilizing two different pin diameters. Current CANDU designs do not need enriched uranium to achieve criticality (due to their more efficient heavy water moderator), however, some newer concepts call for low enrichment to help reduce the size of the reactors.

A number of less common nuclear fuels are:

- *Magnox* nuclear fuel is a type of fuel used in reactors, which are pressurized, carbon dioxide cooled, graphite moderated reactors using natural, unreached uranium as fuel, and magnox alloy as fuel cladding.

- *Tristructural-isotropic* (TRISO) nuclear fuel is a type of micro fuel particle. It consists of a fuel kernel composed of uranium compounds (UOX, UC, or UCO) in the center, coated with four layers of three isotropic materials. The four layers are a porous buffer layer made of carbon, followed by a dense inner layer of pyrolytic carbon (PyC), followed by a ceramic layer of silicon carbide SiC to retain fission products at elevated temperatures and to give the TRISO particle more structural integrity. There is a dense outer layer made out of PyC.

- *Quad-structural-isotropic* (QUADRISO) nuclear fuel particles consist of europium oxide, erbium oxide, or carbide layer surrounding the fuel kernel of ordinary TRISO particles, which helps to better manage the excess reactivity.

- *Реактор Большой Мощности Канальный* (*RBMK*) nuclear fuel was used in Soviet designed and built RBMK type reactors. This is a low enriched uranium oxide fuel. The fuel elements in an RBMK are 3 m long each, and two of these sit back-to-back on each fuel channel, pressure tube. Reprocessed uranium from Russian Water-Water Energetic Reactor (VVER) reactor spent fuel is used to fabricate RBMK fuel. As a result of the Chernobyl accident, the enrichment of nuclear fuel was changed from 2.0% to 2.4%, to compensate for control rod modifications and the introduction of additional absorbers.

- *CerMet* nuclear fuel consists of ceramic fuel particles (usually uranium oxide) embedded in a metal matrix. It is thought that this type of fuel is what is used in United States Navy reactors. This fuel has high heat transport characteristics and can withstand a large amount of expansion.

- *Plate type* nuclear fuel is commonly composed of enriched uranium sandwiched between metal cladding. It is used in several research reactors where a high neutron flux is desired, for uses such as material irradiation studies or isotope production, without the high temperatures seen in ceramic, cylindrical fuel.

- *Sodium bonded* nuclear fuel consists of fuel that has liquid sodium in the gap between the fuel pellet and the cladding. The sodium bonding is used to reduce the temperature of the fuel. It is often used for sodium cooled liquid metal fast reactors, and has been used in EBR-I, EBR-II, and the FFTF type reactors. The fuel pellets may be metallic or ceramic.

Plutonium

The second-most used fissile isotope is plutonium-239. It can also fission on absorbing a thermal neutron, with the end product being plutonium-240 (Pu-240). Pu-240 makes up a large proportion of reactor-grade plutonium used today, which is plutonium recycled from spent fuel that was originally made with enriched natural uranium and then used once in a light water reactor (LWR).

Note: Current light water reactors (LWR) make relatively inefficient use of nuclear fuel by fissioning only the very rare and expensive uranium-235 isotope. Nuclear reprocessing can make this waste reusable, and more efficient reactor designs allow better use of the available resources.

Pu-240 decays with a half-life of 6,561 years into U-236. In a closed nuclear fuel cycle, most Pu-240 will be fissioned (after more than one neutron capture) before it decays. However, Pu-240 discarded as nuclear waste will decay over thousands of years.

Thorium

Recently, several countries have experimented with using thorium as a substitute nuclear fuel in nuclear reactors. The growing interest in a thorium fuel cycle is due to its abundance in some areas (3-4 times more abundant than uranium), its safety benefits, and absence of non-fertile isotopes.

Thorium is a naturally occurring radioactive chemical element with the symbol Th and atomic number 90. In nature, virtually all thorium is found as Th-232, which has a half-life of about 14.05 billion years. Other isotopes of thorium are short-lived intermediates in the decay chains of higher elements, and only found in trace amounts.

Thorium is mostly refined from monazite sands as a by-product of extracting rare earth metals. Thorium undergoes a complete combustion in specialized nuclear reactors, vs. only 1% for standard uranium reactors using natural uranium.

Thorium reactors are popular in India (albeit still in development) and will become more popular around the world when global supplies of uranium ore are near depletion. Thorium reactors generate 3.6 billion kWh of heat per ton of thorium at 40% efficiency, which means that a 1 GW reactor uses about 6 tons of thorium per year.

Worldwide thorium resources are estimated at 2 million tons, so the thorium supply (theoretically) could power the world for several centuries India's three-stage nuclear power program is possibly the most famous, well funded, and most advanced thorium nuclear process development effort.

Metal Nuclear Materials

Metal nuclear materials have much higher heat conductivity than oxide types, but cannot withstand high temperatures. Metal fuels have the highest fissile atom density, and are normally alloyed, made with pure uranium metal.

Uranium alloys include uranium aluminum, uranium zirconium, uranium silicon, uranium molybdenum, and uranium zirconium hydride. Any of these can be made with plutonium and other actinides as part of a closed nuclear fuel cycle. Metal fuels have been used in water reactors and liquid metal fast breeder reactors, such as EBR-II.

Some of the metal nuclear materials are:

• *Uranium dioxide* (UO2) nuclear fuel is a black solid material, which is prepared by reacting uranyl nitrate with ammonia to form a solid (ammonium uranate). It is then heated (calcined) to form U3O8 that can then be converted by heating in an argon/hydrogen mixture at 700°C to form UO2. Thus obtained UO2 is mixed with an organic binder and pressed into pellets, which are then fired at a high temperature again in argon/hydrogen gas mixture to sinter the pellets into a solid material with few pores.

• *Mixed oxide*, or MOX, is a blend of plutonium and natural or depleted uranium which behaves similarly to the enriched uranium feed, for which most nuclear reactors were designed. MOX fuel is an alternative to low enriched uranium (LEU) fuel used in the light water reactors (LWR) which are used in the global nuclear power generation.

• *TRIGA* nuclear fuel is used in TRIGA (training, research, isotopes, general atomics) reactors, which use uranium-zirconium-hydride (UZrH) fuel, which has a built-in safety, where as the temperature of the core increases, the fuel reactivity decreases. This pretty much eliminates the possibility of a meltdown. Most cores that use this fuel are "high leakage" cores where the excess leaked neutrons can be utilized for research.

TRIGA fuel was originally designed to use highly enriched uranium, however in 1978 the U.S. Department of Energy launched its Reduced Enrichment for Research Test Reactors program, which promoted reactor conversion to low-enriched uranium fuel. A total of 35 TRIGA reactors have been installed at locations across the USA. Anotherr 35 reactors have been installed in other countries.

• *Actinide* nuclear fuel is a by-product of fast neutron reactors, where minor actinides produced by neutron capture of uranium and plutonium can be used as fuel. Metal actinide fuel is typically an alloy of zirconium, uranium, plutonium and the minor actinides. It can be made inherently safe as thermal expansion of the metal alloy will increase neutron leakage.

Note: The minor actinides include neptunium, americium, curium, berkelium, californium, einsteinium, and fermium. The most important isotopes in spent nuclear fuel are neptunium-237, americium-241, americium-243, curium-242 through -248, and californium-249—252.

Ceramic Nuclear Materials

Ceramic nuclear materials, in addition to the oxides, also have high heat conductivities and melting points, but they are more prone to swelling than oxide fuels and are not understood as well.

Some of the ceramic nuclear materials are:

• Uranium nitride (UN) is used in NASA reactor designs, because it has a better thermal conductivity than UO2, since it has a very high melting point. UN fuel has the disadvantage that a large amount of 14C would be generated from the nitrogen by the (n,p) reaction. As the nitrogen required for such a fuel would be so expensive it is likely that the fuel would have to be reprocessed by a pyro method to enable the 15N to be recovered. It is likely that if the fuel was processed and dissolved in nitric acid that the nitrogen enriched with 15N would be diluted with the common 14N.

• Uranium carbide was used in the form of pin-type fuel elements for liquid-metal fast breeder reactors during their intense study during the 60s and 70s. Recently there has been a revived interest in uranium carbide in the form of plate fuel and most notably, micro fuel particles (such as TRISO particles). The high thermal conductivity and high melting point makes uranium carbide an attractive fuel. In addition, because of the absence of oxygen in this fuel, as well as the ability to complement a ceramic coating, uranium carbide could be the ideal fuel candidate for certain Generation IV reactors such as the gas-cooled fast reactor.

Liquid Nuclear Materials

Liquid nuclear fuels are basically liquids that contain some percentage of dissolved nuclear fuel. Liquid-fueled reactors generally have large negative feedback mechanisms and therefore are particularly stable designs. Liquid fuels have the disadvantage of being easily dispersible in the event of an accident, such as a leak in the primary system.

Molten salts nuclear fuels have nuclear fuel dissolved directly in the molten salt coolant. Molten salt-fueled reactors, such as the liquid fluoride thorium reactor (LFTR), are different than molten salt-cooled reactors that do not dissolve nuclear fuel in the coolant. Molten salt fuels were used in the LFTR known as the Molten Salt Reactor Experiment, as well as other liquid core reactor experiments.

The liquid fuel for the molten salt reactor was a mixture of lithium, beryllium, thorium and uranium fluorides: LiF-BeF2-ThF4-UF4 (72-16-12-0.4 mol%). It had a peak operating temperature of 705°C in the experiment, but could have operated at much higher temperatures, since the boiling point of the molten salt was in excess of 1400°C.

Aqueous solutions of uranyl salts are used in the aqueous homogeneous reactors (AHRs) in a solution of uranyl sulfate, or other uranium salt, in water. Historically, AHRs have all been small research reactors, not large power reactors. An AHR, known as the Medical Isotope Production System is being considered for production of medical isotopes.

Now, we make a great jump from the futuristic world of sophisticated and futuristic nuclear fuels to the pre-industrial revolution, low-tech—and usually ignored by the uninitiated—world of firewood. Actually, this is not such a great leap of faith, because firewood—believe it or not—provides many times the energy generated by nuclear power plants. Hard to believe, yes. Read on:

Nuclear Power Generation

Nuclear fission is what 100% of the global nuclear power industry uses to generate electricity today. There is no better way to use nuclear power at present, so a number of variations of nuclear fuels, reactors, and methods—all used in fission nuclear reactions—have evolved.

There is significant effort in developing the more advanced and more efficient, *fusion* nuclear reaction, but the results from this effort thus far has been unsatisfactory.

Although nuclear power plants are considered as some of the most efficient in the world, they are far from being 100% efficient, mostly due to the heat to electricity conversion process, which is similar to tjhat of coal-fired power plants.

Table 6-3. Efficiency in electricity generation

Energy Source	A	B
Nuclear power	0.32	0.33
Hydroelectric	0.33	1.00
Biomass	0.33	0.53
Wind and solar	0.33	1.00
Geothermal	0.16	0.10

A and B in Table 6-3 represent different ways of calculating the efficiency of power plants. This simply means that the efficiency (calculated as primary energy) of different power plants varies depending on who is doing the calculations. Regardless of the calculation methods, however, nuclear plants are surprisingly low in their overall efficiency. Although they are very efficient in converting nuclear power into steam, they lose the advantage when the steam (heat) is converted to electricity. At this point, they use 19th century technology similar to that used by any fossil-fired power plant. And with that, their overall conversion efficiency advantage is lost, bringing them close to the efficiency of fossils-fired power plants.

Since 3 units of energy are needed to generate each 1 unit of electricity, 2/3 of the energy introduced is lost.

Note: Conversion efficiency is not to be confused with capacity factor. The conversion efficiency of a power plant depends solely on its ability to convert the energy source into electricity, which is a set number, as 0.32 in Table 6-3. The capacity factor, on the other hand, is the total time that the power plant is available for operation. Most nuclear plants operate 24/7/365 and are shut down only for periodic maintenance and refilling. Their capacity factor is about 80-90%.

Below we take a brief look at the major products and processes used in today's fission nuclear power generation.

The Nuclear Process

In a nuclear reactor, a nuclear chain reaction is usually initiated and conducted under normal production regime, but it is strictly controlled at all times. Proper and efficient control is the key to safe nuclear power plant operation.

> *Redundant controls and safety mechanisms ensure that the nuclear reaction is under complete control at all times.*

Complete control are the words here, because any type of control less than *complete* means disaster is ready to happen. The reaction is controlled (it can be slowed down and quickly stopped) via *neutron poisoning;* a process that absorbs some of the free neutrons that cause acceleration of the reaction.

This is seemingly simple to do, but there are many pieces of equipment and systems involved in the control sequence; all driven by humans. When humans err like at Chernobyl, or when Mother Nature misbehaves as at Fukushima, the nuclear reactors go out of control. When that happens the devastation is *complete.*

The nuclear power generation process starts by pilling the control rods in the nuclear reactors up to initiate a nuclear reaction. The control rods (or better said neutron absorbers), made of special materials, control the neutron poisoning.

When the nuclear reaction starts, bombarding uranium-235 isotopes in the uranium rods with slow neutrons, cause the uranium atoms to divide into two smaller nuclei, releasing nuclear energy (which was stored in the atoms and used for the binding). Large quantity neutrons are released at this point as well, and since uranium has the ability to absorb thermal neutrons, the reaction may go one of two ways:
- Over 80% of the time it will fission, or
- 18% of the time it will not fission, and will instead emit gamma radiation, thus yielding U-236.

When many neutrons are absorbed by uranium-235 nuclei, a nuclear chain reaction occurs. If properly controlled, the reaction produces heat, which heats the liquid in the reactor, sending it as steam into the steam turbine where it generates electricity. The steam is then condensed in the cooling towers and sent back into the reactor to repeat the heating-steam-power generation cycle.

To find the actual amount of energy released by uranium (U-235) we could use the famous formula:

$$E = mc^2$$

where
  E = energy released
  c = speed of light (3.0×10^8 m/s)
  m = nuclear mass

The fission of 1 kg of uranium fuel produces 9×10^{16} Joules of energy compared to 3×10^7 Joules of energy produced, when burning 1 kg of coal. Under real world conditions, approx. 8 kWh of heat can be generated from 1 kg of coal, approx. 12 kWh from 1 kg of mineral oil, and around 24,000,000 kWh from 1 kg of uranium-235.

> *Uranium contains about 3 million times the energy in the same amount of coal.*

If a fission reaction releases 100 MeV of energy, then the different components of the released fission energy (assuming complete capture and accounting of the species in the reaction) are as follow:
  85.00 MeV of kinetic energy of fission fragments.
  2.50 MeV of kinetic energy of neutrons.
  7.50 MeV of energy beta particles and gamma rays.
  5.00 MeV as energy of antineutrinos.

In practice, however, we must consider interferences and losses in many places during the entire length of the nuclear power generating processes. These are different for the different fuels, equipment, and processes, so we end up with different numbers for each combination and permutation.

One kilogram of uranium-235 can produce (theoretically) about 45 MWh of electric energy, assuming complete fission and no losses. In practice, considering the losses, we would get much less than 45 MWh, maybe 25-30 MWh, which is still a lot of energy. It is as much as contained in several thousand tons of coal and thousands of barrels of crude oil.

This is a tremendous amount of power, which, if properly controlled, could be harnessed to generate a lot of electricity. If not properly controlled, however, it could do a lot of damage.

> *A nuclear reaction out of control creates a chain reaction that gets so violent that it cannot be stopped. Instead, it propagates; overheating and even exploding the reactor.*

This is what happened at Chernobyl reactor #4 in April, 1986, and then the process was repeated at Fukushima's nuclear plant in March of 2011. These were different reactors, operating under different conditions by different (more qualified) people. The results were also different, but similarly very tragic. Although the magnitude of overheating, and/or the explosion conditions, depends on the reactor type, and the type and quantity of nuclear materials involved, the results never vary.

Loss of control brings problems requiring special handling. Complete loss of control brings utmost devastation!

Commercial nuclear power plants use fuel that is typically enriched to around 3% uranium-235. The CANDU and Magnox reactor designs are the only commercial reactors capable of using un-enriched uranium fuel. Fuel used in United States Navy reactors is typically highly enriched uranium-235 (the exact values of this fuel are classified, but is believed to be well over 3%).

In the military, uranium is used in nuclear reactors to power ships and submarines. It is also used for making atomic bombs. Fifteen pounds of uranium-235 is all that is needed to make a small atomic bomb with huge destructive powers.

The first nuclear bomb used in war, the Little Boy, was based on uranium fission, while the very first nuclear explosive (The Gadget) and the bomb that destroyed Nagasaki (Fat Man) was plutonium based.

Different bombs, different materials, different locations. The results, however, just like those from an out-of-control nuclear reactor, were similar; bringing 100% devastation to the local area, wounding and mercilessly killing thousands of innocent people.

Regardless of all the good reasons and convincing justifications, this is still the most inhumane act committed by the U.S. military machine ever.

We can only hope that this will never happen again.

Nuclear Fission Reactions
Now we know all there is to know about the nuclear fuels, so let's see what happens to them in the nuclear reactor. There are two types of nuclear reactions; fission and fusion. Presently fusion nuclear reactions are in a development stages, and only fission is used in nuclear power plants,. Because of that, it is fission that we will consider in more detail in this text.

During uranium fission reaction neutrons are captured by the nucleus of the U-235 atoms and are absorbed within it. This briefly turns the nucleus into a highly excited U-236 atom, and very quickly it splits into two lighter atoms, Ba-141 and Kr-92, plus two or three neutrons. The number of ejected neutrons depends on the conditions under which the U-235 atom splits. This process (neutron capturing and splitting) is very fast and energetic.

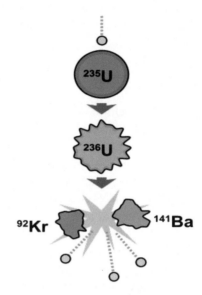

Figure 6-9. Nuclear fission reaction

The decay of a single U-235 atom releases approximately 200 million electron volts (MeV). Since there are billions of atoms undergoing this reaction in a uranium fuel source, the energy release is enormous.

For example, the energy released by just a kilogram of highly enriched uranium undergoing a fission reaction is equal to the energy released by nearly a million gallons of gasoline burning. In addition to energy (heat) released during the splitting of the atoms, a large amount of gamma radiation (radiation made of high-energy photons) is released.

The two atoms that result from the fission of each atom go in their own way releasing beta radiation (super-fast electrons) and gamma radiation of their own. These particles creates the radioactivity of nuclear fuels, and they are what makes nuclear accidents so very dangerous.

For all this to work most efficiently, uranium used for nuclear fuel must be enriched so to contain 2 to 3% additional U-235. Three-percent enrichment is sufficient for nuclear power plants, but weapons-grade uranium requires at least 90% U-235.

Note: It is important to point out here that one neutron impinging onto the U-235 nucleus creates three neutrons, which can in turn impinge on three other U-235 nucleus, thus propagating and escalating the reaction ad infinitum. This spontaneous, exponential generation of neutrons is why nuclear reactions must be controlled to avoid overheating the vessel.

The fission process is accompanied by large amounts of kinetic (heat) energy that is generated during the U-236 fission. When a large amount of U-235 material is bombarded by neutrons in a containment vessel (as in a nuclear power plant) an enormous amount of heat is generated and consequently used to make electricity.

Nuclear Fusion

Nuclear fusion is a futuristic concept that promises to have great impact on the energy sector…in the distant future, if ever. Fusion power is the energy generated by nuclear fusion processes—not to be confused with nuclear fission processes, which we reviewed above. In nuclear fusion, two or more atomic nuclei collide at a very high speed and join to form a new type of atomic nucleus.

During this process, no matter is conserved because some of fusing nuclei are converted to photon energy. This is the opposite of the fission power, where neutrons are captured by the nucleus and are absorbed within it, which excites it and causes it to split into two lighter atoms and emit neutrons.

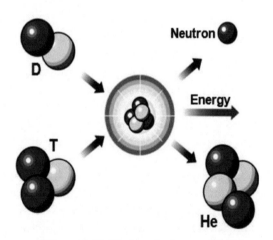

Figure 6-10. Nuclear fusion reaction

The fusion reaction needs a lot of energy to reach the operating state, at which point it releases a huge amount of energy—much more than the input energy—from overcoming the binding energy of the powerful nuclear forces. This results in a rapid and very large increase in temperature at the reaction site.

Fusion power is primarily an area of research in plasma physics, with only a few universities and government labs working on it full-time. The term is commonly used to refer to potential (future) commercial production of net usable power from a fusion source.

The major designs for controlled fusion reaction use magnetic (tokamak) design, or inertial (laser) confinement of a plasma ignition. The heat released by the fusion reaction can be used to run a steam turbine and electrical generator setup, very similar to those in conventional power stations. Both approaches are many years away from success in the labs, and even many more away from commercial application.

The temperature developed by full fusion reaction is so high that it makes the modern fission reaction seem like a toy. This is a big problem, because, according to some estimates, there are no materials known to man that can withstand these temperatures. And if we have a problem with the present day fission (relatively low temperature) reactors, just imagine the magnitude of problems with the fusion high temperature monsters.

In our estimate, if the Fukushima power plant's nuclear reactors were fusion based, the damage would've been many times faster, longer lasting, more serious, and with much more devastating consequences in terms of environmental damage and human suffering.

Nuclear Power Plant Life Cycle

In addition to the regular permits and authorizations, as needed for the construction and operation of coal, oil and gas power plants, a nuclear power plant requires a number of additional analyses, permits, certifications, and licenses. These are needed to comply with safety rules and regulations, specific to the nuclear industry, and/or as required by local, state and federal authorities. This is truly a complex and expensive process.

A nuclear power plant's cradle-to-grave process includes:

- Nuclear plant site search
- Proposed area survey
- Environmental studies
- Local rights and agreements
- Federal and state permits
- Legal concerns and related effort
- Plant site preparation
- Power plant construction
- Initial tests, certification and licensing
- Reactor(s) ramp-up
- Power generation, process optimization
- Daily power plant O&M sequence
- Fresh and spent nuclear fuel logistics
- EOL plant decommissioning, and
- Local area reconstruction

The difficulties and expense at every step of the way are enormous, which is one of the reasons why there have not been any new nuclear plants built in the US since the mid 1970s. The problems start on day one; locating an area to construct a nuclear plant is a difficult task. The locals usually evoke their NIMBY (not in my back yard) rights. NIMBY demonstrations of local activists and neighbors accompany all plans for nuclear plant construction. We have seen these all around the country and the world since the beginning of nuclear power generation, and the movement is not going away.

Obtaining the needed local, state and federal permits, certifications and authorizations is another enormous task. It is actually the major reason of the stoppage and/or delay in new nuclear plants construction in the U.S. and most other countries.

Due to changes in regulations, safety and environmental requirements, it takes forever (and a bank full of money) to complete the permitting process. Most planned nuclear plant plans are on hold presently for this reason alone.

Once a site for the new nuclear plant is chosen, the opposition is kept at a respectable and safe distance, and there is enough assurance of successful permitting that the site area survey and analysis can start. These consist of the usual geological analyses and historical data collection, as needed to determine if the site can support large structures, deep excavations, roads, and infrastructure as needed for the construction activities.

Several stages of environmental tests and studies are conducted, to determine the present environmental conditions, and as needed for the required reports. A long chain of events in the never ending permitting and licensing processes follows.

Once the necessary permits and licenses have been issued, after many years into it, plant construction starts in a fashion similar to that of coal, oil, and gas-fired power plants...although somewhat differently, as we will see below.

Since no new nuclear power plants have been built in the U.S. since 1974, we are unclear how the construction of the new plants would be handled today. Most of them are now stuck in permitting and licensing stages, so we must wait and see how the situation develops in the near future.

Note: The above sub-chapters are actually a very condensed summary of what is needed to design, build, and commission a modern nuclear power plant. A complete description of the entire process and the related procedures would require an entirely new, and quite thick, book. We believe, however, that our text is enough to give the reader a good idea of the complexity and grandeur of a nuclear plant site construction and use.

Nuclear Power Plant Construction

Nuclear power plant (using fission nuclear processes for now) design and construction is a complex and expensive undertaking. It is different from a coal- or gas-burning power plant by the different (nuclear) fuel it uses, and the special requirements for efficient and safe use of that fuel.

<u>*Safe*</u> here must be underlined, emphasized, and over-emphasized, for the future of the new plant, and the entire nuclear energy industry depends on the safe operation of all plants at all times. This is true now more than ever! This serious requirement makes a big difference in designing and building the front-end of the plant (the nuclear reactors). The back-end (steam generation, turbines and electric generators) and other support infrastructure and equipment are quite similar in design to that of coal- and gas-fired power plants.

Another big difference between nuclear and any other power plant are the waste materials. Coal- and gas-fired power plants emit a lot of air emissions and some solid and liquid waste, which could be stored and disposed safely. In contrast, nuclear waste materials cannot be disposed at any time and at any cost. Instead, they are locked in special coffins and stored in special safe places forever. This is (one of) the Achilles heels of the nuclear industry, as discussed in more detail later on in this text.

The fission nuclear plant consists of a number of key elements, each of high level of complexity, size, and safety requirements.

For example:
- Reactor Pressure Vessels (RPV) are complex and very large pieces of equipment, over 20 feet inside diameter by 90 feet high, that weigh up to 1,200 tons. Each complete GEN II+ unit has one RPV and one RPV head, so with everything included, the total weight of the entire unit is well over 2,000 tons.

- Steam Generators are nearly 80 feet tall with an 18-foot-diameter upper section and a 14-foot-diameter lower section, weighing about 730 tons. Moisture separator re-heaters are up to 100 feet long and 13 feet in diameter, weighing around 440 tons. Each GEN III+ unit uses either two steam generators or two to four moisture separator re-heaters;

another set of gigantic, albeit less complex, equipment.

- Control Rod Drives and Fuel Elements are key elements of a nuclear power plant and consist of 200 accurate fine-motion control rod drives per reactor. Over 1,000 fuel elements are used per reactor. Steam Turbine Generators (STG) and condensers are about 1,500 MVA and have low pressure (LP) turbine with last-stage blades that are about 52 inches long. The high pressure steam turbine weighs over 500 tons. Several low pressure rotors are used, each weighing about 250 tons. The generator stator weighs 500 tons and the generator rotor another 250 tons. The STG condenser's lower sections each weigh over 650 tons, with dimensions of 57 x 31 x 34 feet. Each STG has up to three condensers.

- Pumps are used to move liquids and gasses around the plant. There are ten reactor coolant pumps, two turbine-driven feedwater pumps and two motor-driven feedwater pumps for each reactor. Each unit also has several very large (>400HP) safety-related pumps, and over 100 smaller pumps. Some reactors have "passive safety" features and do not require that many safety related pumps.

- Valves are used to control the flow of liquids and gasses around the nuclear power plant. Each reactor unit has over 2,000 valves. Over 1,000 motor operated valves (MOVs) and air operated valves (AOVs) are used in each unit, with 700 valves of 3" size and larger. For example, different GEN III+ units have a total of 9,000-18,000 valves, with over 2,000 values used in the reactors alone.

- Class 1E Switchgear and Equipment are used to assist the power plant operations and usually consist of two or three medium-voltage switchgear panels, three 5MW emergency diesel generators, nine 480V motor control centers, four 125VDC uninterruptible power supply systems, and three 120VAC uninterruptible power supply systems. Some reactors have "passive safety" features and do not require many, if any, emergency diesel generators.

- Control Equipment is used to control the power plant operations, and consists of 2,000-3,500 instruments, digital plant control systems, main control panels, reactor protection panels, local panels, and a plant simulator.

This short review reveals that thousands of tons of metals and other materials are used to make the different components of a nuclear power plant. These materials are processed one way or another, then machined before assembly, loaded on special transport vehicles and delivered to the plant for final assembly and installation.

Some of the plant equipment pieces used in nuclear reactors and the other plant components are prefabricated. There are 500-600 prefabricated modules used in the assembly of each reactor, half of which are fabricated by third-party contractors, usually at their own offsite facilities and then shipped to their final destination. These units include electro-mechanical equipment modules, piping, pipe supports, valves, controls and measurement instrumentation, tubing, conduit, cable tray, junction boxes, structural bases, and structural supports.

The maximum size of a module or sub-assembly fabricated off-site is 12 x 12 x 80 feet, to allow shipment by rail or truck. Larger structural and equipment modules are assembled (welded and bolted on) onsite from scratch or from multiple sub-assemblies.

Over 250 reinforcing steel modules and piping assemblies and over 140 mechanical equipment modules are required for each plant of the most modularized design. In addition over 60 structural modules and 20 electrical equipment modules are required for each unit.

The largest structural modules consist of numerous factory preassembled sub-modules that weigh up to 800 tons each. Some of the structural modules include leave-in place formwork for concrete placement. Upon delivery of the different parts and components, they undergo final assembly onsite and are mounted in place, to complete the plant construction.

Construction Labor

From the first shovel to initial plant testing, construction work requires hundreds, sometimes thousands, of engineers, technicians and general labor personnel. The construction labor force is usually temporary and is different for different power plant designs.

The construction labor, and on-site labor support personnel includes craft supervision, warehouse personnel, clerical staff, security personnel, quality control inspectors, EPC contractors, engineers, schedulers, start-up personnel, and the ever-present Nuclear Regulatory Commission inspection staff.

All these people are highly trained and have specific daily tasks and overall goals. The labor force is divided into groups according to their specialty and work location. All jobs and tasks are managed, monitored and inspected.

Each unit site requires 130-150 administrators, engineers, and loss-control personnel during the peak construction period. This does not include the work of supervision, quality control and system start-up personnel associated with vendors and subcontractors since these are different specialized groups that are counted separately.

During the construction phases, 40-50 quality control inspectors are assigned to different areas of the plant. At the same time, NRC is also represented by 10-20 NRC inspectors on-site and at different off-site locations during construction, start-up, and testing periods. This is in addition to over 60 highly trained specialists onsite during the critical initial start-up phases.

EPC contractors are involved and are usually in charge of plant construction. They need a staff of over 100 specialists at each unit site during different phases of construction. This includes project managers, engineers, schedulers, and other specialized and clerical personnel.

Usually the plant owner is represented by an operations and maintenance (O&M) staff of 200 specialists who support the commissioning, start-up, and maintenance of unit systems during the construction and start-up phases. The O&M staff amount to over 650 for a single unit and 400 for the second unit of a twin unit plant, during normal plant operation.

Most of the permanent plant O&M personnel and plant management, engineering, and security staff are also present onsite during the construction period, but are not included in most labor efforts. They undergo hands-on training for the duration, instead.

Craft labor and on-site labor supporting construction and start-up during the peak construction period totals 2,400 people of different occupations and specialties. Basically, 800 onsite managers and specialists support 1600 personnel during daily labor activities for the duration of plant construction and start-up.

Some of these people are gradually transferred to, or interchangeably used in, the construction of other units. Others are trained for long-term positions at the plant.

A five-year nuclear power plant construction schedule includes 12 to 18 months for site preparation, 36 to 42 months for construction from first concrete to fuel load, and 6 to 12 months for commissioning and testing.

There are 8-10 million man hours of craft-labor estimated for the construction of one unit of an average nuclear plant in the United States. Peak-construction craft labor is estimated at 1,500-1,700 personnel working on each unit. Most construction schedules have 12 to 18 months between the commercial operation dates of units in multiple-unit plants.

The cost of the consecutive units within the staggered 12-18-month construction period is usually significantly lower than the construction of the first unit for several reasons. There are preliminary estimates that additional GEN III+ units will be built after 2014, at the rate of eight units under construction at any time.

By staggering the construction effort, two units could be entering commercial operation every year. This would require a substantial number of sustained total labor force during the deployment period. This means that in case of full deployment of nuclear power, the world could add several nuclear power plants annually to its power generation mix. This scenario, however, is highly unlikely, due to resistance from the public and nuclear safety concerns.

Nuclear plant design, engineering, equipment set, construction, testing and initial start-up are some of the world's most complex, sophisticated, efficient and safe operations. The same is true for the plants' O&M procedures, which are a model of precision and accuracy that simply cannot even be achieved in most other industrial operations.

Unfortunately, even though hundreds of nuclear plants have been operating safely all over the world since the 1950s, it took only 2-3 accidents to put the nuclear industry on the 'guilty until proven' bench. As things are going, it will be sitting on that bench for a long time to come, praying that no other accidents follow. If, God forbid, we witness another Chernobyl or Fukushima type nuclear accident, the global nuclear industry might shrink and even disappear in some places.

The Construction Process

On top of the regular permits and authorizations needed for the construction and operation of coal, oil and gas power plants, nuclear power plants require additional analyses, permits, certifications, and licensing.

These are needed to comply with safety rules and regulations specific to the nuclear industry, as required by local, state and federal authorities. This is a complex and expensive process.

A nuclear power plant cradle-to-grave process includes:

• Nuclear plant site search

- Proposed area survey
- Environmental studies
- Local rights and agreements
- Federal and state permits and legal work
- Plant site preparation
- Plant construction
- Initial tests, certification, and ramp-up
- Daily plant O&M operations
- Fresh and spent fuel transport and storage
- EOL decommissioning and area reconstruction

The difficulties in every step of the way are enormous, which is the reason there have not been any new nuclear plants built in the US since the mid 1970s. Problems start with locating an area to construct a nuclear plant. Locals usually evoke their NIMBY rights. Not in my backyard demonstrations of local activists and neighbors accompany any plans for nuclear plant construction.

We have seen these around the country since the beginning of nuclear power generation, and the movement is not going away. Obtaining the needed local, state and federal permits, certifications and authorizations is another enormous task. Due to changes in regulations and technologies, it is almost impossible to complete the permitting process. Many nuclear plant plans are on hold for this reason alone.

Once a site is chosen, and the opposition is kept at a safe distance, the area survey and analysis can start. These consist of the usual geological analysis and historical data collection to determine if the site can support large structures, deep excavations, roads and infrastructure for construction activities.

Several stages of environmental tests and studies are conducted to determine the present environmental conditions, and as needed for the required reports. A long chain of events in the never-ending permitting and licensing processes follow.

Once the necessary permits and licenses have been issued, plant construction proceeds in a fashion similar to that of coal, oil, and gas-fired power plants. Since no new plants have been built in the U.S. since 1974, we are unclear how the construction of new plants will be planned and executed. Most of them are now stuck in the permitting and licensing stages, so we must wait and see how the situation develops.

Nuclear Plant Operation

A nuclear power plant is a 24/7, 365 days a year operation. It never sleeps, and is only shut down from time to time for refueling, and for major scheduled maintenance. Sometimes it is shut down in emergencies to prevent accidents, which in many cases is a complex undertaking that requires skilled operators and some luck.

In all cases, a nuclear power plant is fully staffed by O&M personnel around the clock, regardless of the condition or stage it is in. Each unit has a shift supervisor, several control room operators, and additional auxiliary operators whose job is to operate the equipment. Multi-unit power plants do have a general shift manager who is responsible for the entire site.

During normal operation, the operators perform many tasks, such as supporting O&M activities, testing safety and emergency equipment, performing minor maintenance, and processing radioactive liquids and gases.

During refueling, which is conducted at least every 2 years, O&M personnel handle the old fuel removal, and transfer new fuel into the reactor, following carefully designed and strictly supervised fuel replacement and safety procedures.

The shift manager and supervisors are highly trained and licensed by the national regulatory agency. Mid- and high-level managers and supervisors are licensed as senior reactor operators, for which they have undergone special training and tests, as needed to show superb accident assessment, supervisory, and team management abilities. They spend a lot of time working in plant simulators, where real-life situations can be reproduced, followed by written and oral tests. In addition, supervisory personnel usually have many years of hands-on experience as plant operators or engineers, prior to assuming the higher responsibility.

Control room operators are licensed as reactor operators by NRC, and are in charge of and control different areas of the plant from the control room. For example, while one operator watches and controls the operating parameters of the turbine, generator, circulating water and related systems, another is in charge of the reactor, reactor cooling, and emergency systems. The different plant systems are usually controlled from different areas of the control board in the room, but the operators communicate with each other and make coordinated decisions, with plant efficiency and safety as a priority.

Operators are also trained on simulators, where they learn and test the basic duties by practicing daily activities, some of which they would be rarely required to perform (usually in emergency siotuations).

Plant startups, shutdowns, and emergencies are included in simulator training, and are needed to provide operators with the maximum understanding and comfort level of managing the plant under all possible situations. In addition, plant personnel undergo constant on-the-job training and re-qualification, via plant simulator training and testing, as needed to maintain their skills and demonstrate competence.

Note: The plant simulator is a functional (but not operational) copy of the control room, with an exact duplicate of gauges, alarms, operational controls, etc. gadgets that function as these in the real plant. The simulator is driven by computers and can be programmed to simulate any possible situation—from a refueling to a critical accident. It is an integral part of plant operation and, in addition to operator training, it allows process improvements and provides solutions for unusual situations, ultimately optimizing the plant's efficiency and safety.

Spent Nuclear Fuel

As in any power generation plant, nuclear plants generate waste materials. Three percent of spent-fuels mass consists of fission products of $235U$ and $239Pu$, which comprise the radioactive waste. These can be processed and separated further for various industrial and medical uses. Fission products include the second transition metals row, Zr, Mo, Tc, Ru, Rh, Pd, Ag, and the next in the periodic table, I, Xe, Cs, Ba, La, Ce, and Nd.

Many fission waste products are either non-radioactive or only short-lived radioisotopes, while a considerable number of these are medium- to long-lived radioisotopes such as $90Sr$, $137Cs$, $99Tc$ and $129I$.

Fission waste products can modify the thermal properties of the uranium dioxide, where the lanthanide oxides tend to lower the thermal conductivity of the fuel, while the metallic nanoparticles slightly increase its thermal conductivity.

Different types of nuclear fuels produce different kinds of spent nuclear fuel (nuclear waste). Spent low enriched uranium nuclear fuel, for example, is a type of a nanomaterial. In the uranium oxide spent fuel, intense temperature gradients cause fission products to migrate. The zirconium tends to move to the center of the fuel pellet where the temperature is highest, while the lower boiling fission products move to the edge of the pellet.

The pellet is likely to contain a lot of small bubble-like pores which form during the fission cycle. Fission xenon migrates to these voids, and some of it decays to form cesium, C-137 ($137Cs$), so many of the bubbles contain a large concentration of $137Cs$.

In MOX, the xenon tends to diffuse out of the plutonium-rich areas of the fuel, and it is then trapped in the surrounding uranium dioxide, while the neodymium is not very mobile. Metallic particles of an alloy of Mo-Tc-Ru-Pd also tend to form in the fuel, while other solids form at the boundary between the uranium dioxide grains. The majority of fission products remain in the uranium dioxide as solid solutions.

Successful efforts have been led to segregate the rare isotopes in fission waste including the "fission platinoids," Ru, Rh, Pd, and Ag, as a way of offsetting the cost of reprocessing. Although this is possible, it is not economical enough, so it is not done commercially.

In all cases, spent nuclear fuel has only two possible paths: stored as a waste or ship for reprocessing. Reprocessing is uneconomical in most cases, but some fuels are showing promise, so efforts in recycling, reprocessing, and reusing spent nuclear fuels will continue and intensify.

Most spent nuclear fuels, however, are stored in large onsite water pools at the nuclear plant, where they are left to cool down for a long time. Eventually the cooled fuel rods and bundles are transferred into special canisters and shipped to an offsite storage facility.

Waste Disposal

All spent nuclear fuel is considered to be nuclear waste. Nuclear waste and its transport, storage, and disposal in particular have been of most concern to all involved—from the power plant owners to the investors, and the customers. Because of that, special attention is paid to the technology and procedures used in this process.

Low-level Radiation Nuclear Fuels

Low-level and intermediate-level wastes (LLW and ILW respectively) are generated throughout the nuclear fuel cycle and from the production of radioisotopes used in medicine, industry and other areas. The transport of these wastes is commonplace, and they are safely transported to waste treatment facilities and storage sites.

Low-level radioactive wastes are a variety of materials that emit low levels of radiation, slightly above normal background levels. They often consist of solid materials, such as clothing, tools, or contaminated soil. Low-level waste is transported from its origin to waste treatment sites, or to an intermediate or final storage facility.

A variety of radio-nuclides give low-level waste its radioactive character. However, the radiation levels from these materials are very low and the packaging used for the transport of low-level waste does not require special shielding.

Low-level wastes are transported in drums, often after being compacted to reduce their total volume. The drums commonly used contain up to 200 liters of material. Typically, 36 standard, 200-liter drums go into a 6-meter transport container. Low-level wastes are moved by road, rail, and internationally by sea. However, most low-level waste is only transported within the country where it is produced.

Intermediate-level Radiation Nuclear Fuels

The composition of intermediate-level wastes is broad, but they all require shielding. Much ILW comes from nuclear power plants and reprocessing facilities. Intermediate-level wastes are taken from their source to an interim storage site, a final storage site (as in Sweden), or a waste treatment facility. They are transported by road, rail and sea.

The radioactivity level of intermediate-level waste is higher than low-level wastes. The classification of radioactive wastes is decided for disposal purposes, not on transport grounds. The transport of intermediate level wastes take into account any specific properties of the material, and requires shielding.

In the US there had been 9,000 road shipments of defense-related trans-uranic wastes for permanent disposal in the deep geological repository near Carlsbad, NM, by October 2010, without any major accident or any release of radioactivity. Almost half the shipments were from the Idaho National Laboratory. The repository, known as the Waste Isolation Pilot Plant (WIPP), is about 700 meters deep in a Permian salt formation.

High-level Radiation Fuels

Used fuel unloaded from a nuclear power reactor contains 96% uranium, 1% plutonium, and 3% fission products (from the nuclear reaction), and trans-uranics. It emits high levels of both radiation and heat, so it is stored in water pools adjacent to the reactor to allow the initial heat and radiation levels to decrease. Typically, used fuel is stored on-site for at least five months before it can be transported, although it may be stored there long-term.

From the reactor site, used fuel is transported by road, rail or sea to either an interim storage site or a reprocessing plant where it will be reprocessed. Used fuel assemblies are shipped in Type B casks which are shielded with steel, or a combination of steel and lead, and can weigh up to 110 tons when empty. A typical transport cask holds up to 6 tons of used fuel.

Since 1971 there have been some 7,000 shipments of used fuel (over 80,000 tons) over millions of miles

with no property damage or personal injury, no breach of containment, and very low dose rate to the personnel involved (e.g. 0.33 mSv/yr per operator at La Hague). This includes 40,000 tons of used fuel shipped to Areva's La Hague reprocessing plant, at least 30,000 tons of mostly UK used fuel shipped to UK's Sellafield reprocessing plant, 7,140 t used fuel in 160 shipments from Japan to Europe by sea (see below) and 4,500 tons of used fuel shipped around the Swedish coast.

Some 300 sea voyages have been made carrying used nuclear fuel or separated high-level waste over a distance of more than 6 million miles. The major company involved has transported over 4,000 100-ton casks each carrying 8,000 tons of used fuel or separated high level wastes. A quarter of these have been through the Panama Canal.

In Sweden, more than 80 large transport casks are shipped annually to a central interim waste storage facility called CLAB. Each 80 ton cask has steel walls 30 cm thick and holds 17 BWR or 7 PWR fuel assemblies. The used fuel is shipped to CLAB after it has been stored for about a year at the reactor, during which time heat and radioactivity diminish considerably. Some 4,500 tons of used fuel had been shipped around the coast to CLAB by the end of 2007.

Shipments of used fuel from Japan to Europe for reprocessing used 94-ton type B casks, each holding a number of fuel assemblies (e.g. 12 PWR assemblies, total 6 tons, with each cask 6.1 meters long, 2.5 meters diameter, and with 25 cm thick forged steel walls). More than 160 of these shipments took place from 1969 to the 1990s, involving more than 4,000 casks, and moving several thousand tons of highly radioactive used fuel—4,200 tons to the UK and 2,940 tons to France. Within Europe, used fuel in casks has often been carried on normal ferries e.g. across the English Channel.

Following reprocessing, plutonium is transported as an oxide powder since this is its most stable form. Plutonium oxide is transported in several different types of sealed packages, and each can contain several kilograms of material.

Risk of exposure is reduced by the design of the package, limiting the amount within and the number of packages carried on a transport vessel. Special physical protection measures apply to plutonium consignments too. A typical transport consists of one truck carrying one protected shipping container. The container holds a number of packages with a total weight varying from 80 to 200 kg of plutonium oxide. A sea shipment may consist of several containers, each of them holding between 80 to 200 kg of plutonium in sealed packages.

Nuclear Waste Recycling

One of most plausible solutions to reducing nuclear waste is recycling it. To this purpose, in 2000 the US and Russia signed a bilateral agreement, committing to eliminate 34 metric tons of surplus military plutonium produced during the Cold War by recycling it as fuel for civil nuclear applications. In 2008, the Department of Energy made an agreement with a joint venture created by the AREVA and SHAW groups for the construction of a MOX fuel production plant.

The effort consists of two parts:

1. Construction and operation of a pit disassembly and conversion facility (PDCF), where nuclear warheads are dismantled and where the recovered metal is converted into plutonium oxide, and 2. Construction and operation of a fuel fabrication plant, mixed-oxide (MOX) fuel fabrication facility (MFFF), where plutonium oxide is mixed with uranium oxide to make MOX assemblies.

The 600,000 ft^2 MFFF plant is currently under construction at the Savannah River Site near Aiken, SC, and is on track to be completed by its target date of 2016. Its purpose is to reduce the surplus weapons grade plutonium and provide fuel for commercial plants. If successful, the MOX process might be the ultimate answer both politically and environmentally for a safer nuclear power industry.

It is estimated that when the MFFF operation is complete, enough electricity could be generated to power all households in South Carolina for up to 20 years. The project is overseen by the National Nuclear Security Administration (NNSA), via third-party contractors. In addition, there will be a waste solidification building, and a pit disassembly and conversion unit that are pivotal in the attempt to shrink the waste plutonium mountain.

At the MFFF facility, surplus plutonium will be processed and blended with depleted uranium oxide to make mixed oxide fuel that will be used as new fuel for nuclear plants. The waste solidification building and the pit disassembly and conversion unit are vital cogs in the wheel of using the plutonium as a resource. The waste unit is forecast to treat 150,000 gallons of transuranic waste, and approximately 600,000 gallons of low-level radioactive waste from the MFFF and pit disassembly buildings.

Note: As an additional benefit to national and international security, once the MOX fuel has been irradiated by the commercial reactors, the plutonium can no longer be used for nuclear weapons activity. Most im-

portantly, MOX facilities would provide environmental safety for future generations by converting potentially dangerous radioactive materials into safe commercial nuclear fuel.

Nuclear Waste Storage

Ideally, a nuclear power plant runs at the maximum allowed power level from one refueling to the next. It must be shut down for several hours, or days, during each refueling, and then restarted and brought up to the maximum power output. Used nuclear fuel, which is removed from the reactor after a year or two of service, is a solid material that must be stored safely, usually at nuclear plant sites. This temporary storage is only one component of an integrated used fuel management system in use at all nuclear plants, addressing all facets of storing, recycling and disposal.

The integrated used fuel management approach mandates that used nuclear fuel will remain safely stored at nuclear power plants for the near term, with safety as the key word. Hopefully, the government will find a way to recycle it eventually, and place the unusable end product in a deep, long-term repository.

Low-level wastes are byproduct remaining from uses of a wide range of radioactive materials produced during electricity generation, medical diagnosis and treatment, and various other medical processes. These could be liquids or solids, and are treated in a number of ways prior to disposal as hazardous waste, burning, or storage in temporary containers.

Recycling of waste nuclear fuels is a program of the federal government, which includes plans to develop advanced recycling technologies to take full advantage of the unused energy in used fuel, and to also reduce the amount and toxicity level of byproducts requiring disposal.

Transportation of waste nuclear fuel is the responsibility of the U.S. Department of Energy, which will transport used nuclear fuel to the repository by rail and road, inside massive, sealed containers that have undergone safety and durability testing.

Repository for long-term storage is under review, and under any used fuel management scenario, disposal of high-level radioactive byproducts in a permanent geologic repository is necessary.

Note: The criticality of nuclear waste storage is reflected in the roughly 53 million gallons of nuclear waste stored in 177 large underground tanks at DOE's Hanford Nuclear Reservation in Washington State. Of these, 149 are more than 40 years beyond their expected 25-year design life. One-third of them are known or suspected to be leaking, releasing roughly 1 million gallons of waste to Hanford's surrounding soils.

Hanford lacks the storage capacity to retrieve the waste from these tanks until the waste treatment and disposal process is underway. Washington's $12.3 billion Waste Treatment Plant (WTP) continues to be designed and constructed to meet standards specific to the Yucca Mountain facility. Design and engineering for the WTP is 78% complete and construction is 48% complete.

In 2002, Congress designated Yucca Mountain as the nation's sole current repository site for deep geologic disposal of high-level radioactive waste and spent nuclear fuel. At that time, the Secretary of Energy concluded that, "The amount and quality of research the DOE has invested...done by top-flight people...is nothing short of staggering...I am convinced that the product of over 20 years, millions of hours, and four billion dollars of this research provides a sound scientific basis for concluding the site can perform safely."

Congress then directed DOE to file a license application for the Yucca Mountain site with the Nuclear Regulatory Commission (NRC) and thereby commence a formal evaluation and licensing process overseen by the NRC. But...in January 2010, President Obama, Secretary Chu, and DOE determined that they would withdraw with prejudice the application submitted by DOE to the NRC for a license to construct a permanent repository at Yucca Mountain, NV, for high-level nuclear waste and spent nuclear fuel.

Also that month, President Obama, Secretary Chu, and DOE chose to unilaterally and irrevocably terminate the Yucca Mountain repository process mandated by the Nuclear Waste Policy Act, 42 U.S.C. §§ 10101-10270. Several law suites were filed as a result, but these will take years to resolve, during which time the US has no place to store its large pile of nuclear waste produced during power generation and nuclear weapons production.

The Promised Land...

In Finland, a radical solution is planned to build an enormous bunker for permanent storage of the dangerous radioactive waste. It consists of burying the stuff deep underground, sealing the depository and throwing the key away. Literally. The intent is to keep everyone—including future civilizations—away by hiding the nuclear waste somewhere so unremarkable and unpleasant that nobody would ever think to go there, let alone dig into it.

On Olkiluoto Island, just off Finland's southwest coast, the underground facility known as Onkalo will hold all of the country's 5,500 tons of nuclear waste—all

that is expected to be produced by the end of the century. It is designed to keep that waste secure for at least 100,000 years, taking measures and counting on making humans forget that Onkalo was ever there.

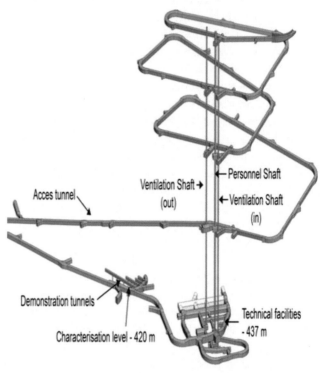

Figure 6-11. Onkalo nuclear waste depository

Onkalo is intended to permanently and safely store high-level waste (HLW) which consists of spent nuclear fuel and some equally dangerous decay products. This residue emits dangerous types and levels of radiation for tens of thousands of years. Over 300,000 tons of this stuff now exist around the world and about 12,000 more tons are produced annually. It is also expected that numbers will increase significantly in the years to come. Onkalo boasts a maze of deep underground bunkers carved from impermeable rock, in geologically stable zones, where the waste can be redundantly sealed and then permanently buried.

At a $3 billion price tag thus far, Onkalo will start accepting nuclear waste in 2020, while construction of new tunnels will continue as needed until the facility is shut down and sealed forever.

Decommissioning of Nuclear Plants

The U.S. power plants shutdowns to date include:
19,772 MWe coal-fired power plants,
12,167 MWe natural gas-fired,
6,793 MWe oil-fired, and
3,554 MWe nuclear power plants,

Many power plants, including most of the existing nuclear, are over half a century old, and needs to be refurbished or decommissioned. Refurbishing has been done several times on most of them, and as the safety requirements get tighter, decommissioning seems more likely.

Yes, decommissioning of nuclear (and all other types of) power plants is shaping as a huge market. It is expected to grow at a CAGR of 14.1% by 2018. This is a lot of growth and a lot of decommissioning work. Decommissioning costs of nuclear plants are significantly higher and much different than those of other power plants. This is due to the complexity of the technology, and also because of the added danger of nuclear radiation at every step of the way. This means that the demolition crew must be well trained in and thoroughly familiar with radioactive components and materials. Every action could result in the unexpected release of radiation, which could be damaging and even fatal.

Decommissioning involves a number of administrative and technical actions, including inspection, clean-up of radioactivity, demolition of the plant, and removal and disposal of components. There are contaminated materials, parts and components that must be handled with utmost care.

Engineering services and labor used in decommissioning of nuclear plants can be described as specialized services, starting with concrete and metal cutting and removal, including:

- Man (hole) access cuts
- Equipment hatch enlargement
- Containment wall penetrations
- Dog house cutting for SGRPs
- Elevated platform removals
- Fuel pool and canal segmentation
- Fuel transfer canal segmentation

There are a number of tasks performed by experienced professionals during decommissioning of a nuclear power reactor, such as:

- CRDM cutting
- Reactor nozzle cutting
- Heat exchanger cutting
- Ion guide tube removals
- Monitoring line cutting
- Steel plate stabilizer cutting
- Carbon containment liner cutting

Hands-on demolition services require highly specialized and experienced workers, thoroughly trained in understanding nuclear plant components, their func-

tion, and safe handling. Only a well-trained, committed workforce attuned to personal radiation and toxic materials exposure and waste minimization awareness could be allowed in such an undertaking. Anything less might result in environmental and personal damage and even death.

Decommissioning costs represent about 10% of the nuclear plant's initial capital investment cost.

When discounted, the numbers go down and contribute only a few percent to the investment cost and even less to the generation cost. For example, in the US, decommissioning costs are estimated at average 0.1-0.2 cent/kWh, which is about 5% of the cost of the electricity produced. Decommissioning and land remediation is a complex and expensive undertaking.

The dismantling process could cost from $5 million to $10 million. Handling and removal of the fuel could cost another $5-10 million. The entire decommissioning could be in the range of $25 million to $1 billion. In some cases there is an additional long term charge for upkeep of the area, which could amount to over $10 million annually. And yet, these are only some of the costs…if everything goes well. Additional and unforeseen, costs could pile up very quickly too, amounting to many billions of dollars.

One can only hope that all these costs were foreseen and estimated in the overall operating budget of the plant. Once a nuclear facility is decommissioned properly, there is no longer danger of radioactivity. After official inspection and certification, the area is released from regulatory control, and the owner of the plant is no longer responsible for the nuclear safety of the area.

NUCLEAR POWER MARKET

First, let's remember that the term *Nuclear Power Market* includes anything and everything related to nuclear power; starting with nuclear materials mining and processing; nuclear plants design, construction, and O&M; as well as the distribution and sale of electricity coming from nuclear plants. This also includes anything and anybody participating in this long process; consultants, advisors, lobbyists, accountants, lawyers, insurance agents, law makers, regulators, etc., etc. people, companies, and entities.

There are roughly 1100 nuclear reactors operating today around the world. Of these, 430 are large reactors of around 1.0 GW of electricity generation capacity each.

Table 6-4. Nuclear power generation (in Mtoe)

USA	22.7
France	10.5
Japan	4.6
Russia	4.3
Germany	2.9
S. Korea	2.8
Canada	2.6
Ukraine	2.2
China	2.0
Sweden	1.7

The USA has the largest installed nuclear power capacity—over one third of the total global capacity. Our nuclear fleet consists of 100 power plants with over 100,000 MWe of nuclear power capacity currently in operation. France is close behind, with over 60,000 MWe of capacity.

The difference is that while nuclear in the U.S. accounts for less than 20% of the total power generation mix, in France, nuclear holds a total of 75% of the electric market share. On top of that, a significant amount of thus generated nuclear power is exported to Italy, Germany and other countries, as well. Of all countries, France has most to lose in case of total nuclear power shut down in the future.

Increasing concerns about the use of conventional energy sources are forcing developed economies to depend on nuclear energy for power generation in order to satisfy their power demands. Due to its energy efficiency, relatively low price, and sufficient availability, nuclear power is used for electricity generation across the world. Compared to other power generation technologies, nuclear power has the minimum impact on greenhouse gas emission, which adds to its attractiveness.

The nuclear power industry has been in the headlines and going through hard times recently, but there are signs of growth in the past few years. Significant technological development in reactor technology, emphasis on safety measures, and zero carbon emissions from nuclear power plants are some important factors driving the growth of this market.

Nuclear disasters in the past few years have questioned the safety of using nuclear power. Nevertheless, the global nuclear power market is expected to grow at the compounded annual growth rate of 4% until 2020.

The market segmentation based on different technologies is as follow:

Generation II Technology
- Pressurized Water Reactor (PWR)
- Boiling Water Reactor (BWR)
- Pressurized Heavy Water Reactor (PHWR)
- Canada Deuterium Uranium (CANDU) Reactor
- Fast Breeder Reactor (FBR)
- Advanced Gas-Cooled Reactor (AGR)

Generation III Technology
- Advanced Boiling Water Reactor (ABWR)
- Advanced Pressurized Water Reactor (APWR)
- Advanced Heavy Water reactor (AHWR)
- European Pressurized Water Reactor (EPR)
- Fast Neutron Reactors
- Pebble Bed Modular Reactor (PBMR)

Generation IV Technology
- Gas-cooled Fast Reactors
- Lead-cooled Fast Reactors
- Molten Salt Reactors
- Sodium-cooled Fast Reactors
- Supercritical Water-cooled Reactors
- Very High-temperature Gas Reactors

Generation V Technology
- Fusion technology, *and*
- Future developments

The nuclear power market is spread over the major global geographies. The key players in it—equipment manufacturers, construction contractors, power generators, etc.—today shown in Table 6-5.

The key nuclear energy market driver today is the growing energy demand, while the key market challenges are the high initial cost and the threat of nuclear accidents.

Nuclear power—like the other energy markets—is a truly international undertaking. What makes it unique in this, and many other respects, is its awesome powers that cross national and international boundaries. Literally!

Raw materials and components are imported and exported, huge amounts of nuclear power are traded, especially in Europe. And, of course, we all share the effects of occasional (but devastating) nuclear accidents. The combination of the best vs. the worst nuclear power has to offer is in the foundation of the nuclear energy markets and one does not go without the other.

The international aspect of nuclear power was demonstrated also after the Chernobyl nuclear accident,

Table 6-5. Major nuclear power companies

— Ameren Corporation
— Areva Inc.
— Atomic Energy of Canada
— Bharat Heavy Electricals
— BKW FMB Energie AG
— Bruce Power Inc.
— China National Nuclear Corp. (CNNC)
— China Power Investment Corporation
— Chubu Electric Power Company
— Chugoku Electric Power Company Inc.
— Constellation Energy Group Inc.
— Duke Energy
— EDF Electricite de France SA
— Entergy Corporation
— Eskom Holdings Limited
— Exelon Corporation
— GE Hitachi
— General Electric Company, USA
— JSC Atomstroyexport
— KEPCO
— Larsen and Toubro Ltd.
— Mitsubishi Heavy Industries
— ROSATOM
— Siemens AG
— Westinghouse Electric Co. LLC, and others

when plumes of nuclear radiation drifted over Western Europe in its wake. We surely need its cheap, clean, and reliable power for our daily routines, but at the same time we fear this same power that can easily terminate our existence in a split second. This makes the nuclear energy markets quite complex and full of controversies, excess regulations, and the need for special materials, equipment, and procedures.

Nuclear Renaissance

A pending nuclear renaissance—still using fission reactors—has been in the news for as long time. This trend (dream for now) describes ways for a possible return to the nuclear power industry's glory days. Global economics and rising fossil fuel prices would be the drivers, while rising concerns about greenhouse gas emission limits and global warming would provide the support for unlimited nuclear progress. Sounds good on paper. Unfortunately, there are a number of unresolved issues on the way to a full nuclear renaissance.

Some of the key issues slowing down the nuclear progress are:

• Political currents and tight regulations

• Serious delays in permitting and siting of new power plants,

• Unfavorable economics and safety record as compared to other sources of energy,

• Industrial bottlenecks and personnel shortages in the nuclear sector,

• Uncertainty about what to do with nuclear waste and spent nuclear fuel,

• Lingering national security issues related to nuclear plants, fuels, and weapons,

• The increasing threat of nuclear terrorism,

• The increasing threat of nuclear weapons proliferation. and

• The fear of additional nuclear accidents, such as Chernobyl,

Memories of the nuclear accidents at Fukushima I Nuclear Power Plant, Chernobyl, Three Mile Island and other nuclear facilities linger in people's minds and restrict the introduction of new nuclear programs around the world. These recent developments raise serious questions about the future of nuclear power, and the nuclear renaissance is on hold for now, with some evidence of a nuclear reversal underway in some places.

Germany led the nuclear reversal trend when, following the 2012 Fukushima nuclear disaster, it decided to shut down all nuclear power plants in the country by 2020, while reviewing the safety of the nuclear energy in general. Several countries followed Germany's example to a certain degree, but for a short time. Several new nuclear reactors planned for construction in European countries (which were paving the way toward nuclear renaissance) have been delayed or cancelled.

There are also estimates that 30 nuclear plants would be closed worldwide in the near future. Those located in seismic zones, or close to national boundaries, are the most likely to shut down first. Switzerland, Israel, Malaysia, Thailand, United Kingdom, Italy and the Philippines are all reviewing the safety of their nuclear power programs.

At the same time, Australia, Austria, Denmark, Greece, Ireland, Latvia, Lichtenstein, Luxembourg, Portugal, Israel, Malaysia, New Zealand, and Norway remain unshakably opposed to nuclear power.

Indonesia and Vietnam are the exception, and are still planning to build nuclear power plants in the near future. China is an exception, with 27 new nuclear reactors under construction. A number of new nuclear power plants are being built also in South Korea, India, and Russia.

The IEA reduced its estimate of additional nuclear generating capacity built by 2035 by more than 50%.

On top of that, about a hundred older reactors will need to be decommissioned by 2025 around the world due to age, and many nuclear power programs are running over-budget and out of time. Even a pro-nuclear country like France is planning to close at least two reactors to demonstrate political action and restore the public acceptability of nuclear power.

As a confirmation of the upcoming debacle, Siemens (one of the giants in the industry) recently decided to withdraw entirely from the nuclear industry.

In the U.S., Exelon Corporation, the nation's largest nuclear operator, threw in the towel on a planned twin-reactor project in Victoria County, TX, in the face of fierce opposition from people who claim that there is not enough water in the area and that the ground was subject to subsidence that could wreck a cooling pond. The Nuclear Regulatory Commission might have approved the site over these objections, so Exelon finally admitted that, in addition to the resistance, the economics were not favorable, so it would be better to give up on the project.

Two years ago the Unistar consortium fell apart when it could not obtain a loan guarantee from the Department of Energy on terms acceptable to Constellation Energy, which was later on bought by Exelon. Two new nuclear plants, each with two reactors, are underway, one in Georgia and one in South Carolina, but permitting and financial issues make us believe that this will be another long and expensive process.

At the same time, a panel of administrative law judges ruled that Électricité de France (EDF) could not proceed with a plant in Calvert Cliffs, Maryland. That plant was originally a joint venture between Constellation Energy, which owned the adjacent Calvert Cliffs 1 & 2, and the French EDF.

These developments show how weak and vulnerable the nuclear industry is. Its future is equally fragile, for it's determined by many factors. It could grow very fast, if things line up in its favor. Or it could be blown away like a paper tower, if and when the winds of change blow against it. A tsunami, much bigger than the one that devastated Fukushima, could hit the global nuclear industry if and when another accident occurs. Let's hope that this won't happen, because it might be the end of nuclear power as we know it today.

The Beginnings… U.S.

In the U.S., commercial nuclear power development started after WWII, when Westinghouse designed the first fully commercial pressurized water reactor (PWR). It was very small to today's standards, of merely 250 MWe capacity. Westinghouse's Yankee Rowe reactor started operation in 1960 and operated successfully until 1992, when it was decommissioned as unsafe and obsolete.

The next step up in the nuclear development process was the boiling water reactor (BWR), which was developed by the Argonne National Laboratory. A prototype BWR reactor, named Vallecitos, was installed in 1957 and ran until 1963. Its first cousin was the first commercial nuclear power plant in the U.S. and the world. Dresden 1 was a small nuclear power plant of only 250 MWe. It was designed by General Electric and the construction was completed in 1960.

Following the success of the prototypes, the construction of several larger PWR and BWR reactors started by the end of the 1960s, and many more were planned. Most of these units were 1.0 GWe capacity, or larger. A major nuclear power plants construction program got under way in the USA in the 1970s. During a short period of time, the U.S.' nuclear industry was born. The reactors build during that time are still in operation are the foundation of the U.S. nuclear industry.

All was well and nuclear was taking over the country by a storm…until the 1979, when the Three Mile Island nuclear accident shook the country and world to the reality of the dangers behind nuclear power. People were shocked that nuclear reactors are that vulnerable to accidents, caused by natural forces and human error. Although no one was injured or even exposed to harmful radiation during and after the accident, the perception of the safety (or lack thereof) of nuclear power was born. It persisted and grew with each accident that followed.

Shortly thereafter things changed for the worst for the nuclear industry. Most orders for new nuclear plants and projects were cancelled or suspended. During the 1980s and 1990s, the nuclear construction industry went into the doldrums for two decades.

Nevertheless, as a result of the many construction projects underway since the 1970s, over 100 commercial power reactors were commissioned and under full operation in the U.S. by 1990. Nuclear power regulations and increasing concerns put the U.S. nuclear industry in a corner, where it is still sitting and sulking today. The costs of running the nuclear plants increased with time, too but the regulated utilities who own most of the nuclear plants simply increased their rates gradually to cover the expenses. Customers are not very happy with that and the wave of discontent overflows from time to time.

A lot of water has run under the bridge since the 1970s, and the US nuclear industry dramatically improved the safety and efficiency of its operations. By 2000, it was among the world leaders in safety and efficiency. There have not been any nuclear incidents in the U.S. since 1979, with all safety indicators constantly exceeding the targets.

Most U.S. nuclear power plants run at average net capacity factor of over 90%, and their safety is estimated at 100%

90% capacity and 100% safety factors are the highest possible expected from any power plant today. The U.S. energy industry continued the path of deregulation, which begun with the passage of the Energy Policy Act in 1992. Changes accelerated after that, included mergers and acquisitions affecting the ownership and management of most nuclear power plants. The trend continues, but the future is still unclear.

The U.S. Nuclear Industry Today

Today, the U.S. is the world's largest producer of nuclear power, accounting for more than 30% of worldwide nuclear generation of electricity. The country's nuclear reactors produced 789 billion kWh in 2013, nearly 20% of the total electrical output of the country. There are presently 100 units operable and five are under (delayed) construction.

Following a 30-year period, during which a few new reactors were built, six new units are expected to come on line by 2020. Four of those are the result of 16 license applications made since mid-2007 to build 24 new nuclear reactors.

Today, however, excess of low price energy sources (natural gas in particular) are putting the economic viability of some existing and proposed nuclear reactors in doubt. This, in addition to increased safety concerns since Fukushima's nuclear disaster, have a great negative effect on the U.S. nuclear industry and represent great obstacles in its future development.

Government policy changes somewhat since the late 1990s, mostly in support of nuclear power generation, which have helped to pave the way for significant growth in nuclear capacity. The developments in the nuclear industry are a testament of government and industry working closely on all aspects of the business. This

cooperation—from expedited approval for construction to implementing new plant designs, as well as finding new ways to store nuclear waste—is a good example of a successful business model.

Note: Without government support, the U.S. nuclear industry could not have achieved even half of what it is today. As long as this support continues, the nuclear industry will be successful. Government support in key areas of the industry; raw materials procurement, regulations, safety equipment and procedures development and implementation, and waste transport and storage, are fundamental and indispensible. Without government support in these areas, the U.S. nuclear industry would not be what we know it as today, and might even fall on its face.

Figure 6-12. U.S. nuclear fleet

Presently, the U.S. operates 100 nuclear power reactors in 31 states, run by 30 different power companies. Seven of the power plants in Figure 6-12 are expected to be in full operation by 2020.

Since 2001, the U.S. nuclear plants have achieved an average capacity factor of over 90%, generating over 800 billion kWh per year and accounting for 25% of total electricity generated. Capacity factor has risen from 50% in the early 1970s, to 70% in 1991, and it passed 90% in 2002, remaining at around this level since. In 2013 it reached 91%. The industry invests about $7.5 billion per year in maintenance and upgrades of the operating power plants.

The U.S. nuclear industry operates 65 pressurized water reactors (PWRs) with combined capacity of about 64 GWe, and 35 boiling water reactors (BWRs) with combined capacity of about 34 GWe–for a total capacity of 98,951 MWe.

The Government's Support and Control

Commercial nuclear power in the U.S. is a brain child of the Cold War's race for more powerful and larger arsenal of nuclear weapons. A lot of research and effort was spent by the government's labs and military installations on understanding the nature of the nuclear reaction, its application, and the related positive and negative effects.

The negative effects, in terms of mass destruction, were packed into A-bombs and ballistic missiles armed with nuclear charge, which could obliterate an entire city upon impact. During all this time (1940-1960) thousands of people were involved in, and billions of dollars were spent on, intensive nuclear research and development effort. In addition to figuring out the destructive side of nuclear power, a significant effort was undertaken to understand and implement the positive effects of nuclear power, in form of electric power generation for civilian use.

No other sector has benefitted so thoroughly by government's political, technical, and financial support, which continues even today and which will surely continue in the distant future.

Many steps of the nuclear power cycle are still under government control, or are actively supported one way or another. Research into new and more efficient nuclear technologies, for example, continues, while nuclear waste disposal is almost completely under government control. These and other activities require billions of dollars to be spent by the government in support of the nuclear industry.

The U.S. government is still responsible for, and provides solutions to, all nuclear issues as needed to run the nuclear power industry.

The Problems

Almost all of the U.S. nuclear generating capacity comes from reactors built between 1967 and 1990. There had been no new construction starts from 1977 until 2013, largely because construction schedules during the 1970s and 1980s had frequently been extended by opposition, compounded by heightened safety fears following the Three Mile Island accident in 1979.

Another reason of late, for the nuclear power construction slow down, is the fact that natural gas generation is now considered more economically attractive. While there are plans for a number of new U.S. reactors, the prospect of low natural gas prices continuing for several years has dampened these plans and only four new units have a real chance of coming on line by 2020. The first of these, the new PWR, Watts Bar 2, is expected to be operational in 2016.

Despite a near halt in new construction of more than 30 years, US reliance on nuclear power has continued to grow. In 1980, nuclear plants produced 251 billion kWh, accounting for 11% of the country's electricity generation. In 2008, that output had risen to 809 billion kWh and nearly 20% of electricity, providing more than 30% of the electricity generated from nuclear power worldwide. Much of the increase came from the 47 reactors, all approved for construction before 1977, that came on line in the late 1970s and 1980s, more than doubling U.S. nuclear generation capacity. The US nuclear industry has also achieved remarkable gains in power plant utilization through improved refueling, maintenance and safety systems at existing plants.

All this is fine, but all good things eventually come to an end. The U.S. nuclear fleet is no exception. It is getting old and cannot be expected to remain safe and efficient forever. And so, in February 2013 Duke Energy's 860 MWe Crystal River PWR in Florida was decommissioned due to damage to the containment structure sustained when new steam generators were fitted in 2009-10, by the previous owner. Its 40-year operating license was due to expire in 2016 and some $835 million in insurance coverage was claimed as a result of the closure.

Dominion Energy's 566 MWe Kewaunee PWR in Wisconsin was decommissioned in May 2013, after 39 years operation. Then in June 2013, the two 30-year old PWR reactors (1,070 & 1,080 MWe) at San Onofre nuclear plant in California were retired permanently, due to regulatory delay and uncertainty following damage in the steam generators of one unit. In August 2013 Entergy announced that its 635 MWe Vermont Yankee reactor would be closed down at the end of 2014 as it had become uneconomic. Ten other nuclear plants, with total of 13 reactors, are considered for closure today too. Most of these are located in deregulated states in the northeast USA.

There are lots of factors giving rise to the uncertainty surrounding nuclear power, but the major ones are:

• High construction costs and low power price returns,

• Complex and expensive to deal with regulatory issues, and the most important

• Ever present concerns with safety of nuclear power plants.

Table 6-6. U.S. Power generation (in GW, 2014)

Coal	1,510
Gas	1,236
Nuclear	770.1
Hydro	276.2
Renewables	240.4
Oil	22.2

Table 6-6 tells us that nuclear power generation is almost half of that generated by natural gas, and that the renewables (including hydropower) are approaching the amount of electric power generated by nuclear. The difference here is awesome; while it took nuclear power over 50 years to achieve this capacity, the renewables grew this fast in less than a decade. Also, they (the renewables) used less than 1% of the support (subsidies, etc.) that nuclear has been, and still is, using.

Coal still occupies the lion's share of the U.S. electricity generation mix, and is expected to remain so until 2035, even though about 50 GWe of coal-fired capacity is expected to be retired by 2020. The main reasons for the shut downs are growing environmental constraints and coal's low efficiency, coupled with a continued drop in natural gas prices. Natural gas-fired capacity will increase proportionally during this time to replace coal. The renewables will also grow, but at a much slower pace for a number of reasons, as we discuss in detail in this text.

The nuclear power's future in the country is still uncertain. It depends on many factors, as discussed above, so we cannot even start guessing. In best of cases, it will increase by 10-12 GW by 2030. In worst of cases, it would decrease by that amount during the same time. There is also the doomsday scenario, where all nuclear power plants will be shut down, in case of another nuclear accident, as it happened in some countries following the Fukushima nuclear accident.

Global Competition

Globally, around 1/7th of all electricity is generated by nuclear power plants.

Table 6-7. Installed nuclear capacity

Location	ktoe	GWh
All Europe	4.6	53.5
North America	3.2	37.2
Asia	2.6	30.3
Latin America	.08	0.93
Africa	.05	0.58
MENA	.03	0.35

Table 6-7 illustrates the discrepancy among the developed and the developing nations, with MENA being the exception (a fairly developed country with very little nuclear power). Recently, however, we see a new trend; that of serious competition among the nuclear and the nuclear-wannabe countries. Increased costs for today's new generation of nuclear plants, for example, are due in large part to fierce worldwide shortage of, and competition for, the resources and manufacturing capacity needed in the design and construction of new power plants. The competition, due to lack of expertise of this dying art, and double digit annual increases in the costs of key power plant commodities such as steel, copper and concrete, is hindering the full expansion of nuclear power.

Worldwide demand is straining the limited capacity of EPC (Engineering, Procurement, and Construction) firms, equipment manufacturers, and other firms specialized in nuclear products and services.

There is a limited number of nuclear equipment manufacturers and consumables suppliers today.

This causes bottlenecks in construction projects when there are multiple orders for new power plants in a certain location. There are, for example, only two companies in the entire world that have the heavy forging capacity to create the largest equipment and components in new nuclear plants—Japan Steel Works and Creusot Forge in France. The demand for heavy forgings is significant because it is on demand by many other industries too. As such, the nuclear industry will be waiting in line for the materials and equipment, alongside the steel and petrochemical industries as a result of the increasing demand for new equipment in different production facilities and refineries.

Get this: Twenty years ago there were about 400 suppliers of nuclear plant components and 900 so-called nuclear stamp, or N-stamp, certifications from the American Society of Mechanical Engineers. Today there are fewer than 80 suppliers in the U.S. and fewer than 200 N-stamp certifications. Five-fold decrease of available services in two decades pretty much describes the state of the nuclear industry in the country.

It appears that the U.S. nuclear industry will be forced to rely on overseas companies to manufacture plant systems and components. This will complicate things, as the NRC would need to make a great effort to inspect the quality of the manufacturing programs in foreign firms, in order to ensure that substandard materials or equipment don't end up installed in the new U.S. nuclear plants. This would require more time to inspect foreign-made components than it would to check quality control of U.S.-manufactured components. The heavy reliance on overseas suppliers will also lead to significant delays and cost increases, due to added services and the continuing weakness of the U.S. dollar relative to other currencies.

Political uncertainty, regulatory changes, environmental pressure, and other such factors are overwhelming the industry as well, and it needs to come up with proper response soon, if it is to show progress in the 21st century.

Standardization would be very helpful in this area too, for it will force all global manufacturers and providers of commodities and services to operate in a uniform and reliable manner. This, however, is far in the future, so for now we must do things the old and proven ways.

Above we only scratch the surface of potential problems and difficulties in front of the U.S. nuclear industry, which, in addition to the technical issues, are making the already complex situation even more so.

China's Nuclear Power

In comparison to the rest of the world, China is in the midst of an largest energy revolution ever, led by a strong nuclear renaissance. China is quickly becoming self-sufficient in reactor design and construction, as well as in other aspects of the fuel cycle. It is quickly shaping up as the emerging giant of the nuclear industry.

Newly planned nuclear plants in China include some of the world's most advanced reactors, which (according to Party plans) will give a five- or six-fold increase in nuclear capacity to 58 GWe by 2020, then possibly 200 GWe by 2030, and 400 GWe by 2050. This, if implemented in practice, would make China the undisputed global leader in nuclear power generation.

The Chinese made AP1000 reactor units have substantially lower costs, which is mostly due to significantly lower labor rates. For example, the cost for the first two AP1000 units under construction in China was $5.3 billion. An additional four AP1000 reactors constructed in China were estimated to cost a total of $8 billion. This is about $2,000 per kW installed capacity, which is much lower than that in the U.S.

Construction costs in China are expected to fall even further once full-scale mass production is underway. Another domestic CAP1400 reactor design, based on the AP1000 model, started construction in 2013 with scheduled completion in 2017. Once the CAP1400 de-

sign has been installed and proven, work is scheduled for a CAP1700 design with a target construction cost of $1000 per kW installed. Wow…

Mainland China presently has 16 nuclear power reactors in operation, 27 under construction, and even more in planning stages, or about to begin construction. This includes China's most ambitious nuclear project, started in the fall of 2012, when China resumed construction on a "fourth generation" nuclear power plant. It was actually started in 2011, but the construction was suspended in the wake of the 2011 Fukushima disaster.

The Shidao Bay nuclear plant in Rongcheng, a city in eastern China's Shandong province, resumed in November, 2012, and is planned to be China's biggest nuclear project. The 6.6 GW reactors will be cooled by high-temperature gas and will also become the world's first successfully commercialized 4th generation nuclear technology, designed as the safest and most cost effective plant in the world.

Official plans, however, do not specify the level of safety of this overgrown nuclear giant. Nuclear safety information, like all important information in semi-communist China, is kept secret for as long as possible. Human life is undervalued in China, and quality is not always up to par, so we would not be surprised if we see some of the greatest nuclear accidents there in the future.

Is China going to bring the nuclear industry out of its slumber, or would it bring us Fukushima on steroids accidents instead?

The Major Issues

Nuclear power reactors have a number of notable advantages. They use much less raw materials than any other competitor. In a 1 GW nuclear power station, for example, 25 tons of uranium would produce as much electricity in a year as around 2.3 million tons of coal. This is a factor of about 25 to 2.3 million, or almost 1 to 230,000. When everything considered, however, only small part of the uranium ore is U-235 (fissible), so the true weight factor of uranium over coal power generation is about 10,000. Still a huge number; energy materials for 1 nuclear plant vs. that for 10,000 coal-fired such.

The key nuclear power advantages:
- Nuclear plants are very efficient as far as conversion of small amount of material into useful energy is concerned. This makes their O&M expenses much lower than conventional power plants.

- Nuclear plants have much lower emissions than any of the competitors. This includes air, soil, and water pollution. One ton of high level radioactive waste (after reprocessing) produced by a nuclear plant equals about seven million tons of carbon dioxide and sulfur dioxides, plus around 150,000 tons of ash and sulfur solids. To this we must also add the emissions due to mining, transport and milling. Here again, we see a ratio of about 1 to 10,000 in favor of uranium.

But nuclear power also has a number of disadvantages too:

- Nuclear waste is highly radioactive and although it is in much smaller quantities than coal, its handling, storage and processing and processing are very cumbersome, dangerous, and expensive. Today radioactive waste is stored in special canisters, which are left for eternity in "temporary" storage facilities. The problem is that these facilities are actually becoming permanent, since no decision has yet been taken on what to do with the growing waste piles. The most accepted plans are to bury the nuclear waste deep underground inside steel canisters, or to encapsulate it in corrosion resistant metals vessels made out of copper or lead.

- Nuclear proliferation is a threat to the global safety and peace. Spent nuclear fuel contains large quantities of weapons-grade plutonium, which can be used to make dirty bombs. Each reactor produces 500-600 lbs. of this material annually, which is enough to produce about 50-60 bombs. This material can be used by terrorist groups or rogue governments to position themselves in developing nuclear weapons capability for achieving their objectives.

- Centrifuge plants are compact and readily concealed, and a centrifuge plant can switch quickly from the production of reactor-grade to bomb-grade fuel. According to one expert, "Since the entire reactor can be emptied within days, by the time you arrive to verify the declared inventory of fuel elements, which power the reactor, all the evidence of illicit irradiations to obtain plutonium could be covered up."

Considering that most of the world's enrichment plants are not safeguarded, and seem highly unlikely to come under IAEA (International Atomic Energy Agen-

cy) safeguards for the foreseeable future, the threat is real and present.

- Uranium mining is very dangerous work, which results in millions of tons of radioactive tailings or waste materials. Uranium tailings retain about 85% of the radioactivity of the ore and pose a major waste management problem. The majority of the in-situ leaching projects for obtaining uranium are using sulfuric acid in the process, with the residual leaching solutions from uranium mines migrating in the local environment.

In Bulgaria and the Czech Republic, these solutions have led to contamination of good quality groundwater systems feeding local towns. Similarly, there are large areas in the U.S. that are contaminated by nuclear waste products from past nuclear research projects.

- Indirect CO_2 emissions from mining, transport, and refining of nuclear materials are estimated to release 4-5 times more CO2 than equivalent power production from renewable sources. So, nuclear power is clean at the last stages of its life-cycle, but is quite dirty in the beginning stages.

- Nuclear plants have a history of financial boon-doggles. They require a much higher level of precision during construction, operation, and maintenance than that of most other types of power generating facilities. Cutting corners during financial hard times is a concern that the nuclear industry cannot justify, let alone control.

- Safety issues are part of the nuclear industry and its Achilles Heel. The Chernobyl nuclear accident, for example, contaminated 160,000 square kilometers of land in Europe, and displaced nearly half a million people. The radioactive plum floating over the land led to the illness and premature deaths of incalculable numbers of people.

The claims that the 'safer' western-style, supposedly more advanced and safer, reactors are not prone to Chernobyl-like disasters were disproven by the near disasters at Windscale-Sellafield in the UK, Three Mile Island in the U.S., and at the Monju reactor in Japan. That clam was totally proven wrong once and forever by the Fukushima nuclear accident, the damage from whose western technology is even greater than that from the inferior Russian technology used in Chernobyl.

- Finally, a team of researchers found in 2015 that the federal government had never fully revealed the true toll of what working around nuclear materials has done to men and women at nuclear facilities. In 2001, the U.S. government set up a compensation fund for some of the workers hurt by nuclear exposure. It appears now that more than 33,000 of them have died from nuclear exposure and related illnesses. This is a significant number—more than four times the number of Americans killed in the Iraq and Afghanistan wars—who died heros. Instead of being given medals of honor, they were hidden from the public and quickly forgotten.

In addition, more than 100,000 Americans, contractors and consultants, working at nuclear facilities have been diagnosed with cancer and other diseases. These people working at over 300 nuclear related facilities around the country did everything from pipe fitting and production work. There were also white-collar work, nuclear physicists and scientists, and even some executives who ran contracting companies that managed the work at the plants.

And of course, we must remember the multitude of people killed, hurt, and displaced after the Chernobyl and Fukushima nuclear disasters.

In summary, there have been significant advances in reactor design in the past, but a number of questions still remain unanswered. The economics of nuclear power plants, as well as the impacts of uranium mining, nuclear waste storage, overall process safety, and nuclear proliferation must be answered and resolved before nuclear is considered safe and welcomed. Until then, we will have to keep our eyes wide open for another Fukushima, hoping that the next one is far away from home.

Nuclear Plants Standardization

Evaluating all issues discussed above, we consider global standardization as one of the major tasks ahead of the nuclear industry. Presently, nuclear plants construction and operation is much like a well hidden black box. There are good reasons for this, but these reasons combined are hindering the progress of the nuclear industry.

Because of the 'black-box' mentality, we now have multiplicity of specialized and customized reactor designs, governed by different regulatory approaches and licensing requirements. As in many other cases, this lack of uniformity in equipment and procedures has the effect of increasing cost and uncertainty of operations, and is far from being optimally conducive to nuclear plant efficiency and safety.

Also affected are investment and political decisions, which depend on the level of risk manageability. Future standardization, resulting in transparent and predictable licensing processes and oversight, would contribute significantly to a stable investment framework and contribute to a rapid, efficient and orderly expansion of nuclear power worldwide. The concept of standardization need not cover every single detail in a nuclear plants' operation.

To begin with, all that is needed is sufficient standardization to:

1. Enable the owner to prepare standardized specifications for the procurement of new plant equipment, and
2. Establish standardized regulations in determining the adequacy of a nuclear facility's efficiency vs. its safety.

This would limit the degree of individual nuclear power plant adaptation, but would allow enough flexibility to meet site-specific conditions and other local factors. In more detail, nuclear plant standardization would mean taking the nuclear power plant's equipment and procedures out of the black box and making them uniform.

The Future

A successful effort along these lines would eventually lead to:

- Developing much higher levels of detail for standard reactor designs, where special reactor designs will be the exception rather than the rule, and will need special, and different requirements;

- Harmonizing the nuclear industry's standards and requirements, focusing on convergence of codes and standards applicable to key components affecting efficiency and safety;

- Clarifying and expanding the existing feedback sharing among utilities and the participating parties, during power plant construction and operation;

- Enhancing the cooperation between vendor and utilities by establishing efficient mechanisms for long-term design knowledge management, such as training materials, operator certification, and plant operation procedures;

- Information and expertise sharing among governments, vendors and regulators, which will eventually lead to wide adaptation of the standards.

There are presently efforts to identify the differences and develop international codes and standards in different areas, such as mechanical codes and instrumentation and control (I&C). Organizations like ASME and AFCEN, and IEEE and IEC are fully involved in this effort. Some general utility requirements for new reactor designs have been developed already by EPRIURD in the U.S. and EUR in Europe.

In addition, a number of multinational regulatory initiatives have been created, such as the Multinational Design Evaluation Program (MDEP) with its main objective as establishing convergent reference regulatory practices. Regional initiatives have also been taken by regulators and utilities, such as the Western European Regulators' Association (WENRA) and the European Nuclear Installations Safety Standards (ENISS) initiative in Europe). WENRA has established common reference levels for reactor safety to be implemented in member countries, which will lead to further harmonization.

The International Atomic Energy Agency's (IAEA) Integrated Regulatory Review Service (IRRS) provides reviews of national regulatory systems to identify and spread best practices in licensing and oversight. It also provides a reference point for states seeking to establish a nuclear infrastructure. IAEA Safety Standards specify safety requirements and guides representing best/good practices, which are increasingly used as a reference for review of national safety standards and as a benchmark for harmonization in all countries utilizing nuclear energy for peaceful purposes.

There is much more work to be done in the standardization area, though, and the steps taken thus far are the baby steps that are needed to develop a strong and safe international nuclear power generating industry.

Nuclear power is a viable, efficient and clean energy source, which is badly needed to support our life style and industrial progress, so we need to support and expand these efforts. Nevertheless, all these efforts must be done very carefully, because once the sleeping Genie in the bottle is released, it could cause immeasurable damage to people and property.

ITER

The international collaboration for a new source of energy, ITER (International Thermonuclear Experimental Reactor) is one of the few good Cold War creations. It was when Presidents Reagan and President Gorbachev agreed during a meeting in Geneva in 1985 to initiate and pursue an international effort to develop *fusion* energy in order to benefit of all mankind. It was nearly 30 years ago, when a group of industrial nations joined the effort and agreed on a project to develop a new, cleaner, sustainable source of energy.

At that Geneva Superpower Summit, Presidents Mitterand of France and Thatcher of the United Kingdom also agreed with the proposal to collaborate in this new international project aimed at developing fusion energy for peaceful purposes. Thus, The ITER project was born. Then other joined; the European Union (via EURATOM), Japan, China, and South Korea in 2003, and India in 2005. This is not a small thing, because Together, these six nations plus Europe represent over half of the world's population and most of the global nuclear generation.

The conceptual design work for the fusion project began in 1988, followed by increasingly detailed engineering design phases until the final design for ITER was approved by the Members in 2001. Further negotiations established the Joint Implementation Agreement to detail the construction, exploitation and decommissioning phases, as well as the financing, organization and staffing of the ITER Organization.

ITER was one of the largest and most ambitious international science projects ever conducted. ITER actually means "the way" in Latin, which requires unparalleled levels of sophisticated international scientific collaboration. The power plant design and operation components were provided by the members, where each member established a domestic ITER-related agency, in order to manage its contributions to the project.

The ITER members agreed to share every aspect of the project: theory, hands-on scientific work, materials procurements, finance, labor, etc., The goal here was for each Member to share in the know-how in order to eventually construct and operate its own fusion energy plant(s).

The location for ITER was a chosen in 2005, when representatives of the ITER members unanimously agreed on the site proposed by the European Union— the research facility was to be built at Cadarache, near Aix-en-Provence in Southern France. The ITER Agreement was officially signed at the Elysée Palace in Paris in November 2006 by representatives from the seven ITER members. And so, the new ITER Agreement established a legal international entity to be responsible for construction, operation, and decommissioning of the ITER effort.

Another, "Broader Approach" for complementary research and development was signed in 2007 between the European Atomic Energy Community (EURATOM) and the Japanese government. It established a framework for Japan to conduct research and development in support of ITER over a period of ten years. The new Broader Approach included three projects that focused on: a) materials testing, b) advanced plasma experimentation and simulation, and c) the establishment of a design team to prepare for DEMO.

Note: The DEMO was the demonstration power plant, which was the next step after ITER effort was completed. The Broader Approach projects were considered of great import and absolutely necessary for the advancement of fusion energy, as outlined by ITER agreements. In the fall of 2007, all members officially ratified the ITER Agreement and officially established the ITER Organization.

During its operational lifetime, ITER will test key technologies necessary for the next step: the demonstration fusion power plant that will prove that it is possible to capture fusion energy for commercial use.

The ITER fusion reactor was designed on a principle to produce more output energy than is needed for the operation. The present research model is designed to produce 500 MW of output power while needing less than 50 MW to operate. The aim of producing more energy from the fusion process than is used to initiate it, is something that has not yet been achieved in any fusion reactor.

Construction of the ITER Tokamak complex started in 2013, and the building costs now are estimated at about $16 billion. This is already over 3 times the original figure of about $5 billion.

If ITER becomes operational, it will become the largest magnetic confinement plasma physics experiment in use, surpassing the Joint European Torus and everything else designed and built by man. The first commercial demonstration fusion power plant of the ITER project, named DEMO, will follow the successful demonstration of the fusion process, most likely around 2030…if everything goes according to plans and expectations.

Update:

The construction phase of the ITER prototype is expected to be completed in 2019, when the actual commissioning of the reactor will be initiated. The plasma experiments will start in 2020 with full deuterium-tritium fusion experiments expected in 2027. But that's on paper… As of 2013 the construction effort has run into many delays and budget overruns. The facility is now not expected to begin operations until the year 2027, which is 11 years after initially anticipated.

And then in February 2014, the ITER Management issued a report, listing 11 essential recommendations to the continuation of the project. Some of these were unexpected and even disturbing, while some of them made no sense, for example: "Create a Project Culture," "Instill a Nuclear Safety Culture," "Develop a realistic ITER Project Schedule" and "Simplify and Reduce the IO Bureaucracy." This did not sound like a sophisticated, efficient, and even well thought of attempt to make this right.

Later on the same year, the US Senate published its report stating that: "the Committee directs the Department of Energy to work with the Department of State to withdraw from the ITER project." And so, here we go, international collaboration in the nuclear industry.

But even worse, if ITER management has been more proactive and looking more carefully around, they would've certainly notice that similar efforts for the design and operation of fusion reactors have been conducted in other countries. The most notable is the U.S. based National Ignition Facility fusion experiment.

U.S. Fusion Research

The National Ignition Facility (NIF) is a large laser-based research facility, located at the Lawrence Livermore National Laboratory in Livermore, California, USA. The inertial confinement fusion (ICF) process at NIF uses lasers to heat and compress a small amount of hydrogen fuel in order to initiate and maintain *nuclear fusion* reaction.

The goal is to achieve fusion ignition with high enough energy gain, which is many times greater than the input power needed to start and continue the process. NIF, of course, had other missions, such as to support nuclear weapon maintenance and design by studying the behavior of matter under the conditions found within nuclear weapons.

Construction of the NIF facility started in 1997 but management problems and technical delays slowed progress for several years. All in all, NIF was completed five years behind schedule and well above the initial budget estimates. Construction was completed in March 2009, with the U.S. Department of Energy officiating a dedication ceremony later on that year.

The first large-scale laser target experiments were performed around that time too, and the first "integrated ignition experiments" (which tested the laser's power) were completed in the fall of 2010, when the most important milestone towards achieving fusion reaction was achieved by a fuel capsule that produced more energy than was applied to it.

This, however, is step one in a long, long process,

and still very far from satisfying the Lawson Criterion of an efficient and sustainable nuclear reaction. It is, however, a major step forward it. But what are the problems?

Note: The author was a Quality Control manager of a contractor company serving NIF in the mid-, late 2000s, and has first hand experience with the activities of the supply chain at the facility.

From this perspective, the issues are many. The initial idea to create a fusion reaction was and still is a mirage. And many more billions of dollars will be needed to implode the pea-size target and release its enormous energy. But that—even if achieved some day—would be only the beginning of a practical fusion process. Billions more dollars will be needed to design a commercial reactor that could withstand the enormous temperature, and the accompanying problems, created during the fusion of a sizeable chunk of fusionable material.

As demonstrated by NIF's 500 TW laser blast, we need as much energy as the entire country uses at the time to initiate the fusion reaction in a small ~2 mm diameter target.

Please note that at one split moment, NIF's lasers used as much energy as the entire U.S. at the time.

This situation—the monstrous size and the uncertainty of what to be expected—reminds us of the 2013 TV mini-series, *Eve of Destruction. A* nuclear accelerator operating at similarly high energy levels misbehaves and creates a dark energy black hole (whatever that means) the magnitude of which threatens to destroy the entire planet several times over.

With that in mind and after everything else we know about fusion, there are still a number of unanswered questions:

1. How much energy could be produced by a 2 mm target during its fusion process, and how long would it last?
2. How much of the energy released by the target can be captured and used?
3. Is it possible to use larger targets for practical power generation with the NIF setup?
4. How much energy is needed for the fusion of a 4 mm target? 4 cm? 4 foot?
5. How big of a target, and how much power, do we need to produce 1 GW of electric power via fusion reaction?
6. What would be the maximum practical amount of energy generated by a larger target?
7. Could the fusion reaction of a larger target be controlled?
8. What materials would be used to build the vessels needed to contain the enormous heat generated during thermonuclear fusion of larger targets?
9. Would a practical fusion power plant be as safe, or safer, than today's fission power plant of the same size?
10. What would happen if a fusion power plant of the Fukushima size experiences similar problems?

Judging from how things stand today, the answers to most of these questions are unknown, and work is underway to answer them theoretically. So, it seems that there is a lot of work to be done before fusion becomes a practical nuclear power generator. One doesn't have to be a laser or nuclear scientist to figure out that NIF is the model T of the fusion industry, and that much more time and money are needed to build a practical, functioning and safe fusion reactor on Earth. So, we see many thousands of hours and billions of dollars will be spent by NIF, and other such enterprises, on tests, equipment modifications, and upgrades until the NIF dream becomes a reality…if ever!

Update:

The NIF's plan was that upon successful completion, the fusion reaction triggered in the fuel cell, would release net energy many times greater than the energy the 192 lasers use. The fuel cell made up of the two hydrogen isotopes tritium and deuterium was bombarded by the lasers on several occasions and produced enormous pressures and temperature for less than a billionth of a second.

Alas, this was still not enough to trigger the expected sustained fusion reaction. This is step one in the process of several steps, which ultimately requires that the target (fuel capsule) produces more energy than the input for the duration.

Note: Presently, NIF facility contains world's largest and most energetic ICF device ever built, comprised of almost 200 of the most powerful lasers ever designed, built, and tested successfully. Equally famously, the NIF facility was used as the set for the starship Enterprise's warp core in the movie Star Trek Into Darkness. Is this the end of the Fusion Dream and deviating into other areas of interest, or is it just a short break before more intense work starts. The answer is hiding somewhere in NIF's lasers.

We must always remember that even if and when fusion is achieved and sustained, the reaction temperature is so high that it would be impossible to contain it. The Sun's core temperature is over 28 million degrees F, which is what would be expected in a large scale fusion reactor. Such high temperature could melt any presently known material and drill a hole in the ground all the way to China. We simply have no materials on Earth today that can withstand even a fraction of this temperature.

So, provided that somebody develops a fusion reaction, they should also know what materials would be used to build the reactors with. Having a one-shot ignition is just not enough for any practical purposes.

After all the hoopla about the potential of the NIF's amazing technology, in 2014 the management quietly announced—as if this was not such a big deal—that the initial goal of igniting a fusion reaction was officially abandoned…for now!

Effective immediately, NIF's $10 billion facility will be used for more mundane basic research and possibly more futuristic movies. This is analogous to converting a rockets manufacturing building facility into a donut shop. Another analogy comes to mind here, that of Solyndra, where poorly designed solar equipment was pitched to the U.S. government as the technology of the future. After half a billion taxpayer dollars were wasted, Solyndra went belly up overnight. Similarly so, after $10 billion wasted, NIF's high purpose was changed to a daily maintenance routine operation, employing hundreds of people.

The biggest difference here is that Solyndra went away quickly, while NIF is planning to continue operations for many years, and spend many additional billions. The most likely result of these efforts would be hundreds of theoretical papers justifying NIF's existence. As with many grandiose ideas in the past, NIF did not even come close to achieving the goal of demonstrating a complete fusion reaction. Instead, it wasted about $10 billion on worthless hardware and finally gave up on the idea.

"The experiments at NIF are laying the groundwork to provide the nation with abundant clean energy by using lasers to ignite fusion fuel," was how the management sold the NIF project. Sounds good, and those who do not know what fusion is would fall for it. The U.S. Congress did!

Today, NIF touts the amazing power of its 192 lasers, and there are plans to shoot at and break a lot of things with them over time. At this price, we could've fed half of the world instead of dumping so much money in a hole in the ground…literally.

Without going into the details, we can positively say that if fusion experiments are to be financed again with public funds, the money should go to private companies, not to government entities.

A number of institutions and private firms are claiming that they are working on "the last stages of a fusion reaction." What they are saying, with no exception, is that they are close to creating a split-second fusion reaction in a target as big as a pea. The questions they need to answer are:

- What would it take to expand the small-scale fusion reaction to large commercial scale?
- What materials would be used to contain the out of this world heat in the reactors?
- What safety measures would be implemented to prevent nuclear fusion accidents?

Until all these questions are answered plausibly, and all problems resolved in practice, nuclear fusion will remain just a pipedream and a way for institutions and companies to make money from government subsidies and private investment.

Nuclear Pros and Cons

There are two trends of thought today, as far as nuclear power is concerned; pro and con nuclear. The pro-nuclear activists and supporters are optimistic about the prospects of nuclear revival and growth in the near future. They have a good reason to be optimistic; nuclear is an amazing force second to none.

The more pessimistic group of con-nuclear lobby, is not willing to risk another Chernobyl, or Fukushima disaster, for it is too much to pay for daily comforts. They are also concerned with the long term availability of uranium and other radioactive fuels.

The pro-con balance has been tittering from side to side throughout the decades, tilted to one or other side according to the developments of the time. One thing that is not debatable, however, is the fact that no matter how good or bad nuclear turns out to be, the uranium and other radioactive materials, which are used to generate power, are in limited quantity.

No matter how much uranium we have today, its availability on Earth is limited.

Today the world needs about 70,000 tons of uranium ore annually to fabricate fuel for the currently operating nuclear power reactors. According to the nuclear power authorities and the UN's International Atomic

Energy Agency, the world's present known economic resources of uranium, exploitable at a reasonable price (below $80 per kilogram of uranium), are approximately 3.5 million tons. At the present rate of use, this amount is will last the world 50 years.

The estimates of the possible and expected uranium resources, including all that are not yet economically produced or are not properly quantified, is about 12 million tons, or enough for about 150 years of exploitation at today's rate of usage. If the rate of exploitation increases, as planned by China, India, Russia and other countries, the available uranium would be exhausted much faster.

Whether it would take 50 or 150 years, uranium will be depleted sometime in the next century.

And we have no solution for that problem. At that time, according to our calculations, the world would be already in the midst of a fossil-less existence, so the depletion of nuclear materials would be a great blow. Something like the proverbial straw that broke the camel's back. What a sad, dark, and horrible day that would be!

Here we must also emphasize that the above uranium estimates provide only some measure of the future uranium availability. We must also remember, however, that the presently known resources of any and all minerals are a poor indicator of to what is actually in the Earth's crust. And even less indication of what is potentially extractable for practical use.

In the case of current economic resources of uranium, the 50-year quantification is no more than a rearview mirror perspective on supply. During future consumption of these resources, the dynamics of supply and demand will produce price signals that will inevitably trigger effects involving several "resource-expanding factors." Predictions of availability or lack thereof of mineral reserves do not stand up to close scrutiny because they fail to take adequate account of these key "resource-expanding factors"

Some of these factors are already clearly evident in today's uranium market. Uranium, unlike the other fossils which have been around for a long time, has a very short history on the energy markets, because it had no direct use until the 1940s, when it started supplying limited quantities for military purposes. Later on, its use was greatly expanded into nuclear power generation. Nevertheless, today's uranium market is not much different from that of other metals, including the other energy sources. Just like them all, it is subject to price changes, cycles of exploration, and discovery and production issues.

The only big difference here is that uranium has experienced only one such cycle thus far. After the initial discoveries and use, uranium ore prices declined, but following the price spike in the late 1970s, we saw a significant exploration boom and price drop. This one cycle offers some reassurance in meeting the global reactor requirements during the last half a century, while also providing 3.5 million tons of known uranium resources that are available and awaiting recovery.

So, it might be too early to talk about uranium scarcity, but on the long run the bucket will get empty eventually no matter what the economists and scientists say or do.

The Nuclear Future at a Glance

Nuclear power is amazing in what it can do, both; in helping and hurting humans. The prevailing sentiment, however, is that we will proceed with using it… until another Fukushima-like disaster hits. At that point we will have to re-evaluate everything…again!

As a result, there are a lot of cautious expectations, and even more possibilities, awaiting the nuclear industry. Our advice is for the risks ahead to be thoroughly considered, and the expectations lowered to the very minimum.

Here is a summary of the nuclear market conditions with the information available to date. Keep in mind that everything can change overnight…as it did after Chernobyl and then after Fukushima.

• The nuclear energy industry (the nuclear market's back end) consists of the net generation of electricity by nuclear power plants all around the world. (The volume of the market is calculated as the net volume of electricity produced via nuclear power in gigawatt hours (GWh). The market value is calculated according to an average of annual domestic and industrial retail prices $/kWh, inclusive all applicable taxes, etc. charges.)

• The nuclear energy market is segmented by the total volume of energy generated through nuclear power, as a percentage of net generation for the relevant country or region.

• The global nuclear Industry production volumes declined with a CAGR of -2.1% between 2009 and 2013, reaching a total of 2.3 TWh power generation in 2013-2014.

- The global nuclear energy industry had total revenues of over $300 billion annually during the same time.

- The performance of the nuclear industry is expected to accelerate with an anticipated CAGR of 4.8% by 2018, which is expected to drive the industry to a value of $395 billion (up from about $300 billion in 2013 and 2014).

- The estimates today are that the global nuclear power *equipment (its manufacturing alone)* market will grow to $70 billion by 2020.

Summary

Nuclear power is a huge industry with even huger market. There are dozens and hundreds of government bodies and private, companies, factories, and organizations (technical, environmental, legal, financial, etc.) involved in the entire life cycle of the nuclear power generation, distribution, and use.

Thousands of skilled, unskilled, and specialized personnel are involved in all stages of nuclear plant equipment manufacturing, transport, installation, and operation. Other thousands are involved in the production, transport, and processing of raw materials. Numerous other people are involved in the manufacturing and transport of primary and secondary materials, supplies, and consumables; needed for the day-to-day operations. And of course, many more are involved in the transport, storage, disposal, and re-processing of waste nuclear materials.

We must always keep in mind the undeniable importance of the utility companies and their personnel, which are involved day and night, and are responsible for delivering the right amount of high quality, cheap electric power to our homes and businesses.

Designing, directing, and coordinating these steps is a very important job, which the marketing people of the different companies are heavily involved in. They are responsible for ensuring the uninterrupted and profitable market for their materials, components, supplies, electricity, etc.

All these companies, factories, organizations—as well as the thousands of people working in them—are elements of the national and global nuclear energy market. Without even one of these elements, the entire energy market structure won't function properly. If several of these elements are missing or fail, then the entire structure could collapse. At the very least, it won't be the same as we know it today!

We witnessed the result of nuclear industry failures during the aftermath of the Fukushima nuclear accident. Several countries hesitated in their response and shut down some of their nuclear plants off, while others stopped nuclear power generation all together. The effect on the separate countries, and the global, energy markets was profound. We saw drastically increased fossil-fired power generation, with the accompanying pollution and the related over-spending.

One thing became clear at that time; the nuclear industry's importance cannot be underestimated. There is no shadow of a doubt left that we will be in dire straights if we at any point we are forced to shut down our nuclear power plants. That would be a very sad and dark day for many of us.

Weather the industry grows or shrinks in the future, however, depends on how all the companies, organizations, and people involved in the nuclear industry will ensure its safe and accident free operation. These people and companies ultimately determine its direction and pace of its development.

Recently, the customers and regular people in general, have been getting more involved, and have more say so in the direction and pace of the nuclear power development in general. The world governments are also on their toes as far as nuclear power plants operation and safety is concerned.

This trend is expected to grow, and so the decisions on the overall direction and pace of development (or not) of the nuclear industry will be made on many levels. This is good, and we can only hope that as the trend grows so will the safety of nuclear power plants. We need our nuclear power plants. They are one of our gifts to the future generations too. We just have to be more careful how we design, build, and use them. This is something that has to be done. Period!

HYDROPOWER

Hydropower is the cleanest power generating technology to date. It is also the most reliable... if and when Mother Nature decides.

We must clarify hydropower's uniqueness, which is the reason we consider it in a class of its own among the *other* energy technologies. Hydropower is a major power generator, which has been, and still is, a significant contributor to our economic development. It is very safe and clean—the safest and cleanest in its finished form—of all energy sources. It is also renewable, albeit not officially recognized as such for a number of reasons.

The problems with hydropower, and its major flaws, are,

- Massive devastation of large land mass and human misery during its initial construction, and
- Extended, severe draughts, floods, earthquakes, and other natural events are putting a big question mark on its future renewability, reliability, and expansion across the globe.

Global warming is changing the world climate drastically to where a greater number and even more severe draughts, earthquakes and other natural disasters are expected in many areas of the world in the near future. This could put hydropower in a very bad place. No water from the sky, no hydropower...it is that simple! Floods and earthquakes complete the list of bad things that happen to any hydropower installation.

These abnormalities overemphasize the fact that construction of large hydro dams is also very expensive, extremely painful for the local population, and destructive for the local environment. For these reasons, we prefer to put hydropower in the special class of the *other* energy sources.

Background

Water is the liquid gold of the Earth, and is abundant over much of the world. Unfortunately, the type of water that is necessary for human consumption and other practical uses—including hydropower—is limited and becoming more so each year. The reduction of water quality and quantity in many places is becoming critical, and, of course, there are large areas that simply have no water. Our liquid gold is becoming more scarce and expensive, raising a number of questions as to its use in general and for power generation in particular.

The lack of water is becoming a huge problem for the entire energy sector, because most power plants use a lot of water for cooling. Hydropower power plants do not use water for cooling their turbines, since falling water turns them, but lots of water is wasted nevertheless. Hundreds and thousands of tons of water evaporate every hour from the millions of acres of water storage behind the world's dams.

And yet, hydropower is a major part of our energy generating mix.

Hydropower refers to the huge potential energy contained in water (dams, rivers, and streams).

It is energy created by the hydrologic (water) cycle

which is ultimately driven by the sun, thus making it an indirect form of solar energy. The water cycle is driven by the energy contained in sunlight, which evaporates water from the oceans and other water bodies and then deposits it on land in the form of rain and snow. This never ending cycle of evaporation and precipitation forms and maintains thousands of streams, rivers, and lakes around the globe.

Some rain water is absorbed by the ground, where it feeds the ground water tables. The part that is not absorbed runs off the land and into the oceans by means of streams and rivers. Some of the water in the long journey to the ocean is lost to evaporation, irrigation, refilling the water table etc., while the water reaching the ocean repeats the evaporation cycle time after time.

It is funny to think that we now drink the same water in which dinosaurs bathed millions of years ago.

Most often, hydroelectric plants are built along streams and rivers, where they generate power by using either:

- Running river water to turn the wheels of their turbines, or
- Falling water from a large lake behind a dam wall.

In the first case, no dam is needed, since the blades are submerged in the running river water and turn by the pressure applied by the river water current. In the latter case, however, water is stored behind concrete dams in a form of a large lake, built across the river and turns the turbine blades. The energy released by the water falling through a turbine is converted into mechanical power. The mechanical energy of the rotating turbines then drives generators to produce electricity.

Hydropower from dams accounts for approximately 75% of the nation's total renewable electricity generation.

This fact makes hydropower the leading renewable energy source of power. Yet, hydropower is NOT recognized officially as a renewable energy source in the U.S.

The U.S. annual hydropower output is equivalent to the energy produced from burning over 200 million barrels of heating oil

There are also estimates that more than 200 million tons of CO_2 emissions are avoided in the U.S. annually because of hydropower generation (as compared with coal, oil, and gas power generation).

Hydropower is the most efficient energy source.

Hydropower turbines are capable of converting more than 90% of the available energy into electricity, making it the most efficient form of electricity generation. By comparison, fossil fuel and nuclear plants are only about 30-40% efficient.

Hydropower is also the cheapest power generator.

In addition to providing low-cost electricity, multipurpose dams and storage reservoirs provide water for irrigation, wildlife, recreation, barge transportation, and flood control benefits.

Hydropower plants range in size from large power plants that supply many consumers with electricity, to small and micro plants operated for individual needs or to sell power to utilities.

Hydropower plants vary in size, but according to the U.S. DOE:

1. Large hydropower plants are facilities that have a capacity of more than 30 megawatts.

2. Small Hydropower plants are facilities that have a capacity of 100 kilowatts to 30 megawatts, and

3. Micro hydropower plants are facilities with capacities of up to 100 kilowatts.

Large and some small hydropower plants use dams to store water and release it when needed to generate electricity. Most small and all micro-hydroelectric power systems use rivers or streams current to generate electricity. They are also usually designed to produce enough electricity for a home, farm, ranch, small business or village. In some cases, the excess power is sold to the local utility.

Large (Dam) Hydropower

Dams on rivers and streams have been around for a long time, and are still present and in use in almost all regions of the globe. They have played a key role in the development of human activities, and are used for irrigation, water supply, flood control, electric power generation, and improvement of navigation. They also provide recreation, such as fishing and swimming, and become refuges for fish and birds, and they slow down streams and rivers so that the water does not carry away soil, thereby preventing erosion.

During the last century, dams have also been used to generate electricity, and some of them are the largest installations on Earth, generating huge amount of electricity.

Unfortunately, the water that fuels hydro-power generation is not considered (officially) a renewable power source. Because of that, hydropower does not benefit from the assistance and subsidies the renewables get. To remedy that, the issue was recently brought up before the US Congress for a vote.

Water's renewability is being debatable, since it means different things to different people. So, regardless of the Congressional decision, hydropower will remain debatable power generation alternative. This is mainly because there are many ways to look at global water production, storage, preservation, and use, including the damage and waste that come with these activities.

Hydroelectric plants range in size from several kW to many GW. For example, the Three Gorges Dam on the Yangtze river, the largest hydroelectric dam in the world, has installed capacity of over 20 GW, which is nearly twice the power of the next largest hydropower plant, the Brazil-Paraguay Itaipu dam, at 14 GW. It is also many times the power generated by the largest fossil, or nuclear plants.

There are about 35,000 large dams in existence around the world today.

Not all dams are equipped to generate electricity, but many are. There is also a drive to equip many existing dams for electricity generation.

The importance of hydropower in boosting economic development is undeniable, and is even increasing in recent decades. There is a trend today—most significantly in developing countries—of increased plans for new dams construction. Since there is not much land mass (or river water) available for large dams, hydropower development in most countries is now focused on building smaller dams, or refurbishing and upgrading existing hydroelectric plants, and on retrofitting dams constructed for other purposes.

The feasibility of small hydropower plants depends on the availability of a back-up source of electricity, since most of the smaller plants often do not have a reservoir for storage, thus depending on the variable river flow. Large- and intermediate-size dams, however, will continue to be very important in developing countries, and in particular in Russia, China, and in some industrialized nations, such as Canada.

Case Study: Hoover Dam

Building a hydropower plant on a large dam is a huge undertaking, usually of national proportions. This cannot be done by one company alone. Instead a number of companies, under the direction and supervision of government bodies are usually involved in the project.

A good example, illustrating the magnitude of issues related to large hydropower dams is the construction of Hoover dam on the Arizona-Nevada border.

Since about 1900, the Black Canyon and nearby Boulder Canyon had been investigated for their potential to support a dam that would control floods, provide irrigation water and produce hydroelectric power. In 1928, the U.S. Congress authorized the project. The winning bid to build the dam was submitted by a consortium called Six Companies, Inc., which began construction on the dam in early 1931.

The project was approved by the U.S. Congress and executed under the direction of the Bureau of Reclamation. A decision on the massive concrete arch-gravity dam structure was made and the design was overseen by the Bureau's chief design engineer John L. Savage.

The Bureau issued bid documents available to interested parties, where the government was to provide the materials, but the contractor was to prepare the site and build the dam. The dam was described in minute detail, covering 100 pages of text and 76 drawings. A $2 million bid bond by the contractors accompanied each bid, and the winner posted a $5 million performance bond, allowing him seven years to build the dam, or else.

There were three valid bids, and a bid of $48,890,955 was the lowest. It was amazingly only within $24,000 of the confidential government estimate of what the dam would cost to build. The best bid was also $5 million dollars lower than the next lowest bid. This is truly amazing, considering the fact that the estimates were done in the late 1920s and early 1930s. There were no computers, nor advanced construction equipment at that time, and yet they show the amazing professionalism and high level of expertise of the American engineers at the time.

Before the work on the dam started, an entire city was built in the desert near the dam site, which is still there as we know it today as Boulder City, Nevada. There was also a special railroad line constructed to join Boulder City to Las Vegas for transport of people and materials.

The dam building began by diverting the Colorado River away from the construction site. Four diversion tunnels (56 ft. in diameter and nearly 3 mi total length) were drilled through the canyon walls, on both sides of the proposed dam structure. This work had to completed quickly in late fall and winter, when the water level in the river was low enough.

When the tunnels were completed, they had to be lined with concrete, using Gantry cranes running on rails through the entire length of each tunnel. Then the sidewalls were poured, using movable sections of steel forms, to create 3 ft. thick concrete lining.

The river was then diverted into the two Arizona tunnels, by exploding the temporary cofferdam protecting the Arizona tunnels, while at the same time dumping rubble into the river until its natural course was blocked and it started flowing through the two tunnels. The Nevada side tunnels were kept in case of high water floods. Upon completing the dam, the entrances to the diversion tunnels were sealed at the opening and halfway through the tunnels with large concrete plugs, while the downstream halves of the tunnels following the inner plugs are now the main bodies of the spillway tunnels.

Two cofferdams were constructed to facilitate the river's diversion, each 96 ft high, and 750 feet thick at the base, which is actually thicker than the dam base. Each contained 650,000 cubic yards of rock material and cement.

The site was then drained of water and the accumulated erosion soils and other loose materials in the riverbed were dredged until sound bedrock was reached for the dam foundation. This required the excavation and removal off-site of over 1,500,000 cubic yards of river bed material. Since the dam was an arch-gravity type, the side-walls of the canyon would bear the force of the impounded lake. The side walls of the surrounding rock channel were excavated too, to reach virgin rock as needed for the load-bearing side walls and eliminate water seepage.

The dam foundation was reinforced with grout and holes were driven into the walls and base of the canyon, as deep as 150 feet into the rock, and all cavities were filled with grout, in order to stabilize the rock. This would also prevent water from seeping past the dam through the canyon rock and limit the upward pressure of water seeping under the dam.

After the dam base and sides were secured, the pouring of concrete into the dam structure was initiated. This is a complex undertaking in such an enormous structure, because concrete heats and contracts for a long time while it cures. The potential for uneven cooling and contraction of the concrete is a serious problem, and so to avoid a very long curing process, the dam was built in sections. The ground where the dam was to rise was marked with rectangles, and concrete blocks in columns were poured. These were 50 ft square and 5 ft. high, strengthened by a series of 1-inch steel pipes through which first cool river water, then ice-cold water from a refrigeration plant was run.

Once each individual block had cured and had stopped contracting, the pipes were filled with grout. Grout was also used to fill the hairline spaces between columns, which were grooved to increase the strength of the joins.

Huge steel buckets (7 ft. high by 7 ft. in diameter, and weighing 18 tons when full), suspended from aerial cableways above the construction site, were used to pour the concrete for each block. The concrete was prepared at two large concrete plants on the Nevada side, and were delivered to the site in special railcars. A team of men worked the newly poured cement in each block throughout the form until achieving the desired uniformity.

A total of 3,250,000 cubic yards of concrete were used in the dam. Additional 1,110,000 cubic yards concrete were used for the construction of the power plant and other works. More than 582 miles of cooling pipes were placed within the concrete. It was estimated that there is enough concrete in the Hoover dam and the surrounding structures and infrastructure to pave a two-lane highway from San Francisco to New York.

Although the dam was completed in 1935, concrete cores removed from the dam for testing in 1995 showed that the concrete has continued to slowly gain strength. It was also confirmed that the dam is composed of a durable concrete having a compressive strength exceeding the range typically found in normal mass concrete. Hoover Dam concrete is not subject to Alkali-Silica Reaction (ASR) as the Hoover Dam builders happened to use nonreactive aggregate, unlike that at downstream Parker Dam, where ASR has caused measurable deterioration.

The huge monolithic dam is now 660 ft. thick at the very bottom and getting thinner as it goes up, ending with a 45 ft. wide road, connecting Nevada and Arizona. It has a convex face towards the water level above the dam. It was estimated that the curving arch would transmit the water's force into the abutments of the rock walls of the canyon.

Following an upgrade in 1993, the total gross power rating of the Hoover hydropower plant, including two 2.4 MW Pelton turbine-generators that power Hoover Dam's own operations, is a maximum capacity of 2.1 GW. The annual power generation varies, according to water conditions and other factors. The maximum annual generation of 10.3 TWh/y was recorded in 1984, and the minimum was 2.6 TWh in 1956. The average has been about 4.2 TWh/year.

The dam reservoir has been very low since the mid 2000, and although still far from minimum level, it is watched closely and preventative measures are planned.

The upstream picture is much more impressive

Figure 6-13. Upstream side of Hoover dam before filling the reservoir

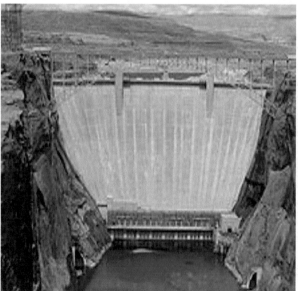

Figure 6-14. Downstream of Hoover dam.

and even dramatic. 250 square miles (160,000 acres, or 28,537,000 acre/feet) in the Arizona/Nevada desert were permanently flooded, thus causing a radical change of the local environment. Good or bad, this change has to be taken seriously and evaluated for what it is, which what we will do in the next chapters.

Labor

Working on the dam and the power plant construction was very dangerous and exhausting job. 5,000 people on the average were involved in the daily dam construction. The most dangerous job was that of the "high scalers." These were people suspended from the top of the canyon with ropes, who climbed down the canyon walls every day. Suspended on the vertical rock wall, they removed the loose rock with jackhammers and dynamite.

There were falling rocks and other that hurt and killed workers. One high scaler was able to save a government inspector, who lost his grip on a safety line and began tumbling down a slope towards a certain death, when the high scaler intercepted him, risking his own life, and pull him into the air and into safety.

The workers were under severe time constraints too, because the concrete pour was due, and at times negligently ignored seepage and cavities in the wall. As a result, many holes were incompletely filled. This eventually caused unacceptable leaks in the completed dam, and the Bureau decided to fix the problem by drilling new holes from inspection galleries inside the dam into the surrounding bedrock. This work was done in secrecy, and at additional cost, taking over nine years to complete after the dam was already in full operation.

Although there are myths that men were caught in the pour and are entombed in the dam to this day, each bucket only deepened the concrete in a form by an inch, and Six Companies engineers would not have permitted a flaw caused by the presence of a human body.

Nevertheless, there were over 110 deaths during the construction of the Hoover dam. One of the first victims was a surveyor who drowned on in 1922, while looking for an ideal spot for the dam. Incidentally, his son was the last man to die working on the dam's construction, 13 years to the day later. Ninety-six of the deaths occurred during construction at the site, and 91 of these were contracted employees. The rest were helpers and visitors.

In addition to the official fatalities due to accidents and incidents, there were a number of deaths due to illness, such as pneumonia. There were allegations, however, that this diagnosis was a cover for death from carbon monoxide poisoning of people overexposed to the gasoline-fueled vehicles in the diversion tunnels, which diagnosis was used by the contractor to avoid paying compensation claims.

This is a good place to take a close look at the entire cradle-to-grave life cycle of hydropower plants.

Hydropower Generation

Hydroelectric power plants produce electricity by using a power source (water in this case) to turn a turbine, which then spins an electric generator that produces electricity. While a coal- or gas-fired power plant uses steam to turn the turbine blades, the hydroelectric plant uses falling water to turn the turbine and generate electricity. The results are the same; electric power flows from the facility into the grid. A distinguishable exception in the case of a coal-fired plant is the significant emission of toxic gasses, liquid and solid waste, in addition to the electrical flow.

Hydroelectric power is generated by the gravitational force of falling water, so the capacity to produce energy is dependent on both the amount of flow and the height from which it falls. The water behind the dam accumulates huge amounts of potential energy which is transformed into mechanical energy when the water rushes down the sluice and strikes the turbine's rotary blades.

The power available from falling water can be calculated from the flow rate and density of water, the height of fall, and the local acceleration due to gravity, as follows:

$$P = \eta \varrho Q g h$$

where

P is power generated by falling water in watts

η is the dimensionless efficiency of the turbine

ϱ is the density of water in kilograms per cubic meter

Q is the flow in cubic meters per second

g is the acceleration due to gravity

h is the height difference between inlet and outlet

The turbine's rotation spins electromagnets which generate current in stationary coils of wire. Finally, the current is put through a transformer where the voltage is increased for long-distance transmission over power lines.

Hydropower Plants Cost

Hydropower plants (especially these with large

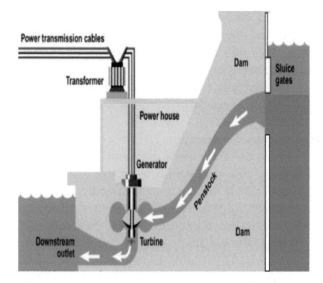

Figure 6-15. Hydroelectric power plant at a dam

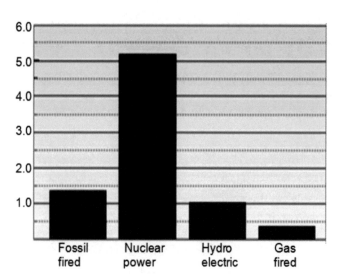

Figure 6-16. Power plants construction costs (in $/watt)

water storage reservoirs) are not cheap or easy to construct. As a matter of fact, the actual cost of a large hydropower plant design and construction are astronomical. While the Hoover Dam's construction in 1936 was estimated at less than $50 million ($838 million today), the total cost of the Three Gorges dam in 2012 was estimated at $28 billion. Yes, with a B.

Large dams are things of the past, however, due to lack of suitable land and reduced water availability. So now we can talk mostly about smaller dams and power plants. These are much cheaper, but still in the billions of dollars range. The cost of the proposed Marvin Nichols reservoir in northeast Texas, for example, has been estimated at $2.2 billion, with no power plant included. It would flood 30,000 acres of rare bottomland hardwood forest, and another 42,000 acres of mixed forest, family farms and ranches along the Sulphur River in northeast Texas.

This is the other thing to consider, when discussing hydropower; the external cost of hydropower, especially during the construction stages. Thousands of acres of productive land are lost, accompanied by the suffering of relocated families, due to flooding of their local areas. The outcry of the affected residents cannot be ignored. Long lasting bad memories and negative publicity surround most large hydro projects.

The actual cost of power plant construction (as related to the amount of produced power) varies from plant to plant. The main variable in the power generation cycle is always the size of the plant. For example, it takes almost as many people to operate and maintain a power plants with small one-unit generator as it would to operate and maintain two larger generators. This

means that the cost of operation and maintenance per kilowatt produced would be maybe twice as high for the smaller plant.

This is only one of the factors that determine the final cost of the produced electricity. A number of other variables also affect the final cost of the generated energy. The type, availability, quantity, and price of the fuels used in different power plants are another example of critical components of the cost structure.

Hydropower is the only major power source that needs no fuel to operate.

This is because water is always available (if the external effects are not considered) and is free. Recently, however, water has been becoming a precious commodity, so this issue is becoming more important, and it is quite likely that eventually water used by hydropower plants will come at a cost.

Maintenance and operation variables are also significant and must be considered. Again, hydropower requires little daily maintenance, and only scheduled periodic maintenance and replacement of turbines and other components add cost.

Operational costs of hydropower plants are also quite low, so the entire fuel-maintenance-operation regime is quite stable, efficient and cheap—provided there is enough water.

Table 6-8 shows average hydropower plant costs and power generation expenses.

In general, larger hydroelectric plant have higher initial coast (land flooding and plant construction), but enjoy cheaper cost per kilowatt of produced electricity. When compared to other means of producing electricity,

Table 6-8. Hydropower plant costs and expenses

Activity	Cost/Expense
Capital cost	$1700-2300/kW
Operation cost	0.4¢/kW
Maintenance cost	0.3¢/kW
Capacity factor:	60-80%
Operating life:	50+ years
Average power plant size:	30 MW
Large power plant size:	over 1.0 GW

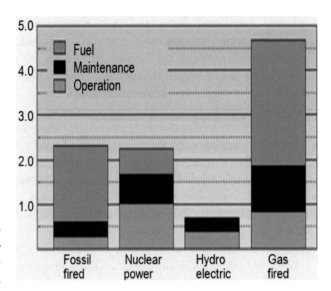

Figure 6-17. Power generation expenses in cents/kWh

hydroelectric production costs run about one third those of either fossil-fueled (coal or oil) or nuclear power plants, and about half of natural gas turbine electricity production. The main contributing factor for the difference in this cost of production is the fuel costs of the other means of producing electricity.

The original plant cost for a small hydroelectric plant is somewhat lower than either fossil fuel or nuclear plants (not including the support infrastructure—water reservoir, etc.). On the other hand, gas turbine plants are the cheapest to build but the most expensive to operate. As long as there is sufficient water to run the turbines, electricity can be produced very cheaply. Compared even to mature nuclear plants, hydropower costs less than half as much to produce, at under a penny per kWh.

Thus generated power is transmitted to the national power grid and sold to the wholesale and retail markets at the same prices as electricity generated by other means, complete with premiums for peak demand production. This provides significant profit for the middle man and the utilities.

In summary, hydropower is a great energy source, which will continue to provide a major part of the world's energy, but there are the unanswered issues of land and water availability. These must be properly addressed and resolved once and forever, if we are to rely on hydropower in the 21st century.

As Figure 6-17 suggests, hydropower is the only one of the major power generators that does not need input fuel. This is a great advantage as long as we can ensure that enough water runs across the turbines in the hydro-power plants. That, however, is debatable issue that depends on a number of extraneous factors.

River Hydropower

Hydropower plants based on the *run of the river* generate electricity on a smaller scale by simply using the water flow of a local stream or river. One way is to generate electricity is to channel a certain amount of water through a pipe from the rive to a turbine at the outlet. The water flow makes the turbine blades spin, which in turn spins a generator to produce electricity. This method is preferred to damming the river because it is much cheaper, and causes much less environmental damage by flooding large areas (as in damming the rivers).

Electricity can also be generated from turbines installed directly in the river. While this seems as a most practical and economic way to use the water flow, this method has not yet been commercially used, although there are a number of tests underway in the U.S. and Europe.

The problem here is that the water flow is usually low, so the power generator relies solely on the density and mass of the running water. The slow speed limits significantly the amount of generated electricity, while on the positive side of things it also reduces the risk to fish life.

River dams of the *run of the river* type are the new trend in hydropower generation. One of the most remarkable examples is the joint Romanian-Yugoslavian mega hydro project on the river Danube that started operation in 1972. The first version, Iron Gate I dam, had two dams with power plants each, containing 6 turbines at approximately 1.0 GW power generation each. Iron Gate II dam extension followed in the late 1990s by an upgrade and 10% increase of the total power, and two additional smaller power plants are still to be built as well.

The construction of these dams created large reservoir in the Danube shoreline, and raised the water level of the river near the dam by 35 meters. Six villages, totaling a population of 17,000, were evacuated and the villages, together with the local Orşova island were flooded. The locals were relocated and the settlements have been lost forever to the Danube. The dam construction had a major impact on the local environment, especially on the fauna. The spawning routes of several species of sturgeon were permanently interrupted.

The good news is that during post-dam construction, the geo-morphological, archaeological and cultural historical artifacts of the Iron Gates have been under protection by both nations. In Serbia the Dzerdap National Park was created in 1974 on 245.59 sq mi., and in Romania, the Porţile de Fier National Park was created in 2001, with 446.55 sq mi., both officially protected territories.

There are a number of similar examples of river hydropower, but nevertheless, a lot of work remains to be done on this technique, before its efficiency and cost-effectiveness can be assessed, and all problems resolved, before it finds wider commercial applications.

Small Hydropower Plants

Building a small hydropower plant on an existing lake, spillway or a river, is a straight forward process, which, just like any other power plants, would go through the normal channels of development. The only difference here is the fact that there is no fuel to be procured and transported, nor is there any waste or byproducts to be stored and disposed of. And of course, with that the emissions coming from the hydropower power during full operation are close to nil too.

A typical cradle-to-grave development process for a small hydropower plant includes:

• Hydropower plant site search
• Proposed area survey
• Environmental studies
• Local rights and agreements
• Permits and legal issues
• Plant site preparation
• Plant construction
• Daily plant O&M operations
• Decommissioning and area reconstruction

All steps of this process must be considered, in order to obtain a complete picture of the project, as needed to assess properly its technical and financial feasibility. As with all power plants, the survey, environmental studies and permitting processes are complex undertakings, in need of expert personnel. They also might take months, but most likely years, of intense work and negotiations with local and federal authorities and regulators.

After all studies have been completed, and the necessary permits secured, the construction of the plant can begin. Here we see again large pieces of equipment moving in, leveling the site and digging holes in the ground. Cement trucks dump their loads in the foundation and construction workers erect the buildings and complete the other elements of the plant infrastructure.

In most cases, small to medium size hydropower plants are built in remote locations—far away from populated centers and/or the power grid. In such case, a substation has to be built nearby in order to transform the generated electricity into voltage suitable for transmission to a connection point in the national power grid, or to remote customers. In all cases, right of way for the transmission lines is needed, which is another great obstacle in building remote power plants.

After all the work has been completed, the power plant goes through a series of start-up, testing and certification procedures. If everything checks OK, then the plan is given a clean bill of health and is switched into full operation, at which point it starts sending electricity to the grid, and normal O&M procedures take place on daily, usually 24/7, basis.

Case study:

In the 1980s the author was part of a team contracted for the design and implementation of small hydropower plants in Eastern Europe. The permitting and financing of the projects was the greatest hurdle, which had to be considered first. Due to the complexity of these procedures, several sites were rejected. Another set of potential sites had to be rejected because of their remoteness, which made connecting to any customers or the grid impossible (or at least prohibitively expensive). The remote sites also bore the risk of vandalism, which was not justifiable from the investors' point of view.

After the preliminary evaluation of dozens of proposed sites, only two were approved for development, and only one of these was eligible for government support. And so, the team focused on the proposed site, located on a large stream in a mountain region, 1.5 miles away from a large resort, which was the customer.

The stream is full and water flow is fast 8 months out of the year, while for the other 4 months the level and flow vary from year to year. The stream also freezes occasionally for 2-3 months in the winter too, during which time the turbine is frozen and shut down.

A natural curve on a hill, where the stream zigzags for about a mile, as in Figure 6-18, was the proposed hydropower plant site. An intake at the top of the curved section was to be built and the water diverted from the

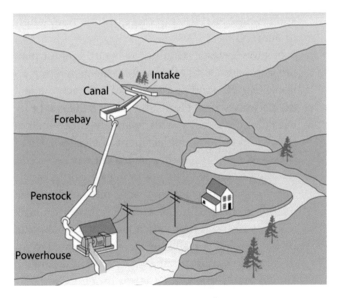

Figure 6-18. Small hydropower plant

natural stream. A catch box at the mouth of the intake was to filter out floating debris and fish, using an array of bars to keep large debris out, and a large mesh screen to remove smaller objects. An inspection gate was installed to divert the water for inspection and maintenance of the box.

A large bore cement pipeline was run ¼ mile from the intake to the turbine building. This was a major challenge, due to the mountainous terrain. An electro-mechanical control was installed in the turbine building to regulate the speed of the turbine and the electric generator attached to it. Thus generated power was sent to a transformer and then to the main breakers of the building for use on demand.

The power plant was rated at 150 kW and was equipped with a German turbine/gen set. The plant operation was quite simple, and consisted of periodic cleaning and inspection of the catch box, and normal periodic maintenance of the turbine/gen stack.

Figure 6-19. Pumped energy storage.

The only anticipated long term problem was the availability of water in the stream during the low flow months in the summer and freeze in the winter. Also the late night use of power at the resort was drastically reduced, and since the power grid was far away, the turbine was idling several hours every night. The team could not come up with a better (economically feasible) solution for using the excess power generated during the night time.

The total cost from concept to final inspection was about $1.2 million, or $8,000/kW. This somewhat higher cost was due to extensive works needed for the ¼ mile pipe on the mountain hills.

Pumped Storage

There are several types of water storage for hydropower generation. *Pumped water* storage is one of them. It is basically storing water and reusing water for peak electricity demand. Demand for electricity is not "flat" and constant. Demand goes up and down during the day, and overnight there is less need for electricity in homes, businesses, and other facilities. For example, in Phoenix, Arizona on a hot August day, the demand for electricity to run thousands of large air conditioners is huge But later on at midnight, it is drastically reduced.

Some hydroelectric plants use "pumped storage" to generate power on demand any time of day by reusing the same water more than once.

Pumped energy storage is a method of keeping water in reserve for peak period power demands by pumping water that has already flowed through the turbines back up a storage pool above the power plant *at a time when customer demand for energy is low*, such as during the middle of the night. The water is then allowed to flow back through the turbine-generators at *times when customer demand is high*, like at noon on hot August day in Arizona. This reduces the heavy load placed on the electric system at peak hours, when the air conditioners are cranked up.

The reservoir acts much like a battery, storing power in the form of water when demands are low and producing maximum power during daily and seasonal peak periods. An advantage of pumped storage is that hydroelectric generating units are able to start up quickly and make rapid adjustments in output. They operate efficiently when used for one hour or several hours. Because pumped storage reservoirs are relatively small, construction costs are generally low compared with conventional hydropower facilities.

Compressed air hydro is used when a plentiful head of water can be made to generate compressed air directly without moving parts. In these designs, a falling column of water is purposely mixed with air bubbles generated through turbulence at the high level intake. This is allowed to fall down a shaft into a subterranean, high-roofed chamber where the now-compressed air separates from the water and becomes trapped. The height of falling water column maintains compression of the air in the top of the chamber, while an outlet, submerged below the water level in the chamber allows water to flow back to the surface at a slightly lower level than the intake. A separate outlet in the roof of the chamber supplies the compressed air to the surface. A facility on this principle was built on the Montreal River at Ragged Shutes near Cobalt, Ontario in 1910 and supplied about 4.5 MW electric power to nearby mines.

Another method of compressed air storage uses excess electric energy during low power demand hours to run compressors, which compress air in special storage tanks. During high power demand hours the compressed air can be released to run turbines connected to generators. Thus produced power is sent into the grid to reduce the power generated by peaking power plants.

Compressed air storage suffers from low efficiency (less than 50%), so there are designs to improve the efficiency by using a mist of water sprayed into the air storage tanks, which acts to absorb and store the heat generated from the compression (and release it on expansion). This allows the improved system to achieve a 80-90% thermodynamic efficiency, and a total 70% overall efficiency.

Hydropower Characteristics

Hydropower has its own up- and downsides.
The major benefits of hydropower generation are:
- The water resources are widely spread around the world. Potential exists in about 150 countries, and about 70% of the economically feasible potential remains to be developed. This is mostly in developing countries. Here we must consider the many factors affecting hydropower; impending global warming causing draughts, construction and infrastructure huge costs, etc.
- It is a proven and well advanced technology (more than a century of experience), with modern power plants providing the most efficient energy conversion process (>90%), which is also an important environmental benefit.
- The production of peak load energy from hydropower allows for the best use to be made of base load power from other less flexible electricity sources, notably wind and solar power. Its fast response time enables it to meet sudden fluctuations in demand.
- It has the lowest operating costs and longest plant life, compared with other large-scale generating options. Once the initial investment has been made in the necessary civil works, the plant's life can be extended economically with relatively cheap maintenance and periodic replacement of electromechanical equipment. Typically a hydro plant in service for 40-50 years can have its operating life doubled.
- The "fuel" (water) is renewable, and is not subject to fluctuations in market. Countries with ample reserves of fossil fuels, such as Iran and Venezuela, have opted for a large-scale program of hydro development, recognizing environmental benefits. Hydro also represents energy independence for many countries.

Note: Can water used for hydropower generation be considered a renewable resource? This question is becoming a focal point in U.S. politics. A resolution in Congress calls for officially recognizing it as such, but its destiny is uncertain.

Hydropower provides unique benefits, rarely found in other sources of energy. These benefits can be the electricity itself, or those associated with reservoir development. Despite recent debates, few would disclaim that the net environmental benefits of hydropower are far superior to fossil-based generation. In 1997, for example, it was calculated that hydropower saved GHG emissions equivalent to all the cars on the planet (in terms of avoided fossil fuel generation).

While development of all the remaining hydroelectric potential could not hope to cover total future world demand for electricity, implementation of even half of this potential could have enormous environmental benefits in terms of avoided generation by fossil fuels.

Carefully planned hydropower development can also make a vast contribution to improving living standards in the developing world (Asia, Africa, Latin America), where the greatest potential still exists. Approximately 2 billion people in rural areas of developing countries are still without an electricity supply.

As the most important of the clean, renewable energy options, hydropower is a benefit of a multipurpose water resources development project. As hydro schemes are generally integrated within multipurpose development schemes, they can often help subsidize other vital functions of a project.

Typically, construction of a dam and its associated

reservoir results in a number of benefits associated with human well-being, such as secure water supply, irrigation for food production and flood control, and societal benefits such as increased recreational opportunities, improved navigation, the development of fisheries, and cottage industries. This is not the case for any other source of energy.

Hydropower, as an energy supply, also provides unique benefits to an electrical system.

• First, when stored in large quantities in the reservoir behind a dam, it is immediately available for use when required.

• Second, the energy source can be adjusted to meet demand instantaneously.

There are other benefits too, known as ancillary services. They include:

• Spinning reserve—the ability to run at a zero load while synchronized to the electric system. When loads increase, additional power can be loaded rapidly into the system to meet demand. Hydropower can provide this service while not consuming additional fuel, thereby assuring minimal emissions.

• Non-spinning reserve—the ability to enter load into an electrical system from a source not on-line. While other energy sources can also provide non-spinning reserve, hydropower's quick start capability is unparalleled, taking just a few minutes, compared with as much as 30 minutes for other turbines and hours for steam generation.

• Regulation and frequency response—the ability to meet moment-to-moment fluctuations in system power requirements. When a system is unable to respond properly to load changes its frequency changes, resulting not just in a loss of power, but potential damage to electrical equipment connected to the system, especially computer systems.

Hydropower's fast response characteristic makes it especially valuable in providing dependable regulation and frequency.

• Voltage support—the ability to control reactive power, thereby assuring that power will flow from generation to load.

• Black start capability—the ability to start generation without an outside source of power. This ser-

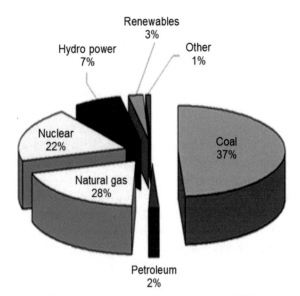

Figure 6-20. Power generation in the U.S.

vice allows system operators to provide auxiliary power to more complex generation sources that could take hours or even days to restart. Systems having available hydroelectric generation are able to restore service more rapidly than those dependent solely on thermal generation.

U.S. Hydropower

Hydro power generation is probably the oldest method of producing (mechanical) power. We can even imagine the amazement of the first caveman who noticed that moving stream water can do work. Then he figured out how to make the water spin a wheel that crushed grain, feeding the entire village with very little effort. This, as a matter of fact, is one of the first customer service enterprises, later on used widely by both; the capitalist and communist systems.

People have used moving water in increasing sophistication to facilitate their work throughout history, while presently people make use of moving water mainly to produce electricity. However, most of the electric power in the U.S. is produced by fossil-fuel and nuclear power, and only 7% of total power is produced by hydroelectric plants. Most of the hydropower comes from huge power generators placed inside dams.

The greatest limitation of this type of power generation is the fact that there is no more room (area and water) to build new hydropower plants, since both; the available land and water are in short supply. Increasing populations need more land, while frequent draughts and other natural events make hydropower's reliability questionable on the long run.

Most U.S. states use some type of hydroelectric

power, but states with low topographical relief, such as Florida and Kansas, produce very little hydroelectric power. Some states, such as Arizona, Idaho, Washington, and Oregon use hydroelectricity as one of the main power sources. As a matter of fact, most of Idaho and Washington states' electric power comes from hydroelectric plants.

In 2006, Texas had only 23 dams with hydroelectric power plants out of the state's hundreds of medium to large dams built for purposes other than power generation. The 23 dams have a total generating capacity of 673 MW, but the amount of electricity they actually produce annually is well below their maximum generation potential. This is due to water level and demand variations, and other factors that reduce the plants' up-time and efficiency.

For example, in 2004 that time, Texas' hydropower plants operated at an average of 22% capacity factor, while two years later (in 2006) the capacity factor averaged only 11%; mostly due to water availability issues. This confirms the fact that hydropower production is limited most drastically by droughts or other factors that affect surface water flows.

Officially, there are six types of hydropower plants in the U.S.:

• Municipal and other non-federal public
• Private utility
• Private non-utility
• Industrial
• Federal
• Cooperative

There are 2,388 licensed U.S. hydroelectric plants, assigned to one or more of the six owner classes. These plants represent the bulk of the U.S. hydroelectric capacity of approximately 75 GW in 1996. Of these, 69% of the plants are owned by private owners, as follow: private utility 31%, and private non-utility 27%, while industrial hydropower plants represent 9%, and cooperative 2%.

Table 6-9. Large U.S. dams

Dam name	Year built	MW
Grand Coulee Dam	1942	6,809
Chief Joseph Dam	1958	2,620
Robert Moses Niagara	1961	2,515
John Day Dam	1949	2,160
Bath County PSP	1985	2,100
Hoover Dam	1936	2,080
The Dalles Dam	1981	2,038

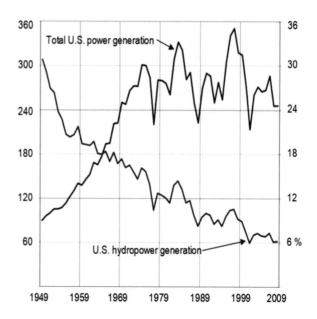

Figure 6-21. U.S. total- vs. hydro-power generation
- Left hand side is GW and
- Right hand side is in %

Nearly 75% of the total hydropower capacity is owned by federal and public owners, as follow: federal 51% and public 22%. Different federal agencies are considered owners of different dams and hydro plants. Since an owner may own plants in more than one owner class (e.g., private non-utility and industrial) the total number is referred to as "presence" rather than owners.

The total number of owners, therefore, is 1,134, while the total number of presences is 1,152. As with the distribution of the plant population by owner class, the distribution of the plant ownership shows that approximately 70% of the plant ownership is in the private sector.

No doubt, we use all the hydropower we generate since it is clean and cheap, and we want more too. Hydroelectric power is the best and cheapest power available. It is reliable (to a point), and is totally clean at the customer end. So the question is why don't we use more of it to produce a lot (or all) of the electric power we need? The answer is quite simple—there is just not enough of it, and not much more can be produced due to lack of hydro resources.

There are no plans to build any large dams in the U.S, while small dams construction faces fierce opposition.

Hydropower is good, but the beginnings of each hydropower plant on a dammed river are bad and even ugly. This is due mainly to the fact that lots of water and a lot of land are needed to build a dam and create a large lake (reservoir) at the expense of the local environment and the people living in it. The local environment covered by the lake is lost forever, and once flooded it can never be restored to its original condition.

There are also a lot of controversies with existing dams, where fishing and other traditional activities of the locals have been disturbed or destroyed by hydro dams and river flow changes.

Large dam hydropower projects also require a lot of money, time, construction materials and effort. Today, most of the good locations for large dams and hydropower plants in the U.S. have been taken, so only smaller plants can be built.

While in the early part of the 20th century hydroelectric plants supplied about half of the nation's power, that number is down to about 7% today and falling, as gas and nuclear use increases. The only possible trend for the future is to build small-scale hydro plants on rivers and streams that can generate electricity for single communities. Each step in this direction, however, encounters opposition from the locals, so we do not expect to see a large number of small dams popping up in the U.S. ant time soon.

Global Hydropower

Worldwide hydropower represents about 20% of total global electricity production. Hydro power supplies more than 50% of national electricity in about 65 countries, more than 80% in 32 countries and almost all of the electricity in 13 countries.

China has been building large hydroelectric facilities in the last decade and now leads the world in hydroelectricity use. Canada and Brazil follow. In many countries it is the most important widely used renewable energy source. In addition to large hydro dams, which are no longer an option in most countries, hydropower is produced by using the power of large rivers and natural drops in elevation.

While large-scale hydropower is almost fully developed in developed countries, untapped hydro resources which are still abundant in Latin America, Central Africa, India and China.

The world's total technically feasible hydropower (the amount of hydropower that can be achieved if no other conditions were considered) is estimated at approximately 15,000 TWh/year, of which nearly 13,000 TWh/year is currently considered economically feasible for development. North America generates about 700 out of 3100 TWh/year already in operation around the world, with an additional 100+ GW under construction. Most of the remaining Hydropower potential is in Africa, Asia and Latin America.

Table 6-10. Global hydropower potential vs. use

Continent	Technically feasible	Present use
Africa	1700 TWh/year	300 TWh/year
Asia	6800 TWh/year	1600 TWh/year
N. America	1600 TWh/year	700 TWh/year
S. America	2600 TWh/year	500 TWh/year

Although there are significant natural hydropower resources (rivers, stream, etc.), as seen in Table 6-10, using them is not that easy. Countries, such as China India, Iran and Turkey, are undertaking large-scale hydro development programs, and there are projects under construction in about 80 other countries around the world. According to the recent world surveys, a number of countries see hydropower as the key to their future economic development. Some of these, in addition to the above mentioned, are: Sudan, Rwanda, Mali, Benin, Ghana, Liberia, Guinea, Myanmar, Bhutan, Cambodia, Armenia, Kyrgyzstan, Cuba, Costa Rica, and Guyana.

Nevertheless, building a large hydro dam is not an easy, cheap, nor a quick thing to do. While planning is easy, the actual construction—including the land area clearing and modifications—is a huge undertaking with many social, political, and economic consequences and ramifications. Funding such large projects of national size is another barrier.

Addition of smaller hydropower projects on rivers and streams is easier, but the generated power is much less and the economics often do not jive. Climate warming is another factor, since it is changing the weather patterns, which affect the rain and snow fall. And so, it remains to be seen how many, if any, of the grandiose plans do materialize.

In all cases, we must remember that water coming from above is needed to operate hydropower plants, and that we have no control of its timing or quantity. This is a serious limiting factor that plays a major role now, and will play even greater role in the future, as far as hydropower generation is concerned.

The world's oceans, however, contain unlimited amount of water. It is not good for human consumption, but can be used for power generation. Not easy to do, but possible, and so the efforts continue as follow:

OCEAN ENERGY

The world's oceans contain the largest amount of potential energy on Earth. Using this energy is a daunting task that may never be completed.

The world's oceans contain immeasurable amount of water, which contain equally immeasurable and never ending, renewable, and clean (potential) power. The key here is *potential*, because although the energy is there for the taking, taking it out of the ocean and using it is not easy. As a matter of fact, it is extremely difficult, very expensive, and usually nearly impossible. This is the main reason we put ocean energy in the *other energy sources* category.

Nevertheless, there are many efforts underway—large and small—to capture, tame, and use the (potential) energy contained in the world's oceans.

There are basically three different ways to look at ocean energy. These are:

- Wave power,
- Tidal power, and
- Thermal energy phases

Ocean Wave Energy

Kinetic energy (movement) exists in the moving waves of the ocean and can be used to power a turbine, which in turn can generate electric power.

In Figure 6-22, the wave rises into a chamber, where the rising water forces the air out of the chamber. The moving air spins a turbine, which can turn a generator. When the wave goes down, air flows through the turbine and back into the chamber through doors that are normally closed.

This is only one type of wave-energy system. Others actually use the up and down motion of the wave to power a piston that moves up and down inside a cylinder.

That piston can also turn a generator. Most wave-energy systems today are very small for the purpose of investigating their work function and the future possibilities. Many of them can be used to power a warning buoy or a small light house. Larger systems in the future could supply power to harbors, and coastal towns.

Ocean Tidal Energy

Another form of ocean energy is the energy created and contained in the periodic tides. When tides come into the shore, usually on regular intervals, they can be trapped in reservoirs behind dams. Then when the tide drops, the ocean water behind the improvised dam can be let out just like in a regular hydroelectric power plant.

In order for this to work well, however, a large increase in tides height is necessary. An increase of 16 feet between low and high tide is the minimum needed for significant power generation. There are only a few places around the world where such large tide changes can be found, and some power plants are already operating on this approach. One such power plant in France, for example, makes enough energy from tides to power 240,000 homes.

Another, more practical for small applications, method for tidal power generation is depicted in Figure 6-23. Here the tidal waves are directed into a channel, where they speed one way when the tide comes and, and back the other way when the tide goes out. A turbine is located in the channel and its blades turn with the tides coming in and out. The rotational energy is converted into electricity by an electrical generator, and is used in the power grid or for directly powering electrical devices.

Ocean Thermal Energy

The ocean thermal power generating method uses temperature differences in the ocean. If you have ever

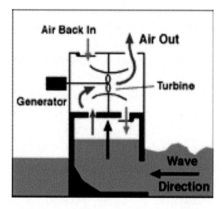

Figure 6-22. Ocean wave engine

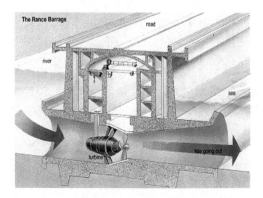

Figure 6-23. Ocean tidal engine

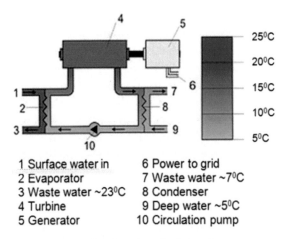

1 Surface water in   6 Power to grid
2 Evaporator   7 Waste water ~7°C
3 Waste water ~23°C   8 Condenser
4 Turbine   9 Deep water ~5°C
5 Generator   10 Circulation pump

Figure 6-24. Closed cycle thermal system

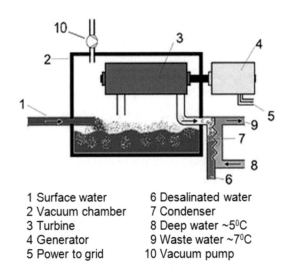

1 Surface water   6 Desalinated water
2 Vacuum chamber   7 Condenser
3 Turbine   8 Deep water ~5°C
4 Generator   9 Waste water ~7°C
5 Power to grid   10 Vacuum pump

Figure 6-25. Open cycle OTEC system

swam in the ocean, diving deep below the surface you would have noticed that the water gets colder the deeper you go. It's warmer on the surface because sunlight warms the surface water. But below the surface, the ocean wager gets very cold. That is why scuba divers wear wet suits when they dive down deep. The wet suits trap the divers' body heat to keep them warm.

Power plants can be built that use the difference in temperature to make energy, but a temperature difference of about 20.5 degrees Celsius (38 degrees Fahrenheit) is needed between the warmer surface water and the colder deep ocean water.

There are several types of ocean thermal energy systems:

Closed cycle system

Closed-cycle systems use fluid with a low boiling point, such as ammonia, to power a turbine to generate electricity. Warm surface seawater is pumped through a heat exchanger to vaporize the fluid. The expanding vapor turns the turbo-generator. Cold water, pumped through a second heat exchanger, condenses the vapor into a liquid, which is then recycled through the system.

The first mini ocean thermal experiment, done in 1979, achieved the first successful at-sea production of net electrical power from closed-cycle thermal system. The mini ocean thermal energy generating vessel was moored 1.5 miles (2.4 km) off the Hawaiian coast and produced enough net electricity to illuminate the ship's lights and run its computers and television set.

Open cycle system

Open-cycle ocean thermal energy generator uses warm surface water directly to make electricity. Placing warm seawater in a low-pressure container causes it to boil. In some schemes, the expanding steam drives a low-pressure turbine attached to an electrical generator. The steam, which has left its salt and other contaminants in the low-pressure container, is pure fresh water. It is condensed into a liquid by exposure to cold temperatures from deep-ocean water. This method produces desalinized fresh water, suitable for drinking water or irrigation as well.

In other schemes, the rising steam is used in a gas lift technique of lifting water to significant heights. Depending on the embodiment, such steam lift pump techniques generate power from a hydroelectric turbine either before or after the pump is used.

The first vertical-spout evaporator to convert warm seawater into low-pressure steam for open-cycle plants was developed in 1984. Conversion efficiencies were as high as 97% for seawater-to-steam conversion (overall efficiency using a vertical-spout evaporator would still only be a few percent). In May 1993, an open-cycle ocean thermal plant at Keahole Point, Hawaii, produced 50,000 watts of electricity during a net power-producing experiment. This broke the record of 40 kW set by a Japanese system in 1982.

Hybrid thermal energy

A hybrid cycle combines the features of the closed- and open-cycle systems. In a hybrid system, warm seawater enters a vacuum chamber and is flash-evaporated, similar to the open-cycle evaporation process. The steam vaporizes the ammonia working fluid of a closed-cycle loop on the other side of an ammonia vaporizer. The vaporized fluid then drives a turbine to produce electricity. The steam condenses within the heat exchanger and could be used for desalination of sea water at the same time as well.

Similar ocean thermal energy conversion systems are being used in Japan and in Hawaii presently in demonstration and small-scale pilot projects.

Water is becoming a huge problem in many areas of this world. There are countries in the Middle East and Africa that don't have enough water for drinking and basic necessities, let alone using it for hydropower. Some countries, like Kuwait and Israel are desalinating huge quantities of ocean water, thus converting it into potable water. Desalinating millions of gallons of salty ocean water every day is technically complex, expensive, and ecologically problematic process, but what is their choice? So let's take a close look at water desalination:

Water Desalination

Please note that the reason we include *water desalination* in the energy markets discussion at all is because:

- The desalination process uses huge amounts of energy (electricity or some sort of heat). In Israel, for example, the five water desalination plants along the coast line consume about 3.5% of the total electric power used in the country. This energy, however, provides 40% of the county's life-giving potable water. The desalination plants are so efficient that a significant amount of water is also exported to neighboring countries as a humanitarian gesture.

- Another well known, and even more important, fact is that fresh water supplies have been drastically reduced in some parts of the world due to global warming and draughts.

A little know fact is that over 1% of the world's population do not have access to potable water.

This 1% represents millions of people around the world. And the number of these unfortunate people continue to rise. These millions dependent on water supplies trucked from distant locations, or on desalinated water to meet their daily needs. As climate warming affects water availability, the situation is growing worse by the day.

The UN expects over 15% of the world's population to be lacking adequate water supply by 2020.

Get this: in only 5 years, every 7th person on Earth would be lacking clean potable water. There are many reasons for this anomaly, but the choices are limited.

Unless people figure out and implement a radically better water conservation measures, transport from distant places, or water desalination is the only option many of these unfortunate people have.

Water desalination (or desalinization) is a process that removes some (most) of the sand, salt, and other minerals and contaminants from ocean water in order to make it useful by humans and for irrigation. The modern process, known as reverse osmosis, involves forcing seawater at high pressure through a membrane that screens out the salt, leaving behind a heavily brackish residue.

Salt water desalination is not a new process. It has been used on many seagoing ships and submarines for decades. Most of the modern interest in desalination, however, is focused on developing quicker and cheaper ways of providing fresh water for human use. Along with recycled wastewater, this is one of the few rainfall-independent water solutions to the needs of a thirsty world with diminishing natural water resources.

Australia, which is also poor on natural water supplies traditionally have relied on collecting rainfall behind dams to provide their drinking water supplies. But as the rains are unpredictable and are also diminishing, Australia may have to take a close look at the Israel's example and start building water desalination plants.

The International Desalination Association reports that in 2011 there were 15,988 desalination plants operating worldwide, producing 66.5 million cubic meters of fresh water daily. This amount is enough to supply the needs of 300 million people. In 2013, these plants produced 78.4 million cubic meters, or 57% more than 5 years prior.

The largest desalination project in the world is

Figure 6-26. $1.0 billion Carlsbad water desalination plant

Ras Al Khair in Saudi Arabia, which produces over 1.0 million cubic meters of fresh water per day. The largest desalination plant in the U.S. is the new Carlsbad facility, expected to produce about 50 million gallons of fresh water daily, which is enough to supply 7% of the San Diego County residents.

So what is the future of water desalination? And why is it not used more extensively in California, for example, which is in the midst of an unprecedented draught that promises to get even worse in the coming years. Building several large ocean water desalination plants on the Pacific Ocean cost of California would resolve all its water problems…forever. Or would it?

Case Study. Water Desalination in California

The persisting and increasingly severe California droughts bring calls to build desalination plants up and down the Pacific seashore. With so much ocean water, how can the state be dying thirsty? All that has to be done, is install some equipment and run the ocean water through it. When done, pump it into the water supply pipelines, thus solving our water supply problems instantly and permanently. Easy, no? Nope, it is not so easy.

A $1.0 billion desalination project is scheduled to start operating in Carlsbad in 2016 and is attracting lots of attention. The new desalination plant is the largest of its kind in the U.S., and upon completion will provide San Diego County with about 50 million desalinated gallons a day, which is about 7% of its total water needs.

Note: In order to provide desalinated water to its entire population, San Diego County must spend over $14 billion for the construction of additional processing plants. And of course, the annual expenses of the O&M of all these plants, running non-stop up and down the California coast line would add several more billion dollars of annual expenses. Not a small thing, but there are other problems too,

People are watching the Carlsbad desalination experiment, since the future decisions will be based on its success.

Very few expect desalination to be a major component of the state's overall water supply in our lifetime, although there is a proposal to build a second desalination plant, in Huntington Beach. That plant is still awaiting approval from the California Coastal Commission and other authorities.

But some already think that this is a big mistake. Enthusiasm cannot replace the high costs of construction and the enormous energy demand. But the biggest resistance seems to originate from serious environmental footprint of the desalination process.

Southern California has become more dependent on fossil-fueled electric generation since the shutdown of the San Onofre nuclear power plant. The Carlsbad desalination plant will be moderating the effects of climate change on the region while contributing to the GHG emissions that cause it. So the plant is expected to buy carbon credits and restore local wetlands to offset its negative environmental impact.

While all agree that there are definite advantages to seawater desalination, since it is a reliable supply, independent of weather conditions like drought. But unfortunately it's still among the most expensive, energy intensive and polluting fresh water supply options.

There are, however, a number of serious issues related to desalination of large quantities of water needed by the nearly 40 million people living and visiting the great state of California. Some of the major issues related to large scale water desalination are:

Permitting

Desalination plants are huge facilities that require years of environmental and other studies in order to be issued a permit. This is because the construction and daily operations would destroy the coastal qualities that attract people to the shoreline. But the plants need to be close to customers, with enough room for their equipment and pipelines.

Three locations were studied for the new Carsbad desalination plant, before settling on the Carlsbad site. It happens to be next to the Encina Power Station, which allows it to share the warm seawater used by the power station, which reduced its cost and its impact on marine life. And yet, the new plant had to install a lot of extra equipment in order to reduce its marine impact during periods when Encina power plant is idling.

And in worst case scenario, if and when Encina power plant is decommissioned, which will happen eventually, the new desalination plant will have to submit a new environmental impact report and await issuance of a new permit.

Costs

Due to relatively high energy consumption, the costs of desalinating sea water are generally higher than the alternatives (fresh water from rivers or groundwater, water recycling, and water conservation), but alternatives are not always available and rapid overdraw and depletion of reserves is a critical problem worldwide.

Yet, initial cost and operational expense are major issues. The San Diego County Water Authority (SD-

CWA) is going to purchase the new desalination plant's entire output for 30 years. This arrangement was crucial for financing of the plant, and sets a price of about $2,100-$2,300 per acre-foot, accounted for inflation. An acre-foot is 325,851 gallons, which is about what an average five-member family would use.

This comes to about $110 million a year, which SDCWA will be paying, even if does not need the water. With this, the San Diego water bills will be raised by an average of $5-$7 a month per customer to cover the cost. The county estimates that it might pay about that much in the future for water imported from out of state. So, this commitment is a long-term hedge against a continuing water crisis. But if the draught ends, the county and its customers are stuck with the bills, which is a great waste of money.

In any case, desalinated water is far more expensive than the other options. San Diego, for example, pays $923 per acre-foot for imported water. It could also obtain recycled water for as little as $1,200 per acre-foot. So, desalinated water is about twice the price of the other options.

Santa Barbara is an example of a failed desalination effort, which began by building a $34-million desalination plant during the drought-stricken 1980s. When it was completed in 1992, the draught had ended, and although the new facility went through a few weeks of pilot testing, it was mothballed and dismantled. Recently the city contemplated restarting the plant at a cost of $40 million, plus $5 million a year in operating costs. That would place the cost of desalinated water at about $3,000 an acre-foot. It will also raise the average monthly household water bills to $110 from today's average of $75 per month.

Australia has replicated this failure on a much larger scale by investing more than $12 billion in six desalination plants in 2006. The largest of these plants was twice the capacity of Carlsbad's, but four of these were mothballed in 2012, after the drought ended. Abundant, and free rains overfilled the country's water reservoirs at the time, making the desalination plants obsolete.

Waste Salt

The least visible cost, of course, is environmental damage. The natural ocean inflows suck up and kill larval marine organisms and maintains the natural cycle. One potential byproduct of desalination is salt, produced at the end of the desalination cycle. This salt, extracted from seawater, produces a heavy brine during the desalination process, and has to be pumped back into the ocean. Some settles on the ocean floor, while

the rest increases the ocean salinity. The huge amount of salt have a potentially destabilizing effect on the ecology around the plant's outflow.

As a result, some native marine organisms can't survive, while outsiders move in and change the marine environment. The environmental groups are protesting with the argument that the ocean isn't a big garbage can.

Air pollution

While desalinated water might be a necessity in many locations of the near future, we must always consider their environmental impact. Especially the large amount of energy needed to operate the pumps and other equipment in these huge facilities. Desalination plants use about 15.0 MWh of electric power for every million gallons of fresh water produced. In comparison, reusing wastewater uses almost half—8.0 MWh of power for the same volume of fresh water. At the same time, importing a similar amount of water into Southern California from out of state sources requires 14.0 MWh of electricity. Desalination, therefore, is more energy intensive and expensive than any of the options.

The new Carlsbad desalination plant would require 750 MWh electricity daily (50 million gallons x 15 MWh). Or about 274 GWh annually are needed to run the plant non-stop, year around. It needs its own power plant, but it doesn't have one, so where does this electricity come from?

Note: 750 MW is the power a medium to large size power plan generates. In order to supply 100% of the California water needs, 14 such power plants are needed for the grand total of 10.5 GW additional electric power generation. This means that California needs to add a dozen, or more, large power plants (1.0 GW in size each) to desalinate enough water for its population. So, we need to add another $20 billion (for the power plants) to the $14 billion cost of new desalination plants. Expensive proposition, no doubt, but as things are going, some version of it will be implemented in the future.

The new San Diego desalination plant would use the electric energy generated by gas- and coal-fired power plants, located mostly out of state. These are the power plants which emit the large amount of GHGs, we are all talking about. Not helping the global warming, are we? As a result, Santa Cruz and several other Northern California water districts, after taking a close look at the desalination technology, decided to cancel any such plans, mostly because of its huge expense and unpredictable environmental implications.

In summary, we need water to live and thrive. Drinking water comes mostly from snow and rain, and when it is not enough, we experience severe draughts. Water desalination is one option to dealing with fresh water scarcity, but it comes at a great expense. While most of us in the U.S. cannot relate to this problem, because all we have to do is turn the water faucet on and abundant, clean water comes out. The U.S. is also a rich country, so in worst of cases, we could afford water desalination.

This is not the case with the developing countries. And as their population and needs increase, the drinking water problem could grow quickly into a global disaster. Water desalination might be a solution for some of these people, but it requires a lot of energy, so let's think long and hard, before we waste billions of dollars, millions of GWh of electricity and pollute further the environment by building large desalination plants.

The solution is not easy, but we have many options today; including different energy technologies (including renewables) for water transport, sterilization, and desalination.

GEOTHERMAL POWER

Geothermal power represents the largest concentration of power on Earth, since it is fed by its overheated core deep underground.

The term *geothermal* comes from the Greek words *geo* (earth) and *therme* (heat), which put together mean earth's heat. And this is exactly what geothermal power generation is—using the earth's interior, which is naturally and (almost) eternally very hot. This heat, fueled by a number of intensive nuclear and other types of reactions, create geothermal energy that we are learning to capture and use.

This thermal energy originates from the Earth's creation over 6 billion years ago. At the center of the Earth is the Earth's *core,* which is over 4,000 miles deep, and where ferocious thermo-nuclear reactions are at play, at temperatures of over 9,000°F.

Just thinking that there is such an inferno below our feet should make us humble and appreciative of the immense amount of energy available on our small planet.

This energy, however, is not there for the grabbing. It is available only at certain places and requires special equipment for its capture and use. Like any heated body, the heat in the Earth's core is continuously emitted and flows towards cooler bodies in the surroundings. In some

places the heat flows outward towards the surface, heating the rocks and earth as it cools down on its way up.

The nearby layer of rock, the *mantle, is heated red hot, and w*hen temperatures and pressures become high enough, some mantle rock melts, becoming *magma.* During this process the magma becomes less dense and lighter than the surrounding rock, it rises upward. This motion, or convection, makes it move slowly up toward the earth's crust, still hot but cooling down with every inch of the way.

If and when the hot magma reaches the surface, via volcanic explosions or through cracks in the ground, it flows down the hills like molten metal, which we know as *lava.* Most often, however, the magma remains contained in pools well below earth's crust. The pools of magma are heating nearby rocks and water in the water table or rainwater seepage. The rocks can get very hot—nearly 1000°F—and can keep the nearby water very hot too.

Some of the heated water flows up to the surface through cracks, where it forms *hot springs* or *geysers.* Most often, however, it stays in pools deep underground, where trapped in cracks and porous rock it forms natural formations called *geothermal reservoirs.*

Geothermal Electricity Generation

So, here we have a totally different type of energy, ready to be extracted and used. People have used the hot water on the Earth's surface for relaxing and therapeutic activities since the beginning of time. Today, this free energy is used in other much more creative and practical ways; mostly for heating buildings, and generating electric power.

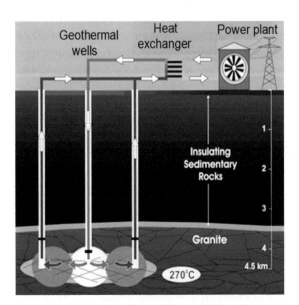

Figure 6-27. Geothermal power generation

The process of generating electricity from geothermal reservoirs is quite simple. We simply drill deep wells into the hot geothermal reservoirs (where available close to the surface), pump the hot water to the surface. Here we can use it (and the *geothermal power in it*) to heat buildings or to make steam. Thus produced steam is sent into a turbine, where it provides the force to spin the *turbine blades* and the *generator*, which produces electricity. The used, and cooled, geothermal water, exiting the turbine is then returned down the *injection well* into the reservoir to be reheated, thus the cycle can be repeated ad infinitum.

There are three kinds of *geothermal power plants,* depending on the temperatures and pressures of the underground geothermal reservoir, which can be divided into.

Dry steam reservoir

A "dry'" steam reservoir produces a lot of steam but very little water. The steam can be piped directly into a *"dry" steam power plant* to provide the force to spin the turbine generator. There are, however, only a few known dry steam reservoirs around the globe. The largest dry steam field in the world is The Geysers, about 90 miles north of San Francisco. Production of electricity started at The Geysers in 1960, at what has become the most successful alternative energy project in history.

Hot Water Reservoir

A geothermal reservoir that produces mostly hot water is called a "hot water reservoir" and is used in a *"flash" power plant*. Water ranging in temperature from 300-700°F is brought up to the surface through the production well where, upon being released from the pressure of the deep reservoir, some of the water flashes into steam in a 'separator.' The steam can be then used for heating or to power electric power generating turbines.

Binary Systems

A reservoir with temperatures between 250 and 360 degrees F is not hot enough to flash enough steam but can still be used to produce electricity in a *"binary" power plant*. Here, the hot geothermal water is passed through a *heat exchanger*, where its heat energy is transferred to a second (binary) liquid, such as isopentane, that boils at a much lower temperature than water. When heated, the binary liquid flashes to vapor, which, like water steam, expands upon release and can spins electric power generating turbine's blades. The vapor is then re-condensed to a liquid and is reused repeatedly. One advantage of such a closed loop cycle is that there

are no harmful emissions into the air.

The geothermal electricity generation is most prevalent where Earth's large oceanic and crustal plates collide and one slides beneath another. These are called subduction zones, with the best example being; the Ring of Fire bordering the Pacific Ocean. the South American Andes, Central America, Mexico, the Cascade Range of the U.S. and Canada, the Aleutian Range of Alaska, the Kamchatka Peninsula of Russia, Japan, the Philippines, Indonesia and New Zealand.

Active geothermal energy sources are also found where these plates are sliding apart, such as in Iceland, the rift valleys of Africa, the mid-Atlantic Ridge and the Basin and Range Province in the U.S. There are also places called "hot spots," which are fixed points in the mantle that continually produce magma to the surface. In these cases the plate is continually moving across the hot spot, forming strings of volcanoes, such as the chain of Hawaiian Islands.

The countries currently producing the most electricity from geothermal reservoirs are the United States, New Zealand, Italy, Iceland, Mexico, the Philippines, Indonesia and Japan, but geothermal energy is also being used in many other countries.

Worldwide there are about 10 GW of geothermal power generated in over 20 countries. Of these, about 1/3 are generated in the U.S., which is equivalent to *not* burning over 60 million barrels of oil each year.

Other Uses of Geothermal Energy

Geothermal power can be used even when the water is not hot enough, for applications other than generating steam and electricity. The main non-electric ways for using low temperature geothermal energy are; a) direct use, and b) geothermal heat pumps.

Direct Use

Geothermal waters ranging from 50 degrees F to over 300 degrees F, are used directly from the earth in domestic and industrial applications, such as; soothing aching muscles in hot springs, and health spas (balneology); growing flowers and vegetables in greenhouses (agriculture); growing fish, shrimp, abalone and alligators to maturity (aquaculture); pasteurizing milk, drying onions and lumber and washing wool (industrial uses);

Space Heating

Heating individual buildings and entire population centers is the most common and oldest direct use of nature's hot water. Geothermal district heating systems pump geothermal water through a heat exchanger, where it transfers its heat to clean city water that is piped to buildings in the district. There, a second heat exchanger transfers the heat to the building's heating system. The geothermal water is injected down a well back into the reservoir to be heated and used again.

The first modern district heating system was developed in Boise, Idaho. In the western U.S. there are 271 communities with geothermal resources available for this use. Modern district heating systems also serve homes in Russia, China, France, Sweden, Hungary, Romania, and Japan. The world's largest district heating system is in Reykjavik, Iceland. Since it started using geothermal energy as its main source of heat Reykjavik, once very polluted, has become one of the cleanest cities in the world.

Community Uses

Geothermal heat is being used in some creative ways. For example in Klamath Falls, Oregon, geothermal water is piped under roads and sidewalks to keep them from icing over in freezing weather. In New Mexico and other places, rows of pipes carrying geothermal water have been installed under soil, where flowers or vegetables are growing. This ensures that the ground does not freeze, providing a longer growing season and overall faster growth of agricultural products that are not protected by greenhouses.

Geothermal Heat Pumps

We all know that deeper into the ground the temperature is relatively stable compared to the surface air temperature. Using this free energy, geothermal heat pumps (GHPs) are used to take advantage of the stable earth temperature (about 45-58^{0}F) just a few feet below the surface. Thus obtained heat can be used to keep households and business buildings' indoor temperatures comfortable.

The GHPs simply pump and circulate water or other liquids through pipes buried in a continuous loop (either horizontally or vertically) underground, next to a building to be heated. GHPs use very little electricity and are very easy on the environment. In the U.S., the temperature inside over 300,000 homes, schools and offices is kept comfortable by these energy saving systems, and hundreds of thousands more are used worldwide. The U.S. Environmental Protection Agency has rated GHPs as among the most efficient of heating and cooling technologies.

Depending on the weather, the GHP system could be used for heating or cooling.

- *Heating uses* Earth's heat (the difference between the earth's temperature and the colder temperature of the air) to be transferred through pipes into the buildings to be heated.
- *Cooling is used d*uring hot weather seasons, where the continually circulating fluid in the pipes 'picks up' heat from the building—thus helping to cool it—and transfers it into back to reservoirs deep in the earth.

Shallow ground heat is one of the most cost-effective from initial capital point of view. The earth's temperature a few feet below the ground surface is relatively constant everywhere in the world (about 45-58 degrees F), while the air temperature can change from summer to winter extremes. Shallow ground temperatures are not dependent upon tectonic plate activity or other unique geologic processes, thus geothermal heat pumps can be used to help heat and cool homes anywhere and very cheaply so.

Geothermal Power Production

United States leads the world in geothermal electricity production with about 4.0 GW of installed capacity generated over 80 geothermal power plants around the country. The largest concentration of geothermal power plants in the world is located at The Geysers geothermal field in California. The Philippines is the second highest producer of geothermal power in the world, with close to 2.0 GW of capacity, which is nearly 20% of the country's electricity generation.

There are at least 1.5 million geothermal wells is Texas alone, most of them small and on private lands.

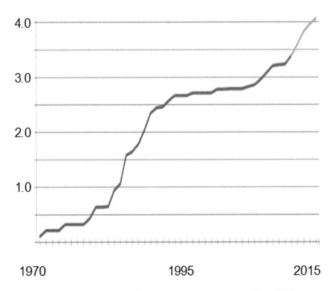

Figure 6-28. U.S. Geo-power generation (in GW)

The state looks like a Swiss cheese from that point of view. How many more holes can be drilled, and what is their use is uncertain, but geothermal power is increasing in importance, so expect to hear more good news about geothermal power in Texas in the coming years.

There are nearly 11 GW of geothermal power generated in 24 countries today, which generate over 65 GWh of electricity annually. This is about 20% increase in geothermal power online capacity since 2005, and which is expected to grow to nearly 20 GW by 2020. There are a number of projects presently under consideration, often in areas previously assumed to have little geothermal value, so these numbers seem plausible and reachable.

Al Gore, the guy who invented the Internet and discovered global warming, said that Indonesia is on its path to become a super power in geothermal electricity production in the near future. It is not exactly clear what 'superpower' means in this case, but Al has never been wrong before, so we will watch for the rise of Indonesia' superpower.

Several countries, El Salvador, Kenya, the Philippines, Iceland, and Costa Rica presently generate more than 15% of their electricity from geothermal sources. Enhanced geothermal systems, several kilometers in depth are operational in France and Germany and some are being developed or evaluated in several other countries around the world. The advance of new types of technologies might bring geothermal power to places that were not considered possible until recently.

Iceland is an example of efficient use of geothermal energy.

Case Study: Iceland

Iceland has one of the world's most productive geothermal installations and most geo-power use per capita. This is because Iceland has the world's largest number of volcanoes per area, and population of only about 325,000 people, or about that of an average U.S. city. Considering this, and other factors, such as the fact that almost half of the country's population lives in the capital, Reykjavik, and year-round inclement weather, the energy data has to be translated in Icelandic terms.

The five geo-power plants, Hellisheiði Power Station (303 MW), Nesjavellir Geothermal Power Station (120 MW), Reykjanes Power Station (100 MW), Svartsengi Power Station (76.5 MW), and Krafla Power Station (60 MW) generate less than 600 MW electric power and heat. Nevertheless, geothermal power heats nearly 90% of the houses in Iceland and over 50% of the primary energy usage comes from geothermal sources.

Geothermal power has many applications in Iceland, but the majority, almost 60% of the energy is used for space heating in buildings, while 25% is used for electricity, and the remaining amount is used in many other areas, such as; fish farms, greenhouses, and residential applications.

The total potential for electricity production from existing high-temperature geothermal fields in Iceland is estimated at about 1500 TWh, so with a little more effort and expense, Iceland could become energy independent thanks to geothermal power.

Does this mean that Iceland has thousands of years supply of geothermal energy at the present level of use? Maybe, or maybe not… It all depends on many factors, some of which are not under the control of the Icelanders, and some of which might change the picture entirely and very quickly…if and when Mother Nature decides to intervene.

As the world's climate patterns and weather conditions change unpredictably and irreversibly, so could the geothermal energy availability and accessibility in Iceland. So, we must remind them that it is not advisable to put all their eggs in one basket.

Land on Fire

The PBS documentary, *Land on Fire*, takes a very close look at the volcanic nature of Iceland; an island created by volcanoes, many of which are still active. With so much fire underneath, Iceland has a lot of geothermal energy for the taking, but with that come problems. Sure enough, Iceland has a huge problem. Earth's geothermal sources pack immense amount of power. Just like nuclear power, geothermal power can bring a lot of benefits, but it can also kill. It does not ask for permission, and does not hesitate to do so at a whim.

Geothermal power is similar to nuclear power in bringing enormous amount of free energy, while at the same time causing a lot of damage.

Iceland is an island of many active volcanoes, spread across its landscape and even extending beneath the sea around the island. Most of these are along the Mid Atlantic Ridge, which runs in the middle of the island. This is a divergent tectonic plate boundary, separating the Eurasian and North American Plates, along the floor of the Atlantic Ocean.

In Iceland's most recent history the massive eruption of the Laki volcano in 1783 caused one of the greatest disasters in living history. Laki volcano erupted in the June of 1783 killing many thousands and spreading a massive plume of smoke and haze that covered most of Europe and parts of North America, and as far as Asia and North Africa. The impact was unprecedented weather patterns with violent thunderstorms and hailstorms, killing cattle in the fields and destroying crops.

Benjamin Franklin referred to Laki's eruption during a lecture in 1784 by saying, '...when the effect of the sun's rays to heat the earth in these northern regions should have been greater, there existed a constant fog over all Europe, and a great part of North America...' There was simply no summer that year over major parts of the world. This was followed by several years of unusual and extreme weather, due mostly to the prolonged sunlight obstruction.

Another famous event occurred in the spring of 2010, when Eyjafjallajökull volcano erupted and spit millions of tons of smoke, dirt, and particles in the air high above it. Although the explosion was relatively small in volcanic terms, the plume it emitted caused enormous disruption to air travel across western and northern Europe that lasted a week. Dust particles spread and suspended high up in the atmosphere are dangerous for jet engines. If swallowed by the turbines they could melt, clump together, and restrict the air intake, which would bring down any jet engine powered aircraft.

The main explosion was followed by additional localized disruptions during the next month. The eruptions finally stopped in October 2010, when snow on the glacier did not melt. Ash covered large areas of northern Europe for the duration and 20 countries closed their airspace to commercial jet traffic, which affected about 10 million travelers. The total cost of this event to the air transport, tourism, fishing, and agricultural sectors of many countries is in the billions of dollars.

Today, Iceland's volcanoes are getting loaded and ready for another explosion. The only question is which will go first and when. At that time, the land will and the sky will get black again and people will suffer the aftermath, as they have done for generations.

Living in Iceland might have its advantages, but just like living on top of San Andreas Fault in San Francisco, it has its disadvantages. People living in these dangerous areas are constantly watching and evaluating the situation, looking for the signs of trouble and hoping that the *next big one* does not harm them, their families, and properties.

Cost of Geothermal Power Generation

The cost of geothermal power installations consists of the cost of:

- Exploration,
- Site planning and design,
- Financing,
- Site and power plant permitting,
- Wells drilling,
- Power plant construction,
- Power plant operation and maintenance, and
- Site decommissioning

The cost of building a geothermal power plant heavily weighs toward the initial expenses. This is different from other power plants, where fuel supplies, transport and disposal, as needed to keep them running are major factors. Not so in geo-power's case. Here, fuel is free and non-polluting. Win-win situation.

Well drilling starts the actual project development. It is followed by resource analysis of the drilling information. Next is design of the actual power plant, be it for thermal use, electricity, or both. If the drilling information shows acceptable conditions for geo-power generation, such as; large quantity high temperature rocks at a reasonable depth, and the location is favorable (preferably close to a population center or large industrial enterprise), then plans for constructing a new geo-power plant can be made.

Power plant construction is usually considered concurrent with entire field development. The initial cost for the field and power plant is around **$2500 per installed kW** for large installations (over 1 MW installed capacity) in the U.S. It is somewhat higher—$3000 to $5000/kW—for smaller power plants and in other countries that need to import expertise and technology.

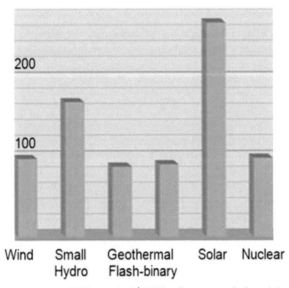

Figure 6-29. LCOE cost in $/MWh of generated electricity

Operating and maintenance costs are among the lowest in the energy field—ranging from $0.01 to $0.03 per kWh. Most geothermal power plants can run at greater than 90% availability, which means that they are fully operational (and at 100% capacity) 90% of the time. This is remarkable, and should be looked at as the very best the power generation industry has to offer thus far.

Running at higher availability of 97% or 98%, however, is also possible but it can increase maintenance costs. Because of that, only higher-priced electricity justifies running the plant 98% of the time, as needed to recover the higher maintenance costs.

Recent reports suggest that geothermal power may actually be cheaper than all other fuels, including coal, natural gas, nuclear, hydro, and all renewables. The U.S. stimulus package that includes $28 billion in direct subsidies for renewable energy, and additional $13 billion for research and development, gave renewable energy sources—geothermal power included—a shot in the arm recently, and so we expect some great things to happen soon in this area of the energy industry.

Estimates from Credit Suisse estimate geothermal power costs at 3.6 cents per kilowatt-hour, versus 5.5 cents per kilowatt-hour for coal generated power. This, however, does not include the elevated risks and a number of assumptions that have to be made in case of a new geothermal power plant development. And there are many serious concerns.

Nevertheless, the financial basics in the U.S. are there; starting with the tax incentives, which save about 1.9 cents per kilowatt-hour. Unfortunately, they won't last long, even if and when extended further into the future.

Table 6-11. Installation cost of conventional geothermal power plant (in $.kW installed)

Stage	$/kW	%
Exploration	14	0.4
Permitting	50	1.4
Steam Gathering	250	6.8
Exploratory Drilling	169	4.6
Production Drilling	1,367	37.5
Plant & Construction	1,700	46.6
Transmission	100	2.7
TOTAL	**3,650**	

The development of geothermal energy requires the consideration and evaluation of a number of factors, such as site (geography), geology, reservoir size,

geothermal temperature, and plant type. The majority of the overall cost is typically attributed to construction of the power plant, due to the high cost of raw materials including steel. The second highest cost intensive processes are the exploratory and production drilling stages, which put together cost as much as the plant construction.

Geothermal power production is relatively capital-intensive with high first-cost and risks, but it has fairly low operating and maintenance (O&M) costs and a high capacity factor. This makes it one of the most economical baseload power generation options available. A number of factors contribute to the cost of developing a geothermal power plant, with the power conversion technology having some, but nor great effect on the final cost.

Low-temperature reservoirs typically use binary power plants, while moderate- to high-temperature reservoirs employ dry steam or flash steam plants, based on whether the production wells produce primarily steam or water, respectively. The different technologies show similar construction costs and O&M expenses.

One thing to keep in mind here is the fact that the significant up-front capital requirements, pervasive resource and development uncertainty, and long project lead times bring greater risk-related mark-up, as compared to other renewable and traditional energy alternatives. These factors, combined with current economic conditions, mean private firms seeking to develop geothermal projects may face greater difficulties in obtaining the requisite capital for exploration and development. Industry analysts suggest that although financing is still available, the terms will be less attractive to investors and developers.

Most cost estimates rely on the *levelized cost of energy* (LCOE), which represents the total cost to produce a given unit of energy, usually in $/MWh. This figure assumes that the money to build a new geothermal plant is available at reasonable interest rates.

This, however, is not the case today. Although there is interest in geothermal energy generation, the general consensus is that it's difficult to get to and that it's expensive to produce. Therefore, only very low risk and outstanding in their profitability projects can get even credit card interest rates, according to insiders. This means that the up-front costs are too high for a profitable project. So, utilities and other companies prefer to spend their money on lower front-end costs, like natural gas powered plants, which are cheap to build but relatively expensive to operate on the long run, due to the cost of fuel needed to run them.

Natural gas LCOE is averaged at $0.052 per kWh (vs. $0.036/kWh for geothermal power) but is finding greater use because it can be deployed anywhere. This is a great advantage over most other location-specific technologies—including geothermal power.

Only 13 U.S. states have identified significant geothermal resources, which severely limits its usefulness, since it is not possible to transport it like gas or oil. Nevertheless, there are estimates that the U.S. has over 30 GW of geothermal energy supplies that could be exploited using today's technology. These supplies are also estimated to last indefinitely, but, of course, all this is to be tested, verified, and proven.

Advantages of Geothermal Energy

There are a lot of advantages of using geothermal heat for making electricity, and cooling/heating homes and businesses. Geothermal power plants, like wind and solar power plants, do not burn fuels to make steam to turn the turbines. Generating electricity with geothermal energy helps to conserve nonrenewable fossil fuels, and by decreasing the use of these fuels, we reduce emissions that harm our atmosphere. There is no smoky air around geothermal power plants, and in fact some are built in the middle of farm crops and forests, and share land with cattle and local wildlife.

For over ten years, Lake County California, home to five geothermal electric power plants, has been the first and only county in the U.S. to meet the most stringent governmental air quality standards in the U.S.

The land area required for geothermal power plants is smaller (per megawatt) than for almost every other type of power plant. Geothermal installations don't require damming of rivers or harvesting of forests—and there are no mine shafts, tunnels, open pits, waste heaps or oil spills.

Geothermal power plants are designed to run 24 hours a day, all year, providing consistent and more reliable power than any other technology. A geothermal power plant sits right on top of its fuel source, so there is no need for mining, processing, transport, etc. operations. It is resistant to interruptions of power generation due to weather, natural disasters or political rifts that can interrupt mining, processing and transportation of fuels. Geothermal "fuel"—unlike the sun and the wind—is always on and the economic benefits remain in the region, and no fuel price shocks are expected.

Geothermal power plants can have modular designs, with additional units installed in increments when needed to fit growing demand for electricity.

And finally, geothermal projects, with all their benefits could help developing countries grow their energy system efficiently and without pollution. Installations in remote locations can raise the standard of living and quality of life by bringing electricity to people far from "electrified" population centers.

And if this is such a wonderful energy source, then why is it not used more frequently around the world? There are a number of reasons, some of which are; unavailability of equipment, experienced staff, and infrastructure issues. All these hinder the proper and efficient location, installation and operation of geothermal plants across the globe. Not enough skilled manpower and availability of suitable to build locations (in addition to funding barriers) have been identified as the most serious problems in adopting geothermal energy globally.

Getting geothermal energy requires installation of power plants equipped with expensive equipment, as needed to pump up hot water from deep within the earth and converting it into useful electric power. All this requires huge one time investment, a certified installer, and the relocation of experienced staff to the remote power plant location. Then, thus generated electric power has to be delivered to the customers, who are usually far away, so another great expense is needed for adding piping and/or electricity infrastructure that has to be set up in order to move the steam and electric power. This alone often makes the entire project prohibitively expensive and limits its financing opportunities.

Geothermal energy is only practical;
a) in regions with hot rocks (not too deep) below the Earth's surface, and
b) if it can produce steam over a long period of time.

To identify such places, great research is required which is done by the companies before setting up the plant. This initial investment (time and money) is significant and only a few can afford it, since it increases the total expense in setting up a geothermal power plant.

There is also a potential of higher level of corrosion in some sites, due to dissolved minerals in the water used in the pipes and the power plant. This might increase the maintenance cost significantly and reduce the profits accordingly.

Since geothermal sites penetrate miles below the Earth's surface, the drill holes and hot rock reservoirs often contain toxic gases which can escape to the surface and there is a fear of toxic substances getting released into the local atmosphere. The toxic gasses can also travel deep within the Earth and contaminate the water table, all of which is a liability to the investors and owners.

There is also a real danger that any geothermal site can run out of steam over a period of time. This could be due to drop in temperature, as a result of shifts in the Earth layers, and/or if too much water is injected to cool the rocks. Since these are unpredictable events in most cases, the risk of financial loss for the investors is high.

In summary, many thousands of megawatts of heat and electric power could be generated from already-identified geothermal resources. The major obstacle is the remoteness of the geothermal sites, which increases the initial cost of power plant construction and maintenance. The remoteness also requires additional and equally expensive pipelines and transmission lines to bring the power to populated centers or the grid.

Usable geothermal resources are not limited to the "shallow" hydrothermal reservoirs at the crustal plate boundaries. Much of the world is underlain (4-6 miles below the surface), by *hot dry rock* which in most cases contains no water, but packs a lot of heat. This heat is there for the taking, as difficult the undertaking and unpredictable the results it might be. New technologies allow work at these depths, but here again, cost is a major issue.

The U.S., Japan, England, France, Germany and Belgium are experimenting with piping water into very deep hot rock layers to create more hydrothermal resources for use in geothermal power plants. As drilling technologies improve, allowing us to drill much deeper, geothermal energy from hot dry rock could be available anywhere. They will also get cheaper with time, which will allow us to tap the true potential of the enormous heat resources in the earth's crust and beyond.

BIO-MASS

Bio-mass in this text refers to any and all vegetable species that can be used for generating heat or electric power. The resulting heat and electricity, on the other hand, are considered to be bio-energy. So, although the terms *bio-mass* and *bio-energy* are often used interchangeably, there is a difference. Bio-mass is the material, while bio-energy is the final product of bio-mass use and conversion into some sort of energy.

Bio-mass is bio-energy resource that contains large amount of energy that can be harvested at regular time periods.

Efficient, cheap, and safe use of biomass for heat or electricity, however, requires the creation of new ap-

proaches, technologies, and even new global socio-economic and political systems. Bio-energy generation has been shown to interfere with the global food supply, so at this time it is the most controversial of all energy technologies. The scale of this controversy goes wider and deeper than any other such related to energy sources.

This is because although bio-energy materials are plentiful and renewable, they are subject to resource overuse and abuse. In many cases, the demand outstrips the supply, which causes distorted and even criminal responses on part of suppliers or customers. And since food is concerned, the controversies are overwhelming.

Also, contrary to some opinions, serious greenhouse gas (GHG) emissions are produced during the cultivation and use of bio-materials. The problem is that bio-energy materials (forests and crops) are grown in one area, but transported to and used in other. This way, the environment in the growing areas remains relatively clean, while that of areas that use biomass in large quantities gets increasingly polluted.

The major components of the bio-energy category are: biomass (natural woods, vegetation, and agricultural crops), bio-fuels and their sub-components, as we will see below in this text. Bio-energy, in the general meaning of the word, is renewable energy that is contained in, and can be extracted from, materials created by natural biological sources (generally referred to as bio-mass). In more specific terms, bio-mass is any organic material which has converted and stored sunlight energy in its structure in the form of chemical energy. This energy can then be extracted and converted into heat or electricity.

Wood, wood waste, straw, manure, sugarcane, and by-products from many agricultural crops and processes are considered bio-energy sources.

Presently, bio-energy is the world's largest renewable energy resource. It provides about 10% of world primary energy supply, and plays a crucial role in many developing countries. Here it provides basic energy for cooking and space heating, often at the price of severe health and environmental impacts.

The deployment of advanced biomass cook stoves, clean fuels and additional off-grid biomass electricity supply in developing countries, are key measures to improving the current situation and achieving universal access to clean and safe energy facilities by 2030.

Bio-energy is also used to generate electricity for the grid, mostly in developed countries. There are about 45 GW of globally installed bio-energy capacity for electricity generation, of which 12 GW was in the

U.S. Its role as a transportation fuel is expected also to increase from about 2-3% today to 25-30% by 2050. This way, bio-fuels play an important role in enhancing our energy security and reducing the CO_2 emissions of the transport sector. That role is expected to grow fast, so we just have to hope that it does not damage or destroy the already fragile global energy market. The danger of damage to the socio-economic systems in some developing countries, due to overuse of bio-mass, is even greater; with significant and longer lasting consequences.

In its most narrow sense, bio-energy is a synonym to bio-fuel, since it is fuel derived from biological sources. In its broader sense, it includes biomass, the biological material used as a bio-fuel, as well as the social, economic, scientific and technical fields associated with using biological sources for energy. This leads to a common misconception about these terms, so we need to remember that bio-energy is the energy extracted from bio-mass.

Bio-mass is the fuel, while bio-energy is the energy contained in the fuel.

In many cases, the terms *bio-mass*, *bio-energy* and *bio-fuel* are used interchangeably. We must remember, however, that the major difference between these is that:

- Bio-mass is the natural organic matter, which can be used *as is* to produce *bio-energy*, by burning it directly, without any processing, or any modifications.
- Bio-fuels, and their sub-components, on the other hand, are usually derived from processing bio-mass and other organic or inorganic materials.

In other words, *bio-mass* is the raw material from which bio-fuels are produced. Looking from another vintage point, *bio-mass* consists of mostly naturally grown materials, which require little, if any, human assistance. *Bio-fuels*, on the other hand, need 100% human intervention, a lot of equipment and energy, in a long chain of complex processes.

Now, that we have thoroughly cleared the *bio-mass* vs. *bio-energy. vs. bio-fuels* controversies and all bio-mass related misconceptions, we will take a closer look at this important, but often ignored or misunderstood, energy source.

Background

The official definition of *biomass* is any organic, decomposable, matter derived from plants or animals available on a renewable basis. This includes wood and agricultural crops, herbaceous and woody energy crops, municipal organic wastes as well as manure.

Traditional biomass use refers to the use of wood, charcoal, agricultural resides and animal dung for cooking and heating in the residential sector. It has very low conversion efficiency (10% to 20%), unsustainable supply, and is a source of significant GHG emissions.

Biomass, for the purposes of this text, is any organic material classified as renewable energy source, such as; wood, crops, and all types of animal, vegetation and other biological wastes. Biomass is a type of renewable energy source, and while it is replenished by nature in most cases, when used it is by no means green, or clean, as far as the environment and the fight against climate change are concerned, as we will see in this text.

The amount of energy released when a given unit of fuel is combusted is known as its energy content. For example, the energy content of wood is generally in the range of 6 to 18 megajoules per kg (MJ/kg) of wood, depending on the type and moisture content of the wood.

Freshly cut wood could have as much as 60% moisture and would have a relative low energy content (e.g., 6 MJ/kg), whereas oven-dried wood with close to zero moisture content could have up to 18 MJ/kg. An average commonly used value for wood is 15 MJ/kg.

The energy content in some common energy sources are approximately:

Table 6-12. Energy content in different materials

Material	Energy	
	MJ/kg	kW
Crude oil	42	11.7
Coal	27	7.5
Natural gas	18	5.0
Paper	17	4.7
Dung (dry)	16	4.4
Straw (dry)	15	4.2
Wood	14	3.9
Domestic waste	9	2.5
Grass (fresh cut)	4	1.1

The biomass production rate in terms of the quantity of biomass that can be grown on a parcel of land per unit time, is generally given as kilograms of biomass per hectare per year (kg/ha/yr). The production rate varies greatly depending on the crop, soil type, availability of water, and moisture content of the crop.

Some representative yields of biomass products in kilogram/hectare/year are:

Table 6-13. Yield of energy crops

Wood	20,000 kg/ha/yr.
Wheat, rice, and sorghum	15,000 kg/ha/yr.
Sugar cane	35,000 kg/ha/yr.

* **Note:** 1 hectare (ha) is 10,000 m² or about 2.5 acres.

For liquid biofuels, the energy content is usually given in terms of megajoules per liter (MJ/L). Representative values for the energy content of crops grown for biofuels are:

Gasoline	35 MJ/L
Sunflower oil	33 MJ/L
Castor oil	33 MJ/L

The biomass production rate for plants to be grown for biofuels, given as liters per hectare or L/ha varies greatly too. Some representative values for plants grown as biofuels are:

Sunflower	952 L/ha
Soybean	446 L/ha
Castor oil	1,413 L/ha

Examples:

Assuming one crop per year, a hectare of castor beans would produce the equivalent of approximately 1,400 L of gasoline. Given an average crop yield of 15,000 kg/ha/yr, and an energy content of 15 MJ/kg, the amount of thermal energy that can be expected to be available from cultivated wood production would be about 225,000 MJ/ha/yr, which would be about the same amount of energy contained in approximately 8 tons of coal.

The biomass power generating industry in the U.S. consists of approximately 11 GW of summer operating capacity actively supplying power to the grid, which produces about 1.4% of the U.S. electricity supply.

The 140 MW New Hope Power Partnership is the largest biomass power plant in North America. It uses mainly sugar cane fiber (bagasse), and recycled urban wood as fuel to generate enough power for large industrial operations, and to supply electricity for nearly 60,000 homes.

The power generated by bio-mass reduces the import of 1 million barrels of oil annually, and by recycling sugar cane and wood waste, preserves landfill space.

We will now take a close look at the different types of biomass and bio-fuels:

Solid Biomass

Biomass is the common name for organic materials used as renewable energy sources such as; wood, crops, and waste. Solid biomass include a number of natural and man-made products, such as: wood, sawdust, grass trimmings, domestic refuse, charcoal, agricultural waste, nonfood energy crops, and dried manure.

Again, biomass is not to be confused with biofuels, which are a product of the organic material in the biomass. Biomass, then, refers only to the organic matter, which can be used as a renewable energy source in a number of different ways.

Raw biomass in a suitable to use form such as firewood can burn directly in a stove or furnace, as needed to provide heat or generate steam. When raw biomass is in a different form, such as sawdust, wood chips, grass, urban waste wood, agricultural residues, the typical process is to process it in order to increase the density of the as is biomass.

This process sometimes includes grinding to an appropriate particulate size to produce hogfuel, which, depending on the densification type can be from 1 to 3 cm in size, and which is then concentrated into a fuel product. The current processes produce wood pellets, cubes, or pucks of condenses biomass materials. The pellet process is common in Europe, and is typically a pure wood product. The other types of densification, used in commercial operations, are larger in size, and are compatible with a broad range of input feedstocks. The resulting densified fuel is easier to transport and feed into thermal generation systems, such as boilers.

Table 6-14. Energy value of bio-mass materials

Bio Fuel	Energy content (MJ/kg)
Coal (reference)	28
Commercial wastes	16
Domestic refuse	9
Dung, dried	16
Biogas	55
Newspaper	17
Oil waste	42
Straw, harvested, baled	15
Sugar cane residues	17
Wood, green	6
Wood, air-dried	15
Wood, oven-dried	18

One of the advantages of solid biomass fuel is that it is often a byproduct, residue or waste-product of other processes, such as farming, animal husbandry and forestry. This means that this type of fuel does not compete for resources with food production, although this is not always the case as we saw in 2008 with the ethanol production in the U.S.

More specifically, different biomass materials contain different energy levels.

1 ton dry wood	=	18 GJ = 5 MWh = 500 liters oil = 500 m^3 natural gas
1 ton air-dry wood fuel	=	11 GJ = 3 MWh
1 m^3 wood chip (.7 ton)	=	3.6 GJ = 1.0 MWh
1 barrel of crude oil	=	6.1 GJ = 1.7 MWh
1 barrel of crude oil	=	1.7 m^3 wood chip = 0.5 m^3 dry wood

There are a number of materials that qualify as sold biomass. Below, we will take a look at several of these materials, and their use as energy sources.

Although biomass is classified as a renewable energy source, it heavily depends on the whims of Mother Nature and many other factors.

Its environmental impact, and role in the fight against climate change are also debatable. In order to produce energy from biomass, the organic matter must be burned in some way. This releases carbon dioxide into the air, unlike the use of solar, wind, and other renewable energy sources.

Although the processing of biomass emits carbon dioxide, it is classified as carbon neutral fuel. This is due to the total cradle-to-grave carbon cycle of the biomass materials. The thinking is that while the crop grows, it absorbs carbon dioxide from the air, thus reducing the total GHG count in the affected area. On the other side of the cycle, however, GHGs are released back into the atmosphere when biomass is burned as is or in the form of biofuels. So, in practice lots of carbon dioxide is emitted when burning biomass, which affect the environment in the immediate area.

This way, some areas of the world are affected positively, while others are affected negatively. Firewood is one example of this effect on the environment. The forests, where firewood is grown enjoy clean air most of the year, while populated centers where firewood is burned, suffer from heavily polluted air.

And speaking of firewood...

Firewood

Until the middle of the nineteenth century, the major source of energy were coal and firewood. Firewood is one of the oldest energy sources that has been used for a number of purposes—cooking, heating, etc.—since the beginning of time. Even when coal became a major energy source, firewood was the next in line, mostly due to its unlimited supply from forests. It accounted for over 90% of U.S. energy generation, mostly for home heating and cooking.

An average of 18 cords (stack of 4x4x8 feet) of wood per year were needed to heat an American house in the mid 1800. This is quite a waste of energy, since it is the heat energy equivalent of 2.5 times the amount of heat energy used to warm a typical home today. This is mostly due to the inefficiency of the old fireplaces, which had less than 10% efficiency rating, and which swallowed three quarters of all wood used at the time (approximately 75 million cords of wood). Stoves are about four times as efficient, but were not popular in the Americas at that time for convenience reasons. Wood stoves became fashionable in North America in the mod 20th century, and the wood consumption increased with time.

Wood also powered the growing industries of the 19th century too. This included steamships and trains running on wood fire, and burned nearly 8 million cords of wood a year at the time. The iron and steel industry were the major industrial users of wood, although there was a twist. Most of the energy for iron smelting came from charcoal, which was made out of 1.5 million cords of wood per year.

The charcoal used in iron smelting totaled some 700,000 to 750,000 tons. This compared to estimated 750,000 tons of charcoal are used every year by outdoor chefs in the back yard barbeques in this country.

Charcoal is still important worldwide. Brazil uses charcoal in 45% of its iron smelting. 3.6 million tons of charcoal are used in Brazil annually, and consumption is still increasing. Charcoal is also a major fuel in less developed countries such as Ghana and Kenya, where 250-300,000 tons are used every year.

Firewood stoves are fashionable today, and are selling at a rate of 200,000 per year in the United States. Over 50 million cords of firewood were are used annually for home heating alone, and wood and wood residue provided almost 2 quads of total energy that year, most of it in the wood products and paper industries. In these industries these "biofuels" served about half of the total energy needs too.

Consumption of fuel wood in this country reached

Figure 6-30. Daily supply of firewood.

a peak of almost 150 million cords in 1870, since at that time wood was the primary fuel. In comparison, the 100 million cords used today in the US account for only a few percent of the total primary energy input.

Even if the use of wood and biomass increases to a maximum, the total annual potential of biomass energy generation, could provide only about 1/4 of the U.S. primary energy input.

Firewood pollution

Firewood is the best kept secret of the energy markets, so we must point to the amazing fact that it's the world's major energy source. It is number one in terms of number of people using it, Btus produced, and amount of pollution emitted every day. Yes, firewood in form of brushes, branches, twigs and stumps of trees that are burned by millions of people in the developed and developing countries for a number of reasons, is the largest source of energy around the globe.

While burning wood in the developing countries is mostly by choice, over 600 million people in Sub-Saharan Africa, 800 million in India, 500 million in China and 100 million in South America are using firewood for their daily needs. Firewood use in these cases is by necessity, for these people have no other choice.

Deadwood is collected around the roadways and forests, and many live trees are cut for this purpose as well. As time goes on, the deadwood gets used, and more and more trees are cut from the forest, thus helping in the propagation of the deforestation phenomena in some parts of the world.

Consumption of firewood is estimated at approximately 800 kg. per person per year). This comes to only 2-3 lbs. per person per day, but it is a large amount of wood—especially in areas that have been cleared from trees and vegetation for years. This creates great problems, including deforestation and tensions, especially in the more densely populated areas in a number of countries.

Let's assume for the purposes of this rough calculation that 500 kg. firewood are used per person annually by approximately 2 billion poor people worldwide. Firewood has energy content equivalent to 0.35 tons of oil per ton of firewood, and assuming that firewood stoves and fire pits burn at maximum 15% efficiency, we can conclude that to replace the firewood used worldwide, we would need over 150 million tons of oil a year.

This is nearly a quarter of U.S. oil consumption, which means that the U.S.' 300 million inhabitants use 4 times the amount of oil used by all 2 billion poor around the world. Or, the U.S. uses over 20 times more energy per person than the entire developing world.

While this is not very surprising, the fact that these 2 billion people emit more pollution than any other group would come as a surprise to many. With every wood fire there is smoke. Lots of it. Toxic gasses belch in the air, in sometimes enclosed quarters, poisoning the inhabitants, and causing a number of health problems. This is one of the great challenges in front of the social reforms and global warning planners.

Firewood is also used in large quantities in the rural areas of developing countries. The 1973 oil crisis brought out an entire firewood burning industry in the US, complete with most efficient wood burning stoves and processes. There are also many fireplaces in the US cities that use natural wood to heat the houses during the cold months.

The firewood use in the developed countries is not of necessity. In most cases, it is a type of luxury, or convenience. Cuddling with a book by a roaring wood fire is part of the American Dream, and people very seldom pass the opportunity to do just that. Fire wood use has increased lately, so now there are many winter days with local wood burning ban, due to increased particulate or CO_2 pollution.

Wood burning is one of the largest energy sources and polluters in the world today.

It is the least known and discussed subject in the energy field, and is also one that most energy principles prefer to ignore. It is of no consequence to them, or their business, so why bother? And because of that, the entire issue has fallen between the cracks and has become a non-issue.

The majority of the wood burning people happen to live in the poorest segment of the world' population, and since there are not much business opportunities in this area due to lack of money, the wood burning is left alone and will continue this way for a long time.

It is hard to say if the extensive wood use is creating a problem big enough to classify wood burning as a non-renewable energy. In any case, we hesitate to call it *renewable*, simply because it leads to uncontrollable and extensive destruction of forests and vegetation in entire regions. Plus, it is simply not something we should encourage. Instead, we should find a way to replace it with a more sustainable and less damaging energy systems.

As a matter of fact, just think that one simple 100 Watt solar panel, installed on a hut's roof might solve the energy problems of an entire Sub-Saharan family. A hundred PV modules in the village might prevent a hundred families from wasting 2-3 hours gathering wood every day. This will save a lot of wood too, and will eliminate a lot of smoke emission in the process.

Multiplied by the millions of wood burning families, this might save millions of tons of wood, and that many toxic gasses from been released in the atmosphere. Doing this would be the greatest environmental success ever! This, however, is not going to happen any time soon, for a number of reasons; the main of which is indifference.

Note: Flying over some areas of Asia or Africa at night, one can easily notice hundreds and thousands of fires on the ground below. Even large cities, like Taipei, Beijing and others are marked by the fires and plums of smoke coming from their suburbs at night. Thousands of people burn brushes and tree trunks to prepare their supper, or to clear land.

In other places around the world one can see large areas of bare land, where forests and woodlands were before. These bare areas create additional problems, such as soil erosion, and are becoming increasingly troublesome in a number of countries.

Firewood burning cannot be eliminated completely, for it has some useful purposes, but providing alternative energy sources for millions of people in places such as India and Sub-Saharan Africa, who are forced to burn wood daily, would help their development, and contribute significantly to reducing deforestation and cleaning the environment from harmful gasses and the related effects.

Sawdust

Sawdust, or wood dust, is a by-product of commercial wood processes: cutting, grinding, drilling, sanding, or otherwise pulverizing wood. It is therefore composed of fine particles of wood. It is also the byproduct of certain animals, birds and insects which live in wood, such as the woodpecker and carpenter ant. It can present a hazard in manufacturing industries, especially in terms of its flammability. Sawdust is also the main component of particleboard materials, which are used in great quantities around the world.

A major use of fine sawdust is for particleboard, while coarser sawdust is used for wood pulp and paper making. Sawdust has a variety of other practical uses, including serving as a mulch, as an alternative to clay cat litter, or as a fuel in commercial processes, heating, and electric power generation.

Until the advent of refrigeration, it was often used in icehouses to keep ice frozen during the summer. It has been used in artistic displays, and as scatter. It is also sometimes used to soak up liquid spills, allowing the spill to be easily collected or swept aside. As such, it was formerly common on barroom floors, and used to make Cutler's resin. Mixed with water and frozen, it forms pykrete, a slow-melting, much stronger form of ice.

Sawdust is also used in the manufacture of charcoal briquettes, the invention of which is credited to Henry Ford who made some from the wood scraps and sawdust produced by his automobile factories.

Burning sawdust to generate electricity is not very common, but is still used in some special situations, mostly in developing countries.

In addition to the pollution problems described above, sawmill sawdust burners generate a lot of toxic gasses. Sawdust is also stored in large outdoor piles where it causes harmful leachates in local water systems, creating additional environmental hazard. This has become a serious problem for small sawyers and environmental agencies, putting them in a deadlock.

Of larger, global, concern are substances such as lignin and fatty acids that protect trees from predators while they are alive. These compounds leach in mass quantities into water as the trees decay. When sawyers process a large volume of wood in one place, large concentrations of these materials permeate into the runoff, and the toxicity they cause is harmful to a broad range of organisms.

Domestic and agricultural refuse

Also called garbage, domestic refuse is a waste type consisting of everyday items that are discarded by the public and end up in the municipal dumps, where they rot and emit all kinds of gasses. Large part of our waste consists of bio-mass materials, such as grass, leaves, brush and branches from back yard maintenance. These materials comprise a significant amount of the domestic waste, which also end up in the municipal dumps to add to the emissions with time.

Agricultural waste, consisting of dead vegetation and small animals is added in large amounts to the pollution emitters on daily basis. Manure, of course is inevitable sign of large animal herding, and can be found in large amounts in the fields and in the cow and pig feeding farms.

Large part of thus generated refuse can be used for biogas generation, but some has to be disposed of in municipal dump sites, landfills, and land spreading (in special circumstances). In many cases, the municipal solid waste can be used to generate energy. Several technologies have been developed that make the processing of solid waste for energy generation cleaner and more economical than ever before. This includes landfill gas capture, combustion, pyrolysis, gasification, and plasma arc gasification methods.

Older waste incineration plants emit high levels of pollutants, recently new technologies have significantly reduced this type of pollution. For example, EPA regulations in 1995 and 2000 under the Clean Air Act reduced emissions of dioxins from waste-to-energy facilities by more than 99% from 1990 levels. At the same time mercury emissions have been reduced by over 90% as well. These improvements allowed waste-to-energy source to have less environmental impact than almost any other source of electricity.

Table 6-15. Husbandry biofuels (potential)

Animal	Manure		Biogas	
	Kg/day	MJ/day	m³/day	MJ/day
Cow	40	62	1.2	26
Hen	0.19	0.9	0.18	3.8
Pig	2.3	6.2	0.18	3.8

Agricultural refuse, mostly in form of manure and feed scraps, is mostly organic matter that can be used as organic fertilizer in agriculture, where they contribute to the fertility of the soil by adding organic matter and nutrients. Large quantity of manure can be processed in digesters to produce methane gas, in a process similar to that of producing biogas.

Charcoal

Charcoal is a type of man-made fuel, produced from wood or vegetation and animal matter, via special process, where the raw material is burnt in a controlled presence of oxygen. The main reason for producing this type of fuel was, and still is, to convert unusable materials (even waste materials) into useful fuels that are easy to transport and use.

Charcoal is soft, brittle, light, with a dark grey to black appearance, very similar to coal. It consists mainly of carbon and ash, which remain after the water and other volatile compounds were removed from the raw materials. Its combustion properties are also close to these of coal.

In the past, charcoal was produced by piling wood logs, or other combustible materials, in a conical pile, which was covered to restrict contact with the air. A small openings at the base was used to control the intake of air, which was directed through a central air shaft serving as a flue. The fire was started at the bottom of the flue, and let to gradually spread upwards, which process took days at a time to complete.

Slow combustion is the key to producing quality charcoal. Small scale production yields 50% by volume, or 25% by weight, of charcoal, which produces heat equivalent to the heat produced by the starting raw materials. Today charcoal is produced by a special carbonization process, where small pieces of wood, sawdust, or other waste materials are burnt under controlled conditions in cast iron retorts, which produce charcoal of different quality and for different purposes.

When charcoal is made at 300°C it is brown, soft, friable, and readily inflames at 380°C. When it is made at higher temperatures it is hard and brittle, and does not fire until heated to about 700°C. Charcoal production is extensively practiced for the recovery of byproducts that have some useful heat content, such as tree branches, wood shavings, sawdust, etc.

There was massive production of charcoal during the last several centuries, which supported a large industry employing hundreds of thousands of workers in Central Europe and the UK. Thousands of acres of forests were cut, which created a major deforestation in those areas. Large wooded areas were cut and regrown cyclically, in order to provide a steady supply of charcoal at all times. Over-exploitation and lack of new supplies, and the increased demand for charcoal created a supply and demand unequilibrium, which facilitated the switch to coal and later on to fossil fuels for domestic and industrial use.

Biochar

Biochar is charcoal, which is used for particular purposes, especially as a soil amendment. It is a possible source of carbon sequestration as it produces negative carbon dioxide emissions, so it has the potential to help mitigate climate change, via carbon sequestration.

When mixed with soils, biochar can increase soil fertility, increase agricultural productivity and provide protection against some foliar and soil-borne diseases. Biochar is a stable, solid, rich in carbon material that can endure in the soil for thousands of years.

Biochar made from agricultural waste can substitute wood charcoal. As wood stock becomes scarce, this alternative is gaining ground. In some parts of Africa, for example, biomass briquettes are being marketed as an alternative to charcoal to prevent deforestation associated with charcoal production.

Bagasse

Bagasse is the dry, fibrous matter that remains after crushing sugarcane, blue agave, or sorghum stalks, and extracting their juice. It is mostly used in the production of biofuel, paper products, and some special building materials.

Sugarcane and other plants are taken to a processing plant, where the juice contained in them is extracted for use in foods.

Sucrose accounts for little more than 30% of the chemical energy stored in the mature sugarcane plant, while about 35% is in the leaves and stem tips, which are left in the fields during harvest. Another 35% of the energy in the plant is in the fibrous material (bagasse) left over from crushing and pressing the sugarcane stalks, which can then be used for making fuel gas.

Bagasse is often used as a primary fuel source for sugar mills, where it is burned in large quantity to produce enough heat to run an entire sugar mill. The energy in biogas can be also used to provide both; heat energy, used in the mill, and electricity, which is typically sent into the electricity grid. This allows the plants to be energetically self-sufficient and even sell surplus electricity to utilities.

An average sugar- or ethanol-producing plant could produce 500 MW electricity for self-use, and 100 MW for sale. The sale of power is expected to boom as new regulations force the utilities to pay "fair price." This type of power is also especially valuable to utilities because it is produced mainly in the dry season when hydroelectric dams, and the electricity produced by them are running low.

Estimates of the potential power generation from bagasse in Brazil range from 1,000 to 9,000 MW depending on technology. Higher estimates assume gasification of biomass, replacement of current low-pressure steam boilers and turbines by high-pressure ones, and use of harvest trash currently left behind in the fields.

Presently, it is economically viable to extract about 288 MJ of electricity from the residues of one ton of sugarcane, of which about 180 MJ are used in the plant itself. Thus a medium-size distillery processing 1 million ton of sugarcane per year could sell about 5 MW of surplus electricity. At current prices, it would earn $18 million from sugar and ethanol sales, and about $1 million from surplus electricity sales. Not bad.

With advanced boiler and turbine technology, the electricity yield could be increased to 648 MJ per ton of sugarcane, but current electricity prices do not justify the necessary investment. Presently the World Bank would only finance investments in bagasse power generation if the price were at least $0.068/kWh.

In many other countries (such as Australia), sugar factories significantly contribute 'green' power to the electricity supply. In the U.S., for example, Florida Crystals Corporation, one of America's largest sugar companies, owns and operates the largest biomass power plant in North America. The 140 MW facility uses bagasse and urban wood waste as fuel to generate enough energy to power its large milling and refining operations as well as supply enough renewable electricity for nearly 60,000 homes.

Researchers are also working with cellulosic ethanol, in order to optimize the extraction of ethanol from sugarcane bagasse and other plants viable on an industrial scale. The cellulose-rich bagasse is being widely investigated for its potential for producing commercial quantities of cellulosic ethanol. For example, Verenium Corporation is building a cellulosic ethanol plant based on cellulosic by-products like bagasse in Jennings, Louisiana.

Bagasse is being sold for use as a fuel (replacing heavy fuel oil) in various other industries too, including citrus juice concentrate, vegetable oil, ceramics, and tire recycling. The state of São Paulo, Brazil uses 2 million tons, saving about $35 million in fuel oil imports.

Bagasse burning is environmentally friendly compared to other fuels like oil and coal. Its ash content is only 2.5%, vs. 30–50% of coal-fired power plants, and it also contains very little sulfur. Since it burns at relatively low temperatures, it produces little nitrous oxides, which has a dual effect of; a) reducing some of the worst pollutants (acid rain especially), and allowing to introduce ways to reduce nitrous oxides generation (which is not possible at high sulfur levels).

The resulting CO_2 emissions are equal to the amount of CO_2 that the sugarcane burnt in the power plant absorbed from the atmosphere during its growing phase, which makes the process of cogeneration greenhouse gas-neutral.

All in all, sugarcane and bagasse are shaping as an integral part of our energy future.

BIO-FUEL

Bio-fuels play a major part in the transportation sector, and are expected to be even more important in the future.

Biofuels are types of fuel derived from biomass and other organic raw materials. Biofuels can be divided into several categories; solid, liquid and gas fuels.

• Solid biofuels are usually found in nature, and include wood, sawdust, grass trimmings, domestic refuse, charcoal, agricultural waste, nonfood energy crops, and dried manure. We took a close look to some of these above, so it suffices to say that the current commercial solid biomass processes are used worldwide to make a number of products from biomass that are convenient for use. These include; wood pellets, cubes, or pucks, which are most common in Europe, and are typically a pure wood product processed in some form and shape for commercial distribution and use.

• Other types of densification produce larger in size products, as compared to a pellet, and are compatible with a broad range of input feedstocks. The resulting densified biofuels (in form of logs and other shapes) are easier to transport and feed into thermal generation systems, such as boilers and cooking stoves.

• *Liquid biofuels* include bioethanol, biodiesel, etc. organic compounds in liquid form that have a high heat value. These fuels are produced from different raw biomass materials using special processes.

• *Gaseous biofuels* are often byproducts from the processing of solid or liquid biofuels. These include a number of bio-gasses that can be burnt onsite or delivered to other locations for burning.

We reviewed biomass, which is considered solid bio-fuel, above, so here we will take a close look at the liquid and gas-phase biofuels, which can be divided into first, second, and third generation biofuels:

• *First generation biofuels* are made from the sugars and vegetable oils found in energy crops (sugarcane, soy, corn, etc.), which can be easily extracted using conventional technology.

These are also called conventional biofuels, and include well-established processes that are already producing biofuels on a commercial scale. These biofuels, commonly referred to as first-generation, include sugar- and starch-based ethanol, oil-crop based biodiesel and straight vegetable oil, as well as biogas derived through anaerobic digestion. Typical feedstocks used in these processes include sugarcane and sugar beet, starch-bearing grains like corn and wheat, oil crops like rape (canola), soybean and oil palm, and in some cases animal fats and used cooking oils.

• *Second- and third-generation biofuels* are made from ligno-cellulosic biomass or woody crops, agricultural residues or waste, which makes it harder and more expensive to extract the required fuel.

These are also called advanced biofuels, and are basically conversion technologies, which are still in R&D, pilot or demonstration phase. They could be of the second- or third-generation biofuels, and some are ready for reclassification to the first generation category.

These include hydro-treated vegetable oil (HVO), which is based on animal fat and plant oil, as well as biofuels based on ligno-cellulosic biomass, such as cellulosic-ethanol, biomass-to-liquids (BtL)-diesel and bio-synthetic gas (bio-SG).

This category also includes novel technologies that are mainly in the R&D and pilot stage, such as algae-based biofuels and the conversion of sugar into diesel-type biofuels using biological or chemical catalysts. The lines between the second- and third-generation technologies are fading, as the state-of-the-art changes and provides more options.

Recently the boundaries between the biofuels of the first-, second- and third-generation have been getting even hazier, with many gaps and overlaps contributing to the confusion. In some cases, the same fuel might be classified differently depending on its use, different technology specifics and level of maturity, as well as heating value, GHG emission balance, and feedstock used in making the distinction:

In more practical terms, today's biofuels could be divided into the following categories.
1. Commercially available biofuels:
   a. Ethanol from sugarcane and corn
   b. Biodiesel, via trans-esterification, and
   c. Biogas, via anaerobic digestion (biogas)

2. Biofuels in early commercial stages:
   a. Hydro-treated vegetable oils
   b. Bio-methanol
   c. Bio-gas reforming (H_2 gas)

3. Biofuels in demonstration stages:
   a. Cellulosic ethanol
   b. BtL diesel from gasification
   c. Biobutanol and DME,
   d. Pyrolysis-based fuels, and
   e. Bio-SG (biogas)

4. Biofuels in some type of R&D stage:
   a. Biodiesel from microalgae,
   b. Furanics (novel biofuel),
   c. Hydrogen, novel bio-generation
   d. Sugar based hydrocarbons, and
   e. Gasification with reforming (H_2 gas)

For the purposes of this text, we will review the biofuels that have most practical applications today. To start with, we'll take a look at some of the biofuels currently in use:
- Bioethanol
- Biobutanol
- Biodiesel
- Bioethers
- Biogas
- Syngas

Biofuels made from processing biomass are considered much cleaner than petrol/diesel alternatives. Theoretically they could be considered carbon neutral, since the biomass they were made from absorbs roughly the same amount of carbon dioxide during growth, as when burnt. While this is true, the fact that vegetation is grown in one place, biofuels are processed in another, and burnt at a third place, contributes to the regional inequality of global pollution. The air at the first place (the agricultural fields) is usually clean, while that at the processing plants and the burning site are badly contaminated.

On top of that, biofuels are responsible for other environmental inequalities. In many cases, large areas of forest are cut down to make space for the plantation of biofuel suitable crops. This deforestation not only harms the carbon cycle, but also harms surrounding civilizations/tribes who live off the forest.

Many environmentalists argue that biofuel is a disaster in the making, and doesn't offer a significant positive long-term environmental impact. No doubt, biofuels have some drawbacks, which—together with the benefits—must be thoroughly understood, carefully evaluated, and efficiently and safely implemented in the global energy future.

Biofuels can contribute to global warming as a result of "carbon leakage," an example of which is large scale deforestation taking place in some areas of the world. It is a cause of carbon leakage, as we are reducing the worlds carbon absorption capacity, disturbing the natural equilibrium of carbon dioxide between the; atmosphere, biosphere, geosphere, and hydrosphere.

We must also take into account the energy involved during the planting, maintaining, harvesting, transporting, and manufacturing of the crops. And let's not forget the mountains of fertilizer, pestisides, and other chemicals that are used during the crop planting and growing cycles.

Water usage is another troublesome issue, especially large scale biofuel crops growing in desert-like areas, These crops require significant amounts of water several months at a time year after year. Water is becoming a precious commodity in many areas of the world, so it remains to be seen how much of it could be dedicated to growing biofuel crops.

The processes involved in biofuels production are energy guzzling operations, which also emit large amount of GHGs.

Some of these process are:
- *Hydrothermal processing.* is a chemical process where biomass can be processed in a liquid media (typically water) under pressure and at temperatures between 300-400°C. The reaction yields oils and residual solids that have a low water content, and a lower oxygen content than oils from fast pyrolysis. Upgrading of the so-called "bio-crude" is similar to that of pyrolysis oil.

- *Pyrolysis oil* can be produced by fast pyrolysis, a process involving rapidly heating the biomass to temperatures between 400-600°C, followed by rapid cooling. Through this process, thermally unstable biomass compounds are converted to a liquid product. The obtained pyrolysis-oil is more suitable for long-distance transport than for instance straw or wood-chips.

As a by-product of this process, bio-char can be used as solid fuel, or applied on land as a measure of carbon sequestration and soil fertilization. The oil can be processed in ways similar to crude oil, and several research efforts are currently undertaken to upgrade pyrolysis oil to advanced biofuels.

• Dimethylether (DME) is another biofuel that can be produced from methanol through the process of catalytic dehydration or it can be produced from syngas through gasifying ligno-cellulosic and other biomass feedstocks. Production of DME from gasification of biomass is in the demonstration stage, and the first plant started production in September 2010 in Sweden (Chemrec, 2010). DME is the simplest ether and can be used a s a substitute for propane in liquefied petroleum gas (LPG) used as fuel, and it is a promising fuel in diesel engines, due to its high cetane number.

• Biobutanol is used as a fuel in a number of applications, including unmodified internal combustion engines. It has a greater energy density (29.2 MJ/l) and is more similar to gasoline than ethanol, and could thus be distributed through existing gasoline infrastructure.

   Biobutanol can be produced by fermentation of sugar via the acetone-butanol-ethanol (ABE) process using bacteria such as Clastridium aceto-butylicum. Demonstration plants are operating in Germany and the US and others are currently under construction. Biobutanol can be produced from the same starch and sugar feedstocks that are used

for conventional ethanol. In addition sugars can also be derived from lignocellulosic biomass, using the same biochemical conversion steps required for advanced ethanol production. This underlines the need for enhanced research into the biochemical conversion of biomass to sugars.

• *Solar bio-fuels* are produced by processing biomass into syngas using heat generated by a concentrating solar plant, thus potentially improving the conversion efficiency and providing higher GHG emission savings. More demonstration plants and further research are needed to make the process more efficient and allow for commercial-scale operation. Upon full development, solar bio-fuels production method has the potential to become the best, most efficient, and cheapest fuel production in the country.

We will now take a close look at the key biofuels, remembering that some of them are well established and widely used, while others are still in R&D or small scale production phases.

Bio-ethanol

   Bioethanol, ethanol, or ethyl alcohol (C_2H_5OH) is a clear colorless liquid. It is biodegradable, low in toxicity and causes little environmental pollution if spilt. When burning, ethanol produces carbon dioxide and water. Bioethanol is actually ethanol, but produced from bio materials; instead of from petrochemicals. It is the fuel substitute mixed with gasoline, used in the U.S. and other countries for passenger and commercial vehicles.

   Ethanol is a high octane fuel and has replaced lead as an octane enhancer in petrol. Ethanol oxygenates the gasoline fuel mixture so it burns more completely and reduces polluting emissions. Ethanol fuel blends are widely sold in some states, mostly in the summer

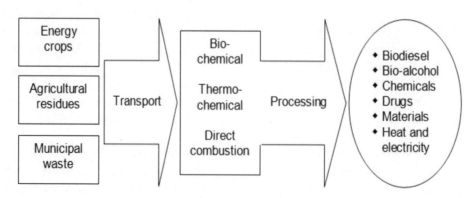

Figure 6-31. The bio-mass cradle-to-grave process

months, with the most common mix being 10% ethanol and 90% petrol, also called E10. Vehicle engines run well on this mixture, without the need of any modifications. Some hybrid vehicles can run on up to 85% ethanol and 15% gasoline blends, called E85.

In Brazil 25% ethanol is added to all gasoline mixes, resulting in E25 fuel mixture. Millions of flex-vehicles in Brazil can also run on pure, 100%, ethanol, or E100. As a matter of fact, Brazil has been running on ethanol since the 1970s, and why this practice is not widely implemented in other countries—including the U.S.—is a mystery, which we have been trying to figure out for decades now. We have large areas around the U.S. that have climate favorable for sugarcane and similar crops, which need only water and sunlight to grow. And yet, we produce very little of this precious crop.

Ethanol can be produced also by fermenting other biomass matter (types of vegetation and agricultural crops), or from petrochemicals by a reacting ethylene in these with steam. The so called *energy crops, however,* are the main sources of sugar required to produce bioethanol commercially. These crops are grown specifically for bioethanol production, and include; sugar cane, corn, maize and wheat crops, waste straw, willow and popular trees, sawdust, reed canary grass, cord grasses, Jerusalem artichoke, myscanthus and sorghum plants. Solid municipal wastes are another possible source of ethanol fuel stock, but these are still in the R&D or small pilot plant stages.

Bioethanol is produced from biomass by the hydrolysis and sugar fermentation processes of biomass wastes which contain a complex mixture of carbohydrate polymers from the plant cell walls known as cellulose, hemi cellulose and lignin. It is also produced from the energy crops, containing starch, or other carbohydrates.

In order to break these into sugars, the raw materials are treated with acids or enzymes. This initially reduces the size of the feedstock and opens up the plant structure, thus making it more susceptible to the fermentation process.

Where

A is kilograms of material needed to produce one liter ethanol, and

B is square meters of land needed to produce one liter ethanol,

It is easy to see from Table 6-16 that sugarcane produces the most ethanol from unit mass and that wood products need the least land area for the same purpose (trees are very tall, you know). The big difference is that it is much easier (much less energy intensive) to produce ethanol from sugar than it is from wood products.

Note: Some biomass types contain mostly cellulose $(C_6H_{10}O_5)n$, while the most used (and profitable) energy crops contain starch $(C_6H_{10}O_5)n$. Even better—easier to convert into energy and most profitable—is bio-mass containing sugar (CnH_2nOn). Sugarcane is the best example of this type of biomass, although there are other sugar-containing species that can be used with similar success.

Note that the chemical formulas of cellulose and starch are basically the same, and that of sugar is close enough, and yet their behavior is totally different. Starch lends easily to chemical treatment when converted into sugars, while cellulose requires much more aggressive and expensive processing. Sugar in bio-mass, on the other hand is ready to be used without much effort or expense.

Corn is the most used crop for bio-fuels today. Its molecule is actually one long chain of glucose molecules held together loosely, so adding enzymes is all that is

Table 6-16. Ethanol production

Material	Ethanol Yield	
	A	B
Cassava roots	0.18	0.05 - 0.4
Maize grain	0.36	0.03 - 0.2
Sugar cane stalks	0.07	0.04 - 1.2
Sweet potato roots	0.12	0.1 - 0.5
Wood products	0.16	0.02 - 0.4

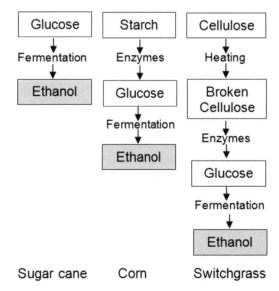

Figure 6-32. Bioethanol processing from different biomass materials

needed to break the chains and separate the individual glucose molecules (sugar). Releasing sugar from sugarcane is even easier, since it is already in the stalk and can be easily squeezed out of it. Nothing else is needed.

Also note in Table 6-16 that it takes much less sugar cane stalks (0.07 kg) to produce 1 liter of ethanol, verses more than twice (0.16 kg) the amount of wood that is needed for the production of the same amount of ethanol. This is mostly due to the fact that cellulosic materials (wood is composed mostly of cellulose) contains less glucose per unit mass.

Cellulose is made out of similar long chains of glucose molecules, but the bonds between these are what make the big difference. The bonds (links) holding the glucose molecules in the chain together are very different in orientation and behavior, and so fewer enzymes are capable of breaking down the long chains in the cellulose molecule. Enzymes work in a *lock and key* system, where each enzyme is effective exclusively with a particular molecule, so the right enzyme is needed to build or degrade an organic molecule biologically.

Sugarcane, sugar beets and several other energy crops, on the other hand, are quite unique in that they contain *pure sugars* in their molecule, which are very easily extracted my mechanical or chemical treatment. The big difference here is that it requires much less energy and chemicals to convert sugarcane into ethanol.

As can be seen in Figure 6-32, the processing of sugar cane is much simple and cheaper than that of cellulosic materials, such as Switchgrass, which is becoming a favorite. In all cases, however, the goal is to convert the raw biomaterials into sugar—whatever it takes.

The cellulose and starch components are hydrolyzed (broken down) by enzymes or dilute acids into sucrose (type of sugar) in a solution. With time and with the help of heat and yeasts the mass is then fermented into ethanol. The lignin and other waste products in the biomass are normally separated, and some are burnt as a fuel for the ethanol production plants boilers.

There are three main commercial methods of extracting sugars from biomass and fuel crops;
• Concentrated acid hydrolysis,
• Dilute acid hydrolysis, and
• Enzymatic hydrolysis.

Concentrated Acid Hydrolysis Process

The Arkanol process works by adding 70-77% sulfuric acid to the biomass that has been dried to a 10% moisture content. The acid is added in the ratio of 1.25 acid to 1 biomass, and the temperature is controlled to 50C. Water is then added to dilute the acid to 20-30%

and the mixture is again heated to 100C for 1 hour. The gel produced from this mixture is then pressed to release an acid sugar mixture and a chromatographic column is used to separate the acid and sugar mixture. The sugar mixture is then fermented with the help of enzymes or acids and the resulting dilute alcohol is distilled to pure bioethanol.

Dilute Acid Hydrolysis

The dilute acid hydrolysis process is one of the oldest, simplest and most efficient methods of producing ethanol from biomass. Dilute acid is used to hydrolyze the biomass to sucrose. The first stage uses 0.7% sulfuric acid at 190C to hydrolyze the hemi cellulose present in the biomass. The second stage is optimized to yield the more resistant cellulose fraction. This is achieved by using 0.4% sulfuric acid at 215C. The liquid hydrolates are then neutralized and recovered from the process.

Enzymatic Hydrolysis

In this process, instead of using acid to hydrolyse the biomass into sucrose, enzymes are employed to break down the biomass in a similar way. However this process is very expensive, and is still in R&D and in early stages of development.

Bioethanol in the U.S. is produced mostly from corn, via
• Wet milling, and
• Dry milling process

Wet Milling Process

In this process, the corn kernels are first steeped in warm water (thus the name) to break down the skin, and soften the kernel and the proteins, in order to release the starch locked in. The corn is then milled in a mechanical mill, where germ, fiber and starch are produced. The germ is extracted and used for the production of corn oil, while the starch fraction is centrifuged and then left for saccharifcation, which produces a wet cake of gluten material.

The ethanol is extracted from the gluten by a fractional distillation process. The wet milling process is used in large scale bioethanol production plants, which produce hundreds of millions gallons of ethanol every year.

Dry Milling Process

The dry milling process involves cleaning and breaking down the corn kernel into fine particles using a hammer mill process. This creates a powder with a course flour type consistency. The powder contains the corn germ, starch and fiber. In order to produce a sugar solution the mixture is then hydrolyzed or broken down into sucrose sugars using enzymes or a dilute acid. The mixture is then cooled and yeast is added in order to ferment the mixture into ethanol. The dry milling process is used in lower volume factories, which on the average produce less than 50 million gallons of ethanol every year.

Bioethanol from sugarcane is the easiest to process, since the sugarcane contains free sugars, which only need to be squeezed out from the stalks and fermented into alcohol.

<u>Sugar Fermentation Process</u>

The hydrolysis process breaks down the cellulosic part of the biomass or corn into sugar solutions that can then be fermented into ethanol. Yeast is added to the solution, which is then heated. The yeast contains an enzyme called invertase, which acts as a catalyst and helps to convert the sucrose sugars into glucose and fructose (both $C_6H_{12}O_6$).

The chemical reaction is:

$$C_{12}H_{22}O_{11} + H_2O = C_6H_{12}O_6 + C_6H_{12}O_6$$
Sucrose Water Fructose Glucose

The fructose and glucose are produced in the presence of invertase (a catalyst). These sugars then react with another enzyme called zymase, which is also contained in the yeast to produce ethanol and carbon dioxide.

The chemical reaction of this process is:

$$C_6H_{12}O_6 = 2C_2H_5OH + 2CO_2$$
Sugars Ethanol Pollution

The fermentation process takes around three days to complete and is carried out at a temperature of between 250C and 300C. All we have to do after that is separate the alcohol from the mixture.

Note: We see here that a significant amount of CO_2 is emitted during this process. Two molecules of CO_2 are emitted for every molecule of sugar entering the reaction. This is in addition to the pollution emitted during planting and growing the crops, their transport, and additional energy used in their conversion to ethanol process.

Fractional Distillation Process

The ethanol, which is produced from the fermentation process, described above, still contains a signif-

icant quantity of water, which must be removed. This is achieved by using the fractional distillation process. The distillation process works by boiling the water and ethanol mixture. Since ethanol has a lower boiling point (78.3C) compared to that of water (100C), the ethanol turns into the vapor state before the water and can be condensed and separated.

Since bioethanol is a primary source of biofuels in North America, many organizations are conducting research in that area. The National Corn-to-Ethanol Research Center (NCERC) is a research division of Southern Illinois University Edwardsville dedicated solely to ethanol-based biofuel research projects. On the federal level, the USDA conducts a large amount of research regarding ethanol production in the United States. Much of this research is targeted toward the effect of ethanol production on domestic food markets. A division of the U.S. Department of Energy, the National Renewable Energy Laboratory (NREL), has also conducted various ethanol research projects, mainly in the area of cellulosic ethanol.

Glucose (a simple sugar) is created in the plant by photosynthesis.

$$6CO_2 + 6H_2O + light \rightarrow C_6H_{12}O_6 + 6O_2$$

During ethanol fermentation, glucose is decomposed into ethanol and carbon dioxide.

$$C_6H_{12}O_6 \rightarrow 2C_2H_5OH + 2CO_2 + heat$$

During combustion ethanol reacts with oxygen to produce carbon dioxide, water, and heat:

$$C_2H_5OH + 3O_2 \rightarrow 2CO_2 + 3H_2O + heat$$

After doubling the combustion reaction because two molecules of ethanol are produced for each glucose molecule, and adding all three reactions together, there are equal numbers of each type of atom on each side of the equation, and the net reaction for the overall production and consumption of ethanol is just: light $\rightarrow$ heat

The heat of the combustion of ethanol is used to drive the piston in the engine by expanding heated gases. It can be said that sunlight is used to run the engine (as is the case with any combustion-based energy source, as sunlight is the only way energy is added to the planet, and geothermal energy, which comes from the heat already present inside the earth).

Glucose itself is not the only substance in the plant that is fermented. The simple sugar fructose also undergoes fermentation. Three other compounds in the plant can be fermented after breaking them up by hydrolysis into the glucose or fructose molecules that compose them. Starch and cellulose are molecules that are strings of glucose molecules, and sucrose (ordinary table sugar) is a molecule of glucose bonded to a molecule of fructose. The energy to create fructose in the plant ultimately comes from the metabolism of glucose created by photosynthesis, and so sunlight also provides the energy generated by the fermentation of these other molecules.

Ethanol may also be produced industrially from ethylene. Addition of water to the double bond converts ethene to ethanol:

$$C_2H_4 + H_2O \rightarrow C_2H_5OH$$

This is done in the presence of an acid which catalyzes the reaction, but is not consumed. The ethene is produced from petroleum by steam cracking.

When burning in pure oxygen environment, ethanol produces large amounts of CO_2 and H_2O

$$C_2H_5OH + 3O_2 \rightarrow 2CO_2 + 3H_2O$$

When ethanol is burned in the atmosphere rather than in pure oxygen, however, other, much different and dangerous chemical reactions take place, since there are different gasses in the atmospheric air, such as nitrogen (N_2).

$$C_2H_5OH + 5O_2 + N_2 \rightarrow 2CO_2 + 2NO_2 + 3H_2O$$

During burning in this air mixture, ethanol produces lots of nitrous oxides, which is also a major air pollutant. It is actually 300 times more dangerous than CO_2, which makes burning large quantity of alcohol a very dangerous affair, as far as the environment and global warming in particular are concerned.

It appears that equal amounts of CO_2 and NO_2 gasses are generated during the bioethanol burning process, which gasses are known to produce unwanted environmental effects.

Example: Starting with, let's say 100 lbs. of starch or cellulosic materials, we get about 120 lbs. of glucose after adding the enzymes to the mixture. After fermentation we measure about 60 lbs. of bioethanol and 60 lbs. of CO_2.

Note that there is almost as much CO_2 produced during this process as there is ethanol. Still, the process

is considered CO2 neutral, because the released during the conversion process CO_2 was actually absorbed by the plants during their growth in the fields.

But since the nitrogen in the air is also oxidized during the ethanol burning process, we now have 30 lbs. of CO_2 and 30 lbs. of NO_2. Considering the fact that NO_2 is 300 times more damaging than CO_2, we arrive to the conclusion that thus burned alcohol does at least 150 times more damage than burning equal amount of other fuels that produce only CO_2. Not a good thing…

Another problem, related to environmental issues, that we need to consider here is the actual *location, where* the different operations take place. The CO_2 in question was absorbed in the fields where the biomass was grown, which usually are several hundred miles away from the processing plants. During their growth, the plants absorbed a lot of CO_2, while at the same time releasing a lot of oxygen, thus making the air clean and fresh in that location. And, of course, there is no any trace of NO_2 in these fields.

Fast forwarding to the place where bioethanol is burnt; most likely by the cars cruising the roadways of large populated centers, where the entire amount of CO_2 stored in the biomass (plus equal amount of NO_2) are released, both of which pollute the air at that location. If you happen to be in Los Angeles, or Beijing, in a late summer afternoon, you'll see and smell a lot of these polluting gasses in the air. At times you cannot even see more than a few feet ahead, nor take your breath, from the dense smog in these places, some of which is caused by ethanol burning.

The effects of this local gas pollution un-equilibrium are open to debate, as is the entire effect of CO_2 on the global environment, but it is a proven fact is that some areas have totally different air quality than others.

It is possible to minimize the CO_2 and NO_2 emissions, and significantly improve the greenhouse gas profile of ethanol, but we need to be aware of, and deal squarely with, the importance issue. It is a serious one and should not be ignored. So basically speaking, we have a good thing in bioethanol that comes with some bad consequences, which we just need to take a close look at, and deal with, properly and efficiently.

Case study 1: Sacramento bio-ethanol production and power cogeneration plant.

Sacramento ethanol and power generation plant was one of the first attempts for large scale bio-ethanol production from cellulosic materials in the U.S. It was initiated around 1992, but it took 2 years to obtain construction permit. At the time it was not only the first

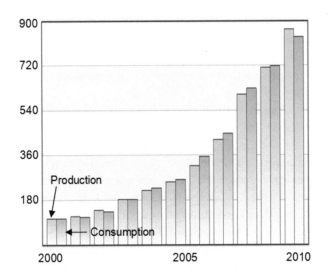

Figure 6-33. U.S. Bio-ethanol production and use (in thousand barrels/day)

commercial cellulosic biomass-to-ethanol plant, but also was a key part of the solution to the rice straw disposal problem facing California's rice growing industry in the face of regulations banning most field burning of such agricultural residues. This process would eliminate the post-harvest open field burning of rice straw on some 40,000 acres, approximately 15% of the total Sacramento Valley rice under cultivation.

Win-win situation, no doubt...at least on paper. In practice things are much more difficult—from political, regulatory, and technical points of view. The plant was to be built on a 90 acres land in order to convert over 400 tons per day of rice straw and other cellulosic agricultural residue into approximately 35,000 gallons per day of fuel grade ethanol, using patented concentrated acid hydrolysis technology. This is 12 million gallons bio-ethanol obtained by processing 132,000 tons of rice-straw per year. Thus produced bio-ethanol would be used in tandem with natural gas to generate about 150 MW electricity as a base-load power plant.

The plant was essentially two separate yet linked projects with different owners united by a contractual arrangement, sharing a site and various operational synergies, including process heat and power supplied to the ethanol plant by the cogeneration plant, shared water supply and waste disposal provisions, etc. It's complicated...

Partnership issues flared up, delayed the project and finally caused the permit to expire. Five years later the owner applied for a new permit, which was granted in 2000. The struggles continued, but the plant was never built.

The plant construction did not go forward mostly

due to complexities with the joint venture and multiple parties, who could not divide the pie to everyone's satisfaction. The technological approach was also unproven and somewhat too advanced for the time, which complicated things further.

Nevertheless, there were a number of lessons (albeit expensive) learned from this experience. It was an early test case of the California regulatory process for permitting a bio-refinery facility, where bio-alcohol production would be combined with electricity generation. As such, the project could be considered as a partial success story.

Some of the successfully executed tasks and lessons learned have been:

- Extensive environmental analysis was conducted and complex mitigation measures considered on a full range of issues, including air quality, water supply and water quality, hydrology, and biological resources. Issuance of an air quality permit for the entire project was based on emission offsets to be obtained via the discontinuation of rice straw burning resulting from use of rice straw as the ethanol plant feedstock.

- Flood plain concerns resulted in modifications to the facility site plan. Original plans to use groundwater wells were changed to use of surface water; water supply arrangements included mitigation measures at the Sacramento River water intake to protect salmon. Various other mitigation measures were adopted involving several different endangered species found on the site.

- The unique features of the project, combining rice straw to ethanol production and electricity cogeneration, posed a number of considerations not previously encountered in the California regulatory proceedings.

- Reliability, or lack thereof, of the unproven cellulosic ethanol production process affected both the cogeneration performance and emission offset viability of the power plant. Various issues associated with the feedstock supply plan based on the yet-to-be-demonstrated use of rice straw were addressed as well.

Case study 2. Gridley Ethanol Plant

The Gridley Ethanol Project was another attempt, designed as a potential solution to the rice straw disposal problem in the Sacramento Valley region of California, which became acute with legislative mandates to significantly reduce the amount of rice straw burning after the fall rice harvest.

The Rice Straw Burning Reduction Act of 1991 (AB 1378) mandated a reduction in rice straw burning by the year 2000 to no more than 25% of the planted acreage. The California rice straw burning phase down has proceeded as required by the statute, with growers burning less than the statutory limitations. Other open-field burning laws and regulations further limit the actual rice straw acreage burned annually.

Despite the ongoing reduction of rice straw burning, no alternative market or disposal option sufficient to handle the quantities of rice straw being produced was available at the time. As a result, very large volumes of this material continue to accumulate without a viable market alternative to dispose of the rice straw. This could eventually render useless thousands of acres of rice lands, since in these hard clay-pan soils, no other crops have been successful. Production of bio-ethanol from rice straw was seen as a potential solution.

The original concept of the Gridley plant involved application of an unproven enzymatic hydrolysis process to produce ethanol. Lignin remaining from the hydrolysis process was to be utilized as combustion fuel for firing the facility's boiler for the production of steam and electricity to be used on site, with excess steam potentially used by adjacent facilities. Excess electricity would be supplied to the municipal utility and/or sold to the grid.

The actual Gridley plant planning began also in 1994 and was authorized in 1996 by NREL. Phase 1 started with initial screening of the technical and economic feasibility of a commercial rice straw-to-ethanol facility in the Gridley area. Phase II was to acquire financial and site commitments, perform pilot plant studies of the technology at NREL, prepare a preliminary engineering package, evaluate the economics and risks, and finally to prepare an implementation plan to commercialize the process. Phase II was to lead to a "go/no go" decision regarding the construction of the plants.

Here again, there were many partners, and sure enough, in 1997 the original conversion technology developer withdrew from the project. Since Phase I tasks had been completed and a rice straw-to-ethanol facility appeared feasible, a new partner was chosen to provide the conversion technology, which was basically also acid hydrolysis and fermentation, with lignin as a co-product.

During the progress of Phases I and II, it was determined that project economics with the then-current state of conversion technology would be enhanced by converting the bio-ethanol production plant into a cogeneration facility. It was then to be sited next to an existing biomass power plant in the region, which uses orchard prunings and forest wastes as feedstock. This co-location would reduce the costs and improve the efficiency of both the power plant and the proposed ethanol facility.

Construction of the new plant was projected to commence in early 2002 with operations to begin in late 2003. The collection and processing of rice straw became a paramount consideration, since infrastructure to harvest rice straw for use in the plant was virtually nonexistent. Processing of the rice straw for use as feedstock (i.e., grinding) presented technical challenges due to the high silica content of rice straw.

Rice straw supply studies indicated that the rice straw would cost over $30.00/bone dry ton (BDT) to be delivered to the facility. This did not include the grinding and processing of the rice straw at the facility. To produce the 23 million gallons of ethanol would require 300,000 dry tons of rice straw (some of which could be provided by orchard and forest wood wastes).

On top of that, there were environmental permitting and impact assessment studies that indicated some potentially higher costs than originally anticipated. Wastewater from the plant would have to be discharged to the local municipal wastewater treatment plant, which alone would cost several million dollars. Also, in order to discharge to the wastewater plant, an expensive wastewater pretreatment adding several million dollars to the operating cost, would be necessary. Oh, yea, additional air emission control equipment would be needed to complete the environmental safety, which was not anticipated previously.

This, combined with the technical uncertainties of the two stage dilute sulfuric acid conversion technology, led a conclusion that the acid hydrolysis technology was not financially viable for use. A decision was made to investigate the use of a gasification technology to create syngas that could be converted to ethanol or other fuels. This evaluation, done in 2002, indicated that switching from the dilute sulfuric acid process to a gasification process could have a number of advantages, such as, a) increased yields of ethanol, with associated reductions in feedstock and other operating costs per gallon of ethanol produced, b) lower capital investment cost, c) fewer air emissions and wastewater effluents, and d) reduced feedstock requirements, which better fit the initial needs of the area for disposing of a critical mass of rice straw The plant was also to be moved again to its original location, due to a new industrial site availability, shorter transportation hauling distances from the rice fields, significantly reduced wastewater disposal costs and available infrastructure to better support the proposed facility.

NREL continued to finance and support the plant and its new gasification technology. A pilot facility was used for a proof-of-concept. The results were encouraging, and the Gridley plant was able to get funding from the U.S. Department of Energy. In December 2006 the plant management issued a Request for Proposals to construct and operate a thermo-chemical conversion system using rice straw to produce bio-ethanol and electricity.

The plant was awarded also a CEC grant in April 2007 to demonstrate an integrated biofuels and energy production system. The project was geared to support the construction, demonstration and validation of a cost effective and energy efficient biomass conversion system. But the conflicts among the partners, shifting locations, and changing processing technologies took their toll and, when the farmers refused to participate due to high transportation costs, the plant was put on hold. Indefinitely...after millions of taxpayers money was spent in the process.

In summary, there were a great number of failed bio-ethanol conversion plants in the U.S. during the last 2-3 decades too, but nevertheless, in the summer of 2013 there were 211 bio-ethanol operational plants in the U.S., producing the grand total of 13.5 billion gallons of bio-ethanol every year, mostly from corn.

What does the future hold for bio-ethanol is anyone's guess, but we venture say that it won't go much further for awhile—and until natural gas prices are so low. It will, however, be the fuel of choice for electricity generation and transportation in the distant future, when oil and gas are gone. Make no mistake about it !!!

Biobutanol

Bio-butanol is a type of a 4-carbon butyl alcohol (C_4H_9OH), while ethanol is an alcohol with two carbon atoms (C_2H_5OH). Both of these alcohols are types of fuel that can be used for many different purposes. Butanol is more efficient than ethanol, since it is able to produce about 110,000 Btus per gallon, while ethanol can only produce 84,000 Btus per gallon.

Butanol can be used to replace gasoline without modifying the car's fuel system. It does not evaporate as quickly as ethanol. Anther big difference is that while ethanol must be transported by rail or truck, butanol can be delivered through existing gas lines. Butanol could become the next fuel of the future entering the fray, thanks to a new technologies that could solve many of the problems associated with ethanol. Industry stakeholders, such as BP and others, have been seeking ways to make a cost-efficient transition to butanol, the "advanced biofuel," and a scientific breakthrough could make this possible.

Butanol trumps ethanol in several ways:
- Adding ethanol to gasoline reduces fuel mileage, but butanol packs almost as much energy as gas, meaning fewer fill-ups.
- Butanol doesn't damage car engines like ethanol, so more of it can be blended into gas, and
- butanol doesn't separate from gasoline in the presence of water, so it can be blended with gasoline right at the refinery, while ethanol has to be shipped separately from gas and blended closer to the filling station. Big advantage for the future of this promising fuel.

The problem is that even with all its advantages and new technological advancements, producing butanol in the U.S., with its 200 ethanol-producing plants, could cost roughly $15-20 million to modify each of those facilities to produce butanol...so the future of butanol looks bright, but it may be delayed in coming at a later date.

Bio-butanol is produced from a number of biomass feedstocks via fermentation. Large variety of biomass types can be used in this process; corn grain, corn stovers, and many other feedstocks. Like in other processes, these are processed into sugars, and the special microbes of the Clostridium acetobutylicum species, are introduced to the sugars, which are then broken down into various alcohols, including butanol.

Biobutanol can be produced also by fermentation of sugar via the acetone-butanol-ethanol (ABE) process using bacteria such as *Clastridium acetobutylicum. It can also* be made using Ralstonia eutropha H16, which process requires the use of an electro-bioreactor, and the additional input of carbon dioxide and electricity.

The difference from ethanol production is primarily in the fermentation of the feedstock and minor changes in the distillation setup and process parameters. The feedstocks are the same as for ethanol: energy crops such as sugar beets, sugar cane, corn grain, wheat and cassava. Prospective non-food energy crops such as switchgrass and even guayule in North America, as well as agricultural byproducts such as straw and corn stalks are being investigated.

In addition sugars can also be derived from ligno-cellulosic biomass, using the same biochemical conversion steps required for advanced ethanol production. According to the industry experts, existing bioethanol plants can cost-effectively be retrofitted to biobutanol production, but at a high price.

Unfortunately, high alcohol concentration makes the butanol mixture toxic to some needed microorganisms. This condition made the fermentation process expensive and impractical when compared to the petroleum costs. New technological advances have improved the efficiency and reduced the cost of the fermentation process. Today, genetically engineered processes are making it possible for the most efficient microbes to withstand even the highest alcohol concentrations. This allows large quantity of biobutanol to be produced commercially.

As with the case of bi ethanol, biobutanol can be prepared easier and directly from a ready source of sugar, such as sugarcane, but production from crop wastes and energy grass is possible too. It can also be made entirely with solar energy, from algae, called Solalgal Fuel, or special diatom materials.

Butanol has total of four different isomers, but only three are used commercially: n-butanol, isobutanol, tertbutanol. They all have multiple uses in industrial and consumer products. The market for n-butanol and isobutanol is over 7 billion lbs. annually, and the producers bio-butanol have a captive market already.

The different types of butanol are:

N-butanol finds applications as production intermediate for a number of chemicals, such as: butyl acrylate, butyl acetate, dibutyl phthalate, and also as extractant for antibiotics, hormones, vitamins. It is also ingredient in perfumes, degreasers, repellents, and cleaning solutions

Iso-butanol finds applications as; paint solvent, ink ingredient, gasoline additive, derivative ester precursor, viscosity reducer in paint, automotive polish, and paint cleaner additive

Tert-butanol finds application as; perfume ingredient, gasoline octane booster, paint remover ingredient, solvent, as well as a synthesis intermediate of: methyl tert-butyl ether (MTBE), ethyl tert-butyl ether (ETBE), and tert-butyl hydroperoxide (TBHP).

Biobutanol has a greater energy density (29.2 MJ/l) and is more similar to gasoline than ethanol, so it is used in internal combustion engines, primarily as a gasoline additive, or a fuel blend with gasoline. The energy content of biobutanol is only 10% less than that of regular gasoline, while the energy density of ethanol is over 40% lower. Biobutanol is also more chemically similar to gas-

oline than ethanol, so it can be integrated into regular internal combustion engines much easier than ethanol.

Additionally, biobutanol has the potential to reduce the carbon emissions by 85% as compared to gasoline, which makes it superior alternative to gasoline and to the gasoline-ethanol blended fuels. It also can be produced from feedstocks which do not compete with food. For example, algae biomass and waste wood particles can be converted to biobutanol, with the advantage that some of these require only a 10th of the land resource needed to grow corn.

Nevertheless, most of these achievements are derived in small scale or R&D lab settings for now. Because of that, the commercial future of biobutanol is…still in the future.

Biodiesel

Biodiesel is a substance of pure or somewhat modified vegetable oil, or animal based fat, that is used to power diesel engines. It has a technical definition of *mono-alkyl* (methyl, propyl, or ethyl) *ester* of long-chain fatty acids, which are produced from vegetable oils or animal fats. Soybean methyl ester is one of the pure biodiesel varieties. It is made from soybeans, has an average molecular weight of 292.2 and is comprised of:

- Palmitic acid $C_{15}H_{31}CO_2CH_3$,
- Stearic acid $C_{17}H_{35}CO_2CH_3$,
- Oleic acid $C_{17}H_{33}CO_2CH_3$,
- Linoleic acid $CH_3(CH_2)_4CH=CHC_2CH=CH(CH_2)_7CO_2CH_3$, and
- Linoleic acid $CH_3(CH_2CH=CH)_3(CH_2)_7CO_2CH_3$.

Who knew that diesel is nothing but a large acid concoction. Since biodiesel is used to power expensive diesel engines in passenger and commercial cars and trucks, it has to meet the strictest quality specifications of ASTM D 6751, and has to be compatible in any blend with petroleum diesel fuels.

Biodiesel is typically made by chemically reacting (trans-esterification) of vegetable oil or animal fat feed-

Table 6-17. Energy content of key fuels

Fuel	Energy density	Air-fuel ratio	Specific energy
Gasoline and biogasoline	32 MJ/L	14.6	2.9 MJ/kg air
Butanol fuel	29.2 MJ/L	11.1	3.2 MJ/kg air
Ethanol fuel	19.6 MJ/L	9.0	3.0 MJ/kg air
Methanol fuel	16 MJ/L	6.4	3.1 MJ/kg air

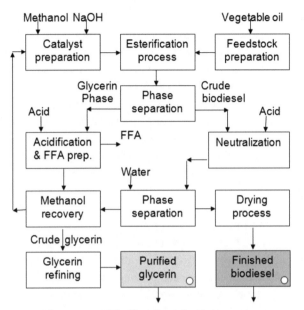

Figure 6-34. Biodiesel production process

stock. lipids, vegetable oil, and animal fat (tallow) with an alcohol producing fatty acid esters. Recycled oil can be used after processing it to remove impurities from cooking, storage, and handling, such as dirt, charred food, and water. Virgin oils are refined to commercial grade purity (not to a food-grade level.) Degumming process step is used when needed to remove phospholipids and other plant matter via different refinement processes.

Excess water is removed to prevent triglycerides from hydrolysis during base-catalyzed trans-esterification process, thus preventing the formation of salts of the fatty acids (soaps) instead of producing biodiesel. The acids present in the oil are either; esterified into biodiesel, esterified into glycerides, or removed through neutralization with bases.

Trans-esterified biodiesel is a mix of mono-alkyl esters of long chain fatty acids, with the most common form being methanol converted to sodium methoxide to produce methyl esters. These are commonly referred to as fatty acid methyl ester, or FAME.

Methanol is used to produce fatty acid ethyl ester, or FAEE biodiesel, but other alcohols such as isopropanol and butanol can also be used. Using alcohols of higher molecular weights improves the cold flow properties of the resulting ester, at the cost of a less efficient trans-esterification reaction.

A lipid trans-esterification production process is used to convert the base oil to the desired esters. Any free fatty acids (FFAs) in the base oil are either; converted to soap and removed from the process, or they are

esterified, which yields more biodiesel, via an acidic catalyst. After this processing, unlike straight vegetable oil, biodiesel has combustion properties very similar to those of petroleum diesel, and can replace it in most current uses.

Glycerol is a by-product of the trans-esterification process, where every 1 ton of biodiesel also produces 100 kg of glycerol. Crude glycerol contains 20% water and catalyst residues and has no practical use today. Research is underway to find use for it as a chemical building block in some products, such as epoxy resins.

Diesel consists of primarily stable molecules such as $C_{12}H_{22}$, $C_{13}H_{24}$ and $C_{12}H_{24}$. Chemists sometime describe diesel as having an average chemical composition of $C_{12}H_{23}$. This, however, is only a very general chemical description, since it does not accurately reflect diesel's spatial chemical composition.

Organic chemists will shrug at this, but if we are to condense diesel's chemical formula, we would get something like this ($C_{15}H_{31}CO_2CH_3$-$C_{17}H_{35}CO_2CH_3$-$C_{17}H_{33}CO_2CH_3$-$CH_3(CH_2)_4CH=CHCH_2CH=CH(CH_2)_7CO_2CH_3$-$CH_3(CH_2CH=CH)_3(CH_2)_7CO_2CH_3$). This is complicated, no doubt.

Biodiesel's ozone (smog) forming potential, its CO, particulate, and total hydrocarbons content, are about 50% less than that of conventional diesel fuels.

There are contaminants in any diesel fuel, but biodiesel is a cleaner overall alternative to standard diesel fuel. Sulfur is one of the main contaminants and pollution emitters in standard diesel fuels, but it is essentially eliminated with pure biodiesel. And so are the unburned hydrocarbons, carbon monoxide, and particulate matter, all of which are typical for standard diesel fuels.

The bad news is that NO_x emissions from B100 biodiesel are 10% higher than those from standard diesel, which (NO_x)—as we saw above—is 300 times worse for the environment than same amount of CO_2. This is a serious issue that needs to be resolved, if biodiesel is to become accepted as environmentally safe alternative to standard, petro-diesel.

The good news is that biodiesel's lack of sulfur content allows the use of NO_x control technologies that cannot be used with conventional diesel engines. There are also some additives developed lately that can reduce NO_x emissions in biodiesel blends.

Basically speaking, biodiesels reduce the health risks associated with standard diesel fuels.

Biodiesel emissions also show decreased levels of polycyclic aromatic hydrocarbons (PAH) and nitrated polycyclic aromatic hydrocarbons (nPAH), which have been identified as potential cancer causing compounds. In recent health effects testing, PAH compounds were reduced by 75 to 85%, while benzo(a)anthracene was reduced by roughly 50%. Targeted nPAH compounds were also reduced dramatically with biodiesel, with 2-nitrofluorene and 1- nitropyrene reduced by 90%, while the rest of the nPAH compounds are reduced to minute (trace) levels.

Biodiesel is extensively used in European and other countries around the world. Pure biodiesel, used in standard diesel engines, is different from waste oils (used for cooking) that are sometimes used to fuel *converted* diesel engines. Note the term *converted*, because using waste vegetable oil in a standard diesel engine would cause damage that might require expensive repairs.

Biodiesel can be used pure, or blended with petrodiesel. It can be also used as a low carbon alternative to heating oil. Blends of biodiesel and conventional hydrocarbon-based diesel are most commonly used in the retail diesel market.

The "B" factor system is usually used to reflect the amount of biodiesel in the respective fuel mix, as follow: 100% biodiesel is labeled B100, 20% biodiesel in 80% petrodiesel is labeled B20, 5% biodiesel in 95% petrodiesel is labeled B5, and 2% biodiesel in 98% petrodiesel is labeled B2.

Blends of 20% biodiesel and lower are used in standard diesel equipment with only minor modifications, but in some cases it can violate the manufacturer's warranty. Biodiesel used in its pure form (B100) usually require major engine modifications.

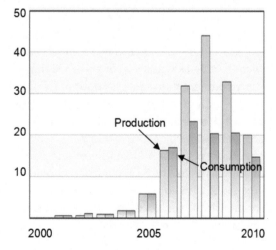

Figure 6-35. U.S. biodiesel production and use (in thousand barrels/day

Advanced biodiesel

Several processes are under development that aim to produce fuels with properties very similar to diesel and kerosene. These fuels will be blendable with fossil fuels in any proportion, can use the same infrastructure and should be fully compatible with engines in heavy duty vehicles. Advanced biodiesel and bio-kerosene will become increasingly important to reach this roadmap's targets since demand for low-carbon fuels with high energy density is expected to increase significantly in the long term.

Advanced biodiesel includes:

• Hydrotreated vegetable oil (HVO) is produced by hydrogenating vegetable oils or animal fats. The first large-scale plants have been opened in Finland and Singapore, but the process has not yet been fully commercialized.

• Biomass-to-liquids (BtL) diesel, also referred to as Fischer-Tropsch diesel, is produced by a two-step process in which biomass is converted to a syngas rich in hydrogen and carbon monoxide. After cleaning, the syngas is catalytically converted through Fischer-Tropsch (FT) synthesis into a broad range hydrocarbon liquids, including synthetic diesel and bio-kerosene.

Advanced biodiesel is not widely available at present, but could become fully commercialized in the near future, since a number of producers have pilot and demonstration projects underway.

Biodiesel has a great potential as a plentiful, clean and cheap fuel. With some more improvements it might be one of the energy sources that will carry us through the energy transition gap of the 21st and 22nd centuries.

Bioethers

Bio-ether is a derivative from bioethanol, which is obtained from the distillation of energy crops and sugar beet. The best known and widely used fuel ethers are MTBE (methyl-tertiary-butyl-ether) and ETBE (ethyl-tertiary-butyl-ether).

Bioethers are a class of ethers that are produced from biomass materials, energy crops, and such. They are classified as organic compounds that are characterized by an oxygen atom bonded to two alkyl or aryl groups. Ethers are similar in structure to alcohols, and both ethers and alcohols are similar in structure to water...somewhat...

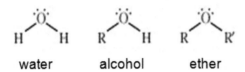

Figure 6-36. Chemical formulas of basic fuels

The difference here is that in the alcohol molecule, one hydrogen atom of a water molecule is replaced by an alkyl group, while in the ether molecule, both hydrogen atoms are replaced by either an alkyl or aryl groups. Just like water, but a little bit different…which makes a huge difference.

In the presence of acid, two alcohol molecules may lose water to form an ether. In practice, however, this bimolecular dehydration to form ether competes with uni-molecular dehydration to give an alkene. Bimolecular dehydration produces useful yields of ethers only with simple, primary alkyl groups such as those in dimethyl ether and diethyl ether. Dehydration is used commercially to produce diethyl ether.

The most practical method for making ethers is the Williamson ether synthesis, which uses an alkoxide ion to attack an alkyl halide, substituting the alkoxy (–O–R) group for the halide. The alkyl halide must be unhindered (usually primary), or elimination will compete with the desired substitution.

Ether fuel can be produced from a mixture of both petrochemical and agricultural feedstocks. In all cases, the building blocks for fuel ethers are isobutylene or isoamylenes compounds, reacted with methanol or ethanol. A complete bio-chemical process involves using *bio-ethanol*, which is derived by a fermenting process from wheat, sugar beet and other agricultural products. It is also the major feedstock for the production of ETBE or TAEE (tert-amyl-ethyl-ether).

Bio-methanol, which is also derived from biomass, is the second feedstock used in the production of MTBE or TAME (tert-amyl-methyl-ether). Isobutylene is yet another feedstock used in both MTBE and ETBE production, but it is derived from fossils; byproduct from natural gas or petroleum refining.

Similarly, isoamylenes used in the production of TAME or TAEE are by-products of petroleum refining.

Bioether production facilities are typically located close to a refinery with a fluid catalytic cracker unit, or in a chemical plant with a steam cracker. Large-scale "stand alone" units are also in operation lately, but these are based mostly on fossils, using either; butane isomerisation/dehydrogenation technology (where both the butane and the methanol are derived from natural gas),

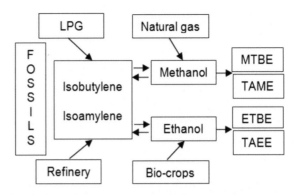

Figure 6-37. Bioether production scheme

or by dehydration of tertiary butyl alcohols.

Ethers are pleasant-smelling, colorless liquids. They are less dense than alcohol and are less soluble in water. Ethers are relatively un-reactive and have lower boiling points than their cousins, the alcohols. They are used widely as solvents for fats, oils, waxes, perfumes, resins, dyes, gums, and many hydrocarbons. Vapors of certain ethers are used as insecticides, miticides, and fumigants for soil.

Due to its lower ignition point, ethyl ether is also used as a volatile starting fluid for diesel engines and gasoline engines in cold weather. Dimethyl ether is used as a spray propellant and refrigerant. Methyl *t*-butyl ether (MTBE) is a gasoline additive that boosts the octane number and reduces the amount of nitrogen-oxide pollutants in the exhaust. Its chemical formula is CH_3CH_2—O—CH_3CH_3.

Bioethers are one of the biofuels of choice today in Europe, with over three quarters of all bioethanol being used as bio-ether (ETBE). European energy policy is promoting the use of bio-fuels for transportation. Bioethers and bioalcohols are used as blending agents for enhancing the octane number. By changing the properties of common fuels, they make gasoline work harder, help the engine last longer and reduce air pollution. Development of renewable fuels needs both knowledge of new thermodynamic data and improvement of clean energy technologies. In this context, the use of ethanol of vegetable origin in its manufacture process, increases the interest in ETBE or bio-ETBE as an oxygenated additive.

Used Cooking Oil

There is a lot of oil—of vegetable and animal origin—used for cooking in restaurants and fast food chains in the U.S. and around the world. Options for disposal of used cooking oil and grease are limited in most places. Disposal is difficult, and expensive when available, because used cooking oil is usually in a liquid, or semi-solid, form, and most waste disposal regulations restrict the disposal of liquids in landfills.

Other disposal methods can also be problematic, for example; open burning of used cooking oil causes black smoke, which is prohibited. Using it as fuel in most standard heating systems will cause black smoke and soot too, and may even damage the system. Pouring used cooking oil down the drain can clog pipes and damage wastewater or septic systems, and so it is a major no-no.

So, in addition to using it in pet food preparations, one of the best ways of disposing of and actually using cooking oil is to burn it in some approved incinerators. Another way is to use it as a fuel in modified diesel engines. Since the modifications and the supply might be questionable, the best way is to refine it into biodiesel for use in standard diesel engines.

For example, McDonald in the UK is recycling 100% of its used cooking oil for biodiesel to be used as fuel in fuel delivery trucks. This is emission savings of more than 3,500 tons of CO_2, or the equivalent of 1,500 cars being removed from the road each year.

In the USA, McDonald's is implementing a bulk cooking oil delivery and retrieval program, which includes collection of waste cooking oil in a separate tank, which is periodically taken back to a distribution facility where it is sold for re-use to a variety of vendors, including biodiesel companies. Nearly 9,000 U.S. McDonald restaurants are enrolled in this program, and additional franchises are being converted for this service.

This means that the average participating U.S. restaurant recycles nearly 1,500 gallons of used cooking oil per year, which together with eliminating large amounts of plastic and corrugated paper packaging, which eliminates the pollution created during the manufacturing of the packaging, to begin with, and then keeps additional packaging waste from going into the landfill.

Today there are small, portable, biodiesel generators which can process used cooking oil into biodiesel. These can be used at home, or commercially, and can produce 20 to 200 gallons of biodiesel per day.

Additionally, there are recipes for home-made biodiesel, in which vegetable oil (including used cooking oil) is mixed with sodium hydroxide and methanol in a blender. After blending, the mixture is left to separate, and the top layer can be used as biodiesel. Easy to do but the quality of the biodiesel is questionable, and this process cannot be used for large scale fuel production.

Another problem is that biodiesel has a limited shelf life. Some oils contain the antioxidant tocopherol, or vitamin E (e.g., rapeseed oil), and they remain usable longer than biodiesel from other types of vegetable oils. Biodiesel's stability can diminished after a week or two, and is usually unusable after 1-2 months. Higher temperature also affects negatively the fuel stability by denaturing it.

Biogas

Biogas is a type of gas-fuel produced by the breakdown of organic matter (biomass and other organic materials) in the absence of oxygen. It is considered a renewable energy source, which can be produced from regionally available raw materials, recycled vegetation, and animal waste. Biogas is considered environmentally friendly and CO_2 neutral fuel.

Anaerobic digestion or fermentation of biodegradable materials used in the bio-gas production such as; manure, sewage, municipal waste, green waste, plant material, and many crops, has also the advantage of reducing the overall amount of final waste from these materials.

Biogas generated from digestion of organic matter is comprised primarily of ~60% methane (CH_4), 30% carbon dioxide (CO_2), with small amounts of nitrogen (N_2), hydrogen (H_2), oxygen (O_2), hydrogen sulfide (H_2S), moisture, and siloxanes.

Biogas is most often produced as landfill gas (LFG) or digested gas in *biogas plants. These are simple* anaerobic (oxygen-free) digesters, designed to treat farm wastes or energy crops, such as maize silage. Some can process biodegradable wastes including sewage sludge and food waste into biogas as well.

The process requires air-tight tanks, where the biomass waste is *digested*, or transferred, into methane gas, thus producing renewable energy that can be used for heating, electricity, and many other operations, including fuel for car and truck engines.

Large-scale *landfill gas* is produced by wet organic waste decomposing under anaerobic conditions, where it is covered and mechanically compressed by the weight of the material that is deposited above. This material prevents oxygen exposure thus allowing anaerobic microbes to thrive and the gas builds up with time. It is slowly released into the atmosphere if the landfill site has not been engineered to capture the gas, or in containers for later use, if the design allows.

Landfill gas is hazardous and polluting gas because,

• It is explosive when mixed with oxygen. Even small amount, about 5% methane can create an explosion,

• Methane from landfill gas is 20 times more potent

Material	Typical weight %
Ash	25
Carbon	25
Water	20
Oxygen	18
Hydrogen	3
Chlorine	0.7
Nitrogen	0.6
Sulfur	0.003

Table 6-18. Approximate content of household wastes

as a greenhouse gas than carbon dioxide, so if landfill gas is allowed to escape into the atmosphere, it contributes to the global warming, and

• Landfill gas contains volatile organic compounds (VOCs) which contribute to the formation of photochemical smog; another bad environmental polluter.

In most cases, thus produced gas mixture is not good enough for use as fuel gas for car engines and other machinery. I.e., H_2S in it is corrosive enough to quickly destroy the internal components of a car engine or power plant. Because of that, thus generated raw biogas must be purified, in order to remove the contaminants. When the contaminants are scrubbed out, more methane per unit volume of gas is available for burning, and less harmful gasses are released.

There are several methods of refining and upgrading biogas:

• water washing,
• pressure swing absorption,
• selexol absorption, and
• amine gas treating.

The most practical of these is water washing, where high pressure gas flows into a process column in which the carbon dioxide and other impurities removed by cascading water running counter-flow to the gas. This method produces over 98% pure methane, but some 2% of the total methane content is loss in the system. It also takes about 5% of the total energy content in the gas to run the scrubbers.

Thus produced upgraded biogas can be compressed, like natural gas, and used to power cars and trucks, and to fuel a number of important applications anywhere natural gas is used. Biogas, for example, has

the potential to replace around 15-20% of vehicle fuel in some European countries, and since it also qualifies for renewable energy subsidies in some of these countries, the interest in it is increasing.

Another major use for biogas is the so called gas-grid injection, where biogas is mixed into the natural gas distribution grid network. This way the gas mixture is delivered to customers, where it can be used in domestic or commercial power generation (heat or electricity).

One thing to note here is the fact that the typical energy losses in natural gas transmission systems are in the 1–2% range, while the energy losses from a large electrical system range from 5–8%. The problem is that methane gas, escaping from the distribution system is many times more environmentally damaging than any of the losses in the electrical system. This leads us to the conclusion that we need to separate and use different measures and calculations for a) the *energy conversion efficiency losses,* and b) the actual *physical loss* of physical materials in the energy generation networks. Since we are talking about huge quantities of matter and energy, the exercise is well worth it.

Syngas

Syngas is a mixture of carbon monoxide, hydrogen and other hydrocarbons, which is produced by partial combustion of biomass. It is emitted during processes, such as the making of charcoal, where the biomass combustion is done with restricted (controlled) amount of oxygen. The gasification process usually proceeds at temperatures greater than 700°C, but due to insufficient oxygen, it does not convert the biomass completely to carbon dioxide and water. Instead, it removes enough of these to convert the raw biomass into a lighter combustible fuel, accompanied by release of syngas.

The resulting gas mixture, syngas, is more efficient than direct combustion of the original biomass, since more of the energy contained in it is extracted and contained in the resulting biogas.

The wood gas generator is a wood-fueled gasification reactor, that can feed syngas directly in an internal combustion engine. It can also be used to produce methanol, DME and hydrogen, or converted via the Fischer-Tropsch process to produce a diesel substitute. A mixture of alcohols, used for blending into gasoline can be produced too.

Lower-temperature gasification, as that used normally for co-producing biochar, results in syngas polluted with tar. After refining, syngas may be burned directly in internal combustion engines, turbines or high-temperature fuel cells.

There are a number of biogases available commercially today, depending on the raw materials and processes used in their production. They all, however, have similar composition and energy value, and all have to be refined and otherwise processed for convenient storage, transport and use.

A number of other biogases are in R&D mode, and/or are produced on small scale. These are the second- and third-generation biogases, which we review below.

Second-generation (advanced) biofuels

Second generation biofuels are made from ligno-cellulosic biomass or woody crops, agricultural residues or waste, which makes it harder and more expensive to extract the required fuel. These are produced from biomass materials, which are available in large quantity for mass biofuel production, and whose impact on GHG emissions, on biodiversity and land use, are well known and acceptable...to a point.

Most second-generation biofuels are under development in universities, private and government labs and other facilities. Cellulosic ethanol, Algae fuel, biohydrogen, biomethanol, DMF, BioDME, Fischer-Tropsch diesel, biohydrogen diesel, mixed alcohols, and wood diesel are prime examples of fuels under development.

Cellulosic ethanol production uses nonfood crops or inedible waste products and does not divert food away from the animal or human food chain. Ligno-cellulose is the "woody" structural material of plants. This feedstock is abundant and diverse, and in some cases (like citrus peels or sawdust) which present a significant disposal problem before, during and after processing.

Producing ethanol from cellulose is a difficult technical problem to solve. In nature, ruminant livestock (such as cattle) eat grass and then use slow enzymatic digestive processes to break it into glucose (sugar). In cellulosic ethanol laboratories, various experimental processes are being developed to do the same thing, and then the sugars released can be fermented to make ethanol fuel. In 2009, scientists reported developing, using "synthetic biology," "15 new highly stable fungal enzyme catalysts that efficiently break down cellulose into sugars at high temperatures," adding to the 10 previously known.

The use of high temperatures has been identified as an important factor in improving the overall economic feasibility of the biofuel industry and the identification of enzymes that are stable and can operate efficient-

ly at extreme temperatures is an area of active research. In addition, research conducted at Delft University of Technology by Jack Pronk has shown that elephant yeast, when slightly modified, can also produce ethanol from inedible ground sources (e.g. straw).

The recent discovery of the fungus *Gliocladium roseum* points toward the production of so-called my-co-diesel from cellulose. This organism (recently discovered in rainforests of northern Patagonia) has the unique capability of converting cellulose into medium-length hydrocarbons typically found in diesel fuel. Scientists also work on experimental recombinant DNA genetic engineering organisms that could increase biofuel potential.

Scientists working with the New Zealand company Lanzatech have developed a technology to use industrial waste gases, such as carbon monoxide from steel mills, as a feedstock for a microbial fermentation process to produce ethanol. In October 2011, Virgin Atlantic announced it was joining with Lanzatech to commission a demonstration plant in Shanghai that would produce an aviation fuel from waste gases from steel production.

Scientists working in Minnesota have developed co-cultures of *Shewanella* and *Synechococcus* that produce long-chain hydrocarbons directly from water, carbon dioxide, and sunlight. The technology has received ARPA-E funding.

Author's note: The conversion of cellulosic matter into alcohol has been a major effort of global proportions since the 1970s. The author had a chance to work at a South American cellulose pilot plant in those days, experimenting with Brazilian jungle biomass. The results were promising, but the processes were never efficient, nor the final product profitable enough, for commercial application.

Fast forward 30 years and we still have problems with producing bioethanol from cellulosic materials. As a matter of fact, this was confirmed recently by the Appeals Court, which ruled in 2013 that the EPA's blending targets for a the type of advanced biofuel—cellulosic ethanol—were simply infeasible. The EPA had demanded that between 2010 and 2012, 20 million gallons of cellulosic ethanol be produced. But to date almost no cellulosic ethanol has been blended into commercial fuel, and cellulosic ethanol startups have been unable to provide significant stock to blenders.

The EPA claims it has the authority to enforce blends based on the 2007 Energy Act passed under the Bush Administration, which promoted biofuels (and corn ethanol) growth. But the Appeals court rejected that argument calling the decision to enforce targets on refiners—customers of the fuel producers—rather than the producers themselves as a bizarre and unprecedented government effort. Wow!!!

It is apparent that although significant technologically advances of late have introduced some improvements, a steady, fully efficient, and profitable large scale commercial *cellulosic bioethanol* conversion process is still in the distant future.

Algae biofuel

Algal biofuel is a fuel derived from processing bio-materials as an alternative to fossil fuel. This process uses different types of algae as its raw materials and/or catalysts. There is an ongoing effort by several companies and governmental agencies to develop the process in order to reduce capital cost and operating expense as needed to make algae fuel production commercially viable.

Algae materials release significant amount of CO_2 when burnt, but that amount is compensated by the CO_2 taken out of the atmosphere during the algae growing process. The world food crisis is increasing the interest in algae-culture in the production of vegetable oils, and biofuels on land unsuitable for the usual agricultural crops.

Algal fuels can be grown with minimal impact on fresh water resources, and some can even be produced by using ocean and wastewater. Most algae products are biodegradable and harmless to the environment in any form and shape.

Algae cost more per unit mass, about ~$5000/ton, mostly due to high capital and operating costs. This is compensated by the fact that they can yield up to 100 times more fuel per unit area than most other crops. There are estimates that for algae fuel to replace all petroleum fuel used in the United States, we would require total of 15,000 square miles of land and a lot of construction and processing equipment.

But, as with many other potential energy sources, these estimates remain on paper and far from any real and sizeable commercial application. And like most other renewables, biofuels production relies heavily on government support in form of grants, income tax, and production tax credits. This shows the immaturity of this technology, and as promising it sounds, we still have to wait awhile to see it compete shoulder to shoulder with the big guys—coal, oil, and natural gas—and even with the renewables (solar, wind, and geothermal).

In the 80s and 90s U.S. scientists experimented with algae as a biofuels source. Algae grown in ponds of wastewater treatment plants can be harvested and processed into biofuels and ethanol. After all the hoopla, the efforts died off, and the potential of harvesting oil for biofuels from waste pond algae is still far from any practical large scale commercial, application.

Recently, there have been isolated attempts to revive the interest in using algae to make biofuels, and there are a number of reported success stories. Still, lots work remains to be done in order to bring this process to market.

One great advantage, in addition to its projected high yield, algae-culture, unlike crop-based biofuels, does not interfere with food production, which led to the crop-ethanol demise in 2008-2009. This is because no agricultural products can be grown in municipal waste plants, and other waste lands. Since algae does not requires farmland nor fresh water for its growth, it could be the solution to several problems at once.

A number of companies are looking into algae bioreactors for various purposes, including scaling up biofuels production to commercial levels, but again; the work is still confined to small test sites and university lab benches.

In recent years, several novel biofuel conversion routes have been announced, such as the conversion of sugars into synthetic diesel fuels. These include: a) use of a micro-organisms (yeast, heterotrophic algae or cyanobacteria that turn sugar into alkanes), b) transformation of a variety of water-soluble sugars into hydrogen and chemical intermediates using aqueous phase reforming, and then into alkanes via a catalytic process, and c) the use of modified yeasts to convert sugars into hydrocarbons that can be hydrogenated to synthetic diesel. Unfortunately, these processes, as promising as they sound, have not been able to produce commercial products.

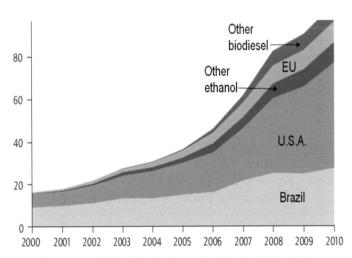

Figure 6-38. Global biofuels production in billion liters

Bio-synthetic gas

Bio-SG (or biomethane) is actually methane derived from biomass via thermal processes. The first demonstration plant producing biomethane thermochemically out of solid biomass started operation in late 2008 in Güssing, Austria, and another plant is operating in Gothenburg, Sweden).

With the increased use of natural gas vehicles (NGV) during the last decade, their share is reaching over 25% and more of the total vehicle fleet in some countries like Armenia and Pakistan. These vehicles can also be run on biomethane, derived from anaerobic digestion or gasification of biomass.

Several routes to fuels and additives at different commercialization are available, including hydrothermal processing, pyrolysis oil, dimethylether (DME), biobutanol, and "solar" fuels (these produced with the help of solar as energy input).

Biofuel Markets

Biofuels are increasing in volume and importance, as the environmental issues take increasingly front stage. Biofuels are considered by some as the solution to all of these problems (to one extend or another), and so their use has increased from almost zero, 20 years ago, to over 3.0% of the total road transport fuel use in some European countries. For example, in the UK, 400 million gallons of biofuels are produced and sold annually.

At the same time, the U.S. produces over 5 billion gallons of ethanol, and about 2 billion gallons of biodiesel. Brazil also produces over 5 billion gallons of ethanol, and China follows with 1 billion. India is next with 500 million, followed by Germany, France, and Russia with about 200 million each. Canada and South Africa are next with about 100 million each annually. Many other countries also produce biofuels, but in much smaller volume.

It is expected that by 2020 over 10% of the energy used in road and rail transport in European countries will come from renewable sources. This is equivalent to replacing 4.3 million tons of fossil oil each year. Conventional biofuels are likely to produce between 3.7 and 6.6% of the energy needed in road and rail transport, while advanced biofuels could meet up to 4.3% of theUK's renewable transport fuel target by 2020

The world leaders in biofuel development and use are Brazil, the United States, France, Sweden and Germany. At the same time Russia, which has over 20% of world's forests is one of the largest biomass (solid biofuels) supplier, and is also looking into the possibility to increase its share in the biofuels production.

The transport sector presently is the largest consumer of fuels—fossil and biofuels—and will most likely remain in the lead for the foreseeable future. It is expected that the biofuels in 2050 will be used mostly for road transport too (over 50%), with 25% used for jet fuel, and the rest for shipping and other commercial purposes.

International organizations, such as IEA Bioenergy, were established by the OECD International Energy Agency (IEA), in order to improve the cooperation and information exchange between countries with bioenergy programs, and assist in the research, development and deployment of biofuels. The UN International Biofuels Forum, formed by Brazil, China, India, Pakistan, South Africa, the United States and the European Commission has similar agenda and goals.

The global bioethanol industry continues to be a bright spot in the world economy, and continues to grow. It is presently supporting nearly 1.5 million jobs and contributes over $300 billion to the global economy. The future looks bright for biofuels, and bioethanol in particular.

Bio-energy Costs

The diversity of raw materials and processes for making biofuels is staggering. We can divide the sector into biomass and biofuels, although there is some intermingling possible.

Table 6-19. Global fuel use by 2050 (estimate)

Fuel types	2050
Biofuels	27%
Diesel	23
Gasoline	13
Jet fuel	13
Electricity	13
Hydrogen	7
Heavy fuel oil	2
CNG, LPG	2

Table 6-20. Global bioethanol productio
(in million liters)

Area	2008	2010	2012
N. America	36,000	51,600	54,600
S. America	24,500	26,000	21,300
Europe	2,900	4,250	5,000
Asia/Pacific	2,750	3,100	4,000
Africa	65	130	235
World total	**66,000**	**85,000**	**85,000**

Table 6-21. Biomass origins and uses

Forest biomass				
Wood			Residues from wood based industry	
Forest residues	Stem wood	Whole trees	Dry residues	Wet residues
Pretreatment				
Solid biofuels				
Wood chips			Pellets	Residues
Bio-heating and bio-electricity systems				
District heat & power	Micro grids		Central heat & power	Stove

Table 6-22. Bulk weight and heating values of biomass products

Wood products	Bulk weight of dry matter in ton / m^3	Heating value in MW/h per ton dry matter.
Pellets	0.70	4.80
Wood chips	0.45	4.70
Forest residues	0.40	4.40
Stem wood	0.45	4.40
Sawdust	0.40	3.00
Shavings	0.25	4.40

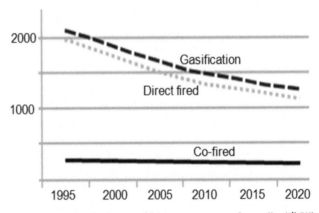

Figure 6-39. Capital cost of biomass power plants (in $/kW)

Biomass costs

Biomass can be sub-divided into wood products from forests, and other (crops, waste materials etc.). The forest-wood for use as energy generator and other purposes can be sub-divided into:

a. Direct use of wood (as forest tree trunks and branches), and
b. Wood and wood processing byproducts.

As it can be seen in Table 6-21, there are a number of products and uses of wood and wood byproducts. Putting a value on each product and use is beyond the scope of this text, but in all cases the price of products and uses are determined by the market supply and demand dynamics. The availability of the wood products also has a determining role in the final price structure.

The harvesting and marketing of forest biomass in Kenya, for example, would be different from that in Sweden. In Kenya there are no laws, nor is there any significant enforcement of wood products harvesting and sale. Because of that, wood harvesting and use in forests

and fields is basically free for the locals, but is sold at a certain minimal price in the large cities.

In Sweden, on the other hand, the wood harvesting process is heavily regulated and enforced, which will determine the market price, which varies with location and the seasons. And no doubt, wood products market prices in Sweden are much higher than these in Kenya.

A significant factor in determining the use of biomass for energy production is the capital expense. It has been going down since the 1990s, and is expected to be competitive in the near future.

The price of biomass products is also determined by its specific heat value in relation to mass. For example, wood pellets are the densest wood product with the highest heating value per ton of material. On the other side of the spectrum, wood shavings are the least dense material and although the heat value per ton is almost as good as that of the wood pellets, it will take almost 4

Table 6-23. Average yield from different energy crops per unit area

Crop	Liters per hectare
Sugarcane	4,900
Sugar beet	4,000
Palm oil	3,600
Cellulose	3,100
Corn	2,600
Rapeseed	1,700
Soy beans	700

Table 6-24. Biofuel prices

Biofuel	$/ton
Bio-crude	167
Corn oil	802
Cottonseed oil	782
Crude palm oil	543
Peanut oil	891
Rapeseed oil	824
Soybean oil	771
Tea seed oil	514
Maize	179
Sugar	223
Wheat	215
Waste oil	224
Poultry fat	256
Yellow grease	412

times the volume to generate equal heat value.

The final cost of biomass materials (delivered on site) depends on a number of factors, based on type of materials, labor and transport conditions, etc. Further from the source the processing facility and the power plant are, higher the cost of transport would be. This, of course, will increase the cost of the final product as well.

Another factor, determining price is the biomass quality; especially that destined for the generation of bioenergy. The European Committee of Standardization (CEN) is working on standards for fuels and has established three main categories for biomass quality:

- *Pure biomass*, which comprises pure wood and other biomass;
- *Contaminated biomass*, which is fuel that cannot be burnt without meeting the demands of waste combustion set the EU Government, and
- *Hazardous biomass*, e.g. CCA impregnated wood, which can only be burnt using a special combustion- and cleansing technology.

Obviously, the price will be different for pure vs. contaminated biomass, while hazardous biomass will have a totally different price schedule, which might be on the negative side.

The price of the different energy crops also depends on the type and location of the crops. For example, the price of a ton of sugarcane would be different from that of a ton of corn, and very different from a ton of bagasse. One of the reasons for this is the fact that sugarcane yields almost twice the amount of fuel from a hectare than corn, and 5 times that of soy beans. The cost of the land and amount of effort is reduced by the fact that sugarcane requires less land to produce equivalent energy.

Where the crop was grown and processed also matters in price setting calculations. A ton of bagasse grown in the U.S. would cost much more than if it is grown in Brazil. The conversion to bio-fuel would also more expensive in he U.S., due to higher labor and ma-

terials other costs.

Another important factor to consider in the price formation scenario of energy crops used for biofuels is the ease of processing. Sugarcane is the easiest and cheapest to process into biofuel, because it is pure sugar, which requires fewer process steps. This would raise its price significantly, as compared to the rest.

Due to the scarcity of wood and wood products in many places around the globe, wood products import-export activities have increased during the last decade. Typical CIF prices for sea transport, including loading costs, ocean freight and insurance, are in the range of 100 to 120 Euro per ton for wood pellets with heating value around 17 GJ/ton.

At this price, wood pellets are imported in the EU from Canada, United States, the Baltic States and Russia. Large scale in freight volumes from 10,000 to 50,000 tons makes wood products import profitable, since the cost for ocean fright is estimated to approximately $7.5/MWh.

The cost of energy produced by biomass materials is very near that of the fossils. Presently, it is estimated at $0.04-$0.06 per kWh. Not bad.

Biofuels Costs

The average international prices for the common bio-fuels are as shown in Table 6-24.

The biofuels prices, however, vary significantly with time, location, and socio-economic developments. This is because the production of biofuels varies significantly with location, weather, seasons, and raw materials types and availability. A municipal waste biogas generating plant, for example, would have a different price tag and production cost than an algae biofuel power plant.

As an example, a small production plant of 200,000 tons of biodiesel per year has been estimated at approximately $500 million, or approximately at the cost of $2,500 per produced ton of biofuel. This is about $2.50 per liter, or $10 per gallon of biodiesel. $150 million of the total, however, would go towards the construction of an oxygen generating facility. So, if the plant is located near an oxygen generator, the construction costs would be significantly reduced, as would the price of the biodiesel it produces.

In contrast, the average production cost of a large biodiesel production facility today is estimated at $500-600 per ton, or about $0.50-0.60 per liter of petro-diesel equivalent. This is about $2.00-2.40 per gallon of biodiesel. There are other costs too, so the price would go up by the time it gets to the consumer.

Biofuel Markets

The new U.S. Renewable Fuels Standard (RFS), signed into law in the summer of 2015 as part of the revised Energy Bill, sets high goals for the U.S. biofuels industry. It mandates the annual production of 36 billion gallons of biofuels (ethanol and biodiesel primarily) to be achieved by 2022. 21 billion gallons of the total is supposed to come from "advanced biofuels." These can be produced using a variety of new feedstocks and technologies. 16 billion gallons of the advanced bio fuels total is expected to come from "cellulosic biofuels," which are derived from plant sources such as trees and grasses.

But some of these technologies are in testing stages, and others have not yet been developed. So, are these targets realistic? And if pushed hard, would they have a serious negative impacts on the nation's farmlands, forests, waterways, and populated areas? There are no answers to any of these questions, but fortunately the Bill contains a few conditions that can be used to stop the escalation if things go wrong anywhere in the system.

The entire U.S. corn crop would produce only 34 billion gallons of bio-ethanol.

So, unless there are some secret factors we are not aware of, the U.S. must stop consuming corn all together in order to get even close to the mandate. Our educated guess is that this is logistically impossible, regardless of the technologies and methods used. So, undoubtedly, the goals will have to be lowered to a more practical level.

Presently there are a number of technical, social, economic, and environmental issues, related to the production and use of biofuels. Most of these have been discussed in the media and the scientific journals, and include: the effect of moderating oil prices, the *food vs. biofuels* debate, poverty reduction potential, carbon emissions levels, sustainable biofuel production, deforestation and soil erosion, loss of biodiversity, impact on water resources, as well as energy balance and efficiency, just to mention a few.

Scientists warn that not all biofuels perform equally in terms of their impact on climate, energy security, and ecosystems, and suggest that environmental and social impacts need to be assessed throughout the entire life-cycle, prior to making a decision on which biofuel to use, when and where.

And then, there is the land conflict, which is growing bigger by the day. In Africa—the poorest continent ion Earth—66% of land deals, that have been cross-referenced by researchers recently, are intended for biofuel production. The continents, which houses 90% of the world's poorest people is dedicating more land to biofuels production that food. 60% of new land deals are for bio-crops production, versus 15% of the total for food crops. The sad news is that most of these bio-crops will be exported to developed countries.

This, at a time when millions in sub-Saharan Africa go hungry to bed every day, where 1 out of 4 children is badly malnourished, and where over 70% of the rural population does not have access to electricity. Go figure!

The fast development of new biofuel crops and second-generation biofuels is introducing new variables, which at times circumvents the common knowledge. We are well aware of the current issues with biofuel production and use, and are looking into new ways to handle the most pressing problems. The good news is that a new trend today points to the development of biofuel crops that require less land and use fewer resources than current biofuel crops do.

Algae, as a source for biofuels, is one of the solutions, since it uses unprofitable land and wastewater from different industries. Different algae are able to grow in wastewater, which does not affect the land or freshwater needed to produce current food and fuel crops. Also, algae are not part of the human food chain, so do not take away food resources from humans.

Nevertheless, the effects of the biofuel industry on food and the environment in general are still being debated. A recent study shows that biofuel production accounted for up to 30% increase in food prices in 2008. Market-driven expansion of ethanol in the US increased corn prices by more than 20% in 2009, as compared with the prices of ethanol production at 2004 levels. This changed the direction of the basic research, forcing the development of biofuel crops and technologies that will reduce the impact of a growing biofuel industry on food production and cost.

This prompted some drastic reactions, such as the 2012 decision of the United States House Committee on Armed Services to put language into the 2013 National Defense Authorization Act that would prevent the Pentagon from purchasing biofuels for any reason, even if they offered improved performance for combat aircraft and other uses. Developing biofuel crops that are optimized for the local climate and market is the best solution to this problem.

Using local biofuel crops voids the need to import fossil fuels from faraway places.

The problem is that many areas of the globe are unsuitable for producing the most efficient energy crops, which require large amounts of water and nutrient-rich soil. So biofuel crops, such as corn, sugarcane and others are impractical in these places and must be grown in different environments and regions of the globe. That complicates the picture, and creates a several-fold un-equilibrium, where rich nations produce the crops to be used by poor nations. This increases the poor countries import-export balance and also creates an environmental un-equilibrium, where CO_2 absorbed in one part of the world is released in another part.

In summary

Solid biomass use around the world is projected to continue at the present levels at least for the foreseeable future. The negative effects of that use are well known, but there is very little that can be done to reduce them due to the fact that the major use of solid biomass (in form of firewood) is for providing basic, and often only, human comforts, which cannot be denied to the needy people of the world.

Liquid and gaseous biofuels fuels' production is expected to rise significantly in the future. Biofuels derived from organic matter will play an ever increasing important role in providing fuel and reducing CO_2 emissions in the transport sector, thus enhancing our energy security and helping to clean the environment.

Biofuels could provide 20% of total transport fuel by 2025 and 30% by 2050. They would play a particularly significant role in the replacement of diesel, kerosene and jet fuel.

Sustainably produced and efficiently used biofuels could avoid the generation of around 2.1 gigatonnes (Gt) of CO_2 emissions per year. For this to happen, however, most conventional biofuel technologies need to improve conversion efficiency, cost and especially their overall sustainability.

There are other requirements too:

- Advanced biofuels have to be deployed on a large scale, which requires additional and quite substantial investment in their future development and demonstration. Special attention to and support for commercial-scale advanced biofuel plants is a must to the timely success of advanced biofuels.

- Support policies should be developed and focused on incentivizing the most promising and efficient biofuels in terms of GHG avoidance. The policy framework must ensure that food security and biodiversity are not compromised, and that only social impacts are the result of the expansion efforts.

- Sustainable land-use management and proper certification schemes, together with support measures that promote "low-risk" feedstocks and efficient processing technologies, are the key factors in ensuring fast and efficient development of large scale biofuels production,

Meeting the biofuel demand by 2050 would require around 65 exajoules of biofuel feedstock, occupying around 100 million hectares. This is in direct competition with the need for land and feedstocks to meet the rapidly growing demand for food and fiber around the world. Additional 80 exajoules of biomass will be needed by 2050 for generating heat and power. The required 145 exajoules of total biomass for biofuels, heat and electricity from residues and wastes, along with sustainably grown energy crops could be achieved only by the implementation of sound policy framework and by the application of efficient equipment and processes.

Trade in biomass and biofuels will become increasingly important as well, especially in the effort to supply biomass to areas with very high levels of production and/or consumption. Scale expansion and efficiency improvements will reduce biofuel production costs over time. In a low-cost scenario, most biofuels could be competitive with fossil fuels by 2030. In a scenario in which production costs are strongly coupled to oil prices, they would remain slightly more expensive than fossil fuels.

The total biofuel production costs from 2010 to 2050 are estimated in the range between $11 trillion to $13 trillion, while the marginal savings or additional costs compared to use of gasoline/diesel are in the range of only +/-1% of total costs for all transport fuels.

United States produces 48 billion liters of biofuels annually. Of this amount only 20 million is produced from cellulosic materials (trees, grass, corn stalks, etc.) There is an estimate of 150 billion liters of bio-ethanol to be produced by 2025, 60 billion liters of which would be from cellulosic-ethanol.

The increased biofuels production, transport and use will also generate millions of jobs around the world, which will contribute to the economic development of a number of developing countries.

Note: The author was part of an R&D team in a South American government R&D lab, which specialized in the production of cellulosic products (paper and biofuels). The results of this effort, after several years of intense work, led to the conclusion that it is *technologically feasible* to produce cheap bio-ethanol and other bio-fuels, as demonstrated recently by some companies.

Technological feasibility, however, is one thing. Getting the technology down to practical mass production is another. The big problem with biofuels production today is the huge amount of water, energy and chemicals used in these types of processes, which makes the energy input-output balance questionable. Also, there would be mountains of waste solid byproducts generated during the processing of raw biomass.

Since the useful content of the cellulosic materials is extremely low, the remaining waste comprises 90-95% of the raw materials input. This means that very large quantity of raw materials are hauled in, processed at a great expense of energy and chemicals, and then dried and stored in large storage bins, or hauled away again as waste.

This is a very large amount of solid waste products that are basically useless for any type of human or animal use. If the production increases to the estimated levels, we will eventually end up with mountains of waste cellulosic materials. Some of the wastes could be burned for heat generation at the biofuel processing plants, but the huge amount of smoke, soot, and GHGs would make this process one of the dirtiest and therefore objectionable from an environmental point of view. This, in addition to rivers of chemicals used in the process and disposed later on, would create a nightmarish environmental scenario at any large biofuels production facility.

A number of not easy to deal with problems—especially at the large scale of production anticipated in the next several decades. Biofuels, however, are very promising and important part of our energy future, so solutions must be found.

In any case, bio-energy crops are needed, no doubt, and if properly implemented can produce significant social and environmental benefits. Proper management of the biofuel crops and processes can help us get our energy independence quicker, and could also provide excellent carbon reduction at increased energy production ratios, which in turn would result in reduction in pollution and greenhouse gas emissions.

The key to success here is the implementation of *proper and balanced* soil and crop management techniques, which must be adapted to the local conditions. This way, greater crop efficiency and environmental benefits could be achieved. Land selection is critical too, where the best case scenario is to use land that is abandoned or in some way is in degradation and/or useless. Energy crops compete for land with food production, livestock grazing, and firewood gathering practices, so engaging and educating the local communities in the proper development and use of energy crops is another key to success.

Lots of uncertainties lay ahead, and lots of battles will be fought on different fronts, but one thing is for sure; energy crops have an important role to play in our energy and environmental futures. But as we saw during the uncontrolled boom-bust of 2007-2008, they can also create large scale problems.

Using the lessons of the past, we must make sure that future energy crop growing and processing methodologies are properly designed, and implemented in a manner that would benefit the environment and those who are affected.

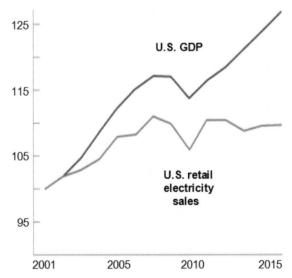

Figure 6-40. U.S. GDP vs. retail electricity prices

ENERGY EFFICIENCY

Yes; energy efficiency (EE) is also part of the energy markets, but in reverse. It is not generating power, but voiding its use. So, yes, it is part of the energy formula;

Energy Efficiency equals the amount of energy used yesterday minus the energy used today. It is how much energy we save by not using it.

Its potential is almost unlimited, mostly because we here in the U.S. waste more energy than we need to use. The U.S.' EE market (or rather money spent on EE) has been growing fast in recent years, from $1.0 billion in 1990 to over $14 billion in 2014. EE spending has more than tripled from 2005, with broad based growth, covering all sectors of the energy markets.

The difficulty here is that to quantify EE, we need to measure the energy that was avoided by not using it, or replacing it with more efficient systems. The growth in EE investment correlates with the trend of slowing demand growth for electricity for the present period of time. It is uncertain if this downward trend will continue or if demand growth will level out at a little less than 1%. If the trend in EE deployment if sustained, would suggest that electricity demand growth could continue to slow in the coming years, even if the population grows.

For a closer look at the relationship between energy use and the U.S. economy, we can consider the GDP and total retail electricity sales. It shows that retail electricity sales diverging from GDP starting in 2003, with the gap growing over time. Following the "Great Recession," as the economy began to recover, retail electricity sales stayed essentially flat, and as of 2014, they were still lower than the peak in 2007. Still, the growth rate in retail sales is slightly negative between 2010 and 2014, even as the economy grew by 9% in real terms. Part of this is due to successful implementation of energy efficient measures and programs.

Energy efficiency is a critical part of the energy markets, since it represents the most efficient use of energy—the simple act of using less today than yesterday.

What this means is that we are:

- Using less electricity, which makes us more like the rest of the world,
- Saving a lot of fossil fuels from being burnt, which in turn is
  - Helping to preserve some of the fossils for the future generations, while at the same time
  - Emitting less GHGs in the air and less solid and liquid pollution in the soil.

The overall growth in EE innovation and deployment is clearly one of the factors driving the energy markets. There are, however, other variables, including shifts in economic activity (from production to service) and the growth of distributed (rooftop) solar.

Nevertheless, we foresee the electricity consumption remaining flat or even declining even more as population and economy growth continue. EE, on the other hand, has no limits. Our imagination is the only boundary.

Notes and References

1. Nuclear Power International. http://www.nuclearpowerinternational.com/index.html
2. Oak Ridge, http://nhaap.ornl.gov/HMR/2014
3. DOE, http://www.energy.gov/hydropower-market-report
4. NRGExpert, http://www.nrgexpert.com/energy-market-research/global-hydropower-report/
5. DOE, http://energy.gov/science-innovation/energy-sources/nuclear
6. WNA, http://www.world-nuclear.org/info/Country-Profiles/Countries-T-Z/USA—Nuclear-Power/
7. *Photovoltaics for Commercial and Utilities Power Generation*, Anco S. Blazev. The Fairmont Press/CRC, 2011
8. *Solar Technologies for the 21st Century*, Anco S Blazev. The Fairmont Press/CRC, 2013
9. *Power Generation and the Environment*, Anco S. Blazev. The Fairmont Press/CRC, 2014
10. Energy Security for the 21st Century, Anco S. Blazev. The Fairmont Press/CRC, 2015

Chapter 7

Electricity and
Transportation Markets

Over 90% of the primary energy resources are converted into electric power and transportation fuels, which drive the world's economies. Without them we would be back in the stone age.

—Anco Blazev

BACKGROUND

Electric power and transportation (fuels) are part of the "secondary energy" markets, since they are derived from the primary energy sources; the fossils and renewable. They, however, are the most important part of the energy markets, because they represent the largest, most practical, and most useful side of the energy resources. They are integral part of out lives. Without them, life as we know it would not exist, or at least it would be quite different and very difficult.

Electricity and fossil fuels are the largest part of the energy markets. They literally drive our daily personal and business lives

Electric power generation started with a crazy man standing in the midst of a severe storm, trying to light a light bulb with a lightning strike. Over a century later, electricity is now one of the hugest businesses on the planet. Power generation, distribution, and use is a day and night non-stop activity, which moves mountains and runs the global economies.

The transportation sector, which also had humble beginnings, consisted of thousands of horses pulling carts on unpaved roads, is now the lifeblood of our lives. Millions of 18-wheelers and other trucks, ships, planes, and trains loaded with tons of products speed down the highways, waterways, and airways day and night. Additional millions of cars, light trucks, and other vehicles bring billions of people to and from work and anywhere in between daily.

99.9% of all vehicles are powered by fossil fuels.

Yes, we all know that most hybrid and electric car owners are very proud of their contribution to energy conservation and clean environment. We must congratulate and encourage them for the good thoughts and their contribution. At the same time, however, we must reveal to them that fossils are still driving their vehicles. Natural gas, used in some *'green'* vehicles, is 100% fossil, while the electricity used to charge them is at least 75% from fossils too. So, no, these vehicles are not the solution to our energy and environmental problems. They are only part of it. Very small part.

Nope, we still rely in fossils much more than we think we do. Just imagine for a moment a day when the power plants sit idle and all gas stations are closed. Imagine all houses and businesses dark, all vehicles parked, and all daily activities stopped. A total economic paralysis...without any way to get to work, produce, or transport anything. What a dark, cold, and desperate day this would be!

We console our selves that this disaster cannot happen here today, or tomorrow. While this is true, we have to always remember that we have been saying this for a century now, and that someday we would get surprised. It is inevitable, because no matter how large the fossils bucket is, it has a bottom.

The transformation from candles to electric lights, and from horse buggies to motorized vehicles was not simple, quick, nor cheap. It took thousands of engineers, technicians, and workers many years and billions of dollars to generate and deliver power to the consumers and convert the horse-carts to fast moving, powerful vehicles. The 21st century power generation and transportation sectors are products of their ingenuity and hard work.

And as with any successful business in the capitalist society, power generation and transportations have developed into huge markets, which is what we are looking at in this chapter. They are very important part of the energy sector. Electric power is one of the key drivers of today's society, where transportation provides the mobility needed for the proper and efficient function and development of our economy. Both markets, working together, are responsible for the amazing economic and technological achievements of the 19th and 20th centuries. They are still on the driver seat of our society and its economy.

The *energy sector*, and the *related energy markets*, consist of thousands of electricity, oil, natural gas, and renewable energy assets. These are geographically dispersed and connected into systems and networks spread all over the country and the globe. Interdependency within the energy sector and across the Nation's critical infrastructure sectors is complex and critically important. The energy infrastructure provides electric power and fuels to the Nation, and in turn depends on the Nation's transportation, information technology, communications, finance, and government infrastructures for its proper function. One cannot exist without the other.

A complex picture that cannot be reviewed in its minute details in one book, so we try to cover the major points and inter-relations that drive the energy markets.

Below, we take a closer look at the electricity and transportation markets:

ELECTRIC POWER MARKETS

Ben Franklin may have discovered electricity, but it is the man who invented the electric meter who made the money.

—Earl Warren

Actually, it is Thomas Edison, with the help of his business-partner Samuel Insull, who brought up electricity from the stormy night experiments into the homes and the energy markets. Thomas Edison took the undeveloped concept of electric power generation and put it to practical use, in form of light bulbs, wires, electric motors, meters, and such.

While Edison was the technical genius, Samuel Insull, his business partner, was the engine behind Edison's commercial success. He figured a way to manufacture and distribute electric appliances to the masses and make piles of money. Insull is the lesser known of both, but he alone developed a successful business model to bring electricity and appliances into homes, stores, offices, and industry at affordable prices.

It all started in 1881, when 34-year-old Edison (the technical genius) and 21-year-old Insull (the business guru) became partners in a new company that is still around and still known as General Electric (GE). For 10 years GE grew under their capable technical and business partnership, but then the 32-year-old Insull left GE and moved to Chicago with another grandiose plan in mind.

He was planning to turn electricity use from selling luxury items and services (which GE delivered in those days) to mass-produced, affordable necessity for the masses. Insull applied Edison's model of central-power generation, and redesigned it to serve large areas. This was to replace small, isolated plants serving a local area. Instead of making electricity, the masses were encouraged to buy it from the central-power plants.

Soon "Insullization" became a model for successful electricity generation and distribution business, soon to be one of the largest in the U.S., covering 39 states. The business then expanded from electricity to coal and gas distribution, and urban transportation as a major user of power of the time. Soon, Insull enterprises were leading the energy field, even eclipsing the Rockefeller's achievements.

Thomas Edison is the father of electricity, while Samuel Insull is the father of the electric industry.

Since most miracles last only a day or two, Insull (the clever businessman he was) kept on expanding and diversifying his operations during the Great Depression. As a result of the changing times, his holdings became worthless within months. Insull's brilliant ascent ended with a tragic twilight, during which period he died almost penniless.

Amazing business trajectory, that very few have been lucky, or unlucky, to duplicate. Except Enron. Very much like Insull, Enron went from a leader in the energy field to a shadow of a company, with its leaders making all the wrong decisions and ending in prison, or dead.

Insull's holding companies controlling 85% of the U.S. electric power market, started failing under the weight of the Great Depression. Insull, blinded by his success, ignored the surroundings and unable to turn down new deals brought him to his knees.

The Insull Empire bankruptcy filings started in 1932, and over half million Americans lost most of their investment. Company loans carried Insull's personal guarantees, which brought his personal net worth from $150 million to zero, and even negative, value within days.

Under Congressional criticism, accompanied by state and federal lawsuits after his bankruptcy, Insull pleaded innocent. Then he put himself and his brilliant career on public trial, which cost him everything he had, and ended up in court. At the end he was acquitted, after the jury felt sorry for this failed businessman, and did not consider his dishonesty during the trials.

They saw good intentions behind the manipulated accounting, the clumsy efforts, and fruitless results. Samuel Insull lived the last years of his life as broken man; an icon of celebrity and success fading away abroad amidst a modest, self-imposed exile.

In a similar fashion, the Enron's bosses were faced with the wrath of the people and the courts. During the Enron trial, they blamed each other—each guilty in their particular area for the Enron demise—but each throwing the blame at the other. With defiance and arrogance on the witness stand, they only exposed their real nature of predators in disguise. Many were found guilty, unlike Insull's verdict some decades before.

The Enron bosses might have considered Insull's case as a beacon for their faith, since the past can inform the present, but in this case they simply ignored the details of the lessons of history and paid for their criminal actions and negligence with prison sentences.

Still, Insull and Enron should be both given credit for creating a number of innovative products and services, and addressing efficiency, price, reliability, regulatory compliance, and other basic needs of the energy marketplace.

Today we just have to hope that the new leaders in the energy industry will look carefully at, and develop further, these concepts without straying away from the established and proven methods of operation and norms of conduct.

Background

Electricity is a commercial product—a physical entity with unique properties. It is actually very different from most other forms of energy and products. It is the flow of electrons that cannot be seen, so what we are buying is invisible to the naked eye. With special equipment, however, electricity can be made to bring light, heat, or do work.

Note, please, that for the sake of convenience and efficiency we use the terms: *electricity, electric power, power, electric energy,* and *energy* interchangeably in this chapter. All these terms refer to the physical concept of electricity (flow of electrons through wires or appliances), but there are differences that must be acknowledged, as we will see below.

Electricity is a *secondary energy source*, since it is generated by the conversion of other energy forms such as oil, natural gas, coal, uranium, or the energy inher-

ent in wind, sunshine, or the flow of water turning a hydropower turbine. Although electricity is invisible and difficult to measure, it can be turned on and off and measured by special, dedicated devices.

As unusual and hard to understand electricity might be, it is the foundation of our modern lives. It provides comforts at home and drives the economy.

Today, the electric power market is an important, well organized and coordinated effort of each nation to ensure enough, cheap electric power, and to secure the energy infrastructure for the populace. Thus generated electric power is then transferred and distributed to millions of homes and businesses across all countries and the continents.

The electrical systems and the related electric energy generators and networks are huge businesses that cross national borders. This makes international collaboration a necessary component of the national and global energy markets.

Electricity is a physical product–the actual flow of electrons at a given time is the actual product for sale here. The electricity market is a system created around the generation and use of electricity, which enables purchases, through bids to buy; sales, through offers to sell; and short-term trades, generally in the form of financial or obligation swaps. Bids and offers use supply and demand principles to set the price. Long-term trades are contracts similar to power purchase agreements and generally considered private bi-lateral transactions between counterparties.

Wholesale bids and offers in electricity are typically cleared and settled by the market operators, or, in some cases by a special-purpose (or independent) entity. Market operators use their knowledge and experience to maintain generation and load balance in the system by all means available to them.

The practical aspects of electricity can be also looked at as consisting of two different physical entities: these of power and energy. In technical terms:

- Power is the rate of electric power transfer through a measured point at a given moment. It is an instantaneous unit, measured in kilowatts (kW), megawatts (MW), gigawatts (GW), etc.

Power in this case refers to the size of a power plant, which could be 200 kW, 200 MW, or 2 GW. This also means that at any given moment, the plant generates 200 kW, 200 MW, or 2 GW of electricity.

- Energy, on the other hand is that same power measured over time. The amount of electricity that flows through a metered point for a given period and is measured in kilowatt hours (kWh) megawatt hours (MWh), gigawatt hours (GWh), etc.

Energy in this case refers to the amount of electricity a power plant generates in one hour. This also means that at any given moment, our 200 kW, 200 MW, or 2 GW power plants generate 200 kWh, 200 MWh, or 2 GWh of electricity respectively.

So, using the terms *power* and *energy* interchangeably is not the right thing to do, since they mean totally different things. Yet, this approach is accepted by the masses, and must be taken as such with a grain of salt.

The national electric energy markets trade net generation output during a certain interval of time, usually in increments of 5, 15 and 60 minutes. There are also (financial) markets for power-related commodities, which are required and managed by the market operators, in order to ensure reliability of the entire system. These are considered *ancillary* services, which include such specialized entities and events like: installed capacity, spinning and non-spinning reserve, operating and responsive reserve, regulation up and down, etc.

Some operators use special techniques for managing transmission congestion, including trading of electricity derivatives, such as electricity futures and options. These are special markets, developed as a result of the restructuring of the electric power systems in some countries, often in parallel with the restructuring of the natural gas markets.

The electricity market has also retail and wholesale components:

- Retail markets involve the sales of electricity to residential and commercial consumers;
- Wholesale markets typically involve the sales of electricity among electric utilities, electricity traders (and even countries) before it is eventually sold to consumers.

Much of the wholesale market and certain retail markets are competitive, with prices set separately and competitively. Other prices are set based on the service provider's cost of service. For the U.S. wholesale markets, FERC either authorizes jurisdictional entities to sell at market-based rates or reviews and authorizes cost-based rates.

In competitive markets, prices reflect the factors driving supply and demand, which comprise their physical fundamentals. In markets where rates are set based on costs, these fundamentals matter as well. Electricity supply incorporates generation and transmission, which must be adequate to meet all customers demand simultaneously, instantaneously, and most importantly for the customers; reliably.

The key supply factors affecting electricity prices include; fuel prices, capital costs, transmission costs, transmission capacity and constraints, and the overall operating characteristics of power plants and power transmission networks.

Sharp up-down changes in demand, as well as extremely high levels of demand (peak periods), affect prices as well, especially if less-efficient, more-expensive power plants (peakers) must be turned on to serve the load. This condition is becoming increasingly important as new large scale wind and solar power plants are plugged into the national grid.

Wind and solar are variable energy sources—on, when wind and sun are available, and off, when wind and sun are not available. This forces many peaker plants to start and shut-down many times during the day or night. The cost of doing this at times exceeds the savings from the renewable power plants.

In the United States and other developed countries, consumers expect high quality electricity to be available whenever they need it...right now and for as long as needed.

Note: Here we need to mention that *quality* has always been a requirement for the U.S. electric supply, since we don't want to live with flickering electric lights, or TV sets. Lately, however, the increased number of computers, communications, and the related electronic hardware, has made quality an absolute necessity. Computer equipment is quite sensitive to voltage, frequency, and other electric power variations, so high quality electric power today is on the list of priorities of all power companies.

Electricity use has also grown enormously as consumers in the developed world today consider not only refrigerators, TVs, and hair dryers but also computers, iPods and other electronic devices as absolute necessities. Consumers also expect to pay reasonable prices for the electricity they use. Meeting these customer expectations is challenging, since the electric power quality is expected to be higher than ever before.

One of the difficulties in providing constant supply of high quality electric power is the fact that it cannot be stored in any appreciable quantities. Instead, it is produced on the fly; as needed and when needed. Unlike most other markets, electricity's historical inelastic demand does not move with prices. To provide electricity on demand, electric system operations have to be planned and conducted with that goal in mind.

Lacking energy storage and responsive demand, operators must plan and operate power plants and the transmission grid so that demand and supply exactly match, every moment of the day, every day of the year, in every location. Not an easy nor cheap thing to do.

This may sound easy to most of us, since we are used to it, but in actuality is a very difficult task. There are thousands of people involved in accomplishing this task, and it is these people we must thank for the reliable electric power we use daily.

The Discrepancies

Generating and consuming a lot of electric power and fossil fuels is the norm today. We don't even blink when prices go higher, and continue using as much energy as we need. Yet, changes are coming our way.

Table 7-1. U.S. energy additions

New construction in the U.S. since 1980	
Coal-fired plants	0
Nuclear power plants	0
Hydropower plants	0
Refineries	0
LNG export terminals	3
Coal export terminals	4

Obviously, the U.S. is not adding any new energy generation facilities. Instead, we are gearing up for export of different types of energy. What it does NOT tell us, is that while there are no new power plants built in the U.S. for 35 years, many coal-fired plants have been retrofitted for burning natural gas. Also, it does not reflect the many gigawatts of renewable power generation added to our power generation mix. All of which is good, of course…to a point.

What Table 7-2 also tells us is that the global energy demand and GHG emissions are increasing rapidly. On global level, we have to wonder where would all this energy come from. In three decades the world has to come up with almost twice of what we have today. This simply sounds unreasonable, since all indicators point to decreased in energy sources production and much higher prices. So, unless new breakthrough technologies are invented and implemented, we don't consider such increase of energy possible.

Table 7-2. Energy statistics

Global energy demand increase by 2050e	
Oil	25%
Gas	45%
Coal	50%
Electricity	75%
Total demand	40%

Global GHG emissions	
500,000 BC	380 ppm
1800	280 ppm
2010	380 ppm
2015 420 ppm	
2050(e)	490 ppm
2100(e)	550 ppm

At the same time, we see the increase of GHGs about 25% above present day levels. This means that the global warming will continue and even accelerate. The ocean levels will swallow low lying land masses and many islands, and the weather patterns will bring more devastation by storms, floods, and draughts.

This is enough of a warning to wake us up from our indifference, and prompt us into action. The problem is that the developed nations don't see this as a serious issue and continue the status quo. The developing nations, on the other hand, are busy catching up at all cost and whatever it takes. The future generations are the losers in this mad race. They will have to figure out a way to live with less energy, because we are not planning to leave them much.

Electricity Market Types

The major components of the day-to-day electricity markets are; retail and wholesale. Retail markets involve the sales of electricity to consumers, while wholesale markets typically involve the sales of electricity on a larger (even national) scale. It is done by selling and buying electric power among electric utilities and electricity traders before it is eventually sold to final residential or business consumers.

• Retail energy markets are run by demand, which varies with time, seasons and location, which strongly influences the wholesale markets. The wholesale market and certain retail markets are competitive, with prices set equally competitively. Prices are set, among other things, on the service provider's cost of service. For wholesale markets, FERC either authorizes jurisdictional entities to sell at market-based rates or reviews and authorizes cost-based rates.

- The traditional U.S. wholesale electric markets exist primarily in the Southeast, Southwest and Northwest. About 40% of all retail customers are in traditional wholesale markets where utilities are responsible for system operations and management, and, typically, for providing power to retail consumers.

Utilities in these markets are frequently vertically integrated; they own the generation, transmission and distribution systems used to serve a block of local consumers. These may include also federal entities, such as the Bonneville Power Administration, the Tennessee Valley Authority, and the Western Area Power Administration.

More than half of the U.S. population is served by electricity markets run by regional transmission organizations (RTOs) or independent system operators (ISOs). The main distinction between RTO and ISO markets and their predecessors; the vertically integrated utilities, municipal utilities and co-ops, is that RTO and ISO markets deliver electricity through competitive market mechanisms.

Competitive markets' prices reflect all factors driving supply and demand, which are the markets' physical fundamentals. These fundamentals matter also in markets where rates are set based on costs, The supply component here includes generation and transmission of electric power, meeting customers' demand instantaneously and reliably. The key supply factors affecting prices include the fuel prices, capital costs, transmission capacity and constraints, as well as the O&M characteristics of the different power plants.

Changes in demand, and/or very high levels of demand, affect prices across the board as well. This is especially true when the less-efficient and more-expensive peaker power (which are usually idling) have to be turned on and off to meet the increased load demand.

All customers in the U.S. and the other developed countries expect—no; demand—electricity to be available whenever needed…right now, immediately, in unlimited quantity, and at the lowest price possible. Meeting these customer expectations is an increasing challenge for the power companies, which they have to deal with on daily basis. And amazingly enough, they have been quite successful in doing just that.

How do they do it?

ELECTRIC POWER GENERATION

Electricity is a necessity for most consumers, who depend on the electricity system but who have very little say about its structure and pricing. Since customer cannot pack up and move to another vendor, they are forced to live by the local utilities rules with no say so at all. Until now. Recently, the electric market started paying attention to the customers and things are changing, albeit slowly

Unlike many other products, electricity cannot be stored in any appreciable quantities, but is used continuously by all and at all times. Electric equipment and appliances are tuned to a very specific standard of power, measured as voltage. Deviations in voltage can cause devices to operate poorly or may even damage them.

Consequently, the supply side of the electric market must provide and deliver exactly the amount of power customers want at all times, at all locations. This requires constant monitoring of the grid and close coordination among industry participants. Electricity service relies on a complex system of infrastructure that falls into two general categories: generation and the delivery services of transmission and distribution.

Figure 7-1 shows the basic electric system setup. Electric power generated by the generating station (power plant) is sent into high voltage transmission lines, which transfer it long distance near the customers. Here, the high voltage is stepped-down to user-friendly levels by special equipment in a 'substation'. From there the power is distributed to the customers, some of which—the industrial enterprises—use higher voltage. Most of us—residential and small businesses—use the low voltage of 120 and 240 volts for everyday needs.

Together, the power generation and high-voltage transmission lines that deliver power to distribution facilities constitute the bulk power system of *power generation and transmission*. The *power distribution* lines and facilities deliver power of a much lower voltage to the end user.

Note: For the purposes of this writing, and when referring to *electric power*, we intermingle the words power and "energy" simply because they mean one and the same for all practical purposes related to electric power generation and use.

We will, however, remind the esteemed reader, again, that when talking about the electric system "power" is not exactly the same as "energy," since:

- Power is the rate at which energy is converted and used at a given moment, measured in kW, MW,

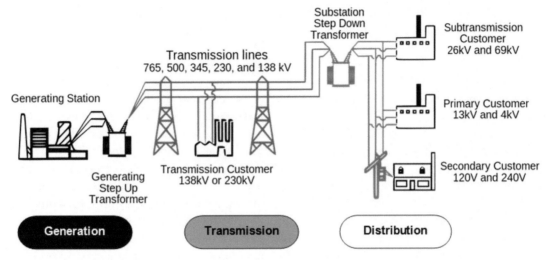

Figure 7-1. Power generation, transmission, and distribution

and GW while
- *Energy* is that same power generated during some amount of time (usually one hour). It is measured in kWh, MWh, and GWh

And so, energy is power as referred to electric power generation, transmission, and use.

- Electricity is measured in terms of watts, and commercially in kilowatts (kW=1,000 watts) or megawatts (MW=1,000 kilowatts).
  — A kilowatt (or watt or megawatt) is the amount of energy used, generated or transmitted at a point in time—a split second, or as long as it takes to take a quick measurement.
  — The aggregation of kilowatts possible at a point in time for a power plant is its *capacity*.
  — The aggregation of kilowatts used at a point of time is the demand at that time period.
- The number of kilowatts used in an hour is expressed as kilowatt-hour (kWh) and is the amount of electricity a customer uses or a power plant generates over a period of time. Kilowatt-hours are also used by the power companies to bill customers.

In any case, the energy we refer to when discussing power generation and use (be it electric, heat, or vehicle fuels) is energy and power all in one. This is the reason we do not make much of a distinction between these words in this text, unless absolutely necessary.

Power generators are typically categorized by the fuel they use and subcategorized by their specific operating technology. The U.S. has more than 1,000 GW of electric power generating capacity. Coal, natural gas, nuclear, and hydropower and dominate the power generation market. The renewables, solar and wind in particular, are a growing part of it too.

Power plants have different operational and cost characteristics, both of which determine when, where and how plants will be built and operated. Plant costs fall into two general categories:
- Capital cost (investment cost), which is amount of money spent to build the plant, and
- Operational costs, which is the amounts of money spent to run and maintain the plant.

In general, there is a trade-off between these expenses, where the most capital intensive plants are usually the cheapest to run, since they have the lowest variable costs. Conversely, the least capital intensive are more expensive to run, since they use the cheapest equipment and have the highest variable cost.

For example, nuclear plants produce vast amounts of power at low variable costs, but are quite expensive to build. Coal-fired combustion turbines are far less expensive to build, but are more expensive to run, since they require enormous amount of coal to be used as fuel. Solar and wind power plants fall into a different category all together, because they have no fuel expenses.

Power Quality

The quality of the electric power determines its fitness for proper, efficient, and safe use by consumers' electrical equipment, devices and appliances. Synchronization of the voltage frequency and phase is a must for the electrical system, if they are to function as intended—without significant loss of performance, efficiency, or life span.

Without proper power quality, an electrical device (or load) may malfunction, fail prematurely, or not operate at all.

There are many ways in which electric power can be of poor quality and there are many causes of poor quality power.

Electric power is generated (as AC power), which then goes into the high voltage power transmission lines. Finally, it goes into the low voltage electric power distribution lines before coming into our houses and businesses. Here, it is sent into a electricity meter located at the premises of the end user. From here, the electricity is fed through the wiring system of the end user's building into its final destination; the consumer's load, which could be a computer, electric stove, or an air conditioner.

Even if *high quality* power is generated, the complexity of the system to move electric energy from the point of production to the point of consumption provide many opportunities for the quality of supply to be compromised.

The power quality is affected by variations, such as: location, time of day, seasons, weather, demand and many other factors.

Power quality is actually a general term used by all, while from technical point of view it is the *quality of the AC voltage* and frequency that actually determine the quality of the electric power we use. Power is simply the flow of energy at a given moment, which together with the current demanded by a load are largely uncontrollable entities.

Voltage variations affect the quality of the electricity supply the most.

Another factor in assessing electric power quality is its frequency, which affects some devices and appliances. U.S. made devices use 50 hertz electric power, while these in most European countries use 60 hertz.

From a technical point of view, the quality of electrical power can be described as a set of parameters, such as:
• Continuity of service
• Variation in voltage magnitude
• Transient voltages and currents
• Harmonic content in the waveforms for AC power

Ideally, AC voltage is supplied by a utility in a si-

nusoidal wave with an amplitude and frequency given by national standards with an impedance of zero ohms at all frequencies. However, no power source is ideal and its power quality can deviate in a number of ways.

Each of the power quality problems has a different cause, but most problems are as a result of the shared infrastructure. Any fault on the network side, for example, may cause a voltage dip that affects some customers. The higher the level of the fault, the greater is its seriousness and the number of customers affected. Also, a problem on one site may cause a transient that affects other customers on the same subsystem.

Problems of synchronization, such as voltage or current harmonics, arise within the customer's own installation and may propagate onto the network and affect other customers. Harmonic problems can be dealt with by a combination of good design practice and well proven reduction equipment.

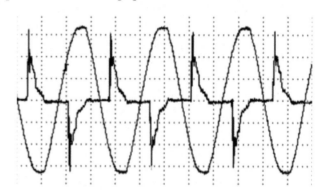

Figure 7-2. Distorted voltage and current (harmonics)

Power conditioning is used in many cases, as needed to modify the power and improve its quality. An uninterruptible power supply (UPS), for example, can be used to switch off the mains power, if a temporary transient condition is detected on the line.

Cheap UPS units could create poor-quality power themselves, which could impose a higher-frequency and lower-amplitude square wave atop the sine wave. High-quality UPS units utilize a double conversion topology which breaks down incoming AC power into DC, charges the batteries, then creates a proper AC sine wave. This new sine wave is highly controlled on-site, so it is of much higher quality than the original AC power feed.

A *surge protector* or simple *capacitor or varistor* can also protect against most overvoltage conditions, while a *lightning arrester* protects against severe spikes. In many cases electronic filters are also used to remove undesirable harmonics in the power fed to sensitive devices.

Modern systems use sensors called *phasor measurement units* (PMU) installed throughout the network to monitor power quality. In some cases they are set to respond automatically to the quality variations. These are components of the smart grids and feature rapid sensing and automated self healing of anomalies in the network. These systems promise to bring higher quality power and less downtime in the future.

In the near future, they will also support power delivered by variable power sources (such as solar and wind power plants) and distributed generation (such as rooftop solar installations), which degrade power quality, if injected in the grid untreated.

The manufacturing of devices designed to ensure the quality of electric power, or somehow modify it and its behavior, is a big and growing business. It is one of the peripheral markets of the energy markets.

Electric Power Generation

Electricity powers 90% of the developed world, including most residential and commercial applications. It is produced primarily by means of burning fossils (coal, oil, and natural gas), nuclear reactors, hydro power, and lately through the renewable energy generators, solar, wind and bio-fuels.

We will review these different energy sources in the following chapters, in order to understand completely their structure and operation. We will then translate into practical terms, their usefulness and size, while estimating the environmental damage these cause now, or are capable of causing during their entire cradle-to-grave life span.

The key energy conversion and power generation methods today can be divided into: a.) conventional (coal, gas, and oil), b.) renewable (solar and wind power), and c.) the other power sources (nuclear, hydro-power, ocean-, bio-, and geo-power).

Fossil fuels (coal, crude oil, and natural gas) are mostly used to generate electricity usually by burning them (converting their potential energy into heat). Thus generated heat is used to boil water into steam, which then turns its energy into mechanical energy by turning a steam turbine and a generator, which finally generates electricity.

Coal Power Generation

After coal is mined, it is transported to a coal burning power plant by trains, barges, and trucks. A conveyor belt carries the coal to a pulverizer, where it is ground to a very fine powder, similar to talcum powder.

The powdered coal is then blown into a combustion chamber of a boiler, where it is burned at around 1,400°C. Surrounding the walls of the boiler room are pipes filled with water. Because of the intense heat, the water vaporizes into superheated high-pressure steam.

The steam passes through a turbine (which is similar to a large propeller) connected to a generator. The incoming steam causes the turbine to rotate at high speeds, which in turn rotates the generator rotor.

The rotation creates a magnetic field inside the wound wire in the stator coils in the generator. This action creates, and pushes, an electric current through the wire coils out of the power plant and into the transmission lines.

After the steam passes through the turbine chamber, it is cooled down in cooling towers and is then returned into the water/steam cycle.

Fossil Power Plants

Fossil fuels (coal and natural gas mostly) are used for power generation via three major technologies, each with its distinct set of market advantages and limitations.

The major fossil-fired power generators today are:
- Steam boilers,
- Gas turbines,
- Combined cycle generators, and
- Oil and diesel power plants

Coal-fired power plants generate more than a third of the electricity in the United States, while natural gas is catching up with, and even exceeding, the coal-fired capacity. The fossil-fired power plants tend to be large units, which run continuously day and night. These power plants provide most of our nation's baseload (constant power). They usually have high initial capital costs and are also somewhat complex in their design, operations, and maintenance. The cost of fuel is also an important factor in their operational and cost structures.

Note: Baseload power plants are capable of running at full power non-stop 24/7/365. The power these generate can be also ramped up and down as needed by the demand side of the power grid. But in most cases, they run continuously and stop only for periodic or emergency maintenance.

Coal-fired power plants have low marginal costs and can produce substantial amounts of power. Most of the coal-fired plants in the United States are owned by traditional utility companies located in the Southeast and Midwest.

The key technologies used today for coal-, or gas-fired power generation are as follow:

- *Steam boiler* technology is an older design that burns gas in a large boiler furnace to generate steam at both high pressure and a high temperature. The steam is then run through a turbine that is attached to a generator, which spins and produces electricity. Typical plant size ranges from 300 MW to 1,000 MW. Because of their size and the limited flexibility that is inherent in the centralized boiler design, these plants require fairly long start-up times to become operational and are limited in their flexibility to produce power output beyond a certain range.

Furthermore, these plants are not as economical or easy to site as newer designs—which explains why none has been built in recent years.

- *Gas turbines* are small, quick-start units similar to an aircraft jet engine. These plants are also called simple cycle turbines or combustion turbines. Gas turbines are relatively inexpensive to build, but expensive to operate being relatively inefficient, providing low power output for the amount of gas burned, and having high maintenance costs.

They are not designed to run on a continuous basis and are used to serve the highest demand during peak periods, such as hot summer afternoons. Gas turbines also run when there are system wide shortages, such as when a power line or generator trips offline.

Gas turbines typically have a short operational life due to the wear-and-tear caused by cycling. The typical capacity of an average gas turbine is 10-50 MW and they are usually installed in banks of multiple units in order to provide larger amount of power.

- *Combined cycle* power plants (CCPPs) are a hybrid of the gas turbine species and steam boiler technologies. Specifically, this design incorporates a gas-combustion turbine unit along with an associated generator, and a heat recovery steam generator along with its own steam turbine. The result is a highly efficient power plant. They produce negligible amounts of SO_2 and particulate emissions and their NO_x and CO_2 emissions are significantly lower than a conventional coal plant. CCPPs, on average, require 80% less land than a coal-fired plant, typically 100 acres for a CCPP versus 500 acres for comparable coal plant, and CCPPs also use modest amounts of water, compared to other technologies.

- *Oil- and diesel* fired power plants

These play a minor role in U.S. power markets, but are nevertheless very important, because of their logistic roles. For example, most of the power in Washington DC, Hawaii, and several other locations is generated by oil fired power plants. Just imagine Washington DC or Hawaii without power for whatever reason.

These facilities are expensive to run and also emit more pollutants than gas plants. These plants are frequently uneconomic and typically run at low capacity factors. Like gas-fired generators, there are several types of units that burn oil; primarily, these are steam boilers and combustion turbines.

Generally, two types of oil are used for power generation: number 2 and number 6 (bunker) fuel oil. Number 2 is a lighter and cleaner fuel. It is more expensive, but because it produces fewer pollutants when burned, it is better for locations with stringent environmental regulations such as major metropolitan areas. Conversely, number 6 fuel oil is cheaper, but considered dirty because of its higher emissions.

It is highly viscous (thick and heavy) and it comes from the bottom of the barrel in the refining process.

For fairness sake, we must mention the other power plants here too:

- *Nuclear plants* provide roughly 20% of the nation's electricity. There are 104 operating plants with a total capacity of about 100 GW. These plants are used as baseload units, meaning that they run continuously and are not especially flexible in raising or lowering their power output.

Nuclear plants have high capital and fixed O&M costs, but low variable costs, which includes fuel cost. They typically run at full power for 18 month, which is the duration of a unit's fuel cycle. At that point, they are taken off-line for refueling and maintenance. Outages typically last from 20 days to significantly longer, depending on the work needed.

After the Three Mile Island plant accident in 1977, there was a cessation in the development of new plants. Most projects under construction in 1977 were finished, albeit with tremendous cost overruns. The last unit built in the United States came online in 1996. The future o9f nuclear power depends on if and when other nuclear accidents would occur.

- Hydropower plants provide over 15% of the national electric power today, but just like the nuclear plants, their future is uncertain. The problem here is the fact that climate changes are limiting the amount of water available in many parts of the world. Without steady water refill, most hydro dams would dry up, and not many more could be built.

• Renewable energy power plants are these that use solar, wind and other renewable energy sources (ocean, geo-, and bio-energy). Their contribution is still small today, but is expected to grow significantly in the future.

Heat-to-electricity Conversion Process

The first step of the energy conversion takes place in the furnace where coal is burned. Most of the potential energy in coal is converted to heat energy. Only those carbon atoms that don't join up with two oxygen atoms and form GO_2, carbon dioxide, are not effectively used. This wasted raw energy goes up the stack as carbon or as partially burned carbon, carbon monoxide, CO. Inside the furnace the usable energy is in the form of the thermal energy of hot combustion gases, GO2, formed when coal is completely burned, and a host of other chemical combinations.

But some of the heat energy is lost as friction and other process inefficiencies, while some goes up in smoke through the smokestack. But the smokestack is essential for efficient burning process, for its draft pulls oxygen into the furnace. It also carries away some of the unburnable pollutants, together with close to 10% of the total heat energy generated by the coal's burning.

The 90% that remains in the furnace is absorbed by the pipes lining the furnace walls in which water is converted into high-pressure, high-temperature steam. The steam is then transported through insulated pipes to the turbine. But because insulation is never perfect, some of the steam's heat travels out from the pipe, through the insulation, warms the air around it, and is also lost from its intended use.

The high-pressure steam is directed to the steam turbine and sets it spinning. This conversion of heat energy (in the steam) to mechanical energy (in the spinning turbine propeller) lowers both the pressure and the temperature of the steam, as energy is transferred from it to the spinning turbine.

In most cases, a number of turbines are arranged in a row of multiple stages. Each one in the sequence is designed to work with lower-pressure steam, as the steam moves down the line.

Finally the pressure becomes too low to be useful and the steam is now ready to be exhausted. Much energy, 50-60% of the total, still remains in it and its temperature has fallen from 700°C to about 100°C. The steam then exits the turbines area and enters the "condenser," but at this point it has lost all its energy.

At this point the steam has lost most of its energy, and can no longer be used for any work functions, but is still hot enough and needs to be cooled down for return into the process cycle. Water from a river or lake is usually pumped through coils in the "condenser" to cool the steam down. As its cooled, the steam changes state from gaseous to liquid. It thus condenses back to warm water, which is then pumped back to the furnace boiler. The cooling water, which is now also warm can be discharged back into the river or lake where it came from, or is recycled after post-cooling treatment (expensive process).

The turbine is connected to an electrical generator and its mechanical energy (spinning generator rotor) is converted into electrical energy. Some energy is lost here too due to friction of the spinning parts and electrical resistance. As the electric current runs through the various conductors in the generator and out through wires to the transmission lines, it heats these wires and the heat is lost into the environment.

In each of these conversion steps we can see significant loss of energy from the flow. Almost all of this energy escaped in the form of waste heat energy—by either direct transition into the surrounding air, or by mechanical friction. There is a significant "heat tax" on any type of energy conversion. This loss of energy is actually a loss of the final output. The "lost" energy still exists, but is now in different forms and outside the system. It is lost forever, as far as the related process is concerned.

The energy losses in these conversions fall into several different categories; in the furnace, the steam pipes, the turning turbine and generator, and the generator's electrical losses, the amount of loss is subject to some control. We can control the speed and quality of some of the different steps of the process by, for example, blowing more air into the furnace to improve burning, letting the hot gases heat water as they go up the smokestack, doubling the insulation in the steam pipes, using oil to reduce friction, and increasing the conduction effectiveness of the wires to reduce their heating. By doing this we can reduce the "heat lax" at these conversions significantly.

One major loss that is different, however, and which we have no control over is that at the condenser. As long as the steam is hot, it contains energy, and there is no practical way to get the temperature of the exhaust steam below that of its surroundings (the air or the cooling water from the river). While the other losses are limited only by human ingenuity and a willingness to spend money to improve the equipment, the loss at the condenser is obligatory, controlled by the laws of Mother Nature—and we can only sit and watch…while wasting energy and water in cooling the hot steam.

The heat-to-electric process of steam turbines is 19th century technology. It suffers from a number of inefficiencies, the largest of which is the need for cooling the low-temperature steam, in order to convert it back to water, as needed to use it again to create more steam. In addition to heat loss, we also need and waste lots of water due to evaporation.

Note: Just imagine a set of large turbines operating at a solar thermal (CSP) power plant in the desert. Yes, CSP solar plants also heat water to generate steam and run it through turbines, just like described above. But where is the cooling water going to come from in the desert? And since each turbine uses millions of gallons of water daily, the immediate question is, how could we justify using, and wasting, so much water in the desert?

Note: A 250 MW solar power plant (CSP technology) uses about a million gallons of cooling water daily. Significant part of this water is lost due to evaporation, leaks, etc.

All steam turbines create a great waste, but this is the state-of-the-art today (amazingly primitive, we must say), so until a new way is found lots of calories and millions of gallons of precious water will be lost for cooling the steam in coal, gas and solar power plants.

There were over 2,300 coal-fired power plants globally in 2012. 620 of these were in China and 550 in the U.S.

The remarkable thing here is that while the coal-fired plants in the U.S. have been shut down or converted to burning natural gas, this is not the case in China. On the contrary; Chinese coal-fired plants are increasing at the rate of 3-4 monthly.

The problem is that, in addition to generating electricity, coal burning power plants also generate over 2.0 billion tons of CO_2 every year in the U.S. alone. That much is generated by the Chinese power industry and more by the rest of the world.

Extrapolating this number to global level, we see that coal burning is a major polluter.

Over 8.0 billion tons of CO_2 are emitted in the atmosphere annually by coal power generation.

We take a close look at this problem in our book titled, *Power Generation and the Environment,* published by Fairmont Press in 2014. In it, we review in detail the present issues related to power generation, and what can be done to reduce the damages. The overwhelming conclusion is that the world—the developing countries in particular—are not willing to consider reduction of the use of fossil fuels, so we will continue on the path to extreme global warming in the future.

Waste and pollutants

Several by-products, including solids and gases, are created in the electricity generation process, when using coal as a fuel. A substance called "clinker" or bottom ash (glassy particles of melted coal ash) settles at the base of the furnace. This material is periodically removed and disposed of. Fly ash, the noncombustible minerals found in coal (including ash, dust, soot, and cinders) travels upward with gaseous by-products. Fly ash can be captured in an electrostatic precipitator and then transported by pipes to a holding pond, where it settles. Over 98% of all waste solids are captured in the plant.

Gaseous by-products include carbon dioxide (CO_2), sulfur oxides (SO_x), and nitrogen oxides (NO_x). These are air polluters, or green house gasses (GHGs) that are held responsible for the present day climate warming

Sulfur oxides can be controlled by the installation of scrubbers at coal-fired power plants. Scrubbers allow high-sulfur coals to be used because they remove sulfur dioxides out of the gas stream in the stacks (a process called desulfurization).

Scrubbers work by spraying limestone slurry directly in the path of the materials leaving the boiler chamber. The limestone reacts with the sulfur in the gases within the stacks. The combination of carbonate (limestone) and sulfur forms the mineral gypsum. Gypsum is a solid, which falls to the bottom of the stacks, where it can be collected, removed and disposed of. Many tons of it, where the by-product gypsum created in this process can be used to make drywall, bowling balls, and other practical uses.

Nitrogen oxides are managed by careful control of the furnace temperature., while current technology is also available to control carbon dioxide emissions. Using high-efficiency coals (such as those found in Kentucky) also helps reduce the CO_2 output without much equipment modification.

Coal is also used in industrial process heating, as needed to heat commercial boilers and ovens. Cement production, which represents the biggest worldwide industrial use of coal, glass, ceramic, and paper industries all use large amounts of coal. In some North American states, industrial process heating accounts for 10% of the annual coal usage, all of which contributes to high levels of GHG emissions.

Similarly polluting are the other fossil fuels—oil and natural gas—albeit natural gas does not pollute much during burning. It, however, is a great polluter during its fracking, transport and storage process steps.

Natural Gas Power Generation

Natural gas also can be burned in order to generate steam to power a turbine-generator set to generate electric power. The process is similar to that described in coal burning above, so we will skip a repetition. Natural gas, however, is preferred for power generation today because it is abundant, cheap, easier to transport, emits less gasses, and generates no solid waste.

As a matter of fact there is a growing trend today in the U.S. and other developed countries of converting coal-burning to natural gas-burning power plants. This move is justified mostly by the 60% reduction of GHG emissions when burning natural gas. Yet, when the entire life cycle of coal and natural gas is taken into consideration, the GHG savings are much less.

Diesel Power Generation

Crude oil and its derivatives can be used to generate electricity instead of coal or natural gas as well. The oil, in form of heating oil, gasoline, or diesel fuels, are injected into the furnace, where they are burnt in a similar way and to similarly high temperature to generate steam. The rest of the process—including the energy and water wasted to cool the used steam—is identical to that employed in coal and gas-fired power plants.

Diesel engine generator sets are often used for power generation in peaker plants, and for remote communities power generation. In addition, emergency (standby) power systems may use reciprocating internal combustion engines operated by fuel oil or natural gas. Standby generators may serve as emergency power for a factory or data center, or may also be operated in parallel with the local utility system to reduce peak power demand charge from the utility.

Gas or diesel engines can produce strong torque at relatively low rotational speeds, which is generally desirable when driving an alternator, but diesel fuel in long-term storage can be subject to problems resulting from water accumulation and chemical decomposition. Rarely used generator sets may correspondingly be installed as natural gas or LPG to minimize the fuel system maintenance requirements.

Spark-ignition internal combustion engines operating on gasoline (petrol), propane, or LPG are commonly used as portable temporary power sources for construction work, emergency power, or recreational uses.

Reciprocating external combustion engines such as the Stirling engine can be run on a variety of fossil fuels, as well as renewable fuels or industrial waste heat, but installations of Stirling engines for electric power production are relatively uncommon.

There are over 1020 oil/diesel power plants worldwide, of which over 740 are located and operational in the United States. Just imagine the amount of oil and diesel that goes through these oil guzzlers on daily basis.

Nuclear Power Generation

Nuclear power generation consists of capturing the energy of nuclear reactions in nuclear power plants, and converting it into electricity. Solid nuclear fuel rods are activated in the nuclear reactors, which raises the temperature within enough to boil water. The water is then sent into steam turbines very similar to these used in coal and gas-fired power plants.

Nuclear power is a fancy, and very efficient, way of boiling water, used to generate electricity.

Nevertheless, nuclear power is very different from all other energy sources. It is an immense force—thousands of times more powerful than any competitor. It can also deliver many times the power of the competitors when under complete control. At the same time, however, it is a merciless killer that spares nothing and no one in its path when control is lost. Just ask the survivors of the Chernobyl and Fukushima nuclear accidents.

Because of this anomaly, we see two distinct, and very different, paths for nuclear power development in the future. It could become a leading base-load power generator if all goes well, and no other nuclear accidents occur. It could, however, be completely removed from operation in case of more accidents.

Table 7-3. Global nuclear power share

1980	10%
2010	20%
2030	25%
2050	30%, or 0…?

If another Chernobyl- or Fukushima-like disaster occurs anywhere in the world, nuclear power might be shelved for good. As unlikely this might be, the chance exists simply because people make mistakes. While a mistake (human error) in a coal-fired power plant might cause a local problem, an accident at a nuclear power plant could destroy the plant and the entire region around it—including people, animals, and vegetation.

Also, nuclear power depends on the availability of materials (uranium and other radioactive ores), which are in limited quantities. They are estimated to last no more than 100 years, at the present level of nuclear power generation, but this number could go up or down with time.

For these reasons alone, the future of nuclear around the world is uncertain for now.

Hydro Power Generation

Hydroelectric power is produced by the force of falling water, which is determined by the mass, height and speed of the water flow. In the past, flowing or falling water was channeled through a water wheel, making it spin, thus providing mechanical power for some activities like turning grinding wheels.

Today, the most important use of water power is storing it behind a high dam, where it builds a lot of potential energy waiting to be released. When the water is allowed to flow down the penstock pipes into the spinning turbine blades, it releases the potential energy, converting it into kinetic, and in more practical terms; mechanical) energy at the moment when it strikes the turbine blades.

The turbine's axle is attached to the rotor of a large electromagnet which spins and generates current in stationary coils of wire arranged around it. The current is fed into a transformer and conditioning devices where the voltage is increased for long distance transmission over power lines.

There are many energy losses in this energy conversion process. To start with, the sunlight falling on the lake behind the dam converts many tons of water into vapor, which evaporates in the atmosphere. When the water starts going down the penstock its potential (storage) energy is converted into kinetic and with some losses in the way due to friction with the walls of the pipes. Thus acquired kinetic energy is soon enough converted into mechanical energy as soon as the water hits the turbine blades and make them turn.

The turbine bearings cause some resistance and some of the mechanical energy acting upon them is lost when converted into heat. There are also losses in the

wires from the generator to the transformer and to the point of use, where the electrical energy traveling in them encounters resistivity and part of it converted into heat energy.

Hydro energy generation depends on the amount of snow falling at a certain period of time, and rain water running into the lake. Since these amounts are unpredictable, and are heavily dependent on the climate conditions, we cannot rely on them fully under the increasing environmental uncertainties. Also, the number of locations suitable for building hydropower plants around the world is reaching its limits and not many more hydro plants can be build.

Damming also causes extensive damage in the up- and down stream areas. Entire populations have been moved in some cases, and the extent of the damage over millions of acres of good land is questionable on the long run. Because of that, hydropower is not welcome in most places.

The Renewables

Capturing the energy of the sun, wind, the oceans etc. and converting it to electricity is considered renewable, green and clean. These energy sources and the related technologies are called renewable because the respective sources do not need any fuels, since their energy input is renewed constantly, and because we don't fear them getting depleted ever.

They are also considered green and clean because compared with the conventional energy sources, they do not emit significant quantity of GHGs, liquid, or solid waste during operation. There is, however, some pollution during the production cycle of the equipment needed for their conversion into useful energy. Additional pollution is generated during the transportation, installation and decommissioning of the equipment in the power plants as well.

Let's take a look at these promising energy sources.

Solar Energy

Why solar energy? The answer seems obvious, but solar energy conversion equipment has been around for over half a century without showing much progress until recently. What's the catch? We will take a brief look at solar energy generation here, leaving the more detailed portion of the answer for the following chapters.

Solar energy, and its energy components; radiant light and heat in form of photons are generated during the fusion reactions on the Sun's surface. The part that arrives on Earth has been used by humans since ancient times by means of different devices; usually for solar heating of water and drying of fruits and vegetables.

Present day solar energy technologies include; solar heating, solar photovoltaics, solar thermal electricity and solar architecture. These disciplines are considered promising in solving our energy problems, and are expected to contributes to solving some of the urgent environmental problems as well.

Solar technologies can be divided into passive and active solar devices, according to the way they capture, convert and distribute solar energy.

An example of passive solar technique is a building that is designed and oriented to use the available sunlight most efficiently, in order to save maximum amount of energy needed for heating and cooling the living spaces in the building. By selecting proper materials with favorable thermal mass, and/or light dispersing properties, and by designing living and working spaces that naturally circulate air within the building, we can maintain comfortable living temperatures with minimum outside energy input.

Active solar techniques include the use of photovoltaic (PV) modules and solar thermal collectors to harness the energy and eventually convert it to electricity. An example of an active solar technology is a solar water heater mounted on a housetop. It consists of a heating plate with pipes, where water running through pipes in it gets heated and can then be used for residential and commercial purposes.

The main component in the PV modules is the solar cell. It can be made by special processing of silicon material, or some other semiconductor materials like Ga, As, Ge and others. Sometimes the solar cells are made by depositing other types of semiconductor materials on a piece of glass. In all cases, the resulting product (solar cells) cannot be used by themselves, so they are encapsulated in a special package called PV module.

When sunlight hits the solar cell surface, the photons (radiation energy) energize the electrons to the point where they exit the device via wires attached to it. In this case the radiation energy of the photons is converted directly into electric energy. This is the photovoltaic (PV) type of solar energy conversion

Here we also have a number of significant energy losses. To start with, when the sunlight strikes a surface—any surface—part of it gets reflected and part of it gets converted to heat, Only a part, approximately 10-20%, of the sunlight is presently converted to electricity by the available solar conversion devices. And then, we still have to familiar losses in the electric wires, in the inverters and transformers.

The design and manufacturing processes of the so-

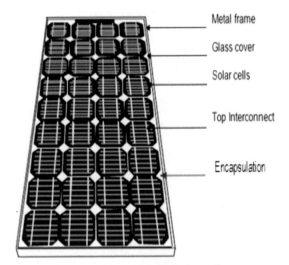

Figure 7-3. PV module (panel)

lar cells and modules are quite complex, so we will take a closer look at these in the following chapters.

On the commercial side of things we use PV modules and systems to generate electric energy by *direct conversion* of sunlight to electricity. We can also use solar thermal conversion technologies for *indirect conversion*, where water is heated to high temperature and is then run through a steam turbine (just like in the coal power plants discussed above). The steam energy spins the turbine propellers, which in turn spin a generator to produce electric energy.

There are a number of other solar technologies, such as thin film PV, thermal solar collectors and other solar energy generators, which will take a very close look at in the following chapters.

Wind Energy

Wind power generation is defined as the conversion of wind energy into mechanical energy and then into electric power, which can then be used as needed. Wind turbines have been used for a long time to directly provide mechanical power, such as rotating a grinding wheel, water pumping and many such activities.

A large wind farm may consist of several hundred individual wind turbines which are connected to the electric power transmission network. Offshore wind farms can harness more frequent and powerful winds than are available to land-based installations and have less visual impact on the landscape but construction costs are considerably higher. Small onshore wind facilities are used to provide electricity to isolated locations and utility companies increasingly buy surplus electricity produced by small domestic wind turbines.

A wind turbine is the basic device used to converts

Figure 7-4. Wind turbines

wind's kinetic energy into mechanical and then electric energy; a process known as wind power. If the mechanical energy is used to produce electricity, the device may be called wind turbine or wind power plant. If the mechanical energy is used to drive machinery, such as for grinding grain or pumping water, the device is called a windmill or wind pump.

A generator attached to the blades of the wind turbine spins the rotor of an electric power generator, and thus generated electricity is transferred to the point of use by electric wires.

The energy losses in the wind-to electricity conversion processes in wind turbines are significant. First, a major part of the incoming wind power is lost by the inefficiency of the blades. Only 30-40% of it is converted into mechanical energy that is used to rotate the blades. Some energy is lost also during the rotation of the blades and the generator, due to the friction in the axle bearings, where it is converted into heat. And the ever present resistivity in the electrical wires robs us of some additional energy, which is also converted into heat.

Today's wind turbines are manufactured in a wide range of sizes and types with vertical and horizontal axis of rotation of the blades. Large wind farms consist of several dozen to hundred individual wind turbines which are connected to the electric power transmission network. Offshore wind farms harness more frequent and powerful winds than those at land-based installations, but construction costs are higher.

Ocean and Geothermal Energy

Electric power can be generated in the world's oceans by using the wave or tidal motion of the waters, and/or their temperature differential. Different types of equipment can be designed and installed on the ocean surface or bottom, where they capture the energy available in the ocean waters, and deliver it into the grid.

Geothermal power is obtained by drilling deep wells into the ground until reaching a hot spot. Hot water, or steam escaping from the well is captured and used to heat buildings, or is converted into electric power in steam turbines.

The biggest problem with these technologies is the fact that they are usually located in remote areas, far from the national grid. Because of that, extensive and expensive transmission systems have to be built to deliver the power into the grid. This, and other technical issues are limiting the use of these technologies in most countries.

Bio-energy

Bioenergy is energy generated by materials obtained from biological sources. Biomass is any organic material which has stored sunlight in the form of chemical energy. As a fuel biomass might refer to wood, wood waste, straw, manure, sugarcane, and many other by-products from a variety of agricultural processes.

In a practical sense bio-energy is a synonym to biofuel, which is derived from biological sources. In its broader sense it is referred to as biomass, which is the biological material used as a biofuel itself, as well as for the production of biofuels like bio-alcohols and bio-diesel. Bio-energy basically encompasses a number of social, economic, scientific and technical fields associated with using biological sources used for generating energy.

There is some confusion in terminology, where the word *bioenergy* is used in Europe, while *biofuel* is used in North America to mean one and the same. In both cases, some clarification must be given as of the type of bio-technology and its use, in order to get a clear and final picture of the final product and its use.

The energy conversion process here consists in converting the chemical energy contained in the vegetation materials into heat, light, electricity and other forms of energy. There are many types and sizes of bio-energy conversion equipment, but in most cases direct generation of heat is the most efficient energy product from any type of bio-mass.

Liquids and gaseous fuels generated from biomass are more convenient to use, but a lot of energy is wasted in their production process. Lately, the generation of gasses is becoming very important and a lot of effort is spent in that area. The energy losses in these processes are confined mainly in the production cycle, during the burning process where some heat is lost in the surrounding environment as well.

The major, and most practical, bio-fuels today are *bio-ethanol, bio-diesel and bio-gas*.

- Bio-ethanol is a man-made alcohol, which is usually made by fermentation of carbohydrates containing vegetation. It is produced by sugar or starch crops such as corn or sugarcane. Cellulosic biomass, contained in non-food vegetation like trees and grasses can also be developed as a feedstock for ethanol production. Ethanol can be used as a fuel for vehicles in its pure form, but it most cases it is used as a gasoline additive to increase octane and improve vehicle emissions. Bio-ethanol is used in great quantities the USA and in Brazil.

- Biodiesel is also made from vegetables containing oils, and/or from animal fats. Biodiesel can be used as a fuel for vehicles in its pure form, but it is usually used as a diesel additive to reduce levels of particulates, carbon monoxide, and hydrocarbons from diesel-powered vehicles. Biodiesel is produced from oils or fats using trans-esterification. It is a common type of biofuel in Europe, but is finding its place in the US energy markets as well.

- Bio-gas can be produced by breakdown of different types of organic matter in the absence of oxygen. Dead plants and animal material, animal feces, and kitchen waste can be converted into a gas that can be used as fuel. Man-made bio-gas is produced by the anaerobic digestion or fermentation of biodegradable materials such as biomass, manure, sewage, municipal waste, green waste, plant material, and crops. Biogas comprises primarily methane (CH_4) and carbon dioxide (CO_2) and may have small amounts of hydrogen sulphide (H_2S), moisture and siloxanes. Biogas can be used also as a fuel for heating purposes and cooking, or in anaerobic digesters where it is typically used in a gas engine to generate electric or heat power.

Author's note: Although biomass and its products can be considered as renewable form of energy, we must remember that biomass comes from agricultural vegetation or animals which quantity is variable and sometimes (like in extreme draughts) simply unavailable.

Nevertheless, the global biomass production is a truly renewable resource—something we can count on every day. And like all other fuels, its production and use must be done in a systematic fashion, so we don't end up with another disaster like the ethanol overproduction and global food crops deficiency of 2008-2010. The mad cow disease around that time also reminds us that biomass is a fragile, variable and unreliable, energy source, so we need to be very careful when planning its production and use in the 21st century.

Below we review the key components of electric power distribution.

ELECTRIC POWER DISTRIBUTION

Thus far we discussed the different energy sources and the way electric energy is generated. Once generated, the electric power has to be delivered to homes, offices and factories. The power distribution in the US is done by the National power grid, as reflected in Figure 7-5.

The economic significance of generating, transmitting, distributing and using electricity today is staggering. It is one of the largest and most capital-intensive sectors of the economy. Total asset value is estimated to exceed $800 billion, with approximately 60% invested in power plants, 30% in distribution facilities, and 10% in transmission facilities.

Annual electric revenues—the nation's electric bill—are about $247 billion, paid by America's 131 million electricity customers, which includes nearly every business and household. The average price paid is about 8-15 cents per kilowatt-hour, although prices vary from state to state depending on local regulations, generation costs, and customer mix.

There are more than 3,300 traditional electric utilities in the US:

- 213 stockholder-owned utilities provide power to about 73% of the customers
- 2,000 public utilities run by state and local government agencies provide power to about 15% of the customers
- 930 electric cooperatives provide power to about 12% of the customers

Additionally, there are nearly 2,000 non-utility power producers, including both independent power companies and customer-owned distributed energy facilities.

The entire national transmission and distribution network (the national grid) consists of:

- Over 30,300 substation, located in special locations all around the country, maintaining the connectivity and ensure the power quality of the distribution system.

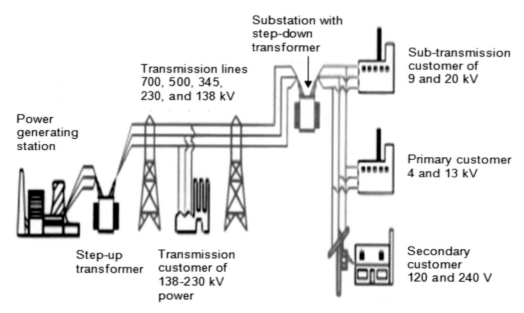

Figure 7-5. Power generation, transmission and distribution

- About 210,000 miles high voltage AC transmission lines,
- Over 6,200 miles of high voltage DC transmission lines, and
- 145 million customers are served daily by the national power grid

The bulk power transmission system consists of three independent networks: Eastern Interconnection, Western Interconnection, and the Texas Interconnection. These networks incorporate international connections with Canada and Mexico as well. Overall reliability planning and coordination is provided by the North American Electric Reliability Council, a nonprofit organization formed in 1968 in response to the Northeast blackout of 1965

Power Generation

America operates a fleet of about 6,500 power plants, mostly thermal (coal and natural gas), generating an average of 1,000 GW of electric power. The average efficiency of our power generating fleet is around 33%, which has not changed much since 1960. This is mostly due to slow turnover of the capital stock and the inherent inefficiency of central power generation that cannot recycle heat. Nuclear and hydro plants are more efficient, but much more expensive as well.

Power plants are generally long-lived investments, with the majority of existing US capacity 30 years old or older. They can be divided into:

a. Baseload power plants, which are run all the time to meet minimum power needs,

b. Peaking power plants, which are run only to meet power needs at maximum load (known as peak load), and

c. Intermediate power plants, which fall between the two and are used to meet intermediate and emergency power loads.

The roughly 5,600 distributed energy facilities typically combine heat and power generation and achieve efficiencies of 40% to 55%, accounting for about 6% of US power capacity in 2001 and several times more today.

A shift in ownership is occurring from regulated utilities to competitive suppliers. The share of installed capacity provided by competitive suppliers has increased from about 10% in 1997 to about 35% today. Recent data suggest, however, that this trend is slowing down. Also, cleaner and more fuel-efficient power generation technologies are becoming available. These include combined cycle combustion turbines, wind energy systems, advanced nuclear power plant designs, clean coal power systems, and distributed energy technologies such as photovoltaics and combined heat and power systems.

Because of the expected near-term retirement of many aging plants in the existing fleet, growth of the information economy, economic growth, and the forecasted growth in electricity demand, America faces a significant need for new electric power generation. In this transition, local market conditions will dictate fuel and technology choices for investment decisions, capital markets will provide the financing, and federal and state policies will affect siting and permitting. It is an enormous challenge that will require a large commitment of technological, financial and human resources in the years ahead.

Power Transmission

Adequate electric generation in the US is hindered by bottlenecks in the transmission system, which interferes with reliable, efficient, and affordable delivery of electric power to the customers. America operates about 215,000 miles of high voltage electric transmission lines.

Construction of transmission facilities has decreased about 30%, and annual investment in new transmission facilities has declined significantly over the last 25 years. The result is grid congestion, which can mean higher electricity costs because customers cannot get access to lower-cost electricity supplies, and because of higher line losses. Transmission and distribution losses are related to how heavily the system is loaded. U.S.-wide transmission and distribution losses were about 5% in 1970, grew to 9.5% in 2001, and are even higher today, due to heavier utilization and more frequent congestion.

Congested transmission paths, or "bottlenecks," now affect many parts of the grid across the country. In addition, it is estimated that power outages and power quality disturbances cost the economy up to $150-200 billion annually. These costs could soar if outages or disturbances become more frequent or longer in duration. There are also operational problems in maintaining voltage levels.

America's electric transmission problems are also affected by the new structure of the increasingly competitive bulk power market. Based on a sample of the nation's transmission grid, the number of transactions have increased substantially recently. For example, annual transactions on the Tennessee Valley Authority's transmission system numbered less than 20,000 in 1996. They exceed 250,000 today, a volume the system was not originally designed to handle. Actions by transmission operators to curtail transactions for economic reasons and to maintain reliability (according to procedures developed by the North American Electric Reliability Council) grew from about 300 in 1998 to over 1,000 in 2000 and is much higher today.

Additionally, significant impediments interfere with solving the country's electric transmission problems. These include opposition and litigations by different groups against the construction of new facilities,

uncertainty about cost recovery for investors, confusion over whose responsibility it is to build and maintain, and jurisdiction and government agency overlap for siting and permitting. Competing land uses, especially in urban areas, leads to opposition and litigation against new construction facilities.

In Figure 7-5 above, we get a glimpse into the complexity of the generation-transmission-distribution scheme of electric power transfer. The generator (coal or nuclear power plant) might be miles away from the point of use (POU)—residential or industrial customer.

The power generated at the power plant is sent to a step-up transformer (substation), where it has to be transformed into higher voltage, as needed for long distance transfer. Some of this power is used by larger users who have their own sub-stations as needed for power use in their facilities. The rest of the power (most of it) is transported via the national power grid to step-down substations all over the country, where it is converted to lower voltage and is sent via the distribution power lines for use by residential and commercial customers, who also have their own sub-stations or transformers for converting the power to exact voltage they can use.

Power Distribution

The "handoff" from electric transmission to electric distribution usually occurs at the substation. America's fleet of substations takes power from transmission-level voltages and distributes it to hundreds of thousands of miles of lower voltage distribution lines. The distribution system is generally considered to begin at the substation and end at the customer's meter. Beyond the meter lies the customer's electric system, which consists of wires, equipment, and appliances—an increased number of which involve computerized controls and electronics operating on DC.

There are two types of distribution networks in the U.S.; radial or interconnected.

- A radial network leaves the power generating station to its final destination with no connection to any other supply in the network. This is typical of long rural lines with isolated load areas.

- An interconnected network is generally found in more urban areas and will have multiple connections to other points of supply. These points of connection are normally open but allow various configurations by the operating utility by closing and opening switches.

The distribution system supports retail electricity markets. State or local government agencies are heavily involved in the electric distribution business, regulating prices and rates of return for shareholder-owned distribution utilities. Also, in 2,000 localities across the country, state and local government agencies operate their own distribution utilities, as do over 900 rural electric cooperative utilities. Virtually all of the distribution systems operate as franchise monopolies as established by state law.

The greatest challenge facing electric distribution is that of responding to rapidly changing customer needs for electricity; i.e., increased use of information technologies, computers, and consumer electronics has lowered the tolerance for outages, fluctuations in voltages and frequency levels, and other power quality disturbances. In addition, rising interest in distributed generation and electric storage devices is adding new requirements for interconnection and safe operation of electric distribution systems.

Finally, a wide array of information technology is entering the market that could revolutionize the electric distribution business. For example, having the ability to monitor and influence each customer's usage, in real time (part of the smart grid solution), could enable distribution operators to better match supply with demand, thus boosting asset utilization, improving service quality, and lowering costs. More complete integration of distributed energy and demand-side management resources into the distribution system could enable customers to implement their own tailored solutions, thus boosting profitability and quality of life.

Power Grid Operation

Transmission and distribution facilities are also referred to as the power grid, also referred to as the national grid, the electricity or electric grid. It is an interconnected network of wires and equipment for delivering electricity from suppliers to consumers. It consists of power generating stations that produce electrical power, high-voltage transmission lines that carry power from distant sources to demand centers, and distribution lines that connect individual customers.

The grid is coordinated and operated by a grid coordinator. Nationally, the grid is split into three main sections—the Western, Eastern and Texas Interconnections. These sections operate independently and have limited interconnections between them.

The transmission network moves power long distances, sometimes across international boundaries, until it reaches the local company (utility) that owns, operates, and services the local distribution network.

Upon arrival, the high voltage coming from pow-

er plants, is fed into a substation, where the power is stepped down from a transmission level voltage to a much lower distribution level voltage. From the substation, the lower voltage power enters the local distribution network and is delivered to the customers. Here, it is stepped down again from the distribution voltage to the required service voltage.

The U.S. territory, along with Canada and a small part of Mexico, is also divided into regional grid entities. The regional reliability entities fall under the purview of the North American Electric Reliability Corp. (NERC), which was designated by the Federal Energy Regulatory Commission (FERC) as the nation's energy reliability organization and which develops standards, among other things, to ensure the grid's reliability.

The standards, once issued by FERC, must be met by all energy industry and market participants. These standards are mandatory and enforceable. Consequently, the grid is designed and operated to meet these standards.

NERC's regions include:
- Florida Reliability Coordinating Council (FRCC),
- Midwest Reliability Organization (MRO),
- Northeast Power Coordinating Council (NPCC),
- Reliability First Corporation (RFC),
- SERC Reliability Corp. (SERC),
- Southwest Power Pool (SPP),
- Texas Reliability Entity (TRE) and
- Western Electricity Coordinating Council (WECC)

The grid functions synchronously, which is "interconnection" of a group of distribution areas all operating with alternating current (AC) frequencies. These are synchronized so that the peaks of the AC current occur at the same time. This allows transmission of AC power

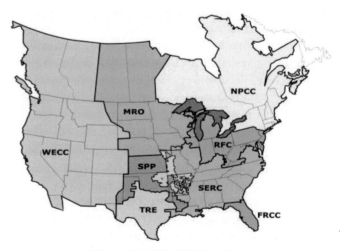

Figure 7-6. U.S. NERC regions

throughout the area, connecting a large number of electricity generators and consumers. This synchronization also allows more efficient functioning of the electricity markets and redundant power generation and distribution.

In a synchronous grid, all power generators run at the same frequency and at the same phase. Here, each generator is maintained by a local governor that regulates the driving torque by controlling the steam supply to the turbine driving it. Generation and consumption are thus balanced across the entire grid, because energy is consumed almost instantaneously as it is produced. Energy is stored in the immediate short term by the rotational kinetic energy of the generators.

Note: High-voltage direct current (DC) lines, or variable frequency transformers, are sometimes used to connect two alternating current interconnection networks which are not synchronized with each other. This provides the benefit of interconnection without the need to synchronize an even wider area.

A large failure in one part of the grid (unless quickly compensated for) can cause the current to reroute itself. It then could over flow from the remaining generators to transmission lines of insufficient capacity, thus causing further failures. A widely connected grid, therefore, is vulnerable to cascading failures and widespread power outage.

A central authority is usually designated to facilitate communication and develop protocols to maintain a stable grid. For example, the North American Electric Reliability Corporation gained binding powers in the United States in 2006, and has advisory powers in the applicable parts of Canada and Mexico. The U.S. government has also designated National Interest Electric Transmission Corridors, where it believes transmission bottlenecks have developed.

Some areas, for example rural communities in Alaska, do not operate on a large grid, relying instead on local diesel generators. These run independently and the generated power is distributed in the local area without concerns about other parts of the grid.

U.S. Grid's Growing Pains

The U.S. electrical grid, and its power delivery infrastructures especially, are old and continue aging, which brings a number of bad consequences. Aging power equipment have higher failure rates, leading to higher power interruption rates, which affects the economy and the society as a whole. Older assets and facilities also lead to higher inspection maintenance costs and further repair/restoration costs.

Obsolete system layout require serious additional substation sites and rights-of-way that cannot be obtained in current area and are forced to use existing, insufficient facilities. Outdated engineering and tools for power delivery planning and engineering are ineffective in addressing current problems of aged equipment. The old cultural values of planning, engineering, and operating the system by using concepts and procedures that used to work in vertically integrated industry, today augment the problems of working in a deregulated industry

The U.S. utilities are under pressure to modify their classic approaches as needed to accommodate distributed generation. Rooftop solar is shaping a way to erase the differences between distribution and transmission grids. This presents a problem for the utilities, since they end up with the rising costs of grid operation and maintenance.

Demand response is a growing management technique where larger customers are requested to reduce their load at certain periods of time. Until now, transmission grid operators used the demand response approach to request load reduction only from very large energy users. The new demand response increases the number of companies that can participate in this effort.

The utilities are forced to allow and even encourage distributed generation. Small power generators, can be brought on-line to help supply the need for power, thus providing more options to the utility and selling more electricity. This is also a new opportunity to increase profits, since the utility can, for example, buy DG solar power for \$0.02/kWh and sell it for \$0.10/kWh. Not bad profit.

The new and upcoming "smart grid" is another factor that the utilities are considering. In the U.S., the Energy Policy Act of 2005 and Title XIII of the Energy Independence and Security Act of 2007 are providing funding to encourage smart grid development. This would to enable utilities to better predict their needs, and provide better service. In some cases this involves consumers in some form of time-of-use based tariff. Funds have also been allocated to develop more robust energy control technologies.

Various planned and proposed systems to dramatically increase transmission capacity are known as super, or mega, grids. The expected benefits include enabling the solar and wind energy industries to sell electricity to distant markets, as well as the ability to increase usage of intermittent energy sources by balancing them across vast geological regions. This might also contribute to eliminating congestion that hinders electricity markets' proper and efficient function

These developments, however, are facing financial barriers and increasing local opposition. The construction of new transmission lines often violates land owner rights, and the added expense is often paid by the tax payers. These are major obstacles in the development of new smart and super grids.

1,000 miles, 3.0 GW HVDC Trans-Canada project was priced recently at $3 billion and would require a wide land corridor. This is at least 3-4 million per mile, if the right-of-way is obtained easily and cheaply.

The energy markets deregulation continues, forcing the utilities to sell their assets as the energy market follows in line with the gas market in use of the futures and spot markets and other financial arrangements. Even globalization with foreign purchases are taking place. The UK's National Grid, the largest private electric utility in the world, bought the U.S based New England's electric system for $3.2 billion. Scottish Power also purchased Pacific Energy for $12.8 billion.

Local electric and gas firms begin to merge operations as they see advantage of joint affiliation especially with the reduced cost of joint-metering. Technological advances taking place in the competitive wholesale electric markets include fuel cells used in space flight, aero-derivative gas turbines used in jet aircrafts, solar engineering and photovoltaic systems, off-shore wind farms, and the communication advances spawned by the digital world particularly with microprocessing which aids in monitoring and dispatching.

All in all, electricity is expected to see growing demand in the future. The Information Revolution is highly reliant on electric power. Other growth areas include emerging new electricity-exclusive technologies, developments in space conditioning, industrial process, and transportation, such as hybrid vehicles, super-locomotives, etc.

The Utilities

The utilities dictate, directly or indirectly, the way things are done in the energy sector. They make and break the rules, and operate in a way that is much less than democratic or fair, and yet we need them. Without the utilities there will be no power every time we flip the light switch, or air conditioning unit on. Life would be quite different without the utilities running the power plants and distribution lines that deliver this precious commodity to our front doors.

So while we admit our dependence on electric power—and the utilities—we also need to protect our rights. It all starts with understanding how the utilities operate, which we have dedicated large portion to in our previous books on the subjects at hand, see Reference 20 and 21.

The U.S. Utilities Status

The utilities industry (yes, there is such a thing as utility industry and is a big one) has been changing lately; faster in some areas and much slower in other, as dictated by the regulatory changes, supply-demand factors, price volatility, and more importantly the new competition. The new competition is from energy sources that were but ignored in the past; solar, wind, bio-, and geo-powered systems and installations.

Until recently, the large regional monopolies ran the whole show without any interference. They controlled everything from the way power was generated all the way down to retail supply. Recently, however, the industry structures that once led the electricity industry has been going through a slow disintegration. It is slowly breaking down into several separate (supplier related) segments.

The new utility industry segments are:

- *Power Generators*, are the electric power systems and power plants that generate electrical power. Large utilities continue still build and operate such plants, but a growing number of independent "merchant generators" build power capacity. This development is done on speculative basis or is based on acquired utility-divested plants. These independent companies market their output at competitive rates in unregulated markets.

- *Grid Operators* determine the way power is distributed via regional and distribution networks. They also sell access to their networks to retail service providers. This aspect of the industry is heavily regulated and very expensive, so these network operators form an oligarchy of natural monopolies. Breaking into this business and duplicating the existing network operations is simply not possible, because it would be very expensive, redundant, and extremely risky. No takers thus far…

- *Energy Traders* buy and sell energy futures and other derivatives. These traders and marketers have created a complex system of "structured product." One of the useful function of these participants in the energy market is that they help the utilities and businesses that use a lot of power to secure a dependable supply of electricity at a stable and somewhat predictable price. Traders usually wager on the direction of power prices in order to increase their returns.

- *Energy Retailers* are choice of services is now available in most U.S. states. Now consumers can choose from a number of retail service providers. In a number of states the power grid is open to third party providers, so many such companies are entering the market and buy power at competitive prices from transmission operators and energy traders. The energy retailers can then sell thus acquired power to the local end users. These services are often bundled in competitive packages with gas, water and even financial services included.

- *Energy Price Regulation* is also changing today, where wholesale electricity prices are no longer set by regulatory agencies. Instead, they are mostly free to fluctuate with supply and demand, which introduces a variable that is no longer under the control of regulators or the utilities (which was basically one and the same in many cases in the past). This in turn increases significantly the risk of uncontrollable price increases. And finally we see the new group with a voice in the utilities operations:

- The customers. An additional new and large factor in the utilities operations is the 21st century customer. Instead of being a silent participant, and obedient bills-payer, the new customer is participating actively in the utilities operations. Millions of customer generate electricity today and send it into the grid; unheard of thing until recently. As the number of these new type of customer-generators increase, so does their say-so of how the utilities operate.

The Costs

Average electricity costs in the U.S. are in the $0.08 to $0.40 per kWh, depending on location and season. In Arizona, for example, power bills vary from $0.05/kWh during normal hours in the winter months, to over $0.35/kWh during summer peak hours.

Under the right conditions, the power bills could very quickly go sky high as dictated by the supply-demand market conditions. This creates a price risk that is a full-time headache for the utilities. For assuming some control over pricing in the increasingly deregulated energy market, forwards and futures options provide the tools to help hedge against unexpected price swings.

As with anything else, however, these tools can help avoid short term risks and losses, but they are helpless on the long run. A major disaster, such as shutting down all nuclear power plants in the U.S., for example, could jack up energy prices permanently to levels that are even hard to imagine today.

As the use of electric power increases, expecting 355 GW of new electric generating capacity (more than 40% currently supplied) will be needed by 2020 to meet growing demand. Upward consumption growth is almost guaranteed over the long run, but the short-term fluctuations of the energy market increase the risk. This is because the demand for electricity fluctuates on a daily and seasonal basis. An unusually mild winter, for instance, can moderate consumption and reduce revenues. This makes estimating the appropriate level of investment in power generation capacity a difficult task.

The Players

Several agencies are in charge of regulating the activities of the publicly owned utility companies, but very little regulation is forced on the independent operators.

The publicly owned utilities are regulated by:

- *The Federal Energy Regulatory Commission (FERC)* regulates the U.S. electricity industry, which oversees rates and service standards as well as interstate power transmission, and most any activities related to power generation, distribution, and use.

- *California Public Utilities Commission (CPUC)* regulates privately owned electric, natural gas, telecommunications, water, railroad, rail transit, and passenger transportation companies in California. The CPUC main goal is to protect consumers and ensure the provision of safe, reliable utility service and infrastructure at reasonable rates, with a commitment to environmental enhancement and a healthy California economy.

- *Public Utility Holding Company Act* was enacted during the Great Depression and was designed to prevent industry consolidation. The act's repeal would evoke a wave of mergers among publicly traded utility companies, and so for now it is set in cement.

The Tools of the Trade

- *Power Purchase Agreements (PPA)* is a contract entered into by a power producer and its customers, and/or independent power generators. PPAs require the power producer to take on the risk of supplying power at a specified price for the life of the agreement—regardless of price fluctuations.

- *Load* is the amount of electricity delivered or required at any specific point or points on a system.

The load of an electricity system is affected by many factors and changes on a daily, seasonal and annual basis. Load management attempts to shift load from peak use periods to other periods of the day or year.

- *Megawatt Hour (MWh)* is the basic industrial unit for pricing electricity, equal to one thousand kilowatts of power supplied continuously for one hour. One kWh equals 1,000 watt hours. One kWh = 3.306 cubic feet of natural gas. An average household uses 0.8 to 1.3 MWh/month, depending on location and season. The use is much higher in the Northwest during cold winters, and the Southwest during hot summers

The Fight

As a confirmation of the changing situation, and the continuing efforts of utility companies to control the situation by manipulating their customers, Arizona Public Service (APS) recently proposed a drastic overhaul of net energy metering (NEM) policy in its service areas.

Note: Current NEM policy credits solar homeowners at a full retail rate for the energy they send back to the grid when panels produce more than a solar home consumes, which helps Arizona residents stabilize their utility costs.

The plan is to replace consumer-friendly NEM practices with more favorable to APS options.

APS has proposed two options:
- Add a charge of $50-$100 to solar owners' monthly bills for—get this—'use of the grid', and/or
- Credit solar homeowners for the electricity they generate at the wholesale rate*.

Note: Wholesale rate is about third to a quarter of retail.

Both options mean that any financial benefits for solar owners would be eliminated. And as usual, those who would be hurt the most is the middle class, whose sole purpose of installing solar on their roofs was to save money on rising monthly electric bills.

If the proposal is approved, it would signal an end to rooftop solar installations in Arizona, so the battle is just now starting in the Southwestern desert. The proposal is so important to APS that they are using tactics that have not been seen in Arizona for a long time. APS is using fake grassroots movement, consisting of front groups 60 Plus and Prosper HQ, to launch TV ads in support of the APS demands. Using outdated scare tactics and financial figures that have been publicly denounced, the groups appear to be blatantly misrepresenting the truth to the public.

There is also an anti-APS proposal resistance movement, consisting of the solar advocacy group Tell Utilities Solar won't be Killed (TUSK). TUSK has launched their own ad in opposition to the proposed changes to the existing NEM policy. TUSK claims that the proposal is a slap in the face to APS' own customers. There are also other Arizona NGO and civic groups who are getting involved in the fight by applying pressure directly to a number of government agencies.

By ignoring the benefits of rooftop solar for all of its customers, APS sees it only as a threat to its bottom-line and is determined to eliminate it at any cost. This is so important to APS, that even if they don't succeed this time, they will try again and again until finally getting what they want. From previous experience, we know that the utilities always win, so we would be surprised if APS does not win on the long run.

In another example, Maryland's regulators are working on a provision that allows state utilities to bill customers—get this—for an electricity outage, even if it is not their fault. So imagine for a moment; the power goes out, all customers are in the dark for days and week and when it is all over, they are forced to pay for the black out.

In response, the Maryland Public Service Commission (PSC) deemed the charge applicable only for the first 24 hours of an outage. The fee kicks in for any storm with more than 100,000 outages, or 10% of a utility's Maryland service territory, whichever is less.

A severe storm stuck numerous Maryland electric customers in the summer of 2012, resulting in many long-lasting outages. The affected people suffered in stifling heat and misery living without power, on top of spoiled food in the hot refrigerators. And when it was over the customers got a bill for living without power for several days or weeks.

Even if the amount was quite small (this time), it is still hard to explain to people why do they have to pay for suffering without power. Or who is going to reimburse them for the suffering and spoiled food.

The utilities' explanation is that since they cannot control the weather, they deserve to be reimbursed for unexpected revenue losses. While this makes sense, the principle of paying for a service that is not working is not accepted in the U.S. and this alone should be enough to prohibit the fee from being assessed during an outage. Customers should not be punished for damages not caused by them.

ELECTRICITY MARKETS

Electricity is a key resource in modern societies. We use it continuously to satisfy our basic household needs It is the backbone of our economy, as it is drives all manufacturing and service industries. Due to the utmost importance of electricity in our society, it is essential that the entire chain of events from the generation to the delivery of power to the end consumers is managed in a reliable, cost-efficient, and safe manner.

These activities are currently being performed in the framework of the *electricity markets.*

Background

The top level electricity market is a system designed for executing (usually large size) purchases through bids to buy, or sales through offers to sell. This could include long or short-term trades, generally in the form of financial instruments, or obligation swaps. Electric power bids and offers use supply and demand principles to set the price.

Long-term trades are contracts similar to power purchase agreements and generally considered private bi-lateral transactions between counterparties.

Wholesale transactions (bids and offers) in electricity are typically cleared and settled by the market operator or a special-purpose independent entity charged exclusively with that function. Market operators do not clear trades but often require knowledge of the trade in order to maintain generation and load balance.

The products and services sold and bought on the electric market generally consist of two distinct types: power and energy.

- Power is the metered net electrical transfer rate at any given moment and is measured in kilowatts (kW), megawatts (MW), and gigawatts (GW).

- Energy, on the other hand, is electricity that flows through a metered point for a given period of time and is measured in kilowatt hours (kWh), megawatt hours (MWs), and gigawatt hours (GWh). The concepts of power and energy are important, because they are in the foundation of the electricity markets, so they have to be well understood.

For example, a power plant has a capacity to generate 500 MW electric power. The actual production capacity can change from zero to 500 MW during the day by ramping up and down the steam turbines. The generated power is then sent into the grid and every drop is measured, recorded, and charged to the customers in units of kW, MW, or GW.

Note: Nevertheless, the popular convention in electricity use—right or wrong—is to use the terms *energy and power* interchangeably. And so, we also use these words indiscriminately in this text.

The Developments

Until the 1980s, power systems worldwide were centralized. State-owned, vertically-integrated utilities were in charge of the entire electricity life cycle: generation, transmission, distribution, and retail. The utilities were monopolies in all these fields and ruled the electricity market.

These state monopolies remained unchallenged for almost a century. The first steps towards the creation of modern electricity markets were taken in Chile in 1982. It happened during the Pinochet dictatorship, when separation of generation and distribution activities (unbundling of electric power) was introduced, bringing with it competition between producers. At the same time, pricing and trading of electricity based on the production cost (marginal pricing) was also implemented.

UK was the first country in Europe to break the monopoly with the creation of an electricity market in England and Wales in 1990. Norway followed in 1991 and the rest is history. Australia and New Zealand were among the first movers in that part of the world by following the EU example in 1996. The USA started the liberalization of the electricity markets in California (CalPX), New York (NYISO). and Pennsylvania. New Jersey, and Maryland (PJM) in the 1990s too.

As time went by, the electricity market developed in a variety of ways, which are too complex for detailed review in this text. There are, however, a number of common features, which characterize the trend:

- The most important feature in the new electricity markets is the separation of generation, transmission, distribution, and retail activities. While the electricity markets promote competition in generation and retail distribution, the power transmission remains largely as a monopoly managed by non-commercial organizations (system operators).

- Trading the bulk of electricity is organized in pools or exchanges, with the preferred method of short-term transactions being the "day-ahead" market. It is a forward market in which hourly location-based marginal prices are calculated for the next operating day based on generation offers, demand bids and scheduled bilateral transactions.

This market is organized as a two-sided auction where producers, retailers, and large consumers submit offers and bids for electricity delivery to and withdrawal from the grid, respectively, throughout the following day. Market participants must usually submit 24 selling offers or purchasing bids; one for each hour of the following day. In New Zealand, for example, electricity market offers are submitted on a 30-minute basis.

Each offer-bid is specified as a set of price in quantity pairs, indicating the amount of energy the participant is. willing to sell or purchase at a given price. In order to clear the market, the market operator determines aggregate sale and purchase curves by sorting the sale offers according to increasing prices, and the purchase bids in the inverse order.

If the transmission grid is not considered in the market-clearing procedure, the intersection between the two curves sets the system price and this price applies for all market participants. More specifically, die sale (purchase) offers whose price is not greater (lower) than this price arc accepted, and this determines the day-ahead schedule.

If the transmission network is considered in the market-clearing procedure, instead of a single system price, a location-based marginal price is associated with each node of the power system and may differ due to line congestion and line losses. Later adjustments of day-ahead contracts are possible in intra-day markets, also known generically as adjustment markets.

- The balancing market, which is also called regulation market. is a last-resort market that ensures that production equals consumption at any point in time. This implies that all unwanted deviations from the production and consumption plans resulting from the day-ahead and intra-day markets are balanced by the activation of regulating power from other market participants. These unwanted energy deviations, or imbalances, are typically settled ex post according to the metered production and consumption of the market participant that deviates from its forward schedule.

- Day-ahead, intra-day, and balancing markets are energy markets, where the payment to or from the market operator is proportional to the amount of energy actually delivered to or withdrawn from the grid.

- In addition to energy markets, reserve capacity markets are in place in some countries to guarantee the availability of sufficient balancing power during the real-time operation of the power system. Producers participating in reserve capacity markets are paid proportional to the available capacity. Here the price is determined using rules similar to those employed to compute the energy price.

- Most electricity markets also provide clearing services for long-term financial contracts (forward, options, and derivatives), which are part of the financial markets reviewed later on in this text.

Electricity Use

The U.S. total electric power generating capacity—from coal, gas, oil, nuclear, hydro, and renewable power plants—is about 1,100 GWe or about 4,000 TWh annually.

This is a tremendous amount of energy, of which:
- 1,600 TWh (38%) is from coal-fired plants,
- 1,300 TWh (30%) from natural gas,
- 800 TWh (19%) from nuclear,
- 300 TWh from hydro power, and
- 140 TWh from wind and solar.

Annual electricity demand is projected to increase to 5,000 billion kWh annually by 2030, even though it is depressed presently and not expected to recover to previous levels until 2017-2018.

The annual electricity consumption in the U.S. is estimated at 13-14 MWh per capita.

This is a lot of energy, and compared to that used by people in other countries, we see the annual use of electricity in Haiti to be 25 kWh per capita, or 500 times less than that used in the USA. At the same time Iceland's population uses 51-52 MWh per capita annually, which is 4 times that used in the U.S.

Upon generation, the electric power of each power plant is sent into the grid, where it enters the wholesale electricity market. Here, competing generators offer their electricity output to a number of retailers, who reprice the electricity and take it to the retail market.

Until recently, the wholesale pricing was the exclusive domain of large retail suppliers, now the markets are beginning to open up to end-users. Larger end-users seeking to cut out unnecessary overhead in their energy costs have recognized their power to negotiate and take advantage of it. This is a relatively recent phenomenon, which is expected to grow fast.

There are, of course, drawbacks in buying whole-sale electricity. There are market uncertainties, membership costs, set up fees, collateral investment, and organization costs, in addition to the added hassle of buying electricity on a daily basis. Because of that, only large users could benefit from such activity. The larger the end user's electrical load, the greater the bargaining power and benefits.

The electricity wholesale market can succeed only if a number of criteria are met. The most important in this case is the existence of a coordinated spot market that has "bid-based, security-constrained, economic dispatch with nodal prices." These criteria have been thus far successfully adopted in the U.S., Australia, New Zealand and Singapore

Today's electricity markets also include a huge system of additional products and services, which are part of the electricity life-cycle. These are a large number of: mines, gas wells, transport vehicles, power plants, transmission and distribution networks, and millions of customers. These are all interconnected and interwoven in a complex system of products and services, which of which affects the It would be impossible to squeeze the details of each these segments in this text, so we focus on the key elements of the electricity markets function.

On the technical level, today's electricity markets (retail and wholesale) consist basically of two quite different and almost separate entities: power generation and power distribution.

The key elements of these sectors, and the related markets, are:

Electric Power Generation
- Exploration and siting services
  — Engineering design services
  — Power plant equipment manufacturing
  — Power plant construction
  — Transport services
  — Power plant operation and maintenance

Electric Power Distribution
- Power grid O&M
- Rail, road, and water transport
  — Pipelines
- Retail sales

In more detail:

Electric Power Life Cycle

The entire cradle-to-grave energy production, distribution, and use process, which determines and drives the electricity energy markets, is long and complex, with many different steps, such as:

Energy Source Production
- Need and request for energy
- Initial investment and legal work
  — Private banks and investors
  — Government finance
- Energy source locating and discovery
  — Equipment and services
  — Operations
- Energy source development
  — Mines and wells construction
  — Equipment and services
  — Operations
  — Risks

Electric Energy Generation
- Energy source procurement
- Energy generation
  — Power plants construction
  — Equipment and services
  — Power plants operation
  — Power plants decommissioning

Electric Energy Distribution
- The National Electric Grid
- Equipment and services
- Operations
- The smart grid
- Risks

The Power Grid
- Operation and maintenance
- Smart Grid
- The challenges

Energy Storage
- Importance
- Market development

Financial Aspects
- Private investments
- Government finance and subsidies
- Energy stocks, options, and futures trading
- Risks

Political and Regulatory Aspects
- Government direction
- Political decisions
- Regulatory compliance

Legal Aspects
- Policy challenges
- Lobbying issues
- Lawsuits

Environmental Aspects
- Environmental markets
- CO_2 trading
- Government and legal action
- Risks

Future Energy Concerns
- Fossil-less lifestyle options
- The role of the renewables
- New technologies and approaches

It is worth noting here that the concept of *energy* and *energy markets* in this text encompasses everything and everybody involved in the production, distribution, and use of electric power, vehicle fuels, and different types of petrochemicals, and other products derived from energy materials, or somehow generating electricity or heat.

And, of course, included herein are the thousands of employees, contractors, consultants, and government employees, who are in one or another way involved in the energy markets.

Demand Drivers

In general, the amount of electricity demanded on regular basis is relatively insensitive to the price of electricity in the short-term. It is basically inelastic and irresponsive to the prices, with very few exceptions.

- Smaller–residential and small business—customers do not get price signals (on time, or ever) to which they can respond. Most residential customers are billed monthly on a preset rate structure. They are basically locked into a one-way relation, where the local utility makes the rules and executes the actions without any input from the customers.

- Large industrial customers have more influence and more options, since they receive real-time price signals to which they may adjust accordingly, if possible and/or if they chose to do so. They also have more leeway to negotiate with the utility and make changes in their consumption as needed to either increase efficiency, or reduce energy costs.

The most import factor in the demand side of the equation is the fact that electricity is an absolute necessity for all of us.

Electricity is indispensable to most people and businesses in the developed countries, and is a basic necessity in the developing world. There are no valid options of not-using electricity in the U.S. and the other developed nations. As a matter of fact, a hot Arizona summer day without electricity would be as close to Hell as we here on Earth could get to. Similarly so, a freezing cold winter day in Maine without electricity would be equally deadly to those affected.

While we, the customers, like to complain about availability or price of electricity, we are stuck with it. The only option we, the customers, have in controlling our electric bills is to reduce its use in the short-term by turning down the thermostat or turning off lights. Yet, most of us find it difficult to do so, and, of course, living without electricity is not an option at all for the vast majority. And so we complain, but still use as much as we want, and even more than we need some of the time.

There is little options for electricity storage and few realistic substitutes—now and in the near future. Consequently, customer demand tends to drive prices, especially when the system is stressed during some time periods. In the long-term, options for reducing electricity use include switching to natural gas, installing insulation and implementing other energy efficiency measures. Larger consumers may consider building their own generation facilities, or use alternative energy sources.

Governments and businesses are well aware of the problems at hand and are developing demand response programs, using energy efficiency measures, or programs in which customers agree to reduce load in exchange for compensation.

The Key Factors

In addition to the customers' demand influence, there are a number of other factors that drive the electricity supply and demand cycles.

Some of the most important *other* factors are:
- Demographics,
- Climate and weather,
- Economic activity, policies and regulations, and
- Other factors, as follow:

Demographics

Population levels affect demand, with greater population levels tending to increase electric consumption. Shifts in population also affect regional demand. Population flight in the 1980s from northern industrial regions—the Rust Belt—to warmer climates in the South, for example affected residential consumption patterns.

In the 1990s. As a result, consumption in the South surpassed that in the Midwest, making it the region with the greatest electric use and highest prices.

Climate and Weather

Weather is the biggest factor driving demand in some areas—especially these with greatest temperature extremes. General climatic trends drive consumption patterns and therefore the infrastructure needed to ensure reliable service. Cold weather and short days drive winter demand in northern regions.

Southern regions rely more on electric space heating, and, thus, see demand rise in the winter, although demand typically peaks in the summer with air conditioning load. In the winter, lighting contributes to the occurrence of peaks during the seasonally dark early morning and early evening hours.

Weather also can have extreme short-term effects on electricity usage. A sudden cold snap can drive heating use up quickly and a heat wave can push air conditioning loads.

Other, less obvious weather patterns affect demand, such as rain and wind, for example, may result in sudden cooling, affecting heating or air conditioning.

Economic Activity

The pattern of socioeconomic life affects the cycle of electric use, with weekends and holidays showing a different pattern than weekdays. Demand typically rises as people wake up and go to work, peaking in the afternoon. The health of the United States and regional economies also affects power demand. During periods of robust activity, loads increase. Similarly, loads drop during recessions.

These changes are most evident in the industrial sector, where business and plants may close, downsize or eliminate factory shifts. In addition to reducing overall demand, these changes may affect the pattern of demand; for example, a factory may eliminate a night shift, cutting baseload use but continuing its use during peak hours. In some cases these effects can be significant.

Energy Policies and Regulations

State regulatory agencies set prices and policies affecting retail customer service. Some states are considering changes that would enable customers to receive more accurate price signals. They include, among other things, changing rate structures so that the rate varies with the time of day, or is even linked to the cost of providing electricity. Efforts to reduce overall demand by improving energy efficiency are underway through several governmental and utility venues.

Government policies are a major driver of the U.S. energy markets. The present trend of increased drilling for oil and gas is overshadowing the U.S. energy markets. The regulations and the low oil and gas prices are basically slowing down the development of renewables and other energy sources.

According to the US Energy Information Administration forecast in 2014, the country will lose 10,800 MWe of nuclear generation by 2020. This is mostly due to lower prices of natural gas and stagnant growth in electricity demand. This will have significant implications for CO_2 emissions, which will increase by 500 million tons annually by 2040.

The situation is even more serious in the developing countries. Here, coal-fired power plants are being built as fast as possible, in order to provide a better life style for the population.

The Customers...again

There is no doubt that the electricity market is dominated, and driven mostly, by the number of customers and their power use preferences and patterns. So, a even closer look is needed, as follow:

Retail Customer Mix

Most electric utilities serve different types of customers: residential, commercial and industrial. Each class uses electricity differently, resulting in a differing load profile, or the amount that each customer class uses and the daily shape of the load. If a consumer uses electricity consistently throughout the day and seasons, his load shape is flat, and the load will be baseload. Another consumer may use more at some times than others, resulting in baseload and peaks.

Greater variability in demand is typically more expensive to serve, especially if the peak occurs at the same time other customers' use peaks. Consequently, the mix of customer types affects a region's overall demand. Residential consumers form the largest customer segment in the United States at approximately 38% of electricity demand. Residential consumers use electricity for air conditioning, refrigerators, space and water hearing, lighting, washers and dryers, computers, televisions, cell phones and other appliances. Prices for residential service are typically highest, reflecting both their variable load shape and their service from lower-voltage distribution facilities, meaning that more power lines are needed to provide service to them.

Commercial use is the next largest customer segment at approximately 36%, and includes hotels and motels, restaurants, street lighting, retail stores and wholesale businesses and medical religious, educational and social facilities. More than half of commercial consumers' electric use is for heating and lighting. Industrial consumers use about 26% of the nation's electricity. This sector includes, for example, manufacturing, construction, mining, agriculture and forestry operations. Industrial customers often see the lowest rates, reflecting their relatively fl at load structure and their ability to take service at higher voltage levels.

Transportation demand for electricity stems primarily from trains and urban transportation systems. This, however, is less then 1% of electricity demand.

Load Forecasting

Demand is constantly changing, challenging grid operators and suppliers responsible for ensuring that supply will meet demand. Consequently, they expend considerable resources to forecast demand. Missed forecasts, where actual demand differs significantly from the forecast, can cause wholesale prices to be higher than they otherwise might have been.

Forecasts are necessary as well for the variety of actions that must occur if sufficient supply is to be available in the immediate or long term: planning the long-term infrastructure needs of the system, purchasing fuel and other supplies and staffing, for example. Load forecasts are also extremely important for suppliers, financial institutions and other participants in electric energy generation, transmission, distribution and trading.

Load forecasting uses mathematical models to predict demand across a region, such as a utility service territory or RTO footprint. Forecasts can be divided into three categories: short-term forecasts, which range from one hour to one week ahead; medium forecasts, usually a week to a year ahead; and long-term forecasts, which are longer than a year. It is possible to predict the next-day load with an accuracy of approximately 1%-3% of what actually happens.

The accuracy of these forecasts is limited by the accuracy of the weather forecasts used in their preparation and the uncertainties of human behavior. Similarly, it is impossible to predict the next year peak load with the similar accuracy because accurate long-term weather forecasts are not available. The forecasts for different time horizons are important for different operations within a utility company.

Short-term load forecasting can help to estimate transmission system power flows and to make decisions that can prevent overloading of transmission systems. Timely implementation of such decisions leads to the improvement of network reliability and to the reduced occurrences of equipment failures and blackouts. Forecasted weather parameters are the most important factors in short-term load forecasts; temperature and humidity are the most commonly used load predictors.

The medium- and long-term forecasts, while not precise, take into account historical load and weather data, the number of customers in different customer classes, appliances used in the area and their characteristics, economic and demographic data, and other factors. For the future peak forecast, it is possible to provide an estimated load based on historical weather observations.

Long-term forecasts are used for system infrastructure planning and are meant to ensure that there are sufficient resources available to meet the needs of the expected future peak demand. These forecasts are made for periods extending 10 to 20 years into the future.

Demand Response

Electric demand is generally insensitive to price, meaning that demand does not typically fall significantly when wholesale prices rise. However, some utilities and grid operators are developing ways to stimulate a response from consumers through demand-response programs. Demand response (DR) is the ability of customers to respond to either reliability or price triggers by forgoing power use for short periods, by shifting some high energy use activities to other times or by using onsite generation.

These programs may use price signals or incentives to prompt customers to reduce their loads. The signals to respond to electric power system needs or high market prices may come from a utility or other load-serving entity, a regional transmission organization (RTO) or an independent DR provider. These programs are administered by both retail and wholesale entities.

DR has the potential to lower system wide power costs and assist in maintaining reliability. It can be used instead of running power plants or to relieve transmission congestion There can also be environmental benefits because peaking units tend to be costly—and dirty—to run. Demand response rewards consumers for reducing load during certain market conditions and at specific times.

However, it is difficult to measure and quantify this reduction. Measuring and verifying the reduction requires development of consumers' baseline usage, against which their actual use is measured to determine the reduction in the event they are called to lessen their load. An accurate measure of their typical usage is important to prevent (or detect) gaming by participants.

Electricity Market Deregulation

Today our economies are driven by computers, other digital products, gadgets, and processes. Our daily existence is device-driven, in which the devices are powered exclusively by electricity from the grid. Recently, electric vehicles—cars, trucks, and trains of all sorts—are becoming increasingly important. And so, stabile, quality, riskless, and cheap electricity supply is the key to our economic progress and comforts at home and business.

In the U.S., electric black-out—even a very short-lived one—is as rare as a Lunar eclipse. Electricity is the foundation upon which we build our lives and in many cases there are no substitutes, nor compromises. So any lack of electricity is a not an option here. Period!

We walk in a room, flip a switch and presto; light come on. Press the computer ON button and it comes alive. The same with the water heater, clothes washer, radio, TV, AC, and all other gadgets we have at home, our garages, and our businesses. And the credit goes to our incredibly efficient electrical system. Try to do the same in Syria, Iran, Iraq, and many other developing countries. Nope, it would not work most of the time. Even if there is a switch on the wall—and often there is none—flipping it won't make any change. The power is off for the duration, and non-existent all together for a long periods of time.

Deregulation

Originally the U.S. power system function and operation was quite simple. It was based on one-way transport of electricity from the power plants to our homes and businesses through transmission and distribution lines. The power companies sold their power to the local utility company, which then send it via their own wires to the local customers. Customers paid the utility for the power they used, the utility paid the power company, and the cycle repeated—no problems, and few complaints.

In time, numberd of problems and issues arose, eventually initiating the structural changes started in the 1980s. At that time, research showed that the original belief of economies of large-scale power generation was no longer valid.

The increased awareness and change in attitude led to two major changes:

• Deregulation of power generation and retail marketing, and

• Unbundling of generation, transmission, and distribution.

Deregulation and liberalization of the energy systems brought up a new concept, where physical delivery of electricity (by the power company) is to be differentiated from its buying and selling (by the local utility). So now, (some of) the consumers are allowed to act as their own agents; able to negotiate and enter into contracts with generators, no matter where these might be located. This way, the competition among the generators is increased, since they now can compete for a market share.

During the past two decades we have seen comprehensive electricity sector deregulation in many countries. The situation is different in the U.S. since we have no mandatory federal restructuring and competition laws. Instead, the individual states are left to decide on their own. As a result, many states have introduced only a partial reforms mostly affecting the wholesale markets. As a matter of fact, there is a trend now, where some states with these reforms in place are planning to re-regulate the industry.

As a result, implementing and evaluating the reforms may take different forms at different times and places. In some cases it means permission for independent generators to enter the market, in others it is the creation of a power pool, and sometimes the horizontal separation of incumbent generators. It could also refer to the vertical disintegration of state-owned monopolies into generation, transmission, and distribution businesses.

In its most practical form, deregulation usually means the sale of state-owned assets, either completely or partially, to the private sector. This usually improves the economic efficiency of the enterprise and the entire industry.

In Arizona, for example, we have one regulated and one unregulated (private) utility companies. Which one is better depends on whom you talk to, but they both have one thing in common; they need a lot of improvement in the customer relations and support areas of their activities. In their defense, we must admit that their job is not easy. Letting the old, comfortable way of life and jumping in this new, brave, and revolutionary world is not easy. And so, they are trying their best to adapt to the new times. It is a slow process, but progress is noted and appreciated.

While unbundling and regulation of generation, transmission, and distribution of electric power is a good idea, since it allows competition, the ownership and use of power distribution networks is still a natural monopoly of sorts. This has created a number of issues, which are increasing in importance as the renewable technologies tend to use the power distribution networks too.

We take a closer look at this problem elsewhere in this text.

Electricity Pricing

The price sensitivity of electric power consumers historically has been quite low. We are just glad to have electric power any time we need it. The rest is just a matter of complaining about high electric bills. But we like to complain about all bills; it is our favorite pass time, so we do…every month.

But price of electricity, or rather the possibilities for price adjustment, have not been a matter of serious concern or discussion…until recently. Also, the contracts have traditionally been based on flat rates so that the volatility of the real-time price has little, or no direct effect on the monthly bill.

In power generation, however, the cost of generation varies greatly depending on a number of factors, the main of which is demand levels. The merit order of the power system is based on using most cost-efficient equipment and approaches. More expensive methods such as the firing of peaker power plants, are used during the demand peak hours only, when the prices are much higher than during the low demand periods. These basic generation cost-differences occur daily, and even hourly, and yet they do not reach the consumers directly in most cases. At least not until recently…

The unbundling of marginal costs from current prices has always been problematic, and technological constraints have prevented the use of efficient price setting methods. The development and installation of intelligent automated metering technologies and devices, however, is changing the picture. Many real-time pricing (RTP) contracts with residential and business customers have been signed, and this development is increasing as we speak. What kind of long-term effect the RTP method has on generation and consumption, and consequently on emissions, is still an open question.

Based on the ever evolving growth of new technologies such as the smart grid, micro grids, and renewable energy production, these and other new pricing strategies are being considered to capture the dynamics of supply and demand in the electricity market. The electricity pricing strategy development is one of the most important and complex aspects of the present-day energy debate.

The dynamic pricing strategies that are used to offer customers shifting prices depend on several internal and external factors. Dynamic pricing makes value and cost of energy use transparent to consumers which enables them to determine when cost exceeds value, which in turn enables the adjustment of energy consumption and production needs.

With the advent of smart grid and other hi-tech solutions emerging daily, the necessity of new and efficient means of energy pricing become more urgent. There has been an array of attempted pricing and forecasting models by utility companies and regulators—usually designed to increase end use efficiency, energy conservation, capacity utilization, savings, in the short term, and to ensure energy security, in the long run. Many trial and error efforts have been documented; some have been successful, some not so much. It is work in progress…

Price Structure

Presently, a price-based energy network is structured mainly on:
* Marginal costs,
* Consumer supply and demand,
* Competition trends, and
* The effects of deregulation

A number of different factors impact electricity pricing, which change with changes in the socio-political system and technological advances. These factors affect pricing schemes no matter which pricing model is implemented. Prices fluctuate by location—city, state, country, region—and by input source (petroleum, oil, coal, nuclear, solar, etc.).

Key factors that affect the price of electricity and energy include:
* Peak and demand rates are vehicles that help the power companies pay their bills and make capital investments to meet the present and anticipated peak demands (by charging the customers as needed).

* Alternative energy technologies can help reduce the rates (especially the peak rates), but despite advances in recent years their use is limited due to high implementation costs, uncertainty in their long-term operation, legal constraints, and recently due to conflicts with utilities operations (it costs too much to operate the grid with many renewables hooked into it).

- The state and local taxes imposed on utilities change from state to state and from time to time, affecting energy rates.

- The power generation or input source determines the price of energy delivered to the customers. A mix of energy sources is ideal for trimming peak rates, where, for example, base-load (gas-fired power plant) is assisted by a solar power plant, which delivers maximum power during the peak hours.

- Environmental considerations complicate the situation and limit the choice of the utilities, forcing them to use technologies with a smaller carbon footprint (wind and solar), which are more expensive to use.

- The capital expense and maintenance of the transmission network is of great importance, as is the distance from the energy source. For example, greater losses are experienced over long-distance power transmission, also making rates higher. Many factors affect energy rates. The overall picture is complex, and the expenses for producing and maintaining energy supplies are great.

- The utilities are caught in the middle between the customers (who demand lower prices) and the regulators (who order lower prices). So, the utilities are forced to kick back by increasing prices in whichever area of their activities they can. Recently, for example, the utilities have been trying the increase the prices charged to residential solar power owners to compensate for those owners' free use of the power grid.

The utilities claim that the increased use of the power grid by thousands of new residential solar installations is forcing new capital investment and additional maintenance. True, but the customers do not like the utilities' dictatorship and are rebelling against it. Another work in progress.

Electricity Market Update

Electric power sector coal consumption was largely unchanged in 2014. Power sector coal consumption is projected to increase by 0.3% in 2015, despite a 0.8% increase in electricity demand, as comparative natural gas prices decline and retirements of coal power plants rise in response to the implementation of the Mercury and Air Toxics Standards. The full effect of the coal plant retirements is felt (or projected to be felt) in 2016, as projected electric power sector coal consumption declines by 1.4%.

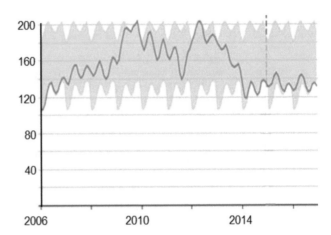

Figure 7-7. U.S. coal-fired electricity (in million tons)

At that time, 2017-2017, the U.S. solar industry will lose its main incentive—the investment tax credit (ITC). This will change the dynamics of the renewables energy market, and will affect the entire U.S. energy market.

Electricity Exports

The U.S. is a major electricity importer. With all the power we generate, we import electricity from our neighbors; the equivalent of power generated by a dozen power plants. But we also export some power. Our major (if not only) electricity import and export are our neighbors, Canada and Mexico.

According to Table 7-4, the U.S. used about 59 million megawatt hours of imported electricity. The imports came from Canada, with slightly under 52 million MWh balance, and Mexico, with over 7 million MWh.

Canada's power exporters are the provinces of Quebec, Ontario, Manitoba and British Columbia. The main U.S. customers are the states of New England, New York, with some power going to the Midwestern and Pacific Northwest states.

Most of the power trade is north-south, because the distances between power plants and markets in the U.S. are shorter than the east-west distances. The value of the American dollar is also a contributing factor in the north-south trade trend.

Some electricity sales to the U.S. are settled in advance through fixed supply contracts between provinces and states. There are also electricity markets, where blocks of energy are traded at prices established by supply and demand.

Table 7-4.
U.S. Electricity trade balance (in MWh)

Year	Imports	Exports
2003	30,394,551	23,974,703
	34,210,063	22,897,863
2005	43,929,314	19,150,968
	42,691,310	24,271,335
2007	51,395,702	20,143,592
	57,019,381	24,198,159
2009	52,190,595	18,137,984
	45,083,186	19,106,180
2011	52,299,710	15,048,552
	59,257,069	11,995,649
2013	70,355,069	11,352,850

Electricity Trade

The main characteristic (and a main problem) of electricity is that it must be used as soon as it is generated. This is because there is no reliable method for storing directly large amounts of power (as yet). Instead, utilities have to use some the generated power to make and store other types of energy that can be converted back into electricity when needed. This is usually in the form of water pumped up and held behind dams, storing extra coal or natural gas on-site, etc.

This way, the utilities have extra capacity that they bring online to meet periods of higher demand, thus avoiding blackouts. When demand increases and creeps towards the limit of supply, the utilities switch to this stored (backup) power, including thermal generating stations or hydroelectric plants, if they are available.

In some cases, this is not enough power to keep the grid going in peak periods of electric use. In such cases utilities are forced to buy quantities of power on the electricity spot market; which is where electricty is sold and bought. The extra power is provided (sold) by neighbouring provinces or states, where excess generation is available.

Because prices fluctuate with time and location, companies will either generate or purchase based on whatever is least expensive. This way, the market is given more chance to operate in a balanced manner, thus avoiding fluctuations and blackouts. Big electricity producers also methodically review weather conditions and customer consumption profiles to determine the best times to sell their electricity. The estimates are accurate to within an hour, or even within minutes, and allow some planning.

The potential buyers evaluate the offers that are made and choose the best one if and when they need extra power. In addition to balancing the energy flow, this also determines the electric market prices. Hydropower is the best and cheapest method of storing and delivering electric power, and Canada has a lot of it.

The electricity trade (coming mostly from hydropower) accounts for only a fraction of the value of Canadian energy exports, which are ruled by oil and gas. And in the best years, Canada's electricity exports account for maximum of 2% of total U.S. power consumption.

Canada makes about $2-2.5 billion from electricity exports to the U.S. It, however, could be making much more money from this export, if it was not for a serious stumbling block. It's not quotas or tariff barriers imposed by the U.S., that hinder the Canadian exports of electricity to the U.S. Nope; it is the poor electric grid that hinders the export capacity of Canada.

The Canadian transmission system is the step child of the market deregulation, and saw no significant investments. Because of that, it is the same as it was 20-25 years ago. So, the key to increasing Canadian electricity exports is developing and expanding high-voltage transmission lines. This effort, however, takes a lot of time and money, so it would be awhile before Canada is able to upgrade its grid and potentially increase electricity exports to the U.S.

For now, the electricity exports lags far behind these of oil and natural gas, which generate fossil-fuel export revenue over 20 times higher than sales of electrical power.

Case Study: Hydropower Advantages

Electricity must always meet or exceed demand, so Canadian provinces which have dams with large water reservoir and hydro power stations have a big advantage when it comes to exports. This is because water stored behind a dam is like a car battery; it has stored potential energy, which can be released at any time (when needed) to generate extra electricity by opening the floodgates as much as needed and when needed.

This cannot be done as easily, if at all, with nuclear power plants or thermal generating stations, so they are considered fixed (base) load electricity generators. We also cannot adjust the wind speed of wind farms, or the sun intensity of solar power plants, and because of that they are considered variable source of electricity.

Here the variability is uncontrollable, so in some cases they cause fluctuations and other problems in the power grid.

On the other hand, it is quite easy and quick to start and stop hydroelectric plants, and with that provide as much power as needed. Canadian utilities use this unique feature to provide power when needed, thus earning profits at home and at the export markets.

Some Canadian provinces have a lot of hydro-power capacity, where the dams can be wide open during peak periods and closed when the grid has an abundance of energy or when prices are low.

When the power plant is idling, more water collects in the reservoir, ready to be released. This is like putting money in the bank, so in periods of domestic demand increase, or when export prices on the spot market increase, utilities simply open the floodgates to generate power. Cha-ching, cha-ching...

Because of the advantage of mass hydropower generation, Canadian provinces make several billion dollars in profit by using their hydropower resources. For example, the Canadian Hydro-Québec's net exports account for only 10% of its total sales, but constitute over 22% of net profits. Not bad for using water to make money.

As mentioned before, the Canadian electricity exports to the U.S. could be several fold higher if and when the power grid infrastructure is upgraded. This is a good news—win-win situation—for both; Canada and the U.S. It means that in the future, pending the grid upgrade, Canada could ramp up its hydropower production at will, and send the excess to the U.S., where it could be used to maintain the balance in the power grid, and for providing more power to homes and businesses.

TRANSPORTATION MARKETS

Transportation is the "endangered species" of the energy markets. It is also the Achilles Heel of our energy and national securities.

The fossils—coal, natural gas, and oil—run the global economies, where coal and natural gas provide electric power, while crude oil powers the transportation sector. We have enough coal and gas, but not enough crude oil. Because of that, and since only crude oil can run the millions upon millions of vehicles around the world, it is considered a critical commodity of exceptional importance to our economy, and our energy and national security.

Without crude oil derived fuels, powering billions of vehicles, the global economy will stop, or at the very least it will slow down to a snail's pace.

True, there are some alternatives to crude oil for use as transportation fuel. LNG, fuel cells, bio-fuels, electric cars and many other innovations are getting ready to jump in the transportation sector. We, however, are still to see any of these fuels lift off a 777 superliner, power a cruise ship, or move fully loaded eighteen wheeler on a cross country trip. Nope! Not yet...maybe some day, but we should not hold our collective breath until then, because it is very dangerous to hold one's breath for a long period of time.

Note: Sorry Tesla, only the very rich and some extravagant entrepreneurs can afford the luxury of owning your slick, sophisticated and expensive electric cars. But even they cannot cross the country without a fuel stop, and since there are not many of these yet, it will take a long time before electric cars become a plausible answer to the fossil powered beasts.

And let's not forget that the electric—all slick, pollution-free EVs—need to charge their batteries with electricity generated by polluting power plants using quickly depleting fossil remains somewhere in the country. The super-rich have an option to install a special solar charging station at their estates dedicated to charging their fancy EVs, but most of us will think twice, or more times, before going that route. Maybe some day things will change, but for now, most of us will continue driving the fossil guzzlers for a long time to come. Probably until all fossils are gone.

We must always remember that electricity is as dirty as the energy source that generated it.

Another important fact we must never lose track of is that 90% of the electricity generated in the U.S. comes from fossils. So, when considering the efficiency losses of the, a.) fossils-to-electricity, and b.) electricity-to-miles conversion processes, each mile driven by an EV emits as much pollution as a gasoline fueled vehicle. In some cases Evs emit even more pollution than gas guzzling monsters, because of the inefficiency of the equipment.

Nevertheless, the experiments and R&D efforts to replace the fossils continue, and a lot of progress will be made in the future. We actually already see LNG and biofuels at the gas pumps, albeit in very small numbers. With all this in mind, we must also remember that none of the transportation fuel wanna-be alternatives have the thermal and kinetic energy, nor the convenient small size, of liquid fossil fuels—gasoline, diesel, and jet fuels. This is a facts of physics that cannot be changed, no matter how much work goes into it.

Because of that, we must admit and agree that the fossil fuels will be around for a long time. As a matter of fact, and as things are going today, it is obvious that they will be a major vehicle power until every drop is burned and no more is left on the face of the Earth. This is also a fact that cannot be changed.

Maybe in this futuristic fossil-less age, people will focus on developing new transportation fuels. For now, however, such efforts are few and between and are not given a priority in most cases.

Transportation is especially, directly, and heavily dependent on fossils—crude oil especially. Just think that 99% of all vehicles and moving equipment on Earth run on gasoline, diesel, jet fuels, and different fuel blends, all of which include crude oil and crude oil products.

The energy markets encompass, and in one or another way affect almost every area of our daily lives. One must live in a cave in a far, faraway place to not feel the presence of, and the need for, energy in his/her life. Amazingly, crude oil is the driving force behind it all. It is what brings the other energy sources and makes the entire energy market work. Everything around us—directly or indirectly—needs crude oil or its derivatives to function.

Without crude oil, the global economy will stop, and life as we know it, will change drastically.

This is not that hard to imagine. Empty highways, waterways and skies…Nothing and nobody can move anywhere at any time. What a dark and miserable day that would be… And yes, it can happen. Some day in the not-so distant future…unless we do something (or many things) that we are not doing today.

Transportation Fuels

"Over the past four decades, eight presidents tried to reduce America's oil dependence but, despite some tactical progress, were unsuccessful, and each passed to his successor a country that was more, rather than less, dependent on oil. One of the reasons for this failure is that they all focused on reducing the level of oil imports when the real problem is oil's status as a strategic commodity which stems from its virtual monopoly over transportation fuel."
—*The United States Energy Security Council. 2014.*

We can add a lot to this statement to make things more clear and poignant, but it says it all; we use a lot of crude oil and allowing it to monopolize the transportation sector, we have allowed it to affect essentially any and all aspects of our personal and business lives. No exception.

Crude oil powers everything; our personal cars, commercial trucking, trains, ships, factories, and most military vehicles.

Transportation is especially, directly, and heavily dependent on crude oil.

Just think for a moment that 99.9% of all vehicles and moving equipment on Earth runs on gasoline, diesel, jet fuels, and different fuel blends, all of which are derived from crude oil and its byproducts.

Crude oil also affects the energy markets, which in turn encompass, and in one or another way affect, almost every area of our daily lives. One must live in a cave in a far, far away place not to feel the presence of, and the need for, energy in his/her life.

Amazingly enough, crude oil is the driving force behind it all. It is what brings the other energy sources to market or to the users, and what makes the entire energy market work. Everything around us—directly or indirectly—need crude oil or its derivatives to function and operate properly and efficiently

The transportation sector is driven mostly by crude oil, which is getting very expensive and won't last forever.

It is quite obvious from the developments of late that crude oil is a big problem. In addition to getting more expensive by the day, most of it comes from shady countries, which governments are not concerned with our well being…on the contrary. We can clearly see behind the politically correct words and actions of the U.S. government, that we have problems with most oil exporters.

Except Canada and Mexico, not one single oil exporter has any respect towards—let alone real friendship with—the U.S. As a result, we are forced to be friends with people who are willing to harm and kill us. Literally. 9/11 is only one reminder of that fact, and we must always remember the truth behind it.

It would be also unreasonable and even ridiculous to think that we could have crude oil forever. So, no doubt, new approaches are needed for the long term survival of the transportation sector and our own well-being. We need to put together and sort through the lessons of the past and use the information to figure out the best ways to proceed in the future.

So let's start from the beginning of the energy markets as we know them today. Nope, they did not start with the Big Bang, but developed with time and were dormant for many centuries. Until finally the energy balance was disturbed and they were brought up to life only a century ago. And within a single century we managed to burn more than half of the entire fossil supply in Mother Earth's bosom.

In the Beginning...

The global energy markets have been around—in one form or another—for centuries. As recently as the beginning of the 20th century, however, have they been identified as a critical part of the global economy. Today, energy is a major preoccupation for most people, companies, organizations, and governments.

One thing that is obvious from Table 7-5; there have NOT been a lot of improvements in the land, water, air, or space transportation sectors since the 1980s. No major inventions or developments in these areas have been introduced for over two decades. Does this mean that we are at the end of an era, and that things need to change? Or does this signifies a still before a huge transportation revolution explosion.

Although energy products and services have been sold, bought, and traded since the beginning of time, the beginning of the energy markets, as we know them today, started in the 1970s with a big bang.

The U.S. and most of the other developed countries were asleep on the wheel and misjudged the global energy situation until 1970s. Then, one beautiful fall day in 1973 we saw ourselves cornered by an oil embargo that we did not expect and which cost us dearly.

It opened page one of our Energy Security book, assuring us that we no longer control the type and amount of energy we use. Factories shut down, gas pump lines went around the block, and the country's economy was faced with a serious risk of collapse. A lesson we will never forget and one that serves us quite well until today. We seem better prepared now and such a disaster would not happen again...maybe!

Almost overnight, in the fall of 1973 the energy markets were clearly defined and energy security became the main preoccupation and goal of all developed countries.

The new reality became painfully obvious. While we were enjoying the cheap oil supplies, most of them were controlled by foreign states. The wicked OPEC

Table 7-5. Energy markets history

3500 BC	Wheels for moving objects and carts.
	Oar powered river boats
2000 BC	Horses are domesticated and used for transportation.
200-250	The wheelbarrow.
770	Iron horseshoes improve transportation reliability
1492	Leonardo da Vinci dreamt about flying machines
1620	Human oared submersible
1662	Horse-drawn bus
1740	A clockwork powered carriage
1783	A practical wheel paddle steamboat
1784	Hot air balloon
1789	Self-propelled road vehicles
1790	Modern bicycle
1801	Steam powered locomotive (roadster)
1807	Hydrogen gas powered vehicle
1809	Steamboat with regular passenger service
1814	Steam powered railroad locomotive
1862	A gasoline engine automobile
1867	A motorcycle
1868	Compressed air locomotive brakes
1871	Cable car
1885	The internal combustion engine
1899	The dirigible
1903	Engine powered airplane
1907	The helicopter
1908	Mass production automobile manufacturing line
1908	Hydrofoil boat
1926	Liquid propelled rocket
1940	Modern helicopter
1947	Supersonic jet flight
1956	Hovercraft
1964	Bullet train
1969	Space mission Apollo
1970	Jumbo jet production
1981	Space shuttle
1990s	Computer revolution
2000s	Internet driven life

cartel had control over the global energy markets and our energy security was in the hands of cartel members, which included some enemy states.

This was not a new thing; it was something that the politicians simply ignored, thinking that since they control the purchasing end of the deal, they can also control the supply end. Wrong! Wrong! Wrong! Lesson learned? Partially so, but good enough...for now!

The "Seven Sisters " term was used since the 1950s to describe the seven oil companies of the "Consortium for Iran" cartel which dominated the global petroleum industry at the time. The Anglo-Persian Oil Company (now BP); Gulf Oil, Standard Oil of California (SoCal), Texaco (now Chevron); Royal Dutch Shell; Standard Oil of New Jersey (Esso) and Standard Oil Company of New York (Socony) (now ExxonMobil) were the seven pillars of the oil cartel at the time.

The Seven Sisters controlled around 85% of the world's petroleum reserves, but since the 1960s the dominance of these companies had declined due to the increasing influence of the OPEC cartel and state-owned oil companies in emerging-market economies, like Venezuela and Nigeria.

Everything changed in 1970s as the influence of OPEC grew substantially and the repercussions of the 1973 oil crisis shook up the global energy markets. The U.S. and EU politicians woke up from their decades long slumber and started talking about energy issues, energy security, and developing energy market plans. They started creating a national energy policies that encourages the development of an energy industry in a safe and competitive manner. This included the development of renewable technologies like solar and wind power generation.

The global energy markets were born, albeit under duress. They then developed as the major vehicle ensuring the achievement of the energy security goals. Unfortunately, the renewables were put back in the closet, when the oil crisis subsided. They were brought out several times since the 1970s, only to be shoved back in the closet following the original pattern.

Today most countries, including the U.S., have given up on the idea of achieving complete energy independence. This is mainly due to the fact that they lack one or another energy source needed for their ensuring 100% of their energy supplies. Instead, they are focusing on ensuring some resemblance of energy security.

Even that, however, is not an easy task...even for the all mighty U.S. of A. Just take a close look at the daily problems in ensuring a safe passage of the oil tankers headed to the U.S., while going through the world's choke points. And the worse is still ahead...with fossil fuel price hikes first on the global energy horizon.

The Present Situation

The transportation sector is driven 99.9 % by crude oil, so it is directly dependent on crude oil's availability and pricing.

The transportation sector and its sub-sectors consist of:

Vehicle Fuels
- Energy source procurement
- Energy source processing
  - Refineries construction

  - Equipment and services
  - Operations
- Fuels transport
  - Pipelines
  - Trains, trucks, barges
- Retail distribution and sales
- Risks

Petrochemicals
- Energy source procurement
- Petrochemicals processing
  - Production facilities construction
  - Operations
  - Final products transport
    - Pipelines
    - Trains, trucks, barges
- Retail distribution and sales
- Risks

Unconventional energy sources
- Firewood
- Coal products

Energy market transactions
- Domestic energy markets
  - Production issues
  - Transport issues
- International energy markets
  - Trade
  - Global shipping
- Security
- Risks

Financial Aspects
- Private investments
- Government finance and subsidies
- Energy stocks, options, and futures trading
- Risks

Political and Regulatory Aspects
- Government direction
- Political decisions
- Regulatory compliance

Legal Aspects
- Policy challenges
- Lobbying issues
- Lawsuits

Environmental Aspects
- Environmental markets

- CO$_2$ trading
- Government and legal action
- Risks

Future Energy Concerns
- Fossil-less lifestyle options
- The role of the renewables
- New technologies and approaches
- Risks

Looking at the above list, we see one persistently reoccurring factor; *Risk*. This is not a coincidence, nor a mistake. It is a fact of life! Crude oil is risky business no matter what point, which category, or from what of view we look at it.

We, however, are fed only positive news. For example, in 2014, U.S. EIA praised the rise in oil and gas production over the last decade: The "proved reserves of U.S. oil and natural gas rose by the highest amounts ever recorded since 1977. This is stunning news, considering the general trend of decreasing production between 1980 and 2000, which now has being reversed.

U.S.' proved reserves of crude oil are now at their highest levels ever.

True, but this sounds like new oil has been created constantly, and does not say how long the new oil bonanza would last. Could it last 10 years, 50, 100? No matter how long it lasts, it is unlikely to last forever, so what shall we do? How about just starting to consider the options and talk about them. That would be a good start.

Proved reserves are estimates of oil and natural gas deposits that may reasonably be recovered from known reservoirs under existing economic and operating conditions. The estimates change from time to time, of course, and depends on factors such as discoveries of new fields, reassessments of existing reserves, and the availability of new detection and recovery technologies. The overall extraction prices, which depend on the new technologies, play the most important roles in determining what's reasonably recoverable as well.

According to EIA: "The expanding application of horizontal drilling and hydraulic fracturing in shale and other 'tight' (very low permeability) formations, the same technologies that spurred substantial gains in natural gas proved reserves in recent years, played a key role." This pretty much explains what accounts for the record growth in reserve estimates of late.

Rising oil prices are to be credited in part for that

phenomena, too, since they provided incentives for producers to explore and develop additional resources. This is a natural self-correction, since when the energy markets are allowed to work unobstructed, the new resources are usually developed in response to price signals. For awhile, both technology and market's price signals combined to boost proved reserves of crude oil by more than 12%, and natural gas by nearly as much.

There were already some impressive reserves estimates issued in 2009; a record 22.3 billion barrels in crude oil proved reserves, which went to 25.2 billion barrels in 2010 and kept increasing with time.

These are truly extraordinary developments, worth celebrating especially when considering the economic benefits. Thousands of jobs, affordable and abundant energy supplies and new revenues for government through taxes and royalties. Those are achievements, which we all must applaud, no doubt.

All this helps in furthering our energy security goals too, but we must keep the distant future in mind too. We must never, ever, forget that no matter how much oil and gas we have, these are commodities with finite reserves. Sooner or later, they will be depleted, so we must be mindful of the way we exploit and use them.

But even before we get there, we must consider the external factors in play in the energy markets. A good example of their profound influence on our lives is the Saudi Arabia's manipulation of the oil market, when oil prices dropped under $50 in 2014-2015. This caused all U.S. oil producers to reduce production and many to file Chapter 11.

How could it be that while having so much oil, we cannot afford producing it? Well, this is the role of the energy markets, and where the different variables kick in with full force. The global energy markets determine the supply and demand, which dictate the pricing levels, but in this case (global oil price drop of 2014-2015) the global pricing dictated the supply and demand levels.

This is just another way to manipulate the energy markets. It is another variation of the Arab 1973 oil crisis, brought upon us compliments of our Arab friends. This simply means that we still have the wrong countries for friends, which also means that we are still very far from energy independence.

California Car Dreams

California is symbolized by sunlit beaches, girls in bathing suites, and cars—all kind of cars. No other part of the world has embraced the automobile as passionately, nor is any other state defined as much by the car, as California. Presently, there are over 26 million cars and 1 million trucks on California roads and highways.

> *About half of all energy used in California fuels cars and trucks, or nearly 20 billion gallons of gasoline and diesel fuel annually. This results in the estimated emission of over 250 million metric tons of GHGs, or about 40% of all state's GHG emissions.*

California is determined to keep the wheels rolling, but due to changes in the global energy markets and the environment, it must make some changes in its car-loving ways. The rolling resistance of tires, for example, has been blamed for a leading role in waste of fuel. Solving that problem could lead to substantial improvements in vehicle fuel economy. So, a consumer bill AB 844 was issued in 2003, directing the California Energy Commission to develop and implement a Fuel-Efficient Tire Program that included standards for tires on passenger vehicles and light duty trucks.

Other such programs are underway as well, with the 2007 Bill 118, Alternative and Renewable Fuel and Vehicle Technology Program (ARFVTP) leading the way to fossil fuels reduction. The Bill created an annual $100 million public investment fund to promote development and deployment of advanced technologies, low carbon fuels and vehicles that will help the state achieve its greenhouse gas reduction goals.

Alternative fuels supported by ARFVTP include electricity, hydrogen, biofuels, natural gas and renewable natural gas, plus the fueling infrastructure needed to deliver these new fuels to California consumers.

Alternative technology vehicles supported by ARFVTP include: electric drive cars, shuttles and trucks; hydrogen fuel cell cars, trucks and buses; natural gas trucks, buses and cars; as well as trucks and cars that can use biofuels such as ethanol and biodiesel.

ARFVTP also provides funding support for workforce training and development in order to prepare the next generation of mechanics and technicians to support new technology vehicles and fuels.

How much these programs contribute towards reduction in fossil fuels use remains to be seen. Nevertheless, they are a good step forward in the battle for energy and environmental responsibility. We can only hope that other states and countries will follow California's example soon enough and before it is too late.

Vehicles and Equipment Manufacturing

For its proper function, the transportation sector needs a lot of fuel to power all the millions of vehicles

running on land, water, and air. We discussed the fuels that power the transportation, so now we need to look at the vehicles these fuels power.

Shortly after their manufacturing by the automotive industry, these millions of vehicles become part of the transportation sector. So directly and indirectly, the automotive, ship, and planes manufacturing industries are part of the transportation markets.

In general terms, the global automotive industry consists of a wide range of companies and organizations involved in the design, development, manufacture, marketing, and selling of motor vehicles. It is one of the world's largest and most important economic sectors by size and revenue.

Table 7-6. Global vehicles production

Year	Vehicles Produced	% Change
1999	56,258,892	6.2
2000	58,374,162	3.8
2001	56,304,925	-3.5
2002	58,994,318	4.8
2003	60,663,225	2.8
2004	64,496,220	6.3
2005	66,482,439	3.1
2006	69,222,975	4.1
2007	73,266,061	5.8
2008	70,520,493	-3.7
2009	61,791,868	-12.4
2010	77,857,705	26.0
2011	79,989,155	3.1
2012	84,141,209	5.3
2013	87,300,115	3.7
2014	??????	????

As Table 7-6 points, the number of vehicles produced every year is staggering. If we allow 15 years useful life for the majority of vehicles, with 10% reduction rate, then we have nearly a billion cars and trucks running on the world's roads. This number would double, if we add all other motor vehicles: motorcycles, scooters, buses, boats, trains, planes, etc. And, according to the experts, it will double again by 2030, if the present trend continues.

The motor vehicle industry employs millions of workers at its cars and trucks assembly facilities. Millions of additional people work at third party factories, where they manufacture different parts. These parts are shipped to the assembly facilities for final installation on the cars and trucks.

Nearly 5.0 million people in the U.S. are employed in some area of the motor vehicle industry. This includes vehicle parts manufacturing transport, assembly, retail, maintenance, etc. This number represents many times more people than those working in the entire power generation and distribution sector.

Now, here is another question, and one that ties the transportation sector and the automotive manufacturers even closer. It is the question of how much energy (in terms of electricity, fuels and lubricants) are used every year by the vehicles manufacturers? There are over 10,000 parts (minimum) per car, and each of these parts has to be manufactured, shipped, and installed. Each of these operations requires a lot of energy, but they are so many and so complex that we just cannot answer the question. Nevertheless, the amount of energy used is significant, and when added to the total energy (fuels and such) a car uses during its life cycle, we come with another set of staggering numbers.

Americans drive an average of 10,000-15,000 miles annually. This requires (another average) of 500-750 gallons per car per year. With about 70 million cars and trucks on the roads, we are looking at 35-50 billion gallons of fuel used annually by the U.S. transportation sector. This number represents about a billion barrels of oil that are burned in motor vehicles every year. And, of course, there are millions of additional commercial vehicles that use almost that much fuel too. Most of these fuels are derived from crude oil.

So how much energy are we talking about? A lot! Much more than what we could consider sustainable on the long run.

Fuel Transport

Fuel transportation includes moving all types of energy products. The energy products and fuels that are transported from place to place include:
- Coal
- Compressed and liquid natural gas
- Gasoline (regular, midgrade, premium, racing fuels)
- Diesel (LSD, ULSD)
- Fuel oil (type 0, 1, 2, 4, 5, 6)
- Kerosene (K1, K2, K3)
- Marine (distillate, residual)
- Aviation fuels (JP-4, JP-5, JP-8, Jet-A, Jet-A1, D 910, 100 LL, Grade 100, Grade 80)
- Liquid Petroleum Gas (LPG)
- Butane
- Bio-diesel
- Ethanol
- Lubricants
- Asphalt
- Oilfield products transport
- Refineries products transport

Coal and crude oil transport is a big business. See for yourself:

Coal Transport

After the coal is dug out from the ground and taken up to the surface, it is processed per consumer specifications of size, cleanliness, etc. before being shipped to the customer for burning. Transportation is a major step in the overall coal production-use process.

The coal transport system will depend on site-specific and project-specific factors and could be a conveyor system within the mine site to the coal preparation plant or a rail system. A system of haul roads is also likely to be present. Transporting coal off-site may be accomplished by rail, truck, barge, or some combination thereof. A coal-slurry pipeline also may be used to send coal off-site.

Usually, coal transport includes, a.) loading on railroad cars and moved to a central point for preparation and upgrading—cleaning, sorting, refining, etc., and b.) loading and shipping to a power plant where it is unloaded and burnt for electricity generation.

A brief look at the coal transportation systems shows that:

Coal mines are growing larger and their production increases exponentially. There were 316 mines in the mid-1970s with an annual output of above 500,000 tons. These large mining complexes execute all coal preparation operations onsite. From there the coal moves, usually by train, to the end use site—utility power plant or a large industrial consumer.

Most of the 4325 coal mines in the U.S., however, produce less than 50,000 tons per year. It is impractical, and/or economically unfeasible to process the coal at the smaller mines, so it is usually trucked to a central processing facility. Upon arrival it is unloaded, cleaned, sorted by size and otherwise processed to customer specs before shipping to the point of use (POU).

Most of the coal in the U.S. is shipped by "unit train." These are special train compositions, containing 100 or more coal hopper cars. The unit trains shuttle back and forth between large coal mines and coal-fired power plants non-stop, and sometimes on dedicated railroads lines

The typical coal train is 100-120 hopper cars long, with each of the cars holding 100-115 tons. This is almost a mile of coal, which can feed a large coal burning power plant operating about a day, or two maximum. The larger surface mines load two or three unit trains of coal a day.

There are approximately eighty trains leaving Wyoming mines every day, or about 26,000 trains annually. This is 26,000 miles of coal, or more than the circumference of the earth. If the unit trains from the other coal producing states are added, they can be wrapped around the Earth several times.

Figure 7-8. Unit trains

There is a trend of building "mine mouth" plants, in order to avoid the transportation costs. These power plants are built and operated right at, or very close to the mines, because the coal is cheap at the mine (about $5 per ton). The transportation makes 60-80% of the total cost of coal. Environmental and capital expense considerations, however, make this practice unprofitable in some locations, such as the West Coast.

There are several large complexes in the Four Corners area (where Colorado, New Mexico, Arizona, and Utah join) and in the Dakotas and Montana, which are good example of mine mouth plants.

Coal can be shipped by water—by barge on the nation's inland waterways, or by ocean freighters, to coal's export customers. A huge transoceanic coal export facility is located at Norfolk, Virginia. Coal is currently shipped also from sea ports of Baltimore, Philadelphia, New York City, New Orleans, and Los Angeles.

Coal can be shipped by another water method,

via *slurry pipelines*. In this method, a mixture of powdered coal and water is mixed at the mine and pumped through a long, large diameter, pipeline to the coal burning power plant. An example of that is/was the 273-mile-long pipeline from the Peabody mine on the "Black Mesa" in Arizona to the 1500-MW Mohave power plant near Page, Arizona. The Black Mesa slurry pipeline delivered about 8 tons of coal per minute along with 2700 gallons or 11 tons of water. At the plant the coal was dried and burned.

Note: This slurry pipeline is an example of lack of consideration to the natural resources and the local environment. Built on Indian land in the harshest and driest desert on Earth, it used 2700 gallons of water a minute, or close to 4 million gallons of water per day. This is 1.5 billion gallons—an entire lake—taken from one area of the desert and pumped to another, where it is discharged as waste water after the coal has been filtered out.

What brilliant mind conceived such a monstrosity? And where did they get these lakes of water to waste in the middle of one of the most arid deserts in the world? Did they think of the consequences?

Similar slurry pipeline are proposed for the North Central coal fields of the Dakotas, Montana, and Wyoming to feed the power plants of the Midwest (Chicago, St. Louis, etc.). Although studies show that slurry transport is less expensive than rail—after the pipeline is installed—environmental and economic controversies have so far frustrated this pipeline projects.

Coal Transportation Cost

The cost of transport via unit train today is approximately $0.020 per ton/mile, or 1,000 mile trip would cost approximately $20.00 per ton of coal transported from the mine to the power plant. This multiplied by 100 tons per car and 100 cars per unit train gives us the grand total of $200,000, which is the amount a large scale power plant pays *every day* for coal transport alone. Remember that at the mine coal started with a humble price of $5 per ton at some mines.

The transportation cost is even higher when considering other methods, such as barges, trucks, etc. This, compared with the cost of mining the coal, which is approximately $5-15 per ton in the U.S. (but much higher in other countries), is a true highway robbery. There is, however, no way around it; if you want it you have to ship it. Except in the case of "mine mouth" versions, where the power plant is located at the mine site, so the coal goes directly into the furnace to be converted into electricity.

There is a loss of power during the transmission of the generated electric power from the remote power station to the populated centers, but this is a small price to pay, compared with shipping millions of tons of coal across country.

Crude Oil Transport

Huge quantities of oil have to be transported from the point of production to the refineries, and then to the point of use. After the crude oil is pumped from the ground, the next stop is to transport it to a refinery where the desired products are produced. It gets to the refinery by pipeline typically, but the pipeline may either be connected to the wells or filled at a tanker port. And there are many other modes of transport as well.

The transport of oil is big business. The world tank ship fleet numbers more than 5000. On any given day as much as 750 million barrels of crude oil and products may be in tankers on the world's oceans. The U.S. pipeline system totals about 223,000 miles. From the refineries, gasoline, fuel oil, and so on, go by truck or train tank car to wholesalers or large consumers. The U.S. tank truck fleet numbers about 160,000, and the number of rail tank cars' is slightly larger, over 165,000.

And so, let's take a look at the different oil transport methods; pipelines, marine vessels, tank trucks, rail tank cars and so forth are used to transport crude oils, compressed and liquefied hydrocarbon gases, liquid petroleum products and other chemicals from their point of origin to pipeline terminals, refineries, distributors and consumers.

We'll take a closer look at the main components of this important chapter of the oil's cradle-to-grave life cycle.

Pipelines

At one time or another all crude oils, natural gas, liquefied natural gas, liquefied petroleum gas (LPG) and petroleum products flow through some type and length of a pipelines, at one or another point of their migration from the well to a refinery, gas plant, a terminal, or a consumer.

There are aboveground, underwater and underground pipelines, varying in size from several inches to a several feet in diameter. These move vast amounts of crude oil, natural gas, LHGs and liquid petroleum products around the US and most other countries.

The first successful crude-oil pipeline, a 2 inch diameter wrought iron pipe, 6 miles long with a capacity of about 800 barrels a day, was built in Pennsylvania in 1865. During WWII large pipe networks were built in the US, as needed to move oil from coast to coast. A large pipeline project, planned to deliver Canadian syncrude oil to the Gulf Coast Refineries is stalled in the permitting and political debate stages.

Figure 7-9. The Alaska pipeline

The 48 states receive 1.5 to 1.6 million barrels of oil per day, courtesy of the Alaskan pipeline. This pipeline, 789 miles of 48-inch steel pipe, curving up and down and around hills and valley, carries great quantities of oil from the rich North Slope field to the port of Valdez in southern Alaska. Here the oil is loaded on tankers for shipment south. The Alaskan pipeline, was built in controversy, which is still unresolved and is a subject of high level debates and legal battles.

Today liquid petroleum products are moved long distances through pipelines at speeds of up to 6 miles per hour, assisted by large pumps and compressors along the way at intervals ranging from 60 miles to 200 miles. Pipeline pumping pressures and flow rates are controlled throughout the system to maintain a constant movement of product within the pipeline. Sensors, control mechanism and safety devices are installed along the length of the major pipelines to prevent spills, fires, etc. disasters.

Pipelines run from the frozen tundra of Alaska and Siberia to the hot deserts of the Middle East, across rivers, lakes, seas, swamps and forests, over and through mountains and under cities and towns. There are over 95,000 mile network of petroleum product pipelines in the United States. This network delivers finished petroleum products to the end customers. It is separate from the network of crude oil pipeline, and balances the demand and supply conditions in each region.

The initial construction of pipelines is difficult and

expensive, but once they are built, properly maintained and operated, they provide one of the safest and most economical means of transporting these products.

Types of Pipelines

There are four basic types of pipelines in the oil and gas industry; flow lines, gathering lines, crude trunk pipelines and petroleum product trunk pipelines.

Flow lines move crude oil or natural gas from producing wells to producing field storage tanks and reservoirs. Flow lines may vary in size from 5 cm in diameter in older, lower-pressure fields with only a few wells, to much larger lines in multi-well, high-pressure fields. Offshore platforms use flow lines to move crude and gas from wells to the platform storage and loading facility. A lease line is a type of flow line which carries all of the oil produced on a single lease to a storage tank.

Gathering and feeder lines collect oil and gas from several locations for delivery to central accumulating points, such as from field crude oil tanks and gas plants to marine docks. Feeder lines collect oil and gas from several locations for delivery direct into trunk lines, such as moving crude oil from offshore platforms to onshore crude trunk pipelines. Gathering lines and feeder lines are typically larger in diameter than flow lines.

Crude trunk pipelines move natural gas and crude oil along long distances from producing areas or marine docks to refineries and from refineries to storage and distribution facilities by 1- to 3-m-diameter or larger trunk pipelines.

Petroleum product trunk pipelines move liquid petroleum products such as gasoline and fuel oil from refineries to terminals, and from marine and pipeline terminals to distribution terminals. Product pipelines may also distribute products from terminals to bulk plants and consumer storage facilities, and occasionally from refineries direct to consumers. Product pipelines are used to move LPG from refineries to distributor storage facilities or large industrial users.

Batch Shipments and Interface

Although pipelines originally were used to move only crude oil, they evolved into carrying all types and different grades of liquid petroleum products. Because petroleum products are transported in pipelines by batches, in succession, there is commingling or mixing of the products at the interfaces. The product intermix is controlled by one of three methods: downgrading (derating), using liquid and solid spacers for separation or reprocessing the intermix.

Radioactive tracers, color dyes and spacers may be placed into the pipeline to identify where the interfaces occur. Radioactive sensors, visual observation or gravity tests are conducted at the receiving facility to identify different pipeline batches.

Petroleum products are normally transported through pipelines in batch sequences with compatible crude oils or products adjoining one another. One method of maintaining product quality and integrity, downgrading or derating, is accomplished by lowering the interface between the two batches to the level of the least affected product. For example, a batch of high-octane premium gasoline is typically shipped immediately before or after a batch of lower-octane regular gasoline. The small quantity of the two products which has intermixed will be downgraded to the lower octane rating regular gasoline.

When shipping gasoline before or after diesel fuel, a small amount of diesel interface is allowed to blend into the gasoline, rather than blending gasoline into the diesel fuel, which could lower its flashpoint. Batch interfaces are typically detected by visual observation, gravitometers or sampling.

Liquid and solid spacers or cleaning pigs may be used to physically separate and identify different batches of products. The solid spacers are detected by a radioactive signal and diverted from the pipeline into a special receiver at the terminal when the batch changes from one product to another. Liquid separators may be water or another product that does not commingle with either of the batches it is separating and is later removed and reprocessed. Kerosene, which is downgraded (derated) to another product in storage or is recycled, can also be used to separate batches.

A third method of controlling the interface, often used at the refinery ends of pipelines, is to return the interface to be reprocessed. Products and interfaces which have been contaminated with water may also be returned for reprocessing.

Marine Transport

The majority of the world's crude oil is transported by tankers from producing areas such as the Middle East and Africa to refineries in consumer areas such as Europe, Japan and the United States. Oil products were originally transported in large barrels on cargo ships. The first tanker ship, which was built in 1886, carried about 2,300 SDWT (2,240 pounds per ton) of oil. Today's supertankers can be over 300 m long and carry almost 200 times as much oil.

Gathering and feeder pipelines often end at marine terminals or offshore platform loading facilities, where the crude oil is loaded into tankers or barges for transport to crude trunk pipelines or refineries. Petroleum products also are transported from refineries to distribution terminals by tanker and barge. After delivering their cargoes, the vessels return in ballast to loading facilities to repeat the sequence.

Marine Vessels

Oil tankers and barges are vessels designed with the engines and quarters at the rear of the vessel and the remainder of the vessel divided into special compartments (tanks) to carry crude oil and liquid petroleum products in bulk. Cargo pumps are located in pump rooms, and forced ventilation and inerting systems are provided to reduce the risk of fires and explosions in pump rooms and cargo compartments. Modern oil tankers and barges are built with double hulls and other protective and safety features required by the United States Oil Pollution Act of 1990 and the International Maritime Organization (IMO) tanker safety standards. Some new ship designs extend double hulls up the sides of the tankers to provide additional protection. Generally, large tankers carry crude oil and small tankers and barges carry petroleum products.

Oil tankers are smaller vessels, which in addition to ocean travel can navigate restricted passages such as the Suez and Panama Canals, shallow coastal waters and estuaries. Large oil tankers, which range from 25,000 to 160,000 SDWTs, usually carry crude oil or heavy residual products. Smaller oil tankers, under 25,000 SDWT, usually carry gasoline, fuel oils and lubricants.

Barges carrying oil products operate mainly in coastal and inland waterways and rivers, alone or in groups of two or more, and are either self-propelled or moved by tugboat. They may carry crude oil to refineries, but more often are used as an inexpensive means of transporting petroleum products from refineries to distribution terminals. Barges are also used to off-load cargo from tankers offshore whose draft or size does not allow them to come to the dock.

Supertankers are the least expensive way for long-distance shipment of oil and oil products. These modern ocean-going vessels are huge floating oil tanks that make their slow cumbersome way from the giant oil fields of the Mideast to ports in the industrial countries. But there are no supertanker ports in this country, so they are usually unloaded off-shore into smaller tankers. Supertankers are huge. The largest ones presently in service have a DWT (deadweight tonnage, i.e., cargo and fuel capacity) of more than 546,000 tons. They are up to 400 meters long (the length of about five football fields laid end to end) and can carry about 40 million barrels of oil.

Ultra-large and very large crude carriers (ULCCs and VLCCs) are restricted by their size and draft to specific routes of travel. ULCCs are vessels whose capacity is over 300,000 SDWTs, and VLCCs have capacities ranging from 160,000 to 300,000 SDWTs. Most large crude carriers are not owned by oil companies, but are chartered from transportation companies which specialize in operating these super-sized vessels.

Their "draft" or how deep they set in the water is very important for their navigation. A fully loaded draft of over 90 feet prevents such a large ship from going into any of the U.S. ports. Our deepest port, Los Angeles, cannot handle ships of more than 100,000 deadweight tons. As a result supertankers now unload offshore in the Caribbean and transfer their cargo to smaller tankers for delivery to the United States.

This off-shore transfer method costs a lot of additional energy and money and for this purpose an alternative LOOP facility was built. *LOOP* is The Louisiana Offshore Oil Port (LOOP), which is a deepwater port in the Gulf of Mexico off the coast of Louisiana, near the town of Port Fourchon. It provides offloading and temporary storage services for crude oil transported on some of the largest tankers in the world, since most of them cannot enter US ports.

LOOP presently handles 13% of the nation's oil imports, or about 1.2 million barrels a day. It is connects by a pipeline to the refineries, thus feeding 50% of the U.S. refining capability.

Tankers offload at LOOP by pumping crude oil through hoses connected to a Single Point Mooring (SPM) base. Three SPMs are located 8,000 feet from the Marine Terminal. The SPMs are designed to handle ships of up to 700,000 deadweight tons. The crude oil then moves to the Marine Terminal via a 56-inch diameter submarine pipeline.

The Marine Terminal consists of a control platform and a pumping platform. The control platform is equipped with a helicopter pad, living quarters, control room, vessel traffic control station, offices and life support equipment. The pumping platform contains four 7,000-hp pumps, power generators, metering and laboratory facilities. Crude oil is only handled on the pumping platform where it is measured, sampled, and boosted to shore via a 48-inch diameter pipeline.

Railroad Tank Cars

Railroad tank cars are constructed of carbon steel or aluminium and may be pressurized or unpressurized. Modern tank cars can hold up to 171,000 liters of compressed gas at pressures up to 600 psi. Non-pressure tank cars have evolved from small wooden tank cars of the late 1800s to jumbo tank cars which transport as much as 1.31 million liters of product at pressures up to 100 psi.

Non-pressure tank cars may be individual units with one or multiple compartments or a string of interconnected tank cars, called a tank train. Tank cars are loaded individually, and entire tank trains can be loaded and unloaded from a single point. Both pressure and non-pressure tank cars may be heated, cooled, insulated and thermally protected against fire, depending on their service and the products transported.

All railroad tank cars have top- or bottom-liquid or vapor valves for loading and unloading and hatch entries for cleaning. They are also equipped with devices intended to prevent the increase of internal pressure when exposed to abnormal conditions. These devices include safety relief valves held in place by a spring which can open to relieve pressure and then close; safety vents with rupture discs that burst open to relieve pressure but cannot reclose; or a combination of the two devices.

A vacuum relief valve is provided for non-pressure tank cars to prevent vacuum formation when unloading from the bottom. Both pressure and non-pressure tank cars have protective housings on top surrounding the loading connections, sample lines, thermometer wells and gauging devices. Platforms for loaders may or may not be provided on top of cars.

Older non-pressure tank cars may have one or more expansion domes. Fittings are provided on the bottom of tank cars for unloading or cleaning. Head shields are provided on the ends of tank cars to prevent puncture of the shell by the coupler of another car during derailments.

Tank Trucks

Petroleum products and crude oil tank trucks are typically constructed of carbon steel, aluminium or a plasticized fiberglass material, and vary in size from 1,900-l tank wagons to jumbo 53,200-l tankers. The capacity of tank trucks is governed by regulatory agencies, and usually is dependent upon highway and bridge capacity limitations and the allowable weight per axle or total amount of product allowed.

There are pressurized and non-pressurized tank trucks, which may be non-insulated or insulated depending on their service and the products transported. Pressurized tank trucks are usually single compartment, and non-pressurized tank trucks may have single or multiple compartments. Regardless of the number of compartments on a tank truck, each compartment must be treated individually, with its own loading, unloading and safety-relief devices. Compartments may be separated by single or double walls. Regulations may require that incompatible products and flammable and combustible liquids carried in different compartments on the same vehicle be separated by double walls. When pressure testing compartments, the space between the walls should also be tested for liquid or vapor.

Tank trucks have either hatches which open for top loading, valves for closed top- or bottom-loading and unloading, or both. All compartments have hatch entries for cleaning and are equipped with safety relief devices to mitigate internal pressure when exposed to abnormal conditions. These devices include safety relief valves held in place by a spring which can open to relieve pressure and then close, hatches on non-pressure tanks which pop open if the relief valves fail and rupture discs on pressurized tank trucks.

A vacuum relief valve is provided for each non-pressurized tank truck compartment to prevent vacuum when unloading from the bottom. Non-pressurized tank trucks have railings on top to protect the hatches, relief valves and vapor recovery system in case of a rollover. Tank trucks are usually equipped with breakaway, self-closing devices installed on compartment bottom loading and unloading pipes and fittings to prevent spills in case of damage in a rollover or collision.

Transport Fuel Pricing

Gasoline and diesel fuel prices at the pumps are calculated on the basis of the following components:

- Crude Oil prices determine the monthly average of the composite refiner acquisition cost, which is the average price of crude oil purchased by refiners. Price of crude oil is the main contributing factors to the substantial changes in gasoline prices the U.S. has experienced in recent years. Crude oil prices are affected by levels of supply relative to actual and expected demand for the petroleum products made from crude oil.

Crude oil prices contribute about 62% to the total fuel price at the pumps.

- Distribution and Marketing costs and profits are the difference between the average retail price of gasoline or diesel fuel as computed from EIA's weekly survey and the sum of the other 3 components.

Distribution and marketing contribute about 17% to the total fuel price at the pump.

- Taxes are added according to a monthly national average of federal and state taxes applied to gasoline or diesel fuel.

Taxes contribute about 15% to the total fuel price at the pump.

- Refining costs and profits reflect the difference between the monthly average of the spot price of gasoline or diesel fuel (used as a proxy for the value of gasoline or diesel fuel as it exits the refinery) and the average price of crude oil purchased by refiners (the crude oil component).

Refining contributes about 6% to the total fuel price at the pump.

Note: All of the above factors vary from time to time, but the refining and distribution components are most unpredictable. They depend fully on the time when they are being calculated. Since there is typically a lag between when the spot price changes to when the retail price changes, the refining costs & profits component and the distribution & marketing costs & profits component can vary from month to month.

For example, as prices increase on the spot market, often the retail prices take time to adjust. Thus, at this point in the cycle, the refining costs & profits component (assuming no corresponding increase in crude oil prices) would be relatively large while the distribution & marketing costs & profits component would be relatively small.

However, later on, as retail prices "catch-up" with the previous spot price increases, the distribution & marketing costs & profits component would increase while the refining costs & profits component would decrease. Quick and frequent changes would introduce havoc in the system, which increases the unpredictability of fuel pricing.

All things considered, 2014 was an year full of surprises in the energy sector. The transportation fuels—starting with crude oil in particular—were the biggest surprise.

Note: Fuel price data is collected by EIA daily, using Form 878, "Motor Gasoline Price Survey." The survey is designed to collect and publish data on the cash price (including taxes) of self-serve, unleaded gasoline, by grade of gasoline. The data are used to calculate average gasoline prices at the national, regional, and select State and city levels across all gasoline grades and formulations

The information on U.S. fuel prices is collected every Monday, by calling by telephone a sample of approximately 800 retail gasoline outlets. The prices of all fuel types and grades are published around 5:00 p.m. ET every Monday, except on government holidays. The data is released on Tuesday morning, but still reflects Monday's prices.

The reported price includes all taxes and is the pump price paid by a consumer as of 8:00 A.M. Monday. This price represents the *self-serve* price except in areas having only full-serve. The price data are used to calculate weighted average price estimates at the city, state, regional and national levels using sales and delivery volume data from other EIA surveys and population estimates from the Bureau of Census.

Table 7-7 shows the gradual increase in prices of all fuels from 1990 to 2000, the steep increase by 2014 and the sudden drop in prices in 2015. The gradual increase until 2014 reflects the gradually changing status of the global energy markets. We could call this increase "normal," if there was anything normal in the energy markets.

Table 7-7. U.S. fuel prices 1990-2015

Date	1990	2000	2014	2015
Regular	1.20	1.40	3.40	1.90
Mid-grade	1.25	1.50	3.70	2.00
Premium	1.30	1.60	3.90	2.10
Diesel	1.10	1.30	4.20	3.10

The sudden drop of prices in 2015, however, is anything but normal. It reminds us of the sudden change in prices in 1973—except that in 1973 the prices went the other way. In both cases, however, the changes were driven by Saudi Arabia.

The greatest mystery here is in the fact that while in both cases, the emirate reduced the amount of oil it produces, the results were opposite. The global fuel price increase in 1973 could be attributed to a large extent to the unpreparedness of the U.S. and most other countries to handle energy problems. The sudden decrease in fuel availability caused mass panic, which resulted in long gas pump lines and drastic price increases.

In contrast, the reduced oil production by Saudi Arabia in 2014 was designed to hurt the oil and gas producers in the U.S. and other producing countries. It achieved that goal, but the domestic fuel prices stayed low, because the oil imports prices were low.

The long term lesson here is that we must never sleep, and should always watch carefully what our "partners" in the energy markets do today and plan to do tomorrow.

Crude Oil Storage

Crude oil is stored in thousands of facilities around the country. From small storage sites at processing and delivery facilities, to huge storage facilities at government installations. They all use specially designed steel storage tanks, equipped to handle all types of weather and emergency situations.

There are a number of different types of vertical and horizontal aboveground atmospheric and pressure storage tanks in tank farms, which contain crude oil, petroleum feedstocks, intermediate stocks or finished petroleum products. Their size, shape, design, configuration, and operation depend on the amount and type of products stored and company or regulatory requirements.

- Aboveground vertical tanks may be provided with double bottoms to prevent leakage onto the ground and cathodic protection to minimize corrosion.

- Horizontal tanks may be constructed with double walls or placed in vaults to contain any leakage.

Types of Storage Tanks

There are a number of different types and sizes of oil storage tanks. The major types are

Atmospheric cone roof tanks are aboveground, horizontal or vertical, covered, cylindrical atmospheric vessels. Cone roof tanks have external stairways or ladders and platforms, and weak roof to shell seams, vents, scuppers or overflow outlets; they may have appurtenances such as gauging tubes, foam piping and chambers, overflow sensing and signaling systems, automatic gauging systems and so on.

When volatile crude oil and flammable liquid petroleum products are stored in cone roof tanks there is an opportunity for the vapor space to be within the flammable range. Although the space between the top of the product and the tank roof is normally vapor rich, an atmosphere in the flammable range can occur when product is first put into an empty tank or as air enters the tank through vents or pressure/vacuum valves when product is withdrawn and as the tank breathes during temperature changes. Cone roof tanks may be connected to vapor recovery systems.

Conservation tanks are a type of cone roof tank with an upper and lower section separated by a flexible mem-

brane designed to contain any vapor produced when the product warms up and expands due to exposure to sunlight in the daytime and to return the vapor to the tank when it condenses as the tank cools down at night. Conservation tanks are typically used to store aviation gasoline and similar products.

Atmospheric floating roof tanks are aboveground, vertical, open top or covered cylindrical atmospheric vessels that are equipped with floating roofs. The primary purpose of the floating roof is to minimize the vapor space between the top of the product and the bottom of the floating roof so that it is always vapor rich, thus precluding the chance of a vapor-air mixture in the flammable range. All floating roof tanks have external stairways or ladders and platforms, adjustable stairways or ladders for access to the floating roof from the platform, and may have appurtenances such as shunts which electrically bond the roof to the shell, gauging tubes, foam piping and chambers, overflow sensing and signaling systems, automatic gauging systems and so on. Seals or boots are provided around the perimeter of floating roofs to prevent product or vapor from escaping and collecting on the roof or in the space above the roof.

Floating roofs are provided with legs which may be set in high or low positions depending on the type of operation. Legs are normally maintained in the low position so that the greatest possible amount of product can be withdrawn from the tank without creating a vapor space between the top of the product and the bottom of the floating roof. As tanks are brought out of service prior to entry for inspection, maintenance, repair or cleaning, there is a need to adjust the roof legs into the high position to allow room to work under the roof once the tank is empty. When the tank is returned to service, the legs are readjusted into the low position after it is filled with product.

Aboveground floating roof storage tanks are further classified as external floating roof tanks, internal floating roof tanks or covered external floating roof tanks.

External (open top) floating roof tanks are those with floating covers installed on open-top storage tanks. External floating roofs are usually constructed of steel and provided with pontoons or other means of flotation. They are equipped with roof drains to remove water, boots or seals to prevent vapor releases and adjustable stairways to reach the roof from the top of the tank regardless of its position. They may also have secondary seals to minimize release of vapor to the atmosphere, weather shields to protect the seals and foam dams to contain foam in the seal area in case of a fire or seal leak. Entry onto external floating roofs for gauging, maintenance or other activities may be considered confined-space entry, depending on the level of the roof below the top of the tank, the products contained in the tank and government regulations and company policy.

Internal floating roof tanks usually are cone roof tanks which have been converted by installing buoyant decks, rafts or internal floating covers inside the tank. Internal floating roofs are typically constructed of various types of sheet metal, aluminum, plastic or metal-covered plastic expanded foam, and their construction may be of the pontoon or pan type, solid buoyant material, or a combination of these. Internal floating roofs are provided with perimeter seals to prevent vapor from escaping into the portion of the tank between the top of the floating roof and the exterior roof. Pressure/vacuum valves or vents are usually provided at the top of the tank to control any hydrocarbon vapors which may accumulate in the space above the internal floater. Internal floating roof tanks have ladders installed for access from the cone roof to the floating roof. Entry onto internal floating roofs for any purpose should be considered confined-space entry.

Covered (external) floating roof tanks are basically external floating roof tanks that have been retrofitted with a geodesic dome, snow cap or similar semi-fixed cover or roof so that the floating roof is no longer open to the atmosphere. Newly constructed covered external floating roof tanks may incorporate typical floating roofs designed for internal floating roof tanks. Entry into covered external floating roofs for gauging, maintenance or other activities may be considered confined-space entry, depending on the construction of the dome or cover, the level of the roof below the top of the tank, the products contained in the tank and government regulations and company policy.

Tank farms

Tank farms are groupings of storage tanks at producing fields, refineries, marine, pipeline and distribution terminals and bulk plants which store crude oil and petroleum products. Within tank farms, individual tanks or groups of two or more tanks are usually surrounded by enclosures called berms, dykes or fire walls. These tank farm enclosures may vary in construction and height, from 45-cm earth berms around piping and pumps inside dykes to concrete walls that are taller than the tanks they surround.

Dykes may be built of earth, clay or other materials; they are covered with gravel, limestone or sea shells to control erosion; they vary in height and are wide enough for vehicles to drive along the top. The primary functions of these enclosures are to contain, direct and divert rain water, physically separate tanks to prevent the spread of fire in one area to another, and to contain a spill, release, leak or overflow from a tank, pump or pipe within the area.

Dyke enclosures may be required by regulation or company policy to be sized and maintained to hold a specific amount of product. For example, a dyke enclosure may need to contain at least 110% of the capacity of the largest tank therein, allowing for the volume displaced by the other tanks and the amount of product remaining in the largest tank after hydrostatic equilibrium is reached. Dyke enclosures may also be required to be constructed with impervious clay or plastic liners to prevent spilled or released product from contaminating soil or groundwater.

The U.S. Transportation Market

The transportation market basically consists of the supply and use of vehicle fuels of all sorts. Most of these, however, are fossil-based, so we will focus on them herein.

Vehicle Fuels
- Energy source procurement
  — Crude oil and natural gas wells
  — Renewables
- Energy source processing
  — Refineries construction
  — Equipment and services
  — Operations and maintenance
- Fuels transport
  — Pipelines
  — Trains, trucks, barges
- Fuels distribution and sales
  — Commercial fueling
  — Retail distribution
- Risks and solutions

And, of course, there are the indirect products and services that are also part—albeit indirectly—of the vehicle fuels markets. They affect the function of the transportation sector in a number of ways, as we will see below in this text.

The indirect products and services are:
- Fossil fuels and petrochemicals production
- Unconventional energy sources and fuels
- Equipment (and vehicles) manufacturing
- Energy market transactions
- Financial Aspects
- Political and Regulatory Aspects
- Legal Aspects

- Environmental Aspects
- Future Energy Concerns

U.S. Transportation Sector

According to the U.S. EPA, the U.S. transportation sector includes air, ground (non-automobile), water transportation and transportation cleaning equipment. The sector is now classified under the 2007 North American Industry Classification System (NAICS) as:

Air Transportation (EPA Section 481)

Industries in the Air Transportation subsector provide air transportation of passengers and/or cargo using aircraft, such as airplanes and helicopters. This subsector distinguishes scheduled from nonscheduled air transportation. Scheduled air carriers fly regular routes on regular schedules and operate even if flights are only partially loaded. Nonscheduled carriers often operate during nonpeak time slots at busy airports. These establishments have more flexibility with respect to choice of airport, hours of operation, load factors, and similar operational characteristics. Nonscheduled carriers provide chartered air transportation of passengers, cargo, or specialty flying services. Specialty flying services establishments use general-purpose aircraft to provide a variety of specialized flying services.

Scenic and sightseeing air transportation and air courier services are not included in this subsector but are included in a special Subsector 487, Scenic and Sightseeing Transportation and in Subsector 492, Couriers and Messengers.

Although these activities may use aircraft, they are different from the activities included in air transportation. Air sightseeing does not usually involve place-to-place transportation; the passengers flight (e.g., balloon ride, aerial sightseeing) typically starts and ends at the same location. Courier services (individual package or cargo delivery) include more than air transportation; road transportation is usually required to deliver the cargo to the intended recipient.

Rail Transportation (EPA Section 482)

Industries in the Rail Transportation subsector provide rail transportation of passengers and/or cargo using railroad rolling stock. The railroads in this subsector primarily either operate on networks, with physical facilities, labor force, and equipment spread over an extensive geographic area, or operate over a short distance on a local rail line.

Scenic and sightseeing rail transportation and street railroads, commuter rail, and rapid transit are not

included in this subsector but are included in Subsector 487, Scenic and Sightseeing Transportation, and Subsector 485, Transit and Ground Passenger Transportation, respectively.

Although these activities use railroad rolling stock, they are different from the activities included in rail transportation. Sightseeing and scenic railroads do not usually involve place-to-place transportation; the passengers trip typically starts and ends at the same location. Commuter railroads operate in a manner more consistent with local and urban transit and are often part of integrated transit systems.

Water Transportation (EPA Section 483)

Industries in the Water Transportation subsector provide water transportation of passengers and cargo using watercraft, such as ships, barges, and boats. The subsector is composed of two industry groups: (1) one for deep sea, coastal, and Great Lakes; and (2) one for inland water transportation. This split typically reflects the difference in equipment used.

Scenic and sightseeing water transportation services are not included in this subsector but are included in Subsector 487, Scenic and Sightseeing Transportation. Although these activities use watercraft, they are different from the activities included in water transportation. Water sightseeing does not usually involve place-to-place transportation; the passengers trip starts and ends at the same location.

Truck Transportation (EPA Section 484)

Industries in the Truck Transportation subsector provide over-the-road transportation of cargo using motor vehicles, such as trucks and tractor trailers. The subsector is subdivided into general freight trucking and specialized freight trucking. This distinction reflects differences in equipment used, type of load carried, scheduling, terminal, and other networking services. General freight transportation establishments handle a wide variety of general commodities, generally palletized, and transported in a container or van trailer. Specialized freight transportation is the transportation of cargo that, because of size, weight, shape, or other inherent characteristics require specialized equipment for transportation.

Each of these industry groups is further subdivided based on distance traveled. Local trucking establishments primarily carry goods within a single metropolitan area and its adjacent nonurban areas. Long distance trucking establishments carry goods between metropolitan areas.

The Specialized Freight Trucking industry group includes a separate industry for Used Household and Office Goods Moving. The household and office goods movers are separated because of the substantial network of establishments that has developed to deal with local and long-distance moving and the associated storage. In this area, the same establishment provides both local and long-distance services, while other specialized freight establishments generally limit their services to either local or long-distance hauling.

Transit and Ground Passenger Transportation (EPA Section 485)

Industries in the Transit and Ground Passenger Transportation subsector include a variety of passenger transportation activities, such as urban transit systems; chartered bus, school bus, and interurban bus transportation; and taxis. These activities are distinguished based primarily on such production process factors as vehicle types, routes, and schedules.

In this subsector, the principal splits identify scheduled transportation as separate from nonscheduled transportation. The scheduled transportation industry groups are Urban Transit Systems, Interurban and Rural Bus Transportation, and School and Employee Bus Transportation. The nonscheduled industry groups are the Charter Bus Industry and Taxi and Limousine Service. The Other Transit and Ground Passenger Transportation industry group includes both scheduled and nonscheduled transportation.

Scenic and sightseeing ground transportation services are not included in this subsector but are included in Subsector 487, Scenic and Sightseeing Transportation. Sightseeing does not usually involve place-to-place transportation; the passengers trip starts and ends at the same location.

How many vehicles are we talking about here? Many millions, to be sure! And this does not even include the private mopeds, motorcycles, cars, SUVs, trucks, vans, off-road, racing, and many other types of ground, water, and air vehicles Americans use daily. Nor does it include millions of private business vehicles; rental cars, executive cars, ambulances, industrial self propelled equipment, etc.

The U.S. has largest passenger vehicle market of any country in the world. There are over 275 million passenger vehicles registered in the country.

This number has increased steadily every year since 1960, indicating a growing number of vehicles per capita. All vehicles on the U.S. roads put together come to over one vehicle registered to every American.

The number of commercial and business vehicles, combined with these used for special purposes (military, police, emergency services, etc.) is another large number, so we are looking at a huge number of vehicles that move many miles every day on their different tasks and destinations. They all have one thing in common; their thirsty engines need a lot of crude oil based fuels every day. Without it, the army of vehicles will just stop.

The transportation sector consists of billions of tons of metal, plastics, and glass that move and operate as needed to provide us with all we need in our daily lives. And all, or most, of this movement requires crude oil and its derivatives; gasoline, diesel, jet fuels, different mixes, and lubricants. No oil, no movement. The consequences of immobility would be awesome and in some cases, life threatening.

The fuel we depend so much on come from Saudi Arabia, Iraq, Venezuela, Nigeria, and several other similarly unstable and/or unfriendly countries around the world.

So, since we import half of the crude oil we use, at least half of our transportation sector depends totally on imported fuels coming from shaky political regimes. In case of a problem overseas, at least half of our transport network would stop moving. Which one would it be? There would be enough for the military, police, and emergency services vehicles, but how about the rest? How would we proceed with our daily life?

One good thing to consider in such a case would be that if we don't have enough crude oil to run the millions and billions of vehicles around the world, the global environment would take a deep breath of clean air for a first time since the 1950s. Good for the environment, but devastating to our energy and national security, with the world economy taking a great hit.

U.S. Military Transport

The U.S. military is the world's largest transportation agency. The U.S. Army has a special division, Transportation Corps, which manages the transport of thousands of troops and provides logistics and other support for thousands of vehicles on land, water, and air. It is a Force Sustainment branch of the U.S. Army headquartered at Fort Lee, Virginia. It is responsible for the movement of personnel and materiel by truck, rail, air, and sea. The Transportation Corps provides a full spectrum of transportation capabilities at the tactical, operational, and strategic levels of war.

The Transportation School trains soldiers how to operate and maintain Army tactical trucks, material handling equipment and watercraft. The school trains Transportation Operations, Traffic Management, Convoy Operations, Cargo Transfer, Cargo Documentation, Movement Control, Operation of Heavy Material Handling Equipment, Sailing and Maintenance of Army Vessels, and Unloading Aircraft, Ships, Railcars and Trucks.

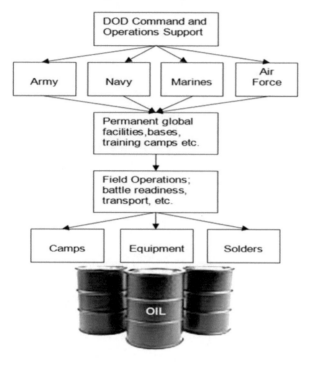

Figure 7-10. The U.S. military machine

The U.S. Army is huge business, that uses millions of gallons of fossil fuels and lubricants. It is one of the largest consumers of fuel in the country, using over 100 million barrels crude oil annually. This amount cannot be reduced, since F-22 jets, M1A2 Abrams tanks, and all other heavy equipment cannot run easily on renewables.

DoD is one of the largest consumers of energy in the world too and uses more crude oil than most countries in this world. The military uses 93% of all U.S. government energy, at the cost of $4.0 billion/year, as needed to power its 300,000 bases and facilities worldwide

The military is involved with, and responsible for, many different technologies and operational strategies. As huge and important as it is, all its technologies and strategies depend fully on, and are driven 100% by—crude oil.

We take a very close look at the U.S. military use of energy sources in our book titled, *Energy Security for the 21st Century*, published by The Fairmont Press in 2015.

So it suffices to say here that the U.S. military depends 100% on crude oil for its field operations, and 90% on other fossils for power generation at its bases and training facilities.

The remarkable scenario, 'no crude oil, no movement' says it all.

In such case, all vehicles sit still, and all military operations are cancelled. How ridiculous and dangerous this is, remains to be figured out, but it is a scary picture. This alone should make us consider the alternatives and look for ways to reduce the use of fossils and crude oil in particular.

We must give credit to the U.S. military in this respect, because they have implemented a number of programs efficiency programs intended to reduce fossil energy use—including crude oil. Reducing crude oil use in any significant numbers, however, would be very difficult, if not impossible for the foreseeable future.

Fuel Subsidies and Taxes

Fossil and renewable fuels have benefitted from influx of enormous subsidies through the years, which continue presently as well. On the other hand, the fuels have been exposed to taxation that varies from type to type and from location to location.

The problem is that there is lack of clarity on the subject, because the entire financial aspect of the different fuels is veiled in thin layers of foggy information or lack of information.

Fuels Subsidies

Recent estimates of the direct and indirect fossil fuel subsidies in the U.S. range from $14 to $50 billion annually. This includes military spending on defending the international oil routes. Fair? Without that spending for patrolling the world's oil routes, we might find ourselves in an oil conundrum.

The breakdown of overall energy related U.S. government subsidies of several years ending at 2008 was as follow:

• $72.5 billion of subsidies were allocated to the fossil industry, with $70.2 billion going to the traditional fossil fuels, and $2.3 billion spent on carbon capture and storage,

$16.8 billion of fuel subsidies were allocated to corn ethanol development, production, and use projects, and

$29.0 billion of fuel subsidies were allocated to the

renewable energy sources, with $12.2 billion spent on the traditional renewable energy sources, and the rest used for R&D of new technologies.

The amount of government fossil fuels subsidies around the globe in 2012 alone was an astounding $775 billion (yes, with a B).

The breakdown of fossil fuels subsidies around the globe is as follow:
- $630 billion of fossil fuel subsidies were allocated to consumption subsidies in the developing countries alone. This number fluctuates widely with the price of oil from $409 billion in 2010, and $557 billion in 2008.
- $45 billion of fossil fuel subsidies were allocated to consumption subsidies in the developed countries, which is considered to be an average annual consumption subsidies,
- $100 Billion of fossil fuel subsidies were allocated to the global producers, as cited in the June 2010 Report for G-20 leaders from OECD, IEA, World Bank and OPEC. Here again, lack of total transparency is hindering getting a clear picture of the fossils subsidies.

In 2009, G20 leaders pledged to phase out fossil fuel subsidies, which was an historic commitment. Put on paper this was a great beginning, but which did not go very far, partially because it is not clear how much money the governments provide in fossil fuel subsidies. But the major problem is that politicians change their minds with the political winds, so whatever is promised today is no longer valid tomorrow…usually justified by unforeseen change in economic and political conditions.

Efforts to remove even small portions of those subsidies are constantly defeated in Congress, because it is generally assumed that keeping fuel prices low by subsidizes is beneficial to the economic development of the country. The low fuel price is credited with increasing employment, invigorating the general business activities, reducing property values, and increasing tax revenues. Another benefit credited to low fuel prices is helping the poor people, as an extra benefits that offsets some of the external costs.

These assumptions, however, are very often wrong. No doubt, that low fuel prices and subsidies do stimulate some—mostly fuel related—activities, such as fuel purchases and transport. These gains are small comparatively speaking, and the fuel subsidies often cause greater economic harm by:

a) Transferring wealth from fuel consumers to producers, since the producers benefit from receiving payment for their regular prices, while the consumers are eventually paying the price of the subsidies. And by

b) Reducing the overall transport system efficiency, which often lead to traffic congestion, accidents, higher road expenses, sprawl, and pollution emissions increases, as compared with higher fuel price conditions.

Low fuel prices also cause increased vehicle travel which we should remember increases among the wealthy and much less in urban areas. Adding up to 5,000 annual vehicle-miles per person of the wealthy class only increases some of the above mentioned issues, and does not do anything to help the economy, especially that of the urban areas.

Economic benefits increase with higher fuel prices in some cases, particularly in oil consuming regions, where the fuels are imported. This is because higher fuel prices encourage people and businesses to think and operate outside the box by using resource efficient fuel consuming and transport options. This results in less wealth to be transferred to the oil producers, thus leaving more money in the local economy. This also could reduces transportation costs, increase employment, and invigorate the local business activity.

This can happen even in oil producing regions, as for example Norway, which is a major petroleum producer, and yet maintains high fuel prices. The high fuel prices in turn force the implementation of energy conservation policies, thus leaving more oil to export. As a result, Norway has one of the world's highest incomes, a competitive and expanding economy, a positive trade balance and the world's largest legacy fund.

On the other hand, Saudi Arabia, Venezuela and Iran subsidize the fuel in order to keep its price low and keep the masses happy, which, however, results in inefficient resource consumption, wasteful behavior, and relatively anemic economic development. In such and most other cases fuel expenditures are regressive, since they increase with income levels, and so the overall equity impacts depend on how revenues (and subsidies) are used and the quality of transport options available.

Complicated picture these subsidies, no doubt, the magnitude of which brings an obvious question, "Why are the world governments involved so closely in the fuels business, and why are they trying to control it?" The official answer is also obvious, "Fossils are the blood of the world' economies." But the more important question that remains unanswered is, "What will happen if the governments leave the energy sector alone and let it operate on its own in a true capitalist manner—with no outside assistance and control, and complete with competition, etc.?" And the answer is…silence.

Fuels Taxation

Many countries tax fuels directly for some applications; the UK, for example, imposes direct taxes on vehicle gasoline and diesel fuels, which is adjusted to equivalent levels among the different fuels. The direct tax sends a clear signal to the consumer, but its use as an efficient mechanism to influence consumers' fuel needs to be revised for a number of reasons.

Some of the uncertainties about fuel taxes are related to the fact that in many countries fuel is already taxed to influence transport behavior and to raise other public revenues. The fuel taxes in most cases are used as a source of general revenue but since the price elasticity of fuel is low, increasing fuel taxation has a very slight impact on their economies. Under these circumstances, the role of the carbon tax is unclear.

A tax on vehicle fuel may counterbalance the "rebound effect" that has been observed when vehicle fuel consumption has improved through the imposition of efficiency standards. Rather than reducing their overall consumption of fuel, consumers have been seen to make additional journeys or purchase heavier and more powerful vehicles.

There is also enough evidence that consumers' decisions on fuel economy are not entirely aligned to the price of fuel. This can deter manufacturers from producing efficient vehicles, which have lower sales potential and bring lesser profits. Other broader efforts, such as imposing efficiency standards on manufacturers, or changing the income tax rules on taxable benefits, may be at least as significant.

If fuel taxes are used to finance services that benefit lower-income households, such as improved public education or transit, and if lower-income people have fuel efficient transport options, then high fuel taxes are not necessarily regressive.

This reasoning is complicated by the diversity of the population, and the different socio-economic policies, but it basically leads to the basic conclusion that higher fuel taxes can be good and bad, depending on a number of factors. They can depress or support economic development depending on the combination of these factors. If implemented gradually and predictably, they could help to create more equitable transport systems. They can be even more beneficial if combined with

policies that increase transport system efficiency and diversity, such as improved walking, cycling and public transit service.

But wait, walking, cycling and using public transit service is not part of the American Dream? Is that even possible? How many people living in the suburbs would get off their comfortable SVUs and do walking and cycling to work or to a concert, even if it was possible? Nope, not in America. Not today! We love our cars and trucks, so we will stand by them, and in them, until the last drop of oil, or until death do us apart; whichever comes first.

Vehicle Aftermarket

Officially, the automotive industry does not count industries dedicated to *after-sale* maintenance of automobiles following delivery to the end-user. So the thousands of automobile repair shops, and motor fuel filling stations, so parts manufactured and services provided after the sale are classified as *aftermarket*. They, however, are an important part of the transportation sector.

> *Without aftermarket products and services, many of the vehicles cruising the world's highways and city streets would be sitting on the roadside in need of repair, or with empty gas tanks.*

With more than 250 million vehicles traveling across the highways, byways, and city streets throughout the U.S., the vehicle aftermarket is booming. It includes repair facilities, car washes, towing companies, aftermarket parts manufacturers, and many other businesses that help keep the millions of cars, trucks, boats, trains, and planes running. Its mission is to keep America on the road by ensuring that all vehicle owners and operators have continued access to quality and affordable parts and service.

By 2020, the experts expects over 275 million vehicles on the road in the U.S. alone. Travel on all roads by all vehicles is increasing by about 1.0%, to 18.1 billion vehicle miles annually in 2013. Gas prices are a factor, and in 2013 gasoline was about $3.50 a gallon, so in 2015 with an average price of $1.75 per gallon, we will drive even more.

And so, with people driving more miles and owning vehicles longer, these vehicles will need more frequent unscheduled maintenance and repairs. The millions of vehicles in need of periodic and emergency maintenance (a major part of the vehicle aftermarket) is a significant sector of the overall U.S. economy. It employs over 4.0 million people and encompasses all products and services purchased for light, medium and heavy duty vehicles after the original sale.

The U.S. vehicle aftermarket employs more American workers than any other industry, except construction, and the federal government.

Today the U.S. vehicle aftermarket consists of over 27,000 repair and body shops, 35,500 automotive parts stores, and 1,400 warehouse parts distributors. The aftermarket includes: replacement parts, accessories, lubricants, appearance products, tires, collision repairs as well as the tools and equipment necessary to make repairs. The aftermarket consists of hundreds of small, medium and large companies manufacturing parts and providing services for vehicle maintenance and repairs.

In 2013-2014, the U.S. aftermarket industry provided services for over 250 million vehicles. It brought in about $310 billion in parts and services in the U.S. economy. This was about 2% of total U.S. GDP at the time, and accounts for more sales than all other industries, except construction, hotels, and food service industries.

Military Fleet Maintenance

Since 2000 the demand for military vehicles in the U.S. has been driven by the extended overseas deployment of U.S. armed forces. The counterinsurgency in Iraq and Afghanistan have taken a toll on the vehicle fleets, with IUDs destroying or damaging thousands of vehicles. The new situation has prompted the development of new vehicle repair platforms to fit the asymmetric warfare scenarios.

All these factors have driven the global market for motor vehicle maintenance to unprecedented growth. Surprisingly enough, even though most of the U.S. and allied forces have been withdrawn from Iraq and Afghanistan the surge of maintenance services continued. Experts estimate this new market of military vehicles maintenance in 2013-2014 to have been around $6.0 billion.

As combat operations are decreasing, the large U.S. military vehicle maintenance market is also decreasing. With the largest military vehicle maintenance market (in the U.S.) in decline, the experts predict its strong expansion in other countries. Russia, China, India and many other countries will continue the trend.

Wrecking Yards

Why in the world would we include wrecking yards in the energy markets? What do they have to do with energy? Actually, a lot. Wrecking yards (junk, recy-cling, or salvage, yards) are a big part of the transportation markets in all countries around the world. They keep old vehicles running more efficiently and cleanly, and save money for people and businesses who can use the recycled parts.

These facilities are large sites (yards) specialized in purchasing disabled, wrecked, or decommissioned vehicles for the purpose of dismantling them. The vehicles are brought in and their usable parts are sold for use in operating vehicles, while the unusable parts and car frames (scrap), are sold to metal-recycling companies.

The major types of wrecking yards in the U.S. are:
* Aircraft boneyard,
* Shipbreaking yard,
* Train dismantling, and
* Cars and trucks wrecking yards.

About 75% of all vehicle's parts can be recycled and reused.

A wrecking yard in Phoenix, AZ has a plaque on the wall that says, '*All cars on the road run on used parts.*' How true, wise, and to-the-point summary of the role and importance of wrecking yards for the wellbeing of the transportation sector.

The size of the wrecking yard facilities vary from a few acres of a mom and pop wrecking yard, to thousands of acres of a aircraft bone-yard; all filled with disables vehicles awaiting their turn to be dismantled.

Many thousands of people are employed in the dismantling and parts sales business. Many more thousands of people visit these yards daily to purchase used parts, which usually are a fraction of the price of new parts and function equally well.

Most wrecking yards operate on a local level, purchasing local vehicles direct from from owners, or dealers. They also buy all wrecked, derelict and abandoned vehicles that are sold at auction from police impound storage lots, and often buy vehicles from insurance tow yards as well.

The vehicles are usually towed from the location of purchase to the yard, unless they can be driven. Here, the vehicles are typically arranged in rows, and often stacked on top of one another to save space. Some yards keep extensive parts inventories in a warehouse or their office, keeping track of the usable parts, and the vehicle location in the yard. Most U.S. wrecking yards have computer inventory systems and communicate with the other yards.

The more sophisticated yards use web-based and satellite services to find parts by contacting multiple salvage yards. This is far cry from the 20th century set-up, when wrecking yards used call centers to get leads for customers by paying premium rates. There are call centers today too—web based, service oriented enterprises—whose role is to locate a specific part among the thousands of wrecking yards and offer a choice to the customers.

Once the vehicles have been stripped with no more usable parts remaining, the hulks are sold to a scrap-metal processor, who usually crushes the bodies on-site at the yard's premises, using a mobile baling press, Shredder, or flattener, with final disposal occurring within a hammer mill which literally smashes the vehicle remains into fist sized chunks. These chunks are then sold by the ton for melting in steel mills.

There are over 17,000 wrecking yards in full operation in the U.S. today.

There are also hundreds of aircraft bone-yards, ship break-yards, and train scrap-yards around the U.S. The car and truck wrecking yards are the majority of the total, and deal with all types and kinds of road based vehicles. This is a big business in terms of volume of parts sold, people employed, and money transactions executed daily.

What is hard to estimate is the money saved by thousands of customers, who use recycled parts or vehicles. It is also hard to estimate, let alone put a number on, the overall impact the wrecking yards have on the U.S. economy.

The impact in terms of energy and environmental savings, and the effect on the transportation sector in particular, are also unknown. Nevertheless, wrecking yards are alive, active, and very important part of life in America. While there are vehicles on the roads, sky and water, there will be wrecking yards.

What will happen in the distant future is uncertain, but join us in this improvised walk in the inevitable fossil-less future.

Walk in the Future...

It is not very hard to imagine a world without crude oil, natural gas, and coal since they drive us and everything around us. Crude oil crisis, however, would have immediate and disastrous effect on our society and economy. Just remember the fall of 1973 and the long lines at the gas stations.

Let's recall those days and imagine for a moment a catastrophic event that cuts all global crude oil supplies. No oil comes from the oil wells and no oil products come from the refineries. The world will not end, but it will feel like it, at least during the initial stages, and until a way around it is found.

No oil, no cars and trucks on the highways, no trains on the railroads, no boats in the water, no airplanes in the sky. If there is no gasoline and diesel, people could not get around and even drive to work. Nope, electric cars would not move either, because they depend on the power generated by power plants, which are driven by fuel (coal mostly) delivered by diesel driven vehicles. No diesel, no fuel, no electricity, since coal from the mines cannot get to the power plants without diesel to drive the unit trains.

Those who get to work won't be able to do much work, because there would be no lights, no air conditioning, no heat, no power for the computers and the electric motors driving the machinery. Agricultural production would stop dead in its tracks, because it is totally dependent on diesel driven equipment in the fields, for transport, and to run the food processing plants.

Renewable energy will keep the lights on and off, but mostly off. It is the nature of the beast! The operating solar and wind power fields need replacement parts and human intervention, none of which would be available, so they would start failing one by one. The entire power grid would be shut down, because the intermittent (solar and wind) power generation cannot keep it going. Natural gas and hydropower plants could continue operating (at least for awhile) but they cannot keep the entire power grid operational. After awhile, they would be forced to shut down too, due to lack of replacement parts.

Sooner or later the nuclear power plants will also shut down due to lack of uranium (which is transported overseas) and the situation would get only worse. As a result, the grid malfunctions will increase, causing massive blackouts and myriad of electrical malfunctions.

But the most dangerous part of such a scenario is that the entire U.S. military machine will stall. All trucks, tanks, fighter boats and planes, drones and everything else that moves would be grounded. Iraq, Iran, Ukraine, and all other places would become unreachable and the U.S. would stop being the global policeman and peacemaker. The consequences of such development on global scale cannot be predicted, but no good can come out of it.

Internally, the police force would be grounded too, since their cars and SWAT vehicles need a lot of diesel and gasoline. The fire engines could not even With reduced counter-terrorism capabilities, and lack of quick emergency response, the country would become much more vulnerable to internal terrorist attacks. From here on things go from bad to worse ad infinitum, with the end result being anyone's guess.

Agreed; this entire thing is a fantasy; waste of words and a highly unlikely scenario (today), figment of a sick imagination. Yes, it is hard to see the forest from the trees, especially today, when we are awash in oil and gas. So not to worry, all is well, the experts say…today! The present fossil fuels bonanza will keep us going for several decades, so don't pay attention to the doomsayers and their absurd scenarios.

Maybe so, but common sense dictates that crude oil is here today and gone tomorrow. It is that simple, and one doesn't have to be a scientist to figure out that simple fact. There is no way around it; if we continue using the fossils the way we use them today, a day like the one described above is going to arrive. And that would be the darkest and saddest day humanity has ever experienced. That pretty much equates to a dead society. Back to the 18th century we go.

Because of that, even remote danger to the national and global energy markets—and those of crude oil especially—deserve our full and undivided attention.

Notes and References

1. Electrioity Markets, https://emp.lbl.gov/research-areas/renewable-energy
2. FERC, http://www.ferc.gov/market-oversight/mkt-electric/overview.asp
3. EPSA, http://www.epsa.org/industry/primer/?fa=wholesaleMarket
4. MIT, http://ocw.mit.edu/courses/engineering
5. NREL, http://www.nrel.gov/electricity/transmission/electricity_market.html
6. EMRF, http://emrf.net/
7. *Photovoltaics for Commercial and Utilities Power Generation*, Anco S. Blazev. The Fairmont Press, 2011
8. *Solar Technologies for the 21st Century*, Anco S Blazev. The Fairmont Press, 2013
9. *Power Generation and the Environment*, Anco S. Blazev. The Fairmont Press, 2014
10. *Energy Security for the 21st Century*, Anco S. Blazev. The Fairmont Press, 2015

Chapter 8

Peripheral Energy Markets

The global economy is driven by the energy markets, which proper,
efficient, and safe operation is ensured by the numerous global peripheral energy markets.

—Anco Blazev

BACKGROUND

In the previous chapters we took a fairly close look at the different energy markets and their operation. In order for these to function properly, however, there is another tier of third party suppliers of materials, products, components, equipment, and all kinds of services. Most of these businesses and entities are secondary to, and only support the activities in, the principal energy markets. In that role, they are *peripheral* to, but very important part of, the direct energy production and use.

The term *peripheral* has to be clarified here, because logistically speaking there is nothing peripheral about a construction company, for example, that builds a power plant. Its role is glaringly important; no construction, no power plant. Period. And yet, that construction company, and its products and services, are not a direct part of the act of power generation. Nor are they part of the operation of the energy markets per se.

Nevertheless, we cannot imagine a power plant without the construction company activities, thus it—and the rest of the peripheral products and services related to the energy markets function—are reviewed herein in the necessary detail, under the special heading peripheral.

We call the products and services discussed in this chapter 'peripheral,' but at the same time we underline their importance and give them due credit as a critical part of the energy markets. We must emphasize again that the companies providing peripheral products and services are not directly involved in the energy markets function, but only ensure their proper, efficient, and safe operation.

And because of that, the 'peripherals' deserve a close look as a special category of especial importance to the energy markets in general. The energy markets simply will not function well, or at least not as we know it today, if it was not for the peripheral energy markets.

The great importance of the peripherals is further emphasized and reflected in the fact that this is the longest chapter of this book. Here is a brief list of the peripheral energy market products and services:

Table 8-1. Partial list of the 'peripherals'

- Political and regulatory control
    - Federal and state energy policies
    - Regulatory enforcement
- Legal Actions
    - Operational support
    - Risk management
- Financial Activities
    - Funding and Investment
    - Energy and energy products trading
- Equipment manufacturing
    - Production equipment
    - Power generation and distribution equipment
    - Transportation equipment
- Contracting and Consulting services
    - Technical and logistics
    - Lobbying
- Physical and Cyber security
- Energy Efficiency (buildings and vehicles)
- Energy Storage
- Smart Grid
- Specialty Products and Materials
- Transportation Services
- Maintenance Services
- Waste Management
- Decommissioning services
- Environmental services

The 'peripherals' list above is quite impressive, but still incomplete. It gets even more impressive when one gets deeper into the labyrinth of materials, products, services, and people involved in all stages and disciplines of their activities. That opens new disciplines and niche markets. So much so, that an entire book could be written about this aspect of the energy sector. Because of the limited space allocation herein, we will only scratch the surface by taking a look at some of the most important parts of the 'peripherals' in this text.

Note: A better title for this chapter would be 'Secondary' Energy Sector, but the danger with this is that the word 'secondary' could be interpreted to mean 'less important.' Since there is nothing less important here, we prefer the description 'peripheral' instead.

Let's start with a closer look at the most important 'peripheral' entity; the political and regulatory system that determines the direction and growth of the energy markets. Politicians dictate the level and direction of development of the different energy markets via policies and rules. Regulators enforce the policies and rules established by the politicians.

POLITICS AND REGULATION

Figure 8-1 shows that the primary concern of any government is its national security. The importance of the other priorities vary and the rhetoric on the subjects changes according to current events and developments. Politicians make appealing speeches, company officials promise to put energy prices, personal safety, and the environment on top of the list of their priorities.

In the end, however, compromises are usually the final result. Making quick buck dictates the actions in the capitalist reality. Quicker, the better. The governments agree and design their energy policies accordingly. Their role is indisputably important. This is why the word 'government' is one of the most repeated in this text.

There is actually some progress in these areas, but the government and corporate priorities on making money at times overshadow the interests of the energy markets and the masses. Even worse, many companies pollute the soil we walk on, the water we drink, and the air we breath without blinking an eye. They justify these actions as, "done for the good of the majority," often with the support of the governments.

"All politics are local," so the common good usually stops at the door steps of the locals.

In many cases, the locals pay with their properties, health and lives, in order to benefit the majority and/or those bent on making a quick buck at all cost. Capitalism at its best...

When a sinkhole full of black, bubbling liquids shows up in my backyard my water is contaminated, and the air is full of poisons, then all politics—including concerns with energy and environment—fade away and-matter not.

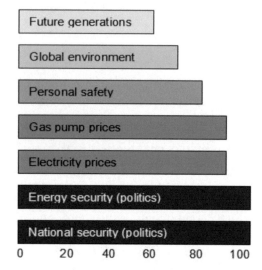

**Figure 8-1. The government concerns
(100 is most concerned)**

The politicians and corporate executives often sit on the other side of the issues, since they prefer to see the big picture from that angle only, and look for quick solutions to maintain the status-quo. They, for example, look at mass production of energy as the major, if not only, solution to our energy and national security problems. And so, "drill baby, drill and burn baby, burn" dominate the global energy scene today. The rest matters little, or none.

The U.S. government is not an exception. Although there is a lot of talk about environmental issues, energy efficiency, etc. good things lately, there is very little done in practice. In the end, the fossil industry is encouraged to expand, while the renewables' role has been diminished for most part. Again.

The race to dig out and pump up as much fossils as possible is on, until the last drop of oil and chunk of coal are burned up. Whatever the consequences...

The related environmental and health issues are discussed on all levels, and some action is taken here and there, but coal burning—in China, India, and even environment conscious Japan, Germany and other countries—is on the increase.

How long can the Earth's atmosphere take the millions and billions of tons of GHG gasses sent down its throat (and that of its population) is uncertain, but nothing good is expected to come from this race. And the future generations? Well, they can wait...we have too many problems to worry about now.

The unintended consequences of the Fukushima disaster point the way to solving the energy problems in anticipation of the many weird things to come (nuclear plants accidents and fossils depletion in particular.) The Japanese government, many companies, and the public are looking to replace nuclear power and the fossils by diversifying their energy portfolio.

The island of Japan is poor on fossils, its nuclear energy is misbehaving, and there is not enough land for bio-fuels and other alternative energy sources. But Japan has a huge coast line, which can be used for wind power generation. It also has many unutilized residential and commercial roofs most of which can be used for mounting solar panels.

Although the sun and wind are not as reliable as nuclear energy, they are much less dangerous. They are also less polluting than coal-fired power too, and will never be depleted. Because of that, the country is headed in the direction of the alternative energies, but it will be a long and difficult road. Nevertheless, for Japan and many other countries, the renewables are the only feasible long-term solution to the energy, environmental, and health problems.

The U.S. and the world should pay close attention to the developments in Japan, because what happens there is a reflection of the energy crisis coming our way in the future. If Japan does the right things, it might outline the shortest path to successful long-term energy future for the entire world.

National Energy Policy

The energy policy of the United States has been, and still is, work in progress that changes with time, depending on current events. Looking 40-50 years back, it seems to take a sharp turn after each major technological and global political development. It all started over 40 years ago with the 1973 Arab oil embargo, which was a slap in the face that woke up the American public and policy makers to the fact that our energy supplies, which determine our energy and national security, are vulnerable.

For a first time since WWII, our national interests were attacked directly, mercilessly and successfully by foreign powers.

October 16, 1973 was the 9/11 of the era. Oil prices went 400% up in a split second. The U.S. was surprised, stunned and paralyzed. It became quite obvious that we can be hit where it hurts most—by depriving us of life giving energy supplies—and that there is very little we could do in response. We just stood dazed and numb, watching the long lines at the gas pumps in disbelief. The American Dream seemed to be turning into a nightmare.

Lesson learned? Partially so! The shock did not last long. Soon after the Arabs decided to let us of the hook, we went back to our old ways of life and doing business as usual—including using as much gasoline we want. Except that now we would sleep with one eye open and watch for another Arab oil embargo, adjusting our national energy policy according to the developments.

Some examples of what happened and didn't happen through the years;

- Nixon vowed to make sure that gasoline and diesel will never exceed $1.00/gallon. Instead, we paid $6.00 per gallon of diesel in Bakersfield, California in 2008.
- Carter beat himself in the chest, proclaiming that the U.S. will never again import as much oil as it did in 1977, but no comprehensive long-term energy policy has been proposed, albeit the discussions continue. The result is obvious; we almost doubled the 1977 oil import volume 30 years after Carter's proclamation.

Figure 8-2. U.S. oil imports (in billion bbls/year)

- The U.S. now imports several times the amount of crude oil it imported in the mid-1970s.
- The use of crude oil produced in the U.S. has increased significantly as well, so the oil consumption in the country is well above anytime in the past.

- The federal energy policies have been dominated by crisis-management thinking, and jerk-knee reaction and actions that favor expensive quick fixes and single-shot solutions—a la Solyndra style—which for the most part focus on masking the problem by the "band-aid" approach, while ignoring market and technology realities.

Government's role is to provide solid, stable rules in support of basic research, while leaving the American enterprise machine alone

Things work best when the government provides and supports the technological foundation by supporting basic research and transferring the know-how to the different industrial sectors. Letting the American ingenuity take over after that is proven to bring the best results.

Entrepreneurship and innovation drive the U.S. economy and should not be restrained, or limited by government rules and regulations.

Instead, Congress and administrations have time after time introduced energy policies that seem politically expedient, but with a short and inefficient life in the energy markets. The renewable energy technologies have suffered most through the years by the changing government policies.

The trend was started by President Nixon, who called for urgent "energy independence." This was a wake-up call for the U.S. policy makers, and a number of measures were taken in order to save energy and to put energy production and use under some resemblance of control.

Some of these measures were successful, some not so much:

- In 1974, the national maximum highway speed limit was reduced to 55 MPH through the Emergency Highway Energy Conservation Act. Soon after the "energy emergency" was gone, the speed limit was restored back to 65 and 75 MPH, and even higher in some states. Soon, everything went back to "normal"; large cars, trucks, and SUVs were and still are the predominant vehicles in the U.S.
- Year-round daylight saving time began in Jan. 1974, in an attempt to save some energy, but due to wide public protests, pre-existing daylight savings rules were restored in 1976. It was later on restored and is now the law of the land.

- President Ford signed a bill creating the first fuel efficiency standards, which required auto companies to double fleet-wide fuel use savings averages by 1985.
- The Energy Policy and Conservation Act (EPCA) of 1975 were enacted and expected to achieve a number of goals; to increase energy production and supply, reduce energy demand, provide energy efficiency, and give the executive branch additional powers to respond to disruptions in energy supply.

EPCA established the Strategic Petroleum Reserve (SPR) in 1975 for the storage of up to 1 billion barrels of oil for emergency use only, which is still part of our energy policy today.

EPCA sought to roll back the price of domestic crude oil. Old oil was to be priced at the May 15, 1973 price plus $1.35 per barrel. New oil and stripper oil prices were set at the September 30, 1975 new oil price less $1.32 per barrel. That didn't last long.

- The Corporate Average Fuel Economy (CAFE) standards was established at that time too. It mandated increases in average automobile fuel-economy. The standards set a corporate sales-fleet average of 18 MPG beginning with the 1978 model year, and established a schedule for attaining a fleet goal of 27.5 miles per gallon by 1985.

CAFE standards apply separately to domestic and imported sales fleets. That program has been relatively successful, but its effect is minimal due to the extensive use of low MPG vehicles (large trucks and SUVs) in the country.

- In 1977, President Carter called the energy crisis the "moral equivalent of war" and established the U.S. Department of Energy (DOE), which consolidated several energy-related entities, and functions of the Federal Government into a single, Cabinet-level organization. Until the 1970s, energy and fuel-specific programs were handled by several Federal departments: The Department of the Interior managed most Federal programs affecting the coal and oil industries, the Federal Power Commission regulated natural gas prices, and the Atomic Energy Commission (AEC) oversaw the development of nuclear power. Post 1970 events accelerated the reorganization of the energy-related programs of the Federal government. AEC was replaced with two new agencies; the Federal Energy Administration (FEA) was created in 1974 to administer programs that included crude oil price and allocation, and the SPR replacement of natural gas and oil with coal, and energy conservation.

One of the key functions of DOE originally was to formulate comprehensive energy policy, but energy management in general became the primary role of the agency in the early 1980s. Energy shocks and public sensitivity during the last 2-3 decades have molded DOE into the country's energy manager. Energy conservation, alternative fuels, oil consumption reduction, national security, and energy prices are now DOE's major concerns.

- The second oil crisis hit the U.S. and the world in 1979 when the Shah of Iran was overthrown and the Iranian revolution began. The "tractorcade" (encircling of the U.S. Capitol by 3,000 tractors) called the attention of policy makers to the need for ethanol production and use. Now ethanol is part of the energy mix and its role is becoming increasingly important.

- The Three Mile Island nuclear plant accident around that time changed the plans for building more nuclear plants, and the U.S. nuclear industry has been in decline ever since.

- President Carter showed his support to renewables by installing a bank of solar panels (hot water heaters) on the roof of the White House in 1977. That experiment did not last very long and the panels were removed during the next presidency.
  Another set of solar panels was installed on the roof of the White House (PV modules this time) in 2013 by President Obama. These events demonstrate the on again, off again fate of solar development in the U.S.

- Three Energy Policy Acts were passed later on in 1992, 2005, and 2007, which include many provisions for conservation, such as the Energy Star program.

- A set of energy development programs, complete with grants and tax incentives for both renewable energy and non-renewable energy were also enacted. Most of these are now terminated, or changed to a lesser function.

- The developments, starting with the financial crisis in the U.S., and the subsequent global economic slow-down during 2008-2012 brought up the same jerk-knee reaction from the federal and state administrations. Billions of dollars were thrown at energy companies and projects, many of which were proven to be waste of time, effort, and money. What were the government officials and their advisers at DOE thinking is above and beyond our comprehension, but we saw the similar awkward decisions and money waste during renewables revival of the late 1970s, so the latest frenzy was not totally surprising.

- Recently, the plans to build mega-power solar and wind installations in the U.S., although still in the news, are on the government's back burner. Fossil energy—crude oil and natural gas—are abundant in the U.S., so we will use every last chunk and drop of these while we can, instead of leaving some to the next generations.

- And then, the 1973 lesson was repeated, albeit in a different form and shape in 2014. This time the Arabs kept the production levels and the oil prices ridiculously low, in order to kill our increasing oil and gas production. With friends like these, who needs enemies?

Because of the artificially low prices, the U.S. energy market is going through unprecedented complications. Hundreds of oil and gas drilling operations, that popped up during the last several years, and were thought to be signs of the new energy reality, are now struggling to survive the relentless attack of lower oil prices.

The prolonged energy crisis brought many small oil and gas producers close to bankruptcy, and some went out of business. How many would be able to withstand the direct attack is uncertain, but the new situation clearly shows that our energy future still depend heavily on unreliable foreign powers.

Today's Energy Policies

Officially speaking, today the U.S. energy policy is determined by federal, state, and local entities. As fragmented and temporary as it might be, it is moving in the direction of ensuring our energy security whatever it takes. It is geared to address issues of energy production, distribution, and consumption, such as; building codes, gas mileage standards, pollution, etc. issues. It also includes legislation, international treaties, subsidies and incentives to investment, guidelines for energy conservation, taxation and other public policy approaches and techniques. State-specific energy-efficiency incentive programs also play a significant role in the overall U.S. energy policy.

All good things considered, the U.S. refused to endorse the Kyoto Protocol, preferring to let the market drive CO_2 reductions as needed to mitigate global warming by means of CO_2 emission taxation. The major (unofficial) reason for this is that any emissions restrictions might hurt the related industries and cripple the U.S. economy.

President Obama has proposed an aggressive energy policy reform, including the need for a reduction of CO_2 emissions. This includes a cap and trade program, which could help encourage more clean renewable, sustainable energy development. The new energy policy, however, is partially viable now due to more urgent issues in the drastically changing global energy markets, and the dysfunctional U.S. Congress.

President Obama also set a goal of 20% renewable energies to be implemented in the U.S. government installations (including DOD) by 2020. The US government is the largest energy consumer in the country, so the energy use reduction targets (which only apply to electricity consumption) will be phased-in gradually.

According to it, the agencies must draw no less than 10% of their electricity from renewables by 2015, 15% by 2017, 17.5% by 2019, and no less than 20% by 2020. The different departments are required to do so by funding and installing their own generation through on- or off-site renewables. They can also purchase from third-party owned clean energy plants built at their request, by purchasing renewable power from the grid or by paying for renewable energy certificates. All power also must be generated by renewable sources less than 10 years old.

The US government is also looking to facilitate the work of developers as needed to install renewable energy generation projects on federal land. A recent government auction for available federal land in Colorado in 2013 did not result in any bids, however, so we will have to see if this development will have any effect in the future.

While all this is a good step in the right direction, it is limited to government installations and cannot be expanded to the entire U.S. economy. A larger, national energy policy is badly needed, in order to eliminate the uncertainties and provide a solid direction for the energy sector in the country.

Update 2015

The U.S. installed 1,330 MW of solar PV in Q1 2014, up 79% over Q1 2013, making it the second-largest quarter for solar installations in the history of the market. Cumulative operating PV capacity stood at 13,395 MW, with 482,000 individual systems on-line as of the end of Q1 2014. Growth was driven primarily by the utility solar market, which installed 873 MWdc in Q1 2014, up from 322 MWdc in Q1 2013.

Q1 2014 was the first time residential PV installations exceeded non-residential (commercial) installations nationally since 2002. For the first time ever, more than 1/3 of residential PV installations came on-line without any state incentive in Q1 2014. During the same time we saw school, government, and nonprofit PV installations add more than 100 MW for the second straight quarter.

74% of new electric generating capacity in the U.S. in Q1 2014 came from solar. PV installations are up about 40% over 2013 and nearly double the market size in 2012.

Q1 2014 was the largest quarter ever for concentrating solar power due to the completion of the 392 MWac Ivanpah project and Genesis Solar project's second 125 MWac phase. With a total of 857 MW (AC) expected to be completed by year's end, 2014 will likely be the largest year for CSP in history.

In 2015, the progress in the sector continued. With nearly 3.0 GW installed capacity—about 50% higher than 2014—it is pumping the grand total of nearly 25 GW of electricity in over 5 million American homes and businesses.

Great achievements, no doubt, but remember that this was possible primarily due to federal and state governments' regulatory and financial support.

Woops, when ITC is gone in 2017, the U.S. solar industry is projected to collapse. At least for awhile and until it finds its sea legs, to walk and run on its own.

In all cases, solar and wind power generation are still very small part of the energy mix in the country. Combined, they provide about 3-4% of the total power used on daily basis. And most importantly, they depend on steady government's support.

The bigger problem facing the renewables in the near future is the waning federal government support. The renewables—wind and solar in particular—are no longer considered critical components in the feds' energy programs. They are left on their own, and without any significant support are faced with unfair competition and insurmountable barriers, which will curtail their progress in the near future.

The Renewables' Share

As Table 8-2 shows, the renewables, which category includes all types of renewable energy technologies (hydro, solar, wind, bio-, and geo-power), are about

15% of the total energy mix, but solar and wind are only about 3-4% of the total power generation.

Table 8-2. U.S. U.S. power generation by technology (2013)

Source	# Plants	Capacity (GW)	Annual energy (billion kWh)	% of annual production
Coal	557	336.3	1,514.04	36.97
Natural Gas	1,758	488.2	1,237.79	30.23
Nuclear	66	107.9	769.33	18.79
Hydro	1,426	78.2	276.24	6.75
Renewables	90	218.33	5.33	5.33
Petroleum	1,129	53.8	23.19	0.57
Misc	64	2	13.79	0.34
Storage	41	20.9	-4.95	-0.12
Imp-Exp	-	-	47.26	1.15
Total	6,997	1,168	4,095	100

Note the important discrepancy here. The renewables have a total capacity of over 218 GW, but generate only 5.3 billion kWh of power, due to their variability. A similar capacity coal plant would generate over 1,000 billion kWh of electric power during the same time. This is not twice, or 20 times difference, but 200 times.

This says it all; the renewables (except hydro) cannot be used for base load power generation.

We discussed some of the issues that are stopping the full development of solar and wind power, such as variability, lack of storage, high price, increasing competition and resistance, etc. This means that these renewable technologies have a long ways to go before they can compete with the conventional (fossil) fuels.

One of the major obstacles in the renewables' progress is the fact that the fossil energy sources are abundant and cheaper. This is causing a catch 22 scenario in the U.S., where we are using a lot of fossil fuels, because they are abundant and cheap. Because of that, they are also supported by the federal government as a solution to our energy problems.

This is stifling the progress of the renewables, while at the same time we are emitting a lot of GHGs which are choking the environment and causing global warming.

And to top it off, we are using such great quantity of fossil fuels (and even exporting significant quantities) that they will not last very long. So it seems like we are living in a temporary bonanza, indulging in energy gluttony in total disregard of tomorrow.

Regardless of the total amount of available fossil reserves we must remember that they are finite and will

be depleted soon; well before we have found a suitable replacement.

Table 8-3 points the way to a sustainable future. The only problem is that the decision makers do not see it this way, let alone follow the path to sustainability. They are, instead, gearing the economy for using and exporting more fossils in the near future.

Table 8-3. Sustainable National Energy Plan (s)

Year	Coal	NG	Oil	Ren	GHG
2010	100	100	100	2	100
2015	95	95	95	5	95
2020	90	90	90	10	90
2025	85	85	85	15	85
2030	80	80	80	20	80
2035	75	75	75	25	75
2040	70	70	70	30	70
2045	65	65	65	35	65
2050	60	60	60	40	60
2055	55	55	55	45	55
2060	50	50	50	50	50
2065	45	45	45	55	45
2070	40	40	40	60	40
2075	35	35	35	65	35
2080	30	30	30	70	30
2085	25	25	25	75	25
2090	20	20	20	80	20
2095	15	15	15	85	15
2100	10	10	10	90	10

Where:

NG = natural gas
Ren = renewables
GHG = green house gasses (pollution)

But what is really needed? Very simple. We need to ensure sustainable energy future for the U.S., which can be easily achieved by:
*   50% reduction in fossil fuels use NOW,
*   To be replaced by a 50% increase of renewables, which will result in
*   50% reduction in GHG emissions, and
*   If all goes per plan, then by 2100 we will be on the path to a sustainable and clean future.

But first, we need to start this process, and achieve the goals set for 2050-2060. If this plan does not go as scheduled in Table 8-3 by then, and instead we continue the present rate of fossils exploitation and GHG emissions, then it might be too late to go back.

The U.S federal government is the only entity that has the power to outline a feasible plan to a sustainable energy future for the nation. If and when the policy makers stop bickering about incidentals and focus on solving the problems in the energy sector, they may come up with a plan of attack. Until then, we will continue wandering in the uncertainty of the global energy markets.

Energy Strategy

The link between energy strategy and energy production and use is critical. It is the most important aspect of developing new methods for the electrical grid and ensuring our energy and national security in general. Government energy strategy and the related policies have the power to regulate the types of energy sources utilized in the market through various directives and regulations. Policies can also influence consumer preferences based on different initiatives.

Government policies can be negatively enforced or positively enforced, as needed to help prevent or promote different results. For example, taxes and regulations could be used to reduce pollution and encourage the development of new technologies.

The political landscape certainly influences the types of energy used within a country as well as any advancements to the grid through economic incentives. These advancements, due to economic incentivizing is called *induced innovation*, which has a substantial role in reducing costs for energy and increasing research into different sources of production.

In more general terms, government policy has consequences on pricing and production, since changes in the capacity and generation of energy are driven mostly by price. These changes are ultimately shaped by government policy and initiatives that greatly influence not only electricity producers but also consumers.

The impact of government policy and directives such as the NetZero program, for example, is significant and drives social behaviors, business practices, and energy initiatives for the national energy grid; yet, it is possible for the relationship to be symbiotic. The energy sector can also help to influence government policy at times creating an interactive, rather than a reactive relationship.

Note: This influence is not to be confused with the negative effects of excessive lobbying and cronyism on government levels, which we have observed in the past.

Considering the possibility of energy crisis hitting at anytime, the looming depletion of fossil resources,

and pending threats of natural disasters or terrorist attacks, it is clear that there ought to be a focus on altering the energy policies in order to mitigate these concerns. The electric grid is also of concern, but there are several different alternatives to answer the concerns with the objective of building a stronger, smarter, more efficient, and cleaner electric grid.

The basis of pricing as a means to influence consumer preferences and energy source investment also factors into most government policies. By influencing behavior in the social and business sectors it helps to reinforce government policies focused on advancing the energy markets. On the other hand, the economic inefficiencies, and other factors in the market, may put pressure on politicians to alter their ideology that significantly influence policy proposals with regards to energy. Recently, we see this happening in the U.S. time after time and with increasing frequency.

Pricing policy and strategy can also provide incentives for producers to invest in research and development into new technology which in turn alters the markets by replacing old buying practices for consumers with more choices to meet their preferences. The relationship between government policies, initiatives, directives, and the energy markets is a critical factor in developing a methodology to optimize efficiency and mitigate many of the concerns facing the U.S. now.

The real and perceived threats to the power grid, and the energy markets in general, will continue to be a pressing policy issue for many years to come. As the threats increase in number and seriousness, understanding the relationships between policy, energy sources, and outputs is imperative to building successful economic solutions on national and global levels.

Proper and timely introduction and implementation of such solutions should be a priority, and is not to be hindered by third party (political, personal, or corporate) interests, which might twist the decision makers thinking and influence their actions. The wish list for improvements in the energy sector is arm long, and the government has the ultimate responsibility to provide the proper solutions. These must be then incorporated in one all-encompassing, comprehensive National energy policy.

U.S. GOVERNMENT AND ENERGY

The U.S. government is a large energy user. It is actually the largest user of electricity and fuels in the country. It accounts for nearly 2% of the nation's annual energy consumption, accompanies by the emission of a huge chunk of GHG emissions.

The amount of energy consumed by all U.S. government entities is larger than that used by most countries today.

Recently, the federal government decided to take steps to reduce its energy use and consequently reducing its heat-trapping emissions by 28% by 2020, compared with 2008 levels. If met, these measures would save over $10 billion in energy expenses over the next decade. This cumulative reduction over 10 years would be equivalent to saving 205 million barrels of oil, or taking 17 million cars off the road for a year.

"As the largest energy consumer in the United States, we have a responsibility to American citizens to reduce our energy use and become more efficient" according to the President. "Our goal is to lower costs, reduce pollution and shift federal energy expenses away from oil and towards local, clean energy."

Federal agencies are pursuing measures to reduce energy consumption by making buildings and vehicles more fuel efficient or switching to renewable sources of energy. The Central Intelligence Agency, for example, opened two new high-efficiency buildings in Virginia that will consume 20% less energy than existing structures. At the same time, the U.S. Army is installing a 500-megawatt solar plant in the Mojave Desert at its Fort Irwin training grounds.

The Veterans Affairs Department has contracted for a wind turbine generation system to provide 15% of the electricity for its hospital in St. Cloud, Minn. The DOD, as the federal government's largest consumer of energy, set its emissions reduction goal at 34% by 2020, although that figure excludes energy consumed by combat forces. Many of the near-term costs of the program would be covered by the federal stimulus package. An added benefit will be the creation of thousands of new jobs.

The U.S. governments is involved in all aspects of the American private and business activities. From the White house to the last house on a remote Hawaiian island, the Feds are watching and controlling the daily activities of all people and companies. This is sometimes good and sometimes bad, but no matter what it is, we have accepted it and are willing to live with it.

The only problem the government has is its inability to control fully the fact that all these activities require a lot of energy. And that energy is getting more expensive by the day, and that eventually there will be not enough of it. At that point the government activities

Figure 8-3. U.S. government issues, projects, and tasks

would be seriously crippled, so it does everything possible to prevent this from happening. This is done by subsidizing energy companies and projects. In recent times, energy efficiency measures have been on the forefront of the U.S. government's efforts to postpone the inevitable.

Below is a narration of some of the government's solutions of late:

Executive Order 13514

Executive Order (E.O.) 13514, Federal Leadership in Environmental, Energy, and Economic Performance, was signed in October, 2009 by President Obama. It expanded upon the energy reduction and environmental performance requirements of the previous E.O. 13423. Surely, the Administration sees a strong relation between energy production and use, and the environment. This is a good step ahead toward energy efficient and environment friendly future.

E.O. 13514 states that the U.S. government is the largest consumer of energy in the country, with over 500,000 large, old, and mostly energy-inefficient buildings. E.O. 13514 sets numerous Federal energy requirements, starting with a mandate for 15% of existing federal buildings and leases to meet Energy Efficiency Guiding Principles by 2015.

This means that at least 75,000 buildings are to be converted to energy efficient structures by 2015. A tall order. The annual progress is to be made toward 100% conformance of all federal buildings, where 100% of all new federal buildings are to achieve zero-net-energy status by 2030.

"Zero-net-energy building" according to E.O. 13514 is "a building that is designed, constructed, and operated to require a greatly reduced quantity of energy to operate, meet the balance of energy needs from sources of energy that do not produce greenhouse gases, and therefore result in no net emissions of greenhouse gases and be economically viable."

The executive order also states the general direction the Administration is taking the country, "...the federal government must lead by example...increase energy efficiency; measure, report, and reduce their greenhouse gas emissions from direct and indirect activities...design, construct, maintain, and operate high performance sustainable buildings in sustainable locations; strengthen the vitality and livability of the communities in which federal facilities are located; and inform Federal employees about and involve them in the achievement of these goals."

In October 2013, President Obama also signed a Memorandum under his Climate Action Plan, directing the Federal Government to consume 20% of its electricity from renewable sources by 2020. This is more than double the current level. Another tall order.

As part of this effort, Federal agencies will also identify formerly contaminated lands, landfills, and mine sites that might be suitable for renewable energy projects, in order to return those lands to productive use with minimal fossil energy use.

In order to improve the Federal agencies' ability to manage energy consumption and reduce costs, the Memorandum also directs them to use Green Button (a tool developed by the industry) which provides utility customers with easy and secure access to their energy usage information in a consumer-friendly format. This will provide transparency and will help further in managing the effort.

This is a great step ahead for the U.S. energy policy makers. Unfortunately, it is relevant only to the government complex. Similar patchwork of energy and environmental guidelines exists here and there across the national economy too, but they are much more sporadic, and not as clearly defined.

The final decision in executing the civilian energy policies in most cases, is left to the individual states and local authorities, which makes it hard to predict how the entire program will develop in the future. A more unified, country-wide federal level energy program is needed, but we will not hold our collective breath waiting for it.

The American Recovery and Reinvestment Act of 2009

The American Recovery and Reinvestment Act of 2009 (ARRA) (Pub.L. 111–5), commonly referred to as the Stimulus or The Recovery Act, was an economic stimulus package enacted by the 111th United States Congress in February 2009 and signed into law by President Obama.

Its intent was to respond to the recession by saving and creating jobs immediately. Secondary objectives were to provide temporary relief programs for those most affected by the recession and invest in infrastructure, education, health, and renewable energy.

The cost of ARRA activities is estimated at $831 billion between 2009 and 2019.

The Act included direct spending in infrastructure, education, health, and energy, federal tax incentives, and expansion of unemployment benefits and other social welfare provisions. It also created the President's Economic Recovery Advisory Board.

The rationale for ARRA was derived from the Keynesian macroeconomic theory. It argues that, during recessions, the government should offset the decrease in private spending with an increase in public spending in order to save jobs and stop further economic deterioration. Critics of the law claim it is too weak, since it covered only one third of the spending gap at the time.

Other Key Federal and State Energy Policies and Regulations
- Energy Policy Act of 2005, and Energy Independence and Security Act of 2007
- Executive orders 13221, 13423, 13514, and 13693
- Renewable Portfolio Standard (RPS)
- Renewable Fuels Standard (RFS), 2015
- Financial incentives; grants, loans, rebates, and tax credits
- Public benefits funds for renewable energy
- Output–based environmental regulations
- Interconnection standards (WIP)
- EPA regulations affecting renewable energy, 2015
- EPA programs in support of renewable energy;
  — Green Power Partnership (GPP)
  — Methane Outreach Programs, LMOP and Ag-STAR
  — RE–Powering America's Lands

We take a closer look at some of these programs elsewhere in this text.

Public Energy Programs
In addition to its own energy saving programs, as discussed above, the U.S. government has established a number of public energy programs. One of these is the Low Income Home Energy Assistance Program (LIHEAP). It is designed to help keep families safe and

healthy through initiatives that assist families with energy costs.

The Energy Start program, established in 2009, gave consumers nationwide a chance to take advantage of a federal "cash for appliances" program offering rebates on purchases of a wide array of home appliances certified as energy-efficient by the EPA's Energy Star program. Backed by an initial $300 million in funding from the American Recovery and Reinvestment Act, the state-run rebate program is intended to help make American homes more energy-efficient while further stimulating the economy.

"Appliances consume a huge amount of our electricity, so there's enormous potential to both save energy and save families money every month," according to DOE "These rebates will help families make the transition to more efficient appliances, making purchases that will directly stimulate the economy and create jobs."

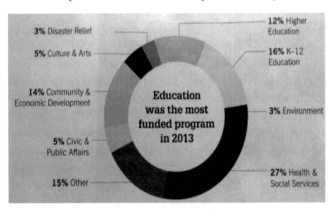

Figure 8-4. U.S. government spending in 2013.

Another significant program was established in January, 2015, where the government awarded $300 million to U.S. states, tribes and territories of funding available under the Consolidated and Further Continuation Appropriations Act, 2015. The program determines which household's income qualifies for the program, and requires households to meet additional eligibility criteria in order to receive energy assistance.

Government Organizations

Energy, in its different forms—especially electricity, natural gas, and heating oil—are indispensable for our homes and businesses. Air conditioning, heating, lighting, and most industrial applications depend on it. Electricity as a commercial product joined the energy markets late in the game, but quickly became favorite for lighting, household appliance, and industrial applications during the early decades of the twentieth century. It is now indispensible part of our lives. So much so

that we cannot even imagine our life without electricity.

From the very beginning of oil, gas, and electricity markets in the U.S., the customers have been concerned about their vulnerability to the machinations and manipulations of powerful producers, trans-porters, and buyers and sellers of energy.

Policymakers and regulators recognize the critical importance of energy to consumers, and have been trying to balance the desire to harness the benefits of competition with the risks unbridled markets pose to vulnerable consumers. These concerns produce different regulatory responses aimed at ensuring an adequate supply of energy at reasonable prices.

In oil markets, regulators rely primarily on anti-trust law as a tool of control, using lawsuits to punish the exercise of market power and push markets toward greater competition. In electric and gas markets, by contrast, government regulators chose the public utility model. That model accepts the absence of competition on the premise that these network industries were natural monopolies, granting to electric and gas companies monopoly service rights. In exchange, electric and gas companies ceded control over their rates and certain other terms of electric and gas services.

With the increased emphasis on environmental issues and global warming, the U.S. government is revising its policies and regulations. The changes in this area are needed, but they are adding another layer of complexities in the already complex government regulatory system.

All in all, these are complex issues in need of proper and effective action,. For that purpose, the U.S. government enlists the help of several organizations to guide its efforts in managing its energy sector and the issues related to energy generation, distribution, pricing, and efficient use.

DOE

The U.S. Department of Energy ((DOE) is the premier energy expert, watch dog, and a guide for all past, present, and future energy related programs conducted by the U.S. government. It is a Cabinet-level department of the U.S. government, in charge of all policies related to energy and safety in handling nuclear material. Its responsibilities include the nation's nuclear weapons program, nuclear reactor production for the United States Navy, energy conservation, energy-related research, radioactive waste disposal, and domestic energy production.

DOE also directs the Human Genome Project, which originated as a DOE initiative. DOE sponsors research in the physical sciences, the majority of which is conducted through its system of National Laboratories. The agency is led by a Secretary of Energy from its headquarters in Washington, D.C.

As its primary purpose is to ensure energy security and maintain the safety of the nuclear weapons stockpile, it manages the clean-up of environmental damage left from 50 years of nuclear defense activities that impacted two million acres in communities across the country.

DOE also operates 24 preeminent research labs which employs 30,000 scientists and engineers, including the Lawrence Livermore Lab, NREL, the Argonne National Lab and three accelerator labs. It has four power marketing administrations market electricity, some of which are from hydro-electric dams built by the U.S. Army Corps of Engineers.

DOE also creates and supervises the national energy policies. It helps to develop and execute the President's Management Agenda via eight program offices, which are responsible for: disposal of radioactive waste, clean up of land, buildings and material contaminated by nuclear weapons research and production facilities, and maintain the closed facilities, support safe use of nuclear energy, monitor and ensure the reliability of the nation's electric grid, promote use of renewable energy sources, monitor the nation's petroleum usage and supplies, and fund research in high-energy physics, nuclear physics, and fusion energy sciences.

DOE Financial Programs

DOE has funded a number of renewable projects to date. Among these are the 290 MW thin film technology Agua Caliente project, built by First Solar, the 250 MW NextEra Energy CSP project of Genesis Solar. and the 392 MW Ivanpah CSP tower plant, to mention a few.

One of the latest DOE' crown jewel is its famous investment in Elon Musk's Tesla Motors electric vehicle large-scale manufacturing efforts. Tesla repaid its loan with interest in 2013, nine years earlier than scheduled.

DOE data released in 2014 shows more than $5 billion of profit, exceeding greatly the many millions in losses from defaulting companies. As a matter of fact, the losses make up only about 2.5% of over $35 billion DOE has loaned to different companies and projects by 2013.

The US Department of Energy loans guarantee programs have brought in more from interest payments than was lost on failed companies such as Solyndra and others.

Although Solyndra is still a symbol of government's mis-direction and poor judgment in investing in clean tech firms, and a valid argument from technical and logistics point of view, it does not apply to the DOE's bank accounts. Some companies and technologies might be affected by poor management, assisted by inappropriate DOE intervention, but the taxpayers have not lost any money. On the contrary.

FEMP

The U.S. Department of Energy's Federal Energy Management Program (FEMP), under the DOE coordination, is a new program that plays a critical role in reducing energy use and increasing the use of renewable energy at federal agencies. FEMP is designed to work with key individuals and entities to coordinate energy production and use change within federal organizations by bringing expertise from all levels of project and policy implementation. This would enable federal agencies to meet energy-related goals and provide energy leadership to the country and the world.

As the nation's largest user of energy, the U.S. federal government has a great opportunity and even greater responsibility to lead by example in saving energy. With the technical assistance of FEMP, the energy intensity of federal facilities has decreased by roughly 45% since 1975 and much is still left to do.

FEMP also helps federal agencies with funding mechanisms for their projects, such as energy-saving performance contracts (ESPCs), under which a contractor pays the up-front cost of improvements and is repaid through a portion of the energy savings. Thus far FEMP has arranged a number of ESPCs that have saved taxpayers more than $5 billion in federal energy costs.

FERC

The Federal Energy Regulatory Commission (FERC) is a federal government agency, which mission is to "assist consumers in obtaining reliable, efficient and sustainable energy services at a reasonable cost through appropriate regulatory and market means."

As an independent agency, FERC regulates the interstate transmission of electricity, natural gas, and oil. It also reviews proposals to build liquefied natural gas (LNG) terminals and interstate natural gas pipelines as well as licensing hydropower projects. The Energy Policy Act of 2005 gave FERC additional responsibilities as outlined and updated Strategic Plan.

As part of that responsibility, FERC also regulates the transmission and wholesale sales of electricity in interstate commerce; reviews certain mergers and acquisitions and corporate transactions by electricity companies; regulates the transmission and sale of natural gas for resale in interstate commerce; regulates the transportation of oil by pipeline in interstate commerce; approves the siting and abandonment of interstate natural gas pipelines and storage facilities; reviews the siting application for electric transmission projects under limited circumstances; ensures the safe operation and reliability of proposed and operating LNG terminals; licenses and inspects private, municipal, and state hydroelectric projects; protects the reliability of the high voltage interstate transmission system through mandatory reliability standards; monitors and investigates energy markets; enforces FERC regulatory requirements through imposition of civil penalties and other means; oversees environmental matters related to natural gas and hydroelectricity projects and other matters; and administers accounting and financial reporting regulations and conduct of regulated companies.

A long list of responsibilities, and yet FERC is not involved in most day-to-day functions of the energy markets. In more detail, FERC's responsibility in the different energy sectors is as follow: for:

- Electric power. FERC is responsible for regulating interstate transmission rates and services, wholesale energy rates and services, corporate transactions, mergers, and securities issued by public utilities.
- Hydropower. FERC is responsible for licensing of nonfederal hydroelectric projects, overseeing related environmental matters, and inspecting nonfederal hydropower projects for safety conditions and compliance with license terms and conditions.
- Natural Gas. FERC is responsible for regulating interstate transportation rates and services for natural gas pipelines, the construction of natural gas pipelines, and overseeing related environmental matters.
- Crude oil. FERC is responsible for regulating interstate transportation rates and services of crude oil and petroleum products.

EPA

EPA's mission is to protect human health and the environment. Period. In more detail, EPA aims to ensure that: all Americans are protected from significant risks to human health and the environment where they live, learn and work; the national efforts to reduce environmental risk are based on the best available scientific information; the federal laws protecting human health and the environment are enforced fairly and effectively;

environmental protection is an integral consideration in U.S. policies concerning natural resources, human health, economic growth, energy, transportation, agriculture, industry, and international trade.

All these factors are similarly considered in establishing environmental policy; all parts of society—communities, individuals, businesses, and state, local and tribal governments—have access to accurate information sufficient to effectively participate in managing human health and environmental risks; environmental protection contributes to making our communities and ecosystems diverse, sustainable and economically productive; and that the United States plays a leadership role in working with other nations to protect the global environment.

Simple, no? In order to do this, EPA must develop and enforce appropriate and efficient regulations. Every time the U.S. Congress issues an environmental law, EPA must implement it by writing detailed and enforceable regulations. EPA also sets the national standards that states and tribes enforce through their own regulations. EPA then helps them to meet the national standards by training responsible state and company officials, and enforcing the regulations.

For these purposes, EPA also issues grants to state environmental programs, non-profits, educational institutions, and others, which amounts to nearly half of its budget. This money is used for a wide variety of projects, from scientific studies that help to make decisions to community cleanups and other activities. The grants are a major part of the EPA activities, since they help it achieve its overall mission, which is to protect human health and the environment.

Military Energy Markets

The country of the U.S. Department of Defense (DoD), in terms of money, technology, and energy use, is larger, much more advanced, and many times more powerful than any other country in this world. As such, it is also held responsible for its actions. In response to the government's energy efficiency and environmental impact management efforts, the DOD has established its own energy plans with respective goals and accountability measures.

The U.S. DoD energy security goals are based on:
- Reducing energy consumption
- Increasing energy efficiency across platforms and facilities
- Increasing use of renewable and alternative energy sources

- Ensuring access to sufficient energy supplies
- Reducing adverse impacts on the environment, and
- Changing the overall energy culture

The energy management approaches in these plans are excellent and deserve to be reviewed and adopted by corporations and governments alike. Some of these should be urgently implemented in the overall U.S. economy, but there is not much chance of that happening anytime soon. The U.S.' economy is simply too large and diverse, but more importantly, it lacks the discipline needed to execute some of the basic energy use restrictive measures in a coordinate way.

Since the DoD is smaller and much more organized than the larger U.S government system, its energy plans are easier to design and implement. It also enjoys a huge advantage of operating under strict discipline at

all ranks. This way, once the plans are agreed upon by the top commanders, they are quickly and immaculately executed through the ranks. No questions asked, no debates, no protests, no filibusters, and no delays!

Some of the recent results from the DoD's energy efficiency improvement tactics are reflected in Table 8-4. It is obvious here that DOD takes the energy management effort seriously. There are still unreached goals and targets to meet, to be sure, but the effort is well underway and DOD is determined to succeed in completing its goals.

In more detail:

- At the end of FY 2009, DoD was managing 1.93 billion square feet of facility space (EO 13423 defined goal-subject facilities) and had spent $3.6 billion on facility energy alone. In addition, DoD spent $9.6 billion on fuel for vehicles (non-fleet and fleet) and other equipment. This included jet fuel,

Table 8-4. U.S. Department of Defense Energy Plan execution summary (2011)

GOALS	UNITS	ENTITY	FY2011	
			Achieved	Target
Reduce facilities energy use intensity per FY 2003 baseline Per EI&SA*, 2007, #431	BTUs of energy consumed per gross square foot of facility space in use.	DOD	-13.30%	-18%
		Army	-11.80%	
		Navy	-16.90%	
		Marines	-9.40%	
		Air Force	-16.30%	
Consume more electric energy from renewable sources. Per EPA**, 2005, #203	Total renewable electricity consumed as a percentage of the total facility electricity use.	DOD	3.10%	5%
		Army	0.50%	
		Navy	1.00%	
		Marines	1.20%	
		Air Force	6.00%	
Reduce potable water water use intensity relative to FY2007 baseline. Per Exec. order 13423	Gallons of water used per square foot of facility space in use.	DOD	-10.70%	8%
		Army	-10.30%	
		Navy	-4.00%	
		Marines	-22.70%	
		Air Force	-13.10%	
Produce or procure more energy from renewable sources. Per U.S. Code Title 10, #291(e)	Total renewable energy (electric or non-electric) produced or consumed as a percentage of total facility electricity use.	DOD	8.50%	25% by 2025
		Army	4.30%	
		Navy	20.60%	
		Marines	0.60%	
		Air Force	7.10%	
Reduce petroleum use in non-tactical vehicles relative to FY2005 use. Per EI&SA, 2007, #142, Exec. Order 13514 #2(a)	Gallons of gasoline equivalent of petroleum fuel consumed.	DOD	-11.80%	12%
		Army	-10.30%	
		Navy	-26.00%	
		Marines	-16.60%	
		Air Force	-8.30%	

* Energy Independence & Security Act
** Energy Policy Act

aviation gasoline, Navy-special fuel, automobile gasoline, diesel-distillate and liquefied petroleum gas (LPG)/propane.

- DoD delivered more than 209,000 billion British thermal units (BBtus) of energy to its EO13423 goal-subject facilities during FY 2009. This was a 1.3% increase over the FY 2008 amount. About ninety-four percent of the energy was to the military departments.

- About 79% of the delivered energy to DoD facilities in FY 2009 was from electricity and natural gas. In addition to using fuel oil, coal and purchased steam, DoD facilities also used a small percentage of LPG (Liquefied Petroleum Gas)/propane and renewable energy sources. In FY 2009 DoD used renewable energy for 3.6% of its delivered electricity.

- In FY 2009 all DoD facilities had an overall energy intensity level of 104,527 Btus per gross square foot (GSF), which was a 1.1% increase over the 103,692 Btus/GSF level for FY 2008

- When renewable energy is measured per 10 USC section 2911(e), DoD procured or produced 6.8% of its electric consumption from electric renewable sources in FY 2009.

- The Army had 67 active renewable energy projects operating in FY 2009. Of the total, 42 were generating electricity qualifying for credit toward the renewable energy goal and nearly all the energy produced was used on-site in Federal Army facilities. The Army consumed 2.1% of its electricity from renewable sources in FY 2009.

Figure 8-5. Mid-flight refueling of U.S. fighter jets.

How many gallons of jet fuel does this operation take? How many times a day is it preformed across the globe? Keeping in mind that a fighter jet burns from 1,000 to 3,000 gallons an hour, and adding all other fuels

used 24/7/365 around the globe, we see an ocean of fuel floating out of the U.S. refineries in support of the daily military operations.

Because of that, and in order to keep up with the times, the U.S. DoD is taking serious measures to reduce fossils use for fuels or other uses. Here are some of the fact:

- The Air Force purchases Renewable Energy Certificates (RECs) to help achieve its renewable energy goals and continues to pursue the development and installation of renewable energy, with 5.8% of total electric consumption from renewable sources in FY 2009.

- The Navy consumed 0.6% of its electricity from renewable sources in FY 2009. These sources include wind and solar electric generation, but do not include the Navy's Naval Air Weapons Station (NAWS) China Lake geothermal site, whose 270 megawatts (MW) production capacity is not directly consumed by the Navy.

- The Navy uses thermal energy from the waste heat of six cogeneration systems to further meet energy intensity reduction goals. Cogeneration credits account for six percent of the energy intensity reduction, the largest single technology contribution. The most recent addition, a 39 MW cogeneration plant in Yokosuka Japan, came on line in November 2008 to contribute 2.5 percent of the reduction to the Navy's overall energy intensity.

- DoD continued efforts in FY 2009 to acquire alternative fuel capable vehicles and provide the necessary supporting infrastructure. DoD acquired 105 neighborhood electric vehicles; received 863 low-speed or mini-utility vehicles, 150 hybrid electric vehicles, and 1,485 E-85 alternative fuel capable vehicles; and ordered 800 low-speed electric vehicles (LSEVs). DoD also completed the infrastructure for 16 E-85 and/or B-20 alternative fueling stations.

- DoD continues to make progress installing cost effective renewable energy technologies and purchasing electricity generated from renewable sources (solar, wind, geothermal, and biomass). In FY 2009, 3.6% of DoD's electrical consumption came from renewable electricity, exceeding the EPAct 2005 goal of 3% and improving on the 2.9% achieved in FY 2008.

- DoD produced or procured 6.8% of all electricity from renewable sources. This is three percent less than the 9.8% reported by DoD for FY 2008 because of a change in the calculation method. In FY 2008 and earlier, the calculation compared all renewable energy (electric and non-electric) produced or procured to total electricity consumed. Had FY 2008 been reported limiting renewable energy to only renewable electricity sources, the FY 2008 achievement would have also been reported as 6.8%.

- The Army purchased 148,000 MWH of electricity qualifying toward the renewable energy goal. A large portion of the electricity was a direct purchase from a two megawatt photovoltaic array at Fort Carson, Colorado. Other sources included Renewable Energy Certificates (RECs) purchased by Fort Lewis, Washington; Fort Carson, Colorado; and the Pennsylvania Army National Guard. The Army also purchased a substantial amount of energy from renewable municipal solid waste plants at Redstone Arsenal, Alabama, and Aberdeen Proving Ground, Maryland.

- The Air Force acquires the most economical renewable energy available on the market, either by acquiring bundled renewable electricity, or through RECs. Bundled renewable electricity purchases represent 39% of the renewable purchases; RECs represent the remaining 61%.

- In 2010, U.S. armed forces consumed more than five billion gallons of fuel in military operations. The number one factor driving that fuel consumption is the nature of today's defense mission. Twenty-first century challenges to U.S. national security are increasingly global and complex, requiring a broad range of military operations and capabilities, which means large, steady supplies of energy.

Wow. Impressive! And this is only part of the long list of energy related projects and operations undertaken by the Pentagon. There are surely many other applications of energy around the globe that we are not aware of and which information can not be released to the general public. In any case, try as it may, the mighty Pentagon cannot replace, nor even reduce, the use of fossil fuels for its vehicles.

Since we don't see any plausible solutions for this problem, the fossil fuels remain the Achilles Heel of the U.S. Armed Forces. Their cost is the biggest problem at the present, and steps have been taken to reduce it.

The long-term problem, however—the imminent and finite depletion of the fossil fuels—is like a sharp dagger hanging over the military. The generals know very well that without fossil fuels the mighty, agile military machine will become an immobile piece of rust-

ing junk. The consequences for the U.S. and the world would be devastating…

Operational Energy Strategy

In focusing on energy for the Warfighter, the goal of the "Operational Energy Strategy" is to ensure that the armed forces will have the energy resources they require to meet 21st century challenges.

> *"Our dependence on foreign oil reduces our international leverage places our troops in dangerous global regions, funds nations and individuals who wish us harm, and weakens our economy."*
> From: *Risks to National Security, U.S. Military Advisory Board report*

The DoD energy strategy outlines three principal ways to a stronger and more efficient armed forces:

- *More fight*, less fuel: Reduce the demand for energy in military operations. Today's military missions require large and growing amounts of energy with supply lines that can be costly, vulnerable to disruption, and a burden on Warfighters. The Department needs to: reduce the overall demand for operational energy; improve the efficiency of military energy use in order to enhance combat effectiveness; and reduce military mission risks and costs.

- *More options*, less risk: Expand and secure the supply of energy to military operations.

Most military operations depend on a single energy source, petroleum, which has economic, strategic, and environmental drawbacks. In addition, the security of the energy supply infrastructure is not always robust.

This includes the civilian electrical grid in the United States, which powers some fixed installations that directly support military operations. The Department needs to diversify its energy sources and protect access to energy supplies in order to have a more reliable and assured supply of energy for military missions.

- *More capability*, less cost: Build energy security into the future force. Current operations entail more fuel, risks, and costs than are necessary, with tactical, operational, and strategic consequences. Yet the Department's institutions and processes for building future military forces and missions do not systematically consider such risks and costs.

Fuel convoys like that in Figure 8-6 haul thousands of gallons of gasoline, diesel, jet fuel, and lubricants

Figure 8-6. U.S. Army fuel supply convoy.

to U.S. military bases and field operations around the world. Each fuel convoy in war time operations (like during the Iraq war) supplies fuel to move the military hardware around the battle arena for several days or weeks. The same amount of energy could supply power to half a dozen Sub-Saharan countries for a month or more. This number multiplied several times is what the entire military machine needs to operate efficiently.

The Department of Defense still needs to integrate operational energy considerations into the full range of planning and force development activities. Energy will be, in itself, an important capability for meeting the missions envisioned in the QDR and the National Military Strategy

Military installations should also have the ability to "island" themselves from the power grid in order to support their strategic mission. Ideally a military base can switch off of the energy grid, and still manage to power the essential operations of the installation necessary to conduct their mission. Note that the social implications of islanding will not be addressed but they are profound.

This is important in the event of a physical or cyber attack on either the energy grid or our energy resources. Should this happen, the Army, must be able to keep key functions alive until power is reestablished. Currently the military is mainly buying its energy commercially having outsourced most of its power production facilities.

However, there is no current plan or funding to harden these outside facilities and the supplying grid. In order to island itself, an installation would almost certainly have to move some kind of energy source onto the installation itself, whether that is solar, wind, biomass, a nuclear generator, or any other kind of renewable energy source.

Critical national security and homeland defense missions are at risk of extended outage from failure of the grid. Currently the Army and the Department of Defense have implemented goals and policies to facilitate energy security and other NetZero energy initiatives.

Note: Net Zero Energy Installation (NZEI) is an installation that produces as much energy on site as it uses, over the course of a year. To achieve this goal, installations must first implement aggressive conservation and efficiency efforts while benchmarking energy consumption to identify further opportunities. The balance of energy needs then are reduced and can be met by renewable energy projects.

From a solely ROI perspective these alternative energies are not cost efficient when compared to fossil fuels. The argument must be made here that the value of alternative energy is directly related to our energy security and good environmental stewardship—regardless of price and any other difficulties.

Renewables might be expensive and a hassle, but are the only viable energy sources on the long run.

The problem with the military energy use is that the huge machinery used in field operations require petroleum products to move efficiently. It is not possible (at least for now) to run Abrams tanks or fighter jets with solar power, or any other fuel. Using biofuels is the best (if not only) option for replacing the fossil fuels in the future, but that in itself is problematic and requires much more effort.

So, DoD is faced with a real dilemma; while reducing energy use, or using renewables, on the military bases is possible, fueling the U.S. war machine is not that easy. It is a thirsty beast that is built to use a lot of fossil fuels—crude oil derivatives mostly. Replacing these on mobile units would not be easy nor fast…

So, for now DoD is doing all that is in its power to save energy and fuels; whatever and whenever possible.

DoD Net Zero Energy Initiative

The DoD Net Zero Energy Installation Initiative is an effort to increase the energy independence of installations by offsetting total annual energy use through on-site energy production. The Army goals for this initiative are for five Army installations to be net zero (outside energy independent) by 2020, 25 installations by 2030, and all Army installations by 2050.

Some of Army's renewable energy efforts in support of this initiative are: a large concentrated solar system at Fort Irwin, California, a geothermal steam resources at Hawthorne Army Depot, Nevada, replacing 800 petroleum-fueled non-tactical on-post vehicles with neighborhood electric vehicles, and develop consolidated waste to fuel projects at several locations. McGuire Air Force Base, New Jersey is implementing four measures (including day lighting) to render a 30,774-square-foot facility the first energy-neutral facility in the Air Mobility Command.

Metering of Electricity Use

DoD has identified 37,493 buildings requiring either standard or advanced metering. The Navy accounts for 70% of this total (26,311), with meters installed in 64% (16,929) of its buildings in FY 2009. This effort, combined with that of the Air Force (93% complete), Army (44% complete) and the other ten DoD components (44% complete), brought the total number of metered buildings in DoD to 23,674, or 63% of all identified buildings in FY 2009.

Buildings Energy Efficiency Standards

In FY 2009, 99% of DoD's new building designs include provisions to make them 30% more energy efficient than the American Society of Heating, Refrigerating and Air Conditioning Engineers (ASHRAE) 90.1-2004 standard.

The Army Corps of Engineers continues work with the DOE and the Office of the Army Assistant Chief of Staff for Installation Management to develop design guides for implementing building efficiency standards mandated by the EPAct 2005. The Corps has completed prescriptive design guides for battalion headquarters buildings, permanent party barracks, training barracks, and tactical equipment maintenance facilities.

Four of the most prevalent types of buildings were constructed in conjunction with Army troop stationing actions. Use of these design guides will help new building designs be 30% more energy efficient than ASHRAE standards without having to model each individual project.

The Air Force initiated 98 new building designs in FY 2009. 100% will be life-cycle cost effective and meet the goal to exceed the ASHRAE efficiency standards by at least 30%.

The Navy is expecting 99% of the FY 2010/2011 military construction facility designs to meet or exceed Federal standards and achieve life cycle cost effective sustainable designs.

The Defense Logistics Agency is redesigning a Child Development Center project in Columbus, Ohio, to ensure that it will be at least 30% more efficient than ASHRA standards and that it will be eligible for LEED Silver status.

The Missile Defense Agency is presently planning three military construction building projects (Redstone Arsenal, Alabama and Fort Belvoir and Dahlgren, Virginia). These buildings are designed in accordance with EPAct 2005, EO 13423 and the Whole Building Design Guide, and each building will be LEED certified.

The TRICARE Management Agency's Bureau of Medicine and Surgery (BUMED) is working with Naval Facilities Engineering Command (NAVFAC) to ensure that all new design work will exceed the ASHRAE efficiency standard by at least 30% where achievable. Ninety-three percent of new BUMED building designs started since the beginning of FY 2007 are expected to exceed the ASHRAE standard by 30%.

Although the above figures are somewhat outdated, they clearly show the trend towards energy efficiency improvements in the U.S. military. These measures are saving a lot of energy and are diverting tons of fossil fuels from use in the military bases.

Water Conservation

In FY 2009 DoD facilities had an overall water intensity (consumption) level of 57.1 Gallons per gross square foot (GSF), which was a 1.1% decrease over the FY 2008 level of 58.1 GSF. Although the Army had an increase in water intensity compared to FY 2008 (58.2 compared to 54.0), the other military departments and DoD components (particularly the Defense Commissary Agency) had lower water intensity levels, which resulted in DoD meeting its water intensity reduction goal of 4.0% for FY 2009

All of the military departments continue to install water conserving toilets and urinals, low flow faucets and showerheads. Some installations have instituted aggressive leak detection surveys and followed up with repair programs for leaky valves and damaged pipelines, which have significantly reduced water consumption (as much as 20% at one location). All facility projects executed by the Army Corps of Engineers follow the International Plumbing Code, which prescribes water conserving fixtures.

The Air Force is implementing stronger conservation measures to reduce landscape irrigation, encourage low-water plantings, and repair leaking water and steam lines. The Defense Commissary Agency requires low-flow toilets and urinals with electronic flush sensors for new and renovated commissaries. Proposed landscaping for new DeCA facilities is closely reviewed during all phases of the design for low maintenance and watering requirements and includes requirement for xe-riscaping and drip versus sprinkler irrigation systems.

Energy Innovation

A letter signed by more than 350 U.S. veterans, including retired generals and admirals, issued in 2012 urges Congress to support the Pentagon's initiative to diversify its energy sources, limit the demand, and lower overall energy costs. This effort is needed, if the military is to be able to deploy clean energy technology, as needed to reduce its dependence on fossil fuels, while at the same time contributing to our energy security and strengthening the national security.

The concern here is that some Congressmen are working to restrict the department's ability to introduce energy innovation by using advanced biofuels programs. On the long run, this would hurt DoD's capacity to reduce its dependence on oil prices, which might affect negatively its operational effectiveness.

Some congressional amendments, if adopted, would limit and even bar the DoD from purchasing alternative fuels, which could also affect the fuels used to power all (including unmanned) vehicles for training and military operations.

The U.S. military is the world's largest consumer of liquid fuels, according to the experts, which needs over 22 gallons of fuel per soldier per day to support combat operations.

Every $10 increase per barrel of crude oil, costs DoD additional $1.4 billion yearly.

The Pentagon uses nearly 400,000 barrels of crude oil daily, at the cost of over $16 billion in 2013. During 2011 and 2012, the department had additional $3 billion in unanticipated fuel costs, which additional budget burdens adversely affect the military readiness and put troops overseas at imminent risk.

So, most people agree that both the DoD and the nation as a whole must reduce their expenses and dependence on foreign oil. In response, DoD is trying to become more energy efficient by testing different advanced biofuels in its ships, planes, and land vehicles.

Alternative fuels are seeing as the energy products to be used when they become readily available and cost-competitive with conventional fuels. The so called "second generation" biofuels can be produced from domestic biomass sources, for "as is" use. It will require no changes to current engine design, and would provide the same or better performance than conventional fuels.

The development to date are as follow:

- Since 2007 a provision of the Energy Security and Independence Act, or Section 526, has allowed the Pentagon to develop advanced biofuels and other clean and renewable transportation alternatives. These are domestically produced fuels from non-food agricultural feedstocks, such as camelina and sawgrass, that could be used in existing engines in fighter jets, land vehicles, and ships. They are, theoretically, supposed to provide similar power and performance as the traditional fuels used today in these areas.

The Navy and Air Force have successfully tested and certified some of these fuels, along with similar efforts by some commercial airlines, but the new defense authorization bills would repeal or weaken the provision for such fuels. It would also send a negative signal to the biofuels industry, which could result in adverse impacts to U.S. job creation, competitiveness, and overall rural development efforts.

- The military's energy policy and investments have encouraged a domestic industry that is creating jobs and businesses in America. For these and other reasons, groups such as Airlines for America, the Farm Bureau, the Advanced Biofuels Association, the American Security Project, and others support the defense department's renewable energy programs and policies.

Table 8-5 shows the progress of reducing energy consumption at DOD.

DOD and The Environment

One significant achievement of DoD's policies is the full recognition of the direct connection between energy today and tomorrow, and the environment. Secretary of DoD, Mr. Panetta drew a clear line between environmental, energy use and our national security, since their relationship is important, as recently established in the Pentagon strategy.

Table 8-5. Energy consumption per square foot of Federal buildings (as compared to 2003 energy consumption)

Fiscal Year	Percentage Reduction
2006	2
2007	4
2008	9
2009	12
2010	15
2011	18
2012	21
2013	24
2014	27
2015	30

"In the 21st century, reality is that there are environmental threats that constitute threats to our national security," Mr. Panetta stated in 2013, stopping short of naming the individual threats, which are many. Instead, Mr. Panetta lays out a strategic framework for how the military thinks about and acts on long-term environmental and energy issues. All this at a time of austerity, and when the military's alternative energy investment and achievements are under assault by members of Congress.

Never mind the hard times. Mr. Panetta confirmed the military services' commitment to add 3 GW of renewable energy in the coming years, emphasizing the military's history of anticipating changing trends in the global markets. Tomorrow is knocking on the energy door and we must answer.

Security implications of climate change have been a special interest of Mr. Panetta's policies too. Rising sea levels, extended droughts, and more frequent and severe natural disasters around the world are seen by him as a signal to increase the energy conservation effort by the U.S. military and a new and increasing demand for humanitarian assistance from the U.S. military.

Note: Melting polar ice caps are prompting competing claims by different nations in the mineral-rich Arctic, so Panetta calls for the U.S. to ratify the U.N. Convention on the Law of the Sea, which is the international treaty governing activity in the world's waters. It has been endorsed by U.S. military leaders, business groups and environmentalists, but has been stalled in the Senate for years by opponents who claim that it is a threat to U.S. sovereignty.

The U.S. is the only industrialized nation that has not approved that treaty as yet. There are many reasons for that, but the reality is changing daily and needs a clear and decisive action, and soon.

DoD Energy Roadmap

The U.S. DoD has a seemingly firm grasp on the energy reality in its ranks, and its operations at home and abroad, as follow:

- Energy security and climate change goals are clearly integrated into national security and military planning processes.
- DoD aims to design and deploy systems to reduce the burden that inefficient energy use place on our troops as they engage overseas.
- DoD understands its use of energy at all levels of operations. It should also know its carbon footprint (or bootprint) and strive to reduce it.
- DoD is attempting to reduce its use of energy at all installations through aggressive pursuit of energy

efficiency, smart grid technologies, and use of electrification and renewables in its vehicle fleet.

All this sounds good, and the U.S. DoD is on the right path thus far, but the implementation of all these plans and ideas requires a lot of money. So, the only question here is how and when will all this happen in light of the decreasing DoD budget and the ongoing political constipation in Washington DC.

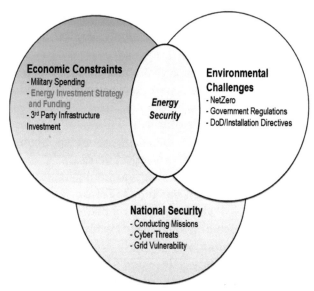

Figure 8-7. Energy security a la DoD

Of course, as far as DoD is concerned, national security is its long-term goal number one, with the national energy security in the center of its short-term concerns.

Our national security depends heavily on energy, which is why our energy security is a high priority to DoD and U.S. politicians.

Lack of energy is a direct and serious threat to our national security and our way of life. Simple as that. Hence the extra effort in ensuring energy supplies at all cost. It takes a lot of energy to get energy. For that purpose we police the world at all times, making sure parts of the world don't blow up, and that the energy supply import-export channels are clear of obstructions.

The economic constrains and the environmental challenges are another set of significant factors that affect our energy security. They are aggressively addressed because they also affect negatively our national security. So, as DoD is moving forward to a new era of energy dependency (or rather less dependency), we must remember that tanks, jet planes, patrol boats, and super aircraft carriers do not run on words and promises. Nor can they run on solar or wind power. Not now, and not anytime soon. They, instead, require thousands and millions gallons of fossil fuels...every single day... day after day. These fuels—crude oil and natural gas mostly—transport our solders and power most military hardware.

> *Crude oil ensures the proper function of the military machine which ensures our energy security, which ensures our national security. No oil; no security; energy, national, or otherwise!*

So, the DoD is planning to save some energy, which it is doing quite successfully thus far. Nevertheless, its world wide operations will still swallow a significant portion of our national (and global) fossil fuels in the process. There is no way around it for now.

In summary, we clearly see that the U.S. DoD understands and appreciates the role energy plays in its operations. The U.S. generals feel the pain of high fuel prices, and are worried about fuel supply disruptions. Both of these are potential barriers in the drive to a more powerful, mobile, and versatile global fighter units.

In order to anticipate and control the threats to its energy budget and supplies, the DoD is actively pursuing a well designed short-term plan for ensuring its energy supplies and the national energy security. The long-term plan of achieving a full energy independence is also in the planning stages, but it is too far fetched to become a reality anytime soon. Certainly not in our life time, since there are some serious problems in its way, which even the mighty U.S. military cannot overcome.

The proverbial "giant with feet of clay" comes to mind here, except that the feet of the U.S. military are made out of oil. It (the giant) runs fast and tirelessly on his feet of oil, but as soon as the oil runs out, the giant would not be able to move an inch. No more running! Instead, the giant will be laying in the dust on its back, and the era of U.S. military might and total superiority would come to an abrupt end. All because of lack of oil...Unless drastic measures are taken to anticipate and prevent this from happening.

But for now, the U.S. military machine's energy supply is secure, and all the DoD generals have to worry about is the energy supplies of tomorrow. And this is where things might get very complicated and even dangerous. If DoD does not anticipate and control the upcoming energy debacle; that of global demand increase and pending depletion of the fossil energy sources, we might find ourselves sitting motionless and mightily

vulnerable on our massive island. The only hope then would be that our enemies would also lack fuel needed to cross the ocean in order to attack us. That would be a dark and scary day.

Case Study—Afghanistan

We were, and still are (albeit partially), engaged in bloody wars in Iraq and Afghanistan, the latter so-called "Just War." We sent hundreds of thousand young American men and women to fight a deadly force across the world. First, it was Saddam Hussein and his weapons of mass destruction. We run Bagdad overnight and killed Saddam, but did not find his WMDs. Instead, we found the Taliban and Al Qaeda fighters and the battles were transferred from the city streets into the cliffs and caves of the Bora-Bora mountains.

This was the beginning of a new and long war geared to eliminate "Islamic terrorism" and instate Western style democracy upon the people of these countries. After 15 years of fighting and democracy building we are still fighting. The nature of the battles changes daily, and so do the strategies, but the end is still elusive.

After thousands of killed and wounded solders and many hundreds of billions of taxpayers dollars, we are still not sure where these countries are going. Democracy, however, is not in the cards...it never was. The counter-terrorism campaigns have been also used to obfuscate some other objectives of the wars in that part of the world.

Could it be that we have a profit (access to cheap crude oil) driven agenda? A war for resource? Iraq has a lot of oil, which we are very interested in keeping control over? Afghanistan is a strategic hub in Central Asia, bordering on the former Soviet Union, China and Iran. It is also at the crossroads of pipeline routes and oil reserves. And then, there is its huge mineral wealth and untapped natural gas reserves.

Afghanistan is a land bridge, in the midst of the strategic trans-Afghan transport corridor, which links the Caspian sea basin to the Arabian sea. Several oil and gas pipeline projects have been contemplated in Afghanistan, including the planned crucial-transit-corridor; the $8.0 billion Turkmenistan, Afghanistan, Pakistan, India pipeline project. It consists of 1500 miles of pipelines transporting Turkmenistan's natural gas across Afghanistan to export points. Turkmenistan possesses the third largest natural gas reserves after Russia and Iran, so controlling the transport routes would be of great importance to the global energy markets.

Afghanistan also has huge and yet untapped natural gas, coal, and oil reserves. Recent estimates, confirmed by U.S. EIA in 2008, place Afghanistan's proved gas reserves at about 5 trillion cubic feet. Are we planning to convert the war gains into a colonial policy in order to control the natural resources rich country? Are we planning to take possession of Afghanistan's natural gas exports, in order to prevent the Russia and other neighbors from getting them first?

For now, however, narcotics is the main natural resource production in Afghanistan. The opium trade has increased over 35 times, with about 7,000 tons produced and exported annually at about $2.5 billion annual income. This, compared to less than 200 tons produced in 2000 cannot be counted as a success after a decade of bloody invasion.

And now we se ISIS taking hold in that part of the world. Experts believe the increase of Taliban attacks is related to ISIS attempting to gain a foothold in the country. ISIS is a growing threat inside of Afghanistan, where disaffected Taliban people are looking for a more radical cause. Taliban commanders are worried about the rise of ISIS and the fracturing in their own ranks inside of Afghanistan, and how it affects their ability to take over the country following the completion of the U.S. army drawdown.

All this means that our mission in that part of the world is not finished. It also suggests that many additional millions of gallons of crude oil will be wasted in conducting never ending operations there.

LEGAL ENERGY MARKETS

Laws are spider webs through which the big flies pass and the little ones get caught, stated Honore de Balzac a long time ago. It seems that, as far as the legal profession is concerned, Mr. Balzac's statement still holds true. Just looking back—not very far back—at our energy markets, we see Exxon, BP, and a number of other big flies passing through the legal spider webs. And surely enough, the smaller ones, Solyndra, Enron, etc., and many other, much smaller, got caught.

> *As far as the energy markets are concerned, the activities usually start with financial, and end up with legal, actions.*

The spider web of financial activities and legal actions is affecting the energy markets around the globe. As the energy markets grow and change, the legal profession finds itself more and more involved with their daily activities.

Background

Traditionally, the American energy markets have been regulated by antitrust and utility laws, predominately in; a) crude oil, b) natural gas, c) coal, and d) electricity energy markets respectively. Today, however, the energy markets have got very complex and competitive. The change is due to ever changing dynamics in the world's socio-political structure, its energy needs, as well as the advent of new technologies in the energy sector. Changes in policies and regulation in different countries have contributed significantly to the drastic changes of late.

The increased competitiveness in the energy markets have increased also the number and severity of risks for the energy companies. Because of that, many energy companies are turning to energy derivatives and other financial instruments to help hedge the risk. The changes have also led regulators to look for new approaches. For that purpose, they borrow from securities regulation and focus on actions of "manipulation and deceit" by some major energy market participants.

The securities model does not fit the energy markets well, because it exposes consumers to price risks associated with the exercise of market power by the sellers. We saw a glimpse of what could happen in 2008, when global oil prices shot to over $180 per barrel.

The increased risks are a new phenomena, that was unknown under the traditional energy regulations. The problem is that the new securities regulation model overlooks important ways in which sellers exert market power at the expense of consumers in the absence of fraud or deceit enforcement. But more about that later...

The Legal Context

The curriculum of some law schools—for students interested in obtaining a Certificate in Natural Resources Law, as needed to practice law in the natural resources, energy, or environmental fields—includes a number of disciplines. It starts with introductory classes in Natural Resources Law and Environmental Law, designed to familiarize students with the law and the related issues.

The studies then progress to Sustainable Energy, which focuses on the legal implications of policies and technologies that seek to minimize carbon emissions in the development and delivery of energy.

Additional studies include Oil & Gas Law, which examines correlative rights of surface and mineral owners, and the rights to explore, mine and extract, develop, and transportation. There are also short courses, like the Law of Renewables. These courses must be short because the solar industry law is still in its infancy, and the legal activities are still not as numerous or profitable as in the mature energy markets. This confirms the fact that technology usually progresses much faster than the law. In this case, the law is at least several decades behind the development in the solar energy markets.

In more detail, the legal curriculum of some legal schools includes:

Introduction to Natural Resources, which introduces students to various natural resources law offerings, geared both to provide a broad base of knowledge to interested students and to inform students who may be considering the natural resources law certificate.

Natural Resources Law examines the specialized property rules governing estates in natural resources, the correlative rights of surface and mineral owners, and the rights to explore, mine and extract, develop, and transport natural resources, with primary emphasis on "hard" minerals.

Sustainable Energy Law explores the significant challenges facing the energy industry, including climate change, energy independence and security, traditional pollution, regulatory burdens, jobs, energy prices, "peak" supply, and increased energy demand.

Oil and Gas Law applies property law and contract law principles to a complex natural resource, and evaluates resource rights from the perspective of the developer, the property owner, and the regulator.

Environmental Law examines selected topics in the law governing the protection of air, water, and land from pollution.

Mineral Title Search and Examination provides the students with an overview of the process of examining mineral titles and rendering legal opinions on title in the context of mineral production and development.

Regulation of Energy Markets and Utilities provides insight to state and federal utility law and regulation, that governmental power over electric, natural gas and oil markets. Students will explore and study administrative law issues, regulatory agencies, and the role of regulation.

Water Resources Law examines regulation of water systems by states and the federal government. Water is arguably our most important natural resource, so the course explores increasing water scarcity, degraded water quality, stresses to watersheds, and public water supply issues stemming from aging infrastructure, global issues like international trade, management of waters shared with Mexico or Canada, including the effects of global warming.

Coal and Mineral Law examines the legal, business and environmental side of the coal and hard mineral law. The focus here is on the nature of ownership of subsurface minerals; methods of transferring ownership; property rights; partition among co-owners; analysis of leasehold estates, rights and duties; coal mining rights and privileges; regulatory and environmental issues; and administrative processes.

The Law of Renewables examines the laws and policies designed to promote renewable energy development, where students review existing renewable energy technologies and the practical limitations on their development, siting and integration into the U.S. electricity grid.

Environmental Dispute Resolution Practicum explores the characteristics of environmental and natural resource disputes, how they arise, and how we choose to resolve them.

Real Estate Transactions Practicum focuses on how commercial and residential real estate is conveyed. Lectures discuss legal theories of titles, title transfer, and present day ownership issues.

This is a long list of legal subjects and issues to be put into the right perspective and eventually resolved. It also shows how deep into the matter, and how involved with the issues at hand, the legal profession is. One is left wondering what would the energy markets be without the legal professionals help...?

We can only guess that since there are billions of dollars involved in energy related cases, we will see many lawyers heavily involved and working hard in the energy markets. We rest assured that while there are energy markets, there would be lawyers taking care of, and resolving, the problems in them.

Legal Aspects of the Energy Markets

All power generation products, facilities, processes, and services carry some liability. The type and size of the liability varies with type of fuel (coal, solar, etc.) and the type and size of the facility (large or small coal-fired power plant, or solar fields).

For the sake of this text we can divide the different power generators into:

• The power generators that do not require any fuel to operate, or fuel-less power generators such as solar, wind, hydro, and ocean-wave technologies. They do not need a special effort to be made, or expense accrued, to obtain fuel for their operation. Sunlight, wind, river and dam water, and ocean tides and waves are always present for free and the power contained within them is used to generate heat and/or electric energy as is.

• Power generators that require fuel for power generators, such as coal, oil, gas, nuclear, geothermal steam, and biofuels, do require fuel to be produced and delivered to the plant for burning. So, this is a special effort needed to obtain the respective fuels. This effort might consist of digging a mine shaft, drilling a well, or plowing the fields to grow energy crops. And of course, this comes at an additional expense and liabilities.

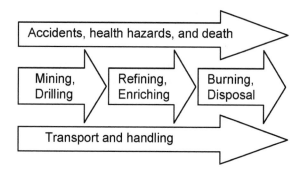

Figure 8-8. Cradle-to-grave cycle of fossil fuels.

As seen in the previous chapters most of the steps in the fuels production cycle (mining, drilling, refining, enriching, burning and disposal) are dirty, dangerous, and expensive. There is always a possibility of a work-related accident affecting a worker, or a group of workers. These could range from a mine shaft ceiling collapse, or flooding, to a an explosion in a oil refinery. The extent of the damages would depend on a number of factors, but in any case it is a lawsuit waiting to happen.

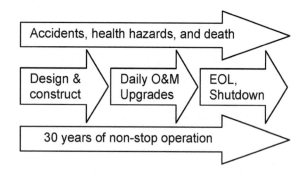

Figure 8-9. Cradle-to-grave cycle of power plants

Each minute of every step of the cradle-to-grave operations cycle of a power plant is an accident waiting to happen, with a lawsuit to follow soon after. Although safety procedures and equipment have progressed immensely lately, and so we don't hear about power plant accidents every day, they do happen. From a single injury of a worker falling from scaffolding, to a major

Chernobyl-like disaster; accidents do happen and will continue happening in the future.

And, of course, there are the long-term damages and illnesses. Black lung disease, caused by mining operations has affected millions of miners and local people. Many other have been hurt by hydro-dam reservoirs construction and collapses, oil trains derailments, nuclear accidents, etc. disaster brought up by the numerous steps of energy production, transport, and use.

This is fact of life. It is the way the beast works, and there is not much we can do to eliminate accidents from the power generating industry. The only way to protect people and the environment is to design better protection equipment and safer operation methods. And of course, the legal profession is waiting; able and willing to help sort out and resolve all issues related to the energy markets.

The Watchdogs

All processes related to power generation are complex, expensive and labor intensive, and are usually accompanied by significant environmental effects, including damage to human health and life. Government bodies try, with varying level of success, to ensure a level of safety and environmental control into these fuels production activities.

The key agencies in charge of ensuring the health of the people and the environment in the U.S are:

- The U.S. Environmental Protection Agency (EPA), as the name suggests, is the agency in charge of ensuring that all effort is made to protect the environment—including some very serious damages to the land surface and the underground layers during mining, well drilling, and all such activities related to fuel production.

- The U.S. Occupational Safety and Health Administration (OSHA) is an agency established under the U.S. Occupational Safety and Health Act signed by President Nixon in 1970. Its mission is to "assure safe and healthful working conditions for working men and women by setting and enforcing standards and by providing training, outreach, education and assistance." In addition, OSHA is also responsible for enforcing a variety of whistleblower statutes and regulations.

- The National Institute for Occupational Safety and Health (NIOSH) is responsible for research of, and making recommendations for, the prevention of work-related injury and illnesses. NIOSH is part of the Centers for Disease Control and Prevention within the U.S. Department of Health and Human

Services, with research laboratories and offices in Alaska, Colorado, Georgia, Ohio, Pennsylvania, Washington State, and W. Virginia. NIOSH' professionals specialize in a number of disciplines including epidemiology, medicine, industrial hygiene, industrial safety, psychology, and statistics.

These, and a number of other federal and state agencies, are responsible for setting and enforcing the rules for the work conditions, and the overall business environment in the above mentioned fuel producing operations. A number of universities, R&D labs, environmental groups and many private citizens are involved in watching, reporting, and analyzing the activities at different work sites.

The key concerns that are regulated in one way or another during construction, operation and decommissioning of fuel products and power generation facilities, which are under constant watch and scrutiny, are as follow:

• Acoustics
• Air Quality
• Cultural Resources
• Ecological Resources
• Environmental Justice
• Hazardous Materials/Waste Management
• Health and Safety
• Land Use
• Paleontological Resources
• Socioeconomics
• Soils and Geological Resources
• Transportation
• Visual Resources
• Water Resources

The Labor Problems...

Making money as a priority #1 is a principle that the energy industry has been using since its inception. It all started with making a quick buck by abusing coal miners...

In 1883, a railroad from Pocahontas to Tazewell County, Virginia opened a gateway to the untapped coalfields of southwestern West Virginia, which brought a dramatic population increase virtually overnight. Thousands of European immigrants and a large number of African Americans migrated to the area.

Thus a new economic system controlled by the coal industry was created. Miners worked on company terms, which controlled everything from salaries to rentals the cost of items from the company store which was the only place they can buy food. The prices of everything were over-inflated, since there was no other alternative. Miners were paid tokens which could be used only at the company store, so this way even when wages were increased, the companies simply increased prices at the company store to balance what they lost in pay. Capitalism at its best. Or as Merle Travis famously put it in his 1946 song Sixteen Tons, "...I owe my soul to the company store."

Miners were paid based on tons of coal brought from the mines in cars holding 2,000 pounds. The owners altered the cars to hold more coal, so miners would be paid much less than what they produced. On top of that, the miners were docked pay for slate and rock mixed in with the coal in each car. Docking was a judgment call on the part of the supervisors so miners were regularly and intentionally cheated.

Safety in mines was of the greatest concern. In 1907, an explosion Monongah, Marion County, WV, killed 361. There were suggestions at the time that during World War I, a U.S. soldier had a better chance of surviving in battle than did a coal mine worker. So the United Mine Workers of America (UMWA) was formed in Columbus, Ohio, in 1890. The UMWA successfully organized miners in Pennsylvania, Ohio, Indiana, and Illinois.

In 1912 UMWA miners on Paint Creek in Kanawha County, WV demanded higher wages, but the request was rejected and the miners walked off the job. This was the beginning of one of the most violent strikes in the nation's history. The owners brought in mine guards whose primary responsibility was to break the strike by making the lives of the miners as miserable as possible. Evicting miners and their families from company houses was one of the measures, so the evicted miners set up tent colonies on site. These events became known, national labor leaders got involved. As the strike was becoming more violent, the Governor imposed martial law and sent 1,200 state militia men to disarm both the miners and mine guards. The violence increased nevertheless, until on the night of February 7, 1913 when an armored train, the "Bull Moose Special," run over a miners' tent colony which sparked a gun fight during which sixteen people died.

A series of attempts to settle the strike failed, and after additional violence it was settled in July, 1912 without answering the two primary grievances: the right to organize and the removal of mine guards. Nevertheless, this strike produced a number of labor leaders who would play prominent roles in the years to come. Corrupt UMWA leaders were ousted and a group of young rank-and-file miners were elected as leaders of the union.

Many other strikes followed, the most famous of which is the 1921 Blair Mountain strike. During several months of violence a number of policemen were killed, which forced the dispatch of federal troops. Those who surrendered were placed on trains and sent home, but the leaders were held accountable for the actions of all the miners. Special grand juries handed down 1,217 indictments, including 325 for murder and 24 for treason against the state.

The Chief Villains

One of the most violent, and most famous, incidents in the U.S. mining history is the Ludlow Massacre. In the spring of 1914 the Colorado National Guard and camp guards opened a random fire on a tent colony of 1,200 striking coal miners and their families at Ludlow, Colorado.

The strike was organized by the United Mine Workers of America (UMWA) against coal mining companies in Colorado against the Rockefeller family-owned Colorado Fuel & Iron Company (CF&I), the Rocky Mountain Fuel Company (RMF), and the Victor-American Fuel Company (VAF).

When the owners decided to put an end of the strike, the guards opened fire that killed several men. As the battle intensified, 2 women and 11 children were killed too. Some of the victims were asphyxiated and burned to death under a collapsed tent.

In retaliation, the Ludlow miners attacked several mines during the next ten days, destroying property and engaging in several battles with the Colorado National Guard. In the end, nearly 200 people lost their lives in the skirmish, making it deadliest strike in the history of the United States.

John D. Rockefeller Jr., who was the chief mine owner at the time was blamed for the violent deaths of over 20 people. He made amends with the workers and the American public by going unescorted to the mine and spending time with the workers, promising better work and living conditions and things at the mine got back to normal for everyone…except for those who had to be buried.

Nevertheless, even when forgiven by the miners, this incident put the Rockefellers on the list of most notorious U.S. villains, since that was not the first or last time they have been violating workers' rights. John D. Rockefeller Sr. got into the oil business in the mid 1800s, and started Standard Oil Company in 1870. His success is attributed to inventing new operating practices and using different forms of manipulation and exploitation of workers to achieve his goals.

Using the lack of legal structure in the new oil business as a crutch in achieving his goals, Mr. Rockefeller was able to push the competition aside and exploit the workers for the benefit of his bank accounts. A number of approaches used by him and his cohorts were later on found to be illegal, and even greater number are today considered immoral.

As a result, the U.S. Supreme Court of the United States found Standard Oil Company of New Jersey in violation of the Sherman Antitrust Act. The court ruled that the trust originated in illegal monopoly practices and ordered it to be broken up into 34 new companies. Rockefeller benefitted personally from the breakup, which during the next 10 years brought him close to $1.0 billion in profits from the 34 new companies.

John D. Rockefellers, Sr. was an example of a successful American entrepreneur-capitalist at the turn of the 19th century. Even today, he is held as an example of what should and should not be done in business. He enriched himself upon rivers of sweat from thousands of workers, exploited carelessly natural resources, and blatantly used unfair government influence. He invented procedures, organizations, plans, and what not with one and only goal in mind—to make more and more money. He claimed that making money is his gift, and is what God has called him to do on this Earth. A great man as far as capitalism is concerned, but a low class person in human terms.

He succeeded, no doubt, and reached his goal of becoming the richest man in America, but God works in mysterious ways, and made Mr. Rockefeller's life quite miserable in 1890, giving him a gift of alopecia; a rare condition that resulted in the loss of all his body hair. Thus marked, and quite embarrassed, Mr. Rockefeller would've given half (or more) of his fortune for a piece of hair on his head.

And he did just that, when shortly before his death in 1937, giving away half of his fortune to churches, medical foundations, universities, and many other organizations. Was that act driven by good will, conscience, or his devout faith in God is still debated, but John D. Rockefeller will be always remembered for all good and bad he has done to America, the American business and the people.

Although oil companies cannot follow his example while operating on U.S. territory, they do use Rockefeller-like illegal and immoral practices in other countries. A number of oil companies—including U.S. based such—are exploiting and poisoning people in Africa, Asia, and South America. They don't hesitate to use unfair MOs in order to achieve their goal of making a profit, even if that means destroying the local environment and people's lives.

We can't help it but wonder if Mr. Rockefeller, Sr. is turning in his grave, or if he is cheering them on.

But let's go back to the present day energy markets; starting at the very beginning of the energy sources production cycle:

Mineral Rights

Usually all mineral resources, including valuable rocks, minerals, coal, oil or gas found on, or within the Earth, belong to the government. Organizations or individuals in many countries cannot legally extract and sell any mineral commodity without first obtaining a permit from the government.

In the U.S. and a few other countries ownership of the mineral resources is granted to the owners of the surface area. The property owners had both "surface rights" and "mineral rights," which complete private ownership is known as a "fee simple estate," which is the most basic type of ownership.

Basically, the owner controls the surface, anything under the surface and the air above it. The owner has unrestricted access to these commodities, and a right to sell, lease, or give away each of these rights, or the entire package.

With the expansion of commercial mineral production, the ways in which people own and manage their properties became much more complex. Today complex leases, sales, and gift that were executed in the past have created an environment of people or companies who have partial or full ownership of, or rights to, many land parcels intended for different commercial applications.

Different states have different laws that govern mining and drilling, as well as the transfer of mineral rights. Companies usually pay money for using the underground commodity (coal, gas, etc.) to an owner who retains his right to use the surface land for whatever he needs. This fee simple owner does not have interest or the ability to extract the minerals beneath his property but a large company does. And so the company pays the owner for the right to do just that. Win-win situation.

An agreement is basically made to share the property, which is usually very complex to begin with, and could get even more so as time goes on. The transaction can involve all unknown mineral commodities that exist beneath the property, or it can be limited to a specific mineral commodity.

Buying a coal seam, for example, is much more complex than buying a house where you simply pay for it, and file a title transfer. With mineral rights, the buy-er basically moves in (under your house) and is given certain rights to exploit the property. And here is where things get hairy, because digging under ones house could bring a number of unwanted consequences.

When a company buys mineral rights it also buys the right to enter the property and remove the resource at will. The surface owner has no say in when or how the mining or well digging will be done. Most disagreements between buyers and sellers occur at the time of mining, and/or after some unspecified and unexpected consequences surface. The only way to have some control at these cases is to anticipate what might happen and put it in the contract. Otherwise the dispute may grow too big for resolution and the case might end up in court.

Mineral rights in general also include the rights to any oil and natural gas deposits that might be in the contract area, and these rights can be sold or leased to others. In most cases, oil and gas rights are leased, simply because the there is usually uncertainty if, or how much, oil or gas is under ground, so it makes sense to pay a small lease amount.

Most lease agreements also include a signing bonus, and the owner gets a share of the oil or gas sales. The customary royalty percentage is 12.5% of the value of the oil or gas at the wellhead, which in some states is a state law. Some owners can get 15-25% or more.

Coal Mining Laws and Regulations

The coal industry is very likely the most regulated in the energy sector. Federal legal requirements apply to all specific activities associated with coal mining, transport, and use. A number of federal and state laws, regulations, and Executive Orders apply to coal mining activities.

The extent to which the federal requirements apply to specific coal mine projects on tribal lands depends upon the nature of the project, its location, and size.

Note: For the most part, state laws and regulations do not apply to coal mining on tribal lands.

Some of the federal laws and/or requirements that apply to specific activities associated with coal mining are in the areas shown in Table 8-6.

Just imagine the breadth of knowledge and understanding the legal professionals involved in the coal industry must have to handle any—let alone all—of these activities and the related issues. One thing is for sure looking at this extensive list; thousands of lawyers will be busy for a long, long time solving the issues that arise in different sectors of the coal industry. Can't live with them, can't live without them...so we must agree that

Table 8-6. Coal related activities and regulations

Acoustics
- Noise Control Act

Air Quality
- Clean Air Act

Cultural Resources
- American Indian Religious Freedom Act
- Antiquities Act
- Archaeological and Historic Preservation Act
- Archaeological Resources Protection Act
- Executive Order 11593: Protection and Enhancement of the Cultural Environment
- Executive Order 13007: Indian Sacred Sites
- Executive Order 13175: Consultation and Coordination with Indian Tribal Governments
- Executive Order 13287: Preserve America
- Historic Sites, Buildings, and Antiquities Act (Historic Sites Act)
- Illegal Trafficking in Native American Human Remains and Cultural Items
- National Historic Preservation Act
- Native American Graves Protection and Repatriation Act
- Theft and Destruction of Government Property

Ecological Resources
- Bald and Golden Eagle Protection Act
- Clean Water Act
- Endangered Species Act
- Executive Order 11988: Floodplain Management
- Executive Order 11990: Protection of Wetlands
- Executive Order 12996: Management and General Public Use of the National Wildlife Refuge System
- Executive Order 13112: Invasive Species
- Executive Order 13186: Responsibilities of Federal Agencies to Protect Migratory Birds
- Federal Insecticide, Fungicide, and Rodenticide Act
- Fish and Wildlife Coordination Act
- Migratory Bird Treaty Act
- National Wildlife Refuge System Administration Act
- Noxious Weed Act
- Rivers and Harbors Act
- Wild Free-Roaming Horses and Burros Act

Energy Resource Development
- Surface Mining Control and Reclamation Act
- Tribal Energy Resource Agreements

Environmental Justice
- Executive Order 12898: Federal Actions to Address Environmental Justice in Minority Populations and Low-Income Populations

Hazardous Materials & Waste Management
- Comprehensive Environmental Response, Compensation, and Liability Act
- Emergency Planning & Community Right-to-Know Act
- Executive Order 12856: Federal Compliance With Right-to-Know Laws and Pollution Prevention Requirements
- Federal Insecticide, Fungicide, and Rodenticide Act
- Hazardous Materials Transportation Act
- Pollution Prevention Act
- Resource Conservation and Recovery Act
- Toxic Substances Control Act

(Continued)

Table 8-6. (*Continued*)

Health & Safety
- Emergency Planning & Community Right-to-Know Act
- Executive Order 13045: Protection of Children From Environmental Health Risks and Safety Risks
- Federal Mine Safety and Health Act
- Occupational Safety & Health Act

Land Use
- Air Commerce and Safety Act
- Farmland Protection and Policy Act
- Federal Land Policy and Management Act
- National Trails System Act
- Rivers and Harbors Act
- Soil and Water Resources Conservation Act
- Surface Mining Control and Reclamation Act
- Wild and Scenic Rivers Act
- Wilderness Act

National Environmental Policy Act
- National Environmental Policy Act

Paleontological Resources
- Antiquities Act
- Paleontological Resources Preservation
- Theft and Destruction of Government Property

Soils & Geological Resources
- Farmland Protection and Policy Act
- Soil and Water Resources Conservation Act

Water Quality
- Clean Water Act
- Safe Drinking Water Act

---

the legal profession is an important, albeit peripheral, part of the energy markets.

Note: Some of the peripheral energy markets, as we reviewed in this text, are more peripheral than others. Although the legal profession seems quite remote and far from the actual energy markets operation, its importance and influence is hard to deny.

Just consider the impact of a $100 million dollar lawsuit on a small drilling company. It might make the difference between its further growth or immediate demise. Or the lawsuit filed by a NIMBY group against the construction of a new nuclear power plant in their neighborhood. The lawyers, fighting the issues on both sides, will determine if the multi-million plant will be constructed, or that it would die in its inception stages.

Operational Safety

Safety during design, installation and operation of energy sources production and power production is a major concern to all involved. It is also a major business for the legal profession. In the U.S., a lot of effort and money is poured in safety measures and initiatives on

different levels, geared to ensure the maximum safety of personnel, equipment, and facilities. Even more money is spent on resolving lawsuits, resulting from different O&M related accidents and incidents.

Figure 8-10 shows the safety levels during the different stages of production and use of the major power generating sources. Coal, for example, causes large number of injuries and fatalities during the production (mining) operations, mostly in foreign countries. Nuclear, on the other hand, is dangerous during use, where it could cause major damage—including environmental damage, illness and death—during a nuclear accident like the one at the Fukushima nuclear plant in 2011.

Finding the right balance here is critical, because it means not only cheaper electricity, but it could also make the difference between life and death for many people. Japan's population, for example, was getting cheap electricity from the Fukushima nuclear power plant, but the locals paid for it with their lives, properties, and well being. And now, the nuclear plant owners and operators will pay a significant amount of money to mitigate the land and compensate for health damage claims resulting from the disaster.

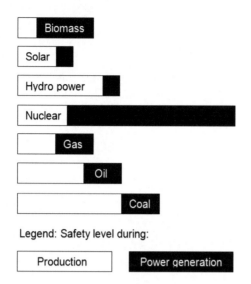

Legend: Safety level during:

Figure 8-10. Safety during production and use

Damages and Royalties

Underground activities in the U.S. and around the world are a big business. Thousands of different mining and drilling operations are conducted every single day. While coal, as the only solid fossil energy source, stays put in its place forever, oil and gas move through the rocks in a random and unpredictable patterns. This movement is even more unpredictable when prompted by high pressure chemicals during hydrofracking operations.

The high pressure chemicals travel freely through the maze of rocks and within tiny cracks created in them. When the pressure is increased enough, a well can connect to and drain oil or gas from adjacent wells in neighbors' lands. The states have recognized this problem and have produced regulations that govern the fair sharing of oil and gas royalties. The drilling companies are required to specify how oil and gas royalties will be shared among adjacent property owners, based upon preliminary analysis.

This procedure, known as "unitization" is used to regulate the process and is part of the permitting procedures in many states. The situation is more complicated in states that do not have unitization rules. In Pennsylvania, for example, there are different rules for natural gas sharing. The rules are different for different depths, as well as at certain positions in the stratigraphic column.

No matter what the particular situation, and regardless of how well the contracts might be written, disagreements erupt in many cases at some time during the life of any project. The problems arise usually as soon as the surface owners notice damage to the land or

any property on it. This is where a well written contract could save a lot of hassles. It is also where lawyers get involved and take over the off-, or in-court dispute negotiations.

In worst of cases, land and/or other property damages appear after the drilling operation has been completed. Some of the worst damages also appear months and years after the mine or well are shut down. Land collapse, or fracking liquids leaks nearby could destroy a property. If the mining or drilling companies are already gone and the contract is complete, then there would be no one to hold responsible. In such cases, even more, and higher paid, lawyers get involved.

A broadly spread damage occurring in mining or drilling operations, could damage large areas of public lands or water supplies. Entire communities relying on these water sources could be affected. This situation might also affect the health of the locals, and the value of their properties. A rural property without a water supply is close to worthless, and a number of lawsuits have been filed for that reason as well.

The amounts of money that change hands in mineral property transactions in most cases is huge.

The total yield from the sum of land lease, signing bonus, and royalties often exceeds the value of the surface rights.

Oil and natural gas transactions involve large sums of money but the true value can be difficult to estimate, especially in areas where very little drilling has occurred in the past, or where deep rock units are being tested for the first time. The profits derived from any underground exploration depend on many other factors and variables as well.

As an example, a 200 acre property situated on top of a shallow, eight feet thick coal seam resulted in a royalty agreement on the time of contract signing of $5 per ton to be paid to the owner. During the 20 years coal mining operations and a coal recovery rate of 90% the owner would be paid nearly $8 million, or $400,000 annually.

The same 200 acre property also has natural gas reserves underneath and property owner will receive a 12.5% royalty based upon the wellhead value of the gas. At the time of production, gas is $6 per thousand cubic feet at the well, and the well produces 2 million cubic feet of gas per day, then the property owner would receive over $80,000 dollars for one year of gas production. This, however, is five times less than the coal mine royalties paid for a similar energy value.

Hydrofracking Legal Market

A large number of laws and regulations address the process of hydrofracking. Directly, or indirectly, they focus the attention of the law professionals on the issues at hand—mainly the dangers of air, water, and soil contamination and their direct and indirect effects on humans and properties.

Due to its size and importance of this issue, and since we are no legal experts, we cannot possibly get deep into the legal aspects. Instead, we can only provide a partial list of laws and legislations containing provisions for use or not use of hydraulic fracturing techniques and materials.

Some of the key laws and initiatives related to, and/or applicable to hydrofracking are:

- Safe Drinking Water Act;
- Federal Role in Regulation of Underground Injection;
- Energy Policy Act of 2005: A Legislative Exemption for Hydraulic Fracturing;
- EPA Guidance for Permitting Hydraulic Fracturing Using Diesel Fuels;
- Clean Water Act;
- Clean Air Act;
- Resource Conservation and Recovery Act;
- EPA's Regulatory Determination, Comprehensive Environmental Response, Compensation, and Liability Act;
- National Environmental Policy Act;
- Toxic Substances Control Act;
- Emergency Planning and Community Right-to-Know Act;
- Emergency Release Notification and Hazardous Chemical Storage Reporting Requirements; + + State Tort Law; etc.

Where do we start? One does not have to be a lawyer, or any type of specialist to see the labyrinth of laws and issues that the hydrofracking industry, and those affected by it, have to deal with. And, of course, there is an army of lawyers ready to jump and defend both sides of each issue at hand.

How much does the legal profession helps in the resolution of the hydrofracking issues is questionable at best, but this is the system we live in. The legal profession did not create the problems; they are product of the hydrofracking industry's enormous and very quick expansion. It is a large peripheral market, created by thousands of drillers and most oil companies.

Nuclear Power and the Law

Laws and legal decisions relating to the use of nuclear energy in the United States are difficult, because they are complicated by the technological complexity and uncertainties in some aspects of nuclear power generation

The Nuclear Power Laws

The use of nuclear energy for both military and peacetime applications is strictly regulated by a number of federal, state, and local laws. Here in we focus on the civilian applications of nuclear energy, with special emphasis on the conditions under which nuclear power plants may be built, operated, and dismantled, as well as the environmental restrictions placed on such plants.

Some of the most important laws related to nuclear power are:

Atomic Act Energy Act, 79-585 of 1946

The Atomic Energy Act of 1946 was the first law passed relating to the control of nuclear energy in the United States. At the conclusion of World War II, the question arose as to the fate of the Manhattan Engineering District (MED), the program under which the first nuclear weapons were built. The preponderance of opinion among lawmakers appeared to be that the MED should be continued, probably in some modified form, such that control over the development and use of nuclear energy would remain in the hands of the U.S. military.

In May 1945, Representative Andrew J. May and Senator E.C. Johnson introduced a bill into the U.S. Congress aimed at such an objective. The May-Johnson bill would have assigned control over all nuclear energy development to the War Department.

At first, the bill seemed assured of passage, especially when President Harry S. Truman announced that he favored its provisions. The May-Johnson bill, however, aroused a considerable amount of concern among scientists, many of whom had worked on the Manhattan Project and appreciated the horror inherent in nuclear weapons.

The scientists were worried that the future development of nuclear energy would remain in the hands of the military. Although traditionally a largely nonpolitical group, the scientific community rapidly organized to oppose the May-Johnson bill and eventually threw its support to a competing bill introduced by Senator Brien McMahon in December 1945.

The McMahon bill proposed the creation of a civilian agency that would take over responsibility for the development and promotion of nuclear energy in the United States. After a congressional debate that had lasted nearly a year, the McMahon bill was passed. It became the Atomic Energy Act of 1946.

The Atomic Energy Act of 1946 provided the fundamental structure under which nuclear energy was to be controlled in the United States, a structure that remains in place to the present day. The act authorized the establishment of two agencies, the Atomic Energy Commission (AEC) and the Joint Committee on Atomic Energy (JCAE) of the U.S. Congress.

The role of the JCAE was to provide oversight on the activities of the AEC. The AEC was assigned six major responsibilities:

1. Assisting and fostering private research and development of nuclear energy,
2. Providing for the open and free dissemination of scientific information about nuclear energy,
3. Sponsoring of research on nuclear energy,
4. Controlling the production, ownership, and use of nuclear materials,
5. Studying the social, political, and economic effects of nuclear energy, and
6. Keeping Congress informed about developments in the field of nuclear science.

ACE was also to be responsible for the design, development, construction, and maintenance of all nuclear weapons in the nation's arsenal. In order to carry out its functions, the Atomic Energy Commission was to be organized into four major sections: divisions of research, production, materials, and military applications.

In spite of its important accomplishments, the Atomic Energy Act of 1946 contained some serious defects. For example, an amendment offered by Senator Arthur Vandenberg (R-Mich.) gave veto power over AEC decisions to the committee's Board of Military Advisors, essentially limiting to some extent its scope of operations.

Also, the monopoly on nuclear materials given to the committee by the act was a matter of serious concern to private industry, which had great hopes for the use of such materials in the development of many peacetime applications of nuclear science.

Atomic Energy Act, 83-703 of 1954

The Atomic Energy Act of 1954 was adopted primarily to remedy one of the defects that many people saw in the original Atomic Energy Act of 1946, namely, the prohibition against private ownership of nuclear materials.

With the election of President Dwight D. Eisenhower and a Republican controlled Congress in 1953, corporate interests received more attention than they had under earlier Democratic-controlled administrations and Congresses.

On February 17, 1954, President Eisenhower asked Congress to consider revisions in the Atomic Energy Act of 1946 that would make it easier for the federal government to share information about nuclear energy and to assist private corporations in the development of nuclear facilities than had earlier been the case. In response to this request, Congress passed the Atomic Energy Act of 1954, and the president signed the bill into law on August 30 of that year.

The major purpose of the act was to provide for "the development, use, and control of atomic energy [in such a way] as to promote world peace, improve the general welfare, increase the standard of living, and strengthen free competition in private enterprise." The Atomic Energy Commission was directed to provide information about nuclear science and technology to private industry and to cooperate with private corporations in the development of peacetime applications of nuclear science.

The act also instructed the AEC to develop regulations and standards for the design, construction, and operation of nuclear power plants for the protection of human health and the environment, and to establish methods by which these regulations and standards were to be enforced.

Detailed instructions about the licensing required for nuclear power plants and other nuclear facilities were provided. Overall, the act significantly expanded the authority and responsibilities of the ACE and the Joint Committee on Atomic Energy. Finally, the AEC was authorized to expand its efforts to work with other nations and international agencies in the development of peacetime applications of nuclear energy.

Rice-Anderson Act, 85-256 of 1957

Enthusiasm for the development of nuclear power plants in the 1950s was tempered by a number of problems that made private industry reluctant to become involved in such construction. Among the most important of these problems was the legal liability a company would face in case of an accident at a nuclear facility. A study conducted by researchers at the Brookhaven National Laboratory in 1956 (WASH-740) concluded that, in a worst-case scenario, 3,400 people would be killed and 43,000 injured in case of a nuclear accident. In addition, property damage could reach as much as $7 billion.

Few, if any, private companies were willing to accept this level of risk in the construction of a nuclear reactor. In addition, the amount of insurance from private companies (even high-risk takers, like Lloyd's of London) available to cover a company's liability in case of a nuclear accident was far too low. The best terms available capped limits at $60 million for liability and an additional $60 million for property damage.

Under these circumstances, the U.S. Congress became convinced that the development of nuclear power in the United States was going to be possible only if the federal government itself assumed all or most of the financial responsibility for accidents that might occur at a facility.

As a result, it passed a piece of legislation authored by Senator Clinton Anderson and Representative Melvin Price that absolved private companies from all legal liability for any accidents that might occur at a nuclear power plant.

In addition, the Price-Anderson bill allocated $500 million to a fund designed (along with the $60 million available from private insurers) to pay the victims of any such accident. The legislation included an expiration date of 1967, 10 years after its adoption.

As expiration of the original Price-Anderson Act approached, Congress adopted an extension to the legislation in 1966. This extension maintained liability protection for nuclear power plants in essentially the same form as the original act, although it did make some minor adjustments in provisions for liability and payments in case of an accident. The act was amended again in 1975 (for 12 years) and 1988 (for 14 years).

As the latest extension was about to expire, Congress took up yet another amendment to the act in 2 001. Although the latest extension has not yet been approved, a temporary continuation of Public Law 85-256 was passed by the Congress in 2003.

One of the most significant changes included in the 1975 and 1988 amendments was a shift of primary liability in case of an accident from the federal government to private industry. A fund to cover costs of such an accident was created and paid for by a tax on nuclear power plant owners.

Over time, the value of that fund has increased; today it amounts to more than $10 billion, an amount that would be increased to nearly $11 billion if the 2001 amendment passes.

Private Ownership of Special Nuclear Materials Act, 88-489 of 1964.

One of the most serious concerns of legislators in their early debates over the peacetime applications of nuclear energy was the ownership of nuclear materials. With the experience of the first two fission bombs fresh in their minds, public officials worried that these materials might fall into the wrong hands and be used for weapons production.

Under both the Atomic Energy Act of 1946 and the Atomic Energy Act of 1954, therefore, ownership of nuclear materials was restricted to the U.S. government. As reasonable as this policy may have been from a security standpoint, it proved to be a blessing and a curse in the development of nuclear power plants. The industry began lobbying almost immediately after World War II, then, for the right to own nuclear materials on their own.

Slowly, government officials and legislators were won over to this position and, in 1964, the Congress passed the Private Ownership of Special Nuclear Materials Act, which allowed private companies to purchase and own nuclear materials. The government still tracks and has the ultimate say so over the procurement, handling, and use of these materials.

Energy Reorganization Act, 93-438 of 1974

One of the fundamental criticisms that had long been aimed at the Atomic Energy Commission was the inherent conflict in two of its major responsibilities: promoting the use of nuclear power in the United States while adopting and enforcing standards that ensured the safety (and, hence, the cost) of nuclear power plants. In trying to carry out these two somewhat conflicting roles, the AEC was often accused of siding too often with the nuclear industry and too often ignoring the safety problems associated with the reactors.

In 1974, the U.S. Congress dealt with this problem by abolishing the AEC and reassigning its responsibilities to two new agencies: the Nuclear Regulatory Commission (NRC) and the Energy Research and Development Administration (ERDA).

The former agency was charged with the regulatory functions previously carried out by the AEC. It was assigned the task of regulating the:
1. Design, construction, and operation of nuclear reactors;
2. Research on nuclear materials; and
3. Safety and safeguard functions related to nuclear energy.

The latter agency was given the task of promoting research and development on nuclear power. ERDA remained in existence for only three years. In 1977, it was abolished as a separate agency and its functions were transferred to the new Department of Energy, created by the Department of Energy Organization Act of 1977. Some critics have suggested that the goal of the 1974 Energy Reorganization Act was never adequately achieved since the NRC continued to have relationships too closely tied to the nuclear power industry to allow it to carry out its regulatory tasks adequately.

Nuclear Waste Policy Act, 97-425 of 1982

The third in the federal government's trio of acts designed to deal with the nation's nuclear waste problems was the Nuclear Waste Policy Act of 1982 (NWPA), fashioned to deal with high-level wastes. The act charged the Department of Energy with responsibility for developing a plan by which the federal government would develop an underground repository for the permanent storage of high-level wastes.

The act also set out a time table for the selection, testing, approval, and opening of a site for the nuclear waste depository. According to that timeline, the first wastes were to be delivered to the storage site no later than January 31, 1998.

As with the LLWPA, the NWPA has been largely unsuccessful in solving the problem it was drafted to handle. The selection of a site (Yucca Mountain, Nevada) was accomplished, and extensive research on the site has been conducted. But, largely as a result of unexpected environmental problems and the fervent opposition of the state of Nevada and a number of environmental groups, progress in construction of the waste repository has gone forward only very slowly. In 2004, development of the site was delayed once again when the Federal Appeals Court for the District of Columbia ruled that the Department of Energy's plans for ensuring the safety of stored wastes for a period of 10,000 years was inadequate.

The court ordered DOE to modify its plans to extend the period of time during which the buried wastes could be considered to be safely entombed. Unfortunately the Yucca Mountain plan was abandoned in 2012.

What all this shows is that since its very inception, the U.S. nuclear energy industry has been supervised, directed, and propped (technologically and financially) every step of the way by the U.S. government. All kinds of technical, regulatory, and financial assistance were provided to the nuclear industry. It could have not developed on its own. As a matter of fact, it cannot even

function today without the governments' non-stop daily intervention and assistance.

Some people would disagree with the above statement, but taking a close look at the nuclear waste materials disposal procedures and accident insurance provisions, we clearly see the helping hand of the U.S. government. Without this help, these essential activities might prove impossible for the industry to handle on its own. The disposal of nuclear materials, would be an enormous task that the nuclear industry would stumble over, if left on its own recourses.

There are many reasons for this preferential treatment, but the final result is the fact that nuclear and fossil power generators have benefited tremendously from the government's assistance, and are now the dominating energy sources in the country. This shows the enormous power the U.S. government has in directing the energy markets.

This also means that if and when the U.S. government decides that the renewables are worth supporting—as we saw happening (albeit briefly) during the 2008-2012 boom and bust of the renewable technologies—then the U.S. and global energy markets would change profoundly too.

But such development is still far in the future. For now, most governments are focusing on providing maximum amount of power via fossils and nuclear energy sources, while taking their sweet time in deciding what to do with the renewables.

And speaking of renewables, the most promising, and the one that needs most help today, is the solar power generation sector.

Solar Power and the Law

Some of the key legal elements involved in solar power generation are:

Transaction

Solar contracts are usually signed after negotiating new solar projects. These are structured in several different ways that have their own benefits, costs and complexities.

The major contractual structures for new solar projects are:
- Power purchase agreements (PPAs), which are agreements of the energy consumer (utility company usually) to purchase the solar power generated by a solar system which is owned, operated and maintained by a third party,
- Full solar system ownership, where a consumer purchases, owns and maintains a solar system and

uses the generated solar electricity for his/her own power needs, and

- Solar system leasing, which is one of the financing alternative to owning a solar system for residential or commercial use.

Regulations

Building a new solar project, which is designed to generates electricity for public consumption converts a private individual or a commercial entity into a regulated mini-utility. The status of the project depends on whether the power it generates is used on site or is sold to third-parties. The sale can be done via power purchase agreement (PPA), or directly to the public through an authorized connection into the grid.

The different cases bring different jurisdiction-specific questions, so the power provider must be able to evaluate and resolve all regulatory issues prior to starting the construction. Connecting into the grid is usually highly regulated, and the owners must understand and be quite familiar with the regulatory framework, the local and regional interconnection rules, and especially with the requirements of the local power utility.

Permitting

This is the start (and sometimes the end) of solar projects. Ensuring that the design and construction is in compliance with all applicable federal, state, and local laws and regulations in the areas of environmental protection, land use, and zoning requirements is a complex and expensive undertaking. Ensuring that all necessary permits can be obtained and all other requirements to construct and operate the solar system is to be done first and before any other work can be started.

Liability protections from environmental regulatory agencies needs to be assessed in most cases, as well as lease, easement and other access rights need to be secured, as needed for the successful completion of the project, and have to be done before the actual construction work is started.

Engineering, Procurement and Construction (EPC)

The cost of large-scale solar energy systems is in the tens and hundreds of millions of dollars. Performance contracts are signed with the EPC parties should be properly drafted and thoroughly negotiated with all parties involved, in order to clearly set forth the respective obligations and responsibilities. This is a critical step of the effort, because it determines the completeness and fairness of the undertaking. A well done EPC

contract can make the difference between success and failure of the project.

Financing

Solar projects' financing is usually a mix of existing and new equity and debt, which varies from case to case since there is no one-size-fits-all financing solution. In most case, the solar projects are completed using a combination of all options, usually both; equity and debt. The ITC fate will challenge the financing industry and will determine the future of solar in the U.S.

Operations & Maintenance

System owners have the option to take care of the system themselves, or to hire third-party contractor for all operation and maintenance services. The costs for such services must be factored in the owner's calculations and financial models. Long-term maintenance of a solar system comes with a number of obvious and some not so obvious risks, which must be taken into account at the very beginning of the project design stages.

Many small and large projects have failed because of small details were overlooked at the design stage. Even small detail can grow very big during 25-30 years of non-stop operation, so we cannot over-emphasize the need for thorough familiarity with the solar project's function, O&M, and other requirements.

Risk Assessment

If a solar system fails to perform as predicted, the consequences can have economic impacts, and instead of saving money the owners could easily find themselves in a losing territory. If, for example, the regulatory policies and programs, or the federal and state incentives are modified or revoked, the initial profit-loss calculations will have to be seriously modified.

These uncertainties of the present vs. future operational and regulatory risk cannot be eliminated entirely, the participants in the solar energy industry must be well informed and able to identify and evaluate the real and perceived risks. They should be able to also allocate them accordingly and reflect them as needed in the calculations and contracts in order to ensure the success of the project.

Long-term Performance

Solar and wind power generating systems are supposed to operate non-stop for 20, 25, 30 years, or more. As with any other technology, such a long period of operation cannot be expected to be totally free of problems. Equipment in solar and wind power plants can fail on day one as easily as during year #30. It depends on initial quality of the different components, quality of installation, and the particulars of long-term O&M.

A solar power plant, consisting of millions of solar panels cannot possibly be expected to function 100% of the time at 100% capacity during 30 years of non-stop operation. Neither can a wind power plant be 100% efficient during such a long time.

So what happens if and when a part of the solar pant malfunctions, due to a batch of failing solar panels, for example? Or a wind power plant loses several wind turbines during a storm. In most cases, there is insurance that would take care of the problem. But even then, the plants will have reduced outputs and lesser profit margin.

How about a massive failure, with 50% or more of the power plant being taken out by bad weather, flood, or other natural events? In such case, we can only hope that the owners have good insurance and good lawyers to help them overcome the problems.

In all cases, however, we must remember that sensitive equipment is left to the mercy of Mother Nature, who misbehaves from time to time. Because of that, the initial plant design must always include 'what-if' responses to reduced performance, and handling severe incidents.

Legal Issues

There are a number of legal cases involving solar projects; small and large alike. On the residential side the most common complaint and a reason for legal action is roof access, where neighboring trees or structures are shading a planned or existing solar installation. New laws passed on the matter have resolved some of the issues, but there are still disputes on the subject.

For example, California bill SB1399 signed into law Governor Arnold Schwarzenegger took effect on January 1, 2009, amending the 1979 Solar Shade Control Act by making disputes over trees and shrubs that cast shade on a neighbor's solar panels civil matters rather than crimes. The law applies also to existing trees and shrubs that later grow big enough to shade the solar panels.

The law was prompted by a Sunnyvale, California couple who ordered to top their redwood trees to allow sunlight to fall on their neighbors' solar panels. The trees grew to 40 feet tall and shade more than 10% of their neighbor's solar panels between the hours of 10 a.m. and 2 p.m.

The court has determined that 2 of the 8 trees must be cut. The redwood tree owners decided to fight the law by appealing the decision.

The greatest and most important battles have been raging in the deserts, where large and very large solar power plants were planned and constructed. The desert looks open and empty, but it is not. It is a delicate habitat to birds, insects, and plants; some of which are endangered.

There are over 80 large solar projects planned or in construction stages in the California deserts alone. That many more are planned for the deserts in the other Western states—Arizona, Utah, New Mexico, Colorado, and Nevada. The California projects cover over 1,000 square miles, located mostly on pristine BLM publicly-owned lands, and most of the projects have already been given the right-of-way grants.

The impact on rare plants, vegetation, animals and majestic landscapes is hard to comprehend, let alone evaluate in terms of numbers and money. Solar construction severely impact the area by modifying the surface and damaging thousands of acres of pristine desert land. And there is always the indirect impact caused by the building of access roads, running new power lines, and brining in invasive plants and animals, which eventually expand the amount of land modified or destroyed above and beyond the initial project size.

There are recommendation by the California Native Plant Society (CNPS) and other organizations that government agencies need to discuss their intentions in advance and negotiate alternative locations for solar plants, especially these for new solar thermal plants. The anticipated project impacts have to be well understood and appropriately mitigated, after the regulators fully evaluate the impact of the new installations on the local ecosystem.

Using millions of gallons of fresh water for cooling of the steam generated at solar thermal plants is another issue that needs a complete evaluation. The desert cannot be qualified as a reliable water source and any use of water in these amounts will eventually lead to depletion of the water table, which brings additional hardships for the local plants, animals and humans.

In response to all these issues, there have been a number of law suites filed by private citizens and organizations, attempting to stop or at least modify planned solar projects. Some of these have been successful, some not so much.

There are also lawsuits by owners and developers against PV module manufacturers, solar companies against homeowners associations, home owners associations against solar companies and a number of additional combinations and permutations. These actions complicate the already complex situation in the solar sector, eventually making it more expensive.

And so the legal battle continues; awaiting more efficient policies and new product and services standards and regulations.

Wind Power and the Law

Since the 1800's, farmers in the United States have been using millions of windmills across the Midwest and the Plains for water pumping and later on for generating electric power. These windmills fit nicely into the existing landscape and have not created any significant problems.

Today, however, the wind energy industry is building large- scale wind power fields, where their monstrous 200-300 feet tall structures with waving arms, a la Don Quixote, have a tremendous impact on the visual and noise landscape, and many other aspects of the rural culture. In many cases, wind energy development has raised issues among neighbors and private landowners on one hand, and wind energy development companies on the other. There are also notable conflicts between local officials and wind power development companies.

Many farmers and other rural landowners have entered into long-term agreements with wind energy companies for the placement and operations of wind turbines on their property. Generally, those agreements are drafted in favor of the wind energy company and require negotiation and modification of numerous provisions to make them fair from the landowner's perspective.

Unfortunately, and regardless of its elegancy and good intentions, wind power expansion has roughed up lots of feathers, and has brought up a number of serious lawsuits. Some potential liability concerns associated with wind energy development have been raised through the years. Although some of them could be classified as a nuisance, the legal system takes them seriously, so they must be properly addressed. Of especial concern is the fact that many gigantic wind turbines have been installed on private land, and/or adjacent to residential areas Some of the legitimate concerns, and issues that have brought up legal action, are:

- Damages to adjacent property caused by alteration or damage of the flow of surface water due to the construction of the wind power field and the related access roads
- Aesthetic damage is claimed by such, which although considered a nuisance is a liability,
- Damages and/or injury caused by ice throws by the blades have been reported and lawsuits filed,

- Stray voltage from the wind generators have the potential (no pun intended) to hurt people and animals,
- Interference with electromagnetic fields is a concern of to people living in nearby residential areas.
- Fire caused by wind turbine malfunction, or as a result of a lightning strike could cause damage to people and animals,
- Interference with television and radio signals of nearby residents,
- Death of protected birds and/or bats,
- Adverse health impacts on residents living or working nearby.

These are serious issues, which have caused a number of NIMBY (not in my back yard) anti-wind power actions on part of the locals living and working near large wind power fields.

For example:
- In 2007, a wind power station with 200 turbines was to be constructed close to a residential development. The locals sued to permanently enjoin the construction and operation of the wind power station. They cited increased noise, aesthetical impact on the view-shed, flicker and strobe effect of light reflecting in their houses from the turbine blades. Even more serious were the claims for potential danger from broken blades, ice throws and potentially reduced property values. The court's decision was in favor of the locals, stating that the wind power turbines represent a nuisance. This was enough to conclude that the local' claims were sufficient to enjoin a nuisance. In addition, the court ruled that even though the State had previously approved the facility, their approval did not abrogate the common law of nuisance.
- In 2008, a landowner sued the county commission which approved the construction of a large-scale wind farm adjacent to his property. The landowner also claimed that he was physically attacked by a county commissioner for his public opposition to the siting of the wind turbines. In addition, the landowner claimed that the wind turbines were a nuisance, because his land was completely surrounded by the turbines, the turbines caused a "powerful strobe light effect," were loud and contributed to the loss of equity and marketability of his home and the loss of view and quiet enjoyment of his property. The Federal District Court for the Western District of Missouri dismissed the

case, but noted that the plaintiff could amend his complaint to replace the county commission with a private party as the defendant

- In 2008 a wind plant development in England, was denied due to noise concerns and errors that the developer made in assessing the project's noise impact. At least one academic study has concluded that background noise does not effectively "mask" the thumping sounds produced by blades, and that such sounds are perceptible only at a certain distance from the wind turbine. Regardless, living with a constant thumping noise in the background is considered intolerable.

- In 2008 a Texas Court of Appeals upheld a trial court ruling that dismissed a nuisance lawsuit filed by property owners that complained about the "aesthetical impact" of a large-scale, 421-turbine wind farm. The court refused to expand nuisance law to cover actions for aesthetical impact that causes emotional injury. The court found that the common-law doctrine of nuisance in Texas had never recognized a nuisance claim based on aesthetical impact.

- In 2008, the Federal Aviation Administration (FAA) was ordered to reconsider its decision to allow the construction of a wind farm near the site of the new Las Vegas Airport. The evidence presented indicated that the turbines would interfere with the airport's radar systems. The Federal district court determined that the FAA's determination was irresponsible, arbitrary and capricious.

The potential legal troubles have not affected wind power developers in a significant way as yet. As the number of cases grows, however, it will become more and more important for the issues to be resolved one way or another.

In addition to the legal issues, and/or because of them, there are additional problems that are affecting the wind power developers.

Some of the additional problems faced by wind power developers are:

- Difficulty in negotiating viable wind energy power purchase and off-taker agreements is the biggest challenge facing the wind power industry today. The current economic chaos and abundance of cheap energy, such as natural gas, has created an unfavorable pricing which makes the successful negotiation of purchase contracts difficult. That also makes it difficult to secure the important long-term revenue streams needed to fund new investments.

- Lack of viable, and/or inefficient, Federal Renewable Energy Standards is another major issue before the wind power industry. Strong Federal Standards are needed for wind owners and developers to get assurance of a viable and growing market for their wind energy now and in the future.

- Outdated and undeveloped electric power transmission infrastructure is another major industry challenge. Lack of efficient regional planning and effective federal and state transmission policies hinders the secure transmission investments needed to support new wind power generation projects.

As a result, there are lately difficulties with either, a) the builder's capability to secure project bonding, b) power transmission and interconnection issues, and/or, c) financial failure of the general contractor or subcontractor(s) managing the project. In either case, these are serious issues affecting the growth of one of the most promising power sources in the world.

In summary: Wind power is a proven power generator. Besides the issues outlined above, it is renewable and can be used to supplement the basic load of the national grid. There are a number of issue remaining to be resolved, as we saw above, but wind power is here now, its capacity is increasing around the world, and it will be a major power generator for the future generations.

Update

A number of federal laws, regulations, and Executive Orders apply to wind energy development activities. For the most part, state laws and regulations do not apply to wind energy development on tribal lands. The extent to which the Federal requirements will apply to specific wind projects on tribal lands depends upon the nature of the project, its location, and size.

Here again, the technology developments are much faster than the law. Legal issues regarding wind rights appear when evaluating who has the right to capture wind freely, are just now arising. The question here is simple, "Who owns the wind?" and its simplicity, or lack thereof, will make many lawyers rich.

Modern day wind turbines are also known to create wind disturbances or "wakes" for hundreds of yards downwind. If an upwind property owner installs a wind turbine on his property, the wind reaching the less-lucky downwind neighbors would affect them and their wind turbines. This may cause the unlucky down-wind owner to lose out on potential earnings. But there are no clear legal rules related to property owner's "wind" rights. This anomaly will surely bring conflicts among potential wind turbine owners, and to court they will go.

An attempt to bring a legal solution to the "wind ownership" problem could use the Cathedral Model; a legal model used to analyze allocation of scarce resources. It proposes that in rural areas, zoned for commercial wind energy development, wind turbines can be located anywhere, within ordinary safety restrictions. This precludes the up-wind neighbors from any liabilities, but is not acceptable to the down-wind property owners, since it might curtail the efficient use of wind resources on his property.

And so, the down-wind neighbors must be notified of the up-wind landowner's plans to install a wind turbine. There must be also a provision for a legal right to pay the upwind owner in order to keep him from installing the turbine. This value, however, is difficult to establish and could be as much as the value of the turbine site itself.

This way, the down-wind property owner would chose to exercise this legal option only if he considers his wind power much more valuable than the up-wind neighbor. This would allow the disputing neighbors to discuss and settle the conflict in a way that promotes the optimal use of the wind resources in the area.

The wind ownership issue sound like a huge area for legal work, which we will certainly see expanding in the next several decades across the good, 'ol U.S. of A.

Decommissioning Impacts

Decommissioning and site reclamation activities are a major issue in a number of states, and a frequent visitor in the U.S. courts during recent years. These activities have been blamed for causing environmental impacts, including filling in the mine, or well, removal of related infrastructure and materials, as well as land de-contamination, re-contouring, and re-vegetation, etc., etc.

The renewables are no exception. They are subject to the same laws and are expected to go through the same procedures during the last stage of their lifecycle as the rest. Even though decommissioning a solar power plant might seem easier, faster, and cheaper than that of a coal-fired such, the work has to be done properly and thoroughly in all cases.

There are numerous real, potential, and perceived impacts from these activities, depending on the type of technology used.

The following potential impacts are expected to result from decommissioning and site reclamation, especially at coal mines, drilling sites, and coal- and gas-fires power plants:

- Acoustics (Noise) is the noise during decommissioning, which is similar to that during construction and mining. It includes equipment (rollers, bulldozers, and diesel engines) and vehicular traffic noise. The acceptable noise levels are established by the EPA and local ordinances. The distance to the nearest residence or business is very important as well.

- Air Quality (including Global Climate Change and Carbon Footprint) are emissions from decommissioning activities, which include vehicle tailpipe emissions; diesel emissions from large construction equipment and generators. There is also the possibility of excess dust generated during backfilling, dumping, restoration of disturbed areas (grading, seeding, planting) operations, and equipment traffic.

- Cultural and Paleontological Resources are not a serious concern during decommissioning, since any such resources would have been removed, or destroyed by prior activities. Artifacts collection could be a problem if the surface area was damaged and was left unmonitored. Poor planning, however, could be an obstacle to a complete recovery of related artifacts during decommissioning.

- Visual impacts from mining or drilling activities would be mitigated if the site were restored to its preconstruction state. In most cases, however, despite the attempts to remove all surface facilities, the scarred landscape remains and might be objectionable by the locals.

- Ecological Resources impacts during decommissioning activities are similar to those during construction and mining or drilling operations. Negligible effect on wildlife habitat would be expected, with minor injury and mortality rates of vegetation and wildlife. Acid mine drainage, land collapse, and other effects however, could continue if not properly managed.

- Environmental Justice impact is a significant effect that occurs in any resource area and which disproportionately affects a minority group or low-income populations. This impact might be damage to air or water quality, loss of employment and income, and visual impacts.

- Hazardous Materials and Waste Management deals with the removal of industrial wastes, such as lubricants, hydraulic fluids, coolants, solvents, cleaning agents, etc. chemicals and contaminated materials from the site. These are usually classified and treated as hazardous wastes that require special handling, packaging, transport, and disposal. Further damage to the environment and human health is possible if these wastes were not properly handled and removed.

- Human Health and Safety is utmost import for protecting workers' and public health and safety during the decommissioning and reclamation process. The rules and procedures here are similar to those during construction.

Added risk may be the reclamation of underground mines due to the potential for mine collapse. Health and safety issues also include working in potential weather extremes and possible contact with natural hazards, such as bad weather, uneven terrain and dangerous plants, animals, or insects.

- Land use damages often result from underground and strip mining, and well drilling. Some of the damages could be reversed, but delayed collapse of underground mines is a serious long-term issue, with no plausible solutions. Open pit mines could have lasting land-use impacts since the land is usually irreversibly altered and reclamation to pre-development condition is not possible. In such cases alternate land uses are the best, if not only, solution.

- Socioeconomics are directly impacted by decommissioning, due to permanent jobs (miners) and revenue (royalties) loss. There are, however, temporary jobs created during this activity. A number of indirect impacts are to be expected too, mostly related to jobs coming and going, and the related changes in the local economy. Proper site reclamation might actually result in increase of the value of residential properties adjacent to the site.

- Soils and Geologic Resources impacts during the decommissioning/reclamation phase include removal of access and on-site roads, and heavier (or different) than usual vehicle traffic. Surface disturbance, heavy equipment traffic, and changes to surface runoff patterns can cause soil erosion, which could lead to soil nutrient loss and reduced water quality in nearby surface water bodies.

- Transportation activities are increased during this stage, and consist of increased use of local roadways, including overweight and oversized loads when removing heavy equipment from the site. These are known to cause temporary disruptions to local traffic.

- Visual Resources impacts would be similar to those from the initial construction activities. Restoration of a site to pre-construction conditions entails re-contouring, grading, scarifying, seeding and planting, and otherwise stabilizing the disturbed surfaces. The work creates visual contrasts that would persist for a number of seasons or even years, before the new vegetation starts disguising the scars. Invasive species are known to move in and take over the area, which further changes its visual appearance.

- Water Resources (Surface Water and Groundwater) are affected all though the decommissioning process. Water is a key element in maintaining the environment and ensuring healthy population. Its importance have been increased lately with extended draughts in the South West of the U.S.

During mining and drilling operations, water is pumped up from local groundwater wells, or is trucked in from distant off-site such. A lot of water is used for dust control, for road traffic, surface mine filling, as well as for consumptive use by the hydrofracking operations, and finally during decommissioning and site reclamation efforts. Water quality could be affected by continued acid mine drainage, or hydrofracking fluids remaining in the ground.

If the water transport, use and disposal are not properly managed, these activities could cause soil erosion, or contamination, and weathering of newly exposed soils leading to leaching and oxidation that could release chemicals into the water, discharges of waste or sanitary water, and pesticide applications.

Upon completion of decommissioning, disturbed areas would be contoured and revegetated to minimize the potential for soil erosion and water-quality-related impacts.

Surface and groundwater flow systems would be affected by withdrawals made for water use, wastewater and storm-water discharges, and the diversion of surface water flow for access road reclamation or storm-water control systems. The interaction between surface water and groundwater could also be affected if the two resources are hydrologically connected, potentially resulting in unwanted dewatering or recharging of any of these water resources.

Criminal Activities

The global energy markets provide ample opportunities for criminal activities. Some of the most notorious are as follow:

Nigerian Bunkering

Nigeria's Oil companies are battling against rising theft costing the country over 250,000 barrels of crude a day. The loss is roughly 8% of Nigeria's total GDP output. As militant attacks on oil installations in the southern Niger Delta region have slowed down in recent years, the oil theft has surged.

Bunkering, as the oil theft activities are known in Nigeria, consists of siphoning crude from pipelines into makeshift vessels. Bunkering is the downstream business in the maritime sector, which includes fueling of ships of all kinds in the high seas, inland water ways, and within the ports.

Figure 8-11. Nigeria bunkered oil transport

It is a massive, well organized "business" that, in addition to damaging the national economy, fosters crime and corruption among the locals. Stolen oil is a lucrative black market in Nigeria, Africa's top oil exporter.

The locals see it as one way to get out of poverty and do not hesitate to do most outrageous and dangerous things to achieve their goals. The oil crime wave is spilling in the neighboring countries as well.

In the recent past, attacks on oil installations by militants, slashed Nigeria's oil production by about a million barrels a day. In 2009 the government offered amnesty to the militants, which largely ended the violence and the oil production increased, but is still below the height of the pre-militancy levels.

Nigeria has over 20-30 billion barrels of oil in proven reserves, with a potential to produce 3-5 million barrels every day, but production is kept at 2.0-2.5 million barrels. The oil and gas industry accounts for around two-thirds of government revenue and more than 90%

of Nigeria's exports.

Efforts to clean up the corruption-ridden energy sector forced the government to scrap subsidies on locally consumed fuel in 2012. That more than doubled pump prices in a country where nearly two-thirds of the population of 160 million survive on less a dollar a day. Nationwide strikes and protests forced the government to partially abolish the subsidies.

Update 2015

As part of efforts to make Nigeria the hub of oil and gas business, as well as marine activities, the federal government has approved the legalization and resuscitation of bunkering operations in Nigerian territorial waters, a few years after the operations were suspended.

The resumption of bunkering operations in Nigeria is expected to reduce the theft and generate revenue for the government and create employment opportunities for the growth of the economy. These actions will also divert oil, gas and marine operators from going to Senegal, Cape Verde and Ivory Coast to fuel vessels that operate in the Nigerian territorial waters.

The operators are now required present to the Nigerian Navy Headquarters an application for bunkering clearance, detailing: the vessel involved; location of bunkering operation or discharge point; quantity of bunker fuel and duration of the operation. The operators must obtain license from the Nigerian Maritime Administration and Safety Agency on the quality of the products and the vessels.

All documents and facts will be verified by the government agencies for validity prior to approval of the bunkering activities. This is to make sure that the vessels to be used for bunkering must meet certain obligations required by the Customs. Vessels operating in national waters must show proof that the duty is fully paid too, while leased foreign vessels must obtain temporary importation permit from the Nigerian Customs. The bunkering vessels will not be allowed to leave the Nigerian territorial waters, regardless of their registry, and must obtain approval from the nearest Area Command when moving from one Nigerian port to another.

These actions, if properly implemented, will not stop the criminal bunkering activities completely, but will reduce the incentives of the criminal organizations, and the number of those willing to take the risk.

Solar in Spain

The solar energy battle in Spain is raging on several fronts. The Spanish government is faced with legal action over its proposal to cap profits for solar power plants at a prohibitively low rate. The solar industry is challenging the retroactive 7.5% pre-tax profit cap placed on solar investments with financial backers likely to explore their legal options too. This promises to be a long process, complicated by the fact that legal challenges have been ongoing since the first cuts in 2010.

There is another side of this battle too, where local and foreign investors have taken Spain to international courts and there are international investment funds that will look to do the same. The strongest challenge, however, could come from the investors rather than the Spanish solar industry itself. The proposed profit cap, which works out at around 5.5% after tax, applies not to the electricity generated but to the initial investment. If the rate of borrowing underpinning a project is higher than this rate, it becomes impossible to generate any returns at all.

International investors see the government unfairly changing the rules, and their returns getting cut from 6% to 1%. What this means is that an investment in a project a year ago the numbers that you used are now worth nothing.

The Spanish government is faced with dire economic circumstances, compounded by a deficit in its energy budget of $34 billion that was estimated to grow by $5.9 billion in 2013 alone.

The new legislation also includes some unusual measures that make the latest changes far more radical than previous alterations to subsidy rates. This opens the door for incentives to close down renewable energy projects, which is equal to the first time ever that money is actually paid to shut down renewable technologies.

On top of the that a new levy applied to customers consuming their own electricity will also make it cheaper to buy from the grid than to use your own solar panels, effectively killing the residential solar market. This is at a time when solar is close to reaching grid parity and needs financial support. The Spanish government, however, is making it hard to impossible by its resistance to supporting distributed energy. The government-forced reduction in income since 2012 amounts to over 50%, while suggesting that it is going to reduce the costs of electricity. Instead of some cuts, however, it imposed severe reductions. Go figure.

As a result, a large part of the Spanish PV sector in trouble and going broke. This also means that as things are going on in Spain, the government will not be able to reach its target of sourcing 20% of its primary energy demand from renewable energy. Investing in renewables will be hard to impossible in Spain now at least

for the for the next several years; until and unless the internal and international lawsuits are settled, and the confidence in the government support for renewables is restored. Good luck with that, Spain!

The Czech Republic

The situation facing the beleaguered renewable energy industry in the Czech Republic has taken another twist. In the fall of 2013, the Czech Energy Regulatory Office (ERU), claimed that over 1,500 solar plant owners may have submitted falsified electricity production data in order to receive more subsidies than they were entitled to.

Public media sources claim that some solar plants have recorded more sunlight than would be expected in the Czech Republic annually. Allegedly, more than 1,500 owners of PV power generating plants recorded over 1,200 hours of sunlight in a year. According to the television reports, this would be closer to the amount of sunshine expected in California, but not even close to that in most local areas.

The Czech renewable energy and photovoltaic industry associations are questioning the strength of evidence and even logic behind the accusation. They argue that compared against data prepared by the Czech Hydro-meteorological Institute, recorded periods of sunlight fall well within expected levels and that the accusation is false.

They also argue that the number of hours of sunlight expected in the country each year is between 1,200 and 1,800, and that critical factors have not been considered in the above said estimates. This includes diffuse sunlight on cloudy days and the role of the electricity distribution companies in coordinating measurements of energy production with the market operator.

Czech industry reps clams that it is unclear where the data of the accusations came from, and that the data seemed "mixed up" at best. The industry is also unhappy with the high public profile of the regulator accusations, arguing that equivalent bodies in other countries were not constantly in the press influencing opinion.

The ERU has asked the State Energy Inspection Authority to become involved in investigating energy producers, and so in August, 2013 the Czech government began dramatically slashing subsidies for renewable energy producers, following a wave of negative publicity and artificially inflated energy prices for consumers.

Soon after that it was reported that Alena Vitaskova, chairwoman of the ERU could face criminal charges along with nine other colleagues for allegedly allowing solar power plants built in 2011 to be fraudulently registered as completed in or before 2010, in order to receive higher subsidy rates.

And then in September, 2013 the Czech lower house of parliament voted to end feed-in tariff payments all together as of January, 2014, and to apply a 10% rate of solar tax for future installations on top of that.

The change to the law will signify a major shift in the Czech solar power industry, and if fully implemented it might just slow it down to a crawling pace.

Fuel Theft in the U.S.

Fuel theft sounds like something we would expect to see in Nigeria and some other developing countries. But fuel theft in the U.S. sounds unreal. And yet, the rate of fuel theft in this country is accelerating when fuel prices climb. The cost of gas and diesel tempts more employees to steal fuel.

Some employees either siphon fuel from fuel tanks, or use special devices making it hard to detect that fuel would gradually be leaving the fuel tank. This employee-related crime is costing the U. S. trucking industry over $2.0 billion annually.

Gasoline theft (sometimes known colloquially as fill and fly, gas and dash and drive-off) is the removal of gasoline from a station without payment. The thief will usually use some form of decoy to prevent nearby witnesses from noticing the lack of payment until they have left the station. Common decoys include pretending to press the wrong buttons after swiping the credit card, or having multiple people get gas at the same time with one paying for another person and the other running off with both cars.

With typical gas thefts costing station owners in the range of $50 per incident, many stores have fought back by installing better video equipment and requiring pre-payment.

Since the oil price increases after 2004, the surge in fuel theft has gone up,[citation needed] which has included license plate thefts (when gasoline is stolen in the case that the vehicle has the original tags, the vehicle tags will ID the registered owner).[citation needed]

A single operation in a Texas city, operating 6 trucks outfitted with bladder gas tanks steals: 6 trucks x 600 gallons x 6 days x 56 weeks = 1,209,600 gallons annually.

This represents over $3.0 million financial loss for the fuel retailers in a single city.

Fuel prices rising by as much as 20% a year have spurred an increase in fuel theft from, and waste of fuel, from commercial vehicles. The employee-related fuel crimes cost the U S. trucking industry $2.5 billion a year during 2012-2013, according to the insiders.

The rate of fuel theft is accelerating with climbing fuel prices, which tempt more employees to steal fuel. Fuel waste also is a growing concern as transportation companies struggle to find a cost-effective answer to curb what is threatening their bottom line.

Latest figures from the American Trucking Associations (ATA) indicate that the U.S. trucking industry faces a $75 billion fuel cost annually. Taxpayers foot most of the bill when public funds are used to compensate for misuse of state-purchased fuel. This is a double blow for the economy, but government authorities and private firms are taking steps to stop the waste and theft of trucking fuel by using cutting-edge RFID (radio frequency identification) technology to combat this problem.

Using RFID to monitor rolling assets such as trucks and company vehicles, dramatically reduces fuel theft. RFID tracking is geared to monitor and communicate activities of trucks and operators, so trucking companies can tailor each RFID product to suit their needs, without unnecessarily going over their current budgets. This saves the company, the taxpayer, and the US economy billions of dollars of resources and revenue.

Industry reports show an increased use of RFID in monitoring truck activity in construction and mining operations as well, which ensures that loads are properly transported and accounted for. RFID locate assets of transport management systems.

This is the same sensor-based technology that is used in refineries and chemical facilities to measure and monitor flow of liquids. It is now being applied to monitoring truck fuel consumption and other parameters of fuel use via remote sensors. RFID coupled with GPS could provide real-time tracking of the truck movements, as well as idle time when the engine consumes fuel instead of shutting it down.

Another new development of RFID technology allows a fuel tank cap to be matched with an authorized fuel pump. In case that an operator opens the fuel cap away from an authorized fuel pump, the RFID transponder unit logs the event. The trucking company, through reporting and messaging, receives a record of the event. RFID-able fuel pumps also match drivers and trucks through closed-circuit TV images with the truck RFID transponder. Much like EZ-Pass, if the truck doesn't match the transponder, the trucking company is alerted and appropriate action is immediately taken.

Traffic jams waste fuel at high rates too, so RFID can re-route drivers away from congestion, thus saving them time and fuel costs. This way RFID controls fuel usage waste, in addition to making theft harder to do. Because RFID works seamlessly within the trucking industry, and creates a mutually beneficial working environment, the savings and enhanced efficiency more than cover its initial cost. With such a costly activity as fuel waste and theft reduced, RFID's cost-saving results will give companies the edge over trucking competitors and competing transportation modes.

The higher the price of fuel, higher the value of the technology used to reduce fuel consumption and reduce fuel theft. U.S. trucking companies are using RFID with considerable positive results, while RFID-enabled smart cards at the fuel pump are making fuel dispensing easier and quicker.

Case Study: Fleet Fuel Theft

In 2014, at a truck rental center employees interrupted thieves in the act of stealing 10,000 gallons of diesel. The thieves abandoned a cube van outfitted with plastic tanks, electric pump and generator, especially equipped to steal diesel unnoticed.

Earlier, the police has suspected these and other thefts in the area stealing over 50,000 gallons of diesel in one occurrence, and 75,000 gallons in another.

In Edmonton, Canada and surroundings alone, there were hundreds of incidents in 2013-2014, the majority of them occurring at cardlocks. This theft happens when a thief finds a working gas card, and fills storage containers until the card gets to its maximum, or until the thief runs out of storage containers.

The B.C. Forest Safety Council issued an alert in 2014 that thieves were also showing up at overnight truck stops, long-term parking lots, work sites where there was unattended equipment and fuel storage tanks. They then proceed to fill storage containers from parked trucks and other idling and/or unattended vehicles and other equipment.

A truck company reported thefts anywhere from $16,000 to $30,000 in one month alone, plus there were about a dozen reports of fuel theft, and probably lots more, that were unreported and undocumented. The reason for that is that many people do not want to draw attention to themselves and/or the steps they are taking to secure their assets. There is an element of embarrassment in the response too.

In many cases trucks were broken into and ransacked by thieves looking for gas card PIN numbers. Entire tankers full of gasoline or diesel are stolen and

emptied at times. Fuel theft is a wide-spread problem, and at times it is easier to ask who is it not being stolen from? Employees steal, sometimes outsiders steal. Some drivers sell it, others buy it. The activities take place when the drivers are at a break, or often with their co-operation. Very few get caught.

The vast quantities of fuel being sucked from cardlocks is most puzzling. One would think that a single very large transaction, enough to fuel 10-20 trucks, or multiple purchases in different areas in one night, would ring the alarm bells. Today cardlock thefts should be completely preventable. One problem has been identified as carelessness on part of the truck drivers. The thieves find sticky notes with PINs on the backs of the gas cards laying in the truck cabs, or PINs written on the cards, which then makes stealing easy.

Also, some fleets do not have daily purchase limits on their cards, which explains the amazing amount of stolen fuels. Setting daily limits, changing PIN numbers, pick and choose which locations drivers can fuel at, and other measures are easy to do today, so gaps are hard to understand. This is a simple matter for fleet management program that monitor fuel mileage at the truck level to look for trucks with unusual fuel consumption.

Cameras and improvements in gates and fences of commercial filling stations are other ways to combat the fuel theft. All these measures, combined with appropriate legislation and effective legal action, could reduce and even eliminate fuel theft in North America.

Energy Markets Manipulations

Manipulation of the physical and financial (commodities trading) markets is not a new phenomena. As of 2008, however, the accusations of mass manipulations in the energy markets have grown as never before. When oil and natural gas prices skyrocketed in mid-2008, allegations of manipulation increased and were believed to be the main reason for the price discrepancy.

Prices have gone down lately, but the allegations, the pain from, and fears of repeating that event, remain. These allegations have caused the U.S. and the world to look for more effective market regulatory system to reduce its prevalence. Unfortunately, the issues and the respective responses have been confusing, to say the least. And the worst part is that the U.S. legislators and regulators are part of the confusion.

But what is energy market manipulation? Amazingly enough, precise and clear, official legal definition does not exist. According to the unofficial sources, "Market manipulation is a deliberate attempt to interfere with the free and fair operation of the market and create artificial, false or misleading appearances with respect to the price of, or market for, a security, commodity or currency."

According to the FTC official definition, "…market fraud includes any action, transaction or conspiracy for the purpose of impairing, obstructing or defeating a well-functioning market. Fraud is to be determined by evaluating all circumstances of each case."

Market manipulation is prohibited in most countries, including the U.S. It is prohibited here under Section 9.a of the Securities Exchange Act of 1934. But the exact values of "interfere," and "free and fair operation" above are still not well defined by the legislators, regulators, and the courts.

The U.S. Congress has been puzzled and undecided on the issue. According to one U.S. Senator, "The difficulty, as I understand it, is that [the acts that constitute manipulation] are various and perhaps impossible of direct definition. I do not know how we could draw a definition to bring it home to the individual."

The U.S. courts have also been unable to define market manipulation per se. In a famous court case, the U.S. Eighth Circuit declared that "The methods and techniques of manipulation are limited only by the ingenuity of man." According to the court then, market manipulation has no limits, and therefore is illusive and impossible to judge in real life situations. Because of that, the law can be twisted by the lawyers to mean whatever they feel fit.

As a result, the U.S. energy market, and the energy market manipulations in particular, have been and still are object of intense criticism and scrutiny. The wild rise of energy prices in 2008 fueled the accusations further. It clearly showed that the laws and regulations are just not efficient enough to prevent a major price runoff. What happened in 2008 could easily happen today…and tomorrow.

Using different legal and illegal techniques, the energy financial market traders have been, and still are, capable of moving the markets. By doing this, they have been able to create abnormal conditions, which affect the global energy markets. So now there are legislators and regulators set on taking additional action to reduce the severity of market manipulation.

Renewables Patents

The different renewable energy technologies continue to grow, which is evidenced by the number of patents filed and issued during the last several years.

The granting of patents by the United States Patent

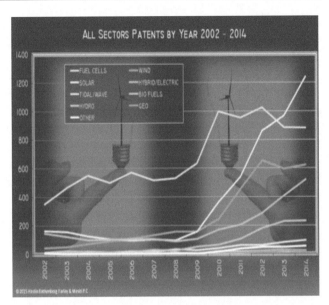

Figure 8-12. Patent filing for different energy technologies

and Trademark (PTO) is often cited as a measure of the inventive activity and evidence of the effectiveness of research & development investments. Patents are considered to be such an indicator, because to be awarded a patent, it requires not only the efforts of inventors to develop new and non-obvious innovations but also successful handling by patent counsel to shepherd a patent application through the PTO. Thus, the granting of a patent is an indicator that efforts at innovation have been successful and that an innovation had enough perceived value to justify the time and expense in procuring the patent.

The Clean Energy Patent Growth Index (CEPGI), published quarterly by the Cleantech Group at Heslin Rothenberg Farley & Mesiti P.C. provides an indication of the trend of innovative activity in the Clean Energy sector from 2002 Solar patents topped all other technologies, having 358 patents more than nearest competitor Fuel Cell technologies. There were 1238 Solar patents granted in 2014 which gave Solar technologies its second annual win and being over 200 higher than the previous annual technology record set by Fuel Cells in 2012 at 1024 granted patents.

The fall of longtime annual technology winner (until 2013) Fuel Cells by 6 patents to 880 compared to 2013 further cemented Solar technology's ascendancy. Patents in Wind technologies (623) trailed Fuel Cell patents by over 250 and jumped about 30 compared to the previous year.

Patents granted in Hybrid/Electric Vehicle technologies were about 100 less than in Wind at 521. HEV patents were up 112 compared to the year before. Biomass/Biofuel patents (230) were up four compared to 2013 while Tidal/Wave energy patents were up 12 to 94 granted patents.

Hydroelectric patents fell two to 24 while Geothermal patents jumped two to 23. There were 55 clean energy patents granted in 2014 in other technologies.

Environmental Lawsuits

There have been thousands of environmentally based lawsuits field in the U.S. since the 1980s, and many are being filed as we speak. They cover all areas of the energy markets, from mining to power generation and use. And there are many more around the world too.

Case Study: Oregon Environmental lawsuit

A large crowd was seen in the spring of 2015 outside the courthouse in Eugene, Oregon. Kids, teachers, parents, grandmas; all kinds of people trying to draw attention to the environmental issues debated in court. A 19-year-old was suing the State of Oregon, for not doing enough to stop climate change and mitigate its effects. Lack of action is bringing irreversible, catastrophic crises in the country and the world, the young plaintiff claims.

The unusually young age of the complainer is drawing attention, all right, but there is another unusual thing in this case too. It is the legal approach that's being used. Worried that the environmental law does not address many of the most significant problems, the Public Trust Doctrine was pulled out of storage, dusted out, and put in front of a judge.

The Public Trust Doctrine is a legal theory that claims that the government should hold certain natural resources in trust for the public. It can be traced back to ancient Roman law and to the English common law. It has been used in the U.S. too, mostly to guarantee public access to waterways. It has been part of American case law since 1890's Supreme Court ruling that private developers in Chicago couldn't prevent public access to Lake Michigan.

As early as the 1970's, environmental lawyers were arguing that the Doctrine should be extended to wildlife and even the air. Governments can and should protect these resources, which today includes protection of the atmosphere and the environment. It is easy to make the connection, since the atmosphere controls the climate system human life depends on.

And so, if the atmosphere is considered a natural resource covered by the Public Trust Doctrine, then the courts could force the government to take steps needed to protect it. Whatever it takes; be it reduction of greenhouse gas emissions and other pollution. The legal proponents called the idea, "atmospheric trust litigation" and enlisted the help of famous environmentalists and climate scientists.

And now we see this idea becoming into a movement. Many kids backed by an army of attorneys and scientists working pro bono, have brought lawsuits against the federal government in 14 other states. The movement is catching on globally, with similar lawsuits filed in Ukraine, Uganda, the Philippines and the Netherlands.

The idea is to force the governments to reduce harmful emissions by at least six percent a year by means of remedies like carbon tax and such.

Similar lawsuits, including a suit in U.S. federal court, have been dismissed in the past. The exception are a couple of state judges who have been receptive to the idea that the public trust could include the atmosphere. None of this has made any difference, or forced any state government to take action thus far.

The legal problems are many. They go well beyond what any state court has done thus far to start with, thus lack of precedence. But even more importantly, the atmosphere is global and greenhouse gases travel globally. So no state alone can solve the problem, even in its proper territory...let alone globally.

Then imagine the courts agreeing that action is needed immediately. Then what? The courts do not dispute the existence of a climate problem, but cannot provide and enforce a remedy? Only the political branches can prescribe and enforce some action in this area.

So, it all boils down to asking the court to do its job in interpreting the public trust rights and prompt the legislature in upholding its duty in making laws that protect people's rights. Far fetched, yes, but it is what the people want.

The Doctrine proponents aren't going to slow down. They will file three more state and federal lawsuits. New lawsuits are also expected in other countries, including new groups in India and Pakistan.

We have to agree that a court decision in Oregon or India—even in best of cases—cannot fix the climate problems. The movement, however, can. State after state, country after country. It all starts with the people uniting around an idea. And this one is worth while the effort. It is the air we breath and the water we drink that are in real and present danger. It is the wellbeing of the future generations that is at stake too.

FINANCIAL ASPECTS

There are two major types of trading in the global energy markets: a) trading the physical (actual) energy commodities, and b) using financial instruments as investment or trading tools

The different types can be briefly described as follow:

- *The physical* energy markets can be divided into a) wholesale, and b) retail—The wholesale energy markets, is the major part of the global energy market. Here, thousands of tons of coal, millions of barrels of oil, or cubic feet of natural gas are traded (bought and sold) between companies and countries. These trades consist of actual deliveries of physical goods, where trains and ships are loaded with the energy commodity and sent their way across country, or the globe. The winners here are always the producing countries, while the losers are the importers. On which side of the trade a country sits, usually determines how rich or poor it is.
  - The retail energy markets consist of the distribution (sales) of small quantities of energy to homes and businesses. This includes the sale of coal in some countries, natural gas via pipes or storage tanks, and gasoline to fill cars and trucks.

Note: We review the wholesale and retail energy markets structure and function in the previous chapters, so below we will focus on their financial aspects.

- *The financial* energy markets are where companies and individuals go when they need to raise or invest money. No physical commodities change hands in these markets, and yet, they are very important in the development of the physical (coal, crude oil, natural gas, and electric) markets, by providing:
  - Access to capital needed for business start-up, or operation, and
  - Protection against market variations and future loses, and/or a platform for speculations, or trading of derivative financial products in the general financial markets

Let's start with the basics of capital investment.

Capital Finance and Investment

Capital finance and investment is what banks, different financial institutions, VCs, and wealthy individuals do to start, or support. In most cases, their ultimate, if not only, goal is to make profit from the venture. Every project starts with capital finance and investment consideration. No project can start or properly develop without it.

The quick rise of the renewables in the energy markets recently has introduced a new and serious variable in the market's function. This is affecting the way financing and investment work too, so we will take a closer look at the new energy markets, starting with the basics.

Energy Economics

When talking about economics, we must remember that there are people with titles such economists, financial experts, lawyers, etc., whose sole (professional) goal in life is to put a value on all products and services. This includes evaluation and pricing of energy and environmental products and services...as well as ways to make profit from it all.

Since we are not as qualified as most of these professionals to discuss such complex subjects (just like they are not qualified to discuss the technical aspects of the energy markets), we will meet in the middle and will view the subjects at hand through the prism of our long and wide technical expertise.

Energy generation and use is a huge and growing in importance business. We all depend increasingly on technology—computers run our lives—which in turn depend on reliable energy. Energy brings problems, such as inconveniently high prices, and is usually accompanied by a lot of pollution, which is creating another huge business in its wake; that of environmental protection.

Today investment in the fastest growing information and telecommunications sectors total ~$100 billion, while energy and utilities total over ~$300 billion.

Adding environmental technologies and the related business transactions to the energy markets, doubles and triples the dollar amount, which makes the energy sector a major part of the U.S. economy. Recently, the global energy sector has been seeing more capital expenditures than any other sector. The energy and environmental markets are also among the few with undeniable great future growth potential...for better or worse!

So like it or not, energy, environment, and investment are interrelated, and very large global business enterprises. And this is how it is seen by many economists too. As an example, just think of the BP Deep Horizon accident in the Gulf of Mexico. How many billions of dollars has BP made from its Gulf oil wells (part of its energy business) before and after the accident, and how many billions it is still paying (part of its environmental business). And then, think how many economists, financial and legal experts were, and still are, involved in these business operations.

Note: In the summer of 2015, BP agreed to settle the 2010 blowout accident in the Gulf of Mexico for $17.5 billion, but this is not the end of the case. We will be hearing about the developments for many years to come from the environmentalists, economists, and the army of legal experts.

So, let's clarify who the economists and legal experts are and what makes them tick. To be sure, they think and function differently than most of us "normal and average people." They see the world as a large ball made out of numbers and dollars, where; a) everything has a value, and b) everything and everybody operates (supposedly) under the law.

If for some reason things go astray, as with the BP Gulf accident, then the economic and legal experts jump into action, trying to find a way to:

- Value each action in every step of the way in order to put a dollar number on it, and then
- Figure out which of these actions were outside the law and what would that cost the violators and the violated.

Energy and environmental issues to these experts are simply market-based instruments represented by factors such as environmental cost, human life cost, physical and psychological damages, carbon taxes, cap-and-trade systems, fuel-economy standards, government subsidies, etc.; each of which has value and price. It is all brought out in the open as numbers—including numbers for human life.

It is actually much better this way, because if one gets too involved in the action, emotion and sentiment might take over and fog up the uninhibited reasoning. Such fog would inevitably hurt the victims and the underprivileged, so we must thank our esteemed economists, financiers, and lawyers for their contribution in the process of clarifying, untangling and (partially) resolving the energy-environmental dilemma accompanying the energy markets.

The economists and lawyers evaluate the activities in the different areas, including the political actions and regulatory policies on the basis of pragmatic consequences and in terms of the corresponding cost vs.

benefit framework, although the right vs. wrong debate (and its ideology) is not a direct part of the economic, financial, and legal systems.

There is a joke out there that economists are against the death penalty primarily because it is too expensive and financially unjustifiable...

This is another complex subject, which only confirms that to them it is much better to not get involved in the action, and only judge it for what it is from a healthy distance. This is done by assigning dollar numbers to the activities related to the main subject. Unemotionally and professionally.

Different lawyers have different views on the issues at hand, which vary from person to person, but usually boils down to twisting the law (to the breaking point) in order to protect their own (and that of their clients) interests. And who can blame them; the capitalist system is set that way; it allows and encourages it.

In principle, most lawyers are not against the death penalty per se, because it provides an alternative to chose from in some complex cases, and offers a carrot and stick solution in others. But from a pragmatic point of view, it is just business as usual. Nothing personal.

Regardless of how imperfect these people (economists and lawyers) and their actions might be, and how many bad examples we could find in history, economists, and financial and legal experts are the gatekeepers of the capitalist system. Their major contribution to the preservation of our capitalist system functionality is their ability and drive to assign dollar value to everything, as well as defending and preserving the laws of the Land. Right or wrong, this is the system we have chosen and live our lives by.

These people have assumed the responsibility to provide clarity, as well as protect and keep everything moving in the right direction. This includes clarification (by the economists), financial responsibility (by the financial experts), and protection (by the legal experts) of the energy and environmental products and services, and the related infrastructures and markets.

Economists, and financial and legal experts are the gatekeepers of the capitalist system

Now that we know who is who, we will take a close look at the economics and finances of the different power generating products, technologies, and processes dominating the energy markets. For this, it is imperative that we understand the technologies and processes involved in the energy markets.

Let's start with a quick look at the energy products and sources today:

Energy Generation and Use

As another clarification, in this text we use the terms energy, energy generation, power, and power generation interchangeably when describing the act of electricity or heat generation and use. By definition, however, energy is only contained in the energy sources (fossil fuels) which we burn to generate electric power. Because of that, it would be more appropriate to say that we are releasing the energy in the energy sources (the fossil fuels) by burning them as needed to generate electric power.

With all this said, we will continue using these terms interchangeably, except in cases where distinction is absolutely necessary.

In Figure 8-13 we compare the overall (lifetime) cost of construction and operation of power plants using different energy sources. It is obvious from the first glance that there is a big difference between the cost of coal and solar power plants, for example.

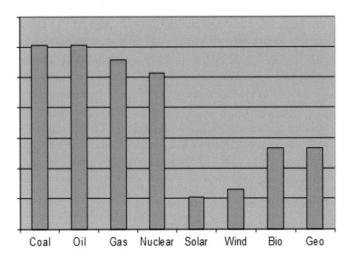

Figure 8-13. Lifetime expenses of power generation.

Please note that Figure 8-13 reflects the "lifetime" cost of using the respective technologies. In case of coal, for example, it includes everything from mining, transport, and burning coal, to generating, delivering, and using the electric power. Some, but not all, externalities, are also included in this estimate.

Note: Including all externalities would change the dynamics significantly in favor of the renewable technologies. This concept and the related processes, however, have not been integrated in the energy markets as yet.

Comparing the energy of the different energy sources to that of crude oil (a standard 42 gallon barrel) containing 1.7 MWh of energy, we see that:

- 1,000 m3 natural gas contain 10.5 MWh of energy, while
- 1 ton of coal contains 6.5 MWh energy equivalent, and
- 1 kilogram of uranium contains the energy of 3,000 tons of coal, or 12,000 barrels of oil.

Thus, a $90 barrel of oil contains only 1/4 of the energy contained in $60 ton of coal, 1/6 of the energy contained in 1,000 cubic feet natural gas worth $40, and only a small fraction of the energy contained in 1 kilogram of uranium-235 worth several hundred dollars.

Note: It is important to understand here that the above numbers represent the theoretical content of energy contained within the different energy sources. The actual amount of electricity, or heat, produced while burning any of these depends on a number of additional factors. The final energy output, or the efficiency of the heating capacity of the different fuels, depends (in addition to their own qualities and energy content) also on the type and efficiency of the equipment, the facilities and conversion processes involved, etc. factors.

The energy prices also reflect not only the actual energy value of the energy sources, but some other factors, such as ease and convenience of use. It is, for example, much easier, cleaner, and convenient to burn a kilogram of uranium than it is to transport and burn several thousand tons of coal; just to get the same amount of heat or electric energy. This is because uranium has much greater energy content than coal, which makes it much more convenient to use.

Also, it looks on the surface that solar power beats all other energy sources hands down, as far as initial construction and subsequent operation are concerned. In reality, however, solar has a number of serious problems that simply disqualify it as a major power source—even if the power generated by it is totally free.

Surprisingly, coal generates nearly 50% of the electricity used in the U.S. and even more around the world; even in the midst of all this talk about the harm it causes the environment. Coal is not that cheap either, but it is always there, and always available and ready for use. This is its big advantage that will never go away...until all coal is gone, of course.

Today, coal use is actually increasing around the world, and with it more and more GHG gasses are emitted in the atmosphere daily. Why is that? Are we all blind, or stupid? Far from it! After considering all the options, analyzing all steps of the processes, and running the financial calculations, we have figured out that coal—regardless of its cost and damages—is always there when needed, while sunshine and wind are here now, and gone in a minute or two.

We also have figured out that we have a lot of coal, while oil and other fuels are not as available, which also contributes to the reliability of coal as an energy source. Lately, however, natural gas has been replacing coal at a quick pace as large deposits are being discovered. Gas prices have the potential of going down gradually, which is further affecting the energy dynamics in the U.S. and the world.

But this is still only the tip of the iceberg. Economists (the good ones) are knowledgeable enough to understand how the different technologies work, and are familiar with their cradle-to-grave advantages and disadvantages. Some have realized that there is a large number of variables that need to be considered, so the lifetime cost, as important is it might be, is not the most important factor in their calculations. There are many other factors to consider, which we aim to reveal and cover in this text.

Let's look at what the economists and financial experts see in the energy markets

The Costs

Comparing the costs of the different power generating technologies is not easy, but we have to start somewhere, so Table 8-7 will serve this purpose for now. It shows the costs of the different technologies as far as their initial cost (capital cost), the cost of operation and maintenance (O&M cost), and the levelized cost of energy (LCOE) are concerned.

The numbers in Table 8-7 show that there is a big difference in constructing a conventional combined cycle (CCC) gas-fired power plant vs. a solar CSP plant. As a clarification, this is because coal power is a well established technology, using readily available conventional equipment, while CSP is a relatively new technology, using specially designed and made equipment. There are a also a great number of other factors at play as reviewed in this text, and as detailed in our earlier books on the subject, see references.

- While a coal power plant is cheaper to construct than most solar plants of equivalent power, it uses millions of tons of coal, which makes it expensive to operate. Also, the new EPA GHG standards make it impossible to meet by any coal technology,

Table 8-7. Power generation costs (in $/MWh)

Energy source	Capital cost	O&M cost	LCOE cost
CCC gas	18	2	**68**
Natural gas	40	3	**80**
Hydro	77	4	**90**
Coal	65	4	**100**
Wind*	83	10	**100**
Geo	77	11	**100**
Nuclear	90	10	**115**
CCS coal	95	10	**140**
Solar PV	145	8	**160**
Solar CSP	210	40	**250**

so no new coal plants will be build anytime soon in the U.S., regardless of the price and anything else. This is a huge game changer for the coal and all other industries.

- There is another important factor that deserves a special attention and evaluation: it is the variability of the renewable energy sources. It is important to understand that the conventional technologies (coal, gas, and oil) operate at 85-90% capacity, or basically they are available 24/7/365. At the same time, the renewables (solar and wind) operate at 20-30% capacity, or they are available only 1/4 to 1/3 of the time.

- Another cost related factor is land use. While a 500 MW coal-fired power could be built on 20-30 acres piece of waste land, a 500 MW solar plant would require nearly 10,000 acres of sun bathed land close to populated centers. Not an easy, nor cheap undertaking.

- It is also important to note here that:

- Solar is most efficient during the day, when electric power is most needed, and

- Combining solar and wind at some location provides much higher capacity factor, thus their availability rate can be increased up to 75%. The special locations, however, where enough wind and sunshine co-exist are rare, and so we cannot expect much development in this area.

- Solar and wind, however, have one huge advantage; they will never be exhausted. Coal, on the other hand will last one more century at best. And then, there will be none. How do you put a dollar number on this fact? Not to worry; our good friends economists, and financial and legal experts are working on it.

Only if and when the above factors and effects are fully understood, analyzed and calculated will we be able to see the full advantage of the variable power sources (wind and solar). Even then, however, due to their variability, they would not be able to keep up with the rest without energy storage. Some day, when reliable and cheap energy storage becomes available, wind and solar will then, and only then, become major sources of base load power in the U.S. and the world.

Presently coal and gas energy prices are the lowest (with externalities* excluded) and are the standard of the power generating industry.

***Note:** Externalities in this case are the negative effects of coal use on the economic activity on a third party. When coal is mined, transported and burned to generate power, external costs include the impacts of; air, soil and water pollution, toxic coal waste, the long-term damage to local and global ecosystems, and most importantly; their effect on human health and life.

If the total cost of externalities is included in the energy prices, however, then coal, oil and gas-fired, as well as hydro and nuclear power prices would be increased greatly—several fold in some cases.

The Initial Costs

Table 8-8 shows the relative cost to install, operate and decommission a solar power plant compared with a coal-fired one. Using coal's cost as a reference, or 1.0 unit, the relative levels of the cost and expenses for the solar technologies are derived from a breakdown of the different steps in their cradle-to-grave life cycle:

Table 8-8. Relative costs and expenses

Task	Coal	Solar
Land & permits	1.0	0.6
Construction	1.0	0.2
Equipment	1.0	0.8
Installation	1.0	0.6
Operation	1.0	0.1
Maintenance	1.0	0.5
Supplies	1.0	0.1
Emissions	1.0	0.1
Solid waste	1.0	0.1
Carbon tax	1.0	-1.0
Decommission	1.0	0.2
Recycling	1.0	0.1
Waste disposal	1.0	0.2
TOTAL		13.02.6

It is obvious from Table 8-8 that the relative costs and expenses of coal power plants (taken as 1.0) are much higher when compared with solar power generation. This ratio, however, becomes irrelevant, and is even reversed, when some practical factors are considered, the most important of which are; overall power availability, quantity, variability, and the reliability of the power generated by the different energy sources.

Solar and the other renewables are simply not readily available in large quantities, and/or as reliable as coal, oil, and natural gas power sources. Even with the glaring cost disadvantages shown in Table 8-8, coal-fired power generates the lowest price electricity, and is much more practical as a reliable power generator.

There are a number of factors that make it the preferred power generator world-wide. The most important factors that make coal so important as a fuel must be well understood and considered in order to get the entire picture, as follow:

- Coal is abundant, readily available, and relatively cheap,
- Coal has a high energy value, which produces efficiently large quantity of heat and electricity,
- Coal-fired power is constant, since the power plant can operate non-stop 24/7/365.
- Coal-fired plants can also generate variable power if and when needed

These factors do make coal the preferred power source for large-scale grid power generation, and because of that, coal is the backbone of the power generating system in the U.S. and most other countries. This is also the reason for its mass global deployment today.

Coal provides the base-load; the one major and constant value any power network depends on. And since it can be produced in very high quantities, the coal-fired power prices are much lower and more stable than most other power generating technologies.

The negative externalities of coal-fired electricity generation, however, are estimated at several times the actual price of the electricity generated by coal itself. If this is put in dollar figures, coal-fired power generation would be the most expensive of all other technologies. Although this is a major consideration, new coal-fired power plants are built every day around the world. With that the environmental issues increase by the day too

Energy vs. Environment

It is easy to see the growing discrepancy, arising from growing use of coal around the world. So how do we justify this inequality? Actually, we cannot, mostly because it is a fairly new subject, so we don't have a firm grasp on its intricacies. We are in fact just now starting to delve in that process, and here is where our economists, financial and legal experts, and their tools of the trade come in to help.

There are thousands of these professionals feverishly trying to find the best fit for all possible scenarios and calculate the best ways to approach this huge problem that is growing and becoming more urgent by the day.

Regardless of what happens, we won't have a complete, standardized, and supported by all people (and countries) solution anytime soon. Presently, a number of financial instruments, such as carbon taxes etc. are being tried as equalizers. Some of these are actually showing results in reducing the price difference among the different fuels and power generating technologies, but a number of controversies have developed around the strategies that do not allow a clear view and/or solution.

The energy vs. environment equation is very complex and far from being completely understood. Americans are split on the subject almost 50:50, so the debate continues and could go in any direction, depending on the global developments and domestic circumstances.

There won't be a solution anytime soon, because some major players (China, India, Russia and other developing countries) are determined to go their own way. They will continue to build large number of new coal-fired power plants regardless of what the international community decides, or the consequences thereof.

These countries have their justifications, the major of which is that they demand the same level of development we have here in the developed West. And how can we argue? We have emitted most of the poisonous GHG gases that are causing the present climate changes during our 20th century furious industrial development and relentless chase of the American Dream.

And just recently other countries, some well developed like Germany, have also made a decision to ramp up their coal-fired plants. They also have a justification, mostly based on economic principles, and which totally ignores the environmental problems created by excess coal burning.

This makes it obvious that coal power will be around for a long time—regardless of the high (environmental) price we have to pay. So, hang in there, solar and wind power; your time is coming, but it will take awhile. And even then, you are still the new and weak kids on the block, so the older and proven technologies (the fossils) will dominate until the last piece of coal is mined, the last whiff of natural gas, the last drop of oil, are pumped out of the ground.

All these abnormalities complicate the economic plans and mess up the financial calculations. Cost of fuels, transport, operating and maintenance (O&M), gas emissions, waste disposal, mines and power plants decommissioning, etc., etc. factors make the picture very complex. Some of these factors are changing with time as well, which brings further complications.

In the end, however, all these factors must be understood, sorted out, finalized, standardized and implemented globally. Then, and only then will we have achieved the ultimate balance of energy vs. cost vs. environment. We rely on our economists, and financial and legal experts to come up with the results and prescribe solutions.

Our goal in this text is to shed as much light on the technical aspects of the complex energy-cost-environment scenario, in order to bring us all to a common denominator and allow us to start an intelligent discussion on the issues at hand. Since economics and legal matters is the subject of this chapter, we will be looking at the energy technologies and issues through these lenses as well.

Let's start with the basics. In determining the economics of different energy sources, we use a number of financial formulas and techniques.

The major components of the energy economic picture are:
- Capital cost,
- Operating cost and,
- LCOE.

In more detail:

Capital Cost

The capital cost of any energy or environmental project consists of the specific construction cost (SC), the equipment cost, and the related labor etc. tasks as needed to build the necessary infrastructure and start operation.

Specific construction cost is the cost of building a new power plant, in $ per kW of capacity installed.

Specific construction costs decreased from the beginning of the industry until about 1970, then doubled between 1970 and 1987. The specific construction cost is a very part of the overall cost system, and could easily be the subject of a separate decomposition study unto itself. Such a study would have a different scope from our intent to present the overall picture. Nevertheless, we can discuss the determinants of specific construction cost and in some cases provide numerical estimates of their impact.

The most important factors are: Economies of scale, Add-on environmental controls, Thermal efficiency, and Construction inputs in terms of prices and quantities

- Economies of scale (the size of the project) usually lower the construction costs as unit capacity grows. This is due mostly to the fact that larger projects are better coordinated, and also because advanced ordering of large quantities of materials and equipment lowers the costs significantly.

- Add-on environmental controls raise plant costs by an amount proportional to the type and size of additions. These costs may decrease over time with increased engineering knowledge and technological advances, but are still a major part of the plants costs. Estimates show that sulfur scrubbers and cooling towers add about 15% and 6%, respectively, to the cost of plants. Besides raising plant costs directly by requiring new equipment, there is evidence that pollution controls also raised costs indirectly by increasing the complexity of the plant, which now required greater planning and longer construction times. For example, adding new CCS emission controls will add major costs to new coal power plants and might contribute to 60-80% increase of the total price tag.

- Thermal efficiency can raise or lower the specific construction cost of a plant. This is because increasing the efficiency typically raises materials and construction costs, but may also increase the capacity of the plant. As the technologies mature, it may be possible to obtain a higher thermal efficiency for the same materials and building costs, which effectively reduces the final costs in the long run.

- Varying price or quantity of construction materials and components during ordering affects plant costs accordingly. The quantities of materials and labor may decrease over time with the new technology developments, while changes in the national and global economies may also alter prices up or down, according to the developments at hand at the moment of ordering materials and components.

More specifically, capital costs of energy sources and power generators are considered to be these costs and expenses related to mines, drilling rigs, and power plants construction, equipment purchases and setup, etc. costs needed to put a project together and in operation.

As an example, the future of new coal-fired power plants is uncertain today, due to the increased EPA emission standards, which new coal-fired power plants cannot meet. But even if EPA lowers the standards, which is possible under a new administration, the capital costs of any coal-fired plant are extremely high, and have been best described as "soaring," "skyrocketing," and "staggering."

As recently as 2005, proposed coal-fired power plants were estimated to cost $1,500/kW, but the estimated construction costs of new coal plants have risen significantly since then. Today, the estimated costs of building new coal plants have reached $3,500 per kW, without financing costs, and are still rising. This means that a new 500 MW power plant would cost of well over $2 billion after including the financing, insurance, etc. costs.

The cost increases have been driven by a worldwide competition for power plant design and construction resources, commodities, equipment and manufacturing capacity. And since there is no reason to expect that this worldwide competition will end anytime in the foreseeable future, the costs are expected to rise even more.

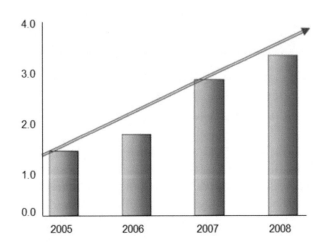

Figure 8-14. U.S. coal power plant capital cost increases (in $ billion)

To make things worse, in order to meet the new EPA rules, new coal plants must be equipped with new carbon capture and sequestration (CCS) equipment, which is very expensive. And to make things even close to impossible, this new CCS equipment is not even commercially available as yet. When it becomes available—most likely around 2025-2030—it will add 60-80% to the cost of building new coal-fired power plants.

It would be fair to say, then, that there will be no new coal-fired power plants build in the U.S. until then…if then. That, however, will not stop China, India and Russia in building hundreds of new coal plants on their territories, which totally defeats the purpose of the U.S. restrictions.

Many coal-fired power plants are being converted into gas-fired operation today, and many more are expected to be converted in the near future as well.

Nuclear plants in the U.S. are in a similar state of limbo since the 1980s, and their status is even more unclear since the 2010 Fukushima accident. There are plans for some new plants construction, but these are mired in political and regulatory debacles and hesitation.

The SC costs for wind and solar power plants, on the other hand, have been dropping significantly during the last several years.

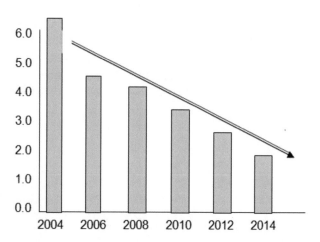

Figure 8-15. Solar power plants cost decrease (in $/Watt)

The solar (PV) installation costs have been falling drastically lately, mostly due to millions of cheap Chinese made PV modules flooding the world markets. Serious stagnation in the solar markets is also contributing to the low installation costs.

Due to the turbulence in the global solar markets, following the boom-bust cycle of 2007-2012, the PV equipment and installation costs might go in either direction.

Cost Breakdown

The capital cost breakdown (installation, or specific construction cost), of the different power generating technologies can be broken down as follow:

The cost of design, construction, installation, and O&M of the different power plants contain complex

variables that depend on a number of factors. The production equipment—capital cost—is a significant part of the equation.

While there is no consensus on the initial price (the capital cost) of each type of power generation plant, we have summarize these in Table 8-9 for general information purposes.

The estimated cost of the existing and most promising power generating technologies today is as follow:

Table 8-10 shows how different the initial costs of the different technologies are. These numbers do not reflect some peripheral, but yet important, factors, as ease of obtaining permit, environmental resistance, etc. They also do not indicate the added cost and complexities due to externalities, such as accidents, negative effects on human health and the environment, etc.

O&M Expense

The operation and maintenance (O&M) costs of an energy system, or technology, reflect the day-to-day operations of a mine, drill rig, or a power plant. These costs are as significant as, and in some cases even more so than, the capital cost. O&M represents only about 10–15% of the total power generation costs of an average power plant, and there are other components included in the final equation that can increase the cost of O&M and the entire power production process significantly.

Operation and maintenance costs of any mine, drill rig, or power plant are actually part of a larger group, which we call production costs, and which can be separated in several sub-groups to encompass all cost elements of running the operation.

The operating expenses of a coal, oil, or natural gas power plant operation are shown in Table 8-11.

No doubt, fuel (thousands tons of coal, thousands gallons oil, or thousands cubic feet of gas) used on continuous basis is the most expensive part of the operation.

As an example, the costs of a coal-fired power plant vary with a number of factors, the major of which are:

a)   The cost of coal is a major factor, because an average coal-fired power plant goes through 8-10,000 tons of coal daily, so any increase in the price affects the bottom line,

b)   The cost of labor and the related expenses, and

c)   How much power cycling (power reduction and increase cycles) the plant goes through daily. Power cycling is an unwelcome, but necessary part of power plant operations. It is needed to match the load demand, and is a major expense, because it is

Table 8-9. Cost breakdown of different technologies

Nuclear power plant

Components	% of total
Equipment	17.5
Buildings	47.6
Engineering	15.9
Other costs	19.0
SC cost	**$6,100/kW installed**

Combined gas cycle power plant

Components	% of total
Gas turbine	14.0
Steam turbine	5.0
Buildings	58.0
Engineering	6.0
Other costs	17.0
SC cost	**$1,230/kW installed**

Combustion gas turbine plant

Components	% of total
Turbine	40.0
Buildings	40.0
Engineering	3.0
Other costs	17.0
SC cost	**$651/kW installed**

Pulverized coal-fired power plant

Components	% of total
Turbine	5.0
Boiler	9.0
Buildings	61.0
Engineering	8.0
Other costs	17.0
SC cost	**$2,890/kW installed**

Carbon capture and sequestration addition

Components	% of total
CC equipment	100.0
CC cost	**$3,750/kW installed**

Note: Carbon capture and sequestration equipment is to be added to the total cost of all new combined cycle fossil-fired power plants, as required to meet the new EPA requirements for drastically reduced GHG emissions.

Standalone biomass power plan

Components	% of total
Turbine	17.0
Boiler	23.0
Buildings	26.0
Engineering	15.0
Other costs	19.0
SC cost	**$3,830/kW installed**

Geothermal power plant

Components	% of total
Well	26.0
Piping	8.0
Heat exchanger	2.0
Turbine	13.0
Buildings	26.0
Engineering	8.0
Other costs	17.0
SC cost	**$5,940/kW installed**

Note: An enhanced geothermal power plant would cost additional $4,000/kW installed.

Hydropower plant

Components	% of total
Reservoir	26.0
Tunnels	14.0
Turbines	16.0
Buildings	14.0
Engineering	7.0
Other costs	23.0
SC cost	**$3,500/kW installed**

Ocean wave power plant

Components	% of total
Hydro-absorber	34.0
Power takeoff	28.0
Controls	2.0
Fixation	8.0
Engineering	10.0
Other costs	18.0
SC cost	**$9,240/kW installed**

Ocean tidal power plant

Components	% of total
Hydro-absorber	15.0
Power takeoff	18.0
Controls	6.0
Fixation	27.0
Engineering	18.0
Other costs	16.0
SC cost	**$5,880/kW installed**

Solar (PV) power plant (2012)

Components	% of total
PV modules	49.0
Structures	29.0
Inverters	8.0
BOS	7.0
Engineering	2.0
Other costs	5.0
SC cost	**$2,830/kW installed**

Solar trough (CSP) power plant with energy storage (2012)

Components	% of total
Solar field	40.0
HTF system	9.0
Turbines	18.0
Heat storage	9.0
Engineering	8.0
Other costs	16.0
SC cost	**$7,060/kW installed**

Table 8-9. (*Continued*)

Solar tower (CSP) power plant with energy storage (2012)	
Components	**% of total**
Solar field	38.0
Receiver	10.0
Tower	2.0
Turbines	14.0
Heat storage	6.0
Contingency	7.0
Engineering	8.0
Other costs	15.0
SC cost	**$7,040/kW installed**

Wind power plant, onshore	
Components	**% of total**
Wind turbines	68.0
Distribution	10.0
Construction	13.0
Engineering	4.0
Other costs	5.0
SC cost	**$1,980/kW installed**

Wind power plant, offshore	
Components	**% of total**
Wind turbines	50.0
Distribution	12.0
Construction	27.0
Engineering	5.0
Other costs	6.0
SC cost	**$3,310/kW installed**

Wind power plant, floating platform	
Components	**% of total**
Wind turbines	45.0
Distribution	13.0
Construction	30.0
Engineering	6.0
Other costs	6.0
SC cost	**$4,200/kW installed**

Table 8-10. Initial cost of different power generating technologies (in $/kW installed)

Power plant type	$/kW
Combustion gas turbine	651
Combined gas cycle	1,230
Wind power, onshore	1,980
Solar (PV) power	2,830
Pulverized coal	2,890
Wind power, offshore	3,310
Hydropower plant	3,500
Standalone biomass	3,830
Wind power, floating	4,200
Ocean tidal power	5,880
Geothermal	5,940
Nuclear power	6,100
Coal with CC addition	6,640
Solar tower w/ES	7,040
Solar trough w/ES	7,060
Enhanced geothermal	9,010
Ocean wave power	9,240

Table 8-11. O&M expenses of a fossil power plant

•	Fuel costs,	60% of total cost
•	Day-to-day O&M,	10%
•	Capital costs,	10%
•	Depreciation	10%
•	Miscellaneous	10%

very expensive to shut-down and restart a coal- or gas-fired power plant.

The boilers must be kept at a constant temperature in order to maintain efficient and cheap operation. A lot of time and fuel are needed to bring them back to normal operating temperature, after cycling (shut down).

Power Cycling

Power cycling is a condition that is forced onto the power plant operators by the demand-supply ratio in the power grid. When the demand is low, some power plants are forced to reduce the power, or shut down all together. Doing this is easy on the front end; just flip a switch or two and the entire plant goes down. But of course, the actual process is much longer and more complicated.

Bringing the plant to normal operating condition, is even more complex and very expensive procedure. In addition to the wear and tear on the equipment, and the long time needed to restart (heat) the boilers, power cycling is also dangerous, since it is associated with risks due to equipment malfunction and human errors.

The risk of personnel errors due to power cycling is multifaceted. It could create risks to human health and life conditions, such as: implosions, explosions, low boiler water level conditions, water Induction into the turbines, low load instability, improper valve alignment, and other man/machine/electronics/and control interface problems during turbine restarts and power ramp-ups.

The factors affecting cycling cost vary, but can be generalized as: type of unit and its particular design, operator training and level of care, type of upgrades for cycling efficiency, overall O&M costs, system marginal energy and capacity costs, cost of new capacity, annual maintenance issues, and number and type of cycles in the past.

Excess power cycling is hard on the equipment, since it causes excess wear and tear, which require additional repair and replacement expense. It is especially hard on the boilers, which experience a number of premature failures due to excess cycling, such as: seals degradation, tube rubbing and damage, boiler hot spots, drum humping and bowing, down-comer to furnace sub cooling, excess corrosion, expansion joint stress, and other expensive to fix failures.

The turbines are also affected by excess cycling, which results in a number of failures, such as: water Induction to turbine, increased thermal fatigue due to steam temperature mismatch, steam chest fatigue cracking, steam chest distortion, bolting fatigue distortion and cracking, blade and nozzle blockage, solid particle erosion, rotor stress and defects, etc. Any of these conditions require a turbine shut down and expensive repairs and parts replacements.

The total cost of power generation is usually significantly increased during each cycle, due to: excess maintenance/overhaul costs for parts and labor, forced outage rates, plant performance (efficiency) loss, overall system production cost increase, reduction in unit life expectancies, long-term capacity cost increase, and overall increase of emissions per kWh generated.

Nuclear power plants are very similar to any other power plants, but with some additional, and equally expensive, issues to deal with. Due to the overall complexity of nuclear plants design, and the ever-present threat of nuclear accidents, we need to have a better understanding of their function.

Although cost is not the most important factor when analyzing a nuclear power plant operation, it is always taken into consideration in order to be compared with the competing technologies. as needed to extract best value of the particular situation.

Nuclear power is a very important factor in the global energy markets. It is also the most volatile and interesting of the participants, so we will take a closer look at its specifics herein.

Nuclear Power Specifics
Nuclear power plants are the most complex (and most dangerous) of all man-made and managed operations on Earth. They pack enormous power that can do miracles as far as power generation is concerned, but can also cause unspeakable damage and complete devastation—including loss of human life—equal to no other man-made contraption. Just ask the survivals of Chernobyl and Fukushima; they could tell you some horror stories of the aftermath of these nuclear accidents.

The world stands still after each of these accidents and considers the options. Some measures, albeit temporary, are taken shortly thereafter, but after a short hesitation after each nuclear accident, nuclear power plants are ramped up to full power. This is because we need a lot of electricity to conduct our daily lives routines, and to run our energy-hungry commercial operations. Only nuclear power can deliver so much power, so we won't give it up…no matter what.

So, here are some of the specifics of running a nuclear power plant:

Fuel Costs
This is the total annual cost associated with the "burnup" of nuclear fuel resulting from the operation of the unit. This cost is based on the amortized costs associated with the purchasing of the nuclear fuel: uranium material, its conversion, enrichment, and fabrication services, along with storage and shipment costs, and inventory (including interest) charges less any expected salvage value.

A typical 1.0 GW BWR or PWR nuclear power plant, the approximate cost of fuel for one reload (replacing one third of the core) is about $40 million. This amount of fuel and the resulting value is base on an 18-month refueling cycle.

The average fuel cost at a nuclear power plant in

2012 was 0.75 cents/kWh. Nuclear plants refuel every 18-24 months, which allows efficient advanced planning, so they are not affected by fuel price volatility as much as coal, gas and oil power plants.

Fuel costs make up 30% of the overall production costs of nuclear power plants. At the same time, fuel costs for coal, natural gas and oil make up about 80% of the production costs.

O&M Expenses

The O&M annual cost of running a nuclear power plant is associated with the day-to-day operation, maintenance, administration, and support of the non-stop activities. All costs related to labor, material & supplies, contractor services, licensing fees, and other costs such as insurance, accidents, employee expenses and regulatory fees, as well as all other expenses are included.

The average non-fuel O&M cost for a nuclear power plant in 2012 was 1.65 cents/kWh. This cost might increase significantly in a case of a minor accident. A major accident can put the entire plant out of commission... for good. Just ask the plant operators at the Fukushima nuclear plants.

Production Costs

Production costs are the combined O&M, fuel and all other costs at a power plant put together. Production costs are important in comparing the different technologies, because they reflect all steps and materials involved in the power generation. It has been estimated that nuclear power plants have achieved the lowest production costs as compared with the fossils; coal, natural gas and oil.

There are, however, several components in the production costs which are not reflected in the general O&M expenses of the plant. Some of these are waste management and plant decommissioning costs that have a high price tag, but yet are hidden from view until the last moment.

Another hidden cost is that of additional safety and precautions taken on local, state, and Federal levels for each nuclear plant. On top of that there is the lingering threat of nuclear accidents that incurs a number of additional expenses, the worst of which is actual nuclear accident, as we saw recently at Fukushima.

Waste Management Costs

In a addition to the tons of GHG emissions, all fossil-fired power plants generate liquid and solid waste, which has to be stored and/or disposed one way or another. The fossil-fired power plants deal with this by creating on-site lagoons and piles of wasted materials, which are removed and disposed from time to time.

Spent fuel and waste management (or lack thereof) in nuclear power plants is the Achilles heel of the nuclear industry. Presently, some of the waste fuel is transported to several outside storage facilities around the U.S. Most spent fuel, however, is stored on-site, with plans for long-term storage in limbo.

For this purpose, the nuclear power plants contribute regularly to the Nuclear Waste Fund, which has about $35.8 billion, $10.8 billion of which have been already spent on different (mostly failed) initiatives. The contributions amount to about 0.1 cent per kWh of electricity generated at nuclear power plants plus interest. These payments to the Nuclear Waste Fund are included in the fuel costs.

However, with all this money on hand and with all the good intentions in mind, the U.S. nuclear industry has no reliable, permanent nuclear waste storage, nor there are any plans for building one. After Yucca mountain was dropped, after spending billions, the nuclear waste issue hangs in thin air, where it will be for a very long time to come.

Decommissioning Costs

After running without an accident for 40 years or more (or after a major accident) any power plant is shut down and decommissioned. Coal and gas power plants are simply demolished and the land is returned to its pre-construction status as much as possible.

The nuclear power plants are also demolished, but special efforts are required to decontaminate the equipment and land in order to return it to its previous condition. The estimates for this effort are about $300-600 million per plant.

Decommissioning and remediation of a nuclear plant includes efforts such as:
• Remediation of any and all radiological effects,
• Used fuel disposal, and
• Overall site restoration costs.

These costs average $300 million, $100-150 million and $50 million, respectively.

The estimates for the entire nuclear industry are in excess of $30-40 billion, or about $300-400 million per reactor. The catch here is that the decommissioning costs are not included in the production, or any other, costs. Of the total $30-40 billion estimated to decommission all eligible nuclear plants at an average cost of $300 million, about two-thirds have already been funded.

The remaining over $10 billion will be funded in

the future, so not to worry: our nuclear future is free from additional costs from nuclear plants decommissioning. We just have to make sure that we don't allow any Fukushima-like accidents, which could alter significantly the spread-sheet of the local area, the country, and even the entire world.

Levelized Cost of Energy

Levelized cost of energy (LCOE) is the price at which electricity generated from a specific source breaks even over the lifetime of the project. Keeping in mind that the break even price is a moving target, LCOE is just an economic entity—an assessment of the cost of the energy-generating system over its lifetime cycle, including; capital investment, installation, O&M, fuel, and all other cost for the duration. With some externalities (but not all) included.

The inclusion of ALL externalities would require a brand new and much more complex equation. In this new equation nuclear power, for example, would have to be taxed with billions of dollars of past, present and future damages to property, human health and death.

For now, however, LCOE is the best tool for calculating, and comparing, the costs of power generation by different sources.

LCOE can be defined in a single formula as:

$$LCOE = \frac{\sum_{t=1}^{n} \frac{I_t + M_t + F_t}{(1+r)^t}}{\sum_{t=1}^{n} \frac{E_t}{(1+r)^t}}$$

where

$LCOE$ = Average lifetime levelized electricity generation cost
I_t = Investment expenditures in the year t
M_t = Operations and maintenance expenditures in the year t
F_t = Fuel expenditures in the year t
E_t = Electricity generation in the year t
r = Discount rate
n = Life of the system

Typically LCOE of a power generating plant is calculated over 20 year lifetime, and is given in dollars per kilowatt-hour, or megawatt-hour. The LCOE for a given energy source is highly dependent on the available (accurate) information and any (accurate) assumptions, such as financing terms and technological characteristics. For example, the assumption of the value of the capacity factor has significant impact on the calculation of LCOE.

A solar PV power plant may have a capacity factor as low as 10% depending on location, which would affect the LCOE negatively. At the same time a nuclear plant could be in the 80-90% range. This is a big difference, and requires using reliable data and justified assumptions for achieving accurate LCOE results.

Table 8-12 reflects the conventional wisdom on the matter. Now, it is quite obvious that LCOE is a very complex variable, for which common wisdom would suggest using different measures and variables for the different power generating technologies. And yet it is applied equally and uniformly to all energy sources and situations; as different as coal in the depths of the Earth is different from the sunshine blasting at the Earth from above.

Here are some of the problems:

* Electricity at different places and from different power generators sells at different prices, depending on the location, time of day and season, type of electricity (base-load, peak following, intermittent) etc., etc. variables,

* LCOE cannot account for complex mechanisms, such as power plants that co-produce energy (elec-

Table 8-12. LCOE of energy generation (estimates)

Plant Type	LCOE in $/kWh		
	Min.	Avg.	Max.
Conventional Coal	0.090	0.097	0.116
Advanced Coal	0.104	0.112	0.126
Advanced Coal w/CCS*	0.130	0.141	0.162
Natural Gas Fired			
Conventional CC**	0.062	0.069	0.881
Advanced CC	0.059	0.066	0.083
Advanced CC w/CCS	0.083	0.093	0.111
Combustion Turbine	0.095	0.132	0.164
Nuclear	0.108	0.113	0.120
Geothermal	0.085	0.100	0.114
Biomass	0.102	120.2	142.8
Wind, onshore	0.078	0.097	0.114
Wind, offshore	0.307	0.330	0.350
Solar, PV modules	0.122	0.157	0.246
Solar Thermal (CSP)	0.183	0.251	0.401
Hydro	0.058	0.089	0.148

*CCS—carbon capture and sequestration
**CC—combined cycle

tricity and heat) and/or chemicals and other products or services,

- There are a number of discount rates, subsidies, and such that are (sometimes) not properly stated and included in the calculations. Some of these vary from time to time too, so a LCOE today is not the same as LCOE tomorrow,

- The units in LCOE calculations, usually currencies, are problematic since it makes it harder to compare between different countries, or between past cases. LCOE is divided by the price of electricity, so the units of normalized LCOE and the rate of return are not the same; one is dimensionless, and the other has units of time.

- The LCOE calculations hide the underlying nature of economics. In case of a LCOE value higher than the current electricity costs, the actual price of electricity would have to increase. In such a case, the calculation of the fuel costs, the labor costs, and the maintenance cost estimates will increase also, which will cause the LCOE to increase. And going back and forth like that we get into a vicious cycle. The same is true if we add more plants, in which case the cost of electricity will keep increasing according to the amount of generated power, getting us into another vicious circle.

- But perhaps the most significant variable, which is not even part of the LCOE calculations, is that of externalities. We continue ignoring that variable as if it does not exist, or as if it is going away tomorrow. But taking a close look at Fukushima and talking to the locals would reveal the true price of power generation. It would reveal a horrific situation that is difficult to describe in human terms, let alone put in dollar numbers or economic projections. It is a situation that is getting even worse with time, and will be with us for many years and maybe centuries.

How and when would we be able to add this cost the LCOE? Is it even possible? And yet, leaving it unaddressed is not the solution.

Alternative Cost Analysis

Nevertheless, LCOE is a valuable metric, because it allocates the costs of an energy plant across its useful life. This can give us an effective price per unit of energy

(kWh). This is like averaging the up-front costs across production over the entire life of the power plant.

LCOE gives a single metric that can be used to compare different types of systems; renewables, where the up-front capital cost is high and the "fuel" cost is near zero, to a coal power plant, where the capital cost is lower, but the fuel cost is higher. LCOE of different power generators can even be compared against the utility bills in $/kWh.

Simpified LCOE-like Analysis

LCOE is too complex for some purposes, so instead, analyzing and comparing different levels of detail might be more useful in matching the level of complexity we are trying to accomplish.

Starting with the most basic assumptions is very useful in giving us a first glance idea of the difference between different energy sources and technologies. We can then expand by adding more detail and complexity until we get the answer that is most appropriate for our case.

In more detail:

1. We start with the minimum amount of data and analysis that is absolutely needed to get a simple LCOE-like numbers to compare the technologies. Here we need only to know the:
   - The system size, or the 'nameplate capacity' of the power system in question. This tells us how much power the system could produce when running at full power.
   - The system cost, which is the cost to install the system expressed in $/Watt.
   - Watt-hours per watt-peak (Wh/Wp) is an extension of the nameplate (which only tells us how much power can be generated.) The Wh/Wp number gives us is how much energy we actually get out of our system over a period of time as compared to the nameplate number. I.e. 100/90 means that 100 MW nameplate is producing 90 MW during a certain period of time. From this number we can also calculate the number of hours in a 24 hr. cycle that the system was under full operation.
   - Years of operation reflect the time from installation to decommissioning of our power plant. Most solar power plants are expected to operate and are warranted for 20-25 years of non-stop operation.

Using the above analysis we can get a good idea as of what to expect from a given power system, regardless of what technology is used. We just compare the system size to the system cost and then add the variable of how much power is expected for how many years.

Let's take for example two power plants; one is coal and the other solar power plant.

Starting with the minimum level analysis:

Table 8-13a. Minimum analysis example

Variables	Coal	Solar
System size	100 MW	100 MW
System cost	$2/Watt	$2/Watt
Wh/Wp	100/90	100/25
Years	25	25

In this example, with everything equal, we see that the coal plant provides 90% of its maximum power, while the solar plant generates only 25% of the 100 MWp nameplate capacity, due to the variability of sunlight and night time power down state.

Also, the solar plant requires more O&M expenses, which reduces our profits further, so if given a choice we would chose a coal plant for more reliable and profitable operation.

2. An added level of detail could include another set of assumptions and variables that are significant, but not always needed for a number of reasons:
    *   Derating reflects the actual power generation, due to efficiency losses in the system: turbines, condensers, inverter, wire, etc., which reduce the power by the time it enters the grid.
    *   Discount rate is the theoretical future value discounted against today's actual value of the system and the money used to pay for it. This number would be the same for any and all power systems, with some rare exceptions.
    *   Government incentives drive the energy industries, and matter a lot. For example, there is 30% tax credit for renewable energies and dozens of state and municipal incentives, but much less for the conventional energies.

So, adding these numbers to our minimum level exercise above, gives us:

Now we see, with everything considered in 8-13.b above, that solar benefits from a 30% tax break. This amounts to millions of dollars savings during the plants construction, so now we have to decide if the long-term inefficiency of the solar plant (expressed in dollars 25

Table 8-13b. Added level of detail

Medium level analysis:

Variables	Coal	Solar
System size	100MWp	100MWp
System cost	$2/Watt	$2/Watt
Wh/Wp	100/90	100/25
Years	25	25
Derating	20%	20%
Discount rate	5%	5%
Incentives	5%	30%

years down the road) is more or less than the 30% worth of savings that can be realized today.

In order to do make the right choice we will need another set of variables, as follow:

3. The final variables needed for completing the analysis are:
    *   Components and system' degradation is expected to affect all components of any power plant. PV modules degrade much faster (0.5-1.0% per year), due to their physical characteristics. Coal plants degrade too, but at a much slower rate, most of which is absorbed by the maintenance cost.
    *   Operations costs include everything that is needed to run the plans for the duration—including fuel—and are different for different plants. In this case, coal plants require a lot of coal, while solar plants do not need any fuel.
    *   Maintenance costs include the equipment and labor needed to fix and replace power plant components. Coal power plants have a number of complex and expensive components, such as furnaces, turbines, generators, etc., that need to be repaired and replaced periodically. Solar plants also need fixing and replacement of equipment—PV modules, inverters, wiring, etc., but at a much lower level of complexity and expense

Adding these variables to the medium level exercise, gives us:

Complete Analysis

Now the picture gets more complicated with all this data piled up. Note that the units here are 0 to 1, which has no real value and is used only for compara-

Table 8-14. All power plant details included

Variables	Coal	Solar
System size	100MWp	100MWp
System cost	$2/Watt	$2/Watt
O&M	$0.04/Watt	$0.08/Watt
Wh/Wp	100/90	100/25
Years	25	25
Derating	20%	20%
Discount rate	5%	5%
Incentives	5%	30%
Degradation	0.1	1.0
Operation cost	1.0	0.1
Maintenance	1.0	0.5

tive purposes.

With the degradation and operations cost awash, the maintenance costs are to be considered as a priority in this scenario, but will require full knowledge of the equipment and procedures involved and some guess work too.

Combining this last choice with the choices in 1. and 2. above, adds a level of complexity to the final decision. All variables need to be well understood and properly presented. Nevertheless, the number of choices is reduced to only 3—one in each category, and although these are complex in nature, they can be resolved by simple individual evaluation and calculations, thus arriving at a plausible final choice and decision at each level of the analysis, as needed for the project design.

In conclusion, while LCOE is a valuable tool that covers the basics of power generation and its costs, it is not an ideal tool as far as comparing different technologies used under different conditions, and at different locations are concerned. Using the simplified cost calculation method above is also not perfect, but it reduces the complexity and eliminates some of the uncertainties contained in LCOE calculations.

We will, however, leave it to the economists to resolve these problems and come up with a more efficient method.

Case Study: Wind Power O&M

Recent estimates by leading data providers show that the North American wind industry will increase from nearly 70.0 GW in 2014 to over 105.0 GW in 2020. About 30% increase of new installations means that the other 70% (older facilities) will need increased need maintenance and repairs.

There are 450 companies that will be competing for

the slice of the wind power operations & maintenance (O&M) market. At this growth rate, the U.S.' wind O&M market is expected to reach new highs of over $5.0 billion by 2020, including major corrective operations and upgrades. At this level, the O&M market has a clear market visibility going forward, and so many components suppliers and service providers are expected to compete for a piece of the pie.

O&M revenues and profit opportunities vary significantly from company to company. The final success is based on the ability to provide high quality and cheap scheduled maintenance, remote monitoring, major and minor repairs, technical support, spare parts, and component upgrades.

In the end, the LCOE of a wind or solar power plant depends heavily on its O&M expenses during the 30 years of field operation. These expenses increase from zero during the initial months o operation, to millions of dollars during the last years of operation.

Since none of the modern wind and solar installations has completed a full life-cycle (30 years) of non-stop operation, we don't know where the breaking point is. This leaves a big gap in our understanding of the usefulness of these installations. It also makes it very hard to estimate their life-time LCOE. We may be faced with some disappointments during the next 10-20 years, when some wind and solar power plants fail to deliver enough power, and/or their components fail prematurely.

Environmental Economics

Please note that environmental economics is closely related to energy, but only partially because there are many other factors and variable involved when talking about environmental issues. As a matter of fact, at certain point and time the two part ways. In some instances they even contradict each other. This is where and when the energy and environmental experts of the legal industry get involved.

It is obvious, for example, that a coal-fired power plant operator would be penalized when the plant generates more power. It then emits more pollutants, which leads to a higher carbon tax. At the same time, a large solar power generator is increasingly rewarded as it increases the power generation, which generates no pollutants at the point of use. We usually ignore the fact that a lot of energy has been used and a lot of pollutants have been emitted at different times and locations during the manufacturing, transport and installation of the solar equipment.

So, is it fair to consider rewarding the solar plant on the basis of what it is producing now at this location, while totally disregarding the previous energy use and environmental damages at a different location? And then, is it fair to penalize the coal plant for emitting pollution today and ignoring the good it does by providing us with useful energy day after day? If so, then why and how much?

These are subjects that energy and environmental economists and legal experts are dissecting and analyzing. Not an easy thing to do, so let's see who, when, and how this is done...

Basically speaking, environmental economics is the study of the environment and the related responses, uses and abuses evaluated through the lens of economics. Market failures, externality, or valuation methods are applied to environmental topics such as; air pollution, fuel consumption, and alternative forms of energy.

One way of energy vs. environment type of evaluation today is based on the fact that we are taking too much from the earth (fossils), and hurting the environment in the process. The environmental economics is stepping in to find a balance between the usage and the damage on one hand, and obtaining the greatest societal benefit on the other.

The solutions and measures are often based on dollar amounts which one party has to pay for the benefit of another, but ultimately benefitting the society as a whole. The legal profession is involved in all this in order to ensure that the law is properly interpreted and followed, and that justice is done at the end.

Environmentalists and economists are trying (usually independently from each other) to determine the amount and value of the existing natural resources, and how to make sure that we all benefit from them without posing great risks to life on Earth.

Most natural resources used in every day life, have been assigned a monetary value based on their availability, usefulness and use. The task now is to determine and establish a value on what we need to do in order to live comfortably and at the same time preserving the Earth for the future generations.

The problem is that putting universal value on this key concept, in order to measure everything with it, is almost impossible because it has a great value to some, while it is meaningless to others. And to make things worse for economists, some of the natural events such as the value of clean air or water (which are taken for granted by most people) are intangible, and no one has figured out how to put a price tag on them.

And this is causing a great divide in the society. For example, we rely heavily on fossil fuels, which produce pollution that is harmful to all. But they are also a neces-

sary part of life and are needed for our basic needs and the advancement of our society.

Different people look at this complex situation from different points of view. The environmentalists, for example, focus on the problems and insist on reducing the carbon footprint to the minimum. The pollution emitters (power plants and such) insist that this is a necessary evil which helps the economy and should not be excessively punished or eliminated.

The energy and environmental economists are stuck in between, performing cost benefit analyses in order to find the middle ground and provide solutions. In some cases, they are willing to settle for mild environmental damage that brings great economic benefits, while other times they figure out that the damages outweigh the benefits.

In most cases, there is always somebody, or some group, that will disagree, and this is where the legal experts step in to resolve the differences via court proceedings, arbitration, counseling, etc. methods.

Environmental economists usually try to find an acceptable compromise, in order to ensure economic and societal advancement, keeping in mind the effect of all activities on the environment.

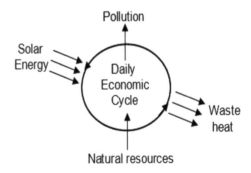

Figure 8-16. The energy-environment cycle

Figure 8-16 shows the daily cycle of economic activities and the related environmental impact, where the positive effects are represented by the input of solar energy and natural resources into the system. The negative effects of the economic cycle are represented by pollution and waste heat generation.

The lasting negative effects in this scenario are,

- Gradual depletion of natural resources which is only indirectly impacting us now, but which will become a major factor soon, unless we find a way to control the complete depletion of the fossil fuels.

- Increase in pollution, which damages the natural environment and all life forms in it, including human life, and

- Increase in global temperatures, which is causing cataclysmic changes in some parts of the world.

The environment has a several-fold function:

1. It is a source of raw materials and services for human use, which can be interrupted by:
   a) Depletion of natural resource as humans use them quicker than they could be regenerated, and
   b) increase of pollution affecting human activities to the point of extreme events and disasters. In case that any of the above points is reached, human activities will have to change drastically or cease all together.
2. The second function is that of a life giver and supporter. Without our environment there would be no life. Period. And then it supports life by providing and controlling the proper type, quality and quantity of critical life elements: air, light, water, and food to all living organisms, including humans. This function is the most important factor in any and all discussions of environmental issues and actions. It is also taken for granted some times, and ignored others.
3. The environment also functions as a sink—a self-contained garbage disposal and storage bin in one—by absorbing and storing the natural and man-made poisons and waste by-products in the air, soil, and water. The sink and storage bin combination has its limits and stops normal function when the types and amount of pollution exceed certain levels in a given time period, and/or when the wastes exceed certain level of toxicity.

At such a point, the most critical components of the environment; soil, water, and air on which we depend daily could become damaged, polluted, and/or poisoned. This is a critical point, which we are approaching fast today, the issues of which must be fully addressed and resolved soon, and before it is too late.

Modern Environmental Economics

Modern environmental economists take a broader view and perspective in framing environmental questions by incorporating laws derived from the natural sciences. A fundamental principle of environmental economics is that:

Human economic activity is limited by the environment's carrying capacity and its overall tolerance to change.

The carrying capacity of the environment is defined as the level of human and animal consumption of natural (non-renewable) resources. This is basically what the available natural resource base can sustain without complete depletion on one hand, and the capacity of the environment to absorb and store waste products on the other.

The environment's carrying capacity is the natural Supply-Demand-Waste balance

Please note here that while the energy markets depend only on the Supply and Demand ratios, the environmental markets have an added variable; this of the type and amount of Waste in the environment. This waste could be tons of coal ash, liquids, and/or air pollution.

Environmental economists emphasize the importance of the different energy resources and their use, and especially the fossil fuels, as related to the current economic systems. They also focus on the pollution the different energy sources generate in different areas, as an unavoidable factor in all energy related calculations, and look for solutions and alternatives.

The rapid growth of economic production during the twentieth century required enormous energy inputs, and global economic systems will make even greater energy demands in the twenty-first century and beyond. This means that energy availability and environmental implications of energy use are central issues for environmental economics.

For the human population, the issues are getting more complex by the day too. The main issue of food supplies is certainly relevant as the world population grows toward a projected 10 billion by 2050. Ecological economists also point to energy supplies, scarce natural resources, and cumulative environmental damage as contradicting conditions and constraints on economic growth.

One problem is that the standard economic theories and practices are not aware of, or simply ignore, many of the key energy and environment related factors, and also that major structural changes in the nature of economic activity are required to adapt to the upcoming energy and environmental limits.

Another problem is that the environmental economists are focusing their attention on providing better understanding and valuation of the complex energy-environment dilemma of the 21st century, which most of us have a hard time understanding, or accepting. In this text, we are trying to clarify some of these discrepancies and point to ways of solving the related issues.

So what is the proper way of accounting for environmental costs?

Valuation Methods

Direct or indirect values of energy and environment do not provide precise dollar amount from losses of properties and services. We must agree, however, that the assignment of "some" number is better than no number at all.

The assignment of "some" number, however, is still a complex and controversial undertaking. A lousy job, but somebody has to do it, and our tireless environmental economists are best suited for that effort. So what would they do? Where would they start?

Fortunately, they have an arsenal of tools and weapons that can be successfully used for this purpose. Successfully is highly subjective in this case, because no matter what number they assign, it will be questioned and disputed by one or another group of interested entities. A thank-less job, to be sure, so we do appreciate their efforts.

Some of the tools used for valuation of hard to value energy and environmental targets are:

Cost-Benefit Analysis and Valuation

Valuating the energy sector and the related social benefits and environmental effects is difficult but needed endeavor. Different groups see it from different points of view, which makes the effort even harder. While some see the environment as priceless gift from above, and reducing it to monetary values is absolutely unacceptable. The other extreme insists on putting values on all ecological functions before the economic system gives them a zero value.

No matter what, we all agree that it is extremely difficult, and in some cases impossible, to conduct proper cost-benefit analysis of energy-environmental systems. This is due to lack of sufficient reliable information, and also because certain elements, such as social, moral, and spiritual values are impossible to present in percentages or dollar numbers.

Nevertheless, when properly used, cost-benefit analysis is a useful tool; always keeping in mind that we are barely scratching the surface of the immense area environmental responsibility and accountability. We should also realize that the methods are not set in cement and will slowly undergo gradual modifications and optimization, before becoming standards. If they ever become such.

Presently there a number of approaches that re-

quire polluters to pay dollar amounts corresponding to the amount of pollution emitted. These approaches are well received by some groups and rejected by others. Only time will tell who is right, and who is wrong, and which approach will win the battle of universal acceptance.

- *Contingent valuation* is basically a survey to determine how much people would be willing to pay to preserve their community and its environmental functions. The resulting estimate is their willingness to pay (WTP) that can be included in the cost-benefit analysis (CBA). The only problem here is that this method is almost always challenged by a group of people on one side of the issues at hand.

The hypothetical nature of the assumptions, when no actual money is involved often leads to un-realistic (very high or very low) valuations and unrealistic assumptions. Since respondents have limited knowledge of the actual science and technology of energy generation, or environmental function, they tend to under-, or over-estimate, the real problems.

- The *estimated willingness to accept* (WTA) is another approach, quite different from the more recognized willingness to pay (WTP) method. For example, when asked how much is an adequate compensation for the loss of picnicking, hiking, and fishing activities, most peoples' dollar estimates are much higher than the WTP estimates for the same items. This leads to criticism as of the usefulness of the contingent valuation approach.

The disparity between WTP and WTA valuation is a problem which has particular importance in the economic valuation of environmental issues. Unlike a more ordinary economic transaction such as the loss of land use, pricing the environmental effects involves serious ethical and philosophical considerations, in addition to the usual socio-economic considerations. This usually leads to the overwhelming question if the true value of environmental preservation can really be expressed in dollar terms.

For example, the public's WTP as needed to avoid the Valdez spill's wildlife damages was estimated at $2.8 billion, or about 75% of total cleanup and compensation costs. In other cases it was much lower, which means that the society's willingness to pay to avoid environmental damages, is significantly lower than the actual cleanup and compensation expenses.

Not a precise science, and yet, the estimates, the valuation, and the related cleanup and restoration have to be done, because this is the only way for our capitalist system to function properly. And in fact, money helps a lot, as in the numerous cases of restoration of abandoned mining sites, cleaning water supplies, etc. Without that money, the mining area would be lost forever, and the water supply would make many more people ill.

So we must thank environmental economists for making the effort to put a money value on things that we have a hard time putting a value on, but which we nevertheless need in order to survive and move on.

Demand-side Valuation Methods

This valuation method is closely related to the way actual market decisions are made by valuing environmental products and services as they relate to the value of marketed goods and their availability. The price of a farm located next to a controlled conservation area is usually much higher than that of a comparable piece of land situated next to a dusty and congested road leading to a dirty and stinky oil rig.

In many cases, however, there is a bias in demand valuations—unless and until actual money is to change hands. Surveyed people often say one thing, when in reality they are not willing to put up real money in support. The inflated estimate is an example of strategic bias, where people attempt to bias the survey upward by stating an amount that exceeds the actual willingness to pay.

It has been determined that the disparity between WTP and WTA is usually found in non-marketed goods such as land preservation, where willingness to accept (WTA) exceeds willingness to pay (WTP) by a factor of 7. Or with other words, people are often willing to place value on an environmental product or service that is seven times higher than what they are actually willing to pay for it.

Supply-side Valuation Methods

In contrast, the valuation on the supply side of the market is based on production function and engineering cost methods. Here more precise science plays a significant role, since the goal is to evaluate the situation and determine the cost of restoring or preserving the environment via established products and services.

Building a waste-ash storage pond close to a coal mine would affect the forest and its pollution-absorption capabilities. If the forest is preserved, and building and operating the storage pond is unnecessary, thus avoided cost can provide an estimate of the value of pollution control services provided by the forest. We can easily make technical estimates of the facility and calculate the associated costs, which will give us the approximate value of the forest.

If an oil spill destroys a wetland, the responsible will be required to pay to restore it to its prior state, or to create a comparable wetland nearby. The cost estimates generated in such cases are typically significantly higher than those based solely on demand-side studies.

A combination of demand and supply side methods should be used to determine the best dollar value for environmental products and services. The usefulness of these valuations depends on our basic concept of "value," which consists of, a) the purely economic value, and b) the broader concept of ecological value, which is very difficult to define.

In all cases value refers to what is important to humans, and is defined by our willingness to pay, or to provide substitutes. Some key environmental issues, such as the climate change, are not possible to define in such terms. No matter how much money we pay, we cannot reverse the global warming, because we don't even know what would work, nor we have enough money to pay for the effort, even if we knew what to do—this is how big this problem is.

The potential and the limits of economic valuation of environmental costs and benefits is in question. There is a long list of environmental effects and issues that cannot be valued, and some of them fall through the gaps. This makes the environmental economics models incomplete.

But when, how and who would come up with the exact valuation of a human limb, or life. How do we value the future events that may, or may not happen, even if we knew their magnitude and severity? How can we value the costs of tomorrow, if we are having problems valuing the costs of today? Lots of questions in need of answers, so we thank our dedicated economists for working on all these issues.

Human Life Valuation

Just talking about valuating human life, sounds like an inhumane way to deal with the most sacred thing on Earth. We all well know that human life is not for sale and that no price that can justify hurting, or taking it away. So how do economists value it, and why? Isn't anything sacred left? The environmental cost-benefit analyses force these questions and demand answers.

It might be easy for most of us to turn a blind eye on the subject, but ask a miner's wife whose husband has been killed in a mine accident. She will tell you that funeral expenses have to be paid on time, and that rent, utilities, and food bills are due every month in a household where income has been slashed to zero.

Some valuation is needed here to right at least some of the wrong. How much would she accept as a recompense for her husband's life? How much would the mining company be willing to pay? And how and who determines the actual amount?

On a broader scale, tighter control of coal-fired plant's smoke stacks may reduce the amount of cancer-causing chemicals and save some of the locals. The tighter controls, however, inevitably cause increased production costs. So the benefits are reduced cancer cases to the locals, who end up paying higher electricity rates. And the question again is, how and who will determine if the tighter controls are needed, and determine the rate increase if so needed.

The evaluation in this case requires reliable estimate of the reduction in cancer cases, in which case we would multiply that number by the cost of hospital care in each case to obtain a measure of total medical costs.

But if the evidence indicates that several deaths per year could be avoided, then how do we balance the lives saved with the increased production costs? A simple economic estimate of what an individual contributes in terms of production of goods and services to society might be one (albeit unfair) way. This approach assigns a high dollar number on the life of a CEO whose salary is in the millions, versus that of an hourly worker who is struggling to pay the rent.

Another approach is to consider the amount of life insurance and safety precautions taken in some cases, which will derive an estimated value in the range of several million dollars for a human life. No matter how look at it, however, we cannot possibly capture the full value of human life. Instead what we are talking about is a compromise, which indicates an arbitrary value of what economists call a "statistical life." It is based on a demonstrated willingness and ability to pay certain amount of money in order to avoid risks to life.

Again, not a perfect way to do things, but the option is to set the value of life at zero, which is wrong, or make it infinite, which is impossible. Taking the middle road, economists are deriving different solutions for different cases, which help the miner's wife to pay the bills, forces the coal plants to enforce safety procedures and install pollution control devices.

Financial Valuations

It seems that in the capitalist system almost everything can be equated to money. From the smallest trin-

ket, to a complicated gadget, or important service, and even human life; they all have a price attached to them. The energy markets are no exception. Here everything has a price too. Everything from raw materials, to buildings, equipment, labor, services, and people's lives have a price tag.

How are these prices established in the energy markets is the question we are tackling below.

Power Plants Valuation

Power plants have an initial value based on the cost of materials and services during the design and construction phases. As the system's components get old, the value of the entire plant decreases until the end of its operation life cycle (30-40 years down the road) when the value is down to zero. And even worse; millions of dollars might have to be spend to demolish the plant's structures, and decontaminate and restore the land to its original state.

There are several tools available to analyze and valuate the power plant during different stages of its operation, most of which are theoretical exercises. What is most important to owners and investors is the final number reflecting their return on investment (ROI).

ROI estimates are done during the plant's design stages and are basically the difference between all projected income and all foreseen expense for the duration, or

$$ROI = IT - ET.$$

where:

IT is the total income, and
ET the total expenses

The total income IT includes:

- Total amount of sales (electric power, heat, and other products)
- Total amount of subsidies and incentives (during the construction and operation)
- Total amount of other money and goods received for the duration.

The total expense ET includes components such as:

- Total cost of design and construction,
- Total O&M expenses
- Total amount of other expenses (law suites, insurance increases, interest payments, etc.)

So far everything seems clear and transparent, except that there are items that are buried deep into the above formula assumptions. Some are hidden behind complex technical operations, while some are purely financial functions and manipulations. Each of these could change the outcome and make the ROI go in whichever direction.

For example, solar modules failing at a higher rate than initially calculated, or increase in coal price and/or insurance premiums 5-10 years into the project that were not foreseen and included in the budget during the design stages, could cripple the project and even force it into bankruptcy.

Here is where solid technical understanding of the technology on one hand, and thorough financial analysis have to be applied. Expert technical and financial specialists working hand in hand during the design stages could and should foresee anything and everything to be expected to happen during the plant's lifetime. One missed, or miscalculated, variable could become the rock that overturned the car.

The Future Discount Rate

Power plants operate for many years, so comparing the real value of money now and at the end of its lifecycle is a convenient valuation method. It should not be used alone, but instead should be included in the comparative valuations of different technologies, locations, and timing schemes.

Discounting over time is used to value cost or benefit expected in the future, based on its present value and the present value of money. The discount rate is the annual rate at which dollar values are considered to change over time. It is a known fact that a dollar today is worth more than a dollar tomorrow after correcting for inflation.

For example, at a 10% discount rate, $1.00 today becomes $1.10 next year and so on ad infinitum. Looking from the other side of the formula, $1.00 at this discount rate to be spent 10 years from now is worth only about $0.50 today.

Use of a discount rate in valuating environmental costs and benefits is a bit different and much more complicated. For longer time periods, the discount factor becomes much more dramatic. The present value of $1,000 to be received 50 years from now is only $87.20 at a 5% discount rate. And the present value of $1,000 to be received in 100 years is only $7.60 today. This would mean that, according to present value theory, it is not worth spending more than $7.60 today to avoid $1,000 worth of damages 100 years from now.

The problem is that no matter how we do it, it seems that we are basically justifying serious damages to future generations as less important than moderate costs today that might prevent these damages.

Discounting is essential if we consider the economics of, for example, taking a mortgage to buy a house or a loan to finance a business investment. The benefits of being able to own and live in the house starting today may well outweigh the future costs of paying interest on the mortgage over the next 20 years. Similarly, the income generated by the business investment can be compared to the annual payments on the loan.

In this case it makes sense to use the commercial discount rate, determined in current markets, to compare present benefits and future costs. Even so, can we say that a GNP gain today, or in the near future, outweighs major damage in the next generation? How should we evaluate broader environmental impacts that will continue over long periods of time?

In some cases, like that of a cost-benefit evaluation of a new coal-mine project, the choice of discount rate may play a crucial role in the relative weighting of present and future costs and benefits.

There are actually three time periods of the mine cradle-to-grave life that must be considered: the mine construction period, the period of active mine operation; and the period after the mine ceases to provide benefits.

During the construction period there are heavy costs, no little benefits (except temporary job creation and local revitalization). Once the mine starts operation, the benefits are tons of coal sold to power plants. At this period we pay two types of costs: operating and maintenance costs, and external environmental and social costs. As a matter of fact the external costs begin long before the mine operation starts. During the 30-40 year expected lifetime of the mine, both actual benefits and costs are to be considered.

After the coal supply is depleted and the mine ceases operation, the environmental and social costs continue for a long time, and in some cases indefinitely. Who pays for these depends on many factors. If the original company or property owner are out of business, then the ongoing costs are absorbed by state and federal entities.

The evaluation of net present value includes all of the above components, which must be discounted back to year 0. A number of complex formulas would show significant benefit during the 30-40 coal-production years, which are, however, overshadowed by the long lasting damages to the local environment, and especially the wild life and humans living nearby.

Ecosystem Valuation

How do we put a value on the environment we live in and depend on? Nevertheless, we need to try, because if we don't things might get even worse. Valuation can be a useful tool that aids in evaluating different options to managing natural resourced.

Our ecological resources and services are varied in their composition, so it is often difficult to examine them on the same level. If we, however, assign a value to each environmental resource or service (as hard as this might be) it can then be compared to any other similar resource or service.

The Basics

- Value is generally defined as the amount of alternate goods (money usually) a person is willing to give up in order to get one "additional unit" of the good in question. An individual's preference for certain goods may either be stated or revealed. In the case of stated preferences, the amount of money a person is willing to pay for a good determines the value because that money could otherwise be used to purchase other goods.

  Value may also be determined by simply ranking the alternatives according to the amount of benefit each will produce. Revealed preferences can be measured by examining a person's behavior when it is not possible to use market pricing. Marketing techniques of many products today are based on revealed preferences, which serve as a base for price structure and market posture.

  There are typically two ways to assign value to environmental resources and services; use and non-use. There are approaches to measuring environmental benefits based on these defined values.

  — When environmental resources or services are being used, it is easier to observe the price consumers are willing to pay for the conservation or preservation of those resources.

  — Non-use cases are more complex to resolve, due to uncertainties and a guessing factor effects. Non-use value is usually evaluated by employing contingent valuation through the use of surveys that attempt to assess (assume) an individual's willingness to pay for a resource that they do not consume.

- Market cost pricing is used when there are tangible products to measure, such as the amount of fish caught in a lake.
- Replacement cost can be calculated based on any expenses incurred to reverse environmental damage.
- Hedonic pricing will measure the effect that neg-

ative environmental qualities have on the price of related market goods.

- A cost-benefit analysis requires the quantification of possible impacts of a proposed project. The impacts could be physical or monetary, but both must be calculated and included since a financial analysis that requires assigning dollar values to every resource evaluated is also performed.

The process of environmental resource or service valuation provides a way to compare alternative proposals, but it is not without problems.

All valuation techniques encompass a great deal of uncertainty: flaws can exist in the methods of assigning value accurately due to a wide number of variables and it is difficult to compartmentalize and measure environmental and natural resources and/or services within an ecosystem that functions as an interconnected web.

Ecosystem Valuation

Ecosystem valuation is a method of assigning a monetary or any other value to environmental resources, including the production, use and/or services provided by those resources. Environmental resources and/or services are particularly hard to quantify due to their intangible benefits and multiple value options.

It is almost impossible to attach a specific value to some of the experiences we have in nature, such as viewing a beautiful sunset. What is the value of the sunset itself, and how do we value the pleasure we get from viewing it. Or how do we valuate a forest that provides timber, fresh air, animal shelter and even beneficial environmental services by preventing hill sliding and downstream flooding.

Resource that can be used for multiple purposes, such as a forest has to be valued in all aspects of its usefulness. The timber quantity has a certain value, while the prevention of the hill sliding onto the neighborhood houses has a totally different value. Both can be expressed in terms of dollar amounts, but the quantity would vary according to the usefulness of the forest in each case.

The quantity of the different resource must also be taken into consideration because value can change depending how much of a resource is available. A small power plant generates a little smoke and the environment might not be harmed…at least not for a long time. A huge power plant, however, could destroy the local environment and hurt the locals in a very short time.

The small power plant's emissions are not valued very highly because the environment can easily recover, even if it is damaged. However, if the pollution from the large emissions generator continues until the air is becoming toxic to its surroundings; the value of preserving clean air by preventing additional pollution is going to be increasingly valued.

In summary, ecosystem valuation is a complex process by which economists and regulators attempt to assign a value to natural resources, their production, use and the related services as provided by those resources. This is not easy, but it is the only way for policymakers to make decisions based on specific comparisons, typically monetary, rather than some other arbitrary basis. In recent years, the U.S. government has placed increasing emphasis on cost-effective environmental laws and projects, in an attempt to establish a common measure by which to evaluate the alternatives.

The Environmental Externalities

The externality concept is a fairly new development, as far as the general public is concerned, but which has been gaining popularity quite quickly recently. It is basically a cost or benefit incurred by someone other than the buyer or seller (a third party), and the effect of which is not accounted for in the price the buyer paid initially. Additional costs paid later on (insurance, medical bills, etc.) might pay for part of the externalities eventually.

A positive externality provides a net benefit, such as improving the economic and social life in the affected areas. A negative externality results in a cost to a person or the society in general. Groundwater contaminated by hydrofracking is a great example of a negative externality since the locals are deeply affected and suffer from the effects. The cost of damaged property and health issues is not included in the price the buyers paid.

A major problem with the fossils is that their full costs are not reflected in the market price.

Looking at the greater picture, we see that although coal prices are low today, in the long run we are paying a much higher cost. The impacts on human and environmental health not reflected in the price of coal are the "externalities."

Most of us benefit from the seemingly cheap electricity but we eventually end up paying for the externalities in the form of medical bills, environmental cleanups, special taxes, etc.

A full cost accounting for the life cycle of coal, taking these externalities into account, show the externali-

ties and their effects as:

- Particulates and gasses causing air pollution
- Climate change from greenhouse gas emissions
- Toxic liquids and solid wastes causing soil and water pollution
- Increased illness and mortality due to mining pollution
- Loss of biodiversity around the country and the globe
- Damages from mudslides as result of mountaintop removal
- Surface and infrastructure damages from mine blasting
- Acid rain damage resulting from coal combustion byproducts
- Decreased property values
- Government coal, gas, and oil subsidies
- Cost to taxpayers of environmental monitoring and cleanup

The majority of the externality costs come from reduction in air quality, contribution to climate change, accidents, and other impacts to public health. The total cost of these externalities ranges from approximately 15 to 25 cents per kilowatt-hour (kWh) of electricity generated. This is a conservative estimate, because there are other externalities that are hard to impossible to account for—such as the added stress to the life, and discomfort of, the locals confronted by the externalities.

An important fact here is that most of these external factors do not apply to most non-fossil fuel energy sources. On the other hand, some of the renewables have their own externalities, which are a new development and have not been fully evaluated and accounted for. Some of these are the noise and flickering causing health problems to wind turbine neighbors, and/or the land grabbing in Africa due to the expansion of the use of biofuels.

Some solar PV modules, for example, contain significant amounts of toxic materials, and since there are millions upon millions of them spread all over the world's deserts, we can expect a large toxicity issue in the next decades. These and many other externalities are unaccounted for, but they should not be ignored. Addressing them soon, and putting them into the proper perspective, would be the right thing to do.

Quantifying Externalities

A closer look at the energy-environment externalities dilemma, reveals that if, for example, a hydrofracking well pollutes a town's water supply, the cost of water treatment provides an estimate of real environmental damage that can be expressed in dollar figures. This, however, does not include less tangible factors such as damage to a local lake or river, parks, or the wild-life ecosystems.

The diagnosed health problems resulting from the local air or water contamination and the related medical expenses provide additional monetary damage estimate, which cannot account for some externalities like excess pain or the aesthetic damage to the local. Contaminated water and dirty air are unpleasant, regardless of the health effects or payment thereof, but it is not possible to put an exact dollar value on that aspect of life.

All the complexities presented by the above mentioned externalities are not easily converted into monetary rewards, but we need to assign some value to them and the environmental and health damage they cause. If we don't assign any value, then the society and market will automatically assign a 0 (zero) value of zero, simply because it is the easiest thing to do, and because none of these issues are directly reflected in quantifiable consumer products and services.

Different groups of specialists (environmental, energy producers, energy users, economists, legal experts) will describe and suggest different ways to put a numeric, or monetary value on the externalities. They will all agree, however that no matter how many methods can be used, it all starts with obtaining the proper information. Such information is the foundation onto which the house of truth would be built. And since no house can stand up for too long on poor foundation, we must make sure that the information we are working with is precise, accurate, and sufficient.

Once that is done, we need to execute cost-benefit analysis (CBA), which in this case involves integration of economics into environmental facts and decisions through the use of economic techniques as needed to properly appraise the related projects, actions, policies, and their effects.

Actually CBA is successfully used to sort and analyze the information for proper decision and policy making. If a town's water supply is contaminated by fracking activities, the water assets become scarce, and the locals' attitude towards the drilling will change. They will start expressing their discontent and preferences for further action. These changes in people's attitude and behavior can be measured and expressed in monetary value, according to what is needed to restore the environmental quality and/or prevent further damage.

CBA is just a tool that is as useful as the master's skill allows, and since this is a new tool the master has

to be careful. There are always gaps in the information, which have to be filled with not-so-perfect data and values. Although it is better to input some imaginary values than to leave gaps unfilled, the input data must be based on some scientific principles. It also has to be reasonable and open to discussion and changes.

In the power generation sector, the environmental goods (and damages) must be given the same weight as all other goods and services (power generation and use). We must always remember that when excess resources are allocated to protection of the environment, the resources spent on production of other goods and services will become more expensive or their number or quality will be reduced.

The principle that economic efficiency is fundamental criterion for public investment and policy-making is fundamental in the use of CBA for economic analysis and decision making tools. Although energy-related CDA methods are based on the principle that the benefits from the use of the scarce resources are maximized net of the costs of using them, we must also acknowledge that economic efficiency should not be the sole criterion in decision making.

Distributional and competitive considerations can be rational justifications for deviation from the principles of economic efficiency as an absolute criterion for maximizing the welfare of society. A big obstacle here are the gaps that still exist in reliable environmental information and the accompanying uncertainty about prices and valuation estimates that can make the result of CBA tentative, to say the least.

This is so because there are no ways to measure some environmental pollution sources and their after effects, so economists use alternative methods of valuation. This is complicated process and often ads to the uncertainty and reduces the reliability of the outcome of valuation efforts.

Nevertheless, since something has to be done, we rationalize that this imperfect "something " is better than nothing. So, we must embrace the short-term discomfort, while searching for perfection, for the benefit of having a standardized analysis and valuation system on the long run.

There are several ways to look at, and take care of, the environmental externalities, as follow:

- Government intervention in environmental externalities decisions is at times used as justification for imposing new taxes and additional charges on energy companies and projects, so it has to be reduced or eliminated.

- Incentives allow flexibility in responding to problems rather than forcing a singular approach on all individuals, so their use in reducing environmental externalities must be reevaluated and increased.

- Obstacles that prevent the energy market from functioning freely must be eliminated, if we are to achieve an optimal level of environmental protection and resource use.

Different people, corporations, entities, groups, and nations have embraced one or another of these approaches, insisting that theirs is the only way. While there is true in any of the above factors, none are the solution on their own. An integrated approach is needed, where all groups and nations unite and decide to do the right thing for the good of humanity. This might happen some day, but not in the foreseeable future. We now are too busy making life more comfortable by burning more and more fossils.

Use Values

There are various established ways to estimate values of goods and services, but there are no established ways to determine environmental benefits and damages, let alone assign a monetary value. And so we use the established estimation and valuation techniques, at times in a modified form, for that purpose.
Use value is one of the types of valuation in use presently. In the case of hydrofracking, for example, we see two types of use value;

- *Direct use value*, such as loss (due to contamination), or loss of use, of producing farm land. Here we see a direct relation between the loss of the land and loss of economic value. The farmer(s) simply will not get the crops they usually get, which will result in a monetary loss. This loss can be easily and accurately determined by comparing with the previous years income. If the average past income is A, and the income from the contaminated land is B, then the loss is determined by simply subtracting B from A. The resulting dollar number is a fair account of the loss that can be put in dollar amount. This amount, plus additional penalties, would be a fair remuneration for the farmer(s) for losing use of their farm land.

- *Indirect use value*, such as loss (contamination) of a community park, or lake, that has no economic value per se. In this case, the loss cannot be represented as direct monetary loss, simply because there was no past income to show. In such case, different groups would use different criteria to measure and

value the loss. The locals will remind us that they used to go to the park for relaxation, and swim and fish in the lake—the use of which is now lost due to the contamination. The environmentalists will emphasize on the effect of the loss on the entire local flora and fauna, but ion all cases a dollar number equal to the loss would be some random value assigned by an environmental economist.

- *Non-use value*, such as loss (extermination) of local species of birds that also have no economic value. Here again, we will see several different groups stepping in, insisting on recompense for the loss. Although the birds have not contributed to any economic growth, they are part of the environment, therefore their loss will be felt. The locals might claim that the birds have helped in crop pollination, and their loss means crop reduction. The environmentalists would expand the loss to a loss on national and international level, but in both cases no actual dollar value can be assigned to the loss. And so, here again we look up to the esteemed environmental economists to assign some dollar number—some, any, number. Any number is still better than no number at all.

In Figure 8-17 we have depicted the environment and the use values within it. No doubt, this is not a perfect way of evaluating the environment and the effects of human activities in it. Even if we ever get to a perfect and standardized way of doing these evaluations, there will still be gaps left in between the different methods and approaches, as shown in Figure 8-17.

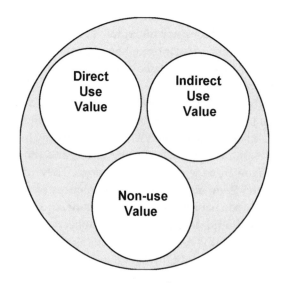

Figure 8-17. The use-value method of valuation

The environment is too large and complex to be covered completely by our human ways of valuing the most invaluable thing we have.

Nevertheless, it is what it is, and we need to do something, so here we are. We are trying. The environmental sciences are young, and as with all youngsters, it is trying to walk before learning how to run and do all kind of fancy things. Until then, we expect to see a lot of trial and error efforts.

One of the earlier steps in this direction is the carbon tax.

Carbon Tax

Figure 8-18. A coal-fired power plant

Carbon is present in every fossil fuel (coal, petroleum, and natural gas). It is released as carbon dioxide (CO_2) and other gasses, when the fossils are burnt. Some of these gasses are toxic, while some are green house gases (GHGs), which are blamed for the increasing global climate warming.

In contrast, non-fossil energy sources—wind, sunlight, hydropower, and nuclear—do not convert hydrocarbons to CO_2 and have no emissions at the energy generation stages. They do, however, leave a heavy GHG footprint during the manufacturing and transport of their equipment, and during the construction of the power plants.

CO_2 is a heat-trapping "greenhouse" gas. Scientists have theorized the effects on the climate system of releasing GHGs into the atmosphere. Since GHG emissions caused by the combustion of fossil fuels are closely related to the carbon content of the respective fuels, a tax on these emissions can be levied by taxing the carbon content of fossil fuels at any point in the product cycle of the fuel.

Carbon taxes offer a potentially cost-effective means of reducing greenhouse gas emissions. From an economic perspective, carbon taxes are a type of Pigovian tax. They help to address the problem of emitters of greenhouse gases not facing the full (social) costs of their actions. Carbon taxes can be a regressive tax, in that they may directly or indirectly affect low-income groups disproportionately.

The regressive impact of carbon taxes could be addressed by using tax revenues to favor low-income groups. However, there are about USD $550 billion in fossil fuel subsidies annually worldwide, so their importance needs no other explanation.

A number of countries have implemented carbon taxes or energy taxes that are related to carbon content. Most environmentally related taxes with implications for greenhouse gas emissions in OECD countries are levied on energy products and motor vehicles, rather than on CO_2 emissions directly.

Opposition to increased environmental regulation such as carbon taxes often centers on concerns that firms might relocate and/or people might lose their jobs. It has been argued, however, that carbon taxes are more efficient than direct regulation and may even lead to higher employment. Many large users of carbon resources in electricity generation, such as the United States, Russia, and China, are resisting carbon taxation and are looking for other ways to correct the increasing emission problems.

Carbon tax is a charge for amount of CO_2 gas emitted from any power plant.

It is a form of pricing the amount of carbon we emit in the atmosphere. Since carbon is present in every fossil fuel (coal, petroleum, and natural gas) when burned some of it is released as carbon dioxide (CO_2). At the same time the renewable technologies, solar, wind, and hydropower, do not emit much CO_2.

CO_2 is causing some global climate effects, so a tax on these emissions can be levied by taxing the carbon content of fossil fuels at any point in the product cycle of the fuel. If properly designed and implemented the carbon tax could be a potentially cost-effective means of reducing greenhouse gas emissions on the long run.

It is still not clear if these taxes would help the environment, but at the very least they help to remind us that the emitters of greenhouse gases must pay for the social costs of their actions. Until we find a permanent solution.

Carbon tax is also a regressive tax, because it directly or indirectly affects mostly the low-income groups. The power generators pay the carbon tax by increasing their rates across the customer base, so the low income class is disproportionately penalized.

A carbon tax is a tax on a transaction, and not a direct tax, which taxes income. A carbon tax also sets a direct price for carbon dioxide emissions, and usually results in higher power rates. In economic theory, pollution is considered a negative externality, a negative effect on a party not directly involved in a transaction, which results in a market failure.

Prices of fossil fuels are expected to continue increasing as more countries industrialize and add to the demand on fuel supplies. This would increase the amount of pollutants pumped in the atmosphere daily. The tangible consequences of carbon taxes is that they indirectly create incentives for energy conservation. Carbon taxes are also helping put renewable energy sources such as wind, solar and geothermal on a more competitive footing, thus stimulating their growth.

Social Cost of Carbon

The social cost of carbon is a new concept, which defines the marginal cost of emitting one extra ton of carbon dioxide at any point in time. To calculate the social cost, the atmospheric residence time of carbon dioxide must be estimated. The impact of this ton of CO_2 on climate change must be then calculated and converted to the equivalent impact during the exact time this ton of carbon dioxide was emitted.

This time shift requires calculating the discount rate, which determines the weight placed on impacts occurring at different times and different locations. Under perfect conditions, the social carbon tax should be equal to the social cost of carbon and to the value of emission permits.

In reality, however, we live in an imperfect world, so the social carbon cost estimates are not complete. Presently, the amount of CO_2 pollution is measured by the actual weight (mass) of the gas emissions, or the actual weight of the carbon dioxide molecules in a ton of carbon dioxide. Alternatively, the pollution's weight can be measured by adding up only the weight of the carbon atoms in the pollution, ignoring the oxygen atoms. This unit is called a ton of carbon and is equivalent to 4 tons of CO_2.

A fully encompassing carbon tax is one that compensates for the social cost of carbon by fuel source, and which can be expressed as the carbon dioxide production of the fuel source per unit mass or volume, multiplied by the social cost of carbon.

For example, gasoline, diesel and jet fuels produce around 20 lbs of CO_2 per gallon of burned fuel. An es-

timate for the total carbon tax would be in the order of $0.11-0.13 per gallon for each of these fuels.

Natural gas used for power generation would be penalized about $0.007 per each kWh electricity produced, while the carbon tax for coal burned at coal-fired power plants would be about $0.012 or almost twice the tax amount charged for natural gas use. This, combined with the lower production cost of natural gas, and the higher efficiency of gas-fired power plants, explains the recent increase of its use for power generation in the U.S.

Estimates of the social cost of carbon are highly uncertain. There are some estimates of $40-80 per ton of carbon, which is actually more than the total cost of coal in most countries. The wide range here is due the uncertainties in the science of climate change and control. The climate sensitivity, the different choices of discount rate, different valuations of economic and non-economic impacts, treatment of equity, and how potential catastrophic impacts being some of the key uncertainties.

In all cases, the true value of the social cost of carbon is expected to increase over time as the population and use of fossils are increasing rapidly. The rate of that increase will be proportional to those increases.

Carbon Leakage

This is also a new concept, introduced recently to describe how the emissions regulation in one country (or geographic location) affects the emissions in other countries (or geographic location) that are not subject to the same regulation, but are yet affected. Carbon leakage could result in the increase in carbon dioxide emissions in one country as a result of an emissions reduction by a second country with a stricter climate policy.

Carbon leakage occurs when the emissions policy of one country raises local emission costs, then a neighboring country with a more relaxed policy may have a trading advantage. If demand for these goods remains the same, production may move offshore to the cheaper country with lower standards, and the total global emissions will not be reduced.

At the same time, if environmental policies in one country add a premium to certain fuels or commodities, then the demand may decline and their price may fall. Countries that do not place a premium on those items may then take up the demand and use the same supply, thus negating any environmental benefit.

The type and amount of leakage effects, therefore, could be negative by reducing overall emissions, or positive by increasing overall emissions. Also, short-term leakage effects are different from, and must be compared with, long-term leakage effects. An emission policy that sets carbon taxes in developed countries might lead to leakage of emissions into developing countries.

Border Adjustments, Tariffs and Bans

A number of policies have been suggested to address concerns over competitive losses due to one country introducing a carbon tax, while another does not. Such policies have been considered in an attempt to encourage some countries to introduce carbon taxes. These policies include border tax adjustments, trade tariffs and trade bans.

The border tax adjustments are basically additional taxes on imports from nations without a carbon tax. In extreme cases, these actions could result in trade bans or tariffs applied to countries refusing to implement carbon taxes.

Cross-location Leakage

Another 'leakage' that is even harder to estimate in terms of social cost, is that of the effects of emissions in location A, vs. those in location B. Assuming that a coal-fired power plant is located in location A, its local environment and population will see the greatest negative effects of the emissions.

But then, the wind takes the smoke and pollution to location B, which is far away; different city, state, or country. Day after day, the environment and people living at location B suffer the consequences of high emissions coming from the power plant at location A. Although they do not benefit from the proximity of the power plant (jobs, subsidies, etc.), they pay for its operation with their property values and health.

In the end, the social costs are different in both locations, but are much more difficult to estimate and mitigate at location B. Here, properties could be damaged, and people hurt, but proving that the reason is the cross-location leakage of pollutants from a distant power plant is not easy. This is the next complex task to be tackled by lawyers and environmental scientists.

Energy and carbon taxes have been implemented in some countries in response to commitments under the United Nations Framework Convention on Climate Change. In most cases, however, where an energy or carbon tax is implemented, the tax is implemented in combination with various forms of exemptions.

Subsidies

There are a number of negative impacts of fossils' mining, pumping, and burning, all of which have direct and indirect economic impact and cost. These range from the jobs lost by fishermen and farmers downstream

of a coal mine, the health care costs of the people sickened by coal-fired power plant pollution, and to the cost of cleaning up spills of toxic coal waste and hydrofracking chemicals.

Federal support and subsidies for fossil energy production have been, and still are, a major part of the energy markets. The U.S. fossil energy production subsidies totaled over $72 billion between 2002 and 2008. During the same period of time, the renewable energy industry received about $29 billion. Less than half. But this is only part of the difference.

The big difference here is that the renewables received significant (one time) subsidies during just that time period, when they were needed for whatever reasons. Those subsidies are now largely gone, and are not expected to come back anytime soon. So the renewables are left to fend on their own

The fossils have been, and continue, receiving large amount of subsidies.

These include foregone tax revenues due to special tax provisions, under-collection of royalty payments, as well as direct spending on R&D and other such programs. Fossil fuel subsidies benefit also from tax breaks, such as the Foreign Tax Credit (for taxes paid in other countries) and the Credit for Production of Nonconventional Fuels.

Note: Nonconventional fuels include; oil produced from shale or tar sands, synthetic fuels produced from coal, and gas produced from either pressurized brine, Devonin shale, tight formations, or biomass and coal bed methane. The credit is about $6.00 per barrel of liquid fuels, and over $1.00 per thousand cubic feet for gaseous fuels. The production credit phases out when oil prices fall within a certain range.

Direct government subsidies to nuclear energy companies and projects equaled $115 billion by 1999, plus $145 billion in indirect subsidies, such as R&D and other activities not directly related to the power generation cycle.

U.S. federal energy sector subsidies affected approximately 75 programs/tax breaks, in the following major direct and indirect energy subsidies categories:

- Defending Persian Gulf and other crude oil shipping lanes.
- Subsidized water infrastructure for coal, gas, and oil industry use.
- Federal spending on energy research and development.
- Accelerated depreciation of energy-related capital assets.

- Under-accrual for reclamation and remediation at coal mines and oil and gas wells.
- The ethanol exemption from the excise fuel tax.

The IEA's, World Energy Outlook 2008 annual report estimated that energy subsidies to oil, gas, and coal totaled $557 billion annually. The IEA estimated also that eliminating those subsidies would reduce global GHG emissions over 10% by 2050. The subsidies, however, are not going away, and as a matter of fact they are on the increase in some countries, like China and India.

At the same period of time, the European energy subsidies totaled EUR 29 billion, mostly for coal production. These included direct grants; preferential tax treatments, regulations and loans; trade restrictions; infrastructure investments; and uncompensated security and environmental costs.

The Canadian government subsidies for the oil and gas industry totaled CA$1.5 billion, or about CA$50 per capita. These analysis includes federal grants, tax benefits (such as the Resource Allowance and the Accelerated Capital Cost Allowance for oil sands), and government expenditures that directly support oil, gas and oil sands industries.

Note: During 2002-2008, the U.S. taxpayer subsidized the energy sector at the tune of nearly $300 per person, or $50 annually. Of this amount, the taxpayers paid (one time payment of) about $80 each for renewables, or $14 per capita annually for the duration. This, however, does not include the subsidies for the nuclear industry, which if added, alone would more than double this number to about $750 per capita by 1999. The major expenses here were during the initial stages of the nuclear power program, which dwindled since the 1980s.

A UN study concluded that energy subsidies are widespread, but vary depending on definitions, analysis methodologies, fuel type and location. It concludes that producer subsidies, usually in the form of direct payments or support for research and development, are most common in OECD countries. Most subsidies in developing and transition countries go to consumers, usually through price controls that hold end-user prices below the full supply costs.

Fossil-fuel and nuclear industries receive the majority of such subsidies in most cases. OECD countries significantly increased their support for renewables alternative energy technologies renewable during the 2008-2013 boom and bust cycle. This, one time life line, however, did not last long. It was recently withdrawn, so that the renewables and the alternative energy tech-

nologies are left on their own. They are presently choking in the stormy waves of the energy markets, fighting for their lives.

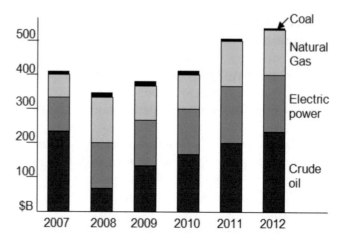

Figure 8-19. U.S. subsidies of different energy sources (in $ billion/year)

Common sense tells us that if the renewable energy sector can't stand on its own in the free market, then it doesn't deserve to exist. This is also what we hear from the experts. But wait…as can be seen in Figure 8-19, fossil fuel's historical subsidies are like skyscrapers next to the renewables humble shack that is about to be demolished.

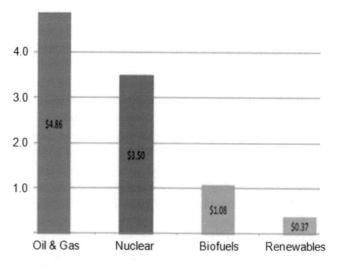

Figure 8-20. Global energy subsidies (in $ billion)

The lion's share of fossils in the subsidies jackpot is remarkable.

The fossil subsidies do not include the massive indirect subsidies in unchecked and unaccounted for externalities.

Adding all externalities to the bottom line would make the financial and social responsibilities of the fossil energy industries overwhelming, while putting some out of business. This is because their negative effects wreak havoc on our health, and our quality of life. Not to mention the potential liability and viability of the human species, after the results from the increasing climate change are accounted for.

For a long time oil and gas have taken the lion's share of subsidies allocated to the energy sector, and the financial support hasn't gone away yet. In fact, in many ways, it still trumps the fragile support for renewables. Recently, the government support has been increasing in some aspects of the U.S. and global energy markets.

For example, in addition to the other subsidies, the oil and gas industries are benefiting tremendously from Master Limited Partnerships (MLPs) and Real Estate Investment Trusts (REITs), which represent two low-capital-cost ways of financing infrastructure, and which are now rapidly expanding in the financial services world.

None of these are available to renewable energy investors, and both cost less than the tax equity funds derived from solar Investment Tax Credit (ITC) and wind's Production Tax Credit (PTC). The U.S. government also heavily subsidized natural gas production and use, starting with the combustion turbine which was developed for aircraft and later reapplied to the gas-fired energy generation sector.

There is a talk in the U.S. Congress about including the renewables in the MLP and REIT financing schemes, but talk is cheap. Today Washington is dysfunctional, so the implementation of these acts into reality might take a long while.

Global Pre-tax Energy Subsidies

Consumer subsidies include two components: a) pre-tax subsidy (if the price paid by firms and households is below supply and distribution costs) and b) tax subsidy (if taxes are below their efficient level.

Most economies impose consumption taxes to raise revenue to help finance public expenditures. Efficient taxation requires that all consumption, including that of energy products, be subject to this taxation. The efficient taxation of energy further requires corrective taxes to capture negative environmental and other externalities due to energy use (such as global warming and local pollution).

- The pre-tax subsidy estimates capture both those that are explicitly recorded in the budget and those that are implicit and off-budget. The evolution of energy subsidies closely mimics that of international energy prices. Although subsidies declined with the collapse of international energy prices, they have started to escalate since 2009. In 2011, global pre-tax subsidies reached $480 billion (0.7% of global GDP or 2% of total government revenues).

Petroleum and electricity subsidies accounted for about 44% and 31% of the total respectively, with most of the remainder coming from natural gas. Coal subsidies are relatively small at $6½ billion.

Energy subsidies comprise both consumer and producer subsidies. Consumer subsidies arise when the prices paid by consumers, including both firms (intermediate consumption) and households (final consumption), are below a benchmark price, while producer subsidies arise when prices received by suppliers are above this benchmark.

Where an energy product is internationally traded, the benchmark price for calculating subsidies is based on the international price. Where the product is mostly non-traded, such as electricity, the appropriate benchmark price is the cost-recovery price for the domestic producer, including a normal return to capital and distribution costs. This approach to measuring subsidies is often referred to as the price-gap approach, and is used widely in analyses by other international agencies.

For example, the average pre-tax subsidies of some oil exporting countries in 2011 were as follow:

Table 8-15. Pre-tax subsidies for oil exporters

MENA	8.5 % of GDP
Sub-Africa	5.5 % of GDP
CIS	4.5 % of GDP
Asia	4.0 % of GDP
Latin America	2.0 % of GDP

At a closer look the subsidies in Table 8-15 can be detailed as follow::

- The Middle East and North Africa region accounted for about 50% of global energy subsidies. Energy subsidies totaled over 8½% of regional GDP or 22% of total government revenues, with one-half reflecting petroleum product subsidies. The regional average masks significant variation across countries. Of the 20 countries in the region, 12 have energy subsidies of 5% of GDP or more. Subsidies are high in this region for both oil- exporters and importers.

- Countries in Emerging and Developing Asia were

responsible for over 20% of global energy subsidies. They amounted to nearly 1% of regional GDP or 4% of total government revenues, with petroleum products and electricity accounting for nearly 90% of subsidies. Energy subsidies exceeded 3% of GDP in four countries (Bangladesh, Brunei, Indonesia, and Pakistan).

- The Central and Eastern Europe and CIS accounted for about 15% of global energy subsidies, including the highest share (at nearly 36%) of global natural gas subsidies. Energy subsidies amounted to over 1½% of regional GDP or 4½% of total government revenues, with natural gas and electricity accounting for about 95%. They exceeded 5% of GDP in four countries (Kyrgyz Republic, Turkmenistan, Ukraine, and Uzbekistan).
- The Latin America and Caribbean made up over 7½% of global energy subsidies (approximately ½% of regional GDP or 2% of total government revenues), with petroleum subsidies accounting for nearly 65%. Energy subsidies exceeded 5% of GDP in two countries (Ecuador and Venezuela).
- Sub-Saharan Africa accounted for about 4% of global energy subsidies. Energy subsidies amounted to 1½% of regional GDP or 5½% of total government revenues, with electricity subsidies accounting for over 70%. Total subsidies exceeded 4% of GDP in three countries (Mozambique, Zambia, and Zimbabwe).
- The only advanced economy where energy subsidies were a non-negligible share of GDP was Taiwan at 0.3% of GDP spent on electricity development.

In most economies, there are elements of both producer and consumer subsidies, although in practice it may be difficult to separate the two. The advantage of the price gap approach is that it also helps capture consumer subsidies that are implicit, such as those provided by oil-exporting countries that supply petroleum products to their populations at prices below those prevailing in international markets. The price gap approach does not capture producer subsidies that arise when energy suppliers are inefficient and make losses at benchmark prices.

Pre-tax subsidies are pervasive and impose significant fiscal costs in most developing and emerging regions. They are most prominent in Middle East and North Africa, especially among oil exporters. Given that energy consumption can be expected to rise as incomes grow, the size of subsidies could climb in regions where

they currently account for a small share of the global total, such as sub-Saharan Africa.

Global Post-tax Energy Subsidies

These are much larger than pre-tax subsidies, amounting to $1.9 trillion in 2011—about 2½% of global GDP or 8% of total government revenue. Virtually all of the world's economies provide energy subsidies of some kind when measured on a tax-inclusive basis, including 34 advanced economies.

For some products, such as coal, post-tax subsidies are substantial because prices are far below the levels needed to address negative environmental and health externalities. The fact that energy products are taxed much less than other products also contributes to the high level of post-tax subsidies. In MENA, for example, applying the same rate of VAT or sales taxes to energy products as other goods and services would generate 75% of GDP.

Globally, pre-tax subsidies account for about 25%, and tax subsidies account for about 75%. The advanced economies account for about 40% of the total.

The top three subsidizers are the U.S. ($502 billion), China ($279 billion), and Russia ($116 billion).

The Effect of Subsidies

Basically speaking subsidies have a number of effects, some positive, some negative.

The major effects of subsidies are as follow:

- Subsidies can discourage investment in the energy sector. Low and subsidized prices for energy can result in lower profits or outright losses for producers, making it difficult for SOEs to expand energy production and unattractive for the private sector to invest both in the short and long run. The result is severe energy shortages that hamper economic activity.
- Subsidies can crowd out growth-enhancing public spending. Some countries spend more on energy subsidies than on public health and education. Reallocating some of the resources freed by subsidy reform to more productive public spending could help boost growth over the long run.
- Subsidies diminish the competitiveness of the private sector over the longer term. Although in the short-run subsidy reform will raise energy prices and increase production costs, over the longer term there will be a reallocation of resources to activities that are less energy and capital intensive and more efficient, helping spur the growth of employment.

- Subsidies create incentives for smuggling. If domestic prices are substantially lower than those in neighboring countries, there are strong incentives to smuggle products to higher-priced destinations. Illegal trade increases the budgetary cost for the subsidizing country while limiting the ability of the country receiving smuggled items to tax domestic consumption of energy.

Note: Fuel smuggling is a widespread problem in many regions around the world including in North America, North Africa, and the Middle East, parts of Asia, and Africa. For example, Canadians buy cheap fuel in the United States; Algerian fuel is smuggled into Tunisia; Yemeni oil is smuggled into Djibouti; and Nigerian fuel is smuggled into many West African countries

- Energy subsidies exacerbate the difficulties of both oil importers and exporters in dealing with the volatility of international energy prices. The balance of payments of many energy-importing countries is vulnerable to international price increases. The adverse impact can be mitigated by passing through international price increases and by providing greater incentives for improving energy efficiency and lowering energy consumption.
- Removing energy subsidies helps prolong the availability of non-renewable energy resources over the long term and strengthens incentives for research and development in energy-saving and alternative technologies. Subsidy reform will crowd-in private investment, including in the energy sector, and benefit growth over the longer term.
- The volatility of subsidies also complicates budget management. For oil exporters, energy subsidies accentuate macroeconomic volatility by increasing subsidies during periods of international price increases. Allowing domestic prices to rise with international prices can help cool off domestic demand during commodity booms and build up fiscal buffers for use during periods of declining prices. To offset concerns about the transmission of high international price volatility to domestic prices, some smoothing of price increases can be considered (see paragraph.
- The negative externalities from energy subsidies are substantial.
  - Subsidies cause over-consumption of petroleum products, coal, and natural gas, and reduce incentives for investment in energy efficiency and renewable energy. This over-consumption in turn aggravates global warming and worsens local pollution.
  - The high levels of vehicle traffic that are encouraged by subsidized fuels also have negative externalities in the form of traffic congestion and higher rates of accidents and road damage.
  - The subsidization of diesel promotes the over-use of irrigation pumps, resulting in excessive cultivation of water-intensive crops and depletion of groundwater.
  - The subsidization of electricity can also have indirect effects on global warming and pollution, but this will depend on the composition of energy sources for electricity generation.

- The over-consumption of energy products due to subsidies can also have effects on global energy demand and prices. The multilateral removal of pre-tax fuel subsidies in non-OECD countries, under a gradual phasing-out, would reduce world prices for crude oil, natural gas, and coal by 8%, 13%, and 1% respectively in 2050 relative to the no-change baseline. The reduction would be substantially larger if prices were raised to levels that eliminated subsidies on a post-tax basis. These spillover effects suggest that non-subsidizers would share the gains from subsidy reform, as well as extending the availability of scarce natural resources.
- Energy subsidies are highly inequitable because they mostly benefit upper-income groups. Energy subsidies benefit households both through lower prices for energy used for cooking, heating, lighting and personal transport, but also through lower prices for other goods and services that use energy as an input.
- On average, the richest 20% of households in low- and middle-income countries capture six times more in total fuel product subsidies (43%) than the poorest 20% of households (7%). The distributional effects of subsidies vary markedly by product, with gasoline being the most regressive (i.e., subsidy benefits increase as income increases) and kerosene being progressive. Subsidies to natural gas and electricity have also been found to be badly targeted, with the poorest 20% of households receiving 10% of natural gas subsidies and 9% of electricity subsidies.

Negative Subsidies Effects

Some of the simplest, and quite significant economic costs of coal come in the form of subsidies and tax breaks which are not reflected in the market price of coal. Also not include here are the estimated $4.8 billion in coal-related subsidies in the 2009 stimulus package. Coal mining and combustion projects require major investments, which are accompanied by a number of uncertainties and risks, the costs of which are often passed on to taxpayers via infrastructure subsidies and loan guarantees.

One example of extreme waste of money is the Healy Clean Coal Plant (HCCP), which has cost the State of Alaska tax payers nearly $300 million since the mid 1990s. Ant the plant is still to produce power in return. There are estimates that coal pays only 5% of its market value to the state of Alaska, even though the nominal rates are much higher.

A recent study in Kentucky determined that the government spends $115 million more on subsidies for the coal industry than it receives in taxes or other related benefits. In addition, the taxpayers also pay the costs of cleaning up environmental disasters caused by the coal industry. Cleanup of the recent coal ash spill in Tennessee is estimated to cost up to $1 billion, not including the costs of pending litigations. And since the cleanup at this site has been taken over by the EPA under the Superfund law, most of this cost will be borne by the US taxpayer, and very few would even know the total price they will end up paying.

The health impacts of coal pollution also have enormous economic costs, due to increasing health care costs and lost productivity. An Ontario government study estimated these costs as billions of dollars within Ontario alone. A similar recent study in West Virginia found that the cost associated with premature death due to coal mining was five times greater than all measurable economic benefits from the mining. How do you put a dollar number on these statistics?

Another growing issue is the fact that other major industries depend on the ecosystems coal mining destroys. Industries such as recreational fishing, commercial fishing, and tourism that are particularly relevant in a number of the affected states.

In Alaska, almost 55,000 direct jobs on full time equivalent basis are closely linked to the health of Alaska's ecosystems. These jobs make up more than a quarter of Alaskan employment base and produce almost $2.6 billion of income for Alaska workers. These 55,000 ecosystem-dependent jobs are threatened by oil production and transport, and mining operations, such as the

Chuitna Coal strip mine. which could generate about 350 jobs. So how do we justify this project—and many similar such all over the U.S.—if the job-to-job ratio, and/or the economic benefits cannot justify the environmental damage to be done.

Negative effects on the economy, such as coal and oil projects present with their emissions, usually lead to worse health in the population. This in turn has a great impact on health care costs, compounding the negative economic impact.

On the other hand, coal, no doubt, provides some benefits to society. We must give it credit for powering the 19th and 20th century global economic development. It still provides the majority of reliable and cheap electricity, which is a boon to the economy, where health is improved, and health care costs are lowered. While this additional health effect should indeed be considered, it should be applied after the economic impacts of its externalities are considered and calculated (some time in the future).

Once the costs of pollution, global warming, and habitat destruction are added to the benefits of cheap electricity, the economic impact of coal may no longer be positive—especially in the most affected locations and among the most affected people. This will also become clear in the near future.

We, however, must be fair and keep all these numerous and complex facts in retrospective when discussing and analyzing the different pros and cons of fossils use and their effects.

Renewable Subsidies

Demonizing the fossils is easy. They are very convenient and defenseless target for many politicians and company executives. Like sitting ducks on one end of the energy spectrum, they just sit, take, and endure whatever is thrown their way.

The new, shiny, and mistreated renewables are on the other end. Words of criticism are not allowed, for the poor, weak renewables might just crumble. With the fear of being politically incorrect, we must remind the esteemed reader that:

• Without subsidies, the renewables (solar and wind especially would not even be here. Period, and

• Crying, 'unfair, why are the fossils subsidized?' would not help. It is what it is, and we all must do what we must do. This includes all renewables.

Until recently the renewables were the pampered, even spoiled, baby of the energy markets. Wrapped up in comfortable government blanket of subsidies and billions of dollars of perks, they grew in comfort, even luxury, growing slowly while making mess after mess.

Today, things are changing. The comfortable government blanket of subsidies and special privileges is being slowly unwrapped, and the different renewable energy markets are getting exposed to the cold and cruel reality. Cry as they may, the government support in the U.S. most countries is gone and the renewables must learn to walk and grow on their own, or else…

Let's look at the past and present, where subsidies and other such instruments help the renewables development, and especially the solar and wind energy markets.

PPAs

Power purchase agreement (PPA) was, and still is, one of the major instruments in, and drivers of, the recent solar and wind power development in the U.S. and other countries. The definition of a PPA is; a financial agreement where a developer arranges for the design, permitting, financing and installation of a solar energy system on a customer's property at little to no cost. The developer sells the power generated to the host customer at a fixed rate that is typically lower than the local utility's retail rate. This lower electricity price serves to offset the customer's purchase of electricity from the grid while the developer receives the income from these sales of electricity as well as any tax credits and other incentives generated from the system.

PPAs typically range from 10 to 25 years and the developer remains responsible for the operation and maintenance of the system for the duration of the agreement. At the end of the PPA contract term, a customer may be able to extend the PPA, have the developer remove the system, or choose to buy the solar energy system from the developer.

PPA prices are determined by the utility company buying the power, after negotiations with the solar plant owners. Recent PPAs range from $0.05 to $0.25 per AC kW generated, depending on location and size of the solar installation. PPA prices also vary with time of day, which is where solar power generators make most of their profits.

The solar or wind power customers under PPA benefit from

- No or low upfront capital costs, since the developer handles all upfront costs of sizing, procuring and installing the solar PV system. Without any upfront investment, the host customer is able to adopt solar and begin saving money as soon as the system becomes operational.

— Reduced energy costs: Solar PPAs provide a fixed, predictable cost of electricity for the duration of the agreement and are structured in one of two ways. Under the fixed escalator plan, the price the customer pays rises at a predetermined rate, typically between 2%-5%. This is often lower than projected utility price increases. The fixed price plan, on the other hand, maintains a constant price throughout the term of the PPA saving the customer more as utility prices rise over time.

— Limited risk: The developer is responsible for system performance and operating risk.

— Better leverage of available tax credits: Developers are typically better positioned to utilize available tax credits to reduce system costs. For example, municipal hosts and other public entities with no taxable income would not otherwise be able to take advantage of the Section 48 Investment Tax Credit.

— Potential increase in property value: A solar PV system has been shown to increase residential property values. The long-term nature of these agreements allows PPAs to be transferred with the property and thus provides customers a means to invest in their home at little or no cost.

RECs

Renewable Energy Certificates (RECs) are another major tool that was used successfully in growing the solar and wind industries. RECs, also known as Green tags, Renewable Energy Credits, Renewable Electricity Certificates, or Tradable Renewable Certificates (TRCs), are tradable, non-tangible energy commodities. One REC basically represents a proof that 1 megawatt-hour (MWh) of electricity was generated from an eligible renewable energy resource.

RECs can be sold and traded or bartered, and the owner can claim to have purchased renewable energy. According to the U.S. Department of Energy's Green Power Network, RECs represent the environmental attributes of the power produced from renewable energy projects and are sold separately from commodity electricity.

While traditional carbon emissions trading programs use penalties and incentives to achieve established emissions targets, RECs simply incentivize carbon-neutral renewable energy by providing a production subsidy to electricity generated from renewable sources.

It is important to understand that the energy associated with a REC is sold separately and is used by another party. The consumer of a REC receives only a certificate.

If the Renewable Energy Credit (REC) is to be used for the purposes of participation in programs such as LEED-certified green building projects, ensuring compliance with these program requirements is a must. Basically speaking, participating in REC programs and monetizing their values requires understanding of both business and legal issues.

SRECs

Solar renewable energy credits (SRECs) are RECs, generated by a solar power system. RECs are used to show how much electricity was produced using solar energy. They are typically bought and sold mostly by regulated utilities as needed to meet their obligations associated with state-level renewable energy standards.

SRECs are also used by consumers who voluntarily purchase them for marketing claims or other use. Most often in PPAs and/or SRECs are owned by the developer.

Note: When entering into a PPA, the customer must clearly understand who owns and who/how can sell the SRECs generated from the PV system in question. Full understanding of the risks associated with SREC ownership, and the tradeoffs with respect to PPA price, is utmost important for the profitable operation of any solar power generating system.

ITC

The federal investment tax credit (ITC) is the most important and widely used financial incentive today. It provides qualifying entities with a tax credit equal to a certain percentage of the eligible costs of a project, and up to 30% of the total. In many states there are also grants, rebates and attractive loan programs for such qualifying projects. In most cases, a combination of ITC and other financial incentives are used to increase the monetary returns from each solar project.

Note: The present ITC expires in December 2016, so after a flurry of activities, in 2017 the solar industry would fall of the financial cliff, dug out by the ITC withdrawal. Unless, of course, the ITC is extended. If that happens, then well; solar will continue its ascend, at least for the duration. If not...then it would be on its knees for some time to come, and until it learns how to walk on its own.

The U.S. solar industry is facing some serious issues in the coming years. The solar energy market will be full of flurry of political, financial, and legal activities. It will be a miserable life for those involved, but it will make a very good telling. It would be amazingly interesting to watch...from the outside.

Hi-tech Subsidies

Starting with the billions of dollars spent on nuclear research since the 1940s, we see the U.S. government spending enormous amount of money in the energy sector. Most of this money and effort is focused on developing and optimizing high tech equipment and processes, used in the energy industry. In addition to nuclear, the coal, oil, and gas sectors have been beneficiaries of generous government subsidies, which are flowing unrestricted to this day.

While most of these subsidies produced some good results, some of the money was wasted. Solyndra comes to mind as the most advertised example of hastily spent tax payers money by bureaucrats who have no idea what energy does. But even worse is the advise given these politicians by scientists (at DOE, NREL and Sandia) who should know better, but for some reason prefer to ignore their common sense for the sake of political ideas and favors.

How much of the subsidies go to waste—as in the now famous Solyndra and many other such cases—is unknown. Most of the waste does not get publicized, and although there were many Solyndras since 2008, we just don't hear about them.

Note: We have described dozens of these cases in our book on the subject, *Solar Power for the 21st Century*, published by Fairmont/CRC Press in 2013. In it, we have an entire chapter describing in detail the failures of renewable energy companies during the 2008-2013 boom-bust period. Interesting reading, with many lessons to be learned. Unfortunately, it seems that the lessons have not helped much.

Many government projects, developed with good intentions become victim of poor planning, design, execution, and most often than none, fall victim to a combination of these factors. One such case is the multi-billion dollar NIF facility at the Lawrence Livermore National Laboratories, which was hoped to demonstrate the next phase of nuclear power generation by producing a fusion reaction.

After spending billions of dollars during 15 years of non-stop work by thousands of scientists and engineers, the goal was not achieved. Lots of excuses, of course...but the goal has been pushed into the distant future. The dream of nuclear fusion has been replaced by basic research projects, justified as needed to pay the operating bills and the salaries. So the waste continues...

All this money given to private companies, would bring miracles, but no use of crying over spilled milk. Such day is not in the cards.

Energy (Financial) Trading

Energy trading (on the financial markets) is different from actual trading of energy products and services.

Energy trading is activity similar to trading any and all other commodities.

The techniques are the same, the only difference being that the commodity is some sort of energy product. In addition to stock trading, many people are involved in futures trading.

While the level of complexity of energy markets has increased rapidly, practitioners have often adapted to it, using approaches and methods inherited directly from the other asset classes, with some modifications.

The energy markets, however, have their own peculiarities, which at times present difficulties to the participants.

There are two basic features of energy that together result in price behavior that can be dramatically different from that of other asset classes.
- The first is that for a variety of reasons, including weather events and infrastructure failure, fluctuations in supply and demand for many energy commodities can change rapidly on daily or even hourly time scales.
- The second is that it costs real money to move a commodity through time or between locations. One cannot electronically transfer ownership and delivery of natural gas from Louisiana to Boston on a cold winter day as one could, for example, title to a bond.

The facilities required to store or transport energy also vary in cost, with coal and crude oil relatively easy to store and transport natural gas much more challenging and costly. Electricity itself is also expensive to move and cannot be stored in meaningful quantities at any cost.

In each of these markets storage capacity by region, as well as transport capacity between regions, are finite. This sometimes tests the infrastructure limits, be it through mechanical failure, unusually high supply or demand, or extreme levels of inventory. In all cases the final price is affected.

All disruptions bring appropriate type and level of consequences:
- The first consequence of these fundamental aspects of energy markets is that energy prices are significantly more volatile than benchmarks in equities, rates, and currencies. Higher volatility affects an array of commercial considerations, including stability of pricing during the negotiation of potential transactions, investment decisions relating to asset development or acquisition, and the capital required to support trading operations and collateral requirements.
- Another consequence is that energy markets exhibit a much higher degree of "specialness," that is to say, negative forward yields or rapid changes in relative pricing between closely related commodities. As a result, the liquidity in energy markets, as in most asset classes, is usually concentrated in a few benchmark commodities.

The practical consequence of chronic specialness is that energy portfolios often have risk profiles that are very high dimensional—not in the sense of a large number of closely related tradables such as a few hundred Treasury bonds outstanding on any given day, but in the sense of distinct regional and temporal delivery that can dislocate rapidly and unpredictably from the benchmarks most commonly used to hedge. Specialness, and the resulting high dimensional attributes of typical portfolios, is the fundamental challenge in energy risk management.

Many seasoned risk managers in other asset classes consider energy trading as simply more of a 'white knuckle" experience than other businesses. This view is often based in part on the empirical observation, easily gleaned from any screen with west Texas Intermediate or Brent oil futures prices, that energy commodities can exhibit exceptionally high volatility—within time frame and/or region. The relatively frequent blowups of energy trading desks, and other anomalies, reinforce the volatility image.

Noteworthy instance of abuse of market mechanics, is the case of Enron and other power marketers in California in the early 2000s. These were accompanied by index manipulation in natural gas, and culminated recently with the Amaranth spectacle and FERC actions against several power marketers. This added credence to the notion that financial energy markets can be challenging environments in which to operate.

Much of this perception is based on very high-level views of the more newsworthy mishaps, with only cursory knowledge of the commercial realities of energy markets and the risk-management practices that are required to run an energy trading operation. Small and large traders are involved in these operations, all having different approaches, but a similar goal—to make money somehow.

The financial energy markets, however, have a much higher purpose than providing large hedge funds an opportunity to speculate

Most trading activity involves balancing variations in supply and demand across time and between locations. This is needed by many companies to reduce or eliminate future losses. The variations occur both on short and long-term scales.

- The short-term, or so-called cash desks, move commodities from supply centers to demand centers on a daily basis.
- The long-term scales involved in the construction of a new refinery, pipeline, or generator, are activities that often require significant hedging programs to support the sizable financing required.

The distinguishing features of energy price dynamics affect market participants in different ways, an understanding of which is central to the development of useful risk-management and valuation frameworks. What this means is that energy trading is not Las Vegas casino, where chance rules the game. Instead, energy has its own rules, which are many and very complex, and which determine the market direction.

In more detail:

Financial Energy Markets

The trading of energy on the financial markets consist of energy commodities trading, where investors and speculators "buy and sell" imaginary amounts of energy and energy sources. Here, however, nobody loads or ships anything. Instead, only papers and money gets traded. This type of trading is similar to gambling, since nobody can predict the trends accurately all the time. This situation also creates losers and winners, but they cannot be easily separated into categories, because Mother Luck has no favorites.

Amazingly enough, the financial markets can have significant impact on the physical energy markets. As a matter of fact the impact could be extraordinary,

such as the oil prices reaching $150/bbl in 2008, which were blamed on financial markets' traders frenzy.

Where there are markets, there will be those who will attempt to manipulate them for their own benefit.

How true, and how well was this illustrated during the 1973 oil embargo, the 2008 financial crisis, and the following artificially low global oil prices of 2014-2015.

Crude oil is the favorite trading instrument on the financial energy markets. Finding oil on Wall Street, however, is as likely as finding a lake in the middle of the Sahara desert. And yet, there was a fad that came from nowhere a decade ago about the great opportunity of "finding oil on Wall Street." Only fools would rush to Wall Street with buckets in their hands to scoop oil from the Street. Poor idiots!

Equally, though, thousands of people who surely do not think of them as fools, do throw a lot of money on the Street, where Las Vegas type of chance to win awaits them. Most lose, but just like those who win the Vegas jackpots or the weekly Lotto, Wall Street also creates millionaires too…with oil.

Energy (oil included) has been traded in one form or another for almost a century. The modem energy markets, however, emerged in the early 1980s with the launch of the West Texas Intermediate (WTI) crude oil contracts in 1983.

Subsequent growth has been significant, thanks to an expanding array of physical and financial instruments traded across almost all energy commodities. There has been also a trend toward deregulation in markets such as natural gas and power which until relatively recently were fully regulated.

The past decade has also witnessed growing interest among institutional investors in energy, and commodities more broadly, as a new component of investment portfolios.

And then, there are other, even more important, instruments in the financial energy markets.

Energy Futures Trading

Energy trading, where energy is treated as a commodity, is an important part of the energy markets function. It is not to be confused with actual energy products selling and buying. Energy commodities, futures, and stocks trading are financial transactions, where no products change hands. Instead, only money is traded by the rules of the financial markets.

There are two main types of futures traders: the

speculators and the hedgers.

- The speculators

The word "speculator" usually conjures a harsh images of a greedy millionaire manipulating prices for own gains, which cause the rest of us pain at the stores and at the pump. While this is true to an extent, there is much more to it. Most commodity speculators usually take a long or short position In a commodity with the only goal in mind; that of making a profit and nothing else.

They don't really care about the commodity itself, or the market in general. All they are concerned with is making a buck. For that purpose they are willing to take a gamble, which could go either way. That alone makes them not as evil as one might think, and is even beneficial for the commodity markets.

As a matter of fact, commodity trading speculators play a critical (and balancing) role in the futures market. They are not trying to manipulate the market as if it were a puppet on strings. Nope; their only goal is to identify (properly) a specific trend and profit from it. Nothing more or less, since many of them don't even know, nor care, how the commodity markets function, and/or what is good or bad for them.

Most speculators simply take on the risk that other people stay away from.

This, believe it or not, stabilizes the markets and makes them more fluid.

- The hedgers

Hedgers are individuals, groups, or companies already involved with buying, selling, or producing some commodity. By hedging, they are only trying to minimize their price risk by establishing a short or long hedge in their commodity. The hedgers' goal is to basically manage, and reduce or eliminate, the risk of future price volatility.

For example, oil and natural gas companies routinely lock in a price for future production. If a company has 1.5 billion cubic feet per day (Bcf/d) natural gas production, it would hedge using fixed price contracts at the New York Mercantile Exchange (NYMEX) at a price, which is well above the average spot price projected for the year. This way, regardless of what the future of this commodity is, the company would be able to get the price it wants.

There are two hedging techniques:
- Short hedge is a strategy producers can use to secure a price of the commodity for future delivery. The hedger essentially establishes a short futures position while still owning the commodity. If the price of the commodity falls, the value of the short position offsets some of the revenue lost upon delivery.
- Long hedge involves taking a long futures position to protect against future price increases.

In addition to these specific categories, the style a company chooses for trading futures contracts directly affects the trading of contract futures.

Futures traders typically fall under one of two classes:
- Day trading is conducted by executing several trades throughout the day. The focus here is on getting in and getting out in a short period of time. The ultimate, and only, goal here is to make some profit at the end of the day. This was a method used mostly by small traders, but today corporate traders with fast computers, using sophisticated trading programs corner the market and execute millions of transactions within seconds, thus skimming profits from the markets.
- Position trading is the typical buy-and-hold strategy trading. It's the method used by larger organizations. Just like in day trading, the traders buy a position in an energy commodity, hoping that the prices will go up and bring them some profit on the long run.

Traded energy products include
- Coal
- Natural Gas
  - Natural gas (gas)
  - Liquefied Natural Gas (LNG)
- Oil Products
  - Crude Oil
  - Light Products
  - Heavy Products
  - Natural Gas Liquids (NGLs)
  - Distillates
  - Fuel Oil
- Biodiesel
- Ethanol
- Electricity
- Weather risks
- Emissions

Leading market makers in exchange-traded and over-the-counter (OTC) environmental products include: the European Union Allowances (EUAs), and Certified Emission Reductions (CERs). These commodities are traded as swaps, options and different investor structures as needed to manage the participants exposure to environmental risk across all regions.

Energy Futures

Energy futures are relatively new in comparison to other commodity markets. They grew almost overnight out of a need to control price volatility with calculated risk management. Futures and options have been integral in keeping prices in check, while also letting the market move naturally with geopolitical events. Like with most commodity markets, supply and demand drive the energy markets.

Unlike other commodity markets, supply and demand are greatly affected by international politics. To protect prices from quick dramatic swings, futures and options regulate how much prices can change in a given timeframe. This is to protect consumers and producers alike.

Popular Contracts

*   Crude Oil Is an essential commodity that is used in nearly all aspects of modern society—gasoline, plastics, paint, detergent, medicine, make-up, etc. Because of this, crude oil and its price are always in high demand. Crude oil is not a uniform product and it is common practice to classify crudes based on three characteristics: gravity, sulfur content, and field of origin. The names of respective crude oils are indicative of these characteristics, such as West Texas Intermediate and Brent.

*   Natural Gas, although traded internationally, is primarily a domestic commodity. Oklahoma, Louisiana, Texas, and New Mexico account for eighty percent of the natural gas production in the U.S, while Mexico and Canada account for a small portion of the total consumed. Natural gas is measured by volume and heating quality. Demand peaks with winter's heating needs and summer's air conditioning usage.

*   RBOB (Reformulated Gasoline Blendstock for Oxygen Blending) is mainly used for transporting ethanol, RBOB is a refined crude oil product. RBOB prices are heavily dependent on crude oil prices; this product is imported when crude oil producers do not have the means to refine it into gasoline. Though a crude oil producer makes more money off of the crude oil they sell, the producer must still pay higher prices to import gasoline.

Forward Contracts

A forward contract is an agreement between two parties to exchange at some fixed future date a given quantity of a commodity for a price defined when the contract is finalized. The fixed price is known as the forward price. Such forward contracts began as a way of reducing pricing risk in food and agricultural product markets, because farmers knew what price they would receive for their output.

Forward contracts for example, were used for rice in seventeenth century Japan.

Futures Contract

Futures contracts are standardized forward contracts that are transacted through an exchange. In futures contracts the buyer and the seller stipulate product, grade, quantity and location and leaving price as the only variable.

Agricultural futures contracts are the oldest, in use in the United States for more than 170 years.[27] Modern futures agreements, began in Chicago in the 1840s, with the appearance of the railroads. Chicago, centrally located, emerged as the hub between Midwestern farmers and east coast consumer population centers.

Swaps

A swap is a derivative in which counterparties exchange the cash flows of one party's financial instrument for those of the other party's financial instrument. They were introduced in the 1970s.

Case Study: SandRidge Petroleum

Saved by the hedge...or almost? It was becoming clear that in 2015 many independent mid-cap E&Ps were hedged. One of these was a small company, SandRidge Petroleum (SD). At $80/barrel this action looked good, but the continued drop in oil has sent oil prices down through the 'floor' of most of the company's three-way collars.

Note: A three-way collar is a hedging strategy employed by oil and gas producers which, like a regular collar, has a 'floor' price and a 'ceiling' price for a barrel of oil. But unlike a regular collar, a three way collar also has another ceiling which is much lower, but a ceiling that the company would have to accept if oil prices dropped significantly. This third 'way' is often done to reduce the cost of the collar itself.

SD's three way collar is a good example of a typical planning that did not work well, but prevented even bigger mishap. Here the 'floor' was $90, the ceiling $103.50, but the third ceiling about $76.50. In other words, if oil prices fell below $76.50, which was deemed highly unlikely until recently, SD would have to take that lower price.

Unfortunately, the 'unthinkable' happened, and now SD is on the hook for an additional 4.6 million barrels of oil. This makes SD's hedge book considerably less extensive, and it also exposes a good chunk of the company's oil production to the downward volatility of today's oil prices. Many small oil drilling companies are in the same shoes, waiting for better days...errr...higher oil prices.

So, can the company make it through the spell of low oil prices? As of the second quarter of 2014, SandRidge had about $919 million in cash on hand. Cash operating costs, which include lifting, production taxes and G&A, are quite low at just $16.75 per barrel, so capex needed to keep production flat was $500 million.

If SD produces just the 5.6 million hedged barrels of oil and 15.4 million hedged cfe, it would then earn $518 million and $69 million in revenue, respectively. It's difficult to assume that the rest of production would be economically viable, and because these oil wells have first-year decline rates of 80%, most production does have to come from drilling. In other words, there isn't much 'base production' to work with, and if drilling is thoroughly uneconomical, so SandRidge will not do it unless the company absolutely has to.

So, with that production we get $587 million in revenue and $135 million in operating costs. As per the second quarter's financial statement, annual interest expense was another $250 million. Therefore, SandRidge would have only $200 million or so in 'cash flow' by which to apply to capex. This does not include royalties or transportation expenses.

In such a scenario, management may want to spend $500 million in order to keep production steady. While a lot of that drilling may be uneconomical right now, further drilling and well design improvements in 2015 may help with some of that. In any case, if SandRidge wants to maintain production levels. It has $919 million in cash on hand, so SandRidge should be just fine through 2015.

How about 2016? Could SD make it two-three more years? Hedges in 2016 are virtually nonexistent, so this becomes more difficult. It all depends on where oil prices settle for the long-term. If prices stay low through 2016, SandRidge might have to look at some other options.

In order for the SD's plans to work in the long run, drilling needs to make economic sense. It all boils down to a minimum of $75 WTI as a break even point. Less than that is no go, and the SD pumps will stay idling for the duration. Higher, $85 WTI, price would help SD to start delivering oil at a profit and recover financially. In worst of cases, the company has no outstanding balance on its credit lines and no senior notes due until 2020, so there's still plenty of flexibility there.

But the global oil prices are still the driver in the U.S. oil production. While large companies can endure the punishingly low oil prices, the small guys, like SD, are hanging on a thread—the minimum $75 WTI price thread. If the low oil prices continue, SD and many others will have to shut down shop and send everybody looking for new jobs.

In September, 2015, SD stock went down to a record $0.25 per share. This is way down from the record high of near $13 in 2011. What a difference time makes...especially when the market is against you.

Oil and Gas Trading

- Crude oil, in all its different forms, accounts for about 1/3rd of the entire global energy consumption. This includes crude oil, shale oil, oil sands, and natural gas liquids. As a result, we see different numbers depending on whether the report accounts for oil only, or total liquids.

Oil is traded by the barrel, containing 42 U.S. gallons. Oil and gasoline are traded in units of 1,000 barrels (42,000 US gallons.). WTI crude oil is traded through NYMEX under trading symbol CL and through Intercontinental Exchange (ICE) under trading symbol WTI. Brent crude oil is traded in through ICE under trading symbol B.

Gulf Coast Gasoline is traded through NYMEX with the trading symbol of LR. Gasoline(reformulated gasoline blendstock for oxygen blending or RBOB) is traded through NYMEX via trading symbol RB. Propane is traded through NYMEX, a subsidiary of IntercontinentalExchange since early 2013, via trading symbol PN.

For many years (WTI) crude oil, a light, sweet crude oil, was the world's most-traded commodity. WTI is a grade used as a benchmark in oil pricing. It is the underlying commodity of Chicago Mercantile Exchange's oil futures contracts. WTI is often referenced in news reports on oil prices, alongside Brent. WTI is lighter and sweeter than Brent and considerably lighter and sweeter than oil from Dubai or Oman, for example.

Crude oil can be light or heavy. Oil was the first form of energy to be widely traded. Some commodity market speculation is directly related to the stability of certain states, e.g., Iraq, Bahrain, Iran, Venezuela and many others. Most commodities markets are not so tightly tied to the politics of volatile regions as crude oil.

- Natural gas accounts for just under one-quarter of global energy consumption. You can measure natural gas by volume, usually in cubic meters or feet, or by the amount of heat it can produce in therms, or British thermal units (Btus).

Natural gas is traded through NYMEX a subsidiary of Intercontinental Exchange (ICE) in units of 10,000 million Btus (MMBtu) with the trading symbol of NG. Heating Oil is traded through NYMEX, a subsidiary of ICE, under trading symbol HO.

The Exchanges
- Exchange-traded commodities (ETCs) is a term used for commodity exchange-traded funds, or commodity exchange-traded notes. These instruments track the performance of each commodity index including the total return indices based on a single commodity. The trading structure is similar to ETFs, where the trade and settlement is done exactly like with stock funds. ETCs have market maker support with guaranteed liquidity, thus enabling investors to invest in commodities by evaluating the risks of each transaction.

ETCs were introduced partly in response to the tight supply of commodities in 2000, combined with record low inventories and increasing demand from emerging markets such as China and India. Starting in 2003, only professional institutional investors had access to ETC trading, but today online exchanges opened some ETC markets to almost anyone.

- Over-the-counter (OTC) commodities derivatives originally involved two parties without an exchange. Exchange trading offers greater transparency and regulatory protections. In an OTC trade, the price is not generally made public. OTC are higher risk but may also lead to higher profits.

Between 2007 and 2010, global physical exports of commodities fell by 2%, while the outstanding value of OTC commodities derivatives declined by two-thirds

as investors reduced risk following a five-fold increase in the previous three years.

Money under management more than doubled between 2008 and 2010 to nearly $380 billion. As a result, the inflows into the sector totaled over 72 billion in 2009. The bulk of funds went into precious metals and energy products. The growth in prices of many commodities in 2010 contributed to the increase in the value of commodities funds under management.

The financial energy markets trade energy and energy related products and services. One of the newest is the carbon trading (not to be confused with Carbon Tax.)

Carbon (Emissions) Trading

Carbon emissions trading is a form of financial trading that targets carbon dioxide (CO_2), calculated in tons of carbon dioxide equivalent (tCO_2e). This type of transaction constitutes the bulk of emissions trading at the present. It is a form of permit trading that is a method used by different companies and countries as needed to meet their environmental obligations.

These obligations were specified by the Kyoto Protocol and are geared to reduce carbon emissions in an attempt to mitigate climate change. Using this method, any country emitting more carbon than allowable, can purchase (or 'trade') carbon credits with a country having less emission.

Carbon trading is purchasing the right to emit more GHGs than maximum allowed.

More money one has, more emissions can be bought and emitted in the atmosphere, the common wisdom goes. The idea, however, is that since it costs real money to pollute over the limits, carbon emitting countries and companies will try to keep close to the allocated limit of carbon emissions.

The problem is that the excess emissions of GHGs are not felt directly by the emitters. Nor are they fully affected by these actions. This is because, a) they pass the increases to their customers, and, b) the emissions spread all over the area, and sometimes cross political boundaries. So, a city, or country, downwind from large polluting center might be negatively affected by the pollution. In such case, regardless of the money paid by the polluters, the downwind population suffers the consequences.

Here we enter into the gray area of externalities caused by energy generation, which is another huge subject, we review in other areas of this text.

Note: There is a difference between carbon tax, which we reviewed previously in this chapter, and carbon trading. Some economists suggest that carbon taxes should be preferred to carbon trading as a more fair distribution of wealth and responsibilities. The counter-arguments to this are based on the preference politicians have for emissions trading vs. taxation.

One of these arguments is that emission permits can be freely distributed to polluting industries, rather than the revenues going to the government. In some cases, companies may successfully lobby to exempt themselves from a carbon tax. This then means that with emissions trading polluters have an incentive to cut emissions, while exempted from a carbon tax, they have no such incentive.

On the other hand, distributing emission permits in unfair way could create corrupt corporate behavior, and make the situation even worse. It is obvious here that more work is needed in this area.

In all cases; carbon tax and trading practices cannot fix the big problem. This is because no matter how much a polluter pays, or by what method, the additional money does not help the environment. Polluters usually have a good reason to pollute and will continue doing so no matter what. They would pay money to pollute, and might even reduce the pollution, but will not stop. We see this happening increasingly in the developing world, which leads us to the conclusion that other approaches are needed to reduce GHG pollution.

2015 Update

The 2014-2015 oil prices crash exposed new serious risks for many U.S. shale oil drillers. Oil prices went so low that a weakness in the insurance that some U.S. shale drillers bought to protect themselves against a crash was revealed.

A number of companies are using the three-way collar strategy, which does not guarantee a minimum price if crude falls below a certain level. This type of insurance policy is much cheaper, than other hedge instruments, but the extremely low global oil prices left these drillers exposed to steep losses.

Oil drillers are usually bullish. How can you lose drilling for oil? Oil business is a profitable business 99.9% of the times. But in the summer-fall of 2014 the world entered that rare 0.1% of the times, when oil prices kept falling to unprecedented levels.

West Texas Intermediate crude, the U.S. benchmark, dropped over 50% since the summer of 2014 amidst a worldwide oil glut. The OPEC held produc-

tion steady in order to competes for market share against U.S. shale drillers that have the highest ever domestic oil output.

Shares of oil companies in 2015 dropped 40, 50, and even 80% since the summer of 2014. Since the oil drilling in the U.S. was driven by the high oil prices and low-cost financing during 2008-2013, many companies spent $1.30 for every dollar earned selling oil and gas.

Financing was the way to get in the business and to expand operations. Now, however, financing costs are rising as prices sink. The average borrowing cost for energy companies in the U.S. high-yield debt market has almost doubled since the all-time low of 5.60% in the summer of 2014.

Locking in a minimum price for crude reassures investors that companies will have the cash to keep expanding and lenders that debt can be repaid. Pioneer Oil, for example, uses the three-ways method to cover 85% of its projected 2015 output. The strategy capped the upside price at $99.36 a barrel and guaranteed a minimum (the floor) of $87.98. This by itself would mean heavy losses at the very low oil price levels.

Pioneer, however, added a third element to its hedge by selling a put option, sometimes (subfloor) at $73.54, which gives the buyer the right to sell oil at that price by a specific date. This strategy ensures that the bulk of Pioneer's production will earn more than the low market price. The three-ways will also prove valuable if oil rises above the subfloor.

Still, as prices move lower and lower, the floor and subfloors would collapse and the oil drillers end up with a lot of red ink on their balance sheets, simply because none of them foresaw oil prices getting as low as $40/bbl. And very few could stay afloat at that price…for a long time

Many anomalies in the energy markets cost companies and governments billions of dollars in losses. In order to reduce their financial losses, many companies and governments use different types of insurance vehicles. Below we take a look at some of the new insurance trends.

INSURANCE

Insurance of energy companies, projects, and other activities is a big—and growing exponentially—business.

The insurance industry has been, and still is, heavily involved in the energy sector. It provides the

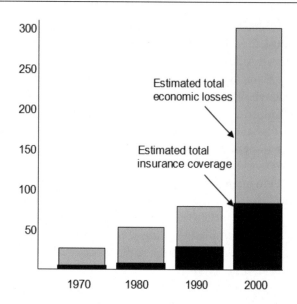

Figure 8-21. Energy sector weather damages (in $ billions)

usual and standard coverage for all kinds of risks and accidents at mines, power plants, and transport routes. These insurance vehicles are well established and used daily, so we won't spend much time on them.

Similar coverage is offered to the renewable technologies, but with a twist. Or several twists, which the insurance industry has to deal with today, as we will see below. But now, we need to take a look at the basics:

The Insurance Basics

Insurance companies world-wide are in the business of allocation of risk as their primary concern. People buy insurance to protect against financial damages or loss of property or persons. Insurance underwriters provide insurance to protect against potential damage or loss based on the associated risks.

Risk, as far as the insurance companies are concerned, is: "The chance of loss to the person or entity that is insured." Simple...but in each situation, the actual risk depends on a multitude of factors, such as what is being insured, what events might occur, and how likely each occurrence might be.

The insurance companies assess the total risk they are covering for any particular situation and determines the likelihood that they will have to make a payment if that risk becomes an actual loss. The annual insurance premium payment is determined based on the potential for the risks to become actual losses. For insurance companies to be profitable, the premiums they receive for insuring against losses must be greater than the actual payouts in the event of losses occurring.

There are three main types of insurance companies:

a.   Insurance brokers
b.   Insurance underwriters, and
c.   Reinsurers

Brokers, just like the name suggests, are the intermediaries (salesmen) between the underwriters and the customers. As agents of the underwriters, their role may sound trivial, but they are responsible for finding the customers, negotiating the deals and in the process they come up with new and unique products, which are useful to all involved.

Industry-specialized brokers work with renewable energy product manufacturers, facility developers, and project finance clients to create unique product offerings and the necessary underwriting support that are not well covered by traditional insurance. The goal of an industry-specialist or consultative insurance broker is to strategically partner with clients and provide innovative, responsive, and cost-effective solutions that mitigate the risk and uncertainty of renewable energy systems.

Insurance underwriters, on the other hand, are the companies that actually have the money to pay the customers when legitimate claims are made. They approve the deals negotiated by the brokers, and formulate the payment estimates designed to cover their risk. All terms of the deal are outlined in a binding contract which is approved by the underwriters. They also define the price and risk allocation terms under these contracts through a multitude of processes, including research, engineering analysis, risk modeling, contractual negotiations, and modification of pre-existing underwriting forms.

Insurance underwriters are specialized in different areas of the economy, which allows them to gain expertise in a particular area, which leads to a better understanding of the risks involved. This allows customers to receive best possible insurance premiums and faster turnaround on the creation of an insurance policy for one or both; liability and/or property, depending on the project location, size and other conditions.

Working together with the insurance brokers, underwriters are gaining more understanding of the intricacies of the renewable energy business, since it is quickly becoming one of the major areas of their business, with great growth potential in the near future, as well.

Basically speaking, the process of establishing insurance policies begins with developers or project owners identifying the risks that they need or want to cover and looking for an insurance broker to put their needs into an insurance package.

The insurance broker (industry-specialist preferably) might have suggestions for what to cover, how and how much. After a preliminary agreement on the type and size of insurance coverage needed, the broker presents the underwriting company s/he is working for with an outline of the insurance products that interest this particular client.

The client, on the other hand, is presented with the range of policies, premiums, and terms of coverage from which to choose. Once the customer decides which policy it wants, the broker returns to the underwriter to complete the policy formation process, a contract is designed and signed and the insurance policy, and all its conditions, takes effect. The project and the people are therefore protected by the terms of insurance policy for the duration of the contract.

Note: We did not discuss the role of reinsurers, because these are specialized insurance companies that provide risk management, and even insurance, for the insurance contract issued by the insurance underwriters. With other words, they insure the insurers.

Today, the energy markets are bringing new meaning to insurance.

The New Energy Dimension

The insurance industry is being significantly affected by the current economic downturn, new regulation requirements, and increased losses to policyholders. In the context of these large-scale forces shaping the insurance industry as a whole, the insurance market for PV systems is evolving and maturing rapidly.

The addition of the renewable energy market insurance options to the portfolio of insurance products, have not made the insurance company executives lives any better. Needless to say investments in PV products and projects are often viewed by underwriters as quite risky for two main reasons:
1. The technologies are newer (i.e., most systems do not have a long history of operational data), and,
2. There are fewer installations relative to other technology deployments (e.g., the real estate or the automobile).

Insurers use the "law of large numbers," which says that "the larger the group of units insured, the more accurate the predictions of gain and loss will be." And although the solar technologies have been in existence since the 1950s, their deployment at a significant level occurred much later, so the numbers are not that large.

Since 2007 new commercial on-grid installations increased to more than 1,500, which statistically speaking is not a large number. Not enough solar installations have been in place for long enough for underwriters to feel they can accurately predict what the losses associated with them would be. The experiences with some of these fields have not been very reassuring either, so all this fuels the uncertainty of the insurance industry.

On top of that, data of the different projects, products, installation and operation is not readily available, which complicates the process. Insurers have access to data on only projects that they insure, so the more insurers there are, the less data each insurer has.

Another major factor that adds to the complexity of insuring solar installations is the typical involvement of many parties, including installers, developers, investors, lenders, and insurance companies in each project. And typically, each party to the transaction attempts to minimize the risks it assumes while defining the recourse available in case a risk event occurs and leads to actual losses. In many cases, insurance products can be used to mitigate and manage the allocations associated with complex contractual arrangements.

Until the recent growth in the PV market, demand for PV system coverage was not adequate to encourage insurance underwriters to develop PV-specific products. If the upward trend of new PV installations continues, there will be a great demand for insurance products for PV. This growing market of new solar installations, which is being driven by state policies, federal incentives, and corporate responsibility, represents a possible market opportunity for insurance underwriters.

So let's look at the present day solar energy insurance from our vintage point:

The Insurance Issues

In this text we've discussed the major technical, administrative, and financial issues facing the solar industry. We determined that long-term performance and longevity are some of the most important, albeit least discussed, issues in any PV undertaking with any of the existing PV technologies. And looking at the data in some long-term desert tests we can't help but wonder how anyone could put a positive spin on the results and accept the risks associated with 20-30 years on-sun operation of these modules…especially in the world' deserts. We obviously don't know much about the different possibilities during such a long time exposure to such unforgiving climates. At the very least, we should take a very close look and estimate the extremes.

The long-term field tests are conclusively assuring us that the risks are real and are warning us of the potential and serious long-term physical deterioration, power degradation, failures and other dangers.

The lack of convincing test data, plus negative data from a number of field test studies, point to major obstacles in the early stages of the PV projects' financing, because investors are concerned with the lack of reliable scientific or test data that can convince them that the risks associated with long-term on-sun operation are well-known and manageable.

Enter the insurance industry…again!

A number of insurance companies do offer conditional performance guarantees designed to put the risk factors on a level ground which would make potential customers and investors more comfortable. PV performance insurance is a new field and like the PV industry it is equally un-standardized and unregulated thus far. In some cases, coverage is available to all participants: manufacturers and distributors of PV modules, integrators, utilities, customers and investors, while in other cases it covers only certain segments of the PV cycle, such as PV modules' efficiency and performance only.

Different insurance companies cover different types and percentages of the risks related to PV products and services. That might be just enough to eliminate the majority of risks, but we'd advise customers and investors to take a close look at the guidelines and definitions of the coverage in each policy before making a final decision. This is new thing and we all need to be careful.

Nevertheless, this is an encouraging development that might be just what is needed to bridge the gap of financing PV projects today and in the near future. Proper performance insurance fills the gap between quality, longevity and profitability of PV modules and projects, so it might be the key to providing acceptable levels of risk management, thus increasing the customers' and investors' confidence in solar installations.

One serious catch: the insurance companies would favor the more established manufacturers (at least for now) who produce better quality PV products. But there are a lot of PV project owners and investors who prefer to use the "cheap" Asian-made PV modules. So how

would the insurers determine which of these have acceptable performance risk? How do insurers know what a better performance is to begin with?

What do they know about longevity, let alone assessing the risks associated with it? Because of its importance, these questions and the entire solar performance insurance sector needs special attention, so let's take a closer look at the characteristics and conditions of providing "performance" insurance.

In order to mitigate the performance and longevity risks of PV modules, insurers must have a good idea what these are. Since they are not specialists in the solar field, and since there are many risks in all areas of the cradle-to-grave life span of modules, they will need to create a "PV products performance insurance" technical team, which would consist of a number of technical specialists with varying areas and degrees of responsibilities.

Expertise

As an example of a team of experts, as needed for the success of both; the insurer and insured, we would suggest the following additions to the insurance company's professional team:
1. Materials specialist, or expert-consultant, in charge of evaluating the type and quality of the all process materials and chemicals used for making the modules in each of the insured projects.
2. Front end process engineer, in charge of determining the quality of each step of the solar cells manufacturing operations.
3. Back end process engineer, in charge of evaluating the PV modules assembly process.
4. Quality control engineer, in charge of determining the quality procedures of each process step and the final product.
5. In addition, a group of other specialists is also needed to ensure the integrity of the PV project's installation and operation. These specialists should be able to evaluate the project's design, structure, land, interconnect and other aspects and issues related to the modules' installation and operation.

Note: Actually two separate technical teams are needed to cover all aspects of PV modules manufacturing and use—one team consisting of crystalline silicon (c-Si) solar cells and modules manufacturing and use specialists, and the second of thin film PV (TFPV) modules manufacturing and use specialists. This diversification is needed because the materials, processes, installation and operation of these different types of

PV technologies are quite different and require special knowledge and experience to fully address and resolve the potential issues. Because of that, the different tasks and responsibilities in the different technologies are not interchangeable.

The Tasks

These specialists' and engineers' tasks and responsibilities would be to, a) obtain the most recent and reliable information on the manufacturer, the materials and products to be used, and, b) analyze and summarize the risks in the respective areas.

The goal is to arrive to conclusions in the key areas of interest, as follow:
1. Supply chain management and quality
2. Manufacturing process execution
3. Quality control verification
4. Efficiency evaluation under local climate conditions
5. Performance evaluation under local climate conditions
   — a. Temperature coefficient verification
   — b. Annual degradation verification
6. Longevity under the local climate conditions
   — a. Early mortality estimate
   — b. Useful life estimate
   — c. End-of-life measures
7. Toxicity and safety evaluation
8. Land and interconnect evaluation
9. Evaluation of political and regulatory issues in the local area

After completing these tasks, the technical team members would make a thorough and detailed report on each material and process step in the cradle-to-grave life span of the modules and estimate the level of quality, efficiency, performance and longevity of the final product and the entire project.

With all this done, the insurers will have a very good idea what they are dealing with and will have a better chance to calculate and mitigate their risks. It is possible that within several years of this thorough due diligence, a pattern would emerge that can lead to standardizing the entire insurance process. Actually, it is most likely that several patterns would emerge—patterns related to different manufacturers (established vs. new companies), different locations (hot desert vs. cloudy climates), different technologies (c-Si, vs. TFPV, vs. HCPV), and several combinations of the above.

Short of that, the insurers are working with their eyes closed, so rolling a die would be the only worse thing they could do. While the insurance business is full of uncertainties and risks, mitigating the risks in the chaotic solar frenzy is only logical.

Sorting and sifting the low quality players and technologies would prevent disasters and save money for all involved. This is not an easy task and some serious knowledge, professionalism and hard work is needed before obtaining a good handle on the situation.

We hope that the insurance companies involved in providing performance insurance to the solar products business will be able to understand and successfully tackle this fascinating but risky business venture.

"The insurance industry doesn't work this way," some might say. Maybe, but the insurance industry has no experience with long-term PV installations either, and is far from being a specialist in all areas of PV products manufacturing and use, so it needs to take a very close look at all options (including the above suggestions) before jumping on the solar bandwagon.

Or, it might end up in a mess of its own and up to its neck in risks, problematic issues and unresolved cases. Just saying...

Peace of Mind

If everything is properly set, solar project installers, owners and investors would have a choice of insurance protection, which could vary from full, unconditional coverage to a minimal coverage, and whatever else is required by law.
1. Manufacturers who use their own PV modules to build PV power plants may not need or want performance insurance*, although they still might choose a level of conventional insurance coverage to protect the plant from natural and manmade disasters.
2. Another class of customers using the highest quality PV modules available may also decide to skip insurance coverage altogether for obvious reasons.
3. Most of the other players, however, especially those using cheap PV modules will be able to choose a level of coverage as they see fit. This way the gap between uncertain quality, deteriorating performance and reduced longevity will be at least partially closed and the PV plant will survive unpredictable events, including excess performance degradation and failures.

Note: There is a difference between "performance" and regular insurance coverage required by state or federal law. Performance insurance in this text refers only to covering the amount of energy lost within a time period due to poor product performance, or other factors affecting the overall performance and yield of the power plant.

Nevertheless, there is some overlapping of conditions and issues when talking about any type of insurance for power plants, so utmost care must be used when designing and signing insurance contracts.

Captive Insurance Option

There are still major challenges in the newly developed (or rather developing) solar and wind energy fields, and so not many insurance companies would be willing to provide adequate long-term insurance to projects using these technologies.

As an alternative, mid- to large-size renewable energy developers and manufacturers (or projects) concerned about the overall cost of insurance and product warranty conditions and premiums may have the option of establishing a captive insurance company (hereafter referred to as "captive"). Such enterprises are common with fortune 500 companies and have been successfully used for a long time in other industries.

A captive subsidiary established with the parent company (or project) explicitly as a risk mitigation arm is intended to maintain control of the operation—including any accidents, underperformance and other issues.

Like traditional insurance companies, the insured (a solar company or project) pays a premium in exchange for coverage as outlined in the policy. And like any insurance company, the captive should be managed in such a way that it does avoid, and somehow do not absorb all losses of the insured.

This could be an attractive alternative to buying insurance from a traditional underwriter, since insurance premiums for plant construction and operation are limited in scope and way too high in price. Insurance premiums of 25% of a system's annual operating expense, and/or 0.25%-0.5% of the total installed cost of the facility, have been reported.

Based on NREL's System Advisor Model (SAM), a 1MW system located in Phoenix, AZ with a total installation cost of $4.5 million would have an annual insurance cost between $11,500 and $22,500. This means that our 100 MWp at the same location, and insured under the same conditions, might pay half a million dollars annually in insurance fees alone. This is a large chunk of the profits no matter how you look at it, so a better alternative is needed.

The higher premiums in most cases are due to the solar industry's turbulent childhood; with its maturing

technologies, imperfect business models, and evolving applications—most of which are not fully established and/or standardized.

A major issue for solar projects is the fact that none of them has completed a full life cycle—30 years of successful non-stop field operation. Thus, they all lack a reliable track record. Because of that, it is hard to impossible for insurers to assess the risk levels of the different locations and technologies.

Solar power lacks a track record of one full life-cycle, which leaves a large gap in their reliability record.

Most insurance companies lack expertise and investment strategies as needed to evaluate and manage renewable companies and projects. So the combination of all these factors increases the (real and perceived) risks, which drive prices unreasonably high. At least for now...

The Challenges

The challenges of this captive enterprise, therefore, are numerous and serious, as follow:

a. The captive may lack the actuarial expertise needed to assess risks, especially for the uncertainties associated with innovative technologies, their financial structures, etc. conditions and issues

b. Setting up a captive includes, among other things, transaction costs (feasibility study costs, organizational costs, actuarial fees, business plan costs, permitting and PPA costs, management fees, legal fees, audit fees, and taxes). These costs could be staggering—in the millions of dollars for significant size projects.

c. Costs to establish a captive range from about $50,000 to over $125,000, depending on the captive's complexity. This means that developers would need to determine the minimum project risk (and optimum operating conditions) that would make a captive worthwhile.

d. Determining where to establish the captive is tricky too, because not all states have regulations conducive to establishing such a venture. Developers must research state and local code to determine possible locations. Vermont is the most common state for captive domiciles, and foreign countries may also be used.

e. A major claim event or series of claims could be costly, so the possibility of such an event must be entered into the equation in order to make a proper decision.

Solar Insurance Details

We realize that this is a brand new and unexplored area, so jumping into it is full of risks—it is just like jumping in a swamp full of hungry crocodiles. We, however, have been living in this swamp for many years, have been attacked many times and have learned the tricks, so hopefully will be able to deal with the dangers ahead without getting swallow.

The View from the Top

The insurance executives view of the solar industry in general is somewhat limited for a number of reasons—some of which we discussed above. Their view of the solar business in general is as follow:
1. Manufacturers cannot ensure the long-term efficiency and reliability of their product
2. Developers do not know exactly what coverage they need.
3. Broker's understanding of the solar technology ensures proper coverage.
4. The solar PV market is a maturing industry, but it is not yet mature.
5. Lack of established standardization and regulations is of great concern
6. The solar PV industry has excellent fundamentals (e.g., strong product demand, declining input costs), but there are still uncertainties and risks.

The View from the Bottom

On the other side of the equation, the opinion of the solar industry about the insurers is also somewhat skewed, as follow:
1. Insurance policies are not affordably priced
2. Insurance companies sometimes lack the background knowledge of solar PV technologies
3. Insurance premiums could go down if the insurance industry had better data and understanding about system operation and historical data
4. Insurance brokers are considered the 800-pound gorilla in the room
5. If brokers aren't educated on the technologies and risks, then the underwriters won't be either.
6. Some brokers pretend they understand solar technologies and place policies that do not fully cover what needs to be covered.

I.e., in a famous case, one insurer asked about the use of molten salt in a PV system to be insured. Molten salt is relevant to CSP, but not for PV, thus revealing the severe lack of understanding of the business in this particular case.

The Negotiations

Solar project insurance negotiations usually go along the lines of the general insurance business, as follow:
1. General Liability—covers for death or injury to persons or damage to property owned by third parties (i.e., not the policyholder).
2. Property Risk Insurance covers damage to or loss of policyholder's property, such as theft, natural disasters, transit of goods etc.
3. Additional policies, such as Environmental Risk Insurance, Business Interruption, Contractor Bonding etc., AND
4. We will add here Performance Insurance. The long-term performance of the equipment is a major issue we are concerned with during 30 long years of non-stop operation. We must have some assurance (and insurance) that whatever happens to or with our investment (power generation) is protected against any unforeseen events—including and above the manufacturers' warranties.

Summary

Project performance (and any other type) insurance might be the solution to successfully using any type of PV modules and products, including those of unknown quality—as is the case quite often today with Chinese, Indian, and other imports.

The European and Japanese solar markets are more mature than the U.S. solar market, with most of the growth in PV capacity in Europe being in Germany, which was the world leader in cumulative installed capacity in 2010-2011. Spain and Japan also have robust PV markets

The maturity of the European PV market gives European insurance companies more experience underwriting renewable energy generation systems than the U.S. market. Because of Europe's greater experience with solar PV installations, some U.S. insurance companies use performance and other data from European solar fields to project probabilities of future losses and risk of similar installations in the United States.

Provided that the insurance companies are well aware of the risks and accept them (for a price, of course) most PV power plants with adequate performance insurance coverage will survive the test of time. Without adequate performance insurance, however, many PV plants may experience serious problems in the deserts and other harsh climate areas during long-term exposure to the elements.

These add some fun to doing business in the sector.

The newest of these, most important, and interesting is the brand new and yet unexplored area of performance insurance.

Performance Insurance

This is a new concept, which albeit discussed in the past, and even partially applied in some cases, was never fully implemented in one single project. It is today, however, becoming a prerequisite for starting (and for sure financing) new solar projects.

And so, we are now looking at the newest branch of the US insurance industry, *performance insurance of solar and wind power plants*.

It is like the knight in shining armor coming to save the PV industry from its problems. Or is it...? This type of insurance is too new for us to assign it an exact place and role in the 21st century, but our long experience shows that it is one (if not the only) way to fill the gaps between quality, performance, and profits.

We know what to expect from the PV manufacturers, installers and operators, so before jumping into the abyss of insuring PV fields—and especially large-scale installations—we need to meet the other part of the equation; the insurance companies.

Solar Risk Insurance Examples

As a confirmation of the positive developments in the area of performance insurance coverage, and to free us from our worries and hypothesis about field failures, several companies are actively looking into it, and some are taking immediate action as follow:

LDK Solar Risk Insurance Package

LDK Solar and Solarif introduced a new type of insurance package for PV systems in the summer of 2012. It is a full insurance package, including "all risk insurance," with the related warranty and inherent defect insurance.

The all risk insurance is for all components of PV systems (including BOS), while LDK Solar power and product warranty cover their PV modules.

LDK Solar's manufacturing process was audited by Solarif and the insurance is covered by German insurer HDI Gehrling. The insurance package consists of full coverage of all products sold on the European continent, including compensation in case of production loss during warranty exchange, transportation and all other costs related to module replacements per warranty conditions.

This, to our best knowledge, makes LDK the very first PV company to introduce such innovative, complete, and very much needed insurance solution, which sets new standards for the PV insurance industry.

Nevertheless, we see a number of problems with this first attempt to cover the long-term performance of PV components, starting with the financial health of the participants.

Solar manufacturers have varying degrees of short- and long-term debt as well as cash levels, and these vary from day to day with the market fluctuations and according to companies' activities, so it is hard to get a good picture of the situation, let alone predict the future.

But at a quick glance of the financial results of 2011 wee see that LDK Solar had (and still has) $1.2 billion long-term debt, and $2.2 billion short-term debt. This is a lot of money owed to banks, suppliers, governments etc. Other Chinese solar companies, like Trina Solar and Yingli Green Energy show much better results. During the same period they had $383 million and $0 in long-term debt, and $343 million and $1.1 billion in short-term debt, respectively.

So there is a big difference in the bottom line for sure. And although it is hard to predict what will happen, we don't see LDK as a healthy company. And things are not looking any better for them in the near future either. On top of that LDK is a Chinese company, who is dependent on its communist government whims, and so if they decide to shut down LDK (as it happened to 50% of the Chinese solar manufacturers) nobody can stop them. So what will happen then to the full insurance package? We don't know what the small print says, so we have to wait and see.

Nevertheless, this is a good example and a good start in the right direction. This type of full insurance coverage, we believe, is what is needed badly today, and is a solution for securing investments in PV projects. It is the make-it-or-break-it for large-scale solar installations in the US and Europe.

Not withstanding the high price of such insurance, we see it as the missing link that hindered development of solar projects thus far in the US in particular. It is also absolutely the best (if not only) way to ensure power plants' long-term profit. It is how business will be done in the near future, and for sure during the latter part of the 21st century. There is simply no other, or better, way!

In another case, Assurant, Inc. and GCube Insurance Services, Inc., both insurance industry veterans, have partnered and have created a new insurance for commercial-scale solar projects. The insurance package covers solar installations between 100 kW and 3.0 MW in size. In addition to the standard property and liability insurance, the new insurance package offers an unique warranty component at a project-specific level.

The new warranty concept addresses the concern that the manufacturers typically offer 25 years insurance, usually handled by a company that has been in business for at least 4-5 years. The new package brings experience and capital to reassure the financiers that it can stand behind the warranty. Also, the project developer has the option of warranty management coverage as long as the project's property and liability coverage remains with the original insurer.

This new insurance instrument is a huge value-added, because it is really the only one with warranty management available in the market. The warranty management will be done through the manufacturers and the O&M providers, similar to the way auto insurance providers use mechanics and body shops.

Note: No question, this is the path of the future. And while we do not question the ability of these insurance companies to make money, we only wonder how much expertise they actually have in the solar power field, and if they have full understanding of the risks they are taking.

On the other hand, we wonder if the potential insured parties would be able to afford the insurance premiums (which won't be cheap, for sure) without breaking the bank.

No question, there is a large gap between the insurers and the insured right now, which again points to the immaturity of the solar industry. We do hope that with time the gap will be narrowed down to an acceptable level for the benefit of all parties involved.

Time will tell, but we clearly see the companies in the above described cases clearing a path into the future, which everyone will follow during the 21st century. The path is not going to be short, nor easy, but is one that provides most security and reliability to the solar power generation industry.

ReneSola 25-year Performance Insurance

As another confirmation of the need and feasibility of performance insurance, ReneSola, a China-based manufacturer of solar wafers and modules, signed a solar products performance insurance agreement with PowerGuard Specialty Insurance Services, who specializes in insurance for solar and other alternative energy companies and products.

The new coined agreement, PowerGuard, provides conditional insurance and warranty-related coverage for ReneSola's PV products for 25 years. This insurance will provide an edge for ReneSola to offer its products with high degree of confidence and to increase its mar-

ket share to include customers with varying risk-management profiles.

This type of insurance addresses the customer needs for reliable long-term operation, and reduces the difficulties that are often encountered by solar project developers, installers and operators. It will also allow ReneSola to focus on their area of expertise, thus increasing the quality of their products.

PowerGuard is a good example of the things to come. The PowerClip warranty product has been particularly popular, with the likes of China Sunergy, Lightway Green New Energy and SunEdison having all adopted the protective cover during 2012.

The Practical Solar Insurance Aspects

With the economic and technical fiascos of solar companies and projects of late, investors and customers alike are getting more and more suspicious and careful about putting money in solar companies and projects. The fact that the world's PV modules manufacturers Have been (and still are) sitting on 50% surplus is one indicator of the issues at hand. The spontaneous failure of solar companies and projects of late has increased the level of uncertainty, which similar to what happened several times in the past, is threatening the very existence of the solar industry.

Things are getting so bad that customers and investors are asking PV modules installers to provide an insurance from reputable insurance companies. Several such companies are involved in the solar insurance game today.

These companies cover warranty claims even if the solar panel maker goes out of business. And this is becoming the accepted modus operandi, if not the only way that solar companies, even the large ones, can sell their products on the US and EU markets.

Warranty Claims

Warranty claims resulting from manufacturing defects appear to be one of the biggest expenses facing PV modules makers. There were recently a number of news stories of such claims eating up profits at solar companies. I.e. First Solar has received more than 5,000 warranty claims since 2008, mostly due to manufacturing defects, field failures, rejects, and other quality deficits. Over 5,000 unhappy customers. and that many projects to spend resources and money at.

Industry experts report that First Solar has paid around $150 million in warranty claims thus far, which have resulted from defects that cause the thin film PV modules to fail in hot desert climate. Additional $35 million have been allocated as a reserve for covering future failures.

While $150 million is drop in the bucket for a company that can get $4.5 billion from Uncle Sam, the failed modules cause chaos in the solar fields, resulting in lost revenue and embarrassment for the owners.

Did First Solar officials and their friends at DOE and the White House, know that the CdTe PV modules are not proven reliable for operation in hot desert climate? Did they hear the official statements that CdTe PV modules degrade and fail prematurely under extreme heat?

Just think, the CdTe PV modules have been in mass production only 4-5 years, with most of the larger installations in cloudy Europe. The large-scale desert installations of CdTe PV technology in the US deserts started just couple of years back. Proven track record? Nope. None! But it appears that the "new normal" of only a couple of years lab testing is good enough for DOE to issue billions of dollars in loans and loan guarantees (taxpayers money is cheap) for product that may, or may not, live up to the expectations.

There are also other similar reports of major field failures of PV modules made by other world-class manufacturers. Most of these, however, have been taken care of quickly and silently, thus not allowed to spread throughout the media. Such problems are expected to grow in the future, and get worse as the manufacturers down-scale, or go into bankruptcy under the uncertainty of the present day economic situation.

Most manufacturers do not use a third-party insurance, the major reason for which abnormality is that most insurers don't have enough understanding and confidence in the PV products. Some also lack sufficient cash reserves to cover larger claims, as expected in many cases in the near future.

So, in our opinion, the warranty claims will keep coming in ever increasing number, but the solutions will get fewer and far between. The imbalance will be too much to handle for some companies and projects, so we won't be surprised if some of them fail and go bankrupt.

Quality of the Manufacturers

What can we expect when buying from different quality manufacturers, while considering adding performance insurance to our large PV project? Let's see:

Poor Quality Manufacturers

Buying poor (or uncertain at best) quality Chinese PV modules made by government subsidized, hastily put together, flight-by-night operations, using cheap

materials and unqualified labor was wide spread in the US and Europe until recently. Not a wise move, and we are still to see the full range consequences in the near future.

This is the worst case scenario, where unpredictable quality of the PV products (modules and BOS components) will shock us with low performance and prematurely failing modules and projects.

In such cases, the risks are enormous and extremely expensive to mitigate (if at all possible). We dare say there are many such cases, because during 2007-2011 most wanna-be solar installers and specialists purchased and installed the cheapest PV junk they could find.

Driven by greed and blinded by lack of expertise, they bought large quantity of PV modules, the only requirement for which, in addition to the lowest price, was that they look shiny enough. They then slapped these toys as hastily as they could on every roof they could find, Now, how will these modules perform? How long would they last? These questions were very seldom asked and are now becoming irrelevant, because most of the wanna-be installers are gone and never to be found.

So we have to wish best of luck to the PV installations owners, who were gullible enough to allow their roofs to be covered with untested Chinese junk made and installed by flight-by-night operators.

And we feel even more for the large-scale solar power fields owners who used the same untested Chinese modules. Many of them will be put through the stress test of their lives, together with the modules. We hope that, for their sake, we are wrong and the modules will perform efficiently and flawlessly for 30 years.

Medium Quality Manufacturers

In case of below-average or intermittent quality of the PV products (PV modules and BOS components) the risks are lower, but still unpredictable. So the risks still need to be evaluated and mitigated, and the problems corrected; be it through warranty, negotiations with the manufacturers or via insurance claims.

Since in most cases the manufacturers are of some reputation, they will agree to cooperate. This makes things much better, but some additional expense (lost power generation revenue. module replacement modules transport and installation, etc.) is expected to be paid by the customer.

Excellent Quality Manufacturers

The quality of the PV products (PV modules and BOS components) in this case is good to excellent, which is ensured via thorough understanding, and or direct inspections and tests, of the manufacturing process. Here some of the risks, inherent to PV products are still present, but they are monitored and controlled by the customer and/or a third party, with the full cooperation of the manufacturer.

In all cases some long-term performance insurance will save us a lot of effort and money on the long run. And so, how much could the insurance industry help the suffering of the solar industry? This is yet an unknown (or rather still developing) factor, but we suspect that there will be some good examples, as well as some "quick buck" deals that will give another black eye to the solar industry and its insurance counterparts.

Nevertheless, the evidence of late points to the fact that solar power plants of the 21st century will be either self-insured, or insured by a third party.

One way or another, customers and investors are awaking to the fact that the impressive, big-name labels on the shiny PV modules and inverters do not ensure their long, trouble-free life (especially in the desert), and do not guaranty profits during the long and difficult trip. And so, the insurance industry, in our opinion, will have a decisive, if not the last, word in the world's solar energy game in the 21st century.

Certification (Performance) Insurance

And speaking of insurance, here is a new, different, more elaborate and reliable way to test and certify PV modules. It is a brand new service offered by a few companies, who claim to thoroughly check and test the three key aspects of the modules' quality, and namely; a) overall performance, b) individual cells' quality, and c) level of workmanship.

These factors basically do reflect the performance and reliability of PV modules, so the new tests would determine how good a PV module is, but also how a manufacturer is able to produce good PV modules on a large scale in mass production mode.

The tests are done via the conventional test methods, but a more thorough check and evaluation is also undertaken. And what is even more important here, and something that is simply not done in any enterprise thus far, is that 100% of the modules are tested in some way in order to avoid gaps in individual modules, and/or batch performance.

The results are then sorted and reported as follow:

Performance [Weight 34%]

Performance shows mainly the power output of a module, but power alone is not entirely representative of a module performance.

A module with higher efficiency will reduce the cost of the BOS (Balance Of System) by reducing the land use and the number of structure elements.

Under this category is evaluated the power output of each module, based on the result of a flash test at the Standard Testing Conditions (STC) but also at low irradiance to evaluate the power output in less sunny day.

In addition, the lot tested is evaluated as a whole or lot homogeneity as PV modules in the field will be grouped together to produce electricity and the weaker one will affect the power output of all the other PV modules.

Also evaluated under this category is the ability for the manufacturer to provide PV modules with positive tolerance on power, appreciated by the buyers.

Cells Quality [Weight 30%]

The performance is evaluated at the time of testing, but users of PV modules want to get a high power output from their modules for the longest time possible.

A solar cell containing cracks may perform correctly at the moment, but with transportation, mechanical stress and climate constraints, cracks and other cell defects will spread in the cell with time and result in a cell that does not produce any more electricity.

Under this category, the quality of the cells themselves is evaluated, or more exactly their electroluminescence picture.

Workmanship [Weight 36%]

Under this category is grouped the quality related to the making of the PV modules.

Assessment includes the visual appearance of the PV modules but also the safety to the user and the packaging as well as the quality of the sealing, construction and other parameters.

The workmanship score indicates the quality of the work that has been performed to produce the PV modules.

Additional Performance Insurance Factors

The mark given to each PV panel by the leading testing labs is the result of a thorough and detailed testing, during which the following factors are checked and tested step by step:

Positive Tolerance [Weight 5%]

A higher score is given to manufacturers providing PV modules with a positive tolerance for their PV modules. This seems to be easy, but the output will later be evaluated against this positive tolerance.

Output against Nominal Value [Weight 12%]

The power output is determined by flash test under STC (25°C, AM=1.5, 1000W/m2). To show the ability of the manufacturer to produce quality PV modules on a mass production scale, we take into account the average power of the lot, the minimum and the consistency (homogeneity in power) of the whole lot.

Efficiency [Weight 6%]

In order to save cost of system and space, the PV modules need to provide the maximum of power on the smaller area. A higher score is given to modules with a higher efficiency.

Efficiency at Low Light [Weight 6%]

Modules are expected to perform well in sunny weather but also during less sunny days. The power output is therefore determined under low irradiance (25°C, AM=1.5, 400W/m2) to evaluate its performance under these conditions.

Inefficient Area [Weight 10%]

Appearing as black area under EL test, these areas of the cells do not produce power and reduce the power output of the module. Another risk is heating of these areas resulting in hotspot (burns on the modules)

I-V Curve Appearance [Weight 6%]

When natural sunlight is simulated through a flash test machine, the response from the module in terms of current and voltage is well defined. Bumps and accident in that curve show some problems in the function of the solar cells or the module itself.

Cracks [Weight 10%]

Crack in solar cells are a threat to the reliability of the PV modules with time. Cracks in cell are identified under electroluminescence test.

Pollution [Weight 5%]

Processing issues during solar cells manufacturing might result in less efficient area which in turn can results in hotspot problem or lower output of the cells. These defects are identified under electroluminescence test.

Soldering Defects [Weight 5%]

Processing issues during solar cells manufacturing might result in less efficient area which in turn can results in hotspot problem or lower output of the cells. These defects are identified under electroluminescence test.

Packing [Weight 5%]

If packing is not always considered as a major item by manufacturers, it crucial to protect the PV modules during long transportation and, of course, user-friendly.

Marking [Weight 3%]

Required by IEC standards and mandatory to enter European or US market, marking must include specific information and be legible.

Construction [Weight 3%]

Under this category are taken into account both the design of the PV modules construction and the quality of the assembly.

Dimensions [Weight 5%]

Installers are basing their structure and installation on the dimensions of the PV modules. Dimensions such as length, width, skew and compliance of mounting holes are measured.

Frame Assembly [Weight 5%]

Poor docking of the frame elements can result in sharp edges putting the user at risk. The presence of these sharp edges and quality of the work is evaluated.

Sealing [Weight 5%]

In order for the PV module to perform for more than 20 years, the encapsulation of the cells and the sealing must be waterproof to prevent moisture from entering and corrode the inner elements.

Cells layout [Weight 3%]

Spacing between cells and between strings is important to avoid any risk of short circuit inside the modules. A minimum distance between the electrical conductors and the frames must be respected as well to avoid electrical shock risks.

Stringing [Weight 2%]

In order to perform correctly, the cells must be correctly connected to each other. The conductors have to be well soldered and properly laid to ensure good conductivity in time.

Component Appearance [Weight 3%]

As end users are more and more concerned with the visual appearance of the products they buy, cosmetic defect are taking a higher importance in the choice of the PV modules. Components such as glass, frames, junction box and cables are evaluated.

Components Usage and Function [Weight 2%]

When tested against IEC standards, the PV modules are a combination of different components.

Taken separately, some components might seem equivalent, however, the combination of all these components might not work properly. This is the reason why it is mandatory for the manufacturers to respect the use of the specified components. The function of these components such as connecting function of connectors is also evaluated.

In conclusion, the test labs involved in this type of thorough checks and tests are looking for a set of product issues related to the overall performance and reliability, as follow:

Performance Estimates

One of the major aspects of the solar products and projects is their ability to deliver the expected power for the duration. We can only estimate what will happen during year #20, allowing for an ample room of error, which gets even larger by year #30.

The maximum power determination is performed according to IEC 60904. Even though, IEC 61215 requires the output power determination to be performed on class B sun simulators, AAA Class equipments should be used to obtain the most accurate and reliable results.

The tests are performed in a controlled environment, to guarantee that the tests are performed in conditions as close as possible to the Standard Test Conditions (STC). During this test we not only control the output power, but also check each electrical characteristic as well as the shape of the IV curve to identify any intermittent malfunction of some solar cells.

Individual Cells Quality

One guarantee that the modules will perform well in time, is that the solar cells are free from defects such as cracks, pollution and so on. Those defects are invisible to the naked eye.

The most efficient way to detect this type of defect is to use the principle of electro luminescence: In practice, the solar cells are transforming the light (photons) that they receive into electricity. The solar cells also have the reverse property.

When subject to a current, the semi conductors will emit light through the emission of photons. The electro luminescence equipment is using this property to perform an "x ray " like picture of the module where the dark areas show the less efficient part of the cells.

Workmanship

As a lot of operations remain manual during the production, the quality of workmanship can vary considerably from one module to another.

The quality of workmanship will not only affect the appearance of the module which has become a criterion of choice for the end users, but it can also affect the safety of people installing the products or the reliability of the products in time.

We perform a thorough inspection of the construction, focusing on the dimensions the quality of sealing, the components used and the packaging.

Note: The final performance tests (flash tests) are done via solar simulation tests. IEC 60904-9 defines 3 categories of sun simulators A, B, C (A being the best class) based on 3 different criteria:

Temporal Instability

In order to be accurate, the measure has to be performed with a stable irradiance (light sent to the module by the simulator) during the whole period of testing (the time to take the entire I-V curve).

The temporal instability refers to the difference between the maximum and the minimum irradiance in percentage. The lower the instability, the better.

Spatial Non Uniformity

Equally, the irradiance provided by the sun simulator shall be the same all over the testing area.

The spatial non uniformity refers to the difference of irradiance between different points of the testing area. The lower the non uniformity, the better.

Spectral Match

As the natural sunlight is composed of different wavelengths, in order to provide a good evaluation of the behavior of the module in the field, the sun simulator shall be able to recreate the composites of the natural light.

In nature, different wavelengths of light will participate in a certain proportion of the irradiance. The spectral match is the ability of the sun simulator to provide irradiance similar to the natural light for given wavelength ranges or its deviation from the reference irradiance for different wavelengths.

Summary

In summary, lab tests done by qualified labs using thorough, sophisticated, 100% checks and testing of the product, using the best equipment and procedures available today, is a step forward in the battle for efficient and reliable product, which must withstand 30 years of extreme heat and humidity.

If all PV modules coming to the energy markets are tested this way, we will have to take back most of what has been said thus far about suspected poor quality of modules and PV installations, and the related future PV modules failures.

Unfortunately, this type of testing is,

a)   not done by 99.9% of the manufacturers, and
b)   although 100% thorough check and testing routine, as described above, would increase significantly the long-term reliability and durability of the PV modules, there is still a chance that many modules would fail with time, due to latent problems created during the manufacturing process—many of which are not noticeable at first glance.

This is simply because there are dozens (if not hundreds) of different materials and process steps with many combinations and permutations of different conditions, which affect the performance and quality of the modules. Not ALL of these can be checked and tested even with the most vigorous and thorough test methods available today. The key word here is, TODAY.

Today we simply do not have the technology, the know how, and even the will, needed to take a very close look at the inner workings of PV modules—especially during their actual under-load field operation, which is the most relevant part of their function. If we are able to do that, we would be able to see the actual performance issues, and even predict their long-term behavior and potential failure mechanisms.

This, however, is today, and we just cannot do that. We are, however, absolutely sure that it will be a routine operation later on in the 21st century, when engineers, equipped with special equipment, will be able to see *in real-time* exactly what is going on *inside* of each and every module operating in their power plants. This way they would be able to correct any and all problems in a timely fashion and avoid expensive and embarrassing surprises in the solar fields down the road.

This is a dream today, of course, so until then we would at least hope that more modules are checked and tested as described above.

The Future of Performance Insurance

A report on the subject was issued by Bloomberg New Energy Finance in August 2013, which estimated that the renewable energy industry could be spending three times as much on performance insurance by 2020. If and when fully implemented across the entire energy

industry, such insurance would mitigate risks to projects and convince investors that the renewables are here to stay and fighting for a equal place in the energy markets.

The report, looked at six of the world's leading markets for solar and wind; Australia, China, France, Germany, the United Kingdom and the United States. Based on different factors and scenarios, Bloomberg estimates that insurance premium volumes in these markets will increase from $850 million today to $1.5-$2.8 billion by 2020. The likely drivers for this increased demand is the growing industry scale and growing interest in the sector from more risk-averse investor groups.

New global renewable power capacity by 2030 will grow to more than $2 trillion in total investment, of which 75% (900 GW) of capacity additions will be in the solar and wind sectors. Most of these developments are expected in the above mentioned major global markets, where the demand for risk management is growing as owners and developers of renewable energy projects are looking into new financing options. Of special interest are large groups of institutional investors such as pension funds and other large money managers.

Wind and solar projects are technological complex, and are exposed to adverse weather effects through the duration. Physical damage to the projects infrastructure, unplanned delays, and unavoidable downtimes are the norm, which can significantly reduce the returns on investment.

There are also construction and operational risks as well, and changing market conditions, where the government subsidies and incentives going away. The new risks are many, including balancing charges, curtailment due to grid constraints, counterparty risk, possible retroactive policy changes, and power price volatility.

Project owners are finding ways to manage these risks internally, but are also looking for third-party services to assist them in their choices as new insurance products are developed.

Convincing them of the advantages of renewable energy requires also a new and better way to assure them that all risks of wind and solar are accounted for all through the life of the projects. According to the new Bloomberg report, this is doable and is happening already.

"The analysis conducted for this report shows that the demand for risk management solutions will grow, partly because the renewable energy sector will simply get bigger, but also because of increasing uncertainty affecting power markets in general. As the renewables mature and become part of the mainstream energy industry, they will need to evolve from an innovative sector where risks are taken on the chin, to one where returns are predictable and there are fewer surprises" according to Bloomberg New Energy Finance, which issued the report

"Insurance is not a silver bullet. But by mitigating the risk in the construction phase and improving the consistency and surety of revenues during operation, insurance can help improve the return on investment for renewable energy projects. This, in turn, would allow the sector to attract the scale of investment necessary to put the world's energy mix on a more sustainable footing.," according to the Swiss Re Corporate Solutions, which commissioned the report.

In more practical terms, all the talk about performance insurance is directly related, and boils down to ensuring the quality of the final product by strict control of the down-stream materials, processes and labor during the manufacturing, construction, and O&M stages.

The first step in ensuring quality performance in the field during 25-30 years of non-stop operation is making sure that the renewable energy products are of the highest quality possible.

The quality battle starts at the beginning of the manufacturing process with obtaining and maintaining the quality of all incoming process materials and components used in every production step. Most of these commodities come from third party suppliers, we call the supply chain, which is a critical part of ensuring the quality of the final product.

Then we follow the product through all stages of its life cycle, all the way until it ends up in the garbage dumps. Any gaps in, and between, the different steps represent risks and can affect the bottom line. To minimize the damages, proper measures—such as having adequate performance insurance—must be taken.

Summary

Insurance companies run a great risks in insuring any renewable energy projects. There are many reasons for this anomaly, which vary from location and technology types. Below we summarize the major issues insurers encounter in evaluating and assigning risk on renewable companies and projects:

Hydro Power
- Damage to turbines due to water ingress
- Electrical failure
- Flood

- Impact on river flow and ecosystems
- Land use and displacement of communities

Geothermal Energy
- Pump and controllers failure
- Heat exchanger failure
- Generator failure
- Some risks if land use not well managed
- Use of chemicals, but risks managed by closed loop systems

Wave and Tidal Power
- Fatigue failure of structure
- Water ingress and damage
- Power cable failure
- Impact on marine environment
- Impact on shipping
- Sensitive siting

Bio-Fuels
- Agricultural land use
- Effects on produce prices
- Limited volume
- Questionable environmental benefits

Photovoltaics (PV)
- Variable power generation
- Inverter failures
- Moisture ingress
- High temperature
- Storm damage
- Premature failure
- Hazardous materials
- EOL recycling

Concentrated Solar Power
- Variable power generation
- Fire danger
- Heat exchanger failure
- Receiver failure
- Turbines failure
- Storm damage
- Lack of water in desert areas

Onshore Wind Power
- Fatigue failure of main rotating parts
- Electrical failure
- Cable damage
- Potential biodiversity impacts
- Visual impact
- Sensitive siting

Offshore Wind Power
- Fatigue failure and corrosion damage
- Electrical failure
- Blade failure due to lightning strike
- Potential biodiversity impacts
- Potential impact on shipping
- Sensitive siting

EQUIPMENT MANUFACTURING

Many different pieces of equipment are used in the production of fossil fuels and renewables, their transport and subsequent conversion into heat, or electric power. The equipment is usually designed and manufactured by a number of companies, which have a large supply chain of parts and consumables that are used daily in their operations.

The equipment we are talking about is
- Production equipment (equipment and spare parts manufacturing)
- Transportation equipment
- Power generation equipment, and
  — Power transmission and distribution equipment

There are many companies and industries involved in manufacturing the equipment in the above list. We cannot possibly cover every single piece of equipment or company that makes it, so we would just cover the key segments of this peripheral market.

Fossils Fuels Production Equipment

Coal mining is the most intensive fuel producing business, in terms of size and volume, today. Presently there are about 2,500 coal-related businesses in nearly 500 categories globally, generating approximately $90 billion in annual revenue. While the coal mining in the U.S. is slowing down, thus reducing the demand for mining equipment, China, India and other developing countries (along with some developed) are projected to account for over 75% of all mining equipment purchases in the next several decades.

Mining drills and breakers are the fastest growing equipment segment, which reflects their universal use for breaking through the ground and subsurface materials. The second strongest demand is posted by crushing, pulverizing and screening equipment, which are widely used in both surface and underground mining. Although demand for surface mining machinery is growing at a slower rate, they will continue to account

for the largest single share of equipment demand in the future.

While the major part of the coal, oil and gas markets are in the production and sale of the fuels (by coal, oil and gas companies), there are many other industries involved in the process of bringing these products to market. Some of the other key players are equipment manufacturers and other supply chain participants, as well as the power producers (who use the fuels to generate power) and power distributors (who sell the power to the public).

The equipment manufacturers are usually large companies that design and make the equipment and/or the supplies used in coal, oil, and gas production operations, which heavily depend on specialized equipment for efficient and profitable operation. The equipment manufacturers, on the other hand, depend on the fuel producers to buy their equipment and services, and replacement parts for subsequent maintenance work. A perfect win-win situation…

A number of construction and engineering companies are also involved in the different stages of field exploration, construction and maintenance of coal mines and drill sites. Although most of these companies are not directly involved in the mining or drilling process, together, they represent a large chunk of the U.S. and global economies and also use a lot of heavy duty equipment.

As we saw in the previous chapters, mining, drilling, and transport equipment for the most part consists of very large pieces of machinery and structures, such as cranes, oil tankers, trucks, drag lines, diggers, train cars, etc. used in these operations. The equipment used in power generating plants is also very large and very expensive.

The fossils fuel production, transport, and use supports the large equipment manufacturing industry, consisting of a very large number of equipment and parts manufacturers. The variety and volume of products made by these companies (in size, weight and money) is tremendous, and is the foundation of one of the largest global niche markets.

Thousands of tons of steel, plastics, and other materials, miles of wires, and a lot of electro-mechanical components are used to make energy production equipment. The energy used, and the pollution emitted, during manufacturing and transport of these huge pieces of machinery are significant as well. We cannot look at a coal mine, an oil rig, or any power plant without thinking of what it took to make, transport, install, and maintain all the pieces of equipment used in these operations.

Mining Equipment

Mine operations require a number of specialized pieces of equipment, as needed for digging, loading and transport of coal from the mining site to the surface and beyond.

Table 8-16. Mining equipment specs and prices

Equipment	Specifications	Weight in lbs.	Cost in $
Dragline	55 cu yd bucket, 250 ft. dump height	16 million	$100 million
Shovel, hydraulic	5.2 cu yd bucket 23.6 ft (7.2 m) dump height	131,000	$925,000
Loader, wheel	9.0 cu yd bucket, 12'1" dump height	114,000	$720,000
Truck, rear-dump	60 ton, 46 cu yd, mechanical drive	22,000	$120,000
Drill, rotary (crawler)	5.13" to 7.88" hole, 25 ft drill length	30,000	$600,000
Tractor, crawler (dozer)	13.7' maximum blade width	39,100	$300,000
Grader, road	14 ft blade width	52,200	$450,000
Truck, water	5,000 gallon water tank	20,000	$255,000
Truck, service	Off-road tire service truck	15,000	$55,000
Truck, shot loader	1,000 per minute capacity	15,000	$75,000

Mining equipment is usually very large, really large, and not cheap. When all additional expenses related to equipment transport, assembly, maintenance, disassembly, EOL disposal, etc. are added, we get many more billions of dollars spent on mining equipment (after the initial purchase) through its long life in the mines.

Well Drilling and Power Plant Equipment

We describe in detail the equipment used for well drilling and power generation in the previous chapters, so it suffices to say that it is important part of the energy sector. In most cases, these are large, complex, and expensive pieces of equipment. Usually, it takes several equipment manufacturers and significant amount of time and effort, to design and make the pieces.

Thus manufactured components are then loaded on tracks, trains, or ships for transport to the power plant. Here, they are unloaded and installed with great care. The entire process requires the participation and full cooperation of thousands of people.

Renewables Production Equipment

Most of the renewable do not require any fuels (except biofuels, which need biomass). Instead, the renewable companies manufacture products that are used for generating electricity. For example, wind power companies manufacturer large wind turbines, while solar energy companies manufacture PV modules, or troughs for thermal energy conversion.

In all cases, these companies purchase the production equipment that is used to make their products from third party vendors. This equipment is then used to make the final products—wind turbines, solar modules, etc.

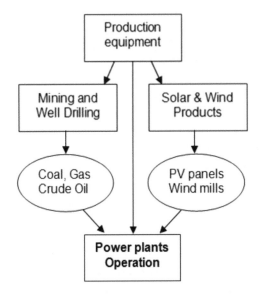

Figure 8-23. Production Equipment Use

Figure 8-22. Mining equipment…try it for size.

The production equipment in most renewable energy manufacturing operations is not that large, albeit it is quite complex and expensive. One exception, where very large pieces of equipment are used is in the large furnaces and earth moving equipment needed for producing silicon material from sand, and that used in mining operations. Wind mills manufacturing equipment is also formidable in size, where the blades manufacturing, for example, huge (up to 200 feet) machinery.

In addition to silicon, large quantity of glass, aluminum, steel, silver, copper, cadmium, tellurium and other minerals and chemicals are also needed in the renewables' manufacturing processes, all of which are mined and processed using large pieces of machinery. The transport of the materials and the finished product also requires large vehicles.

All in all, manufacturing the equipment and vehicles needed for energy production is a big business, spread over several market niches and general market areas. The production equipment manufacturing industry, serving the energy sector is quite large and provides work for thousands of people.

Many of the production equipment manufacturers are specialized and totally dependent on orders from coal and oil and gas producers, or from wind and solar energy manufacturing companies. Others specialize in power plants equipment or transportation vehicles.

We are taking a close look at the solar modules and wind mills manufacturing in the previous chapters, so here we will only touch on the basics of production equipment used in these operations.

Solar Equipment

Solar cells and modules are manufactured in special facilities, where modern production equipment is installed and operated, usually on 24/7 basis. The production equipment is made by different companies around the world, transported to, and installed in, the production facility.

Millions of solar cells and modules are then manufactured using the newly installed, state-of-the-art production equipment. Different raw materials; different ores, metals, chemicals, and gasses are brought to the plant to be processed and converted into cells and modules.

Half of the cost of installation of PV power plants today is the hardware; solar modules and the related equipment. The solar (PV) module prices fell significantly recently, from about $3/Wp in 2008 down to $0.50/Wp in 2015. This phenomena is due mostly to pressure from falling Chinese PV module prices, which are heavily subsidized by the Chinese government.

The solar power installation costs vary with different technologies and locations, but the average residential PV installation in Arizona was $8.0/Wp in 2008, dropping to $6.0/Wp in 2010, leveling at approximately $5.0/Wp in 2012, and down to $4.50/Wp in 2015. The prices for large-scale installations are lower, and are expected to fall even lower by 2020.

Price of Chinese PV modules contributed significantly to the overall PV installations price reduction, and there is a chance that prices might go up, if and when Chinese panels get more expensive, due to the recent DOC restrictions, and/or due to natural price leveling as the global PV markets stabilize.

One of the major factors in the reduced price of solar and wind products coming from China is the fact that the production equipment used in these operations is made in China. It is heavily subsidized by the Chinese government, which means very low prices, compared to those on the global markets.

In addition, the majority of the Chinese made production equipment was copied (illegally in many cases) from U.S. and European models. This allows the Chinese manufacturers to bypass the most expensive and risky product development stages and keep their prices lower than the competition.

Note: More details on production equipment used in these operations can be found in the chapter on Renewable Energy. We would also suggest our books on the subject: Photovoltaics for Commercial and Utilities Power Generation, 2011, and Solar Technologies for the 21st Century, 2012, published by The Fairmont/CRC Press, where we take a very close look at solar cells and modules, and the production used in their manufacturing.

Wind Turbines

Wind turbines components are manufactured by original equipment manufacturers (OEMs) who design, assemble, and brand the different parts and assemble them in sub-assemblies, which are shipped to final assembly at the wind turbine manufacturer plant, or for assembly in the field.

Wind turbine manufacturers operate similarly to automobile assemblers that assemble car or truck with parts that are partially made by the OEMs and partially in house. So wind turbine manufacturers are system integrators. They must bring together an estimated 8,000 precision made parts and components to produce a wind turbine.

One supplier might roll large plates of steel into the towers that support the turbine. Another company makes the turbine blades from special carbon fiber materials, and a third might manufacture the electronic computerized control systems.

Each of these components might be produced and assembled domestically from imported inputs, or might be imported as an assembled product. In all cases, the wind turbine manufacturer puts together as many sub-systems as possible in their production plants. The assemblies are then shipped to the field for final assembly into a wind turbine.

The OEMs and specialty firms are part of a complex global supply chain. Tier 1 suppliers make large components such as towers, hubs, blades, or gearboxes. They include firms such as LM Wind (blades), SKF (bearings), and Winergy (gearboxes). Tier 2 suppliers produce subassemblies such as ladders, fiberglass, control systems, hydraulics, power electronics, fasteners, resin, machine parts, or motors. They include companies such as American Roller Bearings (power transmission bearings), Cardinal Fasteners (structural fasteners), and Timken (power transmission bearings).

A wind turbine is a significant investment. Price quotes range from as low as $900/kW to a high of $1,400/kW, so an average 2 MW turbine would cost between $1.8 million and $2.8 million, plus transportation, construction, installation, testing, insurance, operation, etc. costs.

Each wind turbine assembler uses different sourcing strategies and levels of vertical integration. Some produce almost all major components internally or through subsidiaries, while others outsource many of their critical components. For instance, some manufacturers produce blades, generators, or gearboxes in-house, while others opt for outside suppliers. Hundreds of smaller companies make specialized parts such as clutches, rotor bearings, fasteners, Sensors, and gears for the wind industry.

Very high levels of expertise and specialization are required of wind turbine suppliers, with the level of precision similar to that of the aerospace industry. Turbine manufacturers often establish relationships with suppliers in the interest of quality, as a failure in a turbine part can be very expensive to fix. Wind turbines are expected to survive largely unattended in extreme climactic conditions for a design life of as much as 20-25 years of non-stop operation under any weather conditions. Product quality is also of concern to wind farm operators, as a malfunctioning turbine can reduce operating revenue and even cause a project to fail.

U.S. Wind Turbine Manufacturing Facilities

At the end of 2011 there were more than 470 wind turbine manufacturing facilities in the United States, up substantially from the 30-40 wind-related manufacturing facilities nationwide 5-6 years earlier. Over that period, the number of tower plants increased from 6 to 18; blade facilities rose from 4 to 12; and, nacelle assembly facilities grew from 3 to 14.

Today there are more than 550 such manufacturers in the U.S., an increase due to the fact that overseas manufacturers cannot compete with the locals. The transport of the huge structures over long distances is too expensive, which allows the U.S. wind industry to flourish.

In addition to the cost savings related to transportation, and concern about the risks associated with currency fluctuations, greater demand for wind turbines is a big reason for the increase of wind turbine and component manufacturers in the U.S.

Even with increased domestic production capacity, wind turbine assemblers source parts and components from global suppliers, which reflects the industry's global supply chain nature. Many wind manufacturers with production facilities in the United States also produce elsewhere, typically in Europe and Asia.

Total investment in facilities to manufacture for the wind industry in the United States has exceeded $2.0 billion.

ENERGY SECURITY

Providing physical and cyber security of energy generation, installation, distribution and use networks is opening a huge peripheral marketplace for the related products and services. What we see today in this area is the tip of the iceberg. As time goes on, cyber security will become a major business enterprise.

Multi-billion dollar markets are opening in everything related to security of energy production and use.

Since producing and using energy includes practically any and every industry, business, and person in the global economy, physical and cyber security issues affect each one of us. To start with, we see billions of dollars spent by governments on maintaining armed forces, one of the main goals of which is to keep the energy supply lines open. There are also a myriad of contractors and consultants assisting in the design, preparations, and execution of security activities, designed to keep the energy cycle undisturbed.

Physical and cyber security affect everything in our economy, with energy being the most important sector.

This is because without energy, all sectors would be affected and even paralyzed. Who knew? But it is a fact! Just think of one single industry that can run without energy input. Nope, there no exceptions. We all need energy (gasoline, diesel, jet fuel, and electricity mainly) to live a decent live and conduct normal business.

So, let's take a close look at what this hoopla is all about. What are we afraid of? What are the risks that we see in the energy sector that are so important, and why? The simple answer is that energy security is the key to our being able to turn the lights on, power our computers and drive our cars.

This is a huge subject. So much so, that we have published a 800 pages book on the subject titled, *Energy Security for the 21st Century*, by the Fairmont/CRC Press. Because of that and the limited space in this text, we will cover only the basics of energy security, as follow.

We divide energy security into two major sectors: physical security, and cyber security. Although these are totally different things, they are intermingled at times and depend on active intervention.

Physical and cyber security risks can also be divided into internal and external.

Internal Risks

There are a number of internal risks to our energy security. They are internal, because they are based, and conducted. on our national territory. The internal risks can also be divided into natural (caused by natural disasters and such) and man-made (caused intentionally or unintentionally by human activities).

Unforeseen natural events and man-made accidents in mines, transport routes, refineries, and power plants threaten to disrupt the energy supply chain. Mine accidents, triggered by natural events (earthquakes, floods, etc.) are still happening in the U.S. and around the world. They take lives and shut down mining operations temporarily or permanently.

Railroad cars and transport ships accidents and spillage can close a transport route indefinitely and do significant damage to the environment at the same time. Power plants are vulnerable to internal sabotage and terrorism, and the risks of this happening in the U.S. are increasing.

Many accidents of the past have resulted in property damage and loss of human life as well. Hurricanes damage and shut down refineries and power plants every year, which cause another set of problems.

As the state-of-the-art of the energy technologies matures, and we learn how to handle the natural and man-made accidents, some of the internal risks get more manageable. With that, the number of mine, railroad and transport ships accidents has been significantly reduced. Refineries and power plants, because of their complex and relatively fragile infrastructure, are vulnerable to the force of hurricanes and other natural events, and not much can be done to improve that situation.

The worst and most dangerous risks of all in the energy sector are these caused by nuclear power plant incidents, accidents, and failures. Just mentioning Chernobyl and Fukushima brings gruesome memories and forces us to imagine what would happen if a similar accident happens nearby. Nuclear accidents are the worst type of internal risk. Although they are an extremely rare occurrence, when they happen, the local devastation is total and permanent, and their effects are felt over broad areas—even across countries and continents.

Nuclear power safety is a critical component in considering our energy security and safety.

One single nuclear plant accident can not only disrupt our national energy supply chain, but could even destroy the entire area and kill and sicken thousands. This is a serious issue that needs to be taken into consideration as we consider our energy options for the 21st century.

Internal terrorism is the most dangerous types of internal terrorism, since it is one way to gain control and damage a refinery, or a power plant. There are a number of reported incidents, and even a greater number of unreported such.

Computer Hacking

Today computers run everything—including our energy infrastructure. Every step of our energy supply chain, from mining operations to transporting, processing, refining, and using energy sources is also monitored and controlled by computers. That creates a serious problem of computer hacking and terrorism activities, especially in nuclear power plants.

In 2012, a backdoor (unauthorized computer access) in a piece of industrial software used to control power plants, allowed hackers to illegally access a New Jersey power company's internal heating and air-conditioning system. One of the viruses was accidentally discovered after an employee called in IT technician to troubleshoot the USB drive. A simple virus check discovered sophisticated malware, capable of doing a lot of damage to the plant equipment.

In this case, the hackers could have gained access to the control of the boiler, disrupt the cooling water supply, and trick the system to think that everything is OK, until the boiler overheats and explodes. Such an accident could cost dozens of lives and billions of dollars in repairs.

While this is bad enough, if the hackers gain access to a the cooling system in a nuclear power plant, the overheated boiler might blow up like those in Chernobyl's power plant. Such an event would be followed by a disaster similar to Chernobyl too. In this case, it would be humans causing the damage intentionally, instead of unintentionally as in Chernobyl, or Mother Nature, as in Fukushima.

Since the computer safety technology is fairly new, the defense mechanisms have not been fully developed and/or deployed, as was in the above mentioned New Jersey, and many other cases. The workstations, in this case, lacked adequate backup systems, so the operators were lucky in discovering the malware before it was activated. We are sure that the software system at the plant has been modified to prevent any such incidents. The question is how long before the hackers find another back door?

Computer hacking is a full time business for many people and governments, so we must be aware of the threats

Another intentional malware attack, spread by a USB drive, affected 10 computers in a steam turbine control system of another power plant. The incident resulted in downtime for repair of the impacted systems, which delayed the plant restart by three weeks.

The Stuxnet worm and the Flame malware were developed by the US and Israel to spy on, and even control, critical systems in power plants of our enemies. In the summer of 2012, these programs successfully disabled an entire enrichment facility in Iran. These programs relied on USB drives to store the commands, propagate attack codes, and carry intercepted communications over computer networks. Microsoft has patched these vulnerabilities on Windows computers, but there are additional steps to be taken by power plant and other energy sector operators.

And of course, we cannot leave out the humbling experience of the massive computer security breach in December 2013 by foreign hackers who stole the identity and other private information of several hundred thousand Target and other U.S. companies' customers.

If hackers can get into one of America's largest and most prestigious chain-stores' computers, getting into utilities' control systems is only a step away.

As a matter of fact, hackers were able to get into several other chain stores and even some banks. Also during that time, critical control systems in two U.S. power plants were found infected with computer malware, spread intentionally by computer virus brought in the plants via USB drives. The infected computers controlled critical systems controlling power generation and cooling equipment.

Intentionally planted malware poses a real threat by allowing the attackers to disable key equipment, thus disrupting its normal operation, or destroying it all together. What a disaster that would be, if the virus is allowed to shut down the cooling system of a nuclear plant!

Physical Attacks

But it is not just computers that are the internal terrorists' target. In 2013, for example, a well organized and executed sniper attack on power grid components shut down the power in a large part of the Silicon Valley area, and threw residences and businesses into darkness for several hours and even days in some areas.

The sniper shot at several transformers in a local high voltage substation which malfunctioned as a result and caused the power grid in a large part of the San Jose-Fremont area to shut down. The repairs were started almost immediately, but it took some time to figure out what is happening and to repair the damages.

These examples are cautionary tales and lessons in safety procedures, as well as a call to action by owners and operators of critical energy infrastructure. New, 21st century security policies must be developed and implemented, in order to maintain up-to-date antivirus mechanisms, manage system patching and the use of removable media.

Other, site-specific protection measures must be implemented for ensuring the energy infrastructure's physical safety. There are weak points all through the power system—from power generation to delivery—which need to be evaluated and protected from physical and cyber attacks. This is not simple, not cheap task, but it has to be done, if we don't want to risk waking up some day in darkness, or even worse; in the midst of a nuclear disaster.

On the overall, internal risks are unavoidable, but we are well positioned to anticipate and respond in most cases. The internal security situation is basically under control for the most part, and we just have to add the necessary cyber and physical protection to key components of our energy infrastructure, be more careful in all we do, and not let our guard down.

The external risks are more serious, have greater consequences, and are much more complicated.

External Risks

While we have some control over internal physical and cyber attacks, these coming from outside our national territory are even worse. This is mostly due to the fact that the U.S. has made many enemies through the years.

Traveling around the world, one can see the signs of animosity towards the U.S. in almost every country. Some hate us because we have done bad things in the past (and God know we have). Others hate us because they are jealous of what we have achieved, and yet others hate us for the sake of hating.

Many groups have their own reasons and ways to inflict damage and pain to the mighty USA its energy infrastructure, and the people.

Figure 8-24. Terrorist attack on French oil tanker Limburg in 2002

Terrorist attacks targeting oil facilities, pipelines, tankers, refineries, and oil fields are referred to as "industry risks," because they are part of daily life in the energy sector. The energy market (crude oil in particular) is extremely vulnerable to external sabotage, due to extensive global oil transportation, and its exposure at the five oceans and their chokepoints.

Crude oil transport is on top of the external risks list.

The Iranian controlled Strait of Hormuz is a prime example of a chokehold, where one attack on a Saudi oil field, or on tankers in the Strait of Hormuz, could disturb the oil supply. A prolonged conflict in the area, would surely throw the entire world energy market into a chaos.

External risks have been always, and will continue to be, a big problem in the future.

Foreign terrorists spend a lot of time planning attacks on U.S. properties and personnel around the world and at home. This is the price we pay for being the world's greatest power and a shining example of a functioning democracy.

International terrorism affecting the world's energy reserves is of great concern lately as well, as evidenced by NATO leaders meeting in Bucharest in 2008, where international terrorism against energy resources was one of the key subjects. The group discussed the possibility of using military force to ensure the energy security of the region. One of the possibilities discussed include strategic placement of NATO troops in the Caucasus energy fields to police and protect oil and gas pipelines from terrorist attacks.

The U.S. energy supply and energy infrastructure are in the terrorists eyesight too, and attacks are, no doubt, planned daily in the terrorists' dens around the world. But how would they do that? We are too far away, and too powerful, to invade by sea or air. Instead, they do it by terrorism, which could be a physical attack on people or structures, but that doesn't work very well either, so recently they have been choosing different weapons and changing the battlefield tactics

The Risk Levels

So let's take a look at the risk levels of the different energy sources.

Figure 8-25 is a rough summary of the level of risk the different energy sources present to our energy security. Going from left to right, we see a decreasing level of risk-free operation of the different energy sources.

• Solar, wind, and geo-power are relatively risk-free energy sources, simply because the fuels (sunshine and wind) are readily available and so these technologies, do not depend on any additional external factors. There are, of course, some inherited problems, such as the variability of sunshine and wind currents. There are also other risks, such as materials availability and cost, but these are few and between and are easily manageable for the most part.

Terrorism on large-scale solar or wind power plants is possible, but not easily done, which

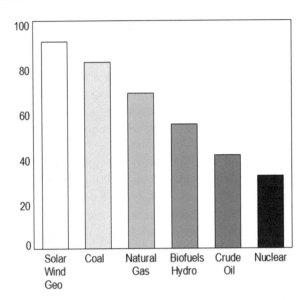

Figure 8-25. Energy security risks and dependencies (100 is risk-free)

makes these energy sources virtually risk-free and highly reliable ad infinitum. We foresee them becoming major power generators in the future, as their technologies mature and the fossils get more expensive. They are the best (if not the only) bet for the future generations.

• Coal is also relatively risk-free as far as its production, distribution, and use are concerned. Apart from its excess pollution, and mine accident problems, coal is, and will be, readily available and cheap enough for reliable and risk-free operation for a long time.

Its depletion in the more distant future, however, is imminent, so it is only a temporary solution to the present-day energy issues. Let's not forget that some day soon—most likely during the next century—there will be no more coal on this Earth. Terrorists can not make coal disappear, but we are doing it quite successfully ourselves.

• Natural gas presents an increased level of risks, due to its nature and the related production and distribution issues. The transport of natural gas, via pipelines and different land and ocean based vehicles increases the risks of natural accidents and criminal activities.

But even more importantly, the U.S. natural gas distribution infrastructure—the thousands of miles of gas pipelines running under our streets and homes—is old. In some places the pipelines are over 100 years old, and are developing leaks and creating all sorts of problems. The incidents

of gas line failure are increasing and getting more violent. The last and most revealing example is the gas explosion that destroyed two apartment complexes is New York city and killed 8 people.

Natural gas is also a temporary solution to our energy problems. No matter how exiting the latest natural gas bonanza in the U.S. might be, we should not forget that Its quantity is limited and that sooner of later—and surely sometime this century—it will be completely depleted.

• Biofuels and hydro power share the same unfortunate dependency on natural events—extended draughts and uncontrolled climate changes and weather patterns threaten to the reliability of these energy sources. Biofuels also compete with the world's food supply and as the global population increases, the competition between crops for biofuels and food will grow too.

Cellulosic biomass has a great potential as a future energy source, but it is too expensive for now. It, however, will become more reliable and cheaper in the future, as the technology is improved and optimized.

• Crude oil, and its imports in particular, present a major risk to our energy security. 50% of the oil used in the U.S. is imported. It comes from some of the most volatile and unfriendly areas of the world, and is transported through some of the most dangerous chokepoints today. There is a constant threat to the production levels in some countries, in addition to the risks of the distribution channels, thus its import is a risky proposition.

But most importantly, let's not forget that crude oil production is entering a period of decline. The prices will keep rising, and the quantities will diminish steadily. One thing is for sure—there will be no more crude oil some day—and surely sometime this century.

• Nuclear power presents serious dangers and risks to our energy security and personal safety. Computer warfare is the name of the game today, so a terrorist sitting in a cave in Afghanistan, or in a high rise in Beijing, could gain access to a the computer controls at a power plant or refinery and simply put them out of commission, or worse.

Nuclear reactors are potential terrorist targets, since they are not designed to withstand attacks by large aircraft, rockets and other air-born weapons. A well-coordinated attack, using powerful weapons, could damage the reactors in a nuclear plant, which will in turn have severe consequences for human health and the environment.

A recent study concluded that such an attack on the Indian Point Reactor in Westchester County, New York, could result in high radiation within 50 miles of the reactor, and cause 44,000 deaths from acute radiation sickness. Additional 500,000 long-term deaths from cancer and other radiation-caused illnesses would be expected as well.

Terrorists could also target a spent fuel storage facility by using high explosives delivered by ground or air vehicles. This would also results in radiation contamination of the immediate area. A terrorist group may infiltrate the personnel of a nuclear plant and sabotage it from inside. They could, for example, disable the cooling system of the reactor core, or drain water from the cooling storage pond. An internal attack is perhaps the most likely, and most dangerous, terrorist attack on a nuclear-power reactor.

There is also an inextricable link between nuclear energy and nuclear weapons, which pose the greatest danger related to nuclear power. The problem is that the same process used to manufacture low-enriched uranium for nuclear fuel, also can be used for the production of highly enriched uranium for nuclear weapons.

Expansion of nuclear power generation could, therefore, lead to an increase in the number of rogue states with nuclear weapons, produced at their "civil" nuclear programs. This is exactly what we are seeing being played out in Iran and possibly North Korea. The use of nuclear power would, at the same time, increases the risk that commercial nuclear technology will be used to construct clandestine weapons facilities, as was done by Pakistan in the recent past.

Today, the majority in Japan is against nuclear power. As a result, Japan's nuclear plants are shut down and the GHG emissions increased by 15% as more fossils are used for power generation in Japan since Fukushima.

One more, even much smaller, nuclear accident on the island of Japan will surely put an end to nuclear in Japan forever. In order to prevent that from happening, Japan's nuclear plants spent billions of dollars for upgrades with the latest hardware and security measures, as needed to fight Mother Nature and eliminate any man-made accidents.

But even without any additional accidents, nuclear power in Japan will be under constant and intensive watch, for a long time. Its destiny around the globe is also uncertain, and will be determined mostly by its safety record during the years to come.

Price Manipulation

Starting with the 1973 oil embargo, the world has been subjected to terrorism of another kind—that of intentional manipulation of quantity and price of crude oil on the global markets. Saudi Arabia and other OPEC countries have been trying from time to time to control world events by fluctuating price and availability of crude oil exports.

In 1973 OPEC nations reduced the oil imports, which brought up high prices. Today they are doing the opposite; flooding the market with low-price oil.

OPEC has been bent on controlling the global crude oil prices since its creation, and will continue doing so until its disintegration. In all cases of abnormal pricing, the U.S. economy takes a beating. High crude oil prices increase the cost of doing business, and cause price increases of most commodities. Low oil prices help the economy on one hand, by providing cheap gasoline and diesel fuel for people and industries. On the other hand, however, low prices stifle the domestic oil and gas production sector.

While, the U.S. has learned how to handle the oil price and availability fluctuations, some countries cannot. Falling oil prices are of particular concern to Russia and other oil producing (non-OPEC) countries. Russia, for example, needs high oil prices to keep its economy afloat. Recently, its economic performance has been slow due to sanctions over Ukraine and weakening domestic demand. The Russian Central Bank forecasts growth in 2014 was estimated at a meager 0.4%, improving marginally to about 1.0% in 2015.

Russia's biggest problem is that its budget requires global oil prices to average at least $100 a barrel in order to cover the government's spending promises. The government already needs to borrow around $7 billion from foreign investors next year and as much as 1.1 trillion rubles ($27.2 billion) from domestic investors. Given the country's sanctions-imposed isolation from international bond markets, any additional borrowing would be a big concern for policymakers in Moscow.

Even OPEC countries need oil prices of well over $100 a barrel to survive. Venezuela, for example, needs $121 a barrel to break even, according to the experts. Because of that, its economy is in a state of free fall at $60 oil prices. Venezuela is asking its larger OPEC partners to help by increasing the oil prices, but Saudi Arabia is prepared for at least two more years of keeping the global oil prices low. How low, remains to be seen, but regardless, this is energy price manipulation of global proportions that serves one country's interests. Is this the new form of terrorism?

ENERGY EFFICIENCY

Energy efficiency is the new game in town and a welcome addition to the peripheral energy markets. It is saving millions of tons of coal and natural gas, while reducing the GHG emissions. It is also creating new industries and providing thousands of job nationwide.

Energy efficiency refers to reducing the amount of energy usually required to provide essential energy products, services, while not affecting much the comfort we enjoy in homes and businesses. This is done by different means; from more efficient appliances to insulating houses and buildings.

Making power plants more efficient in terms of using less fuel is another major effort that brings greatest results. It, however, is the hardest and most expensive part of the energy efficiency effort.

Insulating a home, on the other hand, is easy and cheap to do, and reduces the energy use by requiring less heating and cooling to maintain a comfortable temperature inside. Using compact fluorescent lights (CFL),

and/or natural skylights reduces the amount of energy required to light a room by 75%, as compared to using traditional incandescent light bulbs. CFLs also last 5-10 times longer than their old incandescent cousins.

Great energy efficiency increases are made possible and affordable also by the introduction of new and more efficient technologies and production processes, and/or the application of new and proven methods of eliminating energy losses.

The key motivations to improve energy efficiency are: reducing energy costs to the owner, and cost saving to customers. Energy efficiency is also one of the solutions to reducing carbon dioxide emissions.

Energy efficiency is also seen as a pillar in national energy security, because it reduces the energy imports from foreign countries. And equally importantly, it also slows down the rate at which domestic energy resources are used and depleted.

Figure 8-26 needs some clarification. The U.S. is perched way high up on top of the world as far as its high productivity is concerned, but at the same time it is near the bottom of the global energy efficiency. Could it be that Bangladesh is more energy efficient than the

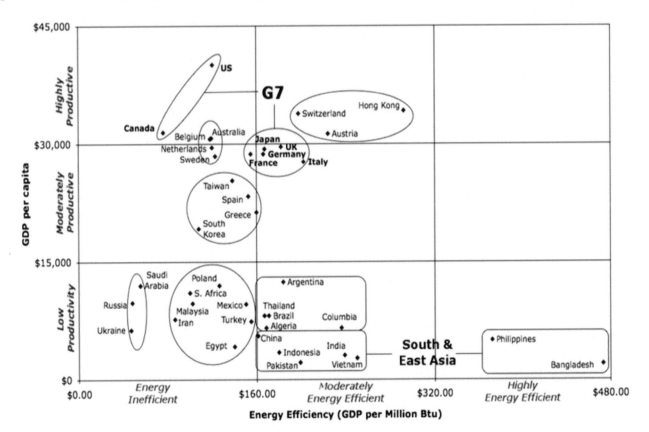

Figure 8-26. Energy efficiency vs. GDP

mighty U.S. of A.? Yes; about three times more efficient. Why?

It is the way life in this great country is structured. We are very smart and productive, but do not know how (and/or don't care) to conserve energy at home, or at work. Bangladesh, on the other hand, has no energy reserves and people are not used to, nor can they afford (even if they had), wasting energy like we do.

> *Optimized energy efficiency in buildings, industry and transportation could reduce the world's energy needs by 1/3 by 2050. It could also help control global GHG emissions by an equal amount.*

So, yes, energy efficiency is an important part of the global energy markets and as such deserves a closer look.

The U.S. Energy Efficiency Syndrome

Although the energy debate in the U.S. is escalating, and there are some good results to show, we don't expect any major developments on that front in the near future. Energy efficiency is an effective word, thrown around by politicians and industry experts alike, mostly when it is convenient to do so. We all agree that energy efficiency is one of the keys to our energy future, and yet there are several obstacles in our way to an energy bliss.

This is due to two main factors:

• *Infrastructure.* The way the U.S. roads and transport infrastructure are designed and built is very differently than those in most other countries. The vast territory of the 48 continental states is scarcely populated, so public transport is not the best option. Although there are trains, planes, and buses running all over the place, they account for a very small percentage of the human transport. Instead, most people prefer to drive from place to place, and even from state to state.

Mass transport systems exist in a few major cities, but in most U.S. cities and rural areas it is non-existent. Because of that, people drive everywhere around town in their own cars.

The interstate highway system is so well developed that one can make 100-200 miles drive in a car faster than in a train, bus, and in many cases even faster than a plane. Not to mention that this is a much more convenient and pleasant way to travel. And so millions of people do just that; drive and drive!

Nevertheless, it would not be that difficult or expensive to modify the national mass transportation system, so that more people use it. Trains and busses crisscross the cities and the country half-empty, so filling them would boost our energy efficiency. This, however, is unlikely to happen anytime soon, because of the other—even more important—factor; our life style.

• Our Way of Life is very different from that of any other humans on Earth. We are used to having our cake and eat it too. Like spoiled children, we want everything to be done our way, and done now or sooner. And amazingly enough, we usually get it that way too.

These high expectations and the resulting high standard of living come at the cost of long hours of stressful work. However, the culture has become so entrenched in this system that most people feel there is no alternative. After all, it is a small price to pay for our almost unlimited freedoms and comforts.

So, considering the above issues, we all must agree that a major change is needed.

Major Change…?

For generations, Americans have enjoyed living in large houses with large appliances, and large brightly lit rooms maintained at constant temperature all through the year. Since the 1950s, Americans have been driving large cars, trucks, and SUVs without regard to price or anything else. The automobile is a symbol of freedom and independence which are two of the highest values for Americans so that any thought of altering our transportation system is viewed as a threat to our most cherished values.

> *Americans insist on doing whatever, whenever at a whim. It is unreasonable to expect Americans to live any other way unless something drastic happens to force a major change.*

But what major changes could we expect? Even during the Great Depression many people—even the poorest of the poor—drove cars, so we see cars as a key indicator of our way of life. It is how it is. This means that we will hang on the car driving habit as a necessity until the last drop of oil. Therefore, only if and when oil is totally exhausted could we talk about a major change in the energy sector.

Also during the Depression, FDR authorized electric power subsidies, which indicates the importance of this energy source. Today we can't even imagine life without electricity, so it is to be expected that we will burn coal and natural gas until they are also gone for good. Then, and only then will we take steps towards implementing the major change.

Europeans and Asians are accustomed to living in small, crowded, dimly lit rooms, with shoe box size refrigerators, air-dry clothes lines all over the place, and many ride bicycles to work. Most have accepted the idea of living with ever increasing energy sacrifices, while others are even contemplating life in strictly regulated eco-villages.

Well, that's perfectly is OK, if this is what they want to do. We will let them. This, however, is not going to happen on any large scale in the U.S. Not now, not ever! Or at least not until the major changes are upon us

The conservative thinkers of today must understand that "our way of life" cannot be changed, and work around it.

After all is said and done, this is what most Americans think. It matters not what our opinions of the status quo are, or how much better the alternatives might be; it is what it is and making any significant changes—at this time and under the present socio-political conditions—is not an option. It is the American way of life and the American people will fight for it, regardless of what is right or wrong, and the energy and environmental issues at hand.

While a small number of free thinkers in Portland, OR, might consider living in an eco-village, going to Tyler, TX, with such an idea might get you in big trouble with the locals. Imagine a Texas farmer living on a 500-acre farm in a 3,000-square-foot house, with a walk-in refrigerator/freezer and a 5-ton AC unit on the roof.

Then take a look at his even larger equipment barn, full of huge tractors and other agricultural machinery, and a 40 foot Allegro Bus parked in one corner. Then, in the three-car garage we see F-150 truck, Escalade luxury SUV, a Softail Harley Davidson, and in a distance is a twin engine fishing boat tied to the lake pier.

Now imagine this same guy riding a bicycle to the local bar at night, or to church on Sunday morning. Or how about moving out of the 4 bedroom home into a studio apartment with a shoebox-size refrigerator, 25 W lights, no air conditioner, and air-dry clothes line in the bathroom. Not a chance! This is not going to happen. Not in America. Not in Tyler, Texas. Not now, not later, not ever! Unless and until the big major change comes. But what change?

Perhaps we should focus on greater energy efficiency instead? Yes, it is good to talk about it, so don't

stop. You, however, cannot take my four-bedroom 3,000 square foot home with two full-size refrigerators/freezers, and 5 ton air conditioner on the roof, away from me. And I will also continue driving my 8 cylinder SUV, my 1200 cc. motorcycle, and 40 foot diesel motor home anytime and everywhere I want.

Our way of life, the society, the economy, and even the government, encourage this type of behavior and only a major change could make a difference. At this time, we see no significant change coming our way—seeing ourselves awash in oil in gas for a long time to come—so the American Way will remain untouched and undisturbed for a long duration.

But seriously, let's continue the energy efficiency talk; it is a good thing, if it is not pushed down our throats. Instead, it has to be somehow woven in the American fabric without many sudden changes of our daily routines.

The politicians know this very well, since they live in even bigger houses and drive even bigger cars and campaign busses. They are very careful not to step on any toes too, because one wrong step could cost them their political future, together with their expense accounts, large homes, and luxury cars.

And so, the energy efficiency, environmental protection and the need for major change debate in this country continues, but the changes that are planned or expected will have to be carefully crafted and even more carefully implemented. They must come slowly and without much sacrifice on part of the people and their way of life. How far we can go that way is uncertain, but there are few alternatives. And that's the honest (albeit politically incorrect) truth!

Energy Intensity

Energy intensity measures the energy efficiency as compared to the economy of a nation, calculated as units of energy per unit of GDP. High energy intensities indicate a high cost of converting energy into GDP, while low energy intensity means lower cost of converting energy into GDP.

For example, 1 million Btus consumed with an energy intensity of 8,553 produced $116.92 of GDP for the U.S. in the mid-2000s. At the same time, 1 million Btus of energy consumed in Bangladesh with an Energy Intensity of 2,113 produced $473 of GDP. This is over four times the effective US rate. This, however, is an extreme case that does not set a precedent.

Energy intensity can be used as a comparative measure between countries; whereas the change in energy consumption required to raise GDP in a specific country over time is described as its energy elasticity.

Energy intensity is different in different world regions. For example, it is three times higher in CIS than in European countries. Energy intensity in OECD Asia and Latin America are about 15% above those in Europe, while North America stands 40% higher, but remains below the world average. India is on a par with the world average, with energy intensity levels 60% higher than in Europe.

The high energy intensity in the CIS, the Middle East, China and other Asian developing countries is due to the predominant use of energy-intensive industries and relatively low energy prices.

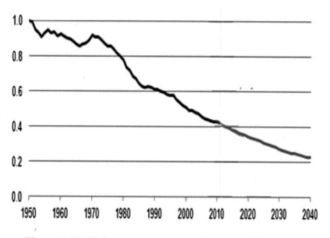

Figure 8-27. U.S. energy consumption per dollar GDP

Figure 8-27 shows that the U.S. energy intensity is in decline (a good thing). From 1950 to 2010, the energy intensity in the U.S. decreased by almost 60% per GDP dollar. There was a period of sharp increase around the 1970s, although energy prices fluctuated only about 3% from year to year.

After the 1973 Arab oil embargo, energy prices rose significantly above the previous years, which lead to changes in the national energy policy. New vehicle efficiency standards were established, and the consumer attitudes began to change as well. Since 1973, the U.S. energy intensity has declined at a rate closer to 2% per year, and is expected to continue through 2040.

Surge of the knowledge-based economy, expressed in the rise of computer hardware, software and digital technology from 1991 to 2000, global economic productivity increased without parallel increases in energy use. New energy production technologies and use improved energy efficiency in almost every aspect of the economy. At the same time, the energy-intensive industries in the U.S. declined as well, with continuing structural changes in the economy.

Global energy intensity declined at an average annual rate of almost 1% in the 1980s and 1.40% in the 1990s. From 2001 to 2010, the rate dropped to 0.03%. This was a period of decline, when the developed countries restructured their economies with energy-intensive heavy industries accounting for a shrinking share of production. Global energy intensity increased 1.35% in 2010, reversing a broader trend of decline over the last 30 years.

In contrast, China's energy intensity declined 4.52% annually between 1981 and 2002, and a staggering 15.37% between 2005 and 2010. But that fell short of the government's incredible goal of 20%. The main reason for this shortfall was that over half of China's $630 billion stimulus plan was invested in infrastructure development, which drove up energy consumption.

The latest economic recovery has led to an increase in total energy consumption per unit of GDP, for the first time in more than 20 years (+ 0.5%) in the developed countries.

The European Union accounts for the highest increase in energy intensity, about 2.5% as compared to a 1.7% average annual decrease before the crisis. The poor EU performance was due to the industrial sector, where energy consumption did not decrease at the same pace as the value added segment due of lower efficiency.

The key drivers of the present and expected decline patterns in the U.S. include:

- Residential energy intensity, measured as delivered energy used per household, is expected to decline about 27% by 2040.
- Commercial energy intensity, measured as delivered energy used per square foot of commercial floor space, is expected to decline about 17% by 2040.
- Industrial sector's energy intensity, measured as delivered energy per dollar of industrial sector shipments, rises above its 2005 level initially owing to the 2007-09 recession but is expected to decrease 25% below its 2005 level by 2040.
- Transportation sector's energy intensity is more difficult to measure because of the multiple modes of transportation. Light-duty vehicles are by far the largest energy consuming part of the sector. Light-duty vehicle energy intensity, which is measured as their consumption divided by the number of vehicle-miles traveled, is projected to decline by more than 47% by 2040 from their 2005 value.

Things are looking good in the good 'ol US of A from that perspective at least.

Energy Use Intensity

Energy use intensity (EUI) is a measure of energy use per unit area of a building or business. As one of the key metrics in the efficiency mix, EUI basically expresses a building's energy use as a function of its size, function, and other characteristics.

Note: EUI is not to be confused with the (national) energy intensity we discussed above. EUI is a measure of a single unit—energy use in building, business, or home.

For most property types, the EUI is expressed as energy per square foot per year. It's calculated by dividing the total energy consumed by the building in one year (measured in kBtu or GJ) by the total gross floor area of the building. This can also be calculated from the energy use information in the monthly utility bills.

Generally speaking, a low EUI signifies good energy performance. An church campus, for example, uses much less energy compared to a hospital. The intensity of energy use, and the time periods are what makes the difference. Certain property types, however, will always use more energy than others.

It is obvious from Figure 8-28 that supermarkets and hospitals (usually 24 hrs./day operation) are using much more energy than churches with their 2 hour weekly use, or warehouses with dim and partial lighting. Because of that, the efforts to increase the energy efficiency must be focused on the large users.

Why a bank branch would use that much energy is a mystery to us, but the figures are right, so we have to assume that banks have some secret operations (like money printing and such) that require a lot of energy. No wonder…with the amount of money we need to print today just to keep the boat afloat.

The U.S. Department of Defense (DoD) provides a good example of properly designing and enforcing building energy efficiency standards at its properties.

Case Study: DoD Building Energy Efficiency

In FY 2009, 99% of DoD's new building designs include provisions to make them 30% more energy efficient than the American Society of Heating, Refrigerating and Air Conditioning Engineers (ASHRAE) 90.1-2004 standard.

The Army Corps of Engineers continues work with the DOE and the Office of the Army Assistant Chief of Staff for Installation Management to develop design guides for implementing building efficiency standards mandated by the EPAct 2005. The Corps has completed prescriptive design guides for battalion headquarters buildings, permanent party barracks, training barracks, and tactical equipment maintenance facilities.

Four of the most prevalent types of buildings were constructed in conjunction with Army troop stationing actions. Use of these design guides will help new building designs be 30% more energy efficient than ASHRAE standards without having to model each individual project.

The Air Force initiated 98 new building designs in FY 2009. 100% will be life-cycle cost effective and meet the goal to exceed the ASHRAE efficiency standards by at least 30%. The Navy is expecting 99% of the FY 2010/2011 military construction facility designs to meet or exceed Federal standards and achieve life cycle cost effective sustainable designs.

The Defense Logistics Agency is redesigning a Child Development Center project in Columbus, Ohio, to ensure that it will be at least 30% more efficient than ASHRA standards and that it will be eligible for LEED Silver status. The Missile Defense Agency is presently planning three military construction building projects (Redstone Arsenal, Alabama and Fort Belvoir and Dahlgren, Virginia). These buildings are designed in accordance with EPAct 2005, EO 13423 and the Whole Building Design Guide, and each building will be LEED certified.

The TRICARE Management Agency's Bureau of Medicine and Surgery (BUMED) is working with Naval

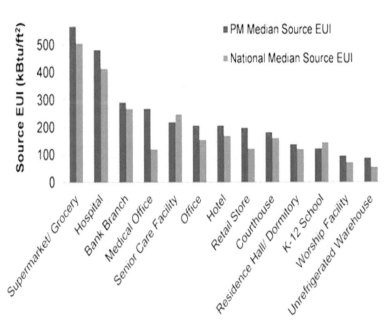

Figure 8-28. EUI of different building/business types

Facilities Engineering Command (NAVFAC) to ensure that all new design work will exceed the ASHRAE efficiency standard by at least 30% where achievable. Ninety-three percent of new BUMED building designs started since the beginning of FY 2007 are expected to exceed the ASHRAE standard by 30% in the near future.

Although the above figures are somewhat outdated, they clearly show the trend towards energy efficiency improvements in the U.S. military. These measures are saving a lot of energy and are diverting tons of fossil fuels from use in the military bases.

Widening this effort on a national level would save billions of dollars in fuels and will reduce the GHG pollution significantly.

Energy and GDP

The amount of energy that a nation uses is related to its productivity, as expressed in its GDP and its population, which of course drives GDP. Therefore, in order to compare nations with different outputs and populations, it seems wise to divide both consumed energy and GDP by population.

Figure 8-29 shows unequivocally the wide range in power use (in kW per capita) between the lowest consumer (India) and the highest (USA). The GDP per capita is directly proportional to the energy use per capita. With other worlds; more money people make, more energy they use.

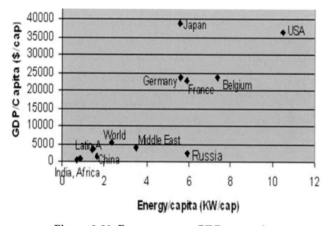

Figure 8-29. Energy use vs. GDP per capita

Half of the rural areas in India don't even have access to electric power, so the locals do with whatever they can; burning wood, coal, and many businesses generate electricity via diesel generators instead. In stark contrast, each U.S. citizen needs an average more than 10 kW at any moment to keep our way of life going. This includes all types of energy use in our daily lives: transportation, work, food, housing, leisure, etc. This brings

the question of how much of this energy is wasted...?

There is a rough overall correlation between GDP/capita and energy/capita. This is because higher productivity per individual will cause a higher energy need for each. But comparing a developing country with a highly developed one is like comparing apples with oranges. A more fair way to do this is compare the U.S. and Japan, where the productivity per capita is comparable and are the best in the world from a productivity point of view. Yet Japan uses only half of the power used in the U.S.

Because of that, Japan emits 9.5 tons of CO_2/capita, vs. 19.7 tons per capita in the U.S., which roughly corresponds to the 2:1 energy consumption ratio of the two countries. Remember that the global per-capita CO_2 emission is only 4 tons per year—2.5 times less than the Japanese and almost 5 times less than the U.S. For some developing countries this ratio goes up to 20 times less than the U.S.

If we use energy as frugally and efficiently as the Japanese, we would not need any oil imports.

This sounds so simple, and yet we are so far from the Japanese in everything we do. Most of us realize that energy efficiency is one (perhaps the only for now) way to obtain energy security and energy independence, while significantly lowering the total GHG emissions at the same time. Still, the present day energy conditions in the U.S. do not favor any intensive efforts in the energy efficiency areas. Maybe some day...

Note: After the 2011 Fukushima disaster, however, Japan shut down its nuclear plants and replaced the lost power by coal and natural gas fired such. This, of course, increased the air pollution significantly, so the emission ratio has changed, but is still lower than the U.S. The energy efficiency measures in the country, however, remained in place and were even reinforced.

The U.S. energy use per capita was constant from 1990 to 2007 and began to fall after 2007. Energy use per capita continues to decline as a result of improvements in energy efficiency, reflected in the use of new more-efficient appliance and CAFE standards. We have been also slowly introducing changes in the way we use energy in form of energy conservation. As a result, under the present trend, the U.S. population is expected to increase 21% by 2040, but the energy use will increase only 12%. The energy use per capita is expected to decline 8% during the same time.

The U.S. CO_2 emissions (in 2005 dollar of GDP) have tracked closely with energy use. With lower-carbon fuels accounting for a growing share of total energy use, CO_2 emissions per dollar of GDP are expected to decline more rapidly than energy use per dollar of GDP, or about 2.3% per year for a total of 56% below their 2005 level by 2040.

This is good and probably the most that can be expected from the U.S. consumers, which is actually much better than the rate of energy use and CO_2 emissions expected from other, especially the developing countries. It is, however, unfair to compare these countries, due to the much different (even opulent) lifestyle in the U.S., which makes a huge difference in every area of daily life.

The Opulence Effect

There is no official measure of the level of opulence (or extreme life styles) for different countries, but we don't have to dig too deep to see the striking difference. That of the haves and the have nots. Just one example would illustrate the great life style chasm that exists between being and American and people in other parts of the world.

During the 2014 Ebola epidemic events in Africa, thousands of people, including many doctors, were infected and thousands died. In best of cases, the infected people—the lucky ones—got a bed and some food, but not much more than that. There were simply no adequate facilities and services to help them adequately. They were basically left to fight the vicious Ebola on their own with very little outside help. Many of the infected died, and the rest were left struggling in their misery for weeks and months.

At the same time, three American doctors were also infected by the Ebola virus. At that news, the world stopped and held back its collective breath. Literally. The news were brimming with concerns about the American doctors status and following their progress step by step. An intense and very expensive effort to save the lives of the Americans at all cost started. No amount of effort or expense was going to delay, let alone stop, the operation.

A private jet was equipped with special isolation tents and specialized equipment, attended by dozens of specialists on board and on the ground. The doctors were carefully isolated in special suites, placed in the isolation tents in the jet and flown to specialized hospitals in the U.S., which are equipped for, and specialized in, handling acute infectious cases. Several dozes of a brand new experimental vaccine were administered with the hope of stopping the virus and speeding the recovery period. The evacuation and treatment operations took several weeks at the cost of millions of dollars and ended up successfully with saving the Americans' lives, who were up and well within a week or two.

Looking at the two different scenarios we clearly see that Americans have enormous advantage over the locals in any underdeveloped African country. Money, expertise and preferential treatment is always available when our wellbeing is at stake. This includes the advantages we have in the energy sector, where many African villages don't even have access to electricity. During the 2014 Ebola disaster in Africa, many (mostly make-shift) hospitals didn't have the basic necessities to keep the patients alive. Some didn't even have electricity and/or running water. Thousands of people died from this lack of the most basic patient care needs.

At the same time, each and every American has access to unlimited electric power, water, health care, and all the physical and financial benefits that come with life in America; including the availability of unlimited resources in the effort to save American lives overseas. No wonder the developed world is trying to catch up with us at all cost. And who can blame them?

Sustainability

Sustainability is a fairly new concept that means different things to different people. As such, we see it thrown around as a boomerang, often ending back in the lap of the thrower. The word sustainability can be seen printed on the packaging of all kinds of products, but lately companies are trying to cash on it. They proclaim their products or services to be sustainable of one way or another.

Entire cities and even countries are now claiming to be 100% sustainable. The city of Burlington, VT, for example claims that 100% of its electric power is generated from renewable resources and is, therefore, 100% sustainable. Denmark claims that it is the most sustainable country in the world, due to its huge wind mill power generation and use of energy saving approaches like bike lanes, pedestrian walkways, etc.

So if this is so easy to do, and is so beneficial for all involved, why aren't we all living in sustainable cities and countries? This is a good reason to take a closer look at what sustainability is and how it applies to the energy markets in particular.

What is sustainability? In ecological terms, sustainability is the way for biological systems to remain diverse, productive, and live in harmony. For this to happen, healthy ambient; wetlands, deserts, oceans, and forests need to be ample and remain healthy for the long run.

Sustainability, therefore, is the long-term endurance and life in prosperity, of systems and processes.

Please note the key words; long-term endurance, and life in prosperity. Without one of these terms, we have either a) temporary (unsustainable) sustainability, or b) life in misery caused by forced, or wrongly applied, sustainability. As we take a close look around us, very often we see one or the other, but very seldom do we see a complete, 100% efficient and practical sustainability.

Sustainability is usually achieved by sustainable development of energy, food, and all other life style systems. This process usually includes several interconnected domains, the main of which are the local: a) ecology, b) politics, c) economics, d) culture, and e) the way these are interwoven in the local and global environment.

Basically speaking, sustainable ecosystems and environments are the foundation of healthy and prosperous human life. But humans have chosen ways which are hurting the very ecosystems and environments we live in.

We know exactly what is happening, and the ways to reduce the negative impacts, but have chosen temporary comforts, and less painful solutions to the problems, instead. Because of that, sustainability is becoming a social challenge that entails international and national law, urban planning and transport, local and individual lifestyles and ethical consumerism.

Looking for ways to return to sustainably requires drastic changes in our life styles, some of which are simply unacceptable to the American psyche. Eco-villages, eco-municipalities and sustainable cities have been suggested and tried. Restructuring of the economic sectors by the introduction of concepts like perma-culture, green building, sustainable agriculture etc. also sound attractive, but has not been proven successful. Using new technologies, such as solar and wind, and other renewable energy sources, is the new approach taken lately, but it seems that it has issues that will take a long time to resolve.

And so, although the term sustainability is very popular, and everyone agrees plausible today, the possibility of achieving global environmental sustainability is illusive. Instead, we see increasing waste, overconsumption, and pursuit of economic growth, all of which brings additional environmental degradation, climate change, and ecosystems destruction. All this boils down to suffering of Earth's population and life forms on the short term, and disastrous events on the long run.

Sustainable Living

Sustainable living is a lifestyle that also attempts to reduce the carbon footprint and minimize the use of Earth's natural resources. This is most often done by alternative methods of transportation, energy consumption, and special diets.

Sustainable living aims to structure human life in ways that are consistent with sustainability, or making sure that biological and all other life-supporting systems remain diverse and productive. This ensures their natural balance and preserves humanity's symbiotic relationship with the Earth's natural resources, ecology and cycles. Sustainable and ecological living is interrelated with the overall principles of sustainable development, which is a form of sustainable sustainability.

The concept of sustainable living is changing in the twenty-first century. Now it is focused—in addition to reducing carbon footprint, etc.—on shifting to a renewable energy-based, reuse/recycle economy and a diversified transport system.

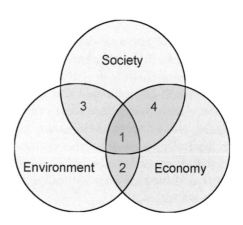

Figure 8-30. The sustainability model
1—True sustainability
2—Energy efficiency and environmental incentives
3—Natural resources overuse
4—Social justice and fair trade

The sustainability model in Figure 8-30 reflects the reality of today's interaction between the three basic elements of the capitalist society. It is driven by economic and environmental concerns and the resulting combinations and permutations of their intersections.

We clearly see that true sustainability, shown in the segment marked with number 1 in Figure 8-30, is achieved only in case when all three segments of today's life intersect at equal shares. The intersections of the other segments, 2, 3, and 4 above, are critical parts of the sustainability equation, but cannot achieve it on their own. When all are working together, however, sustainability is a perfect result.

Sustainable Living Example

One example and a trend in sustainable living today are the so called eco-villages. Actually, the concept of intentional, planned communities living is not new, but today it is focused on becoming socially, economically, and ecologically sustainable, and/or perhaps independent from the conventional societal life support systems.

The new era eco-villages usually vary in size from 20 to 200 people, and some are much smaller. Larger eco-villages of up to 2,000 individuals exist as networks of smaller sub-communities. Some eco-villages have grown by the addition of individuals, families, or other small groups who are not necessarily members settling on the periphery and participating in the eco-village community's activities.

Much larger eco-villages are also being developed in the midst of large metropolitan areas. These function as independent patches of alternative living in the midst of a conventional life style. Parts, or even entire large cities in the Europe and Asia are using the eco-village principals and trending towards sustainable living.

One example of a sustainable living patch (mini eco-village) in the midst of a large city is Christiania, or Freetown Christiania—a self-proclaimed autonomous neighborhood of about 850 residents—in the Christianshavn neighborhood in Copenhagen, Denmark. Christiania covers about 85 acres within the city limits and is officially regarded as a large commune with its own and unique legal status. The commune is regulated by the Christiania Law of 1989, which transfers parts of the supervision of the area from the municipality of Copenhagen to the state.

Christiania has been a source of controversy since its creation in a squatted military area in 1971. Its cannabis trade was tolerated by authorities until 2004, which is when the conflicts escalated. Since then, measures for normalizing the legal status of the community have led to frequent police raids and negotiations which are ongoing.

During the height of the conflict, the neighborhood was closed by residents in the spring of 2011, while discussions continued with the Danish government as to its future. It is now open again and has resumed its normal daily operations. Conflict with the official powers is not a good, sustainable, nor exemplary lifestyle, so we just have to wait and see what will happen at Christiania on the long run.

The practical, 21st century eco-village lifestyle requires a shift to renewable technologies, which approach is complex and expensive. This means that the concept can be successful only if the new lifestyle and the surrounding environment are attractive to (and can be afforded by) the local culture. Only then the eco-village concept can be maintained on the short run and adapted as best and necessary life style by the generations to come.

Case Study: Ten Thousand Villages

Ten Thousand Villages is all together different concept of collective sustainable living on global scale. It is a nonprofit fair trade organization that produces and markets handcrafted products around the world. It consists of disadvantaged artisans from more than 120 artisan groups in more than 35 countries.

As one of the world's largest and oldest fair trade organizations, Ten Thousand Villages cultivates long-term buying relationships in which artisans receive a fair price for their work and consumers have access to gifts, accessories and home décor from around the world. As part of the "global village" this concept opens new opportunities for people who are looking for a way out of the 20th century job market.

Large or small, the eco-village concept is growing and pointing towards a way of life, which actually reminds us of the communist system, which was designed and developed in a similar, self-sustaining, closed-loop environment. The concept actually was good—equality and opportunities for all—and the system could've been successful, but human greed, ignorance, and stupidity took over. After a short period of misery and agony for the masses, while the rulers twisted the rules and piled wealth, communism fell with a great crash. The effects of the 70 years of life in a closed society still reverberate around Russia and Eastern Europe, so we must be careful when considering new closed loop systems. And so, while the new "eco-village" concept might be a good one, we must remember the lessons of the past.

The eco-village concept, also, is contrary to the American way of life, where freedom from any organized bondage, rules, and regulations is priority number one for most free-thinking Americans. The American Dream is contrary to living in closely regulated communities. Because of that, we don't expect any great movement in that area in the U.S. anytime soon.

Case Study: City of Burlington, VT

Burlington, Vermont's largest city with population of about 42,000 people, has achieved a great socially conscious feat. The city now boasts that 100% of its electricity is generated by renewable sources; mostly wind, water, biomass, and hydropower.

About a third of Burlington's renewable energy is produced at this biomass facility, which uses wood chips to produce energy. Truck after truck haul in scrap wood from across Vermont and use it to make steam and generate electricity.

20% or so is sourced from wind turbines like these on the hills of a neighboring town, where a 12.5 MW wind farm went online in 2009, and is producing electricity in excess of that consumed by the local community of 850, according to the city officials. The excess power is sold to other locals.

There are solar arrays at the airport, and some residences, but that is a small amount of the total.

The biggest portion of the city's renewable production, however, comes from hydropower. Some of it is sourced from other places, like the older hydroelectric dams in other states, while some is generated locally by the 7.4 MW Winooski 1 hydroelectric power plant on the Winooski River at the city's edge.

All this is what accounts for the city's ability to produce as much energy from renewables as it uses in a year.

The Burlington's example is a part of a broader movement that includes a statewide goal of generating 90% of Vermont's energy from renewable resources by 2050. This includes all kinds of energy, including electricity, heating, and even transportation.

On even a larger scale, this movement is growing across the country too. For a number of reasons most state governments and businesses want to use power produced by environmentally friendly sources.

Here are the issues with the developments in Burlington, and why this strategy would be hard to impossible to expand across the entire state.

1. Vermont has no sun to speak of, and there are other weather related problems, so with all considered, solar has no great future in the area. Plus, the small amount of solar arrays in Burlington were installed thanks to Obama's generous solar subsidies, which are now gone; not to be seen for many decades to come.

2. Wind is available at some remote spots, usually far from the power grid, which makes capital investment for wind power quite expensive, in addition to the other issues related to it. The existing 12.5 MW wind farm provides only 1/50th of the power needed for the city, and only part of the time.

3. Wood chips for the Burlington bioenergy power plant come from all over the state. A lot of wood chips are needed to keep the power plant working day and night, so we wonder if there are enough

wood chips to power the entire state? Not very likely. Also, wood chips emit a lot of GHGs during the burning process, which makes the city's "clean" status somewhat less than 100%

4. And most importantly, the city's 50% of "renewable" energy comes from hydropower. But Winooski 1 hydropower plant is only 7.4 MW in capacity, which is about 1/60th of the total power needed by the city.

So, here is the catch; in order to optimize its "renewable energy" status, the city of Burlington buys power from other hydropower facilities and sells renewable energy credits for the renewable power they produce to utilities in southern New England, where their value is highest. Then the city electric utility buys less expensive credits from other sources to offset the credits they have sold. Selling high and buying low is a new gimmick of using accounting tricks to show more clean and renewable energy use.

So could the entire state of Vermont follow Burlington's example? Maybe, but not easily and not very soon.

Case Study: Denmark

The official Denmark website claims that, "Creating a green and sustainable society is one of the key goals for Denmark. More than 20 per cent of Denmark's energy already comes from renewable energy, and the goal is to reach 100 per cent by 2050. Much of the renewable energy comes from wind turbines, where Denmark is a world leader when it comes to developing new technology.

The Danish cycling culture is another example of a green and sustainable society and Copenhagen alone has around 400 km of cycle paths, and about 40 per cent of the capital's population commute to work by bicycle.

So, granted, good things are happening in Denmark and a good example to follow. But how do we do that here, in the good 'ol U.S. of A.? There is no way to obtain 100% power from wind or any other renewable sources, so that idea is out.

And how about 40% of the population going to and fro work on bicycles? Now, we see so may problems with this option, that some of them are even ridiculous. Just imagine the thousands of people living in the San Fernando Valley commuting to Los Angeles and the suburbs daily? Now imagine all cars absent from the bumper-to-bumper highways during traffic hours and instead, men and women pedaling bicycles.

At 10 mph average speed, the 22 miles (35 km.)

distance would take minimum of two hours to, and two hours from, work. This amounts to about 4 hours of travel (pedaling) time daily, plus dressing, undressing, and taking a shower after each trip, we will need about 5-6 hours every day to get to work and back...in good weather. During rainy days, this trip might turn into a full day adventure.

But how many of these people can even make the trip...once? Some of them won't even fit on regular bikes, so a special industry has to be developed for oversized bicycles. Absurd, is one way to put it. And even impossible, if Phoenix, Arizona is considered, because here we need to add an air-conditioning to each bicycle, else most people would just die in the 150 degree heat on the highways asphalt top during the 6 months of blistering summer heat.

We also reviewed the eco-village concept above, including the Christiania eco-village by Copenhagen in Denmark. We don't need to look very long at life in the U.S. to determine that this concept also does not fit at least 99.999% of our present day society.

So we will just have to congratulate Denmark, and consider their sustainability achievements as unachievable under the present conditions in the U.S. We just have to find some other, more American-like, solutions to deal with the energy problems of today and tomorrow.

Energy efficiency is one effective tool we have to reducing our energy dependence and environmental destruction.

Energy efficiency is something that doesn't cost much and we all can participate in for the good of the society and the world. Government participation and encouragement are needed for its full implementation in our lives.

ENERGY STORAGE MARKETS

Energy storage is another serious effort underway today. The new kid on the (energy) block is becoming the talk of the town.

Energy storage is the key to making solar and wind energy base load electric power providers.

This is so, because energy storage is the only way to balance the variable output from these renewable energy sources. Energy storage is also needed in case of emergencies, when it could prevent massive blackouts.

Although there are many forms of storing energy, they all suffer from side effects, such as loss of efficiency during the conversion process of converting electricity into storage media and then converted back to electricity. There are also a number of reliability, safety, and pricing concerns that must be resolved before energy storage becomes

Types of Energy Storage

- Batteries are certainly the most discussed technology, and are the greatest hope of the solar and wind industry to compete with the conventional fuels. Batteries use chemical reactions with two or more electrochemical cells to enable the flow of electrons. Examples include lithium-based batteries (ex: lithium-ion, lithium polymer), sodium sulfur, and lead-acid batteries (used usually in cars).

- Pumped hydropower storage systems utilize elevation changes to a) store off-peak electricity in an elevated water reservoir for later use, and then use it to generate electricity when needed. In this case, water is pumped from a lower reservoir to a reservoir at a higher elevation during off-peak periods. When more power is needed, thus stored water is allowed to flow back down to the lower reservoir, generating electricity via conventional hydropower plant.

- Molten salts are solid at room temperature and atmospheric pressure, but undergo a phase change when heated. This liquid salt is frequently used to store heat in CSP facilities for subsequent use in generating electricity.

- Underground thermal energy storage systems pump heated or cooled water underground for later use as a heating or cooling resource. These systems include aquifer and borehole thermal energy storage systems, where this water is pumped into (and out of) either an existing aquifers or man-made boreholes.

- Compressed air energy storage (CAES) systems use off-peak electricity to compress air, storing it in underground caverns or storage tanks. This air is later released to a combustor in a gas turbine to generate electricity during peak periods.

- Pit storage systems use shallow pits, which are dug and filled with a storage medium (frequently gravel and water) and covered with a layer of insulating materials. Water is pumped into and out of these pits to provide a heating or cooling resource.

- Thermochemical storage uses reversible chemical reactions to store thermal energy in the form of chemical compounds. This energy can be discharged at different temperatures, dependent on the properties of the thermochemical reaction.

- Chemical-hydrogen storage uses hydrogen as an energy carrier to store electricity, for example through electrolysis. Electricity is converted, stored, and then re-converted into the desired end-use form (e.g. electricity, heat, or liquid fuel).

- Flywheels are mechanical devices that spin at high speeds, storing electricity as rotational energy. This energy is later released by slowing down the flywheel's rotor, releasing quick bursts of energy (i.e. releases of high power and short duration).
  — Supercapacitors store energy in large electrostatic fields between two conductive plates, which are separated by a small distance. Electricity can be quickly stored and released using this technology in order to produce short bursts of power.

- Superconducting magnetic energy storage (SMES) systems store energy in a magnetic field. This field is created by the flow of direct current (DC) electricity into a super-cooled coil. In low-temperature superconducting materials, electric currents encounter almost no resistance, so they can cycle through the coil of superconducting wire for a long time without losing energy.

- Solid media storage systems store energy in a solid material for later use in heating or cooling. In many countries, electric heaters include solid media storage (e.g. bricks or concrete) to assist in regulating heat demand.

- Ice storage is a form of latent heat storage, where energy is stored in a material that undergoes a phase change as it stores and releases energy. A phase change refers to transition of a medium between solid, liquid, and gas states. This transition can occur in either direction (i.e. from a liquid to a solid or vice versa), depending on if energy is being stored or released.

- Hot- and cold-water storage in tanks can be used to meet heating or cooling demand. A common example of hot water storage can be found in domestic hot water heaters, which frequently include storage in the form of insulated water tanks.

Since wind and solar are the greatest beneficiaries of energy storage—as a matter of fact their future depends on it—we will take a closer look at the available options in this area today:

Solar and Wind Energy Storage

Variable power output of solar and wind power plants' output is unavoidable. Solar power is available only when the sun shines, and wind power only when there is enough wind. Since this is always the case, solar and wind power plants idle most of the time.

Solar and wind power output is available only about 20-30% of the time. And most of the time, it is not the full power the plant is capable of generating.

Solar power can be combined with other energy sources (i.e., wind) to smooth the variations and even match the peak load, but even that is not a complete solution, simply because there is not enough sun and wind to provide 24/7 power generation.

During cloudy or rainy days, for example, the solar plant output will be limited to near zero and wind alone cannot provide the missing component. The same is true for wind; there are hours and even days when the wind is just not enough, or too strong for safe wind power generation. In these cases the wind power is limited, or none. In all cases, the total power output is variable and unreliable, thus unsuitable as load power.

The only way to rectify the variability problem is using energy storage as a supplemental power generation.

Energy Storage Types

Excess energy produced during the times of maximum power generation by wind and solar power plants can be stored and later on used during periods of low energy generation. There are a number of potential energy storage solutions for use with solar and wind power plants, as follow:

- Electrochemical storage
  Batteries
  Flow batteries
  Fuel cells

Electrical
Capacitor
Supercapacitor
Superconducting magnetic energy storage (SMES)

- Thermal storage
  Steam accumulator
  Molten salt
  Cryogenic liquid air or nitrogen
  Seasonal thermal store
  Solar pond
  Hot bricks
  Fireless locomotive
  Eutectic system
  Ice Storage

- Mechanical storage
  Compressed air energy storage (CAES)
  Flywheel energy storage
  Hydraulic accumulator
  Hydroelectric energy storage
  Spring
  Gravitational potential energy (device)

- Chemical storage
  Hydrogen
  Biofuels
  Liquid nitrogen
  Oxy-hydrogen and Hydrogen peroxide

- Biological storage
  Starch
  Glycogen

- Electric grid storage

Here we review only the most developed energy storage options to date:

Battery Energy Storage

No doubt, this is the most direct and efficient way to store a large amount of electricity generated by PV power plants. The generated DC electric energy is stored as DC power in batteries for later use.

There are several types of batteries, the most commonly used as follow:

1. Lead Acid batteries

These are the most common type of rechargeable batteries in use today. Each battery consists of several electrolytic cells, where each cell contains electrodes of elemental lead (Pb) and lead oxide (PbO_2) in an elec-

trolyte of approximately 33.5% sulfuric acid (H_2SO_4). In the discharged state both electrodes turn into lead sulfate ($PbSO_4$), while the electrolyte loses its dissolved sulfuric acid and becomes primarily water. During the charging cycle, this process is reversed.

These batteries last a long time, and can go through many charge-discharge cycles, if properly used and maintained. They are affected, however, by high temperatures, when the electrolyte can boil off and destroy the battery. Since there is water in the cells, the electrolyte can freeze during winter weather, which could destroy the battery as well.

2. Lithium batteries

These are a mature technology, having been used widely for a long time in consumer electronics. They are actually a family of different batteries, containing many types of cathodes and electrolytes. The most common type of lithium cell used in consumer applications uses metallic lithium as anode and manganese dioxide as cathode, with a salt of lithium dissolved in an organic solvent.

A large model of these can be used to store large amounts of electric power generated by a PV power plant, and due to their highest known power density they could be quite efficient—70-85%. They are suitable for smaller PV installations, too, because scaling up to large PV plants would be a very expensive proposition.

3. Sodium Sulfur batteries

These are high temperature, molten metal, batteries constructed from sodium (Na) and sulfur (S). They have a high energy density, high efficiency of charge/discharge (89–92%) and long cycle life. They are also usually made of inexpensive materials, and due to the high operating temperatures of 300-350°C they are quite suitable for large-scale, grid energy storage.

During the discharge phase, molten elemental sodium at the core serves as the anode, and donates electrons to the external circuit. The sulfur is absorbed in a carbon sponge around the sodium core and Na+ ions migrate to the sulfur container. These electrons drive an electric current through the molten sodium to the contact, through the electric load and back to the sulfur container.

During the charging phase, the reverse process takes place. Once running, the heat produced by charging and discharging cycles is sufficient to maintain operating temperatures and usually no external source is required.

There are, however, a number of safety and corrosion problems, due to the sodium reactivity, which need to be resolved before full implementation of this technology takes place.

4. Vanadium redox batteries

These are liquid energy sources, where different chemicals are stored in two tanks and pumped through electrochemical cells. Depending on the voltage supplied, the energy carriers are electrochemically charged or discharged. Charge controllers and inverters are used to control the process and to interface with the electrical source of energy.

Unlike conventional batteries, the redox-flow cell stores energy in the solutions, so that the capacity of the system is determined by the size of the electrolyte tanks, while the system power is determined by the size of the cell stacks. The redox-flow cell is therefore more like a rechargeable fuel cell than a battery. This makes it suitable as an efficient energy storage for PV installations.

A number of additional types of batteries are under development, and some show great potential for use in larger PV installations in the near future. Most batteries have problems with moisture, high temperature, memory effect, and use of scarce and toxic exotic materials, all of which cause longevity problems and abnormally high prices. If and when all these problems are resolved, the energy storage problems of PV power plants will be resolved as well.

Case Study: Younicos

Together with regional green utility WEMAG, Younicos is currently demonstrating the business case for stand-alone commercial storage. Since September 2013 the company has been building a battery park in the north-east German city of Schwerin. In September 2014 the battery park was inaugurated and now uses its 5 MW rated power to stabilize grid frequency in wind-swept West Mecklenburg.

WEMAG operates the battery park commercially by competing with conventional generators in the so-called primary control power market. To date, in this market, conventional energy producers receive money in return for quickly boosting on lowering the performance of their power plants by a few percent in order to bring supply and demand of electricity into equilibrium. As explained earlier, batteries can do this much faster, more accurately as well as cheaper—allowing us to outcompete fossils plants.

A second lighthouse project of Younicos, Europe's largest commercial storage trial in Leighton Buzzard northeast of London, showcases even more of what is possible today: Just like in Schwerin the battery will mainly provide frequency regulation—however it will also automatically respond to price signals and then be used to shift peak loads or provide other important system services.

Such multipurpose use also opens new and additional revenue streams for storage. For instance, by peak-shifting just for a few weeks in the year, the project will save over £6 million in investment cost for transformers, underground cables or power lines, which would otherwise have been necessary.

Case Study: Hawaii Battery Storage

Battery energy storage connected to a 6.0 MW solar power plant in Hawaii uses lead carbon batteries made by Xtreme Power, which went bankrupt and was consequently bought by Younicos. A second storage system is to be installed by the French battery maker SAFT at a 12.0 MW solar power plant at Kauaii.

Recently, the island's main utility, Hawaii Electric Company, issued a request for proposals for over 200 MW of energy storage at its solar plants. A brave move forward in the unproven field of energy storage on part of the Hawaii utilities, especially in the aftermath of the failed 6 MW energy storage facility by Xtreme Power site mentioned above.

The accelerated capacity fade, which took place at the site and brought it down, is part of the learning process, and not a show stopper, according to the experts. It actually represents a setback for the brand of lead carbon batteries, especially after the exit of Xtreme Power. Younicos, the German battery storage maker which purchased Xtreme Power in 2014 dropped the lead carbon technology, so it is now dead. Lead carbon is a very small component of the energy storage marketplace, so no major consequences are expected.

The batteries used by Younicos are actually Samsung made lithium-ion batteries. Similarly, SAFT also uses containerized lihtium-ion batteries at the newest of Kauai's utility-scale storage systems.

Nevertheless, the failure (accelerated capacity fade) at the 6.0 MW Kauai project opens a can of worms that grid-connected energy storage has to deal with. It is the very important question about the lifetime of any type of batteries exposed to the stresses of daily field operation. There is no solid answer thus far, so buyers and developers ate taking a very close look at the specs and claims made by manufacturers.

The batteries performance and lifetime question will not go away, and since the object of owners and developers paying for the systems want them to last as long as the power plant. This is 20-30 years exposed to the daily torture of the elements. Since no battery we know of lasts that long, the question remains wide open.

Although we agree that the lead carbon batteries failing is not the end of the battery energy storage, and yet assurances given by manufacturers as far as life time has grey areas which need to be addressed. This, in addition to the fact that there is also a lack of available data in this relatively new industry, means that energy storage needs to be proven reliable and long-lasting.

So, failures or not, Hawaii utilities are planning to integrate more PV using battery energy storage as part of the learning curve. It is part of the wider aims of integrating more PV on the island, with the absolutely necessary help of battery storage.

Note: Safety considerations have been a major issue with any and all battery systems. There have been a number of accidents already—even though the technology has limited use. Here, in addition to the usual risks of electrical equipment that may exist due to high voltages or high currents, there are many safety-critical characteristics of the battery itself. These properties have to be considered to ensure a safe operation of the systems.

Lead-acid, and most other, batteries are very heavy and awkward to handle, so they pose a potential danger during installation and operation. The electrolyte used in some batteries is dilute acid or some other dangerous chemical. Non-encapsulated lead-acid batteries, for example, require the electrolyte to be refilled manually at certain periods of time, which is a dangerous operation.

During the charging process hydrogen gas is formed, and eventually which comes out of the battery. Above a certain concentration in the ambient atmosphere an explosion hazard can result. The rechargeable Lithium based batteries can, under certain conditions, cause a fire in the battery in a process known as thermal runaway.

For the purpose of identifying and preventing accidents, the UN has established a test certificate procedures, UN 38.3 (Transportation Test). The test is geared to provide minimum safety during batteries transport, but has no significance for their installation and operation.

DKE, the German Commission for Electrical, Electronic & Information Technologies working group was established to draft a STD 1000.3.1. This is an application rule for stationary electrical energy storage for low-voltage grid applications. The aim of the standard is to summarize the relevant standards, guidelines and regulations, thus providing safety help, advice, and general rules for batteries storage handling and operation.

China

The battery storage work is just now beginning in China. And yet, the Chinese battery manufacturer BYD has already connected a 40.0 MWh large-scale battery storage system to the grid. The system supports the electricity supply of the company's production site near Hong Kong. A much larger-scale battery systems is planned for installation at the same site in the near future.

The world's largest commercially used energy storage system, connected to the to the grid is installed at BYD's headquarters in Shenzhen, China: It is primarily used to mitigate load spikes in the energy demand, and to ensure the frequency stability of the local grid.

This system consists of over 60,000 single lithium-iron-phosphate cells, with 230 Ah capacity each. This capacity is equal to about 500,000 laptop batteries, or the battery packs of 2,000 electric cars. A specially designed safety system ensures safe operation of the lithium iron phosphate batteries.

BYD has also plans for even larger-scale battery systems. Working together with its German partner FENECON, they have targeted a large market for these projects in Germany.

Japan

Japanese energy service provider Eneres is planning to substitute the unpopular FiT changes with a huge quantity of battery storage systems from Toshiba IT & Control Systems. In view of recent issues with Japanese FiT structure, which caused withdrawal from new contracts for renewable energy purchases by some electric power utilities, Eneres expects that the new battery storage systems will gain awareness in the domestic market.

Eneres plans to supply about 10,000 such systems to their domestic customers. The systems are based lithium-ion battery technology and have storage capacity of 9.9 kWh each.

Singapore

Singapore's Energy Market Authority (EMA) has established a $25 million Energy Storage Program (ESP) to develop technologies that enhance the overall stability and resilience of the power system in the country. The program is designed to support the development and integration of large-scale cost-effective energy storage systems for the power grid. New funding for the program was announced in late 2014.

Battery technologies capable of storing electricity on a large scale, like electrochemical batteries and flywheels, will be used to reduce demand during peak periods in Singapore. These systems could be used for frequency regulation and in support of the deployment of intermittent generation sources like solar energy.

Germany

Recent survey revealed the willingness of Germany's residential PV customers to invest in energy storage systems (batteries). A survey of over 4,000 German PV end customers show that about 20% of participating PV owners already have an energy storage system.

Especially the German manufacturers Deutsche Energieversorgung and SMA have gained the lion share of the new marketplace. This is 60% increase over the 8% energy systems owners a year ago. It seems that the only obstacle is the high investment costs. This keeps many interested end customers away from buying a new battery storage installation.

Another reason for many PV owners, who have installed their system several years ago, not to invest in a storage system at the moment, are their guaranteed feed-in-tariffs. These FiT guarantees will last more than a decade, at which point the energy storage market might get a boost.

Summary

The U.S. manufacturer, Tesla, must be given credit for bringing battery storage for use in solar installations to the media attention recently. There are plans for a giant factory to make millions of Li-ion batteries, and contracts were signed with major solar companies to install these batteries in people's homes.

We will have to wait and see how this is going to play, but an expert analysis of the battery energy market shows that it is too crowded for any one company to be able to claim a large portion of it. Another big problem for battery storage is that it will be limited to home roof tops and small business installations. Large solar power plants will not benefit from battery energy storage anytime soon for a number of reasons:

- High price,
- Lack of long-term performance data, and
- Lack of reliability data of long-term desert operations.

The deserts—where most large-scale solar power plants are installed—with their extreme heat and cold and other natural phenomena, promise merciless treatment, which batteries would not be able to withstand.

Nevertheless, Tesla and many others are trying. We will watch carefully.

Thermal Energy Storage

Thermal energy storage (TES) technologies operate with a goal of storing energy for later use as heating or cooling capacity. Individual TES technologies operate in the generation and end-use steps of the energy system and can be grouped by storage temperature: low, medium, high. Thermal storage technologies are well suited for an array of applications including seasonal storage on the supply-side and demand management services on the demand-side portion of the energy system.

As heating and cooling requirements represent 45% of the total energy use in buildings, these demand-side services can represent significant value to the energy system.

Some thermal energy storage technologies have already realized significant levels of deployment in electricity and heat networks, including underground thermal storage (UTES) systems and ice storage systems for residential cooling. Further, some end-use technologies that have already been deployed to meet other societal requirements include TES capabilities, though this potential is not currently being fully realized (e.g. residential hot water heaters). Today's R&D in thermal energy storage is primarily focused on reducing the costs of high-density storage, including thermo-chemical process and phase-change material (PCM) development.

The thermal storage technologies are divided mainly to low, medium, and high temperature storage systems.

Low-Temperature (<10°C) Energy Storage

Cold-water storage tanks in commercial and industrial facilities are already installed around the world to supply cooling capacity. Larger UTES systems, including aquifer thermal energy storage (ATES) and borehole thermal energy storage (BTES), have been successfully commercialized in order to provide both heating and cooling capacity in countries such as the Netherlands, Sweden, Germany, and Canada.

Due to the higher energy storage densities seen with PCMs compared to sensible heat storage, the United States and Japan have already installed significant amounts of thermal storage that uses ice for cooling applications. In the United States, an estimated 1.0 GW of ice storage has been deployed to reduce peak energy consumption in areas with high numbers of cooling-degree days.

Beyond water, significant R&D activities have been dedicated to developing other PCMs for the transportation of temperature sensitive products. Thermo-chemical storage, where reversible chemical reactions are used to store cooling capacity in the form of chemical compounds, is currently a focus in thermal storage R&D projects due to its ability to achieve energy storage densities of five to 20 times greater than sensible storage.

Medium temperature (10°C-250°C) Energy Storage

Distributed thermal energy storage has been around for decades in countries such as New Zealand, Australia and France that use storage capabilities in electric hot water storage heaters. By allowing the heater system to be controlled by the local utility (or distribution company), the demand from these systems is used to manage local congestion and has reduced residential peak demand. In France, for example, thermal storage capabilities in electric water heaters are used to achieve a 5% annual peak reduction.

BTES and aquifer UTES systems have been successfully deployed on a commercial scale to provide heating capacity in the Netherlands, Norway and Canada. These systems utilize holes drilled deep into the ground to store and release energy for heating. Pit storage—where hot water is stored in a covered pit—is used throughout Denmark's district heating networks.

Thermo-chemical storage systems can be designed to discharge thermal energy at different temperatures, making them an appealing option for medium temperature thermal energy storage applications. As with low-temperature applications, this storage mechanism's relatively high energy density potential has prompted significant R&D efforts.

High-temperature (>250°C) Energy Storage

Perhaps the most well-known form of thermal energy storage for high-temperature applications is found in molten salts. This material increases the dispatchability of power from CSP facilities by storing several hours of thermal energy for use in electricity generation (IEA, 2010). Heat storage with PCMs, thermo-chemical energy storage, and waste heat utilization methods offers many potential opportunities. However, these technologies will need to overcome containment vessel design and material stability challenges at very high temperatures before achieving widespread deployment.

Solar Thermal Energy Storage.

Presently, thermal storage is the most widely used method of energy storage in residential solar heating and in concentrating solar power (CSP) plants. Here, heat is transferred to a thermal storage medium in an insulated reservoir during the day, and withdrawn for power generation at night. Thermal storage media used today include pressurized steam, a variety of phase change materials, and molten salts such as sodium and potassium nitrate.

The thermal energy storage systems in CSP plants used today can deliver electric power 6-8 hours after the sun has gone down. They, however, depend fully on the availability of sun in the day before, during which time the thermal reservoirs are replenished. If there is no sun, the thermal storage would stay idle.

In case of prolonged sun-less days, a cogenerator system might be used. Here a gas-fired furnace could provide power when needed, and/or heat water for energy storage. Due to excess losses, however, this is not very efficient use of fossil fuels, so its application is limited.

The most widely used heat transfer methods today are:

- Pressurized steam energy storage. Some thermal solar power plants store heat generated during the day in high-pressure tanks as pressurized steam at 50 bar and 285°C. The steam condenses and flashes back to steam, when pressure is lowered. Storage time is short—maximum one hour. Longer storage time is theoretically possible, but has not yet been proven.

- Molten salt energy storage. A variety of fluids can be used as energy storage vehicles, including water, air, oil, and sodium, but molten salt is considered the best, mostly because it is liquid at atmospheric pressure, it provides an efficient, low-cost medium for thermal energy storage, and its operating temperatures are compatible with today's high-pressure and high-temperature steam turbines. It is also non-flammable and nontoxic, and since it is widely used in other industries, its behavior is well understood and the price is cheap.

  Molten salt is a mixture of 60% sodium nitrate and 40% potassium nitrate. The mixture melts at 220°C, and is kept liquid at 290°C (550°F) in insulated storage tanks for several hours. It is used in periods of cloudy weather or at night using the stored thermal energy in the molten salt tank to generate steam and turn a turbine, which in turn generates electricity. These turbines are well established technology and are relatively cheap to install and operate.

- Pumped heat storage. Pumped heat storage systems are used in CSP power plants and consist of two tanks (hot and cold) connected by transfer pipes with a heat pump in between performing the cold-to-heat conversion and transfer cycles. Electrical energy generated by the PV power plant is used to drive the heat pump with the working gas flowing from the cold to hot tanks. The gas is heated and pumped into the hot tank (+50°C) for storage and use at a later time. The hot tank is filled with solids (heat absorbing materials), where the contained heat energy can be kept at high temperature for long periods of time.

The heat stored in the hot tank can be converted back to electricity by pumping it through the heat pump and storing it back in the cold tank. The heat pump recovers the stored energy by reversing the process. Some power (20-30%) is wasted for driving the heat pump and during the transfer and conversion cycles, but the technology can be optimized for use in large-scale PV plants.

In all cases, large heat energy losses accompany the energy storage processes. At least one third of the energy is lost during the conversion of stored heat energy into electricity. More is lost during the storage and following cooling cycles.

Compressed Air Energy Storage

Compressing air into large high-pressure tanks is one of the most discussed and most promising energy storage methods for use with PV power plants today. It is quite simple way of energy storage, using a compressor powered by the electricity produced by the PV plant compressing air into the storage tank.

A lot of energy is lost by activating the compressor and, heat is wasted during the compression process, so there are several compressing methods that treat generated heat so as to optimize conversion efficiency.

Some of these are as follow:
1. *Adiabatic storage*
Adiabatic storage retains the heat produced by compression via special heat exchangers, and returns it to the compressed air when the air is expanded to generate power. Its overall efficiency is in the 70-80% range, with the heat stored in a fluid such as hot oil (300°C) or molten salt solutions (600°C).
2. *Diabatic* storage
Here extra heat is dissipated into the atmosphere as waste, thus losing a significant portion of the generated energy. Upon removal from storage, the air must be re-heated prior to expansion in the turbines, which requires extra energy as well. The lost and added heat cycles lower the efficiency, but simplify the approach, so it is the only one implemented commercially these days. The overall efficiency of this method is in the 50-60% range.
3. *Isothermal compression and expansion*
This method attempts to maintain constant operating temperature by constant heat exchange to the environment. This is only practical for small power plants, which don't require very effective heat exchangers, and although this method is theoretically 100% efficient, this is impossible to achieve in practice, because losses are unavoidable.

There are a large number of other methods using compressed air, such as pumping air into large bags in the depths of lakes and oceans, where the water pressure is used instead of large pressure vessels. Pumping air into large underground caverns is another approach that is receiving a lot of attention lately.

Pumped Hydro Energy Storage

Pumped hydro energy is a variation of the old hydroelectric power generation method used worldwide, and is used quite successfully by some power plants. Energy is stored in the form of water, pumped from a lower elevation reservoir to one at a higher elevation. This way, low-cost off-peak electric power from the PV power plant can be used to run the pumps for elevating the water.

Stored water is released through turbines and the generated electric power is sold during periods of high electrical demand. This way the energy losses during the pumping process are recovered by selling more electricity during peak hours at a higher price. This method provides the largest capacity of grid energy storage—limited only by the available land and size of the storage ponds.

Flywheel Energy Storage

Flywheel energy storage works by using the electricity produced by the PV power plant to power an electric motor, which in turn rotates a flywheel to a high speed, thus converting the electric energy to, and maintaining the energy balance of the system as, rotational energy.

Over time, energy is extracted from the system and the flywheel's rotational speed is reduced. In reverse, adding energy to the system results in a corresponding increase in the speed of the flywheel. Most FES systems use electricity to accelerate and decelerate the flywheel, but devices that use mechanical energy directly are being developed as well.

Advanced FES systems have rotors made of high-strength carbon filaments, suspended by magnetic bearings, and spinning at speeds from 20,000 to over 50,000 rpm in a vacuum enclosure. Such flywheels can come up to speed in a matter of minutes—much quicker than some other forms of energy storage.

Flywheels are not affected by temperature changes, nor do they suffer from memory effect. By a simple measurement of the rotation speed it is possible to know the exact amount of energy stored. One of the problems with flywheels is the tensile strength of the material used for the rotor. When the tensile strength of a flywheel is exceeded, the flywheel will shatter, which is a big safety problem.

Energy storage time is another issue, since flywheels using mechanical bearings can lose 20-50% of their energy in 2 hours. Flywheels with magnetic bearings and high vacuum, however, can maintain 97% mechanical efficiency, but their price is correspondingly higher.

DOE Energy Storage Program

One of the advantages of electric power generation is that the amount of electricity flowing across the national power grid is relatively stable (fixed) over short periods of time. This, despite of the fact that the actual demand for electricity fluctuates throughout the day. Yet, there are periods of time when the grid is pushed to its limits and new technology to store electrical energy is needed so it can be available to meet demand whenever needed.

Such development would represent a major breakthrough in electricity distribution, where electricity storage devices would manage the amount of power required to supply customers at times when need is greatest (the peak load periods). Energy storage is also needed to help make the variable solar and wind energy output stable in order to be controlled by grid operators.

Energy storage can also balance microgrids to achieve a good match between generation and load. Storage devices can provide frequency regulation to maintain the balance between the network's load and power generated, and they can achieve a more reliable power supply for high tech industrial facilities. Thus, energy storage and power electronics hold substantial promise for transforming the electric power industry.

High voltage power electronics, such as switches, inverters, and controllers, allow electric power to be precisely and rapidly controlled to support long distance transmission. This capability will allow the system to respond effectively to disturbances and to operate more efficiently, thereby reducing the need for additional infrastructure.

The U.S. Department of Energy (DOE), through its Office of Electricity, has developed an Energy Storage Program, which ultimate goal is to develop and reduce the cost of energy storage technology and power electronics and to accelerate their market acceptance.

Since energy storage technology can be applied to a number of areas that differ in power and energy requirements, the Energy Storage Program performs research and development on a wide variety of storage technologies. This broad technology base includes batteries (both conventional and advanced), flywheels, electrochemical capacitors, superconducting magnetic energy storage (SMES), power electronics, and control systems.

The Energy Storage Program works closely with industry partners, and many of its projects are highly cost-shared. It also collaborates with utilities and State energy organizations such as the California Energy Commission and New York State Energy Research and Development Authority to field major pioneering storage installations that are several-megawatts in size. It also supports analytical studies on the technical and economic performance of storage technologies.

Enhanced energy storage can provide multiple benefits to both the power industry and its customers. Among these benefits are:

• Improved power quality and the reliable delivery of electricity to customers;

• Improved stability and reliability of transmission and distribution systems;

• Increased use of existing equipment, thereby deferring or eliminating costly upgrades;

• Improved availability and increased market value of distributed generation sources.

The Energy Storage Program also seeks to improve energy storage density by conducting research into advanced electrolytes and nano-structured electrodes. In Power Electronics, research into new high-voltage, high power, wide-band-gap materials such as silicon-carbide composites, diamonds, and diamond-graphite composites is underway.

For the purposes of implementing the energy storage programs, DOE is also working on new standards for safe and efficient energy storage systems. These are the initial steps in a long way towards mass energy storage, but we know that all great things start with small steps.

Electricity Storage

Based on the types of services that they provide, electricity storage technologies can be grouped into three main time categories:

- Short-term,
- Distributed battery storage, and
- Long-term energy storage

These systems include a number of technologies in various stages of development. Broadly speaking, PSH, CAES, and some battery technologies are the most mature, while flow batteries, SMES, super-capacitors and other advanced battery technologies are currently at much earlier stages of development.

Major R&D efforts exist for many electricity storage technologies. In particular, battery and hydrogen technologies have received significant funding in support of research, development, and demonstration projects in regions including the United States, Japan, and Germany. The primary technology characteristics used in assessing a technology's potential for use in specific applications include storage and operation properties (including energy and power capacity, density, efficiency, scale, discharge capacity, response time, and lifetime or cycling performance), and cost (Inage 2009).

- *Short-term* (several seconds or minutes) storage applications

Super-capacitors and SMES technologies use static electric or magnetic fields to directly store electricity. Flywheels store and then release electricity from the grid by spinning and then applying torque to its rotor to slow rotation. These technologies generally have high cycle lives and power densities, but much lower energy densities. This makes them best suited for supplying short bursts of electricity into the energy system. Modern technologies struggle in today's energy markets due to high costs relative to their market value.

- *Distributed* battery storage

Batteries use chemical reactions with two or more electrochemical cells to enable the flow of electrons (e.g. lithium-based, NaS, and lead-acid batteries). The battery is charged when excess power is available and later discharged as needed. This storage technology can be used for both short- and long-term applications (i.e. both power and energy services) and benefits from being highly scalable and efficient. Furthermore, it can be installed throughout the energy system and has already achieved limited deployment in both distributed and centralized systems for mobile and stationary ap-

plications at varying scales. Widespread deployment, however, is hampered by challenges in energy density, power performance, lifetime, charging capabilities, and costs.

- *Long-term* (hours) storage applications

PSH are currently the most mature and widespread method for long-term electricity storage. In addition, two CAES facilities have been successfully used by utilities in the United States and Germany for several decades. These technologies face high upfront investment costs due to typically large project sizes and low projected efficiencies for non-adiabatic CAES design proposals. In the case of pumped hydro and CAES, geographic requirements can lead to higher capital costs.

Today, there are two CAES systems in commercial operation, both of which use natural gas as their primary onsite fuel and are equipped with underground storage caverns. The larger of these two facilities is a 321 MW system in Huntorf, Germany. Commissioned in 1978, this system uses two caverns (300 000 m3) to provide up to 425 kilograms per second (kg/s) of compressed air (pressure up to 70 bars) produce efficiencies up to 55%. The other system, in McIntosh, Alabama, uses flue gas from its natural gas power plant for preheating to increase overall power plant efficiency (US DOE, 2013).

- *Hydrogen storage*

Hydrogen storage can be used for long-term energy applications. Electricity is converted into hydrogen, stored, and then re-converted into the desired end-use form (e.g. electricity, heat, synthetic natural gas, pure hydrogen or liquid fuel). These storage technologies have significant potential due to their high energy density, quick response times, and potential for use in large-scale energy storage applications. However, these technologies struggle with high upfront costs, low overall efficiencies and safety concerns, as well as a lack of existing infrastructure for large-scale applications (e.g. hydrogen storage for fuel-cell vehicles).

Of all these types, battery energy storage seems the technology of choice today. Because of that, there are many companies involved in R&D of advanced battery energy storage devices, and/or in manufacturing of such.

Electric Grid Storage

Grid energy storage is large-scale storage of electrical energy, using the resources of the national electric grid, which allows energy producers to send excess electricity over the electricity transmission grid to temporary electricity storage sites that become energy producers when electricity demand is greater. Grid energy storage is a very efficient storage method, playing an important role in leveling and matching electric power supply and demand over a 24-hour period.

There are several variations of this method, one of which is the proposed grid energy storage called vehicle-to-grid energy storage system, where modern electric vehicles that are plugged into the energy grid can release the stored electrical energy in their batteries back into the grid when needed. Far fetched, yes, but the future will demand many such ingenious approaches, if we are to be energy independent.

According to the Energy Storage Subcommittee of the Electricity Advisory Committee, "Our vision is that there will be multiple viable energy storage options for competitive and regulated markets, and for different applications and regions, which will yield positive outcomes with high confidence under a wide range of economic, regulatory, climate, and energy scenarios (sometimes called "no regrets" scenarios).

Sounds good, right? "However," the Committee continues, "storage differs from conventional resources in some critical ways that limit its adoption for existing market services. For example,

- Inverter based storage technologies do not meet current North American Electric Reliability Corporation (NERC) definitions for inertial response due to the lack of rotating machinery.
- Many storage technologies are unable to cost effectively meet market rules for serving some applications over a long duration.
- Storage is inherently a zero net energy device so that either the system operator or the resource operator must provide for the restoration and maintenance of storage energy levels as explicit features of new product definitions or as implicit behavior in the market by the resource operator

Grid energy storage presents a number of opportunities, while bringing new issues and problems in the energy markets.

- *Regulatory problems*: The National Association of Regulatory Commissioners (NARUC), FERC, and NERC all have regulatory oversight of aspects of storage. Storage brings challenges to all three groups as well as crossing boundaries in some cases, which complicates things further. This problem cannot be resolved quickly, nor completely.

- *Planning problems*: NERC, RTOs, and regional groups all are responsible for assessing capacity requirements, transmission plans, and regional reliability requirements. Storage can play a role in many of the issues these groups address, which could also present another set of problems.
- *Market participants* on the load and generation sides have opportunities to apply storage in their businesses. That, however, could impact regulated transmission and distribution (T&D) utility operations and reliability if not coordinated.
- *Demand and supply side* provides T&D utilities with opportunities to apply storage to improve reliability and asset utilization. This could impact the wholesale markets and complicate the function of the energy markets.

So it seems that, in addition to the serious technological problems, the existing policies and regulations are inadequate for if and when the electric grid energy storage technologies start coming online.

Conclusions

There are a number of efficient energy storage methods such as fuel cells, new types of batteries, superconducting devices, super-capacitors, hydrogen production, and many others under development, and some in mass production, so the future looks bright in this area.

In practice, however, there are a limited number of energy storage installations around the world, totaling only about 3.0 GW, with the major technologies being:

- Thermal energy storage is over 1,500 MW
- Batteries energy storage is over 500 MW
- Compressed air energy storage is over 500 MW
- Flywheels energy storage is over 200 MW

Energy storage has many advantages, in addition to its unique potential to transform the electric utility industry by improving wind and solar power variability, availability and utilization. It can contribute to the overall energy independence and environmental cleanup by avoiding the building of new power plants and transmission and distribution networks.

Experts consider energy storage to be the solution to the electric power industry's issues of variability and availability, opening new opportunities for wind and solar power use along the way. Complexity, safety, price and other restrains, however, will have to be worked out well before any of the energy storage methods become accepted reality across the U.S. and the world in large.

Energy Storage Finance

In 2015, venture capital (VC) funding for smart grid, storage and efficiency sectors reached well over a billion dollars. Smart grid investment accounting for half of the investments, while battery/storage followed, but large VC funding deals were few and between.

The energy storage gained momentum with the California Public Utilities Commission's (CPUC) passage of the first energy storage mandate in the U.S., requiring the state's largest utilities, Pacific Gas and Electric, Southern California Edison, and San Diego Gas and Electric, to buy the total of 1,325 MW of energy storage by 2020.

Solar residential and commercial funds raised about $4 billion in 2014 and that much in 2015, so a game changer for energy storage companies is in the making. The energy storage market could explode if and when these companies gain access to the large pools of capital through partnerships, mergers and acquisitions. SolarCity, the largest U.S. solar installer and third-party finance company, for example, obtained $1 billion to finance commercial solar energy systems, including such with battery storage.

Solar plus energy storage systems can receive the investment tax credit (ITC) for both energy storage and solar on top of any local incentives, at least until the end of 2016. We foresee a lot of money and effort in the solar plus energy storage area in a rush to meet the deadline. After that, things in the residential and commercial solar and energy storage market get somewhat murky.

Generally speaking, the U.S. battery storage sector is where the solar industry was in the mid 2000s. At that time, the VCs were trying to decide in which solar technology to invest. There were a lot to chose from: c-Si, CIGS, CdTe, a-Si, CSP, CPV and others. In the end, the market chose the c-Si as the most reliable and cheaper on the long run.

Today, the development of the c-Si technology continues, which is evidenced by dropping prices and increasing efficiencies and orders. With this, the attention of the VCs is on the downstream participants; project developers, integrators, and companies adapting innovative financing solutions to gain access to large pools of capital and lower the cost of capital to make economics work for end-users.

The U.S. battery energy storage industry is at the same cross-roads today, c-Si was short time ago. During the next several years we will see the good and bad sides of battery energy storage. We will see if it is suitable for U.S. residential, commercial, and/or utilities type installations.

Case Study—Tesla and Solar City

As an example of things to come, Tesla and SolarCity have joined forces to bring solar energy storage (batteries) systems to the masses by slicing current prices by 60%. This is to be done by bringing new equipment, approaches, and business models for power generation and management.

The price of the new Tesla Powerwall stationary energy storage systems for homes, businesses, and off-grid communities starts at $3,000 for the 7kWh, and $3,500 for the 10kWh models. This new energy storage system, to be packaged and sold by SolarCity, is about a third of the price of similar products available on the market today. It also comes with 9 year optional warranty, which will add to the final cost of the product.

Only a few hundred of these batteries have been deployed at residential and commercial buildings by SolarCity, but regardless, Tesla is expanding its storage batteries manufacturing operations, complete with some technological, business, and marketing innovations.

The Tesla Energy's range of products and services includes DemandLogic, a commercial storage product which has been on the U.S. market since 2014, and GridLogic, a new micro-grid battery energy storage system.

The variety of systems will enable SolaCity to "…offer fully-integrated and affordable solar battery back-up systems for homes, businesses and utilities," according to the officials. Tesla's management is sure that their battery business will grow as the solar market grows in accepting the new technology, with low prices driving the success of the new products.

This all sounds good and doable, especially if the successful Tesla company is involved. Tesla is surely in a position to offer low prices and financing to a large number of customers, which could spur spontaneous acceptance and adoption of the new technology. But, of course, there is a big IF…or several IFs.

The success of this new and unproven technology depends on a number of factors:
- If the technology performs as specified,
- If there are no failures and fires, as with similar batteries in the past,
- If the batteries can withstand the desert heat,
- If the customers can afford the initial cost and O&M expense.

Note: Remember, the battery packs have to be changed at an additional expense 9-10 years down the road.

- If the lapse of ITC in 2017 does not affect negatively the U.S. solar industry.

Note: In 2017, the 30% ITC for residential solar power

drops to zero, and down to 10% for commercial installations. This will have a ripple effect on the U.S. solar industry, and battery energy storage will be affected by it. We just don't know in what way and how much.

Nevertheless, the latest addition to the booming battery storage market, Tesla's Powerwall battery is a new, elegant, and exiting product. It is designed for use in residential and small commercial solar installations, in order to bridge the gap between solar energy supply and the actual power demand.

Nicely said, but is it easily done? The Powerwall battery pack comes in 10 kWh weekly cycle and 7 kWh daily cycle models. They are guaranteed for 9 years of non-stop operation, as needed to provide sufficient power during peak evening hours. Multiple batteries—up to 90 kWh total for the 10 kWh battery and 63 kWh total for the 7 kWh battery—can be installed in homes or businesses with greater energy usage.

Let's see:

"...two Powerwalls installed together could support the majority of household loads, including heating and cooling..." according to SolarCity (Tesla's partner) official. But wait, the two batteries can support heating and cooling? Our 5 ton air conditioning unit might require 20 of these batteries to just turn over once, so what are we missing here? Is the official misinformed, or is this just another example of poor marketing of the 'make a quick buck' type?

The other great sales point, according to the same official, is that, "When the battery is depleted, it can be recharged by solar power even if the outage continues for multiple days." But wait again. When was the last time we had a power outage? About four years ago. And how many days did it last? Just about half an hour. So how do we justify the additional expense for battery storage, if we have half an hour outage every 4-5 years?

But then, America is a rich country, with a lot of rich people, who don't mind spending $3,000 every 9-10 years on a new shiny gadget, that has not been proven efficient or reliable on the long run.

Simple calculations show that battery storage adds about $50 of additional cost to our monthly bill, so this technology might be hard to sell to people with limited finances. It would be even harder to sell in other, poorer, countries, so we just have to wait and see where all this is going.

Using batteries of any kind for large-scale energy storage is even more complicated and many times more expensive undertaking. 1.0 MW solar power plant would require about 200 Powerwall packs. At $3,000 each, the energy storage option would add about $600,000, or $0.60 per watt—and battery replacement every 9 years. 100 MW solar plant would need 20,000 Powerwall packs equivalent, or $60,000,000 every 9 years. Not to mention that this is a best case scenario, and that the batteries might give up after 2-3 years of non-stop operation under the extreme desert conditions.

One needs not to be a solar or energy storage specialist to see that this is enormous amount of money that is simply unjustifiable under normal business conditions. There may be special cases, where energy storage is a must, and where money is not a factor, but we see very few of these around us.

The additional expense would add large financial burden to the already burdened solar power plants, which in most cases makes it simply financially unjustifiable. Not to mention the added risk by using unproven technology to the already uncertain future of the solar industry in the U.S.

In conclusion, if and when the new Tesla battery storage technology is proven efficient, reliable, and profitable in the U.S., its proliferation around the globe is also ensured. Albeit, only developed countries could afford this luxury for the foreseeable future, so it would be a long time before those who need it most—people in the developing countries, where power outages are the norm—will have a chance to enjoy the benefits of battery energy storage.

SMART GRID MARKETS

'Smart grid' is another hot item of late in the energy markets. It is the way of things to come in the near future. We are seeing a lot of activities in this area already, but this is just the beginning. All we are seeing right now is a baby in diapers, laying helplessly in its crib, screaming loudly for attention. This baby will grow fast, but we can only imagine what it would look like when fully matured. We can only imagine the wonders it will bring to the energy sector.

Smart grid means different things to different people and organizations. In general terms, it is an electricity supply network that uses digital communications technology to detect and react to local changes in usage. In everyday terms, it is a class of technology (or a compilation of many technologies) designed to bring utility electricity delivery systems into the 21st century. This will be done by modernizing and automating end-user appliances, interfaces, and communications.

In simplistic terms, the Smart Grid is computer-based remote control and automation of the electrical network.

In technical terms, the Smart Grid is a network of specially designed, complex system of different types of equipment, electronic gear, computers, and software. All components are designed to use two-way communications and computer data processing technologies, as needed to ensure efficient and economically feasible use of electric power by end-users.

The Smart Grid concept includes all parts of the electric power networks; from the power plants, through the transmission and distribution lines, all the way to the consumers of electricity and their appliances in homes and businesses. The benefits to utilities and consumers are mostly in terms of improvements in energy management and efficient operation of their electrical devices and the electricity grid in general. This in turn facilitates and optimizes power use efficiencies, thus resulting in reduced power use and lower monthly bills. Sounds good on paper, right? In reality, however, it is a dauntingly complicated and expensive undertaking.

The smart grid is one way of bringing the outdated power grid and utility companies into the 21st century.

True, the Smart Grid will take us out of the century old way of doing business and bring us a new world of efficiency and comfort. This change is badly needed, because for almost a century now, utility companies have been dictating the rules of the electric generation and usage. Since the 19th century, for example, they have been gathering customers data manually; which was done by meter readers going from house to house. The utilities rely on these workers to also look for broken equipment and detect other problems. And this is just one example of the outdated methods by which utilities deliver and manage electricity, and which the smart grid aims to automate and computerize.

There is no shortage of options and products available to the power industry to help modernize it, but the big problem right now is that the grid itself is not suitable for modernization. The existing "power grid" consists of different networks in different locations, that carry electricity from the power plants to the consumers. This includes millions of electrical poles, many thousands of miles of wires, thousands of substations, transformers, switches, and other complex and expensive equipment. With one word, the existing power grid is an outdated, rigid, frigid, and not-very-smart animal. Some call it stupidly inept and inefficient.

To make it more friendly and smart, the grid operators need to understand and adopt the two-way digital communication language, in order to communicate with devices that could help control the grid operation. All devices hooked into the grid network, such as power meters, voltage sensors, fault detectors, etc., would have sensors to gather data. They would also have two-way digital communication capabilities, providing life updates from the devices to the field and vice versa.

Today, the communications are fed into, and controlled by, the utility's network operations center, where operators makes decision and execute commands to the devices and the grid equipment, in an attempt to optimize the electricity distribution and use. The new Smart Grid, when fully implemented will use automation technology that would basically allow the utility to adjust and control the individual devices and the grid response automatically.

Some of the decisions will be made from computers in a central location, while others will be made at different locations across the grid and at final destination points.

The number of benefits from smart grid operations—in addition to increased power use efficiency and reduced power use—include:
• Enhanced cyber-security,
• Handling power generation from variable wind and solar energy sources, and
• Integrating electric vehicles onto the grid.

The companies involved in developing smart grid technology products and services include technology giants, established communication firms, and many firms developing the new technologies.

The role of the government is the most powerful tool in the Smart Grid development. New policies and technical and financial support measures are facilitating the smart grid growth in the U.S. and other developed countries.

Legislation

In 2007, the U.S. Congress passed Title XIII of the Energy Independence and Security Act of 2007 (EISA). 2007 EISA provided the legislative support for DOE's smart grid activities and reinforced its role in leading and coordinating national grid modernization efforts.

Key provisions of Title XIII include:
• Section 1303 establishes at DOE the Smart Grid Advisory Committee and Federal Smart Grid Task Force.
*Section 1304 authorizes DOE to develop a "Smart Grid Regional Demonstration Initiative."

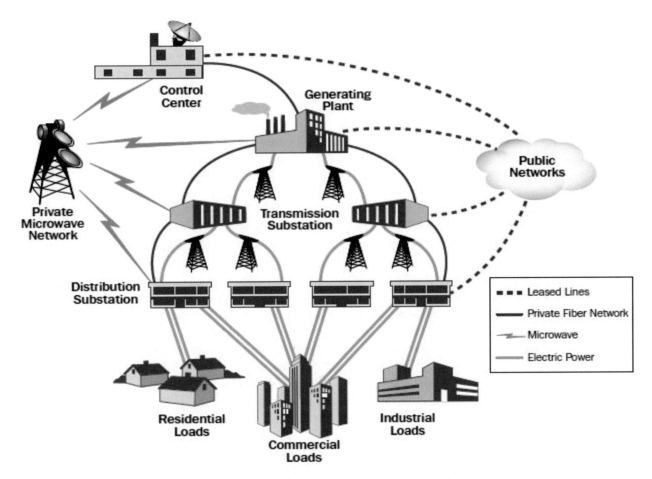

Figure 8-31. The Smart Grid concept (top tier).

- Section 1305 directs the National Institute of Standards and Technology (NIST), with DOE and others, to develop a Smart Grid Interoperability Framework.
- Section 1306 authorizes DOE to develop a "Federal Matching Fund for Smart Grid Investment Costs."

The Leaders

The major contributors and leaders in the U.S. Smart Grid development effort are:

- The U.S. Office of Electricity (OE) is the national leader in this area. OE has partnered with key stakeholders from industry, academia, and state governments to modernize the nation's electricity delivery system. OE and its partners identify research and development (R&D) priorities that address challenges and accelerate transformation to a smarter grid, supporting demonstration of not only smart grid technologies but also new business models, policies, and societal benefits.

  OE has demonstrated leadership in advancing

this transformation through cooperative efforts with the National Science and Technology Council (NSTC) Subcommittee on Smart Grid and the Federal Smart Grid Task Force.

- The National Science and Technology Council Subcommittee on Smart Grid: Chaired by the Assistant Secretary for OE and the National Director for Smart Grid at NIST, the Subcommittee is promulgating a vision for a smarter grid including the core priorities and opportunities it presents; facilitating a strong, coordinated effort across federal agencies to develop smart grid policy; and developing "A Policy Framework for the 21st Century Grid" which describes four goals the Obama administration will pursue in order to ensure that all Americans benefit from investments in the Nation's electric infrastructure.

  These include: better alignment of economic incentives to boost development and deployment of smart-grid technologies; a greater focus on standards and interoperability to enable greater innovation; empowerment of consumers with enhanced information to save energy, ensure privacy, and shrink bills; and improved grid security and resilience.

• Federal Smart Grid Task Force: Directed by OE, the Task Force includes experts from the departments of Agriculture, Commerce, Defense, Homeland Security, and State; the Federal Energy Regulatory Commission (FERC); the Environmental Protection Agency; and the Federal Communications Commission. The mission is to ensure awareness, coordination, and integration of the diverse federal activities related to smart grid technologies and practices.

The Task Force implements Administration policies articulated by the NSTC Subcommittee on Smart Grid while coordinating federal research, development and demonstration; international activities; and outreach and education efforts.

The Activities

The key activities that comprise DOE's OE smart grid strategy are as follow:

• Smart grid demonstrations and deployment activities take advantage of the catalytic effect of substantial investments in the manufacturing, purchasing and installation of devices and systems. These activities leverage efforts under way in the research and development activity area and will help develop critical performance and proof-of-concept data. This activity area is also developing a framework for analyzing smart grid metrics and benefits, which is necessary to help build the business case for cost-effective smart grid technologies.

• Research and development activities advance smart grid functionality by developing innovative, next-generation technologies and tools in the areas of transmission, distribution, energy storage, power electronics, cybersecurity and the advancement of precise time-synchronized measures of certain parameters of the electric grid.

• Interoperability and Standards activities ensure that new devices will interoperate in a secure environment as innovative digital technologies are implemented throughout the electricity delivery system, advancing the economic and energy security of the United States. The ongoing smart grid interoperability process promises to lead to flexible, uniform, and technology-neutral standards

that enable innovation, improve consumer choice, and yield economies of scale. Interoperability and standards activities are not limited to technical information standards; they must be advanced in conjunction with business processes, markets and the regulatory environment.

• Interconnection planning and analysis activities create greater certainty with respect to future generation, including identifying transmission requirements under a broad range of alternative electricity futures (e.g., intensive application of demand-side technologies) and developing long-term interconnection-wide transmission expansion plans.

• Workforce development intends to address the impending workforce shortage by developing a greater number of well-trained, highly skilled electric power sector personnel knowledgeable in smart grid operations. An example of this is OE's involvement with the Consortium for Electric Reliability Technology Solutions (CERTS), a consortium of national laboratories, universities and industry that performs research and develops and disseminates new methods, tools and techniques to protect and enhance the reliability of the U.S. electric power system and the efficiency of competitive electricity markets.

• Stakeholder engagement and outreach activities identify R&D needs for planning, sharing of lessons learned for continuous improvement, and exchanging technical and cost performance data. Information is provided on www.smartgrid.gov to inform decision makers about smart grid technology options and facilitate their adoption.

• Monitoring national progress activities establish metrics to show progress with respect to overcoming challenges and achieving smart grid characteristics.

The Developments

The major efforts and developments in the U.S. Smart Grid technology are as follow:

Advanced Metering Infrastructure

Advanced metering infrastructure (AMI) is one of the most feasible and urgent components of the Smart Grid technology. AMI encompasses smart meters, the communications networks that transmit meter data to the utility at regular intervals (hourly or shorter), and the utility office management systems (such as meter data management systems) that receive, store, and process the meter data.

Usage data from AMI systems can also be sent directly to building energy management systems, customer information displays, and smart appliances. About 46 million smart meters are in place in the United States today. An estimated 65 million smart meters will be installed nationwide by 2015, accounting for more than a third of the approximate 145 million U.S. meters (of all types) in use today.

AMI enables a wide range of capabilities that can provide significant operational and efficiency improvements to reduce costs, including:

- Remote meter reading and remote connects/disconnects that limit truck rolls.
- Tamper detection to reduce electricity theft.
- Improved outage management from meters that alert utilities when customers lose power.
- Improved voltage management from meters that convey voltage levels along a distribution circuit.
- Measurement of two-way power flows for customers who have installed on-site generation such as rooftop PV systems.
- Improved billing and customer support operations

Some real benefits, such as improved operational efficiencies, are being observed where AMI is deployed. For example, Central Maine Power Company has deployed smart meters to its 625,000 customers and has reduced its meter operations costs by more than 80% with annualized savings of about $6.7 million. This is due largely to fewer service calls, resulting in about 1.4 million fewer annual vehicle miles traveled.

Projects under ARRA estimate operational cost savings from 13% to 77%, depending on the nature of legacy systems, the particular configuration of the utility service territory, system integration requirements, and customer densities per line mile.

Nearly 75% of AMI installations to date have occurred in only 10 states and D.C., where on average more than 50% of customers now have smart meters. AMI investments have been driven largely by state legislative and regulatory requirements for AMI, ARRA funding, and by specific cost recovery mechanisms in certain regions.

AMI requires significant investment, and adoption barriers remain for utilities where the business case for AMI is not clear and where prior investments in older metering technology (such as automated meter reading) may present stranded costs. Concerns over meter safety, costs, and consumer privacy protections are being addressed, and enhanced consumer education is a key part of the solution.

Transmission System

The old U.S. transmission system need modernization, if it is to be compatible with the pending Smart Grid upgrades. These include the application of digitally based equipment to monitor and control local operations within high-voltage substations and wide-area operations across the transmission grid.

Synchro-phasor technology, which uses devices called phasor measurement units (PMUs) to measure the instantaneous voltage, current, and frequency at substations, is being deployed to enhance wide-area monitoring and control of the transmission system. Synchro-phasor data are delivered in real time to sophisticated software applications that permit grid operators to identify growing system instabilities, detect frequency and voltage oscillations, and see when the system exceeds acceptable operating limits. This allows them to correct for disturbances before they threaten grid stability.

Additionally, synchro-phasor data enable improved coordination and control of generators, including renewable resources (e.g., wind power plants), as they interact with the transmission grid.

Since the 2003 Northeast blackout investigation revealed inadequate situational awareness for grid operators, utilities have increasingly deployed synchro-phasors to provide real-time, wide-area grid visibility.

Synchro-phasors can provide time-stamped data 30 times per second or faster, which is 100 times faster than conventional supervisory control and data acquisition (SCADA) technology. Technology deployments includes phasor data concentrators that combine, time-align, and verify data from multiple PMUs; communication networks that deliver synchro-phasor data; and information management, visualization, and other analytical tools to process synchro-phasor data and support new data applications for grid operators.

The ARRA projects include a total public-private investment of about $330 million that will increase U.S. synchro-phasor coverage from 166 networked PMUs in 2009 to more than 1,000 networked PMUs deployed by 2014-2015 time frame.

As PMUs are deployed, transmission owners and reliability coordinators are working to develop suitable applications, build out high-speed data networks, improve data quality, and share synchro-phasor data between transmission owners and operators across large regions

Power Grid

Parts of our national power grid are over 100 years old and in need of renovation and modernization. Since we are talking about implementing Smart Grid technologies in it, the renovation effort includes the deployment of sensors, communications, and control technologies that, when integrated with field devices within circuits, permit highly responsive and efficient grid operations.

Smart distribution technologies enable new capabilities to automatically locate and isolate faults using automate d feeder switches and reclosers, dynamically optimize voltage and reactive power levels, and monitor asset health to effectively guide the maintenance and replacement of equipment.

Industry analysts indicate that investments in distribution automation technology are now exceeding those in smart metering and will continue to grow. More than half of the American Recovery and Reinvestment Act of 2009 (ARRA) projects were in deploying distribution automation technologies across 6,500 circuits, representing about 4% of the estimated over 160,000 U.S. distribution circuits.

U.S. government (ARRA based) projects have invested about $2 billion as of 2013 in distribution automation to deploy field devices, such as automated feeder switches and capacitors, and to integrate them with utility systems that manage data and control operations.

The utilities are also beginning to upgrade and integrate their computer systems for managing distribution grid operations including meter operations and customer support, outage management, automated operations within substations and distribution circuits, and as set management.

The impetus for advancing and integrating distribution management systems comes from the significant inflow of new data from field devices, such as smart meters and sensors on equipment and lines that provide utilities with enhanced understanding of grid status and new capabilities for planning and operations.

As utilities begin to apply this information, increased coordination between departments is becoming possible along with greater collaboration between field operations and business processes, including customer interactions. In addition, advanced distribution systems allow greater degrees of automation, including both centralized and distributed control schemes.

Communications

The application of smart grid technologies, such as AMI, distribution automation, customer systems, and synchro-phasors, poses increased data communication challenges for legacy utility systems. To meet these challenges, utilities are investing in a range of technologies with varying bandwidth, latency, reliability, and

security characteristics. Each smart grid application has unique bandwidth and latency requirements, often requiring utilities to use a combination of different communications technologies.

These technologies can be deployed over either an existing public network (e.g., cellular and radio frequency [RF] mesh), which is often economical and readily available, or a licensed private network (e.g., communication over fiber, licensed RF mesh, or microwave links). Cost, reliability, performance, and technology longevity impact a utility's decision-making on communications technologies.

While some utilities implement private communications networks, lower costs and increased technical support are causing public networks to gain momentum for utilities. Recently, public cellular carriers have lowered the per-megabyte cost of AMI communications, making wireless broadband technology (e.g., 2G/3G and 4G LTE networks) more popular with utilities. However, certain applications, such as feeder switches and synchro-phasors, require higher speeds than what cellular networks can offer.

RF-based mesh networks have emerged as the leading technology for AMI and distribution automation deployments in North America, although fiber-optic cable is also used. Many U.S. municipal utilities also use microwave or Wi-Fi wide area communications for AMI backhaul and distribution applications.

To meet the high-speed, high-security communication needs of its utilities, the Western Electricity Coordinating Council is using a secure, fiber-optic, wide-area network, built to the same standard as the nation's air traffic control network, that sends PMU data in less than 30 milliseconds to grid control centers.

The Costs

All this work and the resulting developments come at a cost. Figure 8-32 shows that the U.S. government is serious about implementing the Smart Grid concept in our daily lives and is spending a lot of money to start the effort.

Although the initial steps are encouraging, the overall spending required to fully implement the smart grid is many times higher than what has been spent thus far. The estimates vary from $350 -$500 billion to be spent over a 20 year period to fully implement the smart grid. This includes preliminary estimates of $80-$90 billion for new transmission systems and substations, and upgrades, $250-$350 billion for distribution systems, and $30-$50 billion for consumer systems.

Others estimate that total transmission and distri-

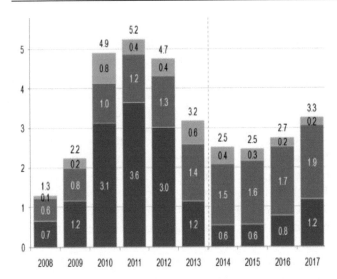

Figure 8-32. U.S. Smart Grid spending (in $ billion)

bution investment needed is nearly $900 billion if the forecast electricity demand is to be met by 2030. Knowing how engineering estimates work; with the final costs usually being 2-3 times higher than the initial estimates, we should allow $2-3 trillion for the completion of the U.S. Smart Grid effort. This includes repairs and updates of the old electrical transmission and distribution systems.

The U.S. is a rich country, and could (probably) afford several trillion dollars to be spent in 2 decades on Smart Grid additions, and we think that it will happen eventually. Maybe much later than planned, but it is inevitable. Many other countries, however, cannot even dream of such luxury. So, how far are we into the smart grid implementation process, depends on what part of the world we live in.

In the U.S. we are talking about it, and obviously, quite a bit of work is under way, but the completion date is not set yet. On the other hand, China, Japan, and some EU countries are at a similar level of development, but are more affected by, and interested in, Smart Grid implementation. The developments of late in the smart grid area, show that some of these countries might get ahead of the U.S. in the near future.

California's Smart Grid Roadmap

Southern California Edison (SCE), which serves about 13 million people in the Golden State, is expected to make more changes to the grid and its overall power infrastructure in the next ten years than in the previous century. It is a good example of a big investor-owned utility that's taken a leading role in the smart grid.

There is already an impressive list of smart grid accomplishments, which mark the beginning of a trans-

formation that is going to bring many changes. That, however, comes with a long list of tasks and projects to be undertaken as needed to meet the myriad challenges in the coming years.

The Issues

One of the top concerns of the California utilities is how to manage the growing share of intermittent renewables (solar and wind) powering its grid. The recent rise of distributed energy resources (solar roof systems) have complicated the situation even further.

State mandates call for utilities to get 20% of their power from green energy today, and that figure is set to grow to 33% (and even 50%) by 2020. Some of that power comes from stable sources like geothermal or small hydropower, an increasing share is coming from solar and wind power. And much more renewable power is to be added during the next 2-3 decades.

The variable power sources are creating problems for the utilities that have traditionally delivered power from central generation plants to end-customers in a number of ways.

The present day issues and solutions can be summarized as follow:

- Solar and wind power are different from the base load sources (fossil power plants) because they cannot be dispatched when needed. Instead they force the utilities to react to the whims of Mother Nature which drives the solar and wind power generation.
- Many of the generation resources are connected directly to the distribution networks, thus bypassing the transmission lines all together. The distribution network, however, was not designed to handle two-way power flow. This prevents the central grid control structure from managing this extra power, so a new system control must be developed and implemented.
- Plug-in electric vehicles are another headache for the utilities. As their number grows, they still use enough power when plugged, which at the grid looks like an entire household power use.
- Millions of new smart meters deployed recently represent another disrupting factor, because they force the utilities to collect and process more data than usual. The new meters also enable customers to sign up for time-of-use rates, instead of the old way of paying flat fees for power used. This is forcing the utilities to base the energy use on real-time feedback from home energy management platforms.

- The old traditional customers are now sophisticated "transactive energy" players, who generate electricity and, a) either send it into the energy markets, or b) shift energy use to take advantage of price fluctuations.
- Customers now have a chance to make energy decisions based on economic choices. This bring a new level of unpredictability into the utilities equation. And as the renewables increase in number, so will the utilities headaches.
- Battery-based, grid-scale energy storage is presently on the drawing boards as one way of balancing the grid. Two Department of Energy smart grid stimulus grant-funded projects are fueling the drive. The energy storage effort is focused simultaneously on the large-scale and small-scale energy storage systems.
- On the large scale, one California utility is working with A123 Systems on an 8-megawatt, 32-megawatt-hour lithium-ion battery storage option in the Tehachapi mountains. The goal is to stabilize and integrate large-scale wind power generation into the grid. For that purpose, SCE has undertaken a number of projects, including a California Energy Commission (CEC)-funded partnership with GRIDiant and New Power Associates, which analyzes transmission and distribution systems for a grid serving about 275,000 customers. The California utilities are also leading players in a DOE stimulus grant-funded project to deploy synchrophasors, which are devices that monitor transmission lines by collecting data in sub-second intervals, across the entire western United States.
- On the small scale, the utilities are integrating several different configurations of batteries, ranging from a 2-megawatt substation battery to smaller residential energy storage units (RESUs), into its smart grid demonstration project in Irvine, California. This is a $80 million, Department of Energy grant-funded project designed to test the interplay of energy-smart homes, solar panels, grid batteries, plug-in car chargers, grid voltage management, self-healing circuits, and communications and controls networks in a one single neighborhood. The list of partners, and potential designers and manufacturers of the related equipment includes; Boeing, General Electric, SunPower and Space-Time Insight, to name a few.
- The utilities are also testing a variety of "smart inverters" and other devices that could help integrate the roof solar power into the grid by allowing each rooftop solar system to better manage its interaction with the grid. This has been successfully tried in Germany, where solar inverter regulations and requirements were put in place recently, in order to manage the massive share of distributed solar power and the grid interactions.

Ultimately, using digital devices on the grid can help record and even manage the variable flow of distributed energy resources and customers' energy use. The challenges, however, are great; from hooking up smart meters to distribution management systems, to integrating transmission grid measurement units into the utility's operations.

- The future role of energy storage is undeniable, but the type and size of the systems will depend on the respective economics, reliability, and efficiency factors. CPUC recently decided that it would require at least 50 megawatts of energy storage for the Los Angeles region power mix by 2020. This may or may not happen, but if it is not done right it might distort the market without solving the energy storage problems.
- At the same time, the utilities are spending a lot of effort and money on keeping the old grid humming. Replacing old power poles and wires, transformers and other components of the crumbling electric infrastructure cost SCE over $3-billion-per-year in capital investment. The number is tripled when the efforts of the other California utilities is added. Utilities in other states are in the same situation, and like it or not must dedicate a lot of effort and money to fixing the old grid and its components. This alone makes any investment in smart grid initiatives involving the old grid imprudent.

With all this considered, we must conclude that:
- The U.S. power grid is not ready for some smart grid additions and improvements,
  — The most advanced and most promising smart grid technologies are not ready for mass-deployment, and
  — Many of the new and advanced Smart Grid technologies that are commercial-ready and available are not approved by the California Public Utilities Commission, since they requires a lot of testing and verification of the cost-benefit equations involved with deploying them.

And so, the smart grid is a great idea and would bring miracles to the grid, the utilities, and their customers, but it its full implementation is still far way in the distant future. Partial additions and improvements re underway, so we have to wait and see how these work out and how they affect the overall energy situation in the country.

U.S. DOE

One example of good intentioned intervention of the government is the Department of Energy's Smart Grid Investment Grant Program. Grants ranging from $500,000 to $200 million are to be issued for deployment of smart grid technologies, most of the recipients of which seem to be working on smart metering solutions (since it is the easiest, and most accepted by the public, smart grid related technology).

In addition, IRS issued a guidance in 2010 providing a safe harbor, under which the $3.4 billion of federal Smart Grid Investment Grants (SGIGs) issued under the American Recovery and Reinvestment Act of 2009 (ARRA) will not be taxable to corporate recipients.

Great news, right? But here is the catch; energy conservation, derived from the full implementation of smart meters (as encouraged by the government grant and tax programs) means lost revenue to utilities at a time when the utilities are in the midst of expensive changes and are not ready for additional revenue losses. The lost revenue, forced by the government programs is simply untimely and is not aligned with the objectives of most utilities and their shareholders.

In addition, long-established regulations have favored supply-side resources over energy conversation, so utilities have been encouraged to add new generation because they earn a rate of return on investments on their assets; and mainly the power generation, transmission and distribution infrastructure. And now, without proper warning and preparation, the utilities must abandon the profitable business model and gear for energy conservation, which simply translates to loss of revenue.

This misalignment might cause delays and even failure of some energy-efficient technologies and services in the U.S. and reflects the disconnect and fragmentation in the sector. Some of the disconnect is caused by the inadequacy of the government bureaucrats, who talk the talk, but seem more interested in their agendas than anything else. This anomaly is fueled by the ignorance of the regulators as well, which completes the circle of incompetency, and which will make the smart grid implementation that much harder.

The bureaucrats do not take into account the fact that the technologies that promote energy conservation have evolved much faster than the regulations that govern utilities and other suppliers. This creates an imbalance, which will continue until new policies and regulations promoting the implementation of demand response and other energy conservation programs are designed and implemented. To cross the divide seamlessly, the new regulations and policies should be timely and structured to justify and encourage full utility participation and investments in demand response.

International Smart Grid

The work on the Smart Grid around the world has started, no doubt, but it is still full of gaps, fragmentation and misdirection. In order to coordinate and encourage the activities, new legislation must be approved, and regulation must be implemented in different countries, as needed to create incentives for the participants to consider investing in the effort.

A lot of work still remains to be done on the power generation, transmission and distribution assets, in addition to promoting energy efficiency and conservation. There must be a balance between the approaches, and the participants must be encouraged to use all these vehicles as part of the power generation and use cycle.

Table 8-17. Smart grid investment

Country	$ Million
China	7,500
USA	7,000
Japan	850
S. Korea	850
Spain	800
Germany	400
Australia	360
UK	300
France	260
Brazil	200

Europe

Europe's Smart Energy Collective is a sector-transcending cooperation involving a wide range of companies working for smart energy and smart grid implementation. Members include: ABB, Alliander, APX Endex, BAM,DELTA, DNV KEMAEnergy & Sustainability, Efficient Home Energy, Eaton, Eneco, Enexis, Essent, GEN, Gemalto, Heijmans, IBM, ICT Automatisering, Imtech, KPN, Nedap, NXP Semiconductors, Philips, Priva, Siemens, Smart Dutch, Stedin and TenneT. —Eco-Seed Staff

In May 2012, members of the Smart Energy Collective in Europe have approved the second phase of the smart grid initiative, which involves the development of five large-scale smart grid demonstration projects in the Netherlands.

Schiphol airport, ABB and Siemens offices, and several residential districts were the chosen sites for smart grid implementation and tests. In order to make this possible, an intelligent energy system is designed that uses a combination of innovative technologies and services. It will enable the owners to keep the costs of energy low with a comparably high level of reliability.

This effort would be an important step towards standardization, in addition to testing the smart grid operation in actual use. A survey during the first phase showed that there are more than 6,000 relevant standards that play a role in the new technologies, in which according to the group, will be introduced to the market over the coming years.

Working groups have been established for standardization, market mechanisms, services and business cases, smart grids, privacy and security, and ICT infrastructure, in order to establish a solid foundation for the design of the five demonstration projects, as well as for future projects.

Asia

Asia is quickly becoming a major player, and the center of global smart grid activity. The combined smart grid market in China, Japan and South Korea is estimated at over $10.0 billion, with estimated increase to over $30.0 billion by 2020, according to industry specialists.

As Asian countries are irreversibly becoming the predominant smart grid markets, the competition and the positioning of different vendors is increasing as well. Lack of clear understanding of the energy scenarios (present and future) in the major Asian countries (China, Japan, and South Korea) is an obstacle that needs to be resolved first.

It is widely expected that the smart grid markets in Asia will move forward at a breakneck pace during the next decade or two. The developed countries are already positioned for the race with over $45 billion in funding, earmarked by the respective governments and utilities across China, Japan and South Korea. The majority of funding and related opportunities, of course, are located in China and its booming energy market.

A level of uncertainty is still distorting the smart grid vision, so determining the trends and establishing a clear path of energy policies and currents will allow much faster implementation of smart grid technologies

in these countries. Once the uncertainties of today are lifted, the a meaningful entry in the Asian smart grid markets will proceed very fast.

China

The growth of the smart grid market is characterized by the special needs of each country's energy demands, as well as the local utilities and existing grids specifics. For example, the smart grid investment in China is focused on transmission and distribution automation, as needed to support the plans for a new power grid (planned to be developed) and robust renewable energy (planned to be built.)

China is aiming to become a world leader in smart grid technologies in the next decade. The "Strong and Smart Grid," an 11-year plan, revealed in 2009, outlines the ambitious steps to get there. It involves all aspects of the power grid, including; increase of power generation capacity, implementation of smart meter programs, emphasis on large-scale renewable energy, and a large transmission lines and substation build-out.

Today, the plan is still alive and well. The State Grid Corporation of China, which is actually one of the largest utilities in the world, and the executor of China's smart grid plans, is already in phase two of the three-phase program. It is the Construction Phase, which is scheduled for completion in 2015.

New transmission lines are a major focus for State Grid in the Construction Phase, which is struggling to meet the growing energy demands of the rising middle class in the East and South. Most coal, hydro, wind, and solar load sources are over 1,000 kilometers away from the populous east and south. High voltage (HV, under 300 kilovolts), extra-high voltage (EHV, 300 kilovolts to 765 kilovolts), and ultra-high voltage (UHV, 765 kilovolts and up) lines are being installed currently, with at least one 1,000-kilovolt UHV AC or DC line installed annually until 2015. Overall transmission line investments for 2015 are approximately $269 billion, equivalent to the combined market cap of ABB, GE, and Schneider Electric as of May 21, 2012.

China is adding so much new transmission capacity and so many new power lines that it could build the equivalent of 3/4 the length of the American transmission grid in just five years. When the dust settles, there will be over 200,000 kilometers of new 330-kilovolts-and-up transmission lines built, for a total of 900,000 kilometers of transmission lines, compared to 257,500 kilometers of transmission lines presently in the U.S.

At a cost of $1.05 million per mile for UHV transmission line and equipment, each UHV line requires billions of dollars to build, and State Grid put in a staggering $80 billion investment into 40,000 kilometer of UHV lines for the 2011 to 2015 Construction Phase. The business case is readily apparent: a 2,000-kilometer, 800-kilovolt UHV DC line has an incredibly low 3.5% line loss rate per 1,000 kilometer and a high 6.4-gigawatt transmission capacity, all the while being 30% cheaper than a 500-kilovolt EHV DC or 800-kilovolt UHV AC line of the same length. By 2020, UHV lines will have 300 gigawatts of transmission capacity, roughly split 60% AC and 40% DC.

The competitive business environment seen in the transmission grid build-out is indicative of the rest of the smart grid market in China—high-quality goods, competitive costs, and a well-built relationship with State Grid all go a long way toward winning a contract. Fierce vendor competition exists, due in part to State Grid's competitive construction procurement process. All projects costing over $300,000 to build are required to go through an open bidding process that aims to enforce fairness and transparency, but State Grid still holds the reigns tightly on choosing project developers. In the process, State Grid has the final say and does a rough 45/45/10 split when evaluating meters, based on quality, cost, and bankability of the company.

With the promise of power shortages disappearing and a stable energy supply base, the build-out of the transmission grid is ushering in the next era of smart grid opportunities in China. Smart meters and renewable integration are already big businesses, and new substation infrastructure has brought with it a vibrant and growing substation automation market. The need for better monitoring equipment has risen as China is keen on decreasing its system average interruption duration index (SAIDI) and improving power quality to its customers.

State Grid has earmarked over $40 billion toward these smart grid technologies between 2011 and 2016, with smart meters alone being a $2.5 billion to $3 billion annual market. State Grid has paid special attention to substation automation technologies, and plans on installing 74 new digital substations for 63 kilovolts to 500 kilovolts by 2015.

While this number is small compared to the existing 40,000+ substation base, State Grid has stated it intends to include digital technology in all new substations built. Companies such as BPL Global have been expanding their substation operations in China, which has been met with stiff domestic competition. The substation market offers promising growth over the next ten years.

The transmission grid build-out also has an impact on technologies at the distribution level and downward. China is building 36 million new urban homes between 2011 and 2015, and modern building automation and smart meter technologies will be utilized. The coming years promise to create a new and vibrant building automation market, but for the time being, the market continues to focus on meeting demand shortfalls and other key infrastructure challenges.

Expect to see an exciting shift toward technologies at the distribution level and downward in the next five to ten years, as China's grid solidifies its transmission grid and generation sources. If the past three years have been any indication of future progress, expect to see China become a leading smart grid market for the next five to ten years. The distribution grid build-out and digitization will be the next major indicator of China's smart grid prowess.

The Rest of the World…

In Japan, the shutdown of many nuclear plants created a need for demand response, energy management and smart meter deployments. Smart grid is on the agenda. Money is available, the need is there, the customers are willing, so only time separates Japan from the full implementation of efficient smart grid concepts.

We would not be surprised if Japan leads the world in this area in the near future, and even become the first country to claim complete smart grid deployment.

The South Korean market is quite different. As the country with the most reliable power grid in the world, South Korea is looking into developing the next-gen smart grid technologies and components (hardware and software) across all segments; primarily for global export.

So, in summary, the global smart grid future is bright. It has the potential to develop into one of the most important parts of the world' energy markets, thus bringing us closer to the overall goals of achieving energy independence and cleaning the environment.

The 21st century will bring a lot of developments in this area, and the Asian countries will be the first to claim smart grid implementation and use.

SPECIALTY MATERIALS AND PRODUCTS

There are a number of energy materials and products today that fall under the "specialty materials" category. They are "special" because they are not part of the top-tier energy materials, but are somehow related to these, and/or are used at some stage of the energy cycle.

Their participation and use in the energy sector, in order of practical importance, can be classified as:

- Practical and very important
- Practical, but not absolutely necessary
- Practical, but too rare or too expensive,
- Show-offy, and
- Just because

Some of the most important specialty materials and products, which are responsible for the proper function and progress of the energy markets, are discussed below:

Petrochemical Products

The petrochemical industry produces a large variety of products and chemicals—in addition to gasoline and diesel. We include these in this section, because they are not an integral part of the energy markets, but are byproducts derived from processing of crude oil, coal, and natural gas. Yet, they are extremely important part of our lives, and we all, and our entire economy depend on them.

The petrochemical industry is undergoing major changes that now affect everything and everyone, including the energy markets. Starting with the 19th century, refineries in the U.S. processed crude oil primarily to recover the kerosene, which was used for heating and lighting purposes. There was no market for the more volatile fractions, including gasoline, which were considered waste and were often dumped directly into the rivers and lakes.

The invention of the automobile shifted the demand from kerosene to gasoline and diesel, which remain the primary refined products today, and which revolutionized the industry. Number of huge refineries popped-up all over the country, hustling and bustling to meet the ever increasing demand for automotive fuels and lubricants. This trend continued until the mid-1970s.

Today, however, federal and state legislation requires refineries to meet stringent air and water cleanliness standards. Obtaining a permit to build a modern refinery today is so difficult and costly that no new refineries have been built in the U.S. since 1976. One exception is the small Dakota Prairie Refinery in North Dakota, which started operation recently.

More than half the refineries that were operating in 1980 are now closed now for a variety of reasons. In 1982, the US. operated 301 refineries with a combined capacity of about 18 million barrels of crude oil each calendar day. In 2010, the number of refineries dropped to 149, with a combined capacity of about 18 million barrels per calendar day.

Got that? Today the 149 U.S. refineries produce as much as the 300 in 1980s.

Truly amazing development! Although the total number of refineries was reduced by more than half, the total production capacity has been fairly constant until now. This is mostly due to the fact that new technologies, increases in facility size, and improvements in efficiencies have offset much of the lost physical capacity.

During 2008-2012 time period of financial crisis, revenue streams in the oil business dried up and profitability of oil refineries fell due to lower demand for product. At the same time high reserve levels prior to the economic recession, forced oil companies to close or sell some refineries. As a result, today we have the fewest refineries on record, and yet their capacity is enough to satisfy the U.S. market and offer significant amount of petrochemicals for export.

Some of the products coming out of crude oil refineries (in alphabetical order) include:

Some of these products are used 'as is,' while others are sent to different facilities for further processing into medicines, cosmetics, fertilizers, etc. useful products. The complete list of petrochemicals and the finished products derived from them is too long to fit in this text, so we mention only the most important products and those that we are most familiar with.

The partial list of petrochemical products in Table 8-18 includes some products that we cannot even live without. Gasoline, diesel, and jet fuels, for example, are primary energy sources produced in huge amounts daily and are responsible for moving people and products around the country and the world. They power our cars, trucks, ships, and planes, without which we all our activities would be paralyzed.

Paints, fertilizers and medications, on the other hand, are part of the 'specialty' materials, and part of the 'peripheral energy' markets, as defined in this text. They are used in huge amounts every day to provide us with comfort and basic necessities. Many of the above materials also appear in other products of practical use, such as cosmetics, medications, paints, plastics, etc., which are used in many different applications every day around the world.

Rare and Exotic Materials

Here we take a closer look at some specialty materials, used in different areas of the energy sector, because

Table 8-18. Key refinery products

Acid gas	LPG
Alkylation	Lubricants
Amine gas	Medication starters
API oil-water separation	Naphtha
Aromatic petrochemicals	Natural-gas
Aromatics	Olefins
Asphalt	Oxygenates
Butanol fuel	Paint bases
Coker gas oil	Pentenes
Diesel	Petrochemical feedstocks
Ethanol fuel	Petroleum coke
FCC gasoline	Propane fuel
Fertilizer feedstock	Sour gas
Gas oil	Solvents
Gasoline	Stripped water
H2S	Sulfur
H-Bio	Sulfuric acid
Hydrogen (MCFD)	Tar shipping
Jet fuels	Wax (paraffin)
Kerosene	

Figure 8-33. Some petroleum products

some of them are truly rare, thus only very limited quantities are available.

Rare and exotic 'specialty' materials are hard to obtain, which increases their prices and that of the technologies that use them.

All energy generating technologies use one or another type of specialty materials in their turbines, generators, etc. Because of that, we need to keep this important factor in mind. when talking about the future of different energy technologies; especially these that use large amount of specialty materials.

Some examples of the growing specialty materials market are:

Rare Earth Oxides and Metals

• Rare earth oxides (REO) is a group of chemical elements which is sometimes referred to as just the Rare Earths or sometimes as REO. This is an informal group of minerals, as most official mineral groups are related first by structural similarities and secondly by chemistry. Since there are no structural characteristics to this group as a whole, the reason for its existence is because of the chemical and general physical property similarities of its members.

The importance of rare earth oxides is on the rise with their use in the energy sector. Rare earth oxides (REOs), such as neodymium oxide, are used in wind turbines, in addition to many other modern devices such as catalytic converters, LCD screens, rechargeable batteries, etc. The wind generators using REOs are more compact, more efficient, and usually require less maintenance, which is especially important in off shore applications.

• Rare earth metals (REM) are part of a series of elements called the Lanthanides that runs from atomic number 57 to 71. In addition, the elements yttrium 39, thorium 90 and scandium 21 are also considered to be rare earth metals because they share similar properties. Uranium, although often associated with rare earth metals, is not technically a rare earth metal.

REM materials are used in nuclear power and nuclear weapons industries for both practical and experimental purposes. Some rare earth metals have found more down to earth applications in metallurgy, ceramics, glass making, dyes, lasers, televisions and other electrical components. Yttrium, for example, has been used for decades in a diamond simulant called YAG (yttrium aluminum garnet) used in lasers, which has physical properties similar to natural garnets but with a brilliance and fire more similar to diamond.

Lithium

Lithium is a chemical element with symbol (Li) and atomic number 3. It is a soft, silver-white metal belonging to the alkali metal group of chemical elements.

Under standard conditions, it is the lightest metal and the least dense solid element.

Li is highly reactive and flammable, so it is typically stored in mineral oil. Because of its high reactivity, lithium only appears in compounds, and is most frequently isolated electrolytically from a mixture of lithium chloride and potassium chloride.

Lithium is used extensively in lithium-ion batteries (Li-ion or LIB). It is a member of a family of rechargeable battery types in which lithium ions move from the negative electrode to the positive electrode during discharge and back when charging. Li-ion batteries use an intercalated lithium compound as one electrode material, compared to the metallic lithium used in a non-rechargeable lithium battery. The electrolyte, which allows for ionic movement, and the two electrodes are the consistent components of a lithium-ion cell.

Li-ion batteries are the most popular types of rechargeable batteries for portable electronics. They have high energy density, exhibit no memory effect, and only a slow loss of charge when not in use. Li-ion batteries are becoming a common replacement for the lead acid batteries that have been used historically for golf carts, EVs, and utility vehicles. Lightweight lithium-ion battery packs provide the same voltage as lead-acid batteries, so no modification to the vehicle's drive system is required.

Li-ion and related batteries are finding increasing use in the energy storage systems today too.

Cadmium, Tellurium, Arsenic, Selenium, Gallium

Cadmium, tellurium, arsenic, selenium, gallium are a series of special materials used in the new thin film solar (PV) panels. Most notably, they are used in the manufacturing of CdTe and SIGS solar panels. Although these materials are not rare, nor very expensive, they are usually found in remote corners of the world, and/or are produced as a bi-product of zinc, copper, lead and other mining operations. In addition, for the most part these chemicals are toxic, so special handling and manufacturing procedures are used.

Summary

The major concern with all these materials today is that scarcity of rare earth (and other specialty) materials may impede the future large-scale deployment of wind and solar power and other energy technologies, or at least make them more expensive.

The total known rare earth materials reserves are estimated to represent over 500 years of supply at current consumption levels, according to U.S. Geological Society. At the same time, the prices for neodymium oxide used to produce magnets, for example, dropped from \$195/kg to \$80/kg in 2012 and the trend continues, which does not suggest imminent scarcity. On top of that, the estimates show that the global wind power industry uses less than 1% of the global demand, so in worst of cases, it would be the last affected in case of global rare-earth materials crisis.

But there is still a big problem on the rare earth materials and the solar and wind energy horizon. It is the fact that over 95% of the current global rare earth and other special materials production occurs in China. As a result, China, for example, controls most of the global rare earth materials exports. Nevertheless, 70% of the world's known (albeit undeveloped) reserves are located in other countries.

As the world demands more of these materials, many mining projects are currently under consideration in more than 20 countries. Advanced research for finding alternative materials in many applications, including solar and wind power generation, is underway as well.

Today we see a combination of a) increased use of specialty materials with the increase of the renewable energy markets, and b) development of new (man-made) materials to replace the natural ones that are in limited quantity. Although the ratio is still in favor of using natural specialty materials, we count on the discovery and use of man-made such in the near future. This is the only way to ensure the continuing growth of the energy markets.

CONTRACTING AND CONSULTING SERVICES

All companies try to do as much work as possible by using their internal talent and resources. 'Vertical integration' is the dream of every company official, because it means end of dependence on third party suppliers. No matter how small or large a company is, however, it often lacks the necessary expertise or resources to tackle the very large and complex projects on its own.

A company in charge of building a nuclear power plant, for example, cannot provide all the equipment, expertise, and services needed to complete the project. In most cases, the company (owner or developer), no matter how large, rich and/or diversified s/he might be, needs to buy some equipment and parts from outside manufacturers and vendors.

Sooner or later s/he also has to hire third party specialized contractors and consultants in order to ensure the timely and quality completion of the project. The outside consultants and contractors bring expertise that the primary developer or owner do not have, so their participation is required in many cases.

And so, an army of outside contractors and consultants start work on the project from day one. Then they descend on the construction site, providing help in many areas of the project development. This could be expertise in the areas of; architectural design, permitting facilitation, legal advice, pouring the foundation, installing the equipment of the future power plant, designing the O&M procedures, ensuring plant's safety and other compliances, planning the decommissioning steps, etc., etc.

During the operation stages, more materials, parts, components, and consumables are purchased from third party vendors. Many services, like transport, maintenance, accounting, waste disposal, personnel hiring, etc. are also executed by third parties. And that process continues for the duration of the project.

At the end of its useful life, the power plant, mine, or oil/gas well has to be decommissioned and the land returned to its original condition. At that point another army of specialists in demolition, waste treatment and disposal, transport services, personnel management, etc. professionals start work on the last stage of the project until the last traces of its existence are removed.

Table 8-19 is a partial list (in alphabetical order) of possible outside manufacturers, vendors, contractors, and consultants in the energy sector. These people and companies participate in all areas and stages of the life cycle of the different energy technologies; their products and services.

Table 8-19 is part of a long list of individuals and companies specialized in some areas, who participate in some stages of large and complex energy related projects design and development. Looking at the Equipment Manufacturers and Concrete Contractors category alone, for example, we see hundreds of companies involved in these disciplines, and thousands of people providing billions of dollars worth of equipment and services to mines, power plants, and other energy related projects.

As another example, going through all steps of a power plant's location search, design, and construction, we see hundreds of other third party companies and individuals specialized, and involved, in every step of the way. Surveying, site evaluation, permit negotiations, site design and development, construction, equipment ordering and delivery, equipment installation, plant operation and maintenance, waste disposal, and finally decommissioning at EOL, step require knowledgeable and reliable people and companies.

Table 8-19. Partial list of third party products and services

Accident Prevention Specialists	Landscape Architects & Landscape Designers
Accounting Specialists	Landscape Contractors
Architects & Building Designers	Lawn & Sprinklers
Audio Visual Systems Specialists	Legal Experts and Consultants
Backyard and Courts Design	Lighting Specialists
Bath and Kitchen Design	Machine Shops
Building Supplies	Marketing Advisors
Cabinets & Cabinetry	Mechanical Engineers
Carpet & Flooring	Media Design & Installation
Chemical Engineers	Mining Specialists
Chemicals Manufacturers	Nuclear Power Consultants
Closet & Storage Designers	Outdoor Enclosures and Lighting
Concrete Contractors	Paint & Wall Coverings
Decks and Patios	Paving Services
Demolition Services	Personnel Trainers
Design-Build Firms	Photographers
Doors Makers and Installers	Plumbing Contractors
Driveways & Paving	Pools Construction and Maintenance
Electrical Contractors	Power Plant Design and Construction
Electronics Specialists	Real Estate Agents
Emissions Specialists	Roofing & Gutters
Environmental Services & Restoration	Rubbish Removal
Equipment Manufacturers *	Septic Tanks & Systems
Equipment Moving & Installs	Siding & Exterior Contractors
Fencing & Gates	Solar Energy Contractors
Financial Specialists	Seismographic Specialists
Furniture & Accessories	Specialty Contractors
Garage Doors	Staircases & Railings
Garden & Landscape Supplies	Stone Works
General Contractors	Storage Tanks and Facilities
Health Advisors	Thermodynamic Engineers
Heavy Transport Vehicles	Tile, Stone & Countertops
HVAC Contractors	Transport Services
Insurance Specialists	Tree Services
Interior Designers & Decorators	Upholstery Specialists
Ironwork Specialists	Waste Treatment and Disposal
Kitchen & Bath Designers	Welders
	Wind Power Design & Installs
	Windows Installs & Treatment

All these 'peripheral' companies and individuals work on their specific tasks, the proper and efficient completion of which determines the final success of the entire project. A failure at any step of the project would delay, make more expensive, and/or even bring the entire project to a complete failure.

Solar power plant contractors and consultants, for example, would work with the owners and principals of the project to provide a variety of services, which are not the core competency of the principals. These contractors and consultants—specialists in different areas of the project's disciplines—provide their specialized services and assist with whatever is needed to complete the project on time and within budget.

In more detail, some of the 'peripheral' products and services provided by third party specialists in the solar energy sector are:

The Solar Sector

The quick development of solar power installations in the country and the world, brought to life many new industries. The quick development in this field required the assistance of thousands of contractors and consultants of different types.

Some of the key services contractors and consultants provide to an existing or planned solar power installation are shown in Table 8-20.

Each of these tasks requires thorough understanding of the technical and logistics aspects of the critical elements involved in making proper decisions. Each of these tasks can also involve several individual consultants, contractors, and companies. Taking appropriate action, and supervising the proper completion and operation are on top of the list of priorities of the contractors and consultants, and usually prove to be critical for the success of the project.

Once the construction of the solar plant has been completed, it goes into full operation, Here, another team of specialists would advise and/or provide hands-on assistance with the safe and efficient operation and maintenance of the plant.

O&M Services

As power generation (solar and wind mostly) spreads over huge areas, it brings new and unprecedented challenges. The newly installed wind and solar power plants operate without much human intervention... until something breaks. Then the company has to send its own employees, or hire an outside service company, to do the work.

Assistance with operation and maintenance (O&M) of solar and wind power plants is a new and mostly unexplored area of the energy markets, where the traditional concepts and approaches apply only to an extent. As a result, as the solar and wind power plants grow and multiply, so does the maintenance industry. The third-party specialists and companies bring their expertise and know how, in addition to trying new and improved methods for more efficient, cost effective, and safe O&M procedures.

Solar Power Plants Maintenance

By investing in a PV system, we are setting the course for a sunny, energy independent future. Sure enough, a brand new solar installation with its shiny modules looking at the sun is a symbol of that promise.

Table 8-20. Partial list of third party specialists involved in solar power generation

- Site evaluation and overall design
- Budget and financial assistance
- Design of a racking system to fit the site
- Design the electrical system for the site
- Provide electrical schematics and racking certifications for the permits
- Assist with planning, permitting, interconnection, and PPA issuance
- Assist with inspectors and plan checkers via NABCEP certified engineers
- Legal research and assistance
- Marketing research and assistance
- Suggest manufacturers who provide quality equipment at the right price
- Suggest contractors able and willing to do the job properly
- Quote the parts and services needed to complete the project
- Assist with subsidies and rebates paperwork
- Provide and verify the solar equipment
- Oversee the installation of the solar equipment
- Assist with, and coordinate, long-term O&M procedures
- Design the EOL disassembly and waste disposal procedures

But wait…as time goes on, these modules get abused by the weather and people and need help.

Improperly operated and maintained solar system can drop up to 30% in efficiency, thus reducing investment payback. For this and other reasons, periodic and emergency maintenance procedures have to be considered as part of the regular O&M work load.

Some power field developers and owners have their own maintenance teams, but as the power fields increase in size and number this becomes a daunting task. Because of that, they often prefer to hire local service companies to do the maintenance work.

The local service companies usually offers customized service and excellent support which ensures the proper and safe system operation over its entire operating time. These companies perform commissioning, remote monitoring, regular and emergency maintenance that the owners can rely on from the very start to the end of the project.

The outside services used most often to optimize performance and reduce failure rates, are

Real Time Monitoring

To ensure maximum return on investment (ROI), it is important to have a watchful and proactive eye on the system performance. Solar monitoring services usually consist of advanced, real-time monitoring equipment and capabilities, able to track and analyze the performance and detect potential issues. These issues are then quickly resolved remotely, or by dispatching field service engineers to get the system back on track. This is the best and most flexible way to manage performance, maintenance schedules, as well as tracking and reporting warranty claims and providing complete financial information.

Preventative Maintenance

Preventative maintenance plans include manufacturer preventative maintenance and a visual inspection of the arrays and BoS components, such as: panels, combiner boxes, disconnects, and data acquisition systems (DAS) systems. Thermal scans on all electrical DC wiring, inverters, e-housing, pad-mounted transformers, and other electrical equipment identify anomalies. Torque checks and DAS maintenance must also be included. Expert advice is provided through remote system monitoring with 24/7 customer support. A written maintenance summary report after each visit, and an annual maintenance report are used to provide complete documentation.

Proactive Maintenance

Proactive (aka preventative) maintenance plans expand upon the essential service components of the Preventative plan. The Proactive plan adds in a 24-hour emergency response time and written report for all emergency services. Spare parts inventory management is an integral part of the plan, and simplifies maintenance planning and preserves uptime in the event of an unforeseen mechanical issue. Warranty administration allows for greater focus on revenue-generating activities. System performance is improved through regularly scheduled sub-system testing and verification, real-time monitoring, electrical, thermography, and full calibration services.

Performance Maintenance

Performance maintenance plans builds upon the Proactive plan scope of services, and delivers the ultimate in full 24/7 O&M support. It consists of a complete system design review and 30-day validation test, annual performance ratio guarantees, and enhanced management and reporting. This service provide system owners with the optimum level of performance and long-term security. All aspects of the project are considered and all deviations (except for weather pattern shifts) are taken care of immediately. By leveraging the collective expertise of the service company, the power field owners benefit from a superior plan that encompasses all necessary service components to generate bankable guarantees, which virtually eliminating risk altogether.

PV Panels Cleaning Services

After months of operation, solar panels get covered with dust, dirt, bird droppings, leaves and other residuals. Any and all foreign objects on the solar panel surface hinder its efficiency. This causes reduced power output and may even lead to electrical failure of the affected solar cells.

Solar panels cleaning is therefore recommended on scheduled period of time, and after some weather caused anomalies such as dust storms. Solar panels are a big investment, and like all investments, they must be taken care of.

For this purpose, the developers or owners of the solar projects have specially trained maintenance teams that specialize in solar panels cleaning. As the fields are growing in size and number, more and more developers and owners use local third party service companies for this task.

In all cases, the first step of the cleaning process is a thorough inspection and documentation of the solar field and its components. The foundation, the supports and other peripherals must be free of damage before starting the cleaning process. If all is well, then the entire system is inspected from top to bottom. The search is focused on looking for broken panels, loose wires/connections, wear/tear on the panels, seals and frames, and broken zip ties. Any problems must be corrected before the cleaning begins.

After all the inspections and repairs are complete, the cleaning can start. Full size truck/trailers equipped with large water tanks are most often used. Pure de-ionized water is the preferred cleaning media since it leaves no residue. Special solar panel brushes, made specially for solar panels cleaning are used. These brushes have soft bristles for safe and gentle removal of dirt and debris. This allows thorough cleaning and virtually spot free drying of the panels' surface.

Our conclusion thus far is that when compared with other more established engineering based industries, O&M operations in the solar and wind energy industries still have a long ways to go up the learning curve.

Waste Management

In February, 2014, the European Waste Electrical and Electronic Equipment Directive (WEEE) came into force. It mandates PV companies to ensure the collection and recycling of their discarded end-of-life products and to guarantee the financial future of their PV waste management.

In 2015, Germany introduced the Waste Electronics and Electrical Equipment (WEEE) legislation (ElektroG). According to it, solar (PV) modules are classified as household equipment.

Therefore:

• Manufacturers are responsible to finance the collection, transport, processing, and disposal of their end-of-life modules. The actual collection of the waste materials can be performed by a municipal waste-collection network, a subcontracted industry scheme, or by the manufacturer.

• In order to keep track of the use and disposal of PV modules, the manufacturers must provide a legally binding financial guarantee on their annual sales of PV modules in the country.

Note: The new law applies to Germany-based companies, which manufacture or import PV modules for the German market after October 24, 2015.

• The official register Stiftung ElektroAltGeräte (EAR) will coordinate communications between the municipal waste network and the manufac-

turers. EAR will also issue mandatory Producer Identification Numbers, which gives permission for firms to market PV modules in Germany. All manufacturers must now have an EAR-accredited financial guarantee before being given a Producer Number.

There are limitations to disposing of waste modules at municipal collection points already, so in such cases, the waste is managed by an industry-led compliance and waste management scheme. This could be quite expensive and might cause excessive financial burden to many companies.

Such dedicated, industry-focused scheme, however, has the relevant market and product knowledge to collect PV modules when they become waste, and recycle or dispose of them as needed. This with time will allow to bring order in the entire life cycle, and control costs, while maintaining high environmental standards.

Compliance with WEEE regulations is crucial for allowing a level playing field in the PV industry where compliant companies do not have to pay for their free-rider peers. Free-riders bring a tremendous disadvantage to compliant companies, especially in countries where PV modules are classified as household products, such as France, the United Kingdom, the Netherlands or Germany under its imminent new law.

Those who are compliant pay double the bill. Companies pay for their WEEE-compliance and waste operations but they may also end up carrying the burden of poorly pre-financed waste at municipal collection points. If the pre-financing through compliant companies does not match the actual market volumes, non-compliance is a clear game changer.

Japan is also headed in this direction, and will soon begin implementing recycling rules for PV modules, having recognized that the vast growth in solar deployment under the feed-in tariff could lead to an accumulation of huge PV modules waste in landfills.

Wind Power Plants Maintenance

60-70% of the U.S. wind power capacity are slowly coming off their new product warranties, new service opportunity is opening for many technical companies. The question now is how the new customers/owners/operators will handle the scheduled and emergency maintenance needs

There are over 70 GW of wind power projects in the U.S. alone and 117.3 GW of installed wind capacity in Europe. This includes a mix of 110.7 GW of onshore projects and 6.6 GW of offshore projects.

With so much equipment working day and night in the field, getting the right strategy of operating and maintaining wind power plants has been increasing in importance.

The European wind energy operations & maintenance (O&M) industry is one of the most dynamic in the world with many other markets looking to Europe for inspiration on contracting structure, innovation in service optimization and cost reduction strategies.

The O&M of wind projects worldwide accounts for one quarter of the total world wind industry's levelized cost of energy (LCOE).

The preliminary results show that the transition from 'end of warranty' into the long-term phase of operations and maintenance of the facilities is a very important one. The Operators of wind farms around Europe are increasingly assessing which O&M model to opt for, with the final goal of taking greater control of their assets than ever before. This is echoed by the huge focus on O&M strategy and data analysis lately.

The supply chain working in wind energy O&M are being called on to help asset operators get the most out of their farms through services that are easily monitored and increasing the energy yield of their assets.

Table 8-21. Wind industry's major contract areas

O&M strategies	27.1%
Data analysis	16.5
Blades	11.9
Gearboxes	11.3
Health & safety	9.0
Electrical, wires	7.4
CMS	7.1
Offshore work	5.1
Other	4.6

Traditionally, owners and operators of power plants have had a choice of two types of maintenance: Reactive and Preventative. It is easy to deduct that reactive maintenance is the least effective and most costly strategy in the long run. In spite of the disadvantages, however, it is still relatively common across the industry.

This is due mostly to lack of understanding and planning on part of the owners. As a result, instead of spending a little money in the beginning, they continue spending a lot of money for the duration of the project. Reactive maintenance in the operation of a nuclear plant, for example, is unthinkable and even criminal, so the accepted and tried method is that of Preventative maintenance.

And so, the most established method of preventative (PM), or scheduled, maintenance is used predominantly in the energy industry. Here, a PM schedule is agreed upon and specified to be executed at certain periods of time following engineering specifications for electro-mechanical and safety checks and remediation.

In recent times, however, a new, even more efficient and cost-effective method—that of predictive maintenance—has been taking root in the industry. This method allows the use of the latest technologies, of wireless communications, field sensors, weather data etc. to be entered in and analyzed by a computer, which then identifies and reports potential problems. On the basis of this information, combined with the hands-on experience of the maintenance engineers and technicians, teams are dispatched to the field to verify and correct the problems BEFORE any incidents occur.

The advantages and disadvantages of the different maintenance approaches are as follow:

Reactive Maintenance
Advantages:
- Little or no upfront cost
- Relatively low staff overhead
- May be the only reasonable approach to dealing with unforeseen failures

Disadvantages:
- Increased losses from unplanned downtime
- Increased staff costs if expensive overtime is needed
- Increased capital costs of replacement parts (which may themselves be charged at a premium for short-notice supply), plus related logistics
- Additional costs from damage due to knock-on impact of failure on other components
- Higher insurance premiums, or inferior policy terms due to higher risks

Preventative (scheduled) Maintenance
Advantages:
- Cost effective for many capital intensive processes and equipment
- Provides flexibility for adjusting maintenance frequencies
- Lengthens component lifespan
- Reduces equipment and/or process failures
- Results in cost savings over reactive methods

Disadvantages:
- Labor-intensive approach
- Increases risk of failure due to incorrect reassembly or through accidental contamination of lubricant systems
- Does not take into account condition or actual rate of deterioration of components

Predictive Maintenance
Advantages:
- Increased component operational life and availability
- Allows for pre-emptive corrective actions
- Results in decreased downtime, minimizing lost income from generation
- Lowers costs for parts and labor
- Reduces risk associated with catastrophic failure
- Typically results in cost savings over preventative maintenance

Disadvantages:
- Increases investment in diagnostic equipment
- Increases investment in staff training
- Savings potential can be a 'hard sell' to management.

Recent surveys of different U.S. and European power plants show that 27% of the respondents use reactive maintenance, 23% predictive, and 50% have switched, or are switching, to predictive maintenance.

Regardless of the official response, however, over third of respondents claim to still use reactive maintenance in some cases. No doubt, that there non-critical situations where this kind of approach is acceptable; but the experience shows clearly that moving away from a reactive approach improves reliability and reduces unnecessary financial losses over the longer run.

The owners have a choice to contract an outside company for the O&M operations, or to do it themselves (DIY). The contractors could be original equipment manufacturer (OEM), or a third party independent service provider (ISP) and in both cases they charge much more than self-O&M.

Wind Power O&M Trend

The 7th Annual Wind Energy Operations and Maintenance Summit, USA was held in April, 2015 in Dallas, Texas. The previous summits have clearly outlined the necessity and importance of figuring out the operations and maintenance (O&M) procedures and expenses of wind power plants.

The global O&M expenditure for wind turbines, according to the insiders, is expected to rise from $ 9.25 billion in 2014 to over $18 billion by 2020. This unprecedented rise of expenditure for wind turbines O&M is driven mainly by:
- Sharply increasing numbers of new wind installations, and
- Increasing number of aging turbines.

The above O&M market drivers require sophisticated power performance and reliability centered maintenance software and hardware, as turbine owners look to decrease the maintenance expenses of the aging fleets, and maximize profits.

The owners now have to make decisions involving the type of O&M they would prefer to use on the short and long term. The choices are limited, and a wrong decision could bring a project to its knees.
It all comes down to choosing between:
- OEM, ISP, Self-Perform, and/or a combination of O&M approaches, and
- If this O&M approach would be used temporarily, mid-, or long-term.

The right choice would determine the future of the project and its profits. Making the right choice requires owners and operators to evaluate and understand fully the risks of the operation. For that purpose, there are commercially available models that deal with every available approach, which take all risks and trends, in order to provide the right O&M choice, or combination of choices.

It is important to understand that over 30% of the U.S. wind turbines will be at least decade or more old by 2020. In wind turbine language, this is considered quite old. Making plans for end of life (EOL) procedures is (or better be) on the agenda of the wind power plant owners. Many parts on the old wind turbines are worn out, and others are wearing down fast.

40 GW of GE and 20 GW of Vestas and Siemens wind turbines would be off-warranty by 2020 in the U.S.

The turbines are tired and break more often, but they have to run another decade or two, so the cost of their upkeep is increasing. Keeping in mind that most wind turbines are financed for 20 to 25 years, and could last 30-40 years, they must be kept in perfect operational order for the duration. And for that purpose, a new 'peripheral' service industry is developing in the U.S. Here, the latest trends in O&M practices employed by the US industry at large are applied to wind turbine support.

The wind power market is already quite dynamic and evolving, with the latest in the O&M area being the trend of owners and operators to self-perform the maintenance of their turbines. The main reason for that is control over the quality of the operations and reduced cost of DIY modus operandi. This is not always the case, but as the trend develops and matures, it might prove to be the right thing to do.

A wind farm's O&M is a complex and very expensive undertaking. The expenses can reach 10-15 % of the total cost of power generation in an onshore wind farm. In the U.S., the O&M expense was slightly less than $1.00 billion in 2014, and is expected to increase to $4.00 billion by 2020. This will increase its share in the overall wind market from 10% to nearly 22% by that time.

This cost is even higher, over 25%, in the case of offshore wind farms. Such costs might wipe out all hopes for profit, especially as the fleets age and the O&M expenses increase. Offshore O&M is several times more difficult and expensive to execute than onshore O&M for obvious reasons. For these reasons, in 2014, the global offshore wind power accounted for about 2.5 % of the world's cumulative wind power capacity.

In 2015, the global PV market showed the strongest growth since 2008. Demand increased by 30% and wind installations reached 57 GW, which is over 10 times increase during the same time. The global onshore wind O&M market is estimated to grow fom $8.34 billion in 2014 to over $14 billion by 2020, or compound annual growth rate (CAGR) increase of well over 9.0%.

At the same time, the offshore wind O&M market is expected to increase at even faster rate of 26% CAGR. The trend of quick offshore projects development is due to broad experience gained in onshore projects, and favorable renewable energy policies, mostly in Europe.

With all these wind turbines in installation and operation stages, O&M becomes a critical component of the equation. Since O&M of large-scale wind power plants is a new industry, we expect to see a lot of changes in the days to come. Even if the trend of self O&M (O&M done by the owner) persists, the complexity of the operation suggests that a number of supporting industries would be created or expanded to support these efforts.

ENERGY INFRASTRUCTURE

The production of different energy sources (coal, natural gas, and crude oil mainly) and energy types (LPG, gasoline, electricity) requires extensive network of pipes, storage vessels, transport vehicles, and transmission lines. One could argue that this important and complex network is not 'peripheral' but rather part of the energy markets, and it would be hard to argue with such an argument. For the purposes of this text, and for more clarity, however, we consider it thus.

The power grid is the largest and most important part of our energy system. It drives our economy and provides comfort in our homes. The U.S. is what it is thanks to this amazing compilation of power plants, substations, control centers, transmission wires, and thousands of people working day and night to ensure the proper and safe function of each pieces of equipment.

The Power Grid

Our national power grid is very old—with some portions in operation more than 50 years and showing signs of deterioration. Its 200,000 miles of metal and wooden poles, wires, and electrical and electronic equipment, are in bad need of upgrades.

The sharp increase of renewable energy sources (solar and wind) with their variability puts additional stress on the aging network. The ongoing implementation of Smart Grid elements into the national grid complicates its function further.

We hear talk about China and other countries having the capability to shut down the U.S. power grid and other critical infrastructure through cyber attacks. While this seems unlikely, due to the immensity of the power grid, there is no doubt that parts of it can be disabled by physical or cyber attacks.

The U.S. national power grid is a model T, which four cylinder engine is being replaced by a F-35 jet engine

It is hard today to find anyone who has anything good to say about the U.S. power grid. Bill Richardson, ex-energy secretary during the Clinton administration, called the grid 'a third-world grid.' That was over 20 years ago, and still summarizes the opinion of most Americans.

The Report Card for America's Infrastructure, issued by the American Society of Civil Engineers, gives the U.S. electric grid a D rating, and summarizes it as follow, 'The U.S. power transmission system is in urgent need of modernization. Growth in electricity demand and investment in new power plants has not been matched by investment in new transmission facilities. Maintenance expenditures have decreased 1% per year since 1992. Existing transmission facilities were not designed for the current level of demand, resulting in an increased number of "bottlenecks," which increase costs to consumers and elevate the risk of blackouts.'

Edwin D. Hill, president of the International Brotherhood of Electrical Workers, says, 'The average age of power transformers in service is 40 years, which also happens to be the average lifespan of this equipment. Combine the crying need for maintenance with a shrinking workforce, and we may find that the 2005 blackout that affected parts of Canada and the northeastern United States might have been a dress rehearsal for what's to come. Deregulation and restructuring of the industry created downward pressure on recruitment, training and maintenance, and the bill is now coming due.'

The Federal Energy Regulatory Commission chairman Joseph Kelliher says, 'The U.S. transmission system has suffered from underinvestment for a sustained period. In 2005, the expansion of the interstate transmission grid in terms of circuit miles was only 0.5 percent. At the same time, congestion has been rising steadily since 1998.'

Continuous underinvestment in the national grid is a problem, which will affect the future of the U.S. energy sector and the entire economy. A drastic national solution is needed. How this would be done, however, remains to be seen. Many billions of dollars are needed to do a complete overhaul and add improvements, as needed to make it 100% reliable and safe.

Gas Pipelines

There is a lot of talk lately about the ageing infrastructure of most large cities. In some cases, like New York, the underground system of electric cables, and gas and water pipes is over 100 years old. More than half of the gas mains are made of cast iron, wrought iron, or unprotected steel. These are materials that are vulnerable to corrosion and cracking, especially in cold weather.

As time goes on, moisture, pressure, and rodents do a lot of damage to the defenseless wires and pipes, so that many are failing and in need of repair. This opens a large marketplace for maintenance services in most American cities.

The rusting and rotting gas pipes present a real danger hidden beneath the streets.

There are over 1.2 million miles of gas pipes across the U.S. in 2014 there were about 12 leaks per 100 miles of gas pipes.

New York City has over 6,000 miles of pipes transporting natural gas, large part of which has leaked.

Figure 8-34. Underground gas pipe replacement.

Some have caused explosions that hurt and even kill people. As a matter of fact, these accidents are startlingly common, numbering in the thousands every year, according top federal records.

In 2012 alone, there were nearly 10,000 leaks reported in New York city and Westchester County. Many of these were considered hazardous because of the dangers they posed to people or property. Most of the leaks proved harmless, simply dissipating into the soil or air. But when gas finds an ignition source, the results can be deadly.

Several separate episodes in Queens in recent years killed people, and a half-dozen others in the city left people injured, according to federal records dating back 10 years. A rupture in a major pipeline in San Bruno, California caused an explosion that killed 8 people. A leak from an 83-year-old cast-iron main in Allentown, Pa., caused an explosion that killed 5 people.

Utilities across the country have been struggling to replace thousands of miles of these old, metal pipes with pipes made of plastic or specially coated steel that are less prone to leakage. This is not an easy, nor cheap undertaking. To replace all of the old mains, the city of New York alone is looking at a cost of over $10 billion.

The present replacement schedule calls for 50-70 miles of pipe per year, so it would take nearly three decades to replace what regulators have identified as the most leak-prone pipes. In the meanwhile, there could be more explosions expected in New York and most other places in the U.S. that use gas.

The most aggressive state in old gas pipes replacement efforts in the country is Ohio. Utility executives and state regulators were concerned that the original 40-year schedule to replace pipes that were already 50 to 75 years old was too slow. As a result, the state's major utility, Duke Energy, serving the Cincinnati area, was granted approval in 2002 to begin a 10-year, $700 million program to replace about 1,200 miles of cast-iron and bare-steel gas pipes. The number of leaks per miles for Duke Energy now ranks among the lowest in the country. In 2007, Dominion East Ohio, which serves Cleveland and northeast Ohio, initiated a 25-year, $2.7 billion program to replace 4,000 miles of pipe.

The work in New York cannot proceed at such fast speed, because of heavy traffic disruptions. As a result, it cost over $2,000 a foot, or well over $10 million a mile, to replace a gas main in the city. This means that the gas pipe replacement effort in New York and the suburbs will be going for decades at a cost of many billions of dollars.

Oil Pipelines

Transporting oil via pipes is the safest, albeit not 100% safe, method of moving oil from one location to another—even if these are thousands of miles a part. The U.S. pipeline system totals about 223,000 miles. From the refineries, gasoline, fuel oil, and such, go by truck or train tank car to wholesalers or large consumers. The U.S. tank truck fleet numbers about 160,000, and the number of rail tank cars is slightly larger, over 165,000.

The risks related to pipe transport are few, but serious in nature if and when the pipe system fails. Oil spills are the most common result from pipeline failure, and result in contamination of large land areas. These spills are difficult to clean and are the cause of a number of law suites and protests from locals and environmental organizations—some of which have contributed to shutting down pipeline sections and prevented others from being built.

Background

Pipelines are widely used for transport of crude oil from the oil wells to the refineries for processing. Thus processed products are also often delivered via pipelines to the point of use. There are aboveground, underwater and underground pipelines, varying in size from several inches to several feet in diameter. These pipelines move vast amounts of crude oil, natural gas, LHGs and liquid petroleum products across the US and most other countries.

The first successful crude-oil pipeline, a 2-inch-diameter wrought iron pipe, 6 miles long with a capacity of about 800 barrels a day, was built in Pennsylvania in 1865. During WWII large pipe networks were built in the US, to move oil from coast to coast. The Keystone XL pipeline, planned to deliver Canadian syncrude oil to Gulf Coast refineries, is stalled in the permitting and political debate stages. If and when completed, it will transport millions of gallons of crude oil and petrochemicals across the U.S.

Many locals and a number of environmental groups are fighting the project, because of oil spill dangers, but it is unlikely that they can stop it. Oil is too important to our energy and national securities; so, if not this one, the next administration for sure, will make the Keystone XL pipeline happen.

The lower 48 states receive 1.5 to 1.6 million barrels of oil per day, courtesy of the Alaskan pipeline. This pipeline, 789 miles of 48-inch steel pipe, curving up and down and around hills and valleys, carries great quantities of oil from the rich North Slope field to the port of Valdez in southern Alaska. Here the oil is loaded on tankers for shipment south. The Alaskan pipeline was built in controversy which is still unresolved and is a subject of high-level debates and legal battles.

Today, liquid petroleum products are moved long distances through pipelines at speeds of up to 6 miles per hour, assisted by large pumps and compressors along the way, at intervals ranging from 60 miles to 200 miles.

Pipeline pumping pressures and flow rates are controlled throughout the system to maintain a constant movement of product within the pipeline. Sensors, control mechanisms, and safety devices are installed along the length of the major pipelines to prevent spills, fires, and other possible disasters.

Pipelines run from the frozen tundra of Alaska and Siberia to the hot deserts of the Middle East, across rivers, lakes, seas, swamps and forests, over and through mountains and under cities and towns.

There is a network of over 95,000 miles of petroleum product pipelines in the United States. This network delivers finished petroleum products to the end customers. It is separate from the network of crude oil pipelines, and balances the demand and supply conditions in each region.

The initial construction of pipelines is difficult and expensive, but once they are built, properly maintained and operated, they provide one of the safest and most economical means of transporting these products.

Types of Pipelines

There are four basic types of pipelines in the oil and gas industry—flow lines, gathering lines, crude trunk pipelines and petroleum product trunk pipelines.

- Flow pipelines move crude oil or natural gas from producing wells to producing field storage tanks and reservoirs. Flow lines vary in size from 5 cm in diameter in older, lower-pressure fields with only a few wells, to much larger lines in multi-well, high-pressure fields. Offshore platforms use flow lines to move crude and gas from wells to the platform storage and loading facility. Lease lines carry all of the oil produced on a single lease to a storage tank.

- Gathering and feeder pipelines collect oil and gas from several locations for delivery to central accumulating points, such as from field crude oil tanks and gas plants to marine docks. Feeder lines collect oil and gas from several locations for delivery directly into trunk lines, such as moving crude oil from offshore platforms to onshore crude trunk pipelines. Gathering lines and feeder lines are typically larger in diameter than flow lines.

- Crude trunk pipelines move natural gas and crude oil long distances, from producing areas or marine docks to refineries and from refineries to storage and distribution facilities by 1- to 3-m-diameter or larger trunk pipelines.

- Petroleum product trunk pipelines move liquid petroleum products such as gasoline and fuel oil from refineries to terminals and from marine and pipeline terminals to distribution terminals. Product pipelines may also distribute products from terminals to bulk plants and consumer storage facilities, and occasionally from refineries direct to consumers. Product pipelines are used to move LPG from refineries to distributor storage facilities or large industrial users. Batch Intermix and Interface Although pipelines originally were used to move only crude oil, they evolved into carrying all types and grades of liquid petroleum products.

Because petroleum products are transported by pipelines in successive batches, there is co-mingling or mixing of the products at the interfaces. The product intermix is controlled by one of three methods: downgrading (derating), using liquid and solid spacers for separation, or reprocessing the intermix. Radioactive tracers, color dyes and spacers may be placed into the pipeline to identify where the interfaces occur. Radioactive sensors, visual observation, or gravity tests are conducted at the receiving facility to identify different pipeline batches.

Petroleum products are normally transported through pipelines in batch sequences with compatible crude oils or products adjoining. One way to maintaining product quality and integrity, downgrading or derating, is by lowering the interface between the two batches to the level of the least affected product.

For example, a batch of high-octane premium gasoline is typically shipped immediately before or after a batch of lower-octane regular gasoline. The small quantity of the two products which has intermixed will be downgraded to the lower octane rating regular gasoline. When shipping gasoline before or after diesel fuel, a small amount of diesel interface is allowed to blend into the gasoline, rather than blending gasoline into the diesel fuel, which could lower its flashpoint.

Batch interfaces are typically detected by visual observation, gravitometers or sampling. Liquid and solid spacers or cleaning pigs may be used to physically separate and identify different batches of products. The solid spacers are detected by a radioactive signal and diverted from the pipeline into a special receiver at the terminal when the batch changes from one product to another. Liquid separators may be water or another product that does not co-mingle with either of the batches it is separating and is later removed and reprocessed. Kerosene, which is downgraded (derated) to another product in storage or is recycled, can also be used to separate batches.

A third method of controlling the interface, often used at the refinery ends of pipelines, is to return the interface to be reprocessed. Products and interfaces which have been contaminated with water may also be returned for reprocessing.

Here's a closer look at the most controversial pipeline in the U.S., the Keystone XL pipeline.

Case Study: Keystone XL Pipeline Debate

There is a big question before the American consumer—especially the people in the states that are affected by the Keystone XL pipeline. Carrying oil from the Canadian oil sands to Texas—crossing the U.S. from border to border—has a lot of pros and cons. Those who care about energy security support the pipeline, and those who care about the environment oppose it. In that number are the locals, closest to the pipeline route.

And if you care about both, then you pick a side, or design your own battlefield. Maybe there is a middle ground: that of supporting the Keystone XL pipeline as long as its construction and use are regulated by strict environmental protections which are enforced. Here are the problems from different points of view:

The Pros

As far as our energy supply and national security are concerned, increasing domestic oil production instead of importing petroleum from the Middle East is a win-win situation. It is in our best interests, and for our best friend and neighbor, Canada, too, while imports of petroleum from regions like the Middle East not so much. By buying their oil, we support governments and philosophies that are contrary to our values and our democracy.

Pumping dollars into the Middle East and other volatile regions of the world also supports unfriendly currents and terrorist organizations in those areas. Some of these groups are fighting against us and would not hesitate to attack us and our allies. The Keystone XL pipeline helps mitigate this problem.

The Keystone XL pipeline also has a number of economic advantages to the United States and the state of Texas in particular. Canada is our largest trading partner, so the richer Canada gets, the more we benefit too, since Canadians buy a lot of U.S.-made goods.

The new pipeline would also reinforce the technical cooperation with Canada, since we have the advanced technologies and skills that are needed to get oil from the oil sands. Increasing the oil imports from stable Canada, would decrease imports from unstable countries like Nigeria and Venezuela. This will have a leveling effect on oil prices too.

The Cons

Water supply protection is a major issue that won't go away no matter what happens now or later. Like all oil and gas pipelines, the Keystone XL pipeline poses some real and present risks and dangers. The construction and operation of a large-diameter pipeline, with thousands of gallons of oil flowing every day cannot possibly be free of accidents.

The following land contamination, property devaluation, and in particular poisoning of local water supplies are unavoidable. The only question is when, where, and who will be affected and how? For those who live near the line, this becomes a life and death issue...thus the persisting battles.

These major risks are at, a) the sand oil production sites in Canada, and b) along the pipeline's path from Canada to Texas. The risks in Canada are under the control of Canada's environmental agencies. The risks on this side of the border are under the control of EPA and other government bodies.

Oil leakage along the pipeline in water bodies such as the Ogallala aquifer are a reality that is under consideration and a subject of lively discussion. We already have tens of thousands of miles of pipelines carrying oil and gas across the U.S. over sensitive aquifers and other environmentally delicate areas. The new pipeline increases the risk of leaks proportionally, which is actually a small number comparatively speaking. Looking from the eyes of the locals potentially affected by accidents, that number is huge.

Opposing the pipeline won't do much if and when a decision is taken to proceed. A better approach would be to raise the standards of pipeline integrity and safety inspections. Imposing a fee on pipeline owners and operators would ensure regular inspections, repairs and mitigation of risks.

The other big problem under evaluation is the additional carbon emissions that the sand oil project brings. Due to process complexity, oil produced from oil sands is much more energy-intensive. This, in turn, makes it much more carbon-intensive than conventional oil production.

Therefore, while we increase oil production from the Canadian oil sands, we also proportionally increase the carbon emissions at the production sites. We cannot change the minds of Canadian companies and government, who are determined to develop the process to its maximum. Because of that, the oil sands will be exploited and oil will be shipped somewhere no matter what. If it is not shipped to Texas, it will be surely shipped to China, which will use much more energy (and emit much more GHGs) than shipping it to Texas.

Also, the oil shipped to China will be refined in outdated refineries, which emit many times the GHGs of comparable facilities in the U.S. All in all, sending the oil to the U.S. will produce fewer carbon emissions, in addition to the other benefits like jobs, stable energy prices, etc. So, it is obvious that Canada oil sands and the Keystone XL pipeline offer a serious economic and national security advantage, provided that we design and implement proper protection procedures to ensure the safety of the environment and the people living nearby.

Oil-Tank Trucks

From the wells, or refineries, oil, gasoline, fuel oil, and other petrochemical products go by truck or train tank car to wholesalers, or to large consumers. The U.S. tank truck fleet numbers about 160,000, and the number of rail tank cars is over 165,000.

Tank trucks are normal heavy-duty trucks, modified to carry large metal oil tanks. The oil tanks are typically constructed of carbon steel, aluminum, or plasticized fiberglass material, and vary in size from 1,900 liter tank wagons to jumbo 53,200 liter capacity. The optimum capacity of tank trucks is governed by regulatory agencies, and usually is dependent upon highway and bridge capacity limitations and the allowable weight per axle or total amount of product allowed along the scheduled routes.

There are pressurized and non-pressurized tank trucks, which may be non-insulated or insulated depending on their service and the products transported.

Pressurized tank trucks are usually single-compartment, and non-pressurized tank trucks may have single or multiple compartments. Regardless of the number of compartments on a tank truck, each compartment must be treated individually, with its own loading, unloading and safety-relief devices. Compartments may be separated by single or double walls. Regulations may require that incompatible products and flammable and combustible liquids carried in different compartments on the same vehicle be separated by double walls. When pressure testing compartments, the space between the walls is also tested for liquid or vapor.

Tank trucks have either hatches which open for top loading, valves for closed top- or bottom-loading and unloading, or both. All compartments have hatch entries for cleaning and are equipped with safety relief devices to mitigate internal pressure when exposed to abnormal conditions.

These devices include safety relief valves held in place by a spring which can open to relieve pressure and then close, and hatches on non-pressure tanks which pop open if the relief valves fail and rupture discs on pressurized tank trucks.

A vacuum relief valve is provided for each non-pressurized tank truck compartment to prevent vacuum when unloading from the bottom. Non-pressurized tank trucks have railings on top to protect the hatches, relief valves, and vapor recovery system in case of a rollover. Tank trucks are usually equipped with breakaway, self-closing devices installed on compartment bottom loading and unloading pipes and fittings to prevent spills in case of damage in a rollover or collision.

Railroad Tank Cars

Railroad tank cars are constructed of carbon steel or aluminum and may be pressurized or unpressurized. Modern tank cars can hold up to 171,000 liters of compressed gas at pressures up to 600 psi. Non-pressure tank cars have evolved from small wooden tank cars of the late 1800s to jumbo tank cars which transport as

much as 1.31 million liters of product at pressures up to 100 psi.

Non-pressure tank cars may be individual units with one or multiple compartments, or a string of interconnected tank cars, called a tank train. Tank cars are loaded individually, and entire tank trains can be loaded and unloaded from a single point. Both pressure and non-pressure tank cars may be heated, cooled, insulated and thermally protected against fire, depending on their service and the products transported.

All railroad tank cars have top- or bottom-liquid or vapor valves for loading and unloading and hatch entries for cleaning. They are also equipped with devices intended to prevent the increase of internal pressure when exposed to abnormal conditions. These devices include safety relief valves held in place by a spring which can open to relieve pressure and then close; safety vents with rupture discs that burst open to relieve pressure but cannot reclose; or a combination of the two devices.

A vacuum relief valve is provided for non-pressure tank cars to prevent vacuum formation when unloading from the bottom. Both pressure and non-pressure tank cars have protective housings on top surrounding the loading connections, sample lines, thermometer wells and gauging devices. Platforms for loaders may or may not be provided on top of cars.

Older non-pressure tank cars may have one or more expansion domes. Fittings are provided on the bottom of tank cars for unloading or cleaning. Head shields are provided on the ends of tank cars to prevent puncture of the shell by the coupler of another car during derailments.

Oil Tanker Transport

A great portion of the world's crude oil and about half of the oil headed to the U.S. is transported via ocean tankers. Oil transport is very big business. The world tank ship fleet numbers more than 5000.)n any given day as much as 750 million barrels of crude oil and products may be in tankers on the world's oceans. These undergo major risks every step of the way as they traverse the world's oceans, and especially in the most vulnerable places—choke points.

The issues related to security of oil transport could be summarized as follow:

* Global problems, such as political changes and turmoil, can affect the oil routes and the safety of the tankers and the people working on them.
* Weather related problems, such as storms and hurricanes on the tanker route could cause serious damage to the vessels and their personnel.

- Terrorist attacks are the most threatening part of global oil transport routes. This is the worst form of damage to the oil transport tankers and their personnel. There are several "choke points" around the world, where the ocean transports are most vulnerable.

Oil tankers and barges are vessels designed with the engines and quarters at the rear of the vessel and the remainder of the vessel divided into special compartments (tanks) to carry crude oil and liquid petroleum products in bulk. Cargo pumps are located in pump rooms, and forced ventilation and inerting systems are provided to reduce the risk of fires and explosions in pump rooms and cargo compartments.

Modern oil tankers and barges are built with double hulls and other protective and safety features required by the United States Oil Pollution Act of 1990 and the International Maritime Organization (IMO) tanker safety standards. Some new ship designs extend double hulls up the sides of the tankers to provide additional protection.

Generally, large tankers carry crude oil and small tankers and barges carry petroleum products.

- Oil tankers are ocean traveling vessels, which in addition to ocean travel can navigate restricted passages such as the Suez and Panama Canals, shallow coastal waters and estuaries. Large oil tankers, which range from 25,000 to 160,000 SDWTs, usually carry crude oil or heavy residual products. Smaller oil tankers, under 25,000 SDWT, usually carry gasoline, fuel oils and lubricants.

- Barges carrying oil products operate mainly in coastal and inland waterways and rivers, alone or in groups of two or more, and are either self-propelled or moved by tugboat. They may carry crude oil to refineries, but more often are used as an inexpensive means of transporting petroleum products from refineries to distribution terminals. Barges are also used to off-load cargo from tankers offshore whose draft or size does not allow them to come to the dock.

- Supertankers are the least expensive way for long-distance shipment of oil and oil products.

These modern ocean-going vessels are huge floating oil tanks that make their slow cumbersome way from the giant oil fields of the Middle East to ports in the industrial countries. But there are no supertanker ports in this country, so they are usually unloaded off-shore into smaller tankers.

Supertankers are huge pieces of complex machinery. The largest ones presently in service have a DWT (deadweight tonnage, i.e., cargo and fuel capacity) of more than 546,000 tons. They are up to 400 meters long (the length of about five football fields laid end to end) and can carry about 40 million barrels of oil.

- Ultra-large and very large crude carriers (ULCCs and VLCCs) are restricted to specific routes of travel by their size and "draft" (depth of water to which a ship sinks according to its load). ULCCs are vessels whose capacity is over 300,000 SDWTs, and VLCCs have capacities ranging from 160,000 to 300,000 SDWTs.

Most large crude carriers are not owned by oil companies, but are chartered from transportation companies which specialize in operating these super-sized vessels. Their draft is very important for their navigation. A fully loaded draft of over 90 feet prevents such a large ship from going into any U.S. ports. Our deepest port, Los Angeles, cannot handle ships of more than 100,000 deadweight tons. As a result supertankers now unload offshore in the Caribbean and transfer their cargo to smaller tankers for delivery to the United States. The ocean transfer method costs a lot of additional energy and money, and for this purpose an alternative facility was built. The Louisiana Offshore Oil Port (LOOP) is a deepwater port in the Gulf of Mexico off the coast of Louisiana, near the town of Port Fourchon. It provides offloading and temporary storage services for crude oil transported on some of the largest tankers in the world, since most of them cannot enter US ports.

LOOP presently handles 13% of the nation's oil imports, or about 1.2 million barrels a day. It is connected by pipeline to the refineries, thus feeding 50% of the U.S. refining capability. Tankers offload at LOOP by pumping crude oil through hoses connected to a single point mooring (SPM) base. Three SPMs are located 8,000 feet from the marine terminal.

The SPMs are designed to handle ships of up to 700,000 deadweight tons. The crude oil then moves to the marine terminal via a 56-inch diameter submarine pipeline. The marine terminal consists of a control platform and a pumping platform. The control platform is equipped with a helicopter pad, living quarters, control room, vessel traffic control station, offices and life-support equipment. The pumping platform contains four 7,000-hp pumps, power generators, metering and laboratory facilities. Crude oil is handled only on the pumping platform where it is measured, sampled, and boosted to shore via a 48-inch-diameter pipeline.

The Issues

Oil transport is a huge business, and as with any business; bigger it is, bigger the problems it encounters. The majority of the world's crude oil is transported by tankers from producing areas such as the Middle East and Africa to refineries in consumer areas such as Europe, Japan and the United States.

Oil products were originally transported in large barrels on cargo ships. The first tanker ship, which was built in 1886, carried about 2,300 SDWT (2,240 pounds per ton) of oil. Today's supertankers can be over 300 meters long and carry almost 200 times as much oil.

Gathering and feeder pipelines often end at marine terminals or offshore platform loading facilities, where the crude oil is loaded into tankers or barges for transport to crude trunk pipelines or refineries. Petroleum products are also transported from refineries to distribution terminals by tanker and barge. After delivering their cargoes, the vessels return in ballast to loading facilities to repeat the sequence.

Oil tankers also pose a significant danger of oil spills due to accidents or terrorist attacks. Millions of tons of crude oils are transported across world's maritime channels every day by tankers which are exposed to considerable risks.

Oil spills from transport tankers account for less than 8 percent of oil spills in the ocean. Although infrequent, these are usually large spills caused by accidents, and the public is well acquainted with most of them.

For example, everybody knows about the Exxon Valdez spill in Alaska, the effects of which are still visible. Recently published list of oil spill effects by the International Tanker Owners Pollution Federation shows that the most harmful result of oil-related incidents is the negative effect these have on marine animals.

Toxic chemical components in the oil smother and kill marine life. Less-than-lethal levels of toxicity have long-term effects on marine animals' ability to feed and reproduce. In the long run, oil spills contaminate the entire marine food chain, causing a crippling and even deadly domino effect on all species.

Aquatic birds and mammals are particularly affected. Even a slight, or momentary exposure to oil products is fatal for many of these animals, and ingested oil poisons them. Birds with oil in their feathers lose their ability to fly, and the waterproof coating that protects their vital organs is damaged, which causes health problems

and even death. Older and sick mammals often suffer hypothermia due to oil-related complications.

The *Exxon Valdez* spill in 1989, for example, killed more than 40,000 birds and thousands of other sea mammals after the spill and during the clean-up efforts. Following the *Exxon Valdez* spill, increased government regulations were implemented. One of these is the International Safety Management Code of 1998, which requires tankers to conform to new, much tougher, standards of quality and their owners are held fully accountable for all damages. Individual states have crafted their own laws and methods of preventing oil-related accidents. In California, for example, the regulatory agencies require oil tankers and transport ships to have active contingency plans for handling oil spills, as well as an additional $300 million in accident insurance.

Terrorism

The total world oil production amounts to approximately 80-90 million barrels per day. About one-half of this quantity is moved by oil tankers on fixed maritime routes. The global economic downturn of 2008 reduced the world oil demand, and the volumes of oil shipped to markets via pipelines and along maritime routes.

There are several chokepoints around the world through which large quantities of oil float daily and which are under constant threat of terrorism. By volume of oil transit, the Strait of Hormuz leading out of the Persian Gulf, the Strait of Malacca linking the Indian and Pacific Oceans, Bab-el-Mandeb at the entrance to the Red Sea, and the Strait of Gibraltar at the exit from the Mediterranean Sea, are the world's most strategic and dangerous chokepoints.

Oil tankers crossing those chokepoints are exposed to political unrest in the form of wars or hostilities and are vulnerable to theft from pirates and terrorist attacks. All this can lead to shipping accidents, resulting in disastrous oil spills and or blockage of the waterways. The blockage of a chokepoint, even temporarily, can lead to substantial increases in total energy costs.

Over half of America's oil is imported, so terror organizations like al-Qaeda and its affiliates see the disruption of oil transportation routes as one of the ways to hurt the U.S. and its allies. And sure enough, disruption of scheduled oil flow through any of those routes could impact negatively the global oil prices.

Several attacks against oil tankers in the Arabian Gulf and Horn of Africa have been planned in the past, starting with June 2002, when a group of al-Qaeda operatives suspected of plotting raids on British and American tankers passing through the Strait of Gibraltar were arrested by the Moroccan government. In October of the same year, a boat packed with explosives rammed and badly damaged a French supertanker off Yemen. al-Qaeda claimed responsibility for that attack, and for plans to seize a ship and crash it into another vessel or into a refinery or port.

The terrorists' goal is to cut the economic lifelines of the world's industrialized societies, but such attacks would also weaken and perhaps topple some of the Gulf oil monarchies. This would have an even greater effect on the world's oil supplies and transport routes.

There are approximately 4,000 tankers crossing the world's oceans every day. Each of those slow and vulnerable giants can be attacked at the narrow passages, where a single disabled or burning oil tanker and its spreading oil slick could block the route for other tankers. This constant threat and the increased risks contribute to further increase in oil prices. Insurance carriers are raising the premiums to cover tankers in risky waters, as evidenced by the fact that the insurance for oil tankers entering Yemeni waters tripled since the attack in Yemen a decade ago.

A typical supertanker with two million barrels of oil on board would pay about $450,000 instead of the usual $150,000 for insurance of the ship alone. The cargo requires a separate and equally expensive insurance policy each trip. This adds about 15-25 cents a barrel to the cost of the oil. Even the best insurance cannot ensure complete safety and reliability of our energy supply under those conditions. The only way to achieve energy independence, therefore, is to eliminate the need to import and transport oil across the globe. This, however, is not going to happen anytime soon, so the new situation has created a new niche market for individuals and companies that provide insurance and ensure the physical safety of the oil tankers.

Some oil tank owners are resorting to placing security personnel on board, especially when crossing pirate-infested international waters. This measure, in addition to equipping the tankers with special equipment, is one of the services that is expected to grow. The number of complexities of those services, as well as their costs, will increase as the pirate attacks are on the rise, at least until we figure a way around it.

WASTE STORAGE, DISPOSAL, AND REUSE

Humans generate a lot of waste. What is amazing here is that all waste is related to some amount of energy used in its initial production, transport, and disposal.

Food waste, for example, started as growing crops, which were taken care of, harvested, transported to processing facilities, processed and packaged, transported to the markets, stored, and refrigerated.

Finally, the food products were purchased by individuals, transported home, cooked, eaten, and the leftovers turned waste were thrown in the garbage can. The waste from the garbage can was loaded on a garbage truck, which transported it to the waste disposal site, where it was buried, or was sorted and transported to another site for final disposal or reprocessing. The reprocessed and recycled items require even more energy.

Each step of this long process requires energy—mostly fossil fuels—to run the agricultural and processing equipment, the transport vehicles, and operate the waste sites. Energy, energy, energy.

Note: Mines and power plants also generate a lot of waste materials. In addition to the air emissions, there are huge amounts of ashes, slurry, and other waste materials generated and disposed of. We take a close look at these operations, including the waste materials, in the previous chapters, so we will not cover them here.

For the purposes of the energy markets, the important facts are that:

• Waste products are responsible for consuming some energy during their life cycle,

• They all contain some energy, which can be extracted and used with the help of appropriate equipment and methods, and

• They all require a lot of energy for transport to, and disposal at, their final destination.

Since there is so much waste around the world, this subject deserves our undivided attention. So, let's take a close look at the different types of waste and how they fit in the energy and environmental markets.

Animal Waste

In a 2012 report on U.S. greenhouse gas emissions, the U.S. Environmental Protection Agency (EPA) reported that the U.S. cattle fleet emits over 140 million metric tons of methane annually.

Nope, this is not a typo; cattle fleets emit as much GHGs (mostly methane gas released unintentionally from both ends of the animals) as over a dozen coal-fired power plants put together.

Methane is energy source extraordinaire, so treating it as 'waste' is just not right. It also so happens that methane is 23 times more potent than carbon dioxide as a greenhouse gas.

The U.S. agricultural sector (not just the cattle fleet) is responsible for over 8% of our total GHG emissions.

So, no joke; this is a serious issue. By not capturing the waste methane we not only waste large quantity of energy, but contaminate the environment at the same time.

Enteric fermentation from ruminant animals is the largest anthropogenic source of GHG (methane) in the U.S. The natural gas systems* are the second largest source with 127 million metric tons annual GHG emissions.

Note: The natural gas systems are oil/gas wells, pipelines, and all kinds of equipment involved in the production, transport, and use of natural gas. This fact is important to note and understand, because we are talking about huge amounts of pollution emitted even before the natural gas has been burned in the power plants or in vehicles engines.

It is also important to know that raw natural gas contains methane and other GHG elements that are many times more dangerous to our environment than CO_2.

In 2006, the United Nations' Food and Agriculture Organization released a report on environmental issues related to livestock which recommended cutting in half the "environmental costs" of livestock, which includes greenhouse gas emissions, and land and water degradation caused by these activities.

According to the UN environmental specialists: "When emissions from land use change are included, the livestock sector accounts for 9% of CO_2 deriving from human-related activities, but produces a much larger share of even more harmful greenhouse gases (such as methane). It also generates 65% of human-related nitrous oxide, which has 296 times the Global Warming Potential (GWP) of CO_2. Most of this comes from manure."

And as the economies of developing nations improve, and their people get richer, they eat more meat. Meat comes from animal farms, so the UN projects meat production to double from the 240 million metric tons in the early 2000s to nearly 500 million metric tons by 2050. The milk output is expected to almost double too; from 580 million tons to over 1,000 million tons during the same time period.

The sharp increase of demand for meat and milk will require more GHG emitting cattle population around the world, so we can expect an equal increase of GHG gasses. So, the present 140 million tons of methane emitted annually by the U.S. cattle fleet will double and triple during the next several decades.

This, added to the total world cattle fleet, means that we are to expect over 1 billion tons of methane to hit the Earth's atmosphere every year from 2050 on.

Presently 21% of global fossil-related GHG emissions come from natural gas, and 44% from coal—both burned for power generation. 35% of the global GHG emissions come from crude oil, most of which is used in transportation and some for power generation.

An average U.S. coal-fired power plant emits about 2,000 lbs. of CO_2 per each MWh, so an average 1.0 GW power plant at 80% operational capacity will emit about 14 billion lbs. of CO_2, or about 7 million tons annually from an average coal-fired power plant.

The volume of GHGs emitted by the global cattle fleet equals that emitted by 150 coal-fired power plants.

But now, remember that methane is 23 times (and nitrous oxide is 296 times) more potent than CO_2 as far as its GHG potential is concerned. So if we multiply the above numbers by 23 and 296 respectively we come up with an astronomically high GHG amount.

Everything considered, cattle emitted GHGs are equal to those of 3,200 average coal-fired power plants, or close to the total GHGs emitted by the power generation around the world.

Not a small thing, and since the world will not slow down—let alone stop—growing and increasing the meat and milk consumption, we have to brace for a significant growth of the global cattle fleet. With it, we will see a huge influx of GHGs in our air, which, no doubt, will do significant damage to the quality of our environment and will contribute to serious deterioration of the quality of air we breath.

But, on the good side of things, there are increasing efforts to convert the gas, liquid, and solid waste from animal feeding lots into useful energy. Methane emitted by the manure lots can easily be captured, burned, and converted into electricity. This way we can kill two birds with one stone; we produce electricity and reduce global GHGs. And considering the fact that we are talking about huge amounts of energy and GHGs, this could create an equally huge industry in the near future.

Human Waste

People, just like all animals, create a lot of waste. In addition to the regular kitchen waste—excess food and leftovers—we also generate a lot of liquid and solid waste. These waste streams contain useful energy, which is wasted by burying the waste products in garbage dumps, where they rot and contaminate the environment.

Not to mention the important fact that a lot of energy has been used throughout the entire life cycle of the waste stream; producing, transporting, processing, and marketing. One doesn't have to be a genius to figure out that there is a lot of energy wasted during this long process, and that there must be a better way to handle the increasing human waste…

Let's take a look at the man-generated waste streams.

Packaging

Packaging is actually one of the most visible and noticeable symbols of energy waste and environmental impact of the 21st century. Plastic bags, containers, wrappers of all sorts, and sizes are all around us. We see them around the house, the park, in the rivers and the oceans.

Packaging has to be credited with heavily contributing to our economic development and raising our standard of living. Just one look at the items in a U.S. supermarket, as compared with the street market in most African and Asian countries, gives you a good idea of the stark difference that packaging makes.

> In the U.S., with our sophisticated packaging, storage, and distribution systems only 2-3% of the food is wasted before it gets to the consumer. In developing countries, on the other hand, 40-50% of food shipments are spoiled because of inadequate packaging, storage, and distribution systems.

On the consumer side, however, this ratio is reversed; 30-50% of the food from U.S. consumers ends in the waste baskets, while in the developing countries only 2-5% of the food total is wasted by the consumers.

So how can people without efficient cooling and packaging of food items be so efficient in their use? How can they do without packaging, and why are we so wasteful having good packaging options?

Packaging plays a major and growing role in the global economy, especially in the developed countries. Packaging in the developing countries is not only for protection of goods from contamination, but also to convey information about their contents, preparation, and use, but in most cases it also keeps would-be tamperers from poisoning or otherwise hurting the customers.

At the same time, however, packaging—its manufacturing, transport, use, and disposal—use enormous amount of energy and is causing an environmental disaster. It is the largest and fastest growing contributor to one of the most troubling environmental problems: garbage. In the U.S., packaging accounted for more than 30% of the municipal solid waste stream in 1990.

Where is all this packaging going? In this country, most packaging and other waste is buried in landfills. But even with its abundance of open land, America is running out of room for its garbage. One quarter of the country's municipalities have already exhausted their landfill capacities, so that more than half the population lives in regions with lack of landfill capacity.

Even though the environmentally sound alternatives to burying garbage—recycling, reuse, and energy recovery—are well developed, our disposable and throwaway society is still generating more garbage from packaging than the entire world combined.

Packaging is not the only culprit in the solid waste crisis, but it is the most highly visible component in a garbage dump. It also directly involves the customers, who use it for a very short time. Its short lifetime exacerbates the problem, since the useful lives of most packages is very limited.

While some packaging materials may be as long as several years (i.e. paint and food cans, and reusable canisters), the useful lives of most others can be as fleeting as a few minutes, or few hours. Think of the hamburger wrapper, or a new gadget in a large cardboard box received in the mail.

We deal with a huge volume of different types of packaging daily going in the solid waste stream. This is a big problem, which also presents an opportunity, since even very small improvements in packaging, due to its huge amount, can make a real difference in the magnitude of the garbage crisis. So packaging offers this unique opportunity for individuals and companies to assume a leadership role in environmental responsibility.

The industry's response to the environmental challenge of excess packaging has been focused on recycling and packaging materials reduction. But this is just not enough, because these are complex issues that demand a systemic, integrated approach based on comprehensive analysis. Long-term vision and implementation of innovative solutions are also some of the tools that we must use.

Life-cycle analysis is another one of these tools, which needs to be applied to each and every product we consume. Think of millions of penny-worth gadgets coming from China enclosed in large packages made out of heavy-duty plastics. How much materials and energy was use to make these packages? How much energy was used to manufacture, use, and dispose of them? What is their overall benefit to our well-being?

Life-cycle analysis is an important step toward understanding the full energy and environmental implications of packaging choices.

Using this technique, we consider the energy use and environmental implications as integral part of each product. Considering the energy used during the entire product cycle—from raw material acquisition, design, manufacturing, transportation, to final use and disposal of the packaging materials—will allow us to see the other side of packaging. The side that affects our economy and contributes to great energy waste. This will allow us to take a closer look at the problem and make better packaging choices.

Our energy security will not be achieved by solving the packaging problems (even if could), but the present-day packaging is one of the elements of our way of life that is indicative of our societal problems in need of fixing. Less waste means more and cheaper energy for all.

Food Waste

Food is what we humans cannot live without. Because of that, it is on the priority list of our government and a lot of effort (and energy) goes in ensuring the food supply on the American tables. This, however, comes at a heavy price.

Getting food from the farm to our dinner table eats up 10% of the total U.S. energy budget, uses 50% of U.S. land, and swallows 80% of all freshwater consumed in the country.

Yet, 40% of food in the United States today is unused and/or wasted for one or another reason. This is over 20 pounds of food per person every month ending up in the waste baskets and eventually in the municipal garbage dumps. Amazing numbers…

Today, the average American consumer wastes 10 times as much food as people in Southeast Asia and Africa, and 50% more than what we used to waste in the 1970s.

This means that Americans are throwing out the equivalent of $165 billion of different food items each year. It also means that the uneaten foods have to be collected, transported, buried, and left rotting in landfills. Food waste is the single largest component of U.S. municipal solid waste, where it also accounts for a large portion of U.S. methane emissions.

And of course, huge amount of energy has been used to produce, transport, and prepare the foods, which then end up in a garbage dump. What a waste of energy and source of contamination.

Organic matter (mostly food leftovers) in municipal waste dumps accounts for 16% of U.S. methane emissions, and contaminates 25% of all fresh water supplies in the country.

Not to mention the huge amounts of energy, chemicals, energy, and land that are wasted in the garbage dumps burial process too.

There is work under way in Europe in search of better understanding of the drivers of the problem of, and identify potential solutions for, food waste. In 2012, the European Parliament adopted a resolution to reduce food waste by 50% by 2020. It also designated 2014 as the "European year against food waste."

An extensive UK public awareness campaign called "Love Food Hate Waste" has been conducted over the past five years too. As a result, over 50 of the leading food retailers and brands have adopted a resolution to reduce waste in their own operations, as well as upstream and downstream in the supply chain. As a result of these initiatives, during the last five years, household food waste in the United Kingdom has been reduced by 18%.

The situation is somewhat different in different countries, where America is in class of its own. Here, food represents a small portion of most Americans' budgets, which makes the financial cost of wasting food low enough to be considered a regular mode of life, where time and convenience are the priorities of most people.

Then there is the basic economic truth of advanced capitalist society, that more food consumers waste, more food the food industry is able to sell. This affects the entire supply chain—from the fields to the supermarkets. Here any waste downstream translates to higher sales for the upstream companies.

Overcoming these and other related challenges is not an easy thing, and will require the total cooperation of the government, consumers, and businesses.

But this effort can start only after everyone understands and agrees that this is a problem. Then, and only then, the problem of reducing food waste to a significant level can be raised to a higher level of priority it deserves.

Reducing waste food losses in the U.S. by just 15% could feed 25-30 million people in Africa every year. It could also save millions of barrels of oil wasted during the food production, storage, transportation, preparation, and disposal.

One simple solution (one of many that the government can undertake) is to standardize and clarify the meaning of expiration date labels on food packages. All foods have recommended use periods stamped on their packages. These stamps are hard to read and are easy to misunderstand and/or misinterpret. This is causing tons of foods to be thrown out due to misinterpretation.

Expiration date clarification on food packages could prevent about 15-20% of wasted food in U.S. households.

Businesses must understanding the extent of their own waste streams and adopt best practices. For example, an U.S. food chain saves over $100 million annually after an extensive analysis of the freshness, shrink, and customer satisfaction in the perishable foods department.

The consumers can reduce their waste by getting better educated on the basics of foods; shelf life, when and how the different types of food go bad, as well as buying, storing and cooking food with food waste reducing in mind.

Implementing efficient food waste reduction strategies, can bring tremendous social and economic benefits. It can reduce hunger, save energy, and reduce environmental pollution. Are we ready for it?

China's Poo Energy

A growing portion of China's toilet waste is converted into fertilizer and biogas. Over the past decades, millions of rural workers have fled into its cities in the largest migration in human history. This has caused overcrowding conditions, where the number of urban dwellers is now over 750 million. This is more than 100 million more than the rural population.

In the city of Beijing alone, about 7,000 tons of human excrement are produced and treated each day. City planners and energy experts are getting together to solve the problem. They intend to re-purpose feces into energy resources or fertilizer, hoping that their effort can be duplicated across China and other developing countries.

There is a taboo around reusing fecal matter in developed countries, so it is up to the Chinese to develop, implement and prove the science for its safe and efficient reuse.

A variety of techniques are used for treating human waste in China, where some is dumped in rivers, some is incinerated, and some is buried. While there are parts of the world that also attempt to harness fecal sludge into resources, China's opportunity in this area is unmatched as the cities become more crammed and solutions is urgently needed.

Presently, a model used in rural China consists of a holding tank for human and animal waste, which is partly sanitized by oxygen starvation. The solids are then converted to fertilizer for the farms. This method cannot be used in Beijing, where the average annual amount of human waste processed is increasing by 200 to 300 tons every day. So the human waste pile needs urgent attention.

And so, Beijing has several human waste processing plants; usually two story, red tiled buildings. Here the process begins on the ground floor, where the odor of human excrement in the air is overwhelming. At each plant, bright yellow trucks unload 800-1,000 tons of fecal sludge every day, making 200-300 trips.

The stinky human waste is pumped through a pipe into the processing equipment. Here, unrecyclable solid material like tissue paper and plastic bags are separated and sent for disposal into a garbage dump. The rest of the sludge then goes for further separation, and the liquid material is routed into tanks to generate biogas, after which it pumped to water-treatment plants.

The solid waste is pumped into compost polls, where it remains fermenting at 600 C for 10-14 days. During this step, harmful bacteria and parasitic roundworms are killed and the solids turn nutrients rich fertilizers for field applications.

Presently, only small part of Beijing's waste is converted into energy and other resources. Most of it is still sent to treatment plants, or is simply dumped in rivers or landfills. Proper waste treatment is still not the norm, but China has entered a human waste revolution, which promises to grow quickly in volume in the near future.

There are efforts to harness energy from fecal sludge in other parts of the world too. Recently, the Bill & Melinda Gates Foundation invested $1.5 million in a project in Ghana aimed to explore biodiesel production from human waste. The Foundation has been also investing in small biogas generating units and special septic tanks that process human waste in Thailand and India.

United Utilities Group Plc, Britain's largest publicly traded water company, is working on a project to process sewage from 1.2 million people in Manchester at a sludge recycling plant. The plant generates enough electricity to power over 25,000 homes. Who is next...?

Nuclear Waste

Nuclear power plants generate a lot of solid waste. Every 12-18 months the nuclear reactor is shut down and part, or all, of its contents emptied and stored in a special storage. Usually this is a pool of water, where the used rods stay for awhile and until are loaded for transport to their more permanent storage place, which might be the back yard o the power plant or a far away storage facility.

Basically, 3% of the spent (waste) nuclear fuels mass consists of fission products of 235U and 239Pu, which comprise the radioactive waste. These can be processed and separated further for various industrial and medical uses. The fission products include the second transition metals row; Zr, Mo, Tc, Ru, Rh, Pd, Ag, and the next in the periodic table; I, Xe, Cs, Ba, La, Ce, and Nd.

Many of the fission waste products are either non-radioactive or only short-lived radioisotopes, while a considerable number of these are medium to long-lived radioisotopes such as 90Sr, 137Cs, 99Tc and 129I.

The fission waste products can modify the thermal properties of the uranium dioxide, where the lanthanide oxides tend to lower the thermal conductivity of the fuel, while the metallic nanoparticles slightly increase its thermal conductivity.

Different types of nuclear fuels produce different kinds of spent nuclear fuel (nuclear waste).

Spent low enriched uranium nuclear fuel, for example, is a type of a nanomaterial. In the uranium oxide spent fuel, intense temperature gradients cause fission products to migrate. The zirconium tends to move to the centre of the fuel pellet where the temperature is highest, while the lower-boiling fission products move to the edge of the pellet. The pellet is likely to contain lots of small bubble-like pores which form during the fission cycle. Fission xenon migrates to these voids, and some of it decays to form cesium, C-137 (137Cs), so many of the bubbles contain a large concentration of 137Cs.

In MOX, the xenon tends to diffuse out of the plutonium-rich areas of the fuel, and it is then trapped in the surrounding uranium dioxide, while the neodymium is not very mobile.

Metallic particles of an alloy of Mo-Tc-Ru-Pd also tend to form in the fuel, while other solids form at the boundary between the uranium dioxide grains. The majority of the fission products remain in the uranium dioxide as solid solutions.

Successful efforts have been led to segregate the rare isotopes in fission waste including the "fission platinoids," Ru, Rh, Pd, and Ag as a way of offsetting the cost of reprocessing. Although this is possible, it is no economical enough, for now, so it is not done commercially today. But things might change soon enough...

In any and all cases, spent nuclear fuel has only two possible paths; stored as a waste, or shipped for reprocessing. Most of the spent nuclear fuels are stored in large onsite water pools at the nuclear plant, where they are left to cool down for a long time. Eventually the cooled fuel rods and bundles are placed into special canister and shipped to an offsite storage facility.

Nuclear fuel stored as waste remains radioactive for hundreds and thousands of years, and the fact that there are no permanent storage areas is becoming a huge controversy. There are only a few options around the world, so this unresolved issue will continue throw a shadow of danger and uncertainty over the future of nuclear power.

Reprocessing is an expensive and uneconomical process with the present state-of-the-art, so it is done infrequently, while the R&D efforts continue. Some fuels are showing promise, so the efforts in recycling, reprocessing, and reusing spent nuclear fuels will continue and even intensify in the future.

Nuclear Materials Transport

Uranium production and processing are dangerous undertakings, and so is the transport of the radioactive materials from place to place. The transport of nuclear materials is a complex process that involves a number of sophisticated equipment pieces, many different regulations, and equally numerous and complex procedures.

Unlike the other fuels, which can be dumped on any old truck, uranium ore from the mines, or spent nuclear fuel from the plant, is transported using special equipment and procedures, and following strict regulations. No exceptions!

This has created a special industry that deals with nuclear products transport exclusively.

These special precautions are needed because uranium and its products are radioactive, and people must be protected. Additionally, uranium presents security issues, so transport has to be done in such a manner as to prevent it falling in the hands of terrorists and rogue nations.

Uranium ore is transported from the mines to the milling plant in order to produce the yellow cake. The yellow cake is then transported to the enrichment facility to make UF6 which is transported to the fuel fabrication plant, where the final nuclear fuel is produced. Thus enriched UF6 (nuclear fuel) is finally transported to the power plant to be used for generating electricity. Once used at the plant, the nuclear fuel is transported to used fuel storage facility, from where it is transported to a nuclear waste disposal site or to a reprocessing facility.

From the reprocessing plant, the transportation cycle repeats. These procedures are full of dangers, precautionary and safety measures, expert technical personnel, and specialized security personnel.

There are estimated 20 million shipments of radioactive material from one place to another every year around the globe. These vary from a single small package, or a number of packages sent from one location to another via common carriers, to large containers shipped on special trucks or trains.

Radioactive materials are used in many other areas (not nuclear fuel related) too, such as medicine, agriculture, university research, special manufacturing, non-destructive testing, and minerals' exploration. As a matter of fact, only 5% of these annual shipments are nuclear fuel cycle related, so shipping of this commodity is a big business.

The shipment of any radioactive materials is governed by international regulations for the transport of radioactive materials, established as far back as 1961. These regulations control shipment of radioactive material in a manner that is independent of the material's intended application.

Nuclear fuel processing facilities are located in various parts of the world and materials of many kinds are transported between them. Many of these are similar to materials used in other industrial activities. However, the nuclear industry's fuel and waste materials are radioactive, and it is these 'nuclear materials' about which there is most public concern.

In the U.S. only 1% of the 300 million packages of hazardous materials shipped each year contain radioactive materials. This is still 3 million radioactive packages on the U.S. roadways every year.

Of these 3 million packages, about 250,000 contain radioactive wastes from US nuclear power plants, and only 50 to 100 shipments contain actual used fuel.

Nuclear Waste Storage

Ideally, a nuclear power plant runs at the maximum allowed power level from one refueling to the next. It has to be shut down for several hours, or days, during each refueling procedure, and then restarted and brought up to the maximum power output

Used nuclear fuel, which is removed from the reactor after a 6 or 12 months of service, is a solid material that has to be stored safely; which is done usually at the proper nuclear plant site. This storage is only 'temporary.' This is one component of the integrated used fuel management system in use at all nuclear plants, that addresses all facets of storing, recycling and disposal of nuclear materials.

• The integrated used fuel management approach mandates that used nuclear fuel will remain safely stored at nuclear power plants for the near term, with safely as the key word. Eventually, the hope goes, the government will find a way to recycle it, and place the unusable end product in a deep long-term repository.

Figure 8-35. Airtight containers for nuclear waste storage

• Low-level waste are byproduct remaining from uses of a wide range of radioactive materials produced during electricity generation, medical diagnosis and treatment, and various other medical processes. These could be liquids or solids, and are treated in a number of ways prior to disposal as hazardous waste, burning, or storage in temporary containers.

• Recycling of waste nuclear fuels is a program of the federal government, which includes plans to develop advanced recycling technologies in order to take full advantage of the unused energy in the used fuel, and to also reduce the amount and toxicity level of byproducts requiring disposal.

• Transportation of the waste nuclear fuel is the responsibility of the U.S. Department of Energy, which will transport used nuclear fuel to the repository by rail and road, inside massive, sealed containers that have undergone safety and durability testing.

• Repository for long-term storage is under review, and under any used fuel management scenario, disposal of high-level radioactive byproducts in a permanent geologic repository is necessary.

Note: The desperation of the nuclear waste storage is reflected in the roughly 53 million gallons of nuclear waste stored in 177 large underground tanks at DOE' Hanford Nuclear Reservation in Washington State. 149 of these are over 40 years beyond their expected 25-year design life. And of these more than one-third are known or suspected to be leaking, releasing roughly 1 million gallons of waste to Hanford's surrounding soils.

Hanford lacks the storage capacity to retrieve the waste from these tanks until the waste treatment and disposal process is underway. Washington's $12.3 billion Waste Treatment Plant (WTP) continues to be designed and constructed to meet standards specific to the Yucca Mountain facility. Design and engineering for the WTP is 78% complete and construction is 48% complete.

And then, in 2002, Congress designated Yucca Mountain as the nation's sole current repository site for deep geologic disposal of high-level radioactive waste and spent nuclear fuel. At that time, the Secretary of Energy concluded that, "The amount and quality of research the DOE has invested…done by top-flight people…is nothing short of staggering…I am convinced that the product of over 20 years, millions of hours, and four billion dollars of this research provides a sound scientific basis for concluding the site can perform safely."

Congress then directed DOE to file a license application for the Yucca Mountain site with the Nuclear Regulatory Commission (NRC) and thereby commence a formal evaluation and licensing process overseen by the NRC.

But…in January 2010, President Obama, Secretary Chu, and DOE determined that they would withdraw with prejudice the application submitted by DOE to the NRC for a license to construct a permanent repository at Yucca Mountain, Nevada, for high-level nuclear waste and spent nuclear fuel.

Also that month, President Obama, Secretary Chu, and DOE chose to unilaterally and irrevocably terminate the Yucca Mountain repository process mandated by the Nuclear Waste Policy Act, 42 U.S.C. §§ 10101-10270. Several law suites were filed as a result, but these will take years to resolve, during which time. the US of A has no place to store its large pile of nuclear waste produced during power generation and nuclear weapons production.

Case Study—Onkalo Nuclear Waste Storage Site

Now there are real plans to build an enormous bunker for permanent storage of the dangerous radioactive waste; not in the U.S., but in Finland, where this radical new solution is planned. It consit of burying the nuclear waste deep underground, sealing the depository and throwing the key away. Literally. The intent is to keep everyone—including future civilizations—away by hiding the nuclear waste somewhere so unremarkable and unpleasant that nobody would ever think to go there, let alone dig into it.

Located on Olkiluoto Island, just off Finland's southwest coast, the underground facility known as Onkalo will hold all of the country's 5,500 tons of nuclear waste—al that is expected to be produced by the end of the century. It is designed to keep that waste secure for at least 100,000 years, taking measures and counting on making humans forget that Onkalo was ever there.

Onkalo is intended to store permanently and safely high-level waste (HLW), which consists of spent nuclear fuel and some equally dangerous decay products. This residue emits dangerous types and levels of radiation for tens of thousands of years. Over 300,000 tons of this stuff now exists around d the world and about 12,000 more tons are produced annually. It is also expected that numbers will increase significantly in the years to come.

A maze of deep underground bunkers carved from impermeable rock, in geologically stable zones, where the waste can be redundantly sealed and then permanently buried.

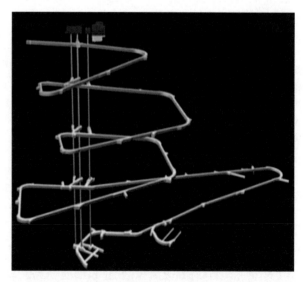

Figure 8-36. The Onkalo' maze of underground tunnels

At a $3 billion price tag thus far, Onkalo will start accepting nuclear waste in 2020, while construction of new tunnels will continue as needed until the facility is shut down and sealed forever.

Would this grandiose plan work? Who knows… it is too big, expensive, complex, and ambitious to be properly assessed with the available information, so we just have to wait and see.

Case Study—Mixed Fuel Production Plant

One of the best and most plausible solutions to reducing nuclear waste is its recycling and reuse. To this purpose, in 2000 the United States and Russia signed a bilateral agreement, committing to eliminate 34 metric tons of surplus military plutonium produced during the Cold War by recycling it as fuel for civil nuclear applications.

In 2008, the Department of Energy made an agreement with a joint venture created by the AREVA and SHAW groups for the construction of a mixed oxide (MOX) fuel production plant. The effort consists of:

1. Construction and operation of a pit disassembly and conversion facility (PDCF), where nuclear warheads are dismantled and where the recovered metal is converted into plutonium oxide, and
2. Construction and operation of a fuel fabrication plant, Mixed-oxide Fuel Fabrication Facility (MFFF), where plutonium oxide is mixed with uranium oxide to make MOX assemblies.

The 600,000 ft2 MFFF plant is currently under construction at the Savannah River Site near Aiken South Carolina, is on track to be completed by its target date of 2016. Its purpose, upon completion, is to reduce the surplus weapons grade plutonium and provide fuel for commercial plants. If successful, the MOX process might be the ultimate answer both politically and environmentally for a safer nuclear power industry.

It is estimated that when the MFFF operation is complete, enough electricity could be generated to power all households in South Carolina for up to 20 years. The project is overseen by the National Nuclear Security Administration (NNSA), via third party contractors.

In addition, there will be a waste solidification building, and a pit disassembly and conversion unit that are pivotal in the attempt to shrink the waste plutonium mountain.

At the MFFF facility, surplus plutonium will be processed and blended with depleted uranium oxide to make mixed oxide fuel that will be used as new fuel for nuclear plants.

The waste solidification building and the pit disassembly and conversion unit are vital cogs in the wheel of using the plutonium as a resource, the waste unit is forecasted to treat 150,000 gallons of transuranic waste, and approximately 600,000 gallons of low level radioactive waste from the MFFF and pit disassembly buildings.

Note: As an additional benefit to national and international security, once the MOX fuel has been irradiated by the commercial reactors, the plutonium can no longer be used for nuclear weapons activity.

But most importantly, MOX facilities would provide environmental safety for the future generations, by converting potentially environmentally dangerous radioactive materials into safe commercial nuclear fuel.

DECOMMISSIONING SERVICES

So, after 20, 30, 40 years of non-stop operation, all energy facilities—mines, oil and gas wells, power plants, etc—are decrepit enough for the owners to start thinking about putting them out of their misery. A day comes when the operations are discontinued and demolition equipment and personnel arrive to start the last stage of their life-cycle: decommissioning.

Here we take a brief look at the decommissioning operations, which are usually performed by third party service companies, which create a sizeable peripheral industry within the energy market.

Oil Rigs Decommissioning

Oil production has problems that start with the initial stages of discovery, planning, and drilling of oil reservoirs, and continue all through their productive year. After 20, 30, or 50 years of non-stop operation, the oil deposits are depleted, and/or the oil rig is getting too old for reliable and safe operation. At that time the oil rigs go into the last phase of their lifespan—the decommissioning process.

Decommissioning of oil rigs is a complex and expensive effort. It can cost $5-$10 million per rig in the shallow water Gulf of Mexico (GOM), and several times that much for decommissioning of deep water rigs.

The U.S. Department of the Interior, Bureau of Ocean Energy Management, Regulation, and Enforcement (BOEMRE) Gulf of Mexico' OCS Region, issued a new decommissioning regulation in September 2010, which tightens further the requirements and increases the costs.

The new NTL 2010-G05 requires oil and gas wells that have not been used for the last five years to be classified as: permanently abandoned, temporarily abandoned, or zonally isolated by Oct. 15, 2013.

If wells are zonally isolated, operators have 2 additional years to permanently or temporarily abandon the wellhead. Platforms and supporting infrastructure that have been idle for five or more years must be removed within 5 years as of the same date.

This means that the new NTL on top of the typical volume of decommissioning work in the GOM, will increase demand for contractors and, in turn, the expenses

The Process

There are 10 steps to the decommissioning process:
- Project Management,
- Engineering and Planning;
- Permitting and Regulatory Compliance;
- Platform Preparation;
- Well Plugging and Abandonment;
- Conductor Removal;
- Mobilization and Demobilization of Derrick Barges;
- Platform Removal;
- Pipeline and Power Cable Decommissioning;
- Materials Disposal; and finally
- Site Clearance and final test

Project management, engineering and planning for decommissioning an offshore rig usually starts three years before the well runs dry. The process involves review of contractual obligations engineering analysis, operational planning, and derrick barges contracting.

Due to the limited number of derrick barges, many operators contract these vessels two to three years in advance. In addition, much of the decommissioning process requires contractors who specialize in a specific part of the process. Most operators contract out the project management, hardware cutting, civil engineering, and diving services.

Permitting and Regulatory Compliance requires obtaining permits to decommission an offshore rig, which can take up to three years to complete. Often, operators will contract a local consulting firm, which is familiar with the regulatory framework of their region, in order to ensure that all permits are in order prior to decommissioning.

An execution plan is one of the first steps in the process. The plan includes environmental information and field surveys of the project site and describes a schedule of decommissioning activities and the equipment and labor required to carry out the operation.

An execution plan is required to secure permits from Federal, State, and local regulatory agencies. The

BOEMRE (a Federal agency) will also analyze the environmental impact of the project and recommend ways to eliminate or minimize those impacts.

Federal agencies often involved in decommissioning projects include BOEMRE, National Marine Fisheries Service, US Army Corps of Engineers, US Fish and Wildlife Service, National Oceanic and Atmospheric Administration, US Environmental Protection Agency, US Coast Guard, the US Department of Transportation, and the Office of Pipeline Safety.

Platform Preparation

To prepare a platform for decommissioning, all holding tanks, and processing equipment and piping must be flushed and cleaned, and the waste waters and solid materials have to be disposed of.

Then the platform equipment has to be removed which includes cutting pipe and cables between deck modules, separating the modules, installing padeyes to lift the modules; and reinforcing the structures in order to withstand the disassembly and transport efforts. Underwater workers usually prepare the jacket facilities for removal, which includes removing marine growth.

Well Plugging and Abandonment

Plugging and abandonment is one of the major costs of a decommissioning project and can be broken into several stages.

The planning phase of well plugging includes:
- Data collection
- Preliminary inspection
- Selection of abandonment methods
- Submittal of an application for BOEMRE approval

In the GOM, the rig-less method, which was developed in the 1980s, is primarily used for plugging and abandonment jobs. The rig-less method uses a load spreader on top of a conductor, which provides a base to launch tools, equipment and plugs downhole.

Well actual abandonment involves:
- Well entry preparations
- Use of a slick line unit
- Filling the well with fluid
- Removal of downhole equipment
- Cleaning out the wellbore
- Plugging open-hole and perforated intervals(s) at the bottom of the well
- Plugging casing stubs
- Plugging of annular space
- Placement of a surface plug
- Placement of fluid between plugs

Plugs must be tagged to ensure proper placement or pressure-tested to verify integrity.

According to BOEMRE, all platform components including conductor casings must be removed to at least 15 ft below the ocean floor or to a depth approved by the Regional Supervisor based upon the type of structure or ocean-bottom conditions.

To remove conductor casing, operators can chose one of three procedures:

- Severing, which requires the use of explosive, mechanical or abrasive cutting
- Pulling/sectioning, which uses the casing jacks to raise the conductors that are unscrewed or cut into 40 ft-long segments.
- Offloading, which utilizes a rental crane to lay down each conductor casing segment in a platform staging area, offloading sections to a boat, and offloading at a port. The conductors are then transported to an onshore disposal site.

Mobilization/Demobilization and Platform Removal

Mobilization and demobilization of derrick barges is a key component in platform removal. According to BOEMRE, platforms, templates and pilings must be removed to at least 15 ft below the midline.

First, the topsides are taken apart and lifted onto the derrick barge. Topsides can be removed all in one piece, in groups of modules, reverse order of installation, or in small pieces.

If removing topsides in one piece, the derrick barge must have sufficient lifting capacity. This option is best used for small platforms. Also keep in mind the size and the crane capacity at the offloading site. If the offloading site can't accommodate the platform in one piece, then a different removal option is required.

Removing combined modules requires fewer lifts, thus is a time-saving option. However, the modules must be in the right position and have a combined weight under the crane and derrick barge capacity. Dismantling the topsides in reverse order in which they were installed, whether installed as modules or as individual structural components, is another removal option and the most common.

Topside can also be cut into small pieces and removed with platform cranes, temporary deck mounted cranes, or other small (less expensive) cranes. However, this method takes the most time to complete the job, so any cost savings incurred using a smaller derrick barge will likely be offset by the day rate.

Removing the jacket is the second step in the demolition process and the most costly. First, divers using explosives, mechanical means, torches or abrasive technology make the bottom cuts on the piles 15 ft below the mudline. Then the jacket is removed either in small pieces or as a single lift. A single lift is possible only for small structures in less than 200 ft of water. Heavy lifting equipment is required for the jacket removal as well, but a derrick barge is not necessary. Less expensive support equipment can do the job.

Pipeline and Power Cable Decommissioning can be done in place if they do not interfere with navigation or commercial fishing operations or pose an environmental hazard. However, if the BOEMRE rules that it is a hazard during the technical and environmental review during the permitting process, it must be removed.

The first step to pipeline decommissioning in place requires a flushing it with water followed by disconnecting it from the platform and filling it with seawater. The open end is plugged an buried 3 ft below the seafloor and covered with concrete.

Site Clearance and Materials Disposal

Clearance and disposal of platform materials is used to ensure that all materials are refurbished and reused, scrapped, recycled or disposed of in specified landfills.

To ensure proper site clearance, operators need to follow a four-step site clearance procedure.

- Pre-decommissioning survey maps the location and quantity of debris, pipelines, power cables, and natural marine environments.
- Post decommissioning survey identifies debris left behind during the removal process and notes any environmental damage
- ROVs and divers target are deployed to further identify and remove any debris that could interfere with other uses of the area.
- Test trawling verifies that the area is free of any potential obstructions.

How much does each step of this effort costs depends on the location and size of the rig. In all cases, many millions dollars is a good estimate…for each step of the process. The total cost can be a mind boggling number, which the oil companies are usually reluctant to release.

In case of an accident, such as a rig collapse, or human injury and fatality, the final total costs can be even more unbelievable. This is another reason why crude oil will never be cheap, and why its cost is only going to go up in the future—regardless of how much more oil we may discover.

Mines Decommissioning

After many years of exploitation, the coal in any mine would get depleted and the mine abandoned. A large hole in the ground and significant contamination in the surrounding area is all that remains. The law, however, requires that the owners clean up the mess before leaving permanently. The land used during the mining operations must be brought back, as close as possible, to its original condition. Restoring the land 100% is very seldom possible, and the process is very expensive, so enforcing the law is subjective at best.

Land reclamation is the preferred post-strip mining method for bringing it back to original conditions. It, however, is expensive and at a cost of about $10,000 per acre, it would cost $3-5 billion to reclaim all damaged land in the U.S. This is possible but improbable to happen anytime soon.

And so, what happened in the past more often than none is that mine companies would just abandon the mine, file for bankruptcy, and hope for the best. In many cases, they look for other ways to evade the expensive land restoration process—even if it involves legal procedures.

There are, however, examples of great success in this area. For example, in the Rhineland, Germany, coal fields they store both the topsoil and the subsoil during excavation. It is then used to cover the damaged areas, thus bringing the land back to where it was on a first place...literally. The land is fully refilled, graded and then fertilized and seeded. Drains carry the water away before it can get contaminated by forming sulfuric acid. The acid in the leachate is sometimes neutralized with limestone to prevent further damage.

The acid can be also neutralized with ashes, which are a serious solid waste from coal burning. Using these ashes to fill the trenches and neutralize the leachate acidity would, however, be practical only at mine mouth plants (power plants operating close to the mine). Transport of large quantity of ashes to mines that are far away from the power plant would be prohibitively expensive and has never been tried.

In other cases, sewage sludge and even liquid sewage from the water reclamation plants can be spread over strip-mined areas to help restore the fertility and soil texture. The land then can be revegetated, thus slowly returned to its original state.

When the land is restored, a long-term land management plan kicks in, which the owners finance and supervise.

This is a critical phase of the process of recovering damaged land mass. This is because even in best of cases, the initial efforts of land reclamation do not meet fully the requirements. And also because, there are secondary damaging processes that could continue to damage the land for years.

There are some areas that are very difficult to reclaim. One of the most difficult reclamation is in the North Central U.S. regions, where grassy ranch land is being destroyed in large numbers. Restoring this arid land to usable status would be almost impossible from economics point of view. As a result, no coal company in Montana has ever gotten its reclamation performance bond (money deposited at the beginning of the mine process.

Meeting the requirements of the federal surface mining controls adds to the cost of the coal. The cost of reclaiming Western surface mines, for example, is $1,000 to $5,000 per acre. A more useful comparison is provided by the additional cost per million Btu (MBtu) of energy obtained from coal. It, however, varies from location to location and from mine to mine. In the mid-1980s coal energy was available to electric utilities at an average cost of about $1.50 per MBtu. It is estimated now that federally mandated reclamation adds about $0.02/MBtu to Western coal (where the seams are deep, but the energy content is relatively low) about $0.05/MBtu to Midwestern coal, and $0.11/MBtu to Appalachian coal.

Existing requirements already cost an additional $0.10 MBtu for Appalachian coal. Added to these costs is a tax, equivalent to $0.02/MBtu, assessed on the coal companies for the reclamation of abandoned strip-mined land.

Thus, the highest total of additions, Appalachia's $0.23/MBtu, raises the cost of coal to the utility, and ultimately of electricity to the consumer, by over 15%.

About 1.1 million acres of coal mined land currently needs reclamation and new land is being disturbed at a rate of about 65,000 acres/year. At this pace, in 20 years we would have an additional 1.2 million acres of pit mining land in need of surface restoration and reclamation.

The present day reclamation laws are not up to par and cannot keep with the fast pace of proliferation of coal draglines around the country. They also need careful enforcement, which has not been done in the past. The situation is even worse in many other countries.

Power Plant Decommissioning

After several decades of operation, every power plant ends up on the chopping block and the land it once stood on is returned to its original state. At least this is what needs to happen. Sometimes it does happen, but very often it does not. At least not fully.

The major steps in the coal burning power plant decommissioning and reclamation are:

- Plant shut down and disassembly
- Turbine and generator dismantling and removal
- Smoke stack demolition and dismantling
- Rigging and removal of support equipment
- Construction and equipment waste disposal
- Surface land decontamination and reclamation

Coal burning power plant decommissioning is similar to that of any power plant, and basically consists of safe removal of the buildings by demolition, dismantling and disposal of their components.

The decommissioning effort usually starts with disassembly and removal of key electro-mechanical components, including; breaker boxes, transformers, conduit, wring, etc. electrical components. The rest of the equipment; furnace components, piping, turbine rotor and blading, generators, static exciters, frequency changers, heat exchangers, and other related mechanical equipment is disassembled and removed from the facilities as well.

This effort might also involve selective dismantling and rigging of some power equipment pieces, which are in good condition and could be reused or sold. When all equipment has been removed from the facilities, the buildings are demolished and the waste, together with the demolished equipment, are loaded on dump trucks fro removal and disposal.

Last, but not least, the land that was occupied by the plant buildings has to be decontaminated, leveled, and/or otherwise brought to its original state. This is usually also a very complex, lengthy, and expensive undertaking.

All decommissioning, rigging and dismantling work is done in accordance to all safety rules and regulations and is quite labor intensive and expensive.

Decommissioning of Nuclear Plants

Decommissioning costs of nuclear plants are very high, and significantly different than that of other power plants. This is due to the complexity of the technology, but also because of the added serious danger of nuclear radiation at every step of the way. This means that the demolition crew must be well trained and thoroughly familiar with, and trained in, dealing with radioactive components and materials. Every action could result in the unexpected release of radiation, which could be damaging and even fatal.

Decommissioning involves a number of administrative and technical actions, including inspection, clean-up of radioactivity, demolition of the plant, and removal and disposal of components. There are contaminated materials, parts and components that have to be handled with utmost care.

Engineering services and labor used in decommissioning of nuclear plants can be described by specialized services, starting with concrete and metal cutting and removal, including:

- Man access cuts
- Equipment hatch enlargement
- Containment wall penetrations
- Dog house cutting for SGRPs
- Elevated platform removals
- Fuel pool and canal segmentation
- Fuel transfer canal segmentation

And in more detail, there are a number of tasks performed by experienced professionals during decommissioning of a nuclear power reactor, such as:

- CRDM cutting
- Reactor nozzle cutting
- Heat exchanger cutting
- Ion guide tube removals
- Monitoring line cutting
- Steel plate stabilizer cutting
- Carbon containment liner cutting

Hands-on demolition services experience is needed here, but it has to be supplemented by thorough understanding of the nuclear plant components and their function. Only well trained and committed workforce, well attuned to personal radiation and toxic materials exposure concerns, and waste minimization awareness, could be allowed and successful in such undertaking. Anything less might result in environmental and personal damage and even death.

Note: The above is a very sketchy and superficial description of a very complex, dangerous, and lengthy operation. It is intended to give the esteemed reader just a brief glimpse of the entire nuclear fuel life cycle. This is because going into intimate technical detail would require another thick book.

Decommissioning costs represent about 5-10% of the initial capital cost of a nuclear power plant, but when discounted, they contribute only a few percent to the investment cost and even less to the generation cost. For example, in the US the decommissioning costs average to 0.1-0.2 cent/kWh, which is about 5% of the cost of the electricity produced.

Decommissioning and land remediation is a complex and expensive undertaking. The dismantling could cost from $5 million to $10 million. Handling and removal of the fuel could cost another $5-10 million alone. The entire plant decommissioning could be in the range of $25 million to a $1 billion. In some cases there is an additional charge for upkeep of the area, which could be up to $ 10 million annually.

One can only hope that all these costs were foreseen and estimated in the overall operating budget of the plant. Else the owner might be faced with seriously uncertain financial future, or worse.

Once a nuclear facility is decommissioned properly, there is no longer danger of radioactivity. At that time, and after official inspection and certification, the area and the owner of the plant are released from regulatory control, and the owner is no longer responsible for the nuclear safety of the area.

Renewable Power Plants Decommissioning

The renewables, like geo- and bio-energy, use power plant buildings and processes similar to the conventional power plants described above, so we will not spend time on these here.

Other renewables, like solar, wind, and ocean power generators, use stand-alone structures that do not require buildings. Their decommissioning, therefore, is straightforward and consists of removal of the equipment and reconstructing the land to its original state. This operation is similar to that described in the Power Plants Decommissioning sections above.

It is important to note that solar and wind power plants do not damage the land as extensively as any of the fossil power plants. This, therefore, means that the expense would be much less. There might be some exceptions in the future, where toxic materials used in these plants contaminate the land.

Cadmium Telluride PV panels, for example, contain toxic, carcinogenic, heavy metal cadmium. While locked within the panels cadmium is harmless, but if and when released, in case of a major accident where thousands of PV panels are broken, then the land would be heavily contaminated. In such case, the broken PV panels would be handled as toxic waste, the top soil layer would be removed and new soil laid in its place. This could be a very expensive undertaking, for which the owners and manufacturers would be held responsible for financing, supervising, and completing.

ENVIRONMENTAL SERVICES

We cannot possibly discuss the global energy markets without considering the global environmental balance and its history. These facts and variables are closely related, so they need to be considered, understood, and properly managed as part of the big picture.

There is so much talk today about the changing climate and weather patterns. While we cannot argue that man has been abusing the Earth—and especially its natural resources—during the last century, we must also consider the fact that the pre-industrial condition of Mother Earth was anything but stable. Like a drunk sailor, she went from very hot to very cold periods—see Figure 8-37. There were good reasons for these changes every single time, and we think that we understand most of them.

Figure 8-37 shows dramatic temperature fluctu-

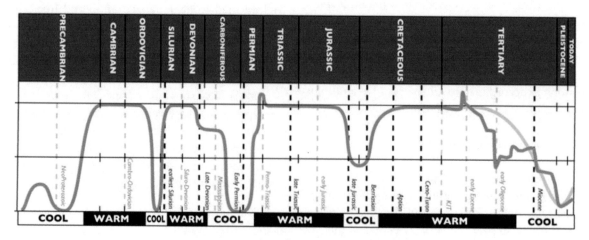

Figure 8-37. Historical global temperatures

ations through the centuries past. Starting 550 million years ago with a cooler climate, it went up and down on several occasions, lasting many centuries every time. During the mid-Cretaceous period (about 100 million years ago) the earth went through the hottest climate ever. It is a time of total absence of ice on the planet. But then, by the end of the period, the global temperature plummeted to the present levels, which are actually (and even with the present increase) some of the lowest on record.

So what is different today? Yes, we have a slight rise in temperature. Yes, we see changes all around us. Yes, we know that man has a major influence on the Earth's environment, and we know well what the final results might be. But as Figure 8-37 shows, the Earth is a dynamic structure with many facets and uncontrolled behavior. The man-made effects complicate the picture further, and so we are having a hard time superimposing them over the natural such in order to distinguish the difference these make on the long run.

A closer look at the entire picture might shed light on the uncertainties:

The Participants

There are willing and unwilling participants in the global environmental markets. The willing ones are the government and private environmental organizations. The unwilling are those who profit from, and most of us; unsuspecting indirect participants in, the energy cycle. And so now we see the globe enveloped by the 'energy production and use vs., the environmental damages' debacle.

Who are the major participants in this debacle?

U.S. Environmental Watch Dogs

The U.S. is blessed by a number of environmental friends, who watch over our skies, waters, and soil. Sometimes they provide solution, some times they create problems. Since the U.S. has no viable energy policy, in many cases the environmental protection procedures and actions are questionable.

The principal environmental watchdogs in the U.S. are:

EPA

The overall mission of Environmental Protection Agency (EPA) is to protect human health and the environment. In more detail, EPA aims to ensure that all Americans are protected from significant risks to human health and the environment where they live, learn and work. Environmental protection is an integral consid-

eration in U.S. policies concerning natural resources, human health, economic growth, energy, transportation, agriculture, industry, and international trade, and these factors are similarly considered in establishing environmental policy.

EPA leads national efforts to reduce environmental risk are based on the best available scientific information. Makes sure that federal laws protecting human health and the environment are enforced fairly and effectively.

EPA also ensures that all parts of society—communities, individuals, businesses, and state, local and tribal governments—have access to accurate information sufficient to effectively participate in managing human health and environmental risks.

EPA's environmental protection efforts contribute to making our communities and ecosystems diverse, sustainable and economically productive, and that the United States plays a leadership role in working with other nations to protect the global environment.

Office of Environmental Markets

The Office of Environmental Markets (OEM) was established in response to the Food, Conservation, and Energy Act of 2008, America's Farm Bill. Section 2709 of the Conservation Title directs the Secretary to facilitate the participation of America's farmers, ranchers, and forest landowners in environmental markets.

OEM supports the Secretary of USDA in the development of emerging markets for water quality, carbon sequestration, wetlands, biodiversity, and other ecosystem services. Environmental markets have the potential to become a new economic driver for rural America.

Office of Air and Radiation

The Office of Air and Radiation (OAR) develops national programs, policies, and regulations for controlling air pollution and radiation exposure. OAR is concerned with pollution prevention and energy efficiency, indoor and outdoor air quality, industrial air pollution, pollution from vehicles and engines, radon, acid rain, stratospheric ozone depletion, climate change, and radiation protection.

OAR is responsible for administering the Clean Air Act, the Atomic Energy Act, the Waste Isolation Pilot Plant Land Withdrawal Act, and other applicable environmental laws.

Energy and Environmental Markets Advisory Committee

The Energy and Environmental Markets Advisory Committee was created in 2008 to advise the U.S.

Commodity Futures Trading Commission on important new developments in energy and environmental futures markets that may raise new regulatory issues, and the appropriate regulatory response to ensure market integrity and competition, and protect consumers.

Sierra Club

The Sierra Club, founded in 1892, is one of the oldest conservation organizations in existence.) With over 1.3 million members, this organization is one of the most effective and powerful at effecting changes in government and corporate America. Fighting for the preservation of land and forest, clean air and water, and a host of other issues, the Sierra Club is well-known and respected.

Other

There are many other organizations that are focused on promoting market-based solutions for environmental challenges through revision and implementation of public policies, industry best practices, effective education and training, and member networking. These organizations usually represent a diverse membership; including large utilities, emissions brokers and traders, exchanges, law firms, project developers, consultants, academics, NGOs and government agencies.

Their efforts consist of reviewing and commenting on the developments at hand and taking some (limited) action on the issues at hand.

Global Environmental Organizations

And of course, there are many environmental organizations in different countries. Some operate on local level, while others are concerned with and active on global level. The most significant of these are as follow:

Intergovernmental Panel on Climate Change

The Intergovernmental Panel on Climate Change (IPCC) is a scientific intergovernmental body under the auspices of the United Nations, set up at the request of member governments. It was first established in 1988 by two United Nations organizations, the World Meteorological Organization (WMO) and the United Nations Environment Program (UNEP), and later endorsed by the United Nations General Assembly through Resolution 43/53. Membership of the IPCC is open to all members of the WMO and UNEP.

The IPCC produces reports that support the UN Framework Convention on Climate Change (UNFCCC), which is the main international treaty on climate change. The ultimate objective of the UNFCCC is to "stabilize greenhouse gas concentrations in the atmosphere at a level that would prevent dangerous anthropogenic [i.e., human-induced] interference with the climate system." IPCC reports cover the scientific, technical and socio-economic information relevant to understanding the scientific basis of risk of human-induced climate change, its potential impacts and options for adaptation and mitigation.

Greeneace

The Greenpeace movement began in 1971 when a group of activists put themselves directly in harm's way in order to protest nuclear testing off the coast of Alaska. Believing that concerted action from ordinary people is the best way the organization has helped to stop whaling, nuclear testing, as well as leading efforts to protect Antarctica. Greenpeace has over 2.5 million members worldwide.

Global Environment Organization

The Global Environment Organization is a non-advocacy, non-partisan, non-for-profit organization with a mission to assist in solving environmental problems in Andhra Pradesh state. the organization fulfils this mission by encouraging co-operation among non-governmental organizations, governments, businesses and other environmental stakeholders, by supporting the exchange of information and by promoting public participation in environmental safeguard, protection and improve human well-being and social equity, while significantly reducing environmental risks and ecological scarcities.

The Unwilling Participants

There are many companies and private individuals—probably most of us at one time or another—who do not agree with what is going on around us. We often object—for whatever reasons—the utilities and energy companies' plans to build new facilities or expand existing.

In some cases we fear that people and businesses could be negatively affected by a nearby power plant, or energy project. This has brought the almighty NIMBY (not in my back yard) movement in existence.

NIMBY

We know that personal safety and comfort have a price and a limit. NIMBY is a powerful movement in a number of countries—including the U.S.—that is trying to establish and enforce the limits of what energy companies can and cannot do, when and where. In the U.S.

it started with opposition to nuclear power, which part of this movement has been considerable since day one. Starting with the first reactor, Fermi 1 to be built close to Detroit in 1957, where the United Autoworkers Union led a fierce anti-nuclear opposition, resulting in cancellation of the plans.

Then came the decision of PG&E to build the first large, commercially viable nuclear power plant in the U.S. at Bodega Bay, close to San Francisco. This was a very controversial project, and serious conflict with local citizens began in 1958. After several years of non-stop protests and strikes, the NIMBY movement won in 1964. Plans for a power plant at Bodega Bay were abandoned. Another set of attempts to build a nuclear power plant in Malibu, CA, failed under similar circumstances.

The awareness about the danger of nuclear power, which started the NIBMY movement was fueled by a number of little known nuclear accidents during the 1960s. Starting with a small test reactor exploding at the Stationary Low-Power Reactor Number One in Idaho Falls in January 1961, and followed by a partial meltdown at the Enrico Fermi Nuclear Generating Station in Michigan in 1966.

The environmentalists see the advantages of nuclear power in reducing air pollution, but are in general critical of nuclear technology on the grounds of its safety record. The major concerns with nuclear power were and still are about the possibility of nuclear accidents, nuclear proliferation, high cost of nuclear power plants, the possibility of nuclear terrorism, and the problems with radioactive waste disposal.

Because of that, nuclear power has been unable to face the opposition successfully thus far, and as things are progressing, it has a long ways to go before achieving mass acceptance in the U.S. and the world. The success (or not) of the ongoing work at the destroyed Fukushima nuclear power plant in Japan will have a lot to do with determining the future of global nuclear power, so we will be watching carefully.

The NIMBY movement is now spreading and growing as a mass opposition to the expansion of hydrofracking activities. This strong opposition has led companies to adopt a variety of public relations measures to educate, explain, and reduce fears about hydraulic fracturing. The admitted use of military tactics to counter drilling opponents and other such abnormal behavior have been in the center of the media reports.

An executive of a large oil company plans to download the U.S. Army "Counterinsurgency Manual" to deal with the NIMBY insurgency.

Some companies admittedly use psychological warfare operations tactics and veterans in dealing with the NIMBY crowd. They use the experience learned in the Middle East when dealing with emotionally charged township meetings and advising townships on zoning and local ordinances dealing with hydraulic fracturing and other energy projects.

Police departments are also increasingly forced to deal with intentionally disruptive and violent opposition to oil and gas development in different parts of the country. In March 2013, ten people were arrested during an "anti-fracking protest" near New Matamoras, Ohio. They illegally entered a well development zone and locked themselves to drilling equipment.

A drive-by shooting at a well site in Pennsylvania was recorded, in which an individual shot a rifle at the drilling rig, shouting profanities at the site and fleeing the scene. There were cases of gas pipeline workers finding pipe bombs at pipeline construction sites, which would cause a disaster if not discovered and detonated on time.

Surprisingly, this paradox can work out much better, if handled properly. European countries have learned how to deal with the locals successfully. Where wind power developers are building on a decentralized scale, they've found ways to work around NIMBY locals, by giving them a stake in the power plant that's being built nearby.

Denmark, for example, has a lot of wind power but not much of an issue with local anti-wind protests and NIMBY incidents. In most cases, to build a wind power plant, a developer legally has to give the nearby community at least 20% of the proceeds. A multiturbine wind farm developer usually also offers the community at least one of the turbines' output as a reward for good behavior. This keeps the locals happy and wind power development continues uninterrupted.

The Carbon Markets (The New Carbon Age)

Historically, we have lived through many significant ages, such as the discovery of fire, invention of the wheel, adoption of agriculture, the development of various technologies, and the development of new kind of political systems.

The most significant criteria in all cases has been the physical material with which most technologies of a specific era were manufactured. Officially, there have only been three such "ages." The stone age was the first, followed by the bronze age, and eventually the iron age. The transitions between these ages occurred at different times throughout the world, but represent a common pattern in societal and archeological development.

A number of other similar ages might have occurred as well, but they have not left a heavy footprint to be remembered by. Since the stone, bronze, and iron ages mark significant transitions in the technologies that can be created, any new material that has the effect of transforming a large fraction of a civilization's technologies should count as well.

We know well how stone, bronze, and iron use contributed to the development of humanity. Similarly so, we could easily consider the plastics age in the beginning of the 20th century, followed by the silicon age during the latter half. Plastics truly changed many aspects of the world we live in, while silicon introduced us into a totally different world—that of electronic gadgets and computers.

Many other materials have also had a tremendous effect on our way of life. Glass, rubber, cotton, paper, copper, steel, and of course, everyone's modern favorite, silicon. Each of these new materials has contributed to changes and a transformation of our civilization. Some of the changes have been so drastic that we have a hard time imagining life without them.

Today, however, we are entering a threshold of a major paradigm shift, which is very special because it has only happened a few times in human history.

*While we are developing numerous new materials in our modern era, one that really stands out is **Carbon**.*

What is most significant about carbon is the enormous amounts of it, which we are intentionally and unintentionally producing. Human body, for example is 80% water, while most of the rest (bones, flesh, skin) contains some sort of carbon derivative.

What we are most interested here, however, are the billions of tons of carbon related materials on Earth—coal, crude oil, natural gas—and their products and byproducts. The products we use on daily basis, some of which emit byproducts like CO_2, CH4 and other GHG related carbon containing solids, liquids, and gasses.

Some think of the 20th century as the electronics and plastics age, while what we are going through now in the 21st century is the "carbon age." The concerns with carbon actually started in the beginning of the 20th century, and have been growing in size like a snowball running down a hill.

This is because the use of fossil energy sources; coal, natural gas, and crude oil, in our daily lives have been increasing exponentially. And with that, the air, water, and land pollution have increased at the same

astronomical rate as well.

Today we have a new dimension to the carbon age; where we find new uses for, and build new and fancy materials and structures, out of carbon. These materials, such as zero dimensional bucky-balls, one dimensional carbon nanotubes, two dimensional carbon sheets, and possibly three dimensional carbon solids or shells are "cool" and useful in many ways for our future development as intelligent species. That trend is projected to increase this century as well.

What is not cool, however, is the fact that the carbon emissions are increasing proportionally by the day too.

*The new **carbon age** has been blamed for the drastic temperature and climate changes around the globe.*

So what can we do to prevent the millions of tons of carbon GHGs from entering our atmosphere and choking it, and us, to death? There are several possible solutions, the implementation of which is not easy nor cheap. Nevertheless, the work on the new technologies and processes is intensifying and that has created new 'peripheral' energy related markets. Carbon capture and storage is one such market.

Carbon Capture and Storage

Global energy-related CO_2 emissions continue to rise. In 2011 they increased by 3.2% from 2010, reaching a record high of 31,000 million tons. If this trend continues, by 2050 we will have increased the GHG emission over 120%, or nearly 70,000 million tons of GHGs in one year alone.

This increase will put carbon emissions on a trajectory corresponding to an average global temperature increase of around 6°C by 2050. The greater the emissions of greenhouse gases (GHGs), such as CO_2, the greater the warming and severity of the associated consequences.

The worst effect, at least on the short term, is the quick rise of Earth's temperature. This brings unintended and serious consequences, which we already see in many areas of the world. These include a rise in sea levels, causing dislocation of human settlements, as well as extreme weather events, including a higher incidence of heat waves, destructive storms, and changes to rainfall patterns, resulting in droughts and floods affecting food production, human disease and mortality.

A major fear here is the presumption that as global warming increases, the accompanying events will increase proportionally. At some point they might accel-

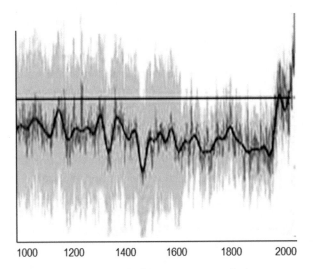

Figure 8-38. Global temperature variations

erate uncontrollably, thus bringing the entire world to its proverbial knees. How realistic this presumption is remains to be seen, but we must do everything possible to get even close to such dooms day scenario. This, however, can be done only by controlling and restricting the global carbon emissions.

In 2013, global CO_2 emissions due to fossil fuel use (and cement production) were 36 gigatonnes ($GtCO_2$); this is 61% higher than 1990 (the Kyoto Protocol reference year) and 2.3% higher than 2012. In 2014, global CO_2 emissions increased by an additional 3% over the 2013 level. CO_2 emissions were dominated by China (28%), the USA (14%), the EU (10%), and India (7%). There was increase of emissions in all of these states, except for a 2% decline in the EU 28 member states.

The breakdown of carbon dioxide emissions from fossil fuels and cement production only is as follow: coal (43%), oil (33%), gas (18%), cement (5.5%) and gas flaring (0.6%).

In order to restrict global temperature increase to 2 °C max, only 880 Giga tons of CO_2 can remain in the atmosphere by 2050.

This means that during the 40 years from 2010 to 2050, the average global GHG emissions cannot exceed 22 Giga tons annually. Not 36, and surely not 70 Giga tons, as emitted presently, but 22 Giga tons annually for the duration. Otherwise, we would be living on a hot and smoky Earth, with many serious and unpredictable climate changes bringing untold human suffering.

We can't even imagine how accurate these huge numbers are and what they mean, so we would point to another, more reliable method of measuring the GHGs

and their effects around us. It is the actual count of particles per million (ppm) of contaminants floating in the air we breath. While in the past there were 200-300 ppm counts, today we see over 400 in some places. The ppm count is much higher around power plants, cement factories, and other large industrial complexes.

These are not imaginary or insignificant numbers. They represent actual poisons in the air we breath. The worst part is that their number is increasing steadily.

Carbon capture and storage (CCS) technologies might be one of the solutions to reducing air pollution. These technologies are developed and widely implemented to assist with the problem. CCS would serve our climate objectives, while investing in development and deployment of CCS is an important risk management response for companies and governments who derive significant income from fossil fuels. It is perhaps the best way of hedging against the uncertainties, fossil fuels price increases, and eventual global GHG annihilation.

CCS is the best way to preserve the economic value of the global fossil reserves, the associated infrastructure, and reduce climate change effects.

While the different forms of crude oil use contribute to the global emissions significantly, coal and natural gas power plants are the greatest GHG emitters. There is no intention to reduce their use, at least for now, and as a matter of fact their number is increasing. Because of that, we must at least consider doing something to reduce their GHG emissions. CCS to the rescue, as the most efficient way to curb GHG emission releases in the atmosphere.

Note: CCS is not the final and permanent solution to the environmental problems, because sooner or later we would run out of storage space. It is only one way to avoid a great disaster, while looking for alternative ways to reduce GHG emissions in the (hopefully near) future.

What is CCS?
In essence, CCS consists of:
- CO_2 separation from exhaust gasses
- CO_2 compression and clean-up
- CO_2 transport by pipeline or container
- CO_2 Underground storage and monitoring

where
1. Saline formations and aquifers
2. Coal and coal-methane seams
3. CO_2 use in enhanced oil recovery
4. Depleted oil and gas reservoirs

Geological storage of CO_2 involves the transport and injection of CO_2 into appropriate geologic formations that are typically located between one and two miles underground. It also involves the subsequent monitoring of injected CO_2. Suitable geologic formations include saline aquifers, depleted oil and gas fields, oil fields with the potential for CO_2-flood, and coal seams that cannot be mined with potential for enhanced coal-bed methane recovery. Storage in other types of geologic formations (e.g. basalts) and for other purposes, such as enhanced gas recovery or geothermal heat recovery, are active topics of investigation as well.

CO_2 capture can be done by a number of methods, according to the production process and how it needs to be modified to enable CO_2 separation and transport to the storage site or facility. In some cases, these approaches can be combined to create hybrid routes to capture.

The Present CCS State

The main CCS processes today are:

- Post-process capture, which is done by separation of CO_2 from a mixture of combustion gases at the end of the power generation process. This is referred also to as post-combustion capture in power generation applications.

- *Syngas/hydrogen capture* could be used for production of syngas (a mixture of hydrogen, carbon monoxide and CO_2, which can be generated from fossil fuels or biomass.) The CO_2 is then removed, leaving a combustible fuel, reducing agent, or feedstock. In some cases, where either pure hydrogen or additional emission reductions are required, the syngas can be reduced to hydrogen while converting the carbon monoxide to separable CO_2. This method is also referred to as *pre-combustion capture* in power generation applications.

- *Oxy-fuel combustion* is obtained when nearly pure oxygen is used in place of air in the combustion process to yield a flue gas of high-concentration CO_2. While in oxy-fuel combustion a specific CO_2 separation step is not necessary, there is an initial separation step for the extraction of oxygen from air, which largely determines the energy penalty.

- *Inherent separation* is the generation of concentrated CO_2 as an intrinsic part of the production process (e.g. gas processing and fermentation-based biofuels). Without CO_2 capture, the generated CO_2 is ordinarily vented to the atmosphere.

For all applications where CO_2 separation is an inherent part of production, CO_2 capture processes are commercially available and in common use. In other applications, such as coal-fired electricity generation, however, CO_2 separation processes are less advanced or require considerable redesign of traditional processes.

Since coal-fired CO_2 is the largest component of the GHG emissions, a larger emphasis on its CCS is needed, if and when governments undertake ambitious measures to combat climate change. The current trends of increasing global energy sector CO_2 emissions and the dominant role that fossil fuels continue to play in primary energy consumption, means that the urgency of CCS deployment is only increasing.

> *The present trend of using fossil fuels will raise global temperature over 6°C by 2050. Properly implemented CCS could reduce the GHG emissions by 15-20% by that time.*

> *In all cases, CCS is obviously only one, but very important, part of the total solution.*

The individual components of the technologies required for CO_2 capture, transport, and storage are gen-

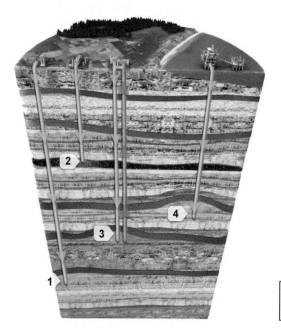

Figure 8-39. CO_2 storage methods

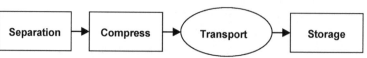

Figure 8-40. Basic CCS process

erally well understood and, in some cases, technologically mature. For example, capture of CO_2 from natural gas sweetening and hydrogen production is technically mature and commercially practiced, as is transport of CO_2 by pipelines.

While safe and effective storage of CO_2 has been demonstrated, there are still many lessons to gain from large-scale projects, and more effort is needed to identify viable storage sites. However, the largest challenge for CCS deployment is the integration of component technologies into large-scale demonstration projects. Lack of understanding and acceptance of the technology by the public, as well as some energy and climate stakeholders, also contributes to delays and difficulties in deployment.

Governments and industry must ensure that the incentive and regulatory frameworks are in place to deliver upwards of 30 operating CCS projects by 2020 across a range of processes and industrial sectors. This would be equivalent to all projects in advanced stages of planning today reaching operation by that time. Co-operation among governments should be encouraged to ensure that the global distribution of projects covers the full spectrum of CCS applications, and mechanisms should be established to facilitate knowledge sharing from early CCS projects.

It is important to understand also that CCS is not only about electricity generation. Almost half (45%) of the CO_2 captured between 2015 and 2050 would be from industrial applications. In this scenario, between 25% and 40% of the global production of steel, cement and chemicals must be equipped with CCS by 2050. Achieving this level of deployment in industrial applications will require capture technologies to be demonstrated by 2020, particularly for iron and steelmaking, as well as cement production.

As Figure 8-41 shows, the list of possible CCS technologies is long, and encompasses a number of industries. The commercially feasible CCS technologies today are listed in Figure 8-41. Of these, Sweetening, Coal-to-liquids, Hydrogen production, Ammonia/methanol, and Ethanol fermentation are presently used on commercial basis, while the rest are either in lab, pilot, or demo mode.

Given the rapid growth in energy demand, the largest deployment of CCS will need to occur in developing countries. By 2050 they need to account for 70% of the total cumulative mass of captured CO_2, with China alone accounting for one-third of the global total of captured CO_2 between 2015 and 2050.

This would be hard to impossible to do, unless OECD governments and multilateral development banks work together with non-OECD countries to ensure that support mechanisms are established to drive deployment of CCS in the developing countries in the coming decades.

This decade is also critical for moving deployment of CCS beyond the demonstration phase. Mobilizing the large amounts of financial resources necessary will depend on the development of strong business models for CCS, which are so far lacking. Urgent action is required from industry and governments to develop such models and to implement incentive frameworks that can help them to drive cost-effective CCS deployment. Moreover, planning and actions which take future demand into account are needed to encourage development of CO_2 storage and transport infrastructure.

Transporting CO_2 is the most technically mature step in the CCS process. Transport of CO_2 in pipelines is a known and mature technology, with significant experience from more than 6 000 km of CO_2 pipes in the U.S. There is also experience, albeit limited, with transport of CO_2 using offshore pipelines in the Snøhvit project in Norway. Guidance for the design and operation of CO_2 pipelines that supplements existing technical standards for pipeline transport of fluids was released in 2010.

CO_2 is also transported by ship, but in small quantities. Understanding of the technical requirements and conditions for CO_2 transport by ship has improved recently, but it still very expensive.

To achieve CCS deployment at the scales envisioned in the ETP 2012 2DS, it will be necessary to link CO_2 pipeline networks across national borders and to shipping transportation infrastructure (i.e. temporary storage and liquefaction facilities) to allow access to lowest-cost storage capacity.

Even though the transport technologies are best understood, they remain a major challenge. The main and most expensive problem is developing long-term strategies for CO_2 source clusters and pipeline networks that optimize source-to-sink transport. Government-led national or regional planning exercises are required for these technologies to become reality.

The Future

While geological storage of CO_2 is the preferred CCS method today, there are other methods under evaluation.

Some of the CCS methods in research mode are:

• Stimulating Earth's terrestrial ecosystems to take up more carbon dioxide has a huge potential,

		Syngas-hydrogen capture	Post-process capture	Oxy-fuel combustion	Inherent separation
First-phase industrial applications	Gas processing	-	-	-	Sweetening
	Iron and steel	direct reduced iron (DRI)*, smelting (*e.g.* Corex)		-	DRI*
	Refining	-	-	-	Coal-to-liquids; synthetic natural gas from coal
					Hydrogen production
	Chemicals	-	-	-	Ammonia/methanol
	Biofuels	-	-	-	Ethanol fermentation
Power generation	Gas	Gas reforming and combined cycle	Natural gas combined cycle	Oxy-fuel combustion	Chemical looping combustion
	Coal	Integrated gasification combined cycle (IGCC)	Pulverised coal-fired boiler	Oxy-fuel combustion	Chemical looping combustion
	Biomass	IGCC	Biomass-fired boiler	Oxy-fuel combustion	Chemical looping combustion
Second-phase industrial applications	Iron and steel	Hydrogen reduction	Blast furnace capture	Oxy-fuel blast furnace	-
	Refining	Hydrogen fuel steam generation	Process heater and combined heat and power (CHP) capture	Process heater and CHP oxy-fuel	-
	Chemicals	-	Process heater, CHP, steam cracker capture	Process heater and CHP oxy-fuel	-
	Biofuels	Biomass-to-liquids	-	-	Advanced biofuels
	Cement	-	Rotary kiln	Oxy-fuel kiln	Calcium looping
	Pulp and paper	Black liquor gasification	Process heater and CHP capture	Process heater and CHP oxy-fuel	-

Figure 8-41. The CCS technologies today

but there are unresolved issues, and the overall impacts are difficult to assess. To start with how would we deliver the CO_2 from China to the Brazilian jungle, where it is most needed?

Also, these systems are not very likely to be able to absorb all the millions of tons of extra CO_2. generated daily around the world. In best of case, this method may help to stabilize carbon emission rates for awhile. But even then, a stabilization of GHG emission rate is not likely to help the Earth to recover to pre-industrial carbon levels. In worst case scenario, biological feedbacks to global warm-

ing is expected to increase the incidents of forest fires, drying soils, rotting permafrost, etc, which may actually accelerate carbon emissions. The final result of such developments may actually bring massive carbon de-sequestration to our already overloaded environment.

• *Phytoplankton growth* is another physically plausible scheme that is under investigation. Here, phytoplankton growth on the ocean surface is stimulated by seeding the world ocean surface at certain locations with iron particles. Iron is the limiting nutrient in the phytoplankton growth, so this will cause an increased uptake of carbon by the plankton. This would then increase phytoplankton population several fold, most of which would eventually find its way to the ocean floor. Large fishing companies are particularly interested in this method, as it would increase fish harvests and garner credit for carbon sequestration.

There are, however, a number of uncertainties in this approach. Ecological disruptions could occur in unpredictable manner and force, and could be especially devastating if this method is conducted on a very large scale.

- *Ocean bottom storage* is a CCS method under investigation with the possibility of depositing waste CO_2 into large pools on the world oceans' bottom. It is, however, unlikely that such pools to be stable, since they will most likely erupt to the surface.

    Even more likely, and unavoidable, consequence is their diffusion into the ocean waters, where they form H2CO3 (carbonic acid) and alter the oceans pH.

So, with such a great number of possibilities and uncertainties, we see a best case scenario, where the enormous amount of CO_2 we are presently emitting will force us to develop and implement some sequestration approaches for the short run. This may allow us to fill the gap to curbing global emissions, while working on the transition to renewable, carbon-free, energy sources.

Under the worst case scenario we see most carbon sequestration efforts failing, which can be used (by companies and governments) as a tool to delay and postpone as long as possible the transition to renewable energy sources. This scenario will certainly prolong the existing problems and may even create many additional and very serious such.

CCS is a new technology. which is now making its first steps and looking forward to great things in the future. Figure 8-42 shows that the major emphasis in CCS deployment in this century is at coal powered power plants, followed by that in bioenergy. The rest of the industries in diminishing size are as follow: Iron and steel, cement, gas power, chemicals, gas processing, refining, and pulp and paper.

The next decade is critical to the accelerated development of CCS, which is necessary to achieve low-carbon stabilization, as needed to limit the long-term global average temperature increase to 2°C maximum. The key actions listed below are necessary to lay the foundation

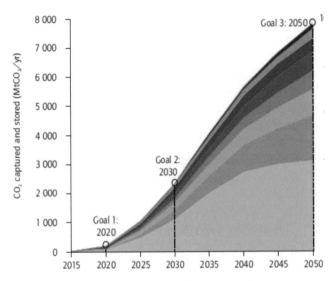

Figure 8-42. CCS technology goals

for scaled-up CCS deployment by 2020. They require serious dedication by governments and industry, but are realistic and cover all three elements of the CCS process.

These actions are:

- Introduce financial support mechanisms for demonstration and early deployment of CCS to drive private financing of projects.
- Implement policies that encourage storage exploration, characterization and development for CCS projects.
- Develop national laws and regulations as well as provisions for multilateral finance that effectively require new-build, base-load, fossil-fuel power generation capacity to be CCS-ready.
- Prove capture systems at pilot scale in industrial applications where CO_2 capture has not yet been demonstrated.
- Significantly increase efforts to improve understanding among the public and stakeholders of CCS technology and the importance of its deployment.
- Reduce the cost of electricity from power plants equipped with capture through continued technology development and use of highest possible efficiency power generation cycles.
- Encourage efficient development of CO_2 transport infrastructure by anticipating locations of future demand centers and future volumes of CO_2.

Water Depletion

What does water has to do with the energy markets, you ask? Well, try to imagine Hoover Dam without water to run the water turbines, or a large nuclear power station without cooling water.

Without enough water, any and all thermal power plants will come to a screeching halt.

On a broader perspective, we know that electricity is very important for our wellbeing and comforts. Crude oil is critical for our transport and commodities manufacturing. Water, however, is simply indispensible for all aspects of life for humans and wild life on Earth. We could survive without electricity and crude oil, just like our ancestors lived without these commodities for centuries.

Life without water is unthinkable in more ways than one.

While living without electricity or crude oil would make life very difficult, uncomfortable, and unsafe, life without water is not possible at all. If we don't have water, we would not have to worry about the energy markets, because there would be none. Most animals, including all mammals and humans, get ill and die within weeks without water. In extreme cases, the vegetation follows within several more weeks, and if the water scarcity expands, the entire Earth could become parched dry desert, totally void of life.

The world has a total of about 332 million cubic miles of water—including all the water in the oceans, ice caps, lakes, and rivers, as well as groundwater, atmospheric water of freshwater—or all water we can account for.

Fresh water is about 1/110th of the total amount of water on the Earth surface. This is about 2.5 million cubic miles of water that can be used for drinking, cooking, and other human needs. Over 23 thousand cubic miles of the fresh water is contained in our lakes, rivers and streams. The rest is contained in underground water sources.

Groundwater aquifers contain over 95% of this water, while rain, rivers, and lakes make up the remaining 5%. Approximately 1,700 m3 of water exist for every person on the planet, which is an alarming low number, and which is actually decreasing. According to the World's Water Stress Index, a region with less than 1,700 m3 per capita is considered "water stressed." The number of these regions is great and growing.

In the last century alone, the global water usage per person doubled.

The increasing use of water is creating a situation today where many areas of the world are consuming water at an unsustainable rate. At the same time, water to some population centers has to be brought up from far away places. How long could this abnormality and inequality last?

The global water supply is not distributed evenly around the planet, nor is water equally available at all times throughout the year. Many areas of the world have seriously inadequate access to water, and many places with high annual averages experience alternating seasons of drought and monsoons.

The global climate change is adding to the uncertainty in a negative way. So far it seems that the regions that are prone to draughts are getting drier, while the wet ones are getter wetter.

Water usage in developing countries is also different from that in the developed countries. The difference is striking. While developing countries use 90% of their water for agriculture, 5% for industry, and 5% for urban areas, the developed countries use 45% of their water for agriculture, 45% for industry, and 10% for urban areas.

Case Study: Groundwater Pumping

Today we are faced with a serious problem of water depletion in some of the most populated and productive areas of the U.S. and the world. 4-5 years severe and unprecedented draught in California is bringing serious water shortages, which are damaging private properties, agriculture crops, and industrial enterprises.

Due to the draught, farmers and ranchers are forced to pump underground water in many places in California, which is causing a number of problems. Pumping groundwater at a faster rate than it can be recharged can have significant negative effects of the environment and the people who make use of the water.

Some of the negative effects of groundwater depletion are: drying up of wells, reduction of water in streams and lakes, deterioration of water quality, increased pumping costs, and land subsidence

In more detail, excess groundwater pumping causes:

- *Lowering of the water table* is the most common and most severe consequence of excessive groundwater pumping. As the water level drops, the water table, below which the ground is saturated with water, can be lowered. If groundwater levels decline too far, the well owner might have to deepen the well, or drill a new well.

Also, as water levels decline, the rate of water the well can yield may decline too. For example, some the level wells in California have dropped 200-300 feet during the last 4-5 years alone. If the draught continues, then that number might go as high as 500-600 feet, which is unheard of and which might bring even more dangerous consequences to the Golden State.

Low water levels usually mean increased costs for the users, since the water must be lifted higher to reach the land surface, and if pumps are used to lift the water (as opposed to artesian wells), more energy is required to drive the pump. Using the well under these conditions can become prohibitively expensive. Sometimes new wells have to be drilled, which is very complicated and expensive undertaking. New well drilling process costs upwards of $200,000 each.

- Reduction of the water table is usually accompanied by *reduction of the streams and lakes levels*. The interaction between the water in lakes and rivers and groundwater is an important phenomena, most of us are not even aware of. Often a great deal of the water flowing in rivers comes from seepage of groundwater into the streambed. Groundwater also contributes to streams in most physiographic and climatic settings, and since the proportion of stream water that comes from groundwater inflow varies according to a region's geography, geology, and climate, with the reduction of the water table bringing chaos in this delicate system.

- Excessive groundwater pumping alters the way water moves between an aquifer and a stream, lake, or wetland. This is usually done by intercepting groundwater flow that discharges into the surface-water body under natural conditions, or by increasing the rate of water movement from the surface-water body into an aquifer. A related effect of groundwater pumping is the lowering of groundwater levels below the depth that streamside or wetland vegetation needs to survive. The overall effect is a loss of riparian vegetation and wildlife habitat.

- *Land subsidence* is often caused by loss of support below ground, so when water is taken out of the soil, the soil collapses, compacts, and drops lower. This depends on factors such as the type of soil and rock below the surface. Land subsidence is most often caused by human activities, mainly from the removal of subsurface water, which is exactly what is happening in California now.

- *Deterioration of water quality* is another real threat to our water supply. Here, due to dropping water table level, fresh groundwater supplies are contaminated from saltwater via saltwater intrusion.

Under natural conditions the boundaries between the freshwater and saltwater tends to be relatively stable, but excess pumping of underground water can cause saltwater to migrate inland and upward, resulting in saltwater contamination of part of the water supply. Once that happens, the affected fresh water is no longer useful for human consumption; including irrigation or other agricultural applications.

In response to the water issues, new peripheral activities, businesses, and markets have been created. Some of these are as follow:

Water Well Drilling

During the present 5-year draught, California farmers and other enterprises have drilled more water wells than all other U.S. states put together. California's case is a reflection of what can be expected in the future in other states and countries.

Pumping Equipment Manufacturing and Service

The increased water wows are bringing increased profits to the manufacturers of water drilling, pumping, and treatment equipment. As the California draught continues, for example more and more farmers are drilling for water, and buying more and more powerful pumps. Since fresh water is in diminishing quantities in many states and countries, desalination equipment is at increased demand too.

Land Subsidence

Inelastic (or permanent) land subsidence is basically sinking (collapse) of ground surface, caused by excess groundwater extraction. In most cases, the damage to the ground surface is permanent. This is happening mostly because recently in many parts of California, more water is extracted from underground reservoirs than is being replenished.

This is a major problem in California, which is exacerbated by additional and unprecedented demand for groundwater during the present drought period. Since spring 2008, groundwater levels are at all-time historical lows (for the period of record) in most areas of the state, especially in the southern San Joaquin Valley, portions of the Sacramento Valley, and portions of the San Francisco Bay, South Lahontan, and South Coast Hydrologic Regions.

The California Department of Water Resources have found that:

- In general, the areas with a higher estimated potential for future subsidence are in the southern San Joaquin, Antelope, Coachella, and western

Sacramento Valleys. These areas match well with the areas of subsidence documented by Luhdorff & Scalmanini Consulting Engineers (LSCE), Borchers and Carpenter (2014).

- Most of the groundwater basins with a higher estimated potential for future subsidence are also ranked as high or medium priority by the CASGEM Basin Prioritization Process.

- Subsidence is occurring in many groundwater basins in the state especially in the southern San Joaquin River (4 sub-basins) and Tulare Lake (7 sub-basins) Hydrologic Regions. Subsidence is also occurring in the South Lahontan (5 basins), South Coast (4 basins), Sacramento River (3 sub-basins), Central Coast (2 basins), and Colorado River (2 basins) Hydrologic Regions.

- Since the spring of 2008, groundwater levels are at all-time historical lows (for period of record) in most areas of the state especially in the southern San Joaquin Valley, portions of the Sacramento Valley, and portions of the San Francisco Bay, South Lahontan, and South Coast Hydrologic Regions— these areas exhibit groundwater levels more than 50 feet below previous historical lows experienced sometime prior to 2000. There are many areas of the San Joaquin Valley where recent groundwater levels are more than 100 feet below previous historical lows and correspond to areas of recent subsidence.

- In the Central Valley, groundwater levels in 36% of the long-term wells (216 of 599) in the Sacramento Valley and 55% of the long-term wells (1,718 of 3,124) in the San Joaquin Valley are at or below the historical spring low levels.

And this seems to be the beginning...

Now—five years into the draught—the Golden State is a parched piece of land that is sinking at the same time.

Experts estimate that if the draught continues for several more years, the land subsidence will continue to increase, with many areas—Central and Southern California—threatened with massive and very serious water table drops and large areas of land subsidence.

The effects of this abnormality are numerous and serious. Land subsidence is reducing the State's land and is causing safety issues. The new situation is forcing the state and private farmers and individuals to take drastic measures, the final results of which are uncertain at this point.

Land subsidence is also affecting the road infrastructure. Many canals, bridges, and roads in the country are falling apart, due to old age and land subsidence as discussed above. Additional money and effort will have to be allocated to fix these before major accidents occur.

Ground Water Contamination

Adding insult to the injury in the midst of the severe drought, the oil industry has injected about 3 billion gallons of fracking chemicals into central California half-depleted aquifers, that are supposed to be used for drinking-water and farm-irrigation.

California is experiencing a severe drought, of the most serious type, and the ongoing fracking operations are not helping. Each fracking site requires as much as 140,000-150,000 gallons of water daily. This a lot of water that is taken from the aquifers and cannot be used for human consumption or in farming operations. But on top of that, the same fracking sites can generate that much waste water, full of dangerous chemicals every day. Some of that contaminated water is entering the aquifers, and making then useless for human consumption or for any other use.

The water aquifers are protected by state law, in addition to the federal Safe Water Drinking Act, and since they are supposed to supply quality water in the midst of the current unprecedented drought, they are treated as endangered species.

As a result of a series of inquiries, the California State Water Resources Board has found that nine of the 11 fracking sites that were shut down in the summer of 2014 for suspicion of contamination were in fact full of toxic fluids, which are used to fracture the rocks and pressurize gas and oil deep underground. 19 more injection wells may have also contaminated the sensitive, protected aquifers.

At the same time, many more similar wells are operating in the state, and have been dumping wastewater into the aquifers with unknown results

The Central Valley Water Board reports show that high levels of toxic chemicals—including arsenic, thallium, and nitrates—were found in water-supply wells near the fracking and wastewater-disposal sites. Arsenic and thallium are extremely dangerous chemicals, and the fact that high concentrations are showing up in multiple water wells close to wastewater injection sites raises major concerns about the health and safety of nearby residents.

Despite these damning findings, the extent of wastewater pollution is still undetermined, as the Central Valley Water Board has thus far only tested eight water wells of the more than 100 in the area, according to the documents. Half of those tested came up positive for containing an excessive amount of toxic chemicals. Fracking has been linked to groundwater contamination in other states, and has been also linked to an uptick in earthquakes, exacerbation of drought conditions, and a host of health concerns for wild life and humans.

The Center for Biological Diversity noted that the contamination of water sources could be much worse in another regard, as flowback water that comes from oil wells in the state can contain levels of benzene, toluene, and other toxic chemicals that are hundreds of times higher than legally allowed. Flowback fluid is then released back into wastewater storage wells. Chemicals like benzene can take years to eventually find their way to water sources.

Clean water is one of California's most crucial resources, and these documents make it clear that state regulators have utterly failed to protect our water from oil industry pollution," according to attorneys for the Center for Biological Diversity.

So what is California going to do? Stop fracking? Not likely! We just have to wait and see.

Recycling of Sewage Water

California uses a lot of water, most of which is dumped down the drain after use. Now the state is looking into using—or rather reusing—this waste water. In the fall of 2014, the San Diego City Council votes to regenerate recycled water for thirsty residents in exchange for giving a "pass" to the city's sewage plant, which has never met federal Clean Water Act standards.

The Point Loma Wastewater Treatment Plant is part of the region's sewage system, that pumps about 180 million gallons of sewage every day. It was started in 1963, as a small facility on a cliff top in Point Loma. The plant has later on expanded to keep up with demand. More digesters were added, and a bio-solids holding tank and several sedimentation basins were built in as well.

The waste treatment facility serves about 2.2 million people in 16 San Diego county municipalities. Here, sewage is treated and discharged offshore via a four-mile long underwater pipe. After all the improvements, the facility does not meet the clean water standards, which is a prerequisite for sewage plants in the U.S. It, in fact, it is the only plant in the United States that doesn't meet secondary standards, but it meets the regulations of the Clean Water Act; only because of EPA waivers allowing it to operate under special rules.

According to the Clean Water Act rules, waste treatment plants have to meet three benchmarks: pH levels, suspended solids, and biochemical oxygen demand (BOD). San Diego's plant meets only the first one, pH levels.

Now, after all these years, the plant will be retrofitted to a drought-proof water supply. Under this deal, the Point Loma discharge permit is modified to where a waiver is no longer needed. Instead, the city agreed to upgrade the plant so that it will generate 50 million gallons a day of potable water by 2023, and 83 million gallons a day by 2035.

This would be done by diverting as much as half of the sewage before it even reaches the Point Loma sewage plant. Then, three additional (newly built) treatment plants would clean the water multiple times before being distributed to homes.

And so, San Diego, as well as several other counties in California, will no longer discharge most of the liquid waste, but process it for human consumption. That does not sound pretty, but under the draught conditions is the only thing the water-deprived state can do to survive the pending and any future draughts.

Water Desalination

Fresh water supplies in California are dwindling, but there is an entire ocean full of water—salt water that is—that is not good for any human use. With the 5 year draught underway, and in anticipation of things to come (increasing drought conditions), the city of Carlsbad, California received final approval to build a $1.0 billion desalination plant. It took twelve years of planning and over six years of permitting battles with the California Coastal Commission, State Lands Commission and the Regional Water Quality Control Board, but now the city has a 30-year Water Purchase Agreement with the San Diego County Water Authority for the entire output of the plant and is ready to go.

Construction on the plant and pipeline is already under way and the plant will be delivering water to the businesses and residents in San Diego County by 2016. When operational, the plant will use huge amount of electricity for the desalination process, as needed to pump, process and deliver 50 million gallons of drinking water every day, which will be then sent to local homes and businesses to use as needed.

This amount, however, is only 5-6% of the total water demand of the service area, so it is a little bit more than the proverbial drop in a bucket. The new desalination project will also increase the water bills to each household by $5-6 every month. Is it worth it? With the world of a city official, "When you turn the water tap on and no water comes out, price becomes insignificant."

True, or false? This is the question California have to answer spoon, because as the draught continues additional desalination plants are scheduled to be built.

Obviously, California cannot exist on desalinated water, so the government is taking all feasible steps to reduce water use, including curbing outdoor water use, all through the drought crisis. At the same time, Governor Brown called on all Californians to (voluntarily) reduce their water use by 20%, or more, and to eliminate all kinds of water leaks and waste.

All this is fine and dandy, and will help with managing the drought, but what if it continues for 7 more years? What if it never goes away? What will happen to the Golden State? What will the Silicon Valley crowd do? Where would the San Francisco and Los Angeles millions of people go?

There are a number of other questions that can be asked here, but there would be no clear answers either. Just like there are no answers what to do when the BIG ONE hits along the San Andreas fault line.

California is an amazing place with versatile environmental conditions and natural disasters waiting to happen…! And yet, people are going around their daily tasks as if nothing is if nothing can happen. The author was a part of this crowd for a long time, so the feeling is mutual, and mutually unexplainable. One has to live in that place for awhile to understand the emotions of the locals.

Sooner or later the present draught will end, so the state can recover and get back to normalcy. But climate experts predict that other, even more severe and of longer duration, draughts are expected in the future. It is not certain how much the state can take and how long it could fight, but the choices are few.

Water Conservation and Use Efficiency

Water conservation is one of the tools the state government is using to control (or reduce) the draught consequences. Urban communities are cooperating for the most part, and reached 11.5% statewide in the fall of 2014. This is a significant jump from the 7.5% reported a month before and 4%, as compared to a year ago. This is approximately 27 billion gallons of fresh water saved in one month, up from 18 billion gallons saved during the previous month.

This success is due in part to the fact that 81% of the local water agencies have instituted outdoor water use restrictions. Nevertheless, about the 19%, or 76 water suppliers have not implemented any water shortage

plans to mandate outdoor use restrictions.

The outdoor water use restriction is a key action for urban water suppliers under the emergency water conservation regulation because outdoor watering comprises a large percentage of urban water use, as much as 80% in some areas. Water districts are now required to report "residential gallons per capita per day" in an effort to determine average water consumption per person.

The adequacy of compliance of those submitting plans is also checked in order to assure that they meet the letter and intent of the regulations. "Many more California communities are taking the drought seriously and making water conservation a priority—and residents are responding. Increasing urban water conservation is definitely a good thing. The trend is terrific. However, while we can hope for rain, we can't count on it, so we must keep going. Every gallon saved today postpones the need for more drastic, difficult, and expensive action should the drought continue into next year." according to the State Water Board. "At a time when hundreds of thousands of acres of farmland lie fallow, communities across the state are running out of water, fish and wildlife are suffering, and no one knows whether it will rain or snow much this winter, it makes good common sense for urban Californians to do what they can to use less water."

In Southern California, water conservation through turf removal is underway, with about 4.5 million square feet from almost 3,000 front yards been removed from residential customers in 2014. In the commercial sector, 9 million square feet, or 150 football fields' worth of turf, was removed too. In 2014 about $45 million in rebates for turf removal were requested, showing that Southern Californians are interested in permanently reducing their water use.

Case Study: Government Water Use

In 2013, the National Security Administration brought a huge building in Bluffdale, Utah, and converted it into a monster data center. Bluffdale was chosen to take advantage of the cheap local power and a cut-rate deal for using millions of gallons of fresh water. The water is used to cool the 1-million-square-foot building and banks upon banks of monster servers storing and processing the data.

Good deal for the Feds, no doubt, but now Utah's law makers are considering a bill that could shut off the water spigot to the massive data center. The reasons are several-fold, ranging from protest against collecting data on U.S. citizens to refusing to let the Federal government take advantage of the cheap rates at the expense of the local tax payers.

And so, the bill will deny renewal of the current contract come 2021. At that time, the city will cut the water flow to the NSA building. Depending on the final version of the bill, the Feds will have no choice but move the data center, or will be forced to pay higher water rates.

Regardless of the outcome of this situation, one thing is clear; water is more precious than good relations with the government, and Utah's law makers are not hesitating to put it in writing.

We see similar situations developing in other parts of the country and with other industries. For example, a number of concentrated solar power (CSP) plants were installed in the South West deserts, where they are most efficient. This type of solar power plants use millions of gallons of fresh local water for cooling their turbines. But wait, the deserts don't have much water, and whatever water is available is getting harder to get since the water table is sinking, making the locals unhappy. And the water depletion brings another set of problems as discussed above.

These examples show the increasing importance and value of water in the U.S., and we can find many similar examples overseas too.

Government Policies

Governments have an important place in controlling water use. For example, in January 2014 Governor Brown issued an Emergency Drought Proclamation, calling for all Californians to voluntarily reduce their daily water use by minimum of 20%. The trend of increasing water use reductions shows that many California communities have met and exceeded the call to conserve.

Are these measures enough to meliorate the fierce draught is not clear. But since half of the fresh water in the state is used by agriculture, where most water conservation measures cannot be implemented, we see the water problems in the state continuing. Here, the state government may have to figure out other measures that would ensure enough water without affecting the farmers negatively.

The water crisis in the Golden State—the fruit and vegetable basket of the U.S.—might get even worse, if the draught continues at the same level too long. The federal and state governments are expected to step in and provide some additional assistance to the farming community, but it is just not clear what that assistance might be, and if it would come on time.

International Water Issues

Above we considered the water issues that we deal with in the U.S., but they are even more serious in other countries.

Food Scarcity

According to the International Food Policy Research Institute (IFPRI), if current water consumption trends continue, by 2025 the agricultural sector will experience serious water shortages. The IFPRI estimates that crop losses due to water scarcity could be as high as 350 million metric tons per year, slightly more than the entire crop yield of the U.S. This massive water crisis will be caused by water contamination, diverting water for industrial purposes, and the depletion of aquifers. Climate change may also play a part. The Himalayan glaciers, which feed the rivers that support billions of people, are shrinking in size every year. Their disappearance would cause a major humanitarian disaster.

The greatest danger to global food security comes from aquifer depletion. Aquifers are an essential source of water for food production, and they are being overdrawn in the western U.S., northern Iran, north-central China, India, Mexico, Australia, and numerous other locations. Additionally, many aquifers are contaminated each year by pollution and seawater intrusion.

Despite their importance, data on underground water reservoirs remains imprecise. There is little evidence regarding how many aquifers actually exist, and the depth of known aquifers is often a mystery. However, it is clear that water from these sources takes centuries to replenish, and that they are being consumed at a highly unsustainable rate.

International Conflicts

According to the UNEP, there are 263 rivers in the world that either cross or mark international boundaries. The basins fed by these rivers account for 60% of the world's above ground freshwater. Of these 158 have no international legislation, and many are the source of conflict.

For example, water has always been a central issue in Arab-Israeli situation. Ariel Sharon once said the Six Days War actually began the day that Israel stopped Syria from diverting the Jordan River in 1964. Decades later, the Egyptian military came close to staging a coup against Egyptian president Anwar Sadat, who had proposed diverting some of the Nile's water to Israel as part of a peace plan.

The Nile River, which runs through Ethiopia, Sudan, and Egypt, exemplifies the potential for future water conflicts. The banks of the Nile River support one of most densely populated areas on the planet. In the next fifty years the number of people dependant on the Nile could double, creating a serious water crisis in the region. The Nile is not governed by any multilateral treaties, and Egypt would not shrink from using military strength to guarantee its future access to water.

The potential for water conflicts are less likely outside the Middle East, but never the less there are many problematic areas. The Mekong River is the lifeblood of South East Asia, but it begins in one of the most water poor countries on Earth: China. The Indus River separates Pakistan and India, and aquifer depletion by Indian farmers has one of the highest rates in the world.

U.S.-Mexican relations are already strained over water use on their mutual border. The Niger River basin in West-Central Africa runs through five countries. Surging populations coupled with decreasing rainfall in the region seriously threaten water security for millions of people.

Although the specter of international water wars can seem very real, in the last 50 years there have only been 7 conflicts over water outside the Middle East. While a global water crisis has the potential to tear international relations at the seams, it also has the potential to force the global community into a new spirit of cooperation. The question is which will come first in each case.

Case Study—Qatar

Qatar's water consumption is among the highest in the world. It is estimated that each Qatari uses about 350-400 gallons each day. There are many rich people in the country who demand all their comforts and privileges to be met daily, which requires large volumes of water. But Qatar is a desert, and what is underground is oil—a blessing and curse. It is a blessing, because Qatar is also among the richest nations on Earth. It is curse, because water in the underground reservoirs has been replaced by oil. Oil and money, unfortunately, cannot be drunk, so the Qataris are very rich and thirsty people.

Qatar has only two days of water supplies available at any given time.

If the unthinkable happens, within several days Qatar would become a dead zone, full of dying of thirst rich people. Qatar is a very dry country and has very little fresh water, so the country's burgeoning population depends on desalinated water—water taken from the ocean and processed via complex and expensive desali-

nation plants. Desalination accounts for about 99.9% of drinkable water in the country.

Got that? 99.9% of all water used in the country is produced by desalination of ocean water. And even more surprisingly, as a bonus for being good citizens, Qatari water is subsidized, so the people feel entitled to use and waste as much as they want. Just like their American counterparts, the Qataris have bigger houses than most people in the world, complete with gardens and swimming pools. They also like grass, and like to see large green places and many fountains.

But as they say, there is no free lunch, so the expensive price of the desalination process is also becoming increasingly apparent in the country's deteriorating marine ecosystems. By processing so much ocean water, the Qatar desalination plants put back into the Gulf very warm and excessively salty brine water…day after blessed day…for decades now.

For Qatar's approximately 2 million inhabitants (about 300,000 citizens and 1.6 million expatriates, each requiring 400 gallons of desalinated water each day), this is 2,000,000 x 400 = 800 million gallons of water is desalinated every day. This is over 290 billion gallons annually, and about 3 trillion gallons in 10 years. So by the end of the century, Qatar will desalinate the total of 25 trillion gallons of ocean water.

The largest percent of desalinated water used in any country is in Israel which produces about 40% of its domestic water by seawater desalination.

But back to Qatar…when the hot salty water (millions of gallons every day) is released in the ocean, the area around that release creates a "plume" of modified "dead" water, which affects negatively everything in its proximity. The harmful effects of this artificial water—in addition to using huge amounts of energy and chemicals used during the desalination process—are already evident in the Gulf environment.

As the amount of warm, briny water discharged in the Gulf daily continues and increases, so will its negative environmental effects. And so does it go day after day ad infinitum. The estimated 25 trillion gallons of desalinated water

Note: Keep in mind that fresh water contains less than 1,000 ppm of solids (salt and other contaminants), slightly saline water contains 1,000 ppm to 3,000 ppm, moderately saline water 3,000 ppm to 10,000 ppm, and highly saline water 10,000 ppm to 35,000 ppm.

Desalinated water falls in the latter category, so the Gulf is receiving 35 times the salt and particulate concentration it usually contain. Life forms cannot exist in this severely modified water, so guess what? It disappears by dying off, or moving further in the ocean. In all cases, the Gulf is overwhelmed and changed to a new type.

Note: 35,000 ppm (brine concentration) is 3.5% in weight of dissolved salt in the natural ocean water. This is a lot, and as the discharges continue, the concentration increases even further, to where ocean life can no longer exist in it.

The energy used, and the overall cost of the desalination process are significant. The average cost is about $75 per 250 gallons, or about 30 cents. Not an amount that can break the average Qatar citizen, but with the millions of gallons used every day, it amounts a great sum of money. With Qatar's large petroleum deposits, however, the cost of energy is not even considered…for now. But things might change in the future!

Qatar's residents also pay (personally) the price for consuming this water, since it is not natural water and is treated with different chemicals When taking showers, the chemicals used to treat desalinated water come into direct contact with people's skin and hair and many people bold because of it. Drinking thus treated water day after day is also causing other negative effects on the population as well. Today, the Qatar government has taken several measures to educate its people on water scarcity, and its problems, and help secure its water supply for the future.

In order to provide some more water for emergencies, the government plans to build five "mega" reservoirs on the outskirts of Doha by 2016. But despite all these efforts, desalination is seen as the only way to proceed, so Qatar has plans to build more desalination plants for the future, with some already under construction. And as the number of desalination plants increase in Qatar, their harmful impacts will become even more pronounced in the future.

A more reasonable effort for efficient water use at homes, farms, the industry, and government offices is essential for the future of this unnatural life style in Qatar. Fresh clean water is critical for all life, and has to be used with moderation and efficiency in mind—especially in the perpetually dry regions.

Qatar is not the only country that uses desalination. The most important users of desalinated water, in addition to Qatar, are in the Middle East, mainly Saudi Arabia, Kuwait, the United Arab Emirates, and Bahrain. These countries combined use about 70% of worldwide desalination capacity. North African countries, like Libya and Algeria use about 6% of worldwide capacity.

In the developed world, the United States is one of the most important users of desalinated water, recently especially in California and Florida.

There are about 12,800 desalination plants around the world in 120 countries today. These combined produce over 15 million cubic meters freshwater daily, which is about 1% of total world fresh water consumption. Not much, but it is all for countries with limited fresh water supplies.

The side effects of water desalination are seen in the world's oceans and agriculture, where estimates are that about 1/3 of all irrigated areas around the suffer from salinity problems. Since remediation of these areas is very costly, no changes are foreseen in the future, so the problem will only escalate.

What does water has to do with the energy markets?, you may ask. How does all this affect the energy markets? The simple answer is, profoundly.

In more detail:

- A lot of energy is used (wasted) for pumping and processing water. Entire states (California) and countries (Qatar) are increasingly using processed sea water to replace lost or inadequate fresh water supplies. Most of the energy used for the water processing is generated by fossil fuels, which activities, in addition to emitting a lot of GHGs, will bring the fossils depletion sooner than expected.

- Water is essential to many energy activities.
  — Most power plants need water for cooling their turbines.
  — Hydropower plants need water to turn the turbines.
  — Hydrofracking operations use zillions of gallons of fresh water every day, and without water the natural gas bonanza would just fizzle away.
  — Bio-fuels need water to grow the energy crops. Without water these industries will not function properly, if at all. If the water crisis continues too long, the entire global energy infrastructure might crumble piece by piece.

- On the extreme end of things, if the global water problems increase enough, a day might come when we will be forced to chose between using water for energy production and generation, or drinking. The present day water conflicts, as described above, might escalate into real wars between neighboring nations, bringing chaos to the entire world.

For the most part, the energy markets are ignoring, or not taking into consideration, is the growing water problems and the related issues. How the energy markets will be affected in the future, if the water issues grow into a serious global crisis, is anyone's guess, but we must keep water (and its relation to energy) in mind every time we talk, or make decisions, about energy production and generation.

Earthquakes

In the spring of 2014 an earthquake of 4.4 magnitude was centered about 5 miles northwest of Westwood, California. Coincidentally, or not, the quake center happened to be 8 miles from a large fracking site. California is a quake country, no doubt, so is the fracking causing more problems than the state can handle. Water contamination is one thing, and there might be a solution to it. But triggering the Big One could end the state of California as we know it. This is serious!

Seismologists described the 6:26 a.m. earthquake as the strongest to hit directly under the Santa Monica Mountains in the 80 years since seismic record-keeping began in the area. And so, Los Angeles City Council members called on city staff to investigate whether oil and natural gas drilling methods like fracking caused the magnitude temblor. The staff works directly with several regional, state and federal agencies to produce a report looking into whether a link exists between the hydraulic fracturing activities and the quake.

This new earthquake is about the size that seems to be happening in other states, close to fracking sites, so looking into it is justified. There are states that historically have not had earthquakes, they've been getting earthquake after the fracking activities began. Texas is one of these states. So a parallel between man's activities and Mother Nature's plans can be drawn, which is especially important for a state like California, sitting on top of the active San Andreas fault.

Experts in the field—the U.S. Geological Survey officials—think there has been a dramatic rise in earthquakes that exceed 3.0 magnitudes in the central and eastern U.S. in recent years. The agency also found that some quakes happening in Oklahoma, Arkansas, Texas and Ohio were caused by activities related to fracking.

So fracking and earthquakes go hand in hand in states that are not known to have earthquakes. How about California, which is earthquake prone, and where people are waiting the arrival of the Big One every day? Would you sleep well at night, living in Los Angeles, knowing that somebody is drilling under the San Andreas fault, bringing the inevitable closer by the day?

Fuel Waste

We live in time and age of wastes. From disposable everything to leisurely Sunday drive (averaging 50-100 miles) Americans are wasting fuel on daily basis.

Some examples:

Car Travel

Let's take a look at this symbol of modern civilization, the reflection of the American Dream; the large 300 horse power, 3,000 pounds iron horse with a one or two people riding in it.

How much power does a regular car-user consume?

$$Ed = \frac{Dd}{Df} \times Ef$$

Where:
  Ed   is energy used per day
  Dd   is distance travelled per day
  Df   is distance travelled per gallon of gasoline
  Ef   is the energy contained in unit of gasoline (about 30 kWh/gallon).

$$Ed = \frac{90 \text{ miles/day}}{20 \text{ miles/gallon}} \times 30 \text{ kWh} = 135 \text{ kWh/day}$$

So our leisurely drive wasted about 4. 5 gallons of gasoline, which could've generated 135 kWh (135,000 watt/hour) of electric power. This is more than what most people in the underdeveloped countries in this world use in a month…if fuel or electricity are available to begin with.

Add to this the energy-cost and pollution emitted during the production of gasoline. There are estimates that making each unit of petrol requires an input of 1.4 units of crude oil and other primary fuels. So during our leisurely Sunday drive we have actually wasted 1.4 times more fuel of the global supply that what we've burned. Our total waste is then about 6.3 gallons, which could generate about 190 kWh electricity.

Now consider the GHG gasses emitted during the production of the gasoline fuel, its transport to the gas stations and the related evaporation and additional emissions and pollution from storage tanks, diesel trucks, etc.

Airplane Travel

We all fly from place to place; some more than others. Let's assume one intercontinental flight per person in the developed world. How much energy does that use and what the energy and pollution costs are?

Let's take as an example a Boeing 747-400 that carries 65,000 gallons of fuel and 416 passengers. A round trip to Europe and back would be approximately 10,000 miles, during which flight about 2 x 60,000 gallons, or 120,000 gallons are used to propel the plane. Jet fuel has calorific value is about 38 kWh per gallon, so the energy cost of one full-distance roundtrip on this plane would be equal to:

$$120,000 \text{ gallons} \times 38 \text{ kWh} = 4.5 \text{ GWh}$$

This is as much power as an average nuclear plant generates in two-three hours of full power operation.

If divided among all passengers, then we get:

$$\frac{120,000 \text{ gallons}}{416 \text{ passengers}} \times 38 \text{ kWh/gallon} = 10,900 \text{ kWh/person}$$

This means that just one international trip per year for each person on that plane has an energy equivalent greater than leaving 1.0 kW electric stove on at full blast non-stop, 24 hours a day, all year long. Add to this the energy-cost and pollution emitted during producing and transporting the jet fuel.

There are estimates that making each unit of jet fuel requires an input of 1.4 units of crude oil and other primary fuels. So during our international flight we actually wasted 1.4 times more fuel of the global supply that what we the plane burned. Our total energy waste is then the grand total of about 15,000 kWh resulting from just one trip. And similarly so, the pollution footprint we left is larger too.

And now add to this the energy of manufacturing the plane, airport maintenance, and the pollution emitted during the entire port-to-port process. This will most likely double our energy cost and pollution estimates of our flight.

Note: Flying creates other greenhouse gases in addition to CO_2, such as water and ozone, and indirect GHGs, such as nitrous oxides. The actual carbon footprint in tons of CO_2-equivalent from our flight can be obtained by getting the actual CO_2 emissions and multiply them by two or three. Then multiply this amount by the thousands of flights every day around the world; the numbers you get will amaze you.

Cruise Ships

The most unjustifiable waste of our precious energy resources, and clear indication of the unjustifiable

extreme opulence we live in, is the cruise ship industry. Most adults in the developed countries take at least one 3-5 day cruise during their lifetime. This amounts to millions of people boarding monstrous ships to nowhere, circling the world's oceans aimlessly and burning millions of gallons of fossil fuels.

The worldwide cruise industry is estimated at $29.34 billion for 2011, which is a 9.5% increase over the previous years. It is the industry with the fastest growth in the last decades, and with that its energy and environmental impacts grow too. Host communities usually bear some of the economic, environmental and socio-cultural effects deriving from ships and passengers' presence and the related activities.

While there are some positive results from these activities, the negative environmental externalities are significant: large amounts of waste, erosion and degradation of vegetation, deprivation of historical and geological sites, which are caused mainly by conduct producing physical and visual impacts.

Further negative socio-cultural externalities can be produced too, since cruise passengers tend to "invade" the destination just for a few hours in a single day. This effect is particularly visible in small locations where cruisers compete for roads with the residents.

But the most obvious and constant negative result is waste of millions of gallons of diesel every year for an activity that is counter-productive and absolutely unjustifiable.

A large cruise ship, of the Royal Caribbean type, burns close to 3,000 gallons of diesel per hour under full sail. Assuming 50% lower speed, we get 2,000 gallons per hour are needed to keep the ship moving and power supplied throughout its dozen of stories high platform and thousands of apartments, venues, and halls. A 3 day cruise will then burn over 140,000 gallons, while a seven day cruise will waste another 335,000 gallons.

With 3,000 people on board we come up with a grand total of 112 gallons wasted by each person on a trip to nowhere. A trip that we make just because we can, and because everybody does it.

Multiplying these 112 gallons wasted by millions of people aimlessly cruising the oceans every year, we come up with a staggering number of millions of gallons of precious crude oil wasted. This number alone will tell us that we are quickly cutting the branch we are sitting on for no good reason at all.

Case Study: Cruise to Nowhere

Recently this author decided to check out the cruising phenomena personally by taking a week-long cruise from San Diego, California to Tijuana, Mexico. The actual air-distance port-to-port is 85 miles, with one stop at the mid-point port of Catalina Island.

And so, we departed on a huge cruise ship with 3,000 passengers on board on a week-long voyage that covered the 85 miles distance from San Diego to Tijuana (via Catalina Island). The enormous luxurious ship offered untold number of amenities and out-of-this-world comforts, It traveled at a medium speed most of the time, which was decreased only when nearing a port. It is still not clear how it managed to navigate 185 miles, traveling non-stop between point A and B, separated by 15 miles distance.

5 days at sea, or 120 hours sailing (with about 10 hrs. stop total at Catalina and Tijuana) at an average speed of about 10 MPH. At the end, the ship was under sail for about 110 hrs., which at 10 MPH would mean that we actually travelled 1,100 miles from start to end of the voyage. This also means that in order to travel the 1,100 miles between point A, B, and C only 85 miles apart, the great ship was going up and down the ocean expanse to pass the time.

At one occasion a passenger asked the captain where the ship is and how fast it was going. The captain's answer was, "I don't know exactly, because the ship is on autopilot." So it seemed that arriving at port A, B, or C was not the point. Instead, filling the time in between the ports was.

And so all 3,000 souls on board spent a fun-filled week on a ship headed to nowhere, on which nobody—even the captain—knew where it is exactly, nor how fast it was going. All this was done at the expense of several thousand gallons of diesel, which emitted that much more pollution into the air and the ocean.

The cruise ship industry is here to stay, and millions of people will be wandering aimlessly the world's oceans until the last drop of diesel has been burned. Or maybe at that point the great cruise ships will be equipped with electric engines, or more likely retrofitted to burn natural gas. In any case, the cruise industry will figure it out, because there is a lot of money to be made in this business.

Emissions vs. Income

There are a number of theories on the relation between pollution and economic development. One of the most famous is the so-called Kuznets curve, see Figure 8-43, which represents the theory that economic inequality increases as a country develops industrially.

As seen in the Figure 8-43, the SO_2 gas emissions rise as the income of the population increases. After a

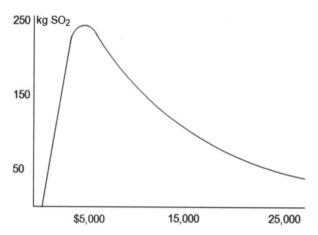

Figure 8-43. SO_2 emissions levels according to GNP per capita (in kg SO2 vs. $ income)

while, however, the inequality begins to decrease as the economy matures. During their initial economic development, the developing countries generate large amounts of pollution, the amount of which shrinks as they approach economic maturity.

This phenomena occurs because at low incomes people tend to value development over environmental quality, but as they achieve greater wealth they are willing to devote greater resources to environmental quality improvements. So, the developing countries tend to use highly polluting heavy industry in the race to catch up with the rest. Developed economies, on the other hand, rely on more advanced (albeit more expensive) and relatively clean technologies and services.

This is logical, because clean air and water provide benefits and an enjoyment that are income elastic. As income increases above a certain threshold, which is different for different countries, individuals and society are willing and able to spend a larger share of their incomes on these goods. This reduces the pollution per unit of output and even decreases the total pollution with time.

We see the events of China today supporting this theory. China and India are now at the highest point of the curve, where economic development and acquiring goods is the top priority. This is why the Chinese are planning to build hundreds of new coal-fired plants. They need the energy to continue their rise as an economic power, regardless of the effects this might have on the environment. Hopefully, they will realize the problem an slow down with the coal burning.

This theory, however, cannot be applied for every single country and so it is contested by ecological economists, who believe that good environmental stewardship is possible at all times and at different levels of development.

Studies done on this subject are complex and contradictory, and the relevance of this curve seems to vary from case to case. One great factor is the location, because so much depends on whether the pollutants in question are local and are essentially fouling their own environment, or are coming from far away as externalities generated by some far away source.

In the case of SO_2 and particulates pollution, the damage is more evident, the pollution prices are near their marginal social costs, and so turning points have been observed at relatively low-income levels. On the other hand, turning points are found at much higher income levels, or not at all for pollutants such as CO_2—the damages from which are less immediate and less evident.

Other analysis show that rich and poor countries alike tend to reduce pollution relative to economic output over time, but that since developing countries grow at a rapid rate their overall pollution grows rapidly with it. In rapidly growing middle income countries the scale effect, which increases pollution and other degradation overwhelms the time effect. In wealthy countries, therefore, economic growth is slower, and pollution reduction efforts can overcome the scale effect. And so, obviously, a fast-growing economy will produce more pollution despite technological advances.

Another recent study, using a number of different methods, finds evidence that the relation between GNP and pollution exists in many countries, but that it still a fragile concept. Only in a few cases, growing GNP and environmental damage evidence are absolutely undeniable.

As an example, a number of Latin American countries show significant evidence of the GNP-pollution relationship demonstrated by heavy deforestation which is a result consistent with the theory that highly visible local conditions are the best evidence of this relation. There is also an element of externalities here as well, since the large-scale deforestation is affecting neighboring countries and the entire world.

In all cases, the local environmental policies and institutional arrangements need to be accounted for in order to get a complete and accurate picture in each case. Scale, specific situation, and type of environmental degradation are the other set of variables that must be carefully examined when evaluating the relation. Still it's clear that, particularly in China and India, that economic growth is far outpacing improved environmental efficiency. Simply allowing development to take its course is insufficient, so other measures for environmental protection must be taken before it is too late.

Economy and Environment

The relationship between the local and national economy on one hand, and the global environment on the other is two-directional. It all depends on which side of the equation we sit. If we are concerned with the ways environmental protection activities impact the overall economic performance, we see one thing. But if we are mostly concerned with how economic growth impacts environmental quality, then the picture looks quite different.

In this text we are looking from both sides of environmental protection and economic development activities as related to the energy markets. Since we all agree that economic growth impacts environmental quality, we take a close look at this phenomena in this text.

If we have to summarize it all, we'd say; the impact is significant. More specifically, as people and nations get richer over time, the environmental quality in the area (directly), and around the globe (indirectly) change. While the pattern follows some logic, the answers are not immediately obvious.

Richer people usually use more energy and other resources, and produce more waste and pollution. At the same time, however, a richer nation can afford to invest in renewable energy, install expensive pollution control equipment, and implement effective environmental policies.

Environmental quality is widely accepted as a normal good, which means that rich people will use more of it as their income increases, in order to protect the environment and themselves.

This brings the question if environmental quality is also a luxury good, since spending on it increases disproportionately as income grows. While this may be so over some income levels, it is merely a normal good at other income levels.

It seems logical that economic growth eventually provides the resources to reduce the negative environmental impacts, since there is a clear evidence that although economic growth usually leads to environmental deterioration in the early stages of the process, in the end the best—and probably the only—way to attain a decent environment in most countries is to become rich.

The notion that the negative environmental impacts tend to increase initially as a country becomes richer, but then eventually decrease with further income gains, is known as the Environmental Kuznets curve (EKC) hypothesis.

EKC hypothesis proposes that the relationship between income and environmental impacts is an inverted-U shape, as in Figure 8-44 is based on actual data on sulfur dioxide emissions in the U.S. and Europe. The data shows that per capita SO_2 emissions increase sharply with income up to a per-capita income of around $4,000, but then as the income level increases, SO_2 emissions per-capita decline steadily. See Figure 8-44.

This is an encouraging result because the "turning point" occurs at a relatively modest income level. Thus a moderate amount of economic growth can lead to substantial SO_2 emission reductions.

The Problem

The major problem here is that the EKC hypothesis while accurate to SO_2 emissions, does not apply to all environmental impacts. Most importantly, the EKC hypothesis does not match the data on carbon dioxide emissions, which is the primary greenhouse gas. Any attempts to fit an inverted-U trend line through the CO_2 data fail. No turning point is to be found and per-capita CO_2 emissions continue to rise as per-capita income increases. See Figure 8-44.

A more sophisticated statistical analysis tested the EKC hypothesis for carbon emissions and concluded that, while it may be valid for some air pollutants such as SO_2, particulate matter, and nitrogen oxides, it does not seem to apply more broadly to other environmental impacts. Despite the new statistical] approaches, there is no clear-cut evidence supporting the existence of the EKC for total carbon emissions. There is no relation between economic growth and total pollution.

There is, however, inverted U-shaped relation between urban ambient concentrations of some pollutants and income, but EKC is only a partial model of some types of emissions and concentrations.

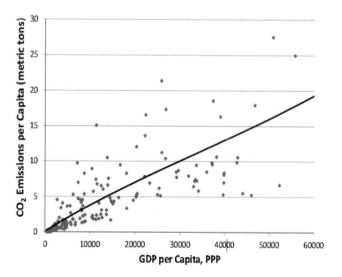

Figure 8-44. CO_2 emissions vs. income per capita

In any case, improvement of the environment is not automatic with income growth. It depends on a number of other factors, such as the types of policies and institutions. GDP growth creates the conditions for environmental improvement by raising the demand for improved environmental quality and makes the resources available for supplying it.

So, promoting economic growth as a vehicle for environmental remediation does not appear to be a means to solve the global warming, or most other environmental problems.

Whether environmental quality improvements materialize with economic growth or not, and when and how, depend mostly on proper government policies, the activities of interested social institutions, and the functions of the energy, transport, and other pollution prone markets.

But what is all that matter if we, humans, get to the point that we cannot even live on a planet that is changing drastically and out of control day after day in front of our eyes. A planet that makes life harder for millions as a consequence of our mindless behavior, excessive depletion of fossils, and uncontrolled pollution of the environment to the point of no return.

Table 8-22 clearly shows what will happen if we do not change our ways soon! All this talk about finances, regulations, markets, etc. will become irrelevant. Once we reach the point of no return, all that would matter is basic survival—finding fresh water to drink and clean air to breath. And that may be very hard to do and getting harder as unforeseen forces are unleashed and bring disaster after disaster to our fragile world.

Maybe that will lay the foundation of another economy, and another financial system. A system where a breath of fresh air would cost a lot of money, and where a glass of fresh water would be the new measure of wealth.

We should not ignore the possibility of a future global economic system where a glass of clean water would be worth its weight in gold…or where we would have to pay for each breath of fresh air.

If and when such a day arrives, the affected people would look back at us in disbelief and anger. How could such obvious crimes of energy and environmental abuse be committed in the 20th and 21st century? And why these premeditated crimes, committed intentionally by millions of violators go unpunished? How could we obliterate the fossil fuels from the face of the Earth, and contaminate the environment to the point that it is unsuitable for human life?

What excuse would we give them?

Table 8-22. The effects of global warming

IMPACT	Temperature rise of 1⁰C would cause:	Temperature rise of 3⁰C would cause:	Temperature rise of 5⁰C would cause:
Fresh water supplies	• Glaciers in the Andes and other areas around the globe shrink and disappear. • Over 50 million people are left without fresh water. • Water supplies decrease 30% in Southern Africa and Mediterranean countries.	• Serious draughts occur in Southern Europe every 10 years of even more often. • 2-4 billion people world-wide are left without water. • Water supplies decrease 50% in Southern Africa and Mediterranean countries.	• Large glaciers in the Himalayas disappear all together. • Quarter of the population in China, and surrounding countries, is left without fresh water supplies. • Disease and death increase.
Food and agriculture	• A modest increase in crop yields is possible in some temperature regions. • Crop yields in the tropics decrease significantly. • 5-10% crop decrease in Africa.	• Crop yields increase at higher altitudes. • Half a billion people are at risk of hunger. • 20-30% crop yield decrease in Africa.	• Significant increase of ocean acidity and toxicity is blamed for reducing fish stocks. • Warmer ocean waters change fish population habits and reduce fishing.
Human Health	• Reduction in mortality in high latitude countries. • Quarter million die yearly from climate-related disease • 50-60 million more people in Africans are exposed to malaria and other diseases.	• 5-6 million more people world-wide suffer from malnutrition, and 2-3 million more die from it. • 80 million more people are exposed to malaria in Africa and many die from it.	• The number of climate-related diseases and death increase, putting additional burden on the health care and services systems, especially in the world's poorest countries and regions.
Coastal areas	• Increased damage from flooding in coastal areas. • Flooding and inundation of low-lying counties, and permanent change of coastal areas. • Forced evacuations of 10-20 million people from coastal areas.	• 200 million people subject to forced evacuation from low-lying coastal areas • The numbers of people getting sick due to changing climate and living conditions increases.	• Ocean level rise is threatening major population centers in the U.S. and the West. • New York, Tokyo, London, and others are on the evacuation list.
Ecosystems	• 10-15% of land species are facing extinction. • 20-30% of all species face some change, or extinction • Wild life risk is increased significantly.	• 40-50% of all wild life species face extinction. • Half of the Arctic tundra is lost. • Amazon forest collapse. • Significant coral reefs loss.	• Uncontrolled world-wide wild-life extinction, coral reefs and glaciers loss. • The entire human race is threatened by irreversible negative changes.

International Environmental Planning

The UN has a number of programs designed for and implemented in developing countries as needed to help them achieve the goals of the international community. Some of these programs are:

• Multilateral Fund for the Implementation of the Montreal Protocol on Substances that Deplete the Ozone Layer is a financial mechanism established in 1991 to assist developing countries to meet Montreal Protocol compliance targets. The Fund facilitates the transfer of ozone friendly technologies and financial assistance to developing countries so that they can phase out ozone depleting substances

(ODS). The Montreal Protocol, adopted in 1987, has contributed to reversing the damage done to the ozone layer and is recognized as a global success story of global environmental protection. This is demonstrated by the massive reductions in the use of ODS worldwide since it came into force. The Montreal Protocol had helped phase out over 95% of the ozone depleting substances it set out to control.

• The Environment and Energy Thematic Trust Fund gives seed money to innovative and catalytic initiatives which allow the United Nations to receive and quickly deploy resources into new business areas, thereby responding to new global or partner country demands. It supports the work within Key Result Areas of the United Nations' Strategic Plan that guides the work in the area of environment and sustainable development, such as: 1) environmental mainstreaming, 2) environmental finance, 3) climate change, and 4) local capacity for service delivery.

- The Global Environment Facility provides grants and concessional financing to assist countries to meet the costs of achieving their commitments under the global environmental conventions in biodiversity, climate change, and desertification, as well as supporting action on international waters and the phase out of persistent organic pollutants. As a partnership between the Implementing Agencies of the United Nations and the World Bank, with the cooperation of seven Executing this program works to integrate action to generate global environmental benefits into the mainstream of country led development.

- The United Nations' Green Commodities Facility is working to scale up existing initiatives with companies, governments and others intended to help shift markets to drive the production and sale of green commodities. Green commodities are those produced and supplied in a manner which minimizes negative environmental impacts, and includes bulk traded products as well as more specialized niche varieties.

- Reducing Emissions from Deforestation and Forest Degradation)is an effort to value the carbon stored in standing forests as a way to create incentives for developing countries to protect forests. Financial flows resulting from this program will not only significantly reduce carbon emissions, but it can also benefit developing countries, support poverty reduction and help preserve biodiversity, as well as other vital ecosystem services. Further, maintaining resilient forest ecosystems can contribute to adaptation to climate change.

Renewable Energy Country Attractiveness Index

Ernst & Young established a global quarterly publication in 2003 that ranks 40 countries by comparing the level of attractiveness of their renewable energy investment and deployment opportunities. It is called Renewable Energy Country Attractiveness Index. The ranking in the Index is based on several factors related to the global energy market and a number of technology-specific indicators, and several factors, including incentives, opportunities for projects and other.

In 2013, the 10th annual Renewable Energy Country Attractiveness Index determined that the U.S. is the most attractive place in world for renewable energy, replacing China as the index leader in previous years.

The global investment in clean energy grew to $269 billion in 2012, a five-fold increase since the Renewable Energy Index inception. This shows that the renewable energy sector has moved from the fringes of the energy markets close to the mainstream.

"The sector now competes for investment with more traditional energy sources, where solar, biomass boilers, and mini wind turbines are allowing people to produce their own electricity, which changing the way businesses and consumers think about energy," according to Ernst & Young.

"Market fundamentals, such as energy demand growth, security of energy supply and the affordability of renewable energy (relative to other sources), feature as some of the most prominent drivers of renewable energy growth today. Our revised methodology allows us to analyze each market's investment attractiveness much more effectively by considering these factors and weighting them accordingly."

The U.S. achieved the highest overall score of 71.6 on the Index, where China and Germany received slightly lower scores of 70.7 and 67.6, respectively. The U.S. success is due to factors such as; economic "macro stability," ease of doing business, and the bankability of renewables.

Obama's support for renewables and a permanent extension of the Production Tax Credit (PTC) were some of the factors that raised U.S. ranking in the Index. In his 2012 State of the Union Speech, President Obama called for doubling the renewable energy generation in the country by 2020, while at the same time reducing oil imports to half of present levels.

He also called for a $2 billion Energy Security Trust as needed to support clean energy research initiatives, in addition to the Department of the Interior drive to improve its permitting process for clean energy projects.

Another great factor for the renewables' success in the U.S. is the involvement of "big business," which is involved and increasingly involved and supporting the development and implementation of renewables energy sources.

U.S. business leaders are looking towards a smooth transition to unsubsidized renewable energy sources that can compete with the conventional such. Their competitiveness is based on extending while reducing, and ultimately eliminating, the PTC over time, which makes business sense and is supported by all parties, including the renewable companies and organizations.

Congress is considering extension of the master limited partnerships (MLPs) to renewable energy projects, which would make it much easier for the public and financial institutions to invest in the renewable projects.

In China, renewables are still a priority, but high barriers to entry for external investors, and other unfavorable financial factors placed China into second place. And yet, China's GDP is growing, the energy demand is increasing, and renewable energy manufacturing and use is a key to the local economies. These factors make renewables a good market for internal renewable energy investment. If and when China opens the doors to foreign investment, the growth renewables might bring China back to first place in the Index.

The real winner in such a case would be the global environment, but will not hold our breath for this to happen anytime soon. China is very busy today build-ing coal-fired power plants and ramping up electric power generation and distribution as fast as possible. This trend is estimated to last at least a decade or two, after which China might slow down to consider the alternatives.

Notes and References

1. EU legal, http://www.theguardian.com/business/2011/feb/20/carbon-emissions-trading-market-eu
2. DOE Legal, http://energy.gov/gc/services/laws-legal-resources
3. USDA, http://energy.gov/gc/services/laws-legal-resources
4. WSJ, http://www.wsj.com/articles/is-it-time-to-invest-in-energy-stocks-1422028657
5. DOE, http://energy.gov/downloads/doe-clean-energy-investment-center-fact-sheet
6. EERE, http://energy.gov/oe/mission/energy-infrastructure-modeling-analysis/state-and-regional-energy-risk-assessment-initiative
7. *Photovoltaics for Commercial and Utilities Power Generation*, Anco S. Blazev. The Fairmont Press, 2011
8. *Solar Technologies for the 21st Century*, Anco S Blazev. The Fairmont Press, 2013
9. *Power Generation and the Environment*, Anco S. Blazev. The Fairmont Press, 2014
10. *Energy Security for the 21st Century*, Anco S. Blazev. The Fairmont Press, 2015

Chapter 9

Energy Markets' Risks

*The amount, price, and availability of **risk-free energy sources**
determine the status and level of economic development of
each country, and predict its long term economic progress.*

—Anco Blazev

BACKGROUND

The global energy market is a big ocean of products, services and the related activities. A close look reveals a mix of complex and expensive equipment, processes, procedures and all kinds of projects driven by different types of people and entities of different nationalities, backgrounds, with varying professional and personal interests and motives.

Entire countries economies depend on energy supplies—some selling and some buying—all trying to keep things moving and their people happy, which boils down to energy availability and prices. Saudi Arabia, for example, needs to produce and export a lot of crude oil in order to afford the excess subsidies that support the opulent and decadent life of its population. At the same time, most African countries—some across the border from the Saudis' palaces—are scrambling to import enough energy to provide the very basics of life for their population.

Energy imbalance and struggle exist on all levels, all around the world. So why should we worry so much about energy in general and the energy markets in particular? What and where are the major risks? Surely, there are some, as is the case with all other markets.

To start with, we must understand that the energy market is a huge and dynamic enterprise, which has been undergoing some major, unforeseen, and quite rapid changes lately. Some of the changes are good, some are bad, but they are so many, fast developing, and complex on the good and bad sides of the equation, that it is hard to keep track, let alone understand, and predict the trends and the associated risks.

Nevertheless, we will attempt to do just that, keep-

Electricity	Petroleum	Natural Gas
· Generation	· Crude Oil	· Production
– Fossil Fuel Power Plants	– Onshore Fields	– Onshore Fields
» Coal	– Offshore Fields	– Offshore Fields
» Natural Gas	– Terminals	· Processing
» Oil	– Transport (pipelines)	· Transport (pipelines)
– Nuclear Power Plants	– Storage	· Distribution (pipelines)
– Hydroelectric Dams	· Petroleum Processing Facilities	· Storage
– Renewable Energy	– Refineries	· Liquefied Natural Gas Facilities
· Transmission	– Terminals	· Control Systems
– Substations	– Transport (pipelines)	· Gas Markets
– Lines	– Storage	
– Control Centers	– Control Systems	**Coal**
· Distribution	– Petroleum Markets	
– Substations		· Mining
– Lines	**Environmental**	· Transport
– Control Centers		· Processing
· Control Systems	· Air, soil, water pollution	· Control systems
· Electricity Markets	· Climate change	· Coal markets
	· Global warming	

Figure 9-1. The energy sector and the related markets

729

ing in mind that a thorough analysis of all risks involved in all day-to-day energy markets operations is a daunting exercise, and that, even in best of cases, whatever the situation might be today, it will change tomorrow. Different countries and regions present different difficulties when dealing with energy, and the related risks vary accordingly too.

Our goal is to reveal and discuss briefly as many of these risks as possible, focusing on those with the greatest and longest-lasting consequences.

As Figure 9-1 suggests, the different components of the global energy sector, and the related energy markets, are many and diverse. They undergo periodic and unscheduled changes according to local and global events, so let's see if we could sort them out in light of the new order in the global energy system. In the process, we will also discuss the risks associated with the different components of the energy markets.

Supply and Demand

Energy is the most precious commodity on Earth. We could not live our lives as we know them without enough and cheap energy.

Demand drives the energy markets, while the supply response shapes them

It all starts with somebody wanting something. In this case, energy of some sort. If there is no demand, the market would stagnate and idle until enough demand is available for the normal activities to start. Fortunately, in the energy markets the demand is always there and growing as we speak.

When the demand becomes apparent in some area of the world, the supply chain (usually in another part of the world) is activated, or created, and the market activities begin. The market then grows in depth and peripherally according to the supply and demand ratios.

Presently the demand for energy is at all times high, but the supply chain is facing difficulties and is nearing a break down. This is the most significant threat facing our energy marketplace. The law of supply and demand works always, but not always in predictable ways. In cases when one side of the supply-demand equation fails, the entire market responds accordingly. Sometimes the response is erratic and in some cases it brings the market close to collapse.

Today the demand is increasing, while the supply is trying to catch up. In the long run, the demand will continue increasing, but supply will decrease.

The increasing demand-to-supply ratio is clearly visible today. The beginning stages of this abnormality were observed in the energy markets in the near past. Growing energy consumption resulted in tight markets and an upward pressure on prices. The economic slowdown of 2008 dampened the worldwide energy demand temporarily, but as of 2013, the world's demand for energy has been increasing exponentially. This unprecedented and uncontrollable increase of demand is most noticeable in developing countries.

This marks a new trend in increasing excess energy consumption around the globe, driven by the developing countries, who insist on achieving better life style. And who can blame them? This leads to the conclusion that the long-term outlook for energy is questionable at best. Demand will increase continually and supply will not be able to keep up and the energy markets will respond with high prices and/or shortages.

Global energy demand is projected to increase 30% by 2030. It is impossible to increase energy output by 30% over the same period of time so something has to give. What will be it?

With all that said, we must remember that there are exceptions of all rules. In this case, we know for sure that the world demand for oil and other energy sources is growing beyond the supply capabilities, and yet we see exceptions all around us. This is exactly what we saw in 2014-2015, when oil prices dropped almost 50%. There are many reasons for this drop, but it is an exception of the rules, nevertheless.

No matter how big and important these exceptions may be, however, we must remember that the general rule of supply and demand still apply. As long as we have increasing world population with ever increasing energy needs, and ever decreasing energy reserves, the demand threatens to catch up with and exceed the supply on the long run and bring new balance in the energy markets. We may also witness a new world order, when oil prices rise above the tolerance threshold of the oil importing nations.

The Exceptions

The oil supply-and-demand game is stable from day to day and quite predictable well in advance… except when it is not. At times things go wrong and the global markets respond with unexpected actions, such as shortages, prices increases, etc. Most often, these abnormalities are short lived and the markets get back to normal in a matter of days. At other times, however, these abnormalities have catastrophic long term effects.

For example, the oil supply and demand dynamics were abruptly interrupted in 1973, when the Arab oil producing nations decided to punish the U.S. and the entire Western world by sudden reduction of oil exports. The energy market was disrupted suddenly and without a warning, resulting in panic, long lines at the gas pumps, and unreasonably high gasoline prices.

The Arab embargo was the first significant event since WWII, which brought panic and misery to millions, lined up at the gas pumps. Gas use restrictions and high prices forced a number of measures by the federal and state governments to avoid this happening in the future.

Later on in 2008 oil prices jumped to over $150 per barrel, but this time the reasons were different. This was another reminder of how much we depend on oil, and how many things could cause problems in the energy markets. In 2008, the oil problem was created by the global financial chaos and traders' manipulations of energy commodities.

In another twist of events, with the latest one starting in the summer of 2014, all of a sudden oil prices started dropping. By December 2014, oil prices have dropped about 50% on the global markets. Gasoline and diesel prices followed the down trend too.

Arab Embargo in Reverse

The oil crisis of 2014-2015 was different, and on the other end of extremes from that of 1973. Instead of increasing oil prices, it reduced them well below break even point. The only similarity is that as every time before, nobody saw what was coming until it was too late. The U.S. was caught again by a surprise.

The reasons for the 2014-2015 oil price anomaly that brought unexpected global oil price reductions are several fold:

• The world was awash in oil. U.S. oil production in the Gulf of Mexico, Alaska, Canadian oil sands, and the use of fracking on land almost doubled the oil production. Domestic oil production has been increasing steadily since 2008, which contributed to a slowly growing national and world supply. At the same time, the U.S. refineries were buying fewer barrels of more expensive foreign crudes.

The increased oil supply in the U.S., combined with weakening expectations for the global economy and world oil consumption, pushed oil prices lower. This then led to lower price of petroleum products like gasoline and diesel.

• In 2014, OPEC (Saudi Arabia mostly) decided to

NOT restrict oil production, which it usually does to balance the supply and demand ratio. Supply was abundant, so oil process went down below $50/barrel.

• There were no major natural disasters in the Gulf of Mexico in 2014. Since most U.S. refineries are located in that area, and usually suffer losses during the hurricane seasons, they were unscathed in 2014, thus no production shortages and no price increases.

• The gasoline prices in the U.S. actually started falling in the fall of 2014. This was partially due to, in addition to the above mentioned factors, to winter season price reductions. Around September of every year, the U.S. government eases off clean-air standards for oil refineries, thus allowing them to produce cheaper gasoline by using cheaper hydrocarbons like butane. Refineries don't benefit much from the cost savings, because petroleum products traders know how to buy gasoline at lower prices without hurting the refinery's profits, and so prices drop in response until late spring of the next year.

The Lesson…

The combination of these four factors is the reason we saw the drastic, albeit temporary and on-and-off gasoline price drop in 2014 and 2015. Our long-time experience dealing with the oil exporters tells us that another surprise is behind the corner, with global oil prices jumping sky high. As soon as the supply and demand ration increases in favor of demand, we expect oil prices to increase proportionally. The petroleum product prices will follow like a well trained puppy.

The overall lesson of the past is that the energy markets depend on many factors and many players, so anything can happen at any time.

The unpredictable volatility of the oil market is not new, nor is it a secret. There are armies of researchers, analysts, bankers, lawyers and many other specialists looking very carefully at the energy markets. Their job description is very simple; find a way to predict the future, reduce the risks, and prevent major disturbances in the energy markets. Simple goal, but very difficult to impossible to achieve.

It appears that the global energy markets puzzle is too complex for the armies of professionals to solve. They, however, have learned the lessons of the past, so future anomalies would be easier to predict and handle. At the very least, we would be better prepared for them.

But now back to the supply and demand controversy, which is the major problem and dictates the markets. Taking a closer look at the various components of the energy mix, we must remember that energy is the foundation of our lives. And that it is getting even more so with time.

The global oil consumption today is 85 million barrels per day, increasing to 120-130 million barrels by 2030.

Similar increase is expected in the production and use of coal and natural gas. Producing such huge amount of fossils, however, is becoming harder to achieve as time goes on. There are many reasons for this, but the major one is the fact that the existing fossil deposits are getting depleted fast. Their prices are expected to increase accordingly, until all fossils are completely depleted.

The end-times of the fossils is evident by the decreasing quality of crude oil coming from deep-wells. Drilling miles underground, and pumping this lesser quality oil, which also requires more energy and expense to produce and refine, makes the oil prices to go up. Reduced quantity and quality also means that the oil companies need more new production, which would cost more too.

Peak Oil…again

Here we need to refer to our old friend, the "Peak Oil" theory. *Peak oil* is the point at which we will have exhausted half the world's resources, after which production will inevitably go down. This has actually already happened in some U.S. states, where oil production peaked in the 1970s and beyond. It was actually this peak in output, vs. increasing demand that brought up the oil shortage of the 1970s. We were able to recover for a while after opening the Alaska's Prudhoe Bay and other new oil sources, but overall our oil production continued to decline until recently.

Today things have changed. We benefit from the use of new technologies, like hydrofracking, that allow squeezing oil from rocks deep under ground. These technologies, however, cannot create more oil, and only help explore these "new reservoirs," which, however, are difficult and more expensive to exploit, which in turn makes production costs higher than most conventional oil production.

The new fracking sites production also usually decline 60-80% during the first year of operation, which increases the long term pricing further. So, we consider this as a temporary exception in the Peak Oil theory, as

are the Canadian tar sands and other such major new developments in North America.

The bigger question today, therefore is if the Peak Oil theory is valid on national and global level? Some of the trends of late are controversial. The main phase of global oil discovery actually peaked in the 1960s-1970s. The peak in discovery is supposed to precede the peak of oil production by about 40 years. So, we are in the midst of the global oil peak, or very close to it. Some experts think that the global production from conventional oil sources has peaked in 2006. But around that time, huge unconventional sources such as the Athabasca tar sands in Alberta and sharply increased hydrofracking production changed the scenario.

As a consequence, we must now modify the Peak Oil theory to state that higher prices will always bring forth new—more difficult to produce and more expensive—oil supplies.

We must agree that the cheap oil peak is over.
Crude oil will only get more expensive from now on.

The short lived cheap oil of 2014-2015 is a temporary event, that cannot last long. Production costs are going to go up on the long run, and crude oil is going to get more expensive with time. Much more expensive…

No matter how look at the global energy markets, however, we must be aware of, and watch for, the many risks at play at different times and locations.

The Energy-Related Risks

In order of importance and priority, we can categorize the major risks to the proper function of the energy markets as follow:
- Short-to-mid-term price fluctuations
- Pending and inevitable depletion of fossil energy sources,
- Increasing global socio-political unrest and criminal activities, and
- Increasing environmental problems

Price fluctuations

Starting with the 1973 Arab oil embargo, crude oil prices have been going up and down (mostly up) and creating havoc in the energy markets and affecting the global economy. The oil price depends on many factors, including supply and demand conditions. The demand end of the equation has been slowly, and steadily increasing with time, while the suppliers have been varying their output, in order to control the prices and achieve all sorts of other self-serving goals.

It all started in 1973, when the Arabs decided to punish the U.S. and the Western world by reducing the oil exports. They succeeded then, but the element of surprise—their biggest and most effective weapon—was no longer a factor. The U.S. and the world took measures to eliminate the oil supply surprises.

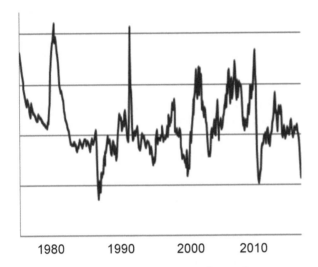

Figure 9-2. Global oil price fluctuations

Oil prices surged on several occasions and for different reasons during the 1980s and 1990s due isolated and short-lived events, like the Iran revolution, and the Iraq war. The price variations were significant for awhile and so were their consequences.

And then came the 2008 global economic meltdown. At that time we saw oil prices jump up to $180/barrel for a time...albeit for a very short time. The 2008-2012 time period was a bonanza for the oil suppliers, which, however did not last long. In 2014, Saudi Arabia manipulated its exports in order to shock the world with low oil prices. This was another surprise that the world did not expect, and was not ready for.

The booming U.S. oil and gas business came to a halt and many companies went into idle mode, or filed bankruptcy. Most oil producing countries were also negatively affected. They were forced to operate on very thin profit margins, which created economic problems in some of the smaller countries.

Russia was also in oil hell for the duration. At $50/barrel global price, the Russian oil drillers were losing money, because they need $80-100/barrel to break even. The situation was similar in most other oil producing countries.

In times of higher prices, everybody in the importing countries is unhappy. The damage the high prices cause is constantly in the news, and on the minds of policy makers, and the public. With this increased attention, political rhetoric regarding (possible) governmental regulations and market manipulations emerge in the search of ways to lower the prices.

In times of very low prices, as we saw during 2014-2015, there are different type of problems in different places. Politicians and oil company executives worry about the effect of low prices on their business affairs. They look for ways to survive the effects, knowing that this is a temporary situation and that low oil prices cannot last forever.

The most interesting, and of great importance in such cases is the fact that the consumers are happy. Paying less at the gas pump is always a good thing. This creates sort of a drift in the socio-political life of the affected countries, but its effect, if any, is minimal and short lived.

Oil prices eventually go up, and the up-down cycle repeats.

Socio-Political Unrest

Another significant factor in the new global international energy order is the old fact that countries can be divided into energy-surplus and energy-deficit categories. This is not a new characteristic, but some major changes within some major countries and globally have changed the way we look at the energy markets.

Saudi Arabia and Venezuela are in the energy surplus (energy producers) category, which controls the oil price by manipulating the volume of oil they produce. The U.S. and China are on the other end of the equation; the energy deficit (energy importers) category, which is undergoing major changes discussed in detail in this text.

On the extreme end of things, we see unrest in many countries around the world—both in oil producers and importers categories. Since many of these are important producers and users of energy, the internal socio-political turmoil is reflected in the function of energy markets.

Socio-political unrest in important areas of the world increases the uncertainty in the global energy markets.

The Environment

We cannot possibly talk about anything energy without considering the environment. These two are intimately interwoven, which actually represents an important risk to the energy markets. It is the fact that the increasing global pollution, due to overuse of fossil fuels for power generation and transport, is becoming a centerpiece for energy production and us reforms in many government and private efforts.

Increased use of fossil fuels increase the global pollution, which brings rapid climate change and more energy markets uncertainties.

As governments and companies all around the world are trying to combat the environmental pollution, they introduce changes that make energy more expensive. Because of that, we see shifts in energy use, such as the switch from coal to natural gas power generation in the U.S. and other countries. This is helping the renewables to an extent, but the focus on these is still marginal, thus allowing them to proceed, but at a slow pace, with many uncertainties awaiting them in the near future.

We also see a trend of switching transportation fuels to natural gas and other non-fossil alternatives. It is much faster and wider spread success story, where entire city bus fleets and other vehicles are converted to use LNG instead of gasoline and diesel.

Although we can clearly see some progress in many areas of the energy markets, we don't know where all this is leading, so let's go back to today's reality and analyze today's energy markets.

The changes in the U.S. energy sector of late were driven by a number of developments. The most important of these are the new oil and natural gas bonanza in the U.S., Canada, and other major energy producers and users. The amazing quantity of oil and gas coming from thousands of new wells and tar pit sites is unprecedented and is shaping the energy markets in a way no one thought possible a short time ago.

As a result, Saudi Arabia decided to control the energy markets by causing a slump in global oil prices, where in the beginning of 2015, a barrel of oil was 60% less than the same time in 2014.

Artificially low oil prices cannot last long, nor can it be repeated often, because it hurts everyone involved—including the instigators.

Because of that, we will consider the abnormal 2014-2015 slump in oil prices as a glitch in the otherwise stable oil markets. This temporary event showed that Saudi Arabia and its OPEC cronies still have huge influence over the global energy markets. Their influence, however, is decreasing slowly, as new energy sources and capabilities (hydrofracking oil and gas, sand oil etc.) are brought up all the time. And so, the temporary 2014-

2015 oil glut and price slump does not deserve much attention, because it will not be repeated any time soon, if ever.

On the other side of the ocean; Japan's wows after the Fukushima nuclear disaster, increasing Middle East turmoil, Russia's rogue behavior, Europe's ongoing financial problems, and China's aggressive attitude are also affecting and reshaping the global energy markets. We address these factors in more detail in different parts of this text, so we will not waste time to repeat them here.

Fossils Depletion

A major factor to consider in analyzing the global energy markets is the overall status of the global energy sources—and especially their availability in the past, present and the future. Energy sources are the foundation of the energy markets, so their availability and price are of utmost importance.

Figure 9-3 is showing the availability of the different energy sources world-wide from 1900 to 2200 (estimate). The fossils were untouched in 1900, thus they were 100% available for exploitation and use. The renewable energy sources were available as always, but the technologies to produce and use them were not around, so their use was limited to none.

Today, we are entering a period of excessive production and use of all kind of energy sources. Because of that, the risks have changed and are increasing. At the present level of energy production and use, as well as the technological developments now and in the future, we see that by 2200 the fossils will be gone, and the renewables will have taken their place in the global energy markets.

All renewables will be available until then and used 100% by then, so all that the future generations will have to generate electric power and heat is renewable energy. Hydro- and nuclear power would be around, but are facing some serious problems.

Hydropower has a chance for a quick, 100% development around the globe, but it might be limited by draughts caused by global warming and other climate change related phenomena.

Nuclear power will be developed 100%…IF there are no more serious nuclear accidents. It is the only major energy resource the future of which is totally uncertain, and which fortunes could change within a blink of an eye. In best of cases, by 2200 nuclear might be the major power supplier around the world. In worst of cases, it might be history due to increasing number of nuclear incidents.

Global Energy Resources	% Availability			
	1900	2000	2100	2200
Coal	100	80	20	0
Natural gas	100	80	10	0
Crude oil	100	60	10	0
Hydropower	0	50	80	100
Nuclear power	0	50	75	100
Biofuels	0	1	50	100
Geo power	0	0.2	50	100
Ocean power	0	0.1	50	100
Solar power	0	0.1	50	100
Wind power	0	1	50	100

Figure 9-3. Energy resources availability (in % of total needed)

Because of these uncertainties, the future of hydro- and nuclear power can not be clearly defined, but in best of cases (barring major disasters) they would be used 100% by 2200.

In all cases,

The fossils are getting more expensive and nearing their depletion, while the renewables are getting cheaper and increasing quickly.

The renewables are becoming a major factor in the global energy markets, and even driving some areas of their developments. Their future looks bright, but the present not so much. There are serious socio-economic changes around the world, where the most important factor—and one driving the energy markets right now—is the exponential increase of fossil fuels use around the world.

A number of major players in the energy markets, mostly developing countries like; China, Russia, India, and even some developed countries in the EU have increased the use of fossils for a number of reasons. The plans for the future also call for ramp up in the use of fossils, which dwarfs the increase of renewables.

The New Global Order

The new developments brought up a new order in the way countries rank in the general global pecking order. Under the old order, a nation's ranking in the global hierarchy was determined by a number of criteria, such as its stockpile of nuclear warheads, its warships at sea, and the number of its armed forces. The superpowers had a priority, due to their superiority measured by their power to destroy other armies and countries.

In the new global energy order, which is just now being established, a nation's rank is determined by:

- The amount of available energy (fossil reserves mostly),
- The amount of energy it can produce, deliver, and use when needed,
- The ability to acquire additional (emergency) energy resources, if and when needed, and
- The ability to purchase the energy resources of the energy-rich countries at will

In the near future, this list will be further expanded to include:

- The availability of renewable energy resources (sunshine, wind, ocean access, etc.), and
- The ability to develop and produce renewable energy in order to sustain economic growth.

The above factors are quite complex, and the picture is further complicated by the fact that the socio-political status of many countries is changing quickly. All this has brought up a serious realignment of the energy markets and the different countries' global status in general.

At a quick glance, we see a rough alignment of the global ranking today as follow:

- Powerful and energy-rich countries (USA, Canada)
- Powerful, but energy-deficient countries (China)
- Weak, but energy rich countries (Venezuela, Nigeria), and
- Weak and energy poor countries (Eastern European countries)

The global socio-political and economic changes are ongoing, so there is no way to make a precise and accurate classification of the present day global ranking. One thing is for sure, though; the new realignment of the global power lineup into energy-capable vs. energy-impotent states introduces new and very significant socio-political and economic implications.

There are few powerful and energy-rich countries, except the U.S. and Canada, but there are quite a few examples of powerful, but energy-deficient states like China, Japan, Russia, and most West European countries, who are forced to pay ever higher prices for imported fuels to weak, but energy-rich states.

Note: When almost 100 years ago Einstein said, *'Everything Is energy and that's all there is to it'* he was speaking of the overwhelming importance of energy on cosmic scale. He was not aware of the problems we will have with fossils, environment, renewables, etc. It seems, however, that he was/is right in that energy is all that matters today. Without enough and cheap energy, our lives would be quite miserable, and the entire world would be faced with a great number of insurmountable problems and disasters.

In order to avert energy deficiency, all countries are competing on the energy markets, and the race is approaching a critical stage. The competition among the importers for the limited quantity of energy materials that the energy-rich states can and are willing to supply is increasing, which brings another unknown in the energy markets equation—what will happen in the future!

With all this going on, the energy-surplus states are becoming richer as they parcel out their increasingly valuable commodities at whatever prices the market will bear. In 2006 alone, for example, oil-exporting countries pocketed almost a trillion dollars from the oil-importers. This amount almost doubled in 2007 and 2008, when the oil prices jumped up. Today (2014-2016) the low oil prices are devastating some of the small oil producing countries.

Saudi Arabia, Abu Dhabi, Dubai, and the other OPEC nations are benefiting immensely from all this. They control the output levels and with that, the global oil prices. Oil also brought up immense wealth to Russia since 2000, as result of its oil and natural gas exports to Europe and CIS countries.

Most oil-rich states are weak (as far as military and other power is concerned), so they have been using their other power—the newly acquired wealth from oil to purchase U.S. banks, corporations, and other enterprises. As an example, the Abu Dhabi Investment Authority acquired a $7.5 billion stake in Citigroup, America's largest bank holding company, which later on sold an even larger share, worth $12.5 billion to Kuwait Investment.

So, the weak militarily oil exporting states don't need military power; all they need is to acquire most of the powerful nations' wealth. Then they can control them and the world.

Billions of U.S. money (paid to purchase oil) have been used to purchase American properties?

But it is not just buildings, that the Saudis are buying with our money. They are also aiming to own our water and other resources in the future too. The beginnings of this trend are reflected in a large Saudi owned agricultural enterprise in northern Arizona.

Scarce underground water of the Arizona deserts will be used by Saudi sheiks to water hay crops for export to Saudi Arabia.

Isn't that like pulling the rug from under our feet? The conspiracy theories abound in this and many such cases, but the facts of life remain: the most precious commodity of the desert is water. Without it, there will be no population centers, no solar power plants, and no life in any part of the Arizona deserts.

Other oil exporters made substantial investments in Advanced Micro Devices (AMD), a major U.S. semiconductor company, and the Carlyle Group, a private equity giant. The AMD affair didn't go very well, as the company was restructured, consequently losing most of its markets share and overall value. Or, maybe it was an intentional inside activity that caused AMD's restructuring and demise.

The level of oil imports and exports determine the economic gradations of the surplus and deficit in the nations. Russia's wealth, for example, has been growing lately, because of its vast crude oil, natural gas, and coal reserves. This has allowed it to, in addition to providing cheaper energy internally, to export huge quantities of energy (oil and gas) to neighboring countries. Saudi Arabia and other Middle East states have been getting rich off oil exports for many years, and this trend will continue for many years to come.

Some of the smaller oil suppliers have a history of producing oil for a couple of decades, generating huge fortunes for the privileged few, but then running dry and returning to the previous nation-wide misery. A similar diversity is documented in the case of the energy deficit countries, except for the wealthy countries, like China, Japan, and the United States, who continue to buy their way out of scarcity. In this never-ending process they damage their economies on the long run. The poor countries lack such advantages, so they will continue to suffer long time in the future.

The Presentday Risks

The above mentioned examples are only a few of the long list of risks plaguing the energy markets. How about the other risks?

The risks in the energy markets today could be summarized as short-, mid-, and long-term:

Short Term; Environmental Issues

Right now and without any delays we must consider the increasing risk of overwhelming environmental pollution, climate warming, weather changes, ocean rising, etc. disasters headed our way as a result of the increased use of fossils.

This is a great and imminent risk, which cannot be left unattended. Tomorrow might be too late...at least for some (millions) of us!

Mid-Term; Prices

On the mid-term, we must worry—and do something—about the simple, but extremely important, law of energy supply and demand. In recent years, growing energy consumption has resulted in tight markets, which leads to an upward pressure on energy prices. The economic slowdown of 2008 dampened worldwide demand temporarily, but this did not bring a new trend in energy consumption. It was a glitch, which is now history and the world is back at the all-you-can-eat energy buffet.

Some specialists of the optimistic kind see the long-term outlook for energy (and energy prices) as dim, while others, more pessimistic, see it as desperate...if the present trend of energy exploitation and use continues unhindered. Demand is increasing continually, and drastically in some cases, while the global energy supplies are flat-lined and will have a very hard time keeping up with the growing demand. Increased demand usually means increased prices, which is what we expect to see in the future, and which should be our major concern for the next several decades.

Global energy demand is projected to increase 45% by 2030—a staggering number that makes it even impossible to even imagine how this would happen, and what this world would look like at that time. It is not possible to imagine, realistically speaking, that the world is capable of increasing energy output almost double the present values over such a short period of time. And if so, then at what price?

Long Run; Availability

The challenges ahead become even more eminent when we take a close look at the various components of the energy mix on the long run—by the end of this century and during the next.

Just think:

The U.S. consumes a staggering 85 million barrels oil every day. By 2030, experts expect this to rise to 100-110 million barrels per day and this amount doubling by 2050. Is this even possible?

The same is true for the 50% increase in global use of coal and natural gas expected by that time too. Pro-

ducing this amount—technologically, logistically, and financially speaking—would be unthinkable today.

And—get this—technologically speaking, it is getting harder and harder to produce the fossil fuels with every passing day. Deeper we dig and drill, harder it and more expensive it is to produce fossil fuels. This is especially true for the oil industry, where, crude oil well and refinery production will likely have to add new more expensive technologies and [processes, and use much more energy to produce equivalent amounts of fossil fuels. This, in turn, will drive the energy process up and up to new, staggering, and unsustainable levels.

We are inventive species, so we might be able to find a way to increase production, but wait!

There is finite amount of fossils on Earth, so more we dig and pump out, quicker we get to their end.

And then what?

Many production wells and mines are near depletion, and many will be completely depleted over the next several decades.

The Energy Circle

The most significant, irreversible, and even catastrophic threat facing the global energy markets on the long run is the eventual depletion of the fossils. Especially serious is the unavoidable running out of oil by the end of this century.

Crude oil drives our societies, businesses, and armies, so life without it is unthinkable.

We've said enough in this text (and even more in the previous books—see references below) about fossils depletion, so it suffices to mention here that everything and anything in this world has beginning and an end; including the fossil energy reserves.

There are still significant supplies left but they are going fast. So the bare fact is that no matter what, some day soon they will be gone. We must keep this in mind at all times and be ready to either take action, or suffer the consequences of a fossil-less existence.

We have been talking about the pending "peak oil" for many years now. This is the point at which we will have exhausted half the world's crude oil resources. From that point on, production will inevitably go down. This is not just another fantasy of a sick genius. It has already happened in the 48 states, where oil production peaked in the 1970s.

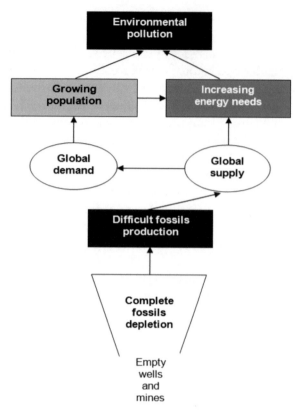

Figure 9-4. The wicked energy circle

The discovery of oil worldwide actually peaked in the 1960s, which according to the theory is supposed to precede the peaking of oil production by about thirty-seven years. Since the early part of this decade some experts have been saying that world production is near a peak. And according to the records, global production from conventional sources does seem to have peaked in 2006.

Fortunately, the difference was quickly made up by unconventional sources such as the Athabasca tar sands in Alberta and increased hydrofracking activities. Weather we have reached oil peak or not is questionable under the new circumstances, but we have surely reached the limits of cheap oil at the levels to which we've become accustomed during the last century.

Energy costs are going to go up, energy will get very expensive, while the competition for energy will increase sharply with time.

The 1970s marked the peak of easy and cheap oil production at high output, which when matched against demand kept on going upward, and which eventually helped precipitate the oil shortages of the 1970s. We recovered shortly after that, albeit for a while, with the help of the Alaska's Prudhoe Bay and other new re-

serves. Still, our production continued to decline until recently, which is why we import 60% of our oil today… even with the new oil discoveries and new production techniques.

Today the conditions seem ripe for the 1970s disaster to happen again on a global level, and as a matter of fact, some of the trends point in that direction. Just look how Saudi Arabia controlled the oil prices in 2014-2015, and how Russia controls the availability and prices of gas exports to Europe.

In the broadest sense of the word, we could divide the risks to the energy markets as internal and external

Internal Risks

We have the lessons of history, where we were betrayed, attacked, killed, and maimed within the U.S. by a number of American citizens. Charles Manson, Ted Bundy, Ted Kaczynski, Aldrich Ames, John Walker, Jr., Tom Horn, Andrew Kehoe, Jim Jones, John Gacy, Robert Hanssen, Timothy McVeigh, Nidal Hasan, James Holmes, Adam Lanza, Edward Snowden, just to mention a few who did not hesitate to hurt the country they were born in and its people.

These people have several things in common; they were all pure blood Americans—some real patriots and some soldiers and FBI agents; they were well educated and came from good families; they did not hesitate to betray and kill Americans on American soil; they did not care what will happen to themselves and; they intended to inflict the maximum damage possible under their particular situation.

Internal risks are as inevitable as human nature is unpredictable.

History teaches us that what happened then could, and most likely will, happen again and again. The criminals among us today have access to superior technology and given a chance will do maximum damage possible. With maximum damage as a goal and self-sacrifice as a price, stopping them and preventing a disaster is almost impossible.

Just imagine a Tim McVeigh wanna-be criminal with access to the controls of a nuclear reactor… The results would be devastating. Such scenario is possible in many other critical infrastructure points, with similarly disastrous results.

There are many types of internal risks and threats to the energy sector, and it would take an entire book to list all of them, so here we only mention the most important. A much more detailed list and a complete description of the different risks can be found in our book on the subject, titled, *'Energy Security for the 21st Century.'* Published by the Fairmont/CRC Press in 2015.

Corporate Terrorism

Terrorism comes in many shapes, forms, and colors. Serious damage to the energy markets can be done from the 25th floor of a Houston, Texas based oil company building, as easily as blowing up a power substation in San Jose, California. Enron is a good example of corporate intentions turned into terrorist-like actions, affecting millions. Enron's executives were given the opportunity and did not hesitate to do the unthinkable in the beginning of the 21st century.

By betraying the trust of the regulators and customers alike, Enron brought up a new era and new thinking in the energy industry. It also brought up a number of organizational failures, including that of the accounting firm Arthur Andersen, which was Enron's most important partner in mischief. On the larger scale, there was also breakdown of Big Government via political capitalism. Here, Big Business brought out the worst in our Government, while the Government reciprocated by bringing the worst in Big Business. Enron started the trend, which Solindra and others continued in a similarly tragic way.

Smart guys led companies into disarray and economic suffering due to poor advice and misdirected investments. In many cases, instead of liquidating the bad investments, New Deals by Government misguided gurus in power propped the bad investments with more investment via deficit spending and other means.

In the end, the 2008 recession turned into a semi-Great Recession, just as government activism transformed a depression into the Great Depression of the 1920s and 1930s.

The Enron problem became universal issue, for it discovered the "smartest-guys-in-the-room" syndrome as the best and fastest route to personal and organizational failure.

"Another lesson from Enron is that corporate behavior is fundamentally a product of the culture of the company," wrote Franklin D. Raines, CEO of Fannie Mae, in 2002. "At Fannie Mae we take pride in the tone we set at the top, in our risk management focus, in our commitment to integrity and intellectual honesty and in the values of our people." A lot of good words, and yet Mr. Raines and Fannie Mae were instilled insular culture. They invented new ways of accounting tricks; denial, arrogance, all covered by political capitalism.

The fall of Edison's partner, Samual Insull and his enterprises, predates Enron by a century, but it had a similar effect on the energy industry. It brought up the Securities Act of 1933, the Securities Exchange Act of 1934. and the Public Utility Holding Company Act of 1935. A century later, the fall of Enron inspired the Sarbanes-Oxley Act of 2002 and the Bipartisan Campaign Reform Act of 2002. It also prompted the Financial Derivatives Act of 2002, which still hangs unresolved.

The overwhelming lesson of all these events is that we need less, not more, government intervention. It teaches us that government direction, albeit unintentional, often leads into the wrong direction. It creates false safe harbors, promotes prudence in place of opportunism, and brings artificial booms that usually turn into corrective busts. Government bureaucrats and politicians often lack the technical and managerial skills and training needed to think and act like business executives.

Many times, what looks good to them is not what is good for business. Just take a close look at the U.S. solar industry, which was propped by government subsidies in the 1970s, just to be dropped and shoved in a closet in the 1980s—before the industry had a chance to step on its feet. Like a toddler learning to walk, it fell on its butt, not being able to get up for a long time. Corporate terrorism? Not exactly, but the tragic results of the solar industry past remind us of the aftereffects of terrorist actions.

If the solar industry had full and uninterrupted government support since the 1970s, it would've been a major power generator by now.

Instead, the on-and-off trend was repeated several times since then, ending with the last boom-bust cycle of 2007-2013. Now the government is pulling out its support to the solar industry—again—so we just have to see if it had stepped solidly on its feet to be able to walk and run on its own.

The lessons from Insull, Enron, and Solyndra show a trend, which we must be fully aware of and watch for. Having this rich background of successes and failures, we should be able to make better decisions today.

NIMBY

Not in My Backyard (NIMBY) movement participants are far from being terrorists. Although their actions at times cause damage, and the overall results remind us of terrorist actions, they mean well. In most cases their only demand is to protect themselves, their neighbors, and/or the environment.

These people are our friends, neighbors, and relatives—common people—who are exercising their rights to object the building of a wind farm, oil well, or a nuclear power plant in their back yards. The final result of these actions in most cases, however, is impeding the development of energy resources.

The NIMBY movement reflects the simple fact that most of us want affordable energy without experiencing any of the bad consequences.

We love and demand using energy, but don't want to have anything to do with its production, or distribution—especially if it affects us negatively in any way. Usually, NIMBY opposition is a local matter, often bringing great national and international consequences.

It is natural to object if someone wants to build a nuclear plant in our immediate vicinity, even if the imposition is very slight. At times, the opposition is just a matter of principle differences, without major consequences. In other cases, however, it has been proven detrimental to the projects at hand.

The Cape Wind Project off the coast of Nantucket Island and Martha's Vineyard, for example, was delayed for years, and may never be built as planned, because the locals object to the fact that the tall windmills would be visible from the high-priced real estate properties on the shoreline. Although opponents have tried to characterize their objections as environmentally-based, many of Cape Wind's opponents are prominent, influential, and well-funded, which has proven pivotal and decisive in the project's development.

An important difference here is the fact that the wind mills in the far distance are not posing any real danger to the life style of the locals. Unless they bump into the wind mill pillars with their motor yachts, after a night of drinking and partying, the wind mills are just a change in the scenery.

The opposition from the citizens of the state of Nevada to the Yucca Mountain Nuclear Waste Repository is another, albeit entirely different case. The site is ninety miles north of Las Vegas in the middle of an old air force bombing range and nuclear testing site. But local people resented the idea that their state was going to become the nation's nuclear waste dump. In response to this political pressure induced by the locals, the U.S administration cancelled the project in 2009.

The simple and common fact here is that nobody wants to be associated with the heavy burden and consequences of producing energy. The easiest thing to do is to export the hard work and problems to other countries; out of sight, out of mind. Let somebody else deal with the problems.

On a larger scale, environmental concerns (by NIMBY and government) have played a significant role in limiting the scope of offshore wind and oil energy production in the U.S. We have forbidden offshore drilling on the Atlantic and Pacific Coasts for fear of oil spills and have stopped offshore wind installations. But other countries are not handicapping themselves in the same way, which gives them an advantage in the energy markets. How fair this is, is a different question and a matter of different discussion, but the facts remain.

Big environmental and NIMBY battles will occur when extracting shale oil in the Rockies gets on the agenda. Western Colorado has the largest concentration of hydrocarbons in the world, which promise to be much messier than the Alberta tar sands.

One of the most inflammatory projects, full of environmental concerns is the Arctic National Wildlife Refuge (ANWR) in Alaska. According to the U.S. Geological Survey estimates, there are over 16 billion barrels of oil in that location. But the environmentalists have used ANWR as the prime example of a fragile ecosystem, which is too important to be used for energy production.

Figure 9-5. Start of the Alaska's anti-Shell revolt

Advocates for greater U.S. energy independence have called for opening up a small portion of ANWR for production, just to try it. Despite the very modest footprint of the proposed drilling site, opposition continues to prevent this project from moving forward. Nevertheless, some work has already started in the area, so we will just have to wait and see how far it gets under the environmental and NIMBY opposition fire.

The Keystone Pipeline System, a long oil pipeline system from Canada to the United States was commissioned in 2010, but is going nowhere. It is supposed to run from the Western Canadian Sedimentary Basin in Alberta to oil refineries in Illinois and all the way down to the Texas coast. It will also serve oil tank farms and oil pipeline distribution center in Cushing, Oklahoma.

Three phases of the project are in operation, and the fourth is awaiting U.S. government approval. NIMBY actions and other issues delay the final decision. In best of cases, the line will not be completed before 2020.

And so, NIMBY—in its local and national forms—is a strong movement, which has a lot to do with the energy supply availability and prices in the country and the world. It represents one of the risks (or opportunities) in the energy markets that shape our energy future.

The U.S. Power Infrastructure

The U.S. is no more energy independent now that it has been in the four decades past, even though we are witnessing a boom in cheaper clean energy sources. New natural gas and crude oil deposits were discovered, and new advanced technologies were used to produce a lot of oil and gas. Megawatts of new wind and solar power plants were constructed and are in operation.

The national infrastructure—power grid, roads, bridges, and railroads is the weak link that presents serious challenges.

Those are the conclusions of the federal government's first Quadrennial Energy Review, a report President Obama directed the Department of Energy to prepare.

The report cites a laundry list of current and future threats to the stability of the U.S. infrastructure network. Much of the problems relate to aging infrastructure, or infrastructure that was simply not designed to handle the increased use, and/or counter the threats now facing it.

For instance, 59% of the pipeline systems that transport natural gas were constructed in the 1960s or earlier.

Table 9-1. Age of U.S. gas pipelines

1920s	4%
1940s	8%
1950s	23%
1960s	24%
1970s	11%
1980s	10%
1990s	11%
2000s	9%

The gas pipeline system is vulnerable to attacks and is falling apart due to old age. The country's transportation, storage and distribution system is also increasingly vulnerable to extreme weather events like hurricanes, flooding and wildfires. Changes in the geography of domestic energy production stress the ability of existing infrastructures to move both liquid fuels and electricity from supply regions to demand centers. Also, increased congestion in the nation's ports, waterways and rail systems affect the timing and cost of moving not just energy products, but all commodities as well.

The only upside of the plans to revamp the national infrastructure is that the effort will necessarily create up to 1.5 million jobs. The downside is that it won't be cheap. The Department of Energy estimates that a program designed to improve maintenance of natural gas pipelines would run $2.5-$3.5 billion over 10 years.

Additional plan to promote innovative responses to threats to the stability of the electric grid would cost additional $3-$5 billion. And a comprehensive modernization of the national power would cost an additional $3.5 billion.

About $8-12.0 billion and 10 years are needed to bring our decrepit and outdated power lines, roads, and railroads into the 21st century.

The report also calls on the federal government to address the growing vulnerabilities posed by climate change, the evolving energy mix, cyber and physical threats, growing interdependencies, aging infrastructure and workforce needs.

To get the entire U.S. infrastructure to a state of good repair by 2020, the American Society of Civil Engineers estimates spending $1.6 trillion more than what we now spend. This is a lot of money, but the sad state of our infrastructure already costs the economy close to $200 billion annually. So the logical choice is to, a) continue patching the holes at $200 billion annually, or b) fix the entire system at $1.6 trillion cost.

In the winter of 2015, a $1.0 trillion bill was introduced in the U.S. Senate to fix the infrastructure over a five year period of time. It was later on reduced to $478 billion bill for the same infrastructure repairs, but was promptly blocked by the Senate. So, while we all know where and what the problems are, we don't know where the money to do all this work would come from…and/or when?

The Regulators

Can't live with, and can't live without them, is the short version of the relation between the government policies and regulations, on one hand, and the energy companies and the public, on the other. The energy companies and the public are not on the same page most of the time either, but they are both not very happy with the regulators all the time, albeit for different and sometimes controversial reasons. So, the regulators are stuck between a hammer and a hard place.

Who are the U.S. energy regulators and what do they do? Below is a list of the major ones and their function:

FERC

The Federal Energy Regulatory Commission (FERC) is an independent agency that regulates the interstate transmission of electricity, natural gas, and oil. FERC also reviews proposals to build liquefied natural gas (LNG) terminals and interstate natural gas pipelines as well as licensing hydropower projects.

The Energy Policy Act of 2005 gave FERC additional responsibilities as outlined and updated Strategic Plan. As part of that responsibility, FERC: regulates the transmission and wholesale sales of electricity in interstate commerce; reviews certain mergers and acquisitions and corporate transactions by electricity companies; regulates the transmission and sale of natural gas for resale in interstate commerce; regulates the transportation of oil by pipeline in interstate commerce; approves the siting and abandonment of interstate natural gas pipelines and storage facilities; reviews the siting application for electric transmission projects under limited circumstances; ensures the safe operation and reliability of proposed and operating LNG terminals.

FERC also licenses and inspects private, municipal, and state hydroelectric projects; protects the reliability of the high voltage interstate transmission system through mandatory reliability standards; monitors and investigates energy markets; enforces FERC regulatory requirements through imposition of civil penalties and other means; oversees environmental matters related to natural gas and hydroelectricity projects and other matters; and administers accounting and financial reporting regulations and conduct of regulated companies.

NARUC

The National Association of Regulatory Utility Commissioners (NARUC) was founded in 1889, as a non-profit organization dedicated to representing the State public service commissions who regulate the utilities that provide essential services such as energy, telecommunications, water, and transportation.

NARUC's members include all fifty States, the District of Columbia, Puerto Rico, and the Virgin Islands. Most State commissioners are appointed to their positions by their Governor or Legislature, while commissioners in 14 States are elected. For a complete breakdown, click here.

NARUC's mission is to serve the public interest by improving the quality and effectiveness of public utility regulation. Under State law, NARUC's members have an obligation to ensure the establishment and maintenance of utility services as may be required by law and to ensure that such services are provided at rates and conditions that are fair, reasonable and nondiscriminatory for all consumers.

NARUC's objectives are to advance commission regulation through the study and discussion of subjects concerning the operation and supervision of public utilities and carriers; to provide a forum for the exchange of information and ideas and between members and other organizations; to act as an effective advocate for members by coordinating activities and increasing members' influence with federal and State decision-makers; to provide and support education and training opportunities for members; and to promote coordinated action by State commissions to protect the public interest

EPA

The U.S. Environmental Protection Agency (EPA) has the goal of protecting human health and the environment. Its purpose is to ensure that: all Americans are protected from significant risks to human health and the environment where they live, learn and work; national efforts to reduce environmental risk are based on the best available scientific information; federal laws protecting human health and the environment are enforced fairly and effectively.

Usually Congress writes the environmental laws, which are then by EPA via new regulations. EPA sets the national standards that states and tribes enforce through their own regulations. If they fail to meet the national standards, EPA works with them on enforcing the regulations by helping them understand the requirements.

Environmental protection is an integral consideration in U.S. policies concerning natural resources, human health, economic growth, energy, transportation, agriculture, industry, and international trade, and these factors are similarly considered in establishing environmental policy; all parts of society—communities, individuals, businesses, and state, local and tribal governments—have access to accurate information sufficient to effectively participate in managing human health and environmental risks; environmental protection contributes to making our communities and ecosystems diverse, sustainable and economically productive.

EPA also ensures that the United States plays a leadership role around the world, for which purpose it works with other nations to protect the global environment.

BLM

The U.S. Bureau of Land Management (BLM) manages all public lands in the nation. In the energy sector, it has authority over lands that have the potential to make significant contributions to the Nation's renewable energy portfolio.

BLM is responsible to provide land for large-scale solar, wind, geothermal, and biomass energy sources deployment. BLM also manages the Federal onshore oil, gas and coal operations as the Nation transitions to a clean energy future.

BLM manages about 245 million surface acres as well as 700 million sub-surface acres of mineral estate. These lands are increasingly tapped to develop clean, renewable energy, in order for the U.S. to lessen its dependence on foreign oil. BLM also manages public lands that are needed for new transmission lines and facilities.

BLM reviews and approves permits and licenses from companies to explore, develop, and produce both renewable and non renewable energy on Federal lands. The BLM's goal is to ensure that proposed projects meet all applicable environmental laws and regulations. The bureau then works with local communities, the states, the energy industry, and federal agencies in this approval process from four Renewable Energy Coordination Offices and five oil and gas Pilot Offices to facilitate reviews.

BLM also participates in a Cabinet-level working group tasked to develop a coordinated federal permitting process for siting new transmission projects that would cross public, State and private lands.
BLM is then responsible for ensuring that developers and operators comply with use authorization requirements and regulations. BLM also approves and supervises mineral operations on Indian lands.

BLM collects royalty, rentals, and bonus payments from its lands, which vary from year to year. In fiscal year 2008, over $5.5 billion were collected for Federal onshore energy projects leasing. For oil and gas, half of this money goes to the States and half goes to the U.S. Treasury. Distribution of revenue from renewable energy varies from state to state and depends on the authority used.

SSP

The U.S. DoE, as the lead sector-specific agency for the U.S. energy sector, works closely with dozens of government and industry partners on the execution of many programs. One of these, of especial importance to the energy sector is its Energy Sector-Specific Plan (SSP). The SSP anticipates the hazards in the energy sector, with natural disasters as a key focus of the efforts.

Note: Although DOE is not officially a 'regulator' it works with, and has the power to regulate, the regulators and other energy related enterprises by a number of technological and financial means.

The SSP program addresses pandemic events and highlights potential cyber security activities, as needed to protect and improve the resilience of the U.S. energy sector in the face of both manmade and natural disasters. The regulators then translate the SSP findings and recommendations into policies and procedures to be followed by different units, or the entire energy sector.

The most valuable aspect of the SSP development has been the ongoing development of a trusted relationship and true partnership between government, regulators, and industry. This has enabled the development of a unified vision for the energy sector that enhances the national effort to implement the sector's critical infrastructure and key resources protective programs.

Some of the results are as follow:

- The North American Electric Reliability Corporation, acting as the federally authorized electric reliability organization, has developed several reliability standards for the power grid. These standards have been approved by the Federal Energy Regulatory Commission.
- The Electricity and the Oil and Natural Gas subsectors have developed enhanced approaches to plan for and counter cyber security threats to the U.S. energy infrastructure operations.
- The Energy Sector working closely with the Chemical Sector implemented new rules regarding safety and security at chemicals facilities, many of which are energy-related facilities.
- The Oil and Natural Gas Sector Coordinating Council (SCC) developed an Emergency Response Working Group, which aim to promote an integrated private sector and government response during natural disasters and terrorist incidents.
- The sector established a working group under the Critical Infrastructure Partnership Advisory Council to develop sector-specific approaches to metrics in order to better track and report on sector advances.

Internationally

On international level, the Energy Regulators Regional Association (ERRA) is an organization comprised of independent energy regulatory bodies. It started as a cooperative exchange among 12 energy regulatory bodies, mostly from the Central European and Eurasian region. ERRA also has affiliates agencies in Africa, Asia the Middle East and even the USA.

An ad-hoc Steering Committee, formed in 1999, began development of the structural and operational elements of the future association, including its purpose and objectives. In December 2000, a Constitution was signed by 15 members that established the ERRA in Bucharest.

To date ERRA lists 24 Full, 5 Associate and 7 Affiliate Members. The Association was legally registered in Hungary in April 2001 and its Secretariat operates in Budapest. NARUC and USAID have been providing continuous support for the operation of the Association.

ERRA's main objective is to increase exchange of information and experience among its members and to expand access to energy regulatory experience around the world—and especially among its member countries.

The Founding Members identify the purpose of ERRA as the need to improve national energy regulation in member countries, and foster development of stable energy regulators with autonomy and authority and to improve cooperation among energy regulators.

In addition, the Association strives to increase communication and the exchange of information, research and experience among members and increase access to energy regulatory information and experience around the world and promote opportunities for training.

So, all things considered, the U.S. energy sector is well protected from internal risks, some of which we discussed above. We also agree that some of the internal risks are controllable, and the rectification of others is work in progress.

Now, let's take a look at the external risks, which are also numerous, and perhaps more serious

External Risks

While the internal risks are numerous and serious, most of them are controllable, and some predictable… to a point. The external risks to the energy markets are even more numerous and serious. And most importantly, we have no way to predict the risks, let alone control the events these bring.

Here are some of the most serious external risks to the present and future of the energy markets:

Geopolitical Risks

The new global order is affected by numerous and complex geopolitical developments. Under the old order, military power was the principal factor of the global ranking. The major nuclear powers, the U.S. and the Soviet Union were the unchallenged super-powers and used their influence to manipulate the behavior of other, lesser, powers.

Military power concerns today are #2 in importance (after energy concerns, which proudly occupies # 1 place). The U.S. military power still conveys an advantage in today's world, but is increasingly overshadowed by the clout of energy abundance or lack thereof. The military machines defends the global pecking order and the prompt delivery of energy, but without energy (fuels), the military machine stops.

Catch 22—we need a lot of energy to get enough energy.

The U.S. has the mightiest military complex in the world. In order for it to function properly and efficiently, however, we need millions of gallons of oil and oil products…daily. Without these products, the machine freezes and cannot take one single step forward.

The U.S. military needs a lot of energy to run its operations. Without energy, all activities stop.

Where do we get the energy products needed for our daily lives? At least half of what we use daily comes from shady and unfriendly places all over the world.

One of these places is Saudi Arabia; oil rich country with a negligible military power. And yet, it has incredible leverage in world affairs. The only reason for this virtual power is that the country is sitting on top of the world's largest known petroleum reserves.

Saudi Arabia also enjoys the direct and indirect protection by the mightiest of them all—the USA military. We are BFFs with Saudi Arabia, not because we have anything in common, since in reality we have absolutely nothing in common. We just need each other… for now.

It would be quite interesting to see how this relation will develop (or not) in the future, when the oil supplies decrease and get more expensive. It will be even more interesting when crude oil is all gone. The Saudi royalties will most likely go back to live in their tents in the deserts, letting the country join the third world, where it came from not very long ago. That would be the end of the close Saudi-U.S. relations too. Sparks will fly sky high, the day we no longer need each other.

There are other, albeit smaller oil producing countries, such as Azerbaijan, Kazakhstan, Angola, Nigeria, Sudan, and others with nonexistent military that are enjoying influence that is several times greater than their size and military or economic capabilities. All because of oil...

Many of these energy-rich countries have been able to maneuver their privileged status to obtain, in addition to billions of dollars, many other concessions from their customers. Some of these are in the form of political support and military assistance. Why would Saudi Arabia, for example, need a large army, even if it could buy the most powerful weapons from its friends in the U.S. and UK? The Saudi sheiks know very well that the U.S. would jump in their defense at the slightest threat, so they sleep well at night.

And on top of that, the Western world looks the other way when human rights in these countries are frequently violated. Fair? Nope; it is only another confirmation that crude oil—and those who have a lot of it—benefit tremendously and even run this world. Make no mistake about it.

Lately, both oil-rich and oil-poor nations have been getting increasingly aggressive in their desire to enhance their respective competitive positions. This has led to the formation of questionable relations and associations, such as the proposed "natural gas OPEC," which resembles the old oil OPEC cartel. On the other side of the equation, we have the International Energy Agency (IEA), which defends the interests of the importers.

One of the most obvious and questionable alliances today, based solely on protecting energy interests, is that between China and Russia. The principal goal of this formation is to reduce the U.S.' influence in Asia, while benefiting from the closeness of the two countries. It is not clear how all this will develop in the future, but we clearly see an unprecedented global political realignment, and new energy market trends based on energy, being constantly developed in unexpected places.

Energy—and the fossils in particular—is very important today, and while the ownership of untapped petroleum reserves is concentrated in the hands of national oil companies (NOCs) their actions are carefully watched and analyzed by the world. The major NOCs are: Saudi Aramco, National Iranian Oil, Iraq National Oil Co., Kuwait Petroleum Corp., Abu Dhabi National Oil, Petróleos de Venezuela S.A., National Oil Corp. of Libya, Nigerian National Petroleum Corp., Qatar Petroleum, Gazprom (Russia), Pemex (Mexico), Petrobras (Brazil), and China National Petroleum Corp.

There are also independent oil and gas companies, the major of which are: Lukoil (Russia), Chevron (USA), Shell (USA) BP (UK), and many other large and small oil companies all over the globe.

The oil companies control most of the available global oil supplies. The NOCs hold the majority of oil supplies, and are mostly driven by their own governments' interests. These include the redistribution of national wealth, the creation of jobs, support of national economic development, and as a tool of foreign policy. And so, the global NOCs hold the key to the world's energy resources, and are pushing aggressively the buttons of global politics.

Energy is a major element in national politics, and the NOCs represent power (in terms of energy and political power), which is creating frictions. Because of the increased competition for energy, we now see a degree of militarization of the energy markets.

Militarization of the Energy Markets

A significant change in the world economy has been reflected in a radical transformation over the past half century; from significant government intervention in the form of regulation and planning to one based more and more on market forces, including these of the global energy markets.

The global oil market is the best example of this trend, as it moves away from contracts or government relationships between specific buyers and producers, to a versatile global market system based on competitive bidding and price discovery through the commercial dealings of a wide number of players.

The United States, as a major world power in energy production and consumption, favors an open, transparent and competitive global market. If given a chance, this would be an ideal market, where no seller or group of sellers can dominate the market and threaten the access to oil supplies as needed to conduct normal and everyday consumer, business and military operations.

The current global energy trading system that has been in operation during the last half a century is led, and to some extent controlled, by the U.S. Since WW II, the United States has taken the lead in many rounds of international negotiations to reduce tariffs, open markets to unrestricted capital flows and establish rules for protecting investments and intellectual property.

At times, the United States has relied on multilateral negotiations, but when multilateral talks were not promising, multi-track negotiations were used. This system has worked well and is supported by most countries, albeit there are complaints at times by the international community.

The U.S. is the world's military superpower, and backs up the operation of the global energy markets by policing the seas and international commerce from attack by hostile nations or non-governmental groups. The threat is increasing, and with that the role of the U.S. military as a global policeman is increasing too. With that, the energy markets get increasingly polarized and militarized.

The militarization of the global energy markets is entering a new stage today. We no longer count on casual opportunistic attacks on our infrastructure. Instead, we are in full battle readiness—at home and overseas—to fend terrorist and unfriendly nations attacks on daily and hourly basis.

24/7 watch over the 5 seas is what the U.S. military has to do to preserve the passage of energy sources (crude oil mostly) from one place to another.

Some of the major external risks to our energy future are:

Risk #1. MENA

Middle East and North America region (MENA) is a blessing and a curse to the world's energy markets. On one hand, we must be thankful for the huge amount of oil they have. On the other hand, however, MENA countries—as the world's largest oil producers—have locked us and the entire world in a $50-150/barrel price, which they control as they wish. Their actions are defined by whichever part of the supply-demand balance proves to be most advantageous to them at any given time.

The political unrest of 2011-12 in the area continues, and is increasing and spreading to other parts of the world. The Gulf Monarchies are also affected and are making everything possible to retain the power and the status-quo. Political succession is one of the key points to watch for in the future. Bahrain, UAE, Qatar, Kuwait and Saudi Arabia would be in trouble if the old succession model of passing power among family members is broken.

At such a point, global oil prices could explode and the unrest could spread into these monarchies as well. At that point, a myriad of events could lead these countries, and the global energy markets, into a disastrous down spiral with no end in sight.

Risk #2. Future Oil Prices

In order to stay in power and benefit to the max, the ruling class in the producing countries have to maintain high oil prices. OPEC nations clear over $1.0 trillion annually, so they have enough money to keep their people happy and buy themselves more time. OPEC no longer considers $100/barrel enough. It is instead a minimum price needed to ensure the survival of the ruling class.

The 50% drop in oil prices during 2014-2015 was an intentional move to control the global oil production on the long run. Most smaller producing states, like Algeria, Nigeria, Venezuela, Russia, Iran, and even the Gulf States, are starting to feel the low price pinch. The geopolitical cost of survival is seriously out of sync with the geological cost of production, which dictates minimum of $100/barrel. Period.

So how long can the Saudis live on $50/barrel remains to be seen, but not too long, we dare guess. Oil prices will get at $100 very soon, will stay high, and will even increase significantly with time. Oil is also nearing its end. There is just no way around it. Law of nature stipulates that what goes up must go down. This is true as far as the depletion of the global oil reserves is concerned too.

Risk #3: Iran

Iran's nuclear ambitions reminds us of an old European saying, "Can't afford a pair of pants, but has a sword on his belt." Just like this poor, but proud European peasant, Iran is enduing the sanctions with the U.S. turning the screw a little harder every day. In response, the Iranians increase their nuclear barking and bluster around the Strait of Hormuz. At the same time they continue vying for regional influence in the Gulf. No matter what Washington and Tehran do and say, the nuclear barking will continue adding to the uncertainty in the region.

Iran might be able to cause a major international accident, but that would be the last chance they get. Soon after, all their nuclear installations would be permanently disabled and buried in sand and rock by U.S. and Israeli bombers. The entire Iran military would be then sterilized by fighter jets. The Iranians know this very well, so they bark as loud as they can, and bite as much as they dare, but always back down just on time. Because of that, we must expect a lot of provocations and even minor conflicts in the Strait, which will keep the U.S. Navy ships busy for many years to come.

Risk #4: Russia

The energy markets, especially those in Europe depend on reliable and cheap gas and oil supplies from Russia. But Russia is in the midst of achieving its own regional conquers at the expense of its energy industry.

One additional factor that needs to be taken into consideration here is that Russia is also in the midst of internal wars between the major energy providers; Rosneft and Gazprom. Rosneft is the new national champion of choice. It produces over 4.6 million barrels daily, which is over 50% of the total daily output of the country.

Gazprom failed in pursuing a number of political goals, such as the Nord Stream and South Stream pipelines, Gazprom now needs to spend over $45 billion annually for the next several years to stay viable. And then there is the competition, but not from overseas companies. It is coming from Rosneft. The battle is intensifying, and under the present low energy prices, the Russian energy production is suffering.

Risk #5. The Renewables

We all agree that the renewables are our hope for a decent energy and environmental future. The problem is that the over $200 billion in global investments thus far have been allocated, or driven, by government tax incentives and subsidies, thus inflating the green tech bubbles. Cost curve basics are just now entering the equation, and as government props are being reduced or eliminated only a few countries will remain active in the renewables market in 2016-2020.

China is one of these countries. There is enough money allocated by the government to keep the solar and wind industries expansions at least for the next several years. Nevertheless, coal is still the *fuel of the future* in Asia and even Europe. And as the global energy prices are in a slump, the renewables will have to wait several more years.

Risk #6. The U.S. Military

The U.S. has 700, or 1,000, or even 1,900 operating military bases around the world. These numbers were plucked out of the author's hair, since no one really knows the real number. One number we have seen claims that there are 4,900 military related outposts in 148 countries, which includes embassies and other non-military operations with attached military units.

These operations include a total of nearly 850,000 buildings and pieces of equipments. The land surface area is estimated at about 2.2 million hectares, which makes the Pentagon one of the largest landowners in the world.

The primary mission of these outposts is to make sure that the global economy and its financial and energy markets are under control and function properly. The U.S. is also making sure that the natural resources (primary resources and nonrenewable sources of energy) are distributed as planned, including the daily oil flow to the U.S. ports. The latter activity is the cornerstone of the U.S.' power, which is achieved through the activities of its multinational military operations.

All these bases and military units have dozens and hundreds of cars, trucks, planes, ships, and other heavy military equipment that requires energy (fossil fuels mostly) to move around and otherwise function properly. Although it is hard to even estimate the amount of gasoline and diesel these vehicles use, their number being in the thousands suggests that hundreds of thousands of gallons are burned every day during the completion of their missions.

Some of these missions are designed to ensure the flow of crude oil and energy to the States and around the world. So, the first question that pops in mind here is, 'What would happen with all these bases and military units when oil price gets sky high, which they will sooner or later?' 'And the second, even more important question is, 'What will happen to the flow of crude around the world when these bases and military units don't have fuel to run around ensuring the oil flow.

This is a Catch 22 situation of sorts, that might seem ridiculous right now, but which becomes more realistic when we consider life by the end of the 21st century. Well, yea, we don't worry about that now. We will let those who are alive then to figure out the solutions. Yet, we must stare the facts in the face and recognize the huge storm brewing on our otherwise sunny horizon.

Risk #7. Terrorism

Saudi Arabia is shaping as increasingly attractive terrorist target. It's the only country with excess supply, still pumping up to 12 mb/d of crude oil. A terror attack could take some of the producing capacity out and the energy markets would collapse, since there is no way to cover any oil supply gaps. This alone would cost easily $200-300 billion in repairs and lost revenue.

One of the potential targets, Abqaiq, a 5-6 mb/d processing plant was unsuccessfully attacked in 2006. Another prized target, would be Ras Tanura terminals, where 8-9 million barrels oil is shipped out of the Kingdom every day.

If Saudi Arabia is the major oil producer in the world, Qatar is the largest LNG producer. 77 million tons of LNG are shipped from Qatar, which is about 25% of the total global supply. That allows it to set spot prices in Europe, and increasingly so in Asia. Ras Laffan plays would be a key target, since half of Qatar's LNG exports come from it.

Such an event might not be fatal for the global economy, and would have zero effect on American Henry Hub benchmarks, but a serious uptick for gas (and even oil) prices can be expected. In such a case, the energy markets would fear the security of the other Gulf producers, which might drag the world into a long energy crisis.

And, of course, well coordinated and executed terrorist attacks on the global maritime choke points would be devastating. The Strait of Hormuz is one of the core target, since about 17.5 million barrels of oil and 35 billion cubic feet of gas are transported through it daily.

Such an attack could wipe out the bulk of Gulf/Persian energy exports for the duration. Regardless of how long such an attack could sustain outages is unclear, but in all cases it would have a major price effect.

The major problem for the U.S. is the transport of oil across the world's oceans. It is an unending task, that has been repeated day after day during the last several decades. It is expected to continue to be a major preoccupation of the U.S. government and its armed forces.

Transoceanic Transport

Energy sources—crude oil, coal, liquefied natural gas (LNG), petrochemicals, and other energy products and byproducts—are transported day and night in huge quantities around the world oceans. The slow moving oil tankers and other commercial products transport ships with their precious cargo, are an attractive target for international; terrorists. As a result, a number of container ships and oil tankers have been attacked and even taken hostage in the past.

There are several very dangerous places around the world, some of which have the potential of interrupting the oil transport routes for a long while. This, however, is not going to happen as long as the U.S. patrol boats and jets have fuel to complete their missions. So, the U.S. military is at a battle readiness 24/7 in order to ensure the flow of crude oil is uninterrupted.

Let's take a look at the worst and most dangerous places our oil tankers cross daily:

The Chokepoints

Chokepoints are narrow channels along widely used global sea routes, some so narrow that restrictions are placed on the size of the vessel that can navigate through them. They are a critical part of global energy security due to the high volume of oil traded through their narrow straits.

Global crude oil production is at approximately 85 million barrels per day, half of which oil has to be

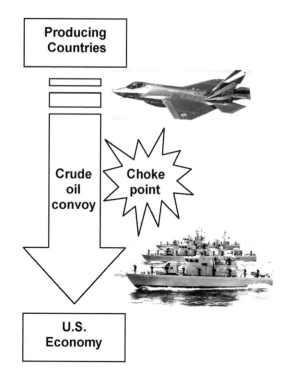

Figure 9-6. The choke points defense

transported by oil tankers all over the world. Oil tankers traveling along these fixed routes fixed has to cross some of the especially dangerous chokepoints, which present especially high risk for piracy that might result in hazardous oil spills.

Even a temporary blockage of a strategic chokepoint leads to substantial economic damage to the affected countries, and overall increase in energy costs.

In 2011, total world oil production amounted to approximately 87 million barrels per day (bbl/d), and over one-half was moved by tankers on fixed maritime routes. By volume of oil transit, the Strait of Hormuz, leading out of the Persian Gulf, and the Strait of Malacca, linking the Indian and Pacific Oceans, are two of the world's most strategic chokepoints.

The international energy market is dependent upon reliable transport The blockage of a chokepoint, even temporarily, can lead to substantial increases in total energy costs. In addition, chokepoints leave oil tankers vulnerable to theft from pirates, terrorist attacks, and political unrest in the form of wars or hostilities as well as shipping accidents that can lead to disastrous oil spills. The seven straits highlighted in this brief serve as major trade routes for global oil transportation, and disruptions to shipments would affect oil prices and add thousands of miles of transit in an alternative direction, if even available.

World oil chokepoints for crude oil transport are a

critical part of the global energy security dilemma, since most of the world's crude oil and petroleum production move on maritime routes.

Table 9-2.
The major choke points (in million barrels per day)

Choke Point	Volume
Panama Canal	1.6
Danish Straits	3.0
Suez Canal	3.0
Bab el Mandab	3.5
Turkish Straits	3.8
Strait of Malacca	15.0
Strait of Hormuz	18.0
Total	**47.9**

Due to the importance of oil supplies to the U.S. economy, and since about half of the crude oil used daily is imported, some of the world chokepoints are of especial interest.

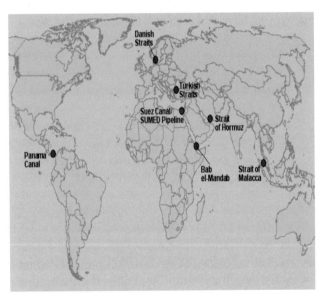

Figure 9-7. The major oil transport choke points.

The Strait of Hormuz

About 18 million barrels of crude oil are transported along the Strait of Hormuz in the Persian Gulf every day. Located between Oman and Iran, the Strait of Hormuz connects the Persian Gulf with the Gulf of Oman and the Fabian Sea. Flow through the Strait account for 30-40% of all seaborne oil. This is almost 20% of all oil traded worldwide.

More than 85% of these crude oil exports go to

Asian countries, with Japan, India, South Korea, and China as the largest customers.

Qatar also exports about 2 trillion cubic feet of liquefied natural gas (LNG) through the Strait of Hormuz annually, which is about 20% of all global LNG trade. Kuwait, on the other hand, imports huge LNG volumes that also travel through the Strait of Hormuz. This is a total of about 100 billion cubic feet per year passing through one single spot every year.

At its narrowest point, the Strait is 21 miles wide, but the width of the shipping lane in either direction is only two miles, separated by a two-mile buffer zone. The Strait is deep and wide enough to handle the world's largest crude oil tankers, with about two-thirds of oil shipments carried by tankers in excess of 150,000 deadweight tons.

The Strait of Hormuz is the world's most important oil chokepoint due to its daily oil flow of about 18 million barrels per day—which is up from 16 million barrels per day flowing during 2010.

Even temporary closure of this chokepoint would immediately and seriously affect the U.S.' economy and national security.

There are no options for oil tankers to bypass Hormuz presently. Iraq, Saudi Arabia, and UAE, however, have pipelines for shipping crude oil outside of the Gulf, thus bypassing the Strait at least partially. Only Saudi Arabia, and UAE have enough pipeline capacity to bypass Hormuz completely. The total available, but not utilized, pipeline capacity from these two countries combined is approximately 1 million bbl/d, which is to be increased to 4.3 million bbl/d, as both countries try to increase their pipeline capacity in order to bypass the Strait if and when needed.

Iraq' major crude oil pipeline, the Kirkuk-Ceyhan (Iraq-Turkey) pipeline transports oil from the North Iraq to the Turkish Mediterranean port of Ceyhan. This pipeline carries has a nameplate capacity of 1.6 million bbl/d, but has been the target of sabotage attacks and is operating at much lower levels. It also cannot send additional volumes to bypass the Strait of Hormuz unless it receives oil from southern Iraq via the Strategic Pipeline, which links northern and southern Iraq. Portions of that pipeline have been closed on and off during the war activities, and renovations are under way.

Saudi Arabia' 745-mile-long Petroline (the East-West Pipeline), runs across Saudi Arabia from its Abqaiq complex to the Red Sea. The Petroline system consists oft two pipelines with a total nameplate capacity of about 5.0 million bbl/d. This means that Saudi Arabia's spare oil pipeline capacity—able to bypass the Strait of Hormuz—has a capacity of 3.0 million bbl/d at full capacity.

The UAE constructed a 1.5 million bbl/d Abu Dhabi Crude Oil Pipeline in order to bypass Hormuz. It runs from the Habshan collection point (serving Abu Dhabi's onshore oil fields), to the shipping port of Fujairah on the Gulf of Oman. The pipeline is capable of carrying nearly 2.0 million bbl/d, which is over half of UAE's oil exports at full capacity

The Strait of Malacca

The Strait of Malacca, located between Indonesia, Malaysia, and Singapore, links the Indian Ocean to the South China Sea and Pacific Ocean. This is the shortest route between the Persian Gulf and the Asian markets. Most oil tankers headed to China, Japan, South Korea, and the Pacific Rim pass through Malacca.

Crude oil makes up about 90% of the daily cargo transiting Malacca, while petroleum products make the remainder of the traffic.

The Strait of Malacca is the key chokepoint in Asia, with over 15 million bbl/d headed to China and Indonesia—the major Asian oil consumers. This flow is up from 13 million bbl/d in the mid-2000s.

A temporary closure of this chokepoint would not affect the U.S. much. A permanent closure, however, might result in a global chaos and overall energy markets instability.

At the Phillips Channel of the Singapore Strait (its narrowest point), Malacca is only 1.7 miles wide, which creates a natural bottleneck. It also presents a great potential for collisions, grounding, and oil spills. Over 60.000 vessels navigate through the Strait of Malacca every year, where piracy, including attempted theft and hijackings, are a constant threat. The number of attacks has dropped since 2005, due to the increased patrols and interventions.

In case of a blockade, half of the world's fleet will have to be rerouted around the Indonesian archipelago and through Lombok Strait, between Bali and Lombok. Another long route is through the Sunda Strait, between Java and Sumatra.

Options to bypass Malacca have not materialized, so China is working on the 400,000 bbl/d Myanmar-China Oil and Gas Pipeline bypass. These are two parallel

(crude oil and natural gas carrying) pipelines from the ports in the Bay of Bengal, Myanmar to the Yunnan province of China. This addition will transport crude oil imports from the Middle East in order to bypass the Strait of Malacca, thus ensuring steady oil supply to China and the neighboring countries.

Turkish Straits

The Turkish Straits are located at the Bosporus and the Dardanelle, which divide Asia from Europe. The Bosporus is a 17-mile long passage, between the Black Sea and the Sea of Marmara. The Dardanelles is a 40-mile long waterway, linking the Sea of Marmara with the Aegean and Mediterranean Seas. Both straits are located on Turkish territory and serve tankers supplying Europe with oil from the oil producers of the Caspian Sea Region.

Almost 3 million barrels of crude oil flow through the Turkish Straits daily. The ports of the Black Sea are one of the primary oil export routes for Russia and other former Soviet Union republics. Oil shipments through the Turkish Straits varies as Russia shifts crude oil exports toward the Baltic ports and as crude production and exports from Azerbaijan and Kazakhstan rise.

Oil exports from the Caspian Sea region to Western and Southern Europe. make the Turkish Straits one of the busiest and most dangerous chokepoints in the world

A temporary closure of this chokepoint would affect the U.S. much. A permanent closure might result in a global chaos and overall energy markets instability.

At its narrowest point, the Turkish Straits are only half a mile wide. This makes them one of the world's most difficult waterways to navigate due to their peculiar geography. 50,000 vessels, including 5,500 oil tankers, pass through the straits annually, so it is also one of the world's busiest chokepoints.

Commercial shipping has the right of free passage through the Turkish Straits in peacetime, but Turkey is concerned about over the navigational safety and environmental threats to the Straits and reserves the right to impose regulations for safety and environmental purposes.

Bottlenecks and heavy traffic present dangers and create problems for oil tankers in the Straits, but there are no alternate routes from the Black and Caspian Sea region westward. As a solution to the problem, there are several pipeline projects near completion and others in various phases of development underway. These will help reduce the congestion and dangers of oil tankers passing through the Straits.

Bab el-Mandab

Bab el-Mandab is located between Yemen, Djibouti, and Eritrea, and connects the Red Sea with the Gulf of Aden and the Arabian Sea. It is a major chokepoint of oil shipments between the Horn of Africa and the Middle East.

It is also a strategic link between the Mediterranean Sea and Indian Ocean. Tankers from the Persian Gulf crossing the Suez Canal and SUMED Pipeline also pass through the Bab el-Mandab.

Four million barrels of oil flow through Bab-el-Mandab annually, headed to ports in Europe, the United States, and Asia. The flow varies with time; from 4.5 million bbl/d in 2008, down to 2.9 million bbl/d in 2009 as a result of the global economic downturn and the decline in northbound oil shipments to Europe.

Decline in northbound traffic through the Suez Canal and SUMED Pipeline in 2009 also reflect the adverse affects of the latest global economic crisis. Over half of the northbound oil shipments through Bab el-Mandab, about 2.0 million bbl/d, move en route to the Suez Canal and SUMED Pipeline.

Closure of the Bab el-Mandab could keep tankers from the Persian Gulf from reaching the Suez Canal and Sumed Pipeline, with the only alternate route being around the southern tip of Africa.

Even a temporary closure of this chokepoint would immediately and seriously affect the U.S.' economy and national security.

Bab el-Mandab is 18 miles wide at its narrowest point, which is limited to two 2-mile-wide channels in the inbound and outbound directions, which is not much for the large amount of the huge oil tankers crossing the chokepoint.

Restriction in the Bab el-Mandab traffic would restrict the flow of oil entering the Red Sea from Sudan and other countries, since it will have no access to the most direct route to the Asian markets. Instead, the tankers have to go into the Mediterranean Sea and through the other chokepoints of the Suez Canal and SUMED Pipeline.

A large French oil tanker was attacked off the coast of Yemen by terrorists in 2002, causing damage to the vessel and chaos around the world. Since that event, se-

curity is a major concern of tanker owners, operating in the region. During the last several years Somali pirates (off the northern Somali coast in the Gulf of Aden and southern Red Sea, including the Bab el-Mandab) have attacked a number of tankers, capturing their crews and demanding large sums of ransom money.

The Suez Canal

The Suez Canal is located in Egypt and connects the Red Sea and Gulf of Suez with the Mediterranean Sea. Crude oil and LNG shipments account for 30% of the total annual Suez Canal business. The Canal is too small for Ultra Large Crude Carriers (ULCC) and also for fully laden Very Large Crude Carriers (VLCC) crude oil tankers. In 2010 when the Suez Canal Authority extended the depth to 66 feet to allow more tankers to go through, so now over 60% of all world tankers can use the Canal.

About 3 million bbl/d of oil tankers transit the Canal in both directions. This is about 6-7% of the total worldwide seaborne oil. The majority of the oil goes north toward the European and North American markets. The remainder flows south toward Asian markets.

The oil volume varies according to economic and other conditions. For example, it increased in 2012 due to the restart of oil production in Libya, following the civil war. At the same time, southbound oil flow from Libya quadrupled.

SUMED pipeline

A 200-mile long SUMED Pipeline, or Suez-Mediterranean Pipeline, was constructed to serve ULCC and VLCC type vessels that are too large to transit through the Canal. These vessels are unloaded at Ain Sukhna terminal along the Red Sea coast, and the oil is transported via pipelines to the Sidi Kerir terminal on the Mediterranean. Here it is loaded on other tankers for shipment to its destination points.

Some fully laden VLCCs trying to cross the Suez Canal also use the SUMED Pipeline for reducing their weight and draft by offloading some of their cargo. A portion of the crude is offloaded at the SUMED Pipeline at the Ain Sukhna terminal and pumped to the Sidi Kerir terminal. The lighter VLCC now can go through the Suez Canal, and then picks up the offloaded portion of its load at the other end of the pipeline at Sidi Kerir terminal.

The crude oil is pumped through two parallel, 42 inch diameter, pipelines with a total pipeline capacity of around 2.5 million bbl/d. The SUMED pipeline is owned by the Arab Petroleum Pipeline Co., a joint venture between the Egyptian General Petroleum Corporation (EGPC), Saudi Aramco, Abu Dhabi's National Oil Company (ADNOC), and Kuwaiti companies.

In 2012, around 1.54 million bbl/d of crude oil was transported through the SUMED pipeline. SUMED crude flows decreased somewhat in 2012, the total crude oil transited northbound from Suez and SUMED combined increased to 2.44 million bbl/d in 2012 from 2.20 million bbl/d in 2011.

The SUMED Pipeline is the only alternative route to transport crude oil from the Red Sea to the Mediterranean, for tankers too large to navigate through the Suez Canal.

Closure of the Suez Canal and the SUMED Pipeline would force oil tankers to travel around the Cape of Good Hope, or 2,700 miles additional miles from Saudi Arabia to the United States, adding time and cost to the transport.

Even a temporary closure of this passage would immediately and seriously affect the U.S.' economy and national security.

Navigating around the southern horn of Africa adds 15 days of travel time to Europe bound tankers, and adds 8-10 days to the tanker going to the United States. The oil cost goes up accordingly too.

LNG also flows through the Suez Canal in both directions, amounting to about 1.5 trillion cubic feet in 2012, or around 13% of total LNG traded worldwide.

Southbound LNG transit originates in Algeria and Egypt and is headed mostly for the Asian markets. The northbound transit originates mostly from Qatar, destined for European markets.

U.S. LNG imports from Qatar fell by around 63% in 2012 compared with 2011, mostly due to sharp increase in domestic supply in the U.S., and a decrease in LNG demand in some European countries. This is also an indication of increased complexity and competition in the global LNG marketplace.

Danish Straits

The Danish Straits is 300 miles long waterway passage that connects the Greenland Sea (an extension of the Arctic Ocean) and the Irminger Sea (a part of the Atlantic Ocean). It is 180 miles wide at its narrowest point, and 625 feet deep at the shallow end, so it does not present nearly as much danger as other narrower chokepoints around the world.

Except that the cold East Greenland Current passes through the strait and carries icebergs south into the North Atlantic. The icebergs are the only danger in the path of oil tankers in the straits, but we are far from the Titanic days, so the danger is minimal.

The Danish Straits are a major route for Russian oil exports to Europe, where about 3 million bbl/d flow mostly westwards. Russia is increasing its oil production and is using its Baltic ports for crude oil exports to Europe.

The new port of Primorsk accounts for nearly half of the exports through the Danish Straits.

Additional 500,000 bbl/d of crude oil from Norway flow eastward to the other Scandinavian markets. About one-third of the westward shipments through the Straits are of refined petroleum products coming from Baltic Sea ports of Tallinn, Venstpils, and St Petersburg.

Panama Canal

The United States' is the top user of the Panama Canal (as country of origin and destination) for all commodities going to/from the U.S. It is, however, not a significant route for U.S. crude oil or petroleum transports.

The Panama Canal is an important route connecting the Pacific Ocean with the Caribbean Sea and the Atlantic Ocean. It is 50 miles long, and only 110 feet wide at its narrowest point, Culebra Cut, on the Continental Divide.

Over 14,000 vessels transit the Canal annually, of which more than 60% {by tonnage) represent United States coast-to-coast trade, along with United States trade to and from the world that passed through the Panama Canal.

Closure of the Panama Canal would greatly increase transit times and costs adding over 8,000 miles of travel. Vessels would have to reroute around the Straits of Magellan, Cape Horn and Drake Passage under the tip of South America.

The Panama Canal is not a significant route for U.S. petroleum imports, so a closure would not affect the U.S. much. A permanent closure might result in a global chaos and overall energy markets instability

Roughly one-fifth of the traffic through the canal is oil tankers. 755,000 bbl/d of crude and petroleum products were transported through the canal in 2011, of which 637,000 bbl/d were refined products, and the rest crude oil. Nearly 80% of total petroleum, or 608,000 bbl/d, passed from north (Atlantic) to south (Pacific).

The relevance of the Panama Canal to the global oil trade has diminished, as many modern tankers are too large to travel through the canal. Some oil tankers can be nearly five times larger than the maximum capacity of the canal.

The largest vessels that can transit the Panama Canal are the PANAMAX-size vessels—ships ranging from 60,000 to 100,000 dead weight tons in size, and no wider than 108 ft.

In order to make the canal more accessible, the Panama Canal Authority began an expansion program to be completed by the end of 2014. Many larger tankers will be able to transit the canal after 2014, and yet some ULCCs will still be unable to make the transit. These vessels will have to find an alternative route.

The Trans-Panama Pipeline

The Trans-Panama Pipeline (TPP) is located outside the former Canal Zone; near the Costa Rican border. It runs from the port of Charco Azul on the Pacific Coast, to the port of Chiriquie Grande in Bocas del Toro on the Caribbean.

The pipeline' original purpose was to facilitate crude oil shipments from Alaska's North Slope to refineries in the Caribbean and the U.S. Gulf Coast. In 1996, the TPP was shut down as oil companies began shipping AJaskan crude along alternative routes.

In 2009, TPP completed a project to reverse its flows in order to enable it to carry oil from the Caribbean to the Pacific. The pipeline's current capacity is about 600,000 bbl/d.

Today BP is leasing storage located on the Caribbean and Pacific coasts of Panama and uses the pipeline to transport crude oil to the U.S.' West Coast refiners. BP has leased 5.4 million barrels of PTP's storage and committed to east-to-west oil shipments through the pipeline averaging 100,000 b/d.

This route reduces significantly transport time and costs of tankers navigating around Cape Horn at the tip of South America on their way to the U.S. West Coast.

The Oil Exporters

While worrying about the safe ways to transport oil across the world, we also need to take a look at the global oil producers. One thing most of them have in common is that they have nothing in common with us. As a matter of fact most of them hate America and all it stands for. This makes them unwilling partners in this extremely important market. And they show their colors from time to time.

We are also unhappy with many of them, since their political, religious, and social systems are often far from what we consider, and value as, democratic and fair. In many cases, these are countries ruled by monarchs, who are shamelessly imposing their will on the people, which is as undemocratic as undemocratic can be.

Some of these monarchs are as anti-American as they could be. Many of them support terrorist organizations, and/or are openly declaring their animosity towards America and everything it stands for. And yet, we bite our thongs, tolerate them, even call many of these people 'friends', and engage in 'friendly' negotiations with them, fully knowing that behind our backs their ultimate goal is to hurt us. And all this in the name of crude oil... It is a game of broken promises, desperate moves, conflicts, and wars, all of which have been going for a century now and will continue going like this for awhile longer.

The world's energy, financial, and political systems change. The change is sometimes very fast and in unpredictable manner, so we have to be ready to react and counter-react in order to protect our own, and the global, energy markets. This is vital for our survival as the most powerful nation in the world, so we make it a priority.

The final dreadful result would be that without enough energy—and crude oil in particular—our military strength would be diminished, our economic growth crippled, and our safety compromised.

The national strategic petroleum reserves were established in the 1970s as another level a defense against energy crisis, in case of crude oil shortages or extreme price increases. The reserves were created in response to the 1973 Arab oil embargo, but have proved to be a very useful tool in ensuring uninterrupted energy supplies under any critical and emergency situations.

Strategic Petroleum Reserves

The U.S. energy politics and programs started with President Carter, who put his right foot forward (and his neck on the line) to design and implement a plausible, comprehensive, all encompassing national energy policy.

Unfortunately, this was too early in the game, and his efforts brought little practical results. Nevertheless, he is given credit for starting the effort, which was to be changed several times by others, and which is still to be completed.

The 4th and most important principle of President Carter's 1977 energy policy was that we must reduce our vulnerability to potentially devastating oil embargoes.

The policy, as questionable as it might look today, was to be implemented by 1985. President Carter' policy was designed to protect ourselves from uncertain foreign supplies by reducing our demand mostly for oil, by,

* Making the most of our abundant resources, such as coal, and
* Developing a strategic petroleum reserves.

Note: After the 1973 Arab oil embargo, reducing oil use and looking for other energy sources was on the national agenda for awhile, and on a number of times after that. Every time the crisis was over, however, it was put on the back burner…again. It is so today too, mostly due to the new crude oil and natural gas bonanza we are experiencing. It is back to normal, acting as if there is no tomorrow.

Reducing oil use does not seem to be a priority now, and the alternatives are too expensive and bothersome. The good old happy days of 8 cylinder cars, SUVs, and trucks are back again. Only somewhat higher prices at the gas pump keep us from getting full blast into the 1950-1960s energy oblivion. Happy days are here again, so we are taking a full advantage of the situation…but for how long.

The strategic petroleum reserves (SPR) plans were implemented as scheduled at the tune of many billions of dollars. Now the U.S. strategic petroleum reserve is a federally owned stockpile of oil with a capacity of nearly 750 million barrels stored at different strategic locations.

The national strategic oil reserves are actually crude oil and petroleum products stored in different underground storage reservoirs and/or above ground tanks at different areas of the country. Thus, great quantity of crude oil and petroleum products sit in storage, but are available for use only in case of emergency, or sudden disruption in oil supply.

In more detail, the storage facilities vary in type and size as follow:

Oil Storage Tanks

There are different types of vertical and horizontal aboveground atmospheric and pressure storage tanks in tank farms, which contain crude oil, petroleum feedstocks, intermediate stocks or finished petroleum products. Their size, shape, design, configuration, and operation depend on the amount and type of products stored and company or regulatory requirements. Aboveground vertical tanks may be provided with double bottoms to prevent leakage onto the ground and cathodic protection to minimize corrosion.

Horizontal tanks may be constructed with double walls or placed in vaults to contain any leakage.

* *Atmospheric cone roof tanks* are aboveground, horizontal or vertical, covered, cylindrical atmospheric vessels. Cone roof tanks have external stairways or ladders and platforms, and weak roof-to-shell seams, vents, scuppers or overflow outlets; they may have appurtenances such as gauging tubes, foam piping and chambers, overflow sensing and signaling systems, automatic gauging systems and so on.

When volatile crude oil and flammable liquid petroleum products are stored in cone roof tanks there is an opportunity for the vapor space to be within the flammable range. Although the space between the top of the product and the tank roof is normally vapor rich, an atmosphere in the flammable range can occur when product is first put into an empty tank or as air enters the tank through vents or pressure/vacuum valves when product is withdrawn and as the tank breathes during temperature changes. Cone roof tanks may be connected to vapor recovery systems.

* *Conservation tanks* are a type of cone roof tank with an upper and lower section separated by a flexible membrane designed to contain any vapor produced when the product warms and expands due to exposure to sunlight and to return the vapor to the tank when it cools and condenses at night. Conservation tanks are typically used to store aviation gasoline and similar products.

* *Atmospheric floating roof tanks* are aboveground, vertical, open top or covered cylindrical atmospheric vessels that are equipped with floating roofs. The primary purpose of the floating roof is to minimize the vapor space between the top of the product and the bottom of the floating roof so that it is always vapor rich, thus precluding the chance of a vapor-air mixture in the flammable range.

All floating roof tanks have external stairways or ladders and platforms, adjustable stairways or ladders for access to the floating roof from the platform, and may have appurtenances such as shunts which electrically bond the roof to the shell, gauging tubes, foam piping and chambers, overflow sensing and signaling systems, automatic gauging systems and so on. Seals or boots are provided around the perimeter of floating roofs to prevent product or vapor from escaping and collecting on the roof or in the space above the roof.

Floating roofs have legs which may be set in high or low positions depending on the type of operation. Legs are normally maintained in the low position so that the greatest possible amount of product can be withdrawn from the tank without creating a vapor space between the top of the product and the bottom of the floating roof. As tanks are brought out of service prior to entry for inspection, maintenance, repair or cleaning, there is a need to adjust the roof legs into the high position to allow room to work under the roof once the tank is empty. When the tank is returned to service, the legs are readjusted into the low position after it is filled with product.

- *Aboveground floating roof storage tanks* are further classified as external floating roof tanks, internal floating roof tanks or covered external floating roof tanks.

External (open top) floating roof tanks are those with floating covers installed on open-top storage tanks. External floating roofs are usually constructed of steel and provided with pontoons or other means of flotation. They are equipped with roof drains to remove water, boots or seals to prevent vapor releases, and adjustable stairways to reach the roof from the top of the tank regardless of its position.

These may also have secondary seals to minimize release of vapor to the atmosphere, weather shields to protect the seals, and foam dams to contain foam in the seal area in case of a fire or seal leak. Entry onto external floating roofs for gauging, maintenance or other activities may be considered confined-space entry, depending on the level of the roof below the top of the tank, the products contained in the tank, and government regulations and company policy.

- *Internal floating roof tanks* usually are cone roof tanks which have been converted by installing buoyant decks, rafts or internal floating covers inside the tank. Internal floating roofs are typically constructed of various types of sheet metal, aluminum, plastic or metal-covered plastic expanded foam, and their construction may be of the pontoon or pan type, solid buoyant material, or a combination of these.

Internal floating roofs are provided with perimeter seals to prevent vapor from escaping into the portion of the tank between the top of the floating roof and the exterior roof. Pressure/vacuum valves or vents are usually provided at the top of the tank to control any hydrocarbon vapors which may accumulate in the space above the internal floater. Internal floating roof tanks have ladders installed for access from the cone roof to the floating roof. Entry onto internal floating roofs for any purpose should be considered confined-space entry.

- *Covered (external) floating roof tanks* are basically external floating roof tanks that have been retrofitted with a geodesic dome, snow cap, or similar semi-fixed cover or roof so that the floating roof is no longer open to the atmosphere. Newly constructed covered external floating roof tanks may incorporate typical floating roofs designed for internal floating roof tanks. Entry into covered external floating roofs for gauging, maintenance or other activities may be considered confined-space entry, depending on the construction of the dome or cover, the level of the roof below the top of the tank, the products contained in the tank, and government regulations and company policy.

Tank Farms

Tank farms are groupings of storage tanks at producing fields; refineries; marine, pipeline and distribution terminals; and bulk plants which store crude oil and petroleum products. Within tank farms, individual tanks or groups of two or more tanks are usually surrounded by enclosures called berms, dykes or fire walls. These tank farm enclosures may vary in construction and height, from 45-cm earth berms around piping and pumps inside dykes to concrete walls that are taller than the tanks they surround.

Dykes may be built of earth, clay or other materials; they are covered with gravel, limestone or sea shells to control erosion; they vary in height and are wide enough for vehicles to drive along the top. The primary functions of these enclosures are to contain, direct and divert rainwater, physically separate tanks to prevent the spread of fire from one area to another, and to contain a spill, release, leak or overflow from a tank, pump or pipe within the area.

Dyke enclosures may be required by regulation or company policy to be sized and maintained to hold a specific amount of product. For example, a dyke enclosure may need to contain at least 110% of the capacity of the largest tank therein, allowing for the volume displaced by the other tanks and the amount of product remaining in the largest tank after hydrostatic equilibrium is reached. Dyke enclosures may also be required to be constructed with impervious clay or plastic liners to prevent spilled or released product from contaminating soil or groundwater.

The History of Strategic Oil Storage

Following the 1973 Arab oil embargo, the U.S. started looking into ways to protect itself from similar disasters. In 1975, the U.S. founded the strategic petroleum reserves (SPR) in order to mitigate future temporary supply disruptions.

In July, 1977, approximately 412,000 barrels of Saudi Arabian light crude was delivered for a first time to the SPR caverns in the Gulf. This was the beginning, and the flow continued until total of about 600 million barrels were completed by the end of 1994.

But then direct purchase of crude oil was suspended during 1995-1999 due to budget resources being redirected to refurbishing the SPR equipment and other work intended to extend the life of the complex through the first quarter of the 21st century.

The filling activities were resumed in 1999 using a joint initiative between the Departments of Energy and the Interior to supply royalty oil from Federal offshore tracts to the Strategic Petroleum Reserve, known as the Royalty-in-Kind (RIK) program. It continued in phases from 1999 through 2009, when the Department of the Interior discontinued the RIK program for good.

The first direct purchase of crude oil from the SPR since 1994 was in January 2009 using revenues available from the 2005 Hurricane Katrina emergency sale. DOE purchased 10.7 million barrels at a cost of $553 million.

In 2012, the SPR conducted an emergency exchange with Marathon Oil following the Hurricane Isaac disruptions to the commercial Gulf Coast oil production, refining, and distribution systems. Marathon Oil repaid the one million barrels in 3 months.

Today, the U.S.' SPR is the largest emergency supply in the world, with the capacity of over 700 million barrels, worth some $20-30 billion. Depending on how much oil is used, the U.S. SPR represents 60-90 day supply of oil.

But the maximum total withdrawal capability from the SPR is only slightly over 4.0 million barrels per day, so it would take nearly 200 days to use the entire inventory. During this time, provided that there are no other fuel supplies, the U.S. transportation sector would function on 50% capacity.

The U.S. oil reserves are stored at four sites on the Gulf of Mexico, located near a major center of petrochemical refining and processing. The storage areas are actually a number of artificial caverns created in salt domes locations below the surface. These structures were built at a cost of $4 billion by drilling into the salt deposits and then dissolving the salt with water.

Each cavern can be up to 3,500 feet below the surface, and capable of holding 10-35 million barrels of oil. The caverns-concept was used in order to reduce costs, since it is 10 times cheaper to store oil below surface, vs. in large above ground tanks. The added advantages are these of no leaks, and there is also a constant natural churn of the oil due to a temperature gradient in the caverns, which keeps it fresh.

Figure 9-8. U.S. SPR above ground delivery system

Decisions to withdraw crude oil from the SPR are made by the U.S. President under the authorities of the Energy Policy and Conservation Act. In the event of an energy emergency, SPR oil would be distributed by competitive sale.

The SPR has been used under these circumstances only three times, most recently in June 2011 when the President directed a sale of 30 million barrels of crude oil to offset disruptions in supply due to Middle East unrest.

The United States acted in coordination with its partners in the International Energy Agency (IEA). IEA countries released all together a total of 60 million barrels of petroleum. The SPR's formidable size (capacity of well over 700 million barrels) is a significant deterrent to oil import cutoffs and a key tool of foreign policy.

The U.S., however, has no significant gasoline or diesel reserves. So, while we are somewhat protected from disruptions in oil supplies, we depend on other sources in emergencies in case of major disruption to refinery operations. This is a problem, since no new refineries have been constructed in the US for thirty years, thus there is little excess refining capacity. Hurricane Katrina made that point, when many Gulf coast oil refineries were disrupted for the duration.

Update 2015

There have been some more than dramatic developments in the U.S. energy sector recently, marked by an unprecedented increase in domestic crude oil production. This has brought significant changes in the overall national energy system, which include pipeline expansions, and construction of new energy infrastructure.

New methods of handling our energy sources (crude oil in particular) have been developed, including reversed flow of existing pipelines, and increased and diversified use of the domestic crude oil terminals.

The government is also assessing the strategic petroleum reserves system' capabilities and evaluating the appropriate responses in the event of emergencies and oil supply disruptions. For that purpose, In March 2014, the U.S. Department of Energy initiated a test of the system by drawing and selling about 5 million barrels of sour crude oil from its strategic petroleum reserves. This exercise was done solely for operational purposes, and was designed to have a relatively minimal market impact. The release was a "test sale," according to DOE, designed to evaluate how the reserve system works when releasing and selling the oil.

Refineries that were interested in buying crude oil from the reserve did so in preparation for the summer driving season. This is when the U.S. refineries switch over to summer grade gasoline, which is more expensive than the winter types.

As a result, the oil markets responded by extending the losses, due to unexpected flood of crude-oil supplies, and the accompanying announcement of selling significant amount of oil from the U.S. strategic petroleum reserve. Light, sweet crude for April delivery fell $2.05, or 1.5%, on the New York Mercantile Exchange. Brent crude on ICE Futures Europe slipped 33 cents, or 0.3%.

The last time the U.S. DOE conducted a similar test sale was in August 1990 when 4 million barrels of crude oil were put on sale. Such test sales are required by law and are done periodically, but there is no specific order or schedule of when and how these tests are conducted. Crude oil was also released from the U.S. strategic oil reserves in 2011, as discussed above.

Our strategic oil reserves system can keep the U.S. economy going at full speed for only 60-90 days. Longer with some modifications.

It, however, cannot keep it going much longer than four months under any conditions, let alone forever. And under many circumstances, things can get even much tighter in case of a more serious, unexpected, problems. We can easily see how such problems could occur, but we are well prepared to handle most of them. And yet, there is another, even bigger, problem that even the mighty U.S. of A. cannot handle.

The biggest problem of the energy markets is the unavoidable depletion of the global fossil reserves, and crude oil will be the first fossil fuel to go extinct.

The so called Peak Oil, if it is not already here, is approaching fast and there is not much we can do about it. As a matter of fact, most of us—including oil companies, regulators, and politicians—are misinformed about, or turn a blind eye to, its existence. It is the Achilles Heel of the energy sector, which is still largely ignored by the world' energy markets. It will, however, no doubt, become a major factor in the near future, most likely when oil prices increase, and/or oil supplies decrease to certain levels by the end of this century.

And speaking of Achilles Heels, there are quite a few around us. Another one that no one dares talk about is the situation with rare earth elements.

Technological Barriers

The energy markets are driven by the type, quality, and quantity of energy available sources. We know that the quantities of fossil fuels are limited, that the prices are going up, and that there are other risks associated with their production, transportation, and use. The technologies related to these steps multiply and develop quickly, which also increases the risks? What risks? Oh, yes, there are several, and some are quite serious. And this is what we are discussing here.

The major energy sources today are: coal, natural gas, crude oil, nuclear power, hydropower, ocean wave power, geo-power, bio-energy, wind, and solar power.

We took a close look at the fossil fuels—coal, oil, and gas—and we saw that no major new technological developments are expected in these areas any time soon. With the exception of implementing the new and expensive carbon sequestration and storage (CSS) methods, we will rely on the 19th century power generation the fossils offer.

Because of that, we will now focus our attention on the other energy sources that have the potential of replacing the fossils someday soon. Nuclear power and renewables are the energy sources with the greatest potential to take us into a bright energy future. But how and at what cost?

Nuclear Power

For a number of reasons, nuclear power is one of the most promising power generating technologies. Although it can also be classified as a fossil power, since uranium is mined out like coal, the global supplies are large enough, so we can ignore that fact...for now.

> *From a technical point of view, nuclear power can be considered a reliable, renewable, clean, green, and cheap energy source.*

Nuclear power generation has a huge advantage over the other energy sources because it packs a mighty punch within very small space. It also has very low pollution emissions and other wastes. It can produce amazing quantity of electric power, when under control. This is actually its major problem too, because when out of control, the huge energy contained within the reactors poses a great threat in case of a failure.

> *Like the Genie in the bottle, nuclear power does miracles when controlled. But when out of control, it is a vicious, merciless killer, and destroyer.*

After what happened during the Chernobyl and Fukushima nuclear accidents, these possibilities cannot, and should not, be ignored. We never thought that a seemingly simple negligence, as at Chernobyl, could cause so much damage. Similarly, we never thought that a well designed, build, and operated power plant like Fukushima, could fall victim so easily to Mother Nature's whims.

And yet, these disasters happened, many people were killed and made sick in the process, so we cannot discount the possibility of this happening somewhere else...and in even greater dimension.

Yes, nuclear power is the cleanest, cheapest, and most reliable power source known to man, but we cannot and should not ignore the risks it presents. We must be always on the watch, when dealing with nuclear power. This way we will not get caught with our pants down as Japan did after the Fukushima nuclear accident. So, we only hope that everything goes well during the next several decades and nuclear power gets safer and safer to the point where we no longer feel threatened by its increase worldwide.

But just in case; let's assume a worst case scenario, where all nuclear power plants in the U.S. shut down for good. We must be ready for such a day too, and ask ourselves what can replace them?

What can provide the 800 terawatt of electric power the 104 U.S. nuclear power plants generate before the shutdown? This is about 20% of the electric power we use day and night, and not having it is just not an option.

In response, coal power plants could crank up their turbines to squeeze some more power capacity. But as most of them are already old enough and ready for retirement, we cannot expect them to deliver the additional 20% base load. We saw this scenario play in Japan since Fukushima. The coal power plants were cranked up, which led to enormous increase of coal imports and GHG emissions.

New gas-fired power plants could be built in a hurry, but that would take several years and a lot of money, so at least for awhile, the country would be crippled by power shortages and the related disasters.

Such a doomsday scenario promises to bring a lot of headaches to all involved. We—all of us—will be affected one way or another. From the politicians in Washington and the states' capitals, to the utilities, and the customers, we all will be faced with difficult decisions.

The country would enter a state of emergency never experienced before that might end up in an economic melt down and social unrest.

The Alternatives; The Renewable

How about the renewables as an alternative? Could they fill the gap? Probably, but that would require some advanced thinking and action. Solar and wind are the most promising of them all, so let's see what would it take to generate 800 terawatt of electric power annually? Can they fill the gap left by the nuclear power plants by pumping the missing 20% of electricity in the national electric power grid?

Let's see:

- 104 U.S. nuclear power plants = 100 GWe electric power generation capacity
- 100 GWe = 800 TWh of electric power sent into the grid annually in a 24/7 operation.
- 800 TWh generated by 100 GW power plants = 8,000 Wh generated per watt installed power.
- (8,000 Wh = 8,000 hours of 24/7 non-stop power generation in one year)

One year = total of 8,760 hours, so on the basis of these numbers we conclude that the U.S. nuclear power plants fleet operates at over 80% of the time at full capacity.

This means that the nuclear power plants, although fully operational 95-99% of the time, are actually idling a little. The idle is mostly because of electricity demand during night hours, weekends, holidays and many other days, is somewhat reduced.

But now, under our doomsday scenario, this huge 800 TWh generating capacity is lost. What can wind and solar do to replace it? We will ignore the other renewables, like bio-, geo-, ocean and other renewables for now, because they are not easily, quickly, or cheaply ramped up.

Let's look first at what wind power can do:

Ramping-up Wind Power

Replacing 100 GWe of lost power generation by wind turbines is not impossible, but would be logistically complex, quite expensive, and lengthy undertaking.

Such an effort would require:

- Manufacturing and installing 50,000 wind turbines (2 MW each) quickly
- Finding an appropriate location for each installation,
- Allocating thousands of acres of land for the new installations
- Obtaining the necessary financing, insurance, etc.
- Obtaining the necessary permits for the new installation,
- Constructing transmission lines and substations
- Manufacturing and assembly of parts for all turbines
- Transporting the huge parts to the installation site,
- Installing each of these giants, testing, and certifying them all.
- Designing and implementing proper O&M procedures

Technically speaking, in best of cases:

- Total installed cost would be about $300 billion
- It would take at least 10 years, or more, to manufacture so many parts,
- Locating sites with enough wind would be a great challenge,
- Permitting and siting would take 2-3 years per site
- Connecting the new sites to the power grid would be even more challenging, due to their remoteness, which would add another $30-50 billion and several years.
- Maintenance of the huge forests of giants spread all over the country would be a very expensive

logistics nightmare.

- Unexpected and unforeseen incidents and accidents must be prevented

The worst part is that unlike the conventional power generators, which provide power 24/7, wind comes and goes as it wishes, so wind turbines idle major part of the time. Because of that, the grid operators cannot rely on wind solely, so additional base load capacity (renewable or fossil-fired) must be introduced even in best of cases. In best of cases, due to the variability of wind currents, the 100 GWp (gigawatt at peak) wind fleet would provide only a fraction of the power requirements at unpredictable periods of time.

And so, what we end up is a discrepancy of great proportions and with great consequences. While the 100 GWp wind power fleet can reach 100 GW output at times, it does so only at certain peak periods of time; only when the wind is at optimum speed. Since wind is uncontrollable, it introduces a great variability in the output.

We usually think of wind power plants as operating at about 20-25% capacity. This means that our new 100 GWp wind power installations would provide only 1/5 to 1/4 of the total power needed to fill the gap. This also means that we need to add more wind turbines—4-5 times more—to generate the 800 TWh of power. But that won't help much either, because often the wind blows at times when we don't need power, so the turbines would be idle at the best wind conditions.

And so, although the total generated power in this case is 800 GWe, only small part of it is useful...unless, in addition to increasing the number of wind mills from 100,000 to 400,000, we also add huge amount of energy storage. This would cost even more than the entire wind power capacity of the country.

So, no matter how we look at this, and what we do to correct the issues at hand, wind alone cannot replace the lost capacity properly.

In this case we are looking at the alternatives, which are to:

- Combine wind with conventional power generators,
- Combine wind with solar power, and/or
  — Combination of the above.

Case Study: Wind power Terrorism

Wind turbines and solar systems are vulnerable to physical and cyber attacks.

German security researcher Maxim Rupp has discovered numerous security flaws with solar lighting systems and wind turbines which, if maliciously exploited by an attacker, could result in disrupting energy supplies.

Rupp recently reported numerous flaws in the web controls for the following systems, the XZERES 442SR Wind Turbine, the Sinapsi eSolar Light and the RLE Nova-Wind Turbine, with the ICS-CERT subsequently issuing public warnings on all three of these.

One of these flaws is a cross-site scripting (XSS) request forgery vulnerability affecting the XZERES turbine, could potentially be used by an attacker to change the administrator password for the web management interface, and then gain complete control of the turbine.

A 'black hat' hacker could get access and change the wind vane correction, or the network settings to lock access to the web interface, and damaging the equipment in some other way. The ICS-CERT has ranked this security issue as 10 of 10 on the standard Common Vulnerability Scoring System (CVSS). This flaw is dangerous due to the ease of remote access and exploitation.

There are several reasons for the security bridges, including lack of experience with designing secure IT solutions, and/or shipping, installing, and operating utility components that are part of critical control infrastructure.

The utility companies usually run remote, in-house control centers, from where they control power plants, distribution facilities, etc. Wind power plants are smaller scale installations, which in most cases are not integrated into existing internal networks and control rooms. In many cases, their controls are placed on the Internet, where they are monitored and supervised by a third party.

Often micro-generation systems, such as a wind and solar power plant, have a substandard firewall, where cheap SOHO equipment is connected to an external Internet connection (via a 3G or 4G connection). Behind the firewall, which in some cases can be bypassed by built-in vulnerabilities, are web-based management interfaces that control the power plant. These often have standard passwords, which makes unauthorized access easy.

In addition, some utilities use the same network connection that connects the power plant to the outside word for other services, such as field network video cameras, or shared network connection for the physical security, such as access control systems, CCTV, burglar alarms, fire alarms, etc.

So, we will not be surprised if there is an increase in the problems associated with these types of micro-generation plants and facilities, if they continue to be used with little or no security built-in.

Solar Power

Replacing 100 GW of lost nuclear power generation by solar (PV or thermal) is not impossible, but not easy, quick, nor cheap.

Note: We do not consider solar thermal (CSP) power generation as a feasible alternative, because of the need for extensive maintenance of the moving parts, and most importantly, due to the need for a lot of fresh water. Lots of water is needed to generate and cool the steam, which is used in the steam turbines. The deserts, where this technology is most efficient, do not have excess water, so this technology has limited use at least for now, So, until solutions (such as efficient and cheap dry cooling) are found, CSP will sit on the side lines.

Because of that, for now at least, photovoltaic (PV) power generation is the only alternative, which is what we look at below.

100 GWe solar PV power generation would require:
- Manufacturing of 500 million PV modules (200 W each)
- Manufacturing of 100,000 inverters (1 MW each)
- Manufacturing the supports and related equipment
- Finding an appropriate location for each installation,
- Allocating over 500,000 acres of land for the new installations
- Obtaining the necessary financing, insurance, etc.
- Obtaining the necessary permits for the new installations,
- Transporting the parts to the installation site,
- Installing, testing, and certification of each site.

Technically speaking, in best of cases:
- It would take 5 years, or more, to manufacture all these parts;
- Locating sites with enough solar intensity is limited to the U.S. Southwest;
- Connecting the new remote sites to the power grid would be challenging;
- Maintenance of the millions of PV modules would be a logistics nightmare.

But then, the real problems start. Solar energy (sunlight) is variable even in the deserts, so the grid operators cannot rely on it solely. In all cases additional base load capacity (fossil-fired) must be introduced even in best of cases. The worst part is that—due to the variability of sunlight—the 100 GWp (gigawatt at peak) of solar power would provide only a fraction of the power requirements at unpredictable quantities and periods of time.

And so, the 100 GWp solar fleet power fleet could generate 100 GW of power, but only at short periods of time, mostly around the peak periods; only when the sun is high up in the sky during the summer months of the year. Sunlight is uncontrollable, which introduces a great variability in the final output.

We usually think of solar power plants as operating at about 20-25% capacity. This means that our new 100 GWp wind power installations would provide only 1/5 to 1/4 of the total power needed to fill the gap...and only at peak hours.

This then also means, that we need to add more PV modules—4-5 times more—to generate the total of 800 TWh of power. Manufacturing and installing 2-3 billion PV modules, and that many acres of available land, is a logistical nightmare, and we just can't even consider it as a real possibility.

And even then, the sun do not shine when needed at times and so solar power plants idle even at the height of the peak hours in the desert and at night. And so, even if the total generated power by the new PV fleet reaches 800 TWh, only small part of it is of practical use. Unless we add energy storage, which, for the most part, would allow uninterrupted use, but would also double and triple the already astronomically high initial price and O&M expenses.

No matter how we look at this, solar power alone cannot replace the lost capacity properly.

The alternatives are to:

- Combine solar with conventional power, or
- Combine solar with wind power, and/or
- Combine solar, wind, and conventional power generators.

There are indications that wind and solar can be successfully combined to produce a less variable total output. Let's take a look at how we can do this:

Wind, Solar, and the Grid

The major excuse for not adding more wind and solar into grid operations has been that they variable nature. Their power output depends on, and varies with the ups and downs of wind and sun availability. Because of that, they could not be used for on demand (dispatched) power when additional power is needed by the grid operators.

Solar at the rescue... There are efforts today to demonstrate how the unique characteristics of wind and solar working together can, much like traditional generation, perform Active Power Control (APC) to support the grid. The effort is focused on two areas of APC; the regulating reserves and the primary frequency control.

Grid operators would like to have more support (excess power) for transmission line overload, voltage drops, and frequency variations, but don't typically ask that renewables to provide it. On the other hand, the regulators have not awaken yet, and have not imposed the requirement to use renewables in that way.

And so, wind turbine manufacturers can provide the capability to provide APC services, but see no demand for it from the customers and developers. The developers don't see any point in driving up costs by building in the capability to provide a service for which there is no demand from grid operators. The regulators see no point in requiring grid operators to buy a service from renewables they may get for free from traditional generators. And the utilities are hesitating for now, not knowing which path to take—renewables or no renewables is the question before them today. A truly catch 22 situation on a large scale.

And so, if all parties work together, a complete solution would be found through APC, combined with other means, such as energy storage and combining wind and solar. While the future of large scale energy storage is still uncertain, combining wind generated at certain locations especially chosen for this purpose, with solar power generated nearby is one such solution that can be implemented immediately.

At such locations PV power is complementary to the output of wind generation, since it is usually produced during the peak load hours when wind energy production may not be available. Variability around the average demand values for the individual characteristic wind and solar resources can fluctuate significantly on a daily basis. However, as illustrated by Figure 9-9, solar and wind power plant profiles—when considered in aggregate—can be a good match to the load profile and hence improve the resulting composite capacity value for variable generation.

In the example illustrated in Figure 9-9, the average load (upper line) is closely followed during the day by the average output from the combined wind and solar generators (the second from top line) during the same time. This average is created regardless, and because, of the fluctuations of the individual wind and solar power generators. This is a marriage made in heaven, and this combined power generating combination will work quite well if wind and solar power outputs can be matched as closely as the one represented here.

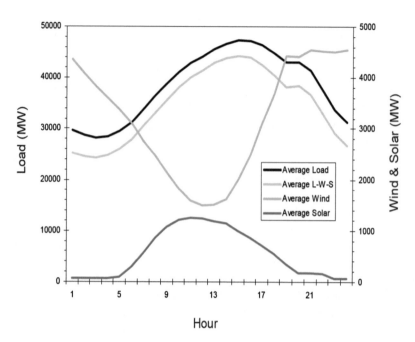

Figure 9-9. Simultaneous wind and solar generation

There are areas in the US and abroad where this scheme is hard to execute because the best places for wind and solar are at different locations (often miles apart). The lack of infrastructure at the most suitable locations is another factor to be considered. Because of that, it will require a great and very expensive effort to implement large-scale "wind-solar load matching" schemes anytime soon with existing technologies.

Nevertheless, having as a goal matching wind and solar power outputs will force us to look for and find the most suitable locations and appropriate technologies for this match. This is not going to happen overnight, but if we approach this solution seriously and intelligently, we will have a large-scale power output—nationally—that matches the grid power loads.

A Game Changer: Solar-Wind

A brand new development is making energy specialists to pause and wonder. The problems with wind projects lately, and falling solar panel prices, have prompted a number of wind companies to look into the competing technologies, and solar in particular.

The wind industry added more than 13 GW of capacity in 2012, but as a result of the fiscal cliff delays and hesitation, only part was installed in 2013. At the same time solar was surging. U.S. solar installations would surpassed new wind additions in 2013 for the first time ever.

So what is a wind guy to do? Sitting on their idling turbines and sucking their thumbs won't do any good, so some of them are jumping head first into solar technologies and projects. Most of them are looking into it, and some have hired solar specialists to help them evaluate and get into the solar business.

This is a new trend of adding or switching to solar, which is extending throughout the entire supply chain, from consultants to construction and engineering firms. In some extreme cases, wind companies are getting off the wind business all together.

Some of the more practical companies, like EDF Renewable Energy in San Diego, are developing solar-wind hybrid technologies and installing them in California's Mojave Desert. For example a 140 MW wind farm was constructed close by to 143 MWp solar power plant, which is actually one of the best ways to level the energy output of the combined power generation. Solar and wind work together very well, so this might be the best, if not only, way to a sustainable power generation.

And why not? Wind companies already have many of the in-house expertise and field skills needed to develop energy projects. Many of the disciplines used in wind power installation are similar to that used in solar. GIS mapping, real estate contracts, transmission issues, and even the customers, are the same for both types of energy.

And since wind and solar go hand in hand so well, the future for the wind-solar hybrids is just now starting. Smaller-scale solar and storage projects are shaping as source of future growth. Wind and solar generation can compliment each other on the grid, since the wind blows even when the sun isn't shining, and vice versa.

> *Solar is posed to become a close and reliable partner of the wind power industry by complimenting it, which would help both sectors to grow more than on their own.*

One of the greatest challenges we see in the solar industry is the fact that the utilities have the resources and know-how to design, install, and operate solar projects. So instead of contracting these projects, some U.S. utilities are building their own solar fields and taking over distributed generation (DG) projects. So, with other words, the solar field is getting crowded, which may be good on the long run.

In the short term, however, so many players coming in the solar field so quickly might create more problems than solutions.

And then we have the regulators, who at times create more problems than they solve:

The Energy (Commodity) Market

The energy (commodity) markets are part of the overall commodity (financial) markets scheme, which are driven by statutes that deal with financial markets operation and manipulation. These are the Commodity Exchange Act (CEA), the Energy Policy Act of 2005, and the Energy Independence and Security Act of 2007.

There are also regulations issued on the basis of these statutes by the Commodity Futures Trading Commission (CFTC), the Federal Energy Regulatory Commission (FERC), and the Federal Trade Commission (FTC).

Trading of energy stocks and commodities is a major driver in the function of the energy markets and contributes to their stability. The problem is that, as with any business, or any opportunity to make a quick buck, there are people who will try to take advantage of the situation. This is a problem that we must watch for, because it could cause a chaos in the energy markets.

For that purpose The Energy Policy Act of 2005 added anti-manipulation provisions to the Federal Power Act, 16 U.S.C. § 824v (2006), and the Natural Gas Act, 15 U.S.C. § 717c-1 (2006), which the Commission implemented in Order No. 670. This rule has been codified as 18 C.F.R. § 1c (2011) (Anti-Manipulation Rule).

The Anti-Manipulation Rule prohibits anyone from,

- Using a fraudulent device, scheme or artifice, or making any untrue statement of a material fact or omitting to state a material fact necessary to make a statement that was made not misleading, or engaging in any act, practice or course of business that operates or would operate as a fraud or deceit upon any entity;
- With an intentional or reckless state of mind, and
- In connection with a transaction subject to FERC jurisdiction.

The Anti-Manipulation Rule applies to any person, entity or form of organization, regardless of its legal status, function or activities.

In Order No. 670, the FTC models its Anti-Manipulation Rule on Securities and Exchange Commission (SEC) Rule 10b-5 in an effort to prevent (and where appropriate, remedy) fraud and manipulation affecting the markets the Commission is entrusted to protect, while providing a level of certainty to market participants that is beyond that which the Commission would be otherwise required to provide. Like SEC Rule 10b-5, FERC's Anti-Manipulation Rule is intended to be a broad anti-fraud catch-all clause.

The problem is that the majority of these rules and regulations are not-well-defined, imprecise, and in part fundamentally flawed. This has led to even more confusion in the energy markets.

For example, the CEA's anti-manipulation provisions (according to the courts' interpretation) are full of loopholes, which allow manipulation and/or allow the manipulators to escape punishment. On top of that, a recent court decision challenges the very Constitutionality of these provisions in criminal manipulation prosecutions. It also raises serious questions about the efficacy of the CEA as an anti-manipulation tool all together.

The FERC and FTC anti-manipulation rules have not been tested in litigation, and according to the experts, they are not well suited to address and punish manipulators of the energy financial markets. There are also a number of proposals on the legislative agenda, intending to address anti-manipulation measures in energy markets, but these also fall short of the goal of stopping criminal manipulations.

As a result of these inefficiencies, even blatant manipulators have a chance of escaping punishment, while the only option for the innocent is litigation in expensive and ineffective court action.

The existing judicial and regulatory decisions on manipulation, however, are like a sitting on a rickety three-legged stool. Each leg; artificial price, causation, and intent is shaky at best and ready to collapse when least expected.

Fixing markets manipulation law is not easy. The

process must start by recognizing the fact that there are different kinds of manipulation. These based on fraud must be clearly defined and separated from the market power activities. The respective statutes must be designed accordingly to address each individual activity.

Addressing manipulations that rely on fraud and deceit is a totally different matter than that of market power manipulations. Since this is the major point of confusion, the statute language must identify clearly the different types of trades. The emphasis should be on market power manipulators vs. fraudulent activities, clearly identifying the criminal points in each. This must be followed by appropriate punishment of the respective criminal intent.

Case Study: "Bang the Close"

In May, 2008 amidst record oil prices and calls from legislators to crack down on speculative oil trading and market manipulation, the Commodity Futures Trading Commission (CFTC) announced a wide-ranging probe into oil price manipulation. To that end, the agency initiated dozens of investigations of oil trading companies.

In July 2008, Optiver Holding, two of its subsidiaries and three employees were accused by the CFTC with manipulation and attempted manipulation of crude oil, heating oil and gasoline futures on the New York Mercantile Exchange.

The company acquired large trading positions shortly before the market closing, and then quickly executed the trades at a profit. This is a classic *"Bang the Close"* type of market manipulation, designed to bully the markets and generate huge profits. But according to CFTC this is Illegal trading.

A complaint was filed against the company's chief executive. the head trader, and the head of trading at a subsidiary. According to CFTC these people orchestrated the "Bang the Close" move by buying large positions just before markets closed, thus forcing prices up and then selling them quickly to drive prices down and pocket the difference.

A "Bang the Close" move was attempted 19 times during 11 days in March 2007. In five of those 19 times, traders succeeded in driving oil prices higher twice and lower three times. And yet, CFTC and other experts insist that there is no evidence that speculators are to blame for the rising oil prices. Trading information shows no correlation between investment activity and price swings, according to them.

However, there must be some correlation between a four-fold increase of investment money into oil futures and a four-fold increase in oil prices since 2004, which

should not be ignored. Many lawmakers, consumer rights advocates and oil industry analysts insist that market speculation and manipulation is at least partly to blame for the 2008 oil price crash.

Over the years, the CFTC has found isolated incidents of price manipulation, usually when oil producers attempt to control production and product moves in order to influence prices, or other cases of wrongdoing. Since 2002, the agency has charged 66 companies with some sort of energy market violations, including manipulation.

More recently, BP settled a civil lawsuit that alleged that the company tried to corner the propane market to inflate prices in 2003 and 2004. In response, BP agreed to pay a $303 million settlement, thus avoiding official court proceedings. Just another day on the global energy markets…

ENERGY SECURITY

Energy security, in general terms, is the effort to ensure reliable delivery of energy sources, as needed to run the national economy efficiently and consistently. In more detail, *energy security* means that the quantity, speed of delivery, and price of the energy supplies must be reasonable, thus maintaining a healthy, growing economy, and happy populace.

Note: Strictly speaking, energy security is *peripheral* to the energy markets, because it neither produces energy sources, nor generates or distributes any energy products. It, however, is essential to the function of the energy markets, because it adds a level of stability and control over national and international developments.

The energy security tasks include; securing the supply of energy sources from overseas suppliers, for which purpose, the U.S. military patrols the world's oceans and ensures the flow of energy sources (crude oil, natural gas, etc.) across the globe. This is a major effort that costs billions of dollars and uses millions of barrels of oil annually. We took a closer look at these activities in the previous sections.

Equally important is the within-country distribution of energy products. It won't do us much good to have a ship-load of crude oil on the docs, if we cannot deliver it to the refinery. Similarly, a large power plant won't be of much use, if the power lines are down as a result of a terrorist attack.

We distinguish two different types of energy security today; a) physical energy security, and b) cyber energy security, as we will see below. The physical security protection consists of ensuring proper and efficient function of the different elements of the hardware of national energy infrastructure. The cyber-security, on the other hand, is focused on ensuring proper and uninterrupted function of the software and other non-hardware assets that assist, or work in tandem with, the national energy infrastructure.

Physical Security

As with any tangible and intangible things in the capitalist society markets, energy security cost money and could even be considered *marketable* commodity. Although it might seem illogical to put a price on security, any security—including energy security—the experts have estimated the world's energy security market at over $50 billion, and expected to reach nearly $100 billion by 2020.

How did they reach these conclusions and derived these huge dollar numbers is a question that we cannot answer, nor dispute, since it involves a number of specialized disciplines and complex decisions. So, we just have to accept their estimate as a fact—$100 billion is a lot of money, which translates into a lot of activities, products, and services.

According to these same experts, North America is expected to own over 30% of the global share of the technology-based energy security market, by offering a myriad of new, innovative, and mostly untested products and services.

Our modern society relies on energy supplies to fuel the basic requirements; our comfort living, running businesses, providing transportation, communications, healthcare, personal security and public safety. Energy has become the backbone for the growth of the worlds greatest economies.

Many developing countries are also focusing on optimum usage of their energy resources for the entire nation development. But due to increasing threats across the countries these energy resources have become more vulnerable.

To protect the power plants and energy resources, most countries are focusing on securing them against any attacks; including physical and/or cyber attacks. These attacks could be very sophisticated in nature, and are often carried out by terrorists groups, cyber criminals, and entire counties—all using modern and well refined tools.

Different protective measures and solutions need to be designed and executed to protect the energy providers and resources against all kinds of threats, such as attacks on supply infrastructure, accidents, natural and unnatural disasters and rising terrorism and cyber attacks.

These are some of the drivers of the energy markets security principals. Other drivers are: political pressure, government regulations, pending terrorist threats, and other factors related to ensuring the security of the energy sector.

Increased number and level of threats of late are contributing to rapid growth of the North America's energy security market.

Power generating facilities and power infrastructure have been classified by government agencies as primary vulnerable targets of terrorists and cyber attackers. This has led to industry-wide push to deploy technology and methods in answering the call, which is fueling the growth of the market.

Governments in the developed countries, including the U.S., have developed laws and regulations that define security standards, which need to be deployed by the owners and operators of power plants and the related infrastructure. This includes facilities that develop, generate, and store nuclear energy and materials, and all oil and gas energy, thermal and hydropower and renewable energy entities.

Ensuring the physical security of these facilities requires a number of sophisticated technologies and approaches. These include, but are not limited to: microwave intrusion detection, perimeter fencing, IR isolation fields, video surveillance, motion detectors and other access control devices, air and underwater surveillance, fire detection and alarm systems, building management systems, and remote hazardous and nuclear materials detection technologies.

The rising investments in energy resources and infrastructure around the globe are triggering a sharp demand for providing their long term security. This, however, is not an easy thing to do, and is often a very challenging proposition that requires a lot of specialized effort and most advanced technological solutions.

The effort of providing security to the entire energy system is complicated by the fact that there are many different components, most of which are spread over large and often remote areas. These are the elements of the national power grid (high power transmission lines) with thousands of miles of wires stretched around the country and thousands of substations installed far away from populated centers.

There are also thousands of miles of pipelines, collection and distribution centers, etc. gas and oil transport infrastructure crisscrossing the country—mostly in remote areas. In order to ensure the safety of these installations, and avoid terrorist attacks, surveillance along with remote monitoring technologies have been increasingly deployed.

Video cameras, intrusion detection, supervisory control and data acquisition (SCADA), aerial and underwater surveillance are being used by the power companies to protect their infrastructure from attacks and natural disasters.

New and emerging technologies have been focused on in recent years, such as unmanned air and water vehicles, and underwater surveillance systems have been deployed to secure the perimeters in the remote areas of their installations.

The energy security market is projected to grow, but political instability and economic stagnation might reduce the estimates in the near future. Nevertheless, it is a needed measure in the fight of protecting our energy sources and way of life.

Note: The experts forecast the global energy security market to grow 8-10% by 2020. One of the key factors contributing to this market growth is the increased occurrence of cyber threats.

The global energy security market has also been witnessing the increase in the construction of smart grid projects. The thread and unpredictability of cyber threats poses a real threat to the growth of this market, so a lot of work is done in that area too.

The energy security is one of the new comers in the energy field, and as such it is immature and in many cases the results are uncertain. Nevertheless, the energy security field is growing in importance, which will force the energy security market to mature fast too.

The Principals

The number of participants on all levels is growing with time; the effort is becoming very important and of increased visibility. Various intelligent security solutions are under development, which enable the providers to integrates collect and analyze different scenarios through the data generated by their Supervisory Control and Data Acquisition (SCADA) networks and grids and are becoming a must.

The key companies dominating this space are: ABB Group, Aegis Defense Services Ltd., BAE Systems PLC, Cassidian SAS, Elbit Systems Ltd., Ericsson A.B., Flir Systems Inc., Honeywell International Inc., and Siemens AG.

Other companies, directly and indirectly involved in the sector are: Acorn Energy Inc., Agiliance Inc., Anix-

ter, Inc., HCL Infosystems Ltd., Intergraph Corp., Lockheed Martin Corp., Mcafee Inc., Moxa Inc., Northrop Grumman Corp., Qinetiq Group, Raytheon Co., Safran S.A., Symantec Corp., Thales Group, Tofino Security, and Tyco International Ltd.

Energy Security Types

In more detail, the global *energy security* market can be summarized in the sub-categories shown in Table 9-3.

Energy Security Risks

There are uncertainties and risks galore in the energy sector, which we cannot detail in this text. We actually wrote a whole book on the subject, *Energy Security for the 21st Century*, published in 2015 by the Fairmont/CRC Press, which details the number and degree of risks that the U.S. and global energy sectors are facing today.

The energy risks can be separated into risks created by, and/or affecting, the different sectors of our lives. Residential, commercial, industrial, transportation, governmental, etc. users of energy depend on reliable and cheap power and transport. Making the two major energy markets—power generation and transport—less variable and cheaper are the keys to ensuring well being to the people and economic prosperity for the country. And yet, we see a number of risks today and many more on the horizon, which threaten our energy and national security.

*The major risk for the energy markets is **uncertainty**, fueled by ignorance, negligence, greed, and misinformation*

Uncertainties

There are a number of internal issues in the energy sector that need better understanding:
- *Customers and Investors:*

One energy report, published in 2014 states that the consumers and investors alike have very little confidence in the future price and availability of energy supplies. The painful memories of what happen, as during the 1970s and 1970s the 1990s are still fresh in our minds. Large-scale investments were made at the time only to have the viability completely lost later on, when North American energy prices spiked form the lowest to the highest in the world almost overnight.

- *Energy policies*

The lack of confidence is exacerbated by the lack of proper energy policies to address the key environmental and energy issues. Rules and regulations are required

Table 9-3. The major energy security elements

Power Infrastructure
— Nuclear power plants
— Oil and gas power plants
— Thermal and hydro power plants
— Oil wells and rigs
— Refineries
— Renewable energy power plants
— Distribution lines
— Substations
— Fuel storage depots
— Pipelines
— Fuel transport vehicles

Physical security and safety
— Microwave intrusion detection
— Perimeter fencing and IR fields
— Secured communications
— Video surveillance and CCTV
— Transportation security
— Detectors and access control
— Air and ground surveillance
— Costal and water surveillance
— Fire detection and alarm systems
— Personnel tracking and RFID
— Identity and access management
— Building management systems
— Scanning systems
— CBRNE/HAZMAT
— Plant and personnel safety
— Biometrics and card readers

Network security
— Distributed denial of service (DDoS) protection
— Firewall
— IDS/IPS
— UTM
— SIEM
— Disaster recovery
— Antivirus/malware
— Incident management SCADA

Professional and specialized services
— Risk management services
— Armed protection
— Security system design
— Compliance, policy, and procedure consulting
— Auditing and reporting
— Integration and consulting, and
— Management and consulting services;

World regions
— North America
— Europe
— Middle East Africa
— Asia-Pacific
— Latin America

to direct us to the economic and environmental future. Consumers and investors must have the assurance that that fracking, for example, can be done safely under well-defined and enforced rules. And even more importantly, a renewables-focused energy mix, with the help of natural gas, is indispensable for the medium term.

Mergers

Government Accounting Office reported in 2008 that more than 1,000 mergers occurred in the petroleum industry between 2000 and 2007. These mergers were mostly between firms involved in crude oil exploration and production, and were generally driven by the challenges associated with producing oil in extreme physical environments such as offshore in deep water and increasing concerns about competition with large national oil companies. Other mergers took place in the segment of the petroleum industry that refines and sells petroleum products.

Most mergers were generally driven by the desire for greater operational efficiencies and cost savings. While mergers could help oil companies overcome some of these challenges, they also have the potential to increase their market share, sometimes at the expense of the consumers.

The U.S. Shale Revolution

We cannot ignore this monumental development in the U.S. energy sector; a trail blazer showing huge gains over the past decade. The sharply increased U.S. shale gas production has turned natural gas into a true buyers' market. The only suffering victim is the Henry Hub gas prices, which are basically speaking a victim of their own success.

Recently, the low energy prices have put the breaks on this progress. Now we wonder if America will continue producing 650 billion cubic meters of natural gas annually (bcm/y), and how much of it will be ex[ported as LNG? 125 bcm/y of LNG trains are awaiting FERC approval, but only half of that is realistic on the short term.

America is pushing towards 12 million barrels/day production levels, which gets us closer to energy independence. That, together with (possible passing of) the Keystone XI pipeline, is a chilling news for OPEC countries. mostly because of the damage these huge developments will do to oil prices. But maybe the worst for OPEC is the U.S. will use this bounty as a production hedge to send tankers to China, Japan, and Europe. This would basically undermine the controlling role of OPEC in the global energy markets.

Educational Reform

Educational issues have been receiving increased attention lately. We see educational specialists, government officials and all kinds and types of people describing the problems and projecting their opinions. While this all is necessary, talk is cheap, as they say, and a coordinated action is needed. Unlike the other risks, this one is easier to solve, since it is on a smaller scale and within the domain of the federal and state governments.

Just because it is easier to solve does not mean that it will be solved soon. There is, however, a good chance that good things will happen soon enough. Well educated engineers and work force are essential to the development and expansion of the U.S. energy markets, and ensuring their security. We, therefore, need to have an educational system that produces this type of professionals.

The Divisions

The greatest threats are from within our society. The United States is so powerful because it was *united*. Today, we are a nation divided. There are divisions among us that are stopping our progress as the most advanced—technologically and economically—nation in the world. These divisions generate hate and result in conflicts, including violent encounters.

The divisions are among all layers of our society, and are in many pieces.

> *Republicans and democrats; white and black, brown and yellow; Anglo-American, African-American, Mexican, Puerto Rican, Jewish, and other ethnicities; Christian, Muslim, Buddhist, and other religions; heterosexual and bisexual orientation; physical and psychological disabilities; male and female gender groups; educated and less-educated; with and without criminal record; and other divisions prevail in our society.*

Each of these groups thinks that their issues are the most important. That their cause has been shelved for too long, and must be brought up to the national discussion table and resolved ASAP. The rest; those who don't belong to any of these groups, are required to be politically correct, keep their mouths shut, say nothing and do nothing. The penalty for being politically incorrect is isolation, labeling, and even persecution. The free speech in the U.S. has a built in restriction—you can say whatever you wants as long as it is politically correct.

This colorful, heterogenic and immiscible mix of values and interests is causing havoc and turmoil in the socio-political and economic lives in the country.

Energy Divisions

As if this was not enough, we now have a set of new serious divisions, based on our personal interests and understanding of energy production and use, and the related environmental, safety and security issues.

Climate warming vs. environmental pollution, fossils vs. solar and wind, biomass and bio-fuels vs. hydrofracking, and on and on the list of energy issues goes. Different beliefs, combined with misunderstandings and misrepresentation of the issues at hand divide the society and complicate the pertinent issues even further. We now have many groups of people divided along the lines of energy and energy related issues and projects.

And the battles on the different fronts rage non-stop, weakening day after day the fabric of our society, converting the mighty United States into the Un-United States. So, now we have to deal with and manage these and many other internal divisions, on top of the pressing external risks and threats; some of which are even more serious as far as our energy and national security are concerned.

A nation divided is a nation in trouble... Internal divisions bring conflicts, which is what brought the mighty empires of the past to ruin. Are we headed towards the inevitable too?

Power Grid Security

The power grid, with its exposed, unprotected, and vulnerable infrastructure is an attractive target for vandals and terrorists. The attacks on the national power grid can be divided into two distinct categories:

- Physical attacks on the power grid infrastructure (distribution lines, substations, etc.), geared towards physical damage and destruction, and
- Cyber attacks on the grid infrastructure, conducted from remote computer terminals that are geared to interfere with the proper and safe power generation and distribution.

Cyber attacks on power plants have been successfully attempted on several occasions in the past with varying results. In all cases, they have shown that a remote attack, resulting in a great damage is possible. While still a real threat, cyber attacks seem to be properly addressed and well within our ability to control them...at least for now.

Physical attacks on the grid, however, are not so common and are not so easy to control. The grid is spread over thousands of miles across the entire country. There are over 2,000 power plants, 4,000 substations, connected by thousands of miles of high and low voltage distribution lines. How do you protect all of them non-stop day and night?

There is also reluctance on part of some of the principals to even discuss the issues. Thus, there is clear and present danger of physical attacks on the power lines, power plants, and sub-stations around the U.S.

The potential of great damage as a result of physical attacks on the grid became reality in the spring of 2013 after the attack on a major substation in California. The April, 2013 attack on PG&E's Metcalf transmission substation in the Silicon Valley started when somebody entered an underground vault and cut telephone cables leading to the substation. Several attackers then fired more than 100 shots at the substation wires and equipment, causing millions in damage.

Quick response from PG&E personnel was able to avert a massive blackout, but it took almost a month to repair the damages.

Note: The Metcalf substation is one of the several extremely critical, strategically positioned, power distribution centers in the country. Located in the middle of the all-important Silicon Valley in California, it serves several million people and many businesses, some of which are major names in the hi-tech industry. The selection of this particular target and the precision of the attack shows that the attackers are well trained and determined to do most harm. This was not a midnight target practice by a hand full of teenagers. It was premeditated and well planned attempt to disrupt the power supply to the area.

According to the insiders, this was the most significant incident of domestic terrorism resulting in a successful attack on the U.S. power grid infrastructure ever. It was done by well trained and well coordinated group of people determined to do most harm possible. Energy experts fear that the 2013 attack was a dress rehearsal, and that it is only a glimpse to what could happen in the future.

Amazingly enough, the attack was (intentionally or unintentionally) hidden from the public until it was finally publicized by *The Wall Street Journal* almost a year later. Several senators sent a letter to the Federal Energy Regulatory Commission and the North American Electric Reliability Corporation, stating, their concern that voluntary measures may not be sufficient to provide a reasonable response to the risk of physical attack on the electricity system.' The senators also warned that this attack is a wake-up call to the risk of physical attacks on the grid.

The senators' letter was followed by a meeting of several other senators, and industry and government officials in order to further discuss the adequacy of voluntary measures in protecting the national power grid.

The lawmakers may know something we don't, but apart from isolated vandalism, this is the first major attack on a key substation that got close to throwing part of the Silicon Valley into darkness. It is also the only one, to our knowledge, that has led to such serious damages to the infrastructure, which required major and lengthy repairs.

The utilities, and the national grid in extension are already heavily regulated as is, so it is not clear how additional regulations will add to the national power grid safety. The House Energy and Commerce Committee insist that protecting the grid remains a top priority and that the work on mitigating all emerging threats is ongoing.

The substation attack had the potential to cause a lot of damage, and the perpetrators have not been identified yet, so the threat cannot be ignored. These people may strike at any time anywhere in the country, so something has to be done. But what?

Smart Grid Cyber Security

Smart grid activities are shaping as one of the most promising as far as ensuring and improving the future grid functionality is concerned. From a technical stand point, this is doable, exciting, and profitable effort, but there is a big problem—the new smart grid is computer driven, which opens the new smarter power grid to cyber-attacks. Since the smart grid will grow no matter what, ensuring its security is going to be a very big business.

Unfortunately, this big business will not make money for the utilities that will end up paying for a major portion of it. Nope, the majority of cyber-security investments will go mostly to third party (peripheral) companies and enterprises, as needed to meet regulations and make the investors happy.

A major effort and expense will go into justifying the extra costs for preventing grid attacks, and the related incidents.

There are estimates that the $120 million spent on cyber-security products and services in 2011 will double by 2015. This number will grow over $6 billion spent mostly by the utilities and the U.S. government by 2020 to secure the grid and avert the consequences of a successful cyber-intrusion. World-wide, this number could grow to $80 billion for the same time frame.

The energy security expenses will exceed the total amount of money allocated for the smart grid itself.

This is truly big business that is taken very seriously by all involved. Nevertheless, it is a very risky business too, since the grid must be protected from any eventuality. Any and all cases of minor data theft to taking over grid assets, causing blackouts, or damaging substations and power plants equipment must be prevented, and/or dealt with successfully. A lot of research and trials and error will go into making sure this is done right.

A number of cyber-attacks and hacks at major U.S. organizations and installations have elevated the issue at the White House level. As a consequence, the Obama administration has issued an executive order demanding that all industries involved in the smart grid and power generation come up with a plan to deal with the threats.

How far and how fast this effort will develop remains to be seen, but it will not be a transparent process, since one of the key tenets of cyber-security is that cyber-security is most secure when nobody knows how secure it is. We surely will not learn much about the specifics of discovering, isolating, and eliminating and new intrusions and attacks. And since the methods of the intruders change daily, there will be a lot of changes and adjustments done to the cyber-security infrastructure as well.

The utilities are interested in, and required to secure their infrastructure by all means available and necessary. The effort will spread on all levels from installing fences and locking gates and doors to ensuring the safety of the smart meters and computer networks, and to implementing the latest cyber-intrusion detection software and schemes. All this will require new products and support services, similar in size and form to what the banking and telecommunications industries have been doing for the last 20 or more years.

The Regulations

The cyber-security spending in North America is controlled by the North American Electric Reliability Corporation (NERC)'s Critical Infrastructure Protection (CIP) requirements. In the U.S. and Canada these rules are implemented and enforced by stiff fines. There are cases of up to $1 million per day fines for utilities and other entities that do not meet the security guidelines.

The Department of Energy is dangling a $4.5 billion carrot in front of the cyber-security principals and wannabes. The DOE stimulus grants for a number of projects require cyber-security compliance, as outlined and coordinated by the U.S. National Institute of Standards and Technology.

Around the world, governments are working on their own regulations for ensuring the safety of their respective power grids. The European Union's cyber-security agency, ENISA, for example, is building a framework that is focused on a risk-based approach with a certain degree of freedom for the utilities and all companies involved, in order to do their job more promptly and efficiently.

Securing the new IT infrastructure of the power grid against cyber-attack is going to be big business, but that's not because it makes money for the utilities that are buying it. Instead, today's smart grid cyber security investments are mostly about meeting regulations, satisfying shareholders, and trying to justify the costs based on what it's trying to prevent.

Putting a dollar figure on averting the consequences of a successful cyber-intrusion is a risky business, of course. Utilities have to worry about everything from data theft to a full-blown attempt to take over grid assets in order to cause blackouts or overload generators.

Reports of cyber-attacks on U.S. infrastructure by foreign actors, as well as hacks at the Defense Department and Department of Energy, have brought the issue into focus this year, and the Obama administration has issued an executive order demanding that the industries involved come up with a plan to deal with the threats.

Recent research forecast that the trends can lead to market opportunity at a pretty big chunk of change. U.S. utilities are expected to spend a cumulative $7.25 billion in security from now until 2020, with distribution automation assets as the core focus.

Other estimates reach $14 billion through 2020, with about 63% of that spending aimed at the industrial control systems and SCADA networks used to control today's grid assets. Even wilder claims exist of an exuberant $80-90 billion global market by 2020, which exceeds total smart grid spending figures in most of the world.

North America accounts for over 40% of the global market presently, with Europe at 30% and Asia-Pacific at 17%. China is expected to overtake the U.S. as the largest market by 2020, leaving the Asia-Pacific region with 35% of the market. China, Japan, South Korea, and Russia will see growth rates of 40% and up in the next decade.

How far along the industry is toward that vision is hard to say, since one of the key tenets of cyber-security is that it is secretive and you just don't talk about cyber-security. And even less we know about the specifics of how the principals are discovering, isolating, eliminating and building new protections against new intrusions and attacks that change from day to day.

In North America, much of that spending is being driven by the North American Electric Reliability Corporation (NERC)'s Critical Infrastructure Protection (CIP) requirements. Covering the U.S. and Canada, these rules come with stiff fines of up to $1 million per day for utilities that can't prove they're meeting security guidelines, and newer versions add a lot more serial-connected smart grid assets to their purview.

The Department of Energy's $4.5 billion in stimulus grants also came with cyber-security strings attached, as outlined by the ongoing government-industry work being coordinated by the U.S. National Institute of Standards and Technology, or NIST.

Other parts of the world have their own regulations in place for how to secure the grid. The European Union's cyber security agency, ENISA, is building a framework that, in "contrast to the U.S.' strict regulatory path," is aimed at a risk-based approach that allows "a certain degree of 'freedom'" to the utilities and technology partners involved.

U.S. Grid Cybersecurity

Cybersecurity issues affecting the power grid remain as a critical challenge, for which government and industry are actively developing tools, guidance, and resources. These are necessary to develop robust cybersecurity practices within utilities and the related entities.

In response to Executive Order 13636, in 2014 NIST released the Framework for Improving Critical Infrastructure Cybersecurity. It offers a prioritized, flexible, repeatable, and cost-effective approach to manage cyber risk across sectors. This effort is built upon NIST's collaborative work with industry to develop the NISTIR 7628 Guidelines for Smart Grid Cyber Security.

In the same year, DOE released a second version (1.1) of the Electricity Subsector Sybersecurity Capability Maturity Model (ES-C2M2), which uses a self-evaluation methodology to help grid operators assess their cybersecurity capabilities and prioritize actions and investments for improvement. The ES-C2M2 provides a complementary, scalable tool for NIST Framework implementation.

To date, 104 utilities covering 69 million customers have downloaded the ES-C2M2 toolkit. Combined with the Risk Management Process that DOE released in 2012, and upcoming cybersecurity procurement language, utilities now have a holistic view of cybersecurity

best practices across business processes.

In addition, DOE required each recipient of SGIG funding under ARRA to develop a Cybersecurity Plan that ensures reasonable protections against broad-based, systemic failures from cyber breaches. DOE followed up with extensive guidance on plan implementation, annual site visits to the 99 recipients, and two workshops to exchange best practices.

As a result, recipient utilities are instituting organizational changes and leveraging new tools to strengthen organization-wide cybersecurity capabilities. Advanced technologies with built-in cybersecurity functions are now being developed and deployed across the grid.

For example, research funded by DOE has led to advancements in secure, interoperable network designs, which have been incorporated into several products, including a secure Ethernet data communications gateway for substations, a cybersecurity gateway (Padlock) that detects physical and cybersecurity tampering in field devices, and an information exchange protocol (SIEGate) that provides cybersecurity protections for information sent over synchrophasor networks on transmission systems.

In addition, the University of Illinois developed NetAPT, a software tool to help utilities map their control system communication paths, allowing utilities to perform vulnerability assessments and compliance audits in minutes rather than days.

There are many other companies and entities working on optimizing the cyber security of the U.S. power grid and its components. Our enemies and terrorists, however, are close behind, developing new tools in responses to our efforts to secure the power infrastructure. The race is on…

NPP

For the purpose of strengthening our energy security in response to events of the early 2000s, in 2006, the U.S. Department of Homeland Security (DHS) announced the National Infrastructure Protection Plan (NIPP). This is a comprehensive risk management framework that defines critical infrastructure protection (CIP) roles and responsibilities for all levels of government, private industry, and other sector partners. The NIPP builds on the principles of the President's National Strategy for Homeland Security4 and strategies for the protection of critical infrastructure and key resources (CIKR). The NIPP was revised and reissued in January 2009.

The NIPP fulfills the requirements of the Homeland Security Act of 2002, which assigns DHS the responsibility to develop a comprehensive national plan for securing CIKR, as well as Homeland Security Presidential Directive 7 (HSPD-7), which provides overall guidance for developing and implementing the national CIP program. In accordance with HSPD-7, the national infrastructure is divided into 18 distinct CIKR sectors, and CIKR protection responsibilities are assigned to select Federal agencies called Sector-Specific Agencies (SSAs).

The U.S. Department of Energy (DOE) has been designated the Energy SSA. In this role, it collaborates with dozens of government and industry partners to rewrite and revise the 2007 Energy Sector-Specific Plan (SSP). DOE also conducted formal review and comment periods for the draft 2010 Energy SSP.

CIKR protection and resilience are not new concepts to Energy Sector asset owners and operators. The Electricity and Oil and Natural Gas subsectors have faced challenges from both natural and man made events well before September 11, 2001. Since that time, the Energy Sector has made significant progress in developing plans to protect the energy CIKR and to prepare for restoration and recovery in response to terrorist attacks or natural disasters.

More recently, potential threats from cyber penetration have created new concerns and industry responses. Through the Energy SSP process, government and industry have established unprecedented cooperation and close partnership to develop and implement a national effort that brings together all levels of government, industry, and international partners. An updated, 2010 Energy SSP, was later on issued, which is a reflection of the partnerships and the achievements of the national security sector over the last three years.

Since energy runs the entire economy, our energy security is an integral part of the NIPP. We have written an entire book on the subject—Energy Security Now and Later, published by the Fairmont Press in 1915—in which we detail the different components of the energy production and distribution systems and outline the developments and risks within the energy sector.

A healthy energy infrastructure is one of the defining characteristics of a modern global economy. Any prolonged interruption of the supply of basic energy—electricity, petroleum, or natural gas—would do considerable harm to the U.S. economy and the American people.

Numerous characteristics of the Nation's energy infrastructure, including the wide diversity of owners and operators and the variety of energy supply alternatives and delivery mechanisms, make protecting it a real challenge.

One of the greatest challenges is the fact that the energy production and distribution infrastructure assets and systems are geographically dispersed. Thousands of mines, drilling sites, power plants, and millions of miles of electricity lines and oil and natural gas pipelines and many other types of assets exist in all 50 States and the Territories. In many cases these assets and systems are interdependent and the entire Energy Sector is subject to regulation in various forms.

ENVIRONMENTAL RISKS

For more than a decade, and especially since Al Gore made his movie *An Inconvenient Truth*, in 2006, our attention has been focused on the problem of global warming. Al Gore, the same guy who invented the Internet, had another vision at the time, that of 100% environmental apocalypse. As far fetched as the events in the movie might've seemed at the time, the movie caught the attention of the masses and even our leaders.

Only 10 short years later, the same events seem more real. This is because within the last decade we have witnessed some of the Al Gore's predictions come to pass. The movie all of a sudden takes a life of its own, and serves a good purpose as a measure of real present life events and many others to come.

Now our eyes and minds are open to see clearly the environmental events around us. The movie initiated a broad public fervor for regulating large portions of our economy, which now are becoming the law of the land, as needed to prepare us for the upcoming environmental crisis.

Figure 9-10. Just a reminder...

The environmental awareness is growing today to where the U.S. public demands are heavily affected by global events. Even though the majority of the disasters are happening across the world's oceans and may not reach this continent for many decades, we realize that we are affected.

As a result, several years back the Obama administration proposed cap-and-trade (CAT) legislation, which is a complex system where carbon emitters must buy permits to trade among themselves. This, according to some, would force us to sort out the issues and decide who will spend for expensive clean-up methods in order to find the cheapest way of reaching the desired results.

The opponents, however, consider this approach a straight tax on the U.S. economy—the greatest tax increase ever, as a matter of fact. The energy sector is affected more than most by this rule and is looking for ways to reduce the burden.

The final CAT bill was watered down to where it would not accomplish much, and finally stalled in the Senate. So all the environmental hoopla finally ended up in the waste basket…for now. Sooner or later, however, someone, somewhere will take it out, dust it off and put it on the agenda again.

Today the major environmental battles in the U.S. are focused on a number of projects, one of the major of which is bringing tar oil from Canada, with the Keystone XL pipeline rejected by Obama. Another one is extracting shale oil in the Rockies, which promises to be even messier to extract than the Alberta tar sands. The other huge environmental concern is the debate about drilling for the 16 billion barrels of oil in the Arctic National Wildlife Refuge (ANWR). The battles rage on and off, but will not stop…until the last drop of oil is extracted.

Most other countries are not even considering the environmental problems. China is talking about it, but continues building several coal-fired plants every week. Cuba is planning to drill for oil close to our offshore territory, and even Europe is firing its coal power plants. This way, even though we (the U.S.) clean our power sector, we get increased air pollution from far away China, and other environmental risks from our back yard neighbor, Cuba. With friends like these…

So why are we so worried about it? Oh, right, the environment, and its problems:

- *Global Warming* is the increase of Earth's average surface temperature due to effect of greenhouse gases (GHG), such as carbon dioxide. These gasses come from many places, but burning fossil fuels have been blamed for the greatest GHG emissions. Another culprit deforestation, which trap heat that would otherwise escape from Earth, is also contributing the climate warming via greenhouse effect.

Global warming is caused by both; natural events and human activity. Earth's climate is mostly influenced by the first 6 miles or so of the atmosphere which contains most of the matter making up the atmosphere. This is a very thin layer, which, viewed from space, would only be about as thick as the skin on an onion. The onion is the Earth, of course.

Yet, this onion skin has a killer effect on us; little ants crawling on the onion's surface. Realizing this makes it more plausible to suppose and demand that human beings can and should change the climate. We may not be responsible for the entire problem, but we cannot deny our participation in the atmospheric changes. A brief look at the amount of greenhouse gases we are spewing into the atmosphere (see below), confirms that we have affected the atmosphere.

- *Greenhouse Gases.* Green House Gasses (GHG) from coal and gas-fired power plants have been blamed for contaminating the atmosphere and causing the climate change. While this cannot be denied, we must also remember that there are other factors.

One of the major GHGs on Earth is the large amount of water vapor from the oceans and other water bodies.

Huge amounts of water vapor are created daily by evaporation from the world's water bodies, which occupy 2/3 of the Earth's surface. This is a natural phenomena, dating from the beginning of time, and not something produced directly by humankind. The problem is that while water vapor is constant, even slight increases in atmospheric levels of carbon dioxide (CO_2) can destroy the environmental balance and cause a substantial increase in temperature.

This is because the concentrations of the GHG gases is significant (albeit not nearly as large as that of oxygen and nitrogen (O_2 and N_2), which are the main constituents of the atmosphere). While oxygen and nitrogen are not GHGs, but are key elements of the environment and life on Earth, water vapor, CO_2 and other gasses are considered GHGs.

O_2 and N_2 have only two atoms per molecule, thus they lack the *internal vibrational modes* of molecules with more than two atoms. Water (H2O) and CO2, however, have large number of atoms, which have internal vibrational modes.

The vibrational modes of large-atom molecules like H_2O, CO_2, CH_4, etc. cause absorption and re-radiation of infrared radiation, which causes the greenhouse effect.

Water vapor is constantly present over the water bodies. It moves around with the wind currents, and undergoes changes with time, while CO_2 remains higher in the atmosphere for a very long time (hundreds of years) without much change. Water vapor easily condenses, depending on local conditions, usually taking down to Earth some of the air contamination in the area.

Water vapor is, therefore, a *reactive factor*, which tends to adjust quickly to the prevailing conditions, such that the energy flows from the Sun and re-radiation from the Earth achieve a balance, while at the same time changes the concentration of pollutants.

In all cases, water vapor has been a constant variable through the, while the CO_2 amount has been increasing through the last decades. Since the global warming has been increasing sharply during the recent decades, we make the conclusion that man-made CO2 emissions are causing it.

While this cannot be denied, we must also remember the combined effect of increasing water vapor, and melting glaciers, as the global temperature rises. This only aggravates the already serious situation, adding more water vapor to the atmosphere and accelerating the global warming further.

These are unpredictable and even controlling, rather than reacting, factors. More CO_2 we emit, and more water evaporates, higher the temperature required for a balance get, and as things are going now, we are quickly approaching the upper limits of this equation...

We have increased the atmospheric CO_2 concentration by over 30%, mostly during the 20th century, which is a significant increase on any time scale. Man-made activities are now considered responsible for most of this increase, because it is almost perfectly correlated with increases in fossil fuel combustion.

While water vapor and other natural emissions are still fairly constant, and are largely ignored for now, there are measured changes in the ratios of different carbon isotopes in atmospheric CO_2 that are consistent with "anthropogenic" (human caused) emissions.

'Doing business as usual', we will soon reach CO_2 concentrations higher than ever seen before, and which we assume will bring some horrible events to happen on Earth.

The total worldwide GHG emissions are estimated at 22 billion tons of CO_2 equivalent added to the atmosphere each year. About a third of this comes from electricity generation, and another third from transportation. The other third is from all other sources.

22 billion tons of anything dumped in the atmosphere in one single year is enormous amount to deal with. What it means is that in 10 years we dump 220 billion tons, and in hundred years, we have dumped 2,200 billion tons of CO_2 in the air we breathe.

The man-made emissions stay in the atmosphere, causing the total atmospheric levels of CO_2 to rise quickly and well beyond the safe levels. The current emissions level, and projected increase over the next hundred years, go out of the charts. This is startling and disturbing news, which is reflected in changes in the Earth's average surface and air temperatures.

While there are many reasons for global pollution and the related global warming, some of which are natural fluctuations, there is the undeniable effect of man's activities:

- *Rising Seas*, or inundation of fresh water marshlands (the everglades), low-lying cities, and islands with seawater are now evident in many places in the USA, Europe, and Asia.
- *Changes in rainfall patterns* are evident in increasing droughts and fires in some areas, while flooding persists in other areas.
- *Increased likelihood of extreme events*, such as flooding, hurricanes, etc. has been predicted by scientists all over the globe.
- *Melting of the ice caps* and loss of habitat near the poles has been also recorded. Polar bears are now endangered by the shortening of their feeding season due to dwindling ice packs.
- *Melting glaciers* have been increasingly observed in areas where significant melting of old glaciers is predominant.
- *Vanishing animal populations* is logical after widespread of their habitat loss.
- *Wide spread of diseases* and/or migration of diseases, such as malaria to new, warmer (recently) regions.
- *Bleaching of coral reefs* has been also recorded, caused by warming seas and acidification due to excess carbonic acid formation. Over one third of the world's coral reefs appear to have been severely damaged by warming seas.
- *Loss of plankton* is caused by warming seas. The 900 mile long Aleution island ecosystems of orcas, sea

lions, sea otters, sea urchins, kelp beds, and fish populations, is disappearing mostly due to loss of plankton. This leads to urchin explosions, loss of kelp beds, and their associated fish populations.

- *Warmer oceans and other large bodies* evaporate more water every single day. As the global temperatures increase, the amount of evaporated water from the large water bodies and the oceans increase too. The additional water vapor in the atmosphere adds to the green house effect and together with increasing CO2 volumes, further accelerate the global warming.

Is the problem fixable?

There is urgency in "fixing" the excess emissions problem by near term reduction of coal-fired electricity. Substituting coal with natural gas, solar and wind power can make substantial cuts in these emissions in the near term. Even then, we still would be above the acceptable limits of GHG emissions, because of natural gas burning and transportation vehicle emissions.

Energy efficiency is another means of emission reduction, which potential impact must be revisited and re-calculated. The U.S. National Academy of Sciences determined, as far back as 1991, that the U.S. could reduce current emissions by over 50% at a minimum cost to the economy by full implementation and use of practical and cost-effective efficiency improvements nationwide.

In conclusion:

- The producers of CO_2 emissions, and all interested parties, must agree that there is a problem, since the first step to progress in correcting a problem, is admission that there IS a problem.
- Then they must agree that their emissions are a major part of the problem, and that something has to be done to solve it.
- The GHG emissions data and its effects thus far are very convincing and hard to argue with, so we ALL need to agree that we have a problem, or continue arguing until we choke to death.
- Recently, the observation of record-warm years is statistically significant. Thus, it is dangerous (and statistically unlikely) for the events of late to be a random fluctuation.
- And most importantly, it is a proven fact that the CO_2 levels are rising dramatically. So, at the present level of fossil fuels use, the amount of GHGs in the atmosphere will soon exceed the amount of carbon in the living biosphere. This will certainly

have very serious, negative effects on life on Earth, as mentioned in several places above,

And there are other problems too:

Energy Technologies Risks

How clean and safe are the energy producing technologies? Let's clarify first what is safety in the energy sector? It may be best to explain with several examples of present day energy production.

Fossils Pollution

- Fossils production and use—especially coal mining and burning—emit huge quantities of gasses, many of which are toxic and/or considered green houses gasses (GHG). These gasses, emitted in huge amounts affect the global climate and contribute to its warming of late. The consequences of this warming are detrimental to some areas of the world, as discussed elsewhere in this text.
- Soil pollution is another serious consequence of fossils production, where thousands of acres of mountain tops and surface land have been erased from then face of the Earth, and the surrounding areas heavily contaminated with toxic liquids of mining and well drilling operations.
- Water table pollution is probably the most ignored, most disputed, and most urgent phenomena of today's energy markets, where rivers, streams, and drinking water reservoirs have been flooded with toxins from ongoing fossils extraction projects. People and animals have been suffering from the effects and some have even died.

Hydrofracking

Once a hydrofracking well is drilled and start producing, all that is visible on the surface is a small hole, or several holes, with a pipes on top. Clean and safe, no doubt. What happens underneath the surface, however, is anyone's guess. According to the well operators, the chemicals are pumped into the rock formations and diligently do their work by behaving according to engineering principles and within design parameters.

But a closer look underground reveals a maze of tunnels, cracks, voids, and all kind of other formations along the miles of pipe belching toxic chemicals all over the place. Large caverns could be filled with the chemicals, which after awhile could cause collapse of its ceiling.

A large hole could open in the ceiling all the way to the surface, with the chemicals blowing out of it, as it has been seen in many occasions. When many such cavities are enlarged, the land will experience a shock, which could result in an earthquake.

The used chemicals are stored in large waste chemical pit's dug into the ground. With time the retaining materials of the pits' walls and bottom disintegrate and the chemicals could spill into the ground and the water table. With chemicals coming from below and from above, the local soil and water table are under constant attack from all sides.

Although this does not happen often, the results of these attacks—whenever and wherever they get out of control—could be devastating for the local environment and people's properties and health.

Nuclear Accidents

It goes without saying that just mentioning the words Chernobyl and Fukushima brings images of misery, illness, and death. The source of these images is the nuclear power gone berserk. When that happens, people suffer and die...in the name of energy and energy production. This always brings the question, is it worth it? And the answer, my friend, is blowin' in the wind...

The Renewables

The renewables are proclaimed clean and safe, which is true to a point. Wind turbines are known to kill birds by the thousands, and people have lost their lives by falling off the tall structures. Wind mills have also collapsed in the past, causing additional property damage and taking lives.

Solar might be the cleanest and safest of them all, and yet we can quote many accidents involving solar panel installers falling off house roofs, or through skylights, solar technicians being electrocuted, workers being hurt by lifting solar panels, etc., etc.

There have been also incidents, albeit isolated, when solar activities have caused workers' death. One such case is the explosion at Mitsubishi Materials Yokkaichi polysilicon plant, in Mie Prefecture, on the main Japanese island of Honshu.

At 2:05 local time on January 8, 2014, during a scheduled maintenance of the plant's heat exchangers a loud explosion was heard and a fire started nearby. The explosion was caused by a chemical reaction involving trichlorosilane materials, while the heat exchangers were removed for cleaning, and resulted in the deaths of 5 workers and 12 injured. Three of the dead were directly employed by Mitsubishi Materials, and the other two were subcontracted workers. One of the injured, a 39-year-old man, was critically injured and was rushed to hospital by helicopter.

Following the accident, the complex was closed for any extended period of time and suspension of other work at the plant was necessary. The materials that caused the blast were removed from the site and taken outside the facility.

This was not the first explosion at the Mitsubishi Materials polysilicon plant, and the troublesome fact is that there are no guarantees that it will be the last.

The Mitsubishi plant supplies polysilicon for solar cells, which implicates the solar power industry in a deadly accident. But solar does not mix well with hurting and killing people, so this accident would have to be remembered as the way things should NOT be done.

The Solar Materials

Since we are on the subject of the safety of solar materials we need to expand the issues at hand and see how renewable, green, and safe the materials used in the production of renewable technologies really are.

In more detail this claim means that:

- Renewable: there are no supply chain problems, and no such are expected any time in the future. As we will see below, this is not so for most of the "renewable" technologies.
- Green: there are no negative environmental effects during the cradle-to-grave lifespan of these technologies. That their components are nearly perfectly clean to manufacture, transport and use, which as we will see below is not so, for most of the "green" technologies.
- Safe: there are no serious safety issues related to the manufacturing, transport, and use of these technologies and their products. This is also far from the truth, as we will see below.

Silicon

80% of all solar products contain silicon wafers. The production of silicon raw material starts with digging, transport, melting and processing of large quantities of sand. The bi-products of these processes, most of which generate, leak and release tons of dust, liquids and gasses into the soil, water table and the atmosphere, are difficult to access. The huge size of these undertakings alone is a good indication that surely they have impact (with a measurable negative component) on the environment, and the health of the workers and the locals.

So how renewable, green, and safe is the silicon material?

- *Renewable?* Maybe, but only special type and quality of sand can be used, and there is a limited number and amount of these sources on the Earth's surface. So, at the present level of use, the quality silicon will eventually be depleted. Lower quality silicon costs more to process, so prices will go up, until it becomes unprofitable to continue its production.

- *Green?* Not so much. Due to the extremely large quantity of raw and process materials, and the huge amounts of energy used to produce silicon, excessive air, soil and water contamination during these processes.

- *Safe?* Maybe, if properly manufactured and used. The concern here is with the mining and refining operations, where many people are involved (and some exploited) in the daily operations. Many of these people are exposed to unsafe working conditions, where accidents cost people's health and life.

Note: Regardless of how safe or unsafe silicon might be, it is much more so than most of the newer solar technologies. Some of these use toxic materials in their production and operation. Concentrating solar power (thermal) solar plants, for example, use hot, toxic, liquids to transfer heat to the boilers. Any accident with or around these liquids could prove fatal.

Other solar technologies use toxic and carcinogenic materials in the panels. Cadmium, arsenic, and others are widely used in SIGS and CdTe solar panels. Since these technologies are fairly new, we don't know what would happen with the poisons locked in the PV panels during 30 years of non stop on-sun operation.

Solar Cells Production

In the solar cells manufacturing facilities, all these materials, silicon, silver, all kinds of chemicals, and gasses are mixed, boiled, baked, sifted, crushed, liquefied, gasified, and solidified in a never-ending action. There are a number of liquid chemicals that are quite expensive in this process, so they are recycled and otherwise reused via complex distillation, filtration, and other chemical process, all of which use huge amounts of electric power, cooling water, and additives.

Millions of cubic tons of CO_2 and other toxic liquids and gasses are the by-products of these processes. Some are difficult to recycle or capture, so they are just freely exhausted or disposed of into the environment without any treatment, as in many Asian polysilicon plants.

Organic chemicals such as silane, dichlorosilane (DCS), trichlorosilane (TCS), silicon tetrachloride (STC), and many others are mixed, heated, evaporated, con-

densed, and transported, along with many inorganic liquids and gasses such as HCL, HF, HNO_3, H_2O_2, etc. Some of these are also dumped in the soil or vented into the air, where they mix within the soup of other gasses. The resulting mixture of organic and inorganic compounds sometimes stagnates over population centers as it becomes part of the atmosphere to accelerate global warming.

Hazardous liquid by-products, some dangerously corrosive and even pyrophoric (self-igniting), are created, transported, and processed along the complex process sequence. All of these require special handling, placing chemical and fire safety on the top of management's list of priorities in most facilities.

- *Renewable?* Maybe, if we don't consider some of the materials that are in short supply. See below.

- *Green?* Not so much, due to the extremely large quantity of processing materials and energy used to produce it, as well as excessive air, soil and water contamination during transport and processing.

- *Safe?* Maybe, if properly manufactured and used. The concern here is that in some countries, people are exploited in the daily operations. They are also exposed to unsafe working conditions, where accidents are daily occurrence.

Here is a description of some the materials used in the solar industry...you be the judge!

Silver, Copper and Other Metals

Huge quantities of silver, aluminum, copper and other metals, as well as plastics and many chemicals are needed to produce millions of PV modules. These metals are also dug out of mines on the earth's surface or deep into it. Here again, massive amounts of dirt are moved and processed, in order to get the pure metals out.

Here again, the dirt is dug out and processed via heat and chemicals, which processes emit great amounts of poisonous gasses and liquids, which then contaminate the air, soil and water table.

Large amounts of silver metal are used to provide good ohmic contact between the metal grid on top of the cell and the interconnecting wires. Silver is also used for reflective backing for mirrors in thermal solar plants.

According to the VM Group in London, over 1,000 tons of silver have been projected for making PV modules in 2011. This is over the 1.0 million kilograms of silver used today, and the amount is projected to triple by 2016, to nearly 3,000 metric tons of silver used for making PV modules worldwide every year. If we add that much more silver metal for coating heliostat mirrors, we end up with some very large numbers.

So the question here is, "How much silver will be left in the world in 10, 50 and 100 years if we use it at this pace?"

Silver is a precious metal, the price having gone from $4.00/ounce several years ago to over $40.00 today. Prices are expected to go higher with time. What will that do to PV module prices?

Similarly, prices of copper and aluminum metals have sky rocketed lately, and although there are large deposits of these left on Earth, the increased prices will play a significant role in PV manufacturing operations and cost.

Although there are significant amounts of these metals around the world, they cannot be considered "renewable," simply because they are in limited quantities.

Although they are non-toxic they could be considered "green" to use, but the mining and refining operations, with their excessive air, soil and water contamination, are far from green.

Safe? Yes, if we ignore mining labor exploitation and safety violations.

Cadmium

Cadmium (Cd) is a toxic, carcinogenic heavy metal, used in some PV modules, known as cadmium telluride (CdTe) thin film PV modules. Raw cadmium is generally recovered as a byproduct of zinc mining and its concentrates. Zinc-to-cadmium ratios in typical zinc ores range from 200:1 to 400:1. It is used for making NiCd batteries, electroplating, lasers, electronics, paints, and most recently in thin film PV modules.

In January 2010 USGS estimated ~600,000 tons of cadmium reserves worldwide (calculated as a percentage of available zinc reserves), of which the world mines and uses ~19,000 tons annually. If USGS is correct, in 32 years there will be no more cadmium.

Cadmium is a toxic heavy metal, which displaces zinc in many metallo-enzymes in the body, so cadmium toxicity can be traced to a cadmium-induced zinc deficiency. It concentrates in the kidneys, liver and other organs, and is 10 times more toxic than lead or mercury. Inhaling cadmium-laden dust leads to respiratory tract and kidney problems which can be fatal. Ingestion of significant amounts of cadmium causes immediate and irreversible damage to the liver and the kidneys. Japanese agricultural communities consuming Cd-contaminated rice developed itai-itai disease and renal abnor-

malities, including proteinuria and glucosuria.

Cadmium is one of several substances listed by the European Union's Restriction on Hazardous Substances (RoHS) directive, which bans certain hazardous substances in electrical and electronic equipment but allows certain exemptions and exclusions from the scope of the law. In February 2010 cadmium was found in an entire line of Wal-Mart jewelry, which was subsequently removed from the shelves. In June 2010 cadmium was detected in paint used on McDonald's tumblers, resulting in a recall of 12 million glasses.

Compared to other serious toxins, cadmium is more dangerous as it accumulates, and is not dissipated. Even a negligible dose of cadmium in the air or water, inhaled or ingested daily, will accumulate to eventually reach toxic levels, causing cancer or organ failure.

USGS says, "Concern over cadmium's toxicity has spurred various recent legislative efforts, especially in the European Union, to restrict its use in most end-use applications. If recent legislation involving cadmium dramatically reduces long-term demand, a situation could arise, such as has been recently seen with mercury, where an accumulating oversupply of by-product cadmium will need to be permanently stockpiled."

So how "renewable," "green" and "safe" are the cadmium-based PV technologies? From the above facts we see that cadmium is not a "renewable" commodity per se, but we don't foresee a shortage during the next 3 decades. It is, however, far from "green" or safe."

Tellurium

Tellurium (Te) metal is produced by refining blister copper from deposits that contain recoverable amounts of tellurium. Relatively large quantities of tellurium are also found in some gold, lead, coal, and lower-grade copper deposits, but the recovery cost from these deposits is too high to be worthwhile.

Tellurium is used mostly in making steel and copper alloys to improve machinability, in the petroleum and rubber industries, and for making catalysts and some chemicals. One of the rarest elements in the Earth's crust, it is found in considerable quantities as a secondary metal in mining operations.

The world produces 100-200 tons of tellurium annually, and while its total availability is uncertain, we do not foresee a shortage anytime soon. Tellurium is used as cadmium telluride in manufacturing thin film PV modules. It is mildly toxic material; nevertheless, utmost precaution must be taken when handling its pure form or its basic compounds as contained in the PV modules.

Although there is significant amount of tellurium around the world, it cannot be considered "renewable," although we do not foresee a shortage anytime soon. Due to its toxic properties it is not "green," nor "safe." Tellurium and its compounds are known to cause sterility in men working with tellurium-containing materials, even under strict monitoring conditions such as in semiconductor fabs and hard disk manufacturing operations.

Selenium

Selenium (Se) is a non-metal that is chemically related to sulfur and tellurium. It is obtained by mining sulfide ores, and is used in glassmaking, metallurgy, and pigments. While toxic in large amounts, trace amounts of selenium are needed for cellular function in most animals.

Selenium toxicity was noticed first by doctors who found increased sickness among people working with it. A dose as small as 5 mg per day can be lethal, causing selenosis. Symptoms include a garlic odor on the breath, gastrointestinal disorders, hair loss, sloughing of nails, fatigue, irritability, and neurological damage. A number of cases of selenium poisoning of water systems were attributed to agricultural runoff through normally dry lands.

Selenium quantity and price depend on mining operations of other metals and minerals, but we don't foresee a shortage anytime soon. Its use is estimated at 1,500- 2,000 tons annually. Though there is a significant amount of selenium worldwide, it cannot be considered renewable, green, or safe.

Arsenic

There are over 1 million tons of arsenic (As) worldwide, of which 54,000 tons are extracted annually, mostly in the form of arsenic sulfur compounds. Another 11 million tons might be recovered from copper and gold ores. The main use of metallic arsenic is for strengthening alloys of copper and especially lead used in automotive batteries. Arsenic has proven toxic and poisonous qualities. Although there is significant amount of arsenic around the world it cannot be considered "renewable." Due to its toxic properties it is not "green," or "safe."

Gallium and Indium

These and other hard-to-find-and-isolate mildly toxic metals are also presently used in significant amounts in thin-film PV modules manufacturing processes. These elements are rare, so they cannot be considered "renewable." They are mildly toxic but could be qualified as "green," or "safe," if properly produced and used.

EVA and Other plastics

A number of organic materials (plastics) are used throughout the entire PV module manufacturing process. They are too varied and complex to qualify or quantify in this text. Since their production is based mostly on fossils (extracts from crude oil, coal and such) they are not "renewable," but we don't foresee shortage anytime soon.

Because they are manufactured with the help of poisonous solvents and other toxic materials, we'd have a hard time classifying them as green or safe. They are, however, safe enough to work with, following basic precautions like wearing gloves and masks.

Due to outgassing, their safety in long-term operations has been questioned and needs more research—especially in light of the large-scale PV installations in the deserts, where organics are most vulnerable and unpredictable.

Glass and Aluminum

The top surface of the silicon PV modules is covered with glass, while most thin film PV modules take two glass sheets—top and bottom. This is a lot of glass. Entire mountains of sand are melted and refined in the glass manufacturing process, where many other chemicals and a lot of energy is used as well. Many tons of toxic gasses, liquids, and solids are produced as well. So, although glass is clean and green during use, its production leaves a long and deep footprint of environmental damage.

The back surface of the silicon PV modules is usually covered with a thin aluminum sheet. The production of aluminum metal also requires digging deep into the ground in aluminum ore mines, where thousands of tons of dirt are moved to get some of the aluminum ore contained in it. The ore is then purified chemically and the aluminum metal is separated electrochemically from the impurities.

This aluminum production process requires enormous amount of electric energy. As a matter of fact, it is one of the most energy intensive metal working processes today. And again, tons of toxic gasses, liquids and solids are produced along each step of the aluminum mining, refining, and shaping process.

Aluminum is also clean and green during operation, but we should not ignore its long and deep environmental footprints.

One can easily see that the complications here are endless, and interwoven so much so that we could write an entire book on the properties and effects of the different combinations and permutations of the availability, mining dangers, toxicity, adverse short- and long-term environmental and health effects, improper field use, etc. of the materials and components used in making and using PV cells and modules.

Many questions remain to be answered in these areas, so we must insist that these issues are brought out for honest discussion by all parties involved, including manufacturers and proponents of the PV technologies.

Note: At the time of this writing, several manufacturers have applied for "non-toxic" certification of their PV modules. The trend will continue until it becomes an industry-wide standard, and we envision different types and levels of "toxicity" assigned to different types of PV modules in the not-so-distant future.

The full implementation of this process will take awhile. It will be ignored by some, and resisted by others, but it is unavoidable on the long run. It is a step in the right direction, because the consumers must be fully aware of how renewable, green, and safe the products they use are and what to expect in the long run.

The Externalities

Using coal, natural gas, oil, nuclear power and other energy products and services comes at a price. Actually we pay two prices for these commodities: one we pay with dollars for their use, and the other we pay with our health and even our lives for being around them.

There is not much open discussion, and not much public information on the matter, but what is available is surprising and terrifying in its magnitude.

Coal Externalities

This is the skeleton in the closet of the coal industry. In economic terms, an external cost, or externality, is a negative effect of an economic activity on a third party or the society in general. When coal is mined and used to generate power, for example, the external costs include the impacts of water pollution, toxic coal waste, air pollution, and the long-term damage to ecosystems and human health.

The list of such damages is long, and we cannot possible look at all items and cases in that list, so we will limit this writing to the very basics, as needed to underline the importance of this subject:

• Greenpeace released an analysis in 2008 showing that the global total cost of coal, including the ex-

ternalities, was at least $450 billion annually and growing. This number was later on updated to $500 billion in 2011 by a Harvard study group. Today, it is most likely much higher, because although the U.S. is reducing coal use, other countries (China and India in particular) are adding two-three coal-fired plants to their power mix every week.

The researchers arrive at the externalities figures by looking at CO_2 and other pollution damages, resulting in health and property costs from mining and power plant activities—including accidents, air, soil, water pollution, etc. The first report was released at the time when Industry Ministers from 20 big GHG emitting countries met in Warsaw with the world's climate-polluting industries. It was agreed then that the relentless expansion of the coal industry is the single greatest threat to increasing the damages of the global climate change.

Coal is presently the most environment-polluting fossil fuel, responsible for one third to one-half of all CO_2 emissions, which are projected to increase to 60% of the total by 2030.

Thus, reducing coal burning will benefit not only the global climate, but will also reduce the other external impacts and the related costs which everybody else has to pay for. In calculating the $450 billion annual global cost of coal figure, the focus was on the cost for rectifying damages attributable to climate change, human health impacts from air pollution and fatalities due to major mining accidents, and other factors for which reasonably reliable global data is currently available.

The above dollar number was derived by taking into account 90% of the global emissions and looking at the damages, which are projected to rise significantly, due to the impacts of climate change. With these changes, the total number of global cost of coal is likely to increase sharply...unless its use is reduced and/or the climate change is stopped soon enough.

The projected impacts of climate change include billions of people who will face water shortage, while others will be flooded by rising ocean levels. This also includes the hundreds of millions who will be threatened by food insecurity and hurt by exceedingly extreme weather events, such as storms, earthquakes, tsunamis, etc.

• There are, however, scientists and specialists on the matter who insist that it is already too late now. There are climate models showing that no matter what we do now—even if we stop all CO_2 and other GHG emissions immediately—we have already triggered a massive global warming and other environmental changes that can be reversed only by reducing the GHG gasses already in the atmosphere. Doing this is simply impossible to achieve, so no matter what we do, we cannot stop, let alone reverse the environmental damages.

- The negative impacts of coal are not only related to global climate change. Large geographic areas are affected and changed forever by blowing apart entire mountain tops in order to make room for mines and processing plants. Coal and its gaseous, liquid, and solid byproducts from these pollute the local soil, agricultural fields, water resources. This, in addition to contaminating the local air which causes black lung disease in the vicinity.

A dramatic increase of air pollution in many Chinese cities is a good example of such large scale damages to environment and humans alike. Entire river systems and lakes are lost to pollution from mining operations in China and other countries as well.

In the spring of 2015, torrential rain in Vietnam caused sludge from mine storage areas to spill onto local communities, creating immediate and ongoing health dangers and environmental hazards. Operations, in the 4,000 miners coal mine were suspended, but the source of the deluge was not stopped for days. The flow of electricity in the affected areas was disconnected or reduced. The pace, large scale, and long term effects of this disaster will be evident for years to come.

In the fall of 2015, a similar collapse of the dam of a mine sludge storage pool, swamped an entire town in Brazil. Over a dozen people were killed after a wall of sludgy mining spoils cascaded down a valley, wiping out everything in its wake and burying people alive.

- The new 'clean coal' technologies have the potential to sharply reduce CO_2 emissions from coal-fired power plants, but the industry experts http://www.greenpeace.org/international/PageFiles/31030/false-hope.pdfclaim that the newly developed Carbon Capture and Storage (CCS) techniques have their own problems that could lead to another set of dangers. The CCS technology is immature, unproven, and contains inherent risks. In addition, it comes with an enormous price tag and a long process and equipment development time period.

One thing everyone who is well informed on the CCS technology agrees is that the global GHG emissions must start declining soon, and in best of cases the new CCS approach is in no position to play a significant role in making this happen any time soon for the above mentioned reasons.

- Greenpeace insists for the world governments to see and agree on a "climate vision" that is to address immediately the threat of serious global emissions peaking by 2015. Developed countries must agree on immediate reduction targets in the 30-40% range, if any significant results are to be achieved before it is too late. How this is to be done and how much will be done, if any at all, is still uncertain, but the talks continue… Yes, we are still in the talking stages.

Note: Talk is cheap, while the actions show a totally different scenario. While the U.S. is reducing the use of coal for power generation, China seems to be racing to the finish line and get all world records in this area. There is an astounding 155 coal-fired power plants that received a permit in 2015 alone.

The total capacity of the new fleet is close to 40% of all coal-fired power plants in the U.S.

If all permitted coal-fired power plants in China are built, which is not a given, they would increase the global pollution levels by 10-15%. Fortunately, there are indicators that point to some reason and order coming in China's relentless drive to increase fossil fuels use.

China's economic slowdown of late, and the government's pledges to use more renewable and nuclear energy make some of the country's existing plants (including the 155 new ones) too expensive and even unnecessary for now. We, however, never know what will happen in China next. As any communist (or semi-communist in this case) system, it does not always comply with the world's standards. So, new the coal-fired power plants are ready for construction…just in case.

There are a number of environmental changes of late, some of which are attributed to coal mining and burning. We cannot list them all, let alone discuss each in detail in this text, but here is a sample of what is happening and what could be expected to happen in the future.

Habitat Destruction

The majority of the coal mining in the U.S. is surface mining, which includes strip mining and mountaintop removal. In these cases, the original ecosystem which once was on the surface is destroyed and removed in the process of removing the surface dirt and then mining the coal underneath.

Mining destroys fish and wildlife habitat, which has rippling effects not only on their populations, but on the local population that rely on them. In addition to the mine site itself, coal mining affects the surrounding areas too. Air pollution, wide spread coal dust, mine runoff discharge into nearby water bodies, etc. damages are the usual results from coal mining operations.

Coal dust particles, chemicals, and sediment from coal mining discharged in the local rivers and lakes can reduce life expectancy of fish, damage their immune systems, and suffocate fish eggs.

After the coal in the mine is depleted coal companies usually restore the surface to a certain level and plant vegetation in an attempt to restore the original ecosystem. This, however, is difficult and sometimes impossible, especially in wetlands areas.

It is very unlikely that any amount of effort or money could restore the environment and the wild life in it exactly the way it was during pre-mining times. Because of that we must accept the fact that mining operations, if and when approved, will leave a permanent scar in the Earth's surface, which might take centuries to heal completely.

The worst part of these disasters is unplanned and uncontrollable release of contaminants in the environment, as in the examples below.

U.S. Case Studies

In the summer of 2000, a coal slurry impoundment owned by Massey Energy in Martin County, Kentucky, broke into an abandoned underground mine below. The slurry came out of the mine openings, sending an estimated 306,000,000 gallons of toxic slurry down two tributaries of the Tug Fork River. Wolf Creek was oozing with the black waste. The Coldwater Fork, a 10-foot wide stream, became a 100-yard expanse of thick slurry.

The spill was over five feet deep in places and spilled into residents' backyards. It polluted 400 miles of the Big Sandy River and its tributaries, and spread to the Ohio River. The water supply for over 27,000 residents was contaminated, and all aquatic life in Coldwater Fork and Wolf Creek was killed.

The spill was 30 times larger than the Exxon Valdez *oil spill, and was considered one of the worst environmental disasters ever in the southeastern U.S., according to the EPA.*

But then in 2008, the record was broken by the Kingston Fossil Plant coal fly ash slurry spill. An ash dike ruptured at an 84-acre solid waste containment area at the Tennessee Valley Authority's Kingston Fossil Plant in Roane County, Tennessee. 1.1 billion gallons of coal fly ash slurry was released into the surroundings.

The coal-fired power plant, located across the Clinch River from the city of Kingston, uses slurry storage ponds to dewater the fly ash, a byproduct of coal combustion. The slurry is stored in wet form in dredge cells, and when the cells were broken, the mixture of fly ash and water traveled across the Emory River and its Swan Pond embayment, on to the opposite shore, covering up to 300 acres of the surrounding land, damaging homes and businesses.

The slurry then flowed in the nearby waterways; the Emory River and Clinch River, both tributaries of the Tennessee River, contaminating everything its wake. The cleaning effort and the damages remain to be finalized.

Then, in the summer of 2015, we saw another example of large scale coal-related contamination. A work crew broke the containment basin full of sludge from old coal mining operations in Durango, Colorado. The Entire river system in several states was contaminated with sludge, containing, arsenic and other toxic materials.

The estimates are that over two million gallons of this toxic waste found its way into the river. The spill was caused when the EPA was investigating the abandoned Gold King Mine, during which they accidentally breached a debris dam that had formed inside the mine. This triggered the release of the waste sludge from the containment into Cement Creek, a tributary of the Animas River in San Juan County.

Ironically, the goal of the EPA effort was to find a safe way to pump out the wastewater from the mine, which was abandoned in 1923 and treat it for safer use. Good intentions turned into evil disaster... This only confirms that we are dealing with mighty forces—gaseous, liquid, and solid—which we know precious little about, and which we cannot expect to control.

Health Effects

Mining and burning coal in coal-fired power plants release a number of toxic pollutants, some of which rise in the air, and remain behind as liquid or solid waste materials and chemicals. These pollutants are responsible for a large number of illnesses and premature deaths, both to people directly involved in the industry, people who live nearby, and people worldwide.

• *Coal dust* in mines and near storage and transport facilities contributes to serious respiratory illnesses such as asthma and Pneumoconiosis (black lung). Solid combustion wastes such as fly ash pollute groundwater near storage facilities, contaminating individual and community water supplies.

Airborne pollutants have a larger footprint. Despite air pollution regulations, toxic emissions (soot, sulfur dioxide, nitrous oxides) from coal-fired power plants are estimated to be responsible for thousands of deaths due to lung disease each year in the US and Canada. A government study in Ontario found that the coal-fired plants in that province alone were responsible for an average annual total of about 660 premature deaths, 920 hospital admissions, 1090 emergency room visits, and 331,000 minor illness.

• *Coal combustion emissions* released into the atmosphere contain nitrous oxides which are responsible for industrial and urban smog, sulfur dioxide which is the primary reactive agent behind acid rain, mercury which accumulates in the food chain, and large amounts of carbon dioxide which is the most important greenhouse gas contributing to climate change. Coal mining itself also releases significant amounts of methane, another extremely potent greenhouse gas. Coal mining is responsible for over 25% of the energy-related methane emissions in the US.

Coal combustion wastes (CCW) include ash, sludge, and boiler slag left over from burning coal to make electricity. These wastes (120 million tons/year in the US) concentrate toxins such as arsenic, mercury, chromium, cadmium, uranium and thorium. In addition, these wastes create an expensive storage problem "in perpetuity."

Coal-fired power plants are also a major source of atmospheric mercury, which accumulates in the food chain and can damage the developing nervous systems of human fetuses, as well as leading to reduced immune function, weight loss, reduced reproduction rate, mental defects and other neurological problems.

There are a number of health effects and illnesses attributed to the pollution released during coal mining, preparation, transport, combustion, and waste removal and storage.

Some of the negative coal-related health effects and illnesses in the U.S. are:

• Respiratory illnesses like asthma OCPD and such, caused by particulates, ozone, and sulfur dioxide emissions.
• Black lung from coal dust
• Congestive heart failure, caused by particulates and carbon monoxide emissions
• Non-fatal cancer, osteroporosia, ataxia, renal dysfunction, caused by benzene, radionuclides, heavy metals, etc. emissions and waste materials
• Chronic bronchitis, asthma attacks, etc., caused by particulates and ozone emissions
• Loss of IQ, caused by excess air and water pollution
• Nervous system damage, caused by mercury releases in the drinking water.
• Terminal cancer, caused by air and water pollution
• Reduction in life expectancy, caused by particulates, sulfur dioxide, ozone, heavy metals, benzene, radionuclides, etc. emissions and releases.
• Degradation and soiling of buildings that can effect human health, caused by excess sulfur dioxide, acid deposition, and particulates emissions
• Ecosystem loss and degradation with negative effects on human health and quality of life, caused by excess emissions and releases of gasses, liquids, and solids.
• Global warming, caused by large quantity of carbon dioxide, methane, nitrous oxide, and sulfur dioxide emitted   by coal-fired power plants.

Coal Related Illnesses

Coal-fired power plants emit that may act as triggers to a number of medical conditions in humans and animals. These pollutants include sulfur dioxide, nitrogen oxides, and particulate matter. In addition, the carbon dioxide emissions from coal accelerate global warming, which is likely to increase the concentration in air of pollen from some plants, such as ragweed, and thereby contribute to the development of additional asthma and other conditions.

Chronic inhalation of coal dust causes several lung disorders, including simple coal-workers' pneumoconiosis, progressive massive fibrosis, chronic bronchitis, lung function loss, and emphysema.

We will review some of the most important illnesses related to coal mining and use, such as:

Asthma

Coal dust is one of the primary causes of asthmatic conditions for coal miners, people living and working next to coal mines, or living in areas with high pollution levels. There are a number of cellular actions, interactions, and conditions related to coal dust toxicity on human tissues, and the lungs in particular. The dust particles are known to cause lung disorders on cellular level like; macrophages and neutrophils, epithelial cells, and fibroblasts.

Inhaling coal dust particles induces cellular and non-cellular activities in the lungs, and may be involved in the damage of lung cells as well as some macromolecules including α-1-antitrypsin and DNA.

Recent studies with coal dusts show its effects on important leukocyte recruiting factors, such as: Leukotriene-B4, Platelet Derived Growth Factor, etc. Coal dust particles also stimulate the macrophage production of various factors with potential capacity to modulate lung cells and/or its extracellular matrix

In more practical terms, asthma is a chronic disease of the lungs characterized by inflammation and narrowing of the airways. Airway inflammation in asthmatics causes swelling in the throat that narrows a bronchial tree that has been previously sensitized to inhaled irritants, including many air pollutants. Exposure to an inhaled irritant causes further narrowing of the airways and the production of mucus that makes airways even narrower. A vicious and very dangerous process for all affected.

Patients with asthma experience recurrent episodes of dyspnea (shortness of breath), a sensation of tightness in the chest, wheezing, and coughing that typically occurs at night or early in the morning. This can lead to hypoxia (low blood oxygen level), hypercarbia (high blood carbon dioxide level), and respiratory acidosis (acidification of the blood caused by carbon dioxide retention) that may, in turn, cause cardiac arrhythmias and other medical emergencies.

During severe attacks, the lungs also fail to perform their task of exchanging carbon dioxide, produced by metabolic processes in the body, for live-giving oxygen. This, combined with other conditions, and the resulting panic, could prove fatal in some cases.

There are about 25 million asthmatics in the U.S., including 6-7 million children. The Centers for Disease Control and Prevention report that the number of persons with asthma increased by 84% from 1980 to 2004. More than half of the U.S. states report that 8.6% or more of their inhabitants have asthma. These high-asthma states are clustered in the northeast and Midwest.

Oxidative stress

"Oxidative stress" is the term used to describe the physiological state characterized by an excessive concentration of oxidizing free radical molecules. Oxi-

dative stress, or the possibility that oxygen, or reactive forms of oxygen, to be toxic to certain cellular functions has been known over half a century. Research on the importance of highly reactive forms of oxygen, known as oxygen free radicals, in biological systems is ongoing, but we now know that some of these free radicals exert critical controls over normal cellular metabolic process and cellular signaling. This abnormality can cause tissue damage and illness

Oxidative stress is usually created when oxygen ions, free radicals, or other reactive species are produced in excess of the body's ability to fight with remove these molecules. Genetic polymorphisms responsible for controlling the inflammatory response also increase an individual's susceptibility to the respiratory effects of ozone.

Studies have shown differences in the susceptibility to ozone (O_3—a form of reactive oxygen) that are due to polymorphisms (subtle differences in genes that control the expression of a trait) in the genes responsible for dealing with oxidative stress.

Thus, the probability that an individual will develop an oxidative stress symptoms or illness depends on exposure to a trigger, such as ozone, and the individual's susceptibility to that trigger, i.e., a complex combination and interaction between genetic and environmental factors.

Our current understanding is that oxygen free radicals are important contributors to the initiation and propagation of a several diseases, including; cardiovascular and pulmonary disease, atherosclerosis, hypertension, rheumatoid arthritis, diabetes mellitus, neurodegenerative disorders such as Alzheimer's disease and Parkinson's disease

Since coal-fired power plants emit large quantitiea of reactive oxygen, or ozone, we need to be aware of the consequences of living close to coal-fired power plants, or in cities with high ozone pollution levels.

COPD

Chronic obstructive pulmonary disease (COPD), unlike asthma which is a reversible condition, is permanent damage to the airway blamed on inhaling coal dust. It consists of a scary chronic obstructive airway condition characterized by narrowing of the airway passages. Smoking tobacco is the most important risk factor for the development of COPD, and approximately 85% of all cases of COPD are attributed to this single, preventable cause.

Data that have emerged during the past several years have shown that there is a smaller but important link between air pollution, including pollutants produced by burning coal, and the subsequent development of COPD exacerbations.

Cancer

The National Cancer Institute estimates that in 2008 alone there were 215,020 new cases of lung cancer, the leading U.S. cancer killer in both men and women, with 161,840 deaths annually. Smoking tobacco, radon and other radioactive gases, second hand smoke, asbestos, arsenic, nickel compounds, and other airborne organic compounds have been identified as risk factors for developing lung and stomach cancer.

New data from several large epidemiological studies show that air pollution may also be a risk factor in these cases. Since air pollution is often as a result from coal-burning emissions, it is obvious that something has to be done quickly to reduce it.

Coal dust has also been blamed for large number of lung and stomach cancer occurrences. Since coal miners are most affected by the coal dust and other coal-related illnesses, NIOSH' *Criteria for a Recommended Standard—Occupational Exposure to Respirable Coal Mine Dust*, issued in 1995 recommends:

1. Exposures to respirable coal mine dust should be limited to 1 mg/m^3 as a time-weighted average concentration for up to a 10 hour day during a 40 hour work week;
2. Exposures to respirable crystalline silica should be limited to 0.05 mg/m^3 as a time-weighted average concentration for up to a 10 hour day during a 40 hour work week;
3. The periodic medical examination for coal miners should include spirometry;
4. Periodic medical examinations should include a standardized respiratory symptom questionnaire;
5. Surface coal miners should be added to and included in the periodic medical monitoring

This is one of the tightest health specs, and best controlled mining work environment, in the world. Unfortunately, the GHG emissions from coal-fired power plants are not this easily controlled, so there are still many cases of coal-emissions related illnesses and death among miners and others affected persons in some areas of the U.S.

The situation is much worse in other countries, especially the developing nations, where the occupational exposure limits are too lax, or non-existent. The air pollution in China, India and other major countries is above any and all acceptable limits, and where coal-mining and burning-related illnesses and death are the norm.

The Economics of Coal Externalities

All activities surrounding coal mining, transport, and burning have a direct and indirect economic cost; ranging from the health care costs of the people sickened by coal-fired power plant pollution, to the cost of cleaning up spills of toxic coal waste, and the jobs lost by fishermen downstream of a coal mine. And there are a number of other consequences, the list of which is too long to list here.

Some of the indirect economic costs of coal are in the form of subsidies and tax breaks which are not reflected in the market price of coal. For example about $4.6 billion in coal-related subsidies was introduced in the 2009 stimulus package. And on top of that, in the state of Kentucky for example, the government spends $115 million more on subsidies for the coal industry than it receives in taxes or other benefits.

Coal mining and power plant operation are expensive projects, and often require major investments. The risks and costs of those investments are often passed on to taxpayers via different financial vehicles, such as: infrastructure subsidies and loan guarantees. For example, the Healy Clean Coal Plant (HCCP) cost the State of Alaska and the Federal Government nearly $300 million since its construction started in the mid 1990's and it is yet to produce any power in return. In addition to the other abnormalities, the Alaskan coal industry pays only 5% of its market value to the state, even though the nominal rates are much higher.

Taxpayers also pay the costs of cleaning up environmental disasters caused by the coal industry. These cases and the related expenses are not well publicized but many millions of dollars are spent on these activities every year.

Cleanup of the recent coal ash spill in Tennessee is estimated to cost up to $1 billion, not including pending litigation. Now that the cleanup at this site has been taken over by the EPA under the Superfund law, most of this cost will be borne by the US taxpayer. Congratulations, taxpayers; here is another gift from the coal industry, in which cases we have no say so.

The most serious impact of the coal industry is the air pollution, which has enormous economic costs through health care costs and lost productivity. Recent study by the government of Ontario estimated these costs in the billions of dollars within the state alone.

Similarly, the cost associated with premature death due to coal mining in the state of Virginia is five times greater than all measurable economic benefits from the mining industry. Measuring human death in terms of economic benefits is not easy, but our esteemed economists find the ways, and we agree that it is the best way to reflect the reality in our capitalist society.

A recent study found out that coal mining in the state of Illinois resulted in a net cost to the state of almost $20 million in 2011, even before including any externalities. Including the externalities would bring this number in the hundreds of millions dollars.

Another direct effect of coal mining is that it also hurts the ecosystems that other industries and population use and depend on. Recreational fishing, camping, commercial fishing, and tourism are important for the economy of most states. For example, there are over 55,000 full time jobs in Alaska that are closely related to, and depending on, the health of the state's fragile ecosystems. These jobs are over a quarter of Alaskan employment and produce over $2.5 billion of income in the state.

At the same time, the proposed Chuitna Coal strip mine in the Cook Inlet, Alaska, would create 350 jobs, which will impact the 55,000 ecosystem-dependent jobs in a negative way. These effects will affect negatively the economy by leading to worsening the health of the population, which in turn impacts the health care costs, thus compounding the economic impact.

It is undeniable that coal provides cheap electricity, which helps the economic development and contributes to improving the health of the population, thus lowering the total health care costs. While this is true in theory, it has to be applied in the same equation with the negative economic impacts. When the total of all costs of air, soil, and water pollution, global warming, habitat destruction, and human health are added and compared with the benefits of cheap electricity, the overall economic and social impact of coal is negative by almost 2 to 1 ratio.

$$\text{Economic impact} = \frac{\text{Negative effects}}{\text{Positive effects}} = \frac{2}{1}$$

Could it be that the negative effects of coal mining and burning are twice as many and as serious as the positive? The coal industry proponents will surely argue with the above numbers, and the overall conclusions herein, but even if the negative effects are not that great, the fact that coal is continuously emitting more and more air pollution is undeniable. The pollution in turn is causing more and more illness and climate warming, which is also undeniable.

The other fact that needs to be considered in this context is that China, India and other large developing countries are planning to increase significantly—even double and triple—their use of coal-mining and burning in the coming decades. This will cause an increase in coal pollution to levels we have never seen before.

What would that increase in GHG emissions will do to the health of the people and to the global environment is anyone's guess, but we venture guess that it won't be good…and might even have detrimental consequences to the people and Mother Earth's tortured environment.

On the other hand, shutting all coal mining and power generating operations, if it is even possible, would throw us and most of the world into darkness and will bring unprecedented global economic chaos.

So, like it or not, coal is here to stay and we just have to deal with it. One choice, the renewables, are still immature and their full implementation is far in the future. The other, more plausible choice, natural gas, will continue to replace coal, but there is a limit to that too, so coal is not going anywhere. We have to expect its use to continue increasing with time, and until suitable alternatives are found.

Natural Gas Externalities

Natural gas has been hailed as the solution to our energy and environmental problems. Is it? Although gas drilling and exploitation have a number of positive effects on the local economy by creating jobs and wealth, it contains hidden and at times questionable indirect and induced effects with their associated costs.

These natural gas externalities include the perceived and actual damages done at the gas well site, during gas transport, gas burning, and the related activities, to:
- the local environment (land, water, and air)
- the local property values
- the human health, and
- the local wild life,
- the negative global effects

- The direct effects are obvious and clearly observable damages caused by the gas well drilling and operation. Examples of these are air and water contamination, and human illnesses as a result of these effects.
- The indirect effects are represented by a number of additional economic activities of (or effects upon) the value chain network, which are mainly caused by the gas well drilling and operation. These are different from, and in addition to the direct effects. The results from these effects can be positive and/or negative.

- The induced effects are a different set of additional economic activity of (or effect upon) other, unrelated, firms and households that are also (directly or indirectly) caused by the gas well drilling and operation and its direct and indirect impacts. The effects from these effects can also be positive and/or negative.

For example, 2009 estimates of spending on the Marcellus Shale industry in Pennsylvania show direct expenditures of approximately $3.8 billion. This includes the direct costs of well drilling, O&M etc. expenses that are reflected in the books.

Additional $1.6 billion of expenses have been attributed to *indirect* spending by other industries on goods and services. The combined direct and indirect effects generate approximately $1.8 billion in *induced* effects, which result in economic benefits for the local area. The overwhelming conclusion in this case is that each $1 spent by the gas industry generates $1.90 in gross output and sales in the state and the country.

For example, gas well owners pay royalty payments for the land they are renting or leasing. This way, the local land owners could get bonanza payments for their, sometime otherwise useless land, of over $50,000 annually for leasing each 100 acres to the drilling operations.

What is much harder to estimate is the latent negative effects, which are either hidden or which presence is delayed by months or years.

The External Costs

One of the biggest problems with natural gas—and especially its production and transport today—is the accidental release of contaminants in the air, soil, and water. The size and cost of these types of contamination have not been determined, but a fierce battle is raging across the U.S. and the rest of the world, intended to stop, or at least tightly regulate, hydrofracking.

The cost of this battle to the gas industry, and the entire U.S. industrial complex is unknown, but it is significant. Legal action, delaying or stopping well drilling projects, and other related anomalies cost money and will have to be accounted for sooner or later.

Another great cost in the natural gas production and transport is from legal action resulting from accidents. Just like spilled oil, spilled (accidentally vented) natural gas can cause serious damage. The difference is that in some cases, the damage is immediate. Spilled (leaked) natural gas mixed with air and exposed to an open flame will explode and burn very hot. It would destroy anything in its path within minutes.

This is the case of the 2010 natural gas pipeline explosion in San Bruno, California that killed eight people, injured 66, and destroyed 38 houses. Property damage of about $220 million was assessed, but how about the other losses—including the loss of human life. The $2.5 litigation is the largest penalty ever levied by a state regulatory body in the U.S.

The largest penalty under federal pipeline safety laws thus far is $101.5 million for an August 2000 explosion on the El Paso Natural Gas pipeline in New Mexico. Twelve people camping under a concrete-decked steel bridge that supported the pipeline across the river were killed instantly and their three vehicles destroyed. Two nearby steel suspension bridges for gas pipelines crossing the river were extensively damaged. Property and other damages or losses totaled nearly a million dollars, and resulting in $101.5 million in punitive damages awards.

The Hydrofracking Facts...

Fact 1. The toxic fracking liquids enter into the Earth's depth under very high pressure and strictly via controlled process. But then, they are released into porous ground without much, if any, control from above, because at this point any control the process crew had is lost. The chemicals are free to act on their own and can travel in whichever direction they (not the ground crew) want.

Fact 2. At the point of release from the delivery pipes the chemicals are on their own and move around according to the particularities of the local geological formations. This way they can enter into the water table, and/or eventually flow up and break through the surface. There is no way to control their movement under the ground. This fact cannot be disputed!

Fact 3. What we know for certain is that there are increasing numbers of cases of fracking liquids and gasses showing up in places where they don't belong and where they have not been present before. This fact cannot be disputed!

Fact 4. There is also mounting evidence of increased earthquake activities in areas near hydrofracking sites. This phenomena could be explained by the increased water and chemicals damage deep underground, which dissolves and moves large amount of soil and rocks out of their natural places. Thus created soil movement disturb the natural formations, causing earthquakes.

Fact 5. Any intentional or unintentional contact with these liquids and gasses by humans or animals would lead to some negative health effects. Common sense tells us that the effects of any contact with toxic fracking liquids, or any petroleum products, would range from a skin rash to cancer, depending on the chemicals type, concentration, and length of exposure. This fact cannot be disputed!

Fact 6. A number of researchers in veterinarian medicine have looked into this problem and a number of cases in several shale states have been investigated. The conclusions are that livestock had often experienced neurological, reproductive and gastrointestinal problems after being exposed to fracking fluids either accidentally or incidentally.

Note: Some of the most dramatic cases thus far have been the death of 17 cattle within an hour after being exposed directly to hydraulic fracturing fluid. In another case 140 cattle exposed to fracking wastewater after an impoundment was breached, of which 70 died and the remainder produced only 11 calves with only three survivals. There are also many other deaths reported as a result of poisoning of animals accidentally getting into the fracking fluids.

The exposure symptoms usually result in neurological symptoms and weight loss in horses, problems with dairy goats reproducing, and abnormalities in dogs. In some farms, adult animals suffered abnormally high miscarriages or stillborn births after the animals had been accidentally exposed to fracking fluid.

The health issues and death of animals as a result of exposure to fracking fluids only confirms that the fracking chemicals:
a)   are uncontrollable,
b)   are found where they are not supposed to be,
c)   are causing property damage, and
d)   are dangerous to all life forms, including humans.

The above facts are supported by a large body of research and it all leads to a very real conclusion; that there is a danger brewing in the numerous hydrofracking fields of the U.S. and abroad, and that we must take all possible precautions to ensure the safety of the locals and the environment.

Nevertheless, the fracking industry, supported by government authorities, claims that any and all accidents and incidents related to fracking, and any and all assumptions made as a result of these are not based on any scientific merit. Instead, they claim, the opposition is using unverifiable, anecdotal information to fit predetermined set of results all of which conflicting a number of studies on the same issue done by the industry.

This brings to mind the claims of the lead, asbestos,

tobacco industries some years ago, when they blamed the users of their products for the high incidence of deteriorating health and cancer cases. Just like then, the present day coal, oil and gas industries attributes the health problems to a mass hysteria that has swept the nation and the world. According to fossil industry representatives, most of the claims are anecdotal, since no study has been definite enough to offers any clear evidence supporting the notion that fracturing is making people sick.

Representatives of oil and gas companies usually claim that they take health claims seriously, and we have to believe that they mean that…to an extent. They claim that they would respond to any medical reports, or scientific data showing a direct link between fracking and human health, but…they would not respond to anecdotal accounts.

Because all claims are "anecdotal," according to their measure of anecdotal vs. non-anecdotal accounting, there is no real problem for now, and as far as they are concerned there would be no real action; probably until a very large disaster is documented showing a direct link between fracking and human health.

In one of the affected states, the state oil and gas commission plans to study whether oil and gas pollution affects health, but it will take years before any reliable information is available. The health-risk assessment phase of the study won't begin until sometime in 2016. It will probably take another 2-3 years to complete the study, so by 2020 we might have some results. Until then…tough!

Many people are not willing to wait any longer, so as a result a number of law suites have been filed; blaming the fracking operations and their chemicals and byproducts for health related abnormalities and death, and blaming the drilling companies for the property damage and the related health problems. For more info on the subject visit the FrackCheck WV website (2).

We can only hope that we don't have to witness another tobacco-, or asbestos-like gruesome large scale disaster, and the end of such a high visibility issue of national and international importance. But as things are going, we would not be surprised much…

Update 2015. Radioactivity
The hydraulic fracturing process is designed to extract oil and natural gas, which it does well. Together with these useful products, however, the process brings to the surface millions of gallons of water at each well. In addition to hundreds of different chemicals, the oil/gas/water mixture also contains small amounts of radi-

um. Where does radioactivity comes from?

Most materials deep underground—including oil, gas, and ground water—are slightly radioactive as a by-product of uranium, and other sediments in the Earth's depths. Low radiation levels in small quantities of materials do not present a real danger to people and animals, but large quantities and/or increased radiation levels do.

There are nearly 700,000 underground waste and injection wells in the U.S., with over 150,000 of them injecting chemicals and water solutions thousands of feet below the surface.

Some of the water is pumped back to the surface together with the oil and gas mixture. Once on the surface, the water/oil/gas mixture is separated, and the water is pumped back into the ground. Before that, however, the water undergoes a filtering process in order to remove sand, pebbles and other materials that could potentially clog injection wells. The filtering is done with the help of *filter socks*, large quantities of which are used at every drill site.

Thus used filter socks and the filtered material are collected and disposed of periodically. But they cannot be disposed as regular waste because they are slightly radioactive. Some states only accept waste with very low radiation—about 5 picocuries radioactivity in a gram of material.

Some fruits, like bananas, have on average 3-4 picocuries of radiation per gram. And as such, the filtered material, which usually contains much higher radiation levels, has to be subjected to special processing to eliminate the radiation, or is trucked to other states with higher radioactive waste restrictions. The officials of the affected states are looking to either increase the waste sites radiation level restrictions, or find other ways to dispose of the radioactive waste.

On the radioactive scale, the radioactive materials coming from the drilling sites is usually low, but the quantities of material are quite large, so this presents a problem for the dump site operators. And lately, people who live near these waste sites in a number of states are concerned about the growing quantity of radioactive waste in their back yard.

This, in addition to increased seismic tremors near drilling and hydrofracking sites, that have been connected to the disposal of fracking wastewater into the ground, is bringing a rise in the anti-fracking NIMBY movement. Now the battle is expanding to include concerns of increased radioactivity around fracking wells.

The low oil prices are putting even more pressure on oil and gas drilling companies to reduce spending, so there is not much room for negotiations. Because of that, we expect the battles over drilling and waste materials disposal to become even more intense.

Crude Oil Externalities

From economics point of view, the **externalities of any product or service** reflect the cost (of damage or benefit in dollar amount) which results from an activity or transaction which affects a third party that is *not involved directly* in the activity, and who did not choose to incur that cost or benefit.

The cost of externalities is not easy to calculate, because they usually contain hidden costs; costs that have not been valued yet, or are simply hard to value in dollar terms. Nevertheless, there are a number of examples, which could be used as guidelines, or at least to compare similar cases in the future.

Some examples of external costs resulting from crude oil production, transportation, and use are:

- Estimates for oil spill cleanup and damage costs in the U.S. average approximately $16 per gallon, or $670 per barrel.

The 1979 Ixtoc-I spill in the Gulf of Mexico cost $7 per gallon, or $300 per barrel. The 1980 Exxon Valdez spill in Alaska cost $630 per gallon, or $25,000 per barrel.

These numbers reflect only the compensation costs, while many other damages (such as long term effects on people, wild life, and the environment in large) cannot be calculated. These are long term effect that affect communities, large land areas and all life in them. The cost of this damage exceeds any and all compensations to date.

These *hidden* damages are usually not compensated due to the complexity and the inherited difficulty to attribute costs to some damage types, or locations. This is particularly true for some small spills, spills in international waters, and when assessing damages to ecological areas and services that lack legal status.

Some of the costs of the externalities are paid by the companies involved, but a significant part is paid directly by the taxpayers, and indirectly by the locals that are affected by the damages. In all cases, the locals are the most affected and suffer from the damages the longest.

The major annual costs of externalities in the petroleum production and use in the U.S. are:

- Surface transportation $91 billion
- Defense spending   83 billion
- National security   24 billion
- Climate change   16 billion
- Production externalities   1 billion

Total petroleum externalities   **215 billion**

If applied to the production levels, these numbers would average about $14 of new taxes per barrel of crude oil. This breaks down to $0.33 per gallon for transport expenditures, and $2.96 per gallon of external costs needed to pay for oil production and import services. This is another $3.00 per gallon that needs to be added to the gas pump price to pay for all externalities. These numbers are increasing by the day...so who pays for them all? You guessed it! *YOU do!*

- In the mid- to late 2000s a federal report estimated U.S. petroleum import external economic costs, *excluding* military expenditures, to a total of $13.60 per barrel, with a range of $6.70 to $23.25, or about $54 billion annual cost to the U.S. economy. This phenomena was further qualified as "a measure of the quantifiable per-barrel economic costs that the U.S. could avoid by a small-to-moderate reduction in oil imports."

Adding the military expenditures needed to ensure the flow of oil across the world's oceans, might double this number.

The external costs per gallon and vehicle-mile from the extraction, distribution, and consumption of various fuels and vehicles for various time periods were estimated at:

- Passenger car health damages, $36 billion
- All vehicles damages, $56 billion.

Note: Studies of electric vehicles and grid-dependent hybrid vehicles show higher environmental damage than many other technologies; if and when electricity used to recharge them is generated by fossil fuels. This is due to efficiency conversion losses, which waste energy and emit additional pollution during the different steps of conversion.

The total externality, as a result of the U.S. economy's oil dependence, has been estimated at $200-$250 billion annually in the mid 2000s, and a total of $10 to $12 trillion between 1970 and 2005. These costs are evenly divided between direct transfer of wealth (purchase money) from the U.S. to oil producing countries, the loss of economic potential due to oil prices elevated above competitive market levels, and disruption costs caused by sudden and large oil price movements.

These estimates do not include military, strategic or political costs associated with U.S. and world dependence on oil imports. They also do not include the constant risk of oil flow disruptions and/or price variations, which affect the U.S. economy and markets.

The cost of the U.S. Middle East military intervention, as needed to maintain U.S. access to petroleum resources, and their transport average about $500 billion annually. These costs are in addition to comparable magnitude economic costs (as discussed above).

The total U.S. oil dependency cost is about $1 trillion annually, significant part of which are external costs.

Vehicular Externalities

Crude oil today is used mostly for transportation; fueling cars, tracks, trains, planes, and ships. Let's take for an example the intense use of cars and trucks that results in a number of real costs at one time or another (property damage, accidents and air pollution), most of which are usually not included in the initial manufacturer's cost, or the cost of the fuels.

Some of these costs are included later on in our taxes, insurance premiums and court costs, all of which cover some of the damages, accidents, etc. casualties. Nevertheless, all costs must be considered in the cradle-to-grave life cycle of the car or truck in question, in order to get a full picture of the reality.

A major component in the environmental cost are its externalities, which must be brought out, accounted for and included in all economic analysis. This often missing component, reflecting a major damage done to the environment by smoky cars and trucks exhaust pipes, has been ignored until recently for a number of reasons; mostly a combination of ignorance, negligence, and greed.

The first problem we face in attempting to do this is assigning a true and accurate monetary value to the environmental damages caused by the multitude of vehicles. Reducing complex environmental effects to a single dollar value is nearly impossible—especially at this early stage of development of the environmental economics sciences. There is no clear-cut answer to this question. In some cases, economic damages may be identifiable, but not in others.

Annually, the U.S. motor vehicles fleet emits about 50 million tons of carbon monoxide, 7 million tons of nitrogen oxides, and thousands of tons of other toxins including particles, formaldehyde and benzene.

This is the amount of pollution emitted by a half dozen coal-fired power plants. In addition, car and truck accidents kill more than 40 thousand people and injure more than 3 million every year in the U.S. alone as well. These are serious external costs, most of which are not covered by manufacturer's warranty, personal insurance, or any other financial instruments, nor are they properly accounted in the costs of vehicles and fuels.

Additional external costs include natural habitat destruction from building roads and parking lots, as well as from activities related to manufacturing vehicles, vehicle parts, filling stations, repair shops, oil and waste disposal etc. And then, even more abstractly, there are other costs associated with national security in securing petroleum supplies, noise pollution damage, and on and on the list goes.

A number of studies attempt to monetize the external costs of motor vehicle use by estimating the costs associated with global air pollution, crop losses, reduced visibility, national security, noise pollution, and so forth. The public sector pays some of these costs; others become external social losses, paid for by the government.

Assumptions of the total annual monetary and non-monetary externality costs of automotive externalities in the U.S. vary from $250 billion to $900 billion. Annual public sector costs for motor vehicle infrastructure and services are valued at $120 billion to $240 billion. Free-parking subsidies and cross-subsidies caused by congestion amounted to another $50-$100 billion per year.

An expert estimate of the public sector externalities costs related to vehicles use is given in Table 9-4.

An additional gasoline tax of $2.50-$3.50 per gallon might be needed to internalize all the social costs, and even that won't be enough in many cases. To get closer to the reality, air pollution externalities should be internalized based on a vehicle's emissions levels rather than on gasoline consumption, or any other parameters.

And then, there is another great number related to health related externalities brought upon by car and truck accidents around the country. There is an estimate from the 2000s of $400-450 billion spent on health issues and death from road accidents annually in the U.S. If added to the cost of gas these numbers would represent additional $2.10 per gallon on top of the price at the pump.

Table 9-4. Externality costs of vehicle use

Air pollution, human mortality	$300 billion
Highways and parking	200 billion
Productivity loss	150 billion
Imposed travel delays	100 billion
Accident costs (pain, death)	50 billion
Global warming	20 billion *
Noise pollution	15 billion
Air pollution, crops damage	15 billion
Water pollution	10 billion
Pollution regulation and control	5 billion
Highway patrol and safety	8 billion
National security costs	7 billion
Other externality costs	35 billion
Other public sector costs	25 billion

No matter how we look at it and what we decide to do, we will not provide a complete, let alone perfect estimate and solution. Doing nothing, however, sends the wrong signal to the users and abusers—that the external costs are irrelevant, or that do not exist. This is why the efforts of our esteemed economists and legal experts in this area are so important.

The trend of increased electric vehicles use is not the solution, as hailed by some. As a matter of fact, the global warming effects, which are already increasing exponentially, could be raised sharply by mass use of electric (charged by electric power) vehicles. There are many reasons for this, the major of which is the fact that:

• Electric vehicles are charged with electric power generated mostly by fossil-fired power plants, and

• The inefficiency of the energy conversion process—from burning coal to generate power, its transmission, and charging batteries—is nearly 75%. Or ¾ of the energy contained in coal is lost before the wheels of the electric vehicles hit the road.

This translates in more waste of energy and additional pollution in the future.

Medical Externalities

The energy industry has been operated in a dangerous territory since the very beginning. Mining, oil and gas drilling, fuel transport, etc. activities can be classified as dangerous—and in some cases extremely dangerous—operations. A lot of improvements have been made recently to reduce the environmental damage and avoid human illnesses and death resulting from these activities, and we do give the necessary credit to the industry for those efforts.

The case of the newly expanded gas fracking, however, stands out in its large scale land use and huge amount of toxic chemicals used. The worst part of this is that we don't know the exact behavior and the negative effects of the millions of gallons of toxic chemicals pumped deep under the ground.

While the pumping itself might create problems by leaking toxic chemicals through poorly cemented well walls, the biggest problem is the uncertainty of what happens after that. Once the chemicals are pumped underground, the high pressure forces them in every crack and cranny that they can get into.

Immediately after the hydrofracking chemicals leave the pressurized pipe and are released deep underground, they are on their own. There is no control over them whatsoever. They can go in any direction they want—usually choosing the path of less resistance. Since the underground mass is porous they easily find and follow a path of minimum resistance. And although their main job is to make new cracks in order to push the gas out, they can make new paths and travel through them under the high pressure—again without any control or guidance from the ground crews.

Injecting huge amount of pressurized harsh chemicals underground, and controlling them, is a gambling game.

The well drillers' claims that they can control the action of the fracking chemicals is only partially true. While they can control the type, amount, pressure, and initial location of the injectors, as soon as the chemicals leave the injecting pipe, other—totally uncontrollable—forces take over.

Deep underground, the rock structure, gravity, and reduced non-directional pressure control the action and direction of the fracking chemicals.

So, it is quite easy to see that the fracking chemicals can travel up or down, left or right at will. Controlling their activity by the ground crew is not an option. It is equally impossible to predict the exact direction of their travel underground. If so, then what would stop them from taking the path of less resistance, following a crack in the rocks that leads in Joe Farmer's back yard? Once there, they can cause extensive damage to properties, the environment, people, and animals.

And of, course the ground crew activities are accompanied by extensive and ever increasing use of vehicles in personal and commodities transport, which causing additional externalities. Entire communities have been changed as a result of increased hydrofracking activities. The final effects of these changes are still to be accounted for.

Cars, trucks, and all other vehicles are another great contributor to medical externalities. They use a lot of energy and emit a lot of pollution during their life time. It starts from the manufacturing process, and continues with the emission of huge amount of GHG gasses on the roadways, waterways and the air. The exhaust gases, while contributing to abnormal climate phenomena and global warming, also harm people and animals.

As a matter of fact, most of the crude oil used today goes for fueling cars and trucks. And yet, most of the people in the world insist on having more cars and driving more miles. The car manufacturers around the world are ramping their production volumes and encourage car driving in their flashy media commercials. Every single person in the world desires to possess one of these fast, comfortable, and elegant machines, Since there is no mechanism to limit the trend, the number of cars and trucks around the globe is going to increase proportionally in the future.

Cruise companies put huge ships, some as big as a small cities, in route to nowhere, pointlessly circling the world's oceans and burning millions of gallons of fossil fuels in the process.

Thousands of airplanes lift off every day crisscrossing the country and the world, burning many more millions of gallons of fuel.

And there is no end in sight. We all like driving a lot of miles, take cruises to nowhere, and go for an airplane ride whenever possible. We are basically acting as if there is no tomorrow. As if the fossils will last forever, which is absolutely not true, because we already see the bottom of the fossil's bucket. And in the process, we ignore the fact that the fossils are sickening and killing us every day.

The Costs

In economic terms, all these activities are costing us billions in health costs annually.

Researchers have estimated that the health impacts (medical externalities) by fuel type are :
- 25 to 45 cents per kilowatt-hour for coal,
- 10 to 20 cents per kilowatt-hour for oil, and
- 5 to 8 cents per kilowatt-hour for natural gas.

Note: Nuclear power externalities are much more complex and more difficult to calculate. Taking into account research activities, proper liability insurance, waste disposal, and other factors, its cost per kilo-watt-hour could get close to that of oil.

Keeping in mind that the average retail price for electric power in the U.S. is about 9-10 cents per kilowatt-hour (kWh), then the coal and oil costs with added externalities are several times higher than the typical retail price of electricity. The difference points to a very large cost of the respective externalities, which corresponds to, and accounts fully for, the adverse environmental, climate and health impacts associated with fossil fuel usage.

Combining the average retail cost of electricity and the health impacts from fossil fuels brings the average electric bill 24-45 cents/kWh higher than it is today. This difference, applied to using alternatives such as energy efficiency and renewable technologies, could be enough to limit the use of fossil fuels.

A paper recently published by researchers at the U.S. EPA finds that full accounting for the health impacts of burning fossil fuels would bring the retail electricity price an average of 14-35 cents per kWh higher than the actual retail cost of electricity. This amount could be 24-55 cents depending on type of fuel and location of its production and use.

EPA scientists estimate that the hidden fossils-related health costs add up to $886.5 billion of externalities expenses annually, which is about 6% of the U.S.' GDP.

European NGO released a report recently, which found that emissions from Europe's coal-fired power plants cost the EU citizens up to $55 billion in health costs annually. This includes costs associated with medical expenses; hospitalizations, medications, premature deaths, and reduced activity including work hours, and lost work days due to illness.

In 2013, the International Monetary Fund (IMF) released a report calling for the end of $1.9 trillion in annual global energy subsidies. That includes $480 billion in direct subsidies to consumers and $1.4 trillion in the "mispricing" of fossil fuels. Mispricing is due to the fact that polluters are not forced to pay the full costs associated with fossil fuel combustion on climate change and public health.

According to the IMF, the biggest offender was by far the United States, with $502 billion in subsidies to energy companies, followed by China with $279 billion. Russia was third at $116 billion of subsidies to energy companies. Offsetting the impact of the energy subsidies in the U.S. would require new fees, or taxes totaling over $500 billion per year.

And as mentioned above, there is also a great number of health related externalities brought upon by car and truck accidents around the globe. This amount is estimate at $400-450 billion spent on health issues and death from road accidents annually in the U.S. alone. If added to the cost of gasoline these numbers would represent additional $2.10 per gallon on top of the price at the pump.

And, worst of all, we keep on using the fossils as if there is no tomorrow. Some of us don't even know that we are running out of fossils. Others deny it, but most of us don't care. And so, we entered the 21st century with our eyes closed to the possibility of a fossil-less and car-less future. The talk volume is increasing, but not much is done to change that any time soon.

Drill, baby drill is still the predominant call in the global energy sector. As things are going, that call will only get louder. While more drilling will supply us with enough oil and gas to power the economy and drive our cars for awhile longer, it does not answer the fundamental questions of fossils depletion, and environmental destruction.

Actually, there are calls of reason in the U.S., which demand reduction of fuel use and GHG emissions. As a matter of fact there is a noticeable reduction in oil imports and GHG emissions lately, as hydrofracking is providing more oil and as coal power plants are converted to gas.

The problem is that even if this trend continues in the U.S., which is not certain, it is still a drop in the global bucket as far as GHG emissions and global warming are concerned. This is because China, India and a number of other major polluters have no intentions of reducing the use of their coal-fired power plants and vehicles. The overall effect is that of increased energy (fossils mostly) use and much more GHG emissions.

Nuclear Power Externalities

The U.S. nuclear power plants produce over 50 billion kilowatt-hours of electricity annually, which is over 20% of the total utility-generated power in the country. This means that U.S. is heavily dependent on nuclear power, regardless of its enormous expense, radiation pollution, the threat of a nuclear accident, nuclear terrorism, and the nuclear waste debacle.

Nuclear power is where it is today mostly because of government's technical assistance, and huge subsidies and incentives in the past. It is also cheap today for the same reasons. It costs residential consumers an average of 8.0 cents per kWh of electricity, while industrial users pay almost twice less. Not bad, you say and we agree, but the actual cost of nuclear power would be much higher, if it was not for the federal subsidies, which also pay for the extra costs, including these of the "externalities."

We know by now that externalities are hidden charges where some people bear costs that they are not paid or compensated for. Here, the decision-makers do not pay for the additional costs, and doesn't even take them into consideration in deciding how resources shall be allocated.

The power plant owners have no reason to produce benefits that he doesn't get, nor to cut back on costs that he doesn't pay for. So, if there are 'external' costs of any sort we can expect to get charged for things we are not responsible for, all of which makes the energy markets to be inefficient.

Externalities in the nuclear power business have been estimated by the Greenpeace': *The Economic Failure of Nuclear Power*, a 1993 study, which estimated the federal expenditures for nuclear power until 1990 to over $97 billion. This is many billions of dollars spent on research, development, regulation, construction costs, uranium enrichment, and for contributions to the nuclear waste fund. These funds, however, did not pay for the environmental destruction caused by nuclear accidents and other real and perceived damages.

There are many other externalities and costs that we don't hear much about. The Shoreham Nuclear Generating Station in New York was finished in 1988 at a cost of $7 billion only to be decommissioned shortly thereafter without ever producing a single watt of electricity. And because one of the reactor cores was turned on for a test, the decommissioning cost an extra $400 million. Who paid for it? You and I, of course.

And then, many nuclear plants in the past were completed with 500-700% cost overruns, while others were just abandoned because they were too expensive. Who paid for these? You guessed it.

Figure 9-11 says it all; we all see the enormous amount of energy and other benefits nuclear power plants provide day and night. At the same time, we are all painfully aware of the even greater, but less visible, side of nuclear power; uncontrollable nuclear accidents.

For now, the benefits of nuclear power outweigh the dangers. The big question is how long?

All is well with nuclear power now and until another Fukushima devastates thousands of acres and kills hundreds. One doesn't have to be a nuclear scientist to

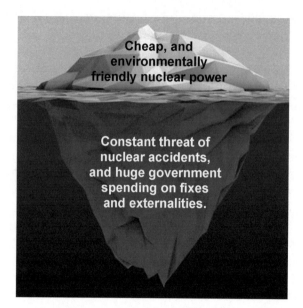

Figure 9-11. The nuclear power iceberg.

figure out that we, the taxpayers, are paying billions upon billions of dollars for on-going nuclear research, waste disposal, accident prevention, and remediation of old nuclear sites.

The industry is heavily subsidized, so what we see on the surface—the cheapest and most environmentally friendly electric power—is only the tip of the iceberg. What lies below are accounts of deadly accidents, mixed with government policies, and regulations. Most of us do not see the related activities and expenditures, and yet we live with, and pay for, them. There is no choice for now, so we need to live with it. It is what it is!.

The subsidies, however, hide the true cost of nuclear power and create an underestimation of its true cost. But why are the feds so much into covering up for nuclear to such a great extent; something they do not do for any other industry? A study by the Nuclear Information and Research Service and U.S. Public Interest Research Group found out that by 1992 only three members of the Senate and seven members of the House have not benefited from money donated by the nuclear industry's Political Action Committee campaign contributions. Any questions? But this is only one, and probably minor, reason. Nuclear power has Uncle Sam's support in many other areas and for many other reasons.

When General Electric was unable to obtain accident insurance for their reactors, for example, Congress responded with the Price-Anderson Indemnity Act, which limited the liability of the nuclear industry in case of a major nuclear accident to $12.6 billion. Any amount above that will be paid by the U.S. taxpayer. The same law also indemnifies nuclear plants if an accident is caused by operator error, even it is by negligence.

Now, this is a big deal, because recent estimates of the losses from the Fukushima nuclear accident, and the subsequent clean up and remediation of the surrounding area, come to a trillion dollars. And even then, when all that can be done is done, neither the multibillion dollar nuclear power plant, nor the area around it, will be the same. Nor will the local people be allowed to return to their homes. Ever! How much worse can a man-made disaster get?

Note: Thus far, the remediation cost of the Fukushima's nuclear accident is estimated at $150 billion, plus $50 billion for decontamination. These amounts do not include rebuilding the power plant and decontaminating the land around it, if it is even possible. Pending and future legal action payments are also not included.

According to some experts, the work at Fukushima will continue for decades to come, and if and when it is done, it would have cost the Japanese taxpayers $1.0 trillion, or more.

The U.S. nuclear industry is obligated to maintain only $7 billion of insurance under the Indemnity act, while the official estimates that a nuclear accident could cost many times that amount. A meltdown of one single reactor could cost over $500 billion for remediation and recompense, not including the cost of the facility itself. And this, in light of NRC' estimated of a 45% chance of having a meltdown in the near future.

50% chance means that the question is no longer if, but where and when, the next accident would happen.

And of course, the growing pile of nuclear waste requires large subsidies as well. In addition to the daily transport and storage of nuclear waste from active power plants, we also have to deal with the old and abandoned nuclear sites. The nuclear repository at Hanford, Washington has been in the news with a third of its 177 underground storage tanks leaking toxic chemicals and radioactive waste in the local rivers and water table. In response, the locals have filed a number of lawsuits.

This brings the question of the externalized cost of nuclear power' effects on people's health. Uranium mining, enrichment, nuclear accidents, waste disposal sites, etc nuclear-related activities affect our health by contaminating our environment with cancer-causing radioactive particles and chemicals. This is the costs we pay for having cheap electricity, and this is how it will be for many years to come.

ExternE

ExternE is a major European study of the external costs of various fuel cycles, focusing on coal and nuclear, ExternE, shows that in clear cash terms nuclear energy incurs externalities amounting only about one tenth of the costs of coal. This is before the costs of any major accidents are accounted for. In these cases—such as Chernobyl and Fukushima nuclear disasters—the externalities well exceed any other.

The *external* costs are defined as those actually incurred in relation to health issues and environmental damage. These costs are quantifiable, but are presently *not* built into the cost of the electricity. If these costs were in fact included, the EU price of electricity from coal would double and that from gas would increase 30%. And even these costs do not include the external costs of accidents, global warming and other factors.

Note: In the nuclear plant case, things are complicated further by a very important factor that is not included in the external costs. It is the external cost of environmental damage, and pain and suffering caused by nuclear plants accidents. These costs are hard to even estimate, let alone putting a harder number on their effects. How do you estimate the cost of human health damages and death caused by the Chernobyl accident. Or Fukushima?

These accidents caused a complete devastation of a large land mass, and its animal life, crops, water sources etc. And unlike other accidents caused by power generators, due to nuclear radiation, which lasts long time, the area is rendered useless for many, many years; maybe forever. other

But the most important and irreconcilable damage is the thousands of animals and people that have been made ill and killed as a result. How do you estimate the cost of a human life? Thousands of them? Or the pain and suffering of over a million people after Chernobyl.

If the enormous externalities costs are spread over the entire global nuclear industry, it may prove to be the most expensive and dangerous energy source ever!

So, while the impacts of global warming can be classified as "inter-generational," the impacts of the nuclear accidents and waste cycle debacles are better described as "inter-civilizational" in addition to, and on top of, being inter-generational.

The nuclear power industry points global warming as a good reason for its renaissance. The nuclear power environmental benefits are real, true, and without question the best possible...only assuming that the problems of, a) nuclear accidents, and b) disposing of spent nuclear fuel will be solved for good sometime in the near future.

Unfortunately, such assumptions have been made by nuclear energy proponents for the last fifty years or more, and yet we see serious accidents from time to time. Even one of these is one too many, so if we haven't solved these problems during the past half century, what makes us sure that we will be more likely to do so now or any time.

There are many technological advances, but people still make mistakes as they did 2,000 years ago, and Mother Nature is still unpredictable as it was millions of years ago. And sometimes this is all that it takes to blow up a nuclear reactor. Look at Chernobyl's reactor, which was praised as the safest ever, where one wrong move by a negligent operator blew up the reactor and everything in 20 miles radius around it. Or Fukushima, which was built with the U.S. best and latest equipment, and was operated and maintained by the best trained professionals. It took Mother Nature seconds to turn this technological marvel into twisted pile of metal and contaminate the entire region.

And the U.S. is not immune against nuclear accidents. We all remember Three Miles Island scare of the not-so-distant past. And even today, we deal with nuclear problems, like leakages at abandoned nuclear waste storage in some sites, and other nuclear mishaps.

Case Study: WIPP

The Waste Isolation Pilot Plant (WIPP) in New Mexico is one of several U.S. facilities for storing nuclear waste from previous nuclear experiments. It is part of the Los Alamos National Lab (LANL), nuclear power process development programs, and is, therefore, one of the most sophisticated such facilities in the U.S. and perhaps the world.

Note: Nuclear research and the related activities are done by a number of U.S. universities and government labs. Most of the research is done in secrecy, so information is not readily available. Unless an accident forces the release of some information.

Working on a project to clean its facilities from all nuclear waste by June, 2014, LANL workers with the help of numerous contractors, would package the waste and ship it to WIPP for further processing and storage. After shipping over 1,000 loads successfully, workers at the site came across a batch that was extraordinarily acidic, thus unsafe for shipping. Following the rules, the work had to stop and undergo a rigid set of reviews as needed to determine the proper procedures to follow. The work, however, did not stop, most likely due to complications and time waste that jeopardized the lab's goal of meeting the June deadline.

Instead, the lab and the contractors decided to treat the acidic nuclear waste by neutralizing the excess acidity, using organic kitty litter to absorb the excess liquids. Wrong decision! The mix turned into a bomb, a type of slow-acting plastic explosive.

The mix was put in a 55-gallon drum and shipped 330 miles to WIPP, which is actually the nation's only underground repository for nuclear waste in the southeast. The accompanying documents, which must include a detailed description of the contents, were doctored up, so they didn't even mention the dangerous procedural deviation at LANL.

The organic materials used were mischaracterized as *clay-based*, without mentioning the *combustible organic variety* in them. Given a chance and proper information, any chemist at WIPP would have recognized the hazards of the mix and possibly taken the needed precautions. Alas, they were not given a chance, because the information was intentionally hidden.

So, in February 2014, with close to 90% of the campaign to clear the nuclear waste from LANL completed, the drum with the vicious mixture cracked open, and the "Genie in the bottle" was released. Nuclear radiation leaked into the air and contaminated part of the facility. The temperature in the underground cavern increased to over 1,600 degrees. It could've gone even worse, if dozens of drums stored nearby have been damaged by it, thus making things even worse.

Regardless of the details, the accident was caused by a human error. Ignorance, negligence, and glaring lack of discipline by one or several people played major role in the disaster. Just like Chernobyl, WIPP was a victim of a man-made accident.

As a consequence, during the time of the accident and before full evacuation of the personnel took place, 21 workers were contaminated and treated for nuclear radiation. Shortly thereafter, the site was immediately and indefinitely closed, due to excess contamination.

WIPP is a very important facility, costing billions of dollars to construct, maintain, and run. How could one wrong action blow up such a big hole in the reputation of all involved and the entire nuclear industry? This, after all, is the 21st century, where we rely on sophisticated equipment to do most of the work. Obviously not enough, since the equipment is told what to do by people who can make mistakes. We, humans, are fallible, no doubt.

The saying, "To err is human, to forgive divine" is valid now as it was when it was first spoken, centuries ago. Knowing that we are entitled to forgiveness is quite reassuring and comforting, and it usually works in real life. Except nuclear power mistakes. While we can expect forgiveness for everything, even after big mistakes, there is no forgiveness with nuclear power. It punished immediately, cruelly, and irrevocably any and all mistakes.

Since the nuclear process allows no faults, it will not be totally safe until and unless man is totally removed from any nuclear activities.

The 2014 WIPP accident confirms this conclusion. One single mistake, due to ignorance or negligence of one single person, is all that it took to hurt many people and shut down the entire facility. It also added more expense during the shut down and maintenance procedures for the duration. The planned efforts to re-open WIPP, and a possible set of legal procedures, will add even more expenses.

And then, as if this is not enough, a group of engineers discovered in January, 2015 that portion of the ceiling in Panel 3 of WIPP's facility had collapsed. Panel 3 contains the access drift of the WIPP underground, which is an area that is restricted due to low levels of radiation contamination close by.

It has been estimated that it would cost at least half-billion dollars, on top of everything else, to re-open WIPP. As is usually the case in such complex situations, however, it would most likely cost twice that much; a billion dollars or more. If and when that will happen, however, is unknown at this point.

As a result of the WIPP accident, the state of New Mexico outlined more than 30 state-permit violations at the local WIPP. The state accuses LANL and WIPP of mixing incompatible nuclear waste streams, treating hazardous waste without a permit, and failing to notify regulators about changes in the way waste was being handled.

The New Mexico Environment Department has filed charges for $54 million in fines against the U.S. DOE, LANL and WIPP. Then, the U.S. Congress cut $40 million from cleanup programs at the LANL, but at the same time adding $100 million to help reopen WIPP. Not to worry, the government is committed to nuclear power and will do whatever necessary to prop it up it up when it stumbles.

One thing worth mentioning here is the arrogance and inappropriate behavior of LANL top managers who for two years preceding the incident refused to allow state inspectors inside the facility. Only following the accident has the New Mexico Environment Department demanded and was allowed inside the facility for inspections.

After the accident, these same top managers most likely sat back in their comfortable office chairs, washing their hands of any responsibility, while blaming and punishing the workers for the accident. But they are accountable to us, the people, and we must find a way to make them step up to the plate, do their job better, and assume the responsibility when the job is not done right.

Would annual state inspections have prevented the accident is debatable, but the hidden lesson here is that a second pair of eyes and opinions is always preferable when working with nuclear materials and processes.

All in all, the new estimates are that WIPP might reopen by 2018 at an additional cost, but the uncertainty is too great for more precise estimates. The problem here, and in similar cases, is that once radiation is released in significant quantities, the area cannot be cleaned easily or quickly. As a matter of fact, it might not be even possible to clean it, depending on the conditions it is in. In all cases, it takes a lot of effort, money, and time to bring it to the original condition.

So, we wish WIPP, LANL, and DOE good luck, and do not doubt for a moment that they will be able to correct the problem with time…this time. They will do whatever it takes, and spend as much as they have to, in getting WIPP back on-line, come hail or high water.

One thing they would not be able to fix, however, is the public's perception about the dangers of nuclear power, confirmed by the fact that some of the problems we saw at Chernobyl and Fukushima were repeated here.

Nuclear power, just like the Genie in the bottle, can provide a lot of power when properly managed. Once released by mistake, however, it has a mind of its own and could destroy anything and anybody in its way. And it follows that we have not learned how to keep it in the bottle…yet!

The Achilles Heel

Nuclear power has an Achilles Heel, or two, which will determine its future.

- Nuclear power is an awesome source of power when under control, but it is even more awesome when out of control. In such cases, it devastates anything and everyone in its path. No excuses, no exceptions, no mercy! Full and permanent devastation of large areas is left behind, accompanied by loss of life, in addition to heavy contamination of even larger areas. Chernobyl and Fukushima are the unwilling witnesses to this awesome power, and what it is capable of doing within very short span of time.

• Nuclear fuel wastes remain hazardous for hundreds of thousands, and as much as a million years. Since human history goes back only about 10,000 years, this means that the radioactivity of the spent nuclear fuels and waste materials will be dangerous at least 100 such periods of time.

Thus far no successful implementation of permanent disposal option for nuclear waste is available, although some plausible solutions have been proposed. Thus we will leave the unresolved externality of nuclear energy production for future generations to worry about, and a task for the future civilizations to solve. This is the most blatant transfer of the most expensive part of the cost of the nuclear fuel cycle to future generations and civilizations to deal with. We must pause and wonder what would they think of us…

• There is also a real damage done by the nuclear power plants to the environment presently, where most significant of these are the aquatic impacts associated with cooling water withdrawals and heat discharges common to most steam electric generating facilities.

Aquatic impacts from even one facility can be staggering. According to the EPA, "…impingement and entrainment at individual facilities may result in appreciable losses of early life stages of fish and shellfish." Three to four billion could be affected, in addition to serious reductions in forage species and recreational and commercial landings; estimated 23 tons were lost annually in one case. There are also extensive aquaria losses over relatively short intervals of time. For example, one million fish were lost during one three-week study period.

These impacts can be reduced and even eliminated by implementing closed cycle cooling systems, but this is expensive and not on the industry's agenda…yet.

• In addition to the risks and potential dangers of accidental reactor meltdown and release of radioactivity during normal operation, there is also a real and present danger of nuclear proliferation and terrorism. Rogue nations, as well as many terrorists and criminal organizations are working towards obtaining nuclear weapons and using them against the civilian population of large cities in the U.S. and Europe. These risks are inseparable from any scheme of nuclear energy production and waste disposal, and must be entered in the power generation and use equations.

• There are also other environmental and public health costs of nuclear materials, facilities, spent nuclear fuel and radioactive wastes that the nuclear energy industry prefers to keep hidden, let alone fully accounting for and externalizing them. All this is done with the full knowledge and monetary support of the world's governments, so it is not something that will change anytime soon.

The big question here is, "Who is going to pay for the piling externalities?" And, unless and until it is answered in a way that is satisfactory to all parties involved, the long pending nuclear renaissance seems to be just a mirage. Thus, nuclear power will be viewed by many of us as a dangerous, and even unwelcome, member of the power generation family. This is a pity, because nuclear power has the greatest potential than any other power generating technology…

Renewables Externalities

All power generation has some external cost, which is a cost (or benefit) resulting from an activity or transaction that affects a third (otherwise uninvolved) party who was not part of the activity, and did not choose to incur that cost or benefit.

The everyday consumer is bombarded by information telling him/her how clean and green the renewables are. And at first look, they are as promised—clean and green.

A closer look in the renewables' life-cycle reveals a number of not-so-clean-and-green events. Who knew?

The "externalities" costs of solar and wind equipment, and the related power plants' setup and operation, are defined by monetary quantification of the socio-economic and environmental effects (benefits and damages) they inflict to a local area, a nation, and the world as a whole during their cradle-to-grave life cycle.

Although we think of solar and wind power as "green," "clean" and "safe" they are far from it. There are serious damages to land and the surrounding environment in all cases, where real estate values, and even human health and well-being, are affected.

The externalities could be expressed in $ per kWh generated, for lack of better way, and should account for all materials, procedures, and events from cradle to grave. Included in these calculations are: environmental effects; human health; materials production; effects of use and disposal (waste materials, gasses, chemicals, etc.) on agriculture; noise; audio and visual pollution; ecosystem effects (acidification, CO_2 damage); and all other effects.

Thusly obtained numbers could be used to provide scientific basis for legislative and regulatory policies, energy taxes and incentives, global warming policy adjustments, etc.

Basically, PV systems must be designed and manufactured with long-term environmental concerns and considerations in mind. This includes the effects produced at each step of their life-cycle:

- Manufacturing materials and procedures, including the production and use of MG and SG Si, thin-film metals and chemicals, glass panes, metals frames, gasses, chemicals, etc. process supplies
- Solar wafers and cells manufacturing materials and processes
- Hazardous materials and chemicals used in the manufacturing process
  - Module encapsulation and framing materials and processes
  - Evaluation of direct and indirect processing energy (including transport, storage, etc.)
  - Gross energy requirements of input materials (supply chain and internally generated products and byproducts)
- Allocation schemes used in the calculations
- Separate thermal energy, electrical energy, and "material energy" calculations
- Land use and local environment and wild life protection
- End-of-life recycling and disposal activities (including transport, recycling, and disposal)

Looking closely at each step of the renewables' life cycle, we see a lot of raw materials, chemicals, and waste products. There is also enormous amount of energy used at each step.

It takes several years of operation for solar and wind power plants to generate the amount of energy that was used during their production.

There are a number of dislocated efforts in estimating the benefits and damages in the above areas. Thus far, however, no uniform, standardized method exists that is capable of capturing all energy and environmental factors into one, all-encompassing methodology. Such a methodology must be developed, in order to account for the effects of the above concerns on the PV manufacturing processes and long-term use of the related PV products.

At first glance, the negative effects of solar and wind power generation are much less than these of the fossils. The big difference is that the new renewable technologies occupy huge land areas. Each power plant needs thousands of acres of land, so we need to be fully aware of, anticipate, and take the necessary measures to counteract any negative effects that could affect this huge land mass and its inhabitants.

Case Study. ExternE project

The "ExternE project, was introduced and funded by the European Commission in order to develop a methodology to calculate the external costs caused by energy consumption and production. This is another far-thinking program of the EU community, which, along with their RoHS program, makes them far superior in this area, when compared to the almost vacuum of such considerations in the US and other developed and developing countries.

ExternE defines costs as the monetary quantification of the socio-environmental benefits and damage, expressed in eurocents/kWh.

The goal and real possibility here is providing a scientific basis for policy decisions and legislative proposals. The data can then help decision makers to justify subsidizing cleaner technologies and introduce energy taxes to internalize their external costs.

ExternE looks at all energy-production technologies using a methodology developed for this project that allows the various fuel cycles to be compared. An outline of the initial results for the solar (PV) fuel cycle, starting with a very small sample of PV systems, shows that the results are not consistent and require more work. There has been pressure from the PV industry for the PV cycle estimates to be re-done using a larger sample of more representative systems, but this is still in the making.

One of the earliest publications from this project by Baumann et al. is still valid. In this paper the author outlines the basic assumptions and requirements of the methodological framework for the quantification of external costs and compares the environmental effects of different energy technologies, including renewables and the conventional technologies of coal, oil, nuclear, and natural gas.

While this is a valid effort and deserves further funding and development, it is our opinion that no conclusive, 100% reliable, data could be obtained until we get at least *one full cycle*—30 years—of the different types of renewable power generating plants field operation.

A complete—30 years—life cycle data is needed for reliable evaluation of the renewables' externalities

This means that we have to wait at least 10-15 years to get fully documented picture of the reliability and efficiency of the new renewable technologies. Without this data, any and all estimates would be just that; estimates of incomplete guesswork.

The U.S. Environmental Movement

Environmental and conservation organizations in the United States are usually established to protect the environment, land, water, flora, and fauna on government, or private land, coastal areas, conservation areas, parks and municipalities

Most environmental organizations operate as nonprofit companies. The revenue is used mostly to achieve their goals, instead of making profit and paying owners and shareholders. The nonprofit environmental organizations usually enjoy tax exempt status, which obligates them to maintain certain level and type of operation as required by the law. Failure to do so, results in loss of the tax exempt status. Different states and localities offer nonprofits additional exemptions from taxes, such as sales or property tax. The tax exempt environmental organizations are required to file annual financial reports, which are available to the public.

There are a number of environmental groups in the U.S., who influence the direction and speed of the development of different energy technologies.

As Table 9-5 shows, some of these groups have been around for a long time, so they have established methods of operation, some of which are very effective. This, however, is a partial list, since the number of active environmental organizations in the U.S. is well over 100.

The U.S. environmental movement is a huge and mighty entity, which has the power to change the direction of the energy sector and affect the energy markets one way or another. The different environmental groups focus on different subjects, that deal with important national or local issues, such as:

Table 9-5. The oldest U.S. environmental organizations

Sierra Club (1892)
National Audubon Society (1905)
National Parks Conservation Association (1919)
Izaak Walton League (1922)
Wilderness Society (1935)
National Wildlife Federation (1936)
Defenders of Wildlife (1947)
The Nature Conservancy (1951)
WWF-US (1961)
Environmental Defense Fund (1967)
Friends of the Earth (US) (1969)
Natural Resources Defense Council (1970)
Greenpeace USA (1972)

Animal Rights
Conservation
Deep Ecology
Eco-feminism
Eco-preservation
Eco-spiritualism
Environmental Health
Environmental Justice
Green/Anti-Globalization
Nuclear Safety
Reform Environmentalism
Wildlife Management
NIMBY, etc.

The major environmental issues, addressed by these groups are:
— Land use and impact
— Human heath and lifestyle impact
— Water use and protection
— Wild life protection
— GHG emissions

The different environmental organizations use different methods of operation, which range from attending meetings on environmental issues, organizing protests, etc. If all peaceful attempt fail to produce the desired results, most of these groups use the state and federal courts to enforce environmental and conservation regulations and laws. Greenies Dictatorship

Like any good thing, environmental consciousness and the movement around it, could grow into a cancer similar to that of the communist dictatorship. Not that environmentalism is a bad thing, nor that environmentalists are bad people, but when people—any people—see themselves empowered, they tend to take advantage of the situation.

Empowered people usually look for ways to benefit their cause—and their cause only—sometimes at the expense of others.

This trend eventually ends up with a focus on benefiting their own cause and themselves at all cost. A full blown dictatorship regime follows close behind. This never fails. It is our human nature that makes us do such unreasonable things. The capitalist system supports, and even rewards, such behavior.

The Time Machine is a novel written by H. G. Wells in 1895 as a sort of a sci-fi horror scene. A scientist-wanna-be young man builds a time machine that carries him into the future. Upon arrival some centuries in the future, he finds a beautiful, pastoral world populated with sickly looking, passive, willowy people; totally ignorant of machinery, or any kind of industrial development. They are non-confrontational and terrified of the dark and many other things.

The young hero soon discovers that another race of semi-human beings below the surface of the world has established a terrifying dictatorship. The ape-like, violent group of creatures runs the world from the dark underground, and lives off of the blood and flesh of the surface people.

At the end of the novel, the young man travels even further into the future, where he observes the gradual end of the human race and its devolution into crustaceans drowning into primordial slime.

H.G. Wells was a sort of a socialist and his novel was an allegory for the control that the rich and powerful capitalists had over the workers at the time. He believed that the exploitation of workers by big industrialists destroyed the worker class by creating weak people which leads to total dependence, while making the industrialist owners bloodthirsty for profits.

Wells believed that a *one world government* is needed to provide for all of the needs of the population. He went as far as believing that sterilization of inferior, sub-standard people would be a betterment for society as a whole.

The Time Machine is just a figment of a sick imagination in the late 1800, you say. But what followed during the years of violent communist oppression of millions of people in the Soviet Union and Eastern Europe is very close to what Wells saw in the future. Mass genocide, followed by relentless persecution and exploitation of the ordinary people. This is what the communist dictatorship brought us, under the disguise of, "To every one according to their abilities, and to everyone according to their needs."

Today, the environmental movement wants to change the world for the better too. We have messed up, so step away and let us show you the way. All you have to do is move to the deserts, live in tents, ride bicycles, do not use any electricity or fuels, and have 0.2 children per family, or less. This will allow the world environment to recover and then we would have our beautiful world back.

The lesson of the past is that if any minority class is successful in imposing its will upon the majority, it eventually creates dictatorship. From the same lessons of the past we also know that dictators usually focus on pushing their agendas down people's throats. In this case, the environmental dictators would most likely luxuriate in the new and better world by living in large houses and driving luxury cars, while leaving the rest of us to live in our desert tents, riding government issued bicycles.

A confirmation of the possibility of such Wells-like scenario is the United Nations' Agenda 21 document, adopted at the 1992 United Nations Conference on Environment and Development (UNCED) in Rio de Janeiro.

Agenda 21 reflects a perceived international consensus that worsening poverty and growing stresses on the environment require greater integration between environmental and development concerns. Such a comprehensive approach to development, according to Agenda 21, is necessary for the common well-being. In order for countries to be able to continue to meet the basic needs of the world's population, improve living standards, and manage the planet's natural resources in an efficient manner.

While this is all good and dandy, this is pretty much what the communists aimed to achieve too. Or rather, this is the motto they hid behind, while trying to justify their inefficiencies and greed.

Agenda 21 seeks to promote **world government** *through the creation of a* **centralized planning agency***, that would be* **responsible for oversight into all areas of our lives***.*

But wait...this sounds so very familiar. These words sound like they were copied from the Communist Manifesto. Wasn't this what the role of KGB in the communist system was? Once in power, the communists, through cruel and merciless KGB actions, started making and enforcing their own, more often than none awkward and even stupid, rules and regulations.

With some notable exceptions, the new social system was acceptable at first and many people supported it. Then, the dictators felt the awesome power of their new system, and kept on modifying it until by the end it was a totally perverted, self-serving form of blatantly inefficient dictatorship. And look at the results! The signs of the damages, the communist system dictatorship caused to the people in the old Soviet Union and Eastern Europe, will be quite noticeable for many decades.

The lesson here is to NOT let any one person, or group of people, tell us what is best for all of us as a society. We are free, thinking human beings, fully entitled to chose our destiny. And this includes our energy and environmental future.

It is better to live free in a polluted world, than under any dictatorship—including environmentalism.

So, Agenda 21 is a good start of talking about the problems, but it is not the solution. By no means! There are many other moving parts that need to be included in the final solution. Environmentalism is only one part of the dynamic global society and our energy markets. It is an important part, but let it remain just that; a part; not the one and only solution.

Case Study: Paris 2015

In December 2015 the world gathered in Paris to discuss and solve the climate changes caused by man-made gas emissions. An agreement was signed by all participants—great success. The first step in the right direction was made.

As a result of the Climate Change summit;
- The world will look for ways to limit global warming by 2100 to, or below, 2 degrees Celsius (about 3.6 degrees Fahrenheit), as compared to pre-industrial levels. It also lists 1.5 degrees Celsius (2.7 degrees Fahrenheit) as an 'aspirational' goal, but leaves the actual decision-making for that goal until a later date.

- According to climate models, the national climate pledges submitted by countries before the Paris summit will only limit warming to between 2.7 and 3.5 degrees Celsius (about 4.9 and 6.3 degrees

Fahrenheit respectively), which means that further, and very serious, action must be taken soon if the 2 degree Celsius reduction is to be achieved.

- To that end, the agreement formally asks the Inter-governmental Panel on Climate Change (IPCC), the U.N. climate science and research body, to issue a special report in 2018 detailing steps needed to reach the 2- and 1.5-degree Celsius targets.

- The agreement also requests that countries re-submit their pledges by 2020, reflecting the conclusions of the IPCC report and new developments in technology. A similar review is scheduled to take place every five years starting in 2020.

- With the worldwide economic crisis still lingering, most countries were slow to finance the Green Fund. Industrialized countries promised in 2009 to provide $100 billion per year to help poor countries adapt to the impacts of climate change by 2020. Only $10 billion is available today.

- The delegates also could not agree on a finance target after 2020. Instead they agreed to use the $100 billion figure as a floor for subsequent years. More specific goals were left to be decided in the future.

- The agreement provides a big indirect push for the development of renewable energy sources, but few direct incentives. Some key initiatives calling for "de-carbonization" or "zero emissions" at some future date has been left out. These will be discussed at the following sessions.

- The experts agreed that the long-term temperature goals, along with individual national climate pledges that often include goals for clean energy, would act indirectly as incentives for a big ramp-up in investments in renewable energy.

The final decisions, as far as the type of technologies to be used and the size of their implementation, were left to the different countries.

- Paris 2015 summit put climate change in the headlines and made it an urgent issue to be resolved by the world as a whole. When the Kyoto Protocol was 'agreed' to back in 1997, it was seen as an attempt to reaching an initial global environmental agreement.

After Paris 2015, it is a top geopolitical issue, that attracted 147 heads of government at the opening sessions, and resulted in a unprecedented and unforeseen consensus.

• Great start, no doubt. But only a start, with a lot of work left to be done. The 'voluntary' aspects of the agreement was supported by the U.S. delegation, which knew well that Congress will no approve any binding commitment.

This makes the full and timely implementation of the agreement questionable. So now, the nations will gather again and again every five years to calculate the progress and outline the new steps to be taken.

NATURAL RISKS

The major risks affecting the energy markets can be divided into a) natural, these that are caused by Mother Nature, and b) man-made risks, these caused by human activities.

We will take a close look at these, starting with the natural

Draughts

Hydropower generation is only one of the many uses of water. Its importance for the survival of the species cannot be overemphasized. In recent years, however, we have been witnessing some serious abnormalities related to the natural water cycle. There are increasing incidents of draughts and floods, most likely caused by global climate changes. These events affect the way water is delivered and used around the world.

Droughts around the globe are affecting the climate (or is it the other way around—hard to tell today), and in the process, agricultural crops, animals, and many people's lives are affected. It is so bad, in fact, that even the Global Drought Monitor web site crashed in 2013, displaying a morbid sign, "IMPORTANT NOTICE: 19th November 2013. The server running the UCL Global Drought Monitor has malfunctioned beyond repair. We are currently exploring possibilities and will provide an update on the future of this service as soon as possible. Please accept our apologies for any inconvenience." A sign of the times?

This sounds like the end of the world, but is actually just a reminder of how futile human efforts to control Mother Nature are at times. That incident only emphasizes that fact. But seriously, periodic droughts are not uncommon, and we often hear the "seven year drought" theories, and while there is something to that, Mother Nature is very complicated and well beyond our ability to predict her actions.

The "seven-year" drought in Arizona, for example, is going on over ten years now. And is actually getting worse, with no signs of change for the better any time soon. At the same time California seems to be entering its "seven-year" drought period, that seems to be worse than any seen before. And yes, there were many such droughts in its past, so we should not be surprised, nor attribute it to one single factor. Mother Nature won't like that.

Droughts today are caused by global warming (at least partially, according to the scientists). Elevated temperatures increase the evaporation of water (evapo-transpiration) from land and water surfaces and from plants due to evaporation and transpiration.

This leads to increased drought in many areas, which felt worst in the drier regions of the world. There, evapo-transpiration produces long periods of drought, which are basically lower water levels of rivers, and lakes. Eventually, the groundwater level drops too, which brings reduced soil moisture in the agricultural areas.

As a result, there has been a significant decline in precipitation in the tropics and subtropics since the early 1970s. At the same time, Southern Africa, the Sahel region of Africa, southern Asia, the Mediterranean, and the U.S. Southwest, are getting noticeably drier. And even the relatively wet areas are increasingly experiencing longer dry conditions.

And even worse, there is a noticeable expansion of dry areas around the world. And not only in the Sahara desert sands. There are estimates of the amount of land affected by drought to grow, and water resources in these areas to decline over 30% by 2050.

These changes are thought to occur, at least partly, due to expanding atmospheric circulation pattern known as the Hadley Cell. Here, warm air in the tropics rises and loses moisture to tropical thunderstorms, which usually dissipate over the ocean or over wet areas which don't need that much water—sometimes causing huge floods.

The dry air from the tropics descends in the subtropics as more dry air, that only creates more drought conditions. As jet streams continue to shift to higher latitudes, and storm patterns shift along with them, semi-arid and desert areas are expanding steadily.

Once flourishing gardens are being converted into parched lands with little hope for recovery. The big problem is that droughts in key states usually mean reduction of agricultural production. This results in higher food prices and deficit of agricultural products. Form energy security point of view, extended droughts mean reduction of the amount of hydropower and biofuels we can produce. And, of course, the energy prices will increase accordingly too.

One of the best examples of the negative effects of extended droughts on the energy supplies can be seen at the sugar cane the fields of Brazil. Brazil runs on ethanol, which comes from sugar cane. Sugar cane is the most efficient biofuel raw material, but it needs a lot (with capital L) irrigation water.

Brazil, fortunately, has a lot of water (mostly from rain), which allows it to grow a lot of sugar cane. This in turn has allowed it to process it into ethanol, which fuels 90% of its transportation sector since the 1970s...until now.

The persisting global climate change, accompanied by extended drought periods threatens to put Brazil's transport system back on the crude oil life support system. One of the worst droughts in three decades slashed Brazil's northeastern sugar cane crop in 2013-2014 by as much as 30% in some areas and even more in other.

Brazil's north and northeast regions account for only about 10% of national cane output, but the crop is an important source of sugar and ethanol at home and abroad when the main center-south crop is idle between harvests. The Alagoas region, which begins harvest in September, typically crushes cane to produce sugar and ethanol through March, or about six months, just as the center-south starts up harvest of the new crop.

But this time, cane processing operations ended sooner simply because there was not enough cane to crush. The northeast produced only 52 million tons in 2013, down 15% from the record 62 million tons in 2013. Sugar output from the region also fell by 15% from the over 4 million tons sugar production in 2013.

Some states saw a 20-25% drop in cane output as direct result of the fact that the cane fields received only 25-30% of the normal rainfall in 2013. This was the worst crop since 1983. When the cane and sugar production is down, the ethanol production suffers and gas pump prices go up accordingly. If the drought continues, Brazil may have to import ethanol (and more crude oil) in order to supply its gas stations.

Case Study: California drought

California is in the midst of a severe, unprecedented, drought. The state is not a stranger to dry conditions, but in terms of severity and length, the present 7-year drought is unprecedented. The new estimates show that the drought is so severe that it's causing the ground to rise. What?

Scientists estimate that 63 trillion gallons of water have been lost since 2013 alone. So what happens when 63 trillion gallons of water disappear? A lot. This is about 240 billion tons of water, that rests on the ground beneath the waters of lakes and rivers. As the water evaporates, the ground shifts...in upward direction. There are measurements in the California's mountains, where huge ex-lake bottom areas have been lifted as much as half an inch.

California's water supply is in three largest reservoirs, all of which are at roughly 30% capacity. Other, smaller, water reservoirs are doing better. The statewide lowest average of 41% was recorded in 1997, when another devastating drought struck the state. The present drought is nearing the 1997 record and as things are going, it will set new records.

Figure 9-12. Lake Oroville...now stream Oroville.

Lake Oroville is (was) one of California's largest water reservoirs. Now it is a skinny steam flowing where the lake bottom was...a near-record low. So what that does to the state's energy production and the overall economy?

What is a state running out of water to do?

In August, 2014 The California Public Utilities Commission (CPUC) ordered all state water companies under its jurisdiction to provide direct notice to their customers of mandatory water use restrictions. These are accompanied by potential fines for non-compliance with the State Water Resources Control Board's Emergency Regulation for Statewide Water Conservation.

Starting in February, 2014, CPUC adopted drought

procedures for water conservation, rationing, and service connection moratoria for regulated water utilities. In April, 2014, Governor Brown issued an Executive Order to strengthen the state's ability to manage water in drought conditions.

And then in July, 2014 the Water Board adopted Emergency Regulation that prohibits the use of drinking water for outdoor landscapes in a manner that causes runoff; the use of a hose without a shut-off nozzle to dispense drinking water to wash a motor vehicle; the application of drinking water to driveways and sidewalks; and the use of drinking water in a fountain or other decorative water feature, except where the water is part of a recirculation system.

Violation of the prohibited actions brings a fine of up to $500 for each day in which the violation occurs. Additionally, all CPUC jurisdiction water utilities are ordered to comply with the Water Board's requirements in implementing either a) mandatory outdoor irrigation restrictions or alternatively, b) mandatory water conservation measures.

Utilities must include notice of the implementation of either the mandatory outdoor irrigation restrictions or mandatory water conservation measures as required and as part of the required customer notification the CPUC required in its last order.

The CPUC-regulated water utilities are to implement water conservation measures consistent with the Water Board's mandate. CPUC will closely monitor the Water utilities' progress in encouraging water conservation and consider further action if warranted. CPUC-regulated water utilities also will take steps to monitor and promptly inform customers about leaks and to work with customers to stop leaks and water waste, consistent with best practices being implemented by other California water utilities.

Case Study: Produced Water

Pumping oil and gas out of the ground also produces large volumes of water with undesirable quality known as produced water. In many cases there is much more water than oil pumped every day from the wells. Safely disposing of this highly saline water and mitigating the effect of past disposal practices is a national concern for environmental officials, land managers, petroleum companies, and land owners.

Produced water is by far the largest volume byproduct stream associated with oil and gas exploration and production. Approximately 21 billion barrels of produced water are generated each year in the U.S. from its over 900,000 oil and gas wells. This is equivalent to

a volume of 2.4 billion gallons per day pumped up and disposed in some way.

This amount of water is enough to meet 4.5 million people's daily needs…if it was clean enough. But it is not. The physical and chemical properties of produced water vary considerably depending on the geographic location of the field, the geological formation from which it comes, and the type of hydrocarbon product being produced. Produced water properties and volume can even vary throughout the lifetime of a reservoir.

The major contaminants in produced water are: salt content. oil and grease, inorganic and organic chemicals, and naturally occurring radioactive materials. Storing, purifying and properly disposing this water is expensive, which is forcing some operators to take short cuts. The costs to properly dispose produced water vary from a few pennies to $5 and more per barrel. Multiplied by the millions of barrels produced daily, produced water has created a significant market within the energy market.

Intentional and unintentional leaks and dumps of produced water are a common occurrence in the oil fields. The issues related to produced water are not new, but have been coming under increased scrutiny lately, due to reported leaks and dumps into the water table and the streams in some localities.

In draught stricken California, state officials are taking a closer look at the effects of oil well operations on the quickly depleting water table. A number of wells have been shut down to protect the water supplies.

In response, some oil well operators in the state have figured out a way of purifying produced water and pumping it back into the water table, or piping it directly to nearby farm fields. Some wells produce 9 barrels of water for each barrel of oil, which comes to many millions of barrels of water added to the California empty water reservoirs. Yet, this is still just a drop in the ocean of parched fields across the state.

And while some areas are dying from thirst, there are areas in the U.S. and around the globe where periodic floods are devastating properties and hurting people.

Floods

On the other end of the water shortage extreme are the periodic floods that cause enormous damages. Globally, floods are becoming more frequent and dangerous. In the U.S. floods have been a way of life for people in some flood-prone areas since the beginning of time (or the Union). Some of the floods in the past were exceptionally damaging, so in response to heavy flood damages in the 1960s, the National Flood Insurance

Program (NFIP) was established. It is a federal program created by the U.S. Congress through the National Flood Insurance Act of 1968.

The program enables property owners in participating communities to purchase insurance protection from the government against losses from flooding. This insurance is designed to provide an insurance alternative to disaster assistance to meet the escalating costs of repairing damage to buildings and their contents caused by floods.

Participation in the NFIP is based on an agreement between local communities and the federal government that states that if a community will adopt and enforce a floodplain management ordinance to reduce future flood risks to new construction in Special Flood Hazard Areas (SFHA), the federal government will make flood insurance available within the community as a financial protection against flood losses.

The SFHAs and other risk premium zones applicable to each participating community are depicted on Flood Insurance Rate Maps (FIRMs). The Mitigation Division within the Federal Emergency Management Agency manages the NFIP and oversees the floodplain management and mapping components of the Program.

The intent was to reduce future flood damage through community floodplain management ordinances and provide protection for property owners against potential losses through an insurance mechanism that requires a premium to be paid for the protection.

The NFIP is meant to be self-supporting, though in 2003 the GAO found that repetitive-loss properties cost the taxpayer about $200 million annually. Congress originally intended that operating expenses and flood insurance claims be paid for through the premiums collected for flood insurance policies. NFIP borrows from the U.S. Treasury for times when losses are heavy, and these loans are paid back with interest. Since 1978, the National Flood Insurance Program has paid more than $38 billion in claims (as of 2011). More than 40% of that money has gone to residents of Louisiana.

The Biggert-Waters Act of 2012 mandated that the NFIP charge actuarial rates, resulting in a large rate increase for consumers. As of November 2013, legislation is pending in both the House and Senate that would delay implementation of these rate increases for four years.

As of April 2010, the program has insured about 5.5 million homes, the majority of which were in Texas and Florida. And now, as the frequency and intensity of floods—be it from river overflow, or ocean hurricanes—is increasing, the NFIP is expected to undergo a number of alterations.

As the global climate changes, and the flood risks increases, the national budget simply would not be able to support the expected flood of floods predicted to hit our rivers and shore communities in the future. The amount of money needed is just not available to take care of all the damage predicted to be caused by future floods.

So what do people in other, poorer, countries do when their homes and fields are flooded? They usually start from scratch., or move to higher ground. So maybe the U.S. national flood program should adapt this method too. It would save a lot of money and grief on yearly (and in most cases predictable) floods.

Earthquakes

Earthquakes are a real threat to the energy infrastructure. Earthquakes have occurred on Earth's surface since the beginning of time. Thousands of earthquakes of different magnitudes occur every year, but most of them go unnoticed as they are either too weak on the Richter scale or happen far from populated centers.

Large earthquakes, however, almost always result in loss of life, property, and environmental damage. On May 22, 1960 the world witnessed Valdicvia, Chile's strongest earthquake ever, with a magnitude of 9.5 on Richter scale. It caused 20,000 fatalities. The second strongest earthquake on record occurred on December 26, 2004, with a magnitude of 9.3 on Richter scale. The ocean floor of west Sumatra and Indonesia were the epicenter of this earthquake that caused over 300,000 causalities and the disastrous tsunami in the Indian Ocean.

On March 27, 1964, Prince William Sound, Alaska, faced the third strongest recorded earthquake. It measured 9.2 on the Richter scale and caused a great deal of damage in Anchorage.

The quake in Kamchatka, 1952 having 9.0 magnitude, is the fourth strongest earthquake on record. After this comes the 2011 Japan earthquake, measuring 8.9 on the Richter scale.

On 26 December, 2004, the Sumatra-Andaman earthquake (9.2 on the Richter scale) with an epicenter off the west coast of Sumatra, Indonesia, caused a large tsunami (called 2004 Indian Ocean tsunami, South Asian tsunami, Indonesian tsunami, or the Boxing Day tsunami). The earthquake had the longest duration of faulting ever observed, close to 10 minutes. It caused the entire planet to vibrate as much as 0.4-05 inches and triggered other earthquakes as far away as Alaska. 200,000 people lost their lives in the tsunami, which was accompanied by massive infrastructure destruction and environmental damage.

Other large earthquakes are the Tangshan, China, earthquake of 1976, in which at least 255,000 were killed; the earthquake of 1927 in Xining, Qinghai, China, with 200,000 dead; the Great Kanto earthquake which struck Tokyo in 1923 with 143,000 victims); and the Gansu, China, earthquake of 1920 which killed 200,000 people.

The deadliest known earthquake in history occurred in 1556 in Shaanxi, China, with an estimated death toll around 800,000. Of course, we know about the 2012 earthquake off the coast of Japan that caused a large tsunami. The combined effect of these two natural disasters was total devastation of portion of Japan's coastline, and the destruction of the Fukushima nuclear power plant.

The radiation emitted by the disabled nuclear reactors added a third and equally devastating component to the tragedy and completed its totality. Hundreds of square miles of coastland are uninhabitable and will remain so for decades to come.

The Fukushima, and the above listed disasters are a perfect picture of Nature at its worst, hitting man without warning and where we are most vulnerable. They are also a reminder of how small we are, as compared with Nature. We cannot provide 100% protection for our energy infrastructure, but following the Fukushima lesson, we will pay more attention to earthquake dangers.

Sun Flares

The sun is a giant nuclear reactor, which never sleeps, and is ever changing. From time to time it explodes, like billions Chernobyls and Fukushimas, sending electromagnetic pulses and charged particles in all directions. These are known as sun storms, or flares, which travel at the speed of light, or around 186,000 miles a second and could reach Earth in as little as eight minutes.

In 2015, the solar storm activities are their maximum, but the solar flare activity on sun's surface have been unusually quiet. The sun's first X-flare of the year, large enough to damage electrical grids and all kinds of electronic equipment—including satellites orbiting the Earth—erupted in March, 2015 in the direction of Earth. It impact was felt several days later.

The unusually large flare was observed by NASA's Solar Dynamics Observatory satellite. It was estimated that it contains the energy of several million hydrogen bombs exploding all at once. Luckily, most of this energy is directed away from the Earth, so what we receive is only a minute part of the entire force of the flares.

Many of the flares are more 20-30 times the size of the Earth in size, and fly fast, spewing electromagnetic

energy in form of gamma rays and such. They have been categorized as A, B, C, M, and X type flares, with X being the most powerful. A is respectively the weakest, followed by B, C, and M. Each of these letters represents a 10-fold increase in energy from the flare. An X-flare, for example, is 10 times as powerful than M, 100 times more powerful than C, and so on. In addition, there is also a scale of 1 to 9 within each letter class, with 9 being the most powerful flare.

A direct hit from an X-flare, as the most powerful category, would be catastrophic for our life-sustaining power infrastructure. It could also have a major impact on all modes of communications, electrical grid systems, and all kinds of electronics and control systems.

The giant flare that spewed from the Sun in March, 2015 was an X2.1, which was loaded with megatons of highly charged particles, known as coronal mass ejections (CMEs). These can cause geomagnetic storms, which actually could cause the massive disruptions of Earth to communications and electrical grids. While the electromagnetic radiation travels fast, the accompanying particles are much slower. Because of that, they might arrive on Earth days after the flares radiation hits.

The initial effect of the March X.2.1 flare impacted the local communications, resulting in a radio blackout in the 15-26 MHz frequency range. It lasted about a half hour on some areas of the sunny side of the Earth at the time.

The most affected areas on Earth, affected by solar flares and CMEs, extend from the Northeastern U.S., to the East Coast, into the Gulf of Mexico, and into Central and South America. The U.S. and other Western countries are technologically advanced societies, with complex and critical infrastructures run by electronics, so the space events bring another risk element. This danger takes on an added importance, since our national grid system and its components (all of which most of us take for granted and cannot live without) are already quite vulnerable. A large electromagnetic pulse (EMP), can fry the unprotected grid, and its substations, transformers, thousands of electrical components and automated control systems.

A major hit could damage a large part of the system which could take weeks, months, and even years, to replace. According to NASA, a direct hit to Earth from one of these enormous flares would have a catastrophic, cascading impact on the nation's critical infrastructures over a very wide geographical area.

As many as 350 of the large, customized transformers, which maintain the steady power supply across the nation would be destroyed. Since utilities do not keep spare inventory (expensive components) and these are produced and imported from abroad, it would be weeks and months before repair of the grid would be even possible. And if a the producing countries are also affected by a direct solar flare or CME impact, then our grid would stay immobilized for much longer time.

Such a disaster could cost the U.S. upward of $2 trillion, which would take 5 to 10 years to recover from, if at all. It would also affect the lives of some 150 million people, who will be threatened by starvation, thirst, illnesses, and even death.

The experts fear that such a catastrophic event could drastically affect life in the urban centers. Life in these is totally dependent on critical infrastructures for electricity, communications, food and water delivery, oil and gas, transportation, automated banking and financial institutions, and even emergency services. Grocery stores, for example, would be cleared of food within hours, due to the total panic. They cannot be restocked, due to lack of fuel for delivery trucks, so food would become scarce within days, which will fuel the panic further.

Automated controls that regulate the flow of oil and natural gas through the national pipeline system would be disabled. This would cause widespread malfunctions, fires and explosions in remote locations and under city streets, houses, and businesses. Fire and medical emergency vehicles would be grounded to, which would escalate the disastrous scenario.

And finally, and most dangerously for large populated centers, fresh water delivery would be discontinued too, since all filtering and sewage systems in the urban setting would be shut down. Water shortage would lead quickly to cholera and dysentery. If hospitals and emergency equipment are shut down, many people would get sick and die.

Yes, this is a nightmarish extreme, would may never happen, but the point is that there is a good possibility that it might happen…and we are not prepared for it.

The Carrington Event of 1859

At 11:18 AM on the cloudless morning of Thursday, September 1, 1859, 33-year-old Richard Carrington—widely acknowledged to be one of England's foremost solar astronomers—was in his well-appointed private observatory. Just as usual on every sunny day, his telescope was projecting an 11-inch-wide image of the sun on a screen, and Carrington skillfully drew the sunspots he saw.

On that morning, he was capturing the likeness of an enormous group of sunspots. Suddenly, before his eyes, two brilliant beads of blinding white light appeared over the sunspots, intensified rapidly, and became kidney-shaped. Realizing that he was witnessing something unprecedented and "being somewhat flurried by the surprise," Carrington later wrote, "I hastily ran to call someone to witness the exhibition with me. On returning within 60 seconds, I was mortified to find that it was already much changed and enfeebled." He and his witness watched the white spots contract to mere pinpoints and disappear. It was 11:23 AM. Only five minutes had passed.

Just before dawn the next day, skies all over planet Earth erupted in red, green, and purple auroras so brilliant that newspapers could be read as easily as in daylight. Indeed, stunning auroras pulsated even at near tropical latitudes over Cuba, the Bahamas, Jamaica, El Salvador, and Hawaii.

Even more disconcerting, telegraph systems worldwide went haywire. Spark discharges shocked telegraph operators and set the telegraph paper on fire. Even when telegraphers disconnected the batteries powering the lines, aurora-induced electric currents in the wires still allowed messages to be transmitted.

What Carrington saw the day before was a white-light solar flare—a magnetic explosion on the sun's surface. Now we know that solar flares happen frequently, especially during solar sunspot maximum. Most betray their existence by releasing X-rays (recorded by X-ray telescopes in space) and radio noise (recorded by radio telescopes in space and on Earth). In Carrington's day, however, there were no X-ray satellites or radio telescopes. No one knew flares existed until that September morning when one super-flare produced enough light to rival the brightness of the sun itself.

A modern solar flare was recorded Dec. 5, 2006, by the X-ray Imager onboard NOAA's GOES-13 satellite. The flare was so intense, it actually damaged the instrument that took the picture. Researchers believe Carrington's flare was much more energetic than this one. The explosion produced not only a surge of visible light but also a mammoth cloud of charged particles and detached magnetic loops—a "CME"—and hurled that cloud directly toward Earth. The next morning when the CME arrived, it crashed into Earth's magnetic field, causing the global bubble of magnetism that surrounds our planet to shake and quiver. Researchers call this a "geomagnetic storm." Rapidly moving fields induced enormous electric currents that surged through telegraph lines and disrupted communications.

Another huge solar flare was observed on August 4, 1972. It knocked out long-distance telephone communication across Illinois and caused AT&T to redesign its power system for transatlantic cables. A similar flare on March 13, 1989, provoked geomagnetic storms that disrupted electric power transmission from the Hydro Québec generating station in Canada, blacking out most of the province and plunging 6 million people into darkness for 9 hours; aurora-induced power surges even melted power transformers in New Jersey.

In December 2005, X-rays from another solar storm disrupted satellite-to-ground communications and Global Positioning System (GPS) navigation signals for about 10 minutes. That may not sound like much, but we would not have wanted to be on a commercial airplane being guided in for a landing by GPS, or on a ship being docked by GPS during these 10 minutes.

Another Carrington-class flare would dwarf these events, but fortunately they appear to be rare and in the 160-year record of geomagnetic storms, the Carrington event is the biggest.

Examining arctic ice shows that energetic particles leave a record in nitrates in ice cores. Here again the Carrington event sticks out as the biggest in 500 years and nearly twice as big as the runner-up. These statistics suggest that Carrington flares are once in a half-millennium events. The statistics are far from solid, however, and we stand warned that we don't understand sun flares well enough to rule out a repeat in our lifetime. Let alone preventing potential damages.

And on top of that, we must be aware of the fact that electronic technologies have become more sophisticated and are driving our lives. With time, they have also become more vulnerable to outside influences, including solar activities. In case of a large sun flare attack, power lines and long-distance telephone cables might be affected by the auroral currents, as happened in 1989. Radar, cell phone communications, and GPS receivers could be disrupted by solar radio noise.

Experts claim that there is little we can do to protect earth installations and satellites from a Carrington-class flare. In fact, a recent paper estimates potential damage to the 900-plus satellites currently in orbit could be significant and cost $50-$70 billion. The only solution in case of such a catastrophic event is to have backup to maintain earth operations and a fleet of satellites ready for launch.

What would be the effect of another "Carrington Effect" on our energy supplies and the energy markets is hard to estimate. Nevertheless, we know that electronics run every step of the energy lifecycle, so we must assume that any damage to the electronic systems would affect the energy markets negatively.

If the national power grid (or part of it) is knocked down, the immediate effect would be chaos in the electric power supply. How far and deep the chaos will spread is anyone's guess, but it would be a mistake to ignore such a possibility.

A direct hit on the U.S. would most likely damage, or wipe out, significant part of our satellite receivers, which would cause another shock that would reverberate throughout the entire country. Such an event is unprecedented and unlikely, but not impossible. And we are totally unprepared.

Just like millions of people living on top of active earthquake faults, all of us on Earth are living in a massive sun flare destruction zone.

We can also easily see millions of solar panels and related electronics installed on thousands of acres desert land seriously damaged. A direct hit could affect the wiring and put the panels on fire, evaporating many of them, and disabling entire areas of affected solar power fields. Not likely, but entirely possible. We must never say *never*, when Mother Nature is giving us danger signals.

NASA is working with other space agencies around the world on a study intended to understand better this phenomena and allow them to predict sun flare activities. For that purpose, a fleet of spacecraft is positioned above the Earth and is monitoring the sun activities. Their instruments are gathering data on all flares in order to eventually figure out the sun flares' creation mechanism and their behavior.

Spacecrafts SOHO, Hinode, STEREO, ACE and several others are already doing this work deep in space, while others like the Solar Dynamics Observatory will continue the effort.

Although no amount of research would be able to affect, let alone stop, sun flares, with time it might give us a better understanding of what we are dealing with, and at the very least eliminate the element of surprise and the fear of the "unknown." This might then lead to designing and implementing response and defense mechanisms to use in case of powerful sun flares headed our way.

Lightening

Lightning is another natural event that can cause devastating damage to solar panels, wind turbines, substations, and all types of power plants. Appropriate lightening protection must be considered and included in the design of the facilities, if damage is to be limited.

Direct or indirect lightening strikes, or strong sun emissions, can damage and even destroy unprotected equipment. In some cases of extraordinary power discharges, the damage can extend over protected equipment and facilities too. In all cases, the expense in identifying and replacing damaged components could be quite great. Direct strikes hitting a solar power plant could destroy the PV panels, inverters, etc. equipment by melting critical components, burning wires, and causing fires and shorts. A direct strike hitting a wind turbine, could damage the body, the blades, or the electrical equipment, as well as burn the wires and cause fires in the local area.

Indirect strikes are encountered more often, and could be equally damaging by inducing high voltage surges and creating sparks, which could burn wires, PV panels and other components. Any of this damage would cause loss of power and additional expense in repairs.

Participants in such projects—owners, financiers, insurers and operators, are increasingly aware of the risks and are looking into lightning protection. This is especially important in areas of high lightening incidents.

PV farms installed in open areas where there are no high buildings or trees—like in the deserts—have a very high probability of being hit by lightning. Lightning protection for such installations is a must. Likewise, roof-mounted solar panels on high buildings are also a target and should be protected.

Requirements for lightning protection of solar panels can be found in NFPA780, the standard for the installation of lightning protection systems.

The fundamental principle of lightning protection is based on placing the solar panels within a zone of protection so that downward lightning leaders attach to streamers emanating from air terminals or other strike termination devices and is directed to ground rather than hitting the panel and/or other components of the solar system and causing damage.

Critical to an effective lightning protection system are surge protection devices (SPDs) to suppress lightning induced high voltages to ground and avoid damage to the PV system. Most damage occurs from nearby lightning strikes to ground, usually within a few hundred feet. A near-strike can induce thousands of volts onto the PV system if not protected.

Underwriters Laboratories (UL) certified lightning protection systems for solar panels verifies that the system complies with NFPA780 and will give owners, investors, inspectors and insurance companies the assurance that the solar panels will be protected. As an independent third party, UL will inspect the system and determine compliance, which results in the issuance of a safety certificate.

Dark Energy

There is another form of energy that we are just now getting to know, or rather recognizing that we don't know well. As a matter of fact, the experts, can't even describe it properly, except for assuring us of its ominous presence. It is the so-called 'dark energy,' which we are told is all around us. As a ghost of unknown origin or behavior, dark energy is lurking in all corners of the Universe and we don't even know its powers and/or what to expect, when, and where.

During the 1990s, we were certain that the Universe might have enough energy density to stop its expansion and start going in reverse—and eventually collapsing. But in 1998, the Hubble Space Telescope (HST) observations of very distant supernovae showed that a long time ago, the Universe was actually slowing down, but for some reason it is now expanding much faster today.

This totally unexpected discovery was confirmed by a number of scientific organizations around the world. And yet, no one expected this, no one understood it, and no one knew how to explain it. Something was causing this "abnormality," but what was it?

Finally the experts came up with several explanations, ranging from a version of the long-discarded version of Einstein's theory of gravity, the one that called a "cosmological constant." The suggestion was that some strange kind of energy-fluid filled and ruled the space.

Some suggested that something was wrong with Einstein's theory of gravity and that a new theory is needed now. It would include some kind of energy field (or matter) that creates this cosmic acceleration. We still don't know what the correct explanation is, but have given the newly discovered phenomena a name. It is a mysterious one...*Dark Energy*.

And so, now we see ourselves travelling on our spaceship Earth, surrounded by this mysterious, but evidently energy filled *Dark Energy* stuff. This gives us the uneasy nightmare-like the feeling of being on a boat floating on something—some matter not like water—that we are not familiar with. How long would this matter allow us to float in/on it? What if decides to not allow us to float anymore? Lots of unknowns…

What we know now is that our voyage amidst this matter started 13.7 billion years ago, with unknown final destination or time of arrival. We now also see changes in the rate of expansion of the universe, starting about 7.5 billion years ago when objects in it began flying apart at a faster rate. This means that the universe instead of compressing is expanding due to increased energy.

The astronomers still don't know what this energy is, but speculate that the faster expansion rate is due to the mysterious Dark Force that is providing the energy to pull planets, starts and entire galaxies apart.

We can calculate (approximately) the amount of Dark Energy around us, because we know (somewhat) how it affects the Universe's expansion. But this is the extent of our knowledge on the strange matter. And beyond that, it is a complete mystery. But, oh, how important of a mystery it is, since it is all around us!

So, get this, 68% of the Universe is filled with Dark Energy that we have no idea what it is.

And then there is the *Dark Matter—vs. Dark Energy—which is* another thing that we can only speculate about. Dark Matter, we are told, makes up about 27% of the total, with the rest—including everything on Earth and around it—as *normal matter.*

Only less than 5% of the total mass and space of the Universe. Is occupied by normal energy and matter.

If this is so, then how *normal* is our *normal*, if it comprises only 5% of the totality of *abnormal stuff* all around us?

We don't know much about the rulers of the Universe, and their majesties; Dark Energy, Dark Force and Dark Matter.

Oh, yes, the scientists now agree that space is not empty and that Dark Energy is a property of space. There is no such thing as *empty space*, which is actually something that Albert Einstein realized almost a century ago. Space has amazing properties, most of which (almost 95%), we know nothing about, and which are just now beginning to be slowly revealed, but not completely understood.

The quantum theory of matter provides an explanation for how space acquires energy. According to it, "empty space" is actually full of temporary, *virtual*, particles that continually form and disappear with time. Physicists used this theory to calculate how much energy this *virtual stuff* would give empty space, the answer came out wrong. Miserably wrong—about 1 with 120 zeros after it, times wrong. How do you even get an answer that bad, unless everything you are working with is wrong? So the mystery only gets more mysterious.

So now we think of Dark Energy as a new kind of *dynamical energy*. Some sort of a fluid or energy field, that fills all of space. The only problem is that its effect on the expansion of the Universe is the opposite of that we know about *normal* matter and energy. Some have named it *quintessence*, after the fifth element of the Greek philosophers. But no matter what we call it, we still don't know what it is like, what it interacts with, or how and even why it exists. Mystery upon mystery!

And lastly, we now blame Einstein for giving us a wrong theory of gravity. That affects not only the expansion of the Universe, but also the way that our *normal* matter in galaxies and clusters of galaxies behaved.

This brings the possibility of another Einstein coming up with a new gravity theory. We only have no idea what it would be…for now, since to correctly describe the motion of the bodies in the Solar System, as Einstein's theory does, it should also give us different prediction for the Universe's behavior than what we would expect with our *normal* way of thinking. Maybe our thinking is that *normal* either…

In conclusion, we are on a long journey onboard our old spaceship Earth. We are traveling at a high speed through space—which is not empty as we thought until not very long ago. Instead, it is filled of type of energy and matter; stuff we have no idea what it is. Is our spaceship going to be swallowed by it, or just go up in flames? Or, would it be something more benign, like a mild electric storm, or electrical discharge like a large scale lightening?

So, at least for now, and until we get a deeper understanding, we would classify this new Dark Energy and Dark Matter phenomena as most capable of complete destruction of our Earth. The other potentially dangerous events, such as a large meteor colliding with our Earth, huge solar storms, nuclear accidents, and others are also on the back of our minds, and cause some of us to lose sleep.

The effect of the Dark majesties coming in contact with us, however, is billions of times more powerful and dangerous than anything else we know of. A collision with their Darkness might convert our spaceship Earth into a golf ball and hurl it across the expanse. Or it might shred it into small pieces and splash them all over the map of the Universe. But, most likely it would shift us into another world, or a sixth or seventh dimension. Much like *Alice in Wonderland* on steroids.

The seriousness, totality, and finality of an encounter with Dark Energy stuff, with such magic properties and of such magnitude, are unprecedented and unthinkable.

And yet, this is something most of us are not even aware of. Those who are, know that there is nothing they can do about it and so life goes on. Just like the people of San Francisco don't worry about the Big One coming their way at any moment. What can they do? Why worry about it, if nothing can be done to change things and make us safer?

So, we keep going about our daily affairs; worrying about the energy markets, the potential of nuclear accidents, or the depletion of the fossils. These are everyday worries that preoccupy us and absorb all our attention and energy. Dark Energy and Dark Matter don't matter much today. We will deal with them when time comes... if we are given a chance at all.

On a more practical level, we see problems with the supply chain of some natural materials. We discuss the pending depletion of the fossils—and crude oil in particular—elsewhere in this text, so will not spend more time on it. There are, however, a number of other very important materials that need to be mentioned here. These are the rare earth metals.

Rare Earth Metals

The rare earth metals are a special group of elements, number 57 through 71 in the Periodic Table of Elements. Number 21 and 39 are also included in this group because of their similar properties. Many of these elements, also called lanthanides, are critical components in a large number of everyday items, such as cell phones, vehicles, wind turbines, and even jet fighters, rockets and satellites. Life would be quite different without these, and the U.S. military would just lose its superiority on the battlefield.

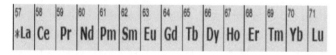

Figure 9-13. The rare earths (lanthanides)

The name "rare" is actually misleading, because they are found in many places around the world. They are, however, dispersed in small quantities all over the place, which makes them difficult and expensive to mine in large quantities.

Decades ago, the U.S. had the world's only rare earth elements industry, the strategic value of which was not fully appreciated at the time. It all started in the mid-1990s, when China purchased the critical technology that the U.S. had developed, together with America's largest rare earth magnet company, Magnequench. With the factory went the intellectual property, patents,

equipment, and even people. That was a big mistake, but at the time, the technology was not appreciated for what it is. And so, the rare earth industry was gradually shifted to China, where today 80-90% of these crucial metals are mined, processed, and export to the rest of the world at the Chinese whim.

Prices soared in 2010 after Beijing cut rare earth export quotas by 40%, or just over 30,000 tons. The restrictions were necessary, according to the officials, to reduce the environmental damage which mining rare earths creates. The restrictions bolstered Chinese industry at the expense of other countries by forcing them to pay two-three times more than what Chinese companies paid for the same earths.

The U.S., which is almost totally dependent on China for the metals, filed a complaint in 2012, which was joined by the European Union and Japan soon after. They challenged the export restrictions on 17 rare earths, as well as two metals used in steel alloys: molybdenum and tungsten. An interim report by the W.T.O. panel last October had indicated that the panel would rule against China. The Chinese reversed their decision soon thereafter and even apologized for it.

Soon thereafter, in 2014, Chinese and Japanese ocean vessels collided, and in the ensuing international chaos the Chinese government imposed export sanctions to Japan. As a result, for several months Japan did not receive the contracted rare metals. And although there are other factors that play a significant role in this case, the fact that China has the power to control exports of critical commodities is quite troubling.

The problem today, therefore, is not that there are not enough of these materials to go around, but rather the increasingly troubling trend of the Chinese government manipulating their exports. The danger of China monopoly of the industry became even more threatening by news of the largest rare earths mine and processing facility outside China being in financial trouble.

Molycorp is a Colorado-based company; the only rare earths mine and processing facility in the U.S. and one of very few outside China. Molycorp's was warned by its auditors that it might not be able to stay in business much longer. Chinese competition works...

In the early 1990s, China's government saw great potential in the rare earths industry. The former Chinese leader Deng Xiaoping indicated, in a little known speech, that rare earths could provide additional strategic power. "The Middle East has oil," he said, "China has rare earths." And so, China now has more power to pull the international strings of the rare earths markets.

Labor in China is cheap, the environmental destruction by large factories is a non-issue, and China has a lot of money. All this put together, explains why the U.S. exported the industry to China. It also explains why no one—the U.S. included—can compete with China.

Rare earths are metals that, among other uses, are bedrock of the miniaturization and militarization processes. They enable electronic gadgets to be more compact and effective. Tiny, super-powerful magnets made from these metals have become crucial in the consumer electronics, green energy and defense industries.

Wind mills, steam turbines, power plant controls, and many other energy related components and systems contain significant number of rare earth metals.

The U.S. Air Force jet fighters contains significant amounts of rare earths. Some say that F-35 jet fighter contains almost 1,000 lbs. of rare earths on its wings, in the engine, and navigation components.

But this is only a small sample of a great number of advanced weapons systems that rely on these elements. Weapons guidance systems, smart bombs, satellites, and many other sophisticated electronic devices contain rare earths.

We would go as far as saying that anything hi-tech made today contains some type of rare earths.

So how does this affect the global energy markets. Simple; without rare earths we would be stuck in the 19th century and have to revert to analog, or manual, weapons guiding and GPS systems. This in turn, means that our military would not be effective enough to police the world's dangerous places. At such time, the flow of crude oil and other energy sources, to the U.S., and around the world, would be interrupted.

Extreme case scenario, yes, but it is quite easy to derive from it that even more moderate cases would bring us a lot of misery and dangers.

MAN-MADE RISKS

In addition to the internal and external natural risks, man has created conditions, which pose risks to humans, wildlife, and the environment, all of which affect the energy markets too.

The Human Factor

Everything man does on Earth has primary and secondary consequences. Mining coal, for example, produces energy materials, which are then transported and burned. This results in energy, which is the primary goal

and consequence of the coal mining. These activities, however, are accompanied by a number of risks and dangers.

The energy sector is basically a dangerous place to work in. Here, people deal with large equipment pieces that have to be moved from place to place, huge structures have to be erected. People are involved in all these dangerous activities, which can hurt and even kill people.

Oil Field Risks

Unprecedented oil and gas drilling rush is underway in North Dakota, Pennsylvania, Texas and other places. Oil and gas production has risen several-fold recently there and with it, the business has become that much more dangerous too. There have been 74 reported deaths and many injuries in the Bakken oil fields alone during the last decade. Many more have happened in the other locations as well.

It is a dangerous business. Workers deal with large pieces of equipment, heavy cables, climb tall structures, and perform heavy physical labor. All the while, underneath their feet is a massive inferno of liquid and gas combustible materials.

Blowouts and other accidents are rare, but when they occur, those people who happen to be nearby suffer the consequences.

And oil companies have found a way to evade accountability from deadly accidents. It has become the norm for the owner and operator of the well to be separate companies, which shields both from accountability. Most of the workers in the oil fields are considered outside contractors, so this way the regulators cannot cite the company directly for any accidents.

On top of that, the energy companies hire what are known as company men. They're site supervisors, and as independent contractors they are outside of OSHA's purview. Company men are officially representatives of the oil companies. They are usually people with a lot of experience in the field, and are ultimately in charge of all activities on the sites. This is another layer of insulation for the owner and operator to further distance themselves from any direct responsibility in case of an accident.

And finally, but very importantly, workers sign oil field contracts upon hire. These contracts are designed to shield the companies even when they are responsible for accidents. A number of attempts to pass anti-indemnification bill for North Dakota's oil wells, similar to that passed in Texas, were outgunned by the oil lobby. This way a system was created, where the oil companies are legally allowed to shift the responsibility for their own negligence onto other parties.

And so, with several tiers of protection, the oil companies in parts of the Bekken oil field are free to rush into the black gold deposits, drill fast and make profit at mind boggling speed...sometimes at the expense of the workers' health and lives. Is this fair? Chasing the mighty dollar doesn't mean running over dead bodies. Or does it?

Man's activities are underway all over the Earth. One of the most widely spread and despicable is the deforestation of large pristine forests and areas around the world.

Deforestation

Deforestation is the new, and largest, plague on Earth, where huge forests are cleared on a massive scale. This results in waste and often damages the land and the local environment.

—Forests still cover about 30% of the world's land area, but swaths of forests the size of Texas are lost every year.—At the current rate of deforestation, the world's rain forests could completely vanish by the end of this century.

Forests are cut down for many reasons; people in developing countries, for example, gather wood every day for heat and cooking. This is one of the least known facts of the energy sector; millions upon millions of people depend on daily wood supply to run their homes and businesses. This has already caused deforestation of large parts in Africa and Asia.

But most of the forests are cut down for making money one way or another. The biggest driver of deforestation is clearing land for new agricultural fields. Farmers cut forests to provide more room for planting crops or grazing livestock. Small farmers clear a few acres each to feed their families by cutting down trees and burning them in a process known as "slash and burn" agriculture. Large farming operations are responsible for uncontrolled deforestation of South America's forests that is shrinking by the day as a result of these activities.

Deforested lands are used for developing different projects, like the construction of new populated centers, and commercial enterprises. Many thousands of acres, however, are destined for some types of energy production. Huge deforested areas are used to grow crops for bio-fuels, while others are used for the construction of solar and wind power plants.

Large part of the deforested land is used for renewable energy projects.

How good are the renewables then, if we are killing the precious forests on Mother Earth? This is extremely serious matter that needs to be addressed before it is too late.

Logging operations, which provide the world's wood and paper products also cut millions of trees each year. Loggers, many of whom are acting illegally, clear forests to build roads as needed to access more and more remote forests. This leads to further deforestation. And finally, large forested areas are cleared to accommodate the growing urban sprawl, mostly in developing countries.

Some deforestation is unintentional; caused by natural forces, or a combination of human and natural factors. Wildfires (natural and man-made) consume millions of forested acres every year, while overgrazing of the lands prevents the growth of young trees.

Removing trees deprives the forest of portions of its canopy, which blocks the sun's rays during the day and holds in heat at night. This leads to more extreme temperatures swings that can be harmful to plants and animals. Trees also play a critical role in absorbing the greenhouse gases that fuel global warming. Fewer forests means larger amounts of greenhouse gases entering the atmosphere—and increased speed and severity of global warming.

Deforestation has many other negative effects on the environment. The most dramatic impact is a loss of habitat for millions of wild life species.

Over 70% of Earth's land animals and plants live in forests. Deforestation destroys their habitat and kills many of them.

Deforestation is a major driver of climate change as well. Forest soils are moist, but after deforestation they dry and become unfertile without protection from sun-blocking tree cover. Trees help the proper function of the natural water cycle too, by returning water vapor back into the atmosphere. Without trees to process the water vapor, deforested lands become barren deserts within short time period.

Managing forest resources by reducing (and even eliminating) clear-cutting of large areas is one way to ensure that the world's forests, and their environment, remain healthy. Planned and coordinated cutting is a must, in order to ensure planting of young trees to replace the old ones.

The tree cutting today exceeds the growing by a large margin, so the world's forests are themselves endangered species.

The world's forests represent a huge market of wood production and use. Here, private individuals, small and large enterprises, and entire governments are involved in the wood business. Each of these entities has their own reasons and goals, but most are interested mostly in cutting trees.

It is hard to think of wood as a renewable energy source, if we continue demolishing the world's forests at the present rate.

Desertification

In a similar way, during the last century the world has been undergoing another large environmental disaster, called *desertification*. It is a type of land degradation in which a relatively dry land region becomes increasingly arid and useless. This is due to losing its bodies of water as well as vegetation and wildlife. This phenomena is caused by either climate change and/or human activities. In any and all cases, it is a man-made phenomena. We are making the bed we have to lay in eventually.

There is an evidence of serious and extensive land deterioration occurring several centuries ago in arid regions had three epicenters: the Mediterranean, the Mesopotamian Valley, and the loessial plateau of China with its previously dense population.

Desertification has contributed to the collapse of several large empires of the past, such as Carthage, Greece, and the Roman Empire. Recently, it has been blamed for causing displacement of huge local populations.

Desertification is presently a significant global ecological and environmental problem, equal in size and importance to deforestation. As a matter of fact, deforestation and desertification go hand in hand in many areas around the globe. A major factor today is over-grazing, where animals are allowed to roam uncontrolled and turning previously dense vegetation areas into desert lands.

A major effort is underway to control, stop, and reverse the desertification in several countries. China's State Forestry Administration, for example, has implemented a nationwide land control program, which aimed to get a basic control of desertification by 2010. The second phase aims then to reduce the area of desertification year by year until 2030; and the third phase aims to raise the nation's forest cover and bring all desertification sources under effective control by the year 2050.

Similar efforts, designed to control the deforestation, are needed all over the world too. There are some local attempts to do that, but a wide-spread, coordinated effort is needed for best results. Only then, with deforestation and desertification under control, will we be able to keep the planet from further and out of control excess erosion.

Coal Mining Risks

There are many health risks linked to coal mining operations. There are obvious workplace hazards associated with working in a coal mine, but these are not the only risks. People who live within proximity of a coal mine can also suffer a variety of health problems.

The miners, however, are faced directly by a plethora of dangers in the cramped, unsafe facilities. Injuries of all kinds are common occurrences. Miners get injured by falling objects from the shafts' ceilings, small and large equipment use, and at times by roof collapse.

The threat of physical injury is a real concern for many, it is only one of the threats miners face daily. They run the risk of respiratory damage through the high levels of dust and other chemicals and particulates present in the coal mining facilities.

The disorders caused by these particulates include COPD, coal worker's pneumoconiosis (CWP, also known as black lung), and progressive massive fibrosis (PMF). Damage to hearing, as a result of the high noise level in the enclosed areas, is a constant concern.

Both types of mining (underground and surface) pose their own set of problems to the nearby communities too. Some of these are as follow:

- The communities in proximity of underground mines include; increased rates of lung cancer, respiratory disease, and low birth weight. Some illness and disease rates (COPD and hypertension) are directly related to the amount of coal extracted from the mine. These, and other, health issues pose a significant risk to all people located within proximity of underground coal mines.

The communities in proximity of surface, or mountaintop, mining sites also face a number of health problems. Explosive charges, used at these sites cause loud noise and huge dust clouds, which drift into the residences and affect the respiratory health of the locals.

The explosives emit chemical gasses, which have been linked to poisoning local area residents. Explosions have been also known to fracture underground water tables, and lead to the contamination of drinking water by heavy metals, mine drainage, and methane gas. Explosions have also caused flying debris that have hit homes, cars, and people, and causing property damage and death.

Several of the illnesses associated with underground coal mining are also present in communities located within proximity of surface mining sites. In a word: coal business is dangerous business. Not only for the miners, but also to those that happen to live nearby.

Gas Line Risks

There are thousands of miles of natural gas pipelines under the streets in our cities. As the pipeline get old, they start leaking. Most of the leaks are very small, and very hard to detect without specialized equipment.

This was confirmed by a study published in January by Duke University and Boston University researchers in 2014, who spent two months driving the streets of Washington, D.C., with mobile monitoring equipment. They found 5,893 leaks of methane, the main component of natural gas coming from pipes in different areas of the city. In a similar study of Boston, which like Washington has aging natural gas infrastructure, the team found 3,356 leaks under the city's streets.

One dozen of the leak sites had underground methane concentrations as high as 500,000 parts per million, which is ten times the threshold at which an explosion can occur. "If there's a telecommunications manhole there, and you get a spark or a short, the air can ignite," according to the researchers. "If someone drops a cigarette butt down a manhole, it can do the same thing."

It's not known how many such leaks exist nationwide, since there has been no systematic effort to search for them. But the gas explosion accidents continue.

In 2011, for example, a natural gas explosion in a residential neighborhood in Allentown, Pennsylvania, killed five people. A state investigation discovered that the cause was a crack in a main that had been installed in 1928.

The big problem are the cast-iron and wrought-iron pipes that gas companies used to build their distribution networks in many parts of the nation back in the early to mid-20th century. These materials are very strong, but the iron pipes are vulnerable to corrosion, and their rigidity makes them susceptible to stresses. Pounding from construction and vehicle traffic above increases the possibility of leaks opening in the stressed areas.

Although most of the iron pipes have been replaced, about 46,000 miles of iron mains and service lines to individual homes and businesses are still in place.

Only about 2.5% of the nation's gas lines are made of cast iron, 11% of the major incidents have involved cast-iron pipes

In a separate incident, a natural gas explosion in New York's East Village residential area caused two apartment buildings to collapse and burst into flames. Two others caught fire, and injured a dozen people. The explosion occurred shortly after 3 p.m., sending flames into the sky, with bloodied victims running out of the burning buildings, and collapsing on the street. Only three people were critically injured.

Preliminary evidence indicated it was a gas-related explosion, caused by a private gas and plumbing upgrade work that had been going on in one of the buildings. Con Edison utility inspectors had been on the scene about an hour before the explosion and found the work did not pass inspection and needed to be reworked. The explosion made the rework unnecessary.

In California, a pipeline explosion occurred in San Bruno' Crestmoor residential neighborhood in 2010. A 30-inch steel natural gas pipeline owned by Pacific Gas & Electric exploded into flames and loud roar. The noise and heavy shaking made some residents and the first responders to believe that it was an earthquake, or that a large jetliner had crashed.

It took nearly an hour to determine that the cause of the explosion was a gas pipeline leak. Eight people died and several were injured in this incident. The explosion and the resulting shock wave were recorded as a magnitude 1.1 earthquake, with a wall of fire more than 1,000 feet high.

Man-Made Earthquakes

Man-made earthquakes…is this even possible? Yes, possible! It is a phenomena discovered and documented recently. A wave of brand new earthquake centers were observed in some areas of the U.S. which were suspect to be caused by man's activities.

Man-made earthquakes are caused by excess mining and hydrofracking activities.

The general consensus, confirmed by scientific studies, is that hydrofracking has caused small to medium sized earthquakes in a number of states in the U.S. and abroad.

The general theory behind the hydrofracking-caused earthquakes is that the injected deep into the ground fluids can free loose rocks and other formations, thus shifting large land mass. It is also possible that the fraking liquids lubricate locked (by gravity and friction) fault lines, causing them to prematurely break apart or slide to one side. This movement of large land mass could then cause tremors and earthquakes.

It is also possible that the increased stresses caused by the injected fluids (mechanical pressure, hydro-erosion, and chemical action) disturb the balance of the underground layers and influence certain faults to prematurely break.

This phenomena is not new, since it was first identified as early as the 1960s around Denver, CO. Toxic chemicals from the Rocky Mountain military arsenal were injected at great depth to dispose of them, but in doing that, it was observed that small earthquake activity increased as a result. Some people have thought the releasing energy in this matter via small to medium sized earthquakes might be a good idea so as to prevent or reduce the size of larger earthquakes. But the issues are very complex and not well understood, and no one is confident enough to undertake such an effort given the complexities, uncertainties and liabilities present.

Case Study. Oklahoma Earthquakes

Most recently, a statistically significant increase of number of earthquakes in Oklahoma brought the issue in the open. For the first time in the state's history, Oklahoma's state government officially recognized the long held scientific consensus linking the hydrofracking activities and the disposal of oil and gas wastewater with the record number of earthquakes plaguing the adjacent areas in recent years.

In 2015, the Oklahoma Geological Survey released a statement, declaring that it is "very likely that the majority of recent earthquakes, particularly those in central and north-central Oklahoma, are triggered by the injection of produced water in disposal wells."

The statement coincided with the launch of a website produced by Oklahoma's Energy and Environment Cabinet, featuring an interactive map and links to expert studies detailing the scientific evidence behind the link between Oklahoma's earthquakes and the disposal of oil and gas waste water.

This is the first time, after years of skepticism, that the state of Oklahoma officially recognized the link between earthquakes and hydrofracking activities.

Hydraulic fracturing, including the controversial gas extraction process that involves injection of waste water into deep underground wells, has boomed in Oklahoma. There are more than 30,000 oil wells across the state, according to the U.S. Energy Information Administration.

In the early 2000s there were only about one and a half earthquakes exceeding magnitude of 3.0 on the Richter's scale annually. 3.0 is the minimum level at which humans can detect an earthquake without scientific instruments.

In 2014, there were over 550 quakes of 3.0 or larger, which is about 600 times greater than the background seismic rate. A recent report by the U.S. Geological Survey claimed that the 5.6 magnitude quake that struck Prague, OK, in 2011, resulting in several injuries and damage to more than a dozen homes, appears to have been "hydrofracking waste water disposal induced."

The oil industry is denial about the issue, "There may be a link between earthquakes and disposal wells, but we—industry, regulators, researchers, lawmakers or state residents– still don't know enough about how waste water injection impacts Oklahoma's underground faults."

This, in the face of all the evidence is all we can expect for true oil people, closing their eyes to the well established theory, and numerous laboratory tests, and field experiments. The danger is actually increasing. The earthquake rate in Oklahoma has increased so much that it raises the risk of a larger damaging earthquake.

Oklahoma is not the only state experiencing a dramatic rise in earthquakes. Large parts of central and eastern parts of the U.S. have recorded an uptick in seismic activity during the last several years of the oil and gas boom.

During 1973 and 2008 there were about 20 earthquakes annually, of magnitude 3.0 or larger in the area. Recently, there were over 600 quakes of magnitude 3.0 or larger. This increase in seismic activity coincides with the increase in hydrofracking activities and the injection of wastewater in deep disposal wells in Colorado, Texas, Arkansas, Ohio, and Oklahoma, according to the U.S. Geological Survey.

This way, the oil and gas industry is introducing a new, clear and present danger to the local population. The friction between the locals and the industry will continue to increase as a result too. There will be many NIMBY demonstrations and lawsuits filed in the near future.

Power Plants Risks

Conventional (coal and natural gas fired) power plants do not pose a significant danger to their workers.

There are isolated incidents of health and live threatening situation, but these are to be expected. Like in any large work setting, where large pieces of equipment and huge structures are involved, there are isolated incidents of workers getting hurt and even paying with their lives.

While this is regrettable, and deserves the attention of the authorities, there is a much bigger problem that conventional power plants pose today. It is the air, water, and soil pollution they cause to their surroundings during their normal operation.

Studies issued since 2005, quantifying the deaths and other adverse health affects attributable to smog and fine particle air pollution resulting from power plant emissions, show about 8,000-10,000 deaths each year in the U.S. These are attributable mostly to the fine particle pollution from power plants operating all over the country.

Although this is still a lot of people perishing thanks to the power industry, it still represents a dramatic reduction from the previous studies, done 10 years previous, which show about 24,000 deaths by 2004. This three-fold decrease in mortality is due mostly to improvements of the power plant technologies, and the introduction of a variety of federal and state laws, regulatory, and enforcement initiatives. This includes the Mercury and Air Toxics Rule (MATS) and the Cross-State Air Pollution Rule (CSAPR). Increased awareness and better enforcement of existing regulations such as New Source Review (NSR) has also contributed to the good news—if 8,000 people dying every year can be considered as such.

Recently, the new regulatory measures have dropped sulfur dioxide (SO_2) pollution by 68% and nitrogen oxide (NO_x) by 55%, which are the leading components of fine particle pollution. This was achieved by doubling of the amount of scrubbers used to reduce pollution, which were installed at power plants. But even greater contributor to clean air is the retirements of large coal-fired power plants that were replaced by natural-gas fired such.

Updated studies show that strong regulations that require stringent emission controls can have a dramatic impact in reducing air pollution across the country, saving lives, and avoiding a host of other adverse health impacts.

The studies, however, also show that some areas of the country still suffer from unnecessarily large levels of pollution from power plants that could be cleaned up with the application of proven emission control technologies. Some segments of the power industry and entire states resist this trend and continue trying to overturn the MATS and CSAPR regulations in court and reverse the life-saving trend brought up by them. We cannot blame them, because they are trying to preserve jobs and the livelihood of thousands of miners and power plant operators.

There are also serious issues of water and soil contamination by power plants activities. Runoff from containment tanks and other spills do occasionally contaminate local rivers and streams and cause environmental and financial damage to the local communities.

Although federal and state regulations are strict enough, major accidents have happened and continue to happen. Coal-fired power plants are mostly to blame for these accidents. As they are being slowly but surely replaced by gas-fired power plants, the environmental accidents are reduced in number and severity too.

Gas-fired power plants will not solve all the environmental and health problems of the nation, but will contribute significantly to cleaner and healthier environment. The rapid increase of renewable energy technologies will further expand this trend and bring us a better and even healthier future.

Nuclear Power Risks

The two major, Earth-shaking and unforgettable nuclear accidents—Chernobyl and Fukushima—define the nuclear industry and confirm its status as the most powerful and most destructive force on Earth. This immense power is a blessing and a curse at the same time. It is amazingly efficient in providing huge quantities of electricity from small amount of raw materials.

The competitors can't get even close in this respect, but neither can any accident at these facilities be compared to that at a nuclear reactor. While major accidents at the competitors facilities do damage equipment and hurt a few people, major nuclear accidents erase entire cities and regions from the map. The local area becomes uninhabitable for ever, together with the properties located on it, and taking the lives and health of thousands of innocent bystanders in the process.

Let's take a closer look at these ominous events:

Chernobyl

Just like the Genie in the bottle, nuclear power possesses awesome powers. When contained within the constrains of a nuclear reactor, it is a mighty force, equal to none. It can generate more and cheaper electric power than any other energy source with minimal waste and environmental impact. When released intentionally or unintentionally, however, it is a monster, which unbridled and uncontrollable power mercilessly devastates anything in its path.

And we have several examples of its fearsome behavior.

The Chernobyl nuclear accident is the most "unusual" of all nuclear accidents to date in all aspects of the word. During that faithful night of April 1986, several factors combined to cause a monstrous explosion in reactor #4 that blew up the roof of the structure, contaminated massive land areas, and hurt and killed people thousands of miles around the site.

A wicked combination of several factors played decisive role in the disaster:

- A faulty socialist-communist system designed and built the reactors in a hurry, while disregarding some basic safety measures in materials selection, and construction and operating and procedures. The reactors were installed in an even greater hurry, in order for the party bosses to receive extra bonuses. During this stage, non-standard materials and procedures were used, which left gaps in the reactors' overall safety.

- A faulty set of control systems in the reactors, and substandard construction materials, contributed to the reactors overheating, which led to total loss of control and caused the subsequent explosion.

- A poorly trained and badly behaving management team was in charge at the time of the accident, where some of the engineers did not have the necessary experience, and some were at their posts only because they were good Communist Party members. As such, they followed blindly the Party rules, and other communist system behavioral quirks, while at the same time disregarding some basic nuclear power principles and procedures.

- A faulty set of operational and emergency procedures, executed during the last minutes before the explosion, reveal the prevailing Soviet culture of arrogant ignorance, blunt negligence, never ending cover-ups, and lack of respect for human life.

Faced with all these obstacles, Chernobyl reactor #4, and many people working in control room #4, had no chance of survival that night. The explosion was inevitable under these circumstances, so many died and many more suffered burns from excess nuclear radiation. The local area was contaminated and evacuated as a result.

It, however, took the local communist leaders 14 hours to notify the population of the nearby town of the pending disaster caused by the nuclear radiation that was emitted from the reactor. And it took them over a week to hesitate and cover up, as was the norm of the communist system at the time, to evacuate all 160,000 people away from the radiation zone.

But then, at the same time 6,000 people were brought in to work on the destroyed reactor and its structure. These people worked in the zone of active radiation with very little to none safety equipment. Many more of these people also suffered the consequences of exposure to excess nuclear radiation. As a consequence many got sick and some died.

And then, instead of continuing the clean-up, the Soviet authorities just buried the reactor under thousands of tons of cement. To top it off, while the radiation from reactor #4 spread around the area and the world, the nearby nuclear reactor #3 was kept active for a long time, while reactors #1 and 2 were never shut down and operate even now at full power as if nothing happened.

Update:

A decision was made some time ago to attempt a new way of burying the blown nuclear reactor, using a New Safe Confinement (NSC) structure. This arched structure-some 300 feet high, 700 feet wide and 600 feet long is to be assembled on concrete rails and slid into place over building #4 at Chernobyl, which was destroyed by the steam and hydrogen explosions that followed the unstoppable power excursion in April 1986. 20,000 tons of steel and that much cement were used in the structure's construction.

The hermetically sealed NSC will allow engineers to remotely dismantle the hastily constructed cement and metal 'sarcophagus' that was built to shield the remains of the reactor from the weather, shortly after the accident. The instability of the sarcophagus was one of the major risk factors at the site, and its potential collapse threatened to emit more radioactive materials in the surrounding air and land mass.

Another half-baked project to shore up the structure was completed in mid-2008, but it was never believed to be of any consequence. The NSC would reduce the danger of structural collapse, while also allowing the sarcophagus to be taken apart under controlled and relatively safe conditions. The new structure will reduce the emissions from the buildings during the next 100 years. At the same time it will also stop the ingress of water, which increases the risk that nuclear fuels scattered inside the building could potentially see sustained and even increased fission reactions.

The huge building is also designed to facilitate the removal of materials containing nuclear fuel and accommodate their characterization, compaction and packing for disposal. This task represents the most important step in eliminating nuclear hazard at the site. It will also signal the initiation of the real decommissioning efforts. The NSC will allow remote operation and handling of the dangerous materials, and using as few personnel as possible.

In 2014, Ukrainian President Viktor Yanukovych laid flowers at the Chernobyl memorial near the plant's north gate that commemorates the work of those that took early action during the 1986 accident. Then he took part in a ceremony to launch the assembly of the NSC structure.

Yanukovych told those gathered at the site, "We are witnesses to a historic event in the transformation of the Object Shelter into an ecologically safe system: the construction of the arch of the Chernobyl nuclear power plant." He added, "This project's successful implementation will ensure that the surrounding area becomes environmentally safe and that there are no radiation leaks into the atmosphere."

A historic event, that feels like putting new garments on a dead body, giving it a new name and meaning.

Yet, the effort is appreciated by the 29 donor countries of the Chernobyl Shelter Fund, which was set up in 1997. The fund is administered by the European Bank for Reconstruction and Development. The necessary funds for the $621 million project were secured in July 2011 and the entire project is to be completed in 2016.

Figure 9-14. The new grave stone over Chernobyl reactor #4

The NSC structure is being built by the French-led Novarka consortium, which includes Bouygyes and Vinci, as well as German and Ukrainian firms. It's a lot of work and money only to bury the giant, which has been dead for over 30 years. Imagine that!

There should be an engraving on the new multi-million dollar NSC headstone, reading, "Here rests the worlds most powerful serial killer, let lose by the communist system."

And while all this sounds quite bad, Fukushima is thousands of times worse. That giant is even bigger and still alive, causing enormous damage even now; five years after the disaster. The Fukushima recovery effort is even more complicated and many times more expensive than Chernobyl. The radiation in the local area and the ocean around the hurt giant continue to increase, so it will take much more than one multi-million NSC to bury that monster and stop the spread of radiation all around the blown reactor.

Fukushima

A strong earthquake and tsunami combination shook Japan in March 2011. The Fukushima Daiichi nuclear power plant on Japan's eastern coast was also surprised and its major support systems failed. Their failure resulted in nuclear reactors running out of control, and used fuel storage tanks evaporating, triggering three meltdown incidents in a short time period. Assisted by hydrogen explosions at the site, the meltdowns caused massive escape of radioactive material in the environment.

Initially several workers were severely injured or killed by the disaster conditions. Some drowned, and some were hurt by falling equipment, mostly as a result from the earthquake. There were no immediate deaths due to direct radiation exposures, but at least six workers have exceeded lifetime legal limits for radiation and more than 300 have received significant radiation doses.

Mass evacuations, tests and checks followed, but the damage was already done and now we could only watch, sympathize, and expect to see more disasters and negative consequences. The plant was totally destroyed within a short time period and the radiation was reduced to minimum.

Several years later, predicted future cancer deaths due to accumulated radiation exposures in the population living near Fukushima are inconclusive. But it is certain that it would take decades to decontaminate the surrounding areas, to decommission the plant, and to get life in the area somewhat near normal.

The locals were affected the most, but the radiation spread quickly to distant areas too. Over 3,000 individuals from a town over 15 miles away from the disabled nuclear plant also bore detectable levels of radioactive cesium in their bodies. Their total dose of less than one milli-Sievert is considered safe, and no radiation sickness was observed. Nevertheless, the exposed men, women and children need to be watched for the long-term effects of the radiation for the rest of their lives.

The damages are noticeable in the local wildlife as well. One example of the continuing damage is the pale grass blue butterfly, a delicate local insect, famous for the way it changes the colors and patterns of its wings in response to environmental changes. Butterflies caught in the Fukushima region six months after the meltdowns (a generation in butterfly lifetime terms) had different colors and patterns than before. Some of the butterflies also had badly deformed legs, antennae, wings and even eyes.

The deformities persisted and got even worse in the second generation of offspring as well. The same deformities were found in butterflies collected from the wild away from the failed plant. Such results are found in research labs, where normal butterflies are exposed to radiation from cesium particles like those escaping from Fukushima Daiichi. This leads to only one answer; the damage is due to radiation.

How many more changes to wildlife and the human population will be encountered in the future remains to be seen. The affected, however, will live in fear of developing cancer, or some other radiation-related illness, for the rest of their lives, asking, "Is it worth it?"

Note: These are the largest and most damaging nuclear power accidents since the inception of the commercial nuclear technology. But here is where the similarities end. The Chernobyl accident happened during the reign of communist awkwardness, and was caused by inferior Russian made technology, clumsily operated by poorly trained technicians.

The Fukushima nuclear accident was caused by natural events, triggering an earthquake, followed by a large tsunami. Faced with that awesome force, even the superior made in the U.S. technology, and U.S. trained technicians and engineers could not prevent, nor contain, the nuclear meltdown. The nuclear plant was dead in the water—literally. What followed was a disaster that no one had ever even thought possible.

So, here we have two different accidents; Chernobyl and Fukushima. One is manmade and the other nature-made (with some help from imperfect humans). But the results are the same: complete, thorough and merciless devastation of environment and human life. We were not able to prevent neither disaster, even though we were aware of the dangers and took measures (albeit inefficient) to prevent them.

These two accidents open a large hole in the safe-ty track of nuclear power use, a hole so large that the entire world fears that a third accident could easily slip through the same hole—even if we were advised ahead of time of the possibility. The fear was/is so great that some governments decided to shut down all their nuclear facilities as the only way to prevent another Chernobyl, or Fukushima.

There is a nuclear power plant close to many large population centers in the Western world, so the burning question—fueled by the memories from the two greatest and totally different nuclear disasters we discussed above—is, who is next? And when? But even more importantly, it is time for the different governments—and the world community as a whole—to take much closer looks at the pros and cons of using nuclear power and come up with a long-term plan that minimizes the environmental and health risks and yet provides enough power for normal life. Not an easy task, but it is one that is urgent in nature and that we cannot/should not ignore.

Update

The Fukushima disaster is a never-ending nightmare, unlike Chernobyl, which was packed in cement and left to brew silently for ages to come. Out of sight and out of mind; almost forgotten in its silent grave, Chernobyl is now a fading memory. Fukushima, however, is very much alive and continues to create serious problems on a daily basis. Water draining from the local mountains runs through or close to the plant, becomes contaminated with radioactive materials and runs into the ocean. Local fisheries are contaminated with ever increasing doses of radioactivity, spreading, threatening the globe.

Fukushima cannot be entombed like Chernobyl, so other alternatives to prevent further damage to the environment are sought. The latest one, approved by the Japanese government, calls for building a wall of iced soil around the plant, and containing the radiation within a "frozen wall."

Figure 9-15 shows the location of the proposed "frozen wall" to be built around reactors 1, 2, 3, and 4. The idea is simple; a frozen soil barrier deep underground would contain the contaminated water from the plant's leaking infrastructure, while at the same time not allowing the local runoff and groundwater to enter the plant perimeter, This would reduce the amount of contaminated water and allow full containment of radiation coming from the site. At least this is how the theory goes. Things are usually much more complicated, and this case is n exception of the rule.

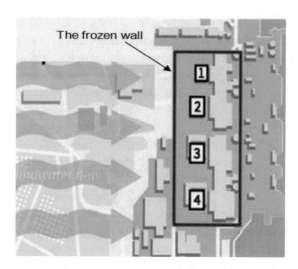

Figure 9-15. Fukushima's grave in the 'frozen wall'

The actual implementation of the containment project is designed as follows:

- First a row of deep shafts will be drilled into the ground around the wrecked reactors.
- A dedicated refrigeration plant will be then built nearby.
- A network of "freeze pipes" will be laid from the refrigeration plant, or plants, all around the "frozen wall" perimeter and into the underground shafts.
- Special coolant, kept at subzero temperature will circulate in the freeze pipes, freezing the soil in the deep shafts around them.
- The cryogenic action will eventually create a deep and thick layer of frozen soil in the shafts along the "frozen wall."

If successful, the new underground wall would prevent some 300-400 tons of radioactive water from running into the ocean every day. Although the frozen soil method has been used in other, albeit much smaller projects in the past, it has never been attempted on such a large scale and under such complex conditions.

This is a huge construction project, expected to last 2-3 years (if no problems are encountered), at the cost of half a billion dollars. The project is faced with a number of technological, safety, and economic problems, and nobody knows to what extent it would be able to contain the radiation contamination of the ocean. The biggest unknown is how long the "freeze wall" has to be operated and maintained. Radiation contamination can last hundreds of years. Can the "freeze wall" last that long too? And at what expense?

So now imagine 1 or 2 similar nuclear disasters in the next several years. Would the world continue think-

ing of nuclear power as safe, or would it bury all nuclear power plants once and forever? Good question...

Solar Power Risks

'Solar power' and 'risk' usually don't go in the same sentence, but this is only because solar power, in the large scale development of today, is a new industry. We just don't have enough experience with the different types, sizes, and configurations of solar modules and field systems.

It is not difficult to imagine, however, that solar power plants exposed to the elements would develop problems with time. This is especially true for the solar equipment installed and operating under extreme desert heat during the day, freezing at night, UV radiation, sand and rain storms, and other surprises the deserts have keep hidden in their sleeves.

Figure 9-16. Apple/First Solar PV installation on fire

Although the solar industry insists that the risks of operating solar power plants are negligible, we must disagree, since there are a number of examples of the risks solar installations pose to the environment and the people who live in them.

The latest of these examples, and a confirmation of the worst to come, is a fire of a rooftop solar installation on an Apple-owned facility in Mesa, Arizona. The PV panels just caught fire one beautiful sunny spring morning in 2015, lighting up the roof of one of Apple's data processing centers in the making. These solar panels were part of a deal to provide 14,000 local homes with electricity.

Apple is planning to spend $2 billion converting the Mesa facility into a "command center" for the company's global operations. The plant is expected to create 500 construction and 150 full-time jobs.

Ironically, the fire comes as Apple is in the midst of a campaign to promote its "green" image. The company hired former EPA administrator Lisa Jackson as head of "Environmental Initiatives" and has promised to use 100 percent green power at all of its data centers.

Apple already claims to run 100 percent of its "U.S. offices, data centers, and retail stores with renewable energy. Apple's green image, however, has come into question recently. Critics have pointed out that much of the "green" energy used by Apple is obtained from carbon offsets—where companies pay extra on their energy bills to "offset" their carbon dioxide emissions by funding green energy projects.

Carbon offsets, however, don't mean Apple is only using green energy. The company could still be using coal, natural gas or oil, but then simply paying more to have green power theoretically produced elsewhere. A great PR trick, but it does little to actually "green" up a company.

It ain't easy being green. It all starts with shedding a lot of greenbacks to get thing started, then you shed more greenbacks to keep things going, after which you shed even more greenbacks to prop up the aging hardware, and finally, you put some greenbacks in a safe place in case of a storm, fire, or other disaster headed your way.

Even then, when a major disaster happens, as it happened in this case to the most unlikely, green-wanna-be partners, all planning goes down the drain. And in this case, the CdTe PV modules made by First Solar contain significant amount of the notorious toxic, carcinogenic heavy metal cadmium.

About 10 grams of cadmium poison are locked in each CdTe PV module.

About 100 CdTe PV modules were ignited and burned for over an hour on Apple's warehouse rooftop. What happened with the cadmium in the burning PV modules is anyone's guess, but it is most likely that most of the CdTe and CdS compounds inside the burning modules were converted into some volatile gaseous form of cadmium poison. These toxic gasses were then taken up in the air by the smoke particles of the burning structure; carried away and spread over the nearby residential areas.

Some of the toxins would remain in the ashes on the rooftop, awaiting cleanup by hopefully careful and knowledgeable professionals. Cadmium gas or solids

dangers are not to be ignored, because any unsafe handling would result in illness, or worse.

This is a small size accident, the after-effects of which would not be that great, but imagine this fire occurring in a 10,000 acres power plant covered with CdTe PV modules. Millions of these modules burning in flames and belching toxic fumes would create the accident of the century. Or imagine another set of several million cadmium poison laden PV modules broken and buried in the soil by a large earthquake. The moisture in the soil will disintegrate and dissolve the cadmium and carry it into the water table. Not possible, yes possible! Just ask the people living close to the recent large earthquakes in Chile.

So how green and risk-safe solar is depends on whom you talk to, where, and when.

False Certification

This is a new dimension in the solar energy market. In an isolated case, the failed module manufacturer QSolar used fake TÜV SÜD certificates on some its products. A desperate move by a dying giant, or business as usual? How many other companies have had, are, or will be using false certifications to trick us in believing that we are buying quality product? Haw many such products have been already installed? And how many of these will fail?

The false certificates were discovered after a number of complaints from customers have been documented. Customers have shared photos of a TÜV SÜD mark (a little kite) on the back of badly degraded 175W PV modules made by QSolar at their Calgary, Canada address.

TÜV SÜD (a German testing lab) had confirmed that the PV modules in the photos had never been certified. "Our internal research has shown that QSolar had one certification which was declared invalid in April 2014. Within this certification there was no product QS 175 Rainbow and no model with an indicated performance of 175W. Furthermore certification owner was QSolar Photovoltaic in Shanghai. An address in Calgary, Alberta, Canada never owned a certification. This means that the TÜV SÜD certification mark on the panel is faked." Wow! This smells like a criminal offense. But it is too late to do much about it now, since the company is out of business now.

But this is not all of QSolar's criminal activities. After selling off his own shares, QSolar's CEO Andres Tapakoudes, claiming that he is putting the money back into the company, the company's share price took a sharp dive. QSolar's share price was just under $1.00

in May 2014, around $0.20 in January 2015 and less than $0.05 in March. At this point the trading of QSolar stock was halted.

Shortly after, the CEO Tapakoudes, along with CFO Preston Maddin and the rest of the board, resigned without issuing a press release, in breach of Canadian securities rules. The board members went in hiding—another criminal offense—and the court-appointed board has not been able to make contact with any member of the QSolar's former management.

The court-appointed directors were trying to salvage something from the company for the shareholders' benefit, but encountered a number of barriers. One was the fact that the previous management left just three boxes of paperwork representing two years' worth of company activity.

"The logical operation from the former board and management is a huge impediment from us trying to conduct normal due diligence. The fact there was no information to conduct due diligence on leaves you to think the worst," according to the court appointed representatives.

So, QSolar is no more, the management is in hiding, but their products are still perched on many roofs with false certification. What if they start failing? The unlucky owners if the QSolar PV modules will be left with the bill and hassles. And finally, how many QSolar-like companies are pulling the same trick. It is getting harder to do so, but the crooks of the world have always been able to stay ahead of, and find ways to, beat the system.

Solar Installations

We need to add another 'risk' here that we have noticed in a number of places of the global solar energy installations. It is actually the act of installing solar modules and other equipment. Due to poor training, hurried up schedules, poor supervision, and lack of standards, many solar installations are installed wrongly, and/or even damaged during installation.

Just take a look at Figure 9-17, and see if you can guess what the installer is doing wrong. Oh, yes, he is doing something very, very wrong.

What is the guy in Figure 9-17 doing wrong? Firstly, he is not well trained and has not been told that to not step on the glass modules. Even if the glass doesn't break, 100 lbs/ft2 pressure on a small area of the 1/8" glass surface would bend it enough to damage the solar cells underneath. They may not break, but the bending would cause micro cracks, and/or add more stress to the already stressed soldered joints and other critical func-

Figure 9-17. How *not* to install solar modules, part 1

tional points in the affected cells.

Although the modules might not exhibit any problems at the time, the excess stress often creates so called *latent failure* issues. In this case this is structural damage that remains undetected at first, but which eventually causes performance deterioration, and could even trigger a massive failure several years down the road. Joints and other areas under excess stress cannot handle the extreme heat-freeze cycling very well, and eventually give up—usually when least expected.

Second, this installer is not supervised, or is supervised by a supervisor who also lacks the experience to assess the situation and prevent damage to the PV modules. The lack of international standards for installation and use of solar energy equipment adds to this abnormality.

Figure 9-18. How not to install PV modules, part 2

Another problem we see often in solar installations is the way the modules are stacked on roofs. As Figure 9-18 shows, the modules in the middle of the stack are not reachable. If so, then how would they be tested and serviced in case of a failure in the future?

It is quite obvious that there are limited options for reaching and servicing the modules in the middle of the stack. Option 1 would be to walk on the adjacent modules, which would create a big mess. Option 2 is to remove a row of the adjacent modules, which requires a lot of time and effort, and in some cases is not even possible. There is no other option, so we just have to wonder what the designers were thinking when working on this project?

So, the obvious maltreatment of PV modules, and the unserviceable arrangement of modules, depicted above beg the question, 'How many modules has been installed this way by these groups, and how many such groups are they that design and operate this way?'

We have actually seen quite a number of such abnormalities in the field, so we expect to hear a lot about problems created by poor design and awkward installation procedures in the future. Customers and investors must be aware of such problems, which pose serious future performance and failure risks.

The Solar Process

The shortcomings of the solar installations in Figures 9-17 and 18 above are clearly visible and documented. It seems like speed of installation is all that matters in these cases. Future maintenance does not seem to be part of the design, because there is no access to the panels in the middle of the array. This makes it very difficult to test and/or replace any of the modules in that area.

This, however, is only one, and one of the easy to correct, problems facing the solar industry. There are dozens of steps in the solar panels' life cycle that are impossible to detect and correct. Starting with the manufacturing process, the fragile solar cell or module assemblies could be easily damaged by careless handling. There is no way to observe and document every mistake in every step of the process, so the operators' mistakes usually go unnoticed...until the components show signs of poor performance, or fail prematurely and has to be replaced.

It doesn't take much to damage a solar cell or module. One wrong move, one improperly executed operation by a distracted or negligent operator, is all that it takes to damage their delicate structure. For example, 'cold welds' are quite common during solar cells assembly into panels. During this step, several critical manipulations are critical to the long term performance of the finished module.

The cleanliness of the surfaces to be bonded (soldered), the temperature of the solder materials, and the time it is applied to each joint are absolutely critical. A single deviation from the specs could lead to defective joint, which could fail at any time due to increased resistance and excess heat. Thus damaged solar panel may pass the final tests before shipping, but the defective joint will continue deteriorating.

This is only one example of what can happen, but there are many steps of the process; each in need of accurate and precision operation. Since many of the operations in today's solar industry are manual, the possibility of introducing defects in the final product are numerous. Once in the field, the defective solar modules will start performing poorly and/or fail prematurely in the field at the owner and investor's expense.

It is impossible to observe, account for, and document any and all mistakes in the dozens of steps of the long manufacturing process. The subsequent in-house or certification tests (using today's standard protocol) cannot detect all damaged cells and modules. So, in the end, most of the damaged components end up in the field. From there on, under the present day modus operandi, it is anyone's guess...and luck has a lot to do with it in such cases.

Just recently, there have been attempts to introduce another tier of certification tests, which would test the long term durability of PV modules. These tests introduce new and more vigorous testing methods that simulate long term sun exposure in the field. The new, more rigorous, test procedures might be able to detect defective cells and modules and alert the manufacturers of potential problems lurking in their production lines.

But even then, regardless of how thorough the tests are, they cannot detect all problems, and certainly cannot test every single module shipped to the customers. Manufacturers also have a way to avoid certification failures, by simply sending hand-picked

So, for now at least, installing and using PV modules has elements of surprise and risks. A gambling game of sorts.

It is a different sort of 21st century gambling game. Since no commercial PV module has a complete, proven 30 years of non-stop successful field operation, we have no track record to compare products and companies. Instead, we are putting our money and trust in the hands of assembly line operators in China, who sit on their chairs all they long stringing and soldering solar cells into modules. We bet on these people not making any mistakes, or at least not great mistakes that might cause latent failure to our product.

How much of a gamble buying solar cells and modules is today, depends on the type of product, who the manufacturer is, how much we know and understand the particular PV process and products, and how much money we put down. The combination of these factors would eventually determine our bottom line.

Wind Farm Risks

Most traditional energy sources have serious impact on the environment in form of solid, liquid, or gas pollution. Compared with these, the impact of wind power is relatively minor. Wind power turbines consume no fuel, and do not emit any kind of pollution. While a wind farm may cover a large area of land, the land under the turbines can be used for agricultural and husbandry purposes, since only small areas of the land are occupied by the turbine foundations.

There are, however, issues related to using wind turbines. One of the greatest concerns is with bird and bat mortality rates today. Many of these flying creatures are being killed by the large wind turbine blades upon impact.

The extent of the overall ecological impact, if any, depends on specific circumstances, but until more data is generated or better ways to estimate the impacts are found, the prevention and mitigation of wildlife fatalities. This is important, because the protection of natural resources is playing a significant role in the public's perception and affects the actual permitting, siting, and

Figure 9-19. Bird killing machines

O&M of wind turbines.

In the summer of 2015, environmentalists prevailed and the federal court took notice of the indiscriminate killing of birds in the U.S. A US District Court in San Jose, California ruled invalid a Department of the Interior regulation allowing wind energy and some other companies to kill eagles and other birds for 30 years, in the name of "clean energy.

With that, any new wind power plants would be given only a 5 years license.

Five years license to kill, as some call it, will be renewed following a revision of the casualties and other environmental effects. This is a major move, which would force the plant operators to obtain a separate Environmental Impact Statement and a new permit to extend operations. What this move does is bring the problem under the spot light and focus the attention of potential wind farm owners and operators on it. It is doubtful that any new wind projects, limited to five years permit, could obtain a quick financing. At a minimum, there will be a pause and more concern with wild life when designing and operating wind power plants in future. This might bring a radical change in the way the wind industry operates in the U.S.

Another type of negative effects, of concern to the locals, are the noise and the unsightly appearance of the giant wind turbine. These factors also play a significant role in the siting and permitting processes.

On top of that, a closer look reveals that the cradle-to-grave cycle of wind power plants also has significant effect on the environment as follow:

Equipment Manufacturing Impact

It is generally accepted that the energy consumed to manufacture and transport the materials used to build a wind power plant is equal to the new energy produced by the wind plant within a few months or years, depending on location. While this is true, we must also take into account the pollution generated during the manufacturing and transport of the large wind turbines and the support infrastructure and construction materials.

Imagine a 500 foot giant metal tube sticking up in the air. It is 15-20 ft. wide at the bottom, and is cemented a dozen or more feet into the ground. Inside the tube, spiraling metal stairs lead to the top of the structure, where another giant is mounted. It is the structure, as big as a house, perched 500 feet in the air, housing electrical generator, complete with its own transmission box, electrical components and controls. But the most spectacular piece of the entire structure are the 3 or 4 blades, each 150-200 feet in length, swinging their long arms non-stop, as if challenging Don Quixote to attack again.

This massive structure contains many tons of steel, aluminum, copper, plastics, and fiberglass. These materials were not blown in by renewable winds. Nope, they were dug out of the ground as metal ore, or pumped out of it crude oil, from which the final components were molded and machined into their final shape.

A lot of solid, liquid and gas pollution was created during the mining, pimping, machining and transport operations. Making a 500+ foot long steel tube is not a small task. Molding and machining 200 foot blades is another enormous task, all of which use a lot of energy and emit tons of pollution.

The manufacturing of the transmission, the generator, and the controller parts and the support infrastructure (spiral stairs, cement, rebar, etc.), together with the site construction work are also major polluting activities.

Figure 9-20. A wind turbine blade transport

We also need to mention the transport of these huge components. Entire train composition, or dozens of 18-wheelers would be needed to load and transport all the parts of one single wind mill. Not a small undertaking, where tons of fuel are used and that much pollution emitted.

On the other hand, transport and installation of wind mills is a big business, which have created entire industries in the U.S. and around the world.

O&M Impact

Some pollution is also emitted during the operation and maintenance of wind turbines. It is simply not possible to operate any type of electromechanical equip-

ment of this complexity and size without having some sort of pollution and other negative effects.

During transport and installation, many vehicles crisscross the area, leaving trails of destruction and pollution. During operation, dozens of operators, maintenance workers, and contractors drive to the site periodically for checks, testing, and periodic maintenance work. Periodic and emergency maintenance work is done at times, where parts are hauled along the country roads, and hoisted up and down the giant structure. All these activities, including the manufacturing, transport, and installation of the replacement parts also use more energy and generate some additional pollution.

Still, wind turbines are considered pollution neutral pieces of equipment, because the pollution generated during the equipment manufacturing operations is compensated by the pollution which is avoided by using wind power (instead of fossils) during long term operation. It is estimated that a 2 MW wind turbine would compensate its pollution contribution within a year or two of normal operation. How true this remains to be seen, but we must be aware of the fact that there is no such thing as free, or pollution-free, energy.

Note: We can describe a similar scenario for solar power plants, which pollution footprint is similar, and in some cases greater than that of wind turbines. We have, for example, allowed unproven solar technologies, such as toxic cadmium telluride and SIGS containing thin film PV modules, to be installed in huge numbers in the U.S. deserts.

Safety

Working with, or on, the monstrous wind giants 200-300 feet up in the air is a dangerous business. Skilled and well trained technicians and experienced engineers are in charge of the different operations during the entire cradle-to-grave process of the wind turbine. Training is a major part of the wind farm installation and O&M process, which helps prevent excess casualties.

Safety is a priority in most wind power plants operations, but the dangers are constantly there; awaiting an opportunity to do damage. And since it is human to err, and equipment pieces malfunction from time to time, a number of accidents are reported during all stages of the operation.

Falling from the high towers is the most frequent cause of people getting hurt or killed. This opens the doors to law enforcement, insurance, and legal experts to get involved on either side of the case. And so it goes non-stop in the wind power plants of the U.S. and all around the world...

Ocean Power Risks

The power generation via tidal and wave power generating devices usually proceeds mostly without generating carbon dioxide, or any other harmful gasses or liquids. But as the number of wave and tidal projects in the world's oceans increases, the attention is focusing on the negative effect power generating devices might have on marine life.

Ocean Environment

Any man-made equipment placed into the oceans would create some sort of interference with the marine environment and life in it. At the very least, the metals would have some chemical reactions, while the moving parts would affect it in another way.

The presence of any electro-mechanical devices in the world's oceans would introduce a number of new risks. Some of the anomalies and risks associated with ocean power generation are

- Static and dynamic effects
- Chemical effects;
- Acoustic effects;
- Electromagnetic effects;
- Energy changes; and
- Many cumulative effects.

These effects would result in some interference in or with the:

- Near and far-field physical environment;
- Habitat;
- Invertebrates;
- Migratory fish;
- Resident fish;
- Marine mammals;
- Seabirds; and
- General ecosystem interactions.

Noise, electromagnetic fields, mechanical damage etc. possible effects are suspected to affect marine life.

Figure 9-21. Ocean power generation

Noise, for example, is known to confuse marine mammals, causing them stress and loss of orientation. Other marine species—sea turtles, crabs, sharks, skates, salmon and other fish—rely on Earth's magnetic fields for migrating and searching for food. The wave, tidal and hydrokinetic power devices, and the cables that bring the electricity they generate to shore, produce similar electromagnetic fields, which might confuse them.

And then, one thing that is known for sure is that fish swimming close to these devices might get hurt or killed by the moving parts—propellers, pistons, turbines, etc. Precautions have been taken to avoid collisions by installing special devices that emit noise to alert the potential victims, but it is not clear what other effects such a noise makers would have.

We really don't know if the animals will be affected by all this or not. There's surprisingly little comprehensive research to tell us one way or the other. But besides that (all of which needs to be verified and taken care of) the ocean power generating technologies are clean as a whistle. They do not emit any emissions, nor do they cause any damage to the environment during operation.

There is also considerable amount of disturbance to the ocean floor caused during the exploration and installation phases. Large pieces of equipment are brought in to dig deep holes in the ocean floor for the support structures. Large amounts of cement and other materials are then dumped in these holes to fix the structures in place.

These activities might go for days, during which time all marine life in the area would be killed or chased away. It cannot be even estimated how much damage is done during that time, but it is clear that larger the project, more damage would be inflicted to the ocean environment and local life in it.

Finally, large ocean power plants might obstruct maritime traffic, and even cause accidents. All these issues are important and we must consider them thoroughly before we start building huge structures in the world's oceans.

Geothermal Power Risks

The environmental problems caused by building a geothermal energy plant start while looking for a suitable site. During exploration, researchers will do a land survey, which may take several years to complete. To extract profitable heat generating site, we have to find certain hot spots within the Earth's crust, which are most often around volcanoes and fault lines. But who wants to build their geothermal energy plant next to a volcano? So away we go in search of similar conditions somewhere else.

Some areas of land may have the sufficient hot rocks to supply hot water to a power station, but many are in harsh areas of the world; near the poles or high up in mountains. And so, the search continues for a long time, with people and machinery crawling all over the place, digging and drilling deep holes, and causing some environmental damage in the process.

And of course some damage to the local environment and life would be done during the construction of the production facility. Large pieces of equipment are brought in to level the place and drill more and deeper holes in the ground. Lots of pipes and other hardware are delivered also and laid in the exploration and production holes.

During the well drilling and operation stages, toxic gasses can escape from the site, since it is hard to control what's happening deep under the surface. Things also change down there, so that although there are no problems today, tomorrow things might change and toxic gas leakage could be the result. These gasses can escape into the local air or penetrate deep into the ground where they could contaminate the water table. In all cases, the environment could be damaged, and all life in it—including people—could be hurt.

Another danger of geothermal energy is the likelihood of recurring earthquakes. People who live in areas of geothermal production wells report that there is an increasing number of earthquakes in the area. Although they are low-level quakes, no stronger than magnitude 4, they are damaging homes and foundations.

Increased volcanic activities could bring the eruption of an old or a new volcano during hot rock drilling. Such an eruption could devastate the local flora and fauna in an instant. This, of course, is not as dangerous as a nuclear power plant explosion, nevertheless, those unfortunate to be nearby during the eruption will pay with their health or lives.

And finally, after years of generating power in many cases the site "cools down" and production stops. At that time more environmental damage will be done by the shutting down and decommissioning of the power plant. Another batch of large equipment arrives at the site and starts demolishing buildings, digging trenches and levelling the ground. This all is at the expense of the local environment.

The good thing is that in most cases, the geothermal site—unlike coal mines and gas and oil fields—can be restored to near-perfect (original) condition.

Note: Most importantly, looking at the larger picture, we can easily see that injecting large amounts of cold water into any hot body (including Mother Earth's bosom) day after day has limits. There are quantity and time related limitations of how much heat we can extract before a major event occurs.

Where these limits lie is uncertain, but we should not be as naïve as to think that we can continue doing this indefinitely, and even increase the volume of cold water injected in the Earth's hot rocks, without any negative effects sometime in the future.

Like any other technology, when geothermal is applied in small quantities—as it is today—it cannot do much harm for a long time. When deployed in large and very large scale, however, things will surely change and negative effects will follow...as they always do in such cases!!!

It is great mistake to take examples of old, small, good, and reliable production geothermal wells, and extrapolate their results to large and very large scale power generation.

Doing this would be like comparing apples and oranges...or worse!

There is a serious issue to consider here. The energy in Earth's core is huge, no doubt! but it is not unlimited. Cooling it down by pumping lots of cold water in/on it, without stopping to even consider the possible consequences, might be the dumbest and most dangerous thing we've ever done. It is like permanently cutting the branch we are sitting on, in order to warm up our hands temporarily. Just think...!

And this is why, we have to reconsider putting geo-thermal power together with the renewable energies. It is renewable while it lasts, yes, but then it is gone, which happens sooner or later at each particular location.

Biomass Risks

As with all energy sources, biomass has its own problems and risks. Biomass goes through several life cycles, starting with agricultural and forestry operations. The biomass materials (agricultural or forest waste) destined for conversion into energy, have to be collected, loaded on huge trucks or trains, and transported to their final destination.

Here, the organic matter in the biomass products must be burnt, or processed, one way or another in order to make or extract energy. These processes use a lot of energy and release carbon dioxide and other gasses into the air, unlike the use of solar, wind, and other renewable energy sources.

The difference is that although processing biomass emits carbon dioxide, it is still classed as a carbon neutral fuel—which in some respects is better than wind and solar. The reason behind this unique classification is the carbon life-cycle completion, which means that while the crop grows it will absorb carbon dioxide, releasing it back into the atmosphere when burnt, or processed.

The theoretical biomass carbon cycle is considered to be simple:

CO_2 in – CO_2 out = zero (or near zero)

There is only one problem; it is that in most cases the biomass is grown in one place, but transported, processed and/or burned in another place. This creates distinct zones of environmental benefits at the growing sites, and damage at the burning, or processing sites. This and other factors contribute to the conclusion that extensive use of biomass for energy is actually serious contributor to climate change.

Although biomass is a carbon neutral fuel, other, somewhat unrelated, factors play a significant role, and can disturb this aspect. For example, biomass can contribute to global warming as a result of "carbon leakage." Deforestation is an example of carbon leakage, since we are reducing the world's carbon absorption capacity, and disturbing the natural equilibrium of carbon dioxide between the atmosphere, biosphere, geosphere, and hydrosphere.

There is also large amount of different forms of energy (diesel, fossil electricity, fertilizers) used, and the related pollution taking place, during the planting, maintaining, harvesting, transporting, and processing of biomass and energy crops. With other renewable energy sources such as solar, wind, and geothermal, the only carbon based energy used will be to manufacture, transport, and construct the system, and the amounts are fairly low.

Continuous energy use and pollution emissions are present throughout the biomass life cycle.

And so, keeping in mind that biomass is still much cleaner than the fossil fuels, it is not perfect in that respect, and so improvements and new alternatives must be sought.

The major culprit in using biomass is the fact that in most cases it involves burning, or processing which emits considerable amounts of pollutants, such as particulates and polycyclic aromatic hydrocarbons. Even modern pellet boilers generate much more pollutants than oil or natural gas boilers. Pellets made from agricultural residues are usually much worse than wood pellets, producing much larger emissions of dioxins and chlorophenols.

Nevertheless, estimates show that biomass fuels (when considering the entire growing-to-burning process) have significantly less impact on the global environment than fossil based fuels, due to the CO_2-in-and-CO_2-out concept.

We must understand, and keep in mind, however, that power generation with these fuels emits significant amounts of greenhouse gases (GHGs), mainly CO_2, in the local environment. Sequestering CO_2 from the power plant flue gas is possible and can significantly reduce the released GHGs. The CO_2 capture and sequestration, however, consumes additional energy, thus lowering the plant's fuel-to-electricity efficiency and increasing its operating costs. To compensate for this, more fossil fuel must be procured and consumed to make up for lost capacity, which in turn produces more harmful gasses. Catch 22...

Taking this into consideration, the global warming potential (GWP), which is a combination of CO_2, methane, and nitrous oxide emissions, and the energy balance of the system, need to be examined using a life cycle assessment. This takes into account the upstream processes which remain constant after CO_2 sequestration, as well as the steps required for additional power generation.

Black carbon—a pollutant created by incomplete combustion of fossil fuels and biomass—is the second largest contributor to global warming.

A recent study of the giant brown haze (containing black carbon) that periodically covers large areas in South Asia, determined that it had been principally produced primarily by biomass burning and to a lesser extent by fossil-fuel burning. Researchers have also measured a significant concentration of carbon associated with recent plant and biomass life-cycles, rather than with fossil fuels.

Extensive use of forest-based biomass has recently come under fire from a number of environmental organizations, including Greenpeace and the Natural Resources Defense Council, for the harmful impacts it can have on forests and the climate. Greenpeace recently released a report entitled "Fuelling a BioMess, which outlines their concerns about forest-based biomass.

Because any part of the tree can be burned, the harvesting of trees for energy production encourages the so called 'Whole-Tree Harvesting', which removes more nutrients and soil cover than regular harvesting, and can be harmful to the long-term health of the forest. In some jurisdictions, forest biomass is increasingly consisting of elements essential to functioning forest ecosystems, including standing trees, naturally disturbed forests and remains of traditional logging operations that were previously left in the forest.

Recent scientific research indicates that it can take many decades for the carbon released by burning biomass to be recaptured by re-growing trees, and even longer in low productivity areas. Also, logging operations may disturb forest soils and cause them to release stored carbon. In light of the pressing need to reduce greenhouse gas emissions in the short term in order to mitigate the effects of climate change, a number of environmental groups are opposing the large-scale use of forest biomass in energy production.

Another very serious issue that needs to be discussed and resolved, is the location of the different phases of the seed-to-burn of plant and animal life cycles. While the fields and the forests that are used for making the biomass are relatively green, clean, and safe places, the actual processes of transport, burning and otherwise using the biomass fuels suffer from pollution—quite serious at times. How do we assign a value to the contribution and the damage in the different areas?

Money is paid some times (carbon tax) for excessive pollution and the related damages to those who are hurt by it? But is it possible, or fair, to assign monetary value to environmental damage and human suffering? These cases are no accidents, so can they be considered as planned and premeditated violation of environmental and human safety laws?

And finally, we must mention the increasing discrepancy of uncontrolled land-grabbing in many countries, where unorthodox methods are used to secure land for energy-related biomass growth. This situation is causing unspeakable hardship to the locals and is also contributing to food price increases.

All these issues remain to be understood, discussed, and resolved.

Environmental Effects

Due to its increasing importance and wide-spread use, the life cycle of biomass and biofuels deserves a close analysis, as needed to sort out the positive and negative effects.

In-out energy (energy ratio)

Biomass contributes 30% of the energy in the U.S. and even more in other countries.

Who knew? If this is so, then biomass is a major fuel, and a close look is more than urgent. It is undoubtedly a *renewable* source of energy, although its renewable nature is heavily dependent on location, weather, economic, and socio-political conditions. Its seed-to-energy cycle inevitably involves the consumption of some fossil fuels at one or another point and time of the cycle. It is also a source of significant pollution.

How much of it is used depends on the location, type of biomass, and the production methods. A close look usually includes fuels consumed by farm machinery in land preparation, planting, tending, irrigation, harvesting, storage, and transport.

There are also great quantities of fossil feedstocks for chemical products such as herbicides, pesticides, and tons of fertilizers used during the growth cycle. And, of course, a lot of energy, in form of fuels for tractors and other machinery, is required for plowing, seeding, growing, harvesting, and processing the bioenergy crop into a usable biofuel.

Energy requirements are generally higher for some crops because that need greater use of machinery and a higher level of water and chemical inputs (fertilizers and such). For many energy crops, the energy ratio, or the amount of useful bioenergy the particular crop can produce, compared to the fossil fuel consumed for feedstock production, could be very high.

Table 9-6 shows that hydropower is the most energy efficient power generator, producing 250 units of electric power for each unit of fossil fuel input. Biomass produces 30 units, while biofuel crops produce only 5 units of energy for each unit of fossil fuels used during their life-cycle.

Table 9-6. Energy ratios for different technologies

Technology	Ratio
Hydropower	250
Wind power	35
Biomass waste	30
Nuclear	15
Bio-fuel crops	5
Coal	5
Solar PV	5
Natural gas	5
Fuel cells	3
Crude oil	3

Some crops, such as poplar, sorghum, and switchgrass grown in temperate climates have energy ratios of up to 30. This simply means that each BTU used in the bio-crop cycle generates up to 30 BTUs of energy from the respective biomass crops. In tropical climates with good rainfall these ratios could even be considerably higher, due to higher yields and less labor and energy-intensive agricultural practices.

The energy ratios are much lower for crops that require much labor and mechanization, or yield a relatively small proportion of usable bioenergy feedstock per unit of plant matter produced. Some oil crops (like soy bean), however, could have an energy ratio close to 1, which means that the amount of energy this crop produces is equal to the energy it required during its entire seed-to-fuel life cycle. For some crops, and in some cases, the energy ratio is so low that we need to think twice about using them as energy sources.

Another factor to consider in the energy ratio estimates is the pollution created during production, transport, and processing of the crops. In most cases heavy machinery is used to cultivate the bio-crops, which uses a lot of fossil fuels and emits a lot of GHGs. After harvesting, the crops are loaded on large trucks or trains to be delivered to the processing facility. Even if the processing facility is a mile away, which is rarely the case, the crops still have to be loaded on vehicles and transported to the intake elevator.

At the processing facility some more energy is used during the different steps of the conversion process and more polluting gasses, liquids and solids are generated. In all cases, massive amounts of fossil fuels are burnt, which further reduces the energy in/energy out ratio. And, of course, significant pollution is created in this process as well.

Emissions

Bioenergy is hailed as carbon neutral, and although at a first glance this is so, this generalization needs a closer look.

The net carbon benefit is calculated by comparing the reality with what would have happened if fossils were used instead. The amount and type of fossil fuel that would otherwise have been consumed, as well as the land use that would otherwise was to be used, must be entered into the calculations.

The relative carbon intensity must be assessed on the basis of the emissions associated with the biofuel crop production and the efficiency of the energy technology in which the biofuel is used.

As can be seen in Table 9-7, all biomass materials have fairly large carbon content. In most cases, the carbon is released in one or other form, but most likely as

Table 9-7. Carbon content in biomass materials

Type of Material	Carbon-Nitrogen Ratio
Alfalfa hay	18 : 1
Bagasse	150 : 1
Cichen manure	25 : 1
Clover	2.7 : 1
Cow dung	25 : 1
Cow urine	0.8 : 1
Grass clippings	12 : 1
Kitchen refuse	10 : 1
Lucerne	2 : 1
Pig droppings	20 : 1
Pig urine	6 : 1
Potato tops	25 : 1
Sawdust	200 : 1
Seaweed	80 : 1
Straw	200 : 1
Sewage sludge	13 : 1
Silage liquor	11 : 1

CO and CO_2 during the digestion, burning, or otherwise processing of these materials.

Table 9-8 shows natural gas having the highest efficiency, or 45% of its energy input is converted into electric power. Biomass averages 20% efficiency, or only 20% of the available energy contained in it (and everything else considered) can be converted into useful electric power.

Table 9-8.
Efficiency and CO_2 emissions of different fuels (per kWh)

Technology	% Efficiency	Gr. CO_2 emitted
Diesel generator	20%	1,320
Coal steam cycle	33%	1,000
LNG combined cycle	45%	410
Biogas digester and diesel generator	18%	220
Biomass steam cycle	22%	100
Biomass gasifier and gas turbine	35%	60

At the same time, diesel power generators seem to be the most polluting, while biomass is the cleanest; assuming that the bioenergy crop is harvested in a carbon-neutral manner, or that there is no carbon net change in/on the crop fields and in the soil over the course of the complete crop growth and harvesting cycle.

In practice, the carbon in and on the land (in the soil) changes significantly, depending on how the biomass is produced and what would have happened in other cases. Taking as an example clearing the jungle forests, which leaves a bare land for growing energy crops, we see that the bare site has lost its carbon-reducing value and cannot be readily regenerated. The carbon emissions from site preparation alone (trucks, bulldozers and such emissions) could exceed the savings from biofuel use, and could be greater than the carbon emissions from a fossil-fuel cycle generating same amount of power.

The justification for the land clearing activities, from an environmental perspective, is that long term use of biofuels produced at the cleared land will compensate for the increase of CO_2 at the initial stages. This is a frequently used model for the production of energy and non-energy biomass, under which millions of acres have been deforested around the world. The overall global effect of this massive land clearing is yet to be figured out, let alone precisely calculated and extrapolated into the future.

Clearing large areas of natural forests, some of which are used for energy crops planting, means that the CO2 sequestered in the natural forest will be released eventually during burning of the biomass materials at an amount that depends on the type of the forest.

About 125 tons of carbon are released per acre of biomass

As biomass feedstock is grown and harvested in cycles, carbon will be held on the land, partly compensating for the carbon released when the natural forest was cut down. Averaged over a growth cycle, a typical amount of carbon sequestered on the land might be 15 tons per acre, which leaves about 110 tons balance uncounted for.

If the purpose of a bioenergy crop in this case is to displace fossil fuels in order to reduce carbon emissions, it will achieve this (by compensating for the 110 tons difference derived from the above estimate) over a period of roughly 40-45 years.

When environmental and social considerations (preserving habitat and protecting watersheds), are taken into account, the total of all considerations might increase the number of years and even outweigh any and all carbon saving benefits. This is the negative side of biomass.

On the positive side, if biomass or bioenergy crops are grown on unproductive land (waste, degraded, or abandoned land), then both; that land and the local environment could benefit from their re-vegetation. The degraded land is most likely carbon-starved, so the crop field will store some CO_2 in the soil and other be-low-ground biomass. This would compensate for some of the initial carbon emissions.

And so, the overall effect of using degraded or waste-land for energy crops show definite benefits in:
- Providing immediate measurable carbon and other local ecosystem benefits, and
- Displacing fossil fuels for power generation on the long run.

Carbon emissions from biofuels in general can be reduced by:
- Providing energy that can displace fossil fuel energy, and
- Changing the amount of sequestered carbon.

Soil

Biomass crops are very similar to other crops, as far as managing soil, water, agrochemicals, and biodiversity are concerned. Following good crop managing practices, while taking into consideration the specific technical and environmental challenges, are the key to success.

One difference worth noting here is the fact that biomass plants are often harvested completely—in their entirety. They are usually chopped down to the roots, and sometimes are harvested with the roots. This MO leaves little organic matter or plant nutrients in the soil of the harvested field. This would create soil problems especially in the developing world where the soil depends on recycling crop wastes and manure, rather than on the use of expensive fertilizers. In such cases, biomass production could contribute to dramatic deterioration of soil quality.

In order to maintain the soil quality, clever farmers would choose to keep sufficient plant matter on the land for the soil's sake, at the expense of reduced crop yields. Or, as another alternative, farmers might allow some nutrient-rich parts of the plant, like small branches, twigs, and leaves to decompose on the field.

Some feedstock nutrient content can be also recovered from the conversion facility in the form of ash or sludge, and trucked back to the fields to be used for soil improvement, instead of being put in a landfill. This, however, is expensive undertaking, and is rarely used. Nevertheless, there are options for those who are determined to produce biomass and biofuels.

Hydrology

Bioenergy crops usually consume more water than many food crops, and some energy crops like sugarcane consume immense amount of water, which creates problems with the local irrigation systems and water wells. Sugarcane and other such thirsty crops are known to lower the water table, or reduce stream flow, all of which makes the local irrigation system less reliable.

An amazing fact to remember is that three quarters of our fresh water, which is only 5% of all the total Earth's water supply, is used for crops growth and animals raising. For this reason alone, many agricultural communities have resisted the introduction of some energy crops and tree plantations.

A number of practices, such as; growing tree crops without undergrowth, or planting species that do not generate adequate litter, reduce the ability of rainfall to replenish groundwater supplies, thus increasing the local water overconsumption.

Water scarcity and the desertification of our agricultural lands are proceeding at a frightening pace. Irrigation, for example is quickly depleting our Ogallala Aquifer on the Great Plains. The annual water consumption is about 1.3 trillion gallons faster than it can be replenished by rainfall.

The ongoing draught in California is putting additional stress on the forests and agricultural fields in the state. Longer it lasts, more changes in the status of the biomass and biofuels in the state would be expected. The final outcome of this devastation is still pending, but it cannot be good.

Biodiversity

All farm crops have significant effect on the local ecosystem by enhancing or suppressing the biodiversity of the region. Energy crops have a similar effect, but provide even more biodiversity, which is closer to a natural habitat than other crops. By enhancing biodiversity some gaps between fragments of natural habitat can be filled. For this purpose, in Brazil for example, environmental regulations require 25% of the plantation area to be left in natural vegetation. This helps to preserve biodiversity and provides other ecosystem benefits, where bioenergy crops can also serve as corridors between adjacent natural habitat areas for the benefit of all wildlife.

Energy crops, like many industrial crops have been escaping from the cultivated area and growing uncontrollably in the natural areas at the expense of the local species. There are many such examples, where crops in various regions have reproduced widely beyond plantations to harm the local species. Growing energy crops at the same field for extended periods of time (monoculture) could have the effect of an incubator for pests or disease, which can then spread into natural habitats.

Meat production

A lot of water, fuels, and other resources are used to grow meat for human consumption. Watering and cleaning the animals and the facilities consumes huge amounts of water, which usually ends up in a stinky cesspool somewhere nearby.

Raising one pound of meat requires 2,000 gallons of water and produces 58 times more greenhouse gasses than growing 1 pound of potatoes. It takes 7,000 pounds of grain to feed the animals that produce 1,000 pounds of meat. It takes sixteen times more energy (mostly fossil generated) to produce six ounces of meat than to produce a cup of broccoli, a cup of vegetables, or a cup of rice. These are significant numbers that deserve our attention and understanding

Each year, ten billion bushels of corn are grown in the U.S. alone, 60% of which is used to feed cows and other livestock. The meat of these animals is the major ingredient in processed and fast foods, which in turn are the major cause of obesity and diabetes.

Beef meat also produces a lot of greenhouse gases, because of the methane and nitric oxide produced by cow flatulence. Grains and grass ferment in the cows' stomachs and produces large amounts of methane gas, which is one of the largest contributors to global warming. And that gas has to go somewhere...yes, you guessed it.

Just think, if we all skipped one processed meat meal each week, this would be the equivalent to taking half a million cars off the road. But the Exxons and BPs of the meat industry would not allow that. When the USDA suggested "Meatless Mondays," the National Cattleman's Beef Association lobbied the government to retract their recommendation. And, yes, you guessed right again; there is no Meatless Monday in the U.S., or anywhere else...for now.

Biofuels' Risks

It could be argued that biofuel is a product of solar energy, as the sun is needed to grow the biomass crops, which can then be manufactured into usable fuel. It is also true that biofuels are carbon-neutral, simply because the CO_2 emitted during their burning was absorbed during their growth cycle. It is also undeniable that the use of biofuels could help reduce the costs associated with the purchasing of mainstream fuels such as petrol and diesel.

Although *biofuel* sounds like a modern day invention, we have used this type of fuel since the discovery of fire. If firewood can be classed as a biofuel, as wood is a biomass product, then biofuels are the oldest type of energy used for domestic and industrial applications. It is the material that was burnt to released energy in the form of heat since man walked on this Earth.

Note: The word *biofuels*, as used this text (biodiesel, biogas etc.) relate to the modern day uses of biomass as (liquid) fuel energy source.

We, however, should not get carried away and exaggerate the importance of biomass and biofuels as future energy sources. To provide biofuel for every car, truck, bus, plane, boat, and factory across the globe would require a colossal amount of land to be used for the plantation of renewable biomass crops. This would also result in the deforestation of the world, as a direct result of biomass plantation becoming out of hand. This is unrealistic and unsustainable development. Instead moderation and careful expansion of biofuels must be adapted to ensure its safe participation in the energy mix.

Electric cars powered from the electricity generated by renewable energy sources would be the best option, however, implementing this on a global scale will be a very challenging task. Instead, we need to be looking into all possible methods for powering vehicles and machinery—including the use increased of biofuels.

Biofuels Externalities

The production and use of biofuels have been advertised as clean, green, and safe economic engine for most global economies. Is this so?

Consider this in light of the facts that:

- In Mexico, the corn for fuel programs caused loss of agricultural workers' income, consolidation of the tortilla industry, excess urban migration, and increased poverty.
- In East Asia, increased use of corn for fuels caused growth of the livestock industry, accompanied by more meat consumption, which in turn leads to a number of health complications, and
- In Africa, corn for biofuels was transformed into a battle for land. Land grabbing for expansion of the biofuel crops by governments and multi-national companies caused forced resettling of entire populations. The diversion of corn crops from the kitchen table of the masses to the gas tanks of the rich also created untold misery. This was a huge global problem that resulted in riots, hunger for millions, and in death from starvation in some cases.

The fact that the same resources are used for production of both food and biofuel crops, creates a disequilibrium in the global food-energy systems.

If the demand for one of the products (biofuel crops in this case) is rising, the availability and price of other (food crops in this case) will be negatively affected.

During the last boom-bust biofuels cycle of 2007-2009, there was a sharp increase in the demand for biofuel crops. As a result, the food crops acreage decreased quickly and with that the food became scarce, forcing its price to rise accordingly.

A bushel of corn produces approximately 2.8 gallons of ethanol.

Filling a 25-gallon gas tank with pure ethanol requires biofuels made from over 450 pounds of corn, which contains enough calories to feed one person in Mexico or Africa for an entire year.

The production and use of biofuels is classified as renewable, clean, and green technology. Is it?

In addition to the alcohol, a bushel of corn also produces 16 lbs. of high-protein distiller grains, which can be used for animal feed, or for protein enrichment of foods. This is the good thing...

The bad part is that that same bushel is also responsible for the emission of 17 lbs. of CO_2 during its life cycle. This amount, multiplied by the billions of bushels of corn produced world wide every year, puts corn—and corn ethanol especially—among the major GHG polluters.

During the recent ethanol boom and bust cycle, price of corn increased from \$2.25 per bushel in summer 2006 to \$4.25 in February 2007, and has been up and down (mostly up for a number of reasons) ever since. Corn prices have risen and fallen many times in the past, but now the bio-ethanol pressure will keep them high for an extended period of time, this time around.

Substituting corn for other commodities such as soy is also causing an increase in the price of these commodities. In developing countries, where food versus fuel trade-off is even more critical and the battles are expanding in other areas of the affected economies.

There are winners and losers in the battle for more energy sources, but regardless of who is winning, the problems will only grow with time. On the long run, the price of biofuels for the rich people will go down, while the food prices for the poor will increase.

INTERNAL RISKS

There are a number of additional risks created and propagated within the energy markets themselves.

Some of these are borderline criminal, as we will see below.

False Advertising

There has been a lot of erroneous, or flatly *false* advertising in the energy sector. Oil and gas companies are paying millions of dollars for TV ads that show how wonderful their products are. Lately, they have been sneaking in the idea that natural gas is clean and beneficial to the environment. We argue elsewhere in this text how true (or not) this statement is, so here we will only say that calling natural gas 'clean' is distortion of the facts.

The oil and gas companies have been also trying to diminish and discredit the claims of increasing global warming. They especially argue the claim that the fossils are a major contributor to this abnormality. For many years they have been presenting 'research data' that show that the fossils are a minor contributor to the recent environmental and climate problems.

Some oil companies have gone as far as consistently lying about the results of their research. One large oil company was recently taken to court to answer the allegations of misleading the public by presenting intentionally modified research data. How many such legal actions will follow is anyone's guess, but we are sure that most oil and gas companies have at one time or another presented twisted in their favor data.

The new kid on the block, the solar industry is no exception. One solar company, for example, claims that their 50 GWp of solar farms are capable of supporting 1/3 of the U.S. population. We are sure that the advertisers know that this statement is far from the truth, but we doubt that they know how far it is. So let's help them:

1/3 of the population would be roughly 110 million people. These people live in groups of 3-4 per household, so we would estimate 1/3 of the people (about 110 million today) live in about 30 million households. The average U.S. home uses 12 MWh electricity annually, so 30 million households x 12 MWh = 360,000,000 MWh, or 360,000 GWh.

Assuming that 1 kW power plant would generate 8,760 kWh annually (1 kW x 24 hrs x 365 days) under full load (100% non-stop operation), we come up with 41,100 MW, or 41 GW of power needed for running this number of households on 24 per day basis. This is the electricity generated by 80 mid-size power plants.

Here are the problems:

• The 50 GWp solar power, advertised to power 1/3 of American homes, is available ONLY when the sun shines.

• This means that we can get electricity only 6-8 hours during the 24 hour cycle, and then ONLY if and when the sun shines.

• the 50 GWp (the *p* after GW stands for peak) would produce that much power only at noon. During the rest of the days, the power output diminishes as the sun angle increases. There are also other losses from such things as:

a) Temperature coefficient (0.5% loss of power per each degree C above 25 C). In the desert, this is a loss of 20-30% at hot noon hour, when temperature soar well above 100 degrees C, and

B) Annual power output degradation, which means that the total output of the plant decreases by 1.0% every year. By year 10, the plant would have lost 10% of its total output, by year 20%, the loss would increase to 20%, by year 30 to 30% and so on.

Considering all this, in best of cases, any solar power plant in Arizona would produce maximum 20-25% of its nameplate (peak power) capacity. A similar solar power plant in the state of Maine would produce half of this power, due to weather (clouds, rain, snow, etc.) weather obstructions.

So now our 50 GWp becomes about 15 GW average output during the first year—which is a best case scenario, assuming 365 days of full sunshine. Since this is not probable even in the deserts, the output would drop by another 15-20%, due to cloudy, dusty, or rainy days. Now, our solar power plants generation is down to 12 GW of total power production.

Due to the annual degradation phenomena, by year #10 the output would drop further to 11 GW, 10 years later to 10 GW, and 10 years after that to 9 GW and so on. All in all, this is quite far from the 41 GW of power needed to power 1/3 of American homes.

On top of that, similar solar power plants in areas that have less sunshine than Arizona (most of the USA) would produce even less power, depending on weather conditions.

We must conclude, therefore, that the advertisers in this, and many similar cases, either have no idea what they are talking about, or count on the customer's ignorance and negligence to promote their product and services. Such mass misleading approaches have been used since day one of the capitalist enterprise system, but we did not expect it to be continued by the new, more sophisticated, green, and clean renewable technologies.

We have seen similarly wrong assumptions made

for the performance of wind power plants too. Although there is a lot of data available on wind currents and other weather conditions, it is hard to estimate the throughput of a large wind power plant. This is due to the fact that here are too many variables that affect the final results.

Another problem, which we must always keep in mind is the fact that no modern, large solar or wind power plant has completed a full life cycle (30 years) of non-stop operation. This simply means that we have no track record to use as a basis of our assumptions.

Because of that, when we hear that a 100 MW wind farm will produce X amount of power annually, we must pause and consider the variables in the overall estimates.

Energy Market Manipulation

The conduct that results in (accusations of) manipulation, according to the legal cases in this area, show that the confusion is based on the fact that there are several types of manipulative acts without clear boundaries.

The most common (financial) market manipulation techniques are,

- Market power manipulation, and
- Market fraud manipulation.

Market *power* manipulation

Market *power* manipulation is the most common and broadly used type of manipulation of the energy market. It is also called a 'long market power manipulation', a 'corner', or 'squeeze.' It is usually executed by a large (or rich) trader (squeezer) purchasing a very large number of futures contracts. Here, the rich squeezer acquires energy futures position, thus demanding delivery of more oil or gas than is currently available at a competitive price.

For example, a large squeezer may acquire futures of 20 million barrels of oil in Texas, while there are only 10 million barrels available at the time. Additional oil supplies are available in other states, of course, but at a much higher cost. The squeezer then can exploit the discrepancy by giving the futures sellers a choice to either, a) pay the higher cost of the additional quantity, or b) buy back their futures positions at a lower price.

Win-win situation for the squeezer and lose-lose for the sellers or hedgers. This way the squeezer could corner the market and squeeze more money from the unfortunate sellers. In essence, this is monopoly position (albeit temporary) in futures contracts, which could distort prices in the larger markets, introduce chaos on the trade floor, and even disrupt the actual flow of oil in the affected area for the duration.

Such price distortions undermine the effectiveness of the futures contract as a hedging mechanism. Some hedgers incur heavy losses because the squeezer causes the price of their futures to rise relative to the hedged prices. This makes the markets more volatile and unpredictable than they should be. Instead of providing stable competitive supply and demand conditions, the manipulated market prices reflect a volatile market, driven by monopoly power.

On the other hand, there are also *short* market power manipulations. Here, a large short trader deliveries excess quantity of the commodity in order to drive down the futures price. This move allows repurchasing of the futures positions at a lower price.

Another method, used to manipulate energy prices is *withholding*. This is done by the removal of energy supply (oil or gas) from the market in order to create a supply shortage. This classic manipulation of a market corner involves taking a long contract position in a deliverable commodity and stockpiling physical supply to force those who have taken a short position to buy back those positions at an inflated price. It is one of the oldest forms of commodities manipulation, and preferred method of quick-buck making for some large traders.

Exploiting the *imperfect liquidity* of the market is another type of power manipulation. The futures market trades are limited, thus reflecting liquidity and steady orders flow. Here, the relation between bid/offer prices and commodity quantities is usually respectively downward/upward sloping.

A trader submitting a sell or buy order for a quantity larger than the quantity of outstanding bids or offers currently outstanding at the highest bid or lowest offer, could cause the price to move above or below the current best bid or offer. By submitting a very large order, a trader can cause a large price movement, especially in a somewhat illiquid market where the quantities of bids and offers are small. This way, these trades create unreasonable prices, and could cause a long term impact on prices in the energy market.

Market *Fraud* Manipulation

Market *fraud* manipulation is the other major type of market manipulation. While the market *power* manipulation is unethical and border-line legal, market *fraud* manipulation is criminal, since it always involves some sort of fraud or other criminal activity.

For example, a trader can spread a false rumor before or during trading sessions in order to panic the markets and force prices to move in his favor. Sometimes called pump and dump schemes, these activities were prevalent in the U.S. futures market of the 1980s and 1990s.

Note: The author was a victim of such a get-rich-quick scheme in the mid 1990s, when the solar company he was working for was shut down by FTC due to suspicion of illegal pump and dump stock manipulation. The brokers, who were authorized to manage and trade the company's stock were issuing false advertising, full of exaggerations as of the company's projects and potential future growth.

Excess stock purchases by innocent and ill-informed investors, driven by the ads and their false promises, drove the company's stock prices up overnight. Next day, the brokers would dump large quantity of stock, and get a part of the profits. The problem is that you can fool FTC once or twice, but not all the time. And so, the company—together with its employees and plans for bright future—went down the drain.

Misreporting the transaction prices, used to determine the settlement price of a derivatives contract, is another example of criminal manipulation. Here, a trader may execute a trade, or series of trades, at an inflated price in order to mislead the market as of the actual price, which affects the traders' buying or selling strategies.

Figure 9-22. Commodity markets complexity

As Figure 9-22 shows, the commodity markets are very complex, and a fertile field for quick-buck attempts. Improper and even criminal market manipulations are part of it. In addition to the types discussed above, there are also see other types, such as, ***market churning, pools, stock bashing, runs, wash trades, quote stuffing, etc.***

As mentioned above,

Only people's imagination is the limit to cheating on the marketplace.

In all cases, these activities bring confusion and disrupt the proper function of the energy markets. They also hurt honest traders financially, while the crooks end up making money and get away with it.

The Green Corruption

Money corrupts, the saying goes. Big money corruption, we know by now, goes to the very top of our society. We often see politicians and other 'big fish' caught in the justice nets, paraded in the evening news. Time after time, we see people on high posts trying to cheat the system in order to help themselves or their crony buddies. And we are seldom surprised at the new faces we see getting caught.

But what has the new, clean and green, renewable energy market has to do with corruption? Well, prior to 2005, the U.S. renewable sector was dormant and practically non-existent. Then came the flood of money.

In 2008-2009, the Obama administration opened the gates and flooded the renewable energy sector with money. Billions of dollars went into it, with a good chunk going down the drain and even bigger part was wasted.

It happened very fast and the insiders were able to capture the moment...and the money. So much money has never been available so fast to any industry, and many renewable companies benefited by generous amounts thrown at them.

The industry was not ready for this sudden influx, so the corporate principles and wiz-kids had to improvise and pull strings in order to capture as much as possible of the bonanza. It mattered not that what they intended to use the money for made no sense. It mattered not that the money would be wasted. They grabbed the chance and filled their pockets with tax-payers money.

In our 2013 book, Solar Technologies for the 21st Century, published by the Fairmont/CRC Press, we have dozens of pages full of failed renewable energy companies during the 2008-2012 global renewable energy boom. Abound Solar, BP Solar, ECD, Evergreen Solar, LDK Solar, Q-Cells, REC Solar, Solar World USA, Solyndra, Amonix, SpectraWatt, Miasole, Stirling Systems, Solar Day, Solon, etc. companies made the list and are an example of wasted taxpayers money.

Billions of dollars went into these companies, which in most cases had no viable product and/or real strategy. One thing in common many of them had was the fact that they had powerful connections in the federal and state governments, Ex-Im Bank, DOE, and many private organizations.

There is no compromise when it comes to corruption. You have to fight it, said A.K. Anthony, and we agree. But how do we do it? It appears that corruption is woven into the socio-political fiber of our society and the energy markets are no exception. Capable and slippery lobbyists and insiders, representing Abound Solar, Solyndra, First Solar, GE, Amonix, and other companies were able to secure billions (yes with a B) for their sponsors.

These companies were helped by their buddies high up in the government to secure grants, loans, and other perks for their technologies and projects that in many cases made no sense. Sure enough, many of these companies and projects failed, giving black eye after black eye to the renewable energy sector.

In the end, a lot of money was wasted and is still being wasted for financing worthless technologies and projects. But who cares, the U.S. is a rich country, so several billion dollars is just a drop in the bucket.

There is no point in crying over spilt milk, another saying goes, so we turn our sights to the future. How do we learn from the lessons of the past, in order to make sure the errors and disasters are not repeated. Not an easy thing to do in such a crowded field, with so many sharks waiting and looking for the next opportunity.

First, we must look around, identify the problems and the culprits, and wipe the slate clean. We still see money being thrown at companies and projects that have very little to show, except their large bank accounts and crony buddies at high posts.

Rich helping the rich is an ongoing trend n the renewables energy sector. It is the new and very profitable game in town, often operating in the gray areas of the energy markets, which offers new opportunities for the insiders to profit.

In some cases, it is the *'first come, first serve'* principle that counts most. And who can be more first than the insiders, so like the early bird, they get the early worm. This way, more billions of dollars of taxpayers money are piped through secret channels and flowing down the drain. It is an amazing development, which is summarized in several articles by Christine Lakatos, The Green Corruption Files (3, 4). We use some of the information and data from the Corruption Files in the following sections.

The problem with this type of corruption is that it is way high up in the political and society circles, of the rich and powerful, where no outsiders are allowed. Any attempt to penetrate, or God forbid, expose the circle's evil deeds is punished by discrediting, or threatening the violators. Most of us cannot afford the punishment, so we stay out; watching the insiders doing their stuff and shaking our heads in disbelief.

And so it goes under the capitalist umbrella; the rich get richer…any way they can. The rest of us just watch and wonder.

DOE's Secrets

DOE is the energy tsar of the U.S. energy sector, and trusted technology advisor to the President. Because of that, it is hard to determine what exactly happened, how it happened, and who is responsible for the billions of dollars wasted during the 2008-2012 financial crisis and beyond.

The most unexplainable and damaging of the period were some of the 27 loan guarantees under the 1705 program, of which the DOE doled out in excess of $16 billion. 23 of these loans were rated "Junk grade" due to their poor credit quality, while the other four were rated BBB, which is at the lowest end of the 'investment' grade of categories." Good business, or political cronyism? That is the question.

The DOE's 1705 loan portfolio overall average was BB, according to the financial experts. This is like lending $100,000 for a new house to an unemployed person. So why did the DOE back so many high-risk investments, particularly at a time when we were drowning in a tsunami of debt?

Twenty-one energy firms were behind the 27 projects found in the House Oversight investigation, and 18 of them are politically connected to President Obama and the Democratic Party. This is over 85% of the lot. 13 were bundlers, donors, and supporters for Obama's 2008 campaign, and two are members of the president's Job Council, while three are allied to Senator Harry Reid—all with ample other Democrat links in the mix.

According to a report by Christine Lakatos in the summer of 2015, the beneficiaries were:

- General Electric with two projects, worth over $1.4 billion:
  - 1366 Technologies Inc, rating B by Fitch, 09/2011—$150 million
  - Caithness Shepherds Flat, LLC, rating BBB- by Fitch, 10/2010—$1.3 billion)
- Abound Solar (financial issues), rating B by Fitch in 2010—$400 million
- Beacon Power Corporation (gone bust), rating CCC+ by S&P in 2010—$43 million
- BrightSource Energy, Inc with three projects, total

of $1.6 billion
— Ivanpah I and Ivanpah III, rating BB+ by Fitch
— Ivanpah II, rating BB by Fitch, 2011

- Cogentrix of Alamosa, LLC, rating B by Fitch, 2011—$90.6 million
- Exelon (Antelope Valley Solar Ranch), rating BBB- by Fitch, 2011—$646 million
- Granite Reliable Power, LLC, rating BB by Fitch, 2011,—$135 million
- Kahuku Wind Power LLC, rating BB+ by Fitch, 2010—$117 million
- NRG with two projects, grand total of over $2.0 billion
  — NRG Solar, LLC (Agua Caliente), rating BB+ by Fitch, 2011—$967 million
  — NRG Energy (California Valley Solar Ranch), rating BB+ by Fitch, 2011— $1.2 billion
- NextEra Energy Resources, LLC with two projects, total of well over $ 2.0 billion
  — Genesis Solar (environmental issues), rating BBB+ by S&P, 2010—$852 million
  — Desert Sunlight, rating BBB- by Fitch, 2011—$1.46 billion
- Record Hill Wind, LLC, rating BB+ by S&P, 2011—$102 million
- SolarReserve Inc, LLC (Crescent Dunes), rating BB by Fitch, 2011—$737 million
- SoloPower Inc., rating CCC+ by S&P, 2011—$197 million
- Solyndra, Inc, rating BB- by Fitch, 2009—$535 million
- U.S. Geothermal, Inc (Malheur County, Oregon), rating BB by S&P, 2011—$97 million
- Nevada Geothermal Power Company Inc., rating BB+ by Fitch, 2010—$78.8 million
- Ormat Nevada, Inc., rating BB by S&P, 2011—$280 million
- SolarReserve Inc., rated BB by S&P, 2011—$877 million
- LS Power (Transmission Line project), rating BB+ by Fitch, 2011—$343 million
- Mesquite Solar I, LLC (Sempra Mesquite), rating BB+ by Fitch, 2011—$337 million
- Prologis (Project Amp), rating BB by Fitch, 2011 —$1.4 billion
- Abengoa with three projects, total of nearly $3.0 billion
  — Abengoa Bioenergy Biomass of Kansas LLC, rating CCC by Fitch, 2010—$132.4 million
  — Abengoa Solar, Inc (Solana), rating BB+ by Fitch, 2010—$1.45 billion

— Abengoa Solar, Inc (Mojave Solar), rating BB by Fitch, 2011—$1.2 billion

Note: Abengoa is a Spanish company. So how is it that a foreign company was able to get so much money from the U.S. taxpayers to build solar plants in the U.S.? U.S. solar companies have been building similar plants in the U.S. deserts since the 1980s. Local effort to bring the abnormality to the lawmakers failed.

The author was part of that effort, and can testify that Abengoa did not introduce any new or unusual equipment, materials, processes, methods or approaches. Similar materials and processes have been used since the 1980s in the U.S. The only explanation here is that Abengoa was able to get to convince the local utility and other powerful people and politicians, who in their ignorance and negligence authorized billions of dollars to be used by the clever Spanish company. The story doesn't end here, and we take a close look at the technical, financial and other issues elsewhere in the text.

Looking at the above list, we see lot of money spread in thick layers among companies with questionable financial abilities and technological capabilities. And all this, in the wake of the failed mortgage debacle, where similar tactics were used to lend money to people who could not otherwise afford the high prices.

The key points of this taxpayers money massacre of 2008-2012 are that:

- The majority, 23 of the 27 large loans, were junk grade. The other four were considered at the lowest end of the investment grade categories.
- There were documented, special connections and ties between most of these projects and some layers of the political establishment, including DOE and the White House.
- Two of the DOE junk loans were considered bailouts, which was a clear violation of DOE's basic contract regulations.
- Many of the listed companies and projects are having serious problems, and some went out of business all together; sinking billions of dollars in the process.

First Solar

First Solar is an U.S. company that deserves a special place in the risk assessment part of our text. It is Solyndra on steroids, which is alive and well, thank you, mostly due to government assistance. With over $3.0 billion (with capital B) of DOE approved, and over $1.0 billion from Ex-Im bank, worth of loans, loan guarantees, and other financial and administrative assistance, to be exact.

Republican politicians have been criticizing federal loans that First Solar Inc. received for its massive power plants in Arizona and California since day one. The allegations were that DOE officials broke the rules in order to secure money for the company's projects, but the response has been muted at best.

- First Solar did not qualify, and should not have been given, a $967 million loan guarantee for the 290 MW Agua Caliente power plant in Yuma County. Nor a $646 million loan for the 230 MW Antelope Valley power plant in Los Angeles County, or a $1.46 billion loan guarantee for the 550 MW Desert Sunlight power plant in Riverside County, according to a report from a House committee.

- There is nothing innovative in these projects, which is required for the type of federal loan they got, according to the report from the House Committee on Oversight and Government Reform. Internal DOE e-mails reveal that officials questioned whether Agua Caliente and Antelope Valley met the 'innovative' aspect of the funding requirements, since all First Solar power plants use the same, old, flat solar panels it has been making since the mid-2000s, and which others have been making long before that too.

- DOE claimed that a special type of inverter would be used in Agua Caliente that would make it innovative. But an e-mail from DOE staff said the inverters were not innovative because they have been used before by other companies and projects.

- The Antelope Valley project was planned to use a special tracking system on at least 20% of the solar panels, in order to provide higher output and be qualified as 'innovative.'. But what is innovative about trackers; a simple technology have been used since the 1970s. An e-mail from a DOE employee stated the trackers did not constitute innovation and complained that documents regarding the project were continually edited to make that claim.

- DOE also factored in its multi-billion dollar decision the fact that First Solar's planned to build a large production plant in Mesa, AZ. The factory was never built, even though the loans were approved on that premise. The plans were changed, the plant was never started, and the building was sold recently, giving the appearance that it was just a gimmick used to secure the loans.

Another set of very important facts were also overlooked by DOE officials at the time—be it by blatant ignorance, or negligence, shows that:

- First Solar's CdTe thin film solar modules have very low efficiency as compared to other solar technologies. At the time of these loans, First Solar's PV panels were about 2 times less efficient than some of the competition's silicon panels. There were other U.S. based companies that have more efficient, and proven by long term tests, modules that did not get even close to the level of preferential treatment First Solar got. Who can explain that anomaly?

- First Solar's thin film technology is relatively new—only 5-6 years in large field exposure at the time—so how were DOE experts convinced that it is efficient, durable, reliable, and safe enough for 30 years non-stop operation in the U.S. deserts?

- And finally, there is another major issue (the elephant in the room nobody saw) that was simply swept under the rug, even though DOE engineers and officials were notified time after time by specialists and were fully aware of the risks.

The cadmium telluride and cadmium sulfide (CdTe and CdS) in First Solar's PV panels contain significant amount of the notorious toxic, carcinogenic heavy metal cadmium.

About 8-10 grams of this poison is locked in each First Solar CdTe solar panel. Although this is not a great amount, when multiplied by the millions upon millions of poison laden PV panels, we see thousands of tons of toxic, carcinogenic materials spread thinly over thousands of acres in the U.S. deserts. Sometimes these potentially poison laden fields are located close to populated centers.

A major earthquake, wild fire, or flood in the area, could damage many of the fragile (glass) PV panels. Cadmium compounds would be easily dispersed in the air and/or dissolved in the soil and water table. The possibility of poisoning the environment and creating a large scale environmental disaster is quite obvious, and yet not taken into consideration. All DOE and local officials simply refused to discuss the matter, while the company flatly denied the toxicity risks.

How could it be that hundreds of scientists and engineers at DOE and its National labs, who were familiar with the CdTe technology and were ultimately responsible for reviewing and authorizing the loan guarantees, did not acknowledge that fact? It is not that they did not know or forgot about it. Nope! As a matter of fact, many of them were intimately familiar with the CdTe technology, since they worked on it for years. They were also reminded of it by a number of emails and letters sent to DOE and personally to many of the principals at the time.

We find it difficult to believe that responsible people—scientists and engineers, experts in the field no less—would allow many millions of untested and unproven for desert operation solar panels loaded with cadmium poison to be installed in the U.S. deserts. And yet, this is exactly what happened. So we just have to wonder, how would they justify a large scale toxic contamination in these cadmium poison-laden fields sometime in the future?

Was Dr. Chu—the man in charge of DOE and the loan authorizations at the time—aware or warned about the problems? Or was he just pressured by higher ups to authorize the loans at the very last moment, which is exactly what he did? Either way, just like the captain of the Titanic, who did not know about the icebergs waiting to sink his ship, Dr. Chu is ultimately responsible for this negligence, arrogance, and money waste that still might lead to an even larger disasters in the U.S. deserts sometime in the future.

"From the very inception of the Obama administration's $14.5 billion loan program, warning signs pointed to a likely loss of taxpayer dollars," said Rep. Darrell Issa, a chairman of the committee, which held hearings on the matter. "There appears to be a significant amount of evidence indicating that DOE manipulated analysis and strategically modified evaluations in order to get loans out the door."

Should we bring Dr. Chu back from cushy retirement for additional explanation of what happened? He left the DOE, just like the captain of the sinking Concordia did around that time, so we wonder, if he would even have the guts to come back and tell us the truth now. Not very likely; no matter how nice and brave he might be. At least not during this administration. Maybe later on we will find out the truth behind his actions.

And so, for now at least, First Solar's and other unjustifiable loan files are locked in the U.S. government safe of questionable projects of the times past, and the key was thrown away. No point in crying over spilled milk. Looking forward is all we can do now.

Update 2015.

Another loan-guarantee application filed by First Solar fell through recently only because it missed the dead line. But not to worry, because First Solar got more

loans from Ex-Im bank that went to Canadian company to use the money to buy First Solar panels. But wait. That company was First Solar subsidiary. So, in fact, First Solar bought panels from First Solar with taxpayers' money.

Then, First Solar got more loans from the U.S. Export-Import (Ex-Im) bank to fund solar plant installations in India and elsewhere. That's another long story, made into a joke, when the Indian government turned down the request of the Ex-Im bank, to be recognized as a "Financial Institution."

So, it seems that under these conditions, Ex-Im bank can lend money for projects in India, but it is not a financial institution, thus not able to collect the debt as and when needed. How is this even possible now day and age?

In another series of amazing events, in 2014, Ex-Im bank added James Hughes, the CEO of First Solar, to represent "the environment" in the Bank's deals. We can't help it but wonder which part of the environment is he going to represent; that which is to be damaged by First Solar's toxic panels, or the one that is not? That remind us of the wolf in sheep's clothing fable… What would his advice be, you think, if asked about First Solar panels' toxicity issues?

According to Ms. Christine Lakatos at the Green Corruption Files blog, Ex-Im Bank's activities, and its deals with First Solar in particular, during the period between March 2011 and July 2012, total about $632.3 million, as follow:

- March 2011, $19 million loan for a 15 MW project in Gujarat, India.
- July 2011, $16 million loan for a solar project in Rajastan, India:
- August 2011, $84.3 million for a 40 MW solar project in Rajastan, India.
- September 2011, $455.7 million for exports to solar projects in Ontario, Canada:

Note: This is in addition to $17.3 million in grants and $15 million in government loans. This transaction also enabled First Solar to sell solar panels to itself. Surprised? Naahh, it is just another small wheel in the U.S. corporate-political machine.

- July 2012, $57.3 million for other solar plants in Rajastan, India

Note: Ex-Im bank justifies these loans by bragging that "these transactions will support 200 U.S. jobs at First Solar's manufacturing facility in Perrysburg, Ohio." So, it seems, according to Ex-Im bank's statement, that it takes ¾ billion dollars to support 200 workers for several months work.

"First Solar is expected to be one of the green energy firms that receives loans and loan guarantees through Ex-Im's *$1 billion partnership with India,*" according to an Ex-Im bank official. But with ¾ of the money pot given to First Solar, not much remains for any other U.S. companies. We must ask also, how green are the cadmium toxins, while partnership with India is another question in need of clarification.

It is, however, late to ask Ex-Im bank anything, for it is no more. Logging on the ex Ex-Im bank web site, you will see: 'AUTHORITY HAS LAPSED. Due to a lapse in EXIM Bank's authority, as of July 1, 2015, the Bank is unable to process applications or engage in new business or other prohibited activities.'

Now things are changing, since DOE has no more money to lend for such projects, and Ex-Im bank was shuttered by U.S. Congress. It would be interesting to see what avenue First Solar, and others like it, will take now with the biggest supporters gone.

And there is another big name, we need to mention here:

Abengoa Solar

In December 2015, the official translation of the word Solyndra into Spanish was Abengoa Solar.

Just like Solyndra, Abengoa Solar fooled the U.S. utilities, regulators, and the government to give it over $1.5 billion (yes, with capital B) for Solana and additional $1.2 billion (this one is with capital B too) to build several other smaller projects.

Note: Mr. Obama personally hailed Abengoa as a model solar company in 2010, while attempting to justify committing $1.5 billion to the Solana project. But wait, he did the same thing with Solyndra, so it is obvious that he has to go back to school. Talking Business and solar Energy 101 would do him a lot of good, and would've saved taxpayers billions of dollars.

Just like Solyndra Abengoa fooled everybody that it has a proven and viable technology. And just like Solyndra, its technology is full of problems. Just ask some of the engineers that worked on the Solana solar project in Arizona. The $2.7 billion of federal loan guarantees (which is actually better and safer than cash) was basically wasted on showing us how NOT to do business in the U.S.

So, in December, 2015, Abengoa Solar begin insolvency proceedings, which is the first step toward a possible bankruptcy. Many U.S. government officials are beginning to feel just like they felt during Solyndra's conundrum. They, however, would not show it, and will find ways to defend the huge losses and the embarrassment.

But Abengoa's news also chose the worst possible timing—only a week away from the Paris (potentially) historic climate change summit. It is here, where U.S. officials call loudly for the world leaders to reject fossil fuels. Instead, they will promote use of renewable energy, and mainly solar power. What would they say, 'Don't do business like we do, because if you do, you will waste a lot of money.'

The White House would not reveal how much money the U.S. taxpayers will loose when Abengoa is no more. This is just another reminder that the U.S. government does not know how to handle the U.S. energy sector. In the end, the taxpayers get burdened by paying billions of unpaid loans, and the solar industry gets another black eye.

There is something wrong with any company that exists on government subsidies, and it is expected, sooner or later, to run into financial trouble. This is because their business model is not based on sound economics, but rather it relies on political favors. Looking at First Solar's many loans and other benefits listed above, we have to wonder in which direction it is headed and what will happen in the end. Many other solar companies, who benefitted by billions of dollars in loans, come to mind here too, and we fully expect to see them on the bankruptcy bench in the near future as well.

Note: In our book on the subject, *Solar Technologies for the 21st Century,* published by The Fairmont Press in 2013, we have an entire chapter, describing the demise of many energy companies in the U.S. and abroad, during the renewables boom-bust cycle of 2008-2013. Many of these companies had benefitted from government loans and handouts. In many of cases, the free money caused them to deviate from the proper business model and crash in the bottomless pit of negligence and arrogance.

Abengoa is one of these companies, who got what's coming to it, albeit much later than the rest. It's stock fell over 60% on the news, and many international bankers got hit by the realization that a full Abengoa bankruptcy would cost them over $21 billion (with capital B again). Abengoa doesn't seem to bother with small amounts, and works only with billions of dollars.

Spain is now bracing for the aftereffects of what could turn out to be largest bankruptcy in the kingdom's history. Not to mention that it comes with a large doze of international embarrassment. It is also a huge black eye for the Spanish solar industry, which is already dealing with a number of serious problems.

The bankruptcy announcement, however, was

good and welcomed news for some large Spanish investors, who were planning to pour hundreds of millions of dollars in Abengoa's deep pockets. They can now keep their money, or pour it in another company, which would hopefully have a better luck.

Abengoa started the negotiating process with its creditors, aiming to reach a compromise that would guarantee some financial viability under Article 5 of the Bankruptcy Act. We have to wait and see how this case will develop, but some seriously negative aftereffects are expected.

But while Abengoa's saga continues, we need to move on...

The Media

There are a number of media organizations that dominate the energy market news. Some of these specialize in presenting (their version) of news and developments in the fossils energy sector. Others, focus on news the renewables energy sector. As such, they present daily news of events and developments around the U.S. and the world.

While we must appreciate the effort of their reporters, we must also look very carefully between and behind the lines. At times, they reveal the ignorance of the authors, but in many cases they show their bias. Some of these people have favorite subjects and companies, and don't hesitate to tilt the information in whichever direction they, or their favorites want.

This became quite obvious during the turbulent period 2008-2013, when lots of companies were struggling. Looking for ways to help their desperate situation, they used the media (and some biased reporters) to promote ideas and present situations that would distract from their problems and help their cause.

There are numerous examples of this happening today too, and one just have to take a careful look at the media channels for the evidence. One example is the BP TV commercials that show how beautiful the Gulf Coast is now, after BP took care of it. It is like, BP is the savior of the environment, for which we must be forever thankful. What the commercials don't show is that the marine environment is still disturbed and polluted.

Another example of the media role in misinformation-turned-brainwashing are the frequent TV commercials that show how clean and useful natural gas is. This is, however, only partially true, because in some cases the air contamination from natural gas is proving to be even worse than that of coal. And there are increasing number of serious incidents of water and soil contamination from gas well operations.

Nevertheless, watching these commercials night after night one starts believing that BP is doing us a great favor and that natural gas is the solution to all our problems. Most of us live far away from the Coast and the gas wells, so we could care less, but for the millions who live near and are exposed to the damages, things look differently.

The renewables sector has similar, albeit less pronounced, problems. We can clearly identify several sources and individual reporters who have favorite companies. These companies and their projects are presented in the most favorable light possible...often bordering misinformation.

Since most of us do not have time, and/or enough understanding on the subjects at hand, we do believe what we are told. This way, the reporters are able to promote their own (and that of their favorites) agendas. Because of that, we must not take the word of the reporters—regardless of how good or bad they may be—as the ultimate word on any subject. Instead, we must think of it as just another opinion, do our own research, and evaluate the facts from several points of view before making a final conclusion or decision.

The media can help, or hurt, energy projects, companies, and the related energy markets. It is up to us to determine and decide how much effect we would let them have on us and our society.

The Rich, The Famous, and the Powerful

Tich, famous, and powerful groups are among the most important components of the energy markets. Their actions, or lack thereof, often affect the energy market conditions, and even determine its direction.

Recently, Warren Buffet and many other rich and famous people started spending millions on renewable companies and projects. By doing this they are sending a signal to the investors and the general public that they like what they see. This way they, intentionally or unintentionally, can bend the public opinion towards favoring the companies, technologies, and projects involved.

We doubt that Warren Buffet, and any of his rich friends, know anything about solar or wind energy. We also doubt that they care much about the companies, technologies, and projects involved. So we cannot help it but wonder how they make purchasing decisions. It is very likely a combination of things, such as; taking technical and marketing advise from paid consultants, conducting analysis of the companies and projects involved, analyzing stock market performance etc.

In any and all cases, the final objective is making money. There are no other more compelling reasons for doing what they are doing—this is capitalism, after all. Including concerns for the environment does not hurt—it even embellishes their humanitarian side.

The final result of this is that when rich people get involved in any aspect of the energy markets, they send a signal; this company, technology or project is a money maker. Since they have a high visibility social profile, many people see their decisions and imitate their actions. After all, being rich and famous means that you are smart and knowledgeable on the subject at hand. And since you know how to make money, we would follow you.

The crowds follow the decisions of these people, hoping to benefit from their experience and achieve some level of success. How many succeed in doing this is uncertain, but it affects the market and drives it in one or another direction.

Case Study. Warren Buffett's Solar Adventure

Warren Buffett invested over $2.0 billion recently on purchasing several solar projects in California. Good thing, right? Yes, it is a good thing for the developers that they were able to sell their problems to the man who can afford to deal with them.

We are, however, not sure that he was told that none of the modules in his newly acquired power fields have a proven track record for 30 years non-stop operation in the deserts.

Without a full-life cycle track record, at least 30 years on-sun operation, PV modules represent great risk.

Most, if not all, PV modules in today's solar power plants lack such track record, because most of the companies involved—from the manufacturers to the installers—have been in business less than 10 years. Because of that, we already see problems with modules underperforming and foresee even more problems in the future with modules failing prematurely.

Note: An argument that deserves attention here is the fact that some solar module manufacturers have been in business much longer than 10 years. The determining factor here is the fact that while they have been making PV modules, the materials, production processes, and assembly procedures have been changing. Because of that, a PV module made mostly by hand in 1990 by BP, is totally different from one made today at First Solar's fully automated production lines, or by SolarCity/Silevo brand new manufacturing process.

Comparing any of these is like comparing the proverbial apples and oranges.

Each manufacturer and production process produce different types and quality products.

The difference could be quite serious, so each of these different types and quality PV modules must be properly identified and tested separately.

Was Warren told that all PV modules lose about 1% of their efficiency annually. That their output drops 0.5% for every degree Celsius above 25°C, which means 30% reduced power output during the noon hours, when the plant should be most productive.

And we also doubt that Warren is fully aware of the fact that PV modules in some of the projects he bought contain considerable amounts of toxic, carcinogenic heavy metal cadmium. We also doubt that he understands the full impact of huge amounts of cadmium thin films baking under the merciless sun, disintegrating and finding their way out into the environment in.

With millions of these modules installed in the U.S. deserts, this might signal the beginning of a great environmental disaster. Warren was surely advised that at their end-of-life, the PV modules have to be disassembled and discarded. We are not sure, however, that he knows that since some of these modules contain heavy metals, they have to be handled with extreme caution. This means that they require special hazmat crews with special hazmat equipment to be brought in to remove, load, and transport the useless (toxic heavy metals containing) modules to a special hazardous materials waste recycling and disposal sites.

Obtaining special permits for this work, bringing specially trained and equipped people to the site, loading the hazmat modules on a special train, and taking them to a special facility for special processing would require a lot of time and even more expenses.

In our estimate, the hazmat effort done under today's regulatory conditions would be more expensive than the original installation. By 2040-2050, when this work will have to be done, the related expenses might double. With millions of these toxic modules baking on the deserts sun, the effort would be monumental and cost billions of dollars. But Warren can afford it, so we won't worry about it. On top of it, Warren and most of us will most likely be gone by then, so why worry about it now.

The Most Powerful

The most powerful people in all aspects of life in the U.S., however, are the politicians, heads of different government departments, and many corporate CEOs. The politicians—on federal, state, or local political posts—determine the general direction the energy sector and the energy markets take. The corporate CEOs and other top officials, on the other hand, drive their companies in the general direction, as outlined by the politicians, but add their own touch and input as needed to maximize the company profits.

These different groups often see things differently, which helps balance the market dynamics, dictate the market conditions, and determine the final market direction. A good example of this dynamic is the government subsidies to the solar energy industry; the famous ITC (investment tax credit.

The ITC was introduced in 2006, and provides 30% tax credit for all residential and commercial solar installations. This is basically 30% cash back, based on the initial project cost. With other words, one-third of the cost of every solar installation in the U.S. since 2009 has been paid by the government. Significant portion of it went to Chinese, Spanish, and other foreign companies, but that is a different subject, which we discuss elsewhere in this text.

The ITC is solely responsible for the boom in solar and wind power generation.

This was/is a political move of unprecedented size and long-lasting reverberation across the U.S. and global energy markets. During the last several years, many companies and people benefited from the generous ITC, and the entire renewable energy sector went into a frenzy of activities. A lot of new renewable products and processes were introduced and prices were slashed. Record after record of number of wind and solar installations were broken on regular basis.

The future will tell, but we foresee some very interesting developments ahead of us in the U.S. renewable energy markets.

Summary

The issues and risks related to energy production and use can be boiled down to a number of points, as follow:

National Issues
• The developing countries are increasing the use of fossil energy, which increases the global fossil energy consumption. This trend is expected to continue for several decades.
• In developed countries, existing systems, policies, and infrastructure limit the adoption and growth of the renewables.

• In many countries, the national and local policies, and the preferred market-based mechanisms conflict, instead of help building, a sustainable energy future.
• U.S. utilities have not decided how to handle the renewable energy sources, so we expect to see some surprises in the future.

Global Issues
• There are serious disagreements between the developed and developing world on the overall on responsibility for, and the way to transform, the way energy is produced and used.
• Lack of global standardization and unified energy policies co-ordination among countries is affecting the regional stability. This presents obstacles to the regional energy markets planning and function.
• Energy externalities are not well defined and handled properly. A global consensus is needed, in order to bring these into the lime light and account for them properly.
• High environmental degradation related to fossil fuel extraction and use is recognized, but disagreements between developed and developing countries continue, which slow overall progress.
• The increasing environmental problems are not a sufficient deterrent to the increase of energy use. And so, the world continues in the path of increased fossils use.
• Socio-political changes in some regions are affecting the global energy markets by bringing significant changes in energy distribution and use.

General
• Economic progress and personal well-being depend on, and are driven by, energy use.
• There are unresolved issues and insufficient mitigation of short and long term risk associated with nuclear power generation, waste disposal, accidents response and responsibilities, etc. All this delays nuclear power developments—technologically and logistically speaking—which affect its future.
• The renewables are not proven reliable, safe, cost-effective, or efficient enough to compete, let alone replace, the current carbon intensive energy sources. There are no proven technologies this far, that can change the 'second-class citizen' role of the renewables.
• There is misinformation, miscommunication and basic lack of public awareness, all of which create

barriers to the long-term development of sustainable energy markets.

- Energy efficiency is a powerful tool in the direction of sustainable energy markets and environmental solutions, but it is not handled properly. There are political, financial, and cultural drivers that hinder its full implementation, despite significant benefits to sustainable energy and environment.

Notes and References

1. FrackCheck WV, http://www.frackcheckwv.net/
2. Inside Job: Export-Import Bank financing themselves, their friends and Obama-Clinton wealthy 'green' cronies
3. Green Corruption: Department of Energy "Junk Loans" and Cronyism, Part One and Two
4. Risk.net, http://www.risk.net/energy-risk
5. DOE, http://www.energy.gov/sites/prod/files/2015/06/f22/TX_Energy%20Sector%20Risk%20Profile.pdf
6. UoP, http://physics.isu.edu/radinf/np-risk.htm
7. EERE, http://www1.eere.energy.gov/geothermal/pdfs/geothermal_risk_mitigation.pdf
8. DOE, http://science.energy.gov/~/media/opa/pdf/processes-and-procedures/doe/g4133-Risk_Management.pdf
9. NREL, http://www.nrel.gov/docs/fy13osti/57143.pdf
10. Energy, http://www.cii.co.uk/media/4043851/ch12_energy.pdf
11. NRDC, http://www.nrdc.org/energy/frackingrisks.asp
12. The Green Corruption Files, http://greencorruption.blogspot.com/
13. *Photovoltaics for Commercial and Utilities Power Generation*, Anco S. Blazev. The Fairmont Press, 2011
14. *Solar Technologies for the 21st Century*, Anco S Blazev. The Fairmont Press, 2013
15. *Power Generation and the Environment*, Anco S. Blazev. The Fairmont Press, 2014
16. Energy Security for the 21st Century, Anco S. Blazev. The Fairmont Press, 2015

Chapter 10

Energy Markets' Future

It is up to us to shape the path to fair and sustainable energy future by diversification and efficient use of the available energy resources.
—Anco Blazev

BACKGROUND

Life on Earth has been changing slowly throughout the centuries. From living in caves, hunting animals with bare hands, and cooking them on camp fire, we (humans) went to living in comfortable homes, raising animals for food, and cooking on wood stoves. From bow and arrow we went to rifles and fishing nets. Candles were replaced by kerosene lights. Much later, the horse carts were replaced by self-propelled automobiles...and now technology is all around us. It even runs our daily lives.

The fastest progress ever recorded, started by the end of 19th century and accelerated even more during the last half of the 20th century. It was marked by incredible successes in all areas of the industry, transport, and agriculture. This fast development continues today, and promises to bring us to even higher levels of technological advances in the future.

The progress of the last century was made possible largely by the amazing properties of the fossil energy sources; coal, natural gas, and crude oil.

The remarkable thing here is that without the energy sources—their amazing properties and abundance—we would still be living in the horse-cart age.

Humanity's progress from smoke signals to satellite signals can be attributed to our ingenuity, powered by fossil fuels.

Note the "fueled by fossil fuels" statement above and think of its significance and importance. We owe all our achievements to the abundance and ease of use of the fossils. Not one single advancement would've been possible without the fossils. And even now, energy provided by the fossils drives the majority of the nations'

economic development. Without coal, gas, and crude oil, economic development is simply impossible. It is that simple and finite. Period!

The global energy markets regulate the prices and the distribution of energy sources around the world.

The energy markets are the engine of the global economic development. They coordinate and control the distribution of the fossil, and other energy sources, around the globe.

Today, the rich developed nations, and the nations blessed with a lot of fossil reserves, control the energy markets. They determine the amount of different fossils to be produced and used, and this way control their prices.

While on the surface, we see some resemblance of rule and order in the global energy markets, deep beneath there is an ocean of disagreements, manipulations, confrontations, and even wars. These abnormalities bring uncertainties in many areas of our lives today, and promise to bring even more in the future.

The global population, and its needs for energy increase daily, while at the same time the global fossil fuels reserves get more expensive, and decrease in volume. This creates an imbalance in the supply and demand chain, which in turn increases the prices proportionally. As global prices increase so will the conflicts between nations fighting for fossil energy sources.

Of course, we also have other energy sources—the renewables—which are awaiting their turn to become major players in the global energy markets. They, however, have long ways to go before being able to compete with the fossils.

Today's renewables will not be able to compete with the fossils until these are nearly depleted and when their prices rise astronomically high.

So, let's take a close look at the present day energy markets.

Energy Market Segments

Once again, we would like to clarify that in this text we look at energy and the energy markets from the broadest possible point of view. We consider anything that takes part in the development, production, distribution, and function of energy, and anything related to energy, as an integral part of today's energy markets.

This is to say that a large mining operation in Montana is equally important as the manufacturer of shoes for the miners. This is simply because the miners would not be able to enter the mine barefooted. But even if they enter the mine, they would not be as productive and/or as safe as having high quality shoes made by a mom and pop shop in El Paso, Texas.

There is not enough space in this text to detail the contribution of the Texas shoe shop to the mine operation, let alone its global effect, but we do keep in mind the army of such people and companies behind the scenes, who are involved in the supply chain of the energy industry.

Even the last woman, gathering wood for cooking the family dinner in Ghana, is a significant contributor to the energy-environment cycle. Not taking her, the Texas shoe shop, and millions of such small contributors into account, would paint a distorted picture of the global energy markets.

Actually, it is hard to think of even one sector of the economy (and even a single person) that is not involved—one way or another—in the energy business.

It's a complex picture, for sure, so let's see if we can see where all this is going.

We took a close look at the conventional, renewable, and peripheral energy markets above, so here we will only summarize their function. In general, the fossils still rule the global energy markets, but the battle is expanding on all fronts.

The core segments of the energy markets today are shown in Table 10-1.

The basic peripheral products and services, related to the energy markets (as we see and present them in this text), are shown in Table 10-2.

Quite a list, for sure! Far from complete too, but representative of the major segments of the energy sector and the related energy markets. We take somewhat superficial look at each of these sectors here, simply because the matter is overwhelmingly complex and requires many pages for a complete disclosure.

Table 10-1. Energy Markets' Core Segments

- Raw materials production (coal, gas, silicon, metals)
- Equipment manufacturing
  - Construction equipment
  - Transportation vehicles
  - Production equipment
  - Power generation equipment
  - Safety equipment
- Production of finished goods (fertilizers, solar cells, etc.)
- Refining and processing of fossils (crude oil, natural gas)
- Transport, distribution, and use of energy goods and fuels
- Power generation, distribution, and use, and
- Renewables' development, production, and use

Table 10-2. Energy Markets' Peripheral Segments

- Political and regulatory control
  - Federal and state energy policies
  - Regulatory enforcement
- Legal Actions
  - Operational support
  - Risk management
  - Financial Activities
  - Funding and Investment
  - Energy and energy products trading
  - Peripheral equipment manufacturing
  - Production equipment
  - Power generation and distribution equipment
  - Transportation equipment
- Contracting and Consulting services
  - Technical and logistics
  - Lobbying
- Physical and Cyber security
- Energy Efficiency (buildings and vehicles)
- Energy Storage
- Smart Grid
- Specialty Products and Materials
- Transportation Services
- Maintenance Services
- Waste Management
- Decommissioning services
- Environmental services
- Energy commodities and futures trading

We, however, take a much more close look and get into the details of the different segments of the energy sector in our other books on the subject: *Photovoltaics for Commercial and Utilities Power Generation, Solar Technologies for the 21st Century, Power Generation and the Environment,* and *Energy Security for the 21st Century,* all published by The Fairmont/CRC Press.

In this chapter, we will attempt the hardest part of the assignment—analyzing the factors that play role in, and even predicting, our energy future. Since we don't have a crystal ball, we had to consult the experts in different areas, compare their opinions and ideas (yes, they differ quite a bit across the spectrum of energy disciplines), and make the necessary conclusions. On the basis of these conclusions, and our experience in the field, we could then make some (unbiased) educated guesses and assumptions.

Due to the complexity and dynamics of the subjects at hand, most of our analysis and conclusions would be easy to argue with. Because of that, we suggest that they are taken for what they are; opinions based on available information, filtered through our technical experience in the respective areas. From here on, the esteemed reader can form his/her own opinions and make the necessary conclusions on the different subjects.

Below, we review the fossils first, and then look at the other energy sources, keeping their future in focus.

COAL'S FUTURE

Coal is largely responsible for bringing our civilization to the level of technical and economic development we enjoy presently. Coal is also the fuel that might help us with the transition to fossil-less energy future too. But today, coal is blamed for excess pollution...a catch 22 that we will have to deal with for a long time—hopefully in a responsible manner.

While we are thankful for all the good things coal has done for us and for our economic progress, we are at the same time blaming it for great environmental evil, property damages, as well as for making people sick and even killing them. It is as if we want to have our cake and eat it too.

Before we delve deep into the future, let's see what we are doing with coal today:

Present-day Coal Use

Coal is still the primary energy source for a number of countries worldwide, and provides about 1/3 of the world's primary energy and over half of the power generation. Coal is the main fuel for the generation of electricity since its price is low, compared to other fuels. Unfortunately, it is also the highest polluting source of electricity.

Other major uses of coal are in the production of steel and synthetic fuels. Bituminous coal is also used to produce coke for making steel and other industrial process heating. Coal gasification and coal liquefaction (coal-to-liquids) are used also to produce synthetic fuels.

The primary use of coal, however, is to generate electricity or heat, with electric generation surpassing any other use. Coal generates over $1.0 trillion annual revenue, which its use is rapidly growing around the world.

In contrast, coal use in the U.S. is decreasing, driven mainly by the conversion of coal-fired power plants to natural gas fuel.

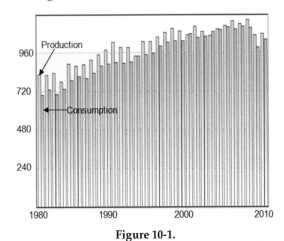

Figure 10-1.
U.S. coal production and use (in millions of tons/year)

Obviously, we produce more coal than we need for internal use. So, we export some of the excess. Only 4-5% of the coal mined in the U.S. was exported in the past, mostly for making steel. The export plans, however, call for a major jump in the coal exports in the near future.

Large quantity U.S. coal will be exported to Asia and Europe mostly for electric power generation.

How that will help our energy and environmental problems remains to be seen. The combination of increased coal exports, leading to increased GHG emissions overseas, however, promises more problems coming our way in the near future. Once again we see the U.S. caught in a huge controversy; proudly telling the world that we are reducing coal use in the country, while quietly exporting coal for polluting the global environment by other countries.

Significant quantities of coal are mined commercially in over 50 countries. Over 7,000 million tons annually (Mt/yr) of hard coal is currently produced, a substantial increase over the past 25 years. Recently, the world production of brown coal and lignite was slightly over 1,000 Mt, with Germany the world's largest brown coal producer at 194.4 Mt, and China second at 100.6 Mt.

Coal production has grown fastest in Asia, while Europe has declined. The top coal mining nations in 2010 were, unsurprisingly, China and the U.S.

Table 10-3. Global coal production and use (millions of tons).

Country	Annual production	% of world production	% of world use
China	3,050	39.4	38.6
US	960	19.3	18.4
India	557	6.8	7.7
Australia	415	6.6	1.7
S.Africa	250	4.7	3.0
Russia	310	4.9	3.6
Indonesia	250	4.7	3.2
Poland	150	2.2	1.9
Kazakhstan	125	1.8	1.1
Japan	—	—	3.9
S.Korea	—	—	1.8

Note the obvious discrepancies, where countries like Australia produce much more coal than they use, while other countries, like Japan, do not produce any coal, and yet use a huge amounts. Obviously, Australia would export its extra coal, while Japan would import it, hence the global coal market in action.

Most coal production is used in the country of origin, but some of it is being exported. U.S. coal production in 2011, for example, increased slightly from 2010, driven by export demand, to roughly 1.1 billion short tons. Production in the Western Region, which includes Wyoming, totaled 587.6 million short tons, a 0.7% decline from 2010, and yet totaling over half of the entire U.S. coal production.

Productive capacity of coal mines increased by 2.5 million short tons to 1.3 billion short tons. At the same time, the average number of employees in U.S. coal mines increased 6.3% to 91,611.

Domestic coal consumption of metallurgical coal by the coking industry rose 1.6% to 21.4 million short tons. The average sales price of coal increased 15.2% to $41.01 per short ton.

The average coal use in the U.S. is estimated at 18 lbs. per person per day (p/d). The global average is around 6.4 lbs p/d, while in Australia it is over 33 lbs. p/d.

We can't help but wonder what are the Aussies doing with so much coal? Well, they export a lot of it, of course? At the same time, most of the coal in the U.S. is used for power generation and industrial processes. Using 18 lbs. of coal per person per day (6.3 billion lbs., or 2.5 million tons), every day, requires a lot of coal to be mined, transported, and burned. This huge amount reflects our comfortable and even opulent life style too.

Global Coal Reserves

Coal reserves are available in almost every country worldwide, with significant recoverable reserves in around 70 countries. At current production levels, the global proven coal reserves are estimated to last over 150 years. However, production levels are by no means *level*, and are in fact increasing. This brings the question of what to expect in the future.

Again, we don't have a crystal ball, so we have to rely on the estimates of insiders and experts, which show that *peak coal* could arrive in many countries such as China and America around 2030.

Total global coal production is about 7,000 Mt/yr, and is expected to reach 12-13,000 Mt/yr by 2030.

But wait, aren't we in agreement that coal pollution is the source of climate change? Didn't we agree—including China—to reduce coal emissions? Yet, China accounts for most of the present use, and future increase of coal burning. So, how can we reduce pollution, if we increase the coal use?

We also know that coal has a beginning and end. We know the beginning, but are not sure of the end. Actually, we are ignoring the fact that coal is a finite energy resource. And that it is in danger of depletion, just like its fossil cousins; crude oil and natural gas.

Currently, about 2.24 trillion short tons, or 2/3, of the U.S.' 3.68 trillion tons of remaining (potential) coal resource inventory are classified as hypothetically undiscovered. Much additional unknown coal may be concealed in the central parts of coal basins and is not as yet included in the Nation' s coal inventory. The estimation of hypothetical coal resources in areas where geologic, thickness, rank, and areal size data are sparse or absent is a key to promote exploration for poorly known and undiscovered coal areas.

This additional unknown coal must be identified, as must the 2.24 trillion tons currently remaining in the inventory, because knowledge of the quantity, quality, and rank of the unknown and hypothetical coal could influence the Nation's energy plans. Or would it...?

Coal reserves are usually stated as either:

1. "Resources," which consist of the total of all "measured" + "indicated" + "inferred" resources,
2. "Run of Mine" (ROM) reserves (or the actually recoverable coal), which is usually much less than the general "reserves" estimates, and
3. "Marketable reserves," which after sorting of the produced coal may be only 60% of ROM reserves.

The standards for reserves are set by stock exchanges, in consultation with industry associations. For example in ASEAN countries, reserves standards follow the Australasian Joint Ore Reserves Committee Code used by the Australian Securities Exchange.

Because of increased pollution concerns, coal has fallen on hard times lately and the world-wide coal industry is forced to adapt. The aftereffects of this are quite obvious in the U.S. The axe is mercilessly falling on many coal-related businesses around the country. EPA regulations issued in the last several years limit the emissions from power plants to a point where coal can no longer compete with natural gas and the renewables.

The last round of EPA regulations put the coal power generation in absurdly awkward position. It simply made coal-fired power plants thing of the past. No new plant can be build and/or expected to operate at a profit. Period. The same is true for the existing coal-fired power plants too, and many of them are operating at a break even point.

The coal mining is in trouble, but it seems to be a temporary situation. While in-country use of coal is decreasing, coal exports are expected to reach unprecedented levels. The U.S. coal mines will most likely increase their output for the foreseeable future to meet the export demands. Most of the coal will be exported to Asia and Europe for burning in coal-fired power plants. Hello climate change.

Case study: The Contrast

Loy Yang coal mine in the Latrobe Valley in Victoria, Australia, is the biggest coal mine in the southern hemisphere, excavating over 30 million tons of brown coal annually. At the same time, it burns 65,000 tons of brown coal every day for power generation, releasing the same amount of carbon dioxide as 60 million cars. This is about 50-75% of the mine's daily coal production, so the excess coal is transported to other power plants across the country.

The estimates are that there is enough brown coal easily accessible in the Loy Yang mine to continue present-day levels of operation for the next 1,200 years. In contrast, official estimates claim that, at the present rate of exploration and exploitation, China has only about 50 years of coal reserves left. Others insist that the coal reserves would last over a century, even if the coal use increases and outpaces production, as planned.

And this brings the BIG questions: Are these estimates accurate? Are any of the estimates accurate? And what is the real situation?

Unfortunately, there is no answer to this question. The only thing that is for certain is that China and many other countries, which already relay on coal for most of their electricity and heating needs, will increase its use in the years to come. Where the coal is going to come from is irrelevant. They will find it and burn it.

China has the third largest coal reserves in the world, after the U.S. and Russia. But even then, the reserves are limited, so how long would they last, at the ever increasing rate of use, is anyone's guess. At the rate of about 2.5 billion tons of coal used annually in China alone, and increasing by the day, one doesn't have to be a genius, or a coal expert, to guess that the end of coal will eventually come sometime in the future.

And the other important question, begging for an answer is, "What are the Chinese people going to do for electricity and heat when the inevitable coal-less future arrives someday?" There is no way around it; any bucket full of water, with a hole in the bottom of it—no matter how small the hole—will eventually run dry. China's coal reserves are leaking fast, so it is only a matter of time before the bucket gets empty. And no amount of money or politics can change that fact.

Oh, wait, China and many other countries in a similar situation can import a lot of coal from the Aussies and the Americans for many years to come. This way, they can continue the unrestricted coal burning and polluting the air, while depleting the global coal reserves many times faster. Wise solution...

U.S. Coal

Coal is a big business world wide. Very big and getting bigger by the day. With nearly 25% of the world's reserves located in the U.S., it is, and will remain, big business in this country for a long, long time as well. No matter how many regulations and restrictions are introduced on the use of coal, mining and transport of coal (and the profits from its export) will only increase in the near future. At least this is Plan A now. Plan B calls for somewhat reduced export volume.

Coal in the U.S. is found in great quantities in several regions, such as the:

- Appalachian region in the states of PA, OH, WV, KY, TN, and AL;
- Midwest region in the states of IL, IN, and KY;
- Gulf Coast region in the states of TX, LA, AR, and MS; and in
- West in the states of UT, CO, AZ, NM, WY, SD, ND, and MT.

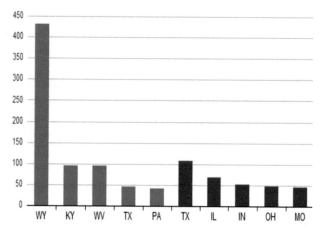

Figure 10-2. Coal producing states (in millions of tons)

To begin with, we need to emphasize that there are different types and quality of coal, which determine the design and operation of the coal mine, as well as the use of the coal produced at each mine.

Coal vary in composition from deposit to deposit, and sometimes even within the same deposit.

There are four major and very different types of coal. Each of these is characterized by a differences in appearance, abut more importantly in energy output. These are determined by the different pressure, heat, and time the different coal reserves have endured.

The different types of coal are:

- *Lignite* is a brownish-black coal with high moisture and ash content, which has the lowest heating value of the four types of coal. It is considered an "immature" coal that is still soft. It is used for generating electricity.
- *Sub-bituminous* coal is a dull black coal with a higher heating value than lignite, and is used principally for electricity and space heating.
- *Bituminous coal* is the most common type in the U.S., accounting for over 50% of the demonstrated reserve base. It is the most commonly used type of coal for electric power generation in the U.S. It is a dark, hard coal that has a higher heating value than lignite and sub-bituminous coal, but a lower heating value than anthracite.

- *Anthracite* is also known as "hard coal" that was formed from bituminous coal under increased pressures in rock strata during the creation of mountain ranges. In the U.S., it is located primarily in the Appalachian region of Pennsylvania. It is very hard and shiny. This type of coal is the most compact and therefore, has the highest energy content of the four levels of coal. It is used for space heating and generating electricity. It makes up only 1.5% of the demonstrated reserve base for coal in the U.S.

Different types of coal can be found in different areas, as follow:

- *Lignite* is found in the states of MT, TX, and ND;
- *Sub-bituminous coal* is found in the states of MT and WY;
- Bituminous coal is found in the states of IL, KY, and WV; and anthracite is found predominantly in the state of Pennsylvania.

There were 1,325 mines in the U.S. in 2011, with total coal production of roughly 1.1 billion short tons annually. This is over three times less coal production that than in China.

While the U.S. is steadily reducing the use of coal, China is increasing its use.

Production in the Western Region, which includes Wyoming, totaled 587.6 million short tons. The productive capacity of the U.S. coal mines was approximately 1.3 billion short tons.

The average number of employees in U.S. coal mines at the same time was 91,611. Domestic coal consumption of metallurgical coal by the coking industry was 21.4 million short tons. The average sales price of coal was $41.01 per short ton.

Approximately 25% of the world's recoverable coal deposits are in the US; approximately 270 billion tons, with Russia following at 176 billion tons, except that most of the deposits are in areas that are very difficult to mine.

There are estimates of over 1.0 trillion tons of coal available globally, with the largest coal reserves located in the U.S., Russia, China, Australia and India.

The global coal reserves are equivalent to about 120 years of non-stop production at the present day rate of use.

The major coal producers today are China, the U.S., India, Australia, Russia, Indonesia and South Africa. The global coal production lately has been about 7.5 billion tons annually. China accounts for nearly 50% of world's coal consumption, followed by the U.S., India, Japan and Russia as major coal consuming nations.

In the U.S., a number of regulations have changed the rate and the way coal has been produced and used. So, let's take a brief look at the regulations that affect coal production and use:

The Regulations

The problem is that coal is not clean by any means, and the attempts to clean it are few and far in between. This is mostly because coal cleaning requires complex and expensive technologies/and processes. Since the cost of cleaning coal's pollution is very high, we see no way to change the present situation. As a result, the U.S. government agencies in charge of the energy sector, while encouraging coal cleaning, have given up and are not waiting any longer.

Instead, they decided to tighten the GHG emission regulations for power plants, which in fact means, a) no new coal-fired plants can be build in the U.S. ever, and b) many old such were, or will be, forced to shut down (or retrofitted to gas-burning) soon.

There have been numerous sweeping regulations, affecting the U.S. coal mining and coal power generation industries. Some of the recent and more important ones are as follow:

CSAPR

In July, 2011, the EPA issued a rule intended to protects the health of Americans by helping states reduce air pollution and attain clean air standards. This rule, known as the Cross-State Air Pollution Rule (CSAPR), requires states to improve air quality by reducing power plant emissions that contribute to ozone and/or fine particle pollution in other states.

In a related, regulatory action, EPA finalized a supplemental rulemaking on December 15, 2011 to require five states (Iowa, Michigan, Missouri, Oklahoma, and Wisconsin) to make summer time NO_x reductions under the CSAPR ozone season control program.

CSAPR requires a total of 28 states to reduce annual SO_2 emissions, annual NO_x emissions and/or ozone season NO_x emissions. The goal is to attain the 1997 ozone and fine particle and 2006 fine particle National Ambient Air Quality Standards (NAAQS).

In February and June, 2012, EPA also issued two adjustments to the CSAPR, which replace EPA's 2005 Clean Air Interstate Rule (CAIR).

Note: A court decision in 2008 kept the requirements of CAIR in place temporarily but directed EPA to issue a new rule to implement Clean Air Act requirements concerning the transport of air pollution across state boundaries. This action is in response to the 2008 court's decision.

MATS

In December 2011, EPA issued the first national standards for mercury pollution from power plants (MATS). This is the first national standard intended to limit power plant emissions of mercury, arsenic, acid gas, nickel, selenium, and cyanide. The new standards also slash emissions of these dangerous pollutants by relying on widely available, proven pollution controls that are already in use at more than half of the nation's coal-fired power plants.

EPA estimates that the new safeguards will prevent as many as 11,000 premature deaths and 4,700 heart attacks a year. The standards will also help America's children grow up healthier—preventing 130,000 cases of childhood asthma symptoms and about 6,300 fewer cases of acute bronchitis among children each year.

All these actions are done in good faith, no doubt. And while these may be justifiable measures, they also mean that nearly 30,000 workers are losing their jobs, millions of consumers will be paying more for their electricity and the reliability of our electricity supply is being compromised.

The coal related problems were (partially) eliminated overnight by not burning more coal. As a result, many coal-fired power plants were shut down or converted to natural gas fuel. What is not clear yet, however, is how much GHG savings these moves bring, since natural gas is also a polluter.

Total of 24.5 GW of coal power generation have been eliminated thus far and much more will be eliminated in the future as a result of the new EPA emission rules. Over 60 power plants were shut down in the recent years—most of them to never be restarted. Thousands of jobs were lost in the process, with the accompanying financial devastation for the local economies.

The amount of lost power generation thus far is more than the total power generation of most countries in this world. The future loss of power generation will be even greater.

The Environmental Protection Agency (EPA), however, seems to have grossly underestimated the impact of its mandates on coal-based electricity generation. Instead of 4.8 to 9.5 GW of electric plant retirements, as

Table 10-4. U.S. coal-fired power plants shut down due to tightened EPA regulations

State	Plant	Owner	MW	State	Plant	Owner	MW
CO	Clark	Black Hills	42	OR	Boardman	Portland	601
GA	Harlee Branch	Georgia Pwr	581	PA	Armstrong	FirstEnergy	326
IL	Hutsonville	Ameren	150		New Castle	GenOn	330
	Meredosia	Ameren	203		Shawville	GenOn	597
KY	Green River	LG&E	189		Titus	GenOn	243
	Cane Run	LG&E	618		Portland	GenOn	401
	Big Sandy	AEP	278		Elrama	GenOn	460
	Tyrone	LG&E	135	SC	Jefferies	Santee	306
ME	R. Paul Smith	FirstEnergy	110		McMeekin	SCE&G	294
MA	Salem Harbor	Dominion	806		Urquhart	SCE&G	100
	B.C. Cobb	CMS	312		Canady's Sta.	SCE&G	490
	D.E. Karn	CMS	515		Grainger	Santee	170
MI	J.R. Whiting	CMS	345	SD	Ben French	Black Hills	25
MN	Black Dog	Xcel	294	TX	Monticello	Luminant	1,186
MO	Meramec	Ameren	924		Welsh	AEP	528
NM	Four Corners	APS	633		J.T. Deely	CPS	897
NC	LV Sutton	Progress	604	VA	Yorktown	Dominion	376
	Cape Fear	Progress	323		Chesapeake	Dominion	813
	Weatherspoon	Progress	171		Clinch River	AEP	235
	HF Lee	Progress	397		Glen Lyn	AEP	335
OH	Miami Fort	Duke	163	WV	Albright	FirstEnergy	278
	WC Beckjord	Duke	1,222		Kammer	AEP	630
	Muskingum	AEP	840		Rivesville	FirstEnergy	110
	Conesville	AEP	165		Willow Island	FirstEnergy	213
	Bay Shore	FirstEnergy	499		Kanawha	AEP	400
	Avon Lake	GenOn	732		Phillip Sporn	AEP	600
	Lake Shore	FirstEnergy	300	WI	Alma	Dairyland	45
	Niles	GenOn	217	WY	Neil Simpson	Black Hills	22
	Ashtabula	FirstEnergy	256		Osage	Black Hills	35
	Eastlake	FirstEnergy	1,289			**Total**	**24,459**
	Pickway	AEP	100				

predicted by EPA recently, 63 power plants with over 24 GW of generating capacity are already shut down, or on the chopping block, due to the new regulations.

The EPA rules also resulted in higher monthly bills for most electric power customers. Arizona, for example, the Solar capital of the world, gets 40% of its electricity from coal, 22% from natural gas, 29% from nuclear, and 9% from renewables. It is now the 18th most expensive state for energy prices in the country.

The average retail price of electricity in Arizona is less than 10 cents per kWh. During the hot summer months, the price during peak hours (1-8 PM) rises to 25 cents per kWh. Go figure! Additional EPA regulations might send the prices even higher.

As things have been going, the worst for the U.S.

coal mining and power generation industries is yet to come. In the summer of 2014, EPA came out with new rules, directing the states to cut greenhouse gas (GHG) emissions from power plants. The new rules call for 25% reduction of GHGs by 2020 and 30% by 2030 from the 2005 emission levels. Considering the fact that the 2005 emission standards were already too low to be met by coal, the new standards basically mark the end of coal-fired power plants era in the U.S.

According to the EPA, this move will provide the country with cleaner air and $90 billion in climate and health benefits, in addition to avoiding hospitalizations due to health issues such as asthma. And so, the environmentalists are welcoming the new rules as potential gain for the American customers. The hope is that the new rules will force people and companies to look into alternative power generation and introduce efficiency measures. The major gains could come from improved energy efficiency, as well as the health and environmental benefits of curbing climate change.

The already impossible to meet GHG standards, have made it very hard for coal-fired plants to operate profitably. As a result, many coal power plants have been closed down or converted to natural gas. and we expect many more to shut down or get converted. This also means that no new coal plants will be built in this country ever.

In May 2014, the U.S. Chamber of Commerce released a paper estimating that the new expected rules would cost the U.S. economy $50 billion a year. This move would also eliminate a quarter—224,000—of the total 800,000 coal industry jobs. This is in stark contrast with the EPA estimate of over $90 billion saved in climate and health benefits. Who is right, who is wrong…?

The new EPA standards give the states full flexibility in deciding how to meet the new limits—as long as they cut the GHG emissions 30% by 2030. Period. But how? Power plants can cut emissions directly, by switching to a fuel source with lower carbon emissions, such as from coal to natural gas. They can also make upgrades to equipment performance, and improve the process efficiency.

The states also have a choice to meet the new standards by increasing the amount of energy generated by renewable sources such as solar, wind, biofuels, or hydropower.

There are already successful programs based on similar principles already in place in some locals. The northeast's Regional Greenhouse Gas Initiative is a good example of this effort. It is a nine-states agreement of 2005 that uses a cap-and-trade system to reduce emissions. These states have been able to make greater cuts in GHG emissions at much lower costs than was initially expected.

This achievement have been made possible mainly by the increased use of natural gas, with some help from expansion of, and reduced costs for, renewables and other innovations in the energy market.

The new EPA mandate puts coal at a great disadvantage and makes natural gas officially the leader of the U.S. power sector.

Natural gas is winning simply because there is no other fuel or technology—abundant, clean, and cheap enough—that can replace coal. At least for now. Thing

will change as time goes by, but for now, natural gas rules the U.S. energy market. It is gaining in popularity in both; electric power generation and transportation.

Natural gas is replacing both; coal and crude oil in the U.S. energy markets

Obama's 2015 Environmental Plan

The proverbial 'last straw that broke the camel's back' was also the strongest action ever taken in the U.S, to combat climate change. It was the President Obama's unveiling of a set of environmental regulations devised to sharply cut GHG emissions from the nation's power plants. The biggest victim are the coal power plants, and the coal industry at large. If and when these regulations become the law of the land, they will ultimately transform America's entire electric power industry.

The new rules are the final, and much tougher versions of the proposed EPA regulations of 2012 and 2014. The legal challenges are just starting, but when the dust settles, these new regulations will bring sweeping policy changes that would certainly shut down hundreds of coal-fired power plants. They will also stop the construction of any new coal plants, while at the same time creating a boom in the natural gas, solar and wind and other renewable energy industries.

The Issues

Here are the major issues with the new regulations, as we see them:

- The U.S. coal industry won't give up. On the contrary; plans are made right now to produce more coal than before. If it is no longer needed in the U.S., then it will be exported to China, India, Europe and other energy hungry places around the globe.

- It is obvious from Figure 10-3 that U.S. produces much more coal than it uses. Most of the excess coal will be exported. The increase in U.S. coal exports will contribute to increased emissions in other countries, because nothing can stop China, India, and the other developing, and some developed, countries from using more coal as planned presently. In the end, we are only moving the pollution overseas—and the global emissions will continue rising…with our help.

- The U.S. is giving a good example of what can and should be done to reduce GHG emissions before it

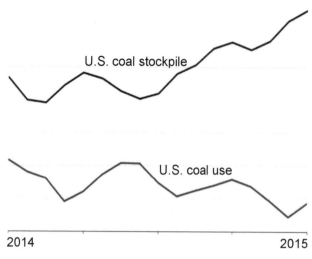

Figure 10-3. U.S. coal production and use.

is too late—starting by reducing coal use. The U.S. is a rich country and can afford all these new rules and regulations. Most of the world is poor, however, and cannot afford such drastic changes, and therefore will not follow our example. Period!

• The new regulations put us in disadvantage as compared to most other countries, where coal production and use are unrestricted. The disadvantage can be expressed in terms of lost jobs, and a number of future problems—read on.

• The very worst part of the new rules is that the mighty U.S. of A. will now rely on natural gas for most of their power generation needs. Our economy would become 70-80% natural gas based. This is one big disadvantage that can profoundly affect our national security, since any serious interruptions in natural gas supplies would cripple the economy that relies exclusively on it. EPA regulators apparently have never heard that putting all eggs in one basket could lead to a disaster.

• Another disadvantage is for the general public, which has to pay more every month while all these changes take place.

• The natural gas bonanza is good for the short term, but how long could it last? A decade, two, three? The production from existing and any new gas deposits will eventually decrease to a point where gas will become more expensive than coal. Would the EPA issue new, less strict, rules so that we can

start using coal again?

• In best of cases, at the present rate of use, the gas deposits will be drained completely sooner or later. What regulations would the President and EPA issue then?

Clean Carbon

Clean carbon is a misnomer, since coal is inherently dirty material. Period. It would be more appropriate to say *'cleaner'* carbon is the goal. Carbon cleaning is a new concept that is taking a new meaning as we enter the era of quickly rising CO_2 levels, accompanied by serious climate and weather changes. We fully understand and agree that, on one hand, coal is not going away, since it is a most important part of our energy supply, while on the other hand, we also know that it is polluting the air we breath.

We all, as a society, have accepted these facts of life, which we must deal with. But what if there is a way to clean coal…at least partially, so it stops killing us with these noxious fumes emitted at high volume? The clean coal technology is here and it is just a matter of implementing it…but that requires putting lots of money into it.

Yes, it might be somewhat more expensive than the conventional processes, but it is important that we stop, or at least reduce, the CO_2 and other GHG gases produced during coal burning from killing the environment, and all life in it. It can be done, IF the regulators and the coal industry agree on a path to a compromise. Yes, compromise on both sides of the issue is needed.

Here are some not-so-new ideas for cleaning coal's pollution, which time has come:

IGCC Process

Integrated gasification combined cycle (IGCC) is a technology that uses special equipment, gasifier, to process coal and turn it into gas. Thus produced synthesis gas, or syngas is then further purified before it is sent to the power plant for burning.

Some pollutants found in coal and released during the IGCC process, such as sulfur, can be turned into re-usable byproducts. This brings additional income to the processing facility, and reduces the emissions of sulfur dioxide, small particulates, and mercury.

Further processing of syngas via the water-gas shift process, shifts the carbon in it to hydrogen, which results in even cleaner; nearly carbon free fuel. The reaction produces carbon dioxide, which can be compressed and stored.

Excess heat from the primary combustion and syngas fired generation can be re-used in a combined cycle gas turbine, which results in improved efficiency compared to conventional pulverized coal burning.

The IGCC process consists of several major steps:

- Coal and oxygen are combined in the gasifier and converted into syngas and steam

- Hot syngas is sent into a refining unit to remove sulfur and other contaminants, such as mercury and particulate matter.

- Thus processed clean syngas is piped into a combustion turbine where it is burned to produce steam.

- The steam in turn rotates the blade of a turbine connected to an electric power generator.

- The generator spins, during which action the its rotor and stator generate electric power.

- Thus generated electric power is used locally, or is sent into the national grid.

- Large amount of steam, generated in the combustion turbine is recovered and used to generate more steam.

- The steam recovered in the different steps of the process, is sent into a heat recovery steam generator, where the steam is conditioned and sent into another steam turbine attached to a generator to generate additional electric power.

Case Study: Edwardsport IGCC Power Plant

The Edwardsport coal-fired power plant in the southwest corner of the state of Indiana has powered the state since the 1940s. It was recently retired, to give place to a new modern coal-fired power plant. The old power plant could not meet today's air quality standards, which require clean fuel, and in a much more efficient way.

The new 618 MW Edwardsport IGCC was designed as a model of the coal-fired power plants of the future. The new base-load plant also has space for carbon capture and storage (CCS) if and when Duke Energy decides to add this option.

The problem is that the original plant design called for the CCS option, but it was determined later on that the site is not geologically suitable for CCS. So, Duke Energy needs get a permit and construct a pipeline to transport the exhaust gases to a more suitable site nearby.

The new IGCC plant actually turns coal into a cleaner-burning fuel.

Sounds complicated and expensive, yes, but this is what we thought of computers half a century ago. And look what happened. Regardless, focused on the future, Duke Energy started the construction of the new IGCC power plant. This was the first of its kind and size ever.

The IGCC process at the new plant converts coal

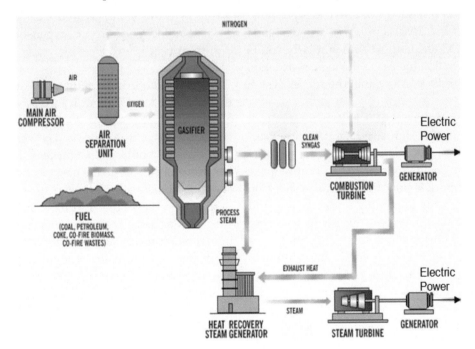

Figure 10-4. IGCC process

into a combustible *syngas*, which is then fed into the combustion turbine to generate steam. The exhaust gases are routed through a large heat recovery steam generator and are then used in a steam turbine to further increase the plant's efficiency.

The pollution emitted by the plant will be gradually with time, as particulate matter and GHGs will be removed from the syngas. The coal-derived synthetic gas is much cleaner than conventional coal, and is also much easier to clean it further. Coal gasification allows carbon capture to occur before the fuel is combusted, which is easier and economically feasible operation than carbon capture of exhaust gases.

The plant was completed and started operation in 2013 at a cost of about \$3.5 billion, which is up 75% from the initial estimates of less than \$2 billion in 2007. The added costs were blamed on the added "scale and complexity" of the project. But money talks, and we are watching, but what we see doesn't look that good:

618 MW at a cost of \$3.5 billion, puts the plant capital cost at \$5.60 per installed watt. This is about three times the cost of a natural gas-fired power plant, so the additional cost has to be justified and figured out in future undertakings of this nature.

\$2.6 billion of this amount will be recovered from Duke's 790,000 Indiana taxpayers. This also means that each customer will have to pay about \$2,700 in addition to the present charges, for the new sophisticated plant. And this is if everything goes well, and no additional expenses are incurred.

Another similar project, the 582 MW Kemper power plant in east-central Mississippi, was also designed to use similar IGCC technology. And similarly so, its initial cost of \$2.0 billion increased during construction to more than \$4.2 billion, or double the initial design estimates.

Price is not everything, but is always a significant part of the equation. The competition in the energy sector is steep. Today low, and decreasing, natural gas prices are changing the energy markets. They simply make it uneconomic to upgrade pollution-control systems on the old coal-fired plants to comply with the new and much stricter strict U.S. EPA rules. This is why we see so many old coal-fired power plants shutting down.

The new IGCC technology might be the solution to coal-fired power generation, since it is much more efficient than the conventional coal-fired power plants. And even more importantly, IGCC reduces the GHG (sulfur dioxide, nitrogen oxide and particulates) emissions by about 70%. It also reduces CO_2 emissions by about 50%, as compared to the old coal-fired power plants.

The reduction of pollution gases, and the possibility of CCS, are very attractive options for the old and tired coal-fired power plants, but capital cost has to be reduced significantly before any great developments are to be expected.

CCS Process

Carbon capture and storage, or sequestration (CCS) process is used to capture waste carbon dioxide (CO_2) from fossil fuel power plants, and storing it in special underground reservoirs. This way, large quantities of CO_2 are prevented from entering the atmosphere, in an attempt to mitigate the contribution of fossil fuel emissions to environmental damage. CCS can also be used in a similar manner to remove large amounts of CO_2 from ambient air, but power plants CCS of CO_2 and GHG other gases is the focus today.

Storage of the CO_2 can be done either as gas or liquid in deep geological formations, or in the form of mineral carbonates, but natural geological formations are the most promising sequestration sites for now. North America has enough storage capacity for more than 900 years worth of carbon dioxide at current production rates.

The Major Problems with CCS:

• *Long-term predictions about submarine or underground storage security are uncertain, and*

• *There is always the risk of uncontrollable, excess CO_2 leakage into the atmosphere.*

• The more sophisticated CCS methods (such as liquefied CO_2) are too expensive for all practical purposes

Note: Capturing, compressing and liquefying CO_2 is one method of CCS, but it increases the fuel needs of a coal-fired CCS plant by 30–40%, which ultimately increase the cost of the generated power.

Another sophisticated process, deep ocean storage, has been considered in the past too, but has been largely abandoned, because it brings further uncertainty to the problem of excess ocean acidification.

The industry has an experience with injecting CO_2 into geological formations for different purposes, such as enhanced oil recovery. The huge amount of CO_2 released by today's power plants, and the need for it to be stored indefinitely are new concepts that need to be proven.

CCS applied to a modern conventional power plant could reduce CO_2 emissions to the atmosphere by approximately 80–90% compared to a plant without

CCS. The economic potential of CCS could be between 30% and 50% of the total carbon mitigation effort until year 2100.

Applying the CCS technology to existing power plants would be even more expensive since they are usually far from a sequestration site.

There is no way to change the existing power generating infrastructure to alternative renewable sources of energy, so the best approach is to work with the existing technologies, while working on the alternatives until they become economically feasible. That process, however, could take many decades, so storing CO_2 and reducing greenhouse gas emissions might be one way to do this painlessly.

Coal-fired power generation plants produce a rather dilute flue gas, so coupling coal-fired power generation with CO_2 geological storage through carbon capture and storage would produce green electricity, since there are much less GHG emissions emitted during power generation.

With successful research, development and deployment, and considering the related laws and regulation, CCS coal-based electricity generation may cost less than un-sequestered coal-based electricity generation by 2030.

Case Study—German CCS Projects

One of first commercial CCS projects was started in 2000, where a 200 miles long pipeline transfers $CO CO_2$ gas from the Weyburn Sask oil fields to a coal gasification plant in Beulah, North Dakota. Tons of carbon dioxide are transported through it daily to be injected deep underground for a long-term storage.

This effort alone is equivalent to removing 8 million cars off the road every year, over the 35 years of operation of this project. Over a million tons a year of $CO CO_2$ are being injected into the reservoir and kept in there as long as possible. This method of carbon capture and storage is seen as a practical way of reducing CO_2 emissions in the short term.

Another pilot-scale CCS power plant began operating in 2008 in the eastern German power plant Schwarze Pumpe, hoping to answer some of the pertinent questions about the technological feasibility and economic efficiency of CCS methods.

At a total investment of $96 million, the project consists of a steam generator with a single 30 MW top-mounted pulverized coal burner and the subsequent flue gas cleaning equipment (electrostatic filter/precipitator), wet flue gas de-sulfurization, and a flue gas condenser.

The $CO CO_2$ purification and compression plant is downstream of the flue gas condenser to produce liquid $CO CO_2$ and gaseous oxygen with 99.5% purity. This high level of purity is required for combustion is supplied by a cryogenic air separation unit.

In 2009 the project achieved nearly 100% $CO CO_2$ capture, and in the middle of 2010, Schwarze Pumpe reported 6500 hours operation during 18 consecutive months without any major problems.

The pilot plant is designed to have flexibility in terms of construction and the ability to exchange key components, like the burners and such. The plant has undergone two remodeling/rebuilding periods as needed to enlarge the project. During those times changes were made to three different burners for testing and optimization purposes.

Schwarze Pumpe, also known as the 'CO CO_2-Free Power Plant Project', is scheduled to be operated for a ten year period (until 2018), delivering base knowledge and validation data. The project boasts a complete process from fuel input to delivery of liquid clean $CO CO_2$ ready for storage. Life cycle assessment of the entire CCS process, including the upgrades, and upon completion will provide information for future practical applications of the CCS technology.

Case Study: IGCC-CCS Combo

How about using both: IGCC and CCS at the same time, as was the original plan for the above mentioned Edwardsport IGCC power plant in Indiana? A good example of this new and promising technology is the new GreenGen power station near Beijing, China. It is a 400 MW coal burning plant that is using a coal-to-gas conversion and gas burning process, which is much cleaner than solid-coal burning.

The plant is also equipped with efficient gas sequestration equipment, which ensures that the gases that are released in the atmosphere are 90-95% clean. Thus generated CO_2, which is the major component in the waste gases, is trapped, segregated, purified and stored. It is then sold for bottling of soft drinks and other commercial needs.

GreenGen demonstrates multiple Integrated Gasification Combined Cycle (IGCC) technologies that can be scaled to simultaneously address several environmental challenges in response to climate change and energy security concerns. The $1.0 billion GreenGen plant features pre-combustion technology that will strip pollutants such as SO_2 and particulates from the coal burning process.

The second phase will implement fuel cell power

generation and carbon capture and sequestration (CCS) technology for nearly zero-emission power generation.

The third phase of the project is planned for completion by 2016, when the plant would produce a total of 650 MW and 3,500 tons of syngas per day.

When completed and optimized, this model will be duplicated in many such IGCC-CCS plants. And with China continuing to build about 30 power plants a year, these technologies could have a significant impact on the air quality and greenhouse gas emissions not only in China but all around the world.

So far so good, so what is stopping us? The answer is money. Lots of money is needed, as well as some more work on, and improvements in some areas of the CCS/IGCC technologies as a whole.

In any case, even if and when all these processes work, they are not the final solution to all our energy problems. The problem is that only a few rich countries could afford the new technologies, and most wouldn't care to. And then, we still have to deal with increasing global pollution and the depletion of the fossils.

Nevertheless, we have to agree that during this energy (fossil-less) transitional period, which is upon us already, we must look into any and all possible solutions; no matter how expensive, or temporary they might be today. Things may change tomorrow so that what is impossible or impractical today, becomes a valuable solution tomorrow.

Little by little, we might just be able to add another drop to the bucket of emergency energy and environmental solutions. Any measure that reduces environ-mental pollution must be given a priority and implemented immediately—it is a matter of life and death at this point in time.

Coal Liquefaction

Coal liquefaction is the process of converting coal into liquid fuels. Although not simple nor cheap, it might be one of the most important uses of coal, as far as our energy security and independence are concerned.

As discussed above, we have enough domestic fuels for power generation and heating, but are importing half of our transportation fuel—crude oil—from abroad.

Making enough liquid fuels via coal liquefaction would reduce or even eliminate the oil imports.

Coal liquefaction seems as a reasonable alternative, so this is one process/product which we need to watch in the future. This is one technology, or rather a number of different technologies, that are capable of producing liquid oils and fuels using hydrogenation and/or carbon-rejection processes. This is necessary, because liquid hydrocarbons have a much higher hydrogen-carbon molar ratio than coal. So the coal molecules must be changed in order to become more hydrocarbon-like, in order to be used as liquid fuels.

The coal liquefaction technologies can be divided into:

* *Direct coal liquefaction (DCL) and*
* *Indirect coal liquefaction (ICL).*

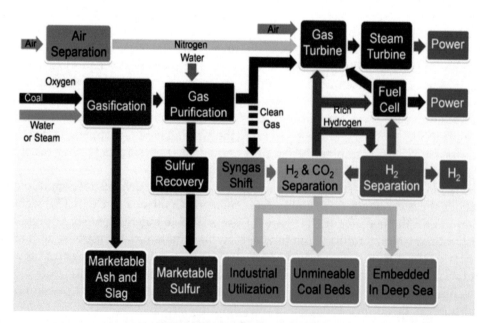

Figure 10-5. Green Gen function

Coal liquefaction is usually high-temperature and high-pressure process, which requires a lot of energy input, and which increases its cost often well above the crude oil price. Also, the large-scale industrial coal liquefaction operations require multi-billion dollar capital investments. Because of these factors, coal liquefaction is only economically viable when crude oil is unavailable, and/or its prices are very high.

This presents a high investment risk, so we won't see many coal liquefaction operations around until either,

a) crude oil price jumps well above $150/bbl, and/or,
b) international or national disaster forces us to use coal liquefaction for transportation and other applications dominated by crude oil.

In more detail, the major coal liquefaction technologies in use today are:

Direct Coal Liquefaction (DCL) Process

Direct coal liquefaction (DCL) process converts coal into liquid oils and fuels directly without the intermediate step of gasification as used in the ICL process. This is done by breaking down coal' organic molecule via solvents and catalysts at very high pressure and temperature.

Some of the key DCL processes are:

• *The Bergius Hydrogenation process* is a DCL process for coal liquefaction by hydrogenation, developed by Friedrich Bergius in the beginning of the 20th

century. Here, dry coal is mixed with heavy oils that has been used previously and recycled. The mixture (slurry) is pumped into a reactor, where catalyst is added to the mixture. The catalyst enriched mixture is exposed to high temperature in the 800-950°F range and up to 10,000 PSI pressure in hydrogen atmosphere.

The resulting reaction is of the type:

$$xC + (x+1)H_2 \rightarrow C_xH_{2x+2}$$

or

$$6C + 7H_2 \rightarrow C_6H_{14} \text{ (hexane)}$$

The mixture is sent into a flash drum, where the liquid and solids are separated, and the liquid is vaporized. It is then sent into a distillation column, where the different fractions of oils and fuels are separated and collected individually. These liquids are then stored, and shipped to customers for use as lubricants and liquid fuels.

Figure 10-6 depicts a generic *coal* hydrogenation process, involving reactors and distillation columns, which is all that is needed to convert the single minded carbon molecule in coal into much more versatile hydrocarbon molecules in liquid fuels.

This relatively simple (but very expensive) petrochemical process amplifies the uses of coal from that of power and heat generation to thousands of applications that crude oil is dominating now. But most importantly,

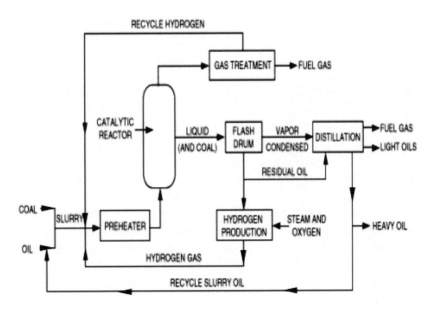

Figure 10-6. Coal liquefaction via hydrogenation.

thus liquefied coal can replace the huge, expensive, and insecure crude oil imports.

There are several variations of this process:

- *The H-Coal process,* developed in the 1960s uses pulverized coal mixed with recycled liquids, hydrogen and catalyst in an ebullated bed reactor. The dissolution and oil upgrading takes place in the single reactor at a fast pace and the products have high H/C ratio. The main disadvantages of this process are high gas yield, since it is basically a thermal cracking process, high hydrogen consumption, and limitation of the produced oil usage as a boiler oil because of impurities.

- *The NEDOL process* was developed in the 1970s, where coal is mixed with a solvent and a synthetic catalyst and the mixture is heated in a tubular reactor to 800-900°F and 3,000 PSI in H2 atmosphere. The produced oil has low quality and requires expensive refining before regular use.

- *The Solvent Refined Coal (SRC-I and II)* processes were developed in the 1960 and 70s. They use dried, pulverized coal mixed with molybdenum catalyst. The hydrogenation is conducted under high temperature and pressure via syngas, produced in a separate gasifier. This process yields synthetic liquid product Naphtha, some C3/C4 gas, light-medium weight liquids (C5-C10) suitable for use as fuels, small amounts of NH3 and significant amounts of CO_2.

There are also a number of two-stage direct liquefaction processes; however, after the 1980s only the Catalytic Two-stage Liquefaction Process, modified from the H-Coal Process; the Liquid Solvent Extraction Process by British Coal; and the Brown Coal Liquefaction Process of Japan have been developed.

Historically, there have been many small and large size coal liquefaction attempts, with the most famous dating back to WWII, when several plants were built in Germany to supply the German army with fuel and lubricants from coal.

The Kohleoel Process was used in the demonstration plant with the capacity of 200 ton of lignite per day, built in Bottrop, Germany during1981-1987. This process was also explored by SASOL in South Africa around the same time.

The NEDOL process was used by Japanese companies Nippon Kokan, Sumitomo Metal Industries and Mitsubishi Heavy Industries in the 1980s.

Although the different approaches and processes show promise, most of them proved to bee too expensive for all practical purposes. Some of the prototype and demo plants are still in operation, but, with few exceptions, the majority of the effort in this area is limited to small-scale R&D.

Case Study: Shenhua DCL Plant

The Chinese coal mining company Shenhua built a DCL plant in Inner Mongolia in the 2002. The process includes a slurry of pulverized coal in recycled (heavy coal-derived oil) which is premixed and pumped through a preheater along with hydrogen and catalyst into the first stage reactor.

The effluent from the first stage undergoes separation to remove gases and light ends with the heavier liquid stream flowing to the higher temperature second stage. Efluent from the second stage, joined with overhead from the inter-stage separator, flows to the fixed bed in-line hydro-treater for enhanced upgrading to very clean fuels.

The effluent from the hydro-treater is the major liquefaction product, mostly diesel, naphtha, and a jet fuel fraction. Bottoms product from the second-stage separator is flashed, and the overheads are pumped to the in-line hydro-treater for upgrading.

The atmospheric bottoms stream containing solids is used as recycle with a portion going to a vacuum still and to solvent solids separation, with the resulting bottoms going to partial oxidation and the overheads to recycle.

The plant coal liquefaction line was designed for 12,000 tons of coal use daily, and produced around 3,000 MTD of gasoline, 18 tons ammonia, and about 53 MTD sulfur every day.

The economics of the coal liquefaction plant show 18.5% discounted-cash-flow rate of return, with 33.3% equity financing and a I0-year debt carrying an interest rate of 10.5%. The process economics can be improved significantly by; decreasing the investment cost, lowering the coal price, using cheaper natural gas for hydrogen production, extending the plant' operating lifetime.

But most importantly, the economic feasibility of a direct coal liquefaction plant depend heavily on the prevailing price of gasoline and diesel fuel products. Since these vary wildly, any attempt of coal liquefaction comes with a heavy risk.

At the present day abundance and price of crude oil below \$50/bbl, coal liquefaction is economically unfeasible.

Indirect Coal Liquefaction

The indirect coal liquefaction (ICL) process, on the other hand, involves gasification of the coal solids, where coal is reacted in order to produce a mixture of carbon monoxide and hydrogen (syngas) first. These are then treated via special processes, such as Fischer-Tropsch, which convert the syngas mixture into liquid oils and fuels.

The major ICL processes operate in two stages. First, coal is converted into syngas (a purified mixture of CO and H2 gas). In the second stage, the syngas is converted into light hydrocarbons using one of three main processes: Fischer-Tropsch synthesis, Methanol synthesis with subsequent conversion to gasoline or petrochemicals, and methanation.

The first stage of the process—conversion of coal into syngas—or gasification, uses a gasifier, a cylindrical pressure vessel about 40 feet high by 13 feet diameter. Feedstocks (coal in this case), water and oxygen are fed into the top of the reactor and when the temperature and pressure produce are high enough, syngas and steam are produced. Recycled black water and slag are also produced and discharged from the bottom of the reactor.

Any kind of carbon-containing material can be a feedstock, but coal gasification, of course, requires coal. A typical gasification plant could use 16,000 tons (14,515 metric tons) of lignite, a brownish type of coal, daily.

Coal gasifiers operate at high temperature and pressures—about 2,500 F and 1,000 PSI respectively. This causes the coal to undergo a number of thermo-chemical reactions, starting with partial oxidation, where the carbon in coal releases heat feeds further the gasification reaction.

Pyrolysis reactions occur under these conditions, during which coal's volatile matter decomposes into several constituent gases. This process leaves char, a charcoal-like substance, as a main byproduct.

Reduction reactions following the pyrolysis step, convert the remaining carbon in the char to a gaseous syngas mixture, with carbon monoxide and hydrogen as its two primary components.

The raw syngas is then run through a gas-cleaning process in a cooled chamber, where the various components accompanying it are separated. During this step, harmful impurities, including sulfur, mercury and unconverted carbon, are removed, separated and used in different steps of the process, or for other purposes.

Carbon dioxide can be also separated at this step, to be either stored underground or used in ammonia, methanol, or other chemicals production.

At this point, hydrogen and carbon monoxide only

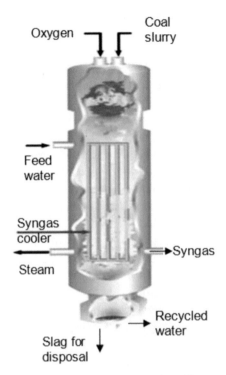

Figure 10-7. A typical coal gasifier

remain in the syngas, and it can now which be burned cleanly and efficiently in gas turbines as needed to produce electricity. Syngas can also be converted into natural gas-like substance by passing the cleaned gas over a nickel catalyst. This causes carbon monoxide and carbon dioxide to react with free hydrogen and form methane gas.

Thus produced gas has the same properties and behaves just like regular natural gas, so it can be used in gas turbines to generate electricity, heat homes and businesses, or for fueling vehicles.

The second step of the ICL process could be the Fischer-Tropsch reaction, which converts the syngas into liquid hydrocarbons. The simplified form of this reaction is as:

$$CO + 2H_2 \rightarrow CH_2 + H_2O$$

This process was discovered in the 1920's and has been used for the production of liquid fuels and chemicals from syngas ever since. Fischer-Tropsch process was first used on large technical scale in Germany during WWII, and is currently being used by Sasol in South Africa, and other companies around the world.

The Fischer-Tropsch process works by synthesizing hydrocarbons via chain reactions, where the length of the chain depends on the catalyst properties and the reaction conditions.

There are two types of Fischer-Tropsch conversion steps in use today. The first one uses a Slurry Phase Reactor to produce substances like waxes and distillate fuels, and the other uses the Advanced Synthol Reactor to produce light olefins, gasoline, and other liquid fractions.

In the Fischer-Tropsch process, syngas is preheated and fed into the reactor. Here it is spread into a slurry made out of liquid wax and catalyst particles. The gas is allowed to travel through the slurry, where it diffuses into it and is subsequently converted into even more wax by the Fischer-Tropsch reaction.

The heat released by the reaction is removed via cooling coils mounted inside the reactor, which results in steam generation. The wax is separated from the slurry containing the catalyst particles via special process. At the same time, the more volatile gas fractions and water leave the reactor as a gas stream and can be captured for later use by cooling the steam.

Thus separated hydrocarbon streams are collected and sent for upgrading to their final state as liquid oils and fuels. The water stream is recycled by treatment in the water recovery unit.

- *Methanol synthesis processes* use the reaction of converting syngas into liquid methanol, which is then polymerized into alkanes in special reactors in the presence of zeolite catalysts. This process, known as MTG (Methanol To Gasoline) method, was developed in the early 1970s, and was tested at a demonstration plant by Jincheng Anthracite Mining Group (JAMG) in Shanxi, China.

- *Methanation reaction* converts thus produced syngas into synthetic natural gas (SNG). This process is used at the Great Plains Gasification Plant in Beulah, North Dakota, which is a coal-to-SNG facility, known to produce 160 million cubic feet SNG per day. The plants has been in operation since the early 1980s.

The Problems

As we see in any fossil energy product and process, coal liquefaction processes emit significant CO and CO_2 emissions. These result from the gasification process and from the generation of heat and electricity needed for the proper operation of the reactors.

There is also the need of huge amount of energy to run the different reactors. Electric power used in these processes comes from fossil power plants, which increases further the pollution footprint of these fuels. And then, there is high water consumption and waste

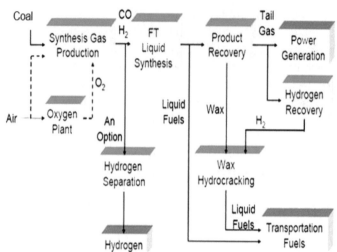

Figure 10-8. Fischer-Tropsch ICL process

water discharge rates in water-gas shift or methane steam reforming reactions, which also cause adverse environmental effect.

Nevertheless, properly designed and implemented synthetic fuels produced by coal liquefaction processes are proven to be less polluting than naturally occurring crudes, at least a far as sulfur pollution is concerned. This is because hetero-atom (sulfur in this case) compounds are not synthesized during these processes, and/or are always separated from the final product before its final use.

In conclusion, we must admit that a full integration of coal liquefaction in the U.S. energy supply system for use in the overall economy won't happen overnight. The DCL and ICL technologies require a lot of energy to operate, the infrastructure and operation are expensive undertakings, so it will take a while before any significant large-scale DCL or ICL plants are in production.

We need to remember, however, that this technology can bring us the very much needed relief from expensive and unreliable foreign crude oil imports. We also must remember that as soon as crude oil prices hit the $150 mark, DCL and ICL will become instantly economically feasible, so we must be ready to scale up their production in order to avoid another economic disaster.

Coal Wars

Internally, coal is a subject to power plays. It started with President Obama initiating a major overhaul of the coal mining regulations recently. The House of Representatives did not go along with the administration plans, and the coal war started.

Bill HR 3409, "Stop the War on Coal," was introduced which in effect prohibits the Secretary of the interior from issuing regulations under the Surface Mining Control and Reclamation Act. This measure was taken in order to avoid adverse affect on the production of coal, the delivery of coal, and to preserve jobs in the coal industry. Also, the legislation compiled several bills already passed by the House and stalled in the Senate, intended to halt the onslaught of regulations intended to severely curtail the use of coal for electricity generation in the U.S.

At the same time, layoffs and mine closures were announced, mostly due to the tight regulations, which forced coal-power plants shut down. Since less coal is burned across the country, less coal would be needed from the Wyoming mines, that could lead to layoffs in the Powder River Basin.

Wyoming is a major coal-producer. About 40% of all the coal used in the U.S. comes from that one state. The new regulations affect the state' coal mining operations, the workers and their families.

In response to the administration move, congress-woman Cynthia Lummis (R-WY), issued the following statement after the bill's passage: "There is a clear anti-coal agenda in the President's so called 'all-of-the-above' energy plan. In casting coal aside, this Administration is casting aside domestic energy security, high-paying jobs, and affordable abundant energy. President Obama has turned a blind eye to the real-life consequences of these actions in Wyoming and across the nation. We cannot sacrifice the jobs or energy supply these regulations take away, period. It's time the Administration takes the choke hold off America's most abundant energy supply and lives up to their promise of domestic energy security and job creation."

The coal war in the U.S. is on, and it promises to intensify as time goes on. So, looking at this newly created (potential) coal-less situation, we see a number of complex scenarios at play:

- On one hand, reducing the coal used for power generation would benefits the environment, but we just don't know how much. The problem is that coal is being replaced by natural gas, which also emits significant—if not equal, and by some estimates larger—amounts of GHG during its entire cradle-to-grave life cycle,
- On the other hand, we see abandoned mines, lost jobs, and most importantly making abrupt change that might affect our energy supplies and security.
- Natural gas production and delivery is not as stable as coal, so the reduction of the amount of coal used for power generation will result in increased energy costs, at least in the short term.

- Converting the entire U.S. power generating fleet to natural gas might have a significant impact on our energy security. A temporary—and God forbid long lasting—Interruption in the natural gas supply would reduce the power generation and even shut down parts of the sector.
- In the long run, we see exports of large quantity of coal to China and other countries, which in effect nullifies the effect of reduced coal use in the U.S. Instead of eliminating the coal-related GHG emissions, we are simply transferring them from the U.S. to China. Is this a smart thing to do? How does that benefit the global environment?

This is a serious subject that is taking new turns every day with some surprising results at each turn. Here is one of these traitorous turns of events.

Coal Cost

Coal is a predominantly domestic fuel—at least 85-90% is produced and used locally. Domestic energy markets are more stable, since they are not usually exposed to international prices and the related problems with transport etc. Yet the prices can vary significantly because of quality, geographic, contractual and regulatory aspects. Different types of coal and purchase conditions, including time and point of delivery, create a great variety of pricing schemes too.

We must consider the not-so-obvious fact that coal is not a single product. Instead, it is a family of many types of different materials. There are many coal classifications, but the main price determining types are:
a) non-coking (steam or thermal coal and lignite), and
b) coking coal.

In more detail:
- Coking coal, which is used in iron making, is of a higher quality, since it has more desirable higher caking properties and strength, than non-coking coal. Because of that, its price is usually higher too. But this is oversimplification. Coking coal is also not a 'homogenous' product, since there are a variety of qualities and other factors that determine its final price.
- Hard coking coal represents the highest quality, while the other types, such as semi-soft or high-volatile coking coal are usually sold at much lower prices.
- Some high-quality non-coking coals are also used in metallurgy for pulverized coal injection, which reduces total coke consumption in blast furnaces.

The price of these types is related to coking coal, although still at some discount rate.

- And then, there are the different market niches. For example, high-grade anthracite can be used for a number of application and would follow the usual prices in each of these. Ultra-high-grade anthracite can replace coke in the blast furnaces, where it approaches coke price at a slight discount.

- The non-coking coals, such as steam and thermal coals are used for heat and power generation, where calorific value is the main factor in defining performance and price. Lower calorific value generally means a lower price, where the price falls faster than the energy value.

- The calorific values can also be referred to as 'gross' or 'net'.

- The price-reporting agency Argus lists five different price indexes for one single source—Indonesian coal. Each index reflects the actual kilocalorie counts of 6500, 5800, 5000, 4200 and 3400 kilocalories per kilogram of different coals.

- Supply and demand is another determining factor of price formation.

- The different physical conditions—air dry, wet, etc.—of the as received coal also determines the price per ton.

- Different geographic markets, although basically well integrated, depend on many factors, such as transportation.

- The different importing and exporting regions also have different prices, determined mostly by availability and production and transportation costs.

- Seaborne coal' price is determined mainly by the freight and insurance costs. Here, terms such as free-on-board (FOB), cost insurance freight (CIF) or cost freight (CFR) determine the final price.

- The price of coal also depends on the type of purchase instrument; contracted coal or spot purchase. Generally, coal is traded internationally on a spot basis, so most price markers and indexes refer to spot purchases. Some Japanese utilities are the exception. They buy most of their thermal coal through one-year term contracts, while coking coal is bought on a quarter or monthly contracts.

- Coal futures, forwards, as well as swaps for different dates, different types of coal and different locations are also common on the coal marketplace.

And so, it is quite obvious that our old, seemingly simple friend coal is a complex commodity. So much so that we really have to know very well what we need and expect in order to get the best coal quality and value.

In the U.S. coal prices vary from state to state, from mine to mine, and also depend heavily on the seasons and the overall market conditions.

Table 10-5. Average sales price of coal (in $/ton)

STATE	Under ground	Surface mine
Alabama	100.17	108.71
Illinois	51.43	46.60
Indiana	51.77	44.91
Illinois	51.43	46.60
Indiana	51.77	44.91
Kentucky Total	63.38	64.01
East Kentucky	78.63	70.86
West Kentucky	47.87	38.93
Ohio	47.86	43.41
Pennsylvania	78.67	82.89
Tennessee	66.27	77.27
West Virginia	89.40	77.39
North W. Virginia	60.91	65.74
South W. Virginia	114.25	78.15

These are only examples of the purchase costs of coal from mines in different U.S. states. The price would vary according to quality of coal, location, transport charges, and other fees. Compare these prices with the average cost of $15 per ton in 1950.

The official estimates show coal prices in the $40-60 per short ton (2,000 lbs.). At the same time, Powder River Basin coal costs only $8.75 per ton. The variation is due to the quality of coal and the amount of heat as measured in Btu generated.

Figure 10-9 shows the relative prices of the fossil fuels during the energy crisis of 2008-2012. The fluctuating prices were caused and driven faulting economy and rising crude oil prices and production uncertainties.

Note that while we can clearly see the price fluctuations in Figure 10-9, the actual dollar per unit comparison among the different energy sources is very difficult to do for a number of reasons.

- The physical properties of the different energy sources are so very different—oil is liquid, coal is solid and natural gas is gaseous, then

- The heating values of the different fuels are quite different, and to top it off;

- The official market prices are expressed in different units; oil is given in $ per barrel (42 gallons), while the price of coal is in $ per ton (1,000 kilograms), and that of natural gas in $ per thousand cubic meters.

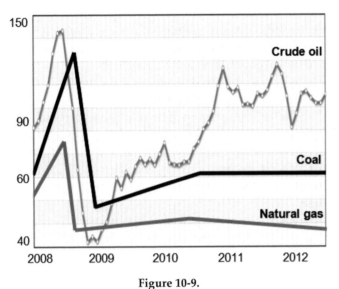

Figure 10-9.
U.S. fossil fuels, comparative prices in ($/unit)

If we equate the energy of the different sources to that of crude oil (42 gallons/bbl), which contains approximately 1.7 MWh of power, we see that 1,000 m³ natural gas contains 10.5 MWh of energy, while 1 ton of coal contains 6.5 MWh energy equivalent.

This information reveals that a $90 barrel of oil (2012 prices) contains only 1/4 of the energy contained in $60 ton of coal, and 1/6 of the energy contained in 1,000 cubic feet natural gas, worth $40. When the price of crude oil fell down to $40/bbl in 2014-2015, then the price ratios changed too. At that price, crude oil is unbeatable, mostly due to its versatility of use in different applications.

We also must remember that the above numbers represent only the theoretical content of energy in different energy sources. The actual amount of final (practical) energy—electricity, or heat, produced when using the different fuels in different applications—depends on a number of additional factors. Here, the location, the type and efficiency of the energy conversion equipment, and the actual quality of the input materials—for example, all coal is not the same—make a big difference.

So, the conclusion is that the market prices reflect not only the type and theoretical energy value of the different energy sources, but other factors, such as the convenience of using them in different applications. For example, no matter how cheap coal is, we cannot pour it in our gas tanks and drive away in our gas guzzlers. Crude oil is replaceable that way. Similarly so, natural gas is very convenient for use in power generation, since it is easy to transport (via pipelines) and does not require much processing before burning.

Coal Workers

There is one large group of Americans for whom the new 'environmental protection', implemented by reducing coal use, means losing jobs and benefits. It is a very large group too—that of workers in the different branches of the coal industry. Thousands of these people working in the surface and underground mines, railroads, and coal-fired power plants all across the U.S. are the latest victims of the 'green' revolution.

During the last 4-5 years the products and services they have been providing for over a century have been labeled 'evil' and have been targeted for extermination. As a result of changing regulations, coal mines and coal-fired power plants are shutting down. By no fault of their own, the working people are left without jobs and their families suffer the consequences.

While the issues brought out as a reason for the shut downs are legit, the way this was done is questionable. President Obama' re-election campaign slogan was "all of the above" energy strategy. "We need an energy strategy ... for the 21st century that develops every source of American-made energy," Obama said in March, 2012. Voters in the coal states of Virginia, Ohio, and Pennsylvania, believed that this means fairness.

A year after his re-election, the President pushed for tighter environmental restrictions on GHG emitting power plants and businesses. As a result, EPA issued CO_2 regulations, which, in fact, made impossible building new coal-fired power plants. The new regulations also make the operation of most existing such plants more expensive, so they operate near break-even point. Many—63 to be exact—coal-fired power plants were shut down all together as well.

During his 2014 State of the Union address, Obama confirmed that he meant what he said in 2013, and that he intends to finish what he has started—reduce GHGs to the maximum possible levels. Since coal is in the cross-hairs of the anti-CO_2 movement, it is logical to conclude that coal will have to go away one way or another, sooner or later. Sooner seems to be in the administration plans right now.

With other words, coal mines and coal-fired power plants, and their workers, have no place in the America's energy future. As a result, many of them simply have no future at all. How does that serve our energy security and independence, if some of us are badly hurt in the process?

And things can get even worse with the new EPA mandate on GHG emissions, which will surely affect the coal industry as dozens of coal-fired power plants are shut down and more will follow.

It is obvious that President Obama intends to make his final years in office an example of how to deal with climate emissions. The cancellation of the Keystone XL pipeline being the last confirmation of the administration plans. He has decided to proceed—on his own if he has to—and do whatever is needed to make a drastic change in the way we do business. It will be his legacy: "When our children's children look us in the eye and ask if we did all we could to leave them a safer, more stable world, with new sources of energy, I want us to be able to say yes, we did," he declares.

This is actually a noble act, no doubt, but the results would not match the initial intent. While we are shutting down the coal mines and power plants in the U.S., we are making plans to export more coal overseas, which will be burned in China and India, producing the same, and even worse, GHG pollution, but somewhere else around the globe. This means that the overall effect of sharply cutting coal production and use in the U.S. on the global environment would be a big round zero, or worse.

So, instead of such a drastic change—reducing the emissions overnight—the EPA should have implemented a gradual reduction. This would have given the coal producing and using industries enough time to take measured actions, instead of going into panic mode.

How fair or unfair all this is remains to be debated by the experts, but what we can clearly say is that abrupt changes like this are not a good way to do business. They hurt the industry, which has powered, and is still powering, most of our economy. And those who suffer most and immediately are the workers in the coal industry, as a consequence of these changes. At least 30,000 of them are to lose their jobs.

Our energy security is also compromised by such drastic and sudden changes. The widely spreading use of natural gas is not a proven alternative and any problems in the plans might cost us dearly. Energy security should not mean abundant, safe, and cheap energy for some, on the expense of others. But it is too late to cry over spilt milk, so we just need to find solutions to the issues at hand before we end up with a bigger national and international problems.

Coal Externalities

This is the skeleton in the closet of the coal industry. While we might not agree with Obama's approaches, we must agree that coal has problems. Big problems, as discussed in this text. In economic terms, an external cost, or externality, is a negative effect of an economic activity on a third-party or the society in general. When coal is mined and used to generate power, for example,

the external costs include a number of negative impacts, expressed as: water pollution, toxic slurry and solid waste, air pollution, and the long-term damage to local and global ecosystems and human health.

The list of such damages is long, and we cannot possibly look at all items and all cases.

Note: For more details on coal and its externalities, we would recommend our books on the subject, *Power Generation and the Environment*, and *Energy Security for the 21st Century*, published by The Fairmont/CRC Press. In these books we take a very close look at the coal industry.

Here, we take a look at the very basics, as needed to underline the importance of this subject.

- Greenpeace released an analysis in 2008 showing that the global total cost of coal, including the externalities, was at least $450 billion annually and growing. This number was later on updated to $500 billion in 2011 by a Harvard study group, and now is most likely much higher.

The researchers arrive at these figures by looking at CO_2 and other pollution damages, and the damages, resulting in health costs from mining and power plant accidents, air, soil, and water pollution.

- The first report was released at the time when Industry Ministers from 20 big GHG emitting countries met in Warsaw with the world's climate-polluting industries, where was agreed that the relentless expansion of the coal industry is the single greatest threat to increasing the damages of the global climate change.

- Coal is presently the most environment-polluting fossil fuel, responsible for one third to one-half of all CO_2 emissions, which are projected to increase to 60% of all emissions by 2030. Thus, reducing coal burning will benefit not only the global climate, but it also will reduce the other external impacts and the related costs which everybody else has to pay for.

- In calculating the $450 billion annual global cost of coal figures, the focus was on the external costs of coal. This includes: the cost for rectifying damages attributable to climate change; human health impacts from air pollution; fatalities due to major mining accidents; and other factors for which reasonably reliable global data is currently available. The above dollar number was derived by taking into account 90% of the global emissions and looking at the damages, which are projected to rise significantly, due to the impacts of climate change.

With these changes, the total number of global cost of coal is likely to increase sharply—unless its use is reduced and/or the climate change is stopped soon enough.

• The projected impacts of climate change include billions of people who will face water shortage, while others will be flooded by rising ocean levels. This also includes the hundreds of millions who will be threatened by food insecurity and hurt by exceedingly extreme weather events, such as storms, earthquakes, and tsunamis.

U.S. Coal Future

Thus far, most coal produced in the U.S. has been consumed domestically at a fairly steady pace. At the same time, the U.S. coal exports have been going up and down. Now, with decreased domestic demand for coal, as discussed in more detail above, there are serious plans to export increasing amounts of coal to China and other countries.

The new regulations are forcing the coal companies to look at exporting their production in order to keep the mines open. And so, as a result we foresee export volumes of coal to increase in recent years.

The proportion of coal production going toward exports has also increased, doubling from 5% in 2009 to 10% in 2011, sharply increasing for awhile in 2012, and projected to take off in the near future.

Several new shipping terminals have been planned in the U.S. and the coal companies are gearing up for exporting as much coal as they can produce—to whomever wants to buy the stuff.

But what does that mean to the U.S. energy security? Especially that of the future generations? What would exporting so much coal do to their energy supplies? Isn't that similar to the proverbial 'shooting oneself in the foot?'

The majority of the present day coal production consists of thermal (steam coal) used mostly for electricity generation. Metallurgical (coking coal), although in smaller quantities, is also important for iron and steel production.

70% of global steel production depends on this type of coal. In 2011, metallurgical coal exports accounted for 77% of U.S. coal exports in terms of trade dollars and 65% in terms of volume.

In 2010, exports to Asian markets increased 176% from 2009 levels, primarily because of a surge in exports of metallurgical coal to China, Japan, and South Korea.8 Metallurgical coal exports accounted for 83% of the growth in 2010 export volumes.

U.S. coal producers increased both export volumes and prices in 2010 because there were huge demands from China and India, due to reduced world coal supply caused by heavy rains and flooding in major coal exporting countries.

In 2012, export coal prices turned down sharply, due to decreased domestic demand, a slowdown in economic growth in both China and India, and the continuing weakness in some European countries, which promise to be a bottomless coal market.

Oil and natural gas prices in Europe are, and will remain, quite high for a long time, so relatively inexpensive U.S. coal will become even greater demand by European utility companies. This, combined with the decline of carbon emissions permits in the European Union, makes coal even more attractive to the end users.

Today the U.S. coal export volume and prices are influenced by the difference between domestic and international demand and gas prices, as well as by the coal use in developing countries, and the Asian coal production levels.

And while all indicators point to a favorable trade winds for the U.S. coal exports, the nagging question persist, 'Is this the best we can do with such a key commodity, that is in limited supplies?' Is this the best for our long-term energy security? The answer is also blowing in the wind…

China's Coal Future

China has seen amazing growth in all industrial sectors. This requires a lot of energy. So, in order to continue its progress, the economy requires even more energy. This is forcing the Chinese leaders to make grandiose plans for its energy sector, and coal power generation in particular:

Table 10-6 tells us that China's plans to almost double its electric power production and use by 2020. The increase is based mostly on increase of use of coal, which while decreasing in proportion is increasing in total GW power generation.

Since 2020 is only 4-5 years away, we expect some spectacular developments in China's energy sector.

The developments would be most visible in the coal-fired power generation. Presently, about 70% of China's primary energy consumption is in the form of coal and 80% of its power generation is from some type of thermal power generation. To achieve the 2020 goals, the number of coal-fired power plants will be almost doubled by 2020. For that purpose, several coal-fired

Table 10-6. China's electric power sector

Technology	2010 GW	2020 GW
Coal (all)	646.60	1030.00
Gas	26.42	58.90
Nuclear	10.82	80.83
Hydro	198.21	340.00
Pumped storage	17.84	50.00
Wind	29.57	150.00
Solar	0.26	24.00
Biomass	1.70	15.00
Other	36.92	36.92
Total	968.34	1785.65

power plants are added to China's energy portfolio every week. Every week...?

China's leaders talk the environmental talk, but are far from walking the real anti-pollution walk.

Coal pollution is visible and thick all around large population centers. Spending just a day breathing the thick Beijing smog would tell you the real story; that of making a lot of energy today without any thought about tomorrow.

To be fair, during the past five years, China has devoted a lot of attention to developing clean-coal technologies, mainly focusing on coal processing, high efficiency clean coal burning, coal conversion, pollution control and waste processing.

In the new five year plan (FYP) China plans to continue along this line but will no longer rely solely on technological research and development and application. Instead, it will place gasification of coal through poly-generation at the heart of its clean-coal applications. This can increase the overall efficiency and reduce pollution by 10–15%. At least, this is what the theory is. Small-scale tests continue, so it remains to be seen, if the process can be as efficient as to save energy and reduce pollution, or if this is just another round of talking the environmental talk.

More than half of the thermal power plants generation capacity in the coal-rich regions in central and west China were built to meet the electric loads in the East. In the new FYP, China would like to optimize the coal-fired power profile for sustainable development of the power industry by reinforcing the inter-provincial transmission capability. In addition, strategic measures are to

be taken to ensure sufficient and timely coal supplies so as not to threaten the security and stability of power systems as well as to contribute to sustainable economic development.

Meanwhile, chemical products can be sold to earn extra revenues for power producers, and the balance between chemical production and power generation can be adjusted in accordance with demand. In addition, new technologies, such as the smart grid, renewable energy sources, efficiency measures, distributed energy expansion, and other will help reduce the reliance on fossils. And yet, fossils are still over 75% of the Chinese energy budget.

Come hail or high water, however, China is moving forward with its energy plans, which include major increase in use of coal power, regardless of the thick poisonous air in the major cities, and despite all international environmental agreements. Adding to that the crude oil used, and GHGs emitted, by millions of cars and trucks driven by newly grown middle class on China's ways and byways, we see a picture of a gluttonous and even reckless developments in China.

How long would China be able to support this unsustainable life style? And what are the Chinese people going to do some day soon, when the global energy supplies dwindle and prices rocket sky-high? This is not something they are worried about now, though. Living is good today, which is what every Chinese is dreaming of...whatever the ultimate cost.

NATURAL GAS

Total natural gas deposits, estimated to exist in a particular areas around the U.S., show 2,170 trillion cubic feet (Tcf) of future gas supply in the country as of 2010. This included 273 Tcf of proved dry gas reserves and 1,898 Tcf of total potential resources at the time. These numbers have increased lately, due to new discoveries and the use of hydrofracking.

The five major shale plays in the U.S. include Barnett, Fayetteville, Woodford, Haynesville, Eagle Ford, and Marcellus. The large gas sites are seeing heavy exploration activity and are quickly becoming major contributors of natural gas supply.

The shale plays are widely distributed through the entire country, which has the added advantage of putting production closer to demand centers, thus reducing transportation bottlenecks and costs. Together, they are now estimated at 3,420 Tcf of shale gas resources, according to the experts. Over 30% increase from the 2010 estimates. How accurate these estimates are, remains to be seen.

The indications of late, however, are that although the natural gas estimates might be close to reality, their extraction is another issue that has not been resolved. Very often, new gas wells with great estimated quantity of gas dry-up within months of exploration. This is because, hydrofracking in its present form is not capable of extracting every last drop of gas in the ground.

Presently, the natural gas markets have three major segments:

1. Supply, which includes exploration, development, and production of natural gas resources and reserves (this includes drilling, extraction and gas gathering),

2. Midstream distribution, which uses small-diameter gathering pipeline systems to transport the gas from the wellhead to natural gas processing facilities, where impurities and other hydrocarbons are removed from the gas to create pipeline-quality dry natural gas.

3. Bulk transportation, which includes large diameter intrastate and interstate pipeline systems that move natural gas to storage facilities and a variety of consumers, including power plants, industrial-facilities and local distribution companies, which deliver the gas to retail consumers.

Natural Gas Market

The U.S. natural gas markets are divided into several areas: West, Midwest, Gulf Coast, Northeast and Southeast regions. These regions have differing supply, transportation and demand characteristics, resulting in different prices. Within these regions are hubs—the interconnection of two or more pipelines—which are also market hubs for buying and selling gas.

The key hub used to reflect the U.S. natural gas market as a whole is the Henry Hub, in Louisiana. Prices at other locations are frequently shown as Henry Hub plus or minus some amount. The regional differences in supply and demand result in different prices for natural gas at various locations.

Prices are usually lowest in areas with low-cost production, ample infrastructure and limited demand, such as in the the Opal Hub in Wyoming. They are highest where production or transportation is limited and demand is high as in Algonquin, Massachusetts, for example.

The U.S. natural gas market is accommodated by extensive infrastructure, consisting of roughly 306,000 miles of wide-diameter, high pressure inter- and in-tra-state pipelines. These make up the mainline pipeline transportation network, run by more than 210 companies.

More than 1,400 compressor stations maintain pressure on the natural gas pipeline network with more than 5,000 receipt points, 11,000 delivery points, and 1,400 interconnection points implement the flow of gas across the U.S. Nearly three dozen hubs or market centers provide addition interconnections, serviced by 400 underground natural gas storage facilities that increase the flexibility of the system.

There are 49 import-export locations in the U.S. serviced by pipelines.

There are also over 100 LNG peaking facilities across the country that use stored natural gas held for electric power generation during peak demand periods.

More than 1,300 local distribution companies deliver natural gas to retail customers as well.

Natural gas markets in the U.S. have been undergoing some major changes. Within the last decade, various factors have shifted the dynamics of supply and demand. The major factor has been the development of new production technologies—hydrofracking—which have revolutionized the industry.

The demand for natural gas for electric power generation, however, is the major driver for the rapid increase of hydrofracking around the country.

Electric Power

Natural gas is a major fuel for electric power generation that is becoming more important by the day. The U.S. produces a significant amount of natural gas, which is increasing exponentially as we speak. Some of that amount is exported, while most of the rest is used for power generation.

There is a trend in the U.S. presently, where old coal-fired power plants are converted to gas-fired such. The trend is increasing as the benefits of cheap gas are added to the list. EPA has also tightened the GHG emissions specs so that the operation of existing coal-powered plants is getting more expensive. At the same time, no new coal-power plant can be build, because they cannot meet the new GHG specs.

Because of that, and as a sign of the times, natural gas provides a major part of the electric power used in the country, and especially in California. While there are a lot of discussions, disagreements, and even fights over the renewable energies, natural gas is quietly taking the driver's seat of the California energy industry and other sectors.

Figure 10-10 points to the fact that California is leading the country in using natural gas for power generation. Over three times more natural gas-fired capacity than nuclear is used in the state—four times more than hydro, and four times more than all other energy sources put together. The trend continues, and now California is on the path to double the use of natural gas by increasing its use in this and other sectors of its economy—transportation primarily.

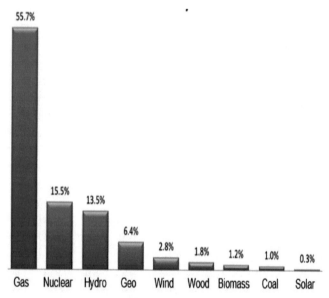

Figure 10-10. California electricity generation

Gas usage for electric power generation is also growing in many other states due to its abundance, low price, and (relative) environmental friendliness. Natural gas is literally taking over the electric power generation in the U.S. and a number of countries are following closely.

LNG Market

Natural gas is sold in gaseous form to domestic users, such as refineries and power plants. It is also distributed as liquefied natural gas (LNG) to domestic users and for export in specially designed pressurized vessels. Many shale reservoirs contain natural gas liquids (NGL), which are sold separately, and which augment the economics of producing natural gas. Good business, no doubt…if price is right.

Natural gas imports play an important role in regional U.S. markets, accounting for about 3,800 Bcf, or 11%, of the natural gas used in the U.S. in 2011. The natural gas pipeline systems of the U.S. and Canada are integrated, and about 90% of imports came from Canada, while 10% was imported as liquefied natural gas

(LNG). The import-export balance changes according to local demand and pricing.

Imported natural gas flows into the U.S. via pipelines at numerous points along the U.S. border with Canada. Imports from Canada have been of strategic importance in the Northeast and the West, which were traditionally far from the major domestic production centers. However, Canadian exports to the U.S. have been declining as U.S. shale production has increased.

Net U.S. gas imports have declined from a recent high of 3,785 Bcf in 2007 to 1,948 Bcf in 2011 and are going down with time. The estimates are that imports will continue to decrease as shale-gas production increases.

The U.S. also exports natural gas to Canada and Mexico. We also export significant quantity of LNG to Japan. These exports are expected to increase significantly in the near future.

In order to prepare for export, natural gas has to be converted to LNG. For this purpose, it is sent to liquefaction facilities, which are major industrial complexes, typically costing $2 billion each, with some costing as much as $20 billion each to construct and even more to operate.

Once liquefied, the LNG is typically transported by specialized ships with cryogenic, or insulated, tanks. Once LNG reaches an import (regasification) terminal, it is unloaded and stored as a liquid until ready for sendout. The regasification terminal warms the LNG to return it to a gaseous state and then sends it into the pipeline transportation network for delivery to consumers.

Currently, more than 80 Bcfd of regasification capacity exists globally, more than double the amount of liquefaction capacity. Excess regasification capacity provides greater flexibility to LNG suppliers, enabling them to land cargoes in the highest-priced markets. This flexibility has fostered a growing spot market for LNG.

The cost of the LNG process is $2-$4 per million British thermal units (MMBtu), depending on the costs of natural gas production and liquefaction and the distance over which the LNG is shipped. Liquefaction and shipping form the largest portion of the costs. Regasification contributes the least cost of any component in the LNG supply chain. The cost of a regasification facility varies considerably; however, the majority of these costs arise from the development of the port facilities and the storage tanks. A 700-MMcfd terminal may cost in the range of $500 million to $800 million. LNG in the U.S.

The U.S. is second to Japan in LNG regasification

capacity. As of 2011, there were 11 LNG receiving or re-gasification terminals in the continental U.S., with approximately 13 Bcfd of import capacity and 79 Bcf of storage capacity. All of these facilities are on the Gulf or East coasts, or just offshore. In addition, the U.S. imports regasified LNG into New England from the Canaport LNG terminal in New Brunswick, Canada, and into Southern California from the Costa Azul LNG terminal in Mexico's Baja California.

Between 2003 and 2008, the U.S. met 1-3% of its natural gas demand through LNG imports. LNG imports peaked at about 100 Bcf/month in summer 2007. Growth in relatively low-cost U.S. shale gas production has trimmed U.S LNG imports, affecting Gulf Coast terminals the most.

Today, most LNG enters the U.S. under long-term contracts (about half of the total) coming through the Everett (Boston) and Elba Island (Georgia) LNG terminals. The remainder of the LNG enters the U.S. under short-term contracts or as spot cargoes. LNG prices in the U.S. generally link to the prevailing price at the closest trading point to the import terminal. During 2011-12, the growth in shale gas production led to proposals to export significant volumes of domestically produced LNG. Since 1969, small quantities of LNG have been shipped from Alaska to Pacific Rim countries.

Drilling Regulations

As part of the effort to reduce climate change, in 2015 EPA announced a new goal to cut methane emissions from the oil and gas sector by 40–45% by 2025, and a put forward a set of actions to put the U.S. on a path to achieve this ambitious goal.

Methane—the primary component of natural gas—is a potent greenhouse gas, with 25 times the heat-trapping potential of carbon dioxide over a 100-year period. Methane emissions accounted for nearly 10% of U.S. greenhouse gas emissions in 2012, of which nearly 30% came from the production transmission and distribution of oil and natural gas.

Emissions from the oil and gas sector are actually down 16% since 1990 and current data show significant reductions from certain parts of the sector, notably well completions. At the same time, however, emissions from the oil and gas sector are projected to rise more than 25% by 2025 if no additional steps are taken to lower them. For these reasons, the new strategy for cutting methane emissions from the oil and gas sector is thought to be an important component of efforts to reduce the effects of climate change.

Reducing methane emissions also means less

waste by capturing valuable fuel and reducing other harmful pollutants in the process. This would save up to 180 billion cubic feet of natural gas in 2025, enough to heat more than 2 million homes for a year and continue to support businesses that manufacture and sell cost-effective technologies to identify, quantify, and reduce methane emissions.

Fracking Regulations

The newest and most important development in the oil and gas drilling industry is the first major federal regulations on hydraulic fracturing, announced in the spring of 2015. It is focused on drilling safety of fracking, which led to a huge surge in the production of oil and gas. That also brought up a lot of environmental concerns and complaints from neighbors.

The fracking boom of late has brought the U.S. on the global energy scene as the world's largest oil and gas producer. Unfortunately, the expansion of drilling sites comes with fears and complaints and fears that injecting a soup of hundreds of different chemicals deep underground, as needed to break up the rocks around oil and gas deposits, could escape and contaminate the water supplies, kill wildlife, and harm humans.

As fracking expanded, so did the controversy and fights over the regulations. Presently, the states have jurisdiction over drilling on private and state-owned land, which is where the vast majority of fracking is done. So, the new federal regulations cover about 100,000 oil and gas wells drilled on public lands only—they do not affect drilling on state lands.

So, although the federal rules will not solve the problem, they will serve as a guidance and a standard for the different states to tighten up their own regulations. Easier said than done! Different states have different priorities, so an uniform action is not expected any time soon.

Nevertheless, the federal well-drilling regulations are very old—over 30 years old—so new regulations, complying with the technical complexities of today's hydraulic fracturing operations were overdue. The regulators spent over four years developing the new rules, working closely with oil and gas companies, state authorities and environmental groups. They have also reviewed more than 1.5 million public comments and complaints.

Yet, they are standards, no matter how good or bad they may be, cannot be enforced on state lands. The new regulations, therefore, would be most useful in states that have no regulations, and where the federal regulations can be used as a template. States with existing regulation will have the option to integrate (or not) the new regulations in their system.

The key points in the new regulations:

- Allow government officials to inspect and validate the safety of fracking wells. In particular interest here is the integrity of the concrete barriers that line fracking wells.
- Drilling sites will be also required to disclose the chemicals used in the fracturing process, and
- Companies must follow a set of safety standards for storing used fracking chemicals around well sites.
- Drillers will be required to submit detailed periodic reports on the chemicals and the well geology to the Bureau of Land Management.

And, of course, the Independent Petroleum Association of America filed a lawsuit shortly thereafter, challenging the regulations. They think that it is unnecessary and a reaction to unsubstantiated concerns. Thus, the official request to set the regulations aside doe fear that the new mandates would add new burdensome costs on the independent producers.

The environmental groups, on the other hand, consider this as one initial step of a multi-step process. They see too many concessions to the oil industry groups, which have to be considered eventually.

The best news in all this is that many drillers in states that have fracking regulations in place, are already compliant with most of the new regulations, which add about $5,500 to the cost of fracking. This is less than 1% of the total cost of well drilling, so it is a small price to pay for piece of mind—for the drillers and the locals. Another good new is that the regulations are also expected to be revised and adjusted periodically.

Natural Gas Vehicles

This is a new, and very important application for natural gas. Replacing crude oil with natural gas in the transportation sector would be a major win for all involved. And so, many companies jumped into the compressed natural gas (CNG) game, with Honda leading the effort for awhile…and until recently.

For some reason, however, Honda's effort to get consumers to buy natural gas cars was stuck in neutral for years. And finally in 2014 it shifted into reverse. Only 751 Civic GXes, CNG powered cars were sold in 2014, a 65% drop from the 2,198 sold in 2013. Since 2008, when the car was introduced, Honda has sold around 1,000 to 3,000 of these vehicles annually.

Honda has been the only major manufacturer to ever release a CNG car for consumers, although General Motors has released flex fuel cars that can be powered by CNG. But the sales of Honda GX clearly are falling fast.

What's the problem? CNG vehicles come with most of the drawbacks of electric vehicles, mostly: range anxiety and higher sticker price. The GX, for instance, costs more than a regular Civic and gets slightly worse mileage. It also lacks power in demanding situations, loaded with a few passengers, and especially under hilly terrain conditions.

The worst part is that, gasoline prices are declining, which saps the urgency for consumers to give CNG cars a second look. Thus the slowdown is sales.

CNG is still a viable transportation fuel in commercial markets, particularly with heavy-duty vehicles, he said. A large truck or bus might drive 100,000 miles a year and get only 1 or 2 miles per gallon. In that situation, CNG vehicles make economic sense because the comparatively higher cost of diesel. Commercial fleet owners can also invest in their own commercial-grace filling stations, which can cost $500-750,000. One of the pains of getting alternative fuels to market has been the capital requirements they can put on general-purpose gas station owners.

Sales of natural gas heavy duty vehicles are expected to climb in the coming years because of the benefits they provide to fleet owners. Shipments are expected to grow from 2.3 million annually in 2014 to 3.9 million in 2024 worldwide across all segments

CNG cars have also been more popular in Europe, where gas prices are higher and more station owners have invested in CNG pumps for consumers. China could also become a market for general-purpose, where CNG cars as part of future initiatives to reduce emissions. The general concept of CNG transportation has its benefits, and we expect to see companies trying to turn natural gas into a feedstock for liquid fuel to be used in regular gasoline engines.

But in the meantime, Honda GX and most other CNG cars will continue gathering dust in the car lots.

Reality Check

At a closer look, we see the U.S. increasingly depending on natural gas for power generation, heating, vehicular and commercial use. So, let's see…

Fact. The estimated amount of natural gas, based on analysis of geologic and engineering data gathered though drilling and testing, that can be reasonably projected to be recoverable under existing economic and operating conditions totals 273 Tcf.

Fact. The U.S. economy uses about 23 Tcf annually, which in simplistic terms means that we have only 10 years left of proved gas supplies.

Assumption. Proved gas reserves have been growing every year since 2010, due to new discoveries and drilling/extraction technologies, so assuming that they have doubled, we have 20 years of gas supplies left.

Assumption. At the same time, gas use is increasing in the U.S., as we depend more and more on gas for power generation and other uses, so we are back to maximum of 10-15 years of reliable supplies.

Assumption. Assuming best of cases, that we can extract all proved and unproved gas supplies on the U.S. territory, which by now is about 2,200 Tcf, we have natural gas for about 100 years total. A more realistic assumption would be that half of these supplies can be recovered. Also, gas use in the country and gas exports will double later on this century, so realistically speaking, we have 25-30 years total natural gas supplies.

Within the next 25-30 years the U.S. will run out of natural gas supplies. What are we going to use for power generation and vehicle fuel then?

We agree that these numbers are arbitrary, and that we might have more (or less) gas supplies as assumed above. We also must agree, however, that even the natural gas bonanza is extended by several decades, the gas supplies are finite. They will be depleted sooner than later, and we must be aware of that possibility and get ready for the time when the natural gas pipes will be rusting empty in the ground.

The great danger we see with this scenario is that we are switching quickly from coal to gas-fired power generation. This means that we will increasingly rely on gas for our electricity, but its end is not that far in the future, so we need to keep that in mind.

CRUDE OIL

Crude oil is a critical commodity for any nation, and even more for the U.S. Without oil our economy would stop, and the military paralyzed. Not a good thing!, so there is a major orchestrated effort and billions of dollars spent on making sure that we have enough and cheap as possible crude oil. Not an easy, not cheap task, but one of utmost importance, and sure enough; we have not missed a beat…thus far.

The global oil market is getting more complicated and more difficult to navigate in as time goes on.

2014 was a reminder of that fact. It brought up a lot of troubles and heartaches to the U.S. oil and gas drillers. Global oil prices went well below $50/barrel for quite a awhile, which made the gas pump customers happy, but at the same time it put most drillers in a perilous situation. Oil rigs were reducing and even shutting down, production due to the low oil prices, which made them operate at a loss. The overall oil and natural gas production in the country declined, in response to the extremely low oil prices, and the reduced demand.

A new Arab boycott was on the horizon. This time it was in reverse of the 1973 oil embargo. Since today OPEC cannot hurt us by reducing the oil production and exports, they flooded the market with cheap oil, thus stifling the oil production in many countries—including the U.S.

The Oil Conundrum

2010-2014 marked a new era for the U.S. and global crude oil present and future development. Many important events and mishaps occurred on the national and global energy markets, with 90 million barrels of oil produced around the world daily.

Today OPEC member countries produces about 20 million, with Saudi Arabia accounting for half of that, or 10 million barrels daily. During the financial crises, the demand went down, the surplus increased, and the Saudis held production levels steady. On purpose.

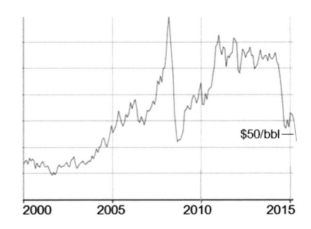

Figure 10-11. Global crude oil price variations

As a result, oil prices went way down and stayed down during the last half of 2014 and most of 2015. During the summer months of 2015 we saw oil prices fall down to nearly $40/barrel, something that has not happened since the good 'ol days of the 1980s. The U.S. drillers paid the price. Most cut down production, and many went bust.

National Developments

With oil prices in free fall, the fear in 2014-2015 was that hunting for new shale in the U.S. may just not be that profitable for the time being. ConocoPhillips was the first major U.S. oil company to reveal that it is slashing spending for 2015. Many other energy companies followed.

Oil prices dropped 60% since June 2014. And OPEC's decided in November of the same year not to scale back on production, which sent prices even lower. This was clearly an attempt by the oil cartel to choke off the U.S. shale boom. America, however, is defiant and did not avoid the 'oil war' with OPEC

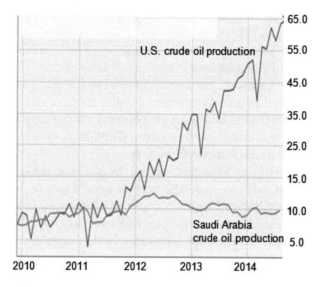

Figure 10-12. Crude oil production in percent change

Lower oil prices caused immense and quite unexpected turmoil in the energy industry. Nobody was prepared to see such low oil prices that lasted so long. Russia, Iran and Venezuela as well as higher-cost producers in North America need higher prices to generate a profit. So, pain was the norm in these countries, accompanied by major cuts across the board to U.S. capital spending in the oil industry.

ConocoPhillips' decision to slash its budget for 2016 by 20% is one of the indications of things to come. The new target reflects lower spending on major projects as well as on "unconventional plays." The U.S. shale oil production is not dead. It is just taking a break. The surge in spending like in recent years is not going to be seen for awhile, but the potential is still there.

All this is happening at oil prices stabilized at $40-50 per barrel. Chevron, ExxonMobil, and other major U.S. oil producers also announces "some spending cuts," though not as deep as Conoco. This is mostly because Chevron and Exxon are bigger players with more financial flexibility. Exxon thinks that it could survive oil as low as $40 per barrel, at least for some time. We will see how that works out for them and the rest.

The small well drillers and oilfield servicing companies are hit the worst. They rely on heavy investment from big oil companies, or investors, which is just not available under the present conditions. What are they going to do? For now, they are tightening the belts, and getting ready for more storms to hit their tents. Many are giving up and folding the tents.

International Developments

Until recently, there was a steady worldwide crude oil production growth that averaged nearly 2.0% per annum on a compounded basis. Among the world's largest producers were the U.S. and Canada as major market share gainers, exceeding the world's overall production growth rate by a very wide margin. Recently North America's oil production grew despite major price realization discounts (as wide as $20-$25 per barrel at some points) relative to waterborne oil benchmarks due to infrastructure bottlenecks.

Middle East producers as a group produced essentially in lockstep with the worldwide production. A closer analysis of individual production volumes by country reveals a major market share loss by Iran that occurred beginning in 2011 (a consequence of the trade sanctions) and, most recently, Syria. Saudi Arabia and Iraq, on the other hand, picked up market share over the same period.

Russia and China barely managed to keep pace with the world's production growth.

Europe saw one of the largest production declines (-25%) which reflected mature asset base in the North Sea. Mexico also posted a significant decline (-8%) as the country's business environment remained difficult for foreign investment.

Africa's production decline was driven primarily by Libya (that lost more than half of its volumes from January 2010 to September 2014) and, to a lesser degree, Angola.

The very strong incremental volumes from the U.S. and Canada, reversal of the earlier production cuts by Saudi Arabia, and Iraq's gradual rebuilding of its capacity were the primary drivers behind the recent global production growth. These factors kept the market adequately supplied in the past five years.

The biggest market share losers were those countries that experienced major political turmoil or international sanctions (Libya, Syria, Iran). Countries and regions with mature asset bases (Europe, Mexico, Russia to some degree, and several others). While previously they struggled to keep up with the global demand growth, during the $100 per barrel oil era, now at $40-50/bbl they are losing money with every barrel of oil produced.

Since the Saudi Arabian-forced drastic drop in oil prices, the trends of the past five years will likely continue in the future, but some new factors have been added to the equation that may play a major role in the oil market. Some of the most important are:

- Iran, Iraq and Libya are the major uncertainty in the global oil market. Combined, these three nations hold very large low-cost production capacity that is currently interrupted or requires investment and development. If brought online, this capacity has the magnitude to upset global supply/demand balance for a number of years and cause a prolonged depression in oil prices. Production increases from these three nations have already been a source of supply pressure on the market.

During 2014, Iraq and Libya added about 1.1 million barrels per day of combined incremental production. This contributed significantly to the current oversupply situation and oil price collapse.

- Producer nations, like Russia, Venezuela, Nigeria, and others, that are unable to sustain profitable production at less than $100 per barrel are unlikely to gain market share in an environment where oil prices are half that much. Their production will

likely continue to lag for the duration.

- The economic sanctions against Russia have not made a big difference in the overall economy, but the deterioration of the overall investment and political climate in the country, and the continued flight of capital will likely render major new development projects non-viable.

Note: Even without the sanctions of late, Russia has been a relatively high-cost, high-risk oil province for new energy projects. The estimate is that the Russian crude oil production breaks even at $79 per barrel.

Given the continued maturation of the legacy West Siberia production base, natural declines may become increasingly difficult to offset. Given Russia's large share of the global supply, a relatively small decline in the country's production in percentage terms could make a noticeable difference for the world's overall supply/demand balance, especially in case of a sustained decline.

- North American unconventional oil and gas plays are considered a marginal source of oil supply. At the end of the day, however, if a new supply source takes a large market share gain in a short period of time, while the other suppliers are losing market share, then this new source of supply is not all that marginal. Nevertheless, the North American unconventional plays may indeed be pushed to the margin, at least temporarily, if and when the low-cost producers of Iran, Iraq and Libya bring significant volumes to the market.

In the near future, however, flexible, low-cost sources of supply that can benefit from existing infrastructure will likely come out as strategic winners in the current oil industry downturn. The U.S. and Canada unconventional energy sources are well positioned for the fight, while mega-projects in technically challenging or politically unstable environments are not. At least during this period of time.

In the end, the potential return to the market of Iran, Iraq and Libya's production capacities might disrupt all plans and predictions. Significant output from these major producers would create an "all bets are off" scenario for the price of oil that may change the global energy order for many years to come. The game is on and is getting more interesting by the day. We just have to keep in mind that oil won't last forever. This is a key factor that we seem to ignore in times of great excitement, like these we are going through today.

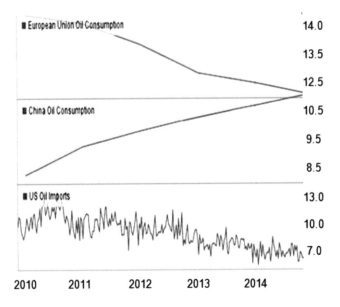

Figure 10-13. Global crude oil consumption (in Mbbl/day)

Case Study: Indonesia

Indonesia is unique case of gasoline monopoly. For decades price-fixing by the government has kept foreign oil companies out of the huge fuels market in the country. But finally, in 2015 things changed. The gasoline subsidies were finally removed and for a first time now people can fill up at gas pumps owned by foreign oil companies.

And with more than one million new cars and almost 8 million motorcycles sold yearly, Southeast Asia's biggest economy is a potentially lucrative market forecast to become the world's largest importer of gasoline by 2020.

State-oil firm Pertamina has dominated sales of gasoline, because subsidies meant it sold fuel cheaper than anyone else, but the move to float prices in January changes this. The new situation is opening a lot more opportunities for foreign players.

Soon after Indonesia liberalized its downstream market in 2001, Shell, Total and Malaysia's Petronas opened filling stations. But because of Pertamina's cheap (subsidized) prices, foreign firms' sales were limited to rich drivers, prepared to shell out for higher quality fuel that was less likely to clog their engines. Foreign-run gas stations had to partly rely on sales from shops on their forecourts to top up earnings. So, in 2012, after nearly a decade of disappointment, Petronas sold its filling stations.

Indonesia is the world's eighth-largest consumer of motor fuel. In Asia, only China and Japan use more. Pertamina accounts for about 90% of the 70 billion liters of fuel sold annually, while Shell and Total account for barely 10% between them.

Despite the removal of subsidies, Pertamina's low-grade, RON88 fuel, known locally as Premium, is still the cheapest. Motorists now pay a much smaller premium for buying foreign firms' RON92 fuel, and more of them are choosing it.

Crucially, market pricing means they won't suddenly find their fuel becoming relatively more expensive if oil rallies, because the state firm should raise its prices too. Also, phasing out sales of low-grade Premium altogether in a bid to end smuggling of cheap fuel is in the cards too, and as that happens, the market will become even larger for the foreign players.

Diesel Fuel Conundrum

Diesel is the preferred fuel for use in large engines. It is, therefore, the commercial side of the transportation fuels. As such, it is primarily used for powering transport vehicles; ships, trains, and eighteen wheelers. Since these vehicles move most of the commerce around the U.S. and the world, the availability and price of diesel on the retail level are of utmost importance.

In August 2008, a gas station in Bakersfield, CA, sold diesel at $6.75 a gallon. In December 2014, the average price of diesel in the U.S. was about $3.50 a gallon. In the spring of 2015, diesel went down to $2.25 and even lower at some gas stations around the U.S. Three times lower than the price of 2008 diesel fuel. Why?

A break-down of the diesel price reveals that taxes are 13% of the total, distribution and marketing 21% (vs. 15% for gasoline), refining is 11% (vs. 8% for gasoline), and crude oil components is 55% (vs. 63% for gasoline). Not a big difference, and yet gasoline price is, and has been for a long time, at least a dollar per gallon lower than that of diesel.

One reason why diesel fuel today is higher priced than gasoline is because of the unintended consequences of the 2007 EPA mandated ULSD (Ultra Low Sulfur Diesel) fuel—not necessarily because it costs more to produce. It all started in October of 2006, when the new U.S. ULSD regulations were implemented. Current U.S. ULSD is regulated to contain no more than 15-parts per million sulfur. In actual practice, U.S. ULSD contains just 7 or 8-ppm, which perhaps not coincidentally allows our ULSD to meet even the somewhat stricter, 10-ppm, sulfur regulations in Europe.

So, ULSD produced here in the U.S. has, for the first time, become acceptable for use in Europe. And that, according to the experts, allowed U.S. diesel to be exported in large quantity. U.S exports of diesel fuel average 350-400,000 barrels per day, up almost seven-fold from 60,000 barrels a day exported in 2007.

The supply and demand economics do not apply to the U.S. diesel fuel market. No matter what American truckers do, under the present situation, the price of diesel fuel won't change in the near future. In 2008, for example, U.S. oil companies were exporting more than 1.8 million barrels of crude oil, gasoline, diesel fuel, jet fuel and other refined products per day. The top five buyers of U.S. petroleum products were Mexico, Canada, the Netherlands, Chile and Singapore.

And then get this: Venezuela owns three CITGO refineries in the U.S., and about 30,000 barrels of refined products per day are shipped back to Venezuela, where government-subsidized gas/diesel is sold for a whopping $0.19 per gallon. What's wrong with this picture?

If we don't export diesel fuel, there would be a surplus, which could result in parity between gas and diesel fuel prices.

Right! But we do export and even want to increase the exports. On top of that, we have the U.S. government policies, starting with Reagan's Surface Transportation Assistance Act (STAA). Entering office in 1981, when Reagan recognized the need to increase funding for the country's highway system: Four thousand miles of highway needed resurfacing and 23,000 bridges needed repair or replacement. That need gave birth to the STAA of 1982. And diesel fuel prices started going up.

Heavy duty trucks cause more damage to roads than passenger cars, so the bill included a higher user tax on heavy trucks, along with a $.05 per gallon increase on gasoline and diesel taxes—equally taxed at the time.

Yet, that fee was significant, with a trucking industry official estimating that these fees alone for each truck could jump from $1,700 to $2,300 annually. That fear was confirmed in a 1984 report from the Government Accountability Office (GAO), which estimated taxes on very heavy trucks (70,000 pounds and up) would jump from $1,506 to $1,742, or from $1.40 to $1.56 per mile driven. At least a quarter of that would have to be paid at the start of the fiscal year, adding an extra financial burden.

The American Trucking Associations (ATA) was "appalled" by the proposal and asked Congress to vote against the bill and to start over again in the next session in establishing new truck use fees. That didn't happen, and ATA lost the battle for good. Reagan signed the bill into law in 1983, saying, "...as this bill becomes law, America ends a period of decline in her vast and world-famous transportation system." Weather Reagan was right or not remains to be seen, but there are still thousands of miles and many bridges in need of repair all around the country.

In January 1984 the issue became violent and the Independent Truckers Association led a strike by drivers around the country. In addition to striking, opponents of the diesel fuel fee began attacking drivers who continued on their routes. One trucker was shot through the neck by a sniper and killed while driving. Seven other North Carolina drivers were shot at during that time too.

The strike did not last long and was never widespread, but the violence it kicked up highlighted the passion among truckers over the use fees. It didn't take long for popular opinion to shift against the use fee too. The month the bill became law, the Department of Transportation signaled it would be willing to see the truck fees replaced with an increased tax on diesel, the fuel used by large trucks and a few passenger cars.

The idea behind taxing diesel is that it's used mostly by trucks, but few passenger cars. That way, trucks still pay their fair share, in accordance with how much time they spend on the road, and are not hit with large fees that they must pay up front. And so, they are heavily taxed at the diesel pump.

In 1984, Reagan signed the Deficit Reduction Act (raising taxes by $50 billion) into law, sharply cutting the use fee and raising the tax on diesel from $.09/gallon to $.14/gallon. Higher rates scheduled to take effect under the Highway Revenue Act of 1982 would have imposed a large tax on trucking operations which did not necessarily relate to the amount of business they might do, and that an alternative form of highway excise taxation should be devised which is more definitely correlated with the use of trucks. Therefore, Congress decided to substitute a high diesel fuel tax for a lower use tax.

And so, now diesel is taxed more heavily than gasoline, where the federal excise tax on gasoline is $0.184/gallon, while that for diesel is $0.244/gallon. The 1984 law included a tax credit to help out owners of diesel-powered light vehicles. That tax credit no longer exists, though farmers get a break for using diesel on their farms.

The Re-Fracking Phenomena

The slump in oil prices during 2014-2015 was a very bad news for the U.S. oil industry. Many oil and gas wells shut down production and some companies went broke. This, however, gave a chance for the oil industry to pause and take a second look at about 50,000 old (low producers or abandoned) wells in the U.S. as candidates for a second wave of fracking, using techniques that didn't exist when they were first drilled decades ago.

New wells cost $6-$8 million, while re-fracking costs of existing wells are about $1-$2 million.

This is significant savings in times as these, when the price of crude is hovering close to, and even below, $50 a barrel. The only question is how much oil and gas can be obtained from old wells. Historically, re-fracking offered mixed results, thus the nickname "pump and pray." But now, technologies have improved and with oil prices so low, companies are forced into drastic measures in the name of cheap oil production. So, analyzing data from older wells has become the key to finding a solution to increasing production without increasing costs. Re-fracking old wells seems to fit the times.

While fewer new wells would seem to mean less total fracking, the re-fracking phenomenon means there won't be as big a reduction as some had expected. Local communities, however, will continue to feel the impact from increased traffic and pollution, and more natural resources being used,

The number of wells fracked in the U.S. in 2014 climbed to 18,200, the total number of fracking stages—the holes punched in the rock—more than doubled. That means there are a lot of older wells with primitive frack work that are prime candidates for a fresh workover, using superior technology and know-how. And the timing is perfect for this opportunity to be tried. When oil prices go up, re-fracking will most likely be abandoned, and left for later time.

The hardest part about re-fracking is pumping new fracturing fluid down the length of a well running horizontally for 5,000 feet (over 1,500 meters), and isolating the spots that need to be blasted. That's more difficult than fracking a new well, and that's why operators in the past have dubbed the technique "pump and pray." Now, however, drillers have perfected the technique to send fracking fluid into old wells and direct it to the best targets, and so most of them are looking into re-fracking.

The low oil price period is giving oil companies more time to further develop the new technology, since when oil prices are high, there's no one available to look into such experimentation. And so, for now experimentation is alive and well and is expected to take place during the next couple of years. The incentives and the money are there to help the re-fracking experiments.

What this also means is that when someday the oil and gas bonanza starts slowing down for good—marking the end of the oil as we know it—the number of new wells will decrease dramatically, and the re-fracking will be the major approach. At that point—decades down the road—new techniques will allow drillers to pump more oil and gas from decades old wells, thus continuing the flow of fossils into the U.S. economy for a long time to come. How long…no body knows.

Reality Check

The bulk of known, proven, and remaining crude oil reserves in the world today are located in the Middle East. There are over 750 billion barrels of oil in reserve in that area alone, which is far more than what is available anywhere else on Earth.

For example, the known reserves in Central and South America are an estimated 100 billion barrels, in Africa an estimated 90 billion barrels, in the former Soviet Union nations, an estimated 80 billion barrels, and in Western Europe and China there is an estimated 20 billion barrels in each location. Mexico has an estimated

15 billion barrels of oil in reserve, and India an estimated 5 billion barrels of oil in reserves.

Currently, the U.S. ranks 12th among the top 20 countries by proven crude oil reserves. That places us just ahead of China and Qatar, but below Libya and Nigeria. By contrast, Russian oil reserves are estimated to be less than 80 billion barrels.

According to the U.S. Energy Information Administration (EIA), our proved reserves of crude oil and lease condensate (ultra-light crude oil) rose by about 4.5 billion barrels to 33.4 million barrels in 2013, which is the highest level since 1976.

The main factor in the rapid growth in oil reserves exploitation is the explosion of activity in Texas' Eagle Ford shale and Permian Basin and North Dakota's Bakken shale. Improvements in drilling technologies made it technically possible and economical to exploit previously inaccessible resources within these and many other fields.

Note: It is important to understand that the fact that the oil reserves are NOT *'growing.'*

> *Crude oil deposits in the earth's bosom are fixed. New oil is not made now, nor will it be made ever.*

The new crude oil bonanza is simply discovering old oil reservoirs that have been hidden or impossible to exploit until now. This is still a good reason for optimism. These reserve additions aren't due to luck, but rather due to heavy investments in technology made by U.S. energy companies over the past several years. Without those technological improvements, it wouldn't have been possible to economically harvest these 'new' reserves.

Somebody said that dealing with crude oil is like trying to catch falling knives. While it is quite amusing to watch the falling knives and the changing strategies of the players catching them, one very important thing is missing in this picture. It is the fact that no matter how many knives are caught and by whom, there are limited number of knives and eventually they will stop falling. And then, the game will just stop.

Not today, not tomorrow, but soon enough there would be no more knives falling—and no more oil flowing from the oil wells. We will be then caught by surprise, since we had too much fun with the game and nobody warned us that it will end some day. As a result, dire consequence will follow us for the rest of our lives.

This allegory applies directly to the oil market. We are so busy producing, transporting, processing, and using oil that we totally ignore the fact that it is limited in quality and quantity. While prices could go up and down today according to the whims of the producing countries, eventually oil reservoirs will be depleted to the point where only very deep and hard to obtain oil remains under the Saudi Arabian sands. This oil is of very low quality, it is very viscous, hard to work with, and is full of impurities and contaminants.

Between the extreme expense of pumping oil from deep underground, and the added expense of processing low quality oil, the global prices will start climbing up. $150, 200, 250, 500 per barrel… And finally the cost would become so great, and oil so impossible to use, that we would give up on it all together.

Life without crude oil is just not possible under the present day societal structure.

Crude oil is absolutely needed in all areas of the economy and our private lives. Especially those of us living in developed countries will experience first hand the dire consequences of life without crude oil. The Saudi sheikhs will go back in their grandparents' tents in the sand, while we will go back to using candles for light and wood for heat…

Not possible? Yes, possible! And not only possible, but a reality, which cannot be avoided. Unless we do something right now to extend the oil's presence long enough to figure out what to do without it, or replace it with something else. This doesn't seem to be happening today, so the only bright scenario is that the future generations would be so smart and ingenious as to be able to *replace oil before it is depleted*. We just have to hope that this is so, else there would be a lot of unhappy and suffering people on Earth, come the 22nd century.

But let's take a closer and more realistic look at the situation, as we see it today.

In 2013, the annual crude oil production averaged just under 7.5 million barrels per day, which comes out to a little over 2.7 billion barrels for the entire year. Assuming we deplete reserves by 2.7 billion barrels each year, the current reserves (33.4 billion barrels total) would run dry in less than 13 years.

Assuming that we find more oil, and decrease its use at the same time, which is very possible, we might end up with 25-30 years of oil supplies at prices and conditions we are used to. In best of cases, however, we will see very high oil prices by mid-century and the end of oil soon thereafter. The signs are obvious and we must not ignore them.

The question is how long would our oil supplies last? 25, 30, 40 years? More? And how much more?

Sounds funny and even scary thinking that crude oil's end is approaching, and for now one can blame the oversimplification of our calculations. Nevertheless, the point is that even though the U.S. oil production now exceeds every other country except Saudi Arabia and Russia, and the oil bonanza might last good deal more than our estimates, the size of our reserves is much smaller than our future needs dictate. And it will get even smaller with time. In the end, we are looking at empty oil reservoirs and idling oil fields. It is just a matter of time. Are we ready for that time?

NUCLEAR POWER

We must give credit to nuclear power as the most powerful energy source on Earth. Compared pound to pound, no energy source gets even close to the awesome power of uranium.

It takes thousands of tons of coal and crude oil to generate the power of a single pound of uranium.

With all due respect to its positive side, we must also agree, however, that nuclear power is the most destructive and vicious power known to man. Only mentioning Chernobyl and Fukushima, brings to mind vivid pictures of nuclear power gone wild and bringing terror to anything and everyone in its path.

And so, keeping in mind all this, we must consider and evaluate fairly the good and bad points of nuclear power as an energy source of the future.

Let's start by looking at the myths about nuclear power:

The Myths

There are a number of myths about nuclear power, circulating around. Let's put them to rest:

Myth 1. Nuclear Power Is Green and Clean.

Although nuclear power plants do not emit much GHGs, there is a lot of emissions during the mining and transport of nuclear materials and nuclear waste. Also, nuclear power plants use a lot of water for cooling, which is heated and released back in the water bodies, where it heats the ambient waters, which changes the hydrological environment.

In addition to that, the huge reactors and other pieces of equipment in the plant required a lot of energy and emitted a lot of GHGs during the steps of their manufacturing, transport, and installation. And so did the buildings during their construction. It would take a nuclear plant a year or two to generate power equivalent to what was used to make its equipment and buildings.

Nevertheless, compared to the mess created by coal-fired power plants, nuclear power must be given credit of being green and clean under normal operation.

Myth 2. Nuclear Power is Safe.

How safe it is would depend on whom you ask. A nuclear power official could provide a lot of information and data on safe performance of the hundreds of nuclear plants around the world. A Chernobyl survival, however, would show you pictures of his kids, who died painful deaths as a result of the accident and the following radiation exposure.

Safe, yes! If and when operating normally. But when out of control, nuclear power is a merciless destroyer of properties and killer of wild life and humans who happened to be nearby.

Myth 3. Nuclear Power is Cheap

Nuclear power generation is relatively cheap, but the construction of nuclear power plants is not. Neither is the waste disposal, which is still subsidized by governments. As is, the externalities cost—that of potential major disaster—is also largely subsidized at the tune of many billions of dollars annually.

Nuclear power is cheap, as far as the monthly electricity bills go, but the actual costs of creating, operating, and maintaining nuclear plants are enormous. Significant portion of these costs are paid by the taxpayers too.

Myth 4. Nuclear Power is Good Investment

Yes, it is. When it works right. When it doesn't, however, the investors wish they were not involved. Investing in a nuclear plant is a gamble at best, and a huge risk at its worse.

Investing in nuclear power is not for everyday investors. And this is why, it is governments, who are usually heavily involved in the financing of nuclear power plants construction and O&M.

Myth 5. Nuclear Power Is the Energy of the Future

Yes, it truly is…or at least it has the capacity to become so, *IF* a safe operation is ensured and no additional accidents occur. If we find a sure way of preventing 100% any and all major nuclear accidents, nuclear power could bring us energy galore, like no other energy source. This day, unfortunately, is still long way in the future.

So, in all cases we see that nuclear power is a good thing, IF… There is always this big IF that throws its shadow over nuclear power. Looking at the dark side of things, there are varying estimates of the damages and deaths attributable to nuclear power. Ultimately, it all comes down to the three major nuclear energy disasters: Three Mile Island, Chernobyl, and Fukushima. In the wake of events such as those, it's very difficult to know exactly which cancers and other health effects, spread out over geographic locations and the course of time, were caused by the accidents, and how many wouldn't have happened anyway. Even where there's correlation, causation is still difficult to prove.

In reality, the same thing is true about health problems and deaths caused by air and particle pollution from mines coal and coal-fired power plants; but there's much more information and less of a public debate about those cases and the related numbers.

In both situations, scientists analyze historical data on public health and make the best estimations that they can make. Different scientists don't always agree, but many of them come down on the side of coal being more deadly (at least in numbers) than nuclear power.

For instance, if we take a close look at the time period between 1987 and 2004, the only way the number of deaths in Europe caused by nuclear-power exceed the number of deaths caused by coal-powered electricity is if we use the very highest estimates for deaths in the related cases.

However, the striking difference here is that the coal-related health and death numbers come from many hundreds and even thousands of cases. The nuclear power related health and death numbers come from one or two isolated accidents. Not statistically sufficient data for any meaningful comparison.

Either way, nuclear power is here to stay (for now) and is a valuable contributor to the present day power systems. Barring some new and unprecedented disasters, it will be growing slowly around the globe.

Nuclear-Solar Future

There is a good opportunity for nuclear and solar to work together hand in hand. Since neither nuclear, nor solar can provide the entire power load of the country, combining forces might be the best approach for the future. Doing it all with nuclear would be exceedingly risky, doing it with solar alone would be foolish and simply impossible. So, how about combining these two power sources?

For example, Palos Verdes nuclear plant located 50 miles from Phoenix, Arizona, is one of the largest nuclear plants in the country. It hums happily non-stop during the entire year, until the broiling summer days invade the valley. In most of the exceedingly hot afternoons, May through October, electric demand in the area exceeds the nuclear power plant's capacity.

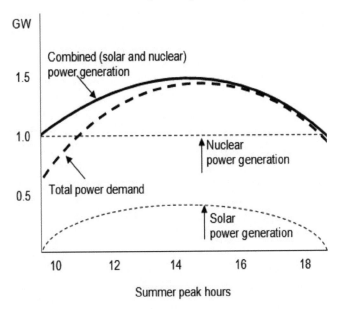

Figure 10-14.
Combined solar and nuclear power peak power generation

A number of peaker power plants, powered by diesel or natural gas, kick in during the hot summer afternoons to help the nuclear power plants keep up with demand. The peakers are usually idle until called into action by hot weather or in emergencies at a great expense. Even with their help, however, the power grid is still close to the breaking point during most summer afternoons, and the prices jump up 2-3 times the regular rates.

So obviously, nuclear power needs help at some times of the day in the summer months. Quite conveniently so, solar is most productive during the same times; the sunniest (and hottest) summer days.

Properly designed and implemented nuclear-solar power cogeneration could play a major role in providing enough and cheap electric power during the peak summer hours.

The additional power provided by the solar power plants would prevent the peakers from starting up and burning extra quantities of fossils. The power grid would take a break, and the electric power prices won't go up so high.

Figure 10-14 is a hypothetical scenario, showing a 1.0 GW nuclear power plant operating at maximum capacity on a summer day in the Arizona desert. Close by a smaller, 500 MW, solar plant is cranking some power too. The nuclear power plant alone is capable of keeping the power demand met most of the time, but around noon the residential and commercial air conditioners kick in at full power and start consuming megawatts of power in anticipation of another blistering hot summer day in the Valley of the Sun.

With temperatures approaching 110°F in the shade, the total power demand increases steadily, and eventually surpasses the maximum power level provided by the nuclear plant. At 1 p.m. the local utility starts charging peak rates, which are 2-3 times higher than off-peak rates, but that would not reduce the demand by much.

Life without air conditioning during Arizona's summer afternoon is a death wish, so every single home and business building has a large air conditioning system. Some of the commercial air conditioning systems consist of multiple units, some bigger than a truck and consuming more power in a day than entire villages in Africa use in a year.

All this cooling and fanning requires a lot of energy, so the power grid reaches full capacity around noon. It would shut down at this point, if only power from the nuclear plant were available. Instead, it starts sucking electric power from other states. In most cases, even that is not enough, so around 1 p.m. the local peakers (diesel or natural gas power plants on standby for emergency use only) fire up and start generating the additional 500 MW electric power to keep the grid operating and thousands of air conditioning units running.

The local solar power plant, which has been ramping up slowly, reaches its maximum output at this time (maximum solar insolation) and provides the additional 500 MW needed to keep the locals cool and happy. Now the peakers can slow or even shut down, saving money and fuel, and sending less GHGs into the atmosphere.

Figure 10-14 shows combined solar-nuclear power generation as an almost perfect fit, but it's a far-fetched as a standard approach. Too much controversial politics and conflicting private interests are involved in this process, so although we think this is the best way to build a bridge to the fossil-less future, it is not happening—not on a large enough scale to make a difference.

Still, common sense dictates that the intense summer heat, caused by too much sunlight could be successfully counteracted by generating equal amounts of electric power from the same sunlight. Common sense and future generations are asking, "How can you ignore such a winning combination?" But for now, common sense and the future generations are absent for all practical purposes, so the decisions are often based on political and corporate interests. Because of that, the energy markets are unbalanced, and the customers are paying for the consequences.

HYDROPOWER

Hydropower has the potential of delivering electric energy for the centuries to come, except that there are several problems, the major of which are:

- Most of the best sites for large hydro-power dams around the world have already been used as such. Those that remain are much less attractive from efficiency and economic points of view.

- Increasing severity and duration of draughts around the world is impacting the hydro-power industry. Since this is something we have no control over, it is becoming a major obstacle in counting on hydropower as a steady power generator.

- The global climate changes and other natural events are intensifying with time, and could reduce hydropower's output and limit its growth in the future.

- Hydropower is not officially recognized as a renewable energy source in the U.S., although scientifically and practically speaking it is. This is limiting its progress in the country.

While we are running out of sites suitable for large-scale hydropower deployment, there are still many sites suitable for smaller hydropower projects. The big problem—which we have some control over—has been the heavy regulation of hydropower projects in the country; especially new and planned such.

For that purpose, the U.S. Congress passed the Hydropower Regulatory Efficiency Act of 2013 (H.R. 267). This is one of the few bills that the divided Washington politicians were unanimous about. It passed the House by a vote of 422-0. A record bar none. This alone is an incredible accomplishment in the present polarized political atmosphere.

The new Hydropower Act aims to making it easier to develop new, smaller-size, hydropower plants.

Current regulations are inadequate since they treat small hydro projects the same way as the huge-size such. The new Act alters those regulations to make it easier for smaller plants to get approval quickly. This

way, by lifting a significant hurdle, we now can expect to create many small to medium size hydropower plants around the country.

The legislation also requires the Federal Energy Regulatory Commission (FERC) to study how to further improve the hydropower regulatory process. The Bill also amends the Public Utility Regulatory Policies Act of 1978 (PURPA) and the Federal Power Act. Currently, hydropower projects that produce 5,000 kilowatts or less of power do not require certain licenses, now the limit is raised to 10 MW. This further facilitates the speed at which smaller hydropower projects could be built.

All in all, instead of the average five years required for hydropower projects to get approval, the new regulations would reduce the permit time to months. This way, we can hope for the U.S. to add over 60 GW of new hydropower capacity by 2025. This would revive the hydropower industry in the country, and create over 700,000 jobs in the process.

So, the future of hydropower in the U.S. is not building more large dams, like the Hoover dam. Although it is the best use of hydropower, we just don't have many places left for such large projects. Instead, we need to focus on building small to medium size hydro-power plants.

Small-Size Hydropower Plants

With the passing of the new hydropower Act, we expect to see many small hydropower projects popping all over the country. We can divide these small projects into groups, according to their size and designation.

Small Hydro

Small hydro is any hydropower plant that can serve a small local community or business. Presently, small hydro is any hydro project with generating capacity of up to 10 MW as an upper limit. In contrast, many hydroelectric projects are of enormous size, such as the generating plant at the Hoover Dam of 2.1 GW, or the vast multiple projects of the Tennessee Valley Authority.

The small hydropower type of power generation is usually considered as *distributed power generation*—similar to the solar power generation from private roof tops.

There is a chance the *small hydro* upper limit to be stretched up to 30 MW in the U.S., and up to 50 MW in Canada. in the near future.

Small hydro can be further subdivided into:
- Small hydro is any hydropower plant of less than 10 MW
- Mini hydro is defined as less than 1.0 MW, and
- Micro hydro is usually less than 100 kW.

Micro Hydro

Micro hydro is a hydroelectric power that typically produces up to 100 kW of electricity, usually by using the natural flow of water. Micro hydro is usually the application of hydroelectric power sized for smaller communities, single families or small enterprise. They are sometimes connected to electric power networks. There are many of these installations worldwide, particularly in developing nations, as they can provide an economical source of energy without the purchase of fuel.

Micro hydro systems complement photovoltaic solar energy systems because in many areas, water flow, and thus available hydro power, is highest in the winter when solar energy is at a minimum. Micro hydro is frequently accomplished with a Pelton wheel for high head, low flow water supply. The installation is often just a small dammed pool, at the top of a waterfall, with several hundred feet of pipe leading to a small generator housing.

Mini and Small Hydro

These types of hydro plants can be distributed, or connected to conventional electrical distribution networks as a source of low-cost renewable energy. Small hydro projects may be built in isolated areas that would be difficult, or uneconomic, to connect to the grid, or in areas where there is no national electrical distribution network.

Since small hydro projects usually have minimal infrastructure, such as water storage reservoirs and extensive construction work, they are seen as having a relatively low environmental impact compared to large hydro. This decreased environmental impact depends strongly on the balance between stream flow and power production.

Hydro plants with reservoirs, like small storage, or small pumped-storage hydropower plants, can contribute to distributed energy storage and decentralized peak and balancing electricity. Even more importantly, such plants can be built to integrate the intermittent renewable energy sources, wind and solar.

During peak hours, wind and solar power plants can pump up and store water at a higher elevation, for later use. When wind and solar do not produce enough power, the stored water can be released and used to generate electricity.

The future of hydropower is uncertain at best, but we see clearly the advantages of using it in combination with renewable energy sources—wind and solar in particular. This way we increase the efficiency of power generation, thus making it more practical and cost effective for use by all.

WIND POWER

Onshore (inland) wind energy is a tried and tested technology that is already cost competitive with conventional power in some parts of the world. In Australia, Brazil, and in parts of the U.S., wind power is already used extensively.

In the midst of the economic downturn which started in 2008, during 2012, nearly $80 billion were invested in the wind energy sector globally. And while the global economic downturn continued to act as both a direct and indirect drag on investment, wind energy continued nevertheless to become a significant global industry in its own right.

The wind power revolution continues, albeit at a slower pace. Wind energy plants are being widely deployed wherever economic conditions are conducive. Wind energy can already claim to be a major source of electricity: in 2012, wind energy provided about 30% of electricity consumption in Denmark, 20% in Portugal, 18% in Spain, 15% in Ireland, 8% in Germany, nearly 4% in the U.S. and 2% in China.

The benefits of wind energy are numerous and varied. Wind energy:
• Provides opportunities to diversify a nation's supply mix,
• Reduces reliance on fossil fuels, and
• Provides clean and green energy

Deployment of wind power at scale can reduce dependence on imported fuels,3 and reduce exposure to price volatility of those fuels. Additionally, wind power can generate significant value for a country's economy through supply chain investment and job creation. More broadly, there is increasing recognition of the ability of wind energy, along with other renewables, to help spur innovation and thus stable, long-term economic growth.

Wind energy began to emerge in the 1970s, partly in response to the oil crisis, and particularly in countries exposed to fossil fuel price inflation with limited reserves of their own, such as Denmark. However, up until the 1990s, global wind power capacity remained at low levels: only 1.7 gigawatts (GW) in 1990. It was not until the end of that decade that the market for wind energy really began to accelerate, reaching a global installed capacity of over 282 GW in 2012 (GWEC, 2013).

Figure 10-15 demonstrates how the global wind energy market has grown cumulatively from 1995 to 2012, and provides a breakdown for the top ten global markets. The data show how rapidly growth has accelerated in the last decade, and that growth is forecast to continue.

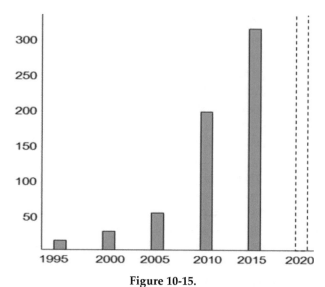

Figure 10-15.
Global wind power generation (in GW installed)

Figure 10-15 shows the rapid growth of wind power around the world. Its future, however, depends on many things. Due to drastic changes in the world's political, economic, and technological developments, it is not certain what will happen in the wind energy sector in the next several years.

Modern wind turbines and related technology have evolved rapidly over recent decades. This evolution has had several aspects, of which two are particularly relevant.

*   Turbines have continued to increase in size, from an average of around 1 megawatt (MW) in 2002 to 2 MW in 2012, and
*   Average capacity factors have risen: each MW installed now produces more electricity than in the past.

The improvements are due mainly to numerous technological achievements of late. Larger rotors and improved electro-mechanical design are the major contributors. To some extent, improved siting, installation, and O&M procedures have helped wind power to become more efficient and cost-effective too.

Such improvements will continue to support the deployment of wind power in more remote locations, will further expand the offshore market, and will support the repowering and replacement of older turbines in existing wind power plants.

Offshore Wind Energy
The offshore wind energy market is younger and much less developed than its onshore cousin. By the end of 2012, only 5.4 GW of onshore wind power had been installed, 90% of that in northern European waters. This is because offshore projects present higher risks during project development, and through construction and in operation, as well as greater average costs and higher complexity.

The major offshore wind energy markets presently face political uncertainty in the longevity and extent of support mechanisms for the technology. This is likely to impact investor confidence and national ambitions, slowing development. In spite of the uncertainties, the offshore wind energy industry is still expected to grow at a compound annual growth rate (CAGR) in excess of 15% per year in Europe until 2020.

By 2020 offshore wind energy could reach 35 GW globally, delivering a total of over 100 GWh of electric power.

The challenges of developing offshore wind energy are such that capital costs can be, at present, up to three times higher than for onshore, although this is offset to a degree by the higher capacity factors experienced offshore. Depending on a number of factors, including distance from the shore and water depth, recent studies indicate that investment costs for offshore wind energy span from $3.6 million/MW to $5.6 million/MW.

In comparison, the investment cost for land-based wind power generation ranges from $1.1 million/MW in China, to a high of $2.6 million/MW in Japan. Mid-range prices are found in the U.S.—about $1.6 million/MW, and Western Europe, $1.7 million/MW. This is still several times higher than the cost of onshore power.

The future of both onshore and offshore wind power depends on many factors. First, however, we need to address and solve some of the problems related to their development into main stream energy technologies.

To start with, we need to understand the present day barriers, as reflected in Table 10-7.

System Integration
Wind energy, like most renewable energy, is strongly dependent on weather and geography, since wind electricity output fluctuates with the changing wind speed. Although this variation is not discernible in the range of seconds to minutes, in the space of one day the aggregated production of a country can, on occasion, ramp from near zero to near maximum, and vice versa, of the total installed wind energy capacity. Because of that, wind power systems that incorporate a large share of wind power need sufficient flexibility to respond to the variability.

Table 10-7. Wind power barriers

Barrier	Details	Action options
Connection to grid is constrained	• Transmission and/or distribution grid owner may not wish (or lack capacity) to connect • Offshore connection costs may be prohibitive • Connection fee may be inappropriate • Local opposition prevents construction of new grid connection • Point of connection may be disputed among developers or with transmission owner • Long distance between potential site and grid node can be a barrier due to cost or existing rights of way	• Regulate monopoly control to allow access for Independent Power Producers (IPP) • Educate local population on benefits of wind power (GHG reduction, green jobs) • Consider underground power lines • Regulate system operators to ensure rates reflect costs • Distinguish connection costs from grid reinforcement costs and assign appropriately • Engage with local stakeholders to manage trade-off between new grid infrastructure and benefits of wind power
Operational aspects	• Wind turbines present health and safety challenges (e.g. ice throw) • Assignment of decommissioning costs • Repowering demands grid upgrade • Shortage of qualified personnel for the operations and maintenance (O&M)	• Ensure interface with planning process to avoid conflicts and provide contact point for local residents • Ensure wind energy policy addresses end-of-life issues (decisions regarding recycling or decommissioning equipment versus repowering) • Ensure that O&M training programmes exist at national or regional level that are consistent with the desired level of wind energy deployment

Such flexibility comes from both: the available resources and from institutional arrangements with regulatory and market entities. Flexible resources differ from case to case, ranging from dispatchable plants (such as reservoir hydropower and gas plants, pumped hydro storage, demand-side management and response), to trade with neighboring systems through interconnectors. This includes the variable wind and solar power plants themselves, as is the case with some offshore wind energy plants in Denmark.

It is, therefore, not enough that such flexible resources exist: they must be available and able to respond to demand, while being given incentive to do so, as and when required. Wind power is well on its way to taking its place of a major power generator, but the gaps—some of which were mentioned above—need to be filled by different means.

Costs and Expenses

The wind power equipment and installation investment costs for onshore wind energy, including turbine, grid connection, foundations, infrastructure and installation, ranged from $1.45/W to $2.60/W in the past. Today, the range today is even larger, spanning from the low $1.10/W in China to the high $2.60/W in Japan. Mid-range prices are found mostly in the U.S. at about 1.60/W, and Western Europe at $ 1.70/W.

Following a period of steady decline, investment costs rose considerably in 2004-09, and doubling in the U.S. This increase was due mostly to supply constraints on turbines and components (including gear boxes, blades and bearings), as well as higher commodity prices, particularly for steel and copper (the increase in commodity prices also affected conventional power production).

Since 2009, investment costs have fallen along with commodity costs and the reversal of supply constraint trends as well as increased competition among manufacturers. All factors considered, investment price declined by 33% or more since late 2008

Investment costs for offshore wind is usually a factor of 2-3 higher than onshore wind installations. Limited data on offshore costs make it difficult to calculate accurate estimates. Presently, the cost of off-shore installed turbines represent less than half of the total investment cost. This, compared to 75% turbine cost element for onshore projects.

Offshore projects are more expensive, because they incur additional expenses for foundation, electric infrastructure and installation costs. The costs vary with location, distance from shore, and sea water depth. In 2008, for example, offshore investment costs ranged from $3.10/W to $4.70/W.

Total costs increased during the 2010-13 period, spanning from $3.60/W to $5.60/W.

Note: Some of the low investment numbers are from Denmark, and does not include the expense for grid connection to the shore.

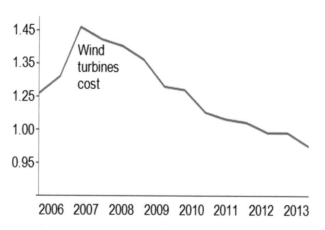

Figure 10-16.
Cost trend of land-based wind turbine prices (in $/Watt)

Some of the investment costs of offshore wind in the United Kingdom have significantly increased since the first commercial-scale wind power plants were deployed in the early 2000s. This results from underlying cost increases, reliability concerns and deeper water sites: while earlier plants were in relatively shallow waters, most new plants since 2010 are located in water depth exceeding 65 feet. Recently announced wind power plants for similar sites show that capital costs have leveled off at $4.00/W to $4.40/W, including transmission capital costs. This reflects several factors including

a better understanding of the key risks in offshore wind construction and larger projects leading to greater economies of scale.

O&M

The operation and maintenance (O&M) costs of wind turbines represent an important component—15% to 25%—of the total cost of wind power. O&M activities typically include scheduled and unscheduled maintenance, spare parts, insurance, administration, site rent, consumables and power from the grid. Low availability of data makes it difficult to extrapolate general cost figures, as does the rapid evolution of technology: O&M requirements differ greatly, according to the sophistication and age of the turbine.

Problems with electrical and electronic systems are the most common causes of wind turbine outages, although most of these faults can be rectified quite quickly. Generator and gearbox failures are less common, but take longer to fix and are more costly.

From 2009 to 2013, O&M expenses show a 44% decrease in average prices in $/MW/year. With a capacity factor of 25%, the 2013 costs for land-based wind power plant would thus be $10.25/MWh. The span, however, can be large, ranging from $5.0/kWh $16.0/kWh. Offshore wind power O&M costs could be as high as $70-$80/MWh. This is an increase from about $50/MWh in 2007

LCOE

The leveraged cost of investment (LCOE) of wind energy can vary significantly according to the quality of the wind resource, the investment cost, O&M requirements, the cost of capital, and also the technology improvements leading to higher capacity factors.

Turbines recently made available with higher hub heights and larger rotor diameters offer increased energy capture. This counterbalances the decade-long increase in investment costs, as the LCOE of recent turbines is similar to that of projects installed during the 2002-2003 time period.

For some sites, LCOEs of less than $50/MWh have been announced; as at the recent Brazil auctions and some private-public agreements signed in the U.S. Technology options available today for low wind speed—tall, long-bladed turbines with greater swept area per MW—reduce the range of LCOE across wind speeds.

More favorable terms for turbine purchasers, such as faster delivery, less need for large frame agreement orders, longer initial O&M contract durations, improved warranty terms and more stringent performance guarantees, have also helped reduce costs.

Higher wind speeds off shore mean that plants can produce up to 50% more energy than land-based ones, partly offsetting the higher investment costs. However, being in the range of \$136/MWh to \$218/MWh, the LCOE seen in offshore projects constructed in 2010-12 is still much higher compared to land-based. This reflects the trend of locating wind power plants farther from the shore and in deeper waters, which increases the foundation, grid connection and installation costs. Costs of financing have also been higher for larger deals at new sites, as investors perceive higher risk related to weather and other natural events.

The Future

Wind energy technology is proven and mature; making serious progress all over the world. No single element of onshore turbine design is likely to cause any problems, or increase dramatically the cost of energy in the years ahead. Nevertheless, the turbines' design and reliability can be improved in many areas. If and when put together, these factors will reduce both cost of ener-

gy and the uncertainties that stifle investment decisions.

Greater potential for cost reductions, or even technology breakthrough, exists in the offshore sector. Actions related to technology development fall into three main categories:

- Wind power technology: turbine technology and design with corresponding development of system design and tools, advanced components, O&M, reliability and testing;
- Wind characteristics: assessment of wind energy resource with resource estimates for siting, wind and external conditions for the turbine technology, and short-term forecasting methods;
- Supply chains, manufacturing and installation issues.

In light of continually evolving technology, continued efforts in standards and certification procedures will be crucial to ensure the high reliability and successful deployment of new wind power technologies. Mitigating environmental impacts is also important to pursue.

Table 10-8. Wind equipment optimization efforts

System design	Time frames
Wind turbines for diverse operating conditions: specific designs for cold and icy climates, tropical cyclones and low-wind conditions.	Ongoing. Commercial-scale prototypes, 2015.
Systems engineering: to provide an integrated approach to optimizing the design of wind plants from both performance and cost optimization perspectives.	Ongoing. Complete by 2020.
Wind turbine and component design: improve models and tools to include more details and improve accuracy.	Ongoing. Complete by 2020.
Wind turbine scaling: 10 MW to 20 MW range turbine design to push for improved component design and references for offshore conditions.	Ongoing. Complete by 2020-25.
Floating offshore wind plants: numerical design tools and novel designs for deep offshore.	Ongoing. Complete by 2025.
Advanced components	**Time frames**
Advanced rotors: smart materials and stronger, lighter materials to enable larger rotors; improved aerodynamic models, novel rotor architectures and active blade elements.	Ongoing. Complete by 2025.
Drive-train and power electronics: advanced generator designs; alternative materials for rare earth magnets and power electronics; improved grid support through power electronics; reliability improvements of gearboxes.	Ongoing. Complete by 2025.
Support structures: new tower materials, new foundations for deep waters and floating structures.	Ongoing. Complete by 2025.
Wind turbine and wind farm controls: to reduce loads and aerodynamic losses.	Ongoing. Complete by 2020-25.

Table 10-9. Cumulative investment in $ billion

Area	by: 2020	2030	2050
OECD Europe	256	337	831
OECD Americas	209	455	628
OECD Asia Oceania	32	69	120
Africa and Middle East	42	173	194
China	305	385	839
India	36	38	158
Latin America	25	12	74
Other developing Asia	53	105	279
Other non-OECD	22	61	185
TOTAL	980	1 635	3 308

Cost reduction is the main driver for technology development but others include grid compatibility, acoustic emissions, visual appearance and suitability for site conditions (EWI, 2013). Reducing the cost of components, as well as achieving better performance and reliability, thus optimizing O&M procedures and resulting in reduced cost of energy.

The long-term target is to cut the cost of wind power to $125-$150/MWh. All parts of the supply chain play major roles in building the industry and bolstering innovation to drive down the cost of energy, along the lines of:

- Introduction of larger turbines with higher reliability and energy capture and lower operating costs;
- Greater competition in key supply markets (e.g. turbines, foundations and installation) from within the United Kingdom, Europe and East Asia;
- Greater activity at the front end of projects, including early involvement of suppliers and improved wind farm design;
- Economies of scale and standardization;
- Optimization of installation methods;
- Mass-produced, standardized deep water foundations;
- Lower costs of capital through de-risking construction, and O&M.

As cost reductions require a larger market, predictability and permanence of the market is needed to achieve maximum results. Wind farm developers and suppliers must work together to deliver continuous, end-to-end cost and risk reduction. Managing a pipeline of projects, rather than working project by project, will help to drive down cost.

The cost of capital is a key driver of LCOE for offshore wind plants. A drop of one percentage point in the weighted average cost of capital (WACC) reduces LCOE by about 6%. As the offshore industry gains experience, key risks (e.g. installation costs and timings, turbine availability, and O&M costs) will be better managed, and the overall risk profile of offshore projects will decline, thereby lowering the returns sought by capital providers.

Moving to products specifically designed for offshore wind, and industrializing and standardizing the supply chain and procedures will provide multiple opportunities to reduce capital and operating costs while increasing power generation.

The future of wind power is not clear of obstacles, yet we foresee a steady increase of wind power generation in the U.S., most developed, and some developing, countries in the future.

One very promising technology of the future is using a combination of wind and solar plants to compensate for their variability.

Solar/Wind Power Cogeneration

Solar energy generation could be combined with other energy sources when constant output is the goal. Addition of solar power to conventional power sources (power plants) is one approach. Using energy storage devices (batteries or water storage) is another. A more natural and most efficient way, however, is combining PV power with wind at locations especially chosen for this purpose. At such locations PV power is complementary to the output of wind generation, since it is usually produced during the peak load hours when wind energy production may not be available.

Variability around the average demand values for the individual characteristic wind and solar resources can fluctuate significantly on a daily basis. However, as illustrated by Figure 10-17, solar and wind plant profiles—when considered in aggregate—can be a good match to the load profile and hence improve the resulting composite capacity value for variable generation.

In the Figure 10-17 example, the average load (upper line) is closely followed during the day by the average output from the combined wind and solar generators (the second from top line) during the same time. This average is created regardless and because of the fluctuations of the individual wind and solar power generators. This is a marriage made in heaven, and this combined power generating combination will work quite well if wind and solar power outputs can be matched as closely as this one.

Although there are areas in the US and abroad that match this wind and solar profile, the combined effort is usually hard to execute, because the best places for wind and solar are at different locations, often miles apart, and also because of lack of infrastructure at the most suitable locations.

Truthfully, it will require a great effort and expense to implement large-scale "wind-solar load matching" schemes anytime soon with existing technologies.

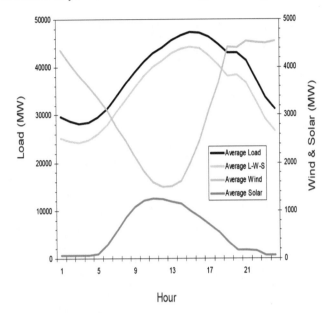

Figure 10-17. Simultaneous wind and solar generation

Still, having as a goal matching wind and solar power outputs will force us to seek the most suitable locations and appropriate technologies for this match. This won't happen overnight, but if we approach the solution intelligently, we will have a large-scale power output—nationally—that matches the grid power loads.

SOLAR POWER

Solar power is the power of the future. We must understand and agree that the only free energy we have left on Earth is solar energy coming from the Sun. Non-stop, every single day, 365 days of the year. It is the only constant, uninterrupted, energy source on Earth. We just have to find, capture, and use it properly and efficiently. Not an easy task, but this what it is, and the options are getting few and between.

Background

The brutal shakedown of the global solar manufacturing industry during 2011-2013 is not finished. Instead, it will continue and even increase in the near future. Industry insiders predict over 70% of the solar manufacturers worldwide will shrink, get bought, disappear all together, or otherwise exit the market by 2020. Not an encouraging news, but fact of life, and a confir-

mation of the immaturity of the global solar industry.

The world's total number of PV modules manufacturers is expected to go down to 125-150 companies at that time, compared to 500 in 2012 and 650 in 2011. The 2010-2012 worldwide glut of PV modules slashed prices in recent years, driving many companies like Fremont's Solyndra into bankruptcy. The price plunge has been great for consumers, helping drive up the number of solar installations. But for photovoltaic manufacturers, the bloodletting continues.

Consolidation was rampant in the PV solar industry in 2012, one example of which was the consolidation (acquisition) of MiaSole, the thin-film solar manufacturer headquartered in Santa Clara, by China's Hanergy Holding Group. MiaSole, like many other solar companies was getting ready to make a big dent in the solar energy markets during the boom years, or at least the media hoopla promised so. It, however, was not able to take-off. Instead, its wings were clipped by technological glitches, the ongoing economic downturn, and the plunging PV module prices.

MiaSole is now at the mercy of its new Chinese bosses. Hanergy took over and will keep MiaSole's Sunnyvale plant open (for awhile at least), retaining the 100-plus people employed there and maybe even hiring more. Maybe!

The new website proudly announces MIASOLE (and down below with smaller letters it clarifies, *a Hanergy Company*). MiaSole's fledgling technology benefited from more than $550 million in U.S. venture capital over the years, but its sale price to Hanergy was just $30 million. This is 1:20 ROI ratio, which is hard to justify under any conditions. China 1, U.S. 0.

Today, MiaSole/Hanergy's flexible SIGS technology is still under development, and still under actual 10% field efficiency. This is limiting the marketing share of the company, which is selling small quantity of flexible solar rooftop materials to pay the bills. The market share is actually near zero, according to the experts.

In 2012-2013 things got progressively worse for the major PV products suppliers too, with most upstream PV supply operations simply ceasing to exist, rather than being acquired by other companies. Many of the suppliers—most of them in Asia—actually have already stopped production, and will never restart it

Dozens of new solar factories in China, working non-stop 24/7, created the PV modules glut, and have driven many U.S. solar companies out of business. But most of the newly established Chinese wanna-be solar manufacturers companies won't survive the shakeout. The Chinese government will prop up some of the bigger companies, but most of the smaller ones will be left on their own. They will most likely shut their doors, or change the line of business and switch to manufacturing some other gadgets. Many of these companies switched to solar from some other business, so going back where they came from won't be too difficult.

A huge bumper crop of solar panels already had caused a sharp decline in the prices and bankrupted many manufacturers worldwide over the past two years. New reports say another 180 solar panel makers will likely go out of business or be bought by the end of 2014. Nearly half of them–88 companies, according to the experts' estimates–will shut down factories in countries that have become too expensive for producing solar panels, namely the U.S., Europe and Canada.

The prognosis is not only shocking, but it also answers a perennial question, at least for now, about whether solar manufacturing can thrive in some labor-expensive countries, like the U.S. China, which has used state owned banks and utilities to finance solar factory expansion and create domestic demand for solar panels, will continue to dominate solar manufacturing, though the government is reportedly working on rescuing only 12 large companies and forcing mergers in others. Even with the low labor costs, it is estimated that over 50 solar panel makers in China will not survive over the next 3-5 years.

Given where the industry is right now, and how committed China is to its solar manufacturing industry, it's very difficult for the European and U.S. manufacturers to compete. China's rise as the world's epicenter for solar manufacturing has elicited resentment from rivals who believe Chinese companies haven't competed fairly. The U.S. Department of Commerce recently sided with petitioners of such a trade complaint and imposed tariffs after finding that Chinese solar companies have indeed received illegal government subsidies and sold solar panels at below cost.

Signs of trouble began to show in early 2011, when changes in solar incentive policies in key European markets prompted solar panel makers' customers—distributors and project developers—to delay purchasing decisions. Prices for solar panels began to fall faster than what manufacturers had expected.

The prices dropped 50-60% recently and have continued to decline slowly. At the same time, many manufacturers had built up massive factories and were counting on a huge surge in demand in the global market. In fact, they continued to churn out solar panels to keep their factories running and workers employed even though demand wasn't keep pace.

First Solar, the wonder-kid of the solar industry, a giant in the field and the investors' darling (at the time), saw flat revenues and lower earnings since the first quarter of 2011. Then, its stock plummeted to unprecedented and unforeseen lows—from nearly $300 a share down to $12 in 2012. The stock recovered somewhat soon after, hanging in the $40-50 range. The future of this, and many other companies hang in the balance.

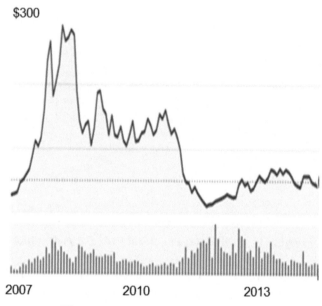

Figure 10-18. First Solar stock plunge

Life pretty much went downhill from there for many other solar panel makers and their suppliers. Young companies that were entering mass production of their technologies in order to compete effectively with larger rivals went bust, following the examples of Solyndra, Abound Solar, and dozens other solar- manufacturers wanna-bes.

GE, which once embarked on an ambitious plan to build a 400-megawatt thin film modules factory in Colorado, decided to shelve that project in 2013. First Solar, long the king of low-cost manufacturing, decided to gradually shut down its big factory in Germany and put on hold its plans to build new factories in Vietnam and Arizona. This, after both companies pocketed over $3.5 billion in loan guarantees and other perks from the U.S. taxpayers.

Other solar panel makers in the U.S., Europe and Asia made similar decisions to shutter factories and/or file bankruptcy. SunPower was one of them. It suspended production at six of the 12 solar cell production lines and cut solar panel production by 20% in the Philippines, and laid off about 900 employees there. It is still stumbling in an effort to find the path to success.

Still, some solar manufacturers have proceeded with their plans to build new factories for a variety of reasons. Some thought the oversupply problem would be over soon; others needed to scale up their production to cut costs. As a result, the world will likely see 35 GW of excess solar panels for sale per year over the next 5 years. China is absorbing a good amount of the surplus by increasing the number and size of its solar installations. Even then, the surplus conditions are here to stay for awhile.

The plummeting prices for solar panels are good news for installers and solar power plant developers—and ultimately consumers. An increasing number of manufacturers are entering the business of developing solar energy generation projects since they are not making enough money by selling solar panels only.

Some U.S. utilities are also looking at becoming solar projects developers and owners. All these, and other pending, developments in the solar sectors promise that the game is changing quickly, and no one knows what to expect.

Solar Then and Now

The U.S. solar industry was born in the 1970s as a result of a government action. It was propped by multimillion dollar financing and other assistance geared to develop solar technologies as a last resort against rising oil and energy prices. The trend actually started with the Arab oil embargo in 1973. It is when U.S. politicians realized that oil is not the solution. So, solar was going to be our secret weapon, with everyone hoping that it will pull a miracle or two, to bail us out of the energy doom the Arabs were cooking for us.

Solar was encouraged to show what it can do by billions of dollars thrown into R&D, production optimization, and new facilities construction. The industry giants at the time, Motorola, Shell, and others, together with a number of universities and national labs were the leaders and the beneficiaries of countless grants, tax breaks, and different subsidies. But try as it may, solar was not able to provide even a glimpse of hope for replacing the fossils. The technology was not well developed, and was simply incapable of providing much help to the desperate situation at the time.

And then the miracle everyone was looking for came; the Arabs dropped the oil price once again. Solar? Who cares about solar now? And back in the closet the embarrassingly weak and impotent solar stepchild went. Some people made money in the solar frenzy of the 1970s, but most of those who got caught in the frenzy lost some money and hope in solar.

A number of similar situations occurred through the years past, with the latest solar boom-bust being the best example of the knee-jerk reaction of Washington politicians. Lacking long-term plans, they panic at every oil price fluctuation and jump into action.

During the last financial crises, during which oil prices jumped to nearly $180/bbl, things were quite different. The entire world was panicking. Solar was seen the solution to our energy problems by the U.S. and EU politicians. Billions upon billions were thrown…again… into worthless technologies and wasted on mismanaged companies and projects. This was another desperate, albeit poorly executed, effort to use solar, and the renewables, as a solution to our energy problems.

Soon enough we all figured out that solar is not THE solution, but only one small piece of the total solution. Nevertheless, Germany jumped in solar with both legs; leading the 2007-2012 solar boom-bust cycle with the most PV installations in the world with the most generous subsidies and benefits for installing and using solar. Ever!

But relying on solar energy in Germany is like looking for water in the middle of the Sahara desert. Yes, some water is found in the Sahara desert too, but you cannot build a city around it…in most cases. Or can you? In the same way, there is some sunshine in some places in Germany, but you cannot realistically build a sustainable national energy plan around it. There are many more rainy and cloudy days in most of Germany, than sunny days.

At maximum of 15% efficiency at full sunlight, PV modules drop down to less than 5% under the cloudy, drizzly, and gloomy German skies. At that rate, you need an area 10 times the size of the entire German territory to provide even half of the country's energy needs on constant and sustainable basis. And even the, there would be gaps in the generated power.

But Germany has no extra land, or excess money, so reality hit like a loaded truck hitting a brick wall. The brick wall was the extended world financial crisis that slowly extinguished the German solar dreams.

Germany's solar industry went bust with a great bang, dragging the entire world solar industry down with it. The subsidies were cut down, the benefits were reduced and solar became a losing proposition. This woke up most PV cells and modules manufacturers, giving them a chance to stop relying on quantity, but to look into providing options to create successful technological propositions within the PV industry in order to keep the ball rolling and the bills paid.

Actually, the technology aspect of the PV industry

per se was most often on the back seat. Instead, most manufacturers adapted the *brute force* modus operandi, where technology is no longer the driving force. The technology is expected to meet only the very minimum industry requirements to satisfy module power ratings and—very importantly—price.

Increasing production volume and lowering base price were the drivers of that period of wild, unsustainable solar energy progress.

Within this approach, the key (and maybe the only) requirement at the time was credit availability with *market-share at any cost* as the prevailing line of operation in the U.S. and the developed countries. The frenzy was fed by often inflating and pushing the limits of the global markets with strong government, corporate and other types of financial back-up.

Other technology-agnostic methods included domestic market-share protectionism, which is achieved by pursuing trade barrier routes, and/or the reliance on government sponsorship and assisted marketing, in order to create perceived brand superiority.

As the frenzy increased so did the number of questionable companies and products. But who has time for such incidentals. Volume and low price is all that matters.

Case Study: CdTe modules

Good example of the frenzy of the moment and unjustified government sponsorship was the wide deployment of cadmium telluride (CdTe) thin film PV technology in the U.S. CdTe PV panels were in mass production about 5-6 years old at the time, and had absolutely no track record for long-term use in the U.S. deserts.

Regardless, the untested and unproved for large-scale desert use CdTe PV technology was awarded several billion dollars of U.S. DOE loan guarantees, which allowed many millions of CdTe PV modules to be installed in the U.S. deserts.

This expansion was allowed regardless of the fact that this technology is not proven efficient or reliable for long-term desert operation. And now thousands of acres U.S. desert land is covered with PV modules that no one knows what to expect from. But what is the problem?

In addition to low efficiency, CdTe PV modules contain significant amounts (8-10 grams each) of the toxic, carcinogenic, heavy metal cadmium in each module.

Not much, yes, but there are millions of these tox-ic-laden modules on each of the thousands acre desert land in the U.S. So, 8-9 grams multiplied by several million modules means that tons of toxic materials with potentially disastrous behavior are spread over thousands of desert acres. And some are very close to populated centers.

The top and bottom glass-covers of the CdTe modules will survive the climate extremes in the desert during 30 years of non stop broiling by the blistering desert sun. The vulnerable plastic edge seals and encapsulation in the frame-less modules, however, will not. Not even close. They would be lucky to remain intact 5-10 years.

Once the edge seals and lamination are damaged, the fragile thin films inside will undergo major changes too. Heat, moisture and the elements will finish the job of destroying the insides of the modules. So the question is not if, but when, the plastic edge encapsulation will be disintegrated by the merciless UV and IR radiation coming down non-stop all day long, day after day. And how long would the disintegrating system be able to retain the toxic cadmium films from escaping in one or another form in the air, soil, and the water table.

Was this a wise move, or a one-time rush judgment by ignorant and negligent politicians and DOE bureaucrats, remains to be seen. One thing is for sure, though; solving temporarily our energy problems by poisoning the environment and ourselves is not the type of solid technological solution we need in the long run.

Now that the government money has run out, there would be less cadmium spread in the U.S. deserts, but we still need to watch for the results in the thousands of acres covered with cadmium ridden PV modules. This is needed in order to detect, and hopefully reduce, the damages in the years to come.

But back to the world's solar developments! Now, that the dust is settling, we can clearly see the aggressive Chinese invasion, and the short lived expansion of the new and unproven technologies. One of the conclusions we can draw from this experience is that only the superior PV technologies—not the cheapest, or these backed by the German, Chinese or U.S. Government subsidies—will pave the road to efficiency and safety enhancements, and move the PV industry forward in the long run.

Only efficient, cheap, and reliable solar technologies will survive the test of time.

We learned during this time also that technological progress cannot exist in an oversupply environment, which allowed wanna-be Chinese pop-and-mom shops to market untested, below-spec, "low cost" products, giving the appearance as though something new and superior has been achieved on the production lines.

Also, setting up R&D labs and producing limited runs of high-performing solar wafers, cells, and PV modules, and publishing datasheets with amazing test results—as many companies did in the recent past—does not count as progress and will create problems in the long run.

This time we also learned that lowering direct and indirect production costs does not denote technological progress. Cost reduction has nothing to do with squeezing the suppliers on price, and must be kept separate from technology innovation. Product quality, reflected in high efficiency and long-term reliability is the most important factor in assessing PV products, and they cannot be replaced by anything.

Technology roadmaps set by academic or R&D labs, including U.S. DOE and its subjugate National Labs, were typically biased to their respective favorite technologies and the respective funding schemes, and were mostly harmful to normal private technology progress.

Technology roadmaps were used by influential institutions as a vehicle to encourage funding schemes, and we have a number of good examples of PV roadmaps that were created as investor relations soundbites, and intentioned to convince the world that a credible technology path is under control. It suffices to mention Solyndra here…

There were too many ulterior motives influencing PV technology roadmaps this time around, and many times in the past. Fortunately, the PV industry has largely ignored any and all PV technology roadmaps, although the consequences have been a lot of wasted government funds.

A large number of heavily government subsidized thin film PV companies failed during the 2007-2011 solar boom-bust cycle. They were encouraged by interest groups, including the National Labs, into believing that although their technology is inferior and needs a lot of work, it somehow can compete and survive. Not so! Many of them—hundreds—are now history. And many are still struggling with the hope of a successful comeback.

The global PV industry was heavily subsidized during 2007-2011. The trend started with the overly generous European feed-in-tariffs (FITs), which started the ball rolling by propping up great, but unsustainable and temporary, demand. Many companies got in the solar

Figure 10-19. Solar art...its time has come!?

business, and most of them had no idea what they were getting into. Even at the $3-4/W price for PV modules in 2007-2008, any manufacturer could make lots of money, and many got in the business attracted by the easy money it promised.

And when the bust time came in 2010-2011, most of the new comers were not able to adjust to the over-supply and extremely low prices, and so most of them are history now. PV module prices dropped below $1.0/W, which put an end to the prevailing strategy of just adding more and more production capacity in order to make more money. Now, only a fraction of the former technologies and manufacturers would survive in the long run.

This time around we know better. We know that survival in the solar business is based solely on technological advantage…now that all other schemes of easy money and variations of improvised technological progress were tried, were proven wrong and went out the window, never to be seen again! And so we expect to see a new, more robust and more technologically aware PV industry, where technology advantage is the only measure of success.

We consider the time of the 2008-2012 solar boom-bust as transitional for the PV industry. It is a period when the industry survivals no longer rely on the thinking that their own brand of technology is the only thing they should worry and concentrate on.

This is a new era of team work within the framework of the PV industry, where helping others mean increasing your own prospects for survival and success. Hard to swallow concept, but it will prove to be the right path to success for those who understand and accept it.

There are several dozen different next-generation PV technologies (mostly thin film based), which tried to make it but failed during the last boom cycle. Bad timing, or marketing hype…who knows? But regardless, this is not a good time for any of the next-generation to compete on the open market, so they may have to wait until they mature, or for the next boom cycle. Another hard to swallow concept, but if any of them violate it, they will be history too.

Note: The U.S. DOE supported a number of solar and wind projects during the last boom-bust cycle, through its loan guarantee program. The program offered DOE' guarantees for loans from private lenders to energy projects, in which the government promising to step in to cover the loan in the event of default.

The DOE loan guarantee program was actually created by a 2005 law, passed under former US President George Bush. The program was expanded, changed and funded through the 2009 American Recovery and Reinvestment Act ("the stimulus package"), a key piece of legislation for the Obama Administration.

During the last phase of the program, which basically run out of money in 2010, 18 large wind and solar projects (representing 87% of the program) were financed at the tune of several billion dollars. Five large utility-scale solar photovoltaic (PV) and concentrating photovoltaic (CPV) plants supported by the US DOE's loan guarantee program went online in 2012.

The beneficiary projects were; Agua Caliente, Alamosa Solar, Antelope Valley Solar 1, Mesquite Solar and California Valley Solar Ranch plants, which currently represent about 631 MW-DC of new capacity. As the largest of the five, the Agua Caliente PV reached 287 MW-DC, with more capacity scheduled to come online later on. The Alamosa Solar is the largest CPV plant in the world at 30 MW-DC, nearly four times as large as the next-largest plant, and the only CPV plant among the five.

Some of the 18 projects were under construction until recently, including the 370 MW Ivanpah concentrating solar power (CSP) project, which is the largest single-site CSP project in the world when completed. Ivanpah, however, is failing to reach even half of the design capacity, so the project is under scrutiny. This is a bad news for the CSP sector of the solar industry. It seems that CSP is not the best solar technology for use in the U.S. deserts.

The other bad news, affecting the entire solar sector, is that the era of large solar projects is over. Without government subsidies, private investments of billions of dollars for solar farms are just not available. Investors see increasing risks in the solar fields, where most technologies have no completed a full 25-30 year life cycle. So, with a few exceptions, we are entering a new era; that of smaller solar projects.

All solar technologies need to survive one full life-time cycle, of 25-30 years minimum, in order to determine all their inherited risks.

Thus far not one single solar technology has completed a full life-cycle reliability test. As much as we'd like to think that the solar industry has stepped on its feet, and as much as PV equipment suppliers and manufacturers would like to think that their unique brand of technology will be a roadmap driver with global ramifications, we have a news for them. Their technologies are too immature and unproven; not having completed a full life-cycle test period.

This is a confirmation that they all—no exceptions—need to grow up before establishing themselves as reliable technologies. They are also at the mercy of supply/demand and many other global financial and socio-political currents, so that nobody (even governments with their large subsidies) can change that. Capitalism works, and no one can change its ways?

The customers will continue to decide the best technologies based on quality and price, and select the winner products and companies. End-users know enough about solar technologies, but they are not specialists so they usually buy not what is offered on the spec sheets, but what they think they need. Within the solar world, that means specific power in Watts that can be produced for many years, and real dollars spent for each unit. Simple, and so the solar industry continues to grow for now.

These are interesting times we live in. They are also very dynamic from all points of view, so it is hard to predict what will happen later on in this decade.

SOLAR GROWING PAINS

Global solar installations were near 60 GWp in 2015. China alone added over 15 GW, followed by Japan with 10 GW, and the U.S. with 9 GW. These three countries account for about 60% of all installations for the period. Asia-Pacific region accounts for more than half of the installations in 2015, while Europe is crawling way behind. Once the solar power leader of the world, today Europe is stumbling, but is expected to start another upswing soon.

2015 and 2016 are considered to be the best years for the U.S. solar industry, The global solar market earns about $60-70 billion annually, which is expected to more than double to nearly $140 billion by 2020. The annual demand is expected to reach more than 100 GW by that time too, according to the insiders' predictions.

China installed only 15 GW of the government's target of almost 18 GW, set in January 2015. This is due to the fact that the provincial allocations came out in March, so they started signing PPAs in the second quarter of the year. Most of the development in the first quarter of 2015 was overhang from the previous year. Similarly so, 2014 target of 14GW was not met, with only 10.6 GW installed in the country.

Reduction in the coal-fire tariffs due to lower coal prices, as well as long delays in FiT payments might be one of the major obstacles in the Chinese solar installations presently. It is known that FiT payments have been delayed by six to nine months on the average, while recently they have been delayed even longer than previously, for a number of reasons.

Europe accounts for 21% of the global market in 2015, with a grand total of around 12 GW of installed solar power capacity. Europe bottomed out and is expected to begin growing in 2016, with 42 GW installed by 2020, or over 30% of the total global market.

The UK has installed almost 2 GW in the first quarter of 2015, well ahead of the ROC regime's termination. This alone might guaranty a lot of new installations in the UK. Germany, the leader in the field until recently added only slightly over 1.0 GW of new solar power in 2015. The solar market in Germany will continue to grow slowly as in 2014 and 2015, especially in the self-consumption market, which offsets the declines in the other areas. France has a tender process, which guarantees a stable amount of PV installations. Italy has a growing number of solar projects—most without subsidy support—operating successfully under the net-metering program.

The solar equipment supply abnormalities are expected to subside in the near future, as annual growth rates of PV markets stabilize at around 20% annually from 2018 on and as the global solar market expands.

The emerging markets in Latin America, Middle East, Africa, and Southeast Asia are more difficult to predict. They are expected grow fast, but it is anyone's guess how fast is fast in these areas of the world. We already see fast development in Latin America, but a number of unpredictable socio-political and economic factors are at play in these countries, so anything can happen there.

The same is unfortunately true for China and Japan, where things are changing fast too. We would not be surprised much, if we see these countries on the top of the list of global solar installations by 2020, nor would we be surprised if they are at the very bottom. We will

be watching…

The global solar industry, and the related markets, are young, immature, and going through serious growing pains. The situation is dynamic and ever changing, which would require an entire book to describe fully, so here we can only provide a snap-shot of what is going now.

Note: For people who need more information, we have written a book on the matter, titled, *Solar Technologies for the 21st Century.*, published by The Fairmont Press in 2013. In it we review all solar technologies in use today, focusing on these that are most likely to become important in the future.

Trade Wars…Update 2015…

Several rounds of formal complaints against alleged breaches of trade agreements, and the initiation of circumvention investigations in the U.S., Europe, and China, demonstrate that the multidimensional trade conflict in global PV markets is just now starting. The trade dispute is largely focused on the import of downstream products (c-Si wafer, cells and modules) in current and prospective high-volume markets, mostly in the EU, the U.S., and India.

These nations have implemented, or have proposed to implement, anti-dumping and countervailing duties, predominantly targeted against Chinese downstream solar producers. New rounds of investigations might lead to existing tariffs being extended to Taiwanese manufacturers that directly or indirectly import into the U.S., while the EU might scrap a previous quota and minimum price system and revert to tariffs.

More than 70% of the global downstream solar products manufacturing capacity is located in China and Taiwan, so the manufacturers in these regions have no choice but to export. And export of huge quantities all over the world they do. In response, the U.S. Department of Commerce made its final ruling in 2014, in the China solar trade case, levying trade duty rates of up to 165% on some firms.

Trina Solar faces anti-dumping tariffs of 26.71% and anti-subsidy rates of 49.79%. Yingli Solar faces 52.13% rate, and the China wide anti-subsidy of 38.72%. The penalty rate for companies not assigned an individual level of tariff will be 165.04%.

The Taiwanese producers have been given an increase of 19.5% with the exceptions of Motech and Gintech who were assigned 11.45% and 27.55% respectively.

"These remedies come just in time to enable the domestic industry to return to conditions of fair trade. The tariffs and scope set the stage for companies to create new jobs and build or expand factories on US soil," according to the solar industry experts.

Many are unhappy with the decision, of course. "Unfortunately, today's ill-advised and unprecedented decision will harm many and benefit few. We remain steadfast in our opposition because of the adverse impact punitive tariffs will have on the future progress of America's solar energy industry," according to one of the insiders.

The fear is that the ruling might have a negative effect on solar jobs within the US. High level U.S. and Chinese trade officials continue meeting periodically to discuss the issues and find level ground. "We hope to continue to work with China on both the solar and the polysilicon issue to create a stable environment in which, our trade laws are being enforced, secondly that we are able to see the further deployment of clean technology to fight climate change, and thirdly, that we're able to maintain world-class industries in both our countries to manufacture these important products," according to one U.S. government representative.

After all said and done, the trade wars continue. China will not give up easily, because solar is a major part of its economy. The Chinese solar industry is well developed, and fits well within China's future plans for global expansion. Solar also helps the country meet some of its energy and environmental goals, so it is not going away—nor the issues related to it—any time soon.

China Quality Syndrome

While discussing the issues of solar cells and modules, we cannot overlook the fact that millions of PV cells and panels are made in, and exported from, China; much more than from any other country in the world. What we have seen, and know, about the Chinese solar industry in general is that it is not much different than any other industry in that vast country.

Like any other product, solar cells and panels can be made well, or not so well. It all depends on the company, its management, and the production and marketing techniques used. China is famous for its huge labor force, its controversial labor laws and production procedures.

We saw during the solar frenzy of 2007-2012, millions of Chinese made solar panels ordered by U.S. installers enter the U.S. packed in large containers. What was the quality of these panels is anyone's guess. That issue was not even discussed in most cases. Or if it was, there was very little information to go by. Price was the main criteria of most U.S. installers at the time.

The installers didn't have much time, nor did most

of them have the experience needed to perform thorough testing in order to verify the quality of each panel. Instead, some did spot checks using a voltmeter to see that a panel or two in a batch generate some power. Some didn't even bother, or didn't know that they are supposed, to do that.

Weather the panels will last a day, month, or years was not a concern at the time. Installing many panels quickly and cheaply, while the subsidies last, was the goal. In any case, short of controlling the quality at the Chinese production lines, there was nothing they could do. And so, with blind faith in Chinese quality, we installed millions upon millions of solar panels in fields and on house roofs all over the U.S.

Looking back, we see the millions of Chinese made solar panels installed during 2007-2012 in the U.S. as the largest invasion of a product of uncertain and unverified quality ever.

There have been other large-scale quality problems coming from China in the recent past. There were millions of lead tainted paint cans and drinking cups imported and used even at seemingly quality oriented companies like McDonalds and others. We also remember contaminated Chinese baby formula being sold in American stores, making babies sick.

And recently we saw hundreds of thousands of American homes dressed with defective floor laminates and wall panels from China. The formaldehyde emissions from these laminates were dozens of times higher than those allowed in this country. Many people were made sick and forced to remove the defective laminate products.

And mind you; these are products with obvious defects. How many millions of other products have been used without noticing the defects?

The product quality in China is a variable based on cultural phenomena. Most Chinese suppliers believe that what the customer doesn't know can't hurt him. Thus they allow themselves to change product specifications at a whim and without notifying the customers. The Chinese managers believe that it is better to *beg forgiveness later than to ask permission first.*

Profitability is the main goal for most Chinese companies, and so quality is usually seen as a barrier to greater profit making.

Due to the drive to greater profits, quality issues

are not openly discussed, or documented. Many of the Chinese companies that exported solar panels to the U.S. and Europe fall in this category. They had to produce and export large quantities finished product, in order to justify the subsidies and qualify for more.

On top of that, many of these companies were startups, so they could not have been up on all quality control procedures—even if they were available at the time. How did they learn to make solar panels this quickly? What was the quality of the panels they produced during the first months of their operation? How did they check and verify their quality? We will never know the answers to these questions, until someday in the near future the solar panels start failing during field operation.

Sure, all Chinese solar panels imported in the U.S. had some sort of testing and certification stamped on them. But so did the contaminated drinking cups and floor laminates. And yet they failed and even hurt people.

The Chinese manufacturers have learned how to play the game, and would not hesitate to manipulate the system. They would, for example, send a good quality batch for certification, but then feel free to switch materials and/or modify the production procedures as needed to speed up the production. They would then make and ship many solar panels with the certification of the previous models; without concern of quality and with maximized profit as a goal number one.

The problem in the solar cells and modules manufacturing process is that it doesn't take much to introduce defects in solar panels. All it takes is one poor material type, or one careless execution of a key step. Low quality starting materials, mechanical stress in the silicon wafers, contaminated surfaces, contaminants in the diffusion media, low temperature soldering of the electrical contacts, and poor execution at any other step of the process have been known to create a defect, or many defects.

The wicked part of the solar products is that most of the defects introduced during the manufacturing process would not show on the production line. They would usually even pass the final test before shipping to the customer. But then, once in the field, under the long-term effects of extreme temperatures and the elements, many of these defects (what we know as latent defects) would eventually grow and cause serious failures.

And so, we expect to see significant field failures of Chinese made solar modules in the U.S. and abroad during the next 10-15 years.

Recently, international test procedures, tighter certifications, and exposure to public opinion have forced the Chinese solar panels manufacturers to introduce more stringent quality control procedures. This is good, but even then, we cannot be sure of the quality of any products coming from China. The outward behavior can be easily changed, but cultural changes take long time. And these are slow to come in China under its present leadership.

India's Quality

Intersolar, India, 2015 was an uneventful event, except for a revelation of national and global proportions. The executive panel of experts expressed concern that many solar projects built in India have used PV modules of questionable quality. To make bad things even worse, the performance of builders and the execution of many solar projects has been determined to be well below the expectations as well.

Poor quality PV panels and shoddy construction practices prevail in the Indian solar installations.

The executive panel also expressed concern about the extremely low priced bids quoted in the state solar auctions in 2015. This condition may exacerbate the issue of quantity and efficiencies being a priority over the final quality and long-term reliability of the solar power projects. The issues at hand have been noticed by, and are putting off, the international investors. Without these investors, India's planned 100 GW of solar by 2022 target might also be compromised.

The only good news here is that today's quality problems might turn into a good opportunity for some EPCs to get involved in retrofitting and optimizing the solar plants' efficiency in the near future. Such 'fix', however, is not what India's solar industry needs. Fixing problems at a later date is very expensive, and might overburden and even break many solar projects. Instead, the issues must be fixed first, and before any large projects are undertaken. Otherwise, this might result in some owners going bankrupt and selling solar installations cheaply, just to get them off their shoulders.

According to the managing director of an India-based module manufacturer, "There is a total disregard for quality of execution and sustainability of these large projects, which are eventually going onto the grid and possibly making the grid unstable. These are very significant issues which nobody is willing to talk about."

The Intersolar, India panel of experts also concluded that, "People have compromised in evaluating the modules suppliers." Wow! The truth coming directly from the horse's mouth. It appears that while some companies execute a final flash test at the production floor, before shipping to India, other are skipping this essential test. In addition, many developers who are working on solar projects, have used questionable practices for transport and installation of the support structures.

This situation cannot continue for long, because after a period of time the developers will not be able to obtain financing for their projects. This would cause rise of tariffs, and a market correction would be likely to follow. The correction, however, will not be seen in the next six months as there are still some companies looking to secure market position by bidding at very low tariffs.

One example is SunEdison's recent bid on a 500 MW of solar in Andhra Pradesh. It was at a ridiculously low tariff of less than 7 cents per kWh. This, and other such projects were a major talking point at Intersolar India, 2015 as the solar industry is trying to justify a 20% drop in tariffs in 2015.

The conclusion is that if everything continues as is, a major correction will take place within the next 18-24 months. This is unavoidable, because thus far the solar projects' execution risk in India is huge. In many cases, the developers also face challenges with the local society, land acquisition, right of way, and obtaining transmission permits.

The hope of the market's self-correcting mechanism to take over is fading fast, because there are many bidders looking to secure their market share in India's solar market. So, it might be a long time before normalcy prevails India's solar fields. For now, many international players without appropriate experience are winning the low bids and are then failing to deliver projects. This is hurting the projects and all developers. The government can make it difficult for such companies or speculators to enter in India's solar sector, but that will take a long time too.

All in all, the conclusion at Intersolar India, 2015 is that the developments of late are harming India's solar program. It needs to mature, and be self-sufficient, which would require full support of the domestic solar manufacturing sector. However the current tariff-based bidding does not favor the domestic solar program since the winning PV developers are not using local products and services.

What this means is that India's solar industry is going through growing pains. It is in the midst of a trial-and-error phase, which will have to transition quickly—within the next 24 months—into a mature and self sustaining industrial complex. Else, India will be

dealing with a lot of problems in their solar installations for a long time to come.

Australia

Australia's energy ministry ordered an inquiry into the quality of solar system installations in 2014, in response to an investigation by Fairfax Media, which claimed that PV systems have been failing after just one or two years of operation. The Clean Energy Council (CEC) has been ordered to open an investigation into the matter ASAP. The CEC is responsible for accreditation of PV systems installers, which gives them eligibility to the government subsidies and support.

The solar industry is unhappy, of course, and has launched a personal attack on key government officials and agencies. But at the same time, a survey by consumer watchdogs has found high incidences of consumer dissatisfaction with the quality of their solar systems. In January 2014, for example, it found that 32% PV owners had an issue with their installer, and 25% had experienced problems with the systems. 10% of people in the survey claim that they had had to replace the inverters of their new systems.

These figures are far higher than those found by the government's own Clean Energy Regulator, which carried similar surveys in 2013. It found that about 3.3% of almost 4,000 tested systems were unsafe and 9.7% were substandard.

Over 13% of all inspected solar projects in Australia are unsafe or substandard.

All things considered, there is a problem that needs to be addressed. And so, the country has been embroiled in a bitter debate on renewable energy since the new prime minister Tony Abbott came to power. Funding through the Australian Renewable Energy Agency was curtailed and so the country's Renewable Energy Target faces an uncertain future.

The new open investigation is quite unique, because although we are aware of many other failures in the U.S., Germany, Italy, Span, and other country, official investigations usually do not deal with the quality of products and services. Or if they do, their focus is so narrow that it is difficult to assign numbers as in the investigation above.

In the U.S., neither the government, nor the members of the solar industry want to know about the problems customers have. At least this is how it has been until now. Recently, some local utilities have been taking a closer look at the quality of solar products and services, and this promises to become a major trend, as it should have been since day one.

But since there is not much use in crying over spilled milk, we will cut our losses, lick our wounds, and keep on keeping down the solar path; much better informed and watchful of the problems on the way.

Japan

In 2014, Kyushu Electric Company (KEC), the regional utility for the southern region of Kyushu, started suspension of applications for solar installations. Small islands of land on the KEC territory started showing signs of overcapacity, where that adding more solar was impossible for the local grid infrastructure to handle. Other regional utility companies were also affected in a similar way, and followed KEC's example of slowing down the solar installations growth on their territory.

This led the Japanese government to establish a working group focused on grid connection problems. The group reviewed the capacity issues at the affected utilities and calculated that the slightly over 17.0 GW of solar projects approved to date would go over the available capacity limits of the affected utilities. But the group also found out that total of over 50.0 GW of capacity remained across the affected regions.

As a result, new rules on output from, and grid connection to, solar farms were implemented in the winter of 2015, in an attempt to resolve a situation that forced half of Japan's electric utility companies to stop accepting applications for new solar projects in 2014. The affected utilities were cleared to resume considering applications for solar projects. It is, however, still up to the individual utilities to make decisions on whether or not to grant grid connection to individual projects.

At the same time, the government launched another investigation in the country's feed-in tariff (FiT). The object of this investigation is to help further in resolving the grid connection issues and reduce the large number of complaints about rejected applications. The investigation also addressed the cost of renewable energy, which (both the renewable energy installations and their cost) have been growing in importance.

Although the new rule changes are quite strict and introduce a level of uncertainties, there is one positive aspect. The rule changes do not apply retrospectively on solar power plants granted FiT approval before the ruling. Also, the effort confirmed that there is enough spare capacity on the grid for adding more power generation by all main electric companies.

The biggest problem thus far, that cannot be easily solved, is that land is scarce and remains at a premium in Japan. Developers are developing new solutions to the issue, such as floating PV plants on reservoirs, and / or reclaiming landfill sites and old golf courses, but even then the land is severely limited on the island.

The new rules in effect enable utilities to exercise more controls on the output from solar farms. They can stop accepting power generated from solar farms at any point, when supply exceeds demand. They can also restrict power output, or deny payment for up to 30 days of power production annually. So, one certain risk for the solar power plants owners is that 1/12 of their annual power production could go without compensation.

Here go all the profits... Fortunately, this new rule does not apply to existing smaller, residential solar power systems, below 10 kW generation capacity. All new projects, however, will be affected by it.

In addition, Japan's Agency for New and Renewable Energy plans to set new targets for renewable energy generation as a proportion of the national energy mix in 2016. This is a good news for the solar industry, because thus far there has been lack of defined targets for renewables, which has been a cause of many of the recent problems in Japan's solar industry.

Grid connection issues are present in most other countries, including the U.S., although the problem here is quite different. Unlike space-confined, crowded, and cloudy Japan, the U.S. has vast stretches of land in the sunny South West deserts that are perfect for almost unlimited amount of solar installations.

The problem is that the huge desert areas usually lack grid connection points, so most of the solar installations are in the areas close to the grid. In order to develop solar power generation in the entire desert area, the U.S. has to spend a lot of money on new transmission lines and substations. This will happen, no doubt, but slowly.

Spain

Spain is one of the sunniest European countries, and because of that and with government's help Spain had most advances in the development of solar energy. Royal Decree 436/2004, issued March 2004, leveled the field for large-scale solar thermal and photovoltaic plants and guaranteed feed-in tariffs. It simply removed economic barriers to the development and connection of renewable energy technologies to the electricity grid and the race was on.

In 2008, Spanish government set another ambitious goal of achieving a 12% of primary energy from renewable energy by 2010, and 10 GW by 2020. Spain was the fourth largest manufacturer of solar power technology in the world at the time. It was exporting 80% of this output to Germany and other countries, and set a world record in 2008 by adding 2.6 GW of solar power, for a total solar capacity to 3.5 GW.

Solar power in Spain increased slightly to 3.9 GW by the end of 2010 and 4.5 GW by 2013. Spain was the world's leader in concentrated solar power (CSP), with over 2.0 GW by 2013.

Then came the 'big one' in 2008 and things started slowing down. In the wake of the 2008 financial crisis, the Spanish government drastically cut its subsidies for solar power and capped future increases in capacity at 500 MW per year, which effected the industry entire global solar industry. A series of corruption and cheating scandals rocked the industry since 2010 and Spain's solar industry went into a free fall.

In 2014, Spain installed only 22 MW of PV power, and that much in 2015. A weak reflection of the gigawatts of solar power installed just a few years earlier.

As if that was not enough, in the summer of 2015 the Spanish government added insult to the injury. It proposed a new "sun tax" on use of batteries for residential self-consumption of solar energy in the country. Only owners of off-grid PV systems will be allowed to install battery backup without paying the new tax. The new tax applies to grid-connected solar PV installations of up to 15 kW. Depending on the size of the installation, the new tax will be between $10 per kW for domestic consumers, and up to $41 per kW for medium size businesses. Obviously, the Spanish government sees the addition of battery storage to any solar installation as its enemy.

This is like beating a dead horse. A significant increase, which in effect doubles the estimated payback time of a solar installation from around 15 years to over 30 years. Fines for infringement, capped at about $68 million, are part of the new change.

This is the only self-consumption law in the world, which seems to be created only to slow down self-consumption. The new law also puts a cap on installations at 100kW, where owner and consumer must be the same person.

Projects already approved are also affected, and those not fitting within the new rulings will be considered a "very serious" infringement, with fines of up to $68 million. This is an astronomic number, since it is almost twice the fine for releasing nuclear waste.

UNEF and others call the new tax "discriminatory" as no other energy generation or energy efficiency measure has to pay such tax. Cogeneration is free of the tax until 2020, while, nuclear, coal and gas, which self-consume at least 8% of their energy, are not subject to the tax at all.

According to UNEF, if the government applied the same "sun tax" to traditional energy plants, it could gain about $260 million. Applying the tax to cogeneration would bring an additional $113 million, while the 100 MW of PV self-consumption installed in Spain currently would bring only about $7 million from the tax.

Generally speaking, most people feel that the terms and policies of the Spanish government during the last several years have been anti-renewables. It appears as if there is an ideological campaign against solar, triggered by the utilities, who have a dominant position in the market.

All in all the new laws would have a very negative impact on the Spanish solar industry. The new rules are on top of the moratorium against new utility-scale PV in the last three years and strong anti-self-consumption rhetoric from the government in the last two years.

SOLAR'S FUTURE

Above we reviewed some of the key problems observed in the solar sector of late. They are many and some are very serious. But remember that the modern solar industry is still very young and immature. It is growing fast and so are the growing pains. Nevertheless, PV technologies are shaping as major energy generators, and will provide some, if not most, of the electric power of the future generations.

In our opinion, the present and near-term future of the solar energy sector and the related markets is as follow:

• Concentrated solar power (CSP) technologies were the darling of the solar industry during the last decade. It became clear soon after, however, that there are problems with the CSP installations that cannot be easily overcome. For example; CSP power plants have to be very large—over 100 MW—in order to make economic sense. They also use steam turbines, which are expensive to install, operate, and maintain, and the conversion efficiency of which is quite low. But most importantly, the turbines require a lot of water for cooling. Since CSP power plants are usually installed in the deserts, where water is a premium, the future CSP power plants development is questionable.

   This makes the photovoltaic (PV) solar power generation the preferred technology of the solar industry.

- *PV is an ideal energy source for our changing needs.* There are several reasons why PV solar power is such an attractive energy source. To begin with, it is clean for the most part of its operational cycle. Solar panels are typically guaranteed to perform for as long as 25 years, produce electricity without pollution or emissions that can contribute to global warming.

  The process of creating solar panels is also clean, compared with the fossil fuels, but a lot of energy is used in it, which we discuss in this text. Nevertheless, PV modules have the lowest carbon footprint, and use the lowest amount of energy during production.

- *There is a lot of solar energy available around the globe,* which helps to make PV power generation a prime choice within the global mix of energy sources. Solar panels require nothing more than sunlight for the electricity generation process. This means that PV solar technology can provide energy security for virtually every nation in the world, helping to reduce the tension and uncertainty that carbon-based energy sources often create.

- *PV is location specific, but solar energy still can be generated virtually anywhere,* even right in the heart of densely populated urban centers, can help reduce the load on long-distance electricity transmission networks. Here PV solar is a significant improvement from both carbon and hydro power generation sources, which are often located far from where they are consumed.

- PV modules generate electricity during peak summer energy demand hours in most global markets This makes PV well-suited to replace carbon-based sources of peak energy demand, such as gas-fired turbines, which have a much larger impact on emissions.

- *PV is scheduled to provide over 10% of EU and U.S. energy demand by 2020.* Current developments in Europe and the U.S. underline both the strength and growth potential of PV solar energy. The European Union (EU) has already set a goal of meeting 20% of energy demand through the use of renewable sources by 2020. This might be too optimistic, so taking everything under consideration the European Photovoltaic Industry Association (EPIA) estimates PV power to meet as much as 10-12% of EU electricity demand by 2020.

- Although solar energy makes up just 1% of total installed electricity generation capacity in the EU, it accounted for 10% of the newly installed capacity for 2008. At such growth rates, which are lower than the astronomic growth that the industry experienced over the past decade, the 10-12% goal of solar power in Europe by 2020 is a realistic one.

- The situation is similar in the U.S. too. Several states have mandate to add 20-30% renewable energies to their energy mix by 2020, so the share of PV installations is estimated to be in the 10% of the total by then.

- *PV has room to grow,* since the global solar industry has enough capacity to deliver enough PV solar products to make a difference in the energy mix in the short to medium term. To put this in perspective, in 2011 the global PV solar energy industry made enough solar modules to satisfy the entire electricity demand of California, which is one of the world's largest energy markets. With ample production capacity and an ideal clean energy product, the PV solar industry is more than capable of meeting even the most aggressive growth scenarios, both within Europe and globally.

Generally speaking, the future looks very bright for PV solar power generation.

Table 10-10 shows that the total U.S. PV market in 2050 is expected to grow to 135 GW, while the available capacity is almost 25 times that much. With other words, everything considered and in best of cases, the U.S. has the capability to install and operate over 3,000 GW of solar power plants by 2050. Another pipedream, true, but totally feasible…under certain circumstances.

This option will be considered some day, no doubt, but the reality today is different.

Table 10-10.
U.S. PV markets; present vs. total capacities (in GW)

Markets	Present / Capacity				
	2010	2020	2030	2040	2050
Residential	4 / 18	20 / 120	50 / 450	55 / 950	55 / 1400
Commercial	1 / 3	5 / 25	15 / 100	20 / 250	25 / 400
Utility	2 / 6	10 / 50	30 / 250	40 / 550	50 / 1000
Off-Grid	1 / 2	5 / 25	15 / 110	20 / 275	25 / 460
Totals	8 / 29	40 / 220	110 / 910	135 / 2025	135 / 3260

Figure 10-20 shows the present and near-future status of the U.S. energy markets, with crude oil soon to disappear, simply because we are using so much of it. It will be then followed by natural gas, for the same reasons. Coal will keep the energy markets going for awhile longer, but eventually it will be exhausted too.

Figure 10-20. The near-future energy markets' status

When all fossils are gone, the energy bridge to the future will be seriously damaged. Even possibly beyond repair. The renewables to the rescue...

As we mentioned above, concentrated solar power (CSP) is having problems, which make its future questionable. PV solar technologies are moving in to fill the gaps left by their CSP cousin, which is causing frictions in the solar industry. Since this is very important subject, we would like to clarify it by expanding on the issues at hand.

PV vs. CSP Wars

We have been witnessing official and unofficial wars inside the solar industry. The fights are usually between companies, and entire countries. But recently, we are witnessing wars among the different technologies. c-Si vs. CdTe thin film PV modules. a-Si vs. SIGS PV modules, etc.

The most intensive one, and that with the greatest impact on the future of the solar industry, is the war between the PV technologies and their thermal (CSP) cousins. All these technologies have advantages and disadvantages, so the war is getting quite fierce with time.

Due to the nature of things, however, PV modules seem to be winning for now. Their major advantages are that they:

a) are modular and can be deployed in very small or very large quantities,

b) do not have any moving parts,

c) do not need any outside materials input (like cooling water), and

d) are virtually maintenance free, except for washing the solar panels periodically and after some major storms.

Concentrated solar power (CSP) has one major advantage, which is its ability to store energy for later use, thus limiting its variability issues. On the other hand, the CSP power plants are usually very large, and cannot be downscaled for residential solar power use. Thus they require very large capital investment, which is not easy to come by. They also have a lot of moving parts, which complicates their operation and maintenance, and reduces the profits.

But most importantly, CSP power plants use steam-water cycles and wet cooling (of the steam), which processes use about 6 acre feet of water per each MW annually.

CSP power plants use 6 acre feet of water per each MW annually

This means that a 1 GW CSP power plant would need 6,000 acre feet of water annually. This is an entire lake of about 2 billion (yes with capital B) gallons of water. While this may not be a lot for a rain soaked tropical lands, taking 2 billion gallons every year from the desert floor is no joke. And it is not sustainable—especially in today's increasing draught conditions.

This is simply unsustainable, since there is not enough water in the U.S. deserts to support large-scale CSP development.

A significant portion of the cooling water evaporates during the (evaporative) cooling process, when steam is condensed. This process also requires large quantity of make-up and blow down water, which in turn requires Zero discharge systems. The condensate, boiler feed water also needs chemical treatment, while make up water needs de-mineralization.

The boiler needs to be operated and maintained by highly qualified and certified boiler operators 24/7 in order to avoid accidents and ensure efficient operation...all of which means money and more money spent on equipment and procedures.

Dry cooling—similar to air conditioning—can be used instead of water cooling, but it does not totally eliminate cooling water use and is also very expensive to install and maintain. Dry cooling uses complex and expensive equipment, adds operating costs and impacts performance of the Rankine cycle.

A new development, a Cascading Closed Loop Organic Rankine cycle (CCLC), patented by an U.S. company, totally eliminates water use. It can be operated remotely, un-attended, and can be built as small as 5.0 MWe up to 25 MWe, or multiples thereof. Smaller systems can be located closer to the actual load centers, which eliminates transmission losses. This is one, albeit unproven for long-term desert operation, alternative to saving water in the deserts while generating solar power, and leaving more water for use by our thirsty planet.

The efficiency, reliability, and initial and operating cost of such systems remains to be proven in large-scale, long-term operations, but at the very least they open a new door for the fledgling CSP technology. Without new, and similar, approaches the CSP's future in the U.S. deserts is less than promising...even when their energy storage capability, and all other benefits, are considered.

Case Study: Ivanpah CSP Plant

Solar power plants can underperform and even fail for a number of reasons. Poor design, materials, installation, and operation are the major reasons. The interrelation among these is complex and not easy to understand, but as time goes on, we get more insights of the technological issues at hand.

We looked at some problems with PV solar installations above, so now we will look at one of the examples of a competing technology—concentrated solar power (CSP)— and the Ivanpah Solar Electric Generating System in California in particular.

This "power tower" solar plant, rated at 392 MW power generation capacity, owned by NRG Energy, Google and BrightSource Energy, is the largest in the world,. It was built in the Mojave Desert, at a cost of $2.2 billion, with the help of a $1.6 billion federal loan guarantee, and started commercial operation at the end of 2013.

The tower power technology has been in existence in the U.S. since the early 1980s, and a number of such plants are in operation in the U.S. and abroad.

Even then, the Ivanpah's three units generated about 450,000 MWh of electricity during 2014. That's less than one-half of the annual over 1 million MWh power generation, the original plant design called for. The reasons given were bad weather and equipment problems. Excess dust collected on the mirrors, probably

due to poor location selection, and/or design, was also determined to be a major issue

Because of the reduced power production, the owners are now forced to use much more natural gas—about 60% more—to run their steam turbines, in order to meet their power generation obligations. The steam boilers also use a lot of water, which is quite scarce in the deserts, so this problem will grow in importance as time goes on.

In addition, or maybe because of the above problems, the owners are having difficulties meeting their financial obligations, so loan extensions and other financial instruments have been used recently to keep the plant financially viable.

But with 50% loss of solar power production, and 60% more natural gas use, it would be hard to operate the mega-facility economically, so other problems might appear in the near future as a result too. Ivanpah, however, is paid mostly by taxpayers money, so some of the capitalist laws do not apply here.

The U.S. government is expected to prop-up the facility until it becomes too embarrassing and/or prohibitively expensive to do so.

And speaking of embarrassing problems, in addition to the gross underperformance, the Ivanpah power plant, with its 175,000 mirrors blasting concentrated sunlight high into the sky, is becoming a symbol of mass murder of birds. Workers at the plant, have come up with a name for this new phenomena, occurring when different birds fly across the plant's concentrated sun rays.

Many local birds, unaware of the dangers ahead, ignite in midair and like WWII airplanes shot down in an air fight, fall to the ground with a spectacular smoke trail left behind. "Streamers" the workers call these unfortunate feathered friends, for the smoke plume they leave behind.

Federal wildlife investigators, visiting the Ivanpah power plant watched as birds burned and fell, reporting an average of one "streamer" every two minutes. This comes to about 300 birds "streaming" under the high beam reflectors at Ivanpah every day. This means that the Ivanpah is capable of killing over 100,000 birds annually, although the count thus far is about 20,000 bird kills. Because of this and other reasons, California's officials have been urged to reject the application to build a still-bigger version of the bird killing plant.

Would the innocent bird genocide be enough to curtail and even stop the development of CSP technology in the U.S. and abroad? We don't think that 300 birds a day would be able to convince the regulators to stop CSP projects development. But that, combined with the other issues—such as underperformance, and excess use of natural gas and water—might. We will be watching.

The Future of Silicon

Silicon has been and will remain (at least for the foreseeable future) a key material for the fabrication of solar (PV) cells and modules. A number of materials with similar PV properties are competing, but none comes close to the efficiency, reliability, and price of silicon.

And so, the best approach, short of competing with silicon, is to improve it. This is exactly what a number of R&D teams are doing. There are a number of developments in this area, the most promising of which is the new Si24 allotrope (modified silicon), which has an open-framework structure comprised of 5-, 6- and 8-membered sp3-bonded silicon rings. Small atoms like sodium and lithium, as well as potentially entire molecules like water, can readily diffuse through the channels in the silicon substrate.

This 3rd dimension, so to speak, creates silicon with different and superior performance, and diverse applications. The potential applications, in addition to efficiency improvements of silicon PV cells, are in the areas of electrical energy storage and molecular-scale filtering. Energy storage is of interest, with most experts rating it as a "savior" of the solar and wind technologies of the future. This new material could be one of the keys to efficient and affordable energy storage…or not…it all depends on future research findings.

Note that in Figure 10-21, over 2/3 of the R&D is focused on silicon solar power generation technology. The total R&D expenditures, however, represent only 1-2 % of total product/project sales of the companies involved.

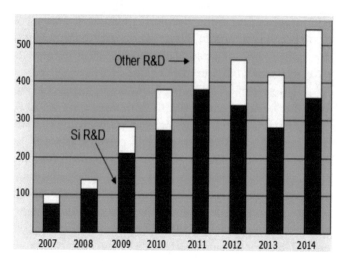

Figure 10-21.
R&D spending in the solar sector (in $ million)

Silicon Cells Progress

The efficiency of the best commercial c-Si modules now exceeds 21% and manufacturers such as US-based SunPower are targeting 23% efficiency, together with significant cost reductions. The major cost in c-Si PV cells is for pure poly-silicon feedstock, which dropped from $67 per kilogram in 2010 to $20/kg in 2012, and has remained below this price since then.

Continued progress in the production processes, and reduction in the use of consumables, will keep the price under $20/kg in the next few years. Cost of ingot growth, wafer (cell precursors) sawing and cleaning will also improve. Efforts to reduce the amount of purified silicon in cells—which is now as low as 5 grams per watt for the best cells—will continue towards 3 g/W or less, with thinner wafers.

Diamond wire sawing and improved slurry-based sawing will reduce losses in slicing c-Si wafers, while kerf-less technologies may or may not offer an alternative to the traditional wafer-based c-Si process.

Manufacturers are also striving to use less silver and other expensive materials (maybe replacing silver with copper), while maintaining or even extending the technical life of cells and modules.

Manufacturing automation is progressing for both cells and modules. For modules, higher throughput could be achieved for the interconnection and encapsulation processes. Energy efficiency improvements over the whole manufacturing process are being sought.

"Mono-like" mc-Si ingots, and reusable ingot moulds, could bring sc-Si performances at mc-Si costs. Back contact and metal wrap-through technologies, which reduce shading and electric losses, have been

Table 10-11. R&D expenditures, 2014

First Solar	144 million
Yingli Green	93 "
SunPower	73 "
RenaSola	53 "
SolarWorld	33 "
JA Solar	23 "
Trina Solar	22 "
REC Solar	18 "
Jinko Solar	17 "
Hanwa Solar One	14 "
Canadian Solar	12 "
Suntech	12"

successfully introduced to markets by various manufacturers. The historical learning rate of 20% could be maintained over the next few years by introducing new double- and single-sided contact cell concepts with improved Si-wafers, as well as improved cell front and rear sides and better module technologies.

The hetero-junction (HTJ) cell design combines two materials—often c-Si wafer and a-Si TF—into one single junction, resulting in higher efficiencies and performance ratios, thanks to a better resistance to high temperatures. The leader in HTJ technology, Sanyo/Panasonic, now develops HTJ cells with back contacts, and has announced a record efficiency of 25.6% in 2014 on a research cell of "practical size" of over 100 cm^2.

Bifacial solar cells offer another emerging option, to be used in glass/glass modules enabling an increase in performance ratio and energy output of up to 15% using the light reflected by the ground or buildings through the rear face. HJT technology, possibly combined with back-contact designs, may be further improved by using alternatives to amorphous silicon, in order to increase the overall spectral response.

The Future

Although silicon is the most common material in today's PV technology, its indirect band gap semiconducting properties limits its efficiency and reliability. This prevents it from being considered for next-generation, high-efficiency applications such as light-emitting diodes, higher-performance transistors and photovoltaic cells.

Metallic substances conduct electrical current easily, whereas insulating (non-metallic) materials conduct no current at all. Semiconducting materials exhibit mid-range electrical conductivity. When semiconducting materials are subjected to an input of a specific energy, bound electrons can move to higher-energy, conducting states. The specific energy required to make this jump to the conducting state is defined as the "band gap." While direct band gap materials can effectively absorb and emit light, indirect band gap materials, like diamond-structured silicon, cannot.

In order for silicon to be more attractive for use in new technology, its indirect band gap needed to be altered. New research shows new forms of silicon, one of which with a quasi-direct band gap that falls within the desired range for solar absorption, something that has never before been achieved.

Thus created silicon is a of the so-called *allotrope* type, which means a different physical form of the same element. This is similar to the way that diamonds and graphite are both different forms of pure carbon molecules.

As mentioned above, unlike the conventional diamond structure, this new silicon allotrope consists of an interesting open framework, called a zeolite-type structure, which is comprised of channels with five-, six- and eight-membered silicon rings.

The new structure is created by using a novel high-pressure precursor process. The first step in this process is the formation of a compound of silicon and sodium, Na_4Si_{24}, under high-pressure conditions. This compound is then recovered to ambient pressure, and the sodium is completely removed by heating under vacuum. The resulting pure silicon allotrope, Si_{24}, has the ideal band gap for solar energy conversion technology, and can absorb, and potentially emit, light far more effectively than conventional diamond-structured silicon. Si_{24} is stable at ambient pressure to at least 842 degrees Fahrenheit.

This type of high-pressure precursor synthesis represents an entirely new frontier in novel energy materials, which promises access to new novel structures with real potential to solve standing materials challenges. These are previously unknown properties for silicon, but by using new methodology entirely different classes of materials can be produced. These new structures remain stable at atmospheric pressure, so larger-volume production is entirely possible.

The future of the new allotrope silicon material is unclear, but it opens a door that was closed until now. What is behind it will be shown by the 21st century scientists in the near future.

n-type or p-type?

In the beginning, and until recently, there were only p-type mono-crystalline solar cells and PV modules. Later on n-doped solar cells were introduced into some PV modules. Then, poly-crystalline silicon solar cells and PV modules, mostly p-doped, took the market by storm. They were cheaper, but less efficient, and since price was a major factor in purchasing decisions making, millions of these modules were sold and installed all over the world.

Much later thin film, organic, and many other solar cells and PV modules were designed and manufactured. They vary in performance and reliability, so their future is not clear as yet. It is very likely, however, that the different types will find different niche markets.

But now the frenzy is fizzling away and the final tally shows that p-type multi c-Si technologies are the winners. They are expected to dominate the solar industry at least until 2020, unless some breakthrough takes over their place.

A number of studies during 2014-2015 show that a dramatic shift to high-efficiency poly-Si, p-doped solar cells and PV modules has been going on since 2012. At the same time, the mono-Si and thin film technologies are falling behind by a great margin.

There is a drive underway in China and Southeast Asia to improve the performance of the cheaper poly-Si solar cells and modules. It seems that these types of cells are more flexible as far as reducing their prices and increasing efficiency are concerned. This is in stark contrast to the previous estimates and predictions of academic and "expert" PV technology roadmaps.

Some manufacturers are stuck with what they already have. For example, SunPower and Panasonic have to use their highest quality, and most expensive, n-type Si ingots to make p-type cells, so switching would be expensive. At the same time, Yingli's 'PANDA' can make n-type cells without the need for high quality and expensive p-type silicon. Luckily, most solar cells and PV manufacturers can switch from mono to poly types and back at will. They only need to secure enough p-type wafer supplies.

Some Asian solar cells manufacturers switched from poly-Si to mono-Si recently, but are finding it difficult to compete, so it is questionable if they can continue this trend very long. Since the Japanese market is one of their market targets, and it is increasingly driven by global price drivers, they must adjust to the new trend and reduce prices, or switch to the cheaper poly-Si technologies.

The p-type poly-Si is seeing increasingly strong market share gains, with the HE poly-Si PV module shipments predominating during the last two years. One of the reasons is improvements in poly-Si ingot growth, resulting in increased efficiency and reduced prices.

Several major PV module suppliers, such as Yingli Green Energy and Trina Solar, have chosen to manufacture exclusively p-type poly-Si. At the same time, poly-Si ingots and wafer suppliers, like GCL, are spending a lot of time and money on developing better and cheaper ingots and wafers to their key Asian customers.

So, on one hand, we see Yingli, Trina and GCL setting a trend towards high efficiency p-type poly-Si technology. On the other hand, we see the poly-Si solar cells and modules potential increased by REC Solar and Hanwha Q-Cells as they ramp production in their Southeast Asia facilities. These two companies are the most experienced in the solar silicon area, and are staunch p-type

poly—Si cell proponents.

Their entire production volume consists of p-type poly-Si solar cells and modules. In addition, they are planning expansion of the p-type poly-Si production lines, featuring higher module power output, mostly as a result of improvements in their poly-Si ingots manufacturing process in the near past.

A number of companies are also looking into adding PERC type solar cells to their line of products, but that might take more time than estimated, since there are some other easier and cheaper improvements in the pipeline. And yet, we see p-doped, n-type poly-Si as the technology of the near term future of the PV industry. At least until a new, more efficient and cheaper process replaces it.

Solar Cells Efficiency

The future belongs to the highest efficiency (and cheap enough) technologies, no doubt. High concentration PV (HCPV), using solar concentration of 500x and more, for example, is approaching 50% efficiency—over 2 times higher than c-Si technologies and over 3 times higher than the thin film ones.

HCPV solar cells are expected to approach 50% efficiency by 2020, and 60% by 2030.

This technology, unfortunately, it is not finding wide application for now. This is due mostly to the fact that there are complex assemblies and moving parts in the HCPV modules and tracker assemblies. Time will come in the distant future, however, when the HCPV-based technologies will be predominant in solar power generation around the world.

The time for switching to the more complex HCPV technology, however, is not today or tomorrow. It will take some time and a lot of money to bridge the technological and economic gaps.

For now, there are other methods and approaches, with solar silicon technologies paving the way to the future. Trina Solar, for example, is producing n-and p-type mono-crystalline and p-type multi-crystalline solar cells with new record conversion efficiencies. For now this work is done exclusively in the R&D labs, but as the processes improve and become more cost-effective, the new PV technologies will eventually become part of the future production roadmap.

The major candidate for the highest efficiency 1x (non-concentrating) solar cell in the near future is the PERC-IBC solar cell.

Best Research-Cell Efficiencies

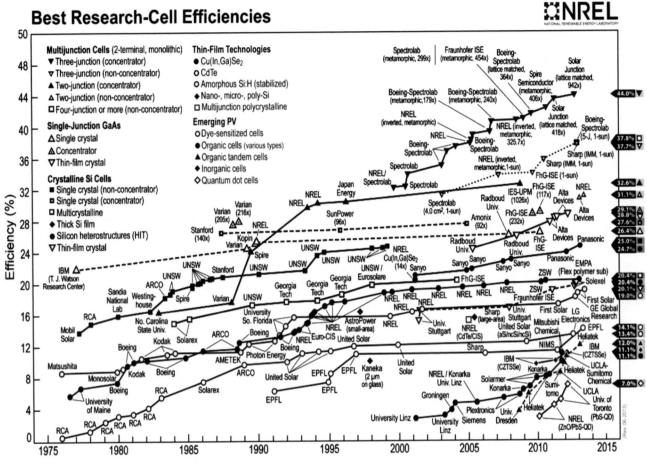

Figure 10-22. Solar cells efficiencies according to NREL.

This is a n-type passivated emitter rear contact (PERC) mono-crystalline solar cell, with Inter-digitated back contact (IBC) structure. Manufactured via proprietary process by companies like Trina Solar and others, this new PV cell achieved a record conversion efficiency of 22.9% on a 156×156 mm2 n-type Cz wafer in 2014. Recently another record was set with similar cells at 24.4%. These independently verified cell efficiency records are the highest so far reported on industrially feasible production processes using the standard (156 x 156 mm) wafer size.

"We are very pleased to announce these new efficiency results achieved by Trina Solar's researchers at the State Key Laboratory of PV Science and Technology. Though these technologies are not currently in production, they will be part of our future commercialized Honey Plus and IBC products. Our aim is to continuously integrate technological developments into our PV products, which are currently commercialized or would be in the future, to further drive down the product cost, strengthen our competitive advantages and provide affordable solar power to the world," according to the

company's chief scientist.

The key here is the fact that some additional, non-standard, equipment and special (quite sophisticated and expensive) process steps were added to enhance the efficiency of the final product. Note also that these cells are made of Cz wafer material. This is also very special and expensive material, manufactured via special equipment and processes. The Cz silicon material has some special properties that are superior to those of the commonly used mono- or poly-silicon (which most standard PV cells are made out of today). The Cz material has also some limitations; mostly the complexity of the manufacturing process and the cost of the finished product.

And so, we are slowly moving from the standard 14-16% efficiency of PV cells to 20-23% efficient such. It might take several years for the PERC-IBC manufacturing processes, using Cz wafers, to get optimized, ramped up, and made cost effective in mass production, but the work is underway and we have no doubt that it will happen sooner or later. This is actually a huge jump of about 30% efficiency increase, which would fill, albeit partially, the gap to the highest efficiency solar technology today; the HCPV tracking technology

The Technology Map

In 2010, the International Energy Agency (IEA) issued a Technology Roadmap for the global solar photovoltaic energy sector, see Table 10-12. IEA has not updated this Roadmap since then, so we feel compelled to point some inadequacies in it and clear some major points of misunderstanding.

The IEA Roadmap in Table 10-12, especially the estimates for 2030 and 2050, are overly-optimistic at best, and misleading, or totally wrong at worst.

This is why:

A. Efficiency. The IEA Roadmap claims that "typical flat-plate module efficiencies" would reach 40% by 2050. Why is this impossible you ask? See below:

- Typical flat-plate module efficiencies of 40% is impossible, because the Shockley-Queisser limit (the *maximum possible* theoretical efficiency of a solar cell) places maximum solar conversion efficiency of any material known to man at around 33.7% assuming a single p-n junction with a band gap of 1.34 eV and AM 1.5 solar spectrum.

- The most popular solar cell material, silicon, has even lower band gap of 1.1 eV, which allows a maximum possible theoretical efficiency of about 28-29%. Theoretical efficiency could be achieved on paper, and in some isolated lab tests, but would never be possible in real-life mass production

- The most efficient solar cells, mono-crystalline silicon solar cells, produce about 22-24% conversion efficiency today, or 10-20% less than the maximum possible, due mostly to losses from reflection off the front surface, light blockage from the wires on

its surface, and other internal resistance and external imperfection factors.

- For these reasons, and with optimized cells structure and manufacturing process, mass produced silicon solar cells could reach 25-26% at the very maximum. But 40% efficiency is absolute impossibility.

- But 25-26% is the efficiency of the solar cell only. When encapsulated in a solar module, the efficiency drops by another 10-20%, so in best of cases a "typical flat-plate" solar module cannot possibly reach even 25% by 2050—let alone the 40% efficiency predicted by IEA.

- Using exotic materials and/or multi-junction cells could increase the efficiency some, but that will no longer be a "typical flat-plate" module, since these are complex and expensive technologies. And there are no materials known to man at this point that can reach this efficiency. So unless IEA knows something we don't, the 2050 estimate needs a serious reality check and justification.

B. Pay-back. The IEA Roadmap estimates the "typical system energy pay-back time (in years)" at 0.5 years by 2050. Why is this impossible? Read on…

- Typical PV systems consist of a) solar cells and b) balance of materials (BOS). Solar cells materials and processes can be optimized both in efficiency and cost with time. This is possible because of the pressure on the solar sector, and because solar cell manufacturer have some control over the price and management of the supply chain, and their own manufacturing processes.

Table 10-12. IEA targets for photovoltaic energy development 2008-2050

Targets (rounded figures)	2008	2020	2030	2050
Typical flat-plate module efficiencies	Up to 16%	Up to 23%	Up to 25%	Up to 40%
Typical maximum system energy pay-back time (in years) in 1500 kWh/kWp regime	2 years	1 year	0.75 year	0.5 year
Operational lifetime	25 years	30 years	35 years	40 years

- BOS, however, is a different ball game all together. Nobody has control over the BOS supply chain. These are common materials aluminum beams, bolts, nuts, washers, wires, inverters, and other electronics and hardware. Most of these are common materials used in many other industries, so their prices fluctuate with the global markets. The solar manufacturers and installers cannot control the availability nor the prices of these materials, because the solar industry is a fairly small segment of these markets.

- Because all materials prices continue rising with time, and with everything else considered, we estimate the pay-back time for our "typical flat-plate" solar (PV) systems by 2050 to be 3-5 years or more.

C. Operational lifetime. Here, the IEA Roadmap estimates 40 years of lifetime for a "typical flat-plate" PV module. This is also a stretch, and a big one at that. Here is why:

- All PV modules—including silicon PV modules (which are the most stable for now)—degrade under the non-stop exposure to IR and UV radiation. Because of that, and many other reasons, they lose about 1% of their output annually. This means that in 40 years they would lose 40%—or usually more—of their initial output. At this point, the low output and increasing failures, would make the operation non-profitable.

- It is also proven that most PV modules start failing faster, and some even drastically, with time.

- In best of cases, all PV modules would be still operational as per contract, and yet would have lost 40% of their nameplate output in 40 years. Running at 60% of nameplate output, or less, while possible cannot be profitable.

- Most likely scenario, however, is that 20-30% of the modules would have either failed completely, or are operating with much less power output. This brings the total solar field output down to 30-40% of nameplate. Would the owner be able to justify this low output will depend on the particular situation, but it is not one we should envy.

- In worst of cases, many, if not most of the modules have failed, which combined with the power output loss of the operational modules, bring the solar field to its knees with 10-20% of nameplate output. This cannot be a profitable operation by no means.

In summary, although we have nothing but respect and admiration for IEA, we must agree to disagree (drastically) in this case. So, unless IEA knows something we don't their Technology Roadmap is but a wish list, and an impossible (at least with what we know today) dream.

Thin Films PV Efficiency and Reliability

Three technologies dominate the thin film PV area.

- The leading company in CdTe technology, First Solar, recently revised upward its efficiency objectives, targeting 25% for research cells and over 19% for commercial modules in three years. The US-based firm also claims that its latest generation of technology reduces degradation of performance to 0.5%/year in all climates.

All these claims come from different test labs and have to be verified, before the data could be considered official. But most importantly, the claims would need to be proven by one full life-cycle (30 years) of non-stop, on-sun field operation. So, we, and First Solar, will have to wait and see how many of these claims are true...

- Copper-indium-gallium-selenide (CIGS) technology, with efficiency of 10-12%, lags behind c-Si but offers a slightly higher performance ratio. The largest supplier, the Japanese firm Solar Frontier, which exceeded 20% efficiency in relatively large research cells produced with mass production technology, aims to increase the efficiency of its commercial modules on this basis. It also aims to halve its costs from end-of-2012 levels by 2017.

- a-Si technology offers traditionally the lowest efficiency among commercial modules, and its deployment has long been impeded by concerns about the longevity of its modules and degradation rates. When these issues were solved, the cost gap with c-Si was no longer sufficient to warrant strong deployment. The combination of amorphous and micro-crystalline silicon (a-Si/μc-Si) allows higher efficiencies, but its cost might be prohibitive at least for now.

International PV Roadmap

From all solar technologies, photovoltaic solar energy generation (PV) is the most important and most promising. There are many reasons for this phenomena, and we address the key reasons and issues in our other books on the subject, *Photovoltaics for Commercial and Utilities Power Generation,* and *Solar Technologies for the 21st Century,* published by The Fairmont Press/CRC Press.

The sixth edition of the International Technology Roadmap for Photovoltaic was released in the summer of 2015. It sets a number of key measures driving the cost cutting of the silicon PV value chain. As always, it identifies the critical objectives of the solar industry, such as reductions in materials costs, improved manufacturing processes and the shift to advanced cell technologies. These are crucial step stones that have to be achieved, if the industry is to remain competitive.

After several years of problems and losses, in 2014 some solar companies started to operate at a profit. Their success was attributed mainly to the continued efforts to increase the quality and reliability, and reduce the cost, of each component along the value chain.

The so-called *price experience curve,* has seen average selling prices decrease by around 21% for every doubling of cumulative PV module shipments. This trend is expected to continue over the next several years. The key drivers are: the introduction of new double and single-side contact cell designs, improved Si wafers, optimized solar cell front and back metallization, and improved module manufacturing technologies.

The roadmap clarifies that silicon remains as the most expensive material of any silicon PV module, and therefore is the focus of present and further cost reductions. Fluidized-bed reactor technology for producing silicon material is expected to increase in relation to the Siemens, and/or any other feasible processes. Processes such as upgraded metallurgical-grade silicon (UMG-Si) and such are not expected to provide significant quality or cost advantages compared to conventional poly-Si technologies, but are still, and will continue to be, available on the silicon PV market.

Wafer production is still focused on thinner wafers, kerf loss reduction, and recycling rates increase. Wire sawing is the technology of choice in support of this trend, and is expected to increase in market share by 2025. This emphasizes the need for thinner wafers and reduced loss of starting material.

In the area of cell production, the roadmap predicts a shift from p-type to n-type wafers within the mono-Si market. This trend is driven by the potential of n-type technology to achieve higher efficiencies compared to p-type as time goes on. Since most manufacturers are set for p-type processes, however, the change might take some time to happen, if ever.

The roadmap also predicts multi-silicon (mc-Si) cells to surpassing 20% efficiency in mass production. It also emphasizes the potential for the major cell technologies to continue increasing in efficiency over time. This will also take some time, since increase of efficiency would make the poly-Si cells more expensive, and incompatible with their monocrystalline cousins.

The advanced technologies trend will be led by double-sided contact cell concepts. An example of this is the success story of PERC cells, which are gaining significant market share over the back surface field and any other solar cells manufactured presently.

Multi-junction cells are expected to increase in market share of up to 10% by 2025. Bifacial solar cells, which can generate power from both sides, are also expected to show gains, but their market application is limited. Yet, according to the roadmap, the percentage of bifacial cells will increase to around 20% of the total by 2025.

The roadmap anticipates production capacity of 65 GW in 2016, rising to 80 GW by 2022 and growing to a peak of 220 GW in 2030. After this, the roadmap sees a reduced demand to 150 GW until 2040, falling even further to a minimum of 100 GW by 2050. While this is possible, the solar energy sector will go through a major change and slow-down during 2017 and beyond, which might have a major negative effect on these figures.

The roadmap suggests that the PV modules market is not endless, and will not continue increasing as expected by some. This, therefore, precludes the need for high production capacity. But the future PV modules market will, nevertheless, be of considerable size. This, of course, depends on the ability of the manufacturers to be able to stay afloat and continue developing efficient, competitive, reliable, and cost effective silicon PV products. Not to mention that changing socio-political conditions will also play a major role.

Future Efforts

Expert opinions vary, but they all agree on the tasks and activities that need to be undertaken in the near future in order to ensure the proper and efficient development and deployment of solar technologies around the world:

By 2020:
- Increase module efficiencies to over 50% (HCPV), 25% (sc-Si), 20% (mc-Si, CdTe, CIGS), 12% (a-Si, uc-Si, organic, dye-sensitized)
- Avoid retroactive changes, which undermine the confidence of investors and the credibility of policies.

- Adopt or update medium and long-term targets for PV deployment.
- Implement or update incentives and support mechanisms that provide sufficient confidence to investors; create a stable, predictable financing environment to lower costs for financing. These may notably include FITs and auctions for long-term PPAs.
- In countries (or smaller jurisdictions, such as islands) with highly subsidized retail electricity prices, progressively reduce these subsidies while developing alternative energy sources and implementing more targeted financial support to help the poor.
- Increase performance ratios and decrease degradation rates.
- Diversify module specifications for variable environments.
- Reduce Si consumption to 3 g/W while increasing module longevity. Reduce silver consumption.
- Enlarge wafer size. Generalize reusable molds.
- Develop low-cost high-efficiency high-output bifacial I-sun tandem cells, and design specific system; around them.
- Develop meteorological PV forecast, with feedback loop from PV power plant online data to weather forecasting.
- Ensure the legal framework authorizes electricity generation by independent power producers at all scales and voltage levels (if not already implemented).
- Streamline permitting and connecting, including permissions on buildings. Phase out unnecessary bureaucratic administrative processes that add costs and waiting time.
- Elaborate and enforce performance standards for PV modules and systems in various climatic environments.
- Facilitate distributed PV generation while ensuring grid cost recovery (in order to keep the utilities happy). Remuneration may be based on net-metering, or self-consumption with remuneration of injection based on a fair assessment of the value of solar. Or "pay all, buy all" remuneration similarly based on a fair assessment of the value of solar.

By 2025
- Develop specific FV materials for building integration, road integration and other specific supports.
- Further reduce Si consumption below 2 g/W and increase efficiencies to 50% (HCPV), 28% (tandem

cells), 22% (mc-Si, CdTe, CICS), 16% (a-Si/uc-Si; organic; dye-sensitized cells)
- In countries with large numbers of people without electricity access, work with stakeholders to develop and implement suitable business models for deploying off-grid and mini-grid PV.
- Facilitate rapid market reactions by shortening gate closure times and trading block length.
- Develop and standardize the training and certification schemes for PV installers.

By 2030:
- Elaborate and enforce grid codes that will drive inverters to provide voltage control and frequency regulation
- Incentivize generation during demand peaks through time-of-delivery payments and/or limitation to instantaneous injection except at peak times.
- Incentivize load management and flexibility from existing generating capacities; ensure fair remuneration of ancillary services.
- Investigate options for new PHS plants; anticipate the need for more flexible power capacities.
- Progressively increase short-term market exposure for PV electricity while ensuring fair remuneration of investment. This may include sliding FiPs and auctions with time-of-delivery and location-specific pricing
- Work with financing circles and other stakeholders to reduce financing costs for PV deployment, in particular developing large-scale refinancing of PV (and other clean energy) loans with private money and institutional investors.
- Implement priority dispatch to jump-start PV deployment in new or nascent markets, until market maturity has sufficiently reduced costs for making greater market exposure compatible with sound deployment.
- Facilitate distributed PV generation, either using feed-in tariffs or net-metering.
- Identify the cost structure of current projects and anomalies by comparison with other jurisdictions, and implement specific actions to reduce anomalous soft costs, depending on country's circumstances.

By 2040:
- Work with financing and other entities to reduce financing costs for PV deployment
- Prevent PV hot spot emergence in ensuring geo-

graphical spread, e.g. through spatial remuneration differentiation.

- Continue work on large-scale energy storage, in order to deploy solar as a base load power.

By 2050

- Increase module efficiencies to over 55% (HCPV), 28% (sc-Si), 22% (mc-Si, CdTe, CIGS), 16% (a-Si, uc-Si, organic, and dye-sensitized)
- Optimize and standardize the long-term performance warranty conditions. The goal is to bring the annual degradation down to 0.5% per annum.
- Develop and implement efficient and safe large-scale energy storage capabilities. The success of this effort will determine the overall development of solar energy around the world. Without adequate energy storage, solar cannot be used as base load power, and its use will remain relatively limited.

U.S. Solar Energy Financing

Solar and wind projects are typically financed through a mix of debt and equity:

- Debt investors get their money back plus interest, via regular payments
- Equity investors get a share of the profits from the project plus tax benefits
- Bonds (low rating, but still investable) with 10-11% fixed return, have been issued lately as well

The biggest barrier in financing new solar and wind projects is the fact that these are new and unproven as asset class technology. Investors need to see at least one full life cycle (25-30 years) of non-stop field operation of a solar and wind power plant, in order to make educated decisions. This is absolutely needed to calculate the long-term risks, which are only a guess right now.

Presently, U.S. solar projects are financed via several methods, as follow:

- Home Equity Loans are obtained through commercial lenders, and are the most common method for homeowners to purchase their solar systems. The repayment is scheduled according to contract, similar to that for buying a car or a major appliance.
- Third Party Financing (PPA) is used when a third party owns and maintains the solar system, and is selling power back to the customer at a reduced rate. It is usually done under a Power Purchase Agreement (PPA), which lowers capital costs and reduces the electric bills. Large-scale (utility) solar projects usually use PPAs.

- Third Party Financing (Solar Leases) is when the solar system is simply rented from a bank or a company, similar to a car lease, while earning the benefits from the electricity the system produces. Solar leases are attractive options for home and business owners that plan to stay at their location for less than five years.
- Property Assessed Clean Energy (PACE) is an option to finance solar installations through the local government. Local governments create property tax finance districts to issue loans for energy efficiency and renewable energy. PACE allows local governments to provide low-cost, 20-year loans to eligible property owners seeking to install solar systems. The customer pays additional tax on property annually.
- California Solar Initiative represents financial incentives for solar installations based on the expected performance of the installation, which is derived from estimating the type and size of the solar array, the angle and location of the system installation, and other performance and risk factors. For larger systems, the incentive is based on the actual performance of the system over the first five years.
- Yieldcos are a new form of finance, where a publicly traded company is formed in order to own the assets (solar or wind power plants) that produce a predictable cash flow through long-term contracts. Separating activities, such as R&D, construction, etc. from less volatile cash flows of the assets can lower the cost of capital.

The above mentioned financial methods are expected to be used in the near future as well. New and more innovative methods are expected in the long term, as needed to replace some of the existing schemes. The most important of these is the ITC.

ITC

Investment Tax Credit (ITC) is the most important federal policy mechanism that is responsible for starting the rapid deployment of solar energy in the U.S. The ITC continues to support the solar industry, which is growing stronger by the day, creating thousands of jobs across the country.

According to SEIA:

- The ITC is a 30 percent tax credit for solar systems on residential (under Section 25D) and commercial (under Section 48) properties.

- The multiple-year extension of the residential and commercial solar ITC has helped annual solar installation to grow by over 1,600 percent since the ITC was implemented in 2006—a compound annual growth rate of 76 percent.
- The existence of the ITC provides market certainty for companies to develop long-term investments that drive competition and technological innovation, which in turn, lowers costs for consumers.

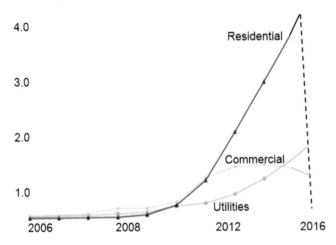

Figure 10-23. U.S. solar annual growth (in GWp).

Figure 10-23 shows the amazing growth of solar energy in the country. Ten years of unprecedented bliss. But there is a big problem on the horizon:

The ITC was extended in 2015, and could bring an additional 55 GW of solar installations by 2020. According to the experts, U.S. solar installations could provide 10% of the electricity used in the country by 2025. Could, would, should…it is given a chance, so now the sky is the limit…

There are still problems ahead, and these will be dealt with accordingly. First, we must make sure that they are well understood and managed.

Warranty Conditions

The rising importance of solar energy is also raising a number of questions about their technical characteristics and performance. Thus far, the only warranty most manufacturers provide is related to annual reduction of output. It is estimated that all PV modules will drop output by 1% annually across the board.

Each customer issues a warranty for its products, which varies widely and usually specifies 3-5% annual degradation during the first year of operation, followed by 1.0% annual degradation up to year 25. In the summer of 2015, Canadian Solar upgraded the long-term warranties on its multicrystalline and high-end monocrystalline PV modules.

The new and improved warranty guarantees a lower first-year power output degradation of less than 2.5% on all its multicrystalline modules. The degradation levels in year two to year 25 are not to exceed 0.7%. This way, by the end of modules life cycle—around year 25—the actual power output from each module in the field would be guaranteed to be no less than 80.7% of the nameplate (initial) power output.

The high-end 'Diamond' double-glass monocrystalline modules are also guaranteed for first year power degradation of no more than 2.5% from initial nameplate power. After that, and every year up to year 30, total power declines would be no more than 0.5% annually. This way, by the end of year 30, the actual power output would be no less than 83% of the nameplate power output.

The new and upgraded warranties are a direct results of intensive investment in solar cell and module R&D efforts, according to the company officials. According to our information, however, Canadian Solar has been consistently ranked near the bottom of the R&D spending of major solar companies.

While we must congratulate Canadian Solar for the bold move, further analysis of the situation brings new questions to mind. The most important of these is, "How did Canadian Solar change the basic property of silicon, and/or the physics of the silicon degradation process to ensure such an advance?"

We also cannot help it but wander; a) how much time and effort has the company invested in actual field research prior to making this decision, b) Why is Canadian Solar making this move now, and c) this move might brings lots of business to the company, so how would it respond if come year 25 its modules do not meet the warranty claims.

In such a case, would all non-complying PV modules be replaced, or would Canadian Solar pay for the difference in performance? In all cases, such failure could easily bring any solar company to its knees, so the logical conclusion is that Canadian Solar either knows something we don't, or is not planning to be around 25-30 years from today.

And speaking of long-term field failures, which is extremely important, and yet unresolved issue, we need to elaborate:

Supply Chain Management

Supply chain is basically the flow of materials from third party manufacturers to the production facility. This includes the ordering, transport, delivery, and proper storage of raw materials, work-in-process inventory, and finished goods from point of origin to point of consumption.

In case of solar modules manufacturing operations, the supply chain consists of purchasing large quantity of glass panes, aluminum sheets, solar cells (unless manufactured in-house), different chemicals and materials. These materials have to be of the highest possible quality, obtained at the best market price, and delivered on time.

Supply chain management is also part of the design and construction of solar, wind, and all other power plants. Here, the ordering, movement, and storage of the different materials is strictly controlled by supply management specialists.

Managing the supply chain is complex, very difficult and often ignored, or improperly managed, undertaking. There is a big difference between *handling supply chain logistics* and *well managed supply chain transactions*. This 'revelation' has transformed the industry, and has made the supply chain management an important part of the path to success. The successful companies employ armies of specially educated and trained supply chain managers, who ensure efficient materials flow.

Most people think that moving trucks from one point to another and moving shipments of materials and products is what supply chain is all about. It is not; it is only logistics. The proper management consists of filling all gaps in the system, resolving issues, and anticipating problems.

> *Due to poor management of the supply chain channels and methods, the global solar industry is losing between $600 and $800 million annually.*

The solar industry's supply chain is a huge business, encompassing several marketplaces. Adding to this the wind energy, and all other sectors, we come up with a huge numbers of specialists and money saved, due to their efforts.

The Solar Supply Chain

The modern solar industry is one of the youngest in the energy sector, and as such is going through growing pains. Supply chain management optimization is one of causes. It, however, has been slowly becoming a focus for the PV industry's ongoing drive to cut costs, according to the experts. Weak supply chain practices are costing the global solar industry many millions of dollars annually, according to the International PV Equipment Association (IPVEA)—close to a billion

dollars, to be exact. The industry is well aware of this problem and is making attempts to get to grip on it by dealing with factors such as delayed or lost shipments, missing inventory, and mis-ordered materials that are causing significant annual losses.

The industry had worked hard to cut costs in manufacturing of panels and balance of system (BOS) items, but the supply chain optimization has been on the back burner and needs to be fixed, in order to reduce costs further. One way to do this is through proper coordination within the company and efficient collaboration with the business partners along the supply chain.

The need for this effort is increasing in step with solar industry's continued expansion worldwide. Solar energy products manufacturing is a complex business, with many materials and the related problems; starting with moving out huge volumes of materials and parts from OEM locations around the world to the solar cells, panels, and BOS production plants.

The semiconductor industry, which business operations are very similar to the solar production, also had similar problems in the years past. These were resolved by establishing proper MOs and introducing international standards. None of this is present in the solar industry, so the solar manufacturers should look at the semiconductor industry as an example of efficiency in supply chain management and optimization.

Optimizing the supply chain of the global semiconductor industry took many years to get to the sophistication it enjoys today. Even then, more work on supply optimization is done on daily basis at all semiconductor facilities. It will take that many years for the solar industry to optimize its supply chain, but there is no way around it. Sooner, the better!

Asset Management

Solar power installations are growing around the world at unprecedented speed. Roughly 6.5 GW of solar photovoltaic power generation came online in 2014, or a growth of 36% over 2013's installations, which marked record installation levels.

Until now, solar owners and operators were focused on getting the power plants up and running, and worry about the rest as it came along. Now, however, the industry's attention is shifting to making sure that the systems are functioning properly, safely and efficiently. This requires *asset management* (AM).

> *Asset management controls all technical, financial, commercial, and administrative activities, as needed to ensure the efficient O&M and maximum ROI.*

Asset management of solar power plants is often divided into *blue and white collar* functions. The *blue collar* function consists of the actual installation, operations, and maintenance efforts. This effort is designed to enhance system performance, warranty enforcement, parts management, mitigate risk and protect asset value.

White collar functions include project design, engineering, accounting, energy production reporting, power purchase agreement (PPA) billing and management, tax filings and financial audits, regulatory and contractual compliance, bank facility management and partnership accounting and insurance processing.

Most solar system managers, which includes commercial and industrial project owners, utilities, banks, private equity firms, and others involved in the day-to-day management of solar projects, manage projects via in-house teams, or through the original project developer. In most cases of late, there is a combination of in-house and contracted management activities.

This is because very few companies can perform all functions well for the simple reason that the designing, constructing, owning, operating and maintaining a solar system in most cases is very complex. Third-party companies usually cover only one aspect of the tasks at hand, which requires the use of several such companies. Since all aspects of asset management are interrelated, this incomplete service leaves gaps in the overall management system and can be a liability.

O&M companies, for example, may not have physical and financial capabilities beyond the basic accounting and reporting, while entities with financial expertise, such as banks and private equity firms, do not have physical asset management skills. And neither has the EPC contractors' skills needed to manage the physical realm.

This could lead to unidentified failures and unaddressed problems to occur and affect the overall efficiency. As a result, decrease in production is to be expected under real-life conditions. In the financial aspect of the projects, the failure to meet regulatory, reporting, and tax functions in a comprehensive and timely manner can also bring liability and affect the bottom line.

The result is that many solar assets are undermanaged, with consequences including lost production and inefficiencies. These are then compounded with each unidentified or unaddressed issue. Only by combining physical and financial asset management, and completely covering all aspects of the solar project, owners and managers can take advantage of the synergies between the teams. This would then allow the creation of additional opportunities, reduce the problems, and increase the ROI.

As solar power installations increase, the awareness for proper asset management services increases respectively. Many owners and managers recognize the need to outsource different functions to entities with the proper experience and knowledge as needed to achieve the best results. But this is not that easy to do, since the provider must be able to integrate efficiently physical and financial asset management services in order to maximize the return on investments—a complex task.

Examples of common asset management problems that are encountered in the solar power sector are:

- *Underserviced assets* scenario occurs when O&M contracts call for annual servicing, when in reality a system may require servicing more frequently. Fast growing weeds, for example, could cover the bottom of the panels of a ground-mounted system. In desert, the PV or CSP panels could get covered with dust and mud after sand storms. In these cases, the result can be a large loss in power generation, unless timely measures are taken. Outsourcing physical asset management to a skilled provider who is monitoring production on a 24/7 basis can alleviate these problems.

- *Emergency preparedness* is required by all well managed operations. Extreme weather events of the past highlight the need for emergency preparedness. A full-service asset management provider can ensure solar assets are protected from damage during the storm and that the electricity keeps flowing once the storm has passed, while many systems owners and managers do not have the expertise or capability to undertake such tasks on their own.

- *Financial reporting* is important requirements, but owners are often reluctant to invest in experienced staff necessary to meet the sophisticated requirements of lenders, the tax laws, etc. Failure to meet deadlines and compliance requirements can lead to a loan default, tax penalties, and such—all of which detract from the bottom line and could lead to some unforeseen and undesirable situations.

- *Year-end audit* is a complex, lengthy and expensive process; a necessary evil, which requires a special set of skills to maintain the books and ensuring compliance with investors and government authorities. If not properly managed, a large project may face some very costly financial penalties.

These are only a few of the many ways in which combining physical and financial asset management can increase productivity and contribute to the bottom line. The biggest benefit of a proper solar asset management program, however, is that it would allow the owners and managers, whose expertise typically lies elsewhere, to spend their time doing what they do best. Instead of handling the day-to-day tasks and the related problems associated with running a solar system, they can focus on optimizing the management system in the respective areas of their expertise.

Long-term Performance

And so, the most important and maybe the only thing that matters for the long-term future of the PV industry is price and a substantial long-term performance guarantee for its products. These are somewhat confusing and conflicting factors, so they need some clarification.

The solar industry priorities today are:
- Affordable product price,
- Reasonable monetary returns.
- High efficiency per unit land,
- Reliable and safe long-term operation.

This is the order of priorities today of many solar manufacturers. With *'reliable and safe long-term operation'* being last in the list of priorities. It is also a black box, which is kept in the background in most cases.

A fresh and closer look at this priority has been undertaken lately, and eventually—after we obtain full 30 years life-cycle data from the different technologies—the priorities will change. Then, the last might become first. Until then, we just have to hope and pray that the solar installation we invested in would last, and perform well, during the 30 long years of non-stop, on-sun operation in the deserts.

One of the greatest risks we see with solar power generation today is the fact that, although solar cells and panels have been around for decades, the industry is relatively new. It grew very fast, driven by financial incentives and goals. As a result, the quantities dramatically increased, and the prices fell, but very little is known about the long-term performance of PV modules under field conditions.

Since none of the present day manufacturers and their products have completed one full life-cycle (minimum of 25-30 years in the field), the actual long-term performance of the millions upon millions of PV modules installed around the world is unknown.

Scary fact: None of the millions of PV modules installed around the world have been proven reliable or safe for 30 year field operation.

How could one be sure that any module would last 30 years blasted by the desert sun, freeze, sand storms, etc extreme conditions, if we have not seen each type survive in the field as expected for the duration (30 long torturous years)? This is over 10,000 days and nights full of extremes, which the otherwise fragile PV modules have to endure without failure.

Knowing what we know about the desert extremes, we would not want to bet that this merciless killer of anything left in there for a long time, would be kind to PV modules. This is a great uncertainty that increases the risk for owners and investors, who might find themselves counting their losses from prematurely failing PV modules.

The current IEC and UL testing procedures and certification protocols for PV modules use only pass/fail criteria for certain key parameters. The only practical usefulness of this test is to identify those modules that would fail immediately in the field. It separates the very bad from the rest, which is good, but it does not give us the entire picture of 30 years field operation.

The existing test procedures only reduce the risk of early field failures (infant mortality), but do not address the issue of successful 30 years operation—especially in very hot and humid climate areas, like deserts and coastal areas.

According to researchers at the Fraunhofer Institute in Germany, 'The existing standard testing procedures do not include a protocol for relative durability assessment of different modules. Hence, financial models rely on a patchwork of methods to forecast relative durability in the absence of these benchmarks. Consequently, the uncertainty in quantifying which solar modules are best suited to a particular installation results in an increase of perceived risk, delayed financing and, ultimately, raises the cost of building PV power plants.'

In order to provide more information on the long-term performance (durability) of PV modules the Institute has undertaken a joint initiative between the Fraunhofer CSE and Fraunhofer ISE; the so called Photovoltaic Module Durability Initiative (PVDI), which is an integrated lifetime assessment initiative to understand and improve PV modules' long-term reliability (durability.)

'The goal of the PVDI is to establish a baseline PV durability assessment program that provides quantitative, independent, third-party assessment on a module's long-term durability. Moreover, the PVDI enables comparison of the relative durability risk between different module designs, which the current test protocols lack.'

The PVDI test sequences are summarized in Table 10-13, and further explanation of the purpose of each group can be found on the Fraunhofer's website.

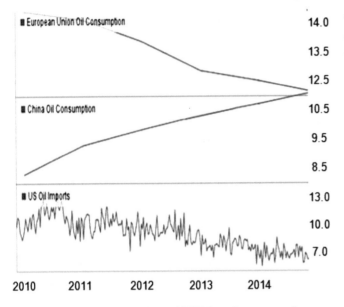

Table 10-13. Overview of PVDI testing protocol

According to the authors, 'PVDI goes beyond the pass/fail criteria of today's testing protocols. Modules are repeatedly characterized as they pass through the assigned test sequence. For example, in Group 4, each module is characterized after every set of two hundred thermal cycles. At each interim test point, electrical performance, electroluminescence and infrared images are collected. In some instances, wet leakage current and insulation resistance are also measured.

Each module is rated based on its performance under PVDI's robust testing protocol to generate a credible rating of PV modules based on their likelihood of performing reliably under different kinds of stress. All the results are summarized in a report that provides solar PV financiers, developers and other industry players with a widely available quantitative dataset to assess long-term durability.'

'PVDI strives to provide accelerated tests to assess a module's reliable (relative) performance over its service life, identify degradation mechanisms and provide information to accelerate innovation in PV modules. PVDI provides an assessment of module's susceptibility

to degradation under each testing group, as well as a relative performance rating to compare against other modules.

This helps manufacturers determine the areas of weakness in their modules and strive to address those, improving their module's performance. The rating system enables financers and investors to compare the relative performance of modules, and making informed decisions.'

This is surely a step in the right direction. Long-term performance and durability are the dark secrets of the industry that no one is prepared, nor willing, to discuss. The only questions remaining unanswered now are, a) why did it take us so long to implement such a procedure, and, b) is even the PVDI protocol enough to ensure 30 years of non-stop, efficient and safe operation?

And speaking about issues, here is one fact that needs to be well understood:

The Solar Reality

We often hear that we have enough solar energy to power the entire Earth. So let's see what would take to do that with today's technology. We will use today's technologies and prices, leaving the potential of future technological developments and extra cost reductions for later.

Roughly speaking, the solar cells needed to power the world would cost about $59 trillion; the mining, processing, and manufacturing facilities to build them would cost about $44 trillion; additional transmission lines and grid upgrades would cost about $30 trillion; and the batteries to store power (which technology is not fully developed as yet) for night use would cost another $40 trillion.

To power the entire world, we will need about $173 trillion initial investment plus about $1.0 trillion per year for operation and maintenance purposes.

Keep in mind that the entire gross domestic product (GDP) of the U.S., which includes all food, rent, industrial investments, government expenditures, military purchasing, exports, and so on, is only about $14 trillion annually. This means that if every American were to go without food, shelter, protection, and everything else while working hard every day, naked, we might just be able to build a photovoltaic array to power the planet in about a decade and a half.

But, unfortunately, these estimations are too optimistic and highly unrealistic. If actual installed costs for solar projects in California are any guide, a global solar program would cost roughly $1.4 quadrillion, about one hundred times the U.S. GDP.

With other words, it would take well 100 years to pay for solar power plants that last 20-30 years.

And then, after 20-30 years of operation, we need to decommission and replace the old solar plants, and build new ones, which would require another 100 years to pay for. So now we are in debt for the next 140-150 years. What's wrong with this picture?

And very importantly, mining, smelting, processing, shipping, and fabricating the solar cells and panels, and the associated hardware, in addition to installation and operation, would yield about 200,000 megatons of CO_2. On top of that, we all would have to move close to the deserts, else the costs of additional transmission infrastructure and losses would make the project even more expensive and even unworkable.

And equally important is the fact, that some of the materials used in the manufacturing of solar cells and modules are in limited quantities, so we might just run out of them after several cycles.

So, many of the above *Ifs* are unreasonable or unsustainable.

What all this tells us is that neither solar, nor any other technology, can solve all our energy problems by itself.

It also tells us that we must not put all our eggs in one basket. Solar is good, but it is not the only solution. Instead, it has to be incorporated in the general mix initially, and slowly increased (to a point) with time.

The Utilities Wars

An increasing problem of late is the change in attitude and behavior on part of some U.S. utilities. They have noticed a number of problems created by the quickly rising number of solar (PV) roof installations. This is causing increase in expenditures, so the utilities are looking into the alternatives. Most of the utilities are waking up to the new reality, which they are not prepared for, nor willing to accept without a fight. For lack of better solution, the utilities are declaring a war to their best customers—those that have, or plan to have, solar systems on their roofs.

Thousands of roof top solar systems are now operational, and the future of many planned thousands, depend on the outcome of the war over use of the power infrastructure, which is presently under the control of the utilities.

The utilities are affected by the quick spread of solar installations, since all power generated by thousands of small and large solar systems is sent into their infrastructure for distribution. All this costs additional money, which utilities are forced to cover, and requires additional resources they don't have. So the utilities are fighting back by imposing extra charges on the solar customers. The battle is intensifying and we are now on the cusp of a new age in power generation, transmission, and distribution.

Recently a $50/month charge was introduced by SRP (one of the two large utilities in Arizona) on all new solar installations. $50 a month is an average saving from an average roof top solar system, so what is the point in getting one?

These were questions asked by existing and potential solar customers during the spring of 2015, during a meeting with SPR's board of directors. There were no satisfactory answers and the directors didn't see to care either way.

As a matter of fact, when asked how many have solar systems on their roofs, the dozen or so board members kept silent.

None of the utility directors has a rooftop solar installation.

This only means that the key players—the decision makers—do not trust, or simply disregard this new technology. Most people at that meeting concluded that SRP (via its directors) is not interested in solar, and even sees it as obstruction to their daily operations and a burden to their pocket books.

SRP also shows reluctance in getting involved in large-scale solar too. Instead of building its own solar power plants, SRP prefers to purchase solar electricity from other states. This is its MO for now and the foreseeable future.

The other local utility, Arizona Public Service (APS)—a regulated utility—joined the trend by trying to impose a similar solar tax in 2014. The customers put up a fight and the Corporation Commission reduced the proposed $50-a-month tax to $5 a month. Sadly, SRP ratepayers have no such protection. SRP is unregulated utility, thus it is not accountable to anyone, including the customers, who have no other choice.

SRP provides extensive lip service to solar, claiming that it is using solar in its energy mix, although it has no significant solar installations. Imposing high taxes on residential solar, however voids any of the utility's good deeds, and makes it look irresponsible.

So, where is solar headed in the sunniest state of the country is anyone's guess, but one thing remains certain; rooftop solar is not on the local utilities' priority list. As a matter of fact, they are fighting against it; looking for ways to reduce its effects on their old, inefficient, but profitable, business model.

Would these types of abnormalities in the energy sector change in the future is anyone's guess, but for now we are convinced that it will take a long time before a compromise is found.

But wait, there is a new trend—a surprise of sorts. It is the decision of several U.S. utilities to jump in the solar game with both feet. As they say, if you can't beat them join them. And so, they are offering a wide range of solar services to their customers. They will compete with the independent solar companies to design, install, and operate rooftop solar systems.

This is a new twist that might lead to a number of new developments in the U.S. solar energy market. It might be the beginning of a new trend, which might elevate rooftop solar to new heights.

Solar 2025

On the bright side of things, the new solar products, methods for harvesting and storing solar energy are getting very advanced and efficient. Because of that, solar is projected to become the primary source of energy on our planet in the near future.

In the summer of 2014, Thomson Reuters analysts tried to identify and predict the rapid progress and emergent trends in scientific research and technological innovation in the solar industry. They concluded that as the world's population grows and global energy needs increase, renewable energy (solar and wind especially) will remain high on the priority lists.

Serious improvements in the PV technology, such as: chemical bonding, photo-catalysts and three-dimensional nanoscale heterojunctions promise to bring solar out of its perception as environmentally conscious undertaking and to wide use by the masses. Solar no longer would be a privilege for the rich people in the rich countries, but to the last villager in Asia and Africa.

The solar equipment will become much more efficient and reliable in the long run. Thus generated energy will be stored and used in cloudy day and at night. Solar thermal and PV technologies from new dye-sensitized and thin-film materials will heat buildings, provide hot water, and electricity at homes, and commercial and manufacturing facilities alike.

Exploring chemical bonds in the PV materials, a new photosynthetic process will make solar energy available when needed; even under low solar insolation. New materials such as cobalt-oxide and titanium-oxide nanostructures, photo-catalysts and 3D nanoscale heterojunctions will dominate the different niche markets.

Fabrication of novel heterostructure, such as CO_3O_4-Modified TIO_2 nanorod arrays with enhanced photo-electrochemical properties, as well as modifying donors behavior in bulk-heterojunction solar cells promise 10-fold increase in efficiency.

At the same time, new production methods using mesoscopic oxide films sensitized by dyes, or quantum dots will increase significantly the present solar conversion efficiency rate of 8-9%.

Could this happen? Yes, sure! When and how... this is the big question that nobody has an answer to. In our humble opinion, however, the new and futuristic technologies look good in the lab (and even better on paper), but they do not belong in large-scale PV installations. We have a hard time seen any of them operating normally under the broiling sun rays in the deserts. Most of these new comers will fry to crisp within hours.

They, however, will find specialized niche markets where they can thrive while being further developed and optimized. Where and what these markets will be is another big subject that needs an entire new book to be fully addressed.

The U.S. and the global solar fields, however, will be equipped with the most efficient and reliable PV technologies. c-Si is the winner today, followed by thin film PV, and that configuration is expected to remain as is in the long run.

In the 22nd century the fossils will be history, and only renewable energy sources would be available for power generation and transport. So, how about a compromise between solar and fossils to fill the gap? The discussions on this subject abound, but the actions do not...thus far.

Yieldcos—The New Wave

When anything solar started look dim, during the last several years, the U.S. finance and investment machinery came to the rescue. As things started leveling during 2013-2015 a new trend, called *yieldco*, hit the solar energy market. It simply consists of bundling the finances of a number of solar power generation assets into one investment fund. Thus created new entity, or yieldco, is expected to lower the cost of renewables in the near future.

This is not an easy concept, and not easy to digest by the uninitiated. It sounds like the formula 1+1=2 has been miraculously changed to 1+1=3. With the 2008 financial fiasco still fresh in our minds, we cannot help it but recall that similar formulas were used to throw us, and the entire world, into one of the greatest economic disasters ever.

The new craze is expected to propel the new yieldco market, and the related companies and projects, into substantial growth during the next few years, according to investment bankers. Bundled securities…again…this time dressed up in new, shiny, solar attire. Or is this similar to the Emperor

Yieldcos are set up as publicly traded companies. Just like them, they pay out periodic dividends. One funny fact is that, although not limited to the solar industry, yieldcos seem to have attracted financiers and investors in the solar, and PV in particular, sector. Yieldcos do not offer huge returns, and instead are expected to offer a stable and long-term platform for investment…here, an image of Bernard Madoff appears to complete the picture.

According to the same bankers, "the yieldco concept has a lot of room to run" and so the yieldco market is on track to quadruple in size from its current total value of around $27 billion. It is not clear what time scale was used for this prediction. Nevertheless, the first steps seem encouraging. A recent report by market analysts confirmed that the growth of yieldcos was one of the drivers for a better-than-average U.S. solar funding in 2015.

The first major yieldco that we know of was launched as an IPO by NRG Energy in the summer of 2013. The new NRG Yield yieldco, represented by over 22.5 million shares, offered 1,324 MW of solar generation capacity from completed renewable energy projects. NRG Yield, however, actually contains a mixture of conventional power generation assets and renewables,

NRG Yield launch was quickly followed by SunEdison's creation of Terraform Power, which is its own, renewables-based yieldco. Others followed soon toot, with the UK's finance and solar industries in particular seeming to be exceptionally enthusiastic about this new investment vehicle. Bundling financial assets has always been profitable…for the bankers…so no surprise there.

The twist here is that while most early yieldcos focus primarily on the utility-scale market, NRG recently said it will also be targeting residential solar sector, through investing in rooftop leasing models.

And then in the spring of 2015, First Solar and SunPower, representing two of the U.S.' biggest names in utility-scale solar, announced the launch of their new yieldco, 8point3 Energy Partners. The new yieldco initially offered investment in a total of 432 M.W. of U.S.

projects. The new company started trading on NAS-DAQ under the symbol CAFD in June, 2015, starting with $21.00 per share. Unfortunately, CAFD lost 20% of its value during the first couple of months. Not a great start, but the near term future still looks bright.

In 2013, just before three major UK yieldcos were about to be launched, one of the principle partners said that the appeal of yieldcos lay largely in the *stability of the investment*. He did not explain how a variable power source, with declining annual efficiency is "stable"… and in the UK, from all places, where a day of full sunshine cannot be seen for months.

This only means that his words are to be taken with a grain of salt (and a big one too), as should anything a banker says today. Nevertheless, the UK institutional investors seem interested in the solar market, and see yieldcos as a particularly attractive instrument for their portfolios.

'Containing only completed and operational assets, the banker clarifies, most yieldcos would probably deter high risk investors seeking an early exit.' Oh, yea, there is some significant risk, we must conclude, so investors would surely look for an exit.

But wait, are we talking about solar performance, or making money as a primary objective here. The name of the new yieldco between First Solar and Sun Power, *'cash available for distribution'* (CAFD) says it all; we have money to burn and will use it as we please. Would any of us; potential customers and conscientious objectors (in the solar sector) care about how much money a solar company is making, how, when, and where? Or would we care only about what quality product and services we would get for our money.

The money making aspect, then, presents a risk for the solar companies involved in these financial schemes.

More money a company makes, more money it is stealing from its customers. Customers beware of the high yields!

Yet, there are other risks for the solar companies— yieldco, or not. The conditions of any project and the related risks vary according to each project's particulars; type of project and equipment used, stage of development, location, etc. The combination of all these factors usually determines what kind of customer and/or investor would be interested.

Since the yieldcos contain only operational assets, the private equity guys who want a higher return on their money must get in the early stages—even at the very beginning stages—of the projects development.

The early bird catches the early fat worm. But with that, as with any high return ventures, there are many uncertainties, which increase the investment risk proportionally.

But yieldco, and all solar assets in general, have weaknesses that must be considered. They all are short-term in nature. While general purpose YieldCo and other businesses are fairly stable with time, with long-lived assets. But solar panels are variable power generators, and degrade slowly over time. Same as wind turbines, which are designed to last around 20 years. Since many solar assets had been operating for some time already—some long time before been acquired—there is a danger of poor performance and major field failures.

Equally importantly, all large-scale solar projects operate under power-purchase agreements (PPAs), under which the generated power is sold to local utilities. The PPAs usually have a 10-20 years duration, at which point they may, or may not, be renewed at favorable rates. Most solar project owners, however, seem to assume that they will sign renewed PPA on similar or even more better terms when the existing PPAs expire. This, however, is not a justifiable assumption, because conditions change, and utilities have the right to grant or deny PPA extensions.

All these uncertainties are not well understood, or are intentionally placed in the small print on the bottom of the page by the solar companies, which is helping the creation of yieldcos as a new investment instrument. It is also providing increasing (albeit full of uncertainties) recognition to the solar energy as a separate financial asset in a class of its own.

The very name of these financial instruments "yieldcos," as interpreted by some media and investment gurus, describes the nature of the asset as capable of paying out much higher regular income to investors than other competing asset classes. Another unjustifiable assumption.

There are concerns that rising interest rates could also affect yieldcos, making them much less profitable as compared to the classic investment instruments—which the financial community prefers in most cases. Some companies' yieldcos, however, are insulated from interest rate fluctuations by holding long-term debt at a fixed rate for the duration.

Yet, a number of official warnings were offered in 2014 year by officials of the solar and investment communities. The major issue for the investment community is that the investment returns of an yieldco can be increased only by adding new projects to the portfolio. How many projects an yieldco can add, and how fast remains to be seen.

Some of the yieldcos already on the market were started in a hurry and might falter and fail since they won't be able to keep up with the growth. This would disappoint the investors and would force the entire investment community to take a much closer look at the participating companies, their assets, track record, and growth potential. Only then would they be able to determine what the companies can deliver and what the yieldco offers.

We would not be surprised to see an increase in such activities in the next several years, and expect to see a lot of new companies not having the experience, the growth potential and whatever else is needed to go into a yieldco.

The new First Solar-SunPower venture, and others large companies yieldcos might be the exception of the rule, but as any new business venture, the risks are there. And sure enough, within several months, the new yieldco, CAFD, fell down 50% from its initial value. It was pronounced 'oversold', which in Wall Street language means, *sell as soon as you can*. What the future hold for this new yieldco is uncertain at best, but the trend is not for the best.

Yieldcos still have a long ways to go before becoming established and proven financial instrument in the solar industry. In all cases, the solar companies must have a solid product, well defined market plans, and show long-term profit before yieldcos can be of any help or benefit to them and the investors. Long term is the key here, because most of these companies have a very short life span records.

Solar and Wall Street

No matter how you look at it, the U.S. solar market has been, and still is making the news day after day. New PV installations are soaring by 30% in 2015, the prices keep going down, and in some locations solar power competes successfully with gas and coal-fired power plants. The solar industry also received a vote of confidence from the White House, when President Obama introduced a new plan to kill coal-fired power generation by reducing further their emissions.

Everybody is excited about solar, but wait...not everybody. We don't see anyone on Wall Street jumping up and down about solar's success. Instead, Wall Street has been dumping solar shares all through 2015. Why is that? Is it because of low crude oil prices, which might reduce demand for renewable energy. But oil is not used to generate electric power, so what is it.

There might be several reasons for this abnormality. Stock prices have been affected by an abundance of new equity issues by companies raising capital to fund their growth. There are also concerns that interest rate hike by the Federal Reserve could reduce the success of the new solar financial instruments, the yieldco units.

The MAC Global Solar Energy index dropped over 35% in 2015, with some of industry giants like SunEdison losing over 50% of its value in a couple of months. SunRun, one of the top residential solar installers, went public in August 2015 $14 a share and promptly lost almost 30% down to $10 a share. First Solar and SunPower yieldco, CAFD, opened at over $21 a share in June, only to drop to below $15 couple of months later. TerraForm Global, another yieldco is now trading more than 20 percent below the IPO price.

Investors are justifiably concerned that many solar companies are trying to achieve quick success by growing too fast and by doing too much. This might make the capital requirements too high, which would shrink the project profit margins.

Solar is now an established segment of the energy markets, but investors still see it as unproven and risky investment.

This only confirms the fact that on Wall Street, the fundamentals, no matter how good, are often powerless in the face of the arbitrary preferences of the traders and the investors.

Despite of the hoopla, solar is actually still more expensive in most U.S. states than conventional power generation. Solar still relies on government subsidies and mandates, some of the major of which will be gone in 2017, which increases the risk factor further. There are also a number of technical and logistics issues with the solar technologies that sill need to be resolved, and which reduces the number of investors able and willing to look at solar objectively and assume the risks.

In summary, many investors stay away from solar, fearing a short-term correction in the making due to rate and tax changes in the field. Long-term investors can be more opportunistic, counting on a 'sunshine after the storm.' How much 'sunshine' will be available in the long run remains to be seen.

THE RENEWABLES' FUTURE

The renewables' presence in the global energy markets is increasing, and with it the problems and risks they bring into the power mix are growing too. Solar and wind are the most promising and trouble-free renewable technologies, but they also come with a number of problems and present risks of different types.

Their variability is the greatest risk for owners, customers, and investors alike.

The variability of solar and wind power is defined as the difference between the total energy demand (at the grid) and the total variable power production.

With other words, 100 MW solar power field idles during the night, and then starts increasing power output with the first sunrays at sunrise. The power level increases slowly and is maximized around the noon hour. If the solar modules are mounted on trackers, then the power can be kept almost constant (at its maximum level allowed by the modules) all through the day.

In all cases, when a cloud covers the sky, all solar power generation simply ceases. It goes near zero and stays there until the sun breaks through the clouds again. The same is true for wind power. In cases of very low, or very high, wind speeds the turbine blades stay motionless, and no power flows into the grid. None, nada, niente!

At the same time, the grid demand level remains steady. This imbalance forces the grid operators must hustle to add power and meet the load demands during cloudy and wind-less days. This is complex and expensive undertaking, which is causing frictions among the players.

As a daily average, the difference caused by variability might not be great, but when measured on moment-by-moment basis, the fluctuations of power generations could be quite significant. The variation—going from full to zero power—causes many problems on the distribution side of the national grid. This is because, while demand remains constant, the decrease of the amount of solar power injected into the grid is reduced.

This abnormality causes the grid operators to scramble to add power to the grid. This is usually done by firing the stand-by peaker power plants. Turning these power plants on and off is very expensive and causes other problems, such as grid instability, power fluctuations, black-outs etc.

These issues can be addressed only by a power system with a higher degree of flexibility, as needed to follow and adjust for the fluctuating net load. To accomplish this, a variety of solutions involving both producers and consumers are being assessed at present and are to be included in the new Smart Grid solution—another complex, expensive, and full of technological and financial challenges and risks undertaking.

One thing is for certain; the variability of solar and wind power generators will not be solved any time soon. The Smart Grid, on the other hand, is still a pipedream, so we must deal with, and solve, the new problems one by one, day after day.

The Power Grid

Our National power grid is quite old and badly outdated. Power network issues and bottlenecks are increasing in number, size, and frequency as new technologies and processes are added to the old decrepit power grid. These new issues are becoming a major problem for the grid operators and an obstacle in the large-scale integration of renewable energy in the national power mix.

On top of everything else, the new issues also force changes in the plans for implementation of the new smart grid concept. The renewables (solar and wind) are not helping much. There are many reasons for this abnormality, but, in addition to the variability issues, it is mainly due to the fact that the renewable energy-sources are location- and time-specific. In many cases, they (large-scale solar and wind power plants) are located in the deserts, and are usually far away from the large populated centers.

These abnormalities force the need for an extended and expensive transmission infrastructure. Solar and wind also generate power in unpredictable bursts, which requires more resources to handle. Some people think that the resources thrown to solve the variability problems could be spent much better in other areas of the energy sector.

Transmission system expansion and modernization involve significant lead-time and huge investment, in addition to fierce public opposition in some cases. So, for optimal allocation of renewable capacity, the power plant owners need to use the existing transmission resources in the short run, while finding a way to expand and optimize it.

Only through adequate transmission expansion projects could the renewables reach their full potential. Here we must add that even that is not enough. Adequate energy storage is a must too, because without it, the solar and wind power transmission lines would stay idle during cloudy days, and/or no-wind conditions.

Power Grid Issues and Risks

Some of the key present day power grid issues and risks, that we must understand and find solutions for in the future, are as follow:

Renewables and the Grid

There is no such thing as a perfect power generation, even if we succeed in mixing and matching the different technologies the best way possible. Generally speaking, the fossil, hydro and nuclear power generators' output is stable and reliable on its own. When renewable energy sources are added, however, the stress on the power system increases exponentially. The power grid is supposed, and actually required, to integrate the new (unpredictably variable) power generation from solar and wind power plants. Their output, however, is strongly dependent on the type, seasons, time of day or night, weather, etc extraneous and uncontrollable factors. This complicates things significantly for the old grid.

Variability on the generation-side of a power system requires variable operation of the conventional power generating units at levels that are much higher or lower than their optimum steady output. This abnormality, coupled with the variability on the demand side, has to be dealt with as needed to provide a steady load.

This is usually done by ramping up or down the base load power generators, which is not an easy or cheap undertaking. The ramping excursions often end up with the start-up or shut-down of conventional units and/or introducing additional units; these of the stand-by peaker power plants, into the mix.

If large variations of renewable generation are to be added to the mix, their variable output will result in conventional power plants operating in an erratic and much less efficient way. Use of peaker power plants to level the load variability, is expensive undertaking from operating point of view. It also reduces significantly the emissions savings brought about by the contribution of renewables to the electricity supply.

While solar and wind power plants contribute to environmental cleanup, their variability introduces additional GHG emissions.

The increasing share of renewable energy sources is also supposed to lower electricity prices, but it is questionable if this will provide adequate economic incentives for electricity producers. The expense related to turning additional fossil power plants on and off, in order to accommodate the variability of solar and wind power plants, is significant. It is a new phenomena, which is just now being looked at, as the number of solar and wind power plants increases and more data becomes available.

Solar and wind power plants are cheaper to operate. but their variability increases the fossil power generation.

Yes, it is a controversial statement, but one that describes the present day conundrum. Solar and wind farms do not use fuel, so their operation is almost free. But their variability (in case of cloudy or low wind days) increases the use of fossil-fired peaker power plants. This, in turn, increases the overall energy costs and GHG emissions.

Why would anyone then invest in variable power generation technologies to fit in larger amounts of renewables, which jeopardize the security of supply, increase the emissions, and bring less profits? Not an easy answer, and yet, there is no turning back now; the renewables with their advantages and disadvantages are here to stay, so solutions must be found. Sooner the better…with the burden falling square on the utility companies shoulders.

The Role of Demand Response

The U.S. power demand is quite steady overall. If the seasonal patterns of renewable production were similar to those of the demand, the variability of renewables would be passively absorbed by consumers with low net load fluctuations. Instead, in most cases power demand and renewable power availability are not perfectly correlated. In many cases, the periods of high or low demand do not coincide with the respective periods of high or low renewable energy production.

And so, in case of low demand and high power generation, like during *windy night time*, or sunny winter afternoon in Arizona (when power is not needed due to balm weather) most of the power generation could be wasted (since it cannot be stored).

In case of high demand and low power generation, like that of a cloudy, hot summer afternoon and no wind to speak of in the desert, the solar power generation cannot match the power peak demand and so expensive peaker power plants are fired to supplement the excess load demand.

A number of approaches have been suggested to remedy this situation. These can be grouped into, a) direct and b) indirect control.
- *The direct control* method allows transmission system operators (TSOs), or other electric market principals, to directly modulate the demand side by rationing consumers.
- The indirect control method uses economic incentives by implementing time-varying power prices. It is also known as *dynamic real-time pricing*.

These approaches, however, require adequate infrastructure, allowing effective communications with the consumers on one hand, and ability of consumer appliances to adapt to the regime. This is what the new Smart Grid, in addition to many other things, is supposed to do too.

The demand response poses challenges also in terms of market design, where price-incentives introduce dynamic properties to the demand, by increasing its cross-elasticity across different time periods. This, however, depends to a large extent on the policy-makers decisions and on the number of distributed generators spread around the particular system.

Further complications are also imposed by the tight capacity constraints of the power grid at the local distribution level. For these, and many other reasons, efficient and cost effective demand response remains one of the great challenges to the power systems operation in the future.

Electricity Storage

It is obvious that the solution to the problems of variable energy sources is by developing large-scale energy storage capabilities. Easily said… There are currently a number of energy storage technologies under development, such as batteries, fly-wheels, electric vehicles, superconducting magnetic energy storage systems, compressed air. and molten-salt or pumped hydro storage systems, etc., and yet none of them seems adequate enough.

Recently batteries have been considered as the best solution to energy storage. Small battery packs for use in residences and remote locations are feasible, but not very popular at least for now. It is difficult to imagine, however, a large battery pack—24 foot container size—baking in the desert for a long time. Batteries of any type do not last very long under the merciless extreme heat of the desert summer and freezing nights in the winter.

Batteries just don't seem like the most efficient and cheapest approach to store large amounts of electricity. New advances and more incentives in this field are needed for large-scale electricity storage to become a reality in the required magnitude!

Note: We take a close look at the batteries energy storage above, so we won't waste more time here. We only need to remember that there are many other forms of energy storage. Because of the great need, the other forms must be given a chance and their use explored in the near future.

This is very important, because the development of energy storage will determine (to a large extent) the development of the renewable energy sources. While batteries seem as the most direct and elegant, albeit not that easily done, solution, we need to look at everything that can be used now.

Without energy storage, solar and wind are not able to compete, let alone becoming base load power generators. This limitation does, and will, keep them at the level of secondary power sources, which progress depends on, and is determined by, the primary energy sources (coal, natural gas, crude oil, and nuclear power.) The only solution to this problem is large-scale energy storage, thus the importance of its deployment NOW, not later.

Distributed Generation

Distributed generation (DG), in this context, is the collective use of solar energy generated by rooftops solar installations. I.e., several houses of a local neighborhood, adjacent businesses, or a combination of the above getting together in order to jointly use the power generated by the entire installation. DG is the fastest growing sector in the U.S. solar market.

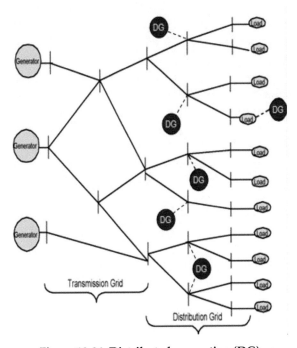

Figure 10-24. Distributed generation (DG)

DG Advantages and Benefits:
- DG allows customers to make energy generation and use decisions
- Property-Assessed Clean Energy (PACE) allows communities to finance energy retrofits for private-property owners which are repaid through property taxes.

- Smart Utility Rate policies encourage energy efficiency and on-site generation by customers while maintaining (or changing) utilities' business model.
- Hybrid power generation—gas, diesel (developing countries, or remote locations)
- DG created peripheral support industries and businesses—hardware and software

DG Issues:
- Possible Fed and State regulations changes scare the investors
- Utilities games add uncertainties:
  — Two-way flow is counter-intuitive for the utilities MO model
  — Customers are now suppliers—and competitors
  — Game changer; forcing the utilities to rethink their MO model
- Solar power is variable, causing problems to the grid and additional expenses
  — Price and regulations manipulation
  — Permitting and interconnect difficulties
  — Take over DG consolidation and the entire DG market—finance, install, operate

Distributed Energy Resources are the new kid on the energy block.

This new kid is different from what we know about electric power, and are also the greatest challenge to the utilities today. This is because they directly affect the day-to-day management of the national power system.

As millions of home and business owners are installing solar panels on their roofs, the power network is suffering from their variability. The old electrical power system was designed to handle limited number of high-voltage, high power output, which is to be distributed to peripheral low-voltage networks.

The AC power grids were designed to allow efficient transmission of electricity over long distances, which added to the economics of scale in electricity generation based on large power plants with lower per unit production costs. This was/is the foundation of our national electric system.

Until now that worked well. and the national electric energy system became a colossal power supply chain. Here electricity is send over long distance to be consumed far away from the large power generation centers. In all cases, the electric power flows in one direction; from the power plant to the transmission and into the distribution systems.

But now, guess what; renewable power generators (solar and wind) and mini smart grids are entering into the power grid in great numbers; some as *distributed energy sources*. Here, millions of roof-top installations are pumping electric power into at millions of contact points. The one-way grid has become a two-way hi-way, with power going in and out of it. What is an utility to do?

Amazing; the electric power in today's power grid flows in both directions.

This totally new situation is changing the electric power management game. This is bringing many (thousands) new power generators and other key-drivers into the mix. The utilities must adapt to this now dynamic and growing industry in distributed generation technology. This is complicating things for the utilities, and is making their work more complex and expensive. It is also posing constraints on the construction of new transmission lines, while customers demand higher quality and highly reliable electricity.

Today, power system operators must deal with the deregulation of the electricity sector and the establishment of electricity markets. This is compiled with concerns about climate change and global warming. The new distributed power generation is further complicating the work of the power grid operators significantly.

While distributed generation may help to provide a more differentiated service to consumers (on the demand side) it is at the same time adding a degree of difficulty in providing a steady power flow (on the supply side). This is increasing the operating and maintenance complexity and costs of the grid functions.

Case Study—New York Style DG

In the summer of 2015, the state of New York approved a bold new community initiative in distributed generation (DG) that would bring solar down to the masses. It will enable millions of New York state residents to access clean and affordable energy for the first time.

This is truly a game-changer; where now everyone—renters, homeowners, low-income residents, schools, businesses, and any and all kinds of facilities will be able to join together in the set up of 'shared renewable energy projects,' most of which will be solar installations on rooftops. This will result in cheaper electricity, less pollution and will build better, healthier and stronger communities.

The so called 'Shared Renewables initiative' is a new way of using solar, bringing it to a level where it has never been before. "This program is about protecting the environment and ensuring that all New Yorkers, regard-

less of their zip code or income, have the opportunity to access clean and affordable power. Together, we will build a cleaner and greener New York," according to Governor Cuomo.

This is good, of course, but New York is not the sunshine capital of the world. On the contrary; it is usually cloudy, foggy, drizzly, snowy, and smoggy place. How much electricity the rooftop solar panels will produce under these reduced sunlight conditions remains to be seen. The excess city dust and smog will make things even more difficult for the solar equipment on rooftops close to the ocean and the roads, to generate enough electricity to justify its existence.

It is an interesting and useful concept, so we hope that it works.

The Future of DG

DG has been in the news for a long time, and are now becoming the focus of the energy industry. DG policy debates are also raging across the country, but mostly at the state level. No federal action has been even mentioned until the summer of 2015.

Sen. Angus King of Main is trying to change that with his Free Market Energy Act of 2015 (FMEA, 2015). It is actually an amendment to the old Public Utilities Regulatory Policy Act (PURPA) of 1978. Its intent is to alter the energy (electricity) market structures by changing the regulations for DERs across the nation. The bill would provide for the right of DG to connect in a reasonable timeframe and only be charged just and reasonable fees

FMEA, 2015 would codify a right of interconnection for DG, in order to change the policy framework to better assess their full value and encourage their proliferation nationwide. With other words, it would establish a clear right for all citizens to generate electricity and interconnect from their homes. It sets of criteria for the states to look at DG in terms of setting reasonable and just interconnection costs and backup charges; including unbundling rates, time-of-use, location and other things that will make DER most valuable to the homeowner and the national grid in general.

FMEA, 2015 would direct states to consider mechanisms such as time-of-use pricing, specific location values, and the societal benefits of DG, among other key components. This would incentivize rational behavior in the energy market and better account for the two-way value of DERs-to-grid and grid- to-DER.

The bill also introduces changes to the concept of a Qualifying Facility (QF), which thus far has been any entity that an electric utility is federally mandated to buy from under the 1978 PURPA act. PURPA's intent was to ensure that the nation's utilities were sourcing their power from a diversity of companies, not just themselves. The lawmakers were basically concerned that the utilities were stifling competition and not allowing other companies, especially minority-owned ones, into the energy market.

So the U.S. utilities are mandated to buy power from any entity with a QF designation at the utility's avoided cost, but the concept was applied only to larger-scale entities like generation companies.

Note: Avoided cost is the dollar amount that the utility would have paid to build, or purchase, the additional energy resource in question, had it not come from a qualified QF. The avoided costs are typically assigned by the state commissions, and that structure has not been challenged by the new bill.

The bill, however, would prohibit state commissions from setting fixed fees for customers higher than $10 per month. Fees increase is used by many U.S. utilities to mitigate the additional cost and lost revenue caused by the proliferation of DG. DG, according to the utilities, is the reason for stagnant power generation load, and increased maintenance expenses.

Small DG sources do not have QF designation presently, so FMEA, 2015 is focusing on this point. Under it, if a state does not consider and implement unbundled rates in an attempt to better value DG, then this state's *DG sources will be given QF status*. In cases where the states refuse to implement the unbundling practices under the new bill, they would be forced to adopt retail net metering instead.

This is a radical shift from the status quo and would provide an incentive for DERs of any size—not just large entities. FMEA, 2015 clarifies and changes the way QFs are designated. It creates a federal obligation for the utilities to buy as much power as any qualified QF wants to sell at the utility's avoided cost.

A big step in the right direction, this bill is. It may not pass this time, but the time is right, so eventually we will see DG with QF designation, as well as many other bills that bring DG to its rightful place in the nation's energy mix. This, and other similar energy bills, would eventually elevate the renewables' status as a *base load* and level the competition field somewhat. The game is just now starting.

Case Study. Project Sunroof

Google and solar go hand in hand. Or rather they have been trying to do so for awhile now, with Google looking into and participating in several energy projects.

The latest effort is for Google to use its extensive slate of mapping data and computing resources, to dive right into the rooftop PV business.

"Project Sunroof" is the latest in Google's arsenal of solar weapons, aimed to assist potential customers that are interested in installing rooftop solar by providing charts of their rooftops and calculating project size, costs, benefits, and savings. Not exactly a new concept, since most solar companies offer similar service for free, but Google is looking for a niche for its technology, so the trials continue.

And so, as of the summer of 2015, the lucky residents in Boston, Fresno, California and the San Francisco Bay Area can participate in the project by logging onto the website and simply entering their address into the program. Google will then analyze the building's rooftop, and provide data (estimates) on how much sunlight is available on the roof over the span of a year and what can be expected from a solar system installed on it.

Project Sunroof can also recommend installation size for a PV system on the roof, with emphasis on providing enough energy to cover close to 100% of the building's electricity costs. Various pricing options will be provided, as needed to help the solar system owners to weight the available finance options, including leasing, loan, or buyout. When the customer is ready to go ahead with one of the installation and finance options, Project Sunroof will connect him/her with local solar providers to negotiate the next step.

Presently, Google would recommend only a few solar installers, in the likes of SunEdison, NRG, and SunPower. Earlier in 2015, SunEdison announced expansions into the rooftop PV market, offering residential power purchase agreements (PPA) in various regions across the US. This, according to the experts is a validation of the rapid growth in the US residential solar market.

So Google is positioning itself between the demand and supply ends of the solar game. While we see the practicality and usefulness of this service, making money from it is not that clear.

No matter what happens in the short term, the renewable energy sector depends primarily on the national and state politics, and the related energy policies and regulations.

Political currents can make or break the renewables.

U.S. Politics

The renewables in the U.S. have an agonizing history, which at times resembled cat and dog relationship.

Driven mostly by government subsidies, the renewables were born in the 1970s after the Arab embargo frenzy. Shocked and looking for quick fixes, the U.S. government poured billions of dollars in developing solar energy alternatives.

By the 1980s, the oil shortages became history, and so solar, wind, bio-fuels and other renewable energy sources were pushed on the back burner. Soon after—when oil prices dropped—they were shoved carelessly in the closet.

The on-again, off-again cycle was repeated several times since 1978.

The last cycle—and one that is quite different from the rest—started in 2008. Exactly 30 years from the first renewables renaissance, the renewables were pulled out of the closet, dusted off, given a huge financial injection, and put to work. The difference is that this time, they grew so fast and furious that there is no way to put them back in the closet.

During the 2008-2012 period of renewable energy renaissance, the renewables were propped up with billions of dollars. They were encouraged to grow very large quickly, and so they did. Now, they are looking for their own place in the energy markets. Success is smiling at them this time.

But regardless of the successes and obvious benefits, U.S. politicians have a different agenda. Now, oil is cheap again, so they are slowly removing the badly needed life support from the renewables. A wave of rulings in several states, designed to tamper with Renewable Portfolio Standards (RPSs) and tax credits for solar and wind installations in the U.S., were introduced in 2014 and 2015. The trend seems to be growing and could eventually prove to be a significant setback for the progress of solar deployment in the U.S.

The RPSs in different states require a different level of energy production from renewable sources, such as wind and solar, to be generated (or purchased) by the local utility companies, thus setting green targets for the individual states.

Texas, for example, has an RPS target of introducing 10% renewable energy by 2015, which was first introduced and agreed upon by the lawmakers way back in 1999. At the 11th hour, the Texas Senate passed a bill, which would scrap the state's RPS as of 2016. It will also undo a billion-dollar competitive renewable energy zones (CREZ) initiative, which many states are still working on. Why is CREZ on the chopping block is a long story, but it suffices to say that it is in conflict with

some local personal and business interests.

Following Texas' example, North Carolina's House leaders also introduced a bill to reduce by more than 50% the state's RPS target too. It is now to be reduced from 12.5% to 6%, which goal is to be reached by 2018 instead of the original target date of 2021.

RPSs have been a driving force to the future development of clean energy in the U.S., and tempering with their original goals and targets is like taking the wind out of the renewable energy's sails. The legislative actions of late are a direct assaults on renewable energy, including solar, by conservative groups funded by oil companies and billionaires.

So what we are looking at now in the U.S., is a trend of changes designed to remove all support from the renewable. Natural gas is the kid on the energy block and all eyes are on it. Solar and wind have to wait their turn again.

On top of that, some states are planning to raise taxes for people who use solar. Such moves put many states out of sync with the national and global movement to adopt solar energy and seriously jeopardizes its further progress.

This simply means that solar, wind and the rest of renewables are at major crossroads and must chose between going back in the closet, or trying to step up on their own feet. Having done that, they must learn how to walk and then run without any support. This is a critical period, and as a result, there is a battle going on many levels in many states. Who is going to win is the big question now.

$4 Billion Investment

In 2015, President Obama announced a $4 billion of private-sector commitments to help fund innovation in clean technologies. These commitments were made by a number of private foundations, private and institution-

Figure 10-25. President Obama

al investors. They are launching several clean energy investment alliances designed to connect would-be investors with clean-energy firms, in order to significantly reduce the transaction costs of investing in new and developing technologies.

The $2 billion set by the White House earlier in 2015, also comes from hundreds of different organizations. To further encourage private-sector investment in clean energy, the White House is also launching a Clean Energy Impact Investment Center at the U.S. Department of Energy. The goal of this new Center is to make information about energy and climate programs more transparent and accessibly by the public and the investors.

The White House has also committed to aid investments by charitable foundations in clean energy technologies, by releasing new Treasury Department guidance on impact investing and by clarifying the role of investing for charitable aims. It will also improve financing options for private investment funds in their search for long-term capital, including early-stage investors in capital-intensive clean energy technologies.

These are steps in the right direction for the embattled renewables sector, which also show the trend of shifting government responsibility to the private sector. This is the first step in weaning the renewables from government funds, and teaching them to walk on their own.

The U.S. Solar Zones

One thing we here in the U.S. have—which is a great advantage for solar power development—is the availability of huge, idling, sun0bathed desert areas. In order to allow the expansion of solar in these areas, the US Bureau of Land Management (BLM) issued a new regulation in the summer of 2013. It is geared to facilitate the development of solar and other renewable energy projects on public land.

Previously, public land included in proposed solar or wind energy right-of-way applications remained open to mining claims (fossil fuels). The new rule is designed to avoid possible conflicting applications to use public land either for renewable energy projects or mining. But the new regulation means the BLM now has the authority to temporarily remove lands included in a renewable energy right-of way application and lands offered for wind or solar energy lease from land appropriations such as mining claims.

Public land segregated under the new regulation will be reserved for solar or wind energy development for up to two years. The new regulation replaces an interim 'final rule' brought in by the BLM in November 2011 that first proposed segregation of public land.

Before the final rule came into effect 216 new mining claims were located within solar energy right-of-way application areas. The segregation will simplify the administration of public lands by making such mining claims impossible, according to BLM.

Table 10-14 lists the new U.S. solar energy zones.

Table 10-14. U.S. solar zones

Arizona
- Brenda (Lake Havasu/La Paz) 3,348 acres/372 megawatts
- Gillespie (Lower Sonoran/Maricopa) 2,618 acres/291 megawatts
  Total 5,966 acres/663 megawatts

California
- Imperial East (El Centro/Imperial) 5,717 acres/635 megawatts
- Riverside East (Palm Springs–South Coast/Riverside) 147,910 acres/16,434 megawatts
  Total 153,627 acres/17,070 megawatts

Colorado
- Antonito Southeast (La Jara/Conejos) 9,712 acres/1,079 megawatts
- De Tilla Gulch (Saguache/Saguache) 1,064 acres/118 megawatts
- Fourmile East (La Jara/Alamosa) 2,882 acres/320 megawatts
- Los Mogotes East (La Jara/Conejos) 2,650 acres/294 megawatts
  Total 16,308 acres/1,812 megawatts

Nevada
- Amargosa Valley (Southern Nevada/Nye) 8,479 acres/942 megawatts
- Dry Lake (Southern Nevada/Clark) 5,717 acres/635 megawatts
- Dry Lake Valley North (Ely/Lincoln) 25,069 acres/2,785 megawatts
- Gold Point (Battle Mountain/Esmeralda) 4,596 acres/511 megawatts
- Millers (Battle Mountain/Esmeralda) 16,534 acres/1,837 megawatts
  Total 60,395 acres/6,711 megawatts

New Mexico
- Afton (Las Cruces/Dona Ana) 29,964 acres/3,329 megawatts
  Total 29,964 acres/3,329 megawatts

Utah
- Escalante Valley (Cedar City/Iron) 6,533 acres/726 megawatts
- Milford Flats South (Cedar City/Beaver) 6,252 acres/695 megawatts
- Wah Wah Valley (Cedar City/Beaver) 5,873 acres/653 megawatts
  Total 18,658 acres/2,073 megawatts

Note that the new solar zones are located in the sunniest states of the country. Grand total of 284,918 acres desert land, capable of generating about 31,658 MW electric power, is presently available in the U.S.

The BLM move basically means that the U.S. is ready to add 31.66 GW of power, which is about the power generated by 20-30 nuclear power plants. Not bad as a start. But keep in mind that this is only a small part of the total land surface available in the above listed states.

It is for obvious reasons that the new energy zones are located in the sunniest (desert like) states of the country. A logical move, since it makes no sense to generate solar power in states with limited solar insolation. As a matter of fact, each of these desert states could easily provide the entire U.S. electric power load, using combination of solar and wind power plants. Adding hydrogen and other energy storage for steady power generation could make the country totally solar powered and energy independent.

Another pipe dream, yes!...but a feasible one; the most plausible large-scale power generation in the U.S. for the next century and beyond. This is the best the future generations could hope for, after we burn all the fossils, which won't take that long.

Case Study—Land Restrictions

In 2010, the US-based solar developer, Iberdrola Renewable, USA, a Spanish energy company, applied for a 200 MW photovoltaic solar project on 1,616 acres about 10 miles north of Baker in the Silurian Valley, San Bernardino County, Southern California. After undergoing a rigorous review process, the project was determined to not be in accordance with the BLM's Western Solar Plan, enacted in 2012. And in the fall of 2014, the US Bureau of Land Management (BLM) rejected the application for the 200 MW project after it was deemed to be "not in the public's interest."

The BLM's Western Solar Plan established 17 solar energy zones across the South Western states of Arizona, California, Colorado, Nevada, New Mexico and Utah, for prioritized large-scale solar development. The plan also allows for some solar development outside of the zones, but such projects must remain in line with the BLM's requirements and require special evaluation and permits by BLM and other authorities.

The Iberdrola Renewable project fell within this category. While evaluating this special case, BLM found that the environmental data and information from public and local, state, federal and tribal governments indicated that Iberdrola Renewable's project would be too negatively impactful to the proposed area of the Silurian Valley, which is known as a largely undisturbed valley full of wildlife. It is also considered an important piece of the Old Spanish National Historic Trail and is well known for its recreational and scenic values.

And so, the 200 MW Iberdrola Renewable solar project is now history. It is the first project to be denied through the variance process, and a lesson to stay within the BLM's approved zones for solar energy development.

When the BLM announced its solar zone plan in 2012, a number of organizations, such as the Sierra Club and others emphasized the positive aspects of developing large-scale solar, while stressing the importance of minimizing the potential environmental and wildlife impact.

The implications of minimizing the environmental effects was reflected in the now infamous Ivanpah solar plant, in California's Mohave Desert. It was initially designed as 440 MW facility, but the final project is 392MW—over 10% reduction—due to land use issues. Some parts of the 4,000 acres were sensitive to wild life. The lesson here is that the desert is a living and thriving portion of our ecology, and must be preserved as such.

Consequently, the BLM has approved 18 solar, wind and geothermal projects on public lands in California since 2010; all within the BLM's approved zones.

In summary, the majority of solar technologies have a short track record, so we really don't know what will happen in the different solar fields after 25-30 years of non-stop operation. This is especially true for installations in the world's worst places; the deserts, coastal, and swampy areas.

The encouraging aspect of the performance thus far is that there are a number of documented tests and long-term installations with flawless record. The overwhelming conclusion from all available documentation, including our own tests through the years, is that the quality, performance and longevity of the solar equipment and systems is directly related to the quality of materials, equipment, and processes used in their production, as well as the performance of the personnel involved in the entire process.

Basically speaking, well-established, reputable companies do have the ability and experience to produce a world-class product, which functions per the manufacturer's specs. At the same time, a number of newcomers and low-cost manufacturers are working hard at getting their manufacturing and customer support processes streamlined, and we expect that many of them will be successful.

The solar energy technologies can be successfully used in many areas of the world, but we'd like to caution the reader that in many cases, it is not the best solution to the energy crisis, as some portray it. Solar is most efficient in certain geographic areas and under special conditions. Due to its temperature and humidity dependence, we'd suggest that all solar technologies could be used without limitations or restrictions, in areas with moderate temperatures and humidity.

Their use in area with extreme climate, such as the deserts and humid areas of the world, however, is not well proven. In our humble opinion, any solar technology operating non-stop in the deserts will have to be tested and proven during 25-30 years operation before considered efficient and reliable. This has not been done until now, simply because the solar industry has not been around this long. Because of that, we expect to see some failures and other surprises in the world's solar fields in the near future. We just hope that they are not big enough to cause harm to the solar industry as a whole.

In all cases, more research is needed to understand the desert, and other weather related, phenomena. This could be done only by appropriate long-term tests, which must be conducted to verify the technologies performance and longevity under these conditions.

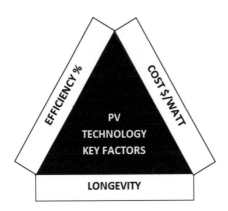

Figure 10-26. PV's key factors

Looking at the present status quo of the solar technologies, see Figure 10-26, we see that cost and efficiency goals are constantly improving, and the work in these areas continues to set new records on daily basis. The progress in longevity (durability and reliability) areas, however, is not so clearly defined. Some tests show good results, including field tests done under normal conditions. The above-mentioned extreme climates, however, do and will continue to present serious problems to all solar technologies.

The emphasis, therefore, should be on optimizing their longevity. The issues must be resolved well before we embark on deploying mega plants in the U.S. deserts and humid areas. Since these areas are the most promising for large-scale PV installations, we all must be aware of the limitations, which must become priority #1 for manufacturers, installers, investors and operators alike. Its not late.

Biofuels

Biofuels are expected to continue to be part of the energy mix, by providing 10-15% of all fuel (bio-ethanol and bio-diesel) used for powering motor vehicles. This is the good news. Looking forward, however, we see other factors playing important role in the production and use of biofuels.

- A lot of energy in form of electricity and heat are used in the production of biofuels. Starting with the crops growing, a lot of diesel fuel is used by the agricultural machinery working the fields.
- Millions of tons of nitrogen-rich fertilizer is used in the corn production too, which also requires a lot of fossil energy to be produced, transported, and spread onto the fields.
- Biofuel crops are also water intensive. Over 30 billion gallons of water are needed annually to grow crops and process them into biofuels. Some countries, like Brazil, are blessed with lots of rains and other water sources, but this is not the case in most other countries,
- In addition to using a lot of fossil fuels and water, the biofuels processes are polluting as well. Carbon emissions analysis for biofuels show that the carbon life cycle assessment (LCA) depends greatly on the type of technology, and types of fossil fuel used.

Considering all these facts we see that:

- Using corn ethanol results in a 20% reduction in CO_2 emissions per MBtu (million British Thermal Units) when substituting for gasoline, and
- The reduction is 40% for biodiesel from U.S. and Europe produced corn biofuel, and 78% for Brazilian made sugarcane such.
- Increasing biofuels production Increases the price of food, by diverting the wheat, soy, and corn production from production of food to filling the gas tanks,
- Given the increased tax relief and other incentives for farmers producing biofuels, a situation might

occur in which more fuel is used than is produced; an obvious misallocation of resources.

These abnormalities could result in some major, unforeseen, and unwanted developments, such as:
- Subsidies for biofuels lead many farmers to switch to corn,
- Fewer food crops (like soybeans and wheat) lead to higher food prices,
- Higher soybean prices lead Brazilian soybean farmers to expand, displacing cattle ranchers, and
- Cattle ranchers are forced to clear new pastures out of the Amazon.
- The deforestation of the Amazon rainforest and other world forests alone most likely outweighs any and all the positive externalities which biofuels were supposed to bring. The removal of millions of trees decrease the plants number and their overall CO_2 consuming potential, thus driving the world's CO_2 to higher levels faster.

There are also positive externalities associated with biofuel production:
- The main positive externality is the fact that the vegetation growth process leads to decrease in the amount of CO_2 in the atmosphere at the growth location
- Increased transport needs can be met by usage of the biofuels as 'green' and clean fuels
- Production of biofuels from sugar beets uses much less fuel than it produces, therefore it could easily meet the increasing demand for fuel.

Good stuff, no doubt. We must however, consider the complexity of each step of the process, and its effect on the overall situation. In all cases, we must proceed carefully with the further increase of biofuel crops as transportation fuels. There future is quite promising, but not easy nor trouble-free.

Europe

European countries need renewable energy sources (RES) as much as, and even more than, most other developed countries. This is because they don't have much natural energy resources, and are forced to import a significant amounts.

RES are one, maybe the only, way to reduce dependence on imported energy and especially the burdensome Russian dominance of the European natural gas markets. And this is why Germany became a leader in solar installations during the solar frenzy in 2007-2012.

Spain, Italy, Portugal, and most other European countries followed their example and were following close.

They (the RES), however are variable in their output, and as such require special attention, since growing variable RES power generation affects the type and quantity of balancing (ancillary) services required to ensure reliability of the electricity system.

Studies indicate that in the near future, it will not be the fastest services that will experience the greatest increase in demand with rising shares of variable RES (i.e., primary reserves or frequency response able to respond in seconds). Instead, the services that will be in greater demand will be those that can respond *reliably and consistently* within minutes, to tens of minutes, to hours.

In order to achieve the task at hand, the RES in Europe (and actually anywhere else) must be able to offer the following capabilities.
- *Flexible, fast start-stop cycling* as needed to shut down and re-start energy generation multiple times within a reasonably short window of time and up to hundreds of times over the course of the year;
- *Regular, dispatchable ramping* is needed to reduce the energy resource to a low level of stable operation and ramp it back up at a specified rate, not in a traditional operating reserve role (contingency conditions) but as a normal-course ramping capability; and
- *Ramping capability* reserved now to be used in the future means a slower type of secondary reserves with flexible ramping capability, a type of ramping service, which can address issues arising in the tens of minutes (e.g., due to forecasting error).

Currently, ancillary services' capacities are usually set aside for a particular trade interval and deployed within that interval when pre-specified conditions are met. But with this ramping service, the ramping capability is set aside for later periods of time when more power is needed on scheduled or unscheduled basis.

These three capabilities are different in nature or degree from the capabilities of services that operators currently procure in balancing/ancillary services markets. These markets can indeed provide some flexibility (often a by-product of gas-fired plant) that is adequate for today's needs, but they would need to be adapted in order to attract the above-listed capabilities.

Also, existing balancing/ancillary services markets are generally short-term in nature with procurement of services just days or weeks ahead, whereas capacity mechanisms generally procure years ahead. Four years has been proposed in Great Britain and the other EU countries are contemplating similar approaches.

To attract investment in capacity with adequate capability, either the services markets or the capacity mechanisms, where they exist, should be respectively adopted and reformed to fit the investors' requirements. Following this path successfully, would bring Europe to a higher level in RES utilization and energy efficiency, while at the same time bringing a new dimension of flexibility to its energy markets.

This also means more independence from foreign energy imports, and less dependence on unreliable Russian natural gas supplies.

If this approach succeeds in Europe, then the entire world will follow in its steps and the renewables will advance a step further towards successful competition with the conventional fuels.

Case Study—UK Solar

All goals and predictions for future growth of solar worldwide are achievable, if we lived in an ideal world. We must, however, remember that there are a number of obstacles in the way. The major ones being the fact that solar energy is variable. It changes from one minute to the other, and—even worse—it is simply not available in some coastal and Northern-most locations.

Also, very importantly, not all countries have unlimited land expanse for large solar power plants. Even smaller number of countries have unlimited quantity of desert areas, which are most suitable for solar power generation. This limits the importance of solar energy as a significant source of electric energy for general use.

Recently, we also see an increase of concerns about using agricultural lands for solar farms. This trend is widely used in the U.S., UK and other countries, where environmental groups expressed their concern with the increase of solar farms on agricultural lands. If this trend continues, their theory goes, it might cause diminishing agricultural production and higher product prices.

The UK, for example, has reduced solar insulation during most of the year, due to clouds and ocean fog. It also has limited unused land mass, suitable for solar installations. This means limited land area for solar use. Not allowing the use of agricultural lands for solar installations would lead to further limitations of solar energy growth. The combination of these factors leads us to believe that solar is not going to be a major power supplier any time soon, or ever, in a place like the UK. The best we can expect in places with UK's conditions is limited growth of solar. As a matter of fact, the UK

environmental ministry is proposing expansion of solar installations mainly on residential and commercial roof tops.

Even in best of cases, however, this is not going to be enough for solar to be come a significant driver in the country's energy mix. Because of that, the UK and countries in its predicament, must look into other energy sources, such as wind, ocean, geo-, and bio-energy.

The situation is quite different in the U.S. Here, we have almost unlimited unused land mass, and huge abandoned agricultural areas. Most of these areas are in, or close to, the deserts. Since the deserts provide the greatest amount of solar insolation, the combination of abundant land and sunshine is quite favorable for the quick growth of solar energy in the country.

China

2015 is the last of China's current 12th Five-Year Plan for Solar Development (2011-2015). The results are staggering. In addition to becoming the world's largest producer of solar cells and modules, China is becoming the largest consumer of solar and wind technologies. The target for installed solar PV to be achieved by the end of 2015 was raised in total four times, from an initial 5GW in 2011, to 10GW the same year, to 20 GW in 2012 to a finally 35 GW in 2013.

But China now already has approximately 45GW of installed solar power capacity, exceeding the national target goals by over 20%. China is overtaking Germany in becoming the world's larger user of solar energy.

China's National Energy Administration (NEA), recently announced plans for 100 GW target of installed solar by 2020, which is now considered as a very minimum. There is a possibility of doubling this amount for the duration.

China aims for 200 GW of solar power by 2020, or as much solar as the entire world has now.

The interesting fact here is that the targets set for Concentrating Solar Power (CSP) are 5 to 10 GW by 2020, which is only 5-10% of the total solar capacity. This reflects the number of problems related to the construction and operation of CSP power plants, as we have observed in the U.S. and documented in this text. This is one of the major reasons we focus on PV as leading solar technology now and in the future, and do not spend much time discussing CSP in this text.

Other significant developments in the Chinese solar sector are a number of key tasks, one of which calls for abolishing the discrimination against renewable energies in dispatch terms, direct trading of power, and establishment of a market-based price mechanism, as needed to intensify the implementation of the renewable energy law and to actively promote distributed generation, including energy storage.

Distributed generation has been identified as a key priority for the coming five-year plan and it is anticipated that a series of related policies will be issued in due course.

Also, the revised Renewable Portfolio Standard (RPS), which was introduced in 2007, calls for a larger share of PV in the renewable energy mix in the respective portfolios. This is expected to drive demand for solar PV, because until now the RPS quota was left to the utilities, which preferred lower-cost hydro and coal power generation.

In addition, the US-China Climate Change Agreement signed in 2014, reflects China's commitment to helping the environment. By 2030, 20% of China's energy consumption would come from renewable sources. This is almost double (up from 11.1%) the 2014 levels. Achieving this ambitious target would require adding annual solar PV installations of up to 40GW in the decades to come. This is well above even the most optimistic expectations, so we just have to see how far China goes with this pledge.

China is also actively promoting its PV industry to not only export its products, and sell them in China, but to also move manufacturing installations abroad. To this end, over the course of the last three to four years China has facilitated solar projects in 40 countries around the world.

All this sounds like a step in the right direction. Yet, due to the rapidly increasing electricity demand, coal will remain king of the energy sector for many decades to come in the country. China's urbanization brings 20 million people each year from rural to urban areas.

By 2030 approximately 1 billion people will live in cities, which creates an enormous demand for on-site power generation and on-site storage solutions. This makes solar PV, in form of distributed generation and energy storage, king #2 in the Chinese energy sector. At least on paper it does.

It remains to be seen how many coal-fired power plants vs. rooftop solar installations will generate China's electric power in the near future. It is also unclear what effect moving manufacturing operations overseas will have on the global energy markets and environment. We just have to watch…

And speaking of energy storage as an indispensable companion of solar energy…

ENERGY STORAGE

Solar energy is here to stay and grow. There are several dark clouds on the solar (PV) horizon that puts a threatening shadow on its near term future; be it n-type, or p-type, or thin films. In addition to the political and financial developments, which promise to bring more difficulties, the wind and solar are still immature and suffer from technical problems.

The biggest problem is the fact that solar and wind power generation—even in best of cases—is variable.

Clouds during the day cause restricted power generation, and no power at night time make the solar fields idle, which is a bad thing. The same is true for wind farms, where winds of variable speed and direction are causing variable, or no output at all. This forces the utilities to crank up power from other sources, which makes their job harder and the generated power more complex and expensive to manage.

Energy storage is the best—albeit still immature and overly expensive—solution to the variability issues.

Energy storage is seen as the one and only solution to solar and wind power variability.

But energy storage is complex, underdeveloped, unreliable, and very expensive, so we don't expect energy storage—especially large-scale such—any time soon. How this will play in the future depends largely on governments' support. Without it, progress will be very slow and painful.

Energy storage technologies are designed to absorb energy whenever possible, store it for a period of time, and then release it whenever needed. This is done in order to provide continuous supply of energy for on-demand services. This way energy technologies can eliminate (or at least reduce) variability and temporal gaps between energy supply and demand.

Theoretically, energy technologies can be implemented in large and small scale, as well is in distributed and centralized manner throughout the energy system. We see energy storage especially difficult, if not prohibitive, in large-scale solar and wind power generation. At least at the present day state-of-the-art.

Although supporters make it sound good on paper, accomplishing these goals is very difficult and extremely expensive. The energy storage equipment and processes are in the early stages of development, which requires a lot of time, effort, and expense to bring them to the energy markets.

Energy storage technologies differ by the type of energy they can store; mainly electric power and thermal energy (heat or cold). These technologies can also serve simultaneously as generators and consumers, which gives them the potential to link the currently disconnected energy sources (wind and sola) on one hand, and the energy markets (power generation, transportation fuels, and local heat/cooling) on the other.

In this respect, the energy storage concept can be viewed as a system integration technology, which, when fully developed will facilitate and optimize the management of energy supply and demand. At its best, a single energy storage infrastructure can provide multiple valuable energy and power services, which would benefit greatly the energy markets.

The greater effort today is focused on energy storage technologies connected to larger energy systems, such as distributed solar, while solar and wind power plants and the electricity grid are next...much later. As a clarification, the efforts to develop and implement distributed and off-grid energy storage technologies, due to their complexity and high cost, seem to be concentrated in opulent countries. The U.S. and some European countries lead the way in this area, but are still far from complete success.

Background

Basically speaking, energy storage technologies are valuable components of the energy markets and could be an important tool in:

- Eliminating the variability issues with solar and wind power generation in the short term,
- Improving the efficiency of the existing power generators and the power grid, in the mid-term, and
- Help with achieving a low-carbon future in the long run.

Presently, energy storage technologies are predominantly installed to take advantage of

a) Dispatchable energy supply resources, and
b) Variable power demand.

The increasing energy prices and emphasis on de-carbonization are augmenting the need for advanced energy storage technologies. Increased resource use efficiency, such as storing waste heat in thermal storage, is supporting the use of the variable solar and wind energy resources. In the future, energy storage will be considered from a system point of view with a focus on the multiple services it is capable of providing in bulk,

and many other applications.

There is intensive R&D underway in the design, manufacturing and implementation of different energy technologies. The primary goals thus far are technological optimization, cost reductions, and improving the performance of existing technologies. New and emerging storage technologies are also developed, but they will take much longer to reach the energy markets.

There are also some non-technical issues, that are being addressed by governments and interested parties. The most important of these; policies, regulations, permitting, interconnect, and other barriers to deployment are on the agenda and many of these will be resolved in the near future.

Thus far, the most important drivers for increasing use of energy storage are:

- Improving energy system resource use efficiency
- Increasing use of variable renewable resources
- Rising self-consumption and self-production of energy (electricity, heat/cold)
- Increasing energy access (e.g. via off-grid electrification using solar PV technologies)
- Growing emphasis on electricity grid stability, reliability and resilience
- Increasing end-use sector electrification (e.g. electrification of transport sector).

The Technology

Energy storage technologies include a large set of centralized and distributed designs that are capable of supplying an array of services to the energy system. Storage is one of a number of key technologies that can support global de-carbonization.

- Energy storage technologies are valuable in (most) energy systems, with or without high levels of variable renewable generation. Today, some smaller-scale systems are cost competitive or nearly competitive in remote community and off-grid applications, although high initial costs keep their growth to the minimum. Large-scale thermal storage technologies are competitive for meeting heating and cooling demand in some regions. Here the problems are mostly of technological nature, since it is.
- Individual storage technologies often have the ability to supply multiple energy and power services. The optimal role for energy storage varies depending on the current energy system landscape and future developments particular to each region.
- To support electricity sector de-carbonization, an estimated 310 GW of additional grid-connected

electricity storage capacity would be needed in the U.S., Europe, China and India. Significant thermal energy storage and off-grid electricity storage potential exists, but additional data is required to provide a more comprehensive assessment at the respective national levels.

• Market design is key to accelerating deployment. Current policy environments and market conditions often cloud the cost of energy services, creating significant price distortions and resulting in markets that are ill-equipped to compensate energy storage technologies for the suite of services that they can provide.

• Public investment in energy storage research and development has led to significant cost reductions in recent times. However, additional efforts (e.g. targeted research and development investments and demonstration projects) are needed to further decrease energy storage costs and accelerate development.

• Thermal energy storage systems appear well-positioned to reduce the amount of heat that is currently wasted in the energy system. This waste heat is an underutilized resource, in part because the quantity and quality of both heat resources and demand is not fully understood and applied in practice.

Application	Output (electricity, thermal)	Size (MW)	Discharge duration	Cycles (typical)	Response time
1. Seasonal storage	e,t	500 to 2 000	Days to months	1 to 5 per year	day
2. Arbitrage	e	100 to 2 000	8 hours to 24 hours	0.25 to 1 per day	>1 hour
3. Frequency regulation	e	1 to 2 000	1 minute to 15 minutes	20 to 40 per day	1 min
4. Load following	e,t	1 to 2 000	15 minutes to 1 day	1 to 29 per day	<15 min
5. Voltage support	e	1 to 40	1 second to 1 minute	10 to 100 per day	millisecond to second
6. Black start	e	0.1 to 400	1 hour to 4 hours	< 1 per year	<1 hour
7. Transmission and Distribution (T&D) congestion relief	e,t	10 to 500	2 hours to 4 hours	0.14 to 1.25 per day	>1 hour
8. T&D infrastructure investment deferral	e,t	1 to 500	2 hours to 5 hours	0.75 to 1.25 per day	>1 hour
9. Demand shifting and peak reduction	e,t	0.001 to 1	Minutes to hours	1 to 29 per day	<15 min
10. Off-grid	e,t	0.001 to 0.01	3 hours to 5 hours	0.75 to 1.5 per day	<1 hour
11. Variable supply resource integration	e,t	1 to 400	1 minute to hours	0.5 to 2 per day	<15 min
12. Waste heat utilisation	t	1 to 10	1 hour to 1 day	1 to 20 per day	< 10 min
13. Combined heat and power	t	1 to 5	Minutes to hours	1 to 10 per day	< 15 min
14. Spinning reserve	e	10 to 2 000	15 minutes to 2 hours	0.5 to 2 per day	<15 min
15. Non-spinning reserve	e	10 to 2 000	15 minutes to 2 hours	0.5 to 2 per day	<15 min

Figure 10-27. Types of energy storage and application characteristics (See legend opposite.)

Non-spinning reserve is at >15 minute response time. Faster response times are generally more valuable to the system. Energy reserve capacity is also referred to as "frequency containment reserve."

U.S. Energy Storage

Energy storage is a fairly newcomer to the U.S. energy sector. It is, however, quickly growing in importance, mostly because of its unique ability to reduce the variability of solar and wind power generators. This variability is what is stopping the wider development of these technologies, so energy storage is considered their "savior." Not an easy task, non the less.

A lot of companies are working on energy storage technologies and options. In the fall of 2014, the U.S. DOE's Sunshot initiative announced $15 million in funding of the development and integration of energy storage technologies into the electrical grid.

The new funding opportunity, branded "Sustainable and Holistic Integration of Energy Storage and Solar PV" (SHINE), is actively looking for projects "enabling the development and demonstration of integrated, scalable, and cost-effective technologies for solar that incorporate energy storage and work seamlessly to meet both consumer needs and the needs of the electricity grid."

The emphasis is on distributed solar power generation, rather than on large-scale, utility-scale plants. This is because inexpensive and flexible solutions to distributed generation energy storage could be deployed quickly and integrated readily into the grid. The power generation, and the related storage solutions are smaller in size, thus more feasible to implement into successful schemes.

Presently, California, Hawaii and New York have the most energy storage, mostly due to favorable policy.

Legend:

1. Seasonal energy storage is the ability to store energy for days, weeks, or months to compensate for a longer-term supply disruption or seasonal variability on the supply and demand sides of the energy system (e.g. storing heat in the summer to use in the winter via underground thermal energy storage systems).

2. Arbitrage (or energy storage trades) represents storing low-priced energy during periods of low demand and subsequently selling it during high-priced periods within the same market. Similarly, arbitrage refers to this type of energy trade between two energy markets.

3. Frequency regulation is the balancing of continuously shifting supply and demand within a control area under normal conditions. Management is frequently done automatically, on a minute-to-minute (or shorter) basis.

4. Load following is the second continuous electricity balancing mechanism for operation under normal conditions, following frequency regulation, is load following. Load following manages system fluctuations on a time frame that can range from 15 minutes to 24 hours, and can be controlled through automatic generation control, or manually.

5. Voltage support is the injection or absorption of reactive power to maintain voltage levels in the transmission and distribution system under normal conditions.

6. Black start is encountered when the power system collapses and all other ancillary mechanisms have failed. In such cases, black start capabilities allow electricity supply resources to restart without pulling electricity from the grid.

7.-8. T&D congestion relief and infrastructure investment deferral is a condition which energy storage technologies use to temporally and/or geographically shift energy supply or demand in order to relieve congestion points in the transmission and distribution (T&D) grids or to defer the need for a large investment in T&D infrastructure.

9. Demand shifting and peak reduction occurs when energy demand can be shifted in order to match it with supply and to assist in the integration of variable supply resources. These shifts are facilitated by changing the time at which certain activities take place (e.g. the heating of water or space) and can be directly used to actively facilitate a reduction in the maximum (peak) energy demand level.

10. Off-grid energy is usually used by consumers who frequently rely on fossil or renewable resources (including variable renewables) to provide heat and electricity. To ensure reliable off-grid energy supplies and to support increasing levels of local resources use, energy storage can be used to fill gaps between variable supply resources and demand.

11. Variable supply resource integration is the use of energy storage to change and optimize the output from variable supply resources (e.g. wind, solar), mitigating rapid and seasonal output changes and bridging both temporal and geographic gaps between supply and demand in order to increase supply quality and value.

12. Waste heat utilization is used by energy storage technology for temporal and geographic decoupling of heat supply (e.g. CHP facilities, thermal power plants) and demand (e.g. for heating/cooling buildings, supplying industrial process heat) in order to utilize heat that is usually wasted.

13. Combined heat and power cycle is used in combined heat and power (CHP) facilities in order to bridge temporal gaps between electricity and thermal demand.

14.-15. Spinning and non-spinning reserve capacity is the capacity of the electricity supply to compensate for a rapid unexpected loss in generation resources in order to keep the system balanced. This reserve capacity is considered spinning at <15 minute response time.

These states are ranked as global leaders in the energy storage sector, although at this early stage of the game it is hard to tell who and what will win at the end. Nevertheless, nine companies in the state of New York will receive $250,000 each to develop prototypes for commercialization of their energy technologies.

There are also existing schemes in support of energy storage in Germany, and recently funding and storage policy announcements were made by the governments of India and Singapore. So energy storage is on the agenda of major governments, so we should expect some good results soon.

It would be simply impossible for the renewables—solar and wind in particular—to compete with the conventional energy technologies without efficient and reliable storage. And this is why the DoE and other government bodies are interested in developing new and better energy storage technologies.

DOE's requirements under SHINE are for energy technologies to be able, in addition to storing energy, to use smart technology, such as metering and smart inverters, all of which must be capable of providing a level of demand response, in order to incorporate solar power generation and forecasting of power loads. The DoE also requests that they feature "easily interoperable hardware, software and firmware technologies." A lot to ask from a toddler technology, but it is what it is.

The SHINE program will fund a number of energy storage related projects, but the awarded companies and projects are require to provide at least 50% of the project cost. Individual awards will range from $500,000 to $5 million. The rewards and the work on the program started in mid-summer of 2015, and the results are expected to be announced in 2016-2017.

The only question here is, "is this enough money for such and important technology?" When fully developed, the energy storage technology is expected to be deployed in thousands of projects world-wide at a cost of billions of dollars. So what can $15 million do to speed up this important development? We don't expect much, but this is nevertheless a good start of an important effort, which we expect to take several decades to develop fully functional, efficient, and cheap energy storage systems.

Note: The Sunshot program is a national effort, funded by DOE, in support of the development of solar energy generation technologies and projects in the country. Sunshot was started in 2011, and has funded over 350 projects thus far. It would benefit tremendously from the introduction of energy (electricity) storage technologies, so the effort is expanding.

One of the major developments in this area is the gigantic effort of Tesla to produce millions of cheap Li-ion batteries for residential and small business energy storage applications. While this endeavor is feasible from a technical point of view, it has to overcome a number of logistics and financial hurdles. Just because we have something that works, does not mean that we all can, or want to use it, so the market will have the last word on this matter.

Try as it may, however, Tesla will not be able to provide efficient, durable, and cheap battery storage for large-scale solar and wind operations any time soon. There are a number of technological issues that have to be resolved first, before the technology can be deployed successfully in the energy market. At that time, it has to fight a set of logistic and financial battles, which its smaller cousin (residential energy storage) is fighting now.

Electric Vehicles

Electric vehicles (EV) are touted as the transportation of choice for the future by some. We don't know if this is so, or not, but we are sure that EVs are *not* the only, nor even the biggest, solution to our future transportation needs. Putting all our eggs in one basket is never a good idea and instead, it is often quite problematic.

Addressing environmental concerns is the major reason given for owning an EV today. Is this simple ignorance, or exuberant extravagance?

We must remember that charging EVs requires electric power, most of which is generated by fossil-fired power plants.

The other benefits EVs are expected to provide is greater flexibility by offering the possibilities of, a) night time charging, and b) energy storage. With that, however, come a number of problems.

- *Night time charging of EVs would be* an advantage for now, since only a limited number of EVs are plugged up in the grid for charging at night. These vehicles utilize the lower rates and unused capacity of the base load power plants, thus contributing to the energy-environmental balance...for now

If and when the number of EVs exceeds that of conventional cars, then the picture will change. If a day arrives when millions of EVs will be sucking power from the grid all night long, then the base load plants might not be able to keep up with the demand. In this case, the auxiliary power plants, the peakers, will have to be fired at an additional expense and increased environmental pollution.

At such distant time, the electric power demand would be almost equal during day and night. In this case, the conventional power plants will be required to generate power at full load 24/7, which would require twice the amount of fossil fuels.

Extra fossil fuels used at increased cost and extra pollution during night time charging defeats the purpose of EVs.

But EVs offer other, yet unexplored, possibilities:

- *Energy storage* is seen by some as a possible collective agreement to charge millions of EVs at night and then release the stored electricity into the grid during the day hours. This might sound as a plausible solution to environmental and socialist minded groups, but we don't see it as a practical solution in the U.S. The major reason is that it would be hard to impossible—short of introducing and enforcing appropriate laws—to get Americans out of their SUVs and RVs, and squeeze them into compact EVs. Let alone make them plug and unplug them at predetermined time intervals, which dictate their travel plans. It is quite hard to imagine such a thing happening any time soon, if ever!

Another option, however, that of re-purposing of EV batteries, sounds more feasible. Li-ion batteries used in EVs lose their efficiency and have to be replaced every 5-7 years. These used batteries are still capable to retain most of the charge, so instead of sending them in the waste disposal sites, they can be re-purposed for use as energy storage systems (ESS) in stationary settings. Albeit temporary, they could serve an useful propose.

This is a promising twist of technology use, capable of supporting improved management of demand and supply of electricity. If and when proper methods and systems for re-using these batteries are developed, we could rely on them since the supply of such (used) batteries in the EVs' (long-term) future would be almost unlimited.

Simulation of residential energy profiles and regulated cost structure have been used by scientists to analyze the feasibility and cost savings from re-purposing EV batteries. In addition to providing cheap initial setup, since these used batteries would be free of charge, thus created power centers would offer possibility of mass energy storage and even peak-shifting; none of which is possible today.

In situ residential energy storage, and larger commercial Li-ion battery based energy storage facilities, can contribute significantly to the implementation of mass energy storage and smart grid solutions by supporting the reduction of demand during typical peak power use time periods.

Although ESS would increase household energy use at low power demand periods, it would shift the peak demand, thus potentially improving the economic effectiveness of electric power and reducing GHG emissions. Recent research efforts also confirm the effective use of financial incentives for Li-ion battery reuse in ESS, which includes lower power rates and auxiliary fees.

Residential Energy Storage

All this might sound like a science fiction now, but imagine a day in the (distant) future, when EVs are the dominant vehicles on city streets and highways. If every household has two EVs, then the huge battery pack, measuring 4 x 4 feet and weighing 1,000-1,500 lbs. has to be replaced every 2-3 years (5-7 years per vehicle).

A shelf in the garage, or a small shed in the backyard, would be used to store the re-purposed batteries, which would be connected to the grid via inverters in a central control center. At 50-75 kWh storage capacity per battery pack, the two-EVs household might have 100-150 kWh energy storage facility in 5-6 years.

The average electric power use in the U.S. is 500-800 kWh per month, and much less in other countries. As the residential energy storage facility grows with each re-purposed battery pack, it won't take long before the household can store enough energy during low demand periods and send it into the grid at high demand (peak) periods. This would generate income for the family (the difference between high and low peak charges), and would also contribute to reduction of GHG emissions.

Eventually, when the number of re-purposed Li-ion battery packs exceeds the residential energy storage capacity, larger commercial energy storage facilities could be constructed. Here, many re-purposed battery packs can be used to store energy for use during peak periods. These mass energy storage facilities would help in reducing the variability of solar and wind power plants, and would also provide additional options to the utility companies.

A feasible, albeit somewhat far-fetched, win-win situation.

Case Study: Daimler Energy Storage

A joint venture of carmaker Daimler, battery-to-grid integrator The Mobility House AG, energy service provider GETEC, and the recycling company REMONDIS was formed in 2015. Its goal is to create the world's first large-scale used batteries energy storage plant. For this purpose, the JV is collecting used lithium-ion batteries from electric and plug-in hybrid vehicles. The idea is to obtain enough batteries to create a large electric power storage system.

The "2nd use battery storage unit" prototype, to be deployed in Lünen, Germany in 2016, consists of over 1,000 batteries, with an estimated 13 MWh of total capacity. This way, in addition to helping the power grid variability, the reused batteries will be kept out of landfills, will help reduce some EV costs by adding an additional revenue stream.

Note: 13 MWh is approximately the annual electric output of a 3.0 kW solar array operating in Arizona. Or a 6.0 kW array operating in Germany. It is, therefore, a fairly insignificant amount of energy compared to the many gigawatts of electricity flowing in the national grid. This means that hundreds and thousands of similar—1,000 batteries each—energy storage plants would be needed to make a meaningful difference in the grid.

According to the JV principals, and depending on the vehicle, new EV batteries have a life-cycle of up to 10 years with at least 80% efficiency. After that, the batteries have to be replaced, because the vehicle's performance is affected. These 'old' batteries, however, are still fully operational, albeit at lower efficiency, which is not critical for stationary applications.

This then gives the already 10 year old batteries a second chance of additional 10 years as a stationary energy storage. Lithium-ion batteries have been becoming the standard for homes energy storage, and now could help electricity generated by solar and wind to be stored on site or used in larger grid-service systems.

Thus stored electric power can be used during peak load hours, thus leveling the variability in the grid. This battery capacity can also be used when solar and wind power is not available; at night and during inclement weather, which in turn reduces grid demand.

Tesla already has home and commercial battery storage systems on the market. The home system, called the Powerwall, is a 7 kilowatt-hour (kWh) system that will retail for $3,000 and a 10kWh unit that costs $3,500. The commercial battery system, called the Powerpack, will store 100kWh of power and retail for $25,000 each.

Daisy-chaining the Powerpacks, allows the creation of large and very large energy storage capacity. At least this is how the theory goes, and we just have to see the results from future large-scale trials. We must remember also that energy storage types and costs, and especially these of large-scale systems, are project and location specific. Thus, a success with one type or location does not mean automatic license for universal use or guaranteed success.

The good thing right now is that the new developments are stirring global competition. SolarEdge Technologies and Enphase, for example, announced plans to bring a lithium-ion home battery to their solar power management systems soon as well.

The U.S. energy storage market will grow 10-fold by 2020, with many new players.

Enphase batteries, manufactured by the Japanese-based ELIIY Power, is a lithium iron phosphate battery, which, they claim, is more stable than lithium-ion. Enphase The batteries are about one cubic foot in size, with 1.2 kWh electric power capacity. It is expected to be available by mid-2016, and might change the energy storage game all together.

Case Study: Tesla EVs

The best example of extreme reliance on battery power solution is Tesla's EV car. It is a fancy creation—Ferrari of EVs—and equally exotic and expensive, so only the *very rich* can afford a Tesla car for now. And only the *extremely rich* can add a large solar installation on their roof, with equally large battery storage in the basement, as needed for night-time charging of their Tesla car—without using the fossil-burning electricity—in order to justify its marginal environmental contribution.

Most of us, however, will continue using the wall sockets, with their dirty (fossil fuel generated) electricity, to charge our EVs, if we can afford them. This means that in all cases we are contributing very little to cleaning the global environment, which is the main selling point of EVs. Or is it?

Tesla—the premier EV manufacturer—thinks that EVs' time has arrived and EVs are justifiable and needed. In anticipation of this futuristic reality, Tesla is building a Gigafactory in the U.S., scheduled to be up and running by 2017. Here, the new Tesla cars would be manufactured complete with battery packs made in house. The Gigafactory is expected to produce about 500,000 of these battery packs annually with the help of Panasonic and others. Some of these battery packs would be also used by solar companies to supply stationary energy storage to residential solar installations.

Tesla's Roadster has a battery pack, called Energy Storage System (ESS). The Tesla ESS contains 6,831 lithium ion cells arranged into 11 sheets, or layers, connected in series, Each sheet contains 9 blocks of batteries connected in series, where each block contains 69 cells connected in parallel.

The battery cells are of the 18650 type, which is found in some laptop batteries. The pack is designed so as to prevent catastrophic cell failures from propagating to adjacent cells due to thermal runaway and other extreme emergencies. A 146 Watt cooling pump delivers coolant continuously through the ESS, when the car is running and when is off—if the pack retains more than a 90% charge.

A full recharge of the battery system requires 3½ hours using a special 70 amp connector, drawing about 17 kW of electricity. The recharge cycles usually starts from a partially charged battery pack, so it requires less time in most cases. A fully charged ESS stores approximately 53 kWh of electrical energy at a nominal 375 volts. The battery pack weighs 992 lbs., or about half a ton

Note: 17 kW electric charge requires a 70, or much better 100 amp breaker, as needed for charging the EV batteries. Since most homes do not have such large breakers, this means additional $2-3,000 for home electric system modification. And some older homes' electrical wires cannot handle so much power. This means that even more of us will be left out of the Tesla car market.

The replacement cost of the ESS presently is about $35,000, with an expected life span of 7 years, or 100,000 miles. Pre-purchase of a battery replacement was offered for $12,000 to be delivered after seven years. Not a small change, which even fewer of us can afford every 7 years.

The ESS is expected to retain 70% capacity after 5 years or 50,000 miles of driving. Tesla also provides a 3 year/36,000 mile warranty on its cars with an optional 4 year/50,000 mile extended warranty available at an "additional cost." A non-ESS warranty extension is available for $5,000 and adds another 3 years/36,000 miles to the coverage of components, excluding the ESS, for a total of 6 years or 72,000 miles of driving.

Note: if the battery pack becomes completely discharged (as in extended driving, or emergency situations), it will be irretrievably damaged and no longer can be used. In such case, the only option is to replace it at a cost of $40,000. No insurance, nor warranty, covers this cost and so the owner is solely responsible for the full cost of the battery pack replacement. Obviously,

this is rich people vehicle, since very few working people would agree to shed $40,000 for several minutes of emergency driving, or any other car related expenses.

With all this in mind, why is Tesla competing with the established major car manufacturers? The three key components in its advantage portfolio (and its goals) are: cost reduction, localized supply chain, and economies of scale. Tesla claims they the car price will be reduced by 30% during mass production. But the numbers don't add up.

Tesla cars are priced in the $100,000 range, which is not your average Joe's price range. So how many cars can Tesla sell, even in the rich USA? And the initial price is only one of the problems at hand for our average Joe car buyer, as we discussed above.

Presently, the cost of battery packs used in the Volt and Leaf EVs is over $300 per kWh of each battery pack. Tesla's price is around $275, and could get down to $245 per kWh, if the Gigafactory is built and functions as planned, if Tesla and its partners succeed in improving the battery chemistry, and if, and if, and if…

And then, we have to deal with the other problems; limited driving range for daily driving, and limited number of filling stations across the country. With that limitation, the number of us, average Joes, willing to deal with this problem dropped significantly. So how Tesla solves these problems would determine the future of its cars and the entire EV industry in the U.S., which is its largest market.

A big potential advantage of Tesla EVs is that they can be used as energy storage, by charging their batteries at night; when rates are low, and sending the extra power into the grid at higher rates. We have to question the wisdom of using $35,000 battery pack with limited charge-discharge cycle number for the purpose of making pennies, but it is a potential use. Although it might end up being a very expensive way to use electric power in the short term, it might have some advantages in the long run.

Tesla Solar Batteries

We don't know how successful the Tesla EV venture would be, nor even how practical and useful it is. There is, however, one good thing about it; it is the fact that Tesla is hoping to use its new facility to produce up to 500,000 battery packs a year by 2020. This would provide the car battery industry with some badly needed experience.

Even more importantly, some of the battery packs will be used for solar energy storage.

This is a good start of a mass battery production for energy storage, which technology is badly needed to reduce the variability of solar power plants.

500,000 battery packs annual production is a large number, but in practical terms, the energy storage of all these batteries together amounts to only about 50 GWh. This is approximately the annual electricity production of a 16 MW power plant, or a small part of the annual electricity generated by an average 500 MW (mid-size) coal-fired power plant.

So the question here is how many battery packs does Tesla need to produce, in order to store the electric power generated by one 500 MW power plant? Here we come up with astronomical numbers of batteries that have to be manufactured, transported, installed, and operated for 5-7 years. At that time they have to be disassembled, loaded on trucks and transported to a specialized hazmat waste disposal facility. The cost of disposal and replacement of the battery packs every 5-7 years might break the project.

Even if justifiable, many Tesla Gigafactories would be needed to manufacture enough batteries to store the energy of our 500 MW power plant. How about storing the energy of all U.S. power plants? Impossible, is the short answer.

The most practical answer, however, is that we don't need to store the electric power of all power plants in the U.S. We only need to store some of the power from some solar and wind power plants. But even in this case of limited use, we would need about 10 Tesla Gigafactories now and many more later on, producing battery packs on full time basis for several years to cover the variability of the U.S. solar and wind power generation.

The problem of today is that large-scale battery storage technology is not proven efficient, reliable, safe, or cost-effective enough for everyday use. So, the above plans for large-scale energy storage do not seem practical, nor even possible, with today's battery energy storage technologies. Rapid developments are necessary to make this badly needed technology a reality. Maybe Tesla with its ambitious plans, would open the doors to future developments in EV and battery storage…

Self-Driven Cars

Just because we can! Why else? This new 'invention' is on the very top of the U.S. list of projects demonstrating our opulence and decadence. For no reason at all, and with no practical purposes in mind, we all of a sudden decided to eliminate all car drivers. Not so fast!

The future of these steering wheel-less vehicles, after millions of dollars are spent on their development,

is full of discussions, debates, and law suites. In the end, thousands of these useless for all practical purposes machines would sit rusting in junk yards around the country.

How else could this experiment to eliminate the drivers in a country of drivers end? We started the car craze over half a century ago, and the car is still our pride and joy. Why let stupid, brainless, emotionless computers drive us around town? Not now, not later, not ever!

The Trend

Energy storage is a critical element of our energy future. It is especially important for the quick development and mass implementation of the variable energy sources like wind and solar. The variability (during temporary unavailability of sunshine or wind) is presently limiting their growth around the world.

Without reliable energy storage, the renewables will continue to be treated as an unreliable energy sources, which limits their use.

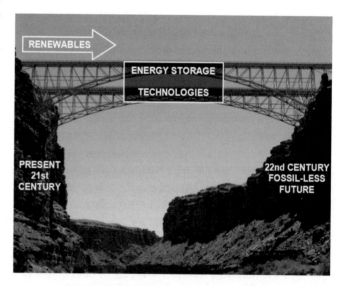

Figure 10-28. The Energy storage bridge.

No joke! The renewables cannot cross the bridge to the future without energy storage.

This is a serious matter that is becoming increasingly important factor in the energy markets. Without large-scale energy storage, solar and wind will be marginal energy providers; it is that simple. The issue is receiving the necessary level of attention and lip service today, but only a few large companies are involved in providing the solutions. The government support is not enough to support escalation of the R&D and deployment efforts—especially in the large-scale energy storage. Without that support, the solar and wind industry will either:

- Wait for very long before adequate energy storage solutions are found, or
- Find themselves prisoners of a single technology provider, who will dictate conditions and prices.

Either of these conditions is unacceptable, and must be avoided. So, it is up to the U.S. government and its technology helpers—DOE, NREL, Sandia, etc.—to come up with the technological solutions before it is too late.

The Issues:

Key issues and actions in front of the international community for the next ten years in the areas of energy storage technologies and markets development are to:

- Determine where near-term cost effective niche markets exist and support deployment in these areas, sharing lessons learned to support long-term development.
- Incentivize the retrofit of existing storage facilities to improve efficiency and flexibility.
- Develop marketplaces and regulatory environments that enable accelerated deployment, in part through eliminating price distortions and enabling benefits-stacking for energy storage systems, allowing these technologies to be compensated for providing multiple services over their lifetime.
- Support targeted demonstration projects for more mature, but not yet widely deployed, energy storage technologies to document system performance and safety ratings. Share information collected including lessons learned widely through storage stakeholder groups.
- Support investments in research and development for early stage energy storage technologies including technology breakthroughs in high-temperature thermal storage systems and scalable battery technologies, and systems that incorporate the use of both electricity and thermal energy storage (i.e. hybrid systems) to maximize resource use efficiency.
- Establish a comprehensive set of international standards in a manner that allows for incremental revisions as energy storage technologies mature.
- Evaluate and broadly disseminate the learning and experience from established installations. Information should include data on both technical aspects (e.g. generation, cost, performance) and contextual

details (e.g. market conditions, energy pricing structures) specific to a region/market.

- Establish international and national data co-operation to foster research, monitor progress and assess the research and development (R&D) bottlenecks. Complete analysis in support of regional assessments to quantify the value of energy storage in specific regions and energy markets, and promote the development and adoption of tools devoted to evaluating energy storage project proposals.

North America

New research shows that North American utilities and companies involved, or seeking to get involved in clean energy, overwhelmingly prefer energy-storage projects. Over 50 proposals, or RFPs, totaling over 4.0 GW of renewable energy capacity were submitted by U.S. utilities and companies, one-third of which included some type of energy-smart technologies including energy storage. This is a strong indicator of the new trend here, where customers and owners are realizing the variability issues of the renewables, and are looking to solve them with the help of the new energy-smart and energy-storage technologies.

Utilities are increasingly targeting solar energy generation, especially in the South West, as confirmed in 2014, which turned out to be a bumper year for utility-scale solar installations. South Western states issued 18 requests, half of which came from the large California utilities, Southern California Edison and PG&E.

At the same time, private companies in the South East requested almost that much solar power generation, with Southern Co. and Duke Energy Corp. leading the pack. Alliant Energy Corporation of Wisconsin issued the largest request for 600 megawatts of wind power.

As the solar and wind power increase in size, so does the importance of energy storage in its role of variability equalizer.

Update:

Solar with batteries for energy storage could become the most economical option for residential and commercial customers in the U.S. by 2040, according to the experts. The Rocky Mountain Institute (RMI) reported in 2015 that grid-connected solar equipped with energy storage will become the most popular choice in some areas.

The RMI report, 'The economics of load defection', examines the potential for customers to choose solar and storage in five key regions of the U.S.; Honolulu,

Figure 10-29. Solar with battery storage

Los Angeles County, San Antonio, Louisville, and New York's Westchester County.

The report analyses the cost and pricing of solar power systems, and forecasts the key drivers of the energy storage market including lithium-ion (Li-Ion) battery production and EV sales. Using software for the modeling work, RMI, puts centrally generated power to as little as 3% of total electric load by 2040.

97% of U.S. homes and businesses will generate their own solar power by 2040.

Parts of New York, Texas, and California, are the exception, where about 25% of central grid power would be still used in these states.

While this statement might be true to a point, we must clarify it by introducing *'partially'* in it. This is because it is simply impossible for all 97% of homes and businesses to generate 100% of their power.

Here are a couple of examples:

*   A summer day in Arizona requires non-stop use of air conditioning systems. A 3.0 ton AC uses approximately 3,500 watts of electricity, which is more than what most rooftop systems can generate. It also requires almost twice that much power to start the compressor. Also, there are some hazy and cloudy days—yes, even during Arizona summer days—so at least 50% power from the grid is needed to power the AC and all house appliances.
*   Any solar (PV) power installation in the states of Rode Island or Maine also has its limitations. Regardless of how much solar panels are on the roof of the richest house in the area, they are simply

useless during the large number of cloudy, rainy, and snowy days. Here again, at least 50%, or maybe more, of the power needed to light and heat homes and businesses has to come from the grid.

Presently, 90%-100% of electric power in the U.S. is provided by the power grid, even in regions with very high level of renewable energy resources.

The revolutionary progress expected in the future is based on the new (and yet to be proven) phenomenon of solar-plus-energy-storage, where batteries, or other technologies, are used to store excess electricity for later use. Reports of the years past have predicted similar defection from the existing grid networks, where individual customers "island" themselves by generating their own power.

Few attempts have been successful, and none have been so optimistic as they are today. Why the optimism? Batteries? Not so fast. Let's consider the facts. Those of us living in the deserts know quite well that batteries do not last very long exposed to the desert extremes. We also know very well that batteries cannot run our air 3-5 ton conditioners—now or ever!

So, while we want to believe that batteries are the answer to energy storage, we don't think that it is possible in the deserts. This is even more true for large-scale solar power generation. The future will show…

Battery Storage Update

To battery, or not to battery? This is the big question in front of the U.S. solar electric power customers. Actually, it is a question for a few of us—the 1% group—that can afford the high cost and maintenance expenses of solar installations with battery storage. Most of us don't have the money, nor a huge space for solar panels and equally huge battery pack.

There are, however, some of us who cannot afford this luxury (yes, it is truly a luxury for now). And there are others, who are interested in this technology for a number of reasons. This is a special group of people, who are willing to take a chance and plunge into the unexplored and untested field of solar energy generation storage, and use.

So, let's congratulate them, and try to figure out the advantages and disadvantages of the new solar energy storage technology.

Unfortunately, the proposed Li-Ion battery energy storage seems overly optimistic to us, practical realists. Here is why.

*   The technology for Li-Ion batteries manufacturing

is still in development. They, therefore, have a long ways to go before becoming efficient, reliable, and cheap enough—especially for use in large installations, where the technology is most needed.

- Yes, we know all about Tesla's innovation and developments in the stationary storage battery field, but its success is to be proven after long-term field tests. Even then, market resistance forces (initial price, reliability, durability, etc. issues) have to be overcome for this product to become a true success. There is a chance for this to happen, but it is not likely until much later in the century.

- i-Ion batteries have one of the best energy-to-mass ratios and a very slow loss of charge when not in use. This is all good, but a very large number of batteries are needed to provide significant amount of energy storage. This increases the chance of failure of individual batteries, connectors, or other components in the packs and the overall system.

- Not all residences and businesses have large unoccupied areas to store and properly maintain a large pack of Li-Ion batteries.

- Sunlight, in many North Eastern U.S. states and European countries, is inconsistent, to say the least. Clouds, rain, and snow dominate the weather in these areas, so it would be difficult to impossible, to generate enough excess solar energy, as needed to store it for later use. Unless, of course, one installs an entire field of solar panels to suck in the reduced and diffused solar radiation reaching them.

- In areas of the SW U.S., where desert sun is a daily visitor, the problem is with enormous power use during summer days. An average air home conditioner sucks 3-6 kWh for 10 hours during the day and half that at night. There is no excess energy to store to begin with, no matter how large the solar array is, so it will only provide part of the needed power. If one is determined to optimize the solar output with the use of energy storage, s/he needs at least 15-30 car batteries at a cost of $3-4,000, or equivalent quantity of Li-Ion batteries at a cost of $20,000-30,000. This setup also requires a larger 8-10 kW solar array at an additional cost of $30-50,000. And even then, power will not be available all the time, since the sun shines on and off during limited number of hours during the day, and none at night.

- The problem with large appliances is that their inductive load requires almost double the rated power to start their motors running. This means that the battery storage capacity has to be doubled, just to start some large appliances, if 100% solar power is the goal. This would require even more space and that much more capital expense. Not an easy thing for most of us.

- And finally, Li-Ion batteries today have about 2,500-3,000 hrs useful operating time. This means that after a couple of years of intensive use, the battery pack has to be replaced at another cost of $20,000-30,000, or more. So, in several years use we have spent $50-60,000 only to get a partial solution to our electric power needs. Too much money for this somewhat marginal performance. This might be acceptable to people, who can justify the extra effort and expense. It, however, is not how the American Dream works, so we have to put this option on the Wish List for later on in this century.

- The combination of above mentioned factors, in our estimate, puts Li-Ion batteries energy storage at 20-30% at best, which is significantly lower than the estimated 75%-97% use by 2040.

Case Study: The Musk Syndrome

Elon Musk has undeniable success in a number of hi-tech areas. Banking on it, in 2015 he jumped with both legs into solar energy storage market. His battery pack, designed for storing solar power generated during the day, can be used later on at night. A brilliant solution to night time use of electricity, no doubt.

Nothing wrong with that, we say. Except that not many of us live in remote cabins, and even less are multi-millionaires like him. While he, and people in his league, can easily afford a remote cabin and a $3,500 battery pack, or two, as required for night time use, most of us do not have remote cabins, nor do we have the excess money to install the new battery packs.

And this is just the beginning of the problems we see:

Once we have shed $3,500—in addition to $20,000 worth of the solar panels installation—we now have to start saving money for replacing the battery pack and inverters, when needed—usually every 5 to 7 years. Although, the estimated life-time of Musk's phosphorous-lithium ion battery pack is 10 years, we challenge any battery *operating in the extreme desert heat* to last more than 5 years at non-stop, full capacity use.

So, let's say, we need to replace the battery pack every 5 years. What does this do to our 25 year cost analysis? It brings the cost of the installation to $17,000 for the battery pack replacement alone, in addition to the initial installation cost of the solar plant.

Translated into layman's terms, this means that we will spend about $40,000 during the 25 years life cycle of the solar power plant with energy storage. This comes to about $1,600 annual expense, which means that we need to save at least $150 every month, in order to pay for the expense of the solar installation and the battery packs.

If any of you has a remote cabin that uses so much power and can afford this expense, then it is the deal waiting for you. Mr. Musk is waiting with open arms, and will take your money without any hesitation. We, on the other hand, wish you best of luck and hope that you will have fun with the new shiny solar installation and the batter pack, something most of us can only dream about.

Note: One of the major sales point of the battery packs is using it in case of black outs. This is a legitimate concern, where black outs are common. In the U.S., however, blackouts are a rarity. We have recorded 2 major blackouts during the last 20 years in Arizona, so very few people would spend money for this purpose.

A much simpler and cheaper alternative for periodic use in the U.S. would be purchasing a $500 gas powered generator from Home Depot. It could run lights and appliances for as long as needed, after which it could be shelved until the next blackout (5-10 years later.) But even that simple and cheap solution is not finding many supporters around here.

Large-scale Energy Storage

The biggest challenge for battery storage is scaling up the technology for use in large-scale solar or wind power plants. So let's see:

A large-scale battery storage system for:
- 1.0 MW solar plant would require 200 Powerwall packs equivalent. At $3,000 each = $600,000 to be spent every 5-7 years
- 100 MW solar plant would require 20,000 Powerwall packs equvalent, or $60,000,000 every 5-7 years.

Is this even feasible from logistics and financial points of view? The ROI in all cases depends on the solar or wind plant location, its operation, efficiency, etc. The battery storage introduces another expensive variable in the already variable equation of renewables power generation.

After all is said and done, the investors need to see one full life cycle (at least 10 years minimum) of field operation in order to figure out and calculate the risks.

Note: We must consider again the *Desert Phenomena* here. It is a fact that most large-scale (utility type) solar power plants are located in the deserts. Here the extreme desert heat and freeze play tricks on anything left in it.

Batteries of any type are one of the most vulnerable victims. Very seldom would a car battery last more than 2-3 years under daily operation in the desert areas. The same destiny awaits any and all battery storage technologies in the desert power fields.

It is hard to imagine any battery technology lasting more than 3-4 years of non-stop operation and daily cycling in the desert. Customers and investors must consider the *'desert phenomena'*, in order to get a realistic estimate of their long-term ROI.

We must also remember that other competing energy storage technologies and methods are available today. For now, sending extra power in the grid is the preferred method for handling excess power. This is because the existing energy storage technologies are still immature. They need to spend some long hours under the desert sun bliss in order to be proven efficient, reliable, or profitable enough.

Renewables-to-Gas Energy Storage

A new *power-to-gas* system, designed to generate hydrogen from the sun and wind during times of excess supply and store it, is under way by Southern California Gas Company (SoCalGas), the National Renewable Energy Laboratory (NREL), and the National Fuel Cell Research Center (NFCRC).

Solar and wind are variable energy sources, so energy storage is the key to their future progress. The new power-to-gas technology intends to rectify this problem by using electrolysis of water to convert excess electricity generated by solar panels and wind mills into hydrogen gas.

The process breaks down the water molecules into hydrogen and oxygen. Thus generated hydrogen can then be stored and used as is. It can also be converted to methane gas and stored in that form for future energy needs.

A great achievement...but wait, this is already existing technology, in use in Germany and other countries, where commercial-scale power-to-gas systems have been functioning for awhile now. It appears that the above mentioned process development group claim to fame is that this is the first such process used in the U.S., so we must give them due credit.

This technology is also in use by ITM Power in the UK, which has two manufacturing and testing sites, equipped with flexible equipment, enabling quick changes of process conditions. The company manufactures integrated hydrogen energy systems for rapid response and high pressure applications, to meet the requirements for grid balancing and energy storage. These systems can also produce clean fuel (hydrogen and methane) for transport, renewable heat and chemicals.

Other companies, such as Hydrogenics, have been working for awhile in the power-to-gas technologies, and have reported significant progress. A tell-tail, or a warning of things to come is the fact that Hydrogenics stock price at its beginning, several years back, was around $200 a share. In 2015, it was slightly over $10. This must tell us something about the progress and the market expectations for this technology.

Taking a closer look at the developments of late, however, we must admit that new power-to-gas energy process offers a larger volume and longer term storage solution than batteries, which are limited to small volume, short-term storage, and are still very expensive. Even with the achievements of late—including the developments at Tesla—and other longer duration battery types such as flow batteries, the facts remain in favor of the power-to-gas energy storage process.

At the Energy Storage Europe conference in Dusseldorf, Germany in March 2015, there was a major strand on power-to-gas technologies. According to the experts at the conference, power-to-gas is likely to become increasingly more practical for large-scale and long duration energy storage applications.

As the technologies improve, power-to-gas energy storage might become the standard.

The grid networks around the world are expected to get well over 10% penetration from renewable energy sources in the near future, which is considered a threshold at which even modern power grids start to suffer stability problems. This is also when power-to-gas storage would become the energy storage of choice; especially in areas with a lot of sunshine.

California, for example, has committed to sourcing one third of its power from renewable sources by 2020 and recently governor Jerry Brown proposed a target of 50% by 2030. This is a lot of power—variable and unpredictable power—which depends on variable sun and wind for power generation.,

Without large-scale energy storage, the goals are simply impossible to achieve.

Using batteries for such large amount of energy storage is absurd at the present day state-of-the-art, so here is where power-to-gas will shine and become the savior of the energy battle in California and other places.

Unfortunately, as with any large-scale technology, there are problems.

- The power-to-gas process development in the U.S. is still in R&D stages, with demonstrations and trials by university and government entities. Major work is underway at the University of California, and in NREL's laboratories in Golden, Colorado, and Irvine, California. The initial results from the demo and test sessions are not expected by 2016. Slow go…but once the process is established and optimized, the private sector will take over and things will go much faster. In any case, we need to hurry, if we are to meet the 2020 and 2030 deadlines.

- Another big problem in the power-to-gas technology is that Western states (with the most solar installations) are running out of water. Unfortunately, the new power-to-gas process requires a lot of water.

Every ton of hydrogen requires the electrolysis of 9,000 tons of water.

Once the water has gone through the process, it is gone forever. Many thousands of tons of hydrogen would be needed to ensure the energy storage of California in achieving its 2020 and 2030 goals. This means that many millions and billions of tons of water (entire lakes) will be needed to feed in the process. Would California be able to provide that much water on top of its other water uses, in the midst of its periodic draughts, remains to be seen?

- And to top it off, there is even a bigger—maybe the biggest—problem of all. It is the fact that (for now) the electric energy required to electrolyze water to hydrogen and oxygen simply exceeds the energy that thus produced hydrogen can generate by a factor of 2.

Here is simple set of formulas, which could be used as a measure of the usefulness and practicality of the power-to-gas process:

One liter of hydrogen contains about 0.50 kWh energy equivalent. It takes slightly over 1.0 kWh to produce a liter of hydrogen, which means that we lose 50% of the initial energy during the conversion.

And it gets even worst from here on, simply because each additional step requires more complex and expensive equipment and more energy:

— Storing hydrogen as gas at 24.8 MPa would take 1.28 kWh per liter.
— Storing hydrogen as gas at 30.0 MPa would take 1.50 kWh per liter.
— Storing hydrogen as liquid at 20.0 K would take 4.72 kWh per liter.
— Storing hydrogen as metal hydride would take 6.36 kWh per liter.

Remember that each liter of H2 contains 0.50 kW of energy equivalent.

Note: The hydrogen energy density in these cases increases proportionally, so that the overall energy loss remains about 50%. This, however, does not include the initial equipment cost, O&M expenses, transport, and storage of the respective hydrogen forms. And let's not forget that a lot of effort and money still remain to be spent on the development and deployment of the new process.

In conclusion, power-to-gas is the most practical and promising of all energy storage technologies. Unfortunately, there are still significant additional expenses and energy losses, where at least 50% of the initial energy is lost during the conversion and storage steps. Not a small thing! So, reducing these significant expenses and energy losses are the keys in the future development of this technology.

Yes, we can make hydrogen, but can we bring it to large-scale energy storage and transportation at a reasonable cost? This question is to be answered before this technology goes to market. Such great expenses and waste of energy are not easily justifiable from technical or financial points of view. We just have to hope that our ingenious energy engineers will come up with the solutions quickly.

Power Combining

'Wedding made in Heaven.' This is what the new technique for "scheduling" flow of energy in the electric grids could be called. It is the future of distributed power generation (via roof top solar) combined with smart grid advances. It bypasses the centralized power management by tapping into the computer assisted distributed power with the help of smart energy devices.

This approach is part of the smart grid concept with a twist.

An important one, since it coordinates the energy being produced, stored, and used by both; conventional and renewable sources. This approach is exceptionally important to the development of the renewable technologies, which, due to their variability, have been treated as the step children of the energy industry.

Currently, the power grid uses centralized scheduling to forecast and manage the power generated at thousands of large power plants around the U.S. Power scheduling is done by centralized controls receiving data from power plants, where the data is crunched and commands are issued to the plants telling them how much power to send in the grid and when.

One big problem here is that If a central control center fails, part of the system it services fails too. If two or more interconnected centers fail, then that could bring the entire system down, at least for a short awhile. If the failure affects even more centers, then the entire area goes black.

As renewable energy sources, such as rooftop solar panels, proliferate and are increasingly incorporated into the power grid, the grid operation and control get even more complicated. Since the infrastructure is not designed to handle variable loads, it fails to handle the load efficiently. It then needs some advanced systems for tracking and coordinating exponentially more energy sources for better handling of the new resources.

A great challenge for the new distributed solar generation is determining how much energy needs to be stored on-site and how much can be sent into the grid. This approach doesn't scale up well either, which is a problem, considering the rapid growth of on-site renewable energy sources.

The rise of on-site energy storage technologies adds to the challenge

This is so, because stored energy is seen by the central control center as 'used', while it is still in the system, but tucked away in storage devices. That energy could then be released at any time, making power scheduling more difficult.

*In many regions of the U.S., we have **two-way flow** of electricity in the grid.*

This unusual and uncontrollable phenomena is due to; a) electricity flow from the central power plants to the grid and the customers, and b) another flow, going in the opposite direction, from the customers' energy storage back into the grid.

This new development is increasing in size and magnitude and presents the central control centers with a dual problem:

- Sudden and unpredictable drop of power output from thousands of rooftop solar installations, due to passing cloud, followed by
- Sudden and uncontrollable release of significant power from thousands of energy storage devices into the grid.

This is like somebody randomly pressing and releasing the gas and break pedals of your car, while you are driving. What do you do? How do you handle the sudden acceleration and breaking? Not an easy situation to deal with, and quite frustrating experience to be sure.

This is exactly what the central power control operators are faced with. Periodic drops and rises of power levels into the grid, which they must adjust in order to keep the grid load constant and balanced. The existing central control technology is not designed for, and is quite limited in, handling the two way flow of power from and into the grid. This new phenomena is causing lots of headaches and problems for the utilities and the control operators.

To address these problems, researchers have developed technology that takes advantage of distributed computing power to replace the traditional control center with a decentralized approach. This is done by tapping into the computers of each distributed energy source.

In this approach, each device is forced to communicate with its neighbors, in order to calculate and schedule how much energy needs to be stored and when. It also calculates and communicates how much power can be sent to the grid, and how much needs to drawn from the grid and when.

This, therefore, becomes a large decentralized/distributed system, which collectively calculates the supply and demand characteristics. It then determines and communicates its optimal schedule to the central controls, which respond accordingly, thus keeping the entire grid's supply and demand load balanced.

The decentralized/distributed method also protects the privacy of home-owners and other power generators, since there is no direct sharing of information with the outside world.

The technology needed to run this new decentralized system is still in R&D mode at several universities and private companies. The expectations are to have a large-scale demo implemented sometime in 2016.

Microgrids

Over a century now, the centralized power grid—a one-way flow of electricity, generated by large, remote power plants and distributed over miles of transmission lines to homes and businesses—succeeded in delivering electricity across continents to billions of people. But in recent years the system's shortcomings have become increasingly evident.

The conventional power grid is largely dependent on outdated power plants generating huge amounts of power, while gobbling even large amounts of fossils and belching enormous amounts of pollutants. Thus generated power goes into equally outdated big, interconnected, and slow to respond power grid.

Due to its age and size, it is not capable of handling the new energy generating and storage technologies, as discussed above. These technologies are here to stay and will increase with time, so the grid has to adapt. The power grid is also vulnerable to massive disruption by natural disasters and susceptible to physical and cyber-attacks, which it handles via old and outdated methods.

For example, in August 2003, 50 million people in parts of Ontario, Canada, and eight U.S. states lost electricity when a sagging power line in a suburb of Cleveland, OH touched an overgrown tree limb and malfunctioned, triggering a cascading sequence of events resulting in the largest blackout in American history. More recently, super-storm Sandy showed us the havoc and extensive damages extreme weather can wreak. This includes bringing the entire local power system down for weeks at a time. It is becoming obvious that our old power grid has done its job well until now, but is not capable of taking us into the 22nd century. Instead a new approach is needed.

Across the globe, regulators, policy makers and businesses are collaborating on creating a new and better electricity delivery system—one that will be more stable and secure, cleaner and cheaper. The new system will be able to accommodate larger shares of variable renewable energy sources.

Local microgrids, including modern renewable energy sources, are the way of the future.

This approach is a departure from the traditional grid model, It is defined by the ability to generate power at or near the point of consumption independent of, or in conjunction with, the other power generators.

"A microgrid is a local energy grid with control capability, which means it can disconnect from the traditional grid

and operate autonomously," according to DOE's official definition.

A microgrid is usually designed to receive, generate, store, and deliver electricity within its boundary, which might be a neighborhood, a military base, university campus, and any other small communities. While maintaining a connection to the larger electricity grid and constantly receiving electricity from it, it also generates its own power and coordinates the usage (two way flow) according to a set of rules. The rules can be set according to local needs and programmed into computerized power control systems.

Its greatest advantage is that in a state of emergency, like grid blackout, fire, or an earthquake, the local microgrid can be safely isolated from the conventional grid. It could then be switched to a different mode and continue to deliver power to the local area it serves.

Today we have models (precursors) to microgrids, which are simple systems consisting of a central power plant that serves a house, business, or a campus with backup power provided by diesel generators or other energy source. Relying on a single generator, however, they are less secure, and because they usually burn fossil fuels to generate electricity, they increase the dependence on these fuels that fluctuate in price and contribute to climate change.

Microgrids under development today benefit from two trends: a) the declining cost of energy storage, and b) the increasing affordability of renewable energy, especially wind and solar power. One of the major components of energy storage, lithium-ion batteries, have been improved and are being deployed across the country. Their prices have fallen by 40% since 2010, and are becoming less expensive by the day. Solar panels are also 80% cheaper than five years ago. Wind turbine prices have fallen down to 35% from their 2008 high.

With affordable energy storage, surplus solar or wind electricity can be stored for later use, enabling microgrids to dispatch power as needed and operate around the clock. Taken together, the price reductions make it possible to bring together variable clean energy and storage into renewable energy microgrids that can support, or operate independently from, the main electricity grid.

Today's sophisticated energy management systems allow the design of microgrids that can bring a diverse mix of distributed energy sources, such as rooftop solar, fuel cells, wind turbines, biomass-fired combined heat and power plants, together. Combined with energy storage, these combined systems result is a small-scale electricity-generating powerhouses, capable of balanc-

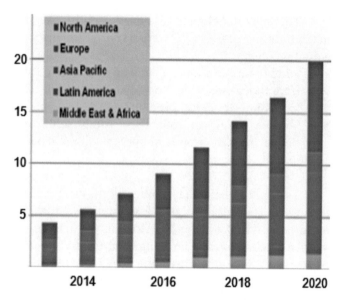

Figure 10-30. Microgrids now and in the future (in $ billion)

ing the variations in energy supply. They can provide auxiliary services, such as voltage and frequency regulation to the conventional grid. They can also send excess electricity into the national grid to make a profit, or to provide additional power during emergencies.

While microgrids are an absolute necessity in developing countries, where power is on and off at random, their price is still too high for mass deployment in most of these areas. Instead, the microgrids' destiny is determined in the U.S. and the developed world, where the efforts are intensifying.

Microgrids have been installed and tried in many areas, already make sense—especially in areas with high energy prices; in remote locations, such as weekend cabins; and areas that use expensive imported diesel fuel for electricity generation. Their most important use for now is at hospitals and military installations, which cannot risk losing power even for a split second.

The U.S. military is actively working on the microgrid concept, since it fits their 'business model' perfectly. One such project is under development on the Hawaiian island of Kauai at the U.S. Navy's Pacific Missile Range Facility. It is located at the end of one branch of Kauai's electricity distribution system, and is its end user. The idea is to clean the power supply and reduce its energy bills, since the average retail price of electricity in Hawaii is 2-3 times higher than the national average.

For that purpose, the Navy installed a number of small rooftop solar arrays. But when trying to connect a large solar array installed on top of a hangar, Kauai Island Utility refused on the grounds that electricity generated by the panels could strain its equipment. It was also afraid that power from the solar array flowing in the opposite direction than usual, could create safety issues.

The National Renewable Energy Laboratory and Honeywell to the rescue. They designed an energy management system (microgrid) used to smooth the surge and slumps in solar output triggered by passing clouds. The system couples fast controls with a small battery to rapidly respond to variations in the solar power supply. There are plans to expand and upgrade the energy management system to a cyber-secure microgrid and use the solution at other U.S. military installations to enable greater flexibility in power use and increase the use of remote renewable energy sources.

Another practical use of microgrids capable of running on renewable energy is in the energy sector itself—including offshore oil and gas drilling platforms. Offshore platforms, in particular, are not tied to mainland power grids and usually generate electricity on-site by burning huge quantities of natural gas in inefficient gas turbines. A well designed microgrid, assisted by solar panels and wind turbines, would help the efficiency of the gas turbines. Energy storage would capture surplus electricity generated by the gas turbines, enabling them to be turned off when not needed, and balance output from the mix of generation sources.

The estimates of the off-grid energy services market, including the so-called "skinny grids" that bundle solar and energy-efficient LED lighting in developing countries, will be worth $12 billion annually by 2030. Remote microgrids will represent an $8.4 billion industry by 2020, with the largest number of deployments occurring in the developing world and activity increasing in North America and Europe.

Global microgrids will become $20-30 billion business in the developing world by 2030.

The developing countries have lesser need for microgrids, which is focused on a number of niche markets, driven by local needs. On the island of Bornholm, Denmark, for example, the local utility Østkraft A/S is building a showcase microgrid demonstration project incorporating a mix of low-carbon solutions, including three-dozen wind turbines, biomass-fired district heating plants and a biogas power plant.

Renewable energy microgrids are also expected to expand in areas with universal electricity access, especially in markets such as California where solar and energy storage have already taken off.

In the U.S., microgrid efforts are focused on another niche market. It is facilities with abnormally high-quality power requirements, where most of the microgrids development effort is at the moment. These are military bases, research facilities, large servers, and data centers. The idea is to ensure constant power under any and all operating conditions. This achieved by the addition of renewable and other energy sources, and sophisticated power controls.

The U.S. microgrid capacity is estimated to approach 2.5 GW by 2020, up from around 1 GW today. Entire cities, communities, and public institutions are looking at the next round of developments in the microgrid sector. The share of microgrids using solar power will grow significantly over the next decade as well.

One unusual fact about microgrids, and one that is helping their development, is that they are not affected or controlled by the bureaucracy in a way the other energy technologies are. Most microgrids can be developed regardless of whether utilities, or regulators, like it or not.

The only thing holding microgrids back is lack of financial incentives. In 2014, the U.S. Department of Energy announced $7 million in funding to advance the design of community-scale microgrids with a capacity as large as 10 megawatts. Recently, it offered additional $6.5 million in matching grants to technologies that address the challenges of integrating renewable energy and storage into the grid.

These token amounts are not enough to boost the development of sophisticated microgrids, so the new microgrid industry has to continue working without government help. This way, it might point the way of the future for all energy technologies. We know from long experience, however, that as soon as the microgrid market increases to a substantial level, government bureaucrats will jump to impose their regulations and restrictions. It never fails...and so it remains to be seen if this future intervention would bring positive or negative results.

Governor Brown, California, signed a bill reauthorizing the Self-Generation Incentive Program, which will provide $415 million in incentives to microgrid components installed on the customer's side of the utility meter. These include solar and wind power, waste heat-to-power technologies, fuel cells, and advanced energy storage systems. This funding will help meet the state's energy storage mandate, which requires investor-owned utilities to add 1.3 gigawatts of energy storage by 2020.

What the future holds for renewable energy microgrids depends on many variables: including regulations, incentives, the future role of the incumbent utilities and many other factors. Judging by the current policy, technology, and pricing trends we see favorable conditions for large-scale adoption of microgrids. These are needed for providing a secure, clean and increasingly affordable alternative to the conventional power grid.

Case Study—First Solar and Caterpillar Microgrids

In a new development, the U.S. companies First Solar and Caterpillar have combined forces to develop projects using solar technologies for establishing microgrids. These can be deployed in remote locations, such as mines and construction sites in developing countries.

First Solar is expected to design and manufacture a standard turnkey package for remote microgrid applications, which would include special *Cat*-brand solar panels. Caterpillar will then sell and manage the packages of microgrid solutions, starting in the Asia Pacific, Africa, and Latin America regions. The marketing of the new microgrid packages is planned to start sometime in 2016-2017.

This type of microgrid packages provide value to diesel and gas fuel burning customers by adding solar power (when the sun shines) to their power mix and saving fuel for running their generator sets.

Recently, Powerhive, which is financed by First Solar, completed a two years field test of off-grid metering and control systems that can be used in the new microgrid package. This would provide tested components to the new technology, thus adding to its quick implementation and ensuring its success.

This is a good step forward for the microgrid sector, and we expect to see many such ventures in the near future. There are many places in this world, that would benefit from the new microgrid technologies.

GLOBAL ENERGY ECONOMY

The world economy is a dynamic and full of ever changing events mechanism, which is affected by many technical and logistics factors and global developments every day and every hour of the day. It, however, has been undergoing some unprecedented and even radical changes and transformations over the last several decades. The changes have been even more dramatic during the last decade.

A century ago, the global economy (if there was such a thing then) was based on significant government intervention in the individual countries. The government dictated regulation and planned all economic activities, often without regard to the overall global economy.

During the 1960s and 1970s the global economy was driven by the oil market, which was dominated by government contracts and special government relationships between buyer and exporter countries. After the oil fiascos in the 1970s and 1980s, the oil market switched to a new; truly global market system.

Now the predominant MO of the market players is that of,
a)   competitive bidding, and
b)   price discovery through the dealings of many players on both sides of the market.

The developments since the 1980s show drastic changes, where the global economy depends more and more on, and is driven by, the global market forces. The oil market is one of the best examples this transformation.

As Figure 10-31 shows, the overall world-wide energy system, and the global energy market, is one complex system of products, services, and entities interwoven in the global economy. Governments and large companies have their say-so in it, but so do customers, investors, and many specialized groups and companies.

The Players

The largest group in the global energy markets are the energy importing countries.

The Importers

Most countries fall into this category, since they all import crude oil. Since oil is the primary imported energy source here too, the U.S., as a world power and major energy consumer, insists on an open, transparent and competitive global oil market. It must be a fair market, where oil exporters (or a group of such) is not allowed to dominate the market, since that threatens the access to the oil supplies needed to conduct normal commerce and military operations.

And so, a new global oil trading system has emerged, which is led and to an extend controlled by the U.S. The role of the U.S. thus far has been that of a negotiator and lead in many international negotiations and conflicts. These range from reducing tariffs, to opening the markets to unrestricted capital flows, including establishing rules for protecting investments and intellectual property. In most cases the U.S. relies on multilateral negotiations and mutual agreements, but at times it relies on multi-track negotiations and approaches, and even military interventions.

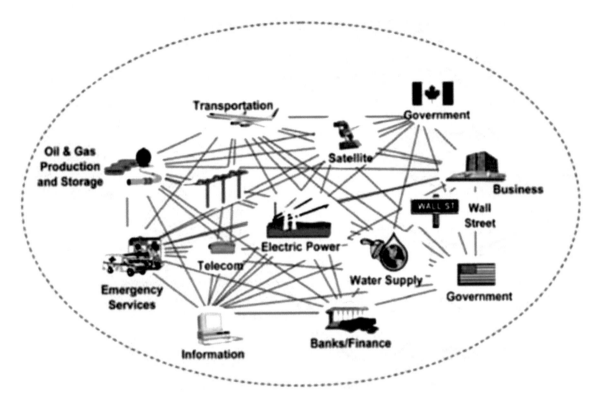

Figure 10-31. The complex North American energy system

This system has worked well and is supported, albeit accompanied by various complaints at different times, by the international community. The U.S., as the world's military superpower, backs up the operation of this global marketplace by policing the seas and international commerce from attack by hostile nations or non-governmental groups.

So far so good! There is a broad support for a liberal international oil trading system among the developed and the developing countries. Most of these countries benefit from high growth rates resulting from their integration into the global economy. As a result, now many of them have a better access to the energy markets and enjoy increased economic growth and abundant foreign investment.

It appears now that the new global energy economy, with more players in each market and more competitive prices, is beneficial for most participants. The energy importers also benefit from a larger diversity of supply sources, where the suppliers are even competing with one another for more customers and better prices.

The Exporters

The new global energy economy, however, is not fully accepted by the exporting countries. There are complains that the energy trading system is biased against the exporters. The major concern of the export-

ers is that while their exports products are based on competitive rates, the goods they import are often not fair, which results in overall unfair trading practices.

As a result, most exporters do not trust the free and competitive international markets and are always looking for ways to cooperate with each other in order to retain control of the export markets. This is especially true when oil prices are falling and the exporters are faced with proportionally growing economic problems at home.

This is exactly what happened during 1997-1998 financial crisis, which coincided with the increase of Iraqi oil exports. At that time, the market share struggle between Venezuela and Saudi Arabia added to the turmoil and caused the global oil prices to collapse below $10 per barrel. This low price reduced the earnings of the oil exporting countries by a great margin, which in turn caused severe pain in their economies.

Reduced revenues in many cases meant that the governments were not able to provide basic services, thus losing authority. Many changes occurred at that time in Indonesia, Iran, Nigeria, Russia, Venezuela, and others. Saudi Arabia, which was going through an internal transition after King Fahd suffered a debilitating stroke in 1995, was heavily affected too.

The 1998 incident forced the oil exporting countries into action, the first step of which was to consider collective actions. The Organization of Petroleum Exporting Countries (OPEC) was committed to protect its members' revenues mostly by controlling (increasing) global oil prices. It was a battle for survival for many of the exporters governments, who feared that falling revenues would bring social and political unrest. After intensive diplomacy, the oil producers agreed to reduce output and bring the oil prices above $22 a barrel

This marked another turning point for OPEC. Oil prices went back over the target of $22 a barrel. OPEC continued to successfully coordinate production cuts as a justification of the dramatically higher prices. The prices jumped to $30 per barrel, then to $50, and finally to a record of $100 a barrel.

Under the new global market dynamics, OPEC focused on its own economic performance, which was inadequate in the 1990s as compared to other nations.

OPEC rejected the thinking that low oil prices are good for everyone—and surely not good for OPEC members.

For them, oil revenues are critical to run the government properly and fill their pockets with money. Since their budgets rely mostly, if not exclusively, on oil exports, falling oil prices would have great and unacceptable impact on their economies. On the other hand, the impact of higher prices on most oil importing countries is not that great, since in most cases the oil imports are only a fraction of the total economy.

This is partially true, but does not include all importing countries. The rich developed countries are those that were able to implement policies in the 1970s and 1980s aimed to diversify their fuels. By doing this, they were able to reduce the demand for crude oil.

The OPEC members were concerned with the new developments and were willing to do anything to reduce the economic disaster caused by low oil prices in 1997-1998. And so, OPEC increased the prices, thus burdening the oil importing community. By 2006 $55 was considered the price floor, not a price target.

OPEC usually responds slowly to rising oil prices, which puts it in the driver's seat, thus providing it with overwhelming market power and higher revenues.

OPEC members control nearly 80% of the global oil reserves, and base their future on a combination of oil demand increase, and reduced growth of non-OPEC oil production

OPEC members are faced with increasing social and economic issues of rising populations and aging in-

frastructure, so they are looking for short-term solution; the fastest of which is raising prices. Focused on that, they are ignoring the need for new and increased oil production capacity, they resolve the problem by reducing output.

All this means that overall, the long-term trend is increased global crude oil prices.

Enjoying increased market power and higher revenues, thanks to increased market power, OPEC members are not looking into investing in new production capacity. OPEC's total production capacity has not risen substantially since 1998. And so, during the last several decades, capacity gains in Iran, Saudi Arabia, Kuwait, Algeria, Qatar and Libya have been less than the capacity losses in Iraq, Venezuela and Indonesia.

The official OPEC aim is to attain a "fair" price for its oil. The unofficial aim is to get the highest price possible with the least effort now and always.

The New Developments

There is a new, and very serious, twist in the global energy markets equation. Some OPEC countries quite aware and even scared of it, and are actively looking into ways to eliminate or reduce the new threat. We are talking about the new developments in the renewable energy sector, and the large fossil reserves discovered in the U.S., Canada, and other non-OPEC countries.

The new developments are threatening the exporters' pockets, so OPEC has warned that a shift to alternative energy inside major oil consuming economies will discourage its own investment in future oil supplies. This could potentially force oil prices to go through the roof, according to OPEC executives.

"If we (OPEC) are unable to see security of demand… we may revisit investment in the long term," warned OPEC officials recently. The U.S. and European biofuels and the other renewables strategies might backfire, because "You don't get the incremental oil and you don't get the ethanol," said the same official, which points to the fact that the biofuels strategies might not prove successful. And then what…?

OPEC is aggressively standing by its price increase and capacity decrease trend expansion, which causing destabilization to the already unstable oil markets. In addition to the market uncertainties, we see increasing energy security concerns in the U.S. and the other major oil importers.

The recent slump in global oil prices, driven by Saudi Arabia's need to control the markets, is. however, a temporary and unprecedented event. It is somewhat outside the OPEC realm of activities and goals, and as such it is not sustainable. Even oil and gold rich Saudi Arabia cannot sustain low oil price conditions for very long, so prices will go up eventually and will stay up for a long time to come. Maybe forever, or at least until the oil reservoirs are totally empty.

All this turmoil is providing more substance to the fears of worsening oil depravity and rising prices in the coming decades. This is in addition to the fears that oil resources are finite and that oil is getting progressively hard to produce and getting more scarce by the day.

And so the energy markets wheel turns in different direction ever day, so it is difficult to predict what will happen tomorrow. We just have to guess what is going and what can be expected in the future of the crude oil sector. It all, however, boils down to the actions of, and the relation between, exporters and importers. We see the problems today, but these are nothing compared with what will happen in the future, when oil prices rise sky high, and oil supplies start to dwindle. Energy wars, global battles, or worse will most likely follow.

The Utilities

The U.S. utilities are a major player in the energy markets. After all, they control the power generation and distribution. Without them, we will be sitting in dark and cold houses, while most businesses would idle and eventually go bankrupt. Really!

The traditional electric power utility regulation was set up in the early 20th century to provide America with universal access to electricity on cost-for-service basis. Thus far, the utilities have been able to maintain the traditional status-quo, but 2015-2016 might mark a drastic change beyond the ability of the utilities to stop it.

We are now witnessing new and emerging technologies, changing customer expectations,
and new energy processes and economics that are forcing changes in the 100 year old business and te related—and equally outdated—regulatory models. The changes are expected to force the utilities to modify the traditional vertically integrated model to accommodate the new distributed, and highly service-based power generation, transmission, and distribution model.

This is a curse and a blessing. The distributed energy resources (solar and wind mainly) are a hassle, but also offer flexibility, and growth opportunities. At the same time they will improve the presently weak to non-existent customer relations, and will even contribute to improvements in the old transmission and distribution networks.

There are huge opportunity in the full implementation of the distributed energy resources, but old habits are hard to break, so for now there is hesitation as of which business model to use. The utilities have no choice in the matter, and are willing to adapt to the changing times, but are just not sure how to do it. As a result, hesitation, and indecision is the utilities MO today.

The utilities have a number of day-to-day challenges, such as aging infrastructure and workforce, in addition to never ending fights with inefficient political and regulatory policies. But with the advent of smart grid improvements and distributed energy resources, the utilities would receive more public support, which will eventually translate in additional political and regulatory models improvement.

Many utilities suffer from stagnant load growth, and some are even seeing reduced power demand in parts of their territories. There are no solutions right now for the depressed electricity sales growth, but as the new technologies and marketing approaches become refined and optimized, the utilities will address that issue by modifying their service model, and/or by implementing new energy marketing approaches.

At the present, most North American utilities are switching from coal to natural gas power generation, in an attempt to reduce GHG emissions. The utilities are also making plans to use more solar, wind, and distributed energy resources during this decade in order to meet their energy mandates. They are also looking seriously at energy efficiency as a vehicle to reduce costs and improve their services during the next decade.

According to a recent survey, 70% of the U.S. utilities are confident in their future growth and success. The different utility managers see the opportunities as follow: 31% in distributed energy resources, 23% in new and improved customer relationship, 14% in new transmission equipment and methods, 8% in centralized power generation, 5% in consolidation of services, and 4% in using the Internet in its daily operations.

A recent survey shows that the utilities clearly see the challenges ahead:

- 47% in aging infrastructure,
- 39% in aging workforce,
- 38% in the regulatory models,
- 28% in stagnant load growth,
- 25% in federal emissions standards,
- 24% in physical and cyber security of power plants and transmission networks,
- 23% in distributed energy resources,
- 21% in coal plant retirement,

- 17% in grid reliability,
- 16% in inappropriate smart grid deployment,
- 13% in renewable energy standards, and
- 9% in energy efficiency mandates.

The customers are looking at the utilities with "hopeful suspicion." The old, ugly image of the uncompromising utility, with its increasing monthly electric bills, was an open gap in customer relations. Now there is an opportunity for the utilities to change it by using new and improved equipment and customer service methods.

According to the public opinion, the utilities should focus on:

- energy storage,
- energy efficiency,
- utility scale renewables,
- demand response,
- distributed solar,
- micro- and smart-grids,
- electric vehicles charging stations,
- natural gas power generation,
- environmental protection,
- carbon capture and storage, and
- coal gasification.

This is a huge list of tasks and responsibilities, which require a lot of effort and money. The (new and improved) utilities seem to be willing to follow in the path of reason, but in most cases they lack the resources. Since many of the new responsibilities the utilities are charged with also represent new opportunities and lucrative profit sources, the utilities will figure a way to capture them, but it might take awhile.

Nevertheless, we are certain that the utilities will step up to the plate and (sooner than later) enter the 21st century as leaders in the renewables energy market.

And sure enough:

Utilities' Solar

The U.S. utilities are taking a closer look at the solar energy business. Very close look! Most of them are clearly and openly not happy with the hassles it presents to their business model, and are putting up a fight. Some are increasing the solar rates, while others are considering the alternatives.

While most other utilities are fighting by increasing rates and imposing new rules, one utility— Southern Company's largest subsidiary, Georgia Power—has decided that instead of fighting the solar invasion they would join it. For that purpose, Georgia Power has launched a residential solar service in direct competition with the local solar companies.

If you can't beat them join them, and so some utilities jump in the renewables game.

And why not? The utilities have the resources, the know-how and everything needed to design, construct, and operate a solar power plant. Their engineers and technicians could handle anything from a small—several PV panels—installation, to many megawatts solar or wind power plant.

Presently, Georgia Power is offering its customers a number of tools to assess the potential for solar on their home—including solar systems design and installation services. Customers who decide to install new rooftop solar have a choice to use a third-party installation, and are not tied to using Georgia Power's own team…for now.

Georgia Power also runs a solar buyback scheme, where rooftop solar system owners are paid a higher rate for electricity they generate, which is higher than the rate at which they would be compensated under the standard Renewable and Non-Renewable program (RNR-8).

Georgia Power basically purchases 100% of the electricity from photovoltaic (PV) systems at a rate of $0.17/kWh, with a current aggregate program capacity limit of 4.4 megawatts (MW). This program, however, gets full quickly and there is no available capacity at times, and Georgia Power has a long waiting list for this program. The expansion of the buyback program depends on the increase of customers purchasing green energy from Georgia Power.

Georgia has one of the fastest growing and most competitive solar markets in the country. The new programs offered by the local utility demonstrate its commitment to provide different energy choices. This is something the customers want in their expectation to get first class service.

As we see it, the synergy and will on part of both; the utility and customers are there, so only the sky is the limit. If this experiment is successful, then many other utility companies will follow Georgia Power example, which might change drastically the way solar business is done in this country.

The Smart Grid

The so called '*smart grid*' is expected to bring a new dimension in the global energy markets. If and when properly and efficiently implemented, it will basically change the way we generate, distribute, and use electric power.

Presently, large population centers consume about three quarters of the world's energy, but contain less than half of its population. These centers are also directly, or indirectly, responsible for 80% of global CO_2 emissions.

It is expected for the global cities' population to increase even further by 2050 and beyond

This, in turn, will increase the energy demand and CO_2 emissions proportionally. The increased urbanization—especially in developing countries—is increasing the strain on all energy, transportation, water, and buildings sectors. The old approaches to solve these problems are no longer working. New, smart, solutions need to be found.

Today we need to work on:
- Resolving the problems by finding highly efficient and sustainable solutions, while
- Generating economic prosperity and social wellbeing in the affected areas.
- One cannot without the other.

Energy is critical in achieving the goal of generating economic prosperity. The energy markets and related infrastructure in most developing countries is just that; developing. The pace and extent of their development determines the wellbeing of the population. Each country has its own particular problems, which are not easy to generalize or summarize. Lack of capital is in the foundation of the slow development of the energy sector, which affects the economic growth in most cases.

At the same time, the energy systems and infrastructure in the developed economies are well developed and serve the population needs properly. They, however, are getting old and decrepit, and need increasing maintenance efforts. The majority of the transmission lines, roads, bridges, canals and other important (and very expensive) segments of the national infrastructure in the U.S. and Europe are over half a century old. The repair and upgrade costs are growing and do not keep up with the demand.

There are plans to upgrade the infrastructure in the UK, where some $1.3 trillion are planned for upgrading the infrastructures of London, and the "Northern Powerhouses" of Leeds, Liverpool, Manchester, Newcastle and Sheffield. Similar plans are in the making in the U.S., but here different states with their different priorities make most of the decisions, so the upgrade work will not be uniform.

Energy drives all this development, so energy management has to be in the forefront of the repair and upgrade plans, in order to improve their energy efficiency and resilience. The Smart Grid technologies are one of the absolutely necessary tools to achieving that goal.

Reliable, affordable, and sustainable energy is the key to economic growth. The Smart Grid is one of the solutions, and is estimated to grow to nearly $200 billion by 2020.

Energy storage is integral part of the Smart Grid solutions, and is expected to play a major role in supporting our future energy needs. Energy storage is vital to solving the variability problem of solar and wind power generation. By storing surplus power and releasing it when needed, energy storage will provide a steady flow from the renewables, thus making them more suitable for base load power generation.

Storage helps the grid to securely balance capacity, and protects it from "stress events," like brown and black outs, etc. The proper introduction of energy storage in substations can decrease the need and cost of traditional reinforcement, such as transformers and cabling.

Europe's largest battery storage project, a 6 MW (10MWh) battery system, for example, is expected to save approximately $9.4 million over 15 years over traditional reinforcement methods. The technology is proven and we are now trialing the financial viability of the project, which could be replicated across all UK network operators. If these plans materialize, the saving would be over $1.0 trillion by 2040, as compared to the presently used approaches.

Advanced Distribution Automation (ADA) is a lesser known Smart Grid technology, which is expected to play an important role in increasing the grid's resilience. It is Internet based technology, capable of detecting power outages caused by storms or other events, using smart switches to "cordon off" affected areas, thus eliminating or minimizing disruption to the rest of the network.

Tennessee Electric Power Board, which serves about 170,000 customers has been losing around $100 million annually due to power outage costs. By using the "self-healing" ADA technology there was a 60% reduction (45 million minutes) in outage duration. When, for example, a major storm caused severe damages to grid infrastructure, there was a 55% reduction in duration of outages, which translated into $1.4 million in operational cost savings from this single event.

As the entire grid gets smarter, there will be more companies developing such technologies and working together to improve the reliability and quality of the grid by harmonizing power flow and voltage, which would result in greater power system stability and balancing.

Smart technologies alone, however, cannot achieve systemic change. Instead, the different projects need to be supported by substantial changes in the way the energy market operates. The difficulty for smart grid technologies (energy storage, ADA, and such) is that the grid was built and still operates on old and tired infrastructure, driven by rules that were created half a century ago.

Like anything that old, the infrastructure and the rules are not adaptable to the new way of doing things. And like anything old, they are out of date and need to be upgraded, giving way to the new technologies and approaches. And that is the hardest part of the battle. Change in any area of the energy market would be expensive and challenging in more ways than one during the 21st century.

Nevertheless, when planning the implementation of new technologies, in an attempt to convert today's populated centers into "smart cities," we must take into account the fact that the old system is usually a hindrance, and find ways to move away from the historic cross-sector approach. Not an easy task, but a vitally important pone, if we are to see a Smart Grid in place of the existing crumbling power network.

On a broader scale, the different resources have been traditionally considered as independent sectors: electric power, transportation fuels, water, waste management, and health have been managed separately. This, however, is inefficient way of managing the resources, so new, more connected, thinking is needed today. The new 'smart grid' is the beginning of the new thinking process.

The transition to 'smart grid' in the context of 'smart cities' is much easier said than done. When properly designed and executed, however, it will increase the reliability and efficiency of, while bringing additional opportunities to, the global cities, their citizens, and economies.

U.S. Smart Grid

The smart grid, although a fairly new development, is finding enthusiastic support in many developed and developing countries. It is complex technically and expensive financially undertaking, and yet there

are takers, with the U.S. as one of the leaders. This is no surprise, because the U.S. leads in technological innovations and capital availability, so the future of the 'smart grid' starts here. We, however, would not be surprised if other countries take the lead in the future.

Figure 10-32 shows the top level of the smart grid operation. It does not show the millions of sensors, control mechanisms, and electronic devices needed to observe, analyze, make decisions, and control every aspect of the energy flow—from the power plants to the last user.

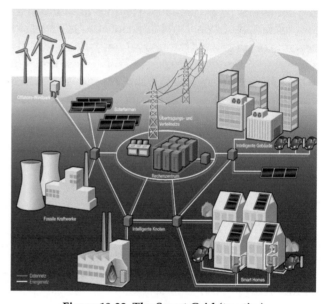

Figure 10-32. The Smart Grid (top tier)

Presently the U.S. Smart Grid effort is focused on:
* Smart meters at homes and businesses
* Electric vehicle integration
* DG systems management
* Communications; generators and users

The renewables integration into the SG will bring:
* Weather forecasting
* Variability management
* Energy storage integration
* Grid load balance (demand-response)

The U.S. power grid consists of 5,000 power plants, 200,000 miles power lines, and 5.5 million miles distribution lines. Most of this infrastructure is very old and outdated, with half of our electricity generated from coal-fired power plants.

The problem is that, due mostly to equipment inefficiencies, only about 50% of the primary energy content in coal, or natural gas is used when generating electricity. And after that more electricity is wasted by the time it gets to the customer.

65% of all electricity generated in the U.S. is wasted while going through the outdated infrastructure.

The power infrastructure is quite inefficient, and is if that was not enough, it is also vulnerable to sudden, uncontrollable blackouts due to natural or criminal activities. Our power grid is basically a Model T car, in which we are trying to install a supercharged engine (the Smart Grid). Can it be done?

The Political Currents

In December 2007, Congress passed, and the President approved, Title XIII of the Energy Independence and Security Act of 2007 (EISA). EISA provided the legislative support for DOE's smart grid activities and reinforced its role in leading and coordinating national grid modernization efforts. Key provisions of Title XIII include:

- Section 1303 establishes at DOE the Smart Grid Advisory Committee and Federal Smart Grid Task Force.
- Section 1304 authorizes DOE to develop a "Smart Grid Regional Demonstration Initiative."
- Section 1305 directs the National Institute of Standards and Technology (NIST), with DOE and others, to develop a Smart Grid Interoperability Framework.
- Section 1306 authorizes DOE to develop a "Federal Matching Fund for Smart Grid Investment Costs"

Office or Electricity (OE) is the national leader, partnered with key stakeholders from industry, academia, and state governments to modernize the nation's electricity delivery system. OE and its partners identify research and development (R&D) priorities that address challenges and accelerate transformation to a smarter grid, supporting demonstration of not only smart grid technologies but also new business models, policies, and societal benefits.

OE has demonstrated leadership in advancing this transformation through cooperative efforts with the National Science and Technology Council (NSTC) Subcommittee on Smart Grid and the Federal Smart Grid Task Force.

"The National Science and Technology Council Subcommittee on Smart Grid: Chaired by the Assistant Secretary for OE and the National Director for Smart Grid at NIST, the Subcommittee is promulgating a vi-

sion for a smarter grid including the core priorities and opportunities it presents; facilitating a strong, coordinated effort across federal agencies to develop smart grid policy; and developing A Policy Framework for the 21st Century Grid which describes four goals the Obama Administration will pursue in order to ensure that all Americans benefit from investments in the Nation's electric infrastructure: better alignment of economic incentives to boost development and deployment of smart-grid technologies; a greater focus on standards and interoperability to enable greater innovation; empowerment of consumers with enhanced information to save energy, ensure privacy, and shrink bills; and improved grid security and resilience."

The above is a long and winding statement, which, in our opinion, reflects the long and winding road in front of the US smart grid. And please note that the statement is calling for a "smarter" grid, while what we need is a totally "smart grid." This might mean that the government bureaucrats are thinking of implementing a "smarter" grid first—a compromise of sorts—before jumping to the "smart" grid version—which only makes the road even longer and more winding.

In the Beginning…

We saw similarly long, winding, confused and even misdirected statements in the 1970s, when solar energy was hailed as the "savior" of the American Dream, and where the US government was taking charge of its development. Solar was hailed as the only way to a sustainable energy future, and so lots of money was thrown at solar companies then (some of it at the wind, as it happened today)…until the government changed, and things went down hill from there on.

The solar energy was brought out and shoved back in the closet several times since then, and it is where it was until recently, when it was taken out, dusted off and hailed again as the only way to energy independence… until the new government takes power and solar is shoved back in the closet…again. That may not be so easy or fast this time around, but the signs unequivocally point in that direction.

Is the same faith expecting the US smart grid? Nobody knows, but due to the great amount of effort and money involved, and judging from the lessons of the past, we cannot, and should not, count on the government bureaucrats—self-proclaimed "smarter" grid saviors—to lead the smart grid effort to a successful end.

Personal interests, changing political winds and everything else that is awkward with politics and politicians today, makes us believe that the government should stay out. Instead, we should let the US capitalist system take over and dictate the rules and the effort. Keeping the government bureaucrats in a sub-serving role, where they belong, and where they can be controlled, is the best, if not the only, way to manage this, and any other, serious effort.

One example of ill-conceived (albeit good intentioned) intervention of the government is the Department of Energy's Smart Grid Investment Grant Program. Grants ranging from $500,000 to $200 million are to be issued for deployment of smart grid technologies, most of the recipients of which seem to be working on smart metering solutions (since it is the easiest, and most accepted by the public, smart grid related technology).

In addition, IRS issued a guidance in 2010 providing a safe harbor, under which the $3.4 billion of federal Smart Grid Investment Grants (SGIGs) issued under the American Recovery and Reinvestment Act of 2009 (ARRA) will not be taxable to corporate recipients.

Great news, right? But here is the catch; energy conservation, derived from the full implementation of smart meters (as encouraged by the government grant and tax programs) means lost revenue to utilities at a time when the utilities are in the midst of expensive changes and are not ready for additional revenue losses. The lost revenue, forced by the government programs is simply untimely and is not aligned with the objectives of most utilities and their shareholders.

In addition, long-established regulations have favored supply-side resources over energy conversation, so utilities have been encouraged to add new generation because they earn a rate of return on investments on their assets; and mainly the power generation, transmission and distribution infrastructure. And now, without a proper warning and preparation, the utilities must abandon the profitable business model and gear for energy conservation, which simply translates to loss of revenue.

This misalignment might cause delays and even failure of some energy-efficient technologies and services in the U.S. and reflects the disconnect and fragmentation in the sector. Some of the disconnect is caused by the inadequacy of the government bureaucrats, who talk the talk, but seem more interested in their agendas than anything else. This anomaly is fueled by the ignorance of the regulators as well, which completes the circle of incompetency, and which will make the smart grid implementation that much harder.

The bureaucrats do not take into account the fact that the technologies that promote energy conservation have evolved much faster than the regulations that govern utilities and other suppliers. This creates an imbalance, which will continue until new regulations, promoting the implementation of demand response and other energy conservation programs are implemented.

To cross the divide seamlessly, appropriate and timely regulations and other policies should be properly structured to justify and encourage full utility participation and investments in demand response.

Presently

The work has started, no doubt, but it is full of gaps, fragmentation and misdirection. In order to coordinate and encourage the activities, new legislation must be approved, and regulation must be implemented to create incentives for utilities to reconsider investing in generation, transmission and distribution assets and promote energy conservation and demand response instead. There must be a balance between the approaches, and the utilities must be encouraged to use demand response as part of the generation portfolio and modus operandi.

There should be also some type of shared-savings program that allows utilities to participate in the savings customers receive from reducing their energy usage. This will persuade them to promote energy conservation because they will benefit, as well. In addition, there should be a penalty if a utility does not encourage customers to reduce their energy usage.

Another way to encourage utility participation would be that a utility gets compensated for a portion of its avoided supply costs obtained through demand response and other related programs.

And so, the future of the smart grid in the U.S. is not very clear and nobody knows what exactly will happen and when, so we can only guess. Looking in the history lessons of the past; solar was the hope for energy independence in the early 1970s. It is still the hope… albeit still in embryonic state and hesitant, at best. After 40 years, we know what solar can do, but we still don't know what to expect from it.

Smart grid technologies are headed in the same direction. Right mow they are changing at an exponential pace—ten years ago we didn't even have smart phones, and the term 'social media' was still in the making. Today, these are everywhere and we cannot even imagine life without them.

So how could we—stuck in the beginning of an era—even imagine where, or what, the smart grid (or anything smart) might be in 2020…2050 and beyond. It is not possible, but we can say with certainty that there will be a progress, with a heavy component of customer (partially virtual) service.

The utilities are well aware of the demographic changes and the pending changes. They see the new generation of informed and active young customers, as well as the large retiring workforce, where by 2020 half of the utility workforce will be retired as well. And they know too much…so nothing can be kept hidden.

Customers are accepting new technologies quickly as evidenced by the fact that, according to some statistics, the volume of text messages now exceeds that of phone calls. But just think; 5 years ago text messaging did not even exist… And so, information is the foundation, customer acceptance is the route, and integration is the vehicle.

The existing data is enormous, and growing exponentially, and all participating systems must be integrated in parallel in order to receive and process the data for maximum efficiency and lowest possible cost.

No matter how you look at it, at least the direction is clear; the smart grid and its smarter virtual applications are here to stay. And while some things that haven't even been invented yet—whatever is coming in the area of smart grids, will have a large virtual component with a high level of self sufficiency and reliability. The utilities must be given a chance to get ready for the new order of things.

Our fully implemented and integrated, smart grid based "green" energy future seems attractive, but we cannot even guess how and when it will arrive. We just have to go step by careful baby step towards its proper and complete development.

Table 10-15. Smart grid investment

Global Smart Grid	
Country	*Total*
China	$7,500 million
USA	7,000"
Japan	850"
S. Korea	850 "
Spain	800"
Germany	400"
Australia	360"
UK	300"
France	260"
Brazil	200"

Europe

Europe's Smart Energy Collective is a sector-transcending cooperation involving a wide range of companies working for smart energy and smart grid implementation. Members include: ABB, Alliander, APX Endex, BAM,DELTA, DNV KEMAEnergy & Sustainability, Efficient Home Energy, Eaton, Eneco, Enexis, Essent, GEN, Gemalto, Heijmans, IBM, ICT Automatisering, Imtech, KPN, Nedap, NXP Semiconductors, Philips, Priva, Siemens, Smart Dutch, Stedin and TenneT.—EcoSeed Staff

In May 2012, members of the Smart Energy Collective in Europe have approved the second phase of the smart grid initiative, which involves the development of five large-scale smart grid demonstration projects in the Netherlands.

Schiphol airport, ABB and Siemens offices, and several residential districts were the chosen sites for smart grid implementation and tests. In order to make this possible, an intelligent energy system is designed that uses a combination of innovative technologies and services. It will enable the owners to keep the costs of energy low with a comparably high level of reliability.

This effort would be an important step towards standardization, in addition to testing the smart grid operation in actual use. A survey during the first phase showed that there are more than 6,000 relevant standards that play a role in the new technologies, in which according to the group, will be introduced to the market over the coming years.

Working groups have been established for standardization, market mechanisms, services and business cases, smart grids, privacy and security, and ICT infrastructure, in order to establish a solid foundation for the design of the five demonstration projects, as well as for future projects.

Asia

Asia is quickly becoming a major player, and the center of global smart grid activity. The combined smart grid market in China, Japan and South Korea is estimated at over $10.0 billion, with estimated increase to over $30.0 billion by 2020, according to industry specialists.

As Asian countries are irreversibly becoming the predominant smart grid markets, the competition and the positioning of different vendors is increasing as well. Lack of clear understanding of the energy scenarios (present and future) in the major Asian countries (China, Japan, and South Korea) is an obstacle that needs to be resolved first.

It is widely expected that the smart grid markets in Asia will move forward at a breakneck pace during the next decade or two. The developed countries are already positioned for the race with over $45 billion in funding, earmarked by the respective governments and utilities across China, Japan and South Korea. The majority of funding and related opportunities, of course, are located in China and its booming energy market.

A level of uncertainty is still distorting the smart grid vision, so determining the trends and establishing a clear path of energy policies and currents will allow much faster implementation of smart grid technologies in these countries. Once the uncertainties of today are lifted, the a meaningful entry in the Asian smart grid markets will proceed very fast.

In Japan, the shut down of many nuclear plants created a need for demand response, energy management and smart meter deployments. Smart grid is on the agenda. Money is available, the need is there, the customers are willing, so only time separates Japan from the full implementation of efficient smart grid concepts.

We would not be surprised if Japan leads the world in this area in the near future, and even become the first country to claim complete smart grid deployment.

The South Korean market is quite different. r the country with the most reliable grid in the world, South Korea is looking into developing the next-gen smart grid technologies and components (hardware and software) across all segments, but primarily for global export.

China

The growth of the smart grid market is characterized by the special needs of each country's energy demands, as well as the local utilities and existing grids specifics. For example, the smart grid investment in China is focused on transmission and distribution automation, as needed to support the plans for a new power grid (planned to be developed) and robust renewable energy (planned to be built.)

China is aiming to become a world leader in smart grid technologies in the next decade. The "Strong and Smart Grid," an 11-year plan,, revealed in 2009, outlines the ambitious steps to get there. It involves all aspects of the power grid, including; increase of power generation capacity, implementation of smart meter programs, emphasis on large-scale renewable energy, and a large transmission lines and substation build-out.

Today, the plan is still alive and well. The State Grid Corporation of China, which is actually one of the largest utilities in the world, and the executor of China's smart grid plans, is already in phase two of the three-phase program. It is the Construction Phase, which is scheduled for completion in 2015.

New transmission lines are a major focus for State Grid in the Construction Phase, which is struggling to meet the growing energy demands of the rising middle class in the East and South. Most coal, hydro, wind, and solar load sources are over 1,000 kilometers away from the populous east and south. High voltage (HV, under 300 kilovolts), extra-high voltage (EHV, 300 kilovolts to 765 kilovolts), and ultra-high voltage (UHV, 765 kilo-volts and up) lines are being installed currently, with at least one 1,000-kilovolt UHV AC or DC line installed annually until 2015. Overall transmission line investments for 2015 are approximately $269 billion, equivalent to the combined market cap of ABB, GE, and Schneider Electric as of May 21, 2012.

China is adding so much new transmission capacity and so many power lines that it could build three quarters the length of a new American transmission grid in just five years. When the dust settles, there will be over 200,000 kilometers of new 330-kilovolts-and-up transmission lines built, for a total of 900,000 kilometers of transmission lines, compared to 257,500 kilometers of transmission lines presently in the U.S.

At a cost of $1.05 million per mile for UHV transmission line and equipment, each UHV line requires billions of dollars to build, and State Grid put in a staggering $80 billion investment into 40,000 kilometer of UHV lines for the 2011 to 2015 Construction Phase. The business case is readily apparent: a 2,000-kilometer, 800-kilovolt UHV DC line has an incredibly low 3.5% line loss rate per 1,000 kilometer and a high 6.4-gigawatt transmission capacity, all the while being 30% cheaper than a 500-kilovolt EHV DC or 800-kilovolt UHV AC line of the same length. By 2020, UHV lines will have 300 gigawatts of transmission capacity, roughly split 60% AC and 40% DC.

The competitive business environment seen in the transmission grid build-out is indicative of the rest of the smart grid market in China—high-quality goods, competitive costs, and a well-built relationship with State Grid all go a long way toward winning a contract. Fierce vendor competition exists, due in part to State Grid's competitive construction procurement process. All projects costing over $300,000 to build are required to go through an open bidding process that aims to enforce fairness and transparency, but State Grid still holds the reigns tightly on choosing project developers. In the process, State Grid has the final say and does a rough 45/45/10 split when evaluating meters, based on quality, cost, and bankability of the company.

With the promise of power shortages disappearing and a stable energy supply base, the build-out of the transmission grid is ushering in the next era of smart grid opportunities in China. Smart meters and renewable integration are already big businesses, and new substation infrastructure has brought with it a vibrant and growing substation automation market. The need for better monitoring equipment has risen as China is keen on decreasing its system average interruption duration index (SAIDI) and improving power quality to its customers.

State Grid has earmarked over $40 billion toward these smart grid technologies between 2011 and 2016, with smart meters alone being a $2.5 billion to $3 billion annual market. State Grid has paid special attention to substation automation technologies, and plans on installing 74 new digital substations for 63 kilovolts to 500 kilovolts by 2015.

While this number is small compared to the existing 40,000+ substation base, State Grid has stated it intends to include digital technology in all new substations built. Companies such as BPL Global have been expanding their substation operations in China, which has been met with stiff domestic competition. The substation market offers promising growth over the next ten years.

The transmission grid build-out also has an impact on technologies at the distribution level and downward. China is building 36 million new urban homes between 2011 and 2015, and modern building automation and smart meter technologies will be utilized. The coming years promise to create a new and vibrant building automation market, but for the time being, the market continues to focus on meeting demand shortfalls and other key infrastructure challenges.

Expect to see an exciting shift toward technologies at the distribution level and downward in the next five to ten years, as China's grid solidifies its transmission grid and generation sources. If the past three years have been any indication of future progress, expect to see China become a leading smart grid market for the next five to ten years. The distribution grid build-out and digitization will be the next major indicator of China's smart grid prowess.

So, in summary, the global smart grid future is bright. It has the potential to develop into one of the most important parts of the world' energy markets, thus bringing us closer to the overall goals of achieving energy independence and cleaning the environment.

The 21st century will bring a lot of developments in this area, and the Asian countries will be the first to claim smart grid implementation and use.

Smart Grid Cyber-Security

Smart grid activities are shaping as one of the most promising, as far as ensuring the future grid safety and functionality are concerned. From a technical stand point, this is doable, exciting, and profitable effort, but there is a big problem

The new smart grid opens the power grid to a new type of unprecedented cyber-attacks.

The sophistication and the number of potential attackers is growing in parallel with the smart grid efforts. Since the smart grid will grow no matter what, ensuring its security is going to be a very big business. Unfortunately, this big business will not make money for the utilities, and other participants, that will end up paying for it. Nope, the majority of cyber-security investments will go mostly to meet regulations and make the investors and customers happy. The major effort and money will go into trying to justify the extra costs for preventing grid attacks, and the related failures, accidents and incidents.

There are estimates that the $120 million spent on cyber-security products and services in 2011 will double by 2016. This number will grow exponentially to over $6 billion spent mostly by the utilities and the U.S. government by 2020, as needed to secure the grid and avert the consequences of a successful cyber-intrusion. Worldwide this number could grow to $80-90 billion during the same time.

Cyber security spending will exceed the total amount of money allocated for spending on the smart grid itself

Ensuring the physical and cyber security of such a large network of power plants, substations, transmission and distribution lines, and the related equipment and personnel is an amazingly complex and expensive undertaking. It is truly big business that is taken very seriously by all involved

Nevertheless, this is also a very risky business. The grid must be protected from any eventuality, such as minor data theft to taking over grid assets. In case of a failure to do so, we will witness blackouts, damaged electric power infrastructure equipment, and even casualties. A lot of research and trials will go into making sure that security on all levels is done right, efficiently and without failures.

This is not the case today, when a number of cyber-attacks and hacks at major U.S. organizations and installations have gone undetected and have caused major damage. The issues have been elevated at very high levels, and as a consequence, the Obama administration has issued an executive order demanding that all industries involved in the smart grid and power generation come up with a plan to deal with the threats.

Such plan, however simple, cannot be done quickly nor implemented cheaply. How far and how fast this effort will develop remains to be seen, but it will not be a transparent process, since one of the key tenets of cyber-security is that cyber-security is most secure when nobody knows how secure it is. We surely will not learn much about the specifics of discovering, isolating, and eliminating most of the new intrusions and attacks. And since the methods of the intruders change daily, there will be a lot of changes and adjustments done to the cyber-security infrastructure as well.

The utilities are interested in, and required, to secure their infrastructure by all means available and necessary. The effort will spread on all levels from installing fences and locking gates and doors to ensuring the safety of the smart meters and computer networks, and to implementing the latest cyber-intrusion detection software and schemes. All this will require new products and support services, similar in size and form to what the banking and telecommunications industries have been doing for the last 20 or more years.

The cyber-security spending in North America is controlled by the North American Electric Reliability Corporation (NERC)'s Critical Infrastructure Protection (CIP) requirements. In the U.S. and Canada these rules are implemented and enforced by stiff fines. There are cases of up to $1 million per day fines for utilities and other entities that do not meet the security guidelines.

The Department of Energy is dangling a $4.5 billion carrot in front of the cyber-security principals and wanna-be companies. The DOE stimulus grants for a number of projects require cyber-security compliance, as outlined and coordinated by the U.S. National Institute of Standards and Technology.

Around the world, governments are working on their own regulations for ensuring the safety of their respective power grids. The European Union's cyber-security agency, ENISA, for example, is building a framework that is focused on a risk-based approach with a certain degree of freedom for the utilities and all companies involved, in order to do their job more promptly and efficiently.

Securing the new IT infrastructure of the power grid against cyber-attack is going to be big business, but that's not because it makes money for the utilities that are buying it. Instead, today's smart grid cybersecurity investments are mostly about meeting regulations, satisfying shareholders, and trying to justify the costs based on what it's trying to prevent.

Putting a dollar figure on averting the consequences of a successful cyber-intrusion is a risky business, of course. Utilities have to worry about everything from data theft to a full-blown attempt to take over grid assets in order to cause blackouts or overload generators.

Reports of cyber-attacks on U.S. infrastructure by foreign actors, as well as hacks at the Defense Department and Department of Energy, have brought the issue into focus this year, and the Obama administration has issued an executive order demanding that the industries involved come up with a plan to deal with the threats.

Recent research forecast that the trends can lead to market opportunity at a pretty big chunk of change. U.S. utilities are expected to spend a cumulative $7.25 billion in security from now until 2020, with distribution automation assets as the core focus.

Other estimates reach $14 billion through 2020, with about 63% of that spending aimed at the industrial control systems and SCADA networks used to control today's grid assets. Even wilder claims exist of an exuberant $80-90 billion global market by 2020, which exceeds total smart grid spending figures in most of the world.

North America accounts for over 40% of the global market presently, with Europe at 30% and Asia-Pacific at 17%. China is expected to overtake the U.S. as the largest market by 2020, leaving the Asia-Pacific region with 35% of the market. China, Japan, South Korea, and Russia will see growth rates of 40% and up in the next decade.

How far along the industry is toward that vision is hard to say, since one of the key tenets of cyber-security is that it is secretive and you just don't talk about cyber-security. And even less we know about the specifics of how the principals are discovering, isolating, eliminating and building new protections against new intrusions and attacks that change from day to day.

In North America, much of that spending is being driven by the North American Electric Reliability Corporation (NERC)'s Critical Infrastructure Protection (CIP) requirements. Covering the U.S. and Canada, these rules come with stiff fines of up to $1 million per day for utilities that can't prove they're meeting security guidelines, and newer versions add a lot more serial-connected smart grid assets to their purview.

The Department of Energy's $4.5 billion in stimulus grants also came with cyber-security strings attached, as outlined by the ongoing government-industry work being coordinated by the U.S. National Institute of Standards and Technology, or NIST.

Other parts of the world have their own regulations in place for how to secure the grid. The European Union's cybersecurity agency, ENISA, is building a framework that, in "contrast to the U.S.' strict regulatory path," is aimed at a risk-based approach that allows "a certain degree of 'freedom'" to the utilities and technology partners involved.

California Edison's Smart Grid Roadmap

Southern California Edison (SCE), which serves about 13 million people in the Golden State, is expected to make more changes to the grid and its overall power infrastructure in the next ten years than in the previous century. It is a good example of a big investor-owned utility that's taken a leading role in the smart grid.

There is already an impressive list of smart grid accomplishments, which mark the beginning of a transformation that is going to bring many changes. That, however, comes with a long list of projects to be undertaken as needed to meet the myriad challenges in the coming years.

The Issues

One of the top concerns of the California utilities is how to manage the growing share of intermittent renewables (solar and wind) powering its grid. The recent rise of distributed energy resources (solar roof systems) have complicated the situation even further.

State mandates call for utilities to get 20% of their power from green energy today, and that figure is set to grow to 33% by 2020. Some of that power comes from stable sources like geothermal or small hydropower, an increasing share is coming from solar and wind power. And much more renewable power is to be added during the next 20-3 decades.

The variable power sources are creating problems for the utilities that have traditionally delivered power from central generation plants to end-customers in a number of ways.

The present-day issues and solutions can be summarized as follow:

- Solar and wind power are different from the base load sources (fossil power plants) because they cannot be dispatched when needed. Instead they force the utilities to react to the whims of Mother Nature which drives the solar and wind power generation.

- Many of the generation resources are connected directly to the distribution networks, thus bypassing the transmission lines all together. The distribution network, however, was not designed to handle two-way power flow. This prevents the central grid control structure from managing this extra power, so a new system control must be developed and implemented.

- Plug-in electric vehicles are another headache for the utilities. As their number grows, they still use enough power when plugged, which at the grid looks like an entire household power use.

- Millions of new smart meters deployed recently represent another disrupting factor, because they force the utilities to collect and process more data than usual. The new meters also enable customers to sign up for time-of-use rates, instead of the old way of paying flat fees for power used. This is forcing the utilities to base the energy use on real-time feedback from home energy management platforms.

- The old traditional customers are now sophisticated "transactive energy" players, who generate electricity and, a) either send it into the energy markets, or b) shift energy use to take advantage of price fluctuations.

- Customers now have a chance to make energy decisions based on economic choices. This bring a new level of unpredictability into the utilities equation. And as the renewables increase in number, so will the utilities headaches.

- Battery-based, grid-scale energy storage is presently on the drawing boards as one way of balancing the grid. Two Department of Energy smart grid stimulus grant-funded projects are fueling the drive. The energy storage effort is focused simultaneously on the large-scale and small-scale energy storage systems.

- On the large scale; SCE is working with A123 Systems on an 8-megawatt, 32-megawatt-hour lithium-ion battery storage option in the Tehachapi mountains. The goal is to stabilize and integrate large-scale wind power generation into the grid. For that purpose, SCE has undertaken a number of projects, including a California Energy Commission (CEC)-funded partnership with GRIDiant and New Power Associates, which analyzes transmission and distribution systems for a grid serving about 275,000 customers. SCE is also a leading player in a DOE stimulus grant-funded project to deploy synchrophasors, which are devices that monitor transmission lines by collecting data in sub-second intervals, across the entire western U.S.

- On the small scale, SCE is also integrating four different configurations of batteries, ranging from a 2-megawatt substation battery to smaller residential energy storage units (RESUs), into its smart grid demonstration project in Irvine, California. This is a $80 million, Department of Energy grant-funded project designed to test the interplay of energy-smart homes, solar panels, grid batteries, plug-in car chargers, grid voltage management, self-healing circuits, and communications and controls networks in a one single neighborhood. The list of partners, and potential designers and manufacturers of the related equipment includes; Boeing, General Electric, SunPower and Space-Time Insight, to name a few.

SCE is also testing a variety of "smart inverters" and other devices that could help integrate the roof solar power into the grid by allowing each rooftop solar system to better manage its interaction with the grid. This has been successfully tried in Germany, where solar inverter regulations and requirements were put in place recently, in order to manage the massive share of distributed solar power and the grid interactions.

Ultimately, using digital devices on the grid can help record and even manage the variable flow of distributed energy resources and customers' energy use. The challenges, however, are great; from hooking up smart meters to distribution management systems, to integrating transmission grid measurement units into the utility's operations.

- The future role of energy storage is undeniable, but the type and size of the systems will depend on the respective economics, reliability, and efficiency factors. CPUC recently decided that it would require at least 50 megawatts of energy storage for the Los Angeles region power mix by 2020. This may or may not happen, but if it is not done right it might distort the market without solving the energy storage problems.

- At the same time, the utilities are spending a lot of effort and money on keeping the old grid humming. Replacing old power poles and wires, transformers and other components of the crumbling electric infrastructure cost SCE over $3-billion-per-year in capital investment. The number is tripled when the efforts of the other California utilities is added. Utilities in other states are in the same situation, and like it or not must dedicate a lot of effort and money to fixing the old grid and its components. This alone makes any investment in smart grid initiatives involving the old grid imprudent.

With all this considered, we must conclude that:
- The U.S. power grid is not ready for serious smart grid additions and improvements,
- The most advanced and most promising smart grid

technologies are not be ready for mass-deployment, and
- Most of the smart grid technologies that are commercial-ready and available are not approved by the California Public Utilities Commission, since they requires a lot of testing and verification of the cost-benefit equations involved with deploying them.

And so, the smart grid is a great idea and would do miracles for the grid, the utilities, and their customers when fully developed. The problem is that its full implementation is still far out in the distant future. For now, partial additions, patches, and marginal improvements are underway. And so, for now we just have to wait and see how these work out and how they affect the overall energy situation in the country.

In the long run, we will most likely continue adding partial smart grid solutions to the mix during the next several decades. Eventually, the entire power system will be upgraded and modernized. This will not happen tomorrow, so for now we can only dream of the day when our entire power system is pronounced 'smart.'

Microgrids

Microgrids are part of the future smart grid. A microgrid is a local energy system (mini-power grid), or a small grid within the large national power grid, that can function independently. If and when needed, it can disconnect totally or partially from the power grid and operate autonomously.

Efficient microgrids can operate while connected to the grid, but can also break off and operate on their own using local (independent from the national grid) power generation. This is especially useful and important in times of crisis like storms or power outages, power price increases, or for whatever other reasons.

Different microgrids are powered by different power sources, but the most efficient operate on distributed generators. They use renewable resources like solar panels to generate power and batteries to store power for later use. Such microgrids can run without outside power input indefinitely.

In most cases, however, microgrids connect to the national power grid via common coupling that maintains voltage at the same level as the main grid. In case of problems on the grid or for any other reason to disconnect, a switch separates the microgrid from the main grid automatically or manually. At that point, it functions as an island within a sea of power around it.

Microgrids have the potential to provide:

- *Security* in times when we are faced with dangers of terrorism, natural disasters, and increasing energy costs. The microgrids are setup to constantly search for ways to isolate disturbances so they do not turn into cascading blackouts. This way, they avoid loss of power to homes and businesses.

Note: In the aftermath of hurricanes Sandy and Irene, for example, there were massive black outs that affected thousands of people. At the same time, however, several microgrids continued to supply electricity and thermal energy without interruption.

- *Self-reliance* is very important in the digital age. It requires high-quality, interruption-free power. Today we don't worry about the cost of power as much as we worry about the cost of *not* having power. This is because power interruptions of business operations are very expensive. Microgrids, equipped with backup power, are the best and most efficient solution to ensuring self-reliance. There are estimates that power outages cost American businesses over $100 billion every year.
- *Power cost reduction* is a major goal of microgrid owners. By disconnecting from the national grid, they can control the costs of power generation and use. This is especially true in peak periods, when the utilities increase the power cost several times above the normal. In Arizona summer afternoons, for example, a well designed and operated microgrid could provide cost savings 2-3 tikes lower than the utility peak rates.
- *Cleaner and more reliable power*. Microgrids can also play a significant and positive role in promoting a cleaner and more resilient energy infrastructure. Growing awareness of these benefits are increasing considerably the interest in deploying microgrids as a primary or supplemental source of heat and power in the country and abroad.

Military U.S. military bases, and even entire countries (like Denmark), are already utilizing microgrids on daily basis. Several U.S. cities and business districts are also using microgrids as a defense against natural and manmade disasters, which is a major motivator to move toward microgrids.

Ownership Models
There are several types of microgrid ownership models:

- Direct Ownership usually retains full control over all aspects of the microgrid's financing, build, own, operate and maintain. It offers greatest potential return, but also creates largest risk for the owner.
- Joint Ownership also retain ownership, but only finances the project. Third party agrees to develop and operate the microgrid. Owners transfers share of returns to the third party in exchange for assuming some of the risk.
- Third-Party Ownership is used when the owners outsource financing, development, O&M and ownership. Significant risk is transferred to third-party, but it also gets primary share of the returns.

Microgrid Problems
The biggest problems with microgrids presently are

- *Standards*. There is a lack of national and international standards for microgrid installation and operation. A lot of work has been done in order to solve the existing interconnect and cost issues via new standards. The CERTS Initiative (a coalition of national labs, universities and others) is developing new, universal specifications for design and operation. NREL is also in the midst of creating national interconnection standards. With time, these standards will solve the interconnection problems while bringing power down costs too.
- *Bankability*. Microgrids require a significant initial capital investment. They are considered to be financially viable or "bankable" when they yield an acceptable return on investment. "Bankability" here means that they create sufficient energy cost savings and/or generate revenues to carry financing costs typically over a period of time. In practical terms, microgrids create financial value by reducing a customer's total cost of energy and/or create additional revenue streams. A robust analysis of the cost savings and/or incomes created by a microgrid is essential to estimating the project's return on investment.

The amount of capital required can vary widely depending on the specific circumstances. The source of the capital may also vary for different projects. Often, third-parties provide the capital needed to construct the microgrid. In other cases, the customer, or more than one source, could provide the financing.

- *ROR*. The rate of return (ROR) expectation could vary for different investors. Depending on the risk profile, large corporations are unlikely to commit funds to support early-stage investments with RORs below 12%. Third-party equity investors typically expect a ROR of at least 15% or more, depending again on the project risk profile. Lower returns are acceptable for investments made later in the development cycle of a project when the project is much more matured and the risks have been substantially mitigated.

- *Standby Rates.* Microgrids interconnected with the conventional grid may be required to pay standby rates to the local distribution utility similar to the stand-by rates that apply to baseload generators.

Note: Standby rates are designed to capture the distribution utility's cost of supplying services to users. These users consume power provided by the conventional grid for supplemental power such as when an on-site generator or microgrid is offline for scheduled or un-scheduled maintenance. Standby rate is often used to describe various rate structures designed for customers with onsite, non-emergency, power generation.

The services provided under standby rates include one or more of the following:

- *Backup service*: serves a customer load that would otherwise be served by onsite generation during unscheduled outages of the onsite generation
- *Supplemental service*: customers whose onsite generation does not meet all of their needs, typically provided under otherwise applicable full service tariff
- *Scheduled maintenance service*: price of grid-sourced power used by microgrid customers during scheduled maintenance outages
- *Economic replacement power*: electricity at times when the cost of producing and delivering it are less than that of the onsite source.

Standby rates include three distinct charges:

- *Customer charges*, designed to recover certain fixed costs (e.g., metering expenses) that do not vary with energy use and shows up on the customer's bill as a fixed monthly charge
- *Contract demand charges* aim to recover variable costs associated with providing electric service to the customer and only applies when the customer's consumption of power from the conventional grid exceeds some pre-determined threshold during a given time period. When consumers exceed their demand threshold, they are subject to penalties in the form of a surcharge equal to between 12 to 24 times the monthly demand charges for all excess usage.
- *Daily-as-used demand charges* are designed to recover the costs of distribution infrastructure required to serve the system's peak demand. The daily-as-used demand charge is based on the customer's daily maximum metered demand during peak-hour periods on the macro system.

Standby rates are different in each utility territory and may make the economic case for a microgrid prohibitively expensive. For example, "demand ratchets" can have adverse financial consequences for onsite generators.

In all cases, the owners must consider these and many other factors in selecting the ownership structure, as follow:

- Long-term uncertainty and risk tolerance
- Standby rates
- Rate of return goals
- Access to low-cost capital
- Financing methods
- Owners' technical expertise
- Operation and Maintenance (O&M)
- Future sustainability goals
- Power Purchase Agreement (PPA), or contracts
- Potential advantages and disadvantages
- Potential future capital constraints
- Accounting services
- Long-term tax implications

Military Microgrids

Traditionally, if the power goes out at a military base, each building within the base will switch to a backup energy source, most often provided by a diesel generator. The military isn't crazy about this setup, in part because this is an expensive solution, and also because generators can fail to start—especially bad news for base hospitals and other critical operations. And if a building's backup power system doesn't start, there is no way to use power from another building's generator.

In addition, most generators are oversized and use a lot of dirty and increasingly expensive fuel. So the military is working on ways to change that by connecting clean energy sources like solar and wind in micro-grids that can function when commercial power service is interrupted.

A new $30 million initiative could transform the way U.S. military bases deal with power failures. Termed SPIDERS—short for Smart Power Infrastructure Demonstration for Energy Reliability and Security—the three-phase project will focus on building smarter, more secure micro-grids on military bases and other facilities, that make use of renewable energy sources. Sandia Labs has been (very appropriately so) chosen as the lead designer and technical support for the SPIDERS project.

Diesel generators are used all the time to provide emergency power to buildings, but they are usually not interconnected with alternative energy sources like solar, hydrogen fuel cells, etc., as needed for a stable and significant energy source. It's a real integration challenge, according to Sandia's engineers working on the SPIDERS program. So, Sandia will work to set up a smart, cyber-secure micro-grid that will allow renewable energy sources to stay connected and run in coordination with diesel generators, which can all be brought online as needed. The new system is expected to not only make the military's power more reliable, it will also lessen the need for diesel fuel and reduce its carbon footprint.

The project is being funded and managed through the Defense Department's Joint Capability Technology Demonstration with the support of the U.S. Department of Energy.

Someday soon, the departments hope to use the SPIDERS plan for civilian facilities like local hospitals and others as well.

The Future of Electricity

Our modern societies run on electricity. Electricity—especially in the developed world—is needed every single minute of the day, and for every step we make. Just look around you. How many lights do you see? The average American house has at least a dozen of ceiling, desk, and table lights. Then there are our refrigerators, heaters, coolers, computers, wi-fi routers, TVs, radios, garage door openers, electric cars, swimming pool pumps. On and on the list goes. Now imagine living without any of these…

U.S. Electricity Divisions

As a nation divided into many different socio-political segments, we are struggling to find our new, 21st century, identity. We are divided along the lines of wealth, sexual preferences, religion, ethnicity, immigration status, and many others. Wealth inequality is the most obvious and controversial in our society.

Why would anyone need a billion dollars? And why after having a billion dollars, most rich people want to have two, three, four billion dollars? It is simply because two and three is more than one. For the same reason, s/he buys a 200 foot yacht? Simply because it is

bigger than the 180 foot yard next to it.

This, at the time when some people can't even pay their rent and have a hard time putting bread on the table. This overwhelming inequity is actually growing, but looking back in history we see that any and all of the societies with similar inequality problems have collapsed. We are no doubt headed in that direction, with the inevitable around the corner. The only question is when?

But back to energy…

As to complicate things, present day energy issues are adding another segment in the already crowded field of divisions in the country and the world.

And now, our society is also divided into power generators and power consumers.

- *Power generators.* The new solar revolution have allowed millions of people and businesses to add solar and wind power generation on their properties. This has created the new class of 'private power generators' This growing class of people now has a voice in, and choice of, the overall electricity generation and the related services.

- *Power users.* Those of us who don't have the money or the proper conditions to add renewable energy on our properties, are just users of energy.

The new energy division is opening another wide gap in our social and economic fabric.

A well-to-do man in Texas, for example, could add solar power and energy storage to his house, thus bypassing the local utility all together. He will also sell extra power to the utility, and will have no problem with the additional regulations, taxation, etc.

Adding solar, on the other hand, is just not possible for a city dweller in New York, living on the 3rd floor of an apartment building; regardless of his, or her, financial or social status. This is even less possible for a mother of three, living on social welfare in a crowded community building. These people will fight against the new 'solar use' charges and utilities add to their monthly bill, since they have nothing to do with solar.

Capitalism, you say. Yes, it is! And as always, the rich get richer… As time goes, we will get even more divided, if we don't get back to the basics, our forefathers built this country upon.

But what does this has to do with energy?, you might ask. Energy, and electricity in particular, is in the very foundation of our lives. As such, we must be able to share and use it responsibly and equally, according to our needs and abilities. Division about energy is something we don't need and which we must avoid.

The advancement of electric power generation via solar panels and energy storage gadgets and toys for the rich in the U.S. is a good thing, since it would pave the way to lower prices (eventually). Unfortunately, for now at least, the new energy model only deepens the gap between the rich vs. the rest of us. The local utilities are stuck in the middle and are looking for ways to get out of the solar and wind trap they see themselves in. And so the electric power battle rages on all fronts and levels.

The Beneficiaries

The energy markets are worth billions of dollars and are continuously generating unprecedented wealth for some. Entire countries depend on the enormous income from oil sales. We all have heard about the wealth of the Arab oil producers, who pile billions of dollars in their coffers. This allows them to live in untold luxury, and even spill some of the bounty over the entire population, which receives monthly benefits without moving a finger. And we know about Venezuela, which entire economy runs mostly on oil proceeds.

We are also periodically informed about the profits of Shell, BP and the other oil giants. They make money when crude oil prices are low, and they make money when crude oil prices are high. They are in a win-win situation, and since we need their products, we can only accept (or not) and envy them. And they are getting clever today by adding numerous services (see our Marketing section above) a la 21st century marketing—making even more money in the process.

Looking further into the energy markets, we see thousands upon thousands of businesses who—directly or indirectly—profit from oil's ups and downs. Many insurance and legal companies depend on the oil's fortunes and misfortunes by providing services. These, and many other large and small financial corporations, credit unions, pension and other funds are heavily invested in the oil stock and futures markets.

Here we also see the thousands of professional and individual oil speculators, who contribute (and at times drive) the energy markets by and cause oil price fluctuations. Some of them make money some lose money, and their influence on the oil markets is expected to increase. There was a fad some time back about "finding oil on Wall Street." Many people took their chance with it, and some became rich during oil's up-and-down excursions since 2008. We take a closer look at this above in this text.

Looking even further, we see some surprises in this mix, where we find the unexpected; U.S. colleges and universities profiting from oil. They use a lot of energy and pay the oil prices like all of us. The more fortunate (or cleverer) schools, however, also have secured cash income from sizable energy endowments. They have invested a block of the total endowment, directly or indirectly, in oil and gas reserves. What a good choice...

Several universities have their energy problem taken care of for the duration (as long as oil rules) Harvard, the University of Texas, Austin and Texas A&M all benefit from revenues from state-owned oil lands. Harvard leads the oil endowments in size, followed by the University of Texas, Austin.

The two Texas universities share income from over 2 million acres in West Texas, with 66% owned by University of Texas, Austin, and 33% by Texas A&M. Not bad investment, so when oil price goes up—which it always does sooner or later—these prestigious institutions receive a handsome recompense for their wisdom and foresight.

University of Texas (UT) opened in 1883, and soon after that an endowment was established by the state of Texas. 2.1 million acres in West Texas were donated to help the University step on its feet. That was a long time before oil craze swept the nation and the world.

Not much was expected of the desolate land at the time, so we even wonder what the donors thought the UT could do with it, besides to perhaps develop it for real estate. Almost 30 years later, some entrepreneurs acquired drilling permits from UT, hoping to strike oil. And that started the huge oil discoveries in Texas, including the oil from the Permian Basin that been very generously provided for the system.

The permanent university fund (PUF) continues to receive royalties from oil and gas production in West Texas, while the available university fund (AUF) continues to receive all surface lease income. Surface lease usually entails "grazing and easements for power lines and pipelines." Together these funds are estimated to be worth about $10 billion; not a small chunk of cash for a couple of universities that also make money from their students tuition.

Globally

The broader, world-wide, picture is somewhat different, but equally unsettling. The projections on global scale are that in about 40 years, the world must find ways to generate enough electricity for additional 3.5 billion people. For this, we must add 50% of the existing power generation capacity in the next 25 years. And most countries are gearing up for the race for more energy production, with China and India leading the pack.

At the same time, there are about 1.5 billion people—almost 20% of the global population—without electricity today, many of whom are determined to live like the rest of us. This means that even more power would be needed than the official estimates. Could we increase the global power generation 50% over the present level? It seems possible, if China and India are used as indicators of things to come.

The effort is quite overwhelming, nevertheless, considering the fact that the global power infrastructure is getting old and decrepit. There are pending requirements to decommission aging nuclear and coal-fired power plants, which added to the above calculations simply means that we must double the existing power generation capacity by building new power plants.

The good thing is that we have new technologies and efficiency procedures, which might help in reducing the need for so much infrastructure increase in such a short time period.

We are in a race between making a fundamental energy transition to more power generation, on one hand, and the growing needs of ever increasing and wealthier population, on the other.

One forecast estimates that, under the present socioeconomic trends, the global energy demand will triple by 2050. This will require a combination of extraordinary demand moderation and extraordinary supply acceleration. Other experts calculate that it will take $60 trillion to meet the very basic energy needs of the world by 2050. It would cost about $50 billion annually to ensure universal access to electricity and access to modern cooking facilities worldwide.

90% of the new demand is expected to come from developing economies. By 2050, China is expected to consume nearly twice the energy used by the U.S., although China's per capita consumption would remain less than half that of the U.S. Did you get that?

China will double its energy use by 2050, but its per-capita consumption would be still two time less than that in the U.S.

What's wrong with this picture?

In best of cases, renewable sources (solar and wind mostly) could meet about 80% of the total global energy demand by 2050, while some (overly optimistic) estimates claim 100% as a real possibility. The costs of the renewables (geothermal, wind, solar, and biomass) are falling, the technologies are improving fast, and could contribute significantly to the energy mix. While such rapid growth is technically possible, it is not very likely under the present day socio-political and economic conditions.

Nevertheless, the innovations in the renewables sector are accelerating. We now see new technological improvements on the horizon: the introduction of concentrated photovoltaics (CPV) dramatically reducing costs; pumping water through solar panels to make them more efficient and making seawater drinkable at the same time; producing electricity from waste heat from power plants; using genomics to create hydrogen-producing photosynthesis; constructing buildings to produce more energy than they consume; solar energy to produce hydrogen; microbial fuel cells to generate electricity; low-energy nuclear reactions (related to cold fusion); light-emitting diodes to significantly conserve energy, nanotubes that conduct electricity, etc., etc.

Renewables are the best, if not the only, answer to our long-term energy future.

Properly designed and implemented carbon emissions standards would increase investment and would help tremendously in the development of the renewables in the energy mix. Taking a close look at, and considering the full financial and environmental costs for fossil fuels—mining, transportation, protecting supply lines, water for cooling, cleanups, waste storage, and so on—shows the renewables as far more cost-effective and easier to manage than the fossils for electric power generation.

Unfortunately, the global energy-related CO_2 missions still increase by about 1.5% annually. And as things look today (lots of talk but little results), the trend will continue ad infinitum. Without major breakthroughs in technologies and drastic behavioral changes, the majority of the world's energy in 2050 will still come from fossil fuels...if there is any left to be had.

Emissions associated with today's fossils use increase, correspond to a long-term average global temperature increase of 3.5°C minimum. Only large-scale carbon capture sequestration (CCS) and reuse (such as reusing waste CO_2 from coal plants to grow algae for biofuels and food, or to produce carbonate for cement) could reduce the emissions significantly.

Such measures could reduce industrial CO_2 emissions by 6Gt annually, if 40-50% of all fossil power plants are equipped with CCS by 2050. This is quite expensive, and is unlikely to happen unless some serious sacrifices are made. This can happen only by the introduction of equalizing carbon taxes, designed to make the transition economically attractive.

Instead, it seems as if we are moving in the opposite direction. For example, the global fossil fuel subsidies remain in the range of $400-500 billion annually. This is many times more than all subsidies to renewables and encourages the inefficient and unsustainable use of the fossil fuels. As things stand now, this trend will continue in the future.

The world is willing to sacrifice everything, even our own health and lives—and those of the future generations—for having enough electricity today

So yes, we will have enough electricity for the foreseeable future, but at what cost? That of polluting the environment, hurting wild life and people in the process, and depleting the fossils until they are finally and irrevocably gone from the face of the Earth. What a heavy price to pay, and what cold and dark this world would be then...

Europe

The European Commission and the Parliament have set a goal to implement "Europe 2020 Strategy" through a secure, competitive and sustainable supply of energy to the economy and the society. The correct transposition of the European electricity and gas legislation in all Member States is still not complete, however, so the Third Internal Energy Market Package was adopted in 2009 to accelerate investments in energy infrastructure to enhance cross border trade and access to diversified sources of energy.

There is still a market concentration on the energy market in the European Union, where a small number of companies control a large part of the market. Together, the three biggest generators of each country hold more than two thirds of the total generating capacity of 840,000 MW. The EU advises three options to weaken the market power of the biggest electricity firms: ownership unbundling, independent system operator (ISO) and independent transmission operators (ITO).

- *Ownership unbundling* is advocated by the European Commission and the European Parliament. It is intended to split generation from transmission of electricity, with the purpose of preventing vertical integration of the energy markets. It remains unclear who can buy the transmission networks, whether such a system will regulate the market-place, and who will pay possible compensation to the energy firms. There is also an argument that the costs will exceed the benefits and possible inequalities arising during the implementation. A better development of a level playing field might be a possible long-term solution.

- *Independent system operators (ISO)* have been put in charge of operation and control of the day-to-day business of the energy groups, according to Art. 13–16 of directive 2009/72/EC, while the transmission networks remain under their ownership. Investments on the network are now accomplished, not only by the owner's funding but also by the ISO's management. This is another form of ownership unbundling, but with a trustee. This would allow transmission and generation to remain under the same owner, but would remove conflicts of interest.

- *Independent transmission operator (ITO)* is a third option, presented by a number of member countries. Under ITO, the energy companies retain ownership of their transmission networks, but the transmission subsidiaries would be legally independent joint stock companies operating under their own brand name, under a strictly autonomous management and under stringent regulatory control. Investment and other decisions would be made jointly by the parent company and the regulatory authority. One prerequisite is the existence of a compliance officer, who is assigned to monitor a specific program of relevant measures against market abuse. This is also a form of legal unbundling.

- ITO Plus is an alternative fourth choice, also called unbundling a la carte. Here, the different States may keep their own system, provided it already existed in 2009 as a vertically-integrated transmission system and it included provisions that ensure a higher independence status for the operation of the system than that of ITO.

The European Community is serious about its energy future. A number of important initiatives are underway and the changes that these will bring promise to be significant. Plagued by financial issues since 2008, however, the continent is moving much slower than anticipated, but the goal is clear and achievable. We will be watching...

Case Study: Standards of Conduct

A new era of energy market fairness is starting in Europe. Ofgem, the Office of the Gas and Electricity Markets, which supports the Gas and Electricity Markets Authority, the regulator of the gas and electricity industries in Great Britain, has mandated new standards geared to reform the retail energy market.

New Standards of Conduct, backed by fines when necessary, require energy suppliers to treat their customers fairly. Consumers can expect all their dealings with energy suppliers to be carried out in an honest, transparent and jargon-free way. Ofgem has also issued directions to suppliers to implement the rest of its reform package including simpler tariffs are due in place by the end of 2015.

The new Standards of Conduct actually came into force in 2013, and are now being enforced. They basically require suppliers and any organizations that represent them, such as brokers or third party intermediaries, to ensure that each domestic customer is well informed and treated fairly.

The Standards cover several broad areas, where suppliers must:

- Behave and carry out any actions in a fair, honest, transparent, appropriate and professional manner.
- Provide information (whether in writing or orally) which is: complete, accurate and not misleading (in terms of the information provided or omitted); communicated in plain and intelligible language; relates to products or services that are appropriate to the customer to whom it is directed; and fair both in terms of its content and in terms of how it is presented (with more important information being given appropriate prominence).
- Make it easy for the consumer to contact them; act promptly and courteously to put things right when they make a mistake; and otherwise ensure that customer service arrangements and processes are complete, thorough, fit for purpose and transparent.
- Make it easier to compare suppliers
- Offer up to four "core" tariffs per fuel (electricity and gas). This will apply to each payment type. Suppliers will be allowed to offer these tariffs to collective switching schemes. They will also be able to offer extra fixed term tariffs into schemes that meet our criteria.
- Introduce new tools to help switching among suppliers.

Suppliers will be also banned from increasing prices on fixed term deals or making other changes to fixed term tariffs (except trackers or structured price increases set out in advance which are fully in line with consumer protection law).

Suppliers will also be banned from rolling forward household customers onto fixed term contracts without their consent.

Suppliers will be required to give all their customers personalized information on the cheapest tariff they offer for them. This information will appear on each bill and on a range of other customer communications.

New rules will be in place requiring all information suppliers send to consumers to be simplified, more engaging and personalized to them.

Ofgem is also looking at ways in which the most vulnerable consumers can be given better information on the cheapest deal across the market. The enforcement component of the Standards is not as clear, but we are sure that Ofgem will figure it out soon enough, and before it becomes a problem.

A new, brighter energy future is rising for the consumers in Great Britain... Will the world follow?

Update

So, customers are given a choice of energy services. But how do they know which service offers what? To answer that question, they go to a number of websites, offering data on the different companies and comparisons of their services and prices. But now, some of them have been accused of "behaving like backstreet market traders" rather than "trustworthy consumer champions."

Some of these energy price comparison websites have been "duping" customers into switching to deals that are not the cheapest on the market. By using misleading language the consumers have been led into options that only displayed deals that earned commissions to the site operators. And others have gone so far as to conceal deals that do not earn them commission behind multiple drop-down web options.

The energy watchdog Ofgem has been called upon to consider requiring price comparison sites to disclose all the data, in addition to the amount of commission received for each switch at the point of sale. These commissions amount to up to £30 every time a customer switches to a participating provider, or up to £60 when a customer switches both their gas and electricity accounts.

Consumers trust price comparison services to help them switch to the best energy deals available on the market, but instead some comparison sites have been behaving more like backstreet market traders than the trustworthy consumer champions they make themselves out to be more like TV advertisers.

As an immediate and essential first step towards rebuilding confidence, the web service companies have been asked to compensate any consumers who have been encouraged to switch to tariffs that may not have been the cheapest or most appropriate for their needs. One of the services, uSwitch has agreed to compensate consumers who had been misled into signing up for an energy tariff that was more expensive than others available.

THE NEW ENERGY REALITY

The 21st century started with a bang in the financial sector, where in 2008 the U.S. and the entire world's economy rolled down the financial cliff. After several years of pain and bruises, it is still recovering. During that time, the U.S. energy sector exploded with major discoveries and significant increases of crude oil and natural gas production in the U.S. and other countries. The renewables also got a boost from the government and grew fast.

The combination of these events is still being analyzed, but whatever it was that happened then (and/or caused the events) is still affecting us profoundly now. While the world is still licking its collective financial wounds, the U.S. energy sector is (was) going gang busters. Gasoline fell to long forgotten levels in 2014-2015, giving us the hope and the illusion that the good times are back again. Maybe so, but whatever it is, unfortunately, is all unsustainable and quite temporary.

The fossils boom in the U.S. is also here today, gone tomorrow, which was proven when the Saudis dropped the global oil prices. The U.S. oil drillers went bust. On-again, off- again the energy sector and the related markets go. We would be fools if we think that we can control them.

We also should not be as naive as to think that we have unlimited resources underground, because we don't. The fossils are finite commodity, whose end is coming; not tomorrow, or even next year, but soon.

In all cases, logic rules that by mid-century there would be very little crude oil and natural gas left in the U.S. and anywhere in the world. Whatever is left, and still possible to extract, would be in limited quantities and very expensive. Sooner we understand and agree that this is the logical truth, sooner would we will look to adapt to the way of things to come.

Unlike other natural disasters, this time there will be no dramatic or apocalyptic events. There would be no

Earth-shaking, or any other spectacular happenings.

The fossil-less future will sneak upon us very slowly and quietly, assisted by oil companies and others, trying to make a buck from the last drop of oil

Instead, just like the proverbial frog that boils to death when the heat is turned up gradually, we are now quite comfortable, as the water we are sitting in heats up. We will, however, find ourselves all of a sudden in the quagmire of the proverbial 'frog in boiling' water sometime this century.

At that time, the energy issues would get to the boiling point and our lives would change dramatically. It is not very hard to imagine what life would be without electricity, without natural gas, without gasoline and diesel. No lights, no heat, no cars…no American Dream.

Is there a way to avoid such hapless day? Of course! If we follow the path of reason. Are we anywhere near to following it? Nope! We are miles away from it, and as a matter of fact in many cases we are headed in the opposite direction.

So what is going on? Why are there so many reasonably smart and experienced people closing their eyes to the reality and pretending that everything is OK. Or, in best of cases, talk about the issues today, but do nothing to solve them tomorrow?

The Present

Energy, Environment, Education, Health, and the Economy are the critical internal issues of the first two decades of the 21st century for the U.S. This, of course, is in addition to a number of external problems and risks, some of which impact us profoundly. Similar internal and external risks seem present, to one or another degree, in most developed countries as well, so at least we are not alone in this debacle.

The worst part of the 21st century is the fact that we, the people of the 'ol mighty US of A, are a nation divided. We are split along political lines, gender, race, ethnicities, religions, sexual preferences, disabilities, educational level, criminal records, etc., etc.

And now, we have a set of new divisions, based on our personal interests and understanding of the subjects of energy and environment. Fossils or no fossils. Solar energy, or no solar energy. Nuclear power, or no nuclear power, fracking or no tracking…and on and on the list goes.

One thing we all agree, however, is that energy is in the foundation of our daily lives, and second in importance after our health.

Health and energy determine our wellbeing and safety.

A profound statement, which is hard to argue with, but how stable and reliable is this foundation?

- We now have many more people living longer and many more are getting sick. All this requires extensive and expensive medical treatments.
- Many are insured under the hastily introduced Obama health plan, which is fair to the less fortunate minority, but burdens the majority. Its future is uncertain, because the efforts to change it continue.
- Increased medical expenses and additional bills, create an unsustainable situation. In the end, if no real and more realistic changes are made, we will wake up one day soon with a collapsed health system that cannot pay its bills.
- Similarly so, our long-term energy future looks dim too, as a consequence of the unsustainable use of energy resources. If this trend continues, soon after the health system collapses, the energy sector will follow and we will be a nation with broken health and energy sectors. What a painful day that will be for all of us.

But while the health plans can be changed overnight by political or voters' will, the present trends in the energy sector cannot be changed. There are many reasons for this, but in the end, we see a number of changes introduced in the U.S. health system, but not much change in the U.S. and global energy markets.

And so, while we might avoid a health system disaster with some clever modifications, we will be forced to jump off the edge of the energy cliff in the future. What will happen then is anyone's guess.

The U.S. Energy Cycle

Taking a closer look at the U.S. energy markets, we see a picture that is less than encouraging. We see huge use of energy resources, which we are borrowing from the future generations. Even more surprising is the fact that large part of our precious and non-renewable fossil resources are wasted due to equipment and processes inefficiencies. And some are exported for piling wealth. Wealth, however, cannot be used to generate electricity, or fuel cars. So we have chosen to end up with piles of money, but empty gas and oil reservoirs.

According to Table 10-16 the expert estimates are that the U.S. used 98.1 quads (quadrillion Btus) of energy in its different forms during 2014. Of this total, 38.4 quads of energy were used for electric power generation

and 27.1 for transportation. The rest (about 1/3 of the total) was used for residential, commercial, and industrial applications.

Table 10-16. U.S. energy use (2014)

Energy Source	BTU Q
Crude Oil	34.8
Natural Gas	27.5
Coal	17.9
Nuclear	8.3
Biomass	4.8
Hydro	2.5
Wind	1.7
Solar	0.4
Geothermal	0.2
Total Use	98.1

What is wrong with this picture?

- The biggest problem is that this is enormous amount of energy, most of which (about 2/3) came from fossils. Since the fossils are in limited quantities, the inevitable question is how long can we rely on them? And what will happen when they are no more?

- Equally important is the fact that crude oil is the largest portion of the import-export imbalance, and the single most important part of the fossils, since it literally drives the wheels of our economy. In addition to the fact that oil is finite commodity and will not last forever, at the present we import half of the oil we use from the darkest corners of the world. How sustainable is this under the increasing world chaos?

- Amazingly enough, most of the total energy used in the U.S.—nearly 60% in fact—is wasted one way or another. We are the most technologically advanced nation in the world, and yet, we waste more energy than we use.

Did you get this?

About 60% of all energy generated in the U.S. is wasted.

There are many reasons for this waste, but the major one is the inefficiency of the existing equipment and processes.

- The transportation sector is the greatest energy

waster, with 1/4 of all energy input converted into useful work, while the rest 3/4 is simply wasted and going up in smoke from billions of exhaust pipes. The internal combustion engine is 100+ years old and is still in use. Although there have been some changes and improvements through the years, it is still horribly inefficient.

Due to its inefficiency, today's internal combustion engine wastes 65% of the fuel we put in its tank.

On top of that, most Americans drive their own car to and from work or anywhere else, which increases the inefficiency of our transportation system millions of times. If all single drivers in the millions of cars and trucks on the roads take a bus or train to work, we may not need to import crude oil anymore. This, however, is not going to happen, because in the U.S. the automobile is part of the American Dream, so we will continue wasting a lot of fossil fuels.

• The electric power generation sector is another great energy waster, due to inefficient generation and use of electric power. Here we see only about 1/3 of all energy input converted into useful work, while the rest 2/3 is wasted at one or another step of its lifetime cycle.

A lot of energy is wasted during the production of the conventional energy sources—mining, well drilling, etc.—and their transport to and from the originating and processing facilities. Similar to the internal combustion engine, the steam turbines, used in power generation, are over 100 year old technology. Their efficiency is similarly low, and so are the other steps of the power production and distribution.

So, if we trace the life cycle of a coal-generated electricity, we see significant losses at every step of the way.

Coal in mine	100%
Mining Losses	2
Power plant	60
Transmission	2
Light bulb	34
Total losses	98%

Only 2% of the total energy input is converted into useful light at the light bulb.

Wow! If this is not waste, what is?

Appliances are somewhat more efficient, but even they waste a lot of energy, so that the actual energy efficiency is still very low. And then, we waste a lot of energy by leaving lights and appliances on when they are not supposed to be on.

There are some changes lately, such as replacing incandescent light bulb with fluorescent lights and LEDs, which would increase the overall energy use efficiency. New gadgets are also introduced to turn lights and appliances remotely and automatically, which helps reduce waste.

Another good news is that the total energy use is expected to decrease some in the future, while the energy efficiency (generation and use) are expected to increase. These changes are quite slow, so we do not expect to see any significant changes in the final efficiency numbers any time soon.

Energy efficiency can be easily increased by changing the way we generate and use electricity.

We took a closer look at the theory behind energy efficiency in the previous chapters, so here we will only touch on its effects on the global energy markets.

Energy Externalities

The costs associated with energy-related activities (and their externalities in particular) differ from most other economic categories. For example, externality costs do not reflect any direct transfers between energy sources and the government, or any mandated monetary transfers. It is also important to recognize that an externality cost, for example that of GHG pollutants, represents a cost that is borne by society as a whole, not simply by the government.

Thus, direct comparisons to government revenues or expenses alone are likely to be misleading. If the costs of the carbon externality, for example, were reflected in government policies designed to "internalize" it, this would affect not only government revenues, but also benefits to consumers and producers across the economy.

The ultimate implications for government revenues would depend on how the demand for carbon emitting products and alternatives responds to changes in their relative prices. One cannot simply assume that if carbon were priced at a level higher than the prices already imposed by existing policies, this would result in lower net revenues to government from every carbon emitting fuel.

Government revenues for individual fuels might stay the same, or decline, or they could even increase, depending on how responsive both demand and supply are to price. Thus one should not simply "net off " the externality costs associated with carbon—or any externality—from government revenues.

With these caveats in mind, we could estimate the implications of different assumed values of the externality cost of carbon. There is significant uncertainty about the cost of the externality per ton of CO_2 (often referred to as the shadow price of carbon), at low, medium and high estimates of €10, 30, and 70/tCO_2.

So, at the medium shadow price of €30/tCO_2, for example, the estimated total externality costs would be: €53 billion for oil, €29 billion for gas, and €35 billion for coal. But money is not all that counts,

Most Affected Industries

Rising temperatures and sea levels, along with increased incidence of extreme weather events, pose a threat to the global economy. Global infrastructure expenses, public food supply, health, and surging demand for energy are among the demographic themes that could come with steep climate-change costs. Energy is actually in the foundation of these sectors, since without energy they would not function properly, or at all.

Energy's importance as a driver of the global economy cannot be overemphasized.

One of the big problems facing us is global warming, which even if leveled (with temperature rise of no more than 2°C from preindustrial levels by 2030, as is the goal today), the damage to the world economy would be significant. Certain regions and industries will be seriously affected—some much more than others—according to the Intergovernmental Panel on Climate Change.

The energy sector has been blamed for the majority of the pending environmental disasters. This creates a controversy between the importance of energy to our wellbeing, and the damages it inflicts on us.

Climate-change skeptics are likely to be proven wrong in the decades to come, when the cost to the society in large would surely rise accordingly. The dollar bill for emergency mitigation efforts would be much higher than timely preemptive efforts (starting now) to keep temperatures from rising.

Certain industries introduce risk management approaches to their operations, which require making assumptions about what is sure in the otherwise unsure future.

The following industries are considered the most at risk in the climate-change era:

Energy

For the energy industry, the risks posed by global warming are reflected in governments' effort to slow it down, and/or change its MO. None of these measures is easy or cheap so the resistance within the energy producers ranks is growing. Regulations on reducing emissions from fossil fuels are increasing and basically changing the oil, gas, and coal industries. That also affects directly the electricity generation, transportation, and the related sectors as well.

The U.S. Environmental Protection Agency's regulations have already shut down dozens of coal-fired power plants. This has contributed to a significant decline in coal production and use in the country. While this is good for the environment, it is killing jobs in the country and complicating the global energy markets. This is because, the coal industry is hurt, but not dead. It is actually very powerful, and will find a way to survive. One of these ways is increased exports of coal to Europe and Asia. With that, the global environment would be polluted even more that before EPA shut down the U.S. coal power plants.

Divestment campaigns by large institutional investors, including university endowments, are also beginning to make an impact. One good example of that is Stanford University's decision to get out of coal.

Another dangerous, though highly speculative, risk is the "stranded assets" situation, where large-scale projects, such as offshore oil rigs and other major fossil fuel deposits were, or would be, abandoned. This leads to huge losses, and even bankruptcies. The Organization for Economic Cooperation and Development (OECD) warned in 2014 that "stranded assets" are not only a major risk to energy companies, but their investors, including sovereign wealth funds, pension funds and university endowments.

Note: OECD governments make total of about $200 billion annually from oil and gas. At the same time, Russia gets about $150 billion a year, which is about 30% of the total government revenues. In comparison, the total OPEC countries oil revenues average $600-$700 billion annually, which is also the great majority of their GDP. These countries live (and quite luxuriously thus far) on pumping oil from the ground. Some of them have gotten so spoiled that they would be crippled economically if and when oil revenues dry up.

Thus far, however, this is a big business, and as is normal in the capitalist enterprise, those who have got the big bone won't let it go easily...environment or no environment.

Agriculture

With climate change on the horizon, many agricultural regions around the world will be affected and some will experience disastrous events. As a matter of fact, almost 30% of the world's population that works in agriculture will be affected one way or another by the rising seas, draughts, water contamination and other negative effects of energy production and climate change.

One beneficial effect is that warmer temperatures help crops grow quicker, but for many major crops, such as the basic grains, faster the growth, the less time seeds have to mature, thus reducing the total yields. This, according to a 2014 United Nations report, may cause food prices to rise up to 85% by 2050, mostly due to reduced crop yields.

Higher prices, caused by reduced crops, are good for producers who can produce, but the climate change may decimate production in certain areas. The Central Valley of California is a good example of such disaster in the making. The ongoing draught caused nearly $2.0 billion loss on 2014 alone. As the draught conditions persist, the losses will increase proportionally. Eventually, in case of much longer draught, the entire agricultural sector in some parts of California will collapse.

Fishing

It is a not-well-known fact that as the sea levels rise, the global fishing industry will be one of the most adversely affected. Salmon and trout, for instance, need cold, free-flowing water, which has been reduced. Habitat loss for these species is expected to rise to nearly 20% by 2030 and 35% by 2060, the global green house effect is not reduced. This would be a loss of about $1.5 billion annually by the fisheries, according to the Natural Resources Defense Council.

Ocean acidification also poses a problem for the fishing industry. Shellfish, such as clams and oysters, find it much more difficult to grow in an increasingly acidic environment.

Insurance

Hurricane Sandy in 2012 resulted in more than $70 billion in economic losses, most of which was flood-related. Only about $26 billion of that total had some sort of insurance, which in most cases was inadequate. Many insurance companies, however, were quite affected and unhappy when faced with such a large bill.

Flood insurance, if and when offered at all, is tricky business for all involved, since floods impact the low-lying areas in different ways and degrees. Since this is a periodic event, it is a well known fact by the insurers and the customers, and so there is a growing discrepancy. Owners of homes above sea level don't buy insurance, and because there are fewer purchasers, flood insurance is significantly more expensive.

Flood insurance is actually not included in most insurance plans of low-lying areas around some rivers, so the government is insuring homes in some of these flood zones. How profitable this business is, is not clear but during periods of severe floods, it seems like it is the worst business to be in.

Rising sea levels and the potential for increased incidences of catastrophic flooding will likely drive up both premiums and payouts, putting a strain on the insurance industry. The National Flood Insurance Program, for example, collects about $3.6 billion in premiums every year. Yet it insures more than $1.25 trillion in total assets.

Moreover, many insurance companies are starting to withdraw altogether from providing insurance to certain catastrophic markets, reveals a report released today by Ceres. The report states that last year just under a third of the $116 billion in worldwide losses from weather-related disasters were covered by insurance, according to data from the reinsurer Swiss Re. After Katrina struck, in 2005, insurance picked up 45% of the bill.

With more than 6.5 million homes in the U.S. at risk of storm damage and a total reconstruction value of nearly 1.5 trillion, according to Corelogic, a global property info and analytics provider, the insurance industry's retreat is an alarming threat to populations and public institutions.

Beverage industry

Increased water shortages are among the biggest threats to the worldwide soft drink and bottled-water market, valued at $247 billion by an IBIS World report. Coca-Cola's 2013 10-K form stated that "changing weather patterns, along with the increased frequency or duration of extreme weather conditions, could impact the availability or increase the cost of key raw materials that the company uses to produce its products."

Coca-Cola's supply of sugarcane, sugar beets and other ingredients looks to be threatened as temperatures rise and extreme weather events occur more often. In 2014, government authorities in India forced the closing of a Coca-Cola bottling plant in the country's north, after the company was accused of extracting too much groundwater.

Winter Sports

Long-lasting draughts in California and other Western U.S. states is causing reduction of rain and snow fall,. This, in turn, is reducing the skiing seasons and damaging the businesses that depend on it. Oddly enough, climate change may actually be good for some skiing Meccas, like Vermont right now. This is because more precipitation is falling with the rising temperatures, so there is more snow during the colder months.

In the long run, however, if winters grow shorter and warmer and the snow precipitations turn to rain, regions that depend on cold-weather tourism will lose a substantial amount of money. There is an estimate that by 2100, two-thirds of European ski resorts may be forced to close.

In the U.S., the $12.2 billion ski and snowmobile industry has already experienced a $1 billion loss brought on by changing weather patterns. The long-term climate and the related winter conditions cannot be estimated accurately, but many changes (mostly negative) are expected. These changes will affect the winter sports industry accordingly.

Wine Production

The production of wine grapes may not be entirely destroyed by climate change, but the geographic landscape of winners and losers in the wine business is already changing. Wine grapes require special environment in which to grow and are very sensitive to even the most subtle shifts in climate. As temperatures rise, there may be a two-thirds drop in production in the traditional wine areas, such as Burgundy and Tuscany. The $10 billion French wine export industry is particularly vulnerable, and is expected to shrink in the near future.

At the same time, certain regions, including southern England and Washington state, could become centers of new and emerging wine industry. These locals have the climate and other conditions, including enough precipitation, that are critical for wine grapes growth.

Wall Street

The impacts of climate change on the financial industry are yet to be determined, but Wall Street is affected by any and all changes, To start with, the location—Lower Manhattan—is in reality a landfill. Until recently, parts of New York, around the Hudson and East rivers, were called the New Amsterdam. Some of this land was flooded in the aftermath of 2012's Hurricane Sandy.

Wall Street is one of the oldest streets in the city, built mostly on old, somewhat elevated land, but is still only inches above sea level. The geography of Lower Manhattan is expected to change drastically in coming years, and if the ocean rise continues as expected, it might find itself under water.

What is Wall Street going to do then? Move? Prolonged shutdown could lead to large losses for the financial services industry and the investors. Over $7 billion were lost in the immediate aftermath of Hurricane Sandy, when the New York Stock Exchange was forced to close for two days. Wall Street under water is not good thing, and many more billions are at stake, if and when the Street sinks under the sea waters.

Energy Efficiency Market

As we saw above, most of the energy produced and used today in the U.S. is wasted.

*Over **60% of all energy** input **is lost** due to inefficient equipment and careless actions during its life cycle.*

Smoke and heat coming out of cars' exhaust, or from power plants' smoke stacks, reflect the huge amount of energy wasted (and pollution emitted) every minute of the day and night. The heat, which residential and commercial water heaters and coolers emit into the surrounding air is another example of wasted energy. Enormous amount of energy is also wasted by heat and cold air escaping from houses and businesses, due to poor insulation and cracks in windows and doors.

These are only few of many examples of wasted energy, due to inefficient or faulty equipment. Keeping in mind that the U.S. is the most technologically advanced country in the world, we can deduct that most other countries waste even more energy.

Energy efficiency, therefore, is a no brainer; a tool available to anyone who is willing and able to use it.

Energy efficiency is the most powerful tool in the fight against depletion of the fossils, and the increasing global pollution.

Energy efficiency is something each one of us can and should consider in our daily lives. It can reduce our monthly energy bills in addition to reducing the pollution. Doing this well is not that simple, but is much more doable than most other approaches available to us, at least on the short run.

There are many ways to increase the efficiency of

our energy production and use, but the major categories are:

- Energy conservation in all its forms, by all participants, all through the life-cycle of the different energy sources,
- Optimizing energy generation and use equipment and processes, and
- World-wide expansion of the renewable technologies, which do not affect our energy balance, thus increasing our energy efficiency.

*The above tasks comprise the top tier of the global **energy efficiency market**.* Getting below the surface, we see billions of people involved in different tasks that could be modified as needed to save energy. We also see millions of pieces of equipment, appliances, and gadgets, which could be also modified, and/or tuned, to use energy more efficiently.

Although at a quick glance the above tasks seem simple and doable, in most cases it is not, and so the progress of implementing some of them is slow and painful. We also see a notable regression in the progress of others in some locations and entire countries.

Recent estimates put the *global energy efficiency market* at over $310 billion annually and growing, according to the International Energy Agency (IEA).

Energy efficiency is the world's largest and easiest to obtain 'fuel supply.'

The above realization is causing the science of energy efficiency to slowly, but surely, penetrate all areas of our lives.

Energy Efficiency Finance

Energy efficiency finance is a new field of the banking business that is quickly becoming an established energy market segment. Here, innovative new products and standards, developed especially for energy efficiency increase, help the customers to overcome risks, and bring stability and confidence to the global energy market. Increased investments in energy efficiency improve our energy productivity, which is the amount of energy needed to produce a unit of GDP.

Energy efficiency is the invisible powerhouse available to the citizens in all states and countries. It is one way to ensure our energy security, lower the energy bills, and help us reach the global climate goals.

Countries that follow this path have shown energy

consumption decrease of 5% in 2011, mostly as a result of investments in energy efficiency. The total avoided energy consumption from energy efficiency improvements in these countries was 1.7 billion tons of oil equivalent, which is more than the annual energy usage of the U.S. and Germany combined.

The IEA claims that energy efficiency is not just a hidden fuel but is actually the "first fuel." It is the fuel that can propel us faster into a cost effective and clean future than any other fuel, or fuels.

Energy efficiency is fuel that we have in unlimited quantities, and the only one that we have complete control over its use.

In Europe, energy efficiency investments to date have avoided more energy consumption than the total annual consumption of the entire European Union. This is reducing the EU's energy demand, in contrast to the increase of energy demand by the developing countries.

Energy efficiency has huge potential world-wide. It is quite effective in developed countries, but it seems most applicable to, and most effective in, emerging economies. The introduction of attractive energy efficiency programs for buildings and equipment, as well as efficient vehicles and transport infrastructure are major opportunities. New efficiency measures, implemented on a global scale, could save up to $200 billion in transport fuel costs alone by 2020. This could also reduce local air pollution and traffic congestion in some countries with rapidly developing urban transport systems.

About 40% of the global energy efficiency market is financed with debt and equity, which simply means that the financial market for energy efficiency can be valued at $120 billion annually. There are a wide spectrum of energy efficiency products and services today, and with that the finance have been expanding in recent years.

Today we have a number of financial tools, such as: common green bonds, corporate green bonds, energy performance contracts, private commitments, carbon, and climate finance. There are also multilateral international development banks and bilateral banks that offer financing of energy efficiency projects Bilateral and multilateral lending alone amounts to $20-25 billion annually

Energy efficiency is no longer a niche market.

It is instead, an established (albeit peripheral) segment of the energy market, with increasing interest from institutional lenders and investors. Energy efficiency is becoming recognized as an essential tool in to meeting our energy and climate goals, while at the same time supporting economic growth. This new market has been expanded by projects and initiatives that contribute to the reduction of barriers and obstacles in its way.

Energy efficiency efforts are also one of the most important parts in the effort to achieve the world's climate objectives. The IEA scenario for limiting the increase in global temperatures to no more than 2 degrees Celsius, attributes the largest share of emissions reductions–about 40%–to successful implementation of energy efficiency laws, standards, and projects.

Energy efficiency—or simply put, limiting the use of energy—is therefore responsible for almost half of our hope to save the world from climate warming and the related disasters.

The big problem today, that is not going away, is the fact that most developing countries are partially or totally ignoring energy efficiency as a tool in their energy and environmental arsenal. Instead, they are rushing toward producing more energy sources, and generating more power, as needed by their increasing population and its needs.

We cannot blame them for trying to achieve what the developed countries have achieved with the help of excess energy usage. We, however, should keep this anomaly in mind and include it in our calculations and estimates. This will force us to consider the consequences and figure out a way to provide the developing countries with the understanding and financing to implement the energy solutions they need—including energy efficiency measures.

Every dollar spent on energy efficiency avoids the burning a lot of coal and its emissions.

Government and Energy

The major technological achievements of late are responsible to a great extent for our economic prosperity. These achievements are based on the knowledge and power platforms, which in turn move our economy.

- The knowledge platform consists of the Internet, the devices that connect to it, and the firms and people that use it. On this platform, the majority of Americans conduct a major part of their business and social lives.
- The power platform comprises the electrical grid, energy consumption, and transport vehicles. Abundant and affordable energy makes modern

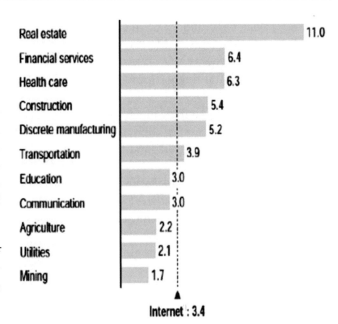

Figure 10-33. Weigh in GDP of different economic sectors vs. the Internet

life possible.

Figure 10-33 shows the contribution of different sectors of the U.S. economy to the total GDP. The Internet—which is actually the youngest of them all—is already high in the list of the major economic sectors. It is also growing faster than most of them and is expected to be among the first three in the list within the next 10-15 years.

The U.S. government has a finger in all sectors of the economy by directly or indirectly directing the activities in these. The Internet, for example was made possible by the efforts, active participation, and continuous support of, the U.S. government.

The development of nuclear energy would've been impossible without the decisive technical and financial support of the U.S. government.

So what does the government do to help the economy and the energy sector in particular?

U.S. Government and Energy

The U.S. government is actively involved in the energy sector, and the related markets, function. Be it by its policies, regulations, subsidies and/or other programs, it coordinates the overall activities in the economy, including the energy sector.

Some of the government activities in the energy sector are:

- Regulate greenhouse gases, waste treatment and disposal through the EPA
- Reform utility regulation to encourage competi-

tion, clean generation, and energy efficiency

- Direct Fannie and Freddie to enable property-assessed clean energy
- Transfer annual Energy grants to state-operated green banks
- Allocate federal land for renewable projects development
- Provide access to public land for renewable projects
- Encourages R&D of new technologies through the DOE and its National labs
- Allow access to its National Labs for energy research by energy companies
- Issues subsidies and loan guarantees to energy companies

These are multi-billion dollar activities, tasks, and projects, which are conducted on daily basis, with a focus on providing abundant and cheap energy to our people and businesses. The final goal is to avoid energy crises and excess pollution, which is a threat to all of us.

President Ronald Reagan, worried about increasing confrontations in the U.S. and abroad ingenuously asked a long time ago, "If all of us in the world discovered we were threatened by a power from outer space, wouldn't we all of a sudden find we didn't have any differences at all?" This question sounds even more true today, since we see clearly the problems around us.

We see the enemy: overuse of energy and out of control GHG emissions.

The U.S., with one quarter of the global economy within its borders, is leading the way with the help of the government. The goal is not only to save the global climate, but also to catalyze a global economic recovery based on the expansion of the power platform in scale and scope at home and around the world.

Compared to the alternatives of dealing with the emergencies of future energy crises and pollution disasters, this could be done (albeit not easily), if the government revised its policies and focused on the key factors of:

- Lowering the cost of capital for clean and renewable energy technologies
- Encouraging achievement of the renewable energy standards at the state level
- Encouraging efficient energy consumption
- Lowering the cost of capital for building, upgrading, and maintaining the power generation and distribution infrastructure

- Reforming and supporting utility regulation to attract private investment
- Changing the tax laws to attract private investment in building the new power infrastructure
- Leading the global battle for energy efficiency and the climate change crisis

Most U.S. government officials are well aware of the situation and what is needed to solve the issues before us. Although some work has been done, and some is under way on all of the above points, we still don't have an uniform, comprehensive National Energy plan. The piecemeal approach used today has its limitations, and is delaying and even stopping the development of efficient long-term solutions.

The Energy Subsidies

Since 2008, many governments had poured billions of dollars in energy companies, technologies and processes. The results of all these subsidies are controversial, to say the least, so IMF has undertaken studies to measuring the effects of energy subsidies, focusing exclusively on fossil fuels. IMF uses a "price gap" methodology for measuring subsidies that is quite similar to that used by IEA.

Note: The definition of *subsidies* distinguishes between *consumer and producer subsidies*, where, consumer subsidies arise when the prices paid by consumers, including both; firms (intermediate consumption) and households (final consumption), are below a benchmark price.

Under the same scenario, the producer subsidies arise when prices received by suppliers are above the same benchmark. In case that the consumer prices are below the benchmark, usually the governments assume the responsibility to cover the difference via subsidies and other financial instruments.

The IMF calculates government subsidies as follow:

- *Pre-tax subsidies* are based on a comparison of pre-tax end-user prices with a particular benchmark price. The benchmark price for most traded products is based on an international benchmark price adjusted according to distribution and transportation costs. Here, the assumption is that margins for distribution and transportation are similar across regions and countries.

For non-traded products, such as electricity import-export activities, the benchmark price is the cost recovery price for the domestic producer, which includes a normal return to capital and distribution costs.

On *pre-tax basis*, the IMF estimates total global fossil fuel subsidies of $492 billion in 2011. However, the majority of this is accounted for by oil exporting countries, with Middle Eastern and North African countries accounting for about 48 per cent of the total. The IMF does not find any significant pre-tax subsidies in advanced economies.

- *Post-tax subsidies* compare end-user prices inclusive of all taxes with a benchmark price that reflects assumptions about a "reference rate" of VAT and allowances for the externalities of greenhouse gas (GHG) emissions. This includes local air pollutants, and some traffic externalities, such as traffic congestions. For VAT, the IMF develops assumptions about what constitutes a reference rate of VAT. In countries where no VAT is paid on the consumption of any product, IMF assumes that the VAT is the same as in countries with similar social structure and income levels.

On *post-tax basis*, the IMF estimates total subsidies of $2.0 trillion in 2011. Externalities account for the most significant share of this total, accounting for about $1.3 trillion. A large proportion of the post-tax subsidy is attributed to the U.S., which according to IMF estimates provides "'subsidies" of $410 billion. In the 28 EU nations, the IMF estimates total post-tax "subsidies" of $113 billion during the same time period.

- *Adjustment for GHG emissions* is included in the benchmark price, and is based on an assumed carbon price of $34/tCO$_2$e (tons of carbon dioxide equivalent), as provided by the U.S. EPA. For local air pollutants, the IMF assumes that countries have similar characteristics to the U.S., using U.S. estimates to quantify the externalities.

The IMF's effort is a good step in the right direction, but it in terms of methodology it has several limitations. For example, the "price gap" approach does not capture subsidies that arise when energy suppliers allow inefficiencies and losses at the benchmark prices.

There are also even more important limitations of this methodology that have to be acknowledged, including:
- Lack of enough data for some products, regions, and/or countries,

- Reliance on a snapshot of prices at a particular point in time and for selected groups only,
- Lack of comparability across countries for some types of fuels because they reflect local characteristics,
- Stylized assumptions about transportation costs, and e) in some cases stylized assumptions for "corrective taxes."

Government Support

Coal, natural gas, oil, nuclear-, hydro-power, and the renewables' energy value chain consists of:
- Upstream: production, infrastructure, land rights, regulations, permits, and taxes,
- Midstream: refining, transportation, storage, power generation and distribution, and taxes
- Downstream: end-use of fuels by homes and industry, and vehicle fuels, and taxes.
- Benefits, damages, and externalities of each stream.

The categories of government revenue and expenses in the energy sector could be summarized as in Table 10-17.

Table 10-17. Government revenues and expenses

Revenue	*Expenses*
Upstream production taxes	Resource extraction support
Corporate taxes	Midstream sector expenses
Excise duties and energy taxes	Consumption support
Value added tax	Liability transfers R&D expenses
Mandated transfers*	Midstream energy expenses Price regulation

Here, the corresponding "revenue" category is shared equally between consumers and other producers. For example, some costs of renewable energy sources are reflected in higher customers bills.

In more detail:
- *Upstream payments* in support of energy extraction activity are mostly focused on production support, which is the largest category of direct government expenditure. It is provided to the coal sector of many countries. This is most notably the case in

Germany and Spain, but these programs are being gradually phased out.

- *Midstream payments* are in support of energy transformation, mostly electricity generation, refining, energy storage, and transportation. There are relatively few of these in the OECD countries, except some Eastern European countries, like Poland in support of losses in coal-fired power generation.

- *Downstream payments* are made to support final consumption, such as consumption grants or price caps for certain types of consumers. Downstream payments are the second largest category of support and are more prevalent in the oil and natural gas sectors. The majority of payments consist of excise duty refunds provided by the government to certain sectors, such as agriculture or public transport. These are distinct from tax expenditures in that the full excise rate is initially paid (and captured within our data). Only after initially paying the full rate of tax can eligible consumers request refunds on this tax payment, which then reduce the initial government revenue.

- *Decommissioning payments* are made when governments cover the cost of asset disposal, such as closing coal mines, and coal and nuclear power plants. This category is applicable mostly to the coal sector and typically is a result of either a lack of provision for decommissioning costs, or the premature closing of mines where the owner has been unable to afford to carry out adequate decommissioning.

- *Compensatory payments* to workers are provided by governments who assume liabilities related to health issues from production activities and unemployment. This category also relates mostly to the coal sector. The closure of mines in certain countries left many otherwise unskilled workers unemployed, while some at the same time facing health problems. Governments in many EU countries, for example, have allocated resources to provide compensation to workers in this category.

- *Environmental damage* is covered by governments who assume the cost of restoring areas of land that suffered from environmental damage as a result of development and production activities. This is a relatively minor expenditure category, in addition to which governments also make contributions to R&D funding in energy the sectors. Significant proportion of this funding is allocated to coal and nuclear power, and energy efficiency activities.

Table 10-18. U.S. government funding

Health and social services	27%
Education, K-12	16
Community development	14
Education, higher	12
Culture and arts	5
Public affairs	5
Disaster relief	3
Environment	3
Other	15

Table 10-18 shows the distribution of U.S. government's subsidies and financial assistance to the different economic sectors. The energy sector is not even figured in this table, although it gets substantial amount every year.

The nuclear, hydro, and oil sectors get the major share of government support in terms of technical and financial assistance. Billions of dollars are spent on these industries all through the value chain. The renewables also get some support, but it is diminishing with time, and will be withdrawn almost completely in the years to come.

The States' Role

The U.S. federal government has set a number of laws and policies in support of the energy sector, but there is still a need for a more comprehensive and unified approach. Instead of waiting for direction from the central government, many states have taken the lead in paving the path for the renewables.

Most notably, the state of California, which leads the nation in many new (good and bad things) has taken the lead in this case by designing and implementing a number of energy measures, the most important of which we took a close look above in this text.

As an example of active state participation in the energy sector, Maine's leadership and attention to detail deserve noting:

Maine

Recently Maine's legislation voted The Maine Solar Act (Sec. 1. 35-A MRSA c. 34-B), in response to which a statutory mandated study was performed in order to evaluate the feasibility and assign value and price to solar energy generation in the state. The Maine Distributed Solar Valuation Study determined that if the customers can get solar for less than $0.33 per kWh; then this is a good deal price-wise with a lot of additional benefits.

The study provides a calculated one-year value and, plus a separate 25-year levelized cost of energy value for the utilities customers. The study is basically made up of 13 cost-benefit factors.

The first eight factors are considered *Avoided Market Costs,* which are quite straightforward, and to each of which corresponding cost and price values can be assigned.

- Avoided Energy Cost,
- Avoided Generation Capacity Cost,
- Avoided Residential Generation Capacity Cost,
- Avoided Natural Gas Pipeline Cost,
- Solar Integration Cost,
- Transmission Delivery,
- Avoided Transmission Capacity Cost,
- Avoided Distribution Capacity Cost, and
- Voltage Regulation,

Then, there are the so-called Societal Benefits, which are regarded as externalized factors, which are emerging as important parts of the solar value equation, but are much more difficult to put in dollar numbers:

- Net Social Cost of Carbon,
- Net Social Cost of SO_2,
- Net Social Cost of NO_x,
- Market Price Response, and
- Avoided Fuel Price Uncertainty.

This is actually a fairly thorough list of factors that are needed to compose a complete picture of the solar power; and its advantages and disadvantages. It could be used as a template by other states in their solar quests.

According to the Maine study, Avoided Market Costs in Maine are valued at $0.09 per kWh for one year and at $0.138 if levelized over 25 years. The Societal Benefits are valued at $0.092 per kWh for one year and at $0.199 per kWh over 25 years.

That puts the one-year total value of solar power at $0.182 per kWh and the 25-year value at $0.337 per kWh.

Remember these values: $0.337/kWh from future solar, vs. $0.18/kWh from present conventional power generators.

So, since solar is a long-term investment, we must take the 25 year estimate ($0.337/kWh) and use it in all our calculations—keeping in mind that this value would be justified (and enough profit realized) only if the weather conditions are perfect and the solar equipment behaves properly.

Plausible undertaking, but not void of problems.

- The typical Maine electricity rate today is $0.13 per kWh, and most of it is imported from fossil-burning power plants in neighboring states. Compare this low rate to $0.337/kWh (2.5 times higher rate) for solar power generated in the state.
- All Maine electric power users (including those without solar on their roofs) will have to pay more for solar generated power, since prices would go up when the utility has to pay several times more for purchasing solar power.
- Solar additions do not affect Maine's environmental conditions, since most of the power used in the state is imported, thus the major pollution emission is out of state.
- But most importantly, Maine is not a "solar" state by any means. It is way up North and smacked into the ocean, where clouds, rain, fog, and snow dominate the weather pattern. Solar installations do not generate much electricity under these conditions, and with all the subsidies gone, it would be even harder to justify adding solar in Maine.

All this considered, it seems that a second look is needed when considering expansion of solar in Maine and in all Northern states. Maybe this is not the place or time for spend so much money on technologies that are not favored by the local conditions. Or maybe it is…

The U.S. DoD

In 2014, the U.S. Secretary of Defense made it official; climate change is real and poses serious risks to U.S. national security. The Pentagon's 2014 Climate Change Adaptation Roadmap issued at that time in a 16-page document details the effects of extreme weather events and rising global temperatures on military training, its operations, acquisitions, and the related infrastructure.

'Climate change is a threat multiplier because it has the potential to exacerbate many of the challenges we already confront." according to Mr. Hagel. And with that the U.S. military is on the forefront of the battle against climate change and the accompanying weather extremes.

This is a good start, since the 2012 and 2013 Roadmap editions labeled climate change as a "future threat." But now it is recognized as a reality that must be dealt with quickly.

The Roadmap begins with, "Climate change will affect the Department of Defense's (DoD) ability to defend the Nation and poses immediate risks," and continues with establishing the Pentagon's changes stance on climate change. This promises to be another show-down between the military and some congressional Republicans, who deny the reality of climate change.

As a matter of fact, just 6 months prior, House Republicans passed an amendment to the annual National Defense Authorization Act, which simply forbids the Defense Department from spending money on any climate-related programs. "This amendment will ensure we maximize our military might without diverting funds for a politically motivated agenda," according to the author, David McKinley.

The democrats defeated the amendment, but climate change is now a new catalyst for conflict, where new concepts and idea are the weapons. Now the Pentagon is on the forefront of the battle, which has been planned for over a decade. In 2004, there were reports about a secret Pentagon document warning about the consequences of climate change. It warned that extreme weather phenomena could push some countries, like China, India, and other, into regional wars over energy sources and fresh water rights.

Note: In our books on the subject, *Power Generation and the Environment*, and *Energy Security in the 21st Century*, published by Fairmont Press in 2013 and 2015 respectively, we detail a number of environmental programs initiated or planned by the U.S. military. Most of these, however, were kept at a low profile partly because the U.S. DoD is financed by Congress, which until recently was officially opposed to any climate change activities and spending.

But that was then! As we saw in the text above, the new DoD Roadmap clearly outlines the Pentagon's position and its decision to wait no more. And so, the U.S. military is going to show us the way to sustainable environmental balance while reinforcing our energy security. This is one of the best news coming from Washington DC lately. It shows that we are aware of the gravity of the problems at hand and that we intend to do something about it.

How much the DoD can do in this area remains to be seen, but at least we all, and the U.S. politicians, now have an example to follow.

Energy Markets Militarization

The Council on Foreign Relations, in its Task Force report. "'National Security Consequences of U.S. Oil Dependency." notes that "the issues at stake affect U.S. foreign policy, as well as the strength of the American economy and the state of the global environment."

The report begins by noting that: "The lack of sustained attention to energy issues is undercutting the U.S. foreign policy and U.S. national security. Major energy suppliers—from Russia to Iran to Venezuela—have been increasingly able and willing to use their energy resources to pursue their strategic and political objectives."

Major energy consumers—notably the U.S., but other countries as well—are finding that their growing dependence on imported energy increases their strategic vulnerability and constrains their ability to pursue a broad range of foreign policy and national security objectives.

Dependence on oil imports also puts the U.S. into proportionally increasing competition with other importing countries, and especially with the rapidly growing economies of China and India. At best, these trends will challenge U.S. foreign policy. At worst, they will seriously strain relations between the U.S. and these countries

In the context of these new concerns about the rising geopolitical power of energy suppliers, and the increasing demand by the world's largest nations, new definitions of energy, energy markets, energy security, and national security are emerging.

The Transition

The world economy has been undergoing a radical transformation over the past half century. Significant government intervention in the form of regulation and planning was replaced by one increasingly based on market forces.

The oil market, for example, slowly moved away from government contracts and relationships between specific buyers and producers, to a global energy market system. This new system is based on competitive bidding and price discovery through uncontrolled commercial dealing of a wide number of players.

The U.S., as a world power and the largest producer and user of energy, favors an open, transparent and competitive global energy market—especially for oil—where no one seller or group of sellers could dominate the market. Any resemblance of market domination threatens the access of the U.S. and its allies to critical oil supplies, as needed to conduct normal and everyday business and military operations.

We are presently dependent on OPEC countries for over half of our crude oil.

The saddest, and most dangerous result of this dependency is the fact that there is nothing we can do to change the threats this situation presents us with daily.

We need a lot of oil to run our huge economy and military machine, so we will get it from wherever we can and at any cost.

The current global energy market system, established after WWII, is led, and in part enforced, by the U.S., which has coordinated many international negotiations to reduce tariffs, open markets to unrestricted capital flows and establish rules for protecting investments and intellectual property.

Often, the U.S. has relied on multilateral negotiations, and when these were not effective, on multi-track such, or even unilateral action. So far, so good. The system works fairly well in most cases, with some glitches and complaints by some countries, in others.

The U.S. military machine enforces the proper operation of the global energy marketplace by policing the seas and protecting international commerce from attack by hostile nations or terrorist groups. This has helped many countries to achieve higher growth rates resulting from integration into the global economy through better access to the energy markets, which also brings increased foreign investment.

All countries benefit from a global economy with many players in the energy market and competitive prices. It is essential for energy importing countries to have a diversity of supply sources and to have suppliers competing with one another.

The energy exporters, however, have a different agenda and often complain about unfair energy trading system. While the products they export are priced at highly competitive rates, the goods they import are typically sold in markets that are oligopolistic, which results is unfavorable trade terms.

Importing countries have a number of complaints, because global commodity price volatility usually translates into local economic instability. And not helping things, exporters often avoid the free and competitive international markets in order to gain some market advantage the export markets—especially in times of energy prices are low, which hurt the exporters' economies.

This is what happened in 1997-1998, during the financial crisis in Asia followed by a ramp up of Iraqi oil exports. There was also a market share struggle between Venezuela and Saudi Arabia at the time; all of which led oil prices to collapse below S10 per barrel. Some oil exporting countries lost over 50% revenue, which intensified the internal economic pain of many oil exporters.

Faced with growing economic problems, the unrest in these countries intensified, and many changes occurred in Indonesia, Iran, Nigeria. Russia, Venezuela, and others. This situation drove the oil exporters to co-operate beyond and above the OPEC areas of influence and authority. Intensive international diplomacy among oil exporters began, with Venezuela and Mexico leading the effort. The goal was to get an agreement among world's oil producers to reduce production and bring oil prices above the $20 per barrel price.

Agreement was achieved, prices went at $22/bbl and that incident marked a major turning point for OPEC. Since then OPEC has been more active and has been successful in coordinating oil production cuts as needed to keep prices high. defend dramatically higher prices, first in the $30 level, then to $50, and then to record levels of $100 a barrel and more.

The Changes

OPEC was able to keep global oil prices well above $100/bbl until 2014-2015 when things changed drastically in the opposite direction. OPEC nations, led by Saudi Arabia, driven by new developments in the global energy markets, took a drastically different approach in an attempt to maintain control over the global energy markets.

After they stood helplessly watching the sharply increased oil and gas production in the U.S., Canada, and other non-OPEC countries, they decided to boycott it. They could not see themselves taking a back seat in the global oil market, so they sprung into action. This time the strategy was quite different from anything OPEC has ever tried.

Instead of pushing oil prices up, they decided to take a hit by keeping production steady, and even increasing it in some cases, in order to keep the world oversupplied. This way, with the oil around $50/bbl, many of the new oil and gas producers could not continue operation. The strategy worked…somewhat. The new well drillers in the U.S. suffered most. Most of them reduced production, some stopped all together, and others headed to the exit.

So now, the global oil market was at a stalemate. Saudi Arabia was willing to lose a lot of money in order to keep others from making money. As they say in the old country, "I don't have, but I'll make sure that you don't have either." But Saudi Arabia is a rich country and could endure many months and even years of low prices.

The U.S. adapted quickly to the new situation, with winners at the gas pump and losers at the drill wells, but the overall effect was positive for the people and the economy. Countries like Russia, Venezuela, Nigeria, and others, however, the low prices were the kiss of death for their economies. They cannot go very long without the vital income coming from oil, and a number of socio-political changes started taking place.

Just as we thought that we have learned all we need to know about OPEC and oil trading, there are a set of major lessons we must learn now from the newly developed scenario during the new, 2014-2015 global oil prices shock:

- Oil over-dependence is U.S.' Achilles Heel, and world-wide cancer, which will not lead to anything good, if left untreated.

- In the short term, oil importers are hurt by the high prices, while the oil exporters are hurt by the low prices. In all cases, oil manipulations bring suffering to innocent people in the most vulnerable places of this world.

- In the long run, oil exporters will run out of oil, and what is considered bad now, with the low oil prices, will become many times worst when these economies—geared to survive on oil exports—will collapse. This includes the oil-mighty Saudi Arabia, which has nothing but oil and sand as economic drivers. And we all know that sand is not an export commodity and does not bring much income.

At that time, the oil importers will have to deal with their oil-less economies, which would simply collapse…unless they have taken measures to avert the consequences of oil-less existence. The Arab sheikhs won't have fuel for their Rolls Royce fleets then, and will have to learn how to ride on camels again.

- For now, however, Saudi Arabia and its cronies control the global oil market, as demonstrated by the 2014-2015 oil boycott. They are not shy about it, and use every occasion to flex their oil muscles, since this is all they have to flex.

So, in 2015 we wondered how long could the low prices scenario continue? After that we wonder how high would the oil prices go, realizing that it all depends (mostly) on what Saudi Arabia and its OPEC partners decide to do. This shows their overwhelming power in controlling the oil market; something that is not good for the U.S. and the world.

While discussing the dangers of over reliance on Middle East oil, former CIA director R James Woolsey wrote in his "Defeating the Oil Weapon": "We had a working partnership with the Saudis for much of the cold war, offering them protection against the Soviets (and Soviet clients states) in exchange for a reliable supply of cheap oil. But in light of the direction taken by the Saudis for nearly a quarter-century now (accommodating extremist views) it is also imperative that we take steps to reduce their hold over us… The wealth produced by oil is what underlies, almost exclusively, the strength of three major groups in the Middle East—Islamists, both Shiite and Sunni, and Ba'athists—that have chosen to be at war with us. Our own dependence on that oil and the effect this has had on our conduct over the past quarter-century, have helped encourage each of these groups to believe that we are vulnerable."

This statement is a reflection of the view of a wide spectrum of political and economic activists in the U.S., including most neoconservatives, mainstream Democrats, and left-wing activists groups. And this opinion is spreading throughout the Western world as well.

Timothy E. Worth, C. Boyden Gray, and John D. Podesta wrote on subject: "The flow of funds to certain oil producing states has financed widespread corruption, perpetuated repressive regimes, funded radical anti-American fundamentalism and fed hatreds that derive from rigid rule and stark contrasts between rich and poor. Terrorism and aggression are byproducts of these realities. Iraq tried to use its oil wealth to buy weapons of mass destruction. In the future, some oil-producing states may seek to swap assured access to oil for the weapons themselves. It is also increasingly clear that the riches from oil trickle down to those who would do harm to America and its friends. If this situation remains unchanged, the U.S. will find itself sending soldiers into battle again and again, adding the lives of American men and women in uniform to the already high cost of oil."

None of this is breaking news. We all know where we stand, but like a deer caught in upcoming car's lights, we are frozen in hesitation and indecision. We cannot move left or right, and the upcoming car cannot stop, so what will happen? We know that too, but are hoping for a miracle, or that it will go away. Miracle of such huge proportions is highly unlikely, and the problems won't go away, so we just have to hope that the collision won't be fatal.

The Future

The high volatility of the global oil markets will continue and will get even worse in the future, as the energy markets get more complicated and the struggle for enough and cheap oil intensifies. After periods of low prices there would be periods of high prices, all orchestrated by OPEC. No matter the direction, extent, and duration these fluctuations are accompanied by significant and lasting economic hardship and uncertainties.

OPEC does not feel responsible for any of this. As a matter of fact, OPEC targets consumer governments within the Organization of Economic Cooperation and Development (OECD), who have been capturing rents via high national Western energy taxes. "People always talk about the revenues of OPEC, but before they point a finger at OPEC, they should probably reduce taxes in their own country," according to one OPEC minister.

So the war of words and deeds is on. And with statements like this, what follows is not hard to guess. In fact, OPEC has openly warned in recent years that a shift to alternative energy inside major oil consuming economies will discourage its own investment in future oil supplies. This potentially will force oil prices to go "through the roof."

In June 2007, OPEC was considering cutting its investment in new oil production, "If we are unable to see security of demand… we may revisit investment in the long term" warned another OPEC minister. He was hinting to the fact that increased emphasis on the U.S. and European biofuels and renewables would backfire.

OPEC's aggressive stance on price and capacity expansion and increasingly tight oil markets have renewed energy security concerns among the major oil importing countries. Western security analysts are warning of the possibility of gradually increasing scarcity of oil in the coming decades.

Taking a broader view off the energy markets, we see that there is a link between the current tightness in the oil market and the geopolitics of OPEC policies. This, in light of the fact that oil resources are finite and are found in smaller and hard to extract deposits, means that the world oil production has either already peaked or will peak shortly, tells us that a serious battle for survival of the fittest (oil-strong) countries is on the way.

Militarization of the energy markets is not new, but it is entering a new phase. The new reality is that of extended and very sophisticated covert operations and smart military actions in different forms are the norm. Due to its military might, the U.S. is presently given a preferential treatment by the oil exporters.

The day when oil runs out, and the U.S. military cannot be deployed freely around the world, would be the end of the mighty US of A. Fully aware of that, the U.S. military will not allow such a day to come without a fight. So ready or not, here we come…

Three Days of the Condor

In the 1975 movie, *Three Days of the Condor*, Robert Redford plays the role of a FBI agent, who accidentally discovers illegal manipulations of global oil resources by U.S. government officials. A fiction, one would say, and it very well might have been. But on the other hand, considering the importance of crude oil in running our economy and the military machine, we have to wonder what would they (the government and the military) stop short of in achieving their goal of getting as much oil as needed for normal operations.

The war in Iraq was considered by many as a conspiracy on global scale that killed thousands of people. Oil rich country was targeted for many reasons, with at least one of the reasons (albeit unofficial) being ensuring the oil flow from the Iraqi oil fields.

How many other covert actions are underway or planned for similar reasons, we will never know, but one thing is for sure: the U.S. military will be the last to run out of oil…whatever it takes…come hell or high water.

In the movie, several people are assassinated by FBI agents in an attempt to cover up the conspiracy. Again, fiction, but would FBI and CIA hesitate to kill, and/or do whatever needed to ensure the flow of energy (crude oil mainly) into the country? They know well that without enough and cheap oil, all military and economic activities would stop. That would make them, and all of us, helpless and vulnerable to terrorist attacks.

The thought of no more planes in the sky, trains, ships, trucks, and cars on the roads is the ultimate nightmare. No more movement, no activities of any sort is un-American and has to be avoided at all cost. The problem is that the dangers are increasing and the energy costs will continue rising. With that, the activities in ensuring the flow of energy (crude oil) will also increase. The theme of the *Three Days of the Condor* might just be a warning of things to come…

Global Economic Failure

According to one of the top minds in the U.S. Intelligence Community, we are nearing the darkest economic period in our nation's history. A 25-year Great Depression barreling down our way. An alarming pattern has caused many in the Intelligence Community to secretly prepare for a "worst-case scenario." Alarmingly, they believe that there is enough evidence that proves this outcome is impossible to avoid.

The CIA's Financial Threat and Asymmetric Warfare division has stepped forward to warn the American people that time is running out to prepare for this $100 trillion meltdown. We have a dangerous level of debt, and the Feds are recklessly printing trillions of dollars. As a result, there are a series of dangerous signals that reveal an economy that has overheated and is reaching a super critical state.

One of the signals, which the CIA financial experts are most concerned with is the Misery Index. This unique warning sign was created many decades ago as a tool to determine how close the U.S. gets to a social collapse.

The U.S. Misery Index simply adds the true inflation rate with the true unemployment rate.

"True" are the key words here, since the numbers the Federal Reserve provides are repeatedly changed, which affects the way the Misery Index is calculated. Different numbers might be used to cover up the true scope of the problem. It is unusual for any government official to talk about the Misery Index openly, so we won't know what they are doing. The truth is that the true Misery Index—according to independent experts—has reached very dangerous levels and there are clear signs pointing to the inevitable.

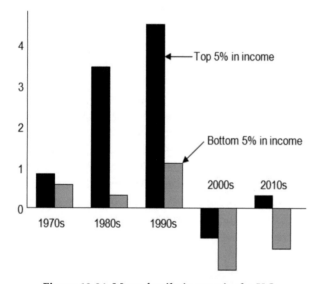

Figure 10-34. Mean family income in the U.S.

Some of the danger signs are similar to the signs we saw prior to the Great Depression. They signal that our complex financial system is about to collapse, and the entire U.S. economy could be flushed down the drain in a matter of days.

There are several dangerous "flashpoints" the intelligence community is closely monitoring, which they believe will eventually unleash this catastrophe.

- An instantaneous 70% stock market crash, which nobody would see coming, would trigger the avalanche. Instead of a flash crash, however, it will cause a systemic meltdown in the economy. An estimated $100 trillion is one conservative estimate of the total damage.

Note: We saw a glimpse of this happening in August 2015, when the stock market went down 1,051 points in early trading. This is the largest drop ever for Dow Jones, which reflects the dependence of our economy on global events. It also shows the vulnerability of our economy, which is affected by things going good or bad across the world.

- The following collapse of the dollar as a global reserve currency would seal the event as the most devastating ever. It would signal the end of the U.S.' reign as the leading superpower. The U.S. would go down in the history books following in the steps of the Roman and British empires. How the mighty fall?
- The nightmarish scenario affecting the U.S. economy would then evolve in a worldwide economic breakdown with an extended period of global anarchy. End of the American Dream, the American global dominance, and end of the world as we know it.

OK, granted; this is an extreme scenario, which might or might not happen as described above any time soon. And if and when it happens, it might be much less damaging and with fewer global consequences.

Yet, we are living in a huge pile of debt, and that energy (oil imports) is a big part of the problem.

Oil imports not only cost us a lot of money every day, but that billions of dollars go to countries who then use that money to fight us around the globe. And even more importantly, the crude uncertainty—global price fluctuations—are driving our economy wild. A good example, is the recent (artificial) drop of oil prices, caused by our good friend, Saudi Arabia. It caused the U.S. energy sector to stumble, and many oil companies (well drillers and such) collapsed.

Trillions of dollars, which we use daily to pay government workers' salaries, buy crude oil and military equipment and such, is not ours. This is money that was lent to our banks by other nations, like China, and which we need to return someday with interest. But most of this money we have spent already and no longer have, so how are we going to pay it back?

As the debt and interest grow, we need even more money to pay the increase. Instead of tightening our belts, we continue spending. In order to cover all our obligations from loans and extra spending, we print more trillions of dollars. We actually print as much as we want, because our currency is the global standard. As such, people worldwide take the dollar believing that the U.S. is financially sound and stand behind its currency.

All this could stop in a split second, if and when the above described worst-case-scenario hits and demolishes our financial empire, which is built on paper money with little value. We saw how this could happen in August 2015, when the stock market collapsed over 1,000 points one morning within minutes of the opening bell. Additional 1,000 point collapse the following day could've resulted in panic, sending the market in uncharted territory with no point of return.

But even a lesser event could damage the U.S. economy and the energy sector. We are presently witnessing a number of variables driving the economy and the energy market. The low oil prices, for example, are badly damaging the U.S. oil and gas industry, where most companies cannot afford to keep operating. Many are slowing down, while others are filing for bankruptcy. The stock market volatility, on the other hand, is not helping either, and if things continue this way, we will witness more oil companies collapsing and more volatility on the stock and energy markets.

Although these are seemingly unrelated elements, the entire economy is directly affected by the behavior of each of its components. If any one of them gets badly out of whack, the entire economy could follow. This may not happen any time soon, but as the global socio-political situation is steadily going into a disarray, it will happen some day. The energy market collapse in the future, might lead the destruction of the world economy as we know it. The upcoming depletion of the fossils, might deliver the first *salvo* in the global economy devastation process.

For now, however things look pretty prosperous and stable...at least in the good, 'ol US of A they do. A major factor in this temporary state of bliss is the low oil prices, but how long? A more permanent reason for optimism is our high level of technological development, which leads the world in many areas. Our technological achievements are based on the knowledge and power platforms, which in turn move our economy, as discussed above.

THE FUTURE ENERGY ALTERNATIVES

Like every good thing that is used excessively, fossils have been (and are getting even more) used and will eventually become extinct. The depletion rate is high,

and although we are not exactly sure what time line each fossil fuel type will follow, we dare guess that the practical use of oil will end around year 2050. Gas, which will become the predominant fuel after that will be mostly depleted by year 2075, and coal will run its course by 2100.

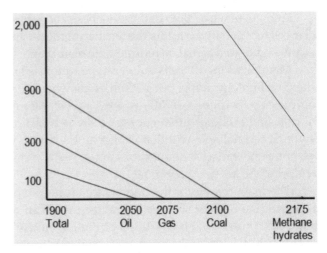

Figure 10-35.
Fossil reserves; billions of tons of fossils vs. depletion time

Figure 10-35 represents the future of the fossil fuels, give or take 25-50 years, What we are looking at some time during the next century is a fossil-less existence, which is hard to even imagine with the present dependence on these critical energy sources.

So, whether it is year 2150, or 2525, the fossils will be exiting the scene one by one; never to be seen again. And unless the alternative energy technologies—the renewables mostly— are fully developed, and/or new energy forms are found and quickly implemented by then, the world will sink in darkness and misery. What a time, what a place, what a life...

Wind and solar technologies are the most promising candidates for replacing the fossils. Wind power generators are a well developed technology, and we see further improvements in their efficiency and price reduction in their future. Solar power technologies are more complex and versatile, so they deserve more attention.

n-type Solar Cells

Crystalline silicon (c-Si) technologies rule now, and will continue so for the foreseeable future. The battle between the c-Si and thin film technologies was decided in favor of c-Si, due to its higher efficiency. There was also a battle among the c-Si technologies, p-type, vs. n-type silicon, which was won by the p-type silicon technologies, which are widely spread today.

There are two c-Si technologies (solar cell manufacturing) concepts with the highest efficiency today. These are:

*   IBC type cells from SunPower, and
*   HIT type cells from Panasonic.

And guess what? Both of these are based on n-type solar cell technology. So is this the revenge of the p-type silicon, and does it signal its return to the main stage?

Manufacturing PV cells out of n-type wafers is not a new technology, nope, but is (still) somewhat more complex and expensive. This is why almost 90% of PV cells and modules production globally is based on p-type Si technology? Will that change in the future is one of the most-discussed questions in the Si PV industry during the last few years now.

The main reason for the dominance of the p-type silicon technology for solar cell manufacturing can be found in the historical background of the PV industry itself. Until the 1980s, the main industrial application of PV was for space applications, mostly for powering satellite. P-type Si proved to be less sensitive to cosmic degradation (high-energy particles bombardment) at the time. Thus from that time on, and for decades after that, all industrial PV cell development and manufacturing was based on p-type silicon.

Since there is no technological difference between the growths of p- and n-type crystals (for making p- and n-type wafers respectively) there is no good reason for the increased manufacturing cost for n-type ingots and wafers. Consequently, it all boils down to economy of scale, where presently over 90% of the worldwide single crystal (Cz-Si) wafers are p-type.

Note: The author was a Process Engineer in one of the oldest PV cells and modules manufacturers in the U.S.—Astra Sciences in Santa Barbara, CA in the late 1970s. At that time, we could only get p-type silicon in 2" rods or wafers. There were very few manufacturers of these materials, and as the industry grew with time, they all continued producing mostly p-type ingots and wafers. It is like the industry froze into the p-type mode, simply because it was what everyone did at the time.

Few attempts were made to produce n-type wafers even later in the 1980s and 1990s, mostly due to the higher cost of low volume production. As a matter of fact, all silicon ingot manufacturers can produce n-type wafers, but do that only by a special order. So here we are today; still frozen into making of p-type wafers mode.

But according to the experts, the n-type technology is now gaining due respect and is growing in impor-

tance. As a result, we see most PV companies looking into it, and some upgrading their lines to produce both; p-type or n-type silicon ingots and wafers. A few are also investing in new capacities for one or both types as well.

And so, now we see the global silicon solar industry divided into two main branches:

*   High-performance (HP) p-type mc-Si cells with better than 22% efficiencies, and
*   n-type cells with over 25% efficiencies.

In the long run, the cost of materials and efficiency of the manufacturing processes will determine which technology is going to win. Nevertheless, today even 0.1% efficiency gain means increased sales, so the battle for efficiency is spreading. Since the two highest efficiencies PV cells in industrial production are based on n-type Cz-Si wafer is a clear indication that n-type wafers might win the battle in the end.

The physical reasons for the superiority of n-type versus p-type can be summarized as follow:

*   The absence of boron, limits the light induced degradation (LID) occurring in p-type Si wafers, which is caused mainly by boron-oxygen complexes.
*   As a result, and very importantly for desert installations, n-type Si cells are less prone to degradation during high temperature exposure to blistering desert sunlight.

Note: One possible solutions for avoiding LID in p-type Si wafers is permanent deactivation of the B-O-complexes. This can be done by a combined heat-illumination treatment. This process is in R&D mode presently, and might prove a challenge for the growth of n-type wafers production.

*   n-type Si is less sensitive to metallic impurities, since the minority carrier diffusion lengths in n-type Cz-Si are significantly higher than these in p-type Cz-Si
*   n-type solar cells can be manufactured from lower grade silicon in the $1.25-1.30 range, vs. $1.00-1.25 range for p-type cells. Fairly insignificant difference, compared to the highest efficiency of the n-type solar cells. In some cases, however—due to low volume production—n-type wafers can be up to 20% more expensive than p-type wafers.
*   As n-type wafer manufacturing volumes increase, their prices would go even lower than these for p-type wafers by 2020, or earlier.

- n-type solar cells could well exceed 21% efficiency, which with the added lower degradation during the high temperature processes and during field operations, will make it the predominant wafer material for solar cell manufacturing in the near future.

Presently, the n-type Si solar cell technologies available on the market, and/or at R&D stage fall into three categories:
- Front side boron emitter H-pattern devices;
- Rear side emitter H-pattern devices; and
- Rear side emitter rear contact devices.

There are many companies working on these types of n-type cell and modules. Some of these are: LG, Neosolarpower, Panasonic, PVGS, SunPower, Yingli, and others. Recently, First Solar, MegaCell, Motech, Nexolon, Silevo, SSNED, etc. have started work on n-type solar cells development as well.

Tempress and Semco have developed new diffusion tube furnaces for BBr3 and BCl3 diffusion processes, and Schmid has developed APCVD B-diffusion equipment for one-sided deposition. AMAT, Kingstone, Intevac and others are working on more efficient and cheaper boron ion-implantation, which, due to its many advantages, is seen as the technology of the future.

And of course, there are challenges, the greatest of which are: the proper execution of the cleaning and passivation of the Boron diffused surface, and the following metallization process. These are tricky, hi-tech processes that determine the final level of Voc losses. Centrotherm, Dupont, Hareaus, Levitech, R&R, RENA, Schmidt, Simgulus, Solaytech and many others are working on the optimization of.

Since Al spiking leads to limited Voc of about 655 mV, it has to be eliminated in future. Surface contamination is another critical issue in n-type solar cell processing, so it has to be fully developed and properly implemented into the cell process. Upon the full development, and implementation in production of these process, n-type Si solar cells are expected to reach mass-production efficiencies in the order of 22-23%.

n-type cells offer higher efficiency, at the cost of introducing additional process steps, which required the introduction of new and advanced solar cell processes and concepts, such as PERC, MWT and PERT. These processes, however efficient their product, require more sophisticated equipment and techniques, which increase the final costs. Especially at this period of low-level production and high R&D expenses.

Nevertheless, n-type solar cells are entering the energy markets by a storm. They are one way of increasing the efficiencies, which have been stagnant for decades, so the ramp for mass production take-off is ready for the first tests. We would not be surprised if n-type solar cells and modules become the preferred energy product by the middle of the 21st century.

III-V Materials and Devices

III-V materials based PV technologies are basically thin films of these materials deposited on substrates of different kinds. These devices are the most promising as far as high efficiency is concerned.
They are divided into single- and multi-junction devices as follow:

Single-junction III-V Devices

A number of compounds such as gallium arsenide (GaAs), indium phosphide (InP), and gallium antimonide (GaSb) have adequate energy band gaps, high optical absorption coefficients, and good values of minority carrier lifetimes and mobility, making them excellent materials for making high efficiency solar cells. These materials are usually produced by the Czockralski or Bridgmann methods, which provide high quality materials with increased efficiency and reliability, but at a higher price.

After silicon, GaAs and InP (III-V compounds) are the most widely used materials for single-junction (SJ) solar cells manufacturing. These materials have optimum band gap values (1.4 and 1.3 respectively) for SJ conversion of sunlight. The construction of solar cells made of these materials is similar to the regular single-junction c-Si solar cells we discussed earlier in this text.

The major disadvantage of using III-V compounds for PV devices is the high cost of producing the materials they are made of and the related manufacturing processes. Also, crystal imperfections, including bulk impurities, severely reduce their efficiencies, so that only very high quality materials could be considered. Too, they are heavier than silicon, which requires the use of thinner cells, but they are weaker mechanically, so their design requires a delicate balance of thickness vs. weight.

The combination of high efficiency, high price, crystal imperfections intolerance, and mechanical weakness makes these devices useful for limited applications, where efficiency and overall behavior is more important than price. Thus they are not widely used in the general PV market, but still can be found in some important niche markets.

Multi-junction Cells and Devices

Single-junction PV devices convert only a portion of the sunlight (with photons just above the band-gap level of the semiconductor material). The problem is that photons with lower or higher energy do not generate electron-hole pairs and are lost as heat, which is also detrimental to the cells, reducing their efficiency and deteriorating them over time.

One way to solve this problem and to increase the efficiency of the PV devices is to add more junctions, thus creating multi-junction (MJ) solar cells. By selecting materials, properties, number and types of junctions, and manufacturing processes designed to capture the majority of sunlight photons, we can reach very high efficiencies.

Thus far, multi-junction solar cells made primarily using the III-V compounds have clearly proven that by minimizing thermalization and transmission losses, large improvements in efficiency can be made over those of single-junction cells. These devices find use in generating power for space applications and in concentrator systems. They show great promise for high efficiency and reliability under harsh climate conditions, such as those in the deserts.

Future development of multi-junction devices using low-cost thin-film technologies is especially promising for producing more efficient and yet inexpensive devices. Cost reductions will also be significant when thin-film technologies are directly produced on building materials other than glass, because many materials such as tiles and bricks can be substantially cheaper than glass and have much lower energy contents.

The major devices in this group are gallium arsenide-based, germanium-based, and CPV solar cells.

- *Gallium Arsenide Based Multi-Junction Cells*

High-efficiency multi-junction cells were originally developed for special applications such as satellites and space exploration, but at present, their use in terrestrial concentrators might be the lowest cost alternative in terms of \$/kWh and \$/W. These multi-junction cells consist of multiple thin films produced via metal-organic vapor phase epitaxy. A triple-junction cell, for example, may consist of the semiconductors GaAs, Ge, and GaInP2.

Each type of semiconductor will have a characteristic band gap energy which, loosely speaking, causes it to absorb light most efficiently at a certain color, or more precisely, to absorb electromagnetic radiation over a portion of the spectrum. Semiconductors are carefully chosen to absorb nearly all of the solar spectrum, thus generating electricity from as much of the available solar energy as possible.

GaAs-based multi-junction devices are some of the most efficient solar cells to date, reaching a record high of 40.7% efficiency under "500-sun" solar concentration and laboratory conditions.

This technology is currently being utilized mostly in powering spacecrafts. Demand for tandem solar cells based on monolithic, series-connected, gallium indium phosphide (GaInP), gallium arsenide GaAs, and germanium Ge p-n junctions is rapidly rising. Prices are rising dramatically as well.

Twin-junction cells with indium gallium phosphide and gallium arsenide can be made on gallium arsenide wafers. Alloys of In.5Ga.5P through In.53Ga.47P may be used as the high band gap alloy. This alloy range allows band gaps in the range of 1.92eV to 1.87eV. The lower GaAs junction has a band gap of 1.42eV.

In spacecraft applications, cells have a poor current match due to a greater flux of photons above 1.87eV vs. those between 1.87eV and 1.42eV. This results in too little current in the GaAs junction, and hampers the overall efficiency since the InGaP junction operates below MPP current and the GaAs junction operates above MPP current. To improve current match, the InGaP layer is intentionally thinned to allow additional photons to penetrate to the lower GaAs layer.

In terrestrial concentrating applications, the scatter of blue light by the atmosphere reduces photon flux above 1.87eV, better balancing junction currents. GaAs was the material of the highest-efficiency solar cell, until recently, when Germanium-based MJ cells capped the world record at 41.4% efficiency.

- *Indium Phosphide Based Cells*

Indium phosphide is used as a substrate to fabricate cells with band gaps between 1.35eV and 0.74eV. Indium phosphide has a band gap of 1.35eV. Indium gallium arsenide (In0.53Ga0.47As) is lattice matched to indium phosphide with a band gap of 0.74eV. A quaternary alloy of indium gallium arsenide phosphide can be lattice matched for any band gap in between the two.

Indium phosphide-based cells are being researched as a possible companion to gallium arsenide cells. The two differing cells may be either optically connected in series (with the InP cell below the GaAs cell), or through the use of spectra splitting using a dichroic filter.

The presence of varying quantities of toxic materials in these devices must be considered when planning their use in large quantities.

• *Germanium-based Single- and Multi-junction Cells*

Germanium (0.86eV band gap) is a semiconductor material, with properties far superior to other substrate materials used for PV cells and modules. It is ~40-50% more efficient than silicon and has a much lower temperature coefficient. It is several times more expensive than silicon, too, but with new superior slicing techniques, it can be cut into very thin wafers, saving a lot of material. This, combined with its higher efficiency and less degradation than silicon, could put it on the competitors' list within the next few years.

Germanium-based solar cells have been used mostly for space applications, but a number of manufacturers have geared up for mass producing them for high concentration HCPV and other high efficiency applications (the record is currently 42.3% efficiency).

CPV Technology

Concentrating photovoltaics (CPV) is a branch of the PV industry, using special cells, optics and tracking mechanisms, developed in the 1970s by several companies under contracts and financing from U.S. Departments of Energy.

Figure 10-36. CPV tracking electric power generator

Early CPV systems used silicon-based CPV cells, which had a problem with elevated temperatures. These were later replaced by GaAs-based multi-junction cells, which have much higher efficiency, but still suffer from the effects of high temperatures. At first GaAs CPV cells were made by using straight gallium arsenide in the middle junction.

Today CPV cells are made by depositing different thin films onto gallium substrate. The combination of efficient multi-junctions with goof lattice match to the base material ensure the devices' high efficiency.

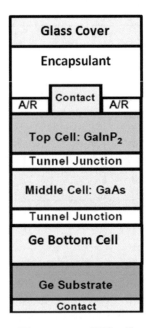

Figure 10-37. CPV cell

Figure 10-37 show CPV cells as complex structures, consisting of many layers (some deposited, some diffused) in, or piled on top of, Germanium semiconductor material, which has more superior process and performance characteristics than silicon.

CPV cells are usually mounted under lenses, which concentrate sunlight onto the cells 100 to 1000 times. This allows high efficiency and reliability, better land utilization, and other benefits. Cell-lens assemblies are mounted on trackers, which track the sun precisely through the day, providing the most power possible.

Efficiencies of 45% on the cell level have been obtained recently. This means that 35% efficiency and more can be obtained by the CPV systems, and sent into the grid. The experts estimate achieving 50% efficiency by 2020 and 60% by 2030. Approaching the theoretical efficiency of 85+% is the goal by the end of the century.

The other solar technologies cannot compare with the potential of efficiency increase of the CPV systems. Because of that, we see the CPV trackers as the leaders in global solar power generation in the near future.

Looking at Table 10-19, we see the obvious advantages of the CPV technology as compared to the usual PV solar panels. CPV efficiency is 2-3 times greater than PV, and so is its potential for further increase, due to the much higher theoretical efficiency of 87%.

Table 10-19. CPV vs. PV

Technology variables	CPV tracking systems	PV conventional systems
Actual system efficiency	36%	12-16%
Theoretical max. efficiency	87%	28%
Full power generation	6-12 hrs./day	1-2 hrs./day
Temperature coefficient	0.015/C⁰	0.5%/C⁰
Annual degradation	0.10%	>1.0%
Capacity per unit area	35%	10%
Land leveling and footprint	5% / 5%	100% / 50%
Local materials and labor	90%	10%

CPV technology uses trackers, which allow it to generate power during most of the day, while PV panels generate maximum power only around the noon hour, when the sun is directly over them. CPV also has significant advantages in the other critical area of solar power generation, such as temperature coefficient, annual degradation, land usage, and land preparation.

Manufacturing of CPV equipment uses over 90% of local materials and labor.

This fact alone, makes CPV the preferred technology for many nations around the world. Because of these advantages, we foresee CPV technologies as the primary choice for installation and use of solar energy in large-scale power plants, especially those in desert regions, in the near future.

So why is CPV not used in greater quantities. This is because, due to the more complex nature of its components it is more expensive to manufacture than the other solar technologies—even if local materials and labor are used. In addition, the trackers onto which the CPV modules are mounted, have moving parts, which also complicates the O&M process and adds to the overall expenses.

So for now, CPV is used only in special projects, but due to its advantages it will became a widely spread technology in the future.

Methane Hydrates

Methane hydrates are the best kept secret of the energy industry. They are well known by many scien-

tists, but some of them are too shy, while others are too scared, to take them out of the closet. And there are some good reasons for that. The science behind this awesome energy source is too complex and overwhelming, while the logistics are even more so. And since the fossils still reign (and pay the bills) most of the professionals prefers to stay on the sidelines when it comes to introducing replacements.

Nevertheless, some of those that are involved in the energy field claim that methane hydrates are the energy source of the future, while others say that it is an expensive exercise in futility, and yet others will warn us that it is the most dangerous effort ever to be undertaken by man.

All this fuss about a single, and otherwise humble, compound, composed of methane (CH_4)—the simplest of all hydrocarbon molecules—buried deep in frozen state under the ocean floor and in the permafrost.

Figure 10-38. Methane

Methane consists of a single carbon atom bonded to four hydrogen atoms. There is nothing special about methane for it is the primary component of natural gas, which we know well and use all the time in many applications. We know now how natural gas was formed, where it is found and how to extract it. These deposits are the traditional natural gas wells we tap into most often, and extract the gas via the controversial hydraulic fracturing (fracking) techniques.

Methane hydrates, however, are quite different from natural gas and the other fossils. They were also formed a long time ago, and in a similar fashion. Methane gas, released deep underground, encountered a frozen soil and became trapped in it. Under the freezing and high pressure conditions of its icy grave, methane gas was literally frozen in place forever, deep under the surface.

Methane hydrate can be envisioned as small balls of ice (frozen water, which gives it the hydrate denomination) with the methane molecules trapped inside the ice balls. See Figure 10-39.

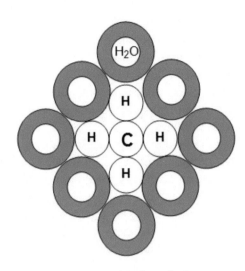

Figure 10-39. Methane hydrate

The conversion from gas to solid methane usually happened deep under the ocean floor in many places around the world, and the permafrost areas. Some of the largest known deposits are found in the permafrost of Siberia and Northern Canada.

The cold water and high pressure in the permafrost or the ocean depth, combined with spending long time under this environment, changed the structure of the methane gas, so that its chemical composition is quite different.

Methane hydrate is somewhat different type of methane or natural gas in properties and behavior. It is chemically and physically the same compound—methane is methane—with the only difference being that its molecules have been trapped and enclosed in ice.

The ice forms a tight ball around each molecule, or a group of molecules, thus immobilizing the methane gas in them, so it no longer acts as gas. It is, instead, a prisoner in a frozen watery icy grave, where it will remain for ever, or at least for long time. At least until the ice cover is broken or melted someday, somehow, and the methane gas is released from the icy clutches around it. At this point, it would act like normal methane gas, going up and going happily things methane gas does.

If we are smart enough, we would figure out ways to release the methane gas from its icy grave and use it for practical purposes.

The estimates of potential methane hydrate deposits in the world' ocean floors are significant—several times larger than all fossils together—but there are no clear estimates of the total quantity. Nevertheless, there is a reliable indication that it is more than the total of all

of the other fossil fuels on Earth.

In order to extract the methane from its ice trap, the entire area has to be treated by:

• Heating, in order to melt the ice around the methane gas molecules, or

• Reducing the pressure it is under in the icy grave, which would have the same effect.

Both processes are, unfortunately, extremely complex, difficult, and expensive. They are well beyond the existing technological capabilities available today. Heating frozen permafrost layers, in addition to being a gigantic undertaking, is unthinkable because the energy (heat) needed for that process would significantly exceed the total energy value of the extracted methane.

Lowering the pressure in the methane hydrate deposits seems more reasonable, but we just don't have the needed technology presently. Maybe in several decades we will get more desperate and focus on it. For now, however, only a few experimental wells to tap into methane hydrate deposits are currently underway in the U.S., Russia, Canada, and Japan.

Note: Obtaining large quantity fuel from the methane hydrates on the ocean floor is not a new concept, but is well hidden from the general public. The main reason for this is that its extraction just does not seem practical, reliable, or cost-effective enough, so it is not included in the calculations of global fossil reserves.

Also, there is a real danger of accidentally releasing large amount of methane gas in the atmosphere, the potential effects of which is unknown. Because of that, methane hydrates are not entered also in the overall global environmental calculation.

And so, our task in the near future is to consider in detail the pros and cons of large quantity methane pumped from the ocean depths. We must be well aware of what we are dealing with, and calculate its potential of:

• Helping us reaching the holy grail of global energy balance,

• Complicate the environmental conditions by excess GHG emissions, or

• Bring us a disaster of unthinkable proportions by accidental release of huge amounts of methane gas during exploration.

There is no doubt that if and when we figure out efficient, practical, cheap, and safe ways to extract methane from the permafrost, it will provide us with a hope for a long term of cheap and fairly clean energy.

Methane hydrates could provide enough energy to power the world for centuries.

But there are numerous, complex, and expensive problems, as we mentioned above. These have to be addressed and solved first, and before any attempts of large-scale extraction are made.

The Issues

Methane hydrates are hidden deep under frozen ground and under ocean floors all over the world. Their cousins—the fossils (coal, oil, and gas)—are quite stable in their present form and state, and sit under ground inactive until taken out by man. Methane hydrates are different in that they are unstable, so much so that even slight changes in temperature or pressure could melt part of their icy grave and release huge amounts of methane gas.

Because of that, we already see problems caused by methane hydrates.

- The immediate problem—even before even touching the methane hydrate deposits—arises from the fact that the oceans and the permafrost are warming up due to the overall global warming. With time, the massive methane hydrate deposits will warm up enough to start releasing huge amounts of methane gas in uncontrolled way.

  Any additional man-made activities in those deposits might accelerate the melting and methane gas release process. We have already observed significant methane gas releases in the arctic areas and other places around the globe.

- Raw methane gas is one of the worst environmental polluters, for it is over 40 times more efficient than CO_2 as a green house gas. A significant methane release from methane hydrate deposits would increase global temperatures which would, in turn, melt more methane hydrate and release more methane gas. The vicious circle would expand as it feeds on itself and accelerate the release of more and more methane.

- There is a great opportunity of great quantity of potential energy in the methane hydrate deposits laying idle on the bottom of the world's oceans. Short of a miraculous technical breakthroughs and global development, however, any of the technical solutions would require a lot of time and money to develop. The methane hydrate deposits are usually deep in the oceans, at a minimum of 1,000 feet and up to miles in depth, so this would be the technological challenge of the century.

- And finally, it will take time for the oil companies and governments to agree on property rights and methods of exploitation. Global consensus will be required on how to proceed with R&D, and equipment and production sites development; all of which requires sustained efforts on all levels and billions of dollars.

- The way politicians and governments cannot agree on the simplest things today, it would be unreasonable to expect any serious action in the short term. So the great methane hydrates deposits will have to wait for the future generations to get to them, which we are sure they will do, after we burn all the fossils during this century.

Another, equally interesting, unexplored, and promising energy source are the huge crude oil and gas deposits deep under the ocean floor.

Deep Ocean Deposits

We have heard people say that today we know more about what is on the surface of the moon and Mars than we know about the deep sea. As a matter of fact, we do have detailed maps of the surface of the moon, Mars and many other planets millions of miles away. We, however, do not have a map of most of the world's ocean floor. This is truly amazing, because we depend on the oceans for so much of our resources—fish, minerals, and even oil and gas.

The oil and gas deposits we are familiar with are usually located just beneath the offshore rigs visible on the horizon. Going further into the ocean, it gets very deep and have made oil drilling very hard to impossible...until now. New technologies will soon allow us to map the ocean floor, discover more oil and gas deposits, and drill into the ocean floor at any depth without the help of surface rigs.

The deep *subsea* wells are changing the concept of offshore oil and gas production. The problem is that these technologies—equipment processes, needed for work miles under the ocean—are still not well developed.

The deep sea is an alien place where the temperature is close to freezing and where every 30 feet of depth the pressure increases by 14.5 pounds per square inch (psi).

A mile beneath the ocean surface, the pressure reaches nearly 2,500 psi. At two miles of dept, the pressure is about 5,000 psi.

This enormous pressure, very few creatures can live in. Keeping in mind that human divers rarely exceed 300-400 feet depth, which is accompanied by great efforts and health dangers, we see that the deep subsea is inaccessible to us mere humans. And so, in order to get to, and work at, a mile or two depths, we have to use robotic or manned submersibles as the only option.

This is a great and very difficult undertaking, but it might be worth the effort and expense. The deep seafloors are loaded with mineral wealth, which we are just now beginning to learn to tap. There are huge reserves of deep sea oil and natural gas, in addition to the seafloor methane hydrates reserves (discussed above). These alone are considered to be at least twice as abundant as all other forms of hydrocarbon deposits combined. Seafloor coal deposits are likewise huge, and well beyond any previous calculations.

The ocean floors are also rich with other minerals. Precious metals, uranium, and large deposits of copper, zinc, iron, sulfur and other metals are stored beneath the deep ocean floor, and are considered by some to be Earth's richest repository of minerals. All we need to do is to learn how to get all these treasures to the surface. When we do, the mineral and fossil deposits on the ocean floor could provide enormous benefits to humanity for centuries.

The minerals, including sulfur, copper, zinc, iron and precious metals, are contained in volcanogenic massive sulfide (VMS) deposits that form on the ocean floor where tectonic plates pull apart and allow magma (molten rock) to invade the Earth's 3.7 mile thick crust. The magma heats seawater to over 650^0 F and moves it through the ocean waters via convection, the seawater then deposits the minerals along the ridge axis.

This is all well, but we still have to worry about the formidable conditions at the deep ocean floor. The answer is subsea wells that produce oil and gas via special installations on the seabed. There are already companies that use this new technology of constructing and operating subsea wells. The industry analysts forecast global subsea hardware capital expenditure totaling $150 billion by 2020. This is a growth of nearly 90% over the preceding decade.

Due to the growth of subsea exploration, the cost of equipment and operations have doubled. The main reason for the higher costs is the fact that the subsea drillers work with custom-made, and very expensive to manufacture and operate equipment, thus costs vary on a project by project basis.

The future success of deep seafloor exploration depends on industry collaboration and global standardization. Several joint industry partnerships are underway, in order to bring together oil and gas producers involved in various stages of the subsea technology. The goal of this effort is to develop internationally agreed industry procedures and standards.

The above discussed energy sources—methane hydrates and subsea oil and gas deposits—sound very promising as far as their quantity is concerned. The problems of extracting them from the depths of the earth and the oceans, however, are overwhelming. As a matter of fact, the present day technologies and just not safe, efficient, and cheap enough to allow mass exploitation of these fossils.

Nevertheless, it is good to know that we have such great potential of energy deposits, if and when the appropriate technologies are developed. And they will be developed some day, no doubt. We just don't know when that day would be.

Another type of massive fossil energy deposits were discovered recently, and which the present day technologies allow safe, efficient, and cost-effective extraction. These are the North American tar sands and shale oil. These reservoirs promise many years of abundant and cheap crude oil and natural gas to flow in the U.S. economy, which is already revolutionizing the global energy markets.

Tar Sands

Tar, or oil sands, also called bituminous sands, are a type of fossil (petroleum) deposits. These are usually loose sandy areas, or sandstone that are saturated with a viscous form of petroleum, or bitumen, which has the odor, color, and other properties of tar. It is so thick, sticky, and viscous that it does not flow at room temperature; it is as thick as molasses. Instead, it has to be heated or diluted with lighter oils to become liquid.

Tar deposits differ in the degree by which they have been degraded from the original conventional oils by bacteria, and the tar oil in these usually has a viscosity greater than 10,000 centipoises, and a gravity of less than 10° API. Since tar is very thick and mixed with sand, it is extracted mostly by strip mining of the sandy deposits, and the tar oil is then forced to flow into wells by reducing its through techniques, such as injecting steam, solvents, and/or hot air into the sands.

These processes use much more water and energy than conventional oil extraction, although most of the conventional oil fields also use a lot of water and energy during daily operations. This process generates a 12% more greenhouse gases per barrel of final product as the "production" of conventional oil.

Natural tar deposits are located in many countries, but in particular are found in extremely large quantities in Canada. Other large reserves are located in Kazakhstan and Russia. Total natural bitumen reserves are estimated at 249.67 billion barrels (39.694×109 m3) globally, of which 176.8 billion barrels (28.11×109 m3), or 70.8%, are in Canada. However, It is estimated that approximately 80-90% of the Canada's oil sands are too deep below the surface to use open-pit mining. Instead, several in-situ techniques have been developed in order to extract these.

The oil sands in the Orinoco Belt in Venezuela are sometimes described as tar sands, but they are not bituminous. Instead they fall in the category of heavy or extra-heavy oil, since their viscosity is even lower. Oil sands reserves are considered as part of the world's oil reserves, but higher costs and technological issues are limiting their use.

Shale Oil

Shale oil, known also as kerogen oil or oil-shale oil, is an unconventional oil produced from oil shale, which is sedimentary rock filled with kerogen. It can be processed by pyrolysis, hydrogenation, or thermal dissolution to convert the organic matter within the kerogen rock into synthetic oil and gas.

The resulting oil products can be used as a fuel or can be refined and upgraded to obtain refinery feedstock by adding hydrogen and removing impurities such as sulfur and nitrogen. The refined products can then be used for a number of purposes, same as these derived from crude oil.

There are estimates of almost a billion barrels of shale oil in the world, or almost half of the available crude oil reserves. A major part of the shale reserves, estimated at over 2 trillion barrels, are in the US. Other countries, like Estonia, Brazil and China also have significant shale oil reserves. Technical issues and high cost are hindering expansion of shale oil, so we estimate that it will be the last of the fossils to go—after all oil is sucked out dry.

Another, still under development, technology is adding to the arsenal of potentially disruptive methods of energy production and use. It is processing of natural gas, via reforming processes, into other versatile products that can be used more efficiently in different areas of the economy.

Natural Gas Reforming

Natural gas, or methane (CH_4), is produced at huge volume all over the world, but due to the remoteness of some drilling sites, significant quantity of methane gas is allowed to escape into the atmosphere, or is burned onsite. This is not only a loss of energy, but it also causes serious environmental problems.

In order to eliminate the waste and reduce the GHG emissions from remote drilling sites and other natural gas facilities, a new, potentially disruptive technology can be used. It uses natural solar energy for natural gas reforming. Thus produced hydrogen gas has about 30-35% higher thermal content than the initial natural gas feed and can be stored, or used for power generation on site.

The commercial process of reforming methane gas is used at the refineries to produce hydrogen (H_2) via thermal processes. The major processes in the commercial conversion of natural gas to H_2 are steam-methane reformation, followed by a water shift reaction. Most hydrogen produced today in the U.S. is made via steam-methane reforming, in which high-temperature steam (700°–1,000°C) is used to produce hydrogen from natural gas.

In the commercial process, the incoming natural gas reacts with steam under pressure in the presence of a catalyst to produce hydrogen, carbon monoxide, and a relatively small amount of carbon dioxide (not reflected below).

$$CH_4 + H_2O \text{ (+heat)} \rightarrow CO + 3H_2$$

Subsequently, in a "water-gas shift reaction," the carbon monoxide and steam are reacted using a catalyst to produce carbon dioxide and more hydrogen.

$$CO + H_2O \rightarrow CO_2 + H_2 \text{ (+ heat)}$$

Finally, in an optional "pressure-swing adsorption" reaction, carbon dioxide and other impurities are removed from the gas stream, leaving essentially pure hydrogen. The advantages of this process are higher energy output, with less GHG emissions.

The major drawback of this method is the relatively high energy input required to initiate and drive the reforming and water-gas shift processes. But what if we replace the high heat input—presently provided by fossil fuels burning—with solar energy. And this is exactly what Los Alamos Solar Energy, LLC, in cooperation with Precision Solar Technologies Corp., both New Mexico companies, are trying to do.

High temperature solar energy is used, as in Figure 10-40, to provide the heat for the natural gas reforming. At the receiver of the solar dish, the gases mix and are converted to hydrogen, which can then be stored, burned onsite to generate electricity, or sent for purifica-

Figure 10-40. Solar dish with receiver at the focal point.

tion and compressing into vehicles fuel.

This new solar methane reforming technology avoids the use of outside energy (usually fossil fuels) by using solar energy for generating high temperature for the reforming reaction. Here, methane reforming is done in the receiver which allows the heat to be deposited in the interior of the device on a specially designed heat exchanger block.

$$2CO_2 + 2CH_4 = 4CO + 4H_2$$

The heat exchange block features very high heat exchange area with a near uniform temperature gradient between the directly heated body and the feed gas. It has also very high energy transfer rates because the heat does not have to be conducted through a wall, but is merely deposited on a top surface where it is expeditiously picked up by the gas.

The system also has a heat exchanger fastened to the boom (the support structure that holds the receiver in position in front of the dish) that transfers heat from the exiting product gas to the incoming feed gas, to increase efficiency.

This new reformed methane (hydrogen mostly) fuel can be then used for:

- Electric power generation at remote drilling sites,
- Electric power generation by conventional boiler/turbine gensets, or could be
- Further purified, compressed, and used as fuel for powering cars and trucks.

As energy and environmental issues grow in importance, so does the sophistication of the related equipment and processes. While the above setup might look too complex or expensive, it is much cheaper and useful than venting thousands of tons of raw natural gas in the atmosphere.

This technology is not the only solution to all our problems, but it offers solution to the problems of wasted energy and methane emissions from some drilling and other operations. Such solutions are what is needed to fill some of the present gaps in the energy production and GHG emissions.

The most promising of all energy technologies in the short term is the promise of recycling waste uranium.

Uranium Recycling

Recycling waste uranium is a truly exciting development in the energy sector. In addition to the fact that uranium is still available and is quite abundant in several areas of the world, it now can be recycled. This changes the energy and logistics equations significantly. Not many, if any, energy materials are recyclable, so if a material is recyclable, then it could fall in its own renewable category. Recycling uranium also eliminates the need of storing dangerous waste radioactive materials.

Unlike coal, where we burn a ton (1,000 kg) to generate power and end up with 950 kg. (95%) ashes, sludge, smog, and other pollutants to deal with for generations, uranium leaves only small amount of solid residue. The residue is packed neatly and stored (safely but temporarily) in special storage yards. It takes several thousand tons of coal to produce the equivalent of one ton of uranium, so uranium waste is of relatively small size.

Either way, waste is waste, and uranium waste brings an additional danger of nuclear radiation, which is very dangerous to all living things. Because of that, special and very expensive packaging and transport are need for every gram of uranium.

But Eureka...today we have the technology and know-how to recycle uranium.

Just think; instead of storing the spent uranium locked into special containers lined up in secure storage yards, we now can just send it to a recycling facility. The used uranium fuel comes out of the nuclear reactors in a hard ceramic form, most of which—95%—is still uranium, accompanied by one percent of other long-lived radioactive elements, called actinides. All of these materials can be processed and converted back into nuclear fuel that is as good as new. Only 4% of the entire nuclear materials are fission products, that are unusable and has to be disposed of.

The uranium recycling process used today, pyro-processing, begins by chopping the used nuclear (ceramic-metal) rods into small pieces and melting the metal chunks. The melt is cooled down and submerged in a vat of molten salts, with an electric current going through it. During this stage, the uranium and other reusable elements are separated via electrolysis, and can be easily shaped back into shape suitable for making fuel rods.

The waste materials are removed from the electrolytic vat and can be cast into stable (much smaller glass discs, which have to be put into permanent storage due to their radioactivity. This is the only waste left from the process, which remains radioactive a few hundred years. This is far less than the thousands of years that untreated used fuel needs to be stored for, and then these disks can be used again in the uranium waste recycling process.

Figure 10-41 shows the new category of energy materials, that opens new possibilities; that of recyclable uranium. Whatever the case may be, and whatever label we give this new development, it is truly remarkable in its significance and magnitude.

So why don't we use this process already, instead of storing hundreds of tons of nuclear waste every year all over the place? The main reason is that it is much cheaper to do what we are doing. Uranium ore is cheap, it is readily available, its processing is cheap, and the waste storage is covered by the governments. So why recycle?

Another plausible reason for limiting the uranium recycling is the fear that when many uranium recycling facilities start operating around the world, the risk of terrorists gaining access to nuclear materials will increase.

Recycling uranium could solve many of our energy problems and those of the future generations.

Be it as it may be, we know that by recycling we have unlimited amount of uranium, which could power the planet for many centuries. So, there is still hope for

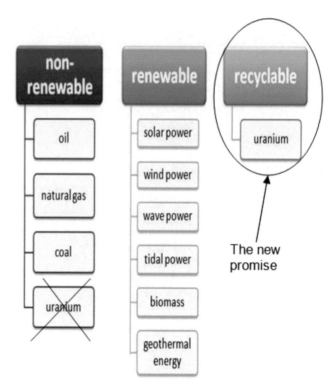

Figure 10-41. The new energy equation

us and the global energy market, provided that we are in full control of the nuclear reactors, and that no new Fukushima accidents would be allowed ever. This is a must! Else, it won't make any sense recycling material that will continue killing us.

OUR FUTURE ENVIRONMENT

This is another major issue of the energy markets, on which subject we wrote entire book titled, *Power Generation and the Environment*, published in 2014 by The Fairmont Press. In it, we dissect the relation between the energy sector and the global environment, so in this text we will limit our analysis to the essentials of today's energy markets and their effects on, and interaction with, the global environment now and in the future.

Current trends in energy supply and use are unsustainable—economically, environmentally and socially. Without decisive action, energy-related greenhouse-gas (GHG) emissions would lead to considerable climate degradation with an average 6°C global warming by 2050.

—IEA Technology Roadmap, 2014

One brief statement summarizes a century long

battle for obtaining fossil fuels and the consequences from their wide spread use. It also simply outlines a doomsday scenario.

According to IEA, *"We can and must change the path we are now on; sustainable and low-carbon energy technologies will play a crucial role in the energy revolution required to make this change happen."*

IEA gives us the solution—a global energy revolution driven by low-carbon technologies. A simple and logical solution, but is it even possible under the present global socio-political conditions?

Energy Efficiency, many types of renewable energy, carbon capture and storage (CCS), nuclear power and new transport technologies will all require widespread deployment if we are to achieve a global energy-related CO_2 target in 2050 of 50% below current levels and limit global temperature rise by 2050 to 2°C above pre-industrial levels.

The formula of achieving this success, according to IEA, is also quite simple: implementing energy efficiency, in addition to ramping up the use of renewable and carbon-limiting technologies.

While we all agree, the IEA Roadmap has an Achilles Heel:

*This will require significant global investment into de-carbonization, which will largely be offset by reduced expenditures on fuels. Nonetheless, this supposes an **important reallocation of capital**.*

Here is the proverbial Catch 22; important **reallocation of capital** is another way of saying, "Increased investment in technologies (solar, wind, bio-fuels and such) that have not been proven to guarantee reasonable return on investment, thus are considered risky." (Author's note) This uncertainty and inequality of renewables, as compared to the boom of profitable (albeit unsustainable and polluting) energy sources (coal, oil, and gas) of late is simply how capitalism works.

Yes, we all agree that IEA is saying the right words, giving us the right solution and the appropriate formula, and yet we prefer to put our money in the proven and profitable technologies—regardless of the consequences of their use.

Making a quick buck today is what the capitalist system and our economy are based on. The fossils are the preferred investment product and so they will continue to dominate the energy markets...pollution or not.

Nevertheless, the dialog has started and there is no stopping it, but the respective actions—the energy revolution, as IEA calls the necessary changes—will be delayed by several decades. Revolutions need time to organize and execute, after all.

A positive development of late is the new U.S.-China agreement to fight carbon emissions. After months of negotiations, the U.S. pledged deeper cuts in greenhouse-gas emissions, while China agreed for the first time to set a target for capping carbon emissions.

This groundbreaking agreement between the world's two biggest economies and biggest polluters is the first step towards a global agreement. Presidents Obama and Xi Jinping outlined the accord, which they hope to help push other nations to agree on a global environmental pact in 2015 during the planned global environmental summit in Paris.

"This is a major milestone in the U.S.-China relationship," Obama said at a news conference with Xi in Beijing. The two nations, which account for more than a third of the global greenhouse-gas emissions, have a "special responsibility" to lead efforts to address climate change, he said.

Obama seems serious in setting new targets for the U.S. to cut greenhouse gas emissions to 26 to 28% below 2005 levels by 2025. This is over 30% higher than the current U.S. target of 17% reduction below 2005 emissions by 2020.

China, as the world's largest greenhouse-gas emitter, agreed to reduce GHG emissions and increase its non-fossil fuel share of energy production to about 20% by 2030 and act on a carbon-dioxide cap soon. China has been taking steps to cut emissions, and plans to start a national carbon-trading market by 2016. China chose Shanghai, Beijing and Guangdong to set regional caps and institute pilot programs for trading rights as part of its initiative to cut emissions by as much as 45% before 2020 from 2005 levels.

Pollution in some large cities, including Beijing, reaches hazardous levels at times. To cut the haze (and reduce the embarrassment) during some special occasions, the government puts limits on the number of cars on the roads and restricted industrial production and construction for the duration of the events.

These temporary measures, however, are the only signs of GHG reduction in China. The new 5 year plan shows clear intent to increase energy generation by fossil fuels. And as a matter of fact, China is building several new coal-fired plants every week. These plants will increase the level of GHGs in the country well beyond and above the ability of any other measures to the contrary. And of course, air pollution does not remain in one place. As a consequence, the poisons emitted by China's coal-fired power plants will be affecting the entire global environment.

Obama's promises are also not a done deal. The majority of Republicans, who control the divided U.S. Senate, and are fighting the Obama administration every step of the way, are promising to fight the administration's efforts to cut GHG emissions from U.S. power plants. They are also looking into ways to reduce or eliminate the administration's pledge to raise $100 billion to help poor nations combat climate change.

And so, environmental talk is easy, and seem to be the politicians' favorite crutch, but actions are much more difficult to implement. We will have to wait and see what will happen in China and the U.S., but all indicators point to a long and difficult way towards significant GHG reductions.

Carbon Capture and Storage

Global energy-related CO_2 emissions continue to rise at a rate of 3-5% annually, reaching a record high every year. Somewhere between 35 and 45 gigatons per year (Gt/yr) are emitted in the air we breath. A gigaton is a million tons multiplied by one thousand, so we are talking about enormous amounts of smoke, particles, and poisons pumped up in the air above us daily.

If this trend continues, the GHG emissions will cause average global temperature increase of 6 °C by 2050.

The greater the emissions of greenhouse gases (GHGs), such as CO_2, the greater the warming and severity of the associated consequences. These consequences include a rise in sea levels, causing dislocation of human settlements, as well as extreme weather events, including a higher incidence of heat waves, destructive storms, and changes to rainfall patterns, resulting in droughts and floods affecting food production, human disease and mortality.

Global warming causes a chain of events, each prompting the next into action, with usually negative or damaging outcome.

Carbon capture and storage (CCS) is one of the solutions to the carbon dilemma. It is becoming increasingly important for countries like China, where the coal-fired mania is increasing. CCS will be a critical component in a portfolio of low-carbon energy technologies if and when governments undertake ambitious measures to combat climate change.

Given current trends of increasing global energy sector carbon dioxide (CO_2) emissions and the dominant role that fossil fuels (coal especially) continue to play in primary energy generation, the urgency of CCS deployment is increasing by the day.

Under the International Energy Agency (IEA) Energy Technology Perspectives 2012, 2°C Scenario (2DS), CCS could contribute one-sixth of CO_2 emission reductions required in 2050, or 14% of the cumulative emissions reductions between 2015 and 2050. This, compared to a business-as-usual approach, which would correspond to a 6°C rise in average global temperature by the same time.

The individual component technologies required for carbon capture, transport and storage are generally well understood and, in some cases, technologically mature. For example, capture of CO_2 from natural gas sweetening, or from hydrogen production, are technically mature and commercially practiced technologies, as is the transport of CO_2 gas by pipelines.

While safe and effective storage of CO_2 has been demonstrated, there are still many lessons to learn from some demo projects, and more effort is needed to identify viable storage (mostly underground) sites.

The major challenge for CCS deployment is the integration of the different components into large-scale (demo for now) projects.

Lack of understanding and acceptance of the technology by the public, as well as some energy and climate stakeholders, contributes to delays and difficulties in deployment. Governments and industry must ensure that the incentive and regulatory frameworks are in place to deliver operating CCS projects by 2020 across a range of processes and industrial sectors.

This would mean that all CCS projects in any stage of development today, reaching 100% operation by that time. Cooperation among governments should be encouraged to ensure that the global distribution of projects covers the full spectrum of CCS applications, and mechanisms should be established to facilitate knowledge sharing from early CCS projects.

One of the well hidden facts of the global heavy industry is that CCS is not only about electricity generation. Almost half (45%) of the CO2 captured between 2015 and 2050 would come from different types of industrial applications.

For this purpose, 25% to 40% of the entire global production of steel, cement, and chemicals producing plants must be equipped with CCS by 2050. Achieving this level of deployment in industrial applications will require capture technologies to be demonstrated by 2020, particularly for iron and steelmaking, and cement production.

Given their rapid growth in energy demand, the largest deployment of CCS will need to occur in non-Organization for Economic Co-operation and Development (OECD) countries. By 2050, non-OECD countries (with China and India leading the pack) will need to account for 70% of the total cumulative mass of captured CO_2.

China alone accounts for one-third of the global total of (to be) captured CO_2 between 2015 and 2050. OECD governments and multilateral development banks must work together with non-OECD countries to ensure that support mechanisms are established to drive deployment of CCS in non-OECD countries in the coming decades. Not an easy, but important and absolutely needed, task for keeping our air clean and healthy.

This decade is critical for moving deployment of CCS beyond the demonstration phase in accordance to the experts estimates. Mobilizing the large amounts of financial resources necessary will depend on the development of strong business models for CCS, which are so far lacking.

Urgent action is required from industry and governments to develop such models and to implement incentive frameworks that can help them to drive cost-effective CCS deployment. Actions which take future demand into account are needed to encourage development of CO_2 storage and transport infrastructure.

Obviously, looking at Figure 10-42, CCS is in its baby stages right now, but IEA has great things planned for it. Starting from measly 200 million tons per year (Mt/yr) in 2020, CCS—in IEA's wish cart—will expand to 2,200 Mt/yr by 2030 and whopping 8 Bt/yr by 2050. This is 8 billion tons, with a B, and great step in the fight for clean air. This is, however, still much lower (about 1/4) than the estimated 35-45 billion tons of annual GHG emissions around the world.

Judging from the way things are going today, this is simply an unrealistic pipedream. But IEA might know something we don't. In any case, we would prefer to think that the next generations will get much more interested and involved in climate corrections, which CCS is integral part of, and will come up with adequate technologies to complete the task at hand.

What We Need to Do?

According to IEA, the next several years are critical to the accelerated development of CCS technologies

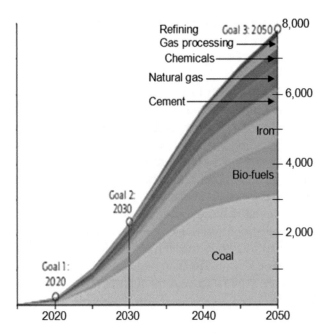

Figure 10-42. CCS goals for 2050 (in Mt/yr CO_2 captured and stored)

and procedures, as needed to achieve the low-carbon stabilization goals of limiting long-term global average temperature increase to maximum of 2 °C.

There are seven key actions, reflected below, which are necessary to be completed by 2020 in order to lay the foundation for scaled-up CCS deployment. They require serious dedication by governments and industry, but are realistic and cover all elements of the CCS process.

- Introduce financial support mechanisms for demonstration and early deployment of CCS to drive private financing of projects.
- Implement policies that encourage storage exploration, characterisation and development for CCS projects.
- Develop national laws and regulations as well as provisions for multilateral finance that effectively require new-build, base-load, fossil-fuel power generation capacity to be CCS-ready.
- Prove capture systems at pilot scale in industrial applications where CO_2 capture has not yet been demonstrated.
- Significantly increase efforts to improve understanding among the public and stakeholders of CCS technology and the importance of its deployment.
- Reduce the cost of electricity from power plants equipped with capture through continued technology development and use of highest possible efficiency power generation cycles.

- Encourage efficient development of CO_2 transport infrastructure by anticipating locations of future demand centers and future volumes of CO_2.

We agree with EIA's conclusions and action plan, and hope that CCS reaches the level predicted above. Even in best of cases, however, CCS alone is not the solution to our problems. Keeping in mind that the world emits 35-45 Gt/y of different GHGs, 8 Gt/yr is only a very small part, about 1/4 to be exact of the total poison pumped into the air above us.

So even if these somewhat incomplete (and thus far unrealistic) projections come to pass, we need much more than CCS to reduce the effects of global warming and the other climate change effects. And so the discussion on the matter, and the fight to a cleaner future, continue.

Carbon-less Future

In the fall of 2014 MIT issued a study that confirms our estimates that the GHG pollution will continue to increase this century. It is expected to double the 2010 levels by 2100, or 70-90 billion tons of GHGs will be added to the air we breath every single year.
At approximately the same time, most of the fossils will be added to the list of endangered species as well.

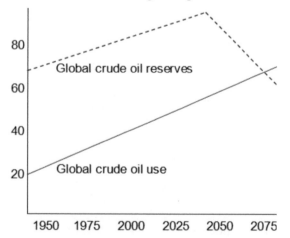

Figure 10-43. Global crude oil reserves vs. use

Figure 10-43 shows that we will most likely continue "discovering" more crude oil during the first part of the 21st century. This 'increase' of crude oil reserves will buy us more time to figure out what to do when oil is gone.

Then, the global oil reserves will start declining quickly as we pass the new (and inevitable) global oil peak. This is inevitable, because the world is using increasingly greater quantities of oil. The decline in global reserves would be exponentially fast with time, because it gets harder and harder to pump oil from the old oil reservoirs, as needed to satisfy the growing needs.

One of the problems is that the newly discovered oil reservoirs that require hydrofracking, usually do not hold much oil, and production gets much more difficult with time. And so, the new ells get depleted much faster than the conventional oil deposits, and the rigs are moved to a new location.

At the present 1:2 ratio of oil discovery vs. use, the U.S. world run out of oil by mid-century.

The entire global production would follow the U.S. example by the end of the century. Short of a miracle, or a drastic change in our habits (which also qualifies as a real miracle) crude oil will be history by New Year's day, 2100, and even long before.

So what is the mankind to do? Reduce the carbon, the specialists say. "Life without carbon," as the media calls it, is not possible, but life with reduced carbon is. It is becoming mandatory, if life on Earth is to continue as we know it. The carbon-less future will occur eventually no matter what we do, since the fossils will be depleted sooner or later. For this purpose, we better get ready, instead of getting caught with our pants down.

A future where electricity and vehicle fuels come from low-carbon, or carbon-free, sources is technically feasible. By gradually implementing it, we would reduce the raw material (fossils) demand, while significantly reducing air pollution. Large-scale solar and wind power generation is feasible and needed for our long-term survival. Combined with reasonable amou7nt of fossil energy, it will take us far into the future. This is what common sense calls us to do, so we don't need the professionals to tell us.

Case Study. Carbon-free Future

Recently, a study published in Proceedings of the National Academy of Sciences investigated this line of thinking. The researchers conducted the first-ever long-term life cycle assessment of the possible global implementation of electricity generation from renewable resources. "This is the first study that has assembled and scaled up the assessment of individual technologies to the whole world and assessed technology implementation to 2050, taking the environmental impacts of production into account," the researchers insist.

The study was done in order to clarify the energy and environmental costs of a widespread global shift to renewable energy technologies. Wind and solar power were considered as the primary energy sources, replacing the fossils. The effect of this materials and technologies shift on energy materials and the environment was then extrapolated.

The main question was, 'how a global shift to low-carbon energy systems would affect the other types of pollution and the world in large.' Previous research in this area have been focused on single issues, single projects, or isolated areas. Some of these studies looked at different energy sources and pollutants, the large land areas use problems, or the need for specialized metals. Since only segments of the puzzle were addressed, the interactions between the different technologies and approaches were not investigated.

And so, the researchers in this study developed an integrated hybrid life cycle assessment model, which allowed the integration of electricity produced by these prospective technologies back into the economic model.

The main subjects of the research were concentrating solar power, photovoltaics, wind power, and hydropower, on the renewables' side. Natural gas and coal-fired power plants with carbon capture and storage (CCS) were evaluated on the fossils side. An assumption was made that the efficiency of the production of raw materials and special metals, such as aluminum, copper, nickel, iron and steel, would improve over time.

Two different energy scenarios, developed by the International Energy Agency on the performance of renewable energy technologies, were used as follow.

- *The Baseline scenario* assumes global electricity production will increase by 134% by 2050. During this time the fossil fuels will remain the major electricity generation mix, accounting for about 75% of the total. Under this scenario, coal-based generation is 149% higher in 2050 than in 2007. It also accounts for 44% of all global power generation.
- *The BLUE map scenario* assumes that electricity demand in 2050 would be about 13% lower than in the Baseline scenario. This is due to a number of factors, such as increased energy efficiency, and reduced pollution as a result increased use of renewables, reduced use of fossils, and the implementation of carbon capture and storage (CCS) technologies.

One problem here is that low carbon technologies usually demand increased use of common and specialized raw materials per unit of power generation than conventional fossil fuel plants. Photovoltaic and wind power systems, for example use up to 40 times more specialized metals than fossil fuel power generators.

The increased demand for such materials is significant and must be included in all estimates and might be unsustainable.

The overall conclusion, however, is that a gradual, but steady, shift to renewable technologies will be beneficial, and absolutely necessary, for achieving the long-term goals of:
1. Reducing the fossils use,
2. Decreasing air pollution, and
3. Extending the life of the fossils.

The latter condition is needed for a smooth transition to a fossil-less energy future. Simple, logical, and true.

Now that we know the truth, all we have to do is act accordingly. Unfortunately, there are barriers in the way to progress in our capitalist society. Regardless of the facts, some of the serious barriers remain. In many, if not most, occasions, logic and reason are trumped by corporate profits and personal interests.

LNG

Liquefied natural gas (LNG) is the energy source of the future for many new applications in countries around the world. It is a significant part of our hope fro cleaner environment. It is actually natural gas that has been compressed into high pressure vessels, where it is in its liquid state. This way, many cubic feet of natural gas, are reduced down to several gallons of fuel.

- The greatest value of gas (LNG included) as an alternative fuel is that it is capable of generating cheaper power,
- It also is capable of replacing crude oil in the transportation sector—is undeniable and very attractive. Most automotive manufacturers are already making, or are planning to make, LNG fueled cars and trucks. The success of LNG as a vehicles fuel is already proven by the fact that many city busses in the U.S. and many other countries already run on LNG.
- The other equally important characteristic is the fact that it is much cleaner burning than coal, gasoline, or diesel, thus solving the environmental problems in addition to its other benefits.

For these reasons alone, natural gas for is becoming more relevant and viable as a clean, efficient, and cheap burning transportation fuel.

NG is the new vehicle fuel of choice; it replaces crude oil imports, while reducing GHG emissions

Across a number of industries LNG is also the preferred energy source, and so its use is increasing daily. The current abundant supply of natural gas is limited in its use by the existing infrastructure and by the reduced applications in the LNG field, for now at least.

The LNG market is segmented and divided among a few end users. The most important LNG market is that of private and mass transportation. In North America about 25% of all GHG emissions come from transportation vehicles, so truck and bus fleets are being converted to LNG. The technology is already here and proven successful and sustainable.

A larger-scale deployment is needed, for which LNG filling stations must be installed all over the country. LNG is the liquid form of natural gas and is the only way to use it in cars, trucks, and busses. Since this is a fairly new form of vehicle fuel, the number of filling stations is severely limited and will remain so regardless of the developments.

But there are many other markets for LNG, the most promising and needed of which are the remote mining and well drilling operations. These sites are huge consumers of diesel and other fuels needed to power their vehicles, pumps, and other equipment. All this equipment–electricity generators, hauling trucks, pimps, etc.-can run on LNG and so the manufacturers are seriously looking into it.

This is promising and equipment manufacturers are modifying theirs to run on LNG by 2017-2018. This is quite different from 5-6 years ago, when there were no such alternatives or promises. Diesel was the only option to produce electricity in remote locations, or to run a large load hauling truck.

Another potentially huge LNG market is the maritime transportation. The International Maritime Organization; a consortium of many countries, has imposed that the amount of sulfur emission from burning liquid fuel for ship transportation is to decrease 10-fold by 2020. Maritime companies must find ways to either buy fuel that has been cleaned of sulfur, or move to alternative fuel, such as LNG, to lower the sulfur emissions. That's a great opportunity for growth of the LNG market share in the U.S.

Nevertheless, multiple challenges remain to be tackled in the existing and potential LNG markets. To start with, permitting, pricing, and environmental assessments need to be handled in a way to achieve social acceptability, which is a key issue. The other major problem is lack of filling stations and other infrastructure as needed for proper delivery and dispensing of LNG fuel into vehicles. And, of course, investment, as with any

new technology is also lagging…for now.

The price of LNG still varies from vendor to vendor and from location to location. That has a great effect on the overall use of LNG fuels, and along with the other issues is an obstacle is getting investment, especially in the U.S. and Canada, for new LNG projects development.

But LNG is not a new technology, and there are very few unknowns about it. Mass LNG development is already under way in Europe and the Far East, and is taking a foothold in the U.S. and Canada too. But most importantly, we are not re-inventing the wheel here. LNG has been around for a long time, so we know that it is safe, efficient, and its cost-effective production and use are proven facts.

Recently, however, natural gas it a brick wall. It was the artificially low price of crude oil, forced on the world by OPEC, and Saudi Arabia in particular. The $40-50 prices forced the U.S. oil and gas drillers to reduce production, and some even went broke. This, however, is a temporary development, which the U.S. oil and gas industry will survive and from which it will emerge larger and better.

LNG Exports

While using LNG in the U.S. and Canada still has ways to go, for reasons we discuss above, it is accepted as a fuel of choice in many EU and Asian countries.

Gas Export Politics

The export of U.S. natural gas (usually as LNG) is one of the most urgent questions on the U.S. energy export agenda. It is surrounded by a number of controversies, which the U.S. politicians and regulators are trying to resolve. Recent congressional votes suggest that the tide may be turning on the issue, as the Energy Department authorizes more companies to sell the fossil fuel overseas. DOE recently also approved multiple applications to export natural gas to non-free trade agreement (FTA) nations, which might have unintended consequences in the near future.

Other major political developments of the gas exports thus far have been as follow:

- On Feb. 15, 2012, the House voted 173-254 to reject an amendment from then-Rep. Ed Markey, D-Mass. that would block oil and gas carried by the Keystone XL pipeline and resulting fuel products made from it from being exported. Twenty-four Democrats joined 230 Republicans in voting against the measure.

- On June 21, 2012, the House voted 161-256 to reject

a Markey amendment that would bar the export of oil and gas produced under new leases covered by an underlying bill. Thirty-one Democrats joined 225 Republicans in voting against the proposal.

- On July 25, 2012, the House voted 158-262 to reject a Markey amendment that would bar companies from exporting any gas resources produced from leases sold that would be sold under the underlying legislation. Thirty Democrats joined 232 Republicans in voting against the amendment.

- In November 2013, the House of Representatives voted 142-276 to reject a plan from Rep. Peter DeFazio, D-Ore., to block export overseas large quantity natural gas produced on public lands. According to DeFazio, foreign sales threaten a resurgence in domestic manufacturing, as companies move operations back to the United States to take advantage of cheap power supplies and feedstocks from surging U.S. natural gas production.

This is because many experts believe that by expanding the marketplace for natural gas harvested inside the U.S., exports could cause the domestic price to rise—but this has not been proven and specialists' opinions vary widely on just how, when, and how much.

Presently many manufacturing companies are bringing production back to the U.S. because of the plentiful and cheap natural gas, but if we begin to export it in great volume, then we are part of the international market, which might mean a dramatic increase in domestic natural gas prices. With this we will instantly lose our competitive advantage for domestic manufacturing.

On the other hand, hoarding U.S. natural gas is not something that will be agreed by most people. The experts believe that increasing the volume of natural gas exports would provide our allies with an alternative and reliable source of energy, helping to strengthen our economic and geo-political partnerships, while restraining U.S. natural gas exports would hurt our abilities to bolster strategic partnerships and create domestic jobs.

Ultimately, 56 Democrats joined 220 Republicans in voting against DeFazio's plan, while 140 Democrats and 2 Republicans voted for it. This is a slight uptick in the number of lawmakers who support energy exports, or at least don't want to ban them outright. This also show growing momentum in Congress for increased natural gas production and exports.

But the Energy Department has already approved several applications to sell natural gas to countries that do not have free-trade agreements with the United States, and increased the planned exports from the Free-

port LNG facility in Quintana Island, Texas.

Does all this mean that we are ignoring our energy and national security in favor of making a quick buck? The results of this development remain to be seen.

California Wows

California has introduced the most aggressive set of targets for emissions reduction ever, anywhere in North America. Governor Brown's executive order revises upward the state target for greenhouse gas (GHG) reduction. Now emissions must be cut down 40% by 2030 below the 1990 levels.

This puts California's climate goals in line with the EU projections, as a precursor for UN talks on climate change in Paris in 2015. The executive order states that climate change "poses an ever-growing threat to the well-being, public health, natural resources, economy, and the environment of California."

California already has in place its own Global Warming Solutions Act, which was enacted in 2006. Under it, the state must reduce GHGs to 1990 levels by 2020, a target the state is currently on track to meet. However the stricter measures would make it possible to reach the 203 target, and aim for 80% emissions reduction by 2050.

A reduction to 80% below 1990 levels is needed for the entire U.S., if we are to play our part in limiting global warming to 2^0C increase by 2030. The California goes even further than the emissions pledge made by President Barack Obama, who set targets for the federal government to reduce emissions by 40%, but from 2008 levels. This would be achieved by the increase of the amount of renewables in the federal energy mix to 30%. Some key federal government suppliers, including IBM and Honeywell, pledged to follow the President's lead.

California is also set to be 50% renewable-powered by 2030. In 2014 the state also extended funding for "critical" energy storage programs, geared to encourage adoption and improve bankability of energy (battery mostly) storage technologies.

Southern California, however, has been in the midst of a serious droughts during the last several years, which has reduced its hydropower generation. This and other factors are playing a major role in the state's energy production, so it is not clear how the new development will affect its future energy and environmental plans.

There are studies of late that suggest that increasing number of solar installations and energy storage systems, including combined distributed generators (large numbers of small-sized commercial or residential batteries) could help. Power generated by these systems in one part of the state could be able to assist the rest of the state in electricity generation and use, thus meeting the energy and environmental goals.

California has always led the country and the world in finding new ways of doing things, so this is not a surprising development. We do believe that the goals set by the State will be achieved…unless…

Unless…is the key word here. Many things could, and most likely will, get in the way to success. These vary from internal problems, like the persistent draughts and earthquakes that threaten to put the state on its knees, to external threats of terrorist actions leading to damage of infrastructure.

Paris 2015

The world is well aware of the problems and gathered at the UN Climate Change Conference, held in Paris in December 2015. This was the hope of the world (last ditch effort, some called it) for taking decisive measures in addressing and resolving the global climate warming.

Nearly 200 countries agreed that something must be done, and ratified a universal pact to slow global warming. Good step forward, which ended several decades-long political stalemate. So, the problem is now been officially recognized and addressed. Now all we have to do is resolve it. And here is where the problems start.

The Paris agreement calls for "rapid reductions" that are "in accordance with best available science." But it also offers a pair of "get out of jail free" cards to heavy polluters.

- The first is the chance to balance man-made emissions with "pollution sinks" such as growing many new trees.

- The second is the potential for new technology to somehow sucks emissions out of the sky, thus cooling the planet without slowing down the burning of fossil fuels.

The questions that these cards raise are, ' How many trees would have to be planted, in order to compensate for the deforestation and the ever increasing air emissions?' And also, 'What new technology, and how much of it, would be needed to make any significant difference?'

So, in the end, the greatest polluter, China, will keep doing what it is doing now until 2030, at which point it may, or may not, start reducing the emissions. The U.S., although showing improvement of late in this area, will also continue to be number 2 in polluting

the global atmosphere for the duration. Between them, these two countries emit more GHGs more than all other countries combined, and will continue doing so for the duration.

And finally, the agreement calls for $100 billion a year in climate finance for developing countries. It does not say who is going to pay this amount. Keeping in mind that only $10 billion has been made available thus far to the Fund, it remains to be seen how much will be raised every year. But even if it becomes reality, it is still not large enough, according to the experts. They have estimated that the cost of adaptation and mitigation will be as much as $800 billion by 2050.

According to one of the experts, the Paris conference was, '…like a very fat man announcing that he's going on a diet, but never changing a thing about the way he eats and lives.' Still much better than the previous conferences, where some delegates walked out in protest and some refused to sign.

So, with Paris as the foundation of a new world order, every five years, the nations of the world will return to the negotiating table, in an attempt to reduce GHG emissions. At that time they would evaluates the world's progress toward zero, or near-zero, net emissions. And all this has to be done within an "enhanced transparency framework for action and support,," which, unfortunately still does not exist.

Paris 2015 was only the beginning of a largely voluntary undertaking. This was the desire of the U.S. delegation, since its members knew very well that they could never get a binding treaty through Congress. We know from past experience that voluntary actions provide marginal results, so at any time, this "historic" agreement could be modified by any nation, and even come undone.

The U.S. Contribution

At the Paris Conference, President Barack Obama supported the negotiators geared to control climate change. "This one trend—climate change—affects all trends," Obama said. "This is an economic and security imperative that we have to tackle now." Well said, Mr. President, if it only was that easy to tackle something so big, hard, and expensive to deal with.

At the same time, U.S. congressional Republicans were doing their best to undermine him. That very same day, the House approved two resolutions aimed at blocking regulations to curb U.S. greenhouse-gas emissions:

- The first would bar the Environmental Protection Agency from enforcing rules aimed at cutting emissions from new power plants, and

- The second would prevent the agency from enforcing rules targeted at existing power plants.

Note: These rules are known as the Clean Power Plan, and they are crucial to the Americans' negotiating position in Paris. This Plan is the pledge the U.S. made in advance of the summit; to cut our emissions by 26% and donate $3 billion to the global Green Climate Fund.

The House votes, which followed Senate's approval of similar resolutions back in November, were explicitly aimed at subverting any agreements during the Paris talks. Our lawmakers sent a message to the climate conference in Paris that there's serious disagreement with the policies of this president—including climate change. They simply put politics before solving world's urgent problems

The new votes would have a minimal impact on the overall situation in the country, and Obama will veto any and all such resolutions, no doubt. But the message sent to the world has a great impact—the greatest polluter is big on words, but short on deeds and is not playing the game fairly.

And this is only the beginning. The congressional Republicans are threatening to block the $3 billion contribution President Obama made to what's known as the Green Climate Fund. The Fund is intended to help developing countries cope with climate change and also to adopt clean-energy systems. As a matter of fact, $3 billion is far less than the U.S.—one of the world's greatest polluters—should be contributing. This move leaves the fund under-financed, and increases the chance of the developing countries to walk away from, and not follow-up on, any proposed accord.

And things may get even worse for the global environment, since the batch of U.S. Presidents-to-be show ignorance on the subject. All received a low mark during a recent poll. "This individual understands less about science (and climate change) than the average kindergartner," according to a renown climate scientist," referring to one of the candidates, "That sort of ignorance would be dangerous in a doorman, let alone a president." This is how our future looks today.

So what can be expected from the U.S. in the future, as far as climate change is concerned? Not much of anything new, is the short answer. 'We have done enough, and will continue the status quo, regardless of the increasing environmental problems,' is the consensus. Similarly so, most of the developing countries are also not willing, or able, to reduce the use of fossils.

Summary

In this text we take a look at some technical, logistics, and financial aspects of the energy markets. We see a lot of good things going on around the world, along with a number of not-so-good such. We also made some predictions based on information available today. How true or not these might turn out to be, depends on many, many factors of national and international importance.

There are, however, several possible and probable scenarios that we need to be aware of, account for, and include in our future plans. The changing global socio-political and environmental conditions are causing unprecedented events, some with great consequences. The experts predict increase of the number and effects of these events, both natural and man-made. Some of these would change the way we live and do things.

The events that we must be aware of, and which are most likely to cause chaos in the global energy markets, and throw the world into despair are:

- ***Armed conflicts*** in Asia, the Middle East, Africa and other places continue and have been increasing in intensity. Atrocities committed by warring factions in these conflicts have killed thousands, displaced millions, and devastated entire regions. There is a possibility of these effects to increase and expand into other areas of the world.

- ***Terrorism*** is the new global cancer. It is spreading all over the world like forest fire, threatening everybody and everything that gets in its way. ISIS, the Taliban, Boko Haram, and others are vowing to destroy Western civilization, and any civilization, as we know it. As if this was not enough, we also deal with internal terrorists, who are planning to do as much damage as possible to people and infrastructure.

- ***Internal terrorism*** is the greatest danger to our country, because, as Abraham Lincoln said, 'America will never be destroyed from the outside. If we falter and lose our freedoms, it will be because we destroyed ourselves.' Lone wolves, clothed in American citizenship, have been trying to harm us, and will continue to do so in the future. The energy, transport, banking, and other important sectors are the targets and we must be ready to face the increasing threats.

- ***Increasing global warming*** is changing the climate in many areas. We see severe storms, hurricanes, and tornadoes hurting people and property. Extended draughts are hurting local agriculture, and flooding of low lying areas are forcing people to leave their homes in search of higher ground. The changing climate is causing a number of other changes, some of which have dramatic negative effect.

- ***Ocean acidification*** is increasing quickly and threatens the oceans flora and fauna. We have been witnessing lately the changes in the coral infrastructure and maritime population. If these changes continue at the present day pace, our oceans would be deserts by the turn of the century.

- ***Desertification and deforestation*** activities are increasing and are becoming way of life in many areas of the world. These conditions are depriving us and the future generations from valuable agricultural and wild life lands and forests.

- ***Earthquakes*** (natural and man-made) are increasing in number and strength as well, and we should ignore the real threat of strong earthquakes hitting large population centers like San Francisco, Los Angeles, Tokyo and many other heavily populated areas.

- ***Nuclear accidents*** are still happening; with Fukushima as the latest confirmation of the fact that we still don't know how to make nuclear 100% safe. Until we do, the threat of mass devastation, a la Fukushima hangs over the regions and people near nuclear power plants.

- ***Depletion of the fossils***, however, will be the most dramatic, irreversible, and permanent devastation human kind has ever experienced. This is not IF, but WHEN scenario. The change will be gradual; with rapidly increasing prices and decreased amount of fossil fuels available for daily use. This anomaly would bring chaos to the global energy markets, and drastic changes to our way of life. The conflicts among countries will increase as the competition for more and cheaper fossils escalates into armed confrontations.

Life without electricity and vehicle fuel is unimaginable. It is a real threat to our lifestyle, which we can avoid.

Amazingly enough, we all see the problems at hand, and even know the ways to eliminate them. We

have even increased the volume of the discussions on the matter, but we are not trying. Or if we are trying, it is not hard enough to make the badly needed difference.

We cannot prevent the gradual disappearance of the fossils, yes, but we can ensure an orderly transition to other energy sources. In order to do that in orderly manner, we must first reduce the use of the fossils. In parallel with that effort, we must accelerate the development of new renewable energy technologies. And lastly, and very importantly, we must figure ways to start replacing the fossils with these newly developed renewable technologies.

Unfortunately, most of us (including entire countries) ignore the grim reality of increasing global pollution and natural disasters, and the even grimmer possibility of life in fossil-less future.

What shall we do? This is the big question that we must answer soon, or else!

In Conclusion

The world's population was less than a billion people in the 19th century. The only energy source these people had at the time was that coming from the sun, which was used mostly to grow and process food crops.

Then came the fossils, which allowed unprecedented progress in all areas of our lives, and so the population doubled, tripled, and is now over 6 times what it was just over a century ago. The fossils use increased proportionally with the population increase.

Within the last century alone, we used more than half of the available global fossil reserves, and there are no solid plans to reduce their use any time soon.

We also increased the environmental pollution to where we are now choking and swimming in our own waste.

This leaves the future generations with two big problems: depleted fossils reserves, and polluted environment.

The renewables are not catching up fast enough and will not be able to replace the amazing power and flexibility of the fossils. On top of that, manufacturing and maintaining renewable energy sources requires huge amounts of fossil fuels, so they would have a hard time staying alive.

The renewables cannot be counted on to substitute the fossils in a fossil-less world.

As the fossil reserves get further depleted, the way of life on Earth will gradually and permanently change. If and when all fossils are gone, the world—living on raw solar energy as the only energy source—would not be able to support more than a billion people for very long.

It is very likely that in the absence of fossils, life on Earth would revert to that of the 19ᵗʰ century

The energy markets would become a history then...

On the other hand, the U.S. is well known for, and is where it is because of the inherited ingenuity, entrepreneurship, and plausible work ethics of our people. There have been many times in the past when serious threats to our wellbeing have been resolved by focusing on these inherited qualities.

As our good friend Churchill said, 'You can always count on Americans to do the right thing—after they've tried everything else.' How true! So for now we are still in the 'trying everything first' stage. After that we will start the real work.

The pending energy and environment problems facing the world are perhaps the most serious ever, but, unlike many times before, they are not hidden, nor are they a big surprise. This then gives us enough time to figure out how to handle them.

And so, although we don't see any serious action coming our way now, or even decades from now, a time will come when all efforts will be focused on solving the major issues of the day. Perhaps as late as the end of this century, all-hands-on-deck approach will be designed and implemented to solve the fossils depletion and environmental pollution issues.

The energy markets would be drastically changed and a new book will be written then, making this one obsolete.

Notes and References

1. Siemens, http://www.nrdc.org/energy/frackingrisks.asp
2. Future Energy Solutions, http://www.future-energy-solutions.com/
3. Green Growth, http://www.thetimes.co.uk/tto/public/greengrowth/article3209304.ece
4. Future Electronics, http://www.futureelectronics.com/EN/COMPANY- INFORMATION/ABOUT- FUTURE-ELECTRONICS/Pages/FutureEnergySolutions.aspx
5. Energy Solutions, https://energy-solution.com/project/distributor-hvac-program/
6. Energy Collective, http://www.theenergycollective.com/gail-tverberg/334631/forecast-our-energy-future-why-common-solutions-don-t-work
7. Union of Concerned Scientists, http://www.ucsusa.org/our-work/energy/smart-energy-solutions/smart-energy-solutions-increase-renewable-energy#.VlpGmTGFM1I
8. Future Technology, http://www.alternative-energy-news.info/technology/future-energy/
9. CEF, http://www.conserve-energy-future.com/causes-and-solutions-to-the-global-energy-crisis.php
10. Photovoltaics for Commercial and Utilities Power Generation, Anco S. Blazev. The Fairmont Press, 2011
11. Solar Technologies for the 21st Century, Anco S Blazev. The Fairmont Press, 2013
12. Power Generation and the Environment, Anco S. Blazev. The Fairmont Press, 2014
13. Energy Security for the 21st Century, Anco S. Blazev. The Fairmont Press, 2015

Appendix

Abbreviations and Glossary

ABBREVIATIONS

′	Minute
	Foot
″	Second
	Inch
α	Angular acceleration
	Alpha rays
α-Si	Amorphous silicon
β	Velocity in terms of the speed of light c (unit-less)
	Beta rays
γ	Lorentz factor
	Gamma ray
	Photon (sunlight)
	Shear strain
	Heat capacity ratio
Δ	A change in a variable
Δt	Temperature change
μc-Si	Micro-crystalline silicon
π	Energy efficiency
	Coefficient of viscosity
λ	Wavelength
ν	Frequency
Σ	Summation operator
σ	Electrical conductivity

A

a	Acceleration
AAO	Antarctic Oscillation
ABE	Acetone-butanol-ethanol (process used in the production of butanol)
ABMA	American Boiler Manufacturers Association
ACC	Arizona Corporations Commission Air Combat Command
ACCS	Assured Combinable Crop Scheme
ACE	Area Control Error
ACER	Agency for the Co-operation of Energy Regulators
AD	Anaerobic digestion
ADB	Asian Development Bank
AEE	Association of Energy Engineers
AEGIS	State-of-the-art radar and missile system
AEP	American Electric Power

AEO	Annual Energy Outlook
AEST	Advanced Energy Storage Trial
AETC	Air Education and Training Command
AEZ	Agro-ecological zoning
AFB	Air Force Base
AFCEE	Air Force Center for Engineering and the Environment
AFCESA	Air Force Civil Engineer Support Agency
AFFEC	Air Force Facility Energy Center
AFIT	Air Force Institute of Technology
AFMC	Air Force Material Command
AFRETEP	African Renewable Energy Technology Platform
AFSO 21	Air Force Smart Operations for the 21st Century
AFSPC	Air Force Space Command
AFV	Alternative Fuel Vehicle
Ag	Silver
AGEB	AG Energiebilanzen e.V.
AGO	Australian Greenhouse Office
AHJ	Authority having jurisdiction
AIC	Amps interruption current
ALCC	Life Cycle Cost Annualized
AMCA	Air Movement and Control Association
AMM	alternative methods and materials
ANL	Argonne National Laboratory
ANGB	Air National Guard Base
AO	Arctic Oscillation
AOGCM	Atmosphere-Ocean General Circulation Model
APCO	Association of Public-Safety Communications Officials International
API	American Petroleum Institute
Ar	Argon
ar	as-received (coal)
AR4	Fourth Assessment Report (of the IPCC)
ARAR	Applicable or Relevant and Appropriate Requirements
ARE	Energy Market Agency, Poland
ARRA	American recovery and reinvestment Act
As	Arsenic
ASHRAE	American Society of Heating, Refrigerating and Air Conditioning Engineers
ASME	American Society of Mechanical Engineers

ASN(I&E)   Assistant Secretary for Installations and Environment
ASQ   American Society for Quality
ASSE   American Society of Safety Engineers
ASTM   American Society for Testing and Materials International
Au   Gold
AU$ Australian dollars
AWEA   American Wind Energy Association

B
B   Boron
BA   Balancing Authority
BASF   Badische Anilin- und Soda-Fabrik, Ludwigshafen, Germany
bbl   Barrels
bbl/d   Barrels per day
BBTU   Billion British thermal units
BC   British Columbia
Bcf   Billion cubic feet (natural gas)
Be   Beryllium
BESS   Battery energy storage system
BIG   Biomass integrated gassifier
Bio-SG   Bio synthetic gas (also referred to as bio synthetic natural gas)
BLM   Bureau of Land Management (USA)
BMDS   Ballistic Missile Defense System
BMS   Battery management system
B-NICE   Biological, nuclear, incendiary, chemical, explosive
BOE   Barrels of oil equivalent
BOP   Balance of Power
BOS   Balance of system
BPA   Bonneville Power Administration
Br   Bromine
BRAC   Base Realignment and Closure
Brent   North Sea Brent
BS   British Standard
BSI   British Standards Institution (UK)
BTL   Biomass-to-liquids
BTU   British thermal unit
BUMED   Bureau of Medicine and Surgery

C
c   Speed of light (300 million meters per second)
C   Carbon
    Celsius (degrees)
Ca   Calcium
CAAA   Clean Air Act Amendments of 1990 (US)
$CaCO_3$ calcium carbonate (limestone)
CCS   carbon (dioxide) capture and storage
CEA   Central Electricity Authority (India)

CAGI   Compressed Air & Gas Institute
CAFE   Corporate average fuel economy
CARB   California Air Resources Board
CAFE   Corporate average fuel economy
CAIR   Clean Air Interstate Rule
CBG   Cleaner Burning Gasoline
CBM   Coalbed Methane
CBRNE   Chemical, biological, radiological, nuclear, explosive
CC   Combined Cycle
CCl_4 Carbon tetrachloride
CCS   Carbon capture and storage
CCSM   Community climate system model
Cd   Cadmium
CdTe   Cadmium telluride
CDC   Central Distribution Center
    China Datong Corporation
CDD   Consecutive dry days
CDM   Clean development mechanism
CDP   Carbon Disclosure Project
CEC   California Energy Commission
CEGB   Central Electricity Generating Board (UK)
CEN   Comité Européen de Normalisation
CEM   Certified Energy Manager
    Clean Energy Ministerial
    Continuous emissions monitoring
CEMT   DFAS Corporate Energy Management Team
CEQA   California Environmental Quality Act.
CER   Certified emission reduction
CERCLA   Comprehensive Environmental Response, Compensation and Liability Act of 1980
CES   Civil Engineering Squadron
    Community Energy Storage (Unit)
CEU   Continuing Education Unit
CEV   Controlled environmental vault
CFC   Chlorofluorocarbons
CFD   Computational fluid dynamics
CFE   National electric utility, Mexico
CFL   Compact fluorescent light
CGA   Compressed Gas Association
CH_2Cl_2 Methylene chloride
CH_3Br Methyl bromide
CH_4 Methane
CHP   Combined heat and power
CIAB   Coal Industry Advisory Board
CIGS   Copper-indium-gallum-selenide
CIKR   Critical Infrastructure and Key Resources (Homeland Security)
cm^2 Square centimeter
CMA   Court of Military Appeals
$CMIP_3$ Coupled Model Inter-comparison Project 3
CMM   Coal Mine Methane

CNG	Compressed natural gas
CNIC	Commander, Navy Installations Command
CNOOC	China National Offshore Oil Corporation
Co	Cobalt
CO_2	Carbon dioxide
CO_2e	Carbon dioxide equivalent
COBRA	Chemical, ordinance, biological, radiological agents
CONUS	Contiguous United States
COP	Conference of Parties
COPD	Chronic obstructive pulmonary disease
CPI	Climate Policy Initiative
CPIC	China Power Investment Corp
CPUC	California Public Utilities Commission
COR	Contracting Officer Representative
CPV	Concentrating photovoltaics (solar PV)
CQEIG	Chongqing Energy Investment Group
Cr	Chromium
CRA	Climate risk assessment
CRM	Climate risk management
Cs	Cesium
CSA	Canadian Standards Association
CSAPR	Cross-State Air Pollution Rule
CSIRO	Commonwealth Scientific and Industrial Research Organization
CSP	Concentrating solar power (thermal solar) Current source converter
CSR	Codes, standards and regulations
CSPGC	China Southern Power Grid Company
CT	Combustion Turbine
CTF	Clean Technology Fund
CTL	Coal-to-liquids
Cu	Copper
CV	Calorific value (also known as heating value)
CWA	CELENEC Workshop Agreement

D

d	Diameter
DASA(E&P)	Deputy Assistant Secretary of the Army for Energy and Partnerships
DASA(I&H)	Deputy Assistant Secretary of the Army for Installations and Housing
DASN(I&F)	Deputy Assistant Secretary Navy for Installations and Facilities
DBCP	Di-bromo-chloro-propane
DCB	Dichlorobenzene
DCMA	Defense Contract Management Agency
DDC	Direct Digital Controls
DDGS	dried distiller's grains with solubles
DDT	An environmentally-persistent insecticide banned for most uses by the U.S. EPA in 1972.
DeCA	Defense Commissary Agency

DECC	Department of Energy and Climate Change (UK)
DEFRA	Department for Environment, Food and Rural Affairs (UK)
DeCAH	Design Criteria Handbook
DESERTEC	"Clean Power from Deserts" foundation
DFAS	Defense Finance and Accounting Service
DFD	Defense Facilities Directorate (WHS)
DHHS	Department of Health and Human Services (USA)
DHS	Department of Homeland Security (USA)
DI	Deionized water
DIA	Defense Intelligence Agency
DIAC	Defense Intelligence Analysis Center
DJF	December, January, February
DLA	Defense Logistics Agency
DME	Di-methyl-ether
DoA	Department of the Army
DoAF	Department of the Air Force
DOC	Department of Commerce (USA)
DoD	Department of Defense
DOD	Depths of Discharge
DOE	Department of Energy (USA)
DOI	Department of Interior (USA)
DOJ	Department of Justice (USA)
DoN	Department of Navy
DOT	Department of Transportation (USA)
DRE	Destruction and removal efficiency
DR	Demand Response
DTSC	Department of Toxic Substances Control
DUSD(I&E)	The Deputy Under Secretary of Defense (Installations and Environment)

E

E	Energy
E85	85 percent ethanol fuel
E/P	Energy/rated power
EACC	Economics of adaptation to climate change
EASA	Electrical Apparatus Service Association, Inc.
ECD	Estimated Completion Date
ECIP	Energy Conservation Investment Program
ECMWF	European Centre for Medium-Range Weather Forecasts
ECPG	East China Power Grid
EDF	Electricite De France
EEE	Electronic equipment enclosure
EH	Euler Hermes
EIA	Environmental impacts assessment Energy Information Administration
EIB	European Investment Bank
EISA	Energy Independence and Security Act
EJ	Exajoule

EKF	Eksport Kredit Fonden (Denmark)		FERC	Federal Energy Regulatory Commission (United States)
EMC	Electromagnetic compatibility		FES	Facility Energy Supervisor
EMCS	Energy Management Control Systems		FGD	Flue Gas Desulfurization
EMSG	Energy Management Steering Group		FFV	Flex-fuel vehicle (multi-fuel use)
ENSO	El Niño-Southern Oscillation		FID	Final investment decision
ENTSO-E	European Network of Transmission System Operators for Electricity		FiP	Feed-in premium
			FiT	Feed-in tariff
EO	Executive Order		FiP	Feed-in premium
EOR	Enhanced oil recovery		FMEA	Failure Mode and Effects Analysis
EPA	Environmental Protection Agency (USA)		FMECA	Failure Mode, Effects and Criticality Analysis
EPAct	Energy Policy Act		FOB2	Federal Office Building #2 (Navy Annex)
ERCOT	Electric Reliability Council of Texas		FSU	Former Soviet Union
EPCRA	Emergency Planning and Community Right-to-Know Act		FT	Fischer-Tropsch
			FTA	Fault Tree Analysis
EPEAT	Electronic Products Environmental Assessment Tool		FTC	Federal Trade Commission
			FY	Fiscal Year
EPRI	Electric Power Research Institute			
EPS	Electric power system		**G**	
EPT	Eagle Picher Technologies		g	Standard gravity acceleration
ERA-40	ECMWF 45 year analysis of the global atmosphere and surface conditions (1957-2002)		G2V	Grid to vehicle
			G8	Group of Eight
ESA	Energy Storage Association		Ga	Gallium (semiconductor)
ESC	Engineering Service Center		GaAs	Gallium Arsenide (semiconductor)
ESCO	Energy Service Company		GAC	Granular activated carbon
ESIC	Energy Storage Integration Council		GBEP	Global Bioenergy Partnership
ESMAP	Energy Sector Management Assistance Program		GBP	Great Britain pound
			GC	Grand Coulee Gram of coal equivalent
ESPC	Energy Savings Performance Contract		GCI	Global Change Institute
ESPP	Energy Savings Performance Program		GCM	General Circulation Model
ESS	Energy storage systems		GCOS	Global climate observing system
ETP	Energy Technology Perspectives		GCV	Gross calorific value
ETSD	Engineering and Technical Services Division		GDP	Gross domestic product
EU	European Union		Ge	Germanium
EUC	Equipment under control		GEF	Global Environment Facility
EUMETNET	European National Meteorological services		GET-CCA	Global Expert Team on Climate Change Adaptation (of the World Bank)
EUR	Euro Estimated ultimate recovery		GFCS	Global Framework for Climate Services
EV	Electric vehicle		GHG	Green house gas (effect)
EWEA	European Wind Energy Association		GIZ	German Development Co-operation Agency
EWI	European Wind Initiative		GJ	Gigajoule
			GMI	Global Methane Initiative
F			GNP	Gross national product
f	Frequency, in Herts or s^{-1}		GPC	Government Purchase Card
F	Fluorine Fahrenheit (degrees)		GPI	Genesis Potential index
FAME	Fatty acid methyl ester		GSA	General Services Administration
FAO	Food and Agricultural Organization		GSF	Gross Square Feet
FAT	Factory acceptance testing		GSHP	Ground Source Heat Pump
FDA	Food and Drugs Administration		Gt	Giga ton
Fe	Iron		GTCO$_2$eq	Giga tons of carbon dioxide equivalent
FEB	Field evaluation body		GTL	Gas-to-liquids
FEMP	Federal Energy Management Program		GTS	Global Telecommunication System

GW	Giga-watt
GWEC	Global Wind Energy Council
GWh	Giga-watt-hour
GWP	Global warming potential

H

h	Height
H	Enthalpy
	as in pH, measure of strength of acidic or basic aqueous solution.
H_2	Hydrogen
H_2O	Water
H_2SO_4	Sulfuric acid
HCFC	Hydro chlorofluorocarbons
HCL	Hydrochloric acid
HDD	Heating degree days
HDI	Human development index
HDV	Heavy-duty vehicle
HEAT	Hands-on Energy Adaptation Toolkit
HF	Hydrofluoric acid
HFC	Hydrofluorocarbons
Hg	Mercury
HGL	Hydrocarbon gas liquids
HHI	Herfindahl-Hirschman Index
HHV	Gross calorific value
HIG	Henan Investment Group
hi-Ren	High renewables (Scenario)
HP	Horse power
HQ	Headquarters
HQCC	Headquarters Command Complex (MDA)
HQDA	Headquarters Department of the Army
hr	Hour
HSPD	Homeland Security Presidential Directive
HTJ	Heterojunction
HVAC	Heating, Ventilating, and Air Conditioning
	High-voltage alternating current
HVDC	High-voltage direct current
HYBLA	Hybla Valley Office Building (WHS)
HVO	Hydrotreated vegetable oil
HWIL	Hardware in the Loop
Hz	Hertz, unit of frequency (one cycle per second)

I

I	Current (in Amps)
	Iodine
IA	Implementing agreement
IAEA	International Atomic Energy Agency
IAPMO	International Association of Plumbing & Mechanical Officials
IAS	International Accreditation Service
IATA	International Air Transport Association
IBC	Inter-digitated back contact

IBEP	International Bio-energy Platform
ICAO	International Civil Aviation Organization
IBC	International Building Code
I-Codes	International Codes
ICC	International Code Council
ICBM	Intercontinental Ballistic Missile
ICPD	International Conference on Population and Development
IEA	International Energy Agency
IEC	International Electrotechnical Commission
IEEE	Institute of Electrical and Electronic Engineers
IESP	Infrastructure Energy Strategic Plan
IEO	International Energy Outlook
IFAD	International Fund for Agricultural Development
IFC	International Finance Corporation
IFES	Integrated Food and Energy Systems
IFI	International financial institution
IFRC	International Federation of Red Cross
IGCC	Integrated combined cycle plant
IGO	International government organizations
IIASA	International Institute for Applied Systems Analysis
IIREA	Irena International Renewable Energy Agency
ILUC	indirect land-use change
IMC	International Mechanical Code
IMCOM	(Army) Installation Management Command
In	Inch
I/O	Input/output
IPC	International Plumbing Code
IPCC	Intergovernmental Panel on Climate Change
IPPC	Integrated Pollution Prevention and Control
IPPF	International Planned Parenthood Federation
IRC	International Residential Code
IRENA	International Renewable Energy Association
ISCC	International Sustainability and Carbon Certification System
ISE	Interconnection system equipment
ISM	industrial, scientific, and medical
ISO	International Standards Organization
ISSA	Inter-Service Support Agreement

J

J	Joule
JJA	June, July, August
JPR	Job performance requirements
JRC	Joint Research Centre
IRR	Internal rate of return

K

K	Potassium
Kelvin	(unit of temperature)

kcal	Kilocalories	Mcf	Million cubic feet (natural gas)
km	Kilometer	MCL	Maximum contaminant level
km²	Square kilometers	MCLB	Marine Corps Logistics Base
kW	Kilowatt	MDA	Missile Defense Agency
kWh	kiloWatt-hour	MDA/DOH	MDA Office of Human Resources
kWh/kW/y	Kilowatt hour per kilowatt and per year	MDMS	Meter Data Management System
kWh/m²/y	Kilowatt hour per square meter and per year	MEA	Multilateral Environmental Agreement
		MEDCOM	Medical Command (DoA)

L

l	Length	MEK	Methyl ethyl ketone
LCA	Life-cycle assessment	MELs	Miscellaneous electric loads
LCC	Life cycle cost	MENA	Middle East and North Africa
LCCA	Life-Cycle Cost Analysis	MEO	Most Efficient Organization
LCOE	Levelized cost of electricity	MEP	Ministry of Environmental Protection
LCPV	Low-concentrating photovoltaic	MESA	Modular Energy Storage Architecture
LDCF	Least Developed Countries Fund	Mg	Magnesium
LDE	Liter of diesel equivalent (energy content 36.1 MJ/litre)		Milligram
		mg/g	Microgram per gram
LDV	Light-duty vehicle	mg/kg	Microgram per kilogram
LED	Light Emitting Diode	mg/l	Microgram per liter
LEED	Leadership in Energy and Environmental Design	mi²	Miles square
		MHA	million hectares
LFD	Lease Facilities Division (WHS)	MILCON	Military Construction
LFF-1	USMC, Facilities and Services Division Facilities Branch	min.	Minute
			Minimum
LER	Light electric rail	MJ	Megajoule
LESTA	Large energy storage test apparatus	mm/yr	Millimeter per year
LEV	Light electric vehicle	MMA	March, April, May
LGE	Liter gasoline equivalent (energy content 33.5 MJ/litre)	MMBtu	Million British Thermal Units
		Mo	Molybdenum
LHPP	Large hydropower	MPa	Megapascal
LHV	Lower heating value, or net calorific value	mpg	Miles per gallon
Li-ion	Lithium-ion	MRI	Meteorological Research Institute, Japan
LMP	Locational marginal price	m/s	Meters per second
LNG	Liquefied natural gas	MSDS	Material Safety Data Sheets
LRS	Load and Resource Subcommittee	mt	Metric ton
LPG	Liquefied natural gas (Propane)	MTBE	Methyl tertiary-butyl ether
LSS	Lean Six Sigma	Mtoe	Million tons of oil equivalent
lt	Long ton	mW	Milliwatt
LTC	Lithium Technology Corp	MW	Megawatt (1,000 kW)
LUC	Land-use change	MWh	MegaWatt-Hour, 1 million Watt-hours
LULUCF	Land use, land use change and forestry		

M

N		
m	Mass	n	number of observations or replicates in a statistical sample
	Meter		
m³/year	Cubic meters per year	N	Nitrogen
M&V	Measurement & Verification	Na	Sodium
MAJCOM	Major Command	NAATBatt	National Alliance for Advanced Technology Batteries
MARPOL	Marine pollution	NABEG	National Grid Expansion Acceleration Act (Germany)
MATS	Mercury and Air Toxics Standards	NAO	North Atlantic Oscillation
mc-Si	Multi-crystalline silicon	NAM	Northern Annular Mode

NAPA	National adaptation plan of action
NaS	Sodium-sulfur
NATO	North Atlantic Treaty Organization
NAVFAC	Naval Facilities Engineering Command
NAVSTA	Naval Station
NAWS	Naval Air Weapons Station—China Lake
NBBPVI	National Board of Boiler and Pressure Vessel Inspectors
NBC	Nuclear, biological, chemical
NBIC	National Board Inspection Code
NCS	National Climate Service
NCV	Net calorific value
NDAA	National Defense Authorization Act
NDRC	China National Development and Reform Commission
NEA	National Energy Administration
NEC	National Electrical Code
NECA	National Electrical Contractors Association
NEM	Net energy metering
NEMA	National Electrical Manufacturers Association
NERC	North American Electric Reliability Corporation
NESC	National Electrical Safety Code
NETCEN	National Environmental Technology Centre (UK)
NETL	National Energy Technology Laboratory (US)
NIST	National Institute of Standards and Technology (US)
NFPA	National Fire Protection Association
NGA	National Geospatial-Intelligence Agency
NGO	Nongovernmental organization
NGPL	Natural gas plant liquids
NGV	natural gas vehicle
NH	Naval Hospital
NiCd	Nickel cadmium (battery)
NIPP	National Infrastructure Protection Plan (Homeland Security function)
NIST	National Institute of Standards and Technology
NMC	Naval Medical Center
NMHS	National Meteorological and Hydrological Services
NNI	Net national income
NNP	Net national product
NOx	Nitrogen oxides
NOOA	National Oceanic and Atmospheric Administration
NPDES	National Pollutant Discharge Elimination System
NPL	National Priorities List
NPP	Net Primary Production
NPV	Net present value

NREL	National Renewable Energy Laboratory
NSA	National Security Agency
NSCOGI	North Sea Countries' Offshore Grid Initiative
NSW	New South Wales
nW	Nanowatt
NWIP	New Work Item Proposal
NWPP	Northwest Power Pool

O

O_2	Oxygen
O_3	Ozone
O&M	Operation and maintenance
OCPD	Overcurrent protective device
OE	Office of Electricity Delivery and Energy Reliability (DOE)
OECD	Organization for Economic Co-operation and Development
OMB	Office of Management and Budget
OPEC	Organization of Petroleum Exporting Countries
OPIS	Oil Price Information Service
OSHA	Occupational Safety and Health Administration
OSP	Outside plant

P

P	Phosphorous
	Power (as in Watts)
	Pressure (as in lbs/in^2)
PADD	Petroleum Administration for Defense District
P/E	Power to energy
Pa	Pascal (unit of pressure)
PACAF	Pacific Air Forces
PADD	Petroleum Administration for Defense Districts
PAH	(or PNA) Polynuclear Aromatic Hydrocarbons,
Pb	Lead
PBMO	Pentagon Building Management Office
PC	Pulveriszd coal
PC1	Planning case
PCS	Power conversion system
PCB	Polychlorinated biphenyls
PCE	Perchloroethylene
PCP	Pentachlorophenol
PCS	Power Conversion System
PEARL	Professional Electrical Apparatus Recyclers League
PENREN	Pentagon Renovation Office
PG&E	Pacific Gas & Electric
PGMA	Portable Generator Manufacturers Association

pH	Level of acidity/alcalinity	RED	Renewable Energy Directive
PH	Pumped hydroelectric	REM	Resource Efficiency Manager
PHEV	Plug in hybrid electric vehicles	REN21	Renewable Energy Network for the 21st Century
PH&RP	Pentagon Heating & Refrigeration Plant		
PHS	Pumped hydroelectric storage	REO	Rare earth oxide
PM	Program Management	RES	Renewable energy source
	Periodic Maintenance	RESS	Renewable energy storage system
PM10	Particulate matter,<10 microns in diameter.	RESU	Residential Energy Storage Unit
PME	Protective multiple earthing	REWP	Renewable Energy Working Party
PNA	Pacific North American Pattern	RFF-PI	Resources for the Future—Policy Instruments
PNNL	Pacific Northwest National Laboratory	RFG	Reformulated gasoline
PPA	Power purchase agreement	RFID	Radio frequency identification
PROMOD	Production cost modeling software by Ventyx	RFP	Request for Proposal
RPS	Renewable portfolio standards	RFS	Renewable fuel standard
PRECIS	Regional climate modeling system, UK Met Office	RI/FS	Remedial Investigation/Feasibility Study
		RIMS	Risk & Insurance Management Society
POP	Persistent organic pollutants	RMB	Renminbi, Chinese currency
PPA	Power purchase agreement	RMCS	Refrigeration Monitoring and Control Systems
ppb	Parts per billion		
ppm	Parts per million	ROI	Return on investment
PPP	Purchasing power parity	RPS	Renewable portfolio standards
Proalcool	Brazilian ethanol fuel program	RRSO	Roundtable on Responsible Soy Oil
PRP	Potentially Responsible Party	RSA	Insurance Group, Munich
PTC	Production tax credit	RSB	Roundtable for Sustainable Biofuel
Pu	Plutonium	RSPO	Roundtable on Sustainable Palm Oil
PUC	Public Utility Commission	RTDS	Real Time Digital Simulation
PV	Photovoltaic	RTU	Remote thermal unit
PVC	Polyvinyl chloride	RVP	Reid vapor pressure
PVES	Photovoltaic energy systems	RX5D	Yearly maximum precipitation in ?? ve consecutive days
pW	Picowatt		
PW	Petawatt	RWQCB	Regional Water Quality Control Board

Q

Q	Heat (BTU, calories)	**S**	
QSR	Quality Surveillance Representative	S	Sulfur
Qual	Quality or abbr. for qualification	SAF/IE	Secretary of the Air Force for Installations, Environment and Logistics

R

r	Radius	SAL	State action level
R	Resistance (in Ohms)	SAM	Southern Annular Mode
R&D	Research and development	SARA	Superfund Amendments and Reauthorization Act of 1986.
RAF	Royal Air Force		
RAP	Remedial Action Plan	SARA Title III—Emergency Planning and Community Right-to-Know Act of 1986	
RCM	Regional climate model		
RCOF	Regional Climate Outlook Forum	SBS	Sick building syndrome
RCRA	Resource Conservation and Recovery Act	Sc-Si	Single-crystalline silicon
RD&D	Research, development and demonstration	SCCF	Special Climate Change Fund
RDD&D	Research, development, demonstration and deployment	SCE	Southern California Edison
		SCEC	Songzao Coal and Electricity Company
		SDD	Sustainable Design and Development
RDF	Remote Delivery Facility (WHS)	SDO	Standards development organization
REAP	Reduced Energy Appreciation Program	sec.	Second
REC	Renewable Energy Certificate	SECNAV	Secretary of the Navy
		SE ITP	Sustainability and Environment Integrated

	Product Team
SERC	State Electricity Regulatory Commission
SERP	Super Efficient Refrigerator Program
SFG	Senior Focus Group
SGIA	Small Generator Interconnection Agreement
SGIRM	Smart Grid interoperability reference model
SHPP	Small hydropower
Si	Silicon
SiO_2	Silicon dioxide (sand)
SIOQ	Quality Assurance Division (NGA)
SIP	State Implementation Plan
SLBM	Submarine Launched Ballistic Missile
SNARL	Suggested No Adverse Response Level
SO_2	Sulfur dioxide
SON	September, October, November
SPC	State Power Corporation
SREC	Solar renewable energy credits
SRES	Special report on emissions scenarios, IPCC
st	Short ton
STC	Standard test conditions
STE	Solar thermal electricity
STLC	Soluble Threshold Limit Concentration
SWC	Surge withstand capability
SWMU	Solid waste management unit

T

t	Time
	Ton
T	Temperature
T&D	Transmission and distribution
TAR	Third Assessment Report (of the IPCC)
TCE	Tetrachloroethylene
	Tons of coal equivalent
Tcf	Trillion cubic feet
TCLP	Toxicity Characteristic Leaching Procedure .
TCP	Tetrachlorophenol
T&D	Transmission and distribution
TEPPC	Transmission Expansion Planning and Policy Committee
TFR	Total fertility rate
TIMES	The Integrated MARKAL (Marketing and Allocation Model)-EFOM.
TF	Thin films
TLV	Threshold Limit Value
TMA	TRICARE Management Agency
TNT	Equivalent measure of energy released during nuclear explosion
TOC	Total organic content
TOD	Time of delivery
toe	Tons of oil equivalent
TOU	Time of use
tpy	Tons per year

TSCA	Toxic Substances Control Act
TSO	Transmission system operator
TSP	Tehachapi Storage Project
TTLC	Total Threshold Limit Concentration
TW	Terawatt (1 billion Watts)
TWh	Terawatt hour (1 billion KWh)
TYNDP	Ten-Year Network Development Plan

U

U	Internal energy (molecular level)
	Uranium
UCI	University of California Irvine
UDS	Ultramar Diamond Shamrock Corporation
UESC	Utility Energy Services Contract
UFRJ	Federal University of Rio de Janeiro
UHV	Ultra High Voltage
UNECE	United Nations Economic Commission for Europe
UNFCCC	United Nations Framework Convention on Climate Change
UK	United Kingdom
UKCIP	United Kingdom Climate Impacts Programme
UL,LLC	UL Limited Liability Corporation (formerly Underwriters Laboratories, Inc.)
UMC	Uniform Mechanical Code
UN	United Nations
UNCCO	UN Convention on Combating Desertification
UNDP	United Nations Development Program
UNEP	United Nations Environment Program
UNESCO	United Nations Educational, Scientific and Cultural Organization
UNFCCC	UN Framework Convention on Climate Change
UK	United Kingdom
UNEP	United Nations Environment Program
UNIDO	United Nations Industrial Development Organization
UNWTO	World Tourism Organization
UPC	Uniform Plumbing Code
UPS	Uninterruptable power supply
U.S.	United States
US$	United States Dollar
USA	United States of America
USACE	US Army Corp of Engineers
USAF	United States Air Force
USAMRIID	United States Army Medical Research Institute for Infectious Diseases
USGBC	United States Green Building Council
USGS	United States Geological Survey
USD	United States Dollars
US DOE	United States Department of Energy

| USEPA | United States Environmental Protection Agency |
| USNC | U.S. National Committee |

V

v	Velocity in m/s, in/s, MPH
V	Potential (in Volts)
	Volume
VAM	Ventilation Air Methane
VAR	Volt-ampere reactive
VAT	Value Added T
VAV	Variable Air Volume
VLA	Vented lead-acid (batteries)
VMT	Vehicle miles traveled
VOC	Volatile organic compounds
VOST	Value-of-solar tariff
vRE	Variable renewables
VRLA	Valve regulated lead acid (batteries)
VSC	Voltage source converter

W

| WACC | Weighted average cost of capital |
| w | Width |

W	Watt
	Work
WACC	Weighted average cost of capital
WBCSD	World Business Council on Sustainable Development
WBG	World Bank Group
WCC-3	World Climate Conference-3
WCRP	World Climate Research Programme
WCSS	World Climate Services Systems
WECC	Western Electricity Coordinating Council
WFP	World Food Program
WG	Working Group
WHO	World Health Organization
WHS	Washington Headquarters Service
W/m^2	Watts per square meter
WMD	Weapons of mass destruction
WMO	World Meteorological Organization
WRAMC	Walter Reed Army Medical Center
WRI	World Resources Institute
WRMA	Weather Risk Management Association
WRMF	Weather Risk Management Facility
WU	Wage units

GLOSSARY OF TERMS

A

Acceptable intake. (as related to sub-chronic and chronic exposure). Numbers which describe how toxic a chemical is. The numbers are derived from animal studies of the relationship between dose and non-cancer effects. There are two types of acceptable exposure values: one for acute (relatively short-term) and one for chronic (longer-term) exposure.

Accumulation. the build up of a particular matter, like pollutants, over time

Acetone: A widely used, highly volatile solvent. It is readily absorbed by breathing, ingestion or contact with the skin. Workers who have inhaled acetone have reported respiratory problems.

Acids. A class of compounds that can be corrosive when concentrated. Weak acids, such as vinegar and citric acid, are common in foods. Strong acids, such as muriatic (or hydrochloric), sulfuric and nitric acid have many industrial uses, and can be dangerous to those not familiar to handling them. Acids are chemical "opposites" to bases, in that they can neutralize each other.

Acid Rain. term applied to acid precipitation formed when emissions of sulfur dioxide (SO_2) and oxides of nitrogen (NO_x) react in the atmosphere with water and other compounds.

Acid Rain Program. created under the Clean Air Act to reduce acid rain; employs a cap and trade framework to achieve SO_2 reductions. Specifically means the limited authorization to emit one ton of SO_2 during a given year.

Act. In the legislative sense, a bill or measure passed by both houses of Congress; a law.

Action level. A guideline established by environmental protection agencies to identify the concentration of a substance in a particular medium (water, soil, etc.) that may present a health risk when exceeded. If contaminants are found at concentrations above their action levels, measures must be taken to decrease the contamination.

Activated sludge. A term used to describe sludge that contains microorganisms that break down organic contaminants (e.g., benzene) in liquid waste streams to simpler substances such as water and carbon dioxide. It is also the product formed when raw sewage is mixed with bacteria-laden sludge, then stirred and aerated to destroy organic matter.

Activity (of a radioactive isotope). The number of particles or photons ejected from a radioactive substance per unit time.

Active solar. Using solar energy with the assistance of external power.

Acute hazards. Hazards associated with short-term exposure to relatively large amounts of toxic substances.

Acute Loading: a term that applies to the short-term build up of a pollutant and which suggests that, in the short-term, significant amounts of a pollutant can accumulate

Adjournment. The end of a legislative day or session.

Adverse health effects. Effects of chemicals or other materials that impair one's health. They can range from relatively mild temporary conditions such as minor eye or throat irritation, shortness of breath or headaches to permanent and serious conditions such as cancer, birth defects or damage to organs.

Advisory level. The level above which an environmental protection agency suggests it is potentially harmful to be exposed to a contaminant, although no action is mandated.

Aeration. Passing air through a solid or liquid, especially a process that promotes breakdown or movement of contaminants in soil or water by exposing them to air.

Aerosol. A suspension of small liquid or solid particles in gas.

Africa. Algeria, Angola, Benin, Botswana, Cameroon, Congo, Côte d'Ivoire, Democratic Republic of Congo, Egypt, Eritrea, Ethiopia, Gabon, Ghana, Kenya, Libya, Mauritius, Morocco, Mozambique, Namibia, Niger, Nigeria, Senegal, South Africa, South Sudan, Sudan, Tanzania, Togo, Tunisia, Zambia, Zimbabwe and other Africa.

Air pollution. Toxic or radioactive gases or particulate matter introduced into the atmosphere, usually as a result of human activity.

Air stripping tower. Air stripping removes volatile organic chemicals (such as solvents) from contaminated water by causing them to evaporate. Polluted water is sprayed downward through a tower filled with packing materials while air is blown upwards through the tower. The contaminants evaporate into the air, leaving significantly-reduced pollutant levels in the water. The air is treated before it is released into the atmosphere.

Alkaline (synonym basic, caustic). Having the properties of a base, a pH greater than 7. Usually used as an adjective, i.e. "alkaline soil."

Allowance. The term generally used to refer to the emission reduction unit traded in emissions trading programs.

Allowance Loan. Transaction wherein an owner of allowances, the lender, allows another party, the borrower, to use the allowances. The borrower customarily promises to return the allowances after a specified period of time with payment for their use, called interest. The allowances returned are not necessarily the exact ones loaned, but are allowances of similar vintage years

Allowance Loan Rate. payment for the lending of allowances over a specified period of time, calculated including the cost-of-carry charge.

Allowance Tracking System (ATS). a computerized system administered by EPA and used to track the allowances and allowance transactions by all market participants

Allowance Transfer Form (ATF). official form used to report allowance transfers to the ATS. The ATF lists the serial numbers of the allowances to be transferred and includes the account information of both the transferor and the transferee.

All-Hazards. A grouping classification encompassing all conditions, environmental or man-made, that have the potential to cause injury, illness, or death; damage to or loss of equipment, infrastructure services, or property; or alternatively causing functional degradation to social, economic, or environmental aspects.

Alluvial deposit. An area of sand, clay or other similar material that has been gradually deposited by moving water, such as along a river bed or shore of a lake.

Alpha particle. A positively-charged particle emitted by radioactive atoms. Alpha particles travel less than one inch in the air and a thin sheet of paper will stop them. The main danger from alpha particles lies in ingesting the atoms which emit them. Body cells next to the atom can then be irradiated over an extended period of time, which may be prolonged if the atoms are taken up in bone, for instance.

Alternative energy. Energy that is not popularly used and is usually environmentally sound, such as solar or wind energy (as opposed to fossil fuels).

Alternative fibers. Fibers produced from non-wood sources for use in paper making.

Alternative fuels. Transportation fuels other than gasoline or diesel. Includes natural gas, methanol, and electricity.

Alternative transportation modes. Travel other than private cars, such as walking, bicycling, rollerblading, carpooling and transit.

Ambient air. Refers to the surrounding air. Generally, ambient air refers to air outside and surrounding an air pollution source location. Often used interchangeably with "outdoor air."

Amendment. A change or addition to an existing law or rule.

Ancient forest. A forest that is typically older than 200 years with large trees, dense canopies and an abundance of diverse wildlife.

Applicable or Relevant and Appropriate Requirements (ARARs). Federal or state laws, regulations, standards, criteria or requirements which would apply to the cleanup of hazardous substances at a particular site.

Apportionment. The process through which legislative seats are allocated to different regions.

Appropriation. The setting aside of funds for a designated purpose (e.g., there is an appropriation of $7 billion to build 5 new submarines).

Aquaculture. The controlled rearing of fish or shellfish by people or corporations who own the harvestable product, often involving the capture of the eggs or young of a species from wild sources, followed by rearing more intensively than possible in nature.

Aqueous. Water-based.

Aquifer. A water-bearing layer of rock or sediment that is capable of yielding useable amounts of water. Drinking water and irrigation wells draw water from the underlying aquifer.

Arbitrage/Storage trades. Storing low-priced energy during periods of low demand and subsequently selling it during high-priced periods within the same market is referred to as a storage trade. Similarly, arbitrage refers to this type of energy trade between two energy markets.

Arms control. Coordinated action based on agreements to limit, regulate, or reduce weapon systems by the parties involved.

Arsenic. A gray, brittle and highly poisonous metal. It is used as an alloy for metals, especially lead and copper, and is used in insecticides and weed killers. In its inorganic form, it is listed as a cancer-causing chemical under Proposition 65.

Artesian well. A well that flows up like a fountain because of the internal pressure of the aquifer.

Asbestos. A general name given a family of naturally occurring fibrous silicate minerals. Asbestos fibers were used mainly for insulation and as a fire retardant material in ship and building construction and other industries, and in brake shoes and pads for automobiles. Inhaling asbestos fibers has been shown to result in lung disease (asbestosis) and in lung cancer (mesothelioma). The risk of developing mesothelioma is significantly enhanced in smokers.

Ash. Incombustible residue left over after incineration or other thermal processes.

Ask. the price a prospective seller is willing to ac-

cept (a.k.a. "offer")

Asset. A person, structure, facility, information, material, or process that has value. In the context of the NIPP, people are not considered assets.

Asthma. A condition marked by labored breathing, constriction of the chest, coughing and gasping usually brought on by allergies.

Atmosphere. The 500 km thick layer of air surrounding the earth which supports the existence of all flora and fauna.

Atomic energy. Energy released in nuclear reactions. When a neutron splits an atom's nucleus into smaller pieces it is called fission. When two nuclei are joined together under millions of degrees of heat it is called fusion.

Average Weighted Price. calculation used to determine price taking into account the quantity of allowances sold

B

Backfill. The word is used in two contexts; a. to refill an excavated area with uncontaminated soils; and b. the material used to refill an excavated area.

Background concentration. Represents the average amount of toxic chemicals in the air, water or soil to which people are routinely exposed. More than half of the background concentration of toxic air in metropolitan areas comes from automobiles, trucks and other vehicles. The rest comes from industry and business, agricultural, and from the use of paints, solvents and chemicals in the home.

Bases. A class of compounds that are "opposite" to acids, in that they neutralize acids. Weak bases are used in cooking (baking soda) and cleaners. Strong bases can be corrosive, or "caustic." Examples of strong bases that are common around the house are drain cleaners, oven cleaners and other heavy duty cleaning products. Strong bases can be very dangerous to tissue, especially the eyes and mouth.

Beach closure. The closing of a beach to swimming, usually because of pollution.

Bear Market. prolonged period of falling allowance prices.

Benzene. A petroleum derivative widely used in the chemical industry. A few uses are: synthesis of rubber, nylon, polystyrene, and pesticides; and production of gasoline. Benzene is a highly volatile chemical readily absorbed by breathing, ingestion or contact with the skin. Short-term exposures to high concentrations of benzene may result in death following depression of the central nervous system or fatal disturbances of heart rhythm. Long-term, low-level exposures to benzene can result in blood disorders such as aplastic anemia and leukemia. Benzene is listed as a cancer-causing chemical under Proposition 65.

Berm. A curb, ledge, wall or mound used to prevent the spread of contaminants. It can be made of various materials, even earth in certain circumstances.

Beta particles. Very high-energy particle identical to an electron, emitted by some radioactive elements. Depending on their energy, they penetrate a few centimeters of tissue.

Bid. price a prospective buyer is willing to pay

Bill. A proposed law, to be debated and voted on.

Billfish. Pelagic fish with long, spear-like protrusions at their snouts, such as swordfish and marlin.

Biodegradable. Waste material composed primarily of naturally-occurring constituent parts, able to be broken down and absorbed into the ecosystem. Wood, for example, is biodegradable, for example, while plastics are not.

Bioaccumulation. The process by which the concentrations of some toxic chemicals gradually increase in living tissue, such as in plants, fish, or people as they breathe contaminated air, drink contaminated water, or eat contaminated food.

Biodiversity. A large number and wide range of species of animals, plants, fungi, and microorganisms. Ecologically, wide biodiversity is conducive to the development of all species.

Biomass. Two meanings. a. the amount of living matter in an area, including plants, large animals and insects; and b. plant materials and animal waste used as fuel.

Bioremediation. A process that uses microorganisms to change toxic compounds into non-toxic ones.

Biosolids. Residue generated by the treatment of sewage, petroleum refining waste and industrial chemical manufacturing wastewater with activated sludge.

Biosphere. Two meanings; a. the part of the earth and its atmosphere in which living organisms exist or that is capable of supporting life; and, b. the living organisms and their environment composing the biosphere.

Biosphere Reserve. A part of an international network of preserved areas designated by the United Nations Educational, Scientific and Cultural Organization (UNESCO). Biosphere Reserves are vital centers of biodiversity where research and monitoring activities are conducted, with the participation of local communities, to protect and preserve healthy natural systems threatened by development. The global system currently includes 324 reserves in 83 countries.

Biota. The animal and plant life of a particular region.

Biotic. Of or relating to life.

Birth control. Preventing birth or reducing frequency of birth, primarily by preventing conception.

Biotransformation. Transformation of one chemical to others by populations of microorganisms in the soil.

Birth defects. Unhealthy defects found in newborns, often caused by the mother's exposure to environmental hazards or the intake of drugs or alcohol during pregnancy.

Birth rate. The number of babies born annually per 1,000 women of reproductive age in any given set of people.

Black start. In cases when the power system collapses and all other ancillary mechanisms have failed, black start capabilities allow electricity supply resources to restart without pulling electricity from the grid.

Bloc. A group of people with the same interest or goal (usually used to describe a voting bloc, a group of representatives intending to vote the same way).

Blood lead levels. The amount of lead in the blood. Human exposure to lead in blood can cause brain damage, especially in children.

Boring. Usually, a vertical hole drilled into the ground from which soil samples can be collected and analyzed to determine the presence of chemicals and the physical characteristics of the soil.

Borrow pit. An area where soil, sand or gravel has been dug up for use elsewhere.

Bottled water. Purchased water sold in bottles.

Broker. person who acts as an intermediary between a buyer and a seller, usually charging a commission

Brownfields. Abandoned, idled, or under-used industrial and commercial facilities where expansion or redevelopment is complicated by real or perceived environmental contamination.

Bubble. a regulatory term which applies to the situation when a company combines a number of its sources in order to control pollution in aggregate; under a bubble facility operators are allowed to choose which sources to control as long as the total amount of emissions from the combined sources is less than the amount each source would have emitted under the conventional requirement

Budget. A formal projection of spending and income for an upcoming period of time, traditionally submitted by the President or Executive for consideration and approval.

Budget reconciliation. Legislation making changes to existing law (such as entitlements under Social Security or Medicare) so that it conforms to numbers in the budget resolution.

Budget resolution. The first step in the annual budget process. This resolution must be agreed to by the House and Senate. It is not signed by the President and does not have the effect of law, but instead sets out the targets and assumptions that will guide Congress as it passes the annual appropriations and other budget bills.

Bull Market. prolonged rise in the price of allowances. Bull markets usually last at least a few months and are characterized by high trading volume

Business Continuity. The ability of an organization to continue to function before, during, and after a disaster.

Bycatch. Fish and/or other marine life that are incidentally caught with the targeted species. Most of the time bycatch is discarded at sea.

Bycatch reduction device (bdr). A device used to cut bycatch while fishing. These gear modifications are most commonly used with shrimp trawls. They are also called "finfish excluder devices" (feds) or, when specifically designed to exclude sea turtles, they are called "turtle excluder devices" (teds).

C

Cadmium. Cadmium is a natural element in the earth's crust, usually found as a mineral combined with other elements such as oxygen. Because all soils and rocks have some cadmium in them, it is extracted during the production of other metals like zinc, lead and copper. Cadmium does not corrode easily and has many uses. In industry and consumer products, it is used for batteries, pigments, metal coatings, and plastics. It is used also to make solar cells and panels. Cadmium salts are toxic in higher concentrations.

Cairo Plan. Recommendations for stabilizing world population agreed upon at the U.N. International

Calendar. In the legislative sense, a group of bills or proposals to be discussed or considered in a legislative committee or on the floor of the House or Senate.

California Environmental Quality Act (CEQA). First enacted in 1970 to provide long-term environmental protection, the law requires that governmental decision-makers and public agencies study the significant environmental effects of proposed activities, and that significant avoidable damage be avoided or reduced where feasible. CEQA also requires that the public be told why the lead public agency approved the project as it did, and gives the public a way to challenge the decisions of the agency.

Call Option. a contract that grants the right to buy, at a specified price, a specific number of allowances by a certain date

Cancer. Unregulated growth of changed cells; a group of changed, growing cells (tumor).

Cancer risk. A number, generally expressed in exponential form (i.e., 1 x 10 -6, which means one in one million), which describes the increased possibility of an individual developing cancer from exposure to toxic materials. Calculations producing cancer risk numbers are complex and typically include a number of assumptions that tend to cause the final estimated risk number to be conservative.

Cap. A layer, such as clay or a synthetic material, used to prevent rainwater from penetrating the soil and spreading contamination.

Carbamates. A group of insecticides related to carbamic acid. They are primarily used on corn, alfalfa, tobacco, cotton, soybeans, fruits and ornamental plants.

Carbon adsorption. A treatment system in which organic contaminants are removed from groundwater and surface water by forcing it through tanks containing activated carbon, a specially-treated material that retains such compounds. Activated carbon is also used to purify contaminated air by adsorbing the contaminants as the air passes through it.

Carbon monoxide (CO). A very poisonous, colorless and odorless gas formed when carbon-containing matter burns incompletely, as in automobile engines or in charcoal grills used indoors without proper ventilation.

Carbon dioxide (CO_2). A naturally occurring greenhouse gas in the atmosphere, concentrations of which have increased (from 280 parts per million in preindustrial times to over 350 parts per million today) as a result of humans' burning of coal, oil, natural gas and organic matter (e.g., wood and crop wastes).

Carbon tax. A charge on fossil fuels (coal, oil, natural gas) based on their carbon content. When burned, the carbon in these fuels becomes carbon dioxide in the atmosphere, the chief greenhouse gas.

Carbon tetrachloride (CCl_4). A colorless, nonflammable toxic liquid that was widely used as a solvent in dry-cleaning and in fire extinguishers. It is listed as a cancer-causing chemical under Proposition 65.

Carcinogens. Substances that cause cancer, such as tar.

Carpooling. Sharing a car to a destination to reduce fuel use, pollution and travel costs.

Catalyst. A substance that accelerates chemical change yet is not permanently affected by the reaction (e.g., platinum in an automobile catalytic converter helps change carbon monoxide to carbon dioxide).

Catalytic Cracker Unit. In a petroleum refinery, the

catalytic cracker unit breaks long petroleum molecules apart, or "cracks" them, during the petroleum refining process. These smaller pieces then come together to form more desirable molecules for gasoline or other products.

Caucus. A meeting of a political party, usually to appoint representatives to party positions.

Caustic. The common name for sodium hydroxide, a strong base. Also used as an adjective to describe highly corrosive bases.

Caustic scrubber. An air pollution control device in which acid gases are neutralized by contact with an alkaline solution.

Chemical Facility Anti-Terrorism Standards (CFATS). Section 550 of the DHS Appropriations Act of 2007 grants the Department of Homeland Security the authority to regulate chemical facilities that "present high levels of security risk." The CFATS establish a risk-informed approach to screening and securing chemical facilities determined by DHS to be "high risk." China (Asia). Bangladesh, Brunei Darussalam, Cambodia, India, Indonesia, Democratic People's Republic of Korea, Malaysia, Mongolia, Myanmar, Nepal, Pakistan, Philippines, Singapore, Sri Lanka, Chinese Taipei, Thailand, Viet Nam and Other Asia.

Chlorinated herbicides. A group of plant-killing chemicals which contain chlorine, used mainly for weed control and defoliation.

Chlorination byproducts. Cancer-causing chemicals created when chlorine used for water disinfection combines with dirt and organic matter in water.

Chlorine. A highly reactive halogen element, used most often in the form of a pungent gas to disinfect drinking water.

Chlorofluorocarbons (CFCs). Stable, artificially-created chemical compounds containing carbon.

Chlorine, fluorine and sometimes hydrogen. Chlorofluorocarbons, used primarily to facilitate cooling in refrigerators and air conditioners, have been found to damage the stratospheric ozone layer which protects the earth and its inhabitants from excessive ultraviolet radiation.

Chlorobenzene. A volatile organic compound that is often used as a solvent and in the production of other chemicals. It is a colorless liquid with an almond-like odor. It is toxic.

Chloroform. Chloroform was once commonly used as a general anesthetic and as a flavoring agent in toothpastes, mouth wastes and cough syrups. It is listed as a cancer-causing chemical under Proposition 65.

Chromated copper arsenate. An insecticide and herbicide containing three metals: copper, chromium and arsenic. This salt is used extensively as a wood preservative in pressure-treating operations. It is highly toxic and dissolves in water, making it a relatively mobile contaminant in the environment.

Chromium. A hard, brittle, grayish heavy metal used in tanning, in paint formulation, and in plating metal for corrosion protection. It is toxic at certain levels and, in its hexavalent (versus trivalent) form, chromium is listed as a cancer-causing agent under Proposition 65.

Chronic exposure. Repeated contact with a chemical over a period of time, often involving small amounts of toxic substance.

CIKR Partner. Those Federal, State, local, tribal, or territorial governmental entities; public and private sector owners and operators and representative organizations; regional organizations and coalitions; academic and professional entities; and certain not-for-profit and private volunteer organizations that share in the responsibility for protecting the Nation's CIKR.

Class I landfill. A landfill permitted to accept hazardous wastes.

Clayoquot Sound. One of the last remaining unlogged watersheds on the west coast of Canada's Vancouver Island.

Clean Air Act Amendments of 1990. reauthorization of the Clean Air Act; passed by the U.S. Congress; strengthened ability of EPA to set and enforce pollution control programs aimed at protecting human health and the environment; included provisions for Acid Rain Program

Clean Air Act. A federal law passed in 1955 and extensively modified in 1970. It is enforced by the California Air Resources Board and the local air quality management or air pollution control districts, as well as by U.S. EPA nationally.

Clean fuel. Fuels which have lower emissions than conventional gasoline and diesel. Refers to alternative fuels as well as to reformulated gasoline and diesel.

Cleanup. Treatment, remediation, or destruction of contaminated material.

Clear cutting. A logging technique in which all trees are removed from an area, typically 20 acres or larger, with little regard for long-term forest health.

Clean Water Act. A federal law of 1977 enforced by U.S. EPA. A key provision is that "any person responsible for the discharge of a pollutant or pollutants into any waters of the United States from any point source must apply for and obtain a permit." This is reflected by the National Pollutant Discharge Elimination System (NPDES), through which the permits are issued by Regional Water Quality Control Boards. Permits are now being required for storm water runoff from cities and other locations.

Cleanup process. A comprehensive program for the clean-up (remediation) of a contaminated site. It involves investigation, analysis, development of a cleanup plan and implementation of that plan.

Clearing Price. price at which a buyer and seller agree to transact a trade

Climate change. A regional change in temperature and weather patterns. Current science indicates a discernible link between climate change over the last century and human activity, specifically the burning of fossil fuels.

Cloture. The formal end to a debate or filibuster in the Senate requiring a three-fifths vote.

Coastal pelagic. Fish that live in the open ocean at or near the water's surface but remain relatively close to the coast. Mackerel, anchovies, and sardines are examples of coastal pelagic fish. commercial extinction—the depletion of a population to the point where fisherman cannot catch enough to be economically worthwhile.

Collar/Zero-cost Collar. set of contracts used to hedge against the risk of prices moving in both directions; involves purchasing a call option and selling a put option. Option premiums in a collar that cancel each other out are "zero-cost" collars

Combined heat and power Electricity and thermal energy storage can be used in combined heat and power (CHP) facilities in order to bridge temporal gaps between electricity and thermal demand.

Combustion gases. Gases produced by burning. The composition will depend on, among other things, the fuel; the temperature of burning; and whether air, oxygen or another oxidizer is used. In simple cases the combustion gases are carbon dioxide and water. In some other cases, nitrogen and sulfur oxides may be produced as well. Incinerators must be controlled carefully to be sure that they do not emit more than the allowable amounts of more complex, hazardous compounds. This often requires use of emission-control devices.

Combustible vapor mixture. The composition range over which air containing vapor of an organic compound will burn or even explode when set off by a flame or spark. Outside that range the reaction does not occur, but the mixture may nevertheless be hazardous because it does not contain enough oxygen to support life, or because the vapor is toxic.

Communities of color. Hispanic, black or Asian people or groups living together or connected in some way.

Community right-to-know. Public accessibility to information about toxic pollution.

Compact fluorescent light (CFL). Fluorescent light bulbs small enough to fit into standard light sockets, which are much more energy-efficient than standard incandescent bulbs.

Compost. Process whereby organic wastes, including food wastes, paper, and yard wastes, decompose naturally, resulting in a product rich in minerals and ideal for gardening and farming as a soil conditioners, mulch, resurfacing material, or landfill cover.

Comprehensive Environmental Response, Compensation and Liability Act of 1980 (CERCLA). Also known as Superfund, authorizes EPA to respond directly to releases of hazardous substances that may endanger public health or the environment.

Conference on Population and Development, held in Cairo in September 1994. The plan calls for improved health care and family planning services for women, children and families throughout the world, and also emphasizes the importance of education for girls as a factor in the shift to smaller families.

Congressional Record. A document published by the government printing office recording all debates, votes and discussions taking place in the Congress; available for free inspection at all government document repositories, as well as in some major libraries.

Consent decree. A legal document, approved and issued by a judge, formalizing an agreement between DTSC and the parties potentially responsible for site contamination.

Confirmation Sheet: formal memorandum from a broker to a client giving details of an allowance transaction.

Consequence. The effect of an event, incident, or occurrence. For the purposes of the NIPP, consequences are divided into four main categories: public health and safety, economic, psychological, and governance impacts.

Containment. Enclosing or containing hazardous substances in a structure to prevent the migration of contaminants into the environment.

Contamination. Pollution.

Contraceptive. Preventing conception and pregnancy.

Copper. Distinctively-colored metal used for electric wiring, plumbing, heating and roof and building construction, and in automobile brake linings. It is known to be toxic at certain levels.

Corrosivity. A characteristic of acidic and basic hazardous wastes. The characteristic is defined by a waste's pH and its ability to corrode steel. A waste is corrosive if it has a pH less than or equal to 2.0 or greater than or equal to 12.5.

Cost-of-carry. out-of-pocket costs incurred while

an allowance holder retains allowances for future transfer.

Counterparty. the party opposite the buyer or seller in a transaction

Control Center. A sophisticated monitoring and control system responsible for balancing power generation and demand; monitoring flows over transmission lines to avoid overloading; planning and configuring the system to operate reliably; maintaining system stability; preparing for emergencies; and placing equipment in and out of service for maintenance and emergencies.

Control Systems. Computer-based systems used within many infrastructures and industries to monitor and control sensitive processes and physical functions. These systems typically collect measurement and operational data from the field, process and
display the information, and relay control commands to local or remote equipment or human-machine interfaces (operators). Examples of types of control systems include SCADA systems, Process Control Systems, and Distributed Control Systems.

Credit Risk. the financial risk that an obligation will not be paid and a loss will result

Creek. A watercourse smaller than, and often tributary to, a river.

Creosotes. Chemicals used in wood preserving operations that are produced by distilling coal-tar. They contain polycyclic aromatic hydrocarbons and polynuclear aromatic hydrocarbons (PAHs and PNAs) and so high-level, short-term exposures may cause skin ulcerations. Creosotes are listed as cancer-causing agents under Proposition 65.

Criteria pollutants. Air pollutants for which standards for safe levels of exposure have been set under the Clean Air Act. Current criteria pollutants are sulfur dioxide, particulate matter, carbon monoxide, nitrogen oxides, ozone and lead. Critical Infrastructure. Systems and assets, whether physical or virtual, so vital that the incapacity or destruction of such may have a debilitating impact on the security, economy, public health or safety, environment, or any combination of these matters, across any Federal, State, regional, territorial, or local jurisdiction.

Critical Infrastructure Information (CII). Information that is not customarily in the public domain and is related to the security of critical infrastructure or protected systems.

Critical mass. The minimum mass of fissionable material that will support a sustaining chain reaction.

Crop dusting. The application of pesticides to plants by a low-flying plane.

Crude Oil. A black, sticky substance used to produce fuels (gasoline, diesel, and jet fuel) and different materials (plastics, fertilizers, medications, etc).

Cryptosporidium. A protozoan (single-celled organism) that can infect humans, usually as a result of exposure to contaminated drinking water.

Cumulative impact. The term cumulative impact is used in several ways: as the effect of exposure to more than one compound; as the effect of exposure to emissions from more than one facility; the combined effects of a facility and surrounding facilities or projects on the environment; or some combination of these.

Cybersecurity. The prevention of damage to, unauthorized use of, or exploitation of, and, if needed, the restoration of electronic information and communications systems and the information contained therein to ensure confidentiality, integrity, and availability. Includes protection and restoration, when needed, of information networks and wireline, wireless, satellite, public safety answering points, and 911 communications systems and control systems.

Cyber System. Any combination of facilities, equipment, personnel, procedures, and communications integrated to provide cyber services. Examples include business systems, control systems, and access control systems.

Cyanide. A highly toxic chemical often used in metal finishing or in extraction of precious metal from ore.

D

Deferred Swap. a trade of one allowance for another in order to exchange the vintage years of the allowances; settlement occurs after more than 180 days

Degrease. To remove grease from machinery, tools, etc., usually using solvents. Aqueous (water-based) cleaners are becoming popular and are required in some parts of the state.

Deionized water. Water which has been specifically treated to remove minerals.

Demand shifting and peak reduction. Energy demand can be shifted in order to match it with supply and to assist in the integration of variable supply resources. These shifts are facilitated by changing the time at which certain activities take place (e.g. the heating of water or space) and can be directly used to actively facilitate a reduction in the maximum (peak) energy demand level.

Demand Side Management (DSM). An attempt by utilities to reduce customers' demand for electricity or energy by encouraging efficiency. A term referring to the need (or demand) for power generation among a utility's customers.

Demersal. Fish that live on or near the ocean bottom. They are often called benthic fish, groundfish, or bottom fish.

De minimis risk. A level of risk that the scientific and regulatory community asserts is too insignificant to regulate.

Department of Toxic Substances Control (DTSC). A department within the California Environmental Protection Agency charged with the regulation of hazardous waste from generation to final disposal, and for overseeing the investigation and clean-up of hazardous waste sites.

Dependency. The one-directional reliance of an asset, system, network, or collection thereof, within or across sectors, on input, interaction, or other requirement from other sources in order to function properly.

Designated Representative. for a unit account, the individual who represents the owners and operators of that unit and performs allowance transfer requests and all correspondence with EPA concerning compliance with the Acid Rain

Destruction and removal efficiency (DRE). A percentage that represents the number of molecules of compound removed or destroyed in an incinerator relative to the number of molecules that entered the incinerator system. A DRE of 99.99 percent means that 9,999 molecules of a compound are destroyed for every 10,000 molecules that enter the system. For some compounds a DRE of 99.9999 is required.

Development. a. developed tract of land (with houses or structures); b. the act, process or result of developing.

Dewater. To remove water from wastes, soils or chemicals.

Diazinon. An organophosphate insecticide. It is used in agriculture, and for home, lawn and commercial uses.

Dibromochloropropane (DBCP). An amber-colored liquid used in a agriculture to kill pests in the soil. Inhalation of high concentrations of DBCP causes nausea and irritation of the respiratory tract. Chronic exposure results in sterility in males. Although not in use as a pesticide in this country since 1979 (until 1985 in Hawaii), it is found as a contaminant at many hazardous substances sites. DBCP is listed as a cancer-causing agent under Proposition 65.

Dichlorobenzene (DCB). A volatile organic compound often used as a deodorizer, and as a moth, mold and mildew killer. It is a white solid with a strong odor of mothballs. It is toxic and is listed as a cancer-causing agent under Proposition 65.

1,1-Dichloroethane. A colorless, oily liquid having an ether-like odor. It is used to make other chemicals and to dissolve other substances such as paint and varnish, and to remove grease. In the past, this chemical was used as a surgical anesthetic, but it is no longer used for this purpose. Because it evaporates easily into air, it is usually present in the environment as a vapor rather than a liquid.

Dieldrin. An insecticide that was used on crops like corn and cotton. U.S. EPA banned its use in 1987. It is listed as a cancer-causing chemical under Proposition 65.

Diesel. A petroleum-based fuel which is burned in engines ignited by compression rather than spark; commonly used for heavy duty engines including buses and trucks.

Diesel engine. An internal combustion engine that uses diesel as fuel, producing harmful fumes.

Dioxins. A group of generally toxic organic compounds that may be formed as a result of incomplete combustion (as may occur in incineration of compounds containing chlorine). RCRA regulations require a higher destruction and removal efficiency (DRE) for dioxins and related furans (99.9999 percent) burned in incinerators than the DRE required for most other organic compounds (99.99 percent). They are rapidly absorbed through the skin and gastrointestinal tract and are listed as cancer-causing chemicals under Proposition 65.

Dispatch. the ordered use of generation facilities by an electric power utility including which units will operate, when they will operate, and at what capacity

Double hulled tankers. Large transport ships with two hulls with space between them, protecting the cargo (in most cases, oil) from spilling in case of a collision.

Downgradient. The direction in which groundwater flows.

Dredge. A fishing method that utilizes a bag dragged behind a vessel that scrapes the ocean bottom, usually to catch shellfish. Dredges are often equiped with metal spikes in order to dig up the catch.

Driftnet. A huge net stretching across many miles that drifts in the water; used primarily for large-scale commercial fishing.

Dump sites. Waste disposal grounds.

E

Ecologist. A scientist concerned with the interrelationship of organisms and their environment.

Ecology. A branch of science concerned with the interrelationship of organisms and their environment.

Ecosystem. An interconnected and symbiotic grouping of animals, plants, fungi, and microorganisms.

Edge cities. Cities bounded by water, usually with

eroding or polluted waterfront areas.

Effluent. Wastewater, treated or untreated, that flows out of a treatment plant, sewer or industrial outfall. Generally refers to wastes discharged into surface waters.

Electric vehicles (EV). Vehicles which use electricity (usually derived from batteries recharged from electrical outlets) as their power source.

Electromagnetic Pulse (EMP). A burst of electromagnetic radiation by deliberate means, such as nuclear attack, or through natural means, such as a large-scale geomagnetic storm. Magnetic and electric fields resulting from EMP have the potential to disrupt electrical and electronic systems by causing destructive current and voltage surges.

Electrostatic precipitator. An air pollution control device that uses electrical charges to remove particulate matter from emission gases.

Emulsifiers. Substances that help in mixing liquids that don't normally mix; e.g., oil and water.

Emissions cap. A limit on the amount of greenhouse gases that a company or country can legally emit.

Endangered species. Species of wild life in danger of extinction throughout all or a significant part of its range.

Endocrine disruptors. Substances that stop the production or block the transmission of hormones in the body. energy conservation—using energy efficiently or prudently; saving energy.

Endosulfan. An insecticide used on vegetable crops, fruits and nuts.

Energy Asset and System Parameters. Six general asset or system characteristics that are important parameters for evaluating the vulnerabilities of energy infrastructure and developing risk management programs. They include: physical and location attributes, cyber attributes, volumetric or throughput attributes, temporal/load profile attributes, human attributes, and the importance of an asset or system to the energy network.

Energy efficiency. Technologies and measures that reduce the amount of electricity and/or fuel required to do the same work, such as powering homes, offices and industries.

Enrolled bill. The final, certified bill sent to the President; House and Senate versions of a bill must match exactly in order to be enrolled.

Equity. In the environmental sense, the planned dispersement of toxic or waste facilities in regions throughout the socioeconomic strata.

Estuary. Areas where fresh water from rivers mixes with salt water from nearshore ocean. They include bays, mouths of rivers, salt marshes and lagoons. These brackish water ecosystems shelter and feed marine life, birds and wildlife.

Ethylene glycol. Used in the manufacture of a wide variety of industrial compounds and in certain cosmetics. It is used most commonly as an automobile antifreeze. It is toxic.

Eugenics. The study of hereditary improvement of the human race by controlled selective breeding.

Everglades. Large and biologically diverse wetland ecosystem in South Florida.

Executive session. A congressional meeting closed to the public (and the media).

Exercise Date (or Expiration Date). last day on which an option can be exercised

Exposure pathways. Existing or hypothetical routes by which chemicals in soil, water or other media can come in contact with humans, animals or plants.

Extraction wells. Wells that are used primarily to remove contaminated groundwater from the ground. Water level measurements and water samples can also be collected from extraction wells.

Exurbia. a. the area of suburbs; b. the region outside a city and its suburbs where wealthier families live.

F

Factory farming. Large-scale, industrialized agriculture.

Factory ships. Industrial-style ships used for the large-scale collection and processing of fish.

Fallout. The radioactive dust particles that settle to earth after the denotation of a nuclear device. It is also used to describe dust particles settling from smoke, etc.

Family planning. A system of limiting family size and the frequency of childbearing by the appropriate use of contraceptive techniques.

Fauna. The total animal population that inhabits an area.

Federal land. Land owned and administered by the federal government, including national parks and national forests.

Feasibility study. An evaluation of the alternatives for remediating any identified soil or groundwater contamination.

Feedlots. A plot of ground used to feed farm animals.

Fertility. The ability to reproduce; in humans, the ability to bear children.

Fertility rates. Average number of live births per woman during her reproductive years, among a given set of people.

Filibuster. A tactic used to delay or stop a vote on a bill by making long floor speeches and debates.

Fiscal year. A financial term referring to any twelve-month period, usually to set a budget. The federal government's fiscal year begins October 1.

Filter cake. A mixture of sediments that results from filtering and dewatering of treated wastewater.

Fisheries. An established area where fish species are cultivated and caught.

Fissile material. Material fissionable by slow neutrons. The fission process and the fissile isotopes are the source of energy in nuclear weapons and nuclear reactors.

Fission. The process whereby the nucleus of a particular heavy element splits into (generally) two nuclei of lighter elements, with the release of substantial amounts of energy.

Flammables. A class of compounds that ignite easily and burn rapidly. The Department of Transportation requires that Vehicles transporting flammables must have special markings (placards).

Flash point. The lowest temperature at which a liquid generates enough vapor to ignite in air. If a waste has a flash point of less than 140° F, then it is an ignitable hazardous waste.

Flora. The total vegetation assemblage that inhabits an area.

Florida Bay. A bay at southern tip of Florida which is bounded by the Florida Keys.

Fly ash. Non-combustible residue that results from burning fuels in an incinerator, boiler or furnace. It can include metal oxides, silicates and sulfur compounds, as well as many other chemical pollutants. It is fine ash carried along by flue gases that must be captured by some means before it reaches the mouth of the chimney.

Footprint. The outline of an area within which hazardous substances are suspected or known to exist.

Forest certification. A process of labeling wood that has been harvested from a well-managed forest.

Forests. Lands on which trees are the principal plant life, usually conducive to wide biodiversity.

Formaldehyde. A water-soluble gas used widely in the chemical industry and in the construction and building industries, largely in wood products and in foam insulation. It is also used in some deodorizing preparations, in fumigants and as a tissue preservative in laboratories. Formaldehyde is listed as a cancer-causing agent under Proposition 65.

Fossil fuel. A fuel, such as coal, oil, and natural gas, produced by the decomposition of ancient (fossilized) plants and animals; compare to alternative energy.

French drain system. A pit or trench filled with crushed rock and used to collect and divert stormwater or wastewater. Most often, perforated piping at the bottom provides easy drainage.

Frequency regulation. The balancing of continuously shifting supply and demand within a control area under normal conditions is referred to as frequency regulation. Management is frequently done automatically, on a minute-to-minute (or shorter) basis.

Fresh Kills. New York City's only operating landfill, located in Staten Island. Infamous as the largest landfill in the world.

Fugitive emissions. Releases of pollutants to the atmosphere that occur when vapors are vented from containers or tanks where materials are stored. They can also be caused by spillage during the unloading of vehicles, leaks from pipes and valves, and through equipment operation.

Function. A service, process, capability, or operation performed by an asset, system, network, or organization.

G

Gamma radiation. A high-energy photon (ray) emitted from the nucleus of certain radioactive atoms. Gamma rays are the most penetrating of the three common types of radiation (the other two are alpha particles and beta particles) and are best stopped by dense materials such as lead.

Gas. Natural gas, used as fuel.

Gasoline. Petroleum fuel, used to power cars, trucks, lawn mowers, etc.

General Accounts. accounts in the SO_2 or NOx ATS's which were created after the initial allocation; general accounts can be opened by any individual and they are not automatically adjusted for compliance

Geophysical logging. A general term that encompasses all techniques for determining whether a subsurface geological formation may be sufficiently porous or permeable to serve as an aquifer. These techniques typically involve lowering a sensing device into a borehole to measure properties of the subsurface formation.

Geothermal. Literally, heat from the earth; energy obtained from the hot areas under the surface of the earth.

Gigawatt (GW). One thousand megawatts, or one billion watts.

Gigawatt hour (GW/h). One thousand MW generated or used in one hour.

Gillnets. Walls of netting that are usually staked to the sea floor. Fish become entangled or caught by their gills.

Global Climate Change. change in the earth's climate; caused by increasing greenhouse gas (GHG) concentrations in the atmosphere; human activities considered to be major new source of GHGs

Global warming. Increase in the average temperature of the earth's surface.

Golden Carrot. An incentive program that is designed to transform the market to produce much

Government Coordinating Council. The government counterpart to the SCC for each sector established to enable interagency coordination. The GCC comprises representatives across various levels of government (Federal, State, local, tribal, and territorial) as appropriate to the security and operational landscape of each individual sector.

Greater energy efficiency. The term is a trademark of the Consortium for Energy Efficiency.

Granular activated carbon (GAC). A form of crushed and hardened charcoal. GAC has a strong potential to attract and absorb volatile organic compounds from extracted groundwater and gases.

Grassroots. Local or person-to-person. A typical grassroots effort might include a door-to-door education and survey campaign.

Grazing. The use of grasses and other plants to feed wild or domestic herbivores such as deer, sheep and cows.

Green design. A design, usually architectural, conforming to environmentally sound principles of building, material and energy use. A green building, for example, might make use of solar panels, skylights, and recycled building materials.

Greenhouse. A building made with translucent (light transparent, usually glass or fiberglass) walls conducive to plant growth.

Greenhouse effect. The process that raises the temperature of air in the lower atmosphere due to heat trapped by greenhouse gases, such as carbon dioxide, methane, nitrous oxide, chlorofluorocarbons, and ozone.

Greenhouse gas. A gas involved in the greenhouse effect. A variety of gases including carbon dioxide, methane, and nitrous oxide; the buildup of these gases in the atmosphere prevents energy from the sun to escape back out into space, creating the "greenhouse effect"

Greenway. Undeveloped land usually in cities, set aside or used for recreation or conservation.

Groundfish. A general term referring to fish that live on or near the sea floor. Groundfish are also called bottom fish or demersal fish.

Ground-level Ozone. the occurrence in the troposphere (at ground level) of a gas that consists of 3 atoms of oxygen (O_3); formed through a chemical reaction involving oxides of nitrogen (NOx), volatile organic compounds (VOC), heat and light; At ground level, ozone is an air pollutant that damages human health, vegetation, and many common materials and is a key ingredient of urban smog.

Groundwater. Water beneath the earth's surface that flows through soil and rock openings, aquifers, and often serves as a primary source of drinking water.

Growth overfishing. The process of catching fish before they are fully grown resulting in a decrease in the average size of the fish population.

H

Habitat. the natural home of an animal or plant; or the sum of the environmental conditions that determine the existence of a community in a specific place.

Half-life. The amount of time that is required for a radioactive substance to lose one-half its activity. Each radioactive substance has a unique half-life. It is also used to describe the time for a pollutant to lose one half of its concentration, as through biological action; and the time for elimination of one half a total dose of a drug from a body.

Halogens. The family of elements that includes fluorine, chlorine, bromine and iodine. Halogens are very reactive and have many industrial uses. They are also commonly used in disinfectants and insecticides. Many hazardous organic chemicals—such as polychlorinated biphenyls (PCBs), some volatile compounds (VOCs) and dioxins contain halogens, especially chlorine.

Harpooning. A surface method of fishing that requires considerable effort in locating and chasing individual fish. Harpoons are hand-held or fired from a harpoon gun and aimed at high-value fish, such as giant tuna and swordfish.

Hazard. A natural or man-made source or cause of harm or difficulty.

Hazardous waste. Waste substances which can pose a substantial or potential hazard to human health or the environment when improperly managed. Hazardous waste possesses at least one of these four characteristics: ignitability, corrosivity, reactivity or toxicity; or appears on special U.S. EPA lists.

Haze. An atmospheric condition marked by a slight reduction in atmospheric visibility, resulting from the formation of photochemical smog, radiation of heat from the ground surface on hot days, or the development of a thin mist.

Health-based remediation targets and levels to which hazardous substances on the site will be cleaned up. These target levels are health-based, meaning that exposure to the hazardous substances at or below the target is not expected to present a significant health risk.

Health risk/endangerment assessment. A study

prepared to assess health and environmental risks due to potential exposure to Hazardous substances.

Hearings. Testimony (sworn statements like those given in court) given before a Congressional committee.

Heavy metals. A group of elements (such as chromium, cadmium, lead, copper and zinc) that can be toxic at relatively low concentrations and tend to accumulate in the food chain..

Hedge. strategy used to offset investment risk. A perfect hedge is one eliminating the possibility of future gain or loss

Heptachlor. An organochloride insecticide once widely used on food crops, especially corn, but has not been in use since 1988. It is listed as a cancer-causing chemical under Proposition 65.

High-Impact, Low-Frequency (HILF). HILF events are occurrences that are relatively unusual, but have the potential to cause catastrophic disruption. Examples include pandemic disease, terrorist attack, and electromagnetic pulse

High seas. International ocean water under no single country's legal jurisdiction.

Highly migratory fish. Fish that travel over great areas.

Homeland Security Presidential Directive 7 (HSPD-7). Homeland Security Presidential Directive 7 establishes a national policy for Federal departments and agencies to identify and prioritize critical infrastructure and to protect them from terrorist attacks. The directive defines relevant terms and delivers 31 policy statements. These policy statements define what the directive covers and the roles various Federal, State, and local agencies will play in carrying it out.

Homeland Security Information Network (HSIN). The Homeland Security Information Network is a comprehensive, nationally secure and trusted web-based platform able to facilitate Sensitive But Unclassified (SBU) information sharing and collaboration between Federal, State, local, tribal, territorial, private sector, and international partners.

Horizontal wells. Extraction and monitoring wells are typically drilled vertically. A horizontal well has the advantage of providing a large area of groundwater capture for a lower overall cost.

Hot spot criteria. Cleanup levels for small areas on the site that have particularly high concentrations of hazardous substances.

Household hazards. Dangerous substances or conditions in human dwellings.

Hydrochloric acid (HCl). Clear, colorless and acidic solution of hydrogen chloride in water often used in metal cleaning and electroplating. Many hazardous

wastes contain chlorine compounds which create small amounts of hydrogen chloride when they are burned. This can contribute to the formation of acid rain. Regulations require that air pollution control equipment remove either 99% of the hydrochloric acid, or that the emissions contain less than four pounds per hour.

Hydroelectric. Relating to electric energy produced by moving water.

Hydrofluorocarbons (HFC). Chemicals used as solvents and cleaners in the semiconductor industry, among others; experts say that they possess global warming potentials that are thousands of times greater than CO_2.

Hydrogeology. The geology of groundwater, with particular emphasis on the chemistry and movement of water.

Hydropower. Energy or power produced by moving water.

Hypoxia. The depletion of dissolved oxygen in water, a condition resulting from an overabundance of nutrients of human or natural origin that stimulates the growth of algae, which in turn die and require large amounts of oxygen as the algae decompose. It was the most frequently cited direct cause of fishkills in the U.S. from 1980 to 1989.

I

Ignitability. A characteristic of hazardous waste. If a liquid (containing less than 24% alcohol) has a flash point less than 140° F, it is a hazardous waste in the United States.

Immediate Settlement. conclusion of an allowance trade in which a party pays for allowances within days of the confirmation of the transaction

Immediate Vintage Year Swap. an trade of one allowance for another in order to exchange the vintage years of the allowances; settlement occurs within days (or at least less than 180 days)

Impoundment. A body of water or sludge confined by a dam, dike, floodgate or other barrier.

In-situ soil aeration. Applying a vacuum to vapor extraction wells to draw air through the soil so that chemicals in the soil are brought to the surface where they can be treated.

Incident. An occurrence, caused by either human action or natural phenomena, that may cause harm and may require action. Incidents can include major disasters, emergencies, terrorist attacks, terrorist threats, wild and urban fires, floods, hazardous materials spills, nuclear accidents, aircraft accidents, earthquakes, hurricanes, tornadoes, tropical storms, war-related disasters, public health and medical emergencies, and other occurrences requiring an emergency response.

Incinerators. Disposal systems that burn solid waste or other materials and reduce volume of waste. Air pollution and toxic ash are problems associated with incineration.

Incompatible wastes. Wastes which create a hazard of some form when mixed together. This could be intense heat or toxic gases, for example.

Indicator chemicals. Chemicals selected from the group of chemicals found at the site and used for a public health evaluation. They are selected on the basis of toxicity, mobility and persistence, and are thought to be the chemicals of the greatest potential risk.

Industrialized countries. Nations whose economies are based on industrial production and the conversion of raw materials into products and services, mainly with the use of machinery and artificial energy (fossil fuels and nuclear fission); generally located in the northern and western hemispheres (e.g., U.S., Japan, the countries of Europe).

Infrastructure. The framework of interdependent networks and systems comprising identifiable industries, institutions, (including people and procedures), and distribution capabilities that provide a reliable flow of products and services essential to the defense and economic security of the United States, the smooth functioning of government at all levels, and society as a whole. Consistent with the definition in the Homeland Security Act, infrastructure includes physical, cyber, and/or human elements.

Interdependency. Mutually reliant relationship between entities (objects, individuals, or groups). The degree of interdependency does not need to be equal in both directions.

Insecticides. Substances used to kill insects and prevent infestation.

Interim remedial actions (IRAs). Also known as Interim Remedial Measures
Cleanup actions taken to protect public health and the environment while long-term solutions are being developed.

International Conference on Population and Development. A conference sponsored by the United Nations to discuss global dimensions of population growth and change in Cairo, Egypt in September 1994. The conference is generally considered to mark the achievement of a new consensus on effective ways to slow population growth and improve quality of life by addressing root causes of unwanted fertility.

International Planned Parenthood Federation (IPPF). An international organization made up of national level affiliates representing every region of the world. IPPF receives and distributes funds from international donor nations to its affiliates, who in turn provide services (prenatal care, contraceptive counseling and service provision, and other reproductive health services) within a country. Some national level organizations provide abortion services, others do not. I P P F sets and supports policies encouraging governmental provision of comprehensive reproductive health care.

Irritant. A chemical that can cause temporary irritation at the site of contact.

J

Joule; unit of energy measurement

K

Key Resources. As defined in the Homeland Security Act, key resources are publicly or privately controlled resources essential to the minimal operations of the economy and government.

Kilowatt (kW). One thousand watts of electric energy

Kilowatt hour (kW/h). 1000 Watts generated or used in one hour.

Kyoto Protocol. an agreement under the UNFCC signed by 84 nations; establishes greenhouse gas targets ("budgets") and framework for implementation; the Protocol has been agreed to and signed by the U.S. and now awaits ratification by the U.S. Senate.

L

Landfill. Disposal area where garbage is piled up and eventually covered with dirt and topsoil.

Landings. The amount of fish brought back to the docks and marketed. Landings can describe the kept catch of one vessel, of an entire fishery, or of several fisheries combined.

Land use. The way in which land is used, especially in farming and city planning.

Law. An act or bill which has become part of the legal code through passage by Congress and approval by the President (or via Congressional override).

Leachate. Typically, water that has come in contact with hazardous wastes. For example: Water from rain or other sources that has percolated through a landfill and dissolved or carries various chemicals, and thus could spread contamination. Current landfills have systems to collect leachate before that can happen.

Lead. A heavy metal present in small amounts everywhere in the human environment. Lead can get into the body from drinking contaminated water, eating vegetables grown in contaminated soil, or breathing dust when children play or adults work in lead-contaminated areas or eating lead-based paint. It can cause damage to the nervous system or blood cells. Children are at highest risk because their bodies are still developing. Lead and its compounds are listed as a reproductive toxic substance for women and men, and a cancer-causing substance under Proposition 65.

Lead agency. A public agency which has the principal responsibility for ordering and overseeing site investigation and cleanup.

Lead poisoning. Damaging the body (specifically the brain) by absorbing lead through the skin or by swallowing.

Least-cost planning. A process for satisfying consumers' demands for energy services at the lowest societal cost.

Leukemia. A form of bone marrow cancer marked by an increase in white blood cells.

Life cycle assessment. Methodology developed to assess a product's full environmental costs, from raw material to final disposal.

Light pollution. Environmental pollution consisting of harmful or annoying light.

Lindane. Lindane (gamma hexachlorocyclohexane) is an insecticide, once used on fruit and vegetable crops. It is still used to treat head and body lice, and scabies. It is highly toxic to humans, freshwater fish and aquatic life. It is listed as a cancer-causing chemical under Proposition 65.

Litter. Waste material which is discarded on the ground or otherwise disposed of improperly or thoughtlessly.

Load following. The second continuous electricity balancing mechanism for operation under normal conditions, following frequency regulation, is load following. Load following manages system fluctuations on a time frame that can range from 15 minutes to 24 hours, and can be controlled through automatic generation control, or manually.

Logging. Cutting down trees for commodity use.

Long. a market position in which a party records (or anticipates recording) emissions less than its yearly emissions allocation, thus it has surplus allowances

Longlines. Fishing lines stretching for dozens of miles and baited with hundreds of hooks. longlines are indiscriminate and unintentionally catch and kill immature fish along with a wide variety of other animals in the Atlantic including tunas, sharks, marlins, sailfish, sea turtles and occasionally pilot whales and dolphins.

Long-term Forward purchase or sale of a specific quantity of allowances, with delivery and settlement scheduled for a specified future date, usually more than one year out

Low-emission vehicles. Vehicles which emit little air pollution compared to conventional internal combustion engines.

Low-impact camping. Camping that does not damage or change the land, where campers leave no sign that they were on the land.

Lumber. Wood or wood products used for construction.

Lung diseases. Any disease or damaging conditions in the lung or bronchia such as cancer or emphysema.

Lymphoma. A tumor marked by swelling in the lymph nodes.

M

Magnesium oxide. Also known as magnesia, magnesium oxide is used medicinally ("Milk of Magnesia"), industrially and in agricultural soil supplements. It is also used to enhance biological processes and to cleanup groundwater.

Magnesium. This light metal and its derivatives are used in aerospace alloys, in incendiary devices such as flares, and elsewhere. When scrap magnesium is thinly shaved or powdered, it is considered to be a hazardous waste, as it ignites easily and burns with an intense, white flame. It is also a nutritionally essential trace metal.

Majority leader. The leader of the majority party in either the House or the Senate.

Malthusian—based on the theories of British economist Thomas Robert Malthus (1766-1834), who argued that population tends to increase faster than food supply, with inevitably disastrous results, unless the increase in population is checked by moral restraints or by war, famine, and disease.

Malathion. Insecticide that, at high doses, affects the human nervous system.

Mammal. An animal that feeds its young with milk secreted from mammary glands and has hair on its skin.

Managed growth. Growth or expansion that is controlled so as not to be harmful.

Manatee. A plant-eating aquatic mammal found in the waters of Florida, the Caribbean, and off the coast of West Africa.

Marbled murrelet. A rare and imperiled bird that nests in ancient forests on the west coast of the U.S.

Marine mammal. A mammal that lives in the ocean, such as a whale.

Market Maker. an individual or company that maintains firm bid and offer prices in allowances by standing ready to buy or sell allowances at market prices.

Mark-up. Action by a Congressional committee to amend and/or approve a bill; following mark-up the bill is "reported" out of committee and is ready for consideration by the entire House or Senate.

Marsh—wetland, swamp, or bog.

Mass transit. Public transportation.

Maximum contaminant level (MCL). A contaminant level for drinking water, established by the California Department of Health Services, Division of Drinking Water and Environmental Management, or by the U.S. Environmental Protection Agency. These levels are legally-enforceable standards based on health risk (primary standards) or non-health concerns such as odor or taste (secondary standards).

Medfly. The Mediterranean fruit fly, a flying insect.

Megalopolis. A large city expanding so fast that city government cannot adjust to provide services (such as garbage disposal).

Megawatt (MW). One million watts.

Megawatt hour (MW/h). One million watts generated or used in one hour

Mercury. Also known as "quicksilver," this metal is used in the paper pulp and chemical industries, in the manufacture of thermometers, and thermostats, and in fungicides. Mercury exists in three biologically important forms, elemental, inorganic and organic. It is highly toxic and affects the nervous system, kidneys and other organs. It also accumulates in animals that are high in the food chain (predators). Organic mercury compounds are the most toxic, and transformations between the three forms of mercury do occur in nature.

Methane (CH_4). An odorless, colorless, flammable gas that is the major constituent of natural gas. It can be formed from rotting organic matter (i.e., trash in a landfill), and seep up through soils or migrate through underground piping to the surface. It also seeps up through the ground in areas that have shallow petroleum deposits or improperly abandoned oil wells, such as certain areas of the Los Angeles Basin. If it collects in a closed space and reaches certain concentrations, a spark can cause an explosion. It can also displace air and cause a suffocation hazard in low, enclosed spaces. This is one of the reasons landfill gas is collected and burned, sometimes for generation of electricity.

Methyl bromide (CH_3Br). The gaseous compound is used primarily as an insect fumigant; found to be harmful to the stratospheric ozone layer which protects life on earth from excessive ultraviolet radiation.

Methyl ethyl ketone (MEK). MEK is a flammable solvent that has many industrial uses, primarily in the plastic industry as a solvent. MEK is also used in the synthetic rubber industry, in the production of paraffin wax, and in household products such as lacquer and varnishes, paint remover, and glues.

Methylene chloride. A colorless liquid that evaporates easily. It has been used as a metal cleaner, paint thinner, in wood stains, spot removers, fabric protectors, shoe polish and aerosol propellants. Mild exposure can cause skin and eye irritation

Microgram per gram (mg/g). A measurable unit of concentration for a solid. A mercury level of 1.0 mg/g means that one microgram (one millionth of a gram) of mercury was detected in one gram of sample. It is equivalent to one part per million.

Middle East. Bahrain, Islamic Republic of Iran, Iraq, Jordan, Kuwait, Lebanon, Oman, Qatar, Saudi Arabia, Syrian Arab Republic, United Arab Emirates and Yemen.

Migration. The movement of chemical contaminants through soils or groundwater.

Milligram per cubic meter (mg/m^3). A unit of concentration for air contaminants. A mercury vapor level of 1.0 mg/m^3 means that one milligram (one thousandth of a gram) of mercury vapor was detected in each cubic meter of sampled air.

Milligram per kilogram (mg/kg). A unit of concentration for a solid. A mercury level of 1.0 mg/kg in fish means that one milligram (one thousandth of a gram) of mercury was found in each kilogram of sampled fish. (A kilogram is 1,000 grams or approximately 2.2 pounds). Also equals one part per million.

Mining. The removal of minerals (like coal, gold, or silver) from the ground.

Minority leader. The leader of the minority party in either the House or the Senate.

Minuteman. An American-made ICBM; 500 Minuteman III ICBMs are deployed currently in the United States.

Mitigation. Actions taken to improve site conditions by limiting, reducing or controlling hazards and contamination sources.

Monitoring wells. Specially-constructed wells used exclusively for testing water quality.

Moratorium. Legislative action which prevents a federal agency from taking a specific action or implementing a specific law.

Mulch. Leaves, straw or compost used to cover growing plants to protect them from the wind or cold.

N

National Ambient Air Quality Standards (NAAQS). health-based standards for a variety of pollutants set by EPA that must be met by states across the country

National Pollutant Discharge Elimination System

(NPDES). A system under the federal Clean Water Act that requires a permit for the discharge of pollutants to surface waters of the United States. In California, NPDES permits are obtained from the Regional Water Quality Control Board.

National Priorities List (NPL). U.S. EPA's list of the top priority hazardous waste sites in the country that are subject to the Superfund program.

Natural long. a party whose allowance allocation is greater than its actual emissions

Natural short. a party whose allowance allocation is less than its actual emissions

National Infrastructure Advisory Council (NIAC). An organization that provides the President, through the Secretary of Homeland Security, with advice on the security of the critical infrastructure sectors and their information systems. The council is composed of a maximum of 30 members, appointed by the President from private industry, academia, and State and local government.

National Recreation Areas. Areas of federal land that have been set aside by Congress for recreational use by members of the public.

Negative Declaration. A California Environmental Quality Act document issued by the lead regulatory agency when the initial environmental study reveals no substantial evidence that the proposed project will have a significant adverse effect on the environment, or when any significant effects would be avoided or mitigated by revisions agreed to by the applicant.

Network. A group of components that share information or interact with each other in order to perform a function.

Neutrals. Organic compounds that have a relatively neutral pH (are neither acid nor base), complex structure and, due to their carbon bases, are easily absorbed into the environment. Naphthalene, pyrene and trichlorobenzene are examples of neutrals.

Nickel. A metal used in alloys to provide corrosion and heat resistance for products in the iron, steel and aerospace industries. Nickel is used as a catalyst in the chemical industry. It is toxic and, in some forms, is listed as a cancer-causing agent under Proposition 65.

Nitrate. Formed when ammonia is degraded by microorganisms in soil or groundwater. This compound is usually associated with fertilizers.

Nitroaromatics. Common components of explosive materials, which will explode if activated by very high temperatures or shocks. 2,4,6-trinitrotoluene (TNT) is a nitroaromatic. Some are listed as cancer-causing chemicals under Proposition 65.

Nitrogen oxides (NOx). Harmful gases (which contribute to acid rain and global warming) emitted as a byproduct of fossil fuel combustion. Gases produced during combustion of fossil fuels in motor vehicles, power plants and industrial furnaces and other sources; is a precursor to acid rain and ground-level ozone

Nitrogen oxides (NOx) Budget Program. a NOx cap and trade program adopted by 13 jurisdictions in the Northeast to address ozone transport in that region

Noise pollution. Environmental pollution made up of harmful or annoying noise.

Non-attainment pollutants. See "Criteria pollutants.» If any of the criteria pollutants exceed established health-based levels in a given air basin, they are identified as "non-attainment pollutants."

Non-OECD (Americas). Argentina, Bolivia, Brazil, Colombia, Costa Rica, Cuba, Curaçao, Dominican Republic, Ecuador, El Salvador, Guatemala, Haiti, Honduras, Jamaica, Nicaragua, Panama, Paraguay, Peru, Trinidad and Tobago, Uruguay, Venezuela and Other Non-OECD Americas.

Non-OECD (Europe and Eurasia). Albania, Armenia, Azerbaijan, Belarus, Bosnia and Herzegovina, Bulgaria, Croatia, Cyprus, Former Yugoslav Republic of Macedonia, Georgia, Gibraltar, Kazakhstan, Kosovo, Kyrgyzstan, Latvia, Lithuania, Malta, Republic of Moldova, Montenegro, Romania, Russian Federation, Serbia, Tajikistan, Turkmenistan, Ukraine and Uzbekistan.

Nuclear energy. Energy or power produced by nuclear reactions (fusion or fission). Alsop called nuclear power.

Nuclear reactor. An apparatus in which nuclear fission may be initiated, maintained, and controlled to produce energy, conduct research, or produce fissile material for nuclear explosives.

Nuclear tests. Government tests carried out to supply information required for the design and improvement of nuclear weapons, and to study the phenomena and effects associated with Nuclear explosions.

O

OECD. Australia, Austria, Belgium, Canada, Chile, Czech Republic, Denmark, Estonia, Finland, France, Germany, Greece, Hungary, Iceland, Ireland, Israel, Italy, Japan, Korea, Luxembourg, Mexico, Netherlands, New Zealand, Norway, Poland, Portugal, Slovak Republic, Slovenia, Spain, Sweden, Switzerland, Turkey, United Kingdom and United States.

Offers. price at which someone who owns an allowance is willing to sell (a.k.a. "Ask")

Off-grid energy consumers are power users who frequently rely on fossil or renewable resources (including variable renewables) to provide heat and electricity. To ensure reliable off-grid energy supplies and to support increasing levels of local resources use, energy storage can be used to fill gaps between variable supply resources and demand.

Oil. A black, sticky substance used to produce fuel (petroleum) and materials (plastics).

Oil spills. The harmful release of oil into the environment, usually in the water, sometimes killing area flora and fauna. Oil spills are very difficult to clean up.

Old growth forests. See ancient forests.

Omnibus spending bill. A bill combining the appropriations for several federal agencies.

Operation Plan. A document submitted to DTSC that gives details about how a permitted hazardous waste facility is built, a detailed description of the hazardous waste operations, the plan to be used in case of emergency, and other plans. A DTSC facility permit requires that the reviewed and approved Operations Plan be followed. It is sometimes referred to as the "Part B" of the hazardous waste facility permit.

Option. a contractual right to buy or sell allowances at an agreed price; the option buyer pays a premium for this right. If the option is not exercised after a specified period it expires

Option Premium. amount per share paid by an option buyer to an option seller for the option Out-of-the-money Call Option term used to describe an call option whose strike price for an allowance is higher than the current market value

Organochlorides. A group of organic (carbon-containing) insecticides that also contain chlorine. These chemicals tend not to break down easily in the environment. DDT, Toxaphene and Endosulfan are all organochlorides.

Organophosphate. A group of organic (carbon-containing) insecticides that also contain phosphorus. Although they do not have a long life, some can be very toxic when first applied. Malathion and Parathion are organophosphates. Malathion is mildly toxic, and parathion is extremely toxic.

Out-of-the-money Put Option: term used to describe a put option whose strike price for an allowance is lower than the current market value

Over-development. Expansion or development of land to the point of damage.

Over-fishing. Fishing beyond the capacity of a population to replace itself through natural reproduction.

Over-grazing. Grazing livestock to the point of damage to the land.

Overpacking. Process used for isolating waste by jacketing or encapsulating waste-holding containers to prevent further spread or leakage of contaminating materials. Leaking drums may be contained within oversized ones as an interim measure prior to removal and final disposal.

Over-the-counter Market: Market in which allowance transactions are conducted through the direction interaction of counterparties rather than on the floor of an exchange

Owners/Operators. Those entities responsible for day-to-day operation and investment in a particular asset or system.

Oxidizers. A group of chemicals that are very reactive, often but not always supplying oxygen to a reaction. Some oxidation reactions can release large amounts of heat and gases, and, under the right conditions, cause an explosion. Others can cause rapid corrosion of metal, damage to tissue, burns and other serious effects. Examples of oxidizers include chlorine gas, nitric acid, sodium perchlorate, and ammonium nitrate.

Ozonation. Ozone reacts with volatile organic compounds (VOCs) to change them into chemicals which pose no potential threat to human health, by breaking them down to form carbon dioxide and water. This is done with an ozonation unit

Ozone. A naturally occurring, highly reactive gas comprising triatomic oxygen (O_3) formed by recombination of oxygen in the presence of ultraviolet radiation. This naturally occurring gas builds up in the lower atmosphere as smog pollution, while in the upper atmosphere it forms a protective layer which shields the earth and its inhabitants from excessive exposure to damaging ultraviolet radiation.

Ozone depletion. The reduction of the protective layer of ozone in the upper atmosphere by chemical pollution.

Ozone hole. A hole or gap in the protective layer of ozone in the upper atmosphere.

Ozone Transport Assessment Group (OTAG): a multi-stakeholder workgroup convened to address problems associated with the long-range transport of ozone and its precursors; encompassed stakeholders in 37 jurisdictions.

P

Pandemic Influenza. Defined by the World Health Organization (WHO) as a global outbreak of influenza, characterized by an emergent strain of the virus, little to no immunity among the general population, rapid and sustained person-to-person transmission, and lack of a vaccine.

Paper. Thin sheet of material made of cellulose pulp, derived mainly from wood, but also from rags and certain grasses, and processed into flexible leaves or rolls. Used primarily for writing, printing, drawing, wrapping, and covering walls.

Paper mills. Mills (factories) that produce paper from wood pulp.

Paper products. Materials such as paper and cardboard, produced from trees.

Parathion and Methylparathion. Toxic insecticides.

Particulates. Small solid or liquid particles, especially those in the emission gases of incinerators, boilers, industrial furnaces or in exhaust from diesel and gasoline engines. Particles below 10 microns (10 one-millionths of a meter, 0.0004 inch) in diameter are considered potential health risks because, when inhaled, they are taken deep into the lungs. Regulations require that an incinerator emit no more than 180 milligrams of total particulates per dry standard cubic meter per minute.

Particulate pollution. Pollution made up of small liquid or solid particles suspended in the atmosphere or water supply.

Parts per million (ppm). A measuring unit for the concentration of one material in another. When looking at contamination of water and soil, the toxins are often measured in parts per million. One part per million is equal to one thousandth of a gram of substance in one thousand grams of material. One part per million would be equivalent to one drop of water in twenty gallons.

Parts per billion (ppb). A unit of measure used to describe levels or concentrations of contamination. A measure of concentration, equaling 0.0000001 percent. For example, One part per billion is the equivalent of one drop of impurity in 500 barrels of water. Most drinking water standards are ppb concentrations.

Passive solar. Using or capturing solar energy (usually to heat water) without any external power.

Pelagic species. Fish that live at or near the water's surface. Examples of large pelagic species include swordfish, tuna, and many species of sharks. Small pelagics include anchovies and sardines.

Pentachlorophenol (PCP). A petroleum-based chemical that is used as a wood preservative because it kills fungus and termites. It is toxic listed as a cancer-causing chemical under Proposition 65.

Perched groundwater. Water that accumulates beneath the earth's surface but above the main water bearing zone (aquifer). Typically, perched groundwater occurs when a limited zone (or lens) of harder, less permeable soil is "perched" in otherwise porous soils. Rainwater moving downward through the soil stops at the lens, flows along it, then seeps downward toward the aquifer.

Perchloroethylene (PCE). A volatile organic compound used primarily as a dry-cleaning agent. It is often referred to as "perc." It is toxic and listed as a cancer-causing chemical under Proposition 65.

Percolation. The downward flow or filtering of water or other liquids through subsurface rock or soil layers, usually continuing to groundwater.

Pesticide. A general term for insecticides, herbicides and fungicides. Insecticides kill or prevent the growth of insects. Herbicides control or destroy plants. Fungicides control or destroy fungi. Some pesticides can accumulate in the food chain and contaminate the environment.

Petrex Method. A method for collecting vapor samples from surface soil.

Petrochemicals. Chemical substances produced from petroleum in refinery operations. Many are hazardous.

Phenols. Organic compounds used in plastics manufacturing, tanning, and textile, dye and resin manufacturing. They are by-products of petroleum refining. In general, they are highly toxic.

Physical Security. The use of barriers and surveillance to protect resources, personnel, and facilities against crime, damage, or unauthorized access.

Piezometers. Small-diameter wells used to measure groundwater levels.

Pilot study. A study of a possible cleanup alternative during the Feasibility Study for a specific site. It is used to gather data necessary for the final selection of the cleanup method.

Plastics. Durable and flexible synthetic-based products, some of which are difficult to recycle and Pose problems with toxic properties, especially PVC plastic.

Plume. A body of contaminated groundwater flowing from a specific source. The movement of the groundwater is influenced by such factors as local groundwater flow patterns, the character of the aquifer in which the groundwater is contained, and the density of contaminants. A plume may also be a cloud of smoke or vapor. It defines the area where exposure would be dangerous.

Plutonium. A heavy, radioactive, man-made, metallic element (atomic number 94) used in the production of nuclear energy and the explosion of nuclear weapons; its most important isotope is fissile plutonium-239, produced by neutron irradiation of uranium-238.

Poison. A chemical that adversely affects health by causing injury, illness, or death.

Poison runoff. See polluted runoff.

Polluted runoff. Precipitation that captures pollution from agricultural lands, urban streets, parking lots and suburban lawns, and transports it to rivers, lakes or oceans.

Pollution prevention. Techniques that eliminate waste prior to treatment, such as by changing ingredients in a chemical reaction.

Polychlorinated biphenyls (PCBs). A group of toxic chemicals used for a variety of purposes including electrical applications, carbonless copy paper, adhesives, hydraulic fluids, and caulking compounds. PCBs do not breakdown easily and are listed as cancer-causing agents under Proposition 65.

Polynuclear aromatic hydrocarbons (PAHs or PNAs). PNAs or Polynuclear Aromatic Hydrocarbons, are natural constituents of crude oil, and also may be formed when organic materials such as coal, oil, fuel, wood or even foods are not completely burned. PNAs are also found in lampblack, a by-product of the historic gas manufacturing process. PNAs are found in a wide variety of other materials, including diesel exhaust, roofing tars, asphalt, fireplace smoke and soot, cigarettes, petroleum products, some foods, and even some shampoos. PNAs tend to stick to soil and do not easily dissolve in water, and generally do not move in the environment. The test method used to analyze for PNAs detects seventeen different compounds. Of the seventeen, seven are suspected of causing cancer in humans.

Polyvinyl chloride (PVC). A plastic made from the gaseous chemical vinyl chloride. PVC is used to make pipes, records, raincoats and floor titles. It produces hydrochloric acid when burned. Health risks from high concentrations of vinyl chloride (not the polymer) include liver cancer and lung cancer, as well as cancer of the lymphatic and nervous systems. Vinyl chloride (not the polymer) is listed as a cancer-causing chemical under Proposition 65.

Population. Two meanings: a. the whole number of inhabitants in a country, region or area; b. a set of individuals having a quality or characteristic in common.

Post consumer waste. Waste collected after the consumer has used and disposed of it (e.g., the wrapper from an eaten candy bar).

Potentially Responsible Party (PRP). An individual, company or government body identified as potentially liable for a release of hazardous substances to the environment. By federal law, such parties may include generators, transporters, storers and disposers of hazardous waste, as well as present and past site owners and operators.

Power plants. Facilities (plants) that produce energy.

Power Pool: a situation where output from different power plants are "pooled" together, scheduled according to increasing marginal cost, technical and contractual characteristics (so-called must-runs), and dispatched according to this "merit order" to meet demand

Preparedness. The activities necessary to build, sustain, and improve readiness capabilities to prevent, protect against, respond to, and recover from natural or man-made incidents. Preparedness is a continuous process involving efforts at all levels of government and between government and the private sector and nongovernmental organizations to identify threats, determine vulnerabilities, and identify required resources to prevent, respond to, and recover from major incidents.

Pretreatment unit. A wastewater treatment unit that is designed to treat wastewater that does not meet the sewage discharge standards so that it meets or exceeds those standards. Pretreatment units usually require a permit from a local agency.

Prevention. Actions taken and measures put in place for the continual assessment and readiness of necessary actions to reduce the risk of threats and vulnerabilities, to intervene and stop an occurrence, or to mitigate effects.

Principal organic hazardous constituents (POHCs). Specific hazardous compounds monitored during an incinerator, boiler or industrial furnace trial burn. They are selected on the basis of their high concentration in the waste feed and the difficulty of burning them.

Proprietary information (trade secret). The Department will classify information as proprietary provided the owner demonstrates the following: the business has asserted a business confidentiality claim; the business has shown it has taken reasonable measures to protect the confidentiality of the information both within the company and from outside entities; the information is not, and has not been reasonably obtainable without the business' consent; no statute specifically requires disclosure of the information; and either the business has shown that disclosure of the information is likely to cause substantial harm to its competitive position, or the information is voluntarily submitted and its disclosure would likely impair the government's ability to obtain necessary information in the future.

Prioritization. In the context of the NIPP, prioritization is the process of using risk assessment results to identify where risk reduction or mitigation efforts are most needed and to subsequently determine which protective actions should be instituted in order to have the greatest effect.

Protected Critical Infrastructure Information (PCII). PCII refers to all critical infrastructure information, including categorical inclusion PCII, that has undergone the validation process and that the PCII Program Office has determined qualifies for protection under the CII Act. All information submitted to the PCII Program Office or Designee with an express statement is presumed to be PCII until the PCII Program Office determines otherwise.

Protection. Actions or measures taken to cover or shield from exposure, injury, or destruction. In the context of the NIPP, protection includes actions to deter the threat, mitigate the vulnerabilities, or minimize the consequences associated with a terrorist attack or other incident. Protection can include a wide range of activities, such as hardening facilities, building resilience and redundancy, incorporating hazard resistance into initial facility design, initiating active or passive countermeasures, installing security systems, promoting workforce surety, training and exercises, and implementing cybersecurity measures, among various others.

Protective Security Advisor (PSA). A field-based liaison between DHS and the State and local CIKR protection community. Using site visits and cross-sector analysis, PSAs identify, assess, monitor, and minimize risks to CIKR assets at the local or district level.

Public estate. Area or plot of public land owned by community or government.

Public health. The health or physical well-being of a whole community.

Public land. Land owned in common by all, represented by the government (town, county, state, or federal).

Public participation plan. A document approved by DTSC that is designed to determine a community's informational needs and to provide a program for public involvement during facility permitting, site investigation and cleanup, or other similar activities.

Public transportation. Various forms of shared-ride services, including buses, vans, trolleys, and subways, which are intended for conveying the public.

Pulp. Raw material made from trees used in producing paper products.

Pump test. A field test by which a well is pumped for a period of time and data are collected for use in assessing characteristics of subsurface water-bearing zones, or aquifers.

Put Option: a contract that grants the right to sell, at a specified price, a specific number of allowances by a certain date

Q

Quench tower. A gas cooling and pollution control device in which heated gases are showered with water. Gases are cooled and particulates "drop out" of the gases. They can generate a waste called "quench tower drop-out."

R

Radiation. The process of emitting energy in the form of energetic particles (such as alpha particles or gamma radiation), light or heat. It also refers to that which is emitted.

Radioactive. Of or characterized by radioactivity.

Radioactive waste. The byproduct of nuclear reactions that gives off (usually harmful) radiation.

Radioactivity. The spontaneous emission of matter or energy from the nucleus of an unstable atom (the emitted matter or energy is usually in the form of alpha or beta particles, gamma rays, or neutrons).

Radionuclides. Radioactive elements, which may be naturally-occurring or synthetic. They emit various types of energetic radiation—alpha and beta particles and gamma radiation. Their half-lives range from a minute fraction of a second to many thousand years. Certain radionuclides have valuable medical and industrial uses. One is used in home smoke detectors at an amount that can cause no harmful effects.

Radium. A radioactive element with a half-life of 1,600 years that emits alpha particles as it is transformed into radon. In the past, radium was mixed with special paints to make watch faces and instrument dials glow in the dark.

Radon. A gaseous, radioactive alpha particle-emitting element with a half-life of about four days. Radon exists naturally in many locations, and may present a serious health risk when it accumulates in basements or crawl spaces beneath homes. A cancer-causing radioactive gas found in many communities' ground water.

Rain forest. A large, dense forest in a hot, humid region (tropical or subtropical). Rain forests have an abundance of diverse plant and animal life, much of which is still uncatalogued by the scientific community.

Ranking member. The lead member of a Congressional committee from the minority party, usually chosen on the basis of seniority.

Reactive. A class of compounds which are normally unstable and readily undergo violent change, react violently with water, can produce toxic gases with water, or possess other similar properties. Reactivity is one characteristic that can make a waste hazardous.

Recess. Ending a legislative session with a set time to reconvene.

Recovery. The development, coordination, and execution of service- and site-restoration plans for affected communities and the reconstitution of government operations and services through individual, private sector, nongovernmental, and public assistance programs that identify needs and define resources; provide housing and promote restoration; address long-term care and treatment of affected persons; implement additional measures for community restoration; incorporate mitigation measures and techniques, as feasible; evaluate the incident to identify lessons learned; and develop initiatives to mitigate the effects of future incidents.

Recycling. System of collecting, sorting, and reprocessing old material into usable raw materials.

Reduce. Act of purchasing or consuming less to begin with, so as not to have to reuse or recycle later.

Redundancy. An energy reliability strategy based on the notion that multiple systems provide needed backup if one system fails or cannot meet demand.

Refrigerants. Cooling substances, many of which contain CFCs and are harmful to the earth's ozone layer.

Regional Clean Air Incentives Market (RECLAIM): initiated in 1993; a set of market initiatives designed address air pollution in the Greater Los Angeles area of California; includes cap and trade programs for NO_x and SO_x.

Regional Water Quality Control Board (RWQCB). Agencies that maintain water quality standards for areas within their jurisdictions and enforce state water quality laws.

Remedial Action Plan (RAP). A plan that outlines a specific program leading to the remediation of a contaminated site. Once the Draft Remedial Action Plan is prepared, and approved by DTSC a public meeting is held and comments from the public are solicited for a period of not less then 30 days. After the public comment period has ended and the comments have been responded to in writing, DTSC may modify the Draft Plan on the basis of those comments before it approves the final remedy for the site (the Final RAP).

Remedial Investigation/Feasibility Study (RI/FS). A series of investigations and studies to identify the types and extent of chemicals of concern at the site and to determine cleanup criteria (Remedial Investigation), and to provide an evaluation of the alternatives for remediating any identified soil or groundwater problems (Feasibility Study).

Remediation. Cleanup of a site to levels determined to be health-protective for its intended use.

Renewable energy. Energy resources such as wind power or solar energy that can keep producing indefinitely without being depleted.

Reservoir. An artificial lake created and used for the storage of water.

Resilience/Resiliency. The ability to resist, absorb, recover from, or successfully adapt to adversity or a change in conditions. In the context of energy security, resilience is measured in terms of robustness, resourcefulness, and rapid recovery.

Resource Conservation and Recovery Act (RCRA). The principal federal law in the United States governing the disposal of solid waste and hazardous waste

Responsible party. An individual or corporate entity considered legally liable for contamination found at a property and, therefore, responsible for cleanup of the site.

Resolution. A formal statement from Congress.

Response. Activities that address the short-term, direct effects of an incident, including immediate actions to save lives, protect property, and meet basic human needs. Response also includes the execution of emergency operations plans and incident mitigation activities designed to limit the loss of life, personal injury, property damage, and other unfavorable outcomes.

Retire (Allowances): to remove a portion of allowances from the market.

Reuse. Cleaning and/or refurbishing an old product to be used again.

Rider. Usually unrelated provisions tacked onto an existing Congressional bill. Since bills must pass or fail in their entirety, riders containing otherwise unpopular language are often added to popular bills.

Riparian. Located alongside a watercourse, typically a river.

Risk. The potential for an unwanted outcome resulting from an incident, event, or occurrence, as determined by its likelihood and the associated consequences.

Risk assessment. A risk assessment looks at the chemicals and all other threats detected at a site, the frequency and concentration of detected chemicals or incidence, the toxicity of the chemicals and how people can be exposed, and for how long.

Risk Management Framework. A planning methodology that outlines the process for setting goals and objectives; identifying assets, systems, and networks; assessing risks; prioritizing and implementing protection programs and resilience strategies; measuring performance; and taking corrective action. Public and private entities often include risk management frameworks in their business continuity plans.

Roadmap to Secure Control Systems in the Energy Sector (Roadmap). The DOE and DHS sponsored document that outlines a coherent plan for improving cybersecurity in the Energy Sector, identifying concrete steps to secure control systems used in the Electricity, Oil, and Natural Gas subsectors over a ten-year period.

Rotary kiln. An incinerator with a rotating combustion chamber. The rotation helps mix the wastes and promotes more complete burning. They can accept gases, liquids, sludges, tars and solids, either separately or together, in bulk or in containers.

Run-off. Precipitation that the ground does not absorb and that ultimately reaches rivers, lakes or oceans.

S

Sagebrush Rebellion. A movement started by ranchers and miners during the late 1970s in response to efforts of the Bureau of Land Management (B.L.M.) to improve management of federal lands. While its announced goal was to give the lands "back" to the western states, its real goal—and the one it achieved—was to force the B.L.M. to abandon its new approach to public land management.

Salvage logging. The logging of dead or diseased trees in order to improve overall forest health; used by timber companies as a rationalization to log otherwise protected areas.

Sanitary landfill. A landfill which does not take hazardous waste, often called a "garbage dump." It must be covered with dirt each day to maintain sanitary conditions. The Integrated Waste Management Board regulates these facilities.

SARA Title III. Or the Emergency Planning and Community Right-to-Know Act of 1986. It requires each state to have an emergency response plan as described, and any company that produces, uses or stores more than certain amounts of listed chemicals must meet emergency planning requirements, including release reporting.

Scrubber: a pollution control technology utilized in power plants to remove pollutants from plant emissions

Seasonal energy storage The ability to store energy for days, weeks, or months to compensate for a longer-term supply disruption or seasonal variability on the supply and demand sides of the energy system (e.g. storing heat in the summer to use in the winter via underground thermal energy storage systems).

Secondary containment. A structure designed to capture spills or leaks, as from a container or tank. For containers and aboveground tanks, it is usually a bermed area of coated concrete. For underground tanks, it may be a second, outer, wall or a vault. Construction of such containment must meet certain requirements, and periodic inspections are required.

Second-growth forests. Forests that have grown back after being logged.

Sector. A logical collection of assets, systems, or networks that provide a common function to the economy, government, or society. The NIPP addresses 18 CIKR sectors, identified by the criteria set forth in HSPD-7.

Sector Coordinating Council (SCC).The private sector counterpart to the GCC, these councils are self-organized, self-run, and self-governed organizations that are representative of a spectrum of key stakeholders within a sector. SCCs serve as the government's principal point of entry into each sector for developing and coordinating a wide range of CIKR protection activities and issues.

Sector Partnership Model. The framework used to promote and facilitate sector and cross-sector planning, coordination, collaboration, and information sharing for CIKR protection involving all levels of government and private sector entities.

Sediment. The soil, sand and minerals at the bottom of surface waters, such as streams, lakes and rivers. Sediments capture or adsorb contaminants. The term may also refer to solids that settle out of any liquid.

Seismic stability. The likelihood that soils will stay in place during an earthquake.

Selenium. This metal is a nutritionally essential trace element that is toxic at higher doses. High levels of selenium have been shown to cause reproductive failure and birth defects in birds.

Semi-volatile organic compounds. Compounds that evaporate slowly at normal temperatures.

Short: a market position in which a party records (or anticipates recording) emissions in excess of its yearly emissions allocation, thus it has a deficit of allowances.

Short-term Forward: purchase or sale of a specific quantity of allowances at the current or spot price, with delivery and settlement at a specified future date, usually within one year

Sick building syndrome. A human health condition where infections linger, caused by exposure to contaminants within a building as a result of poor ventilation.

Silos. Fixed vertical underground structures made of steel and concrete that house an ICBM and its launch support equipment.

Silver. Silver is a metal used in the manufacture of photographic plates, cutlery and jewelry. Silver nitrate is used in an array of industrial chemical processes. It is toxic.

Silvex. A chlorinated herbicide.

Sinkhole. A depression formed when the surface collapses into a cavern.

Site mitigation process. The regulatory and technical process by which hazardous waste sites are identified and investigated, and cleanup alternatives are developed, analyzed, decided upon and applied.

Slurry wall. Barriers used to contain the flow of contaminated groundwater or subsurface liquids. Slurry walls are constructed by digging a trench around a contaminated area and filling the trench with a material that tends not to allow water to pass through it. The groundwater or contaminated liquids trapped within the area surrounded by the slurry wall can be extracted and treated.

Smog. A dense, discolored radiation fog containing large quantities of soot, ash, and gaseous pollutants such as sulfur dioxide and carbon dioxide, responsible for human respiratory ailments. Originally meaning a combination of smoke and fog, smog now generally refers to air pollution; ground level ozone is a major constituent of smog. Most industrialized nations have implemented legislation to promote the use of smokeless fuel and reduce emission of toxic gases into the atmosphere.

SO_2 Allowance Auction: provided for in the Clean Air Act, the SO_2 auction is held annually by the US EPA; the auctions help to send the market an allowance price signal, as well as furnish utilities with an additional avenue for purchasing needed allowances.

Soil borings. Soil samples taken by drilling a hole in the ground.

Soil gas survey. Soil gas or (soil vapor) is air existing in void spaces in the soil between the groundwater and the ground surface. These gases may include vapor of hazardous chemicals as well as air and water vapor. A soil-gas survey involves collecting and analyzing soil-gas samples to determine the presence of chemicals and to help map the spread of contaminants within soil.

Soil vapor extraction. A process in which chemical vapors are extracted from the soil by applying a vacuum to wells.

Solar energy. Energy derived from sunlight.

Solid waste. Non-liquid, non gaseous category of waste from non-toxic household and commercial sources.

Solid waste management units (SWMUs). Any unit at a hazardous waste facility from which hazardous chemicals might migrate, whether or not they were intended for waste management. They include such things as containers, tanks, landfills among others.

Solidification. Mixing additives, such as fly ash or cement, with soil containing hazardous chemicals, especially metals, to make it more stable. This process lessens the risk of exposure to the hazardous chemicals by making it less likely that those chemicals will move into and through surface or groundwater.

Soluble Threshold Limit Concentration (STLC). The limit concentration for toxic materials in a sample that has been subjected to the California Waste Extraction Test (WET), a state test for the toxicity characteristic that is designed to subject a waste sample to simulated conditions of a municipal waste landfill. If the concentration of a toxic substance in the special extract

of the waste exceeds this value, the waste is classified as hazardous in California. This is distinct from the Total Threshold Limit Concentration (TTLC). The California Waste Extraction Test procedure is more stringent than the federal Toxicity Characteristic Leaching Procedure (TCLP).

Solvent. A liquid capable of dissolving another substance to form a solution. Water is sometimes called "the universal solvent" because it dissolves so many things, although often to only a very small extent. Organic solvents are used in paints, varnishes, lacquers, industrial cleaners and printing inks, for example. The use of such solvents in coatings and cleaners has declined over the last several years, because the most common ones are toxic, contribute to air pollution and may be fire hazards.

Soot. A fine, sticky powder, comprised mostly of carbon, formed by the burning of fossil fuels.

Speaker. The leader of the House of Representatives, who controls debate and the order of discussion; chosen by vote of the majority party.

Special Allowance Reserve: roughly 2.8 percent dangerous anthropogenic interference with the of the cap set aside each year to supply the climate system; established a framework for annual allowance auction agreeing to specific actions

Spinning and non-spinning reserve is capacity for the electricity supply when used to compensate for a rapid, unexpected loss in generation resources in order to keep the system balanced. This reserve capacity is classified according to response time as spinning (<15 minute response time) and non-spinning (>15 minute response time). Faster response times are generally more valuable to the system. In some regions, reserve capacity is referred to as "frequency containment reserve."

Spotted owl. Reclusive bird, found in the American West, requiring old-growth forest habitat to survive.

South Coast Air Quality Management District (SCAQMD): the air pollution control agency for the four-county region including Los Angeles and Orange counties and parts of Riverside and San Bernardino counties.

Sprawl. The area taken up by a large or expanding development or city.

Stabilization. Changing active organic matter in sludge into inert, harmless material. The term also refers to physical activities such as compacting and capping at sites that limits the further spread of contamination without actual reduction of toxicity.

State action level (SAL). The maximum concentration of a contaminant in drinking water that The California Department of Health Services considers to be safe to drink. Drinking Water Action Levels (ALs) are health-based advisory levels established by the Department of Health Services (DHS) for chemicals for which primary maximum contaminant levels (MCLs) have not been adopted. There are currently 36 ALs. ALs are usually expressed in parts per billion (ppb) or parts per million (ppm). Drinking water with concentrations of impurities greater than the state action level must be treated to reduce or remove the impurities.

State Implementation Plan (SIP). Mandate for achieving health-based air quality standards. The plan that each state must develop and have approved by the US EPA which indicates how the state will comply with the requirements in the Clean Air Vintage Year: represents the first year in which Act; each State's SIP is amended as they address the allowance can be used for compliance specific or new requirements such as the NO_x reductions required in the NO_x SIP.

State land. Land owned and administered by the state in which it is located.

State parks. Parks and recreation areas owned and administered by the state in which they are located.

Static stability. The likelihood that soils at rest will remain at rest.

Still bottoms. Residues left over from the process of recovering spent solvents in a distillation unit.

Stockpile. Nuclear weapons and components under custody of the U.S. Department of Defense.

Straddling stocks. Fish populations that straddle a boundary between domestic and international waters.

Strangle: sale or purchase of a put option and a call option on the same underlying instrument, with the same expiration, but at strike prices X equally out of the money.

Stratosphere. The upper portion of the atmosphere (approximately 11 km to 50 km above the surface of the Earth).

Strike Price (or Exercise Price): price at which Y the allowance underlying a call or put option can be purchased (call) or sold (put) over the specified period. Z

Strip mining. Mining technique in which the land and vegetation covering the mineral being sought are stripped away by huge machines, usually damaging the land severely and limiting subsequent uses.

Submarine Launched Ballistic Missile (SLBM). A ballistic missile carried by and launched from a submarine.

Subsidence. Sinking or settling of soils so that the surface is disrupted, creating a shallow hole.

Suggested No Adverse Response Level (SNARL). Drinking water standards established by the U.S. EPA, but not enforceable by law. SNARLs suggest the level

of a containment in drinking water at which adverse health effects would not be anticipated (with a margin of safety).

Sulfur dioxide (SO_2). A heavy, smelly gas which can be condensed into a clear liquid; used to make sulfuric acid, bleaching agents, preservatives and refrigerants; a major source of air pollution in industrial areas. A gaseous pollutant which is primarily released into the atmosphere when as a by-product of fossil fuel combustion; the largest sources of SO_2 tend to be power plants that burn coal and oil to make electricity

Sump. A pit or tank that catches liquid runoff for drainage or disposal.

Super Efficient Refrigerator Program (SERP). An organization of 24 U.S. utilities that developed a $30 million competition to produce a refrigerator at least 25% lower in energy use and 85% lower in ozone depletion than projected 1994 models. The winning product, produced by Whirlpool, cut energy use by 40% in 1995.

Superfund Comprehensive Environmental Response, Compensation and Liability Act of 1980 (CERCLA).

Supply-side: a term referring to the generation (or supply) of power by a utility

Surge tanks. A tank used to absorb irregularities in flow of liquids, including liquid waste materials, so that the flow out of the tank is constant.

Sustainable communities. Communities capable of maintaining their present levels of growth without damaging effects.

Swap: an exchange of one allowance for another to exchange the vintage years of the allowances held in accounts Surface water. Water located above ground (e.g., rivers, lakes).

T

T&D congestion relief and infrastructure investment deferral. Energy storage technologies use to temporally and/or geographically shifting energy supply or demand in order to relieve congestion points in the transmission and distribution (T&D) grids or to defer the need for a large investment in T&D infrastructure.

Table. In the legislative sense, an action taken to halt debate on a bill.

Tailings or Mine Tailings. Crushed waste rock deposited on the ground during mining and ore processing, including some of the rock in which the ore is found. Unless they are handled carefully, they frequently release contaminants. As they age under the effects of air, rainfall and bacteria, some oxidize to produce new toxic materials, such as sulfuric acid, that can leach out and poison streams, rivers and lakes.

Tap water. Drinking water monitored (and often filtered) for protection against contamination and available for public consumption from sources within the home.

Tax shift. Replacing one kind of taxes with another, without changing the total amount of money collected. For example, replacing a portion of income taxes with carbon tax or other pollution taxes.

Telecommuting. Working with others via telecommunications technologies (e.g., telephones, modems, faxes) without physically traveling to an office.

Tetrachloroethylene (TCE). Volatile organic compound that is commonly used as an industrial degreasing solvent. TCE affects the central nervous system and is listed as a cancer-causing chemical under Proposition 65.

Tetrachlorophenol (TCP). Tetrachlorophenol is a toxic fungicide.

Thermonuclear. The application of high heat, obtained via a fission explosion, to bring about fusion of light nuclei.

Threatened species. Species of flora or fauna likely to become endangered within the foreseeable future.

Three Gorges. A project along the Yangtze river in China to build the largest hydroelectric dam in the world.

Threshold Limit Value (TLV). Public health exposure level set by the National Institute for Occupational Safety and Health for worker safety. It is the level above which a worker should not be exposed for the course of an eight-hour day, due to possible adverse health effects.

Timber. Logged wood sold as a commodity.

Time Weighting: an investment strategy in which allowance purchases and sales are transacted over an extended period of time and in small increments, thereby eliminating risk associated with highs and lows in the market

TNT Equivalent. A measure of the energy released in the detonation of a nuclear weapon, expressed in terms of the quantity of TNT which would release the same amount of energy.

Toluene. A toxic volatile organic compound often used as an industrial solvent.

Tongass. A national forest in southeast Alaska comprising one of the United States' last remaining temperate rain forests.

Total Threshold Limit Concentration (TTLC). A test for the toxicity characteristic: If the total concentration of a toxic substance in a waste is greater than this value, the waste is classified as hazardous in California. This is distinct from the Soluble Threshold Limit Concentration, or STLC, which is concerned with only the soluble

concentration.

Toxaphene. A chlorinated pesticide insecticide that was widely used to control pests on cotton and other crops until 1982, when it was banned for most uses. (In 1990, banned for all uses.) It was also used to kill unwanted fish in lakes. It is toxic to fresh-water and marine aquatic life and is listed as a cancer-causing chemical under Proposition 65.

Toxic. Poisonous.

Toxic emissions. Poisonous chemicals discharged to air, water, or land.

Toxic sites. Land contaminated with toxic pollution, usually unsuitable for human habitation.

Toxicity. Ability to harm human health or environment, such as injury, death or cancer. One of the criteria that is used to determine whether a waste is a hazardous waste (the "Toxicity Characteristic").

Toxicity Characteristic Leaching Procedure (TCLP). A federal test for the Toxicity Characteristic (TC). If the concentration of a toxic substance in a special extract of a waste exceeds the TC value, the waste is classified as hazardous in the United States (a "RCRA waste"). The extraction procedure is different from that of the California Waste Extraction Test (WET).

Toxic Substances Control Act (TSCA). A federal law of 1976 to regulate chemical substances or mixtures that may present an unreasonable risk of injury to health or the environment.

Toxic waste. Garbage or waste that can injure, poison, or harm living things, and is sometimes life-threatening.

Toxification. Poisoning.

Traffic calming. Designing streets to reduce automobile speed and to enhance walking and bicycling.

Trader: anyone who buys or sells allowances with the intention of making a profit

Transit. See public transportation.

Transportation. Any means of conveying goods and people.

Transportation planning. Systems to improve the efficiency of the transportation system in order to enhance human access to goods and services.

Trash. Waste material that cannot be recycled and reused (synonymous with garbage).

Trawls. Nets with a wide mouth tapering to a small, pointed end, usually called the "cod end." Trawls are towed behind a vessel at any depth in the water column.

Trial burn. A test of incinerators or boilers and industrial furnaces in which emissions are monitored for the presence of specific substances, such as organic compounds, particulates, metals and hydrogen chloride (all specified by agency permits).

Trichloroethane. 1,1,1-TCA, or methylchloroform is used as a cleaning agent for metals and plastics. It is toxic.

Trichloroethylene (TCE). A volatile organic compound that is often used an industrial degreasing solvent. It is toxic and is listed as a cancer-causing chemical under Proposition 65.

Trip reduction. Reducing the total numbers of vehicle trips, by sharing rides or consolidating trips with diverse goals into fewer trips.

Tritium. A radioactive form of hydrogen with a half-life of 12 years. It emits beta particles. It is used to mark chemical compounds so that the structure or chemical activity can be determined. Also used in nuclear weapons research and construction. Small amounts of tritium occur naturally, and some exists as a by-product of previous nuclear testing and nuclear reactor operations.

Trolling. A method of fishing using several lines, each hooked and baited, which are slowly dragged behind the vessel.

Turtle excluder device (TED). A gear modification used on shrimp trawls that enables incidentally caught sea turtles to escape from the nets.

U

Unit Accounts: accounts in the SO_2 or NO_x ATSs which hold allowances initially allocated to those sources required to participate in either the acid rain or OTC NO_x programs; EPA adjusts these accounts for compliance each year

United Nations Framework Convention on Climate Change(UNFCC): a treaty signed in 1992 by 165 countries and ratified by 160 countries (including the U.S.); took effect in March 1994; set a target of stabilizing greenhouse gas concentrations in the atmosphere to a level that would prevent further increase of global climate warming.

Unsaturated zone. Underground soil and gravel that could contain groundwater, but lies above the aquifer. This is in contrast to a saturated zone, where the space between soil particles is filled with water.

Up-gradient. The direction from which water flows in an aquifer. In particular, areas that are higher than contaminated areas and, therefore, are not prone to contamination by the movement of polluted groundwater.

Urban planning. The science of managing and directing city growth.

Uranium (U). A heavy, radioactive metal (atomic number 92) used in the explosion of nuclear weapons (especially one isotope, U-235).

Urban parks. Parks in cities and areas of high population concentration.

Utilities. Companies (usually power distributors) permitted by a government agency to provide important public services (such as energy or water) to a region; as utilities are provided with a local monopoly, their prices are regulated by the permitting government agency.

V

Vanadium. A toxic metal that is both mined and is a by-product of petroleum refining. Compounds of vanadium are used in the steel industry, as a catalyst in the chemical industry, in photography and in insecticides.

Variable supply resource integration. The use of energy storage to change and optimize the output from variable supply resources (e.g. wind, solar), mitigating rapid and seasonal output changes and bridging both temporal and geographic gaps between supply and demand in order to increase supply quality and value.

Veto. A Presidential action rejecting a bill as passed by the U.S. Congress. The President can also effect a "pocket veto" by holding an unsigned bill past the signing period.

Video logging. A method for close-up inspection of the interior of a well or pipe by means of a color camera that can view the well casing and screen at 90 degrees to the well's axis.

Vinyl chloride. Vinyl chloride is widely used in the plastics industry in creating polyvinyl chloride (PVC). It is listed as a cancer-causing agent under Proposition 65.

Virgin forest. A forest never logged.

Viscosity. A measure of the ease with which a liquid can be poured or stirred. The higher the viscosity, the less easily a liquid pours.

Voice vote. A vote where members vote by saying either "yes" or "no" together; individual member's votes are not placed on record.

Void space. The space in a tank between the top of a tank and the liquid level. If the tank is used to store combustible liquids that easily evaporate, this space can fill with vapors which may reach explosive levels.

Volatile. Describes substances that readily evaporate at normal temperatures and pressures.

Volatile organic compounds (VOCs). Organic liquids, including many common solvents, which readily evaporate at temperatures normally found at ground surface and at shallow depths. They take part in atmospheric photochemical (sun-driven) reactions to produce smog.

Volatilization rate. The rate at which a chemical changes from a liquid to gas. It is also known as "air flux."

Voltage support. The injection or absorption of re-active power to maintain voltage levels in the transmission and distribution system under normal conditions is referred to as voltage support.

W

Warhead. The part of a missile which contains the nuclear explosive.

Waste. Garbage, trash.

Waste feed. The flow of wastes into an incinerator, boiler or industrial furnace. The waste feed can vary from continuous to intermittent (batch) flows.

Waste heat utilization. Energy storage technology use for the temporal and geographic decoupling of heat supply (e.g. CHP facilities, thermal power plants) and demand (e.g. for heating/cooling buildings, supplying industrial process heat) in order to utilize previously wasted heat.

Waste site. Waste dumping ground.

Waste stream. Overall waste disposal cycle for a given population.

Waterborne contaminants. Unhealthy chemicals, microorganisms (like bacteria) or radiation, found in tap water.

Water filters. Substances (such as charcoal) or fine membrane structures used to remove impurities from water.

Water quality. The level of purity of water; the safety or purity of drinking water.

Water quality testing. Monitoring water for various contaminants to make sure it is safe for fish protection, drinking, and swimming.

Watershed. A region or area over which water flows into a particular lake, reservoir, stream, or river.

Water table. In a shallow aquifer, a water table is the depth at which free water is first encountered in a monitoring well.

Well. A dug or drilled hole used to get water from the earth.

Wetland. An area that is regularly saturated by surface or groundwater and, under normal circumstances, capable of supporting vegetation typically adapted for life in saturated soil conditions; they are critical to sustaining many species of fish and wildlife, including native and migratory birds. They include swamps, marshes, and bogs, and may be either coastal or inland. Coastal wetlands are brackish (have a certain mixture of salt).

Wilderness area. A wild area that Congress has preserved by including it in the National Wilderness Preservation System.

Wildlife. Animals living in the wilderness without human intervention.

Wildlife refuges. Land set aside to protect certain species of fish or wildlife (administered at the federal level in the U.S. by the Fish and Wildlife Service).

Wind power. Power or energy derived from the wind (via windmills, sails, etc.).

Wise use movement. A loosely-affiliated network of people and organizations throughout the U.S. in favor of widespread privatization and opposed to environmental regulation, often funded by corporate dollars.

Woods Hole. A town on Cape Cod where several important ocean research institutes are located.

Work plan. The site work plan describes the technical activities to be conducted during the various phases of a remediation project.

X

x the horizontal axis in a plot or a chart.

x 8-12 GHz frequency band.

x-ray; a form of electromagnetic radiation with a wavelength in the range of 0.01 to 10 nanometers, and energies in the range 0.1 to 100 keV, which is shorter than the wavelength of UV radiation but longer than that of gamma rays.

Xylene. An aromatic hydrocarbon used in gasoline, paints, lacquers, pesticides, gums, resins and adhesives. It is toxic and flammable.

x-y tracking; frames with PV cells or modules tracking the sun in the x and y directions all day long.

Y

y often used to denote year (i.e., 100 kWh/y means 100 kWh generated yearly)

y the vertical axis in a plot or a chart

y yard

Z

Zoning. The arrangement or partitioning of land areas for various types of usage in cities, boroughs or townships.

Zinc. A metal used for auto parts, for galvanizing, and in production of brasses and dry cell batteries. It is nutritionally essential but toxic at higher levels.

Index